MATHEMATICAL ANALYSIS

Useful for

- Undergraduate and Post Graduate Courses for All Indian Universities.
- UGC-CSIR, JRF/NET, SET, GATE and All Entrance Examinations for Admission in M.Sc., M.Phil. and Ph.D. programmes.

Dr. SUDHIR K. PUNDIR

M.Sc., M.Phil, NET, Ph.D.

Associate Professor
Department of Mathematics
S.D. (P.G.) College,
Muzaffarnagar (U.P.)

CBS

CBS Publishers & Distributors Pvt. Ltd.

New Delhi • Bengaluru • Chennai • Kochi • Kolkata • Mumbai
Hyderabad • Uttarakhand • Nagpur • Patna • Pune • Jharkhand

ISBN: 978-81-239-2667-4

First Edition: 2015
Reprint: 2019

Copyright © Publisher

Published by **Satish Kumar Jain** and produced by **Varun Jain** for

CBS Publishers & Distributors Pvt. Ltd.,
4819/XI Prahlad Street, 24 Ansari Road, Daryaganj, New Delhi - 110002
delhi@cbspd.com, cbspubs@airtelmail.in • www.cbspd.com
Ph.: 23289259, 23266861, 23266867 • Fax: 011-23243014

Corporate Office: 204 FIE, Industrial Area, Patparganj, Delhi - 110 092
Ph: 49344934 • Fax: 011-49344935
E-mail: publishing@cbspd.com • publicity@cbspd.com

Branches:
• *Bengaluru:* 2975, 17th Cross, K.R. Road, Bansankari 2nd Stage,
  Bengaluru - 70 • Ph: +91-80-26771678/79 • Fax: +91-80-26771680
  E-mail: cbsbng@gmail.com, bangalore@cbspd.com
• *Chennai:* No. 7, Subbaraya Street, Shenoy Nagar, Chennai - 600030
  Ph: +91-44-26681266, 26680620 • Fax: +91-44-42032115
  E-mail: chennai@cbspd.com
• *Kochi:* Ashana House, 39/1904, A.M. Thomas Road, Valanjambalam,
  Ernakulum, Kochi • Ph: +91-484-4059061-65
  Fax: +91-484-4059065 • E-mail: cochin@cbspd.com
• *Kolkata:* 6-B, Ground Floor, Rameshwar Shaw Road, Kolkata - 700014
  Ph: +91-33-22891126/7/8 • E-mail: kolkata@cbspd.com
• *Mumbai:* 83-C, Dr. E. Moses Road, Worli, Mumbai - 400018
  Ph: +91-9833017933, 022-24902340/41 • E-mail: mumbai@cbspd.com

Representatives:

• Hyderabad: 0-9885175004   • Nagpur: 0-9021734563
• Patna: 0-9334159340   • Pune: 0-9623451994
• Jharkhand: 0-9811541605   • Uttarakhand: 0-9716462459

Printed at: India Binding House, Noida, UP (India)

Preface

The book entitled 'Mathematical Analysis' is meant for UG and PG students of all Indian Universities. Besides, it will also be very useful for those students preparing for various competitive examinations like CSIR-JRF/NET, SET, GATE and various entrance examinations for admission in M.Sc., MPhil and Ph.D programmes.

This book has evolved the lectures that I have giving to the students for the last sixteen years. Special and conscious efforts have been made to keep the writing style simple. Students who are tired of complex concepts and abstract presentations styles, will find this book simple and straight forward. It is a collection and compilation work from various sources and has been endeavoured to include as much as information could be possible. The book's objective is to provide a conceptual understanding of the fundamentals of real analysis. Different concepts have been explained with the help of examples. A large number of problems with solutions have been provided to assist one get a firm grip on the ideas developed. There is plenty of scope in the form of exercise for the reader to try and solve the problem on his own. To make the book self-contained and competition oriented a chapter review of basic terms, results and questions has been given at the end of each chapter. Also, at the end of each section, graded examples have been given with the name of self assessment test, which help the students to grasp the thing better. These include problems on the entire section and are carefully selected to represent variety.

I express my gratitude to the authors and publishers of various books I consulted.

I wish to sincerely thank **Sh S.K. Jain**, Managing Director, CBS Publishers and Distributors, New Delhi for his encouragement and help in bringing out this publication in a present nice form.

I express my gratitude to the authors and publishers of various books I consulted. My special thanks to Sh. Y.N. Arjuna, Sh. B.M. Singh, Sh. Sunil Dutt and entire team of CBS Publishers and Distributors, New Delhi whose encouragement and unstinted support enabled me to complete my book. Mr. Peeyush Goel, M/s Dreamshapers also deserve special mention for nice type setting.

I must also record my appreciation due to my wife Dr. Rimple, daughter Rijuta and son Shrish for their understanding and love during the long period that I have taken to complete this book.

Above all I am thankful to The Almighty God, without whose grace nothing is possible for any one.

Readers are welcomed to point out errors, if any and send their valuable suggestions for improving the quality of the book.

Dr. Sudhir K. Pundir
email : skpundir05@yahoo.co.in

Contents

Chapter 1

Real Numbers

1.1 INTRODUCTION

The real number system R is one of the most important and beautiful mathematical systems. It is the foundation on which the entire branch of mathematics, known as Real Analysis based.

There are different ways of introducing the real number system, but the most common way is to start with Peano's Axioms for the natural numbers. These axioms for natural numbers discovered by the Italian Mathematician Peano are

(i) 1 is a natural number.

(ii) Each natural number n has a successor $(n + 1)$.

(iii) Two natural numbers are equal if their successors are equal.

(iv) Except 1 each natural number is a successor of a natural number.

(v) Any set of natural numbers which contains 1 and the successor $(k+1)$ of every natural number k whenever it contains k in the set N of natural number.

REMARKS

- Axiom (v) is commonly known as the axiom of induction or principle of finite induction.
- These axioms completely define the set of natural numbers N
  i.e., $$N = \{1, 2, 3, 4, 5, 6,...\}$$
- These axioms can be used to extend the set N of natural numbers to another large system, known as the set of integers, i.e., $$Z = \{... -4, -3, -2, -1, 0, 1, 2, 3, 4,...\}$$
- Integers can be used to construct the set, $Q = \left\{\dfrac{p}{q}, p, q \in Z, q \neq 0\right\}$ known as the set of rational numbers.
- In the definition of rational number, we shall restrict p and q, in such a manner that they may have no common divisor other than 1. In such a case p and q are called relatively prime integers.
- A real number which is not rational i.e., not expressible as $\dfrac{p}{q}$ is called an irrational number.
- The above discussion can be illustrated as follows

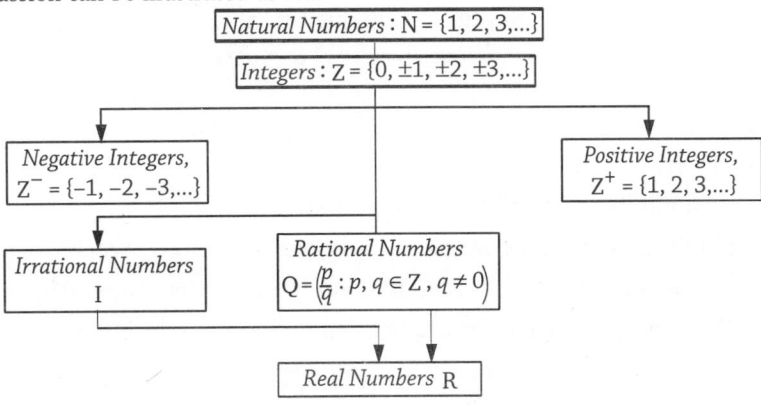

1.2 AXIOMS FOR REAL NUMBERS

The axioms for real numbers are classified as follows :

(i) Extended axioms   (ii) Field axioms

(iii) Ordered axioms   (iv) Completeness axioms

1.2.1 EXTENDED AXIOMS

It states that "the set of real numbers R has at least two distinct members."

1.2.2 FIELD AXIOMS

The set R of real numbers together with two binary operations

$$(+) : R \times R \to R \text{ and } (.) : R \times R \to R$$

respectively called the addition and multiplication of real numbers is called the real number field if the following axioms are satisfied.

(i) Axioms for Addition :

A (1) *Closure property.* The set R is closed under addition operation. This means that the sum of any two real numbers is a real number *i.e.,*

$$a, b \in R \implies a + b \in R.$$

A (2) *Associativity.* Addition operation in R is associative *i.e.* ,

$$a, b, c \in R \implies (a + b) + c = a + (b+c).$$

A (3) *Existence of additive identity.* There must exists a real number 0, (zero) such that

$$a + 0 = 0 + a = a \: \forall \: a \in R.$$

A (4) *Existence of additive inverse.* Corresponding to each $a \in R$, there exists a real number b such that

$$a+b = b + a = 0.$$

(Additive inverses are most commonly known as negatives. The real number b, above, is called the negative of a and is written as $- a$.)

A (5) *Commutative law.* Addition operation in R is commutative *i.e.,*

$$a, b \in R \implies a+b = b + a.$$

(ii) Axioms for Multiplication :

M (1) *Closure property.* The set R is closed under multiplication operation *i.e.,*

$$a, b \in R \implies ab \in R.$$

M (2) *Associativity.* Multiplication operation in R is associative *i.e.,*

$$a, b, c \in R \implies (ab) \: c = a(bc).$$

M (3) *Existence of multiplicative identity.* There must exists a real number 1 (one) such that $\quad a \cdot 1 = 1 \cdot a = a \: \forall \: a \in R.$

M (4) *Existence of multiplicative inverse.* Corresponding to each $a \in R$, other than zero $\exists$ a real number b such that $\quad ab = ba = 1.$

(Multiplicative inverses are most commonly known as inverses. The real number b is called the inverse or reciprocal of a and is written as $1/a$ or a^{-1} .

M (5) *Commutative law.* Multiplication operation in R, is commutative *i.e.,*

$$a, b \in R \implies ab = ba.$$

(iii) Distributive Law :

It states that multiplication is distributive over addition operation

i.e., $a, b, c \in R \Rightarrow a \cdot (b + c) = ab + ac.$

In view of the above properties, the algebric structure $(R, +, .)$ is called a field.

REMARKS

- The set of real numbers R is a field.
- The set of rational numbers Q is a field.
- The set of complex numbers C is a field.
- The set of natural numbers N is not a field.

1.2.3 AXIOMS OF ORDER IN R

The axioms of order in R based on > (greater than) are :

(i) If $a, b \in R$, then one and only one of the following is true

$$a > b, a = b, b > a \hspace{2cm} \text{(Law of Trichotomy)}$$

(ii) If $a, b, c \in R$ and $a > b, b > c$, then $a > c$.   *(Transitivity)*

(iii) If $a, b, c \in R$ and $a > b$, then $a + c > b + c$.   *(Monotone property for addition)*

(iv) If $a, b, c \in R$ and $a > b, \ c > 0$, then $ac > bc$.

(Monotone property for multiplication)

For Example :

(1) The set of real numbers R and set of rational numbers Q are ordered fields.

(2) The set of complex numbers is not an ordered field.

REMARK

- The above axioms can easily be expressed in terms of the less than relation '<' (less than).

1.2.4 PROPERTIES OF REAL NUMBERS

- Division by 0 is not allowed.
- There exists a unique identity element for addition in R.
- There exists a unique additive inverse for each element in R.
- $a + b = a + c \Rightarrow b = c.$
- $a + b = a \Rightarrow b = 0.$
- $a + b = 0 \Rightarrow a = -b.$
- There exists a unique identity element for multiplication in R.
- There exists a unique multiplicative inverse for each non-zero element in R.
- If a and b are any two real numbers, then the equation $x + a = b$ has a unique solution

i.e., $x = [b + (-a)] = b - a$ in R.

- If a, b are any real numbers and $a \neq 0$, then the equation $ax = b$ has a unique solution

$$x = \left(\frac{1}{a}\right) \cdot b = \frac{b}{a} \text{ in R.}$$

- $a \cdot b = 0 \Leftrightarrow a = 0 \text{ or } b = 0.$
- $a \cdot b \neq 0 \Leftrightarrow a \neq 0 \text{ and } b \neq 0.$

THE EXTENDED REAL NUMBERS SYSTEM

We shall denote sets of positive and negative real numbers by R^+ and R^- respectively. Thus

$$R = R^- \cup \{0\} \cup R^+.$$

If the set of real numbers R is extended by the addition of two elements $-\infty$ and ∞, the enlarged set is called extended real number system.

1.2.6 **INTEGRAL POWERS OF A REAL NUMBER**

Let $a \in R$, and n be any positive integer, then we can define $a^n = a \cdot a \cdot a \dots n$ times. In particular

$$a^1 = a$$
$$a^2 = a \cdot a$$
$$a^3 = a \cdot a \cdot a = a^2 \cdot a$$
.... and so on.

If n is any negative integer, then we have $x^{-n} = (x^n)^{-1} = (x^{-1})^n$.

Definition-1. A real number x is called positive if $x > 0$, and the set of all positive real numbers is, denoted by R^+ , is given by

$$R^+ = [x : x \in R, x > 0].$$

Definition-2. A real number x is called negative if $x < 0$, and the set of all negative real numbers, denoted by R^- , is given by

$$R^- = [x : x \in R, x < 0].$$

Definition-3. The order relation 'less than' ($<$), between two real numbers a and b is defined as

$$a < b \text{ if } b > a$$

Definition-4. The symbols $\leq$ and $\geq$ are used for compound statement as follows :

$$a \leq b, \text{ if, either } a < b \text{ or } a = b$$
$$a \geq b \text{ if, either } a > b \text{ or } a = b$$

1.3 INTERVAL

A subset S of R *is called an interval if a, b $\in$ S, x $\in$ R such that a $<$ x $<$ b $\Rightarrow$ x $\in$ S.*
There are following four types of intervals

(1) $a \circ\!\!-\!\!-\!\!-\!\!-\!\!-\!\!-\!\!-\!\!-\!\!\circ b \Rightarrow]a, b[= [x : a < x < b]$

(2) $a \bullet\!\!-\!\!-\!\!-\!\!-\!\!-\!\!-\!\!-\!\!-\!\!\bullet b \Rightarrow [a, b] = [x : a \leq x \leq b]$

(3) $a \circ\!\!-\!\!-\!\!-\!\!-\!\!-\!\!-\!\!-\!\!-\!\!\bullet b \Rightarrow]a, b] = [x : a < x \leq b]$

(4) $a \bullet\!\!-\!\!-\!\!-\!\!-\!\!-\!\!-\!\!-\!\!-\!\!\circ b \Rightarrow [a, b[= [x : a \leq x < b]$

OBSERVATIONS

(1) The set $] a, b[$ in which the end points are not included, is called an open interval.

(2) The set $[a, b]$ also contains both its end points is called a closed interval.

(3) The sets $[a, b[$ and $] a, b]$ are called half open (or half closed) intervals or semi open (or semi-closed) as they contain only one end point.

REMARKS

- If S is any interval and if c and d are two elements of S, then all numbers lying between c and d are also elements of S.
- The proper use of a bracket, for example, parenthesis for open and square bracket for closed interval, itself specifies the interval. As such, to emphasize the nature of an interval, we shall drop the word 'description' and shall simply express the interval by using the appropriate brackets.

1.3.1 LENGTH OF AN INTERVAL

The number $(b - a)$ is called the length of the intervals $]a, b[$, $[a, b]$, $[a, b[$ and $]a, b]$.

If the length of the interval is finite, then interval is said to be finite and if the length is infinite, then it is known as infinite interval.

1.4 FINITE AND INFINITE SUBSET OF R

A subset S of R is said to be finite if either it is empty or there exists, a one to one correspondence or mapping from the set $[1, 2,..., n]$ onto the set S for some $n \in$ N.

Also, a subset, which is not finite is called infinite.

For Example:

(1) The set $[a, b, c]$ is a finite set, because $\exists$ a one to one mapping from the set $[a, b, c]$ onto the set $[1, 2, 3]$.

(2) The set of all prime less than 10^{1000} is a finite set.

1.4.1 ABSOLUTE VALUE OF A REAL NUMBER

The absolute value of a real number, a denoted by $|a|$, is the real number a, $- a$ or 0 according as a is positive, negative or zero, *i. e.*,

$$|a| = \begin{cases} a & \text{if } a \geq 0 \\ -a & \text{if } a < 0 \end{cases}$$

From the definition, we conclude that

(1) $|a| = \max \{a, - a\}$,   (2) $-|a| = \min \{a, - a\}$ and (3) $|a| \geq a \geq -|a|$.

THEOREM 1. *For all $x, y \in$ R, we have*

     (1) $|xy| = |x| \cdot |y|$ (2) $|x + y| \leq |x| + |y|$ (3) $|x - y| \geq ||x| - |y||$.

Proof.   (1)   We know that

$$|xy|^2 = (xy)^2 = x^2y^2 = |x|^2 \cdot |y|^2 = (|x| \cdot |y|)^2$$

$\Rightarrow \qquad |xy| = |x| \cdot |y|$

(2)   We know that

$$|x+y|^2 = (x+y)^2$$
$$= x^2 + y^2 + 2xy \leq x^2 + y^2 + 2|xy| \qquad (\because |xy| \geq xy)$$
$$\leq |x|^2 + |y|^2 + 2|x| \cdot |y|$$
$$= (|x| + |y|)^2$$

$\Rightarrow \qquad |x+y| \leq |x| + |y|$, which is the required result.

(3)   Consider $|x| = |x - y + y| = |(x-y) + y| \leq |x-y| + |y|$ ...(1)

     and $|y| = |(y-x) + x| \leq |y-x| + |x|$ ...(2)

$\Rightarrow \qquad |y| - |x| \leq |y - x|$

or $\quad -(|x|-|y|) \le |x-y|$

$\Rightarrow \quad |x-y| \ge \max\{|x|-|y|), -(|x|-|y|)\}$

$\Rightarrow \quad |x-y| \ge ||x|-|y||$

THEOREM 2. *For any two real numbers x and y*

$$|x-y| \le |x| + |y|.$$

Proof. Consider $\quad |x-y| = |x+(-y)| \le |x| + |-y| \qquad$ (By triangle inequality)

$= |x| + |y|$

$\Rightarrow \quad |x-y| \le |x| + |y|$

THEOREM 3. (i) $\left|\dfrac{x}{y}\right| = \dfrac{|x|}{|y|}$,

(ii) *If $\varepsilon > 0$, then $|x-y| < \varepsilon \Leftrightarrow y - \varepsilon < x < y + \varepsilon$.*

Proof. (i) Consider $\quad \left|\dfrac{x}{y}\right|^2 = \left(\dfrac{x}{y}\right)^2 = \dfrac{x^2}{y^2} = \dfrac{|x|^2}{|y|^2} = \left(\dfrac{|x|}{|y|}\right)^2$

$\therefore \qquad\qquad \left|\dfrac{x}{y}\right| = \dfrac{|x|}{|y|}$

(ii) We have $\quad |x-y| = \max\{(x-y), -(x-y)\} < \varepsilon$

$(x-y) < \varepsilon < -(x-y) < \varepsilon \Leftrightarrow x < y + \varepsilon < (-x+y) < \varepsilon$

$x < y + \varepsilon < (y-\varepsilon) < x \Leftrightarrow y - \varepsilon < x < y + \varepsilon.$

Solved Examples

Example 1. *Show that $\sqrt{2}$ is irrational.*

Solution. Let if possible, $\sqrt{2}$ is not irrational.

$\Rightarrow \sqrt{2}$ is a rational number.

By definition, $\sqrt{2} = \dfrac{p}{q}\,(q \ne 0)$, and p, q have no common factor other than 1.

$\Rightarrow \quad \dfrac{p^2}{q^2} = 2 \Rightarrow p^2 = 2q^2$

Now, since q is an integer, so q^2

and $2q^2$ both are integers.

$\Rightarrow \quad p^2$ is an integer, divisible by 2 $\Rightarrow p^2$ should be even.

$\Rightarrow \quad p$ should also be even.

Let $p = 2k$, then, $\quad 2q^2 = (2k)^2 = 4k^2$ or $q^2 = 2k^2$.

$\Rightarrow q^2$ is even. Therefore, q should also be even.

Let $\qquad\qquad\qquad q = 2l.$

$\Rightarrow \quad p$ and q both have a common factor 2, which is a contradiction.

$\Rightarrow \quad \sqrt{2}$ is not a rational number.

Hence, $\sqrt{2}$ is an irrational number.

Based on the following Results

- Rational numbers can be written as the fraction of two numbers while, irrational are not.

- The set of natural numbers is not a field, while the set of complex numbers, set of real numbers, and set of rational numbers are field.

- The set of real numbers R and set of rational numbers Q are ordered field.

- The set of complex numbers is not ordered field.

- A subset S of R is called an interval if $a, b \in S$, $x \in$ R such that $a < x < b \Rightarrow x \in S$.

- The number $b \sim a$ is called the length of the, interval $[a, b]$.

- $|xy| = |x| \cdot |y|$; $|x+y| \le |x| + |y|$; $|x-y| \ge ||x|-|y||$

Example 2.   *Show that $\sqrt{8}$ is irrational.*

Solution.   Let if possible $\sqrt{8}$ be a rational number

$$\Rightarrow \qquad\qquad \sqrt{8} = \frac{p}{q}, q \neq 0, p, q \in Z.$$

Also p and q have no common factor other than 1.

Clearly, $2 < \sqrt{8} < 3 = 2 < \dfrac{p}{q} < 3$

$$\Rightarrow \qquad 2q < p < 3\,q \Rightarrow 0 < p - 2q < q$$

$\Rightarrow$ $p - 2q$ is positive integer less than q.

$\Rightarrow$ $\sqrt{8}\ (p - 2q)$ is not an integer.   ...(1)

But $\sqrt{8}\ (p - 2q) = \dfrac{p}{q}\ (p - 2q) = \dfrac{p^2}{q} - 2p = \dfrac{p^2}{q^2}q - 2p = 8q - 2p$ is an integer.

$\therefore$ $\sqrt{8}\ (p - 2q)$ is an integer, which is a contradiction.

Hence, $\sqrt{8}$ is not a rational number.

EXERCISE 1.1

1. Prove that

    (i) $|x| \geq 0$

    (ii) $x = 0 \Leftrightarrow |x| = 0$

    (iii) $x \geq - |x|$

    (iv) $|x+y+z| \leq |x| + |y| + |z|$

    (v) $|x+y| \geq ||x| - |y||$

    (vi) $|x-y| = 0 \Leftrightarrow x = y$

    (vii) $\left|\dfrac{1}{x}\right| = \dfrac{1}{|x|}, \ (x \neq 0)$

    (viii) $|x-y| = |y-x|$

2. If $x_1, x_2, ..., x_n$ be any real numbers, then

    (i) $|x_1 + x_2 + ... + x_n| \leq |x_1| + |x_2| + ... + |x_n|$

    (ii) $|x_1 . x_2 . \ .... \ . x_n| = |x_1| . |x_2| . \ .... \ |x_n|$

3. If x, ε be real numbers and $\varepsilon > 0$, then show that $|x| < \varepsilon \Leftrightarrow -\varepsilon < x < \varepsilon$.

4. If x, y be two real numbers, show that
$$|x+y|^2 + |x-y|^2 = 2(|x|^2 + (|y|^2).$$

5. If $y > 0$, then show that

    (i) $x^2 < y^2 \Leftrightarrow -y < x < y$

    (ii) $x^2 > y^2 \Leftrightarrow x > y$ or $x < -y$

    (iii) $xy < 0 \Leftrightarrow x > 0$ and $y < 0$ or $x < 0$ and $y > 0$

    (iv) $x > y$ and $z < 0 \Rightarrow xz < yz$

    (v) $x > y$ and $z > w \Rightarrow x + z > y + w$.

6. Show that π is not a rational number.

1.5 BOUNDEDNESS OF A SUBSET OF REAL NUMBERS

1.5.1 UPPER BOUND OF A SUBSET OF R.

     A subset S of R is said to be bounded above, if there exists a real number u such that $s \leq u \ \forall \ s \in S$. The real number u is said to be upper bound of S.

If there exists no such upper bound, then the set is said to be unbounded above.

For Example:

(1) The set of natural number N $= (1, 2, 3,...)$ is not bounded above or unbounded above.

(2) The set of positive integers Z^+ is not bounded above.

(3) The set $S = [1, 2, 3, 4]$ is bounded above by 4.

(4) The set $\left[\dfrac{1}{n} : n \in N\right]$ is bounded above by 1.

(5) The set of negative integers is bounded above by 0.

1.5.2 LOWER BOUND OF A SUBSET OF R.

A subset S of R is said to be bounded below if there exists a real number l such that $s \geq l \ \forall \ s \in S$. The real number l is said to be the lower bound of S and if there exists no such lower bound, then the set is said to be unbounded below.

For Example :

(1) The set of natural numbers, N is bounded below by 1.

(2) The set $\left[\dfrac{1}{n} : n \in N\right]$ is bounded below by 0.

(3) The set $S = [1, 2, 3, 4]$ is bounded below by 1.

(4) The set of positive real numbers is bounded below.

1.5.3 BOUNDED SET

A subset S of R is said to be a bounded if it is bounded below as well as bounded above i.e., if there exist two real numbers l and u such that

$$l \leq s \leq u, \ \forall \ s \in S.$$

Equivalently, if there exists an interval $I \ (= [l, u])$ such that $S \subseteq I$

For Example :

(1) Every finite set is bounded.

(2) The set $\left[\dfrac{1}{n} : n \in N\right]$ is bounded.

1.5.4 UNBOUNDED SET

A subset S of R, which is not bounded is called an unbounded set.

For Example :

(1) The sets N, Z, Q, R are unbounded sets.

(2) Set of all prime numbers is an unbounded set.

REMARKS

- If a set is bounded above, then it has infinitely many upper bounds in as much as every number greater than an upper bound is also an upper bound.
- If a set is bounded below, then it has infinitely many lower bounds in as much as every number smaller than a lower bound is also a lower bound.
- It is not necessary that lower bounds and upper bounds of a set S are the members of S.
- The null set ϕ is bounded but it is neither possesses lower bound nor upper bound.

1.6 SUPREMUM AND INFIMUM OF A SET

1.6.1 LEAST UPPER BOUND (OR SUPREMUM)

A real number u is said to be a least upper bound of a set S if

    (i) u is an upper bound of S

and (ii) if u' is an another upper bound of S then $u \leq u'$.

i.e., no real number less than u can be an upper bound of S.

1.6.2 GREATEST LOWER BOUND (OR INFIMUM)

A real number l is called a greatest lower bound of a set S if

    (i) l is a lower bound of S

and (ii) if l' is another lower bound of S, then $l' \leq l$

i.e., no real number greater then l can be a lower bound of S.

For Example :

If $S = \left[\dfrac{1}{n} : n \in N\right]$ then $l.u.b. = 1$ and $g.l.b.$ is 0.

REMARKS

- If a real number u is the supremum of a subset S of real numbers then, for every $\varepsilon > 0$, there exists a real number $x \in S$ such that $u - \varepsilon < x < u$.
- If a real number is the infimum of a subset S of real numbers then, for every $\varepsilon > 0$, there exists a real number $x \in S$ such that $l \leq x < l + \varepsilon$.
- Supremum is defined only for the bounded above sets and infimum for the subset, which are bounded below.
- The supremum and infimum of a set may or may not belong to the set.
- If supremum of a set belongs to the set, then supremun is the largest element of the set.
- If infimum of a set belongs to the set, then infimum is the smallest element of the set.
- Supremum and infinium of a bounded subset of R, are unique.
- In case of singleton set $S = [a]$, $a \in R$, supremum and infimum coincide.
- If u and l are the supremum and infimum of a non-empty subset S of R, then $l \leq u$.

THEOREM 1. *The supremum of a set $S \subset R$, if exists, is unique.*

Proof.   Let S be a non-empty subset of R.

Let if possible, s_1 and s_2 be two supremum of S.

To show $s_1 = s_2$.

Since we assume that s_1 and s_2 are the supremums of S.

$\Rightarrow$ s_1 and s_2 are the upper bounds of S.

Let us first suppose s_1 is a supremum and s_2 is an upper bound of S, then

$$s_1 \leq s_2. \hspace{4cm} ...(1)$$

Now, if s_2 is the supremum and s_1 is the upper bound of S, then

$$s_2 \leq s_1. \hspace{4cm} ...(2)$$

From (1) and (2) , $s_1 = s_2$.

Hence, supremum of a set, if exists is unique.

THEOREM 2. *The infimum of a set $S \subset R$, if exists, is unique.*

Proof.   Proof is similar as theorem 1 and left to the reader.

THEOREM 3. *If S is a non-empty subset of R, then a real number s is the supremum of S if and only if*

         (i) $x \leq s$ $\forall x \in S$

and (ii) for each positive real number ε, there exists a real number $x \in S$ such that $x > s - \varepsilon$.

Proof.   **Necessary Condition** (only if part).

Let us first suppose s be the supremum of the set S.

Let s be the supremum of $S \Rightarrow s$ is an upper bound of S.

By definition $x \leq s \ \forall \ x \in S$.

Let $\varepsilon > 0$ be any real number. Then obviously $s - \varepsilon < s$

$\Rightarrow$ $(s - \varepsilon)$ is not an upper bound of S. $\hspace{2cm}$ ($\because s$ is l.u.b. of S.)

Hence, there must exist some $x \in S$ such that $x > s - \varepsilon$.

Sufficient part (If part)

Let us suppose condition (i) and (ii) holds.

Then, to show $\hspace{3cm} s = \sup S$

By condition (i), we have s is an upper bound of S. To show s is the supremum of S, for this, it is enough to show that no real number less than s can be an upper bound of S.

Let s_1 be any real number less than s

$$\Rightarrow s - s_1 > 0.$$

Let us take $\hspace{2.5cm} \varepsilon = s - s_1 \hspace{1.5cm} \Rightarrow \hspace{1cm} \varepsilon > 0.$

Then by condition (ii), these exists $x \in S$ such that $\hspace{0.5cm} x > s - \varepsilon$

$\Rightarrow \hspace{3cm} x > s - (s - s') \hspace{1cm} \Rightarrow \hspace{0.5cm} x > s', x \in S$

$\Rightarrow$ s' is not an upper bound of S.

Hence, we can say that s is an upper bound of S and no real number less than s is an upper bound of S.

$\Rightarrow$ s is the supremum of S.

THEOREM 4. *Let S be a non-empty subset of R, then a real number t is the infimum of S if and only if*

$\hspace{1cm}$ *(i) $x \geq t$ for all $x \in S$ and*

$\hspace{1cm}$ *(ii) For each real number $\varepsilon > 0$, there exists a real number $x \in S$ such that $x < t + \varepsilon$.*

Proof. $\hspace{1cm}$ Proof is similar as theorem 3.

Solved Examples

Example 1. Show that (i) The set R^+ of positive real numbers, is bounded below and unbounded above.

$\hspace{1cm}$ (ii) The set R^- of negative real numbers, is bounded above and unbounded below.

Solution. $\hspace{1cm}$ (i) Since every member of $R^- \cup [0]$ is lower bound of R^+, therefore R^+ is bounded below.

$\hspace{2cm}$ To prove R^+ is unbounded above.

Based on the following Results

- If $l \leq s \leq u \ \forall \ s \in S$. Then S is said to be bounded below by l and bounded above by u. Also, l and u are respectively called lower and upper bound of S.

- least upper bound = supremum; greatest lower bound = infimum.

- supremum and infimum of a bounded subset of R are unique.

- If supremum of a set belongs to the set, then supremum is the largest element of the given set.

- infimum $(S) \leq$ supremum (S).

- A set is singlton $\Leftrightarrow$ supremum = infimum.

Let if possible, suppose u is an upper bound of R^+ we have $u \geq 1$ for $1 \in R^+$. Since $2 \in R^+$, $2 > 0$ and so $u \geq 1$, $2 > 0$ gives $u + 2 > 1+0$ i.e., $u + 1>0$. Thus $(u + 1) \in R^+$ and $(u + 1) > u$ which is a contradiction, that u is an upper bound of R^+.

Hence, R^+ is unbounded above.

(ii) Proof follows in a similar manner.

Example 2. *Show that the set of real numbers, R is an unbounded set.*

Solution. From example 1, we conclude that the set R^+ is unbounded above and R^- is always unbounded below.

Also $R = R^- \cup [0] \cup R^+$

$\Rightarrow$ R is not bounded.

Example 3. *Show that the null set ϕ is neither bounded below nor above, nor unbounded.*

Solution. Since, there is no member in ϕ, we cannot check whether a given real number can be a bound for ϕ or not. Thus, bounds for ϕ do not exists. On the other hand we can as well say that every real number is a lower or upper bound for there is no member in ϕ which does not satisfy the required property of bounds.

Example 4. *Show that every non-empty finite subset of R is bounded.*

Solution. Let S be a non-empty finite subset of R.

$\Rightarrow S$ contains a finite number of elements. Then by the properties of the ordered relation in R, out of these elements, one element $s \in S$, shall be the smallest element of S and another element $b \in S$, shall be the greatest element of S

$\Rightarrow$ $a \leq x \leq b \ \forall x \in S.$

Hence, S is always bounded.

Example 5. *Find the supremum and infimum of the set $S = [x \in Z : x^2 \leq 25]$.*

Solution. Since $S = [x \in Z : x^2 \leq 25]$

$= [- 5, - 4, - 3, - 2, - 1, 0, 1, 2, 3, 4, 5].$

Since S is a finite subset of R, the smallest member of S is $- 5$, which is a lower bound of S, and hence infimum of S is $- 5$. Similarly 5 is the supremum of S.

Example 6. *Find the supremum and infimum, if they exist, of the following sets*

(i) $\left[\dfrac{1}{n} : n \in N \right]$ (ii) $\left[x \in Q : x = \dfrac{n}{n+1}, \ n \in N \right]$

(iii) $\left[1 + \dfrac{(-1)^n}{n} : n \in N \right]$ (iv) $\left[\pi + \dfrac{1}{2}, \ \pi + \dfrac{1}{4}, \ \pi + \dfrac{1}{8}, ... \right]$

Solution. (i) Here we have

$$S = \left[\dfrac{1}{n} : n \in N \right] = \left[1, \dfrac{1}{2}, \dfrac{1}{3}, ... \right].$$

The set S is bounded above by 1, also any member less than 1 is not an upper bound of S, therefore sup $S = 1$.

Also, 0 is a lower bound of S, because $x \geq 0$, $\forall \ x \in S$. Let l be any arbitrary positive small number, then there exists $n \in N$ such that $\dfrac{1}{n} < l$, which shows that l is not an upper bound of S. Thus 0 is a lower bound of S and no other positive real number is a lower bound of S. Therefore, infimum of $S=0 \notin S.$

(ii) Let $S = \left[\dfrac{n}{n+1} : n \in N\right] = \left[\dfrac{1}{2}, \dfrac{2}{3}, \dfrac{3}{4}, ...\right]$

Then, the set S is bounded below by $\dfrac{1}{2}$ and any number greater than $\dfrac{1}{2}$ can not be a lower bound of S, therefore infimum of $S = \dfrac{1}{2}$.

Also, $\left(\dfrac{n}{n+1}\right) < 1, \forall n \in N$, therefore 1 is an upper bound of S, and any number less than 1 not be a upper bound of S.

Therefore supremum of $S = 1$.

(iii) Let $S = \left[1 + \dfrac{(-1)^n}{n} : n \in N\right] = \left[0, \dfrac{3}{2}, \dfrac{2}{3}, \dfrac{5}{4}, \dfrac{4}{5}, \dfrac{7}{6}, \dfrac{6}{7}, ...\right]$

$= \left[\dfrac{0}{1}, \dfrac{2}{3}, \dfrac{4}{5}, \dfrac{6}{7}, \dfrac{8}{9}, ..., \dfrac{2n-2}{2n-1}\right] \cup \left[\dfrac{3}{2}, \dfrac{5}{4}, \dfrac{7}{6}, \dfrac{9}{8}, ..., \dfrac{2n+1}{2n}\right]$.

Here, the proper fraction $\dfrac{0}{1}, \dfrac{2}{3}, \dfrac{4}{5}, \dfrac{6}{7}, ...$ are increasing and tending to 1, and the improper fractions begin with $\dfrac{3}{2}$ are decreasing and tending to 1.

Therefore, infimum of $S = 0$ and supremum of $S = \dfrac{3}{2}$.

(iv) Let $S = \left[\pi + \dfrac{1}{2}, \pi + \dfrac{1}{4}, \pi + \dfrac{1}{8}, ...\right]$.

Here, we have $x \le \pi + \dfrac{1}{2} \forall x \in S$

$\Rightarrow \quad \pi + \dfrac{1}{2}$ is an upper bound for S.

Since, $\pi + \dfrac{1}{2} \in S$, therefore no real number less than $\pi + \dfrac{1}{2}$ can be upper bound for S. Thus, $\pi + \dfrac{1}{2}$ is the least upper bound. Therefore, supremum of $S = \pi + \dfrac{1}{2}$.

Similarly, we can show that π is the infimum of S.

EXERCISE 1.2

1. Find the supremum and infimum, if exist of the following sets.

(i) $\left[\dfrac{1}{5n} : n \in Z, n \ne 0\right]$

(ii) $\left[x : x = 1 + \dfrac{1}{n} : n \in N\right]$

(iii) $[x \in R : x = 2^n : n \in N]$

(iv) $[3, 8, 14, 20]$

(v) $\left[-\dfrac{1}{n} : n \in N\right]$

(vi) $\left[m + \dfrac{1}{n} : m, n \in N\right]$

(vii) $\left[x \in Q : x = \dfrac{(-1)^n}{n} : n \in N\right]$

(viii) $S = \left[\left(1 - \dfrac{1}{n}\right)\sin\dfrac{n\pi}{2} : n \in N\right]$

(ix) $\left[x = (-1)^n\left(\dfrac{1}{4} - \dfrac{4}{n}\right) : n \in N\right]$

2. Find the supremum and infimum of the set

$$S = \left[\dfrac{2n+1}{3n+3} : n \in N\right].$$

3. Prove that every subset of a bounded above (below or both) set is bounded above (below or both).

4. If A and B are subsets of R, then prove that the set $A + B = [x + y : x \in A, y \in B]$ is also bounded and

$$\inf (A + B) = \inf (A) + \inf (B).$$

5. If $A \neq \phi$ is bounded below and $-A$ denotes the set of all $-x$ for which $x \in A$, then prove that $-A \neq \phi$, that $-A$ is bounded above and that $-\sup(-A) = \inf (A)$.

6. If $A \neq \phi$, $B \neq \phi$ and $x \leq y \; \forall \; x \in A$ and $y \in B$, then show that

   (i) $\sup A \leq y \; \forall \; y \in B$

   (ii) $\sup (A) \leq \inf (B)$.

7. If $A \subseteq B$ and B is bounded, then show that

$$\sup B \geq \sup A \geq \inf A \geq \inf B.$$

8. If A and B are two bounded subset of R, then $A \cup B$ and $A \cap B \; (\neq \phi)$ are also bounded and

   (i) $\sup (A \cup B) = \max (\sup A, \sup B)$

   (ii) $\inf (A \cup B) = \min (\inf A, \inf B)$

   (iii) $\sup (A \cap B) = \min(\sup A, \sup B)$

   (iv) $\inf (A \cap B) = \max (\inf A, \inf B)$.

9. Give an example of a set in which supremum is equal to infimum.

10. Show that the set $[x : x \in Q, x > 0$ and $x^2 < 3]$ does not have any supremum in Q.

11. For a real number λ and a subset A of R, let λA be the set defined by $\lambda A = [\lambda x : x \in A]$. Prove that if A is bounded, then λA is also bounded and

$$\inf (\lambda A) = \begin{cases} \lambda \inf (A) & \text{if } \lambda \geq 0 \\ \lambda \sup (A) & \text{if } \lambda \leq 0. \end{cases}$$

12. Give an example of a set which is

    (i) bounded above but not below.

    (ii) bounded below but not above.

    (iii) neither bounded above nor bounded below.

    (iv) both bounded above and below.

13. Find the supremum and infinium of the sets

    (i) $S = [x \in Z : x^2 \leq 16]$

    (ii) $S = \left\{2 + \dfrac{1}{n} : n \in N\right\}$

14. Check the boundedness of the following sets:

    (i) $[-1, -2, -3, ...]$

    (ii) $[1, 2, 3, 4, 5, ...]$

    (iii) $[2, 2^2, 2^3, ..., 2^n, ...]$

    (iv) $\left[1, \dfrac{1}{4}, \left(\dfrac{1}{4}\right)^2, \left(\dfrac{1}{4}\right)^3, ..., \left(\dfrac{1}{4}\right)^n, ...\right]$

    (v) $\left[x : x = (-1)^n\dfrac{1}{n} : n \in N\right]$

    (vi) $[x : x = (-2)^n : n \in N]$

Answers

1. (i) $\inf = -\dfrac{1}{5}$, $\sup = \dfrac{1}{5}$ (ii) $\sup = 2$, $\inf = 1$ (iii) $\inf = 2$, sup, does not exist (iv) $\inf = 3$, $\sup = 20$

   (v) $\sup = 0$, $\inf = -1$ (vi) $\inf = 1$, sup, does not exist (vii) $\inf = -1$, $\sup = 1/2$

   (viii) $\inf = -1$, $\sup = 1$ (ix) $\sup = 15/4$, $\inf = -7/4$

2. $\sup = \dfrac{2}{3}$, $\inf = \dfrac{3}{5}$

9. Singlton set 12. (i) $[-1, -2, -3, ...]$ (ii) N (iii) Z, Q or R (iv) Any finite set.

14. (i) Bounded above but not below   (ii) Bounded below but not above

    (iii) Bounded below but not above   (iv) Bounded below and above, both

    (v) Bounded below and above both   (vi) Neither bounded below nor above.

1.7 COMPLETENESS

1.7.1 COMPLETENESS AXIOMS IN R

Every non-empty subset, which is bounded above has a supremum in R.

In other words, the set of upper bounds of a non-empty subset, which is bounded above has a least member.

REMARKS

- This axioms is also known aa continuity axioms in R. If S is a set bounded below, then by, considering the set. $T = [x : -x \in S]$, we shall state the completeness axioms in the following alternative form. "Every non- empty set of real numbers which is bounded below has an infimum in R".
- Another equivalent form of completeness axioms in R is Dedekind's property, discussed later on.

1.7.2 COMPLETE ORDERED FIELD

The ordered field which satisfying completeness axioms is called complete ordered field.

For Example :

The field of real numbers R, is a complete ordered field while the field of rational numbers Q is an ordered field but not complete.

THEOREM 1. *Any non-empty subset of real numbers which is bounded below has an infimum.*

Proof. Let S be a non-empty subset of R and S be bounded below. To show S has an infimum.

Let us define a set

$$T = [y : y = -x, x \in S].$$

Now, we shall show that T is bounded above.

Since S is bounded below, let l be the lower bound of S so $x \geq l \ \forall x \in S$.

Now, $\qquad\qquad l \leq x \Rightarrow -x \leq -l \Rightarrow y < -l \ \forall y \in T$

$\Rightarrow$ $-l$ is an upper bound of T.

$\Rightarrow$ T is bounded above.

Therefore, by the completeness axioms T has the supremum, say t.

Now, we shall show that $-t$ is the infimum of S,

Since $\qquad\qquad t = \sup T$

$\Rightarrow$ $y \leq t \ \forall \ y \in T \Rightarrow -x \leq t \ \ \forall x \in S$ or $x > -t \ \forall x \in S$.

Therefore, $-t$ is a lower bound of S. Now, if l_1 is a lower bound of S, then $-l_1$ is an upper bound of T. Therefore,

$$t \leq -l_1 \quad \text{or} \quad -t \geq l_1$$

Hence $\qquad\qquad -t = \inf S$.

THEOREM 2. *The set Q is not a complete ordered field.*

Proof. To prove this theorem, we shall show that there exists a non-empty subset of Q which is bounded above but has no rational number as its $l.u.b.$

Let us consider a subset S of Q, such that

$$S = \{\alpha : \alpha \in Q^+, 0 < \alpha^2 < 2\}.$$

$\therefore \qquad\qquad 1 \in S \Rightarrow S \neq \phi.$

Also, S is bounded above by 2. Now, we shall show that no rational number is the l.u.b of S.

Let x be any rational number. Then, there are following cases arise :

Case (i) If $x \leq 0$.

Since every element of S is greater than or equal to zero.

$\therefore$ x is not an upper bound of $S \Rightarrow x$ is not the l.u.b of S.

Case (ii) if $x > 0$ and $0 < x^2 < 2$.

Let $$y = \frac{3x+4}{2x+3}.$$ ....(1)

$\therefore$ $x \in Q^+ \Rightarrow y \in Q^+$.

Also, on simplification

$$y^2 - 2 = \frac{x^2 - 2}{(2x+3)^2}$$ ...(2)

and $$y - x = \frac{2(2 - x^2)}{2x + 3}$$ ...(3)

Now, since $x^2 < 2$ therefore $y^2 < 2 \Rightarrow y \in S$.

But from (3), $y - x > 0 \Rightarrow y > x$

$\Rightarrow$ y is an element of S which is larger than x.

$\Rightarrow$ x is not the l.u.b. of S.

Case (iii) If $x > 0$ and $x^2 = 2$.

This case is not possible, because $x^2 = 2 \Rightarrow x = \sqrt{2}$, which is not rational.

Case (iv) If $x > 0$ and $x^2 > 2$.

Define y as in case (ii).

From (2) $y^2 > 2$ $(\because x^2 > 2)$

From (3) $y - x < 0 \Rightarrow y < x$.

Let z be any element of S

$$0 < z^2 < 2 \Rightarrow 0 < z^2 < y^2 \text{ and } y < x$$

$\Rightarrow$ $0 < z < y < x$

$\Rightarrow$ x and y are both upper bounds of $S \Rightarrow x$ is not the l.u.b. of S.

Hence, in all cases, we have that x cannot be l.u.b. of S. Since, x is any arbitrary rational number, therefore no rational number is the l.u.b. of S. Hence, Q is not a complete ordered field.

THEOREM 3. *The set of natural numbers N is not bounded above.*

Proof.   Let if possible, N be bounded above.

Since $1 \in$ N, N $\neq \phi$, thus N is a non-empty subset of R which is bounded above, therefore by order completeness axiomg N must have supremum say s.

Then, $n \leq s \; \forall \, n \in$ N

$\Rightarrow$ $n + 1 < s \; \forall \, n \in$ N $\Rightarrow n < s{-}1 \; \forall \, n \in$ N

$\Rightarrow s - 1\ (< s)$ is an upper bound of N.

Thus, we have find an upper bound of N, less than the supremum of N, which is a contradiction. Hence, N is not bounded above.

1.8 ARCHIMEDEAN PROPERTY OF REAL NUMBERS

THEOREM 1. *If $x > 0$, then for any $y \in$ R there exists $n \in$ N such that $nx > y$.*

Proof. When $y \le 0$, then theorem is evident. For $y > 0$. Let if possible theorem is not true

$$\Rightarrow \qquad nx \le y, \quad \forall\, n \in N$$

Thus $S = [nx : n \in N\ ]$ is a non-empty set which is bounded above ($\because nx \le y$). Therefore, by completeness axioms, S has the supremum. Let $\alpha = \sup S$. Then $nx \le \alpha\ \forall\, n \in N$

$$\Rightarrow \qquad (n + 1)\,x \le \alpha, \quad \forall\, n \in N$$

$$\Rightarrow \qquad nx \le \alpha - x, \quad \forall\, n \in N.$$

Thus, $\alpha - x$ is also an upper bound of S. But $\alpha - x < \alpha$. This contradicts the fact that $\alpha = \sup S$. Hence, the assumption $nx \le y$, $\forall\, n \in N$ is false and so the statement of the theorem is true.

REMARKS

- For any $x \in$ R, $\exists\, n \in$ N such that $n > x$ (it follows from the above theorem, on replacing y by x and taking 1 for x).
- For any $x \in$ R$^+$, $\exists\, n \in$ N such that $x > \dfrac{1}{n}$.
- For any $x \in$ R, there exists $n, m \in$ Z such that $n > x > m$.
- For any $x \in$ R, there exists $n \in$ Z such that $x \ge n > x - 1$.

1.8.1 ARCHIMEDIAN ORDERED FIELD

An ordered field F is said to be Archimedian ordered field if, $\forall\, x, y \in F,\ y > 0$ there exist some $n \in$ N such that $ny > x$.

1.9 REPRESENTATION OF REAL NUMBERS AS POINTS ON A STRAIGHT LINE

Consider any straight line. Mark a point O on it. The point O divide the straight line into two parts. The part that lies to the right of O, is called positive part and the part that lies to the left of O, is called negative part. Take any point A on the positive part. Let us represent the numbers 0 and 1 by the points O and A respectively.

With OA as unit, we can associate with each real number exactly one point on the line, the positive real numbers being represented by points to the right of O and the negative real numbers being represented by points to the left of O. Also, each point on the line will be associated with one and only one real number. This representation works out to be an important tool in Mathematics. In fact, it serves as a basis of co-ordinate geometry. Due to this representation, we shall frequently talk of a real number as a point on R.

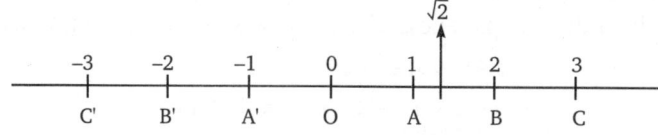

1.9.1 DEDIKIND PROPERTY

If L and U are two subsets of R such that

(i) $L \neq \phi$, $U \neq \phi$

(ii) $L \cup U = R$

(iii) Every member of L is less than every member of U *i.e.*, if $x \in L$ and $y \in U$ then $x < y$.

Then, either L has the greatest number or U has the least member.

1.9.2 DEDIKIND-CANTOR AXIOMS

To every real number, there corresponds a unique point on a directed line and conversely, to every point on a directed line there correspond a unique real number.

REMARKS

- There is a one-to-one correspondance between the real numbers and the points on a directed line. For this reason, the directed line is called the real line or real axis and a real number is called a point of the real line.
- If $a, b \in R$, $a < b$, then the point a lies to the left of the point b.
- The negative numbers lie to the left of 0 and the positive number, lies to the right of 0.
- If $a, b \in R$, then $|a - b|$ is called the distance between the point a and b.

THEOREM 1. *Dedikind's property is equivalent to completeness axioms.*

Proof. By (iii) condition of Dedikind property, the non-empty subset L is bounded above. If L is the greatest member then Dedikind's property is satisfied. If L has no greatest member then by completeness axioms in R, U is the set of upper bounds of L has a least member. Then, Dedikind's property is satisfied. Now, we shall show that the Dedikind's property implies the completeness axioms.

Let S be a non-empty set which is bounded above and let the sets L and U be defined as

$$L = (x : x \text{ is not an upper bound of } S)$$
$$U = (y : y \text{ is an upper bound of } S).$$

Obviously, L and U are non-empty, disjoint and $L \cup U = R$.

Then, by Dedikind's property, either L has the greatest member or U has a least member. If possible, let L have the greatest member say α, then

$$\alpha \in L \Rightarrow \alpha \notin U \Rightarrow \exists \, \beta \in S \text{ such that } \alpha < \beta.$$

Now, $\qquad \alpha < \dfrac{\alpha + \beta}{2} \in U$

$\Rightarrow \dfrac{\alpha + \beta}{2}$ is an upper bound of S.

Also, $\qquad \dfrac{\alpha + \beta}{2} < \beta \in S$

$\Rightarrow \dfrac{\alpha + \beta}{2}$ is not an upper bound of S.

This contradiction implies that L has no greatest member. Therefore, U has the least member. Hence, the set of upper bounds of a non-empty set S which is bounded above has a least member, which is the completeness axiom in R.

$\Rightarrow$ Dedikind's property is equivalent to completeness axioms in R.

THEOREM 2. *Between any two distinct real numbers there always exists infinitely many rational numbers.*

Proof. Let a and b be the two distinct real numbers such that $a < b$

$$\Rightarrow \quad\quad\quad b - a > 0.$$

Then, by Archemedean's property $\exists\, n \in N$ such that

$$n\,(b - a) > 1 \Rightarrow nb > na + 1.$$

Also, for $na \in R$ $\exists$ a unique integer m such that $m > na \geq m - 1$.

Thus, $\quad\quad\quad\quad nb > na + 1 \geq m > na$

which gives $b > \dfrac{m}{n} > a$, where $n \in N$ and $m \in Z$

$\Rightarrow \dfrac{m}{n}$ is a rational number, which lies between two distinct real numbers a and b and is different from a and b. Repeating the argument with b and $m\,/\,n$, we get another rational number say r_1, such that $b > r_1 > m/n > a$. The argument can be continued with $b > r_1$ and so on. Hence, there exist infinitely many rational numbers between any two distinct real numbers.

THEOREM 3. *Between any two distinct real numbers, there always exist infinitely many irrational numbers.*

Proof. Let us suppose a and b be any two distinct real numbers such that $a < b$

$$\Rightarrow \quad\quad\quad b - a > 0$$

Let α be any positive irrational number, then by Archemedean property of real numbers, there exists a positive integer n such that

$$n(b - a) > \alpha \Rightarrow b > a + \frac{\alpha}{n} \quad\quad\quad ...(1)$$

Now, since α is positive, we have

$$a < a + \frac{\alpha}{2n} < a + \frac{\alpha}{n} \quad\quad\quad ...(2)$$

From (1) and (2), we have

$$a < a + \frac{\alpha}{2n} < a + \frac{\alpha}{n} < b.$$

Now $\left(a + \dfrac{\alpha}{n}\right) - \left(a + \dfrac{\alpha}{2n}\right) = \dfrac{\alpha}{2n}$ is an irrational number. Therefore, $\left(a + \dfrac{\alpha}{n}\right)$ and $\left(a + \dfrac{\alpha}{2n}\right)$ cannot be rational number. Let one of them say r be an irrational number. Thus, we have an irrational number r such that $a < r < b$.

Repeating the above argument for a and r and r and b, we get irrational numbers say r_1 and r_2, such that $a < r_1 < r$ and $r < r_2 < b$

$$\Rightarrow \quad\quad\quad a < r_1 < r < r_2 < b.$$

Continuing this process, we get infinitely many irrational numbers between two distinct real numbers a and b.

REMARK

- From theorem 2, we find that between any two distinct rational numbers, there always exist infinite rational numbers. Similarly, theorem 3 gives that between any two distinct irrational numbers, there exist infinite irrationals. Due to these properties the sets of rational and irrationals are said to be dense everywhere in R. Hence, combining the results of theorems (2) and (3), we get **"Denseness property of R"** i.e., *"Between any two distinct real numbers there always exists infinite reals, both rationals and irrationals."*

EXERCISE 1.3

1. Prove that following are equivalent :
   (i) Let R, the set of real numbers be divided into two subsets L and U such that
   (a) $L \neq \phi, U \neq \phi$
   (b) $L \cup U = R$
   (c) $x \in L, y \in U \Rightarrow x < y$

   Then, either L has the greatest member or U has the least member.

   (ii) Every non-empty bounded below subset of R has the greatest lower bound.

2. Assuming as order-completeness axiom for real numbers that every non-empty subset of real numbers which is bounded below has an infimum, prove that every non-empty subset of real numbers which is bounded above has a supremum.

3. Show that the system Q of rational numbers is an Archimedian ordered field.

1.10 NEIGHBOURHOOD OF A POINT

Definition-1 : *A subset N of* R *is said to be a neighbourhood of a point* $p \in$ R. *If there exists a real, however small number* $\varepsilon > 0$ *such that*
$$p \in \,]\,p - \varepsilon, p + \varepsilon\,[\, \subset N.$$
Definition-2 : *A subset N of* R *is called a neighbourhood of a point* $p \in$ R *if* $\exists$ *an open interval I such that*
$$p \in I \subset N.$$

1.10.1 NEIGHBOURHOOD OF A SET

A set T is called a neighbourhood of a set S if T is a neighbourhood of each point of S.

1.10.2 GEOMETRICAL INTERPRETATION OF $\in$-NBD

Any ε-nbd viz. $]\,p - \varepsilon, p + \varepsilon[$, (for $\varepsilon > 0$) of 'p' is the set of all those points which are within an ε distance from p on either side of p.

REMARKS

- For brevity in writing, now onwards, the word 'neighbourhood' is shortened to 'nbd'.
- The open interval $]\,p - \varepsilon, p + \varepsilon[$, $\varepsilon > 0$ is sometimes referred to as ε-nbd of x or simply as nbd of x. An open interval containing a point p is often denoted by I_p. This is also, used for a nbd of p.

1.10.3 DELETED NEIGHBOURHOOD OF p

In case the point p is specially excluded from a nbd N_p of p, then such a nbd is called a deleted nbd of p. Thus, if N is a nbd of p, then the set $N - \{p\}$ is said to be deleted nbd of p.

REMARK

- A symmetric deleted nbd of p is the union of two bounded intervals $]\,p - \varepsilon, p\,[$ and $]\,p, p + \varepsilon\,[$.

☞ ILLUSTRATIONS

(1) *An open interval is a nbd of each of its points.*

Let $]\,a, b[$ be an open interval and let p be any arbitrary point of interval $]a, b[$. To show $]a, b[$ is a nbd of p.

If we take ε as the minimum of two positive numbers $p - a$ and $b - p$ then obviously $\varepsilon > 0$, and
$$p \in \,]\,p - \varepsilon, p + \varepsilon\,[\, \subseteq \,]\,a, b[$$
$\Rightarrow \quad]a, b[$ is a nbd of p.

Since, p is arbitrary point of $]a, b[$ therefore interval $]a, b[$ is a nbd of each of its points.

(2) *The set Q of rational numbers is not the nbd of any of its point.*

Let $p \in Q$. Therefore for any arbitrary positive real number ε, $p - \varepsilon$ and $p + \varepsilon$ are two distinct real numbers contains infinite irrational numbers between them, which are not the members of Q. Therefore $]p - \varepsilon, p + \varepsilon [\not\subset Q \ \forall \varepsilon > 0$

$\Rightarrow$ Q is not a nbd of p.

(3) *A non-empty finite set cannot be a nbd of any of its points.*

Since, a nbd of a point p contains an open interval, containing p. Also, an interval contains infinite points, therefore, for a nbd of a point, the set must contains infinite points. Hence, a non-empty finite set cannot be a nbd of any of its points.

(4) *A closed interval $[a, b]$ is a nbd of each of its points except the end points.*

Since, we know that *"an open interval is a nbd of each of its points"* therefore,

$$x \in]a, b[\Rightarrow]a, b[\text{ is a nbd of } x.$$

Also, $]a, b[\subset [a, b] \Rightarrow [a, b]$ is a nbd of each point of $]a, b[$.

Now, it remains to show that $[a, b]$ is not a nbd of a and b. Take a positive real number ε. Then $a \in]a - \varepsilon, a + \varepsilon [\not\subset [a, b]$

$\Rightarrow$ There is no $\varepsilon > 0$ such that $]a - \varepsilon, a + \varepsilon [\subset [a, b]$

$\Rightarrow [a, b]$ is not a nbd of a.

Similarly we can show that $[a, b]$ is not a nbd of b.

(5) *The null set ϕ is trivally, a nbd of each of its points.*

The null set ϕ is a nbd of each of its points, because there is no point in ϕ, therefore we can say there is no point in ϕ which is not a nbd.

(6) *The set Z^+ of all positive integers or a set of natural numbers is not a nbd of any of its points.*

Use the argument of (2).

(7) *A non-empty subset N of R is a nbd of $p \in R$ iff there exists a positive integer n such that*

$$\left] p - \frac{1}{n}, p + \frac{1}{n} \right[\subset N.$$

Let us first suppose a non-empty subset N of R be a nbd of point $p \in R$, then there exists a positive number $\varepsilon > 0$ such that $p \in]p - \varepsilon, p + \varepsilon [\subset N$.

Now, for a given positive real number ε, we can choose a positive integer n such that $\frac{1}{n} < \varepsilon$

Now, $\qquad\qquad \frac{1}{n} < \varepsilon \Rightarrow p + \frac{1}{n} < p + \varepsilon.$

Also $\qquad\qquad \frac{1}{n} < \varepsilon \Rightarrow -\frac{1}{n} > -\varepsilon \Rightarrow p - \frac{1}{n} > p - \varepsilon.$

Therefore, $\left] p - \frac{1}{n}, p + \frac{1}{n} \right[\subset]p - \varepsilon, p + \varepsilon [$

If N is a nbd of p, then there exists a positive integer n such that

$$\left] p - \frac{1}{n}, p + \frac{1}{n} \right[\subset N.$$

Conversely, let, there exists a positive integer n such that

$$\left] p - \frac{1}{n}, p + \frac{1}{n} \right[\subset N$$

$\Rightarrow \left] p - \frac{1}{n}, p + \frac{1}{n} \right[$ is an open interval containing p and contained in N.

$\Rightarrow N$ is a nbd of p.

(8) *The intersection of two nbds of a point is also a nbd of that point.*

Let N_1 and N_2 be two nbds of a point p. Then, by definition, there exist $\varepsilon_1 > 0$ and $\varepsilon_2 > 0$ such that

$$] p - \varepsilon_1, p + \varepsilon_1 [\subset N_1 \qquad \qquad ...(1)$$

and $\qquad] p - \varepsilon_2, p + \varepsilon_2 [\subset N_2 \qquad \qquad ...(2)$

Take, $\varepsilon = \min\{\varepsilon_1, \varepsilon_2\}$, then from (1) and (2), we have

$$] p - \varepsilon, p + \varepsilon [\subset N_1$$

and $\qquad] p - \varepsilon, p + \varepsilon [\subset N_2$

$\Rightarrow \qquad] p - \varepsilon, p + \varepsilon [\subset N_1 \cap N_2$

$\Rightarrow \qquad N_1 \cap N_2$ is a nbd of p.

(9) *Any superset of a nbd of a point is also nbd of that point.*

Let us suppose N be a nbd of a point $p \in R$ and M is the super set of N i.e., $N \subset M$.

To show M is also a nbd of p.

Since, N is a nbd of p. By definition there exists $\varepsilon > 0$ such that

$$p \in] p - \varepsilon, p + \varepsilon [\subset N.$$

Also $\qquad \qquad \qquad N \subset M$

$\Rightarrow \qquad \qquad p \in] p - \varepsilon, p + \varepsilon [\subset N \subset M$

$\Rightarrow \qquad \qquad p \in] p - \varepsilon, p + \varepsilon [\subset M$

$\Rightarrow M$ is a *nbd of p.*

(10) *On the real line R, for each point $p \in R$, there exists at least one nbd of p.*

If $p \in R$, then for each $\varepsilon > 0$, we have

$$p \in] p - \varepsilon, p + \varepsilon [\subset R.$$

$\Rightarrow R$ is always a nbd of p.

(11) *On the real line R, for each point $p \in R$ and each nbd N of p there exists a nbd M of p such that $M \subset N$ and M is a nbd of each of its points.*

Since, N is a nbd of p, therefore $\exists \varepsilon > 0$ such that

$$p \in] p - \varepsilon, p + \varepsilon [\subset N$$

Take $\qquad \qquad M =] p - \varepsilon, p + \varepsilon [.$

Then, $p \cup] p - \varepsilon, p + \varepsilon [$ is an open interval containing p, so it is a nbd of p and also a nbd of each of its points.

$\Rightarrow \exists$ a nbd M of p such that $M \subset N$ and M is a nbd of each of its points.

Solved Examples

Example 1. *Show that [1, 3] is a nbd of 2 but not of 1.*

Solution. Since $2 \in]1, 3[$ and $]1, 3[\subseteq [1, 3]$, therefore [1, 3] is a nbd of 2.

Also, there exists an open interval $]1, 3[$ such that $1 \notin]1, 3[$ and $]1, 3[\subset [1, 3]$

$\Rightarrow$ [1, 3] is not a nbd of 1.

Example 2. *Which of the following subsets of R are nbd of 3?*

(i) $]2, 4[$ (ii) $]2, 4]$

(iii) $[2, 4[$ (iv) $[2, 4]$

(v) $]3, 8[$ (vi) $]3, 8]$

(vii) $[3, 8[$ (viii) $[3, 8]$

(ix) $[2, 5] - \left[4\frac{1}{2}\right]$

Based on the following Results

- Every non-empty subset, which is bounded above has a supremum in R.

- The set of upper bounds of a non-empty subset, which is bounded above has a least member is called a complete set.

- The field of real numbers R, is a complete ordered field while the field of rational numbers Q is ordered but not complete.

- If $x > 0$, then for any $y \in R$, there exist $n \in N$ such that $nx > y$.

- To every real number there correspond a unique point on the directed line and to every point on a directed line, there correspond a unique real number.

- Between any two real numbers, there always exist infinitely many rational and irrational numbers. (Denseness Property)

- The set of rational numbers is an Archmedean ordered field.

Solution.

(i)   since $]2, 4[$ is an open interval and $3 \in]2, 4[$
$\Rightarrow]2, 4[$ is a nbd of 3.

(ii) Since, there exists an open interval $]2, 4[$ such that $3 \in]2, 4[\subset]2, 4]$
$\Rightarrow]2, 4]$ is a nbd of 3.

(iii) Since, there exists an open interval $]2, 4[$ such that $3 \in]2, 4[\subset [2, 4[$
$\Rightarrow [2, 4[$ is a nbd of 3.

(iv) Since, there exists an open interval $]2, 4[$ such that $3 \in]2, 4[\subset [2, 4]$
$\Rightarrow [2, 4]$ is a nbd of 3.

(v) $]3, 8[$ is not a nbd of 3 as $3 \notin]3, 8[$.

(vi) $]3, 8]$ is not a nbd of 3 as $3 \notin]3, 8]$.

(vii) $[3, 8[$ is not a nbd of 3 since $[3, 8[$ does not contain an open interval containing 3.

(viii) $[3, 8]$ is not a nbd of 3 since $[3, 8]$ does not contain an open interval containing 3.

(ix) $[2, 5] - \left[4\frac{1}{2}\right]$ is a nbd of 3 since there exists an open interval $\left]2, 4\frac{1}{2}\right[$ such that $3 \in \left]2, 4\frac{1}{2}\right[\subset [2, 5] - \left[4\frac{1}{2}\right]$.

Example 3. *Give an example of a set which is a nbd of :*

(i) *each of its points,*   (ii) *not, any of its points,*

(iii) *each of its points, except the end points,*   (iv) *each of its points, except one point.*

Solution.   (i)   The open interval $]a, b[$ is a nbd of each of its points.

(ii)   A finite set is not a nbd of any of its points.

(iii)   The closed interval $[a, b]$ is a nbd of each of its points, except the end points.

(iv)   The right half closed interval $]a, b]$ is a nbd of each of its points except b

Example 4.   *Let* $I_n = \left]-\dfrac{1}{n}, 1+\dfrac{1}{n}\right[$ *be an open interval for each* $n \in N$. *Prove that* $\bigcap\limits_{n=1}^{\infty} I_n$ *is*

not a nbd of each of its points.

Solution.   Since $n \in N$

so, $\dfrac{1}{n} \to 0$ as $n \to \infty \Rightarrow -\dfrac{1}{n} \to 0$ as $n \to \infty$ and $1 + \dfrac{1}{n} \to 1$ as $n \to \infty$.

Therefore, we can find that

$$0 \in \left]-\dfrac{1}{n}, 1+\dfrac{1}{n}\right[\quad \forall\, n \in N$$

$$1 \in \left]-\dfrac{1}{n}, 1+\dfrac{1}{n}\right[\quad \forall\, n \in N.$$

Also, each point lying between 0 and 1 is an element of the open interval

$$\left]-\dfrac{1}{n}, 1+\dfrac{1}{n}\right[\quad \forall\, n \in N.$$

But $-\left(\dfrac{1}{n}\right)$, whatever be the value of n, is not an element of the open interval

$$\left]-\dfrac{1}{n}, 1+\dfrac{1}{n}\right[, \quad n \in N.$$

$\Rightarrow$ All numbers less than 0 are not in $\bigcap\limits_{n=1}^{\infty} I_n$.

Similarly, we can show that all numbers greater than 1 are not in $\bigcap\limits_{n=1}^{\infty} I_n$, $n \in N$.

Now, $[0, 1] = \bigcap\limits_{n=1}^{\infty} \left\{\left]-\dfrac{1}{n}, 1+\dfrac{1}{n}\right[\right\} = \bigcap\limits_{n=1}^{\infty} I_n$

and $[0,1]$ being a closed interval is a nbd of each of the points of the interval $[0, 1]$,

except the end points 0 and 1. Hence, $\bigcap\limits_{n=1}^{\infty} I_n$ is not a nbd of each of its points.

1.11   INTERIOR OF A SET

(i) **Interior point of a set.** A point p is called an interior point of a set S if S is a nbd of p i.e., p is called an interior point of S if there exists an open interval I such that $p \in I \subset S$.

(ii) **Interior of a set.**   The set of all interior points of a set S, denoted by S^o, or S^i is called interior of a set.

(iii) **Open sets.** A subset S of R is said to be open if it is a neighbourhood of each of its points.

or a set $S \subset R$ is said to be open if for each $p \in S$, there exist $\varepsilon > 0$ such that
$$]p - \varepsilon, p + \varepsilon [\subset S.$$

REMARKS

- Every open interval $]a, b[$ is an open set.
- The interval $[a, b[$ is not an open set, because it is not a neighbourhood of a.
- The interval $]a, b]$ is not an open set, because it is not a neighbourhood of b.
- The closed interval $[a, b]$ is not an open set, because it is not the nbd of a and b.
- R is an open set, because if p be any point of R, then the open interval $]p-1, p+1[\subset R$ and consequently R is a neighbourhood of p. Since p is arbitrary, therefore, R is a neighbourhood of each of its points.
- The empty set ϕ is an open set, because there is no point at all in ϕ, and consequently, there is no point in ϕ, of which it is not a neighbourhood.
- The open rays $]a, \infty[$ and $]-\infty, a[$ are open sets.
- The closed rays $[a, \infty[$ and $]-\infty, a]$ are not open sets.
- Every point of an open set is an interior point. Therefore for each open set S, $S° = S$. Thus, S is open iff $S°=S$.

THEOREM 1. *Every open interval is an open set.*

Proof. Let p be any point of the given interval $]a, b[$. Therefore, $a < p < b$.

Consider two numbers c and d such that $a < c < p < d < b$

$\Rightarrow \quad p \in]c, d[\subset]a, b[\Rightarrow]a, b[$ is a nbd of p.

Since, p is arbitrary, therefore $]a, b[$ is a nbd of each of its points. Hence, $]a, b[$ is open.

THEOREM 2. *The union of an arbitrary family of open sets is open.*

Proof. Let G be the union of an arbitrary family $[G_\lambda : \lambda \in \Lambda]$ of open sets in R *i.e.*, $G = \cup[G_\lambda : \lambda \in \Lambda]$. To show that G is open.

Let $p \in G$, since G is the union of members of $[G_\lambda : \lambda \in \Lambda]$ therefore, there must exists an open set $H \in G_\lambda$ such that $p \in H \subset G$. Since, H is an open set and $p \in H$, therefore there must exists $\varepsilon > 0$ such that $] p - \varepsilon, p + \varepsilon [\subset H \subset G$. Again, since $] p - \varepsilon, p + \varepsilon [$ is contained in G, therefore, G is a nbd of p. Now, since p is arbitrary, therefore G is a nbd of each of its points.

$\Rightarrow \quad G$ is an open set.

$\Rightarrow \quad$ Arbitrary union of open sets is open.

THEOREM 3. *The intersection of two open sets is open.*

Proof. Let $G_1 \subset R$ and $G_2 \subset R$ be two open sets. To show $G_1 \cap G_2$ is open.

Case I: If $G_1 \cap G_2 = \phi$, then it is open. ($\because \phi$ is an open set.)

Case II: If $G_1 \cap G_2 \neq \phi$.

Let $p \in G_1 \cap G_2 \Rightarrow p \in G_1$ and $p \in G_2$

$\Rightarrow \quad G_1$ and G_2 are nbds of p ($\because G_1$ and G_2 both are open.)

$\Rightarrow \quad G_1 \cap G_2$ is a nbd of p.

Now, since, p is arbitrary, therefore, it follows that $G_1 \cap G_2$ is a nbd of each of its points. Hence, $G_1 \cap G_2$ is open.

REMARKS

- The intersection of an arbitrary family of open sets is not necessarily open. For example, for each $n \in N$, let

$$G_n = \left] a - \frac{1}{n}, a + \frac{1}{n} \right[$$

Then, each G_n is an open set, but $\cap [G_n : n \in N] = \{a\}$ is not open.

- Every open set is a union of open intervals. This follows from the fact, that, if G is open, then to each $x \in G \; \exists$ an open interval G_x such that $x \in G_x \subset G$ and so

$$G = \bigcup_{x \in G} \{x\} \subset \bigcup_{x \in G} G_x \subset G \Rightarrow G = \bigcup_{x \in G} G_x.$$

- Every non-empty open set consists necessarily infinitely many points.
- Interior of any set S is an open set.
- For any set S, $(S°)° = S°$.
- For two sets S and T, $S \subset T \Rightarrow S° \subset T°$.
- Interior of a set S is the largest open set contained in S.

1.12 LIMIT POINTS OF A SET

1.12.1 ADHERENT POINT

A point $p \in R$ is said to be an adherent point of a set $A \subset R$ if every nbd of p contains a point of A.

The set of all adherent point of A is called the adherence or closure of A, denoted by Adh. (A) or $\overline{A}$.

REMARK

- An adherent point is also called closure point.

1.12.2 LIMIT POINT

Definition (1). *A point $p \in R$ is said to be a limit point of the set $A \subset R$ if every nbd of p contains at least one point of A other than p.*

In symbols, the above definition means that a point $p \in R$ is a limit point of $A \subset R$ iff for each nbd N of p

$$(N \cap A) \sim \{p\} \neq \phi.$$

Definition (2). *A point $p \in R$ is said to be a limit point of the set $A \subset R$ if for each $\varepsilon > 0$, the open interval $] p - \varepsilon, p + \varepsilon [$ contains a point of A, other than p.*

REMARKS

- In order to show that a point p is not a limit point of a set A, it is enough to find a neighbourhood N of p, such that either $N \cap A = \{p\}$ or $N \cap A = \phi$.
- The limit point is also known as accumulation point, limiting point, cluster point or a point of condensation.
- The limiting point of the set A may or may not belong to the set A.
- A finite set cannot have a limit point.
- A set may have no limit point, or a finite number of limit points or infinite number of limit points.

- If A is a non-empty set bounded above and if supremum, say x, does not belong to A, then l is a limit point of A if, for $\varepsilon > 0 \; \exists \, y \in A : l - \varepsilon < y < l$

  $\Rightarrow$ nbd $]l - \varepsilon, l + \varepsilon[$ of l contains a point y of A, other than l,

  If A be a non-empty set bounded below and if its infimum does not belongs to A, then it is a limit point of A.

- A point p is not a limit point of a set A, iff there is a nbd N of p such that $N \cap A$ is finite.

- A point p is not a limit point of a set A, then it is not a limit point of any of its subsets.

- If a point p is a limit point of a set A, then it is a limit point of every subset of A.

- Adh. $A = A \cup D\,(A)$.

- Since $]a, b[\cap \phi = \phi \; \forall \, a, b$ therefore ϕ has no limit point.

- The only limit points of $]a, b[, \;]a, b], \; [a, b[$ or $[a, b]$ are points of $[a, b]$ and every point of R is a limit point of R.

- Every limit point of A is also an adherent point of A, but converse is not always true. For example, 1 is an adherent point of the set $\left[\dfrac{1}{n} : n \in N\right]$ but it is not the limit point of the set.

1.12.3 DERIVED SET

The set of all limit points of a set $A \subset R$ is said to be the derived set and it is denoted by $D(A)$ or A'. Therefore,

$$A' = D(A) = [\,p : p \text{ is a limit point of } A\,]$$

The derived set of A', i. e., $(A')'$, be denote by A'' and so on.

1.12.4 ISOLATED POINT

A point $p \in R$ is said to be an isolated paint of A, if it is not the limit point of A i.e., if there exist a nbd of p which contains no point of A, other than p.

1.12.5 DISCRETE SET

A set A is called discrete, if all its points are isolated points.

REMARK

- Each point of a set A is either an isolated point of A or a limit point of A.

1.12.6 DENSE-IN-ITSELF SET

A subset A of R is said to be dense-in-itself if it possesses no isolated points i.e., every point of A is the limit point of A.

1.12.7 CLOSED SET

Definition (1). *A set A is said to be closed if it contains all its limit points.*

Definition (2). *A set A is said to be closed if its compliment is open.*

For Example :

(1) A finite set is always closed.

(2) The null set ϕ is closed.

(3) The set N, Z and R are closed sets.

(4) The set Q is not a closed set.

(5) The singleton set is always a closed set.

1.12.8 PERFECT SET

A set A is said to be perfect if it is dense-in-itself and contains all its limit points.

or

A set A is said to be perfect if it is dense-in-itself and closed.

REMARK

- A set A is said to be perfect, if it is identical with its derived set.

1.13 CLOSURE OF A SET

The smallest closed set containing A , is called closure of A and denoted by $\overline{A}$.

1.13.1 DENSE SET

(i) A set A is said to be dense or everywhere dense in R if $\overline{A} = R$.

(ii) A set A is said to be no where dense in R if interior of the closure of A is empty *i.e.,* $(\overline{A})° = \phi$.

Definition. *A set A is said to be first species if it has only a finite number of derived sets. It is said to be of second species if the number of its derived sets is infinite.*

REMARKS

- If a set A is of first species, then its last derived set must be empty.
- A set whose n^{th} derived set is a finite set (so that its $(n+1)^{th}$ derived set is empty) is called a set of nth order.
- If every sub-interval of an interval I has a part which does not contain any point of A, then A is called no where dense in I.
- If no sub-interval of I is free from points of A, then A is called everywhere dense in I.

THEOREM 1. *Interior of a set is an open set.*

Proof. Let A be any given set and $A°$ is the interior of A.

Case (i) : If $A° = \phi$, then $A°$ is open.

Case (ii) : If $A° \neq \phi$. Let $p \in A°$

$\Rightarrow$ p is an interior point of A. $\Rightarrow$ $\exists$ an open interval I_p such that $p \in I_p \subset A$.

Since I_p is a nbd of each of its point and $I_p \subset A$

$\Rightarrow$ each point of I_p is an interior point of A *i.e.,* $p \in I_p \subset A°$

Hence. $A°$ is an open set.

THEOREM 2. *A point $p \in R$ is a limit point of a set $A \subset R$ if and only if every neighbourhood of p contains infinitely many points of A .*

Proof. Let us first suppose that every nbd of p contains infinitely many points of A

$\Rightarrow$ every nbd of p contains a point of A, which is different from p

$\Rightarrow$ p is a limit point of A.

Conversely, let p be a limit point of A. To show that every nbd of p contains infinitely many points of A.

Let, if possible, there exists a nbd N of p which contains finitely many points of A. Then there exists a positive number $\varepsilon > 0$ such that the open interval $] p - \varepsilon, p + \varepsilon [$

must contains only finitely many points of A. Now, if p is the only such point which contained in $]\,p - \varepsilon, p + \varepsilon\,[$, then p is not a limit point of A. If $]\,p - \varepsilon, p + \varepsilon\,[$ contains points of A, other than p also, then since we assumed them finite in number, let they be $p_1, p_2, ..., p_n$. Out of these n points of A, let p_i be the point which is nearest to p and let $|\,p_i - p\,| = \varepsilon_1$

$\Rightarrow \qquad \varepsilon_1 = \min[\,|\,p_1 - p\,|,$

$|p_2 - p|, ..., |p_n - p|\,]$

Then, the real number ε_1 is such that the open interval $]\,p - \varepsilon_1, p + \varepsilon_1[$ contains no points of A, other than p.

$\Rightarrow \quad$ there exists a nbd of p which contains no point of A, other than p

$\Rightarrow \quad p$ is not a limit point of A.

Which is a contradiction.

Hence, if p is a limit point of A, then every nbd of p contain infinitely many points of A.

SUMMARY OF THE RESULTS	
• $\phi' = \phi$	• $N' = \phi$
• $Q' = R$	• $R' = R$
• $\bar{\phi} = \phi$	• $\bar{N} = N$
• $\bar{Q} = R$	• $\bar{R} = R$
• $(\bar{\phi})^\circ = \phi$	• $(\bar{N})^\circ = \phi$
• $(\bar{Q})^\circ = R$	• $(\bar{R})^\circ = R$

THEOREM 3. *If a non-empty subset A of R which is bounded above and has no maximum member, then its supremum is a limit point of the set A.*

Proof. Let A be a non-empty subset of R, which is bounded above.

Let $\qquad\qquad \sup A = u$

Now since A has no maximum member $\Rightarrow u \notin A$.

Let $\varepsilon > 0$. Since $\sup A = u \Rightarrow u - \varepsilon$ cannot be an upper bound of A

$\Rightarrow \exists\, x \in A$ such that $x > u - \varepsilon$.

Also, $u + \varepsilon > u$ and $u = \sup A$.

$\Rightarrow \quad u + \varepsilon$ is also an upper bound of A and so $x \in A \Rightarrow x < u + \varepsilon$

$\Rightarrow \quad$ For every $\varepsilon > 0\, \exists\, x \in A$ such that $u - \varepsilon < x < u + \varepsilon$.

Since, $u \notin A$, therefore $x \neq u$.

$\Rightarrow$ every nbd $]\,u - \varepsilon, u + \varepsilon\,[$ of u contains a point x of A which is different from u.

Hence, u is a limit point of A.

THEOREM 4. *If a non-empty subset A of R, which is bounded below has no minimum member, then its infimum is a limit point of the set A.*

Proof. Proof is similar as theorem 3.

THEOREM 5. *The finite set has no limit point.*

Proof. Let A be a finite set. To show that every real number p is not a limit point of the set A. Since A is finite. Therefore, for $\varepsilon > 0$, the open interval $]\,p - \varepsilon, p + \varepsilon\,[$ contains only finitely many points of the set A.

$\Rightarrow \quad]\,p - \varepsilon, p + \varepsilon\,[$ is a nbd of p, which do not contain infinitely many points of A.

$\Rightarrow \quad p$ is not a limit point of A.

Since, p is arbitrary. So, every real number p is not a limit point of A.

$\Rightarrow \quad$ finite set has no limit point.

THEOREM 6. **Bolzano-Weierstrass theorem.** *Every infinite bounded set of real numbers has a limit point.*

Proof. Let $S \subset R$ be any infinite bounded set so that there exist its infimum and supremum k_1 and k_2 respectively. Define a set T as follows :

$x \in T$ *iff it exceeds at most a finite number of members of S.*

Now $T \neq \phi$ $(\because k_1 \in T)$

Also, T is bounded above with k_2 as its upper bound, therefore no number greater than k_2 belongs to T.

Therefore, T is non-empty bounded subset of R.

$\Rightarrow$ T has suprernum in R, say $p = \sup T$ (By ordered completeness property)

Now, we shall show that p is a limit point of S

$p = \sup T \Rightarrow \exists q \in T$ such that $q > p - \varepsilon, \varepsilon > 0$

$q \in T$ $\Rightarrow q$ exceeds at most a finite number of members of S

$\Rightarrow p - \varepsilon$ exceeds atmost a finite number of members of S ...(1)

Also, $p = \sup T \Rightarrow p + \varepsilon \notin T$

$\Rightarrow p + \varepsilon$ exceeds an infinite number of members of S ...(2)

Now, (1) and (2) $\Rightarrow \exists$ a nbd of p, i.e., $]\,p - \varepsilon, p + \varepsilon[$, which contains an infinite number of members of S

$\Rightarrow$ p is the limit point of S.

REMARK

- This theorem gives a sufficient condition for an infinite set to have a limit point.

THEOREM 7. $D(\phi) = \phi$.

Proof. We know that, ϕ, the subset of R is a nbd of each of its point. Therefore

$p \in R$ $\Rightarrow R$ is a nbd of p.

Also $R \cap \phi = \phi$ $\Rightarrow R$ contains no point of ϕ $(\because \phi$ has no points.)

$\Rightarrow p$ is not a limit point of R

$\Rightarrow$ no point of R is a limit point of ϕ

$\Rightarrow D(\phi) = \phi$.

THEOREM 8. *If A and B are subsets of R, then $A \subset B \Rightarrow D(A) \subset D(B)$.*

Proof. Here, $A \subset B$ is given.

To show $D(A) \subset D(B)$.

Let $x \in D(A)$ $\Rightarrow x$ is the limit point of A

$\Rightarrow$ every nbd N of x contains at least one point of A, other than x

$\Rightarrow$ every nbd N of x contains at least one point of B, other than x

$(\because A \subset B)$

$\Rightarrow x$ is the limit point of B

$\Rightarrow x \in D(B)$

Hence, $A \subset B \Rightarrow D(A) \subset D(B)$.

THEOREM 9. *If A and B are any subsets of R, then*

$$D(A \cap B) \subset D(A) \cap D(B).$$

Proof.   Using previous theorem, we have $A \subset B \Rightarrow D(A) \subset D(B)$.

Now $A \cap B \subset A \Rightarrow D(A \cap B) \subset D(A)$ and $A \cap B \subset B \Rightarrow D(A \cap B) \subset D(B)$.

Hence, $D(A \cap B) \subset D(A) \cap D(B)$

REMARK

- $D(A) \cap D(B) \not\subset D(A \cap B)$.

  For Example: let $A = Q$, $B = I$, the set of irrational numbers, then $A \cap B = \phi$ and $D(A \cap B) = \phi$, whereas

$$D(A) = D(B) = R$$

and therefore $D(A) \cap D(B) = R$

Thus, it follows that

$$D(A) \cap D(B) = R \not\subset D(A \cap B) = \phi.$$

THEOREM 10. *If A and B are any subsets of R, then $D(A \cup B) = D(A) \cup D(B)$.*

Proof.   Using Theorem (2), we have

$$A \subset B \Rightarrow D(A) \subset D(B).$$

Since $A \subset A \cup B$ and $B \subset A \cup B$.

Therefore, $D(A) \subset D(A \cup B)$ and $D(B) \subset D(A \cup B)$

$\Rightarrow$ $D(A) \cup D(B) \subset D(A \cup B)$.   ...(1)

Now, to show $D(A \cup B) \subset D(A) \cup D(B)$.

Let x be any point such that $x \notin D(A) \cup D(B)$.

Then, we shall show that $x \notin D(A \cup B)$

Now, $x \notin D(A) \cup D(B) \Rightarrow x \notin D(A)$ and $x \notin D(B)$

$\Rightarrow$ x is neither a limit point of A, nor a limit point of B

$\Rightarrow$ $\exists$ a nbd N_1 of x, which contains no point of A, other than x and $\exists$ a nbd N_2 of x, which contains no point of B, other than x.

Since N_1 and N_2 are the nbds of $x \Rightarrow N_1 \cap N_2$ is a nbd of x (being the intersection of two nbds of x)

$\Rightarrow$ $N_1 \cap N_2$ is a nbd of x, which contains no point of A and B, other than x

$\Rightarrow$ $N_1 \cap N_2$ is a nbd of x, which contains no point of $A \cup B$

$\Rightarrow$ x is not the limit point of $A \cup B$

Therefore, $x \notin D(A) \cup D(B) \Rightarrow x \notin D(A \cup B)$

So, $x \in D(A \cup B) \Rightarrow x \in [D(A) \cup D(B)]$

$$\Rightarrow D(A \cup B) \subset (D(A) \cup D(B)).$$ ...(2)

Now, from (1) and (2), we get

$$D(A \cup B) = D(A) \cup D(B).$$

THEOREM 11. *The derived set of a bounded set is again a bounded set.*

Proof.   Let $A \subset R$ be any set and let m, M the bounds (Infimum, supermum) of A which exist by the ordered completeness property of R

$$A \subset [m, M].$$

Now, we shall show that $D(A)$ is also bounded *i.e.*, no limit point of $D(A)$ can be less than m or greater than M.

Let if possible, $p < m$, be a limit point of A and let $\varepsilon = m - p$ since $m = \inf A$. Therefore, it follows that $\exists$ a nbd $]p - \varepsilon, p + \varepsilon[$, containing no point of A, other than p

$\Rightarrow$ p is not a limit point of A, which is a contradiction

$\Rightarrow$ no limit point of A can be less than m.

Similarly, we can show that no limit point of A, can be greater than M.

Hence $D(A) \subset [m, M]$

$\Rightarrow$ $D(A)$ is bounded.

THEOREM 12. *The derived set of any infinite bounded set attains its bounds.*

Proof.   Let A be an infinite bounded set. Since A is bounded, therefore, there exist $h, k \in R$ such that $A \subset [h, k]$

Now, $A \subset [h, k] \Rightarrow D(A) \subset D([h, k])$

Also, the derived set of the closed interval $[h, k]$ is $[h, k]$, *i.e.*, $D([h, k]) = [h, k]$.

Therefore, $D(A) \subset [h, k]$

$\Rightarrow$ $D(A)$ is bounded.

Since, A is infinite bounded set, therefore, by Bolzano-Weierstrass theorem, A has at least one limit point and so $D(A) \neq \phi$.

$\Rightarrow$ $D(A)$ is non-empty bounded subset of R

$\Rightarrow$ $D(A)$ has supremum and infimum in R.

Let $\inf D(A) = m$ and $\sup [D(A)] = M$.

Now, we shall show that both m and M belongs to $D(A)$ *i.e.*, both m and M are the limit point of A.

Let $\varepsilon > 0$ be given.

Now $m = \inf D(A) \Rightarrow \exists$ some $x \in D(A)$ such that $m \leq x < m + \varepsilon$

$\Rightarrow$ $m - \varepsilon < x < m + \varepsilon$

$\Rightarrow$ $x \in]m - \varepsilon, m + \varepsilon[$

$\Rightarrow$ $]m - \varepsilon, m + \varepsilon[$ is a nbd of some $x \in D(A)$

$\Rightarrow$ $]m - \varepsilon, m + \varepsilon[$ is a nbd of x, which is a limit point of A

$\Rightarrow$ $]m - \varepsilon, m + \varepsilon[$ contains infinitely many points of A

$\Rightarrow$ For every $\varepsilon > 0$, the open interval $]m - \varepsilon, m + \varepsilon[$ contains infinitely many points of A, therefore, m is the limit point of A

Similarly, we can show that, M is the limit point of A

$\Rightarrow$ both m and M *i.e.*, the infimum and supremum of $D(A)$ belongs to $D(A)$

$\Rightarrow$ m is the smallest and M is the greatest member of $D(A)$

$\Rightarrow$ the set A has smallest and greatest limit points.

THEOREM 13. *Finite union of closed sets is closed.*

Proof.   Let $F_1, F_2, ..., F_n$ be n closed sets. Then, to prove that $\bigcup\limits_{i=1}^{n} F_i$ is closed.

Each F_i ($i = 1, 2, ..., n$) is closed $\Rightarrow$ Each F_i' ($i = 1, 2, ..., n$) is open

$$\Rightarrow \quad \bigcap_{i=1}^{n} F_i' \text{ is open} \qquad\qquad (\because \text{ Finite intersection of open sets is open})$$

$$\Rightarrow \quad \left[\bigcap_{i=1}^{n} F_i\right]' = \bigcup_{i=1}^{n} F_i' \text{ is open} \qquad\qquad \text{(By De Morgan's law)}$$

$$\Rightarrow \quad \bigcup_{i=1}^{n} F_i \text{ is closed.}$$

REMARK

- The union of an arbitrary family of closed sets may fail to be a closed set.

    For Example: let $F_n = \left[\dfrac{1}{n}, 2\right]$ for each $n \in N$. Then $\cup F_n = \,]0,2[$, which is not a closed set $\forall\, n \in N$.

THEOREM 14. *The intersection of an arbitrary collection of closed sets is closed.*

Proof.   Let $[F_\lambda : \lambda \in A]$ be an arbitrary family of closed sets. To show $\cap [F_\lambda]$ is closed.

Now, since, each F_λ is a closed $\Rightarrow$ each F_λ' is open

$\Rightarrow \quad \cup [F_\lambda' : \lambda \in \Lambda]$ is open

$\Rightarrow \quad [\cap \{F_\lambda : \lambda \in \Lambda\}]'$ is an open set $\qquad\qquad$ (By De Morgan's law)

$\Rightarrow \quad \cap [F_\lambda : \lambda \in \Lambda]$ is closed.

THEOREM 15. *Let A be a subset of R. Then A is closed if and only if $D(A) \subset A$ i.e., A is closed iff it contains all its limit points.*

Proof.   Let A be any closed subset of R. $\Rightarrow$ A' is open.

Now if $D(A) = \phi$ then $D(A) \subset A$.

Let $D(A) \neq \phi$, let $p \in D(A)$. To show $p \in A$.

Let if possible, $p \notin A$, then $p \in A'$, since A' is open, there exists $\varepsilon > 0$ such that

$$]\,p - \varepsilon, p + \varepsilon[\, \subset A'$$
$$\Rightarrow \quad]\,p - \varepsilon, p + \varepsilon[\, \cap\, A = \phi$$
$$\Rightarrow \quad]\,p - \varepsilon, p + \varepsilon[\ \text{contain no points of } A$$

which contradicting the fact that p is the limit point of A.

Hence, $\qquad\qquad\qquad\qquad p \in A.$

$\Rightarrow$ If A is closed, then $D(A) \subset A$.

Conversely, let us suppose $D(A) \subset A$.

$\Rightarrow$ A contains all its limit points.

To show A is closed.

Let $\qquad\qquad p \in A' \Rightarrow p \notin A \Rightarrow p \notin D(A) \qquad\qquad\qquad (\because D(A) \subset A)$

$\Rightarrow \quad p$ is not the limit point of A

$\Rightarrow \quad \exists\, \varepsilon > 0$ such that $]\,p - \varepsilon, p + \varepsilon\,[$ contains no point of A, other than p

$\Rightarrow \quad \exists\, \varepsilon > 0$ s.t. $]\,p - \varepsilon, p + \varepsilon\,[$contains no point of A $\qquad\qquad (\because\ p \notin A)$

$\Rightarrow \quad \exists\, \varepsilon > 0$ such that $]\,p - \varepsilon, p + \varepsilon\,[\, \subset A'$

$\Rightarrow \quad A'$ contains a neighbourhood of each of its point

$\Rightarrow \quad A'$ is open $\Rightarrow A$ is closed.

THEOREM 16. *Let A be any subset of* R, *then* $\bar{A} = A \cup D(A)$ *i.e.,* $\bar{A}$ *is the set of all adherent points of A.*

Proof.   Firstly, we shall show that $A \cup D(A)$ is closed.

Let p be any limit point of $A \cup D(A) \Rightarrow$ either p is a limit point of A or a limit point of $D(A)$.

If p is the limit point of A then $p \in D(A)$.

If p is the limit point of $D(A)$ then $p \in D(A)$ $(\because D(A)$ is closed.)

$\Rightarrow$ $p \in A \cup D(A)$.

Since p is arbitrary , therefore $A \cup D(A)$ contains all its limit points

$\Rightarrow$ $A \cup D(A)$ is closed.

Now, $A \cup D(A)$ is a closed set containing A, and $\bar{A}$ is the smallest closed set containing A, we have $\bar{A} \subset A \cup D(A)$.   ...(1)

Also, $A \subset \bar{A} \Rightarrow D(A) \subset D(\bar{A})$ ...(2)

Now, since $\bar{A}$ is closed $\Rightarrow D(\bar{A}) \subset \bar{A}$.   ...(3)

From (2) and (3), we have $D(A) \subset \bar{A}$

Now $A \subset \bar{A}$ and $D(A) \subset \bar{A}$ $\Rightarrow A \cup D(A) \subset \bar{A}$ ...(4)

From (1) and (4), we have

$$\bar{A} = A \cup D(A)$$

Solved Examples

Example 1. Find the limit points of the interval $]0,1[$.

Solution.   Let $A =]0,1[$.

Firstly we shall show that every point of the closed interval $[0, 1]$ is the limit point of A.

Let $p \in [0,1]$. Then for $\varepsilon > 0$, the open interval $] p - \varepsilon, p + \varepsilon [$ must contains infinitely many points of A, therefore, it contains at least one point of A, other than p.

$\Rightarrow$ p is the limit point of $]0, 1[$.

Now, we shall show that no points, other than $[0, 1]$ is the limit point of $]0, 1[$ i.e., $p \notin [0, 1]$, then p is not the limit point of $]0, 1[$.

Based on the following Results

- A subset N of R is said to be a nbd of a point $p \in$ R if there exists a real number ε, however small such that $p \in] p - \varepsilon, p + \varepsilon[\subset N$.

- An open interval is a nbd of each of its point.

- A closed interval is a nbd of each of its point except the end points.

- A finite set cannot be a nbd of any of its points.

- A set is said to be open if it is a nbd of each of its points.

- Finite intersection and arbitrary union of open sets is open.

- A point $p \in$ R is said to be an adherent point of a set $A \subset$ R if every nbd of p contain a point of A. Also, if this point is different from p, then it is called limit point.

- The only limit point of $]a, b[,]a, b], [a, b[$ or $[a, b]$ are points of $[a, b]$ and every point of R is a limit point of R.

- $\phi' = \phi$, $N' = \phi$, $Q' = R$, $R' = R$, $\bar{\phi} = \phi$, $\bar{N} = N$, $\bar{Q} = R$, $\bar{R} = R$, $(\bar{\phi})° = \phi$, $(\bar{N})° = \phi$ $(\bar{Q})° = R$, $(\bar{R})° = R$

- Every infinite bounded set of real numbers has a limit point.

- $A \subset B \Rightarrow D(A) \subset D(B)$; $D(A \cap B) \subset D(A) \cap D(B)$; $D(A \cup B) = D(A) \cup D(B)$

Let $\varepsilon > 0$ be such that ε is less than the distance of the point p from each of the end points 0 and 1 of the closed interval $[0, 1]$

$\Rightarrow$ $\varepsilon < |p - 0|$ and $\varepsilon < |p - 1|$

$\Rightarrow$ the open interval $]p - \varepsilon, p + \varepsilon[$ does not contain any point of the set A

$\Rightarrow$ p is not the limit point of A

Therefore, p is the limit point of $]0, 1[$ if and only if $p \in [0,1]$.

Hence, $D(]0, 1[) = [0, 1]$.

Example 2. *Find the limit points of the closed interval $[0,1]$.*

Solution.   Let $A = [0,1]$.

Then, in a similar manner as in Ex. 1.

We have $D([0,1]) = [0,1]$.

REMARKS

- In example (1), we observe that the points 0 and 1 do not belong to $]0,1[$. Also they are limit points of $]0,1[$. Therefore, the set $]0,1[$ is dense-in-itself, which is not perfect.
- In example (2), each point of $[0,1]$ is the limit point of $[0,1]$, therefore the set $[0,1]$ is dense-in-itself. Also $[0,1]$ contains all its limit point, therefore $[0,1]$ is a closed set. Since the set $[0, 1]$ is dense-in-itself and closed, therefore it is perfect.

Example 3. *Find the set of limit points of the set $\left[\dfrac{1}{n} : n \in N\right]$.*

Solution.   Here, we have

$$S = \left[\dfrac{1}{n} : n \in N\right] = \left[1, \dfrac{1}{2}, \dfrac{1}{3}, ..., \dfrac{1}{n}, ...\right]$$

We shall show that the set S has only one limit point, namely 0.

Firstly, we shall show that 0 is the limit point of S.

Let $\varepsilon > 0$ ($\Rightarrow -\varepsilon < 0$). Then, by Archimedean property of real numbers there

exists a positive integer m such that $\dfrac{1}{m} < \varepsilon$ so that

$$-\varepsilon < 0 < \dfrac{1}{m} < \varepsilon.$$

Therefore, for every $\varepsilon > 0$, the open interval $]-\varepsilon, \varepsilon[$ contains a point of S, other

then 0, namely $\dfrac{1}{m}$

$\Rightarrow$ 0 is the limit point of S.

Now, to show that no real number l, other than 0 can be a limit point of S. Now there are following cases.

Case (i) If $l < 0$.

Then, the open interval $]l - 1, 0[$ is a nbd of l which contains no point of S

$\Rightarrow$ l is not the limit point of S.

Case(ii) If $l > 1$.

Then, the open interval $]1, l+1[$ is a nbd of l which contains no point of S

$\Rightarrow$ l is not the limit point of S.

Case (iii) If $0 < l < 1$ but $l \notin S$

$$l < 1 \Rightarrow \frac{1}{l} > 1 \text{ and } \frac{1}{l} \text{ is not an integer.}$$

Therefore, there exists a positive integer m such that $m < \dfrac{1}{l} < m + 1$.

$$\Rightarrow \qquad \frac{1}{m+1} < l < \frac{1}{m}$$

$$\Rightarrow \qquad \left] \frac{1}{m+1}, \frac{1}{m} \right[\text{ is a nbd of } l \text{ which contains no point of } S$$

$\Rightarrow \quad l$ is not the limit point of S.

Case (iv) If $l = 1$.

Then, the open interval $\left] \dfrac{1}{2}, 2 \right[$ is a nbd of 1 which contain no point of S other than 1

$\Rightarrow \quad$ 1 is not the limit point of S.

Case (v) If $l \neq 1$ and $l \in S$.

If $l = \dfrac{1}{m} : m \in N$ and $m \neq 1$ then $\left] \dfrac{1}{m+1}, \dfrac{1}{m-1} \right[$ is a nbd of l, which contains no

point of S, other than l.

$\Rightarrow \quad l$ is not the limit point of S.

Hence, no real number other than 0 is a limit point of S. Therefore, 0 is the only limit point of S.

REMARKS

- Since 0 does not belongs to S. Hence S is not closed.

Example 4. *Find the limit point of the set of natural numbers N.*

Solution. Let p be any element of R. Now, we find a positive integer ε, such that the open interval $] p - \varepsilon, p + \varepsilon[$ contains no point of N, other than p

$\Rightarrow \quad p$ is not the limit point of N.

$$\Rightarrow \qquad D(N) = \phi.$$

Hence, the set of natural number has no limit point.

Example 5. *Find the limit point of the set of irrational numbers.*

Solution. Since, every open interval containing a real number p also contains irrational numbers distinct from p, so every number $p \in R$ is a limit point of the set of irrational number.

Example 6. *Find the limit points of the set S of rational numbers of the form* $\left\{ \dfrac{n}{n+1} : n \in N \right\}$.

Solution. Here, we have

$$S = \left\{ \frac{n}{n+1} : n \in N \right\}$$

Also, $$\frac{n}{n+1} = \frac{n+1-1}{n+1} = 1 - \frac{1}{n+1}$$

Let $\varepsilon > 0$ be arbitrary small positive number, then the nbd $]1 - \varepsilon, 1 + \varepsilon[$ of the point

1 contains a point of S, other than 1, because by taking $n > \left[\dfrac{\varepsilon}{1-\varepsilon}\right]$, we have

$$\frac{n}{n+1} > \frac{\varepsilon/(1-\varepsilon)}{[\varepsilon/(1-\varepsilon)]+1}$$

$\Rightarrow \qquad \dfrac{n}{n+1} > \varepsilon$

$\Rightarrow \quad$ 1 is a limit point of the given set A.

Now, we check whether there is any other limit point of S other than 1.

Let us suppose $p \in A'$, $p \neq 1$.

Now, there are following cases :

Case I. If $p > 1$. then choose $\varepsilon < p - 1$, then the nbd $] p - \varepsilon, p + \varepsilon [$ of p contains no point of S, other than p

$\Rightarrow \quad p$ is not the limit point of S.

Case II. If $p < 1$.

$p \in A'$, implies there exists a point of S, which is nearest to p and let p_r be this element of S, which is nearest to p. Choose a positive integer ε such that $\varepsilon < |p_r - p|$, then the nbd $] p - \varepsilon, p + \varepsilon [$ of the point p contains no point of S and so as before, we conclude that p is not the limit point of S.

Suppose that $p \in S$ and let $p = \dfrac{n}{n+1}$.

Then the point just before p is $\dfrac{n-1}{(n-1)+1}$ $i.e.,$ $\dfrac{n-1}{n}$

and the point just after p is $\dfrac{n+1}{(n+1)+1}$ $i.e.,$ $\dfrac{n+1}{n+2}$

Now, we can find that

$$\frac{n+1}{n+2} - \frac{n}{n+1} = \frac{(n+1)^2 - n(n+2)}{(n+1)(n+2)} = \frac{1}{(n+1)(n+2)}$$

and

$$\frac{n}{n+1} - \frac{n-1}{n} = \frac{n^2 - (n-1)(n+1)}{n(n+1)} = \frac{1}{n(n+1)}.$$

Also $n + 2 > n$

$\Rightarrow \qquad \dfrac{1}{n+2} < \dfrac{1}{n}$

$\Rightarrow \qquad \dfrac{1}{(n+1)(n+2)} < \dfrac{1}{(n+1)n}$

Hence, we have $\dfrac{n}{n+1} - \dfrac{n-1}{n} > \dfrac{n+1}{n+2} - \dfrac{n}{n+1}$

Let us choose a positive number $\varepsilon > 0$ such that $\varepsilon < \left(\dfrac{n+1}{n+2} - \dfrac{n}{n+1}\right)$ then, the nbd

$]p - \varepsilon, p + \varepsilon[$ of p contains no point of S, other than p

$\Rightarrow \quad p$ is not the limit point of S.

Hence, we find that no real number other than 1 is a limit point of A.

Example 7. *Find the limit points of the set*

$$\left[1, -1, 1\frac{1}{2}, -1\frac{1}{2}, 1\frac{1}{3}, -1\frac{1}{3}, \ldots \right].$$

Solution. Let

$$S = \left[1, -1, 1\frac{1}{2}, -1\frac{1}{2}, 1\frac{1}{3}, -1\frac{1}{3}, \ldots \right]$$

Now, let us define the sets A, B and C such that

$$A = \left[1\frac{1}{2}, 1\frac{1}{3}, 1\frac{1}{4}, \ldots \right]$$

$$B = \left[-1\frac{1}{2}, -1\frac{1}{3}, -1\frac{1}{4}, \ldots \right]$$

and $\qquad\qquad C = [1, -1]$, then $S = A \cup B \cup C$

$\Rightarrow \qquad\qquad D(S) = D\,(A \cup B \cup C) = D(A) \cup D(B) \cup D(C).$

Now, since, C is a finite set, it has no limit point

$\Rightarrow \qquad\qquad D(C) = \phi.$

Now, we wish to find $D(A)$ and $D(B)$.

We have, the set A is the sequence $<s_n>$ where

$$s_n = 1 + \frac{1}{n+1} : n \in N \Rightarrow \lim_{n \to \infty} s_n = \lim_{n \to \infty} \left(1 + \frac{1}{n+1} \right) = 1.$$

Therefore, the sequence $<s_n>$ converges to 1.

Now, we shall show that 1 is the limit point of A.

Let $\varepsilon > 0$, then by Archimedean property of real numbers, there exists a positive integer m such that

$$\frac{1}{m} < \varepsilon \Rightarrow \frac{1}{m+1} < \varepsilon \Rightarrow -\varepsilon < 0 < \frac{1}{m+1} < \varepsilon$$

$\Rightarrow \qquad\qquad 1 - \varepsilon < 1 + \frac{1}{m+1} < 1 + \varepsilon.$

Therefore, for every $\varepsilon > 0$, the open interval $]1 - \varepsilon, 1 + \varepsilon[$ contains a point of A, other than 1, namely $1 + \dfrac{1}{m+1}$.

$\Rightarrow \quad$ 1 is the limit point of A.

Also, we can easily show that 1 is the only limit point of A

$\Rightarrow \qquad\qquad D(A) = 1.$

Again, the set B is the sequence $\left[-1 - \dfrac{1}{n+1} : n \in N \right]$

Then, we have

$$D(B) = -1.$$

Hence, $\qquad\qquad D(S) = D(A) \cup D(B) \cup D(C)$

$$= [1] \cup [-1] \cup \phi$$

$$= [1, -1]$$

$\Rightarrow \quad$ 1 and -1 are the only limit points of the given set S.

Example 8. *Find the limit points of the set* $S = \left[\dfrac{3n+2}{2n+1} : n \in N\right]$.

Solution. The given set S is the sequence $< s_n >$ where

$$s_n = \frac{3n+2}{2n+1} : n \in N$$

$$\Rightarrow \qquad \lim_{n \to \infty} s_n = \lim_{n \to \infty} \left(\frac{3+2/n}{2+1/n}\right) = \frac{3}{2}$$

The sequence $< s_n >$ converges to $\dfrac{3}{2}$.

Now, we shall show that $\dfrac{3}{2}$ is the limit point of the set S. Let $\varepsilon > 0$. Since the sequence $< s_n >$ converges to $\dfrac{3}{2}$, therefore for given $\varepsilon > 0$ there exists a positive integer m such that

$$\left| s_n - \frac{3}{2} \right| < \varepsilon \ \forall \ n \ge m \quad \text{i.e.,} \quad \frac{3}{2} - \varepsilon < s_n < \frac{3}{2} + \varepsilon \ \forall n \ge m.$$

For every $\varepsilon > 0$, the open interval $\left]\dfrac{3}{2} - \varepsilon, \dfrac{3}{2} + \varepsilon\right[$ contains infinite number of terms of the sequence $<s_n>$. Therefore, $\dfrac{3}{2}$ is the limit point of S.

Now, to show $\dfrac{3}{2}$ is the only limit point of the set S.

Let if possible l be any other limit point of the set S.

To show $\qquad\qquad l = \dfrac{3}{2}.$

Let $\varepsilon > 0$, Also $s_n \to \dfrac{3}{2}$, therefore $\left| s_n - \dfrac{3}{2} \right| < \dfrac{\varepsilon}{2}, \ \forall \ n \ge p.$

Since, l is a limit point of the set S, therefore, the open interval $\left]l - \dfrac{\varepsilon}{2}, l + \dfrac{\varepsilon}{2}\right[$ contains infinite distinct points of the set S

$\Rightarrow$ there must exists a positive integer $q > p$ such that

$$l - \frac{\varepsilon}{2} < s_q < l + \frac{\varepsilon}{2}$$

$$\Rightarrow \qquad \left| s_q - l \right| < \frac{\varepsilon}{2} \qquad \qquad \qquad ...(2)$$

Now, from (1), we have

$$\left| s_q - \frac{3}{2} \right| < \frac{\varepsilon}{2} \qquad \qquad \qquad ...(3)$$

Consider

$$\left| \frac{3}{2} - l \right| = \left| (s_q - l) + \left(\frac{3}{2} - s_q \right) \right|$$

$$\le \left| s_q - l \right| + \left| \frac{3}{2} - s_q \right| = \left| s_q - l \right| + \left| s_q - 3/2 \right|$$

$$< \frac{\varepsilon}{2} + \frac{\varepsilon}{2}$$

$$\Rightarrow \qquad \left|\frac{3}{2} - l\right| < \varepsilon .$$

Now, since ε is arbitrary, hence letting $\varepsilon \to 0$.

we have $\qquad \left|\frac{3}{2} - l\right| = 0 \Rightarrow \frac{3}{2} - l = 0 \Rightarrow l = \frac{3}{2}$

$\Rightarrow \dfrac{3}{2}$ is the only limit point of S.

Hence, $\qquad D(S) = \left[\dfrac{3}{2}\right]$.

EXERCISE 1.4

1. Show that the set Z^+ of positive integers is not a nbd of any of its points.

2. Show that the right half open interval [2, 3[is a nbd of each of its points except 2.

3. Show that the set [1, 2, 3, 4] is not a nbd of any of its points.

4. Show that the closed interval [2, 4] is a nbd of 3 but not of any of its end points.

5. Find the derived set $\left[\dfrac{1}{m} + \dfrac{1}{n} : m, n \in N\right]$

6. Answer the following:
  (i) Give an example of an infinite set, which does not have any limit point.
  (ii) Show that the set
  $\left[1, -1, \dfrac{1}{2}, -\dfrac{1}{2}, \dfrac{1}{3}, -\dfrac{1}{3}, ...\right]$ is neither open nor closed.
  (iii) Explain why the set [$1^2, 2^2, 3^2, ..., n^2, ...$] has no limit point even though it is an infinite set.

7. Give an example of a set which has
  (i) No limit point
  (ii) Exactly one limit point
  (iii) Exactly two limit points
  (iv) Infinite number of limit points
  (v) Every point of the set as its limit points.

8. Let $S = R - \left[\dfrac{1}{n} : n \in N\right]$. Show that each point of R is a limit point of S.

9. Let $S = \left[3 + \dfrac{1}{n} : n \in N\right]$. Show that the only limit point of S is 3.

10. Find the derived set of each of the following sets.
  (i) The open ray]a, ∞ [
  (ii) The open ray]$-\infty, a$ [
  (iii) [$1 + 3^{-n} : n \in Z^+$]
  (iv) $\left[1 - \dfrac{3}{n} : n \in N\right]$

11. Show that the set of rational number is not an open set.

12. Give an example of each of the following:
  (i) an open set other than interval.
  (ii) a closed set other than interval.
  (iii) a set which is open as well as closed.
  (iv) a set which is neither open nor closed.
  (v) a set where every point is a limit point of the set.
  (vi) a set none of whose point is a limit point of the set.
  (vii) a set which is neither an interval nor an open set.
  (viii) an interval, which is not an open set.
  (ix) an interval which is not a closed set.
  (x) a set which is open but not closed.
  (xi) a set which is closed but not open.

13. Given the set S of numbers
  $S = (1, 1.1, 0.9, 1.01, 0.99, 1.0001, 0.999,...)$
  (i) Is the set bounded ?
  (ii) Determine the supremum and infimum, if they exist.
  (iii) Does the set S have any limit point ? If so, find them ?
  (iv) Is the set S is a closed set ?

14. Prove that a real number p is a limit point of a set S if for each positive integer n

$$\left]p - \frac{1}{n}, p + \frac{1}{n}\right[$$

contains a point of S, other than p.

15. Prove that the subset S of R is open iff for each $p \in S$ there exists a positive integer n such that

$$\left]p - \frac{1}{2^n}, p + \frac{1}{2^n}\right[\subset S.$$

16. Which of the following sets are closed ? Give reason in support of your answer.

(i) The set Z of all integers

(ii) The set [3, 5, 7, 9, 11]

(iii) The null set ϕ

(iv) The set Q of rational numbers

(v) [a]

(vi) $[0, 2] \cup [3, 4]$

17. Which of the following sets are open ? Give reason in support of your answer.

(i) The set Z of integers

(ii) $]0, 1[\cup] 1, 2 [$

(iii) The set [3, 5, 7, 9, 11]

(iv) $\left]-\frac{1}{n}, \frac{1}{n}\right[\quad \forall n \in N$

(v) $[x : 0 < x < 1]$

(vi) The null set ϕ

(vii) The set Q of rational numbers.

═══════════════ Answers ═══════════════

5. $D(S) = \left[\frac{1}{m} : m \in N\right] \cup [0]$ **6.** (i) The set of integers Z   (iii) unbounded above

7. (i) The set of natural numbers N   (ii) The set $\left[\frac{1}{n} : n \in N\right]$

(iii) $S = \left[\frac{1}{2}, -\frac{1}{2}, \frac{2}{3}, -\frac{2}{3}, \frac{3}{4}, -\frac{3}{4}...\right]$ (iv) The open interval $]1, 2[$

(v) The closed interval [1, 2]

10. (i) The closed ray $[a, \infty[$ (ii) The closed ray $] -\infty, a]$

(iii) The singlton [1]   (iv) The singlton [1]

12. (i) $]0, 1[\cup]2, 3[$ (ii) [1]   (iii) ϕ, R

(iv) Q, or finite non-empty set or [0, 1]   (v) R or $[a, b]$ (vi) N or Z

(vii) $[0, 1] \cup]2, 3[$ (viii) $[a, b[\text{ or }]a, b]$

(ix) $]a, b[, [a, b[\text{ or }]a, b]$ (x) $]a, b[$ (xi) N, Z or $[a, b]$

13. (i) Yes   (ii) sup = 1.1, inf = 0.9

(iii) Yes, the only limit point of the set is 1   (iv) Yes

16. (i), (ii), (iii), (v) and (vi) are closed sets.   **17.** (ii), (iv), (v) and (vi) are open sets.

1.14 COUNTABLE SET

A set which can be put in one to one correspondence with the set of natural numbers, is said to be a countable set.

For example, if $A = \{2, 4, 6, 8,...\}$, then $\exists$ an one to one mapping $f : A \rightarrow N$ s.t. $f(x) = x/2$ so that A is a countable set. Further, if $B = \{2, 4, 6, 8, 10\}$, then this set can be put in one to one correspondence with the set $A = \{1, 2, 3, 4, 5\}$, which is a subset of N.

Obviously a countable set can be finite or infinite both.

An infinite countable set is also called *countably infinite* or enumerable set or *denumerable set*. Thus enumerable set always means an infinite countable set.

If a set 'A' is countable then its elements can be put in one to one correspondence with the set N = {1, 2, 3, }. If we denote the elements of A corresponding to the natural numbers 1,2, 3,... , by $a_1, a_2, a_3,$ ... etc., then the set A can be written as $A = \{a_1, a_2, a_3, ... \}$.

Thus a set is countable iff its elements can be written in the form of a sequence.

REMARK

- A set which is not countable, is said to be *uncountable*.

THEOREM 1. *Every subset of a countable set is countable.*

Proof.   Let A be a countable set, then A can be written as a sequence. Let $A = \{a_1, a_2, a_3, ...\}$. If A is finite then its every subset will also be finite and so will be countable. Now consider that A is enumerable. Let B be any subset of A.

If B is finite or empty then the result is trivial, so let $B \neq \phi$ and let k_1 be the least positive integer s.t. $a_{k_1} \in B$. Again, let k_2 be the least positive integer with $k_2 > k_1$ s.t. $a_{k_2} \in B$. Dealing with the elements of B in this way, we reach to the conclusion that B can be written as $\{a_{k_1}, a_{k_2}, ...\}$ which is a sequence and hence B is a countable set.

THEOREM 2. *The union of enumerable collection of enumerable sets is also enumerable.*

Proof.   Let $A = \{A_1, A_2, ... ,\}$ be an enumerable collection of enumerable sets where $A_i = \{a_{i_1}, a_{i_2}, a_{i_3}, ...\}, \forall\, i \in$ N denotes enumerable sets. For the proof of the theorem, we shall construct a progression as given below in which all the elements of A appear.

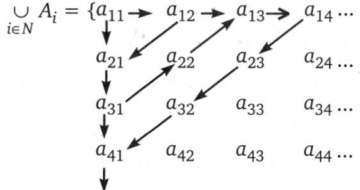

Choose the path as shown above; then every element of A lies somewhere on the path (*i.e.*, each element occupies some particular position say rth position $r \in$ N) and hence a one-one correspondence between $\cup A_i$ and N is implied showing thereby that A is enumerable. Thus we can write $\cup A_i = \{a_{11}, a_{12}, a_{21}, a_{31}, a_{22}, ...\}$ as a sequence.

$\Rightarrow$ set $\cup A_i$ is enumerable.

Besides the above argument, note that the mapping $f : \overset{n}{\underset{}{\bigcup}} A_i \rightarrow$ N, given by

$$f(a_{pq}) = \frac{(p+q-2)(p+q-1)}{2} + p \ \ (= \text{say } n \in \text{N})$$

shows the enumeration of

$$\cup A_i = \{a_{11}, a_{21}, a_{12}, a_{13}, a_{22}, a_{31}, a_{14}, a_{23}, a_{32}, a_{41}, ...\}.$$

$\Rightarrow$ this mapping assigns a unique place to each element of the set $\cup A_i$ in the above sequence and hence that $\cup A_i$ is countable.

REMARKS

- The union of countable collection of countable sets is also countable.
- If A_i is countably infinite set, then $\overset{n}{\underset{i=1}{\bigcup}} A_i$ is also countably infinite.

THEOREM 3. *The set* N × N *is enumerable.*

Proof. We have N × N = $\{(u, v) : u, v \in N\}$; then clearly N × N can be arranged as shown below :

$$N \times N = \{(1,1), (1, 2), (1, 3), (1, 4)$$
$$(2,1), (2, 2), (2, 3), (2, 4)$$
$$(3,1), (3, 2), (3, 3), (3, 4)$$
$$.... \quad .... \quad\quad .... \quad ....\} = \cup A_i,$$
$$\text{where } A_i = \{(i, n): n \in N\}.$$

Now obviously each A_i $(i = 1, 2, 3, ...)$ is enumerable which shows that the set N × N, union of enumerable collection of enumerable sets, is also enumerable by Theorem 2.

THEOREM 4. *The set* Q *of all rational numbers is enumerable.*

Proof. We know that a rational number is written as $p/q : p, q \in Z$ and $q \neq 0$.

Let $A_q = \{p/q : p, q \in Z\}$. Then obviously A_q is equivalent to Z and we know that Z is enumerable and so A_q is enumerable. Also the collection Q = $\underset{q \in Z_0}{\cup} A_q$ is enumerable,

where Z_0 is a subset of Z, because Q is expressed as enumerable union of enumerable sets.

THEOREM 5. (i) *The set of all real numbers in the open interval*]0, 1[*is not enumerable.*

(ii) *Prove that the set of all real numbers in the closed interval* [0,1] *is not enumerable.*

Proof. (i) Let if possible the given interval be enumerable. Then the elements of the interval can be written as a sequence $\{x_1, x_2, x_3, ...\}$. Now using the decimal expansion of these x_m's, we can write as follows :

$$x_1 = \bullet\, a_{11}a_{12}a_{13} \cdots a_{1n}$$
$$x_2 = \bullet\, a_{21}a_{22}a_{23} \cdots a_{2n}$$
$$x_3 = \bullet\, a_{31}a_{32}a_{33} \cdots a_{3n}$$
$$... \quad ... \quad ... \quad ....$$
$$x_n = \bullet\, a_{n1}a_{n2}a_{n3} \cdots a_{nn}$$
$$... \quad ... \quad ... \quad ....$$

where a_{ij}'s may be any integer from 0 to 9, Now let

$$x_n = \bullet\, b_1 b_2 b_3 ...$$

where $b_1, b_2, ...$ are the digits from 0 to 9 s.t. $b_1 \neq a_{11}, b_2 \neq a_{22}, b_3 \neq a_{33}$ etc. In general $b_m \neq a_{mm}$, $\forall\, m \in N$.

Hence $\qquad x \neq x_1, x \neq x_2, ... x \neq x_m,$

Obviously $x \in (0, 1)$ and $x \notin \{x_1, x_2,...\}$; thus we see that besides countable set $\{x_1, x_2,...\}$ there exists elements belonging to interval]0, 1[, showing that the set of all real numbers lying in the interval]0, 1[is not enumerable.

(ii) Open interval]0, 1[is a subset of [0, 1]. Since]0, 1[is not denumerable, the interval [0, 1] is also not denumerable.

THEOREM 6. *The set of all real numbers is not enumerable i.e.,* R *is not enumerable.*

Proof. Let if possible R be enumerable. Then its every subset must be enumerable. Consider the set of all real numbers lying in the interval $]0, 1[$ which is not enumerable by preceding theorem 5; which is again a contradiction since it is the subset of R.

THEOREM 7. *The set of all irrational numbers is not enumerable.*

Proof. Let the contrary be true. Also, by theorem 4, the set of all rational numbers is enumerable. Thus union of rational and irrational numbers would also be enumerable. But we know that R is the union of rational and irrational numbers and hence R would be enumerable which is a contradiction according to theorem 6. Hence the set of all irrational numbers is not enumerable.

1.15 ALGEBRAIC NUMBERS

Let $P_n(x) = \alpha_n x^n + \alpha_{n-1} x^{n-1} + ... + \alpha_1 x + \alpha_0 (\alpha_n \neq 0)$ be a polynomial with α_i's as integral numbers; then we define the algebraic number as a root of the polynomial equation $P_n(x) = 0$ of the above form.

THEOREM 1. *The set of all algebraic numbers is enumerable.*

Proof. Consider the algebraic equation, $\alpha_n x^n + \alpha_1 x^{n-1} + ... + \alpha_n$ of degree n with $\alpha_0 \neq 0$. Now we define the rank of this equation :

$$|\alpha_0| + |\alpha_1| + |\alpha_2| + ... + |\alpha_n| = m.$$

Clearly, rank is a positive number. Also α_i's are integers; so rank is an integer ≥ 1. Obviously for a given rank the roots of the equation will be finite and therefore will be enumerable.

Again we can put a one-one correspondence in the set of natural numbers with the algebraic equation arranged with respect to rank and hence the set of all algebraic equations is enumerable. Now each algebraic equation has enumerable number of roots and so the set of all algebraic numbers is the enumerable collection of enumerable sets and hence enumerable.

THEOREM 2. *The set P of all polynomials*

$$P_n(x) = \alpha_n x^n + \alpha_{n-1} x^{n-1} + ... + \alpha_1 x + \alpha_0 (\alpha_n \neq 0)$$

with integral (rational) coefficients is denumerable.

Proof. If $|\alpha_n| + |\alpha_{n-1}| + ... + |\alpha_0| = m$, then for each pair of natural numbers (m, n), the set P_{mn} of all the polynomials of the form $P_n(x) = \alpha_n x^n + \alpha_{n-1} x^{n-1} + ... + \alpha_1 x + \alpha_0$ is finite and hence countable. Also the sets $P_{(m,n)} = P_k, k = (m, n) \in N \times N$ themselves are countable.

Therefore the set $P = \bigcup_{(m,n) \in N \times N} P_{(m,n)}$ is also countable.

1.16 CARDINALLY EQUIVALENT SET

A set P is said to be cardinally equivalent to a set Q, if there exists at least one- to-one map from P to Q and is denoted by $P \sim Q$.

THEOREM 1. *The relation $P \sim Q$ in the family of sets is an equivalence relation.*

Proof. Here we observe that the given relation is

    (i) Reflexive. Since the identity map $I_p : P \to P$ given by $I_p(a) = a$, $\forall\ a \in P$ is one-one onto *i.e.*, $P \sim P$ for every set P.

    (ii) Symmetric. $P \sim Q \Rightarrow \exists$ a one-one map $f : P \xrightarrow{\text{Onto}} Q$
$f^{-1} : Q \xrightarrow{\text{Onto}} P$ is also one-one *i.e.*, $Q \sim P$.

    (iii) Transitive.

$$P \sim Q \Rightarrow \exists \text{ a one-one map } f : P \xrightarrow{\text{Onto}} Q$$
$$Q \sim R \Rightarrow \exists \text{ a one-one map } g : Q \xrightarrow{\text{Onto}} R$$
$$\Rightarrow g \circ f : P \xrightarrow{\text{Onto}} R \text{ is also one-one.}$$
$$\Rightarrow\ P \sim R.$$

Thus the given relation is an equivalence relation. It is to be noted that an equivalence relation decomposes the set P into equivalence classes, any two of which are either equal or mutually disjoint. The set of these mutually disjoint equivalence classes is said to be the quotient set of the set for the given equivalence relation.

REMARK

- The equivalence relation is also known as equipollent relation.

Example 1. *If A and B are the sets of all real numbers of the intervals $[a_1, a_2]$ and $[b_1, b_2]$ respectively, then show that $A \sim B$.*

Solution. Consider a mapping $f : A \to B$ s.t. if $x \in A$, then

$$f(x) = \frac{b_2 - b_1}{a_2 - a_1} x - a_1 \frac{b_2 - b_1}{a_2 - a_1} + b_1$$

$$= \left(\frac{b_2 - b_1}{a_2 - a_1}\right)(x - a_1) + b_1$$

    (i) f **is one-one.** For, if $f(x) = f(y)$; $x, y \in A$, then

$$\left(\frac{b_2 - b_1}{a_2 - a_1}\right)(x - a_1) + b_1 = \frac{(b_2 - b_1)}{(a_2 - a_1)}(y - a_1) + b_1$$

$$\Rightarrow\quad \left(\frac{b_2 - b_1}{a_2 - a_1}\right)(x - a_1) = \frac{(b_2 - b_1)}{(a_2 - a_1)}(y - a_1) \Rightarrow x = y$$

$$\Rightarrow\quad f \text{ is one-one.}$$

    (ii) f **is onto.** Let $z \in B \Rightarrow b_1 \le z \le b_2$.

If $z = \left(\dfrac{b_2 - b_1}{a_2 - a_1}\right)(x - a_1) + b_1$, then $x = \dfrac{(z - b_1)(a_2 - a_1)}{(b_2 - b_1)} + a_1$

Obviously $a_1 \le x \le a_2$ and hence $x \in A$. Thus for any $z \in B$, $\exists\ x \in A$ s.t. $f(x) = z$ and hence f is onto.

$$\Rightarrow\quad f \text{ is one-one onto and hence } A \sim B.$$

REMARK

- If $A = (a_1, a_2)$, $B = (b_1, b_2)$, then $A \sim B$.

1.17 CARDINAL NUMBER OF A SET

The relation of equivalence, decomposes any collection of sets into equivalence classes containing the equivalent sets. Each equivalent class has a cardinal number which we shall use to represent the property the equivalent sets have in common. It will be in some sense a measure of the points in sets. It is denoted by the card X i.e., cardinal number of X. The basic property of cardinal numbers is that card X = card Y, iff $X \sim Y$.

Definition. *Any set which is equivalent to the set {1, 2, 3, ... , n} is said to have the cardinal number n.* In this way we have defined the cardinal number of finite sets. The null set is said to have the cardinal number zero. Obviously for finite sets, cardinal number is just the number of elements in the sets.

The cardinal number of N (set of natural numbers) is denoted by a and hence all enumerable sets will have the cardinal number a.

The cardinal number of the set of real numbers is denoted by c and all the sets equivalent to the set R are said to have the cardinal number c. The set of all real numbers in the interval [0,1] also has the cardinal number c. Since we have proved that all intervals open or closed are equivalent to [0,1], hence every interval also has cardinal number c. The cardinal number of the set of all real-valued functions defined in the interval [0,1] is denoted by f.

The cardinal number of an infinite set is called a *transfinite cardinal number a* is considered to be the first (smallest) transfinite cardinal. Since every finite set has an enumerable subset which is equivalent to the set of natural number N and so every infinite set has a subset with cardinal number a.

REMARK

- The set of cardinal numbers (0,1, 2, 3,..., a, c, f} is a superset of set of all natural numbers.

1.17.1 SUM OF CARDINAL NUMBERS

Let A and B be any two disjoint sets; then *card A + card B = card A ∪ B.*

In general
$$\sum_{\alpha \in \Delta} \text{card } A_\alpha = \text{card} \left(\bigcup_{\alpha \in \Delta} A_\alpha \right)$$

where $A_\alpha \cap A_\beta = \phi$, $\forall\, \alpha, \beta \in \Delta$ (index set) such that $\alpha \neq \beta$.

1.17.2 PRODUCT OF CARDINAL NUMBERS

We define the product of two cardinal numbers as card P × card Q = card $(P \times Q)$ where P, Q are any sets.

In general,   card P_1 × card P_2 ...= card   [cartesian product of sets P_i $(i = 1, 2, 3,...)$].

1.17.3 COMPARISON OF CARDINAL NUMBERS

Let P and Q be any two sets, then

(i) card P < card Q, if $\exists$ a set $R \subset Q$ s.t. $P \sim R$ and if $P \sim Q$ then card P = card Q.

(ii) card $P \leq$ card Q, if $\exists$ a set $R \subseteq Q$ s.t. $P \sim R$ i. e., there exists a one-one map
$$f : P \xrightarrow{\text{Onto}} R \subseteq Q.$$

(iii) Evidently,   card $(P \cup Q) \geq$ card P or card Q.

(iv) a is the smallest infinite cardinal number.

Solved Examples

Example 1. *Show that card* $(P \times Q) = card\ Q + card\ Q + \ldots P$ *terms.*

Solution. We have

$$P \times Q = \{(x,y) : x \in P, y \in Q\} = \bigcup_{x \in P} \{(x,y): y \in Q\}\ .$$

$$\therefore \qquad card\ (P \times Q) = card\left(\bigcup_{x \in P} \{(x,y): y \in Q\}\right) \qquad \ldots(1)$$

Let $x \in P$ be arbitrary but fixed. Consider the map

$$f : Q \to \{(x,y) : y \in Q\} \text{ defined by } f(y) = (x,y)\ \forall\ y \in Q.$$

This shows that f is one-one onto. Therefore $card\ Q = card\ (\{(x,y): y \in Q\})$. Observing (1) we at once get the required result.

Example 2. *Show that* $\alpha \le card\ P \le \alpha \Rightarrow card\ P = \alpha.$

Solution. We have $a \le card\ P \Rightarrow \exists$ a set Q such that $card\ Q \le card\ P$ where $card\ Q = \alpha$

$$\Rightarrow Q \text{ is equivalent to a subset of } P \qquad \ldots(1)$$

Again $card\ P \le \alpha \Rightarrow card\ P \le card\ Q$

$$\Rightarrow P \text{ is equivalent to a subset of } Q. \qquad \ldots(2)$$

Combining (1) and (2), we get

$$card\ P = card\ Q \Rightarrow card\ P = \alpha \qquad (\because\ card\ Q = \alpha)$$

Example 3. *Let P and Q be any two sets; then show that*

(i) card P + card Q is unique,   (ii) card P . card Q is unique.

Solution. Let $P \sim P_1, Q \sim Q_1, P_1 \cap Q_1 = \phi$. Now

(i) $P \sim P_1 \Rightarrow \exists$ a one-one map $f : P \xrightarrow{\text{Onto}} P_1$

$Q \sim Q_1 \Rightarrow \exists$ a one-one map $g : Q \xrightarrow{\text{Onto}} Q_1.$

Define a function $\psi : P \cup Q \to P_1 \cup Q_1$ s.t.

$$\psi(x) = \begin{cases} f(x), \forall\ x \in P, \\ g(x), \forall x \in Q. \end{cases}$$

Clearly ψ is one-one onto since f and g are one-one onto. Hence $P \cup Q \sim P_1 \cup Q_1 \Rightarrow card\ (P \cup Q) = card\ (P_1 \cup Q_1)$.

This shows that card P + card Q is unique.

(ii) $x \in P, y \in Q \qquad \Rightarrow (x, y) \in P \times Q$

$$\Rightarrow (f(x), g(y)) \in P_1 \times Q_1.$$

Define a function $\psi : P \times Q \to P_1 \times Q_1$ s.t.,

$$\psi(x, y) = (f(x), g(y))\ \forall\ (x, y) \in P \times Q.$$

Again f and g are one-one onto, therefore ψ is one-one onto

$$\Rightarrow P \times Q \sim P_1 \times Q_1 \Rightarrow card\ (P \times Q) = card\ (P_1 \times Q_1).$$

Example 4. *Show that $(P \times Q) \sim (Q \times P)$ i.e., $(P \times Q)$ is cardinally equivalent to $(Q \times P)$.*

Solution. We have $\qquad (P \times Q) = \{(p, q) : p \in P, q \in Q\}.$

Define a map $\qquad f : (P \times Q) \to (Q \times P)$ by the formula $f((p, q)) = (q, p).$

(i) **f is one-one.** Let $f((p_1, q_1)) = f((p_2, q_2))$

$$\Rightarrow (q_1, p_1) = (q_2, p_2) \Rightarrow q_1 = q_2 \text{ and } p_1 = p_2$$
$$\Rightarrow (p_1, q_1) = (p_2, q_2)$$

Obviously f is onto.

THEOREM 1. *Let card $P = p$, card $Q = q$, card $R = r$; then :*

(i) $p+q = q+p$ *i., e., addition of cardinal numbers is commutative.*

(ii) $p \cdot q = q \cdot p$ *i.e., multiplication of cardinal numbers is commutative.*

(iii) $p \cdot (q + r) = p \cdot q + p \cdot r$, *i.e., multiplication is distributive over addition.*

(iv) $p.(q.r) = (p.q).r$, *i.e.,associative law for multiplication holds,*

(v) $p + (q + r) = (p+q) + r$ *i.e., associative law for addition holds.*

Proof.   **(i)** Let $P \cap Q = \phi$. The elements of an arbitrary set may be in any order, therefore

$$P \cup Q = Q \cup P \Rightarrow \text{card } (P \cup Q) = \text{card } (Q \cup P)$$
$$\Rightarrow \text{card } P + \text{card } Q = \text{card } Q + \text{card } P.$$

(ii) We know that $P \times Q \sim Q \times P$

$$\Rightarrow \text{card } (P \times Q) = \text{card } (Q \times P)$$
$$\Rightarrow p \cdot q = q \cdot p.$$

(iii) Let Q, R be disjoint sets. Then

$$p \cdot (q + r) = \text{card } [P \times (Q \cup R)] = \text{card } [(P \times Q) \cup (P \times R)]$$
$$= \text{card } (P \times Q) + \text{card } (P \times R) = p \cdot q + p \cdot r.$$

(iv) $P \times (Q \times R) \sim (P \times Q) \times R$ under the map $f : (x, (y, z)) = ((x, y), z)$,

hence card $[P \times (Q \times R)] = \text{card } [(P \times Q) \times R]$

or $p.(q.r) = (p.q) \cdot r$

(v) Suppose that P, Q, R are pairwise disjoint sets. Also we know that

$$(P \cup Q) \cup R = P \cup (Q \cup R)$$
$$\Rightarrow \quad \text{card } [(P \cup Q) \cup R] = \text{card } [P \cup (Q \cup R)]$$
$$\Rightarrow \quad (p + q) + r = p + (q + r).$$

THEOREM 2.   (i) *If A_i is enumerable set for $i = 1, 2, 3,..., n$, then $\bigcup\limits_{i=1}^{\infty} A_i$ is enumerable and hence*

$$n \cdot a = a.$$

(ii) *If A_i is enumerable set for $i = 1, 2, 3, ...,$ then $\bigcup\limits_{i=1}^{\infty} A_i$ is enumerable and hence*

$$a + a + a + ... \text{ to } \infty \text{ terms} = a.$$

Proof.   (i) Let $A = \bigcup\limits_{i=1}^{\infty} A_i$. We know that countable union of enumerable sets is enumerable, hence A is enumerable.

Deduction. Let $A_i \cap A_j = \phi$ for $i \neq j$; then by definition of sum of cardinal numbers, we have

$$\text{card } A_1 + \text{card } A_2 + ... + \text{card } A_n = \text{card } A \left(\text{since } \bigcup\limits_{i=1}^{n} A_i = A \right) \qquad ...(1)$$

We know that cardinal number of an enumerable set is a and since each A_i $(i = 1, 2, 3, ... , n)$ is enumerable, so we have

$$\text{card } A_1 = \text{card } A_2 = \quad = \text{card } A_n = a.$$

Also A is enumerable so   card $A = a$. So from $\{1\}$, $n \cdot a = a$.

(ii) Let $A = \bigcup\limits_{i=1}^{\infty} A_i$; then A is enumerable being enumerable union of enumerable sets and hence $\qquad$ card A = a.

Deduction. Let $\qquad A_i \cap A_j = \phi$ for $i \neq j$; then card A = a gives
$$\sum\limits_{}^{\infty} \text{card } A_i = a$$
$$\Rightarrow \quad \underset{n}{\underbrace{a + a + a + a + ...}} \text{ to a terms} = a.$$

THEOREM 3. (i) *If A_i is non-enumerable set for $1 \leq i \leq n$, then $\bigcup\limits_{i=1}^{\infty} A_i$ is non-enumerable and hence* $c + c + c + ... $ *to n terms* $= c$

(ii) *If A_i is non-enumerable, $\forall\ i \in N$, then $\bigcup\limits_{i=1}^{\infty} A_i$ is non-enumerable and hence*
$c + c + c + ...$ *to a terms* $= c.$

Proof. (i) Each A_i is non-enumerable for $1 \leq i \leq n$, therefore card A_i = c for $1 \leq i \leq n$.

This implies $A_i \sim [a_i , a_{i + 1}[$ where $a_i, a_{i+1} \in R$ for $i = 1, 2, 3, ..., n$.

Let $a_i < a_{i+1}$ for $i = 1, 2, 3, ... , n$.

Thus, we have
$$A_1 \sim [a_1, a_2[,$$
$$A_2 \sim [a_2, a_3[,$$
$$.... \quad .... \quad ....$$
$$.... \quad ... \quad ....$$
$$A_n \sim [a_n, a_{n+1}[.$$

At first suppose that $A_i \cap A_j = \phi$ for $i \neq j$; then $\bigcup\limits_{i=1}^{\infty} A_i$ is equivalent to some subset of $[a_1, a_{n+1}[.$

$$\therefore \qquad \text{card} \left(\bigcup\limits_{i=1}^{n} A_i \right) \leq c. \qquad\qquad ...(1)$$

Now clearly $\qquad A_i \subseteq \bigcup\limits_{i=1}^{\infty} A_i \Rightarrow \text{card } A_i \leq \text{card} \left(\bigcup\limits_{i=1}^{n} A_i \right)$

$$\Rightarrow c \leq \text{card} \left(\bigcup\limits_{i=1}^{n} A_i \right). \qquad\qquad ...(2)$$

Combining (1) and (2), we have

$$c \leq \text{card} \left(\bigcup\limits_{i=1}^{n} A_i \right) \leq c \text{ i.e., card} \left(\bigcup\limits_{i=1}^{n} A_i \right) = c.$$

$\therefore \bigcup\limits_{i=1}^{\infty} A_i$ is non-enumerable.

Now suppose that $A_i \cap A_j = \phi$ for $i \neq j$, then

$$\bigcup\limits_{i=1}^{\infty} A_i \sim [a_1 , ..., a_{n+1}[\Rightarrow \text{card} \left(\bigcup\limits_{i=1}^{n} A_i \right) = \text{card } [a_1, ..., a_{n+1}[$$

$$\Rightarrow \text{card} \left(\bigcup\limits_{i=1}^{n} A_i \right) = c.$$

Deduction. If we assume that $A_i \cap A_j = \phi$ for $i \neq j$, then as proved above, we have

$$\sum_{i=1}^{n} \text{card}(A_i) = c$$

$$\left(\because \text{card} \bigcup_{i=1}^{n} A_i = \sum_{i=1}^{n} \text{card } A_i \text{ when } A_i \cap A_j = \phi \text{ for } i \neq j \right)$$

i.e., $c + c + c + ...$ to n terms $= c$.

(ii) Let $A = \bigcup_{i=1}^{\infty} A_i$, where card $A_i = c$, $\forall \, i \in \text{N}$. Now

$$\text{card } A_i = c \Rightarrow A_i \sim \left[1 - \frac{1}{2^{i-1}}, 1 - \frac{1}{2^i} \right)$$

Thus, we have

$$A_1 \sim \left[0, \frac{1}{2} \right[, \; A_2 \sim \left[\frac{1}{2}, \frac{3}{4} \right[, \; A_3 \sim \left[\frac{3}{4}, \frac{7}{8} \right[, \; ... \; A_i \sim \left[1 - \frac{1}{2^{i-1}}, 1 - \frac{1}{2^i} \right[...$$

Now assume that $A_i \cap A_j = \phi$ for $i \neq j$.

Then obviously $\bigcup_{i=1}^{\infty} A_i \sim [0,1[$.

$\therefore$ $\text{card} \left(\bigcup_{i=1}^{\infty} A_i \right) = \text{card } [0,1[\text{ or card } A = c$.

Next suppose that $A_i \cap A_j \neq \phi$ for $i = j$, then $\bigcup_{i=1}^{\infty} A_i$ is cardinally equivalent to some subset of $[0,1)$. So

$$\text{card} \left(\bigcup_{i=1}^{\infty} A_i \right) \leq c. \qquad \qquad ...(3)$$

Again $A_i \Rightarrow \bigcup_{i=1}^{\infty} A_i \Rightarrow \text{card } A_i \leq \text{card} \left(\bigcup_{i=1}^{\infty} A_i \right)$

$$\Rightarrow c \leq \text{card} \left(\bigcup_{i=1}^{\infty} A_i \right). \qquad \qquad ...(4)$$

Combining (3) and (4), we get, card $\left(\bigcup_{i=1}^{\infty} A_i \right) = c$.

Hence $\bigcup_{i=1}^{\infty} A_i$ is non-enumerable.

REMARK

- Suppose that $A_i \cap A_j = \phi$ for $i \neq j$, then we have

$$\text{card} \left(\bigcup_{i=1}^{\infty} A_i \right) = c \Rightarrow \sum_{i=1}^{\infty} \text{card } A_i = c$$

or $c + c + c + ...$ to c terms $= c$.

THEOREM 4. *If an enumerable set is added to an infinite set, then cardinal number of the infinite set is unaffected.*

Proof.   Let card $A = \alpha$ *i.e.,* A is an infinite set. Since A is an infinite set and therefore $\exists$ a subset B of A s.t. B is enumerable.

$$\therefore \qquad \text{card } B = a.$$

Now we can write $\qquad A = (A - B) \cup B.$

$$\therefore \qquad A \cup N = (A - B) \cup B \cup N = (A - B) \cup (B \cup N).$$

B and N are enumerable sets $\Rightarrow B \cup N$ is enumerable

$$\Rightarrow B \cup N \sim N$$

Now $\qquad B \cup N \sim N, \ N \sim B \Rightarrow B \cup N \sim B$

$\therefore \qquad (A - B) \cup (B \cup N) \sim (A - B) \cup B$ i.e., $A \cup N \sim A$

or $\qquad$ card $(A \cup N) = $ card A i. e., $\alpha + a = \alpha.$

THEOREM 5. c. c = c.

Proof. Let $\qquad A = \{x : 0 \le x \le 1\}$. Then card $A = c.$

Now, let $\qquad B = \{(0, x): x \in A]$.

Clearly, $\qquad B \subset A \times A \qquad\qquad$ Therefore, card $B \le$ card $(A \times A).$

Again $A \sim B$ under the map f s.t. $f(x) = (0, x), \ \forall \ x \in A,$

$$\text{Therefore, card } A \le \text{card } B.$$

$\therefore \qquad\qquad$ card $A \le$ card $(A \times A).$ $\qquad\qquad\qquad$ ...(1)

Let x, y be any two real numbers in the closed interval [0,1]. Then x and y can uniquely be expanded in the form of infinite decimals which contain non-zero digits. Now define a map $g : (A \times A) \to A$ by writing

$$g(x, y) = 0. x_1 \ y_1 \ x_2 \ y_2 \ x_3 \ y_3$$

Obviously g is one-one. So by definition ,

$$\text{card}(A \times A) \le \text{card } A.\qquad\qquad\qquad ...(2)$$

From (1) and (2), we get

$$\text{card } A \le \text{card } (A \times A) \le \text{card } A; \quad \therefore \text{ card } (A \times A) = \text{card } A$$

or $\qquad\qquad$ c. c = c $\qquad\qquad [\because$ card $A \times$ card $B =$ card $(A \times B)]$

1.18 TRANSCENDENTAL NUMBER

A real number which is not an algebraic called transcendental number. Thus the numbers e and π which are real but not algebraic numbers, are transcendental numbers.

All rational numbers are algebraic number hence every rational number is not transcendental, implying that every transcendental number must be irrational, for R = (rational numbers) $\cup$ (irrational numbers).

There are so many irrational numbers which are algebraic. Therefore every irrational number is not transcendental number.

THEOREM 1. *Every monotonic function in a closed interval is discontinuous at a countable number of points of that interval.*

Proof. Let $f(x)$ be a monotonic function in the closed interval $[a, b]$. Also let it be a monotonically increasing function and be discontinuous at an arbitrary point x.

Then $\qquad\qquad \delta(x) = f(x + 0) - f(x - 0) > 0 \qquad\qquad$ ...(1)

where $\qquad\qquad f(a) = f(a - 0), f(b) = f(b + 0).$

Let $\quad \xi_1, \ \xi_2, \ ..., \ \xi_{m-1}$ be numbers in the intervals $x_k < \xi_k < x_{k+1}$ where $a < x_1 < x_2 < ... x_m < b$. Write $\quad \xi_0 = a$ and $\xi_m = b.$

$$\therefore \qquad f(\xi_k) - f(\xi_{k-1}) \geq f(x_k + 0) - f(x_k - 0) = \delta(x_k) \qquad \text{[by (1)]} \quad ...(2)$$

Therefore $f(b) - f(a) = \sum_{k=1}^{m} [f(\xi_k) - f(\xi_{k-1})] \geq \sum_{k=1}^{m} \delta(x_k).$

Let $\qquad \delta(x_k) > \dfrac{1}{n}, \forall k$

Then, we have

$$f(b) - f(a) > \frac{m}{n} \quad \text{or} \quad [f(b) - f(a)] \, n > m.$$

This shows that m which is the number of points of discontinuity x with $\delta(x) > \dfrac{1}{n}$ is bounded above *i.e.*, the number of points of discontinuity x with $\delta(x) > \dfrac{1}{n}$ are finite in the closed interval $[a, b]$. Since $n \in$ N, we see that the number of points of discontinuity x with $\delta(x) > \dfrac{1}{n}$ are finite in the closed interval $[a, b]$. Since every finite set is countable and for every $x \,\exists\, n \in$ N therefore the number of points of discontinuity in the closed interval $[a, b]$ will be an enumerable union of countable sets and hence countable. Hence the theorem.

THEOREM 2. (Cantor's Theorem)

For any set A, card A < card P(A), P(A) being power set of the set A.

Or

For every cardinal number n, $2^n > n$.

Proof. Let $B^* = \{\{x\}: x \in A\}$; then obviously $B^* \subset P(A)$. Now define a map $f : A \to B^*$ s.t. $f(x) = \{x\}$. Obviously $A \sim B^*$. Hence card $A \leq$ card $P(A)$.

Thus we are only to show that card $A \neq$ card $P(A)$,

Let if possible, $A \sim P(A)$. So $\exists$ a one-one map

$$f : A \xrightarrow{\text{Onto}} P(A).$$

Let $\qquad B = \{x \in A : x \notin f(x)\}.$

Clearly $\qquad B \subset A \Rightarrow B \in P(A).$

Since the mapping f is onto, there must exist $x \in A$ s.t. $f(x) = B$. Now if $x \in B$ then by definition of B, $x \in f(x)$ which is not possible.

Consider the second possibility that $x \notin B$, then $x \in f(x) = B$ which is again impossible. It means that our assumption is wrong. Hence the result.

REMARKS

- From the above theorem, we have $n < 2^n$ where card $A = n$ and
$$\text{card } P(A) = 2^{\text{card } A} = 2^n$$
- Also $\qquad\qquad\qquad 2^a > a.$

THEOREM 3. (Equivalence Theorem). *If $A_1 \subset B \subset A$ and $A \sim A_1$, then $A \sim B$.*

Or

If $A_1 \subset B \subset A$ and card $A = $ card A_1, then card $A = $ card B.

Proof. $A \sim A_1 \Rightarrow \exists$ a one-one map $f : A \xrightarrow{\text{Onto}} A_1.$

As $B \subset A$ so f_B is a one-one onto where f_B is the restriction of f to B. This means that $B \sim B_1 \subset A_1$. Similarly $A_1 \sim A_2 \subseteq B_1$.

Continuing in this way we get equivalent sets

$$A, A_1, A_2, \dots \quad \text{and } B, B_1, B_2, \dots$$

s.t. $\qquad A \supset B \supset A_1 \supset B_1 \supset A_2 \supset B_2 \supset A_3 \supset B_3 \supset \dots .$

Let $\qquad S = A \cap B \cap A_1 \cap B_1 \cap A_2 \cap B_2 \cap \dots$

then we can write

$$A = (A - B) \cup (B - A_1) \cup (A_1 - B_1) \cup \dots \cup S.$$
$$B = (B - A) \cup (A_1 - B_1) \cup (B_1 - A_2) \cup \dots \cup S.$$

Define a map $\psi : A \to B$ s.t.

$$\psi\,(A - B) = A_1 - B_1.$$
$$\psi\,(A_1 - B_1) = A_2 - B_2.$$
$$\psi\,(A_2 - B_2) = A_3 - B_3.$$
$$\dots \qquad \dots \quad \dots \quad \dots$$
$$\dots \qquad \dots \quad \dots \quad \dots$$
$$\psi\,(B - A_1) = B - A_1,$$
$$\psi\,(B_1 - A_2) = B_1 - A_2.$$
$$\dots \qquad \dots \quad \dots \quad \dots$$
$$\dots \qquad \dots \quad \dots \quad \dots$$
$$\psi\,(S) = S.$$

Above definition of ψ makes the mapping ψ one-one and onto showing that $A \sim B$.

THEOREM 4. **(Schorder-Bernstein Theorem).** *If card $A \le$ card B and card $B \le$ card A, then*

$$\text{card } A = \text{card } B.$$

Or   If each of the sets A and B is equivalent to a subset of other, then $A \sim B$.

Proof. Let f and g be one-to-one mappings from A into B and from B into A respectively. Let $f(A) = B_1 \subset B$ and $g(B) = A_2$ and $g(B_1) = A_3$, then we have $A \supset A_2 \supset A_3$. Further $g(B_1) = A_3, f(A) = B_1$ implies $g(f(A)) = A_3$, giving $g \circ f$ is a one-to-one mapping from A to $A_3 \Rightarrow$ card $A = $ card A_3. Hence by the above theorem card $A = $ card A_2. Also existence of g such that $g(B) = A_2$ shows that card $B = $ card A_2.

Hence $\qquad\qquad$ card $B = $ card A.

THEOREM 5. *Show that $2^a = c$.*

Proof. We know that card $[0,1] = c$. On the other hand each $x \in [0,1]$ can be written in the form of binary expansion as $x \in 0.\,x_1 x_2 x_3 \dots$, where each $x_i = 0$ or 1.

But selecting each x_i in two ways (either 0 or 1) we can form at most 2^a numbers. So the card $[0,1] = 2^a$ implying that $2^a = c$.

Solved Examples

Example 1. *Show that every superset of an uncountable set is uncountable.*

Solution. Let if possible B be the superset of an uncountable set A, then B is countable.

But we know that every subset of a countable set is countable and so A must be countable which is a contradiction and so B is uncountable. Hence every superset of an uncountable set is uncountable.

Example 2.   *Show that union of two enumerable sets is also enumerable.*

Solution.   Let A and B be the two enumerable sets.

Case I. When $A \cap B = \phi$.

Let $A = \{a_1, a_2, ...\}$, $B = \{b_1, b_2, ...\}$. Now establish the correspondence
$$f : A \cup B \rightarrow N$$
s.t. $f(a_n) \rightarrow 2n - 1$ (odd positive integer),

and $f(b_n) \rightarrow 2n$ (even positive integer).

Evidently this mapping is one-one from $A \cup B$ onto N.

Case II. When $A \cap B \neq \phi$, then we can write $A \cup B = A \cup (B - A)$. Taking $B_1 = B - A$, we have $A \cap B_1 = \phi$. As already proved $A \cup B_1$ is countable where B_1 is countable being the subset of countable set B and hence $A \cup B = A \cup B_1$ is countable when B_1 is countably infinite but if B_1 is finite say $B_1 = \{e_1, e_2, ..., e_m\}$

Then $A \cup B = A \cup B_1 = \{e_1, e_2, ..., e_m, a_1, a_2, ...\}$

Now set a correspondence $f : N \rightarrow A \cup B_1$ s.t. $f(i) = e_i, 1 \leq i \leq m$,

and $f(m + i) = a_i, \forall i$

$\Rightarrow$ $A \cup B$ is enumerable.

REMARK

- We can generalize the result that union of two countable sets is countable (whether each is countably infinite or finite).

Example 3.   *Show that every infinite set is equivalent to its proper subset.*

Solution.   **Case I.**   When A is countably infinite, then A can be written as a sequence.

Let $A = \{a_1, a_2, a_3, ...\}$. Then the function $f(a_n) = a_{n+1}$ establishes a one to one correspondence between the set A and $A - \{a_1\}$ which is a proper subset of A.

Case II. When A is uncountably infinite, then it has an enumerable subset say B where $B = \{a_1, a_2, a_3, ...\}$. We shall show that
$$A \sim (A - \{a_i\}).$$

Write $C = A - B$; then $A = B \cup C$ and $B \cap C = \phi$. Also $A - \{a_i\} = (B - \{a_i\}) \cup C$. Let $e(x)$ be the identity mapping which associates each $x \in A$ onto itself. Let f be the function such that $f(a_n) = a_{n+1}$. Now define a function h such that
$$h(x) = \begin{cases} e(x) : x \in C, \\ f(x) : x \in B. \end{cases}$$

Then the range of h is $(B - \{a_i\}) \cup C$ which is a proper subset of $B \cup C = A$.

Example 4.   *If α and β are cardinal numbers such that $\alpha \leq \beta$ and $\beta \leq \alpha$, then show that $\alpha = \beta$.*

Solution.   Let card $A = \alpha$, card $B = \beta$. Then
$$\alpha \leq \beta \Rightarrow \text{card } A \leq \text{card } B$$

$$\Rightarrow A \sim B \text{ or } A \sim \text{to a subset of } B, \qquad \ldots(1)$$

$$\beta \leq \alpha \Rightarrow B \sim A \text{ or } B \sim \text{ to a subset of } A. \qquad \ldots(2)$$

(1) and (2) give the required result.

Example 5. *If an enumerable set is subtracted from an enumerable set, then show that the remaining set will be enumerable.*

Solution. Suppose contradiction *i.e.*, $A - B$ is non-enumerable where A and B are enumerable sets. We can write $A = (A - B) \cup B$. Now $A - B$ is non-enumerable gives A to be non-enumerable which is a contradiction. Hence the result follows.

Example 6. *If we subtract an enumerable set from a non-enumerable set, then show that the remaining set is non-enumerable.*

Solution. Let if possible , $A - B$ is enumerable where A is non-enumerable and B is an enumerable set. We can write $A = (A - B) \cup B$. Since both $A - B$ and B are enumerable sets it implies A is enumerable which is a contradiction as A is non-enumerable.

Example 7. *Prove that the set Z of all integers is countable.*

Solution. Write $\qquad \qquad Z^+ = \{1, 2, 3, \ldots\}$ and $Z^- = \{-1, -2, -3, \ldots\}$,

Then, we have $\qquad Z^+ \sim N$ under the mapping $n \rightarrow n$.

Now $\qquad \qquad \quad Z^+ \sim N$ under the mapping $n \rightarrow n$,

$\qquad \qquad \qquad \quad Z^- \sim N$ under the mapping $(-n) \rightarrow n$.

Also singleton set $\{0\}$ is finite, so countable. Thus Z is the countable union of countable sets and hence countable.

REMARK

- An infinite set can be equivalent to its proper subset *e.g.*, $Z \sim N$.

Example 8. *Find the power of an aggregate of numbers given by* $\dfrac{M}{2^m}$, *M and m being positive integers.*

Solution. Let

$$B = \left\{ \frac{M}{2^m} : M, m \in N \right\}$$

Write $\qquad B_M = \left\{ \dfrac{M}{2^m} : m \in N \right\}$

Then we have

$$B_1 = \left\{ \frac{1}{2^1}, \frac{1}{2^2}, \ldots, \frac{1}{2^n}, \right\}$$

$$B_2 = \left\{ \frac{2}{2^1}, \frac{2}{2^2}, \ldots, \frac{2}{2^n}, \right\}$$

$$B_3 = \left\{ \frac{3}{2^1}, \frac{3}{2^2}, \ldots, \frac{3}{2^n}, \right\}$$

$$\ldots \qquad \ldots \qquad \ldots \qquad \ldots$$

$$B_n = \left\{ \frac{n}{2^1}, \frac{n}{2^2}, \ldots, \frac{n}{2^n}, \right\}$$

$$\ldots \qquad \ldots \qquad \ldots \qquad \ldots$$

Evidently,

(i) B_i is enumerable $\forall\, i \in N$ under the mapping $\dfrac{i}{2^n} \to n$,

(ii) B_i 's are pairwise disjoint.

(iii) $B = \bigcup\limits_{i=1}^{\infty} B_i$

Thus B is enumerable being the enumerable union of enumerable sets. Hence card $B =$ a $i.e.$, the power of the given set is a.

Example 9.   *Prove that a < c.*

Solution.   Since $N \subset R \Rightarrow$ card $N <$ card R

$$\Rightarrow \qquad a < c.$$

Example 10.   *Prove that card $(P(A)) = 2^{card\,A}$ for any finite set A*

Solution.   Let   card $A = \alpha$.

Then   card $(P(A)) = 1 + {}^{\alpha}C_1 + {}^{\alpha}C_2 + ... + {}^{\alpha}C_\alpha = (1+1)^\alpha = 2^\alpha = 2^{\,card\,A}$.

Example 11.   *Prove that $\alpha \le \alpha$ for any cardinal number α.*

Solution.   Let   card $P = \alpha$.

Define an identity map $f : P \to P$ written by $f(x) = x,\ \forall\, x \in P$.

Obviously f is one-one. Hence by definition,

$$\text{card } P \le \text{card } P \text{ i.e., } \alpha \le \alpha.$$

Example 12.   *If each α_i, $(i = 1, 2,..., n)$ is a rational number, then the point*

$$a = (a_1, a_2, ..., a_n) \in R^n,$$

is called a rational point.

Show that the set of all rational points in R^n is denumerable.

Solution.   We know that set of rational numbers is countable. So varying $a_1, a_2, ..., a_n$ we can form a^n rational points. But we know that a. a = a and hence a^n = a.

$\Rightarrow$ set of all rational points has the same cardinal number as that of N

$\Rightarrow$ set of rational points is countable.

Example 13.   *Prove that $c^c = 2^c$.*

Solution.   We have $c^c = (2^a)^c = 2^{ac} = 2^c$.

Example 14.   *Let X be any non-empty set and let C be the family of functions $f : X \to \{0,1\}$. Then show that the family of subsets of X i.e., the power set of X is equivalent to C.*

Solution.   Let $A \in P(X)$ where $P(X)$ denotes the power set of X. Also ϕ_A denote the characteristic function of A relative to X.

Now define a map $f : P(X) \to C$ by the formula $f(A) = \phi_A$.

Obviously f as defined above is one-one onto. Hence $P(X) \sim C$.

Example 15.   *Prove that $]0, 1] \sim\,]0, 1[$.*

Solution.   Denote the points of $]0,1]$ by x and of $]0,1[$ by y. Now define a correspondence

$$y = \frac{3}{2} - x \text{ for } \frac{1}{2} < x \le 1; \text{ then } \frac{1}{2} \le y < 1,$$

$$y = \frac{3}{4} - x \text{ for } \frac{1}{4} < x \le \frac{1}{2}; \text{ then } \frac{1}{4} \le y < \frac{1}{2},$$

and so on. $$y = \frac{3}{8} - x \text{ for } \frac{1}{8} < x \le \frac{1}{4}; \text{ then } \frac{1}{8} \le y < \frac{1}{4},$$

From the above correspondence, we see that for every $x \in \,]0,1]$, there corresponds one and only one y of $]0,1[$. Hence by definition $]0,1[\, \sim \,]0,1[$.

Example 16. *Show that for every real number x, the real numbers in the semi-open interval $[x, x+1[$ form an uncountable set.*

Solution.   Let x be any real number. Define a function
$$f : [x, x+1[\,\to\, [0, 1[\text{ given by } f(y) = y - x.$$
Then f is well defined, for obviously,
$$f(x) = x - x = 0, f(x + 1) = x + 1 - x = 1.$$
Again $f(y_1) = f(y_2) \Rightarrow y_1 - x = y_2 - x \Rightarrow y_1 = y_2 \Rightarrow f$ is one-one. Also f is a continuous map which implies that f is an onto map.

Hence $[x, x + 1[\,\sim\, [0,1[$.

∴   card $[x, x + 1[= $ card $[0,1[$.

Now since the set of all real numbers in the semi-open interval $[0,1[$ is uncountable and hence the set of all real numbers in $[x, x+1[$ which is cardinally equivalent to $[0,1[$ is uncountable.

Example 17. *If α is any transfinite cardinal number, then show that $a \leq \alpha$.*

Solution.   Let A be an infinite arbitrary set s.t. card $A = \alpha$.

A is infinite set $\Rightarrow \exists$ an enumerable subset B of $A \Rightarrow$ card $B = a$

$\qquad B \subseteq A \Rightarrow$ card $B \leq$ card $A \Rightarrow a \leq \alpha$.

Example 18. *Show that two enumerable sets are equivalent.*

Solution.   Let P and Q be two enumerable sets. By hypothesis,
$$P \sim N, \ Q \sim N \text{ or } P \sim N, N \sim Q \Rightarrow P \sim Q,$$
since the relation $P \sim Q$ in the family of sets is an equivalent relation.

Example 19. *Show that the set of all transcendental numbers in any interval is non-enumerable.*

Solution.   We know that the set of all algebraic numbers and transcendental numbers is the set of all real numbers which is known to be uncountable. Also we know that the set of algebraic numbers in an interval is enumerable. But we have already proved that if an enumerable set is removed from a non-enumerable set, the remaining set is non-enumerable. Therefore the complement of the set of all algebraic numbers in any interval relative to the set of all real numbers in that interval is uncountable. But this is the set of all transcendental numbers.

Example 20. *Show that the interval $]0,1[$ is equivalent to the set R of all real numbers and hence show that card $]0,1[= $ card R.*

Solution.   Define a function $f : \,]0,1[\,\to\, R$ s.t.
$$f(x) = \begin{cases} \dfrac{2x-1}{x}, & x \in \left]0, \dfrac{1}{2}\right[\\[2mm] \dfrac{2x-1}{1-x}, & x \in \left]\dfrac{1}{2},1\right[. \end{cases}$$

So, that this function is one-one and onto implying that $]0,1[\,\sim\, R$ and hence
$$\text{card } (0,1) = \text{card R} = c.$$
Also since $]0,1[$ is uncountable, the set R is also uncountable.

Example 21. *Prove that a < c < f where a, c and f denote the cardinal numbers of set of all natural numbers, real numbers and the set of all real valued functions defined over* [0,1] *respectively.*

Solution. We have already proved that a < c. It remains to prove that c < f.

Let F be the set of all real valued functions defined over [0,1],

Now consider the mappings $f_k : [0,1] \to R$ defined as $f_k(x) = k$, $\forall\ x \in [0,1]$ and k being a real number in [0,1]. All these functions are real valued and so the set $F^* = \{f_k, : 0 \le k \le 1\}$ is a proper subset of the set F.

We can set up a one-to-one correspondence between [0, 1] and F^*(s.t. $F^* \subset F$).

Hence   card [0,1] = card F^* < card F or c < f.

Example 22. *Show that a countable set is a Borel set.*

Solution. Let $A = \{a_1, a_2, a_3, ...\}$ be a countable set. Note that

$$\{x : x = a_r\} = \bigcap_{n=1}^{\infty} \left\{ x : a_r \le x < a_r + \frac{1}{n} \right\}$$

and $A = \bigcup_{r \in N} \{a_r\}$.

$\Rightarrow A$ is obtained by the formation of countable union and intersection of closed and open sets and hence A is a Borel set.

EXERCISE 1.5

1. (a) Define an enumberable set. Show that the set of all real numbers cannot be enumerable, although the set of rationals is enumerable.

   (b) If $f : A \to B$ and the range of f is uncountable, prove that domain of f is also uncountable.

2. Define cardinal number of a set. Show that $n < 2^n$ for any cardinal number n.

3. Prove that the set of all real numbers in the closed interval [0, 1] is uncountable.

4. Prove that if A and B are enumerable then $A \times B$ is also enumerable.

5. Prove that $\alpha + \alpha = \alpha$ for any infinite cardinal number α.

6. Find the cardinal number of the set $\{x\}$ of those numbers in the interval [0, 1] whose ternary expansion does not have the digit 1.

7. Prove that

   (i) [0,1]–]0,1[,   (ii) [0,1]~[2,5],

   (iii) [0,1[~]0, 1[.

8. If $\{E_n\}$ be a sequence of countable sets and

   $$S = \bigcup_{n=1}^{\infty} E_n ,$$ then prove that S is countable.

9. Let α and β be any two cardinal numbers such that $\alpha \le \beta$ and $\beta \le \alpha$, then prove that $\alpha = \beta$.

10. (i) Prove that the set of all numbers in any interval cannot be enumerable,

    (ii) Show that the set of all characteristic functions on R is uncountable.

11. Set of real numbers is ....

12. Every isolated set of points is ....

13. By an example show that cancellation law does not hold in case of cardinal numbers ?

14. Set of integers (0,1,2, ...) is uncountable.

15. Show that the family of all finite subsets of the natural numbers is countable infinite.

16. Prove that

    (i) a + c = c   (ii) a + a = a

    (iii) c + c = c   (iv) c . c = c

    (v) a . a = a

17. State and prove Schroder-Bernstein theorem.

18. Prove that the set of complex numbers is uncountable.

19. Exhibit a 1-1 correspondence between the points of the closed interval [0,1] of R and the points of the half closed interval (0,1] of R.

20. Show that the set of all polynomial functions with integer (rational) coefficients is countable (or say has the cardinal a).

21. Prove that card $P(A) = 2^{\,\text{card}\,A}$, where A is any finite set.

22. Using the mapping $f : N \times N \to N$ given by $f(x, y) = 2^x (x^y + 1) - 1$. show that set $N \times N$ is countable.

23. Show that the set of points in the closed interval [2,4] and in the open interval]1,2[are cardinally equivalent.

24. If a finite set of elements is added to an enumerable set, then show that executing set is also enumerable.

25. If B is a countable subset of an uncountable set A. Then A– B is ....

26. Show that the set of all sequences whose elements are the digits 0 and 1 is uncountable.

27. If $\{E_n\}$ be a sequence of countable sets and $S = \bigcup_{n=1}^{\infty} E_n$, then prove that S is countable.

===========Answers===========

12. Countable   **14.** False   **25.** Uncountable

CHAPTER REVIEW: A COMPETITIVE APPROACH

SELECTED TERMS AND RESULTS

TERMS

- **Interval:** A subset S of R is called an interval if $a, b \in S, x \in$ R such that $a < x < b \Rightarrow x \in$ S.

- **Finite subset of R :** A subset S of R is said to be finite if either it is empty or there exists a one to one correspondence from the set of natural numbers onto the set S.

- **Absolute value of a Real number :** The absolute value of a real number a denoted by $|a|$ is real number a, $-a$ or 0 according as a is positive, negative or zero respectively.

- **Upper bound :** A subset S of R is said to be bounded above if there exists a real number u such that $s \le u \; \forall \; s \in$ S. Here, the real number u is called upper bound of S.

- **Lower bound :** A subset S of R is said to be bounded below if there exist a real number l such that $s \ge l \; \forall \; s \in$ S. Here, the real number l is called the lower bound of S.

- **Bounded set :** A subset S of R is said to be bounded if it is bounded below as well as above.

- **Supremum :** Least upper bound.

- **Infimum :** Greatest lower bound.

- **Neighbourhood of a point :** A subset N of R is said to be neighbourhood of a point $p \in$ R if there exist a real, however small number $\varepsilon > 0$ such that
  $$p \in \;] \, p - \varepsilon, p + \varepsilon \, [\subset N.$$

- **Interior of a set :** A point p is called an interior point of a set S if S is a neighbourhood of p and set of all interior points is called interior of a set.

- **Open set :** A subset S of R is said to be open if it is a neighbourhood of each of its points.

- **Adherent point :** A point $p \in$ R is said to be an adherent point of a set $A \subseteq$ R if every nbd of p contains a point of A.

- **Limit point :** A point $p \in$ R is said to be a limit point of the set $A \subseteq$ R if every nbd of p contains at least one point of A, other than p.

- **Isolated point :** A point $p \in$ R is said to be an isolated point of $A \subseteq$ R if p is not the limit point of A.

- **Derived set :** Set of all limit points.

- **Discrete set :** Set of all isolated points.

- **Closed set :** A set which contains all its limit points.

- **Dense-in-itself :** A set which possesses no isolated points.

- **Closure of a set :** The smallest closed set containing A is called closure of A.

- **Everywhere dense :** A is said to be everywhere dense if $\overline{A} = $ R.

- **No where dense :** A is said to be nowhere dense if $(\overline{A})° = \phi$.

- **Countable set :** A set which can be put in one-to-one correspondence with the set of natural numbers is called countable set.

- **Cardinal number :** Any set which is equivalent to the set $\{1, 2, ..., n\}$ is said to have the cardinal number n.

- **Algebraic number :** A number which can be written as the root of a polynomial is called algebraic number.

- **Transcendental number :** A real number which is not algebraic is called transcendental number.

RESULTS

- Supremum of a set, if exists is unique.
- Infimum of a set if, exists is unique.
- The set of positive real numbers is bounded below but not above.
- Every non-empty set of real numbers which is bounded above has the supremum (order-complete property)
- If a and b are any two positive real numbers then there exists a positive integer n such that $na > b$. (Archimedean property)
- Order completeness property of a real numbers implies Dedekind's property.
- Dedekind's property implies order completeness property.
- To every real number there corresponds a unique point on the directed line and to every point on a directed line there corresponds a unique real number [Dedekind-Center Axioms.]
- The union of an arbitrary family of open sets is open.
- The intersection of any finite number of open sets is open.
- Every infinite bounded set has a limit point (Bolzano-Weierstrass Theorem).
- The concept of closed and open sets are neither mutually exclusive nor exhausitive. There are some sets which are both open and closed or which is neither open nor closed so, the word 'not closed' should not be considered equivalent to 'open'.
- A set is said to be dense in-itself if every point of A is a limit point of A.
- Union of finite number of closed sets is closed.
- The intersection of an arbitrary family of closed sets is closed.
- A closed set either contains an interval or else is no where dense.
- The derived set of a set is closed.
- The derived set of a bounded set is bounded.
- Every finite bounded set has the smallest and greatest limit points.
- Every subset of a countable set is countable.
- The countable union of countable sets is countable.
- The cartesian product of two countable sets is countable.
- The necessary and sufficient condition for a real number s to be the supremum of a bounded above set is that s must satisfying the following conditions :

  (i) $x \le s \ \forall \ x \in S$ i.e., s must be the upper bound of S.

  (ii) For each positive real number $\varepsilon \ \exists$ a real number $x \in S$ such that $x > s - \varepsilon$.

- The necessary and sufficient condition for a real number i to be the infimum of a bounded below set S is that i must satisfy the following conditions.

  (i) $x \ge i \ \forall \ x \in S$ i.e., i must be the lower bound of S.

  (ii) For each positive real number $\varepsilon \ \exists$ a real number $x \in S$ such that $x < i + \varepsilon$.

- For a bounded set S, $\inf (S) \le \sup (S)$.
- Union and intersection of two bounded sets is again bounded.
- $\sup (A \cup B) = \max [\sup (A), \sup (B)]$
- $\inf (A \cup B) = \min [\inf (A), \inf (B)]$.
- Neither the set of natural numbers nor the set of integers is an ordered field.
- Intersection of a finite number of neighbourhoods of a point is also a nbd of that point.
- Every point has an infinite number of neighbourhoods.
- The closure of a set A is a smallest closed superset of A.

REVIEW QUESTIONS AND PROJECT WORK

1. If A and B are two non-empty bounded sets of non-negative real numbers and $AB = \{xy : x \in A, y \in B\}$

   Then show that

   (i) sup $(AB) \le$ sup A. sup B

   (ii) inf $(AB) \ge$ inf A. inf B

2. If A and B are two bounded sets and $\lambda \in \mathbb{R}$, then prove the following :

   (i) sup $(\lambda A) = \lambda$ sup (A) if $\lambda \ge 0$

   $= \lambda$ inf (A) if $\lambda \le 0$

   (ii) sup $\{x + y : x \in A, y \in B\} =$ sup $A +$ sup B

   (iii) inf $\{x + y : x \in A, y \in B\} =$ inf $A +$ inf B

   (iv) sup $\{x - y : x \in A, y \in B\} =$ sup $A -$ inf B

   (iii) inf $\{x - y : x \in A, y \in B\} =$ inf $A -$ inf B

3. Show that the intersection of a finite family of intervals with at least one point common is either singleton or is an interval.

4. Prove that infimum and supremum of any subset S of $\mathbb{R}$ are also infimum and supremum of $\overline{S}$ and are contained in $\overline{S}$ according as S is bounded below or above.

5. Show that a non-empty bounded open subset of $\mathbb{R}$ has no smallest or greatest member.

6. Show that the set of limit points of an infinite bounded set has smallest and greatest member.

7. Show that there exists no integers n for which $\sqrt{n+1} + \sqrt{n-1}$ is a rational number.

8. Show that the set $\left[\dfrac{r}{2^n}, r = 0, 1, 2, ..., 2^n : n \in \mathbb{N}\right]$ is dense in $[0, 1]$.

9. Show that there exists a bijective mapping between the set of all closed sets of $\mathbb{R}$ and the set $\mathbb{R}$ of real numbers.

10. Let a, b and n be positive integers. Show that if n is not the square of an integer then $\sqrt{n}$ lies between $\dfrac{a}{b}$ and $\dfrac{a + nb}{a + b}$.

11. If there is a mapping from A onto B and B is uncountable show that A is uncountable.

12. Show that every non-empty open set contains both rational and irrational numbers.

13. If r be rational and x be irrational, show that

    (i) $r + x$, $-x$ and $\dfrac{1}{x}$ are irrational.

    (ii) rx is irrational if $r \ne 0$.

SOME IMPORTANT ILLUSTRATIONS

- The set $\mathbb{Z}$, $\mathbb{Q}$ and $\mathbb{R}$ are not bounded.
- Every finite set is bounded.
- The set of rational numbers is not order complete.
- The set of real numbers $\mathbb{R}$ is the neighbourhood of each of its points.
- The set of rationals $\mathbb{Q}$ and set of integers $\mathbb{Z}$ is not a nbd of any of its points.
- Every open interval is a nbd of each of its points.
- Every closed interval is a nbd of each of its points except the end points.
- A non-empty finite set is not a nbd of any of its points.
- The set of real numbers $\mathbb{R}$ is an open set.
- The set of rationals $\mathbb{Q}$ and set of integers $\mathbb{Z}$ are not open.
- The null set ϕ is open.

- $D(\mathbb{R}) = \mathbb{R}$, $D(\mathbb{Q}) = \mathbb{R}$

  $D([a, b]) = [a, b)$, $D\,(]a, b[) = [a, b]$.

- The set $\left[\dfrac{1}{n} : n \in \mathbb{N}\right]$ has only one limit point namely 0, which is not the member of the set.
- The set $\mathbb{R}$ of real number is open as well as closed.
- The $\mathbb{Q}$ of rationals is not closed but dense in itself.
- The set of reals $\mathbb{R}$ is a perfect set.
- The set $\mathbb{Q}$ of rationals and the set of irrationals I are dense in $\mathbb{R}$.
- The empty set is perfect.
- The set of all integers is countable.
- The set of all ordered pairs of integers is countable.
- The set of real numbers is not countable.

TABULAR FORM

Set	Open	Closed	dense-in-itself	Perfect	Countable	dense in R
R	✓	✓	✓	✓	✗	✓
Q	✗	✗	✓	✗	✗	✓
φ	✓	✓	✓	✓	✗	✓
S (finite set)	✗	✓	✗	✗	✓	✗
[a, b]	✗	✓	✓	✓	✗	✓

OBJECTIVE TYPE QUESTIONS

FILL IN THE BLANKS

1. A set is said to be bounded _____, if its lower bound exist.

2. A set is said to be bounded _____ if its upper bound exist.

3. Supremum and infimum of a set _____ exist in the set.

4. Supermum (or infimum) of a set is _____.

5. If supremum = infimum for any set, then the set is _____.

6. A non-empty subset of R which is bounded below has the _____ in R.

7. A non-empty subset of R which is bounded above has the _____ in R.

8. $\sqrt{2}$ is an _____ numbers.

9. The deleted nbd of p does not contain _____ .

10. Every _____ set is a union of open intervals.

11. The interior of a set A is the largest open set contained in _____.

12. Finite intersection of open sets is _____.

13. The set of all limit points is called _____ .

14. A set is said to be closed if it contains all its _____ points.

15. A point p is said to be interior point of a set A, if A is a _____ of p.

16. Every infinite bounded set has a _____ point.

17. The only limit point of the set $A = \left[\dfrac{1}{n} : n \in N\right]$ is _____.

18. The derived set of a finite set is _____.

19. A subset S of real numbers is said to be _____, if it is bounded above as well as bounded below.

20. $D\,(\,]\,2, 3\,[\,) = $ _____ .

21. If $S =]2, 3[$. Then $D(S) = $ _____

22. The infimum of a set $]1, 4[$ is _____

TRUE/FALSE

Write 'T' for true and 'F' for false statement.

1. Every open interval is a nbd of each of its points. **(T/F)**

2. Every closed interval is a nbd of each of its points. **(T/F)**

3. Every closed interval is a nbd of each of its points except the end points. **(T/F)**

4. The set N is a nbd of each of its points. **(T/F)**

5. The set $\left[\dfrac{1}{n} : n \in N\right]$. is not bounded. **(T/F)**

6. Every finite set is a closed set. **(T/F)**

7. The set of real numbers R is bounded. **(T/F)**

8. The set of natural numbers N is bounded below. **(T/F)**

9. A finite set has no limit point. **(T/F)**

10. Every infinite set has at least one limit points. **(T/F)**

11. Every infinite bounded set has at least one limit point. **(T/F)**

12. $D (A \cup B) = D(A) \cup D(B)$. **(T/F)**

13. $D (A \cap B) \subset D(A) \cap D(B)$. **(T/F)**

14. The supremum of the set of natural number does not exist. **(T/F)**

15. The infimum of the set of natural number does not exist. **(T/F)**

16. The empty set ϕ is open as well as closed.

17. The singleton set is always closed. **(T/F)**

18. The set of rational number is a nbd of each of its points. **(T/F)**

19. The supremum and infimum of a set, if exist never coincides. **(T/F)**

20. The only limit point of $]a, b[$ is a and b. **(T/F)**

21. The set $S = \{1, 3, 7, 11\}$ has no limit point. **(T/F)**

22. The finite set $S = \{2, 6, 8, 11\}$ is an open set. **(T/F)**

MULTIPLE CHOICE QUESTIONS

Choose the most appropriate one.

Problem Set- 1

1. The set of natural numbers N is :
   (a) bounded below
   (b) bounded above
   (c) bounded
   (d) none of the above

2. Which statement is not true ?
   (a) The set Z, Q and R are not bounded.
   (b) Every finite set is bounded.
   (c) The supremum and infimum of a set is unique.
   (d) None of the above

3. The supremum and infimum of the set $\left[\dfrac{(-1)^n}{n} : n \in N \right]$ are respectively given by :
   (a) $\dfrac{1}{2}, -1$ (b) $-1, -\dfrac{1}{2}$
   (c) $1, -\dfrac{1}{2}$ (d) none of these

4. Which statement is not true ?
   (a) The set Q, of rational numbers is a field.
   (b) The set Q, of rational numbers is an ordered field.
   (c) The set Q, of rational numbers is a complete ordered field.
   (d) None of the above

5. Which statement is not true ?
   (a) The set C, of complex numbers is a field.
   (b) The set C, of complex numbers is an ordered field.
   (c) The set C, of complex numbers is a complete ordered field.
   (d) None of the above

6. The set is said to be bounded if it is :
   (a) bounded below
   (b) bounded above
   (c) both (a) and (b)
   (d) none of the above

7. The null set ϕ is :
   (a) bounded below only
   (b) bounded above only
   (c) bounded
   (d) none of the above

8. The supremum and infimum of the null set ϕ, are respectively given by :
   (a) 0, 0   (b) 1, 0
   (c) 1, 2   (d) does not exist

9. A non-empty finite subset of R is :
   (a) always bounded
   (b) bounded above only
   (c) bounded below only
   (d) none of the above

10. A subset of an unbounded set is :
    (a) always bounded
    (b) always unbounded
    (c) may or may not be bounded
    (d) none of the above

11. Which is a complete ordered field ?
    (a) The set of integers
    (b) The set of rational numbers
    (c) The set of complex numbers
    (d) None of the above

12. The closed interval $[a, b]$ is a nbd of :
    (a) each of its points
    (b) each of its end points
    (c) each of its point except the end points
    (d) none of the above

13. The set of rational numbers is a nbd of :
    (a) each of its point
    (b) each $n \in N$
    (c) not a nbd of any of its points
    (d) none of the above

14. The number of limit points of set of natural numbers N is :
    (a) 0   (b) 1

(c) 2   (d) infinite

15. The infimum of the set $]4, 8[$ is :
    (a) 1.3   (b) 3.9
    (c) 4.1   (d) 4

16. If $S =]a, b[$, then $D(S) =$
    (a) $[a, b]$ (b) $]a, b[$
    (c) $[a, b[$ (d) $]a, b]$

17. If $S = \left[-\dfrac{1}{n} : n \in N\right]$ then sup $S =$
    (a) -1 (b) 0
    (c) 1   (d) 2

18. If a and b are any two distinct real numbers, then $\exists$ nbds of a and b, which are disjoint. This property is known as :
    (a) housedroff property
    (b) denseness property
    (c) both (a) and (b)
    (d) none of the above

Problem Set-2

1. A subset of real numbers is said to be bounded if it is :
   (a) bounded below
   (b) bounded above
   (c) bounded below as well as above
   (d) none of the above

2. A subset of R is said to be unbounded if it is :
   (a) not bounded above
   (b) not bounded below
   (c) both (a) and (b)
   (d) none of the above

3. The set of positive real numbers is :
   (a) bounded above
   (b) not bounded above
   (c) not bounded below
   (d) none of the above

4. If supremum = infimum for a set, then set is :
   (a) empty   (b) singleton
   (c) can't say   (d) none of these

5. If every non-empty subset S of R which is bounded above has a member of S for its supremum, then S is called :
   (a) complete set   (b) incomplete set
   (c) empty set   (d) singleton set

6. The open interval $]a, b[$ is a nbd of :
   (a) a (b) b
   (c) each of its points   (d) none of these

7. The closed interval $[a, b]$ is a nbd of :
   (a) a
   (b) b
   (c) each of its points
   (d) each of its points except the end points.

8. The set of rational number is a :
   (a) nbd of each of its points
   (b) not a nbd of any of its points
   (c) nbd of a single point
   (d) none of the above

9. The set of irrational number is a :
   (a) nbd of each of its points
   (b) not a nbd of any of its points
   (c) nbd of a single point
   (d) none of the above

10. The set of integers is a :
    (a) nbd of each of its points
    (b) not a nbd of any of its points
    (c) nbd of a single point
    (d) none of the above

11. A finite set is :
    (a) nbd of each of its points
    (b) not a nbd of any of its points
    (c) nbd of its end points only
    (d) none of the above

12. The set of real numbers is a :
    (a) nbd of each of its points
    (b) not a nbd of any of its points
    (c) nbd of a single point
    (d) none of the above

13. A point which is not the limit point is called :
    (a) isolated point   (b) adherent point
    (c) cluster point   (d) none of these

14. The set of all limit points is called :
    (a) derived set   (b) discrete set
    (c) open set   (d) none of these

15. The set of all isolated points is called :
    (a) derived set   (b) discrete set
    (c) open set   (d) none of these

16. A set which contains all its limit points is called:
    (a) closed set   (b) open set
    (c) discrete set   (d) none of these

17. The set which is a nbd of each of its points is called :
    (a) closed set   (b) open set
    (c) discrete set   (d) none of these

18. A finite set is always :
    (a) closed set   (b) open set
    (c) discrete set   (d) none of these

19. If N is a nbd of p, then point p is called a/an :
    (a) interior point   (b) limit point
    (c) adherent point   (d) none of these

20. The set of all adhered points of a set is called :
    (a) closure   (b) interior
    (c) discrete   (d) derived

21. The derived set of empty set ϕ is :
    (a) whole set   (b) ϕ
    (c) $\{\phi\}$ (d) none of these

23. Every point of an open interval is :
    (a) an interior point
    (b) an adherent point
    (c) an isolated point
    (d) a limit point

24. Every singleton is :
    (a) open   (b) closed
    (c) both   (d) none of these

25. If $\overline{A} = A$, then A is :
    (a) closed   (b) open
    (c) discrete set   (d) perfect set

26. A set which is either finite or denumerable is called a :
    (a) countable set   (b) uncountable set
    (c) can't say   (d) none of these

27. Every subset of a countable set is :
    (a) uncountable   (b) countable
    (c) (a) or (b)   (d) none of these

28. Every superset of an uncountable set is :
    (a) uncountable   (b) countable
    (c) (a) or (b)   (d) none of these

29. Union of countable family of countable sets is :
    (a) countable   (b) uncountable
    (c) (a) or (b)   (d) none of these

30. The set of all irrational numbers is :
    (a) finite   (b) countable
    (c) uncountable   (d) none of these

Problem set-3

1. A set A is said to be closed if :
   (a) it contains all its limit points.
   (b) it contains all of its point of closure.
   (c) both (a) and (b)
   (d) none of the above

2. "Every non empty set S of real numbers which is bounded above has a supremum" is known as :
   (a) completeness axioms
   (b) ordered axioms
   (c) field axioms
   (d) none of the above

3. The union of a finite or countable collection of countable sets is :
   (a) uncountable   (b) countable
   (c) infinite   (d) none of these

4. If a is a minimal element of the given set S then :

   (a) $a \in S$

   (b) $a \notin S$

   (c) a is not an upper bound of S

   (d) none of the above

5. For every real number x, there is a positive integer n such that:

   (a) $n > x$ (b) $n < x$

   (c) $n = x$ (d) none of these

6. Every infinite set is :

   (a) uncountable set

   (b) countable set

   (c) both (a) and (b) are true.

   (d) none of these

7. If S is a non-empty subset, then :

   (a) inf (S) < sup (S)

   (b) inf (S) = sup (S)

   (c) inf (S) > sup (S)

   (d) none of the above

8. If $x, y, z \notin R$, the set of real numbers then which one of the following relation does not hold good?

   (a) $x, y > 0 \Rightarrow x - y > 0$

   (b) $x > 0, y > 0 \Rightarrow x.y > 0$

   (c) $x, y > 0 \Rightarrow x + y > 0$

   (d) None of the above

9. If R^* be an extended real number system then inf. R^* is given by :

   (a) ∞ (b) $]-\infty, \infty[$

   (c) 0   (d) none of these

10. If S is a non-empty set of real numbers and S is unbounded below, then :

    (a) inf (S) = $-\infty$ (b) inf. (S) = $+\infty$

    (c) sup (S) = ∞ (d) none of these

11. The half interval $[0,1[$ have :

    (a) minimal element only

    (b) maximal element only

    (c) both (a) and (b)

    (d) none of the above

12. If S is a set of real numbers which is bounded below then inf (S) is :

    (a) prime number

    (b) a point closure to S

    (c) not a point closure to S

    (d) none of the above

13. The closed interval $[0,1]$ is :

    (a) bounded above

    (b) bounded below

    (c) both (a) and (b) are true.

    (d) none of the above

14. The least upper bound (supremum) of the set $S = (0,1)$ is :

    (a) 0   (b) 1

    (c) 2   (d) ϕ

15. The set of negative integers having the least upper bound :

    (a) 1   (b) 0

    (c) -1 (d) ϕ

16. Every open set of real numbers is the union of:

    (a) countable collection of disjoint open intervals.

    (b) uncountable collection of disjoint open intervals.

    (c) countable collection of disjoint closed interval.

    (d) none of the above

17. The empty set ϕ is:

    (a) open

    (b) closed

    (c) both (a) and (b) are true

    (d) none of the above

18. A set is said to be closed if and only if :

    (a) it contains all its limit point.

    (b) it contains all its isolated point

    (c) both (a) and (b) are true

    (d) none of the above

19. The intersection of arbitrary collection of open sets is :

    (a) open   (b) closed

    (c) can not defined   (d) none of these

20. A finite set is always :

    (a) closed

    (b) open

    (c) both (a) and (b) are true

    (d) none of the above

21. The set S is no where dense if :

(a) closure of S contains non empty open sets

(b) closure of S contain no non empty open sets.

(c) closure of S contain empty open sets.

(d) none of the above

22. If a is a maximal element of the given set S then :

(a) $a \in S$ (b) $a \notin S$

(c) a is a lower bound of S

(d) none of the above

23. The closed interval [0, 1] has :

(a) maximal element

(b) minimal element

(c) both (a) and (b) are true

(d) none of the above

24. The set of real numbers R is a/an :

(a) open set   (b) finite set

(c) closed set   (d) none of these

25. If S is a non-empty open sets, then :

(a) its compliment is closed

(b) its compliment is open

(c) its compliment is empty

(d) none of the above

26. The set of real numbers is :

(a) unbounded   (b) bounded

(c) finite   (d) none of these

27. The intersection of finite collection of open sets is :

(a) open   (b) closed

(c) ϕ (d) none of these

28. The intersection of any collection of closed sets is :

(a) closed   (b) open

(c) ϕ (d) none of these

29. If S and T are two subsets of real numbers R and for all $s \in S$, $t \in T$, $s \leq t$, if S and T have infimum then :

(a) $\inf S \geq \inf T$ (b) $\inf S \leq \inf T$

(c) $\inf S = \inf T$ (d) none of these

30. If R* is an extended real number system, then the least upper bound of R* is :

(a) $-\infty$ (b) $+\infty$

(c) 0   (d) none of these

31. The intersection of all sets of the form $\left(-\dfrac{1}{n}, \dfrac{1}{n}\right)$, $n \in$ N is :

(a) an open set   (b) not an open set

(c) {0}   (d) none of these

32. If S is a non-empty set and has no upper bound, then :

(a) $\sup S = \infty$ (b) $\sup S = -\infty$

(c) $\inf. S = -\infty$ (d) none of these

33. The intersection of finite collection of any closed set is :

(a) closed set   (b) open set

(c) null set   (d) none of these

34. The property "Every non-empty set S of real numbers which is bounded below has infimum" is called :

(a) completeness axioms

(b) field axioms

(c) ordered axioms

(d) none of the above

35. The set $R^+ = (0, +\infty)$ is :

(a) bounded above   (b) unbounded above

(c) unbounded below   (d) none of these

36. Let $F_n = \left[-\dfrac{1}{n}, \dfrac{1}{n}\right]$ be a given interval, then $\displaystyle\bigcup_{n=1}^{\infty} F_n$ is :

(a) closed set   (b) open set

(c) both (a) and (b)   (d) none of these

37. If S be a non-empty and bounded from below, then :

(a) $\inf (S) = -\sup(S)$ (b) $\inf S = -\sup(-S)$

(c) $\inf (S) = \sup (S)$ (d) none of these

38. Let $S = [0, 1]$, then maximal element of the set S is given by :

(a) 1   (b) 0

(c) ϕ (d) none of these

39. For the set of negative real numbers $S = [x \in R : x<0]$, then sup $S =$

(a) 1   (b) –1

(c) 0   (d) none of these

40. Which of the following is correct ?

(a) A finite set can not have a limit point.

(b) A finite set must possess a limit point.

(c) An infinite set must have a limit point.

(d) None of the above

41. Which of the following is/are correct?

(a) The empty set is perfect.

(b) The empty set is closed.

(c) The empty set is open.

(d) All are true.

42. If S be the set of points of closure of S, then :

(a) S is closed   (b) S is open

(c) S is empty   (d) none of these

43. Which one of the following is true ?

(a) Null set is an open set.

(b) Null set is semi open set.

(c) Null set is a closed set.

(d) None of the above

44. The property "Between any two real numbers there always exists infinitely many rational and irrational numbers" is called :

(a) Dedkind axiom

(b) denseness property of real numbers

(c) cantor's axioms

(d) none of the above

45. Which is true ?

(a) $\phi' = \phi$ (b) $\bar{\phi} = \phi$

(c) $(\phi)^\circ = \phi$ (d) All are true.

46. Which is/ are true ?

(a) $N' = \phi$ (b) $\bar{N} = N$

(c) $(\bar{N})^\circ = \phi$ (d) All are true

47. Which is/are true :

(a) $Q' = R$ (b) $\bar{Q} = R$

(c) $(\bar{Q})^\circ = R$ (d) All are true.

48. Which is/are true ?

(a) $R' = R$ (b) $\bar{R} = R$

(c) $(\bar{R})^\circ = R$ (d) All are true.

49. The derived set of a bounded set is :

(a) bounded   (b) unbounded

(c) ϕ (d) none of these

50. The number of limit points of the set of natural numbers is:

(a) 1   (b) 2

(c) 0   (d) none of these

Problem set-4

1. A set A is said to be countable if it mapped into natural numbers such that mapping is :

(a) one-one   (b) many one

(c) one-one onto   (d) none of these

2. A set A is called countable if A one to one mapped on the set of :

(a) real nos.   (b) rational nos.

(c) irrational nos.   (d) natural no.

3. A set $A = \{2, 4, 6, 8, ...\}$ is :

(a) countable

(b) not countable

(c) may or may not be countable

(d) none of the above

4. A set $A = \{1, 3, 5, 7, ...\}$ is :
(a) countable
(b) not countable
(c) may or may not be countable
(d) none of the above

5. Finite union of enumerable set is :

(a) countable   (b) uncountable

(c) enumerable   (d) none of these

6. Every infinite countable set is :
(a) enumerable
(b) not emumerable
(c) may or may not be enumerable
(d) none of the above

7. The set of integers is :

(a) enumerable

(b) not emumerable

(c) may or may not be emumerable

(d) none of the above

8. Which set is always countable ?

(a) Finite set   (b) Infinite set

(c) both (a) and (b)   (d) None of these

9. Infinite set is :

(a) countable

(b) uncountable

(c) may or may not be countable

(d) none of the above

10. A set which is not countable is said to be :

(a) enumerable   (b) uncountable

(c) countably infinite (d) none of these

11. Infinite countable set is known as :

(a) enumerable   (b) denumerable set

(c) countably infinite (d) all are true

12. Every subset of a countable set is :

(a) countable

(b) uncountable

(c) denumerable

(d) none of the above

13. Union of two countable sets is :

(a) countable

(b) uncountable

(c) may or may not be countable

(d) none of the above

14. Union of two enumerable sets is :

(a) countable

(b) enumerable

(c) may or may not be countable

(d) none of the above

15. Finite union of countable sets is :

(a) countable

(b) uncountable

(c) may or may not be countable

(d) none of the above

16. The set of natural numbers is :

(a) enumerable   (b) countable

(c) uncountable   (d) none of these

17. The set of all irrational numbers is :

(a) countable   (b) enumerable

(c) uncountable   (d) finite

18. The set of all rational nos. is :

(a) countable   (b) enumerable

(c) uncountable   (d) both (a) and (b)

19. The set of real numbers is :

(a) countable   (b) uncountable

(c) enumerable   (d) finite

20. Transcendental number is :

(a) algebraic

(b) not algebraic

(c) may or may not be algebraic

(d) none of the above

21. The set of all algebraic numbers is :

(a) countable   (b) enumerable

(c) uncountable   (d) none of these

22. The equivalence relation is :

(a) RST   (b) RAT

(c) both (a) and (b)   (d) none of these

23. The difference of two enumerable sets is :

(a) countable   (b) enumerable

(c) not enumerable   (d) uncountable

24. The cartesian product of two uncountable sets is :

(a) countable   (b) uncountable

(c) enumerable   (d) none of these

25. The cartesian product of two enumerable sets is :

(a) countable   (b) enumerable

(c) not enumerable   (d) none of these

26. The difference of an enumerable set and non enumerable set is :

(a) enumerable   (b) not enumerable

(c) countable   (d) none of these

27. If card $A \geq$ card B and card $B \leq$ card A. Then :

(a) card $A \neq$ card B

(b) card $A =$ card B

(c) card $AB =$ card BA

(d) none of the above

28. If each of the sets A and B is equivalent to a subset of other then :

(a) $A = B$ (b) $A \leq B$

(c) $A - B$ (d) $A \geq B$

29. For every cardinal number n :

(a) $2^n = n$ (b) $2^n > n$

(c) $2^n < n$ (d) $2^{n+1} = n$

30. Which set is equivalent to its proper subset ?

(a) Finite set   (b) Infinite set

(c) Super set   (d) Countable set

31. If an enumerable set is substract from an enumerable set, then set thus obtained will be :

(a) countable   (b) enumerable

(c) not countable   (d) not enumerable

32. If $A_1, A_2, ..., A_n$ are enumerable sets then $\bigcup\limits_{i=1}^{n} A_i$ is :

(a) enumerable   (b) not enumerable

(c) countable   (d) none of these

33. Which of the follwoing set is a Borel set ?

(a) Uncountable   (b) Countable

(c) Enumerable   (d) None of these

34. The set of all real numbers between $(0, 1)$ is :

(a) enumerable   (b) not enumerable

(c) finite set   (d) none of these

35. A number which can be written as the root of polynomial is called :

(a) algebraic   (b) transcendental

(c) both (a) and (b)   (d) none of these

Answers

FILL IN THE BLANKS

1. below   **2.** above   **3.** not necessarily   **4.** unique   **5.** singlton   **6.** infimum

7. supremum   **8.** irrational   **9.** point p   **10.** open   **11.** A

12. open   **13.** Derived   **14.** limit   **15.** nbd   **16.** limit

17. 0   **18.** empty   **19.** bounded   **20.** $[a,b]$ **21.** $[2, 4]$

22. 1

TRUE/FALSE

1. T	**2.** F	**3.** T	**4.** F	**5.** F	**6.** T	**7.** F	**8.** T	**9.** T
10. F	**11.** T	**12.** T	**13.** T	**14.** T	**15.** F	**16.** T	**17.** T	**18.** F
19. F	**20.** F	**21.** T	**22.** F					

MULTIPLE CHOICE QUESTIONS

Problem set-1

1. (a)	**2.** (d)	**3.** (a)	**4.** (c)	**5.** (c)	**6.** (c)	**7.** (c)	**8.** (d)	**9.** (a)
10. (c)	**11.** (d)	**12.** (c)	**13.** (c)	**14.** (a)	**15.** (d)	**16.** (a)	**17.** (b)	**18.** (a)
19. (c)	**20.** (c)							

Problem set-2

1. (c)	**2.** (c)	**3.** (b)	**4.** (b)	**5.** (a)	**6.** (c)	**7.** (d)	**8.** (b)	**9.** (b)
10. (b)	**11.** (b)	**12.** (a)	**13.** (a)	**14.** (a)	**15.** (b)	**16.** (a)	**17.** (b)	**18.** (a)
19. (a)	**20.** (a)	**21.** (b)	**22.** (b)	**23.** (a)	**24.** (b)	**25.** (a)	**26.** (a)	**27.** (b)
28. (a)	**29.** (a)	**30.** (c)						

Problem set-3

1. (c)	**2.** (a)	**3.** (b)	**4.** (c)	**5.** (a)	**6.** (b)	**7.** (a)	**8.** (a)	**9.** (b)
10. (a)	**11.** (a)	**12.** (b)	**13.** (c)	**14.** (b)	**15.** (c)	**16.** (a)	**17.** (c)	**18.** (a)
19. (c)	**20.** (a)	**21.** (b)	**22.** (a)	**23.** (c)	**24.** (a)	**25.** (a)	**26.** (a)	**27.** (a)
28. (a)	**29.** (b)	**30.** (b)	**31.** (a)	**32.** (a)	**33.** (a)	**34.** (a)	**35.** (b)	**36.** (a)
37. (b)	**38.** (a)	**39.** (c)	**40.** (a)	**41.** (d)	**42.** (a)	**43.** (a)	**44.** (b)	**45.** (d)
46. (d)	**47.** (d)	**48.** (d)	**49.** (a)	**50.** (c)				

Problem set-4

1. (a)	**2.** (d)	**3.** (a)	**4.** (a)	**5.** (a)	**6.** (a)	**7.** (a)	**8.** (c)	**9.** (a)
10. (b)	**11.** (d)	**12.** (a)	**13.** (a)	**14.** (b)	**15.** (a)	**16.** (b)	**17.** (c)	**18.** (b)
19. (b)	**20.** (b)	**21.** (b)	**22.** (a)	**23.** (b)	**24.** (b)	**25.** (b)	**26.** (a)	**27.** (b)
28. (c)	**29.** (b)	**30.** (b)	**31.** (b)	**32.** (a)	**33.** (b)	**34.** (b)	**35.** (a)	

Mathematical Analysis

SELF ASSESSMENT TEST

Verify each of the following :

1. The set Q of rational numbers and the set of real numbers R are ordered fields while the set N of natural numbers and set of integers Z are not fields.
2. The set Z, Q and R are not bounded.
3. The set N of natural numbers is bounded below but not above.
4. The set of rational numbers is not order-complete.
5. The set R of real numbers has the order completeness property as well as Dedekind's property.
6. The set R of real numbers is the neighbourhood of each of its points.
7. The set Q of rationals is not a nbd of any of its points.
8. The open interval] a, b [is a nbd of each of its points.
9. The closed interval [a, b] is a nbd of each of its points except the end points.
10. A non-empty finite set is not a nbd of any of its point.
11. The interior of the set N, Z or Q is empty but interior of R is non-empty.
12. The set R of real numbers is open as well as closed set.
13. The set Q of rationals is neither open nor closed.
14. The set Z of integers has no limit point.
15. Every point of R is a limit point.
16. Every point of Q is a limit point.
17. The set $\left[\dfrac{1}{n} : n \in N\right]$ has only one limit point say 0.
18. Every point of the closed interval is its limit point and a point not belonging to the interval is not a limit point.
19. The set of real numbers is a perfect set.
20. The set Q of rational numbers and the set of irrationals are dense in R.
21. The set of integers and set of all ordered pairs of integers is countable.
22. The set R of real numbers is uncountable.
23. The set of all polynomial functions with integer coefficient is countable.
24. The set [a, b] = [$x : a \le x \le b$] is a perfect set.
25. The set $S = [x : 0 < x < 1 : x \in R]$ is open but not closed.

Chapter

2 Sequences

George Cantor (1845-1918) is known as the creater of the set theory. He made a considerable contribution to the development of the theory of real sequence, and found a firm base for most of the fundamental concepts of real analysis in the sequence of rational numbers. Though his lay-outs are not convenient in the initial stages, they are quite advantageous while making advanced investigations. The study of many important and advanced concepts becomes easy if the notion of the sequence is employed.

Set of Numbers

We shalll be using capital latters N, Z, Q and R for the set of numbers as specified below:

$N = \{n : n = 1, 2, 3, ...\}$, the set of natural numbers.

$Z = \{x : x = ... - 2, - 1, 0, 1, 2, ...\}$, the set of integers.

$Q = \{x : x \text{ is a rational numbers}\}$, the set of rational numbers.

and $R = \{x : x \text{ is a real numbers}\}$, the set of real numbers.

2.2 SEQUENCES

Let N be the set of all natural numbers and S be any set of real numbers. A function whose domain is the set of natural numbers and range is a subset of S, is called a sequence in S.

Symbolically, if we define a function $f : N \rightarrow S$, then f is a sequence. As in the case of function, we shall denote a sequence in a number of ways.

(i) Usually a sequence is denoted by its images. For a sequence f, the image corresponding to $n \in N$ is denoted by f_n or $f\langle n \rangle$ and is called the nth term of the sequence f.

   For example: $\langle 1, 4, 9, ... \rangle$ is the sequence whose n^{th} terms is n^2.

(ii) Using in order, the first few elements of a sequence, till the rule for writing down different elements becomes clear.

   For example: $\langle 1, 2, 3, ... \rangle$ is the sequence whose n^{th} term is n.

(iii) Defining a sequence by a recurrence formula *i.e.* , by a rule which expresses the nth term by the $(n-1)^{th}$ term.

   For example : Let $a_1 = 1$, $a_{n+1} = 2a_n$, for all $n \geq 1$.

   Above relations define a sequence whose n^{th} term is 2^{n-1}.

REMARKS

- A sequence is represented as $\langle s_n \rangle$ or $\{s_n\}$, where s_n is the nth term of the sequence.
- The set of all distinct terms of a sequence is called the range set of that sequence.
- A sequence whose range, is a subset of R is called a real sequence or a sequence of real numbers.
- Here, we shall study only real numbers. Therefore, the term sequence will be used to denote a real sequence.

☛ ILLUSTRATIONS

(1) $\left\langle \dfrac{1}{n} \right\rangle$ is the sequence $\left\langle 1, \dfrac{1}{2}, \dfrac{1}{3}, \dfrac{1}{4}, ..., \dfrac{1}{n}, ... \right\rangle$

(2) $\left\langle \dfrac{1}{n^3} \right\rangle$ is the sequence $\left\langle 1, \dfrac{1}{8}, \dfrac{1}{27}, ..., \dfrac{1}{n^3}, ... \right\rangle$

(3) $\langle -2n \rangle$ is the sequence $\langle -2, -4, -6, ..., -2n, ... \rangle$

(4) $\left\langle \dfrac{n}{n+1} \right\rangle$ is the sequence $\left\langle \dfrac{1}{2}, \dfrac{2}{3}, \dfrac{3}{4}, ..., \dfrac{n}{n+1}, ... \right\rangle$.

2.2.1 RANGE OF A SEQUENCE

The set of all distinct terms of a sequence is known as its range.

2.2.2 CONSTANT SEQUENCE

A sequence $\langle s_n \rangle$ defined by $s_n = a$ for all $n \in N$, is called a constant sequence.

2.2.3 EQUALITY OF TWO SEQUENCES

Two sequences $\langle s_n \rangle$ and $\langle t_n \rangle$ are said to be equal, if $s_n = t_n; \forall n \in N$.

2.2.4 OPERATION ON SEQUENCES

Since the sequences are real valued functions, therefore, the sum, difference, product etc. of two sequences are defined as follows:

(1) If $\langle s_n \rangle$ and $\langle t_n \rangle$ be any two sequences, then the sequences whose n^{th} terms are $s_n + t_n$, $s_n - t_n$ and $s_n \cdot t_n$ are respectively known as the sum, difference and product of the sequences $\langle s_n \rangle$ and $\langle t_n \rangle$ and are denoted by $(s_n + t_n)$, $(s_n - t_n)$ and $(s_n \cdot t_n)$ respectively.

(2) If $s_n \neq 0$, $\forall n \in N$, then the sequence whose n^{th} term is $\dfrac{1}{s_n}$ is called the reciprocal of the sequence $\langle s_n \rangle$ and is denoted by $\left\langle \dfrac{1}{s_n} \right\rangle$.

(3) The sequence whose n^{th} term is s_n / t_n $(t_n \neq 0, \forall n \in N)$ is known as the quotient of the sequence $\langle s_n \rangle$ by the sequence $\langle t_n \rangle$ and is denoted by $< \dfrac{s_n}{t_n} >$.

(4) The sequence whose n^{th} term is $k s_n$, where $k \in R$ is known as the scalar multiple of the sequence $\langle s_n \rangle$ by k and is denoted by $\langle k s_n \rangle$

2.3 BOUNDED SEQUENCES

2.3.1 BOUNDED BELOW SEQUENCE

A sequence $\langle s_n \rangle$ is said to be bounded below if there exists a real number l such that $s_n \geq l$, $\forall\, n \in N$. The number l is known as the lower bound of the sequence $\langle s_n \rangle$.

2.3.2 BOUNDED ABOVE SEQUENCE

A sequence $\langle s_n \rangle$ is said to be bounded above if there exists a real number u such that $s_n \leq u$; $\forall\, n \in N$. The number u is said to be upper bound of the sequence $< s_n >$.

2.3.3 BOUNDED SEQUENCE

A sequence $\langle s_n \rangle$ is said to be bounded if it is bounded above as well as bounded below.

or A sequence $\langle s_n \rangle$ is bounded if there exist two real numbers l and $u(l \leq u)$ such that $l \leq s_n \leq u$, $\forall\, n \in N$. Equivalently, a sequence is bounded iff there exists a real number $k > 0$ such that

$$| s_n | \leq k, \forall\, n \in N.$$

2.3.4 UNBOUNDED SEQUENCE

A sequence $\langle s_n \rangle$ is said to be unbounded if it is not bounded.

REMARK

- In sequences, terms with equal values can occur. Therefore, a sequence may have more than one term with the smallest value. In such a case any of those is taken for the smallest value. In fact while talking about the smallest value we are interested in the value of the term rather than the position of the term in the sequence. Similar explanation holds for the greatest value. Note that, like sets of real numbers, a sequence bounded below or above may or may not have a smallest or a greatest member accordingly. Clearly, an unbounded sequence cannot have a smallest or a greatest member.

2.3.5 LEAST UPPER BOUND

If a sequence $\langle s_n \rangle$ is bounded above, then there exists a number u_1 such that

$$s_n \leq u_1, \forall\, n \in N. \qquad \qquad \dots (1)$$

This number u_1 is called an upper bound of the sequence $\langle s_n \rangle$. If $u_1 < u_2$, then from (1) we find that $s_n \leq u_2$, $\forall\, n \in N$. Which implies, u_2 is also an upper bound of the sequence $\langle s_n \rangle$. Hence, we can say any number greater than u_1 is an upper bound of $\langle s_n \rangle$.

Hence, a sequence has an infinite number of upper bounds if it is bounded above. Let u be the least of all the upper bounds of the sequence $\langle s_n \rangle$. Then u is defined as the least upper bound (l.u.b.) or supremum of the sequence $\langle s_n \rangle$.

2.3.6 GREATEST LOWER BOUND

If a sequence $\langle s_n \rangle$ is bounded below then there exists a number $l_1 \in R$ such that

$$l_1 \leq s_n ; \forall\, n \in N \qquad \qquad \dots(1)$$

This number l_1 is known as the lower bound of $\langle s_n \rangle$. If $l_2 < l_1$, then from (1) we have

$$l_2 \leq s_n ; \forall\, n \in N \qquad \qquad \dots(2)$$

which implies, l_2 is also a lower bound of the sequence $\langle s_n \rangle$. Hence, we can say any number less than l_1 is a lower bound of $\langle s_n \rangle$.

Hence, a sequence has infinite number of lower bounds, if it is bounded below. Let l is the greatest of all the lower bounds of the sequence $\langle s_n \rangle$. Then l is known as greatest lower bound (g.l.b.) or infimum of the sequence $\langle s_n \rangle$.

☞ **ILLUSTRATIONS**

(1) The sequence $\langle n^2 \rangle$ is bounded below by 1 but not bounded above.

(2) The sequence $\left\langle \dfrac{n}{n+1} \right\rangle$ is bounded as $\dfrac{1}{2} \le \dfrac{n}{n+1} < 1; \forall n \in N$

(3) The sequence $\langle -n^2 \rangle$ is bounded above by -1 but not bounded below.

(4) The sequence $\left\langle \dfrac{1}{n} \right\rangle$ is bounded since $\left| \dfrac{1}{n} \right| \le 1; \forall n \in N.$

(5) The sequence $\langle (-1)^n \rangle$ is bounded since $| (-1)^n | \le 1; \forall n \in N.$

$$[\because | (-1)^n | = 1; \forall n \in N]$$

(6) The sequence $\langle s_n \rangle$ defined by $s_n = 1 + (-1)^n$ for all $n \in N$ is bounded since the range set of the sequence is $\{0, 2\}$, which is a finite set.

(7) The sequence $\langle (-1)^n / n \rangle$ is bounded since $| (-1)^n / n | \le 1$ for all $n \in N.$

(8) The sequence $\langle 2^n \rangle$ is bounded below and has smallest term as 2. Every member of $]-\infty, 2]$ is a lower bound of the sequence and the sequence is unbounded above.

THEOREM 1. *A sequence $\langle s_n \rangle$ is bounded iff there exists a positive integer $a > 0$ and $m, l \in R$ such that $|s_n - l| < a ; \forall n \ge m.$*

Proof. Let $\langle s_n \rangle$ be a bounded sequence. Then there exist two real numbers c_1 and c_2 such that $c_1 < s_n < c_2 \; \forall n \in N$

or $\quad \left(c_1 - \dfrac{c_1 + c_2}{2} \right) < \left(s_n - \dfrac{c_1 + c_2}{2} \right) < \left(c_2 - \dfrac{c_1 + c_2}{2} \right) \forall n \in N$

or $\quad \left(\dfrac{c_1 - c_2}{2} \right) < \left(s_n - \dfrac{c_1 + c_2}{2} \right) < \left(\dfrac{c_2 - c_1}{2} \right) \forall n \in N$

or $\quad -a < (s_n - l) < a \; \forall \; n \in N$ where, $a = \dfrac{c_2 - c_1}{2}$ and $l = \dfrac{c_1 + c_2}{2}$

or $\quad | s_n - l | < a \; \forall \; n \in N$ where, $m = 1 \in N, l \in R$ and $a > 0.$

Conversely, let there exists $l \in R, a > 0$ and $m \in N$ such that

$$|s_n - l| < a \; \forall \, n \ge m.$$

This gives $\quad l - a < s_n < l + a \; \forall \, n \ge m$

Let $\quad c_1 = \min\{s_1, s_2, \dots, s_{m-1}, l - a\}$

and $\quad c_2 = \max\{s_1, s_2, \dots, s_{m-1}, l + a\}$

Then $\quad c_1 \le s_n \le c_2 \; \forall \, n \in N.$

Therefore, $\langle s_n \rangle$ is a bounded sequence.

2.4 LIMIT POINT OF A SEQUENCE

A real number l is called a limit point of a sequence $\langle s_n \rangle$ if every nbd of l contains infinite number of terms of the sequence. Thus $l \in R$ is a limit point of the sequence $\langle s_n \rangle$ if for given $\varepsilon > 0, s_n \in]l - \varepsilon, l + \varepsilon[$, for infinitely many points.

REMARKS

- Limit point of a sequence need not be a member of the sequence.
- A limit point of a sequence may or may not be a limit point of the range of the sequence but the limit point of the range of a sequence is always a limit point of the sequence.
- In the case of set of real numbers, limit points of a sequence may also be called accumulation, cluster or condensation points.

2.4.1 CLASSIFICATON OF LIMIT POINTS

The limit points of a sequence may be classified in two types :
(i) those for which $l = s_n$ for infinitely many values of $n \in N$.
(ii) those for which $l = s_n$ for only a finite number of values of $n \in N$.
But this distinction is not very much needed. As such we do not distinguish the above mentioned two types of limit points of sequences by different titles.

☞ ILLUSTRATIONS

(1) The sequence $\left\langle \dfrac{1}{n} \right\rangle$ has one limit point namely 0.

(2) The sequence $\left\langle (-1)^n \right\rangle$ has two limit points 1 and –1.

(3) The sequence $\langle n \rangle$ has no limit point.

(4) The sequence $\left\langle 1 + \dfrac{(-1)^n}{n} \right\rangle$ has one limit point *i.e.*, 1.

(5) The sequence $\left\langle 1, \dfrac{1}{2}, 1, \dfrac{1}{3}, 1, \dfrac{1}{4} \dots \right\rangle$ has one limit point *i.e.*, 1.

(6) The sequence $\langle n + 1 \rangle$ has no limit point.

2.4.2 SUFFICIENT CONDITIONS FOR NUMBER *l* BE OR NOT TO BE A LIMIT POINT OF THE SEQUENCE $<s_n>$

(1) If for every $\varepsilon > 0$, $\exists \; m \in N$ such that $s_n \in]l - \varepsilon, l + \varepsilon[\; \forall \; n \geq m$ or equivalently $|s_n - l| < \varepsilon \; \forall \; n \geq m$, then l is the limit point of the sequence $\langle s_n \rangle$.

(2) If for any $\varepsilon > 0$. $s_n \in]\, l - \varepsilon, l + \varepsilon$ [for only a finite number of values of n, then l is not a limit point of the sequence $\langle s_n \rangle$. Such a condition is also necessary for a number l not to be a limit point of the sequence $\langle s_n \rangle$.

REMARKS

- Whenever we simply write $\varepsilon > 0$, it is implied that ε may be however small positive number.
- A positive number δ is said to be arbitrary small if given $\varepsilon > 0$, δ may be chosen such that $0 < \delta < \varepsilon$.
- If δ be an arbitrarily small positive number and given any $k > 0$, then $k\delta$ is also an arbitrarily small positive numbers.
- If $\varepsilon_1, \varepsilon_2$, are any two arbitrarily small positive numbers then it follows that l is the limit point of the sequence s_n iff $s_n \in \;] \, l - \varepsilon_1, l + \varepsilon_2$ [for infinitely many values of n.

THEOREM 1. **(Bolzano-Weirstrass Theorem for sequence).** *Every bounded sequence has at least one limit point.*

Proof.   Let $S = \{s_n : n \in \mathbb{N}\}$ be the range set of the bounded sequence $\langle s_n \rangle$. Then S is bounded set. Now there may be two cases :

Case I. Let S be a finite set. Then $s_n = p$ for infinitely many indices n. Here $p \in \mathbb{R}$. Obviously p is a limit point of $\langle s_n \rangle$.

Case II. Let S be an infinite set. Since S is bounded, then by Bolzano-Weierstrass theorem for sets of real numbers, S has a limits point, say p. Therefore every nbd of p contains infinitely many distinct point of S i.e., infinitely many terms of $\langle s_n \rangle$ and hence p is a limit point of the sequence $\langle s_n \rangle$.

2.5 LIMIT SUPERIOR AND LIMIT INFERIOR

The greatest $\overline{\text{limit}}$ point of a bounded sequence is called the upper limit or limit superior and is denoted by $\overline{\lim}\ s_n$ and the smallest limit point of a bounded sequence is called the lower limit or limit inferior and is denoted by $\underline{\lim}\ s_n$.

By definition it is obvious that $\underline{\lim}\ s_n \leq \overline{\lim}\ s_n$

A bounded sequence $\langle s_n \rangle$ for which the upper limit and lower limit coincide with real number l is said to converge to l.

Limit of sequence. A sequence $\langle s_n \rangle$ is said to have a limit l if for a given $\varepsilon > 0$ there exists a positive integer m such that

$$|s_n - l| < \varepsilon, \forall\ n \geq m.$$

2.6 CONVERGENT SEQUENCES

Definition (1) : *A sequence $\langle s_n \rangle$ is said to converge to a number l, if for a given $\varepsilon > 0$ there exists a positive integer m such that*

$$|s_n - l| < \varepsilon\ \forall\ n \geq m.$$

The number l is called the limit of the sequence $\langle s_n \rangle$ and can be written as

$$s_n \to l \text{ as } n \to \infty \ \text{ or } \ \lim_{n \to \infty} s_n = l \ \text{ or } \lim s_n = l.$$

Definition (2): *A sequence $\langle s_n \rangle$ is said to be convergent iff it is bounded and has one and only one limit point.*

In such a case the sequence is said to converge to this limit point 1.

2.7 SUBSEQUENCES

Let $\langle s_n \rangle$ be any sequence. If $\langle n_1, n_2, ..., n_k \rangle$ be a strictly increasing sequence of positive integers *i.e.* , $i > j \Rightarrow n_i > n_j$, then the sequence

$\left\langle s_{n_1}, s_{n_2}, ..., s_{n_k} \right\rangle$ is called a subsequence of $\langle s_n \rangle$.

THEOREM 1. *If $\langle s_n \rangle$ is a sequence of non-negative numbers such that $\lim s_n = l$, then $l \geq 0$.*

Proof.   Let if possible $l < 0$ then $-l > 0$. Now $\lim s_n = l$, therefore, for $\varepsilon = -\dfrac{1}{2} > 0$, there exists a positive integer m such that

$$|s_n - l| < -\frac{l}{2} \ \forall\ n \geq m.$$

In particular

$$|s_m - l| < -\frac{l}{2} \ \Rightarrow l + \frac{l}{2} < s_m < l - \frac{l}{2} \ \Rightarrow s_m < \frac{l}{2} < 0$$

which is a contradiction, because $s_m \geq 0$. Therefore our assumption is wrong. Hence, we must have $l \geq 0$.

THEOREM 2. *A sequence cannot converge to more than one limit point.*

Proof.　　　Let if possible, a sequence $\langle s_n \rangle$ converges to two distinct numbers l_1 and l_2.

Now　　　　　$l_1 \neq l_2$　　$\Rightarrow l_1 - l_2 \neq 0$　　　　$\Rightarrow |l_1 - l_2| > 0$

Let $\varepsilon = \dfrac{1}{2} |l_1 - l_2|$; then $\varepsilon > 0$.

Since $\langle s_n \rangle$ converges to l_1, there must exists a positive integer m_1 such that

$$|s_n - l_1| < \varepsilon, \ \forall \, n \geq m_1 \qquad\qquad\qquad \text{...(1)}$$

Similarly $\langle s_n \rangle$ converges to l_2, there must exists a positive integer m_2 such that

$$|s_n - l_2| < \varepsilon, \ \forall \, n \geq m_2 \qquad\qquad\qquad \text{...(2)}$$

Now, let $m = \max\{m_1, m_2\}$

Then result (1) and (2) hold for all $n \geq m$. So for all $n \geq m$, we have

$$|l_1 - l_2| = |(s_n - l_1) - (s_n - l_2)|$$
$$\leq |(s_n - l_1) + (s_n - l_2)|$$
$$< \varepsilon + \varepsilon \qquad\qquad\qquad \text{[Using (1) and (2)]}$$
$$= 2\varepsilon = |l_1 - l_2|$$
$$\Rightarrow \qquad |l_1 - l_2| < |l_1 - l_2|$$

which is absurd, hence we must have $l_1 = l_2$ *i.e.*, the limit of the sequence is unique.

THEOREM 3. *Every convergent sequence is bounded.*

Proof.　　　Let $\langle s_n \rangle$ be a sequence which converges to l. Take $\varepsilon = 1$. Then there exists a positive integer m such that

$$|s_n - l| < 1, \ \forall \, n \geq m$$

i.e.,　　　　　$(l - 1) < s_n < (l + 1), \ \forall \, n \geq m.$

Let　　　　　$k_1 = \min\{s_1, s_2, ..., s_{m-1}, l - 1\}$

and　　　　　$k_2 = \max\{s_1, s_2, ..., s_{m-1}, l + 1\}$

therefore　　　$k_1 \leq s_n \leq k_2, \forall \, n \in N$

Hence, the sequence $\langle s_n \rangle$ is bounded.

REMARK

- The converse of the above theorem is *not* necessarily true *i.e.*, a bounded sequence need not be convergent. For example $\langle (-1)^n \rangle$ is bounded but not convergent.

THEOREM 4. *If $\langle s_n \rangle$ converges to l, then any subsequence of $\langle s_n \rangle$ also converges to l.*

Proof.　　　Let $\langle s_{n_k} \rangle$ be any subsequence of $\langle s_n \rangle$. Then by definition of subsequence $n_1, n_2, ..., n_k$, are positive integers such that

$$n_1 < n_2 < ... < n_k < ...$$

Now　　　　　$n_1 \geq 1 \ \Rightarrow n_k \geq k$　　　　　　　　[By induction]

Since $\langle s_n \rangle$ converges to l, so given $\varepsilon > 0$, there exists a positive integer m such that

$$|s_k - l| < \varepsilon \ \forall \, k \geq m$$

for $k \geq m$, we have　　$n_k \geq k \geq m$

therefore　　　$|s_{n_k} - l| < \varepsilon$, for all $n_k \geq m$

Hence,　　　　$\langle s_{n_k} \rangle$ converges to l.

REMARKS

- All subsequences of a convergent sequence, converges to the same limit.
- To show that, a given sequence is not convergent, it is enough to show that two of its subsequences converges to different limits.

    (**Ex.** The sequence $\langle s_n \rangle = \langle (-1)^n \rangle$ is not convergent. Since the two subsequences $\langle 1, 1, ... \rangle$ and $\langle -1, -1, -1, ... \rangle$ of the given sequence converges to 1 and -1 respectively, which are not same.)

- If the subsequences $\langle s_{2n-1} \rangle$ and $\langle s_{2n} \rangle$ of the sequences $\langle s_n \rangle$ converges to the same lmit l, then the sequence $\langle s_n \rangle$ converge to l.
- If the subsequence $\langle s_{2n+1} \rangle$ and $\langle s_{2n} \rangle$ of the sequence $\langle s_n \rangle$ converges to the same limit l, then the sequence $\langle s_n \rangle$ converges to l.

THEOREM 5. *The limit of the sum of two convergent sequences is the sum of their limits.*

Proof.   Let $\langle s_n \rangle$ and $\langle t_n \rangle$ be the two given sequences such that

$$\lim s_n = l_1 \qquad \qquad ...(1)$$

and $$\lim t_n = l_2 \qquad \qquad ... (2)$$

Since, $\lim s_n = l_1$, therefore for a given $\varepsilon > 0$, there exists a positive integer m_1 such that

$$|s_n - l_1| < \varepsilon /2, \ \forall \, n \geq m_1.$$

Similarly, $\lim t_n = l_2$, therefore, for a given $\varepsilon > 0$, there must exists a positive integer m_1 such that

$$|t_n - l_2| < \varepsilon /2, \ \forall \, n \geq m_2.$$

Let $m = \max \{m_2, m_2\}$.

Therefore $$|s_n - l_1| < \varepsilon /2, \ \forall \, n \geq m.$$

and $$|t_n - l_2| < \varepsilon /2, \ \forall \, n \geq m.$$

Now, consider $| (s_n + t_n) - (l_1 + l_2) | = | (s_n - l_1) + (t_n - l_2)|, \ \forall \, n \geq m$
$$\leq |s_n - l_1| + |t_n - l_2|, \ \forall \, n \geq m$$
$$< \varepsilon /2 + \varepsilon /2 = \varepsilon, \ \forall \, n \geq m$$

Therefore, the sequence $\langle \, s_n + t_n \, \rangle$ is convergent and

$$\lim (s_n + t_n) = l_1 + l_2 = \lim s_n + \lim t_n.$$

REMARKS

- The converse of the theorem need not be true. For example, the sequence $\langle s_n \rangle = \langle (-1)^n \rangle$ and $\langle t_n \rangle = \langle (-1)^{n+1} \rangle$ are not convergent, but the sequence $\langle s_n + t_n \rangle = \langle (-1)^n + (-1)^{n+1} \rangle$ converges to 0.
- The limit of the differences of two convergent sequence is the difference of the limits. (Proof is similar as Theorem 5).
- If $\langle s_n \rangle$ and $\langle t_n \rangle$ are convergent sequences such that $s_n \leq t_n, \ \forall n$ and $\lim s_n = a, \ \lim t_n = b$, then $a \leq b$.

THEOREM 6. *If $\lim s_n = l_1$ and $\lim t_n = l_2$, then $\lim (s_n t_n) = l_1. l_2.$*

Proof.   We have

$$|s_n t_n - l_1 l_2| = |s_n t_n - l_1 t_n + l_1 t_n - l_1 l_2|$$
$$= |t_n (s_n - l_1) + l_1 (t_n - l_2)|$$
$$\leq |t_n| |s_n - l_1| + |l_1| |t_n - l_2| \qquad \qquad ...(1)$$

The sequence (t_n) is convergent, therefore it is bounded, ($\therefore$ Every convergent sequence is bounded) so there must exists a positive real number c such that

$$|t_n| \leq c, \forall\, n \in N \qquad \qquad \text{...(2)}$$

Since the sequences $\langle s_n \rangle$ and $\langle t_n \rangle$ both are convergent, there must exist positive integers m_1 and m_2 such that

$$|s_n - l_1| < \varepsilon/2c, \forall\, n \geq m_1 \qquad \qquad \text{... (3)}$$

and $\qquad |t_n - l_2| < \varepsilon/2|l|, \forall\, n \geq m_2$

Let $m = \max\{m_1, m_2\}$.

From (1), (2), (3) and (4) we have

$$|s_n\, t_n - l_1\, l_2| < c\,\frac{\varepsilon}{2c} + |l| \cdot \frac{\varepsilon}{2|l|}, \forall n \geq m$$

$$< \varepsilon/2 + \varepsilon/2 = \varepsilon, \forall n \geq m.$$

Therefore $\lim (s_n t_n) = l_1 l_2$.

THEOREM 7. *If $\lim s_n = l_1$, $l_1 \neq 0$ and $s_n \neq 0$, $\forall\, n \in N$ then $\lim \left(\dfrac{1}{s_n}\right) = \dfrac{1}{l_1}$*

Proof. Since $l_1 \neq 0$, there exists a positive number c and positive integer m_1 such that

$$|s_n| > c, \forall n \geq m_1. \qquad \qquad \text{...(1)}$$

Also $\lim s_n = l_1$, therefore, for a given $\varepsilon > 0$, there must exists a positive integer m_2 such that

$$|s_n - l_1| < c|\,l_1\,|\varepsilon, \forall\, n \geq m_2. \qquad \qquad \text{...(2)}$$

Let $m = \max\{m_1, m_2\}$. Then

$$\left|\frac{1}{s_n} - \frac{1}{l_1}\right| = \left|\frac{s_n - l_1}{|s_n| \cdot |l_1|}\right| < \frac{c\,|\,l_1\,|}{c\,|\,l_1\,|}\varepsilon, \forall n \geq m$$

$$= \varepsilon, \forall n \geq m.$$

Therefore, $\qquad \lim \dfrac{1}{s_n} = \dfrac{1}{l_1}$

THEOREM 8. *If $\lim s_n = l_1$ and $\lim t_n = l_2$ ($l_2 \neq 0$), $t_n \neq 0$, $\forall n \in N$ then $\lim \dfrac{s_n}{t_n} = \dfrac{l_1}{l_2}$*

Proof. We have $\quad \lim\left[\dfrac{s_n}{t_n}\right] = \lim\left(s_n \dfrac{1}{t_n}\right)$

$$= \lim(s_n).\lim\left(\frac{1}{t_n}\right) \qquad [\because \text{limit of the product of two}$$

$$\text{sequences is equal to the product of the limits.]}$$

$$= l_1 \cdot \frac{1}{l_2} \qquad\qquad\qquad\qquad\qquad \text{[By previous theorem]}$$

$$\Rightarrow \qquad \lim_{n \to \infty} \frac{s_n}{t_n} = \frac{l_1}{l_2}$$

2.8 DIVERGENT SEQUENCES

Definition (1): *A sequence $\langle s_n \rangle$ is said to diverge to $+\infty$, if for every real number $k > 0$, there exists a positive integer m such that,*

$$s_n > k, \forall\, n \geq m$$

Definition (2): *A sequence $\langle s_n \rangle$ is said to diverge to $-\infty$, if for every real number $k < 0$, there exists a positive integer m such that,*

$$s_n < k, \ \forall \ n \geq m$$

Definition (3): *A sequence is said to be divergent sequence if it diverges to either $+\infty$ or $-\infty$.*

Definition (4): *A sequence which is not convergent, is known as divergent sequence.*

☛ **ILLUSTRATIONS**

(1) $\langle 3, 3^2, 3^3, ... \rangle$ diverges to $+\infty$.

(2) $\langle -2, -2^2, -2^3, ... \rangle$ diverges to $-\infty$.

(3) $\langle 2, 4, 6, ..., 2n, ... \rangle$ diverges to $+\infty$.

(4) $\langle -2, -4, -6, ..., -2n, ... \rangle$ diverges to $-\infty$.

2.9 OSCILLATORY SEQUENCES

A sequence $\langle s_n \rangle$ is said to be oscillatory if it is neither convergent nor divergent.

An oscillatory sequence is said to oscillate finitely or infinitely according as it is bounded or unbounded.

In other words, we can say

(i) A bounded sequence, which is not convergent is said to oscillate finitely.

(ii) An unbounded sequence which does not diverge, is said to oscillate infinitely.

(iii) A bounded sequence which does not converge and has at least two limit points, is said to be oscillate finitely.

☛ **ILLUSTRATIONS**

(1) $\langle 1 + (-1)^n \rangle$ oscillate finitely.

(2) $\langle (-1)^n \rangle$ oscillate finitely.

(3) $\left\langle (-1)^n \left(1 + \dfrac{1}{n} \right) \right\rangle$ oscillate finitely.

(4) $\langle n(-1)^n \rangle$ oscillate infinitely.

__THEOREM 1.__ *If a sequence $\langle s_n \rangle$ diverges to infinity, then any subsequence of $\langle s_n \rangle$ also diverges to infinity.*

Proof. Let $\left\langle s_{n_k} \right\rangle$ be any subsequence of the sequence $\langle s_n \rangle$. Then by definition of subsequence, $\langle n_1, n_2, ..., n_k, ... \rangle$ is a strictly increasing sequence of positive integers

$\Rightarrow \qquad\qquad n_1 \geq 1 \Rightarrow n_k \geq k.$ (By induction)

Take any positive real number c_1.

Now $\langle s_n \rangle$ diverges to $\infty \Rightarrow$ for $c_1 > 0$, $\exists \ m \in N$ such that $s_n > c_1$ for all $n \geq m$ i.e., $s_k > c_1, \ \forall \ k \geq m$ we have $n_k \geq k \geq m$, i.e., $n_k \geq m$.

$\therefore \qquad s_{n_k} > c_1$ for all $c_k \geq m$.

$\Rightarrow \qquad \left\langle s_{n_k} \right\rangle$ diverges to ∞.

REMARKS

- If $s_{2n-1} \to \infty$ as $n \to \infty$ and $s_{2n} \to \infty$ as $n \to \infty$, then $s_n \to \infty$ as $n \to \infty$.
- If $s_n > 0$ for all $n \in N$, then $s_n \to \infty$ as $n \to \infty$, $\Leftrightarrow \dfrac{1}{s_n} \to 0$ as $n \to \infty$.

THEOREM 2. *If the sequence $\langle s_n \rangle$ diverges to infinity and the sequence $\langle t_n \rangle$ is bounded, then $\langle s_n + t_n \rangle$ diverges to infinity.*

Proof.   The sequence $\langle t_n \rangle$ is bounded; therefore for arbitrary positive number k_1 we have
$$|t_n| < k_1.$$
Also, the sequence $\langle s_n \rangle$ diverges to infinity. Therefore for arbitrary positive number k there must exists a positive integer m such that
$$s_n > k + k_1 , \ \forall \, n \geq m$$
Now, for all $n \geq m$, we have
$$s_n + t_n \geq s_n - |t_n| > k + k_1 - k_1 = k.$$
Thus for $k > 0$, $\exists$ a positive integer m such that
$$s_n + t_n > k, \forall \, n \geq m.$$
$\Rightarrow$ The sequence $\langle s_n + t_n \rangle$ diverges to infinity.

THEOREM 3. *If the sequences $\langle s_n \rangle$ and $\langle t_n \rangle$ both diverges to infinity, then the sequences $\langle s_n + t_n \rangle$ and $\langle s_n . t_n \rangle$ diverges to infinity.*

Proof.   Since the sequence $\langle s_n \rangle$ diverges to infinity, therefore for $k_1 > 0$, there must exists a positive integer m_1 such that $s_n > k_1 \ \forall \ n \geq m$. Similarly, the sequence $\langle t_n \rangle$ diverges to infinity, therefore for $k_2 > 0$, there must exists a positive integer m_2 such that
$$s_n > k_2, \forall \, n \geq m_2.$$
Let $m = \max \{m_1, m_2\}$. Then
$$s_n + t_n > k_1 + k_2 = l_1 (\text{say})$$
and $s_n t_n > k_1 . k_2 = l_2 (\text{say}).$
Therefore, sequences $\langle s_n + t_n \rangle$ and $\langle s_n \, t_n \rangle$ diverges to infinity.

Solved Examples

Example 1. *Show that the sequence $\left\langle \dfrac{1}{n} \right\rangle$ converges to 0.*

Solution.   Let $\langle s_n \rangle = \left\langle \dfrac{1}{n} \right\rangle$

Now
$$\lim_{n \to \infty} s_{2n} = \lim_{n \to \infty} \frac{1}{2n} = 0$$
and
$$\lim_{n \to \infty} s_{2n+1} = \lim_{n \to \infty} \frac{1}{2n + 1} = 0$$
Therefore
$$\lim_{n \to \infty} s_{2n} = \lim_{n \to \infty} s_{2n+1} = 0$$

Based on the following Results

- A function $f : N \to S$ whose domain is the set of natural numbers and range is the subset of S is called sequence.

- If for any sequence $\langle s_n \rangle$, $l \leq s_n \leq u$ then l and u are respectively called the lower and upper bounds of $\langle s_n \rangle$

- A sequence $\langle s_n \rangle$ is said to have a limit l if for a given $\varepsilon > 0$ $\exists$ a positive integer m such that
$$|s_n - l| < \varepsilon \, \forall \, n \geq m.$$

- To show that a given sequence is not convergent, it is enough to show that two of its subsequences converges to different limits.

- $\lim (s_n \pm t_n) = \lim s_n \pm \lim t_n$; $\lim (s_n . t_n) = \lim s_n . \lim t_n$; and $\lim \left(\dfrac{s_n}{t_n} \right) = \dfrac{\lim s_n}{\lim t_n}$

$\Rightarrow$ $$\lim_{n \to \infty} s_n = 0, \forall n \in N \cdot$$

Since 0 is a finite quantity. Hence, the sequence $\langle s_n \rangle$ is convergent and converges to 0.

Example 2. *Show that the sequence $\left\langle (-1)^n / n \right\rangle$ is convergent.*

Solution. Let $$\langle s_n \rangle = \left\langle (-1)^n / n \right\rangle.$$

Here $$\lim_{n \to \infty} s_{2n} = \lim_{n \to \infty} \frac{(-1)^{2n}}{2n} = \lim_{n \to \infty} \frac{1}{2n} = 0$$

and $$\lim_{n \to \infty} s_{2n+1} = \lim_{n \to \infty} \frac{(-1)^{2n+1}}{2n+1} = \lim_{n \to \infty} \frac{-1}{2n+1} = 0$$

which gives, $$\lim_{n \to \infty} s_{2n} = \lim_{n \to \infty} s_{2n+1} = 0$$

$\Rightarrow$ $$\lim_{n \to \infty} s_n = 0, \forall n \in N$$

Since 0 is a finite sequence. Hence, the given sequence $<s_n>$ is convergent.

Example 3. *Discuss the convergence of the sequence $\left\langle \dfrac{1}{3^n} \right\rangle$.*

Solution. Let $$\langle s_n \rangle = \left\langle \frac{1}{3^n} \right\rangle$$

Then $$\lim_{n \to \infty} s_{2n} = \lim_{n \to \infty} \frac{1}{3^{2n}} = 0$$

and $$\lim_{n \to \infty} s_{2n+1} = \lim_{n \to \infty} \frac{1}{3^{2n+1}} = 0$$

which implies $$\lim_{n \to \infty} s_{2n} = \lim_{n \to \infty} s_{2n+1} = 0$$

Therefore, $$\lim_{n \to \infty} s_n = 0, \forall n \in N.$$

Since 0 is a finite quantity, hence, the given sequence $\langle s_n \rangle$ is a convergent sequences.

Example 4. *Show that the sequence $\langle s_n \rangle$ defined by $s_n = \left\langle \left(\sqrt{n+1} - \sqrt{n} \right) \right\rangle, \forall n \in N$ is convergent.*

Solution. We have $$s_n = \sqrt{n+1} - \sqrt{n}$$

For any $\varepsilon > 0$, $$| s_n - 0 |= \sqrt{n+1} - \sqrt{n} < \varepsilon$$

$\Rightarrow$ $$\sqrt{n+1} < (\varepsilon + \sqrt{n})$$

$\Rightarrow$ $$n+1 < \varepsilon^2 + 2\varepsilon\sqrt{n} + n$$

$\Rightarrow$ $$1 < \varepsilon^2 + 2\varepsilon\sqrt{n}$$

i.e., if $$\frac{1}{4\varepsilon^2} < n$$

Then, for any given $\varepsilon > 0, \exists m \left(> \dfrac{1}{4\varepsilon^2} \right) \in N$ such that

$$| s_n - 0 |< \varepsilon, \forall n \geq m$$

Therefore, $\quad\quad \lim s_n = 0.$

Since, 0 is a finite quantity. Hence, the given sequence $\langle s_n \rangle$ is convergent.

Example 5. *Show that the sequence $\langle s_n \rangle$ defined by $s_n = r^n$ converges to 0 if $|r| < 1$.*

Solution. If $|r| < 1$. Then

$$|r| = \frac{1}{1+h}, \text{ where } h > 0$$

Since, $\quad\quad (1+h)^n = 1 + nh + \frac{n(n-1)}{2!}h^2 + \ldots + h^n > 1 + nh \,\forall n.$

Now $\quad\quad |s_n - 0| = |r^n|$

$$= |r|^n = \frac{1}{(1+h)^n} < \frac{1}{1+nh} \forall n.$$

Let $\varepsilon > 0$. Then $\quad |s_n - 0| < \varepsilon$ if $\dfrac{1}{1+nh} < \varepsilon$ or $n > \left(\dfrac{1}{\varepsilon} - 1\right) / h.$

Now, if we take a positive integer m such that $m > \left(\dfrac{1}{\varepsilon} - 1\right) / h,$ then, for all $n \ge m.$

$$|s_n - 0| < \varepsilon.$$

Hence, the sequence $\langle s_n \rangle$ converges to 0.

Example 6. *Show that the sequence $\langle s_n \rangle = \dfrac{3n}{n + 5n^{1/2}}$ has the limit 3.*

Solution. Let ε be any positive number.

Consider, $\quad \left| \dfrac{3n}{n + 5n^{1/2}} - 3 \right| = \dfrac{15n^{1/2}}{n + 5n^{1/2}} < \dfrac{15}{n^{1/2}}$

Therefore, $\quad \left| \dfrac{3n}{n + 5n^{1/2}} - 3 \right| < \varepsilon$ if $\dfrac{15}{n^{1/2}} < \varepsilon$ or $n > \dfrac{225}{\varepsilon^2}$

If we choose a positive integer $m > \dfrac{225}{\varepsilon^2}$ then, we get

$$|s_n - 3| < \varepsilon, \,\forall\, n \ge m$$

Hence $\quad\quad \lim_{n \to \infty} s_n = 3$

Example 7. *Show that $\lim\limits_{n \to \infty} \sqrt[n]{n} = 1$*

Solution. Let $\quad\quad \sqrt[n]{n} = 1 + h$, where $h \ge 0$

$\Rightarrow \quad\quad n = (1+h)^n = 1 + nh + \dfrac{n(n-1)}{2!}h^2 + \ldots +$

$\Rightarrow \quad\quad n > \dfrac{n(n-1)}{2}h^2, \forall n$ $\quad\quad\quad\quad\quad\quad\quad\quad\quad\quad$ [$h \ge 0$]

$\Rightarrow \quad\quad h^2 < \dfrac{2}{n-1},$ $\quad\quad\quad\quad\quad\quad\quad\quad\quad\quad\quad\quad\quad\quad$ for $n \ge 2$

$\Rightarrow \quad\quad |h| < \sqrt{\left(\dfrac{2}{n-1}\right)},$ $\quad\quad\quad\quad\quad\quad\quad\quad\quad\quad\quad\quad$ for $n \ge 2$

Let $\varepsilon > 0$ (any positive number, however small) then

$$|h| < \sqrt{\left(\frac{2}{n-1}\right)} < \varepsilon \text{ provided, } \frac{2}{n-1} < \varepsilon^2 \text{ or } n > \frac{2}{\varepsilon^2} + 1$$

If we take $m \in N$ such that $m > \dfrac{2}{\varepsilon^2} + 1$ then $|h| < \varepsilon \ \forall n \ge m$

or $|\sqrt[n]{n} - 1| < \varepsilon \ \forall n \ge m \ \Rightarrow \ \lim_{n \to \infty} \sqrt[n]{n} = 1$

Example 8. *Prove that* $\lim\left(\dfrac{1}{n^p}\right) = 0, p > 0$

Solution. Let ε be any positive number (however small). Then, consider

$$\left|\frac{1}{n^p} - 0\right| < \varepsilon \ \Rightarrow \ \frac{1}{n^p} < \varepsilon \Rightarrow n^p > \frac{1}{\varepsilon} \Rightarrow n > \left(\frac{1}{\varepsilon}\right)^{1/p}.$$

Then, by Archimedean property for $\left(\dfrac{1}{\varepsilon}\right)^{1/p} \in R$, there exists a positive integer

$m > \left(\dfrac{1}{\varepsilon}\right)^{1/p}$. If we take $m \in N$ such that $m > \left(\dfrac{1}{\varepsilon}\right)^{1/p}$, then we have

$$\left|\frac{1}{n^p} - 0\right| < \varepsilon, \forall n \ge m$$

Hence, $\lim \dfrac{1}{n^p} = 0$, when $p > 0$.

Example 9. *Prove that the sequence $\langle n^p \rangle$, where $p > 0$, diverges to infinity.*

Solution. Let $s_n = n^p, p > 0$, then, obviously $s_n > 0, \forall n \in N$, the sequence $\left\langle \dfrac{1}{s_n} \right\rangle = \left\langle \dfrac{1}{n^p} \right\rangle$ exists.

We have, $\dfrac{1}{n^p} \to 0$ as $n \to \infty$ [See example 8]

$\Rightarrow$ $n^p \to \infty$ as $n \to \infty$.

Hence, the sequence $\langle n^p \rangle$ diverges to infinity.

Example 10. *If $\langle s_n \rangle$ is a sequence such that $s_n \ne 0$ for any $n \in N$, and $\dfrac{s_{n+1}}{s_n} \to l$. Then prove that if $|l| < 1$, then $s_n \to 0$.*

Solution. Since $|l| < 1$. Therefore there exist $\varepsilon_1 > 0$ such that $|l| + \varepsilon_1 = h < 1$.

Now $\dfrac{s_{n+1}}{s_n} \to l \ \Rightarrow$ there exists a positive integer m such that

$$\left|\frac{s_{n+1}}{s_n} - l\right| < \varepsilon, \forall n \ge m$$

We have $\left|\dfrac{s_{n+1}}{s_n}\right| = \left|\left(\dfrac{s_{n+1}}{s_n} - l\right) + l\right| \le \left|\dfrac{s_{n+1}}{s_n} - l\right| + |l|$

$$< \varepsilon_1 + |l|, \forall n \ge m$$

i.e., $\left|\dfrac{s_{n+1}}{s_n}\right| < h, \ \forall n \ge m$

Replacing n by, $m, m+1, \ldots, n-1$ successively in the above equation and multiplying the corresponding sides of the resulting $(n-m)$ inequalities, we get

$$\left|\frac{s_{m+1}}{s_m}\right|\cdot\left|\frac{s_{m+2}}{s_{m+1}}\right|\ldots\left|\frac{s_n}{s_{n-1}}\right| < h^{n-m} \Rightarrow \left|\frac{s_{m-1}}{s_m}\cdot\frac{s_{m+2}}{s_{m+1}}\ldots\frac{s_n}{s_{n-1}}\right| < h^{n-m},$$

$$\Rightarrow \quad |s_n| < h^n\left(\frac{|s_m|}{h^m}\right), \quad \text{for } n > m. \qquad \ldots (1)$$

Since, $0 < h < 1$, therefore $h^n \to 0$ and hence, given $\varepsilon > 0$, there exists a positive integer m_1 such that

$$|h^n| < \frac{h^m \varepsilon}{|s_m|}, \forall n \ge m_1. \qquad \ldots (2)$$

Now, let us choose a positive integer p such that $p > \max\{m_1, m_2\}$. From (1) and (2), we get $|s_n| < \varepsilon \; \forall \, n \ge p$.
Hence, $\qquad\qquad\qquad s_n \to 0.$

Example 11. *Show that the sequence* $\left\langle \log\dfrac{1}{n} \right\rangle$ *diverges to* $-\infty$.

Solution. Let $\qquad\qquad s_n = \log\dfrac{1}{n}$

Take any $h < 0$. Then $s_n < h$ if $\log\dfrac{1}{n} < h$

$$\Rightarrow \text{if } (-\log n) < h$$
$$\Rightarrow \text{if } \log n > -h$$
$$\Rightarrow \text{if } n > e^{-h}$$

If we take $m \in \mathbb{N}$ such that $m > e^{-h}$, then

$$s_n < h \text{ for all } n \ge m.$$

Hence, $\qquad\qquad s_n \to -\infty \text{ as } n \to \infty$

2.10 CAUCHY SEQUENCE

A sequence $\langle s_n \rangle$ is said to be Cauchy sequence if, for given $\varepsilon > 0$ there exists $m \in \mathbb{N}$ such that
$$|s_n - s_m| < \varepsilon, \; \forall \; n \ge m$$
or $\qquad\qquad |s_p - s_q| < \varepsilon, \; \forall \; p, q \ge m$
or $\qquad\qquad |s_{n+p} - s_n| < \varepsilon, \; \forall \; n \ge m \text{ and } p > 0.$

REMARKS

- Cauchy sequence is also known as fundamental sequence.
- A sequence cannot converge if even one $\varepsilon > 0$ can be found such that for every positive integer m,
$$|s_{n+p} - s_n| > \varepsilon, \; \forall \; n \ge m \text{ and } p > 0$$
- Here, $|s_p - s_q| < \varepsilon, \; \forall \, p, q \ge m$ means that s_p and s_q are arbitrary close together for large values of p and q.
- The inequality $n \ge m$ in the definition may be replaced by $n > m$.
- If $\langle s_n \rangle$ and $\langle t_n \rangle$ are two Cauchy sequences, then $\langle s_n + t_n \rangle$, $\langle s_n t_n \rangle$ and $\langle s_n/t_n \rangle$ $(t_n \ne 0$ for any $n)$ are also Cauchy sequences.

☞ ILLUSTRATIONS

(1) The sequence $\left\langle \dfrac{1}{2^n} \right\rangle$ is a Cauchy sequence.

(2) The sequence $\left\langle \dfrac{1}{n} \right\rangle$ is a Cauchy sequence.

(3) The sequence $\left\langle \dfrac{1}{n^2} \right\rangle$ is not a Cauchy sequence.

(4) The sequence $\left\langle (-1)^n \right\rangle$ is not a Cauchy sequence.

THEOREM 1. *Every Cauchy sequence is bounded.*

Proof.　　Let $\langle s_n \rangle$ be a Cauchy sequence.

Taking $\varepsilon = 1$, there exists a positive integer m such that
$$|s_n - s_m| < 1, \forall n \geq m$$

$\Rightarrow$ 　　　　$(s_m - 1) < s_n < (s_m + 1), \ \forall n \geq m$

Let 　　　　$k = \min\{s_m - 1, s_1, s_2, ..., s_{m-1}\}$

and 　　　　$K = \max\{s_m + 1, s_1, s_2, ..., s_{m-1}\}$

Then 　　　　$k \leq s_n \leq K, \forall n.$

Hence, the sequence $\langle s_n \rangle$ is bounded.

REMARKS

- Converse of the above theorem is not necessarily true, *i.e.*, a bounded sequence need not be a Cauchy sequence, for example, the sequence $\left\langle (-1)^n \right\rangle$ is bounded, but not a Cauchy sequence.

THEOREM 2. **(Cauchy's General Principle of Convergence).** *A sequence is convergent if and only if it is a Cauchy sequence.*

Proof.　　Let us first suppose $\langle s_n \rangle$ be a convergent sequence. Let, this sequence be converge to l.

$\therefore$ 　for a given $\varepsilon > 0$ these exists a positive integer m such that
$$|s_n - l| < \varepsilon / 2, \forall n \geq m. \qquad \qquad \text{... (1)}$$

In particular, for $n = m$
$$|s_m - l| < \varepsilon / 2. \qquad \qquad \text{... (2)}$$

Now, consider
$$|s_n - s_m| = |s_n - l + l - s_m| \leq |s_n - l| + |s_m - l|$$
$$< \varepsilon/2 + \varepsilon/2, \ \forall n \geq m \qquad \text{(Using (1) and (2))}$$
$$= \varepsilon, \ \forall n \geq m$$

i.e., 　　　　$|s_n - s_m| < \varepsilon, \ \forall n \geq m$

$\Rightarrow$ 　$\langle s_n \rangle$ is a cauchy sequence.

Conversely, let $\langle s_n \rangle$ be a Cauchy sequence.

$\Rightarrow$ 　$\langle s_n \rangle$ is a bounded sequence. 　　　　　　　　[By Theorem 1]

$\Rightarrow$ 　By Bolzano-Weirstress theorem $\langle s_n \rangle$ has at least one limit point, say l. We shall show that sequence $\langle s_n \rangle$ converges to l.

Let $\varepsilon > 0$ be given. Since, $\langle s_n \rangle$ is a Cauchy sequence

∴   ∃ a positive integer m such that

$$| s_n - s_m | < \varepsilon/3, \forall n \geq m. \qquad \qquad ...(3)$$

Since, l is the limit point of $\langle s_n \rangle$.

∴   For above choice of ε and m, ∃ a positive integer $k > m$ such that

$$|s_k - l| < \varepsilon/3. \qquad \qquad ...(4)$$

Since, $k > m$, therefore from (3)

$$|s_k - s_m| < \varepsilon/3. \qquad \qquad ...(5)$$

Now, consider

$$\begin{aligned} |s_n - l| &= |s_n - s_m + s_m - s_k + s_k - l| \\ &\leq |s_n - s_m| + |s_m - s_k| + |s_k - l| \qquad \text{[By Triangle inequality]} \\ &< \varepsilon/3 + \varepsilon/3 + \varepsilon/3 = \varepsilon \end{aligned}$$

i.e., $|s_n - l| < \varepsilon, \forall n \geq m.$

Hence, $\langle s_n \rangle$ is convergent.

REMARK

- Cauchy's general principle of convergence, also termed as "Necessary and sufficient "condition for the convergence ".

Solved Examples

Example 1. If $\langle s_n \rangle$ is a sequence in R, where

$$s_n = 1 + \frac{1}{2} + \frac{1}{3} + ... + \frac{1}{n}$$

evaluate, $\lim\limits_{n \to \infty} |a_{n+1} - a_n|$. Verify, is this sequence satisfy the Cauchy criterion?

Based on the following Results

- A sequence $\langle s_n \rangle$ is said to be Cauchy if, for given $\varepsilon > 0$ ∃ $m \in N$ such that
$$|s_n - s_m| < \varepsilon \ \forall \ n \geq m.$$
- Every Cauchy sequence is bounded.
- A sequence is Cauchy if and only if it is bounded.
- Every bounded sequence is not necessarily Cauchy.
- The importance of the Cauchy general principle lies in the fact that it decides with precision the convergence or non-convergence of a sequence without any idea of the limit of the sequence and involve only the term of the sequences.

Solution.   Here

$$s_n = 1 + \frac{1}{2} + \frac{1}{3} + ... + \frac{1}{n}$$

$$\Rightarrow s_{n+1} = 1 + \frac{1}{2} + \frac{1}{3} + ... + \frac{1}{n} + \frac{1}{n+1}$$

∴ $$s_{n+1} - s_n = \frac{1}{n+1}$$

$\Rightarrow$ $$\lim_{n \to \infty} |s_{n+1} - s_n| = 0 .$$

Also, we have

$$s_{2n} - s_n = \left(1 + \frac{1}{2} + \frac{1}{3} + ... + \frac{1}{n} + \frac{1}{n+1} + \frac{1}{n+2} + ... + \frac{1}{2n} \right)$$

$$- \left(1 + \frac{1}{2} + \frac{1}{3} + ... + \frac{1}{n} \right)$$

$$= \frac{1}{n+1} + \frac{1}{n+2} + ... + \frac{1}{2n} \geq n \left(\frac{1}{2n} \right). \quad \left(\because \frac{1}{n+1} > \frac{1}{2n} \text{ etc.} \right)$$

$$\Rightarrow \qquad |s_{2n} - s_n| > \frac{1}{2} \forall n \in N$$

$\Rightarrow$ There exists a positive integer k such that $|s_n - s_k| \geq \frac{1}{2}$ whenever $n \geq k$

$\Rightarrow$ Cauchy criterion is not satisfied.

Example 2. *Show by applying Cauchy's convergence criterion that the sequence $\langle s_n \rangle$ given by*

$$s_n = 1 + \frac{1}{3} + \frac{1}{5} + \ldots + \frac{1}{2n-1} \text{ diverges.}$$

Solution. Here, we have

$$s_{n+1} = 1 + \frac{1}{3} + \frac{1}{5} + \ldots + \frac{1}{2n-1} + \frac{1}{2(n+1)-1}$$

$$= 1 + \frac{1}{3} + \frac{1}{5} + \ldots + \frac{1}{2n-1} + \frac{1}{2n+1}$$

$$\therefore \qquad s_{n+1} - s_n = \left[1 + \frac{1}{3} + \frac{1}{5} + \ldots + \frac{1}{2n-1} + \frac{1}{2n+1} \right] - \left[1 + \frac{1}{3} + \frac{1}{5} + \ldots + \frac{1}{2n-1} \right]$$

$$= \frac{1}{2n+1} > 0, \forall n \in N$$

Thus, $\qquad s_{n+1} > s_n, \ \forall n \in N$

$\Rightarrow$ The sequence $\langle s_n \rangle$ is increasing sequence.

Also, we have

$$s_{2n} = 1 + \frac{1}{3} + \frac{1}{5} + \ldots + \frac{1}{2n-1} + \frac{1}{2n+1} + \ldots + \frac{1}{4n-1}$$

$$\therefore \qquad s_{2n} - s_n = \left[1 + \frac{1}{3} + \frac{1}{5} + \ldots + \frac{1}{2n-1} + \frac{1}{2n+1} + \ldots + \frac{1}{4n-1} \right]$$

$$- \left[1 + \frac{1}{3} + \frac{1}{5} + \ldots + \frac{1}{2n-1} \right]$$

$$= \frac{1}{2n+1} + \frac{1}{2n+3} + \ldots + \frac{1}{4n-1}$$

$$\Rightarrow \qquad s_{2n} - s_n > n \left(\frac{1}{4n} \right) \qquad \left(\because \frac{1}{2n+1} > \frac{1}{4n} \text{ etc. and there are } n \text{ terms} \right)$$

$$\Rightarrow \qquad |s_{2n} - s_n| > \frac{1}{4}, \forall n \in N$$

$\Rightarrow$ There exists a positive integer k such that $|s_n - s_k| > \frac{1}{4}$ whenever $n \geq k$.

$\Rightarrow$ Cauchy criterion is not satisfied.

$\Rightarrow$ The sequence $\langle s_n \rangle$ cannot converge.

$\Rightarrow$ The sequence $\langle s_n \rangle$ diverges to $+ \infty$.

THEOREM 1. **(Squeeze Principle).** *If $\langle s_n \rangle$, $\langle t_n \rangle$ and $\langle u_n \rangle$ are three sequences such that*

$\quad$ (i) $\ s_n \leq t_n \leq u_n ; \forall n$

and (ii) $\langle s_n \rangle$ converges to l and $\langle u_n \rangle$ also converges to l, then $\langle t_n \rangle$ also converges to l.

Proof. Let $\varepsilon > 0$ be given. Since the sequences $\langle s_n \rangle$ and $\langle u_n \rangle$ converges to l, there must exist positive integers m_1 and m_2 such that

$$|s_n - l| < \varepsilon \; \forall \; n \geq m_1 \qquad\qquad\qquad\qquad \text{...(1)}$$

$$|u_n - l| < \varepsilon \; \forall \; n \geq m_2 \qquad\qquad\qquad\qquad \text{...(2)}$$

Let $m = \max \{m_1, m_2\}$. Then for $n > m$, we have

$$l - \varepsilon < s_n \leq t_n \leq u_n < l + \varepsilon$$

or $\qquad\qquad\qquad l - \varepsilon < t_n < l + \varepsilon$

or $\qquad\qquad\qquad |t_n - l| < \varepsilon, \; \forall \; n \geq m$

$\Rightarrow \qquad\qquad\qquad \lim t_n = l$

Hence, $\langle t_n \rangle$ converges to l.

REMARKS

- If $\langle s_n \rangle$ and $\langle t_n \rangle$ are two sequences such that $|s_n| \leq |t_n| \; \forall \; n \geq m$ where m is a positive integer and $\lim t_n = 0$, then $\lim s_n = 0$.
- The above theorem is also called Sandwitch theorem.

Theorem 2.   **(Cauchy's first theorem on limits).** *If* $\lim\limits_{n \to \infty} s_n = l$, *then*

$$\lim_{n \to \infty} \frac{s_1 + s_2 + ... + s_n}{n} = l.$$

Proof.   Let us define a sequence $\langle t_n \rangle$ in such a way that $t_n = s_n - l$

then $\qquad\qquad\qquad \lim t_n = \lim(s_n - l) = \lim s_n - l = l - l = 0$

and $\qquad\qquad \dfrac{s_1 + s_2 + ... + s_n}{n} = l + \dfrac{t_1 + t_2 + ... + t_n}{n}$

In order to prove this theorem, we have to show that

$$\lim \frac{t_1 + t_2 + ... + t_n}{n} = 0$$

Now, sequence $\langle t_n \rangle$ is convergent ($\because \langle s_n \rangle$ is convergent.), therefore it is bounded and hence there must exists a positive number k such that

$$|t_n| < k, \; \forall \; n \in N$$

Also, $\langle t_n \rangle$ converges to zero. Therefore, for a given $\varepsilon > 0$ there must exists a positive integer m such that

$$|t_n| < \varepsilon/2, \; \forall \; n \in N$$

Now, consider

$$\left| \frac{t_1 + t_2 + ... + t_n}{n} \right| = \left| \frac{t_1 + t_2 + ... + t_m}{n} + \frac{t_{m+1} + ... + t_n}{n} \right|$$

$$\leq \frac{|t_1| + |t_2| + \; + |t_n|}{n} + \frac{|t_{m+1}| + ... + |t_n|}{n}$$

$$< \frac{mk}{n} + \frac{\varepsilon}{2}(n - m), \forall n \geq m$$

Keeping m fixed, we have $\quad \dfrac{mk}{n} < \varepsilon / 2$ if $n > \dfrac{2mk}{\varepsilon}$

Let μ be any positive integer $> \dfrac{2mk}{\varepsilon}$, so that $n \ge \mu$, we have

$$\frac{mk}{n} \le \frac{\varepsilon}{2}$$

Let $\qquad \lambda = \max\{m, \mu\}.$

Therefore, for each $n \ge \lambda$, we have

$$\left| \frac{t_1 + t_2 + \dots + t_n}{n} \right| < \frac{\varepsilon}{2} + \frac{\varepsilon}{2} = \varepsilon.$$

This gives $\qquad \lim_{n \to \infty} \dfrac{t_1 + t_2 + \dots + t_n}{n} = 0$

Hence, we have $\quad \lim_{n \to \infty} \dfrac{s_1 + s_2 + \dots + s_n}{n} = l$

REMARKS

- Here, we can state that the limit of the nth term of the given sequence is equal to limit of the arithmetic mean of first n terms of the sequence.
- The converse of the above theorem need not be true. For example :

Let $\qquad s_n = (-1)^n$

$$s_n = (-1)^n \Rightarrow \frac{s_1 + s_2 + \dots + s_n}{n} = 0, \text{ if } n \text{ is even}$$

$$= -\frac{1}{n}, \quad \text{if } n \text{ is odd}$$

Therefore, $\qquad \lim_{n \to \infty} \dfrac{s_1 + s_2 + \dots + s_n}{n} = 0$

But the sequence $\langle s_n \rangle = \langle (-1)^n \rangle$ is not convergent.

THEOREM 3. (Cauchy's second theorem on limits). *If $\langle s_n \rangle$ is a sequence of positive terms and $\lim_{n \to \infty} s_n = l,$ then $\lim (s_1 \cdot s_2 \cdot \dots \cdot s_n)^{1/n} = l.$*

Proof. Let $\langle t_n \rangle$ be a sequence, such that $\qquad t_n = \log s_n, \forall n \in N$

Now $\qquad \lim s_n = l \Rightarrow \lim t_n = \lim \log s_n = \log l$

$$(\because \lim s_n = l \Leftrightarrow \lim \log s_n = \log l \text{ provided } s_n > 0, \forall n \text{ and } l > 0)$$

Then, by Cauchy first theorem on limits, we have

$$\lim_{n \to \infty} \frac{t_1 + t_2 + \dots + t_n}{n} = \lim t_n = \log l$$

$$\Rightarrow \quad \lim_{n \to \infty} \frac{\log s_1 + \log s_2 + \dots + \log s_n}{n} = \log l$$

$$\Rightarrow \quad \lim_{n \to \infty} \frac{1}{n} \log(s_1 \cdot s_2 \cdot \dots \cdot s_n) = \log l$$

$$\Rightarrow \quad \lim \log(s_1 \cdot s_2 \cdot \dots \cdot s_n)^{1/n} = \log l$$

$$\Rightarrow \quad \log(s_1 \cdot s_2 \cdot \dots \cdot s_n)^{1/n} = l$$

REMARK

- Here, we can state that the limit of the nth term of a sequence of positive terms is equal to limit of the geometric means of first n terms of the given sequence.

- Cauchy's second theorem on limits can also be stated as follows:
  If $\langle s_n \rangle$ be a sequence of positive terms then

$$\lim s_n^{1/n} = \lim_{n \to \infty} \frac{s_{n+1}}{s_n}$$

provided, the limit of right hand sides exists.

2.11 IMPORTANT SEQUENCE $\langle s^n \rangle$

The behaviour of the sequence $\langle s_n \rangle$, where $s_n = s^n$ depends upon the value of 's'. Following cases arise :

(i) $s = 1$. Then $s_n = 1 \ \forall \ n$
                     $s_n \to 1$ as $n \to \infty$

(ii) $s = 0$. Then $s_n \to 0$ as $n \to \infty$

(iii) $s > 1$.

Let $s = 1 + h, h > 0$
        $s_n = (1 + h)^n > 1 + nh$ (By taking first two terms of the binomial expansions)

As $n \to \infty, 1 + nh \to \infty$ Therefore, $\Rightarrow s_n \to \infty$, as $n \to \infty$

(iv) $0 < s < 1$

$\Rightarrow \quad \dfrac{1}{s} = 1$ so $\dfrac{1}{s} = 1 + h$ for some $h > 0$

$\Rightarrow \quad \dfrac{1}{s^n} = (1 + h)^n \qquad \geq 1 + nh$ for $n \geq 1$

$\Rightarrow \quad s^n \leq \dfrac{1}{1 + nh} < \dfrac{1}{nh}$

Since, h is fixed, $\dfrac{1}{nh} \to 0$ as $n \to \infty$.

Hence, $s^n \to 0$ as $n \to \infty$.

(v) $-1 < s < 0$
Let $s = -t$. Then $0 < t < 1$.
By case (iv) $t^n \to 0$ as $n \to \infty$ and hence $s_n = (-t)^n \to 0$ as $n \to \infty$.

(vi) $s = -1$. Then $s^n = (-1)^n$, which is an oscillating sequence.

(vii) $s < -1$
Let $t = -s, t > 1$.
Then by case   (iii) $t^n \to \infty$ as $n \to \infty$
So, $s_n = (-t)^n$ takes values, alternatively negative and positive, numerically greater than any assigned number $i.\ e.,\ \langle s_n \rangle$ oscillate infinitely.

THEOREM 4. *If $\langle s_n \rangle$ is a sequence such that*

$$\lim_{n \to \infty} \frac{s_{n+1}}{s_n} = l \ where \ |l| < 1 \ then \ \lim_{n \to \infty} s_n = 0.$$

Proof.   Since $|l| < 1$, let us choose a positive small number ε such that $|l| + \varepsilon < 1$.

Now, $\lim \dfrac{s_{n+1}}{s_n} = l$, therefore for $\varepsilon > 0$, there must exists a positive integer m such that, for all $n \geq m$

$$\left|\frac{s_{n+1}}{s_n} - l\right| < \varepsilon$$

$$\Rightarrow \quad \left|\frac{s_{n+1}}{s_n}\right| - |l| \le \left|\frac{s_{n+1}}{s_n} - l\right| < \varepsilon$$

$$\Rightarrow \quad \left|\frac{s_{n+1}}{s_n}\right| < |l| + \varepsilon = k \,(\text{say}).$$

Now, putting $n = m, m+1,..., n-1$ in the above inequality and multiplying them, we get

$$\left|\frac{s_n}{s_m}\right| < k^{n-m}$$

or

$$|s_n| < \frac{|s_m|}{k^m}.k^n$$

But $k < 1 \Rightarrow k^n \to 0$ as $n \to \infty$, which gives $\lim s_n = 0$.

REMARK

- The general result obtained here, gives the following important results on limit :

(i) $\lim \dfrac{n^s}{n!} = 0$ (ii) $\lim \dfrac{n^r}{s^n} = 0, |s| > 1$

(iii) $\lim \dfrac{m(m-1)...(m-n+1)s^n}{n!} = 0; |s| < 1$

THEOREM 5. *If $\langle s_n \rangle$ is a sequence such that $s_n > 0$ and $\lim \dfrac{s_{n+1}}{s_n} = l,$ then $\lim \sqrt[n]{s_n} = l$*

Proof.   Let us define a sequence $\langle t_n \rangle$ such that

$$t_1 = s_1, t_2 = \frac{s_2}{s_1},..., t_n = \frac{s_n}{s_{n-1}}$$

Then $t_1.t_2 ... t_n = s_n.$

Also

$$\lim \frac{s_{n+1}}{s_n} = l \Rightarrow \lim \frac{s_n}{s_{n-1}} = l \Rightarrow t_n = l$$

$$= s_n > 0 \Rightarrow t_n = 0, \forall \, n \in N$$

Hence, the sequence $\langle t_n \rangle$ of positive terms and $\lim t_n = l$.
Now, Cauchy's second theorem on limits we have

$$\lim (t_1 \cdot t_2 \cdot.... \cdot t_n)^{1/n} = l$$

or $\lim (s_n)^{1/n} = l$

THEOREM 6.   **(Cesaro's Theorem).** *If $\lim s_n = l_1$ and $\lim t_n = l_2$. Then*

$$\lim \frac{s_1 t_n + s_2 t_{n-1} +...+ s_n t_1}{n} = l_1 l_2 .$$

Proof.   Let us define $s_n = l_1 + u_n$ and $|u_n| = U_n.$

Then $\lim u_n = 0$ and therefore $\lim U_n = 0$.
Now, by Cauchy's first theorem on limits, we have

$$\lim \frac{1}{n}[u_1 + u_2 +...+ u_n] = 0 \qquad\qquad ... (1)$$

Consider, $\dfrac{1}{n}[s_1 t_n + s_2 t_{n-1} + ... + s_n t_1]$

$$= \dfrac{l_1}{n}[t_1 + t_2 + ... + t_n] + \dfrac{1}{n}[u_1 t_n + u_2 t_{n-1} + ... + u_n t_1] \quad ...(2)$$

Since, the sequence $\langle t_n \rangle$ is convergent. Therefore, it is bounded. Hence, there must exists a positive real number k such that $|t_n| < k, \forall n \in N$.

Therefore, $\left| \dfrac{1}{n}(u_1 t_n + u_2 t_{n-1} + ... + u_n t_1) \right| \geq 0$

$\Rightarrow \qquad \dfrac{1}{n}[|u_1||t_n| + |u_2||t_{n-1}| + ... + |u_n||t_1|] \geq 0$

$\Rightarrow \qquad \dfrac{k}{n}[|u_1| + |u_2| + ... + |u_n|] > 0$

$\Rightarrow \qquad \dfrac{k}{n}[u_1 + u_2 + ... + u_n] > 0$

$\Rightarrow \qquad \dfrac{k}{n}[u_1 + u_2 + ... + u_n] \to 0 \text{ as } n \to \infty \qquad$ [By using (1)]

Thus $\qquad \lim \dfrac{1}{n}[u_1 t_n + u_2 t_{n-1} + ... + u_n t_1] = 0$

Since, $\qquad \lim t_n = l_2$, therefore

$$\lim \dfrac{t_1 + t_2 + ... + t_n}{n} = l_2$$

Hence, from (2), we have

$$\lim \dfrac{1}{n}(s_1 t_n + s_2 t_{n-1} + ... + s_n t_1) = l_1 l_2$$

Solved Examples

Example 1. *Prove that*

$$\lim_{n \to \infty} s_n = 1,$$

where $s_n = n^{1/n}$.

Solution. For $n = 1, \quad s_n = 1$

For $n \geq 2, \quad s_n > 1$

Let $s_n = 1 + t_n$,

$n = s_n^n = (1 + t_n)^n$,

$\qquad t_n > 0, n \geq 2$

$= 1 + n t_n + \dfrac{n(n-1)}{2!} t_n^2 + $

$\qquad ... + t_n^n$

(By Binomial Theorem)

$\geq \dfrac{n(n-1)}{2!} t_n^2$

$\Rightarrow 0 \leq t_n^2 \leq \dfrac{2}{n-1}$

- **(Squeeze principle).** If $\langle s_n \rangle$, $\langle t_n \rangle$ and $\langle u_n \rangle$ be three sequences such that

  (i) $s_n \leq t_n \leq u_n \ \forall \ n \in N$

  and (ii) $< s_n >$ converges to l, and also $\langle u_n \rangle$ converges to l then $\langle t_n \rangle$ also converges to l.

- **(Cauchy's first theorem on limits).** If $\lim\limits_{n \to \infty} s_n = l$

  then $\lim\limits_{n \to \infty} \left(\dfrac{s_1 + s_2 + ... + s_n}{n} \right) = l$

- **(Cauchy's second theorem on limits).** If $\langle s_n \rangle$ is a sequence of positive terms and $\lim\limits_{n \to \infty} s_n = l$ then

  $\lim(s_1 . s_2, ..., s_n)^{1/n} = l$ or $\lim(s_n)^{1/n} = \lim\limits_{n \to \infty} \dfrac{s_{n+1}}{s_n}$

- $\lim \dfrac{s^n}{n!} = 0, \ \lim \dfrac{n^r}{s^n} = 0, |s| < 1$

- **(Cesaro's Theorem).** If $\lim s_n = l_1$ and $\lim t_n = l_2$. Then

  $\lim \dfrac{s_1 t_n + s_2 t_{n-1} + ... + s_n t_1}{n} = l_1 l_2$

$$\Rightarrow \qquad 0 < t_n \le \sqrt{\frac{2}{n-1}}$$

Since $\qquad \sqrt{\dfrac{2}{n-1}} \to 0 \text{ as } n \to \infty$

$\Rightarrow \qquad t_n \to 0 \text{ as } n \to \infty.$

Hence $\qquad s_n \to 1 \text{ as } n \to \infty.$

Example 2. If $s_n = \left[\left(\dfrac{2}{1}\right)^1 \left(\dfrac{3}{2}\right)^2 \left(\dfrac{4}{3}\right)^3 \cdots \left(\dfrac{n+1}{n}\right)^n\right]^{1/n}$ then show that $<s_n> \to e$. Show also that

$$\lim_{n\to\infty} \left[\frac{n^n}{n!}\right]^{1/n} = e.$$

Solution. Let $\qquad t_n = \left(\dfrac{2}{1}\right)^1 \left(\dfrac{3}{2}\right)^2 \left(\dfrac{4}{3}\right)^3 \cdots \left(\dfrac{n+1}{n}\right)^n$

so that $\qquad s_n = t_n^{1/n}$

Also, $\qquad \dfrac{t_{n+1}}{t_n} = \left(\dfrac{n+2}{n+1}\right)^{n+1} = \left(1 + \dfrac{1}{n+1}\right)^{n+1}$

$\Rightarrow \qquad \lim_{n\to\infty} \dfrac{t_{n+1}}{t_n} = e$

Hence, by Cauchy's second theorem on limits, we have

$$\lim_{n\to\infty} s_n = \lim_{n\to\infty} t_n^{1/n} = e$$

Also $\qquad s_n = \left[2 \cdot \left(\dfrac{3}{2}\right)^2 \cdot \left(\dfrac{4}{3}\right)^3 \cdots \left(\dfrac{n+1}{n}\right)^n\right]^{1/n}$

$$= \left[\frac{(n+1)^n}{n!}\right]^{1/n} = \left[\frac{(n+1)^n}{n^n} \cdot \frac{n^n}{n!}\right]^{1/n} = \frac{n+1}{n}\left(\frac{n^n}{n!}\right)^{1/n}$$

$\therefore \qquad \lim_{n\to\infty} s_n = \lim_{n\to\infty} \left[\left(\dfrac{n+1}{n}\right)\left(\dfrac{n^n}{n!}\right)\right]^{1/n} = \lim_{n\to\infty} \dfrac{n+1}{n} \lim_{n\to\infty}\left[\dfrac{n^n}{n!}\right]^{1/n}$

$$= 1 \cdot \lim_{n\to\infty} \left(\frac{n^n}{n!}\right)^{1/n}$$

Hence, $\qquad \lim_{n\to\infty} \left(\dfrac{n^n}{n!}\right)^{1/n} = e.$

Example 3. *Show that the sequence* $\langle s_n \rangle$ *when* $s_n = 1 + \dfrac{1}{2} + \dfrac{1}{3} + ... + \dfrac{1}{n}$ *cannot converge.*

Solution. Let us take $\varepsilon = \dfrac{1}{2}$ and $n = k = m$ and apply them in Cauchy's general principle of convergence, we have

$$| s_{n+k} - s_n | = | s_{2m} - s_m | = \frac{1}{m+1} + \frac{1}{m+2} + ... + \frac{1}{2m}$$

$$> \frac{1}{2m} + \frac{1}{2m} + ... + \frac{1}{2m} = \frac{m}{2m} = \frac{1}{2}, \text{ which is a contradiction.}$$

Hence, the given sequence is not convergent.

Example 4. *Show that the sequence* $\langle s_n \rangle$ *where* $s_n = \left\{ \dfrac{1}{\sqrt{n^2+1}} + \dfrac{1}{\sqrt{n^2+2}} + ... + \dfrac{1}{\sqrt{n^2+n}} \right\}$ *converges to 1.*

Solution. Here, we have

$$\frac{n}{\sqrt{n^2+n}} \le s_n \le \frac{n}{\sqrt{n^2}} \implies \frac{1}{\sqrt{1+(1/n)}} \le s_n \le 1.$$

Now the sequence $\langle t_n \rangle, \langle u_n \rangle$ are such that

(i) $t_n \le s_n \le u_n$,

and (ii) $\lim t_n = \lim u_n = 1$

where, $\qquad t_n = \dfrac{1}{\sqrt{1 + \left(\dfrac{1}{n}\right)}}$ and $u_n = 1$

Hence, by squeeze principle, we have $\lim s_n = 1$.

Example 5. *Prove that* $\lim\limits_{n \to \infty} \left[\dfrac{(n+1)(n+2)(n+3) ... (n+n)}{n^n} \right] = \dfrac{4}{e}.$

Solution. Let $\qquad s_n = \dfrac{(n+1)(n+2)...(n+n)}{n^n} = \dfrac{(2n)!}{n^n(n!)}$

Then $\qquad s_{n+1} = \dfrac{(2n+2)!}{(n+1)^{n+1}(n+1)!}$

Therefore, $\qquad \dfrac{s_{n+1}}{s_n} = \dfrac{(2n+2)! n^n (n!)}{(n+1)^{n+1}(n+1)(2n)!}$

$$= \frac{(2n+2)(2n+1)n^n}{(n+1)^{n+2}} = \frac{2(2n+1)n^n}{(n+1)^{n+1}}$$

$$= \frac{2 \times 2n \left[1 + \dfrac{1}{2n}\right] n^n}{(n+1)(n+1)^n} = \frac{4n\left[1 + \dfrac{1}{2n}\right] n^n}{n\left[1 + \dfrac{1}{n}\right](n+1)^n}$$

$$= \frac{4\left(1 + \dfrac{1}{2n}\right)}{\left(1 + \dfrac{1}{n}\right)} \cdot \left[\frac{n}{n+1}\right]^n = \frac{4\left[1 + \dfrac{1}{2n}\right]}{\left[1 + \dfrac{1}{n}\right]} \cdot \frac{1}{\left[1 + \dfrac{1}{n}\right]^n}$$

Now, taking $\lim n \to \infty$, we have

$$\lim_{n \to \infty} \frac{s_{n+1}}{s_n} = \lim_{n \to \infty} \left[\frac{4\left[1 + \dfrac{1}{2n}\right]}{1 + \dfrac{1}{n}} \cdot \frac{1}{\left[1 + \dfrac{1}{n}\right]^n}\right] = \frac{4}{e}.$$

Now By Cauchy's second theorem on limits, we have

$$\lim_{n \to \infty} (s_n)^{1/n} = \lim_{n \to \infty} \left(\frac{s_{n+1}}{s_n}\right) = \frac{4}{e}.$$

$\Rightarrow \qquad \displaystyle\lim_{n \to \infty} \left[\frac{(n+1)(n+2) \dots (n+n)}{n^n}\right] = \frac{4}{e}$

Example 6.   *If $r > 0$, show that $\lim r^{1/n} = 1$.*

Solution.   There are following three cases :

Case I. When $r > 1$.

Let $\qquad s_n = r^{1/n} - 1$, then $s_n > 0 \; \forall \; n \in N$, therefore

$\qquad\qquad r^{1/n} = 1 + s_n$

$\Rightarrow \qquad\qquad r = [1 + s_n]^n = 1 + ns_n + \dots + s_n{}^n$

$\qquad\qquad\qquad \geq 1 + ns_n + \forall \; n \in N$

$\Rightarrow \qquad \dfrac{r-1}{n} \geq s_n, \forall n \in N$

Hence $\qquad 0 \leq s_n \leq \dfrac{r-1}{n}, \forall n \in N$

Then, by Sandwitch theorem, we have $\lim s_n = 0$

Hence, $\quad \lim r^{1/n} = 1$

Case II. When $r = 1$

Here, $\qquad r^{1/n} = 1, \forall \; n \in N$

$\Rightarrow \qquad \lim r^{1/n} = 1$

Case III. When $0 < r < 1$, then $\dfrac{1}{r} > 1$

$\therefore \qquad \lim \left(\dfrac{1}{r}\right)^{1/n} = 1 \; \Rightarrow \; \lim \dfrac{1}{r^{1/n}} = 1$

$\Rightarrow \qquad \lim r^{1/n} = 1 .$

Example 7. *Prove that* $\lim \dfrac{1}{n}[1 + 2^{1/2} + 3^{1/3} + ... + n^{1/n}] = 1.$

Solution. Let $\qquad s_n = n^{1/n}$

$\Rightarrow \qquad \lim s_n = \lim n^{1/n} = 1$

Then, by Cauchy's first theorem on limits, we have

$$\lim \frac{1}{n}(s_1 + s_1 + ... + s_n) = 1$$

$\Rightarrow \qquad \lim \dfrac{1}{n}[1 + 2^{1/2} + 3^{1/3} + ... + n^{1/n}] = 1$

Example 8. *If* $\langle s_n \rangle$ *be a sequence of positive numbers such that*

$$s_n = \frac{1}{2}[s_{n-1} + s_{n-2}], \forall\, n \geq 2$$

show that $\langle s_n \rangle$ *converges, also find* $\lim s_n$.

Solution. **Case I.** If $\quad s_1 = s_2$

Then using $\quad s_n = \dfrac{1}{2}[s_{n-1} + s_{n-2}]$ we find that

$$s_3 = s_1, s_4 = s_1, \quad .... \quad s_n = s_1, \forall\, n \in N$$

$\Rightarrow \quad \langle s_n \rangle$ converges to s_1.

Case II. If $s_1 \neq s_2$

Then $\qquad | s_n - s_{n-1} | = \left| \dfrac{1}{2}(s_{n-1} + s_{n-2}) - s_{n-1} \right|$

$$= \frac{1}{2}| s_{n-1} + s_{n-2} | = \frac{1}{2^2}| s_{n-2} - s_{n-3} | \text{ and so on. } ... (1)$$

For $n \geq m$, we have $| s_n - s_m | = |(s_n - s_{n-1}) + (s_{n-1} - s_{n-2}) + ... + (s_{m+1} - s_m)|$

$$\leq | s_n - s_{n-1} | + | s_{n-1} - s_{n-2} | + ... + | s_{m+1} - s_m)|$$

$$= \frac{1}{2^{n-2}}| s_2 - s_1 | + \frac{1}{2^{n-3}}| s_2 - s_1 | + ... + \frac{1}{2^{m-1}}| s_2 - s_1 |$$

[Using (1)]

$$= \left[\frac{1}{2^{n-2}} + \frac{1}{2^{n-3}} + ... + \frac{1}{2^{m-1}} \right] | s_2 - s_1 |$$

$$= \frac{1}{2^{m-1}}\left[1 + \frac{1}{2} + \frac{1}{2^2} + ... + \frac{1}{2^{n-m-1}} \right] | s_2 - s_1 |$$

$\Rightarrow \quad | s_n - s_m | < \dfrac{1}{2^{m-2}}| s_2 - s_1 |\left[\because \left(1 + \dfrac{1}{2} + \dfrac{1}{2^2} + ... + \dfrac{1}{2^{n-m-1}}\right) < 2 \right]$ ... (2)

Now, let $\varepsilon > 0$ be given. Then choose a positive integer m such that

$$\frac{1}{2^{m-2}}| s_2 - s_1 | < \varepsilon. \qquad\qquad ... (3)$$

Now, from (2) and (3), we have $|s_n - s_m| < \varepsilon, n \geq m$

$\Rightarrow \langle s_n \rangle$ is a Cauchy sequence. Then by Cauchy convergence criterion, the sequence $\langle s_n \rangle$ is convergent.

Let
$$\lim_{n \to \infty} s_n = l.$$

Putting $n = 3, 4, \ldots, k$ in the given relation $s_n = \dfrac{1}{2}(s_{n-1} + s_{n-2}), \forall n \geq 2$

We get

$$\left.\begin{aligned}
s_3 &= \frac{1}{2}(s_2 + s_1) \\[1em]
s_4 &= \frac{1}{2}(s_3 + s_2) \\[0.5em]
&\;\text{-----------------} \\[0.5em]
s_{k-1} &= \frac{1}{2}(s_{k-2} + s_{k-3}) \\[1em]
s_k &= \frac{1}{2}(s_{k-1} + s_{k-2})
\end{aligned}\right\} \qquad \ldots (4)$$

Now, adding relation (4), we have

$$s_k + \frac{1}{2} s_{k-1} = \frac{1}{2}(s_1 + 2s_2)$$

By taking limit as $k \to \infty$ and using (3), we have

$$l + \frac{1}{2}l = \frac{1}{2}(s_1 + 2s_2) \quad \Rightarrow l = \frac{1}{3}(s_1 + 2s_2)$$

$$\Rightarrow \qquad \lim_{n \to \infty} s_n = \frac{1}{3}(s_1 + 2s_2).$$

2.12 MONOTONIC SEQUENCES

(1) A sequence $\langle s_n \rangle$ is said to be monotonically increasing (or non-decreasing).

   if $s_n \leq s_{n+1}, \forall n$

   or $s_n \leq s_m, \forall n > m$

(2) A sequence $\langle s_n \rangle$ is said to be strictly increasing if $s_n > s_{n+1}, \forall n \in N$

(3) A sequence $\langle s_n \rangle$ is said be monotonically decreasing (or non- increasing)

   if $s_n \geq s_{n+1}, \forall n$

   or $s_n \geq s_m, \forall n < m$

(4) A sequence $\langle s_n \rangle$ is said to be strictly decreasing if $s_n > s_{n+1}, \forall n \in N.$

(5) A sequence $\langle s_n \rangle$ is said to be monotonic if it is either monotonically increasing or monotonically decreasing.

☛ **ILLUSTRATIONS**

(1) $\langle 2, 2, 4, 4, 6, \ldots \rangle$ is monotonically increasing.

(2) $\langle 1, 2, 3, \ldots n \ldots \rangle$ is strictly increasing.

(3) $\left\langle 1, 1, \dfrac{1}{3}, \dfrac{1}{5}, \dfrac{1}{5} \ldots \right\rangle$ is monotonically increasing.

(4) $\langle -2, -4, -6, -8, \ldots \rangle$ is strictly increasing.

(5) $\langle 0, 1, 0, 1, \ldots \rangle$ is not monotonic.

REMARKS

- Strictly monotonic sequences are special case of monotonic sequences.
- A strictly monotonic sequence may be monotonic after a certain number of terms.

THEOREM 1. **(Monotone Convergence Theorem).** *Every bounded monotonically increasing sequence converges.*

Proof. Let $\langle s_n \rangle$ be a bounded monotonically increasing sequence.

Let $S = \{ s_n : n \in N \}$ be its range.

Then, obviously S is a non-empty set, which is bounded above. Therefore there exists a number l, which is the supremum of S. We shall show that the sequence $\langle s_n \rangle$ converges to l.

Let $\varepsilon > 0$ be a given number. Since $l - \varepsilon < l$, therefore $l - \varepsilon$ is not an upper bound of S. Hence, there exists a positive integer m such that $s_n > l - \varepsilon$.

Now, since $\langle s_n \rangle$ is monotonically increasing sequence. Therefore

$$s_n \geq s_m > l - \varepsilon \ \forall \ n \geq m. \qquad \qquad ...(1)$$

$$\text{Sup. } S = l \ \Rightarrow \ s_n < l < l + \varepsilon, \forall \ n. \qquad \qquad ...(2)$$

From (1) and (2), we have

$$l - \varepsilon < s_n < l + \varepsilon \ \forall n \geq m$$

$$\Rightarrow \qquad | \ s_n < l \ | < \varepsilon, \ \forall \ n \geq m$$

Hence, $\langle s_n \rangle$ converges to l.

THEOREM 2. *Every bounded monotonically decreasing sequence converges.*

Proof. Let $\langle s_n \rangle$ be a bounded monotonically decreasing sequence. Consider a sequence $\langle t_n \rangle$ such that $t_n = - s_n, \ \forall \ n \in N$. Then, $\langle t_n \rangle$ is bounded monotonically increasing sequence and therefore it converges.   [By Theorem 1]

If $\qquad \lim t_n = l$, then $\lim s_n = \lim (- t_n) = - l$.

THEOREM 3. *A non-decreasing sequence(increasing), which is not bounded above diverges to ∞.*

Proof. Let $\langle s_n \rangle$ be a monotonic non-decreasing sequence, which is not bounded above. Let c be any positive number. Since, the sequence $\langle s_n \rangle$ is unbounded and monotonically increasing, therefore, there must exists a positive integer m such that

$$s_n \geq s_m > c, \forall \ n > m$$

$$\Rightarrow \qquad s_n > c, \ \forall n > m$$

Hence, the sequence $\langle s_n \rangle$ diverges to ∞.

THEOREM 4. *A non-increasing (decreasing) sequence, which is not bounded below diverges to $-\infty$.*

Proof. Proof is exactly on same lines and left as an exercise for the students.

REMARKS

- Every monotonically increasing sequence bounded above converges to the least upper bound.
- Every monotonically decreasing sequence bounded below converges to the greatest-lower bound.

2.13 NESTED SEQUENCE

A sequence of closed intervals $\langle I_n \rangle$ is said to be nested if either

$$I_1 \subseteq I_2 \subseteq I_3 \subseteq \quad \text{or} \quad I_1 \supseteq I_2 \supseteq I_3 \supseteq \dots$$

THEOREM 1 **(Nested Interval Theorem (Cantor's Intersection Theorem).** *If a sequence* $\langle I_n = [a_n, b_n] \rangle$ *of closed intervals is such that each member* $[a_{n+1}, b_{n+1}]$ *is contained in preceding one* $[a_n, b_n]$ *and*

$$\lim_{n \to \infty} [b_n - a_n] = \lim_{n \to \infty} (length \, of \, I_n) = 0.$$

Then $\bigcap_{n=1}^{\infty} I_n$ *contains precisely one point.*

Proof. Since $I_{n+1} \subset I_n, \forall n \in N$, it follows $a_n \leq a_{n+1} \leq b_{n+1} \leq b_n, \forall n \in N$, which implies that $\langle a_n \rangle$ is a monotonic increasing sequence bounded above by b_1 and $\langle b_n \rangle$ is a monotonic decreasing sequence bounded below by a_1.

$\Rightarrow$ Both sequence $\langle a_n \rangle$ and $\langle b_n \rangle$ converges.

Also $\lim (b_n - a_n) = 0 \Rightarrow \lim b_n - \lim a_n = 0$

$\Rightarrow \qquad \lim b_n = \lim a_n = l$ (say).

Obviously, l is the upper bound of sequence $\langle a_n \rangle$ and the lower bound of the sequence $\langle b_n \rangle$. So

$$a_n \leq l \leq b_n, \forall n \in N$$

$\Rightarrow \qquad l \in I_n, \forall n \in N$

Therefore $l \in \bigcap_{n=1}^{\infty} I_n.$

Now we shall show that l is the only point such that $l \in \bigcap_{n=1}^{\infty} I_n$.

Let if possible $l \neq l_1 \in \bigcap_{n=1}^{\infty} I_n$

Then $0 \leq |l_1 - l| \leq |b_n - a_n|, \forall n \cdot$

$\Rightarrow \qquad |l_1 - l| = 0 \qquad\qquad (\because \lim |b_n - a_n| = 0)$

$\Rightarrow \qquad l = l_1$

$\Rightarrow \bigcap_{n=1}^{\infty} I_n$ consist of exactly one point.

REMARK

- The word "closed" in the statement of Cantor's intersection theorem cannot be dropped *i.e*, the intersection of a decreasing sequence of open intervals may be empty.

For Example: let $I_n = \left] 0, \dfrac{1}{n} \right[, n \in N$, then $\bigcap_{n=1}^{\infty} I_n = \phi.$

For, if $x \leq 0$, then $x \notin I_n$ for any n and if $x > 0$, then by Archimedean property of real numbers, there exists a positive integer m such that $m > \dfrac{1}{x}$

$\Rightarrow \qquad \dfrac{1}{m} < x \quad \Rightarrow x \notin I_m \quad \Rightarrow x \notin \bigcap_{n=1}^{\infty} I_n$

Solved Examples

Example 1. *Show that the sequence $\langle s_n \rangle$ defined by*

$$s_n = \frac{1}{n+1} + \frac{1}{n+2} + \ldots + \frac{1}{n+n}$$

converges.

Based on the following Results

- Every bounded monotonically increasing sequence is convergent.

- Every monotonically increasing sequence which is bounded above, converges to its least upper bound.

- Every bounded decreasing sequence which is bounded below converges to the greatest lower bound.

- For a monotonic sequence, there are only two types of behaviour :

  (i) a monotonically increasing sequence is either convergent or divergent to $+\infty$.

  (ii) a monotonicaily decreasing sequence is either convergent or divergent to $-\infty$.

- A monotonic sequence cannot oscillate, it is obviously, bounded below or above according as it is increasing or decreasing.

Solution. The sequence $\langle s_n \rangle$ is defined by

$$s_n = \frac{1}{n+1} + \frac{1}{n+2} + \ldots + \frac{1}{n+n}$$

$\Rightarrow$

$$s_{n+1} = \frac{1}{n+1} + \frac{1}{n+2} + \ldots + \frac{1}{2n+n}$$

Now

$$s_{n+1} - s_n$$

$$= \left(\frac{1}{n+2} + \frac{1}{n+3} + \ldots + \frac{1}{2n+2} \right)$$

$$- \left(\frac{1}{n+1} + \frac{1}{n+2} + \ldots + \frac{1}{2n} \right)$$

$$= \frac{1}{2n+1} + \frac{1}{2n+2} - \frac{1}{n+1} = \frac{1}{2n+1} - \frac{1}{2n+2} > 0, \forall n$$

Hence, the sequence $\langle s_n \rangle$ is monotonically increasing.

Now $\qquad |s_n| = \left| \frac{1}{n+1} + \frac{1}{n+2} + \ldots + \frac{1}{n+n} \right| < \frac{1}{n} + \frac{1}{n} + \ldots + \frac{1}{n} = n \cdot \frac{1}{n} = 1$

i.e,. $\qquad |s_n| < 1, \forall n$

$\Rightarrow$ sequence $\langle s_n \rangle$ is bounded.

Then, by monotonic convergence criterion, the sequence $\langle s_n \rangle$ converges.

Example 2. *Show that the sequences $\langle s_n \rangle$ where $s_n = \sqrt{n^2 + m} - n$ is convergent. Also find its limit.*

Solution. Obviously, $\quad s_n > 0, \forall n \in N$

Further, $\qquad s_n = \frac{(\sqrt{n^2 + m} - n)(\sqrt{n^2 + m} + n)}{(\sqrt{n^2 + m} + n)} = \frac{m}{\sqrt{n^2 + m} + n}$

Also $\qquad s_{n+1} = \frac{m}{\sqrt{(n+1)^2 + m} + (n+1)}$

But $\qquad \sqrt{(n+1)^2 + m} + (n+1) > \sqrt{n^2 + m} + n$

$$\frac{m}{\sqrt{(n+1)^2 + m} + (n+1)} < \frac{m}{\sqrt{n^2 + m} + n}$$

$$\Rightarrow \qquad\qquad s_{n+1} < s_n, \forall n \in N$$

$\Rightarrow$ $\langle s_n \rangle$ is monotonic decreasing sequence and bounded below by 0.

Hence, $\langle s_n \rangle$ is convergent.

Further $\qquad \lim_{n \to \infty} s_n = \lim_{n \to \infty} \dfrac{m}{\sqrt{n^2 + m + n}} = 0.$

Example 3. If $s_n = \sqrt{n}(\sqrt{n+1} - \sqrt{n})$, does $\langle s_n \rangle$ converge?

Solution. We have, $\qquad s_n = \dfrac{\sqrt{n}(\sqrt{n+1} - \sqrt{n})(\sqrt{n+1} + \sqrt{n})}{(\sqrt{n+1} + \sqrt{n})}$

$$= \dfrac{\sqrt{n}(n+1-n)}{(\sqrt{n+1} + \sqrt{n})} = \dfrac{\sqrt{n}}{(\sqrt{n+1} + \sqrt{n})}$$

Also $\qquad s_{n+1} = \dfrac{\sqrt{n+1}}{\sqrt{n+2} + \sqrt{n+1}}$

$\therefore \qquad s_{n+1} - s_n = \dfrac{\sqrt{n+1}}{(\sqrt{n+2} + \sqrt{n+1})} - \dfrac{\sqrt{n}}{(\sqrt{n+1} + \sqrt{n})}$

$$= \dfrac{n+1+\sqrt{n}\sqrt{n+1} - \sqrt{n}\sqrt{n+2} - \sqrt{n}\sqrt{n+1}}{(\sqrt{n+2} + \sqrt{n+1})(\sqrt{n+1} + \sqrt{n})}$$

$$= \dfrac{n+1 - \sqrt{n(n+2)}}{(\sqrt{n+2} + \sqrt{n+1})(\sqrt{n+1} + \sqrt{n})}$$

$$= \dfrac{\sqrt{(n+1)^2} - \sqrt{(n^2 + 2n)}}{(\sqrt{n+2} + \sqrt{n+1})(\sqrt{n+1} + \sqrt{n})}$$

$$= \dfrac{\sqrt{(n^2 + 2n + 1)} - \sqrt{(n^2 + 2n)}}{(\sqrt{n+2} + \sqrt{n+1})(\sqrt{n+1} + \sqrt{n})} > 0, \forall n,$$

$$\text{as } n^2 + 2n + 1 > n^2 + 2n$$

$\Rightarrow$ $s_{n+1} > s_n \Rightarrow \langle s_n \rangle$ is monotonically increasing sequence.

Also $\qquad 1 - s_n = 1 - \dfrac{\sqrt{n}}{\sqrt{n+1} + \sqrt{n}} = \dfrac{\sqrt{n+1}}{(\sqrt{n+1} + \sqrt{n})} > 0$

$\Rightarrow \qquad\qquad s_n < 1, \forall n \in N$

$\Rightarrow$ $\langle s_n \rangle$ is monotonically increasing and bounded (above) sequence.

Hence $\langle s_n \rangle$ is convergent.

Further $\lim_{n \to \infty} s_n = \lim_{n \to \infty} \dfrac{\sqrt{n}}{\sqrt{n+1} + \sqrt{n}} = \lim_{n \to \infty} \dfrac{\sqrt{n}}{\sqrt{n}\left(\sqrt{1 + \dfrac{1}{n}} + 1\right)}$

$$= \lim_{n \to \infty} \dfrac{1}{\left(\sqrt{1 + \dfrac{1}{n}} + 1\right)} = \dfrac{1}{2}.$$

Hence $\langle s_n \rangle$ converges to $1/2$.

Example 4. *Show that the sequence $\langle n/n+1 \rangle$ is a bounded monotonically increasing sequence and convergent too.*

Solution. Let $\langle s_n \rangle = \left\langle \dfrac{n}{n+1} \right\rangle = \left\langle \dfrac{1}{2}, \dfrac{2}{3}, \dfrac{3}{4}, ..., \dfrac{n}{n+1}, ..., \right\rangle.$

Since, $\quad 1/2 < 2/3 < 3/4 < ....$

Now $\quad s_{n+1} - s_n = \dfrac{n+1}{n+2} - \dfrac{n}{n+1} = \dfrac{(n+1)^2 - n(n+2)}{(n+2)(n+1)} = \dfrac{1}{(n+2)(n+1)} > 0$

$\Rightarrow \qquad s_{n+1} - s_n > 0, \forall\, n$

$\Rightarrow \qquad s_{n+1} > s_n , \forall\, n \Rightarrow \langle s_n \rangle$ is monotonically increasing.

Futher, $\qquad n \geq 1 > 0 \;\Rightarrow s_n > 0.$

Also $\qquad 1 - s_n = 1 - \dfrac{n}{n+1} = \dfrac{1}{n+1} > 0 \Rightarrow s_n < 1, \forall\, n$

$\Rightarrow \quad 0 < s_n < 1,\, \forall\, n \Rightarrow \langle s_n \rangle$ is bounded also.

We know that a bounded monotonic sequence is always convergent. Therefore, given sequence is convergent.

Also $\qquad \lim_{n \to \infty} s_n = \lim_{n \to \infty} \left(\dfrac{n}{n+1} \right) = \lim_{n \to \infty} \left[\dfrac{1}{1 + 1/n} \right] = 1.$

Example 5. *Prove that the sequence $\left\langle \dfrac{2^n}{n!} \right\rangle$ is a monotonic decreasing. Also prove that it is bounded.*

Solution. Let $\langle s_n \rangle = \left\langle \dfrac{2^n}{n!} \right\rangle.$

Then $\qquad s_n = \dfrac{2^n}{n!}$ and $s_{n+1} = \dfrac{2^{n+1}}{(n+1)!}.$

$\therefore \qquad \dfrac{s_n}{s_{n+1}} = \dfrac{2^n}{n!} \times \dfrac{(n+1)!}{2^{n+1}} = \dfrac{(n+1)}{2} \geq 1, \forall\, n \in N$

i.e., $\qquad s_n \geq s_{n+1}.$

Hence, the given sequence is a monotonic decreasing sequence.

Putting $\qquad n = 1, 2, 3, 4, ... ,$ in $s_n = \dfrac{2^n}{n}$ we find that

$\langle s_n \rangle = \left\langle \dfrac{2}{1!}, \dfrac{2^2}{2!}, \dfrac{2^3}{3!}, \dfrac{2^4}{4!}, ... \right\rangle.$

It is evident that $\langle s_n \rangle$ is bounded above by 2.

Also, we have

$\dfrac{s_n}{s_{n+1}} = \dfrac{n+1}{2}$ or $\dfrac{s_{n+1}}{s_n} = \dfrac{2}{n+1}$

or $\qquad \lim_{n \to \infty} \dfrac{s_{n+1}}{s_n} = 0$

It means for any given positive ε, $\exists$ a number $m \in N$ such that

$$n > m \Rightarrow \left| \dfrac{s_{n+1}}{s_n} - 0 \right| < \varepsilon \Rightarrow \left| \dfrac{s_{n+1}}{s_n} \right| < \varepsilon$$

$$\Rightarrow \qquad |s_{n+1}| < \varepsilon |s_n| \Rightarrow |s_{n+1}| \to 0$$

$$\Rightarrow \qquad \lim_{n \to \infty} |s_{n+1}| = 0, \forall n \in N$$

$$\Rightarrow \qquad \lim_{n \to \infty} |s_n| = 0, \forall n \in N$$

$$\Rightarrow \qquad \lim_{n \to \infty} s_n = 0.$$

Hence, the sequence $\langle s_n \rangle$ is bounded below by 0.

Example 6. *Show that the sequence $\langle x_n \rangle$ where $x_1 = 1$ and $x_n = (2 + x_{n-1})^{1/2}$ is convergent and converge to 2.*

Solution. We have $x_1 = 1$ and $x_n = \sqrt{2 + x_{n-1}}, \forall n \in N$

$$x_2 = \sqrt{3} \quad \Rightarrow \quad x_1 < x_2.$$

Let us assume that $x_m < x_{m+1}$, then $\sqrt{2 + x_{m-1}} < \sqrt{2 + x_m}$

$$\Rightarrow \qquad x_{m+1} < x_{m+2}.$$

Hence by mathematical induction, we have $x_n < x_{n+2}, \forall n \in N$.

So $\langle x_n \rangle$ is monotonically increasing.

Again $x_{n+1} > x_n \qquad \Rightarrow \quad \sqrt{2 + x_n} > x_n$

$$\Rightarrow \quad 2 + x_n - x_n^2 > 0$$

$$\Rightarrow \quad (2 - x_n)(1 + x_n) > 0 \qquad [\because 1 + x_n > 0]$$

$$\Rightarrow \quad (2 - x_n) > 0$$

or $\qquad\qquad x_n < 2, \forall n \in N$

Hence $\langle x_n \rangle$ is bounded.

Thus $\langle x_n \rangle$ is a monotonically increasing sequence bounded above by 2. Therefore it converges.

Let $\lim_{n \to \infty} x_n = l$, then $\lim_{n \to \infty} x_{n-1} = l$,

Now, $\qquad\qquad\qquad x_n = \sqrt{2 + x_{n-1}}$

$$\Rightarrow \qquad\qquad \lim_{n \to \infty} x_n = \lim_{n \to \infty} \sqrt{2 + x_{n-1}}$$

$$\Rightarrow \qquad\qquad l = \sqrt{2 + l}$$

$$\Rightarrow \qquad\qquad l^2 - l - 2 = 0$$

$$\Rightarrow \qquad\qquad (l+1)(l-2) = 0$$

$$\Rightarrow \qquad\qquad l = -1, 2.$$

But l cannot be -1 since all terms of the sequence are positive. Hence $l = 2$.

Example 7. *Show that $\lim_{n \to \infty} \left(1 + \dfrac{1}{n}\right)^n$ exists and lies between 2 and 3.*

Solution. Let $\qquad\qquad s_n = \left(1 + \dfrac{1}{n}\right)^n$

$$\therefore \qquad\qquad s_1 = 2$$

Now, $s_n = 1 + n\dfrac{1}{n} + \dfrac{n(n-1)}{2!}\dfrac{1}{n^2} + \ldots + \dfrac{n(n-1)}{n!}\cdot\dfrac{1}{n^n}$

[By binomial theorem for positive integral index]

$$= 1 + 1 + \dfrac{1}{2!}\left(1 - \dfrac{1}{n}\right) + \ldots + \dfrac{1}{n!}\left(1 - \dfrac{1}{n}\right)\left(1 - \dfrac{2}{n}\right)\ldots\left(1 - \dfrac{n-1}{n}\right) \qquad \ldots (1)$$

Similarly

$$s_{n+1} = 1 + 1 + \dfrac{1}{2!}\left(1 - \dfrac{1}{n+1}\right) + \ldots + \dfrac{1}{(n+1)!}\left(1 - \dfrac{1}{n+1}\right)\left(1 - \dfrac{2}{n+1}\right)\ldots\left(1 - \dfrac{n}{n+1}\right) \qquad \ldots (2)$$

Comparing (1) and (2), we see that $s_{n+1} \geq s_n\ \forall n$.

$\Rightarrow$ The sequence $\langle s_n \rangle$ is monotonically increasing.

Now from (1), we have

$$2 < s_n < 1 + 1 + \dfrac{1}{2!} + \dfrac{1}{3!} + \ldots + \dfrac{1}{n!}$$

$$\leq 1 + 1 + \dfrac{1}{2} + \dfrac{1}{2^2} + \ldots + \dfrac{1}{2^{n-1}}, \quad \text{which is a G.P.}$$

$$= 1 + \dfrac{1 - \dfrac{1}{2^n}}{1 - \dfrac{1}{2}} = 3 - \dfrac{1}{2^{n-1}} < 3, \forall\, n.$$

$\Rightarrow$ The sequence $\langle s_n \rangle$ is bounded.

Thus, the sequence $\langle s_n \rangle$, being a monotonically increasing sequence bounded above by 3, is convergent.

Since $2 < s_n < 3\ \forall,\, n \Rightarrow 2 \leq \lim\limits_{n\to\infty} s_n \leq 3, \forall\, n.$

$\Rightarrow$ limit of the sequence $\langle s_n \rangle$ lies between 2 and 3.

Example 8. *Show that the sequence $\langle s_n \rangle$ defined by $s_1 = \sqrt{2}, s_{n+1} = \sqrt{(2s_n)}$ converges to 2.*

Solution.   We have $s_{n+1} = \sqrt{(2s_n)}$

For $n = 1$ $\Rightarrow$ $s_2 = \sqrt{(2s_1)}$ $\Rightarrow$ $s_2 = \sqrt{(2\sqrt{2})}$

Since $\Rightarrow$ $1 < \sqrt{2}$ $\Rightarrow$ $2 < 2\sqrt{2}$

$\Rightarrow$ $\sqrt{2} < \sqrt{(2\sqrt{2})} \Rightarrow$ $s_1 < s_2$

Now, let us suppose that $s_m < s_{m+1}$

then $\sqrt{(2s_m)} < \sqrt{(2s_{m+1})}$

$\Rightarrow$ $s_{m+1} < s_{m+2}.$

Now, by the principle of mathematical induction, we have $s_n < s_{n+1},\ \forall\ n \in N$

i.e., $\langle s_n \rangle$ is monotonically increasing sequence.

Now we shall show that $<s_n>$ is bounded.

Since $s_1 = \sqrt{2} < 2.$

Let us suppose that $s_m < 2$. Then $\sqrt{(2s_m)} < \sqrt{2.2} = 2$

$$\Rightarrow \qquad\qquad s_{m+1} < 2.$$

By the principle of mathematical induction, we have $s_n < 2, \forall\, n \in N$

$\Rightarrow \quad \langle s_n \rangle$ is bounded above by 2.

$\Rightarrow \quad \langle s_n \rangle$ is monotonically increasing sequence which is bounded above.

Then, by monotone convergence ctiterion, $\langle s_n \rangle$ is convergent.

Now, let $\displaystyle\lim_{n \to \infty} s_n = l \qquad\Rightarrow\qquad \lim_{n \to \infty} s_{n+1} = l$

Given that $s_{n+1} = \sqrt{(2s_n)} \quad\Rightarrow\quad \lim s_{n+1} = \lim\sqrt{2s_n}$

$$\Rightarrow \qquad\qquad l = \sqrt{2l}$$

$\Rightarrow \qquad l(l-2) = 0$ which gives $l = 2, l = 0.$

But, since $\langle s_n \rangle$ is positive terms sequence with first term $= \sqrt{2}$. Therefore, l cannot be equal to 0. Hence, $\qquad\qquad l=2.$

Example 9. *Show that the sequence $\langle s_n \rangle$ defined by formula $s_1 = 1$, $s_{n+1} = \sqrt{(3s_n)}$ converges to 3.*

Solution. Solution is exactly same as Example 8 and left as an exercise for the students.

Example 10. *Show that the sequence $\langle s_n \rangle$ defined by*

$$s_1 = 1, s_{n+1} = \frac{4+3s_n}{3+2s_n}, \forall n \in N$$

is convergent and find its limit.

Solution. Since

$$s_1 = 1, s_2 = \frac{4+3s_1}{3+2s_1} = \frac{7}{5}$$

$$\therefore \qquad\qquad s_2 > s_1.$$

Now, let us assume that for some positive integer n, $s_{n+1} > s_n$.

We shall show that $\quad s_{n+2} > s_{n+1}$

Consider, $\quad s_{n+2} - s_{n+1} = \dfrac{4+3s_{n+1}}{3+2s_{n+1}} - \dfrac{4+3s_n}{3+2s_n}$

$$= \frac{(4+3s_{n+1})(3+2s_n) - (4+3s_n)(3+2s_{n+1})}{(3+2s_{n+1})(3+2s_n)}$$

$$= \frac{s_{n+1} - s_n}{(3+2s_{n+1})(3+2s_n)} > 0 \qquad\qquad (\because s_{n+1} > s_n)$$

$$\Rightarrow \qquad s_{n+2} > s_{n+1}.$$

Then, by the principle of mathematical induction $s_{n+1} > s_n,\ \forall n \in N$

$\Rightarrow \quad$ The sequence $\langle s_n \rangle$ is monotonically increasing.

Now we have $\quad s_{n+1} = \dfrac{3s_n + 4}{2s_n + 3} = \dfrac{\frac{3}{2}(2s_n + 3) - \frac{1}{2}}{2s_n + 3} = \dfrac{3}{2} - \dfrac{1}{2(2s_n + 3)}$

$$\Rightarrow \qquad s_{n+1} < \frac{3}{2}, \forall n \in N$$

$\Rightarrow$ The sequence $\langle s_n \rangle$ is bounded above by $\dfrac{3}{2}$ $\left(\because s_1 = 1 < \dfrac{3}{2} \Rightarrow s_n < \dfrac{3}{2} \right)$

$\Rightarrow$ The sequence $\langle s_n \rangle$ is bounded, monotonically increasing sequence.

Then, by monotone convergence theorem, $\langle s_n \rangle$ is convergent.

Now, let $\lim s_n = l \Rightarrow \lim s_{n+1} = l$

Consider $s_{n+1} = \dfrac{4 + 3s_n}{3 + 2s_n} \Rightarrow \lim s_{n+1} = \dfrac{4 + 3 \lim s_n}{3 + 2 \lim s_n}$

$\Rightarrow$ $l = \dfrac{4 + 3l}{3 + 2l} \Rightarrow l^2 = 2$ i.e., $l = \pm\sqrt{2}$.

Since $\langle s_n \rangle$ is a positive terms sequence, hence l cannot be negative

therefore, $l = \sqrt{2}$

Example 11. *If $a > 1$, $s_1 = 1$, $s_{n+1} = \sqrt{a + s_n}$, then show that the sequence is bounded and monotonic and converges to the positive root of the equation $x^2 - x - a = 0$.*

Solution.   Since $1 \le s_1 = 1 < a+1$

                      $1 \le s_2 = \sqrt{a+1} < (a+1)$.

Let us suppose $1 \le s_m < a+1$, then

                      $1 \le s_{m+1} = \sqrt{a + s_m}$

                                    $< \sqrt{a + a + 1} = \sqrt{2a} + 1 < a + 1$.

Then, by the method of mathematical induction, we have $1 \le s_n < a + 1$

$\Rightarrow$ $\langle s_n \rangle$ is bounded.

Now, we shall show that $\langle s_n \rangle$ is monotonically increasing sequence.

Consider $s_{n+1}^2 - s_n^2 = (a + s_n) - (a + s_{n-1}) = s_n - s_{n-1}$

$\Rightarrow$ $s_{n+1}^{2(n-1)} - s_n^{2(n-1)} = s_2 - s_1 > 0$

$\Rightarrow$ $a + 1 > s_{n+1} > s_n > 1 \, \forall n \in \mathbb{N}$.

$\Rightarrow$ The sequence is monotonically increasing.

Then, by monotone convergence theorem, $\langle s_n \rangle$ is convergent.

Now, let $\displaystyle\lim_{n \to \infty} s_n = l$. Then $\displaystyle\lim_{n \to \infty} s_{n+1} = \lim_{n \to \infty} \sqrt{a + s_n}$

$\Rightarrow$ $l = \sqrt{a + l} \Rightarrow l^2 = a + l$

$\Rightarrow$ $l^2 - l - a = 0$

By the theory of equation we can say this equation has only one positive root.

$\Rightarrow$ The given sequences $\langle s_n \rangle$ converges to the positive root of $x^2 - x - a = 0$.

Example 12. *If $s_1 > 0$, $t_1 > 0$ and for all $n > 1, s_{n+1} = \dfrac{1}{2}(s_n + t_n)$, $t_{n+1} = \sqrt{s_n . t_n}$. Then prove that $\langle s_n \rangle$ and $\langle t_n \rangle$ are monotonic sequences converges to the same limit.*

Solution.   Since, the arithmetic means of two positive numbers is always greater than or equal to the geometric means between them

$\therefore$ $s_{n+1} \ge t_{n+1}, \forall n$

Now $$s_{n+1} = \frac{1}{2}(s_n + t_n) \le \frac{1}{2}(s_n + s_n) = s_n$$

and $$t_{n+1} = \sqrt{s_n t_n} \ge \sqrt{t_n t_n} = t_n, \forall n$$

$\therefore \langle s_n \rangle$ is a monotonically decreasing sequence and $\langle t_n \rangle$ is a monotonically increasing sequence.

Further, $$s_n \ge t_1 \text{ and } t_n \le s_1, \forall\, n.$$

$\therefore$ Both the sequences $\langle s_n \rangle$ and $\langle t_n \rangle$ are convergent.

Now, to show $\lim\limits_{n \to \infty} s_n = \lim\limits_{n \to \infty} t_n$

Let $s_n \to l_1$ and $t_n \to l_2$

Since $$2s_{n+1} = s_n \to t_n \text{ therefore } \lim_{n \to \infty} 2s_{n+1} = \lim_{n \to \infty}(s_n + t_n)$$

$\Rightarrow \qquad 2 \lim\limits_{n \to \infty} s_{n+1} = \lim\limits_{n \to \infty} s_n + \lim\limits_{n \to \infty} t_n$

$\Rightarrow \qquad 2l_1 = l_1 + l_2$

Hence, $\qquad l_1 = l_2.$

Example 13. *Prove that the sequence $\langle a_n \rangle$ is convergent where*

$$a_n = 1 + \frac{1}{1!} + \frac{1}{2!} + \frac{1}{3!} + \ldots + \frac{1}{n!}.$$

Solution. Since $$a_n = 1 + \frac{1}{1!} + \frac{1}{2!} + \frac{1}{3!} + \ldots + \frac{1}{n!}$$

and $$a_{n+1} = 1 + \frac{1}{1!} + \frac{1}{2!} + \frac{1}{3!} + \ldots + \frac{1}{n!} + \frac{1}{(n+1)!}$$

then $$a_{n+1} - a_n = \frac{1}{(n+1)!} > 0, \forall n \in N$$

Thus, $\langle a_n \rangle$ is monotonically increasing.

Further, $$a_n = 1 + \frac{1}{1!} + \frac{1}{2!} + \frac{1}{3!} + \ldots + \frac{1}{n!}$$

$\Rightarrow \qquad 2 < a_n < 1 + 1 + \dfrac{1}{2^2} + \dfrac{1}{2^3} + \ldots + \dfrac{1}{2^{n-1}}$

$\Rightarrow \qquad 2 < a_n \le 1 + \dfrac{1 - \dfrac{1}{2^n}}{1 - \dfrac{1}{2}} = 3 - \dfrac{1}{2^{n-1}} < 3, \forall n$

$\Rightarrow \qquad \langle a_n \rangle$ is bounded.

Hence, $\langle a_n \rangle$ is convergent.

Example 14. *Show that* $\lim\limits_{n \to \infty} \dfrac{n}{(n!)^{1/n}} = e$

Solution. Let $s_n = \dfrac{n^n}{n!}$, then $s_{n+1} = \dfrac{(n+1)^{n+1}}{(n+1)!}$

$$\therefore \qquad \frac{s_{n+1}}{s_n} = \frac{(n+1)^{n+1}}{(n+1)} \cdot \frac{1}{n^n} = \left(\frac{n+1}{n}\right)^n$$

$$\Rightarrow \qquad \lim_{n\to\infty} \frac{s_{n+1}}{s_n} = \lim_{n\to\infty} (1+1/n)^n = e > 0 \ \forall n \in N.$$

So, by Cauchy second theorem

$$\lim_{n\to\infty} s_n^{1/n} = \lim_{n\to\infty} \frac{s_{n+1}}{s_n} = e$$

Hence $\quad \lim_{n\to\infty} \dfrac{n}{(n!)^{1/n}} = e.$

Example 15. *Prove that the sequence $\left\langle \dfrac{2n-7}{3n+2} \right\rangle$ is monotonically increasing, bounded above and bounded below.*

Solution. A sequence $\left\langle \dfrac{2n-7}{3n+2} \right\rangle$ is said to be a monotonic sequence, if it is either an increasing sequence or a decreasing sequence; *i.e.*, if either $s_{n+1} \geq s_n$ or $s_{n+1} \leq s_n, \forall n \in N$

Since, $\qquad s_n = \dfrac{2n-7}{3n+2} = \dfrac{\dfrac{2}{3}(3n+2) - \dfrac{4}{3} - 7}{3n+2}$

$$= \frac{2}{3} - \frac{25}{3} \cdot \frac{1}{3n+2} \qquad\qquad \text{... (1)}$$

$$\Rightarrow \qquad s_{n+1} = \frac{2}{3} - \frac{25}{3} \cdot \frac{1}{2n+5}$$

$$\Rightarrow \qquad s_{n+1} - s_n = \frac{25}{3}\left[\frac{1}{3n+2} - \frac{1}{3n+5}\right] = \frac{25}{(3n+2)(3n+5)} > 0, \forall n \in N$$

$\therefore \quad s_{n+1} > s_n \ \forall n \in N$ and so the sequence $\langle s_n \rangle$ is a monotonic increasing sequence.

Also from (1), we observe that $s_n < \dfrac{2}{3}, \forall n \in N$ *i.e.*, 2/3 is an upper bound for $\langle s_n \rangle$ and so $\langle s_n \rangle$ is bounded above.

Again $\langle s_n \rangle$ is a monotonic increasing sequence and so

$$s_n \geq s_1 = \frac{2-7}{3+2} = -1, \forall n \in N$$

$\therefore \quad s_1 = -1$ is a lower bound for $\langle s_n \rangle$ and $\langle s_n \rangle$ is bounded below.

Since the sequence $\langle s_n \rangle$ is monotonic increasing and bounded above, therefore by monotone convergence theorem $\langle s_n \rangle$ converges to its supremurn.

From (1), we observe that $\lim_{n\to\infty} s_n = \dfrac{2}{3} - 0 = \dfrac{2}{3}$

Thus the sequence $\langle s_n \rangle$ converges to $\dfrac{2}{3}$ and so sup $\langle s_n \rangle = \dfrac{2}{3}$.

Also $\qquad\qquad\qquad \inf\left\langle s_n \right\rangle = s_1 = -1$

1. Discuss the boundedness of the following sequences $\langle s_n \rangle$, where s_n is given by

    (i) $s_n = 6$ (ii) $s_n = (-1)^n . 4$

    (iii) $s_n = \dfrac{2n+3}{3n+4}$ (iv) $s_n = \left(1 + \dfrac{1}{n}\right)^n$

    (v) $s_n = \dfrac{1}{n^2} + \dfrac{1}{(n+1)^2} + \dots + \dfrac{1}{(2n)^2}$

    (vi) $s_n = n^3$ (vii) $s_n = 1 + (-1)^n$.

2. Discuss the convergence and divergence of sequences in Ques. 1.

3. Give examples of sequence $\langle s_n \rangle$ for which

    $\lim\limits_{n \to \infty} \dfrac{s_{n+1}}{s_n} = 1$ and

    (i) $s_n \to \infty$ (ii) $s_n \to 2$

    (iii) $s_n \to 0$.

4. Verify the following:

    (i) $\lim\limits_{n \to \infty} \dfrac{3n-5}{4-2n} = -\dfrac{3}{2}$

    (ii) $\lim\limits_{n \to \infty} [(n^2+1)^{1/8} - (n+1)^{1/4}] = 0$

    (iii) $\lim\limits_{n \to \infty} \left[\dfrac{1}{n^2} + \dfrac{1}{(n+1)^2} + \dots + \dfrac{1}{(2n)^2} \right] = 0$

    (iv) $\lim\limits_{n \to \infty} \left(1 - \dfrac{1}{n}\right)^{-n} = e$

5. If $\langle s_n \rangle$ diverges to $+\infty$ and $\langle t_n \rangle$ diverges to $-\infty$, then show by example that $\langle s_n + t_n \rangle$ may :

    (i) converge   (ii) diverge to ∞

    (ii) diverge to $-\infty$ (iv) oscillate

6. Prove by definition that the following sequences whose n^{th} term are given below, are Cauchy sequence:

    (i) $\dfrac{1}{n}$ (ii) $\dfrac{1}{n^2}$

    (iii) $(-1)^n . \dfrac{1}{n}$ (iv) $\dfrac{n}{n+1}$

7. Show that the sequences whose n^{th} terms are given below are not Cauchy sequences.

    (i) $(-1)^n$ (ii) n (iii) $(-1)^n . n$.

8. If $\langle s_n \rangle$ and $\langle t_n \rangle$ are two Cauchy sequences, then show that

    (i) $\langle s_n + t_n \rangle$ (ii) $\langle s_n - t_n \rangle$

    (iii) $\langle s_n t_n \rangle$ and   (iv) $\left\langle \dfrac{s_n}{t_n} \right\rangle (t_n \neq 0)$

    are Cauchy sequences.

9. Show that the sequences $\langle s_n \rangle$ defined by

    $s_1 = \dfrac{1}{2}, s_{n+1} = \dfrac{2 s_n + 1}{3}, \forall n \in N$ is convergent.

    Also find its limit.

10. Show that the sequence $\left\langle \dfrac{n^2 + 3n + 5}{2n^2 + 5n + 7} \right\rangle$ converges to $\dfrac{1}{2}$.

11. Show that the sequence $\langle s_n \rangle$ defined by $s_1 = \sqrt{7}, s_{n+1} = \sqrt{7 + s_n}$ converges to the positive root of $x^2 - x - 7 = 0$.

12. Show that the sequence defined by $s_1 = 1, s_{n+1} = \dfrac{4 + 3 s_n}{3 + 2 s_n}, n \in N$ is convergent. Find its limit.

13. Show that the sequence where $s_n = \dfrac{1}{(\log n)^{\log n}}$ is convergent.

14. Show that the sequence $\langle s_n \rangle$ where $s_n = \dfrac{2n^2 + 1}{2n^2 - 1}$ converges to 1.

1. (iii) $s_n = \dfrac{2n+3}{3n+4} \Rightarrow \langle s_n \rangle = \left\langle \dfrac{5}{7}, \dfrac{7}{10}, \dfrac{19}{13}, \dots \right\rangle$

    $\Rightarrow |s_n| \leq \dfrac{5}{7} \Rightarrow \langle s_n \rangle$ is bounded .

4. (i) Let $\varepsilon > 0$ be given. Therefore

    $\left| \dfrac{3n-5}{4-2n} - \dfrac{3}{2} \right| = \left| \dfrac{1}{4-2n} \right| = \dfrac{1}{-4+2n}$

    $< \varepsilon$ if $\dfrac{1}{-4+2n} < \varepsilon$ or $n > \dfrac{4+\varepsilon}{2}$

6. (i) $s_n = \dfrac{1}{n}$. Take $\varepsilon > 0$. If $n \geq m$ then

    $|s_n - s_m| = \left| \dfrac{1}{n} - \dfrac{1}{m} \right| = \left| \dfrac{m-n}{mn} \right|$

    $= \dfrac{n-m}{mn} = \dfrac{n-m}{n} . \dfrac{1}{m} < \dfrac{1}{m}$

If we take m as positive integer, such that

$m > \dfrac{1}{\varepsilon}$ i.e., $\dfrac{1}{m} < \varepsilon$. Then $|s_n - s_m| < \varepsilon$

$\Rightarrow \langle s_n \rangle = \left\langle \dfrac{1}{n} \right\rangle$ is a Cauchy sequence.

9. $s_{n+1} = \dfrac{2s_{n+1}}{3}$

If $\lim s_n = l$.

Then $\lim s_{n+1} = l \Rightarrow \dfrac{2l+1}{3} \Rightarrow l = 1$

10. $s_n = \dfrac{n^2 + 5n + 5}{2n^2 + 5n + 7}$

$\Rightarrow \lim_{n \to \infty} s_n = \lim_{n \to \infty} \left[\dfrac{1 + \dfrac{3}{n} + \dfrac{5}{n^2}}{2 + \dfrac{5}{n} + \dfrac{7}{n^2}} \right] = \dfrac{1}{2}.$

13. $s_n = \dfrac{1}{(\log n)^{\log n}}$

Since we know that $\log(n+1) > \log n, \forall n \in N$

$\Rightarrow [\log (n+1)]^{\log(n+1)} > (\log n)^{\log(n+1)}, \forall n \in N$

$\qquad\qquad\qquad\qquad > (\log n)^{\log n}, \forall n \in N$

$\Rightarrow \dfrac{1}{(\log n)^{\log n}} > \dfrac{1}{[\log(n+1)]^{\log(n+1)}}, \forall n \in N$

$\Rightarrow s_n > s_{n+1} \Rightarrow \langle s_n \rangle$ is monotonically decreasing sequence.

Similarly we can show that $\langle s_n \rangle$ is bounded.

Answers

1. (i), (ii), (iii), (iv), (v), (vii) bounded (vi) unbounded.

2. (i), (iii), (iv), (v) converges (ii), (vii) oscillate (vi) diverges to ∞

3. (i) $s_n = n$　　　(ii) $s_n = \dfrac{2n+1}{n}$　　　(iii) $s_n = \dfrac{1}{n}$

5. (i) $s_n = n, t_n = -n$　　　(ii) $s_n = n^2, t_n = -n$　　　(iii) $s_n = n, t_n = -n^2$.

9. $l = 1$　　　　　　　　**12.** $l = \sqrt{2}$

CHAPTER REVIEW: A COMPETITIVE APPROACH

SELECTED TERMS AND RESULTS

TERMS

- **Sequence:** A function $f : N \to S$ whose domain is the set of natural numbers and range is the set of real numbers is called sequence.

- **Range of a Sequence:** Range set is the set consisting of all distinct elements of a sequence.

- **Bounded above Sequence:** A sequence $<s_n>$ is said to be bounded above if there exists a real number k such that $s_n \le k$, $\forall n \in N$.

- **Bounded below Sequence:** A sequence $<s_n>$ is said to bounded below if there exists a real number k such that $s_n \ge k$, $\forall n \in N$.

- **Bounded Sequence:** A sequence which is bounded below as well as above.

- **Convergent Sequence:** A sequence $<s_n>$ is said to converge to a number l if for given $\varepsilon > 0$ $\exists$ a positive integer m such that $|s_n - l| < \varepsilon$ $\forall n \ge m$.

- **Limit Point of the Sequence:** A real number l is called the limit of sequence $<s_n>$ if every nbd of l contains infinitely many members of $<s_n>$.

- **Oscillate Infinitely:** An unbounded sequence is said to be oscillate infinitely if it diverges neither to $+\infty$ nor $-\infty$.

- **Cauchy's Sequence:** A sequence $<s_n>$ is said to be Cauchy if for given $\varepsilon > 0$ $\exists m > 0$ such that $|s_n - s_m| < \varepsilon$ $\forall n \ge m$.

- **Monotonic Sequence:** (1) A sequence $<s_n>$ is said to be monotonically increasing if $s_{n+1} \ge s_n$ $\forall n \in N$.

  (2) A sequence $<s_n>$ is said to be strictly increasing if $s_{n+1} > s_n$ $\forall n \in N$.

  (3) A sequence $<s_n>$ is said to be monotonically decreasing if $s_{n+1} \le s_n$ $\forall n \in N$.

  (4) A sequence $<s_n>$ is said to be strictly decreasing if $s_{n+1} < s_n$ $\forall n \in N$.

- **Subsequence:** If $<n_k>$ be a strictly increasing sequence of natural numbers then $\left\langle s_{n_k} \right\rangle$ is a subsequence of $<s_n>$.

RESULTS

- Every convergent sequence is bounded.

- A sequence can not converge to more than one limit.

- Every bounded sequence has a limit point (*Bolzano-Weirstrass theorem*).

- The set of limit points of a bounded sequence has the greatest and least members.

- If a sequence is bounded then limit inferior and limit superior are both finite.

- Every bounded sequence with a unique limit point is convergent.

- A sequence is cauchy iff it is convergent.

- The sum, difference, product and quotient under certain conditions of two convergent sequences is convergent. The converse may not be true.

- If $\lim s_n = l$ and $s_n \ge 0$ $\forall n$ then $l \ge 0$.

- If $s_n \le t_n$ then $\lim s_n \le \lim t_n$.

- If $<s_n>, <t_n>$ and $<u_n>$ are three sequences such that

  (i) $s_n \le t_n \le u_n$, $\forall n$ and

  (ii) $\lim s_n = \lim u_n = l$ then $\lim t_n = l$

  (*Sandwitch's theorem*)

- If $\lim\limits_{n \to \infty} s_n = l$ then

  $$\lim_{n \to \infty} \left(\frac{s_1 + s_2 + \ldots + s_n}{n} \right) = l$$

  (*Cauchy's first theorem on limits*)

- If $\lim\limits_{n \to \infty} s_n = l$ then $\lim\limits_{n \to \infty} (s_1 . s_2 , \ldots . s_n)^{1/n} = l$

  (*Cauchy's second theorem on limits*)

- If the sequence $<s_n>$ and $<t_n>$ converges to finite limits l_1 and l_2 respectively, then

  $$\lim_{n \to \infty} \left(\frac{s_1 t_n + s_2 t_{n-1} + \ldots . s_n t_1}{n} \right) = l_1 l_2$$

  (*Cesaro's theorem*)

- If a sequence $<s_n>$ be such that $\lim \dfrac{s_{n+1}}{s_n} = l$, where $|l| < 1$ then $\lim s_n = 0$.

- If $<s_n>$ be a sequence such that $\lim \dfrac{s_{n+1}}{s_n} = l$, then $\lim s_n = \infty$

- A necessary and sufficient condition for the convergence of a monotonic sequence is that it is bounded.

- A monotonic increasing bounded above sequence converges to its least upper bound and a monotonic decreasing bounded below sequence to the greatest lower bound.

- Every monotonic decreasing sequence which is not bounded below diverges to $-\infty$.

- The sequence $<r^n>$ converges to zero if $|r| < 1$.

- If the sequence $<f_n>$ and $<g_n>$ diverges to ∞ then $(f_n + g_n)$ and $(f_n \cdot g_n)$ both diverges to infinity.

- If $<f_n>$ diverges to ∞ and $<g_n>$ is bounded then $<f_n + g_n>$ diverges to ∞.

- If $<f_n>$ diverges to ∞ and $<g_n>$ converges then $<f_n + g_n>$ diverges to ∞.

- $\lim f_n = l$ $\lim |f_n| = |l|$, converse is not necessarily true.

- A necessary and sufficient condition for a sequence $<f_n>$ to be convergent is that to each $\varepsilon > 0$ there corresponds a positive integer m such that $|f_{n+p} - f_n| < \varepsilon, \forall n \geq m, p > 0$ (*Cauchy's general principal of convergence*).

- Sum, difference, product and quotient of two Cauchy's sequences is again Cauchy.

- If a sequence $<s_n>$ converges to l then every subsequence of $<s_n>$ converges to l. Converse of this is not true.

- Every bounded sequence has a convergent subsequence.

- A bounded sequence $<s_n>$ converges to l if and only if $\lim_{n \to \infty} \sup s_n = \lim_{n \to \infty} \inf s_n = l$

- $\lim_{n \to \infty} \sup (f_n + g_n) \leq \lim_{n \to \infty} \sup f_n + \lim_{n \to \infty} \sup g_n$

- $\lim_{n \to \infty} \inf (f_n + g_n) \leq \lim_{n \to \infty} \inf f_n + \lim_{n \to \infty} \inf g_n$

- $\lim_{n \to \infty} \sup (f_n \cdot g_n) \leq \left(\lim_{n \to \infty} \sup f_n \right) \cdot \left(\lim_{n \to \infty} \sup g_n \right)$

- $\lim_{n \to \infty} \inf (f_n \cdot g_n) = \left(\lim_{n \to \infty} \inf f_n \right) \cdot \left(\lim_{n \to \infty} \inf g_n \right)$

- If $\lim \sup s_n = l$ then

  (i) there is a subsequence of $<s_n>$ that converges to l.

  (ii) the limit superior of no subsequence of $<s_n>$ can exceeds l.

  (iii) no subsequence of $<s_n>$ can converge to a number greater than l.

- If $\lim \inf s_n = l$, then

  (i) there is a subsequence of $<s_n>$ that converges to l.

  (ii) the limit inferior of no subsequence of $<s_n>$ can be less than l.

  (iii) no sub sequence of $<s_n>$ can converge to a number less than l.

- If a sequence $<s_n>$ is bounded and there exists $m \in N$ such that $a \leq s_n \leq b \ \forall n \geq m$ then $a \leq \lim s_n \leq \lim s_n \leq b$.

- $\lim_{n \to \infty} n^{1/n} = 1$; $\lim_{n \to \infty} \dfrac{1}{n^a} = 0$

  if $\quad a > 0$; $\lim_{n \to \infty} a^{1/n} = 1$ if $a > 0$

  $\lim_{n \to \infty} n^a r^n = 0$ if $|r| < 1$

  and $a \in R$; $\quad \lim_{n \to \infty} r^n = 0$ if $|r| < 1$

- If non-zero term s_n diverge then $\dfrac{1}{s_n}$ converges to 0.

- Every sequence has a monotonic subsequence which diverges to ∞ or $-\infty$ according as the sequence is unbounded above or below.

- Let $<s_n>$ be a positive terms sequence then

  (i) $\lim_{n \to \infty} \sup s_n = \dfrac{1}{\lim_{n \to \infty} \inf \left(\dfrac{1}{s_n} \right)}$, where the first limit is finite and non-zero.

  (ii) $\lim_{n \to \infty} \sup s_n = 0 \Leftrightarrow \lim_{n \to \infty} \inf \left(\dfrac{1}{s_n} \right) = \infty$

  (iii) $\lim_{n \to \infty} \sup s_n = \infty \Leftrightarrow \lim_{n \to \infty} \inf \left(\dfrac{1}{s_n} \right) = 0$

- Every sequence is a subsequence of itself.

- A non-convergent sequence may have a convergent subsequence.

REVIEW QUESTIONS AND PROJECT WORK

1. Show that the following sequences are bounded.

(i) $\left(1+\dfrac{1}{n}\right)^n$

(ii) $1+r+r^2+....+r^n$, where $-1\le r<1$

(iii) $u_{n+1}=\dfrac{1}{2}\left(u_n+u_{n-1}\right)$ for given $u_1,u_2\in$R.

2. If $s_{n+1}=\dfrac{1}{2}\left(s_n+\dfrac{a}{s_n}\right)$, $a\ge0$ and $s_1\ne0$ then

show that s_n converges to $\sqrt{a}$ or $-\sqrt{a}$ according as $s_1>0$ or $s_1<0$.

3. Show that
$$\lim_{n\to\infty}\frac{1}{\sqrt{n}}\left(1+\frac{1}{\sqrt{2}}+\frac{1}{\sqrt{3}}+...+\frac{1}{\sqrt{n}}\right)=2$$

4. Show that $1<n^{1/n}+1+\dfrac{2}{\sqrt{n}}-\dfrac{2}{n}$ $\forall n>1$ and hence $\lim n^{1/n}=1$.

5. If $s_{n+1}=\dfrac{as_n+bs_{n-1}}{a+b}$ where $a+b>b>0$ then show that s_n converges to $\dfrac{(a+b)s_2+bs_1}{a+2b}$ for any s_1,s_2.

6. If $\lim s_n=l$, show that $\lim_{n\to\infty}\dfrac{1}{2}\left(s_n+s_{n-1}\right)=l$ and $\lim_{n\to\infty}\left(s_ns_{n-1}\right)=|l|$.

7. If $n\to\infty$, show that
$$\left|\frac{s_n-l}{s_n+l}\right|\to0\Leftrightarrow s_n\to l, l\ne0$$

8. Prove the following:

(i) $\displaystyle\lim_{n\to\infty}\frac{1}{n}\left(1+\frac{1}{3}+...+\frac{1}{2n-1}\right)=0$

(ii) $\displaystyle\lim_{n\to\infty}\frac{1}{n}\left(1+2+3^{2/3}+...+n^{2/n}\right)=1$

(iii) $\displaystyle\lim_{n\to\infty}\frac{1}{n}\Big(1.n^{1/n}+2^{1/2}.(n-1)^{1/n-1}$
$$+...+n^{1/n}.1\Big)=1$$

9. If $<s_n>$ and$<t_n>$are non-negative sequences and $\lim(s_n)^{1/n}$ and $\lim(t_n)^{1/n}$ exist, prove that $\lim(s_n+t_n)^{1/n}$ exists and
$$\lim(s_n+t_n)^{1/n}=\max[\lim s_n^{1/n},\lim t_n^{1/n}]$$

10. Show that

(i) $\left(1+\dfrac{1}{n}\right)^n<e<\left(1+\dfrac{1}{n}\right)^{n+1}$ $\forall n$

(ii) $\dfrac{n+1}{(n!)^{1/n}}<e<\dfrac{(n+1)^{1+1/n}}{(n!)^{1/n}}$ $\forall n$.

11. Prove that if $t_n>0$ and $(t_1+t_2+...+t_n)$ diverges to ∞, then
$$\lim_{n\to\infty}\frac{s_n}{t_n}=l\Rightarrow\lim_{n\to\infty}\frac{s_1+s_2+...+s_n}{t_1+t_2+...+t_n}=l$$

(Finite or Infinite)

12. Prove that for a bounded sequence the limit superior and limit inferior are greatest and least limit points respectively.

13. If $<s_n>$ is a sequence of positive terms, then show that
$$\underline{\lim}\,\frac{s_{n+1}}{s_n}\le\underline{\lim}\,s_n^{1/n}$$
$$\le\overline{\lim}\,s_n^{1/n}\le\overline{\lim}\,\frac{s_{n+1}}{s_n}$$

14. If $t_{n+1}>t_n>0$ $\forall n$ and t_n diverges to ∞, show that
$$\underline{\lim}\,\frac{s_{n+1}-s_n}{t_{n+1}-t_n}\le\underline{\lim}\,\frac{s_n}{t_n}$$
$$\le\overline{\lim}\,\frac{s_n}{t_n}\le\overline{\lim}\,\frac{s_{n+1}-s_n}{t_{n+1}-t_n}$$

15. If $<s_n>$ be a sequence of positive terms then show that

(i) $\displaystyle\lim_{n\to\infty}\sup.(s_n)^{1/n}\le\lim_{n\to\infty}\sup.\left(\frac{s_{n+1}}{s_n}\right)$

(ii) $\displaystyle\lim_{n\to\infty}\inf.\left(\frac{s_{n+1}}{s_n}\right)\le\lim_{n\to\infty}\inf.(s_n)^{1/n}$

OBJECTIVE TYPE QUESTIONS

FILL IN THE BLANKS

1. Every convergent sequence is _____ .

2. Every bounded sequence is _____ convergent.

3. The limit of a positive term sequence is always _____ .

4. Limit of the sequence is _____ .

5. A sequence is Cauchy if and only if it is _____ .

6. Every Cauchy sequence is _____ .

7. Every bounded monotonically increasing sequence is _____ .

8. A non-decreasing sequence, which is not bounded above diverges to _____ .

9. Squeeze principle is also known as _____ .

10. Supremum (or infimum) of a sequence is _____ .

TRUE/ FALSE

Write 'T' for true and 'F' for false statements.

1. Every convergent sequence is bounded. **(T/F)**

2. Every bounded monotonically increasing sequence is convergent. **(T/F)**

3. If $\langle s_{n+1} - s_n \rangle$ oscillate finitely, then $\langle s_n \rangle$ oscillate. **(T/F)**

4. If for given k (however large) we can find m for which $a_m > k$ then $s_n \to \infty$. **(T/F)**

5. If $\langle s_{n+1} - s_n \rangle$ oscillate infinitely, then $\langle s_n \rangle$ oscillate. **(T/F)**

6. Every subsequence of a convergent sequence is convergent. **(T/F)**

7. Every subsequence of a bounded sequence is bounded. **(T/F)**

8. Supremum (or infimum) of a given sequence must be the member of the sequence. **(T/F)**

9. Every Cauchy sequence is convergent. **(T/F)**

10. Every convergent sequence is Cauchy. **(T/F)**

MULTIPLE CHOICE QUESTIONS

Problem set-1

Choose the most appropriate one :

1. An oscillatory sequence is :
   (a) always bounded
   (b) may or may not be bounded
   (c) never bounded
   (d) none of the above

2. Formula for s_n, for the given sequences 1, –1, 1, –1, ... is:
   (a) $s_n = (-1)^n \ \forall n \in N$
   (b) $s_n = (-1)^{n+1} \ \forall \ n \in N$
   (c) $s_n = 1$ if n is even
       $= -1$ if n is odd
   (d) none of the above

3. If the sequence $\langle s_n \rangle$ converges to l then the sequence $|s_n|$ converges to :
   (a) l (b) $|l|$
   (c) $-l$ (d) none of these.

4. A sequence $<s_n>$ of real numbers such that $<|s_n|>$ converges but $<s_n>$ does not, is given by :

   (a) $<(-1)^n>$ (b) $\left\langle \dfrac{1}{n} \right\rangle$

   (c) $\left\langle \dfrac{n}{n+1} \right\rangle$ (d) none of these

5. Which of the following sequence is not convergent ?

   (a) $<\sin n\pi>$ (b) $\left\langle \dfrac{1}{n} \right\rangle$

   (c) $\left\langle \dfrac{n}{n+1} \right\rangle$ (d) $\left\langle \sqrt{n} \right\rangle$

6. Let $a<e$, then $\lim_{n \to \infty} n! \left(\dfrac{a}{n} \right)^n$ is given by :
   (a) 0   (b) 1
   (c) –1   (d) indeterminate.

7. Let $a>e$, then $\lim_{n \to \infty} n! \left(\dfrac{a}{n} \right)^n$ is given by :
   (a) 0   (b) $+\infty$
   (c) $-\infty$ (d) 1.

8. The sequence $s_n = \sin n\pi$ converges to :
   (a) 2   (b) 1
   (c) 0   (d) –1

9. If $<s_n>$ and $<t_n>$ are bounded sequence of real numbers such that $s_n \le t_n$, $\forall\ n \in N$, then which of the following is true ?
   (a) $\overline{\lim}\, s_n \le \overline{\lim}\, t_n$ (b) $\underline{\lim}\, s_n \le \underline{\lim}\, t_n$
   (c) $\overline{\lim}\, s_n = \overline{\lim}\, t_n$ (d) (a) and (b) are true

10. If $<s_n>$ and $<t_n>$ are bounded sequences of real numbers then which of the following is true ?
    (a) $\overline{\lim}\,(s_n + t_n) \le \overline{\lim}\, s_n + \overline{\lim}\, t_n$
    (b) $\overline{\lim}\,(s_n + t_n) \ge \overline{\lim}\, s_n + \overline{\lim}\, t_n$
    (c) $\underline{\lim}\,(s_n + t_n) \le \underline{\lim}\, s_n + \underline{\lim}\, t_n$
    (d) $\underline{\lim}\,(s_n + t_n) = \underline{\lim}\, s_n + \underline{\lim}\, t_n$

11. Let $<s_n>$ be a sequence of real numbers, which is bounded above and let
    $$\overline{s}_n = \sup.\left[s_n, s_{n+1} \cdots\right]$$
    then which of the following is true ?
    (a) $\langle \overline{s}_n \rangle$ is non-decreasing sequence.
    (b) $\langle \overline{s}_n \rangle$ is either converges or diverges to $-\infty$
    (c) if $\langle \overline{s}_n \rangle$ converges, then limit superior can be defined.
    (d) All are true.

12. Let $s_n = (-1)^n$, $n \in N$, which of the following is true ?
    (a) $<s_n>$ is bounded below by -1 and bounded above by $+1$.
    (b) $\overline{s}_n = 1$ and $\underline{s}_n = -1$
    (c) $\overline{\lim} s_n = 1$ and $\underline{\lim} s_n = -1$
    (d) All are true.

13. $\lim\limits_{n \to \infty}\left(1 + \dfrac{1}{n}\right)^{n+1}$ is equal to :
    (a) $\dfrac{1}{e}$ (b) e

(c) e^2 (d) 1

14. If $p < 0$ and $c \in R$, then $\lim\limits_{n \to \infty} \dfrac{n^c}{(1+p)^n}$ is equal to :
    (a) 1   (b) 0
    (c) ∞ (d) $-\infty$

15. If $p > 0$, then $\lim\limits_{n \to \infty} \dfrac{1}{n^p}$ is equal to :
    (a) 0   (b) 1
    (c) ∞ (d) $-\infty$

16. Which of the following is not correct ?
    (a) If $<s_n>$ is a sequence of non–negative numbers such that $\lim s_n = l$, then $l \ge 0$.
    (b) The limit of the sequence is unique.
    (c) A Cauchy sequence is convergent but converse is not true.
    (d) Every convergent sequence is bounded.

17. Which of the following sequence oscillate infinitely ?
    (a) $<1^2, 2^2, 3^2 ...>$ (b) $<1,-2,3,-4 ... >$
    (c) $<(-1)^n >$ (d) $<1,2,1,1,4, ...>$

18. Which of the following sequence oscillate finitely ?
    (a) $<1, 3, 5,...>$ (b) $<1,-2,3,-4 ... >$
    (c) $<(-1)^n>$ (d) $<-2,-4,-6,... >$

19. Which of the following is not correct ?
    (a) Every convergent sequence is bounded.
    (b) Every bounded monotonic sequence is convergent.
    (c) A non-decreasing sequence which is not bounded above diverges to $+\infty$.
    (d) A non-increasing sequence which is bounded below diverges to $-\infty$.

20. Which of the following is not a subsequence of $\left\langle \dfrac{1}{n} \right\rangle$?
    (a) $\left\langle \dfrac{1}{2n-1} \right\rangle$ (b) $\left\langle \dfrac{1}{2n+1} \right\rangle$
    (c) $\left\langle \dfrac{1}{n-1} \right\rangle$ (d) $\left\langle \dfrac{1}{2n} \right\rangle$

Problem Set-2

1. The domain of the sequence is always :
   (a) real numbers   (b) integers
   (c) natural numbers   (d) none of these

2. The set of all distinct terms of a sequence is called its :
   (a) range   (b) rank
   (c) domain   (d) none of these

3. The sequence $<s_n>$ defined by $s_n = \dot{a} \; \forall n$ is called :
   (a) null sequence   (b) constant sequence
   (c) unbounded sequence (d) none of these

4. A sequence is said to be bounded if for any $k > 0$:
   (a) $s_n \leq k \; \forall n$ (b) $s_n \geq k \; \forall n \in N$
   (c) $-k \leq s_n \leq k \; \forall n$ (d) none of these

5. Every subsequence of a bounded sequence is:
   (a) unbounded
   (b) bounded
   (c) may or may not be bounded
   (d) none of these

6. The sequence $<n^2>$ is :
   (a) bounded below by 1
   (b) not bounded above
   (c) both (a) and (b) are true
   (d) none of these

7. The sequence $<s_n> = <(-1)^n . n>$ is :
   (a) not bounded below
   (b) not bounded above
   (c) both (a) and (b) are true
   (d) none of these

8. If l be the limit of a sequence of non-negative numbers, then :
   (a) $l > 0$ (b) $l < 0$
   (c) $l \leq 0$ (d) $l \geq 0$

9. If $<s_n>$ converges to l, then any subsequence of $<s_n>$:
   (a) converges to $l_1 < l$ (b) converges to l
   (c) converges to $l_1 > l$ (d) none of these

10. The limit of a convergent sequence is :
    (a) unique
    (b) not unique
    (c) may or may not be unique
    (d) none of the above

11. A sequence $<s_n>$ is called a null sequence if :
    (a) $s_n = 0 \; \forall n$ (b) $\lim s_n = 0$
    (c) $\lim s_n > 0$ (d) none of these

12. A sequence which is neither convergent nor divergent is called :
    (a) bounded sequence
    (b) oscillating sequence
    (c) null sequence
    (d) none of the above

13. If $<s_n>$ diverges to ∞ and $<t_n>$ is bounded then $<s_n + t_n>$:

14. If $<t_n>$ diverges to ∞ and $s_n > t_n \; \forall n \in N$ then $<s_n>$:
    (a) converges to ∞ (b) diverges to ∞
    (c) is bounded   (d) none of these

15. If $\lim s_n = 0$ and the sequence $<t_n>$ is bounded then $\lim (s_n t_n) =$
    (a) 0   (b) 1
    (c) ∞ (d) none of these

16. If $\lim s_n = l$ then $\lim \dfrac{s_1 + s_2 + \dots + s_n}{n}$:
    (a) l (b) $l^{1/n}$
    (c) $-l$ (d) none of these

17. If $<s_n>$ is a sequence of positive terms $\forall n$ and $\lim s_n = l$ then $\lim (s_1 . s_2 \dots s_n)^{1/n} =$
    (a) l (b) $l^{1/n}$
    (c) $-l$ (d) none of these

18. If $<s_n>$ is a sequence such that $s_n > 0 \forall n \in N$ and $\lim \dfrac{s_{n+1}}{s_n} = l$, then $\lim \sqrt[n]{n} =$
    (a) l (b) $l+1$
    (c) $-l$ (d) 0

19. If $\lim s_n = l_1$ and $\lim t_n = l_2$ then $\lim \dfrac{s_1 t_n + s_2 t_{n-1} + \dots + s_n t_1}{n}$:
    (a) l_1 (b) l_2
    (c) $l_1 l_2$ (d) $\dfrac{l_1}{l_2}$

20. Every bounded monotonically increasing sequence is :
    (a) convergent   (b) divergent
    (c) oscillate   (d) none of these

21. A non-decreasing sequence which is not bounded above :
    (a) diverges to ∞ (b) converges to ∞
    (c) diverges to $-\infty$ (d) none of these

22. A non-decreasing sequence which is not bounded below :
    (a) diverges to ∞ (b) converges to ∞
    (c) diverges to $-\infty$ (d) none of these

23. Every bounded sequence has at least :
    (a) 2 limit points   (b) 1 limit points
    (c) 3 limit points   (d) none of these

24. Every Cauchy sequence is :
    (a) convergent

(b) bounded

(c) both (a) and (b) are true

(d) none of these

Problem Set-3

1. The sequence $1, -1, 1 -1, \ldots$ is represented by :

(a) $s_n = (-1)^{n+1} : n \in N$

(b) $s_n = (-2)^n : n \in N$

(c) $s_n = \dfrac{1}{n} : n \in N$

(d) none of the above

2. Let $<s_n>$ be a sequence of real numbers, which is bounded above and let $s_n = [s_n, s_{n+1}, \ldots]$ then which of the following is/are true ?

(a) $\left\langle \overline{s_n} \right\rangle$ is non-decreasing sequence.

(b) $\left\langle \overline{s_n} \right\rangle$ is either converges or diverges to $-\infty$.

(c) If $\left\langle \overline{s_n} \right\rangle$ is converges, then limit superior can be defined

(d) All are true.

3. If $a < e$, then $\lim\limits_{n \to \infty} n! \left(\dfrac{a}{n} \right)^n$ is given by :

(a) 0   (b) $-\infty$

(c) $+\infty$ (d) none of these.

4. If $s_n = (-1)^n : n \in N$ then which of the following is/are true ?

(a) $<s_n>$ is bounded below by -1 and bounded above by $+1$.

(b) $\overline{s_n} = 1$, $\underline{s_n} = -1$

(c) $\overline{\lim} s_n = 1$ and $\underline{\lim} s_n = -1$

(d) All are true.

5. For $p > 0$ the value of $\lim\limits_{n \to \infty} \dfrac{1}{n^p}$ is equal to :

(a) 1   (b) 0

(c) ∞ (d) $-\infty$

6. Which of the following is not true ?

(a) Every bounded monotonic sequence is covergent.

(b) every convergent sequence is bounded

(c) a non-decreasing sequence which is not bounded above diverges to $+\infty$

(d) All are true

7. The sequence $\left\langle \dfrac{1}{n} \right\rangle$ is :

25. If the sequence $<s_n>$ converges to l then sequence $<|s_n|>$ converges to :

(a) $|l|$ (b) l

(c) l^2 (d) $-l$

(a) bounded below by 0

(b) bounded above by 1

(c) bounded

(d) all are true

8. The range set of the sequence $<1, 0, 1, 0, \ldots>$ is :

(a) N   (b) $\{0, 1\}$

(c) Z   (d) none of these

9. The sequence $<s_n = n^2 : n \in N>$ is :

(a) bounded below

(b) bounded above

(c) bounded

(d) none of the above

10. The sequence $<s_n = 1 \ \forall n \in N>$ is :

(a) bounded

(b) unbounded

(c) may or may not be bounded

(d) none of the above

11. The sequence $< 1, \dfrac{1}{2}, \dfrac{1}{4}, \dfrac{1}{8}, \ldots >$ is :

(a) bounded below   (b) bounded above

(c) bounded   (d) unbounded

12. The sequence $< 1, 3, 5, \ldots >$ is

(a) bounded   (b) unbounded

(c) bounded above   (d) none of these

13. Which of the following is a bounded sequence ?

(a) $< -(1)^{n-1} \cdot \dfrac{n}{n+1} >$

(b) $< -(1)^{n-1} \cdot n >$

(c) $< 2, 3, 5, 7, 11, \ldots >$

(d) none of the above

14. The limit of the sequence $\left\langle \dfrac{1}{n} \right\rangle$ is :

(a) 1   (b) 0

(c) ∞ (d) $-\infty$

15. The limit of the sequence $<n>$ is :

(a) 1   (b) 0

(c) exist   (d) does not exist

16. The limit of the sequence $<-1>^n$ is :

(a) 1　　　　　　　(b) −1
(c) 0　　　　　　　(d) does not exist

17. The limit of the sequence $<\dfrac{2n}{n+4\sqrt{n}}>$ is :

(a) 1　　　　　　　(b) 2
(c) 0　　　　　　　(d) does not exist

18. If l be the limit of a sequence of non-negative numbers, then l is :

(a) ≥ 0　　　　　(b) ≤ 0
(c) 0　　　　　　　(d) none of these

19. Which is not true ?

(a) Every convergent sequence is bounded.
(b) Every bounded sequence is convergent.
(c) The sequence $<(-1)^n>$ is not covergent
(d) None of the above

20. If $<a_n>$ is convergent then $<|a_n|>$ is :

(a) convergent
(b) divergent
(c) may or may not be convergent
(d) none of the above

21. The sequences $\left\langle\dfrac{1}{2^n}\right\rangle$ is :

(a) convergent　　　(b) divergent
(c) oscillating　　　(d) none of these

22. The sequence $<r^n>$ is :

(a) always convergent
(b) convergent only for $|r|<1$
(c) divergent
(d) none of the above

23. The sequence $<\dfrac{2n^2+1}{2n^2-1}>$ is :

(a) converges to 0　(b) converges to 1
(c) divergent　　　(d) none of the above

24. Let $<s_n>$ be a sequence such that $s_n \neq 0 \ \forall n$ and $\dfrac{s_{n+1}}{s_n}=l$ if $|l|<1$ then :

(a) $s_n \to 0$　　　　(b) $s_n = 0$
(c) $s_n \to 1$　　　　(d) none of these

25. The sequence $<1+\dfrac{(-1)^n}{n}>$ converges to :

(a) 1　　　　　　　(b) 0
(c) 2　　　　　　　(d) none of these

26. If $\lim\limits_{n\to\infty}\dfrac{s_{n-1}}{s_{n+1}}=0$ then $\lim\limits_{n\to\infty} s_n =$

(a) 1　　　　　　　(b) 0
(c) 2　　　　　　　(d) 3

27. If $<s_n>$ converges to l then $<|s_n|>$ converges to :

(a) $-l$　　　　　　(b) l
(c) $|l|$　　　　　　(d) none of these

28. If the sequence $<a_n>$ converges to 0 and the sequence $<b_n>$ is bounded then $<a_n b_n>$ converges to :

(a) 1　　　　　　　(b) 2
(c) 0　　　　　　　(d) none of these

29. A sequence $<s_n>$ is defined such that $a_n \neq 0$ and $\lim\limits_{n\to\infty}\dfrac{a_{n+1}}{a_n}=l$ if $|l|<1$ then $\lim\limits_{n\to\infty} a_n =$

(a) 0　　　　　　　(b) 1
(c) ∞　　　　　　(d) none of these

30. $\lim\limits_{n\to\infty}\dfrac{x^n}{n!} = \forall x$:

(a) 1　　　　　　　(b) 0
(c) n　　　　　　(d) none of these

31. The limit of the sequence $<\dfrac{3n}{n+5n^{1/2}}>$ is :

(a) 3　　　　　　　(b) $\dfrac{1}{3}$
(c) 5　　　　　　　(d) none of these

32. The sequence $\left\langle\dfrac{n}{n+1}\right\rangle$:

(a) converges to 1　(b) diverges
(c) converges to 0　(d) none of these

33. The value of the $\lim \sqrt[n]{n} =$

(a) 0　　　　　　　(b) 1
(c) ∞　　　　　　(d) none of these

34. The sequence $<(-1)^n>$ is :

(a) convergent
(b) oscillate finitely
(c) oscillate infinitely
(d) none of the above

35. The sequence $< s_n >$:

$$s_n = \begin{bmatrix} \dfrac{n}{2} & ; \text{ if } n \text{ is even} \\ -\left(\dfrac{n-1}{2}\right) & ; \text{ if } n \text{ is odd} \end{bmatrix}$$ is :

(a) convergent

(b) oscillate finitely

(c) oscillate infinitely

(d) none of the above

36. For the sequence $<s_n> = <1, -2, 3, -4,...>$ which is/are not true ?

(a) $<s_n>$ oscillates

(b) not bounded above

(c) not bounded below

(d) All are true.

37. For the sequence $<-1,1,-1,1...>$, which is not true ?

(a) It is bounded below by -1.

(b) It is bounded above by 1.

(c) Oscillates

(d) All are true.

38. The sequence $<n>$ is :

(a) bounded   (b) increasing

(c) decreasing   (d) none of these

39. The sequence $\left\langle \dfrac{n}{n+1} \right\rangle$ is :

(a) strictly increasing

(b) strictly decreasing

(c) monotonically increasing

(d) none of the above

40. A monotonically non-decreasing sequences which is not bounded above :

(a) diverges to ∞ (b) diverges to $-\infty$

(c) converges   (d) none of these

41. The sequence $<n^p>, p > 0$:

(a) diverges to ∞ (b) diverges to $-\infty$

(c) converges   (d) none of these

42. The sequence $< \log \dfrac{1}{n} >$

(a) diverges to ∞ (b) diverges to $-\infty$

(c) converges   (d) none of these

43. The sequence $< \dfrac{1}{n+n} >$

(a) convergent   (b) divergent

(c) oscillating   (d) none of these

44. The value of $\lim\limits_{n\to\infty} \left(1+\dfrac{1}{n}\right)^n$ is:

(a) 2   (b) 3

(c) between 2 and 3 (d) none of these

45. The value of the limit of the sequence

$$s_n = 1 + \frac{1}{1!} + \frac{1}{2!} + ... + \frac{1}{n!} \text{ is:}$$

(a) 2   (b) 3

(c) between 2 and 3 (d) none of these

46. Which of the following is not true ? (For $x>0$)

(a) $\lim\limits_{n\to\infty} \dfrac{1}{n^x} = 0$ (b) $\lim\limits_{n\to\infty} n^{1/n} = 1$

(c) $\lim\limits_{n\to\infty} x^{1/n} = 1$ (d) All are true.

47. Which of the following is not true ?

(a) $\lim\limits_{n\to\infty} \left(1+\dfrac{1}{n}\right)^n = e^2$

(b) $\lim\limits_{n\to\infty} \left(1+\dfrac{1}{n+1}\right)^n = e$

(c) $\lim\limits_{n\to\infty} \left(1+\dfrac{2}{n}\right)^n = 1$

(d) All are true.

48. If $s_n = \dfrac{n}{\lambda^n}$ with $\lambda>1$, then $\lim\limits_{n\to\infty} s_n =$

(a) 0   (b) 1

(c) 2   (d) does not exist

49. Which is not true ?

(a) Every cauchy sequence is convergent.

(b) Every convergent sequence is cauchy.

(c) Every cauchy sequence is bounded.

(d) All are true.

50. For a sequence $<s_n>$ of real numbers which of the following is true ?

(a) $\lim\limits_{n\to\infty} \inf .s_n = \lim\limits_{n\to\infty} \sup .s_n$

(b) $\lim\limits_{n\to\infty} \inf .s_n \le \lim\limits_{n\to\infty} \sup s_n$

(c) $\lim\limits_{n\to\infty} \inf .s_n \ge \lim\limits_{n\to\infty} \sup .s_n$

(d) All are true

Problem Set-4

1. The sequence $\left\langle \dfrac{n}{n+1} \right\rangle$ is :

   (a) decreasing   (b) increasing
   (c) unbounded   (d) none of these

2. The sequence $\left\langle \dfrac{1}{n} \right\rangle$ is :

   (a) bounded
   (b) convergent
   (c) both (a) and (b) are true
   (d) none of the above

3. Which one of the following is not true ?

   (a) $\lim x_n + \overline{\lim} y_n \le \underline{\lim} x_n + \underline{\lim} y_n$

   (b) $\underline{\lim} x_n + \underline{\lim} y_n \le \overline{\lim} x_n + \overline{\lim} y_n$

   (c) $\underline{\lim} x_n + \overline{\lim} y_n \le \overline{\lim}(x_n + y_n)$

   (d) None of the above

4. The sequences $< a + \dfrac{(-1)^n . b}{n} > $:

   (a) bounded   (b) unbounded
   (c) divergent   (d) none of these

5. The sequence $<x_n>$ diverges to $+\infty$ if and only if :
   (a) $\lim x_n \le +\infty$
   (b) $\underline{\lim} x_n = \overline{\lim} x_n = +\infty$
   (c) $\overline{\lim} x_n \le \infty$
   (d) none of the above

6. If $x_n \le y_n$, $\forall\, x_n, y_n \in R$ then :
   (a) $\overline{\lim} x_n \le \lim y_n$
   (b) $\lim x_n \ge \overline{\lim} y_n$
   (c) $\overline{\lim} x_n \ge \overline{\lim} y_n$
   (d) none of the above

7. The set of all distinct terms of the given set is called :
   (a) range set   (b) domain set
   (c) fixed set   (d) none of these

8. If $f : N \to R$ is a sequence then what is $f(x)$, $\forall x \in N$?
   (a) A real number   (b) A natural number
   (c) An integer   (d) None of these

9. The sequence $[1,0,1,0,...]$ is :
   (a) increasing   (b) decreasing
   (c) monotonic   (d) none of these

10. If $<a_n>$ is a decreasing sequence of positive numbers and Σa_n converges, then $\lim\limits_{n\to\infty} na_n = $
    (a) 0   (b) 1
    (c) ∞ (d) $-\infty$

11. If a sequence is not cauchy then it is :
    (a) convergent   (b) divergent
    (c) monotonic   (d) none of these

12. If $<a_n>$ converges to l then :
    (a) $<|a_n|>$ convergs to $+l$
    (b) $<|a_n|>$ convergs to $-l$
    (c) $<|a_n|>$ convergs to $|l|$
    (d) none of the above

13. Every cauchy sequence of real numbers is :
    (a) convergent
    (b) bounded
    (c) both (a) and (b) are true
    (d) none of the above

14. Which of the following statement is not true ?
    (a) Every convergent sequence is bounded.
    (b) Every cauchy sequence is convergent.
    (c) Every convergent sequence is cauchy.
    (d) Every bounded sequence is convergent.

15. A real number l is said to be cluster point of a sequence $<a_n>$ if for any given $\in > 0$:
    (a) $a_n \in]l-\in, l+\in[$ for infinitely many points of N
    (b) $a_n \in [l-\in, l+\in]$ for infinitely many points of N
    (c) $a_n \in]l-\in, l+\in[$ only for one value of $n \in N$
    (d) none of the above

16. Bolzano-Weirstrass theorem states that :
    (a) every unbounded sequence has a limit point.
    (b) every bounded sequence has a limit point.
    (c) a bounded sequence may or may not have a limit point.
    (d) none of the above.

17. The limit superior of the sequence $<(-1)^n>$ is given by :
    (a) 1   (b) -1
    (c) 0   (d) none of these

18. The limit inferior of the sequence $<(-1)^n>$ is given by :
    (a) 1   (b) –1
    (c) 0   (d) none of these.

19. If $\lim\limits_{n\to\infty} \dfrac{s_n}{n} = l \neq 0$ then $<s_n>$ is :

    (a) bounded   (b) unbounded
    (c) convergent   (d) none of these.

20. A real number l is said to be a limit point of the sequence $<s_n>$ if and only if there exists :
    (a) a subsequence of $<s_n>$ converges to l
    (b) a, k such that $k \leq a_n \; \forall n \in N$
    (c) a, k such that $k \geq a_n \; \forall n \in N$
    (d) none of the above

21. A sequence $<s_n>$ converges to l then :
    (a) every subsequence of $<s_n>$ converges to l
    (b) some subsequence of $<s_n>$ converges to l
    (c) no subsequence of $<s_n>$ converges to l
    (d) none of the above

22. The limit superior of the sequence $< \dfrac{(-1)^n}{n} >$ is :
    (a) 1   (b) 0
    (c) –1   (d) none of these

23. The limit inferior of the sequence $< \dfrac{(-1)^n}{n} >$ is :
    (a) 1   (b) 0
    (c) –1   (d) none of these

24. The set of limit points of a bounded sequence is :
    (a) bounded   (b) unbounded
    (c) not necessarily bounded
    (d) none of the above

25. If $x_1 = 1$ and $x_{n+1} = \sqrt{3x_n} \;\; \forall n \in N$ then the sequence $<x_n>$ converges to :
    (a) 1   (b) 2
    (c) 3   (d) none of these.

26. The sequence $<s_n>$ where :
    $s_n = 1 + \dfrac{1}{1!} + \dfrac{1}{2!} + ... + \dfrac{1}{(n-1)!}$ is :

    (a) increasing and bounded
    (b) decreasing and bounded
    (c) decreasing and unbounded
    (d) none of the above

27. If sequence $<s_n>$ where
    $s_n = 1 + \dfrac{1}{2} + \dfrac{1}{2^2} + ... + \dfrac{1}{2^{n-1}}$ is:

    (a) increasing and bounded
    (b) increasing and unbounded
    (c) decreasing and unbounded
    (d) none of the above

28. If sequence $<s_n>$ is a decreasing sequence which converges to 2 then for all $n \in N$:
    (a) $s_n \leq 2$ (b) $s_n \geq 2$
    (c) $s_n = 2$ (d) none of these

29. If sequence $<s_n>$ is a strictly increasing sequence which converges to 3 then for all $n \in N$:
    (a) $s_n > 3$ (b) $s_n \geq 3$
    (c) $s_n \leq 3$ (d) $s_n < 3$

30. $\lim\limits_{n\to\infty} \dfrac{x_1 + x_2 + ... + x_n}{n} = \lim\limits_{n\to\infty} x_n$ only if :

    (a) $<x_n>$ is a convergent sequence of positive terms
    (b) $<x_n>$ is a convergent sequence of non-negative terms
    (c) $<x_n>$ is a convergent sequence of negative terms
    (d) none of the above

31. Let $<s_n>$ be a sequence of positive terms such that $\lim\limits_{n\to\infty} \dfrac{s_{n+1}}{s_n}$ does not exist, then :

    (a) $\lim\limits_{n\to\infty} s_n^{1/n}$ does not exist

    (b) $\lim\limits_{n\to\infty} (s_n)^{1/n}$ exists

    (c) $\lim\limits_{n\to\infty} s_n^{1/n}$ may exist.

    (d) none of the above

32. A sequence contains a convergent subsequence, if it is :
    (a) bounded   (b) bounded above
    (c) bounded below   (d) none of these

33. A cauchy sequence is :
    (a) always convergent
    (b) convergent only in case of real numbers
    (c) divergent
    (d) none of the above

34. A convergent sequence is cauchy if it is a :
 (a) sequence of rational numbers
 (b) sequence of irrational numbers
 (c) sequence of real numbers
 (d) all are true

35. If $<s_n>$ and $<t_n>$ are two sequences which are not convergent, then :
 (a) $<s_n+t_n>$ may or may not be convergent
 (b) $<s_n-t_n>$ may or may not be convergent
 (c) both (a) and (b) are true
 (d) none of the above

36. The sequence $\left\langle \dfrac{1}{2^n} \right\rangle$ is :
 (a) a decreasing sequence
 (b) an increasing sequence
 (c) both (a) and (b) are true
 (d) none of these

37. For a sequence $<s_n>$:
 (a) $\underline{\lim}\left(-s_n\right) = -\overline{\lim}\, s_n$
 (b) $\overline{\lim}\left(-s_n\right) = -\lim s_n$
 (c) both (a) and (b) are true
 (d) none of these

38. Which is a true statement :
 (a) Some cauchy sequences are bounded
 (b) Every cauchy sequence is bounded
 (c) Every cauchy sequence is unbounded
 (d) none of the above

39. Which one of the following is true ?
 (a) A sequence can have more than one limit point.
 (b) A sequence can have at most one limit.
 (c) A sequence can have two limits.
 (d) none of the above

40. If $<s_n>$ is a bounded sequence such that $s_n>0 \; \forall \; n\in N$ then :
 (a) $\underline{\lim}\left(\dfrac{1}{s_n}\right) = \dfrac{1}{\overline{\lim} s_n}$, if $\overline{\lim}\, s_n > 0$
 (b) $\overline{\lim}\left(\dfrac{1}{s_n}\right) = \dfrac{1}{\underline{\lim} s_n}$, if $\underline{\lim}\, s_n > 0$
 (c) both (a) and (b) are true
 (d) none of the above

41. If $<s_n>$ and $<t_n>$ are bounded sequences, then :
 (a) $\underline{\lim} s_n - \overline{\lim} t_n \le \underline{\lim}\left(s_n - t_n\right)$

 (b) $\overline{\lim}\left(s_n - t_n\right) \le \overline{\lim} s_n + \underline{\lim} t_n$
 (c) both (a) and (b) are true
 (d) none of the above

42. Which of the following sequence oscillates infinitely ?
 (a) $<n^2>$
 (b) $<1,-2,3,-4,....>$
 (c) $<(-1)^n>$
 (d) None of the above

43. Which of the following is not a subsequence of $\left\langle \dfrac{1}{n} \right\rangle$?
 (a) $\left\langle \dfrac{1}{n-1} \right\rangle$ (b) $\left\langle \dfrac{1}{2n-1} \right\rangle$
 (c) $\left\langle \dfrac{1}{2n} \right\rangle$ (d) None of these

44. Which of the following sequence oscillates finitely ?
 (a) $<(-1)^n>$ (b) $<1, 3, 5... >$
 (c) Both (a) and (b) are true
 (d) None of the above

45. A sequence $<s_n>$ of real numbers such that $<|s_n|>$ converges but $<s_n>$ does not, is :
 (a) $<(-1)^n>$ (b) $\left\langle \dfrac{1}{n} \right\rangle$
 (c) $\left\langle \dfrac{n}{n+1} \right\rangle$ (d) none of these

46. The sequence $\left\langle \dfrac{2n-7}{3n+2} \right\rangle$ is :
 (a) monotonically increasing
 (b) bounded above
 (c) bounded below
 (d) all are true

47. The limit of $\dfrac{n}{(n!)^{1/n}} = $ (where $n\to\infty$) :
 (a) e (b) $\dfrac{1}{e}$
 (c) e^2 (d) none of these

48. The sequence $\left\langle \dfrac{2^n}{n!} \right\rangle$ is :
 (a) monotonically decreasing
 (b) bounded below

(c) bounded above

(d) all are true

49. Which is not true ?

(a) $<1,2,3...n...>$ is strictly increasing.

(b) $<2,2,4,4,6..>$ is monotonically increasing.

(c) $<1,1,\dfrac{1}{3},\dfrac{1}{5},\dfrac{1}{5},...>$

(d) All are true

50. Which is not true ?

(a) $<3,3^2,3^3...>$ diverges to $+\infty$.

(b) $<-2,-2^2,-2^3...>$ diverges to $-\infty$.

(c) $<2,4,6,...2n,...>$ diverges to $+\infty$.

(d) All are true

Answers

FILL IN THE BLANKS

1. bounded **2.** not necessarily **3.** non-negative **4.** unique **5.** convergent
6. convergent **7.** convergent **8.** $+\infty$ **9.** Sandwitch theorem **10.** unique

TRUE/FALSE

1. T	**2.** T	**3.** F	**4.** F	**5.** F
6. T	**7.** T	**8.** F	**9.** T	**10.** T

MULTIPLE CHOICE QUESTIONS

Problem Set 1

1. (b)	**2.** (a)	**3.** (b)	**4.** (a)	**5.** (d)	**6.** (a)	**7.** (b)	**8.** (c)	**9.** (d)
10. (a)	**11.** (d)	**12.** (d)	**13.** (b)	**14.** (b)	**15.** (a)	**16.** (c)	**17.** (b)	**18.** (c)
19. (d)	**20.** (c)							

Problem Set 2

1. (c)	**2.** (a)	**3.** (b)	**4.** (c)	**5.** (b)	**6.** (c)	**7.** (c)	**8.** (d)	**9.** (b)
10. (a)	**11.** (b)	**12.** (b)	**13.** (b)	**14.** (b)	**15.** (a)	**16.** (a)	**17.** (a)	**18.** (a)
19. (c)	**20.** (a)	**21.** (a)	**22.** (c)	**23.** (b)	**24.** (c)	**25.** (a)		

Problem Set 3

1. (a)	**2.** (d)	**3.** (c)	**4.** (d)	**5.** (b)	**6.** (d)	**7.** (d)	**8.** (b)	**9.** (a)
10. (a)	**11.** (c)	**12.** (b)	**13.** (d)	**14.** (b)	**15.** (d)	**16.** (d)	**17.** (b)	**18.** (a)
19. (b)	**20.** (a)	**21.** (a)	**22.** (b)	**23.** (b)	**24.** (a)	**25.** (a)	**26.** (a)	**27.** (c)
28. (c)	**29.** (a)	**30.** (b)	**31.** (a)	**32.** (a)	**33.** (b)	**34.** (b)	**35.** (c)	**36.** (d)
37. (d)	**38.** (b)	**39.** (a)	**40.** (a)	**41.** (a)	**42.** (b)	**43.** (a)	**44.** (c)	**45.** (c)
46. (d)	**47.** (d)	**48.** (a)	**49.** (d)	**50.** (b)				

Problem Set 4

1. (b)	**2.** (c)	**3.** (a)	**4.** (a)	**5.** (b)	**6.** (a)	**7.** (a)	**8.** (a)	**9.** (d)
10. (a)	**11.** (b)	**12.** (c)	**13.** (c)	**14.** (d)	**15.** (a)	**16.** (b)	**17.** (a)	**18.** (b)
19. (b)	**20.** (a)	**21.** (a)	**22.** (b)	**23.** (b)	**24.** (a)	**25.** (c)	**26.** (b)	**27.** (a)
28. (a)	**29.** (a)	**30.** (a)	**31.** (a)	**32.** (a)	**33.** (b)	**34.** (d)	**35.** (c)	**36.** (a)
37. (c)	**38.** (b)	**39.** (b)	**40.** (c)	**41.** (c)	**42.** (b)	**43.** (a)	**44.** (a)	**45.** (a)
46. (d)	**47.** (a)	**48.** (d)	**49.** (d)	**50.** (d)				

HINTS TO SELECTED PROBLEMS

1. For given $\varepsilon>0$, we have

$$\left|\frac{1}{n}-0\right|<\varepsilon \text{ when } \frac{1}{n}<\varepsilon \ i.e., n>\frac{1}{\varepsilon}$$

Let us choose a positive integer $m>\frac{1}{\varepsilon}$ then

for all $n\geq m$, we have $\left|\frac{1}{n}-0\right|=\frac{1}{n}\leq\frac{1}{m}<\varepsilon$

$\Rightarrow$ the sequence $\left\langle\frac{1}{n}\right\rangle$ converges to 0 i.e., $\left\langle\frac{1}{n}\right\rangle$ has the limit 0.

20. Convergence of $<a_n>$ implies that for given $\varepsilon>0$ there exists a positive integer p such that

$|a_{n+p}-a_n|<\varepsilon \ \forall \ p=1,2,3,....$

But $\pm(|a_{n+p}|-|a_n|)\leq|a_{n+p}-a_n|$

$\Rightarrow \quad ||a_{n+p}|-|a_n||\leq|a_{n+p}-a_n|$

$\qquad\qquad <\varepsilon \ \forall p=1,2,3,...$

21. We have $|s_n-0|=\dfrac{1}{2^n}$

so, for given $\varepsilon>0, |s_n-0|<\varepsilon$ if $\dfrac{1}{2^n}<\varepsilon$

i.e., if $2^n>\dfrac{1}{\varepsilon}$

i.e., if $n>\dfrac{\log\left(\frac{1}{\varepsilon}\right)}{\log 2}$

Let us choose a positive integer m such that

$m>\left(\log\dfrac{1}{\varepsilon}\right)/\log 2$

Then for all $n\geq m, |s_n-0|<\varepsilon$

Hence, $<s_n>$ converges to 0.

22. Let $|r|<1$. Then, we can write $|r|=\dfrac{1}{1+h}, h>0$

Now, since $h>0$, therefore,

$(1+h)^n=1+nh+\dfrac{n(n-1)}{2}h^2+...+h^n\geq 1+nh \forall n$

Now, $|s_n-0|=|r|^n$

$=\dfrac{1}{(1+h)^n}\leq\dfrac{1}{1+nh} \ \forall n$

Let $\varepsilon>0$ be the given. Then $|s_n-0|<\varepsilon$ if

$\dfrac{1}{1+nh}<\varepsilon, i.e., \quad \text{if } n>\left(\dfrac{1}{\varepsilon}-1\right)\Big/h$.

If we take a positive integers $m>\left(\dfrac{1}{\varepsilon}-1\right)\Big/h$

then

for all $n>m$

$|s_n-0|<\varepsilon$

Hence, $<s_n>$ converges to 0.

23. Let $\varepsilon>0$ be the given.

Then $|s_n-1|=\left|\dfrac{2n^2+1}{2n^2-1}-1\right|=\left|\dfrac{2}{2n^2-1}\right|$

$\qquad =\dfrac{2}{2n^2-1}<\varepsilon \text{ if } n>\sqrt{\dfrac{2+\varepsilon}{2\varepsilon}}$

If we choose a positive integer $m>\sqrt{\dfrac{2+\varepsilon}{2\varepsilon}}$ then for all $n\geq m$.

We have $|s_n-1|<\varepsilon$.

Hence, $\lim s_n=1$.

24. Since $|l|<1$ therefore there exists $\varepsilon_0>0$ such that $\qquad |l|+\varepsilon_0=h<1$

Now, $\dfrac{s_{n+1}}{s_n}\to l$ implies $\exists$ a positive integer m

such that $\left|\dfrac{s_{n+1}}{s_n}-l\right|<\varepsilon_0 \ \forall n\geq m$

$\therefore \left|\dfrac{s_{n+1}}{s_n}\right|=\left|\left(\dfrac{s_{n+1}}{s_n}-l\right)+l\right|$

$\qquad\qquad \leq\left|\dfrac{s_{n+1}}{s_n}-l\right|+|l|$

$\qquad\qquad <\varepsilon_0+|l| \ \forall n\geq m$

$\Rightarrow \left|\dfrac{s_{n+1}}{s_n}\right|<h \ \forall n\geq m \qquad ...(1)$

Now, replacing n by $m, m+1, m+2, ...,$ $n-1$ successively in (1) and multiplying the corresponding sides of the resulting $n-m$ inequalities, we get

$\left|\dfrac{s_{m+1}}{s_n}\right|.\left|\dfrac{s_{m+2}}{s_{m+1}}\right|...\left|\dfrac{s_n}{s_{n-1}}\right|<h^{n-m}$

$\Rightarrow \left|\dfrac{s_{m+1}}{s_m}.\dfrac{s_{m+2}}{s_{m+1}}...\dfrac{s_n}{s_{n-1}}\right|<h^{n-m}$

$\Rightarrow \quad |s_n| < h^n \left(\dfrac{|s_m|}{h^m} \right) \forall n \geq m$ ...(2)

Since $0 < h < 1$, therefore $h^n \to 0$ and hence given $\varepsilon > 0 \; \exists \; m_1 > 0$ such that

$|h^n| < \dfrac{h^m \varepsilon}{|s_m|} \forall n \geq m_1$ ...(3)

Choose a positive integer p such that $p > \max\{m, m_1\}$

Then, from (2) and (3), we get

$|s_m| < \varepsilon \; \forall n \geq p$

$\Rightarrow |s_n - 0| < \varepsilon \; \forall n \geq p$

$\Rightarrow \lim s_n = 0 \Rightarrow s_n \to 0$

25. Let $<s_n> = 1 + \dfrac{(-1)^n}{n}$.

Let $\varepsilon > 0$, then we have

$|s_n - 1| = \left| \dfrac{(-1)^n}{n} \right| = \dfrac{1}{n} < \varepsilon \; if \; n > \dfrac{1}{\varepsilon}$

Let us take a positive integer m such that

$m > \dfrac{1}{\varepsilon}$ i.e., $\dfrac{1}{m} < \varepsilon$

Then for all $n \geq m$, we have

$|s_n - 1| = \dfrac{1}{n} \leq \dfrac{1}{m} < \varepsilon$

Hence, for given $\varepsilon > 0 \; \exists \; m > 0$ such that $|s_n - 1| < \varepsilon \; \forall n \geq m$

$\Rightarrow \quad <s_n>$ converges to 1.

26. In view of definition of limit of a sequence, we have the following :

For $\varepsilon > 0$, there exists a positive integer m such that

$n \geq m \Rightarrow \left| \dfrac{s_n - 1}{s_n + 1} - \varepsilon_n \right| < \varepsilon$

Hence we are justified in putting

$\dfrac{s_n - 1}{s_n + 1} = \varepsilon_n \to 0$ with $n \to \infty$

$\Rightarrow (s_n + 1)\varepsilon_n = s_n - 1$

$\Rightarrow s_n(\varepsilon_n - 1) = 1 + \varepsilon_n$

$\Rightarrow \quad s_n = \dfrac{1 + \varepsilon_n}{1 - \varepsilon_n} = \left(1 - \varepsilon_n + 2\varepsilon_n \right) \dfrac{1}{1 - \varepsilon_n}$

$= \left(1 - \varepsilon_n \right) / \left(1 - \varepsilon_n \right) + 2\varepsilon_n / \left(1 - \varepsilon_n \right)$

$= 1 + \dfrac{2\varepsilon_n}{1 - \varepsilon_n}$

$\Rightarrow s_n - 1 = \dfrac{2\varepsilon_n}{1 - \varepsilon_n}$

$\Rightarrow |s_n - 1| = \left| \dfrac{2\varepsilon_n}{1 - \varepsilon_n} \right| < \varepsilon, say$

Thus for $\varepsilon > 0 \; \exists$ a positive integer m.

$n \geq m \Rightarrow |s_n - 1| < \varepsilon$

i.e., $n \to \infty \Rightarrow s_n - 1 \to 0$

i.e., $n \to \infty \Rightarrow s_n \to 1$

$\Rightarrow \quad \lim_{n \to \infty} s_n = 1$

27. $<s_n>$ converges to $l \Rightarrow$ for any $\varepsilon > 0$ there exists a positive integer m such that $n \geq m \Rightarrow |s_n - l| < \varepsilon$

Now $||s_n| - |l|| \leq |s_n - l|$

$\Rightarrow$ for $\varepsilon > 0$ there exists a positive intger m;

$n \geq m \Rightarrow ||s_n| - |l|| \leq |s_n - l| < \varepsilon$

This proves that $< |s_n| >$ convergs to $|l|$.

28. We have $<b_n>$ is bounded $\Rightarrow$ there exists a number M such that $|b_n| \leq M$ for each n. $<a_n>$ converges to '0' $\Rightarrow$ for $\varepsilon > 0 \; \exists$ a positive integer m such that

$\forall n \geq m \Rightarrow |a_n - 0| = |a_n| < \varepsilon / M$

Now $|a_n b_n| = |a_n||b_n|$

$\Rightarrow \quad |a_n b_n| < M . \dfrac{\varepsilon}{M} = \varepsilon \; \forall n \geq m$

$\Rightarrow |a_n b_n - 0| < \varepsilon \; \forall n \geq m$

$\Rightarrow <a_n b_n>$ converges to '0'.

31. Use the reesult

$\left| \dfrac{3n}{n + 5n^{1/2}} - 3 \right| = \dfrac{15 n^{1/2}}{n + 5n^{1/n}} < \dfrac{15}{n^{1/2}}$

32. $|s_n - 1| = \left| \dfrac{n}{n + 1} - 1 \right| = \left| \dfrac{n - n - 1}{n + 1} \right|$

$= \dfrac{1}{n + 1} < \dfrac{1}{n} < \varepsilon$

33. $\sqrt[n]{n} = 1 + h_n, \; h_n > 0$

$\Rightarrow \quad n = (1 + h_n)^n$, now expand by binomial expansion.

41. Let $s_n = n^p, p > 0$.

Then $s_n > 0 \; \forall n \in N, \; p > 0$

$\Rightarrow$ The sequence $< \dfrac{1}{s_n} > = < \dfrac{1}{n^p} >$ exists

We know that $\dfrac{1}{n^p} \to 0$ as $n \to \infty$. Therefore $n^p \to \infty$ as $n \to \infty$. Hence the sequence $<n^p>$ diverges to ∞.

42. Let $s_n = \log \dfrac{1}{n}$.

let us take any $k < 0$.

then $s_n < k$ if $\log \dfrac{1}{n} < k$ i.e., if $-\log n < k$

i.e., if $\log n > -k$

i.e., if $n > e^{-k}$

if we take $m \in N$ such that $m > e^{-k}$, then

$$s_n < k \forall n \geq m$$

Hence, $s_n \to -\infty$ as $n \to \infty$

43. We have $s_n = 1 + 1 + \dfrac{1}{2!} + \dfrac{1}{3!} + ... + \dfrac{1}{n!}$

$\Rightarrow \quad s_{n+1} = 1 + 1 + \dfrac{1}{2!} + \dfrac{1}{3!} + ... + \dfrac{1}{n!} + \dfrac{1}{(n+1)!}$

Consider $(s_{n+1} - s_n)$

$$= \left(\dfrac{1}{n+2} + \dfrac{1}{n+3} + ... + \dfrac{1}{2n+2} \right)$$

$$- \left(\dfrac{1}{n+1} + \dfrac{1}{n+2} + ... + \dfrac{1}{n+n} \right)$$

$$= \dfrac{1}{2n+1} - \dfrac{1}{2n+2} - \dfrac{1}{n+1}$$

$$= \dfrac{1}{2n+1} - \dfrac{1}{2n+2} > 0 \; \forall n$$

$\Rightarrow$ The sequence $<s_n>$ is monotonically increasing.

Now, $|s_n| = s_n = 1 + 1 + \dfrac{1}{2!} + \dfrac{1}{3!} + ... + \dfrac{1}{n!}$

$$< \dfrac{1}{n} + \dfrac{1}{n} + ... + \dfrac{1}{n} (\text{upto } n \text{ terms})$$

$$n . \dfrac{1}{n} = 1$$

$\Rightarrow \quad |s_n| < 1 \; \forall n$

$\Rightarrow$ sequence $<s_n>$ is bounded also.

Hence, by monotone convergence theorem, given sequence is convergent.

45. We have $\quad s_n = 1 + 1 + \dfrac{1}{2!} + \dfrac{1}{3!} + ... + \dfrac{1}{n!}$

$\Rightarrow \quad s_{n+1} = 1 + 1 + \dfrac{1}{2!} + \dfrac{1}{3!} + ... + \dfrac{1}{n!} + \dfrac{1}{(n+1)!}$

Thus $(s_{n+1} - s_n) = \dfrac{1}{(n+1)!} > 0, \forall n.$

This implies that $s_{n+1} - s_n > 0$

$\Rightarrow s_{n+1} > s_n, \; \forall n$

$\Rightarrow <s_n>$ is monotonically increasing.

Now $s_n = 1 + 1 + \dfrac{1}{2!} + \dfrac{1}{3!} + ... + \dfrac{1}{n!}$

$$< 1 + 1 + \dfrac{1}{2} + \dfrac{1}{2^2} + ... + \dfrac{1}{2^{n-1}}$$

$$= 1 + \dfrac{1 - 1/2^n}{1 - 1/2} < 1 + 2 < 3$$

This implies that $s_n \leq 3 \Rightarrow <s_n>$ is bounded above by 3. Hence in view of the above, $<s_n>$ is monotonically increasing and bounded above

$\Rightarrow <s_n>$ is convergent with 3 as its upper bound $\Rightarrow \lim\limits_{n \to \infty} s_n = 3.$

Also $s_n = 1 + 1 + \dfrac{1}{2!} + \dfrac{1}{3!} + ... + \dfrac{1}{n!}$

$$> 1 + 1 = 2, \; \forall n$$

Therefore $2 < s_n \leq 3$.

48. Set $\quad \lambda = 1 + h, \; h > 0$

$\Rightarrow \quad \lambda^n = (1+h)^n$

$$= 1 + nh + \dfrac{n(n-1)h^2}{1.2}$$

$\Rightarrow \quad \lambda^n > \dfrac{n(n-1)}{1.2} h^2$, and therefore,

$$s_n = \dfrac{n}{\lambda^n} < \dfrac{2}{h^2(n-1)} \text{ for } n > 2$$

Thus $n \to \infty \Rightarrow s_n \to 0$.

SELF ASSESSMENT TEST

Verify each of the following:

1. (i) Following sequences are bounded above but not below.
   (a) $<(-1)^n.n+1:n\in N>$
   (b) $<-n:n\in N>$
   (ii) Following sequences are bounded below but not above.
   (a) $<\left(1+\dfrac{1}{n}\right)^{n^2}>$
   (b) $<\dfrac{n^n}{n!}>$

2. Each of the sequence $<n>,<a^n, a>1>$ and $1+\dfrac{1}{2}+...+\dfrac{1}{n}+...$ diverges to ∞.

3. The sequence $<s_n>$ defined by $s_n=1+\dfrac{1}{2^r}+\dfrac{1}{3^r}...+\dfrac{1}{n^r}$ does not converge if $r\leq1$.

4. Each of the sequence $<(-1)^n>,<(-1)^n+\dfrac{1}{n}>$ oscillate finitely.

5. Each of the sequence $<(-1)^n.n>,<a^n,a<-1>$ oscillate infinitely.

6. The sequence $\left\langle\dfrac{n+1}{n}\right\rangle$ is bounded monotonically increasing sequence.

7. $<a^{1/n}>$ is monotonically increasing and bounded when $1>a>0$. Also, it is decreasing or non-increasing for $a=0$ or 1. It decreases monotonically and is bounded when $a>1$.

8. $<a^n>$ is monotonically decreasing if $1>a>0$ and is bounded. It is constant for $a=0$ or 1 and monotonically increasing and unbounded if $a>1$.

9. (i) $\log 2=\lim\left(\dfrac{1}{n+1}+\dfrac{1}{n+2}+...+\dfrac{1}{2n}\right)$

   (ii) $\log 2=\lim\left(1+\dfrac{1}{2}+\dfrac{1}{3}+...+\dfrac{1}{2n}\right)$

10. (i) $\lim\limits_{n\to\infty}\sup.(-1)^n.\dfrac{1}{n}=0$

    (ii) $\lim\limits_{n\to\infty}\sup.(-1)^n n=\infty$

11. (i) $\lim\limits_{n\to\infty}\inf.(-1)^n n^2=-\infty$

    (ii) $\lim\limits_{n\to\infty}\inf n=\infty$

12. (i) $\lim\limits_{n\to\infty}\dfrac{3+2\sqrt{n}}{\sqrt{n}}=2$

    (ii) $\lim\limits_{n\to\infty}\dfrac{5.3^n}{3^n-2}=5$

    (iii) $\lim\limits_{n\to\infty}\left[5-\dfrac{1}{2^{n-1}}\right]=5$

    (iv) $\lim\limits_{n\to\infty}2^{-n}.n^2=0$

13. (i) $\lim\limits_{n\to\infty}\dfrac{1+2+3+...+n}{n^2}=\dfrac{1}{2}$

    (ii) $\lim\limits_{n\to\infty}\dfrac{1+3+5+...+(2n-1)}{n^2}=1$

    (iii) $\lim\limits_{n\to\infty}\left(\sqrt{n+1}-\sqrt{n}\right)=0$

    (iv) $\lim\limits_{n\to\infty}\left(1.\dfrac{2}{1}.\dfrac{3}{2}.\dfrac{4}{3}.....\dfrac{n}{n-1}\right)^{\frac{1}{n}}=1$

    (v) $\lim\limits_{n\to\infty}\dfrac{m(m-1)(m-2)....(m-n+1)}{n!}x^n=0$, $|x|<1$

14. Let $<s_n>=<n(1+(-1)^n)>$ then every neighbourhood of 0 contain an infinite number of terms of $<s_n>$.

15. For any sequence $<s_n>$ if $s_{n+1}-s_n\to1$ then $s_n\to\infty$.

16. If $s_n=2\sqrt{n}-\left\{\dfrac{1}{\sqrt{1}}+\dfrac{1}{\sqrt{2}}+....+\dfrac{1}{\sqrt{n}}\right\}$

    and $t_n=2\sqrt{n+1}-\left\{\dfrac{1}{\sqrt{1}}+\dfrac{1}{\sqrt{2}}+....+\dfrac{1}{\sqrt{n}}\right\}$

    then $<s_n>$ and $<t_n>$ converges to a common limit l such that $1<l<2$.

17. (i) If $s_n^2\to l^2$ then $|s_n|\to|l|$

    (ii) If $s_n^2\to l^2$ and $s_{n+1}-s_n\to0$ then $s_n\to l$ or $s_n\to-l$.

    (iii) If $s_n^3\to l^3$ then $s_n\to l$.

18. (i) $1+t \le e^t \le \dfrac{1}{1-t}$ for $|t| < 1$

(ii) $1 < \dfrac{e^t - 1}{t} < \dfrac{1}{1-t}$ for $0 < t < 1$

(iii) $1 + t < \dfrac{e^t - 1}{t} < \dfrac{1}{1-t^2}$ for $-1 < t < 0$

19. (i) $\displaystyle\lim_{n\to\infty} \left(\dfrac{1}{n+1} + \dfrac{1}{n+2} + \ldots + \dfrac{1}{2n} \right) = \log 2$

(ii) $\displaystyle\lim_{n\to\infty} \left(\dfrac{1}{n+1} + \dfrac{1}{n+2} + \ldots + \dfrac{1}{3n} \right) = \log 3$

20. A sequence $<s_n>$ is defined by the relation $s_1 = \alpha + \beta,\ s_{n+1} = \alpha + \beta - \dfrac{\alpha\beta}{s_n}$ for all $n \ge 1, \alpha > \beta > 0$

then $s_n = \dfrac{\alpha^{n+1} - \beta^{n+1}}{\alpha^n - \beta^n}$

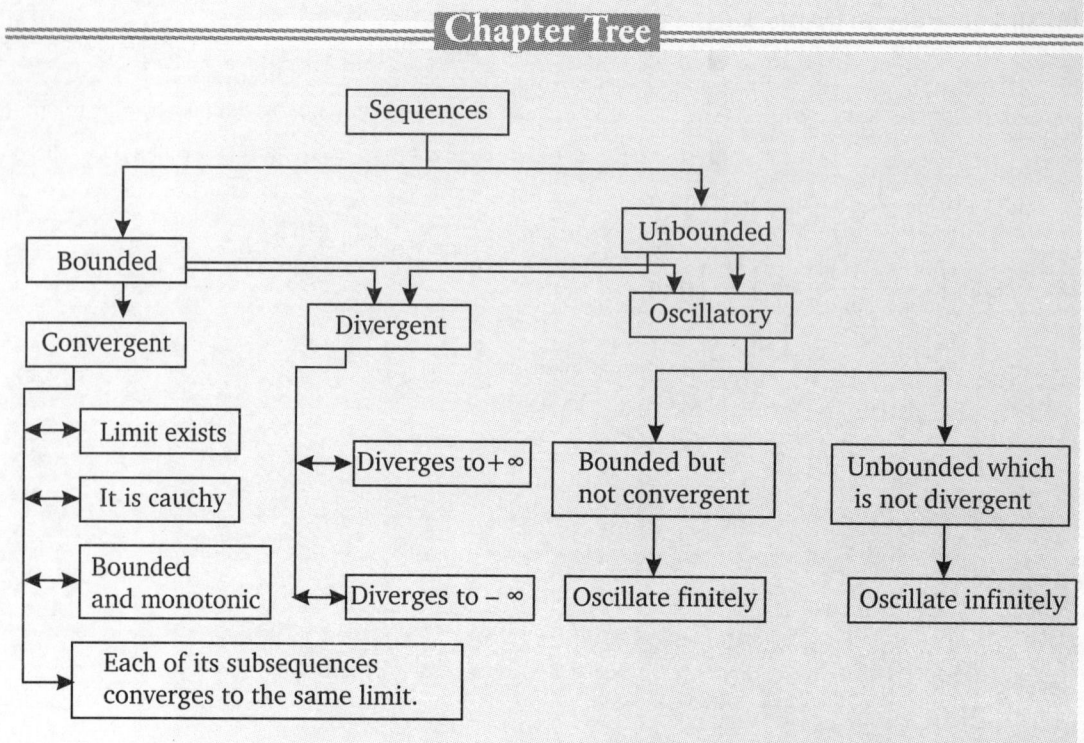

Chapter

3 Infinite Series

3.1 INTRODUCTION

Infinite series are essential in the calculation of values of many functions, and can frequently be used for the evaluation of definite integrals. They can also serve to define new and useful functions that are fundamental in many investigations in advanced mathematics and its applications.

Problems connected with the concept of series attracted the Indian mathematician as early as the third century A.D. Their work on series continued till late fourteenth century, but they never took up a critical study of series. In Europe, it was during the sixth century A.D., that the wider significance of finite and infinite series was realized.

The English mathematicians, Brook Taylor (1685-1731) and James Sterling (1692-1770), and the Scotch mathematician Colin Maclaurin (1698-1746), made important contributions to the study of infinite series. The question of convergent of infinite series was first subjected to rigorious investigation by the German Mathematician Carl Friedrich Gauss (1777-1855).

In this chapter, we are going to discuss the convergence behaviour of infinite series of real numbers and shall obtain a few tests for ascertaining the convergence of the infinite series. Some writer use the word Progression instead of word series. But here the word Series, which is due to the writers of the 17lh century and is most commonly used in preferred.

3.2 INFINITE SERIES

Let $<u_n>$ be a sequence of real numbers, then an expression of the form

$$u_1 + u_1 + \ldots + u_n + \ldots \qquad \ldots(1)$$

is called an infinite series. In symbols it is generally written as

$$\sum_{n=1}^{\infty} u_n \quad \text{or} \quad \sum u_n$$

If all the terms of $<u_n>$ after a certain number are zero then the expression

$u_1 + u_1 + \ldots + u_m$, written as $\sum_{n=1}^{m} u_n$ is called a finite series.

The term u_n is called the nth term or general term of the series (1). The sum of first n terms of the series is denoted by s_n. Thus,

$$s_n = u_1 + u_2 + \ldots + u_n = \sum_{r=1}^{n} u_r$$

3.3 SEQUENCE OF PARTIAL SUM OF AN INFINITE SERIES

An expression of the form $u_1 + u_2 + \ldots + u_n \ldots$ which involves addition of infinitely many terms has in itself no meaning. In order to give a meaning to the value of such as infinite sum, we form a sequence of partial sums. It is the limit of such a sequence which gives meaning to the infinite series.

Let us associate to the infinite series $u_1 + u_2 + \ldots + u_n \ldots$; a sequence $<s_n>$ defined by
$$s_n = u_1 + u_2 + \ldots + u_n.$$
Then the sequence $<s_n>$ is called the sequence of partial sums of the given series
$$u_1 + u_2 + \ldots + u_n \ldots.$$

3.4 CONVERGENCE, DIVERGENCE OR OSCILLATION OF AN INFINITE SERIES

An infinite series $\sum\limits_{n=1}^{\infty} u_n$ is said to be

(i) convergent if the sequence $<s_n>$ of its partial sums converges to a real number l and in that case l is called the sum of the series $\sum\limits_{n=1}^{\infty} u_n$ and we write $\sum\limits_{n=1}^{m} u_n = l$. In this case, we also say that the series is convergent to l.

(ii) converges absolutely if $\sum\limits_{n=1}^{\infty} |u_n|$ converges.

(iii) converges conditionally if $\sum\limits_{n=1}^{\infty} u_n$ converges but $\sum\limits_{n=1}^{\infty} |u_n|$ does not converge.

(iv) diverges to ∞ (or $-\infty$) if the sequence $<s_n>$ diverges to ∞ (or $-\infty$) and in that case
$$\sum\limits_{n=1}^{\infty} u_n = \infty \left(\text{or} \sum\limits_{n=1}^{\infty} u_n = -\infty \right)$$

(v) oscillate finitely if the sequence $<s_n>$ oscillates finitely.

(vi) oscillate infinitely if the sequence $<s_n>$ oscillates infinitely.

(vii) oscillatory if s_n, the sum of its first n terms, neither tends to a definite finite limit nor to $+\infty$ or $-\infty$ as $n \to \infty$.

REMARKS

- Divergent and oscillatory series are often called non-convergent series.
- The value of s_n of oscillate finitely series fluctuate within a finite range as $n \to \infty$.
- The value of s_n of oscillate infinitely series, tends to infinity as $n \to \infty$ and its sign is alternatively positive and negative.

☞ ILLUSTRATIONS

(1) The series $1 + \dfrac{2}{3} + \left(\dfrac{2}{3}\right)^2 + \ldots + \left(\dfrac{2}{3}\right)^{n-1} + \ldots$ is convergent.

(2) The series $\dfrac{1}{2} + \dfrac{1}{2^2} + \dfrac{1}{2^3} + \ldots$ is convergent.

(3) The series $1 + 2 + 3 + \ldots + n + \ldots$ is divergent.

(4) The series $3 - 3 + 3 - 3 + \ldots$ is oscillatory.

THEOREM 1 **(Necessary condition for convergence).** *For a series Σu_n to be convergent, it is necessary that $\lim u_n = 0$.*

Proof.   Let us suppose, the series Σu_n be convergent. Let s_n denote the sum of n terms of the series

$$\left.\begin{array}{l} s_n = u_1 + u_2 + \ldots + u_n \\ s_{n-1} = u_1 + u_2 + \ldots + u_{n-1} \end{array}\right\} \Rightarrow u_n = s_n - s_{n-1} \qquad \ldots(1)$$

The series Σu_n is convergent, therefore s_n and s_{n-1} both will tend to the same finite limit, say l as $n \to \infty$

Now, from (1) $\lim u_n = \lim s_n - \lim s_{n-1} = l - l = 0$

Hence, for a convergent series, it is necessary that $\lim u_n = 0$

REMARKS

- The converse of the above theorem is not necessarily true :

  For example, in the series $\sum u_n = 1 + \dfrac{1}{2} + \dfrac{1}{3} + \ldots$; $u_n = \dfrac{1}{n} \to 0$ as $n \to \infty$

  but the series is not convergent.
- If a series Σu_n be such that u_n does not tend to zero as $n \to \infty$ then the series does not converge.

THEOREM 2. **(Cauchy's General principle of convergence for series).** *A necessary and sufficient condition for a series Σu_n to be convergent is that to each $\varepsilon > 0$, there exists a positive integer m such that $|u_{n+1} + u_{n+2} + \ldots + u_{n+p}| < \varepsilon \ \forall \ n > m, \ p \geq 1$.*

Proof.   Let $<s_n>$ be the sequence of partial sums of the series Σu_n. The series Σu_n will converge if and only if the sequence $<s_n>$ of its partial sums converges. But by Cauchy's general principle of convergence for sequences, we know that a necessary and sufficient condition for the convergence of $<s_n>$ is that for each $\varepsilon > 0$, there exists $m \in N$ such that

$|u_{n+1} + u_{n+2} + \ldots + u_{n+p}| < \varepsilon$, whenever $n \geq m$ and $p \geq 1$.

$$|s_n - s_m| < \varepsilon, \ \forall \ n > m$$

$\Rightarrow \quad |u_{n+1} + u_{n+2} + \ldots + u_{n+p}| < \varepsilon, \ \forall \ n > m$ and $p \geq 1$.

REMARKS

- The nature of a series remains unaltered if :
  (a) the sign of all term are changed.
  (b) a finite number of terms are added or omitted.
  (c) each term of the series is multiplied or divided by the same fixed number k (k $\neq$ 0)
- If Σu_n converges to l and $k \in R$, then $\Sigma k u_n$ converges to lk.
- If Σu_n converges to l_1 and Σv_n converges to l_2 then $\Sigma(u_n + v_n)$ converges to $(l_1 + l_2)$.
- If Σu_n diverges and $k \in R$, $k \neq 0$, then $\Sigma k u_n$ diverges.
- If Σu_n and Σv_n are two divergent series having all terms positive, then $\Sigma(u_n + v_n)$ also diverges.

THEOREM 3. *A series of positive terms is convergent if s_n, the sum of n terms is less than a fixed number for all values of n.*

Proof.   Let $u_1 + u_2 + \ldots u_n + \ldots$ be the series of positive terms.

Then $s_n = u_1 + u_2 + \ldots u_n$.

Obviously if n increases, then s_n increases and may tend to a finite limit or to $+\infty$. The series cannot oscillate. If s_n remains less than a fixed number for all values of n it cannot tend to infinity and so it must tend to a finite limit. Hence the series is convergent.

3.5 FUNDAMENTAL RESULTS FOR THE CONVERGENCE OF POSITIVE TERM SERIES

THEOREM 1. *A series Σu_n of positive term is convergent if and only if the sequence $<s_n>$ (where $s_n = u_1 + u_2 + \ldots + u_n$) of its partial sum is bounded above.*

Proof. Since $u_n > 0 \ \forall \ n$, the sequence $<s_n>$ of partial sums of the series is monotonically increasing. Now the series Σu_n is convergent iff the sequence $<s_n>$ is convergent.

i.e., iff the sequence $<s_n>$ is bounded above.

(∵ a monotonically increasing sequence is convergent iff it is bounded above)

REMARKS

- To show that a series of positive term is convergent it is enough to show the sequence of its partial sum is bounded above and to show that the series of positive term is divergent, we have to show that the sequence of its partial sum is unbounded above.

- A series Σu_n where the terms are not necessarily positive, may fail to the convergent even if the sequence of its partial sums is bounded above.

  For example, consider the series $\displaystyle\sum_{n=1}^{\infty} u_n$ where $u_n = (-1)^n$. Then $s_n = \begin{cases} -1 & if \ n \ is \ odd \\ 0 & if \ n \ is \ even. \end{cases}$

  The sequence $<s_n>$ is bounded above but not convergent and as such, the series is not convergent. Hence, boundedness of the sequence of partial sums of a series Σu_n is only a necessary condition and not a sufficient one. However, it is a sufficient condition for a positive term series.

THEOREM 2. (Pringsheim theorem). *If a series Σu_n of positive monotonic decreasing terms converges then not only u_n tends to zero but also nu_n tends to zero as n tends to infinity.*

Proof. By Cauchy's general principle of convergence, we have for a convergent series, that for given $\varepsilon > 0$ there exists a positive integer k such that

$$|u_{m+1} + u_{m+2} + \ldots + u_{m+p}| < \varepsilon/2 \ \forall \ m \geq k; p \geq 1$$

Choose $m + p = n > 2k$ and $m = \left[\dfrac{n}{2}\right]$ where $\left[\dfrac{n}{2}\right]$ denote the greatest integer not greater than $\left[\dfrac{n}{2}\right]$.

Then $\qquad u_{m+1} + u_{m+2} + \ldots + u_n < \dfrac{\varepsilon}{2}$

But Σu_n is monotonic decreasing sequence of positive terms. Therefore,

$$(n - m)u_n < u_{m+1} + u_{m+2} + \ldots + u_n < \frac{\varepsilon}{2}$$

$\Rightarrow \qquad \dfrac{1}{2}nu_n < \dfrac{\varepsilon}{2} \Rightarrow \qquad nu_n < \varepsilon, \forall \ n \in \mathbb{N}$

$\Rightarrow \qquad \displaystyle\lim_{n \to \infty} nu_n = 0.$

THEOREM 3. *If each term of a series Σu_n of positive terms does not exceed the corresponding terms of a convergent series Σv_n of positive terms, then Σu_n is convergent. On the other hand, if each term of Σu_n exceed (or equals) the corresponding terms of a divergent series of positive terms, then Σu_n is divergent.*

Proof.   Let us suppose $u_n < v_n \ \forall\, n \in N$,

Now let $s_n = u_1 + u_2 + \ldots + u_n$ and $s'_n = v_1 + v_2 + \ldots + v_n$

Since $u_n \leq v_n \ \forall\, n$, therefore $s_n \leq s'_n$

Now Σv_n is convergent, therefore $\lim s_n \leq \lim s'_n = s'$ (a finite quantity).

Thus s_n tends to a finite limit as $n \to \infty$. Hence the series Σu_n is convergent.

Now, if $u_n > v_n$, $\forall\, n$ then $s_n > s'_n$.

But Σv_n is divergent, therefore $s'_n \to \infty$ as $n \to \infty$ and hence $s_n \to \infty$ as $n \to \infty$ which gives that Σu_n is divergent.

THEOREM 4   **(Convergence of geometric series)**. *The geometric series*

$$1 + r + r^2 + \ldots + r^{n-1} + \ldots$$

(i) converges to $\dfrac{1}{1-r}$ *if* $|r| < 1$ *(ii) diverges to* $+\infty$ *if* $r \geq 1$

(iii) oscillate finitely if $r = -1$.   *(iv) oscillate infinitely if* $r < -1$.

Proof.   Here $s_n = 1 + r + r^2 + \ldots + r^{n-1}$

$$= \begin{cases} \dfrac{1-r^n}{1-r} & \text{if } r \neq 1 \\ n & \text{if } r = 1 \end{cases}$$

Now, there are following cases :

Case (I). If $|r| < 1$.

Then $\displaystyle\lim_{n \to \infty} r^n = 0$

so that $\displaystyle\lim_{n \to \infty} s_n = \dfrac{1}{1-r}$

$\Rightarrow$ the series is convergent to $\dfrac{1}{1-r}$.

Case (II). If $r > 1$.

Then $\displaystyle\lim_{n \to \infty} r^n = \infty$

so that $s_n = \dfrac{1-r^n}{1-r} = \dfrac{1}{1-r} + \dfrac{r^n}{r-1} \to \infty$ as $n \to \infty$

Hence, the series diverges to ∞.

if $r = 1$, then $s_n = 1 + 1 + \ldots + 1 + \ldots$ to n times

                            $= n$

Thus, the sequence $< s_n >$ diverges and hence the series diverges.

Case (III). If $r = -1$.

Then, $s_n = \begin{cases} 0 \text{ if } n \text{ is even} \\ 1 \text{ if } n \text{ is odd} \end{cases}$

Therefore the sequence $<s_n>$ oscillate between 0 and 1.

$\Rightarrow$ The series oscillates finitely between 0 and 1.

Case (IV). If $r < -1$

Let $r = -a$ where $a > 1$

Then $s_n = \dfrac{1}{1+a} - \dfrac{(-1)^n \cdot a^n}{1+a}$ so that $s_{2n} \to \infty$ and $s_{2n+1} \to \infty$

Therefore, the sequence $<s_n>$ oscillate infinitely between $-\infty$ and $+\infty$.

Hence, the series oscillate infinitely.

THEOREM 5. *A positive terms series Σu_n either converges to a finite limit or diverges to ∞.*

Proof.   Let $s_n = u_1 + u_2 + \dots + u_n$

$\Rightarrow$ $s_{n+1} = u_1 + u_2 + \dots u_{n+1}$

Therefore, $s_{n+1} - s_n = u_{n+1} > 0$ $\Rightarrow$ $s_{n+1} > s_n, \forall n$

$\Rightarrow$ $<s_n>$ is monotonically increasing sequence.

Since, a monotonically increasing sequence is either convergent to a finite limit or divergent to ∞, the sequence $< s_n>$ of partial sums of the series Σu_n is either convergent a finite limit or divergent to ∞.

Hence, the series Σu_n is either converges or diverges to ∞.

REMARKS

- In view of the above theorem, a positive term series has only two possible behaviours *i.e.* convergence or divergence while a general term has got five behaviour (*i.e.* convergent, divergent to ∞, divergent to $-\infty$, oscillate finitely and oscillate infinitely).
- If, in a positive terms series Σu_n, u_n does not tend to 0 as $n \to \infty$, the series is divergent.
- Similarly, it can be proved that a negative term series either converges to a finite limit or diverges to $-\infty$.

THEOREM 6. $\left(\text{The Auxillary series} \Sigma \dfrac{1}{n^p} \right)$. *The infinite series* $\Sigma \left(\dfrac{1}{n^p} \right) = \dfrac{1}{1^p} + \dfrac{1}{2^p} + \dots + \dfrac{1}{n^p} + \dots$

is convergent if $p > 1$ and divergent if $p < 1$.

Proof.   **Case (I).** When $p > 1$.

Since each term of the given series is positive so that the given series can be written as :

$$\Sigma \left(\frac{1}{n^p} \right) = \frac{1}{1^p} + \left(\frac{1}{2^p} + \frac{1}{3^p} \right) + \left(\frac{1}{4^p} + \frac{1}{5^p} + \frac{1}{6^p} + \frac{1}{7^p} \right)$$

$$+ \left(\frac{1}{8^p} + \frac{1}{9^p} + \frac{1}{10^p} + \frac{1}{11^p} + \frac{1}{12^p} + \frac{1}{13^p} + \frac{1}{14^p} + \frac{1}{15^p} \right) + \dots \quad \dots(1)$$

Since $p > 1$, then

$$3^p > 2^p \Rightarrow \frac{1}{3^p} < \frac{1}{2^p}$$

$\Rightarrow$ $\dfrac{1}{2^p} + \dfrac{1}{3^p} < \dfrac{1}{2^p} + \dfrac{1}{2^p} = \dfrac{2}{2^p} = \dfrac{1}{2^{p-1}}$

Also, $5^p > 4^p, 6^p > 4^p, 7^p > 4^p$

$\Rightarrow$ $\dfrac{1}{5^p} < \dfrac{1}{4^p}, \dfrac{1}{6^p} < \dfrac{1}{4^p}, \dfrac{1}{7^p} < \dfrac{1}{4^p}$

$\Rightarrow$ $\dfrac{1}{4^p} + \dfrac{1}{5^p} + \dfrac{1}{6^p} + \dfrac{1}{7^p} < \dfrac{1}{4^p} + \dfrac{1}{4^p} + \dfrac{1}{4^p} + \dfrac{1}{4^p} = \left(\dfrac{1}{2^{p-1}} \right)^2$

Similarly

$$\frac{1}{8^p} + \frac{1}{9^p} + \frac{1}{10^p} + \dots + \frac{1}{15^p} < \frac{8}{8^p} = \left(\frac{1}{2^{p-1}} \right)^3 \dots \text{and so on.}$$

Now using above inequalities equation (1) becomes

$$\Sigma\left(\frac{1}{n^p}\right) < 1 + \frac{1}{2^{p-1}} + \left(\frac{1}{2^{p-1}}\right)^2 + \left(\frac{1}{2^{p-1}}\right)^3 + \ldots \qquad \ldots(2)$$

The R.H.S. of (2) is a geometric series with common ratio less than 1 as $p > 1$, which is therefore convergent thus the series on L.H.S of (2) is convergent, hence $\Sigma\left(\frac{1}{n^p}\right)$ is convergent, when $p > 1$.

Case (II). When $p = 1$. Then the given series becomes

$$\Sigma\frac{1}{n^p} = 1 + \frac{1}{2} + \frac{1}{3} + \ldots$$

Now, this series may be written as follows

$$\Sigma\frac{1}{n^p} = 1 + \frac{1}{2} + \left(\frac{1}{4} + \frac{1}{4}\right) + \ldots$$

$$> 1 + \frac{1}{2} + \left(\frac{1}{4} + \frac{1}{4}\right) + \ldots$$

$$= 1 + \frac{1}{2} + \frac{2}{4} + \ldots.$$

$$= 1 + \frac{1}{2} + \frac{1}{2}\ldots.$$

Now since $\lim u_n = \frac{1}{2} \neq 0$, the series is divergent.

Case (III). When $p < 1$. Then

$$2^p < 2,\ 3^p < 3,\ 4^p < 4 \text{ and so on.}$$

Hence, the given series reduces to

$$\Sigma\frac{1}{n^p} > 1 + \frac{1}{2} + \frac{1}{3} + \frac{1}{4} + \ldots.$$

Clearly, the series on the right hand side is divergent. [By case (II)]

Hence, the given series is divergent when $p < 1$.

3.6 COMPARISON TESTS

The most important technique for deciding whether a series is convergent or not is to compare it with another suitable chosen series which is already known to be convergent or divergent.

First form. *Let Σu_n and Σv_n be two series of positive terms such that $u_n < kv_n$, $\forall\, n$ Then,*

(i) *Σv_n converges $\Rightarrow \Sigma u_n$ converges*

(ii) *Σu_n diverges $\Rightarrow \Sigma v_n$ diverges.*

Proof. Firstly we shall prove (i) Σv_n convergent $\Rightarrow u_n$ is convergent

Now, $\qquad u_n < kv_n \in N$

$\Rightarrow \qquad (u_1 + u_1 + \ldots + u_n) < k\,(v_1 + u_2 + \ldots\ldots + v_n) \qquad \ldots(1)$

But the series Σv_n is given to be convergent.

$\Rightarrow$ By the fundamental result for positive term series $\exists$ a positive number M such that

$$v_1 + v_2 + \ldots v_n < M\ \forall\, n \in N. \qquad \ldots(2)$$

From (1) and (2), we have

$$u_1 + u_1 + \dots + u_n < kM = k_1 \text{ (say)}, \forall\, n \in \mathrm{N}.$$

$$\Rightarrow \qquad u_1 + u_1 + \dots + u_n < k_1\ \forall\, n \in \mathrm{N}, \text{ where } k_1 = Mk > 0$$

$\Rightarrow \exists$ a positive number k such that $u_1 + u_1 + \dots + u_n < k_1\ \forall\, n \in \mathrm{N}$

So, by the fundamental result for the positive terms series, Σu_n is also convergent. We shall now prove that if Σu_n is divergent, then Σv_n is also divergent.

Since, we are given Σu_n to be divergent.

$\Rightarrow$ The sequence $<s_n>$ of its partial sums is also divergent.

$\Rightarrow \exists$ a positive number k_2 (however large) and positive integer $m \in \mathrm{N}$ such that

$$s_n > k_2\ \forall\ n > m$$

i.e. $\qquad u_1 + u_1 + \dots + u_n > k_2\ \forall\, n > m$ 　　　　　　　　...(3)

From (1) and (3), we have

$$k_2 < u_1 + u_2 + \dots + u_n < k\,(v_1 + v_2 + \dots + v_n)\ \forall\, n > m$$

$$\Rightarrow \qquad v_1 + v_2 + \dots + v_n > \frac{k_2}{k}\ (=k_3)\ \forall\, n > m$$

$$\Rightarrow \qquad t_n > k_3,\ \forall\, b > m$$

where $k_3 = \dfrac{k_2}{k}$ and $\qquad t_n = v_1 + v_2 + \dots v_n$

$\Rightarrow \exists$ a positive number k_3 (however large) and a positive integer m such that $t_n > k_3$, $\forall\, n > m$ and thus t_n is divergent and consequently Σv_n is divergent

Second form. *Let Σu_n and Σv_n be two series of positive terms and let k_1 and k_2 be positive real numbers such that $k_1 v_n \le u_n \le k_2 v_n,\ \forall\, n$*

then series Σu_n and Σv_n converge or diverge together.

Proof. We have

$$k_1 v_n \le u_n \le k_2 v_n,\ \forall\, n \qquad\qquad\qquad ...(1)$$

(i) If the series Σv_n is convergent, then $\Sigma k_2 v_n$ is convergent and hence from second part of form (1) the series Σu_n is convergent.

(ii) If the series Σu_n is convergent, then from first part of the inequality (1), $\Sigma k_1 v_n$ is convergent and hence $\Sigma v_n \left(= \dfrac{1}{k_1}\Sigma k_1 v_n \right)$ is convergent.

(iii) If the series, Σu_n is is divergent, then from second part of inequality (1), $\Sigma k_2 v_n$ is divergent and hence Σv_n is divergent.

(iv) If the series Σv_n is divergent, then $\Sigma k_1 v_n$ is divergent and hence from first part of the inequality (1), Σv_n is divergent.

Third form. *If Σu_n and Σv_n be two given positive terms series such that*

$$u_n \le k v_n,\ \forall\, n > m,\ k > 0 \text{ and } m \in \mathrm{N}$$

Then

(i) *Σv_n is convergent $\Rightarrow \Sigma u_n$ is convergent*

(ii) *Σu_n is divergent $\Rightarrow \Sigma v_n$ is also divergent.*

Proof.

(i) Let us suppose $<s_n>$ and $<t_n>$ be two sequences of partial sums of the two given positive terms series Σu_n and Σv_n respectively.

Therefore, $\quad\quad s_n = u_1 + u_2 + \dots u_n \ \forall \ n \in N$

and $\quad\quad\quad\quad t_n = v_1 + v_2 + \dots v_n \ \forall \ n \in N$

since $\quad\quad\quad u_n \le k v_n \ \forall \ n \ge m \Rightarrow s_n \le k t_n, \ \forall \ n \ge m$

$\Rightarrow\quad\quad\quad s_n - s_m \le k\ (t_n - t_m) = kt_n - kt_m$

$\Rightarrow\quad\quad\quad s_n \le kt_n + (s_m - kt_m) = kt_n + M$ $\quad\quad\quad$ ...(1)

Where $\quad\quad\quad M = s_m - kt_m$, a fixed quantity.

Now, if Σv_n is convergent $\Rightarrow$ $<t_n>$ is convergent and thus is bounded above

$\Rightarrow$ $\ \exists$ a number A such that $t_n \le A, \ \forall \ n \in N$. $\quad\quad\quad$ ...(2)

Now from (1) and (2), we have

$\quad\quad\quad s_n < k.\ A + M = k_1 \ \forall \ n \in N$

and therefore $<s_n>$ is bounded above.

Moreover, $<s_n>$ is a montonically increasing sequence, therefore, $<s_n>$ is monotonically increasing sequence which is bounded above and thus, it is convergent and hence Σv_n is convergent.

(ii) Now if Σu_n is divergent $\Rightarrow$ $<s_n>$ is divergent and therefore $\exists$ a positive number $\varepsilon > 0$ and $m' \in s_n$, then

$\quad\quad\quad s_n > B, \ \forall \ n \ge m.$

Let $\quad m^* = \max \{m, m'\}$ so that $s_n > B \ \forall \ n \ge m^*$.

$\Rightarrow\quad\quad t_n > \dfrac{1}{k}\ (s_n - M) > \dfrac{1}{k}\ (B - M) = C \ \forall \ n \ge m^*, C \ne 0.$

$\Rightarrow\quad\quad <t_n>$ is divergent and hence Σv_n is divergent.

Fourth form. *Let Σu_n and Σv_n be two series of positive terms and let k_1, k_2 be positive real numbers such that $k_1 v_n < u_n < k_2 v_n \ \forall \ n > m$, m being a fixed positive integer. Then the series Σu_n and Σv_n converge or diverge together.*

Proof. Proof immediately follows from the second form of comparison test.

Fifth form. *Let Σu_n and Σv_n be two sereis of positive terms such that*

$$\lim_{n \to \infty} \frac{u_n}{v_n} = l \quad\quad\quad \text{(finite and non-zero)}$$

then both the series converge or diverge together.

Proof. Since $\quad\quad \dfrac{u_n}{v_n} > 0, \forall n \quad \Rightarrow \quad \lim_{n \to \infty} \dfrac{u_n}{v_n} \ge 0 \ i.e.,\ l \ge 0$

But $l \ne 0$ (by assumption) : therefore $l > 0$.

Now, let $\varepsilon > 0$ be chosen in such a way that $l - \varepsilon > 0$.

Since, $\lim\limits_{n \to \infty} \dfrac{u_n}{v_n} = l$ therefore $\exists$ a positive integer m such that

$$l - \varepsilon < \frac{u_n}{v_n} < l + \varepsilon, \ \forall \ n > m. \quad\quad\quad ...(1)$$

Since, $u_n > 0 \ \forall \ n$, therefore, multiplying (1) by v_n, we obtain

$$(l - \varepsilon) < v_n < u_n < (l + \varepsilon) v_n \ \forall \ n > m.$$

Since $l - \varepsilon$ and $l + \varepsilon$ are both positive, therefore applying the fourth form of comparison test, we find that series Σu_n and Σv_n converge or diverge together.

REMARKS

- In the above form of the comparison test, the condition $\lim\limits_{n\to\infty}\dfrac{u_n}{v_n}$ be finite and non-zero cannot be dropped for if $u_n = \dfrac{1}{n}$ and $v_n = \dfrac{1}{n^2}$ then $\lim \dfrac{u_n}{v_n} = +\infty$, Σu_n is divergent and Σv_n is convergent.

  In this case, neither the hypothesis nor the conclusion of the comparison test happens to be true.

- The comparison test is usually applied when the nth term u_n of the given series Σu_n contains the powers of n only which may be positive or negative, integral or fractional

- v_n can be choosen such that

$$v_n = \frac{1}{n^{p-q}}$$

  where p and q are respectively the highest indices of n in the denominator and numerator of u_n when it is in the form of fraction, if u_n can be expanded in ascending powers of $\dfrac{1}{n}$ then to get v_n we should retain only the lowest power of $\dfrac{1}{n}$ the numerical factor being disregarded.

- We always denote the given series by Σu_n and the series which is used for comparison by Σv_n

- The series Σv_n is known as auxiliary series. We select the auxiliary series in such a way that

$$\lim\limits_{n\to\infty}\left(\frac{u_n}{v_n}\right) \text{ exists finitely and non-zero.}$$

Sixth form. Let Σu_n and Σv_n be two series of positive terms and let $\exists$ a positive integer m such that

$$\frac{u_n}{u_{n+1}} \geq \frac{v_n}{v_{n+1}}, \forall n \geq m$$

then Σu_n and Σv_n both converge or diverge together.

Proof. Let us suppose $<s_n>$ and $<t_n>$ be two sequences of partial sum of the series Σu_n and Σv_n respectively, such that

$$s_n = u_1 + u_2 + \dots u_n$$

$$t_n = v_1 + v_2 + \dots v_n \ \forall \ n$$

Now for $n \geq m$, we have
$$\frac{u_m}{u_n} = \frac{u_m}{u_{m-1}} \cdot \frac{u_{m+1}}{u_{m+2}} \dots \frac{u_{n-1}}{u_n}$$

$$\geq \frac{v_m}{v_{m+1}} \cdot \frac{v_{m+1}}{v_{m+2}} \dots \frac{v_{n-1}}{v_n} = \frac{v_m}{v_n}$$

$$\Rightarrow \qquad u_n \leq \frac{u_m}{v_m} \cdot v_n$$

since, m is fixed positive intiger, $\dfrac{u_m}{v_m}$ is a fixed number say k. Thus for $n \geq m$, we have

$$u_n \leq k v_n$$

Hence, Σu_n and Σv_n both converge or diverge together.

Solved Examples

Example 1. If $u_n \geq 0$ for all n and Σu_n converges then show that $\sum \dfrac{\sqrt{u_n}}{n}$ also converges.

Solution. Since $u_n \leq 0$, then for all $n \in N$

or
$$\frac{u_n}{n^2} < u_n^2 + \frac{2u_n}{n^2} + \frac{1}{n^4}$$

or
$$0 \leq \frac{u_n}{n^2} < \left(u_n + \frac{1}{n^2}\right)^2$$

or
$$0 \leq \frac{\sqrt{u_n}}{n} < u_n + \frac{1}{n^2}$$

Since Σu_n and $\sum \dfrac{1}{n^2}$ are convergent, therefore $\sum\left(u_n + \dfrac{1}{n^2}\right)$ is convergent.

Hence by comparison test $\sum \dfrac{\sqrt{u_n}}{n}$ is convergent.

Example 2. If Σu_n is a sereis of positive terms and Σu_n is convergent then show that $\sum \dfrac{u_n}{1+u_n}$ is convergent.

Solution. Since $u_n > 0$, then for all $n \in N$, $1 + u_n > 1$ for all $n \in N$

$\Rightarrow$
$$\frac{1}{1+u_n} < 1 \, \forall n \in N$$

$\therefore$
$$0 < \frac{u_n}{1+u_n} < u_n \, \forall n \in N$$

$\Rightarrow$
$$0 < \frac{u_n}{1+u_n} < \Sigma u_n$$

Since Σu_n is convergent, hence by comparison test $\sum \dfrac{u_n}{1+u_n}$ is convergent.

Example 3. If Σu_n is a convergent series of positive terms, prove that $\sum u_n^2$ is also convergent. Is the converse true?

Solution. Since Σu_n is convergent, then $\lim\limits_{n \to \infty} u_n = 0$

$\therefore$ For given $\epsilon > 0$ there exists a positive integer m such that
$$|u_n| < \epsilon \quad \forall n \geq m$$

$\Rightarrow$
$$-\epsilon < u_n < \epsilon \, \forall n \geq m$$

Since $u_n > 0 \, \forall n$, choose $\epsilon < 1$, then
$$0 < u_n < 1 \, \forall n$$

$\Rightarrow$
$$0 < u_n^2 < u_n \, \forall n \qquad [\because u_n < 1 \Rightarrow u_n^2 < u_n]$$

$\Rightarrow$
$$0 < \Sigma u_n^2 < \Sigma u_n \, \forall n$$

Since Σu_n is convergent, hence by comparison test $\sum u_n^2$ is convergent.

Converse is not always true :

For example if $u_n = \dfrac{1}{n'}$, then $u_n^2 = \dfrac{1}{n^2}$.

Clearly if $\sum u_n^2 = \sum \dfrac{1}{n^2}$ convergent but $\sum u_n = \sum \dfrac{1}{n}$ is divergent.

Example 4. *If $\sum u_n^2$ and $\sum v_n^2$ are both convergent series, prove that the series $\sum u_n v_n$ also convergent.*

Solution. Since $\sum u_n^2$ and $\sum v_n^2$ both are convergent, therefore $\Sigma(u_n^2 + v_n^2)$ is also convergent

$\Rightarrow \quad \sum \dfrac{1}{2}(u_n^2 + v_n^2)$ is also convergent

We know that $\qquad$ G. M. < A. M.

$\Rightarrow \qquad \sqrt{u_n^2 v_n^2} < \dfrac{1}{2}(u_n^2 + v_n^2)$

$\Rightarrow \qquad u_n v_n < \dfrac{1}{2}(u_n^2 + v_n^2)$

Hence by comparison test $\Sigma u_n v_n$ is convergent.

Example 5. *Examine the following series for convergence :*

(i) $\sum \dfrac{1}{(\log n)^{\log n}}$ $\qquad$ (ii) $\sum \dfrac{1}{(\log \log n)^{\log n}}$

Solution. (i) Since we have $\lim\limits_{n \to \infty} \log(\log n) = \infty$

$\Rightarrow \qquad$ There exists a large positive integer n such that $\log(\log n) > 2$

$\Rightarrow \qquad [\log(\log n)] \log n > 2 \log n$

$\Rightarrow \qquad (\log n)[\log(\log n] > \log n^2$

$\Rightarrow \qquad \log[(\log n)^{\log n}] > \log n^2$

$\Rightarrow \qquad (\log n)^{\log n} > n^2$

$\Rightarrow \qquad \dfrac{1}{(\log n)^{\log n}} < \dfrac{1}{n^2}$

since $\sum \dfrac{1}{n^2}$ is convergent hence by comparison test $\Sigma(\log n)^{\log n}$ is also convergent.

(ii) Similarly, $\lim\limits_{n \to \infty} \log(\log \log n) = \infty$

$\Rightarrow$ there exists a positive integer n is so large such that $\log(\log \log n) > 2$

$\Rightarrow \qquad \log n [\log(\log \log n)] > 2 \log n$

$\Rightarrow \qquad \log [\{\log(\log n)\}^{\log n}] > \log n^2$

$\Rightarrow \qquad [\log(\log n)]^{\log n} > n^2$

$\Rightarrow \qquad \dfrac{1}{[\log(\log n)]^{\log n}} < \dfrac{1}{n^2}$

Since $\sum \dfrac{1}{n^2}$ is convergent, hence by comparison test $\sum \dfrac{1}{[\log(\log n)]^{\log n}}$ is convergent.

Example 6. *Test the convergence of the series $\dfrac{2}{1} + \dfrac{3}{4} + \dfrac{4}{9} + ... + \dfrac{n+1}{n^2} + ...$*

Solution. Here $\qquad u_n = \dfrac{n+1}{n^2}$. Take $v_n = \dfrac{n}{n^2} = \dfrac{1}{n}$

Then
$$\frac{u_n}{v_n} = \frac{\frac{n+1}{n^2}}{\frac{1}{n}} = \frac{n+1}{n^2} \cdot \frac{n}{1} = \frac{n+1}{n}$$

Therefore
$$\lim_{n\to\infty} \frac{u_n}{v_n} = \lim_{n\to\infty} \frac{n+1}{n} = \lim_{n\to\infty} \left(1+\frac{1}{n}\right)$$
$$= 1, \text{ which is finite and non-zero.}$$

Thus, by the comparison test two series are either both convergent or both divergent. But the auxiliary series $\sum v_n = \frac{1}{n}$ is divergent, Hence, the given series Σu_n is also divergent.

Example 7. *Test the convergence of the series* $\frac{1}{1\cdot2} + \frac{1}{2\cdot3} + \dots + \frac{1}{n(n+1)} + \dots$

Solution. Here
$$u_n = \frac{1}{n(n+1)} = \frac{1}{n} - \frac{1}{n+1}$$

If s_n is the partial sum of n terms of the series Σu_n, then
$$s_n = u_1 + u_2 + \dots + \dots + u_n$$
$$= \left(1-\frac{1}{2}\right) + \left(\frac{1}{2}-\frac{1}{3}\right) + \dots + \left(\frac{1}{n}-\frac{1}{n+1}\right) = 1 - \frac{1}{n+1}$$

Now,
$$\lim_{n\to\infty} s_n = \lim_{n\to\infty}\left[1 - \frac{1}{n+1}\right] = 1, \text{ which is finite and non-zero.}$$

Hence, the given series is convegent.

REMARK

- This type of series is calld "Telescoping series".

Example 8. *Show that the series* $1 + \frac{1}{2!} + \frac{1}{3!} + \dots$ *is convergent.*

Solution. Since, we $\frac{1}{2!} = \frac{1}{2}$

$$\frac{1}{3!} < \frac{1}{2^2}$$
$$\dots \quad \dots \quad \dots$$
$$\dots \quad \dots \quad \dots$$
$$\frac{1}{n!} < \frac{1}{2^{n-1}}$$

Therefore, $1 + \frac{1}{2!} + \frac{1}{3!} + \dots + \frac{1}{n!} + \dots < 1 + \frac{1}{2} + \frac{1}{2^2} + \dots$

The series on the right hand side is a geometric series with common ratio $\frac{1}{2}$ and hence convergent. So the series on the left hand side will also be convergent.

Example 9. *Test the convergence or divergence of* $\frac{1}{1\cdot2\cdot3} + \frac{3}{2\cdot3\cdot4} + \frac{5}{3\cdot4\cdot5} + \dots + \frac{2n-1}{n(n+1)(n+2)} + \dots$

Solution. Here, $u_n = \frac{2n-1}{n(n+1)(n+1)}$ Take $v_n = \frac{n}{n(n)(n)} = \frac{1}{n^2}$

Then, $\quad \dfrac{u_n}{v_n} = \dfrac{2n-1}{n(n+1)(n+2)} \cdot \dfrac{n^2}{1} = \dfrac{\left(2-\dfrac{1}{n}\right)}{\left(1+\dfrac{1}{n}\right)\cdot\left(1+\dfrac{2}{n}\right)}$

$\Rightarrow \quad \lim_{n\to\infty} \dfrac{u_n}{v_n} = \lim_{n\to\infty} \dfrac{\left(2-\dfrac{1}{n}\right)}{\left(1+\dfrac{1}{n}\right)\left(1+\dfrac{2}{n}\right)} = 2$, which is finite.

Now, the auxiliary series $\sum v_n = \sum \dfrac{1}{n^2}$ is convergent ($\because p = 2 > 1$). Hence the given series is convergent.

Example 10. *Test the convergence or divergence of the series* $1 + \dfrac{1}{2^2} + \dfrac{2^2}{3^3} + \dfrac{3^3}{4^4} + \dots$

Solution. Leaving the first term, we get

$$u_n = \dfrac{n^n}{(n+1)^{n+1}} = \dfrac{1}{n\left(1+\dfrac{1}{n}\right)^{n+1}} = \dfrac{1}{n}\left[1+\dfrac{1}{n}\right]^{-[n+1]}$$

$$= \dfrac{1}{n}\left[1 - \dfrac{(n+1)}{n} + \dots\right] = \dfrac{1}{n} - \left(1+\dfrac{1}{n}\right)\dfrac{1}{n} + \dots$$

Let $\sum v_n = \sum \dfrac{1}{n}$, where $v_n = \dfrac{1}{n}$ be the auxiliary series.

Then $\quad \lim_{n\to\infty} \dfrac{u_n}{v_n} = \lim_{n\to\infty} \dfrac{\dfrac{1}{n}\left[\dfrac{1}{(1+1/n)^{n+1}}\right]}{\dfrac{1}{n}}$

$$= \lim_{n\to\infty} \left[\dfrac{\dfrac{1}{(1+1/n)^n}}{\left(1+\dfrac{1}{n}\right)}\right] = \dfrac{1}{e}, \text{ which is finite and not-zero}$$

Now, since $\sum v_n = \sum \dfrac{1}{n}$ is divergent, therefore by comparsion test the given series is also divergent.

Example 11. *Test the convergence of the series whose general term is* $[n^3 + 1]^{1/3} - n]$.

Solution. Here, we have

$$u_n = (n^3+1)^{1/3} - n$$

$$= n\left[\left(1+\dfrac{1}{n^3}\right)^{1/3} - 1\right] = n\left[\left(1 + \dfrac{3}{3n^3} + \dfrac{\dfrac{1}{3}\left(\dfrac{1}{3}-1\right)}{2!}\cdot\dfrac{1}{n^6} + \dots\right) - 1\right]$$

$$= \dfrac{1}{n^2}\left[\dfrac{1}{3} - \dfrac{1}{9n^3} + \dots\right]$$

Let $v_n = \dfrac{1}{n^2}$, then the auxiliary series $\sum v_n = \sum \dfrac{1}{n^2}$

Now, $\lim \dfrac{u_n}{v_n} = \dfrac{1}{3} - \dfrac{1}{9n^3} + \dots = \dfrac{1}{3}$ which is finite and non-zero.

Since the series $\sum v_n = \sum \dfrac{1}{n^2}$ is convergent ($\because p = 2 > 1$), therefore, the given series is also convergent.

Example 12. _Test the convergence of the series whose n^{th} term is_ $[\sqrt{(n^4+1)} - \sqrt{(n^4-1)}]$

Solution. Here, we have,

$$u_n = \sqrt{(n^4+1)} - \sqrt{(n^4-1)} = n^2\left[\left(1+\dfrac{1}{n^4}\right)^{1/2} - \left(1-\dfrac{1}{n^4}\right)^{1/2}\right]$$

$$= n^2\left[\left(1 + \dfrac{1}{2n^4} + \dfrac{\frac{1}{2}\left(\frac{1}{2}-1\right)}{2!}\cdot\dfrac{1}{n^8} - \dfrac{\frac{1}{2}\left(\frac{1}{2}-1\right)\left(\frac{1}{2}-2\right)}{3!}\cdot\dfrac{1}{n^{12}} + \dots\right)\right.$$
$$\left. - \left(1 - \dfrac{1}{2n^4} + \dfrac{\frac{1}{2}\left(\frac{1}{2}-1\right)}{2!}\cdot\dfrac{1}{n^8} - \dfrac{\frac{1}{2}\left(\frac{1}{2}-1\right)\left(\frac{1}{2}-2\right)}{3!}\cdot\dfrac{1}{n^{12}} + \dots\right)\right]$$

$$= n^2\left[\dfrac{1}{n^4} + \dfrac{1}{8n^{12}} + \dots\right]$$

$$= \dfrac{1}{n^2} + \dfrac{1}{8n^{10}} + \dots$$

Let $v_n = \dfrac{1}{n^2}$, then the auxillary series is $\sum v_n = \sum \dfrac{1}{n^2}$, which is convergent.

Now $\lim\limits_{n\to\infty} \dfrac{u_n}{v_n} = \lim\limits_{n\to\infty}\left[\dfrac{1}{n^2} + \dfrac{1}{8n^{10}} + \dots\right]\Big/\dfrac{1}{n^2} = \lim\limits_{n\to\infty}\left[1 + \dfrac{1}{8n^8} + \dots\right]$

$= 1$, which is finite and non-zero.

Therefore, by comparsion test the given series is also divergent.

Example 13. _Test the convergence of the series_ $\sum \sin\dfrac{1}{n}$.

Solution. Here, we have $u_n = \sin\dfrac{1}{n}$

Let $v_n = \dfrac{1}{n}$, therefore, the auiliary series $\sum v_n = \sum\dfrac{1}{n}$ is divergent.

Now $\lim\limits_{n\to\infty} \dfrac{u_n}{v_n} = \lim\limits_{n\to\infty} \dfrac{\sin 1/n}{1/n} = 1$, which is finite and non-zero.

Therefore, by comparison test the given series is also divergent.

Example 14. _Test the convergence of the series_

$$\dfrac{1}{a\cdot 1^2 + b} + \dfrac{2}{a\cdot 2^2 + b} + \dfrac{3}{a\cdot 3^2 + b} + \dots$$

Solution. Here, we have

$$u_n = \frac{n}{a \cdot n^2 + b} = \frac{n}{n^2 \left[a + \dfrac{b}{n^2}\right]} = \frac{1}{n} \cdot \frac{1}{\left[a + \dfrac{b}{n^2}\right]}$$

Let $v_n = \dfrac{1}{n}$, then the auxiliary series $\Sigma v_n = \Sigma \dfrac{1}{n}$ is divergent.

Now, $\displaystyle\lim_{n\to\infty} \frac{u_n}{v_n} = \lim \frac{\left[\dfrac{1}{n} \cdot \dfrac{1}{(a + b/n^2)}\right]}{\dfrac{1}{n}}$

$$= \lim_{n\to\infty} \frac{1}{a + \dfrac{b}{n^2}} = \frac{1}{a}, \text{ which is finite and non-zero.}$$

Hence, by comparsion test the given series is also divergent.

Example 15. *Test the convergence or divergence of the series*

$$\sum_{n=1}^{\infty} \frac{1}{x^n + x^{-n}}, x > 0$$

Solution. **Case (I)** Let $x < 1$. Take $v_n = x_n$, then

$$\lim_{n\to\infty} \frac{u_n}{v_n} = \lim_{n\to\infty} \frac{1}{x^n + x^{-n}} \cdot \frac{1}{x^n} = \lim_{n\to\infty} \frac{1}{x^{2n} + 1} = 1$$

Since, $x^{2n} \to 0$ for $x < 1$. But the auxiliary series $\Sigma v_n = \Sigma x^n$ is convergent (being a G.P. with common ratio $x < 1$).

Hence, the given series Σu_n is convergent for $x < 1$.

Case (II) Let $x > 1$. In this case take $v_n = x^{-n}$. Then

$$\lim_{n\to\infty} \frac{u_n}{v_n} = \lim_{n\to\infty} \frac{1}{x^n + x^{-n}} \cdot \frac{x^n}{1} = \lim_{n\to\infty} \frac{1}{1 + x^{-2n}} = 1$$

Since, $x^{-2n} \to 0$ for $x > 1$. But the auxiliary series $\Sigma v_n = \Sigma x^{-n}$ is convergent (being a G. P. with common ration $x^{-1} < 1$).

Case (III) Let $x = 1$. In this case $u_n = \dfrac{1}{2}$ (for all n) and so

$$S_n = u_1 + u_2 + \ldots + u_n = \frac{n}{2} \to \infty \text{ as } n \to \infty .$$

Therefore, the given series is divergent when $x = 1$.

Example 16. *Test the convergence of the series whose n^{th} term is $\sqrt{n^3 + 1} - \sqrt{n^3}$.*

Solution. Here, we have

$$u_n = \sqrt{n^3 + 1} - \sqrt{n^3} = n^{3/2} \left[1 + \frac{1}{n^3}\right]^{1/2} - n^{3/2}$$

$$= n^{3/2} \left[1 + \frac{2}{2n^3} - \frac{1}{8n^6} + \ldots\right] - n^{3/2}$$

$$= \frac{1}{2n^{3/2}} - \frac{1}{8n^{9/2}} + \ldots$$

Let us take $v_n = \dfrac{1}{n^{3/2}}$ ($\because$ when u_n is in the form of the series in powers of $1/n$, v_n is taken as the term of lowest power of $1/n$, by ignoring the numerical factor).

Then we have

$$\lim_{n\to\infty}\frac{u_n}{v_n} = \lim_{n\to\infty}\left[\frac{1}{2n^{3/2}} - \frac{1}{8n^{9/2}} + ...\right]\times \frac{n^{3/2}}{1}$$

$$= \lim_{n\to\infty}\left[\frac{1}{2} - \frac{1}{8n^3} + ...\right]$$

$$= \frac{1}{2}, \text{ which is finite and non-zero.}$$

But the auxillary series $\sum v_n = \sum \dfrac{1}{n^{3/2}}$ is convergent ($p = 3/2 > 1$). Hence, the given series is also convergent.

EXERCISE 3.1

Check the convergence of the following series:

1. $\sum u_n = 1 + \dfrac{1}{3} + \dfrac{1}{5} + \dfrac{1}{7} + ...$

2. $\sum u_n = 1 + \dfrac{1}{\sqrt{2}} + \dfrac{1}{\sqrt{3}} + \dfrac{1}{\sqrt{4}} + ...$

3. $\sum u_n = 1 + \dfrac{4}{5} + \dfrac{6}{10} + \dfrac{8}{17} + ... + \dfrac{2n}{n^2+1}...$

4. $\sum u_n = \sqrt{\dfrac{1}{2^3}} + \sqrt{\dfrac{2}{3^3}} + \sqrt{\dfrac{3}{4^3}} + ...$

5. $\sum u_n = \dfrac{1}{2} + \dfrac{\sqrt{2}}{5} + \dfrac{\sqrt{3}}{10} + ... \dfrac{\sqrt{n}}{n^2+1} + ...$

6. $\sum u_n = \dfrac{\sqrt{1}}{1+\sqrt{1}} + \dfrac{\sqrt{2}}{2+\sqrt{2}} + \dfrac{\sqrt{3}}{3+\sqrt{3}} + ...$

7. $\sum u_n = \dfrac{1}{a+b} + \dfrac{1}{a+2b} + \dfrac{1}{a+3b} + ... + \dfrac{1}{a+nb} + ...$

8. $\sum u_n = \dfrac{1}{a(a+b)} + \dfrac{1}{(a+2b)(a+3b)}$
$$+ \dfrac{1}{(a+4b)(a+5b)} + ...$$

9. $\sum u_n = \sum \dfrac{n}{n^2 + \sqrt{n}}$

10. $\sum u_n = \sum \dfrac{n}{(a+nb)^2}$

11. $\sum u_n = \sum \dfrac{\sqrt{n+1} + \sqrt{n-1}}{n}$

12. $\sum u_n = \sum \dfrac{1}{n}\sin\dfrac{1}{n}$

13. $\sum u_n = \sum \tan^{-1}\dfrac{1}{n}$

14. $\sum u_n = \sum \dfrac{n^p}{(n+1)^q}$

15. $\sum u_n = \sum \dfrac{n^2-1}{n^2+1}$

16. $\sum u_n = \sum \dfrac{1}{n}\sqrt{n^2+n+1} - \sqrt{n^2-n+1}$

HINTS TO SELECTED PROBLEMS

1. $\sum u_n = \sum\limits_{n=1}^{\infty}\left(\dfrac{1}{2n-1}\right)$

$\Rightarrow \lim\limits_{n\to\infty} u_n = \lim\limits_{n\to\infty}\left(\dfrac{1}{2n-1}\right) = 0$

Now apply comparison test.

4. $u_n = \dfrac{\sqrt{n}}{(n+1)\sqrt{n+1}}$

Then $\lim\limits_{n\to\infty}\dfrac{u_n}{v_n} = \lim\limits_{n\to\infty}\sqrt{\dfrac{1}{1+\dfrac{1}{n}}}\cdot\dfrac{1}{\left(1+\dfrac{1}{n}\right)} = n \neq 0$

7. $u_n = \dfrac{1}{a+nb}$ Let $v_n = \dfrac{1}{n}$.

Then $\lim\limits_{n\to\infty} \dfrac{u_n}{v_n} = \lim\limits_{n\to\infty} \dfrac{n}{a+nb} = \dfrac{1}{b} \neq 0.$

8. Here, we have

$$u_n = \dfrac{1}{[a+(2n-2)b][a+(2n-1)b]}$$

Let $\qquad v_n = \dfrac{1}{n^2}$

Then $\lim\limits_{n\to\infty} \dfrac{u_n}{v_n} = \dfrac{1}{4b^2} \neq 0$

12. $u_n = \dfrac{1}{n}\sin\dfrac{1}{n} = \dfrac{1}{n}\left[\dfrac{1}{n} - \dfrac{1}{6n^3} + \dfrac{1}{120n^5} - \dots\right]$

Let $v_n = \dfrac{1}{n^2}$. Then $\lim\limits_{n\to\infty} \dfrac{u_n}{v_n} = 1 \neq 0$

14. Here we have

$$u_n = \dfrac{n^p}{(n+1)^q} \quad \text{Let } v_n = \dfrac{n^p}{n^{q-p}}$$

Then $\lim\limits_{n\to\infty} \dfrac{v_n}{u_n} = \lim\limits_{n\to\infty} n^{q-p}\left[\dfrac{n^p}{(n+1)^q}\right]$

$$= \lim\limits_{n\to\infty} \dfrac{1}{\left(1+\dfrac{1}{n}\right)^q} = 1 \neq 0$$

15. $\lim\limits_{n\to\infty} u_n = \lim\limits_{n\to\infty} \dfrac{\left(1-\dfrac{1}{n^2}\right)}{\left(1+\dfrac{1}{n^2}\right)} = 1 \neq 0$

Answers

1. Divergent	**2.** Divergent	**3.** Divergent	**4.** Divergent
5. Convergent	**6.** Divergent	**7.** Divergent	**8.** Convergent
9. Divergent	**10.** Divergent	**11.** Divergent	**12.** Convergent
13. Divergent	**14.** Convergent if $p - q + 1 < 0$ and divergent if $p - q + 1 \geq 0$		
15. Divergent	**16.** Divergent		

3.7 CAUCHY'S ROOT TEST

Let Σu_n be a series of positive terms and let

$$\lim\limits_{n\to\infty} u_n^{1/n} = l.$$

If,

(i) $l < 1$, then Σu_n converges;

(ii) $1 > 1$, then Σu_n diverges;

(iii) $l = 1$, then the test fails and the series may either converge or diverge.

Proof. Case (I) Let $u_n^{1/n} = l < 1$.

Since $l < 1$, we can choose an $\varepsilon > 0$ such that $l + \varepsilon < 1$.

Let $\qquad\qquad\qquad l + \varepsilon < r$ such that $0 < r < 1$.

Since $\lim\limits_{n\to\infty} u_n^{1/n} = l$, therefore there exists a positive integer m_1 such that.

$$|u_n^{1/n} - l| < \varepsilon, \ \forall \ n > m_1$$

$\Rightarrow \qquad\qquad l - \varepsilon < u_n^{1/n} < l + \varepsilon \ \forall \ n > m_1$

$\Rightarrow \qquad\qquad (l - \varepsilon)^n < u_n < (l + \varepsilon)^n \ \forall \ n > m_1$

Since $u_n < r^n \ \forall \ n > m_1$ and since Σr^n converges (being a geometric series with common ratio less than one). Then by comparison test Σu_n converges.

Case (II) Let $u_n^{1/n} = l > 1$.

Since $l > 1$, we can choose an $\varepsilon > 0$ such that $l - \varepsilon > 1$.

Let $l - \varepsilon < R$ then $R > 1$.

Since $R^n < u_n \ \forall \ n > m_2$ and since ΣR^n diverges (being a G.P. with common ratio greater than one). Then by comparison test Σu_n diverges.

Case (III) Let $\quad u_n = \dfrac{1}{n}$

Then $\quad\quad\quad\quad \lim_{n \to \infty} u_n^{1/n} = 1.$

Since $\Sigma\left(\dfrac{1}{n}\right)$ diverges, therefore we find that if $\lim_{n \to \infty} u_n^{1/n} = 1$, then the series Σu_n may diverge.

Again, let $u_n = \dfrac{1}{n^2}$. In this case also $\lim_{n \to \infty} u_n^{1/n} = 1$ but the series Σu_n converges. Thus we find that if $\lim_{n \to \infty} u_n^{1/n}$, then the series Σu_n may converge. The above two examples show that if

$$\lim_{n \to \infty} (u_n)^{1/n} = 1$$

Then the test fail.

REMARKS

- Cauchy's root test can be applied with advantage to series in which the nth term happens to be an exponential fraction of n.
- In this test, it is understood that $u_n^{1/n}$ stands for the positive n^{th} root of u_n.
- The Cauchy's root test can also be stated as follows :
- "A series Σu_n of positive terms is convergent if for every value of $n \geq m$, m being finite, $(u_n)^{1/n}$ less than a fixed number, which is less than unity, and the series is divergent if $(u_n)^{1/n} \geq 1$ for every value of $n \geq m$

3.7 D'ALEMBERT RATIO TEST

Let Σu_n be a series of positive terms and let

(a) $\lim\limits_{n \to \infty} \dfrac{u_n}{u_{n+1}} = l$

Then if,

(i) $l > 1$, the series converges,

(ii) $l < 1$, the series diverges

(iii) $l = 1$, the series may converge or diverge and therefore the test fails.

(b) $\dfrac{u_n}{u_{n+1}} = \infty$ as $n \to \infty$. Then Σu_n *converges*.

Proof. (a) Case(I). When $l > 1$, Let $\varepsilon > 0$ be a positive number such that

$$l - \varepsilon > 1.$$

Now since $\lim\limits_{n \to \infty} \dfrac{u_n}{u_{n+1}} = l$, therefore $\exists$ a positive integer m such that

$$l - \varepsilon < \frac{u_n}{u_{n+1}} < l + \varepsilon, \text{ whenever } n > m$$

Now, putting $n = m + 1, m + 2, \ldots\ldots p - 1$, in succession in the above inequality, we get

$$l - \varepsilon < \frac{u_{m+1}}{u_{m+2}} < l + \varepsilon$$

$$l - \varepsilon < \frac{u_{m+2}}{u_{m+3}} < l + \varepsilon$$

$$\ldots \qquad \ldots \qquad \ldots \qquad \ldots$$

$$l - \varepsilon < \frac{u_{p-1}}{u_p} < l + \varepsilon$$

Multiplying the corresponding sides of the first part of the above inequalities, we get

$$(l - \varepsilon)^{p-1-m} < \frac{u_{m+1}}{u_{m+2}} \cdot \frac{u_{m+2}}{u_{m+3}} \cdots \frac{u_{p-1}}{u_p}$$

$$\Rightarrow \qquad (l - \varepsilon)^{p-1-m} < \frac{u_{m+1}}{u_p}$$

$$\Rightarrow \qquad u_p < u_{m+1}(l - \varepsilon)^{m+1} \cdot (l - \varepsilon)^{-p}$$

$$\Rightarrow \qquad u_p < k(l - \varepsilon)^{-p}, \forall \, p \geq m + 2 \quad \text{and} \quad k = u_{m+1}(l - \varepsilon)^{m+1}$$

Since, the series $\Sigma (l - \varepsilon)^{-p}$ converges (being a geometric series with common ratio $(l - \varepsilon)^{-1}$, which is certainly less than unity), then by comparison test it follows that Σu_n converges,

Case (II) When $l < 1$, let $\varepsilon > 0$ be a positive number such that $l + \varepsilon < 1$.

Now since $\lim\limits_{n \to \infty} \dfrac{u_n}{u_{n+1}} = l$, therefore, $\exists$ a positive intger m such that

$$l - \varepsilon < \frac{u_n}{u_{n+1}} < l + \varepsilon, \, \forall \, n > m$$

Putting $n = m + 1, m + 2, \ldots, p - 1$ in succession in the second part of the above inequality, we get

$$\frac{u_{m+1}}{u_{m+2}} < l + \varepsilon,$$

$$\frac{u_{m+2}}{u_{m+3}} < l + \varepsilon,$$

$$\ldots \qquad \ldots \qquad \ldots$$

$$\frac{u_{p-1}}{u_p} < l + \varepsilon.$$

Multiplying the corresponding sides of the above inequalities, we have

$$\frac{u_{m+1}}{u_p} < (l + \varepsilon)^{p-1-m}$$

$$\Rightarrow \qquad u_p > u_{m+1}(l + \varepsilon)^{m+1}(l + \varepsilon)^{-p}$$

$$\Rightarrow \qquad u_p > A(l + e)^{-p} \, \forall \, p \geq m + 2 \quad \text{and} \quad A = u_{m+1}(l + \varepsilon)^{m+1}$$

Since, $\Sigma (l + \varepsilon)^{-p}$ is a divergent series (being a geometric series with common ratio $(l + \varepsilon)^{-1}$, which is certainly greater than unity), then by comparison test, it follows that Σu_n diverges.

Case (III) Let $l = 1$.

Now, first consider the harmonic series $1 + \dfrac{1}{2} + \dfrac{1}{3} + \dfrac{1}{5} + \ldots + \dfrac{1}{n} + \ldots$

Then $\qquad \dfrac{u_n}{u_{n+1}} = \dfrac{n+1}{n} = 1 + \dfrac{1}{n} \Rightarrow \lim_{n \to \infty} \dfrac{u_n}{u_{n+1}} = 1$

Since, the harmonic series is divergent, we find that if $l = 1$, a series may diverge.

Now, consider the series $\dfrac{1}{1^2} + \dfrac{1}{2^2} + \ldots + \dfrac{1}{n^2} + \ldots$

when $\qquad \dfrac{u_n}{u_{n+1}} = \dfrac{(n+1)^2}{n^2} = \left(1 + \dfrac{1}{n}\right)^2 \Rightarrow \lim_{n \to \infty} \dfrac{u_n}{u_{n+1}} = 1$

since, the series $\sum \dfrac{1}{n^2}$ converges, we find that if $l = 1$, a series may converge.

(b) Let us suppose $\lim\limits_{n \to \infty} \dfrac{u_n}{u_{n+1}} = +\infty$ then there exist positive integers m and p such

$$\dfrac{u_n}{u_{n+1}} > p \,\forall n \geq m, p > 1$$

Replacing n by $m, m + 1, m + 2, \ldots, n - 1$, we have

$$\dfrac{u_m}{u_{m+1}} > p$$

$$\dfrac{u_{m+1}}{u_{m+2}} > p$$

$$\ldots \qquad \ldots \qquad \ldots$$

$$\dfrac{u_{n-1}}{u_n} > p.$$

Multiplying the corresponding sides of the above inequalities, we have

$$\dfrac{u_m}{u_n} > p^{n-m}$$

$\Rightarrow \qquad\qquad u_n < p^{m-n} \cdot u_m,$

$\Rightarrow \qquad\qquad u_n < A.p^{-n} \forall n > m$ and $A = p^m u_m.$

Since Σp^{-n} is convergent, then by comparison test, the series Σu_n is convergent.

REMARKS

- The ratio test is generally applied when the nth term of the series involves factorials, products of several factors, or combination of powers and factorials.
- The ratio test can also be stated as follows:
  "An inifinite series of positive terms is convergent if from and after some terms the ratio of each term to the preceding term is less than a fixed number which is less than unity and series is divergent if the ratio, defined above is greater than or equal to unity,
- The ratio test is easier to apply than the root test. However, the root test is stronger than the ratio test.
- The ratio test does not tell us anything about the convergence of the series Σu_n if we only know that

$$\dfrac{u_n}{u_{n+1}} > 1 \forall n.$$

Example 1. *Test the convergence of the series* $x + 2x^2 + 3x^3 + 4x^4 + ...$

Solution. Here, we have $u_n = nx^n$

$\Rightarrow \qquad (u_n)^{1/n} = n^{1/n} \cdot x$

$\Rightarrow \qquad \lim_{n \to \infty} (u_n)^{1/n} = \lim_{n \to \infty} (x \cdot n^{1/n}) = x.1 = x \qquad\qquad [\therefore \lim_{n \to \infty} n^{1/n} = 1]$

Then, by Cauchy's root test, Σu_n is convergent if $x < 1$ and is divergent if $x > 1$. For $x = 1$, the Cauchy's root test fails.

In this case, the given series becomes

$$1 + 2 + 3 + ...$$

$s_n = $ sum of n terms of the series $= \dfrac{1}{2} n(n+1)$, which is finite

Thus the given series is convergent if $x < 1$ and is divergent if $x \geq 1$.

Example 2. *Test the convergence of the series*

$$\frac{1}{2} + \left(\frac{2}{3}\right) x + \left(\frac{3}{4}\right)^2 x^2 + \left(\frac{4}{5}\right) x^3 + ... \infty, \; x > 0$$

Solution. Omitting the first term of the series (because it will not effect the convergence or divergence of the series),

we have $\qquad u_n = \left(\dfrac{n+1}{n+2}\right)^n \cdot x^n$

Therefore $\quad \lim_{n \to \infty} u_n^{1/n} = \lim_{n \to \infty} \left[\dfrac{\left(1 + \dfrac{1}{n}\right) x}{1 + \left(\dfrac{2}{n}\right)} \right] = x$.

Therefore by Cauchy's root test, the given series Σu_n converges if $x < 1$, divegent if $x > 1$.

For $x = 1$, test fails

$\therefore \qquad\qquad \lim_{n \to \infty} u_n = \lim_{n \to \infty} \dfrac{\left(1 + \dfrac{1}{n}\right)^n}{\left(1 + \dfrac{2}{n}\right)^n} = \dfrac{e}{e^2} = \dfrac{1}{e} > 0.$

$\therefore \quad$ The series Σu_n diverges if $x = 1$.

Hence, the given series is convergent if $x < 1$ and divergent if $x \geq 1$.

Example 3. *Test the series for convergence of the series* $1 + \dfrac{1}{2^2} + \dfrac{1}{3^3} + \dfrac{1}{4^4} + ...$

Solution. Here, we have

$$u_n = \frac{1}{n^n}$$

$\Rightarrow \qquad \lim_{n \to \infty} (u_n)^{1/n} = \lim_{n \to \infty} \dfrac{1}{n} = 0 < 1$

Hence by Cauchy's root test the given series is convergent.

Example 4. *Test the convergence of the series*

$$\left(\frac{2^2}{1^2}-\frac{2}{1}\right)^{-1}+\left(\frac{3^3}{2^3}-\frac{3}{2}\right)^{-2}+\left(\frac{4^4}{3^4}-\frac{4}{3}\right)^{-3}+\dots$$

Solution. Here we have

$$u_n=\left[\frac{(n+1)^{n+1}}{n^{n+1}}-\frac{(n+1)}{n}\right]^{-n}$$

Therefore $\lim_{n\to\infty} u_n^{1/n} = \lim_{n\to\infty}\left[\frac{(n+1)^{n+1}}{n^{n+1}}-\frac{n+1}{n}\right]^{-1}$

$$= \lim_{n\to\infty}\left[\left(1+\frac{1}{n}\right)^{n+1}-\left(1+\frac{1}{n}\right)\right]^{-1}$$

$$= \lim_{n\to\infty}\left(1+\frac{1}{n}\right)^{-1}\left[\left(1+\frac{1}{n}\right)^{n}-1\right]^{-1}$$

$$= (1+0)^{-1}[e-1]^{-1}$$

$$= \frac{1}{e-1} < 1.$$

Hence, by Cauchy's root test the given series is convergent.

Example 5. *Test the convergence of the series* $\sum\left(1+\frac{1}{n}\right)^{-n^2}$.

Solution. Here we have

$$u_n=\left(1+\frac{1}{n}\right)^{-n^2}\Rightarrow(u_n)^{1/n}=\left(1+\frac{1}{n}\right)^{-n}$$

$$\Rightarrow \lim_{n\to\infty}(u_n)^{1/n}=\lim_{n\to\infty}\left(1+\frac{1}{n}\right)^{-n}=\frac{1}{e}<1.$$

Hence, by Cauchy's root test the given series $\sum u_n$ is convergent.

Example 6. *Test the convergence of the series* $\sum\left[\frac{\log n}{\log(n+1)}\right]^{n^2\log n}$

Solution. Here we have

$$u_n=\left[\frac{\log n}{\log(n+1)}\right]^{n^2\log n}$$

$$\Rightarrow u_n^{1/n}=\left[\frac{\log n}{\log(n+1)}\right]^{n\log n}=\left[\frac{\log n\left(1+\frac{1}{n}\right)}{\log n}\right]^{-n\log n}$$

$$=\left[\frac{\log n+\log\left(1+\frac{1}{n}\right)}{\log n}\right]^{-n\log n}$$

$$= \left[\frac{\log n + \dfrac{1}{n} - \dfrac{1}{2n^2} + ...}{\log n} \right]^{-n \log n}$$

$$= \left[1 + \frac{1}{n \log n} - \frac{1}{2n^2 \log n} + ... \right]^{-n \log n} = k \quad (\text{say})$$

Then
$$\log k = \log \left[1 + \frac{1}{n \log n} - \frac{1}{2n^2 \log n} + ... \right]^{-n \log n}$$

$$= (-n \log n) \log \left[\left(1 + \frac{1}{n \log n} - \frac{1}{2n^2 \log n} + ... \right) ... \right]$$

$$= -n \log n \left[\left(\frac{1}{n \log n} - \frac{1}{2n^2 \log n} + ... \right) ... \right]$$

$$= -1 + \frac{1}{2n} - ...$$

$\Rightarrow \qquad \lim_{n \to \infty} \log k = -1$

$\Rightarrow \qquad \lim_{n \to \infty} k = e^{-1}$

$\Rightarrow \qquad \lim_{n \to \infty} u_n^{1/n} = \frac{1}{e} < 1.$ $\qquad\qquad [\because 2 < e < 3]$

Then, by Cauchy's root test the given series is convergent.

Example 7. *Test the convergence of* $\sum \left(\dfrac{n+1}{n+2} \right)^n x^n, (x > 0)$

Solution. The nth term of the given series is
$$u_n = \left(\frac{n+1}{n+2} \right)^n x^n$$

Then we have
$$u_n^{1/n} = \left(\frac{n+1}{n+2} \right) x = \left(\frac{1 + 1/n}{1 + 2/n} \right) x$$

$\therefore \qquad \lim_{n \to \infty} u_n^{1/n} = \lim_{n \to \infty} \left(\dfrac{1 + \dfrac{1}{n}}{1 + \dfrac{2}{n}} \right) x = x$

Now we have the followings cases.

Case I : If $x < 1$ then by Cauhy's root test Σu_n is convergent

Case II : If $x > 1$ then by Cauchy's root test Σu_n is divergent.

Case III : If $x = 1$, the test fails.

Now, when $x = 1$, we have

$$u_n = \left(\frac{n+1}{n+2}\right)^n = \left(\frac{1+\dfrac{1}{n}}{1+\dfrac{2}{n}}\right)^n$$

$$\Rightarrow \qquad \lim_{n\to\infty} u_n = \lim_{n\to\infty} \frac{(1+1/n)^n}{(1+2/n)^n} = \frac{e}{e^2} = \frac{1}{e} \neq 0$$

$\Rightarrow \quad \Sigma u_n$ is divergent.

Hence, the given series is convergent if $x < 1$ and divergent if $x \geq 1$.

Example 8. *Test the convergence of the series* $\displaystyle\sum_{n=1}^{\infty} 3^{-n-(-1)^n}$.

Solution. The nth term of the given series is

$$u_n = 3^{-n-(-1)^n}$$

$$= \begin{cases} 3^{-n+1}, & \text{if } n \text{ is odd} \\ 3^{-n-1}, & \text{if } n \text{ is even} \end{cases}$$

$$\therefore \qquad \lim_{n\to\infty} u_n^{1/n} = \frac{1}{3} \qquad\qquad \left(\because \lim_{\lim\to\infty} x^{1/n} = 1 \text{ if } x > 0\right)$$

$$\Rightarrow \qquad \lim_{n\to\infty} u_n^{1/n} < 1$$

Hence, by Cauchy's root test the given is convergent.

Example 9. *Examine the convergene of the following series :*

(i) $\displaystyle\sum\left(1+\frac{1}{\sqrt{n}}\right)^{-n^{3/2}}$
\qquad\qquad (ii) $\displaystyle\sum\frac{(n-\log n)^n}{2^n.n^n}$

(iii) $\displaystyle\sum\left(1+\frac{1}{n}\right)^{-n^2}$
\qquad\qquad (iv) $\displaystyle\sum_{n=2}^{\infty}\frac{1}{(\log n)^n}$

Solution. (i) We have

$$u_n = \left(1+\frac{1}{\sqrt{n}}\right)^{-n^{3/2}}$$

$$\therefore \qquad \lim_{n\to\infty} u_n^{1/n} = \lim_{n\to\infty}\left(1+\frac{1}{\sqrt{n}}\right)^{-\sqrt{n}} = \lim_{n\to\infty}\left[\left(1+\frac{1}{\sqrt{n}}\right)^{\sqrt{n}}\right]^{-1}$$

$$= e^{-1} = \frac{1}{e} < 1$$

Hence the given series is convergent.

(ii) We have

$$u_n = \frac{(n-\log n)^n}{2^n.n^n}$$

$$\therefore \qquad u_n^{1/n} = \frac{n-\log n}{2n} = \frac{1}{2}\left(1-\frac{\log n}{n}\right)$$

$$\therefore \qquad \lim_{n\to\infty} u_n^{1/n} = \lim_{n\to\infty}\frac{1}{2}\left(1-\frac{\log n}{n}\right)$$

$$= \frac{1}{2}(1-0) = \frac{1}{2} < 1 \qquad \left[\because \lim_{n\to\infty} \frac{\log n}{n} = 0\right]$$

Hence, by Cauchy's root test the given series is convergent.

(iii) We have

$$u_n = \left(1+\frac{1}{n}\right)^{-n^2}$$

∴ $$u_n^{1/n} = \left(1+\frac{1}{n}\right)^{-n}$$

∴ $$\lim_{n\to\infty} u_n^{1/n} = \lim_{n\to\infty} \left(1+\frac{1}{n}\right)^{-n}$$

$$= \lim_{n\to\infty}\left[\left(1+\frac{1}{n}\right)^n\right]^{-1}$$

$$= e^{-1} = \frac{1}{e} < 1$$

Hence, by Cauchy's root test the given series is convergent.

(iv) We have

$$u_n = \frac{1}{(\log n)^n}$$

∴ $$u_n^{1/n} = \frac{1}{(\log n)}$$

∴ $$\lim_{n\to\infty} u_n^{1/n} = \lim_{n\to\infty} \frac{1}{(\log n)} = 0 < 1$$

Hence, by Cauchy's root test the given series is convergent.

Example 10. *Test for convergence the series* $1+\frac{2^P}{2!}+\frac{3^P}{3!}+\frac{4^P}{4!}+...$

Solution. Here, we have

$$u_n = \frac{n^P}{n!} \Rightarrow u_{n+1} = \frac{(n+1)^P}{(n+1)!}$$

Now $$\lim_{n\to\infty} \frac{u_{n+1}}{u_n} = \lim_{n\to\infty} \frac{(n+1)^P}{(n+1)!} \frac{n!}{n^P} = \lim_{n\to\infty}\left[1+\frac{1}{n}\right]^P \cdot \frac{1}{(n+1)}$$

$$= e^P \times 0 = 0 < 1$$

Hence, by ratio test, the given series is divergent.

Example 11. *Test the series* $x+\frac{x^3}{3!}+\frac{x^5}{5!}+\frac{x^7}{7!}+....$ *for convergence, for all positive value of x.*

Solution. Since x is positive. Hence the given series is of positive term series

Here

$$u_n = \frac{x^{2n-1}}{(2n-1)!}, u_{n+1} = \frac{x^{2n+1}}{(2n+1)!}$$

$$\Rightarrow \qquad \lim_{n \to \infty} \frac{u_n}{u_{n+1}} = \lim_{n \to \infty} \frac{x^{2n-1}}{(2n-1)!} \frac{(2n+1)!}{x^{2n+1}} = \lim_{n \to \infty} \frac{2n(2n+1)}{x^2}$$

$$= +\infty, \ \forall \ \text{positive value of } x.$$

Then, by ratio test the given series converges for all positive value of x.

Example 12. *Test for convergence the series* $1 + \dfrac{x}{2^2} + \dfrac{x^2}{3^2} + \dfrac{x^3}{4^2} + \dots$

Solution. Here we have
$$u_n = \frac{x^{n-1}}{n^2}$$

$$\Rightarrow \qquad u_{n+1} = \frac{x^n}{(n+1)^2}$$

Now
$$\frac{u_n}{u_{n+1}} = \frac{x^{n-1}(n+1)^2}{n^2 . x^n} = \frac{1}{x} . \left(1 + \frac{1}{n}\right)^2$$

$$\Rightarrow \qquad \lim_{n \to \infty} \frac{u_n}{u_{n+1}} = \lim_{n \to \infty} \frac{1}{x}\left(1 + \frac{1}{n}\right)^2 = \frac{1}{x}.$$

Hence, by ratio test the series converges if $\dfrac{1}{x} > 1$ *i.e.*, $x < 1$, diverges if $x > 1$ and the test fails if $x = 1$.

For $x = 1$, $u_n = \dfrac{1}{n^2}$. Therefore in the case the series $\Sigma u_n = \Sigma \dfrac{1}{n^2}$ is convergent.

Example 13. *Test for convergence the series* $\dfrac{1}{2\sqrt{1}} + \dfrac{x^2}{3\sqrt{2}} + \dfrac{x^4}{4\sqrt{3}} + \dots$

Solution.
$$u_n = \frac{x^{2n-2}}{(n+1)\sqrt{n}}, \ u_{n+1} = \frac{x^{2n}}{(n+2)\sqrt{(n+1)}}$$

$$\Rightarrow \qquad \frac{u_n}{u_{n+1}} = \frac{x^{2n-2}}{(n+1)\sqrt{n}} . \frac{(n+2)\sqrt{(n+1)}}{x^{2n}}$$

$$= \frac{(1+2/n)}{(1+1/n)} \sqrt{\left(1 + \frac{1}{n}\right)} . \frac{1}{x^2}$$

$$\Rightarrow \qquad \lim_{n \to \infty} \frac{u_n}{u_{n+1}} = \frac{1}{1} . \sqrt{1} . \frac{1}{x^2} = \frac{1}{x^2}$$

Therefore, by ratio test the given series Σu_n is

(i) convergent if $\dfrac{1}{x^2} > 1$ *i.e.*, if $x^2 < 1$.

(ii) divergent if $\dfrac{1}{x^2} < 1$ *i.e.*, if $x^2 > 1$.

and (iii) The test fails if $x^2 = 1$

When $x^2 = 1$, we have $u_n = \dfrac{1}{(n+1)\sqrt{n}}$. Take $v_n = \dfrac{1}{n\sqrt{n}}$. Then $\lim\limits_{n \to \infty} \dfrac{u_n}{v_n} = 1$,

which is finite and non-zero. Hence, by comparison test Σu_n and Σv_n are either both convergent or both divergent.

Since $\Sigma v_n = \Sigma \dfrac{1}{n^{3/2}}$ is convergent as $p = 3/2 > 1$.

Hence, the given series Σu_n is also convergent if $x^2 = 1$.

Example 14. *Test for convergence the series*

$$x + \frac{3}{5}x^2 + \frac{8}{10}x^3 + \frac{15}{17}x^4 + ... + \frac{n^2-1}{n^2+1}x^n + ...$$

Solution. Here, we have

$$u_n = \frac{n^2-1}{n^2+1}x^n, \qquad u_{n+1} = \frac{(n+1)^2-1}{(n+1)^2+1}x^{n+1}$$

$\Rightarrow$
$$\frac{u_n}{u_{n+1}} = \frac{n^2-1}{n^2+1}x^n \cdot \frac{(n+1)^2+1}{(n+1)^2-1} \cdot \frac{1}{x^{n+1}}$$

$$= \frac{1-1/n^2}{1+1/n^2} \cdot \frac{1+2/n+2/n^2}{1+2/n} \cdot \frac{1}{x}$$

$\Rightarrow$
$$\lim_{n\to\infty} \frac{u_n}{u_{n+1}} = \frac{1}{x}$$

Therefore, by ratio test the given series Σu_n is

(i) convergent if $\dfrac{1}{x} > 1$ *i.e.*, if $x < 1$.

(ii) divergent if $\dfrac{1}{x} < 1$ *i.e.*, if $x > 1$.

and (iii) test fails if $x = 1$.

When $x=1$, $u_n = \dfrac{1-1/n^2}{1+1/n^2} \Rightarrow \lim_{n\to\infty} u_n = 1 > 0.$

The given series Σu_n is divergent if $x=1$.

Hence, the given series is convergent if $x < 1$ and divergent if $x \geq 1$.

Example 15. *Test the convergence of the series*

$$\sum_{n=1}^{\infty} \frac{1\cdot 3\cdot 5.... (2n-1)}{2\cdot 4\cdot 6.... (2n)}(1-x^2)^n, \quad 0 \leq x^2 < 1$$

Solution. The nth term of the series is

$$u_n = \frac{1\cdot 3\cdot 5.... (2n-1)}{2\cdot 4\cdot 6.... (2n)}(1-x^2)^n$$

Then
$$u_{n+1} = \frac{1\cdot 3\cdot 5.... (2n-1)\cdot (2n+1)}{2\cdot 4\cdot 6.... (2n)(2n+2)}(1-x^2)^{n+1}$$

$\therefore$
$$\frac{u_n}{u_{n+1}} = \frac{2n+2}{2n+1} \cdot \frac{1}{(1-x^2)}$$

or
$$\frac{u_n}{u_{n+1}} = \frac{\left(1+\dfrac{1}{n}\right)}{\dfrac{1}{2n}} \cdot \frac{1}{(1-x^2)}$$

$$\therefore \qquad \lim_{n \to \infty} \frac{u_n}{u_{n+1}} = \lim_{n \to \infty} \frac{\left(1 + \dfrac{1}{n}\right)}{\left(1 + \dfrac{1}{2n}\right)} \cdot \frac{1}{(1 - x^2)} = \frac{1}{1 - x^2}$$

$$\therefore \qquad \lim_{n \to \infty} \frac{u_n}{u_{n+1}} = \frac{1}{1 - x^2} > 1 \qquad \qquad \because x^2 < 1$$

Hence by D'Alembert Ratio test the given series is convergent.

Example 16. *Test the following series for convergence*

(i) $\quad 1 + \dfrac{2}{5}x + \dfrac{6}{9}x^2 + \dfrac{14}{17}x^3 + \ldots + \dfrac{2^n - 2}{2^n + 1}x^{n-2} + \ldots$

(ii) $\quad \dfrac{x^2}{2\sqrt{1}} + \dfrac{x^3}{3\sqrt{2}} + \dfrac{x^4}{4\sqrt{3}} + \ldots$

Solution.   (i)   Here, we have

$$u_n = \frac{2^n - 2}{2^n + 1}x^{n-2} \quad \Rightarrow \quad u_{n+1} = \frac{2^{n+1} - 2}{2^{n+1} + 1}x^{n-1}$$

$$\therefore \qquad \frac{u_n}{u_{n+1}} = \frac{2^{n+1} + 1}{2^n + 1} \cdot \frac{2^n - 2}{2^{n+1} - 2} \cdot \frac{1}{x}$$

$$= \frac{\left(2 + \left(\dfrac{1}{2}\right)^n\right)}{1 + \left(\dfrac{1}{2}\right)^n} \cdot \frac{\left(1 - \dfrac{2}{2^n}\right)}{\left(2 - \dfrac{2}{2^n}\right)} \cdot \frac{1}{x}$$

so, $$\lim_{n \to \infty} \frac{u_n}{u_{n+1}} = \lim_{n \to \infty} \left[\frac{\left(2 + \left(\dfrac{1}{2}\right)^n\right)}{\left(1 + \left(\dfrac{1}{2}\right)^n\right)} \cdot \frac{\left(1 - \dfrac{2}{2^n}\right)}{\left(2 - \dfrac{2}{2^n}\right)} \cdot \frac{1}{x} \right]$$

$$= \frac{2 + 0}{1 + 0} \cdot \frac{1 - 0}{2 - 0} \cdot \frac{1}{x} = \frac{1}{x}$$

Therefore, by D'Alembert's ratio test the given series is convergent if $x < 1$ and is divergent if $x > 1$.

When $x = 1$, we have

$$u_n = \frac{2^n - 2}{2^n + 1} = \frac{1 - \left(\dfrac{1}{2}\right)^{n-1}}{1 + \left(\dfrac{1}{2}\right)^n}$$

$$\therefore \qquad \lim_{n \to \infty} u_n = \lim_{n \to \infty} \frac{1 - \left(\dfrac{1}{2}\right)^{n-1}}{1 + \left(\dfrac{1}{2}\right)^n} = \frac{1 - 0}{1 + 0} = 1 \neq 0$$

$\Rightarrow \qquad \Sigma u_n$ is divergent.

Hence the given series is convergent if $x < 1$ and is divergent if $x \geq 1$.

(ii) The nth term of the given series is

$$u_n = \frac{x^{n+1}}{(n+1)\sqrt{n}} \qquad \Rightarrow \qquad u_{n+1} = \frac{x^{n+2}}{(n+2)\sqrt{n+1}}$$

$$\therefore \qquad \frac{u_n}{u_{n+1}} = \frac{n+2}{n+1} \cdot \sqrt{\frac{n+1}{n}} \cdot \frac{1}{x} = \left(\frac{1+\dfrac{2}{n}}{1+\dfrac{1}{n}}\right) \cdot \sqrt{1+\frac{1}{n}} \cdot \frac{1}{x}$$

$$\therefore \qquad \lim_{n\to\infty} \frac{u_n}{u_{n+1}} = \lim_{n\to\infty} \left(\frac{1+\dfrac{2}{n}}{1+\dfrac{1}{n}}\right) \cdot \sqrt{1+\frac{1}{n}} \cdot \frac{1}{x} = \frac{1}{x}$$

Therefore, by D'Alembert ratio test the given series is convergent if $x < 1$ and is divergent if $x > 1$.

When $x = 1$, we have $u_n = \dfrac{1}{(n+1)\sqrt{n}}$

Then $\qquad v_n = \dfrac{1}{n\sqrt{n}}$

$$\therefore \qquad \lim_{n\to\infty} \frac{u_n}{v_n} = \lim_{n\to\infty} \frac{n\sqrt{n}}{(n+1)\sqrt{n}} = \lim_{n\to\infty} \frac{1}{\left(1+\dfrac{1}{n}\right)} = \frac{1}{1+0} = 1 \neq 0$$

$\Rightarrow$ By comparison test, $\Sigma v_n = \Sigma \dfrac{1}{n^{3/2}}$ is convergent, then Σu_n is convergent.

Hence, the given series is convergent if $x \leq 1$ and is divergent if $x > 1$.

EXERCIE 3.2

Based on Cauchy's Root Test:

1. Test the convergence of the following series :

(i) $\displaystyle\sum_{n=1}^{\infty} \left(1+\frac{2}{n}\right)^{-n^2}$

(ii) $\displaystyle\sum_{n=1}^{\infty} \frac{n^{n^2}}{(n+1)^{n^2}}$

(iii) $\displaystyle\sum_{n=1}^{\infty} 2^{-n-(-1)^n}$

(iv) $\displaystyle\sum_{n=1}^{\infty} 5^{-n-(-1)^n}$

(v) $\displaystyle\sum_{n=1}^{\infty} (n^{1/n} + x)$ for all positive values of x

(vi) $\displaystyle\sum_{n=1}^{\infty} \frac{n^3}{3^n}$

(vii) $\displaystyle\sum_{n=1}^{\infty} \frac{x^n}{n^n}$, $x > 0$

2. Test the convergence of the following series :

(i) $\displaystyle\sum \left(\frac{n}{n+1}\right)^{n^2}$

(ii) $\displaystyle\sum n^n x^n$, $x > 0$

(iii) $\displaystyle\sum \left(\frac{n+1}{3n}\right)^n$

(iv) $\displaystyle\sum \left(\frac{nx}{n+1}\right)^n$

(v) $\displaystyle\sum \frac{(1+nx)^n}{n^n}$

(vi) $\displaystyle\sum (n^{1/n} - 1)^n$

3. Test the convergence of the following series :

(i) $\dfrac{1^3}{3} + \dfrac{2^3}{3^2} + \dfrac{3^3}{3^3} + \dfrac{4^3}{3^4} + \dots$

(ii) $\dfrac{2}{1^2} x + \dfrac{3^2}{2^3} x^2 + \dfrac{4^3}{3^4} x^3 + \dots + \dfrac{(n+1)^n x^n}{n^{n+1}}$

$\qquad\qquad + \dots$ if $x > 0$

(iii) $\displaystyle\sum q^{n^2} r^2$, $q, r > 0$ (iv) $\displaystyle\sum_{n=2}^{\infty} \frac{1}{[\log(\log n)]^n}$

Based on D'Alembert's Ratio test

4. Test the convergence of the following series :

(i) $\displaystyle\sum_{n=1}^{\infty} \frac{2^{n-1}}{3^n + 1}$

(ii) $\displaystyle\sum_{n=1}^{\infty} \frac{n!}{n^n}$

(iii) $\sum_{n=1}^{\infty} \dfrac{x^n}{n!}, x > 0$ (iv) $\sum_{n=1}^{\infty} \dfrac{x^n}{n^n}, x > 0$

(v) $\sum_{n=1}^{\infty} \dfrac{2^n n!}{n^n}$ (vi) $\sum_{n=1}^{\infty} \dfrac{n^n x^n}{n!}$

(vii) $\sum_{n=1}^{\infty} \dfrac{5^n}{n^2 + 5}$ (viii) $\sum_{n=1}^{\infty} \dfrac{n^3 + a}{2^n + a}$

(ix) $\sum_{n=1}^{\infty} \dfrac{\sqrt{n}}{\sqrt{n^2 + 1}} x^n, x > 0$

(x) $\sum_{n=1}^{\infty} \sqrt{\dfrac{n-1}{n^3 + 1}} x^n, x > 0$

5. Test the convergence of the series with n^{th} term :

(i) $\dfrac{1}{x^n + x^{-n}}$ (ii) $\left[\sqrt{n^2 + 1} - n\right] x^{2n}$

(iii) $\dfrac{1}{2^n + x}, x \geq 0$ (iv) $\dfrac{x^n}{n^2 + 1}$

(v) $\dfrac{a^n}{x^n + a^n}$ (vi) $\sqrt{\dfrac{2^n - 1}{3^n - 1}}$

6. Test the convergence of the following series :

(i) $\dfrac{2!}{3} + \dfrac{3!}{3^2} + \dfrac{4!}{3^3} + \dots + \dfrac{(n+1)!}{3^n} + \dots$

(ii) $\dfrac{1^2 \cdot 2^2}{1!} + \dfrac{2^2 \cdot 3^2}{2!} + \dfrac{3^2 \cdot 4^2}{3!} + \dots$

(iii) $\dfrac{1}{1+2} + \dfrac{2}{1+2^2} + \dfrac{3}{1+2^3} + \dots$

(iv) $1 + 3x + 5x^2 + 7x^3 + \dots$

(v) $1 + \dfrac{x}{2^2} + \dfrac{x^2}{3^2} + \dfrac{x^3}{4^2} + \dots$

(vi) $2x + \dfrac{3x^2}{8} + \dfrac{4x^3}{27} + \dots + \dfrac{(n+1)x^n}{n^3} + \dots$

(vii) $\dfrac{1}{\sqrt{1} + \sqrt{2}} + \dfrac{1}{\sqrt{2} + \sqrt{3}} + \dfrac{1}{\sqrt{3} + \sqrt{4}} + \dots$

(viii) $\dfrac{\sqrt{2} - 1}{3^3 - 1} + \dfrac{\sqrt{3} - 1}{4^3 - 1} + \dfrac{\sqrt{4} - 1}{5^3 - 1} + \dots$

(ix) $\dfrac{1}{2} + \dfrac{2!}{8} + \dfrac{3!}{32} + \dfrac{4!}{128} + \dots$

(ix) $1 + \dfrac{1}{2 \cdot 2^{1/100}} + \dfrac{1}{3 \cdot 3^{1/100}} + \dfrac{1}{4 \cdot 4^{1/100}} + \dots$

7. Test for convergence the following series :

(i) $\dfrac{1}{2 \cdot 3} + \dfrac{1}{3 \cdot 4} + \dfrac{1}{4 \cdot 5} + \dfrac{1}{5 \cdot 6} + \dots$

(ii) $\dfrac{1}{1 \cdot 2 \cdot 3} + \dfrac{3}{2 \cdot 3 \cdot 4} + \dfrac{5}{3 \cdot 4 \cdot 5} + \dots$

(iii) $\dfrac{1 \cdot 2}{3^2 \cdot 4^2} + \dfrac{3 \cdot 4}{5^2 \cdot 6^2} + \dfrac{5 \cdot 6}{7^2 \cdot 8^2} + \dots$

(iv) $\dfrac{1}{3} + \dfrac{1 \cdot 2}{3 \cdot 5} + \dfrac{1 \cdot 2 \cdot 3}{3 \cdot 5 \cdot 7} + \dfrac{1 \cdot 2 \cdot 3 \cdot 4}{3 \cdot 5 \cdot 7 \cdot 9} + \dots$

8. Test the series :

$$1 + \dfrac{x^2}{2} + \dfrac{x^4}{4} + \dfrac{x^6}{6} + \dots$$

for convergence for all positive values of x.

9. Test for convergence the series :

$$\dfrac{x}{1 \cdot 2} + \dfrac{x^2}{2 \cdot 3} + \dfrac{x^3}{3 \cdot 4} + \dfrac{x^4}{4 \cdot 5} + \dots, x > 0$$

10. Show that the series ($\alpha > 0, \beta > 0$)

$$1 + \dfrac{\alpha + 1}{\beta + 1} + \dfrac{(\alpha + 1)(2\alpha + 1)}{(\beta + 1)(2\beta + 1)} +$$

$$+ \dfrac{(\alpha + 1)(2\alpha + 1)(3\alpha + 1)}{(\beta + 1)(2\beta + 1)(3\beta + 1)} + \dots$$

converges if $\beta > \alpha > 0$
and diverges if $\alpha \geq \beta > 0$

11. Test for convergence the series :

$$\dfrac{x}{1 \cdot 3} + \dfrac{x^2}{2 \cdot 4} + \dfrac{x^3}{3 \cdot 5} + \dfrac{x^4}{4 \cdot 6} + \dots$$

12. Test for convergence the following series :

(i) $1 + \dfrac{x}{2} + \dfrac{x^2}{3^2} + \dfrac{x^3}{4^3} + \dots, x > 0$

(ii) $x + 2x^2 + 3x^3 + 4x^4 + \dots$

(iii) $2 + \dfrac{3}{2}x + \dfrac{4}{3}x^2 + \dfrac{5}{4}x^3 + \dots, x > 0$

(iv) $\dfrac{(1+a)(1+b)}{1 \cdot 2 \cdot 3} + \dfrac{(2+a)(2+b)}{2 \cdot 3 \cdot 4}$

$$+ \dfrac{(3+a)(3+b)}{3 \cdot 4 \cdot 5} + \dots$$

(v) $x \log x + x^2 \log 2x + x^3 \log 3x + \dots$

$$+ x^n \log nx + \dots$$

12. Test for convergence the series with n^{th} term :

(i) $\dfrac{n^3-1}{n^3+1}x^n$, $x>0$ (ii) $\dfrac{x^n}{a+\sqrt{n}}$ (v) $\dfrac{x^n}{(2n+1)^p}$ (vi) $\dfrac{3^n-2}{3^n+1}x^{n-1}$, $x>0$

(iii) $\dfrac{x^n}{x+n}$ (iv) $\dfrac{3n+1}{4n+3}x^n$, $x>0$ (vii) $\dfrac{1}{n}\sin\dfrac{1}{n}$

Answers

1. (i) Convergent (ii) Convergent (iii) Convergent (iv) Convergent (v) Divergent (vi) Convergent
   (vii) Convergent
2. (i) Convergent (ii) Divergent (iii) Convergent (iv) Convergent if $x<1$, divergent if $x\geq 1$
   (v) Convergent if $x<1$, divergent if $x\geq 1$ (vi) Convergent
3. (i) Convergent (ii) Convergent if $x<1$ and divergent if $x\geq 1$ (iii) Convergent if $0<q<1$ and divergent if $q>1$, Convergent if $0<r<1$, when $q=1$, divergent if $q>1$ or $q=1, r\geq 1$
   (iv) Convergent
4. (i) Convergent (ii) Convergent (iii) Convergent (iv) Convergent (v) Convergent
   (vi) Convergent if $x<1$, divergent if $x\geq 1$ (vii) Divergent (viii) Convergent
   (ix) Convergent if $x<1$, divergent if $x\geq 1$ (x) Convergent if $x<1$, divergent if $x\geq 1$
5. (i) Convergent if $x>1$ or $x<1$, and divergent if $x=1$ (ii) Convergent if $x<1$, divergent if $x\geq 1$
   (iii) Convergent (iv) Convergent if $x\leq 1$, divergent if $x>1$ (v) Convergent if $x>a$, divergent if $x\leq a$, (vi) Convergent.
6. (i) Divergent (ii) Convergent (iii) Convergent (iv) Convergent if $x<1$, divergent if $x\geq 1$
   (v) Convergent if $x\leq 1$, divergent if $x>1$ (vi) Convergent if $x\leq 1$, divergent if $x>1$
   (vii) Divergent (viii) Convergent (ix) Divergent (x) Convergent
7. (i) Convergent (ii) Convergent (iii) Convergent (iv) Convergent
8. Convergent if $x<1$, divergent if $x\geq 1$
9. Convergent if $x\leq 1$, divergent if $x>1$
11. Convergent if $x\leq 1$, divergent $x>1$
12. (i) Convergent (ii) Convergent if $x<1$, divergent if $x\geq 1$ (iii) Convergent if $x<1$, divergent if $x\geq 1$ (iv) Divergent (v) Convergent if $x<1$, divergent if $x\geq 1$.
13. (i) Convergent if $x<1$, divergent if $x\geq 1$ (ii) Convergent if $x<1$, divergent if $x\geq 1$
   (iii) Convergent if $x<1$, divergent if $x\geq 1$ (iv) Convergent if $x<1$, divergent if $x\geq 1$
   (v) Convergent if $x<1$, divergent if $x>1$, when $x=1$, then convergent if $p>1$ and divergent if $p\leq 1$ (vii) Convergent if $x<1$, divergent if $x\geq 1$ (vii) Convergent.

3.9 RAABE'S TEST

If $\Sigma\, u_n$ be a series of positive terms such that

$$\lim_{n\to\infty}\left\{n\left(\dfrac{u_n}{u_{n+1}}-1\right)\right\}=l.$$

Then, if
(i) $l>1$, the series converges,
(ii) $l<1$, the series diverges,
(iii) $l=1$, the series may either converge or diverge and therefore the test fails.

Proof. Case (I) When $l > 1$. We can write $l = 1 + r$, where $r > 0$. Choosing $\varepsilon = r/2$, we can find a positive integer m such that

$$l - \varepsilon < n\left(\frac{u_n}{u_{n+1}} - 1\right) < l + \varepsilon \ \forall \ n \ge m$$

Now, from the first part of the above inequality, we have

$$(1+r) - \frac{1}{2}r < n\left(\frac{u_n}{u_{n+1}} - 1\right) \forall \ n \ge m$$

$\Rightarrow \qquad \frac{1}{2}ru_{n+1} < nu_n - (n+1)u_{n+1} \ \forall n \ge m$ \hfill ...(1)

Putting $n = m+1, m+2, ..., p - 1$ in succession in (1), we have

$$\frac{1}{2}ru_{m+2} < (m+1)u_{m+1} - (m+2)u_{m+2}$$

$$... \qquad ... \qquad ... \qquad ... \qquad ...$$

$$\frac{1}{2}ru_p < (p-1)u_{p-1} - pu_p.$$

Now, adding the corresponding sides of the above inqualities, we have

$$\frac{1}{2}r[u_{m+2} + u_{m+3} + ... + u_p] < (m+1)u_{m+1} - pu_p,$$

$\Rightarrow \qquad \frac{1}{2}r[u_{m+2} + ... + u_p] < (m+1)u_{m+1},$

or $\qquad u_1 + u_2 + ... + u_p < \dfrac{2(m+1)}{r}u_{m+1} < u_1 + u_2 + ... + u_{m+1}, \ \forall \ p \ge m+2.$

The above inequality shows that the sequence $\langle s_n \rangle$ of the partial sums of the series Σu_n is bounded and therefore Σu_n converges.

Case (II) When $l < 1$. Let us choose $\varepsilon = 1 - l$, then we can find a positive integer m such that

$$l - \varepsilon < n\left(\frac{u_n}{u_{n+1}} - 1\right) < 1 (= l + \varepsilon) \forall \ n \ge m$$

or $\qquad mu_n < (n+1)\ u_{n+1} \ \forall \ n \ge m$

Putting $n = m+1, m+2,..., p-1 \ (p \ge m+2)$, in succession, we get

$$(m+1)u_{m+1} < (m+2)u_{m+2},$$

$$(m+1)u_{m+1} < (m+3)u_{m+3},$$

$$... \quad .... \quad ... \quad ... \quad ...$$

$$(p-1)\ u_{p-1} < pu_p.$$

From the above inequality, we have by transitivity

$$(m+1)u_{m+1} < pu_p \ \forall \ p \ge m+2$$

or $\qquad u_p > k(1/p) \ \forall \ p \ge m+2$ and $k = (m+1)u_{m+1}$.

Now, since the series $\Sigma\left(\dfrac{1}{p}\right)$ diverges, then by comparison test the given series diverges.

Case (III) When $l = 1$. In this case the test fails to give any definite information. For example, consider the series $\Sigma\dfrac{1}{n}$ and $\Sigma\dfrac{1}{n(\log n)^2}$ then, we have

$$\lim_{n\to\infty} n\left[\frac{u_n}{u_{n+1}} - 1\right] = 1.$$

But the former sereis is divergent, while the latter is convergent.

REMARKS

- Raabe's test is to be applied when D'Alembert's ratio test fails.
- Raabe's test is stronger than D'Alembert ratio test.
- It can be shown as in the proof of case (I) above that if

$$\lim_{n\to\infty}\left\{n\left[\frac{u_n}{u_{n+1}} - 1\right]\right\} = +\infty. \text{ , Then, } \Sigma u_n \text{ converges.}$$

- The case in which

$$\lim_{n\to\infty}\left\{n\left[\frac{u_n}{u_{n+1}} - 1\right]\right\} = -\infty.$$

the given series Σu_n diverges.

3.10 LOGARITHMIC TEST

If Σu_n be a series of positive terms such that.

$$\lim_{n\to\infty}\left(n\log\frac{u_n}{u_{n+1}}\right) = l.$$

then Σu_n converges if $l > 1$ and diverges when $l < 1$.

Proof. Case (I) When $l > 1$. In this case, we can choose $\varepsilon > 0$ such that $l - \varepsilon > 1$. Let $l-\varepsilon = p$ (say).

Since $\lim\limits_{n\to\infty}\left(n\log\dfrac{u_n}{u_{n+1}}\right) = l.$ Therefore, we can find a positive integer m such that

$$l - \varepsilon < n\log\frac{u_n}{u_{n+1}} < l + \varepsilon \ \forall \ n \ge m.$$

Consider the first part of the above inequality, we have

$$n\log\frac{u_n}{u_{n+1}} > p \ \ \forall \ n \ge m.$$

$\Rightarrow$
$$\frac{u_n}{u_{n+1}} > e^{p/n} \ \ \forall \ n \ge m.$$

...(1)

Since, $a_n = \left(1 + \dfrac{1}{n}\right)^n$ defines a monotonically increasing sequence converging to e, therefore,

$$e \ge \left(1 + \frac{1}{n}\right)^n \ \forall \ n.$$

...(2)

From (1) and (2), we have

$$\frac{u_n}{u_{n+1}} > \left(1 + \frac{1}{n}\right)^p \ \forall \ n \ge m.$$

$\Rightarrow$
$$\frac{u_n}{u_{n+1}} > \frac{v_n}{v_{n+1}} \ \forall \ n \ge m.$$

...(3)

where
$$v_n = \frac{1}{n^p}.$$

Now since $p > 1$, therefore Σv_n converges and from (3) it then follows by comparison test that Σu_n converges.

Case (II) When $l < 1$. Let the comparison series $\Sigma v_n = \Sigma \dfrac{1}{n^p}$ be divergent, *i.e.*, $p < 1$.

$\therefore \quad \Sigma u_n$ will be divergent if $\dfrac{v_n}{v_{n+1}} > \dfrac{u_n}{u_{n+1}}$

$$\Rightarrow \qquad \frac{u_n}{u_{n+1}} < \left(1 + \frac{1}{n}\right)^p \Rightarrow \log\left(\frac{u_n}{u_{n+1}}\right) < p \log\left(1 + \frac{1}{n}\right)$$

$$= p\left[\frac{1}{n} - \frac{1}{2n^2} + \frac{1}{3n^3} + \ldots\right]$$

$$\therefore \qquad n\log\left(\frac{u_n}{u_{n+1}}\right) = p\left[1 - \frac{1}{2n} + \frac{1}{3n^2} + \ldots\right]$$

$$\therefore \qquad \lim_{n\to\infty}\left[n\log\frac{u_n}{u_{n+1}}\right] = p < 1$$

$\therefore \quad \Sigma u_n$ will be divergent if $l < 1$.

REMARKS

- Logarithmic test is to be applied only when :
  (a) ratio test fails
  (b) the ratio test involves the exponent 'e'
- This test is an alternative to Raabe's test.

3.11 SOME MODIFIED FORMS

Various test of convergence, involving limits can be modified in terms of the upper and lower limits. For example, a few modification are given below :

(i) Cauchy's Root test.

The series of non-negative term Σu_n converges or diverges according as

$$\underline{\lim}\, u_n^{1/n} < 1 \qquad \text{or} \qquad \overline{\lim}\, u_n^{1/n} > 1.$$

(ii) D' Alembert's Ratio test.

The series Σu_n of positive terms converges or diverges according as

$$\underline{\lim}\, \frac{u_n}{u_{n+1}} > 1 \qquad \text{or} \qquad \overline{\lim}\, \frac{u_n}{u_{n+1}} < 1.$$

(iii) Raabe's test.

The series Σu_n of positive terms converges or diverges according as

$$\underline{\lim}\left\{n\left(\frac{u_n}{u_{n+1}} - 1\right)\right\} > 1 \qquad \text{or} \qquad \overline{\lim}\left\{n\log\frac{u_n}{u_{n+1}}\right\} < 1.$$

(iv) Logarithmic test.

The series Σu_n of positive terms converges or diverges according as

$$\underline{\lim}\left\{n\log\frac{u_n}{u_{n+1}}\right\} > 1 \qquad \text{or} \qquad \overline{\lim}\left\{n\log\frac{u_n}{u_{n+1}}\right\} < 1.$$

3.11.1 SOME IMPORTANT LIMITS

(i) $\lim\limits_{n\to\infty}\left(1+\dfrac{x}{n}\right)^n = e^x$

(ii) $\lim\limits_{n\to\infty} n^{1/n} = 1$

(iii) $\lim\limits_{n\to\infty}\dfrac{\log n}{n} = 0$

(iv) $\lim\limits_{n\to\infty}\left(1+\dfrac{x}{n}\right)^p = 1$, if p is finite

(v) $\lim\limits_{n\to\infty}\left(1+\dfrac{x}{n}\right)^{n+p} = e^x$, if p is finite.

3.11.2 SOME OTHER IMPORTANT TESTS

(1) **De Morgan's and Bertrand's tet :** The series Σu_n of positive terms is convergent or divergent according as

$$\lim\left[n\left(\frac{u_n}{u_{n+1}}-1\right)-1\right]\log n > 1 \quad \text{or} \quad < 1.$$

(2) **Alternative to Bertrand's tet :** The series Σu_n of positive terms is convergent or divergent according as

$$\lim\left[\left(n\log\frac{u_n}{u_{n+1}}-1\right)\log n\right] > 1 \quad \text{or} \quad < 1.$$

Solved Examples

Example 1. *Test the convergence of the series*

$$1+\frac{3}{7}x+\frac{3\cdot6}{7\cdot10}x^2+\frac{3\cdot6\cdot9}{7\cdot10\cdot13}x^3+...$$

Solution. After leaving the first term we have

$$u_n = \frac{3\cdot6\cdot9\cdot...\cdot3n}{7\cdot10\cdot13\cdot...\cdot(3n+4)}x^n$$

$$\Rightarrow \qquad u_{n+1} = \frac{3\cdot6\cdot9\cdot...\cdot3n(3n+3)}{7\cdot10\cdot13\cdot...\cdot(3n+4)(3n+7)}x^{n+1}$$

Now $\lim\limits_{n\to\infty}\dfrac{u_{n+1}}{u_n} = \lim\limits_{n\to\infty}\left(\dfrac{3n+3}{3n+7}\right)x = \lim\limits_{n\to\infty}\left(\dfrac{3+3/n}{3+7/n}\right)x = x$

Then, by D'Alembert ratio test the series is convergent if $x < 1$, divergent if $x > 1$ and the test fails if $x = 1$.

For $x = 1$, we have

$$\frac{u_n}{u_{n+1}} = \frac{3n+7}{3n+3}$$

or
$$n\left(\frac{u_n}{u_{n+1}} - 1\right) = n\left(\frac{3n+7}{3n+3} - 1\right) = \frac{4n}{3n+3}$$

$$\Rightarrow \lim_{n\to\infty} n\left[\left(\frac{u_n}{u_{n+1}} - 1\right)\right] = \lim_{n\to\infty} \frac{4n}{3n+3} = \lim_{n\to\infty} \frac{4}{3+3/n} = \frac{4}{3} > 1$$

Therefore, by Raabe's test the series is convergent when $x = 1$.

Hence, the given series is convergent when $x \le 1$ and divergent when $x > 1$.

Example 2. *Test the convergence of the following series*

$$\sum_{n=1}^{\infty} \frac{1.3.5....(2n-1)}{2.4.6....(2n)} \cdot \frac{x^{2n}}{2n}, (x > 0).$$

Solution. Here, we have

$$u_n = \frac{1.3.5....(2n-1)}{2.4.6....(2n)} \cdot \frac{x^{2n}}{2n}$$

and
$$u_{n+1} = \frac{1.3.5....(2n-1)(2n+1)}{2.4.6....(2n)(2n+2)} \cdot \frac{x^{2n+2}}{(2n+2)}$$

$$\Rightarrow \lim_{n\to\infty} \frac{u_n}{u_{n+1}} = \lim_{n\to\infty}\left(\frac{2n+2}{2n+1} \cdot \frac{2n+2}{2n} \cdot \frac{1}{x^2}\right) = \frac{1}{x^2}.$$

∴ By D'Alembert's ratio test, the series is convergent if $x^2 < 1$ and divergent if $x^2 > 1$.

Now since $x > 0$ this gives that the series is convergent if $x < 1$ and divergent if $x > 1$.

If $x = 1$. Then D'Alembert's ratio test fails.

Now consider
$$\lim_{n\to\infty} n\left[\frac{u_n}{u_{n+1}} - 1\right] = \lim_{n\to\infty} n\left(\frac{2n+2}{2n+1} \cdot \frac{2n+2}{2n} - 1\right)$$

$$= \lim_{n\to\infty} \frac{n(6n+4)}{2n(2n+1)} = \frac{3}{2} > 1.$$

Then by Raabe's test, the series is convergent for $x = 1$.

Hence, the series is convergent if $x \le 1$ and divergent if $x > 1$.

Example 3. *Test the convergence of the series* $\frac{a}{b} + \frac{(1+a)}{(1+b)} + \frac{(1+a)(2+a)}{(1+b)(2+b)} + ...$

Solution. Here, we have

$$u_n = \frac{(1+a)(2+a)...(n-1+a)}{(1+b)(2+b)...(n-1+b)}$$

$$\Rightarrow u_{n+1} = \frac{(1+a)(2+a)...(n+a)}{(1+b)(2+b)...(n+b)}$$

∴
$$\lim_{n\to\infty} \frac{u_n}{u_{n+1}} = \lim_{n\to\infty}\left[\frac{n+b}{n+a}\right] = \lim_{n\to\infty}\left[\frac{1+\frac{b}{n}}{1+\frac{a}{n}}\right] = 1.$$

Hence, the D'Alembert's ratio test fails.

Now, consider

$$\lim_{n\to\infty} n\left[\frac{u_n}{u_{n+1}} - 1\right] = \lim_{n\to\infty} n\left[\frac{n+b}{n+a} - 1\right]$$

$$= \lim_{n\to\infty} n\left[\frac{b-a}{n+b}\right] = \lim_{n\to\infty}\left[\frac{b-a}{1+b/n}\right]$$

$$= (b-a).$$

Then by Raabe's test the given series is convergent if $b - a > 1$, *i.e.*, $b > a + 1$ and divergent if $b < a + 1$.

The test fails for $b = a + 1$.

Now, for $b = a + 1$, the given series becomes

$$\frac{a}{a+1} + \frac{1+a}{2+a} + ... = \Sigma\frac{1+a}{n+a}.$$

Taking $v_n = \frac{1}{n}$, by comparison test, we can easily shown that the series is divergent. Hence, the given series is convergent if $b > a + 1$ and divergent if $b \le a + 1$.

Example 4. *Test the convergence of the series*

$$1 + a + \frac{a(a+1)}{1\cdot 2} + \frac{a(a+1)(a+2)}{1\cdot 2\cdot 3} + ...$$

Solution. On leaving the first term we have

$$u_n = \frac{a(a+1)(a+2)...(a+n-1)}{1.2....n}$$

$$\Rightarrow \qquad u_{n+1} = \frac{a(a+1)...(a+n)}{1.2....n(n+1)}$$

$$\therefore \qquad \lim_{n\to\infty}\frac{u_n}{u_{n+1}} = \lim_{n\to\infty}\frac{(n+1)}{(a+n)} = \lim_{n\to\infty}\frac{1+\dfrac{1}{n}}{1+\dfrac{a}{n}} = 1.$$

$\Rightarrow$ The D' Alembert's ratio test fails.

Now $\qquad \lim_{n\to\infty} n\left[\frac{u_n}{u_{n+1}} - 1\right] = \lim_{n\to\infty} n\left[\frac{n+1}{a+n} - 1\right]$

$$= \lim_{n\to\infty} n\left[\frac{1-a}{a+n}\right] = \lim_{n\to\infty}\frac{(1-a)}{(1+a/n)} = (1-a).$$

Hence, by Raabe's test the given series is convergent if $1 - a > 1$, *i.e.*, $a < 0$ and divergent if $a > 0$ and test fails if $a = 0$.

In case $a = 0$, the given series becomes $1 + 0 + 0 + ...$

The sum of n terms is always 1. Therefore, the series is convergent if $a = 0$. Thus the given series Σu_n is convergent if $a \le 0$ and divergent if $a > 0$.

Example 5. *Test the convergence of the series* $\Sigma\dfrac{n!x^n}{3.5.7....(2n+1)}.$

Solution. Here, we have

$$u_n = \frac{n!x^n}{3.5.7....(2n+1)}$$

$$\Rightarrow \qquad u_{n+1} = \frac{(n+1)!\,x^{n+1}}{3.5.7....(2n+1)(2n+3)}$$

Now
$$\lim_{n\to\infty}\frac{u_n}{u_{n+1}} = \lim_{n\to\infty}\left(\frac{2n+3}{n+1}\right).\frac{1}{x} = \lim_{n\to\infty}\frac{2+\frac{3}{n}}{1+\frac{1}{n}}.\frac{1}{x} = \frac{2}{x}.$$

Hence, by D'Alembert's ratio test the series is convergent if $2/x > 1$, *i.e.*, if $x < 2$ and diverges if $2/x < 1$, *i.e.*, if $x > 2$ and test fails when $2/x = 1$, *i.e.*, when $x = 2$.

In case $x = 2$, apply Raabe's test.

When $x = 2$,
$$\frac{u_n}{u_{n+1}} = \frac{(2n+3)}{2(n+1)}$$

$$\therefore \qquad n\left(\frac{u_n}{u_{n+1}} - 1\right) = n\left(\frac{2n+3}{2n+2} - 1\right) = \frac{n}{2(n+1)} = \frac{1}{2(1+1/n)}$$

$$\therefore \qquad \lim_{n\to\infty} n\left(\frac{u_n}{u_{n+1}} - 1\right) = \lim_{n\to\infty}\frac{1}{2(1+1/n)} = \frac{1}{2} < 1.$$

Hence, by Raabe's test Σu_n is divergent if $x = 2$.

Thus, the given series Σu_n is convergent if $x < 2$ and divergent if $x \geq 2$.

Example 6. *Test the convergence of the series* $1 + \dfrac{1}{2}x + \dfrac{2!}{3^2}x^2 + \dfrac{3!}{4^3}x^3 + ...$

Solution. Here, we have $\qquad u_n = \dfrac{(n-1)!}{n^{n-1}}x^{n-1} \qquad \Rightarrow \qquad u_{n+1} = \dfrac{n!}{(n+1)^n}x^n$

$$\therefore \qquad \lim_{n\to\infty}\frac{u_n}{u_{n+1}} = \lim_{n\to\infty}\frac{(n+1)^n(n-1)!\,x^{n-1}}{n!\,x^n.\,n^{n-1}}$$

$$= \lim_{n\to\infty}\left[1+\frac{1}{n}\right]^n.\frac{1}{x} = \frac{e}{x}.$$

Hence, the given series is convergent if $\dfrac{e}{x} > 1$ *i.e.*, if $x < e$ and divergent if $x > e$ and the test fails if $x = e$. In this case

$$\lim_{n\to\infty}\left[n\log\frac{u_n}{u_{n+1}}\right] = \lim_{n\to\infty}\left[n\log\frac{\left(1+\dfrac{1}{n}\right)^n}{e}\right]$$

$$= \lim_{n\to\infty}\left[n^2\left(\frac{1}{n} - \frac{1}{2n^2} + \frac{1}{3n^3} + ...\right) - n\right]$$

$$= \lim_{n\to\infty}\left[-\frac{1}{2} + \frac{1}{3n} - ...\right] = -\frac{1}{2} < 1.$$

Hence, by log test the series Σu_n is divergent if $x = e$.

Thus the given series Σu_n is convergent if $x < e$ and divergent if $x \geq e$.

Example 7 *Test the convergence of the series* $x + \dfrac{2^2 x^2}{2!} + \dfrac{3^3 x^3}{3!} + \dfrac{4^4 x^4}{4!} + ...$

Solution. Here, we have

$$u_n = \frac{n^n x^n}{n!} \qquad \Rightarrow \qquad u_{n+1} = \frac{(n+1)^{n+1} \cdot x^{n+1}}{(n+1)!}$$

Therefore, $\displaystyle \lim_{n \to \infty} \frac{u_n}{u_{n+1}} = \lim_{n \to \infty} \frac{(n+1)! n^n x^n}{(n+1)^{n+1} x^{n+1} \cdot n!}$

$$= \lim_{n \to \infty} \frac{1}{\left(1 + \dfrac{1}{n}\right)^n x} = \frac{1}{ex}.$$

Thus, by D'Alembert's ratio test the series is convergent if $ex < 1$ *i.e.*, $x < \dfrac{1}{e}$, divergent if $x > \dfrac{1}{e}$ and the test fails if $\dfrac{1}{ex} = 1$ *i.e.*, $x = \dfrac{1}{e}$.

In this case

$$\lim_{n \to \infty} n\left[\log \frac{u_n}{u_{n+1}} \right] = \lim_{n \to \infty} n \log \left[\frac{e}{\left(1 + \dfrac{1}{n}\right)^n} \right]$$

$$= \lim_{n \to \infty} n\left[\log e - n \log\left(1 + \frac{1}{n}\right) \right]$$

$$= \lim_{n \to \infty} n\left[1 - n\left(\frac{1}{n} - \frac{1}{2n^2} + \frac{1}{3n^2} - ... \right) \right]$$

$$= \lim_{n \to \infty} \left[\frac{1}{2} - \frac{1}{3n} + ... \right] = \frac{1}{2} < 1.$$

Hence, by Logarithmic test, the series is divergent if $x = \dfrac{1}{e}$.

Thus the given series Σu_n is convergent if $x < \dfrac{1}{e}$ and divergent if $x \geq \dfrac{1}{e}$.

Example 8. *Test the convergence of the series* $1 + \dfrac{2x}{2!} + \dfrac{3^2 x^2}{3!} + \dfrac{4^3 x^3}{4!} + ...$

Solution. Here, we have

$$u_n = \frac{n^{n-1} x^{n-1}}{n!} \qquad \Rightarrow \qquad u_{n+1} = \frac{(n+1)^n x^n}{(n+1)!}$$

Now $\dfrac{u_n}{u_{n+1}} = \dfrac{(n+1)! n^{n-1} x^{n-1}}{(n+1)^n x^n \cdot n!} = \dfrac{\left(1 + \dfrac{1}{n}\right)}{\left(1 + \dfrac{1}{n}\right)^n} \cdot \dfrac{1}{x}$

$\therefore \qquad \displaystyle \lim_{n \to \infty} \frac{u_n}{u_{n+1}} = \frac{1}{ex}.$

Hence, by D'Alembert's ratio test the series is convergent if $\dfrac{1}{ex} > 1$ *i.e.*, $x < \dfrac{1}{e}$, divergent if $x > \dfrac{1}{e}$ and the test fails if $x = \dfrac{1}{e}$.

In this case

$$\lim_{n\to\infty}\left[n\log\frac{u_n}{u_{n+1}}\right]=\lim_{n\to\infty} n\left[\log\frac{\left(1+\dfrac{1}{n}\right)e}{\left(1+\dfrac{1}{n}\right)^n}\right]$$

$$=\lim_{n\to\infty} n\left[\log\left(1+\frac{1}{n}\right)+\log e-n\log\left(1+\frac{1}{n}\right)\right]$$

$$=\lim_{n\to\infty} n\left[\left(\frac{1}{n}-\frac{1}{2n^2}+\frac{1}{3n^3}-...\right)+1-n\left(\frac{1}{n}-\frac{2}{2n^2}+\frac{1}{3n^3}-...\right)\right]$$

$$=\lim_{n\to\infty} n\left[\frac{3}{2}-\frac{5}{6n}+...\right]=\frac{3}{2}>1.$$

Thus, by Logarithmic test, the series is divergent if $x=\dfrac{1}{e}$.

Thus the given series Σu_n is convergent if $x\le\dfrac{1}{e}$ and divergent if $x>\dfrac{1}{e}$.

Example 9. *Test the convergence of the series* $\dfrac{(a+x)}{1!}+\dfrac{(a+2x)^2}{2!}+\dfrac{(a+3x)^3}{3!}+...$

Solution. Here, we have $u_n=\dfrac{(a+nx)^n}{n!}\quad\Rightarrow\quad u_{n+1}=\dfrac{[a+(n+1)x]^{n+1}}{(n+1)!}$

$$\Rightarrow\qquad \frac{u_n}{u_{n+1}}=\frac{\left[1+\dfrac{a/x}{n}\right]^n}{\left[1+\dfrac{1}{n}\right]^n\left[1+\dfrac{a/x}{n+1}\right]^{n+1}}\cdot\frac{1}{x}$$

$$\Rightarrow\qquad \lim_{n\to\infty}\frac{u_n}{u_{n+1}}=\lim_{n\to\infty}\left[\frac{\left[1+\dfrac{a/x}{n}\right]^n}{\left[1+\dfrac{1}{n}\right]^n\left[1+\dfrac{a/x}{n+1}\right]^{n+1}}\cdot\frac{1}{x}\right]$$

$$=\frac{e^{a/x}}{x.e.e^{a/x}}=\frac{1}{ex}.$$

Hence, by D'Alembert's ratio test the series is convergent if $\dfrac{1}{ex}>1$ i.e., $x<\dfrac{1}{e}$, divergent if $x>\dfrac{1}{e}$ and the test fails if $x=\dfrac{1}{e}$.

In this case

$$\lim_{n\to\infty} n\log\left(\frac{u_n}{u_{n+1}}\right)=\lim_{n\to\infty} n\log\left[\frac{\left[1+\dfrac{ae}{n}\right]^n}{\left(1+\dfrac{1}{n}\right)^n\left(1+\dfrac{ae}{n+1}\right)^{n+1}}\right]$$

$$=\lim_{n\to\infty} n\left[n\log\left(1+\frac{ae}{n}\right)+\log e-n\log\left(1+\frac{1}{n}\right)-(n+1)\log\left(1+\frac{ae}{n+1}\right)\right]$$

$$= \lim_{n\to\infty} n\left[n\left(\frac{ae}{n} - \frac{a^2e^2}{2n^2} + \frac{a^3e^3}{3n^3}\cdots \right) + 1 - \left(\frac{1}{n} - \frac{1}{2n^2} + \frac{1}{3n^3}\cdots \right) \right.$$

$$\left. -(n+1)\left(\frac{ae}{n+1} - \frac{a^2e^2}{2(n+1)^2} + \frac{a^3e^3}{3(n+1)^3}\cdots \right) \right]$$

$$= \lim_{n\to\infty}\left[-\frac{a^2e^2}{2} + \frac{1}{2} + \frac{a^2e^2}{2\left(1+\dfrac{1}{n}\right)} + \text{terms containing } n \text{ in the denominator} \right]$$

$$= -\frac{a^2e^2}{2} + \frac{1}{2} + \frac{a^2e^2}{2} = \frac{1}{2} < 1.$$

Thus, by Logarithmic test, the series is divergent.

Hence, the given series Σu_n is convergent if $x < \dfrac{1}{e}$ and divergent if $x \geq \dfrac{1}{e}$.

Example 10. *Test the convergence of the series*

$$1^p + \left(\frac{1}{2}\right)^p + \left(\frac{1\cdot3}{2\cdot4}\right)^p + \left(\frac{1\cdot3\cdot5}{2\cdot4\cdot6}\right)^p +\ldots$$

Solution. Leaving the first term 1^p, we have

$$u_n = \left[\frac{1\cdot3\cdot5\ldots(2n-1)}{2\cdot4\cdot6\ldots(2n)} \right]$$

$$\Rightarrow \qquad u_{n+1} = \left[\frac{1\cdot3\cdot5\ldots(2n-1)(2n+1)}{2\cdot4\cdot6\ldots(2n)(2n+2)} \right]^p$$

Now

$$\frac{u_n}{u_{n+1}} = \left[\frac{(2n+2)}{(2n+1)} \right]^p = \left(\frac{1+\dfrac{1}{n}}{1+\dfrac{1}{2n}} \right)^p$$

$$\Rightarrow \qquad \lim_{n\to\infty}\frac{u_n}{u_{n+1}} = \left(\frac{1}{1}\right)^p \Rightarrow \text{ratio test fails.}$$

Now, applying logarithmic test, we have

$$\log\frac{u_n}{u_{n+1}} = \log\left(\frac{2n+2}{2n+1}\right)^p = \log\left(\frac{1+1/n}{1+1/2n}\right)^p$$

$$= p\left[\log\left(1+\frac{1}{n}\right) - \log\left(1+\frac{1}{2n}\right) \right]$$

$$= p\left[\left(\frac{1}{n} - \frac{1}{2n^2} + \frac{1}{3n^3} -\ldots \right) - \left(\frac{1}{2n} - \frac{1}{2.2^2n^2} + \frac{1}{3.2^3.n^3} -\ldots \right) \right]$$

$$= p\left[\left\{1 - \frac{1}{2}\right\}\frac{1}{n} - \frac{1}{2}\cdot\left\{1 - \frac{1}{4}\right\}\frac{1}{n^2} + \frac{1}{3}\left\{1 - \frac{1}{8}\right\}\frac{1}{n^3} -\ldots \right]$$

$$= p\left[\frac{1}{2n} - \frac{3}{8n^2} + \frac{7}{24n^3} - \dots\right]$$

$$\therefore \qquad n\log\frac{u_n}{u_{n+1}} = p\left[\frac{1}{2} - \frac{3}{8n} + \frac{7}{24n^3} - \dots\right]$$

Therefore $\qquad \lim_{n\to\infty}\left[n\log\frac{u_n}{u_{n+1}}\right] = \frac{p}{2}.$

So that, the series is convergent if $p/2 > 1$ *i.e.*, if $p > 2$, and divergent if $p < 2$ and the test fails if $p = 2$.

EXERCISE 3.3

Test the convergence of the following series

1. $1 + \frac{2}{3}\left(\frac{1}{4}\right) + \frac{2.4}{3.5}\left(\frac{1}{6}\right) + \frac{2.4.6}{3.5.7}\left(\frac{1}{8}\right) + \dots$

2. $\frac{1^2}{4^2} + \frac{1^2.5^2}{4^2.8^2} + \frac{1^2.5^2.9^2}{4^2.8^2.12^2} +$

$$+ \frac{1^2.5^2.9^2.13^2}{4^2.8^2.12^2.16^2} + \dots$$

3. $1 + \frac{1}{2}x + \frac{1.3}{2.4}x^2 + \frac{1.3.5}{2.4.6}x^3, \dots, (x > 0)$

4. $x^2 + \frac{2^2}{3.4}x^4 + \frac{2^2.4^2}{3.4.5.6}x^6 + \dots .$

5. $1 + \frac{1}{2}\frac{x^2}{4} + \frac{1.3.5}{2.4.6}.\frac{x^4}{8} +$

$$+ \frac{1.3.5.7.9}{2.4.6.8.10}.\frac{x^6}{12} + \dots$$

6. $\sum_{n=1}^{\infty} \frac{n!}{(n+1)^n}x^n, x > 0$

7. $\sum_{n=1}^{\infty}\left[\frac{1}{1+\log n}\right]$

8. $1 + \frac{2}{3.5} + \frac{2.4}{3.5.7} + \frac{2.4.6}{3.5.7.9} + \dots$

9. Test the convergence of the series

$$x + x^{1+\frac{1}{2}} + x^{1+\frac{1}{2}+\frac{1}{3}} + x^{1+\frac{1}{2}+\frac{1}{3}+\frac{1}{4}} + \dots$$

10. Test the convergence of the following series:

(i) $\frac{1^2}{2^2} + \frac{1^2.3^2}{2^2.4^2}x + \frac{1^2.3^2.5^2}{2^2.4^2.6^2}x^2 + \dots$

(ii) $1 + \frac{2^2}{3^2} + \frac{2^2.4^2}{3^2.5^2} + \frac{2^2.4^2.6^2}{3^2.5^2.7^2} + \dots$

11. Test for convergence, the following series :

(i) $1 + \frac{x}{1} + \frac{1}{2}.\frac{x^3}{3} + \frac{1.3}{2.4}.\frac{x^5}{5} +$

$$+ \frac{1.3.5}{2.4.6}.\frac{x^7}{7} + \dots.$$

(ii) $\frac{x}{1} + \frac{1}{2}.\frac{x^2}{3} + \frac{1.3}{2.4}.\frac{x^3}{5}$

$$+ \frac{1.3.5}{2.4.6}.\frac{x^4}{7} + \dots (x > 0)$$

(iii) $\sum_{n=1}^{\infty} \frac{1.3.5\dots(4n-5)(4n-3)}{2.4.6\dots(4n-4)(4n-2)}\frac{x^{2n}}{4n}, x > 0$

(iv) $\sum_{n=1}^{\infty} \frac{2.4.6\dots 2n}{1.3.5\dots(2n+1)}$

12. Test for convergence, the following series :

(i) $1 + \frac{x}{1!} + \frac{2^2 x^2}{2!} + \frac{3^3 x^3}{3!} + \dots$ for $x > 0$

(ii) $\frac{1}{2}x + \frac{1.3}{2.4}x^2 + \frac{1.3.5}{2.4.6}x^3 + \dots, x > 0$

(iii) $1 + \frac{2!}{2^2}x + \frac{3!}{3^3}x^2 + \dots, x > 0$

13. Test for convergence, the following series :

(i) $1 + \frac{a(1-a)}{1^2} + \frac{(1+a)a(1-a)(2-a)}{1^2.2^2} +$

$$+ \frac{(2+a)(1+a)a(1-a)(2-a)(3-a)}{1^2.2^2.3^2} + \dots$$

(ii) $\frac{(1+a)(1+b)}{1.2.3} + \frac{(2+a)(2+b)}{2.3.4} +$

$$+ \frac{(3+a)(3+b)}{3.4.5.} + \dots$$

14. Test for convergence the following series :

(i) $1 + \dfrac{\alpha}{1.\beta}x + \dfrac{\alpha(\alpha+1)^2}{1.2\beta(\beta+1)^2}x^2 +$

$\qquad + \dfrac{\alpha(\alpha+1)^2(\alpha+2)^2}{1.2.3\beta(\beta+1)(\beta+2)}x^3 + ...$

(ii) $1 + \dfrac{\alpha.\beta}{1.\gamma}x + \dfrac{\alpha(\alpha+1)\beta(\beta+1)}{1.2.\gamma(\gamma+1)}x^2 +$

$\qquad + \dfrac{\alpha(\alpha+1)(\alpha+2)\beta(\beta+1)(\beta+2)}{1.2.3.\gamma(\gamma+1)(\gamma+2)}x^3 + ...$

15. Test for convergence the following series :

$\dfrac{a}{a+3} + \dfrac{a(a+2)}{(a+3)(a+5)}x +$

$\qquad \dfrac{a(a+2)(a+4)}{(a+3)(a+5)(a+7)}x^2 + ...$

16. Test for convergence the following series :

$\left(\dfrac{1}{2.4}\right)^{2/3} + \left(\dfrac{1.3}{2.4.6}\right)^{2/3} + \left(\dfrac{1.3.5}{2.4.6.8}\right)^{2/3} + ...$

17. Test for convergence the following series :

(i) $\displaystyle\sum_{n=1}^{\infty} \dfrac{1\cdot3\cdot5....(2n-1)}{2\cdot4\cdot6....2n}\cdot\dfrac{1}{n}$

(ii) $\displaystyle\sum_{n=1}^{\infty} \dfrac{4\cdot7\cdot10....(3n+1)}{1\cdot2\cdot3....n}x^n$

(iii) $\displaystyle\sum_{n=1}^{\infty} \dfrac{3\cdot6\cdot9....(3n)}{7\cdot10\cdot13....(3n+4)}x^n, x>0$

(iv) $\displaystyle\sum_{n=1}^{\infty} \dfrac{(2n)!}{(n!)^2}x^n, x>0$

18. Test for convergence the following series :

$\dfrac{1^2}{2^2} + \dfrac{1^2.3^2}{2^2.4^2} + \dfrac{1^2.3^2.5^2}{2^2.4^2.6^2} + ...$

19. Test for convergence the following series :

(i) $\dfrac{1}{(\log 2)^p} + \dfrac{1}{(\log 3)^p} + ... + \dfrac{1}{(\log n)^p} + ...$

(ii) $x^2(\log 2)^p + x^3(\log 3)^p + x^4(\log 4)^p + ...$

Answers

1. Convergent   **2.** Convergent   **3.** $\begin{cases} \text{Convergent if } x<1, \\ \text{Divergent if } x\geq1 \end{cases}$

4. Convergent if $x^2 \leq 1$, divergnet if $x^2 > 1$ **5.** Convergent if $x \leq 1$, divergent if $x > 1$

6. Convergent if $x < e$, divergent if $x \geq e$ **7.** Convergent   **8.** Convergent

9. Convergent if $x < \dfrac{1}{e}$, divergent if $x \geq \dfrac{1}{e}$ **10.** (i) Convergent if $x < 1$, divergent if $x \geq 1$
(ii) Divergent

11. (i) Convergent if $x^2 \leq 1$, divergent if $x^2 > 1$ (ii) Convergent if $0 < x \leq 1$, divergent if $x > 1$
(iii) Convergent if $x \leq 1$, divergent if $x > 1$ (iv) Divergent

12. (i) Convergent if $x < \dfrac{1}{e}$, divergent if $x \geq \dfrac{1}{e}$ (ii) Convergent if $x < 1$, divergent if $x \geq 1$,
(iii) Convergent if $x < e$, divergnet if $x \geq e$ **13.** (i)   Divergent   (ii)   Divergent

14. (i) Convergent if $x < 1$, divergnet if $x > 1$, When $x = 1$, then convergent if $\beta > 2\alpha$, divergent if $\beta \leq 2\alpha$
(iii) Convergent if $x < 1$, divergent if $x > 1$, When $x = 1$, then convergent if $\gamma > \alpha + \beta$, divergent if $\gamma \leq \alpha + \beta$.

15. Convergent if $x \leq 1$, divergent if $x > 1$ **16.** Divergent

17. (i) Convergent (ii) Convergent if $x < \dfrac{1}{3}$, divergent if $x \geq \dfrac{1}{3}$ (iii) Convergent if $x \leq 1$, divergent

if $x > 1$ (iv) Convergent if $x < \dfrac{1}{4}$, divergent if $x \geq \dfrac{1}{4}$.

18. Divergent **19.** (i) Divergent for all values of p, (ii) Convergent if $x < 1$, divergent if $x \geq 1$

3.12 GAUSS'S TEST

If Σu_n be a series of positive terms such that

$$\frac{u_n}{u_{n+1}} = \alpha + \frac{\beta}{n} + \frac{\gamma_n}{n^p},$$

where $a > 0, p > 1$ and $<\gamma_n>$ is a bounded sequence. Then

(i) Σu_n converges for $\alpha > 1$, diverges for $\alpha < 1$, whatever β may be.

(ii) If $\alpha = 1$, Σu_n converges whenever $\beta > 1$, and diverges whenever $\beta \leq 1$.

Proof. We have

$$\lim_{n \to \infty} \frac{u_n}{u_{n+1}} = \alpha.$$

Then by D'Alembert's ratio test Σu_n is convergent if $\alpha > 1$ and divergent if $\alpha < 1$.

For $\alpha = 1$, we have

$$n\left[\frac{u_n}{u_{n+1}} - 1\right] = \beta + \frac{\gamma_n}{n^p},$$

where $p > 1$ and $<\gamma_n>$ is a bounded sequence.

$$\therefore \qquad \lim_{n \to \infty} n\left[\frac{u_n}{u_{n+1}} - 1\right] = \beta.$$

Then, by Raabe's test Σu_n is convergent if $\beta > 1$ and divergent if $\beta < 1$.

Now for $\alpha = \beta = 1$, we compare the series with the divergent series Σv_n where

$$v_n = \frac{1}{n \log n}.$$

Now, consider

$$\frac{u_n}{u_{n+1}} - \frac{v_n}{v_{n+1}} = 1 + \frac{1}{n} + \frac{\gamma_n}{n^p} - \frac{(n+1)\log(n+1)}{n\log n}$$

$$= \frac{\gamma_n}{n^p} - \frac{(n+1)}{n}\left[\frac{\log(n+1)}{\log n} - 1\right]$$

$$= \frac{1}{n^p}\left[\gamma_n - (n+1)\log\left(1 + \frac{1}{n}\right) \cdot \frac{n^{p-1}}{\log n}\right].$$

But $\lim\limits_{n \to \infty} (n+1)\log\left(1 + \frac{1}{n}\right) = \lim\limits_{n \to \infty}\left[\log\left(1 + \frac{1}{n}\right)^n + \log\left(1 + \frac{1}{n}\right)\right]$

$$\lim_{n \to \infty} \frac{n^{p-1}}{\log n} = \infty, p > 1 \text{ and } <\gamma_n> \text{ is bounded.}$$

Therefore, for large value of n, $\gamma_n - (n+1)\log\left(1 + \frac{1}{n}\right)\frac{n^{p-1}}{\log n}$ remains negative.

$$\therefore \qquad \frac{u_n}{u_{n+1}} - \frac{v_n}{v_{n+1}} < 0 \qquad \text{or} \qquad \frac{u_n}{u_{n+1}} < \frac{v_n}{v_{n+1}}.$$

Now, since Σv_n is divergent, by comparison test Σu_n is divergent.

Hence, the series Σu_n is convergent if $\alpha > 1$ or $\alpha = 1$ and $\beta > 1$ and divergent if $\alpha > 1$ or $\alpha = 1$ and $\beta \leq 1$.

3.13 CAUCHY'S INTEGRAL TEST

Let $f(x)$ be non-negative monotonically decreasing integrable function on $[1, \infty[$ then the series $\sum\limits_{n=1}^{\infty} f(n)$ and the improper integral $\int_1^{\infty} f(x)\,dx$ converge or diverge together.

Proof. Let $f(x)$ be monotonically decreasing on $[1, \infty]$.

Then we have

$$f(n) \geq f(x) \geq f(n+1), \text{ where } n \leq x \leq n+1.$$

Also $f(x)$ is non-negative and integrable, we have

$$\int_n^{n+1} f(n)\,dx \geq \int_n^{n+1} f(x)\,dx \geq \int_n^{n+1} f(n+1)\,dx$$

or $$f(n) \geq \int_n^{n+1} f(x)\,dx \geq f(n+1). \qquad \qquad \text{...(1)}$$

Now, putting $n = 1, 2, ..., (n-1)$ in (1) and adding all these, we get

$$f(1) + f(2) + ... + f(n-1) \geq \int_1^2 f(x)\,dx + \int_2^3 f(x)\,dx + ... + \int_{n-1}^{n} f(x)\,dx$$
$$\geq f(2) + f(3) + ... f(n). \qquad \qquad \text{...(2)}$$

Let us suppose

$$s_n = f(1) + f(2) + ... + f(n)$$

and $$I_n = \int_1^n f(x)\,dx.$$

Then (2) can be written as

$$s_n - f(x) \geq I_n \geq s_n - f(1)$$

or $$f(n) \leq s_n - I_n \leq f(1). \qquad \qquad \text{...(3)}$$

Let $$u_n = s_n - I_n \, \forall \, n \in N.$$

Then
$$u_{n+1} - u_n = (s_{n+1} - I_{n+1}) - (s_n - I_n)$$
$$= (s_{n+1} - s_n) - (I_{n+1} - I_n)$$
$$= f(n+1) - \int_n^{n+1} f(x)\,dx$$
$$\leq 0 \qquad \qquad \text{[By using (1)]}$$

Hence, we have $\langle u_n \rangle$ is monotonically decreasing sequence.

Now from (3) $u_n \geq f(n) \geq 0, \, \forall \, n \in N$. Therefore sequence $\langle u_n \rangle$ is bounded below.

Hence $\langle u_n \rangle$ is a convergent sequence and it has a finite limit.

Now, since $S_n = u_n + I_n$, the sequence $\langle S_n \rangle$ and $\langle I_n \rangle$ converge or diverge together.

Hence, the series $\Sigma f(n)$ and the integral $\int_1^{\infty} f(x)\,dx$ converge or diverge together.

3.13 CAUCHY'S CONDENSATION TEST

If $f(n)$ is a monotonically decreasing function of n for all $n \in N$ such that each $f(n)$ is positive, then two infinite series $\sum\limits_{n=1}^{\infty} f(n)$ and $\sum\limits_{n=1}^{\infty} a^n f(a^n)$ converge or diverge together, where a is a positive integer greater than unity.

Proof. Let s_n and r_n be the partial sum of the series $\sum\limits_{n=1}^{\infty} f(n)$ and $\sum\limits_{n=1}^{\infty} a^n f(a^n)$ respectively, then

$$s_n = \sum_{k=1}^{n} f(k) = f(1) + f(2) + f(3) + \dots + f(n)$$

and

$$r_n = \sum_{k=1}^{n} a^k f(a^k) = af(a) + a^2 f(a^2) + a^3 f(a^3) + \dots + a^n f(a^n)$$

The series $\sum_{n=1}^{\infty} f(n)$ can be written as

$$\sum_{n=1}^{\infty} f(n) = [f(1) + f(2) + f(3) + \dots + f(a)]$$
$$+ [f(a+1) + f(a+2) + f(a+3) + \dots + f(a^2)]$$
$$+ [f(a^2+1) + f(a^2+2) + f(a^2+3) + \dots + f(a^3)]$$
$$+ \dots\dots\dots\dots\dots\dots\dots\dots\dots\dots\dots\dots\dots\dots\dots\dots\dots\dots\dots$$
$$+ [f(a^{m-1}+1) + f(a^{m-1}+2) + f(a^{m-1}+3) + \dots + f(a^m)] \quad \dots(1)$$
$$+ \dots\dots\dots\dots\dots\dots\dots\dots\dots\dots\dots\dots\dots\dots\dots\dots\dots\dots\dots\dots\dots$$

The mth group is

$$f(a^{m-1}+1) + f(a^{m-1}+2) + f(a^{m-1}+3) + \dots + f(a^m)$$

Therefore, the number of terms in this mth group $= a^m - a^{m-1} = a^{m-1}(a-1)$

Since $f(n)$ is a decreasing function, then in mth group, we have

$$f(a^{m-1}+1) \geq f(a^m), f(a^{m-2}+1) \geq f(a^m), \text{ and so on}$$

Also $f(a^{m-1}+1) \leq f(a^{m-1}), f(a^{m-2}+2) \leq f(a^{m-1}), \text{ and so on}$

$\therefore$ $a^{m-1}(a-1)f(a^m) \leq f(a^{m-1}+1) + f(a^{m-1}+2) + \dots + f(a^m)$

$$\leq a^{m-1}(a-1)f(a^{m-1})$$

or $a^m \left(\dfrac{a-1}{a}\right) f(a^m) \leq f(a^{m-1}+1) + f(a^{m-1}+2) + \dots + f(a^m)$

$$\leq a^{m-1}(a-1)f(a^{m-1}) \quad \dots(2)$$

Putting $m = 1, 2, 3, \dots, n$ successively in (2), we have

$$a\left(\frac{a-1}{a}\right) f(a) \leq f(2) + f(3) + \dots + f(a) \leq (a-1)f(1) \quad \dots(3)$$

$$a^2\left(\frac{a-1}{a}\right) f(a^2) \leq f(a+1) + f(a+2) + \dots + f(a^2) \leq (a-1)a\, f(a) \quad \dots(4)$$

$$a^3\left(\frac{a-1}{a}\right) f(a^3) \leq f(a^2+1) + f(a^2+2) + \dots + f(a^3) \leq (a-1)a^2 f(a^2) \quad \dots(5)$$

$$\dots\dots\dots\dots\dots\dots\dots\dots\dots\dots\dots\dots$$

$$a^n\left(\frac{a-1}{a}\right) f(a^n) \leq f(a^{n-1}+1) + f(a^{n-1}+2) + \dots + f(a^n)$$

$$\leq (a-1)a^{n-1} f(a^{n-1}) \quad \dots(6)$$

Adding all above n inequalities, we have

$$\left(\frac{a-1}{a}\right)\sum_{k=1}^{n} a^k f(a^k) \leq \sum_{k=2}^{a^n} f(k) \leq (a-1)\sum_{k=1}^{n} a^{k-1} f(a^{k-1}) \quad \dots(7)$$

Adding $f(1)$ throughout the inequality (7), we have

$$f(1) + \left(\frac{a-1}{a}\right)\sum_{k=1}^{n} a^k f(a^k) \leq \sum_{k=2}^{a^n} f(k) \leq f(1) + (a-1)\sum_{k=1}^{n} a^{k-1} f(a^{k-1})$$

$$\Rightarrow \qquad f(1)+\left(\frac{a-1}{a}\right)r_n \leq s_{a^n} \leq f(1)+(a-1)r_{n-1} \qquad \qquad ...(8)$$

Now we have two cases :

Case (I) : If $\sum_{n=1}^{\infty} a^n f(a^n)$ is convergent, then the sequence $<r_n>$ is convergent so it is bounded.

Since $f(n)$ is monotonically decreasing, therefore the sequence $<s_n>$ is increasing.

Since $a > 1$ and a is a positive integer, then

$$a^n > n \; \forall \; n \in N$$

$$\Rightarrow \qquad a_n < s_{a^n}$$

$$\Rightarrow \qquad f(1) \leq s_n < s_{a^n} \leq f(1) + (a-1) \; r_{n-1}$$

$$\Rightarrow \qquad <s_n> \text{ is bounded.}$$

$$\Rightarrow \qquad <s_n> \text{ is convergent.}$$

Hence $\qquad \sum_{n=1}^{\infty} f(n)$ is convergent.

Case (II): If $\sum_{n=1}^{\infty} f(n)$ is convergent, then the sequence $<s_n>$ is convergent so it is bounded.

But $<s_{a^n}>$ is a subsequence of $<s_n>$ therefore $<s_{a^n}>$ is bounded.

Now by virtue of (8), we have

$$f(1)+\left(\frac{a-1}{a}\right)r_n \leq s_{a^n}$$

$$\Rightarrow \quad <r_n> \text{ is bounded above.}$$

$$\Rightarrow \quad <r_n> \text{ is convergent.}$$

$$\Rightarrow \quad \sum_{n=1}^{\infty} a^n f(a^n) \text{ is convergent}$$

Hence $\sum_{n=1}^{\infty} f(n)$ and $\sum_{n=1}^{\infty} a^n f(a^n)$ converge and diverge together.

3.15 REARRANGEMENT OF TERMS

A series Σv_n is said to be rearrangement of a series Σu_n if there exists one-one correspondance between the terms of the two series and if v_n corresponds to u_n then $v_n = u_n$.

In other words, we can say that a series Σu_n is said to be rearragement of a series Σv_n if every term of Σu_n is a term of Σv_n and *vice-versa*.

3.16 ALTERNATING SERIES

A series, whose terms are alternatively positive and negative is called an alternating series. Thus, a series of the form $u_1 - u_2 + u_3 - u_4 +...+ (-1)^{n-1} u_n +...$

where $u_n > 0 \; \forall \; n$, is an alternating series.

3.16.1 ABSOLUTE CONVERGENCE

A series Σu_n is said to be absolutely convergent if the series $\Sigma |u_n|$ is convergent.

3.16.2 UNCONDITIONALLY CONVERGENT SERIES

A series Σu_n is said to be unconditionally convergent if every rearrangement converge to the same sum Σu_n i.e, Σu_n is conditionally convergent iff it is absolutely convergent.

3.16.3 CONDITIONAL CONVERGENCE

A series Σu_n is said to be conditionally convergent if Σu_n is convergent but $\Sigma |u_n|$ is divergent.

REMARK

- The conditional convergence of a series is also known as semi-convergent or non-absolutely convergent.

☞ ILLUSTRATIONS

(1) The series $\Sigma u_n = 1 - \dfrac{1}{2} + \dfrac{1}{2^2} - \dfrac{1}{2^3} + \dots$ is absolutely convergent.

(2) The series $\dfrac{1}{1^2} - \dfrac{1}{2^2} + \dfrac{1}{3^2} - \dfrac{1}{4^2} + \dots$ is absolutely convergent.

THEOREM 1. *An absolutely convergent series is convergent.*

Proof. Let us suppose, the series Σu_n is absolutely convergent. Then by definition $|u_n|$ is convergent.

Now
$$u_n + |u_n| = \begin{cases} 2u_n, & \text{if } u_n \text{ is positive} \\ 0, & \text{if } u_n \text{ is negative.} \end{cases}$$

Therefore, every term of the series $\Sigma(u_n + |u_n|)$ is ≥ 0 and less than equal to the corresponding term of the convergent series $\Sigma 2|u_n|$.

Hence $\Sigma(u_n + |u_n|)$ is convergent. Hence Σu_n is convergent.

REMARKS

- The converse of the above theorem is not necessarily true :

  For example : The series $\Sigma u_n = 1 - \dfrac{1}{2} + \dfrac{1}{3} - \dots$ is convergent, but the series $\Sigma |u_n| = 1 + \dfrac{1}{2} + \dfrac{1}{3} + \dots$

  is divergent. Hence a convergent series need not be absolutely convergent.
- The usefulness of absolute convergence is partly due to the fact that it is often easier to establish absolute convergence than convergence :

  For example : Consider the series $\Sigma \dfrac{a^n}{2^n}$, where $a_n = 1$ if n is prime number and $a_n = -1$

  otherwise. Here, $\Sigma |a_n| = \Sigma \dfrac{1}{2^n}$ is convergent. Accordingly $\Sigma a_n / 2^n$ is absolutely convergent, and hence convergent.

THEOREM 2. *If the terms of a convergent series of positive terms are rearranged, the series remains convergent and its sum is unaltered.*

Proof. Let us suppose Σu_n be a convergent series, and let the terms be rearranged in any manner. Denote the new series by Σv_n, so that every u is a v and every v is a u.

Let $s_n = u_1 + u_2 + \dots + u_n$ and $t_n = v_1 + v_2 + \dots + v_n$.

Then, for any definite value of n, s_n contains n terms each of which occurs, sooner or later, in the v series and so we can find a corresponding m such that t_m contains all the terms of s_n (and possibly other not contained in s_n).

Now, since each term is positive, therefore $s_n \leq t_m$.

Also, suppose that the first m terms of Σv_n are among the first $(n+p)$ terms of Σu_n. Therefore,

$$s_n \le t_m \le s_{n+p}.$$

and m tends to infinity with n.

Let Σu_n converges to s, so that $\lim s_n = \lim s_{n+p} = s$

$\therefore \qquad \lim t_m = s.$

Hence, Σv_n is convergent and has the same sum as Σu_n.

REMARK

- The arrangement fails for a dearrangement such as $u_1 + u_2 + u_5 + \ldots + u_2 + u_4 + u_6 + \ldots$ where Σu_n is broken up into two (or any finite no. of) infinite series.
  Here, we cannot find an m so that the first n terms of Σu_n occur among the first m terms of Σv_n. For instance, u_2 does not occur even if infinitely many of the terms $u_1, u_3, u_5, \ldots$ have been placed.

THEOREM 3. (Dirichlet's Theorem). *If the terms of an absolutely convergent series are rearranged, the series remains convergent and its sum is unaltered.*

Proof. Let Σu_n be an absolutely convergent series, and let its terms be rearranged in a different order. Let, the new series be denoted by Σv_n so that every v occurs somewhere in the u series and every u occurs somewhere in the v series.

Now, we have $u_n + |u_n| = 2u_n$ or 0 according as u_n is positive or negative. Now $\Sigma |u_n|$ is a convergent series of positive terms, so also in the series $\Sigma(u_n + |u_n|)$, because its terms are less than equal to be corresponding terms of the series $\Sigma 2|u_n|$.

Let $\Sigma |u_n| = s$ and $\Sigma(u_n + |u_n|) = s'$ so that $\Sigma u_n = s' - s$.

Also, since $\Sigma |u_n|$ and $\Sigma(u_n + |u_n|)$ are convergent series of positive terms, their sum remains unchanged by any rearrangement of term (By theorem 2).

Accordingly, $\Sigma |v_n| = s$ and $\Sigma(v_n + |v_n|) = s'$.

Hence, $\Sigma v_n = s' - s = \Sigma u_n$.

REMARKS

- If we rearrange the order of terms of a semi-convergent series, we may or may not changed the sum of the series.
- The sum will be changed if we interfere too much with the balance between positive and negative terms.
- By a suitable rearrangement of the terms a semi-convergent series may be made to diverge. The reason is that in a semi-convergent series the positive and negative terms taken separately from two divergent series.

THEOREM 4. (Riemann's Rearrangement theorem). *By a suitable rearrangement of terms of a conditionally convergent series can be made to converge to any number λ or to diverge to ∞ or $-\infty$ even to oscillate.*

In other words, this theorem can be stated as follows

To a given conditionally convergent series and to any given number there corresponds a rearrangement of the given series which is convergent and whose sum is the given number.

THEOREM 5. (Pringsheim theorem). *Let $f(x)$ be a sequence of positive terms which monotonically converges to zero and let the series $\sum_{n=1}^{\infty} (-1)^{n-1} f(x)$ be rearranged so that in the first $p+n$ terms there are p-positive terms and n negative terms, i.e., $\lim_{n\to\infty} n\, f(x) = \lambda$ and $\lim_{n\to\infty} \dfrac{p}{n} = k$ then the sum of the series is increased by $\dfrac{1}{2}\lambda \log k$.*

3.17 LEIBNITZ TEST

If the alternative series $u_1 - u_2 + u_3 - ...(u_n > 0, \forall\, n \in N)$ is such that

(i) $u_{n+1} \le u_n$, $\forall\, n \in N$ (ii) $\lim_{n\to\infty} u_n = 0$

Then the series converges.

Proof. Let $s_n = u_1 - u_2 + u_3 - ... + (-1)^{n-1} u_n$ so that $<s_n>$ is a sequence of partial sums of the given series.

Now for all n

$$s_{2n+2} - s_{2n} = u_{2n+1} - u_{2n+2} \ge 0 \qquad\qquad\qquad \text{[By (i)]}$$

which gives that s_{2n} is a monotonically increasing sequence.

Further,
$$\begin{aligned}
s_{2n} &= u_1 - u_2 + u_3 - ....u_{2n-1} - u_{2n}\\
&= u_1 - (u_2 - u_3) - (u_4 - u_5) - ... - u_{2n}\\
&= u_1 - [(u_2 - u_3) + ... + u_{2n}]\\
&= u_1 - \text{some positive number}\\
&\le u_1.
\end{aligned}$$

Therefore, the monotonically increasing sequence $<s_{2n}>$ is bounded above and consequently it is convergent.

Let
$$\lim_{n\to\infty} s_{2n} = s.$$

Now
$$s_{2n+1} = s_{2n} + u_{2n+1} \quad\Rightarrow\quad \lim_{n\to\infty} s_{2n+1} = \lim_{n\to\infty} s_{2n} + \lim_{n\to\infty} u_{2n+1}$$

$$\left[\because \lim_{n\to\infty} u_n = 0\right]$$

$$\begin{aligned}
&= s + 0\\
&= s
\end{aligned}$$

Thus, the subsequences $<s_{2n}>$ and $<s_{2n+1}>$ both converge to the same limit. Now we shall show that the sequence $<s_n>$ also converges to s.

Let $\varepsilon > 0$ be given. Since, the sequences s_{2n} and s_{2n+1} both converges to s, there exist positive integers m_1, m_2 such that

$$|s_{2n} - s| < \varepsilon \;\forall\, n \ge m_1,$$

and
$$|s_{2n+1} - s| < \varepsilon \;\forall\, n \ge m_2.$$

Let
$$m = \max\{m_1, m_2\}.$$

Then
$$|s_n - s| < \varepsilon \;\forall\, n \ge m$$

which gives that the sequence $<s_n>$ converges to s.

Hence, the given series $\Sigma(-1)^{n-1} u_n$ converges.

REMARKS

- This test gives us a set of sufficient conditions for the convergence of an alternating series.
- If the test does not show a series to be convergent, we may not immediately say that the series is divergent.

Solved Example

Example 1. *Show that* $\lim\limits_{n\to\infty}\left[1+\dfrac{1}{2}+...+\dfrac{1}{n}-\log n\right]$ *exists.*

Solution. Let $f(x)=\dfrac{1}{x},\ x\in[1,\infty]$.

Then $f(x)>0$ and monotonically decreasing on $[1,\infty[$.

Let $S_n=f(1)+f(2)+...+f(n)$

$$=1+\dfrac{1}{2}+\dfrac{1}{3}+...+\dfrac{1}{n}$$

and $I_n=\int_1^n f(x)\,dx=\int_1^n\dfrac{1}{x}\,dx=[\log x]_1^n=\log n.$

It can be easily shown that $f(n)\le S_n-I_n\le f(1)\ \forall\ n\in N$

or $0<\dfrac{1}{n}\le S_n-I_n\le 1\ \forall\ n\in N$

which gives that the sequence (u_n), where $u_n=S_n-I_n$, is bounded below.

Now, it can also be shown easily that the sequence $<u_n>$ is a monotonically decreasing. Therefore it converges.

Hence, $\lim\limits_{n\to\infty}\left(1+\dfrac{1}{2}+...+\dfrac{1}{n}\right)$ exist.

REMARK

- The limit of the above sequence is called Euler's constant and is denoted by γ.

Example 2. *Show by integral test that* $\Sigma\dfrac{1}{n^p}$ *converges if* $p>1$ *and diverges if* $p\le 1$.

Solution. Let $f(x)=\dfrac{1}{x^p},\ p>0$. Then $f(x)$ is positive valued and monotonically decreasing.

Therefore by Cauchy's integral test $\Sigma\dfrac{1}{n^p}$ and $\int_1^\infty f(x)\,dx$ converges and diverges together.

Let $I_n=\int_1^n\dfrac{1}{x^p}\,dx=\int_1^n x^{-p}\,dx$

$$=\begin{cases}\left(\dfrac{n^{1-p}}{1-p}-\dfrac{1}{1-p}\right),&\text{if }p\ne 1\\ \log n,&\text{if }p=1.\end{cases}$$

If $n\to\infty$, $n^{1-p}=\dfrac{1}{n^{p-1}}\to 0$ as $p>1$ and tends to ∞ if $p<1$ and $\log n\to\infty$

$\therefore\qquad \lim\limits_{n\to\infty}I_n=-\dfrac{1}{1-p}=\dfrac{1}{p-1}$, if $p>1$

and $\lim\limits_{n\to\infty} I_n = \infty,$ if $p \le 1.$

Hence, $\int_1^\infty f(x)\,dx$ converges if $p > 1$ and diverges if $p \le 1$. Then by Cauchy's integral

test the series $\Sigma \dfrac{1}{n^p}$ is convergent if $p > 1$ and divergent if $p \le 1.$

Example 3. *Show by Cauchy's integral test that the series* $\displaystyle\sum_{n=2}^{\infty} \dfrac{1}{n(\log n)^p}$ *converges if $p > 1$*

and diverges if $0 < p \le 1.$

Proof.   Let us suppose

$$f(x) = \frac{1}{x(\log x)^p}, \; p > 0$$

and $x \in [2, \infty[$; then obviously $f(x)$ is monotonically decreasing in $[2, \infty[$ and positive
valued.

Let $I_n = \int_2^n \dfrac{dx}{x(\log x)^p}$ Then $I_n = \left[\dfrac{(\log x)^{1-p}}{1-p}\right]_2^n, \; p \ne 1$

$$= \frac{1}{(1-p)}[(\log n)^{1-p} - (\log 2)^{1-p}], p \ne 1$$

and $I_n = [\log\log x]_2^n, \; p = 1$

$$= [\log\log n - \log\log 2], p = 1.$$

Therefore, we have

$$\lim_{n\to\infty} I_n = \lim_{n\to\infty} \int_2^n f(x)\,dx = \infty, \text{ if } p < 1$$

and $\lim\limits_{n\to\infty} I_n = -\dfrac{1}{(1-p)}(\log 2)^{1-p},$ if $p > 1.$

Thus the integral $\int_2^\infty f(x)\,dx$ converges if $p > 1$ and diverges if $0 < p \le 1.$
Hence, by Cauchy's integral test, the series

$$\sum_{n=2}^{\infty} f(x) = \sum_{n=2}^{\infty} \frac{1}{n(\log n)^p}$$

converges if $p > 1$ and diverges if $0 < p \le 1.$

Example 4. *Apply the Cauchy's condensation test to discuss the convergence of the series*

$$\sum_{n=2}^{\infty} \frac{1}{(n\log n)(\log\log n)^p}.$$

Solution.   Here, we have

$$f(n) = \frac{1}{(n\log n)(\log\log n)^p}$$

∴ $a^n f(a^n) = \dfrac{a^n}{(a^n \log a^n)(\log\log a^n)^p} = \dfrac{1}{(n\log a)[\log(n\log a)]^p}$

Since, a is a positive integer greater than 1 and can be chosen that $\log_e a > 1$ so that
$n \log a > n.$

Then $a^n f(a^n) < \dfrac{1}{(n\log a)(\log n)^p}$

Since, the series $\dfrac{1}{\log a}\Sigma\dfrac{1}{n(\log n)^p}$ is convergent when $p > 1$, therefore $\Sigma a^n f(a^n)$ is also convergent and consequently the given series is convergent when $p > 1$.

Now let $p \leq 1$. If we take $a = 2$, then $\log_e a < 1$ so that

$$n\log_e a < n$$

$$\therefore \qquad a^n f(a^n) > \dfrac{1}{(n\log a)(\log n)^p}$$

But the series $\dfrac{1}{\log a}\Sigma\dfrac{1}{n(\log n)^p}$ is divergent when $p \leq 1$ and therefore $\Sigma a^n f(a^n)$

is also divergent. Then by Cauchy condensation test, the given series is divergent when $p \leq 1$.

Example 5. *Test the convergence of the series* $\dfrac{2^2}{3^2} + \dfrac{2^2.4^2}{3^2.5^2} + ...$

Solution. Here, we have

$$u_n = \dfrac{2^2.4^2...(2n)^2}{3^2.5^2...(2n+1)^2}..$$

$$\Rightarrow \qquad u_{n+1} = \dfrac{2^2.4^2...(2n+2)^2}{3^2.5^2...(2n+3)^2}$$

$$\therefore \qquad \dfrac{u_n}{u_{n+1}} = \dfrac{(2n+3)^2}{(2n+2)^2} = \left(1+\dfrac{3}{2n}\right)^2\left(1+\dfrac{1}{n}\right)^{-2}$$

$$= \left(1+\dfrac{3}{n}+\dfrac{9}{4n^2}\right)\left(1-\dfrac{2}{n}+\dfrac{3}{n^2}...\right)$$

(On expanding by binomial expansion)

$$= 1+\dfrac{1}{n}-\dfrac{3}{4n^2}+...$$

$$= \alpha+\dfrac{\beta}{n}+\dfrac{\gamma_n}{n^2}, \text{ where } \gamma_n \to -\dfrac{3}{4} \text{ as } n \to \infty$$

$$\Rightarrow \quad \alpha = 1, \beta = 1. \text{ Then by Gauss test the series } \Sigma u_n \text{ is divergent.}$$

Example 6. *Test the convergence of the series* $\dfrac{1^2}{2^2} + \dfrac{1^2.3^2}{2^2.4^2} + \dfrac{1^2.3^2.5^2}{2^2.4^2.6^2} + ...$

Solution. Here, we have $\qquad u_n = \dfrac{1^2.3^2.5^2...(2n-1)^2}{2^2.4^2.6^2...(2n)^2}$

$$\therefore \qquad u_{n+1} = \dfrac{1^2.3^2...(2n-1)^2(2n+1)^2}{2^2.4^2...(2n)^2(2n+2)^2}$$

$$\therefore \qquad \dfrac{u_n}{u_{n+1}} = \dfrac{(2n+2)^2}{(2n+1)^2}$$

$$\Rightarrow \qquad \lim_{n\to\infty}\dfrac{u_n}{u_{n+1}} = \lim_{n\to\infty}\dfrac{\left(2+\dfrac{2}{n}\right)^2}{\left(2+\dfrac{1}{n}\right)^2} = 1$$

which gives that, the ratio test is fail.

Now, we can easily see that

$$\lim_{n \to \infty} n \left[\frac{u_n}{u_{n+1}} - 1 \right] = 1.$$

$\Rightarrow$ Raabe's test also fails.

Now applying Gauss test,

Consider

$$\frac{u_n}{u_{n+1}} = \frac{(2n+2)^2}{(2n+1)^2} = \left(1 + \frac{1}{n}\right)^2 \left(1 + \frac{1}{n}\right)^{-2}$$

$$= \left(1 + \frac{2}{n} + \frac{1}{n^2}\right)\left(1 - 2.\frac{1}{2n} + 3.\frac{1}{4n^2} \cdots\right)$$

$$= 1 + \frac{1}{n} - \frac{1}{4n^2} + \cdots$$

$$= \alpha + \frac{\beta}{n} + \frac{\gamma_n}{n^2}, \text{ where } \gamma_n \to -\frac{1}{4} \text{ as } n \to \infty.$$

Here, $\alpha = 1$, $\beta = 1$. Therefore by Gauss test the series Σu_n is divergent.

Example 7. *Test the convergence of the series*

$$1 + \left(\frac{2}{3}\right)^p + \left(\frac{2.4}{3.5}\right)^p + \left(\frac{2.4.6}{3.5.7}\right)^p + \cdots$$

Solution. Neglecting first term, we have

$$u_n = \left[\frac{2.4.6\ldots(2n)}{3.5.7\ldots(2n+1)}\right]^p$$

$\Rightarrow$

$$u_{n+1} = \left[\frac{2.4.6\ldots(2n)(2n+2)}{3.5.7\ldots(2n+1)(n+3)}\right]^p$$

$\Rightarrow$

$$\frac{u_n}{u_{n+1}} = \left(\frac{2n+3}{2n+2}\right)^p = \frac{\left(1 + \frac{3}{2n}\right)^p}{\left(1 + \frac{2}{2n}\right)^p} = \left(1 + \frac{3}{2n}\right)^p \left(1 + \frac{1}{n}\right)^{-p}$$

$$= \left[1 + p.\frac{3}{2n} + O\left(\frac{1}{n^2}\right)\right]\left[1 - \frac{p}{n} + O\left(\frac{1}{n^2}\right)\right]$$

$$= \left[1 + \left(\frac{3}{2} - 1\right)\frac{p}{n} + O\left(\frac{1}{n^2}\right)\right] = 1 + \frac{\frac{1}{2}p}{n} + \left(\frac{1}{n^2}\right).$$

Then by Gauss test, the series is convergent if $p/2 > 1$ i.e., $p > 2$ and divergent if $p/2 \le 1$ i.e, $p \le 2$.

Example 8. *If x, α, β, γ are all positive, discuss the convergence of hypergeometric series*

$$1 + \frac{\alpha.\beta}{1.\gamma} + \frac{\alpha(\alpha+\beta)\beta(\beta+1)}{1.2.\gamma(\gamma+1)}x^2 + \frac{\alpha(\alpha+1)(\alpha+2)\beta(\beta+1)(\beta+2)}{1.2.3.\gamma.(\gamma+1).(\gamma+2)}$$

Solution. Since, x, α, β, γ are all positive, the given series is a series of positive terms.

Neglecting first term we have,

$$u_n = \frac{\alpha(\alpha+1)...(\alpha+n-1)\beta(\beta+1)...(\beta+n-1)}{1.2...n.\gamma.(\gamma+1)...(\gamma+n-1)}.x^n$$

$$\Rightarrow \quad u_{n+1} = \frac{\alpha(\alpha+1)...(\alpha+n-1)(\alpha+n)\beta(\beta+1)...(\beta+n-1)(\beta+n)}{1.2...n(n+1).\gamma.(\gamma+1)...(\gamma+n-1)(\gamma+n)}$$

$$\therefore \quad u_{n+1} = \frac{(n+1)(\gamma+n)}{(\alpha+n)(\beta+n)}.\frac{1}{x}$$

$$\Rightarrow \quad \lim_{n\to\infty}\frac{u_n}{u_{n+1}} = \frac{(n+1)(\gamma+n)}{(\alpha+n)(\beta+n)}.\frac{1}{x} = \frac{1}{x}$$

$\therefore$ By ratio test, the series is convergent if $\dfrac{1}{x} > 1$ i.e., $x < 1$ and divergent if $x > 1$.

When $x = 1$, the ratio test is fails.

In this case, consider

$$\therefore \quad \frac{u_n}{u_{n+1}} = \left(\frac{(n+1)(n+\gamma)}{(\alpha+n)(\beta+n)}\right) = \frac{\left(1+\dfrac{1}{n}\right)\left(\dfrac{\gamma}{n}+1\right)}{\left(1+\dfrac{\alpha}{n}\right)\left(1+\dfrac{\beta}{n}\right)}$$

$$= \left[\left(1+\frac{1}{n}\right)\left(1+\frac{\gamma}{n}\right)\right]\left(1+\frac{\alpha}{n}\right)^{-1}\left(1+\frac{\beta}{n}\right)^{-1}$$

$$= \left[1+(1+\gamma)\frac{1}{n}+O\left(\frac{1}{n^2}\right)\right]\left[1-\frac{\alpha}{n}+O\left(\frac{1}{n^2}\right)\right]\left[1-\frac{\beta}{n}+O\left(\frac{1}{n^2}\right)\right]$$

$$= 1+\frac{1+\gamma-\alpha-\beta}{n}+O\left(\frac{1}{n^2}\right)$$

Then by Gauss test, the series is convergent if $1+\gamma-\alpha-\beta > 1$ and divergent if $1+\gamma-\alpha-\beta \le 1$ i.e., the series is convergent if $\gamma > \alpha+\beta$ and divergent if $\gamma \le \alpha+\beta$.

Example 9. *Test the convergence of the series* $1-\dfrac{1}{2^p}+\dfrac{1}{3^p}-\dfrac{1}{4^p}+...(p>0).$

Solution. Since, the given series is an alternating series.

$\therefore$ The nth term $t_n = (-1)^{n-1}u_n$, where $u_n = \dfrac{1}{n^p} > 0,(p>0).$

Now $\quad u_{n+1} - u_n = \dfrac{1}{(n+1)^p} - \dfrac{1}{n^p} = \dfrac{n^p-(n+1)^p}{n^p(n+1)^p} < 0 \, \forall n \ge 1.$

$\therefore \quad u_{n+1} \le u_n \, \forall n \ge N.$ Also $\quad \lim_{n\to\infty}u_n = \lim_{n\to\infty}\dfrac{1}{n^p}, \, p>0.$

Hence, by Leibnitz test the alternating series $\Sigma(-1)^{n-1}\dfrac{1}{n^p}$ is convergent.

Example 10. *Test the convergence of the series*

$$\frac{1}{x}-\frac{1}{x+a}+\frac{1}{x+2a}+...,x>0,a>0.$$

Solution. Since, the given series is an alternating series.

$\therefore$ The nth term $t_n = (-1)^{n-1}u_n$, where $u_n = \dfrac{1}{x+(n-1)a} > 0.$

Now $u_{n+1} - u_n = \dfrac{1}{x+na} - \dfrac{1}{x+(n-1)a}$

$$= \frac{[x+(n-1)a]-[x+na]}{[x+na][x+(n-1)a]}$$

$$= \frac{-a}{[x+ma][x+(n-1)a]} < 0$$

$\therefore \qquad u_{n+1} < u_n.$

Also, $\displaystyle\lim_{n\to\infty} u_n = \lim_{n\to\infty}\frac{1}{x+(n-1)a} = 0.$

Hence, by Leibnitz test, the given series is convergent.

Example 11. *Test the convergence of the series* $\dfrac{\log 2}{2^2} - \dfrac{\log 3}{3^2} - \dfrac{\log 4}{4^2} - \cdots$

Solution. The given series is an alternating series

Here, the nth term $t_n = (-1)^n u_n$, where $u_n = \dfrac{\log(n+1)}{(n+1)^2} > 0$

$$\lim_{n\to\infty} u_n = \lim_{n\to\infty} \frac{\log(n+1)}{(n+1)^2} = \lim_{n\to\infty}\frac{\log(n+1)}{(n+1)} \cdot \frac{1}{(n+1)} = 0.$$

Now, we shall show that $u_{n+1} \le u_n \; \forall\, n.$

Let $\qquad f(x) = \dfrac{\log x}{x^2}$

Then $\qquad f'(x) = \dfrac{x^2 \cdot \dfrac{1}{x} - 2x\log x}{x^4} = \dfrac{1-2\log x}{x^3} < 0$ when $x > e^{1/2}.$

Therefore, the function $f(x)$ is monotonically decreasing for all $x > e^{1/2}$. We know that

$$2 < e < 3 \Rightarrow 2^{1/2} < e^{1/2} < 3^{1/2}$$
$$\Rightarrow 1 < e^{1/2} < 2$$

so $\qquad f(n+2) \le f(n+1)$ for all n.

i.e, $\qquad u_{n+1} \le u_n \; \forall\, n.$

Hence, by Leibnitz test the given series is convergent.

Example 12. *Test the convergence of the series* $\displaystyle\sum_{n=1}^{\infty} \frac{(-1)^{n-1} n}{10n-1}.$

Solution. Here, the given series is an alternating series.

The nth term $t_{n.} = (-1)^{n-1} \cdot u_n$, where $u_n = \dfrac{n}{10n-1}.$

Now, $\qquad u_{n+1} < u_n \; \forall\, n$ if $\dfrac{n+1}{10n+9} < \dfrac{n}{10n-1}$

i.e., if $(n+1)(10n-1) < n(10+9)$

i.e., if $10n^2 + 9n - 1 < 10n^2 + 9n$ which is true.

Now $\displaystyle\lim_{n\to\infty} u_n = \frac{n}{10n-1} \to \frac{1}{10} \ne 0.$

Hence, by Leibnitz test the series does not converge.

Example 13. *Test the absolute convergence of the series* $1 - \dfrac{1}{2\sqrt{2}} + \dfrac{1}{3\sqrt{3}} - \ldots$

Solution. Here $\qquad \Sigma u_n = 1 - \dfrac{1}{2\sqrt{2}} + \dfrac{1}{3\sqrt{3}} - \ldots$

Then series Σu_n is absolutely convergent if $\Sigma |u_n|$ is convergent.

Now $\qquad \Sigma |u_n| = 1 + \dfrac{1}{2\sqrt{2}} + \dfrac{1}{3\sqrt{3}} + \ldots$

$$= 1 + \dfrac{1}{2^{3/2}} + \dfrac{1}{3^{3/2}}$$

$$= \Sigma \dfrac{1}{n^{3/2}}.$$

Hence, the series is convergent ($\because p = 3/2 > 1$).

$\Rightarrow$ The given series is absolutely convergent.

Example 14. *Test the absolute convergence of the series* $\Sigma \left(\dfrac{(-1)^n x^n}{n} \right).$

Solution. Here, we have

$$\Sigma u_n = \Sigma \left[\dfrac{(-1)^n x^n}{n} \right]$$

The series Σu_n is absolutely convergent if the series $\Sigma |u_n|$ is convergent.

Now, first we apply the ratio test, we have

$$\left| \dfrac{u_{n+1}}{u_n} \right| = \left| \dfrac{x^{n+1}}{n+1} \cdot \dfrac{n}{x^n} \right| = \dfrac{|x|}{1 + \dfrac{1}{n}}$$

$\Rightarrow \qquad \lim_{n \to \infty} \left| \dfrac{u_{n+1}}{u_n} \right| = \lim_{n \to \infty} \dfrac{|x|}{1 + \dfrac{1}{n}} = |x|.$

By ratio test the series $\Sigma |u_n|$ is convergent if $|x| < 1$ i.e., $-1 < x < 1$.

Hence, the given series is absolutely convergent if $-1 < x < 1$.

When $x = 1$, the given series become $1 - \dfrac{1}{2} + \dfrac{1}{3} - \dfrac{1}{4} + \ldots$ which converges conditionally by Leibnitz test.

When $x = -1$, the given series $-\left(1 + \dfrac{1}{2} + \dfrac{1}{3} + \ldots \right)$ which diverges to $-\infty$.

When $x > 1$ or $x < -1$ i.e., $|x| > 1$. We have $\lim_{n \to \infty} u_n \neq 0$ and, hence the series Σu_n does not converges. Hence, the given series is convergent if $-1 < x \leq 1$.

For $-1 < x < 1$ it converges absolutely.

Example 15. *Show that the series* $\dfrac{1}{\sqrt{1}} - \dfrac{1}{\sqrt{2}} + \dfrac{1}{\sqrt{3}} - \ldots.$ *is conditionally convergent.*

Solution. The given series is an alternating series.

$\therefore$ The nth term $t_n = (-1)^{n-1} u_n$ where $u_n = \dfrac{1}{\sqrt{n}} > 0.$

Now $\qquad u_{n+1} - u_n = \dfrac{1}{\sqrt{n+1}} - \dfrac{1}{\sqrt{n}} = \dfrac{\sqrt{n} - \sqrt{n+1}}{\sqrt{n}\sqrt{n+1}} < 0.$

$\therefore \qquad\qquad u_{n+1} < u_n. \qquad \Rightarrow \qquad \lim_{n\to\infty} u_n = \lim_{n\to\infty} \dfrac{1}{\sqrt{n}} = 0.$

$\therefore$ By Leibnitz test the given series is convergent.

But the series $\Sigma \left| \dfrac{(-1)^{n-1}}{\sqrt{n}} \right| = \Sigma \dfrac{1}{\sqrt{n}}$ is divergent $\left(\because p = \dfrac{1}{2} < 1 \right)$

Hence, the given series is conditionally convergent.

Example 16. *Discuss the convergence of the series* $1 + \dfrac{x}{1!} + \dfrac{x^2}{2!} + \dots$ *for all values of x.*

Solution. Here, we have

$$u_n = \frac{x^{n-1}}{(n-1)!} \qquad \Rightarrow \qquad u_{n+1} = \frac{x^n}{n!}$$

So, $\qquad\qquad \dfrac{|u_n|}{|u_{n+1}|} = \dfrac{|x|^{n-1}}{(n-1)!} \cdot \dfrac{n!}{|x|^n} = \dfrac{n}{|x|},$ for $x \neq 0$

$\therefore \qquad\qquad \lim_{n\to\infty} \dfrac{|u_n|}{|u_{n+1}|} = \lim_{n\to\infty} \dfrac{n}{|x|} = \infty,$ for $x \neq 0.$

$\therefore$ By the ratio test, the series $\displaystyle\sum_{n=1}^{\infty} |u_n|$ is convergent when $x \neq 0.$

Thus $\displaystyle\sum_{n=1}^{\infty} u_n$ is absolutely convergent. If $x=0$, then the series becomes $1 + 0 + 0 + \dots$ and so is convergent.

Thus the given series is absolutely convergent.

REMARKS

- Since for a convergent series $\displaystyle\sum_{n=1}^{\infty} u_n$, $\lim_{n\to\infty} u_n = 0$. Therefore, $\lim_{n\to\infty} \dfrac{x^n}{n!} = 0$ is a useful result.

3.17.1 MORE ABOUT CONDITIONAL AND ABSOLUTE CONVERGENCE

(i) If Σu_n is an absolute convergent series, then the series of its positive and the series of its negative terms are both convergent.

(ii) The divergence of $\Sigma |u_n|$ does not imply the divergence of Σu_n. For example, if $u_n = \dfrac{(-1)^{n-1}}{n}$ then $\Sigma |u_n|$ is divergent, whereas Σu_n is convergent.

(iii) Since, the series $\Sigma |u_n|$ is of positive terms, therefore, all the tests established for testing the convergence of series of positive terms, will also be the tests for determining the absolute convergence of the series Σu_n.

(iv) If Σu_n is conditionally convergent, then the series of its positive terms and the series of its negative terms are both divergent.

(v) A series with mixed signs cannot converge, if the series of its positive terms is convergent (divergent) and the series of its negative terms is divergent (convergent).

3.17.2 SUMMARY OF THE TESTS

For the guidance of the students we given below a working procedure for determining the convergence of a series.

(i) If in a series of positive terms, n^{th} term does not tend to zero, the series is divergent.

(ii) If n^{th} terms tends to zero, then a comparison test may be applied when its nth term neither involves any power of n nor involve factorials.

(iii) If the nth term is the nth power of some expression, then Cauchy's root test may be applied.

(iv) When the series involves increasing power of x or involves factorials, one should start with the ratio test.

(v) If the $\lim\limits_{n\to\infty} \dfrac{u_n}{u_n+1}$ turns out to be 1, then the ratio test fails and Raabe's test or Gauss's test is applied provided $\dfrac{u_n}{u_{n+1}}$ does not involves e, otherwise logarithmic test is applied.

(vi) For an arbitrary terms series, try with the ratio test for absolute convergence. If the limit turns out to be 1, then try some other tests. When the terms have alternating signs, then Leibnitz's test is suggested.

EXERCISE 3.4

1. Test the convergence of the following series
$$1 - \frac{1}{2} + \frac{1}{3} - \frac{1}{4} + ...$$

2. Prove that the following series is absolute convergent.
$$\left(\frac{\sqrt{2}-1}{1}\right) - \left(\frac{\sqrt{3}-\sqrt{2}}{2}\right) + \left(\frac{\sqrt{4}-\sqrt{3}}{3}\right) - ...$$

3. Show that the series $\Sigma \dfrac{\sin n\theta}{n^2}$ is absolutely convergent.

4. Show that the series
$$\frac{1^2}{4^2} + \frac{1^2.5^2}{4^2.8^2} + \frac{1^2.5^2.9^2}{4^2.8^2.12^2} + ... \text{ is convergent.}$$

5. Examine the convergence of the series
$$1 + a + b^2 + a^3 + b^4 + ...$$

6. Test the convergence of the series
$$\frac{x}{1} + \frac{1}{2}.\frac{x^2}{3} + \frac{1.3}{2.4}.\frac{x^3}{5} + ...$$

7. Show that the series $\Sigma(-1)^{n-1}\sin\dfrac{1}{n}$ is conditionally convergent.

8. Test for convergence the series
$$\Sigma\left(\frac{n^{n-1}.x^{n-1}}{n!}\right).$$

9. Test the convergence of the series
$$1 + \frac{a(1-a)}{1^2} + \frac{(1+a)a(1-a)(2-a)}{1^2.2^2} + ...$$

10. Show that the series $\dfrac{2}{1^2} - \dfrac{3}{2^2} + \dfrac{4}{3^2} - \dfrac{5}{4^2} + ...$ converge conditionally.

11. Show that the series $\dfrac{1}{x+1} - \dfrac{1}{x+2} + \dfrac{1}{x+3} - ...$ is convergent except when x is a negative integer.

12. Show that the series $\Sigma(-1)^n[\sqrt{n^2+1}-n]$ is conditionally convergent.

13. Show that the series $\sum\limits_{n=1}^{\infty} \dfrac{(-1)^{n+1}.n}{n^2+1}$ is not absolutely convergent.

14. Show that the binomial series

$$1 + nx + \frac{n(n-1)}{2!}x^2 + \ldots$$
$$\frac{n(n-1)\ldots(n-r+1)}{r!}x^r + \ldots$$

is absolutely convergent when $|x| < 1$.

15. Test the convergence of the series

$$\sum_{n=1}^{\infty} \left[\frac{1}{n} + \frac{(-1)^{n+1}}{\sqrt{n}} \right].$$

16. Show that the series

$$x + \frac{a-b}{2!}x^2 + \frac{(a-b)(a-2b)}{3!}x^3 +$$
$$\frac{(a-b)(a-2b)(a-3b)}{4!}x^4 + \ldots$$

is absolutely convergent if $|x| < \frac{1}{|b|}$.

17. Show that the series

$$1 - \frac{1}{2^3} - \frac{1}{4^3} + \frac{1}{3^3} - \frac{1}{6^3} - \frac{1}{8^3} +$$
$$\ldots + \frac{1}{(2n-1)^3} - \frac{1}{(4n-2)^3} - \frac{1}{(4n)^3} + \ldots$$

is absolutely convergent.

18. Show that the series

$$2\sin\frac{x}{3} + 4\sin\frac{x}{9} + 8\sin\frac{x}{27} + \ldots$$

converges absolutely for all finite values of x.

19. Discuss the convergence of the series

$$x^2(\log 2)^q + x^3(\log 3)^q + x^4(\log 4)^q + \ldots$$

HINTS TO SELECTED PROBLEMS

1. $u_n = \frac{1}{n}$, $u_{n+1} = \frac{1}{n+1}$

$$u_n - u_{n+1} = \frac{1}{n+1} > 0 \; \forall \; n \in N \Rightarrow u_n > u_{n+1}.$$

Also $\lim u_n = 0$.

3. Since $\Sigma \left| \frac{\sin n\theta}{n^2} \right| \le \Sigma \frac{1}{n^2}$.

The series $\Sigma \frac{1}{n^2}$ is convergent.

7. $u_n = \sin\left(\frac{1}{n}\right) > 0$

$$\sin\left(\frac{1}{n+1}\right) < \frac{1}{\sin n} \Rightarrow u_{n+1} < u_n.$$

Also $\lim_{n\to\infty} u_n = \lim_{n\to\infty} \sin\frac{1}{n} = 0$

8. By D'Alembert's test, series is convergent for

$$x < \frac{1}{e} \text{ and divergent for } x > \frac{1}{e}.$$

Then by logarithmic series, for $x = \frac{1}{e}$ the series is convergent.

18. The given series is convergent if

$$\sum_{n=1}^{\infty} u_n = \sum_{n=1}^{\infty} 2^n \sin\left(\frac{x}{3^n}\right)$$

$$\therefore \quad u_n = 2^n \sin\frac{x}{3^n} > 0, \forall \; n \in \mathbf{N}$$

$$\therefore \quad |u_n| = u_n \Rightarrow \Sigma |u_n| = \Sigma u_n$$

$$u_{n+1} = 2^{n+1} \sin\left(\frac{x}{3^{n+1}}\right)$$

$$\Rightarrow \quad \lim_{n\to\infty} \left| \frac{u_n}{u_{n+1}} \right| = \frac{3}{2} > 1.$$

Then by D'Alembert's ratio test. The given series is absolutely convergent.

Answers

1. Convergent
5. Convergent.
6. Convergent if $x \le 1$ and divergent if $x > 1$.

3.18 MULTIPLICATION OF SERIES

Consider two convergent infinite series given by

$$\sum_{n=1}^{\infty} a_n = a_1 + a_2 + \cdots + \cdots \text{ and } \sum_{n=1}^{\infty} b_n = b_1 + b_2 + \cdots$$

Then
$$\sum_{n=1}^{\infty} c_n = a_1 b_n + a_2 b_{n-1} + \cdots + a_n b_1 = \sum_{r=1}^{n} a_r b_{n-r+1}$$

Is called the Cauchy product or simply product of two given series.

THEOREM 1. **(Cauchy's theorem).** *Let Σa_n and Σb_n be two absolutely convergent series then their Cauchy product Σc_n is also absolutely convergent and sum of the cauchy product series is the product of the sums.*

Proof. By definition of Cauchy product of two infinite series Σa_n and Σb_n we can write

$$\Sigma c_n = \Sigma a_n \cdot \Sigma b_n \text{ where } c_n = \sum_{r=1}^{\infty} a_r b_{n-r+1}$$

Let us suppose

$$\alpha_n = \sum_{r=1}^{n} a_r \ , \ \beta_n = \sum_{r=1}^{n} a_r \ , \ \sum_{n=1}^{\infty} a_n = l \ , \text{and} \ \sum_{n=1}^{\infty} b_n = m$$

Then $<\alpha_n>$ and $<\beta_n>$ are the sequences of partial sums of Σa_n and Σb_n respectively. Also l and m are the sums of Σa_n and Σb_n respectively.

Consider $|a_1 b_1| + |a_1 b_2| + |a_2 b_2| + |a_2 b_1| + \ldots$ to n terms

$$\leq (|a_1| + |a_2| + \ldots |a_n| \, (|b_1|) + |b_2| + \ldots + |\, |b_n|)$$
$$\leq l.m, \forall n \in \mathbb{N}$$

Thus, the series

$$a_1 b_1 + a_1 b_2 + a_2 b_2 + a_2 b_1 + \ldots \qquad \ldots(1)$$

must be absolulety convergent. Then by Dirichlet's theorem, (1) is absolutely convergent. Hence, by grouping, we have

$$\sum_{n=1}^{\infty} \left(\sum_{r=1}^{n} a_r b_{n-r+1} \right) \quad i.e. \quad \sum_{n=1}^{\infty} c_n \qquad \ldots(2)$$

is absolutely convergent.

Finally, since the sum of first n^2 terms of (1) is $\alpha_n \beta_n$ and $\alpha_n \beta_n \to l.m$ as $n \to \infty$, therefore, the sum of the series (1) is lm. Hence, the series (2) must converge to the same sum.

Hence $\Sigma c_n = (\Sigma a_n)(\Sigma b_n)$

THEOREM 2. **(Merten's theorem).**

Let $\sum_{n=1}^{\infty} a_n$ and $\sum_{n=1}^{\infty} b_n$ be two convergent series and $\sum_{n=1}^{\infty} a_n$ converges absolutely. Then

the Cauchy product series and $\sum_{n=1}^{\infty} c_n$ converges to lm where $l = \sum_{n=1}^{\infty} a_n$ and $m = \sum_{n=1}^{\infty} b_n$.

Proof. Let $<A_n>$, $<B_n>$ and $<s_n>$ denote the sequence of partial sums of $\sum_{n=1}^{\infty} a_n$, $\sum_{n=1}^{\infty} b_n$

and $\sum_{n=1}^{\infty} c_n$ respectively. As per given, the series $\sum_{n=1}^{\infty} a_n$ and $\sum_{n=1}^{\infty} b_n$ converges to l

and m respectively, therefore $\lim\limits_{n\to\infty} A_n = l$ and $\lim\limits_{n\to\infty} B_n = m$ ...(1)

Suppose that $p_n = s_n - m \; \forall \; n$ so that $\lim\limits_{n\to\infty} p_n = \lim\limits_{n\to\infty} (B_n - m) = m - m = 0$ ...(2)

Now $\quad s_n = n^{th}$ partial sum of $\sum\limits_{n=1}^{\infty} c_n$

$\therefore \quad s_n = c_1 + c_2 + ... + c_n$

$\quad = a_1 b_1 + a_1 b_2 + a_2 b_1 + a_1 b_3 + a_2 b_2 + a_3 b_1 + ... + a_1 b_n + a_2 b_{n-1} + ... + a_n b_1$

$\quad = a_1(b_1 + b_2 + ... + b_n) + a_2(b_1 + b_2 + ... + b_{n-1})$

$\qquad\qquad + a_3(b_1 + b_2 + ... + b_{n-2}) + ... + ... + a_n b_1$

$\quad = a_1 B_1 + a_2 B_{n-1} + a_2 B_{n-2} + ... + a_n B_1$

$\quad = a_1 (p_n + m) + a_2 (p_{n-1} + m) + a_3 (p_{n-2} + m) + ... + a_n (p_1 + m)$

$\quad = a_1 p_n + a_2 p_{n-1} + a_3 p_1 + ... + a_n p_1 + m (a_1 + a_2 + ... + a_n)$

$\quad = q_n + m A_n$ where $q_n = a_1 p_n + a_2 p_{n-1} + ... + a_n p_1$

$\therefore \quad \lim\limits_{n\to\infty} s_n = \lim\limits_{n\to\infty} q_n + m \lim\limits_{n\to\infty} A_n = \lim\limits_{n\to\infty} q_n + l_m$ ...(3)

Now, we have to show that $q_n \to 0$ as $n \to \infty$.

Here, since $\sum\limits_{n=1}^{\infty} a_n$ converges absolutely $\Rightarrow \sum\limits_{n=1}^{\infty} |a_n|$ converges.

Let $\sum\limits_{n=1}^{\infty} |a_n|$ converges to l'.

From (2) $\lim\limits_{n\to\infty} p_n = 0 \Rightarrow <p_n>$ converges $\Rightarrow <p_n>$ is bounded ...(4)

$\Rightarrow \quad \exists \, k > 0$ such that $|p_n| < k \; \forall \; n$ ...(5)

By definition of convergence of sequence, for a given $\varepsilon > 0 \; \exists$ a positive integer m_1 such that

$$|p_n| < \frac{\varepsilon}{2A' + 1} \quad \forall n \ge m_1 \qquad\qquad ...(6)$$

Since, $\sum\limits_{n=1}^{\infty} |a_n|$ converge, so by Cauchy's general principle of convergence there exists a positive integer m_2 such that

$$|a_{m_2} + 1| + |a_{m_2} + 2| + ... + |a_n| < \frac{\varepsilon}{2k+1} \; \forall \; n > m_1$$

$$\Rightarrow \quad |a_{m_2} + 1| + |a_{m_2} + 1| + ... + |a_n| < \frac{\varepsilon}{2k+1} \; \forall \; n > m_2 \qquad ...(7)$$

Let $m^* = \max\{m_1 + m_2\}$

then (6) and (7) are true for $n > m^*$

When $n > 2m^*$ then $n - m^* > m^*$, we have

$$|q_n| = |a_1 p_n + a_2 p_{n-1} + ... + a_n p_1|$$

or $|q_n| = |a_n p_1 + a_{n-1} p_2 + p_{m+1} a_{n-m} + p_{m+2} a_{n-m-1} + ... + a_1 p_n)$

$$\le |p_1| |a_n| + |p_2| |a_{n-1}| + |p_{m+1}| |a_{n-m}| + |p_{m+2}| |a_{n-m-1}| + \cdots + |p_n| |a_1|$$

$$< k \,(|a_n| + |a_{n-1}| + \cdots + |a_{n-m}|) + \frac{\varepsilon}{2A'+1}(|a_{n-m-1}| + \cdots + |a_1|)$$

$$< k. \frac{\varepsilon}{2k+1} + \frac{\varepsilon}{2A'+1}(|a_1| + |a_2| + \cdots + |a_{n-m-1}|)$$

$$< k. \frac{\varepsilon}{2k+1} + \frac{\varepsilon}{2A'+1} A' \qquad \left[\because \sum_{n=1}^{\infty} |a_n| \models A' \Rightarrow \sum_{n=1}^{n-m-1} |a_n| \not< A' \right]$$

$$< \frac{k\varepsilon}{2k+1} + \frac{\varepsilon}{2A'+1}. A'$$

$$= \frac{\varepsilon}{2\left(1+\dfrac{1}{k}\right)} + \frac{\varepsilon}{2\left(1+\dfrac{1}{A'}\right)}$$

$$\Rightarrow \quad |q_n| < \varepsilon \text{ whenever } n > 2m^*$$

$$\Rightarrow \quad \lim_{n\to\infty} q_n = 0 \Rightarrow \lim_{n\to\infty} S_n = lm$$

$$\Rightarrow \quad \sum_{n=1}^{\infty} c_n \text{ converges to } l.m.$$

THEOREM 3. **(Abel's theorem).** *Let* $\displaystyle\sum_{n=1}^{\infty} a_n$ *and* $\displaystyle\sum_{n=1}^{\infty} b_n$ *be two convergent series such that*

$\displaystyle\sum_{n=1}^{\infty} a_n$ *and* $\displaystyle\sum_{n=1}^{\infty} b_n$ *converge to* l *and* m *respectively. If their Cauchy product* $\displaystyle\sum_{n=1}^{\infty} c_n$

converges, then it converges to $l.m.$

Proof. Let $<A_n>$, $<B_n>$ and $<s_n>$ be the sequences of partial sums of $\displaystyle\sum_{n=1}^{\infty} a_n$, $\displaystyle\sum_{n=1}^{\infty} b_n$

and $\displaystyle\sum_{n=1}^{\infty} c_n$ respectively, then

$$\lim_{n\to\infty} A_n = l \text{ and } \lim_{n\to\infty} B_n = m \qquad\qquad ...(1)$$

Now $s_n = c_1 + c_2 + \ldots + c_n$

$\qquad = a_1 b_1 + a_1 b_2 + a_2 b_1 + a_1 b_3 + a_2 b_2 + \ldots + a_1 b_n + a_2 b_{n-1} + \ldots + a_n b_1$

$\qquad = a_1\,(b_1 + b_2 + \ldots + b_n) + a_2\,(b_1 + b_2 + \ldots + b_{n-1}) + a_3\,(b_1 + b_2 + \ldots + b_{n-2})$

$$+ \ldots + a_n b_1$$

$\qquad = a_1 B_n + a_2 B_{n-1} + a_3 B_{n-2} + \ldots + a_n B_1 \qquad\qquad ...(2)$

which imples

$$s_{n-1} = a_1 B_{n-1} + a_2 B_{n-2} + a_3 B_{n-3} + \ldots + a_{n-1} B_1$$

$$s_{n-2} = a_1 B_{n-2} + a_2 B_{n-3} + a_3 B_{n-4} + \ldots + a_{n-2} B_1$$

$$\cdots\cdots\cdots\cdots\cdots\cdots\cdots\cdots\cdots\cdots\cdots\cdots\cdots\cdots\cdots$$

$$s_1 = a_1 B_1$$

On adding all these we get,

$$s_1 + s_2 + \ldots + s_n = a_1 B_n + (a_1 + a_2) B_{n-1} + (a_1 + a_2 + a_3) B_{n-2} + \ldots + (a_1 + a_2 + \ldots + a_n) B_1$$

$$= A_1 B_n + A_2 B_{n-1} + A_3 B_{n-2} + \ldots + A_n B_1$$

$$\Rightarrow \quad \frac{s_1 + s_2 + \ldots + s_n}{n} = \frac{A_1 B_n + A_2 B_{n-1} + A_3 B_{n-2} + \ldots + A_n B_n}{n}$$

As per given, $\sum\limits_{n=1}^{\infty} c_n$ converges, Let it converges to s, then $\lim\limits_{n\to\infty} s_n = s$

$\Rightarrow \quad \lim\limits_{n\to\infty} \dfrac{s_1 + s_2 + ... + s_n}{n} = s$ (By Cauchy's first theorem on limits)

Since $<A_n> \to l$ and $< B_n> \to m$ then by Ceasaro's theorem

$$\lim\limits_{n\to\infty} \frac{A_1 B_n + A_2 B_{n-1} + ... + A_n B_1}{n} = l.m$$

$\therefore \quad \lim\limits_{n\to\infty} \dfrac{s_1 + s_2 + ... + s_n}{n} = \lim\limits_{n\to\infty} \dfrac{A_1 B_n + A_2 B_{n-1} + ... + A_n B_1}{n} = l.m$

$\Rightarrow \quad s = l.m \qquad \Rightarrow \sum\limits_{n=1}^{\infty} c_n$ conveges to $l.m$.

Solved Examples

Example 1. *Show that the Cauchy product of two divergent series given by*

$$\sum_{n=0}^{\infty} a_n = 1 - \left(\frac{3}{2}\right) - \left(\frac{3}{2}\right)^2 - \left(\frac{3}{2}\right)^3 - ...$$

and $\sum\limits_{n=0}^{\infty} b_n = 1 + \left(2 + \dfrac{1}{2^2}\right) + \dfrac{3}{2}\left(2^2 + \dfrac{1}{2^3}\right) + \left(\dfrac{3}{2}\right)^2 \left(2^3 + \dfrac{1}{2^4}\right) + ...$ *is convergent.*

Solution. Clealy, we can write

$$\sum_{n=0}^{\infty} a_n = 1 - \sum_{n=1}^{\infty} \left(\frac{3}{2}\right)^n \text{ and } \sum_{n=0}^{\infty} b_n = 1 + \sum_{n=1}^{\infty} \left(\frac{3}{2}\right)^{n-1} \left(2^n + \frac{1}{2^{n+1}}\right)$$

If we leave the first term of both the series, we see that remaining series form G.P. with common ratio greater than 1 and therefore, both are divergent.

If $\sum\limits_{n=0}^{\infty} c_n$ be the Cauchy product, then

$$c_0 = a_0 b_0 = 1$$
$$c_n = a_0 b_n + a_1 b_{n-1} + a_2 b_{n-2} + ... + a_n b_0, \quad n \geq 1$$

Therefore

$$c_n = 1.\left(\frac{3}{2}\right)^{n-1}\left(2^n + \frac{1}{2^{n+1}}\right) + \left(\frac{-3}{2}\right)\left(\frac{3}{2}\right)^{n-2}\left\{2^{n-1} + \frac{1}{2^n}\right\} + \cdots + \left(\frac{-3}{2}\right)^n \times 1$$

$$= \left(\frac{3}{2}\right)^{n-1}\left[\left(2^n + \frac{1}{2^{n+1}}\right) - \left(2^{n-1} + \frac{1}{2^n}\right) - \left(2^{n-2} + \frac{1}{2^{n-1}}\right) - \left(2 + \frac{1}{2^2}\right)\right] - \left(\frac{3}{2}\right)^n$$

$$= \left(\frac{3}{2}\right)^{n-1}\left[2^n + \frac{1}{2^{n+1}} - \{2^{n-1} + 2^{n-2} + ... + 2\} - \left(\frac{1}{2^n} + \frac{1}{2^{n-1}} + ... + \frac{1}{2^2}\right)\right] - \left(\frac{3}{2}\right)^n$$

$$= \left(\frac{3}{2}\right)^{n-1}\left[2^n + \frac{1}{2^{n+1}} - 2\left(\frac{2^{n-1} - 1}{2 - 1}\right) - \frac{1}{2^2}\frac{\left(1 - \frac{1}{2^{n-1}}\right)}{1 - \frac{1}{2}}\right] - \left(\frac{3}{2}\right)^n$$

$$= \left(\frac{3}{2}\right)\left[2^n + \frac{1}{2^{n+1}} - 2^n + 2 - \frac{1}{2} + \frac{1}{2^n}\right] - \left(\frac{3}{2}\right)^n$$

$$= \left(\frac{3}{2}\right)^{n-1}\left[\frac{3}{2} + \frac{1}{2^n} + \frac{1}{2^{n+1}}\right] - \left(\frac{3}{2}\right)^n$$

$$= \left(\frac{3}{2}\right)^{n-1}\left[\frac{3}{2} + \frac{3}{2^{n+1}}\right] - \left(\frac{3}{2}\right)^n$$

$$= \left(\frac{3}{2}\right)^n + \left(\frac{3}{2}\right)^{n-1}\cdot\frac{3}{2}\cdot\frac{1}{2^n} - \left(\frac{3}{2}\right)^n$$

$$= \left(\frac{3}{2}\right)^n \frac{1}{2^n} = \frac{3^n}{2^{2n}} = \left(\frac{3}{4}\right)^n$$

$$\Rightarrow \sum_{n=1}^{\infty} c_n = \sum_{n=1}^{\infty}\left(\frac{3}{4}\right)^n \text{ which is a G.P. with common ratio } \frac{3}{4} < 1 \text{ and hence convergent.}$$

Example 2. *Prove that*

$$\frac{1}{1-x}\log\frac{1}{1-x} = \sum_{n=1}^{\infty}\left(1 + \frac{1}{2} + \frac{1}{3} + \cdots + \frac{1}{n}\right)x^n, \text{ for } |x| < 1.$$

Solution. Consider

$$(1+x)^{-1} = 1 + x + x^2 + \cdots = \sum_{n=0}^{\infty} x^n = \sum_{n=0}^{\infty} a_n \text{ where } a_n = x^n, n > 0$$

and $\quad \log\dfrac{1}{1-x} = \log(1-x)^{-1} = -\log(1-x)$

$$= \frac{x}{1} + \frac{x^2}{2} + \frac{x^3}{3} + \cdots + \frac{x^n}{n} + \cdots$$

$$= \sum_{n=0}^{\infty}\frac{x^{n+1}}{n+1} = \sum_{n=0}^{\infty} b_n \text{ (say) where } b_n = \frac{x^{n+1}}{n+1}, n > 0$$

Now, $\lim\limits_{n\to\infty} |a_n|^{1/n} = \lim\limits_{n\to\infty}(|x|^n)^{1/n} = |x|$

Hence, by root test $\sum\limits_{n=0}^{\infty} |a_n|$ is convergent for $|x| < 1$ and therefore $\sum\limits_{n=0}^{\infty} a_n$ is absolutely convergent for $|x| < 1$.

Further $\lim\limits_{n\to\infty} |b_{n-1}|^{1/n} = \lim\limits_{n\to\infty}\left(\left|\dfrac{x^n}{n}\right|\right)^{1/n} = \lim\limits_{n\to\infty}\dfrac{|x|}{n^{1/n}} = |x|$

By root test, $\sum\limits_{n=0}^{\infty} |b_n|$ is convergent for $|x| < 1$ and therefore $\sum\limits_{n=0}^{\infty} b_n$ is absolutely convergent for $|x| < 1$.

If $\sum\limits_{n=0}^{\infty} c_n$ be the Cauchy product of $\sum\limits_{n=0}^{\infty} a_n$ and $\sum\limits_{n=0}^{\infty} b_n$ then

$$c_n = \sum_{k=0}^{n} a_{n-k}.b_k = \sum_{k=0}^{n} x^{n-k}\left(\frac{x^{k+1}}{k+1}\right) = x^{n+1}\sum_{k=0}^{n}\frac{1}{k+1}$$

$$= x^{n+1}\left(1+\frac{1}{2}+\frac{1}{3}+...+\frac{1}{n+1}\right)$$

Now, since both $\sum_{n=0}^{\infty} a_n$ and $\sum_{n=0}^{\infty} b_n$ are absolutely convergent for $|x| < 1$,

so their Cauchy product $\sum_{n=0}^{\infty} c_n$ is convergent for $|x| < 1$.

Also, $\sum_{n=0}^{\infty} a_n.\sum_{n=0}^{\infty} b_n = \sum_{n=0}^{\infty} c_n$

$$\Rightarrow \quad \frac{1}{1-x}\log\frac{1}{1-x} = \sum_{n=0}^{\infty} x^{n+1}\left(1+\frac{1}{2}+\frac{1}{3}+...+\frac{1}{n+1}\right)$$

$$= \sum_{n=0}^{\infty} x^{n}\left(1+\frac{1}{2}+\frac{1}{3}+...+\frac{1}{n}\right)$$

EXERCISE 3.5

1. Show that the Cauchy product of the convergent series. $\sum_{n=1}^{\infty}\frac{(-1)^n}{\sqrt{n+1}}$ with itself is divergent.

2. Show that the Cauchy product of the series $\sum_{n=1}^{\infty}\frac{(-1)^n}{(n+1)^p}, p>0$ with itself converges for $p>\frac{1}{2}$ and diverges for $p\le\frac{1}{2}$.

3. Show that the Cauchy product of two series $3+\sum_{n=1}^{\infty} 3^n$ and $-2+\sum_{n=1}^{\infty} 2^n$ is absolutely convergent, although both the series are divergent.

4. Show that

$$\frac{1}{2}\left(1-\frac{1}{3}+\frac{1}{5}-\frac{1}{7}+...\right)^2 = \frac{1}{2}-\frac{1}{4}\left(1+\frac{1}{3}\right)+\frac{1}{6}\left(1+\frac{1}{3}+\frac{1}{5}\right)...$$

5. Given, $\log 2 = 1-\frac{1}{2}+\frac{1}{3}-\frac{1}{4}+...+\frac{(-1)^n}{n+1}$ prove that

$$\sum_{n=0}^{\infty}(-1)^n\left[\frac{1}{(n+1).1}+\frac{1}{n-2}+\frac{1}{(n-1).3}+...+\frac{1}{1.(n-1)}\right] = (\log 2)^2$$

 **CHAPTER REVIEW: A COMPETITIVE APPROACH**

SELECTED TERMS AND RESULTS

TERMS

- **Infinite series:** An expression of the form

$$\sum_{n=0}^{\infty} a_n = a_0 + a_1 + \ldots + a_n + \ldots$$

where each a_n is a real number, is called an infinite series of real numbers.

- **Sequence of partial sum:** The sequence $<s_n>$ $= a_1 + a_2 + \ldots + a_n$ is called the sequence of partial sum of the series $\displaystyle\sum_{n=1}^{\infty} a_n$

- **Convergent series:** A series Σa_n is said to be convergent if the sequence $<s_n>$ of partial sums of Σa_n is convergent

- **Divergent series:** The series Σa_n is said to be divergent, if the sequence $<s_n>$ of partial sums if Σa_n is divergent.

- **Oscillating series :** The series Σa_n is said to be oscillate if the sequence $<s_n>$ of partial sums of Σa_n oscillate.

- **Positive series:** An infinite series whose all terms are positive is called a positive term series.

- **Absolute convergence:** A series Σu_n is said to be absolutely convergent of the series $|\Sigma u_n|$ is convergent.

- **Conditional convergence:** A series Σu_n is said to be conditionally convergent if (i) Σu_n is convergent and (ii) Σu_n is not absolutely convergent.

RESULTS

- If Σa_n and Σb_n are two convergent series then $\Sigma(a_n \pm b_n)$ are also convergent.

- If Σa_n converges and Σb_n diverges, then $\Sigma(a_n + b_n)$ diverges.

- A necessary condition for the convergence of the series Σu_n is that $\lim_{n\to\infty} u_n = 0$

- **(Cauchy's general principle of convergence).** A necessary and sufficient condition for a series Σu_n to converge is that to real $\varepsilon > 0$ $\exists$ positive integer m such that $|u_{m+1} + u_{m+2} + \ldots + u_n| < \varepsilon \, \forall \, n \geq m.$

- A positive term series Σu_n is convergent if and only if the sequence $<s_n>$ of partial sums is bounded above.

- A positive term series cannot be oscillate.

- If Σu_n is a positive term series such that $\lim_{n\to\infty} u_n \neq 0$ then Σu_n diverges.

- If Σa_n is a convergent series of positive terms then $\sum a_n^2$ is also convergent.

- **Cauchy's n^{th} root test :** If Σu_n be a positive term series such that $\lim_{n\to\infty} (u_n)^{1/n} = l$

Then (i) Σu_n converges if $l < 1$ (ii) Σu_n diverges if $l > 1$ (iii) test fails if $l = 1$

- **D'Alembert's ratio test:** Let Σu_n be a positive terms series such that

$$\lim_{n\to\infty} \frac{u_n}{u_{n+1}} = l$$

Then
(i) Σu_n converges if $l > 1$ (ii) Σu_n diverges if $l < 1$
(iii) test fails if $l = 1$.

- **Raabe' test:** Let Σu_n be a positive term series such that

$$\lim_{n\to\infty} \left(\frac{u_n}{u_{n+1}} - 1 \right) = l$$

Then
(i) Σu_n converges if $l > 1$
(ii) Σu_n diverges if $l < 1$
(iii) test fails if $l = 1$

- Raabe's test is stronger than ratio test.

- **Cauchy's Integral test:** If $u(x)$ is a non-negative monotonically decreasing and integrable function such that $u(n) = u_n \, \forall \, n \in N$ then

the series $\displaystyle\sum_{n=1}^{\infty} u_n$ is convergent if and only

$\int_{1}^{\infty} u(x)dx$ is convergent.

- **Leibnitz test:** If an alternating series
$$\sum_{n=1}^{\infty} (-1)^{n-1} u_n \text{ satisfies}$$

(i) $u_{n+1} \le u_n \ \forall \ n$ (iii) $\lim_{n \to \infty} u_n = 0$

Then the series $\Sigma(-1)^{n-1} u_n$ converges.

- Every absolutely convergent series is convergent but converse is not necessarily true.
- Any rearrangement of a convergent series of positive terms converges to the same sum.
- Cauchy product of two absolutely convergent series is also absolutely convergent.

REVIEW QUESTIONS AND PROJECT WORK

Test the convergence of the following series:

1. $\displaystyle\sum_{n=1}^{\infty} \cos\frac{1}{n}$ **(Ans : Divergent)**

2. $\displaystyle\sum_{n=1}^{\infty} \left(\frac{1}{n^2}\right)^{1/n}$ **(Ans: Divergent)**

3. $\displaystyle\sum_{n=1}^{\infty} \frac{1}{\sqrt{n!}}$ **(Ans: Convergent)**

4. $\dfrac{1}{\sqrt{1\cdot2}} + \dfrac{1}{\sqrt{2\cdot3}} + \dfrac{1}{\sqrt{3\cdot4}} + \cdots$ **(Ans: Divergent)**

5. (i) $\displaystyle\sum_{n=1}^{\infty} \sqrt{n^3+1} - \sqrt{n^3}$ **(Ans: Convergent)**

   (ii) $\displaystyle\sum_{n=1}^{\infty} (\sqrt{n^4+1} - \sqrt{n^4-1}$ **(Ans: Convergent)**

6. $\displaystyle\sum \frac{n^{n^2}}{(n+1)^{n^2}}$ **(Ans: Convergent)**

7. $\displaystyle\sum \left(1+\frac{1}{\sqrt{n}}\right)^{-n^{3/2}}$ **(Ans: Convergent)**

8. $\displaystyle\sum 2^{-n-(-1)^n}$ **(Ans: Convergent)**

9. $\displaystyle\sum_{n=1}^{\infty} \frac{5^n}{n^2+5}$ **(Ans: Divergent)**

10. $\dfrac{\alpha}{\beta} + \dfrac{1+\alpha}{1+\beta} + \dfrac{(1+\alpha)(2+\alpha)}{(1+\beta)(2+\beta)} + \cdots$

   (Ans: Convergent if $\beta > \alpha+1$, divergent if $\beta \le \alpha+1$**)**

11. $\displaystyle\sum_{n=1}^{\infty} \frac{(-1)^{n-1}}{\sqrt{n}}$ **(Ans: Convergent)**

12. $\displaystyle\sum_{n=1}^{\infty} \frac{(-1)^{n-1}}{n\sqrt{n}}$ **(Ans: Convergent)**

13. $\displaystyle\sum_{n=1}^{\infty} \frac{(-1)^n \cos n\alpha}{\sqrt{n^3}}, \ \alpha \in R$ **(Ans: Convergent)**

14. $\dfrac{\sqrt{2}-\sqrt{1}}{1} - \dfrac{\sqrt{3}-\sqrt{2}}{2} + \dfrac{\sqrt{4}-\sqrt{3}}{3} + \cdots$

   (Ans: Convergent)

15. $\displaystyle\sum_{n=1}^{\infty} \frac{(-1)^{n+1}}{\log(n+1)}$

   (Ans: Conditionally Convergent)

OBJECTIVE TYPE QUESTIONS

FILL IN THE BLANKS

1. A series, which contains _____ no. of terms is called infinite series,

2. If sum of the first n terms of an infinite series tends to a finite limit as n tends to infinity, then series is_____.

3. If sum of first n terms of an infinite series tends to $+\infty$ or $-\infty$ as n tends to infinity then series is said to be_____.

4. If sum of first n terms of an infinite series neither tends to a definite limit nor to $+\infty$ or $-\infty$ as n tends to infinity, then series is said to be _____.

5. For every convergent series, it is necessary that_____.

6. The nature of an infinite series remains _____ by addition or removal of a finite number of terms.

7. Every absolutely convergent series is _____.

8. The sum of an absolutely convergent series is_____ of the order of terms.

9. A series whose terms are alternatively positive and negative is called an_____.

10. If Σu_n, is convergent, and $\Sigma |u_n|$ is divergent then series Σu_n is said to be_____.

TRUE/ FALSE

Write 'T' for true and 'F' for false statements.

1. For every convergent series, it is necessary that $\lim u_n = 0$ **(T/F)**

2. The series $\sum \dfrac{1}{n}$ is convergent **(T/F)**

3. If Σu_n is a series of positive terms then $u_n > 0$, $\forall \, n \in N$ **(T/F)**

4. A series of positive terms is divergent if each term after a fixed stage is greater than some fixed positive number. **(T/F)**

5. A series of positive terms is convergent if from and after some terms the ratio of each term to the preceding term is less than a fixed number which is less than unity. **(T/F)**

6. If $\lim u_n > 0$ then series is convergent. **(T/F)**

7. If $\lim u_n = 0$, then the series may or may not be convergent. **(T/F)**

8. If $\lim \ u_n = 0$, then series is always convergent. **(T/F)**

9. Every convergent series is absolutely convergent. **(T/F)**

10. Every conditionally convergent series is convergent. **(T/F)**

MULTIPLE CHOICE QUESTIONS

Choose the most appropriate one.

Problem Set-1

1. A series is defined as :
   (a) the sum of the infinite no. of terms of the sequence
   (b) the product of the infinite no. of terms of the sequence
   (c) same as sequence
   (d) none of the above

2. A series is said to be convergent if (when n tends to ∞):
   (a) the sum of the first n terms tends to a finite limit
   (b) the sum of the first n terms tends to a unique limit
   (c) the sum of the first n terms tends to a finite and unique limit
   (d) none of the above

3. A series is said to be infinite if :
   (a) the sum of the first n terms tends to $+\infty$
   (b) the sum of the first n terms tends to $-\infty$
   (c) the no. of terms are infinite
   (d) none of the above

4. A series is said to be divergent if (when $n \to \infty$):
   (a) the sum of the first n terms tends to $+\infty$
   (b) the sum of the first n terms tends to $-\infty$
   (c) both (a) and (b) are true
   (d) none of the above

5. A series is said to be oscillatory if:
   (a) it is not convergent
   (b) it is not divergent
   (c) neither convergent nor divergent
   (d) none of the above

6. If $\lim u_n = 0$ then :
   (a) series is necessarily convergent
   (b) series is necessarily divergent
   (c) may or may not be convergent
   (d) none of the above

7. If Σu_n converges to l_1 and Σv_1 converges to l_2, then $\Sigma(u_n + v_n)$ converges to :
   (a) l_1 (b) l_2
   (c) $l_1 + l_2$ (d) $l_1 - l_2$

8. If Σu_n and Σv_n are two divergent series having all positive terms, then $\Sigma(u_n + v_n)$ is:
   (a) convergent   (b) divergent
   (c) oscillatory   (d) none of these

9. The nature of the given series will be change if:
   (a) the sign of all terms are changed
   (b) a finite no. of terms are added or omitted
   (c) each terms of the series is multiplied or divided by a non-zero number
   (d) none of the above

10. If Σu_n converges to l and $c \in R$, then $\Sigma c u_n$:
    (a) converges to cl
    (b) converges to l
    (c) may or may not be converge
    (d) none of the above

11. If n^{th} term of the series does not tends to zero as $n \to \infty$ then series is :
    (a) necessarily convergent
    (b) may or may not be convergent
    (c) never convergent
    (d) none of these.

12. A geometric series is convergent if the common ratio is :

(a) less than 1

(b) less than equal to 1

(c) greater than equal to 1

(d) none of the above

13. The auxiliary series $\sum \dfrac{1}{n^p}$ is divergent if:

(a) $p = 1$ (b) $p \le 1$

(c) $p \ge 1$ (d) $p = 0$

14. If $\sum u_n$ be a series of positive terms such that $\lim u_n^{1/n} = l$, then series :

(a) converges if $l < 3$

(b) diverges if $l > 1$

(c) both (a) and (b) are true

(d) both (a) and (b) are false.

15. An infinite alternative series is convergent if :

(a) $\lim u_n = 0$

(b) each term is numerically less than the preceding terms

(c) both (a) and (b) are true

(d) none of the above

16. A series Σu_n is said to be absolutely convergent if :

(a) $\lim u_n = 0$

(b) $\lim u_n = +$ ve

(c) $\lim u_n$ exists only

(d) Σu_n is convergent.

17. A series Σu_n, is said to be conditionally convergent or semi-convergent if :

(a) Σu_n is convergent and $\Sigma |u_n|$ is not convergent

(b) Σu_n and $\Sigma |u_n|$ both must be convergent

(c) Σu_n and $\Sigma |u_n|$ both must be divergent

(d) none of the above

18. Every absolutely convergent series is :

(a) convergent

(b) divergent

(c) oscillatory

(d) may or may not be convergent

19. Every semi-convergent series is :

(a) convergent

(b) absolutely convergent

(c) divergent

(d) none of the above

20. If the terms of an absolutely convergent series are rearranged, then the series :

(a) may or may not be convergent

(b) remains convergent

(c) divergent

(d) may or may not be divergent

Problem Set -2

1. The series $\sum\limits_{n=1}^{\infty} (-1)^n$ is:

(a) convergent   (b) divergent

(c) unbounded   (d) none of these

2. The series $\dfrac{1}{1.3} + \dfrac{1}{2.4} + \dfrac{1}{3.5} + \dots$ is:

(a) convergent

(b) divergent

(c) may be convergent

(d) none of the above

3. The series $1.x + 1.2\,x^2 + 1.2.3\,x^3 + \dots + n!x^n + \dots$ is:

(a) convergent everywhere except at $x = 0$

(b) divergent everywhere except at $x = 0$

(c) divergent for $x = 0$

(d) none of the above

4. The series $\sum\limits_{n=1}^{\infty} \dfrac{n^2}{3^n}$ is:

(a) convergent   (b) divergent

(c) unbounded   (d) none of these

5. The series for $\sum\limits_{n=1}^{\infty} (-1)^n \dfrac{x^n}{n}$ for $|x| > 1$ is:

(a) convergent   (b) divergent

(c) oscillatory   (d) none of these

6. For infinite series $\sum\limits_{n=1}^{\infty} a_n$ and $\sum\limits_{n=1}^{\infty} b_n$, $b_n \ge 0$ for all n and there is a real number N such that for $n \ge N \Rightarrow |a_n| \le b_n$ if $\sum\limits_{n=1}^{\infty} b_n$ converges, then

(a) Σa_n is absolutely convergent

(b) Σb_n is absolutely convergent

(c) Σa_n is divergent

(d) none of the above

7. The series $1 + \dfrac{x}{1^2} + \dfrac{x^2}{2^2} + \dfrac{x^3}{3^2} + \dots$:

(a) converges for $|x| < 1$

(b) diverges for $|x| > 1$

(c) diverges for $|x| = 1$

(d) none of the above

8. The series $1 + x + x^2 + ... + x^n ... +$:
(a) converges for $|x| < 1$
(b) diverges for $|x| > 1$
(c) converges for $|x| = 1$
(d) none of the above

9. Let Σa_n be a convergent series of positive terms and let Σb_n be a divergent series of positive terms converges to 0. Then:
(a) $<a_n>$ converges to 0
(b) $<a_n>$ not converges to 0
(c) $<b_n>$ diverges to 0
(d) none of the above

10. Let Σa_n be a convergent series of positive terms and Σb_n be a divergent series of positive terms. Then:
(a) $\Sigma(a_n + b_n)$ is divergent
(b) $\Sigma(a_n + b_n)$ is convergent
(c) both (a) and (b) may be true
(d) none of the above

11. If for a given series Σu_n

$$\frac{u_n}{u_{n+1}} = \alpha + \frac{\beta}{n} + o\left(\frac{1}{n^p}\right), p > 1, \alpha, \beta \in R. \text{ Then}$$

(a) Σu_n converges if $\beta > 1 \, \forall \, \alpha$
(b) Σu_n converges if $\alpha > 1$ for any β
(c) Σu_n diverges to ∞
(d) none of the above

12. If $n^{1/n} \to 1$ as $n \to \infty$, then the series

$$\sum_{n=1}^{\infty} (n^{1/n} - 1)^n :$$

(a) converges
(b) diverges
(c) neither converges nor diverges
(d) none of the above

13. Which of the following inequalities will be used to show the series $\sum_{n=1}^{\infty} e^{-n^2}$ converges.
(a) $e^n < x$ if $n = 0$
(b) $e^n < x$ if $n > 0$
(c) $e^n > x$ if $n > 0$
(d) none of the above

14. A conditionally convergent series is a series which is
(a) absolutely convergent
(b) convergent but not absolutely convergent
(c) divergent
(d) none of the above

15. For the series $1 + r + r^2 + ...$ $(r > 0)$, which one of the following is true?
(a) It does not converges
(b) It does not diverges
(c) It oscillate
(d) none of the above

16. The series $\frac{1}{3^p} + \frac{1}{5^p} + \frac{1}{7^p} + \frac{1}{9^p} + ...$ converges if :
(a) $p < 1$ (b) $p > 1$
(c) $p = 1$ (d) none of these

17. the series $1 + r + r^2 + ...$ is oscillatory if:
(a) $r < 1$ (b) $r > 1$
(c) $r = 1$ (d) $r = -1$

18. The series $x + \frac{2^2 x^2}{2!} + \frac{3^3 x^3}{3!} + \frac{4^4 x^4}{4!} + ...$ is convergent if:

(a) $0 < x < \frac{1}{e}$ (b) $x > \frac{1}{e}$

(c) $x = \frac{1}{e}$ (d) none of these

19. Let Σu_n be a series of positive terms. Given that Σu_n is convergent and also $\lim\limits_{n \to \infty} \frac{u_{n+1}}{u_n}$ exists then the limit is:
(a) necessarily equal to 1
(b) necessarily less than 1
(c) necessarily geater than -1
(d) none of the above

20. The series $\sum \frac{n!.2^n}{n^n}$ is:
(a) convergent   (b) divergent
(c) oscillatory   (d) none of these

21. Which one of the following test does not give absolute convergence of a series:
(a) comparison test   (b) root test
(c) ratio test   (d) none of the above

22. If $u_n = \sqrt{n+1} - \sqrt{n}, v_n = \sqrt{n^4 + 1} - \sqrt{n^4}$ then:

(a) $\sum\limits_{n=1}^{\infty} u_n$ converges but $\sum\limits_{n=1}^{\infty} v_n$ diverges

(b) $\sum\limits_{n=1}^{\infty} u_n$ and $\sum\limits_{n=1}^{\infty} v_n$ both converges

(c) $\sum\limits_{n=1}^{\infty} u_n$ and $\sum\limits_{n=1}^{\infty} v_n$ both diverges

(d) none of these

23. Ths series:

$$\left(\frac{2^2}{1^2}-\frac{2}{1}\right)^{-1}+\left(\frac{3^3}{2^3}-\frac{3}{2}\right)^{-2}+\left(\frac{4^4}{3^4}-\frac{4}{3}\right)^{-3}+... \text{ is}$$

(a) convergent   (b) divergent
(c) oscillatory   (d) none of these

24. Ths series $\sum\left(1+\frac{1}{n}\right)^{-n^2}$ is

(a) convergent   (b) divergent
(c) oscillatory   (d) none of these

25. Ths series $\sum\dfrac{1}{\sqrt{n}+\sqrt{n+1}+\sqrt{n+2}+\sqrt{n+3}}$ is:

(a) convergent   (b) divergent
(c) oscillatory   (d) none of these

26. Ths series $\sum\dfrac{1}{n^p+n^{-p}}$ converges of:

(a) $p=1$
(b) $|p|>1$
(c) both (a) and (b) are true
(d) none of these

27. The infinite series $\sum[\sqrt[3]{n^3+1}-n]x^n$ converges if
(a) $|x|\le 1$ (b) $|x|\ge 1$
(c) $|x|=e$ (d) none of these

28. Which one of the following is divergent
(a) $\sum\sin\frac{1}{n^2}$ (b) $\sum\sin\frac{1}{n}$
(c) both (a) and (b) are true
(d) none of these

29. If Σu_n is a positive term series such that $\lim\limits_{n\to\infty} u_n \ne 0$ then Σu_n:
(a) converges   (b) diverges
(c) oscillation   (d) none of these

30. The series $\sum\limits_{n=1}^{\infty}(-1)^n.n$ is :
(a) bounded   (b) convergent
(c) divergent   (d) none of these

Answers

FILL IN THE BLANK

1. Infinite   2. Convergent   3. Divergent   4. Oscillatory
5. $\lim u_n=0$ 6. Unchanged   7. Convergent   8. Independent
9. Alternating series 10. Conditionally or semi-convergent.

TRUE OR FALSE

1. T   2. F   3. T   4. T   5. T   6. F   7. T   8. F   9. F
10. T

MULTIPLE CHOICE QUESTIONS

Problem set -1

1. (a) 2. (c) 3. (c) 4. (c) 5. (c) 6. (c) 7. (c) 8. (b) 9. (d)
10. (a) 11. (c) 12. (a) 13. (b) 14. (c) 15. (c) 16. (d) 17. (a) 18. (a)
19. (a) 20. (b)

Problem set -2

1. (b) 2. (a) 3. (b) 4. (a) 5. (b) 6. (a) 7. (b) 8. (a) 9. (a)
10. (a) 11. (b) 12. (a) 13. (c) 14. (b) 15. (a) 16. (b) 17. (d) 18. (a)
19. (b) 20. (a) 21. (a) 22. (b) 23. (a) 24. (a) 25. (b) 26. (b) 27. (a)
28. (a) 29. (a) 30. (a)

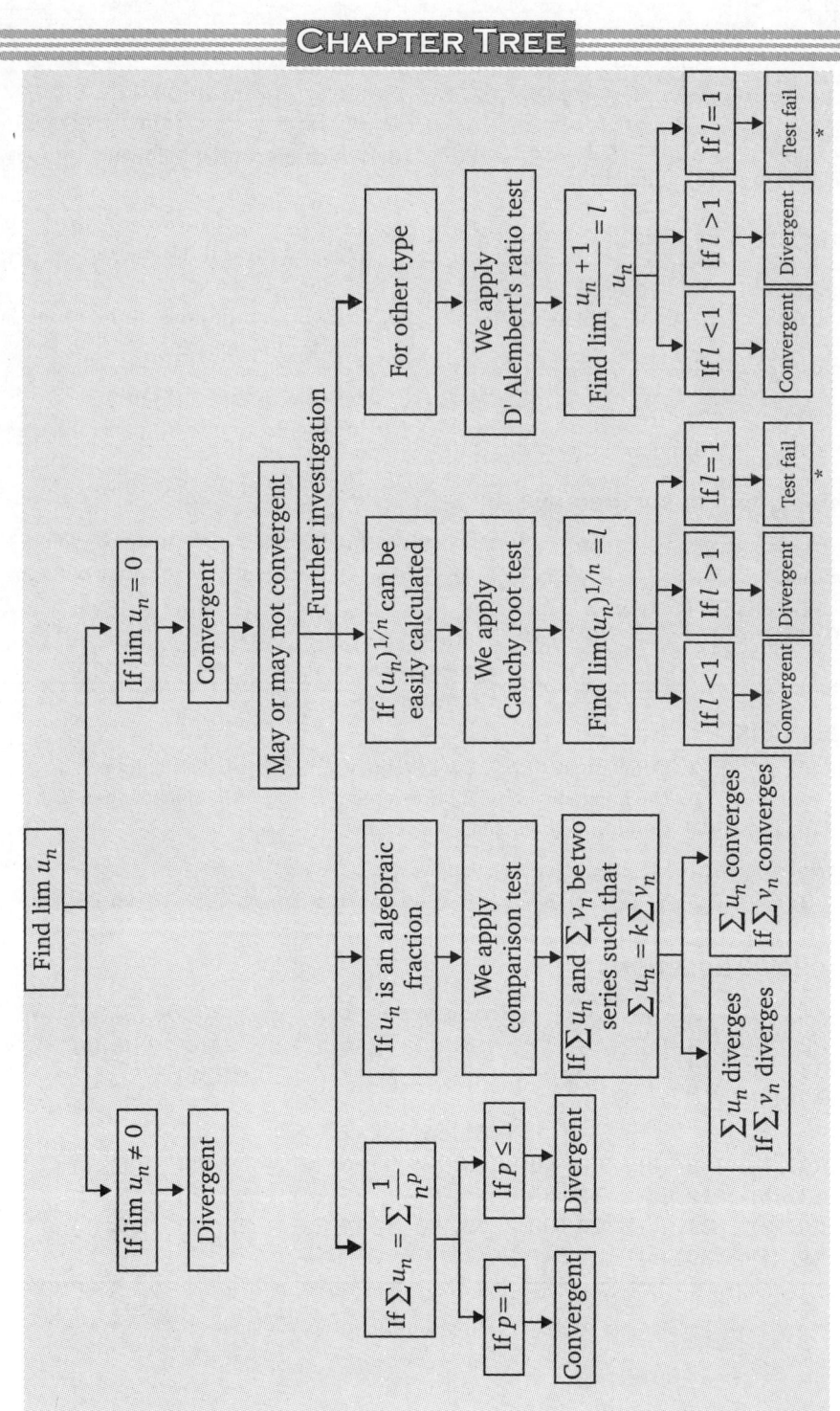

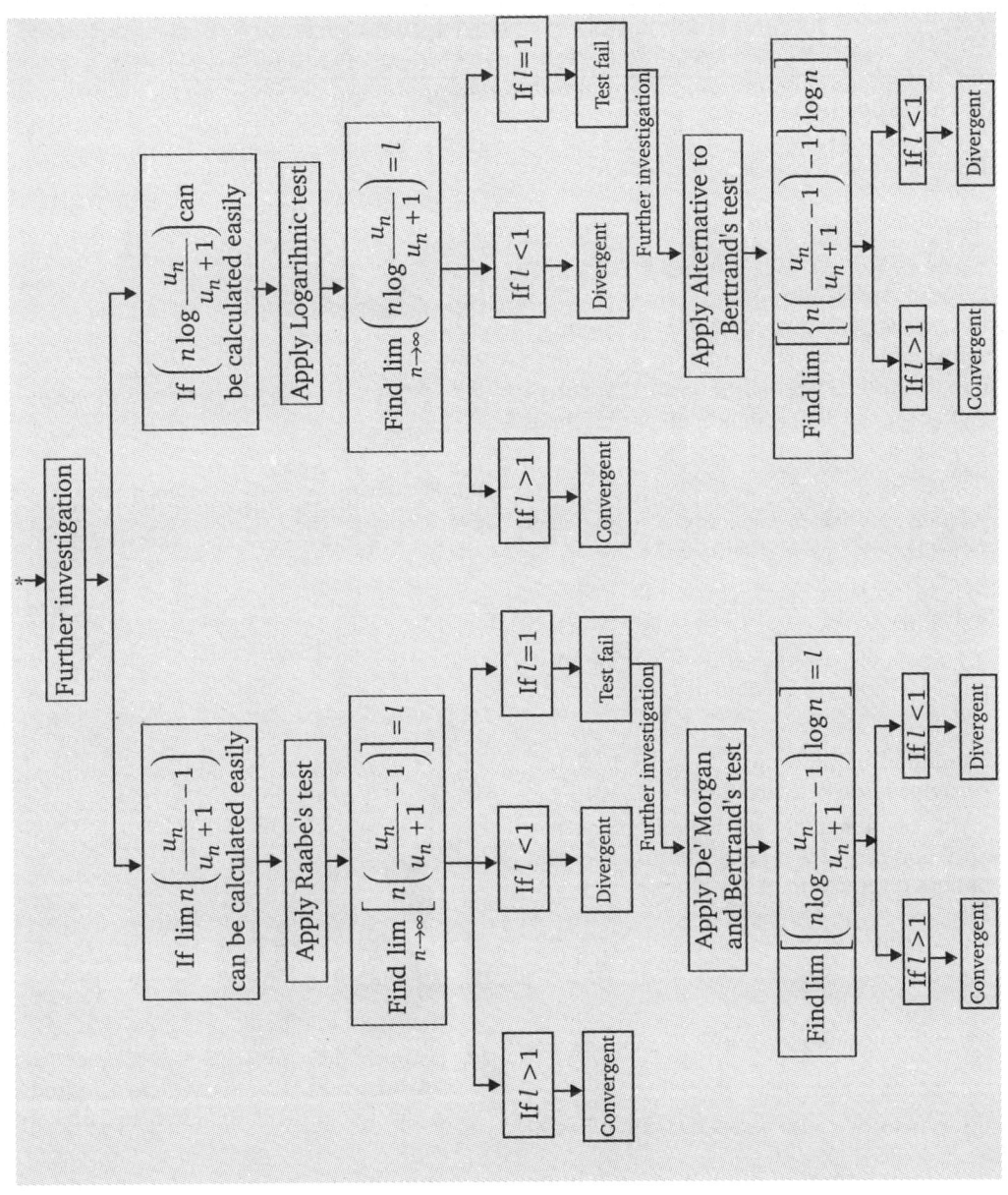

SELF ASSESSMENT TEST

Verify the following:

1. If $|r| < 1$ then the series $\sum_{n=1}^{\infty} r^n$ converges to $\frac{r}{1-r}$.

2. If Σa_n is convergent then $\Sigma a_n x^n$ is absolutly convergent when $|x| < 1$.

3. If $u_n \geq u_{n+1} \geq 0 \ \forall \ n$ then $\Sigma \ (u_n - u_{n+1})$ converges.

4. If the series of positive terms Σa_n diverges and $<s_n>$ be its sequence of partial sums then $\Sigma \frac{a_n}{s_n^2}$ converges.

5. Positive and decreasing terms of convergent series Σu_n implies $n u_n \to 0$.

6. $\sum \left(\sin\frac{1}{n} + \frac{1}{n^2} \right)$ is divergent.

7. If Σu_n is a convergent series of non-negative numbers and v_n is bounded then $\Sigma u_n.v_n$ converges absolutely.

8. The series $\sum a_n^2$ and $\Sigma \ | \ (1+a_n) e^{-a_n} -1 \ |$ converge or diverge together.

9. If $f(n)$ be monotonic and non-negative then $\Sigma f(n)$ and $\Sigma 2^n f(2^n)$ are either both convergent or both divergent to ∞.

10. If $u_n > 0$ and $c > 0$ then $\sum \frac{1}{u_n}$ and $\sum \frac{1}{(c+u_n)}$ converge or diverge together.

11. If $u_n > 0 \ \forall \ n$ then Σu_n and $\sum \frac{1}{u_n}$ shall diverge but cannot converge together.

12. If for a positive term series Σu_n, $\lim n u_n = 0$

then Σu_n may or may not converge.

13. If Σa_n, Σb_n be the convergent series of non-negative terms then Σ min $[\ a_n, b_n]$, Σ max $[a_n, b_n]$ and also $\sum \sqrt{a_n . b_n}$ converges.

14. The series $\sum \frac{n! e^n}{n^n}$ is divergent.

15. The series
$$\sum_{n=1}^{\infty} \frac{2^2.3^2 ... n^2}{(1+2+2^2)(1+3+3^2)...(1+n+n^2)}$$
diverges.

16. The series $\sum_{n=2}^{\infty} \frac{(\log n)^p}{n^q}$ converges if $q > 1$ or if $q = 1$ and $p < -1$ and otherwise diverges.

17. For $x > 0$, the series $\sum x^{1+\frac{1}{2}+...+\frac{1}{n}}$ converges if $x < \frac{1}{e}$ and diverges if $x \geq \frac{1}{e} ... $.

18. The series $1 - \frac{1}{2^s} + \frac{1}{3^s} - \frac{1}{4^s} + ...$ converges but its rearranged series
$$1 + \frac{1}{3^s} - \frac{1}{2^s} + \frac{1}{5^s} + \frac{1}{7^s} - \frac{1}{4^s} + ... \ \text{diverges to}$$
$+ \infty$ if $0 < s < 1$.

19. The series
$$1 - \frac{1}{2} - \frac{1}{4} + \frac{1}{3} - \frac{1}{6} - \frac{1}{8} + \frac{1}{5} - \frac{1}{10} - \frac{1}{12} + ...$$
converges to the sum $\frac{1}{2} \log 2$.

20. Every conditionally convergent series shall be rearranged so as to converge to a desired sum or diverge or oscillate finitely or infinitety.

4 Limit, Continuity and Differentiability of Functions of a Single Variable

4.1 INTRODUCTION

The most important idea in calculus is that of *limit*. The concept of the limit is the foundation of atmost all of mathematical analysis. In this chapter we shall introduce the notion of limits and continuity of a special class of functions whose domain is an interval and range is contained in R. These functions are known as *real valued functions of a single variable*. Since, we shall throughout be concerned with real valued functions only, the word *function* will stand for a real valued function.

4.2 GRAPH OF A FUNCTION

The graph of a function, always play an important role in discussing the nature of a function $f(x)$. It is defined as follows "*If $f : X \to Y$, be a function, then the set of all ordered pair (x, y) in which $x \in X$, appears as a first element and its image appears as its second element is called the graph of f.*

i.e., Graph of a function $f : X \to Y$

is $[\{(x, f(x)) : x \in X, f(x) \in Y\}]$.

For example. Consider the function

$$f(x) = \sin \frac{1}{x}, x \neq 0$$

Then, the graph of $f(x)$ is given.

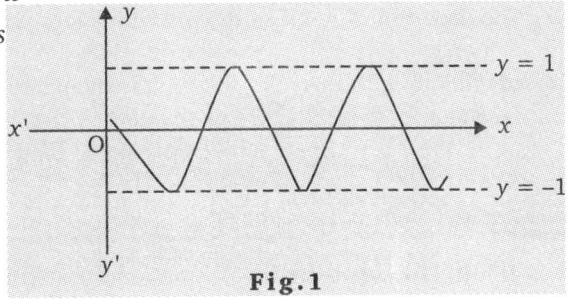

Fig.1

REMARK

- By Dedekind Cantor axiom, we know that to every real number, there correspond a unique point on a directed line and vice versa. Let us consider two mutually prependicular directed straight lines in a plane intersecting at a point O such that the point O represents the real number 0 (zero). We observe that to every ordered pair of real numbers there correspond a point in the plane and vice -versa. Thus a graph of the function can be regarded as a collection of points in the plane.

4.3 LIMIT OF A FUNCTION

Let $f(x)$ be a function defined in some interval I containing a point a, but may or may not be defined at a itself. We consider the behaviour of $f(x)$ as $x \to a$. It may happen that the values of f become closer and closer to a number l as $x \to a$ i.e., the absolute value of the difference $(f(x)-l)$ can be made smaller than any pre-assigned positive number ε, however small, by taking sufficiently close to a. In such a case, we can say that $f(x)$ approaches or converges or tends to the limit l as $x \to a$. We can write

$$\lim_{x \to a} f(x) = l \text{ or } f(x) \to l \text{ as } x \to a.$$

Formally, we define.

Definition. *Let f be a function defined in a neighbourhood of a except possible at a. Then a real number l is said to be the limit of f as x tends to a if given $\varepsilon > 0$, however small, there exists $\delta > 0$ (depending upon ε) such that*

$$|f(x) - l| < \varepsilon \text{ whenever } 0 < |x - a| < \delta$$

i.e., $\qquad\qquad l - \varepsilon < f(x) < l + \varepsilon, \text{ whenever } x \in \,]a - \delta, a[\cup]a, a + \delta[.$

REMARKS

* The definition of limit was first stated By Karl Weierstrass (1850).
* The ε–δ definition does not give the value of l. It just helps to check whether a given number l is the limit of $f(x)$.
* In the above definition, we take any real number $\varepsilon > 0$ and then choose $\delta > 0$ so that $l - \varepsilon < f(x) < l + \varepsilon$, whenever $|x - a| < \delta$ i.e., $a - \delta < x < a + \delta$. Also, $|x - a|$ denote the distance between x and a. Therefore, the limit of the function can also be interpreted as follows:
  "Given $\varepsilon > 0$, we can choose $\delta > 0$ such that if we choose x whose distance from a is less than δ, then the distance of its image from l must be less than ε".
* The number ε is given first and the number δ is to be produced.
* While taking the limit of $f(x)$ as $x \to a$, we are concerned only with the values of $f(x)$ as x takes values closer and closer to a but not when $x = a$.
* The limit of $f(x)$ at a, if exists, will continue to exist and be the same if we change the value of f at a only.
* In order to show that $\lim\limits_{x \to a} f(x) \neq l$, it is enough to produce one $\varepsilon > 0$, such that for each $\delta > 0$ there is some x for which

$$0 < |x - a| < \delta \text{ and } |f(x) - l| \geq \varepsilon$$

4.4 ONE SIDED LIMITS

(i) **Right hand limit.** A function f is said to approach l as x approaches a from right if corresponding to an arbitrary positive number ε, there exists a positive number $\delta > 0$ such that

$$|f(x) - l| < \varepsilon \text{ whenever } a < x < a + \delta$$

It is written as $\qquad f(a+0) \text{ or } \lim\limits_{x \to a+0} f(x) = l$

and $\qquad\qquad f(a+0) = \lim\limits_{h \to 0} f(a+h)$

(ii) **Left hand limit.** A function f is said to approach to l as x approaches a from the left, if corresponding to an arbitrary positive number ε, there exists a positive number $\delta > 0$ such that

$$|f(x) - l| < \varepsilon \text{ whenever } a - \delta < x < a$$

It is written as $\qquad f(a-0) \text{ or } \lim\limits_{x \to a-0} f(x) = l$

If both, right hand limit (RHL) and left hand limit (LHL) of f as $x \to a$ exist and are equal in value, then their common value will be the limit of f as $x \to a$.

REMARK

- If either or both of these limits do not exists, the limit of f as $x \to a$ does not exist. Even if both these limits exist but are not equal in value, then also the limit of f as $x \to a$ does not exist.

WORKING PROCEDURE

(i) To find the limit on right, put $a+h$ for x in $f(x)$ and then take limit as $h \to 0$.

$$\Rightarrow \quad \lim_{x \to a+0} f(x) = \lim_{h \to 0} f(a+h).$$

(ii) To find the limit on left, put $a-h$ for x in $f(x)$ and then take limit as $h \to 0$.

$$\Rightarrow \quad \lim_{x \to a-0} f(x) = \lim_{h \to 0} f(a-h).$$

4.4.1 GRAPHICAL REPRESENTATION OF RHL AND LHL

Let $y = f(x)$ be a function.

If $x \to a$, then for those values of x which greater than a, let l_1 be the limit of $f(x)$

i.e.,

$$\lim_{x \to a+0} f(x) = l_1 \quad \text{or} \quad \lim_{x \to a+0} y = l_1.$$

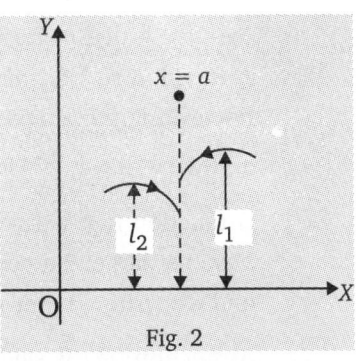

This has been shown in the figure (2) by an arround from the right because for RHL $x \to a$ from the right similarly, the LHL $= l_2$, is shown in the same figure adjoining by an arrow from left.

Fig. 2

4.5 LIMIT AT INFINITY AND INFINITE LIMITS

4.5.1 LIMITS AT INFINITY

(i) A function $f(x)$ is said to tends to a limit l as $x \to \infty$ if for given $\varepsilon > 0$, however small, there exists a positive number δ, such that

$$|f(x) - l| < \varepsilon \; \forall x \geq \delta$$

$$\Rightarrow \quad l - \varepsilon < f(x) < l + \varepsilon \; \forall x \geq \delta$$

and we write

$$\lim_{x \to \infty} f(x) = l$$

(ii) A function $f(x)$ is said to tends to a limit l as $x \to -\infty$ if for given $\varepsilon > 0$, however small, there exists a positive number $\delta > 0$, such that

$$|f(x) - l| < \varepsilon \; \forall x \leq -\delta$$

$$\Rightarrow \quad l - \varepsilon < f(x) < l + \varepsilon \; \forall x \leq -\delta$$

and we write

$$\lim_{x \to -\infty} f(x) = l$$

4.5.2 INFINITE LIMITS

(i) A function $f: A \to R$, where $A \subseteq R$ is said to tend to the limit $+ \infty$ as $x \to a$, if for any given positive number $\delta_1 > 0$, there exists a positive number δ_2 such that

$$x \in A,\ 0 < |x - a| < \delta_2 \Rightarrow f(x) > \delta_1$$

and we write $\qquad \lim_{x \to a} f(x) = \infty.$

(ii) A function $f : A \to R$, where $A \subseteq R$ is said to tend to the limit $- \infty$ as $x \to a$, if for any given positive number $\delta_1, \exists$ a positive number δ_2 such that

$$x \in A,\ 0 < |x - a| < \delta_2 \Rightarrow f(x) < -\delta_1$$

and we write $\qquad \lim_{x \to a} f(x) = -\infty$

(iii) If neither of the above two conditions are satisfied, then the function $f(x)$ is said to oscillate as $x \to a$, if a number δ_1 can possibly be assigned such that

$$|f(x)| < \delta_1 \text{ whenever } 0 < |x - a| < \delta_2$$

then the function f is said to oscillate finitely otherwise infinitely.

(iv) A function $f(x)$ is said to tend to ∞ as $x \to \infty$, if for any given positive number N, however large, $\exists$ a positive number δ such that $f(x) > N\ \forall x \geq \delta$

and we write $\qquad \lim_{x \to \infty} f(x) = \infty$

(v) A function $f(x)$ is said to tend to $-\infty$ as $x \to \infty$, if for any given positive number N, however large, $\exists$ a positive number δ such that $f(x) < -N\ \forall x \geq \delta$

and we write $\qquad \lim_{x \to \infty} f(x) = -\infty$

(vi) A function $f(x)$ is said to tend to ∞ as $x \to -\infty$, if for any given positive number N, however large, $\exists$ a positive number δ such that $f(x) > N\ \forall x \leq -\delta$

(vii) A function $f(x)$ is said to tend to $-\infty$ as $x \to -\infty$, if for any given positive number N, however large, $\exists$ a positive number δ such that $f(x) < -N\ \forall x \leq -\delta$

REMARK

- If a function f does not tend to a finite limit or to ∞ or $-\infty$ then
  (i) if it is bounded in a nbd of a, it is said to oscillate finitely.
  (ii) if it is unbounded in a nbd of a, it is said to oscillate infinitely.

4.6 UNIQUENESS OF LIMIT

THEOREM 1. *The limit of a function, if exists is unique.*

Proof. Let $f(x)$ be a function defined on an interval I. Let $a \in I$. Also, let us suppose

$$\lim_{x \to a} f(x) \text{ exist.}$$

Let if possible, $f(x)$ tends to two different limits l_1 and l_2 as $x \to a$. $(l_1 \neq l_2)$

Take $\qquad \varepsilon = \frac{1}{2} |l_1 - l_2| > 0$

Since $f(x) \to l_1$ as $x \to a$, $\exists\, \delta_1 > 0$ such that

$$|f(x) - l_1| < \varepsilon \text{ whenever } 0 < |x - a| < \delta_1 \qquad \ldots(1)$$

Now, since $f(x) \to l_2$ as $x \to a$, $\exists\, \delta_2 > 0$ such that

$$|f(x) - l_2| < \varepsilon \text{ whenever } 0 < |x - a| < \delta_2 \qquad \ldots(2)$$

Let $\delta = \min\{\delta_1, \delta_2\}$. Then

$$|l_1 - l_2| = |l_1 - f(x) + f(x) - l_2| \text{ whenever } 0 < |x-a| < \delta$$
$$\leq |f(x) - l_1| + |f(x) - l_2| \text{ whenever } 0 < |x-a| < \delta$$
$$< \varepsilon + \varepsilon = |l_1 - l_2|$$

$$\Rightarrow \qquad |l_1 - l_2| < |l_1 - l_2|$$

which is a contradiction.

Hence, $\qquad\qquad l_1 = l_2$

$\Rightarrow$ limit of a function, if exists is unique.

4.7 ALGEBRA OF LIMIT OF FUNCTIONS

THEOREM 1. *If* $\lim\limits_{x \to a} f(x) = l$ *and* $\lim\limits_{x \to a} g(x) = m$, *then*

    (i) $\lim\limits_{x \to a} [f(x) \pm g(x)] = l \pm m$ (ii) $\lim\limits_{x \to a} [f(x).g(x)] = l.m$

    (iii) $\lim\limits_{x \to a} \dfrac{f(x)}{g(x)} = \dfrac{l}{m}$, *provided* $m \neq 0$.

Proof.   (i)   Given that

$$\lim_{x \to a} f(x) = l \text{ and } \lim_{x \to a} g(x) = m$$

By definition, for $\varepsilon > 0\ \exists\ \delta > 0$ such that $|f(x) - l| < \varepsilon/2$

and $\qquad\qquad |g(x) - m| < \varepsilon/2$ for $0 < |x - a| < \delta$

Consider

$$|(f(x) \pm g(x)) - (l \pm m)| = |(f(x) - l) \pm (g(x) - m)|$$
$$\leq |(f(x) - l)| \pm |(g(x) - m)| < \varepsilon/2 + \varepsilon/2$$
$$= \varepsilon \text{ for } 0 < |x - a| < \delta$$

$$\Rightarrow \quad |(f(x) \pm g(x)) - (l \pm m)| < \varepsilon \text{ whenever } 0 < |x - a| < \delta$$

Hence, $\qquad \lim\limits_{x \to a} [f(x) \pm g(x)] = l \pm m$

  (ii) Since $\lim\limits_{x \to a} f(x) = l$, then for $\varepsilon = 1\ \exists\ \delta_1 > 0$ such that

$$|f(x) - l| < 1 \text{ for } 0 < |x-a| < \delta_1$$

or $\qquad |f(x) - l| + |l| < 1 + |l| \text{ for } 0 < |x-a| < \delta_1$

$$\Rightarrow \qquad |f(x)| \leq |f(x) - l| + |l| < 1 + |l| \text{ for } 0 < |x-a| < \delta_1 \qquad \ldots(1)$$

Also we have $\qquad \lim\limits_{x \to a} f(x) = l$ and $\qquad \lim\limits_{x \to a} g(x) = m$

Then, for $\varepsilon > 0\ \exists\ \delta_2 > 0$ such that

$$|f(x) - l| < \varepsilon \text{ and } |g(x) - m| < \varepsilon \text{ for } 0 < |x-a| < \delta_2 \qquad \ldots(2)$$

Now, consider

$$|f(x).g(x) - lm| = |f(x)g(x) - f(x).m + f(x)m - ml|$$

$$= |f(x)(g(x) - m) + m(f(x) - l)|$$

$$\le |f(x)||g(x) - m| + |m||f(x) - l|$$

$$< (1 + |l| + |m|)\varepsilon \qquad \text{[Using (1) and (2)]}$$

$$= \varepsilon_1 \text{ for } 0 < |x-a| < \delta \text{ where } \delta = \min\{\delta_1, \delta_2\}$$

$$\Rightarrow \qquad |f(x)g(x) - lm| < \varepsilon_1 \text{ for } 0 < |x-a| < \delta.$$

Hence, $\lim\limits_{x \to a} f(x).g(x) = l.m$

(iii) Since, $\lim\limits_{x \to a} g(x) = m \ne 0$, then by taking $\varepsilon = \dfrac{1}{2}m$, we can obtain that

$$|g(x)| > \frac{1}{2}|m| \qquad \qquad ...(1)$$

Also, as l, m are the limits of $f(x)$ and $g(x)$ respectively, for $\varepsilon > 0 \ \exists \ \delta_2 > 0$ such that

$$|f(x) - l| < \varepsilon \text{ and } |g(x) - m| < \varepsilon \text{ for } 0 < |x-a| < \delta_2 \qquad ...(2)$$

Now, consider

$$\left|\frac{f(x)}{g(x)} - \frac{l}{m}\right| = \left|\frac{mf(x) - lg(x)}{mg(x)}\right| = \left|\frac{m(f(x) - l) - l(g(x) - m)}{|m|.|g(x)|}\right|$$

$$\le \frac{|m||f(x) - l| + |l||g(x) - m|}{|m|.|g(x)|} < \frac{|m|.\varepsilon + |l|.\varepsilon}{|m|.\dfrac{1}{2}|m|} \quad \text{[Using (1)]}$$

$$= 2\left[\frac{|l| + |m|}{|m|^2}\right]\varepsilon = \varepsilon_1 \text{ for } 0 < |x-a| < \delta, \text{where } \delta = \min\{\delta_1, \delta_2\}$$

$$\Rightarrow \qquad \left|\frac{f(x)}{g(x)} - \frac{l}{m}\right| < \varepsilon \text{ for } 0 < |x-a| < \delta$$

$$\Rightarrow \qquad \lim\limits_{x \to a}\left[\frac{f(x)}{g(x)}\right] = \frac{l}{m}, \text{ provided } m \ne 0$$

REMARK

- $\lim\limits_{x \to a}(f \pm g)(x)$, $\lim\limits_{x \to a}(fg)(x)$ and $\lim\limits_{x \to a}\left(\dfrac{f}{g}\right)(x)$ may exists even if neither of $\lim\limits_{x \to a} f(x)$ and $\lim\limits_{x \to a} g(x)$ exists.

For example. Let f and g be defined as follows :

$$f(x) = \begin{cases} -1 \text{ if } x < a \\ 1 \text{ if } x > a \end{cases} \quad \text{and } g(x) = \begin{cases} 1 \text{ if } x < a \\ -1 \text{ if } x > a \end{cases}$$

Then $\quad (f + g)(x) = 0 \ \forall \ x \ne a$ and $(fg)(x) = -1 = \left(\dfrac{f}{g}\right)(x) \ \forall \ x \ne a$

$$\Rightarrow \quad \lim\limits_{x \to a}(f + g)(x) = 0, \qquad \lim\limits_{x \to a}(fg)(x) = -1 = \lim\limits_{x \to a}\left(\frac{f}{g}\right)(x).$$

But $\qquad \lim\limits_{x \to a} f(x) = -1$ and $\lim\limits_{x \to a+0} f(x) = 1$.

$\Rightarrow \quad \lim\limits_{x \to a-0} f(x)$ does not exist.

Similarly, $\lim\limits_{x \to a} g(x)$ does not exist.

Again, let f and g be defined as follows :

$$f(x) = \begin{cases} 1 & \text{if } x < a \\ -1 & \text{if } x > a \end{cases} \qquad \text{and } g(x) = \begin{cases} -1 & \text{if } x < a \\ 1 & \text{if } x > a \end{cases}$$

Then, $\qquad (f - g)(x) = 0 \quad \forall x \neq a$

$\Rightarrow \quad \lim\limits_{x \to a} (f - g)(x) = 0$, but $\lim\limits_{x \to a} f(x)$ and $\lim\limits_{x \to a} g(x)$ do not exist.

THEOREM 2. *If* $\lim\limits_{x \to a} f(x) = l$, *then* $\lim\limits_{x \to a} |f(x)| = |l|$.

Proof. Given that $\lim\limits_{x \to a} f(x) = l$

Then, by definition, for given $\varepsilon > 0 \; \exists$ a positive number $\delta > 0$ such that

$$|f(x) - l| < \varepsilon \text{ for } 0 < |x - a| < \delta \qquad \qquad \text{...(1)}$$

Also, we have

$$|f(x) - l| \geq ||f(x)| - |l|| \; \forall x \in R. \qquad \qquad \text{...(2)}$$

From (1) and (2), we have

$$||f(x)| - |l|| \leq |f(x) - l| < \varepsilon \text{ for } 0 < |x - a| < \delta$$

$\Rightarrow \quad \lim\limits_{x \to a} |f(x)|$ exists and $\lim\limits_{x \to a} |f(x)| = |l|$.

REMARK

- Converse, of the above theorem need not be true

  For example. Let $f(x) = \begin{cases} -1 & \text{if } x < a \\ 1 & \text{if } x > a \end{cases}$

  Then $\qquad |f(x)| = 1 \; \forall x \neq a$

  $\qquad \lim\limits_{x \to a} |f(x)| = 1$ but $\lim\limits_{x \to a-0} f(x) = -1$ and $\lim\limits_{x \to a+0} f(x) = 1$.

  $\Rightarrow \quad \lim\limits_{x \to a} f(x)$ does not exist.

- Converse of the above theorem is true only if $l = 0$.

THEOREM 3. *If* $\lim\limits_{x \to a} f(x) = l$, *then* $\lim\limits_{x \to a} e^{f(x)} = e^l$.

Proof. Since $\lim\limits_{x \to a} f(x) = l$, then for $e^l > \varepsilon > 0 \; \exists$ a positive number $\delta > 0$

Such that $\qquad \log(e^l - \varepsilon) < f(x) < \log(e^l + \varepsilon)$

$\Rightarrow \quad e^l - \varepsilon < e^{f(x)} < e^l + \varepsilon \Rightarrow |e^{f(x)} - e^l| < \varepsilon$

Hence, $\qquad \lim\limits_{x \to a} e^{f(x)} = e^l$

THEOREM 4. *If* $\lim\limits_{x \to a} f(x) = l$, *then* $\lim\limits_{x \to a} \log f(x) = \log l$.

Proof. If $\lim\limits_{x \to a} f(x) = l > 0$

For $\varepsilon > 0 \; \exists \; \delta > 0$ such that

$$0 < |x - a| < \delta \Rightarrow le - \varepsilon < f(x) < le + \varepsilon$$

$$\Rightarrow \quad -\varepsilon < \log f(x) - \log l < \varepsilon \Rightarrow |\log f(x) - \log l| < \varepsilon$$

Hence, $\lim\limits_{x \to a} \log f(x) = \log l$

THEOREM 5. *If* $f(x)$ *is a function defined on a deleted nbd D of a point a such that* $f(x) \geq 0$, *then* $\lim\limits_{x \to a} f(x) \geq 0$ *provided it exists.*

Proof. Let $\lim\limits_{x \to a} f(x) = l$.

Let if possible $l < 0$.

Setting $\varepsilon = \dfrac{|l|}{2}$, we can find a number $\delta > 0$ such that

$$|f(x) - l| < \frac{|l|}{2} \text{ for } 0 < |x - a| < \delta$$

$$\Rightarrow \qquad l - \frac{|l|}{2} < f(x) < l + \frac{|l|}{2} \text{ for } 0 < |x - a| < \delta$$

$$\Rightarrow \qquad \frac{3l}{2} < f(x) < \frac{l}{2} \text{ for } 0 < |x - a| < \delta \qquad \left(\varepsilon = \frac{|l|}{2} = -\frac{l}{2} \text{ as } l < 0 \right)$$

$$\Rightarrow \qquad f(x) < \frac{l}{2} < 0 \;\forall x \in D, \text{ which is a contradiction as } f(x) > 0.$$

Therefore, $\lim\limits_{x \to a} f(x) \geq 0$.

THEOREM 6. *If f and g are defined on a deleted nbd D of a point a and* $f(x) \geq g(x) \;\forall\; x \in D$, *then* $\lim\limits_{x \to a} f(x) \geq \lim\limits_{x \to a} g(x)$ *provided both limit exist.*

Proof. Let us define a function h on D such that

$$h(x) = f(x) - g(x) \;\forall x \in D.$$

Then $\qquad h(x) > 0$ $\hfill [f(x) > g(x)]$

$\Rightarrow \qquad \lim\limits_{x \to a} h(x) \geq 0$ $\hfill ...(1)$

Now $\qquad \lim\limits_{x \to a} h(x) = \lim\limits_{x \to a} [f(x) - g(x)] = \lim\limits_{x \to a} f(x) - \lim\limits_{x \to a} g(x)$ $\hfill ...(2)$

Now, from (1) and (2), we have

$$[\lim\limits_{x \to a} f(x) - \lim\limits_{x \to a} g(x)] \geq 0$$

$$\lim\limits_{x \to a} f(x) \geq \lim\limits_{x \to a} g(x)$$

THEOREM 7. **(Squeeze principle)** *If functions f, g and h are defined on a deleted nbd D of a point a such that*

$$f(x) \geq g(x) \geq h(x) \ \forall x \in D \text{ and } \lim_{x \to a} f(x) = \lim_{x \to a} h(x) = l$$

then $\lim_{x \to a} g(x)$ *exists and is equal to l.*

Proof. Since $\lim_{x \to a} f(x) = \lim_{x \to a} h(x) = l$, then for any $\varepsilon > 0 \ \exists$ a positive number $\delta > 0$ such that

$$|f(x) - l| < \varepsilon \text{ and } |h(x) - l| < \varepsilon \text{ for } 0 < |x-a| < \delta$$

or
$$l - \varepsilon < f(x) < l + \varepsilon$$

and
$$l - \varepsilon < h(x) < l + \varepsilon \text{ for } 0 < |x-a| < \delta.$$

Therefore, we have

$$l - \varepsilon < h(x) \leq g(x) \leq f(x) < l + \varepsilon \text{ for } 0 < |x-a| < \delta$$

$$\Rightarrow \quad l - \varepsilon < g(x) < l + \varepsilon \text{ for } 0 < |x-a| < \delta$$

$$\Rightarrow \quad |g(x) - l| < \varepsilon \text{ for } 0 < |x-a| < \delta$$

Hence, $\lim_{x \to a} g(x)$ exists and is equal to l.

REMARK

- The Squeeze principle is also known as Sandwitch theorem.

THEOREM 8. *If* $\lim_{x \to a} f(x) = 0$, $g(x)$ *is bounded in some deleted neighbourhood of a, then*

$$\lim_{x \to a} f(x).g(x) = 0.$$

Proof. Since $g(x)$ is bounded in some deleted *nbd* of a, therefore, $\exists$ positive numbers k and δ_1 such that

$$|g(x)| \leq k \text{ whenever } 0 < |x-a| < \delta_1 \qquad \qquad ...(1)$$

Let $\varepsilon > 0$ since $\lim_{x \to a} f(x) = 0$ then $\exists \ \delta_2 > 0$ such that

$$|f(x) - 0| < \varepsilon \text{ or } |f(x)| < \frac{\varepsilon}{k} \text{ whenever } 0 < |x - a| < \delta_2 \qquad \qquad ...(2)$$

Let $\delta = \min\{\delta_1, \delta_2\}$, then $0 < |x - a| < \delta \ \forall x$.

Consider $|f(x)g(x) - 0| = |f(x)g(x)| = |f(x)| \, |g(x)| < \frac{\varepsilon}{k}.k = \varepsilon$ \qquad [Using (1) and (2)]

$$\Rightarrow \quad |f(x).g(x) - 0| < \varepsilon$$

$$\Rightarrow \quad \lim_{x \to a} f(x).g(x) = 0.$$

CERTAIN LIMITS (To be used directly)

(i) $\lim_{x \to \infty} \left(1 + \frac{1}{x}\right)^x = e$

(ii) $\lim_{x \to 0} (1+x)^{1/x} = e$

(iii) $\lim_{x \to \infty} \left(1 + \frac{x}{h}\right)^h = \lim_{x \to \infty} \left(1 + \frac{1}{h}\right) = e^x$

(iv) $\lim_{x \to 0} \frac{\log(1+x)}{x} = 1$

(v) $\lim\limits_{x\to 0} \dfrac{a^x - 1}{x} = \log a \ \forall a>0$

(vi) $\lim\limits_{x\to 0} \dfrac{x^p - y^p}{x - a} = pa^{p-1} \ \forall p \neq 0$ and $a \neq 0$ if $p=0$

(vii) $\lim\limits_{x\to 0} \dfrac{\sin x}{x} = 1$

(viii) $\lim\limits_{x\to 0} \cos x = 1$

Solved Examples

Example 1. *Evaluate* $\lim\limits_{x\to a} \left(\dfrac{x^n - a^n}{x - a} \right)$.

Solution. Here we have

$$f(x) = \dfrac{x^n - a^n}{x - a}$$

$$\Rightarrow f(a+h) = \dfrac{(a+h)^n - a^n}{a+h-a}$$

$$= \dfrac{1}{h}\left[\left\{a^n + na^{n-1}.h + \dfrac{n(n-1)}{2!}\right.\right.$$

$$\left.\left. a^{n-2}.h^2 + ...\right\} - a^n\right]$$

Now, RHL $= f(a+0)$

$$= \lim\limits_{h\to 0} f(a+h) = na^{n-1} \qquad ...(1)$$

Similarly we can find

LHL $= f(a-0)$

$$= \lim\limits_{h\to 0} f(a-h) = na^{n-1} \quad ...(2)$$

Now, from (1) and (2) we conclude that

$$f(a+0) = f(a-0) = na^{n-1}$$

Example 2. *Evaluate* $\lim\limits_{x\to 0} \dfrac{(1+x)^n - 1}{x}$.

Solution. Here we have

$$f(x) = \dfrac{(1+x)^n - 1}{x}$$

∴ $\quad$ RHL $= f(0+0) = \lim\limits_{h\to 0} f(0+h) = \lim\limits_{h\to 0} [(1+h)^n - 1]/h$

$$= \lim\limits_{h\to 0} \dfrac{1}{h}\left[\left\{1 + nh + \dfrac{n(n-1)}{2!}h^2 + ...\right\} - 1\right] = n \qquad ...(1)$$

Also $\quad$ LHL $= f(0-0) = \lim\limits_{h\to 0} f(0-h) = \lim\limits_{h\to 0} [(1-h)^n - 1]/-h$

$$= \lim\limits_{h\to 0} \dfrac{1}{-h}\left[\left\{1 - nh + \dfrac{n(n-1)}{2!}h^2 + ...\right\} - 1\right] = n \qquad ...(2)$$

Based on the following Results

- R.H.L. $= \lim\limits_{x\to a+0} f(x) = \lim\limits_{h\to 0} f(a+h)$

- L.H.L. $= \lim\limits_{x\to a-0} f(x) = \lim\limits_{h\to 0} f(a-h)$

- If $\lim\limits_{x\to a} f(x) = l$, $\lim\limits_{x\to a} g(x) = m$, then

  (i) $\lim\limits_{x\to a} [f(x) \pm g(x)] = l \pm m$

  (ii) $\lim\limits_{x\to a} [f(x).g(x)] = l.m$

  (iii) $\lim\limits_{x\to a} \left[\dfrac{f(x)}{g(x)}\right] = \dfrac{l}{m} \ (m \neq 0)$

  (iv) $\lim\limits_{x\to a} |f(x)| = |l|$

  (v) $\lim\limits_{x\to a} e^{f(x)} = e^l$

  (vi) $\lim\limits_{x\to a} \log f(x) = \log l$

- **Squeeze Principle.** If functions, f, g and h are defined on a deleted *nbd* D of a point a such that

  $f(x) \geq g(x) \geq h(x) \forall x \in D$ and $\lim\limits_{x\to a} h(x) = l$ then

  $\lim\limits_{x\to a} g(x)$ exists and is equal to l.

- If $\lim\limits_{x\to a} f(x) = 0$ and $g(x)$ is bounded, then $\lim\limits_{x\to a} [f(x).g(x)] = 0$

Now, from (1) and (2) we find that

$$\text{LHL} = \text{RHL} = n \Rightarrow \lim_{x \to 0} f(x) = n.$$

Example 3.   *Evaluate* . $\lim_{x \to 0} (1+x)^{1/x}$.

Solution.   Here we have

$$f(x) = (1+x)^{1/x}$$

$$\text{RHL} = f(0+0) = \lim_{h \to 0} f(0+h) = \lim_{h \to 0} (1+h)^{1/h}$$

$$= \lim_{h \to 0} \left[1 + \frac{1}{h} \cdot h + \frac{\frac{1}{h}\left(\frac{1}{h} - 1\right)}{2!} \left(h^2\right) + \ldots \right]$$

$$= 1 + 1 + \frac{1}{2!} + \frac{1}{3!} + \ = e \qquad\qquad \ldots(1)$$

Similarly, $\text{LHL} = f(0-0) = \lim_{h \to 0} (1-h)^{-1/h}$

$$= \lim_{h \to 0} \left[1 - \frac{1}{h} \cdot (-h) + \frac{\left(-\frac{1}{h}\right)\left(-\frac{1}{h} - 1\right)}{2!} \left(-h^2\right) + \ldots \right]$$

$$= \lim_{h \to 0} \left[1 + 1 + \frac{1(1+h)}{2!} + \frac{1(1+h)(1+2h)}{3!} + \ldots \right]$$

$$= 1 + 1 + \frac{1}{2!} + \frac{1}{3!} + \ = e \qquad\qquad \ldots(2)$$

From (1) and (2) we find that RHL=LHL=e

$$\Rightarrow \quad \lim_{x \to 0} (1+x)^{1/x} . = e$$

Example 4.   *Evaluate* $\lim_{x \to 0} \left(x \sin \frac{1}{x} \right)$.

Solution.   Let $f(x) = x \sin \frac{1}{x}$

Now, $\text{RHL} = f(0+0) = \lim_{h \to 0} f(0+h)$

$$= \lim_{h \to 0} (0+h) \sin\left(\frac{1}{0+h} \right) = \lim_{h \to 0} h \sin \frac{1}{h}$$

$$= 0 \times \text{a finite quantity lying between} -1 \text{ and } 1$$

$$= 0. \qquad\qquad \ldots(1)$$

Also, $\text{LHL} = f(0-0) = \lim_{h \to 0} f(0-h) = \lim_{h \to 0} (0-h) \sin\left(\frac{1}{0-h} \right)$

$$= \lim_{h \to 0} h \sin \frac{1}{h}$$

$$= 0. \qquad \qquad \qquad ...(2)$$

Now, from (1) and (2) we conclude that RHL = LHL = 0

Hence, $\lim\limits_{x \to 0} \left(x \sin \dfrac{1}{x} \right) = 0$

Example 5. *Using $\varepsilon - \delta$ definition, evaluate $\lim\limits_{x \to 0} x^2 \sin \dfrac{1}{x}$.*

Solution. Let $\qquad \qquad f(x) = x^2 \sin \dfrac{1}{x}$

then $\qquad | f(x) - 0 | = \left| x^2 \sin \dfrac{1}{x} \right| = |x^2| \left| \sin \dfrac{1}{x} \right|$

Now, since $\left| \sin \dfrac{1}{x} \right| \le 1$ therefore

$$|f(x) - 0| \le |x^2|$$

$\Rightarrow \qquad \qquad |f(x) - 0| < \varepsilon$ whenever $0 < |x^2| < \varepsilon$

i.e., when $\qquad \qquad 0 < |x| < \sqrt{\varepsilon} \qquad$ *i.e.,* when $\qquad \qquad 0 < |x| < \delta (\delta^2 = \varepsilon)$

Hence, by the definition of limit, we have

$$\lim\limits_{x \to 0} x^2 \sin \dfrac{1}{x} = 0$$

Example 6. *Evaluate $\lim\limits_{x \to 0} \left[\left(a^x - b^x \right) / x \right]$*

Solution. Let $\qquad \qquad f(x) = \dfrac{a^x - b^x}{x}$

RHL $\qquad = f(0+0) = \lim\limits_{h \to 0} f(0+h) = \lim\limits_{h \to 0} \dfrac{a^{0+h} - b^{0+h}}{(0+h)} = \lim\limits_{h \to 0} \dfrac{a^h - b^h}{h}$

$$= \lim\limits_{h \to 0} \dfrac{1}{h} \left[\left\{ 1 + h \log_e a + \dfrac{h^2}{2!} (\log_e a)^2 + ... \right\} \right.$$

$$\left. - \left\{ 1 + h \log_e b + \dfrac{h^2}{2!} (\log_e b)^2 + ... \right\} \right]$$

$$\left(\because a^x = 1 + x \log_e a + \dfrac{x^2}{2!} (\log_e a)^2 + ... \right)$$

$$= \lim\limits_{h \to 0} \left[(\log_e a - \log_e b) + \dfrac{h}{2!} \left\{ (\log_e a)^2 - (\log_e b)^2 \right\} + ... \right]$$

$$= \log_e a - \log_e b = \log_e \dfrac{a}{b} \qquad \qquad ...(1)$$

Similarly, we can find

LHL $\qquad = f(0-0) = \lim\limits_{h \to 0} f(0-h) = \log_e \dfrac{a}{b} \qquad \qquad ...(2)$

Thus, we find from (1) and (2) that both RHL and LHL exist and each equal to

$\log_e \dfrac{a}{b}$ hence, $\displaystyle\lim_{x \to 0}\left[\dfrac{a^x - b^x}{x}\right] = \log_e\left(\dfrac{a}{b}\right)$

Example 7. Let $f(x) = \begin{cases} x & \text{if } x \text{ is rational} \\ -x & \text{if } x \text{ is irrational} \end{cases}$. *Check the existence of the limit of* $f(x)$.

Solution. Here, we have

$$f(x) = \begin{cases} x & \text{if } x \text{ is rational} \\ -x & \text{if } x \text{ is irrational} \end{cases}$$

Now, there are following cases:

Case (i) *If* a *is a non-zero rational number* .

Here, $\text{LHL} = f(a - 0) = \displaystyle\lim_{h \to 0} f(a - h)$

$$= \begin{cases} \displaystyle\lim_{h \to 0} & (a - h) = a, & \text{if } (a - h) \text{ is rational} \\ \displaystyle\lim_{h \to 0} & -(a - h) = -a, & \text{if } (a - h) \text{ is irrational} \end{cases}$$

which is not unique.

$\Rightarrow \quad f(a - 0)$ does not exist.

$\Rightarrow \quad \displaystyle\lim_{x \to a} f(x)$ does not exist.

Case (ii) If $\qquad a = 0$.

Here, $\text{LHL} = f(0 - 0) = \displaystyle\lim_{h \to 0} f(0{-}h) = \displaystyle\lim_{h \to 0} f({-}h)$

$$= \begin{cases} \displaystyle\lim_{h \to 0} & (-h) = 0, & \text{if } -h \text{ is rational} \\ \displaystyle\lim_{h \to 0} & h = 0, & \text{if } -h \text{ is irrational} \end{cases}$$

Similarly, $f(0 + 0) = 0$

Hence, $\quad f(0 + 0) = f(0 - 0) = 0 \Rightarrow \displaystyle\lim_{x \to 0} f(x)$ exists and is equal to zero.

Case (iii) *If* a *is an irrational number.*

Here, $\qquad \text{LHL} = f(a{-}0) = \displaystyle\lim_{h \to 0} f(a{-}h)$

$$= \begin{cases} \displaystyle\lim_{h \to 0} & (a - h) = a, & \text{if } (a - h) \text{ is rational} \\ \displaystyle\lim_{h \to 0} & -(a - h) = -a, & \text{if } (a - h) \text{ is irrational} \end{cases}$$

$\Rightarrow \qquad \displaystyle\lim_{x \to a} f(x)$ does not exist.

Hence, we have that $\displaystyle\lim_{x \to 0} f(x)$ exists only when $a = 0$.

Example 8. *Show that* $f(x) = \displaystyle\lim_{x \to 2} \dfrac{|x - 2|}{x - 2}$ *does not exist.*

Solution. Let $f(x) = \dfrac{|x - 2|}{x - 2}$

Now $\text{RHL} = f(2+0) = \displaystyle\lim_{h \to 0} f(2+h) = \displaystyle\lim_{h \to 0} \dfrac{|2 + h - 2|}{(2 + h - 2)} = \displaystyle\lim_{h \to 0} \dfrac{h}{h} = \displaystyle\lim_{h \to 0} 1 = 1$

$$\text{LHL}=f(2-0)=\lim_{h\to0}f[2-h]=\lim_{h\to0}\frac{|2-h-2|}{(2-h-2)}=\lim_{h\to0}\frac{|-h|}{-h}=\lim_{h\to0}-1=-1.$$

Since, $\qquad\qquad\qquad f(2+0)\neq f(2-0).$

Hence, $\lim_{x\to2}\dfrac{|x-2|}{x-2}$ does not exist.

Example 9. *Discuss the existence of the limit of the function f defnied by*

$$f=\begin{cases}1 & if & x<1 \\ 2-x & if & 1\le x\le2 \\ 2 & if & x\ge2\end{cases}$$

Solution. Here, we check the existence of the limit at $x=1$ and $x=2$.

Case (i) At $x=1$

$$\text{RHL}=f(1+0)=\lim_{h\to0}f(1+h)$$
$$=\lim_{h\to0}[2-(1+h)]=\lim_{h\to0}(1-h)=1$$

$$\text{LHL}=f(1-0)=\lim_{h\to0}f(1-h)=\lim_{h\to0}1=1$$

$\Rightarrow\qquad f(1+0)=f(1-0)=1\Rightarrow\lim_{x\to1}f(x)$ exists and is equal to 1.

Case (ii) At $x=2$

$$\text{RHL}=f(2+0)=\lim_{h\to0}=f(2+h)=\lim_{h\to0}2=2$$

and $\text{LHL}=f(2-0)=\lim_{h\to0}f(2-h)=\lim_{h\to0}f[2-(2-h)]=\lim_{h\to0}h=0$

Since $f(2+0)\neq(2-0),$hence $\lim_{x\to2}f(x)$does not exist.

Example 10. *Using $\varepsilon-\delta$ definition,show that $\lim_{x\to2}\dfrac{1}{x}(x\neq0)=\dfrac{1}{2}$*

Solution. Let $\qquad\qquad\qquad\qquad f(x)=\dfrac{1}{x}.$

In order to show that $\lim_{x\to2}f(x)=\dfrac{1}{2}$, we are to prove that for any positive number

ε, we can find a positive number δ, when δ depend upon ε i.e., $\delta=\delta(\varepsilon)$, such that

$$\left|f(x)-\frac{1}{2}\right|<\varepsilon \text{ when } 0<|x{-}2|<\delta.$$

Now, $\qquad\qquad f(x)-\dfrac{1}{2}=\dfrac{1}{x}-\dfrac{1}{2}=\dfrac{2-x}{2x}$

$\Rightarrow\qquad\qquad \left|f(x)-\dfrac{1}{2}\right|=\dfrac{|x-2|}{2|x|}$ ...(1)

Now, choosing $\delta\le1$ and $0<|x{-}2|<\delta$, we find that $0<|x-2|<1$, as $\delta\le1$

i.e., $|x-2|<1$ and $|x-2|>0$

$\Rightarrow\qquad\qquad 2{-}1<x<2+1$ and $x\neq2$

$\Rightarrow$ $1 < x < 3$ and $x \ne 2$

$\Rightarrow$ $\dfrac{1}{1} > \dfrac{1}{x} > \dfrac{1}{3}$ and $x \ne 2$ $\Rightarrow$ $\dfrac{1}{3} < \dfrac{1}{x} < 1$ and $x \ne 2$

$\Rightarrow$ $\dfrac{1}{|x|} < 1$ and $x \ne 2$ $\left(\because \dfrac{1}{x} > \dfrac{1}{3} > 0 \Rightarrow \dfrac{1}{x} = \dfrac{1}{|x|} \right)$

Therefore, from (1), we have

$$\left| f(x) - \frac{1}{2} \right| = \frac{|x-2|}{2} \cdot \frac{1}{|x|} < \frac{\delta}{2} \cdot 1$$

Now, let us choose δ such that $\dfrac{\delta}{2} < \varepsilon$ i.e., $\delta < 2\varepsilon$.

Also $\delta \le 1$, therefore, if we take $\delta = \min\{1, 2\varepsilon\}$, we have

$$\left| f(x) - \frac{1}{2} \right| < \frac{\delta}{2} < \varepsilon \text{ when } 0 < |x-2| < \delta$$

$\Rightarrow$ $\displaystyle \lim_{x \to 2} f(x) = \frac{1}{2}$

Example 11. *If* $\displaystyle \lim_{x \to a} f(x)$ *exists and* $\displaystyle \lim_{x \to a} g(x)$ *does not exist, then show that,* $\displaystyle \lim_{x \to a} [f(x) + g(x)]$ *does not exist.*

Solution.   Let $\displaystyle \lim_{x \to a} f(x) = l$ if exists.

Then by definition of limit of a function, we have

$$\lim_{x \to a+0} f(x) = \lim_{x \to a-0} f(x) = l_1 \qquad \ldots(1)$$

Also, given that $\displaystyle \lim_{x \to a} g(x)$ does not exist. So let

$$\lim_{x \to a+0} g(x) = \lambda_1 \text{ and } \lim_{x \to a-0} g(x) = \lambda_2 \text{ such that } \lambda_1 \ne \lambda_2. \qquad \ldots(2)$$

Now, $\displaystyle \lim_{x \to a+0} [f(x) + g(x)] = \lim_{x \to a+0} f(x) + \lim_{x \to a+0} g(x) = l_1 + \lambda_1$

and $\displaystyle \lim_{x \to a-0} [f(x) + g(x)] = \lim_{x \to a-0} f(x) + \lim_{x \to a-0} g(x) = l_2 + \lambda_2$

Now, since $\lambda_1 \ne \lambda_2 \Rightarrow l_1 + \lambda_1 \ne l_2 + \lambda_2$

$\Rightarrow$ $\displaystyle \lim_{x \to a+0} [f(x) + g(x)] \ne \lim_{x \to a-0} [f(x) + g(x)]$

Hence, $\displaystyle \lim_{x \to a+0} (f(x) + g(x))$ does not exist.

Example 12. *Evaluate* $\displaystyle \lim_{x \to 0} \frac{x - |x|}{x}$.

Solution.   Let $f(x) = \dfrac{x - |x|}{x}$

Now, RHL $= f(0+0) = \displaystyle \lim_{h \to 0} f(0+h) = \lim_{h \to 0} f(h) = \lim_{h \to 0} \frac{h - |h|}{h}$

$$= \lim_{h \to 0} \frac{h-h}{h} = \lim_{h \to 0} 0 = 0$$

and LHL$= f(0-0) = \lim_{h \to 0} f(0-h) = \lim_{h \to 0} f(-h) = \lim_{h \to 0} \frac{-h - |-h|}{-h}$

$$= \lim_{h \to 0} \frac{-h - h}{-h} = \lim_{h \to 0} \frac{-2h}{-h} = \lim_{h \to 0} 2 = 2$$

Since, $f(0+0) \neq f(0-0)$. Hence, $\lim_{x \to 0} f(x)$ does not exist.

Example 13. Find $\lim_{x \to 2} \dfrac{x^2 + 3x + 2}{x - 2}$.

Solution.   Here

$$\text{RHL} = f(2+0) = \lim_{h \to 0} f(2+h) = \lim_{h \to 0} \frac{(2+h)^2 + 3(2+h) + 2}{2 + h - 2}$$

$$= \lim_{h \to 0} \frac{12 + 7h + h^2}{h} = \lim_{h \to 0} \left(\frac{12}{h} + 7 + h \right) = \infty$$

Again LHL$= f(2-0) = \lim_{h \to 0} f(2-h) = \lim_{h \to 0} \frac{(2-h)^2 + 3(2-h) + 2}{2 - h - 2}$

$$= \lim_{h \to 0} \frac{12 - 7h + h^2}{-h} = \lim_{h \to 0} \left(-\frac{12}{h} + 7 - h \right) = -\infty$$

Since $f(2 + 0) \neq f(2-0)$

Hence, $\lim_{x \to 2} f(x)$ does not exist.

Example 14. Find $\lim_{x \to \infty} \dfrac{\sin x}{x}$.

Solution.   Hence, We have

$$\lim_{x \to \infty} \frac{\sin x}{x} = \lim_{y \to 0} y \sin \frac{1}{y} \qquad \left(\text{If } x = \frac{1}{y} \right)$$

Let $f(y) = y \sin\left(\dfrac{1}{y} \right)$

Now,   RHL$= f(0+0) = \lim_{h \to 0} f(0+h) = \lim_{h \to 0} h \sin\left(\dfrac{1}{h} \right)$

$$= 0 \times a \text{ finite quantity lying between } -1 \text{ and } 1 = 0$$

SImilarly,   LHL$= f(0 - 0) = \lim_{h \to 0} f(0 - h) = \lim_{h \to 0} f(-h)$

$$= \lim_{h \to 0} -h \sin (-1/h) = \lim_{h \to 0} h \sin (1/h) = 0.$$

Since $f(0+0) = f(0 - 0) = 0$

Hence, $\lim_{y \to 0} y \sin\left(\dfrac{1}{y} \right) = 0$ i.e., $\lim_{x \to \infty} \dfrac{\sin x}{x} = 0$.

Example 15. *Evaluate the following limit, if exists.*

$$\lim_{x\to 2} \frac{2x^2-8}{x-2}.$$

Solution.   Let $f(x) = \dfrac{2x^2-8}{x-2}$

We have $f(2+0) = \lim_{h\to 0} f(2+h) = \lim_{h\to 0} \dfrac{2(2+h)^2-8}{2+h-2}$

$$= \lim_{h\to 0} \frac{8h+2h^2}{h} = \lim_{h\to 0} \frac{h(8+2h)}{h}$$

$$= \lim_{h\to 0} (8+2h) = 8$$

Again, $f(2-0) = \lim_{h\to 0} f(2-h) = \lim_{h\to 0} \dfrac{2(2-h)^2-8}{2-h-2}$

$$= \lim_{h\to 0} \frac{-8h+2h^2}{-h} = \lim_{h\to 0} \frac{-h(8-2h)}{-h} = 8$$

Since, $f(2+0)=f(2-0)=8$ Hence $\lim_{x\to 2} \dfrac{2x^2-8}{x-2}$ exists and is equal to 8.

EXERCISE 4.1

1. Evaluate the following limits:

(i) $\lim_{x\to 1} \dfrac{x^3-1}{x^2-1}$

(ii) $\lim_{x\to 0} \dfrac{\sin x}{x}$

(iii) $\lim_{x\to 0} \dfrac{a^x-1}{x}$

(iv) $\lim_{x\to 0} \dfrac{e^{1/x}-1}{e^{1/x}+1}$

(v) $\lim_{x\to 0} \dfrac{|\sin x|}{x}$

(vi) $\lim_{x\to 0} \dfrac{e^{1/x}}{e^{1/x}+1}$

(vii) $\lim_{x\to\infty} \left[x\left(a^{1/x}-1\right)\right], a>1$

(viii) $\lim_{x\to 0} \dfrac{\sqrt{1+x}-\sqrt{1-x}}{x}$

(ix) $\lim_{x\to 0} \left[\dfrac{2^x-1}{(1+x)^{1/2}-1}\right]$

(x) $\lim_{x\to 1} \left(\dfrac{\log x}{x-1}\right)$

(xi) $\lim_{x\to 0} \left[\dfrac{e^x-1}{x}\right]$

2. If $f(x) = \dfrac{\sin[x]}{[x]}, [x]\neq 0$ and $f(x)=0, [x]=0$

where [x] denotes the greatest integer less than or equal to x, then find $\lim_{x\to 0} f(x)$.

3. Show that $\lim_{x\to 0} f(x) = \lim_{x\to a} f(x-a)$.

4. Let $f(x) = \begin{cases} x, & 0\le x\le 1 \\ 3-x, & 1\le x\le 2 \end{cases}$

Show that $\lim_{x\to 1+0} f(x)=2$. Does the limit of f(x) at x=1 exists.

5. If $f(x) = a_0 x^n + a_1 x^{n-1} + \ldots + a_n$, then prove that

$$\lim_{x\to a} f(x)=f(a).$$

6. Let $f(x) = \begin{cases} 0, & \text{if } x \text{ is irrational} \\ 1, & \text{if } x \text{ is rational} \end{cases}$

then show that $\lim_{x\to a} f(x)$ does not exist for any $a\in R$.

7. Has the function $e^{-1/(x-a)} \cos\left(\dfrac{1}{x-a}\right)$ a limit when $x\to a$? If so, find it.

8. If $x\to 0$, then does the limit of the following function f exist or not ?

$$f(x) = \begin{cases} x, & \text{if } x<0 \\ 1, & \text{if } x=0 \\ x^2, & \text{if } x>0 \end{cases}$$

9. Find the limit of the function $f(x)$, when $x \to a$, where

$$f(x) = \begin{cases} \left(\dfrac{x^2}{a} - a\right), & \text{if } 0 < x < a \\ 0, & \text{if } x = a \\ a - \left(\dfrac{a^3}{x^2}\right), & \text{if } x > a \end{cases}$$

10. Using the definition of limit, show that $\lim\limits_{x \to 0} f(x) = 1$, where

$$f(x) = \begin{cases} 1 + x^2, & \text{if } x \neq 0 \\ 0, & \text{if } x = 0 \end{cases}$$

Answers

1. (i) $\dfrac{3}{2}$ (ii) 1   (iii) $\log a$ (iv) does not exist   (v) does not exist
                                                                                RHL=1
                                                                                 LHL=-1

(vi) does not exist   (vii) $\log a$ (viii) 1   (ix) $2 \log 2$
     RHL=1 LHL=0

(x) 1   (xi) 1

2. does not exist   **4.** does not exist   **7.** does not exist   **8.** 0   **9.** 0

4.8 CONTINUITY

A continuous process is one that goes on smoothly without any sudden change. Continuity of a function can also be interpreted in a similar way. For better understanding, consider the following figures. The graph of the function in fig. 3(a) has a sudden cut at the point $x = 4$ whereas the graph of the function in fig. 3(b) proceeds smoothly. We say that the function of fig. 3(b) is continuous, while function of fig. 3(a) is not continuous.

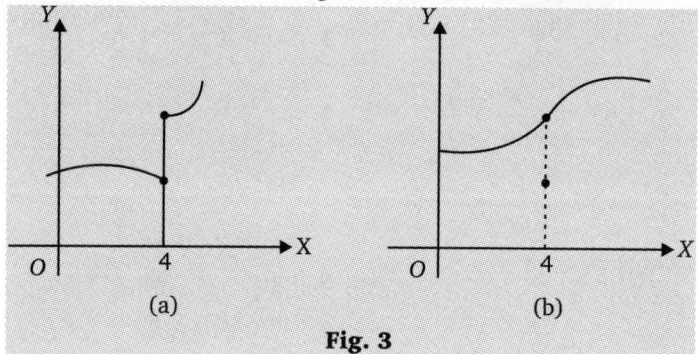

(a)   (b)

Fig. 3

Also, while defining $\lim\limits_{x \to a} f(x)$, the function f may or may not be defined at $x = a$. Even if f is defined at $x = a$, $\lim\limits_{x \to a} f(x)$ may or may not be equal to the value of the function at $x = a$. If $\lim\limits_{x \to a} f(x) = f(a)$, then we say that f is continuous at $x = a$.

REMARK

- Systematic study of the continuous nature of various phenomenon began at the close of the 17th century. The french mathematician G. W. Leibnitz (17th cent.) was a pioneer who first specified the two concepts underlying various physical phenomenon of the universe. The first of these is calculus, which is the natural language of the continuity, and the second is combinatorical analysis which deals with the discrete or the discontinuous function. The study of continuity of functions is the most important aspects of analysis and is based on the notion of limit.

4.8.1 CONTINUOUS FUNCTIONS

A function f, defined on some nbd of a point a, is said to be continuous at a if and only if any one of the following condition is saitsfied.

(i) $\lim\limits_{x \to a} f(x) = f(a)$

(ii) $f(a-0) = f(a+0) = f(a)$

(iii) for $\varepsilon > 0$, $\exists\ \delta > 0$ such that $|f(x) - f(a)| < \varepsilon$ whenever $0 < |x - a| < \delta$.

The above all conditions are equivalent to each other, and being simple, are of common use.

REMARKS

- The definition (iii) is known as Cauchy's definition of continuity.
- When a function f is stated to be continuous at a point a then by definition it is evidently implied that f is defined in some nbd of a.
- A function f is said to be continuous in I if it is continuous at every point of the interval I.
- From definition (iii), we observe that $|f(x) - f(a)| < \varepsilon$ implies that $f(a) - \varepsilon < f(x) < f(a) + \varepsilon$.
- We conclude that from the definition (iii) that the function $f(x)$ will be continuous at $x = a$ in the domain of the function $f(x)$ if the difference between $f(a)$ and the value of the function $f(x)$ at any point in the interval $]a - \delta, a + \delta[$ can be made less than any arbitrary chosen positive small number ε.
- The interval I may be any one of the following forms

  $]a,b[,\ ]-\infty,\infty[,\ ]a,\infty[,\ ]-\infty,b[.$

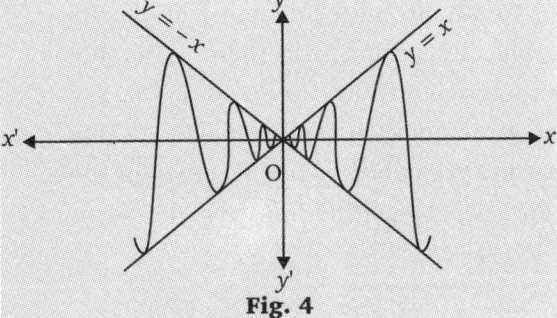

- If a function is not continuous at a point, then it is said to be discontinuous at that point.
- The value of δ depends upon the values of ε and a.
- Checking the continuity of a function from the smoothness of its graph is not a complete method. Consider the graph of the

Fig. 4

function $f(x) = x \sin\dfrac{1}{x}$, then we observe that it has no breaks in the nbd of $x=0$. But this function is not continuous. Observe that the graph oscillate widely near zero.

4.8.2 MORE DEFINITIONS OF CONTINUITY

(i) If $\lim\limits_{x \to a+0} f(x) = f(a)$, then we say that f is continuous to the right of a (or right continuous at a).

(ii) If $\lim\limits_{x \to a-0} f(x) = f(a)$, then we say that f is continuous to the left of a (or left continuous at a).

(iii) A function f is said to be continuous in an open interval $]a, b[$ if it is continuous at every point of $]a, b[$.

(iv) A function f is said to be continuous in a closed interval $[a, b]$ if it is

    (1) right continuous at a

    (2) continuous at every point of $]a, b[$

    (3) left continuous at b.

(v) A function f is said to be continuous in a semi-closed interval $[a,b[$ if it is
    (1) right continuous at a
    (2) continuous at every point of $]a,b[$.
(vi) A function f is continuous in a semi-closed interval $]a,b]$ if it is
    (1) continuous at every point of $]a,b[$
    (2) left continuous at b.
(vii) A function f is said to be continuous at $a \in I$, iff $\lim\limits_{x \to a} f(x)$ exists, finite and is equal to $f(a)$, otherwise the function is said to discontinuous at $x=a$.

4.8.3 SEQUENTIAL CONTINUITY OR HEINE'S DEFINITION OF CONTINUITY

The necessary and sufficient condition for a function f defined on an interval $I \subset R$ to be continuous at a point of interval I is that for each sequence $<a_n>$ in I converges to a, the sequence $<f(a_n)>$ converges to $f(a)$. i.e., f is said to be continuous iff

$$\lim_{n \to \infty} f(a_n) = f(a).$$

4.8.4 GRAPHICAL MEANING OF CONTINUITY OF A FUNCTION

Continuity of a function f at a point a graphically means that there is no break in the graph of the curve $y = f(x)$ at $x = a$ and given however small $\varepsilon > 0 \, \exists \, \delta > 0$ such that the graph of $y = f(x)$ from $x = a - \delta$ to $a+\delta$ lies between the lines $y = f(a) - \varepsilon$ and $y = f(a)+\varepsilon$.

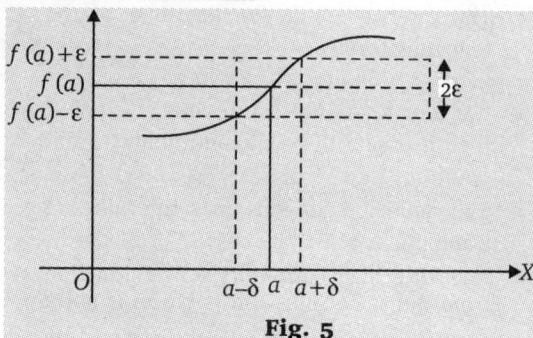

Fig. 5

☛ ILLUSTRATIONS

(1) Every constant function $f : R \to R$ is continuous on R.
    For $\varepsilon > 0$, $a \in R$, $|x{-}a| < \varepsilon \Rightarrow$ $|c - c| = 0 < \varepsilon$
(2) The identity function $f : X \to X \in R$ is continuous on R.
    For $\varepsilon > 0$, $\delta = \varepsilon$ and $|x - a| < \varepsilon \Rightarrow$ $|x{-}a| < \varepsilon \; \forall a \in R$.
(3) The function $f : X \to X^n$, $n \in N$ is continuous on R.
    For any $a \in R$, $\lim\limits_{x \to a} f(x) = a^n = f(a)$.
(4) The polynomial function $f(x) = a_0 + a_1 x + \ldots + a_n x^n$ is continuous on R.
    For any $a \in R$, $\lim\limits_{x \to a} f(x) = f(a)$.

4.9 DISCONTINUITY

(1) A function f which is not continuous at a point a is said to be discontinuous at the point 'a', where 'a' is called the point of discontinuity of f or f is said to have a discontinuity at a.

(2) A function which is discontinuous even at *a* single point of an interval, is said to be discontinuous in that interval.

(3) A function *f* can be discontinuous at a point $x = a$, because of any one of the following reasons :

     (i) $f(x)$ is not defined at $x = a$.   (ii) $\lim\limits_{x \to a} f(x)$ does not exist.

     (iii) $\lim\limits_{x \to a} f(x)$ and $f(a)$ both exist but are not equal.

4.10 TYPE OF DISCONTINUITY

4.10.1 REMOVABLE DISCONTINUITY

A function *f* is said to have a removable discontinuity at a point *a* if $\lim\limits_{x \to a} f(x)$ exists, but is not equal to the function value at *a* ,i.e.,

$$f(a-0) = f(a+0) \neq f(a)$$

REMARK
_____

- A function *f* can be made continuous by assigning some suitable value to *a*, such that
$$\lim\limits_{x \to a} f(x) = f(a)$$

For example. Suppose f is a function defined on $]0, 1[$ as follows :

$$f(x) = \begin{cases} 2, & 0 < x < 1, x \neq \dfrac{1}{2} \\ 1, & x = \dfrac{1}{2} \end{cases}$$

Then, it is clear that *f* is continuous in $]0, 1[$ except at the point $x = \dfrac{1}{2}$. At the point $x = \dfrac{1}{2}$, we have

$$f\left(\frac{1}{2}-0\right) = f\left(\frac{1}{2}+0\right) = 2 \text{ but } f\left(\frac{1}{2}\right) = 1$$

$\Rightarrow$ *f* has a removable discontinuity at $x = \dfrac{1}{2}$.

The discontinuity at $x = \dfrac{1}{2}$ may be removed by choosing $f\left(\dfrac{1}{2}\right) = 1$.

4.10.2 DISCONTINUITY OF FIRST KIND

A function *f* is said to have a discontinuity of first kind at a point *a*, if both the limits $f(a - 0)$ and $f(a + 0)$ exist but are not equal. The point *a* is said to be a point of discontinuity from the left or from right according as

$$f(a-0) \neq f(a) = f(a+0)$$

or $$f(a-0) = f(a) \neq f(a+0)$$

For example. Consider a function *f* defined on $]0, 1[$ as follows

$$f(x) = \begin{cases} 1/2, & 0 < x < 1/2 \\ 0, & x = \dfrac{1}{2} \\ -1/2, & 1/2 < x < 1 \end{cases}$$

Obviously, *f* is continuous over the open interval $]0,1/2[$ and $]1/2,1[$

At the point $x = \dfrac{1}{2}$.

$$f\left(\frac{1}{2}-0\right) = \lim_{h\to 0} f\left(\frac{1}{2}-h\right) = \frac{1}{2} \neq f(1/2)$$

$$f\left(\frac{1}{2}+0\right) = \lim_{h\to 0} f\left(\frac{1}{2}+h\right) = -\frac{1}{2} \neq f\left(\frac{1}{2}\right)$$

$\Rightarrow \qquad f\left(\dfrac{1}{2}-0\right) \neq f\left(\dfrac{1}{2}+0\right)$

$\Rightarrow$ f has a discontinuity of the first kind at $x = \dfrac{1}{2}$.

4.10.3 DISCONTINUITY OF SECOND KIND

A function f is said to have a discontinuity of second kind at a point a if none of the limit $f(a-0)$ and $f(a+0)$ exist at a. The point a is said to be a point of discontinuity of second kind from the left or from the right according as $f(a-0)$ or $f(a+0)$ does not exist.

For example. Consider the function $f(x) = \cos\left(\dfrac{\pi}{x}\right)$ defined on

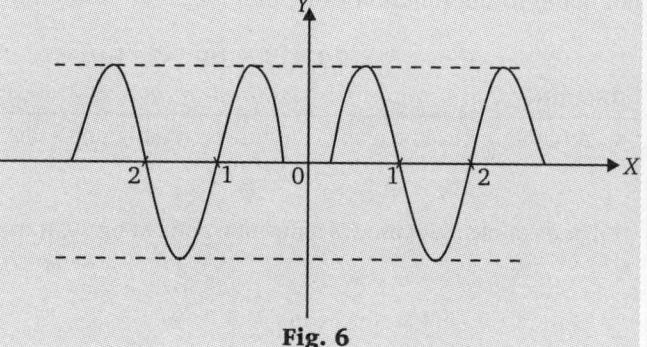

Fig. 6

$]-\infty, \infty[$. The graph of the function is given below :

Obviously, at the point $x = 0$, both the limits $i.e.,$ $\displaystyle\lim_{x\to 0-} \cos\left(\frac{\pi}{x}\right)$ and $\displaystyle\lim_{x\to 0+} \cos\left(\frac{\pi}{x}\right)$ do not exist. Hence, $x = 0$ is a point of discontinuity of the second kind.

4.10.4 MIXED DISCONTINUITY

A function f is said to have a mixed discontinuity at a point a if f has a discontinuity of second kind on one side of a and on the other side, a discontinuity of first kind or may be continuous.

For example. For the function $f(x) = e^{1/x} \sin\dfrac{1}{x}$ then $\displaystyle\lim_{x\to 0-} f(x) = 0$, $\displaystyle\lim_{x\to 0+} f(x)$ does not exist and the function is not defined at $x = 0$.

Therefore, the function has a discontinuity of first kind from the left and a discontinuity of the second kind from the right at $x = 0$. Thus, the function has a mixed discontinuity at $x = 0$.

4.10.5 INFINITE DISCONTINUITY

A function f is said to have an infinite discontinuity at $x = a$ if $f(a+0)$ or $f(a-0)$ is $+\infty$ or $-\infty$. If f has a discontinuity at a and is unbounded in every nbd of a, then f is said to have an infinite discontinuity at a.

For example. Suppose $f(x) = \dfrac{1}{x}$ in $]-\infty, \infty[$.

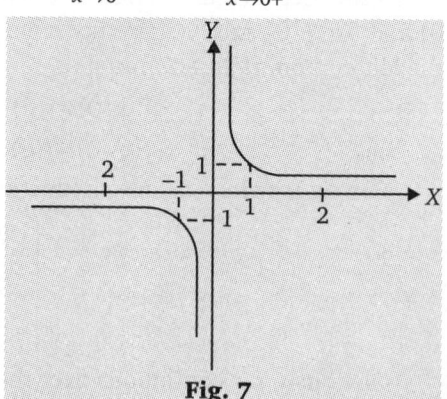

Fig. 7

It is clear that f is continuous on $]-\infty,\infty[$ except at $x=0$. At $x=0$, the limits do not exist but tends to infinity. So, $x=0$ is a point of infinite discontinuity. Hence, a rectangular hyperbola is a curve with one point of infinite discontinuity.

4.10.6 JUMP OF A FUNCTION AT A POINT

If $f(a+0)$ and $f(a-0)$ both exist, but not equal, then the jump in the function at $x=a$ is defined as the non-negative difference $f(a+0)\sim f(a-0)$.

REMARK

- A function having a finite number of jumps in a given interval is called piecewise continuous or sectionally continuous.

4.11 FOR FUNCTIONAL LIMITS

Let us suppose the function $f(x)$ be defined on the closed interval $[a,b]$ and let $x_0 \in [a,b]$.

Let the upper and lower bounds of the function $f(x)$ in the right hand nbd $[x_0, x_0+h]$ of x_0 be denoted by M and m respectively where $M=M(h)$ and $m=m(h)$. Let the sequence of diminishing values $h_1, h_2,...$ be assigned to h, which converges to zero, then $M(h_1), M(h_2), M(h_3)$ ... is a decreasing sequence and so it possesses a lower limit.

Similarly, the sequence $m(h_1), m(h_2), m(h_3)...$ is an increasing sequence and have an upper limit. These lower and upper limits are respectively known as the upper and lower limits of the function $f(x)$ at $x=x_0$ on the right and are denoted by $\overline{f(x_0+0)}$ and $\underline{f(x_0+0)}$.

$$\therefore \qquad \overline{f(x_0+0)} = \lim_{h \to 0} M(h) \text{ and } \underline{f(x_0+0)} = \lim_{h \to 0} m(h)$$

If the right hand upper limits $\overline{f(x_0+0)}$ is equal to the right hand lower limit $\underline{f(x_0+0)}$ common value is known as the right hand limit of the function $f(x)$ at $x=x_0$ and is denoted

by $f(x_0+0)$

i.e., $\qquad f(x_0+0) = \overline{f(x_0+0)} = \underline{f(x_0+0)}$

Similarly, if we consider the left hand nbd $[x_0-h, x_0]$ then the upper limit of $m(h)$ and the lower limit of $M(h)$ are respectively known as the lower and upper limits of the function $f(x)$ at $x=x_0$ on the left and are denoted by $\underline{f(x_0+0)}$ and $\overline{f(x_0+0)}$ respectively.

If the left hand upper limit $\overline{f(x_0-0)}$ is equal to the left hand lower limit $\underline{f(x_0-0)}$, then their common value is known as the left hand limit of the function $f(x)$ at $x=x_0$ and is denoted by $f(x_0-0)$

i.e., $\qquad f(x_0-0) = \overline{f(x_0+0)} = \underline{f(x_0+0)}$

REMARKS

- The four numbers $\overline{f(x_0+0)}$, $\underline{f(x_0+0)}$, $\overline{f(x_0-0)}$ and $\underline{f(x_0-0)}$ are known as four functional limits of the function $f(x)$ at $x=x_0$.
- The four functional limits of the function $f(x)$ at $x=x_0$ are independent of the value of the function $f(x)$ at $x=x_0$.
- At $x=0$, the functional limits are denoted by $\overline{f(+0)}$, $\underline{f(+0)}$, $\overline{f(-0)}$ and $\underline{f(-0)}$.

Solved Examples

Example 1. *Show that* $f(x)=\dfrac{x^2-1}{x-1}$ *is continuous for all values of x except* $x=1$.

Based on the following Results

- A function $f(x)$ defined on some nbd of a point a is said to be continuous at a if and only if either $\lim\limits_{x\to a} f(x)=f(a)$ or $f(a-0)=f(a+0)=f(a)$.

- For removable discontinuity,
$$f(a-0)=f(a+0)\neq f(a).$$

- For discontinuity of first kind,
$$f(a-0)=f(a)\neq f(a+0)$$
or $\quad f(a+0)=f(a)\neq f(a-0).$

- If none of the limit $f(a-0)$ and $f(a+0)$ exist at a. Then $f(x)$ is said to have a discontinuity of second kind.

- If $f(a+0)=\infty$ or $f(a-0)=-\infty$, then $f(x)$ has an inifinite discontinuity at $x=a$.

Solution. If $x\neq 1$, then $f(x)=(x+1)=$ A polynomial $\Rightarrow$ $f(x)$ is continuous for all values of $x\neq 1$.

If $x=1, f(x)$ is of the form $\dfrac{0}{0}$, which is not defined and so the function $f(x)$ is discontinuous at $x=1$.

Example 2. *Show that the function $f(x)$ is defined by* $f(x)=\begin{cases} x^2, & x\neq 1 \\ 2, & x=1 \end{cases}$ *is discontinuous at $x=1$.*

Solution. Here the value of $f(x)$ at $x=1$ is 2.
$$\Rightarrow \qquad f(1)=2$$

Now, $\quad$ RHL $=f(1+0)=\lim\limits_{h\to 0} f(1+h)=\lim\limits_{h\to 0}(1+h)^2=1$

also, $\quad$ LHL $=f(1-0)=\lim\limits_{h\to 0}f(1-h)=\lim\limits_{h\to 0}(1-h)^2=1$

Therefore, we have $f(1+0)=f(1-0)\neq f(1)$
$\Rightarrow$ $f(x)$ is not continuous at $x=1$.

Example 3. *Examine whether or not the function*
$$f(x)=\begin{cases} \dfrac{2\sin x}{x}, & when\, x\neq 0 \\ 2, & when\, x=0 \end{cases}$$
is continuous at $x=0$.

Solution. Given that $f(x)=2$, when $x=0$ $\Rightarrow$ $f(0)=2$

Now, $\quad$ RHL$=f(0+0)=\lim\limits_{h\to 0} f(0+h)=\lim\limits_{h\to 0}\left[\dfrac{2\sin(0+h)}{(0+h)}\right]$
$$=2 \qquad \left(\because \lim\limits_{x\to 0}\dfrac{\sin x}{x}=1\right)$$

and $\quad$ LHL$=f(0-0)=\lim\limits_{h\to 0} f(0-h)=\lim\limits_{h\to 0}\left[\dfrac{2\sin(0-h)}{(0-h)}\right]=2$
$$[\because \sin(-h)=-\sin(h)]$$

Therefrore, we have $f(0+0)=f(0-0)=f(0)=2$
Hence, $f(x)$is continuous at $x=2$.

Example 4. *A function f(x) is defined as follows*

$$f(x) = \begin{cases} \left(x^2/a\right) - a, & \text{when } x < a \\ 0, & \text{when } x = 0 \\ a - \left(a^2/x\right), & \text{when } x > a \end{cases}$$

Prove that f(x) is continuous at x=a.

Solution.   Here, we have

$$\text{RHL} = f(a+0) = \lim_{h \to 0} f(a+h) = \lim_{h \to 0} \left[a - \frac{a^2}{(a+h)} \right]$$

$$\left[\text{By using } f(x) = a - \frac{a^2}{x} \text{ for } x > a \right]$$

$$= \left[a - \frac{a^2}{a} \right] = (a - a) = 0 \qquad \qquad \text{...(1)}$$

and $$\text{LHL} = f(a-0) = \lim_{h \to 0} f(a-h) = \lim_{h \to 0} \left[\frac{(a-h)^2}{a} - a \right]$$

$$\left[\text{By using } f(x) = \frac{x^2}{a} - a \text{ for } x < a \right]$$

$$= \frac{a^2}{a} - a \qquad \qquad \text{...(2)}$$

$$= 0$$

Also $f(x) = 0$ for $x = a$

$\Rightarrow$ $$f(a) = 0 \qquad \qquad \text{...(3)}$$

Now, from (1),(2) and (3), we have $f(a+0) = f(a-0) = f(a) = 0$

$\Rightarrow$ $f(x)$ is continuous at $x = a$.

Example 5. *A function f(x) is defined as follows*

$$f(x) = \begin{cases} 1 + x & \text{if } x \le 2 \\ 5 - x & \text{if } x \ge 2 \end{cases}$$

check the continuity of f(x) at x=2.

Solution.   Here, we have

$$f(2) = 1 + 2 \text{ or } 5 - 2 = 3 \qquad \qquad \text{...(1)}$$

Now, $$\text{RHL} = f(2+0) = \lim_{h \to 0} f(2 + h)$$

$$= \lim_{h \to 0} [5-(2+h)] = \lim_{h \to 0} (3 - h) = 3 \qquad \qquad \text{...(2)}$$

and $$\text{LHL} = f(2-0) = \lim_{h \to 0} f(2-h) = \lim_{h \to 0} [1 + (2-h)] = 3. \qquad \qquad \text{...(3)}$$

Now, from (1), (2) and (3), we have

$$f(2 + 0) = f(2) = f(2 - 0) = 3$$

Hence, the function $f(x)$ is continuous at $x = 2$.

Example 6.　*Show that the function f defined by*

$$f(x) = \begin{cases} 0, & \text{for } x = 0 \\ \dfrac{1}{2} - x, & \text{for } 0 < x < \dfrac{1}{2} \\ \dfrac{1}{2}, & \text{for } x = \dfrac{1}{2} \\ \dfrac{3}{2} - x, & \text{for } \dfrac{1}{2} < x < 1 \\ 1, & \text{for } x = 1 \end{cases}$$

has three point of discontinuity. Find such points. Also draw the graph of the function.

Solution.　Here, we observe that the domain of the function $f(x)$ is closed inverval $[0,1]$ when $0 < x < \dfrac{1}{2}$, the function $f(x) = \dfrac{1}{2} - x$, which is being the polynomial is continuous at each points of its domain.

$\Rightarrow$　$f(x)$ is continuous at each point of the open interval $]0,\dfrac{1}{2}[$ when $\dfrac{1}{2} < x < 1$, $f(x) = \dfrac{3}{2} - x$, which is also a polynomial in x.

$\Rightarrow$　$f(x)$ is continuous in the open interval $]\dfrac{1}{2},1[$.

Now, we check the continuity of $f(x)$ at $x = 0, \dfrac{1}{2}$ and 1.

(i) *At x = 0.*

At $x = 0$,　　$f(x) = 0$

and　　　RHL $= f(0+0) = \lim_{h \to 0} f(0+h) = \lim_{h \to 0} f(h) = \lim_{h \to 0} \left(\dfrac{1}{2} - h \right) = \dfrac{1}{2}$

$\Rightarrow$　　　$f(0) \neq f(0+0)$

$\Rightarrow$　　$f(x)$ is not continuous at $x = 0$.

(ii) *At $x = \dfrac{1}{2}$.*

At $x = \dfrac{1}{2}, f(x) = \dfrac{1}{2}$

$$\text{LHL} = f\left(\dfrac{1}{2} - 0 \right) = \lim_{h \to 0} \left(\dfrac{1}{2} - h \right) = \lim_{h \to 0} \left[\dfrac{1}{2} - \left(\dfrac{1}{2} - h \right) \right] = \lim_{h \to 0} h = 0$$

$\Rightarrow$　　$f(\dfrac{1}{2}) \neq f\left(\dfrac{1}{2} - 0 \right)$

$\Rightarrow$　　$f(x)$ is not continuous at $x = \dfrac{1}{2}$.

(iii) *At x = 1.*

At $x = 1, f(x) = 1$

$$\text{LHL} = f(1-0) = \lim_{h \to 0} f(1-h) = \lim_{h \to 0} \left[\dfrac{3}{2} - (1-h) \right] = \lim_{h \to 0} \left(\dfrac{1}{2} + h \right) = \dfrac{1}{2}$$

$\Rightarrow$　　$f(1) \neq f(1-0)$

$\Rightarrow$ $f(x)$ is not continuous at $x=1$.

Hence, the function $f(x)$ has three points of discontinuity given by $x=0$, $\dfrac{1}{2}$ and 1.

Graph of f(x). The graph of the function consists of the point $(0, 0)$, the segment of the line $y= \dfrac{1}{2} - x$ for $0<x<\dfrac{1}{2}$, the point $\left(\dfrac{1}{2},\dfrac{1}{2}\right)$,

the segment of the line $y = \dfrac{3}{2} - x$ for $\dfrac{1}{2}<x<1$

and the point $(1,1)$. The graph of $f(x)$ is given

as fig. 8.

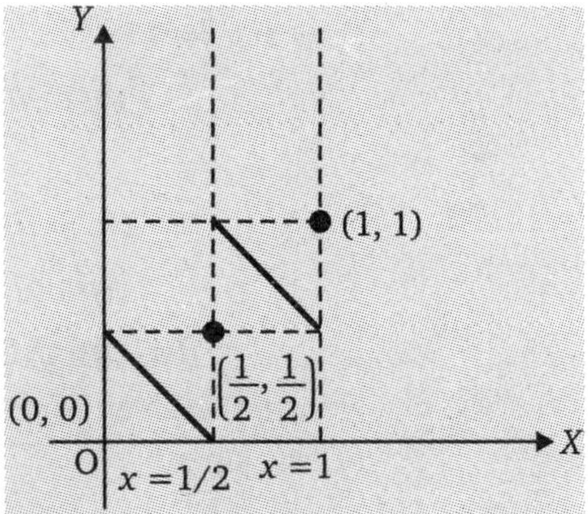

Fig. 8

Example 7. *Test the following functions for continuity*

    (i) $f(x) = x \sin \dfrac{1}{x}$, $x \neq 0$, $f(x)=0$ *at* $x=0$.

    (ii) $f(x)= \dfrac{1}{1-e^{-1/x}}$, $x \neq 0$, $f(x)=0$ *at* $x=0$

Solution.   (i) Here, we have

$$\text{LHL} = f(0-0) = \lim_{h\to 0}f(0-h)= \lim_{h\to 0}f(-h)$$

$$= \lim_{h\to 0}(-h)\sin\left(\dfrac{1}{-h}\right)= \lim_{h\to 0} h\sin\dfrac{1}{h}$$

$$=0\times \text{ (a finite quantity lying between 1 and } -1)=0$$

and $\text{RHL}= f(0+0)= \lim_{h\to 0}f(0+h)= \lim_{h\to 0} f(h)= \lim_{h\to 0} h\sin\dfrac{1}{h} =0.$

Also $f(0)=0$ given

$\Rightarrow$ $f(0+0) = f(0-0) = f(0)$.

Hence, the function $f(x)$ is continuous at $x=0$.

(ii)   Here, we have

$$\text{LHL}= f(0 - 0)= \lim_{h\to 0}f(0-h) = \lim_{h\to 0} f(-h)= \lim_{h\to 0} \dfrac{1}{1-e^{1/h}} =0$$

and $\text{RHL}= f(0+0)= \lim_{h\to 0}f(0+h)$

$$= \lim_{h\to 0} f(h)= \lim_{h\to 0} \dfrac{1}{1-e^{-1/h}} =1$$

Also $f(0)=0$

$\Rightarrow$ $f(0+0)\neq f(0-0)=f(0)$

Hence, $f(x)$ is discontinuous at $x=0$ and this discontinuity is of first kind.

Example 8. *Discuss the kind of discontinuity, if any, of the function.*

$$f(x)= \begin{cases} \dfrac{x-|x|}{x}, & \text{if } x \neq 0 \\ 2, & \text{if } x = 0 \end{cases}$$

Solution. The given function is continuous at all points except possible the origin.

Now at $x = 0$

$$\text{LHL} = f(0-0) = \lim_{h \to 0} f(0-h) = \lim_{h \to 0} f(-h) = \lim_{h \to 0} \frac{-h - |-h|}{-h} = 2$$

and

$$\text{RHL} = f(0+0) = \lim_{h \to 0} f(0+h) = \lim_{h \to 0} f(h) = \lim_{h \to 0} \frac{h - |h|}{h} = 0.$$

Also, $\qquad f(0) = 2$ (given)

$\Rightarrow \qquad f(0-0) = f(0) \neq f(0+0).$

Hence, the given function $f(x)$ is discontinuous at $x = 0$ and this is the discontinuity of first kind.

Example 9. *Discuss the continuity of the function $f(x)$ defined by*

$$f(x) = \begin{cases} x^2 & \text{for} & x < -2 \\ 4 & \text{for} & -2 \le x \le 2 \\ x^2 & \text{for} & x > 2 \end{cases}$$

Solution. Here, we shall check the continuity of $f(x)$ at $x = -2$ and 2.

At $x = -2$

We have $f(-2) = 4$

Now $\qquad \text{LHL} = f(-2-0) = \lim_{h \to 0} f(-2-h) = \lim_{h \to 0} (-2-h)^2 = 4$

and $\qquad \text{RHL} = f(-2+0) = \lim_{h \to 0} f(-2+h) = \lim_{h \to 0} 4 = 4$

$\Rightarrow \qquad f(-2-0) = f(-2) = f(-2+0) = 4$

Hence, $f(x)$ is continuous at $x = -2$.

At $x = 2$

We have $\qquad f(2) = 4$

and $\qquad \text{RHL} = f(2+0) = \lim_{h \to 0} f(2+h) = \lim_{h \to 0} (2+h)^2 = 4$

$\qquad \text{LHL} = f(2-0) = \lim_{h \to 0} f(2-h) = \lim_{h \to 0} 4 = 4$

$\Rightarrow \qquad f(2-0) = f(2) = f(2+0) = 4$

Hence, $f(x)$ is continuous at $x = 2$.

Example 10. *If $[x]$ denotes the positive or negative excess of x over the nearest integer and if exceeds an integer by $\frac{1}{2}$, let $[x] = 0$. Check the continuity of $[x]$. Also, Draw the graph.*

Solution. For any integer n, we can define

$$f(x) = [x] = \begin{cases} x - n & \text{, when } n - \dfrac{1}{2} < x < n + \dfrac{1}{2} \\ 0 & \text{, when} \qquad x = n + \dfrac{1}{2} \\ x - (n+1) & \text{, when } n + \dfrac{1}{2} < x < n + \dfrac{3}{2} \end{cases}$$

At $x = n + \dfrac{1}{2}$ We have $f\left(x + \dfrac{1}{2}\right) = 0$

$$\text{RHL} = f\left[\left(n + \frac{1}{2}\right) + 0\right] = \lim_{h \to 0} f\left[\left(n + \frac{1}{2}\right) + h\right]$$

$$= \lim_{h \to 0}\left[n + \frac{1}{2} + h - (n+1)\right] = -\frac{1}{2}$$

$$\text{LHL} = f\left[\left(n + \frac{1}{2}\right) - 0\right] = \lim_{h \to 0} f\left[\left(n + \frac{1}{2}\right) + h\right]$$

$$= \lim_{h \to 0}\left[n + \frac{1}{2} - h - n\right] = \frac{1}{2}$$

$\Rightarrow$ LHL $\neq$ RHL.

Hence, the function $f(x)$ is discontinuous when $x = n + \frac{1}{2}$, $n \in Z$.

Graph of f(x): We can define the given function as follows :

$$f(x) = \begin{cases} x+1 & \text{when} & -1\frac{1}{2} < x < -\frac{1}{2} \\ 0 & \text{when} & x = -\frac{1}{2} \\ x & \text{when} & -\frac{1}{2} < x < \frac{1}{2} \\ 0 & \text{when} & x = \frac{1}{2} \\ x-1 & \text{when} & \frac{1}{2} < x < 1\frac{1}{2} \end{cases}$$

The graph of $f(x)$ is given below.

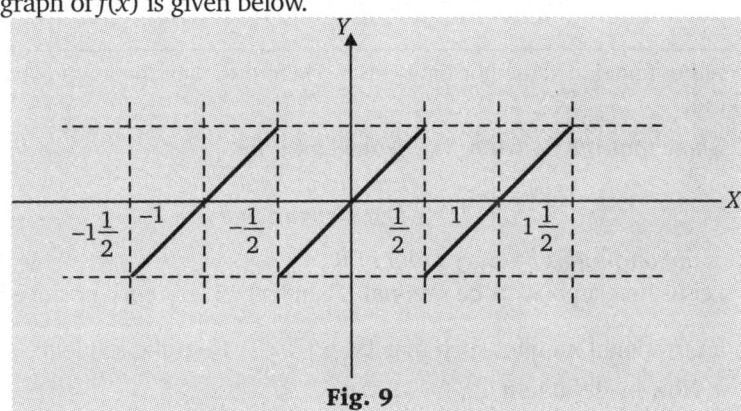

Fig. 9

Example 11. *Determine the discontinuities of the function* $f : R \to R$ *defined by*

$$f(x) = \lim_{n \to \infty}\left[\frac{(1 + \sin \pi/x)^n - 1}{(1 + \sin \pi/x)^n + 1}\right], 0 < x < 1.$$

Solution. For $x = 1, \frac{1}{2}, \dots, \frac{1}{m}, \dots$, we have $\sin \frac{\pi}{x} = 0$. Therefore, at these values of x, we have

$$f(x) = \lim_{n \to \infty} \frac{(1 + 0)^n - 1}{(1 + 0)^n + 1} = 0 \qquad .$$

Now, if $\dfrac{1}{(2m+1)} < x < \dfrac{1}{2m}$, $m \in Z$, then $\sin \dfrac{\pi}{x}$ is positive. Therefore, for these values of

x, we have

$$f(x) = \lim_{n \to \infty} \frac{1 - \dfrac{1}{\left(1 + \sin \dfrac{\pi}{x}\right)^{n}}}{1 + \dfrac{1}{\left(1 + \sin \pi/x\right)^{n}}} = \frac{1 - \dfrac{1}{\infty}}{1 - \dfrac{1}{\infty}} = \frac{1-0}{1+0} = 1$$

Now, for $\dfrac{1}{(2m+1)} < x < \dfrac{1}{2m}$, $\sin \dfrac{\pi}{x}$ is negative and therefore for these values of x,

we have

$$f(x) = \lim_{n \to \infty} \frac{\left(1 + \sin \pi/x\right)^{n} - 1}{\left(1 + \sin \pi/x\right)^{n} + 1} = \frac{0-1}{0+1} = -1$$

Thus, we have

(i) $f(x) = 0, f(x-0) = 1, f(x+0) = -1$ if $x = \dfrac{1}{2m}$.

(ii) $f(x) = 0, f(x-0) = -1, f(x+0) = 1$ if $x = \dfrac{1}{(2m+1)}$.

Hence, the function $f(x)$ has discontinuities of first kind at

$$x = 1, \frac{1}{2}, \frac{1}{3}, \dots, \frac{1}{n}, \dots$$

REMARK

- At $x = 0$, neither function value nor limit exists. Therefore, the function $f(x)$ has discontinuity of second kind.

Example 12. *Show that the function $f(x)$ defined on R by*

$$f(x) = \begin{cases} 1 & \text{when } x \text{ is rational} \\ -1 & \text{when } x \text{ is irrational} \end{cases}$$

is discontinuous at every point of R.

Solution. Let us first suppose, x be rational. Then $f(x) = 1$. For each positive integer n, let x_n be

an irrational number such that $|x_n - x| < \dfrac{1}{n}$. Then the sequence $\langle x_n \rangle$ converges to

x. Now by definition $f(x_n) = -1 \; \forall \; n$.

$\Rightarrow \qquad \lim_{n \to \infty} f(x_n) = -1 \neq f(x)$.

Hence, f is discontinuous at each rational point.

Now suppose x is an irrational number. Then $f(x) = 1$. For each positive integer n, let

x_n be the rational number such that $|x_n - x| < \dfrac{1}{n}$. Then, the sequence $\langle x_n \rangle$ converges

to x. Now $f(x_n) = 1 \; \forall \; n$ so that

$$\lim_{n \to \infty} f(x_n) = 1 \neq f(x).$$

Hence, f is discontinuous at each irrational point.

Therefore, f is discontinuous at every point of R.

REMARKS

- This function is known as Dirichlet's function.

Example 13. *Determine the values of a, b, c for which the function*

$$f(x) = \begin{cases} \dfrac{\sin(a+1)x + \sin x}{x} & \text{for } x < 0 \\ c & \text{for } x = 0 \\ \dfrac{\left(x + bx^2\right)^{1/2} - x^{1/2}}{bx^{3/2}} & \text{for } x > 0 \end{cases}$$

is continuous at x = 0.

Solution.   Here, we have

$$\text{RHL} = f(0+0) = \lim_{h \to 0} f(0+h)$$

$$= \lim_{h \to 0} \frac{\left(h + bh^2\right)^{1/2} - h^{1/2}}{bh^{3/2}} = \lim_{h \to 0} \frac{(1+bh)^{1/2} - 1}{bh}$$

$$= \lim_{h \to 0} \frac{\left[1 + \dfrac{1}{2}bh + ...\right] - 1}{bh} = \frac{1}{2}$$

$$\text{LHL} = f(0-0) = \lim_{h \to 0} f(0-h) = \lim_{h \to 0} \frac{\sin(a+1)(-h) + \sin(-h)}{(-h)}$$

$$= \lim_{h \to 0} \frac{\sin(a+1)h + \sin h}{h} = \lim_{h \to 0} \frac{2\sin\left(\dfrac{1}{2}a + 1\right)h\cos\left(\dfrac{ah}{2}\right)}{h}$$

$$= \lim_{h \to 0} \frac{\sin\{(a+2)/2\}h}{\{(a+2)/2\}h}(a+2)\cos(ah/2) = a+2$$

Now, for continuity at $x = 0$, we have $f(0+0) = f(0-0) = f(0)$.

$$\Rightarrow \qquad \frac{1}{2} = a + 2 = c \Rightarrow c = \frac{1}{2}, a = -\frac{3}{2}$$

Example 14. *Test the continuity at x=0 if* $f(x) = \begin{cases} x\log x, & x \neq 0 \\ 0, & x = 0 \end{cases}$

Solution.   Here we have,

$$f(0) = 0$$

$$\text{RHL} = f(0+0) = \lim_{h \to 0} f(0+h) = \lim_{h \to 0} h\log h = \lim_{h \to 0} \frac{\log h}{(1/h)}$$

$$= \lim_{h \to 0} \frac{1/h}{-1/h^2} = \lim_{h \to 0} (-h) = 0$$

Since $f(x)$ is not defined for $x < 0$, therefore, no need to find $f(0-0)$.

$\Rightarrow \quad f(0+0) = f(0)$

Hence, the function $f(x)$ is continuous at $x=0$.

Example 15. *Prove that the function $f(x) = \dfrac{|x|}{x}$ for $x \neq 0$, $f(0) = 0$ is continuous at all point except $x = 0$.*

Solution.   If $x > 0$, then
$$f(x) = \frac{x}{x} = 1.$$

If $x < 0$, then
$$f(x) = \frac{-x}{x} = -1.$$

Therefore, the function $f(x)$ can be defined as
$$f(x) = \begin{cases} -1 & \text{if } x < 0 \\ 0 & \text{if } x = 0 \\ 1 & \text{if } x > 0 \end{cases}$$

If $x < 0$, $f(x) = -1$ i.e., $f(x)$ is a constant function and a constant function is always continuous at each points of its domain

$\Rightarrow$ $f(x)$ is continuous at each point x where $x < 0$.

Similarly, we can show that $f(x)$ is continuous at each point x where $x > 0$. Now, we shall discuss the continuity at $x = 0$.

$\therefore$ $\text{LHL} = f(0-0) = \lim\limits_{h \to 0} f(0-h) = \lim\limits_{h \to 0} f(-h) = \lim\limits_{h \to 0} -1 = -1$

and $\text{RHL} = f(0+0) = \lim\limits_{h \to 0} f(0+h) = \lim\limits_{h \to 0} f(h) = \lim\limits_{h \to 0} 1 = 1$

Also $f(0) = 0$

$\Rightarrow$ $f(0-0) \neq f(0+0) \neq f(0)$

$\Rightarrow$ $\lim\limits_{h \to 0} f(x)$ does not exist.

$\Rightarrow$ $f(x)$ is not continuous at $x = 0$.

Example 16. *Discuss the nature of the discontinuity at $x = 0$ of the function $f(x) = [x] - [-x]$, where $[x]$ denotes the integral part of x.*

Solution.   Here, we have
$$f(x) = [x] - [-x] \quad \Rightarrow \quad f(0) = f[0] - f[-0] = [0] - [0] = 0 - 0 = 0.$$

Also, $\text{RHL} = f(0+0) = \lim\limits_{h \to 0} f(0+h) = \lim\limits_{h \to 0} f(h) = \lim\limits_{h \to 0} ([h] - [-h])$

$= \lim\limits_{h \to 0} [0 - (-1)] = \lim\limits_{h \to 0} 1 = 1$

$[\because 0 < h < 1 \Rightarrow [h] = 0$ and $-1 < -h < 0 \Rightarrow [-h] = -1]$

and $\text{LHL} = f[0-0] = \lim\limits_{h \to 0} f(0-h) = \lim\limits_{h \to 0} f(-h) = \lim\limits_{h \to 0} \{[-h] - [-(-h)]\}$

$= \lim\limits_{h \to 0} ([-h] - [h]) = \lim\limits_{h \to 0} (-1 - 0) = -1.$

Now, since $f(0-0) \neq f(0+0)$, therefore $\lim\limits_{x \to 0} f(x)$ does not exist and so $f(x)$ is discontinuous at $x = 0$. Also $f(0) \neq f(0-0)$ and $f(0) \neq f(0+0)$ implies that $f(x)$ is discontinuous at $x = 0$ both from the left and from the right. Since $f(0-0)$ and $f(0+0)$ both exists. Hence, the discontinuity is of the first kind, and jump of the function at $x = 0$ is $f(0+0) - f(0-0) = 1 - (-1) = 2$

Example 17. *Let a function f : R→R satisfy the equation*

$$f(x+y)=f(x)+f(y) \quad \forall x,y \in R$$

show that if f is continuous at x=a, then show that it is continuous for all x∈R.

Solution.   Since the function f is continuous at a, we have

$$f(a)= f(a-0) = \lim_{h \to 0} f(a-h)= \lim_{h \to 0} f(a)+ \lim_{h \to 0} f(-h)$$

$$=f(a)+ \lim_{h \to 0} f(-h)$$

$$\Rightarrow \qquad \lim_{h \to 0} f(-h)=0 \qquad \qquad \qquad ...(1)$$

Similarly $\qquad f(a)=f(a+0)= \lim_{h \to 0} f(a+h)= \lim_{h \to 0} f(a)+ \lim_{h \to 0} f(x)$

$$=f(a)+ \lim_{h \to 0} f(h)$$

$$\Rightarrow \qquad \lim_{h \to 0} f(h)=0 \qquad \qquad \qquad ...(2)$$

Now, let x be any arbitrary point of R, then we have

$$f(x-0)= \lim_{h \to 0} f(x-h)= \lim_{h \to 0} f(x)+ \lim_{h \to 0} f(-h)=f(x) \qquad \text{[By using (1)]}$$

and $\qquad f(x+0)= \lim_{h \to 0} f(x+h)= \lim_{h \to 0} f(x)+ \lim_{h \to 0} f(h)=f(x) \qquad \text{[By using (2)]}$

Thus, $\qquad f(x)=f(x-0)=f(x+0)$

$\Rightarrow$ f is continuous at $x \in R$.

Since x is arbitrary. Hence, f is continuous for all $x \in R$.

Example 18. *Discuss the continuity of the function*

$$f(x)= e^x g(x+[x]) \text{ at } x = 0$$

where g(x) is defined as

$$g(x)= \begin{cases} 1 & \text{if} \quad x > 0 \\ 0 & \text{if} \quad x = 0 \\ -1 & \text{if} \quad x < 0 \end{cases}$$

and [x] denotes the integral value of x.

Solution.   Here, we have

$$f(0)=e^0 g(0+[0]) =1.g(0) =1 \times 0=0$$

Now, $\qquad$ LHL$=f(0-0)= \lim_{h \to 0} f(0-h)= \lim_{h \to 0} f(-h)$

$$= \lim_{h \to 0} e^{-h} g(-h+[-h])= \lim_{h \to 0} e^{-h} g(-h-1)$$

$$= \lim_{h \to 0} e^{-h} g(-1)=-1$$

and $\qquad$ RHL$=f(0+0)= \lim_{h \to 0} f(0+h)= \lim_{h \to 0} f(h)$

$$= \lim_{h \to 0} e^h g(h+[h])= \lim_{h \to 0} e^h g(h+0)$$

$$=e^0 \times 1=1 \times 1=1.$$

Thus, we have $f(0) \neq f(0-0) \neq f(0+0)$. Hence, the given function is not continuous at $x=0$.

Example 19. Let $f(x) = \begin{cases} \left[1 + |\sin x|^{a|\sin x|}\right] & , & \text{if} \quad -\dfrac{\pi}{6} < x < 0 \\ b & , & \text{if} \quad x = 0 \\ e^{\tan 2x / \tan 3x} & , & \text{if} \quad 0 < x < \dfrac{\pi}{6} \end{cases}$

Find the value of a and b such that f is continuous at $x=0$.

Solution. Here, we have LHL$=f(0-0) = \lim_{h \to 0} f(0-h) = \lim_{h \to 0} f(-h)$

$$= \lim_{h \to 0} [1 + |\sin(-h)|]^{a|\sin(-h)|}$$

$$= \lim_{h \to 0} [1 + \sin h]^{a|\sin h|} = \lim_{h \to 0} [1 + y]^{a \cdot y} \qquad (\sin h = y)$$

$$= e^a$$

and $\qquad$ RHL$=f(0+0) = \lim_{h \to 0} f(0+h) = \lim_{h \to 0} f(h)$

$$= \lim_{h \to 0} e^{\tan 2h / \tan 3h} = \lim_{h \to 0} e^{2h/3h} = e^{2/3} \qquad \left[\because \lim_{h \to 0} \tan h = h \right]$$

and $\qquad f(0) = b$

The function $f(x)$ is continuous at $x = 0$

$\Rightarrow \qquad f(0-0) = f(0+0) = f(0)$

Therefore, $\qquad e^a = e^{2/3} = b \qquad \Rightarrow a = 2/3,\, b = e^{2/3}$.

Example 20. Find the function defined below for continuity at $x=0$

$$f(x) = \frac{\sin^2 ax}{x^2} \text{ for } x \neq 0$$

and $\qquad f(x) = 1$ for $x = 0$.

Solution. Here, we have $f(0) = 1$.

Now $\qquad f(0+0) = \lim_{h \to 0} f(0+h) = \lim_{h \to 0} f(h)$

$$= \lim_{h \to 0} \frac{\sin^2 ah}{h^2} = \lim_{h \to 0} \left(\frac{\sin ah}{ah}\right)^2 . a^2 = 1. a^2 = a^2$$

and $\qquad f(0-0) = \lim_{h \to 0} f(0-h) = \lim_{h \to 0} f(-h)$

$$= \lim_{h \to 0} \frac{\sin^2(-ah)}{(-h)^2} = \lim_{h \to 0} \frac{\sin^2 ah}{h^2} = a^2.$$

Now $f(x)$ is continuous at $x=0$ if $f(0+0) = f(0-0) = f(0)$.

Hence, $f(x)$ is discontinuious at $x=0$ unless $a=1$.

Example 21. Examine the continuity of the function

$$f(x) = \begin{cases} -x^2 & \text{if} \quad x \leq 0 \\ 5x - 4 & \text{if } 0 < x \leq 1 \\ 4x^2 - 3x & \text{if } 1 < x < 2 \\ 3x + 4 & \text{if} \quad x \geq 2 \end{cases}$$

at $x = 0, 1$ and 2.

Solution. (i) *Continuity at $x = 0$.*

Clearly $f(0) = 0$

Now, $f(0-0) = \lim_{h \to 0} f(0-h) = \lim_{h \to 0} -(-h)^2 = 0$

$$f(0+0) = \lim_{h \to 0} f(0+h) = \lim_{h \to 0} 5(h) - 4 = -4$$

$\therefore$ $f(0) = f(0-0) \neq f(0+0)$

(ii)   *Continuity at x = 1.*

$$f(1-0) = \lim_{h \to 0} f(1-h) = \lim_{h \to 0} 5(1-h) - 4 = 1$$

$$f(1+0) = \lim_{h \to 0} f(1+h) = \lim_{h \to 0} 4(1+h)^2 - 3(1+h) = 1$$

Also $f(1) = 1$

$\therefore$ $f(1-0) = f(1+0) = f(1)$. Hence $f(x)$ is continuous at $x = 1$.

(iii) *Continuity at x = 2.*

$$f(2-0) = \lim_{h \to 0} f(2-h) = \lim_{h \to 0} 4(2-h)^2 - 3(2-h) = 10$$

$$f(2+0) = \lim_{h \to 0} f(2+h) = \lim_{h \to 0} 3(2+h) + 4 = 10$$

Also $f(2) = 10$

$\therefore$ $f(2-0) = f(2+0) = f(2)$.

Hence $f(x)$ is continuous at $x = 2$.

Example 22. *Test the continuity of the function at $x = 0$*

$$f(x) = x \cos\left(\frac{1}{x}\right), \ if \ x \neq 0, \ f(0) = 0$$

Solution.   We have

$$f(x) = x \cos\left(\frac{1}{x}\right), \ x \neq 0, \ f(0) = 0$$

$$\text{LHL} = f(0-0) = \lim_{h \to 0} f(0-h) = \lim_{h \to 0} (-h) \cos\left(\frac{1}{h}\right)$$

$$= 0 \times \text{(some finite value lying between } -1 \text{ and } 1)$$

$$= 0$$

and $\text{RHL} = f(0+0) = \lim_{h \to 0} f(0+h) = \lim_{h \to 0} (h) \cos\left(\frac{1}{h}\right)$

$$= 0 \times \text{(some finite value lying between } -1 \text{ and } 1) = 0$$

$\therefore$ $f(0-0) = f(0+0) = f(0)$

Hence, $f(x)$ is continuous at $x = 0$.

Example 23. *Let f be the function defined on R such that*

$$f(x) = |x| + [x] \ \forall x \in R$$

determine the point of discontinuity of f.

Solution.   Using the definition of modulus and greatest integer function, we can define the given function as follows:

$$f(x) = x + 0 = x \text{ for } 0 \leq x < 1$$
$$f(x) = x + 1 \text{ for } 1 \leq x < 2$$
$$f(x) = x + 2 \text{ for } 2 \leq x < 3$$

$$f(x)=x+3 \text{ for } 3\le x<4$$

$$\dots \quad \dots \quad \dots \quad \dots \quad \dots$$

$$f(x)= -x-1 \text{ for } -1 \le x < 0$$
$$f(x)= -x-2 \text{ for } -2 \le x < -1$$

Therefore, we have

$$\text{LHL}=f(0+0)= \lim_{x\to 0}(-x-1)=-1$$

$$\text{RHL}=f(0+0)= \lim_{x\to 0} x=0$$

Also,

$$f(1-0)= \lim_{x\to 1} x=1 \quad , \qquad\qquad f(1+0)= \lim_{x\to 1}(x+1)=2$$

$$f(2-0)= \lim_{x\to 2}(x+1)=3, \qquad\qquad f(2+0)= \lim_{x\to 2}(x+2)=4$$

$$f(-1-0)= \lim_{x\to -1}(-x-2)=-3, \qquad f(-1+0)= \lim_{x\to -1}(-x-2)=-1$$

Hence, $f(n-0) \ne f(n+0) \; \forall n \in Z$.

Hence, the points of discontinuity of f are $0, \pm 1, \pm 2, \pm 3,\dots$

Example 24. *Examine the continuity of the function*

$$f(x)= \frac{2[x]}{3x-|x|}, \text{ at } x = -\frac{1}{2} \text{ and } x=1$$

Where $[x]$ denotes the greatest integer not greater than x.

Solution.   We know that

$$\left[-\frac{1}{2}\right]=-1 , \qquad [1]=1 , \qquad \left|-\frac{1}{2}\right| = \frac{1}{2} \text{ and } |1|=1$$

Now, for $-\frac{1}{2} < x < 0, [x]=-1$ and for $0\le x <1$, $[x]=0$ etc.

so,

$$f(1-0)= \lim_{x\to 1} \frac{2\times 0}{3x-x}=0$$

and

$$f(1+0) = \lim_{x\to 1} \frac{2\times 1}{3x-x}=1$$

$\Rightarrow$ f is discontinuous at $x=1$.

Now,

$$f\left(-\frac{1}{2}-0\right)= \lim_{x\to -\frac{1}{2}} \frac{2(-1)}{3x-\frac{1}{2}}=-1$$

and

$$f\left(-\frac{1}{2}+0\right)= \lim_{x\to -\frac{1}{2}} \frac{2(-1)}{3x-\frac{1}{2}}=-\frac{1}{2}$$

Also,

$$f\left(-\frac{1}{2}\right)=-1$$

Hence, at $x=-\frac{1}{2}$, $f(x)$ is continuous.

EXERCISE 4.2

1. Discuss the continuity of the following functions :

(i) $f(x) = \cos\left(\dfrac{1}{x}\right)$, when $x \neq 0, f(0) = 0$

(ii) $f(x) = \dfrac{\sin x}{x}$, $x \neq 0, f(0) = 1$

(iii) $f(x) = \dfrac{1}{1 - e^{1/x}}$, when $x \neq 0$, and $f(0) = 0$

(iv) $f(x) = \dfrac{\sin^{-1} x}{x}$, $x \neq 0, f(0) = 1$

(v) $f(x) = \dfrac{e^{1/x} \sin(1/x)}{1 + e^{1/x}}$, $x \neq 0$, and $f(0) = 0$

(vi) $f(x) = \dfrac{e^{1/x} - 1}{e^{1/x} + 1}$, $x \neq 0, f(0) = 0$

(vii) $f(x) = 3x^2 + 2x - 1$ at $x = 2$

(viii) $f(x) = \dfrac{xe^{1/x}}{1 + e^{1/x}} + \sin\dfrac{1}{x}$, when $x \neq 0, f(0) = 0$

(ix) $f(x) = \dfrac{1}{x - a} \sin\dfrac{1}{x - a}$, at $x = a$

(x) $f(x) = \sin x \cos\dfrac{1}{x}$, when $x \neq 0, f(0) = 0$

(xi) $f(x) = \dfrac{e^{1/x}}{1 + e^{1/x}}$, when $x \neq 0, f(0) = 0$

(xii) $f(x) = \begin{cases} \cos x & \text{for } x \geq 0 \\ -\cos x & \text{for } x < 0 \end{cases}$

(xiii) $f(x) = \begin{cases} \dfrac{x^2 - 4x + 3}{x^2 - 1} & \text{for } x \neq 1 \\ 2 & \text{for } x = 1 \end{cases}$

(xiv) $f(x) = \dfrac{1}{x} \cos\dfrac{1}{x}$

2. Examine the following function for continuity at $x = 0$ and $x = 1$

$$f(x) = \begin{cases} x^2 & \text{if } x \leq 0 \\ 1 & \text{if } 0 < x \leq 1 \\ 1/x & \text{if } x > 1 \end{cases}$$

3. Find out the points of discontinuity of the following functions.

(i) $f(x) = \left(2 + e^{1/x}\right)^{-1} + \cos e^{1/x}$ for $x \neq 0$, $f(0) = 0$.

(ii) $f(x) = \dfrac{1}{2^n}$ for $\dfrac{1}{2^{n+1}} < x \leq \dfrac{1}{2^n}$, $n = 0, 1, 2,$ and $f(0) = 0$

4. A function f defined on $[0,1]$ is given by

$$f(x) = \begin{cases} x, & \text{if } x \text{ is rational} \\ \dfrac{1}{3}, & \text{if } x \text{ is irrational} \end{cases}$$

Show that f takes every value between 0 and 1, but it is continuous only at the point $x = \dfrac{1}{2}$.

5. A function $f : R \to R$ is defined as $f(x) = \dfrac{1}{x - 4}$.

Discuss the type of discontinuity which the function $f(x)$ has in $]-\infty, \infty[$.

6. Prove that the function f defined by

$$f(x) = \begin{cases} \dfrac{1}{2}, & \text{if } x \text{ is rational} \\ \dfrac{1}{3}, & \text{if } x \text{ is irrational} \end{cases}$$

is discontinuous everywhere.

7. Show that the function f defined by

$f(x) = \dfrac{xe^{1/x}}{1 + e^{1/x}}$, $x \neq 0, f(0) = 1$ is not continuous

at $x = 0$ and also show how the discontinuity can be removed.

8. Is the function f defined by $f(x) = \dfrac{3x + 4\tan x}{x}$, continuous at $x = 0$? If not, how may the function be defined to make it continuous at this point? Give reason for your answer.

9. Examine the function defined below for the continuity at $x = a$

$$f(x) = \dfrac{1}{x - a} \operatorname{cosec}\left(\dfrac{1}{x - a}\right), x \neq a, f(x) = 0, x = a.$$

10. Let $y = [x]$, where $[x]$ denotes the integral part of x. Prove that the function is discontinuous where x has an integral value.

11. Is the function $f(x) = x\sqrt{x^2 - 1}$ continuous at $x = 0$? Explain it.

===== **Answers** =====

1. (i) Discontinuous at $x=0$ (ii) Continuous at $x=0$
   (iii) Discontinuous at $x=0$ (iv) Contnuous at $x=0$
   (v) Discontinuity of the second kind at $x=0$
   (vi) Discontinuous at $x=0$ (vii) Continuous
   (viii) Discontinuous at $x=0$ (ix) Discontinuous
   (x) Continuous for all x (xi) Discontinuous at $x=0$
   (xii) Discontinuous at $x=0$ (xiii) Discontinuous at $x=1$
   (xiv) Continuous for all x except at $x=0$
2. Discontinuous at $x=0$ and continuous at $x=1$
3. (i) Discontinuous at $x=0$, mixed discontinuity

   (ii) Discontinuous at $x=\dfrac{1}{2^n}$: $n=1, 2,$, discontinuity of first kind.
5. At $x=4$, function has infinite discontinuity and is continuous at all other points in R.
7. We can remove this discontinuity at $x=0$ by taking $f(0)=0$.
8. No, define $f(0)=7$ to make it continuous.
9. Discontinuous at $x=a$, discontinuity is finite at $x=a$.
11. Continuous at $t=0$.

4.12 THEOREMS ON CONTINUITY

THEOREM 1. *If f and g are two continuous function at a point $a \in I$ then the function*

         (i) $f+g$ *(ii) cf*

         (iii) fg   *(iv) $f/g [g(a) \neq 0]$ are also continuous.*

Proof.   Since f and g are continuous at a, then

$$\lim_{x \to a} f(x) = f(a) \text{ and } \lim_{x \to a} g(x) = g(a)$$

(i) By definition, we have $(f+g)(x) = f(x) + g(x) \ \forall \ x \in I$

$\therefore$ $\lim_{x \to a} (f+g)(x) = \lim_{x \to a} [f(x) + g(x)] = \lim_{x \to a} f(x) + \lim_{x \to a} g(x)$

$\Rightarrow$ $(f+g)$ is continuous.

(ii) By definition, we have $(cf)(x) = cf(x) \ \forall \ x \in I$

       Therefore, $\lim_{x \to a} (cf)(x) = \lim_{x \to a} cf(x) = c \lim_{x \to a} f(x) = cf(a) = (cf)(a)$

       Hence, cf is continuous at $x=a$.

(iii) By definition, we have $(fg)(x) = f(x).g(x) \ \forall x \in I$.

       Therefore,

$$\lim_{x \to a} (fg)(x) = \lim_{x \to a} [f(x).g(x)]$$

$$= \left[\lim_{x \to a} f(x) \right].\left[\lim_{x \to a} g(x) \right] = f(a).g(a) = (fg)(a).$$

       Hence, fg is continuous at $x=a$.

(iv) We have

$$\left(\frac{f}{g} \right)(x) = \frac{f(x)}{g(x)} \ \forall x \in I, g(x) \neq 0$$

Therefore, $\lim\limits_{x\to a}\left(\dfrac{f}{g}\right)(x)=\lim\limits_{x\to a}\dfrac{f(x)}{g(x)}=\dfrac{f(a)}{g(a)}=\left(\dfrac{f}{g}\right)(a).$

Hence, $\dfrac{f}{g}$ is continuous.

THEOREM 2. *If f is continuous at $a\in I$, then $|f|$ is also continuous at a.*

Proof.   Since f is continuous at $x=a$ $\Rightarrow$ $\lim\limits_{x\to a}f(x)=f(a)$

We know that $|f|(x)=|f(x)|,\,x\in I$

$\Rightarrow$ $\lim\limits_{x\to a}|f|(x)=\lim\limits_{x\to a}|f(x)|=\left|\lim\limits_{x\to a}f(x)\right|=|f(a)|=|f|(a)$

Hence, $|f|$ is continuous.

REMARK

- The converse of the above theorem need not be true. For example: consider a function f on R defined by

$$f(x)=\begin{cases}1, & \text{if } x \text{ is rational}\\ -1, & \text{if } x \text{ is irrational}\end{cases}$$

then $|f|(x)=1\,\forall x\in R$, therefore $|f|$ is continuous at $x=0$ but f is not continuous at $x=0$.

THEOREM 3. *The necessary and sufficient condition for a function f defined on an interval I to be continuous at a point I is that for each sequence $<a_n>$ of I converges to a, the sequence $<f(a_n)>$ converges to $f(a)$.*

Proof.   **(i) Necessary condition.** Let us first suppose f be continuous at $x=a$ and let the sequence $<a_n>$ in I be such that

$$\lim\limits_{n\to\infty}a_n=a$$

Since, f is continuous at a, therefore for a given $\varepsilon>0$ $\exists$ a positive integer m such that

$$|f(x)-f(a)|<\varepsilon \text{ whenever } |x-a|<\delta \qquad\qquad \text{...(1)}$$

Also, $\lim\limits_{n\to\infty}a_n=a$, therefore, $\exists$ a positive integer m such that

$$|a_n-a|<\delta\,\forall n\geq m \qquad\qquad \text{...(2)}$$

Put $x=a_n$, in (1), we get

$$|f(a_n)-f(a)|<\varepsilon \text{ when } |x-a|<\delta \qquad\qquad \text{...(3)}$$

Now, from (2) and (3), we get

$$|f(a_n)-f(a)|<\varepsilon\,\forall\,n\geq m.$$

Therefore, $\lim\limits_{n\to\infty}f(a_n)=f(a).$

(ii) Condition is sufficient. Let us suppose the sequence $<f(a_n)>$ converges to $f(a)$ if every sequence $<a_n>$ in I converging to a.Then, to show that the function f is continuous at a.

Let if possible, the function is not continuous at a. Then $\exists$ a positive number $\varepsilon>0$ such that for every $\delta>0$ $\exists$ a x such that

$$|a_n-a|<\frac{1}{n}$$

but $|f(a_n)-f(a)|>\varepsilon$ $(\because f \text{ is not continuous.})$

This shows that $\lim\limits_{n\to\infty}a_n=a$. Also $<f(a_n)>$ does not converge to $f(a)$ i.e.,

$$\lim_{n\to\infty} f(a_n) \neq f(a)$$

which is a contradiction.

Hence, f must be continuous at $x=a$.

THEOREM 4. *A function $f : R{\to}R$ is continuous on R iff for each open set $A{\subset}R$, $f^{-1}(A)$ is an open set in R.*

Proof.   **(i) Necessary Condition.** Let us first suppose f be continuous on R and let $A{\subset}R$ be open. To show $f^{-1}(A)$ is open. Let $f^{-1}(A) = \phi$, then $f^{-1}(A)$ is open.

$$(\because \phi \text{ is an open set})$$

If $f^{-1}(A) \neq \phi$, let $a{\in}f^{-1}(A)$, then $f(a){\in}A$. Since A is an open subset of R containing $f(a)$, $\exists\ \delta > 0$ such that

$$]f(a){-}\varepsilon, f(a){+}\varepsilon[\subseteq A$$

Now, f is continuous at $x=a$, $\exists\ \delta > 0$ such that

$$|f(x){-}f(a)| < \varepsilon, \text{ whenever } |x{-}a| < \delta$$

or $\quad x{\in}]a{-}\delta,a{+}\delta[\Rightarrow f(x){\in}]f(a){-}\varepsilon, f(a){+}\varepsilon[$

$\Rightarrow \quad f(a{-}\delta, a{+}\delta) \subset]f(a){-}\varepsilon, f(a){+}\varepsilon[$

$\Rightarrow \quad]a{-}\delta, a{+}\delta[\subset f^{-1}] f(a){-}\varepsilon, f(a){+}\varepsilon[\subseteq f^{-1}(A)$.

Thus for each $a{\in}f^{-1}(A)\ \exists\ \delta > 0$ such that $]a - \delta, a{+}\delta[\subset f^{-1}(A)$

$\Rightarrow \quad f^{-1}(A)$ is open.

(ii) Condition is sufficient. Suppose for each open set A in R, $f^{-1}(A)$ is open. To show f is continuous on R.

Let $\quad\quad\quad a{\in}R \Rightarrow f(a){\in}R$.

For $\varepsilon > 0$, $] f(a{-}\varepsilon), f(a{+}\varepsilon)[$ is an open interval and therefore an open set in R. Then, by our assumption $f^{-1}\ ]f(a){-}\varepsilon, f(a){+}\varepsilon[$ is an open set containing a.

$\Rightarrow \quad\quad \exists\ \delta > 0$ such that $]a{-}\delta, a{+}\delta\ [\subset f^{-1}\{]f(a){-}\varepsilon, f(a){+}\varepsilon[\}$

or $\quad f(a - \delta, a{+}\delta) \subset]f(a){-}\varepsilon, f(a){+}\varepsilon[$.

Hence, for a given $\varepsilon > 0\ \exists\ a\ \delta > 0$ such that $|x - a| < \delta \Rightarrow |f(x) - f(a)| < \varepsilon$

$\Rightarrow \quad f$ is continuous at a.

Since, a is arbitrary. Hence, f is continuous on R.

THEOREM 5. *A function $f : R \to R$ is continuous on R iff for every closed set B in R, $f^{-1}(B)$ is closed in R.*

Proof.   Let us first suppose f is continuous on B, where B is a closed subset of R. To show $f^{-1}(B)$ is closed in R.

Since B is closed $\Rightarrow R - B$ is open.

$\Rightarrow \quad f^{-1}(R{-}B)$ is open and $f^{-1}(R{-}B)=R{-}f^{-1}(B)$

Therefore, we have $R{-}f^{-1}(B)$ is an open set in R.

$\Rightarrow \quad f^{-1}(B)$ is a closed set in R.

Conversely, let $f^{-1}(B)$ be closed in R for every closed set B in R. To show, f is continuous.

Now let A be an open set in R

$\Rightarrow \quad\quad R{-}A$ is closed $\Rightarrow f^{-1}(R{-}A)$ is closed

$\Rightarrow \quad R{-}f^{-1}(A)$ is closed $\Rightarrow f^{-1}(A)$ is open.

Hence, f is continuous.

THEOREM 6. *Let f be a function defined on an interval I_1, $a \in I_1$ and let g be a function defined on an interval I_2 such that $f(I_1) \subseteq I_2$. If f is continuous at a and g be continuous at $f(a)$, then composite function $g \circ f$ is continuous at a.*

Proof.   Since, f is continuous at $a \in I_1$

$$\Rightarrow \qquad \lim_{x \to a} f(x) = f(a)$$

Also, g is continuous at $f(a) \in I_2$ $\Rightarrow \lim_{x \to f(a)} g(y) = g[f(a)]$

By definition, $(g \circ f)(x) = g[f(x)] x \in I_1$ $\Rightarrow (g \circ f)(a) = g[f(a)]$

Now, suppose the sequence $<a_n>$ in I_1 converges to a.

Since, $\lim_{x \to a} f(x) = f(a)$ $\Rightarrow \lim_{x \to a} f(a_n) = f(a)$

Also $f(I_1) \subseteq I_2$ and $<f(a_n)>$ is a sequence in I_2, and

$$\lim_{y \to f(a)} g(y) = g[f(a)]. \text{ Therefore } \lim_{n \to \infty} g[f(a_n)] = g[f(a)]$$

$$\Rightarrow \qquad \lim_{n \to \infty} (g \circ f)(a_n) = (g \circ f)(a).$$

Since, this is true for every sequence $<a_n>$ in I_1 converging to a, therefore,

$$\lim_{n \to \infty} (g \circ f)(x) = (g \circ f)(a).$$

Hence, the composite function $g \circ f$ is continuous at a.

REMARKS

- **Borel's theorem.** If f is continuous function on the closed interval $[a,b]$, then the interval can always be divided up into a finite number of subintervals such that $\varepsilon > 0$. $|f(x_1) - f(x_2)| < \varepsilon$ where, x_1 and x_2 are any two points in the same subinterval.

THEOREM 7. (Boundedness theorem). *If a function f is continuous in a closed interval $[a,b]$, then it is bounded in $[a,b]$.*

Proof.   Let if possible f be unbounded on I. Then for each $n \in N$ $\exists$ $x_n \in I$ such that $|f(x_n)| > n$. The bounded sequence $<x_n>$ in I has a subsequence $\langle x_{n_k} \rangle$ such that it converges to a point $x_0 \in I$

($\because$ every subsequence of a convergent sequence is convergent.)

$\Rightarrow \quad \langle x_{n_k} \rangle \to x_0$ and $\left| f\left(x_{n_k} \right) \right| > n_k \; \forall \; n_k \in N$

$\Rightarrow \quad \langle f\left(x_{n_k} \right) \rangle$ cannot converge to $f(x_0)$.

$\Rightarrow \quad f$ is not continuous at x_0 which is a contradiction.

This contradiction leads to the result that f is bounded on I.

REMARK

- The converse of the above theorem need not be true. For example, the function

$$f(x) = \begin{cases} \sin \dfrac{1}{x} & \text{for } x \neq 0 \\ 0 & \text{for } x = 0 \end{cases}$$

is bounded on $[0,1]$ but not continuous in $[0,1]$.   ($\because$ It is discontinuous at $x = 0$)

THEOREM 8. *If a function f is continuous on a closed and bounded interval [a,b],then, it attains its bounds on [a,b].*

Proof.
Since, the function *f* is continuous on the closed and bounded interval [a,b], therefore, it is bounded.

$\Rightarrow$ supremum *M* and infimum *m* of *f* exist in [a,b].

To show, there exist two points $x_1, x_2 \in [a,b]$ such that $f(x_1)=m, f(x_2)=M$

Then, by definition of supremum $f(x) \leq M \quad \forall x \in [a,b]$.

Let if possible $f(x) \neq M$ for any $x \in [a,b]$, then $f(x) < M \ \forall x \in [a,b]$. Therefore,

$$M - f(x) > 0 \ \forall x \in [a,b].$$

Since, $f(x)$ is continuous on [a,b] and *M* is constant, therefore $M-f(x)$ is continuous on [a,b].

Also $M - f(x) \neq 0$ for any $x \in [a,b]$

$\Rightarrow \quad \dfrac{1}{M-f(x)}$ is continuous on [a, b]

$\Rightarrow \quad \dfrac{1}{M-f(x)}$ is bounded on [a, b]

$\Rightarrow \quad \exists$ a number $k > 0$ such that

$\dfrac{1}{M-f(x)} \leq k \ \forall x \in [a,b]$

$\Rightarrow \quad M - f(x) \geq \dfrac{1}{k} \ \forall x \in [a,b]$

$\Rightarrow \quad f(x) \leq M - \dfrac{1}{k} \ \forall x \in [a,b]$

$\Rightarrow \quad M - \dfrac{1}{k}$ is an upper bound if *f* on [a,b] such that $M - \dfrac{1}{k} < M = \sup \ f(x)$ which is a contradiction

$\Rightarrow \quad \exists$ a point $x_2 \in [a,b]$ such that $M = f(x_2)$.

Similarly, we can show that if $m = \inf f(x) \ \exists$ a point x_1 such that $m = f(x_1)$

SOME COUNTER EXAMPLES

(i) The function $1/x$ is continuous but not bounded on]0,1[and]0,1]. Althought it is continuous on these intervals.

(ii) The function *x* is continuous on R but it is unbounded on every infinite interval.

(iii) The function x^2 is continuous and bounded on]0,1[,[0,1] and]0,1].On each of these it has 0 as the infimum and on]0,l[neither the supremum nor the infimum is attained.

(iv) The function $\dfrac{x^2}{x^2+1}$ is continuous and bounded on R. It attains its infimum 0 but not the supremum 1 on any interval.

THEOREM 9. *If a function f(x) is continuous at x = a and $f(a) \neq 0$ then $\exists$ a number $\delta > 0$ such that f(x) has same sign as f(a) for all values of x in $]a-\delta, a+\delta[$.*

Proof.
Since, *f* is continuous at *x=a*, for a given $\varepsilon > 0$, we can find a number $\delta > 0$ such that

$$|f(x)-f(a)| < \varepsilon \text{ whenever } |x-a| < \delta$$

$\Rightarrow \qquad f(a)-\varepsilon < f(x) < f(a)+\varepsilon \text{ whenever } a-\delta < x < a+\delta.$

Now $f(a) \neq 0 \Rightarrow \quad |f(a)| > 0$. Let us choose $0 < \varepsilon < |f(a)|$, then we have $f(a)-\varepsilon$ and $f(a)+\varepsilon$ having the same sign as $f(a)$

$\Rightarrow \quad f(x)$ has the same sign as $f(a)$ for all *x* in the interval $]a-\delta, a+\delta[$.

THEOREM 10. *If a function f is continuous in [a,b]and f(a), f(b) have opposite signs, then there is at least one value of x for which f(x) vanishes.*

Proof. Since, the function $f(x)$ have opposite signs for a and b i.e., $f(a)<0$ and $f(b)>0$.

Let us define $S=[x:x\in[a,b],f(x)<0]$.

Now, since $f(a)<0$, therefore $a\in S$ $\Rightarrow$ $S\neq\phi$.

Let $u=\sup S$.

Now, to show $a<u<b$ and $f(u)=0$.

First, we shall show that $u\neq a$. Since $f(a)<0$ and f is continuous at a,

$\Rightarrow$ $\exists$ a number δ_1 such that $f(x)<0\ \forall\ x\in\ ]a,a+\delta_1[$.

$\Rightarrow$ $[a,a+\delta_1]\subset S$

$\Rightarrow$ $\sup S$ must be greater than or equal to $a+\delta_1$. Therefore, $u\geq a+\delta_1\Rightarrow u\neq a$.

Now, to show $u\neq b$

Since $f(b)>0\Rightarrow\exists\ \delta_2$ such that $f(x)>0\ \forall\ x\in[b-\delta_2,b]$

$\Rightarrow$ $]b-\delta_2,b[\subset S$

$\Rightarrow$ $u=\sup S\leq b-\delta_2<b\ \Rightarrow\ u\neq b$

Now, we shall show that $f(u)\not>0$. Since $a<u<b$. Therefore, if $f(u)>0$. Then we can find a number $\delta_3>0$ such that $f(x)>0$ for $u-\delta_3<x<u+\delta_3$.

Also, $u=\sup S$. Therefore, $\exists\ x_1\in S:u-\delta_3<x_1<u\Rightarrow\ f(x)>0$.

Also $x_1\in S\Rightarrow\ f(x_1)<0$; which is a contradiction

$\Rightarrow$ $f(u)\not>0$.

Now, we shall show that $f(u)\not<0$. If $f(u)<0$, then we can find a positive number δ_4 such that

$u+\delta_4<b$ and $f(x)<0$ for $u-\delta_4<x<u+\delta_4$.

If x_2 is any other point such that $u<x_2<u+\delta_4$. Then $f(x_2)<0$. But this is a contradiction to the fact that u is the supremum of S consequently $f(u)\not<0$.

Hence, $f(u)=0$.

THEOREM 11. (Intermediate value theorem). *Let f be a function continuous on the closed and bounded interval [a,b].If k be any real number between f(a) and f(b), then there exist a real number c between a and b (a<c<b) such that f(c)=k*

Proof. Let us suppose

$$f(a)<k<f(b). \qquad ...(1)$$

Define a fucntion g such that

$$g(x)=f(x)-k;\ x\in[a,b]. \quad ...(2)$$

Now, since f is continuous on $[a,b]$ and k is constant, g is continuous on $[a,b]$.   ...(3)

From (1), we say that k lies between f (a) and $f(b)$. Therefore, either

$$f(a)<k<f(b)\ or\ f(b)<k<f(a).$$

From (2), $g(a)=f(a)-k<0$ $\Rightarrow$ $g(b)=f(b)-k>0$

$\Rightarrow$ $g(a).g(b)<0$ ...(4)

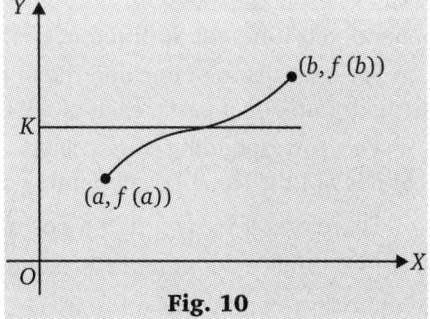

Fig. 10

Now, from (3) and (4) there exists a point $c \in]a,b[$ such that $g(c)=0$

$\Rightarrow$ $\qquad f(c)-k=0$ $\qquad \Rightarrow \qquad f(c)=k$

Hence, these exist a point c such that $a<c<b$ and $f(c)=k$.

REMARKS

- The above theorem can be restated as:

  If a function f is continuous in the closed interval [a,b], then f(x) must take at least once of all values between f(a) and f(b).

- This theorem guarantees only the existence of the number c. It does not tell us how to find it. Also the number c need not be unique.

- The converse of the above theorem is not necessarily true. For example, let f be the function defined by.

$$f(x) = \begin{cases} \sin \dfrac{1}{x} & , x \neq 0 \\ 0 & , x = 0 \end{cases}$$

Then, in the interval $\left[-\dfrac{2}{\pi}, \dfrac{2}{\pi}\right]$ this function takes all values between $f\left(-\dfrac{2}{\pi}\right)$ and $f\left(\dfrac{2}{\pi}\right)$ i.e., between -1 and 1 an infinite number of times as x varies from $-\dfrac{2}{\pi}$ to $\dfrac{2}{\pi}$ but this function is not continuous in $\left[-\dfrac{2}{\pi}, \dfrac{2}{\pi}\right]$. [It is discontinuous at $x=0$].

- If f is continuous on $[a,b]$ and let $k \in [m,M]$ where $m=\inf. f$ and $M= \sup. f$ on $[a,b]$ then there exists $c \in [a,b]$ such that $f(c)=k$.

- If f is continuous on $[a,b]$, then $f([a,b])=[m, M]$. Also, $f([a,b])$ is a closed set.

- If f is a continuous, one to one function on a finite closed interval $[a, b]$, then f is also continuous on its domain.

4.13 UNIFORM CONTINUITY

We know that if a function $f(x)$ is continuous in the closed interval I, then for a given positive number ε, $\exists$ a positive number $\delta>0$ such that

$$|f(x)-f(a)|<\varepsilon \text{ for } |x-a|<\delta, a \in I.$$

Here, we observe that the number δ depends on, besides ε, on the point a as it is a function of a. In general, δ is different at different points in I.

For this, let us consider the figure, where PQ is divided into equal parts, each of length ε.

The corresponding subdivision of $I=[a,b]$ is such that δ is not the same for all points x in $[a,b]$.

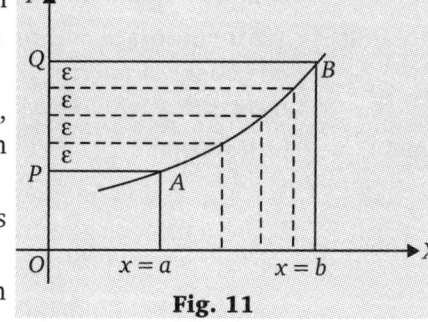

Fig. 11

Therefore, if we can find a positive number δ_0 such that for a chosen ε, $|f(x)-f(a)|<\varepsilon$ for $|x-a|<\delta_0$ where the number δ_0 is independent of the point a, then the function $f(x)$ is said to be uniformly continuous on $[a,b]$.

Definition. *A function f(x) defined on an interval I is said to be uniformly continuous in I if to each $\varepsilon>0$ $\exists$ a positive number $\delta>0$, (depending upon ε) but independent of $x \in I$ such that*

$$|f(x_2)-f(x_1)| < \varepsilon \text{ whenever } |x_2-x_1| < \delta$$

where $x_1, x_2 \in I$.

REMARK

- A function f is not uniformly continuous on I, if there exist some $\varepsilon > 0$ for which no $\delta > 0$ works *i.e.*, for any $\delta > 0 \; \exists \; x_1, x_2 \in I$ such that $|f(x_2)-f(x_1)| \geq \varepsilon$ for $|x_2-x_1| < \delta$.
- The uniform continuity of f on an arbitrary set S can be defined by replacing the interval I by S in the above definition.

THEOREM 1. *If a function f is uniformly continuous on an interval I, then it is continuous on I.*

Proof.   Let us suppose that f is uniformly continuous on I

$\Rightarrow$ given $\varepsilon > 0 \; \exists \; \delta > 0$ such that

$$|f(x_2) - f(x_1)| < \varepsilon, \text{ whenever } |x_2 - x_1| < \delta \; \forall x_1, x_2 \in I$$

In particular, let us take $x_2 \in I$, then we have

$$|f(x) - f(x_1)| < \varepsilon, \text{ whenver } 0 < |x - x_1| < \delta$$

$\Rightarrow$ $f(x)$ is continuous at $x_1 \in I$.

Since, x_1 is arbitrary, consequently $f(x)$ is continuous on I.

REMARKS

- The converse of the above theorem is not true as can be seen in the example, given below:
  Consider the function $f(x) = x^2 \; \forall x \in R$ which is continuous for all $x \in R$ but not uniformly continuous.
- The uniform continuity is a property associated with an interval and not with a single point *i.e.*, the concept of continuity is local in character, while the uniform continuity is global in character.

THEOREM 2. *If a function $f(x)$ is continuous on an closed and bounded interval $I = [a,b]$, then it is uniformly continuous on $[a,b]$.*

Proof.   Since f is given to be continuous in the interval $[a,b]$.

Let $\varepsilon > 0$ be given $\Rightarrow [a,b]$ can be divided into a finite number of subintervals such that $|f(x_2)-f(x_1)| < \dfrac{\varepsilon}{2}$, where x_1, x_2 are any two points of the same subinterval.

Let us divide the whole interval $[a,b]$ into n sub intervals, say

$$[x_0 = a, x_1], [x_1, x_2], [x_2, x_3], ...,[x_{n-1}, x_n = b]$$

$\Rightarrow$ $|f(x') - f(x'')| < \dfrac{\varepsilon}{2}$, where x', x'' belongs to the same subinterval   ...(1)

Let $\delta = \min\{\delta_1, \delta_2 ...\delta_r, ...\delta_n\}$ where δ_r denotes the length of the rth subinterval *i.e.*,

$$\delta_r = |x_r - x_{r-1}|$$

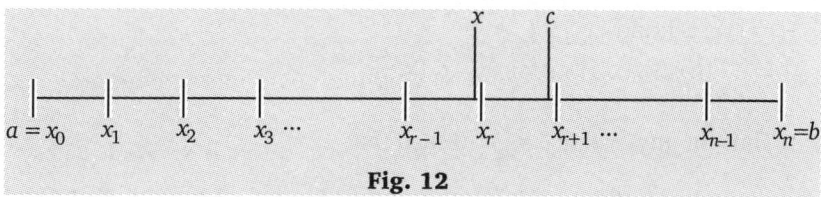

Fig. 12

Let x and c be any two points of $[a,b]$ such that $|x - c| < \delta$.

Since $\delta > 0$, less than the length of each subinterval. Therefore, following two cases may arise:

Case (i) When x and c belongs to same interval:

$\Rightarrow \qquad |f(x)-f(c)| < \dfrac{\varepsilon}{2}$, when $|x-c| < \delta$; where $x, c \in [a, b]$

$\Rightarrow$ function f is uniformly continuous in $[a,b]$.

Case (ii) When x and c belongs to the two consecutive sub intervals say

$$x_{r-1} < x < x_r < c < x_{r+1}.$$

Consider

$$|f(x)-f(c)| = |f(x)-f(x_r)+f(x_r)-f(c)|$$

$$\leq |f(x)-f(x_r)| + |f(x_r)-f(c)|$$

$$< \frac{\varepsilon}{2} + \frac{\varepsilon}{2} \text{ when } |x-c| < \delta < \varepsilon \text{ when } |x-c| < \delta.$$

$\therefore$ Given $\varepsilon > 0 \; \exists \; \delta > 0$ such that $|f(x)-f(c)| < \varepsilon$ where x and c are any two points of $[a, b]$ such that $|x-c| < \delta$

$\Rightarrow \quad f$ is uniformly continuous on $[a,b]$.

Hence, f is continuous on a closed and bounded interval $[a, b]$

$\Rightarrow \quad f$ is uniformly continuous on $[a, b]$.

Solved Examples

Example 1. *Show that the function*

$f(x) = x^2 + 3x, \; x \in [-1,1]$ *is uniformly continuous in* $[-1,1]$.

Solution. Let $\varepsilon > 0$ be given

Let $x_1, x_2 \in [-1,1]$

$\Rightarrow |f(x_2)-f(x_1)|$

$\qquad = |(x_2{}^2+3x_2)-(x_1{}^2+3x_1)|$

$\qquad = |(x_2{}^2-x_1{}^2)+3(x_2-x_1)|$

$\qquad = |(x_2-x_1)(x_2+x_1+3)|$

$\qquad = |x_2-x_1| \, |x_2+x_1+3|$

$\qquad \leq |x_2-x_1| \, (|x_2|+|x_1|+3)$

$\qquad \leq 5|x_2-x_1|$

$\qquad\qquad\qquad (x_1, x_2 \in [-1,1])$

$\Rightarrow |x_1| \leq 1$ and $|x_2| \leq 1$

$\Rightarrow \quad |f(x_2)-f(x_1)| < \varepsilon$ for $|x_2-x_1| < \dfrac{\varepsilon}{5}$

Based on the following Results

- A function $f(x)$ defined on an interval I is said to be uniformly continuous on I if to each $\varepsilon > 0 \; \exists$ a positive number $\delta > 0$ such that

$$|f(x_2) - f(x_1)| < \varepsilon \text{ whenever}$$
$$|x_2 - x_1| < \delta$$

- Every uniformly continuous function is continuous but converse is not, necessarily true.

- If a function is continuous in a closed and bounded interval then it is, uniformly continuous on this interval.

Thus for any $\varepsilon < 0$, $\exists \; \delta = \dfrac{\varepsilon}{5} > 0$ such that

$$|f(x_2) - f(x_1)| < \varepsilon \text{ whenever } |x_2-x_1| < \delta \; \forall \; x_1, x_2 \in [-1,1].$$

Hence, $f(x)$ is uniformly continuous in $[-1, 1]$.

Example 2. *Show that the function f defined by $f(x) = x^3$ is uniformly continuous on $[-2, 2]$.*

Solution. In order to show that the function f is uniformly continuous we have to prove that for a given $\varepsilon > 0 \; \exists \; \delta > 0$ such that

$$|f(x_2) - f(x_1)| < \varepsilon \text{ when } 0 < |x_2 - x_1| < \delta \text{ where } x_1, x_2 \in [-2, 2].$$

Consider $\quad |f(x_2) - f(x_1)| = |x_2^3 - x_1^3|$

$$= |(x_2 - x_1)(x_2^2 + x_1^2 + x_1 x_2)|$$

$$\leq |(x_2 - x_1)|[|x_1^2| + |x_2^2| + |x_1 x_2|]$$

$$\leq 12 |x_2 - x_1|$$

$$(\because x_1, x_2 \in [-2, 2] \Rightarrow |x_1| \leq 2 \text{ and } |x_2| \leq 2)$$

$$\therefore \qquad |f(x_2) - f(x_1)| < \varepsilon \text{ whenever } |x_2 - x_1| < \frac{\varepsilon}{12}.$$

Therefore, given $\varepsilon > 0 \; \exists \; \delta (= \varepsilon/12)$ such that

$$|f(x_2) - f(x_1)| < \varepsilon \text{ whenever } |x_2 - x_1| < \delta, \; x_1, x_2 \in [-2, 2].$$

Hence, f is uniformly continuous on $[-2, 2]$.

Example 3. *Show that the function f defined by $f(x) = \dfrac{1}{x}, \forall x \in]0, 1]$ is not uniformly continuous in $]0, 1]$.*

Solution. In order to show that the function f is uniformly continuous in $]0,1]$ we have to prove that for a given $\varepsilon > 0 \; \exists \; \delta > 0$, independent of the choice of x, $x \in]0,1]$ such that

$$|f(x) - f(c)| = \left| \frac{1}{x} - \frac{1}{c} \right| < \varepsilon \text{ whenever } 0 < |x - c| < \delta$$

i.e., $\qquad |x - c| < \delta \Rightarrow \left| \dfrac{c - x}{cx} \right| < \varepsilon$

i.e., $\qquad x \in]c - \delta, c + \delta[\Rightarrow \left| \dfrac{c - x}{cx} \right| < \varepsilon \qquad \qquad \ldots(1)$

Let us take $c = \delta$, then $]c - \delta, c + \delta[=]0, 2\delta[$.

Since, the condition (1) must hold for all $x \in]0, 2\delta[$

$$\therefore \qquad \text{as } x \to 0, \; \frac{\delta - x}{\delta x} \to \infty \text{ and } x \in]0, 2\delta[$$

i.e., if we choose x close to zero, then condition (1) does not hold.

$\Rightarrow \quad f(x) = \dfrac{1}{x}$ is not uniformly continuous in $]0,1]$.

Example 4. *Show that the function f defined on R^+ as*

$$f(x) = \sin \frac{1}{x}, \; \forall \; x > 0$$

is continuous, but not uniformly continuous on R^+.

Solution. Let $a \in R^+$.

We have $\text{LHL} = f(a-0) = \lim_{h \to 0} f(a-h) = \lim_{h \to 0} \sin \dfrac{1}{a - h} = \sin \dfrac{1}{a}$

$$\text{RHL} = f(a+0) = \lim_{h \to 0} f(a+h) = \lim_{h \to 0} \sin \frac{1}{a + h} = \sin \frac{1}{a}$$

$$\Rightarrow \qquad \qquad f(a) = \sin \frac{1}{a}$$

$$\Rightarrow \qquad f(a+0) = f(a) = f(a-0)$$

$$\Rightarrow \qquad f \text{ is continuous at } a.$$

Since, a is arbitrary point in R^+. Therefore, f is continuous on R^+.

Now, to show f is not uniformly continuous on R^+.

Let δ be any positive number. Take

$$x_1 = \frac{1}{n\pi}, x_2 = \frac{1}{n\pi + \pi/2} = \frac{2}{(2n+1)(\pi)} \quad \text{where } n \in Z^+$$

such that $x_1 - x_2 = \dfrac{1}{n\pi} - \dfrac{2}{(2n+1)\pi} < \delta$

Now, $|x_1 - x_2| < \delta$ but $|f(x_1) - f(x_2)| = \left|\sin n\pi - \sin\frac{1}{2}(2n+1)\pi\right| = 1 > \varepsilon$

which shows that for this choice of ε, we cannot find a $\delta > 0$ such that

$$|f(x_1) - f(x_2)| < \varepsilon \text{ for } |x_1 - x_2| < \delta \ \forall x_1, x_2 \in R^+$$

Hence, f is not uniformly continuous on R^+.

Example 5. *Show that $f(x) = \sqrt{x}$ is uniformly continuous in $[0,1]$.*

Solution. Let $x, y \in [0,1]$, $x > y \geq 0$. Then

$$|f(x) - f(y)| = \left|\sqrt{x} - \sqrt{y}\right| = \left(\frac{\left(\sqrt{x} - \sqrt{y}\right)\left(\sqrt{x} + \sqrt{y}\right)}{\sqrt{x} + \sqrt{y}}\right)$$

$$= \frac{x-y}{\sqrt{x} - \sqrt{y}} \leq \frac{x-y}{\sqrt{x}} \leq \frac{x-y}{\sqrt{x-y}} = \sqrt{x-y}$$

$$\therefore \qquad |f(x) - f(y)| \leq \sqrt{x-y}$$

Now, for given $\varepsilon > 0$, $|f(x) - f(y)| < \varepsilon$ whenever $\sqrt{x-y} < \varepsilon$

$$\Rightarrow \qquad |f(x) - f(y)| < \varepsilon \text{ whenever } |x-y| < \delta (= \varepsilon^2) \ \forall x, y \in [0,1]$$

Hence, $f(x)$ is uniformly continuous in $[0,1]$.

Example 6. *Show that $f(x) = \sin x^2$ is not uniformly continuous on $[0, \infty[$.*

Solution. Let $\varepsilon = \dfrac{1}{2}$ and $\delta > 0$ be arbitrary. Choose a positive integer n such that $n > \dfrac{\pi}{\delta^2}$ ...(1)

Let $\qquad x_1 = \sqrt{\dfrac{n\pi}{2}}, x_2 = \sqrt{\dfrac{(n+1)\pi}{2}} \in [0, \infty[$. Then,

$$|f(x_2) - f(x_1)| = \left|\sin x_2^2 - \sin x_1^2\right| = \left|\sin(n+1)\frac{\pi}{2} - \sin\frac{n\pi}{2}\right|$$

$$= \begin{cases} |0 - (\pm 1)| = 1 & \text{if } n \text{ is odd} \\ |\pm 1 - 0| = 1 & \text{if } n \text{ is even} \end{cases}$$

$$\therefore \qquad |f(x_2) - f(x_1)| = 1 > \varepsilon \text{ and } |x_2 - x_1| = \left|\frac{x_2^2 - x_1^2}{x_2 + x_1}\right|$$

$$= \frac{\pi/2}{\sqrt{(n+1)\frac{\pi}{2}} + \sqrt{\frac{n\pi}{2}}} < \frac{\pi}{2\left[2\sqrt{\frac{n\pi}{2}}\right]} < \frac{\pi}{\sqrt{n\pi}} = \sqrt{\frac{\pi}{n}} < \infty \qquad \text{(By (1))}$$

$\therefore$ $|f(x_2)-f(x_1)|>\varepsilon$ when $|x_2-x_1|<\delta$

Hence, $f(x)=\sin x^2$ is not uniformly continuous on $[0,\infty[$.

EXERCISE 4.3

1. Let $f : \mathrm{R}\to\mathrm{R}$ given $f(x)=x^2$. Show that f is not uniformly continuous on R.

2. Show that the function x^2 and x^3 are not uniformly continuous on $[0,\infty[$.

3. In each of the following cases, show that f is continuous but not uniformly continuous on their respective intervals.

   (i) $f(x)=\sin\dfrac{1}{x}$ $\forall x\in\,]0,1[$

   (ii) $f(x)=\dfrac{1}{2x}$ $\forall x\in[-1,0[$

   (iii) $f(x)=\dfrac{1}{1-x}$ $\forall x\in\,]0,1[$

   (iv) $f(x)=e^x$ $\forall x\in[0,\infty[$

4. If $f(x+y)=f(x).f(y)$ $\forall x, y\in\mathrm{R}$, show that f is continuous on R if and only if f is continuous at least one point of R. If f is continuous at some point $a\in\mathrm{R}$, prove that f is uniformly continuous on every bounded subset of R.

5. Show that the function f defined by

$$f(x)=\begin{cases} x^2\sin\dfrac{1}{x^2} & \text{for } x\neq 0 \\ 0 & \text{for } x=0 \end{cases}$$

   is uniformly continuous in $[-1,1]$.

6. Show that if f and g are bounded and uniformly continuous on an interval I, then the product function fg is also uniformly continuous on I.

4.14 DERIVATIVE OF A FUNCTION

If a function $f(x)$ is defined on nbd of a point a and

$$\lim_{h\to 0}\frac{f(a+h)-f(a)}{h}$$

exist (finitely), then the function $f(x)$ is said to be differentiable at a and this limit is called derivative of the function $f(x)$ at a.

In symbols, this derivative, is denoted by $f'(a)$ and in full read as the derivative of $f(x)$ at $x=a$ with respect to the variable x. The process of evaluating $f'(a)$ is called differentiation.

Graphically, $f'(a)$ means the gradient of the curve $y = f(x)$ at the point $(a, f(a))$.

Quantitatively $f'(a)$ means the rate of change of the function $f(x)$ at a, with respect to the variable x.

4.14.1 LEFT HAND DERIVATIVE

The left hand derivative (regressive derivative) of f at $x = a$ is given by

$$Lf'(a) = \lim_{h\to 0}\frac{f(a-h)-f(a)}{-h}$$

and, is denoted by $Lf'(a)$.

4.14.2 RIGHT HAND DERIVATIVE

The right hand derivative (progressive derivative) of f at $x = a$ is given by

$$Rf'(a) = \lim_{h\to 0}\frac{f(a+h)-f(a)}{h}$$

The derivative $f'(a)$ exists when $Lf'(a) = Rf'(a)$.

4.14.3 DIFFERENTIABILITY IN AN INTERVAL

(i) A function $f : \,]a, b[\to \mathrm{R}$ is said to be differentiable in $]a, b[$ iff it is differentiable at every point of $]a, b[$.

(ii) A function $f : [a, b] \to R$ is said to be differentiable in $[a, b]$ iff $Rf'(a)$ and $Lf'(b)$ exists and f is differentiable at every point of $]a, b[$.

(iii) Let f be a function whose domain is an interval I. If I_1 be the set of all those points x of I at which f is differentiable $i.e.$, $f(x)$ exists and if $I_1 \neq \phi$, we get another function f' with domain I_1. It is called the first derivative of f. Similarly 2nd, 3rd, ...n^{th} derivative of f are defined and one denoted by f'', f''', ..., f^n respectively of course, in order that $f^n(x)$ may be defined, it is necessary (though not sufficient) that $f^{n-1}(x)$ may be defined for all x in some open interval containing a.

REMARKS

- $\lim\limits_{x \to a} \dfrac{f(x) - f(a)}{x - a}$ means the same thing as $\lim\limits_{h \to 0} \dfrac{f(a + h) - f(a)}{h}$

- The derivative of a function at a point and the derivative of a function are two different but related concepts. The derivative of f at a point a is a number while the derivative of f is a function. However, very often the term derivative of f is used to denote both number and function and it is left to the context to distinguish what is intended.

- If $f(x)$ is derivable on internal I then $f'(x)$ at end points of I (if exists) would mean a left or right hand derivative of $f(x)$ according as it is a right or a left hand end point of I. Similar meaning holds for higher order derivatives.

4.15 CONTINUITY AND DIFFERENTIABILITY

THEOREM 1. **(A necessary condition for the existence of a finite derivative).** *Continuity is a necessary but not a sufficient condition for the existence of a finite derivative.*

Proof. Let f be differentiable at a. Then $\lim\limits_{x \to a} \dfrac{f(x) - f(a)}{(x - a)}$ exists and equal to $f'(a)$.

Now we may write

$$f(x) - f(a) = \lim_{x \to a} \frac{f(x) - f(a)}{(x - a)} (x - a) \qquad \text{(If } x \neq a\text{)}$$

Taking limit as $x \to a$, we get

$$\lim_{x \to a} [f(x) - f(a)] = \lim_{x \to a} \left\{ \frac{f(x) - f(a)}{(x - a)} (x - a) \right\}$$

$$= \lim_{x \to a} \left\{ \frac{f(x) - f(a)}{x - a} \right\} \cdot \lim_{x \to a} (x - a)$$

($\because$ limit of the product of two functions is equal to product of their limits)

$$= f'(a) . 0 = 0$$

so that $\lim\limits_{x \to a} f(x) = f(a) \Rightarrow f(x)$ is continuous at $x = a$.

Hence, f is continuous at $x = a$. Thus continuity is a necessary condition for differentiability

REMARKS

- While continuity is a necessary condition for the differentiability, it is not a sufficient condition as it is clear from the following examples :

  (i) Consider the function $f(x)$ defined on R by setting
$$f(x)=0 \quad \text{if} \quad x=0$$
$$f(x)=x \quad \text{if} \quad x \neq 0$$
  f is obviously continuous as also derivative at every point except possibly at $x=0$. At $x=0$, f is continuous but not derivable.

  (ii) Consider the function $f(x)$ such that
$$\begin{cases} x\sin\dfrac{1}{x} & , \ x \neq 0 \\ 0 & , \ x=0 \end{cases}$$
  this function is continuous at $x=0$ but not differentiable at $x=0$.

  (iii) The function $f(x)=|x|$ is a continuous function, but not differentiable at $x=0$.
$$(\because Lf'(0)=-1 \text{ and } Rf'(0)=1)$$

- Continuity of a function even at every point of R has nothing to do with the differentiability of the function at any point.
- Weierstrass is considered as the first mathematician who gave in 1872 examples of functions continuous on R but no where differentiable. The examples given by Weierstrass are:

  (a) $f(x)=\displaystyle\sum_{n=1}^{\infty} a^n \cos b^n \pi x$, where b is an odd integer $1>a>0$ and $ab>1+\dfrac{3}{2}\pi$.

  (b) $f(x)=\displaystyle\sum_{n=1}^{\infty} \dfrac{1}{2^n}\cos(3^n x) \ \forall \, x \in R$.

4.16 ALGEBRA OF DERIVATIVES

THEOREM 1. *Let functions f and g be defined on an interval I. If f and g are differentiable at $x=a \in I$, then $f \pm g$ is also differentiable and*
$$(f \pm g)'(a) = f'(a) \pm g'(a)$$

Proof. Since, the functions f and g are differentiable at a, therefore
$$\lim_{x \to a} \frac{f(x)-f(a)}{x-a} = f'(a) \qquad \qquad \text{...(1)}$$

and
$$\lim_{x \to a} \frac{g(x)-g(a)}{x-a} = g'(a) \qquad \qquad \text{...(2)}$$

Now, consider $\displaystyle\lim_{x \to a} \frac{(f \pm g)(x)-(f \pm g)(a)}{x-a}$

$$= \lim_{x \to a} \frac{[f(x) \pm g(x)]-[f(a) \pm g(a)]}{x-a}$$

$$= \lim_{x \to a} \left[\frac{f(x)-f(a)}{x-a} \pm \frac{g(x)-g(a)}{x-a} \right]$$

$$= \lim_{x \to a} \frac{f(x)-f(a)}{x-a} \pm \lim_{x \to a} \frac{g(x)-g(a)}{x-a}$$

$$= f'(a) \pm g'(a)$$

Hence $f \pm g$ is differentiable at a and
$$(f \pm g)'(a) = f'(a) \pm g'(a)$$

THEOREM 2. *Let a function f(x) be differentiable at a point a and c∈R, then the function cf is also differentiable at a and* $(cf)'(a) = cf'(a)$

Proof. By the defination of the derivative of a function at $x = a$, we have

$$\lim_{x \to a} \frac{f(x) - f(a)}{x - a} = f'(a)$$

Now, consider

$$\lim_{x \to a} \frac{(cf)(x) - (cf)(a)}{x - a} = \lim_{x \to a} \frac{cf(x) - cf(a)}{x - a}$$

$$= \lim_{x \to a} \left\{ c \left(\frac{f(x) - f(a)}{x - a} \right) \right\}$$

$$= c \lim_{x \to a} \left\{ \frac{f(x) - f(a)}{x - a} \right\}$$

$$= cf'(a)$$

Hence, cf is differentiable at a and $(cf)'(a) = cf'(a)$

THEOREM 3. *Let the functions f and g be defined on an interval I. If f and g are differentiable at a∈I, then f. g is also differentiable and*

$$(fg)'(a) = f'(a)g(a) + f(a)g'(a)$$

Proof. Since, f and g are differentiable at a, we have

$$\lim_{x \to a} \frac{f(x) - f(a)}{x - a} = f'(a) \qquad \text{... (1)}$$

and $$\lim_{x \to a} \frac{g(x) - g(a)}{x - a} = g'(a) \qquad \text{... (2)}$$

Consider $\lim_{x \to a} \frac{(fg)(x) - (fg)(a)}{x - a} = \lim_{x \to a} \frac{f(x)g(x) - f(a)g(a)}{x - a}$

$$= \lim_{x \to a} \frac{f(x)g(x) - f(a)g(x) + f(a)g(x) - f(a)g(a)}{x - a}$$

$$= \lim_{x \to a} \left[\frac{f(x) - f(a)}{x - a} \cdot g(x) + f(a) \cdot \frac{g(x) - g(a)}{x - a} \right]$$

$$= \lim_{x \to a} \left[\frac{f(x) - f(a)}{x - a} \right] \lim_{x \to a} g(x) + f(a) \lim_{x \to a} \frac{g(x) - g(a)}{x - a}$$

$$= f'(a)g(a) + f(a)g'(a)$$

Hence, fg is differentiable at a and $(fg)'(a) = f'(a)g(a) + f(a)g'(a)$

THEOREM 4. *If a function f is differentiable at x=a and f(a) ≠ 0, then the function $\frac{1}{f}$ is differentiable at a and*

$$\left(\frac{1}{f} \right)'(a) = -\frac{f'(a)}{[f(a)]^2}$$

Proof. Since f is differentiable at a, therefore, it is continuous also at x=a.

Also, since $f(a) \neq 0$

Consider
$$\frac{\frac{1}{f(x)}-\frac{1}{f(a)}}{x-a}=-\left[\frac{f(x)-f(a)}{x-a}\right]\cdot\frac{1}{f(x)}\cdot\frac{1}{f(a)} \qquad \dots(1)$$

Since f is differentiable at $x = a$, therefore,
$$\lim_{x\to a}\frac{f(x)-f(a)}{x-a}=f'(a) \qquad \dots(2)$$

Also, f is continuous at $x = a$, therefore
$$\lim_{x\to a}f(x)=f(a)\neq 0 \qquad \dots(3)$$

By applying the theorem on the limits of a product to (1), and using (2) and (3), we find that
$$\lim_{x\to a}\frac{\frac{1}{f(x)}-\frac{1}{f(a)}}{x-a}\text{ exist and equal to }-\frac{f'(a)}{[f(a)]^2}$$

THEOREM 5. *Let f and g be defined on an interval I. If f and g are differentiable at $a\in I$, and if $g(a)\neq 0$, then the function f/g is also differentiable at a.*

Proof. Let $F = f/g$. Then
$$F(x)-F(a)=(f/g)(x)-(f/g)(a)$$
$$=\frac{f(x)}{g(x)}-\frac{f(a)}{g(a)}=\frac{1}{g(x)g(a)}[f(x)g(a)-f(a)g(x)]$$
$$=\frac{1}{g(x)g(a)}[f(x)g(a)-f(a)g(a)+f(a)g(a)-f(a)g(x)]$$
$$\therefore\lim_{x\to a}\frac{F(x)-F(a)}{x-a}=\lim_{x\to a}\frac{1}{g(x)g(a)}\cdot\left[\left\{\frac{f(x)-f(a)}{x-a}\right\}g(a)-f(a)\left\{\frac{g(x)-g(a)}{x-a}\right\}\right]$$

or
$$F'(a)=\frac{1}{g(a)g(a)}[f'(a)g(a)-f(a)g'(a)]$$

$$\Rightarrow \quad \left(\frac{f}{g}\right)'(a)=\frac{f'(a)g(a)-f(a)g'(a)}{[g(a)]^2}$$

THEOREM 6. *Let f and g be functions such that the range of f is contained in the domain of g. If f is differentiable at a and g is differentiable at $f(a)$, then gof is differentiable at a and $(g\circ f)'=g'(f(a)).f'(a)$ (This is known as Chain rule).*

Proof. Since, the range of f contained in the domain of g, therefore, gof has the same domain as that of f.
Now, let $y = f(x)$ and $y_0 = f(a)$
Since, f is differentiable at a, we have
$$\lim_{x\to a}\frac{f(x)-f(a)}{x-a}=f'(a)$$
or
$$f(x)-f(a)=(x-a)[f'(a)+A(x)] \qquad \dots(1)$$
where $A(x)\to 0$ as $x\to a$.
Further since g is differentiable at y_0, we have
$$\lim_{y\to y_0}\frac{g(y)-g(y_0)}{y-y_0}=g'(y_0)$$

or $$g(y) - g(y_0) = (y - y_0)[g'(y_0) + B(y)] \qquad \text{... (2)}$$

where $B(y) \to 0$ as $y \to y_0$

Now $(g \circ f)(x) - (g \circ f)(a) = g(f(x)) - g(f(a)) = g(y) - g(y_0)$

$$= (y - y_0)[g'(y_0) + B(y)] \qquad \text{[By (2)]}$$

$$= [f(x) - f(a)][g'(y_0) + B(y)]$$

$$= (x - a)[f'(a) + A(x)][g'(y_0) + B(y)], \qquad \text{[By (1)]}$$

Thus if $x \neq a$, then

$$\frac{(g \circ f)(x) - (g \circ f)(a)}{x - a} = [g'(y_0) + B(y)][f'(a) + A(x)] \qquad \text{...(3)}$$

Also f being differentiable at a is continuous at a and hence $x \to a, f(x) \to f(a)$ i.e.,
$y \to y_0$. $\Rightarrow B(y) \to 0$ as $x \to 0$ and $A(x) \to 0$ as $x \to a$.

Now, taking the limit as $x \to a$, we get from (3)

$$\lim_{x \to a} \frac{(g \circ f)(x) - (g \circ f)(a)}{x - a} = g'(y_0)f'(x_0)$$

Hence the function is differentiable at a and $(g \circ f)'(a) = g'(f(a))f'(a)$

THEOREM 7. **(Derivative of the inverse function).** *If f is differentiable at $x = a$ and is one-one function defined on interval I with $f'(a) \neq 0$, then the inverse of the f is differentiable at $f(a)$ and its derivative at a is $\dfrac{1}{f'(a)}$.*

Proof. Let the domain of f be X and range Y.

If g be the inverse of f, then g is a function with domain Y and range X such that
$f(x) = y \Leftrightarrow g(y) = x$.

Now, let us suppose $y = f(x)$ and $y_0 = f(a)$. Since, f is differentiable at a, we have

$$\lim_{x \to a} \frac{f(x) - f(a)}{x - a} = f'(a)$$

or $$f(x) - f(a) = (x - a)[f'(a) + A(x)] \qquad \text{... (1)}$$

where $A(x) \to 0$ as $x \to a$. Further, we have

$$g(y) - g(y_0) = x - a, \qquad \text{[By definition of g]}$$

$\therefore$ $$\frac{g(y) - g(y_0)}{y - y_0} = \frac{x - a}{y - y_0} = \frac{x - a}{f(x) - f(a)} = \frac{1}{f'(a) + A(x)} \qquad \text{[By (1)]}$$

It can be easily seen that if $y \to y_0$, then $x \to a$.

In fact, f being differentiable at a, it is also continuous at a, which implies that $g = f^{-1}$ is continuous at $f(a) = y_0$ and consequently.

$$g(y) \to g(y_0) \text{ as } y \to y_0 \text{ i.e., } x \to a \text{ as } y \to y_0, \text{ so that } A(x) \to 0 \text{ as } y \to y_0.$$

$\therefore$ $$\lim_{y \to y_0} \frac{g(y) - g(y_0)}{y - y_0} = \lim_{y \to y_0} \frac{1}{f'(a) + A(x)} = \frac{1}{f'(a)}$$

or $$g'(y_0) = \frac{1}{f'(a)} \text{ or } g'(f'(a)) = \frac{1}{f'(a)}$$

THEOREM 8. (Darboux's Theorem or Intermediate Value Theorem). *If f is finitely differentiable in a closed interval $[a, b]$ and $f'(a), f'(b)$ are of opposite sign, then there exist at least one point $c \in\ ]a, b[$ such that $f'(c) = 0$.*

Proof. Let us suppose that $f'(a) > 0$ and $f'(b) < 0$, then there exist intervals $]a, a + h\ [$ and $]b - h, b[, h > 0$ such that

$$f(x) > f(a)\ \forall x \in]a, a + h[\qquad \text{... (1)}$$

$$f(x) > f(b)\ \forall x \in [b - h, b[\qquad \text{... (2)}$$

Now, since f is finitely differentiable, then it is continuous in $[a, b]$ and hence it is bounded on $[a, b]$ and attains its supremum and infimum at least once in $[a, b]$. [$\because$ A continuous function attains its supremum and infimum at least once in $[a, b]$]. Thus if M is the supremum of f in $[a, b]$, then there exist $c \in [a, b]$ such that $f(c) = M$. It is clear from (1) and (2) that the upper bound is not attained at the end points a and b so that $c \in]a, b[$.

Now we shall prove $f'(c) = 0$

If $f'(c) > 0$, then there exist an interval $]c, c + h], h > 0$, such that $f(x) > f(c) = M$ $\forall x \in]c, c + h[$, which is not possible, since M is the supremum of the function $f(x)$ in $[a, b]$.

If $f'(c) < 0$ then there exist an interval $[c - h, c[, h > 0$ such that $f(x) > f(c) = M$ $\forall x \in [c - h, c[$, which is not possible.

Hence, we conclude that $f'(c) = 0$

REMARK

- Darboux's theorem shows that derivative do share an important property of continuous functions. Since the image of an interval under a continuous function is an interval. Darboux's theorem essentially says that the result hold even if a function is not ,continuous, provided of course, it is a derivative. That is, if a function g defined on an interval I is the derivative of some function f, then $g(I)$ is an interval.

THEOREM 9. *Let f be defined and differentiable on $[a, b]$, and if c be any number between $f'(a)$ and $f'(b)$, then there exist a real number k between a and b such that $f'(k) = c$.*

Proof. Let g be the function defined on $[a, b]$ by setting

$$g(x) = f(x) - cx \text{ for all } x \in [a, b]$$

Now, g is differentiable on $[a, b]$ and $g'(a) = f'(a) - c$, and $g'(b) = f'(b) - c$ since c lies between $f'(a)$ and $f'(b)$. Therefore, it follows that $g'(a)$ and $g'(b)$ are of opposite signs.

Since g is differentiable on $[a, b]$, and since $g'(a)\ g'(b) < 0$, therefore there exist a number k between a and b such that $g'(k) = 0$ i.e., $f'(k) = c$.

THEOREM 10. *If f is defined and differentiable on an interval, the range of f' is an interval.*

Proof. Let the domain of f (and therefore, that of f') be an interval X and let the range of f' be Y. Also let p and q be two distinct points of Y. Then there exist two distinct points a and b in X such that $f'(a) = p$ and $f'(b) = q$.

Assume that $a < b$.

Since X is an interval and $a \in X, b \in X$, therefore $[a, b] \subset X$.

Now f is defined and derivable on $[a, b]$. If r be any real number between p and q, then by theorem 9, there exists a real number k between a and b such that $f'(k) = r$, that is $r \in Y$. Thus we find that if p and q are in Y, then every number between p and q is in Y, and this means that Y is an interval.

REMARKS

- If Y does not contain at least two distinct elements, then it is a singleton.
- If f is defined and differentiable on $[a, b]$ and $f'(x) \neq 0$ for any $x \in]a, b[$ then $f'(x)$, retains the same sign, positive or negative in $]a, b[$ i.e., $f(x)$ is either positive or negative for all values of $x \in]a, b[$.

Solved Examples

Example 1. Prove that the function $f(x) = |x| + |x - 1|$ is not differentiable at $x = 0$ and $x = 1$.

Solution. Here, we observe that

(i) $|x| = -x$ and $|x - 1| = 1 - x$
  when $x < 0$.

(ii) $|x| = x$ and $|x - 1| = 1 - x$
  when $0 \le x \le 1$.

(iii) $|x| = x$ and $|x - 1| = x - 1$
  when $x > 1$.

Hence, the given function can be rewritten as

$$f(x) = \begin{cases} -x + 1 - x &= 1 - 2x, & x < 0 \\ x + 1 - x &= 1, & 0 \le x \le 1 \\ x + x - 1 &= 2x - 1, & x > 1 \end{cases}$$

Now, firstly we check the differentiability of $f(x)$ at $x = 0$.

We have $Rf'(0) = \lim\limits_{h \to 0} \dfrac{f(0+h) - f(0)}{h}$

$= \lim\limits_{h \to 0} \dfrac{f(h) - f(0)}{h}$

$= \lim\limits_{h \to 0} \dfrac{1 - 1}{h} = 0$

and $Lf'(0) = \lim\limits_{h \to 0} \dfrac{f(0 - h) - f(0)}{-h}$

$= \lim\limits_{h \to 0} \dfrac{f(-h) - f(0)}{-h}$

$= \lim\limits_{h \to 0} \dfrac{1 - 2(-h) - 1}{-h}$

$= \lim\limits_{h \to 0} \dfrac{2h}{-h} = -2$

Based on the following Results

- $Lf'(a) = \lim\limits_{h \to 0} \dfrac{f(a-h) - f(a)}{-h},$
  $Rf'(a) = \lim\limits_{h \to 0} \dfrac{f(a+h) - f(a)}{h}$

- Every differentiable function is continuous but converse is not necessarily true.

- $(f + g)'(a) = f'(a) + g'(a);$
  $(cf)'(a) = cf'(a), c \in R$

- $(fg)'(a) = f'(a)g(a) + f(a)g'(a)$

- $\left(\dfrac{1}{f}\right)'(a) = -\dfrac{f'(a)}{[f'(a)]^2};$
  $\left(\dfrac{f}{g}\right)'(a) = \dfrac{f'(a)g(a) - f(a)g'(a)}{[g(a)]^2}$

- **(Darboux's Theorem).** If f is finitely differentiable on a closed interval, then there exists at least one point $c \in]a, b[$ such that $f'(c) = 0$.

- If f is defined and differentiable on $[a, b]$ and if c be any number between $f'(a)$ and $f'(b)$ then there exist a real number k between a and b such that $f'(k) = c$.

Thus $Rf'(0) \neq Lf'(0)$ Therefore, the given function is not differentiable at $x = 0$.

Now, we check the differentiability of $f(x)$ at $x = 1$.

We have

$$Rf'(1) = \lim_{h \to 0} \frac{f(1+h) - f(1)}{h} = \lim_{h \to 0} \frac{[2(1+h) - 1] - 1}{h}$$

$$= \lim_{h \to 0} \frac{2 + 2h - 2}{h} = 2$$

and

$$Lf'(1) = \lim_{h \to 0} \frac{f(1-h) - f(1)}{-h}$$

$$= \lim_{h \to 0} \frac{1 - 1}{h} = 0$$

Thus $Rf'(1) \neq Lf'(1)$. Therefore, the given function is not differentiable at $x = 1$.

Example 2. *Prove that the function $f(x) = |x|$ is continuous at $x = 0$, but not differentiable at $x = 0$, where $|x|$ is the absolute value of x.*

Solution. Firstly, we check the continuity of the function $f(x)$ at $x = 0$.

We have

$$f(0) = |0| = 0$$

$$f(0+0) = \lim_{h \to 0} f(0+h) = \lim_{h \to 0} f(h)$$

$$= \lim_{h \to 0} |h| = \lim_{h \to 0} h = 0$$

and

$$f(0-0) = \lim_{h \to 0} f(0-h) = \lim_{h \to 0} f(-h)$$

$$= \lim_{h \to 0} |-h| = \lim_{h \to 0} h = 0$$

$$\therefore \qquad f(0+0) = f(0) = f(0-0)$$

Hence, $f(x)$ is continuous at $x = 0$.

Now, we check the differentiability of the function $f(x)$ at $x = 0$.

We have,

$$Rf'(0) = \lim_{h \to 0} \frac{f(0+h) - f(0)}{h} = \lim_{h \to 0} \frac{f(h) - f(0)}{h}$$

$$= \lim_{h \to 0} \frac{|h| - 0}{h} = 1$$

and

$$Lf'(0) = \lim_{h \to 0} \frac{f(0-h) - f(0)}{-h} = \lim_{h \to 0} \frac{f(-h) - f(0)}{-h}$$

$$= \lim_{h \to 0} \frac{|-h| - 0}{-h} = \lim_{h \to 0} \frac{h}{-h} = -1$$

$$\Rightarrow \qquad Rf'(0) \neq Lf'(0)$$

Hence, the function $f(x)$ is not differentiable at $x = 0$.

Example 3. *Let the function $f(x)$ satisfy the condition*

(i) $f(x+y) = f(x) f(y) \; \forall x, y$ (ii) $f(x) = 1 + x \cdot g(x)$ where $\lim_{x \to 0} g(x) = 1$

Show that the derivative $f'(x)$ exist and equal to $f(x)$ for all x.

Solution. From condition (i), we have

$$f(x + \delta x) = f(x) \cdot f(\delta x)$$

Then $\quad f(x + \delta x) - f(x) = f(x)f(\delta x) - f(x)$

$\Rightarrow \quad \dfrac{f(x + \delta x) - f(x)}{\delta x} = \dfrac{f(x)[f(\delta x) - 1]}{\delta x} = \dfrac{f(x)\delta x \, g(\delta x)}{\delta x}$ $\qquad$ [By (ii)]

$$= f(x)g(\delta x)$$

$\therefore \quad \lim\limits_{\delta x \to 0} \dfrac{f(x + \delta x) - f(x)}{\delta x} = \lim\limits_{\delta x \to 0} f(x)g(\delta x) = f(x).1$

$\therefore \quad f'(x) = f(x)$

Example 4. *If $f(x)$ be an even function and $f'(0)$ exists, then find the value of $f'(0)$.*

Solution. Since $f(x)$ is an even function so $f(-x) = f(x) \forall x$

$$f'(0) \text{exist} \Rightarrow Rf'(0) = Lf'(0) = f'(0)$$

Now $\qquad f'(0) = Rf'(0) = \lim\limits_{h \to 0} \dfrac{f(h) - f(0)}{h}, h > 0$

$\qquad\qquad = \lim\limits_{h \to 0} \dfrac{f(-h) - f(0)}{h}$ $\qquad\qquad [\because f(-x) = f(x)]$

$\qquad\qquad = -\lim\limits_{h \to 0} \dfrac{f(-h) - f(0)}{-h} = -Lf'(0)$

$\qquad\qquad = -f'(0)$

$\Rightarrow \qquad\qquad 2f'(0) = 0 \Rightarrow f'(0) = 0$

Example 5. *Show that the function* $f(x) = \begin{cases} x\tan^{-1}\left(\dfrac{1}{x}\right) &, \quad \text{for } x \neq 0 \\ 0 &, \quad \text{for } x = 0 \end{cases}$ *is not differentiable at $x = 0$.*

Solution. Here

$$Rf'(0) = \lim\limits_{h \to 0} \dfrac{f(0+h) - f(0)}{h} = \lim\limits_{h \to 0} \dfrac{f(h) - f(0)}{0}$$

$$= \lim\limits_{h \to 0} \dfrac{h.\tan^{-1}\dfrac{1}{h} - 0}{h} = \lim\limits_{h \to 0} \tan^{-1}\dfrac{1}{h} = \tan^{-1}\infty = \dfrac{\pi}{2}$$

and $\qquad Lf'(0) = \lim\limits_{h \to 0} \dfrac{f(0-h) - f(0)}{-h} = \lim\limits_{h \to 0} \dfrac{f(-h) - f(0)}{-h}$

$$= \lim\limits_{h \to 0} \dfrac{-h\tan^{-1}\left(-\dfrac{1}{h}\right)}{-h} = \lim\limits_{h \to 0} \tan^{-1}\left(-\dfrac{1}{h}\right)$$

$$= -\tan^{-1}\infty = -\dfrac{\pi}{2}$$

$\Rightarrow \qquad Rf'(0) \neq Lf'(0)$

Hence, $f(x)$ is not differentiable at $x = 0$.

Example 6. Test the continuity and differentiability of the following function in $-\infty < x < \infty$

$$f(x) = \begin{cases} 1 & \text{if} \quad -\infty < x < 0 \\ 1 + \sin x & \text{if} \quad 0 \le x < \dfrac{\pi}{2} \\ 2 + \left(x - \dfrac{\pi}{2} \right)^2 & \text{if} \quad \dfrac{\pi}{2} \le x < \infty \end{cases}$$

Solution.   Firstly, we check the continuity and differentiability at $x = 0$.

(i)   *Continuity of $f(x)$ at $x = 0$.*

$$f(0) = 1 + \sin 0 = 1$$

$$f(0+0) = \lim_{h \to 0} f(0+h) = \lim_{h \to 0} f(h) = \lim_{h \to 0}(1 + \sin h) = 1$$

$$f(0-0) = \lim_{h \to 0} f(0-h) = \lim_{h \to 0} f(-h) = \lim_{h \to 0} 1 = 1$$

$\Rightarrow \qquad f(0+0) = f(0) = f(0-0)$

Hence, $f(x)$ is continuous at $x = 0$.

(ii)   *Differentiability of $f(x)$ at $x = 0$.*

$$Rf'(0) = \lim_{h \to 0} \frac{f(0+h) - f(0)}{h} = \lim_{h \to 0} \frac{f(h) - f(0)}{h}$$

$$= \lim_{h \to 0} \frac{(1 + \sin h) - (1 + \sin 0)}{h} = \lim_{h \to 0} \frac{\sin h}{h} = 1$$

and $$Lf'(0) = \lim_{h \to 0} \frac{f(0-h) - f(0)}{-h} = \lim_{h \to 0} \frac{f(-h) - f(0)}{-h}$$

$$= \lim_{h \to 0} \frac{1 - (1 + \sin 0)}{-h} = \lim_{h \to 0} \frac{0}{-h} = \lim_{h \to 0} 0 = 0$$

$\Rightarrow \qquad Rf'(0) \ne Lf'(0)$

Hence, $f(x)$ is not differentiable at $x = 0$.

Now, we shall check the continuity and differentiability at $x = \dfrac{\pi}{2}$.

(iii)   *Continuity of $f(x)$ at $x = \dfrac{\pi}{2}$*

We have $$f\left(\frac{\pi}{2} \right) = 2 + \left(\frac{\pi}{2} - \frac{\pi}{2} \right)^2 = 2$$

$$f\left(\frac{\pi}{2} + 0 \right) = \lim_{h \to 0} f\left(\frac{\pi}{2} + h \right) = \lim_{h \to 0} \left[2 + \left\{ \left(\frac{1}{2}\pi + h \right) - \frac{1}{2}\pi \right\}^2 \right]$$

$$= \lim_{h \to 0} (2 + h^2) = 2$$

and $$f\left(\frac{\pi}{2} - 0 \right) = \lim_{h \to 0} f\left(\frac{\pi}{2} - h \right) = \lim_{h \to 0} \left[1 + \sin\left(\frac{\pi}{2} - h \right) \right]$$

$$= \lim_{h \to 0} [1 + \cos h] = 1 + 1 = 2$$

$$\Rightarrow \qquad f\left(\frac{\pi}{2}+0\right) = f\left(\frac{\pi}{2}\right) = f\left(\frac{\pi}{2}-0\right)$$

Hence, $f(x)$ is continuous at $x = \frac{\pi}{2}$.

(iv) *Differentiability of $f(x)$ at $x = \frac{\pi}{2}$*

$$Rf'\left(\frac{\pi}{2}\right) = \lim_{h \to 0} \frac{f\left(\frac{\pi}{2}+h\right) - f\left(\frac{\pi}{2}\right)}{h}$$

$$= \lim_{h \to 0} \frac{\left[2 + \left\{\frac{\pi}{2}+h-\frac{\pi}{2}\right\}^2\right] - \left[2 + \left(\frac{\pi}{2}-\frac{\pi}{2}\right)^2\right]}{h}$$

$$= \lim_{h \to 0} \frac{2 + h^2 - 2}{h} = \lim_{h \to 0} h = 0$$

$$Lf'\left(\frac{\pi}{2}\right) = \lim_{h \to 0} \frac{f\left(\frac{\pi}{2}-h\right) - f\left(\frac{\pi}{2}\right)}{-h}$$

$$= \lim_{h \to 0} \frac{1 + \sin\left(\frac{\pi}{2}-h\right) - 2}{-h}$$

$$= \lim_{h \to 0} \frac{-1 + \cos h}{-h} = \lim_{h \to 0} \frac{1 - \cos h}{h}$$

$$= \lim_{h \to 0} \frac{2\sin^2(h/2)}{h}$$

$$= \lim_{h \to 0} \left[\frac{\sin h/2}{h/2} . \sin h/2\right]$$

$$= \lim_{h \to 0} \left[\frac{\sin h/2}{h/2}\right] . \lim_{h \to 0} [\sin h/2] = 1 \times 0 = 0$$

Therefore, $\qquad Rf'\left(\frac{\pi}{2}\right) = Lf'\left(\frac{\pi}{2}\right)$

$\Rightarrow \qquad f(x)$ is differentiable at $x = \frac{\pi}{2}$.

Since, here, we checked the continuity and differentiability at $x = 0$ and $\frac{\pi}{2}$. It is obviously continuous and differentiable at all other points.

Example 7. If $f(x) = \begin{cases} x^2 \sin\dfrac{1}{x} & , \text{ if } x \neq 0 \\ 0 & , \text{ if } x = 0 \end{cases}$

then, show that $f(x)$ is continuous and differentiable everywhere.

Solution. We have

$$f(0+0) = \lim_{h \to 0} f(0+h) = \lim_{h \to 0} (0+h)^2 \sin\frac{1}{0+h}$$

$$= \lim_{h \to 0} h^2 \sin \frac{1}{h} = 0$$

$$f(0-0) = \lim_{h \to 0} f(0-h) = \lim_{h \to 0} (0-h)^2 \sin \frac{1}{0-h}$$

$$= -\lim_{h \to 0} h^2 \sin \frac{1}{h} = 0$$

and $$f(0) = 0$$

$\Rightarrow$ $$f(0+0) = f(0) = (0-0)$$

Hence, the function is continuous at $x = 0$.

Now $$Rf'(0) = \lim_{h \to 0} \frac{f(0+h) - f(0)}{h} = \lim_{h \to 0} \frac{f(h) - f(0)}{h}$$

$$= \lim_{h \to 0} \frac{h^2 \sin \frac{1}{h} - 0}{h} = \lim_{h \to 0} h \sin \frac{1}{h} = 0$$

and $$Lf'(0) = \lim_{h \to 0} \frac{f(0-h) - f(0)}{-h} = \lim_{h \to 0} \frac{f(-h) - f(0)}{-h}$$

$$= \lim_{h \to 0} \frac{(-h)^2 \sin\left(-\frac{1}{h}\right) - 0}{-h} = \lim_{h \to 0} h \sin \frac{1}{h} = 0$$

$\Rightarrow$ $$Rf'(0) = Lf'(0)$$

Hence, $f(x)$ is differentiable at $x = 0$.

Example 8. Let $f(x) = \sqrt{(x)}\{1 + x \sin(1/x)\}$ for $x > 0$, $f(0) = 0$

and $$f(x) = -\sqrt{(-x)}\{1 + x \sin(1/x)\} \text{ for } x < 0.$$

Show that $f'(x)$ exists every where and is finite except at $x = 0$ where its value is $+\infty$.

Solution. We have

$$Rf'(0) = \lim_{h \to 0} \frac{f(0+h) - f(0)}{h} = \lim_{h \to 0} \frac{f(h) - f(0)}{h}$$

$$= \lim_{h \to 0} \frac{(\sqrt{h})\{1 + h \sin(1/h)\} - 0}{h}$$

$$= \lim_{h \to 0} \left[\frac{1}{\sqrt{h}} + (\sqrt{h}) \sin\left(\frac{1}{h}\right) \right] = \infty + 0 = \infty$$

and $$Lf'(0) = \lim_{h \to 0} \frac{f(0-h) - f(0)}{-h} = \lim_{h \to 0} \frac{f(-h) - f(0)}{-h}$$

$$= \lim_{h \to 0} \frac{-\sqrt{[-(-h)]}\left[1 + (-h) \sin \frac{1}{-h}\right] - 0}{-h}$$

$$= \lim_{h \to 0} \left[\frac{1}{\sqrt{h}} + \sqrt{h} \sin \frac{1}{h} \right] = \infty + 0 = \infty$$

$$\Rightarrow \qquad Rf'(0) = Lf'(0) = \infty \qquad\qquad \therefore\; f'(0) = \infty$$

Now, we have

$$f'(x) = \frac{1}{2\sqrt{x}} + \frac{3}{2}\sqrt{x} \sin \frac{1}{x} - \frac{1}{\sqrt{x}} \cos \frac{1}{x} \quad \text{for } x > 0$$

$$f'(x) = \frac{1}{2\sqrt{-x}} + \frac{3}{2}\sqrt{(-x)} \sin \frac{1}{x} - \frac{1}{\sqrt{(-x)}} \cos \frac{1}{x} \quad \text{for } x < 0$$

Hence, $f'(0)$ is finite for all $a \neq 0$.

Example 9. *Show that the function $f : R \to R$ defined by*

$$f(x) = \begin{cases} x\left[1 + \dfrac{1}{3} \sin \log x^2\right] & \text{if } x \neq 0 \\ 0 & \text{if } x = 0 \end{cases}$$

is continuous everywhere but not differentiable at origin.

Solution. Firstly, we check the continuity of $f(x)$ at $x = 0$. We have

$$f(0+0) = \lim_{h \to 0} f(0+h) = \lim_{h \to 0}\left[(0+h)\left\{1 + \frac{1}{3}\sin\log(0+h)^2\right\} \right]$$

$$= \lim_{h \to 0}\left[h + \left(\frac{h}{3}\right) \sin\log h^2 \right] = 0 + 0 \times \text{ a finite quantity}$$

$$= 0$$

Similarly, $\qquad f(0-0) = 0$

Hence, f is continuous at $x = 0$.

Now we shall check the differentiability at $x = 0$. Therefore,

$$Rf'(0) = \lim_{h \to 0} \frac{(0+h)\left\{1 + \dfrac{1}{3}\sin\log(0+h)^2\right\} - 0}{h}$$

$$= \lim_{h \to 0}\left[1 + \frac{1}{3}\sin\log h^2\right]$$

which does not exist, (since $\sin \log h^2$ oscillate between -1 and 1 as $h \to 0$)
Similarly, $Lf'(0) = $ does not exist.

Hence, $f(x)$ is not differentiable at origin.

Example 10. *Draw the graph of the function $y = |x-1| + |x-2|$ in the interval $[0, 3]$ and discuss the continuity and differentiability of the function in this interval.*

Solution. Here, we observe that

$$y = 1 - x + 2 - x = 3 - 2x \quad \text{when} \quad x \leq 1$$

$$= x - 1 + 2 - x = 1 \quad \text{when} \quad 1 \leq x \leq 2$$

$= x - 1 + x - 2 = 2x - 3$ when $x \geq 2$

Hence, the graph consists of the segments of the three straight lines $y = 3 - 2x$, $y = 1$ and $y = 2x - 3$ corresponding to the intervals $[0, 1]$, $[1, 2]$, $[2, 3]$ respectively.

The graph shows that the function is continuous throughout the interval and differentiable at all points of the interval $[0, 3]$ except possibly at $x = 1$ and at $x = 2$.

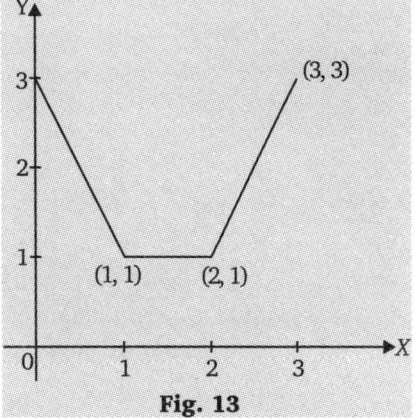

Fig. 13

(i) *Differentiability of $f(x)$ at $x = 1$.*

Here,

$$Rf'(1) = \lim_{h \to 0} \frac{f(1+h) - f(1)}{h} = \lim_{h \to 0} \frac{1-1}{h} = 0$$

and

$$Lf'(1) = \lim_{h \to 0} \frac{f(1-h) - f(1)}{-h} = \lim_{h \to 0} \frac{3 - 2(1-h) - 1}{-h} = -2$$

$\Rightarrow \qquad Rf'(1) \neq Lf'(1)$

$\Rightarrow \qquad f(x)$ is not differentiable at $x = 2$

(ii) *Differentiability of $f(x)$ at $x = 2$*

$$Rf'(2) = \lim_{h \to 0} \frac{f(2+h) - f(2)}{h} = \lim_{h \to 0} \frac{2(2+h) - 3 - 1}{h} = 2$$

$$Lf'(2) = \lim_{h \to 0} \frac{f(2-h) - f(2)}{h} = \lim_{h \to 0} \frac{1-1}{h} = 0$$

$\Rightarrow \qquad Rf'(2) \neq Lf'(2)$.

Hence, $f(x)$ is not differentiable at $x = 2$.

Example 11. *Show that the function*

$$f(x) = \begin{cases} x \left[\dfrac{e^{1/x} - e^{-1/x}}{e^{1/x} + e^{-1/x}} \right], & \text{if } x \neq 0 \\ \qquad 0, & \text{if } x = 0 \end{cases}$$

is continuous but not differentiable at $x = 0$.

Solution. (i) *Continuity of $f(x)$ at $x = 0$.*

We have

$$\text{RHL} = f(0+0) = \lim_{h \to 0} f(0+h) = \lim_{h \to 0} f(h)$$

$$= \lim_{h \to 0} h \left[\frac{e^{1/h} - e^{-1/h}}{e^{1/h} + e^{-1/h}} \right] = \lim_{h \to 0} h \left[\frac{1 - e^{-2/h}}{1 + e^{-2/h}} \right]$$

$$= 0 \times \frac{1 - 0}{1 + 0} = 0 \times 1 = 0$$

and $$\text{LHL} = f(0-0) = \lim_{h \to 0} f(0-h) = \lim_{h \to 0} f(-h)$$

$$= \lim_{h \to 0} -h \left[\frac{e^{1/-h} - e^{-1/-h}}{e^{1/-h} + e^{-1/-h}} \right] = \lim_{h \to 0} -h \left[\frac{e^{-2/h} - 1}{e^{-2/h} + 1} \right]$$

$$= 0 \times \frac{0-1}{0+1} = 0$$

$$\Rightarrow \qquad f(0+0) = f(0-0) = f(0).$$

Hence, f is continuous at $x = 0$.

(ii) *Differentiability of $f(x)$ at $x = 0$.*

Here, we have

$$Rf'(0) = \lim_{h \to 0} \frac{f(0+h) - f(0)}{h} = \lim_{h \to 0} \frac{f(h) - f(0)}{h}$$

$$= \lim_{h \to 0} \frac{\left[h \dfrac{e^{1/h} - e^{-1/h}}{e^{1/h} + e^{-1/h}} - 0 \right]}{h}$$

$$= \lim_{h \to 0} \frac{1 - e^{-2/h}}{1 + e^{-2/h}} = \frac{1-0}{1+0} = 1$$

and $$Lf'(0) = \lim_{h \to 0} \frac{f(0-h) - f(0)}{-h} = \lim_{h \to 0} \frac{f(-h) - f(0)}{-h}$$

$$= \lim_{h \to 0} \frac{\left[(-h) \dfrac{e^{-1/h} - e^{1/h}}{e^{-1/h} + e^{1/h}} - 0 \right]}{-h}$$

$$= \lim_{h \to 0} \frac{e^{-2/h} - 1}{e^{2/h} + 1} = \frac{0-1}{0+1} = -1$$

$$\Rightarrow \qquad Rf'(0) \neq Lf'(0)$$

Hence, the function $f(x)$ is not differentiable at $x=0$.

Example 12. Let $f(x) = \begin{cases} e^{-1/x^2} \sin \dfrac{1}{x} & , \text{ when } x \neq 0 \\ 0 & , \text{ when } x = 0 \end{cases}$

Show that at every point, $f(x)$ is differentiable and f' is continuous at $x = 0$.

Solution. (i) *Differentiability at $x = 0$.*

Here, we have

$$Rf'(0) = \lim_{h \to 0} \frac{e^{-1/h^2} \sin \dfrac{1}{h} - 0}{h} = \lim_{h \to 0} \frac{\sin \dfrac{1}{h}}{h e^{1/h^2}}$$

$$= \lim_{h \to 0} \frac{\sin 1/h}{h \left[1 + \dfrac{1}{h^2} + \dfrac{1}{2!} \dfrac{1}{h^4} + \right]} = \lim_{h \to 0} \frac{\sin \dfrac{1}{h}}{h + \dfrac{1}{h} + \dfrac{1}{2!} \dfrac{1}{h^3} + \ldots}$$

$$= \frac{\text{a finite quantity lying between} -1 \text{ and } 1}{\infty} = 0$$

Similarly, $Lf'(0)=0$

Hence, the function $f(x)$ is differentiable at $x = 0$ and $f'(0) = 0$

(ii) *Continuity of f'*

$$f'(x) = \left(\frac{2}{x^3}\right)e^{-1/x^2}\sin\frac{1}{x} - \left(\frac{1}{x^2}\right)e^{-1/x^2}\cos(1/x)$$

$$= \left\{\left(\frac{2}{x}\right)\sin\frac{1}{x} - \cos\left(\frac{1}{x}\right)\right\}\left(\frac{1}{x^2}\right)\left(\frac{1}{e^{1/x^2}}\right) \qquad \dots (1)$$

Now $\qquad f'(0+0) = \lim_{h\to 0} f'(0+h)$

$$= \lim_{h\to 0}\left(\frac{2}{h}\sin\frac{1}{h} - \cos\frac{1}{h}\right)\cdot\frac{1}{h^2 e^{1/h^2}}$$

$$= \lim_{h\to 0}\left[\frac{2\sin(1/h)}{h^3 e^{1/h^2}} - \frac{\cos(1/h)}{h^2 e^{1/h^2}}\right]$$

$$= \lim_{h\to 0}\left[\frac{2\sin(1/h)}{h^3\left[1+\dfrac{1}{h^2}+\dfrac{1}{2!\,h^4}+\dots\right]} - \frac{\cos(1/h)}{h^2\left[1+\dfrac{1}{h^2}+\dfrac{1}{2!\,h^4}+\dots\right]}\right]$$

$$= \frac{\text{A finite quantity}}{\infty} - \frac{\text{A finite quantity}}{\infty} = 0$$

Similarly, $\qquad f'(0-0) = 0$

Hence f' is continuous at $x = 0$.

Example 13 Let $f(x) = \begin{cases} -x-1 &, & -2 \le x \le 0 \\ x-1 &, & 0 < x \le 2 \end{cases}$ and $g(x) = f(|x|)+|f(x)|$.

Test the differentiability of $g(x)$ in the inteval $]-2, 2[$

Solution. Here, we have

$$|x| = -x, \text{when} \qquad -2 \le x \le 0$$
$$|x| = x, \text{ when} \qquad 0 < x \le 2$$

Therefore, $\qquad f(|x|) = \begin{cases} x-1 &, & -2 \le x \le 0 \\ -x-1 &, & 0 < x \le 2 \end{cases}$

and $\qquad |f(x)| = \begin{cases} 1 &, & -2 \le x \le 0 \\ -x+1 &, & 0 < x \le 1 \\ x-1 &, & 1 < x \le 2 \end{cases}$

so $\qquad g(x) = f(|x|)+|f(x)| = \begin{cases} -x &, & -2 \le x \le 0 \\ 0 &, & 0 < x \le 1 \\ 2x-2 &, & 1 < x \le 2 \end{cases}$

It is obvious that $g(x)$ is differentiable $\forall\, x \in\,]-2, 2[$ except possibly at $x = 0$ and 1.

At $x = 0$. $\qquad Rg'(0) = \lim_{h \to 0} \dfrac{g(0+h) - g(0)}{h} = \lim_{h \to 0} \dfrac{g(h) - g(0)}{h} = \lim_{h \to 0} \dfrac{0 - 0}{h} = 0$

and $\qquad\qquad Lg'(0) = \lim_{h \to 0} \dfrac{g(0-h) - g(0)}{-h} = \lim_{h \to 0} \dfrac{g(-h) - g(0)}{-h}$

$$= \lim_{h \to 0} \dfrac{h - 0}{-h} = -1$$

Thus $\qquad\qquad Rg'(0) \neq Lg'(0)$

Hence, $g(x)$ is not differentiable at $x = 0$

At $x = 1$. $\qquad Rg'(1) = \lim_{h \to 0} \dfrac{g(1+h) - g(1)}{h} = \lim_{h \to 0} \dfrac{2(1+h) - 2 - 0}{h} = 2$

$$Lg'(1) = \lim_{h \to 0} \dfrac{g(1-h) - g(1)}{-h} = \lim_{h \to 0} \dfrac{0 - 0}{-h} = 0$$

Thus $Rg'(1) \neq Lg'(1)$. Therefore $g(x)$ is not differentiable at $x = 1$.

Example 14. Let $f(x) = \begin{cases} \dfrac{x}{1 + e^{1/x}} & , \text{ if } x \neq 0 \\ 0 & , \text{ if } x = 0 \end{cases}$

Show that f is continuous at $x = 0$, but $f'(0)$ does not exist.

Solution. We have

LHL $= f(0-0) = \lim_{h \to 0} f(0-h) = \lim_{h \to 0} f(-h), h > 0$

$$= \lim_{h \to 0} \dfrac{-h}{1 + e^{-1/h}} = \dfrac{0}{1+0} = 0$$

RHL $= f(0+0) = \lim_{h \to 0} f(0+h) = \lim_{h \to 0} f(h), h > 0$

$$= \lim_{h \to 0} \dfrac{h}{1 + e^{1/h}} = 0 \cdot \dfrac{0}{1 + \infty} = 0.0 = 0$$

and $\qquad\qquad f(0) = 0$ (given)

Therefore $\qquad f(0+0) = f(0) = f(0-0)$

Hence, $f(x)$ is continuous at $x = 0$.

Now $\qquad\qquad Rf'(0) = \lim_{h \to 0} \dfrac{f(0+h) - f(0)}{h} = \lim_{h \to 0} \dfrac{f(h) - f(0)}{h}, h > 0$

$$= \lim_{h \to 0} \dfrac{\dfrac{h}{1 + e^{1/h}} - 0}{h} = \lim_{h \to 0} \dfrac{1}{1 + e^{1/h}} = \dfrac{1}{1 + \infty} = 0$$

and $\qquad\qquad Lf'(0) = \lim_{h \to 0} \dfrac{f(0-h) - f(0)}{-h} = \lim_{h \to 0} \dfrac{f(-h) - f(0)}{-h}$

$$= \lim_{h \to 0} \frac{\dfrac{-h}{1+e^{1/h}} - 0}{-h} = \lim_{h \to 0} \frac{1}{1+e^{-1/h}}$$

$$= \frac{1}{1+e^{-\infty}} = \frac{1}{1+0} = 1$$

$\Rightarrow$ $\qquad Rf'(0) \neq Lf'(0)$

Hence, $f'(0)$ does not exist.

Example 15. *Show that the function $f(x) = x|x|$ is differentiable at the origin.*

Proof.   Here, we have

$$Rf'(0) = \lim_{h \to 0} \frac{f(0+h) - f(0)}{h} = \lim_{h \to 0} \frac{f(h) - f(0)}{h}, h > 0$$

$$= \lim_{h \to 0} \frac{h|h| - 0}{h} = \lim_{h \to 0} h = 0$$

and $\qquad Lf'(0) = \lim_{h \to 0} \frac{f(0-h) - f(0)}{-h} = \lim_{h \to 0} \frac{f(-h) - f(0)}{-h}$

$$= \lim_{h \to 0} \frac{-h|h| - 0}{-h} = \lim_{h \to 0} h = 0$$

$\Rightarrow$ $\qquad Rf'(0) = Lf'(0)$

Hence, $f(x)$ is differentiable at $x = 0$.

Example 16. *Let $f\left(\dfrac{x+y}{2}\right) = \dfrac{f(x) + f(y)}{2} \forall x$ and y. If $f'(0)$ exists and equal -1 and $f(0) = 1$, find $f(3)$.*

Solution.   We have

$$f\left(\frac{x+y}{2}\right) = \frac{f(x) + f(y)}{2} \Rightarrow f\left(\frac{x+0}{2}\right) = \frac{f(x) + f(0)}{2}$$

$\Rightarrow$ $\qquad f\left(\dfrac{x}{2}\right) = \dfrac{1}{2}[f(x) + f(0)] = \dfrac{1}{2}[f(x) + 1]$

$\Rightarrow$ $\qquad f(x) = 2f\left(\dfrac{x}{2}\right) - 1$ $\qquad\qquad$ ... (1)

Now $\qquad f'(x) = \lim_{h \to 0} \frac{f(x+h) - f(x)}{h} = \lim_{h \to 0} f\left(\frac{2x + 2h}{2}\right) - f(x)$

$$= \lim_{h \to 0} \frac{\dfrac{1}{2}[f(2x) + f(2h)] - f(x)}{h} = \lim_{h \to 0} \frac{f(h) - 1}{h} \quad \text{[Using (1)]}$$

$$= \lim_{h \to 0} \frac{f(0+h) - f(0)}{h} = f'(0) = -1 \qquad \text{[Given]}$$

$\Rightarrow$ $\qquad f(x) = -x + c$ $\qquad\qquad$ ... (2)

Putting $x = 0$ in (2), we have $c = f(0) = 1$

Therefore, $\qquad f(3) = -3 + \quad -2$

Example 17. *Test the continuity and differentiability* $-\infty < x < \infty$, *of the following function*

$$f(x) = \begin{cases} 1 & , \quad if \quad -\infty < x < 0 \\ 1+\sin x & , \quad if \quad 0 \le x < \pi/2 \\ 2+(x-\pi/2)^2 & , \quad if \quad \pi/2 \le x < \infty \end{cases}$$

Solution. We shall test $f(x)$ for continuity and differentiability at $x = 0$ and $\pi/2$.

(i) *Continuity and differentiability of $f(x)$ at $x = 0$.*

We have $\qquad f(0) = 1+\sin 0 \Rightarrow f(0) = 1$

$$f(0+0) = \lim_{h\to 0} f(0+h) = \lim_{h\to 0} f(h) = \lim_{h\to 0}(1+\sin h) = 1$$

and $\qquad f(0-0) = \lim_{h\to 0} f(0-h) = \lim_{h\to 0} f(-h) = \lim_{h\to 0} 1 = 1$

Since $\qquad f(0) = f(0+0) = f(0-0), f(x)$ is continuous at $x = 0$.

Now $\qquad Rf'(0) = \lim_{h\to 0}\frac{f(0+h)-f(0)}{h} = \lim_{h\to 0}\frac{f(h)-f(0)}{h}$

$$= \lim_{h\to 0}\frac{(1+\sin h)-(1+\sin 0)}{h} = \lim_{h\to 0}\frac{\sin h}{h} = 1$$

and $\qquad Lf'(0) = \lim_{h\to 0}\frac{f(0-h)-f(0)}{-h} = \lim_{h\to 0}\frac{f(-h)-f(0)}{-h}$

$$= \lim_{h\to 0}\frac{1-(1+\sin 0)}{-h} = \lim_{h\to 0}\frac{0}{-h} = 0$$

Hence, $\quad Rf'(0) \neq Lf'(0), f(x)$ is not differentiable at $x = 0$.

(ii) *Continuity and differentiability of $f(x)$ at $x = \pi/2$.*

We have $\quad f(\pi/2) = 2+(\pi/2-\pi/2)^2 = 2$

$$f(\pi/2+0) = \lim_{h\to 0} f(\pi/2+h) = \lim_{h\to 0}[2+\{(\pi/2+h)-\pi/2\}^2]$$

$$= \lim_{h\to 0}(2+h^2) = 2$$

and $\quad f(\pi/2-0) = \lim_{h\to 0} f(\pi/2-h) = \lim_{h\to 0}[1+\sin(\pi/2-h)]$

$$= \lim_{h\to 0}(1+\cos h) = 1+1 = 2$$

Hence, $\quad f(\pi/2) = f(\pi/2-0) = f(\pi/2+0), f(x)$ is continuous at $x = \pi/2$

Now $\quad Rf'(\pi/2) = \lim_{h\to 0}\frac{f(\pi/2+h)-f(h/2)}{h}$

$$= \lim_{h\to 0}\frac{[2+\{\pi/2+h-\pi/2\}^2]-[2+(h/2-\pi/2)^2]}{h}$$

$$= \lim_{h\to 0}\frac{2+h^2-2}{h} = \lim_{h\to 0} h = 0$$

and $Lf'(\pi/2) = \lim_{h\to 0} \dfrac{f(\pi/2-h)-f(\pi/2)}{-h} = \lim_{h\to 0} \dfrac{1+\sin(\pi/2-h)-2}{-h}$

$$= \lim_{h\to 0} \dfrac{-1+\cos h}{-h} = \lim_{h\to 0} \dfrac{2\sin^2 h/2}{h}$$

$$= \lim_{h\to 0}\left[\dfrac{\sin(h/2)}{h/2}.\sin(h/2)\right] = 1\times 0 = 0$$

Hence $Rf'(0) = Lf'(0), f(x)$ is differentiable at $x = \pi/2$.

Example 18. Let $\qquad f(x) = \sqrt{x}\{1+x\sin 1/x\}$ *for* $x \geq 0$

$\qquad f(0) = 0$ *and* $f(x) = -\sqrt{(-x)}\{1+x\sin(1/x)\}$ *for* $x < 0$

Show that $f'(x)$ *exists every where and is finite at* $x = 0$ *where its value is* $+\infty$.

Solution. We have

$$Rf'(0) = \lim_{h\to 0}\dfrac{f(0+h)-f(0)}{h} = \lim_{h\to 0}\dfrac{\sqrt{h}\{1+h\sin 1/h\}-0}{h}$$

$$= \lim_{h\to 0}[1/\sqrt{h}+\sqrt{h}\sin(1/h)] = \infty+0 = \infty$$

and $\qquad Lf'(0) = \lim_{h\to 0}\dfrac{f(0-h)-f(0)}{-h} = \lim_{h\to 0}\dfrac{f(-h)-f(0)}{-h}$

$$= \lim_{h\to 0}\dfrac{-\sqrt{(-h)}[1+(-h)\sin(-1/h)]-0}{-h}$$

$$= \lim_{h\to 0}[1/\sqrt{h}+\sqrt{h}\sin 1/h] = \infty+0 = \infty$$

Hence, $\qquad Rf'(0) = Lf'(0) = \infty \qquad \therefore f'(0) = \infty$

We have $\qquad f'(x) = \dfrac{1}{2\sqrt{x}}+\dfrac{3}{2}\sqrt{x}\sin\dfrac{1}{x}-\dfrac{1}{\sqrt{x}}\cos\dfrac{1}{x}$ for $x > 0$

and $\qquad f'(x) = \dfrac{1}{2\sqrt{-x}}+\dfrac{3}{2}\sqrt{(-x)}\sin\dfrac{1}{x}-\dfrac{1}{\sqrt{(-x)}}\cos\dfrac{1}{x}$ for $x > 0$

Hence, $f'(a)$ is finite for all $a \neq 0$.

Example 19. *Show that the function* $f : R \to R$ *defined by*

$$f(x) = x[1+(1/3)\sin\log x^2], x \neq 0 \text{ and } f(0) = 0$$

is everywhere continuous but has no differential coefficient at the origin.

Proof. Obviously the function $f(x)$ is continuous at every point of R except possible at $x = 0$. Therefore, we have to check the continuity at $x = 0$. Given $f(0) = 0$

Now, $\qquad f(0+0) = \lim_{h\to 0}(0+h) = \lim_{h\to 0}[(0+h)]\{1+1/3\sin\log(0+h)^2\}$

$$= \lim_{h\to 0}[h+(h/3)\sin\log h^2] = 0+0\times \text{a finite quantity} = 0.$$

Similarly, we can show that $\qquad f(0-0) = 0$.

Hence, f is continuous at $x = 0$

Now $$Rf'(0) = \lim_{h \to 0} \frac{(0+h)\{1+(1/3)\sin\log(0+h)^2\}-0}{h}$$

$$= \lim_{h \to 0} \{1+1/3\sin\log h^2\} = \text{which does not exist}$$

$$(\because \sin\log h^2 \text{ oscillates between } -1 \text{ and } 1 \text{ as } h \to 0)$$

and $$Lf'(0) = \lim_{h \to 0} \frac{(0-h)\{1+1/3\sin\log(0-h)^2\}-0}{-h}$$

$$= \lim_{h \to 0} \{1+1/3\sin\log h^2\} = \text{which does not exist as above.}$$

Hence, f has no differential coefficient at $x = 0$.

Example 20. *Let* $f(x) = e^{-1/x^2}.\sin 1/x$ *when* $x \neq 0$ *and* $f(0)=0$. *Show that at every point,* $f(x)$ *has a differential coefficient and this is continuous at* $x = 0$.

Solution. *Differentiability at* $x = 0$

$$Rf'(0) = \lim_{h \to 0} \frac{e^{-1/h^2}\sin 1/h - 0}{h} = \lim_{h \to 0} \frac{\sin 1/h}{he^{1/h^2}}$$

$$= \lim_{h \to 0} \frac{\sin 1/h}{h\left[1+\dfrac{1}{h^2}+\dfrac{1}{2!h^4}+...\right]} = \lim_{h \to 0} \frac{\sin 1/h}{h+\dfrac{1}{h}+\dfrac{1}{2!}\dfrac{1}{h^3}+...}$$

$$= \frac{\text{a finite quantity lying between} -1 \text{ and } +1}{\infty} = 0$$

Similarly, $Lf'(0)=0$

Since $Rf'(0) = Lf'(0)=0$. Hence, the function $f(x)$ is differentiable at $x = 0$ and $f'(0) = 0$.

If x is any point other than zero, then

$$f'(x) = (2/x^3)e^{-1/x^2}\sin(1/x)-(1/x^2)e^{-1/x^2}\cos(1/x)$$

$$= \{(2/x)\sin(1/x)-\cos(1/x)\}(1/x^2)(1/e^{1/x^2}) \quad ... (1)$$

Now $$f'(0+0) = \lim_{h \to 0} f'(0+h) = \lim_{h \to 0} \left(\frac{2}{h}\sin\frac{1}{h}-\cos\frac{1}{h}\right).\frac{1}{h^2e^{1/h^2}}$$

$$= \lim_{h \to 0} \left(\frac{2\sin 1/h}{h^3e^{1/h^2}}-\frac{\cos 1/h}{h^2e^{1/h^2}}\right)$$

$$= \lim_{h \to 0} \left[\frac{2\sin 1/h}{h^3\left(1+\dfrac{1}{h^2}+\dfrac{1}{2!h^4}+...\right)}+\left\{-\frac{\cos 1/h}{h^2\left(1+\dfrac{1}{h^2}+\dfrac{1}{2!h^4}+...\right)}\right\}\right]$$

$$= \frac{\text{some finite quantity}}{\infty} - \frac{\text{some finite quantity}}{\infty} = 0$$

Similarly $f'(0-0) = 0$

Hence, $f'(x)$ is continuous at $x = 0$.

Example 21. If f is differentiable at a point c then show that $|f|$ is also differentiable at c provided $f(c) \neq 0$

Solution. Since f is differentiable at $c \Rightarrow f$ is continuous at c.

If $f(c) \neq 0$ then either $f(c) > 0$ or $f(c) < 0$.

If $f(c) > 0$ then there exists $\delta_1 > 0$ such that $f(x) > 0 \ \forall x \in]c - \delta_1, c + \delta_1[$

If $f(c) < 0$ then there exists $\delta_2 > 0$ such that $f(x) < 0 \ \forall x \in]c - \delta_2, c + \delta_2[$.

Therefore, we have

$$f(x) > 0 \ \forall x \in]c - \delta_1, c + \delta_1[$$

$$f(x) < 0 \ \forall x \in]c - \delta_2, c + \delta_2[$$

$$\Rightarrow \qquad |f(x)| = \begin{cases} f(x) & ; \text{ if } x \in]c - \delta_1, c + \delta_1[\\ -f(x) & ; \text{ if } x \in]c - \delta_2, c + \delta_2[\end{cases}$$

Now since f is given to be differentiable at $x = c$.

Hence, from above $|f|$ is also differentiable at $x = c$.

REMARK

- The above result does not hold if $f(c) = 0$.

EXERCISE 4.4

1. Let $f(x) = \begin{cases} -1 & , & -2 \leq x \leq 0 \\ x - 1 & , & 0 < x \leq 2 \end{cases}$

Test the differentiability of $f(x)$.

2. Find $f'(1)$ if

$$f(x) = \begin{cases} \dfrac{x-1}{2x^2 - 7x + 5} & \text{, when } x \neq 1 \\ -1/3 & \text{, when } x = 1 \end{cases}$$

3. Investigate the following function from the point of view of its differentiability. Does the differential coefficient of the function exist at $x = 0$ and $x = 1$?

$$f(x) = \begin{cases} -x & , & \text{if } x < 0 \\ x^2 & , & \text{if } 0 \leq x \leq 1 \\ x^3 - x + 1 & , & \text{if } x > 1 \end{cases}$$

4. Determine the set of all points where the function $f(x) = \dfrac{x}{1 + |x|}$ is differentiable.

5. Show that $f(x) = |x - 1|$, $0 \leq x \leq 2$ is not differentiable at $x = 1$.

6. Show that $f(x) = \begin{cases} -x & , & \text{when } x < 0 \\ x & , & \text{when } x \geq 0 \end{cases}$ is not differentiable at $x = 0$.

7. Show that the function

$$f(x) = \begin{cases} 2 + x, & \text{if } x \geq 0 \\ 2 - x, & \text{if } x < 0 \end{cases}$$

is not differentiable at $x = 0$.

8. Show that the function $f(x) = |x - 1| + 2|x - 2| + 3|x - 3|$ is not differentiable at the point 1, 2 and 3.

9. Show that the function

$$f(x) = \begin{cases} x & , & 0 \leq x < 1 \\ 2 - x & , & x \geq 1 \end{cases}$$

is not differentiable at $x = 1$.

10. The following limits are derivatives of certain functions at a certain point. Determine these functions and points.

(i) $\displaystyle \lim_{x \to 2} \frac{\log x - \log 2}{x - 2}$ (ii) $\displaystyle \lim_{h \to 0} \frac{\sqrt{(a+h)} - \sqrt{a}}{h}$

11. Let $f(x) = x^2 \sin(x^{-4/3})$ except when $x = 0$ and $f(0) = 0$. Prove that $f(x)$ has zero as a derivative at $x = 0$.

12. Discuss the differentiability of the functions

$$f(x) = \begin{cases} x^2 & \text{for} \quad x < -2 \\ 4 & \text{for} \quad -2 \le x \le 2 \\ x^2 & \text{for} \quad x > 2 \end{cases}$$

13. Discuss the existence of $f'(x)$ at $x = 0, 1, 2$, where $f(x)$ is defined as follows:

$$f(x) = \begin{cases} 1 + x & \text{for} \quad x \le 0 \\ x & \text{for} \quad 0 < x < 1 \\ 2 - x & \text{for} \quad 1 \le x \le 2 \\ 3x - x^2 & \text{for} \quad x > 2 \end{cases}$$

14. Show that

$$f(x) = \begin{cases} x\left[1 + \dfrac{1}{8}\sin(\log x^2)\right] & , \text{for} \quad x \ne 0 \\ 0 & , \text{for} \quad x = 0 \end{cases}$$

is not differentiable at $x = 0$.

15. Show that the function

$$f(x) = \begin{cases} x \sin \dfrac{1}{x} & , \quad x \ne 0 \\ 0 & , \quad x = 0 \end{cases}$$

is not differential at $x = 0$.

16. Define differentiabily of a function at a given point. If a function possesses a finite differential coefficient at a point. Show that it is continuous at this point. Is this converse true? Also, give example.

17. What do you understand by the derivative of a real valued function at the point $a \in R$? Apply your definition, discuss the differentiability of $f(x) = |x|$, $x \in R$ at $x = 0$.

18. Prove that every differentiable function is continuous but the converse is not true? Give an example in support of your answer.

19. Prove that if a function $f(x)$ possesses a finite derivative in a closed interval $[a, b]$ then $f(x)$ is continuous in $[a, b]$.

====== Answers ======

1. Not differentiable

3. Not differentiable at $x = 0$, differentiable at $x = 1$

10. (i) $f(x) = \log x$, point is $x = 2$

12. Not differentiable at $x = -2$ and 2

15. Not differentiable

2. $-2/9$

4. Differentiable in $]-\infty, \infty[$

(ii) $f(x) = \sqrt{x}$, point is $a = 2$

13. Not differentiable at $x = 0, 1, 2$

CHAPTER REVIEW: A COMPETITIVE APPROACH

SELECTED TERMS AND RESULTS

TERMS

- **Continuous function:** A function f is said to be continuous at $x=a$ if f is defined at $x=a$ and $\lim\limits_{x \to a} f(x) = f(a)$.

- **Left continuous :** A function $f : A \to R$ is said to be left continuous at a point $a \in R$ if $f(a-0) = \lim\limits_{x \to a^-} f(x) = f(a)$

- **Right continuous :** A function $f : A \to R$ is said to be right continuous at a point $a \in R$ if. $f(a+0) = \lim\limits_{x \to a^+} f(x) = f(a)$.

- **Continuity in an Interval:**

  (i) A function $f(x)$ is said to be continuous in an open interval (a, b) if f is continuous at every point of (a, b).

  (ii) A function f is said to be continuous in the closed interval $[a, b]$ if it is continuous at each point of the open interval (a, b) and also right continuous at a and left continuous at b.

- **Continuity at a point :** A function $f(x)$ is said to be continuous at $x = a$ if for given $\varepsilon > 0 \; \exists \; \delta > 0$ such that
  $|f(x) - f(a)| < \varepsilon$ wherever $|x - a| < \delta$

- **Discontinuous function:** A function is said to be discontinuous at a point $x=a$ if it is not continuous at $x = a$.

- **Discontinuity of first kind:** If $f(a+0)$ and $f(a-0)$ both exist but not equal, then f is said to have a discontinuity of first kind.

- **Discontinuity of second kind:** If either $f(a+0)$ or $f(a-0)$ or both of them do not exist, then f is said to have a discontinuity of second kind.

- **Mixed discontinuity:** If a function f has a discontinuity of second kind at one side of a and on the other side it may continuous or may have discontinuity of first kind, then f is said to have mixed discontinuity.

- **Removable discontinuity :** If $\lim\limits_{x \to a} f(x)$ exists but not equal to $f(a)$ i.e., $f(a+0) = f(a-0) \neq f(a)$ then the function f is said to have a removable discontinuity.

- **Infinite discontinuity:** A function f has an infinite discontinuity at $x=a$ if either $f(a+0)$ or $f(a-0)$ or both have infinite values.

- **Uniform continuity :** A function f is said to be uniformly continuous on the interval $[a, b]$ if for given $\varepsilon > 0 \; \exists \; \delta > 0$ such that $|f(x_1) - f(x_2)| < \varepsilon$, whenever $|x_1 - x_2| < \delta \; \forall x_1, x_2 \in (a, b)$.

- **Derivative of a function :** Let $f(x)$ be a function defined on nbd of a point a and $\lim\limits_{h \to 0} \dfrac{f(a+h) - f(a)}{h}$ exists finitely then the function $f(x)$ is said to be differentiable at a and this limit is called the derivative of the function $f(x)$ at $x = a$.

- **Left hand derivative:** The left hand derivative (regressive derivative) of f at $x = a$ is given by $Lf'(a) = \lim\limits_{h \to \infty} \dfrac{f(a-h) - f(a)}{-h}$, $h > 0$

- **Right hand derivative** The right hand derivative (Progressive derivative) of f at $x = a$ is given by $Rf'(a) = \lim\limits_{h \to \infty} \dfrac{f(a+h) - f(a)}{h}$, $h > 0$

- **Differentiable function:** A function $f(x)$ is said to be differentiable at $x=a$ if $Rf'(a) = Lf'(a)$.

RESULTS

- If f and g are two continuous functions at $x = a$ then $f+g, f - g, fg, \dfrac{f}{g} \, (g \neq 0)$ are also continuous at $x=a$.

- If f is continuous at $x = a$ then $|f|$ is also continuous at $x = a$. Converse is not necessarily true.

- Composite of two continuous functions is a continuous function.

- Every continuous function is bounded. Converse is not true.

- If a function f is continuous on $[a, b]$ and $c \in \,]a, b[$ such that $f(c) \neq 0$ then there exists some $\delta > 0$ such that $f(x)$ has the same sign as $f(c) \; \forall \, x \in \,]c - \delta, c + \delta[$.

- If a function f is continuous on $[a, b]$ then

  (i) $f(a) > 0 \Rightarrow \exists \, \delta > 0$ such that $f(x) > 0$ $\forall \, x \in [a, a+\delta[$

  (ii) $f(a) < 0 \Rightarrow \exists \, \delta > 0$ such that $f(x) < 0$ $\forall \in [a, a+\delta[$

  (iii) $f(b) > 0 \Rightarrow \exists \, \delta > 0$ such that $f(x) > 0$ $\forall \, x \in \,]b - \delta, b]$

  (iv) $f(b) < 0 \Rightarrow \exists \, \delta > 0$ such that $f(x) < 0$ $\forall \, x \in \,]b - \delta, b]$.

- A function $f : R \to R$ is continuous on R if and only if for every open set A in R, $f^{-1}(A)$ is open in R.

- A function $f : R \to R$ is continuous on R if and only if for each closed set A in R, $f^{-1}(A)$ is closed in R.

- If a function f is continuous on a closed interval $[a, b]$ such that $f(a)$ and $f(b)$ are of opposite sign that there exists at least one point $c \in \,]a, b[$ such that $f(c) = 0$

- If a function f is continuous on a closed interval $[a, b]$ and $f(a) \neq f(b)$ then f assume every value between $f(a)$ and $f(b)$.

- A function f which is continuous on a closed interval $[a, b]$ assumes every value between its bounds.

- Every uniformly continuous function is continuous.

- If a function f is continuous on a closed and bounded interval $[a, b]$ then it is uniformly continuous on $[a, b]$.

- Every differentiable function is continuous.

- Let f be a function defined on $[a, b]$ and $f'(c)$ exists for any point $c \in \,]a, b[$ such that $f'(c) > 0$ then f is increasing at c and if $f'(c) < 0$ then f is decreasing at c.

- Let f be a function defined and derivable on $[a, b]$ such that $f'(a) \, f'(b) < 0$ then there exists some $c \in \,]a, b[$ such that $f'(c) = 0$.

- If f is defined and derivable on $[a, b]$ and $f'(a) \neq f'(b)$ then for each real number k lying between $f'(a)$ and $f'(b) \; \exists$ some $c \in \,]a, b[$ such that $f'(c) = k$.

- If f is differentiable in $[a, b]$ such that $f'(x) \neq 0 \; \forall \, x \in \,]a, b[$ then $f'(x)$ retains the same sign positive or negative in $]a, b[$.

- If f is differentiable at a point c then $|f|$ is also differentiable at c provided $f(c) \neq 0$.

REVIEW QUESTIONS AND PROJECT WORK

1. Show that the function

$$f(x) = \begin{cases} \dfrac{e^{1/x^2}}{1 - e^{1/x^2}} & \text{, when } x \neq 0 \\ -1 & \text{, when } x = 0 \end{cases}$$

is continuous at $x = 0$.

2. Let f be the function defined on $[0, 1]$ such that

$$f(x) = \begin{cases} (-1)^r & \text{if} \quad \dfrac{1}{r+1} \leq x < \dfrac{1}{r} \quad r = 1, 2, \ldots \\ 0 & \text{if} \quad x = 0 \\ 1 & \text{if} \quad x = 1 \end{cases}$$

examine the continuity of f at $x = 1, \dfrac{1}{2}, \dfrac{1}{3}, \ldots$

3. Let f be a function defined on R such that

$$f(t) = \begin{cases} t & ; \quad 0 \leq t < \dfrac{1}{2} \\ 0 & ; \quad t = \dfrac{1}{2} \\ t - 1 & ; \quad \dfrac{1}{2} < t \leq 1 \end{cases}$$

and $f(n+t) = f(t)$, where n is any integer. Determine the point of discontnuity of f.

4. Let f be a function defined on R by setting

$$f(x) = \begin{cases} x - [x] - \dfrac{1}{2} & ; \quad \text{when } x \text{ is not an integer} \\ 0 & ; \quad \text{when } x \text{ is an integer} \end{cases}$$

Show that f is continuous at all points of R– Z and is discontinuous whenever $x \in Z$.

5. Show that the function

$$f(x) = \begin{cases} \dfrac{3x + |x|}{7x - 5|x|} & , \quad x = 0 \\ 0 & , \quad x = 0 \end{cases}$$

is discontinuous at $x = 0$.

6. Show that the function defined by

$$f(x) = \begin{cases} 1 & , \quad \text{if } x \text{ is rational} \\ -1 & , \quad \text{if } x \text{ is irrational} \end{cases}$$

is discontinuous at each point of R.

7. Show that the function
$$f(x) = \begin{cases} 1 & , & \text{if } x \text{ is rational} \\ 0 & , & \text{if } x \text{ is irrational} \end{cases}$$
is discontinuous at every point.

8. Show that the function f defined on R such that
$$f(x) = \begin{cases} x & ; & \text{when } x \text{ is rational} \\ -x & ; & \text{when } x \text{ is irrational} \end{cases}$$
is continuous only at origin.

9. Show that the function
$$f(x) = \begin{cases} x & ; & \text{if } x \text{ is rational} \\ 0 & ; & \text{if } x \text{ is irrational} \end{cases}$$
is continuous only at $x = 0$.

10. Show that the function
$$f(x) = \begin{cases} x & ; & \text{if } x \text{ is irrational} \\ 0 & ; & \text{if } x \text{ is rational} \end{cases}$$
is continuous only at $x=0$.

11. Show that the function f defined on [0, 1] by
$$f(x) = \begin{cases} x & ; & \text{when } x \text{ is rational} \\ 2x & ; & \text{when } x \text{ is irrational} \end{cases}$$

is continuous only at $x = 0$

12. Show that the function $f(x)$ defined by
$$f(x) = \begin{cases} 0 & ; & \text{if } x \text{ is irrational} \\ \dfrac{1}{q} & ; & \text{if } x = \dfrac{p}{q} \text{ where } p \text{ and } \quad q \quad \text{are} \end{cases}$$
positive integers having no common factor is continuous at each irrational point and discontinuous at each rational part.

13. Show that if f and g are continuous on $[a, b]$ and if $f(a) < g(a)$ then there exists some $c \in]a, b[$ satisfying $f(c)=g(c)$.

14. Show that there exists some $x \in [a, b]$ such that $f(x) = x$ where f is continuous from $[a, b]$ to itself.

15. Show that the function
$$f(x) = |x - 2| + |x| + |x + 2| \, \forall x \in R$$
is not differentiable at $x = -2, 0$ and $x = 2$ and is differentiable at every other point.

OBJECTIVE TYPE QUESTIONS

FILL IN THE BLANKS

1. Limit of a function, if exist is _____ .

2. If $\lim f(x) = l$ then $\lim \dfrac{1}{f(x)} = \dfrac{1}{l}$ provided _____ .

3. The limit of the quotient is equal to the _____ of the limits.

4. A function $f(x)$ is continuous at $x=a$ if $\lim\limits_{x \to a} f(x) = $ _____ .

5. $\lim\limits_{x \to 0} (1 + x)^{1/x} = $ _____ .

6. $\lim\limits_{x \to a} \dfrac{x^m - a^m}{x - a} = $ _____ .

7. In the definition of continuity, the value of δ depends upon the value of _____ .

8. A polynomial function is always _____ .

9. $\lim\limits_{x \to 0} \dfrac{a^x - 1}{x} = $ _____ .

10. A function is said is have _____ if $f(a+0)$ $= f(a - 0) \ne f(a)$.

11. If $f(a+0) \ne f(a - 0)$ then $f(x)$ is said to have a discontinuity of _____ .

12. The value of $f(a+0) \sim f(a - 0)$ is known as _____ of a function.

13. If f is continuous then $|f|$ is _____ .

14. If $|f|$ is continuous then f is _____ continuous.

15. Every uniformly continuous function is _____ .

16. Every continuous function is _____ uniformly continuous.

17. Every continuous function is _____ .

18. If f is continuous in $[a, b]$ and $f(a)$, $f(b)$ have opposite sign, then $\exists$ at least one value of x for which $f(x) = $ _____ .

19. A function which is continuous in a closed and bounded interval is _____ continuous.

20. The domain of the function $f(x) = \dfrac{\sin x}{x}$ _____ .

21. Let $f(x) = \begin{cases} 1 & , & x < 1 \\ 2-x & , & 2 \le x < 2 \\ 2 & , & x \ge 2 \end{cases}$

Then $f\left(\dfrac{3}{2}\right) = $ _____ .

22. Every differentiable function is _____ .

23. Every continuous function is _____ .

24. Sum and difference of two differentiable functions is again _____ .

25. The first mean value theorem is also known as _____ .

26. If $f'(x) > 0$ then $f(x)$ is known as _____ .

27. If $f'(x)$ is positive at a point $x = a$, then in the neighbourhood of $x = a$, then function $f(x)$ is _____ .

28. The function $f(x) = x|x|$ is _____ .

29. If f is a function, differentiable on an interval I, than $f'(I)$ is either interval or a _____ .

30. If f is finitely differentiable in a closed interal $[a, b]$ and $f'(a)$, $f'(b)$ are of opposite sign then $f'(c) =$ _____ for at least one value of $c \in]a, b[$.

31. If $f(x)$ is an even function. Then value of $f'(0)$ (if exist) is equal to _____ .

TRUE/ FALSE

Write 'T' for true and 'F' for false statement.

1. Every continuous function in closed interval is bounded. **(T/F)**

2. Every continuous function in open interval is bounded. **(T/F)**

3. For $\lim\limits_{x \to a} f(x)$ to exist, the function $f(x)$ must be defined at $x = a$. **(T/F)**

4. The limit of a products is equal to the product of the limits. **(T/F)**

5. $\lim\limits_{x \to 1} \dfrac{x^3 - 1}{x^2 - 1} = \dfrac{3}{2}$. **(T/F)**

6. For a function $f(x)$ to be continuous at $x = a$, it is necessary that $\lim\limits_{x \to a} f(x)$ must exist. **(T/F)**

7. The function must be defined at the point of continuity. **(T/F)**

8. If a function having a finite number of jumps in a given interval then function is called piecewise continuous. **(T/F)**

9. Sum of two continuous functions is not necessarily continuous. **(T/F)**

10. If a function f is continuous in the closed interval $[a, b]$, then $f(x)$ must take at least once all values between $f(a)$ and $f(b)$. **(T/F)**

11. If f is uniformly continuous on an interval I, then it is continuous on I. **(T/F)**

12. A function which is continuous in an open interval is uniformly continuous. **(T/F)**

13. A function which is continuous in a closed and bounded interval is uniformly continuous. **(T/F)**

14. The function $f(x) = \begin{cases} \dfrac{\sin x}{x} & , \ x \neq 0 \\ 1 & , \ x = 0 \end{cases}$ is continuous at x = 0. **(T/F)**

15. $\lim\limits_{x \to 0} \dfrac{\sin 3x}{x} = 1$ **(T/F)**

16. $\lim\limits_{x \to 3^-} \dfrac{|x - 3|}{x - 3} = 1$ **(T/F)**

17. Every continuous function is differentiable. **(T/F)**

18. Every differentiable function is continuous. **(T/F)**

19. Every differentiable function is bounded. **(T/F)**

20. A function is said to be differentiable if $Lf'(x) = Rf'(x)$. **(T/F)**

21. If $f'(x) > 0$. Then $f(x)$ is an increasing function. **(T/F)**

22. The function $f(x) = |x|$ is differentiable everywhere. **(T/F)**

23. If $f(x) = 0$ at each point in $]a, b[$ then $f(x)$ is a constant function. **(T/F)**

24. If f is differentiable at c and $f(c) \neq 0$ then $\dfrac{1}{f}$ is not necessarily differentiable. **(T/F)**

25. If two functions have equal derivative at all points in (a, b) then they must be equal. **(T/F)**

26. If $f(x)$ is continuous at $x = 0$, then the function $x f(x)$ is differentiable at $x = 0$. **(T/F)**

MULTIPLE CHOICE QUESTIONS

Choose the most appropriate one

Problem Set-1

1. If $\lim\limits_{x \to a} f(x) = l$ and $f(x) \geq 0$, then
   (a) $l = 0$
   (b) $l \leq 0$
   (c) $l \geq 0$
   (d) none of these

2. If $\lim_{x \to a} f(x) = l$, then $\lim_{x \to a} |f(x)| =:$

(a) l (b) $|l|$

(c) 0 (d) 1

3. If $\lim_{x \to \infty} f(x) = l$ and $\lim_{x \to \infty} g(x)$ does not exist, then:

(a) $\lim_{x \to \infty} f(x) \cdot g(x)$ does not exist

(b) $\lim_{x \to \infty} f(x) \cdot g(x)$ exist necessarily

(c) $\lim_{x \to \infty} f(x) \cdot g(x)$ may or may not exist

(d) none of these

4. $\lim_{x \to 2} \dfrac{|x-2|}{x-2} = :$

(a) 0 (b) 1

(c) 2 (d) does not exist

5. The value of $\lim_{x \to 0} \dfrac{\sin x}{x}$ is:

(a) 1 (b) 0

(c) ∞ (d) does not exist.

6. The value of $\lim_{x \to \infty} \dfrac{\sin x}{x}$ is:

(a) 1 (b) 0 .

(c) ∞ (d) 2

7. If $\lim_{x \to a} f(x)$ and $\lim_{x \to a} g(x)$ do not exist, then

$\lim_{x \to a} [f(x) + g(x)]:$

(a) does not exist
(b) necessarily exist
(c) may or may not exist
(d) none of the above

8. The equation $\lim_{x \to 0} f(x) = \lim_{x \to a} f(x - a)$ is:

(a) always true
(b) may or may not be true
(c) always false
(d) depend upon the value of a.

9. The value of k for which

$$f(x) = \begin{cases} \dfrac{\sin 5x}{3x} & , \text{ if } x \neq 0 \\ k & , \text{ if } x = 0 \end{cases} \text{ is continuous at}$$

$x = 0$ is:

(a) $\dfrac{1}{3}$ (b) $\dfrac{3}{5}$

(c) $\overset{.}{0}$ (d) $\dfrac{5}{3}$

10. Let $f(x) = \begin{cases} \lambda x^2 & \text{if } x \leq 2 \\ 3 & \text{if } x > 2 \end{cases}$, then the value of λ is:

(a) 2 (b) 3

(c) $\dfrac{2}{3}$ (d) $\dfrac{3}{4}$

11. The function $f(x)$ is continuous at a point a of its domain if:

(a) $f(x)$ is bounded in the nbd of a

(b) $\lim_{x \to a} f(x)$ exist and finite

(c) $\lim_{x \to a} f(x) = f(a)$

(d) none of the above

12. If $f(x) = \begin{cases} \dfrac{\tan x}{\sin x} & , \ x \neq 0 \\ 1 & , \ x = 0 \end{cases}$ then:

(a) $f(x)$ is monotonic increasing in the nbd of a

(b) $\lim_{x \to 1} f(x) = 0$

(c) $f(x)$ is continuous at $x=0$

(d) none of the above

13. If $f(x) = \begin{cases} 0 & \text{if} & x = 0 \\ 1 - x & \text{if} & 0 < x \leq 1/2 \\ 2x - 1 & \text{if} & 1/2 < x < 1 \\ x & \text{if} & x = 1 \end{cases}$, then:

(a) $f(x)$ is continuous at $x = 0$

(b) continuous at $x = 1/2$

(c) continuous at $x = \dfrac{3}{4}$

(d) none of the above

14. The function $f(x) = x^3$ defined for all values of x is:

(a) increasing in $]0, \infty[$ and decreasing in $]-\infty, 0[$

(b) decreasing in $]0, \infty[$ and increasing in $]-\infty, a[$

(c) decreasing throughout

(d) increasing throughout.

15. If $\lim_{x \to a} f(x) = \pm\infty$, then $\lim_{x \to a} \dfrac{1}{f(x)} =:$

(a) $\pm\infty$ (b) 1

(c) 0 (d) 2

16. A function is said to be piecewise continuous if:

(a) it has a finite no. of jump
(b) it has an infinite no. of jumps
(c) it has no jump
(d) none of the above

17. Jump of a function is defined as:

(a) RHL~LHL   (b) RHL + LHL

(c) RHL/LHL   (d) (RHL) . (LHL)

18. The saltus of f is defined by:

(a) $\sup(f) - \inf(f)$

(b) sup +inf

(c) sup/inf

(d) none of the above

19. If saltus of a function is equal to zero, then function is:

(a) continuous

(b) discontinuous

(c) piecewise continuous

(d) none of the above

20. Continuous image of an open set is:

(a) closed

(b) open

(c) may or may not be open

(d) none of the above

21. If the inverse image of an open set is open then function is:

(a) continuous

(b) discontinuous

(c) piecewise continuous

(d) none of the above

22. The Dirichlet's function f defined as
$f(x) = \begin{cases} +1 & , & x \text{ is rational} \\ -1 & , & x \text{ is irrational} \end{cases}$ is :

(a) continuous at every point of R

(b) continuous at $x = 1$

(c) discontinuous at every point of R

(d) none of the above

23. Let f be a function, which is not continuous, then f^2 is:

(a) not continuous

(b) may or may not be continuous

(c) always continuous

(d) none of the above

24. Which one of the following is not true?

(a) Every constant function is continuous

(b) Every identity function is continuous

(c) Every polynomial function is continuous

(d) None of the above

25. Which one of the following is not true?

(a) The set of points of removable discontinuity of a function is countable.

(b) The set of points of discontinuity of first kind of a function is countable.

(c) The set of point of removable discontinuity

of f is the union of countable number of closed sets.

(d) None of the above.

26. A function $f: [a, b] \to R$ is said to be differentiable if f is:

(a) differentiable at each point of $[a, b]$

(b) differentiable at the ends points only

(c) differentiable at each point of $[a, b]$ except the end points

(d) none of the above

27. A function $f(x)$ is said to be differentiable at $x = a$, if:

(a) right hand and left hand derivative at a exist and equal

(b) only right hand derivative must exist

(c) only left hand derivative must exist

(d) none of the above

28. Every differentiable function is:

(a) necessarily continuous

(b) never continuous

(c) may or may not be continuous

(d) none of the above

29. If f is finitely differentiable in a closed interval $[a, b]$ and $f'(a), f'(b)$ are of opposite sign, then:

(a) $f'(c) = 0 \; \forall \; c \in [a, b]$

(b) $f'(c) = 0$ for at least one $c \in]a, b[$

(c) $f'(c) = 0 \; \forall \; c \in]a, b[$

(d) none of the above

30. Every continuous function is:

(a) necessarily differentiable

(b) never differentiable

(c) may or may not be differentiable

(d) none of the above

31. If $f(x)$ is an even function. Then the value of $f(0)$ (if exist) is equal to:

(a) 1   (b) 0

(c) $+ \infty$ (d) $- \infty$

32. If a function f is continuous on $[a, b]$, differentiable on $]a, b[$ and if $f'(x) = 0$ $\forall \; x \in]a, b[$ then $f(x)$ has a:

(a) constant value throughtout $[a, b]$

(b) constant value only at the end points

(c) constant value through out $]a, b[$

(d) none of the above

33. If $f(x)$ and $g(x)$ are continuous on $[a, b]$ and differentiable on $]a, b[$ and if $f'(x) = g'(x)$ throughout the interval $]a, b[$, then:

(a) $f(x) = g(x) \; \forall \; x \in]a, b[$

(b) $f(x) \neq g(x) \; \forall \; x \in]a, b[$

(c) $f(x)$ and $g(x)$ differ only by a constant

(d) none of the above

34. If f is continuous on $[a, b]$ and $f'(x) \geq 0$ on $]a, b[$, then:

(a) f is decreasing on $]a, b[$

(b) f is decreasing on $[a, b]$

(c) f is increasing on $]a, b[$

(d) f is increasing on $[a, b]$

35. If $y = f(x)$ be an increasing function of x, then:

(a) $f'(x) \leq 0$ (b) $f'(x) = 0$

(c) $f'(x) > 0$ (d) none of these

36. If $f'(x)$ is positive at a point $x = a$, then in the neighbourhood of $x = a$:

(a) $f(x)$ is positive   (b) $f(x)$ is increasing

(c) $f(x)$ is negative   (d) $f(x)$ is decreasing.

37. The function $f(x)$ has equal values at the point $x=a$ and $x=b$, then:

(a) there is a maximum of $f(x)$ between a and b

(b) there is a minimum of $f(x)$ between a and b

(c) there is a maximum or minimum of $f(x)$ between a and b

(d) none of the above

38. If $f''(x) > 0$ at all points in $]a, b[$ then the function f is:

(a) strictly increasing   (b) strictly decreasing

(c) constant   (d) none of the above

39. Let f and g be two functions having the same domain D and $f+g$ or $f.g$ be differentiable at $c \in D$, then:

(a) f and g are necessarily differentiable

(b) f and g are necessarily differentiable

(c) f and g never differentiable at c

(d) none of the above

Problem Set-2

1. If a function f is continuous on $[a, b]$, differentiable in $]a, b[$ and if $f'(x) = 0$ $\forall x \in]a, b[$. Then $f(x)$ has a:

(a) constant value throughout $[a, b]$

(b) constant value only at the end points

(c) constant value throughout (a, b) exist the end points

(d) none of the above

2. If $f(x)$ and $g(x)$ are functions continuous on $[a, b]$ and differentiable on $]a, b[$ and if $f'(x) = g'(x)$ throughout the interval $]a, b[$ then:

(a) $f(x) = g(x) \; \forall \; x \in]a, b[$

(b) $f(x) \neq g(x) \; \forall \; x \in]a, b[$

(c) $f(x)$ and $g(x)$ differ only by a constant

(d) none of the above

3. If f is continuous on $[a, b]$ and $f'(x) \geq 0$ on $]a, b[$, then:

(a) f is decreasing in $]a, b[$

(b) f is decreasing in $[a, b]$

(c) f is increasing in $]a, b[$

(d) f is increasing in $[a, b]$

4. If $f(x) = [x]$, where $[x]$ denotes the greatest integer not greater then x, then :

(a) $f(x)$ is continuous for all values of x.

(b) $f(x)$ is differentiable for all values of x.

(c) $f(x)$ is differentiable for all values of $x \neq m$, where m is an integer.

(d) none of the above

5. Let $f(x) = |x|$. Then:

(a) $f(x)$ is continuous for all values of x.

(b) $f(x)$ is differentiable for all values x.

(c) $f(x)$ is not differentiable at $x = 0$

(d) both (a) and (c) are true

6. The function $f(x) = \begin{cases} x & \text{when } x \text{ is rational} \\ -x & \text{when } x \text{ is irrational} \end{cases}$

(a) $f(x)$ is continuous for all values of x

(b) $f(x)$ is discontinuous for all values of x, except at $x = 0$

(c) The derivative of $f(x)$ is positive for all values of x.

(d) none of the above

7. Let $f(x) = \sin\dfrac{1}{x}, x \neq 0$ and $f(0) = 0$

(a) $f(x)$ is continuous for all $x \neq 0$

(b) $f(x)$ is differentiable for all $x \neq 0$

(c) $f'(x) = 0$ at an infinite no. of points in the neighborhood if $x = 0$

(d) all are true

8. Let $f(x) = \begin{cases} x\sin\dfrac{1}{x} & , & x \neq 0 \\ 0 & , & x = 0 \end{cases}$ then:

(a) $f(x)$ is continuous for all values of x

(b) $f(x)$ is differentiable for all $x \neq 0$

(c) both (a) and (b) are true

(d) both (a) and (b) false

9. Let $f(x) = \begin{cases} x^2 \sin \dfrac{1}{x} &, x \neq 0 \\ 0 &, x = 0 \end{cases}$, then:

(a) $f(x)$ is continuous for all values of x.

(b) $f(x)$ is differentiable for all values of x.

(c) both (a) and (b) are true

(d) both (a) and (b) false

10. Let $f(x) = |x - a| + |x - b|$. Then:

(a) $f(x)$ is continuous for all values of x

(b) $f(x)$ is differentiable for all values of x except a and b

(c) both (a) and (b) are true

(d) both (a) and (b) false

11. Let $f(x) = 1 + |\sin x|$. Then $f(x)$ is:

(a) differentiable everywhere

(b) discontinuous at an infinite no. of points

(c) not differentiable at an infinite number of points

(d) none of the above

12. Let $y = f(x)$ be an increasing function of x. Then:

(a) $f'(x) \leq 0$ (b) $f'(x) = 0$

(c) $f'(x) > 0$ (d) none of the these

13. If $f'(x)$ is positive at a point $x = a$, then in the neighbourhood of $x = a$:

(a) $f(x)$ is positive   (b) $f(x)$ is increasing

(c) $f(x)$) is negative   (d) $f(x)$ is decreasing

14. $f'(x) > 0$ over an interval I, then in this interval:

(a) $f(x)$ is positive

(b) $f(x)$ is a decreasing function

(c) $f(x)$ is stationary

(d) none of the above

15. If $f(x)$ is a differentiable function such that $f(x) < f(2)$; $1 \leq x \leq 3$, then

(a) $f'(2) \leq 0$

(b) $f'(2) = 0$

(c) $f'(2) = 2$

(d) $f'(2)$ does not exist

16. The function $y = x^3$ defined for all values of x, is :

(a) decreasing throughout

(b) increasing throughout

(c) decreasing in $(0, \infty)$ and increasing in $(-\infty, 0)$

(d) increasing in $(0, \infty)$ and decreasing in $(-\infty, 0)$

17. The function $f(x)$ has equal values at the point $x = a$ and $x = b$, then:

(a) there is maximum of $f(x)$ between a and b

(b) there is a minimum of $f(x)$ between a and b

(c) there is a maximum or minimum of $f(x)$ between a and b

(d) none of the above

18. The function $y = x^4$ defined for all real values of (x, ∞):

(a) increasing throughout

(b) decreasing throughout

(c) increasing in $(0, \infty)$ and decreasing in $(-\infty, 0)$.

(d) decreasing in $(0, \infty)$ and increasing in $(-\infty, 0)$.

19. The function $f(x)$ has a maximum at a point $x = a$ if as x passes through a

(a) $f'(x)$ changes sign from positive to negative

(b) $f'(x)$ changes sign from negative to positive

(c) $f'(x)$ has no fixed sign in the neighbourhood of $x = a$

(d) $f'(x)$ has no fixed sign throughout a neighbourhood of $x = a$

20. The functions $f(x)$ has a minimum at a point $x = a$ if x passes through a.

(a) $f'(x)$ changes sign from positive to negative,

(b) $f'(x)$ changes sign from negative to positive.

(c) $f'(x)$ has no fixed sign in the neighbourhood of $x = a$.

(d) $f'(x)$ has no fixed sign throughout a neighbourhood of $x = a$.

21. $f'(x) = 2x^3 - 15x^2 + 36x + 1$ is increasing in the interval

(a) $]\,2, 3\,[$ (b) $]-\infty, 3\,[$

(c) $]-\infty, 2\,[\cup]\,3, \infty[$

(d) none of the above

22. The function $f(x) = x^9 + 3x^7 + 6$ is increasing for:

(a) all positive real values of x

(b) all negative real values of x.

(c) all non-zero real values of x.

(d) none of the above

23. The function $f(x) = x^3 + 3x^2 - 105x + 25$ is decreasing in the interval:

(a) $]-\infty, -7[$ (b) $]-5, \infty[$

(c) $]7, 5[$ (d) none of these

24. The function $f(x) = \cos x - 2px$ is monotonically decreasing for:

(a) $p < \dfrac{1}{2}$ (b) $p > \dfrac{1}{2}$

(c) $p < 2$ (d) $p > 2$

25. The function $f(x) = x^3 + 2x^2 - 1$ is increasing in the interval:

(a) $]-\dfrac{4}{3}, 0[$ (b) $]-\infty, 0[$

(c) $]-\infty, -\dfrac{4}{3}[\cup]0, \infty[$ (d) none of the above

26. If $f(x) = lx^3 - 9x^2 + 9x + 3$ is increasing in each interval, then:

(a) $l < 3$ (b) $l \le 3$

(c) $l > 3$ (d) $l \ge 3$

27. The function $f(x) = (x + 2)e^{-x}$ is:

(a) decreasing for all $x \in R$

(b) decreasing in $]-\infty, -1[$ and increasing in $]-1, \infty[$

(c) increasing for all $x \in R$

(d) decreasing in $]-1, \infty[$ and increasing in $]-\infty, -1[$

28. The function $f(x) = x^x$ decreasing on the interval:

(a) $]0, e[$ (b) $]0, \dfrac{1}{e}[$

(c) $]0, 1[$ (d) none of the above

29. If $k < 0$, then $f(x) = e^{kx} + e^{-kx}$ is decreasing for:

(a) $k > 0$ (b) $k < 0$

(c) $k \ge 0$ (d) $k \le 0$

30. If $f''(x) > 0$ at points in $]a, b[$ then the function f is :

(a) strictly increasing   (b) strictly decreasing

(c) constant   (d) none of the above

31. If $f(x) = 0$ at each points in $]a, b[$ then f is a

(a) strictly increasing   (b) strictly decreasing

(c) constant   (d) none of the above

32. If $f_1(x) = e^x$ and $f_2(x) = e^{-x}$ be two given functions satisfying all conditions of Cauchy mean value theorem in $[a, b]$. Then the value of c is the :

(a) arithmetic mean between a and b

(b) geometric mean between a and b

(c) harmonic mean between a and b

(d) none of the above

33. Let $f(x)$ be a function such that $f'(x)$ exist in $]a, b[$ then:

(a) $f(x)$ can have an ordinary removable discontinuity in $]a, b[$

(b) $f(x)$ can have an ordinary removable discontinuity in $[a, b]$

(c) $f(x)$ cannot have an ordinary or removable discontinuity in $]a, b[$

(d) $f(x)$ cannot have an ordinary or removable discontinuity in $[a, b]$

34. Let $f(x)$ be a function which differentiable in $]a, b[$ and $f'(a) < k < f'(b)$ then there exist a point $x \in]a, b[$ such that :

(a) $f'(x) = k$ (b) $f'(x) \ne k$

(c) $f'(x) > k$ (d) $f'(x) < k$

35. If f is derivable at a point c and $f(c) \ne 0$ then the function $\dfrac{1}{f}$ is :

(a) necessarily derivable

(b) may or may not be derivable

(c) not derivable

(d) none of the above

36. Let f and g be two functions having the same domain and $f + g$ or fg be differentiable at $c \in D$ then:

(a) f and g are necessarily differentiable at c

(b) f and g are not necessarily differentiable at c

(c) f and g never differentiable at c

(d) none of the above

37. The function $f(x) = x|x|$ is:

(a) not differentiable at the origin.

(b) differentiable at the origin.

(c) differentiable at all points except origin

(d) none of the above

38. The function $f(x) = |x| + |x - 1|$ is:

(a) differentiable all points

(b) not differentiable at all points

(c) differentiable at all points except 0 and 1.

(d) none of the above

39. If a function $f(x)$ satisfy the conditions of the mean value theorem and $f'(x) = 0$ $\forall x \in [a, b]$:

(a) $f(x) = 0$

(b) $f(x)$ is increasing function

(c) $f(x)$ is decreasing function

(d) $f(x)$ is constant

40. If two functions have equal derivative at all points $[a, b]$ then they both:

(a) must be equal

(b) must be unequal

(c) differ only by a constant

(d) none of the above

41. If a function f is (i) continuous on $[a, b]$ (ii) differentiable on $]a, b\,[$ (iii) $f'(x) > 0$ $\forall\, x \in]a, b[$ then:

(a) f is monotonically increasing $[a, b]$

(b) f is monotonically decreasing on $[a, b]$

(c) f is strictly increasing on $[a, b]$

(d) f is strictly decreasing on $[a, b]$

42. If f is a function derivable on an interval I, then $f'(I)$ is:

(a) interval only

(b) constant

(c) either interval or a singleton

(d) none of the above

43. Let f be a continuos function on $[a, b]$ and derivable on $]\,a, b\,[$. If $f'(x) > 0\ \forall\, x \in]\,a, b\,[$ then :

(a) $f(x)$ is increasing on $]a, b[$

(b) $f(x)$ is increasing on $[a, b]$

(c) $f(x)$ is decreasing on $]\,a, b[$

(d) $f(x)$ is decreasing on $[a, b]$

44. Let f be a continuous function on $[a, b]$ and differentiate on $]a, b[$ if $f'(x) > 0\ \forall x \in]a, b[$ then:

(a) f is non-decreasing in $]\,a, b[$

(b) f is non-decreasing in $[a, b]$

(c) f is decreasing in $]\,a, b[$

(d) f is decreasing in $[a, b]$

45. Let $f(x)$ is continuous at $x = 0$ then the function $f(x)$ is:

(a) derivable at $x = 0$

(b) not derivable for all x

(c) derivable for all x except $x = 0$

(d) none of the above

46. Let f be a function defined on R by $f(x)$ $= x^2 \cos\dfrac{1}{x}$ when $x \neq 0$ and $f(0) = 0$, is derivable on R then :

(a) $f'(x)$ is continuous at $x = 0$

(b) $f'(x)$ is not continuous at $x = 0$

(c) $f'(x)$ is continuous for all x

(d) none of the above

47. Let $f(x + y) = f(x) + f(y) \forall x, y \in$ R. Then f is differentiable on R if:

(a) f must be differentiable at every point of R

(b) f must be differentiable at one point of R

(c) f must be differentiable at every point of R^+

(d) none of the above

48. Let $f(x, y) = f(x) + f(y) \forall x, y \in R^+$. Then f is differentiable on R if:

(a) f must be differentiable at every point of R

(b) f must be differentiable at one point of R^+

(c) f must be differentiable at every point of R^+

(d) none of the above

49. If functions f and g are such that $f'(x) = g'(x)$ and interval I. Then the difference function $(f - g)$ is:

(a) zero

(b) constant

(c) cannot be obtained

(d) none of the above

50. Let f is continuous on an interval I and differentiable on I^i, then f is non-decreasing on I if:

(a) $f'(x) < 0$ on I^i (b) $f'(x) > 0$ on I^i

(c) $f'(x) \geq 0$ on I^i (d) $f'(x) \leq 0$ on I^i

51. Let f be continuous on an interval I and differentiable on I^i, then f is non-increasing on I if:

(a) $f'(x) < 0$ on I^i (b) $f'(x) > 0$ on I^i

(c) $f'(x) \geq 0$ on I^i (d) $f'(x) \leq 0$ on I^i

52. Let $x \in \left]0, \dfrac{\pi}{2}\right[$, then which one of the following inequality is true?

(a) $\tan x > x > \sin x$

(b) $\tan x < x < \sin x$

(c) $\tan x > x < \sin x$

(d) $\tan x < x > \sin x$

53. Let $x^x (1-x)^{1-x} \geq \alpha^x (1-a)^{1-x} \forall \alpha, x \in]\,0, 1\,[$. Then inequality holds only when

(a) $x \neq \alpha$ (b) $x = 0$

(c) $x = 1$ (d) $x = a$

54. f is twice differentiable and $|f| < \alpha$, $|f''| < \beta$ in the range $x > \alpha$. Then:

(a) $|f'| < 2\sqrt{\alpha\beta}\ \forall\ x > a$

(b) $|f'| > 2\sqrt{\alpha\beta}\ \forall\ x > a$

(c) $|f'| < 2\sqrt{\alpha\beta}\ \forall\ x < a$

(d) $|f'| > 2\sqrt{\alpha\beta}\ \forall\ x < a$

55. If f is twice differentiable real function on $]0, \infty[$ and α, β, γ are the supremum of $|f|$, $|f'|$ and $|f''|$ respectively on $]0, \infty[$. Then:

(a) $\beta^2 < \alpha\gamma$ (b) $\beta^2 > \alpha\gamma$

(c) $\beta^2 \geq \alpha\gamma$ (d) $\beta^2 \leq \alpha\gamma$

56. A function ϕ is defined as follows :

     $\phi(x) = -x$ for $x \leq 0$

            $= x$ for $x \geq 0$ Then $\phi(x)$ is:

(a) continuous for all x

(b) differentiable at $x = 0$

(c) continuous at $x = 0$ not differentiable at $x = 0$

(d) none of the above

57. Let $f(x) = \begin{cases} xe^{1/x} & , & x \neq 0 \\ 0 & , & x = 0 \end{cases}$ then $f(x)$ is:

(a) continuous for all x

(b) differentiable at $x = 0$

(c) continuous at $x = 0$ but not differentiable at $x = 0$

(d) none of the above

58. Let $f(x) = \begin{cases} 2+x & , & x \geq 0 \\ 2-x & , & x < 0 \end{cases}$ then $f(x)$ is :

(a) not differentiable at $x=0$

(b) differentiable at $x = 0$

(c) differentiable for all x

(d) none of the above

59. Let $f(x) = \begin{cases} 1+x & , & \text{if } x \leq 2 \\ 5-x & , & \text{if } x > 2 \end{cases}$ then $f(x)$ is

(a) differentiable for $x = 2$

(b) differentiable for all x

(c) continuous at $x = 2$ but not differentiable at $x = 0$

(d) all are true

60. Let $f(x) = \begin{cases} x^2 & \text{for, } x \leq 0 \\ 1 & \text{for } 0 < x \leq 1, \\ \dfrac{1}{x} & \text{for } x > 1 \end{cases}$ then $f(x)$ is :

(a) differentiable at $x=0$

(b) differentiable at $x=1$

(c) differentiable at $x=0$ and $x=1$ both

(d) not differentiable at $x=0$ and $x=1$

61. Let $f(x) = \begin{cases} x^2 \cos(e^{1/x}) & , & x \neq 0 \\ 0 & , & x = 0 \end{cases}$ then $f(x)$ is :

(a) not differentiable at $x = 0$

(b) not differentiable for all x

(c) not differentiable throughout R

(d) none of the above

62. Let $f(x) = \begin{cases} \dfrac{x-1}{2x^2 - 7x + 5} & \text{when } x \neq 1 \\ -\dfrac{1}{3} & \text{when } x = 1 \end{cases}$ then $f'(1)$ is

(a) $\dfrac{2}{9}$ (b) $-\dfrac{2}{9}$

(c) $\dfrac{3}{9}$ (d) $-\dfrac{3}{9}$

63. Let $f(x) = \begin{cases} x^2 \sin\left(x^{-\frac{4}{3}}\right) & , & x \neq 0 \\ 0 & , & x = 0 \end{cases}$ then $f'(1)$ is

(a) $f(x)$ is not differentiable at $x = 0$

(b) $f(x)$ is not differentiable for all x

(c) $f(x)$ is differentiable at $x = 0$

(d) none of the above

64. The function $f(x) = \log(1+x) - \dfrac{2x}{2+x}$:

(a) monotonically increasing when $x > 0$

(b) monotonically decreasing when $x > 0$

(c) monotonically increasing when $x < 0$

(d) none of the above

65. The function $f(x) = (x^4 + 6x^3 + 17x^2 + 32x + 32)e^{-x}$ is:

(a) monotonically increasing in $[-2, -1]$ and $[0, 1]$

(b) monotonically decreasing in $]-\infty, -2]$ $(-1,]$ and $(1, \infty[$

(c) both (a) and (b) are true

(d) none of the above

66. Let $f : (0, 1] \to R$ be defined by $f(x) = (x-1)^2 + 2$ $\forall$ $x \in [0, 1]$ then the equation of the tangent to the graph of this curve which parallel to the chord joining the points $(0, 3)$ and $(1, 2)$ of the curve is given by:

(a) $4x + 4y = 0$ (b) $4x + 4y = 10$

(c) $4x + 4y = 11$ (d) none of these

67. A function $f(x) = \begin{cases} x\tan^{-1}\dfrac{1}{x}, & x \neq 0 \\ 0 & , & x = 0 \end{cases}$ is

(a) differentiable at $x=0$

(b) differentiable for all x

(c) continuous but not differentiable at $x = 0$

(d) none of the above

Mathematical Analysis

Problem Set-3

1. If $f(x) = \dfrac{\sin[x]}{[x]}, [x] \neq 0,$

$= 0, [x] = 0$

where $[x]$ denotes the greatest integer less than or equal to x, then $\lim\limits_{x \to 0} f(x)$ equals :

(a) 1   (b) 0

(c) –1   (d) none of these

2. $\lim\limits_{x \to \infty} \dfrac{\log[x]}{x}$ where $[x]$ has the usual meaning is :

(a) 1   (b) –1

(c) 0   (d) none of these

3. The left hand limit of

$$f(x) = \left\{ \dfrac{|x|^3}{a} - \left[\dfrac{x}{a} \right]^3 \right\}, (a > 0)$$

where $[x]$ denotes the greatest integer less than or equal to x is

(a) a^2 (b) $a^2 - 1$

(c) $a^2 - 3$ (d) none of these

4. $\lim\limits_{x \to 3} \dfrac{(x-3)}{|x-3|} =$

(a) 0   (b) 1

(c) –1   (d) not exist

5. $f(x) = \begin{cases} \int_0^x \{1 + |1-t|\} dt &, x > 2 \\ 5x - 7 &, x \leq 2 \end{cases}$ then:

(a) $f(x)$ is not continuous at $x = 2$

(b) $f(x)$ is differentiable every where

(c) right hand limit at $x = 2$ does not exist

(d) $f(x)$ is continuous but not differentiable at $x = 2$

6. $\lim\limits_{x \to 1^+} \dfrac{\int_1^x |t-1| dt}{\sin(x-1)} =$

(a) 0   (b) 1

(c) –1   (d) none of these

7. $\lim\limits_{x \to a} \dfrac{x}{x-a} \int_a^x f(x) dx =$

(a) $f(a)$ (b) $a\, f(a)$

(c) 0   (d) none of these

8. If $f(x) = \begin{cases} x & : x < 0 \\ 1 & : x = 0 \\ x^2 & : x > 0 \end{cases}$, then $\lim\limits_{x \to 0} f(x) =$

(a) 0   (b) 1

(c) 2   (d) not exist

9. If $[x]$ denotes the greatest integers less than or equal to x, then the value of $\lim\limits_{x \to 0} (1 - x + [x-1] + [1-x])$ is:

(a) 0   (b) 1

(c) –1   (d) none of these

10. Let $f(x) = \begin{cases} x^2 &, x \in Z \\ \dfrac{k(x^2-4)}{2-x} &, x \notin Z \end{cases}$

Then $\lim\limits_{x \to 2} f(x)$:

(a) exists only when $k = 1$

(b) exists for every real k

(c) exists for every real k except $k = 1$

(d) does not exist

11. Let $f(x) = \begin{cases} x &, x \leq 1 \\ x^2 + px + q &, x > 1 \end{cases}$ and limit $f(x)$ exists finitely for all $x \in R$, then

(a) $p = 1, q = -1$ (b) $p = 1, q = 1$

(c) $p = -1, q \in R$ (d) $p = -1, q = 1$

12. If $f(x)$ is an odd function of x and $\lim\limits_{x \to 0} f(x)$ exists then the limit must be zero. True or False.

13. $\lim\limits_{x \to 0} \dfrac{e^{1/x} - 1}{e^{1/x} + 1}$ is equal to

(a) 1   (b) –1

(c) 0   (d) does not exist

14. $\lim\limits_{x \to 0} (1 + \sin x)^{1/x^2}$ is equal to

(a) 0   (b) ∞

(c) $e^{1/2}$ (d) does not exist

15. $\lim\limits_{x \to 0} \dfrac{\sin[\cos x]}{1 + [\cos x]} =$

([.] denotes the greatest integer function)

(a) is equal to 1   (b) is equal to zero

(c) does not exist   (d) none of these

16. The number of points at which the function

$f(x) = \dfrac{1}{\log|x|}$ is discontinuous is

(a) 1   (b) 2

(c) 3   (d) 4

17. The function $\dfrac{\log(1+ax) - \log(1-bx)}{x}$ is not defined at $x = 0$. The value which should be assigned to f at $x = 0$ so that it is continuous at $x = 0$ is

(a) $a - b$
(b) $1 + b$

(c) $\log a + \log b$
(d) none of these

18. The value of $f(0)$ so that the function

$$f(x) = \frac{\log(1 + x^2 \tan x)}{\sin x^3} (x \neq 0) \text{ is continuous}$$

at $x = 0$ is :

(a) 1
(b) 2

(c) 3
(d) none of these

19. If the function

$$f(x) = \frac{x^2 - (A+2)x + A}{x - 2}, \text{ for } x \neq 2$$

$$= 2 \qquad\qquad , \text{ for } x = 2$$

is continuous at $x = 2$, then A is :

(a) 0
(b) 1

(c) –1
(d) none of these

20. If $f(x) = \dfrac{\cos^2 \pi x}{e^{2x} - 2e^x}, x \neq \dfrac{1}{2}$, the value of $f\left(\dfrac{1}{2}\right)$

so that $f(x)$ is continuous at $x = \dfrac{1}{2}$ is:

(a) $\dfrac{\pi}{2e^2}$
(b) $\dfrac{\pi}{2e}$

(c) $\dfrac{\pi^2}{2e^2}$
(d) $\dfrac{\pi^2}{2e}$

21. The value of b for which the function

$$f(x) = \begin{cases} 5x - 4 & , \ 0 < x \leq 1 \\ 4x^2 + 3bx & , \ 1 < x < 2 \end{cases}$$

is continuous at every point of its domain, is:

(a) –1
(b) 0

(c) 1
(d) 13/3

22. If the function

$$f(x) = (\cos x)^{1/x} \quad , x \neq 0$$
$$= k \qquad\qquad , x = 0$$

is continuous at $x = 0$, then the value of k is

(a) 1
(b) –1

(c) 0
(d) e

23. Let

$$f(x) = \frac{x^3 + x^2 - 16x + 20}{(x-2)^2}, \text{ if } \ x \neq 2$$

$$= k, \qquad\qquad\qquad \text{ if } \ x = 2$$

If $f(x)$ is continuous for all x, then $k = $

(a) 3
(b) 5

(c) 7
(d) none of these

24. The value of $f(0)$, so that the function

$$f(x) = \frac{\sqrt{a^2 - ax + x^2} - \sqrt{a^2 + ax + x^2}}{\sqrt{(a+x)} - \sqrt{(a-x)}}$$

becomes continuous for all x, is given by:

(a) $a\sqrt{a}$
(b) $\sqrt{a}$

(c) $-\sqrt{a}$
(d) $-a\sqrt{a}$

25. The value of $f(0)$, so that the function

$$f(x) = \frac{(27 - 2x)^{1/3} - 3}{9 - 3(243 + 5x)^{1/5}} (x \neq 0) \text{ is continuous,}$$

is given by:

(a) 2/3
(b) 6

(c) 2
(d) 4

26. $f(x) = \begin{cases} \dfrac{\sqrt{(1+px)} - \sqrt{(1-px)}}{x} & , \ -1 \leq x < 0 \\ \dfrac{2x+1}{x-2} & , \ 0 \leq x \leq 1 \end{cases}$

is continuous in the interval $[-1, 1]$, then p is equal to :

(a) –1
(b) –1/2

(c) 1/2
(d) 1

27 $f(x) = (x-1)^{\frac{1}{2-x}}$ is not defined at $x = 2$. If $f(x)$ is continuous, then $f(2)$ is equal to :

(a) e
(b) e^{-1}

(c) e^{-2}
(d) 1

28. The function

$$f(x) = \begin{cases} x^2 / a & , \ 0 \leq x < 1 \\ a & , \ 1 \leq x < \sqrt{2} \\ (2b^2 - 4b) / x^2 & , \ \sqrt{2} \leq x < \infty \end{cases}$$

is continuous for $0 \leq x < \infty$, then the most suitable values of a and b are :

(a) $a = 1, b = -1$
(b) $a = -1, b = 1 + \sqrt{2}$
(c) $a = -1, b = 1$
(d) none of the above

29. If $f(x) = \dfrac{(e^x - 1)^4}{\sin\left(\dfrac{x^2}{\lambda^2}\right)\log\left(1 + \dfrac{x^2}{2}\right)}, x \neq 0$

and $f(0) = 8$ be a continuous function then $\lambda = $

(a) 2
(b) 1

(c) –1
(d) –2

30. Let $f(x) = \dfrac{x(1 + a\cos x) - b\sin x}{x^3}, x \neq 0$

$f(0) = 1$. If $f(x)$ is continuous at $x = 0$ then, a and b are given by:

(a) 5/2, 3/2

(b) –5, –3

(c) –5/2, –3/2

(d) none of these

31. $f(x) = \begin{cases} \dfrac{1 - \cos 4x}{x^2} & , \quad x < 0 \\ a & , \quad x = 0 \\ \dfrac{\sqrt{x}}{\sqrt{[16 + \sqrt{x}]} - 4} & , \quad x > 0 \end{cases}$

If the function is continuous at $x = 0$, then $a =$

(a) 4

(b) 6

(c) 8

(d) 10

32. The function

$f(x) = \begin{cases} x + a\sqrt{2}\sin x & 0 \leq x < \pi/4 \\ 2x\cot x + 6 & \pi/4 \leq x \leq \pi/2 \\ a\cos 2x - b\sin x & \pi/2 < x \leq \pi \end{cases}$

is continuous for $0 \leq x \leq \pi$ then a, b are :

(a) $\dfrac{\pi}{6}, \dfrac{\pi}{12}$

(b) $\dfrac{\pi}{3}, \dfrac{\pi}{6}$

(c) $\dfrac{\pi}{6}, -\dfrac{\pi}{12}$

(d) none of these

33. In order that the function $f(x) = (x+1)^{\cot x}$ is continuous at $x = 0$, $f(0)$ must be defined as:

(a) $f(0) = 0$

(b) $f(0) = e$

(c) $f(0) = 1/e$

(d) none of these

34. Function $f(x) = (\sin 2x)^{\tan^2 2x}$ is not defined at $x = \dfrac{\pi}{4}$. If $f(x)$ is continuous at $x = \dfrac{\pi}{4}$ then $f\left(\dfrac{\pi}{4}\right)$ is equal to:

(a) 1

(b) 2

(c) $\sqrt{e}$

(d) none of these

35. Let $f(x) = \dfrac{\tan(\pi/4 - x)}{\cot 2x}\left(x \neq \dfrac{\pi}{4}\right)$. The value which should be assigned to f at $x = \pi/4$, so that it is continuous everywhere, is:

(a) 1/2

(b) 1

(c) 2

(d) none of these

36. If $f(x) = \dfrac{2 - (256 - 7x)^{1/8}}{(5x + 32)^{1/5} - 2}(x \neq 2)$, then for f

to be continuous every where, $f(0)$ is equal to:

(a) –1

(b) 1

(c) 2^6

(d) 7/64

37. The value of λ that makes the function

$f(x) = \begin{cases} (\cos x)^{1/\sin x} & , \quad x \neq 0 \\ \lambda & , \quad x = 0 \end{cases}$

continuous at $x = 0$ is:

(a) 0

(b) 1

(c) 1/2

(d) none of these

38. Let $f''(x)$ be continuous at $x = 0$ and $f''(0) = 4$. Then value of

$\displaystyle\lim_{x \to 0} \dfrac{2f(x) - 3f(2x) + f(4x)}{x^2}$ is:

(a) 11

(b) 2

(c) 12

(d) none of these

39. The value of $f(0)$ so that the function

$f(x) = \dfrac{2x - \sin^{-1} x}{2x + \tan^{-1} x}$

is continuous at each point on its domain is

(a) 2

(b) 1/3

(c) 2/3

(d) –1/3

40. If $f(x) = \begin{cases} \dfrac{36^x - 9^x - 4^x + 1}{\sqrt{2} - \sqrt{1 + \cos x}} & , \quad x \neq 0 \\ k & , \quad x = 0 \end{cases}$

is continuous at $x = 0$, then k equals

(a) $16 \log 2 \log 3$

(b) $16\sqrt{2}\log 6$

(c) $16\sqrt{2}\log 2 \log 3$

(d) none of the above

41. Let $f(x) = \begin{cases} \dfrac{x - 4}{|x - 4|} + a & , \quad x < 4 \\ a + b & , \quad x = 4 \\ \dfrac{x - 4}{|x - 40|} + b & , \quad x > 4 \end{cases}$

Then $f(x)$ is continuous at $x = 4$, when

(a) $a = b = 0$

(b) $a = b = 1$

(c) $a = -1, b = 1$

(d) $a = 1, b = -1$

42. If $f(x) = \begin{cases} \dfrac{\sin(a + 1)x + \sin x}{x} & , \quad x < 0 \\ c & , \quad x = 0 \\ \dfrac{\sqrt{x + bx^2} - \sqrt{x}}{bx\sqrt{x}} & , \quad x > 0 \end{cases}$

is continuous at $x=0$, then

(a) $a = -\dfrac{3}{2}, b = 0, c = \dfrac{1}{2}$

(b) $a = -\dfrac{3}{2}, b = 1, c = -\dfrac{1}{2}$

(c) $a = -\dfrac{3}{2}, b \in R, c = \dfrac{1}{2}$

(d) none of the above

43. Let $f(x) = \begin{cases} x^p \sin \dfrac{1}{x} , & x \geq 0 \\ 0 , & x = 0 \end{cases}$

Then $f(x)$ is continuous but not differentiable at $x=0$ if :

(a) $p \in (0, 1]$ (b) $p \in [1, \infty[$

(c) $p \in]-\infty, 0[$ (d) $p=0$

44. The value of k which makes

$f(x) = \begin{cases} \sin(1/x) , & x \neq 0 \\ k , & x = 0 \end{cases}$

continuous at $x=0$ is

(a) 8   (b) 1

(c) –1   (d) none of these

45. $f(x) = \begin{cases} -1 & : & x < -1 \\ -x & : & -1 \leq x \leq 1 \\ 1 & : & x > 1 \end{cases}$ is continuous

(a) at $x=1$ but not at $x = -1$

(b) at $x = -1$ but not at $x = 1$

(c) at both $x=1$ and $x = -1$

(d) at none of $x=1$ and -1

46. If $f(x) = \int_{-1}^{x} |t| \, dt, x \geq -1$ then :

(a) f and f' are continuous for $x+1>0$

(b) f is continuous but f' is not continuous for $x+1 > 0$

(c) f and f' are not continuous at $x = 0$

(d) f is continuous at $x = 0$ but f' is not so

47. Given the function $f(x) = \dfrac{1}{(1-x)}$. The point of discontinuity of the composite function, $y = f(f(f(x)))$ are at $x =$

(a) 0   (b) 1

(c) 2   (d) –1

48. If $f(x)$ is defined by

$f(x) = \begin{cases} \dfrac{|x^2 - x|}{x^2 - x} , & x \neq 0, 1 \\ 1 , & x = 0 \\ -1 , & x = 1 \end{cases}$

then $f(x)$ is continuous for all:

(a) x

(b) x except at $x= 0$

(c) x except at $x= 1$

(d) x except at $x =0$ and $x=1$

49. Let $f(x) = |x| + |x-1|$; then:

(a) $f(x)$ is continuous both at $x = 0$ and 1

(b) $f(x)$ is continuous at $x=0$ but not at $x=1$

(c) $f(x)$ is continuous at $x=1$ but not at $x=0$

(d) none of the above

50. The function $f(x) = |x| + |x-1|$ is :

(a) continuous at $x=1$, but not differentiable

(b) both continuous and differentiable at $x=1$

(c) not continuous at $x =1$

(d) none of the above

51. Let $f(x) = \begin{cases} \dfrac{x^4 - 5x^2 + 4}{|(x-1)(x-2)|} , & x \neq 1, 2 \\ 6 , & x = 1 \\ 12 , & x = 2 \end{cases}$

Then $f(x)$ is continuous on the set:

(a) R   (b) R–{1}

(c) R–{2}   (d) R–{1, 2}

52. Let $f(x) = x - |x - x^2|$, $x \in [-1, 1]$. Then the number of points at which $f(x)$ is discontinuous is:

(a) 0   (b) 1

(c) 2   (d) none of these

53. The function $f(x) = [x]^2 - [x^2]$ (where [y] is greatest integer less than or equal to y), discontinuous at:

(a) all integers

(b) all integers except 0 and 1

(c) all integers except 0

(d) all integers except 1

54. On the interval $[-2, 2]$ the function

$f(x) = \begin{cases} (x+1)e^{-\left\{\dfrac{1}{|x|} + \dfrac{1}{x}\right\}} ; & x \neq 0 \\ 0 , & x = 0 \end{cases}$

(a) is continuous for all $x \in Z$

(b) is continuous for all $x \in Z-[0]$

(c) assumes all intermediate values from $f(-2)$ to $f(2)$

(d) has a maximum value equal to $\dfrac{3}{e}$

55. Let $f(x) = \begin{cases} \int_0^x \{5+|1-t|\,dt\} & \text{if } x > 2 \\ 5x+1 & \text{if } x \le 2 \end{cases}$

then

(a) $f(x)$ is not continuous at $x = 2$

(b) $f(x)$ is continuous but not differentiable at $x=2$

(c) $f(x)$ is differentiable everywhere

(d) the right derivative of $f(x)$ at $x=2$ does not exist

56. The function $f(x) = [x] \cos\{(2x - 1)/2\}\pi$, [ ] denotes the greatest integer function, is discontinuous at:

(a) all x

(b) all integer points

(c) no x

(d) x which is not an integer

57. The number of points where $f(x) = [\sin x + \cos x]$ (where [•] denotes the greatest integer function) $x \in (0, 2\pi)$ is discontinuous is:

(a) 3   (b) 4

(c) 5   (d) 6

58. Let $f : R \to R$ be any function. Define $g : R \to R$ by $g(x) = |f(x)|$ for all x. Then g is:

(a) onto if f is onto

(b) one-one if f is one-one

(c) continuous if f is continuous

(d) differentiable if f is differentiable

59. The function f defined as $f(x) = (\sin x^2)/x$ for $x \ne 0$ and $f(0) = 0$ is:

(a) continuous and derivable at $x = 0$

(b) neither continuous nor derivable at $x = 0$

(c) continuous but not derivable at $x = 0$

(d) none of the above

60. If $f(x) = \begin{cases} 1, & x < 0 \\ 1 + \sin x & \text{for } 0 \le x < \pi/2 \end{cases}$

then at $x = 0$, the derivative $f'(x)$ is

(a) 1   (b) 0

(c) infinite   (d) does not exist

61. For a real number y, let $[y]$ denote the greatest integer less than or equal to y.

Then $f(x) = \dfrac{\tan(\pi[x - \pi])}{1 + [x]^2}$ is:

(a) discontinuous at some x

(b) continuous at all x, but the derivative $f'(x)$ does not exist for some x

(c) $f'(x)$ exists for all x but second derivative $f''(x)$ does not exist

(d) $f'(x)$ exists for all x

62. If $f(x) = x[\sqrt{x} - \sqrt{(x+1)}]$, then

(a) $f(x)$ is continuous but not differentiable at $x = 0$

(b) $f(x)$ is continuous and differentiable at $x=0$

(c) $f(x)$ is not differentiable at $x=0$

(d) none of the above

63. The function

$f(x) = \begin{cases} ||x - 3|, & x \ge 1 \\ x^2/4 - 3x/2 + 13/4, & x < 1 \end{cases}$ is

(a) continuous at $x = 1$

(b) continuous at $x = 3$

(c) differentiable at $x = 1$

(d) differentiable at $x = 3$

64. The value of the derivative of $|x - 1| + |x - 3|$ at $x = 2$ is:

(a) – 2   (b) 0

(c) 2   (d) not defined

65. Let [ ] denote the greatest integer function and $f(x) = [\tan^2 x]$. Then:

(a) $\lim\limits_{x \to 0} f(x)$ does not exist

(b) $f(x)$ is continuous at $x = 0$

(c) $f(x)$ is not differentiable at $x = 0$

(d) $f'(0) = 1$

66. If $f(x) = \begin{cases} \dfrac{|x + 2|}{\tan^{-1}(x + 2)} & ; \ x \ne -2 \\ 2 & ; \ x = -2 \end{cases}$

then:

(a) continuous at $x = -2$

(b) not continuous at $x = -2$

(c) differentiable at $x = -2$

(d) continuous but not differentiable at $x = -2$.

67. If $f(x) = \begin{cases} 3x^2 + 12x - 1 & -1 \le x \le 2 \\ 37 - x & 2 < x \le 3 \end{cases}$

then:

(a) $f(x)$ is increasing on $[-1, 2]$

(b) $f(x)$ is continuous in $[-1, 3]$

(c) $f'(2)$ does not exist

(d) $f(x)$ has the maximum value at $x = 2$

68. The set of all points, where the function

$f(x) = \dfrac{x}{(1 + |x|)}$ is differentiable, is

(a) $(-\infty, \infty)$ (b) $(0, \infty]$

(c) $(-\infty, 0) \cup (0, \infty)$ (d) none of these

69. The set of points where the function $f(x) = x\,|x|$ is differentiable is:

(a) $(-\infty, \infty)$ (b) $(-\infty, 0) \cup (0, \infty)$

(c) $(0, \infty)$ (d) $[0, \infty]$

70. The set of all points where the function

$$f(x) = \begin{cases} 0 & , \quad x = 0 \\ \dfrac{x}{1+e^{1/x}} & , \quad x \neq 0 \end{cases}$$

is differentiable is

(a) $(0, \infty)$ (b) $(-\infty, \infty) - \{0\}$

(c) $(-\infty, 0)$ (d) $(-\infty, \infty)$

71. The set of all points of differentiability of the function

$$f(x) = \begin{cases} x^2 \sin(1/x) & , \quad x \neq 0 \\ 0 & , \quad x = 0 \end{cases}$$ is :

(a) $(-\infty, 0)$ (b) $(-\infty, \infty) - \{0\}$

(c) $[0, \infty)$ (d) $(-\infty, \infty]$

72. At the point $x = 1$, the function

$$f(x) = \begin{cases} x^3 - 1 & ; \quad 1 < x < \infty \\ x - 1 & ; \quad -\infty < x \leq 1 \end{cases}$$

(a) continuous and differentiable

(b) continuous and not differentiable

(c) discontinuous and differentiable

(d) none of the above

73. $g(x) = x f(x)$, where

$$f(x) = \begin{cases} x \sin(1/x) & , \quad x \neq 0 \\ 0 & , \quad x = 0 \end{cases}$$

At $x = 0$:

(a) g is differentiable but g' is not continuous

(b) g is differentiable while f is not differentiable

(c) both f and g are differentiable

(d) g is differentiable but g' is continuous

74. Let $f(x) = \begin{cases} 0 & : \quad x < 0 \\ x^2 & : \quad x \geq 0 \end{cases}$ then for all x:

(a) f' is differentiable

(b) f is differentiable

(c) f' is continuous

(d) f is continuous

75. Let $[x]$ denote the greatest integer less than or equal to x. If $f(x) = [x \sin \pi x]$, then $f(x)$ is:

(a) continuous at $x = 0$,

(b) continuous in $(-1, 0)$

(c) differentiable at $x = 1$

(d) differentiable in $(-1, 1)$

(e) none of the above

76. The function $f(x) = 1 + |\sin x|$ is:

(a) continuous nowhere

(b) continuous everywhere

(c) differentiable nowhere

(d) not differentiable at $x = 0$

(e) not differentiable at an infinite number of points

77. If $f(x) = \begin{cases} \dfrac{x \log \cos x}{\log(1+x^2)} & , \quad x \neq 0 \\ 0 & , \quad x = 0 \end{cases}$ then:

(a) $f(x)$ is not continuous at $x = 0$

(b) $f(x)$ is continuous at $x = 0$

(c) $f(x)$ is continuous at $x = 0$ but not differentiable at $x = 0$

(d) $f(x)$ is differentiable at $x = 0$

78. If $x + 4|y| = 6y$, then y as a function of x is :

(a) defined for all real x

(b) continuous at $x = 0$

(c) derivable at $x = 0$

(d) $\dfrac{dy}{dx} = \dfrac{1}{2}$ for $x > 0$

79. If $f'(x_0)$ exists, then $\lim\limits_{h \to 0} \dfrac{f(x_0 + h) - f(x_0 - h)}{2h}$ is equal to :

(a) $\dfrac{1}{2} f'(x_0)$

(b) $f'(x_0)$

(c) $2f'(x_0)$

(d) none of the above

80. The function $f(x) = \begin{cases} |\,2x - 3\,|\,[x] & , \quad x \geq 1 \\ \sin\left(\dfrac{\pi x}{2}\right) & , \quad x < 1 \end{cases}$

(a) is continuous at $x = 2$

(b) is differentiable at $x = 1$

(c) is continuous but not differentiable at $x = 1$

(d) none of the above

81. The function $f(x)$ is defined as under :

$$y = 3^x, \qquad -1 \leq x \leq 1$$
$$= 4 - x, \qquad 1 < x < 4$$

The above function is :

(a) continuous at $x = I$

(b) differentiable at $x = 1$

(c) continuous but not differentiable at $x = 1$

(d) none of the above

82. A function is defined as follows :

$$f(x) = \begin{cases} x^3 & ; \quad x^2 < 1 \\ x & ; \quad x^2 \geq 1 \end{cases}$$

The function is:

(a) continuous at $x = 1$

(b) differentiable at $x = 1$

(c) continuous but not differentiable at $x = 1$

(d) none of the above

83. The left-hand derivative of $f(x) = [x] \sin (\pi x)$ at $x = k$, k an interger, is :

(a) $(-1)^k(k-1)\pi$

(b) $(-1)^{k-1}(k-1)\pi$

(c) $(-1)^k k\pi$

(d) $(-1)^{k-1}k\pi$

84. If the derivative of the function

$$f(x) = \begin{cases} ax^2 + b & , \quad x < -1 \\ bx^2 + ax + 4 & , \quad x \geq -1 \end{cases}$$

is everywhere continuous, then :

(a) $a = 2, b = 3$ (b) $a = 3, b = 2$

(c) $a = -2, b = -3$ (d) $a = -3, b = -2$

85. If $f(x) = \begin{cases} ax^2 + b, b \neq 0, x \leq 1 \\ bx^2 + ax + c, x > 1 \end{cases}$

Then $f(x)$ is continuous and differentiable at $x = 1$ if:

(a) $c = 0, a = 2b$ (b) $a = b, c \in R$

(c) $a = b, c = 0$ (d) $a = b, c \neq 0$

86. Let $f(x) = a|x|^2 + b|x| + c$ where a, b, c are real constants. Then $f'(x)$ exists at $x = 0$ if:

(a) $a = 0$ (b) $b = 0$

(c) $c = 0$ (d) $a = b$

87. Let $f(x) = \min \{x, x^2\}$, for every real number of x. Then:

(a) f continuous for all x

(b) f is differentiable for all x

(c) $f'(x) = 1$ for all $x > 1$

(d) f is not differentiable at two values of x

88. Let $f : R \rightarrow R$ be a function defined by $f(x) = \max \{x, x^3\}$. The set of all points where $f(x)$ in not differentiable is :

(a) $\{-1, 1\}$ (b) $\{-1, 0\}$

(c) $\{0,1\}$ (d) $\{-1, 0, 1\}$

89. The derivative of $f(x) = |x|$ at $x = 0$ is

(a) 1   (b) 0

(c) -1 (d) does not exist

90. For a differentiable function f, the value of

$$\lim_{h \to 0} \frac{[f(x+h)]^2 - [f(x)]^2}{2h}$$ is equal to :

(a) $[f'(x)]^2$

(b) $f(x)f'(x)$

(c) $\frac{1}{2}[f'(x)]^2$

(d) $\frac{1}{2}[[f'(x)]^2 - [f(x)]^2]$

91. If for a continuous function f, $f(0) = f(1) = 0$, $f'(1) = 2$ and $y(x) = f(e^x)\ e^{f(x)}$ then $y'(0)$ is equal to :

(a) 1   (b) 2

(c) 0   (d) none of these

92. The function $f(x) = e^{-|x|}$ is:

(a) continuous everywhere but not differentiable at $x = 0$

(b) continuous and differentiable everywhere

(c) not continuous at $x = 0$

(d) none of the above

93. Let $f(x)$ be defined as

$$f(x) = \begin{cases} \sin 2x & , \quad 0 < x < \frac{\pi}{6} \\ px + q & , \quad \frac{\pi}{6} < x < 1 \end{cases}$$

If f and f' are continuous then (p, q) is equal to:

(a) $\left(\frac{1}{\sqrt{2}}, \frac{1}{\sqrt{2}} \right)$ (b) $\left(1, \frac{1}{\sqrt{2}} + \frac{\pi}{6} \right)$

(c) $\left(1, \frac{\sqrt{3}}{2} - \frac{\pi}{6} \right)$ (d) none of these

94. Let $f(x) = \begin{cases} -\dfrac{1}{|x|} & \text{for } |x| \geq 1 \\ ax^2 - b & \text{for } |x| < 1 \end{cases}$

If $f(x)$ is continuous and differentiable at any point, then:

(a) $a = \frac{1}{2}, b = -\frac{3}{2}$ (b) $a = \frac{1}{2}, b = \frac{3}{2}$

(c) $a = 1, b = -1$ (d) none of these

95. The derivative of $f(x) = |x|^3$ at $x=0$ is

(a) -1 (b) not defined

(c) 0   (d) 1/2

96. If $y = \left| \tan \left(\frac{\pi}{4} - x \right) \right|$, then $\frac{dy}{dx}$ at $x = \frac{\pi}{4}$ is :

(a) −1   (b) 1

(b) does not exist   (d) none of these

97. Which of the following functions is differentiable at $x = 0$?

(a) $\cos(|x|) + |x|$ (b) $\cos(|x|) - |x|$

(c) $\sin(|x|) - |x|$ (d) $\sin(|x|) + |x|$

98. $f(x) = ||x| - 1|$ is not differentiable at :

(a) 0   (b) $\pm 1, 0$

(c) 1   (d) ± 1

99. The number of points at which the function $f(x) = |x - 0.5| + |x - 1| + \tan x$ does not have a derivative in the interval $(0, 2)$ is

(a) 1   (b) 2

(c) 3   (d) 4

100. Consider $f(x) = \begin{cases} \dfrac{x^2}{|x|} & , \ x \geq 0 \\ 0 & , \ x = 0 \end{cases}$ then:

(a) $f(x)$ is discontinuous everywhere

(a) $f(x)$ is continuous everywhere

(c) $f'(x)$ exists in $(-1, 1)$

(d) $f'(x)$ exists in $(-2, 2)$

101. The function

$$f(x) = (x^2 - 1)|x^2 - 3x + 2| + \cos(|x|)$$

is not differentiale at:

(a) −1   (b) 0

(c) 1   (d) 2

102. Consider the following statements S and R:

S : Both $\sin x$ and $\cos x$ are decreasing functions in the interval $(\pi / 2, \pi)$.

R : If a differentiable function decreases in an interval (a, b), then its derivative also decreases in (a, b).

Which of the following is true ?

(a) Both S and R are wrong.

(b) Both S and R are correct, but R is not the correct explanation for S.

(c) S is correct and R is the correct explanation for S.

(d) S is correct and R is wrong.

103. If $f(x) = x^2 + \dfrac{x^2}{(1 + x^2)} + \dfrac{x^2}{(1 + x^2)^2} + \dots$

$$+ \dfrac{x^2}{(1 + x^2)^n} + \dots$$

then at $x = 0$:

(a) $f(x)$ has no limit

(b) $f(x)$ is discontinuous

(c) $f(x)$ is continuous but not differentiable

(d) $f(x)$ is differentiable

104. Let $f(x)$ be a function satisfying

$$f(x + y) = f(x) + f(y) \text{ and } f(x) = xg(x)$$

for all $x, y \in R$, where $g(x)$ is continuous. Then:

(a) $f'(x) = g'(x)$ (b) $f'(x) = g(x)$

(c) $f'(x) = g(0)$ (d) none of these

105. Let $f(x + y) = f(x) + f(y)$ and $f(x) = x^2 g(x)$ for all $x, y \in R$, where $g(x)$ is continuous function. Then $f'(x)$ is equal to:

(a) $g'(x)$ (b) $g(0)$

(a) $g(0) + g'(x)$ (d) 0

106. A differentiable function $f(x)$ satisfies the condition $f(x + y) = f(x) + f(y) + xy$ and $\lim\limits_{h \to 0} \dfrac{1}{h} f(h) = 3$ then f is :

(a) linear   (b) $f(x) = 3x + \dfrac{x^2}{2}$

(c) $f(x) = 3x + x^2$ (d) none of these

107. Let $f(x + y) = f(x)f(y)$ for all x and y. Suppose that $f(3) = 3$ and $f'(0) = 11$ then $f'(3)$ is given by:

(a) 22   (b) 44

(c) 28   (d) 33

108. Let $f(x + y) = f(x)f(y)$ and $f(x) = 1 + (\sin 2x)g(x)$ where $g(x)$ is continuous. Then $f'(x)$ is equal to:

(a) $f(x)g(0)$ (b) $2f(x)g(0)$

(c) $2g(0)$ (d) none of these

109. Suppose the function f satisfies the conditions :

(i) $f(x + y) = f(x)f(y)$ for all x and y

(ii) $f(x) = 1 + xg(x)$ where $\lim\limits_{x \to 0} g(x) = 1$

Then $f'(x) = $

(a) $f(x)$ (b) $f(x)g(0)$

(c) $2f(x)g(0)$ (d) none of these

110. A function $f : R \to R$ satisfies the equation $f(x + y) = f(x)f(y)$ for all values of x and y and for any $x \in R$, $f(x) \neq 0$. Suppose the function is differentiable at $x = 0$ and $f'(0) = 2$, then for all $x \in R$, $f(x) = $

(a) e^x (b) e^{2x}

(c) e^{-x} (d) none of these

111. If f is twice differentiable function such that $f''(x) = -f(x)$, and $f'(x) = g(x)$,

$h(x) = [f(x)]^2 + [g(x)]^2$ and $h(5) = 11$, then $h(10) =$

(a) 22   (b) 11

(c) 15   (d) none of these

112. If $f'(x) = g(x)$ and $g'(x) = -f(x)$ for all x and $f(2) = 4 = f'(2)$ then $f^2(16) + g^2(16)$ is :

(a) 16   (b) 32

(c) 64   (d) none of these

113. Let $f(x + y) = f(x) f(y)$ for all x and y. Suppose $f(5) = 2$ and $f'(0) = 3$, then $f'(5) =$

(a) 0   (b) 2

(c) 6   (d) none of these

114. Let f be a continuous function on $[1, 3]$ which takes rational values for all x. If $f(2) = 10$ then $f(2.5)$ is equal to:

(a) 25   (b) 20

(c) $\dfrac{f(1) + f(3)}{2}$ (d) 10

115. Let f be differentiable $\forall\ x$. If $f(1) = -2$ and $f'(x) \geq 2\ \forall\ x \in [1, 6]$, then :

(a) $f(6) < 5$ (b) $f(6) = 5$

(c) $f(6) \geq 8$ (d) $f(6) < 8$

116. If f is a real valued differentiable function satisfying

$|f(x)| - f(y)| \leq (x - y)^2$, $x, y \in$ R and $f(0) = 0$, then $f(1)$ equals :

(a) 2   (b) 1

(c) -1 (d) 0

117. Suppose $f(x)$ is differentiable at $x = 1$ and $\lim\limits_{h \to 0} \dfrac{1}{h} f(1 + h) = 5$ then $f'(1)$ equals :

(a) 5   (b) 6

(c) 3   (d) 4

Answers

FILL IN THE BLANKS

1. Unique   2. $l \neq 0$ 3. quotient   4. $f(a)$
5. e 6. ma^{m-1} 7. ε 8. continuous
9. $\log_e a$ 10. removable discontinuity   11. First kind
12. jump   13. continuous   14. not necessarily   15. continuous
16. not   17. bounded   18. 0   19. uniformly
20. R$-$[0]   21. 1/2   22. continuous
23. not necessarily differentiable   24. differentiable
25. Lagrange's mean value theorem   26. increasing function   27. increasing
28. differentiable at origin   29. singleton   30. 0   31. 0

TRUE OR FALSE

1. T   2. F   3. F   4. T   5. T   6. T   7. T   8. T   9. F
10. T   11. T   12. F   13. T   14. T   15. F   16. F   17. F   18. T
19. T   20. T   21. T   22. F   23. T   24. F   25. F   26. T

MULTIPLE CHOICE QUESTIONS

Problem Set-1

1. (c)   2. (b)   3. (c)   4. (d)   5. (a)   6. (b)   7. (c)   8. (a)   9. (d)
10. (d)   11. (c)   12. (c)   13. (c)   14. (d)   15. (c)   16. (a)   17. (a)   18. (a)
19. (a)   20. (b)   21. (a)   22. (c)   23. (b)   24. (d)   25. (d)   26. (a)   27. (a)
28. (a)   29. (b)   30. (c)   31. (b)   32. (a)   33. (c)   34. (d)   35. (c)   36. (b)
37. (c)   38. (a)   39. (b)

Problem Set-2

1. (a)   2. (c)   3. (d)   4. (c)   5. (d)   6. (c)   7. (d)   8. (c)   9. (c)
10. (c)   11. (c)   12. (c)   13. (b)   14. (d)   15. (a)   16. (b)   17. (c)   18. (c)
19. (a)   20. (b)   21. (c)   22. (c)   23. (c)   24. (b)   25. (c)   26. (c)   27. (d)
28. (b)   29. (a)   30. (a)   31. (c)   32. (a)   33. (d)   34. (a)   35. (a)   28. (b)
37. (b)   38. (c)   39. (d)   40. (c)   41. (d)   42. (c)   43. (b)   44. (a)   45. (a)
46. (b)   47. (b)   48. (b)   49. (b)   50. (c)   51. (d)   52. (a)   53. (d)   54. (a)
55. (d)   56. (c)   57. (c)   58. (b)   59. (c)   60. (d)   61. (d)   62. (c)   63. (c)
64. (a)   65. (c)   66. (c)   67. (c)

Problem Set-3

1. (d)	2. (c)	3. (a)	4. (d)	5. (d)	6. (a)	7. (b)	8. (a)	9. (c)
10. (b)	11. (d)	12. True	13. (d)	14. (d)	15. (b)	16. (c)	17. (d)	18. (a)
19. (a)	20. (d)	21. (a)	22. (a)	23. (c)	24. (c)	25. (c)	26. (b)	27. (b)
28. (c)	29. (a, d)	30. (c)	31. (c)	32. (c)	33. (b)	34. (c)	35. (a)	36. (d)
37. (b)	38. (c)	39. (b)	40. (c)	41. (d)	42. (c)	43. (a)	44. (d)	45. (d)
46. (a)	47. (a, b)	48. (d)	49. (a)	50. (a)	51. (d)	52. (a)	53. (d)	54. (b,c,d)
55. (b)	56. (b)	57. (c)	58. (c)	59. (a)	60. (d)	61. (d)	62. (b)	63. (a,b,c)
64. (b)	65. (b)	66. (b)	67. (a,b,c)	68. (a)	69. (a)	70. (b)	71. (b)	72. (b)
73. (a,b)	74. (b,c,d)	75. (a,b,d)	76. (b,d,e)	77. (b,d)	78. (a,b,d)	79. (b)	80. (c)	81. (a,c)
82. (a,c)	83. (a)	84. (a)	85. (a)	86. (b)	87. (a,c,d)	88. (d)	89. (d)	90. (b)
91. (b)	92. (a)	93. (c)	94. (b)	95. (c)	96. (c)	97. (c)	98. (b)	99. (c)
100. (b)	101. (d)	102. (d)	103. (b)	104. (c)	105. (d)	106. (b)	107. (d)	108. (b)
109. (a)	110. (b)	111. (b)	112. (b)	113. (c)	114. (d)	115. (c)	116. (c)	117. (a)

HINTS TO SELECTED PROBLEMS

Problem Set-3

1. By def. of $[x]$,

we have $[x] = -1$ when $-1 \le x < 0$

and $\quad [x] = 0$ when $0 \le x < 1$

Hence by def. of f,

$$f(x) = \frac{\sin(-1)}{-1} = \sin 1 \text{ when}$$

$$-1 \le x < 0 \qquad \ldots (1)$$

and $f(x) = 0$ when $0 \le x < 1 \qquad \ldots (2)$

$\therefore \qquad f(0-0) = \lim_{h \to 0} \sin 1 = \sin 1$

and $\quad f(0+0) = \lim_{h \to 0} 0 = 0$

Since $f(0-0) \ne f(0+0)$, the limit of $f(x)$ at $x=0$ does not exist.

2. Let $x = n + k$ where n is integer and $0 < k < 1$

$\therefore \ [x] = n$

$\therefore \quad \lim_{n \to \infty} \dfrac{\log n}{n+k} \qquad \left(\dfrac{\infty}{\infty}\right)$

$= \lim_{n \to \infty} \dfrac{1/n}{1} = 0 \qquad$ (by L' Hospital's Rule)

3. For left hand limit $x < a$ i.e. $x = a - h$, where $h \to 0$ and $a > 0$

$\therefore \ x$ is + ive, $|x| = x$. Also $\dfrac{x}{a} < 1$ but it is +ive,

$\dfrac{x}{a}$ lies between 0 and 1 so that $\left[\dfrac{x}{a}\right] = 0$

$\therefore \lim f(x) = \dfrac{a^3}{a} - 0 = a^2$

4. R.H.L. $= \lim_{h \to 0} \dfrac{3+h-3}{|3+h-3|} = \dfrac{h}{|h|} = \dfrac{h}{h} = 1$

L.H.L. $= \lim_{h \to 0} \dfrac{3-h-3}{|3-h-3|}$

$= \dfrac{-h}{|-h|} = -\dfrac{h}{h} = -1$

Since R.H.L $\ne$ L.H.L.

$\therefore$ Limit does not exist.

5. Let us redefine the function

$$f(x) = \int_0^1 \{1+(1-t)\}dt + \int_1^x \{1-(1-t)\}dt$$

$$= \left[2t - \frac{t^2}{2}\right]_0^1 + \left[\frac{t^2}{2}\right]_1^x = \left(2 - \frac{1}{2}\right) + \left(\frac{x^2}{2} - \frac{1}{2}\right)$$

$\therefore \ f(x) = 1 + \dfrac{x^2}{2} \quad \therefore \ f(x) = \begin{cases} 5x - 7 & ;x \le 2 \\ 1 + \dfrac{x^2}{2} & ;x > 2 \end{cases}$

$L = R = V = 3 \qquad\qquad$ at $x=2$ $\therefore$ continuous

$L' = \lim_{h \to 0} \dfrac{f(2-h) - f(2)}{-h}$

$= \lim_{h \to 0} \dfrac{5(2-h) - 7 - 3}{-h} = 5$

$R' = \lim_{h \to 0} \dfrac{f(2+h) - f(2)}{h}$

$= \lim_{h \to 0} \dfrac{1 + \dfrac{1}{2}(2+h)^2 - 3}{h}$

$= \lim_{h \to 0} \dfrac{2h + \dfrac{1}{2}h^2}{h} = 2$

Since $L' \neq R'$ so the function is not differentiable at $x=2$. Hence at $x = 2$ function is continuous but not differentiable.

6. When $x \to 1$ the given limit is of the form $\dfrac{0}{0}$

$$\lim_{x \to 1+} \frac{\dfrac{d}{dx}\int_1^x |t-1|\,dt}{\dfrac{d}{dx}\sin(x-1)} = \lim_{x \to 1+} \frac{|x-1|.1-0}{\cos(x-1)}$$

Put $x = 1+h$. As $x \to 1^+, h \to 0$

$$= \lim_{h \to 0} \frac{|h|}{\cos h} = 0$$

7. The given limit is of the form $\dfrac{0}{0}$

$$\therefore \quad \lim_{x \to a} \frac{\dfrac{d}{dx}\left[x\int_a^x f(x)dx\right]}{\dfrac{d}{dx}(x-a)}$$

Differentiating numerator by product

$$= \lim_{x \to a} \frac{1.\int_a^x f(x)dx + x.\{f(x).1-0\}}{1}$$

$$= 1.0 + a f(a) = a f(a) \quad \because \int_a^a (fx) = 0$$

8. L.H.L $=0$ and R.H.L. $=0$

9. $\lim\limits_{h \to 0}[h] = 0,\ \lim\limits_{h \to 0}[-h] = -1$

Replacing x by $1+h$ for R.H.L. and by $1-h$ for L.H.L. where $h \to 0$ and and using the above results we have

R.H.L $= \lim\limits_{h \to 0}[-h+[h]+[-h]]$

$\quad = 0 + 0 - 1 = -1$

L.H.L $= \lim\limits_{h \to 0}[h+[-h]+[h]]$

$\quad = 0 - 1 + 0 = -1$

Since R.H.L. $=$ L.H.L. $= -1$

$\therefore$ limit exists and is equal to -1.

10. For, R.H.L. $x = 2 + h$,

For, L.H.L. $x = 2 - h$

When $x \to 2, h \to 0$

$\quad f(x) = -k(2+x),\ x \notin Z$

R.H.L. $=$ L.H.L $= -4k$.

This is true for all real k.

11. $L' = R'$ at $x=1$

$$\lim_{h \to 0} \frac{f(1+h)-f(1)}{h} = \lim_{h \to 0} \frac{f(1-h)-f(1)}{-h}$$

Left hand limt $= \lim\limits_{h \to 0} \frac{(1-h)-1}{-h} = 1$

Right hand limit $= \lim\limits_{h \to 0} \frac{h^2 +(2+p)h+p+q}{h}$

Above limit will be 1 if $p+q = 0$ and $2+p =1$. These two give $p=-1, q =1$.

12. Since $\lim\limits_{x \to 0} f(x)$ exists therefore

$$\lim_{h \to 0} f(0-h) = \lim_{h \to 0} f(0+h)\ i.e., R = L$$

or $\lim\limits_{h \to 0} -f(h) = \lim\limits_{h \to 0} f(h)$

or $2\lim\limits_{h \to 0} f(h) = 0 \quad \therefore \lim\limits_{x \to 0} f(x) = 0$

13. $2 < e < 3 \quad \therefore\ e^\infty = \infty, e^{-\infty} = \dfrac{1}{e^\infty} = 0$

R.H.L. $= \lim\limits_{h \to 0} \dfrac{e^{1/h}-1}{e^{1/h}+1} = \lim\limits_{h \to 0} \dfrac{1-1/e^{1/h}}{1+1/e^{1/h}} = \dfrac{1-0}{1+0}$

$\because\ h \to 0 \Rightarrow \dfrac{1}{h} \to \infty \Rightarrow e^{1/h} \to \infty \Rightarrow \dfrac{1}{e^{1/h}} \to 0$

L.H.L. $= \lim\limits_{h \to 0} \dfrac{e^{-1/h}-1}{e^{-1/h}+1} = \dfrac{0-1}{0+1} = -1$

$\therefore$ R.H.L $\neq$ L.H.L. so limit does not exist.

14. Form 1^∞

$$\therefore\ \lim_{x \to 0}\left[(1+\sin x)^{\frac{1}{\sin x}}\right]^{\frac{\sin x}{x^2}} = e^{\lim\limits_{x \to 0} \frac{\sin x}{x^2}}$$

L.H.L. of $\dfrac{\sin x}{x^2}$

$$= \lim_{x \to 0} \frac{\sin(-h)}{h^2} = -\frac{1}{h}.\frac{\sin h}{h} = -\infty$$

$\therefore$ L.H.L. $= e^{-\infty} = \dfrac{1}{e^\infty} = \dfrac{1}{\infty} = 0$

R.H.L. $= e^\infty = \infty$

$\because$ L.H.L. $\neq$ R.H.L. $\therefore$ Limit does not exist.

15. Both for R.H.L and L.H.L. we have to put $x = 0 \pm h$ and as $x \to 0, h \to 0$

$\lim\limits_{x \to 0}[\cos x] = \lim\limits_{h \to 0}[\cos(0 \pm h)] = \lim\limits_{h \to 0}[\cos h]$

$h \to 0 \Rightarrow \cos h \to 1$ or $\cos h < 1 \Rightarrow [\cos h] = 0$

$\therefore$ Lim is $\dfrac{\sin 0}{1+0} = 0$

16. $f(x)$ is not defined for $\log|x| = 0$ or $|x| = 1$ or $x = +1, -1$. Again by definition of $\log x$, $f(x)$ is not defined for $|x| = 0$ or $x = 0$. Thus we have in all three points of discontinuity.

17. For continuity, limit $=$ value

$$\lim_{x \to 0} f(x) = \left(\frac{0}{0}\right) = \lim_{x \to 0}\left(\frac{a}{1+ax} + \frac{b}{1-bx}\right) = a+b$$

Hence $f(0) = a+b$.

18. For continuity, limit = value = $f(0)$

When $x \to 0$, $\sin x^3 = x^3$ and $\tan x = x$

$$\lim_{x \to 0} \log \frac{(1 + x^2 \cdot x)}{x^3}$$

$$= \lim_{x \to 0} \frac{x^3 - \frac{1}{2}(x^3)^2 + \dots}{x^3} = 1$$

Hence $f(0) = 1$.

19. For continuity, lim = value = 2(given)

$$\lim_{x \to 2} f(x) = \frac{0}{0}$$

Apply *L' Hospital's* rule

$\therefore$　　$\lim = \lim_{x \to 2} \{2x - (A + 2)\} = 2 - A$

$\therefore$　$2 - A = 2 \Rightarrow A = 0$

20. Applying L' Hospital's rule twice, we have

$$\lim_{x \to 1/2} f(x) = \frac{\pi^2}{2e} = \text{value } f\left(\frac{1}{2}\right)$$

21. At $x = 1$, $L = R = V$

$\Rightarrow 1 = 4 + 3b = 1 \therefore b = -1$

22. limit = value = k

But $\lim \log y = \lim \dfrac{\log \cos(x)}{x} \left(\text{form } \dfrac{0}{0}\right)$

$$= \lim\left(-\sin x \cdot \frac{1}{\cos x}\right) = 0$$

$\therefore$　　　$y = e^0 = 1$

23. $R = L = V$, at $x = 2$

24. For continuity, limit = value

Rationalize the given function.

$$\lim_{x \to 0} \frac{-2ax[\sqrt{a + x} + \sqrt{a - x}]}{2x[\sqrt{(a^2 - ax + x^2)} + \sqrt{(a^2 + ax + x^2)}]}$$

$$= -a \frac{2\sqrt{a}}{2a} = -\sqrt{a}$$

25. For continuity, limit = value

$$\lim_{x \to 0} f(x) = \frac{0}{0} \text{ form}$$

Use L' Hospital's rule

$$= \lim_{x \to 1} \frac{1}{3(27 - 2x)^{2/3}} (-2)$$

26. $R = L = V$ for continuity at $x = 0$

$$R = \lim f(0 + h) = -\frac{1}{2}, f(0) = -\frac{1}{2},$$

$L = \lim f(0 - h) = p$, by rationalization or by expansion.

For continuity $R = L = V \therefore p = -\dfrac{1}{2}$

27. For continuity Limit = Value

$\therefore$　$\lim_{x \to 2} (x - 1)^{\frac{1}{2-x}}$　　　　　[Form 1^∞]

$$= \lim_{x \to 2}\left[\{1 - (2 - x)\}^{-\frac{1}{2-x}}\right]^{-1}$$

$$= e^{-1}$$

28. For continuity at $x = 1$.

$$R = V \Rightarrow \frac{1}{a} = a$$

$\therefore$　　$a = 1, -1$

For continuity at $x = \sqrt{2}$

$\therefore$　　$R = L = V \Rightarrow a = b^2 - 2b$

When $a = 1$, $b^2 - 2b - 1 = 0$

$\therefore$　　$a = 1, b = 1 \pm \sqrt{2}$

These values are not given

When $a = -1$, $b^2 - 2b + 1 = 0$

$\therefore$　　$a = -1, b = 1$.

31. $f(0 - 0) = \lim_{h \to 0} \dfrac{1 - \cos 4(0 - h)}{(0 - h)^2}$

$$= \lim_{h \to 0} \frac{1 - \cos 4h}{h^2}$$

$$= \lim_{h \to 0} \frac{2 \sin^2 2h}{h^2} = \lim_{h \to 0}\left(\frac{\sin 2h}{2h}\right)^2 .8 = 8$$

$$\left[\because \lim_{h \to 0} \frac{\sin 2h}{2h} = 1\right]$$

and $f(0 + 0) = \lim_{h \to 0} \dfrac{(0 + h)^{1/2}}{[16 + (0 + h)^{1/2}]^{1/2} - 4}$

$$= \lim_{h \to 0} h^{1/2} \frac{\sqrt{(16 + \sqrt{h})} + 4}{16 + \sqrt{(h)} - 16},$$

on rationalizing

$$= \lim_{h \to 0} \sqrt{[16 + \sqrt{(h)}]} + 4 = 4 + 4 = 8$$

Also　$f(0) = a$ (given)

Since $f(x)$ is continuous at $x = 0$, we must have

$\lim f(0 - 0) = \lim f(0 + 0) = f(0)$

or　　　　$8 = a$

Hence　　　$a = 8$

32. We apply the test of continuity at $x = \pi/2$ and $x = \pi/4$ to get the value of a and b.

At $x = \pi/4$, $L = R = V$

$\therefore \quad \dfrac{\pi}{2} + b = a + \dfrac{\pi}{4}$ or $a - b = \dfrac{\pi}{4}$ ... (1)

At $\qquad x = \pi/2$, $L = R = V$

$$f\left(\dfrac{\pi}{2}\right) = 2 \cdot \dfrac{\pi}{2} \cos \dfrac{\pi}{2} + b = b$$

$$f\left(\dfrac{\pi}{2} - 0\right) = \lim_{h \to 0}\left[2\left(\dfrac{\pi}{2} - h\right)\cot\left(\dfrac{\pi}{2} - h\right) + b\right]$$
$$= b$$

$$f\left(\dfrac{\pi}{2} - 0\right) = \lim_{h \to 0}\left[a\cos 2\left(\dfrac{\pi}{2} + h\right) - b\sin\left(\dfrac{\pi}{2} + h\right)\right]$$
$$= -a - b.$$

$\therefore \qquad b = -a - b$ or $a + 2b = 0$ ... (2)

Solving (1) and (2), we get

$$a = \dfrac{\pi}{6}, b = -\dfrac{\pi}{12}$$

33. For continuity, limit = value

Let $A = \lim_{x \to 0}(x+1)^{\cot x}$

$\therefore \quad \log A = \lim_{x \to 0}\cot x \log(1 + x)$

$$= \lim_{x \to 0}\dfrac{\log(1+x)}{\tan x}$$

$$= \lim_{x \to 0}\dfrac{1/(1+x)}{\sec^2 x} = 1 \text{ by L' Hospital's Rule}$$

Hence $A = e^1 = e$.

$\therefore f(0)$ must be defined as $f(0) = e$

34. For continuity, limit = value

$\therefore \quad f\left(\dfrac{\pi}{4}\right) = \lim_{x \to \pi/4}(\sin 2x)^{\tan^2 2x}$ [Form $(1)^\infty$]

$$= \lim_{x \to \pi/4}(\sin^2 2x)^{\frac{1}{2}\tan^2 2x}$$

$$= \lim_{x \to \pi/4}(1 - \cos^2 2x)^{\frac{1}{2}\tan^2 2x}$$

$$= \lim_{x \to \pi/4}\left[(1 - \cos^2 2x)^{1/\cos^2 2x}\right]^{\frac{1}{2}\sin^2 2x}$$

$$= e^{1/2.1} = \sqrt{e}$$

35. For continuity, $\lim_{x \to \pi/4} = $ value at $x = \dfrac{\pi}{4}$

$$f(x) = \dfrac{1 - \tan x}{1 + \tan x} \cdot \dfrac{2\tan x}{1 - \tan^2 x}$$

where $\tan x = t \to 1$ as $x \to \pi/4$

$\therefore \quad \lim f(x) = \lim_{t \to 1}\dfrac{2t}{(1+t)^2} = \dfrac{2}{4} = \dfrac{1}{2}$

36. For continuity, limit = value

$$\lim_{x \to 0} f(x) = \lim_{x \to 0}\dfrac{2 - 2\left[1 - \dfrac{7x}{256}\right]^{1/8}}{2\left(1 + \dfrac{5x}{32}\right)^{1/5} - 2}$$

$$= \lim_{x \to 0}\dfrac{1 - \left(1 - \dfrac{1}{8}\cdot\dfrac{7x}{256} + ...\right)}{\left(1 + \dfrac{1}{5}\cdot\dfrac{5x}{32}...\right) - 1}$$

$$= \dfrac{\dfrac{1}{8}\cdot\dfrac{7}{256}}{\dfrac{1}{32}} = \dfrac{7}{64}$$

37. For continuity, limit = value = λ

$$y = \lim_{x \to 0}(\cos x)^{1/\sin x} = 1^\infty$$

$$\log y = \lim_{x \to 0}\dfrac{\log(\cos x)}{\sin x} \qquad \left(\dfrac{0}{0}\right)$$

$$= \lim_{x \to 0}\dfrac{\dfrac{1}{\cos x}\cdot(-\sin x)}{\cos x} = 0$$

[L' Hospital Rule]

or $\log y = 0$

$\therefore \qquad y = e^0 = 1 \quad \therefore \lambda = 1$

38. If $f''(x)$ is continuous at $x = 0$, then $\lim f''(0) = f''(0)$ i.e., $\lim = $ value

or $\lim f''(0) = 4$ ... (1)

Now

$\lim_{x \to 0}\phi(x) = \lim_{x \to 0}\dfrac{2f(x) - 3f(2x) + f(4x)}{x^2} \quad \left(\dfrac{0}{0}\right)$

for $x = 0$

$$= \lim_{x \to 0}\dfrac{2f'(x) - 6f'(2x) + 4f'(4x)}{2x}$$

By L' Hospital's rule, $\left(\dfrac{0}{0}\right)$

$$= \lim_{x \to 0}\dfrac{2f''(x) - 12f''(2x) + 16f''(4x)}{2x}$$

$= (1 - 6 + 8)\lim f''(0) = 3.4 = 12$

39. For continuity, $\lim = $ value

$$f(0) = \lim_{x \to 0}\dfrac{2 - \dfrac{\sin^{-1} x}{x}}{2 + \dfrac{\tan^{-1} x}{x}} = \dfrac{2-1}{2+1} = \dfrac{1}{3}$$

By $x = \sin\theta$, $\lim_{\theta\to 0}\dfrac{\theta}{\sin\theta} = 1$

Similarly, $\lim_{\theta\to 0}\dfrac{\theta}{\sin\theta} = 1$

40. $\lim_{x\to 0}\dfrac{(9^x-1)(4^x-1)}{\sqrt{2}\left[1-\cos\dfrac{x}{2}\right]}$

$= \lim_{x\to 0}\dfrac{9^x-1}{x}\cdot\dfrac{4^x-1}{x}\cdot\dfrac{x^2}{\sqrt{2}.2\sin^2\dfrac{x}{4}}$

$= \log 9.\log 4.\dfrac{1}{2\sqrt{2}}\cdot\dfrac{x^2}{(x/4^2)}$

$= \dfrac{16}{2\sqrt{2}}\log 3^2.\log 2^2$

$= 16\sqrt{2}\log 3\log 2$

For continuity, limit = value

$\therefore \qquad 16\sqrt{2}\log 3.\log 2 = k$

41. $\because \quad \dfrac{x-4}{|x-4|} = -1, x < 4$

$\dfrac{x-4}{|x-4|} = 1, \quad x > 4$

$L = R = V$

$\Rightarrow \quad -1+a = 1+b = a+b$

$\Rightarrow \qquad b = -1, a = 1$

42. $f(0-0) = \lim_{h\to 0}\dfrac{\sin(a+1)(0-h)+\sin(0-h)}{-h}$

$= \lim_{h\to 0}\dfrac{\sin(a+1)h+\sin h}{h}$

$= \dfrac{(a+1)h+h}{h} = a+2$

$\because \qquad \lim_{\theta\to 0}\sin\theta = \theta$

and $\qquad f(0+0) = \lim_{h\to 0}\dfrac{(h+bh^2)^{1/2}-h^{1/2}}{bh^{3/2}}$

$= \lim_{h\to 0}\dfrac{(1+bh)^{1/2}-1}{bh}$

$= \lim_{h\to 0}\dfrac{1+\dfrac{1}{2}bh...-1}{bh} = \dfrac{1}{2}$,

which is independent of b and so b may have any real value.
Again by continuity at $x = 0$,

we have $a+2 = \dfrac{1}{2} = c$

$\therefore \qquad a = -3/2$ and $c = 1/2$

43. For continuity, limit = value = 0

$\therefore \quad \lim_{x\to 0}x^p\sin\dfrac{1}{x} = 0$

Now $\sin\dfrac{1}{x}$ is an oscillating function and hence the above limit will be zero if $p > 0$.

44. $\lim_{x\to 0}\sin\left(\dfrac{1}{x}\right)$ does not exist.

Since for continuity, limit = value = k and hence no value of k.

45. At $x=1$, $R=1$, $L=-1$, $V = -1$

At $x=-1$, $R=1$, $L=-1$, $V = 1$

46. Let us divide the interval into two sub-intervals I_1, $-1 \le x < 0$ so that x is $-$ive I_2, $x \ge 0$ so that x is positive

for I_1, $f(x) = \int_{-1}^{x}(-t)dt = -\dfrac{1}{2}(x^2-1)$ ... (1)

for I_2, $f(x) = \int_{-1}^{0}(-t)dt + \int_{0}^{x}t\,dt$

$= -\dfrac{1}{2}[t^2]_{-1}^{0} + \dfrac{1}{2}[t^2]_{0}^{x} = \dfrac{1}{2}(1+x^2)$ ... (2)

Hence the function can be defined as follow:

$f(x) = \begin{cases} -\dfrac{1}{2}(x^2-1) & -1\le x < 0 \\[2mm] \dfrac{1}{2}(x^2+1) & x \ge 0 \end{cases}$

$f'(x) = \begin{cases} -x, & -1 < x < 0 \\ 0, & x = 0 \\ x, & x > 0 \end{cases}$

for, $f, L = R = V = \dfrac{1}{2}$ at $x = 0$ so f is continuous at $x=0$ and for f', $L = R = V = 0$ at $x = 0$ so f' is also continuous at $x=0$. Thus both f and f' are continuous at $x=0$ and hence both are continuous for $x > -1$ i.e., $x+1>0$.

47. The point $x=1$ is clearly a point of discontinuity of the function.

$y = f(x) = \dfrac{1}{1-x}$

If $x \ne 1$, then $v(x) = f[f(x)] = f\left(\dfrac{1}{1-x}\right)$

$= \dfrac{1}{1-[1/(1-x)]} = \dfrac{x-1}{x}$

Hence, the point $x = 0$ is a point of discontinuity of the function v.

If $x \neq 0, x \neq 1$, then
$$w(x) = f[f\{f(x)\}]$$
$$= f\left[f\left(\frac{1}{1-x}\right)\right] = f\left(\frac{x-1}{x}\right)$$
$$= \frac{1}{1-(x-1)/x} = x$$

Hence w is clearly continuous every where. Thus, the points of discontinuity of the composite function.

$f[f\{f(x)\}]$ are $x = 0, x = 1$.

48. $x^2 - x = x(x-1)$ ∴ $x = 0, 1$.

$x^2 - x = +$ive for $x<0, x >1$; $-$ive for $0< x <1$

∴ $|x^2 - x| = x^2 - x$ when $x <0$ or $x >1$

$\qquad = -(x^2 - x)$ when $0<x<1$.

$$f(x) = \begin{cases} 1 & , \quad x < 0 \text{ or } x>1 \\ -1 & , \quad 0 < x < 1 \\ 1 & , \quad x = 0 \\ -1 & , \quad x = 1 \end{cases}$$

∴

At $x = 1, R=1, L= -1, V=-1$

∴ Discontinuous.

At $x = 0, R= -1, L= 1, V= 1$

∴ Discontinuous.

Hence $f(x)$ is continuous $\forall x$ except at $x =0$ and $x=1$.

49. Because of $|x|$ and $|x-1|$ let us redefine the function for

$x < 0, x > 0; x < 1, x > 1$

$$f(x) = \begin{cases} -x - (x-1) = 1 - 2x & , \quad x < 0 \\ x - (x-1) = 1 & , \quad 0 < x < 1 \\ x + (x-1) = 2x - 1 & , \quad x > 1 \end{cases}$$

Consider $x = 0$

L.H.L. $= \lim_{h \to 0} 1 - 2(0-h) = 1$

R.H.L. $= \lim_{h \to 0} = 1$ ∴ continuous at $x=0$

Consider $x=1$

L.H.L. $= \lim_{h \to 0} 1 = 1$

R.H.L. $= \lim_{h \to 0} 2(1+h) - 1 = 2 - 1 = 1$

Hence continuous at $x=1$ also.

50. Redefine the function

$$f(x) = \begin{cases} -2x + 1 & , \quad x < 0 \\ 1 & , \quad 0 \leq x < 1 \\ 2x - 1 & , \quad x \geq 1 \end{cases}$$

It is continuous at $x=1$ but not differentiable at $x =1$, as L.H.D. $= 0$, R.H.D. $= 2$

51. $f(x) = \dfrac{(x^2 - 1)(x^2 - 4)}{|(x-1)(x-2)|}$

Consider continuity at $x=1$. Let us redefine the function for $x<1, 1< x <2$

$x < 1, f(x) = \dfrac{(x^2 - 1)(x^2 - 4)}{(x-1)(x-2)} = (x+1)(x+2)$

$x > 1, f(x) = \dfrac{(x^2 - 1)(x^2 - 4)}{-(x-1)(x-2)}$
$\qquad = -(x+1)(x+2)$

L.H.L. $= \lim_{h \to 0} (1 - h + 1)(1 - h + 2)$
$\qquad = \lim_{h \to 0} (2 - h)(3 - h) = 6$

R.H.L $= \lim_{h \to 0} -[1 + h + 1][1 + h + 2]$
$\qquad = \lim_{h \to 0} -(2 + h)(3 + h) = -6$

At $x = 1, f(x) = 0$

Since $L \neq R$ therefore $f(x)$ is not continuous at $x = 1$. Similarly it can be shown that $f(x)$ is not continuous at $x = 2$

Hence it is continuous on the set $R - \{1, 2\}$

52. $f(x) = x - |x||1 - x|$

Now each of x, $|x|$ and $|1 - x|$ is continuous function and we know that the product and algebraic sum of continuous functions is again a continuous function.

Hence there is no point of discontinuity.

53. Apply $R = L = V$ at $x=0$ and $x =1$

At $x = 1, V = 0, R = 1-1=0, L =0 - 0=0$

At $x = 0, V = 0, R = 0 - 0 = 0,$

$L = (-1)^2 - 0 = 1$.

Hence $f(x)$ is discontinuous at all integers except 1.

54. Rewrite the function for $x < 0, x > 0$

$$f(x) = \begin{cases} (x+1)e^0 & x < 0 \\ (x+1)e^{-2/x} & x > 0 \\ 0 & x = 0 \end{cases}$$

At $x = 0, L=1, R=0$

Hence it is not continuous at $x =0$

It is continuous of all $x \neq 0$ and it assumes all values from $f(-2)$ to $f(2)$ and $f(2) = 3e^{-1} = \dfrac{3}{e}$.

55. We have

$$f(x) = \begin{cases} 1 + 4x + \dfrac{x^2}{2} & x > 2 \\ 5x + 1 & x \le 2 \end{cases}$$

It is easy to show that $L = R = V$ for $x = 2$ and hence continuous but $L'(2) \ne R'(2)$ and hence it is not differentiable at $x = 2$.

57. $y = [x]$, consider the continuity of $f(x)$ at $x = n$, $L = R = V$

$$L = \lim_{h \to 0} [n - h] = n - 1$$

$$R = \lim_{h \to 0} [n + h] = n$$

$$V = [n] = n$$

Since $L \ne R$ the function is discontinuous at all integral values of x.

Here $y = [\sin x + \cos x]$, $x \in (0, 2\pi)$ is an integer at $x = 0, 1, \dfrac{3\pi}{4}, \dfrac{5\pi}{4}, \dfrac{7\pi}{4}$.

Hence at all such five points $f(x) = [\text{Integer } x]$ there will be discontinuity as shown above.

58. f is continuous at $x = a$ if limit = value

i.e., $\lim_{h \to 0} f(a - h) = \lim_{h \to 0} f(a + h) = f(a)$

$$g(x) = |f(x)|$$

$$g(a-0) = \lim_{h \to 0} g(a - h) = \lim_{h \to 0} |f(a - h)| = |f(a)|$$

Similarly $g(a + 0) = \lim_{h \to 0} g(a + h)$

$$= \lim_{h \to 0} |f(a + h)| = |f(a)| = g(a)$$

$\therefore$ g is continuous at $x = a$.

Hence g is continuous iff f is continuous.

59. We have $f(0) = 0$

$$f(0 + 0) = \lim_{h \to 0} \frac{\sin(0 + h)^2}{0 + h} = \lim_{h \to 0} \frac{\sin h^2}{h}$$

$$= \lim_{h \to 0} \frac{\sin h^2}{h} . h = 1 \times 0 = 0$$

Similarly $f(0 - 0) = 0$

Hence $f(x)$ is continuous at $x = 0$.

Now $Rf'(0) = \lim_{h \to 0} \dfrac{\sin(0 + h)^2 / (0 + h) - 0}{h}$

$$= \lim_{h \to 0} \frac{\sin h^2}{h^2} = 1$$

and $Lf'(0) = \lim_{h \to 0} \dfrac{\sin(0 - h)^2 / (0 - h) - 0}{-h}$

$$= \lim_{h \to 0} \frac{\sin h^2}{h^2} = 1$$

Since $Rf'(0) = Lf'(0)$, the function $f(x)$ is derivable at $x = 0$.

60. We have $f(0) = 1 + \sin 0 = 1$, then

$$Rf'(0) = \lim_{h \to 0} \frac{1 + \sin(0 + h) - 1}{h}$$

$$= \lim_{h \to 0} \frac{\sin h}{h} = 1$$

$$Lf'(0) = \lim_{h \to 0} \frac{1 - 1}{-h} = 0$$

Hence $f'(0)$ does not exist.

61. By definition $[x - \pi]$ is an integer whatever x may be and so $\pi[x - \pi]$ is an integral multiple of π.

Consequently $\tan(\pi[x - \pi]) = 0$ for all x.

And since $1 + [x]^2 \ne 0$ for any x, we conclude that $f(x) = 0$.

Thus $f(x)$ is a constant function and so it is continuous and differentiable any number of times, that is $f'(x)$, $f''(x)$, $f'''(x)$,............, $f^n(x)$,............, all exist for every x, their value being 0 at every point x.

62. It is easy to see that

$$f(0 + 0) = f(0 - 0) = f(0) = 0$$

Hence $f(x)$ is continuous at $x = 0$

Now $Lf'(0)$

$$= \lim_{h \to 0} \frac{(0 - h)[\sqrt{(0 - h)} - \sqrt{(0 - h + 1)}] - 0}{-h}$$

$$= \lim_{h \to 0} [\sqrt{(-h)} - (-h + 1)] = 0 - \sqrt{1} = -1$$

and $Rf'(0)$

$$= \lim_{h \to 0} \frac{(0 + h)[\sqrt{(0 + h)} - \sqrt{(0 + h + 1)}] - 0}{h}$$

$$= \lim_{h \to 0} [\sqrt{h} - \sqrt{(h + 1)}]$$

$$= 0 - \sqrt{1} = -1$$

Since $Lf'(0) = Rf'(0) = -1$, the function $f(x)$ is differentiable.

63. Because of $|x - 3|$ we redefine the function as under :

$$f(x) = \begin{cases} x - 3 & , & x \ge 3 \\ 3 - x & , & 1 \le x \le 3 \\ x^2/4 - 3x/2 + 13/4, & x < 1 \end{cases}$$

Now it can be easily seen that $f(x)$ is continuous

at $x = 1$ and $x = 3$, differentiable at $x=1$ but non-differentiable at $x=3$ as shown below.

$$Rf'(3) = \lim_{h \to 0} \frac{f(3+h) - f(3)}{h}$$

$$= \lim_{h \to 0} \frac{h - 0}{h} = \lim_{h \to 0} 1 = 1$$

$$Lf'(3) = \lim_{h \to 0} \frac{f(3-h) - f(3)}{-h}$$

$$= \lim_{h \to 0} \frac{h}{-h} = -1$$

64. For $\quad x = 2 \pm h$

$x - 1 = +$ive but $x - 3 = -$ive

$\therefore \quad f(x) = |x-1| + |x-3|$

$\quad = (x-1) - (x-3) = 2$

Since $f(x) = 2 \quad \therefore f'(x) = 0$ at $x = 2$

65. $\lim_{h \to 0} \left[\tan^2 (0 - h) \right]$

$= \lim_{h \to 0} \left[\tan^2 (0 + h) \right] = \tan^2 0 = 0$

Since L.H.L. = R.H.L. = value
$\therefore \quad f(x)$ is continuous at $x = 0$

66. L.H.L. $= \lim_{h \to 0} \dfrac{|-2 - h + 2|}{\tan^{-1}(-2 - h + 2)}$

$= \lim_{h \to 0} - \dfrac{h}{\tan^{-1} h}$

$= \lim_{h \to 0} \dfrac{-1}{\dfrac{1}{1 + h^2}} = -1$

R.H.L. $= \lim_{h \to 0} \dfrac{|-2 + h + 2|}{\tan^{-1}(-2 + h + 2)}$

$= \lim_{h \to 0} \dfrac{h}{\tan^{-1} h} = 1$

$R \neq L$ and hence the function is not continuous at $x = -2$.

67. $\dfrac{dy}{dx} = 6(x+2)$

and $6(x+2) = +$ive on $[-1, 2] \quad \therefore \phi(x)$ is increasing function on $[-1, 2]$

$$\lim_{h \to 0} f(2 - h) = \lim_{h \to 0} f(2 + h)$$

$$= f(2) = 35$$

continuous at $x = 2$ in $[-1, 3]$

$$R' = \lim_{h \to 0} f(2 + h) = \lim_{h \to 0} \frac{f(2+h) - f(2)}{-h}$$

$$= \frac{37 - (2+h) - 35}{h} = 1$$

$$L' = \lim_{h \to 0} f'(2 - h) = \lim_{h \to 0} \frac{f(2-h) - f(2)}{-h}$$

$$= \frac{3h^2 - 24h}{-h} = 24$$

$\therefore \quad f'(2)$ does not exist.

68. Because of $|x|$ let us redefine the function as under:

$$f(x) = \begin{cases} \dfrac{x}{1-x} & , x < 0 \\ 0 & , x = 0 \\ \dfrac{x}{1+x} & , x > 0 \end{cases}$$

Consider $x = 0$
For continuity

$$\lim f(0-h) = \lim f(0+h) = f(0)$$

$$\lim \frac{-h}{1+h} = 0, \; \lim \frac{h}{1+h} = 0, f(0) = 0$$

Since limit = value $\therefore$ function is continuous.
For differentiability

$$\lim \frac{f(0-h) - f(0)}{-h} = \lim \frac{f(0+h) - f(0)}{h}$$

$$\lim 1/(1+h) = 1, \lim 1/(1+h) = 1$$

Hence both left hand and right hand derivatives exist so the function is differentiable at $x = 0$ and hence for all other real values of x it is differentiable. Therefore it is differentiable in the interval $(-\infty, \infty)$.

69. $f(x) = -x^2, x < 0$

$\quad = x^2, x \geq 0$

The function being algebraic is differentiable for all $x > 0$ and for all $x < 0$. Hence we check its differentiability for $x = 0$

$\therefore$ L.H.D. $= \lim_{h \to 0} \dfrac{-(-h)^2 - 0}{-h} = 0$

R.H.D. $= \lim_{h \to 0} \dfrac{h^2 - 0}{h} = 0$

70. $R = L = V = 0$ at $x = 1$ $\therefore$ continuous
R.H.D. $= 3$, L.H.D. $= 1$
$\therefore$ Not differentiable.

76. If a is any real number, then
$$f(a) = 1 + |\sin a|$$

$$f(a+0) = \lim_{h \to 0} \left[1 + |\sin(a+h)| \right]$$

$$= 1 + |\sin a|$$

and $f(a-0) = \lim_{h \to 0} \left[1 + |\sin(a-h)| \right]$

$$= 1 + |\sin a|.$$

The function is not differentiable at $x = n\pi$, $n = 0, \pm 1, \pm 2, \ldots$ as shown below:

$f(n\pi) = 1 + |\sin n\pi| = 1 + |0| = 1$

Now $Rf'(n\pi)$

$= \lim_{h \to 0} \dfrac{1 + |\sin(n\pi + h)| - 1}{h} \quad (h > 0)$

$= \lim_{h \to 0} \dfrac{|\pm \sin h|}{h} = \lim_{h \to 0} \dfrac{\sin h}{h} = 1$

$[\because h$ is small and > 0 and as such $\sin h > 0] = 1$

and $Lf'(n\pi) = \lim_{h \to 0} \dfrac{1 + |\sin(n\pi - h)| - 1}{-h}$

$= \lim_{h \to 0} \dfrac{|\pm \sin h|}{-h} = \lim_{h \to 0} \dfrac{\sin h}{-h} = -1$

Since $Rf'(n\pi) \neq Lf'(n\pi)$, $f(x)$ is not differentiable at $x = n\pi$, $n = 0, \pm 1, \pm 2, \ldots$

At all other points, $f(x)$ is clearly differentiable. For if a is any real number such that $n\pi < a < (n+1)\pi$.

77. Let us consider differentiability at $x = 0$

$\lim_{x \to 0} \dfrac{f(x+0) - f(x)}{x+0} = \lim \dfrac{\log \cos x}{\log(1+x^2)}$

$= \lim \dfrac{\log\left(1 - \dfrac{x^2}{2!} + \ldots\right)}{\log(1+x^2)}$

$= \lim\left(-\dfrac{x^2}{2!}\right) / x^2 = -\dfrac{1}{2}$

Hence $f(x)$ is differentiable at $x = 0$ and therefore continuous as well.

78. Because of $|y|$ we redefine as under:

$y > 0 \quad \Rightarrow \quad y = \dfrac{1}{2}x, \quad \therefore x \geq 0$

$y < 0 \quad \Rightarrow \quad y = \dfrac{1}{10}x, \quad \therefore x < 0$

Hence we redefine the function as under:

$y = \begin{cases} \dfrac{1}{10}x & , x < 0 \\ \dfrac{1}{2}x & , x \geq 0 \end{cases}$

Also, $f(0+h) = f(0-h) = f(0) = 0$.

$\therefore$ Continuous. It is not differentiable as

$Rf'(0) = \dfrac{1}{2}$ and $Lf' = \dfrac{1}{10}$

79. Subtract and add $f(h)$ in numerator.

$\lim_{h \to 0} \dfrac{1}{2}\dfrac{f(x_0 + h) - f(h)}{h} + \dfrac{1}{2}\dfrac{f(x_0 - h) - f(h)}{-h}$

$= f'(x_0)\left[\dfrac{1}{2} + \dfrac{1}{2}\right] = f'(x_0)$

80. $[2+h] = 2, [2-h] = 1, [1+h] = 1, [1-h] = 0$

At $x = 2$, we will check $R = L = V$

$R = \lim_{h \to 0} |4 + 2h - 3|[2+h] = 2, V = 1.2 = 2$

$L = \lim_{h \to 0} |4 - 2h - 3|[2-h] = 1, R \neq L$

$\therefore$ not continuous.

At $x = 1$

$R = \lim |2 + 2h - 3|[1+h] = 1.1 = 1$

$V = |-1|[1] = 1$

$L = \lim \sin\dfrac{\pi}{2}(1-h) = 1$

Since $R = L = V$ $\therefore$ continuous at $x = 1$.

R.H.D. $= \lim \dfrac{|2 + 2h - 3|[1+h] - 1}{h}$

$= \lim \dfrac{|-1|.1 - 1}{h} = Lt \dfrac{1-1}{h} = 0$

L.H.D. $= \lim \dfrac{|2 - 2h - 3|[1-h] - 1}{-h}$

$= \lim \dfrac{1.0 - 1}{-h} = \lim \dfrac{1}{h} = \infty$

Since R.H.D. $\neq$ L.H.D. $\therefore$ not differentiable.

81. $f(1+h) = f(1-h) = f(1) = 3$

$\therefore$ continuous.

$Rf'(1) = \lim_{h \to 0} \dfrac{f(1+h) - f(1)}{h}$

$= \lim \dfrac{4 - (1+h) - 3}{h} = -1$

$Lf'(1) = \lim_{h \to 0} \dfrac{f(1-h) - f(1)}{h} = \lim \dfrac{3^{1-h} - 3}{-h}$

$= \lim_{h \to 0} \dfrac{3(3^{-h} - 1)}{-h} = \dfrac{3(-3^{-h}\log 3)}{-1}$

$= 3\log 3$

$\therefore Rf'(1) \neq Lf'(1)$ $\therefore$ not differentiable.

82. $x^2 < 1 \Rightarrow x^2 - 1$ is $-$ive.

$\therefore -1 < x < 1, x^2 \geq 1 \Rightarrow (x^2 - 1) \geq 0$

$\therefore$ either $x \leq -1$ or $x \geq 1$.

Hence the given function can be redefined as

$f(x) = x^3, \quad -1 < x < 1$

$= x^2, \quad x \leq -1$ or $x \geq 1$

R.H.L.$= \lim(1+h) = 1$

L.H.L.$= \lim(1+h)^3 = 1$, $V=1$

R.H.L.$=$L.H.L.$=1$

$\therefore$ limit exists. Also limit$=$value

$\therefore$ Continuous at $x=1$.

Also R.H.D.$=1$,L.H.D.$=3$.

It is not differentiable at $x=1$.

83. L' at $x=k$ is

$$\lim_{h\to 0}\frac{[k-h]\sin \pi(k-h)}{-h}$$

$$= \lim_{h\to 0}(k-1)$$

$[\sin \pi k \cos \pi h - \cos \pi k \sin \pi h]$

as $[k-h]$ when k is an integer is $k-1$.

Put $\sin \pi k = 0$, $\cos \pi k = (-1)^k$

$$L' = \lim_{h\to 0}(k-1)\frac{\left[0-(-1)^k \sin \pi h\right]}{-h}$$

$$= (k-1)(-1)^k \pi$$

$$\because \lim_{h\to 0}\frac{\sin(\pi h)}{h} = \pi$$

84.
$$f(x)=\begin{cases} ax^2+b & ,x<1 \\ bx^2+ax+4, & x\geq -1\end{cases}$$

$$\therefore \quad f'(x)=\begin{cases} 2ax & ,x<-1 \\ 2bx+a, & x\geq -1\end{cases}$$

To find a, b we must have two equations in a, b.
Since $f(x)$ is differentiable, it must be continuous at $x=-1$.

$\therefore \qquad R=L=V$ at $x=-1$ for $f(x)$

$\Rightarrow \quad b-a+4=a+b$

$\therefore \qquad 2a=4$ *i.e.,* $a=2$

Again $f'(x)$ is continuous, it must be continuous at $x=-1$.

$\therefore \qquad R=L=V$ at $x=-1$ for $f'(x)$

$-2b+a=-2a$ Putting $a=2$, we get

$-2b+2=-4$ $\therefore$ $2b=6$ or $b=3$

85. For continuity at $x=1$,$R=L=V$

$b+a+c=a+b \quad \Rightarrow \quad c=0$

For differentiability at $x=1$, $R'=L'$

$2a-2b+a \quad \Rightarrow \quad a=2b$.

86. For $f'(x)$ to exist at $x=0$, we must have $L'=R'$ at $x=0$

$$\lim_{h\to 0}\frac{\left\{a|-h|^2+b|-h|+c\right\}-c}{-h}$$

$$= \lim_{h\to 0}\frac{\left\{a|h|^2+b|h|+c\right\}-c}{h}$$

or $-b=b$ or $2b=0$ $\therefore$ $b=0$

87. Let us consider continuity and differentiability for $x=0,1$. Hence we define the function as under

$$f(x)=\begin{cases} x & , & x<0 \\ 0 & , & x=0 \\ x^2 & ,0<x<1 \\ x & , & x>1\end{cases}$$

Clearly $f(x)$ being a polynomial in x is continuous for all x. Also $f'(x)=1 \ \forall \ x>1$

Also both at $x=0$ and $x=1$, f is not differentiable as $L' \neq R'$.

88. Let us redefine the function in $[-1,1]$

$$f(x)=\begin{cases} x & , & x\leq -1 \\ x^3 & , -1<x\leq 0 \\ x & , & 0<x\leq 1 \\ x^3 & , & x>1\end{cases}$$

$y=x$ represents a straight line and $y=x^3$ represents a semicubical parabola. It is clear that $f(x)$ is not differentiable at $x=1$, 0,-1 as L' and R' are different.

89. $\therefore \qquad L'=-1$, $R'=1$.

90. By definition of limit,

$$\text{limit}= \frac{1}{2}\frac{d}{dx}f^2(x) = \frac{1}{2}.2f(x).f'(x)$$

91. $y'(x)=f(e^x).e^{f(x)}.f'(x)+e^{f(x)}f'(e^x)e^x$

Now put $x=0$ and the given data *i.e.,*

$f(0)=0$, $f(1)=0$, $f'(1)=2$

92. $f(x)=e^x$, $x<0$

$\qquad = e^{-x}$, $x\geq 0$

$$L= \lim_{h\to 0} e^{-h} = 1$$

$$R= \lim_{h\to 0} e^{-h} = 1 = V$$

Hence continuous

$$L'= \lim_{h\to 0}\frac{e^{-h}-1}{-h} = \lim_{h\to 0}\frac{(1-h...)-1}{-h} = 1$$

$$R'= \lim_{h\to 0}\frac{e^{-h}-1}{h} = \lim_{h\to 0}\frac{(1-h...)-1}{h} = -1$$

Since $L' \neq R'$ therefore it is not differentiable at $x=0$.

93. Apply $L=R=V$ for continuity of f at $x=\dfrac{\pi}{6}$

$$\lim_{h\to 0}\sin 2\left(\frac{\pi}{6}-h\right)=\lim_{h\to 0}p\left(\frac{\pi}{6}+h\right)+q$$

$$=\sin\frac{2\pi}{6}$$

or $\qquad \dfrac{\sqrt3}{2}=p\,\dfrac{\pi}{6}+q \qquad$ ...(1)

$$L'=\lim_{h\to 0}\frac{f\left(\frac{\pi}{6}-h\right)-f\left(\frac{\pi}{6}\right)}{-h}$$

$$=\lim_{h\to 0}\frac{\sin 2\left(\frac{\pi}{6}-h\right)-\frac{\sqrt3}{2}}{-h}$$

$$=\lim_{h\to 0}\frac{\sin\left(60^\circ-2h\right)-\frac{\sqrt3}{2}}{-h}$$

$$=\lim_{h\to 0}\frac{\frac{\sqrt3}{2}\cos 2h-\frac{1}{2}\sin 2h-\frac{\sqrt3}{2}}{-h}$$

$$=\lim_{h\to 0}\frac{\left(\frac{\sqrt3}{2}-\frac{\sqrt3}{2}\right)-\frac12\sin 2h}{-h}=\frac12.2=1$$

$$R'=\lim_{h\to 0}\frac{f\left(\frac{\pi}{6}+h\right)-f\left(\frac{\pi}{6}\right)}{h}$$

$$=\lim_{h\to 0}\frac{p\left(\frac{\pi}{6}+h\right)+q-\frac{\sqrt3}{2}}{h}$$

$$=\lim_{h\to 0}\frac{p\left(\frac{\pi}{6}\right)+q+ph-\frac{\sqrt3}{2}}{h}$$

$$=\lim_{h\to 0}0+\frac{ph}{h}=p \qquad\qquad \text{by (1)}$$

Now $L'=R' \Rightarrow p=1$ and hence by (1)

$$q=\frac{\sqrt3}{2}-\frac{\pi}{6} \quad \therefore (p,q)=\left(1,\frac{\sqrt3}{2}-\frac{\pi}{6}\right)$$

94. $f(x)=\begin{cases}-\dfrac{1}{|x|}, & |x|\ge 1 \text{ or } x^2-1\ge 0 \\[2mm] & \text{or } x\le -1,\ x\ge 1 \ \ ...(1)\\[2mm] ax^2-b, & |x|<1 \text{ or } x^2-1<0 \\[2mm] & \text{or } -1<x<1 \ \ ...(2)\end{cases}$

Also $|x|=x$ when $x=+$ive $\qquad$ ...(3)

$\qquad |x|=-x$ when $x=-$ive $\qquad$ ...(4)

Hence we redefine the function as under

$f(x)=\begin{cases}ax^2-b & ,x<1 \quad \text{by (2)}\\[1mm] -\dfrac{1}{+x}=-\dfrac{1}{x},x\ge 1 \quad \text{by (1)}\end{cases}$ ...(A)

$f(x)=\begin{cases}-\dfrac{1}{-x}=\dfrac{1}{x},x\le -1 \quad \text{by (1)}\\[1mm] ax^2-b\ ,x>-1 \quad \text{by (2)}\end{cases}$ ...(B)

At $x=1$, for continuity, $R=L=V$

$\therefore \qquad a-b=-1$

For differentiability, $R'=L'$

$\therefore \qquad 2a=1$

Solving $a=1/2, b=3/2$

At $x=-1$, for continuity $R=L=V$

$\therefore \qquad a-b=-1$

For differentiability $R'=L'$

$\qquad -1=-2a$

$\therefore \qquad a=\dfrac{1}{2}, b=\dfrac{3}{2}$

95. $f(x)=\begin{cases}x^3\\ -x^3\end{cases},\ f'(x)=\begin{cases}3x^2,x\ge 0\\ -3x^2,x\le 0\end{cases}$

$Lf'(0)=Rf'(0)=0$

96. L.H.D. $=\lim_{h\to 0}\dfrac{\left|\tan\dfrac{\pi}{4}-\left(\dfrac{\pi}{4}-h\right)\right|-\left|\tan\left(\dfrac{\pi}{4}-\dfrac{\pi}{4}\right)\right|}{-h}$

$\qquad =\lim_{h\to 0}\dfrac{\tan h}{-h}=-1$

R.H.D. $=\lim_{h\to 0}\dfrac{\tan h}{h}=1$

Since $L'\ne R'$, the derivative does not exist.

97. $|x|$ is non-differentiable function at $x=0$ as $L'=-1$ and $R'=1$

i.e., $\lim_{h\to 0}\dfrac{|-h|-0}{-h}=-1,\lim_{h\to 0}\dfrac{|h|-0}{h}=1$

or $y=x,\ x>0,\ y=-x,\ x<0$

But $\cos|h|$ is differentiable.

Now any combination of two such functions will be non-differentiable,

Now consider $\sin|x|-|x|$

$L'=\lim_{h\to 0}\dfrac{\sin|-h|-|-h|}{-h}=\lim\dfrac{\sin h}{-h}=-1$

$R'=\lim_{h\to 0}\dfrac{\sin|h|-|h|}{h}=\lim\dfrac{\sin h}{h}=1$

98. There are two moduli in the problem.

$y=||x|-1|$, Hence we have to consider the cases when $x\ge 0, x<0, |x|\ge 1$ or $|x|\le 1$.

Now $\quad x\ge 1 \Rightarrow x^2-1\ge 0 \Rightarrow (x+1)(x-1)\ge 0$

$\Rightarrow \qquad x>1$ or $x\le -1 \qquad$ ...(1)

$|x| \leq 1 \Rightarrow x^2 - 1 \leq 0 \Rightarrow (x+1)(x-1) \leq 0$

$\Rightarrow -1 \leq x \leq 1$ ...(2)

$\therefore \quad y = |x| - 1, x \geq 1 \text{ or } x \leq -1$ ...(3)

$= -(|x| - 1) = 1 - |x|, -1 \leq x \leq 1$ ...(4)

Now consider $|x| = x, x \geq 0, |x| = -x, x < 0$

Now we redefine the given function

$y = -x - 1 \qquad x \leq -1$

$y = x + 1 \qquad -1 \leq x < 0$

$y = -x + 1 \qquad 0 \leq x < 1$

$y = x - 1 \qquad x \geq 1$

$f(x)$ is not differentiable at $x = -1, 0, 1$.

100. Redefine $f(x) = \begin{cases} -x, x < 0 \\ 0, x = 0 \\ x, x > 0 \end{cases}$

Now clearly $L = R = V = 0$ $\therefore$ continuous

$L' = -1, R' = 1$ $\therefore$ not differentiable.

101. $(x^2 - 3x + 2) = (x-1)(x-2) = +$ive

when $x < 1$ or > 2, $-$ive when $1 \leq x \leq 2$

Also $\cos|x| = \cos x$ $\qquad [\because \cos(-x) = \cos x]$

$\therefore f(x) = -(x^2 - 1)(x^2 - 3x + 2) + \cos x,$

$\qquad\qquad\qquad\qquad\qquad 1 \leq x \leq 2$

$= (x^2 - 1)(x^2 - 3x + 2) + \cos x, x > 2$ ...(A)

Evidently $f(x)$ is not differentiable at $x = 2$ as $L' \neq R'$

102. $\dfrac{dy}{dx} = \cos x$ or $-\sin x$

Both are $-$ive in $(\pi/2, \pi)$. Hence S is correct. Again if a differentiable function $f(x)$ decreases then its derivative $f'(x) = -$ive. It is not necessary that its derivative $f''(x)$ may be $+$ive and hence it may be increasing or decreasing. Hence R is not correct.

103. $f(x) = \dfrac{x^2}{1 - \dfrac{1}{1+x^2}} = 1 + x^2$

$\lim_{x \to 0} f(x) = 1$ and value of $f(x)$ at $x = 0$ is 0.

Since limit is not equal to value $f(x)$ is discontinuous.

104. $f'(x) = \lim_{x \to 0} \dfrac{f(x+h) - f(x)}{h}$

$= \lim \dfrac{f(x) + f(h) - f(x)}{h} = \lim \dfrac{h g(h)}{h}$

$= g(0)$

105. As in last question.

$f'(x) = \lim \dfrac{h^2 g(h)}{h} = \lim h g(h) = 0$

106. Since f is differentiable.

$\therefore f'(x) = \lim_{h \to 0} \dfrac{f(x+h) - f(x)}{h}$

$= \lim_{h \to 0} \dfrac{f(x) + f(h) + hx - f(x)}{h}$

$= \lim_{h \to 0} \dfrac{f(h)}{h} + x = 3 + x$

Integrating $f(x) = 3x + \dfrac{x^2}{2} + k$ ...(1)

Putting $x = 0, y = 0$ in the given relation

$f(0) = f(0) + f(0) + 0$ $\therefore f(0) = 0$

Now from (1) we have $f(0) = 0 + k$ $\therefore k = 0$

$\therefore \quad f(x) = 3x + \dfrac{x^2}{2}$

107. $f'(3) = \lim_{h \to 0} \dfrac{f(3+h) - f(3)}{h}$

$= \lim_{h \to 0} \dfrac{f(3) f(h) - f(3)}{h}$

$f'(3) = 3 \lim_{h \to 0} \left[\dfrac{f(h) - 1}{h} \right]$ ...(1)

$\because f(3) = 3$

Now $f(x+0) = f(x)f(0) \Rightarrow f(x)[f(0) - 1] = 0$

either $f(x) = 0$ or $f(0) = 1$ but

$\qquad\qquad f(3) = 3 \neq 0$

$\therefore$ We must have $f(0) = 1$

Hence from (1), $f'(3) = 3 \lim_{h \to 0} \dfrac{f(h) - f(0)}{h}$

$= 3f'(0) = 3(11) = 33.$

108. $f'(x) = \lim_{h \to 0} \dfrac{f(x+h) - f(x)}{h}$

$= \lim_{h \to 0} \dfrac{f(x) f(h) - f(x)}{h}$

$= f(x) \lim_{h \to 0} \dfrac{f(h) - 1}{h}$

$= f(x). \lim_{h \to 0} \dfrac{1 + (\sin 2h) g(h) - 1}{h}$

$= f(x). 2g(0) = 2f(x) g(0)$

109. Writing δx for y in the given condition (a), we have $f(x + \delta x) = f(x) f(\delta x)$

Then $f(x + \delta x) - f(x) = f(x) f(\delta x) - f(x)$

or $\dfrac{f(x + \delta x) - f(x)}{\delta x} = \dfrac{f(x)[f(\delta x) - 1]}{\delta x}$

$= \dfrac{f(x) \delta x\, g(\delta x)}{\delta x}$

by condition (b)

$= f(x) g(\delta x)$

Hence,

$$\lim_{\delta x \to 0} \frac{f(x+\delta x) - f(x)}{\delta x} = \lim_{\delta x \to 0} f(x) g(\delta x)$$

$= f(x).1$, since by hypothesis $\lim_{\delta x \to 0} g(\delta x) = 1$.

It follows that $f'(x) = f(x)$.

Since $f(x)$ exists, $f'(x)$ also exists and

$$f'(x) = f(x)$$

110. We are given :

$f(x+y) = f(x)f(y)$ for all real x and y ...(1)

For $x = y = 0$, we get from (1)

$f(0) = f(0)f(0) \Rightarrow f(0) = 1$ [$\because f(0) \neq 0$] ...(2)

Now $f'(x) = \lim_{h \to 0} \dfrac{f(x+h) - f(x)}{h}$

$= \lim_{h \to 0} \dfrac{f(x)f(h) - f(x)}{h}$, by (1)

$= f(x) \lim_{h \to 0} \dfrac{f(h) - 1}{h}$

$= f(x) \lim_{h \to 0} \dfrac{f(h) - f(0)}{h}$, by (2)

$= f(x) f'(0) = f(x).2$ [$\because f'(0) = 2$]

Thus $f'(x) = 2f(x)$ $\forall x$.

Again $\dfrac{f'(x)}{f(x)} = 2$. On integrating both sides we have

$\log f(x) = 2x + \log A$

when $x = 0$, $\log f(0) = 0 + \log A$

$\Rightarrow$ $\log 1 = \log A \Rightarrow \log A = 0$.

$\therefore$ $\log f(x) = 2x$ or $f(x) = e^{2x}$

111. Differentiating the given relation with respect to x, we get

$h(x) = [f(x)]^2 + [g(x)]^2$

$h'(x) = 2f(x)f'(x) + 2g(x)g'(x)$ ... (1)

But we are given $f''(x) = -f(x)$ and $f'(x) = g(x)$ so that $f''(x) = g'(x)$.

Then (1) may be re-written as

$h'(x) = -2f''(x)f'(x) + 2f'(x)f''(x) = 0$

Thus $h'(x) = 0$,

whence by integrating, we get

$h(x) = $ constant $= c$, say

Hence $h(x) = c$, for all x.

In particular, $h(5) = c$. But we are given $h(5) = 11$.

It follows that $c = 11$ and we have $h(x) = 11$ for all x.

Therefore $h(10) = 11$.

112. If $h(x) = f^2(x) + g^2(x)$ then

$\therefore$ $h'(x) = 2ff' + 2gg' = 2f(g) + 2g(-f) = 0$

Hence $h(x) = $ constant

$\therefore$ $f^2(16) + g^2(16) = f^2(2) + g^2(2)$

$= f^2(2) + f'^2(2) = 4^2 + 4^2 = 32$

113. We are given

$$f(x+y) = f(x)f(y)$$ ...(1)

for all x and y.

In the identity (1), we put $x = 5, y = 0$. Then

$$f(5) = f(5)f(0).$$

This gives $f(0) = 1$ [$\therefore f(5) = 2 \neq 0$].

Now $f'(5) = \lim_{h \to 0} \dfrac{f(5+h) - f(5)}{h}$

$= \lim_{h \to 0} \dfrac{f(5)f(h) - f(5)}{h}$

by the given identity

$= f(5) \lim_{h \to 0} \dfrac{f(h) - 1}{h}$

$= f(5) \lim_{h \to 0} \dfrac{f(h) - f(0)}{h}$ [$\because f(0) = 1$]

$= f(5) f'(0)$

$= 2 \times 3 = 6.$

114. Since f is continuous function of x on [1, 3]. It must take all real values between $f(1)$ and $f(3)$ but it is given that f takes only rational values for all x. Hence f must be constant function.

$\therefore$ $f(2.5) = f(2) = 10$

115. Given $f(1) = -2$ and $f'(x) \geq 2$ for $\forall x \in [1,6]$

$\therefore$ $\dfrac{f(6) - f(1)}{6 - 1} \geq 2$

$\Rightarrow$ $f(6) + 2 \geq 10 \therefore f(6) \geq 8$

116. We have $\dfrac{f(x) - f(y)}{x - y} \leq |x - y|$

$\therefore$ $\left| \lim_{x \to y} \dfrac{f(x) - f(y)}{x - y} \right| \leq \lim_{x \to y} |x - y|$

$\Rightarrow$ $|f'(x)| \leq 0$

$\therefore$ $f'(x) = 0$

Hence $f(x)$ is a constant function.   ...(1)

Also given $f(0) = 0 \because f(1) = 0$ as well by (1)

117. $\lim_{h \to 0} \dfrac{f(1+h)}{h} = 5 \Rightarrow f'(1) = 5$

SELF ASSESSMENT TEST

Verify each of the following :

1. Let $f(x)=\sin\dfrac{1}{x}, x\neq0$, then $\overline{f(0+0)}=1$,

$\underline{f(0+0)}=-1$, $\overline{f(0-0)}=1$ and $\underline{f(0-0)}=-1$

2. Let $f(x)=[x]+x\sin\dfrac{1}{x}, x\neq0$, then $\overline{f(0+0)}=$

$\underline{f(0+0)}=0$ and $\overline{f(0-0)}=\underline{f(0-0)}=-1$

3. Let $f(x)=\sin\dfrac{1}{x-a}, x\neq0$, then $\overline{f(a+0)}=1$,

$\underline{f(a+0)}=-1$, $\overline{f(a-0)}=1$ and $\underline{f(a-0)}=-1$

4. Let $f(x)=(x-a)\sin\dfrac{1}{x-a}, x\neq0$, then $\overline{f(a+0)}$

$=\underline{f(a+0)}=\overline{f(a-0)}=\underline{f(a-0)}=0$

5. The function $f(x)=\dfrac{1}{1+|x|}$, $x\in$R is continuous

and bounded and attains its supremum at $x=0$ but it does not attains its infimum.

6. The function $f(x)=\dfrac{1}{1+|x|}$ for each real x is

continuous and bounded attains its infimum but does not attains supremum.

7. The function $f(x)=x$ $\forall x\in]0,1[$ is continuous and bounded but it attains neither supremum nor infimum.

8. A function f is defined as follows:

$$f(x)=\begin{cases}\dfrac{x^2-a^2}{x-a},\text{if } x\neq a\\3a\quad,\text{if } x=a\end{cases}$$

then there is a point of removable discontinuity.

But the function $f(x)=\begin{cases}\dfrac{x^2-a^2}{x-a}, x\neq a\\2a\quad,\text{if } x=a\end{cases}$

is continuous at $x=a$.

9. The function $f(x)=\dfrac{\sin x}{x}, x\neq0$ and $f(x)=A$,

$x=0$ has a removable discontinuity at $x=0$ regardless the value of A.

10. The function $\dfrac{\sin(x-a)}{x-a}$ is continuous for

$x\neq a$ and $x=a$ is a point of removable discontinuity.

11. The function $f(x)=|x|+|x-1|$ $\forall x\in$R is continuous but not differentiable at $x=0$ and $x=1$.

12. The function $f(x)=|x-a|$ is continuous but not differentiable at $x=a$.

13. The function $f(x)=\begin{cases}x\sin\dfrac{1}{x} \ ; x\neq0\\0\quad ; x=0\end{cases}$

is continuous but not differentiable at origin.

14. Let $f(x)=\begin{cases}\dfrac{1}{x} \ , x\neq0\\0 \ , x=0\end{cases}$ and $g(x)=-f(x)$ then

$f+g$ is differentiable at origin but f and g are not.

15. The function $f(x)=x|x|$ is differentiable at the origin.

16. The sum function of the infinite series

$x+x(1-x)+x(1-x)^2+...$

is discontinuous at $x=0$.

17. In every nbd of a discontinuity of second kind a function makes on either side infinitely many finite or infinite oscillations.

18. A monotonic function cannot have second kind of discontinuity.

19. The function $f(x)=\dfrac{1}{x}$ is continuous but unbo-

unded on $]0,1[$ and $]0,1]$. Although it is continuous on these intervals.

20. The function $f(x)=x$ is continuous on R but it is unbounded on every infinite interval.

21. The function $f(x)=\begin{cases}x^m\sin\dfrac{1}{x} \ , x\neq0\\0 \ , x=0\end{cases}$

is differentiable at $x=0$ if $m>1$ and $f'(x)$ is continuous if $m>2$.

22. The function $f(x)=\begin{cases}x^2 ,\text{if } x \text{ is rational}\\0 \ , \text{if } x \text{ is irrational}\end{cases}$

is differentiable at 0 and $f'(0)=0$ and f is not differentiable for all $x\neq0$.

Chapter

5 Mean Value Theorems and Expansion of Functions

5.1 INTRODUCTION

In this chapter we shall discuss some important theorems namely, Rolle's, Lagrange's, Cauchy mean value and Taylor's theorem. We shall also discuss Maclaurin's series expansion of some standard functions like e^x, $\log(1+x)$, $\sin x$, $\cos x$ etc.

5.2 ROLLE'S THEOREM

If a function f defined on [a,b] is such that it is

(i) continuous in [a,b],   (ii) differentiable in]a,b[.

(iii) f(a)=f(b),

then there exists at least one vlaue of x, say c,(a<c<b) such that $f'(c)=0$

Proof.   Since, the function $f(x)$ is continuous on $[a, b]$

$\Rightarrow$ $f(x)$ is hounded   [$\because$ Every continuous function is bounded.]

$\Rightarrow$ $f(x)$ attains its bounds   [$\because$ A function, which is continuous on a closed bounded interval $[a, b]$, then it attains its bound on $[a, b]$.]

Let M and m are the supremum and infimum of $f(x)$ respectively.

Now there are two possibilities

(i) If $M=m$, then obviously $f(x)$ is a constant function, and therefore its derivative is zero, *i.e.,* $f'(x)= 0 \; \forall \, x \in \,]a, b[$.

(ii) If $M \neq m$, then at least one of the numbers M and m must be different from the equal values $f(a)$ and $f(b)$.

Let us assume $M \neq f(a)$.

Now, since, every continuous function on a closed interval attains its supremum, therefore, there exists a real number c in $[a,b]$ such that $f(c)=M$. Also since $f(a) \neq M \neq f(b)$. Therefore $c \neq a$ and $c \neq b$, this implies that $c \in \,]a,b[$.

Now, $f(c)$ is the supremum of f on $[a, b]$

$\therefore$ $f(x) \leq f(c) \; \forall x \in [a, b]$ ...(1)

[By the definition of supremum]

In particular, $f(c–h) \leq f(c), \; h>0$.

$\Rightarrow$ $\dfrac{f(c-h)-f(c)}{-h} \geq 0$ ...(2)

Since $f'(x)$ exists at each point of $]a, b[$, and hence, $f'(c)$ exists.

Therefore, from (2)

$$Lf'(c) \geq 0 \qquad \qquad ...(3)$$

Similarly, from (1)

$$f(c+h) \leq f(c) \quad h>0.$$

Then by the same arguments

$$Rf'(c) \leq 0. \qquad \qquad ...(4)$$

Since $f(x)$ is differentiable in $]a, b[\Rightarrow f'(c)$ exist

$$\Rightarrow \qquad Lf'(c)=f'(c)=Rf'(c). \qquad \qquad ...(5)$$

Now from (3), (4) and (5) $f'(c)=0$.

Similarly we can consider the case $M=f(a)\neq m$.

REMARKS

- Converse of Rolle's theorem is not true i.e., $f'(x)$ may vanish at a point $c \in]a, b[$ without $f(x)$ satisfying the three conditions of Rolle's theorem.
- There may be more than one point like c at which $f'(x)$ vanishes but Rolle's theorem ensures the existance of at least one such c.
- Rolle's theorem will not hold good if
  (a) $f(x)$ is discontinuous at some point in the interval $[a, b]$
  (b) $f'(x)$ does not exist at some point in the interval $]a, b[$
  (c) $f(a) \neq f(b)$.
- The hypothesis of Rolle's theorem cannot be weakened.
  For example, if $f(x)=1-|x|, -1\leq x\leq 1$, then $f(-1)=f(1)=0$ and f is continuous on $[-1,1]$. Also if $f'(x)$ exist $\forall x \in]-1, 1[$ except at $x=0$. Then, f satisfies all the condition of Rolle's theorem except that f is not differentiable at $x=0$. For this f, there is no c in $]-1,1[$ for which $f'(c)=0$.

5.2.1 GEOMETRICAL INTERPRETATION OF ROLLE'S THEOREM

Geometrically, Rolle's theorem means that if the curve $y=f(x)$ is continuous from $x=a$ to $x=b$, has a definite tangent at each point of $]a,b[$ and the ordinates at the extremities are equal, then there exists at least one point between a and b at which the tangent is parallel to x-axis.

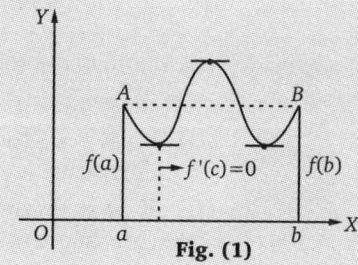

Fig. (1)

5.2.2 ALGEBRAIC INTERPRETATION OF ROLLE'S THEOREM

Algebraically, Rolle's theorem means that if $f(x)$ is a polynomial function in x and $x=a$ and $x=b$ are two roots of the equation $f(x)=0$, then, there is at least one root of the equation $f'(x)=0$ which lies between a and b.

5.3 LAGRANGE'S MEAN VALUE THEOREM

Let f be a function defined on $[a, b]$ such that
(i) f is continuous on $[a, b]$.   (ii) f is differentiable on $]a, b[$.
Then, there exists a real number $c \in]a,b[$ such that

$$\frac{f(b)-f(a)}{b-a} = f'(c)$$

Proof.   Let us define a function $F(x)$ such that

$$F(x)=f(x)+Ax \quad \forall x \in [a,b] \qquad \qquad ...(1)$$

where A is a constant to be suitably chosen such that $F(a)=F(b)$.
Now

(i) Since, f is continuous on $[a,b]$ and Ax is continuous on $[a,b]$ therefore, F is continuous on $[a,b]$

[$\because$ sum of two continuous functions is again continuous.]

(ii) Similarly F is differentiable on (a, b)

(iii) $F(a)=F(b) \Rightarrow -A = \dfrac{f(b)-f(a)}{b-a}$ ...(2)

Hence, we find that F satisfy all the conditions of Rolle's Theorem on $[a,b]$ and consequently, there exists a real number $c \in]a,b[$ such that $F'(c)=0$, this gives

$$f'(c)+A=0$$

$$\Rightarrow \qquad -A=f'(c). \qquad \qquad \text{...(3)}$$

Now, from (2) and (3), we have

$$\frac{f(b)-f(a)}{b-a}=f'(c)$$

REMARKS

- If we take $b=a+h$ and c can be written as $a+\theta h$, where θ is some real number such that $0<\theta<1$. Lagrange's theorem then read as follows :
  " Let f be defined and continuous on $[a, a+h]$ and differentiable on $]a, a+h[$, then for some real number $\theta(0<\theta<1)$

  $$\frac{f(a+h)-f(a)}{h}=f'(a+\theta h).$$

- The hypothesis of the Lagrange's mean value theorem can not be weakened, as it is clear from the following examples :
  " Let f be the function defined on $[-1,2]$ by setting $f(x)=|x|, \quad \forall x \in [-1,2]$.
  Here, f is continuous on $[-1,2]$ and differentiable at all points of $]-1, 2[$ except at $x=0$ (so that second condition is violated)

  Now $\qquad f'(x) = \begin{cases} -1 & \text{if } x \in]-1,0[\\ 1 & \text{if } x \in]0,2[\end{cases}$

  Also $\qquad \dfrac{f(2)-f(-1)}{2-(-1)} \ne f'(x)$ for any x in $]-1, 2[$.

- Lagrange's mean value theorem is known as first mean value theorem.
- The result $f(b)-f(a)=f(b-a)f'(c)$ is also known as the formula for finite increment.
- For $f(a)=f(b)$, the Lagrange's mean value theorem yields Rolle's theorem.

5.3.1 GEOMETRICAL INTERPRETATION OF LAGRANGE'S MEAN VALUE THEOREM

If the curve $y=f(x)$ is continuous from $x=a$ and $x=b$ and has a tangent at each point on the curve between $x=a$ and $x=b$, then, geometrically, the first mean value theorem means that there is at least one point between $x=a$ and $x=b$ on the curve where the tangent to the curve parallel to the chord joining the points $(a, f(a))$ and $(b, f(b))$.

Let ACB be the graph of the function $y= f(x)$ then the co-ordinate of the points A and B are given by $(a, f(a))$ and

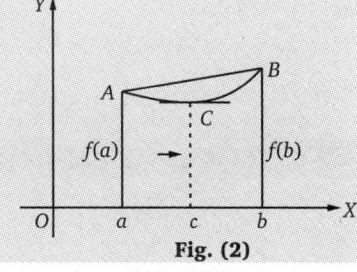

Fig. (2)

$(b, f(b))$ respectively. If the chord AB makes an angle θ with the x-axis, then

$$\tan\theta = \frac{f(b)-f(a)}{b-a} = f'(c), \text{ where } a < c < b.$$

5.3.2 DEDUCTION FROM THE FIRST MEAN VALUE THEOREM

THEOREM 1. *If a function $f(x)$ satisfies the conditions of mean value theorem then*

        (i) $f'(x) = 0 \; \forall \, x \in \,]a, b[\Rightarrow f$ is constant on $[a, b]$,

        (ii) $f'(x) > 0 \; \forall x \in]a,b[\Rightarrow f$ is strictly increasing on $[a,b]$,

   and (iii) $f'(x) < 0 \; \forall x \in]a,b[\Rightarrow f$ is strictly decreasing on $[a,b]$.

Proof. (i) Let x_1, x_2 (where $x_1 > x_2$) be any two distinict points of $[a,b]$, then by Lagrange's mean value theorem,

$$\frac{f(x_2) - f(x_1)}{x_2 - x_1} = f'(c) = 0, \; x_1 < c < x_2 \qquad \qquad ...(1)$$

$\Rightarrow$ $f(x_2) = f(x_1)$.

$\Rightarrow$ function keeps the same value. Therefore $f(x)$ is constant on $[a,b]$.

    (ii) From (1), we have

$$\frac{f(x_2) - f(x_1)}{x_2 - x_1} = f'(c) \text{ for some } c \in \,]x_1, \; x_2[$$

But $f'(c) > 0$ $[\because f'(x) > 0 \; \forall \, x \in [a, b]]$

$\Rightarrow$ $f(x_2) - f(x_1) > 0$.

$\Rightarrow$ $f(x_2) > f(x_1)$.

Thus $x_2 > x_1 \Rightarrow f(x_2) > f(x_1) \; \forall x_1, x_2 \in [a,b]$

Hence, f is strictly increasing on $[a,b]$.

   (iii) Same as (ii).

REMARK

- For a strictly increasing function f, the derivative $f'(x)$ need not be strictly positive. For example, consider $f(x) = x^3$, $x \in \,]-1, 1[$. Here, $f(x)$ is strictly increasing but $f'(x) = 3x^2$, which is zero at $x=0 \in \,]-1, 1[$.

Solved Examples

Example 1. Determine whether $f(x) = \dfrac{1}{x}, -1 < x < 0$ is strictly increasing, decreasing or neither of these.

Solution. Given that $f(x) = \dfrac{1}{x}, \; \Rightarrow f'(x) = -\dfrac{1}{x^2}$

For $-1 < x < 0$ $f'(x) = -\dfrac{1}{x^2} < 0$.

So $f(x)$ is decreasing in $-1 < x < 0$.

5.4 CAUCHY'S MEAN VALUE THEOREM

    Let f and g be two functions defined on $[a,b]$ such that

    (i) f and g are continuous on $[a, b]$,

(ii) f and g are differentiable on $]a, b[$,

and   (iii) $g'(x) \neq 0$ for any point of $]a, b[$.

Then, there exists a real number $c \in]a, b[$ such that

$$\frac{f(b)-f(a)}{g(b)-g(a)} = \frac{f'(c)}{g'(c)}$$

Proof.   Let us define a function

$$F(x) = f(x) + A.g(x) \qquad \qquad ...(1)$$

where A is a constant, to be suitably chosen such that

$$F(a) = F(b) \qquad \qquad ...(2)$$

Now, the function F is the sum of two continuous and differentiable functions. Therefore

        (i) F is continuous on $[a,b]$,

        (ii) F is differentiable on $]a,b[$,

and   (iii) $F(a) = F(b)$.

Then, by Rolle's theorem, there must exists a real number c between a and b such that

$$F'(c) = 0$$

Here, $F'(x) = f'(x) + Ag'(x)$

$$F'(c) = 0 \implies f'(c) + Ag'(c) = 0$$

$$\implies \qquad \qquad -A = \frac{f'(c)}{g'(c)} \qquad \qquad ...(3)$$

Now $F(a) = F(b) \implies f(a) + Ag(a) = f(b) + Ag(b)$

$$\implies \qquad \qquad -A = \frac{f(b)-f(a)}{g(b)-g(a)} \qquad \qquad ...(4)$$

From (3) and (4), we have

$$\frac{f(b)-f(a)}{g(b)-g(a)} = \frac{f'(c)}{g'(c)}.$$

REMARKS

- If we put $b = a + h$, then c can be written as $a + \theta h$, where $\theta \in R$ such that $0 < \theta < 1$, then Cauchy's mean value theorem can be restated as

  "If f and g are continuous on $[a, a+h]$ and are differentiable on $]a, a+h[$ and $g'(x) \neq 0$ for any $x \in]a, a+h[$ then, $\exists$ a $\theta \in R: 0 < \theta < 1$ such that

$$\frac{f(a+h)-f(a)}{g(a+h)-g(a)} = \frac{f'(a+\theta h)}{g'(a+\theta h)}.$$

- If we take $g(a) = g(b)$, then the function g would satisfy all the conditions of Rolle's theorem and consequently for some x in $]a,b[$, we would have $g'(x) = 0$. In view of this we take $g(a) \neq g(b)$.

- In some cases, the Lagrange's mean value theorem is a particular case of Cauchy's mean value theorem (e.g., take $g(x) = k$).

- Cauchy's mean value theorem cannot be deduced by applying Lagrange's mean value theorem to two functions f and g seperately and then dividing. It can be easily seen that the desired result can not be obtained in this manner. In this way, we get

$$\frac{f(b)-f(a)}{g(b)-g(a)} = \frac{f'(c_1)}{g'(c_2)}$$

where $a < c_1 < b$, and $a < c_2 < b$. But, it is not necessary that c_1 and c_2 are equal. Hence, Cauchy's means value theorem is not directly deducable from the first one.

- The conditions in the theorem are sufficient one. The conclusion may still hold even when the function involved do not satisfy the condition on $[a,b]$

5.4.1 GEOMETRICAL INTERPRETATION OF CAUCHY'S MEAN VALUE THEOREM

(1) Under suitable conditions, Cauchy's mean value theorem geometrically means that there is an ordinate $x=c$ between $x=a$ and $x=b$, such that the tangents at the points where $x=c$ cut the graphs of the function $f(x)$ and $\dfrac{f(b)-f(a)}{g(b)-g(a)} g(x)$ are mutually parallel.

(2) The ratio of the mean rates of increase of two functions in an interval is equal to the ratio of the actual rates of increase of the functions at some point within the interval.

Solved Examples

Example 1. *Discuss the applicability of Rolle's theorem in the internal $[-1,1]$ to the function $f(x)=|x|$.*

Solution. Here, we have $\qquad f(x)=|x|$

$$\Rightarrow \qquad \left.\begin{array}{l} f(-1)=1 \\ f(1)=1 \end{array}\right\} \Rightarrow f(1) = f(-1)$$

and

Now the function $f(x)$ is continuous throughout the closed interval $[-1, 1]$ but $f(x)$ is not differentiable at $x=0 \in]-1,1[$. Hence, Rolle's theorem is not satisfied (due to the second condition).

Example 2. *Verify Rolle's theorem the function $f(x) = x^3-4x$ on $[-2, 2]$.*

Solution. The function $f(x)=x^3-4x$ is a polynomial and so it is continuous and differentiable at all $x \in R$. In particular it is continuous in the closed interval $[-2,2]$ and differentiable in the open interval $]-2,2[$. Also $f(-2)=0=f(2)$.

Thus, $f(x)$ satisfies all the three conditions of Rolle's theorem in $[-2,2]$. Therefore, there must exist at least one real number 'x' in the open interval $]-2,2[$ for which $f'(x)=0$.

Also $\qquad f'(x) = x^3-4x$

Now $\qquad f'(x)=0$ gives $3x^2-4=0$ or $x=\pm\dfrac{2}{\sqrt{3}} = \pm 1.55$.

Both these values lie in the open interval $]-2, 2[$ and thus the conclusion of Rolle's theorem is verified.

Example 3. *Discuss the applicability of Rolle's theorem to the function*

$$f(x)= \log\left[\frac{x^2+ab}{(a+b)x}\right],\text{in the interval } [a, b]$$

Solution. Here, we have

$$f(a)= \log\left[\frac{a^2+ab}{(a+b)a}\right] = \log 1 = 0$$

and $\qquad f(b)= \log\left[\dfrac{b^2+ab}{(a+b)b}\right] = \log 1 = 0$

Also, it can be easily seen that $f(x)$ is continuous on $[a,b]$ and differentiable on $]a,b[$. Thus all the three conditions of Rolle's theorem are satisfied. Hence $f'(x)=0$ for at least one value of x in $]a, b[$.

Now
$$f'(x)=0 \Rightarrow \frac{2x}{x^2+ab} - \frac{1}{x} = 0$$

$\Rightarrow \qquad 2x^2-(x^2+ab)=0$

$\Rightarrow \qquad x^2=ab \text{ or } x=\sqrt{ab}.$

Obviously $\qquad \sqrt{ab} \in]a,b[\qquad$ [being the geometric mean of a and b]

Hence, the Rolle's theorem is verified.

Example 4. *Verify Rolle's theorem for the function $f(x)=2x^3+x^2-4x-2$.*

Solution . Since, $f(x)$ is a rational integral function of x, therefore it is continuous and differentiable for all real values of x.

Hence, the first two conditions of Rolle's theorem are satisfied in any interval.

Hence, $f(x)=0$ gives $2x^3+x^2-4x-2=0$

i.e., $\qquad x=\pm\sqrt{2}, -\dfrac{1}{2}$

$\Rightarrow \qquad f\left(\sqrt{2}\right)=f\left(-\sqrt{2}\right)=f\left(-\dfrac{1}{2}\right)=0$

Now take the interval $\left[-\sqrt{2},\sqrt{2}\right]$, then, all the conditions of Rolle's theorem are satisfied in this interval. Then, $\exists$ at least one value of c in $]-\sqrt{2},\sqrt{2}[$, such that $f'(c)=0$

$$f'(x)=0 \quad \Rightarrow \quad 6x^2+4x-4=0$$
$$\Rightarrow \quad x = -1, 2/3.$$

Since, both the points -1 and $2/3$ lies in the open interval $]-\sqrt{2},\sqrt{2}[$, . Hence, Rolle's theorem is verified.

Example 5. *Verify Rolle's theorem for $f(x)=x(x+3)e^{-x/2}$ in $[-3, 0]$.*

Solution . Here, we have
$$f(x)=x(x+3)e^{-x/2}$$

$\therefore \qquad f'(x) = (2x + 3)e^{-x/2} + \left(x^2 + 3x\right)e^{-x/2}.\left(-\dfrac{1}{2}\right)$

$\qquad = e^{-x/2}\left[2x+3-\dfrac{1}{2}\left(x^2+3x\right)\right] = -\dfrac{1}{2}\left[x^2 -x-6\right]e^{-x/2}$

$\Rightarrow$ $f'(x)$ exist for every value of x in the interval $[-3, 0]$. Hence, $f(x)$ is differentiable and continuous in the interval $[-3, 0]$. Also, we have
$$f(-3) = f(0) = 0$$

$\Rightarrow$ All the three conditions of Rolle's theorem are satisfied. So

$\qquad f'(x) = 0 \quad \Rightarrow \quad \dfrac{1}{2}\left(x^2 -x-6\right)e^{-x/2}=0$

$\qquad\qquad \Rightarrow \quad x^2-x-6=0 \qquad \Rightarrow \qquad x=3, -2.$

Since, the values $x = -2$ lies in the open interval $]-3, 0[$, the Rolle's theorem is verified.

Example 6. *Show that there is no real number p for which the equation $x^3-3x+p=0$, has two distinct roots in $]0,1[$.*

Solution. Let, if possible, there are two distinct roots a and b of the given equation in $]0, 1[$, such that $0 < a < b < 1$.

Now, let $\quad\quad\quad\quad f(x) = x^3-3x+p$

Obviously, $f(x)$ is continuous and differentiable for all values of x (being a polynomial)

Also, we have $\quad\quad\quad f(a) = f(b) = 0$

$\Rightarrow$ f satisfies all the conditions of Rolle's theorem in $[a,b]$ hence, $\exists$ a point $c \in]a,b[$ such that $f'(c)=0$.

Now $\quad\quad\quad\quad\quad f'(x) = 0 \Rightarrow 3x^2-3 = 0$

$\quad\quad\quad\quad\quad\quad\quad\quad\quad \Rightarrow x = \pm1$

which is a contradiction $\quad\quad\quad\quad\quad\quad$ $(\because a < c < b$ as $0 < a < b < 1)$

$\Rightarrow$ our assumption is wrong. Hence, there cannot be two distinct roots of $f(x) = 0$ in $]0, 1[$ for any value of p.

Example 7. *Verify the Rolle's theorem for the function $f(x)=x^2$ in $[-1, 1]$.*

Solution. Here, it can be easily seen that the function $f(x)=x^2$ is continuous as well as differentiable on R.

$\Rightarrow$ $f(x)$ is continuous and differentiable in $[-1,1]$.

Also, we have $\quad\quad\quad\quad f(1) = f(-1) = 1$.

Thus, $f(x)$ satisfies all the conditions of Rolle's theorem in $[-1,1]$.

$\Rightarrow$ $\exists$ at least one number, say c, in $]-1,1[$ such that $f'(c)=0$.

Now $\quad\quad\quad\quad\quad\quad f'(x) = 2x$

$\quad\quad\quad\quad\quad\quad f'(x) = 0 \quad\quad \Rightarrow \quad\quad x = 0$.

Since, the root $x = 0$ lies in the interval $]-1, 1[$. Hence, the Rolle's theorem is satisfied.

Example 8. *Verify Rolle's theorem for the function $f(x)= x^2-3x+2$ on the interval $[1,2]$.*

Solution. Here, it can be easily seen that $f(x) = x^2-3x+2$ is continuous as well as differentiable on R (being a polynomial)

$\Rightarrow$ $f(x)$ is continuous in $[1, 2]$ and differentiable in $]1, 2[$.

Also, we have $\quad\quad\quad\quad f(1) = f(2) = 0$.

Thus, $f(x)$ satisfies all the conditions of Rolle' theorem in $[1, 2]$

$\Rightarrow$ $\exists$ at least one number, say c, in $]1, 2[$ such that $f'(c) = 0$.

Now, $\quad\quad\quad\quad\quad\quad f'(x) = 2x-3$

$\quad\quad\quad\quad\quad\quad f'(x) = 0 \quad\quad \Rightarrow \quad\quad x = 3/2$.

Since, the root $x = 3/2$ lies in the interval $(1, 2)$. Hence, Rolle's theorem is verified.

Example 9. *If $a+b+c = 0$, then show that the quadratic equation $3ax^2+2bx+c = 0$ has at least one root in $]0, 1[$.*

Solution. Let us define a function $f(x)$ such that $f(x) = ax^3+bx^2+cx+d$.

Here we have $\quad\quad\quad\quad f(0) = d$ and $f(1) = a+b+c+d = d$ $\quad\quad\quad$ $(\because a+b+c = 0)$

Obviously, $f(x)$ is continuous and differentiable in $]0, 1[$ (being a polynomial).

Thus, $f(x)$ satisfies all the three conditions of Rolle's theorem in $[0, 1]$. Hence, there is at least one value of x in the open interval $]0, 1[$ where $f'(x) = 0$

i.e., $\quad\quad\quad\quad 3ax^2+2bx+c = 0$ has at least one root in $]0, 1[$.

Example 10. *Discuss the applicability of Rolle's Theorem to the function* $f(x)=x^{2/3}$ *in* $(-1,1)$

Solution. We have $f(x) = x^{2/3}$

$\Rightarrow$ $f'(x) = \frac{2}{3}x^{-1/3}$

$\therefore$ $\lim_{x \to 0} f'(x) = \lim_{h \to 0} \frac{2}{3}(0+h)^{-1/3} = +\infty$

Now, $Rf'(0) = \lim_{h \to 0} \left\{ \frac{f(0+h)-f(0)}{h} \right\} = \lim_{h \to 0} \left\{ \frac{h^{2/3}-1}{h} \right\} = +\infty$

and $Lf'(0) = \lim_{h \to 0} \left\{ \frac{f(0-h)-f(0)}{-h} \right\} = \lim_{h \to 0} \left\{ \frac{(-h)^{2/3}}{-h} \right\} = -\infty$

$\therefore$ $Lf'(0) \neq Rf'(0).$

$\therefore f'(0)$ does not exist showing that $f'(x)$ does not exist in the open interval $(-1, 1)$. Hence, Rolle's Theorem is not applicable although $f'(-1) = f(1) = 1$ and $f(x)$ is continuous in the closed interval $(-1,1)$.

Example 11. *Discuss the applicability of Rolle's theorem to the function*

$$f(x) = \begin{cases} x^2 +1 , & when\ 0 \le x \le 1 \\ 3-x , & when\ 1 < x \le 2 \end{cases}$$

Solution.   Here $f(0)=0^2+1$ and $f(2)=3-2=1.$

We shall show that $f(x)$ is continuous for all x in the range $(0,2)$

Also $f(1)=1^2+1=2$

Again, $f(1+0) = \lim_{x \to 1+0} (3-x) = \lim_{x \to 1+h} \left[3-(1+h) \right]$, when $h \to 0$

            $= \lim_{h \to 0} (2-h) = 2$

and $f(1-0) = \lim_{x \to 1-0} \left(x^2 +1 \right) = \lim_{x \to (1-h)} \left[(1-h)^2 +1 \right]$, when $h \to 0$

            $= \lim_{h \to 0} \left(2-2h+h^2 \right) = 2$

Hence, $f(1-0)=f(1)=f(1+0)$ and so the function $f(x)$ is continuous at $x=1$ and the continuous in the whole interval $(0,2)$.

Again, $f'(x) = \begin{cases} 2x , & when\ 0 \le x < 1 \\ -1 , & when\ 1 < x \le 2 \end{cases}$

$\therefore$ $f(x)$ is differentiable in the interval $(0,2)$ except at $x=1$.

Now $Rf'(1) = \lim_{h \to 0} \frac{f(1+h)-f(1)}{h} = \lim_{h \to 0} \frac{\{3-(1+h)\}-2}{h}$

            $= \lim_{h \to 0} \frac{2-h-2}{h} = \lim_{h \to 0} (-1) = -1$

And $Lf'(1) = \lim_{h \to 0} \frac{f(1-h)-f(1)}{-h} = \lim_{h \to 0} \frac{\left[(1-h)^2 +1 \right] -2}{-h}$

            $= \lim_{h \to 0} \frac{2h-h^2}{h} = \lim_{h \to 0} (2-h) = 2$

$\therefore$ Thus $Rf'(1) \neq Lf'(1)$ and so $f'(1)$ does not exist.

Hence, the function $f(x)$ is not differentiable in the entire range $(0, 2)$ and therefore Rolle's theorem is not applicable to the given function $f(x)$ in $(0, 2)$.

Example 12. *Verify Rolle's theorem for the function* $f(x) = x^3 - 6x^2 + 11x - 6$

Solution. Here, we have $f(x) = x^3 - 6x^2 + 11x - 6$, if $f(x) = 0$. Then $x^3 - 6x^2 + 11x - 6 = 0$

$\Rightarrow$ $\qquad (x-1)(x-2)(x-3) = 0 \qquad \therefore \qquad x = 1, 2, 3.$

$\Rightarrow$ $\qquad f(1) = 0 = f(2) = f(3)$

Also $\qquad f'(x) = 3x^2 - 12x + 11$

Now $\qquad Rf'(x) = \lim_{h \to 0} \dfrac{f(x+h) - f(x)}{h}$

$$= \lim_{h \to 0} \frac{\left[(x+h)^3 - 6(x+h)^2 + 11(x+h) - 6\right] - \left[x^3 - 6x^2 + 11x - 6\right]}{h}$$

$$= \lim_{h \to 0} \frac{\left\{(x+h)^3 - x^3\right\} - 6\left\{(x+h)^2 - x^2\right\} + 11\left\{(x+h) - x\right\}}{h}$$

$$= \lim_{h \to 0} \frac{(x+h)^3 - x^3}{h} - 6 \lim_{h \to 0} \frac{\left\{(x+h)^2 - x^2\right\}}{h} + 11 \lim_{h \to 0} \frac{(x+h) - x}{h}$$

$$= 3x^2 - 12x + 11$$

Similarly $\qquad Lf'(x) = \lim_{h \to 0} \dfrac{f(x-h) - f(x)}{-h}$

$$= \lim_{h \to 0} \frac{(x-h)^3 - x^3}{-h} - 6 \lim_{h \to 0} \frac{(x-h)^2 - x^2}{-h} + 11 \lim_{h \to 0} \frac{(x-h) - x}{-h}$$

$$= 3x^2 - 12x + 11$$

Since $Lf'(x) = Rf'(x)$, therefore $f'(x)$ exists for all values of x in $[1, 3]$.

Also $f(x)$ is continuous. Hence, all conditions of Rolle's Theorem are satisfied, and so $f'(x) = 0$ for at least one value of x in $[1, 3]$.

From (1), equating $f'(x) = 0$ where $3x^2 - 12x + 11 = 0$, we get

$$x = 2 \pm \frac{\sqrt{3}}{3}$$

$\therefore \qquad x = 2.577, 1.423.$

Both these above values lie in $[1, 3]$.

Example 13. *Verify Rolle's Theorem for the function* $f(x) = 10x - x^2$.

Solution. Here $\qquad f(x) = 0 \Rightarrow 10x - x^2 = 0 \qquad \Rightarrow x(10 - x) = 0$

$\Rightarrow x = 0, 10.$

Now, $\qquad f(0) = 0, f(10) = 0 \Rightarrow f(0) = 0 = f(10).$

Also, $\qquad f'(x) = 10 - 2x.$ $\qquad\qquad\qquad$ ...(1)

Now $\qquad Rf'(x) = \lim_{h \to 0} \dfrac{f(x+h) - f(x)}{h}$

$$= \lim_{h \to 0} \frac{\left[10(x+h) - (x+h)^2\right] - \left(10x - x^2\right)}{h}$$

$$= \lim_{h \to 0} \frac{\left(10x + 10h - x^2 - 2xh - h^2\right) - \left(10x - x^2\right)}{h}$$

$$= \lim_{h \to 0} \frac{10h - 2xh - h^2}{h} = \lim_{h \to 0} (10 - 2x - h) = 10 - 2x$$

Similarly, $\quad Lf'(x) = \lim\limits_{h \to 0} \dfrac{f(x-h) - f(x)}{-h}$

$$= \lim\limits_{h \to 0} \dfrac{\left[10(x-h) - (x-h)^2 \right] - \left(10x - x^2 \right)}{-h}$$

$$= \lim\limits_{h \to 0} \dfrac{-10h + 2xh - h^2}{-h} = 10 - 2x$$

Thus $Lf'(x) = Rf'(x)$. Therefore $f'(x)$ exists for all values of x in $[0, 10]$. Also $f(x)$ is continuous for all values of x in $[0,10]$.

Now, since every differentiable function is continuous. Hence, all the conditions of Rolle's Theorem are satisfied.

$\therefore \quad f'(x) = 0$ for at least one value of x in $[0,10]$.

From (1), equating $\quad f'(x) = 0 \implies 2x = 10 \implies x = 5$ which lies in $[0, 10]$.

Example 14. *Find 'c' of the mean value theorem, if $f(x) = x(x-1)(x-2); a = 0, b = 1/2$*

Solution. Here, we have $\quad f(a) = f(0) = 0$

$$f(b) = f\left(\frac{1}{2}\right) = \frac{3}{8}$$

$$\therefore \qquad \frac{f(b) - f(a)}{b - a} = \frac{\frac{3}{8} - 0}{\frac{1}{2} - 0} = \frac{3}{4}$$

Now $\qquad f(x) = x^3 - 3x^2 + 2x$

$\therefore \qquad f'(x) = 3x^2 - 6x + 2 \implies f'(c) = 3c^2 - 6c + 2$

Putting all these values in the Lagrange's mean value theorem

$$\frac{f(b) - f(a)}{b - a} = f'(c), (a < c < b)$$

we get $\qquad \dfrac{3}{4} = 3c^2 - 6c + 2$ or $c = 1 \pm \dfrac{\sqrt{21}}{6}$

Hence $c = \dfrac{1 - \sqrt{21}}{6}$ lies in the open interval $]0, \dfrac{1}{2}[$ which is the required value.

Example 15. *If $f(x) = \log x$, find all numbers strictly between e^2 and e^3 such that*

$$f'(x) = \frac{f\left(e^3\right) - f\left(e^2\right)}{e^3 - e^2}$$

Solution. Obviously $f(x) = \log x$ is continuous in $[e^2, e^3]$ and differentiable in $]e^2, e^3[$.

Then by Lagrange's mean value theorem. There exist $c \in]e^2, e^3]$, such that

$$f'(c) = \frac{f\left(e^3\right) - f\left(e^2\right)}{e^3 - e^2} \qquad \implies \qquad \frac{1}{c} = \frac{3 - 2}{e^3 - e^2}$$

$\therefore \qquad\qquad\qquad c = (e^3 - e^2).$

There exist only one value $c = (e^3 - e^2)$ in $]e^2, e^3[$.

Example 16. *Show that any chord of the parabola $y = Ax^2 + Bx + C$ is parallel to the tangent at the point whose abscissa is same as that of the middle point of the chord.*

Solution.   Let a and b (where $a<b$) be the abscissae of the ends of the chord and let $f(x)=Ax^2+Bx+C$. Obviously, $f(x)$ is continuous on $[a,b]$ and differentiable in $]a,b[$ (being a polynomial).

By Lagrange's mean value theorem there exists $c\in]a,b[$ such that

$$\frac{f(b)-f(a)}{b-a}=f'(c)$$

i.e., $Ab^2+Bb+C-Aa^2-Ba-C=(b-a)(2Ac+B)$

which gives $c=\dfrac{1}{2}(a+b)$ i.e., abscissa of the point at which the tangent is parallel to the chord is same as that of the middle point of the chord.

Example 17. *Separate the intervals in which the polynomial $2x^3-15x^2+36x+1$ is increasing or decreasing.*

Solution.   Here, we have $f(x)=2x^3-15x^2+36x+1$

$\therefore$ $f'(x)=6x^2-30x+36=6(x-2)(x-3)$.

Here $f'(x)>0$ for $x<2$ or for $x>3$.

                                         $f'(x)<0$ for $2<x<3$

and $f'(x)=0$ for $x=2,3$

Clearly, $f'(x)$ is positive in the intervals $]-\infty,2]$ and $[3,\infty[$ and negative in the interval $]2,3[$ Hence, the function $f(x)$ is monotonically increasing in the interval $]-\infty,2]$, $[3,\infty[$ and monotonically decreasing in $]2,3[$.

Example 18. *Use the function $f(x)=x^{1/x}$, $x>0$ show that $e^\pi>\pi^e$.*

Solution.   Here $f(x)=x^{1/x}$, $x>0$

$\therefore$ $\log f(x)=\dfrac{1}{x}\log_e x$

Differentiating w.r.t. x, we get

$$\frac{1}{f(x)}f'(x)=\frac{1}{x}.\frac{1}{x}-\frac{1}{x^2}\log_e x$$

$$f'(x)=\frac{x^{1/x}}{x^2}\left[1-\log_e x\right].$$

For $x>e$, $f'(x)<0$ $[\because \log_e x>1 \text{ for } x>e]$

$\therefore$ $f(x)$ is a decreasing function of x for $x>e$.

Hence $\pi>e\Rightarrow f(\pi)<f(e)\Rightarrow \pi^{1/\pi}<e^{1/e}$

$$\Rightarrow \left(\pi^{1/\pi}\right)^{e\pi}<\left(e^{1/e}\right)^{e\pi}$$

$$\Rightarrow \pi^e<e^\pi$$

$$\Rightarrow e^\pi>\pi^e.$$

Example 19. *Show that $\dfrac{x}{1+x}<\log(1+x)<x$, for $x>0$.*

Solution.   Let, $f(x)=\log(1+x)-\dfrac{x}{1+x}$

Obviously, $f(0)=0$.

and $f'(x)=\dfrac{1}{1+x}-\dfrac{1.(1+x)-x.1}{(1+x)^2}=\dfrac{1}{1+x}-\dfrac{1}{(1+x)^2}=\dfrac{x}{(1+x)^2}$

Here, we observe that $f'(x)>0$, for $x>0$.

$\Rightarrow$ $f(x)$ is monotonically increasing in the interval $[0,\infty[$. Therefore

$$f(x)>f(0), \qquad\qquad \text{for } x>0$$

$\Rightarrow$
$$\left[\log(1+x)-\frac{x}{1+x}\right]>0, \qquad\qquad \text{for } x>0$$

$\Rightarrow$
$$\log(1+x)>\frac{x}{1+x}, \qquad\qquad \text{for } x>0 \qquad\qquad ...(1)$$

Now let
$$F(x)=x-\log(1+x).$$

Obviously
$$F(0)=0$$

Then
$$F'(x)=1-\frac{1}{1+x}=\frac{x}{1+x}$$

Here, we observe that $F'(x)>0$, for $x>0$. Hence $F(x)$ is monotonically increasing in the interval $[0,\infty[$.

$\therefore$
$$F(x)>F(0), \qquad\qquad \text{for } x>0$$

$\Rightarrow$
$$[x-\log(1+x)]>0, \qquad\qquad \text{for } x>0$$

$\Rightarrow$
$$x>\log(1+x), \qquad\qquad \text{for } x>0 \qquad\qquad ...(2)$$

Now from (1) and (2), we get

$$\frac{x}{1+x}<\log(1+x)<x, \text{ for } x>0$$

Example 20. *Prove that* $(1+x)<e^x<1+xe^x$, $\forall\ x>0$.

Solution. Let us consider the function $f(x)=e^x$ in $[0,x]$.

Obviously $f(x)$ is continuous as well as differentiable in $]0,x[$.

Then, by Lagrange's theorem $\exists\ c\in\ ]0,x[$, such that

$$f'(c)=\frac{f(x)-f(0)}{x-0}$$

or
$$e^c=\frac{e^x-1}{x} \qquad\qquad ...(1)$$

$$0<c<x \quad\Rightarrow\ e^0<e^c<e^x \qquad (\because\ e^x \text{ is an increasing function})$$
$$...(2)$$

Now, from (1) and (2), we have

$$e^0<\frac{e^x-1}{x}<e^x, \forall x>0$$

$\Rightarrow$
$$1<\frac{e^x-1}{x}<e^x$$

$\Rightarrow$
$$x<e^x-1<xe^x$$

$\Rightarrow$
$$(1+x)<e^x<1+xe^x.$$

Example 21. *Let f be continuous on* $[a-h,a+h]$ *and differentiable in* $]a-h,a+h[$. *Prove that there is a real number* θ *between 0 and 1 such that*

$$f(a+h)-2f(a)+f(a-h)=h[f'(a+\theta h)-f'(a-\theta h)].$$

Solution. Consider the function ϕ defined on $[0,1]$ by $\phi=f(a+ht)+f(a-ht)\ \forall t\in[0,1]$.

Obviously ϕ is continuous on $[0,1]$ and differentiable on $]0,1[$.

Then, by Lagrange's mean value theorem, there is a number θ lying between 0 and

1 such that $\qquad \phi(1)-\phi(0)=(1-0)\phi'(\theta)$

i.e., $\quad f(a+h)-2f(a)+f(a-h)=h[f'(a+\theta h)-f'(a-\theta h)]$.

which is the required result.

Example 22. *Show that Lagrange's mean value theorem does not holds for the function $f(x)=|x|$ in the interval $[-1,1]$.*

Solution . Since $f(x)=|x|$ is a continuous function on $[-1,1]$ but it is not differentiable at $x=0\in]-1,1[$. Hence, Lagrange's mean value theorem does not hold for the function $f(x)=|x|$ in the interval $[-1,1]$.

Example 23. *Verify Lagrange's mean value theorem for the function $f(x)=\sin x$ in $\left[0,\dfrac{\pi}{2}\right]$.*

Solution . The function $f(x)=\sin x$ is continuous and differentiable on R. Hence it is continuous as well as differentiable in $[0, \pi/2]$. Then, by Lagrange's mean value theorem, there must exists at least one c in $]0,\pi/2[$ such that

$$\frac{f(\pi/2)-f(0)}{\pi/2-0}=f'(c) \qquad \qquad ...(1)$$

Here $\qquad\qquad f(0)=0,\ f(\pi/2)=1$

$$f'(x)=\cos x \quad \Rightarrow \qquad f'(c)=\cos c.$$

Put all these values in (1), we have

$$\frac{1-0}{\pi/2}=\cos c \Rightarrow \cos c=\frac{2}{\pi}\Rightarrow c=\cos^{-1}\left(\frac{2}{\pi}\right)$$

Since, $0<2/\pi<1$, therefore the value of $c=\cos^{-1}\left(\dfrac{2}{\pi}\right)$ lies in $\left]0,\dfrac{\pi}{2}\right[$, which is the required value of c. Hence, Lagrange's mean value theorem is verified.

Example 24. *If $f''(x)$ exist for all points in $[a, b]$ and $\dfrac{f(c)-f(a)}{c-a}=\dfrac{f(b)-f(c)}{b-c}$ where $a < c < b$, then, there is a number l such that*

$$a<l<b \text{ and } f''(l)=0.$$

Solution . Since $f''(x)$ exist for all points in $[a, b]$,

$\Rightarrow \quad f'(x)$ is continuous in $[a, b]$

$\Rightarrow \quad f(x)$ is continuous in $[a, b]$.

Now, applying Lagrange's mean value theorem to $f(x)$ in $[a, c]$ and $[c, b]$ respectively, we get

$$\frac{f(c)-f(a)}{c-a}=f'(l_1),\ a<l_1<c \qquad\qquad ...(1)$$

and $\qquad \dfrac{f(b)-f(c)}{b-c}=f'(l_2),\ c<l_2<b \qquad\qquad ...(2)$

Then, from (1) and (2), we get

$$f'(l_1)=f'(l_2) \qquad\qquad \left[\because \ \frac{f(c)-f(a)}{c-a}=\frac{f(b)-f(c)}{b-c}\right]$$

Now $f'(x)$ satisfies all the conditions of Rolle's theorem in the interval $[l_1, l_2]$.

Hence $\qquad\qquad f''(l)=0$ where $l \in]l_1, l_2[$ and $l \in]a, b[$.

Example 25. *If $f(x)=(x-1)(x-2)(x-3)$ and $a=0$, $b=4$, find 'c' using Langrange's mean value theorem.*

Solution .   We have $f(x)=(x-1)(x-2)(x-3)=x^3-6x^2+11x-6$

$$f(a)=f(0)=-6 \text{ and } f(b)=f(4)=6$$

$$\therefore \qquad \frac{f(b)-f(a)}{b-a}=\frac{6-(-6)}{4-0}=\frac{12}{4}=3.$$

Also $f'(x)=3x^2-12x+11$ gives $f'(c)=3c^2-12c+11.$

Putting these values in Lagrange's mean value theorem,

$$\frac{f(b)-f(a)}{b-a}=f'(c) \text{ where } a<c<b$$

we get

$$3=3c^2-12c+11 \text{ or } 3c^2-12c+8=0$$

$$c=\frac{12\pm\sqrt{(144-96)}}{6}=2\pm\frac{2\sqrt{3}}{3}$$

As the value of c lies in the open interval $]0,4[$. Hence both of these are the required values of c.

Example 26.   *Examine if mean value theorem applies to $f(x)=x^3+3x^2-5x$ in the interval $[1,2]$. If it does, then find the intermediate point whose existence is asserted by the theorem.*

Solution .   Given $f(x)=x^3+3x^2-5x.$ ...(1)

$$\therefore \qquad f'(x)=3x^2+6x-5 \text{ and } f'(c)=3c^2+6c-5. \qquad \qquad ...(2)$$

Let $a=1$ and $b=2$, then from (1), we have

$$f(a)=f(1)=1^3+3(1)^2-5(1)=-1.$$

$$f(b)=f(2)=2^3+3(2)^2-5(2)=10.$$

From mean value theorem, we have

$$f(b)-f(a)=(b-a).f'(c) \quad \Rightarrow \quad f(2)-f(1)=(2-1).f'(c)$$

$$\Rightarrow \qquad 10-(-1)=(2-1)f'(c) \quad \Rightarrow \quad 3c^2+6c-5=11 \qquad [\text{using}(2)]$$

$$\Rightarrow \qquad 3c^2+6c-16=0.$$

$$\therefore \qquad c=-1\pm 2.55 \quad i.e., \quad c=-3.55, 1.55.$$

Example 27.   *Verify Cauchy's mean value theorem for the functions $f(x)=x^2-2x+3$, $g(x)=x^3-7x^2+26x-5$ in the interval $[-1, 1]$.*

Solution .   Since $f(x)$ and $g(x)$ are polynomial in x, so these are continuous in the closed interval $[-1, 1]$ and also differentiable and continuous in the open interval $(-1,1)$.

Also $g'(x)=3x^2-14x+26$

$$g'(-1)=3(-1)^2-14(-1)+26=43=+\text{ve}$$

$$g'(1)=3(1)^2-14(1)+26=15=+\text{ve}.$$

Therefore, $g'(x)\neq 0$ for any value of x in $(-1,1)$.

Hence all the conditions of Cauchy Mean Value Theorem are satisfied.

Then, using, $\dfrac{f(b)-f(a)}{g(b)-g(a)}=\dfrac{f'(c)}{g'(c)}$

Putting $a=-1, b=1$ (given), we have

$$\frac{f(1)-f(-1)}{g(1)-g(-1)}=\frac{f'(c)}{g'(c)}$$

$$\frac{\left[1^2-2(1)+3\right]-\left[(-1)^2-2(-1)+3\right]}{\left[1^3-7(1)^2+26(1)-5\right]-\left[(-1)^3-7(-1)^2+26(-1)-5\right]}=\frac{2c-2}{2c^2-14c+26}$$

$$[\because f'(x)=2x-2]$$

or $$\frac{2-6}{15-(-39)}=\frac{2c-2}{3c^2-14c+26}$$

or $$-4(3c^2-14c+26)=54\times2(c-1)$$

or $$3c^2+14c+26=-27(c-1)$$

or $$3c^2+13c-1=0$$

$$\Rightarrow \qquad c=\frac{-13\pm\sqrt{(181)}}{6}=\frac{-13\pm13.454}{6}$$

i.e., $$c=0.076,-4.409$$

Since the value 0.076 lies in [−1,1]. Hence, Cauchy mean value theorem is verified.

Example 28. *Verify Cauchy's mean value theorem for the function x^2 and x^3 in the interval [1,2].*

Solution. Let us suppose $f(x)=x^2$ and $g(x)=x^3$.

Then, obviously $f(x)$ and $g(x)$ are continuous in [1,2] and differentiable in]1,2[.

Also $g'(x)=3x^2\neq0$ for any point in]1,2[.

Then, by Cauchy's mean value theorem there exist at least one real number $c\in]1,2[$, such that

$$\frac{f(2)-f(1)}{g(2)-g(1)}=\frac{f'(c)}{g'(c)} \qquad\qquad ....(1)$$

After solving, we get $c=\dfrac{14}{9}$, which lies in the open interval]1,2[. Hence, Cauchy's mean value theorem is verified.

Example 29. *Use Cauchy's mean value theorem, to evaluate $\lim\limits_{x\to1}\left[\dfrac{\cos\dfrac{\pi x}{2}}{\log(1/x)}\right]$.*

Solution. Let us suppose

$$f(x)=\cos\left(\frac{1}{2}\pi x\right), g(x)=\log x$$

$$a=x \quad\text{and}\quad b=1$$

Putting all these values in Cauchy's mean value theorem

$$\frac{f(b)-f(a)}{g(b)-g(a)}=\frac{f'(c)}{g'(c)}, a<c<b$$

we get $$\frac{\cos\dfrac{\pi}{2}-\cos\dfrac{n\pi}{2}}{\log1-\log x}=\frac{-\dfrac{1}{2}\pi\sin\left(\dfrac{\pi c}{2}\right)}{1/c};x<c<1$$

Now, taking the limit as $x\to1$, which give that $c\to1$, therefore

$$\lim_{x\to1}\left\{\frac{\left[0-\cos\left(\dfrac{1}{2}\pi x\right)\right]}{\log(1/x)}\right\}=\lim_{c\to1}\left\{\frac{\left[-\dfrac{1}{2}\pi\sin\left(\dfrac{1}{2}\pi c\right)\right]}{(1/c)}\right\}$$

or $$\lim_{x\to1}\left\{\frac{\left[-\cos\left(\dfrac{1}{2}\pi x\right)\right]}{\log(1/x)}\right\}=-\frac{1}{2}\pi \qquad\left(\because \sin\frac{1}{2}\pi c\to1\text{ as }c\to1\right)$$

or
$$\lim_{x \to 1} \left\{ \frac{\cos\left(\frac{1}{2}\pi x\right)}{\log(1/x)} \right\} = \frac{\pi}{2}.$$

Example 30. *If in the Cauchy's mean value theorem, we write $f(x)=e^x$ and $g(x)=e^{-x}$, show that 'c' is the arithmetic mean between a and b.*

Solution. Since, we have
$$f(x)=e^x \text{ and } g(x)=e^{-x}$$

$\therefore$
$$\frac{f(b)-f(a)}{g(b)-g(a)} = \frac{e^b - e^a}{e^{-b} - e^{-a}} = -e^a e^b = -e^{a+b}$$

and
$$\frac{f'(x)}{g'(x)} = \frac{e^x}{-e^{-x}} \text{ so that } \frac{f'(c)}{g'(c)} = \frac{e^c}{-e^{-c}} = -e^{2c}$$

After putting all these values in Cauchy's mean value theorem, we get
$$-e^{a+b} = -e^{2c} \quad \Rightarrow \quad a+b=2c$$

$\Rightarrow$
$$c = \frac{a+b}{2}$$

Hence, c is the arithmetic mean between a and b.

Example 31. *If $f(x)$, $g(x)$ and $h(x)$ are functions such that*
  (i) *$f(x)$, $g(x)$ and $h(x)$ are continuous on $[a,b]$*
  (ii) *$f(x)$, $g(x)$ and $h(x)$ are differentiable on $]a,b[$,*

then
$$\begin{vmatrix} f'(c) & g'(c) & h'(c) \\ f(b) & g(b) & h(b) \\ f(a) & g(a) & h(a) \end{vmatrix} = 0 \text{ where } c \in \;]a, b[$$

Solution. Consider the function $F(x)$ such that
$$F(x) = \begin{vmatrix} f(x) & g(x) & h(x) \\ f(b) & g(b) & h(b) \\ f(a) & g(a) & h(a) \end{vmatrix} = 0 \qquad \qquad ...(1)$$

Obviously, $F(x)$ is of the form $A\, f(x) + B\, g(x) + C\, h(x)$, where A, B, C are some real numbers. From the condition (i) and (ii), $F(x)$ is continuous on. $[a,b]$ and differentiable on $]a,b[$.
Also $F(a)=F(b)=0$.
$\Rightarrow$ $F(x)$ satisfies all the conditions of Rolle's theorem. Hence, there exists a $c \in]a,b[$ such that $F'(c)=0$

i.e.,
$$\begin{vmatrix} f'(c) & g'(c) & h'(c) \\ f(b) & g(b) & h(b) \\ f(a) & g(a) & h(a) \end{vmatrix} = 0.$$

Example 32. *Verify Cauchy's mean value for $f(x)=\sin x$ and $g(x)=\cos x$ in $\left[-\dfrac{\pi}{2},0\right]$*

Solution. It can be easily seen that $f(x)$ and $g(x)$ both are continuous on $\left[-\dfrac{\pi}{2},0\right]$ and differentiable on $\left]-\dfrac{\pi}{2},0\right[$.

Also, $g'(x)=-\sin x \neq 0$ for any point in the interval $\left]-\dfrac{\pi}{2},0\right[$.

Then, by Cauchy's mean value theorem, $\exists$ at least one $c \in \left] -\dfrac{\pi}{2}, 0 \right[$ such that

$$\frac{f(0) - f\left(-\dfrac{\pi}{2}\right)}{g(0) - g\left(-\dfrac{\pi}{2}\right)} = \frac{f'(c)}{g'(c)}$$

Putting all the values and after simplification, we have

$$\cot c = -1 \implies c = -\pi/4.$$

Since $c = -\pi/4$ lies in $]-\pi/2, 0[$, hence, Cauchy mean value theorem is verified.

Example 33. *Show that* $\dfrac{\sin\alpha - \sin\beta}{\cos\beta - \cos\alpha} = \cot\theta$

Solution. Let $f(x) = \sin x$ and $g(x) = \cos x$, for $x \in [\alpha, \beta]$, where $0 < \alpha < \beta < \pi/2$.

$\therefore$ $\qquad f'(x) = \cos x$ and $g'(x) = -\sin x$.

It can be easily seen that both the function $f(x)$ and $g(x)$ are continuous in the closed interval $[\alpha, \beta]$ and differentiable in the open interval $]\alpha, \beta[$.

Hence, by cauchy's mean value theorem there exist at least one $\theta \in R$, $\theta \in]\alpha, \beta[$ such that

$$\frac{f(\beta) - f(\alpha)}{g(\beta) - g(\alpha)} = \frac{f'(\theta)}{g'(\theta)}$$

$\implies$ $\qquad \dfrac{\sin\beta - \sin\alpha}{\cos\beta - \cos\alpha} = \dfrac{\cos\theta}{-\sin\theta} = -\cot\theta$

$\implies$ $\qquad \dfrac{\sin\alpha - \sin\beta}{\cos\beta - \cos\alpha} = \cot\theta$, where $0 < \alpha < \theta < \beta < \pi/2$.

Example 34. *Show that the function f and g defined on $\left[0, \dfrac{1}{2}\right]$, by $f(x) = x(x-1)(x-2)$ and $g(x) = x(x-2)(x-3)$ satisfy the condition of Cauchy's mean value theorem.*

Solution. Here, we have

$$f(x) = x(x-1)(x-2) = x^3 - 3x^2 + 2x$$

and $\qquad g(x) = x(x-2)(x-3) = x^3 - 5x^2 + 6x$

$\implies$ $\qquad f'(x) = 3x^2 - 6x + 2$ and $g'(x) = 3x^2 - 10x + 6$

By Cauchy's mean value theorem, we have

$$\frac{f'(c)}{g'(c)} = \frac{f\left(\dfrac{1}{2}\right) - f(0)}{g\left(\dfrac{1}{2}\right) - g(0)}, \quad c \in \left]0, \dfrac{1}{2}\right[$$

or $\qquad \dfrac{3c^2 - 6c + 2}{3c^2 - 10c + 6} = \dfrac{\dfrac{3}{8} - 0}{\dfrac{15}{8} - 0} = \dfrac{1}{5}$

$\implies$ $\qquad 12c^2 - 20c + 4 = 0$

$\implies$ $\qquad c = \dfrac{5 \pm \sqrt{13}}{6}$

The value $\dfrac{5 - \sqrt{13}}{6}$ of c belongs to $\left]0, \dfrac{1}{2}\right[$.

Hence, the Cauchy mean value theorem is satisfied.

Example 35. *Find 'c' of Cauchy's mean value theorem for the functions*

$$f(x) = \sqrt{x}, \ \phi(x) = \frac{1}{\sqrt{x}} \text{ in } [a,b]$$

and show that it is the G.M. of a and b.

Solution .   We have

    (i) $f(x)$ and $\phi(x)$ are continuous in the closed interval $[a,b]$.

    (ii) $f'(x)$ and $\phi'(x)$ exist in the open interval (a,b).

    (iii) $\phi'(x) = -1/2\, x^{-3/2} \neq 0$ for any x in $]a,b[$.

Therefore $f(x)$ and $\phi(x)$ satisfies all the conditions of Cauchy's mean value theorem. Hence there exist a point $c \in]a,b[$ such that

$$\frac{f(b) - f(a)}{\phi(b) - \phi(a)} = \frac{f'(c)}{\phi'(c)} \qquad \qquad \ldots(1)$$

Also here $f'(x) = \dfrac{1}{2}x^{-1/2}, \phi'(x) = -\dfrac{1}{2}x^{-3/2}$

From (1), we get

$$\frac{\sqrt{b} - \sqrt{a}}{\left(1/\sqrt{b}\right) - \left(1/\sqrt{a}\right)} = \frac{1/2c^{-1/2}}{-1/2c^{-3/2}}$$

or $$\frac{\left(\sqrt{b} - \sqrt{a}\right)\sqrt{a}.\sqrt{b}}{\sqrt{a} - \sqrt{b}} = -\frac{c^{3/2}}{c^{1/2}}$$

$\therefore$ $$c = \sqrt{ab} \ .$$

5.5 TAYLOR'S THEOREM

Let $f(x)$ be a single valued function defined on $[a, a+h]$ such that

    (i) all the derivative of $f(x)$ upto $(n-1)^{th}$ order are continuous in $[a, a+h]$, and

    (ii) $f^n(x)$ exists in $(a, a+h)$

then there exists a real number $\theta, 0 < \theta < 1$, such that

$$f(a+h) = f(a) + hf'(a) + \frac{h^2}{2!}f''(a) + \ldots + \frac{h^{n-1}}{(n-1)!}f^{n-1}(a) + \frac{h^n(1-\theta)^{n-p}}{p(n-1)!}f^n(a+\theta h)$$

where p is a given positive integer.

Proof.   Since, f^n exists, all the derivative $f', f'' .. f^{n-1}$ exist and continuous on $[a, a+h]$, consider a function f defined on $[a, a+h]$ such that

$$\phi(x) = f(x) + (a+h-x)f'(x) + \frac{(a+h-x)^2}{2!}f''(x) + \ldots \qquad \ldots(1)$$

$$+ \frac{(a+h-x)^{n-1}}{(n-1)!}f^{n-1}(x) + A(a+h-x)^p$$

where A is a constant to be determined such that $\phi(a+h) = \phi(a)$

Now $$\phi(a) = f(a) + hf'(a) + \frac{h^2}{2!}f''(a) + \ldots + \frac{h^{n-1}}{(n-1)!}f^{n-1}(a) + Ah^p$$

and $$\phi(a) = f(a+h)$$

$\Rightarrow$ $$f(a+h) = f(a) + hf'(a) + \frac{h^2}{2!}f''(a) + \ldots + \frac{h^{n-1}}{(n-1)!}f^{n-1}(a) + Ah^p \qquad \ldots(2)$$

Now

(i) $f, f', f'', ..., f^{n-1}$ being all continuous on $[a, a+h]$ the function ϕ is continuous on $[a, a+h]$,

(ii) Similarly the function ϕ is differentiable on $]a, a+h[$,

and (iii) $\phi(a+h) = \phi(a)$.

Thus, the function ϕ satisfies all the conditions of Rolle's theorem and hence $\exists$ a real number $\theta(0 < \theta < 1)$ such that

$$\phi'(a+\theta h) = 0.$$

Here

$$\phi'(x) = f'(x) + (-f'(x) + (a+h-x)f''(x))$$
$$+ \frac{1}{2!}\left[-2(a+h-x)f''(x) + (a+h-x)^2 f'''(x)\right] + ...$$
$$+ \frac{1}{(n-1)!}\left[-(n-1)(a+h-x)^{n-2} f^{n-1}(x)\right.$$
$$\left. + (a+h-x)^{n-1} f^n(x)\right] - Ap(a+h-x)^{p-1}$$
$$= \frac{(a+h-x)^{n-1}}{(n-1)!} f^n(x) - Ap(a+h-x)^{p-1}$$

[Other terms canceled in pairs]

$$\therefore 0 = \phi'(a+\theta h) = \frac{h^{n-1}(1-\theta)^{n-1}}{(n-1)!} f^n(a+\theta h) - Aph^{p-1}(1-\theta)^{p-1}$$

$$\Rightarrow A = \frac{h^{n-1}(1-\theta)^{n-p}}{p(n-1)!} f^n(a+\theta h), h \neq 0, \theta \neq 1$$

Now, putting the values of A in (2), we get

$$f(a+h) = f(a) + hf'(a) + \frac{h^2}{2!} f''(a) + ... + \frac{h^{n-1}}{(n-1)!} f^{n-1}(a) + \frac{h^n(1-\theta)^{n-p}}{p(n-1)!} f^n(a+\theta h)$$

5.5.1 FORMS OF REMAINDER AFTER N TERMS

(i) The term $R_n = \frac{h^n(1-\theta)^{n-1}}{P(n-1)!} f^n(a+\theta h)$ which occur after n terms, is called the Taylor's remainder after n terms. The theorem with this form of remainder is called Taylor's theorem with Scholomilch and Roche form of remainder.

(ii) For $p=1$, we get

$$R_n = \frac{h^n(1-\theta)^{n-1}}{P(n-1)!} f^n(a+\theta h)$$

Then, R_n is called Cauchy's form of remainder.

(iii) For $p=n$, we get

$$R_n = \frac{h^n}{n!} f^n(a+\theta h)$$

then, R_n is called Lagrange's form of remainder.

5.5.2 ANOTHER FORM OF TAYLOR'S THEOREM

Replacing h by $(x-a)$ in Taylor's theorem, we get

$$f(x) = f(a) + (x-a)f'(a) + \frac{(x-a)^2}{2!} f''(a) + ... + \frac{(x-a)^{n-1}}{(n-1)!} f^{n-1}(a) + R_n$$

The remainder, after n terms can be written as

$$R_n = \frac{(x-a)^n (1-\theta)^{n-p}}{p(n-1)!} f^n(c), a < c < x.$$

Deductions

Putting $a=0$ in second form of Taylor's theorem, we get (Maclaurin's theorem)

$$f(x) = f(0) + x f'(0) + \frac{x^2}{2!} f''(0) + \ldots + \frac{x^{n-1}}{(n-1)!} f^{n-1}(0) + R_n \qquad \ldots(1)$$

(i) If $R_n = \dfrac{x^n (1-\theta)^{n-p}}{p(n-1)!} f^n(\theta x)$, then (1) is known as Maclaurin's theorem with Schlomilch and Roche's form of remainder.

(ii) For $p=1$, $R_n = \dfrac{x^n (1-\theta)^{n-p}}{p(n-1)!} f^n(\theta x)$ is called Cauchy's form of remainder.

(iii) For $p=n$, $R_n = \dfrac{x^n}{n!} f^n(\theta x)$, is called Lagrange's form of remainder.

5.5.3 TAYLOR'S SERIES

Let $f(x)$ possesess continuous derivatives of all orders in the interval $[a, a+h]$, then for every positive integral value of n, we have

$$f(a+h) = f(a) + hf'(a) + \frac{h^2}{2!} f''(a) + \ldots + \frac{h^{n-1}}{(n-1)!} f^{n-1}(a) + R_n$$

where, $\qquad R_n = \dfrac{h^n}{n!} f^n(a+\theta h), (0 < \theta < 1).$ $\qquad \ldots(1)$

Equation (1) can also be written as

$$S_n = f(a) + hf'(a) + \frac{h^2}{2!} f''(a) + \ldots + \frac{h^{n-1}}{(n-1)!} f^{n-1}(a)$$

Then $\qquad f(a+h) = S_n + R_n.$

Let us suppose $R_n \to 0$ as $n \to \infty$, then $\lim\limits_{n \to \infty} S_n = f(a+h)$

i.e., the series $f(a) + hf'(a) + \dfrac{h^2}{2!} f''(a) + \ldots + \dfrac{h^{n-1}}{(n-1)!} f^{n-1}(a) + \ldots$ converges to $f(a+h)$.

Thus,

(i) If f possess a continuous derivatives of every order in $[a, a+h]$.

(ii) The remainder after n terms $R_n \to 0$ as $n \to \infty$, then

$$f(a+h) = f(a) + hf'(a) + \frac{h^2}{2!} f''(a) + \ldots + \frac{h^n}{n!} f^n(a) + \ldots$$

This series is known as Taylor's series for the expansion of $f(a+h)$ as a power series in h.

5.5.4 MACLAURIN'S SERIES

If we put $a=0$ and replace h by x in Taylor's series, we get

$$f(x) = f(0) + x f'(0) + \frac{x^2}{2!} f''(0) + \ldots + \frac{x^n}{n!} f^n(0) + \ldots$$

This series is known as Maclaurin's series for the expansion of $f(x)$ as a power series in x.

REMARKS

- Maclaurin's series is a particular case of Taylor's series.
- Maclaurin's expansions of $f(x)$ fails if any of the functions $f(x), f'(x), f''(x)\ldots$ becomes infinite or discontinuous at any point of the interval $[0, x]$ or if R_n does not tends to zero as n tends to infinity.

5.6 MACLAURIN'S THEOREM

Let $f(x)$ be a function of x which possesses continuous derivatives of all orders in the interval $[0,x]$ and can be expanded as an infinite series in x, then

$$f(x) = f(0) + x f'(0) + \frac{x^2}{2!} f''(0) + \dots + \frac{x^n}{n!} f^n(0) + \dots$$

Proof. Let us define

$$f(x) = A_0 + A_1 x + A_2 x^2 + A_3 x^3 + \dots \qquad \dots(1)$$

Let the expression (1) be differentiable term by term any number of times. Then by successive differentiation, we have

$$f'(x) = A_1 + 2A_2 x + 3A_3 x^2 + 4A_4 x^3 + \dots$$

$$f''(x) = 2.1.A_2 + 3.2.A_3 x + 4.3.A_4 x^2 + \dots$$

$$f'''(x) = 3.2.A_3 + 4.3.2.A_4 x + \dots$$

$$\dots\dots\dots\dots\dots\dots\dots\dots\dots\dots\dots\dots\dots\dots\dots$$

Putting $x=0$, we get

$$f(0) = A_0, f'(0) = A_1, f''(0) = 2!A_2, f'''(0) = 3!A_3 \dots.$$

$$\Rightarrow \qquad A_0 = f(0), A_1 = f'(0), A_2 = \frac{f''(0)}{2!}, A_3 = \frac{f'''(0)}{3!} \dots$$

Substitute all these values in (1), we get

$$f(x) = f(0) + x f'(0) + \frac{x^2}{2!} f''(0) + \dots + \frac{x^n}{n!} f^n(0) + \dots$$

REMARKS

- The Maclaurin's theorem is a particular case of Taylor's Theorem, and can be obtained by replacing $a=0$ and $h=x$ in Taylor's theorem.
- If the function $f(x)$ is denoted by y, then the expansion may be written in the form

$$y = y(0) + x.y_1(0) + \frac{x^2}{2!} y_2(0) + \dots + \frac{x^n}{n!} y_n(0) + \dots$$

where $y(0), y_1(0), y_2(0), \dots, y_n(0)$ etc. denotes values of $y, y_1, y_2, \dots, y_n$ respectively for $x=0$.

5.7 FAILURE OF TAYLOR'S AND MACLAURIN'S THEOREM

(a) *Taylor's theorem fails to expand $f(a+h)$ in an infinite power series in the following cases :*
- If any of the function $f(x), f'(x), f''(x)\dots$ become infinite or does not exists for any value of x in the given interval.
- If R_n does not tends to zero as $n \to \infty$.

(b) *Maclaurin's theorem fails to expand $f(x)$ in an infinite power series in the following cases :*
- If any of the function $f(x), f'(x), f''(x)\dots$ becomes infinite or does not exist in interval $[0, x]$.
- If R_n does not tends to zero as $n \to \infty$.

REMARK

- Before expanding a given function as an infinite Taylor's or Maclaurin's series, it is essential to examine the behaviour of R_n as $n \to \infty$, which is not simple in many cases. We, therefore, generally obtain the expansion by assuming the possibility of expanding it in an infinite series by assuming that $R_n \to 0$ as $n \to \infty$.

5.8 POWER SERIES EXPANSIONS OF SOME STANDARD FUNCTIONS

To find the power series expansion we shall use the following procedure.

Step (1) *Put the given function equal to f(x).*

Step (2) *Differentiate f(x), a number of times and obtain $f'(x), f''(x), f'''(x)$... and so on.*

Step (3) *Put $x = 0$ and find $f(0), f'(0), f''(0)$... and so on.*

Step (4) *Substitute the values of $f(0), f'(0), f''(0), f'''(0),$... in*

$$f(x) = f(0) + x f'(0) + \frac{x^2}{2!} f''(0) + ...$$

We shall now consider Maclaurin's series expansions of the function e^x, sin x, cos x, $(1+x)^m$ and log x.

(i) Expansion of e^x. Let $f(x) = e^x \ \forall x \in R$.

Then $\qquad\qquad\qquad f^n(x) = e^x \ \forall x \in R$.

Thus, for each positive n, f^n is defined in the interval $[-h, h]$.

Writing, Lagrange's form of remainder, after n terms

$$R_n(x) = \frac{x^n}{n!} f^n(\theta x), \ \theta \in R, \ 0 < \theta < 1$$

$$= \frac{x^n}{n!} e^{\theta x}$$

Now, we shall show that $\lim_{n \to \infty} R_n(x) = 0$. Here, it is enough to show that $e^{\theta x}$ is bounded in $[-h, h]$ and $\lim_{n \to \infty} \frac{x^n}{n!} = 0$.

Since, $0 < \theta < 1$ and $x \in [-h, h]$, therefore $|\theta x| < h$ and consequently, $0 < e^{\theta x} < e^h$, hence $e^{\theta x}$ is bounded.

Now, let us write $\qquad\qquad a_n = \frac{x^n}{n!} \ \forall n \in N$.

Then $\qquad\qquad\qquad \frac{a_{n+1}}{a_n} = \frac{x}{n+1} \Rightarrow \lim_{n \to \infty} \frac{a_{n+1}}{a_n} = 0$

$\Rightarrow \lim_{n \to \infty} a_n$ exists and equal to zero.

Now, $\qquad \lim_{n \to \infty} R_n(x) = e^{\theta x} \left[\lim \frac{x^n}{n!} \right] = 0$

Hence, we find that the function $f(x)$ has a Maclaurin's series expansions for each $x \in [-h, h]$. This implies

$$f(x) = f(0) + x f'(0) + \frac{x^2}{2!} f''(0) + ... + \frac{x^{n-1}}{(n-1)!} f^{n-1}(0) + ... \quad \forall x \in R.$$

Substituting $f(x) = e^x, f'(x) = e^x, ..., f^n(x) = e^x$ at $x = 0$, we have

$$e^x = 1 + x + \frac{x^2}{2!} + \frac{x^3}{3!} + ... + \frac{x^{n-1}}{(n-1)!} + ... \quad \forall x \in R$$

(ii) Expansion of sin x. Let $f(x) = \sin x, \quad \forall x \in R$

$\Rightarrow \qquad f^n(x) = \sin\left(x + \dfrac{n\pi}{2}\right), \qquad \forall x \in R$

Writing, Lagrange's form of remainder after n terms, we have

$$R_n(x) = \frac{x^n}{n!} f^n(\theta x), \text{ where } 0 < \theta < 1$$

$$= \frac{x^n}{n!} \sin\left(\theta x + \frac{n\pi}{2}\right)$$

Now, for all $x \in R$,

$$\left|R_n(x)\right| \leq \left|\frac{x^n}{n!}\right|$$

and

$$\lim_{n \to \infty} \frac{x^n}{n!} = 0 \qquad\qquad\qquad\qquad \text{[as in (i)]}$$

$$\lim_{n \to \infty} R_n(x) = 0$$

Thus, we find that, the function $f(x)$ has a Maclaurin's series expansions for each x in $[-h,h]$. Hence, we have

$$f(x) = f(0) + xf'(0) + \frac{x^2}{2!} f''(0) + \dots + \frac{x^{n-1}}{(n-1)!} f^{n-1}(0) + \dots \quad \forall x \in R.$$

Now, substituting $f(x) = \sin x, f^n(x) = \sin \dfrac{n\pi}{2}$, we have

$$\sin x = x - \frac{x^3}{3!} + \frac{x^5}{5!} - \dots \quad \forall x \in R.$$

(iii) Expansion of cos x.

Let $\qquad\qquad f(x) = \cos x, \quad \forall x \in R$

Then $\qquad\qquad f^n(x) = \cos\left(x + \dfrac{n\pi}{2}\right)$

Thus, for each n, f^n is defined in every interval $[-h, h]$.

Writing, Lagrange's remainder after n terms, we have

$$R_n(x) = \frac{x^n}{n!} f^n(\theta x), \text{ where } 0 < \theta < 1$$

$$= \frac{x^n}{n!} \cos\left(\theta x + \frac{n\pi}{2}\right)$$

Now, for all $x \in R$,

$$\left|R_n(x)\right| \leq \left|\frac{x^n}{n!}\right|$$

and

$$\lim_{n \to \infty} \frac{x^n}{n!} = 0 \qquad\qquad\qquad\qquad \text{[as in (i)]}$$

Thus, we find that, the function f has a Maclaurin's series expansions for each $x \in [-h, h]$, which gives

$$f(x) = f(0) + x f'(0) + \frac{x^2}{2!} f''(0) + \dots + \frac{x^n}{n!} f^n(0) + \dots \quad \forall x \in R.$$

Now, substituting $f(x) = \cos x..., f^n(0) = \cos \dfrac{n\pi}{2}$, we have

$$\cos x = 1 - \frac{x^2}{2!} + \frac{x^4}{4!} - \dots \quad \forall x \in R.$$

(iv) Expansion of $(1+x)^m$.

Case (i). Let m is a positive integer, then letting

$$f(x) = (1+x)^m, \quad \forall x \in R.$$

We find that for each $n \in N$, $f^n(x)$ exist for all $x \in R$, and whenever $n > m$, $f^n(x) = 0$ $\forall x \in R$.

$\Rightarrow$ $R_n(x) = 0$, whenever $n > m$.

Hence, $\lim\limits_{n \to \infty} R_n(x) = 0$ and for all $x \in R$, we have

$$f(x) = f(0) + x f'(0) + \ldots + \frac{x^m}{m!} f^m(0), \qquad (\because \text{All other terms must vanish.})$$

Substituting the value of $f(x)$, $f(0), \ldots, f^m(0)$, We have

$$(1+x)^m = 1 + mx + \frac{m(m-1)}{2!} x^2 + \ldots + x^m$$

Case (ii). Let m not be a positive integer (may be a fraction or negative integer).

Here, we find that, if we write

$$f(x) = (1+x)^m, \text{ whenever } x \neq -1$$

then $f^n(x) = m(m-1)\ldots(m-n+1)(1+x)^{m-n}$, whenever $x \neq -1$.

Thus, for each positive integer n, f^n is defined in $[-h, h]$ for each h between 0 and 1.

Now, writing Cauchy's form of remainder after n terms, we have

$$R_n(x) = \frac{x^n (1-\theta)^{n-1}}{(n-1)!} f^n(\theta x), \text{ where } 0 < \theta < 1$$

$$= \frac{x^n (1-\theta)^{n-1}}{(n-1)!} m(m-1)\ldots(m-n+1)(1+\theta x)^{m-n}$$

$$= \frac{m(m+1)\ldots(m+n+1)}{(n-1)!} x^n \left(\frac{1-\theta}{1+\theta x}\right)^{n-1} \cdot (1+\theta x)^{m-1}$$

Now, we observe that

(a) $\lim\limits_{n \to \infty} \dfrac{m(m-1)\ldots(m-n+1)}{(n-1)!} x^n = 0$

If we write $a_n = \dfrac{m(m+1)\ldots(m-n+1)}{(n-1)!} x^n$

Then, we have $\dfrac{a_{n+1}}{a_n} = \dfrac{(m-n)x}{n}$ $\Rightarrow$ $\lim\limits_{n \to \infty} \dfrac{a_{n+1}}{a_n} = -x$

If follows that if $|x| < 1$, then $\lim\limits_{n \to \infty} a_n = 0$

(b) $\lim\limits_{n \to \infty} \left(\dfrac{1-\theta}{1+\theta x}\right)^{n-1} = 0$

In fact, since $0 < \theta < 1$ and $-1 < x < 1$, therefore, $0 < \left[\dfrac{1-\theta}{1+\theta x}\right] < 1$

and hence $\lim\limits_{n \to \infty} \left[\dfrac{1-\theta}{1+\theta x}\right]^{n-1} = 0$

(c)   If $m > 1$, then $(1+\theta x)^{m-1} < (1-|x|)^{m-1}$

For (a), (b) and (c), we find that for all x in $]-1, 1[$ $\lim\limits_{n \to \infty} R_n(x) = 0$

Thus, we find that for each h between 0 and 1, the function f has Maclaurin's series expansion for all $x \in [-h, h]$.

Hence, we have

$$f(x) = f(0) + x f'(0) + \frac{x^2}{2!} f''(0) + \ldots + \frac{x^{n-1}}{(n-1)!} f^{n-1}(0) + \ldots \quad \forall x \in]-1, 1[.$$

Substituting the values of $f(x), f(0), f'(0), ... f^{n-1}(0)$, we have

$$(1+x)^m = 1 + mx + \frac{m(m-1)}{2!}x^2 + \frac{m(m-1)(m-2)}{3!}x^3 + ...$$

$$+ \frac{m(m-1)...(m-n+1)}{n!}x^n + ... \text{whenever } -1 < x < 1$$

(v) Expansion of $\log_e(1+x)$.

Let $\qquad\qquad\qquad f(x) = \log(1+x), -1 < x < 1.$

Then $\qquad\qquad\qquad f^n(x) = \dfrac{(-1)^{n-1}(n-1)!}{(1+x)^n}$, whenever $x > -1$.

Now, we shall consider the following cases :

Case (a) Let $0 \leq x \leq 1$. Writing Lagrange's form of remainder after n terms, we have

$$R_n = \frac{x^n}{n!}f^n(\theta x) = \frac{x^n}{n!}(-1)^{n-1}\frac{(n-1)!}{(1+\theta x)^n} = \frac{(-1)^{n-1}}{n}\left(\frac{x}{1+\theta x}\right)^n$$

Since, $0 \leq x \leq 1, 0 < \theta < 1$, therefore

$$0 < \frac{x}{1+\theta x} < 1$$

$\therefore \qquad\qquad\qquad |R_n| < \dfrac{1}{n}, \text{and } \dfrac{1}{n} \to 0 \text{ as } n \to \infty$

Therefore $\qquad\qquad \lim_{n\to\infty} R_n = 0.$

Case (b) Let $-1 < x < 0$. Since in this case $\left|\dfrac{x}{1+\theta x}\right|$ need not be less than unity, therefore, we may not be able to show easily that $R_n \to 0$ as $n \to \infty$ by considering Lagrange's remainder. Now, writing Cauchy's form of remainder, we have

$$R_n = \frac{x^n}{(n-1)!}(1-\theta)^{n-1}f^n(\theta x)$$

$$= (-1)^{n-1}x^n\left(\frac{1-\theta}{1+\theta x}\right)^{n-1}\cdot\frac{1}{1+\theta x}$$

since $\qquad\qquad |x| < 1$

therefore $\qquad \left|\dfrac{1-\theta}{1+\theta x}\right| < 1$, so that $\left|\left(\dfrac{1-\theta}{1+\theta x}\right)^{n-1}\right| < 1$ and $\left|\dfrac{1}{1+\theta x}\right| < \dfrac{1}{1-|x|}$

Thus $\qquad\qquad |R_n| < \dfrac{|x|^n}{1-|x|}$

This implies that $\lim_{n\to\infty} R_n = 0.$, since $|x| < 1$. Thus we find that if $-1 \leq x \leq 1$, then $\lim_{n\to\infty} R_n = 0.$

$$f(x) = f(0) + xf'(0) + \frac{x^2}{2!}f''(0) + ... + \frac{x^{n-1}}{(n-1)!}f^{n-1}(0) + ... \text{ whenever } -1 < x \leq 1.$$

Substituting the values of $f(x), f(0), f'(0), ..., f^{n-1}(0), ...,$ we get

$$\log(1+x) = x - \frac{x^2}{2} + \frac{x^3}{3} - ..., \text{ whenever } -1 < x \leq 1.$$

Solved Examples

Example 1. *Show that*
$$a^x = 1 + x \log a + \frac{x^2}{2!}(\log a)^2 + ... + \frac{x^{n-1}}{(n-1)!}(\log a)^{n-1} + \frac{x^n}{n!}a^{\theta x}(\log a)^n, 0 < \theta < 1.$$

Solution. Let $f(x) = a^x$ ...(1)

Then $f^n(x) = a^x(\log a)^n \ \forall n \in N$ and $\forall x \in R$ ...(2)

Now, putting $x = 0$, in (1) and (2), we get
$$f(0) = 1, f^n(0) = (\log a)^n \ \forall \ n \in N.$$

From (2) $f^n(\theta x) = a^{\theta x}(\log a)^n$.

Now, by Maclaurin's series with Lagrange's form of remainder after n terms we have
$$f(x) = f(0) + x f'(0) + \frac{x^2}{2!}f''(0) + ... + \frac{x^{n-1}}{(n-1)!}f^{n-1}(0) + \frac{x^n}{n!}a^{\theta x}(\log a)^n \quad ...(3)$$

Now, substituting the above values in (3), we get
$$a^x = 1 + x \log a + \frac{x^2}{2!}(\log a)^2 + ... + \frac{x^{n-1}}{(n-1)!}(\log a)^{n-1} + \frac{x^n}{n!}a^{\theta x}(\log a)^n.$$

Here, Lagrange's form of remainder after n terms
$$R_n = \frac{x^n}{n!}a^{\theta x}(\log a)^n \text{ where } 0 < \theta < 1.$$

Example 2. *Expand $e^{a \sin^{-1} x}$ by Maclaurin's series and find the general term. Hence, show that*
$$e^\theta = 1 + \sin \theta + \frac{1}{2!}\sin^2 \theta + \frac{2}{3!}\sin^3 \theta + ...$$

Solution. Here $y = e^{a \sin^{-1} x}$ ...(1)

Then $y_1 = e^{a \sin^{-1} x} \cdot \dfrac{a}{\sqrt{1-x^2}} = \dfrac{ay}{\sqrt{1-x^2}}$ ...(2)

$\Rightarrow$ $\left(\sqrt{1-x^2}\right)y_1 = ay$

$\Rightarrow$ $\left(1-x^2\right)y_1^2 - a^2 y^2 = 0$ ...(3)

Now, differentiating both the sides, we have
$$\left(1-x^2\right)2y_1 y_2 - 2xy_1^2 - 2a^2 yy_1 = 0$$

$\Rightarrow$ $2y_1\left[\left(1-x^2\right)y_2 - xy_1 - a^2 y\right] = 0$ ...(4)

Since $2y_1 \neq 0$ hence $[(1-x^2)y_2 - xy_1 - a^2 y] = 0$.

Now, differentiating n times by Leibnitz theorem, we get
$$\left(1-x^2\right)y_{n+2} + ny_{n+1}(-2x) + \frac{n(n-1)}{2}y_n(-2) - y_{n+1}x - ny_n \cdot 1 - a^2 y_n = 0$$

$\Rightarrow$ $\left(1-x^2\right)y_{n+2} - (2n+1)xy_{n+1} - \left(n^2+a^2\right)y_n = 0$ ...(5)

Now, we can easily find, (from (1) to (5)) the following values
$$(y)_0 = 1, \ (y_1)_0 = a, \ (y_2)_0 = a^2$$
$$(y_{n+2})_0 = (n^2 + a^2)(y_n)_0. \quad ...(6)$$

Replacing n by $(n-2)$ in (6), we get
$$(y_n)_0 = [(n-2)^2 + a^2](y_{n-2})_0$$
$$= [(n-2)^2 + a^2][(n-4)^2 + a^2](y_{n-4})_0$$

If n is odd, then

$$(y_n)_0 = [(n-2)^2+a^2][(n-4)^2+a^2]...(3^2+a^2)(1^2+a^2)(y_1)_0$$
$$= [(n-2)^2+a^2][(n-4)^2+a^2]...[(3^2+a^2)(1^2+a^2)].a$$

If n is even, then

$$(y_n)_0 = [(n-2)^2+a^2][(n-4)^2+a^2]...(4^2+a^2)(2^2+a^2)(y_2)_0$$
$$= [(n-2)^2+a^2][(n-4)^2+a^2]...[(4^2+a^2)(2^2+a^2)].a^2$$

Hence, $\quad y_n(0) = \begin{cases} a\left(1^2+a^2\right)\left(3^2+a^2\right)...\left[(n-2)^2+a^2\right], & \text{if } n \text{ is odd} \\ a^2\left(2^2+a^2\right)\left(4^2+a^2\right)...\left[(n-2)^2+a^2\right], & \text{if } n \text{ is even} \end{cases}$

Putting $n=1,2,3,4,...$ in (6), we get

$$(y_3)_0=(3^2+a^2)(1^2+a^2)a, (y_6)_0=(4^2+a^2)(2^2+a^2)a^2 \text{ etc.}$$

Now putting all these values in the Maclaurin's theorem

$$y=(y)_0+x.(y_1)_0+\frac{x^2}{2!}(y_2)_0+...+\frac{x^n}{n!}(y_n)_0+...$$

We have $\quad e^{a\sin^{-1}x} = 1+ax+\frac{a^2}{2!}x^2+\frac{a(1^2+a^2)}{3!}x^3+\frac{a(2^2+a^2)}{4!}x^4+...$

The general term is $\frac{x_n}{n!}(y_n)_0$.

Now putting $x=\sin\theta$ and $a=1$, in the above equation, we get

$$e^{\theta} = 1+\sin\theta+\frac{1}{2!}\sin^2\theta+\frac{2}{3!}\sin^3\theta+...$$

Example 3. *Expand log sin$(x+h)$ in powers of h by Taylor's theorem.*

Solution. Let $\qquad f(x)=\log\sin(x)$

$\Rightarrow \qquad f(x+h)=\log\sin(x+h).$

Expanding $f(x+h)$ by Taylor's theorem in powers of h, we have

$$f(x+h)= f(x)+hf'(x)+\frac{h^2}{2!}f''(x)+\frac{h^3}{3!}f'''(x)+... \qquad ...(1)$$

Now $\qquad f(x)=\log\sin x \qquad \Rightarrow \qquad f'(x)=\cot x$

$\qquad f''(x)=-\text{cosec}^2 x \qquad \Rightarrow \qquad f'''(x)= 2\,\text{cosec}^2 x\cot x$ etc.

Substituting all these values in equation (1), we get

$$\log\sin(x+h)= \log\sin x + h\cot x - \frac{h^2}{2!}\text{cosec}^2 x+\frac{2h^3}{3!}\text{cosec}^2 x\cot x+...$$

Example 4. *Expand sin x in powers of $\left(x-\frac{\pi}{2}\right)$ with the help of Taylor's theorem.*

Solution. Let $f(x)=\sin x.$

Since, we want to expand $f(x)$ in powers of $\left(x-\frac{\pi}{2}\right)$, hence, we can write

$$f(x)= f\left[\frac{\pi}{2}+\left(x-\frac{\pi}{2}\right)\right]$$

Now, expanding by Taylor's theorem, we get

$$f(x)= f\left[\left(\frac{\pi}{2}\right)+\left(x-\frac{\pi}{2}\right)\right]$$

$$= f\left(\frac{\pi}{2}\right)+\left(x-\frac{\pi}{2}\right)f'\left(\frac{\pi}{2}\right)+\frac{1}{2!}\left(x-\frac{\pi}{2}\right)^2 f''\left(\frac{\pi}{2}\right)+\frac{1}{3!}\left(x-\frac{\pi}{2}\right)^3 f'''\left(\frac{\pi}{2}\right)+......(1)$$

Now　　　$f(x) = \sin x$　　　　$\Rightarrow$　　　$f\left(\dfrac{\pi}{2}\right) = 1$

　　　　　$f'(x) = \cos x$　　　　$\Rightarrow$　　　$f'\left(\dfrac{\pi}{2}\right) = 0$

　　　　　$f''(x) = -\sin x$　　　$\Rightarrow$　　　$f''\left(\dfrac{\pi}{2}\right) = -1$

　　　　　$f'''(x) = -\cos x$　　　$\Rightarrow$　　　$f'''\left(\dfrac{\pi}{2}\right) = 0$

and so on.

Substituting all these values in (1), we get

$$\sin x = 1 + \left(x - \frac{\pi}{2}\right).0 + \frac{1}{2!}\left(x - \frac{\pi}{2}\right)^2.(-1) + \frac{1}{3!}\left(x - \frac{\pi}{2}\right)^3.0 + \frac{1}{4!}\left(x - \frac{\pi}{2}\right)^4 + \dots$$

$$= 1 - \frac{1}{2!}\left(x - \frac{\pi}{2}\right)^2 + \frac{1}{4!}\left(x - \frac{\pi}{2}\right)^4 + \dots$$

Example 5. If $f(x) = (x-a)^{5/2}$ and $f(x+h) = f(x) + hf'(x) + \dfrac{h^2}{2!}f''(x + \theta h)$ find the value of θ.

Solution . Here, we have

$$f(x) = (x-a)^{5/2} \qquad \Rightarrow \qquad f(x+h) = (x+h-a)^{5/2}$$

$$\Rightarrow \qquad f'(x) = \frac{5}{2}(x-a)^{3/2} \text{ and } f''(x) = \frac{15}{4}(x-a)^{1/2}$$

$$\therefore \qquad f''(x + \theta h) = \frac{15}{4}(x + \theta h - a)^{1/2}$$

Putting all these values in the given relation, we have

$$(x+h-a)^{5/2} = (x-a)^{5/2} + \frac{5}{2}h(x-a)^{3/2} + \frac{15}{4}(x+\theta h - a)^{1/2}\frac{h^2}{2!}$$

Now, taking limit as $x \to a$, we have

$$h^{5/2} = \frac{15}{4}(\theta h)^{1/2}\frac{h^2}{2!}$$

$$\Rightarrow \qquad \theta = \frac{64}{225}.$$

Example 6. Let f is twice differentiable function and $|f| < \alpha$, $|f''| < \beta$, for $x > a$, then show that $|f'| < 2\sqrt{\alpha\beta} \;\; \forall x > a$.

Solution . Let us suppose $x > a$ and $h > 0$, then

$$f(x+h) = f(x) + hf'(x) + \frac{h^2}{2!}f''(x + \theta h), \;\; 0 < \theta < 1$$

$$\Rightarrow \qquad hf'(x) = f(x+h) - f(x) - \frac{h^2}{2!}f''(x + \theta h)$$

$$\Rightarrow \qquad |hf'(x)| = \left|f(x+h) - f(x) - \frac{h^2}{2!}f''(x + \theta h)\right|$$

$$\leq |f(x+h)| + |-f(x)| + \frac{h^2}{2!}|-f''(x + \theta h)|$$

[By using triangular inequality]

$$< \alpha + \alpha + \frac{h^2}{2}\beta = 2\alpha + \frac{h^2}{2}\beta$$

$$\Rightarrow \qquad |f'(x)| < \frac{2\alpha}{h} + \frac{h}{2}\beta = F(h)\,(\text{say})$$

Now, $|f'(x)|$ is independent of h and also less than $F(h)$ for all values of h.

Therefore $|f'(x)|$ must be less than the minimum value of $F(h)$.

For, maxima or minima of $F(h)$, we have

$$F'(h) = 0$$

$$\Rightarrow \qquad -\frac{2\alpha}{h^2} + \frac{\beta}{2} = 0 \qquad\qquad \Rightarrow \qquad h = \pm 2\sqrt{\frac{\alpha}{\beta}}$$

and $\qquad\qquad F''(h) = \frac{2\alpha}{h^3} > 0 \text{ for } h = 2\sqrt{\frac{\alpha}{\beta}}$

Hence $f(h)$ is minimum for $h = 2\sqrt{\frac{\alpha}{\beta}}$,

the minimum value of $F(h)$ is $= 2\alpha . \frac{1}{2}\sqrt{\frac{\alpha}{\beta}} + \frac{\beta}{2} . 2\sqrt{\frac{\alpha}{\beta}} = 2\sqrt{\alpha\beta}$

Hence $\qquad\qquad\qquad |f'(x)| < 2\sqrt{\alpha\beta}$.

EXERCISE 5.1

1. Discuss the applicability of Rolle's theorem of the following functions :
   (a) $f(x) = 2 + (x-1)^{2/3}$ in the interval $[0,2]$
   (b) $f(x) = x^2$ in $2 \le x \le 3$
   (c) $f(x) = \tan x$ in $0 \le x \le \pi$
   (d) $f(x) = x^4 - 3x^2 + 4$ in the interval $[-4,4]$
   (e) $f(x) = 1/(x^2+1)$ in the interval $[-3,3]$
   (f) $f(x) = e^x \sin x$ in the interval $[0,\pi]$
   (g) $f(x) = |x|$ in the interval $[-1,1]$
   (h) $f(x) = (x-2)\sqrt{x}$ in the interval $[0,2]$
   (i) $f(x) = (x-a)^m (x-b)^n, m,n \in Z^+$ in the interval $[a,b]$.

2. Show that between any two roots of $e^x \cos x = 1$, there exists at least one root of $e^x \sin x - 1 = 0$.

3. Let $\dfrac{a_0}{n+1} + \dfrac{a_1}{n} + \dfrac{a_2}{n-1} + ... + \dfrac{a_{n-1}}{2} + a_n = 0.$
   Show that there exists at least one real x between 0 and 1 such that

   $$a_0 x^n + a_1 x^{n-1} + ... + a_n = 0.$$

4. Verify the Rolle's theorem for the following functions:
   (a) $f(x) = x^4 - 1$ on the interval $[-1,1]$
   (b) $f(x) = e^x(\sin x - \cos x)$ in $\left(\dfrac{\pi}{4}, \dfrac{5\pi}{4}\right)$.

5. If $f(x) = \begin{vmatrix} \sin x & \sin\alpha & \sin\beta \\ \cos x & \cos\alpha & \cos\beta \\ \tan x & \tan\alpha & \tan\beta \end{vmatrix}$ where
   $0 < \alpha < \beta < \dfrac{\pi}{2}$. Show that $f'(l) = 0$, where $\alpha < l < \beta$.

6. A function $f(x)$ is continuous in the closed interval $[0,1]$ and differentiable in the open interval $]0,1[$ prove that
   $$f'(x_1) = f(1) - f(0), \; 0 < x_1 < 1.$$

7. Show that the set of all x for which $\log(1+x) \le x$ is equal to $[0,\infty[$.

8. Compute the value of θ in the first mean value theorem $f(x+h) = f(x) + hf'(x+\theta h)$
   if $\qquad f(x) = ax^2 + bx + c$.

9. Show that $x^n - a = 0$ has atmost one real positive root if n is a positive integer.

10. Show that the function f', if it exists in an interval, cannot have an ordinary or removable discontinuity in that interval.

11. Verify the Lagrange's mean value theorem for the following functions :
    (a) $f(x) = x^3$ in $[-1,1]$
    (b) $f(x) = \sin x$ in $[0,\pi/2]$
    (c) $f(x) = x^{2n}$ in $[-1,1], n \in Z^+$
    (d) $f(x) = 2x^2 - 7x + 10, x \in [2,5]$

12. Find the value of c, of mean value theorem, when
    (a) $f(x) = \sqrt{x^2 - 4}$ in the interval $[2,4]$
    (b) $f(x) = 2x^2 + 3x + 4$ in the interval $[1,2]$
    (c) $f(x) = x(x-1)$ in the interval $[1,2]$

13. (a) If $f(x) = \sqrt{x}$ and $g(x) = 1/\sqrt{x}$, then show by Cauchy's mean value theorem that c is the geometric mean between a and b.

(b) If $f(x) = \dfrac{1}{x^2}$ and $g(x) = \dfrac{1}{x}$, then show that c is the harmonic mean between a and b.

14. If f'' exists and continuous on $[a,b]$ and differentiable on $]a,b[$, then prove that

$$f(b) - f(a) - \frac{1}{2}(b-a)\{f'(a) - f'(b)\}$$

$$= -\frac{(b-a)^3}{12} f'''(d)$$

where $d \in R$ such that $d \in]a, b[$.

15. Prove that

$$\sin ax =$$

$$ax - \frac{a^3 x^3}{3!} + \frac{a^5 x^5}{5!} - \dots + \frac{a^{n-1} x^{n-1}}{(n-1)!}$$

$$\sin\left(\frac{n-1}{2}.\pi\right) + \frac{a^n x^n}{n!} \sin\left(a\theta x + \frac{n\pi}{2}\right)$$

16. If $f(x) = f(0) + xf'(0) + \dfrac{x^2}{2!} f''(\theta x)$ find the value of θ as $x \to 1$, $f(x)$ being $(1-x)^{5/2}$.

17. Show that the number θ which occurs in the Taylor's Theorem with Lagrange's form of remainder after n terms approaches the limit

$$\frac{f^{n+1}(a)}{(n+1)}$$ as $h \to 0$ provided that $f^{n+1}(x)$ is

continuous and different from zero as $x \to a$.

18. Show that the function $x^3 - 3x^2 + 3x + 2$ is monotonically increasing in every interval.

19. Obtain by Maclaurin's theorem the expansion of $e^{\sin x}$.

20. If $f(x) = \exp\left[-\dfrac{1}{x^2}\right]$, for $x \neq 0$ and $f(0) = 0$, then show that :

    (i) $f^n(0) = 0$ $\forall n = 0, 1, 2, \dots$

and (ii) The Taylor's series for f about 0 agrees with $f(x)$ only at $x = 0$.

21. Expand "log sec x" by Maclaurin's series expansion, upto the term containing x^6.

22. If $x > 0$, show that

$$x - \frac{x^2}{2} + \frac{x^3}{3(1+x)} < \log(1+x) < x - \frac{x^2}{2} + \frac{x^3}{3}.$$

HINTS TO SELECTED PROBLEMS

1. (a) Since $f'(x)$ does not exist at $x = 1$, the second condition of Rolle's theorem is not satisfied.

2. Let a, b be two distinct roots of $e^x \cos x - 1 = 0$.

Then $e^a \cos a = 1$ and $e^b \cos b = 1$

Define a function $f(x) = e^{-x} - \cos x$.

5. $f(x)$ can be written as

$$f(x) = (\cos \alpha \tan \beta - \cos \beta \tan \alpha) \sin x$$
$$- (\sin \alpha \tan \beta - \sin \beta \tan \alpha) \cos x$$
$$+ (\sin \alpha \cos \beta - \sin \beta \cos \alpha) \tan x$$

Since $\sin x, \cos x, \tan x$ have finite derivatives in $]0, \pi/2[\Rightarrow f'(x)$ exists.

Also, $f(\alpha) = f(\beta)$. Hence, all the conditions of Rolle's theorem are satisfied.

7. Let us suppose $f(x) = \log(1+x) - x$

$$\Rightarrow f(0) = 0$$

$$f'(x) = \frac{1}{1+x} - 1 = \frac{-x}{1+x} \leq 0$$

$\Rightarrow f(x)$ is a decreasing function.

$\Rightarrow f(x) \leq f(0) \ \forall x \geq 0$

$\Rightarrow \log(1+x) - x \leq 0$

$\Rightarrow \quad \log(1+x) \leq x$

9. $f'(x) = nx^{n-1}$. Clearly $f(x)$ is an increasing function.

Let $x_1, x_2 \in]0, \infty[$ and $0 < x < r < x_2$ such that $f(r) = 0$.

Then $f(x_1) < f(r) < f(x_2) \Rightarrow f(x_1) < 0 < f(x_2)$

$\Rightarrow$ If $x \neq r$, $f(x) \neq 0$ on $(0, \infty)$.

$\Rightarrow x^n - a$ has at most one real positive root.

14. Define two functions $g(x)$ and $h(x)$ such that

$$g(x) = f(x) - f(a) - \frac{1}{2}(x-a)$$
$$\{f'(a) + f'(x)\} + A(x-a)^3$$

and $\quad h(x) = \dfrac{1}{2}[f'(x) - f'(a)]^{-1/2}$

$$(x-a) f''(x) + 3A(x-a)^2$$

Clearly, $g(x)$ and $h(x)$ satisfying all conditions of Rolle's theorem. Then use Rolle's Theorem for both the above functions.

18. Since $f'(x) \geq 0$ in $]-\infty, 1]$. Hence, it is monotonically increasing.

22. $f'(x) = \dfrac{1}{1+x} - 1 + x - \dfrac{3x^2 + 2x^3}{3\left(1+x^2\right)}$

$= \dfrac{x^3}{3(1+x)^2} > 0$

f is increasing $\Rightarrow f(x) > f(0) = 0$ for $x > 0$

$x - \dfrac{x^2}{2} + \dfrac{x^3}{3(1+x)} < \log(1+x)$ if $x > 0$

Now, $g'(x) = 1 - x + x^2 - \dfrac{1}{1+x} = \dfrac{x^3}{1+x} > 0$

g is increasing $\Rightarrow g(x) > g(0)$

$\log(1+x) < x - \dfrac{x^2}{2} + \dfrac{x^3}{3}$

Answers

1. (a) Not applicable (b) Not applicable (c) Not applicable (d) Verified
   (e) Verified (f) Verified (g) Not applicable (h) Not applicable
   (i) Verified

4. (a) Verified (b) Verified 8. $\theta = \dfrac{1}{2}$

11. (a) Verified (b) Verified (c) Verified (d) Verified
12. (a) $c = \pm\sqrt{6}$ (b) $c = 3/2$

16. $\theta = \dfrac{9}{25}$ 19. $y = 1 + x + \dfrac{x^2}{2} - \dfrac{x^4}{8} + ...$ 21. $y = \dfrac{x^2}{2} + \dfrac{x^4}{12} + \dfrac{x^6}{45} + ...$

5.9 SOME MORE EXPANSIONS

Example 1. *Expand* $tan^{-1} x$.

Solution. Let $\qquad f(x) = tan^{-1} x \Rightarrow f(0) = 0$

$f'(x) = \dfrac{1}{1+x^2} \Rightarrow f'(0) = 1$

$= (1+x^2)^{-1} = 1 - x^2 + x^4 - x^6 + ...$ (By Binomial expansion)

$f''(x) = -2x + 4x^3 - 6x^5 + ... \Rightarrow f''(0) = 0$

$f'''(x) = -2 + 12x^2 - 30x^4 + ... \Rightarrow f'''(0) = -2$

$f^{iv}(x) = 24x - 120x^3 + ... \qquad \Rightarrow f^{iv}(0) = 0$

$f^{v}(x) = 24 - 360x^2 + ... \qquad \Rightarrow f^{v}(0) = 24$

Put all these values in Maclaurin's series, we get

$$tan^{-1} x = x - \dfrac{x^3}{3} + \dfrac{x^5}{5} - \dfrac{x^7}{7} + ...$$

REMARKS

- To expand an alone inverse function, find its first derivative, expand by Binomial theorem and then find other derivatives.
- The expansion of $tan^{-1} x$ is valid only if $-1 < x < 1$.
- This expansion for $tan^{-1} x$ known as Gregory's series, which is very useful in finding the value of π.
- In a like manner, we may get

$$sin^{-1} x = x + \dfrac{1}{2} \cdot \dfrac{x^3}{3} + \dfrac{1.3}{2.4} \cdot \dfrac{x^5}{5} + \dfrac{1.3.5}{2.4.6} \cdot \dfrac{x^7}{7} + ...$$

Example 2. *If $y = \sin(m \sin^{-1} x)$, then show that*

$$\left(1-x^2\right)\frac{d^2y}{dx^2} - x\frac{dy}{dx} + m^2y = 0$$

Hence, or otherwise expand sin $m\theta$ in powers of sin θ.

Solution . Here, we have

$$y = f(x) = \sin\left(m \sin^{-1} x\right) \qquad \qquad ...(1)$$

$$\Rightarrow \qquad y_1 = \cos\left(m\sin^{-1} x\right).\frac{m}{\sqrt{1-x^2}} \qquad \qquad ...(2)$$

$$\Rightarrow \qquad (1-x^2)y_1{}^2 = m^2\cos^2(m\sin^{-1}x)$$

$$\Rightarrow \qquad (1-x^2)y_1{}^2 = m^2[1-\sin^2(m\sin^{-1}x)]$$

$$\Rightarrow \qquad (1-x^2)y_1{}^2 = m^2(1-y^2) \qquad\qquad [\because y = \sin(m\sin^{-1}x)]$$

$$\Rightarrow \qquad (1-x^2)y_1{}^2 + m^2y^2 - m^2 = 0 \qquad\qquad ...(3)$$

Differentiating w.r.t. x, we get

$$(1-x^2)2y_1y_2 - 2xy_1{}^2 + 2m^2yy_1 = 0$$

$$\Rightarrow \qquad 2y_1[(1-x^2)y_2 - xy_1 + m^2y] = 0$$

$$\Rightarrow \qquad (1-x^2)y_2 - xy_1 + m^2y = 0 \qquad\qquad ...(4)$$

Now, differentiating (4) n times, we get

$$\left(1-x^2\right)y_{n+2} + n.y_{n+1}(-2x) + \frac{n(n-1)}{1.2}y_n(-2) - xy_{n+1} - n.y_n + m^2y_n = 0$$

$$\Rightarrow \quad \left(1-x^2\right)y_{n+2} - (2n+1)xy_{n+1} - \left(n^2 - m^2\right)y_n = 0 \qquad\qquad ...(5)$$

Now, put $x = 0$ in (1), (2), (4) and (5), we get

$$y(0) = 0, y_1(0) = m, y_2(0) + m^2y(0) = 0 \Rightarrow y_2(0) = 0$$

and $\quad y_{n+2}(0) = (n^2 - m^2)y_n(0). \qquad\qquad ...(6)$

Putting $n = 2,4,6,...$ in (6), we get

$$y_4(0) = (2^2 - m^2)y_2(0) = 0$$

$$y_6(0) = (4^2 - m^2)y_4(0) = 0$$

$$y_8(0) = 0$$

.... and so on.

Here, we observe that $y_n(0) = 0$ if n is even.

Now, putting $n = 1,3,5,...$ in (6), we get

$$y_3(0) = (1^2 - m^2)y_1(0) = (1^2 - m^2).m$$

$$y_5(0) = (3^2 - m^2)y_3(0) = (3^2 - m^2)(1^2 - m^2).m$$

....

Putting all these values in Maclaurin's series, we get

$$\sin\left(m\sin^{-1} x\right) = mx + \frac{m\left(1^2 - m^2\right)}{3!}x^3 + \frac{m\left(1^2 - m^2\right)\left(3^2 - m^2\right)}{5!}x^5 + ...$$

Let $\qquad\qquad \theta=\sin^{-1}x \Rightarrow x=\sin\theta$

Then, we get

$$\sin m\theta = m\sin\theta + \frac{m(1^2-m^2)}{3!}\sin^3\theta + \frac{m(1^2-m^2)(3^2-m^2)}{5!}\sin^5\theta + \ldots$$

Example 3. *Expand tan x by Macluarin's theorem as far as x^5 and hence find the value of tan 46°30' upto four decimal places.*

Solution. Let

$$f(x)=\tan x \qquad\qquad\qquad\qquad \Rightarrow\; f(0)=0$$
$$f'(x)=\sec^2 x=1+\tan^2 x \qquad\qquad\qquad \Rightarrow f'(0)=1$$
$$f''(x)=2\tan x\,\sec^2 x=2\tan x(1+\tan^2 x)=2\tan x+2\tan^3 x \quad \Rightarrow f''(0)=0$$
$$f'''(x)=2\sec^2 x+6\tan^2 x\,\sec^2 x=2(1+\tan^2 x)+6\tan^2 x(1+\tan^2 x)$$
$$\qquad\quad =2+8\tan^2 x+6\tan^4 x \qquad\qquad\qquad\qquad \Rightarrow f'''(0)=2$$
$$f^{iv}(x)=16\tan x\,\sec^2 x+24\tan^3 x\,\sec^2 x=8\sec^2 x(2\tan x+3\tan^3 x)$$
$$\qquad\quad =8(1+\tan^2 x)(2\tan x+3\tan^3 x)$$
$$\qquad\quad =16\tan x+40\tan^3 x+24\tan^5 x \qquad\qquad\qquad \Rightarrow f^{iv}(0)=0$$

and $\;\; f^{v}(x)=16\sec^2 x+120\tan^2 x\,\sec^2 x+120\tan^4 x\,\sec^2 x$

$$\qquad\qquad =8\sec^2 x(2+15\tan^2 x+15\tan^4 x) \qquad\qquad \Rightarrow f^{v}(0)=16$$

Now, putting all these values in Maclaurin's series'

$$f(x)= f(0)+xf'(0)+\frac{x^2}{2!}f''(0)+\frac{x^3}{3!}f'''(0)+\frac{x^4}{4!}f^{iv}(0)+\frac{x^5}{5!}f^{v}(0)+\ldots$$

we get $\qquad \tan x = 0+x+\dfrac{x^3}{3!}.2+\dfrac{x^5}{5!}.16+\ldots$

$$\Rightarrow \qquad\qquad \tan x = x+\frac{x^3}{3}+\frac{2}{5}x^5+\ldots$$

Deduction. Here

$$x =46°30'=\left(46\frac{1}{2}\right)^{\!\circ}=\left(\frac{93}{2}\right)^{\!\circ}=\frac{93}{2}\times\frac{\pi}{180}\text{Radians}$$

$$=\frac{31}{120}\times\frac{22}{7}=\frac{31\times 11}{60\times 7}=\frac{314}{420}=0.812$$

Now, putting $\quad x= 46°30'=0.812$ in (1) , we get

$$\tan 46°30'=0.812+\frac{(0.812)^3}{3}-+\frac{2}{15}(0.812)^5 = 0.812+0.1784+0.047 = 1.0374$$

Example 4. *Expand $\log\{x+\sqrt{(1+x^2)}\}$ in ascending powers of x and find the general term.*

Solution. Let $\qquad\qquad y= \log\{x+\sqrt{(1+x^2)}\}$ $\qquad\qquad$ ...(1)

$$\Rightarrow \qquad\qquad y_1= \frac{1}{x+\sqrt{1+x^2}}.\left[1+\frac{2x}{2\sqrt{(1+x^2)}}\right]=\frac{1}{\sqrt{1+x^2}} \qquad\qquad ...(2)$$

$\Rightarrow \qquad y_1^2(1+x^2)-1=0.$

Differentiating again w.r.t. x, we get

$\qquad 2y_1[(1+x^2)y_2+xy_1]=0$

$\Rightarrow \qquad [(1+x^2)y_2+xy_1]=0 \qquad\qquad (\because 2y_1 \neq 0) \qquad\qquad ...(3)$

Differentiating (3) n times, we get

$(1-x^2)y_{n+2}+n.y_{n+1}.2x+\dfrac{n(n-1)}{1.2}y_2.2+y_{n+1}.x+n.y_n=0$

$\Rightarrow \qquad (1+x^2)y_{n+2}+(2n+1)xy_{n+1}+n^2 y_n=0 \qquad\qquad ...(4)$

Putting $x=0$ in (1),(2),(3) and (4), we have

$\qquad\qquad y(0)=0, y_1(0)=1, y_2(0)=0$

$\qquad\qquad y_{n+2}(0)=n^2 y_n(0) \qquad\qquad\qquad\qquad ...(5)$

From (5), we have

$\qquad\qquad y_3(0)=-1^2 y_1(0)=-1^2$

$\qquad\qquad y_5(0)=(-3^2)y_3(0)=(-3^2)(-1^2)=3^2.1^2$

$\qquad\qquad y_7(0)=(-5^2)y_5(0)=(-5^2)(-3^2)(-1^2)=-5^2.3^2.1^2 \;\; ....$ and so on.

Putting $n-2$ for n in (5), we get

$\qquad\qquad y_n(0)=\{-(n-2)^2\}y_{n-2}(0) \qquad\qquad\qquad ...(6)$

$\qquad\qquad\qquad =[-(n-2)^2][-(n-4)^2]y_{n-4}(0).$

Here we observe that

If n is odd, then $\quad y_n(0)=[-(n-2)^2][-(n-4)^2]-...(-5^2)(-3^2)(-1^2).1$

$\qquad\qquad\qquad =[-1]^{(n-1)/2}(n-2)^2(n-4)^2...5^2.3^2.1^2 \qquad\qquad ...(7)$

Also from (5), we get $y_4(0)=-2^2.y_2(0)=0$

$\qquad\qquad\qquad y_6(0)=-4^2.y_4(0)=0 \;...$ and so on.

If n is even.

Then, $\qquad\qquad y_n(0)=0.$

Putting all these values in Maclaurin's series

$$y= y(0)+\dfrac{x}{1!}y_1(0)+\dfrac{x^2}{2!}y_2(0)+\dfrac{x^3}{3!}y_3(0)+...$$

we get $\log\left[x+\sqrt{(1+x^2)}\right]=x-\dfrac{x^3}{3!}.1^2+\dfrac{x^5}{5!}(3^2.1^2)-\dfrac{x^7}{7!}(5^2.3^2.1^2)+...$

General term. The general term $=\dfrac{x^n}{n!}y_n(0)$

where $\qquad\qquad y_n(0)=\begin{cases}(-1)^{(n-1)/2}(n-2)^2(n-4)^2...5^2.3^2.1^2 & \text{, if } n \text{ is odd}\\ 0 & \text{, if } n \text{ is even}\end{cases}$

Example 5.   *Prove by Maclaurin's theorem, that*

$$e^{\sin x} = 1 + x + \frac{x^2}{1.2} - \frac{3.x^4}{1.2.3.4} + ...$$

Solution .   Let $f(x) = e^{\sin x}$ $\Rightarrow$ $f(0) = e^0 = 1$

$$f'(x) = e^{\sin x}.\cos x \quad \Rightarrow \quad f'(0) = e^0 \cos 0 = 1$$

$$f''(x) = e^{\sin x}(-\sin x) + \cos x \, e^{\sin x}\cos x$$

$$= e^{\sin x}[\cos^2 x - \sin x] \qquad\qquad \Rightarrow f''(0) = e^0[1-0] = 1$$

$$f'''(x) = e^{\sin x}[2 \cos x(-\sin x) - \cos x] + e^{\sin x}\cos x.[\cos^2 x - \sin x]$$

$$= e^{\sin x}\cos x[-2\sin x - 1 + \cos^2 x - \sin x]$$

$$= - e^{\sin x}\cos x[3 \sin x + \sin^2 x] \Rightarrow f'''(0) = 0$$

$$f^{iv}(x) = - e^{\sin x}\cos x[3 \cos x + 2 \sin x \cos x]$$

$$+ e^{\sin x}\sin x[3 \sin x + \sin^2 x] \quad -[3 \sin x + \sin^2 x]\cos x \, e^{\sin x}\cos x$$

$$\Rightarrow \qquad f^{iv}(0) = -3.$$

Putting all these values in Maclaurin's theorem, given by

$$f(x) = f(0) + xf'(0) + \frac{x^2}{2!}f''(0) + \frac{x^3}{3!}f'''(0) + \frac{x^4}{4!}f^{iv}(0) + ...$$

we get, $e^{\sin x} = 1 + x + \dfrac{x^2}{1.2} - \dfrac{3.x^4}{1.2.3.4} + ...$

Example 6.   (i) *If $f(x) = x^3 + 8x^2 + 15x - 24$, calculate the valve of $\left(\dfrac{11}{10}\right)$ by Taylor's series.*

(ii) *If $f(x) = x^3 - 2x + 5$, find the value of $f(2.001)$ with the help of Taylor's theorem. Find the approximate change in the value of $f(x)$ when x changes from 2 to 2.001.*

Solution .   (i) By Taylor's Theorem, we have

$$f(x+h) = f(x) + hf'(x) + \frac{h^2}{2!}f''(x) + \frac{h^3}{3!}f'''(x) + ... \qquad\qquad ...(1)$$

We want to find $f\left(\dfrac{11}{10}\right)$ i.e., $f\left(1 + \dfrac{1}{10}\right)$

Put $x = 1$ and $h = \dfrac{1}{10}$, and expand by Taylor's series, we get

$$f\left(\frac{11}{10}\right) = f\left(1 + \frac{1}{10}\right) = f(1) + \frac{1}{10}f'(1) + \frac{1}{10^2}.\frac{1}{2!}f''(1) + \frac{1}{3!}\frac{1}{(10)^3}f'''(1) + ... \,...(2)$$

Now $f(x) = x^3 + 8x^2 + 15x - 24$ $\Rightarrow$ $f(1) = 0$

$\qquad\quad f'(x) = 3x^2 + 16x + 15$ $\Rightarrow$ $f'(1) = 34$

$\qquad\quad f''(x) = 6x + 16$ $\Rightarrow$ $f''(1) = 22$

$\qquad\quad f'''(x) = 6$ $\Rightarrow$ $f'''(1) = 6$

$\qquad\quad f^{iv}(x) = 0$ $\Rightarrow$ $f^{iv}(1) = 0$

Put all these values in (2), we get

$$f\left(1+\frac{1}{10}\right) = 0+\frac{1}{10}.34+\frac{11}{100}+\frac{1}{1000} = 3.4+0.11+0.001 = 3.511.$$

(ii) Here put $x=2$ and $h=0.001$ in Taylor's series, we get

$$f(2.001) = f(2)+(0.001)f'(2)+\frac{(0.001)^2}{2!}f''(2)+\frac{(0.001)^3}{3!}f'''(2)+\dots\dots(3)$$

Now $f(x) = x^3-2x+5$ $\Rightarrow$ $f(2)=9$

$\quad\quad f'(x) = 3x^2-2$ $\Rightarrow$ $f'(2)=10$

$\quad\quad f''(x) = 6x$ $\Rightarrow$ $f''(2)=12$

$\quad\quad f'''(x) = 6$ $\Rightarrow$ $f'''(2)=6$

$\quad\quad f^{iv}(x) = 0$ $\Rightarrow$ $f^{iv}(2)=0$

Put all these values in (2), we get

$$f(2.0001) = 9+(0.001)10+\frac{1}{2!}(0.001)^2(12)+\frac{1}{3!}(0.001)^3.6+\dots$$

$$=9+0.01+0.000006+0.000000001$$

$$= 9.010006001 = 9.01 \text{ approximately.}$$

Approximate value of $f(2.001)-f(2)=9.01-9=0.01$ approximately.

Example 7. *Expand* log $(1+\sin x)$ *by Maclaurin's theorem in ascending power of x upto first five terms.*

Solution . Let $y = f(x) = \log(1+\sin x)$.

By Maclaurin's expansion for $f(x)$, we have

$$y=f(x) = (y)_0 + \frac{x}{1!}(y_1)_0 + \frac{x^2}{2!}(y_2)_0 + \frac{x^3}{3!}(y_3)_0 + \frac{x^4}{4!}(y_4)_0 +\dots\dots(1)$$

Now $y = \log(1+\sin x)$ $\therefore$ $(y)_0=0$

$$y_1 = \frac{\cos x}{1+\sin x} \Rightarrow (y_1)_0 = 1$$

$$y_2 = \frac{-\sin x(1+\sin x)-\cos^2 x}{(1+\sin x)^2} = -\frac{(1+\sin x)}{(1+\sin x)^2} = -\frac{1}{1+\sin x}$$

$\Rightarrow$ $(y_2)_0 = -1$

$$y_3 = \frac{\cos x}{(1+\sin x)^2} = -\frac{\cos x}{(1+\sin x)}\cdot\frac{1}{(1+\sin x)} = -y_1 y_2$$

$\Rightarrow$ $(y_3)_0 = -1(-1) = 1$

$\quad\quad y_4 = -y_1 y_3 - y_2^2$ $\Rightarrow$ $(y_4)_0 = -1.1-(-1)^2 = -1-1 = -2$

$\quad\quad y_5 = -y_1 y_4 - y_2 y_3 - 2y_2 y_3 = -y_1 y_4 - 3y_2 y_3$

$\Rightarrow$ $(y_5)_0 = -1.(-2)-3(-1).1 = 2+3 = 5$ and so on.

Therefore, $\log(1+\sin x) = 0 + \dfrac{x}{1!}.1 + \dfrac{x^2}{2!}.(-1) + \dfrac{x^3}{3!}.1 + \dfrac{x^4}{4!}.(-2) + \ldots$

$$= x - \frac{x^2}{2} + \frac{x^3}{6} - \frac{x^4}{12} + \frac{x^5}{25} \ldots$$

Example 8. Expand $\sin(\pi/4 + \theta)$ in powers of θ.

Solution. Let

$$f(\theta) = \sin(\pi/4 + \theta) \qquad \Rightarrow \qquad f(0) = \sin \pi/4 = 1/\sqrt{2}$$
$$f'(\theta) = \cos(\pi/4 + \theta) \qquad \Rightarrow \qquad f'(0) = \cos \pi/4 = 1/\sqrt{2}$$
$$f''(\theta) = -\sin(\pi/4 + \theta) \qquad \Rightarrow \qquad f''(0) = -\sin \pi/4 = -1/\sqrt{2}$$
$$f'''(\theta) = -\cos(\pi/4 + \theta) \qquad \Rightarrow \qquad f'''(0) = \cos \pi/4 = -1/\sqrt{2}$$
$$f^{iv}(\theta) = \sin(\pi/4 + \theta) \qquad \Rightarrow \qquad f^{iv}(0) = 1/\sqrt{2} \text{ and so on.}$$

The nth derivative of $f(\theta)$ is given by

$$f^n(\theta) = \sin\left(\theta + \frac{\pi}{4} + \frac{n\pi}{2}\right)$$

The Maclaurin's expansion of $f(\theta)$ with Lagrange's from of remainder is

$$f(\theta) = f(0) + \frac{\theta}{1!}f'(0) + \frac{\theta^2}{2!}f''(0) + \frac{\theta^3}{3!}f'''(0) + \ldots + \frac{\theta^{n-1}}{(n-1)!}f^{n-1}(0) + R_n \ldots(1)$$

where $R_n = \dfrac{\theta^n}{n!}f^n(t\theta) = \dfrac{\theta^n}{n!}\sin\left(t\theta + \dfrac{\pi}{4} + \dfrac{n\pi}{2}\right), \; 0 < t < 1.$

Now $|R_n| = \left|\dfrac{\theta^n}{n!}\sin\left(t\theta + \dfrac{\pi}{4} + \dfrac{n\pi}{2}\right)\right|$

$$= \left|\frac{\theta^n}{n!}\right| . \left|\sin\left(t\theta + \frac{\pi}{4} + \frac{n\pi}{2}\right)\right| \le \left|\frac{\theta^n}{n!}\right|.$$

$\therefore \quad \lim\limits_{n\to\infty} |R_n| \le \lim\limits_{n\to\infty} \left|\dfrac{\theta^n}{n!}\right| = 0 \qquad\qquad\qquad \left[\because \lim\limits_{n\to\infty} \dfrac{\theta^n}{n!} = 0\right]$

$\therefore \qquad\qquad \lim\limits_{n\to\infty} R_n = 0$

Thus all the conditions of Maclaurin's series expansion are satisfied. Hence, from (1), the expansion of $\sin(\theta + \pi/4)$ is given by

$$\sin\left(\theta + \frac{\pi}{4}\right) = \frac{1}{\sqrt{2}} + \frac{\theta}{1!}\frac{1}{\sqrt{2}} + \frac{\theta^2}{2!}\left(-\frac{1}{\sqrt{2}}\right) + \frac{\theta^3}{3!}\left(-\frac{1}{\sqrt{2}}\right) + \ldots$$

$$\sin\left(\theta + \frac{\pi}{4}\right) = \frac{1}{\sqrt{2}}\left[1 + \frac{\theta}{1!} - \frac{\theta^2}{2!} - \frac{\theta^3}{3!} + \frac{\theta^4}{4!} + \frac{\theta^5}{5!} - \frac{\theta^6}{6!} - \frac{\theta^7}{7!} + \ldots\right]$$

EXERCISE 5.2

1. Expand the following functions by Maclaurin's theorem :

  (i) Sec x
  (ii) $e^{x\cos x}$
  (iii) $e^x \sec x$
  (iv) $\log_e(1 + e^x)$
  (v) $\log(1 + \tan x)$

2. Apply Maclaurin's theorem to prove that

$$\log \sec x = \frac{1}{2}x^2 + \frac{1}{12}x^4 + \frac{1}{45}x^6 + \ldots$$

3. If $y = \sin^{-1} x = a_0 + a_1 x + a_2 x^2 + \ldots$ Prove that $(n+1)(n+2)a_{n+2} = n^2 a_n$.

4. Show that :

(i) $e^x \cos x =$

$$1 + x - \frac{2x^3}{3!} + \frac{2^2 x^4}{4!} - \frac{2^2 x^5}{5!} + \frac{2^3 x^7}{7!}$$

$$+ \ldots + \cos\left(\frac{n\pi}{4}\right) \frac{2^{n/2}}{n!} x^n + \ldots$$

(ii) $e^x \sin x = x + x^2 - \frac{2x^3}{3!} + \frac{2^2 x^5}{5!} -$

$$\ldots + \sin\left(\frac{n\pi}{4}\right) \frac{2^{n/2}}{n!} x^n + \ldots$$

(iii) $e^{ax} \sin bx = bx + abx^2 + \frac{3a^2 b - b^3}{3!} x^3 +$

$$\ldots + \frac{\left(a^2 + b^2\right)^{\frac{n}{2}}}{n!} x^n$$

$$\sin\left(n \tan^{-1} \frac{b}{a}\right) + \ldots$$

(iv) $e^{ax} \cos bx = 1 + ax + \frac{a^2 - b^2}{2!} x^2 +$

$$+ \frac{a\left(a^2 - 3b^2\right)}{3!} x^3 + \ldots + \frac{\left(a^2 + b^2\right)^{\frac{n}{2}}}{n!} x^n$$

$$\cos\left(n \tan^{-1} \frac{b}{a}\right) + \ldots$$

5. Expand the following :

(i) $\tan^{-1} x$ in powers of $\left(x - \frac{\pi}{4}\right)$.

(ii) $2x^3 + 7x^2 + x - 1$ in powers of $x-2$.

(iii) $\sin^{-1}(x+h)$ in power of x.

(iv) $\log \sin x$ in power of $(x-a)$.

6. Show that

$$\log(x+h) = \log h + \frac{x}{h} - \frac{x^2}{2h^2} + \frac{x^3}{3h^3} - \ldots$$

7. Use Taylor's theorem to prove that

$$\tan^{-1}(x+h) = \tan^{-1} x + h \sin\theta \frac{\sin\theta}{1}$$

$$- (h\sin\theta)^2 \frac{\sin 2\theta}{2} + (h\sin\theta)^3 \frac{\sin 3\theta}{3} +$$

$$\ldots + (-1)^{n-1}(h\sin\theta)^n \frac{\sin n\theta}{n} + \ldots$$

where $\theta = \cot^{-1} x$

8. If $y = e^{\tan^{-1} x}$, show that

$$(1+x^2)y_{n+2} + [2(n+1)x-1]y_{n+1}$$

$$+ n(n+1)y_n = 0.$$

Hence, or otherwise, find out the coefficient of x^5 if $e^{\tan^{-1} x}$ is expanded in powers of x.

9. Expand $(\sin^{-1} x)^2$ in ascending powers of x and deduce that

$$\theta^2 = 2 \cdot \frac{\sin^2 \theta}{2!} + 2^2 \cdot \frac{2\sin^4 \theta}{4!} + 2^2 \cdot 4^2 \frac{2\sin^6 \theta}{6!} + \ldots$$

10. If $y = e^{m \tan^{-1} x} = a_0 + a_1 x + a_2 x^2 + \ldots + a_n x^n + \ldots$, show that $(n+1)a_{n+1} + (n-1)a_{n-1} = ma_n$.

11. If $e^{e^x} = a_0 + a_1 x + a_2 x^2 + \ldots + a_n x^n + \ldots$ show that

$$a_{n+1} = \frac{1}{n+1}\left[a_n + \frac{a_{n-1}}{1!} + \frac{a_{n-2}}{2!} + \ldots + \frac{a_{n-r}}{r!} + \ldots \frac{a_0}{n!}\right]$$

12. Show that

$$f(mx) = f(x) + (m-1)xf'(x)$$

$$+ \frac{(m-1)^2}{2!} x^2 f''(x)$$

$$+ \frac{(m-1)^3}{3!} x^3 f'''(x) + \ldots$$

13. By Maclaurin's theorem find the expansion of $y = \sin(e^x - 1)$ upto and including the term in x. Find also the first non-vanishing terms in the expansion of x as a series ascending powers of y.

14. Prove that $f\left(\frac{x^2}{1+x}\right) = f(x) - \frac{x}{1+x} f'(x)$

$$+ \left(\frac{x}{1+x}\right)^2 \frac{1}{2!} f''(x) - \left(\frac{x}{1+x}\right)^3 \frac{1}{3!} f'''(x) + \ldots$$

15. Calculate the approximate value of :

(i) $\sqrt{17}$ to four decimal places.

(ii) $\sqrt{26}$ to three decimal places

by Taylor's expansion.

HINTS TO SELECTED PROBLEMS

1. (i) $y = \sec x \Rightarrow y_1 = \sec x \tan x = y \tan x$

$$\Rightarrow y_1^2 = y^2 \tan^2 x = y^2(\sec^2 x - 1)$$

$$= y^2(y^2 - 1) = y^4 - y^2.$$

Again differentiating, we get

$$2y_1 y_2 = 4y^3 y_1 - 2yy_1 \Rightarrow y_2 = 2y^3 - y$$

Similarly $\quad y_3 = 6y^2 y_1 - y_1$

$$y_4 = 12yy_1 + 6y^2 y_2 - y_2$$

$$y_5 = 12y_1^3 - 36y_1 y_2 - y_3$$

$$\ldots \qquad \ldots \qquad \ldots \qquad \ldots$$

Now putting $x = 0$ in the above equation and use Maclaurin's series.

2. $y = \log \sec x \Rightarrow y_1 = \tan x$

$$y_2 = \sec^2 x = 1 + \tan^2 x = 1 + y_1^2$$

$$y_3 = 2y_1 y_2$$

$$y_4 = 2y_2^2 + 2y_1 y_3$$

$$y_5 = 4y_2 y_3 + 2y_2 y_3 + 2y_1 y_4$$

$$y_6 = 8y_2 y_4 + 6y_3^2 + 2y_1 y_5$$

$$\ldots \quad \ldots \quad \ldots \quad \ldots \quad \ldots$$

Now putting $x = 0$ in the above equations and use Maclaurin's series.

3. $\qquad y = \sin^{-1} x$

$$\Rightarrow y_1 = \frac{1}{\sqrt{1 - x^2}} \Rightarrow \left(1 - x^2\right) y_1^2 = 1$$

Again differentiating, we get

$$(1 - x^2 y_2 - xy_1) = 0$$

Now apply Leibnitz's theorem to differentiating n times.

4. $\qquad y = e^x \cos x$

$\Rightarrow \quad y_1 = e^x \cos x - e^x \sin x = e^x(\cos x - \sin x)$

$$= re^x(\cos \theta \cos x - \sin \theta \sin x),$$

where $r = \sqrt{2}, \theta = \dfrac{\pi}{4}$

$\Rightarrow \quad y_1 = re^x \cos(x + \theta)$

$\Rightarrow \quad y_2 = r^2 e^x \cos(x + 2\theta)$

$\ldots \quad \ldots \quad \ldots \quad \ldots$

$$y_n = r^n e^x \cos(x + n\theta) = (2)^{n/2} e^x \cos\left(x + \frac{n\pi}{4}\right)$$

Now putting $x = 0$ and use Maclaurin's series.

(ii) Here $y_n = 2^{n/2} e^x \sin\left(x + \dfrac{n\pi}{4}\right)$

(iii) $y_n = \left(a^2 + b^2\right)^{n/2} e^{ax} \sin\left[bx + n \tan^{-1}\left(\dfrac{b}{a}\right)\right]$

5. (i) $y = f(x) = \tan^{-1} x$

$$\Rightarrow y_1 = f'(x) = \frac{1}{1 + x^2}$$

$$\Rightarrow y_2 = \frac{-2.x}{\left(1 + x^2\right)^2}$$

Putting $x = \pi/4$ and find $f(\pi/4)$, $f'(\pi/4)$, $f''(\pi/4)$... and so on.

Now $f(x) = f\left(\dfrac{\pi}{4} + x - \dfrac{\pi}{4}\right)$

Then expand by Taylor's theorem.

6. $y = f(x) = \log x$

$$\Rightarrow f(x + h) = \log(x + h)$$

$$\Rightarrow f'(x) = \frac{1}{x}, f''(x) = -\frac{1}{x^2}, f'''(x) = \frac{2}{x^3} \ldots$$

and so on.

$\Rightarrow$ Now putting $x = h$ in above derivatives and expand $f(x + h)$ by Taylor's theorem.

7. Take $y = f(x) = \tan^{-1} x$

$$\Rightarrow f'(x) = \frac{1}{1 + x^2} = \frac{1}{2i}\left(\frac{1}{x - i} - \frac{1}{x + i}\right)$$

$$\Rightarrow f''(x) = \frac{1}{2i}\left[(-1)(x - i)^{-2} - (-1)(x + i)^{-2}\right]$$

$$\ldots \qquad \ldots \qquad \ldots \qquad \ldots \qquad \ldots \qquad \ldots \qquad \ldots$$

$$f^n(x) = \frac{1}{2i}\left[(-1)^{n-1}(n - 1)!(x - i)^{-n}\right.$$

$$\left. - (-1)^{n-1}(n - 1)!(x + i)^{-n}\right]$$

Put $\theta = \cot^{-1} x$, we get

$$f^n(x) = (-1)^{n-1}(n - 1)! \sin^n \theta \sin n\theta$$

Now use Taylor's series.

8. $y = f(x) = e^{\tan^{-1} x}$

$$\Rightarrow y_1 = \frac{e^{\tan^{-1} x}}{1 + x^2} = \frac{y}{1 + x^2} \Rightarrow \left(1 + x^2\right) y_1 = y$$

$$(1 + x^2) y_2 + (2x - 1) y_1 = 0.$$

Now to find y_n, use Leibnitz's theorem.

9. Let $y = f(x) = (\sin^{-1} x)^2$

$$\Rightarrow \quad y_1 = \frac{2\sin^{-1} x}{\sqrt{1-x^2}}$$

$$\Rightarrow \left(1 - x^2\right) y_1^2 = 4\left(\sin^{-1} x\right)^2 = 4y.$$

Now differentiating n times by Leibnitz's rule.

10. $y = f(x) = e^{m\tan^{-1} x}$

$$\Rightarrow \quad y_1 = \frac{m e^{\tan^{-1} x}}{1+x^2} \Rightarrow \left(1+x^2\right) y_1 = my$$

Now differentiating n times by using Leibnitz's theorem.

11. Let $y = f(x) = e^{e^x} \Rightarrow y_1 = e^x e^{e^x} = e^x . y$

Now to find nth derivative, using Leibnitz's theorem.

12. Write $f(mx) = f[x + (m-1)x]$. Now expand by Taylor's theorem.

13. $y = f(x) = \sin(e^x - 1) \Rightarrow y_1 = e^x \cos(e^x - 1)$,   Now differentiating successively.

14. Write

$$f\left(\frac{x^2}{1+x}\right) = f\left(\frac{x^2}{1+x} - x + x\right)$$

$$= f\left(x - \frac{x}{1+x}\right)$$

Now expand by Taylor's theorem.

15. Let $\quad y = f(17) = \sqrt{17} = \sqrt{x}$

$$y = f(16+1).$$

Now expand by Taylor's theorem.

Answers

1. (i) $1 + \dfrac{x^2}{2!} + \dfrac{5x^4}{4!} + \dfrac{61x^6}{6!} + \dots$

(ii) $1 + x + \dfrac{x^2}{2} - \dfrac{x^3}{3} - \dfrac{11x^4}{24} - \dfrac{x^5}{5} + \dots$

(iii) $1 + x + \dfrac{2x^2}{2!} + \dfrac{4x^3}{3!} + \dots$

(iv) $\log 2 + \dfrac{x}{2} + \dfrac{x^2}{8} - \dfrac{x^4}{192} + \dots$

(v) $x - \dfrac{x^2}{2} + \dfrac{2}{3}x^3 - \dfrac{7x^4}{12} + \dots$

5. (i) $\tan^{-1}\left(\dfrac{\pi}{4}\right) + \left(x - \dfrac{\pi}{4}\right) \bigg/ \left(1 + \dfrac{\pi^2}{16}\right) - \pi \left(x - \dfrac{\pi}{4}\right)^2 \bigg/ \left[4\left(1 + \dfrac{\pi^2}{16}\right)^2\right] + \dots$

(ii) $45 + 53(x-2) + 19(x-2)^2 + 2(x-2)^3 + \dots$

(iii) $\sin^{-1} h + x\left(1 - h^2\right)^{-1/2} + \dfrac{x^2}{2!} h\left(1 - h^2\right)^{-3/2} + \dfrac{x^3}{3!}\left[\left(1 - h^2\right)^{-5/2}\left(1 + 2h^2\right)\right] + \dots$

(iv) $\log \sin a + (x - a)\cot a - \dfrac{(x-a)^2}{2!}\csc^2 a + \dfrac{(x-a)^3}{3!} 2\csc^2 a \cot a + \dots$

8. $\dfrac{1}{24}$ **9.** $2 \cdot \dfrac{x^2}{2!} + \dfrac{2.2^2}{4!}x^4 + \dfrac{2.2^2.4^2}{6!}x^6 + \dots + \dfrac{2.2^2.4^2 \dots (2n-2)^2}{(2n)!} x^{2n} + \dots$

13. $x + \dfrac{x^2}{2!} - \dfrac{5x^4}{24} + \dots, y - \dfrac{y^2}{2} + \dots$ **15.**   (i) 4.123   (ii) 5.099

CHAPTER REVIEW: A COMPETITIVE APPROACH

SELECTED TERMS AND RESULTS

TERMS

- **Continuity in a closed interval:** To test the continuity of $f(x)$ in a closed interval $[a,b]$ we have the following two methods

  Method 1: Here we check following three conditions:

  (i) $f(x)$ is not infinite in the closed interval $[a,b]$.

  (ii) $f(x)$ is not imaginary in the closed interval $[a,b]$.

  (iii) $f(x)$ has no break in the closed interval $[a,b]$.

  Method 2: Use the following results:

  (i) If $f(x)$ is a polynomial function of x, then it is continuous for all real x.

  (ii) If $f(x)$ is not a polynomial function, then we find $f'(x)$. If $f'(x)$ is finite, definite and real in $[a,b]$ then $f(x)$ is differentiable in $[a,b]$ and hence continuous in $[a,b]$.

- **Differentiability in Open Interval:**

  (i) If $f(x)$ is a polynomial function of x then $f(x)$ is differentiable in $]a,b[$ as a polynomial function is always differentiable for all $x \in R$.

  (ii) If $f(x)$ is not a polymomial function, find $f'(x)$. If $f'(x)$ is finite, definite and real in $]a,b[$ then $f(x)$ is differentiable in $]a,b[$.

RESULTS

- **Rolle's Theorem:** If a function f defined on $[a,b]$ is :

  (i) continuous on $[a,b]$.

  (ii) differentiable on $]a,b[$.

  (iii) $f(a)=f(b)$.

  then $\exists\ c \in]a,b[$ such that $f'(c)=0$.

- Geometrically Rolle's theorem states that 'Between two points with equal ordinates on the graph of f, there exists at least one point where the tangent is parallel to x-axis'.

- Algebraically, Rolle's theorem states that 'Between two zeroes of $f(x)$ there exists at least one zero of $f'(x)$'.

- Between two consecutive zeroes of $f'(x)$ there exists atmost one zero of $f(x)$.

- **(Lagrange's mean value theorem)** If a function f defined on $[a,b]$ is

  (i) continuous on $[a,b]$

  (ii) differentiable on $]a,b[$

  then there exists at least one real number $c \in]a,b[$ such that

  $$\frac{f(b)-f(a)}{b-a} = f'(c)$$

- If a function $f(x)$ satisfies the condition of mean value theorem and $f'(x)=0$ for all $x \in]a,b[$ then $f(x)$ is constant on $[a,b]$.

- If two functions have equal derivatives at all points of $]a,b[$ then they differ only by a constant.

- If a function f is continuous on $[a,b]$, differentiable on $]a,b[$ and $f'(x)>0\ \forall x \in]a,b[$, then f is strictly increasing function.

- If f' exists and is bounded on some interval I then f is uniformly continuous on I.

- Geometrically, Lagrange's theorem state that between two points of the graph f there exists at least one point where the tangent is parallel to the chord.

- **Cauchy's mean value theorem:** If two functions f and g defined on $[a,b]$ are

  (i) continuous on $[a,b]$

  (ii) differentiable on $]a,b[$

  (iii) $g'(x) \neq 0$ for any $x \in]a,b[$

  then there exists at least one $c \in]a,b[$ such that

  $$\frac{f(b)-f(a)}{g(b)-g(a)} = \frac{f'(c)}{g'(c)}$$

- Lagrange's mean value theorem can be deduce by Cauchy's mean value theorem as a particular case for $g(x)=x$.

- Geometrically, Cauchy's mean value theorem states that the mean rates of increase of two functions in an interval is equal to the ratio of actual rates of increase of the functions at some points within the interval.

REVIEW QUESTIONS AND PROJECT WORK

1. Show that for any $c \in R$, the polynoimial $f(x) = x^3 + x + c$ has exactly one real root.

2. Show that
$$\frac{\sin\alpha - \sin\beta}{\cos\alpha - \cos\beta} = \cot\theta, \quad 0 < \alpha < \theta < \beta < \frac{\pi}{2}.$$

3. Show that $\dfrac{\tan\theta}{\theta} > \dfrac{\theta}{\sin\theta}$ for $0 < \theta < \dfrac{\pi}{2}$.

4. Apply Lagrange's mean value theorem to the function $\log(1+x)$ to show that
$$0 < \left[\log(1+x)\right]^{-1} - x^{-1} < 1 \quad \forall\, x > 0$$

5. If $f(x) = 0$ has two equal roots, show that $f'(x) = 0$ has one root equal to either.

6. Show that the number θ which occur in the Taylor's theorem with Lagrange's form of remainder after n terms approaches the limit $\dfrac{1}{n+1}$ as $h \to 0$, provided that $f^{n+1}(x)$ is continuous and non-zero at $x = a$.

7. Use Taylor's theorem to show that

(i) $x - \dfrac{x^3}{6} < \sin x < x$ for $x > 0$

(ii) $x - \dfrac{x^3}{6} < \sin x < x - \dfrac{x^3}{6} + \dfrac{x^5}{120} \quad \forall\, x > 0$

8. If $f(x)$ is real valued and differentiable on R and
$$f(x+y) = \frac{f(x) + f(y)}{1 - f(x)f(y)}$$
then show that $f(x) = \tan(x f'(0))$.

9. Show that there exists an $m \in N$ such that $\forall\, n \geq m$ the function
$$f(x) = \frac{x^{2n+2} - \sin x - 1}{x^{2n} + 1}$$
has no zero in $[1,2]$ even though $f(1)f(2) < 0$.

10. If $a = 0$ and $b \geq 2$, show that f is defined by $f(x) = \dfrac{1}{|x-1|}$ when $x \neq 1$ and $f(1) = 24$ do not satisfy the conditions of Lagrange's mean value theorem. However, show that the conclusion holds true if $b > 2 + \sqrt{2}$.

OBJECTIVE TYPE QUESTIONS

FILL IN THE BLANKS

1. The function $f(x) = \dfrac{1}{x}$ is _____ in $-1 < x < 0$.

2. In some cases, Lagrange's mean value theorem is a particular case of _____ .

3. The first mean value theorem is also known as _____ .

4. If $f'(x) > 0$ then $f(x)$ is _____ function.

5. If f is a finitely differentiable in a closed interval $[a,b]$ and $f'(a), f'(b)$ are of opposite sign then $f'(c) =$ _____ for at least one value of $c \in\,]a,b[$.

TRUE/FALSE

Write 'T' for true and 'F' for false statement.

1. Rolle's theorem is a particular case of Lagrange's theorem. **(T/F)**

2. Cauchy theorem is directly deducible from the Lagrange's mean value theorem. **(T/F)**

3. Lagrange mean value theorem is a particular case of Cauchy's mean value theorem. **(T/F)**

4. Maclaurin's theorem is a particular case of Taylor's theorem. **(T/F)**

MULTIPLE CHOICE QUESTIONS

Choose the most appropriate one.

1. If $f(x)$ is an even function then the value of $f'(0)$, if exists is equal to :

(a) 1   (b) 0

(c) 2   (d) ∞

2. If a function f is continuous on $[a, b]$, differentiable on $]a, b[$ and if $f'(x) = 0$ $\forall\, x \in\,]a, b[$ then $f(x)$ has a :

(a) constant value throughout $[a,b]$
(b) constant value only on the end points
(c) constant value throughout $]a,b[$
(d) none of the above

3. If $f(x)$ and $g(x)$ are continuous on $[a,b]$ and differentiable on $]a,b[$ and if $f'(x) = g'(x)$ throughout the interval $]a, b[$ then :

(a) $f(x)=g(x)\ \forall x\in]a,b[$
(b) $f(x)\neq g(x)\ \forall x\in]a,b[$
(c) $f(x)$ and $g(x)$ differ only by a constant
(d) none of the above

4. If f is continuous on $[a,b]$ and $f'(x)\geq0$ on $]a,b[$ then :
(a) f is decreasing on $]a,b[$
(b) f is decreasing on $[a,b]$
(c) f is increasing on $[a,b]$
(d) none of the above

5. If $f(x)$ is an increasing function on x, then :
(a) $f'(x)\leq0$ (b) $f'(x)=0$
(c) $f'(x)>0$ (d) none of the above

6. If $f'(x)$ is positive at a point $x=a$ then in the nbd of a :
(a) $f(x)$ is positive (b) $f(x)$ is increasing
(c) $f(x)$ is negative (d) none of the above

7. The function $f(x)$ has equal values at the point $x=a$ and $x=b$ then :
(a) there is a maximum of $f(x)$ between a and b
(b) there is a minimum of $f(x)$ between a and b
(c) there is a maximum or minimum of $f(x)$ between a and b
(d) none of the above

8. If $f''(x)>0$ at points in $]a,b[$ then the function f is :
(a) strictly increasing (b) strictly decreasing
(c) constant (d) none of the above

9. If a function $f(x)$ satisfy the condition of mean value theorem and $f'(x)=0\ \forall x\in]a,b[$ then :
(a) $f(x)=0$
(b) $f(x)$ is an increasing function
(c) $f(x)$ is constant
(d) none of the above

10. The value of c of Rolle's theorem for the function $f(x)=\sin x$ in $[0,\pi]$ is given by :
(a) $\pi/3$ (b) $\pi/2$
(c) π (d) none of the above

11. The value of c of Lagrange's mean value theorem for $f(x)=x(x-1)$ in $[1,2]$ is given by :
(a) $\dfrac{1}{4}$ (b) $\dfrac{3}{2}$
(c) $\dfrac{5}{4}$ (d) none of the above

12. Lagrange's form of remainder after n terms in Taylor's development of the function e^x in a finite form in the interval $[a,a+h]$ is :

(a) $\dfrac{h^n}{n!}e^{a+\theta h}$ (b) $\dfrac{h^{n+1}}{(n+1)!}e^{a+\theta h}$
(c) $\dfrac{h^n}{n!}e^{\theta h}$ (d) none of the above

13. Let $f:R\to R$ be a differentiable function. Which of the following will follow from mean value theorem ?
(a) for all $a,b\in R$ ∃ some $c\in]a,b[$ such that $\dfrac{f(b)-f(a)}{b-a}=f'(c)$
(b) for all $a,b\in R$ ∃ some $c\in]a,b[$ such that $\dfrac{f(b)+f(a)}{b+a}=f'(c)$
(c) both (a) and (b) are true
(d) none of the above

14. If f is continuous on $[a,b]$ and f' exist at each $x\in]a,b[$ if $f'>0$ in $]a,b[$ then :
(a) f is strictly increasing
(b) f is strictly decreasing
(c) f is constant
(d) none of the above

15. If f is continuous on $[a,b]$ and f' exist at each $x\in]a,b[$ then if $f'=0$, then :
(a) f is increasing (b) f is constant
(c) f is decreasing (d) none of the above

16. Let $f:[a,b]\to R$ be continuous on $[a,b]$ and if f is differentiable on $]a,b[$. If $f(a)=f(b)=0$ ∃ c such that $f'(c)=0$ then :
(a) $c\in]a,b[$ (b) $c\in[a,b]$
(c) $c\in]a,b]$ (d) none of the above

17. If f is continuous on $I(=[a,b])$ then there exists a real number u such that :
(a) $f(x)=u\ \forall x\in I$ (b) $f(x)\leq u\ \forall x\in I$
(c) $f(x)\geq u\ \forall x\in I$ (d) none of the above

18. Let $f:[a,b]\to R$ be continuous on $[a,b]$ and f is differentiable on $]a,b[$. If $f(a)=f(b)=0$ ∃ $c\in]a,b[$ such that $f'(c)=0$. It is known as :
(a) Rolle's theorem
(b) Lagrange's theorem
(c) Cauchy's theorem
(d) none of the above

19. If f satisfies the conditions of Lagrange's mean value theorem and if $f'(x)>0\ \forall x\in]a,b[$, then which one is true ?
(a) f is increasing in $[a,b]$

(b) f is decreasing in $[a,b]$

(c) f is constant

(d) None of the above

20. Which one of the following is correct by mean value theorem ?

(a) $1+x<e^x<1+xe^x \ \forall x>0$

(b) $1+x=e^x=1+xe^x \ \forall x>0$

(c) Both (a) and (b) are true

(d) None of the above

21. Assuming Rolle's theorem for the function $f(x)=\cos x$ in $\left[\dfrac{-\pi}{2},\dfrac{\pi}{2}\right]$. If there exists a real number $c\in \left]\dfrac{-\pi}{2},\dfrac{\pi}{2}\right[$. Then value of c is :

(a) $\pi/2$ (b) $\pi/4$

(c) π (d) 0

22. If f satisfies the conditions of Lagranges's mean theorem and if $f'(x)=0 \ \forall x\in]a,b[$ then

which one of the following statement is true ?

(a) f is constant on $[a,b]$

(b) f is strictly increasing on $[a,b]$

(c) f is strictly decreasing on $[a,b]$

(d) none of these

23. In Lagrange's mean value theorem, the differential function exists :

(a) in $]a,b[$ (b) in $[a,b]$

(c) in $]a,b]$ (d) none of these

24. Does the mean value theorem apply to $f(x)=x^{1/3}; \ -1\leq x\leq 1$:

(a) yes

(b) no

(c) can't say

(d) none of the above

25. The value of c when $ax^2+bx+1=0, \ a\neq 0$ in $[1,p]$ by Lagrange's mean value theorem is :

(a) 5   (b) 15

(c) 18   (d) none of the above

Answers

FILL IN THE BLANKS

   1. decreasing   **2.** Cauchy's mean value theorem   **3.** Lagrange's mean value theorem

   4. increasing   **5.** 0

TRUE/FALSE

   1. T   **2.** F   **3.** T   **4.** T

MULTIPLE CHOICE QUESTIONS

1. (b)	**2.** (a)	**3.** (c)	**4.** (c)	**5.** (c)
6. (b)	**7.** (c)	**8.** (a)	**9.** (c)	**10.** (b)
11. (b)	**12.** (a)	**13.** (a)	**14.** (a)	**15.** (b)
16. (a)	**17.** (a)	**18.** (a)	**19.** (a)	**20.** (a)
21. (b)	**22.** (a)	**23.** (a)	**24.** (b)	**25.** (a)

SELF ASSESSMENT TEST

Verify each of the following :

1. Rolle's Theorem is verified for $f(x) = \sqrt{1 - x^2}$ in $[-1,1]$.

2. Rolle's Theorem is not valid for $f(x) = 1 - (x-1)^{2/3}$ on $[0,2]$.

3. There is no real number for which the equation $x^5 - 45x + t = 0$ possessses two real roots in $[2,5]$.

4. If $f(x)$, $g(x)$ are differentiable on $]a,b[$ and are continuous on $[a,b]$ and $f(a) = f(b) = 0$ then there exists a point $c \in]a,b[$ such that $f'(c) + f(c)g'(c) = 0$.

5. Lagrange's theorem is verified for the function $f(x) = \sqrt{x^2 - 4}$ in $[2,4]$.

6. If f is defined for all $x \in R$ such that

$$|f(x) - f(y)| < (x - y)^2 \quad \forall \, x, y \in R$$

then f is constant.

7. If $a, b \in R$, $a \ne b$ then there exists a real number c between a and b such that $a^2 + ab + b^2 = 3c^2$.

8. If f' and g' exist $\forall \, x \in [a, b]$ and if $g'(x) \ne 0$ $\forall \, x \in]a, b[$ then for some $c \in]a, b[$

$$\frac{f(c) - f(a)}{g(b) - g(c)} = \frac{f'(c)}{g'(c)}$$

9. $f(x) = x^5 - 5x^3 + 100x - 90$ is increasing in every interval.

10. The Cauchy's function f defined on R by

$$f(x) = \begin{cases} e^{-1/x^2} & , \quad x \ne 0 \\ 0 & , \quad x = 0 \end{cases}$$

has derivatives of all orders $\forall \, x \in R$ but it has no Maclaurin's series expansion.

Chapter

6 — Riemann Integrals

6.1 INTRODUCTION

Generally, the process of integration is defined as the inverse of the differentiation. A function F is called an integral of a function f if $F'(x)=f(x)$ on some domain D of f. In view of this definition we may define that "*definite integration is a process of summation*".

The German mathematician G.F.B. Riemann defined the process of integration on certain arithemetical concepts free from dependence on geometrical concepts.

6.6.1 PARTITION OF A CLOSED INTERVAL

Let $I=[a,b]$ be a closed and bounded interval. Then, a finite set of points $P=\{x_0,x_1,x_2,...x_n\}$ such that $a=x_0<x_1<x_2...<x_{n-1}<x_n=b$ is called a partition or division of the interval $I=[a,b]$.

REMARK

- Partition of a set is also called dissection or net.

6.1.2 SEGMENTS OF THE PARTITION

The closed sub-intervals $I_1=[a=x_0,\ x_1], I_2=[x_1,\ x_2]..., I_n=[x_{n-1},x_n=b]$ are called the segments of the partition.

6.1.3 LENGTH OF THE SUBINTERVAL

The length of the subinterval I_r is denoted by Δx_r or δ_r defined by $\delta_r=\Delta x_r=x_r-x_{r-1}$.

6.1.4 NORM OF A PARTITION

The norm of a partition P is the maximum of the lengths of the segments of a partition P, denoted by $\|P\|$, defined by $\|P\|=\max\{\Delta x_r, r=1,2,...,n\}$

6.1.5 REFINEMENT OF A PARTITION

If a partition P^* is a refinement of a closed and bounded interval $[a, b]$ then

$$P^*=P_1\cup P_2 \text{ is called the common refinement of } P_1 \text{ and } P_2.$$

6.1.6 FAMILY OF PARTITIONS

The family of all partitions of the closed interval $[a, b]$ is denoted by $P(a,b)$.

6.2 LOWER RIEMANN SUM, UPPER RIEMANN SUM AND OSCILLATORY SUM

Let f be a bounded real valued function defined on a bounded and closed interval $[a,b]$ and $P=\{a=x_0,x_1,.... x_n=b\}$ be any partition of $[a,b]$. Also, let m_r and M_r denotes the infimum and supremum of the function f on the subinterval $[x_{r-1}, x_r]$ respectively, then the two sums

$$L(P,f) = \sum_{r=1}^{n} m_r \delta x_r \text{ and } U(P,f) = \sum_{r=1}^{n} M_r \delta x_r$$

are respectively called the lower Riemann sum and upper Riemann sum of f on $[a,b]$ with respect to partition P.

Also,
$$U(P,f) - L(P,f) = \sum_{r=1}^{n} [M_r - m_r] \delta x_r$$

$$= \sum_{r=1}^{n} \omega_r \delta x_r \text{ where } \omega_r = (M_r - m_r)$$

Then sum $\sum_{r=1}^{n} \omega_r \delta x_r$ is called the oscillatory sum for the function f with respect to partition P of $[a,b]$.

REMARKS

- The oscillatory sum is denoted as $\omega(P,f)$.
- $L(P,f) \le U(P,f) \; \forall P[a,b]$.

6.3 UPPER AND LOWER INTEGRALS

The infimum of the set of the upper sums is called the upper integral of f over $[a,b]$ and is denoted by $U = \overline{\int_a^b} f(x)dx$. Also, the supremum of the set of the lower sums is called the lower integral of f over $[a,b]$ and is denoted by $L = \underline{\int_a^b} f(x)dx$.

6.4 RIEMANN INTEGRAL

From the above discussion, it is clear that the supremum of the set of upper sums is $M(b-a)$ and the infimum of the set of the lower sums is $m(b-a)$, where M and m be the bounds of f on $[a,b]$ such that for every value of r; $m \le m_r \le M_r \le M$.

Definition. *A bounded function f is said to be Riemann integrable, or simply integrable over $[a, b]$, if its upper and lower integrals are equal; and their common value being called Riemann integral or simply the integral denoted by*

$$\int_a^b f(x)dx$$

REMARKS

- The numbers a and b called the lower and upper limits of integration respectively, while the interval $[a,b]$ is known the range of the integration.
- The concept of integrability defined above, have the following two limitations:
  (i) the function is bounded. (ii) the interval is finite.

THEOREM 1. *Let f be a bounded function defined on $[a,b]$ and let m and M be the infimum and supremum of $f(x)$ in $[a,b]$, then for every partition P of $[a,b]$, we have*

$$m(b-a) \le L(P,f) \le U(P,f) \le M(b-a).$$

Proof. Let $P = \{a = x_0, x_1, ..., x_{n-1}, x_n = b\}$ be any partition of $[a,b]$. Also, let $I_r = [x_{r-1}, x_r], r = 1,2,,..,n$ be the subintervals of $[a, b]$. If M and m be the least upper bound and greatest lower bound of f on $[a,b]$, then we have $m \leq f(x) \leq M \ \forall \ x \in [a,b]$

(By definition of supremum and infimum)

Now let M_r and m_r by the supremum and infimum of f in I_r.

Then, $m \leq m_r \leq M_r \leq M$ for $r \in N$

$\Rightarrow$ $m\delta_r \leq m_r\delta_r \leq M_r\delta_r \leq M\delta_r$ (Multiplying by δ_r)

$\Rightarrow$ $\displaystyle\sum_{r=1}^{n} m\delta_r \leq \sum_{r=1}^{n} m_r\delta_r \leq \sum_{r=1}^{n} M_r\delta_r \leq \sum_{r=1}^{n} M\delta_r$ ...(1)

Now $\displaystyle\sum_{r=1}^{n} m\delta_r = m\sum_{r=1}^{n}\delta_r = m\sum_{r=1}^{n}(x_r - x_{r-1})$ $(\delta_r = x_r - x_{r-1})$

$$=m[(x_1 - x_0)+(x_2 - x_1)+(x_3 - x_2)+...+(x_n - x_{n-1})]$$
$$=m(x_n - x_0)=m(b-a)$$

Similarly, we may find that

$$\sum_{r=1}^{n} M\delta_r = M(b-a)$$

Also, by definition of lower sum and upper sums, we get

$$\sum_{r=1}^{n} m_r\delta_r = L(P,f), \sum_{r=1}^{n} M_r\delta_r = U(P,f).$$

Using all these values in (1), we get

$$m(b-a) \leq L(P,f) \leq U(P,f) \leq M(b-a) \ \forall P \in P[a,b].$$

REMARK

- Clearly, $U(P,f)$ and $L(P,f)$ are bounded if f is bounded.

THEOREM 2. *If f_1 and f_2 are two real valued bounded functions defined on $[a,b]$, then*

(i) $L(P,f_1+f_2) \geq L(P,f_1)+L(P,f_2)$

and (ii) $U(P,f_1+f_2) \leq U(P,f_1)+U(P,f_2) \ \forall \ P \in P\ [a,b]$

Proof.   Let $P=\{a=x_0,x_1,....x_{n-1},x_n=b\}$ be any partition of $[a,b]$. Also, let $I_r=[x_{r-1},x_r]$, $r=1,2,...,n$ be the subintervals of $[a,b]$.

Since, f_1, f_2 both are bounded.

$\Rightarrow$ f_1+f_2 is bounded.   ($\because$ Sum of two bounded functions is also bounded)

Let M_r, m_r, M_{1r}, m_{1r}, and M_{2r}, m_{2r} be the least upper bounds and greatest lower bounds of the functions f_1+f_2, f_1 and f_2 in I_r for $r=1,2,...,n$ respectively.

(i) By definition of infimum, we have

$$f_1(x) \geq m_{1r}$$

and $f_2(x) \geq m_{2r}$ $\forall \ x \in I_r$

Therefore, $f_1(x)+f_2(x) \geq m_{1r}+m_{2r}$

$\Rightarrow$ $(f_1+f_2)(x) \geq m_{1r}+m_{2r}$

$\Rightarrow$ $(m_{1r} + m_{2r})$ is a lower bound of $(f_1+f_2)(x)$ on I_r. But, since m_r to be the greatest lower bound of (f_1+f_2) on I_r, therefore,

$$m_r \geq m_{1r}+m_{2r}$$

$$\Rightarrow \qquad m_r \delta_r \geq m_{1r} \delta_r + m_{2r} \delta_r \qquad \text{(Multiplying by } \delta_r)$$

$$\Rightarrow \qquad \sum_{r=1}^{n} m_r \delta_r \geq \sum_{r=1}^{n} m_{1r} \delta_r + \sum_{r=1}^{n} m_{2r} \delta_r \quad \text{(By summing the above result)}$$

$$\Rightarrow \qquad L[P, f_1 + f_2] \geq L(P, f_1) + L(P, f_2)$$

(ii) By definition of supremum, we have

$$f_1(x) \leq M_{1r} \text{ and } f_2(x) \leq M_{2r}, \forall x \in I_r$$

Therefore, $f_1(x) + f_2(x) \leq M_{1r} + M_{2r}$

$$\Rightarrow \qquad (f_1 + f_2)(x) \leq M_{1r} + M_{2r}$$

$$\Rightarrow \qquad (M_{1r} \text{ and } M_{2r}) \text{ is an upper bound of } (f_1 + f_2)(x) \text{ on } I_r.$$

$$\Rightarrow \qquad M_r \leq M_{1r} + M_{2r}$$

$$\Rightarrow \qquad M_r \delta_r \leq M_{1r} \delta_r + M_{2r} \delta_r \qquad \text{(Multiplying by } \delta_r)$$

$$\Rightarrow \qquad \sum_{r=1}^{n} M_r \delta_r \leq \sum_{r=1}^{n} M_{1r} \delta_r + \sum_{r=1}^{n} M_{2r} \delta_r \quad \text{(By summing the above result)}$$

$$\Rightarrow \qquad U(P, f_1 + f_2) \leq U(P, f_1) + U(P, f_2)$$

THEOREM 3. *Let f be a real valued bounded function defined on [a,b] and $P_1, P_2 \in P(a,b)$ such that P_2 is the refinement of P_1, then*

$$L(P_1, f) \leq L(P_2, f) \leq U(P_2, f) \leq U(P_1, f)$$

Proof. Let $P_1 = \{a = x_0 < x_1 < x_2 ... < x_{r-1} x_r, ..., x_n = b)$ be any partition of $[a,b]$ and let P_2 be any other partition of $[a,b]$ such that

$$P_2 = [a = x_0, x_1, x_2, ... x_{r-1}, \alpha, x_r, ..., x_n = b]$$

contains just one point α more than $P_1 (x_{r-1} < \alpha < x_r)$.

Now let, the least upper bounds of f in the subinterval $[x_{r-1}, x_r], [x_{r-1}, \alpha]$ and $[\alpha, x_r]$ be M_r, M_{1r} and M_{2r} respectively. Then, by the definition of least upper bound it is clear that

$$M_r < M_{1r} \text{ and } M_r < M_{2r} \qquad ...(1)$$

From the definition of lower Darboux sum we find that $M_r(x_r - x_{r-1})$ is the contribution of the closed interval $[x_{r-1}, x_r]$ to $L(P_1, f)$ and $M_{1r}(\alpha - x_{r-1}) + M_{2r}(x_r - \alpha)$ that of the closed interval $[x_{r-1}, x_r]$ to $L[P_2, f]$.

Since α is the only extra point in P_2, which is not in P_1 and $x_{r-1} < \alpha < x_r$

therefore, the contribution of each subinterval except $I_r = [x_{r-1}, x_r]$ to $L(P_1, f)$ and $L(P_2, f)$ is the same. Thus,

$$L(P_2, f) \geq L(P_1, f)$$

$$\Rightarrow \qquad L(P_1, f) \leq L(P_2, f) \qquad ...(2)$$

In a similar manner taking the greatest lower bounds of f in the subintervals $[x_{r-1}, x_r]$, $[x_{r-1}, \alpha]$ and $[\alpha, x_r]$ as m_r, m_{1r} and m_{2r} respectively, we may prove that

$$U(P_2, f) \leq U(P_1, f) \qquad ...(3)$$

Also, we know that

$$L(P_2, f) \leq U(P_2, f) \qquad ...(4)$$

From (2), (3) and (4), we conclude that
$$L(P_1, f) \le L(P_2, f) \le U(P_2, f) \le U(P_1^{\bullet}, f).$$

REMARK

- Here it is clear that, the lower Darboux sum is increased by the introduction of a new point of division where the upper Darboux's sum is decreased by the same reason.

THEOREM 4. *Let f be a real valued function, defined on [a,b] and $P_1, P_2 \in P[a,b]$, then*

(i) $L(P_1, f) \le U(P_2, f)$ and   (ii) $L(P_2, f) \le U(P_1, f)$

Proof. Let P_1 and P_2 be two partitions of the interval [a,b]. Then, it is clear that $P_1 \cup P_2$ is the common refinement of P_1 and P_2. Also $P_1 \subseteq P_1 \cup P_2$ and $P_2 \subseteq P_1 \cup P_2$

Then, from above theorem, we have

$$L(P_1, f) \le L(P_1 \cup P_2, f) \qquad \qquad ...(1)$$

and $\qquad U(P_1, f) \ge U(P_1 \cup P_2, f) \qquad \qquad ...(2)$

Using, theorem (3), equation (1) and (2) gives

$$L(P_1, f) \le L(P_1 \cup P_2, f) \le U(P_1 \cup P_2, f) \le U(P_2, f). \qquad ...(3)$$

Similarly, we may prove that

$$L(P_2, f) \le L(P_1 \cup P_2, f) \le U(P_1, f). \qquad \qquad ...(4)$$

From (3) and (4), we conclude that

$$L(P_1, f) \le U(P_2, f) \text{ and } L(P_2, f) \le U(P_1, f)$$

6.5 LOWER AND UPPER RIEMANN INTEGRALS

If f is bounded on the interval [a,b], then for every $P \in P(a,b)$, $U(P,f)$ and $L(P,f)$ exist and are bounded. Then the lower Riemann integral is defined as

$$\underline{\int_a^b} f = \sup_P L(P, f)$$

and the upper Riemann integral is defined as

$$\overline{\int_a^b} f = \inf_P U(P, f)$$

REMARKS

- $\overline{\int_a^b} f$ and $\underline{\int_a^b} f$ exist uniquely, whenever f is bounded on [a,b].

- $\underline{\int_a^b} (-f) = \overline{\int_a^b} f$ and $\overline{\int_a^b} (-f) = -\underline{\int_a^b} f$

6.6 RIEMANN INTEGRABLE FUNCTION

Definition 1. *A real valued function f(x) is said to be Riemann integrable on [a,b] if and only if their lower and upper Riemann integrals are equal.*

i.e., iff $\qquad \qquad \underline{\int_a^b} f = \overline{\int_a^b} f$

The common value of these integrals is known as the Riemann integral *of f on [a,b] and is denoted by* $\int_a^b f(x)dx$

i.e.,
$$\int_a^b f(x)dx = \int_{\underline{a}}^b f(x)dx = \int_a^{\overline{b}} f(x)dx$$

REMARKS

- The interval [a,b] is known as the range of integration and numbers a and b are known as the lower and upper limits of the integration respectively.
- Riemann integral is abbreviated as R-integral.

Definition 2. *A function f is said to be Riemann integrable over [a,b] if and only if for every* $\varepsilon > 0$ *there exists a positive number* δ *and a number I such that for every partition*

$$P = [a = x_0, x_1, x_2, ..., x_n = b] \text{ with } \|P\| < \delta \text{ and for every } t_r \in [x_{r-1}, x_r]$$

$$\left| \sum_{r=1}^n f(t_r)(x_r - x_{r-1}) - I \right| < \varepsilon$$

Here I is said to be the integral of f over [a,b] and the class of all bounded functions f which are Riemann integrable on [a,b] is denoted by R[a,b].

REMARK

- The equivalence of two definitions of Riemann integrals allow us to write $f \in R[a,b]$ if and only if Riemann integrable over [a,b]. Therefore R[a,b] denotes the class of functions Riemann integrable over [a,b].

THEOREM 1. **(Darboux Theorem)**. *Assume that f is a bounded function defined on [a, b]. Then for every* $\varepsilon > 0$, *there exists* $\delta > 0$ *such that*

$$U(P,f) < \int_a^{\overline{b}} f + \varepsilon \text{ and } L(P,f) > \int_{\underline{a}}^b f - \varepsilon$$

for every partition P with $\|P\| \leq \delta$

Proof. Given that, f is bounded on [a,b], then by definition of boundedness there exist $k > 0$ such that $|f(x)| \leq k \ \forall x \in [a,b]$

Also, since inf $U(P, f)$ is defined as $\int_a^{\overline{b}} f$

$\therefore$ for every $\varepsilon > 0$ there exist a partition $P_1 = [a = x_0, x_1, x_2, ..., x_n = b]$ such that

$$U(P_1, f) < \int_a^{\overline{b}} f + \varepsilon/2 \qquad \qquad ...(1)$$

if $x_0 = a$ and $x_n = b$, then the partition P has $(n-1)$ points. Let $\delta_1 > 0$ be any number such that

$$2k(n-1)\delta_1 = \varepsilon/2 \qquad \qquad ...(2)$$

Now, let P be any partition with $\|P\| < \delta_1$

Also, let $P_2 = P \cup P_1$, then clearly P_2 is a refinement of P and P_1 then P_2 has atmost $(n-1)$ more points than P. Therefore,

$$U(P,f) - 2k(n-1)\delta_1 \leq U(P_2, f)$$

$$\leq U(P_1, f) < \int_a^{\overline{b}} f + \varepsilon/2 \qquad \qquad \text{[using (1)]}$$

$$\Rightarrow \qquad U(P,f) < \int_a^{\overline{b}} f + \varepsilon/2 + \varepsilon/2 \qquad \qquad \text{[using (2)]}$$

$$= \int_a^b f + \varepsilon \text{ for all partition } P \text{ with } \|P\| < \delta_1.$$

Similarly, we may easily shown that there exists a positive number δ_2 such that

$$L(P,f) > \int_a^b f - \varepsilon \text{ for all partition } P \text{ with } \|P\| < \delta_2.$$

Define $\delta = \min\{\delta_1, \delta_2\}$

Then for all partition P of $[a,b]$ with $\|P\| < \delta$, we have

$$U(P,f) < \int_a^b f + \varepsilon \text{ and } L(P,f) > \underline{\int}_a^b f - \varepsilon$$

THEOREM 2. **(Necessary and sufficient condition for integrability).** *A necessary and sufficient condition for R-integrabitity of a bounded function $f:[a,b] \to$ R over $[a,b]$ is that for every $\varepsilon > 0$, there exists a partition P of $[a,b]$ such that*

$$U(P,f) - L(P,f) < \varepsilon \ \forall \ \|P\| \le \delta$$

Proof.   **(i) Necessary Condition.** Let us first suppose f be Riemann integrable on $[a,b]$. Therefore,

$$\underline{\int}_a^b f = \overline{\int}_a^b f = \int_a^b f \qquad \qquad ...(1)$$

Let $\varepsilon > 0$ be given, then by Darboux theorem, there exists $\delta > 0$ such that for every partition P with $\|P\| \le \delta$.

$$U(P,f) < \overline{\int}_a^b f + \varepsilon/2 \qquad \qquad ...(2)$$

and $$L(P,f) > \underline{\int}_a^b f - \varepsilon/2 \qquad \qquad ...(3)$$

Adding inequalities (2) and (3), we get

$$U(P,f) + \underline{\int}_a^b f - \varepsilon/2 < \overline{\int}_a^b f + \varepsilon/2 + L(P,f)$$

which gives $U(P,f) - L(P,f) < \varepsilon$ [using(1)]
which is the required necessary condition.

   (ii) Sufficient Condition. For every $\varepsilon > 0$ and for a partition P of $[a,b]$ with $\|P\| \le \delta$, we have $U(P,f) - L(P,f) < \varepsilon$

By definition of upper and lower integrals, we have

$$L(P,f) \le \underline{\int}_a^b f \le \overline{\int}_a^b f \le U(P,f)$$

$$\Rightarrow \quad \overline{\int}_a^b f - \underline{\int}_a^b f \le U(P,f) - L(P,f) < \varepsilon$$

$$\Rightarrow \quad \overline{\int}_a^b f - \underline{\int}_a^b f \le 0 \qquad \qquad [\varepsilon \text{ is arbitrary}] \qquad ...(4)$$

Also we know that lower Riemann integral never exceed the upper Riemann integral, therefore

$$\overline{\int}_a^b f - \underline{\int}_a^b f \ge 0 \qquad \qquad ...(5)$$

from (4) and (5) we conclude that

$$\underline{\int}_a^b f - \overline{\int}_a^b f = 0$$

$$\Rightarrow \qquad \qquad \underline{\int}_a^b f = \overline{\int}_a^b f$$

Hence, the function f is Riemann integral over $[a,b]$.

ANOTHER FORMS

1. A bounded function f is integrable on $[a,b]$ if and only if for every $\varepsilon > 0$ $\exists$ a partition P of $[a,b]$ such that
$$U(P,f) - L(P,f) < \varepsilon$$

2. A function f is integrable over $[a,b]$ iff there is a number lying between $L(P,f)$ and $U(P,f)$ such that any $\varepsilon > 0$ $\exists$ a parttion P of $[a,b]$ such that
$$|U(P,f) - I| < \varepsilon \text{ and } |I - L(P,f)| < \varepsilon$$

THEOREM 3. *Let f be a bounded function defined on interval $[a,b]$ and P is a partition of $[a,b]$ then*
$$\lim_{\|P\| \to 0} L(P,f) = \int_{\underline{a}}^{b} f \text{ and } \lim_{\|P\| \to 0} U(P,f) = \overline{\int_{a}^{b}} f$$

Proof. Since, f is a bounded function defined on interval $[a,b]$ and P is a partition of $[a,b]$ and $\int_{\underline{a}}^{b} f$ is the supremum of $L(P,f)$ for all partitions P

$\Rightarrow \qquad\qquad L(P,f) \leq \int_{\underline{a}}^{b} f$ \qquad\qquad ...(1)

and $\overline{\int_{a}^{b}} f$ is the infimum of $U(P,f)$ for all partitions P

$\Rightarrow \qquad\qquad U(P,f) = \overline{\int_{a}^{b}} f$ \qquad\qquad ...(2)

Now by Darboux theorem we know that for all $\varepsilon > 0$, $\exists$ $\delta > 0$ such that
$$U(P,f) < \overline{\int_{a}^{b}} f + \varepsilon \qquad\qquad ...(3)$$

and $\qquad\qquad L(P,f) > \int_{\underline{a}}^{b} f - \varepsilon$ $\forall$ partition P with $\|P\| \leq \delta$ \qquad\qquad ...(4)

From equation (1) and (4), we have
$$\int_{\underline{a}}^{b} f - \varepsilon < L(P,f) \leq \int_{\underline{a}}^{b} f$$

$\Rightarrow \qquad\qquad \int_{\underline{a}}^{b} f - \varepsilon < L(P,f) \leq \int_{\underline{a}}^{b} f < \int_{\underline{a}}^{b} f + \varepsilon$

$\Rightarrow \qquad\qquad \int_{\underline{a}}^{b} f - \varepsilon < L(P,f) < \int_{\underline{a}}^{b} f + \varepsilon$

$\Rightarrow \qquad\qquad \lim_{\|P\| \to 0} L(P,f) = \int_{\underline{a}}^{b} f$

Similarly, from (2) and (3) we have
$$\overline{\int_{a}^{b}} f - \varepsilon < U(P,f) < \overline{\int_{a}^{b}} f + \varepsilon$$

$\Rightarrow \qquad\qquad \lim_{\|P\| \to 0} U(P,f) = \overline{\int_{a}^{b}} f$

THEOREM 4. *If $f : [a,b] \to R$ is bounded function then $U(P, -f) = -L(P,f)$ and $L(P,f) = -U(P, -f)$*

Proof. Consider a partition $P = \{x_0, x_1, ..., x_n\}$ in interval $[a,b]$, where $a = x_0$ and $x_n = b$.

Let M_r and m_r be the supremum and infimum of f in I_r.

Since, f is bounded on $[a,b]$ thus $-f$ is also bounded on interval $[a,b]$ and $-m_r$ and $-M_r$ will be supremum and infimum of $-f$ in I_r.

Now, $L(P,-f) = \sum_{r=1}^{n} (-M_r) \delta x_r$ (Lower Riemann sum)

$$= -\sum_{r=1}^{n} M_r \delta x_r$$

$$= -U(P,f) \quad \left\{ \because \sum_{r=1}^{n} M_r \delta x_r = U(P,f), \text{ the upper Riemann sum of } f \right\}$$

Similarly, $U(P,-f) = \sum_{r=1}^{n} (-m_r) \delta x_r$

$$= -\sum_{r=1}^{n} m_r \delta x_r$$ (Upper Riemann sum)

$$\left\{ \because L \text{ is the lower Riemann sum of } f \text{ in} \right.$$

$$= -L(P,f)$$ $$\left. [a,b] \text{ such that } L(P,f) = \sum_{r=1}^{n} m_r \delta x_r \right\}$$

Solved Examples

Example 1. *Find, $L(P,f)$ and $U(P,f)$ if $f(x)=x$ for $x \in [0,3]$ and let $P = \{0,1,2,3\}$ be the partition of $[0,3]$.*

Solution. Let partition P divided the interval $[0,3]$ into the subinterval $I_1 = [0, 1]$, $I_2 = [1,2]$ and $I_3 = [2, 3]$.

The length of these intervals are given by

$$\delta_1 = 1-0 = 1$$
$$\delta_2 = 2-1 = 1$$
$$\delta_3 = 3-2 = 1$$

Let M_r and m_r be respectively the l.u.b. and g.l.b. of the function f in $[x_{r-1}, x_r]$, then we get

$$M_1 = 1, \quad m_1 = 0, \quad M_2 = 2,$$
$$m_2 = 1, M_3 = 3 \text{ and } m_3 = 2$$

Therefore, $U(P,f) = \sum_{r=1}^{3} M_r \delta_r$

Based on the following Results

• A finite set of points $P = \{x_0, x_1, x_2, \ldots, x_n\}$ such that $a = x_0 < x_1 < x_2 \ldots < x_{n-1} < x_n = b$ is called a partition of the interval $I = [a,b]$.

• The length of the subinterval I_r is denoted by Δx_r or δ_r defined by $\delta_r = \Delta x_r = x_r - x_{r-1}$

• $L(P,f) = \sum_{r=1}^{n} m_r \delta x_r$; $U(P,f) = \sum_{r=1}^{n} M_r \delta x_r$ Here $L(P,f)$ is called the lower sum and $U(P,f)$ is called the upper sum.

• $m(b-a) \le L(P,f) \le U(P,f) \le M(b-a)$

• $L(P,f_1 + f_2) \ge L(P,f_1) + L(P,f_2)$

• $U(P,f_1 + f_2) \le U(P,f_1) + U(P,f_2)$

• If P_2 is the refinement of P_1, then $L(P_1,f) \le L(P_2,f)$; $U(P_2,f) \le U(P_1,f)$

• $\int_{\underline{a}}^{b} f = \sup [L(P,f)]$; $\overline{\int_a^b} f = \inf U(P,f)$

• The necessary and sufficient condition for R-integrability is $U(P,f) - L(P,f) < \varepsilon$

$$= M_1 \delta_1 + M_2 \delta_2 + M_3 \delta_3$$
$$= 1.1 + 2.1 + 3.1 = 1 + 2 + 3 = 6$$

and $L(P,f) = \sum_{r=1}^{3} m_r \delta_r = m_1 \delta_1 + m_2 \delta_2 + m_3 \delta_3$

$$= 0.1 + 1.1 + 2.1 = 0 + 1 + 2 = 3$$

Example 2. Let $f(x)=x$, $0 \le x \le 1$ and, let $P=\left\{0,\dfrac{1}{4},\dfrac{1}{2},\dfrac{3}{4},1\right\}$ be a partition of $[0,1]$, find $U(P,f)$ and $L(P,f)$

Solution. Let the partition P divides the interval $[0,1]$ into the subintervals

$$I_1=\left[0,\dfrac{1}{4}\right],\ I_2=\left[\dfrac{1}{4},\dfrac{1}{2}\right],\ I_3=\left[\dfrac{1}{2},\dfrac{3}{4}\right],\ I_4=\left[\dfrac{3}{4},1\right]$$

Clearly, the length of each subinterval is $\dfrac{1}{4}$.

Now, let M_r and m_r respectively be the l.u.b. and g.l.b. of the function f in $[x_{r-1}, x_r]$, then, we get

$$M_1=\dfrac{1}{4},\ M_2=\dfrac{1}{2},\ M_3=\dfrac{3}{4},\ M_4=1$$

and $\qquad m_1=0,\ m_2=\dfrac{1}{4},\ m_3=\dfrac{1}{2},\ m_4=\dfrac{3}{4}.$

Therefore, $\qquad U[P,f]=\sum_{r=1}^{4} M_r\delta_r = M_1\delta_1 + M_2\delta_2 + M_3\delta_3 + M_4\delta_4$

$$=\dfrac{1}{4}\cdot\dfrac{1}{4}+\dfrac{1}{2}\cdot\dfrac{1}{4}+\dfrac{3}{4}\cdot\dfrac{1}{4}+1\cdot\dfrac{1}{4}=\dfrac{1}{16}+\dfrac{1}{8}+\dfrac{3}{16}+\dfrac{1}{4}=\dfrac{5}{8}$$

and $\qquad L(P,f)=\sum_{r=1}^{4} m_r\delta_r = m_1\delta_1 + m_2\delta_2 + m_3\delta_3 + m_4\delta_4$

$$=0\cdot\dfrac{1}{4}+\dfrac{1}{4}\cdot\dfrac{1}{4}+\dfrac{1}{2}\cdot\dfrac{1}{4}+\dfrac{3}{4}\cdot\dfrac{1}{4}=0+\dfrac{1}{16}+\dfrac{1}{8}+\dfrac{3}{16}=\dfrac{3}{8}$$

Example 3. Let $f(x)=x$ on $[0,1]$. Find $\underline{\int_0^1} x\,dx$ and $\overline{\int_0^1} x\,dx$, by partitioning $[0,1]$ into n equal parts. Also, show that $f\in R[0,1]$.

Solution. Let the partition P divides the interval $[0,1]$ into n subintervals such that

$$P=\left\{0,\dfrac{1}{n},\dfrac{2}{n},...,\dfrac{r-1}{n},\dfrac{r}{n},...,\dfrac{n}{n}=1\right\}$$

Clearly, we have

$$m_r=\dfrac{r-1}{n},\ M_r=\dfrac{r}{n}\ \text{and}\ \delta_r=\dfrac{1}{n}\ \text{for}\ r=1,2,...,n$$

Now, by definition, we have

$$L[P,f]=\sum_{r=1}^{n} m_r\delta_r = \sum_{r=1}^{n}\dfrac{r-1}{n}\cdot\dfrac{1}{n}=\dfrac{1}{n^2}\sum_{r=1}^{n}(r-1)$$

$$=\dfrac{1}{n^2}[1+2+3+...+(n-1)]=\dfrac{(n-1).n}{2n^2}=\dfrac{n-1}{2n}$$

and $\qquad U[P,f]=\sum_{r=1}^{n} M_r\delta_r = \sum_{r=1}^{n}\dfrac{r}{n}\cdot\dfrac{1}{n}$

$$=\dfrac{1}{n^2}\sum_{r=1}^{n} r =\dfrac{1}{n^2}[1+2+3+...+n]$$

$$=\dfrac{(n+1).n}{2n^2}=\dfrac{n+1}{2n}$$

Therefore, $\int_{\underline{0}}^{1} x\,dx = \lim_{\|P\|\to 0} L(P,f) = \lim_{n\to\infty} \dfrac{n-1}{2n} = \dfrac{1}{2}$

and $\overline{\int_{0}^{1}} x\,dx = \lim_{\|P\|\to 0} U(P,f) = \lim_{n\to\infty} \dfrac{n+1}{2n} = \dfrac{1}{2}$

From above, it is clear that

$$\int_{\underline{0}}^{1} x\,dx = \overline{\int_{0}^{1}} x\,dx$$

Hence, $\int_{0}^{1} x\,dx = \dfrac{1}{2}$

Example 4. *Give an example of a bounded function which is not R-integrable over the interval [0,1].*

Solution .   Consider the Dirichlet´s function given by

$$f(x) = \begin{cases} 1, & \text{when } x \text{ is rational} \\ -1, & \text{when } x \text{ is irrational} \end{cases}$$

Clearly, the function $f(x)$ is bounded.

Now, consider a partition

$$P = \left\{ \dfrac{0}{n}, \dfrac{1}{n}, \dfrac{2}{n}, ..., \dfrac{r-1}{n}, \dfrac{r}{n}, ..., \dfrac{n}{n} = 1 \right\}$$

of the interval [0,1]. Then, we have $m_r = -1$ and $M_r = 1$ for all $r = 1,2,...,n$. Therefore,

$$L[P,f] = \sum_{r=1}^{n} m_r \delta_r = \sum_{r=1}^{n} (-1)\delta_r$$

$$= -\sum_{r=1}^{n} \delta_r = -(1-0) = -1$$

and $U[P,f] = \sum_{r=1}^{n} M_r \delta_r = \sum_{r=1}^{n} 1.\delta_r$

$$= (1-0) = 1$$

Hence, $\int_{\underline{0}}^{1} f = \lim_{n\to\infty} L[P,f] = -1$

and $\overline{\int_{0}^{1}} f = \lim_{n\to\infty} U[P,f] = 1$

Here it is clear that $\int_{\underline{0}}^{1} f \neq \overline{\int_{0}^{1}} f$

Therefore, $f \notin R[0,1]$.

Example 5. *Find the upper and lower Riemann integrals for the function f defined on [0,1] as follows*

$$f(x) = \begin{cases} \sqrt{(1-x^2)}, & \text{when } x \text{ is rational} \\ (1-x), & \text{when } x \text{ is irrational} \end{cases}$$

Solution .   We have $(1-x^2)-(1-x)^2 = 2x(1-x) > 0 \ \forall x[0,1]$

Therefore, $m_r = 1-x$ and $M_r = \sqrt{1-x^2}$

Now, $\int_0^1 \underline{f} = \int_0^1 (1-x)\,dx = \left[x - \dfrac{x^2}{2}\right]_0^1 = 1 - \dfrac{1}{2} = \dfrac{1}{2}$

and $\overline{\int_0^1} f = \int_0^1 \left(\sqrt{1-x^2}\right)dx = \left[\dfrac{1}{2}x\sqrt{1-x^2} + \dfrac{1}{2}\sin^{-1}x\right]_0^1$

$= \dfrac{1}{2}\sin^{-1}1 = \dfrac{1}{2}\cdot\dfrac{\pi}{2} = \dfrac{\pi}{4}$

Clearly, $\overline{\int_0^1} f \neq \int_0^1 \underline{f}$

Hence, $f \notin R[0,1]$.

Example 6. Let the function f be defined on $\left[0,\dfrac{\pi}{4}\right]$, by $f(x) = \begin{cases} \cos x, & \text{when } x \text{ is rational} \\ \sin x, & \text{when } x \text{ is irrational} \end{cases}$

Show that $f \notin R\left[0,\dfrac{\pi}{4}\right]$.

Solution. Let $P = \left[\dfrac{r\pi}{4n} : r = 0,1,\dots n\right]$ be any partition, such that $\delta_r = \dfrac{\pi}{4n}$

Now, since $\sin x \le \cos x \; \forall\, x \in \left[0,\dfrac{\pi}{4}\right]$, therefore

$m_r = \sin(r-1)\dfrac{\pi}{4n}$

and $M_r = \cos(r-1)\dfrac{\pi}{4n} = \sin\left[\dfrac{\pi}{2} - (r-1)\dfrac{\pi}{4n}\right]$

Consider

$U[P,f] - L[P,f] = \sum_{r=1}^{n}\left[M_r - m_r\right]\delta_r$

$= \sum_{r=1}^{n}\left[\sin\left(\dfrac{\pi}{2} - \dfrac{(r-1)\pi}{4n}\right) - \sin\dfrac{(r-1)\pi}{4n}\right]\dfrac{\pi}{4n}$

$= \sum_{r=1}^{n} 2\cos\dfrac{\pi}{4}\sin\left(\dfrac{\pi}{4} - \dfrac{(r-1)\pi}{4n}\right)\dfrac{\pi}{4n}$

$= 2\cdot\dfrac{1}{\sqrt{2}}\cdot\dfrac{\pi}{4n}\sum_{r=1}^{n}\cos\left(\dfrac{\pi}{4} - \dfrac{(r-1)\pi}{4n}\right)$

$= \sqrt{2}\cdot\dfrac{\pi}{4n}\left[\cos\alpha + \cos(\alpha+\beta) + \dots + \cos(\alpha+(n-1)\beta)\right]$

with $\alpha = \dfrac{\pi}{4}, \beta = \dfrac{\pi}{4n}$

$= \dfrac{\dfrac{\sqrt{2}\cdot\pi}{4n}}{\sin\dfrac{\pi}{8n}}\cos\left[\dfrac{\pi}{4} + \dfrac{(n-1)\pi}{8n}\right]\sin\dfrac{n\pi}{8n}$

$= \dfrac{\dfrac{2\sqrt{2}\cdot\dfrac{\pi}{8n}}{\sin\dfrac{\pi}{8n}}}{\sin\dfrac{\pi}{8n}}\cos\left[\dfrac{\pi}{4} + \left(1 - \dfrac{1}{n}\right)\dfrac{\pi}{8}\right]\sin\dfrac{\pi}{8}$

$$\therefore \quad \lim_{\|P\|\to 0}\left[U(P,f)-L(P,f)\right] = \lim_{n\to\infty} \frac{2\sqrt{2}\cdot\frac{\pi}{8n}}{\sin\frac{\pi}{8n}}\cos\left[\frac{\pi}{4}+\left(1-\frac{1}{n}\right)\frac{\pi}{8}\right]\sin\frac{\pi}{8}$$

$$= 2\sqrt{2}\sin\frac{\pi}{8}\cos\left(\frac{\pi}{4}+\frac{\pi}{8}\right)$$

$$= 2\sqrt{2}\sin^2\frac{\pi}{8} = \sqrt{2}\left(1-\cos\frac{\pi}{4}\right)$$

$$= \sqrt{2}-1 \neq 0$$

Hence, $\qquad\qquad\qquad f\notin R\left[0,\frac{\pi}{4}\right].$

Example 7. *Show that the function $f(x)=\sin x$ is integrable on $\left[0,\frac{\pi}{2}\right]$.*

Solution . Consider the partition

$$P=\left\{0,\frac{\pi}{2n},\frac{2\pi}{2n},....,\frac{(r-1)\pi}{2n},\frac{r\pi}{2n},....,\frac{n\pi}{2n}=\frac{\pi}{2}\right\}$$

which is obtained by dividing $\left(0,\frac{\pi}{2}\right)$ into n equal parts with length of each subinterval

$$=\frac{\pi}{2n}$$

Let $I_r=\left[\dfrac{(r-1)\pi}{2n},\dfrac{r\pi}{2n}\right]$ be the rth subinterval.

Now, since $f(x)=\sin x$ is increasing in $\left[0,\frac{\pi}{2}\right]$, therefore

$$m_r=\sin\frac{(r-1)\pi}{2n} \text{ and } M_r=\sin\frac{r\pi}{2n},r=1,2,...,n$$

Now $\qquad\qquad U(P,f)-L(P,f)=\sum_{r=1}^{n}\left(M_r-m_r\right)\delta_r$

$$=\sum_{r=1}^{n}\left[\sin\frac{r\pi}{2n}-\sin\frac{(r-1)\pi}{2n}\right]\frac{\pi}{2n}$$

$$=\left[\sin\frac{n\pi}{2n}-0\right]\frac{\pi}{2n}=\frac{\pi}{2n}$$

For given $\varepsilon>0$ there exist $m\in N$ such that $\dfrac{\pi}{2n}<\varepsilon$, $\forall n\geq m$. Therefore, for a given $\varepsilon>0$ there exists a partition P of $\left[0,\frac{\pi}{2}\right]$ such that $U(P,f)-L(P,f)<\varepsilon$

Hence, the function $f(x)=\sin x$ is R-integrable.

Example 8. *Let $f(x)=x^2$ on $[0,a]$, $a>0$, show that $f\in R[0,a]$. Also, find $\int_{0}^{a}f$*

Solution . Let $P=\left[\dfrac{ra}{n}:r=0,1,...,n\right]$ be any partition of $[0,a]$. Then, clearly, we have

$$m_r=\frac{(r-1)^2 a^2}{n^2} \text{ and } M_r=\frac{r^2 a^2}{n^2} \text{ with } \delta_r=\frac{a}{n}$$

Now,
$$L(P,f) = \sum_{r=1}^{n} m_r \delta_r$$

$$= \sum_{r=1}^{n} \frac{(r-1)^2 a^2}{n^2} \frac{a}{n} = \frac{a^3}{n^3} \sum_{r=1}^{n} (r-1)^2$$

$$= \frac{a^3}{n^3} \left[\frac{n(n-1)(2n-1)}{6} \right] = \frac{a^3}{6} \left[\left(1 - \frac{1}{n}\right)\left(2 - \frac{1}{n}\right) \right]$$

and
$$U(P,f) = \sum_{r=1}^{n} M_r \delta_r = \sum_{r=1}^{n} \frac{r^2 a^2}{n^2} \frac{a}{n}$$

$$= \frac{a^3}{n^3} \sum_{r=1}^{n} r^2 = \frac{a^3}{n^3} \frac{n(n+1)(2n+1)}{6}$$

$$= \frac{a^3}{6} \left(1 + \frac{1}{n}\right)\left(2 + \frac{1}{n}\right)$$

Hence,
$$\underline{\int_0^a} f = \lim_{\|P\| \to 0} L(P,f) = \lim_{n \to \infty} \frac{a^3}{6} \left(1 - \frac{1}{n}\right)\left(2 - \frac{1}{n}\right) = \frac{a^3}{3}$$

and
$$\overline{\int_0^a} f = \lim_{\|P\| \to 0} U(P,f)$$

$$= \lim_{n \to \infty} \frac{a^3}{6} \left(1 + \frac{1}{n}\right)\left(2 + \frac{1}{n}\right) = \frac{a^3}{3}$$

Therefore, $\underline{\int_0^a} f = \overline{\int_0^a} f$ which implies $f \in R[0,a]$ and $\int_0^a f = \frac{a^3}{3}$.

Example 9. *Let f be a function on (0,1) defined by*
$$f(x) = \begin{cases} 1 & \text{if } x \neq 1/2 \\ 0 & \text{if } x = 1/2 \end{cases}$$

Show that, $f \in R[0,1]$ and $\int_0^1 f = 1$

Solution . By definition of $f(x)$, it is clear that $f(x)$ is bounded in $[0,1]$ $[\because 0 \leq f(x) \leq 1 \ \forall x \in (0,1)]$ Consider the partition P of the interval $[0,1]$ such that point $\frac{1}{2}$ belongs to the interval $[x_{k-1}, x_k]$, then, we have

$$m_r = M_r = 1 \text{ for } r = 1,2....,n \text{ and } k \neq r, m_k = 0, M_k = 1$$

Therefore,
$$U(P,f) - L(P,f) = \sum_{\substack{r=1 \\ r \neq k}}^{n} (M_r - m_r)\delta_r + (M_k - m_k)(x_k - x_{k-1})$$

$$= \sum_{r=1}^{n} (1-1)(x_r - x_{r-1}) + (1-0)(x_k - x_{k-1})$$

$$= x_k - x_{k-1} \qquad \qquad ...(1)$$

Let $\varepsilon > 0$ be given. Then, choose a partition P such that the point $\frac{1}{2}$ is an interior point of one of the subintervals, whose length is less than ε.

Then (1) gives

$$U(P,f)-L(P,f)<\varepsilon \quad \Rightarrow \quad f\in R[0,1]$$

Now,
$$U(P,f) = \sum_{r=1}^{n} M_r\delta_r = \sum_{r=1}^{n} 1.\delta_r = \sum_{r=1}^{n} \delta_r$$

$$=\text{length of the interval } [0,1]=1$$

$$\Rightarrow \qquad \overline{\int}_0^1 f = \lim_{n\to\infty} U(P,f)=1$$

Hence, $f\in R[0,1]$ and $\underline{\int}_0^1 f = \overline{\int}_0^1 f = 1$.

Example 10. *Show that the function* [x], *where* [x] *denotes the greatest integer not greater than* x, *integrable in* [0,3] *and* $\int_0^3 [x]dx = 3$.

Solution. Clearly the given function is bounded and has only three points of discontinuity at 1,2,3. Let $\varepsilon>0$ be given

Consider a partition $P=[0=x_0,x_1,x_2,...,x_i,y_0,y_1,...,y_m,z_0,z_1,...,z_n=3]$ where $y_0=1,z_0=2$.

Now, $U(P,f)=\Sigma 0.\delta x_i+1.(y_0-x_i)+\Sigma 1.\delta y_i+2(z_0-y_m)+\Sigma 2\delta z_i+3(z_n-z_{n-1})$

$$=1+2+\{(y_0-x_i)+(z_0-y_m)+(z_n-z_{n-1})\}$$

Let us choose P such that

$$(y_0-x_i)+(z_0-y_m)+(z_n-z_{n-1})<\varepsilon$$

$$\Rightarrow \qquad U(P,f)<3+\varepsilon$$

Further, $L(P,f)= \Sigma 0.\delta x_i+1.\delta y_i+(z_0-y_m)+\Sigma 2\delta z_i=3$

$$\Rightarrow \qquad U(P,f)-L(P,f)<\varepsilon$$

$$\Rightarrow \qquad f\in R[0,3]$$

Therefore, $\overline{\int}_0^3 [x]dx = \underline{\int}_0^3 [x]dx = 3$

EXERCISE 6.1

1. Show that if f is defined on $[a,b]$ by $f(x)=c$ $\forall x\in[a,b]$ then $\int_a^b c =c(b-a)$.

2. Show that if f is defined on $[0,a],a>0$ by $f(x)=x^3$, then $f\in R[0,a]$ and $\int_0^a f = \dfrac{a^4}{4}$

3. Let f be the function defined on $[0,1]$ by
$$f(x)=\begin{cases} 0, & \text{when } x \text{ is rational} \\ 1, & \text{when } x \text{ is irrational} \end{cases}$$
Show that $f\notin R[0,1]$.

4. If $f(x)=x+x^2$ for rational values of x in the interval $[0,2]$ and $f(x)=x^2+x^3$ for irrational values of x in the same interval, find the upper and lower Riemann integrals of f over $[0,2]$.

5. Show that a constant function is R-integrable.

6. Let $f(x)=x$ on $[0,1]$. Calculate $\underline{\int}_0^1 xdx$ and $\overline{\int}_0^1 xdx$ by dividing $[0,1]$ into n equal parts and hence, show that f is R-integrable.

HINTS TO SELECTED PROBLEMS

1. $U(P,f)= \displaystyle\sum_{r=1}^{n} M_r\delta x_r = \sum_{r=1}^{n} c\delta x_r =c\sum_{r=1}^{n} \delta x_r$

$$=c[\delta x_1+\delta x_2+\delta x_3+...+\delta x_n]$$
$$=c[(x_1-x_0)+(x_2-x_1)+...+(x_n-x_{n-1})]$$

$$=c[x_n-x_0]=c(b-a)$$

Further,

$$L(P,f)= \sum_{r=1}^{n} m_r\delta x_r = \sum_{r=1}^{n} c.\delta x_r = c(b-a)$$

Now find

$$\int_a^b f = \inf U(P,f) \text{ and } \underline{\int}_a^b f = \sup L(P,f)$$

2. Consider a partition

$$P = \left\{ 0, \frac{a}{n}, \frac{2a}{n}, ..., \frac{(n-1)a}{n}, \frac{na}{n} = a \right\}.$$

Since $f(x) = x^3$, therefore $m_r = \dfrac{(r-1)^3 . a^3}{n^3}$ and

$M_r = \dfrac{r^3 . a^3}{n^3}$

Then, $L(P,f) = \sum\limits_{r=1}^{n} m_r \delta x_r = \sum\limits_{r=1}^{n} \left[\dfrac{(r-1)^3 a^3}{n^3} . \dfrac{a}{n} \right]$

$$= \dfrac{a^4}{n^4} \left[1^3 + 2^3 + ... + (n-1)^3 \right]$$

$$= \dfrac{a^4}{4} \left[1 - \dfrac{1}{n} \right]^2$$

which gives

$$\int_0^a f(x)dx = \lim_{n\to\infty} L(P,f) = \dfrac{a^4}{4}.$$

3. Take $m_r = 0, M_r = 1$.

Then

$$L(P,f) = \sum\limits_{r=1}^{n} m_r \delta x_r = \sum\limits_{r=1}^{n} 0.\delta x_r = 0$$

$$U(P,f) = \sum\limits_{r=1}^{n} M_r \delta x_r = \sum\limits_{r=1}^{n} 1.\delta x_r = 1$$

$$\Rightarrow \quad \underline{\int}_0^1 f(x)dx = \lim_{n\to\infty} L(P,f) = 0$$

and $\int_0^1 f(x)dx = \lim_{n\to\infty} U(P,f) = 1$

4. We have $\quad m_r = x^2 + x^3 \quad$ if $0 < x < 1$

$\qquad\qquad\qquad = x^2 + x^2 \quad$ if $1 < x < 2$

$\qquad\qquad M_r = x + x^2 \quad$ if $0 < x < 1$

$\qquad\qquad\qquad = x^2 + x^3 \quad$ if $1 < x < 2$

Then use the following integrals

$$\int_0^2 f(x)dx = \int_0^1 f\left(x + x^2\right)dx$$

$$= \int_1^2 f\left(x^2 + x^3\right)dx$$

and $\displaystyle\int_0^2 f(x)dx = \int_0^2 f\left(x^2 + x^3\right)dx$

$$= \int_1^2 f\left(x + x^2\right)dx$$

5. Do same as (1).

6. Consider the partition

$$P = \left\{ 0 = \dfrac{0}{n}, \dfrac{1}{n}, \dfrac{2}{n}, ..., \dfrac{r}{n}, ..., \dfrac{n}{n} = 1 \right\}$$

and $I_r = \left[\dfrac{r-1}{n}, \dfrac{r}{n} \right]$. Then $\delta I_r = \dfrac{r}{n} - \dfrac{r-1}{n} = \dfrac{1}{n}$

Then take $\quad m_r = \dfrac{r-1}{n}$ and $M_r = \dfrac{r}{n}$

Now,

$$U(P,f) = \sum\limits_{r=1}^{n} M_r \delta x_r = \sum\limits_{r=1}^{n} \left(\dfrac{r}{n} . \dfrac{1}{n} \right)$$

$$= \sum\limits_{r=1}^{n} \dfrac{r}{n^2} = \dfrac{1}{2}\left(1 + \dfrac{1}{n} \right)$$

$$L(P,f) = \sum\limits_{r=1}^{n} m_r \delta x_r = \sum\limits_{r=1}^{n} \left(\dfrac{r-1}{n} . \dfrac{1}{n} \right)$$

$$= \sum\limits_{r=1}^{n} \dfrac{r-1}{n^2} = \dfrac{1}{2}\left(1 - \dfrac{1}{n} \right)$$

Answers

4. $\dfrac{83}{12}, \dfrac{53}{12}$. 　　**9.** $\dfrac{1}{2}, \dfrac{1}{2}$

6.7 INTEGRABILITY OF CONTINUOUS, MONOTONIC FUNCTIONS

THEOREM 1. *Every continuous function is R-integrable.*

Proof.　　Let f be a continuous function on $[a,b]$, then clearly f is bounded.

[∵ Every continuous function is bounded.]

Also f is uniformly continuous on $[a,b]$　　[Being the continuous function in a closed interval.].

Let $\varepsilon>0$ be given. Then there exists a partition

$$P=\{a=x_0,x_1,x_2,...,x_n=b\}$$

of $[a,b]$ such that the oscillation (M_r-m_r) of the partition f in the subinterval (x_{r-1},x_r) is less than $\dfrac{\varepsilon}{b-a}$ for $r=1,2,...,n$. Now, consider

$$U(P,f)-L(P,f)= \sum_{r=1}^{n} M_r\left(x_r-x_{r-1}\right)- \sum_{r=1}^{n} m_r\left(x_r-x_{r-1}\right)$$

$$= \sum_{r=1}^{n}\left(M_r-m_r\right)\left(x_r-x_{r-1}\right)$$

$$< \sum_{r=1}^{n}\frac{\varepsilon}{b-a}\left(x_r-x_{r-1}\right) \qquad\qquad \left(\because M_r-m_r = \frac{\varepsilon}{b-a}\right)$$

$$\Rightarrow\quad U(P,f)-L(P,f)< \frac{\varepsilon}{b-a}\sum_{r=1}^{n}\left(x_r-x_{r-1}\right)$$

$$\Rightarrow\quad U(P,f)-L(P,f)< \frac{\varepsilon}{b-a}\left[\left(x_1-x_0\right)+\left(x_2-x_1\right)+...+\left(x_n-x_{n-1}\right)\right]$$

$$\Rightarrow\quad U(P,f)-L(P,f)< \frac{\varepsilon}{b-a}\left(x_n-x_0\right)= \frac{\varepsilon}{b-a}\left(b-a\right) \qquad [\because x_n=b \text{ and } x_0=a]$$

$$\Rightarrow\quad U(P,f)-L(P,f)<\varepsilon$$

Hence, the continuous function f is R-integrable.

THEOREM 2. *Every monotonic function f is R-integrable.*

Proof.　Let f be a monotonically increasing function on $[a,b]$ i.e., $f(a) \leq f(x) \leq f(b)$ $\forall x \in [a,b]$

Now, for a given positive number ε there exist a partition

$$P=[a=x_0,x_1,x_2,...,x_n=b] \text{ of } [a, b].$$

such that the length of each subinterval is less than $\dfrac{\varepsilon}{\left[f(b)-f(a)+1\right]}$

i.e., $\qquad (x_r-x_{r-1})< \dfrac{\varepsilon}{\left[f(b)-f(a)+1\right]}$ for $r=1,2,..,n$ ...(1)

Now, since the function f is monotonically increasing on $[a,b]$ then it is bounded and monotonically increasing on each subinterval $[x_{r-1}, x_r]$.

Let M_r and m_r be bounds of f on the subinterval $[x_{r-1}, x_r]$ then,

$$M_r=f(x_r) \text{ and } m_r=f(x_{r-1}) \qquad\qquad ...(2)$$

For the partition P, consider

$$U(P,f)-L(P,f)= \sum_{r=1}^{n}\left(M_r-m_r\right)\left(x_r-x_{r-1}\right)$$

$$< \frac{\varepsilon}{\left[f(b)-f(a)+1\right]}\sum_{r=1}^{n}\left[f(x_r)-f(x_{r-1})\right] \quad \text{[Using (1) and (2)]}$$

$$\Rightarrow\quad U(P,f)-L(P,f)< \frac{\varepsilon}{\left[f(b)-f(a)+1\right]}\left[f(x_n)-f(x_0)\right]$$

$$\Rightarrow\quad U(P,f)-L(P,f)< \frac{\varepsilon}{\left[f(b)-f(a)+1\right]}\sum_{r=1}^{n}\left[f(b)-f(a)\right] \qquad [\because x_0 = a, x_n = b]$$

$\Rightarrow \qquad U(P,f)-L(P,f)<\varepsilon$

Therefore, the function f is Riemann integrable on $[a,b]$. Similarly, we may prove that the function f is R-integrable on $[a,b]$ if f is monotonically decreasing function. Hence, every monotonic function f is R-integrable.

THEOREM 3. *A bounded function f is R-integrable in $[a,b]$ if the set of its points of discontinuity is finite.*

Proof. Given that f is discontinuous on $[a,b]$, let $[x_1,x_2,...,x_k]$ be a finite set of points of discontinuity. Also, suppose that M and m be the supremum and infimum of $f(x)$ respectively on $[a,b]$. Let $\varepsilon>0$ be an arbitrary positive number.

Now let the above points of discontinuity of the function f be enclosed in k non-overlapping intervals

$$\left[x_1',x_1''\right],\left[x_2',x_2''\right],...,\left[x_k',x_k''\right]$$

such that the sum of lengths of these subinterval be less than

$$\frac{\varepsilon}{2(M-m)}\text{(with }M-m\ne 0)$$

Since, as in each of these intervals the oscillations of the function f is less than equal to $(M–m)$, therefore, their total contribution to these oscillatory sum

$$\le\frac{\varepsilon}{2(M-m)}(M-m)\ i.e.,\ \le\varepsilon/2$$

Now, consider $(k+1)$ subintervals $[a,x_1'],[x_1'',x_2'],[x_2'',x_3'],...,[x_k'',b]$

The function f is continuous in each of these subintervals. Now, each of the above $(k+1)$ subintervals can be further subdivided so that contribution of each of them separately to the oscillatory sum of these $(k+1)$ subintervals is less than $\dfrac{\varepsilon}{2(k+1)}$.

Therefore there exists a partition of $[a,b]$ such that

$$\text{oscillatory sum }<\varepsilon/2+\frac{\varepsilon}{2(k+1)}.(k+1)$$

i.e., $\qquad$ sum$<\varepsilon/2+\varepsilon/2 = \varepsilon$

Hence, the function f is Riemann-integrable in $[a,b]$.

THEOREM 4. *Let f be a bounded function on $[a,b]$ and let the set of its discontinuities have a finite number of limit points, then $f\in R[a,b]$.*

Proof. Let $[x_1,x_2,...,x_k]$ be the finite set of limit points of the set of discontinuities of f on $[a,b]$ such that $x_1<x_2<...<x_k$

Let $\varepsilon>0$ be given. Now let the above points of discontinuity of the function f be enclosed in k non-overlapping intervals

$$[x_1',x_1''],[x_2',x_2''],...,[x_k',x_k'']$$

such that the sum of their length is $<\dfrac{\varepsilon}{2(M-m)}$ where $M=$supremum of f and $m=$infimum of f.

Now the partition P of $[a,b]$ is given by

$$P = \left[a, x_1', x_1'', x_2', x_2'' \ldots, x_k', x_k'', b \right]$$

which has $(2k+1)$ component intervals of two types.

(i) k intervals. *i.e.,* $\left[x_i', x_i'' \right], i=1,2, \ldots k$ each of which contain a point x_i in its interior.

The total contribution to the oscillatory sum by these intervals is

$$= \sum_{i=1}^{k} (M_i - m_i)(b-a) \le \sum_{i=1}^{k} (M-m)(b_i - a_i)$$

$$= (M-m) \sum_{i=1}^{k} (b_i - a_i)$$

$$< (M-m) \frac{\varepsilon}{2(M-m)} = \varepsilon/2$$

(ii) $(k+1)$ subintervals. *i.e.,* $\left[a, x_1' \right], \left[x_1'', x_2' \right], \left[x_2'', x_3' \right] \ldots \left[x_k'', b \right]$

In the above subintervals, the function f has only a finite number of points of discontinuity. Hence, there exists a partition $P_r : r=1,2,\ldots(k+1)$ respectively of these subintervals such that the oscillatory sum is less than $\varepsilon/2(k+1)$ for $r=1,2,\ldots k+1$. Hence, the total contribution to the oscillatory sum by these subintervals is less than equal to $\dfrac{\varepsilon}{2(k+1)}(k+1)$ *i.e.,* $\le \varepsilon/2$

Therefore, for any partition P the total oscillatory sum $<\varepsilon/2+\varepsilon/2=\varepsilon$

Hence, $f \in R[a,b]$.

6.8 ALGEBRA OF R-INTEGRABLE FUNCTIONS

THEOREM 1. *If f is R-integrable on $[a,b]$ then $|f|$ is also R-integrable on $[a,b]$.*

Proof.   Since the function f is R-integrable on $[a,b]$, therefore f is bounded on $[a,b]$.

[∵ Every integrable function is continuous and hence bounded.]

$\Rightarrow$ $|f(x)| \le \lambda \ \forall \ x \in [a,b]$ for any positive number λ.

Also, since f is R-integrable on $[a,b]$, therefore there exists a partition P of $[a,b]$ such that for any positive number ε.

$$U(P,f) - L(P,f) < \varepsilon \qquad \ldots(1)$$

Let the upper and lower bounds of $|f|$ and f in $\delta_r = [x_{r-1}, x_r]$ be respectively given by M_r, m_r and M_r', m_r'.

Then for all y, z in $[x_{r-1}, x_r]$, we have $\left[\, |f(z)| - |f(y)| \, \right] \le |f(z) - f(y)|$

$\Rightarrow$ $M_r - m_r \le M_r' - m_r'$ (By taking supremum)

$$\Rightarrow \quad \sum_{r=1}^{n} (M_r - m_r)\delta_r \le \sum_{r=1}^{n} \left(M_r' - m_r' \right)\delta_r$$

$$\Rightarrow \quad \sum_{r=1}^{n} M_r \delta_r - \sum_{r=1}^{n} m_r \delta_r \le \sum_{r=1}^{n} M_r' \delta_r - \sum_{r=1}^{n} m_r' \delta_r$$

$\Rightarrow \quad \{U(P,|f|)-L(P,|f|)\} \le U(P,f)-L(P,f)$

$\Rightarrow \quad U(P,|f|)-L(P,|f|)<\varepsilon$ [Using (1)]

$\Rightarrow \quad |f|$ is R-integrable on (a,b).

REMARK

- From above theorem, we conclude that $\left|\int_a^b f(x)dx\right| \le \int_a^b |f(x)|\,dx$

THEOREM 2. *If f_1 and f_2 are R-integrable functions on $[a,b]$ then $f_1 \pm f_2$ is also R-integrable on $[a,b]$.*

Proof. Let f_1,f_2 be two R-integrable functions on $[a,b]$.

If f_1 is R-integerable on $[a,b]$, then for given $\varepsilon>0$ there exists a partition P_1 such that

$$U(P_1,f_1)-L(P_1,f_1)<\varepsilon/2 \qquad ...(1)$$

Also, f_2 is R-integerable.

$\Rightarrow$ For given $\varepsilon>0$ there exists a partition P_2 such that

$$U(P_2,f_2)-L(P_2,f_2)<\varepsilon/2 \qquad ...(2)$$

Define the common refinement P of partitions P_1 and P_2 such that $P=P_1\cup P_2$

Clearly $P\in P[a,b]$, where $P[a,b]$ denotes the family of all partitions on $[a,b]$. Consider

$$U(P,f_1+f_2)-L(P,f_1+f_2)\le[\{U(P,f_1)-L(P,f_1)\}+\{U(P,f_2)-L(P,f_2)\}]$$

$$<\varepsilon/2+\varepsilon/2 \qquad \text{[Using (1) and (2)]}$$

$\Rightarrow \quad U(P,f_1+f_2)-L(P,f_1+f_2)<\varepsilon$

$\Rightarrow \quad f_1+f_2$ is R-integrable on $[a,b]$

Similarly we can show that f_1-f_2 is R-integrable on $[a,b]$.

REMARK

- In the above two theorems, the function f_1 and f_2 must be real and bounded on $[a, b]$.

THEOREM 3. *If f is R-integrable on $[a,b]$ then cf is also R-integrable on $[a,b]$, where $c \in R$*

Also, $\int_a^b cf(x)dx = c\int_a^b f(x)dx$

Proof. Given that the function f is R-integrable on $[a,b]$ therefore, there exists a partition P on $[a, b]$, such that

$$U(P,f)-L(P,f)<\varepsilon \qquad ...(1)$$

Let $c\in R$ be any constant, then we know that

$$(cf)(x)= cf(x)$$

Therefore, $U(P,cf)= cU(P,f)$ and $L(P,cf)=cL(P,f)$

Now, consider

$$U(P,cf)-L(P,cf)=c[U(P,f)-L(P,f)]<c\varepsilon$$

$\Rightarrow \quad cf\in R[a,b]$

Also $\quad U(P, cf)< \int_a^b cf(x)dx + \varepsilon$

and
$$cU(P,f) < \int_a^b cf(x)dx + \varepsilon$$
Now, using (1), we get
$$cU(P,f) = U(P,cf) < \int_a^b cf(x)dx + \varepsilon$$
$$\Rightarrow \qquad c\int_a^b f(x)dx \geq \int_a^b cf(x)dx \qquad \qquad \dots(2)$$
Replacing f by $-f$ in (2), we get
$$c\int_a^b -f(x)dx \geq \int_a^b -cf(x)dx$$
$$\Rightarrow \qquad c\int_a^b f(x)dx \leq \int_a^b cf(x)dx \qquad \qquad \dots(3)$$
From (2) and (3) we conclude that
$$\int_a^b cf(x)dx = c\int_a^b f(x)dx \ .$$

THEOREM 4. *If the function f is R-integrable and if M and m, the supremum and infimum of f on [a,b], then*
$$m(b-a) \leq \int_a^b f(x)dx \leq M(b-a), if\ b \geq a$$
and
$$m(b-a) \geq \int_a^b f(x)dx \geq M(b-a), if\ b \leq a$$

Proof. Let $P[a,b]$ denotes the family of all partitions on $[a,b]$. If $b>a$, then for all $P \in P[a,b]$, we have
$$m(b-a) \leq L(P,f) \leq U(P,f) \leq M(b-a)$$
$$\Rightarrow \qquad m(b-a) \leq L(P,f) \leq M(b-a)$$
$$\Rightarrow \qquad m(b-a) \leq \int_{\underline{b}}^a f(x)dx \leq M(b-a)$$
$$\Rightarrow \qquad m(b-a) \leq \int_a^b f(x)dx \leq M(b-a)$$
$$\left[\because \int_{\underline{a}}^b f(x)dx = \int_a^b f(x)dx\ for\ f \in R[a,b]\right]$$
If $b<a$, then in a similar way, we may get
$$m(a-b) \leq \int_b^a f(x)dx \leq M(a-b)$$
$$\Rightarrow \qquad -m(b-a) \leq -\int_a^b f(x)dx \leq -M(b-a)$$
$$\Rightarrow \qquad m(b-a) \leq \int_{\underline{b}}^a f(x)dx \leq M(b-a)$$

REMARK

- In case of $a=b$, the result is obvious.

THEOREM 5. *If the function f(x) is bounded and R-integrable over [a,b] and f(x) ≥ 0 ∀x ∈ [a,b], then $\int_a^b f(x)dx \geq 0$.*

Proof. Let M and m be the supremum and infimum of f on $[a,b]$. Then by above theorem if $b \geq a$ we have
$$m(b-a) \leq \int_a^b f(x)dx \leq M(b-a) \qquad \qquad \dots(1)$$
Here, it is given that $f(x) \geq 0 \ \forall x\ [a,b]$
Therefore $\qquad \qquad m \geq 0.$
Also $\qquad \qquad b \geq a \qquad \Rightarrow \qquad b-a \geq 0.$

Hence, from (1), we conclude that $\int_a^b f(x)dx \geq 0$

THEOREM 6. **(First Mean Value Theorem).** *If the function f is R-integrable over [a,b] and M,m be supremum, infimum respectively of f on [a,b], then there exists a number k (m ≤ k ≤ M) such that*

$$\int_a^b f(x)dx = k(b-a)$$

Also, if the function f is continuous on [a,b], then there exists c ∈ [a,b] such that

$$\int_a^b f(x)dx = (b-a)f(c)$$

Proof. We know that (From Theorem 4)

$$m(b-a) \le \int_a^b f(x)dx \le M(b-a), \text{if } b \ge a$$

and

$$m(b-a) \ge \int_a^b f(x)dx \ge M(b-a), \text{if } b \le a$$

If $m \le k \le M$, then we conclude that

$$\int_a^b f(x)dx = k(b-a) \qquad\qquad ...(1)$$

Also, if the function f is continuous on $[a,b]$, then there exists a number c in $[a,b]$ such that $f(c)=k$, where $m \le k \le M$.

Hence, from (1), we conclude that

$$\int_a^b f(x)dx = (b-a)f(c)$$

THEOREM 7. *If f and g are R-integrable over [a,b], then fg is also R-integrable over [a,b].*

Proof. Since f and g both are R-integrable over $[a,b]$, therefore f and g both are bounded on $[a,b]$.

$\Rightarrow$ $\exists M>0$ such that $|f(x)| \le M$ and $|g(x)| \le M$, $\forall x \in [a,b]$

Consider $|(fg)(x)| = |f(x).g(x)| \forall x \in [a,b]$

$$\le M^2 \; \forall x \in [a,b] \qquad\qquad ...(1)$$

$\Rightarrow$ fg is bounded on $[a,b]$

Now, let $\varepsilon>0$ be given.

Since, $f \in R(a,b)$, therefore, there exist a partition P_1 of $[a,b]$ such that

$$U(P_1,f)-L(P_1,f)<\varepsilon/2M \qquad\qquad ...(2)$$

Similarly $g \in R(a,b)$, therefore, there exist a partition P_2 of $[a,b]$ such that

$$U(P_2, g)-L(P_2, g)<\varepsilon/2M \qquad\qquad ...(3)$$

Let $P=P_1 \cup P_2$ be the refinement of P_1 and P_2, then we have

$$\left.\begin{array}{l} U(P,f)-L(P,f)<\varepsilon/2M \\ \\ U(P,g)-L(P,g)<\varepsilon/2M \end{array}\right\} \qquad\qquad ...(4)$$

and

Let, $m_r, M_r, m_r', M_r', m_r'', M_r''$ be the infimum and supremum of f, g and $f.g$ respectively over the subinterval $I_r=[x_{r-1},x_r]$. Then for all $x,y \in I_r$ we have

$$|(fg)(x)-(fg)(y)| = |f(x).g(x)-f(y).g(y)|$$

$$= |f(x).g(x)-f(y)g(x)+f(y).g(x)-f(y).g(y)|$$

$$= |g(x)[f(x)-f(y)]+f(y)[g(x)-g(y)]|$$

$$\le |g(x)||f(x)-f(y)| + |f(y)||g(x)-g(y)|$$

$$\le M|f(x)-f(y)| + M|g(x)-g(y)| \qquad\qquad ...(5)$$

Now, $|f(x)-f(y)| \leq M_r - m_r$ ...(6)

and $|g(x)-g(y)| \leq M_r' - m_r'$ ...(7)

$\therefore$ $M_r'' - m_r'' < M(M_r - m_r) + M\left(M_r' - m_r'\right)$ ...(8)

Multiplying both sides of (8) by δ_r and adding on respective sides, we get

$$U(P,fg)-L(P,fg) \leq M[U(P,f)-L(P,f)]+M[U(P,g)-L(P,g)]$$

$$< M.\frac{\varepsilon}{2M} + M\frac{\varepsilon}{2M} = \varepsilon$$

Hence, *fg* is R-integrable.

THEOREM 8. *If f and g are R-integrable functions on [a,b] and $|g(x)| \leq k \ \forall x \in [a,b]$ where k is a positive number then the quotient function f /g is also R-integrable on [a,b]*

Proof.   Since *f* and *g* both are R-integrable on [a,b], therefore, they are bounded on [a,b]. Also, we know that the quotient of two bounded functions is again bounded, therefore *f/g* is also bounded on [a,b].

Let $\varepsilon > 0$ be given. Since $f \in R[a,b]$, therefore, there exists a partition P_1 of [a,b] such that

$$U(P_1,f)-L(P_1,f) < \frac{\varepsilon}{2m}k^2$$ ...(1)

Similarly $g \in R [a,b]$, therefore, there exists a partition P_2 of [a,b] such that

$$U[P_2, g]-L[P_2, g] < \frac{\varepsilon}{2M}k^2$$ ...(2)

Let $P=P_1 \cup P_2$ be a refinement of P_1 and P_2, then from (1) and (2), we have

$$U[P,f]-L[P,f] < \frac{\varepsilon}{2M}k^2$$ ...(3)

and $$U[P,g]-L[P,g] < \frac{\varepsilon}{2M}k^2$$ ...(4)

Now, let m_r, M_r; m_r', M_r'; m_r'', M_r'' be the supremum and infimum of *f,g* and *f/g* respectively over the subinterval $I_r=[x_{r-1}, x_r]$. Then for all $x, y \in I_r$, we have

$$\left|\frac{f}{g}(x)-\frac{f}{g}(y)\right| = \left|\frac{f(x)}{g(x)} - \frac{f(y)}{g(y)}\right| = \frac{|f(x)g(y)-f(y)g(x)|}{|g(x)g(y)|}$$

$$= \frac{|f(x)g(y)-f(y)g(y)+f(y)g(y)-f(y)g(x)|}{|g(x)g(y)|}$$

$$= \frac{|[f(x)-f(y)]g(y)+f(y)[g(y)-g(x)]|}{|g(x)g(y)|}$$

$$\leq \frac{|g(y)||f(x)-f(y)|}{|g(x)||g(y)|} + \frac{|f(y)||g(y)-g(x)|}{|g(x)||g(y)|}$$

$$\leq \frac{M}{k^2}|f(x)-f(y)| + \frac{M}{k^2}|g(x)-g(y)|$$ ...(5)

Now m_r and M_r are the infimum and supremum of *f* respectively over I_r. Therefore,

$$|f(x)-f(y)| \leq M_r - m_r \ \forall \ x, y \in [a,b]$$ ...(6)

Similarly $\quad |g(x)-g(y)| \le M_r{'}-m_r{'} \qquad \forall x,y \in [a,b]$ ...(7)

which implies

$$\left|\frac{f}{g}(x)-\frac{f}{g}(y)\right| \le \frac{M}{k^2}(M_r-m_r)+\frac{M}{k^2}\left(M_r{'}-m_r{'}\right) \qquad \text{...(8)}$$

$\Rightarrow$

$$M_r{''}-m_r{''} \le \frac{M}{k^2}(M_r-m_r)+\frac{M}{k^2}\left(M_r{'}-m_r{'}\right) \qquad \text{...(9)}$$

Multiplying (9) by δ_r and adding on the respective sides, we get

$$U(P,f/g)-L(P,f/g) \le \frac{M}{k^2}\big[U(P,f)-L(P,f)\big]+\frac{M}{k^2}\big[U(P,g)-L(P,g)\big]$$

$$\le \frac{M}{k^2}\frac{\varepsilon.k^2}{2M}+\frac{M}{k^2}\frac{\varepsilon.k^2}{2M}=\varepsilon$$

Hence, $\dfrac{f}{g}$ is R-integrable over $[a,b]$.

THEOREM 9. *If f is R-integrable over $[a,b]$ then f^2 is also R-integrable over $[a,b]$*

Proof. Since, $f \in R[a,b] \qquad\qquad \Rightarrow \quad f$ is bounded on $[a,b]$

$\qquad\qquad\qquad\qquad\qquad\qquad \Rightarrow \quad |f|$ is bounded on $[a,b]$

$\qquad\qquad\qquad\qquad\qquad\qquad \Rightarrow \quad \exists\, m>0$ such that $|f(x)| \le m \,\forall\, x \in [a,b]$

Also $f \in R[a,b] \qquad\qquad \Rightarrow \quad f$ is integrable

$\qquad\qquad\qquad\qquad\qquad\qquad \Rightarrow \quad |f|$ is also integrable

$\qquad\qquad\qquad\qquad\qquad\qquad \Rightarrow \quad$ for $\varepsilon>0\,\exists$ a partition P of $[a,b]$ such that

$$U(P,|f|)-L(P,|f|) < \frac{\varepsilon}{2M}$$

Further since $|f^2(x)| = (f(x))^2 \le M^2 \quad \Rightarrow \quad f^2$ is bounded.

Let m_i, M_i be the bounds of $|f|$ and $M_i{'}$, $m_i{'}$ those of f^2 in δx_i,

then $M_i{'} = M_i^2$, $m_i{'} = m_i^2$

Now, $\qquad U(P,f^2)-L(P,f^2) = \displaystyle\sum_{i=1}^{n}\left(M_i{'}-m_i{'}\right)\delta x_i$

$$= \sum_{i=1}^{n}\left(M_i^2-m_i^2\right)\delta x_i$$

$$= \sum_{i=1}^{n}\left(M_i-m_i\right)\left(M_i+m_i\right)\delta x_i$$

$$\le 2M\left[\sum_{i=1}^{n}\left(M_i-m_i\right)\delta x_i\right]$$

$$= 2M\big[U(P,|f|)-L(P,|f|)\big]$$

$$< 2M.\frac{\varepsilon}{2M}=\varepsilon$$

$\Rightarrow \qquad\qquad f^2 \in R[a,b]$

$\Rightarrow \quad f^2$ is R-integrable over $[a,b]$.

6.9 PRIMITIVE AND INTEGRAL FUNCTIONS

Definition 1. *A function $F(x)$ defined on $[a,b]$ called a primitive of a function $f(x)$, if the function $F(x)$ has $f(x)$ as its derivative at each $x \in [a,b]$ i.e., $F'(x) = f(x) \; \forall \; x \in [a, b]$.*

Definition 2. *Let $f(x)$ be a R-integrable function of the function on $[a,b]$. Then a function $F(x)$ is called the integral function of the function $f(x)$ if $F(x) = \int_a^b f(t)dt$, $\forall x \in [a, b]$*

REMARK

• The primitive of a function is not unique since if $F(x)$ is a primitive of $f(x)$ then $F(x) + c$ is also a primitive function of $f(x)$.

THEOREM 1. *Let $f \in R[a,b]$, then the integral function F of f given by*

$$F(x) = \int_a^x f(t)dt, \; a \leq x \leq b.$$

is continuous on $[a,b]$.

Proof. Let $f \in R[a,b]$, then obviously it is bounded on $[a,b]$. Therefore, there exists a positive number M such that $|f(t)| \leq M \; \forall t \in [a,b]$.

Let $x_1, x_2 \in [a,b]$ such that $x_1 < x_2$. Then, we have

$$|F(x_2) - F(x_1)| = \left| \int_a^{x_2} f(t)dt - \int_a^{x_1} f(t)dt \right|$$

$$= \left| \int_a^{x_2} f(t)dt + \int_{x_1}^a f(t)dt \right|$$

$$= \left| \int_{x_1}^{x_2} f(t)dt \right| \leq M \left| \int_{x_1}^{x_2} dt \right| = M|(x_2 - x_1)|$$

Let $|x_2 - x_1| < \varepsilon/M$ for a given positive number ε. Then, we have

$$|F(x_2) - F(x_1)| < M.\varepsilon/M$$

$\Rightarrow \qquad |F(x_2) - F(x_1)| < \varepsilon$ whenever, $|x_2 - x_1| < \delta \; \forall x_1, x_2 \in [a,b]$

where $\qquad \delta = \dfrac{\varepsilon}{M}$.

$\Rightarrow \quad F$ is uniformly continuous on $[a, b]$. Hence it is continuous on $[a,b]$.

[$\because$ Every uniformly continuous function is continuous]

REMARK

• The above theorem can be restated as "the integral of an integrable function is continuous.

THEOREM 2. *Let f be a continuous function on $[a,b]$ and let $F(x) = \int_a^x f(t)dt$; $\forall x \in [a,b]$.*

Then $\qquad\qquad F'(x) = f(x); \; \forall x \in [a,b]$

Proof. Let $x \in [a,b]$. Then choose $h \neq 0$ such that $x + h \in [a,b]$.

Consider $\qquad F(x+h) - F(x) = \int_a^{x+h} f(t)dt - \int_a^x f(t)dt$

$$= \int_a^{x+h} f(t)dt + \int_x^a f(t)dt = \int_x^{x+h} f(t)dt \qquad\qquad ...(1)$$

Since f is continuous on $[a,b]$, therefore, there exist a number $c \in [x, x+h]$

$$\int_x^{x+h} f(t)dt = hf(c) \qquad\qquad ...(2)$$

Clearly $c \to x$ as $h \to 0$.

From (1) and (2), we conclude that

$$F(x+h)-F(x)=hf(c)$$

$$\Rightarrow \quad \lim_{h\to 0}\frac{F\left(x+h\right)-F\left(x\right)}{h} = \lim_{h\to 0} f(c)$$

$$\Rightarrow \qquad\qquad F'(x)= f(x)$$

Hence, we have $\quad F'(x)= f(x)\ \forall x\in [a,b].$

6.10 FUNDAMENTAL THEOREM OF INTEGRAL CALCULUS

THEOREM 1. *Let f be a R-integrable function on [a,b] and F be a differentiable primitive function on [a,b] such that* $F'(x)=f(x), a \le x \le b$, *then*

$$\int_a^b f(t)dt = F(b)-F(a)$$

Proof. Let f be continuous function on [a,b].

By definition of primitive function, we have $F'(x)= f(x)\ \forall x\in [a,b]$

Also, f is R-integrable function on [a,b]

$\Rightarrow\quad F'(x)$ is R-integrable function on [a,b]

i.e., for a given positive number ε there exists a partition P of [a,b] such that

$$\left|\sum_{r=1}^{n} F'(t_r)(x_r - x_{r-1})-\int_a^b F'(x)dx\right| < \varepsilon\ ;\ \text{where } t_r\in (x_r\text{-}x_{r-1}). \qquad ...(1)$$

By Lagrange's mean value theorem of differential calculus, we find that there exists $t_r\in [x_r\text{-}x_{r-1}]$ such that

$$F(x_r)-F(x_{r-1})=(x_r\text{-}x_{r-1})F'(t_r)$$

$$\Rightarrow\quad \sum_{r=1}^{n}\left[(x_r - x_{r-1})F'(t_r)\right] = \sum_{r=1}^{n}\left[F(x_r)- F(x_{r-1})\right]= F(b)-F(a)$$

Put this value in (1), we get

$$\left|F(b)-F(a)-\int_a^b F'(x)dx\right| < \varepsilon$$

which gives $\quad F(b)-F(a)=\int_a^b F'(x)dx = \int_a^b f(x)dx \qquad [\because F'(x)=f(x)]$

$$\int_a^b f(x)dx = F(b)-F(a)$$

THEOREM 2. *Let $a<c<b$ then $f\in R[a,b]$ iff $f\in R[a,c]$ and $f\in R[c,b]$ then*

$$\int_a^b f(x)dx = \int_a^c f(x)dx + \int_c^b f(x)dx$$

Proof. Let f be bounded on [a,c] and [c,b] iff f is bounded on [a,b].

Let $f\in R[a,b]$. Then for a given $\varepsilon>0$, there is a partition P of [a,b] such that

$$U(P,f)-L(P,f)<\varepsilon \qquad ...(1)$$

Let P^* be a partition of [a, b] and also a refinement of P such that

$$P^*= P\cup\{c\}$$

then $\quad U(P^*,f)-L(P^*,f) \le U(P,f)-L(P,f)<\varepsilon \qquad ...(2)$

Now again P_1 and P_2 be the partitions of [a,c] and [c,b] such that $P^*= P_1\cup P_2$

Now using this result in equation (2), we have

$$U(P^*,f) - L(P^*,f) = [U(P_1,f) + U(P_2,f)] - [L(P_1,f) + L(P_2,f)]$$
$$= [U(P_1,f) - L(P_1,f)] + [U(P_2,f) - L(P_2,f)] < \varepsilon$$

$$\begin{cases} \text{Since } U(P_1,f) - L(P_1,f) < \dfrac{r\varepsilon}{n} \\[2mm] \qquad\qquad U(P_2,f) - L(P_2,f) < \dfrac{(n-r)\varepsilon}{n} \\[2mm] \Rightarrow\quad \{U(P_1,f) - L(P_1,f)\} + \{U(P_2,f) - L(P_2,f)\} < \dfrac{r\varepsilon}{n} + \dfrac{(n-r)\varepsilon}{n} \end{cases}$$

$\Rightarrow \qquad [U(P_1,f) - L(P_1,f)] + [U(P_2,f) - L(P_2,f)] < \varepsilon$

So $\qquad f \in R[a,c]$ and $f \in R[c,b]$

Since, P^* is the union of P_1 and P_2 then

$$U(P^*,f) = U(P_1,f) + U(P_2,f)$$

$\Rightarrow \qquad \int_a^{\overline{b}} f(x)dx = \int_a^{\overline{c}} f(x)dx + \int_c^{\overline{b}} f(x)dx$

$\Rightarrow \qquad \int_a^b f(x)dx = \int_a^c f(x)dx + \int_c^b f(x)dx$

Now we shall prove that the converse part.

Let $f \in R[a,c]$ and $f \in R[c,b]$, then for given $\varepsilon > 0$, $\exists$ partition P_1 and P_2 such that

$$U(P_1,f) - L(P_1,f) < \frac{\varepsilon}{2}$$

$$U(P_2,f) - L(P_2,f) < \frac{\varepsilon}{2}$$

where P_1 and P_2 partition of $[a,c]$ and $[c,b]$ respectively and P is defined as $P = P_1 \cup P_2$

Now $\quad U(P,f) - L(P,f) = [U(P_1,f) + U(P_2,f)] - [L(P_1,f) + L(P_2,f)]$
$$= [U(P_1,f) - L(P_1,f)] + [U(P_2,f) - L(P_2,f)]$$

$$< \frac{\varepsilon}{2} + \frac{\varepsilon}{2} < \varepsilon$$

$\Rightarrow \qquad f \in R[a,b]$

and also $P = P_1 \cup P_2$ on partition $[a,b]$

$\Rightarrow \qquad U(P,f) = U(P_1,f) + U(P_2,f)$

$\Rightarrow \qquad \int_a^{\overline{b}} f(x)dx = \int_a^{\overline{c}} f(x)dx + \int_c^{\overline{b}} f(x)dx$

$\Rightarrow \qquad \int_a^b f(x)dx = \int_a^c f(x)dx + \int_c^b f(x)dx$

THEOREM 3. *Let $f \in R[a,b]$ and let f be continuous at $x_0 \in [a,b]$. If $F(x) = \int_a^x f(t)dt$, $a \leq x \leq b$ then $F'(x_0) = f(x_0)$.*

Proof. $f \in R[a,b] \Rightarrow f$ is continuous at x_0 therefore, for given $\varepsilon > 0$, $\exists$ a $\delta > 0$ such that $|f(x_0 + h) - f(a)| < \varepsilon$ where $|h| < \delta \leq \varepsilon$ and $a \leq x_0 + h \leq b$.

Now $\qquad F(x_0 + h) - F(x_0) = \int_a^{x_0 + h} f(t)dt - \int_a^{x_0} f(t)dt = \int_{x_0}^{x_0 + h} f(t)dt$

$$\int_{x_0}^{x_0 + h} f(x_0)dt = f(x_0)[(x_0 + h) - x_0] = hf(x_0)$$

Now, $\left| \dfrac{F(x_0+h)-F(x_0)}{h} - f(x_0) \right| = \left| \dfrac{1}{h}\int_{x_0}^{x_0+h} f(t)dt - \int_{x_0}^{x_0+h} f(x_0)dt \right|$

$$= \left| \frac{1}{h}\int_{x_0}^{x_0+h} \big(f(t)-f(x_0)\big)dt \right|$$

$$\left[\because \text{ if } f \in R[a,b], |f(x)| \le k\forall x \in [a,b] \Rightarrow \left| \int_a^b f(x)dx \right| \le k\,|\,b-a\,| \right]$$

$\Rightarrow \qquad \left| \dfrac{F(x_0+h)-F(x_0)}{h} - f(x_0) \right| < \dfrac{1}{|h|}\cdot|h|\varepsilon = \varepsilon \ \forall |h| < \delta$

$$\lim_{h\to 0} \frac{F(x_0+h)-F(x_0)}{h} = f(x_0) \Rightarrow F'(x_0) = f(x_0)$$

THEOREM 4. **(The generalised first mean value theorem for Integrals)** *Let f and g be R-integrable functions such that g does not assume both positive and negative values on* [a,b] *and suppose M=sup f and m= inf f on* [a,b] *then* $\int_a^b f.g\,dx = \lambda\int_a^b g\,dx$,$m\le\lambda\le M$. *In particular, if when f is continuous on* [a,b] *then there exists a point t on* [a,b] *such that* $\qquad \int_a^b f.g\,dx = f(t)\int_a^b g\,dx$

Proof. Let us suppose $m=$ inf f and $M=$sup f such that
$$m \le f(x) \le M \ \Rightarrow \qquad mg \le fg \le Mg \text{ for } g>0$$
and $\qquad\qquad\qquad mg>fg>Mg \text{ for } g <0$

$\Rightarrow \qquad \int_a^b f.g\,dx$ lies between $m\int_a^b g\,dx$ and $M\int_a^b g\,dx$

$\Rightarrow \qquad \int_a^b f.g\,dx = \lambda\int_a^b g\,dx \qquad\qquad$ where $m\le\lambda\le M$

Finally continuity of f on [a,b] $\Rightarrow$ $\lambda=f(t)$ for $a \le t \le b$
and hence
$$\int_a^b f.g\,dx = \lambda\int_a^b g\,dx = f(t)\int_a^b g\,dx$$

THEOREM 5. **(Second theorem of mean value)**

(i) **Bonnet's form** *If f and g are continuous on* [a,b] *and g is a positive decreasing function, then there exists a point t in* [a,b] *such that* $\int_a^b f.g\,dx = g(a)\int_a^b f\,dx$

(ii) **Weierstrass form** *If g is a positive monotonic increasing function then there exists a point t on* [a,b] *such that* $\int_a^b f.g\,dx = g(b)\int_t^b f\,dx$

Proof. (i) Let us suppose f and g are continuous on [a,b]

$\Rightarrow \qquad\qquad f, g\in R[a,b]$

$\Rightarrow \qquad\qquad f.\,g\in R[a,b]$

$\Rightarrow \qquad\qquad f(x)g(x) \le f(x)g(a)$

$\Rightarrow \qquad\qquad \int_a^b fg(x)dx \le g(a)\int_a^b f(x)dx$

Now $\int_a^t f(x)dx$ is an increasing continuous function on [a,b].

$\Rightarrow \qquad \int_a^b f(x)g(x)dx = g(a)\int_a^t f(x)dx$;$a \le t \le b$.

(ii) If $g(x)$ is a positive increasing function then similar reason shows that

$\Rightarrow \qquad \int_a^b f(x)g(x)dx = g(b)\int_t^b f(x)dx$;$a \le t \le b$.

THEOREM 6. **(Second Integral theorem of mean value)** *Let $f(x)$ be continuous and $g(x)$ either increasing or decreasing with $g'(x)$ continuous on $[a,b]$ then*

$$\int_a^b f(x)g(x)dx = g(a)\int_a^t f(x)dx + g(b)\int_t^b f(x)dx \; ; a \le t \le b$$

Proof. Let $g(x)$ be monotonically increasing

$\Rightarrow \qquad g'(x) \ge 0$

$\Rightarrow \qquad g(x)$ is continuous.

Let us suppose $F(x)$ is an indefinite integral of $f(x)$ therefore,

$$\int_a^b f(x)g(x)dx = \left[F(x).g(x)\right]_a^b - \int_a^b F(x)g'(x)dx$$

But $g'(x)$ does not change sign in $]a,b[$. Then by first theorem of mean value

$$\int_a^b f(x)g(x)dx = \left[F(x).g(x)\right]_a^b - F(t)\int_a^b g'(x)dx \; , a \le t \le b$$

$$=F(b)g(b)-F(a)g(a)-F(t)g(b)+F(t)g(a)$$

$$=g(a)[F(t)-F(a)]+g(b)[F(b)-F(t)]$$

$$= g(a)\int_a^t f(x)dx + g(b)\int_t^b f(x)dx , a \le t \le b$$

THEOREM 7. **(Second mean value theorem : a particular case)** *If f is monotonic and f, f' and g are all continuous on $[a,b]$ then $\exists \; t \in [a,b]$ such that*

$$\int_a^b f(x)g(x)dx = f(a)\int_a^t g(x)dx + f(b)\int_t^b g(x)dx$$

Proof. Let $G(x) = \int_a^x g(t)dt \; \Rightarrow \; G(a) = \int_a^a g(t)dt = 0$

Also, under the given conditions, $G(x)$ is differentiable and $G'(x)=g(x)$

So, $$\int_a^b f(x)g(x)dx = \int_a^b f(x)G'(x)dx = \left[f(x)G(x)\right]_a^b - \int_a^b G(x)f'(x)dx$$

Now, since G is continuous, therefore, it is integrable.

Also, f is monotonic and continuous on $[a,b]$ then by generalised mean value theorem $\exists \; t \in [a,b]$ such that

$$\int_a^b f(x)g(x)dx = f(b)G(b) - G(t)\int_a^b f'(x)dx$$

$$=f(b)G(b)-G(t)[f(b)-f(a)]$$

$$=f(b)[G(b)-G(t)]+f(a)G(t)$$

$$= f(b)\int_a^b g(x)dx + f(a)\int_a^t g(x)dx$$

THEOREM 8. **(Reimann-Lebesgue Lemma)** *If a function f is bounded and integrable on $[a,b]$ then*

$$\lim_{n \to \infty} \int_a^b f(x)\cos nx \, dx = 0 \;\; and \;\; \lim_{n \to \infty} \int_a^b f(x)\sin nx \, dx = 0.$$

Proof. Let $I_n = \int_a^b f(x)\cos nx \, dx$

Let $\varepsilon > 0$ be arbitrary. Since, f is bounded and integrable on $[a,b]$ $\exists$ a partition $P = [a=x_0, x_1, x_2, ..., x_p = b]$ such that the oscillatory sum

$$U(P,f) - L(P,f) = \sum_{i=1}^{p} (M_i - m_i)\delta x_i < \varepsilon / 2 \; ; \text{ where } M_i, m_i \text{ are bounds of } f \text{ in } \delta x_i$$

Now $$I_n = \int_{x_{i-1}}^{x_i} f(x)\cos nx \, dx$$

$$= \sum_{i=1}^{p} f(x_{i-1})\int_{x_{i-1}}^{x_i} \cos nx \, dx + \sum_{i=1}^{p} \int_{x_{i-1}}^{x_i} \left[f(x)-f(x_{i-1})\right]\cos nx \, dx$$

$$\Rightarrow \qquad |I_n| \le \sum_{i=1}^{p} |f(x_{i-1})| \left| \int_{x_{i-1}}^{x_i} \cos nx\, dx \right| + \sum_{i=1}^{p} \int_{x_{i-1}}^{x_i} \left| [f(x) - f(x_{i-1})] \cos nx \right| dx$$

Now, for all $x \in \delta x_i$, we have $|f(x) - f(x_{i-1})| \le M_i - m_i$

$$\Rightarrow \qquad |[f(x) - f(x_{i-1})] \cos nx| \le M_i - m_i$$

and $\qquad \displaystyle\sum_{i=1}^{p} \int_{x_{i-1}}^{x_i} \left| [f(x) - f(x_{i-1})] \cos nx \right| dx \le \sum_{i=1}^{p} (M_i - m_i)(x_i - x_{i-1}) < \frac{\varepsilon}{2}$

Again $\qquad \displaystyle\left| \int_{x_{i-1}}^{x_i} \cos nx\, dx \right| = \left| \frac{1}{n} [\sin nx_i - \sin nx_{i-1}] \right| \le \frac{1}{n} [|\sin nx_i| + |\sin nx_{i-1}|] \le \frac{2}{n}$

$$\Rightarrow \qquad |I_n| \le \frac{2}{n} \sum_{i=1}^{p} |f(x_{i-1})| < \frac{\varepsilon}{2}$$

Since $f(x_{i-1})$ is a fixed quantity so $\exists$ a positive integer m such that for all $n \ge m$.

$$\frac{2}{n} \sum_{i=1}^{p} |f(x_{i-1})| < \frac{\varepsilon}{2}$$

$$\Rightarrow \qquad |I_n| < \varepsilon \quad \forall n \ge m.$$

$$\Rightarrow \qquad \lim_{n \to \infty} \int_{a}^{b} f(x) \cos nx\, dx = 0$$

Similarly, we may prove that $\displaystyle\lim_{n \to \infty} \int_{a}^{b} f(x) \sin nx\, dx = 0$

Solved Examples

Based on the following Results

Example 1. *Let function f be defined on $[0,1]$ by*

$$f(x) = \frac{1}{n} \text{ for } \frac{1}{n+1} < x \le \frac{1}{n}, n \in N$$
$$= 0 \text{ for } x = 0$$

Show that $f \in R[0,1]$. Also find $\int_0^1 f(x)\, dx$.

Solution. Clearly, the points of discontinuity of f are $\frac{1}{2}, \frac{1}{3}, \frac{1}{4} \cdots$

Also, the set of points of discontinuity of f has only one limit point at $x = 0$.

Then $f \in R[0,1]$

Now

$$\int_0^1 f(x)\, dx = \lim_{n \to \infty} \sum_{r=1}^{n} \int_{1/(r+1)}^{1/r} \frac{1}{r}\, dx$$

- Every continuous function is R-integrable.
- Every monotonic function is R-integrable.
- A bounded function f is R-integrable on $[a,b]$ if the set of its points of discontinuity is finite.
- If f is R-integrable then $|f|$ is also R-integrable.
- Sum and difference of two R-integrable functions is again R-integrable.
- $m(b-a) \le \int_a^b f(x)\, dx \le M(b-a)$ if $b \ge a$.
- $m(b-a) \ge \int_b^a f(x)\, dx \ge M(b-a)$ if $b \le a$.
- If f and g are R-integrable then $f.g$ and $\dfrac{f}{g}$ are also R-integrable.
- A function $f(x)$ defined on $[a,b]$ called a primitive function of $f(x)$ if the function $F(x)$ has its derivatives at each $x \in [a,b]$ i.e., $F'(x) = f(x) \; \forall x \in [a,b]$.
- If f is R-integrable function on $[a, b]$ and F be a differentiable primitive function on $[a,b]$ such that $F'(x) = f(x) \; a \le x \le b$. Then $\int_a^b f(x)\, dt = F(b) - F(a)$

$$= \lim_{n\to\infty} \sum_{r=1}^{n} \frac{1}{r}\left[\frac{1}{r} - \frac{1}{r+1}\right]$$

$$= \lim_{n\to\infty}\left[\left(1 - \frac{1}{2}\right) + \frac{1}{2}\left(\frac{1}{2} - \frac{1}{3}\right) + \dots + \frac{1}{n}\left(\frac{1}{n} - \frac{1}{n+1}\right)\right]$$

$$= \lim_{n\to\infty}\left[\left(1 + \frac{1}{2^2} + \frac{1}{3^2} + \dots + \frac{1}{n^2}\right) - \left(\frac{1}{2} + \frac{1}{2.3} + \frac{1}{3.4} + \dots + \frac{1}{n(n+1)}\right)\right]$$

$$= \lim_{n\to\infty}\left[\left(1 + \frac{1}{2^2} + \dots + \frac{1}{n^2}\right) - \left(\frac{1}{2} + \frac{1}{2} - \frac{1}{3} + \frac{1}{3} - \frac{1}{4} + \dots + \frac{1}{n} - \frac{1}{n+1}\right)\right]$$

$$= \lim_{n\to\infty}\left[\left(1 + \frac{1}{2^2} + \dots + \frac{1}{n^2}\right) - \left(1 - \frac{1}{n+1}\right)\right]$$

The series $1 + \dfrac{1}{2^2} + \dots + \dfrac{1}{n^2} + \dots$ converges to $\dfrac{\pi^2}{6}$, therefore

$$\int_0^1 f(x)\,dx = \lim_{n\to\infty}\left(1 + \frac{1}{2^2} + \dots + \frac{1}{n^2}\right) - 1 + \lim_{n\to\infty}\frac{1}{n+1} = \frac{\pi^2}{6} - 1$$

Example 2. *Let f be a function defined on [0,1] by*

$$f(x) = \begin{cases} \dfrac{1}{2^n} & for \quad \dfrac{1}{2^{n+1}} < x \le \dfrac{1}{2^n}, n = 0,1,2,\dots \\ 0 & for \quad\quad x = 0 \end{cases}$$

Show that $f \in R[0,1]$. Also find the value of $\int_0^x f(t)\,dt$.

Solution. Here, we may defined the function $f(x)$ as follows:

$$f(x) = 1, \text{if } \frac{1}{2} < x \le 1$$

$$= \frac{1}{2}, \text{if } \frac{1}{2^2} < x \le \frac{1}{2}$$

$$= \frac{1}{2^2}, \text{if } \frac{1}{2^3} < x \le \frac{1}{2^2}$$

$$\dots\dots\dots\dots\dots\dots\dots\dots$$

$$= \frac{1}{2^{n-1}}, \text{if } \frac{1}{2^n} < x \le \frac{1}{2^{n-1}}$$

and $\quad f(0) = 0$.

Clearly $|f(x)| \le 1$, $\forall x \in [0,1]$, therefore, $f(x)$ is bounded on $[0,1]$. But f is not continuous on $[0,1]$. The set of points of discontinuity of f in $[0,1]$ is

$$\left\{0, \frac{1}{2}, \frac{1}{2^2}, \dots, \frac{1}{2^n}, \dots\right\}$$

which is an infinite set and 0 is the only limit point of this set. Therefore $f \in R[0,1]$. Also, we have

$$\int_0^x f(x)\,dx = \int_{1/2^m}^x f + \int_{1/2^{m+1}}^{1/2^m} f + \int_{1/2^{m+2}}^{1/2^{m+1}} f + \dots$$

$$= \int_{1/2^m}^x \frac{1}{2^{m-1}} + \int_{1/2^{m+1}}^{1/2^m} \frac{1}{2^m} + \int_{1/2^{m+2}}^{1/2^{m+1}} \frac{1}{2^{m+1}} + \dots$$

$$= \frac{1}{2^{m-1}}\left[x - \frac{1}{2^m}\right] + \frac{1}{2^m}\left[\frac{1}{2^m} - \frac{1}{2^{m+1}}\right] + \frac{1}{2^{m+1}}\left[\frac{1}{2^{m+1}} - \frac{1}{2^{m+2}}\right] + \dots$$

$$= \frac{1}{2^{m-1}}\left[x - \frac{1}{2^m}\right] + \frac{1}{2^{2m+1}} + \frac{1}{2^{2m+3}} + \frac{1}{2^{2m+5}} + \dots$$

$$= \frac{x}{2^{m-1}} - \frac{1}{2^{2m-1}} + \frac{1/2^{m+1}}{1-1/4} \qquad \left[\because \text{Sum of infinite G.P.} = \frac{a}{1-r}\right]$$

$$= \frac{x}{2^{m-1}} - \frac{1}{3.2^{2m-2}}$$

Example 3. *Find $\int_1^2 x^3 dx$, using fundamental theorem of integral calculus.*

Solution. Here, we have $f(x) = x^3$, $\qquad 1 \le x \le 2$
Clearly f is continuous on $[1,2]$

Now if, $\qquad \phi(x) = \dfrac{x^4}{4} \qquad (\because 1 \le x \le 2)$ Then $\qquad \phi'(x) = x^3 = f(x)$

Therefore, by fundamental theorem of integral calculus; we have

$$\int_1^2 x^3 dx = \phi(2) - \phi(1) = \frac{2^4}{4} - \frac{1^4}{4} = \frac{15}{4}$$

Example 4. *Let f be the function defined on $[0,1]$ by*

$$f(x) = \begin{cases} 0 \text{ when } x \text{ is irrational} \\ 1 \text{ when } x \text{ is rational} \end{cases}$$

Show that f is bounded but not R-integrable.

Solution. By definition of $f(x)$, we have $0 \le f(x) \le 1 \ \forall x \in [0,1]$

$\therefore \quad f(x)$ is bounded on $[a,b]$.

Define a partition $P = \{a = x_0, x_1, x_2, \dots x_r, \dots, x_{n-1}, x_n = b\}$ of $[a,b]$
Let $I_r = [x_{r-1}, x_r]$ be any subinterval of P, with length $\delta_r (= x_r - x_{r-1})$. Let M_r and m_r be
respectively the supremum and infimum of f in I_r. Then, we have
$\qquad M_r = 1$ and $m_r = 0$

Now, $\qquad L(P,f) = \sum_{r=1}^{n} m_r \delta_r = \sum_{r=1}^{n} 0.\delta_r = 0$

and $\qquad U(P,f) = \sum_{r=1}^{n} M_r \delta_r = \sum_{r=1}^{n} 1.\delta_r = \sum_{r=1}^{n} \delta_r = [\delta_1 + \delta_2 + \dots + \delta_n]$
$\qquad\qquad = [x_1 - x_0] + [x_2 - x_1] + \dots + [x_n - x_{n-1}] = x_n - x_0 = b - a$

$\Rightarrow \qquad \underline{\int}_a^b f = \sup\{L(P,f)\} = 0 \qquad\qquad \dots(1)$

and $\qquad \overline{\int}_a^b f = \inf\{U(P,f)\} = b - a \qquad\qquad \dots(2)$

From (1) and (2), we conclude that

$$\underline{\int}_a^b f \ne \overline{\int}_a^b f$$

Hence, f is not R-integrable over $[a,b]$.

Example 5. *Show that $f(x) = 3x + 1$ is integrable on $[1,2]$ and $\int_1^2 (3x+1)dx = \dfrac{11}{2}$*

Solution. Here, it is clear that
$$4 \le f(x) \le 7 \ \forall x \in [1,2]$$

$\therefore$ $f(x)$ is bounded on [1,2].

Define a partition $P=\left\{1,1+\dfrac{1}{n},1+\dfrac{2}{n},...,1+\dfrac{n}{n}=2\right\}$ of [1,2]. Let I_r be the rth subinterval of P. Then

$$I_r=\left[1+\frac{r-1}{n},1+\frac{r}{n}\right]$$

and $\delta_r=$ length of $I_r=\dfrac{1}{n},r=1,2,..,n$

Let M_r and m_r be respectively the supremum and infimum of f in I_r,
Since $f(x)=3x+1$ is an increasing function in [1,2], therefore,

$$m_r=3\left(1+\frac{r-1}{n}\right)+1=4+\frac{3(r-1)}{n}$$

and $M_r=3\left(1+\dfrac{r}{n}\right)+1=4+\dfrac{3r}{n}$

Now $L(P,f)=\displaystyle\sum_{r=1}^{n}m_r\delta_r=\sum_{r=1}^{n}\left[\left\{4+\frac{3(r-1)}{n}\right\}\frac{1}{n}\right]$

$$=\frac{1}{n}\sum_{r=1}^{n}\left\{4+\frac{3(r-1)}{n}\right\}=\frac{1}{n}\left[4n+\frac{3}{n}\sum_{r=1}^{n}(r-1)\right]$$

$$=4+\frac{3}{n^2}\left[1+2+...+(n-1)\right]$$

$$=4+\frac{3}{n^2}\cdot\frac{(n-1)n}{2}=4+\frac{3}{2}\left(1-\frac{1}{n}\right)$$

$\Rightarrow$ $\underline{\displaystyle\int_1^2}f(x)dx=\displaystyle\lim_{n\to\infty}L(P,f)=4+\frac{3}{2}=\frac{11}{2}$

Also $U(P,f)=\displaystyle\sum_{r=1}^{n}M_r\delta_r=\sum_{r=1}^{n}\left[\left(4+\frac{3r}{n}\right)\frac{1}{n}\right]$

$$=\frac{1}{n}\sum_{r=1}^{n}\left\{4+\frac{3r}{n}\right\}=\frac{1}{n}\left[4n+\frac{3}{n}\sum_{r=1}^{n}r\right]$$

$$=4+\frac{3}{n}\left[1+2+...+n\right]=4+\frac{3}{n^2}\cdot\frac{n(n+1)}{2}$$

$$=4+\frac{3}{2}\left(1+\frac{1}{n}\right)$$

$\Rightarrow$ $\overline{\displaystyle\int_1^2}f(x)dx=\displaystyle\lim_{n\to\infty}U(P,f)=4+\frac{3}{2}=\frac{11}{2}$

Clearly, $\underline{\displaystyle\int_1^2}f(x)dx=\overline{\displaystyle\int_1^2}f(x)dx$

Hence, $f(x)$ is R-integrable on [1,2] and $\displaystyle\int_1^2f(x)dx=\frac{11}{2}$.

Example 6. *If a function f is defined on $[0,a]$, $a>0$ by $f(x)=x^3$, then show that f is R-integrable on $[0,a]$ and*

$$\int_0^a f(x)dx=\frac{a^4}{4}$$

Solution. Consider a partition $P=\left\{0,\dfrac{a}{n},\dfrac{2a}{n},...,\dfrac{(n-1)a}{n},\dfrac{na}{n}=a\right\}$ of $[0,a]$.

Let I_r be the rth subinterval of P such that

$$I_r=\left[\dfrac{(r-1)a}{n},\dfrac{ra}{n}\right]\text{ with length }\delta_r=\dfrac{a}{n},r=1,2,..,n$$

Now let M_r and m_r be respectively the supremum and infimum of f in I_r, Also since $f(x)$ is an increasing function in $[0,a]$, therefore,

$$m_r=\dfrac{(r-1)^3 a^3}{n^3}\text{ and }M_r=\dfrac{r^3 a^3}{n^3},\ r=1,2,..,n$$

Now $\quad L(P,f)=\displaystyle\sum_{r=1}^{n} m_r\delta_r=\sum_{r=1}^{n}\left[\dfrac{(r-1)^3 a^3}{n^3}\cdot\dfrac{a}{n}\right]$

$$=\dfrac{a^4}{n^4}\sum_{r=1}^{n}(r-1)^3=\dfrac{a^4}{n^4}\left[1^3+2^3+...+(n-1)^3\right]$$

$$=\dfrac{a^4}{n^4}\left[\dfrac{(n-1)n}{2}\right]^2=\dfrac{a^4}{4}\left[1-\dfrac{1}{n}\right]^2$$

$\Rightarrow\quad \underline{\displaystyle\int_0^a} f=\lim_{n\to\infty} L(P,f)=\lim_{n\to\infty}\dfrac{a^4}{4}\left(1-\dfrac{1}{n}\right)^2=\dfrac{a^4}{4}$

Also $\quad U(P,f)=\displaystyle\sum_{r=1}^{n} M_r\delta_r=\sum_{r=1}^{n}\left[\dfrac{r^3 a^3}{n^3}\cdot\dfrac{a}{n}\right]$

$$=\dfrac{a^4}{n^4}\sum_{r=1}^{n}r^3=\dfrac{a^4}{n^4}\left[1^3+2^3+...+n^3\right]$$

$$=\dfrac{a^4}{n^4}\left[\dfrac{n(n+1)}{2}\right]^2=\dfrac{a^4}{4}\left[1+\dfrac{1}{n}\right]^2$$

$\Rightarrow\quad \overline{\displaystyle\int_0^a} f(x)dx=\lim_{n\to\infty}\dfrac{a^4}{4}\left(1+\dfrac{1}{n}\right)^2=\dfrac{a^4}{4}$

Clearly, $\quad \underline{\displaystyle\int_0^a} f=\overline{\displaystyle\int_0^a} f=\dfrac{a^4}{4}$

Hence, f is R-integrable over $[0,a]$ and $\int_0^a f(x)dx=\dfrac{a^4}{4}$

Example 7. If $f(x)=\cos x\ \forall x\in\left[0,\dfrac{\pi}{2}\right]$. Show that f is R-integrable on $\left[0,\dfrac{\pi}{2}\right]$ and $\int_0^{\pi/2}\cos x\,dx=1$

Solution. Since $0\le f(x)\le 1\ \forall x\in\left[0,\dfrac{\pi}{2}\right]$ Therefore, $f(x)=\cos x$ is bounded on $\left[0,\dfrac{\pi}{2}\right]$.

Define a partition $P=\left\{0,\dfrac{\pi}{2n},\dfrac{2\pi}{2n},...,\dfrac{r\pi}{2n},...,\dfrac{n\pi}{2n}=\dfrac{\pi}{2}\right\}$ of $\left[0,\dfrac{\pi}{2}\right]$. Let $I_r=\left[\dfrac{(r-1)\pi}{2n},\dfrac{r\pi}{2n}\right]$

be the rth subinterval of the partition P, with length $\delta_r=\dfrac{\pi}{2n},r=1,2,..,n$

Now let M_r and m_r be supremum and infimum of f in I_r,

We know that $f(x) = \cos x$ is a decreasing function in $\left[0, \dfrac{\pi}{2}\right]$, therefore,

$$M_r = \cos \frac{(r-1)\pi}{2n} \text{ and } m_r = \cos \frac{r\pi}{2n}$$

$$\therefore \quad U(P,f) = \sum_{r=1}^{n} M_r \delta_r = \sum_{r=1}^{n} \left[\left(\cos \frac{(r-1)\pi}{2n}\right).\frac{\pi}{2n}\right]$$

$$= \frac{\pi}{2n} \sum_{r=1}^{n} \left[\cos 0 + \cos \frac{\pi}{2n} + \cos \frac{2\pi}{2n} + \ldots + \cos \frac{(n-1)\pi}{2n}\right]$$

$$= \frac{\pi}{2n} \cdot \frac{\cos\left[0 + \dfrac{(n-1)}{2} \cdot \dfrac{\pi}{2n}\right].\sin \dfrac{n\pi}{4n}}{\sin(\pi/4n)}$$

$$\left[\because \cos A + \cos(A+B) + \ldots + \cos(A+(n-1)B)\right.$$

$$= \frac{\cos\left[A + \dfrac{(n-1)}{2}B\right]\sin \dfrac{nB}{2}}{\sin(B/2)} \Bigg]$$

$$= \frac{(\pi/4n)}{\sin(\pi/4n)} 2\cos\left\{\frac{\pi}{4}\left(1 - \frac{1}{n}\right)\right\}\sin \frac{\pi}{4}$$

$$\Rightarrow \quad \overline{\int_0^{\pi/2}} f(x)dx = \lim_{n\to\infty} U(P,f) = 1.2\cos\frac{\pi}{4}.\sin\frac{\pi}{4} = 1$$

Also $\quad L(P,f) = \sum_{r=1}^{n} m_r \delta_r = \sum_{r=1}^{n} \left[\left(\cos \frac{r\pi}{2n}\right).\frac{\pi}{2n}\right]$

$$= \frac{\pi}{2n}\left[\cos \frac{\pi}{2n} + \cos \frac{2\pi}{2n} + \ldots + \cos \frac{n\pi}{2n}\right]$$

$$= \frac{\dfrac{\pi}{2n}\left[\cos\left\{\dfrac{\pi}{2n} + \dfrac{(n-1)}{2} \cdot \dfrac{\pi}{2n}\right\}\sin \dfrac{n\pi}{4n}\right]}{\sin(\pi/4n)}$$

$$= \frac{(\pi/4n)}{\sin(\pi/4n)} 2\cos\left\{\frac{\pi}{4}\left(1 + \frac{1}{n}\right)\right\}\sin \frac{\pi}{4}$$

$$\Rightarrow \quad \underline{\int_0^{\pi/2}} f(x)dx = \lim_{n\to\infty} L(P,f) = 1.2\cos\frac{\pi}{4}.\sin\frac{\pi}{4} = 1$$

Therefore, $\quad \underline{\int_0^{\pi/2}} f(x)dx = \overline{\int_0^{\pi/2}} f(x)dx = 1$

Hence, $f(x) = \cos x$ is R-integrable on $\left[0, \dfrac{\pi}{2}\right]$.

Also, $\quad \int_0^{\pi/2} f(x)dx = \int_0^{\pi/2} \cos x \, dx = 1$

Example 8. *If the function $f(x)$ defined by*

$$f(x) = \begin{cases} 0, & \text{when } x \text{ is an integer} \\ 1, & \text{when } x \text{ is not an integer} \end{cases}$$

Show that $f(x)$ is R-integrable in every interval.

Solution. Consider an arbitrary interval $[0,a]$, where $a>0, a \in Z$. Then clearly the function $f(x)$ is continuous at all points in the interval excepts at points $x=1,2,3,...,a$, because it is given that $f(x)=0$ when x is an integer and $f(x-0)=1=f(x+0)$

Therefore the given function $f(x)$ has a finite number of discontinuities at $x = 1, 2, 3, ... a$ in interval $[0, a]$. Thus, if each of these a points of discontinuity be enclosed in an interval whose length is less than ε/a, then all these points will be enclosed in a non-overlapping intervals whose total length is less than $\dfrac{\varepsilon}{a}.a$ i.e., less than ε.

Hence, the given function $f(x)$ is R-integrable in the interval $[0, a]$

Example 9. *Verify first mean value theorem for the function $f(x)=\sin x$ and $g(x)=e^x$ for $x \in [0, \pi]$.*

Solution. Clearly, both the functions $f(x)$ and $g(x)$ are continuous on $[0,\pi]$ and $g(x)>0$,

$$\forall x \in \left[0, \frac{\pi}{2}\right] [\because g(x)=e^x \text{ is an increasing function in } \left[0, \frac{\pi}{2}\right]].$$

Then, by first mean value theorem

$$\int_0^\pi f(x)g(x)dx = f(c)\int_0^\pi g(x)dx \ ; \ 0 \le c \le \pi$$

$\Rightarrow$
$$\int_0^\pi \sin x . e^x dx = \sin c \int_0^\pi e^x dx \quad ; \ 0 \le c \le \pi$$
$$= (e^\pi - 1)\sin c \quad ; \ 0 \le c \le \pi \qquad \qquad ...(1)$$

Now
$$\int_0^\pi e^x \sin x \, dx = \left[\frac{1}{\sqrt{2}}e^x \sin\left(x - \frac{\pi}{4}\right)\right]_0^\pi$$

$$= \frac{1}{\sqrt{2}}e^\pi \sin\frac{3\pi}{4} - \frac{1}{\sqrt{2}}e^0 \sin\left(0 - \frac{\pi}{4}\right)$$

$$= \frac{1}{\sqrt{2}}e^\pi . \frac{1}{\sqrt{2}} + \frac{1}{\sqrt{2}} . \frac{1}{\sqrt{2}}$$

$$= \frac{1}{2}\left(e^\pi + 1\right) \qquad \qquad ...(2)$$

From (1) and (2) we conclude that

$$(e^\pi - 1)\sin c = \frac{1}{2}\left(e^\pi + 1\right) \qquad ; \quad 0 \le c \le \pi \qquad \qquad ...(3)$$

$\Rightarrow$
$$\sin c = \frac{1}{2}\left[\frac{e^\pi + 1}{e^\pi - 1}\right]$$

Now $0 < \dfrac{1}{2}\left[\dfrac{e^\pi + 1}{e^\pi - 1}\right] < 1$, therefore, there exists $c \in \left[0, \dfrac{\pi}{2}\right] \subset [0,\pi]$ satisfying (3).

Hence, the first mean value theorem is verified.

Example 10. *Using first mean value theorem to show that*

$$\frac{1}{3\sqrt{2}} < \int_0^1 \frac{x^2}{\sqrt{1+x}}dx < \frac{1}{3}.$$

Solution. Here, we have
$$f(x) = \frac{1}{\sqrt{1+x}} \text{ and } g(x) = x^2$$

Clearly $f(x)$ is continuous on $[0,1]$ and $g(x) > 0$ on $[0,1]$. Also, $g(x)$ is continuous on $[0,1]$. Therefore, by first mean value theorem, we have

$$\int_0^1 \frac{x^2}{\sqrt{1+x}} dx = \frac{1}{\sqrt{1+c}} \int_0^1 x^2 dx, \quad 0 \le c \le 1$$

$$= \frac{1}{\sqrt{1+c}} \left[\frac{x^3}{3} \right]_0^1 = \frac{1}{3\sqrt{1+c}}, \quad 0 \le c \le 1$$

Now, $\quad 0 \le c \le 1 \Rightarrow 1 < (1+c) < 2$

$$\Rightarrow 1 > \frac{1}{(1+c)} > \frac{1}{2} \Rightarrow \frac{1}{\sqrt{2}} < \frac{1}{\sqrt{1+c}} < 1$$

Therefore, $\quad \dfrac{1}{\sqrt{2}} < 3\int_0^1 \dfrac{x^2}{\sqrt{1+x}} dx < 1 \quad$ or $\quad \dfrac{1}{3\sqrt{2}} < \int_0^1 \dfrac{x^2}{\sqrt{1+x}} dx < \dfrac{1}{3}$

Example 11. *Find the value of* $\displaystyle\lim_{n \to \infty} \left[\frac{1}{n+1} + \frac{1}{n+2} + \ldots + \frac{1}{n+n} \right].$

Solution. Here, we have

$$\frac{1}{n+1} + \frac{1}{n+2} + \ldots + \frac{1}{n+n} = \sum_{r=1}^n \frac{1}{n+r} = \frac{1}{n} \sum_{r=1}^n \frac{1}{1+r/n}$$

$$= \frac{1}{n} f\left(\frac{r}{n}\right), \text{where } f(x) = \frac{1}{1+x}, 0 \le x \le 1$$

Since x is continuous on $[0,1]$, therefore $1+x$ is also continuous. Also, $1+x \ne 0$ on $[0,1]$.

$$\Rightarrow \quad \frac{1}{1+x} \text{ is continuous on } [0,1].$$
$$\Rightarrow \quad f(x) \in R[0,1].$$

Now, $\quad \displaystyle\lim \frac{1}{n} \sum_{r=1}^n f\left(\frac{r}{n}\right) = \int_0^1 \frac{dx}{1+x} = \left[\log(1+x)\right]_0^1$

$$= \log 2 - \log 1 = \log 2$$

Hence, $\quad \displaystyle\lim_{n \to \infty} \left[\frac{1}{n+1} + \frac{1}{n+2} + \ldots + \frac{1}{n+n} \right] = \log 2$

Example 12. *Evaluate* $\displaystyle\int_0^1 \left[2x \sin\frac{1}{x} - \cos\frac{1}{x} \right] dx$

Solution. Here the function

$$f(x) = \begin{cases} 2x \sin\dfrac{1}{x} - \cos\dfrac{1}{x}, & x \in]0,1] \\ 0 & , x = 0 \end{cases}$$

is not continuous on $[0,1]$ $\qquad\qquad [\because f(x) \text{ is discontinuous at } x=0]$

but it is bounded on $]0,1]$.

$\Rightarrow$ f is R-integrable on $[0,1]$

Also, the function $g(x) = \begin{cases} x^2 \sin 1/x, & x \in]0,1] \\ 0, & x = 0 \end{cases}$

is differentiable on $[0,1]$ and satisfies

$$g'(x) = f(x) \quad \forall x \in [0,1]$$

$\Rightarrow$ $\int_0^1 \left[2x \sin \dfrac{1}{x} - \cos \dfrac{1}{x} \right] dx = g(1) - g(0) = \sin 1$

Example 13. *Let a function f be defined on $[0,1]$ as follows : If x is irrational, then $f(x)=0$. If x is rational number $\dfrac{p}{q}$ in its lowest terms $\Rightarrow$ $f(x) = \dfrac{1}{q}$.*

and let $f(0)=f(1)=0$. Show that f is integrable over $[0,1]$ and $\int_0^1 f(x)dx = 0$.

Solution. Since f is bounded on $[0,1]$, let $\varepsilon > 0$ then there exist finite number of fractions $\dfrac{1}{q}$ such that

$$\frac{1}{q} > \frac{\varepsilon}{2} \Rightarrow q < \frac{2}{\varepsilon} \qquad \text{(which is finite.)}$$

Now we enclose these exceptional points, in mutually disjoint closed intervals having the sum of lengths is less than $\dfrac{\varepsilon}{2}$

$$[a_1,b_1][a_2,b_2]\ldots[a_m,b_m]$$

Now the remaining subintervals having the lengths less than $\dfrac{\varepsilon}{2}$

$$[b_0=0,a_1],[b_1,a_2],[b_2,a_3]\ldots,[b_m,a_{m+1}]$$

Now for partition

$$P = \{0 = b_0, a_1, b_1, a_2, b_2,\ldots, a_m, b_m, b_{m+1} = 1\}$$

and let $\omega(P,f) = U(P,f) - L(P,f)$

$$\omega(P,f) < \sum_{r=1}^{m} 1.(b_r - a_r) + \sum_{r=1}^{m} \frac{\varepsilon}{2}(a_{r+1} - b_r)$$

$$\left[\because \sum_{r=1}^{m} (b_r - a_r) < \frac{\varepsilon}{2} \text{ and } \sum_{r=0}^{m} (a_{r+1} - b_r) < 1 \right]$$

$\Rightarrow$ $\omega(P,f) < \dfrac{1}{2}\varepsilon + \dfrac{1}{2}\varepsilon$

$\Rightarrow$ $\omega(P,f) < \varepsilon \Rightarrow f \in R[0,1]$

and since for every partition P

$$L(P,f) = 0 \text{ we have } \overline{\int_0^1} f = \underline{\int_0^1} f = 0$$

Example 14. *Show that* $\displaystyle\lim_{n\to\infty}\left[\frac{n}{n^2+1^2} + \frac{n}{n^2+2^2} + \ldots + \frac{1}{2n} \right] = \frac{\pi}{4}$

Solution. L.H.S. $= \displaystyle\lim_{n\to\infty}\left[\frac{n}{n^2+1^2} + \frac{n}{n^2+2^2} + \ldots + \frac{n}{n^2+n^2} \right]$

$$= \lim_{n\to\infty} \sum_{r=1}^{n} \frac{n}{n^2+r^2} = \lim_{n\to\infty} \sum_{r=1}^{n} \frac{1/n}{1+r^2/n^2}$$

Put $\dfrac{r}{n}=x \quad \Rightarrow \quad \dfrac{1}{n}=dx$

Also, when $r=1$, $x=\dfrac{1}{n}\to 0$ as $n\to\infty$ and when $r=n$, $x=\dfrac{n}{n}=1$

$\therefore \quad \lim_{n\to\infty}\sum_{r=1}^{n}\left[\dfrac{n}{n^2+1^2}+\dfrac{n}{n^2+2^2}+...+\dfrac{n}{n^2+n^2}\right]=\int_0^1\dfrac{dx}{1+x^2}$

$$=\left[\tan^{-1}x\right]_0^1=\tan^{-1}1-\tan^{-1}0$$

$$=\dfrac{\pi}{4}-0=\dfrac{\pi}{4}$$

Example 15. *Show that* $\lim_{n\to\infty}\left[\dfrac{1}{n}+\dfrac{n^2}{(n+1)^3}+\dfrac{n^2}{(n+2)^3}+...+\dfrac{1}{8n}\right]=\dfrac{3}{8}$

Solution. L.H.S.$=\lim_{n\to\infty}\left[\dfrac{1}{n}+\dfrac{n^2}{(n+1)^3}+\dfrac{n^2}{(n+2)^3}+...+\dfrac{1}{8n}\right]$

$$=\lim_{n\to\infty}\left[\dfrac{n^2}{n^3}+\dfrac{n^2}{(n+1)^3}+\dfrac{n^2}{(n+2)^3}+...+\dfrac{n^2}{(n+n)^3}\right]$$

$$=\lim_{n\to\infty}\sum_{r=0}^{n}\dfrac{n^2}{(n+r)^3}=\lim_{n\to\infty}\sum_{r=0}^{n}\dfrac{1/n}{(1+r/n)^3}$$

Let $\dfrac{r}{n}=x \quad \Rightarrow \quad \dfrac{1}{n}=dx$ and $r=0$, $x=0$ as $n\to\infty$ and at $r=n$, $x=1$

$\therefore \quad \lim_{n\to\infty}\sum_{r=1}^{n}\left[\dfrac{1}{n}+\dfrac{n^2}{(n+1)^3}+...+\dfrac{1}{8n}\right]=\int_0^1\dfrac{dx}{(1+x)^3}$

$$=\left[\dfrac{(1+x)^{-2}}{-2}\right]_0^1=-\dfrac{1}{2}\cdot\dfrac{1}{4}+\dfrac{1}{2}=\dfrac{3}{8}$$

Example 16. *Prove the inequality*

$$\dfrac{\sqrt{3}}{8}\le\int_{\pi/4}^{\pi/3}\dfrac{\sin x}{x}dx\le\dfrac{\sqrt{2}}{6}$$

Solution. Let $f(x)=\dfrac{\sin x}{x} \quad \forall x\in\left[\dfrac{\pi}{4},\dfrac{\pi}{3}\right]$

$\Rightarrow \quad f'(x)=\dfrac{x\cos x-\sin x}{x^2}=\dfrac{(x-\tan x)\cos x}{x^2}<0 \text{ on }\left[\dfrac{\pi}{4},\dfrac{\pi}{3}\right]$

$$\left[\because x<\tan x \text{ when } x\in\left[\dfrac{\pi}{4},\dfrac{\pi}{3}\right]\right]$$

$\Rightarrow \quad f$ is monotonically decreasing on $\left[\dfrac{\pi}{4}, \dfrac{\pi}{3}\right]$ and it is also bounded.

$\Rightarrow \quad f \in R\left[\dfrac{\pi}{4}, \dfrac{\pi}{3}\right]$

So, $m = \inf$ of f on $\left[\dfrac{\pi}{4}, \dfrac{\pi}{3}\right] = f\left(\dfrac{\pi}{3}\right) = \dfrac{\sin \pi/3}{\pi/3} = \dfrac{3\sqrt{3}}{2\pi}$

and $M = \sup$ of f on $\left[\dfrac{\pi}{4}, \dfrac{\pi}{3}\right] = f\left(\dfrac{\pi}{4}\right) = \dfrac{\sin \pi/4}{\pi/4} = \dfrac{4}{\sqrt{2\pi}} = \dfrac{2\sqrt{2}}{\pi}$

Now we have

$$m(b-a) \le \int_a^b f\, dx \le M(b-a)$$

$\Rightarrow \qquad \dfrac{3\sqrt{3}}{2\pi}\left(\dfrac{\pi}{3} - \dfrac{\pi}{4}\right) \le \int_{\pi/4}^{\pi/3} \dfrac{\sin x}{x}\, dx \le \dfrac{2\sqrt{2}}{\pi}\left(\dfrac{\pi}{3} - \dfrac{\pi}{4}\right)$

$\Rightarrow \qquad \dfrac{\sqrt{3}}{8} \le \int_{\pi/4}^{\pi/3} \dfrac{\sin x}{x}\, dx \le \dfrac{\sqrt{2}}{6}$

Example 17. *Prove the inequality :* $1 \le \int_0^1 e^{x^2}\, dx \le e$

Solution. Clearly e^{x^2} is an increasing function in $[0,1]$.

$\therefore \ 0 \le x \le 1 \qquad \Rightarrow \qquad e^0 \le e^{x^2} \le e^1$

$\Rightarrow \qquad 1 \le e^{x^2} \le e$

$\Rightarrow \qquad \int_0^1 1\, dx \le \int_0^1 e^{x^2}\, dx \le \int_0^1 e\, dx$

$\Rightarrow \qquad [x]_0^1 \le \int_0^1 e^{x^2}\, dx \le e[x]_0^1$

$\Rightarrow \qquad 1 \le \int_0^1 e^{x^2}\, dx \le e$

EXERCISE 6.2

1. Let the function f be defined on $[0,1]$ as follows

$f(x) = 2rx$, when $\dfrac{1}{r+1} < x \le \dfrac{1}{r}$, $r = 1, 2, 3, \ldots$

Show that f is R-integrable in $[0,1]$ and hence show that

$$\int_0^1 f(x)\, dx = \dfrac{\pi^2}{6}$$

2. Let a function f be defined on $[0,1]$ as follows:

If x is irrational, let $f(x) = 0$, if x is rational number p/q in its lowest terms, let $f(x) = \dfrac{1}{q}$ also, let $f(0) = f(1) = 0$. Show that f is integrable over $[0,1]$ and $\int_0^1 f(x)\, dx = 0$

3. Let $f(x) = x$ $(0 \le x \le 1)$. Let P be the partition $\left\{0, \dfrac{1}{3}, \dfrac{2}{3}, 1\right\}$ of $[0, 1]$, compute $U(P, f)$ and $L(P, f)$.

4. Show by definition that $\int_0^1 x^4\, dx = \dfrac{1}{5}$.

5. Find the value of upper and lower integrals for the function f defined on $[0,2]$ as follows:

$$f(x) = \begin{cases} x^2, & \text{when } x \text{ is rational} \\ x^3, & \text{when } x \text{ is irrational} \end{cases}$$

6. Let $f(x) = x^{-1/2}$ on $[1,4]$ consider the partition obtained by dividing $[1,4]$ into n equal parts and hence show that

$$\int_1^4 x^{-1/2}\, dx = 2$$

7. Let $f(x)$ be a function bounded on $[a,b]$ and let P_1 and P_2 be two partitions of $[a,b]$ such that $P_1 \subset P_2$, then show that

$$U(P_1, f) - L(P_1, f) \ge U(P_2, f) - L(P_2, f)$$

8. Give an example of a discontinuous function which is R-integrable on [0,1].

9. Give an example to show that a bounded function is need not be R-integrable.

10. Show with the help of an example that the equation

$$\int_a^b f'(x)\,dx = f(b) - f(a)$$

is not valid.

11. Evaluate $\lim\limits_{n\to\infty}\left[\left(1+\dfrac{1}{n}\right)\left(1+\dfrac{2}{n}\right)\cdots\left(1+\dfrac{2n}{n}\right)\right]^{\frac{1}{n}}$

12. Show that $\int_0^{\pi/4} \log(1+\tan x)\,dx = \dfrac{\pi}{8}\log 2$

13. Show that the integral of an integrable function is cotinuous.

14. Let $a<c<b$, then show that $f \in R[a,b]$ iff $f \in R[a,b]$, either case

$$\int_a^b f(x)\,dx = \int_a^c f(x)\,dx + \int_c^b f(x)\,dx$$

15. If $f(x)=x$, $g(x)=x^2$, show that $\int_0^1 fg\,dx = \dfrac{1}{3}$

16. If $f \in R[a,b]$ and $g \in R[a,b]$ and $f \ge g$ on $[a,b]$, then show that

$$\int_a^b f \ge \int_a^b g$$

HINTS TO SELECTED PROBLEMS

3. Consider the partition P which divides the interval [0,1] into the subinterval

$$I_1 = \left[0,\frac{1}{3}\right], I_1 = \left[\frac{1}{3},\frac{2}{3}\right], I_3 = \left[\frac{2}{3},1\right]$$

with length $\dfrac{1}{3}$, each.

Then, $M_1 = \dfrac{1}{3}, m_1 = 0, M_2 = \dfrac{2}{3}, m_2 = \dfrac{1}{3}, M_3 = 1,$
$m_3 = \dfrac{2}{3}$

$$U(P,f) = \sum_{r=1}^{3} M_r \delta x_r = M_1\delta x_1 + M_2\delta x_2 + M_3\delta x_3$$

$$= \frac{1}{3}\cdot\frac{1}{3} + \frac{2}{3}\cdot\frac{1}{3} + 1\cdot\frac{1}{3} = \frac{2}{3}$$

Similarly, $L(P,f) = \dfrac{1}{3}$

4. Let

$$P = \left\{0, \frac{1}{n}, \frac{2}{n}, \ldots, \frac{n-1}{n}, \frac{n}{n} = 1\right\}$$

and $\quad I_r = \left[\dfrac{r-1}{n}, \dfrac{r}{n}\right]$ with $\delta_r = \dfrac{1}{n}$

Now $m_r = \dfrac{(r-1)^4}{n^4}, M_r = \dfrac{r^4}{n^4}$

which gives

$$L(P,f) = \sum_{r=1}^{n} m_r\delta_r = \sum_{r=1}^{n}\left[\frac{(r-1)^4}{n^4}\cdot\frac{1}{n}\right]$$

$$= \frac{1}{30}\left[\left(1-\frac{1}{n}\right)\left(2-\frac{1}{n}\right)\left(3-\frac{3}{n}-\frac{1}{n^2}\right)\right]$$

and $U(P,f) = \sum_{r=1}^{n} M_r\delta_r = \sum_{r=1}^{n}\left[\dfrac{r^4}{n^4}\cdot\dfrac{1}{n}\right]$

$$= \frac{1}{30}\left[\left(1+\frac{1}{n}\right)\left(2+\frac{1}{n}\right)\left(3+\frac{3}{n}-\frac{1}{n^2}\right)\right]$$

5. $m_r = \begin{cases} x^3 : 0 < x < 1 \\ x^2 : 1 < x < 2 \end{cases} \quad M_r = \begin{cases} x^2 : 0 < x < 1 \\ x^3 : 1 < x < 2 \end{cases}$

7. Let $P_1 \le P_2$ then $U(P_2,f) \le U(P_1,f)$ ...(1)
Again $P_1 \le P_2 \Rightarrow L(P_1,f) \le L(P_2,f)$
$\Rightarrow -L(P_1,f) \ge -L(P_2,f)$ ...(2)
Now using (1) and (2).

8. $f(x) = \begin{cases} \dfrac{1}{2^n} & \text{for } \dfrac{1}{2^{n+1}} < x \le \dfrac{1}{2}, n = 0,1,2 \\ 0 & \text{for } x = 0 \end{cases}$

9. $f(x) = \begin{cases} 1, & \text{when } x \text{ is rational} \\ -1, & \text{when } x \text{ is irrational} \end{cases}$

Answers

4. 2/3,1/3

5. 31/12,49/12

11. $\dfrac{3^3}{e^2}$

CHAPTER REVIEW: A COMPETITIVE APPROACH

SELECTED TERMS AND RESULTS

TERMS

- **Partition:** A finite set P of points $x_0, x_1, \ldots x_n$ where $a = x_0 \leq x_1 \leq x_2 \leq \ldots \leq x_{n-1} \leq x_n = b$ is called a partition of $[a,b]$.

- **Norm of a partition:** The length of the greatest of all the intervals $[x_{r-1}, x_r]$ of the partition P will be called its norm.

- **Refinement of a partition:** Let P_1 and P_2 be two partitions of $[a,b]$ such that $P_1 \subseteq P_2$. Then P_2 is called a refinement of P_1.

- **Upper Riemann sum:** $U(P,f) = \sum_{r=1}^{n} M_r \delta_r$

- **Lower Riemann sum:** $L(P,f) = \sum_{r=1}^{n} m_r \delta_r$

- **Oscillatory sum:** The sum

$$\sum_{r=1}^{n} O_r \delta_r = U(P,f) - L(P,f) = \sum M_r \delta_r - \sum m_r \delta_r$$

is called the oscillatory sum. Here O_r denote the oscillation of the function in subinterval $[x_{r-1}, x_r]$

- **Upper Riemann Integral:**

$$\overline{\int_a^b} f(x)dx = \inf\left[U(P,f)\right]$$

- **Lower Riemann Integral:**

$$\underline{\int_a^b} f(x)dx = \sup\left[L(P,f)\right]$$

- **Riemann Integral:** A bounded function f is said to be Riemann integrable over $[a,b]$ if its upper and lower integrals are equal, the common value being called the Riemann integral.

RESULTS

- We have $\underline{\int_a^b} f \leq \overline{\int_a^b} f$

- The characteristic function of rational numbers, in [a,b] is not Riemann integrable.

- The partition P^* is a refinement of P if $P^* \supset P$ (that is, if each point of P is a point of P^*).

- Assume P_1 and P_2 are two partitions such that $P^* = P_1 \cup P_2$. Then P^* is their common refinement.

- If P^* is a refinement of P, then
$$L(P,f) \leq L(P^*,f)$$
and $\quad U(P,f) \leq U(P^*,f)$

- $U(P^*,f) - L(P^*,f) \leq U(P,f) - L(P,f)$

- The lower R-integral cannot exceed the upper R-integral, equivalently,

$$\underline{\int_a^b} f \leq \overline{\int_a^b} f$$

- A bounded function f on $[a,b]$ is Riemann integrable over $[a,b]$ iff its upper and lower Riemann integrals are equal.

- A necessary and sufficient condition for $f \in R[a,b]$ is that the upper and lower R-integrals be equal.

- Suppose I is a bounded interval of real numbers and f is a bounded real valued function on I. Then we define the oscillation $\omega[f,I]$ as

$$\omega[f, I] = M[f,I] - m[f,I]$$

- A necessary and sufficient condition that $f \in R[a,b]$ is that for $\varepsilon > 0$ there exists at least one partition P for which

$$U(P,f) - L(P,f) < \varepsilon$$

- In order that $f \in R[a,b]$, it is necessary and sufficient that for any pair of positive numbers ω, σ, there exists a partition P of $[a,b]$ such that the sum of lengths of subintervals of P, in the oscillation of $f(x)$ is $\geq \omega, < \sigma$.

- If all the discontinuities of a bounded function f defined on $[a,b]$ are of the first kind, then $f \in R[a,b]$.

- A bounded function f is R-integrable on $[a,b]$ iff the set $D(f)$ of points , where f is discontinuous has Lebesgue measure zero.

- Every continuous function is integrable. Equivalently $f \in R[a,b]$ if it is continuous.

- Every bounded monotonic function is integrable.

- If $f \in R[a,b]$, then $\int_a^b f = -\int_b^a f$.

- If $f \in R[a,b]$, then $\int_a^b dx = b-a$.

- If $f \in R[a,b]$, then $\int_a^b \lambda f = \lambda \int_a^b f$.

- Let $f \in R[a,b]$ and $c \in]a,b[$. Then $f \in R[a,c]$ and $f \in R[c,b]$ and conversely, and in either case
$$\int_a^b f = \int_a^c f + \int_c^b f$$

- If $f \in R[a,b], g \in R[a,b]$, then $f+g \in R[a,b]$ and
$$\int_a^b (f+g) = \int_a^b f + \int_a^b g$$

- If $f \in R[a,b]$ and if $m \leq f \leq M$, then
$$m(b-a) \leq \int_a^b f \leq M(b-a)$$

- If $f \in R[a,b], g \in R[a,b]$, and $f \geq g$ on $[a,b]$, then
$$\int_a^b f \geq \int_a^b g$$

- If $f \in R[a,b]$, then $|f| \in R[a,b]$ and
$$\left| \int_a^b f\, dx \right| \leq \int_a^b |f|\, dx \text{ or simply } \left| \int_a^b f \right| \leq \int_a^b |f|.$$

- Let $\phi(x) = \sum_{j=1}^{n} f_j(x)$. Then
$$\int_a^b \phi = \sum_{j=1}^{n} \int_a^b f_j$$

- If $f \in R[a,b]$ and $g \in R[a,b]$, then $fg \in R[a,b]$.

- If $f \in R[a,b]$ and for $a \leq x \leq b$, put
$$F(x) = \int_0^x f(t)\, dt$$
then $F(x)$ is a continuous on $[a,b]$.

- (The fundamental theorem). Assume $f \in R[a,b]$ and f possesses a primitive (or antiderivative) F. Then $\int_a^b f = F(b) - F(a)$

- A derivative function f, if it exists, with the property that $F = f$ where f is a given function. Then we call F is a primitive of f.

- R-integral of f may exist but not its primitive.

- *First Fundamental Theorem of Integral Calculus* : Assume f is a continuous function on $[a,b]$. Define $F: F(x) = \int_0^x f(t)\, dt$. Then at each x_0 of $[a,b]$, F is differentiable, and $F'(x_0) = f(x_0)$.

- *First mean value theorem* : If $f(x)$ is continuous on $[a,b]$, then there is a point t in $[a,b]$ such that $\int_a^b f = (b-a)f(t)$

- *The generalized first mean value theorem for integrals* : If f and $g \in R[a,b]$, and g does not assume both positive and negative values on $[a,b]$ and suppose $M = \sup f$ and $m = \inf f$ on $[a,b]$. Then
$$\int_a^b fg\, dx = \lambda \int_\alpha^\beta g\, dx, \text{ where } m \leq \lambda \leq M.$$

- Continuity is a stronger condition than integrability but a weaker condition than differentiability.

- If f is continuous on $[a, b]$, then there exists a point t on $[a,b]$:
$$\int_a^b fg\, dx = f(t) \int_a^b g\, dx$$

- *Bonnet's Form of Second Theorem of Mean Value* : If f and g are continuous on $[a,b]$, and g is a positive decreasing function, then there exists a point t in $[a,b]$ such that
$$\int_a^b fg\, dx = g(a) \int_a^b f\, dx.$$

- *Weierstrass form* : If g is a positive monotonic increasing function, then there exists a point t in $[a,b]$ such that
$$\int_a^b fg\, dx = g(b) \int_t^b f\, dx$$

- *The Second Integral Theorem of Mean Value* : Suppose $f(x)$ is continuous and $g(x)$ is either increasing or decreasing with $g'(x)$ continuous on $[a, b]$. Then
$$\int_a^b f(x)g(x)dx = g(a) \int_a^t f(x)dx$$
$$+ g(b) \int_t^b f(x)dx$$

- Suppose f and $\phi \in R[a,b]$. Then
$$\left\{ \int_a^b f\phi\, dx \right\}^2 \leq \left(\int_a^b f^2\, dx \right) \left(\int_a^b \phi^2\, dx \right)$$

REVIEW QUESTION AND PROJECT WORK

1. If $f(x) = |x|$ then show that $\int_{-1}^{1} f\, dx = 1$

2. A function f is defined on $[0,1]$ as follows

$f(x) =$
$\begin{cases} 0 \text{; when } x \text{ is rational} \\ \dfrac{1}{q}, \text{ when } x \text{ is any non-zero rational number } \dfrac{p}{q} \end{cases}$

with least positive integer p and q.

Show that f is integrable on $[0, 1]$ and the value of the integeral is zero.

3. Let f be a non-negative continuous function on $[a,b]$ and $\int_{a}^{b} f\, dx = 0$. Prove that $f(x) = 0$ $\forall\, x \in [a,b]$.

4. If f is continuous and non-negative, show that

$\int_{a}^{b} f\, dx \geq 0$

5. Show that the function f defined on $[0,1]$ as follows :

$f(x) = \begin{cases} 2n & \text{if } x = \dfrac{1}{n}, n \in N \\ 0 & , \text{ otherwise} \end{cases}$

6. If f is bounded and integrable over $[a,b]$, show that $\int_{a}^{b} [f(x)]^2\, dx = 0$ if and only if $f(c) = 0$ at every point c of continuity of f.

7. If a function f is continuous on $[0,1]$. Show that

$$\lim_{n \to \infty} \int_{0}^{1} \frac{n.f(x)}{1 + n^2 x^2}\, dx = \frac{\pi}{2} f(0)$$

8. Show that $\lim_{n \to \infty} \int_{0}^{\delta} \dfrac{\sin nx}{x}\, dx$, $n \in N$ exists and equal to $\dfrac{\pi}{2}$.

OBJECTIVE TYPE QUESTIONS

FILL IN THE BLANKS

1. Partition of a set is also called _____ .

2. The value of $x_r - x_{r-1}$ is called _____ of the interval $[x_{r-1}, x_r]$.

3. Riemann sum is also known as _____ sum.

4. The supremum of the set of lower sums is called the _____ integral.

5. The infimum of the set of upper sums is called the _____ integral.

6. In computing the integral $\int_{a}^{b} f(x)\, dx$, the integral $[a, b]$ is known as _____ of the integration.

7. The function $f(x) = \sin x$ is _____ on $\left[0, \dfrac{\pi}{2}\right]$.

8. Every monotonic function is _____ .

9. Every continuous function is _____ .

10. A bounded function f is R-integrable in $[a,b]$ if the set of its points of discontinuity is _____ .

11. If $f:[a,b] \to$ is a bounded function then $U(P,-f) = $ _____ .

12. If $\overline{\int_{a}^{b}} f\, dx = \underline{\int_{a}^{b}} f\, dx$ then f is _____ .

13. If P_1 and P_2 be any two partitions of $[a,b]$ then $U(P_1, f)$ _____ $L(P_2, f)$.

14. Let $I = [a,b]$ be a closed interval. Then partition of I of real number $P = \{x_0, x_1, ..., x_n\}$ having the property _____ .

15. Let P_1 and P_2 be two partitions of a closed and bounded interval $[a,b]$ then P_2 is called refinement of P_1 if _____ .

16. Let f be a bounded function defined on $[a,b]$ and P_1 be any partition of $[a,b]$. If P_2 is a refinement of P_1 then $L(P^*, f)$ _____ $L(P, f)$.

17. Let f be a real valued function defined on $[a,b]$. Then lower Riemann integral of f over $[a,b]$ is _____ of $L(P, f)$, $\forall P \in P[a,b]$.

18. Let f be a real bounded function on $[a,b]$. Upper Riemann integral of f over $[a,b]$ is _____ of $U(P, f)$, $\forall P \in P[a,b]$.

TRUE/FALSE

Write 'T' for true and 'F' for false statement.

1. Every bounded function is R-integrable.
(T/F)

2. Every R-integrable function is bounded.
(T/F)

3. Every monotone function is not necessarily R-integrable. **(T/F)**

4. A bounded function f is R-integrable in $[a,b]$ if the set of its point of discontinuity is finite. **(T/F)**

5. Lower Darboux sum can not exceed any upper Darboux sum. **(T/F)**

6. There are some integrable function whose set of point of discontinuity on a closed interval has infinitely many limit points. **(T/F)**

7. If f is continuous and non-negative on $[a,b]$ then $\int_a^b f \geq 0$. **(T/F)**

8. If f is continuous and non-negative on $[a,b]$ and $\int_a^b f = 0$ then $f(x)=0\ \forall x\in[a,b]$. **(T/F)**

9. If f has continuous derivative on $[c,d]$ and $a,b\in[c,d]$, then

$$\int_a^b f' = f(b) - f(a).$$ **(T/F)**

10. If $f\in R[a, b]$ then $f^2\in R[a,b]$. **(T/F)**

MULTIPLE CHOICE QUESTIONS

Choose the most appropriate one:

Problem Set-1

1. If P_1 and P_2 be any two partitions of $[a,b]$, then:
   (a) $U(P_1,f) \geq L(P_2,f)$ (b) $U(P_1,f)=L(P_2,f)$
   (c) $U(P_1,f) \leq L(P_2,f)$ (d) none of the above

2. The value of $\lim\limits_{\|P\|\to 0} L(P,f)$ is :
   (a) $\underline{\int_a^b} f$ (b) $\overline{\int_a^b} f$
   (c) $\int_a^b f$ (d) none of the above

3. The value of $\lim\limits_{\|P\|\to 0} U(P,f)$:
   (a) $\int_a^b f$ (b) $\overline{\int_a^b} f$
   (c) $\int_a^b f$ (d) none of the above

4. If $\underline{\int_a^b} f = \overline{\int_a^b} f$ then :
   (a) f may be R-intgrable
   (b) f is not R-intgrable
   (c) f is always R-intgrable
   (d) none of the above

5. The class of bounded function f which is R-integrable on $[a,b]$ denoted by :
   (a) $U(P,f)$ (b) $L(P,f)$
   (c) $R[a,b]$ (d) none of the above

6. Every bounded function is integrable :
   (a) not necessary
   (b) it is necessary
   (c) both (a) and (b) are true
   (d) none of the above

7. Dirichlet's function is :
   (a) R-integrable
   (b) not R-integrable
   (c) may or may not be R-integrable
   (d) none of the above

8. The function
$$f(x)=\begin{cases} 1, & \text{when } x \text{ is rational} \\ -1, & \text{when } x \text{ is irrational} \end{cases};$$
   (a) only bounded
   (b) R-integrable
   (c) bounded but not R-integrable
   (d) none of the above

9. $\int_0^{\pi/2} \sin x\, dx =$
   (a) 1   (b) 2
   (c) 0   (d) $\pi/2$

10. Every monotonic function is :
    (a) bounded   (b) increasing
    (c) R-integrable   (d) none of the above

11. If f_1 and f_2 are two real valued bounded functions defined on $[a,b]$ then for every partition P of $[a,b]$:
    (a) $U(P,f_1+f_2)=U(P,f_1)+U(P,f_2)$
    (b) $U(P,f_1+f_2) \leq U(P,f_1)+U(P,f_2)$
    (c) $U(P,f_1+f_2) \geq U(P,f_1)+U(P,f_2)$
    (d) none of the above

12. A bounded function f is R-integrable in $[a,b]$ if the set of its points of discontinuity is :
    (a) finite   (b) infinite
    (c) zero   (d) none of the above

13. If f is real valued bounded function on $[a,b]$ and m and M are g.l.b. and l.u.b. respectively, then :
    (a) $m(b-a)=M(b-a)$ (b) $m(b-a) \geq M(b-a)$
    (c) $m(b-a) \leq M(b-a)$ (d) none of the above

14. Let f be a bounded function defined on $[a,b]$ and P be a partition of $[a,b]$, P^* is refinement of P, then which is correct :
    (a) $L(P^*,f) \leq L(P,f)$ (b) $U(P^*,f) \leq U(P,f)$
    (c) $U(P^*,-f)=L(P,f)$ (d) none of the above

15. A function f is R-integrable on $[a,b]$ iff :
    (a) only $\int_a^{\overline{b}} f(x)dx$ exists

    (b) only $\underline{\int_a^b} f(x)dx$ exists

    (c) $\underline{\int_a^b} f \neq \overline{\int_a^b} f$

    (d) $\underline{\int_a^b} f = \overline{\int_a^b} f$

16. $f:[0,1] \to R$ such that $f(x) = \begin{cases} 0, & x \text{ is rational} \\ 1, & x \text{ is irrational} \end{cases}$ then :
    (a) the upper and lower integrals of f does not exist
    (b) f is R-integrable
    (c) f is not R-intgrable
    (d) none of the above

17. For function $f(x)=x$ in the interval $[0,3]$ and let $P=\{0,1,2,3\}$ be the partition of $[0,3]$ then the value of $L(P,f)$ is :
    (a) 0   (b) 3
    (c) 6   (d) 9

18. $f:[a,b] \to R$, P and Q are partitions of $[a,b]$ such that $P \subset Q$ then :
    (a) $L(P,f) \leq L(Q,f)$
    (b) $L(P,f) \geq L(Q,f)$
    (c) (a) is true (b) is false
    (d) none of the above

19. Let f be a bounded function defined on the interval $[a,b]$ then f is R-integrable iff for each $\varepsilon > 0$ $\exists$ a partition P of $[a,b]$ if :
    (a) $U(P,f)-L(P,f)<\varepsilon$ (b) $U(P,f)-L(P,f)>\varepsilon$
    (c) $U(P,f)-U(P,f)=\varepsilon$ (d) none of the above

20. If f is R-integrable then $\int_a^b f\,dx + \int_b^c f\,dx$:
    (a) $\int_a^c f\,dx + c$ (b) $\int_a^b f\,dx + c$
    (c) $\int_a^b f\,dx$ (d) none of the above

Problem Set-2

1. The maximum of the length of the partition is called :
    (a) length   (b) norm
    (c) range   (d) none of the above

2. The lower Riemann sum is defined by :
    (a) $m_r \delta x_r$ (b) $\Sigma m_r \delta x_r$
    (c) $M_r \delta x_r$ (d) none of the above

3. The upper Riemann sum is defined by :
    (a) $M_r \delta x_r$ (b) $m \delta x_r$
    (c) $\Sigma M_r \delta x_r$ (d) none of the above

4. If P_1 and P_2 be two partitions of $[a,b]$ such that $P_1 \subseteq P_2$, then :
    (a) $U(P_1,f) \geq L(P_2,f)$ (b) $U(P_1,f)=L(P_2,f)$
    (c) $U(P_1,f) \leq L(P_2,f)$ (d) none of the above

5. $L(P,f+g) \geq :$
    (a) $L(P,f)+L(P,g)$ (b) $L^2(P,f)+L^2(P,g)$
    (c) $U(P,f)+U(P,g)$ (d) none of the above

6. For R-integrability we have :
    (a) $\overline{\int_a^b} f = \int_{\underline{a}}^b f$ (b) $\overline{\int_a^b} f \leq \int_{\underline{a}}^b f$
    (c) $\overline{\int_a^b} f \geq \int_{\underline{a}}^b f$ (d) none of the above

7. The necessary and sufficient condition for R-integrability is given by :
    (a) $U(P,f)-L(P,f)<\varepsilon$ (b) $U(P,f)+L(P,f)<\varepsilon$
    (c) $U(P,f)-L(P,f)>\varepsilon$ (d) none of the above

8. The statement $\int_a^b f$ exists indicates that the function f is :
    (a) bounded   (b) R-integrable
    (c) both (a) and (b)   (d) none of the above

9. If f is continous on $[a,b]$ then f is :
    (a) R-integrable   (b) not R-integrable
    (c) not bounded   (d) none of the above

10. If f is monotonic function on $[a,b]$ then f is :
    (a) R-integrable   (b) not R-integrable
    (c) not bounded   (d) none of the above

11. If f_1 and f_2 are two real valued bounded functions defined on $[a,b]$ then for every partition P of $[a,b]$:
    (a) $U(P,f_1+f_2)=U(P,f_1)+U(P,f_2)$
    (b) $U(P,f_1+f_2) \leq U(P,f_1)+U(P,f_2)$
    (c) $U(P,f_1+f_2) \geq U(P,f_1)+U(P,f_2)$
    (d) none of the above

12. A bounded function f is R-integrable in $[a,b]$ if the set of its points of discontinuity is :
    (a) finite   (b) infinite
    (c) oscillatory   (d) none of the above

13. If f is real valued bounded function on $[a,b]$ and m and M are g.l.b. and l.u.b. respectively, then :

(a) $m(b-a) \le M(b-a)$ (b) $m(b-a)=M(b-a)$

(c) $m(b-a) \ge M(b-a)$ (d) none of the above

14. Let f be a bounded function defined on $[a,b]$ and P be a partition of $[a, b]$, P^* is refinement of P, then which is correct :

(a) $L(P^*,f) \le L(P,f)$ (b) $U(P^*,f) \le U(P,f)$

(c) $U(P^*,-f)=L(P,f)$ (d) none of the above

15. $f:[0,1] \to R$ such that $f(x)=\begin{cases} 0, & x \text{ is rational} \\ 1, & x \text{ is irrational} \end{cases}$ then :

(a) $L(P,f),U(P,f)$ do not exist

(b) f is R-integrable

(c) f is not R-intgrable

(d) none of the above

16. $f:[a,b] \to R$, P and Q are partitions of $[a,b]$ such that $P \subset Q$ then :

(a) $L(P,f) \le L(Q,f)$ (b) $L(P,f) \ge Q(P,f)$

(c) both (a) and (b) (d) none of the above

17. A function f is R-integrable on $[a,b]$ iff :

(a) only $\int_{\underline{a}}^{\overline{b}} f\,dx$ exist (b) only $\int_{\underline{a}}^{b} f\,dx$ exist

(c) $\int_{\underline{a}}^{\overline{b}} f \ne \int_{\underline{a}}^{b} f$ (d) $\int_{\underline{a}}^{\overline{b}} f = \int_{\underline{a}}^{b} f$

Problem Set-3

1. If f is R-integrable then $\int_a^b f\,dx + \int_b^c f\,dx$:

(a) $\int_a^c f\,dx$ (b) $\int_a^b f\,dx + c$

(c) $\int_a^b f\,dx$ (d) none of the above

2. If $f:[a,b] \to R$ is a R-integrable function, then

(a) $\left| \int_a^b f(x)dx \right| \le \int_a^b |f(x)|\,dx$

(b) $\int_a^b |f(x)|\,dx \le \left| \int_a^b f(x)dx \right|$

(c) both (a) and (b) are true

(d) none of the above

3. Let $f:[a,b] \to R$

P and Q are partitions of $[a,b]$ such that $P \subset Q$ then :

(a) $L(P,f) \le L(Q,f)$ (b) $L(P,f) \ge L(Q,f)$

(c) $U(P,f) \le L(Q,f)$ (d) none of the above

4. Which one of the following is true :

(a) $\int_{\underline{a}}^b f(x)dx = \text{l.u.b.}\{L(P,f]\}$

(b) $\int_{\underline{a}}^b f(x)dx = \text{g.l.b.}\{L(P,f]\}$

(c) $\int_{\underline{a}}^b f(x)dx = U(P.f)$

(d) none of the above

5. Which one of the following is/are true ?

(a) Every continuous function is not R-integrable.

(b) If a function is continuous in a closed interval, it is bounded and uniformly continuous in that interval

(c) Continuous function is not necessarily bounded

(d) none of the above

6. Which one of the following is true ?

(a) A constant function is R-integrable

(b) A constant function is not R-integrable

(c) A constant function may or may not be R-integrable

(d) none of the above

7. The statement "$f:[a,b] \to R$ is continuous, then f is R-integrable on $[a,b]$" is :

(a) true (b) false

(c) partially true (d) none of the above

8. Consider the statement :

(A) If f_1 and f_2 are integrable on $[a,b]$ then $f_1.f_2$ is also integrable on $[a,b]$.

(B) If f is integrable on $[a,b]$ then f^2 is also integrable then :

(a) both statements are true

(b) A is true, B is false

(c) A is false, B is true

(d) none of the above

9. If f is R-integrable with respect to α on $[a,b]$,then :

(a) f is increasing and α is bounded function

(b) f is bounded and α is an increasing function

(c) f and α both are bounded

(d) f and α both are increasing

10. A real valued bounded function $f(x)$ is Riemann integrable on $[a,b]$ then :

(a) $\int_{\underline{a}}^b f dx$ and $\int_a^{\overline{b}} f dx$ exist

(b) $\int_{\underline{a}}^b f dx$ and $\int_a^{\overline{b}} f dx$ exist and equal

(c) $\int_{\underline{a}}^b f dx$ and $\int_a^{\overline{b}} f dx$ unequal

(d) none of the above

11. If $f:[a,b] \to R$ is continuous and monotonic function then :
    (a) f is R-integrable on $[a,b]$
    (b) f is not R-integrable on $[a,b]$
    (c) f is R-integrable everywhere
    (d) none of the above

12. For any constant c, which of the following holds :
    (a) $c\int_a^b f(x)dx \neq \int_a^b cf(x)dx$
    (b) $c\int_a^b f(x)dx = \int_a^b cf(x)dx$
    (c) $c\int_a^b f(x)dx \geq \int_a^b cf(x)dx$
    (d) none of the above

13. The value of $\lim_{\|P\| \to 0} L(P,f)$:
    (a) $\int_{\underline{a}}^b f dx$
    (b) $\overline{\int_a^b} f dx$
    (c) $\int_a^b f\,dx$
    (d) none of the above

14. The value of $\lim_{\|P\| \to 0} U(P,f)$:
    (a) $\int_{\underline{a}}^b f dx$
    (b) $\overline{\int_a^b} f dx$
    (c) $\int_a^b f\,dx$
    (d) none of the above

15. Every monotonic function is :
    (a) R-integrable
    (b) not R-integrable
    (c) increasing
    (d) none of the above

16. If f is bounded and integrable over $[a,b]$ and M, m are bounds of f on $[a,b]$, then :
    (a) $m(b-a) \leq \int_a^b f(x)dx \leq M(b-a)$ if $b \geq a$
    (b) $m(b-a) \geq \int_a^b f(x)dx \geq M(b-a)$ if $b \leq a$
    (c) both (a) and (b) are true
    (d) none of the above

17. If f is non-negative function on $[a,b]$ and $\int_a^b f(x)dx = 0 \ \forall x \in [a,b)$, then :
    (a) $f(x)=0 \ \forall x \in [a,b]$
    (b) $f(x)=1 \ \forall x \in [a,b]$
    (c) $f(x)=a \ \forall x \in [a,b]$
    (d) $f(x)=b \ \forall x \in [a,b]$

18. Let c be a fixed point in $[0,1]$. A function $f:[0,1] \to R$ is defined by :
    $$f(x) = \begin{cases} c & ; \text{ if } 0 \leq x \leq c \\ 2c & ; \text{ if } c < x \leq 1 \end{cases}$$

It is given that R-integral $\int_0^1 f(x)dx = \dfrac{7}{16}$ then $c=$
    (a) $\dfrac{1}{2}$
    (b) 1
    (c) $\dfrac{1}{4}$
    (d) does not exist

19. If $\int_{\underline{a}}^b f(x)dx$ and $\overline{\int_a^b} f(x)dx$ are lower and upper R-integrable function on $[a,b]$ then :
    (a) $\int_{\underline{a}}^b f(x)dx \leq \overline{\int_a^b} f(x)dx$
    (b) $\int_{\underline{a}}^b f(x)dx \geq \overline{\int_a^b} f(x)dx$
    (c) both (a) and (b) are true
    (d) none of the above

20. If $f:[a,b] \to R$ is R-integrable function, then :
    (a) $\left|\int_a^b f(x)dx\right| \leq \int_a^b |f(x)|dx$
    (b) $\int_a^b |f(x)|dx \leq \left|\int_a^b f(x)dx\right|$
    (c) both (a) and (b) are true
    (d) none of the above

21. Which of the following is/are true ?
    (a) Every continuous function is integrable.
    (b) Every differentiable function is continuous.
    (c) Both (a) and (b) are true.
    (d) none of the above

22. Which of the following is/are true ?
    (a) Every monotonic increasing function is R-integrable.
    (b) If function is increasing in an interval, it is increasing in each of these interval.
    (c) Both (a) and (b) are true.
    (d) none of the above

23. If f and g are bounded integrable function on $[a,b]$ such that $f \geq g$, then :
    (a) $\int_a^b f\,dx \geq \int_a^b g\,dx$ when $b \geq a$
    (b) $\int_a^b f\,dx \leq \int_a^b g\,dx$ when $b \leq a$
    (c) both (a) and (b) are true
    (d) none of the above

24. The oscillation of a bounded function f on an interval $[a,b]$ is given by :
    (a) supremum of $\{|U(P,f)-L(P,f)| : x_1, x_2 \in [a,b]\}$
    (b) infimum of $\{|U(P,f)-L(P,f)| : x_1, x_2 \in [a,b]\}$
    (c) both (a) and (b) are true
    (d) none of the above

25. Which of the following is/are true ?

(a) Every continuous function is R-integrable.

(b) If a function is continuous in a closed interval, it is bounded and uniformly continuous in that interval.

(c) Both (a) and (b) are true.

(d) none of the above

26. If $f(x)=0$ for all irrationals x and $f(x)=1$ for all rational x then for $a<b$:

(a) $f \in R[a,b]$ (b) $f \notin R[a,b]$

(c) f is continuous   (d) none of the above

27. The characteristic function of rational numbers in $[a,b]$ is :

(a) R-integrable   (b) not R-integrable

(c) continuous   (d) none of the above

28. Each constant function in a closed and bounded interval $[a,b]$ is :

(a) R-integrable   (b) not R-integrable

(c) discontinuous   (d) none of the above

29. Let $f(x)$ be a function defined by

$$f(x)=\sqrt{1-x^2}\ ;\ \text{if } x \text{ is rational}$$
$$=1-x\qquad ;\ \text{if } x \text{ is irrational}$$

The value of $\int_0^1 f$ and $\overline{\int}_0^1 f$ are respectively given by :

(a) $\dfrac{1}{2}, \dfrac{\pi}{4}$ (b) $\dfrac{\pi}{4}, \dfrac{1}{2}$

(c) $\dfrac{1}{2}, \dfrac{1}{2}$ (d) $\dfrac{\pi}{4}, \dfrac{\pi}{4}$

30. If $f(x)=x^2$ on $[0,a]$, $a>0$ then :

(a) $f \in R[0, a]$ (b) $f \notin R[0,a]$

(c) f is discontinuous (d) none of the above

31. The exponential function is :

(a) R-integrable   (b) not integrable

(c) discontinuous   (d) none of the above

32. If $f(x)=x^3$ on $[0,\lambda]$ then :

(a) $f \in R[a,b]$ (b) $f \in R[0,\lambda]$

(c) $f \notin R[0,\lambda]$ (d) none of the above

33. Let $f \in R[a,b]$ then

$$\lim_{n\to\infty}\sum_1^n h\, f(a+rh)=\int_a^b f \text{ if } h=$$

(a) $b-a$ (b) $b+a$

(c) $\dfrac{b-a}{n}$ (d) $\dfrac{b+a}{n}$

34. The value of

$$\lim_{n\to\infty}\left[\frac{1}{n}+\frac{n^2}{(n+1)^3}+\frac{n^2}{(n+2)^3}+...+\frac{1}{8n}\right]=$$

(a) $\dfrac{1}{8}$ (b) $\dfrac{2}{8}$

(c) $\dfrac{3}{8}$ (d) none of the above

35. The value of

$$\lim_{n\to\infty}\left[\left(1+\frac{1}{n}\right)\left(1+\frac{2}{n}\right)...\left(1+\frac{4n}{n}\right)\right]^{1/n}=$$

(a) $\dfrac{5}{e^4}$ (b) $\dfrac{5^4}{e^4}$

(c) $\dfrac{5^5}{e^5}$ (d) $\dfrac{5^5}{e^4}$

36. If $f(x) =$

$$\begin{cases} x+x^2, & \text{if } x \text{ is rational and } x \in (0,2) \\ x^2+x^3, & \text{if } x \text{ is irrational and } x \in (0,2) \end{cases}$$

then value of $\overline{\int}_a^2 f$ and $\int_0^2 f$ are respectively given by

(a) $\dfrac{53}{12}, \dfrac{83}{12}$ (b) $\dfrac{83}{12}, \dfrac{53}{12}$

(c) $\dfrac{7}{12}, \dfrac{83}{12}$ (d) none of the above

37. The Dirichlet's function is :

(a) R-integrable   (b) not R-integrable

(c) continuous   (d) none of the above

38. If f is R-integrable and non-negative on $[a,b]$ then $\int_a^b f$:

(a) non-negative   (b) non-positive

(c) zero   (d) none of the above

39. The function $f(x)=\sin x$ in the interval $\left[0,\dfrac{\pi}{2}\right]$ is :

(a) R-integrable   (b) not R-integrable

(c) discontinuous   (d) none of the above

40. Let $f(x)$ be a function defined on $\left(0,\dfrac{\pi}{4}\right)$ by

$$f(x)=\begin{cases} \cos x\ ;\ \text{if } x \text{ is rational} \\ \sin x\ ;\ \text{if } x \text{ is irrational} \end{cases}$$

then in the interval $\left[0,\dfrac{\pi}{4}\right]$ f is :

(a) R-integrable (b) not R-integrable
(c) discontinuous (d) none of the above

41. If all the discontinuities of a bounded function f defined on $[a,b]$ are of the first kind, then f is :
    (a) R-integrable
    (b) not R-integrable
    (c) may or may not be R-integrable
    (d) none of the above

42. A function f defined on $[0,k]$, k is a positive integer, is as follows :
$$f(x)=\begin{cases}0; & \text{if } x \text{ is an integer}\\1; & \text{otherwise}\end{cases} \text{ then}$$
    (a) $f\in R[0, k]$ (b) $f\in R[a, b]$
    (c) $f\notin R[0, k]$ (d) none of the above

43. A function f difined on $[0,1]$ as follows :
$$f(x)=\frac{1}{a^{r-1}}, \text{for } \frac{1}{a^r}<x\le\frac{1}{a^{r-1}}, r=1,2,...$$
where a is any integer greater than 1. Then value of $\int_0^1 f\,dx =$
    (a) $\dfrac{a}{a-1}$ (b) $\dfrac{a}{a+1}$
    (c) $a+1$ (d) $\dfrac{a+1}{a}$

44. Let $f(x)=2r\,x$ for $\dfrac{1}{r+1}<x\le\dfrac{1}{r}$ on $]0,1[$ then f is :
    (a) R-integrable
    (b) $\int_a^b f=\dfrac{\pi^2}{6}$
    (c) both (a) and (b) are true
    (d) none of the above

45. A function f is defined on $[0,1]$ as follows :
$f(x)=0$ for x is irrational or $x=0$
$=\dfrac{1}{q}$ for x is rational $\dfrac{p}{q}$ such that $(p,q)=1$
then
    (a) $f\in R[a, b]$ (b) $f\in R[0, 1]$
    (c) $f\notin R[0, 1]$ (d) none of the above

46. Define a function f on $[a,b]$ as follows :
$$f(x)=\frac{1}{n} \text{ for } \frac{1}{r+1}<x\le\frac{1}{r}, \text{ then}$$
    (a) $f\in R[a,b]$
    (b) $\int_a^b f=\dfrac{\pi^2}{6}-1$
    (c) both (a) and (b) are true
    (d) none of the above

47. A function g is defined as follows :
$$g(x,t)=\begin{cases}x(t-1) \text{ for } x\le t\\t(x-1) \text{ for } x>t\end{cases}$$
Let f be continuous on $(0,1)$ and
$$f(x)=\int_0^1 f(t)g(x,t)\,dt$$
    (a) $f''(x)=f(x)$ (b) $f(0)=0$
    (c) $f(1)=0$ (d) all are true

48. A derivable function if exist with the property that $F'=f$ where f is a given function then F is called :
    (a) derivative of f
    (b) primitive of f
    (c) both (a) and (b) are true
    (d) none of these

49. "If $f\in R[a,b]$ and f' possess a primitive F then $\int_a^b f=f(b)-f(a)$" is called :
    (a) fundamental theorem of integral calculus
    (b) fundamental theorem of differential calculus
    (c) Bonnet's theorem
    (d) none of the above

50. A bounded function f is R-integrable on $[a,b]$ if the set of its points of discontinuity has only a finite number of:
    (a) Isolated point
    (b) limit points
    (c) both (a) and (b) are true
    (d) none of the above

Answers

FILL IN THE BLANKS

1. dissection or net 2. length 3. Darboux 4. lower 5. upper
6. range 7. R-integrable 8. R-integrable 9. R-integrable 10. finite
11. $-L(P,f)$ 12. R-integrable 13. greater than 14. $a=x_0<x_1<...<x_{n-1}<x_n=b$
15. refinement $P_2\supset P_1$ 16. $\ge$ 17. supremum 18. infimum

TRUE/FALSE

  1. F   **2.** T   **3.** F   **4.** T   **5.** T   **6.** T   **7.** T   **8.** T   **9.** T

10. T

MULTIPLE CHOICE QUESTIONS

Problem Set-1

  1. (a)   **2.** (a)   **3.** (b)   **4.** (c)   **5.** (c)   **6.** (a)   **7.** (a)   **8.** (c)   **9.** (a)

10. (c)   **11.** (b)   **12.** (a)   **13.** (c)   **14.** (b)   **15.** (d)   **16.** (c)   **17.** (b)   **18.** (a)

19. (a)   **20.** (c)

Problem Set-2

  1. (b)   **2.** (b)   **3.** (c)   **4.** (a)   **5.** (a)   **6.** (a)   **7.** (a)   **8.** (c)   **9.** (a)

10. (a)   **11.** (b)   **12.** (a)   **13.** (a)   **14.** (b)   **15.** (c)   **16.** (a)   **17.** (d)

Problem Set-3

  1. (a)   **2.** (a)   **3.** (a)   **4.** (a)   **5.** (b)   **6.** (a)   **7.** (a)   **8.** (a)   **9.** (b)

10. (b)   **11.** (a)   **12.** (b)   **13.** (a)   **14.** (b)   **15.** (a)   **16.** (c)   **17.** (a)   **18.** (c)

19. (a)   **20.** (a)   **21.** (c)   **22.** (c)   **23.** (c)   **24.** (a)   **25.** (c)   **26.** (b)   **27.** (b)

28. (a)   **29.** (a)   **30.** (a)   **31.** (a)   **32.** (b)   **33.** (c)   **34.** (c)   **35.** (d)   **36.** (b)

37. (b)   **38.** (a)   **39.** (a)   **40.** (b)   **41.** (a)   **42.** (a)   **43.** (b)   **44.** (c)   **45.** (b)

46. (c)   **47.** (d)   **48.** (b)   **49.** (a)   **50.** (b)

HINTS TO SELECTED PROBLEMS

Problem Set-3

26. We have

$M_j = 1 \ \forall x$-rational and $m_j = 0 \ \forall x$-irrational.
Suppose P is a partition of $[a,b]$. Then

$$U(P,f) = \sum_{j=1}^{n} M_j \delta_j = \sum_{j=1}^{n} 1.\delta_j$$

$$= \sum_{j=1}^{n} \delta_j = b - a$$

$$\Rightarrow \overline{\int_a^b} f = \inf.(\text{or g.l.b.}), U(P,f) = b - a$$

and $L(P,f) = \sum_{j=1}^{n} m_j \delta_j = \sum_{j=1}^{n} 0.\delta_j = 0$

$$\Rightarrow \underline{\int_a^b} f = \sup(\text{or l.u.b.}); L(P,f) = 0$$

Hence, $\overline{\int_a^b} f \neq \underline{\int_a^b} f$

$\Rightarrow f \notin R[a,b]$, where $b > a$

28. Suppose $f(x) = \lambda$. Then for each partition P on $[a,b]$, we have $M_j = \lambda, \ m_j = \lambda$

and hence $U(P,f) = \sum_{j=1}^{n} M_j \delta_j = \sum_{j=1}^{n} \lambda \delta_j$

$$= \lambda \sum_{j=1}^{n} \delta_j = \lambda (b - a)$$

$$\Rightarrow \overline{\int_a^b} f = \inf.(P,f) = \lambda(b-a)$$

and $m_j = \lambda$

$$\Rightarrow L(P,f) = \sum_{j=1}^{n} m_j \delta_j$$

$$= \sum_{j=1}^{n} \lambda \delta_j = \lambda \sum_{j=1}^{n} \delta_j$$

$$= \lambda(b-a) \Rightarrow \underline{\int_a^b} f = \sup(P,f)$$

$$= \lambda(b-a)$$

Hence, $\overline{\int_a^b} f = \underline{\int_a^b} f = \lambda(b-a)$

where $f(x) = \lambda$
$\Rightarrow f \in R[a,b]$ where f is constant on $[a,b]$.

29. We have

$(1-x)^2 - (1-x^2) = 1 - 2x + x^2 - 1 + x^2$
$\qquad\qquad\qquad = 2x^2 - 2x = 2x(x-1)$

$$\Rightarrow \qquad (1-x)^2 \le (1-x^2) \qquad 0 < x < 1$$

$\Rightarrow m_j = 1-x$ and $M_j = \sqrt{(1-x^2)} \ \forall x \in [0,1]$
Hence,

$$U(P,f) = \sum_{j=1}^{n} M_j \delta_j, L(P,f) = \sum_{j=1}^{n} m_j \delta_j$$

$\Rightarrow \bar{\int}_0^1 f = \int_0^1 \sqrt{(1-x^2)}\,dx$

$= \int_0^{\pi/2} \cos^2\theta\,d\theta$ on putting $x = \sin\theta$

$= \dfrac{\Gamma\left(\dfrac{3}{2}\right)\Gamma\left(\dfrac{1}{2}\right)}{2r(2)} = \dfrac{\dfrac{1}{2}\sqrt{\pi}.\sqrt{\pi}}{2} = \dfrac{\pi}{4}$

$$\left(\because \Gamma\left(\dfrac{1}{2}\right) = \sqrt{\pi}\right)$$

and $L(P,f) = \sum_{j=1}^n m_j\delta_j \Rightarrow \int_{-0}^1 (1-x)\,dx$

$= \left(x - \dfrac{x^2}{2}\right)\Big|_0^1 = \dfrac{1}{2}$

Hence, we find :

$$\bar{\int}_0^1 f = \dfrac{\pi}{4}, \underline{\int}_{-0}^1 f = \dfrac{1}{2} \Rightarrow f \notin R[0,1]$$

32. Suppose $P = \left\{0, \dfrac{\lambda}{n}, \dfrac{2\lambda}{n}, ..., \lambda\right\}$ is a partition of

$[0, \lambda]$ obtained on dividing into n equal parts.
Then for $P \in P[0, \lambda]$, we find

$$U(P,f) = \sum_{j=1}^n f(\xi_j)M_j$$

$= \dfrac{\lambda^2}{n^2}.\dfrac{\lambda}{n} + \dfrac{(2\lambda)^3}{n^3}.\dfrac{\lambda}{n} + \dfrac{(3\lambda)^3}{n^3}.\dfrac{\lambda}{n} + ...$

$= \dfrac{\lambda^4}{n^4}\left(1^3 + 2^3 + ... + n^3\right)$

$= \dfrac{\lambda^4}{n^4}\dfrac{(n+1)^2}{4} = \dfrac{\lambda^4}{n^4}\left(1 + \dfrac{1}{n}\right)^2$

and $L(P,f) = \dfrac{\lambda^4}{n^4}\left[0^2 + 1^2 + 2^2 + ... + (n-1)^2\right]$

$= \dfrac{\lambda^4}{4}\left(1 - \dfrac{1}{n}\right)^2$

Therefore, $\inf U(P) = \dfrac{\lambda^4}{4} = \sup L(P)$ as $\|P\| \to 0$

with $n \to \infty$. This implies that $f \in R[0, \lambda]$ and

$$\int_{[0,\lambda]} f = \dfrac{\lambda^4}{4}$$

33. Let $P = \{a, a+h, a+2h, ..., a(n-1)h, b\}$ be a
partition of $[a,b]$ such that $h = \dfrac{b-a}{n}$. Evidently

$\|P\| = h = \dfrac{(b-a)}{n} \to 0$ as $n \to \infty$. Then

$$\int_a^b f = \lim_{\|P\|\to 0} \sum_{r=1}^n f(a+rh)\delta_r$$

$$= \lim_{\|P\|\to 0} \sum_{r=1}^n hf(a+rh)$$

Suppose $P = \{a, ah, ah^2, ..., ah^n = b\}$ is a partition
of $[a,b]$ belonging to $P[a,b]$. Letting $n \to \infty$
implies that $\|P\| \to 0$. Now

$$\int_a^b f = \lim_{\|P\|\to 0} \sum_{r=1}^n f(ah^r)\delta_r$$

$$= \lim_{\|P\|\to 0} \sum_{r=1}^n f(ah^r)(ah^r - ah^{r-1})$$

34. The $(r+1)$th term of the sum is given by

$$\dfrac{n^2}{(n+r)^3} = \dfrac{1}{n}.\dfrac{1}{\left(1+\dfrac{r}{n}\right)^3}$$

Then letting $f(x) = \dfrac{1}{(1+x)^3}$ implies that
$f \in R[0,1]$, and hence

$$\lim_{n\to\infty}\left\{\dfrac{1}{n} + \dfrac{n^2}{(n+1)^2} + ... + \dfrac{1}{8n}\right\} = \int_0^1 \dfrac{1}{(1+x)^3}\,dx$$

$$= 3/8$$

35. Let

$$P_n = \left[\left(1+\dfrac{1}{n}\right)\left(1+\dfrac{2}{n}\right) + ... + \left(1+\dfrac{4n}{n}\right)\right]^{1/n}$$

Then $\log P_n = \dfrac{1}{n}\sum_{r=1}^{4n}\log\left(1+\dfrac{r}{n}\right)$

Letting $f(x) = \log(1+x)$

$\Rightarrow f \in R[0,4]$ and hence

$$f(x) = \lim_{n\to\infty}\log P_n = \int_0^4 \log(1+x)\,dx$$

$$= (x+1)\log(x+1)\Big|_0^4 - \int_0^4 dx$$

$$= 5\log 5 - 4 = \log(5^5/e^4)$$

Then $P_n = 5^5/e^4$

36. We have

$x + x^2 - (x^2 + x^3) = x - x^3 = x(1-x^2)$

$\Rightarrow x - x^2 > 0$

if $0 < x < 1 \Rightarrow x + x^2 \geq x^2 + x^3\ \forall x \in [0,1]$

$\Rightarrow M_j = x + x^2, m_j = x^2 + x^3\ \forall x \in [0,1]$

Also $x + x^2 - (x^2 + x^3) = x(1-x^2) \leq 0\ \forall\ x \in [1,2]$

$$\Rightarrow M_j = x^2 + x^3,\ m_j = x + x^2\ \forall x \in [1,2]$$

Therefore,

$$\int_0^2 f = \int_0^1 f + \int_1^2 f$$
$$= \int_0^1 (x + x^2) + \int_1^2 (x^2 + x^3)$$
$$= \left(\frac{x^2}{2} + \frac{x^3}{3}\right)\Big|_0^1 + \left(\frac{x^3}{3} + \frac{x^4}{4}\right)\Big|_1^2$$
$$= \frac{1}{2} + \frac{1}{3} + \frac{8}{3} + \frac{16}{4} - \frac{1}{3} - \frac{1}{4}$$
$$= \frac{83}{12}$$

and $\underline{\int_0^2} f = \int_0^1 f + \int_1^2 f$
$$= \int_0^1 (x^2 + x^3) + \int_1^2 (x + x^2)$$
$$= \left(\frac{x^3}{3} + \frac{x^4}{4}\right)\Big|_0^1 + \left(\frac{x^2}{2} + \frac{x^3}{3}\right)\Big|_1^2$$
$$= \frac{1}{3} + \frac{1}{4} + \frac{4}{2} + \frac{8}{3} - \frac{1}{2} - \frac{1}{3}$$
$$= \frac{53}{12}$$

37. Define DIrichlet's function f as follows :
$$f(x) = 1\ \text{if}\ x \neq 0$$
$$= 0\ \text{if}\ x = 0$$
Then

$$U(P,f) = \sum_{j=1}^n M_j \delta_j = \sum_{j=1}^n 1(1-0)/n = 1$$

where $\delta_j = (1-0)/n = 1/n$

$$L(P,f) = \sum_{j=1}^n m_j \delta_j = \sum_{j=1}^n 0(1/n) = 0$$

Hence, $\overline{\int_0^1} f = 1, \underline{\int_0^1} f = 0 \Rightarrow f \notin R[0,1]$

38. $f \geq 0$
$$\Rightarrow L(P,f) \geq 0,\ U(P,f) \geq 0\ \text{for any partition}\ P\ \text{of}\ [a,b]$$
$$\Rightarrow \text{l.u.b.}\ U(P,f) \geq 0\ \text{and g.l.b.}\ L(P,f) \geq 0,$$
Hence $f \in R[a,b] \Rightarrow \int_a^b f \geq 0$

39. Consider a partition P of $\left[0, \frac{\pi}{2}\right]$

$$P = \left\{0, \frac{\pi}{2n}, \frac{2\pi}{2n}, ..., \frac{(r-1)\pi}{2n}, \frac{r\pi}{2n}, ..., \frac{n\pi}{2n} = \frac{\pi}{2}\right\}$$

Here, the length of each subinterval $= \dfrac{\pi}{2n}$

Since, $f(x) = \sin x$ is increasing in $\left[0, \frac{\pi}{2}\right]$, we have

$$m_r = \sin\frac{(r-1)\pi}{2n}$$

and $$M_r = \sin\frac{r\pi}{2n},\ r = 1,2,...,n$$

$$\therefore U(P,f) = \sum_{r=1}^n M_r \Delta x_r$$
$$= \sum_{r=1}^n \left(\sin\frac{r\pi}{2n}\right)\cdot\frac{\pi}{2n}$$
$$= \frac{\pi}{2n}\left[\sin\frac{\pi}{2n} + \sin\frac{2\pi}{2n} + ... + \sin\frac{n\pi}{2n}\right]$$
$$= \frac{\pi}{2n}\left[\frac{\sin\left(\frac{\pi}{2n} + \frac{n-1}{2}\cdot\frac{\pi}{2n}\right)\sin\frac{n\pi}{4n}}{\sin\frac{\pi}{4n}}\right]$$

$$\left[\because \sin a + \sin(a+d) + ... + \sin(a+(n-1)d)\right.$$
$$= \frac{\sin\left(a + \frac{(n-1)}{2}d\right)\sin\frac{nd}{2}}{\sin\frac{d}{2}}\right]$$

$$= \frac{\pi}{2n}\sin\frac{(n+1)\pi}{4n}\cdot\sin\frac{\pi}{4}}{\sin\frac{\pi}{4n}}$$

$$= \frac{\frac{\pi}{2n}\sin\left(\frac{\pi}{4} + \frac{\pi}{4n}\right)\cdot\frac{1}{\sqrt{2}}}{\sin\frac{\pi}{4n}}$$

$$= \frac{\frac{\pi}{2\sqrt{2}.n}\left\{\sin\frac{\pi}{4}\cos\frac{\pi}{4n} + \cos\frac{\pi}{4}\sin\frac{\pi}{4n}\right\}}{\sin\left(\frac{\pi}{4n}\right)}$$

$$= \frac{\pi}{2\sqrt{2}n}\cdot\frac{1}{\sqrt{2}}\left(\cot\frac{\pi}{4n} + 1\right)$$

$$= \frac{\pi}{4n}\left(\cot\frac{\pi}{4n} + 1\right)$$

In a similiar manner, we can find

$$L(P,f) = \frac{\pi}{4n}\left(\cot\frac{\pi}{4n} + 1\right)$$

$\int_{\underline{0}}^{\pi/2} f = \lim_{n \to \infty} L(P, f)$

$= \lim_{n \to \infty} \dfrac{\pi}{4n}\left(\cot\dfrac{\pi}{4n} - 1\right)$

$= \lim_{n \to \infty} \dfrac{(\pi/4n)}{\tan(\pi/4n)} - \lim_{n \to \infty} \dfrac{\pi}{4n} = 1 - 0 = 1$

$\overline{\int_{0}^{\pi/2}} f = \lim_{n \to \infty} U(P, f)$

$= \lim_{n \to \infty} \dfrac{\pi}{4n}\left(\cot\dfrac{\pi}{4n} - 1\right) = 1$

Thus we conclude that

$\int_{\underline{0}}^{\pi/2} f = \overline{\int_{0}^{\pi/2}} \ f = 1$

$= f \in R\left(0, \dfrac{\pi}{2}\right)$ and hence $\int_{0}^{\pi/2} f = 1$

40. Consider the partition P of $\left(0, \dfrac{\pi}{4}\right)$.

$P = \left\{0, \dfrac{\pi}{4n}, \dfrac{2\pi}{4n}, ..., \dfrac{(r-1)\pi}{4n}, \dfrac{r\pi}{4n}, ..., \dfrac{n\pi}{4n} = \dfrac{\pi}{4}\right\}$

Let $\quad I_r = \left\{\dfrac{(r-1)\pi}{4n}, \dfrac{r\pi}{4n}\right\}$

Then $\quad m_r = \sin\left|\dfrac{(r-1)\pi}{4n}\right|$

and $\quad M_r = \cos\left|\dfrac{(r-1)\pi}{4n}\right|$

$\Delta x_r = \dfrac{\pi}{4n}$

$\Rightarrow L(P, f) = \sum_{r=1}^{n} m_r \Delta x_r = \sum_{r=1}^{n} \sin\dfrac{(r-1)\pi}{4n}\cdot\dfrac{\pi}{4n}$

$= \dfrac{\pi}{4n}\left[\sin\dfrac{\pi}{4n} + ... + \sin\dfrac{(n-1)\pi}{4n}\right]$

$= \dfrac{\pi}{4n}\cdot\dfrac{\sin\left(\dfrac{\pi}{4n} + \dfrac{n-2}{2}\cdot\dfrac{\pi}{4n}\right)\sin\dfrac{n\pi}{8n}}{\sin\dfrac{\pi}{8n}}$

$= \dfrac{\dfrac{\pi}{8n}}{\sin\dfrac{\pi}{8n}}\cdot2\sin^2\dfrac{\pi}{8}$

Now

$U(P, f) = \sum_{r=1}^{n} M_r \Delta x_r = \sum_{r=1}^{n} \cos\dfrac{(r-1)\pi}{4n}\cdot\dfrac{\pi}{4n}$

$= \dfrac{\pi}{4n}\left[\cos 0 + \cos\dfrac{\pi}{4n} + ... + \cos\dfrac{(n-1)\pi}{4n}\right]$

$= \dfrac{\pi}{4n}\cdot\dfrac{\cos\left(\dfrac{n-1}{2}\cdot\dfrac{\pi}{4n}\right)\cdot\sin\dfrac{n\pi}{8n}}{\sin\left(\dfrac{\pi}{8n}\right)}$

$= \dfrac{\left(\dfrac{\pi}{8n}\right)}{\sin\left(\dfrac{\pi}{8n}\right)}\cdot2\cos\dfrac{(n-1)\pi}{8n}\cdot\sin\dfrac{\pi}{8}$

$\therefore \int_{0}^{\pi/4} f = \lim_{\|P\| \to 0} L(P, f)$

$= \lim_{n \to \infty} L(P, f)$

$= \lim_{n \to \infty} \dfrac{\left(\dfrac{\pi}{8n}\right)}{\sin\left(\dfrac{\pi}{8n}\right)}\cdot2\sin^2\dfrac{\pi}{8}$

$= 2\sin^2\dfrac{\pi}{8} = 1 - \dfrac{\cos\pi}{4} = 1 - \dfrac{1}{\sqrt{2}}$

and $\overline{\int_{0}^{\pi/4}} f = \lim_{n \to \infty} U(P, f)$

$= \lim_{n \to \infty} \dfrac{\left(\dfrac{\pi}{8n}\right)}{\sin\left(\dfrac{\pi}{8n}\right)}\cdot2\cos\dfrac{\pi}{8}\dfrac{(n-1)\pi}{n}\cdot\sin\dfrac{\pi}{8}$

$= \lim_{n \to \infty} \dfrac{\left(\dfrac{\pi}{8n}\right)}{\sin\left(\dfrac{\pi}{8n}\right)}\cdot2\cos\dfrac{\pi}{8}\left(1 - \dfrac{1}{n}\right)\sin\dfrac{\pi}{8}$

$= 2\cos\dfrac{\pi}{8}\sin\dfrac{\pi}{8} = \sin\dfrac{\pi}{4} = \dfrac{1}{\sqrt{2}}$

Clearly $\int_{\underline{0}}^{\pi/4} f \neq \overline{\int_{0}^{\pi/4}} \ f$

Hence, f is not R-integrable over $\left[0, \dfrac{\pi}{4}\right]$

43. Here $f\left(\dfrac{1}{a^r}\right) = \dfrac{1}{a^r}, f\left(\dfrac{1}{a^r}\right) = \dfrac{1}{a^{r-1}}$

and $f\left(\dfrac{1}{a^r}\right) = \dfrac{1}{a^r}$ .This shows that f has points

of discontinuity of first kind at $x = \dfrac{1}{a^r}$, $r = 1, 2, ...$

The condition for the existence of R-integral is given as below.

The function $f \in R$ on $[0,1]$, then $D(f)$, the set of points of discontinuity is of measure zero. For this

We have $$f\left(x = \frac{1}{a^r}\right),$$

where $$r \to \infty = f(0) = \lim_{r \to \infty} \frac{1}{a^r} = 0$$

and $f(0+0)=0$. This shows that f is continuous on the right of $x=0$. The points of discontinuity

at $x = \dfrac{1}{a^r}$ $(r=1,2)$ of f on $[0,1]$ from a set $D(f)$

with $x=0$ as limit point. $D(f)$ is a closed set. The content of $D(f)$ is zero. In a closed set its content and measure are the same. Therefore, $\mu(D)$ is zero. This shows that $f \in R$ on $[0,1]$.

Now $$\int_0^1 f = \sum_{r=1}^{n=\infty} \int_{1/(a^r)}^{1/(a^{r-1})} \frac{1}{a^{r-1}}$$

Therefore,

if $f \in R[a,b], f = \sum_{n=1}^{\infty} f_n$ for $x \in [a,b]$ converges uniformly on $[a,b]$.

Then, $\int_a^b f = \sum_{n=1}^{\infty} \int_a^b f_n$

Hence,

$$\int_0^1 f = \sum_{r=1}^{\infty} \frac{1}{a^{r-1}}\left(\frac{1}{a^{r-1}} - \frac{1}{a^r}\right)$$

$$= \sum_{r=1}^{\infty} \left[\frac{1}{a^{2r-2}} - \frac{1}{a^{2r-1}}\right]$$

$$= 1 - \frac{1}{a} + \frac{1}{a^2} - \frac{1}{a^3} + \ldots = \frac{1}{1 - \left(-\dfrac{1}{a}\right)}$$

$$= \frac{a}{a+1}$$

SELF ASSESSMENT TEST

Verify each of the following :

1. If f is continuous and positive on $[a,b]$ then $\int_a^b f\,dx$ is also positive.

2. $\lim\limits_{n\to\infty}\left[\dfrac{1}{n}+\dfrac{n^2}{(n+1)^3}+\dfrac{n^2}{(n+2)^3}+...+\dfrac{1}{8n}\right]=\dfrac{3}{8}$

3. If $f(x)=\begin{cases}\sin\dfrac{1}{x};\forall\text{ irrational }x\in[0,1]\\[2mm]0\ \ ;\forall\text{ rational }x\in[0,1]\end{cases}$

then $f(x)$ is not Riemann integrable on $[0,1]$.

4. If $f(x,t)=\begin{cases}x(t-1)\text{ when }x\le t\\ t(x-1)\text{ when }t<x\end{cases}$

and $g(x)$ is continuous on $[0,1]$, then

$F(x)=\int_0^1 g(t)f(x,t)\,dt$ exists and $F''(x)=g(x)$

and $F(0)=F(1)=0$.

5. The sequence

$$a_n=1+\dfrac{1}{2}+\dfrac{1}{3}+...+\dfrac{1}{n}-\log n\ \forall n,$$

monotonically decreasing and bounded between 0 and 1 and converges to a non-zero limit between 0 and 1.

6. The function $f(x)=\begin{cases}\sin\dfrac{1}{x},\text{when }x\text{ is irrational}\\[2mm]0\ \ ,\ \text{otherwise}\end{cases}$

is not Riemann integrable on $[0,1]$.

7. If f is continuous on $[a,b]$ and $\int_a^x f(t)\,dt=\int_x^b f(t)\,dt\ \forall x\in[a,b]$ then $f(x)=0$, $\forall x\in[a,b]$.

8. If f is non-negative Riemann integrable function on $[a,b]$ then $\sqrt{f}$ is R-integrable on $[a,b]$.

9. If $f\in R[a,b]$ then $\lim\limits_{n\to\infty}\int_a^b f(x)\cos nx\,dx=0$

10. If f has continuous derivative on $[c,d]$ and $a,b\in[c,d]$ then $\int_a^b f'=f(b)-f(a)$

Chapter 7
Riemann-Stieltjes Integrals

7.1 INTRODUCTION

German mathematician Riemann was the first to define the process of integration on certain arithemetical concepts which is broad based and free from dependence on geometrical concepts. This concept is known as Riemann integration, which was later on generalised by the Dutch astronomer and mathematician Stieltjes.

In this chapter, we shall discuss the definition of the Riemann-Stieltjes integral, which depends very explicitly on the order structure of real line. We shall discuss also, integrability of vector valued functions on intervals.

7.2 RIEMANN-STIELTJES SUMS

If on an interval $[a, b]$, $\alpha(x)$ be a monotonically increasing function, then by definition of monotonic function, we find that $\alpha(a)$ and $\alpha(b)$ are finite and therefore $\alpha(x)$ is a bounded function in $[a, b]$ corresponding to each partition $P = [x_0, x_1,..., x_n]$ of $[a, b]$.

Let $$\Delta\alpha_r = \alpha(x_r) - \alpha(x_{r-1}) \qquad ...(1)$$

Now, as the function $\alpha(x)$ is monotonically increasing, so we have $\alpha(x_r) \geq \alpha(x_{r-1})$

From (1), we have $\Delta\alpha_r \geq 0$.

Let us now suppose that $f(x)$ is any real valued function defined on $[a, b]$, then evidently $f(x)$ is also necessarily bounded in each subinterval $[x_r, x_{r-1}]$.

Let l.u.b. (or supremum) and g.l.b. (or infimum) of the function $f(x)$ in the subinterval $[x_{r-1}, x_r]$ be denoted by M_r and m_r respectively. Then, the following sums

$$U[P, f, \alpha] = \sum_{r=1}^{n} M_r \Delta\alpha_r$$

and

$$L[P, f, \alpha] = \sum_{r=1}^{n} m_r \Delta\alpha_r$$

are known as upper and lower Riemann-Stieltjes sums or R-S sums with respect to α corresponding to the partition P respectively.

7.3 RIEMANN-STIELTJES INTEGRAL

Definition. *The g.l.b. (or infimum) of all upper R-S sums* $U[P, f, \alpha]$ *is known as upper Riemann-Stieltjes integral w.r.t.* α *over* $[a,b]$ *and is denoted by* $\overline{\int_a^b} f \, d\alpha$

i.e., $$\overline{\int_a^b} f \, d\alpha = \text{g.l.b.} \; [U(P, f, \alpha)]$$

where g.l.b. is taken over all partitions P of $[a, b]$.

Similarly, the l.u.b.(or supremum) of all the lower R-S sums $L[P, f, \alpha]$ is known as lower Riemann-Stieltjes integral with respect to α over $[a,b]$ and is denoted by $\underline{\int_a^b} f \, d\alpha$.

i.e., $\qquad\qquad \underline{\int_a^b} f \, d\alpha = \text{l.u.b. } [L(P, f, \alpha)]$

where l.u.b. is taken over all partitions P of $[a, b]$.

If upper and lower Riemann-Stieltjes integrals with respect to α over $[a, b]$ are equal

i.e., $\qquad\qquad \overline{\int_a^b} f \, d\alpha = \underline{\int_a^b} f \, d\alpha$

then f is called Riemann-Stieltjes integral with respect to α on $[a, b]$.

This common value is known as the Riemann-Stieltjes integral of the function $f(x)$ w.r.t. $\alpha(x)$ on $[a, b]$ and is denoted by

$$\int_a^b f \, d\alpha \quad \text{or} \quad \int_a^b f(x) \, d\alpha(x)$$

The set of all those functions which are Riemann-Stieltjes integrable (or RS-integrable) w.r.t. $\alpha(x)$ is denoted by $R(\alpha)$.

REMARK

- If we take $\alpha(x) = x$, we can find the Riemann integral is a special case of Riemann-Stieltjes integral, therefore, all the theorems which are applicable to Riemann-Stieltjes are also applicable to Riemann integrals.

THEOREM 1. *If P^* is a refinement of P, then*

$\qquad$ (a) $U(P^*, f, \alpha) \le U(P, f, \alpha)$ $\qquad\qquad$ (b) $L(P, f, \alpha) \le L(P^*, f, \alpha)$

Proof. $\quad$ Let $P^* = P \cup \{x^*\}$, where $x_{r-1} < x^* < x_r$, x_{r-1} and x_r being two consecutive points of P.

(a) Let u_1 and u_2 be the supremum of function $f(x)$ in $[x_{r-1}, x^*]$ and $[x^*, x_r]$ respectively. Then, clearly $u_1 \le M_r$ and $u_2 \le M_r$ where M_r is the supremum of the function $f(x)$ in $[x_{r-1}, x_r]$.

By definition, we have

$$U(P, f, \alpha) = \sum_{i=1}^{n} M_i \Delta\alpha_i = \sum_{i=1}^{r-1} M_i \Delta\alpha_i + M_r \left[\alpha(x_r) - \alpha(x_{r-1})\right] + \sum_{i=r+1}^{n} M_i \Delta\alpha_i$$

and

$$U(P^*, f, \alpha) = \sum_{i=1}^{r-1} M_i \Delta\alpha_i + u_1 \left[\alpha(x^*) - \alpha(x_{r-1})\right] + u_2 \left[\alpha(x_r) - \alpha(x^*)\right] + \sum_{i=r+1}^{n} M_i \Delta\alpha_i$$

Now,

$U(P, f, \alpha) - U(P^*, f, \alpha)$

$$= M_r \left[\alpha(x_r) - \alpha(x_{r-1})\right] - u_1 \left[\alpha(x^*) - \alpha(x_{r-1})\right] - u_2 \left[\alpha(x_r) - \alpha(x^*)\right]$$

$$= (M_r - u_2) \left[\alpha(x_r) - \alpha(x^*)\right] + (M_r - u_1) \left[\alpha(x^*) - \alpha(x_{r-1})\right]$$

$$\ge 0 \, (\because M_r - u_1 \ge 0, M_r - u_2 \ge 0, \alpha(x_r) \ge \alpha(x^*) \text{ and } \alpha(x^*) \ge \alpha(x_{r-1}),$$
$$\alpha \text{ being a monotonically increasing function})$$

Therefore, $U(P, f, \alpha) \ge U(P^*, f, \alpha)$ which implies $U(P^*, f, \alpha) \le U(P, f, \alpha)$.

In a similar manner, we may prove the result (b).

THEOREM 2. *Let f be a bounded function and α is a non-decreasing function on $[a,b]$, then lower RS-integral of f relative to α cannot exceed the upper RS-integral*

i.e., $\qquad\qquad \underline{\int_a^b} f \, d\alpha \le \overline{\int_a^b} f \, d\alpha.$

Proof. Let P^* be the common refinement of any two partitions P_1 and P_2 of P^*, then we have

$$L[P_1, f, \alpha] \le L[P^*, f, \alpha] \le U[P^*, f, \alpha] \le U[P_2, f, \alpha]$$

Therefore, $\qquad L[P_1, f, \alpha] \le U(P_2, f, \alpha)$

Now, if P_1 is varied while keeping P_2 fixed, then taking l.u.b. over all P_1, we have $\qquad$ l.u.b. $[L(P_1, f, \alpha)] \le U[P_2, f, \alpha]$

or $\qquad \underline{\int_a^b} f \, d\alpha \le U(P_2, f, \alpha)$

Now, if P_2 is also varied, taking g.l.b. over all P_2, we have

$$\underline{\int_a^b} f \, d\alpha \le \text{g.l.b.} \, U(P_2, f, \alpha)$$

$\Rightarrow \qquad \underline{\int_a^b} f \, d\alpha \le \overline{\int_a^b} f \, d\alpha.$

THEOREM 3. (*Necessary and sufficient condition for RS-integrability*).

Let f be a bounded function and α, a monotonically increasing function on $[a,b]$. Then $f \in RS(\alpha)$ if and only if for every $\varepsilon > 0$ there exists a partitions P of $[a,b]$ such that $U(P, f, \alpha) - L(P, f, \alpha) < \varepsilon$.

Proof. **The condition is necessary.**

Let $f \in RS(\alpha)$ such that

$$\underline{\int_a^b} f \, d\alpha = \overline{\int_a^b} f \, d\alpha \qquad \qquad \dots(1)$$

Since $\underline{\int_a^b} f \, d\alpha$ is the supremum of $L[P, f, \alpha]$ overall partitions P, there exists a partition P_1 such that

$$\underline{\int_a^b} f \, d\alpha < L(P_1, f, \alpha) + \frac{\varepsilon}{2}$$

Similarly, since $\overline{\int_a^b} f \, d\alpha$ is the infimum of $U[P, f, \alpha]$ overall partitions P, there exists a partition P_2 such that

$$U[P_2, f, \alpha] < \overline{\int_a^b} f \, d\alpha + \frac{\varepsilon}{2}$$

If $P = P_1 \cup P_2$, then P is the common refinement of P_1 and P_2, then by definition of refinement, we have

$$\underline{\int_a^b} f \, d\alpha < L[P, f, \alpha] + \frac{\varepsilon}{2} \qquad \qquad \dots(2)$$

and $\qquad U[P, f, \alpha] < \overline{\int_a^b} f \, d\alpha + \frac{\varepsilon}{2} \qquad \qquad \dots(3)$

Adding (2) and (3), we get

$$\underline{\int_a^b} f \, d\alpha + U(P, f, \alpha) < L(P, f, \alpha) + \overline{\int_a^b} f \, d\alpha + \varepsilon$$

Using (1), we have

$$U(P, f, \alpha) < L(P, f, \alpha) + \varepsilon$$

i.e., $\qquad U(P, f, \alpha) - L(P, f, \alpha) < \varepsilon$

The condition is sufficient.

Let for every $\varepsilon > 0$, there exists a partition P of $[a,b]$ such that

$$U(P, f, \alpha) - L[P, f, \alpha] < \varepsilon \qquad \ldots(4)$$

By definition of upper and lower RS-integrals, we have

$$\overline{\int_a^b} f \, d\alpha \le U[P, f, \alpha]$$

and

$$\underline{\int_a^b} f \, d\alpha \ge L[P, f, \alpha]$$

i.e.,

$$L(P, f, \alpha) \le \underline{\int_a^b} f \, d\alpha \le \overline{\int_a^b} f \, d\alpha < U(P, f, \alpha) \qquad \ldots(5)$$

From (4) and (5), we conclude that

$$\overline{\int_a^b} f \, d\alpha - \underline{\int_a^b} f \, d\alpha \le U(P, f, \alpha) - L(P, f, \alpha) < \varepsilon$$

or

$$\overline{\int_a^b} f \, d\alpha - \underline{\int_a^b} f \, d\alpha < \varepsilon$$

Since ε is arbitrary, we have $\overline{\int_a^b} f \, d\alpha = \underline{\int_a^b} f \, d\alpha$.

Hence, the function f is RS-integrable with respect to α.

THEOREM 4.   (i) *If for some partition P and some $\varepsilon > 0$, the inequality*

$$U(P, f, \alpha) - L(P, f, \alpha) < \varepsilon \qquad \ldots(1)$$

holds good, this also holds good for P^*.

(ii) *If (1) holds for $P = (x_0, x_1, \ldots, x_n)$ and if s_i, t_i are arbitrary points in $[x_{i-1}, x_i]$, then*

$$\sum_{i=1}^{n} |f(s_i) - f(t_i)| \Delta \alpha_i < \varepsilon$$

(iii) *If $f \in R(\alpha)$ and the hypothesis of (b) hold, then*

$$\left| \sum_{i=1}^{n} f(t_i) \Delta \alpha_i - \int_a^b f \, d\alpha \right| < \varepsilon.$$

Proof.   (i)   For P^* be the refinement of P, then we have

$$L(P, f, \alpha) \le L(P^*, f, \alpha) \qquad \ldots(2)$$

and

$$U(P^*, f, \alpha) \le U(P, f, \alpha) \qquad \ldots(3)$$

Using (1), (2) and (3), we get

$$U(P^*, f, \alpha) - L(P^*, f, \alpha) \le U(P, f, \alpha) - L(P, f, \alpha) < \varepsilon.$$

(ii)   Given that $f(s_i)$ and $f(t_i)$ lie in $[m_i, M_i]$, so that

$$|f(s_i) - f(t_i)| \le M_i - m_i$$

Therefore, $\displaystyle\sum_{i=1}^{n} |f(s_i) - f(t_i)| \Delta \alpha_i \le \sum_{i=1}^{n} (M_i - m_i) \Delta \alpha_i$

$$= \sum_{i=1}^{n} M_i \Delta \alpha_i - \sum_{i=1}^{n} m_i \Delta \alpha_i$$

$$= U(P, f, \alpha) - L(P, f, \alpha)$$

$$< \varepsilon \qquad \text{[using (i)]}$$

$$\Rightarrow \quad \sum_{i=1}^{n} |f(s_i) - f(t_i)| \Delta \alpha_i < \varepsilon$$

(iii) We have

$$L(P,f,\alpha) \le \sum_{i=1}^{n} f(t_i)\Delta\alpha_i \le U(P,f,\alpha) \qquad \qquad ...(4)$$

and

$$L(P,f,\alpha) \le \int_a^b f\, d\alpha \le U(P,f,\alpha) \qquad \qquad ...(5)$$

From (4) and (5), we have

$$\sum_{i=1}^{n} f(t_i)\Delta\alpha_i - \int_a^b f\, d\alpha \le U[P,f,\alpha] - L[P,f,\alpha] < \varepsilon \qquad \qquad ...(6)$$

and

$$\int_a^b f\, d\alpha - \sum_{i=1}^{n} f(t_i)\Delta\alpha_i \le U(P,f,\alpha) - L(P,f,\alpha) < \varepsilon \qquad \qquad ...(7)$$

From (6) and (7), we conclude that

$$\left| \sum_{i=1}^{n} f(t_i)\Delta\alpha_i - \int_a^b f\, d\alpha \right| < \varepsilon$$

7.4 THE RS-INTEGRAL AS A LIMIT OF SUMS

We know that RS-integral $\int_a^b f\, d\alpha$ is defined by means of the sums $L[P, f, \alpha]$ and $U[P, f, \alpha]$. The numbers m_r, M_r, which appear in these sums are not necessarily the values of f. Now, we shall define $\int_a^b f\, d\alpha$ as the limit of a sequence of sums in which m_r, M_r are replaced by values of f.

Definition. *Let f be a bounded and α is a monotonically increasing function on $[a,b]$. Let $P=\{a=x_0,x_1,...,x_n = b\}$ be a partition of $[a,b]$ and $Q=\{t_1,t_2,...,t_n\}$ an intermediate partition of P so that*

$$x_{r-1} \le t_r \le x_r \ \text{for } r=1,2,...,n$$

Then, the number RS$[P, Q, f, \alpha] = \sum_{r=1}^{n} f(t_r)\Delta\alpha_r$

is defined as the Riemann-Stieltjes sum (or RS-sum) of f relative to α on $[a,b]$ and corresponding to the partition P and the intermediate partition Q , we say that

$$\lim_{\|P\| \to 0} RS(P,Q,f,\alpha) = l$$

iff for every $\varepsilon>0$, there exists a $\delta>0$ such that $|RS(P, Q, f, \alpha) - l| < \delta$ whenever $\|P\| < \delta$.

THEOREM 1. *If $\lim_{\|P\| \to 0} RS(P,Q,f,\alpha)$ exists, then $f \in RS(\alpha)$ on $[a,b]$, then*

$$\lim_{\|P\| \to 0} RS(P,Q,f,\alpha) = \int_a^b f\, d\alpha$$

Proof.   Let us suppose $\lim_{\|P\| \to 0} RS(P,Q,f,\alpha) = \lim_{\|P\| \to 0} \sum_{r=1}^{n} f(t_r)\Delta\alpha_r$ exists and is equal to l.

Let $\varepsilon>0$ be given, then by definition $\exists\ \delta>0$ such that

$$|RS(P, Q, f, \alpha)| < \varepsilon/4 \text{ whenever } \|P\| < \delta$$

$\Rightarrow$ $\qquad l - \dfrac{\varepsilon}{4} < RS(P,Q,f,\alpha) < l + \dfrac{\varepsilon}{4}$ whenever $\|P\| < \delta$ ...(1)

Choosing one such point P, letting the points t_r, range over the intervals $[x_{r-1}, x_r]$ and taking the infimum and supremum of the numbers $RS(P, Q, f, \alpha)$

$$l - \frac{\varepsilon}{4} \le L[P, f, \alpha] \le U[P, f, \alpha] \le l + \frac{\varepsilon}{4} \qquad ...(2)$$

$$U[P, f, \alpha] \le l + \frac{\varepsilon}{4} \text{ and } -L[P, f, \alpha] \le -l + \frac{\varepsilon}{4}$$

On adding, we get

$$U[P, f, \alpha] - L[P, f, \alpha] \le \frac{\varepsilon}{2} < \varepsilon$$

$\Rightarrow$ $\qquad f \in RS(\alpha).$

Therefore, we get

$$\underline{\int_a^b} f\, d\alpha = \overline{\int_a^b} f\, d\alpha = \int_a^b f\, d\alpha$$

But $\qquad L[P, f, \alpha] \le \int_a^b f\, d\alpha \le U[P, f, \alpha].$ ...(3)

From (2) and (3), we conclude that

$$l - \frac{\varepsilon}{4} \le \int_a^b f\, d\alpha \le l + \frac{\varepsilon}{4} \text{ i.e., } \left| l - \int_a^b f\, d\alpha \right| \le \frac{\varepsilon}{4} < \varepsilon$$

Now, since ε is arbitrary, therefore, we have

$$I = \int_a^b f\, d\alpha$$

Hence, $\qquad \lim_{\|P\| \to 0} RS(P, Q, f, \alpha) = \int_a^b f\, d\alpha.$

THEOREM 2. *Let f be continuous and α is monotonically increasing on $[a,b]$ then $f \in RS(\alpha)$ i.e., to every $\varepsilon > 0 \; \exists \; \delta > 0$ such that*

$$\left| \sum_{r=1}^n f(t_r)\Delta\alpha_r - \int_a^b f\, d\alpha \right| < \varepsilon$$

for every partition $P = \{a = x_0, x_1, x_2, ..., x_n = b\}$ with $\|P\| < \delta$ and for every intermediate partitions $Q = \{t_1, t_2, ..., t_n\}$ of P. i.e., $\lim\limits_{\|P\| \to 0} RS(P, Q, f, \alpha) = \int_a^b f\, d\alpha.$

Proof. Let $\varepsilon > 0$ be given, then choose $\eta > 0$ such that

$$\eta |\alpha(b) - \alpha(a)| < \varepsilon \qquad ....(1)$$

Since, the function f is continuous is closed and bounded interval $[a,b]$, therefore, f is uniformly continuous on $[a,b]$, then by definition of uniformly continuous function there exists a $\delta > 0$ such that

$$x, y \in [a, b], |x-y| < \delta \quad \Rightarrow \quad |f(x)-f(y)| < \eta. \qquad ...(2)$$

Now, let $P = \{a = x_0, x_1, ..., x_n = b\}$ be a partition with $\|P\| < \delta$.

The function f is continuous on each subinterval $[x_{r-1}, x_r]$ as it is continuous on $[a,b]$ and therefore, it attains the bounds m_r and M_r on $[x_{r-1}, x_r]$

$\Rightarrow$ there exists points $c, d \in [x_{r-1}, x_r]$ such that $f(c) = m_r, f(d) = M_r.$

Now, from (2), we have $|f(d)-f(c)| < \eta \qquad \Rightarrow \qquad M_r - m_r < \eta$

Therefore, $U(P, f, \alpha) - L(P, f, \alpha)$

$$= \sum_{r=1}^{n} (M_r - m_r)\Delta\alpha_r < \eta \sum_{r=1}^{n} \Delta\alpha_r = \eta \sum_{r=1}^{n} \left[\alpha(x_r) - \alpha(x_{r-1})\right]$$
$$= \eta\left[\alpha(x_1) - \alpha(x_0) + \alpha(x_2) - \alpha(x_1) + \ldots + \alpha(x_n) - \alpha(x_{n-1})\right]$$
$$= \eta\left[\alpha(x_n) - \alpha(x_0)\right]$$
$$= \eta\left[\alpha(b) - \alpha(a)\right]$$
$$< \varepsilon.$$

[Using (1)]

Therefore, $U(P, f, \alpha) - L(P, f, \alpha) < \varepsilon$ ...(3)

$\Rightarrow \qquad\qquad f \in RS(\alpha)$

$\Rightarrow \qquad\qquad \underline{\int_a^b} f \, d\alpha = \overline{\int_a^b} f \, d\alpha = \int_a^b f \, d\alpha .$ ...(4)

Now, from (3), we get

$\qquad\qquad U(P, f, \alpha) - \varepsilon < L(P, f, \alpha)$

$\Rightarrow \qquad\qquad \overline{\int_a^b} f \, d\alpha - \varepsilon < L(P, f, \alpha)$

$\Rightarrow \qquad\qquad \int_a^b f \, d\alpha - \varepsilon < L(P, f, \alpha)$ [Using (4)]

Also, $\qquad\qquad U(P, f, \alpha) < L(P, f, \alpha) + \varepsilon \Rightarrow \qquad U(P, f, \alpha) < \underline{\int_a^b} f \, d\alpha + \varepsilon$

$\Rightarrow \qquad\qquad U(P, f, \alpha) < \int_a^b f \, d\alpha + \varepsilon$

Therefore, $\int_a^b f \, d\alpha - \varepsilon < L(P, f, \alpha) \le U(P, f, \alpha) < \int_a^b f \, d\alpha < \varepsilon$ ...(5)

Also $\qquad\qquad L(P, f, \alpha) \le RS(P, Q, f, \alpha) \le U(P, f, \alpha)$ ...(6)

From (5) and (6), we conclude that

$$\int_a^b f \, d\alpha - \varepsilon < RS(P, Q, f, \alpha) < \int_a^b f \, d\alpha + \varepsilon$$

$\Rightarrow \qquad\qquad \left| RS(P, Q, f, \alpha) - \int_a^b f \, d\alpha \right| < \varepsilon$

$\Rightarrow \qquad\qquad \lim_{\|P\| \to 0} RS(P, Q, f, \alpha) = \int_a^b f \, d\alpha$

THEOREM 3. *Let $f \in RS(\alpha)$ and let α be continuous on $[a,b]$. Then, with the usual notations*

$$\lim_{\|P\| \to 0} RS(P, Q, f, \alpha) = \int_a^b f \, d\alpha \qquad .$$

Proof. Given that $f \in RS(\alpha)$, therefore

$$\underline{\int_a^b} f \, d\alpha = \overline{\int_a^b} f \, d\alpha = \int_a^b f \, d\alpha$$ ...(1)

Let $\varepsilon > 0$ be given. Now, since $\overline{\int_a^b} f \, d\alpha$ is the infimum of $U[P, f, \alpha]$, there exists a partition P^* such that

$$U(P^*, f, \alpha) < \overline{\int_a^b} f \, d\alpha + \frac{\varepsilon}{2}$$

$\Rightarrow \qquad\qquad U(P^*, f, \alpha) < \int_a^b f \, d\alpha + \frac{\varepsilon}{2} .$ ...(2)

Let $M=\sup\{|f(x)| : a \leq x \leq b\}$. Since the function α is continuous on the closed and bounded interval $[a,b]$ therefore, it is uniformly continuous. Then, by definition, there exists $\delta>0$ such that

$$|x-y|<\delta_1 \Rightarrow |\alpha(x)-\alpha(y)|<\frac{\varepsilon}{2pM} \qquad \ldots(3)$$

where, p is the number of subintervals into which $[a,b]$ is divided by P^*. If $P=\{a=x_0,x_1,\ldots,x_n=b\}$ be any partition with $\|P\|<\delta$, then (3) gives

$$\Delta\alpha_r=\alpha(x_r)-\alpha(x_{r-1})<\frac{\varepsilon}{2pM}. \qquad \ldots(4)$$

Divide the subintervals of P into two groups :

(i) those which are contained in a subinterval of P^*.

(ii) those which contain in their interior one or more points of subdivision of P^*.

From (i), we have the sum $U(P, f, \alpha)$ does not exceed $U(P^*, f, \alpha)$. From (ii), we have that p does not exceed $p-1$, and hence their contribution to $U(P, f, \alpha)$ does not exceed $(p-1)kM$, where

$$k=\max\{\Delta\alpha_r, r= 1,2,\ldots,n\}.$$

Therefore, $U(P, f, \alpha) \leq U(P^*,f, \alpha)+(p-1)kM$ $\ldots(5)$

Using (4), we get $k=\dfrac{\varepsilon}{2pM}$ $\ldots(6)$

Now combining (5) with (2) and (6), we get

$$U(P,f,\alpha)<\int_a^b f\,d\alpha+\frac{\varepsilon}{2}+(p-1)M.\frac{\varepsilon}{2pM}$$

$$<\int_a^b f\,d\alpha+\frac{\varepsilon}{2}+\frac{\varepsilon}{2}=\int_a^b f\,d\alpha+\varepsilon$$

Therefore, for all partitions P with $\|P\|<\delta_1$

$$U(P, f, \alpha)<\int_a^b f\,d\alpha+\varepsilon. \qquad \ldots(7)$$

Similarly, it can be shown that there exists $\delta_2>0$ such that for all partitions P with $\|P\|<\delta_2$.

$$L(P, f, \alpha)>\int_a^b f\,d\alpha-\varepsilon \qquad \ldots(8)$$

Define $\delta=\min\{\delta_1,\delta_2\}$. Then, $RS(P, Q, f, \alpha)<\int_a^b f\,d\alpha+\varepsilon$

and $RS(P, Q, f, \alpha)>\int_a^b f\,d\alpha-\varepsilon$

Then $\int_a^b f\,d\alpha-\varepsilon <RS(P, Q,f,\alpha)<\int_a^b f\,d\alpha+\varepsilon$ for all partitions P with $\|P\|<\delta$

Therefore, $\lim_{\|P\|\to 0} RS(P,Q,f,\alpha)=\int_a^b f\,d\alpha.$

7.5 CLASSES OF RIEMANN-STIELTJES INTEGRABLE FUNCTIONS

THEOREM 1. *Let f be continuous and monotonically increasing on $[a,b]$. Then $f \in RS(\alpha)$ on $[a, b]$.*

Proof.   Let $\varepsilon>0$ be given, Since α is monotonically increasing on $[a,b]$ we can choose $\eta>0$ so that

$$\alpha(b)-\alpha(a)<\frac{\varepsilon}{\eta} \qquad \ldots(1)$$

Now, since $f: (a, b) \rightarrow R$ is a continuous mapping in a closed and bounded interval, then f must be uniformly continuous on $[a, b]$, then by definition of uniform continuity, there exists $\delta > 0$ such that for $x \in [a, b], y \in [a, b]$

$$|x-y| < \delta \quad \Rightarrow \quad |f(x)-f(y)| < \eta \qquad \qquad \text{...(2)}$$

Let
$$m_r = \inf_{x \in [x_{r-1}, x_r]} f(x)$$

and
$$M_r = \sup_{x \in [x_{r-1}, x_r]} f(x)$$

and P be any partition of $[a, b]$ such that $\Delta x_r < \delta$ for all r. Then, (2) gives

$$M_i - m_i \leq \eta \qquad \qquad \text{...(3)}$$

From (1) und (3), we conclude that

$$U(P, f, \alpha) - L(P, f, \alpha) = \sum_{r=1}^{n} M_r \Delta \alpha_r - \sum_{r=1}^{n} m_r \Delta \alpha_r$$

$$= \sum_{r=1}^{n} (M_r - m_r) \Delta \alpha_r$$

$$\leq \eta \sum_{r=1}^{n} \Delta \alpha_r = \eta [\alpha(b) - \alpha(a)] < \varepsilon$$

Hence, $f \in RS(\alpha)$.

THEOREM 2. *Let f be monotonic on $[a, b]$ and let α be continuous and monotonically increasing on $[a, b]$, then $f \in RS(\alpha)$.*

Proof.   Let $\varepsilon > 0$ be given small positive number. Given that α is continuous and monotonically increasing function. Therefore, for any positive integer n there exists a partitions P of $[a, b]$ such that $\Delta \alpha_i = \alpha(x_i) - \alpha(x_{i-1})$, $i = 1, 2, \dots, n$

$$= \frac{\alpha(b) - \alpha(a)}{n} \qquad \qquad \text{...(1)}$$

Suppose, f is monotonically increasing. Then, we have

$$M_i = f(x_i) \text{ and } m_i = f(x_{i-1}), \ i = 1, 2, \dots, n \qquad \text{...(2)}$$

From (1) and (2), we conclude that

$$U(P, f, \alpha) - L(P, f, \alpha) = \sum_{r=1}^{n} (M_r - m_r) \Delta \alpha_r$$

$$= \frac{\alpha(b) - \alpha(a)}{n} \sum_{r=1}^{n} [f(x_i) - f(x_{i-1})]$$

$$= \frac{\alpha(b) - \alpha(a)}{n} [f(b) - f(a)]$$

$$= \varepsilon$$

which implies $f \in RS(\alpha)$.

THEOREM 3. *Let f be bounded on $[a, b]$, which has only finitely many points of discontinuity on $[a, b]$ and α is continuous at every point at which f is continuous, then $f \in RS(\alpha)$.*

Proof.   Let $E = [c_0, c_1, \dots, c_n]$ be the finite set of points at which the function f is discontinuous. Since E is finite and function α is continuous at each $c_j (j = 1, \dots, m)$ we can cover E by

m disjoint intervals $[u_i, v_j] \subset [a, b]$ such that

$$\sum_{j=1}^{m} \left[\alpha(v_j) - \alpha(u_j)\right] < \varepsilon.$$

We can construct these intervals in such a way that every point of $E \cap [a,b]$ lies in the interior of some $[u_i, v_j]$

Let $\qquad K = [a,b] - \bigcup_{j=1}^{m} (u_j, v_j)$

Then K is compact, also f is continuous in each subintervals $[a,u_1][v_1, u_2][v_2, u_3]...$ $[v_m, b]$, therefore, f is uniformly continuous on K, then by definition, we have $\delta > 0$ such that for $s \in K, t \in K$.

$$|s-t| < \delta \implies |f(s) - f(t)| < \varepsilon. \qquad \qquad ...(1)$$

Form a partition $P (= \{x_0, x_1, ..., x_n\})$ of $[a, b]$ such that

  (i)   each u_i occurs in P

  (ii)   each v_j occurs in P

  (iii)   no points of any segment (u_i, v_j) occurs in P

  (iv)   if x_{i-1} is not one of the u_j, then $\Delta x_j < \delta$.

Now $\qquad M_i - m_i = |M_i - m_i|$

$$\le |M_i| + |m_i| \qquad\qquad [\because M_i \ge m_i]$$
$$\le 2|M_i|$$
$$\le 2 \sup |f(x)|, \qquad x \in [x_{i-1}, x_i]$$
$$\le 2 \sup |f(x)|, \qquad x \in [a, b]$$

$\implies \qquad M_i - m_i \le 2M$

$\implies \qquad M_i - m_i < \varepsilon$, unless x_{i-1} one of the u_i.

Therefore,

$$U(P, f, \alpha) - L(P, f, \alpha) = \sum_{i=1}^{n} (M_i - m_i) \Delta \alpha_i$$

$$\le [\alpha(b) - \alpha(a)]\varepsilon + 2M\varepsilon$$

Since, ε is arbitraiy, therefore $f \in RS(\alpha)$.

7.6 ALGEBRA OF RS-INTEGRABLE FUNCTIONS

THEOREM 1. *If f and g are RS-integrable on $[a,b]$, then their sum $f+g$ is also RS-integrable on $[a,b]$ and*

$$\int_a^b (f+g)\,d\alpha = \int_a^b f\,d\alpha + \int_a^b g\,d\alpha.$$

Proof. Given that the function f is RS-integrable over $[a,b]$.

Therefore $f \in RS(\alpha)$ on $[a,b]$

Similarly $g \in RS(\alpha)$ on $[a,b]$.

Now, $f \in RS(\alpha)$, implies there exists a partition P_1 such that

$$U(P_1, f, \alpha) - L(P_1, f, \alpha) < \varepsilon. \qquad\qquad ...(1)$$

Similarly $g \in RS(\alpha)$, implies that there exists a partition P_2 such that

$$U(P_2, g, \alpha) - L(P_2, g, \alpha) < \varepsilon \qquad\qquad ...(2)$$

Let P be the common refinement of P_1 and P_2, then (1) and (2) can be written as

$$U[P, f, \alpha] - L[P, f, \alpha] < \varepsilon. \qquad \ldots(3)$$

and $\qquad U[P, g, \alpha] - L[P, g, \alpha] < \varepsilon. \qquad \ldots(4)$

Adding (3) and (4), we get

$$[U(P, f, \alpha) + U(P, g, \alpha)] - [L(P, f, \alpha) + L(P, g, \alpha)] < 2\varepsilon. \qquad \ldots(5)$$

Now let $\qquad h(x) = f(x) + g(x).$

Also, let M_r, M_r', M_r'' be the l.u.b. of $h(x)$, $f(x)$ and $g(x)$ respectively and m_r, m_r', m_r'' be the g.l.b. of $h(x)$, $f(x)$ and $g(x)$ respectively where $x_{r-1} \le x \le x_r$.

For the rth interval, we have $f(x) + g(x) \le M_r' + M_r''$

or $\qquad M_r \le M_r' + M_r'' \qquad \ldots(6)$

Similarly $\qquad m_r \ge m_r' + m_r'' \qquad \ldots(7)$

Therefore, we have

$$U(P, h, \alpha) = \sum_{r=1}^{n} M_r \Delta\alpha_r \le \sum_{r=1}^{n} \left(M_r' + M_r''\right)\Delta\alpha_r$$

$$\Rightarrow \qquad U(P, h, \alpha) \le \sum_{r=1}^{n} M_r'\Delta\alpha_r + \sum_{r=1}^{n} M_r''\Delta\alpha_r$$

$$\Rightarrow \qquad U(P, h, \alpha) \le U(P, f, \alpha) + U(P, g, \alpha) \qquad \ldots(8)$$

Similarly, using (7), we have

$$L(P, h, \alpha) \ge L(P, f, \alpha) + L(P, g, \alpha) \qquad \ldots(9)$$

Therefore, from (5), we have

$$U(P, h, \alpha) - L(P, h, \alpha) < 2\varepsilon$$

$$\Rightarrow \qquad h \in RS(\alpha) \ i.e., \ f + g \in RS(\alpha) \qquad \ldots(10)$$

Now, for partition P, we have $U(P, f, \alpha) < \int_a^b f\, d\alpha + \varepsilon$ and $U(P, g, \alpha) < \int_a^b g\, d\alpha + \varepsilon$.

Adding these, we get

$$U(P, f, \alpha) + U(P, g, \alpha) < \int_a^b f\, d\alpha + \int_a^b g\, d\alpha + 2\varepsilon.$$

Therefore, from (8), we have

$$U(P, h, \alpha) < \int_a^b f\, d\alpha + \int_a^b g\, d\alpha + 2\varepsilon. \qquad \ldots(11)$$

Now, since $\qquad \int_a^b h\, d\alpha < \int_a^b f\, d\alpha + \int_a^b g\, d\alpha. \qquad \ldots(12)$

Replace f by $-f$, g by $-g$ and h by $-h$ in both the sides, we get

$$\int_a^b h\, d\alpha \ge \int_a^b f\, d\alpha + \int_a^b g\, d\alpha. \qquad \ldots(13)$$

Hence, from (12) and (13), we get

$$\int_a^b h\, d\alpha = \int_a^b f\, d\alpha + \int_a^b g\, d\alpha \ i.e., \int_a^b (f+g)\, d\alpha = \int_a^b f\, d\alpha + \int_a^b g\, d\alpha.$$

REMARK

• In a similar manner, we can show that, if f and g are RS-integrable on $[a,b]$, then f–g is also RS-integrable on $[a,b]$ and

$$\int_a^b (f - g)\, d\alpha = \int_a^b f\, d\alpha - \int_a^b g\, d\alpha.$$

THEOREM 2. *If $f \in RS(\alpha)$ on $[a,b]$, then $cf \in RS(\alpha)$ on $[a,b]$, where c is any constant.*

Also $\int_a^b cf\, d\alpha = c\int_a^b f\, d\alpha$.

Proof. Given that $f \in RS(\alpha)$, therefore, by definition of RS-integral there exists a partition P' on $[a,b]$ such that

$$U(P, f, \alpha) - L(P, f, \alpha) < \varepsilon. \qquad \qquad ...(1)$$

Since, for a constant c, $(cf)(x) = cf(x)$, therefore,

$$U(P, cf, \alpha) = cU(P, f, \alpha) \qquad \qquad ...(2)$$

and $\qquad \qquad L(P, cf, \alpha) = cL(P, f, \alpha).$

Thus, $U(P, cf, \alpha) - L(P, cf, \alpha) = c[U(P, f, \alpha) - L(P, f, \alpha)]$

$$< c.\varepsilon. \qquad \qquad \text{[By (1)]}$$

Since ε is arbitrary, therefore $cf \in RS(\alpha)$

Again $\qquad \qquad U(P, cf, \alpha) < \int_a^b cf\, d\alpha + \varepsilon \qquad \qquad ...(3)$

and $\qquad \qquad U(P, f, \alpha) < \int_a^b f\, d\alpha + \varepsilon. \qquad \qquad ...(4)$

Using (2) and (4) we get

$$U(P, cf, \alpha) = cU[P, f, \alpha] < c\int_a^b f\, d\alpha + \varepsilon.c$$

$$\Rightarrow \qquad c\int_a^b f\, d\alpha \geq \int_a^b cf\, d\alpha \qquad \qquad ...(5)$$

Replaced f by $-f$ in (5), we get

$$c\int_a^b f\, d\alpha \leq \int_a^b cf\, d\alpha \qquad \qquad ...(6)$$

From (5) and (6), we conclude that

$$\int_a^b cf\, d\alpha = c\int_a^b f\, d\alpha$$

THEOREM 3. *Let f be RS-integrable function on $[a,b]$, then $|f|$ is also RS-integrabale on $[a,b]$ and*

$$\left|\int_a^b f\, d\alpha\right| \leq \int_a^b |f|\, d\alpha.$$

Proof. Since f is RS-integrable on $[a,b]$, therefore, $f \in RS(\alpha)$.

Then by definition $\exists$ a partition P of $[a,b]$ such that $U(P, f, \alpha) - L(P, f, \alpha) < \varepsilon$

where ε is an arbitirary positive real numbers

$$\Rightarrow \qquad \sum_{r=1}^{n} M_r \Delta\alpha_r - \sum_{r=1}^{n} m_r \Delta\alpha_r < \varepsilon$$

i.e., $\qquad \qquad \sum_{r=1}^{n} (M_r - m_r)\Delta\alpha_r < \varepsilon \qquad \qquad ...(1)$

where M_r and m_r respectively denotes the l.u.b.and g.l.b. of $f(x)$,

$$x_{r-1} \leq x \leq x_r \quad \text{and} \quad \Delta\alpha_r = \alpha(x_r) - \alpha(x_{r-1}).$$

Now, for the partition P of $[a,b]$, let M_r' and m_r' be respectively the l.u.b. and g.l.b. of $|f(x)|$, where $x_{r-1} \leq x \leq x_r$.

If t_1 and t_2 are any two points in Δx_r then

$$\left| |f(t_2)| - |f(t_1)| \right| \leq \left| f(t_2) - f(t_1) \right|.$$

Therefore $$M_r' - m_r' < M_r - m_r$$

i.e., $$\sum_{r=1}^{n} \left(M_r' - m_r'\right)\Delta\alpha_r \le \sum_{r=1}^{n} \left(M_r - m_r\right)\Delta\alpha_r < \varepsilon \qquad \text{[Using (1)]}$$

Hence, $$\sum_{r=1}^{n} \left(M_r' - m_r'\right)\Delta\alpha_r < \varepsilon$$

i.e., $$U(P,|f|,\alpha) - L(P,|f|,\alpha) < \varepsilon$$

$\Rightarrow \qquad |f| \in RS(\alpha) \quad$ i.e., $|f|$ is RS-integrable.

Again as M_r cannot be greater than M_r', therefore,

$$\left| \sum_{r=1}^{n} M_r \Delta\alpha_r \right| \le \sum_{r=1}^{n} M_r' \Delta\alpha_r$$

Therefore, $$\left| \int_a^b f\, d\alpha \right| \le \int_a^b |f|\, d\alpha$$

REMARK

• Converse of the above theorem is not true. For example

Let $$f(x) = \begin{cases} -1 \ , & \text{when } x \text{ is irrational} \\ 1 \ , & \text{when } x \text{ is rational} \end{cases}$$

Here, $\int_a^b |f|\, dx$ exists even if $\int_a^b f(x)\, dx$ does not exist.

THEOREM 4. *If $f(x) \le g(x)$ on $[a,b]$, then $\int_a^b f\, d\alpha \le \int_a^b g\, d\alpha$.*

Proof. Given that $$g(x) \ge f(x) \text{ on } [a,b]$$

$$g(x) - f(x) \ge 0 \text{ on } [a,b]. \qquad \qquad ...(1)$$

Since α is monotonically increasing in $[a,b]$ and so $\alpha(b) > \alpha(a)$, then from (1), we

have $$\int_a^b \left[g(x) - f(x)\right]d\alpha \ge 0$$

$\Rightarrow \qquad \int_a^b (g - f)d\alpha \ge 0$

$\Rightarrow \qquad \int_a^b g\, d\alpha - \int_a^b f\, d\alpha \ge 0 \text{ i.e., } \int_a^b g\, d\alpha \ge \int_a^b f\, d\alpha.$

Hence, $$\int_a^b f\, d\alpha \le \int_a^b g\, d\alpha.$$

THEOREM 5. *If $f \in RS(\alpha_1)$ and $f \in RS(\alpha_2)$, then $f \in RS(\alpha_1 + \alpha_2)$ and*

$$\int_a^b f d(\alpha_1 + \alpha_2) = \int_a^b f d\alpha_1 + \int_a^b f d\alpha_2 .$$

Proof. Since $f \in RS(\alpha_1)$, therefore $\exists$ a partition P_1 such that

$$U(P_1, f, \alpha_1) - L(P_1, f, \alpha_1) < \varepsilon/2 \qquad \qquad ...(1)$$

and $f \in RS(\alpha_2) \Rightarrow \exists$ a partition P_2 such that

$$U(P_2, f, \alpha_2) - L(P_2, f, \alpha_2) < \varepsilon/2. \qquad \qquad ...(2)$$

Let P be the common refinement of P_1 and P_2, therefore (1) and (2) also holds for (1) and (2).

$\therefore \qquad U(P, f, \alpha_1) - L(P, f, \alpha_1) < \varepsilon/2 \qquad \qquad ...(3)$

and $\qquad U(P, f, \alpha_2) - L(P, f, \alpha_2) < \varepsilon/2.$ ...(4)

Now, consider

$$\sum_{r=1}^{n} M_r \left[\alpha(x_r) - \alpha(x_{r-1}) \right] = \sum_{r=1}^{n} M_r \left[(\alpha_1 + \alpha_2)(x_r) - (\alpha_1 + \alpha_2)(x_{r-1}) \right]$$

where $\alpha = \alpha_1 + \alpha_2$

$$= \sum_{r=1}^{n} M_r \left[\alpha_1(x_r) + \alpha_2(x_r) - \alpha_1(x_{r-1}) - \alpha_2(x_{r-1}) \right]$$

$$= \sum_{r=1}^{n} M_r \left[\{ \alpha_1(x_r) - \alpha_1(x_{r-1}) \} + \{ \alpha_2(x_r) - \alpha_2(x_{r-1}) \} \right]$$

$$= \sum_{r=1}^{n} M_r \left[\alpha_1(x_r) - \alpha_1(x_{r-1}) \right] + \sum_{r=1}^{n} M_r \left[\alpha_2(x_r) - \alpha_2(x_{r-1}) \right]$$

Therefore, $\qquad U[P, f, \alpha] = U[P, f, \alpha_1] + U[P, f, \alpha_2].$ ...(5)

Similarly, we obtain

$$L(P, f, \alpha) = L(P, f, \alpha_1) + L(P, f, \alpha_2) \qquad \qquad ...(6)$$

From (4) and (5), we get

$$U(P, f, \alpha) - L(P, f, \alpha) = [U(P, f, \alpha_1) - L(P, f, \alpha_1)] + [U(P, f, \alpha_2) - L(P, f, \alpha_2)] + \varepsilon/2 + \varepsilon/2$$

[Using (3) and (4)]

$\Rightarrow \qquad\qquad U(P, f, \alpha) - L(P, f, \alpha) < \varepsilon.$

Hence, $\qquad f \in RS(\alpha)$, where $\alpha = \alpha_1 + \alpha_2$

$\Rightarrow \qquad\qquad f \in RS(\alpha_1 + \alpha_2).$

Now, from (5), we have

$\qquad$ g.l.b. $U(P, f, \alpha)$ = g.l.b. $[U(P, f, \alpha_1) + U(P, f, \alpha_2)]$

or $\qquad$ g.l.b. $U(P, f, \alpha) \geq$ g.l.b. $U(P, f, \alpha_1)$ + g.l.b. $U(P, f, \alpha_2)$

or $\qquad \int_a^b f \, d\alpha \geq \int_a^b f \, d\alpha_1 + \int_a^b f \, d\alpha_2$ ...(7)

Similarly, from (6), we can find

$\qquad$ l.u.b. $L(P, f, \alpha)$ = l.u.b. $L(P, f, \alpha_1)$ + l.u.b. $L(P, f, \alpha_2)$

$\Rightarrow \qquad \int_a^b f d\alpha \leq \int_a^b f \, d\alpha_1 + \int_a^b f \, d\alpha_2.$ ...(8)

Hence, from (7) and (8), we conclude that

$$\int_a^b d\alpha = \int_a^b f \, d\alpha_1 + \int_a^b d\alpha_2 \,.$$

THEOREM 6. *If $f \in RS(\alpha)$ and c is any positive constant, then $f \in RS(c\alpha)$ and*

$$\int_a^b f d(\alpha) = c \int_a^b f \, d\alpha$$

Proof. Since $f \in RS(\alpha)$, therefore, there exists a partition P such that

$$U(P, f, \alpha) - L(P, f, \alpha) < \varepsilon$$

$\Rightarrow \qquad \sum_{r=1}^{n} M_r \Delta\alpha_r - \sum_{r=1}^{n} m_r \Delta\alpha_r < \varepsilon$

$\Rightarrow \qquad c \left[\sum_{r=1}^{n} M_r \{ \alpha(x_r) - \alpha(x_{r-1}) \} - \sum_{r=1}^{n} m_r \{ \alpha(x_r) - \alpha(x_{r-1}) \} \right] < c.\varepsilon$

$$\Rightarrow \quad \sum_{r=1}^{n} M_r \{c\alpha(x_r) - c\alpha(x_{r-1})\} - \sum_{r=1}^{n} m_r \{c\alpha(x_r) - c\alpha(x_{r-1})\} < \varepsilon'$$

$$\Rightarrow \qquad \sum_{r=1}^{n} M_r \Delta(c\alpha_r) - \sum_{r=1}^{n} m_r \Delta(c\alpha_r) < \varepsilon'$$

$$\Rightarrow \qquad U(P, f, c\alpha) - L(P, f, c\alpha) < \varepsilon'$$

Therefore, $f \in RS(c\alpha)$.

Also, $U(P, f, c\alpha) = \sum_{r=1}^{n} M_r \Delta(c\alpha_r)$

$$= \sum_{r=1}^{n} M_r [c\alpha(x_r) - c\alpha(x_{r-1})]$$

$$= \sum_{r=1}^{n} M_r.c [\alpha(x_r) - \alpha(x_{r-1})]$$

$$= c \sum_{r=1}^{n} M_r [\alpha(x_r) - \alpha(x_{r-1})] = c \sum_{r=1}^{n} M_r.\Delta\alpha$$

$$= c. \, U(P, f, \alpha)$$

Therefore, $\int_a^b f \, d\alpha = \text{g.l.b.} U(P, f, c\alpha) = c[\text{g.l.b.} U(P, f, \alpha)] = c\int_a^b f \, d\alpha.$

THEOREM 7. *Let $f \in RS(\alpha)$ on $[a,b]$ and $m \le f \le M$. If ϕ is continuous on $[m,M]$ and $g(x) = \phi[f(x)]$ on $[a, b]$, then $g \in RS(\alpha)$ on $[a, b]$.*

Proof.   Since ϕ is continuous on a closed interval $[m,M]$ therefore ϕ is uniformly continuous on $[m, M]$ [$\because$ Every continuous function in closed and bounded interval is uniformly continuous.]

Therefore, by definition $\exists$ a positive number $\delta > 0$ ($\delta < \varepsilon$) such that

$$|\phi(y) - \phi(z)| < \varepsilon, \text{ if } |y - z| < \delta \text{ and } y, z \in (m, M).$$

Again, if $f \in RS(\alpha) \Rightarrow \exists$ a partition P of $[a,b]$ such that

$$U(P, f, \alpha) - L(P, f, \alpha) < \varepsilon = \delta^2 \qquad \qquad \qquad ...(1)$$

Let M_r, m_r be respectively l.u.b. and g.l.b. of $f(x)$ and M_r', m_r' be respectively l.u.b. and g.l.b. of $g(x)$, where $x_{r-1} \le x \le x_r$.

Now, let the set $\{1,2,3,...,n\}$ be partitioned into two classes A and B such that

$$M_r - m_r < \delta \Rightarrow r \in A$$

and $M_r - m_r \ge \delta \Rightarrow r \in B$

Then, if $r \in A$, then $M_r - m_r < \delta \quad \Rightarrow \quad |\phi(M_r) - \phi(m_r)| \le \varepsilon$

$$\Rightarrow \qquad |\phi[\text{l.u.b.} \, f(x)] - \phi[\text{g.l.b.} \, f(x)]| \le \varepsilon$$

$$\Rightarrow \qquad |\text{l.u.b.} \, \phi[f(x)] - \text{g.l.b.} \, \phi[f(x)]| \le \varepsilon$$

$$\Rightarrow \qquad |\text{l.u.b.} \, g(x) - \text{g.l.b.} \, g(x)| \le \varepsilon \qquad \qquad [\because \phi(f(x)) = g(x)]$$

$$\Rightarrow \qquad |M_r' - m_r'| \le \varepsilon$$

$$\Rightarrow \qquad M_r' - m_r' \le \varepsilon \text{ since } M_r' > m_r'. \qquad \qquad ...(2)$$

And if $r \in B$, then $M_r - m_r > \delta$

$\Rightarrow$ $\qquad M_r' - m_r' \le 2\lambda$ $\qquad\qquad$ ...(3)

where $\quad \lambda = $ l.u.b. $|\phi(x)|$, $m \le u \le M$.

From (1), we have

$$\sum_{r \in B} (M_r - m_r) \Delta\alpha_r < \delta^2$$

$\Rightarrow \qquad\qquad \sum_{r \in B} \delta\Delta\alpha_r < \delta^2 \qquad\qquad [\because M_r - m_r > \delta \text{ if } r \in B]$

$\Rightarrow \qquad\qquad \sum_{r \in B} \Delta\alpha_r < \delta$. $\qquad\qquad$ ...(4)

Therefore. $U(P, g, \alpha) - L(P, g, \alpha)$

$$= \sum_{r \in A} (M_r' - m_r') \Delta\alpha_r + \sum_{r \in B} (M_r' - m_r') \Delta\alpha_r$$

$$\le \varepsilon \sum_{r \in A} \Delta\alpha_r + 2\lambda \sum_{r \in B} \Delta\alpha_r \qquad\qquad \text{[From (2) and (3)]}$$

$$\le \varepsilon[\alpha(b) - \alpha(a)] + 2\lambda.\delta \qquad\qquad \text{[From (4)]}$$

$$\le \varepsilon[\alpha(b) - \alpha(a) + 2\lambda] \qquad\qquad [\because \delta < \varepsilon]$$

Since ε is arbitrary, therefore, $g \in RS(\alpha)$.

THEOREM 8. *If $f \in RS(\alpha)$ and $g \in RS(\alpha)$ on $[a, b]$, then $f.g \in RS(\alpha)$ on $[a, b]$.*

Proof. $\qquad$ We have $\qquad\qquad\qquad f \in RS(\alpha)$ on $[a, b]$

$\Rightarrow \qquad\qquad\qquad \phi[f(x)] \in RS(\alpha)$ on $[a,b]$ [By previous theorem] $\qquad$ ...(1)

where $m \le f \le M$ and ϕ is continuous on $[m, M]$.

Let $\quad \phi(z) = z^2$ i.e., $\phi(f(x)) = [f(x)]^2$

Then $f \in RS(\alpha)$ on $[a,b] \Rightarrow \{f(x)\}^2 \in RS(\alpha)$ on $[a, b]$ $\qquad$ ...(2)

We also have $f \in RS(\alpha)$ $\quad$ and $\quad g \in RS(\alpha)$, then $f+g \in RS(\alpha)$ and $f-g \in RS(\alpha)$.

Now from (2), we have

$f \in RS(\alpha), g \in RS(\alpha)$ on $[a, b] \Rightarrow (f+g)^2 \in RS(\alpha)$ and $(f-g)^2 \in RS(\alpha)$ on $[a, b]$

$\qquad\qquad\qquad \Rightarrow [(f+g)^2 - (f-g)^2] \in RS(\alpha)$ on $[a, b]$

$\qquad\qquad\qquad \Rightarrow 4f. g \in RS(\alpha)$ on $[a, b]$

$\qquad\qquad\qquad \Rightarrow f. g \in RS(\alpha)$ on $[a, b]$.

THEOREM 9. (First mean value theorem). *If $f(x)$ is continuous, real and $\alpha(x)$ is monotonically increasing on $[a, b]$ then $\exists$ a point t such that*

$$\int_a^b d\alpha = f(t)[\alpha(b) - \alpha(a)], \text{ where } a \le t \le b.$$

Proof. $\qquad$ Let $m = $ g.l.b. $f(x)$ and $M = $ l.u.b. $f(x)$ where $a \le x \le b$.

Then $\qquad\qquad m \le f(x) \le M \; \forall \, x \in [a, b] \qquad$ (By definition of l.u.b. and g.l.b.)

Therefore $\quad m \int_a^b d\alpha \le \int_a^b f \, d\alpha \le M \int_a^b d\alpha$

$\Rightarrow \quad m[\alpha(b) - \alpha(a)] \le \int_a^b f \, d\alpha \le M[\alpha(b) - \alpha(a)]$

Hence, there exists a number λ, $m \le x \le M$ such that

$$\int_a^b f \, d\alpha = \lambda[\alpha(b) - \alpha(a)] \qquad\qquad \text{...(1)}$$

Since, f takes all values between m and M on $[a,b]$.   [$\because f$ is continuous.]

Therefore, $\exists$ a point t on $[a,b]$ for which $f(t)=\lambda$.

Therefore, from (1), we have

$$\int_a^b f\,d\alpha = f(t)[\alpha(b)-\alpha(a)]$$

7.7   RELATION BETWEEN RIEMANN INTEGRAL AND RIEMANN-STIELTJES INTEGRAL

We know that Riemann integral is a special case of Riemann Stieltjes integral. The following theorems shows that under certain conditions, a simple relationship exists between them.

THEOREM 1.   *Let f be R-integrable on $[a, b]$ and let α be a monotonically non-decreasing function on $[a, b]$ such that if derivative α' is R-integrable on $[a, b]$, then $f \in RS(\alpha)$ on $[a,b]$ and*

$$\int_a^b f(x)\alpha'(x)dx$$

i.e., RS-integral of f relative to α on $[a,b]$ is equal to R-integral of $f\alpha'$ on $[a, b]$.

Proof.   Since $f \in R[a, b]$, $\alpha' \in R[a, b]$ so $f\alpha' \in R[a, b]$.

Also $f \in R[a, b] \quad \Rightarrow f$ is bounded on $[a, b]$, therefore, for some constant $M > 0$, we have $|f| \le M$.   ...(1)

Let $\varepsilon > 0$ be given. Since $f\alpha'$ is R-integrable, there exists $\delta_1 > 0$ such that for all partitions P, with $\|P\| < \delta_1$ and for all t_r with $x_{r-1} < t_r < x_r$, we have

$$\left|\Sigma f(t_i)\alpha'(t_r)\Delta x_r - \int_a^b f\alpha'\right| < \varepsilon. \qquad \ldots(2)$$

Again, since α' is R-integrable, $\exists\ \delta_2 > 0$ such that for all partitions P with $\|P\| < \delta$ and $t_r \in [x_{r-1}, x_r]$, we have

$$\left|\Sigma \alpha'(t_r)\Delta x_r - \int_a^b \alpha'\right| < \varepsilon. \qquad \ldots(3)$$

Varying t_r in (3), it is easy to see that

$$\Sigma\left|\alpha'(t_r) - \alpha'(\eta_r)\right|\Delta x_r < 2\varepsilon \qquad \ldots(4)$$

if $\|P\| < \delta_2$ and $t_r, \eta_r \in [x_{r-1}, x_r]$.

Choose $\delta = \min\{\delta_1, \delta_2\}$. Let P be the partition with $\|P\| < \delta$ then using Lagrange's mean value theorem of differential calculus $\exists\ \eta_r \in (x_r, x_{r-1})$ such that

$$\Delta\alpha_r = \alpha(x_r) - \alpha(x_{r-1}) = \alpha'(\eta_r)(x_r - x_{r-1}) = \alpha'(\eta_r)\Delta x_r$$

Therefore, we have

$$\begin{aligned}\Sigma f(t_r)\Delta\alpha_r &= \Sigma f(t_r)\alpha'(\eta_r)\Delta x_r \\ &= \Sigma f(t_r)\left[\alpha'(\eta_r) - \alpha'(t_r)\right]\Delta x_r + \Sigma f(t_r)\alpha'(t_r)\Delta x_r.\end{aligned}$$

Hence,

$$\begin{aligned}\left|\Sigma f(t_r)\Delta\alpha_r - \int_a^b f\alpha'\right| &= \left|\Sigma f(t_r)\left[\alpha'(\eta_r) - \alpha'(t_r)\right]\Delta x_r + \Sigma f(t_r)\alpha'(t_r)\Delta x_r - \int_a^b f\alpha'\right| \\ &\le \Sigma\left|f(t_r)\right|\left|\alpha'(\eta_r) - \alpha'(t_r)\right|\Delta x_r + \left|\Sigma f(t_r)\alpha'(t_r)\Delta x_r - \int_a^b f\alpha'\right|\end{aligned}$$

$$< M.2\varepsilon+\varepsilon \qquad\qquad\qquad \text{[By (1), (2) and (4)]}$$
$$= (2M+1).\varepsilon$$

Therefore,

$$\lim_{\|P\|\to 0}\Sigma f(t_r)\Delta\alpha_r = \int_a^b f\alpha' \qquad\qquad\qquad ...(5)$$

Since $\int_a^b f\alpha'$ exists $\Rightarrow \lim\Sigma f(t_r)\Delta\alpha_r$ exists,

it follows that $\lim_{\|P\|\to 0}\Sigma f(t_r)\Delta\alpha_r = \int_a^b fd\alpha \qquad\qquad ...(6)$

From (5) and (6), we conclude that

$$\int_a^b fd\alpha = \int_a^b f(x)\alpha'(x)dx$$

Solved Examples

Example 1.   Let f be a constant function on $[a,b]$ defined by $f(x)=k$ and α is a monotonically increasing function on $[a,b]$, then show that $\int_a^b fd\alpha$ exists and

$$\int_a^b fd\alpha' = k\big[\alpha(b)-\alpha(a)\big].$$

Solution.   Consider a partition P on $[a, b]$ such that $P = \{a = x_0, x_1,.... x_n = b\}$.

Then $\qquad\qquad M_i = \underset{x\in[x_{i-1},x_i]}{\text{l.u.b.}} f(x) = k$, $i=1, 2,.., n$

$$m_i = \underset{x\in[x_{i-1},x_i]}{\text{g.l.b.}} f(x) = k$$, $i = 1, 2,..., n$

Therefore, $\qquad \int_{\underline{a}}^b fd\alpha = \text{l.u.b.}\ L(P,f,\alpha) = \text{l.u.b.}\sum_{i=1}^n k\Delta\alpha_i$

$$= \text{l.u.b.}\ k\sum_{i=1}^n \big[\alpha(x_i)-\alpha(x_{i-1})\big]$$

$$= \text{l.u.b.}\ k\big[\alpha(x_1)-\alpha(x_0)+\alpha(x_2)-\alpha(x_1)+...+\alpha(x_n)-\alpha(x_{n-1})\big]$$
$$= \text{l.u.b.}\ k\big[\alpha(x_n)-\alpha(x_0)\big] = k\big[\alpha(b)-\alpha(a)\big]$$

Similarly, we can find

$$\int_a^{\overline{b}} fd\alpha = \text{g.l.b.}\ U(P,f,\alpha)$$

$$= \text{g.l..b.}\sum_{i=1}^n k\Delta\alpha_i = k\big[\alpha(b)-\alpha(a)\big].$$

Since, $\qquad \int_{\underline{a}}^b fd\alpha = \int_a^{\overline{b}} fd\alpha$, the integral $\int_a^b fd\alpha$ exists and

$$\int_a^b f\,d\alpha = k\ [\alpha(b)-\alpha(a)].$$

Example 2.   Show that if α is an increasing function on $[a,b]$ and if f is non-negative integral function w.r.t. α on $[a, b]$, then $\int_a^b fd\alpha \geq 0$.

Solution.   Let m be the infimum of f on $[a, b]$.

Also, since $f(x) \geq 0$, $\forall\, x \in [a, b]$ therefore, clearly $m \geq 0$

Now $\qquad\qquad \int_a^b f\,d\alpha \geq m\big[\alpha(b) - \alpha(a)\big]$ ...(1)

Since α is an increasing function on $[a, b]$, therefore $\alpha(b) - \alpha(a) \geq 0$

Now $m \geq 0$ and $\alpha(b) - \alpha(a) \geq 0 \Rightarrow m[\alpha(b) - \alpha(a)] \geq 0.$

Hence, from (1), we conclude that $\int_a^b f\,d\alpha \geq 0$.

Example 3. *Let $f(x) = x$ and $\alpha(x) = x^2$. Does $\int_0^1 f\,d\alpha$ exist? If it exists, find its value.*

Solution. Clearly, we have f is continuous and α is monotonically increasing on $[0,1]$,

therefore $\int_0^1 f\,d\alpha$ exists.

Now consider the partition

$$P = \left\{0, \frac{1}{n}, \frac{2}{n}, ..., \frac{r}{n}, ..., \frac{n}{n} = 1\right\}.$$

Let us take

$$t_r = \frac{r}{n} \text{ in } \left[\frac{r-1}{n}, \frac{r}{n}\right], r = 1, 2, ...,n$$

Then, $\qquad U(P, f, \alpha) = \displaystyle\sum_{r=1}^{n} f\left(\frac{r}{n}\right)\left[\alpha\left(\frac{r}{n}\right) - \alpha\left(\frac{r-1}{n}\right)\right]$

$$= \sum_{r=1}^{n} \frac{r}{n}\left[\frac{r^2}{n^2} - \frac{(r-1)^2}{n^2}\right] = \frac{1}{n^3}\sum_{r=1}^{n}\left(2r^2 - r\right)$$

$$= \frac{2}{n^3}\sum_{r=1}^{n} r^2 - \frac{1}{n^3}\sum_{r=1}^{n} r$$

$$= \frac{2}{n^3}\cdot\frac{n(n+1)(2n+1)}{6} - \frac{1}{n^3}\cdot\frac{n(n+1)}{2}$$

$$= \frac{1}{6n^2}\left[2\left(2n^2 + 3n + 1\right) - 3n - 3\right]$$

$$= \frac{4n^2 + 3n - 1}{6n^2} = \frac{1}{6}\left(4 + \frac{3}{n} - \frac{1}{n^2}\right).$$

Hence, $\qquad \int_0^1 f\,d\alpha = \displaystyle\lim_{\|P\|\to 0} S(P, f, \alpha) = \lim_{n\to\infty}\frac{1}{6}\left[4 + \frac{3}{n} - \frac{1}{n^2}\right]$

$$= \frac{1}{6}[4 + 0 - 0] = \frac{2}{3}.$$

Example 4. *Let α be increasing and f is non-negative and integrable function relative to α on $[a, b]$. Show that if $a \leq c \leq d$, then*

$$\int_c^d f\,d\alpha \leq \int_a^b f\,d\alpha.$$

Solution. Here

$$\int_a^b f\,d\alpha = \int_a^c f\,d\alpha + \int_c^d f\,d\alpha + \int_d^b f\,d\alpha \qquad\qquad ...(1)$$

Now, since f is non-negative on $[a, b]$, therefore, f is non-negative on $[a, c]$ and $[d, b]$.

$\therefore$ $\int_a^c f \, d\alpha \geq 0$ and $\int_d^b f \, d\alpha \geq 0$.

Hence, from (1), we have

$$\int_a^b f \, d\alpha \geq \int_c^d f \, d\alpha \text{ i.e., } \int_c^d f \, d\alpha \leq \int_a^b f \, d\alpha \, .$$

Example 5. *Let $I = [0,1]$ and let f, $\alpha : I \to R$ be the function such that $f(x) = \alpha(x) = x^2$. Then find the value of $\int_0^1 x^2 \, dx^2$.*

Solution. Let P be any partition of $I = [0, 1]$ such that $P = \left\{ 0, \dfrac{1}{n}, \dfrac{2}{n}, \dots, \dfrac{n}{n} = 1 \right\}$

Also, let $I_r = \left[\dfrac{(r-1)}{n}, \dfrac{r}{n} \right]$ be the rth subinterval of $[0,1]$.

Then, $\qquad M_r = \sup_{x \in I_r} f(x) = \dfrac{r^2}{n^2}$

and $\qquad m_r = \inf_{x \in I_r} f(x) = \dfrac{(r-1)^2}{n^2}$

$$\Delta \alpha_r = \alpha(x_r) - \alpha(x_{r-1})$$

$$= \frac{r^2}{n^2} - \frac{(r-1)^2}{n^2} = \frac{2r-1}{n^2}$$

Now, $\qquad U(P, f, \alpha) = \displaystyle\sum_{r=1}^{n} M_r \Delta \alpha_r = \sum_{r=1}^{n} \frac{r^2}{n^2} \cdot \frac{(2r-1)}{n^2}$

$$= \frac{1}{n^4} \left\{ 2 . \sum_{r=1}^{n} r^3 - \sum_{r=1}^{n} r^2 \right\}$$

$$= \frac{1}{n^4} \left[2 . \frac{n^2 . (n+1)^2}{4} - \frac{n(n+1)(2n+1)}{6} \right]$$

$$= \frac{3n^3 + 4n^2 - 1}{6n^3}$$

Therefore, $\displaystyle \overline{\int_0^1} f \, d\alpha = \inf U(P, f, \alpha) = \lim_{n \to \infty} \frac{3n^3 + 4n^2 - 1}{6n^3} = \frac{1}{2}$

Similarly, $\quad L(P, f, \alpha) = \displaystyle\sum_{r=1}^{n} m_r \Delta \alpha_r = \sum_{r=1}^{n} \left[\frac{(r-1)^2}{n^2} \cdot \frac{(2r-1)}{n^2} \right]$

$$= \frac{1}{n^4} \left\{ 2 . \sum_{r=1}^{n} (r-1)^3 + \sum_{r=1}^{n} (r-1)^2 \right\}$$

$$= \frac{1}{n^4} \left[2 . \frac{(n-1)^2}{4} . n^2 + \frac{(n-1).n.(2n-1)}{6} \right] = \frac{3n^3 - 4n^2 + 1}{6n^3} \, .$$

$$\Rightarrow \qquad \int_{\underline{0}}^{1} f\,d\alpha = \sup L(P,f,\alpha) = \lim_{n\to\infty} \frac{3n^3 - 4n^2 + 1}{6n^3} = \frac{1}{2}$$

which gives $\int_{\underline{0}}^{1} f\,d\alpha = \overline{\int_0^1} f\,d\alpha = \frac{1}{2}$

Hence, $\int_0^1 f\,d\alpha$ exists and $\int_0^1 x^2\,dx^2 = \frac{1}{2}$.

Example 6. Find the value of $\int_1^2 x^3 d\left(|x|^5\right)$

Solution. Here, we have

$$\int_1^2 x^3 d\left(|x|^5\right) = \int_1^0 x^3 d(-x)^5 + \int_0^2 x^3 d(x)^5 = -5\int_{\underline{1}}^0 x^7\,dx + 5\int_0^2 x^7\,dx$$

$$= \frac{5}{8} + 160$$

Example 7. Evaluate the following integrals:

(i) $\int_0^2 x^2 d(x)^2$ 　　　　(ii) $\int_0^2 [x]\,dx^2$

Solution. (i) We have

$$\int_0^2 x^2 d(x)^2 = \int_0^2 x^2 . 2x\,dx \qquad \left[\because \int_a^b f\,d\alpha = \int_a^b f(x)\alpha'(x)\,dx\right]$$

$$= \int_0^2 2x^3\,dx = 2.\frac{1}{4}\left[x^4\right]_0^2 = \frac{1}{2}.16 = 8 .$$

(ii) We have $\int_0^2 [x]\,dx^2 = \int_0^2 [x].2x\,dx$

$$= 2\int_0^1 [x].x\,dx + 2\int_1^2 [x].x\,dx$$

$$= 2\int_0^1 0.x\,dx + 2\int_1^2 1.x\,dx \qquad [\because [x]=0 \text{ when } 0\le x\le 1$$
$$\text{and } [x]=1 \text{ when } 1\le x\le 2]$$

$$= 2\int_1^2 x\,dx = 2.\frac{1}{2}\left[x^2\right]_1^2 = 4-1 = 3$$

Example 8. Let $\alpha : [a,b]\to R$ be monotonic increasing on $[a, b]$ and continuous at x', where $a \le x' \le b$. Define $f : [a, b]\to R$ as follows :

$$f(x)= \begin{cases} 0, & \text{when } x \ne x' \\ 1, & \text{when } x = x' \end{cases}$$

Show that $f \in RS(\alpha)$ on $[a, b]$ and $\int_a^b f\,d\alpha =0$.

Solution. Let $P=\{a=x_0, x_1,...,x_n= b\}$ be a partition of $[a, b]$ and let $x' \in [x_{k-1}, x_k]$.

Since α is continuous at x', therefore, for given $\varepsilon>0 \ \exists \ \delta>0$ such that
$$|\alpha(x)-\alpha(x')|<\varepsilon/2, \text{ whenever } |x-x'|<\delta.$$

Also, since, α is continuous on $[a, b]$, we have
$$\alpha(x)-\alpha(x')<\varepsilon/2, \text{ whenever } 0<x-x'<\delta.$$

Then $\qquad \alpha(x_k)-\alpha(x_{k-1})=[\alpha(x_k)-\alpha(x')]+[\alpha(x')-\alpha(x_{k-1})]<\varepsilon/2+\varepsilon/2=\varepsilon.$
$$\text{whenever } x_k-x_{k-1} < \varepsilon$$

Now, let $\|P\| <\delta$, then
$$U(P, f, \alpha)=\Delta\alpha_k=\alpha(x_k)-\alpha(x_{k-1})<\varepsilon$$

and $\qquad L(P, f, \alpha)=0$

Therefore, $\qquad \underline{\int_a^b} f\, d\alpha = \sup\{L(P, f, \alpha) : P \in P[a, b] \text{ and } \|P\| < \delta\}$

$\qquad\qquad\qquad\qquad = 0 \qquad\qquad\qquad\qquad\qquad [\because L(P, f, \alpha) = 0\ \forall\ P]$

Also, $\qquad \overline{\int_a^b} f\, d\alpha = \inf\{U(P, f, \alpha) : P \in P[a, b] \text{ and } \|P\| < \delta\}$

$\qquad\qquad\qquad\qquad = 0 \qquad\qquad\qquad [\because U(P, f, \alpha) < \varepsilon, \text{ and } \varepsilon \text{ is arbitrary.}]$

Thus, $\qquad \underline{\int_a^b} f\, d\alpha = \overline{\int_a^b} f\, d\alpha = 0$

Here, $f \in RS(\alpha)$ on $[a, b]$ and $\int_a^b f\, d\alpha = 0$.

Example 9. Show that $\int_0^3 x\, d([x]-x) = \dfrac{3}{2}$.

Solution . Here x and $([x]-x)$ are of bounded variation on $[0, 3]$ and x is also continuous. Then

$$\int_0^3 x\, d([x]-x) = \Big[x([x]-x)\Big]_0^3 - \int_0^3 [[x]-x]\, dx$$

$$= -\int_0^1 (-x)\, dx - \int_1^2 (1-x)\, dx - \int_2^3 (2-x)\, dx$$

$$= \frac{1}{2} - 1 + \left(\frac{2^2}{2} - \frac{1}{2}\right) - 2.1 + \left(\frac{3^2}{2} - \frac{2^2}{2}\right)$$

$$= \frac{3}{2}.$$

7.8 INTEGRATION AND DIFFERENTIATION

Here, we shall show that integration and differentiation are, in a certain sense, inverse problem.

THEOREM 1. Let $f \in RS(\alpha)$ on $[a, b]$. For $a \le x \le b$, let $F(x) = \int_a^x f(t)\, dt$.

Then F is continuous on $[a, b]$. Also, if f is continuous at a point x_0 of $[a, b]$, then F is differentiable at x_0 and $F'(x_0)=f(x_0)$.

Proof. Since F is RS-integrable, therefore, it is bounded. Then $\exists$ a real number M such that

$$|f(t)| \le M \text{ for } a \le t \le b.$$

If $a \le x \le y \le b$, then

$$|F(y) - F(x)| = \left|\int_a^y f(t)\, dt - \int_a^x f(t)\, dt\right|$$

$$= \left|\int_x^a f(t)\, dt + \int_a^y f(t)\, dt\right|$$

$$= \left|\int_x^y f(t)\, dt\right| \le \int_x^y |f(t)|\, dt$$

$$= M(y-x) \qquad\qquad\qquad ...(1)$$

For given $\varepsilon>0$, we have

$$|y-x| < \frac{\varepsilon}{M} \qquad \Rightarrow |F(y) - F(x)| < \varepsilon$$

$$\Rightarrow F \text{ is uniformly continuous on } [a, b]$$

$$\Rightarrow F \text{ is continuous.}$$

Now, suppose that f is continuous at a point x_0 of $[a, b]$. Given $\varepsilon>0$, we can choose $\delta>0$ such that whenever $a \le t \le b$ with $|t-x| < \delta$

$$\Rightarrow \qquad |f(t)-f(x_0)| < \varepsilon.$$

Therefore, if $x_0-\delta \le s \le x_0 \le t < x_0+\delta$ and $a \le s < t \le b$, we have

$$\left|\frac{F(t)-F(s)}{t-s} - f(x_0)\right| = \left|\frac{1}{t-s}\int_s^t f(u)\,du - f(x_0)\right|$$

$$= \left|\frac{1}{t-s}\int_s^t [f(u) - f(x_0)]\,du\right|$$

$$\le \frac{1}{t-s}\int_s^t |f(u) - f(x_0)|\,du$$

$$< \varepsilon \frac{1}{t-s}\int_s^t du = \varepsilon.$$

Hence, $\qquad\qquad\qquad\qquad F'(x_0) = f(x_0).$

EXERCISE 7.1

1. Let f be integrable on $[a, b]$ in Riemann-Stieltjes sense with respect to a function α, show that $|f|$ is also integrable in the same sense and that

$$\left|\int_a^b f\,d\alpha\right| \le \int_a^b |f|\,d\alpha.$$

2. If $\alpha(x) = [x]$ i.e., the greatest integer not exceeding x, then find the value of the following integrals:

   (i) $\int_0^1 x\,d\alpha$ (ii) $\int_0^2 x\,d\alpha$ (iii) $\int_0^2 e^{2x}\,d\alpha$

   (iv) $\int_0^3 e^{2x}\,d\alpha$ (v) $\int_0^4 e^{2x}\,d\alpha$ (vi) $\int_0^9 e^{-2x}\,d\alpha$

3. If f is monotonic on $[a, b]$ and α is a monotonic continuous function defined on $[a, b]$, then show that f is RS-integrable over $[a, b]$.

4. Find the values of the following integrals :

   (i) $\int_0^2 x^2\,d([x]-x)$

   (ii) $\int_0^2 x\,d([x]-x)$

   (iii) $\int_{-1}^1 e^x\,d\alpha$ if $\alpha(x) = |x|$

   (iv) $\int_{-1}^2 x^5\,d\alpha$ if $\alpha(x) = |x|^3$

5. If f is a real function on $[a, b]$ and

$$f(x) = \begin{cases} 0, & \text{if } x \text{ is rational} \\ 1, & \text{if } x \text{ is irrational} \end{cases}$$

   Show that $f \notin RS[a, b]$.

6. Let f be a bounded real function on $[a, b]$ and $f^2 \in RS[a, b]$. Does it follows that $f \in RS(a,b)$. Does the answer change if we assume that $f^3 \in RS[a, b]$?

7. If f and α are monotonic functions on $[a, b]$ and α is continuous, then show that $f \in RS(\alpha)$.

8. Let f be a real valued function on $[0, 1]$ and $f \in RS[c, 1]$ for every $c > 0$. Define

$$\int_0^1 f(x)\,dx = \lim_{c \to 0}\int_c^1 f(x)\,dx$$

   if the limit exists (and finite)

   (a) If $f \in RS[0, 1]$. Show that the definition agrees with the old one.

   (b) Construct a function f such that the above limits exists, although it fails to exist with $|f|$ in place of f.

9. Suppose $f \in RS(\alpha)$ and $\varepsilon > 0$. Prove that there exists a continuous function g on $[a, b]$ such that $\|f - g\|_2 < \varepsilon$

10. Suppose $f \in RS[a, b]$ for every $b > a$. When a is fixed, define

$$\int_a^\infty f(x)\,dx = \lim_{b \to \infty}\int_a^b f(x)\,dx$$

   if the limit exists (and is finite). Then, we say that integral on the left converges. It is also converges after f has been replaced by $|f|$, it is said to converge absolutely.

   If $f(x) \ge 0$ and f decreases monotonically on $[1, \infty]$, show that $\int_1^\infty f(x)\,dx$ converges if and only if $\sum_{n=1}^\infty f(x)$ converges.

11. If $f(x)$ is bounded and integrable in $[a, b]$ and if $F(x) = \int_a^b f(t)\,dt$ then prove that $F(x)$ is continuous in $]a, b[$.

CHAPTER REVIEW: A COMPETITIVE APPROACH

SELECTED TERMS AND RESULTS

TERMS

- **Partition:** Let $I=[a,b]$ be a closed and bounded interval. Then a finite set of points $P=[x_0,x_1,...,x_n]$ such that
$$a=x_0<x_1<x_2<...<x_{n-1}<x_n=b$$
is called a partition of the interval $[a,b]$.

- **Segment of partition:** The closed subintervals $I_1=[a=x_0,x_1]$, $I_2=[x_1,\ x_2]$ ... $I_n=[x_{n-1},x_n=b]$ are called segments of the partition.

- **Norm of a partition:** The norm of the partition P is the maximum of the length of the segments of a partition P, denoted by $\|P\|$ defined by
$$\|P\|=\max\{\Delta x_r, r=1,2,...,n\}.$$

- **Refinement of a partition:** A partition P^* is called a refinement of a closed and bounded interval $[a,b]$ then
$$P^*=P_1\cup P_2$$
is called common refinement of P_1 and P_2.

- **Riemann-Stieltjes Integral:** The g.l.b. of all upper R-S sums $U[P,f,\alpha]$ is known as upper Riemann-Stieltjes integral w.r.t. α over $[a,b]$ and is denoted by $\overline{\int_a^b} f\, dx.$.

RESULTS

- *Necessary and sufficient condition for RS-integrability.*
Let f be a bounded function and α be a monotonically increasing function on $[a,b]$. Then $f\in RS(\alpha)$ if and only if for every $\varepsilon>0$ there exists a partitions P of $[a,b]$ such that
$$U(P,f,\alpha)-L(P,f,\alpha)<\varepsilon.$$

- Let f be continuous and monotonically increasing on $[a,b]$. Then $f\in RS$ on $[a,b]$.

- If f,g are RS-integrable on $[a,b]$ then their sum $f+g$ is also RS-integrable on $[a,b]$ and
$$\int_a^b(f+g)d\alpha=\int_a^b f\, d\alpha+\int_a^b g\, d\alpha.$$

- If $f\in RS(\alpha)$ on $[a,b]$ then $cf\in RS(\alpha)$ on $[a,b]$, where c is any constant then
$$\int_a^b cf\, d\alpha=c\int_a^b f\, d\alpha.$$

- Let f be RS-integrable function on $[a,b]$ then $|f|$ is also RS-integrable on $[a,b]$ then
$$\left|\int_a^b f\, d\alpha\right|\le\int_a^b|f|\, d\alpha.$$

- *First Mean Value Theorem:*
If $f(x)$ is continuous, real and $\alpha(x)$ is monotonically increasing on $[a,b]$ then $\exists$ a point x such that
$$\int_a^b f\, d\alpha=f(t)[\alpha(b)-\alpha(a)] \text{ where } a\le t\le b.$$

- *Fundamental Theorem of Integral Calculus:*
Let f be R-integrable on $[a,b]$ and let there be a differentiable function F on $[a,b]$ such that $F'=f$ then
$$\int_a^b f(x)dx=F(b)-F(a)$$

REVIEW QUESTIONS AND PROJECT WORK

1. If f is bounded on $[-1,1]$ and
$$\alpha(x)=\begin{cases}0 & ;\text{when } x<0\\ 1/2 & ;\text{when } x=0\\ 1 & ;\text{when } x>0\end{cases}$$
then show that $f\in RS(\alpha)$ on $[-1,1]$ if and only if f is continuous at 0.

2. If $f(x)=\begin{cases}x & ;\text{ when } x \text{ is irrational}\\ 0 & ;\text{ when } x \text{ is rational}\end{cases}$

and $\alpha(x)=\begin{cases}0 & ;\text{when } x<0\\ 1/2 & ;\text{when } x=0\\ 1 & ;\text{when } x>0\end{cases}$

then show that $f\in RS[-1,1]$.

3. Show that $\int_0^2 x^3 d[x^2]=9+2\sqrt{2}+3\sqrt{3}$
where $[x]$ is the greatest integer not greater than x.

4. If f is bounded on $[a, b]$ and

$$\alpha(x) = \begin{cases} 0 & \text{when } x < b \\ 1 & \text{when } x = b \end{cases}$$

Show that $f \in RS\,(\alpha)$ on $[a,b]$ iff $f(b-0)=f(b)$ and then $\int_a^b f\,d\alpha = f(b)$.

5. Prove the following :

(i) $\int_0^3 x^2 d\left([x]-x\right) = -9$

(ii) $\int_0^2 x^2 d[x] = 5$

(iii) $\int_0^1 2x\,d\left(x[x]\right) = n^3 + \sum_1^n r^2$

(iv) $\int_2^3 [5-x]d\left(\log[x]\right) = \log 3$

6. If ΣC_n be a convergent sequence of non-negative terms and $<t_n>$ a sequence of distinct points in $]a,b[$ then for continuous $f(x)$ on $[a,b]$, show that

$$\int_2^3 f(x)\,d\alpha(x) = \sum_{n=1}^{\infty} C_n f(t_n)$$

where $\alpha(x) = \sum_{n=1}^{\infty} C_n \beta(x-t_n)$ and $\beta(x-t_n)=0$ or 1 according as $x \le t_n$ or $x > t_n$.

7. Prove the following :

(i) $\int_0^2 [x]d\left(x^2\right) = 3$

(ii) $\int_\pi^{2\pi} \sin x\,d(\cos x) = \dfrac{-\pi}{2}$

(iii) $\int_{-1}^1 \left(x^2+e^x\right)d(\sin x) = 1$

8. A function α increases on $[a,b]$ and is continuous at t, $a \le t \le b$. Another function f such that

$$f(t) = 1 \text{ and } f(x) = 0 \text{ for } x \neq t.$$

Show that $f \in RS(\alpha)$ over $[a,b]$ and $\int_a^b f\,d\alpha = 0$.

9. Let f be a function bounded on $[-1,1]$ and

$$\alpha(x) = \begin{cases} 0 & x < 0 \\ 1/2 & x = 0 \\ 1 & x > 0 \end{cases}$$

Prove that $f \in RS(\alpha)$ iff f is continuous at $x=0$ then $\int_{-1}^1 f\,d\alpha = f(0)$.

10. If f is continuous and α is monotonic on $[a,b]$ then show that

$$\int_a^b f\,d\alpha = \left[f(x)\alpha(x)\right]_a^b - \int_a^b \alpha\,df.$$

OBJECTIVE TYPE QUESTIONS

FILL IN THE BLANKS

1. $\int_a^b f\,d\alpha = \lambda[\alpha(b) - \underline{\hspace{2cm}}]$.

2. For a bounded function $\overline{\int_a^b}\, f\,d\alpha$ _____ $\underline{\int_a^b}\, f\,d\alpha$.

3. If f is monotonic and α is _____ on $[a,b]$ then $f \in RS(\alpha)$.

4. $\int_0^2 x^2\,dx^2$ _____ .

5. $\int_0^2 [x]\,dx^2$ _____ .

TRUE/FALSE

Write 'T' for true and 'F' for false statement.

1. If f is continuous on $[a, b]$ then $f \in RS(\alpha)$. **(T/F)**

2. If f is monotonic on $[a, b]$ and if α is continuous on $[a, b]$ then $f \in RS(\alpha)$. **(T/F)**

3. For continuous function f, $\lim S(P, f, \alpha)$ exists and equals $\int f\,d\alpha$. **(T/F)**

4. Continuity is a necessary and sufficient condition for integrability. **(T/F)**

5. If f is continuous on $[a, b]$ and α has a continuous derivatives on $[a, b]$ then

$$\int_a^b f\,d\alpha = \int_a^b f\alpha'\,dx.$$ **(T/F)**

MULTIPLE CHOICE QUESTIONS

Choose the most appropriate one.

1. $f \in RS(\alpha)$ exists when:

(a) f is continuous and α is monotonic

(b) f is monotonic and α is continuous

(c) both (a) and (b) are true

(d) none of the above

2. If $\lim S(P, f, \alpha)$ exists as $\|P\| \to 0$ then:

(a) $f \in RS(\alpha)$

(b) $\lim\limits_{\|P\|\to 0} S(P,f,\alpha) = \int_a^b f\,d\alpha$

(c) both (a) and (b) are true

(d) none of the above

3. If $f_1 \in RS\,(\alpha)$ and $f_2 \in RS\,(\alpha)$ over $[a,b]$ then:

(a) $f_1 + f_2 \in RS(\alpha)$ (b) $f_1 - f_2 \in RS(\alpha)$

(c) $f_1.f_2 \in RS(\alpha)$ (d) all are true

4. If $f \in RS(\alpha)$ over $[a,b]$ then:

(a) $|f| \in RS(\alpha)$ (b) $\left|\int_a^b f\,d\alpha\right| \le \int_a^b |f|\,d\alpha$

(c) $f^2 \in RS(\alpha)$ (d) all are true

5. A function f is RS-integrable w.r.t. α on $[a,b]$ if and only if for every $\varepsilon > 0$ $\exists$ a partition P of $[a,b]$ such that:

(a) $U(P,f,\alpha) - L(P,f,\alpha) < \varepsilon$

(b) $U(P,f,\alpha) + L(P,f,\alpha) < \varepsilon$

(c) $U(P,f,\alpha) - L(P,f,\alpha) > \varepsilon$

(d) none of the above

6. For a bounded function f:

(a) $\int_{\underline{a}}^b f\,d\alpha \le \overline{\int_a^b} f\,d\alpha$

(b) $\int_{\underline{a}}^b f\,d\alpha \ge \overline{\int_a^b} f\,d\alpha$

(c) $\int_{\underline{a}}^b f\,d\alpha = \overline{\int_a^b} f\,d\alpha$

(d) none of the above

7. For any two partitions P_1 and P_2:

(a) $L(P_1,f,\alpha) \le U(P_2,f,\alpha)$

(b) $L(P_1,f,\alpha) \ge U(P_2,f,\alpha)$

(c) $L(P_1,f,\alpha) = U(P_2,f,\alpha)$

(d) none of the above

8. $\int_0^2 [x]\,dx^2$:

(a) 4   (b) 3

(c) 0   (d) none of the above

9. $\int_0^2 x^2\,dx^2$:

(a) 8   (b) 16

(c) 4   (d) none of the above

10. $\int_0^x d[t] =$ $\forall x \in R$:

(a) x (b) $|x|$

(c) $[x]$ (d) none of the above

=== Answers ===

FILL IN THE BLANKS

1. $\alpha(a)$ 2. $\le$ 3. continuous   4. 8   5. 3

TRUE/FALSE

1. T   2. T   3. T   4. F   5. T

MULTIPLE CHOICE QUESTIONS

1. (c)   2. (c)   3. (d)   4. (d)   5. (a)

6. (a)   7. (a)   8. (b)   9. (a)   10. (c)

8 Improper Integrals

8.1 INTRODUCTION

The definite integral $\int_a^b f(x)\,dx$ is called improper (or infinite integral) if either any one or both limits are infinite and function $f(x)$ is bounded over the interval or neither the intervals $[a, b]$ is finite nor $f(x)$ is bounded over it.

REMARKS

- The improper integral of a function over any range is defined as the limit of a proper integrals over a part of that range.
- If improper integral having finite value, then it is called convergent.
- Improper integral having no finite value is called divergent.
- If limit is neither finite nor $-\infty$ or ∞, then integral is said to be oscillatory.

8.2 TYPE OF IMPROPER INTEGRALS

By definition of improper Integrals we can divide or categorized it into following three kinds.

8.2.1 FIRST KIND OF IMPROPER INTEGRALS

First kind of improper integral is in which integrand $f(x)$ is continuous but limits are infinite.

Definition. *A definite integral $\int_a^b f(x)\,dx$ in which limits are infinite i.e., $b = \infty$, $a = \infty$ and integrand is continuous is called first kind of improper integral.*

This first kind of improper integral can be further classified into following three categories :

(I) UPPER LIMIT INFINITE

Consider the integral $\int_0^\infty \dfrac{1}{1+x^2}\,dx$. Here it is first kind of improper integral in which upper limit is infinite and $(1/1+x^2)$ is bounded.

(II) LOWER LIMIT INFINITE

Consider the integral $\int_{-\infty}^0 e^x\,dx$. Here, the lower limit of function is infinite.

(III) BOTH LIMITS INFINITE

Let $\int_{-\infty}^\infty \dfrac{dx}{1+x^2}$. It is the example in which both upper and lower limits are infinite.

(i) Consider $\int_a^\infty f(x)\,dx$. Here $f(x)$ is continuous in $[a, \infty[$. There exists a definite number $b > a$ such that $\int_a^b f(x)\,dx$ as $b \to \infty$. This definite integral becomes the improper integral $\int_a^\infty f(x)\,dx = \lim_{b\to\infty} \int_a^b f(x)\,dx.$

If limit is finite, then improper integral $\int_a^\infty f(x)\,dx$ is convergent, otherwise divergent.

(ii) Consider $\int_{-\infty}^\infty f(x)\,dx$, then there exists $a < b$, such that $\int_a^b f\,d\alpha$ as $a \to -\infty$, then

$$\int_{-\infty}^b f(x)\,dx = \lim_{a\to-\infty} \int_a^b f(x)\,dx.$$

If limit is finite, then improper integral is convergent otherwise divergent.

(iii) Consider $\int_{-\infty}^\infty f(x)\,dx$. It is the combination of above 2-procedures so take a constant 'a' between $-\infty$ to ∞ and expressed in the integral in the form of

$$\int_{-\infty}^\infty f(x)\,dx = \int_{-\infty}^a f(x)\,dx + \int_a^{+\infty} f(x)\,dx$$

$\Rightarrow$ $$\int_{-\infty}^\infty f(x)\,dx = \lim_{b\to\infty} \int_a^b f(x)\,dx + \lim_{b\to\infty} \int_a^b f(x)\,dx$$

If both limits are finite then $\int_{-\infty}^\infty f(x)\,dx$ is convergent otherwise divergent i.e., if anyone or both limits are infinite.

8.2.2 SECOND KIND OF IMPROPER INTEGRAL

Second kind of improper integral is in which limits are finite but integrand is infinite. The point at which the integrand is infinite is called a singular point.

Second kind of improper integral is classified into following four categories :

(i) Singular point at right end 'b'. If $x = b$ is only singular point of $f(x)$ then there exists $\varepsilon > 0$ (small positive number) such that

$$\int_a^b f(x)\,dx = \lim_{\varepsilon\to0} \int_a^{b-\varepsilon} f(x)\,dx.$$

Here, $f(x)$ is continuous in $[a, b - \varepsilon]$.

(ii) Singular point at left end 'a'. If $f(x) \to \infty$ as $x \to a$ is only singular point of $f(x)$ then there exists a small positive number $\varepsilon > 0$ such that

$$\int_a^b f(x)\,dx = \lim_{\varepsilon\to0} \int_{a+\varepsilon}^b f(x)\,dx.$$

Here, $f(x)$ is continuous in $[a + \varepsilon, b]$.

If $\int f(x)\,dx = F(x) + c$, then $\int_a^b f(x)\,dx = \lim_{\varepsilon\to0} |f(b) - f(a+\varepsilon)|.$

(iii) Singular point at 'c'. If $f(x) \to \infty$ as $x \to c$ the singular point of $f(x)$ where $a < c < b$, then $\int_a^b f(x)\,dx$ decomposed into following form :

$$\int_a^b f(x)\,dx = \int_a^c f(x)\,dx + \int_c^b f(x)\,dx$$

$$= \lim_{\varepsilon\to0} \int_a^{c-\varepsilon} f(x)\,dx + \lim_{\varepsilon'\to0} \int_{c+\varepsilon'}^b f(x)\,dx.$$

If one or both integrals in R.H.S. be convergent, then $\int_a^b f(x)\,dx$, $a < c < b$ is convergent, otherwise divergent.

(iv) Singular point at both a and b. If a and b are only singular points of $f(x)$ then there exists c such that $a < c < b$, then

$$\int_a^b f(x)\,dx = \int_a^c f(x)\,dx + \int_c^b f(x)\,dx$$

$$= \lim_{\varepsilon \to 0} \int_{a+\varepsilon}^{c} f(x)\,dx + \lim_{\varepsilon' \to 0} \int_{c}^{b-\varepsilon'} f(x)\,dx.$$

If each integral is convergent then $\int_{a}^{b} f(x)\,dx$ is convegent.

8.3.3 THIRD KIND OF IMPROPER INTEGRAL

Third kind of improper integral is in which

   (i) infinite limits   (ii) infinite integrals

"*It is the combination of both first kind and second kind of improper integrals.*"

Let $\int_{a}^{\infty} f(x)\,dx$ is improper integral of third kind when $f(x)$ has a singular point at $x = c$,

   where $a < c < d$ and $c < d < \infty$ then

$$\underset{\text{(I)}}{\int_{a}^{\infty} f(x)\,dx = \int_{a}^{d} f(x)\,dx} + \underset{\text{(II)}}{\int_{d}^{\infty} f(x)\,dx} \qquad \qquad ...(1)$$

Here, $\int_{a}^{\infty} f(x)\,dx$ is convergent if both integrals are convergent otherwise divergent.

8.3 CONVERGENCE OF IMPROPER INTEGRALS

Definition. *The integral $\int_{a}^{\infty} f(x)\,dx$ is said to converge to the value I, if for any arbitrary chosen positive number ε, however small but not zero, there exists a positive number N such that*

$$\left| \int_{a}^{b} f(x)\,dx - I \right| < \varepsilon; \text{ for all values of } b \geq N.$$

If the integral $f(x)$ has a finite limit then improper integral called convergent and if having no finite limit *i.e.*, limt are $+\infty$, $-\infty$ then it is said to be divergent and when having neither finite value nor, 0, $+\infty$ and $-\infty$, the improper integrals is said to be oscillatory.

Solved Examples

Example 1. Discuss the convergence of the integral $\int_{1}^{\infty} \frac{dx}{\sqrt{x}}$ by calculating them.

Solution. Since we have

$$\int_{1}^{\infty} \frac{dx}{\sqrt{x}} = \lim_{x \to \infty} \int_{1}^{x} \frac{dx}{\sqrt{x}} = \lim_{x \to \infty} \int_{1}^{x} x^{-1/2}\,dx$$

$$= \lim_{x \to \infty} \left(\frac{x^{1/2}}{1/2} \right)_{1}^{x} = \lim_{x \to \infty} \left(2x^{1/2} \right)_{1}^{x}$$

$$= \lim_{x \to \infty} (2\sqrt{x} - 2) = \infty.$$

$\Rightarrow$ the limit does not exist finitely.

Hence, the given integral is divergent.

Example 2. Discuss the convergence of the integral $\int_{1}^{\infty} \frac{dx}{x^{3/2}}$ by calculating them.

Based on the following Results

- A definite integral $\int_{a}^{b} f(x)\,dx$ in which limits are infinite *i.e.*, $b = \infty$ or $a = \infty$ and integrand is continuous is called first kind of improper integral.

- Second kind of improper integral is in which limits are finite but integrand is infinite.

- Third kind of improper integral is in which (i) infinite limits (ii) infinite integrand.

- If the limit of the integral $f(x)$ is finite then it is said to be convergent and if having no finite limit i.e, limits are ∞ or $-\infty$ then it is divergent and when having neither finite value nor 0, $+\infty$ and $-\infty$, it is said to be oscillatory.

Solution. Since we have

$$\int_1^\infty \frac{dx}{x^{3/2}} = \lim_{x \to \infty} \int_1^x x^{-3/2} dx = \lim_{x \to \infty} \left[\frac{x^{-1/2}}{-\dfrac{1}{2}} \right]_1^x$$

$$= \lim_{x \to \infty} \left[-\frac{2}{\sqrt{x}} \right]_1^x = \lim_{x \to \infty} \left[-\frac{2}{\sqrt{x}} + 2 \right] = \frac{-2}{\infty} + 2 = 2.$$

⇒   the limit exists and finite.

⇒   the given integral is convergent.

Example 3. *Discuss the convergence of the integral of $\int_0^1 \dfrac{dx}{\sqrt{1-x}}$ by evaluating them.*

Solution. Here given integral is $\int_0^1 \dfrac{dx}{\sqrt{1-x}}$. Clearly, It is not bounded at $x = 1$.

So $$\int_0^1 \frac{dx}{\sqrt{1-x}} = \lim_{\varepsilon \to 0} \int_0^{1-\varepsilon} \frac{dx}{\sqrt{1-x}}$$

$$= \lim_{\varepsilon \to 0} \left[-2\sqrt{1-x} \right]_0^{1-\varepsilon} = \lim_{\varepsilon \to 0} \left[-2\sqrt{\varepsilon} + 2 \right] = 2.$$

which is a finite number.

⇒   the given integral is convergent.

Example 4. *Discuss the convergence of the integral $\int_{-1}^1 \dfrac{dx}{x^2}$.*

Solution. Let $I = \int_{-1}^1 \dfrac{dx}{x^2}$. It becomes infinite at $x = 0$.

So $$\int_{-1}^1 \frac{dx}{x^2} = \lim_{\varepsilon_1 \to 0} \int_{-1}^{-\varepsilon_1} \frac{dx}{x^2} + \lim_{\varepsilon_2 \to 0} \int_{\varepsilon_2}^1 \frac{dx}{x^2}$$

$$= \lim_{\varepsilon_1 \to 0} \left[-\frac{1}{x} \right]_{-1}^{-\varepsilon_1} + \lim_{\varepsilon_2 \to 0} \left[-\frac{1}{x} \right]_{\varepsilon_2}^1$$

$$= \lim_{\varepsilon_1 \to 0} \left[\frac{1}{\varepsilon} - 1 \right] + \lim_{\varepsilon_1 \to 0} \left[-1 + \frac{1}{\varepsilon_2} \right]$$
$$\qquad\qquad \text{I} \qquad\qquad\qquad \text{II}$$

Since (I) and (II) do not exist finitely ⇒ limit does not exist finitely.

Hence given integral is divergent.

Example 5. *Discuss the convergence of $\int_0^{2a} \dfrac{dx}{(x-a)^2}$.*

Solution. The given integral $\int_0^{2a} \dfrac{dx}{(x-a)^2}$ becomes infinite at $x = a$ and $0 < a < 2a$.

So

$$\int_0^{2a} \frac{dx}{(x-a)^2} = \int_0^a \frac{dx}{(x-a)^2} + \int_a^{2a} \frac{dx}{(x-a)^2}$$

$$= \lim_{\varepsilon_1 \to 0} \int_0^{a-\varepsilon_1} \frac{dx}{(x-a)^2} + \lim_{\varepsilon_2 \to 0} \int_{a+\varepsilon_2}^{2a} \frac{dx}{(x-a)^2}$$

$$= \lim_{\varepsilon_1 \to 0}\left[\frac{-1}{(x-a)}\right]_0^{a-\varepsilon_1} + \lim_{\varepsilon_2 \to 0}\left[\frac{-1}{(x-a)}\right]_{a+\varepsilon_2}^{2a}$$

$$= \lim_{\varepsilon_1 \to 0}\left[\frac{1}{\varepsilon_1} - \frac{1}{a}\right] + \lim_{\varepsilon_2 \to 0}\left[\frac{1}{\varepsilon_2} - \frac{1}{a}\right]$$

$$\underset{\text{I}}{} \underset{\text{II}}{}$$

Since the limit of (I) and (II) not exist finitely. Hence, the given integral is divergent.

Example 6. *Discuss the convergence of the integral $\int_0^1 \dfrac{dx}{1-x}$.*

Solution. We have

$$\int_0^1 \frac{dx}{1-x} = \lim_{\varepsilon \to 0}\int_0^{1-\varepsilon} \frac{dx}{1-x}$$

$$= \lim_{\varepsilon \to 0}\left[-\log(1-x)\right]_0^{1-\varepsilon} = \lim_{\varepsilon \to 0}\left[-\log\varepsilon + 0\right].$$

Since $\lim_{\varepsilon \to 0}\log\varepsilon$ is $-\infty$, therefore $\int_0^1 \dfrac{dx}{1-x}$. is meaningless *i.e.,* limit does not exist.
So the integral is divergent.

Example 7. *Test the convergence of the integral $\int_{-\infty}^{\infty} \dfrac{dx}{1+x^2}$.*

Solution. We have

$$\int_{-\infty}^{\infty} \frac{dx}{1+x^2}. \qquad \text{(Both upper and lower limits are infinite.)}$$

So, $$\int_{-\infty}^{\infty} \frac{dx}{1+x^2} = \int_{-\infty}^{0} \frac{dx}{1+x^2} + \int_0^{\infty} \frac{dx}{1+x^2}$$

$$= \lim_{b \to \infty}\int_{-b}^{0} \frac{dx}{1+x^2} + \lim_{b \to \infty}\int_0^{b} \frac{dx}{1+x^2}$$

$$= \lim_{b \to \infty}\left[\tan^{-1}x\right]_{-b}^{0} + \lim_{b \to \infty}\left[\tan^{-1}x\right]_0^{b}$$

$$= \lim_{b \to \infty}[0 - \tan^{-1}(-b)] + \lim_{b \to \infty}[\tan^{-1}b - 0]$$

$$= -\left(-\frac{\pi}{2}\right) + \frac{\pi}{2} = \pi, \text{finite and unique.}$$

Therefore, the given integral is convergent.

Example 8. *Evaluate $\int_1^{\infty} \dfrac{x+2}{x(x+1)}\,dx$.*

Solution. Here, the upper limit of the integral is infinite. So, it is first kind of improper integral.
Now,

$$\int_1^{\infty} \frac{x+2}{x(x+1)}\,dx = \lim_{b \to \infty}\int_1^{b} \frac{x+2}{x(x+1)}\,dx = \lim_{b \to \infty}\int_1^{b}\left(\frac{2}{x} - \frac{1}{x+1}\right)dx$$

(By resolving into partial fraction)

$$= \lim_{b \to \infty}\left[\log\frac{x^2}{x+1}\right]_1^{b} = \lim_{b \to \infty}\left[\log\frac{b^2}{b+1} + \log 2\right]$$

$$= \lim_{b \to \infty} \left[\log \frac{1}{\frac{1}{b} + \frac{1}{b^2}} + \log 2 \right] = \log \infty + \log 2$$

$$= \infty + \log 2 = \infty.$$

Here, the limit of integral is infinite so the given integral is divergent.

Example 9. *Test the convergence of the following integrals:*

(i) $\int_{-\infty}^{0} \sinh x$ (ii) $\int_{-\infty}^{0} \cosh x \, dx$

Solution. We have

(i) $\int_{-\infty}^{0} \sinh x \, dx = \lim_{b \to \infty} \int_{-b}^{0} \sinh x \, dx = \lim_{b \to \infty} \int_{-b}^{0} \frac{e^x - e^{-x}}{2} \, dx$

$$= \frac{1}{2} \left[\lim_{b \to \infty} \int_{-b}^{0} e^x \, dx - \lim_{b \to \infty} \int_{-b}^{0} e^{-x} dx \right]$$

$$= \frac{1}{2} \left[\lim_{b \to \infty} \{e^x\}_{-b}^{0} - \lim_{b \to \infty} \left\{ \frac{e^{-x}}{-1} \right\}_{-b}^{0} \right]$$

$$= \frac{1}{2} \left[\lim_{b \to \infty} \{e^0 - e^{-b}\} + \lim_{b \to \infty} \{e^0 - e^b\} \right]$$

$$= -\infty.$$

Thus, the given integrals is divergent.

(ii) We have

$$\int_{-\infty}^{0} \cosh x \, dx = \int_{-\infty}^{0} \frac{e^x + e^{-x}}{2} \, dx$$

$$= \frac{1}{2} \left[\int_{-\infty}^{0} e^x \, dx + \int_{-\infty}^{0} e^{-x} dx \right] = \infty.$$

Hence, the given integral is divergent.

Example 10. *Test the convergence of* $\int_{a}^{\infty} \frac{4a \, dx}{x^2 + 4a^2}.$

Solution. We have

$$\int_{a}^{\infty} \frac{4a \, dx}{x^2 + 4a^2} = \lim_{b \to \infty} \int_{0}^{b} \frac{4a \, dx}{x^2 + (2a)^2}$$

$$= \lim_{b \to \infty} \left[4a \cdot \frac{1}{2a} \tan^{-1} \frac{x}{2a} \right]_{0}^{b} = 2. \lim_{b \to \infty} \left[\tan^{-1} \frac{x}{2a} \right]_{0}^{b}$$

$$= 2. \lim_{b \to \infty} \left[\tan^{-1} \frac{b}{2a} - 0 \right] = 2.[\tan^{-1} \infty] = 2.\frac{\pi}{2} = \pi.$$

Thus, the limit is finite, therefore the given integral is convergent.

Example 11. *Show that the integral* $\int_{0}^{\infty} \frac{dx}{(1+x)^{2/3}}$ *is not convergent.*

Solution. We have

$$\int_0^\infty \frac{dx}{(1+x)^{2/3}} = \lim_{b\to\infty} \int_0^b (1+x)^{-2/3} dx$$

$$= \lim_{b\to\infty} \left[\frac{(1+x)^{1/3}}{1/3}\right]_0^b = \lim_{b\to\infty} 3[(1+b)^{1/3} - 1] = \infty.$$

The limit is not finite. So integral is divergent *i.e.*, not convergent.

Example 12. *Evaluate* $\int_{-\infty}^\infty \frac{dx}{x^2 + 2x + 2}.$

Solution. We have

$$\int_{-\infty}^\infty \frac{dx}{x^2 + 2x + 2} = \int_{-\infty}^\infty \frac{dx}{(x+1)^2 + 1}$$

$$= \lim_{b\to\infty} \int_{-b}^c \frac{dx}{(x+1)^2 + 1} + \lim_{a\to\infty} \int_c^a \frac{dx}{(x+1)^2 + 1}$$

where c is any real number.

$$= \lim_{b\to\infty} [\tan^{-1}(x+1)]_{-b}^c + \lim_{a\to\infty} [\tan^{-1}(x+1)]_c^a$$

$$= \lim_{b\to\infty} [\tan^{-1}(c+1) - \tan^{-1}(1-b)]$$

$$+ \lim_{a\to\infty} [\tan^{-1}(a+1) - \tan^{-1}(c+1)]$$

$$= \tan^{-1}(c+1) - \tan^{-1}(-\infty) + \tan^{-1}(\infty) - \tan^{-1}(c+1)$$

$$= -\left(-\frac{\pi}{2}\right) + \frac{\pi}{2} = \pi.$$

Thus the limit is finite, therefore the given integral is convergent.

Example 13. *Test the convergent of* $\int_0^\infty e^{-mx} dx \ (m > 0).$

Solution. We have

$$\int_0^\infty e^{-mx} dx = \lim_{a\to\infty} \int_0^a e^{-mx} dx = \lim_{a\to\infty} \left[\frac{e^{-mx}}{-m}\right]_0^a$$

$$= \lim_{a\to\infty} \left[-\frac{1}{m}(e^{-ma} - 1)\right] = \frac{-1}{m}(0-1) = \frac{1}{m}.$$

Thus the limit is finite, therefore the given integral is convergent.

EXERCISE 8.1

Evaluate the following integrals and also discuss their convergence :

1. $\int_1^\infty \frac{dx}{x}$

2. $\int_0^\infty e^{2x} dx$

3. $\int_0^\infty \cos x \, dx$

4. $\int_{-1}^1 \frac{dx}{x^{2/3}}$

5. $\int_{-\infty}^\infty e^{-x} dx$

6. $\int_0^1 \frac{dx}{x^3}$

7. $\int_0^\infty \frac{dx}{(1+x)^{2/3}}$

8. $\int_3^\infty \frac{dx}{(x-2)^2}.$

HINTS TO SELECTED PROBLEMS

2. $\int_0^\infty e^{2x}\,dx = \lim\limits_{a\to\infty} \int_0^a e^{2x}\,dx$

$= \lim\limits_{a\to\infty} \left[\dfrac{e^{2x}}{2}\right]_0^a = \infty.$

3. $\int_0^\infty \cos x\,dx = \lim\limits_{a\to\infty} \int_0^a \cos x\,dx$

$= \lim\limits_{a\to\infty} [\sin x]_0^a = \infty.$

4. $\int_{-1}^1 \dfrac{dx}{x^{2/3}} = \int_{-1}^0 \dfrac{dx}{x^{2/3}} + \int_0^1 \dfrac{dx}{x^{2/3}}$

$= \lim\limits_{\varepsilon\to 0} \int_{-1}^{0-\varepsilon} \dfrac{dx}{x^{2/3}} + \lim\limits_{\varepsilon'\to 0} \int_{\varepsilon'+0}^1 \dfrac{dx}{x^{2/3}}.$

5. $\int_{-\infty}^\infty e^{-x}\,dx = \int_{-\infty}^0 e^{-x}\,dx + \int_0^\infty e^{-x}\,dx$

$= \lim\limits_{b\to\infty} \int_{-b}^0 e^{-x}\,dx + \lim\limits_{a\to\infty} \int_0^a e^{-x}\,dx.$

7. $\int_0^\infty \dfrac{dx}{(1+x)^{2/3}} = \lim\limits_{a\to\infty} \int_0^a (1+x)^{-2/3}\,dx$

$= \lim\limits_{a\to\infty} [3(1+x)]_0^a = \infty.$

Answers

1. ∞, divergent   **2.** ∞, divergent   **3.** oscillates, not convergent

4. 6, convergent   **5.** ∞, divergent   **6.** ∞, divergent

7. ∞, divergent   **8.** 1, convergent

8.4 CONVERGENCE TESTS : FIRST KIND

Recall that, First kind of improper integral is in which limits are infinite and integrand is continuous.

For Example. $\int_a^\infty f(x)\,dx$ or $\int_{-\infty}^b f(x)\,dx$ is the example of first kind of improper integral which cannot be actually integrated. To test its convergence we use the following tests.

8.4.1 COMPARISON TEST

If $\int_a^\infty f(x)\,dx$ and $\int_b^\infty g(x)\,dx$ are positive, continuous (bounded) and integrable in the interval $]\,a, \infty\,[$ and

(i) $f(x) \leq g(x)$, for all x beyond a point $x = c$ and also $\int_b^\infty g(x)\,dx$ is convergent, then

$\int_a^\infty f(x)\,dx$ is convergent.

(ii) $g(x) \leq f(x)$, for all values of x and $\int_b^\infty g(x)\,dx$ is divergent, then $\int_a^\infty f(x)\,dx$ is divergent.

8.4.2 LIMIT FORM OF COMAPRISON TEST

If $\int_a^\infty f(x)\,dx$ and $\int_b^\infty g(x)\,dx$ are such that the integrands are positive and $\lim\limits_{x\to\infty} \dfrac{f(x)}{g(x)} = l$ then

(i) $\int_a^\infty f(x)\,dx$ is convergent when $l = 0$ and $\int_b^\infty g(x)\,dx$ is convergent.

(ii) $\int_a^\infty f(x)\,dx$ is divergent when $l = \infty$ and $\int_b^\infty g(x)\,dx$ is divergent.

(iii) both integrals are either convergent or divergent if l exists but non-zero.

<u>THEOREM 1.</u> *The comparison integral* $\int_a^\infty \dfrac{dx}{x^n}$, *when* $a>0$ *is convergent when* $n>1$, *and divergent when* $n \leq 1$.

Proof.   We have

$$\int_a^\infty \frac{dx}{x^n} = \lim_{x\to\infty} \int_a^x x^{-n}\, dx \qquad \text{(By definition of improper integral)}$$

$$= \lim_{x\to\infty} \left[\frac{x^{1-n}}{1-n}\right]_a^x \qquad \text{if } n \neq 1$$

$$= \lim_{x\to\infty} \left[\frac{x^{1-n}}{1-n} - \frac{a^{1-n}}{1-n}\right] \qquad \qquad \text{...(1)}$$

Now if $n>1$ then $(1-n)<0 \Rightarrow (n-1)>0$ therefore in this case, we have

$$\lim_{x\to\infty} x^{1-n} = \lim_{x\to\infty} \frac{1}{x^{n-1}} = \frac{1}{\infty} = 0.$$

$\therefore$ From (1), we have

$$\int_a^\infty \frac{dx}{x^n} = \frac{a^{1-n}}{n-1} \quad \text{if } n > 1.$$

Hence the given integral is convergent when $n > 1$.

Now, if $n<1$, then $(1-n) > 0$ and $(n-1) < 0$ and $\lim_{x\to\infty} x^{1-n} = \infty$.

$\therefore$ From (1), $\int_a^\infty \frac{dx}{x^n} = \infty.$

Therefore, the given integral is divergent when $n < 1$.

If $n = 1$, then $\int_a^\infty \frac{dx}{x^n} = \int_a^\infty \frac{dx}{x} = \lim_{x\to\infty} \int_a^x \frac{dx}{x} = \lim_{x\to\infty} \left[\log x\right]_a^x$

$$= \lim_{x\to\infty} [\log x - \log a] = \infty - \log a = \infty.$$

$\therefore$ The given integral is divergent if $n \leq 1$.

Hence $\int_a^\infty \frac{dx}{x^n}$ converges when $n>1$ and diverges when $n \leq 1$.

8.4.3 DIRICHLET'S TEST

If $f(x)$, $g(x)$ and $g'(x)$ are all continuous in $[a, \infty[$ and $f(x)$, $g(x)$ satisfy the following three conditions:

(i) $\lim_{x\to\infty} g(x) = 0$ (ii) $\int_a^\infty |g'(x)|\, dx$ is convergent and

(iii) $F(r) = \int_a^r f(x)\, dx$ is bounded i.e., $|F(r)| \leq M$ for some positive constant M.

Then $\int_a^\infty f(x)g(x)\, dx$ is convergent.

8.4.4 THE μ- TEST

Let $f(x)$ be bounded and integrable in the interval $]a, \infty[$ where $a > 0$. Then $\int_a^\infty f(x)\, dx$ is convergent, if there is a number $\mu > 1$, such that $\lim_{x\to\infty} x^\mu f(x)$ exists.

If there is a number $\mu \leq 1$ such that $\lim_{x\to\infty} x^\mu f(x)$ exists and non-zero, then $\int_a^\infty f(x)\, dx$ is divergent and the same is true if $\lim_{x\to\infty} x^\mu f(x)$ is $+\infty$ or $-\infty$.

REMARK

- The value of μ is usually taken to be equal to the highest power of x in the denominator of the integrand minus the highest power of x in the numerator of the integrand.

8.4.5 WEIERSTRASS M-TEST

If there exists a positive continuous function $M(t)$ such that $|f(x, t)| \leq M(t), t \geq a, c \leq x \leq d$, then the improper integral $\int_a^\infty f(x,t)dt$ converges uniformly and absolutely for every x in the interval $[c, d]$ if $\int_a^\infty M(t)dt$ converges.

8.4.6 ABEL'S TEST FOR THE CONVERGENCE OF INTEGRAL OF PRODUCTS

The integral $\int_a^\infty f(x)\phi(x)dx$ is convergent, if $\int_a^\infty f(x)dx$ converges and $\phi(x)$ is bounded and monotonic for $x > a$.

8.4.7 ABSOLUTE CONVERGENCE

The integral $\int_a^\infty |f(x)|dx$ is convergent then the infinite integral $\int_a^\infty f(x)dx$ is said to be absolutely convergent.

REMARK

- The absolute convergence gives a sufficient condition for the convergence of an infinite integral.

8.5 CONVERGENCE TESTS : SECOND KIND

We test the convergence of a definite integral $\int_a^b f(x)dx$ for which limits (intervals) are finite and integrand $f(x)$ is not bounded at one or more points of given interval $[a, b]$.

8.5.1 COMPARISON TEST

If $\int_a^b f(x)dx$ be the given improper integral, whose limits are finite and $f(x)$ is not bounded only at $x = a$. Let $x = b$ be a singular point for both $f(x)$ and $g(x)$ in interval $[a, b]$ and

(i) $0 \leq f(x) \leq g(x)$ everywhere, except at $x = b$ then $\int_a^b f(x)dx$ is convergent if $\int_a^b g(x)dx$ is convergent.

(ii) $f(x) \geq g(x) \geq 0$ everywhere, except at $x = a$ then $\int_a^b f(x)dx$ is divergent if $\int_a^b g(x)dx$ is divergent.

8.5.2 LIMIT FROM OF COMPARISON TEST

(i) If $f(x)$ and $g(x)$ are positive and $\lim_{x \to b} \dfrac{f(x)}{g(x)} = l$, where l is neither zero nor infinite then $\int_a^b f(x)dx$ and $\int_a^b g(x)dx$ either both converge or both diverge at singular point $x = b$.

(ii) If $l = 0$ and $\int_a^b g(x)dx$ converges, then $\int_a^b f(x)dx$ converges.

(iii) If $l = \infty$ and $\int_a^b g(x)dx$ diverges, then $\int_a^b f(x)dx$ diverges.

8.5.3 ABEL'S TEST

If $g(x)$ is bounded and monotonic for $a \leq x \leq b$ and $\int_a^b f(x)dx$ converges. Then $\int_a^b f(x)g(x)dx$ converges.

8.5.4 DIRICHLET'S TEST

If $\int_{a+\varepsilon}^{b} f(x)\,dx$ is bounded and $g(x)$ is bounded and monotonic in $]a, b[$ converging to zero as $x \to a$, then $\int_{a}^{b} f(x)g(x)\,dx$ converges.

8.5.5 INTEGRAND IS BOTH +VE AND –VE

Let the integrand be both positive and negative in $[a, b]$. Let $x = b$ be a singular point of $f(x)$. Now if $\int_{a}^{b} f(x)\,dx$ is convergent then $\int_{a}^{b} f(x)\,dx$ is absolutely convergent, $\int_{a}^{b} f(x)\,dx$ is convergent but $\int_{a}^{b} |f(x)|\,dx$ is divergent then $\int_{a}^{b} f(x)\,dx$ is conditionally convergent.

8.5.6 THE μ-TEST

Let $f(x)$ be not bounded at $x = a$ and bounded and integrable in the arbitrary interval $]a+\varepsilon, b[$, where $0 < \varepsilon < b - a$. If there is a number μ between 0 and 1 such that $\lim\limits_{x \to a+0} (x-a)^{\mu} f(x)$ exists, then $\int_{a}^{b} f(x)\,dx$ is convergent.

If there is a number $\mu \geq 1$ such that $\lim\limits_{x \to a+0} (x-a)^{\mu} f(x)$ exists and is non-zero, then $\int_{a}^{b} f(x)\,dx$ is divergent and the same is true, if $\lim\limits_{x \to a+0} (x-a)^{\mu} f(x) = +\infty$ or $-\infty$.

Solved Examples

Example 1. *Test the convergence of the integral* $\int_{1}^{\infty} \dfrac{dx}{\sqrt{x^3+1}}$.

Solution.   We have

$$f(x) = \frac{1}{\sqrt{x^3+1}} = \frac{1}{x^{3/2}\sqrt{1+\dfrac{1}{x^3}}}.$$

Let us consider

$$g(x) = \frac{1}{x^{3/2}} \text{ and } \lim_{x \to \infty} \frac{1}{\sqrt{1+\left(\dfrac{1}{x^3}\right)}} = 1$$

$\Rightarrow$ limit is finite and non-zero.

$\Rightarrow$ $\int_{1}^{\infty} f(x)\,dx$ and $\int_{1}^{\infty} g(x)\,dx$ are either both convergent or divergent

Now by comparison test

$$\int_{1}^{\infty} g(x)\,dx = \int_{1}^{\infty} \frac{dx}{x^{3/2}} \text{ will be convergent} \qquad \left[\text{Since } n = \frac{3}{2} > 1\right]$$

$\Rightarrow$ $\int_{1}^{\infty} f(x)\,dx$ will be convergent.

Example 2. *Test the convergence of the integral* $\int_{0}^{\infty} \dfrac{1-\cos x}{x^2}$.

Solution.   Consider

$$\int_{0}^{\infty} \frac{1-\cos x}{x^2}\,dx = \int_{0}^{\infty} \frac{2\sin^2\left(\dfrac{x}{2}\right)}{x^2}\,dx$$

$$= \int_{0}^{\infty} \frac{2\sin^2 z}{4z^2} \cdot 2\,dz \qquad \left(\text{let } \frac{x}{2} = z \Rightarrow dx = 2dz\right)$$

$$= \int_{0}^{\infty} \frac{\sin^2 z}{z^2}\,dz = \int_{0}^{a} \frac{\sin^2 z}{z^2}\,dz + \int_{a}^{\infty} \frac{\sin^2 z}{z^2}\,dz.$$

Since $\lim\limits_{z\to\infty}\dfrac{\sin^2 z}{z^2}=1$. Therefore, the integrand $\dfrac{\sin^2 z}{z^2}$ is bounded throughout the interval $(0, a)$.

Here $\int_0^a \dfrac{\sin^2 z}{z^2}\,dz$ is a proper integral. Now we have to check the integral $\int_0^\infty \dfrac{\sin^2 z}{z^2}\,dz$ only.

Now,
$$f(x)=\dfrac{\sin^2 z}{z^2},\ g(z)=\dfrac{1}{z^2}$$

$$|f(x)|=\left|\dfrac{\sin^2 z}{z^2}\right|=\dfrac{\sin^2 z}{z^2}\le \dfrac{1}{z^2},\ \text{since } \sin^2 z\le 1$$

By comparison test $\int_a^\infty \dfrac{\sin^2 z}{z^2}\,dz$ will be convergent because $\int_a^\infty \dfrac{dz}{z^2}$ is convergent.

(since $n=2>1$)

Hence, $\int_a^\infty \dfrac{1-\cos x}{x^2}$ is convergent.

Example 3. Test the integral $\int_0^\infty \dfrac{dx}{1+2x^2+3x^4}$ for convergence.

Solution. By comparison test, we have
$$f(x)=\dfrac{1}{1+2x^2+3x^4}<\dfrac{1}{x^4}=g(x),\ \text{for all } x.$$

Also,
$$I=\int_1^\infty g(x)\,dx=\int_1^\infty \dfrac{1}{x^4}\,dx=\lim_{b\to\infty}\int_1^b \dfrac{1}{x^4}\,dx$$

$$=\lim_{b\to\infty}\left[-\dfrac{1}{3x^3}\right]_1^b=\lim_{b\to\infty}\left[-\dfrac{1}{3b^3}+\dfrac{1}{3}\right]=\dfrac{1}{3},\ \text{which is finite.}$$

$\therefore$ I is convergent.

Hence, $\int_0^\infty f(x)\,dx=\int_0^\infty \dfrac{1}{1+2x^2+3x^4}$ is convergent.

Example 4. Test the convergence of the integral $\int_0^\infty \dfrac{\cos mx}{x^2+a^2}\,dx.$.

Solution. Let $f(x)=\dfrac{\cos mx}{x^2+a^2}$ and $g(x)=\dfrac{1}{x^2+a^2}$.

Here $f(x), g(x)$ are positive in the interval $]0,\infty[$ and $f(x)<g(x)$ for all $x\ge 0$.

Also,
$$I=\int_0^\infty g(x)\,dx=\int_0^\infty \dfrac{1}{x^2+a^2}\,dx$$

$$=\lim_{b\to\infty}\int_0^b \dfrac{dx}{x^2+a^2}=\lim_{b\to\infty}\left[\dfrac{1}{a}\tan^{-1}\dfrac{x}{a}\right]_0^b$$

$$=\lim_{b\to\infty}\left[\dfrac{1}{a}\tan^{-1}\dfrac{b}{a}-0\right]=\dfrac{1}{a}\cdot\dfrac{\pi}{2},\ \text{which is finite.}$$

$\therefore$ $\int_0^\infty \dfrac{dx}{x^2+a^2}$ is convergent. Hence, $\int_0^\infty \dfrac{\cos mx}{x^2+a^2}\,dx$ is also convergent.

Example 5. Test the convergence of the following integrals :

(i) $\int_1^\infty \dfrac{dx}{\sqrt{x^5+1}}$

(ii) $\int_1^\infty \dfrac{x^3\,dx}{(x^2+a^2)^2}$

Solution.   (i) Let $f(x) = \dfrac{1}{\sqrt{x^5+1}}$ and $g(x) = x^{-5/2}$.

So that $\qquad \lim\limits_{x\to\infty} \dfrac{f(x)}{g(x)} = \lim\limits_{x\to\infty} \dfrac{x^{5/2}}{\sqrt{x^5+1}} = 1$, finite,

and $\qquad \int_1^\infty g(x)\,dx = \int_1^\infty \dfrac{dx}{x^{5/2}}$ is convergent. $\left[\because n = \dfrac{5}{2} > 1\right]$

Hence, the given integrals converges.

(ii) $\int_0^\infty \dfrac{x^3\,dx}{(x^2+a^2)^2} = f(x)$ and let $g(x) = x^{-1}$ so that,

$$\lim\limits_{x\to\infty} \dfrac{f(x)}{g(x)} = \lim\limits_{x\to\infty} \dfrac{x^4}{(x^2+a^2)^2} = \lim\limits_{x\to\infty} \dfrac{1}{\left(1+\dfrac{a^2}{x^2}\right)^2} = 1.$$

Since $\int_0^\infty g(x)\,dx = \int_0^\infty \dfrac{1}{x}\,dx$ is divergent therefore given integral is also divergent.

Example 6. *Test the convergence of $\int_0^\infty e^{-x}\dfrac{\sin x}{x}\,dx$.*

Solution.   We can write

$$\int_0^\infty e^{-x}\dfrac{\sin x}{x}\,dx = \int_0^1 e^{-x}\dfrac{\sin x}{x}\,dx + \int_1^\infty e^{-x}\dfrac{\sin x}{x}\,dx.$$

Since $\lim\limits_{x\to\infty} e^{-x}\dfrac{\sin x}{x} = 1$, so that the integrand $e^{-x}\dfrac{\sin x}{x}$ is bounded in finite interval $]0, 1[$.

So, $\int_0^1 e^{-x}\dfrac{\sin x}{x}\,dx$ is a proper integral and therefore it is convergent thus we check the convergence of $\int_1^\infty e^{-x}\dfrac{\sin x}{x}\,dx$. Let $f(x) = e^{-x}\dfrac{\sin x}{x}$ then $f(x)$ is bounded in the interval $]1, \infty[$. Take $g(x) = e^{-x}$ then $g(x)$ is positive in the interval $]1, \infty[$.

We have, $\quad |f(x)| = \left| e^{-x}\dfrac{\sin x}{x}\right| = e^{-x}.|\sin x|.\dfrac{1}{x} \le e^{-x}$. $[\because |\sin x| \le 1 \text{ and } 1/x \le 1]$

therefore $|f(x)| \le g(x)$ throughout the interval $]1,\infty[$.

$\therefore \qquad$ By comparison test $\int_1^\infty f(x)\,dx$ is convergent if $\int_1^\infty g(x)\,dx$ is convergent.

Now, $\qquad \int_1^\infty g(x)\,dx = \int_1^\infty e^{-x}\,dx = \lim\limits_{x\to\infty}\int_1^x e^{-x}\,dx$

$$= \lim\limits_{x\to\infty}\left[-e^{-x}\right]_1^x = \lim\limits_{x\to\infty}\left[e^{-x}+e^{-1}\right] = 0 + e^{-1} = \dfrac{1}{e}\,(\text{finite}).$$

Hence, $\int_1^\infty g(x)\,dx$ is convergent. Therefore $\int_1^\infty f(x)\,dx$ is also convergent.

Also, the sum of two convergent integrals is also convergent. Hence, $\int_0^\infty e^{-x}\dfrac{\sin x}{x}\,dx$ is convergent.

Example 7. *Examine the convergence of $\int_1^\infty \dfrac{dx}{x^{1/3}(1+x^{1/2})}$.*

Solution. Let

$$f(x) = \frac{1}{x^{1/3}(1+x^{1/2})} = \frac{1}{x^{1/3}\cdot x^{1/2}\left(1+\dfrac{1}{x^{1/2}}\right)}$$

$$= \frac{1}{x^{5/6}\cdot\left\{\left(1+\dfrac{1}{x^{1/2}}\right)\right\}}$$

$f(x)$ is bounded in the interval $(1,\infty)$ then by μ-test $\mu = \dfrac{5}{6} - 0 = \dfrac{5}{6}$.

We have

$$\lim_{x\to\infty} x^\mu f(x) = \lim_{x\to\infty} x^{5/6}\cdot \frac{1}{x^{5/6}\left\{1+\dfrac{1}{x^{1/2}}\right\}}$$

$$= \lim_{x\to\infty} \frac{1}{\left(1+\dfrac{1}{x^{1/2}}\right)} = 1 \text{ (finite and non-zero)}$$

Since $\mu = \dfrac{5}{6} < 1$, so by μ-test the given integral is divergent.

Example 8. *Examine the convergence of $\int_a^\infty \dfrac{dx}{x(\log x)^{n+1}}$ where $a > 1$.*

Solution. Let $\log x = t$, $\Rightarrow \dfrac{1}{x}\cdot dx = dt$

$\therefore$

$$\int_a^\infty \frac{dx}{x(\log x)^{n+1}} = \int_{\log a}^\infty \frac{dt}{t^{n+1}}.$$

Let $f(t) = \dfrac{1}{t^{n+1}}$, then $f(t)$ is bounded in the interval $(\log a , \infty)$.

But μ-test, take $\mu = (n+1) - 0 = n + 1$, then

$$\lim_{t\to\infty} t^\mu f(t) = \lim_{t\to\infty} \frac{t^{n+1}}{n+1} = \lim_{t\to\infty} 1 = 1 \quad \text{(finite and non-zero)}.$$

The given integral is convergent if $\mu > 1$ *i.e.*, $n + 1 > 1$ *i.e.*, $n > 0$ and divergent if $\mu \le 1$ *i.e.*, $n + 1 \le 1$ *i.e.*, $n \le 0$.

Example 9. *Examine the convergence of $\int_0^\infty \dfrac{x\,dx}{(1+x)^3}$.*

Solution. We have $\int_0^\infty \dfrac{x\,dx}{(1+x)^3} = \int_0^a \dfrac{x\,dx}{(1+x)^3} + \int_a^\infty \dfrac{x\,dx}{(1+x)^3}$.

The integral $\int_0^a \dfrac{x\,dx}{(1+x)^3}$ is convergent because it is a proper integral. Also, the integrand $\dfrac{x}{(1+x)^3}$ is bounded throughout the finite interval $]0, a[$, we need to check the convergence of $\int_a^\infty \dfrac{x\,dx}{(1+x)^3}$.

Let $f(x) = \dfrac{x}{(1+x)^3}$ then $f(x)$ is bounded in the interval $]a, \infty[$. Take $\mu = 3{-}1 = 2$,

then
$$\lim_{x\to\infty} x^{\mu} f(x) = \lim_{x\to\infty} x^2 \cdot \frac{x}{(1+x)^3} = \lim_{x\to\infty} \frac{1}{\left\{1+\left(\dfrac{1}{x}\right)\right\}^3} = 1.$$

Since $\mu = 2$, *i.e,* > 1, therefore by μ-test the integral $\int_0^\infty \dfrac{x\,dx}{(1+x)^3}$ is convergent.

Example 10. *Test the convergence of integral $\int_0^\infty \dfrac{x^{2m}}{1+x^{2n}}dx$, where m and n are positive integers.*

Solution. We have $\int_0^\infty \dfrac{x^{2m}}{1+x^{2n}}dx = \int_0^a \dfrac{x^{2m}}{1+x^{2n}}dx + \int_a^\infty \dfrac{x^{2m}}{1+x^{2n}}dx.$

$\int_0^a \dfrac{x^{2m}}{1+x^{2n}}dx$ is a proper integral and so it is always convergent.

To test the convergence of $\int_a^\infty \dfrac{x^{2m}}{1+x^{2n}}dx$, let us take $\mu = 2n - 2m$.

We have $\lim_{x\to\infty} x^\mu \cdot \dfrac{x^{2m}}{1+x^{2n}} = \lim_{x\to\infty} x^{2n-2m} \dfrac{x^{2m}}{x^{2n}\left(1+\dfrac{1}{x^{2n}}\right)} = \lim_{x\to\infty} \dfrac{1}{1+(1/x^{2n})} = 1$

$\therefore$ by μ-test, if $\mu > 1$ *i.e.*, $2n - 2m > 1$ then the given integral is convergent.

Also, the given integral is divergent if $\mu \le 1$ *i.e.*, if $2n - 2m \le 1$ *i.e.*, if $n \le m$ (since n and m are positive integers).

Example 11. *Test the convergence of $\int_a^\infty e^{-x}\dfrac{\sin x}{x^2}dx$ where a > 0.*

Solution. We have $\int_a^\infty e^{-x}\dfrac{\sin x}{x^2}dx$.

Let $f(x) = \dfrac{\sin x}{x^2}$ and $\phi(x) = e^{-x}$.

Since $\left|\dfrac{\sin x}{x^2}\right| \le \dfrac{1}{x^2}$ and $\int_a^\infty \dfrac{1}{x^2}.dx$ is convergent, therefore by comparison test

$\int_a^\infty \dfrac{\sin x}{x^2}dx$ is also convergent.

Again e^{-x} is monotonically decreasing and bounded function for $x > a$. Therefore by Dirichlet's test $\int_a^\infty e^{-x}\dfrac{\sin x}{x^2}dx$ is convergent.

Example 12. *Test the convergence of the integral*
$$\int_a^\infty \frac{\sin x}{\sqrt{x}}dx, \text{ where } a > 0.$$

Solution. We have $\int_a^\infty \dfrac{\sin x}{\sqrt{x}}dx$.

Let $f(x) = \dfrac{1}{\sqrt{x}}$ and $\phi(x) = \sin x$

Now, $\dfrac{1}{\sqrt{x}}$ is bounded and monotonically decreasing for all $x \ge a$ and $\lim_{x\to\infty} \dfrac{1}{\sqrt{x}} = 0$.

Also $\left|\int_a^\infty \phi(x)dx\right| = \left|\int_a^\infty \sin x\,dx\right| = |\cos a - \cos\infty| \le 2$

$\because$ For all values of x the value of $\cos x$ lies between -1 and 1.

∴ $\left|\int_a^\infty \phi(x)\,dx\right|$ is bounded for all finite values of x.

Hence by Dirichlet's test the integral $\int_0^\infty \frac{\sin x}{\sqrt{x}}\,dx$ is convergent.

Example 13. *Show that the integral $\int_0^\infty \frac{\sin x}{x}\,dx$ is convergent.*

Solution. We have $\int_0^\infty \frac{\sin x}{x}\,dx = \int_0^a \frac{\sin x}{x}\,dx + \int_a^\infty \frac{\sin x}{x}\,dx$.

Since $\lim\limits_{x\to 0} \frac{\sin x}{x} = 1$, therefore the integral $\int_0^a \frac{\sin x}{x}\,dx$ is a proper integral and

hence convergent. Let $f(x) = \frac{1}{x}$ and $\phi(x) = \sin x$. Now, we check the convergence

of $\int_a^\infty \frac{\sin x}{x}\,dx$.

The function $f(x) = \frac{1}{x}$ is bounded and monotonically decreasing for all $x \geq 0$ and

$\lim\limits_{x\to\infty} \frac{1}{x} = 0$.

Also $\left|\int_a^\infty \phi(x)\,dx\right| = \left|\int_a^\infty \sin x\,dx\right| = |\cos a - \cos\infty| \leq 2$ for all finite values of x.

∴ $\left|\int_a^\infty \phi(x)\,dx\right|$ is bounded for all values.

By Dirichlet's test, the integral $\int_a^\infty \frac{\sin x}{x}\,dx$ is convergent.

Since the sum of two convergent integrals is convergent, therefore $\int_0^\infty \frac{\sin x}{x}\,dx$ is convergent.

Example 14. *Show that the integral $\int_0^\infty e^{-ax} \frac{\sin x}{x}\,dx$, $a \geq 0$ is convergent.*

Solution. We have

$\int_0^\infty e^{-ax} \frac{\sin x}{x}\,dx = \int_0^\alpha e^{-ax} \frac{\sin x}{x}\,dx + \int_\alpha^\infty e^{-ax} \frac{\sin x}{x}\,dx$ where $\alpha > 0$

Since $\lim\limits_{x\to 0} e^{-ax} \frac{\sin x}{x} = 1$, the integral $\int_0^\alpha e^{-ax} \frac{\sin x}{x}\,dx$ is convergent because it

is a proper integral.

Now test the convergence of $\int_\alpha^\infty e^{-ax} \frac{\sin x}{x}\,dx$. Let $f(x) = \frac{e^{-ax}}{x}$ and $\phi(x) = \sin x$.

The function $f(x) = \frac{1}{xe^{ax}}$ is bounded and monotonically decreasing for all values

of $x \geq a$ and $\lim\limits_{x\to\infty} f(x) = \lim\limits_{x\to\infty} \frac{1}{xe^{ax}} = 0$

$\left|\int_\alpha^x \phi(x)\,dx\right| = \left|\int_\alpha^x \sin x\,dx\right| = |\cos\alpha - \cos x| \leq 2$ for all values of x.

∴ $\left|\int_\alpha^x \phi(x)\,dx\right|$ is bounded for all finite values of x.

∴ By Dirichlet's test $\int_\alpha^\infty e^{-ax} \frac{\sin x}{x}\,dx$ is convergent.

Since the sum of two convergent integrals is convergent, therefore $\int_0^\infty e^{-ax} \frac{\sin x}{x}\,dx$ is convergent.

Example 15. *Prove that* $\int_a^\infty \dfrac{\cos \alpha x - \cos \beta x}{x} dx$ *is convergent where* $a > 0$.

Solution. We have

$$\int_a^\infty \frac{\cos \alpha x - \cos \beta x}{x} dx = \int_a^\infty \frac{\cos \alpha x}{x} dx - \int_a^\infty \frac{\cos \beta x}{x} dx.$$

The function $f(x) = \dfrac{1}{x}$ is bounded and monotonically decreasing for all $x \geq a$ and $\lim\limits_{x \to \infty} \dfrac{1}{x} = 0$.

Now, $\left| \int_a^x \cos \alpha x \, dx \right| = \left| \dfrac{1}{\alpha} (\sin \alpha x - \sin \alpha a) \right| \leq \dfrac{2}{|\alpha|}$.

$\therefore \quad \left| \int_a^x \cos \alpha x \, dx \right|$ is bounded for all finite values of x.

Similarly $\left| \int_a^x \cos \beta x \, dx \right|$ is bounded for all finite values of x.

By Dirichlet's test both the integrals $\int_a^\infty \dfrac{\cos \alpha x}{x} dx$ and $\int_a^\infty \dfrac{\cos \beta x}{x} dx$ are convergent.

Hence, the given integral (being the difference of two convergent integrals) also convergent.

Example 16. *Show that* $\int_0^\infty \dfrac{\sin mx}{a^2 + x^2} dx$ *converges absolutely.*

Solution. If $\int_0^\infty \left| \dfrac{\sin mx}{a^2 + x^2} \right| dx$ is convergent, then the integral $\int_0^\infty \dfrac{\sin mx}{a^2 + x^2} dx$ will be absolutely convergence.

Let $f(x) = \left| \dfrac{\sin mx}{a^2 + x^2} \right|$ then $f(x)$ is bounded in the interval $]0, \infty[$

we have $f(x) = \left| \dfrac{\sin mx}{a^2 + x^2} \right| = \dfrac{|\sin mx|}{a^2 + x^2} \leq \dfrac{1}{a^2 + x^2}$, since $|\sin mx| \leq 1$.

$\therefore \quad$ By comparison test, $\int_0^\infty f(x) dx$ is convergent if $\int_0^\infty \dfrac{1}{a^2 + x^2} dx$ is convergent.

But $\int_0^\infty \dfrac{dx}{a^2 + x^2} = \lim\limits_{x \to \infty} \int_0^x \dfrac{dx}{a^2 + x^2}$

$$= \lim\limits_{x \to \infty} \left[\frac{1}{a} \tan^{-1} \frac{x}{a} \right]_0^x = \lim\limits_{x \to \infty} \left[\frac{1}{a} \tan^{-1} \frac{x}{a} - 0 \right] = \frac{1}{a} \cdot \frac{\pi}{2}$$

which is a definite real number.

$\therefore \int_0^\infty \dfrac{dx}{x^2 + a^2}$ is convergent. Hence $\int_0^\infty f(x) dx$ is also convergent and so the given integral is absolutely convergent.

Example 17. *Show that* $\int_1^\infty \dfrac{\sin x}{x^4} dx$ *is absolutely convergent.*

Solution. If $\int_1^\infty \left| \dfrac{\sin x}{x^4} \right| dx$ is convergent, then integral $\int_1^\infty \dfrac{\sin x}{x^4} dx$ will be absolutely convergent. Let $f(x) = \left| \dfrac{\sin x}{x^4} \right|$ then $f(x)$ is bounded in the interval $]1, \infty[$.

Now we have

$$f(x) = \left|\frac{\sin x}{x^4}\right| = \frac{|\sin x|}{x^4} \leq \frac{1}{x^4} \cdot (\because |\sin x| \leq 1)$$

∴ By comparison test, if $\int_1^\infty \frac{1}{x^4} dx$ is convergent then $\int_1^\infty f(x) dx$ is convergent.

But the comparison integral $\int_1^\infty \frac{1}{x^4} dx$ is convergent because here $n = 4$ which is greater then 1.

Hence, $\int_1^\infty f(x) dx$ is convergent and so the given integral is absolute convergent.

EXERCISE 8.2

1. Evaluate the following integrals :

(i) $\int_3^\infty \frac{dx}{(x-2)^2}$

(ii) $\int_0^1 \frac{dx}{1-x}$

(iii) $\int_0^1 \frac{dx}{x^3}$

(vi) $\int_2^\infty \frac{dx}{\sqrt{(x^2 - x - 1)}}$

2. Test the convergence of the following integrals :

(i) $\int_0^\infty \frac{\cos mx}{x^2 + a^2} dx$

(ii) $\int_0^\infty \frac{\cos mx}{1 + x^2} dx$

4. Show that the integral $\int_0^\infty e^{-x^2} dx$ is convergent.

5. Show that the following integrals are convergent :

(i) $\int_0^\infty \frac{x^2}{(a^2 + x^2)^2} dx$ (ii) $\int_1^\infty \frac{dx}{(1+x)\sqrt{x}}$

3. Test the convergence of the following integrals :

(i) $\int_0^\infty \frac{\sin^2 x}{x^2} dx$

(ii) $\int_\pi^\infty \frac{\sin^2 x}{x^2} dx$

(iii) $\int_\pi^\infty \frac{\sin x}{x^2} dx$

(iv) $\int_a^\infty \frac{dx}{x\sqrt{(1+x^2)}}, a > 0$

(v) $\int_0^\infty \frac{x^3}{(x^2 + a^2)^2} dx$

6. Show that the following integral is divergent :

$$\int_b^\infty \frac{x^{3/2} dx}{\sqrt{(x^4 - a^4)}}, \text{ where } b > a.$$

7. Show that the integrals $\int_a^\infty x^{n-1} e^{-x} dx$ is convergent when $a > 0$.

8. Test the convergence of the following integrals :

(i) $\int_a^\infty (1 - e^{-x}) \frac{\cos x}{x^2} dx$, when $a > 0$.

(ii) $\int_a^\infty e^{-x} \frac{\sin x}{x^2} dx$, when $a > 0$.

HINTS TO SELECTED PROBLEMS

2. (i) $f(x) = \frac{\cos mx}{x^2 + a^2}$ and $g(x) = \frac{1}{x^2 + a^2}$

$I = \int_0^\infty g(x) dx = \int_0^\infty \frac{1}{x^2 + a^2} dx$

$= \lim_{b \to \infty} \int_0^b \frac{dx}{x^2 + a^2}$

$= \lim_{b \to \infty} \left[\frac{1}{a} \tan^{-1} \frac{x}{a}\right]_0^b = \frac{\pi}{2a}$,

a finite quantity.

Since $f(x) < g(x)$.

∴ $f(x)$ is convergent.

4. $\int_0^\infty e^{-x^2} dx = \int_0^1 e^{-x^2} dx + \int_1^\infty e^{-x^2} dx$.

Let $f(x) = e^{-x^2}$,

⇒ $g(x) = xe^{-x^2}$ in $]1, \infty[$.
     $|f(x)| \leq g(x)$

Now apply comparison test.

6. Let $f(x) = \dfrac{x^{3/2}}{\sqrt{x^4 - a^4}}$, therefore $f(x)$ is bounded in the interval $]b, \infty[$. Take $\mu = 2 - \dfrac{3}{2} = \dfrac{1}{2}$

$$\lim_{x \to \infty} x^{\mu} f(x) = \lim_{x \to \infty} x^{1/2} \dfrac{x^{3/2}}{\sqrt{x^4 - a^4}}$$

$$= \lim_{x \to \infty} \dfrac{x^{1/2} \cdot x^{3/2}}{x^2 \sqrt{\left(1 - \dfrac{a^4}{x^4}\right)}}$$

$$= 1, \text{ which is finite.}$$

8. Use Abel's test.

Answers

1. (i) $-\infty$ (ii) ∞ **2.** (i) convergent   (ii) convergent   (iii) divergent
3. (i) convergent   (ii) convergent   (iii) convergent   (iv) convergent   (v) divergent   (vi) divergent.

8.6 MORE ABOUT IMPROPER INTEGRALS OF SECOND KIND

We know that an integral $\int_a^b f(x)\,dx$ is said to be of second kind if the range of integration is finite and the integrand $f(x)$ is unbounded at one or more points of the given interval $[a, b]$. Here, it is sufficient to consider the case when $f(x)$ becomes unbounded at $x = a$ and bounded for all other values of x in the interval $[a, b]$.

$\therefore$ We have $\int_a^b f(x)\,dx = \lim_{h \to 0} \int_{a+h}^b f(x)\,dx, \ h > 0.$

Solved Examples

Example 1.   Test the convergence of the integral $\int_0^1 \dfrac{dx}{x^3(1+x^2)}$.

Solution.   Here, it is clear that the integral

$$f(x) = \int_0^1 \dfrac{dx}{x^3(1+x^2)}$$

is unbounded at $x = 0$.

Let $\phi(x) = \dfrac{1}{x^3}.$

$\therefore$ $\lim_{x \to 0} \dfrac{f(x)}{\phi(x)} = \lim_{x \to 0} \dfrac{1}{1+x^2} = 1$, *i.e.*, finite and non-zero.

Then, by comaparison test $\int_0^1 f(x)\,dx$ and $\int_0^1 \phi(x)\,dx$ either both converges or both diverges. But clearly $\int_0^1 \dfrac{dx}{x^3}$ is convergent.

Hence, the integral $\int_0^1 \dfrac{dx}{x^3(1+x^2)}$ is convergent.

Example 2.   Test the convergence of the integral $\int_0^{\pi/2} \dfrac{\cos x}{x^2}\,dx$.

Solution.   Here, the integral $f(x) = \dfrac{\cos x}{x^2}$ is unbounded at $x = 0$.

Let $\phi(x) = \dfrac{1}{x^2}$

Then $$\lim_{x\to 0}\frac{f(x)}{\phi(x)} = \lim_{x\to 0}\left\{\frac{\cos x}{x^2}\cdot x^2\right\}$$

$$\lim_{x\to 0}\cos x = 1, \text{ finite and non-zero}.$$

$\therefore$ By comparison test the integrals $\int_0^{\pi/2} f(x)\,dx$ and $\int_0^{\pi/2}\phi(x)\,dx$ either both converge or both diverge.

But $$\int_0^{\pi/2}\phi(x)\,dx = \int_0^{\pi/2}\frac{1}{x^2}\,dx = \lim_{h\to 0}\int_h^{\pi/2}\frac{1}{x^2}\,dx$$

$$= \lim_{h\to 0}\left[-\frac{1}{x}\right]_h^{\pi/2} = \lim_{x\to 0}\left[-\frac{2}{\pi}+\frac{1}{h}\right] = \infty.$$

$\therefore$ $\int_0^{\pi/2}\phi(x)\,dx$ is divergent.

Hence, the integral $\int_0^{\pi/2}\frac{\cos x}{x^2}\,dx$ is divergent.

Example 3. *Show that the integral $\int_0^1\frac{dx}{\sqrt{\{x(1-x)\}}}$ converges.*

Solution. Here $f(x) = \dfrac{1}{\sqrt{\{x(1-x)\}}}$ is unbounded at $x=0$ and 1.

Let a be any number such that $0 < a < 1$.

Then $$\int_0^1\frac{1}{\sqrt{\{x(1-x)\}}} = \int_0^a\frac{dx}{\sqrt{\{x(1-x)\}}} + \int_a^1\frac{dx}{\sqrt{\{x(1-x)\}}} = I_1 + I_2.$$

In the integral I_1, the integrand $f(x)$ is unbounded at lower limit of integration $x=0$ and in integration I_2, then integrand $f(x)$ is unbounded at the upper limit of integration $x=1$.

To test the convergence of I_1, taking $\mu = \dfrac{1}{2}$, such that

$$\lim_{x\to 0} x^\mu f(x) = \lim_{x\to 0}\frac{x^{1/2}}{\sqrt{\{x(1-x)\}}} = \lim_{x\to 0}\frac{1}{\sqrt{1-x}} = 1.$$

So, the above limit exists.

Since, $0 < \mu < 1$, so I_1 is convergent by μ-test.

To test the convergence of I_2 taking $\mu = \dfrac{1}{2}$, we have

$$\lim_{x\to 1-0}(1-x)^\mu\cdot f(x) = \lim_{x\to 1-0}(1-x)^{1/2}\cdot\frac{1}{\sqrt{\{x(1-x)\}}}$$

$$= \lim_{x\to 1-0}\frac{1}{\sqrt{x}} = \lim_{h\to 0}\frac{1}{\sqrt{1-h}} = 1.$$

Since $0 < \mu < 1$, so I_2 is convergent by μ-test.

Thus, the given integral is the sum of two convergent integrals. Hence, the given integral is convergent.

Example 4. *Show that the integral $\int_0^{\pi/4}\frac{1}{\sqrt{\tan x}}\,dx$ is convergent.*

Solution. Here, then given integrand $f(x)$ is unbounded at $x=0$.

Taking $\mu = \dfrac{1}{2}$, we have

$$\lim_{x \to 0} x^{\mu} \cdot f(x) = \lim_{x \to 0} x^{1/2} \cdot \frac{1}{\sqrt{\tan x}}$$

$$= \lim_{x \to 0} \sqrt{\frac{x}{\sin x}} \cdot \sqrt{\cos x} = 1.1 = 1.$$

Since $0 < \mu < 1$.

Hence, the given integral is convergent by μ-test.

Example 5. *Test the convergence of $\int_0^{\pi/2} \dfrac{\cos x}{x^n} dx$.*

Solution. If $n \le 0$, then given integral is a proper integral and hence it is convergent.

If $n > 0$, then the integrand becomes unbounded at $x = 0$, we have

$$f(x) = \frac{\cos x}{x^n}$$

Then $\qquad \lim_{x \to 0} x^{\mu} f(x) = \lim_{x \to 0} x^{\mu - n} \cdot \cos x = 1$ if $\mu = n$.

Hence, the given integral is convergent if $0 < n < 1$ and the given integral is divergent, if $n \ge 1$.

Example 6. *Test the convergence of the integral $\int_0^1 x^{n-1} \log x \, dx$.*

Solution. Since $\lim_{x \to 0} x^r \log x = 0$ where $r > 0$, the integral is a proper integral if $n > 1$.

If $n = 1$, then we have

$$\int_0^1 \log x \, dx = \lim_{h \to 0} \int_h^1 \log x \, dx = \lim_{h \to 0} \left[x \log x - x \right]_h^1$$

$$= \lim_{h \to 0} \left[-1 - h \log h + h \right] = -1.$$

So the given integral is convergent if $n = 1$.

If $n < 1$ and $f(x) = x^{n-1} \log x$ then, we have

$$\lim_{x \to 0} x^{\mu} f(x) = \lim_{x \to 0} x^{\mu + n - 1} \cdot \log x = 0 \quad \text{if } \mu > 1 - n \qquad \qquad \text{...(1)}$$

$$= -\infty \text{ if } \mu \le 1 - n. \qquad \qquad \text{...(2)}$$

Hence, if $0 < n < 1$, then we can take μ between 0 and 1 and satisfying (1).

Then if $0 < n < 1$ then the integral is convergent by μ–test. Again if $n \le 0$ then, we can take $\mu = 1$ and satisfying (2). Hence if $n \le 0$ then the integral is divergent by μ-test.

Hence, the given integral is convergent if $n > 0$ and divergent if $n \le 0$.

Example 7. *Discuss the convergence of the integral $\int_0^{\infty} \dfrac{\sin mx}{a^2 + x^2} dx$ and show that it will be absolutely convergent.*

Solution. Since given that $I = \int_0^{\infty} \dfrac{\sin mx}{a^2 + x^2} dx$ will be absolutely convergent.

If $\int_0^{\infty} \left| \dfrac{\sin mx}{a^2 + x^2} \right| dx$ is convergent

Let $f(x) = \left| \dfrac{\sin mx}{a^2 + x^2} \right|$ then $f(x)$ is bounded in the interval $(0, \infty)$

$$f(x) = \frac{|\sin mx|}{a^2 + x^2} \le \frac{1}{a^2 + x^2}.$$

$\therefore$ $\int_0^\infty f(x)\,dx$ will be convergent if $\int_0^\infty \dfrac{1}{a^2+x^2}$ is convergent.

$$\int_0^\infty \frac{1}{a^2+x^2} = \lim_{x\to\infty} \int_0^x \frac{dx}{a^2+x^2}$$

$$= \lim_{x\to\infty}\left[\frac{1}{a}\tan^{-1}\frac{x}{a}\right]_a^x = \lim_{x\to\infty}\left[\frac{1}{a}\tan^{-1}\frac{x}{a}-0\right] = \frac{\pi}{2a}.$$

$\Rightarrow$ a finite and non-zero real number

Hence, $\int_0^\infty f(x)\,dx$ will be convergent.

Example 8. *Show that the convergence of given integral is of absolutely convergent where*

$$I = \int_0^\infty e^{-x}\cos mx\,dx.$$

Solution. The given integral $\int_0^\infty e^{-x}\cos mx\,dx$ will be absolutely convergent if $\int_0^\infty |e^{-x}\cos mx|\,dx$ is convergent.

Also, $f(x) = |e^{-x}\cos mx|$ is bounded in the interval $(0, \infty)$

$$= e^{-x}|\cos mx| \le e^{-x}.$$

Now by comparison test, $\int_0^\infty f(x)\,dx$ will be convergent if $\int_0^\infty e^{-x}\,dx$ is convergent.

$$\int_0^\infty e^{-x}\,dx = \lim_{x\to\infty}\int_0^x e^{-x}\,dx = \lim_{x\to\infty}\left(-e^{-x}\right)_0^x$$

$$= \lim_{x\to\infty}(-e^{-x}+1) = 1 \text{ , a finite real number}$$

$$\Rightarrow \int_0^\infty e^{-x}\,dx \text{ is convergent}$$

$$\Rightarrow \int_0^\infty f(x)\,dx \text{ is convergent}$$

$$\Rightarrow \int_0^\infty e^{-x}\cos mx\,dx \text{ is absolutely convergent.}$$

Example 9. *Discuss the convergence of the given integral $\int_0^\infty x^{n-1}e^{-x}\,dx$, if $n > 0$.*

Solution. Here given that

$$I = \int_0^\infty x^{n-1}e^{-x}\,dx = \int_0^1 x^{n-1}e^{-x}\,dx + \int_1^\infty x^{n-1}e^{-x}\,dx.$$

Let

$$I_1 = \int_0^1 x^{n-1}e^{-x}\,dx$$

$$I_2 = \int_1^\infty x^{n-1}e^{-x}\,dx.$$

Here for discuss the convergence of given integral, we use μ-test in I_2 and comparison test in I_1.

$$I_1 = \int_0^1 x^{n-1}e^{-x}\,dx$$

$$f(x) = x^{n-1}e^{-x} \text{ . Clearly, at } x = 0, \text{ it will be unbounded.}$$

Let

$$g(x) = x^{n-1}$$

and

$$\lim_{x\to 0}\frac{f(x)}{g(x)} = \lim_{x\to 0} e^{-x} = 1.$$

By comparison test if $g(x)$ is convergent then $f(x)$ will also be convergent or if divergent then $f(x)$ will be divergent.

$$\int_0^1 g(x)\,dx = \int_0^1 x^{n-1}\,dx = \lim_{\varepsilon \to 0} \int_\varepsilon^1 x^{n-1}\,dx$$

$$= \lim_{\varepsilon \to 0}\left[\frac{x^n}{n}\right]_\varepsilon^1 = \lim_{\varepsilon \to 0}\left[\frac{1}{n} - \frac{\varepsilon^n}{n}\right]$$

$$= \frac{1}{n}, \text{ which is a finite number.}$$

$\Rightarrow \quad \int_0^1 g(x)\,dx$ is convergent. $\quad \Rightarrow f(x)$ will be convergent.

Now $\qquad\qquad I_2 = \int_1^\infty x^{n-1} e^{-x}\,dx$

Here $f(x) = x^{n-1}\, e^{-x}$. It is bounded in the interval $(1, \infty)$

$$\lim_{x \to \infty} x^\mu f(x) = \lim_{x \to \infty} \frac{x^\mu \cdot x^{n-1}}{e^x} = \lim_{x \to \infty} \frac{x^{\mu+n-1}}{1 + x + \dfrac{x^2}{2!} + \dots} = 0$$

For $\mu > 1$, we have $\int_1^\infty x^{n-1} e^{-x}\,dx$ is convergent.

From the above result I will be convergent because I_1 and I_2 both are convergent.

Example 10. *Show that the integral $\int_a^\infty x^{n-1} e^{-x}\,dx$ is convergent where $a > 0$.*

Solution.   For this we proceed same as in I_2 replace 1 by a in above example.

8.7 SOME TYPICAL PROBLEMS

Example 1. *Examine the convergence of the following improper integrals*

(a) $\int_{-\infty}^0 \dfrac{dx}{p^2 + q^2 x^2}$ (b) $\int_{-\infty}^0 \dfrac{x}{1 + x^2}\,dx$

(c) $\int_{-\infty}^\infty \dfrac{dx}{e^x + e^{-x}}$

Solution.   (a) We can write

$$\int_{-\infty}^0 \frac{dx}{p^2 + q^2 x^2} = \lim_{t \to -\infty} \int_t^0 \frac{dx}{p^2 + q^2 x^2}$$

$$= \lim_{t \to -\infty} \int_t^0 \frac{dx}{q^2\left[x^2 + \left(\dfrac{p}{q}\right)^2\right]} = \frac{1}{q^2} \lim_{t \to -\infty}\left[\frac{q}{p}\tan^{-1}\frac{qx}{p}\right]_t^0$$

$$= \frac{1}{q^2} \lim_{t \to -\infty}\left[\frac{q}{p}\tan^{-1} 0 - \frac{q}{p}\tan^{-1}\frac{qt}{p}\right]$$

$$= -\frac{1}{q^2}\cdot\frac{q}{p}\tan^{-1}(-\infty) = \frac{\pi}{2pq}, \text{ which is finite}$$

    Hence, the given integral is convergent.

(b) We can write

$$\int_{-\infty}^0 \frac{x}{1 + x^2}\,dx = \frac{1}{2}\int_{-\infty}^0 \frac{2x}{1 + x^2}\,dx$$

Now $\dfrac{1}{2}\int_{-\infty}^{0}\dfrac{2x}{1+x^2}dx = \dfrac{1}{2}\lim_{t\to-\infty}\int_{t}^{0}\dfrac{2x}{1+x^2}dx$

$$= \dfrac{1}{2}\lim_{t\to-\infty}\,|\log(1+x^2)|_{t}^{0}$$

$$= \dfrac{1}{2}\lim_{t\to-\infty}[\log 1 - \log(1+t^2)]$$

$$= \dfrac{1}{2}(-\log\infty) = -\infty$$

$\Rightarrow$ Given integral is divergent.

(c) We can write

$$\int_{-\infty}^{\infty}\dfrac{dx}{e^x+e^{-x}} = \int_{-\infty}^{c}\dfrac{dx}{e^x+e^{-x}} + \int_{c}^{\infty}\dfrac{dx}{e^x+e^{-x}} \qquad\text{...(1)}$$

Let $\qquad I_1 = \int_{-\infty}^{c}\dfrac{dx}{e^x+e^{-x}}$

$$= \int_{-\infty}^{c}\dfrac{e^x}{e^{2x}+1}dx = \lim_{t_1\to-\infty}\int_{t_1}^{c}\dfrac{e^x}{e^{2x}+1}dx$$

Put $\quad e^x = z \Rightarrow e^x dx = dz$

$\therefore \qquad \int_{-\infty}^{c}\dfrac{dx}{e^x+e^{-x}} = \lim_{t_1\to-\infty}\int_{e^{t_1}}^{c}\dfrac{dx}{1+z^2}$

$$= \lim_{t_1\to-\infty}|\tan^{-1}z\,|_{e^{t_1}}^{e^c} = \lim_{t_1\to-\infty}[\tan^{-1}e^c - \tan^{-1}e^{t_1}]$$

$$= [\tan^{-1}e^c - \tan^{-1}(e)^{-\infty}]$$

$$= [\tan^{-1}e^c - \tan^{-1}0]$$

$$= \tan^{-1}e^c = \text{a finite quantity}$$

Particularly, if $c = 0$, then

$$\int_{-\infty}^{c}\dfrac{dx}{e^x+e^{-x}} = \tan^{-1}1 = \dfrac{\pi}{4} = \text{a finite quantity}$$

Now consider $\quad I_2 = \int_{c}^{\infty}\dfrac{dx}{e^x+e^{-x}} = \lim_{t_2\to\infty}\int_{c}^{t_2}\dfrac{e^x}{e^{2x}+1}.dx$

Put $\quad e^x = z \Rightarrow e^x dx = dz$

Then

$$I_2 = \lim_{t_2\to\infty}\int_{e^c}^{e^{t_2}}\dfrac{dz}{1+z^2} = \lim_{t_2\to\infty}|\tan^{-1}z\,|_{e^c}^{e^{t_2}}$$

$$= \lim_{t_2\to\infty}\left[\tan^{-1}e^{t_2} - \tan^{-1}e^c\right]$$

$$= \tan^{-1}(\infty) - \tan^{-1}e^c$$

$$= \dfrac{\pi}{2} - \tan^{-1}e^c = \text{a finite quantity.}$$

$$= \lim_{t_2\to\infty}[\tan^{-1}e^{t_2} - \tan^{-1}e^c] = \tan^{-1}(\infty) - \tan^{-1}e^c$$

$$= \dfrac{\pi}{2} - \tan^{-1}e^c = \text{a finite quantity}$$

In particular if $c = 0$, then

$$\int_c^\infty \frac{dx}{e^x + e^{-x}} = \frac{\pi}{2} - \tan^{-1} 1 = \frac{\pi}{2} - \frac{\pi}{4} = \frac{\pi}{4}, \text{ (finite)}$$

$\Rightarrow \quad I_1$ and I_2 both integrals are convergent.

Hence, from (1)

$$\int_{-\infty}^\infty \frac{dx}{e^x + e^{-x}} = \tan^{-1} e^c + \frac{\pi}{2} - \tan^{-1} e^c = \frac{\pi}{2}$$

$\Rightarrow$ Given integral is convergent and converges to $\frac{\pi}{2}$.

Example 2. *Test the convergence of the integral* $\int_0^{\pi/2} \frac{\cos x}{\sqrt{1 - \sin x}} dx$.

Solution. Let $\sin x = t \Rightarrow \cos x \, dx = dt$

$$\therefore \qquad \int_0^{\pi/2} \frac{\cos x}{\sqrt{1 - \sin x}} dx = \int_0^1 \frac{dt}{\sqrt{1 - t}}$$

$$= \lim_{\varepsilon \to 0^+} \int_0^{1-\varepsilon} \frac{dt}{\sqrt{1 - t}} = \lim_{\varepsilon \to 0^+} \int_0^{1-\varepsilon} (1 - t)^{-1/2} dt$$

$$= \lim_{\varepsilon \to 0^+} \left[-2(1-t)^{1/2} \right]_0^{1-\varepsilon} = \lim_{\varepsilon \to 0^+} [-2\sqrt{\varepsilon} + 2] = 2$$

$\Rightarrow$ Given integral converges to 2.

Example 3. *Test the convergence of the integral* $\int_0^1 \frac{dx}{x^2 - 3x + 2}$.

Solution. Clearly, 1 is the only point of infinite discontinuity.

Now, $$\int_0^1 \frac{dx}{x^2 - 3x + 2} = \lim_{\varepsilon \to 0^+} \int_0^{1-\varepsilon} \frac{dx}{x^2 - 3x + 2}$$

$$= \lim_{\varepsilon \to 0^+} \int_0^{1-\varepsilon} \frac{dx}{(1 - x)(2 - x)} = \lim_{\varepsilon \to 0^+} \int_0^{1-\varepsilon} \left[\frac{1}{1 - x} - \frac{1}{2 - x} \right] dx$$

$$= \lim_{\varepsilon \to 0^+} \left[-\log(1 - x) + \log(2 - x) \right]_0^{1-\varepsilon}$$

$$= \lim_{\varepsilon \to 0^+} [-\log \varepsilon + \log(1 + \varepsilon) - \log 2]$$

$$= \lim_{\varepsilon \to 0^+} \left[\log \frac{1 + \varepsilon}{\varepsilon} - \log 2 \right]$$

$$= \lim_{\varepsilon \to 0^+} \left[\log \left(1 + \frac{1}{\varepsilon} \right) - \log 2 \right] = \log \infty - \log 2 = \infty$$

$\Rightarrow$ Given integral is divergent.

Example 4. *Test the convergence of the integral* $\int_0^{1/e} \frac{dx}{x(\log x)^2}$.

Solution. Clearly 0 is the point of infinite discontinuity.

$$\therefore \qquad \int_0^{1/e} \frac{dx}{x(\log x)^2} = \lim_{\varepsilon \to 0^+} \int_\varepsilon^{1/e} \frac{1}{x} (\log x)^{-2} dx$$

$$\Rightarrow \quad \lim_{\varepsilon \to 0^+}\left[\frac{(\log x)^{-2+1}}{-2+1}\right]_{\varepsilon}^{1/e} = \lim_{\varepsilon \to 0^+}\left[-\frac{1}{\log e^{-1}}+\frac{1}{\log \varepsilon}\right]$$

$$= \lim_{\varepsilon \to 0^+}\left[\frac{1}{\log e}+\frac{1}{\log \varepsilon}\right] = 1+\frac{1}{\log 0} = 1-0 = 1, \text{ finite}$$

Hence, the given integral converges to 1.

Example 5. Show that the integral $\int_0^{\pi/2} \frac{\sin^m x}{x^n}dx$ exists if and only if $n < m+1$.

Solution. Let

$$I = \int_0^{\pi/2} \frac{\sin^m x}{x^n}dx$$

We have

$$f(x) = \frac{\sin^m x}{x^n} = \left(\frac{\sin x}{x}\right)^m \cdot \frac{1}{x^{n-m}}$$

$$\therefore \qquad \lim_{x \to 0^+} f(x) = \lim_{x \to 0^+}\left(\frac{\sin x}{x}\right)^m \cdot \frac{1}{x^{n-m}} = \begin{cases} 0 \text{ if } n-m < 0 \\ 1 \text{ if } \quad n=m \\ \infty \text{ if } n-m > 0 \end{cases}$$

$\Rightarrow$ For $n \le m$, the given integral I is a proper integral and hence convergent and

0 is the only point of infinite discontinuity of f in $\left(0, \frac{\pi}{2}\right)$ if $n > m$.

Also $f(x) > 0$ in $\left]0, \frac{\pi}{2}\right]$

Let

$$g(x) = \frac{1}{x^{n-m}}, n > m$$

Then $\lim_{x \to 0^+} \frac{f(x)}{g(x)} = \lim_{x \to 0^+}\left(\frac{\sin x}{x}\right)^m = 1 \ne 0, \infty.$

So, the integrals $\int_0^{\pi/2} f(x)dx$ and $\int_0^{\pi/2} g(x)dx$ converge or diverge together.

But the integral $\int_0^{\pi/2} g(x)dx = \int_0^{\pi/2} \frac{dx}{x^{n-m}}, n > m$ is convergent if and only if $n - m < 1.$

Also, $\int_0^{\pi/2} f(x)dx$ is convergent for $n \ge m.$

Hence the given integral is convergent iff $n < m+1.$

Example 6. Show that $\int_0^{\pi/2} \sin x \log(\sin x)dx$ is convergent with the value $\log\left(\frac{2}{e}\right).$

Solution. Let

$$I = \int_0^{\pi/2} \sin x \log(\sin x)dx$$

Here,

$$f(x) = \sin x \log (\sin x)$$

Clearly 0 is the only point of infinite discontinuity of f in $\left[0, \frac{\pi}{2}\right].$

so $\int_0^{\pi/2} f(x)dx = \lim_{\varepsilon \to 0^+}\int_\varepsilon^{\pi/2} \sin x \log(\sin x)dx$

$$= \lim_{\varepsilon \to 0^+}\left[\log(\sin x)(-\cos x)\right]_\varepsilon^{\pi/2} + \int_\varepsilon^{\pi/2} \frac{1}{\sin x}\cos^2 x \, dx$$

$$= \lim_{\varepsilon \to 0^+} \left[\cos \varepsilon \log(\sin \varepsilon)\right] + \int_\varepsilon^{\pi/2} \frac{1 - \sin^2 x}{\sin x} dx$$

$$= \lim_{\varepsilon \to 0^+} \left[\cos \varepsilon \log(\sin \varepsilon) + \int_\varepsilon^{\pi/2} (\cosec x - \sin x) dx\right]$$

$$= \lim_{\varepsilon \to 0^+} \left[\cos \varepsilon \log(\sin \varepsilon) + \left\{\log \tan \frac{x}{2} + \cos x\right\}_\varepsilon^{\pi/2}\right]$$

$$= \lim_{\varepsilon \to 0^+} \left[\cos \varepsilon \log(\sin \varepsilon) - \log \tan \frac{\varepsilon}{2} - \cos \varepsilon\right]$$

$$= \lim_{\varepsilon \to 0^+} \left[\cos \varepsilon \log\left(2 \sin \frac{\varepsilon}{2} \cos \frac{\varepsilon}{2}\right) - \log \sin \frac{\varepsilon}{2} + \log \cos \frac{\varepsilon}{2} - \cos \varepsilon\right]$$

$$= \lim_{\varepsilon \to 0^+} \left[\cos \varepsilon \log\left(2 \cos \frac{\varepsilon}{2}\right) + \cos \varepsilon \log \sin \frac{\varepsilon}{2} - \log \sin \frac{\varepsilon}{2} + \log \cos \frac{\varepsilon}{2} - \cos \varepsilon\right]$$

$$= \log 2 - \lim_{\varepsilon \to 0^+} (1 - \cos \varepsilon) \log \sin \frac{\varepsilon}{2} - 1$$

$$= \log 2 - 1 = \log 2 - \log e = \log\left(\frac{2}{e}\right)$$

Hence, the given integral converges to $\log\left(\frac{2}{e}\right)$.

8.7 CAUCHY'S GENERAL TEST FOR CONVERGENCE AT ∞, WHERE THE INTEGRAND MAY NOT NECESSARILY POSITIVE

The necessary and sufficient condition for the convergence of the improper integral $\int_a^\infty f(x) dx$ *at* ∞ *is that to every* $m > 0 \ \exists$ *a positive number* t_0 *such that*

$$\left|\int_{t_1}^{t_2} f(x) dx\right| < m \ \forall \ t_1, t_2 \geq t_0$$

Proof. We can write

$$\int_a^\infty f \, dx = \lim_{t \to \infty} \int_a^t f \, dx$$

The improper integral $\int_a^\infty f \, dx$ converges if and only if $\lim_{t \to \infty} dx$ exists finitely.

Now let

$$F(t) = \int_a^t f \, dx$$

Then by Cauchy's convergence theorem, we have

$$|F(t_2) - F(t_1)| < m \ \forall \ t_1, t_2 > t_0$$

$$\Leftrightarrow \quad \left|\int_a^{t_2} f \, dx - \int_a^{t_1} f \, dx\right| < m \ \forall \ t_1, t_2 > t_0$$

$$\Leftrightarrow \quad \left|\int_a^{t_2} f \, dx + \int_{t_1}^a f \, dx\right| < m \ \forall \ t_1, t_2 > t_0$$

$$\Leftrightarrow \quad \left|\int_{t_1}^{t_2} f \, dx\right| < m \ \forall \ t_1, t_2 > t_0$$

8.8 FRULLANI'S THEOREM

Let $f(x)$ be a continuous function on $[0, \infty[$ having singularities at 0 and ∞ only, then $\int_0^\infty \dfrac{f(ax)-f(bx)}{x}$ is equal to

(i) 0 if $\int_0^\infty \dfrac{f(x)}{x}dx$ converges to 0 and ∞.

(ii) $f_1 \log\left|\dfrac{a}{b}\right|$ if $\int_0^\infty \dfrac{f(x)}{x}dx$ converges to 0 only.

(iii) $f_0 \log\left|\dfrac{b}{a}\right|$ if $\int_0^\infty \dfrac{f(x)}{x}dx$ converges at ∞ only

(iv) $(f_0 - f_1)\log\left|\dfrac{b}{a}\right|$ if $\int_0^\infty \dfrac{f(x)}{x}dx$ neither converges at 0 nor at ∞.

where $\lim_{x\to 0} f(x) = f_1$ *and* $\lim_{x\to 0} f(x) = f_1$

Proof. Let $1 > \varepsilon > 0$ and $p > 1$, then

$$\int_\varepsilon^p \frac{f(ax)-f(bx)}{x}dx = \int_\varepsilon^p \frac{f(ax)}{x}dx - \int_\varepsilon^p \frac{f(bx)}{x}dx \qquad \ldots(1)$$

Put $ax = t$ and $bx = t$ in first and second integrals on the RHS of (1) we get

$$\int_\varepsilon^p \frac{f(ax)-f(bx)}{x}dx = \int_{a\varepsilon}^{ap} \frac{f(t)}{t}dt - \int_{b\varepsilon}^{bt} \frac{f(t)}{t}dt$$

$$= \int_{a\varepsilon}^{b\varepsilon} \frac{f(t)}{t}dt - \int_{ap}^{bp} \frac{f(t)}{t}dt \qquad \ldots(2)$$

$$= f(\alpha)\int_{a\varepsilon}^{b\varepsilon} \frac{dt}{t} - f(\beta)\int_{ap}^{bp} \frac{dt}{t}, \quad \alpha \in [a\varepsilon, b\varepsilon] \qquad \ldots(3)$$
$$\beta \in [ap, bp]$$

Now, let $\varepsilon \to 0^+$ and $p \to \infty$, $\alpha \to 0^+$, $\beta \to \infty$ respectively and therefore $f(x) \to f_0$ and $f(x) \to f_1$ respectively.

(i) If $\int_0^\infty \dfrac{f(t)}{t}dx$ converges to 0 and ∞ then $\int_0^1 \dfrac{f(t)}{t}dt$ converges at 0 and $\int_1^\infty \dfrac{f(t)}{t}dt$ converges at ∞ and therefore $\int_{a\varepsilon}^{b\varepsilon} \dfrac{f(t)}{t}dt \to 0$ as $\varepsilon \to 0$ and $\int_{ap}^{bp} \dfrac{f(t)}{t}dt$ as $p \to \infty$.

Hence, $\int_0^\infty \dfrac{f(ax)-f(bx)}{x}dx = 0$

(ii) If $\int_0^\infty \dfrac{f(t)}{t}dx$ converges at 0 only, then $\int_{a\varepsilon}^{b\varepsilon} \dfrac{f(t)}{t}dt \to 0$ as $\varepsilon \to 0$, so by (3), we get

$$\int_0^p \frac{f(ax)-f(bx)}{x}dx = -f(\beta)\int_{ap}^{bp} \frac{1}{t}dt$$

$$= -f(\beta)[\log|t|]_{ap}^{bp} = -f(\beta)\log\left|\frac{bp}{ap}\right|$$

$$= -f(\beta)\log\left|\frac{b}{a}\right| = f(\beta)\log\left|\frac{a}{b}\right|$$

Letting $p \to \infty$, we get

$$\int_0^\infty \frac{f(ax) - f(bx)}{x}\,dx = \lim_{p \to \infty} f(\beta)\log\left|\frac{a}{b}\right| = f_1 \log\left|\frac{a}{b}\right|$$

(iii) Suppse $\int_0^\infty \frac{f(t)}{t}\,dx$ converges at ∞ only, then $\int_{ap}^{bp} \frac{f(t)}{t}\,dt \to 0$ as $p \to \infty$,

so by (3), we have

$$\int_\varepsilon^\infty \frac{f(ax) - f(bx)}{x}\,dx = f(\alpha)\int_{a\varepsilon}^{b\varepsilon} \frac{dt}{t} = f(\alpha)\log\left|\frac{b}{a}\right|$$

Letting $\varepsilon \to 0$, we get

$$\int_0^\infty \frac{f(ax) - f(bx)}{x}\,dx = \lim_{\varepsilon \to 0^+} f(\alpha)\log\left|\frac{b}{a}\right| = f_0 \log\left|\frac{b}{a}\right|$$

(iv) Let $\int_0^\infty \frac{f(t)}{t}\,dx$ neither converges at 0 nor at ∞, so by (3)

We have

$$\int_\varepsilon^p \frac{f(ax) - f(bx)}{x}\,dx = f(\alpha)\log\left|\frac{b}{a}\right| - f(\beta)\log\left|\frac{b}{a}\right|$$

$$= [f(\alpha) - f(\beta)]\log\left|\frac{b}{a}\right|$$

Letting $\varepsilon \to 0$ and $p \to \infty$, we get

$$\int_0^\infty \frac{f(ax) - f(bx)}{x}\,dx = (f_0 - f_1)\log\left|\frac{b}{a}\right|$$

8.9 MORE RESULTS

(1) Continuity of the integral

The function $F(\alpha) = \int_a^b f(x, \alpha)g(x)\,dx$ where $f(x, \alpha)$ is continuous in $a \le x \le b$, $\lambda_1 \le \alpha \le \lambda_2$ and $g(x)$ is bounded and integrable on $[a, b]$ is a continuous function of α on $[\lambda_1, \lambda_2]$.

(2) Derivability of the integral [Leibnitz Rule]

The function $F(\alpha) = \int_a^b f(x, \alpha)g(x)\,dx$ where $f(x, \alpha)$, $\frac{\partial}{\partial \alpha} f(x, \alpha)$ are continuous in $a \le x \le b$, $\lambda_1 \le \alpha \le \lambda_2$ and $g(x)$ is bounded and integrable on $[a, b]$, then $F'(\alpha)$ exists and equal to $\int_a^b \frac{\partial}{\partial \alpha} f(x, \alpha)g(x)\,dx$ on $[\lambda_1, \lambda_2]$.

(3) Integrability of an integral of a function of parameter

If $F(\alpha) = \int_a^b f(x, \alpha)g(x)\,dx$ where $f(x, \alpha)$ is continuous in $a \le x \le b$, $\lambda_1 \le \alpha \le \lambda_2$ and $g(x)$ is bounded and integrable on $[a, b]$ then $F(\alpha)$ is Riemann integrable on $[\lambda_1', \lambda_2'] \forall \lambda_1', \lambda_2' \in [\lambda_1, \lambda_2]$ and

$$\int_{\lambda_1'}^{\lambda_2'} F(\alpha)\,d\alpha = \int_a^b \left\{\int_{\lambda_1'}^{\lambda_2'} f(x, \alpha)g(x)\,d\alpha\right\}dx \,\forall\, \lambda_1', \lambda_2' \in [\lambda_1, \lambda_2]$$

Solved Example

Example 1. *Show that* $\int_0^\infty \dfrac{\sin ax \sin bx}{x}\,dx$ *converges to* $\dfrac{1}{2}\log\dfrac{a+b}{a-b}, a>b>0.$

Solution. Let
$$I = \int_0^\infty \frac{\sin ax \sin bx}{x}\,dx$$

Now, $\sin ax \sin bx = \dfrac{1}{2}\cdot 2\sin ax \sin bx$

$$= \frac{1}{2}\big[\cos(a-b)x - \cos(a+b)x\big]$$

So
$$2I = \int_0^\infty \frac{\cos(a-b)x - \cos(a+b)x}{x}\,dx$$

Let us take $f(x) = \cos x$

Since $\int_0^\infty \dfrac{\cos x}{x}\,dx$ is convergent at ∞ and $\lim\limits_{x\to 0^+}\cos x = 1 = f_0$

Then by Frullani's theorem

$$\int_0^\infty \frac{\sin ax \sin bx}{x}\,dx = \frac{1}{2}\int_0^\infty \frac{\cos(a-b)x - \cos(a+b)x}{x}\,dx$$

$$= \frac{1}{2} f_0 \log\left|\frac{a+b}{a-b}\right|$$

$$= \frac{1}{2}\cdot 1\cdot\log\left|\frac{a+b}{a-b}\right| = \frac{1}{2}\log\left|\frac{a+b}{a-b}\right| = \frac{1}{2}\log\frac{a+b}{a-b}$$

Example 2. *Show that* $\int_0^1 \dfrac{x^\alpha - 1}{\log x}\,dx = \log(1+\alpha)\cdot$

Solution. Let
$$F(\alpha) = \int_0^1 \frac{x^\alpha - 1}{\log x}\,dx,\ \alpha > -1$$

Suppose that
$$f(x,\alpha) = x^\alpha - 1, \qquad g(x) = \frac{1}{\log x}$$

Clearly, the function $f(x,\alpha)$ and $\dfrac{\partial}{\partial\alpha}(x^\alpha - 1) = x^\alpha \log x$ are continuous in $[0,1]$

and $\alpha > -1$ and $g(x) = \dfrac{1}{\log x}$ is bounded and integrable.

So,
$$F'(\alpha) = \frac{d}{d\alpha}(F(\alpha)) = \int_0^1 \frac{\partial}{\partial\alpha}(f(x,\alpha))g(x)\,dx$$

But
$$\frac{\partial}{\partial\alpha}f(x,\alpha) = \frac{\partial}{\partial\alpha}(x^\alpha - 1) = x^\alpha \log x$$

$$\therefore \qquad F'(\alpha) = \int_0^1 \frac{x^\alpha \log x}{\log x}\,dx = \int_0^1 x^\alpha\,dx$$

$$\Rightarrow \qquad F'(\alpha) = \left[\frac{x^{\alpha+1}}{\alpha+1}\right]_0^1 = \frac{1}{1+\alpha}$$

On integrating w.r.t. α, we get $F(\alpha) = \log(1+\alpha)+c$

$$\Rightarrow \qquad F(0) = \log 1 + c = c$$

and $$F(0) = \int_0^1 0\, dx = 0 \Rightarrow c = 0$$

Hence, $$F(\alpha) = \log(1+\alpha)$$

Example 3. *Show that* $\int_0^\infty \dfrac{\log(1+\alpha x)}{1+x^2}\, dx,\ \alpha > 0 = \dfrac{1}{2}\log(1+\alpha^2)\tan^{-1}\alpha$.

Solution. Let $$F(\alpha) = \int_0^\alpha \dfrac{\log(1+\alpha x)}{1+x^2}\, dx$$

The function $f(x,\alpha) = \dfrac{\log(1+\alpha x)}{1+x^2}$ and $\dfrac{\partial}{\partial \alpha} f(x,\alpha)$ are continuous.

$\therefore$ $$\frac{d}{d\alpha} F(\alpha) = F'(\alpha) = \int_0^\alpha \frac{\partial}{\partial \alpha}\frac{\log(1+\alpha x)}{1+x^2}\, dx + \frac{\log(1+\alpha^2)}{1+\alpha^2}\frac{d\alpha}{d\alpha} - \frac{\log(1+\alpha.0)}{1+0}(0)$$

$$= \int_0^\alpha \frac{1}{1+\alpha x}\cdot\frac{x}{1+x^2}\, dx + \frac{\log(1+\alpha^2)}{1+\alpha^2} - 0$$

$$= \int_0^\alpha \frac{1}{(1+\alpha^2)}\left[-\frac{\alpha}{1+\alpha x} + \frac{x+\alpha}{1+x^2}\right] dx + \frac{\log(1+\alpha^2)}{1+\alpha^2}$$

$$= \frac{1}{(1+\alpha^2)}\left[-\log(1+\alpha x) + \frac{1}{2}\log(1+x^2) + \alpha \tan^{-1} x\right]_0^\alpha + \frac{\log(1+\alpha^2)}{(1+\alpha^2)}$$

$$= \frac{1}{(1+\alpha^2)}\left[-\log(1+\alpha^2) + \frac{1}{2}\log(1+\alpha^2) + \alpha \tan^{-1}\alpha\right] + \frac{\log(1+\alpha^2)}{1+\alpha^2}$$

$$= \frac{1}{(1+\alpha^2)}\left[-\frac{1}{2}\log(1+\alpha^2) + \alpha \tan^{-1}\alpha\right] + \frac{\log(1+\alpha^2)}{1+\alpha^2}$$

$$= \frac{1}{1+\alpha^2}\left[\frac{1}{2}\log(1+\alpha^2) + \alpha \tan^{-1}\alpha\right]$$

On integrating w.r.t. α, we get

$$F(\alpha) = \frac{1}{2}\int \frac{\log(1+\alpha^2)}{1+\alpha^2}\, d\alpha + \int \frac{\alpha \tan^{-1}\alpha}{1+\alpha^2}\, d\alpha + c$$

$$= \frac{1}{2}\left[\log(1+\alpha^2)\tan^{-1}\alpha - \int \frac{2\alpha}{1+\alpha^2}\tan^{-1}\alpha\, d\alpha\right] + \int \frac{\alpha \tan^{-1}\alpha}{1+\alpha^2}\, d\alpha + c$$

$$= \frac{1}{2}\log(1+\alpha^2)\tan^{-1}\alpha - \int \frac{\alpha}{1+\alpha^2}\tan^{-1}\alpha\, d\alpha + \int \alpha\frac{\tan^{-1}\alpha}{1+\alpha^2}\, d\alpha + c$$

$\Rightarrow$ $$F(\alpha) = \frac{1}{2}\log(1+\alpha^2)\tan^{-1}\alpha + c$$

To find the value of c, putting $\alpha = 0$ in the given function $F(\alpha)$ and in the above equation, we get

$$F(0) = c \quad \text{and} \quad F(0) = 0 \qquad \Rightarrow \qquad c = 0$$

Hence $F(\alpha) = \int_0^\alpha \dfrac{\log(1+\alpha x)}{1+x^2}\, dx = \dfrac{1}{2}\log(1+\alpha^2)\tan^{-1}\alpha$

1. If $a > 0, b > 0$, show that

$$\int_0^\infty \frac{e^{-ax} - e^{-bx}}{x} dx = \log \frac{b}{a}$$

2. Using Frullani's theorem, prove the following:

(i) $\int_0^\infty \frac{\cos ax - \cos bx}{x} dx = \log \frac{b}{a}, a, b \neq 0$

(ii) $\int_0^\infty \frac{\tan^{-1} 10x - \tan^{-1} 5x}{x} dx = \frac{\pi}{2} \log 2$

3. Show that

$$\int_0^\pi \frac{\log(1 + \cos \alpha \sin \alpha)}{\sin x} dx = \frac{\pi^2 - 4\alpha^2}{4}$$

$$\forall \alpha \in] -\pi, \pi [$$

4. Show that

$$\int_0^{\pi/2} \log\left(\frac{\alpha + \beta \sin \theta}{\alpha - \beta \sin \theta}\right) \frac{d\theta}{\sin \theta} = \pi \sin^{-1} \frac{\beta}{\alpha}$$

5. Show that

$$\int_0^\pi \frac{\log(1 + a \cos x)}{\cos x} dx, (|a|) < 1) = \pi \sin^{-1} a$$

6. Show that

$$\int_0^{\pi/2} \log(a^2 \cos^2 \theta + b^2 \sin^2 \theta) d\theta,$$

$$= \pi \log \frac{a+b}{2} \quad (a > 0, b > 0)$$

7. Show that

$$\int_0^\infty \frac{e^{-ax} \sin bx}{x} dx = \tan^{-1} \frac{b}{a}$$

8. Show that

$$\int_0^{\pi/2} \frac{dx}{a^2 \cos^2 x + \sin^2 x} = \frac{\pi}{2a}$$

9. Show that

$$\int_0^{\pi/2} \frac{\log(1 + a \sin^2 x)}{\sin^2 x} dx = \pi(\sqrt{1+a} - 1),$$

$$a > -1$$

10. Show that

$$\int_0^\infty \frac{\tan^{-1} ax}{x(1 + x^2)} dx = \frac{\pi}{2} \log(1 + a), a \geq 0$$

CHAPTER REVIEW: A COMPETITIVE APPROACH

SELECTED TERMS AND RESULTS

TERMS

- **Improper integral:** The integral $\int_a^b f(x)\,dx$ is said to be improper if either a or b or both are infinite or the integrand f becomes infinite (unbounded) on the interval (a, b).

- **Improper integral of first kind:** The integral $\int_a^b f(x)\,dx$ is called improper integral of first kind if either a or b or both a and b are infinite but the integrand f is bounded.

- **Improper integral of second kind:** The integral $\int_a^b f(x)\,dx$ is said to be improper integral of second kind if it has a finite number of points of infinite discontinuities in $[a, b]$.

- **Improper integral of third (mixed) kind:** In case an integral has limit of integration, an infinite interval having points of infinite discontinuity of the integrand f, then such type of integral is called improper integral of third (mixed) kind.

- **Convergence of an improper integral:** If limit of the integral exists as a finite number, then the improper integral is said to be convergent, otherwise divergent.

RESULTS

- If a is the only point of infinite discontinuity of f then we define $\int_a^b f\,dx = \lim_{\varepsilon \to 0^+} \int_{a+\varepsilon}^b f\,dx$ provided the limit exists finitely, then improper integral is said to be convergent, otherwise divergent.

- If b is the only point of infinite discontinuity of f, then we define $\int_a^b f\,dx = \lim_{\varepsilon \to 0^+} \int_a^{b-\varepsilon} f\,dx$. If this limit exists finitely, then improper integral is said to be convergent, otherwise divergent.

- A necessary and sufficient condition for the convergence of improper integral $\int_a^b f\,dx$ at a where f is positive on $]a, b]$ is that there exists a positive number m, independent of $\varepsilon > 0$ such that $\int_{a+\varepsilon}^b f\,dx < m, 0 < \varepsilon < b - a$.

- *Comparison test (I)* : If f and g are two positive functions such that $f(x) \leq g(x)\ \forall x \in\ ]a, b]$ and a is the only point of infinite discontinuity on $[a, b]$ then

  * $\int_a^b g\,dx$ is convergent

  $$\Rightarrow \int_a^b f\,dx \text{ is convergent}$$

  * $\int_b^a f\,dx$ is divergent

  $$\Rightarrow \int_a^b g\,dx \text{ is divergent.}$$

- *Comparison test (II)* : If f and g are two positive functions on $]a, b]$, a being the only point of infinite discontinuity such that

  $$\lim_{x \to a^+} \frac{f(x)}{g(x)} = l\,(\neq, \infty)$$

  Then two integrals $\int_b^a f\,dx$ and $\int_a^b g\,dx$ converges or diverges together.

- Let f and g be two positive functions on $]a, b]$, a being the only point of infinite discontinuity. If

  (i) $\lim_{x \to a^+} \dfrac{f(x)}{g(x)} = 0$ and $\int_a^b g\,dx$ converges,

  then $\int_b^a f\,dx$ converges.

  (ii) $\lim_{x \to a^+} \dfrac{f(x)}{g(x)} = \infty$ and $\int_a^b g\,dx$ diverges,

  then $\int_b^a f\,dx$ diverges.

- The improper integral $\int_a^b \dfrac{dx}{(x-a)^n}$ is convergent iff $n < 1$.

- The improper integral $\int_a^b \dfrac{dx}{(b-x)^n}$ is convergent iff $n < 1$.

- *Cauchy Test* : The necessary and sufficient condition for the convergence of the improper integral $\int_b^a f\,dx$ at a, a being the only point of infinite discontinuity is that to each $m > 0$ $\exists\, \delta > 0$ such that

$$\left|\int_{a+\varepsilon_1}^{a+\varepsilon_2} f\,dx\right| < m,\ 0 < \varepsilon_1, \varepsilon_2 < \delta$$

- The improper integral $\int_b^a f\,dx$ is said to be absolutely convergent if $\int_a^b |f|\,dx$ is convergent.
- Every absolutely convergent integral is convergent.
- A convergent improper integral which is not absolutely convergent is called conditionally convergent integral.
- The improper integral $\int_a^\infty \dfrac{dx}{x^n}, (a > 0)$ converges if and only if $n > 1$ and diverges for $n \le 1$.
- Abel's test : If $\int_a^\infty f\,dx$ is convergent at ∞ and g is bounded and monotonic on $[a, \infty[$ then $\int_a^\infty fg\,dx$ is convergent at ∞.

- Dirichlet's test : If $\int_a^t f\,dx$ is bounded $\forall\, t \ge a$ and g is a bounded and monotonic function for $x \ge a$ tending to 0 as $x \to \infty$, then $\int_a^\infty fg\,dx$ is convergent at ∞.

- If $f(x)$ is a continuous function on $[0, \infty[$ having singularities at 0 and ∞ only then $\int_0^\infty \dfrac{f(ax) - f(bx)}{x}\,dx$ is equal to

  (i) 0 if $\int_0^\infty \dfrac{f(x)}{x}\,dx$ converges at 0 and ∞.

  (ii) $f_1 \log\left|\dfrac{a}{b}\right|$ if $\int_0^\infty \dfrac{f(x)}{x}\,dx$ converges only at 0.

  (iii) $f_0 \log\left|\dfrac{a}{b}\right|$ if $\int_0^\infty \dfrac{f(x)}{x}\,dx$ converges only at ∞.

  (iv) $(f_0 - f_1)\log\left|\dfrac{b}{a}\right|$ if $\int_0^\infty \dfrac{f(x)}{x}\,dx$ neither converges at 0 nor at ∞ when $\lim\limits_{x \to 0^+} f(x) = f_0$ and $\lim\limits_{x \to \infty} f(x) = f_1$.

REVIEW QUESTIONS AND PROJECT WORK

1. Show that $\int_a^b \dfrac{dx}{(x-a)^n (b-x)^m}$ converges iff $n < 1$ and $m < 1$.

2. Show that $\int_0^\pi \dfrac{dx}{\cos\alpha - \cos x}$ is not convergent for any $\alpha \in \mathbb{R}$.

3. Show that the integral $\int_0^\pi x \log\sin x\,dx$ is convergent and its value is equal to $-\dfrac{\pi^2}{2}\log 2$.

4. Show that the integral $\int_0^1 x^{m-1}(1-x)^{n-1}\log\left(\dfrac{1}{x}\right)dx$ converges iff $n > -1$ and $m > 0$.

5. Show that $\int_a^b [f(x) + g(x)]\,dx$ may converge even when $\int_a^b f(x)\,dx$ and $\int_a^b g(x)\,dx$ both do not converge.

6. Show that the integral $\int_0^{\pi/2} x^p \sin^q x\,dx$ converges if $p + q > -1$.

7. Show that $\int_0^1 \dfrac{\left|\sin\dfrac{1}{x}\right|}{\sqrt{x}}\,dx$ converges.

8. Show that $\int_0^1 \dfrac{\sec x}{x}\,dx$ diverges.

9. Show that the integral $\int_0^\infty \left(\dfrac{1}{x} - \dfrac{1}{\sinh x}\right)\dfrac{dx}{x}$ converges.

10. Show that the integral $\int_0^\infty \left(\dfrac{1}{1+x} - e^{-x}\right)\dfrac{dx}{x}$ is convergent.

11. Prove that the integral $\int_0^\infty \dfrac{\sin x}{x}\,dx$ converges but not absolutely.

12. Show that the integral $\int_0^\infty \dfrac{x^p}{1+x^2}\sin^2 x\,dx$ converges iff $2 > p > -3$.

13. Show that $\int_0^\infty \dfrac{x^m}{1+x^n}\,dx$ converges only if $(n-m) > 1$

14. Show that $\int_0^\infty \log(1 + 2\,\mathrm{sech}\,x)\,dx$ converges.

15. Show that $\int_0^\infty \dfrac{1 - \cot x}{x}\,dx$ is absolutely convergent if $1 < t < 3$

OBJECTIVE TYPE QUESTIONS

FILL IN THE BLANKS

1. The definite integral $\int_a^b f(x)\,dx$ is called _____ integral if either any one or both limits are finite and function is bounded over the range of integration.

2. A definite integral $\int_a^b f(x)\,dx$ in which limits are infinite and integrand is continuous is called _____ kind of improper integral.

3. If improper integral having finite value, then it is called _____ .

4. The point at which the integrand is infinite is called _____ point.

5. The integral $\int_0^1 \dfrac{dx}{1-x}$ is _____ .

TRUE/FALSE

Write 'T' for true and 'F' for false statement.

1. The integral $\int_0^\infty \dfrac{dx}{(1+x)^{2/3}}$ is convergent.
(T/F)

2. The comparison integral $\int_0^\infty \dfrac{dx}{x^n}$, when $a > 0$ is convergent when $n > 1$ and divergent when $n \le 1$. **(T/F)**

3. In μ-test the value of μ is usually taken to be equal to the highest power of x in the denominator of the integrand minus the highest power of x in the numerator of the integrand. **(T/F)**

4. If the integral $\int_0^\infty |f(x)|\,dx$ is convergent, then the infinite integral $\int_a^\infty f(x)\,dx$ is said to be absolute convergent. **(T/F)**

5. If $\lim\limits_{x \to \infty} \left\{ \dfrac{f(x)}{g(x)} \right\}$ is a definite number, other than zero, then the integral $\int_a^\infty f(x)\,dx$ and $\int_a^\infty g(x)\,dx$ either both converges or both diverges. **(T/F)**

MULTIPLE CHOICE QUESTIONS

Choose the most appropriate one:

Problem Set-1

1. The integral $\int_a^\infty x^{n-1} e^{-x}\,dx$ is convergent if :

  (a) $n > 0$ (b) $n = 0$

  (c) $n < 0$ (d) none of these

2. The integral $\int_0^\infty \dfrac{\sin x}{x}\,dx$ converges :

  (a) uniformly   (b) conditionally

  (c) absolutely   (d) none of these

3. The integral $\int_0^\infty \dfrac{x^{p-1}}{1+x}\,dx$ converges if :

  (a) $p < 1$ (b) $0 < p$

  (c) $0 < p < 1$ (d) all are true

4. The integral $\int_0^1 x^{m-1}(1-x)^{n-1}\,dx$ converges if :

  (a) $n < 0$ (b) $n > 0$

  (c) $n = 1$ (d) none of these

5. The integral $\int_0^2 \dfrac{1 - \cos x}{x^m}\,dx$ converges for :

  (a) $m > 3$ (b) $m \ge 3$

  (c) $m = 3$ (d) $m < 3$

6. The value of μ in the μ-test of convergence for the integral

$$\int_0^\infty \dfrac{x\,dx}{(1+x)^3} :$$

(a) 1

(b) −1

(c) 2

(d) 3

7. Value of $\int_0^\infty e^{-x^2} dx =$

(a) $\dfrac{2}{\sqrt{\pi}}$

(b) $\sqrt{2\pi}$

(c) $\dfrac{\sqrt{\pi}}{2}$

(d) none of these

8. If $\int_0^\infty |f(x)|\,dx < \infty$ then $\int_a^\infty |f(x)|\,dx$:

(a) converges

(b) diverges

(c) oscillates

(d) none of these

9. By comparison test $\int_0^1 \dfrac{\sec x}{x}\,dx$ is divergent because:

(a) $\dfrac{1}{x}$ is convergent

(b) $\dfrac{1}{x}$ is divergent

(c) $\int_0^1 \sec x\,dx$ is divergent

(d) none of the above

10. The sum of the finite number of improper integrals diverges iff one or more these integrals :

(a) converges

(b) diverges

(c) oscillates

(d) none of these

11. Integral $\int_a^\infty \dfrac{dx}{x^n}, a > 0$ converges if :

(a) $n = 1$

(b) $n > 1$

(c) $n < 1$

(d) none of these

12. If Σv_n is positive monotonic sequence and if Σu_n is convergent series, then $\Sigma u_n v_n$ is also convergent. This criterion is known as :

(a) Dirichlet's test

(b) Abel's test

(c) Weierstrass M-test

(d) none of these

Problem Set - 2

1. The integral $\int_a^b f(x)\,dx$ is said to be proper integral if :

(a) range of integration is finite

(b) $f(x)$ is bounded

(c) both (a) and (b)

(d) none of the above

2. The definite integral $\int_a^b f(x)\,dx$ is said to be improper integral if :

(a) interval (a, b) is not finite

(b) the interval (a, b) is finite and $f(x)$ is not bounded

(c) neither the interval (a, b) finite nor $f(x)$ is bounded over it

(d) all of the above

3. A definite integral $\int_a^b f(x)\,dx$ in which the range of integration is infinite and $f(x)$ is bounded is called:

(a) improper integral of first kind

(b) improper integral of second kind

(c) neither (a) nor (b)

(d) none of the above

4. A definite integral $\int_a^b f(x)\,dx$ in which the range of integration is finite but $f(x)$ is unbounded is called

(a) improper integral of first kind

(b) improper integral of second kind

(c) both (a) and (b)

(d) none of the above

5. The improper integral is called convergent if :

(a) limit is finite

(b) value of the integral equal to the limit

(c) both (a) and (b)

(d) none of the above

6. If the limit is neither a definite number nor $-\infty$ or ∞ then integral is said to be :

(a) convergent

(b) divergent

(c) oscillatory

(d) none of the above

7. If the limit is ∞ or $-\infty$ then integral is said to be :

(a) convergent

(b) divergent

(c) oscillatory

(d) none of these

8. If $f(x)$ is bounded and integrable in the interval $]a, \infty[, a > 0$ and if there is a number $\mu > 1$ such that $\lim_{x \to \infty} x^\mu f(x)$ exists then $\int_a^b f(x)\,dx$ is:

(a) convergent

(b) divergent

(c) oscillatory

(d) none of these

9. If the number μ defined in question 8 is ≤ 1 then integral is:

(a) convergent   (b) divergent

(c) oscillatory   (d) none of these

10. If $\int_a^\infty f(x)\phi(x)\,dx$ converges and $\phi(x)$ is bounded and monotonic for $x > a$.

Then $\int_a^\infty f(x)\phi(x)\,dx$ is :

(a) convergent   (b) divergent

(c) oscillatory   (d) none of these

11. If $\int_a^\infty |f(x)|\,dx$ is convergent then $f(x)$ is said to be:

(a) uniformly convergent

(b) absolutely convergent

(c) divergent

(d) oscillatory

12. If $\lim\limits_{x \to a} \dfrac{f(x)}{g(x)}$ is a definite number, other than zero, the integrals $\int_a^b f(x)\,dx$ and $\int_a^b g(x)\,dx$ both are:

(a) convergent   (b) divergent

(c) either (a) or (b)   (d) none of these

13. The comparison integral $\int_a^b \dfrac{dx}{(x-a)^n}$ is convergent when :

(a) $n < 1$ (b) $n \le 1$

(c) $n > 1$ (d) $n \ge 1$

14. The comparison integral $\int_a^b \dfrac{dx}{(x-a)^n}$ is divergent when:

(a) $n > 1$ (b) $n \ge 1$

(c) $n < 1$ (d) $n \le 1$

15. If $\int_a^b f(x)\,dx$ converges and $\phi(x)$ is bounded and monotonic for $a \le x \le b$ then $\int_a^b f(x)\phi(x)\,dx$ is:

(a) convergent   (b) divergent

(c) oscillatory   (d) none of these

16. By comparison test $\int_0^1 \dfrac{\sec x}{x}\,dx$ is divergent because

(a) $\int_0^1 \dfrac{1}{x}\,dx$ is convergent

(b) $\int_0^1 \sec x\,dx$ is convergent

(c) $\int_0^1 \sec x\,dx$ is divergent

(d) $\int_0^1 \dfrac{1}{x}\,dx$ is divergent

17. The sum of the finite number of improper integral diverges iff one or more these integrals:

(a) converges   (b) diverges

(c) oscillates   (d) none of these

18. $\Sigma\, u_n$ is convergent, then $\Sigma\dfrac{u_n}{n}$ is :

(a) oscillatory   (b) divergent

(c) convergent   (d) none of these

19. "If $< v_n >$ is a positive monotonic decreasing sequence and if Σu_n is convergent series, then series $\Sigma u_n v_n$ is also convergent". This test is called :

(a) Dirichlet's test   (b) Abel's test

(c) Weierstrass test   (d) Cauchy test

Problem Set-3

1. Which of the following is/are true for $\int_0^1 \dfrac{dx}{x(1-x)}$?

(a) 0 is the point of infinite discontinuity

(b) 1 is the point of infinite discontinuity

(c) Both (a) and (b) are true

(d) None of the above

2. For $\int_0^1 \dfrac{1}{x}\,dx$, the point of infinite discontinuity is :

(a) 0   (b) 1

(c) 2   (d) 13

3. The integral $\int_{-\infty}^\infty \dfrac{1+x}{1+x^2}\,dx$ is :

(a) convergent

(b) divergent

(c) may or may not be convergent

(d) none of the above

4. For the integral $\int_{-1}^1 \dfrac{dx}{x^3}$ which is/are true

(a) It is divergent

(b) Its principal value is zero

(c) Both (a) and (b) are true

(d) None of the above

5. For the integral $\int_{-1}^{1} \frac{dx}{x}$ which is/are true ?

(a) Integral is unbounded

(b) 0 is the principal value

(c) General value is

$$\lim_{\varepsilon_1 \to 0} \log \varepsilon_1 - \lim_{\varepsilon_2 \to 0} \log \varepsilon_2$$

(d) All are true

6. For the integral $\int_{-\pi/4}^{\pi/2} \cot x \, dx$, which is/are true ?

(a) General value of the integral is

$$\lim_{\varepsilon_1 \to 0} \log(\sqrt{2} \sin \varepsilon_1)$$
$$- \lim_{\varepsilon_2 \to 0} \log \sin \varepsilon_2$$

(b) Principal value of the integral is $\frac{1}{2} \log 2$

(c) Both (a) and (b) are true

(d) None of the above

7. For the integral $\int_0^2 \frac{dx}{(x-1)^3}$, which is/are true?

(a) 0 is the point of infinite discontinuity

(b) 1 is the point of infinite discontinuity

(c) Both (a) and (b) are true

(d) None of the above

8. The integral $\int_1^\infty \frac{dx}{x^{3/2}}$ is :

(a) convergent

(b) divergent

(c) may or may not be convergent

(d) none of the above

9. For the integral $\int_3^\infty \frac{dx}{(x-2)^2}$, which is/are true ?

(a) $x = 2$ is the point of infinite discontinuity

(b) The integral is convergent

(c) Both (a) and (b) are true

(d) None of the above

10. For the integral $\int_{-\infty}^\infty \frac{dx}{x^2 + 2x + 2}$, which is/are true?

(a) The value of the integral is n

(b) The integral is convergent

(c) Both (a) and (b) are true

(d) None of the above

11. For the integral $\int_{-1}^{1} \frac{dx}{x^{2/3}}$, which is/are true:

(a) $x = 0$ is the point of infinite discontinuity

(b) The value of the integral is 6

(c) Both (a) and (b) are true

(d) None of the above

12. For the integral $\int_0^1 \frac{dx}{\sqrt{1-x}}$, which is/are true?

(a) $x = 1$ is the point of infinite discontinuity

(b) Given integral is convergent

(c) The value of the integral is 2

(d) All are true

13. For the integral $\int_{-\infty}^\infty \frac{dx}{1+x^2}$, which is/are true?

(a) Given integral is convergent

(b) Given integral may be convergent

(c) Given integral is divergent

(d) None of the above

14. The integral $\int_0^\infty \frac{4a}{x^2 + 4a^2} dx$ is :

(a) convergent

(b) divergent

(c) may or may not be convergent

(d) none of the above

15. The integral $\int_0^\infty \frac{dx}{(1+x)^{2/3}}$ is:

(a) convergent

(b) divergent

(c) may or may not be convergent

(d) none of the above

16. The integral $\int_{-\infty}^\infty e^{-x} dx$ is:

(a) convergent

(b) divergent

(c) may or may not be convergent

(d) none of the above

17. The integral $\int_{-\infty}^0 \sin hx \, dx$ is:

(a) convergent

(b) diverges to ∞

(c) diverges to $-\infty$

(d) none of the above

18. The integral $\int_1^\infty \frac{dx}{\sqrt{x}}$ is:

   (a) convergent
   (b) divergent
   (c) may or may not be convergent
   (d) none of the above

19. The integral $\int_1^\infty \frac{dx}{x}$ is:

   (a) convergent
   (b) divergent
   (c) may or may not be convergent
   (d) none of the above

20. The integral $\int_0^\infty e^{-mx}, m > 0$ is:

   (a) convergent
   (b) divergent
   (c) may or may not be convergent
   (d) none of the above

21. The integral $\int_0^\infty e^{2x}\, dx$ is:

   (a) convergent
   (b) divergent
   (c) may or may not be convergent
   (d) none of the above

22. The integral $\int_{-\infty}^0 e^x\, dx$ is:

   (a) convergent
   (b) divergent
   (c) may or may not be convergent
   (d) none of the above

23. The integral $\int_{-\infty}^0 e^{-x}\, dx$ is:

   (a) convergent
   (b) divergent
   (c) may or may not be convergent
   (d) none of the above

24. The integral $\int_{-\infty}^0 \cos hx\, dx$ is:

   (a) convergent
   (b) divergent
   (c) may or may not be convergent
   (d) none of the above

25. The integral $\int_0^1 \frac{dx}{\sqrt{x}}$ is :

   (a) convergent
   (b) divergent

   (c) may or may not be convergent
   (d) none of the above

26. The integral $\int_0^1 \frac{dx}{x^3}$ is:

   (a) convergent
   (b) divergent
   (c) may or may not be convergent
   (d) none of the above

27. The integral $\int_0^1 \frac{dx}{1-x}$ is:

   (a) convergent
   (b) divergent
   (c) may or may not be convergent
   (d) none of the above

28. The integral $\int_{-1}^1 \frac{dx}{x^2}$ is:

   (a) convergent
   (b) divergent
   (c) may or may not be convergent
   (d) none of the above

29. The integral $\int_0^{2a} \frac{dx}{(x-a)^2}$ is:

   (a) convergent
   (b) divergent
   (c) may or may not be convergent
   (d) none of the above

30. The integral $\int_0^\infty e^{-x}\frac{\sin x}{x}dx$ is:

   (a) convergent
   (b) divergent
   (c) may or may not be convergent
   (d) none of the above

31. The integral $\int_a^\infty (1-e^{-x})\frac{\cos x}{x^2}dx, a > 0$ is:

   (a) convergent
   (b) divergent
   (c) may or may not be convergent
   (d) none of the above

32. The integral $\int_1^\infty \frac{\sin x}{x^2}dx$ is:

   (a) convergent
   (b) divergent
   (c) may or may not be convergent
   (d) none of the above

33. The integral $\int_0^\infty \exp\left\{-\left[x-\frac{a}{x}\right]\right\}dx$ is :

(a) convergent

(b) divergent

(c) may or may not be convergent

(d) none of the above

34. The integral $\int_0^\infty \frac{\sin x}{x} dx$ is:

(a) convergent

(b) divergent

(c) may or may not be convergent

(d) none of the above

35. The integral $\int_0^\infty \sin x^2 dx$ is:

(a) convergent

(b) divergent

(c) may or may not be convergent

(d) none of the above

36. The integral $\int_1^\infty \sin x^p dx$, $p > 1$ is:

(a) convergent

(b) divergent

(c) may or may not be convergent

(d) none of the above

37. The integral $\int_0^\infty e^{-a^2x^2} \cos bx \, dx$ is:

(a) convergent

(b) divergent

(c) may or may not be convergent

(d) none of the above

38. The integral $\int_0^\infty e^{-x} \frac{\sin x}{x^2} dx$ is:

(a) convergent

(b) divergent

(c) may or may not be convergent

(d) none of the above

39. The integral $\int_a^\infty \frac{\sin x}{\sqrt{x}} dx$ is:

(a) convergent

(b) divergent

(c) may or may not be convergent

(d) none of the above

40. The integral $\int_a^\infty \frac{\cos ax - \cos bx}{x} dx$ is:

(a) convergent

(b) divergent

(c) may or may not be convergent

(d) none of the above

41. The integral $\int_1^\infty \frac{x \sin x \, dx}{1 + x^2}$ is:

(a) convergent

(b) divergent

(c) may or may not be convergent

(d) none of the above

42. The integral $\int_a^\infty \frac{\sin x}{x} dx$ is:

(a) convergent absolutely

(b) converges but not absolutely

(c) does not converges

(d) none of the above

43. The integral $\int_0^1 \frac{1}{x} \sin \frac{1}{x} dx$ is :

(a) converges absolutely

(b) converges but not absolutely

(c) does not converges

(d) none of the above

44. The integral $\int_0^\infty \frac{\sin^2 x}{x} dx$ is:

(a) convergent

(b) divergent

(c) may or may not be convergent

(d) none of the above

45. The integral $\int_0^\infty \frac{\cos mx}{x^2 + a^2} dx$ is:

(a) convergent

(b) divergent

(c) may or may not be convergent

(d) none of the above

46. The integral $\int_a^\infty \frac{dx}{x\sqrt{1 + x^2}}$, $a > 0$ is:

(a) convergent

(b) divergent

(c) may or may not be convergent

(d) none of the above

47. The integral $\int_0^\infty \frac{x^3}{(x^2 + a^2)^2} dx$ is:

(a) convergent

(b) divergent

(c) may or may not be convergent

(d) none of the above

48. The integral $\int_0^\infty e^{-x} \dfrac{\sin x}{x} dx$ is:

(a) convergent

(b) divergent

(c) may or may not be convergent

(d) none of the above

49. The integral $\int_1^\infty e^{-x} \dfrac{dx}{\sqrt{x^3+1}}$ is :

(a) convergent

(b) divergent

(c) may or may not be convergent

(d) none of the above

50. The integral $\int_0^\infty \dfrac{\cos x}{1+x^2} dx$ is:

(a) convergent

(b) divergent

(c) may or may not be convergent

(d) none of the above

51. The integral $\int_\pi^\infty \dfrac{\sin^2 x}{x^2} dx$ is :

(a) convergent

(b) divergent

(c) may or may not be convergent

(d) none of the above

52. The integral $\int_\pi^\infty \dfrac{\sin x}{x^2} dx$ is:

(a) convergent

(b) divergent

(c) may or may not be convergent

(d) none of the above

53. The integral $\int_2^\infty \dfrac{dx}{\sqrt{x^2-1}}$ is:

(a) convergent

(b) divergent

(c) may or may not be convergent

(d) none of the above

54. The integral $\int_0^\infty \dfrac{1-\cos x}{x^2} dx$ is:

(a) convergent

(b) divergent

(c) may or may not be convergent

(d) none of the above

55. The integral $\int_2^\infty \dfrac{dx}{\sqrt{x^2-x-1}}$ is:

(a) convergent

(b) divergent

(c) may or may not be convergent

(d) none of the above

56. The integral $\int_0^\infty e^{-x^2} dx$ is :

(a) convergent

(b) divergent

(c) may or may not be convergent

(d) none of the above

57. The integral $\int_0^\infty \dfrac{1}{x}\left(\dfrac{1}{1+x}-e^{-x}\right) dx$ is:

(a) convergent

(b) divergent

(c) may or may not be convergent

(d) none of the above

58. The integral $\int_0^\infty \dfrac{1}{x}\left(\dfrac{1}{x}-\dfrac{1}{\sinh x}\right) dx$ is:

(a) convergent

(b) divergent

(c) may or may not be convergent

(d) none of the above

59. The integral $\int_1^\infty \dfrac{dx}{x^{1/3}(1+x^{1/2})}$ is:

(a) convergent

(b) divergent

(c) may or may not be convergent

(d) none of the above

60. The integral $\int_0^\infty \dfrac{x^{2m}}{1+x^{2n}} dx$ is:

(a) converges if $n > m$

(b) diverges if $n \le m$

(c) both (a) and (b) are true

(d) none of the above

61. The integral $\int_0^\infty \dfrac{x^2}{(1+x)^3} dx$ is :

(a) convergent

(b) divergent

(c) may or may not be convergent

(d) none of the above

62. The integral $\int_0^\infty \dfrac{x^2}{(a^2+x^2)^2}\,dx$ is:

(a) convergent

(b) divergent

(c) may or may not be convergent

(d) none of the above

63. The integral $\int_0^\infty \dfrac{x^{p-1}}{1+x}\,dx,\,(0<p<1)$ is:

(a) convergent

(b) divergent

(c) may or may not be convergent

(d) none of the above

64. The integral $\int_1^\infty x^p (\log x)^q\,dx$ is:

(a) converges of $p<-1$

(b) converges if $q>-1$

(c) both (a) and (b) are true

(d) none of the above

65. The integral $\int_0^\infty \dfrac{x}{(1+x)^3}\,dx$ is:

(a) convergent

(b) divergent

(c) may or may not be convergent

(d) none of the above

66. The integral $\int_b^\infty \dfrac{x^{3/2}}{\sqrt{x^4-a^4}}\,dx,\,b>a$ is:

(a) convergent

(b) divergent

(c) may or may not be convergent

(d) none of the above

67. The integral $\int_0^\infty \dfrac{x^{2m}}{1+x^{2n}}\,dx$, m and n are positive integer $(m>n)$ is:

(a) convergent

(b) divergent

(c) may or may not be convergent

(d) none of the above

68. The integral $\int_a^\infty \dfrac{x^{2m}}{x(\log x)^{n+1}}$, $a>1$ is:

(a) convergent for $n>0$

(b) divergent for $n\le 0$

(c) both (a) and (b) are true

(d) none of the above

69. The integral $\int_1^\infty \dfrac{dx}{x^{1/3}(1+x^{1/2})}$ is:

(a) convergent

(b) divergent

(c) may or may not be convergent

(d) none of the above

70. The integral $\int_0^\infty \dfrac{x^2}{(a^2+x^2)^2}\,dx$ is:

(a) convergent

(b) divergent

(c) may or may not be convergent

(d) none of the above

71. The integral $\int_1^\infty \dfrac{dx}{(1+x)\sqrt{x}}$ is:

(a) convergent

(b) divergent

(c) may or may not be convergent

(d) none of the above

72. The integral $\int_1^\infty x^{n-1}e^{-x}\,dx$ is:

(a) convergent

(b) divergent

(c) may or may not be convergent

(d) none of the above

73. The integral $\int_a^\infty x^{n-1}e^{-x}\,dx, a>0$ is:

(a) convergent

(b) divergent

(c) may or may not be convergent

(d) none of the above

74. The integral $\int_1^\infty \dfrac{x^{m-1}}{1+x}\,dx$ is:

(a) convergent

(b) divergent

(c) may or may not be convergent

(d) none of the above

75. Which of the following is/are true?

(a) $\int_2^\infty \dfrac{dx}{x-1}$ and $\int_2^\infty \dfrac{dx}{x}$ are divergent

(b) $\int_0^\infty \left(\dfrac{1}{x-1}-\dfrac{1}{x}\right)dx$ is convergent

(c) Both (a) and (b) are true

(d) None of the above

76. The integral $\int_0^\infty \dfrac{dx}{(1+x^6)^{1/3}}$ is:

(a) convergent

(b) divergent

(c) may or may not be convergent

(d) none of the above

77. The integral $\int_1^2 \dfrac{dx}{\sqrt{x^3-1}}$ is:

(a) convergent

(b) divergent

(c) may or may not be convergent

(d) none of the above

78. The integral $\int_0^1 \dfrac{\sin x}{x^n}dx, n>0$ is:

(a) convergent for $n<2$

(b) divergent for $n\geq 2$

(c) both (a) and (b) are true

(d) none of the above

79. The integral $\int_0^{\pi/2}\log\sin x\,dx$ is:

(a) convergent

(b) divergent

(c) may or may not be convergent

(d) none of the above

80. The integral $\int_0^1 \dfrac{x^{a-1}}{1+x}dx, a>0$ is:

(a) convergent

(b) divergent

(c) may or may not be convergent

(d) none of the above

81. The integral $\int_1^\infty \dfrac{dx}{(1+x)\sqrt{x}}$ is:

(a) convergent

(b) divergent

(c) may or may not be convergent

(d) none of the above

82. The integral $\int_0^{\pi/2}\dfrac{\sin x}{x^{n+1}}dx$ is:

(a) convergent for $0<n<1$

(b) divergent for $n\geq 1$

(c) both (a) and (b) are true

(d) none of the above

83. The integral $\int_0^1 \dfrac{(x^p+x^{-p})\log(1+x)}{x}$, $-1<p<1$ is :

(a) convergent

(b) divergent

(c) may or may not be convergent

(d) none of the above

84. The integral $\int_0^2 \dfrac{\log x}{\sqrt{2-x}}dx$ is:

(a) convergent

(b) divergent

(c) may or may not be convergent

(d) none of the above

85. The integral $\int_0^{\pi/4}\dfrac{1}{\sqrt{\tan x}}dx$ is:

(a) convergent

(b) divergent

(c) may or may not be convergent

(d) none of the above

86. The integral $\int_0^1 \dfrac{dx}{x^3(1+x^2)}$ is:

(a) convergent

(b) divergent

(c) may or may not be convergent

(d) none of the above

87. The integral $\int_0^1 \dfrac{dx}{(x+1)\sqrt{1-x^2}}$ is:

(a) convergent

(b) divergent

(c) may or may not be convergent

(d) none of the above

88. The integral $\int_0^{\pi/2}\dfrac{\cos x}{x^2}dx$ is:

(a) convergent

(b) divergent

(c) may or may not be convergent

(d) none of the above

89. The integral $\int_1^2 \dfrac{dx}{\sqrt{x^4-1}}$ is:

(a) convergent

(b) divergent

(c) may or may not be convergent

(d) none of the above

90. The integral $\int_0^1 \frac{dx}{x^{1/3}(1+x^2)}$ is:

(a) convergent

(b) divergent

(c) may or may not be convergent

(d) none of the above

91. The integral $\int_0^1 \frac{\sec x}{x} dx$ is :

(a) convergent

(b) divergent

(c) may or may not be convergent

(d) none of the above

92. The integral $\int_0^1 x^{n-1} e^{-x} dx, n > 0$ is:

(a) convergent

(b) divergent

(c) may or may not be convergent

(d) none of the above

93. The integral $\int_0^\infty x^{n-1} e^{-x} dx, n > 0$ is:

(a) convergent

(b) divergent

(c) may or may not be convergent

(d) none of the above

94. The integral $\int_a^b \frac{dx}{(b-x)^n}$ is:

(a) convergent if $n < 1$

(b) divergent if $n \geq 1$

(c) both (a) and (b) are true

(d) none of the above

95. The integral $\int_1^\infty \frac{dx}{x}$ is:

(a) convergent

(b) divergent

(c) may or may not be convergent

(d) none of the above

96. The integral $\int_0^1 \frac{1}{x} dx$ is :

(a) convergent

(b) divergent

(c) may or may not be convergent

(d) none of the above

97. The integral $\int_0^1 \frac{dx}{\sqrt{1-x^2}}$ is:

(a) convergent

(b) divergent

(c) may or may not be convergent

(d) none of the above

98. The integral $\int_0^\infty \frac{1}{x^2 + \sqrt{x}} dx$ is:

(a) convergent

(b) divergent

(c) may or may not be convergent

(d) none of the above

99. The integral $\int_0^\infty \frac{1}{x^2} dx$ is:

(a) convergent

(b) divergent

(c) may or may not be convergent

(d) none of the above

100. Which of the following is/are true?

(a) $\int_0^1 \frac{1}{\sqrt{x}} dx$ is convergent

(b) $\int_1^3 \frac{1}{(x-1)^2} dx$ is divergent

(c) Both (a) and (b) are true

(d) None of the above

101. The integral $\int_0^{\pi/2} \log \sin x \, dx$ is:

(a) convergent

(b) divergent

(c) may or may not be convergent

(d) none of the above

102. The integral $\int_0^{\pi/2} \log \cos x \, dx$ is :

(a) convergent

(b) divergent

(c) may or may not be convergent

(d) none of the above

103. The integral

$$\int_0^{\pi/2} \frac{dx}{\sin^m x \cos^n x} (m > 0, n > 0) \text{ is:}$$

(a) convergent for $m < 1, n < 1$

(b) divergent for $m < 1, n < 1$

(c) convergent for $m > 1, n > 1$

(d) none of the above

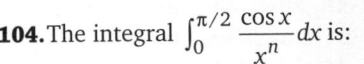

104. The integral $\int_0^{\pi/2} \frac{\cos x}{x^n} dx$ is:

(a) convergent for $n < 1$

(b) divergent for $n \geq 1$

(c) both (a) and (b) are true

(d) none of the above

105. The integral $\int_0^\infty \frac{x}{e^x + 1} dx$ is:

(a) convergent

(b) divergent

(c) may or may not be convergent

(d) none of the above

106. The integral $\int_0^\infty e^{-x^2} dx$ is:

(a) convergent

(b) divergent

(c) may or may not be convergent

(d) none of the above

107. The integral $\int_0^\infty \frac{x^2}{(a^2 + x^2)^2} dx$ is:

(a) convergent

(b) divergent

(c) may or may not be convergent

(d) none of the above

108. The integral $\int_0^\infty \frac{x^3}{(a^2 + x^2)} dx$ is:

(a) convergent

(b) divergent

(c) may or may, not be convergent

(d) none of the above

109. The integral $\int_a^\infty e^{-ax} \cos bx \, dx$ is:

(a) converges but not absolutely

(b) converges absolutely

(c) divergent

(d) none of the above

110. The integral $\int_0^\infty \frac{\cos mx}{q^2 + x^2} dx$ is:

(a) converges but not absolutely

(b) converges absolutely

(c) divergent

(d) none of the above

Answers

FILL IN THE BLANKS

1. improper   **2.** first   **3.** convergent   **4.** singular   **5.** divergent

TRUE/FALSE

1. F   **2.** T   **3.** T   **4.** T   **5.** T

MULTIPLE CHOICE QUESTIONS

Problem set-1

1. (a)   **2.** (b)   **3.** (d)   **4.** (b)   **5.** (d)   **6.** (c)   **7.** (a)   **8.** (a)   **9.** (b)
10. (b)   **11.** (b)   **12.** (b)

Problem set-2

1. (c)   **2.** (d)   **3.** (a)   **4.** (b)   **5.** (c)   **6.** (c)   **7.** (b)   **8.** (a)   **9.** (b)
10. (a)   **11.** (b)   **12.** (c)   **13.** (a)   **14.** (b)   **15.** (a)   **16.** (d)   **17.** (b)   **18.** (c)
19. (b)

Problem set-3

1. (c)   **2.** (a)   **3.** (b)   **4.** (c)   **5.** (d)   **6.** (c)   **7.** (b)   **8.** (a)   **9.** (c)
10. (c)   **11.** (c)   **12.** (d)   **13.** (a)   **14.** (a)   **15.** (b)   **16.** (b)   **17.** (c)   **18.** (b)
19. (b)   **20.** (a)   **21.** (b)   **22.** (a)   **23.** (b)   **24.** (b)   **25.** (a)   **26.** (b)   **27.** (b)
28. (b)   **29.** (b)   **30.** (a)   **31.** (a)   **32.** (a)   **33.** (a)   **34.** (a)   **35.** (a)   **36.** (a)
37. (a)   **38.** (a)   **39.** (a)   **40.** (a)   **41.** (a)   **42.** (b)   **43.** (b)   **44.** (a)   **45.** (a)
46. (a)   **47.** (b)   **48.** (a)   **49.** (a)   **50.** (a)   **51.** (a)   **52.** (a)   **53.** (b)   **54.** (a)
55. (a)   **56.** (a)   **57.** (a)   **58.** (a)   **59.** (a)   **60.** (c)   **61.** (b)   **62.** (a)   **63.** (a)
64. (c)   **65.** (a)   **66.** (b)   **67.** (a)   **68.** (c)   **69.** (b)   **70.** (a)   **71.** (a)   **72.** (a)
73. (a)   **74.** (a)   **75.** (c)   **76.** (a)   **77.** (a)   **78.** (c)   **79.** (a)   **80.** (a)   **81.** (a)

HINTS TO SELECTED PROBLEMS

Problem Set-3

1. We have $\int_{-\infty}^{\infty} \dfrac{1+x}{1+x^2} dx$

$= \int_0^{\infty} \dfrac{1+x}{1+x^2} dx + \int_{-\infty}^0 \dfrac{1+x}{1+x^2} dx.$

Both the integrals on the right are divergent because they are improper integrals of the first kind in view of $\dfrac{1+x}{1+x^2} \geq \dfrac{1}{x}$ for $1 \leq x \leq \infty$

4. We have

$\int_{-1}^1 \dfrac{dx}{x^3} = \lim_{\varepsilon_1 \to 0} \int_{-1}^{\varepsilon_1} \dfrac{dx}{x^3} + \lim_{\varepsilon_2 \to 0} \int_{\varepsilon_2}^1 \dfrac{dx}{x^3}$

$= -\dfrac{1}{2}\left[\lim_{\varepsilon_1 \to 0}\left(\dfrac{1}{\varepsilon_1^2} - 1\right) + \lim_{\varepsilon_2 \to 0}\left(1 - \dfrac{1}{\varepsilon_2}\right)\right]$

and therefore, the given integral does not converge. But

$\varepsilon_1 = \varepsilon_2 = \varepsilon \Rightarrow \int_{-1}^1 \dfrac{dx}{x^3} = \left(\int_{-1}^{\varepsilon} \dfrac{dx}{x^3} + \int_{\varepsilon}^1 \dfrac{dx}{x^3}\right)$

$= \lim_{\varepsilon \to 0}\left[-\dfrac{1}{2}\left(\dfrac{1}{\varepsilon^2} - 1 + 1 - \dfrac{1}{\varepsilon^2}\right)\right] = 0$

the principal value of the given integral is zero.

5. This integral is of second kind, because $f(x) = \dfrac{1}{x}$ is unbounded at $x = 0 \in [-1, 1]$.

Now

$\int_{-1}^1 \dfrac{dx}{x} = \lim_{\varepsilon_1 \to 0} \int_{-1}^{\varepsilon_1} \dfrac{dx}{x} + \lim_{\varepsilon_2 \to 0} \int_{\varepsilon_2}^1 \dfrac{dx}{x}$

$= \lim_{\varepsilon_1 \to 0} \log(x)\Big|_{-1}^{\varepsilon_1} + \lim_{\varepsilon_2 \to 0} \log x \Big|_{\varepsilon_2}^1$

$= \lim_{\varepsilon_1 \to 0} \log \varepsilon_1 - \lim_{\varepsilon_2 \to 0} \log \varepsilon_2$

This is the general value of the integral which does not exist.

But letting $\varepsilon_1 = \varepsilon_2 = \varepsilon$

$\Rightarrow \int_{-1}^1 \dfrac{dx}{x} = \lim_{\varepsilon \to 0} \log \dfrac{\varepsilon}{\varepsilon} = \lim_{\varepsilon \to 0} \log 1 = 0.$

which is the principle value.

6. Here $f(x) = \cot x$, is infinite at $x = 0$
Then

$\int_{-\pi/4}^{\pi/2} \cot x \, dx = \lim_{\varepsilon_1 \to 0} \int_{-\pi/4}^{\varepsilon_1} \cot x \, dx$

$+ \lim_{\varepsilon_2 \to 0} \int_{\varepsilon_2}^{\pi/2} \cot x \, dx$

$= \lim_{\varepsilon_1 \to 0}\left[\log \sin(-\varepsilon_1) - \log \sin\left(-\dfrac{\pi}{4}\right)\right]$

$+ \lim_{\varepsilon_2 \to 0}\left[\log \sin \dfrac{\pi}{2} - \log \sin \dfrac{\varepsilon}{2}\right]$

$= \lim_{\varepsilon_1 \to 0}\left[\log \sin(-\varepsilon_1) - \log\left(-\dfrac{1}{\sqrt{2}}\right)\right]$

$- \lim_{\varepsilon_2 \to 0} \log \sin \varepsilon_2$

$= \lim_{\varepsilon_1 \to 0} \log(\sqrt{2} \sin \varepsilon_1) - \lim_{\varepsilon_2 \to 0} \log \sin \varepsilon_2$

which is the general value of the integral, which does not exist.
Letting $\varepsilon_1 = \varepsilon_2 = \varepsilon$

$\Rightarrow \int_{-\pi/4}^{\pi/2} \cot x \, dx = \lim_{\varepsilon \to 0} \log\left[\dfrac{\sqrt{2} \sin \varepsilon}{\sin \varepsilon}\right]$

$= \log \sqrt{2} = \dfrac{1}{2} \log 2$

which is the principal value.

7. Here the integral $= \dfrac{1}{(x-1)^3}$, infinite at $x = 1$.

8. We have $\int_1^{\infty} \dfrac{dx}{x^{3/2}} = \lim_{x \to \infty} \int_1^x \dfrac{dx}{x^{3/2}}$

$= \lim_{x \to \infty} \int_1^x x^{-3/2} dx$

$= \lim_{x \to \infty}\left[\dfrac{x^{-1/2}}{-1/2}\right]_1^x$

$$= \lim_{x \to \infty} \left[\frac{-2}{\sqrt{x}} \right]_1^x$$

$$= \lim_{x \to \infty} \left[\frac{-2}{\sqrt{x}} + 2 \right] = 2$$

$\Rightarrow$ The limit exists, finite and unique. Hence, the given integral is convergent.

9. We have $\int_3^\infty \frac{dx}{(x-2)^2} = \lim_{x \to \infty} \int_3^x \frac{dx}{(x-2)^2}$

$$= \lim_{x \to \infty} \int_3^x (x-2)^{-2} \, dx$$

$$= \lim_{x \to \infty} \left[\frac{(x-2)^{-1}}{-1} \right]_3^x$$

$$= \lim_{x \to \infty} \left[\frac{-1}{x-2} \right] = \lim_{x \to \infty} \left[\frac{-1}{x+2} + 1 \right] = 1$$

$\Rightarrow$ Limit exists, finite and unique. Hence, the given integral is convergent.

10. We have

$$\int_{-\infty}^\infty \frac{dx}{x^2 + 2x + 2} = \int_{-\infty}^\infty \frac{dx}{(x^2+1)^2 + 1}$$

$$= \lim_{x_1 \to \infty} \int_{-x_1}^c \frac{dx}{(x^2+1)^2 + 1}$$

$$+ \lim_{x_2 \to \infty} \int_c^{x_2} \frac{dx}{(x+1)^2 + 1}$$

$$= \lim_{x_1 \to \infty} \left[\tan^{-1}(x+1) \right]_{-x_1}^c$$

$$+ \lim_{x_2 \to \infty} \left[\tan^{-1}(x+1) \right]_c^{x_2}$$

$$= \lim_{x_1 \to \infty} [\tan^{-1}(c+1) + \tan^{-1}(1-x_1)]$$

$$+ \lim_{x_2 \to \infty} [\tan^{-1}(x_2+1) - \tan^{-1}(c+1)]$$

$$= \tan^{-1}(c+1) - \tan^{-1}(-\infty)$$
$$+ \tan^{-1}(\infty) - \tan^{-1}(c+1)$$

$$= -\left(\frac{\pi}{2} \right) + \frac{\pi}{2} = \pi$$

11. Clearly the integral becomes infinite at $x = 0$ and 0 lies between the given range of integration -1 and 1.

$$\therefore \int_{-1}^1 \frac{dx}{x^{2/3}} = \int_{-1}^0 \frac{dx}{x^{2/3}} + \int_0^1 \frac{dx}{x^{2/3}}$$

$$= \lim_{\varepsilon \to 0} \int_{-1}^{0-\varepsilon} \frac{dx}{x^{2/3}} + \lim_{\varepsilon' \to 0} \int_{\varepsilon'}^1 \frac{dx}{x^{2/3}}$$

$$= \lim_{\varepsilon \to 0} \left[3x^{1/3} \right]_{-1}^{-\varepsilon} + \lim_{\varepsilon' \to 0} \left[3x^{1/3} \right]_{\varepsilon'}^1$$

$$= \lim_{\varepsilon \to 0} [-3\varepsilon^{1/3} + 3] + \lim_{\varepsilon' \to 0} [3 - 3(\varepsilon')^{1/3}]$$

$$= 3 + 3 = 6$$

12. Clearly the integral i.e., $\frac{1}{\sqrt{1-x}}$ becomes unbounded i.e., infinite at the upper limit i.e., $x = 1$.

Therefore, $\int_0^1 \frac{dx}{\sqrt{1-x}} = \lim_{\varepsilon \to 0} \int_0^{1-\varepsilon} \frac{dx}{\sqrt{1-x}}$

$$= \lim_{\varepsilon \to 0} \left[-2\sqrt{(1-x)} \right]_0^{1-\varepsilon}$$

$$= \lim_{\varepsilon \to 0} \left[-2\sqrt{\varepsilon} - 2 \right] = 2$$

which is a definite real numbers. Hence the given integral is convergent and its value is 2.

13. We have

$$\int_{-\infty}^\infty \frac{dx}{1+x^2} = \int_{-\infty}^0 \frac{dx}{1+x^2} + \int_0^\infty \frac{dx}{1+x^2}$$

$$= \lim_{x \to \infty} \int_{-x}^0 \frac{dx}{1+x^2} + \lim_{x \to \infty} \int_0^x \frac{dx}{1+x^2}$$

$$= \lim_{x \to \infty} \left[\tan^{-1} x \right]_{-x}^0 + \lim_{x \to \infty} \left[\tan^{-1} x \right]_0^x$$

$$= \lim_{x \to \infty} [0 - \tan^{-1}(-x)] + \lim_{x \to \infty} (\tan^{-1} x - 0)$$

$$= -\left(\frac{-\pi}{2} \right) + \frac{\pi}{2} = \pi$$

$\Rightarrow$ Limit exists uniquely and finite. Hence, the given integral is convergent.

14. We have

$$\int_0^\infty \frac{4a}{x^2 + 4a^2} \, dx = \lim_{x \to \infty} \int_0^x \frac{4a}{x^2 + (2a)^2} \, dx$$

$$= \lim_{x \to \infty} \left[4a \cdot \frac{1}{2a} \tan^{-1} \frac{x}{2a} \right]_0^x$$

$$= 2 \lim_{x \to \infty} \left[\tan^{-1} \frac{x}{2a} \right]_0^x$$

$$= 2 \lim_{x \to \infty} \left(\tan^{-1} \frac{x}{2a} - 0 \right)$$

$$= 2 \tan^{-1} \infty = 2 . \frac{\pi}{2} = \pi$$

⇒ Limit exists, unique and finite.

Hence, the given integral is convergent.

15. We have

$$\int_0^\infty \frac{dx}{(1+x)^{2/3}} = \lim_{x \to \infty} \int_0^x (1+x)^{-2/3} dx$$

$$= \lim_{x \to \infty} \left[\frac{(1+x)^{1/3}}{1/3} \right]_0^x$$

$$= \lim_{x \to \infty} 3 \left[(1+x)^{1/3} - 1 \right] = \infty$$

⇒ Limit does not exist finitely.

Hence the given integral is not convergent.

16. We have

$$\int_{-\infty}^\infty e^{-x} dx = \int_{-\infty}^0 e^{-x} dx + \int_0^\infty e^{-x} dx$$

$$= \infty + 1 = \infty$$

⇒ Limit does not exist finitely.

Hence the given integral is not convergent.

17. We have $\int_{-\infty}^0 \sinh x \, dx = \lim_{x \to \infty} \int_{-x}^0 \sinh x \, dx$

$$= \lim_{x \to \infty} \int_{-x}^0 \frac{e^x - e^{-x}}{2} dx$$

$$= \frac{1}{2} \left[\lim_{x \to \infty} \int_{-x}^0 e^x dx - \lim_{x \to \infty} \int_{-x}^0 e^{-x} dx \right]$$

$$= \frac{1}{2} \left[\lim_{x \to \infty} \left\{ e^x \right\}_{-x}^0 - \lim_{x \to \infty} \left\{ \frac{e^{-x}}{-1} \right\}_x^0 \right]$$

$$= \frac{1}{2} \left[\lim_{x \to \infty} (e^0 - e^{-x}) + \lim_{x \to \infty} (e^0 - e^x) \right]$$

$$= \frac{1}{2} (1 - \infty) = -\infty$$

Hence, the given integral diverges to $-\infty$.

30. Let $f(x) = e^{-x}$, which is monotonic decreasing

for $x > 0$, and $f(x) = \frac{\sin x}{x} \Rightarrow \int_0^\infty \frac{\sin x}{x} dx$ is

convergent $\Rightarrow \int_0^\infty e^{-x} \frac{\sin x}{x} dx$ converges by

Abel's test.

31. Let $\phi(x) = 1 - e^{-x}$, which is bounded and

monotonic for $x > a$ and $f(x) = \frac{\cos x}{x^2}$.

But $\int_a^\infty \frac{\cos x}{x^2} dx$ is convergent

⇒ the integral converges.

Convergence of $\int_a^\infty \frac{\cos x}{x^2} dx$ is seen by

comparison with $\int_a^\infty \frac{1}{x^2} dx$.

32. Set $\phi(x) = \frac{1}{x^k}, k > 0$ and $f(x) = \sin x$. But

ϕ is monotonic and $x \to \infty \Rightarrow \phi \to 0$ and

$\int_1^\infty \sin x \, dx = \cos 1 - \cos t$

$$\Rightarrow \left| \int_1^t \sin x \, dx \right| < 2 \Rightarrow \int_1^t \sin x \, dx$$

remains bounded $\Rightarrow \int_1^t \sin x \, dx$ converges. By

Abel's test, the given integral is convergent.

33. Set $\phi(x) = e^{-a^2/x^2}$ and $f(x) = e^{-x^2}$

Now ϕ is a monotonic increasing function $\to 1$

as $x \to \infty$, and $\int_0^\infty f(x) dx$ is convergent.

Consequently, the given integral is convergent

by Abel's test.

34. Letting $x \to 0 \Rightarrow \frac{\sin x}{x} \to 1 \Rightarrow 0$ is not a point

of infinite discontinuity.

But $I = \int_0^\infty \frac{\sin x}{x} dx$

$$= \int_0^1 \frac{\sin x}{x} dx + \int_1^\infty \frac{\sin x}{x} dx = I_1 + I_2, \text{ say}$$

I_1 is a proper integral, and hence convergent.

Now it remains to establish convergence of I_2.

We let $\phi(x) = \frac{1}{x}$, which is bounded and

monotonic $\to 0$ as $x \to \infty$, $f(x) = \sin x$ and

$$\left| \int_a^x \sin x \, dx \right| = |-\cos x + \cos 1| \leq |\cos x| + |\cos 1|$$
$$< 2 \ \forall \ x.$$

The conditions for Dirichlet's test hold

$$\Rightarrow I_2 = \int_1^\infty \frac{\sin x}{x} dx \text{ converges by Dirichlet's}$$

test $\Rightarrow I_1 + I_2$ converges.

35. We have $\sin x^2 = \dfrac{1}{2x} 2x \sin x^2$.

But $\displaystyle\int_0^\infty \sin x^2 dx = I$, say

$= \displaystyle\int_0^1 \sin x^2 dx + \int_0^\infty \sin x^2 dx = I_1 + I_2$, say.

I_1 is a proper integral, and hence convergent.

Now $I_2 = \displaystyle\int_1^\infty \dfrac{1}{2x}(2x \sin x^2)\,dx$. Set $\phi(x) = \dfrac{1}{2x}$,

which is monotonic, and $\to 0$ as $x \to \infty$ and

$f(x) = 2x \sin x^2 \Rightarrow \left| \displaystyle\int_1^x 2x \sin x^2 dx \right|$

$= |-\cos x + \cos 1| < 2 \Rightarrow I_2$

is bounded for each $x \geq 1$. The conditions laid down in Dirichlet's test hold $\Rightarrow I_2$ converges

$\Rightarrow I = I_1 + I_2$ converges.

36. We have

$\sin x^p = \dfrac{px^{p-1} \sin x^p}{px^{p-1}}$

Set $\phi(x) = \dfrac{1}{px^{p-1}}$ for $p > 1$,

and $f(x) = px^{p-1} \sin x^p . \phi(x) \to 0$ for $p > 1$

as $x \to \infty$ and $\left| \displaystyle\int_1^\infty px^{p-1} \sin x^p dx \right| \leq 2$ for all

finite x.

$\Rightarrow \qquad \displaystyle\int_1^\infty f(x).\phi(x)dx$

covnerges by Dirichlet's test.

37. Let $\phi(x) = \exp(-a^2x^2)$ and $f(x) = \cos bx$. The given integral converges by Dirichlet's test.

38. We first wish to show that $\displaystyle\int_a^\infty \dfrac{\sin x}{x^2}dx$

converges. Here $f(x) = \sin x$ is bounded and integrable in $[a, x]$ for all $x > a$ and the integral

$\displaystyle\int_a^\infty \sin x\, dx$ oscillates between 0 and 2.

Since

$\displaystyle\int_a^\infty \sin x\, dx = \lim_{x \to \infty} \int_a^x \sin x\, dx$

$= \lim_{x \to \infty} (\cos a - \cos x)$

so that $\left| \displaystyle\int_a^\infty \sin x\, dx \right| < 2.$

Again $\dfrac{1}{x^2} = \phi(x)$ is monotonic decreasing in

$[a, \infty]$ such that $x \to \infty \Rightarrow \phi(x) \to 0$. Hence by

Dirichlet's test, $\displaystyle\int_a^\infty \dfrac{\sin x}{x^2}dx$ converges. Finally

we see that e^{-x} is monotonic and bounded.

Hence by Abel's test, $\displaystyle\int_a^\infty e^{-x} \dfrac{\sin x}{x^2}dx$ converges.

39. Put $\phi(x) = \dfrac{1}{\sqrt{(x)}}$ and $f(x) = \sin x$

Dirichlet's test asserts that the given integral converges.

40. Here $I = \displaystyle\int_a^\infty \dfrac{\cos ax}{x}dx - \int_a^\infty \dfrac{\cos bx}{x}dx.$

Evidently $\dfrac{1}{x}$ is monotonic and $\to 0$ as $x \to \infty$,

and $\left| \displaystyle\int_a^x \cos ax\, dx \right|$ and $\left| \displaystyle\int_a^x \cos bx\, dx \right| \leq 2$, i.e.,

bounded. Hence I is convergent by Dirichlet's test.

41. Put $\phi(x) = \dfrac{x}{1+x^2}$ and $f(x) = \sin x.$

We observe that $\phi(x)$ is monotonic and $\to 0$ as

$x \to \infty$, and $\displaystyle\int_1^x f(x)dx$ is bounded. Therefore

the given integral converges by Dirichlet's test.

42. By second mean value theorem for integrals, we have

$\displaystyle\int_p^q f\phi(dx) = \phi(p)\int_p^t f\, dx + \phi(q)\int_t^q f\, dx,$

for $p \leq t \leq q.$

Hence on letting $\phi(x) = \dfrac{1}{x}$ and $f(x) = \sin x,$

$\displaystyle\int_p^q \dfrac{\sin x}{x}dx = \dfrac{1}{p}\int_p^\varepsilon \sin x\, dx + \dfrac{1}{q}\int_\varepsilon^q \sin x\, dx,$

$0 < p \leq \varepsilon \leq q.$

But $\left| \displaystyle\int_p^\varepsilon \sin x\, dx \right| = |-\cos \varepsilon + \cos p| \leq |\cos \varepsilon|$

$+ |\cos p| \leq 2$

and $\displaystyle\int_\varepsilon^q |\sin x\, dx| \leq 2.$

$\therefore$ for $q > p,$

$\left| \displaystyle\int_p^q \dfrac{\sin x}{x}dx \right| \leq 2\left(\dfrac{1}{p} + \dfrac{1}{q}\right) \leq \left(\dfrac{1}{p} + \dfrac{1}{p}\right) = \dfrac{4}{p}.$

Now for

$$\varepsilon > 0, \left| \int_p^q \frac{\sin x}{x} dx \right| < \varepsilon \; \forall \; p, q > m = \frac{4}{\varepsilon}$$

$\Rightarrow$ the integral converges.

To prove that $\int_0^\infty \left| \frac{\sin x}{x} \right| dx$

does not converge absolutely, for any positive

integral value of r,

$$\int_0^{(r+1)\pi} \left| \frac{\sin x}{x} \right| dx = \int_0^\pi \left| \frac{\sin x}{x} \right| dx$$

$$+ \int_\pi^{2\pi} \left| \frac{\sin x}{x} \right| dx + ...$$

$$+ \int_{r\pi}^{(r+1)\pi} \left| \frac{\sin x}{x} \right| dx$$

$$= \sum_{n=0}^{r} \int_{n\pi}^{(n+1)\pi} \left| \frac{\sin x}{x} \right| dx.$$

Now, for $n \geq 0$,

$$\int_{n\pi}^{(n+1)\pi} \left| \frac{\sin x}{x} \right| dx = \int_0^\pi \frac{\sin y \, dy}{n\pi + y}$$

On putting $x = n\pi + y$.

Note that $(n+1)\pi$ is the maximum value of $n\pi + y$ in $[0, \pi]$. Then we find

$$\int_{n\pi}^{(n+1)\pi} \left| \frac{\sin x}{x} \right| dx \geq \frac{1}{(n+1)\pi} \int_0^\pi \sin y \, dy$$

$$= \frac{2}{(n+1)\pi}$$

Thus $\displaystyle\sum_{n=0}^{r} \int_{n\pi}^{(n+1)\pi} \left| \frac{\sin x}{x} \right| dx$

$$= \frac{2}{\pi}\left(1 + \frac{1}{2} + \frac{1}{3} + ... + \frac{1}{r+1}\right).$$

Hence $\int_0^{(r+1)\pi} \left| \frac{\sin x}{x} \right| dx > \frac{2}{\pi} \displaystyle\sum_{n=0}^{r} \frac{1}{n+1}.$

The series in the right hand member of this inequality diverges as $r \to \infty$. Consequently,

$$\int_0^{(r+1)\pi} \left| \frac{\sin x}{x} \right| dx \to \infty \text{ as } r \to \infty$$

Now if $t \geq (r+1)\pi$, where t is a real number,

$$\int_0^t \left| \frac{\sin x}{x} \right| dx \geq \int_0^{(r+1)\pi} \left| \frac{\sin x}{x} \right| dx$$

Since $t \to \infty$, also $r \to \infty$. Then

$$\int_0^t \left| \frac{\sin x}{x} \right| dx \to \infty \text{ as } t \to \infty.$$

Therefore, $\int_0^\infty \left| \frac{\sin x}{x} \right| dx$

does not converge, i.e.,

$$\int_0^\infty \left| \frac{\sin x}{x} \right| dx$$

is not absolutely convergent.

43. Here $f(x) = \frac{1}{x}\sin\frac{1}{x}$ is unbounded at $x = 0$.

Now by definition

$$\int_0^1 \frac{1}{x}\sin\frac{1}{x} dx = \lim_{\varepsilon \to 0} \int_\varepsilon^1 \frac{1}{x}\sin\frac{1}{x} dx$$

$$= \lim_{\varepsilon \to 0} \int_\varepsilon^1 x \, d\left(\cos\frac{1}{x}\right) dx$$

$$= x \cos\frac{1}{x}\Big|_\varepsilon^1 - \int_\varepsilon^1 \cos\frac{1}{x} dx$$

$$= \cos 1 - \varepsilon \cos\frac{1}{\varepsilon} - \int_\varepsilon^1 \cos\frac{1}{x} dx.$$

Hence

$$\int_0^1 \frac{1}{x}\sin\frac{1}{x} dx = \cos 1 - \lim_{\varepsilon \to 0} \int_\varepsilon^1 \cos\frac{1}{x} dx.$$

Note that $\cos\frac{1}{x}$ is bounded in $[\varepsilon, 1]$ and it is also

continuous in $[\varepsilon, 1]$. Hence $\lim_{\varepsilon \to 0} \int_\varepsilon^1 \cos\frac{1}{x} dx$

exists as a finite number. Therefore, the given integral converges.

Now we wish to show that the integral

$$\int_0^1 \left| \frac{1}{x}\sin\frac{1}{x} \right| dx \text{ diverges.}$$

We consider a sequence of intervals $< [a_n, b_n] >$, where

$$a_n = \frac{1}{\left(2n + \frac{1}{2}\right)\pi}, \; b_n = \frac{1}{\left(2n + \frac{1}{4}\right)\pi}.$$

The sequence of intervals is non-overlapping and is such that

$(b_n - a_n) \to 0$ as $n \to \infty$.

Now for each positive integer m, we find

$$\int_0^1 \left| \frac{1}{x}\sin\frac{1}{x} \right| dx \geq \sum_{n=1}^{m} \int_{a_n}^{b_n} \left| \frac{1}{x}\sin\frac{1}{x} \right| dx$$

$$> \sum_{n=1}^{m} \int_{a_n}^{b_n} \frac{1}{\sqrt{2}} \frac{dx}{x}$$

because $\sin\dfrac{1}{a_n}=1$ and $\sin\dfrac{1}{b_n}=\dfrac{1}{\sqrt{2}}$.

Hence

$$\int_0^1\left|\dfrac{1}{x}\sin\dfrac{1}{x}\right|dx>\dfrac{1}{\sqrt{2}}\sum_{n=1}^{m}\log\dfrac{b_n}{a_n}$$

$$=\dfrac{1}{\sqrt{2}}\sum_{n=1}^{m}\log\dfrac{8n+2}{8n+1}$$

$$=\dfrac{1}{\sqrt{2}}\sum_{n=1}^{m}\log\left(1+\dfrac{1}{8n+1}\right)$$

$$=\dfrac{1}{\sqrt{2}}\log\sum_{n=1}^{m}\left(1+\dfrac{1}{8n+1}\right)$$

$$>\dfrac{1}{\sqrt{2}}\log\sum_{n=1}^{m}\dfrac{1}{8n+1}$$

$$=\dfrac{1}{\sqrt{2}}\log\left\{\dfrac{1}{8}\sum_{n=1}^{m}\dfrac{1}{n+1}\right\}$$

$$=\dfrac{1}{\sqrt{2}}\log\sum_{n=1}^{m}\dfrac{1}{1+n}-\dfrac{\log 8}{\sqrt{2}}$$

But $\lim\limits_{m\to\infty}\sum\limits_{n=1}^{m}\dfrac{1}{1+n}$ diverges.

Hence $\int_0^1\left|\dfrac{1}{x}\sin\dfrac{1}{x}\right|dx$ is divergent.

44. For $a>0$, we can write

$$\int_0^\infty\dfrac{\sin^2 x}{x^2}=\int_0^a\dfrac{\sin^2 x}{x^2}dx$$

$$+\int_a^\infty\dfrac{\sin^2 x}{x^2}dx\quad\dots(1)$$

Now, $\lim\limits_{x\to 0}\dfrac{\sin^2 x}{x^2}=1\Rightarrow$ The integral $\dfrac{\sin^2 x}{x^2}$ is bounded throughout the finite integral $]0,\ a[$. Therefore, $\int_0^a\dfrac{\sin^2 x}{x^2}dx$ is a proper integral, which is always convergent. So we have to check the convergence of the integral

$$\int_a^\infty\dfrac{\sin^2 x}{x^2}dx.$$

Let $f(x)=\dfrac{\sin^2 x}{x^2}$ and let $g(x)=\dfrac{1}{x^2}$.

Clearly $g(x)$ is positive in the interval $]a,\ \infty[$.

Now $|f(x)|=\left|\dfrac{\sin^2 x}{x^2}\right|=\dfrac{\sin^2 x}{x^2}\le\dfrac{1}{x^2}$

$$(\because\ |\sin x|\le 1)$$

Thus, by comparison test $\int_a^\infty\dfrac{\sin^2 x}{x^2}dx$ is convergent if $\int_a^\infty\dfrac{dx}{x^2}$ is convergent. But the comparison integral $\int_a^\infty\dfrac{dx}{x^2}$ is convergent, because here $n=2$ which is >1.

Thus $\int_a^\infty\dfrac{\sin^2 x}{x^2}dx$ is convergent.

Hence, the given integral $\int_a^\infty\dfrac{\sin^2 x}{x^2}dx$ is convergent.

45. Let $f(x)=\dfrac{\cos mx}{x^2+a^2}$.

Also let $g(x)=\dfrac{1}{x^2+a^2}$

$\Rightarrow g(x)$ is positive in the interval $]0,\ \infty[$

Now $|f(x)|=\left|\dfrac{\cos mx}{x^2+a^2}\right|=\dfrac{|\cos mx|}{x^2+a^2}$

$$\le\dfrac{1}{x^2+a^2}\quad(\because\ |\cos mx|\le 1)$$

$\Rightarrow\quad|f(x)|\le g(x)$ when $x\ge 0$.

$\therefore$ By comparison test, $\int_0^\infty\dfrac{\cos mx}{x^2+a^2}dx$ is convergent if $\int_0^\infty\dfrac{dx}{x^2+a^2}dx$ is convergent.

But $\int_0^\infty\dfrac{dx}{x^2+a^2}=\lim\limits_{x\to\infty}\int_0^\infty\dfrac{dx}{x^2+a^2}$

$$=\lim_{x\to\infty}\left(\dfrac{1}{a}\tan^{-1}\dfrac{x}{a}\right)_0^x$$

$$=\lim_{x\to\infty}\left(\dfrac{1}{a}\tan^{-1}\dfrac{x}{a}-0\right)=\dfrac{1}{a}\cdot\dfrac{\pi}{2}$$

a definite real number

$\therefore\ \int_0^\infty\dfrac{dx}{x^2+a^2}$ is convergent.

Hence, the given integral $\int_0^\infty\dfrac{\cos mx}{x^2+a^2}dx$ is also convergent.

46. Let $f(x) = \dfrac{1}{x\sqrt{1+x^2}}$

$\Rightarrow f(x)$ is bounded in the interval $]a, \infty[$.

Let us take $g(x) = \dfrac{1}{x^2}$. Then $g(x)$ is positive in the interval $]a, \infty[$.

Now, $|f(x)| = \left| \dfrac{1}{x\sqrt{1+x^2}} \right| = \dfrac{x}{x^2\sqrt{1+\dfrac{1}{x^2}}}$

$< \dfrac{1}{x^2}$

Thus, by comparison test

$\displaystyle\int_a^\infty \dfrac{dx}{x\sqrt{1+x^2}}$ is convergent if $\displaystyle\int_a^\infty \dfrac{dx}{x^2}$ is convergent.

But the comparison integral $\displaystyle\int_a^\infty \dfrac{dx}{x^2}$ is convergent, because $n = 2 > 1$

Hence $\displaystyle\int_a^\infty \dfrac{dx}{x\sqrt{1+x^2}}$ is also convergent.

47. Let $f(x) = \dfrac{x^3}{(x^2+a^2)^2}$

$\Rightarrow f(x)$ is bounded in the interval $]0, \infty[$.

Let $c > 0$. Then we can write

$\displaystyle\int_0^\infty \dfrac{x^3}{(x^2+a^2)^2}\,dx = \int_0^c \dfrac{x^3}{(x^2+a^2)^2}\,dx$

$+ \displaystyle\int_c^\infty \dfrac{x^3}{(x^2+a^2)^2}\,dx$ ...(1)

Clearly, the first integral of R.H.S. of (1) is a proper integral because the interval of integration $]0, c[$ is finite and the integral $f(x)$ is bounded in this interval. Now, we have to check the convergence of the integral

$\displaystyle\int_c^\infty \dfrac{x^3}{(x^2+a^2)^2}\,dx.$

We can write

$f(x) = \dfrac{x^3}{x^4\left[\left(1+\dfrac{a^2}{x^2}\right)\right]^2} = \dfrac{1}{x\left(1+\dfrac{a^2}{x^2}\right)^2}$

Let us take $g(x) = \dfrac{1}{x}$

We have $\displaystyle\lim_{x\to\infty} \dfrac{f(x)}{g(x)} = \lim_{x\to\infty} \dfrac{1}{\left(1+\dfrac{a^2}{x^2}\right)^2} = 1,$

which is finite and non-zero.

$\therefore \displaystyle\int_b^\infty f(x)\,dx$ and $\displaystyle\int_b^\infty g(x)\,dx$ either both converge or diverge. But the comparison integral $\displaystyle\int_c^\infty g(x)\,dx$ i.e., $\displaystyle\int_c^\infty \dfrac{1}{x}\,dx$ is divergent.

Therefore $\displaystyle\int_c^\infty f(x)\,dx$ is also divergent.

48. We can write

$\displaystyle\int_0^\infty e^{-x}\cdot\dfrac{\sin x}{x}\,dx = \int_0^1 e^{-x}\dfrac{\sin x}{x}\,dx$

$+ \displaystyle\int_1^\infty e^{-x}\dfrac{\sin x}{x}\,dx$

Now, since $\displaystyle\lim_{x\to 0} e^{-x}\dfrac{\sin x}{x} = 1$, thus the integral $e^{-x}\dfrac{\sin x}{x}$ is bounded throughout the finite interval $]0, 1[$. Therefore, $\displaystyle\int_0^1 e^{-x}\cdot\dfrac{\sin x}{x}\,dx$ is a proper integral and therefore, it is convergent. Now, we have to check the convergence of

$\displaystyle\int_1^\infty e^{-x}\cdot\dfrac{\sin x}{x}\,dx.$

Let $f(x) = e^{-x}\dfrac{\sin x}{x} \Rightarrow f(x)$ is bounded in the interval $]1, \infty[$.

Let us take $g(x) = e^{-x}$. Then $g(x)$ is positive in the interval $]1, \infty[$.

Now, $|f(x)| = \left| e^{-x}\dfrac{\sin x}{x} \right| = e^{-x}\,|\sin x|\cdot\dfrac{1}{x}$

$\leq e^{-x} \qquad \left(\because |\sin x| \leq 1, \dfrac{1}{x} \leq 1\right)$

$\therefore$ By comparison test, $\displaystyle\int_1^\infty f(x)\,dx$ is convergent if $\displaystyle\int_1^\infty g(x)\,dx$ is convergent.

Now $\displaystyle\int_1^\infty g(x)\,dx = \int_1^\infty e^{-x}\,dx$

$= \displaystyle\lim_{x\to\infty} \int_1^x e^{-x}\,dx$

$= \displaystyle\lim_{x\to\infty} \left[-e^{-x}\right]_1^x$

$= \displaystyle\lim_{x\to\infty} \left[-e^{-x} + e^{-1}\right] = 0 + e^{-1} = \dfrac{1}{e}$

which is a definite finite number.

Hence $\int_1^\infty g(x)\,dx$ is convergent.

$\Rightarrow \int_1^\infty f(x)\,dx$ is also convergent.

Therefore, the given integral $\int_0^\infty e^{-x}\dfrac{\sin x}{x}$ is convergent because the sum of two convergent integrals is convergent.

49. Let $f(x)=\dfrac{1}{\sqrt{x^3+1}}=\dfrac{1}{x^{3/2}}\cdot\left(\dfrac{1}{\sqrt{1+\dfrac{1}{x^3}}}\right)$.

Take $g(x)=\dfrac{1}{x^{3/2}}$

Then $\lim_{x\to\infty}\dfrac{f(x)}{g(x)}=\lim_{x\to\infty}\dfrac{1}{\sqrt{1+\dfrac{1}{x^3}}}=1$, which is finite and non-zero.

Therefore, $\int_1^\infty f(x)\,dx$ and $\int_1^\infty g(x)\,dx$ are either both convergent or both divergent.

But the comparison integral $\int_1^\infty g(x)\,dx$ i.e,
$\int\dfrac{dx}{x^{3/2}}$ is convergent because $n=\dfrac{3}{2}>1$.

Hence, $\int_1^\infty f(x)\,dx=\int_1^\infty\dfrac{dx}{\sqrt{x^3+1}}$ is also convergent.

57. We have $f(x)=$ integrand

$=\dfrac{e^x-1-x}{x(1+x)e^x}$

$=\dfrac{1+x+\dfrac{x^2}{2!}+\dfrac{x^3}{3!}+\ldots-1-x}{x(1+x)e^x}\to 0$ as

$x\to 0\Rightarrow x=0$ is not a point of infinite discontinuity $\Rightarrow x=\infty$ is the only point of infinite discontinuity.

Now $\lim_{x\to\infty}x^\mu f(x)=\lim_{x\to\infty}x^\mu\dfrac{e^x-1-x}{x(1+x)e^x}$

$=\lim_{x\to\infty}\dfrac{x^\mu}{x^2}\lim_{x\to\infty}\dfrac{e^x-1-x}{e^x}$

$=\lim_{x\to\infty}x^{\mu-2}\lim_{x\to\infty}\dfrac{e^x}{e^x}=\lim_{x\to\infty}x^{\mu-2}=1$

If $\mu=2>1$
$\Rightarrow$ the given integral converges by μ-test.

58. By L' Hospital rule, $\lim_{x\to 0}\left(\dfrac{1}{x}-\dfrac{1}{\sinh x}\right)\dfrac{1}{x}=\dfrac{1}{\infty}$

$\Rightarrow x=0$ is not a point of infinite discontinuity
$\Rightarrow x=\infty$ is the only point of infinite discontinuity

$\Rightarrow \lim_{x\to 0}x^\mu f(x)=\lim_{x\to\infty}x^\mu\dfrac{\sinh x-x}{x^2\sinh x}$

$=\lim_{x\to\infty}\dfrac{x^\mu}{x^2}\lim_{x\to\infty}\dfrac{e^x-e^{-x}-2x}{e^x-e^{-x}}$

$=\lim_{x\to\infty}x^{\mu-2}\lim_{x\to\infty}\dfrac{e^x}{e^x}=\lim_{x\to\infty}x^{\mu-2}=1$ if

$\mu=2>1$
By μ-test, the given integral converges.

59. For, $f(x)=\dfrac{1}{x^{1/3}(1+x^{1/2})}\approx\dfrac{1}{x^{5/6}}$ at $x=\infty$

$\Rightarrow\mu=5/6<1$ the integral diverges by μ-test.

60. We have

$\int_0^\infty\dfrac{x^{2m}}{1+x^{2n}}\,dx=\int_0^a\dfrac{x^{2m}}{1+x^{2m}}\,dx$

$\qquad\qquad +\int_a^\infty\dfrac{x^{2m}}{1+x^{2n}}\,dx,\ a>0$

The first integral on the right is a proper one
$\Rightarrow$ it converges. To examine converges of the second integral,
Consider

$f(x)=\dfrac{x^{2m}}{1+x^{2n}}\le\dfrac{x^{2m}}{x^{2n}}=x^{2m-2n}$

Letting $\mu=2n-2m\Rightarrow$ the second integral
converges if $2n-2m>1$, i.e., if $n>m+\dfrac{1}{2}$
i.e., if $n>m$ and diverges if $n\le m\Rightarrow$ the given integral converges if $n>m$, and diverges if $n\le m$.

62. We have

$=\lim_{x\to\infty}x^\mu f(x)=\lim_{x\to\infty}x^\mu\dfrac{x^2}{(a^2+x^2)^2}$

$\simeq\lim_{x\to\infty}\dfrac{x^{\mu+2}}{x^4}$

$=\lim_{x\to\infty}x^{\mu-2}=1$

if $\mu-2=0$, i.e., if $\mu=2>1$.
By μ-test, the integral converges.

63. We have

$$\int_0^\infty \frac{x^{p-1}}{1+x}dx = \int_0^a \frac{x^{p-1}}{1+x}dx$$

$$+ \int_a^\infty \frac{x^{p-1}}{1+x}dx, \, a > 0$$

The first integral on the right converges unless $p < 1$, since it is the proper one. In case, $p < 1$, the integrand becomes infinite at $x = 0 \Rightarrow$ for its convergence, $1 - p < 1$, i.e., $p > 0$.
For large

$$\frac{x^{p-1}}{1+x} \le \frac{x^{p-1}}{x} = \frac{1}{x^{2-p}} \Rightarrow \int_a^\infty \frac{dx}{x^{2-p}}$$

converges

$2 - p > 1$, i.e., if $p < 1 \Rightarrow \int_a^\infty \frac{x^{p-1}}{1+x}dx$

converges if $p < 1$ by comparison test $\Rightarrow$

$\int_0^\infty \frac{x^{p-1}}{1+x}dx$ converges if $0 < p < 1$ since both

the integrals on the right converge if $0 < p < 1$.

64. The logarithm of a negative number is imaginary. Hence only positive values are to be considered. Letting $x = e^y \Rightarrow$

$$I = \int_1^\infty x^p(\log x)^q dx = \int_0^\infty e^{(p+1)y}y^q dy$$

$$= \int_0^\infty e^{(p+1)y}y^q dy$$

$$+ \int_0^\infty e^{(p+1)y}y^q \, dy, a > 0$$

$= I_1 + I_2$, say.
I_1 is a proper integral if $q \ge 0$, whereas the integrand is infinite of order $-q$, at $y = 0 \Rightarrow$ for its convergence, $-q < 1$, i. e., $q > -1 \Rightarrow$ for any p, I_1 converges if $q > -1$ and diverges if $q \le -1$. I_2 converges if $p + 1 < 0$ for any q and diverges if $p + 1 > 0$ for any q, and if $p - 1 = 0$, it converges for $q > -1$ and diverges for $q \le -1$. Then finally using all the results, we conclude that the given integral converges only if $p + 1 < 0$ and $q > -1$, and diverges in all other cases.
Note that $x = 0$, $x = 1$ and $x = \infty$ are the points of infinite discontinuity.

65. For any $c > 0$, we can write $\int_0^\infty \frac{x}{(1+x)^3}dx$

$$= \int_0^c \frac{x}{(1+x)^3}dx + \int_c^\infty \frac{x}{(1+x)^3}dx \quad ...(1)$$

The first integral of R.H.S. of (1) is convergent

because it is a proper integral.
Now, we have to check the convergence of the

integral $\int_c^\infty \frac{x}{(1+x)^3}dx$

Let $f(x) = \dfrac{x}{(1+x)^3}$

$\Rightarrow f(x)$ is bounded in $]c, \infty[$.
Let us take $\mu = 3 - 1 = 2$ then we have

$$\lim_{x \to \infty} x^\mu f(x) = \lim_{x \to \infty} x^2 \cdot \frac{x}{(1+x)^3}$$

$$= \lim_{x \to \infty} \frac{1}{\left(1 + \dfrac{1}{x}\right)^3} = 1$$

which is a finite real number.
Now, since $\mu = 2$, which is greater than 1, then by

μ-test the integral $\int_c^\infty \dfrac{x}{(1+x)^3}dx$ is convergent.

Hence, being the sum of two convergent

integrals, the given integrals $\int_0^\infty \dfrac{x}{(1+x)^3}dx$

is convergent.

66. Let $f(x) = \dfrac{x^{3/2}}{\sqrt{x^4 - a^4}}$.

Then, $f(x)$ is bounded in the interval $]b, \infty[$.

Let $\quad \mu = 2 - \dfrac{3}{2} = \dfrac{1}{2}$

Then, $\lim_{x \to \infty} x^\mu f(x)$

$$= \lim_{x \to \infty} x^{1/2} \frac{x^{3/2}}{x^2\sqrt{1 - \dfrac{a^4}{x^4}}}$$

$$= \lim_{x \to \infty} \frac{1}{x^2\sqrt{1 - \dfrac{a^4}{x^4}}} = 1,$$

which is finite and non-zero.
Finally, since $\mu < 1$, therefore by μ-test the given integral is divergent.

67. Let $a > 0$

We have $\int_0^\infty \dfrac{x^{2m}}{1+x^{2n}}dx = \int_0^a \dfrac{x^{2m}}{1+x^{2n}}dx$

$$+ \int_a^\infty \dfrac{x^{2m}}{1+x^{2n}}dx \quad ...(1)$$

The first integral of R.H.S. of (1) is a proper integral and so it is convergent. So we have to check the convergence of $\int_a^\infty \dfrac{x^{2m}}{1+x^{2n}}\,dx$

Let us take $\mu = 2n - 2m$

Then

$$\lim_{x\to\infty} x^\mu f(x) = \lim_{x\to\infty} x^{2n-2m} \cdot \frac{x^{2m}}{1+x^{2n}}$$

$$= \lim_{x\to\infty} x^{2n-2m} \cdot \frac{x^{2m}}{x^{2n}\left(1+\dfrac{1}{x^{2n}}\right)}$$

$$= \lim_{x\to\infty} \frac{x^{2n-2m} \cdot x^{2m-2n}}{\left(1+\dfrac{1}{x^{2n}}\right)}$$

$$= \lim_{x\to\infty} \frac{1}{1+\dfrac{1}{x^{2n}}} = 1,$$

which is finite and non-zero.

$\therefore$ By μ-test, the given integral is convergent if $\mu > 0$ i.e., if $2n - 2m > 1$ i. e., if $n > m$.

Hence, the given integral is convergent if $n > m$, otherwise divergent.

68. Let $\log x = t \Rightarrow \dfrac{1}{x}dx = dt$

$$\therefore \int_a^\infty \frac{dx}{x(\log x)^{n+1}} = \int_{\log a}^\infty \frac{dt}{t^{n+1}}$$

Let $f(t) = \dfrac{1}{t^{n+1}}$ then $f(t)$ is bounded in the interval $]\log a, \infty[$.

Let us take $\mu = (n+1) - 0 = n + 1$, then

$$\lim_{t\to\infty} t^\mu f(t) = \lim_{t\to\infty} \frac{t^{n+1}}{t^{n+1}} = \lim_{t\to\infty} 1 = 1,$$

which is finite and non-zero.

Hence by μ-test, the given integral is convergent if

$\mu > 1$ i.e., $n + 1 > 1$ i.e., $n > 0$

and divergent if $\mu \le 1$ i.e., $n + 1 \le 1$ i.e., $n \le 0$.

74. Case I. Suppose $m < 0$. Let $-m = k \ge 0$

$$\Rightarrow \qquad \frac{x^{m-1}}{1+x} = \frac{1}{x^{k+1}(1+x)}$$

$$\Rightarrow \quad \lim_{x\to\infty} x^{k+2} \frac{1}{x^{k+1}(1+x)} = 1 \ne 0$$

Here $\mu = k + 2$. By μ-test, the integral

converges if $\mu = k + 2 = -m + 2 > 1$

Case II. Let $0 < m < 1$. Put $k = 1 - m > 0$

$$\Rightarrow \quad \frac{x^{m-1}}{1+x} = \frac{1}{x^k(1+x)} \Rightarrow \lim_{x\to\infty} x^{k+1}$$

$$\frac{1}{x^k(1+x)} = 1 \ne 0$$

$\Rightarrow$ the integral converges if $\mu = k + 1 > 1$

Case III. Let $m > 1$. Put $m - 1 = k \ge 0 \Rightarrow$

$$\frac{x^{m-1}}{1+x} = \frac{x^k}{1+x} \Rightarrow \lim_{x\to\infty} x^{1-k} \frac{x^k}{1+x} = 1 \ne 0$$

$\Rightarrow$ by μ-test, the given integral converges if $\mu = 1 - k > 1$.

75. Here $\displaystyle\lim_{x\to\infty} \frac{x^\mu}{x-1} = 1$ if $\mu = 1$, and

$$\lim_{x\to\infty} \frac{x^\mu}{x} = 1 \text{ if } \mu = 1$$

By μ-test, $\displaystyle\int_2^\infty \frac{dx}{x-1}$ and $\displaystyle\int_2^\infty \frac{1}{x}dx$ diverges. But

$$\frac{1}{x-1} - \frac{1}{x} = \frac{1}{x(x-1)} \Rightarrow \lim_{x\to\infty} \frac{x^\mu}{x(x-1)} = 1 \text{ if}$$

$\mu = 2$. By μ-test, $\displaystyle\int_2^\infty \left(\frac{1}{1+x} - \frac{1}{x}\right)dx$ converges.

76. We have

$$\lim_{x\to\infty} x^2 \frac{1}{(1+x^6)^{1/3}} = \lim_{x\to\infty} \frac{1}{\left(\dfrac{1}{x^6}+1\right)^{1/3}} = 1$$

the given integral is convergent.

77. We have

$$\lim_{x\to 1} (x-1)^{1/2} \frac{1}{(x^3-1)^{1/2}}$$

$$= \lim_{x\to 1} \frac{1}{\sqrt{(x^2+x+1)}} = \frac{1}{\sqrt3} \ne 0$$

$\Rightarrow$ the integral converges.

78. We have

$$\lim_{x\to 0} \frac{\sin x}{x} = 1 \Rightarrow \lim_{x\to 0} x^\mu \frac{\sin x}{x^n}$$

$$= \lim_{x\to 0} \left(x^\mu \cdot \frac{\sin x}{x} \cdot \frac{1}{x^{n-1}}\right)$$

$$= \lim_{x\to 0} \frac{\sin x}{x} \lim_{x\to 0} \frac{x^\mu}{x^{n-1}}$$

$$= \lim_{x\to 0} x^{\mu-n+1}$$

This limit exists iff $\mu = n - 1 < 1$, *i.e.*, $n < 2$

$\Rightarrow$ the integral converges for $n < 2$ and diverges for $n \geq 2$.

79. We have $\log \sin x = \log \left(x \cdot \dfrac{\sin x}{x} \right)$

$$= \log x + \log \dfrac{\sin x}{x}$$

Hence $\lim_{x \to 0} x^{\mu} \log \sin x$

$$= \lim_{x \to 0} \left(x^{\mu} \log x + x^{\mu} \log \dfrac{\sin x}{x} \right).$$

Evidently $\lim_{x \to 0} x^{\mu} \log \sin x = 0$ for $\mu > 0$ and $\lim_{x \to 0} \dfrac{\sin x}{x} = 1$.

Hence $\lim_{x \to 0} x^{\mu} \log \sin x = 0$ for $\mu > 0$ and therefore the integral converges.

80. $x = 0$ is a point of infinite discontinuity of order 1. Hence for its convergence, $1 - a < 1$, *i.e.*, $a > 0$.

81. $\lim_{x \to \infty} x^{3/2} \dfrac{1}{(1+x)\sqrt{x}} = 1 \Rightarrow$ the integral converges since $\dfrac{3}{2} > 1$ by μ-test.

82. We have $f(x) = \dfrac{\sin x}{x^{1+n}} = \dfrac{\sin x}{x} \cdot \dfrac{1}{x^n}$

$\therefore \lim_{x \to 0} x^{\mu} f(x) = \lim_{x \to 0} x^{\mu - n} \cdot \lim_{x \to 0} \dfrac{\sin x}{x}$

$$= 1 \text{ for } n = \mu.$$

By μ-test, the result follows.

83. Assume $p > 0$ and

$$f(x) = \left(x^p + \dfrac{1}{x^p} \right) \dfrac{\log(1+x)}{x}.$$

Then

$$\lim_{x \to 0} x^p f(x) = (x^{2p} + 1) \dfrac{\log(1+x)}{x} = 1$$

and hence the integral is convergent for $p < 1$ for μ-test.

Similarly

$$\lim_{x \to 0} x^{-p} f(x) = \lim_{x \to 0} \left(1 + \dfrac{1}{x^{2p}} \right)$$

$$\dfrac{\log(1+x)}{x} = 1.$$

The μ-test asserts that the integral converges for $-p < 1$ and hence the given integral converges for $-1 < p < 1$.

84. $\displaystyle \int_0^2 \dfrac{\log x}{\sqrt{(2-x)}} = \int_0^a \dfrac{\log x}{\sqrt{(2-x)}} + \int_a^2 \dfrac{\log x}{\sqrt{(2-x)}} dx$

for $0 < a < 2$.

We observe that $x = 0$ and $x = 2$ are points of infinite discontinuity in the given interval.

Now, $\lim_{x \to 0} x^{\mu} \dfrac{\log x}{\sqrt{(2-x)}} = 0$ for $\mu > 0$.

By μ-test, I_1 converges for $0 < \mu < 1$. Again

$$\lim_{x \to 2} (2-x)^{\mu} \dfrac{\log x}{\sqrt{(2-x)}} = \lim_{h \to 0} \dfrac{h^{\mu} \log(2-h)}{h^{1/2}}$$

$$= \log 2 \text{ for } \mu = \dfrac{1}{2} < 1.$$

Hence I_2 converges by μ-test.

Consequently, $I_1 + I_2 = \displaystyle \int_0^2 \dfrac{\log x}{\sqrt{(2-x)}}$ is convergent.

85. We observe that $x = 0$ is a point of infinite discontinuity of order 1/2.

Now

$$\lim_{x \to 0} x^{\mu} \cdot \dfrac{1}{\sqrt{(\tan x)}}$$

$$= \lim_{x \to 0} x^{\mu} \cdot \sqrt{\left(\dfrac{x}{\sin x} \right)} \cdot \dfrac{\sqrt{\cos x}}{\sqrt{x}}$$

$$= \lim_{x \to 0} x^{\mu - \frac{1}{2}} \lim_{x \to 0} \sqrt{\left(\dfrac{x}{\sin x} \right)} \lim_{x \to 0} \sqrt{(\cos x)}$$

$$= \lim_{x \to 0} x^{\mu - \frac{1}{2}}$$

$$= 1 \text{ for } \mu - \dfrac{1}{2} = 0, \text{ i.e., } \mu = \dfrac{1}{2} < 1.$$

The result follows by μ-test.

86. Let $f(x) = \dfrac{1}{x^3(1+x^2)}$.

Clearly $f(x)$ is unbounded at the lower limit of the integration $x = 0$

Take $g(x) = \dfrac{1}{x^3}$

Then $\lim_{x \to 0} \dfrac{f(x)}{g(x)} = \lim_{x \to 0} \dfrac{1}{1+x^2} = 1$, which is finite and non-zero.

$\therefore$ By comparison test $\displaystyle \int_0^1 f(x) dx$ and

$\int_0^1 g(x)\,dx$ either both converge or both diverge. But the comparison integral $\int_0^1 \dfrac{dx}{x^3}$ is divergent, because we have $n = 3 > 1$.

Hence, the given integral $\int_0^1 \dfrac{dx}{x^3(1+x^2)}$ is divergent.

87. Let $f(x) = \dfrac{1}{(x+1)\sqrt{1-x^2}}$

Clearly $f(x)$ is unbounded at the upper limit of the integration $x = 1$.

Let us take $g(x) = \dfrac{1}{\sqrt{1-x^2}}$

Then $\lim\limits_{x\to 1}\dfrac{f(x)}{g(x)} = \lim\limits_{x\to 1}\dfrac{1}{x+1} = \dfrac{1}{2}$, which is finite and non-zero.

$\therefore$ By comparison test $\int_0^1 f(x)\,dx$ and $\int_0^1 g(x)\,dx$ either both converge or both diverge.

But

$\int_0^1 g(x)\,dx = \int_0^1 \dfrac{dx}{\sqrt{1-x^2}} = \lim\limits_{\varepsilon\to 0}\int_0^{1-\varepsilon}\dfrac{dx}{\sqrt{1-x^2}}$

$= \lim\limits_{\varepsilon\to 0}\left[\sin^{-1}x\right]_0^{1-\varepsilon}$

$= \lim\limits_{\varepsilon\to 0}[\sin^{-1}(1-\varepsilon)] = \sin^{-1}1 = \dfrac{\pi}{2},$

which is a definite real number.

$\Rightarrow \int_0^1 g(x)\,dx$ is divergent.

Hence, the given integral $\int_0^1 \dfrac{dx}{(x+1)\sqrt{1-x^2}}$ is convergent.

88. Let $f(x) = \dfrac{\cos x}{x^2}$

Clearly $f(x)$ is unbounded at the lower limit of the integration.

Let us take $g(x) = \dfrac{1}{x^2}$

Then $\lim\limits_{x\to 0}\dfrac{f(x)}{g(x)} = \lim\limits_{x\to 0}\left\{\dfrac{\cos x}{x^2}\cdot x^2\right\}$

$= \lim\limits_{x\to 0}\cos x = 1,$

which is finite and non-zero.

Therefore, by comparison test, $\int_0^{\pi/2} f(x)\,dx$ and $\int_0^{\pi/2} g(x)\,dx$ either both converge or both diverge.

But $\int_0^{\pi/2} g(x)\,dx$

$= \int_0^{\pi/2}\dfrac{1}{x^2}\,dx = \lim\limits_{\varepsilon\to 0}\int_0^{\pi/2}\dfrac{1}{x^2}\,dx$

$= \lim\limits_{\varepsilon\to 0}\left[\dfrac{1}{x}\right]_0^{\pi/2} = \lim\limits_{\varepsilon\to 0}\left[-\dfrac{2}{\pi}+\dfrac{1}{\varepsilon}\right] = \infty$

$\Rightarrow \int_0^{\pi/2} g(x)\,dx$ is divergent.

Hence the given integral $\int_0^{\pi/2}\dfrac{\cos x}{x^2}\,dx$ is divergent.

89. Let $f(x) = \dfrac{1}{\sqrt{x^4-1}}$

Clearly $f(x)$ is unbounded at the lower limit of integration $x = 1$.

Let us take $g(x) = \dfrac{1}{\sqrt{x^2-1}}$

Then $\lim\limits_{x\to 1}\dfrac{f(x)}{g(x)} = \lim\limits_{x\to 1}\left\{\dfrac{1}{\sqrt{x^4-1}}\cdot\sqrt{x^2-1}\right\}$

$= \lim\limits_{x\to 1}\dfrac{1}{\sqrt{x^2+1}} = \dfrac{1}{\sqrt{2}},$

which is finite and non-zero.

Therefore by comparison test $\int_1^2 f(x)\,dx$ and $\int_1^2 g(x)\,dx$ are either both convergent or both divergent.

But $\int_1^2 g(x)\,dx = \int_1^2\dfrac{dx}{\sqrt{x^2-1}}$

$= \lim\limits_{\varepsilon\to 0}\int_{1+\varepsilon}^2\dfrac{dx}{\sqrt{x^2-1}}$

$= \lim\limits_{\varepsilon\to 0}\left[\log\left\{x+\sqrt{x^2-1}\right\}\right]_{1+\varepsilon}^2$

$$= \lim_{\varepsilon \to 0} \log(2+\sqrt{3}) - \log\left\{1+\varepsilon+\sqrt{\varepsilon^2+\varepsilon}\right\}$$

$= \log(2+\sqrt{3})$, which is a definite non-zero real number.

$\Rightarrow \int_1^2 g(x)\,dx$ is convergent.

Hence, $\int_1^2 \dfrac{1}{\sqrt{x^4-1}}\,dx$ is also convergent.

95. For any integer i, we have

$$\int_1^i \frac{1}{x}\,dx = \sum_{n=1}^{i-1} \int_n^{n+1} \frac{1}{x}\,dx$$

$$\geq \sum_{n=1}^{i-1} \frac{1}{n+1} \int_n^{n+1} 1.dx$$

$$= \sum_{n=1}^{i-1} \frac{1}{n+1} = \sum_{k=2}^{i} \frac{1}{k}$$

But $\displaystyle\sum_{k=2}^{i} \frac{1}{k}$ diverges as $i \to \infty$.

Then we see that $\displaystyle\lim_{t\to\infty} \int_1^t \frac{1}{x}\,dx$ does not exist.

96. For $0 < \varepsilon < 1$, let $F(\varepsilon) = \displaystyle\int_\varepsilon^1 \frac{1}{x}\,dx$

We put $x = \dfrac{1}{u}$.

Then $F(\varepsilon) = \displaystyle\int_{1/\varepsilon}^1 u\left(-\frac{1}{u^2}\right)du = \int_1^{1/\varepsilon} \frac{1}{u}\,du$

$\Rightarrow \displaystyle\lim_{\varepsilon\to 0} \int_1^{1/\varepsilon} \frac{1}{u}\,du$

does not exist. Hence $\displaystyle\lim_{\varepsilon\to 0} F(\varepsilon)$ does not exist, and this proves that the given integral diverges.

97. The integral is improper since $\dfrac{1}{\sqrt{(1-x^2)}}$ is not bounded on $[0, 1]$. We first show that the improper integral

$$\int_0^1 \frac{1}{\sqrt{(1-x)}}\,dx$$

is convergent (and hence absolutely convergent, since $\sqrt{(1-x)} \geq 0$). If $0 < \varepsilon < 1$, we find

$$\int_0^{1-\varepsilon} \frac{1}{\sqrt{(1-x)}}\,dx = 2 - 2\sqrt{\varepsilon}$$

and hence $\displaystyle\lim_{\varepsilon\to 0} \int_0^{1-\varepsilon} dx = 2$

Therefore $\displaystyle\int_0^1 \frac{1}{\sqrt{(1-x)}}\,dx$ converges absolutely.

But, for $0 \leq x < 1$,

$$\frac{1}{\sqrt{(1-x^2)}} = \frac{1}{\sqrt{(1+x)}}\cdot\frac{1}{\sqrt{(1-x)}} \leq \frac{1}{\sqrt{(1-x)}}$$

Hence by comparison test, $\displaystyle\int_0^1 \frac{1}{(1-x^2)}\,dx$ is absolutely convergent.

98. We see that the given integral is an integral over $(0, \infty)$ and $\dfrac{1}{x^2+\sqrt{x}}$ is not bounded for x near 0.

Now $\displaystyle\int_0^\infty \frac{1}{x^2+\sqrt{x}}\,dx = \int_0^1 \frac{1}{x^2+\sqrt{x}}\,dx$

$$+ \int_1^\infty \frac{1}{x^2+\sqrt{x}}\,dx$$

But I_1 is a convergent integral of the second kind since

$$\frac{1}{x^2+\sqrt{x}} \leq \frac{1}{\sqrt{x}} \text{ for } 0 < x \leq 1,$$

and I_2 is a convergent integral of the first kind since

$$\frac{1}{x^2+\sqrt{x}} \leq \frac{1}{x^2} \text{ for } 1 \leq x < \infty.$$

Hence $\displaystyle\int_0^\infty \frac{1}{x^2+\sqrt{x}}$ is convergent.

99. $\displaystyle\int_0^\infty \frac{1}{x^2}\,dx = \int_0^1 \frac{1}{x^2}\,dx + \int_1^\infty \frac{1}{x^2}\,dx.$

The first integral on the right is of the second kind whereas the second integral is of the first kind.

Now $\displaystyle\int_1^\infty \frac{1}{x^2}\,dx$ converges for $n = 2 > 1$.

since $\displaystyle\int_a^\infty \frac{dx}{x^n}\,(a>0)$ is convergent for $n > 1$

and $\displaystyle\int_0^1 \frac{1}{x^2}\,dx$ diverges for $n = 2 > 1$

since $\displaystyle\int_a^b \frac{dx}{x^n}$ diverges for $n \geq 1$

Hence $\displaystyle\int_0^\infty \frac{1}{x^2}\,dx$ diverges.

100. We have $n = \dfrac{1}{2} < 1$ and $n = 2 > 1$.

Applying $\displaystyle\int_a^b \dfrac{dx}{(x-a)^n}$ is convergent for $n < 1$ and

divergent for $n \geq 1$, we have the desired result.

101. We observe that $\log \sin x$ is unbounded at $x = 0$. Hence the only point of infinite discontinuity is $x = 0$.

We know that $\displaystyle\int_0^{\pi/2} \dfrac{dx}{x^\mu}$ converges if $0 < \mu < 1$.

Now $\displaystyle\lim_{x \to 0} x^\mu \log \sin x$, where $\mu > 0$

$$= \lim_{x \to 0} \dfrac{\log \sin x}{x^{-\mu}} \qquad \left[\text{form } \dfrac{\infty}{\infty}\right]$$

$$= \lim_{x \to 0} \dfrac{\cot x}{-\mu x^{-\mu-1}} = \lim_{x \to 0} -\dfrac{1}{\mu} \dfrac{x^{\mu+1}}{\tan x}$$

$$\left[\text{form } \dfrac{0}{0}\right]$$

$$= \lim_{x \to 0} -\dfrac{1}{\mu} \dfrac{(\mu+1) x^\mu}{\sec^2 x} = 0 \text{ if } \mu > 0$$

We set μ such that $0 < \mu < 1$. Then by μ-test, the given integral is convergent.

102. We have

$$f(x) = \dfrac{x^p}{(\sin x)^q} = \left(\dfrac{x}{\sin x}\right)^q . x^{p-q}$$

$$= \left(\dfrac{x}{\sin x}\right)^q \dfrac{1}{x^{q-p}}$$

At $x = 0$, $f(x) \to 0$ if $p - q > 0$ and $f(x) \to \infty$ if $p - q < 0$.

Hence the integral is proper if $p - q > 0$ and improper if $p - q < 0$. Thus $x = 0$ is the only point of discontinuity in the latter case. We take $\mu = q - p$. Then by μ-test the integral converges if $q - p < 1$ or $q < p + 1$.

103. Here $x = 0$ and $x = \dfrac{\pi}{2}$ are points of infinite discontinuity.

Now $I = \displaystyle\int_0^{\pi/2} \dfrac{dx}{\sin^m x \cos^n x}$

$$= \int_0^a \dfrac{dx}{\sin^m x \cos^n x} + \int_a^{\pi/2} \dfrac{dx}{\sin^m x \cos^n x}$$

$$= I_1 + I_2, \text{ say}$$

For I_1, $\displaystyle\lim_{x \to 0} x^m . \dfrac{1}{\sin^m x \cos^n x}$

$$= \lim_{x \to 0} \left\{\left(\dfrac{x}{\sin x}\right)^m . \dfrac{1}{\cos^n x}\right\} = 1$$

Hence I_1 converges for $m = \mu < 1$ only.
For I_2,

$$\lim_{x \to \pi/2} \left\{\left(\dfrac{\pi}{2} - x\right)^n \dfrac{1}{\sin^m x \cos^n x}\right\}$$

$$= \lim_{x \to \pi/2} \left\{\dfrac{1}{\sin^m x} \left(\dfrac{\pi/2 - x}{\cos x}\right)^n\right\} = 1$$

Hence I_2 converges only for $n < 1$.
Therefore, $I = I_1 + I_2$ converges only for $m < 1$, $n < 1$.

104. The integral is a proper one for $n \leq 0$. We observe that $x = 0$ is a point of infinite discontinuity if $n > 0$.
Now

$$\lim_{x \to 0} x^\mu \dfrac{\cos x}{x^n} = \lim_{x \to 0} x^{\mu-n} \lim_{x \to 0} \cos x$$

$$= \lim_{x \to 0} x^{\mu-n} = 1 \text{ if } n = \mu.$$

By μ-test, the integral converges for $n < 1$ and diverges for $n \geq 1$.

105. Here $\displaystyle\lim_{x \to \infty} \dfrac{x}{e^x + 1} \qquad \left[\text{form } \dfrac{\infty}{\infty}\right]$

$$= \lim_{x \to \infty} \dfrac{1}{e^x} = 0$$

Hence $\displaystyle\int_0^\infty \dfrac{x}{e^x + 1}$ is convergent.

106. We have $\displaystyle\lim_{x \to \infty} e^x e^{-x^2} = \lim_{x \to \infty} \dfrac{e^x}{e^{x^2}}$

$$= \lim_{x \to \infty} \dfrac{1 + x + \dfrac{x^2}{2!} + \cdots}{1 + x^2 + \dfrac{x^4}{2!} + \cdots}$$

$$= \lim_{x \to \infty} O\left(\dfrac{1}{x}\right) = 0$$

Hence $|e^x e^{-x^2}| < 0 = \lambda$, and $\mu = 1 > 0$.

Therefore, $\displaystyle\int_0^\infty e^{-x^2} dx$ converges.

Verify each of the following :

1. The integral $\int_0^1 \frac{\log x}{\sqrt{x}}$ is convergent but

$\int_1^2 \frac{\sqrt{x}}{\log x} dx$ is divergent.

2. The integral $\int_0^{\pi/2} \sin x \log \sin x \, dx$ converges to the value $\log 2 - 1$.

3. The integral $\int_0^{\pi/2} \cos 2nx \log \sin x \, dx, n \geq 1$, converges to the value $-\frac{\pi}{4n}$.

4. The integral $\int_0^1 \frac{\sin 1/x}{x^p} dx, p > 0$ converges absolutely for $p < 1$.

5. The integral $\int_0^\infty \frac{x \tan^{-1} x}{(1+x^4)^{1/3}} dx$ is divergent.

6. The integral $\int_0^1 \log \overline{(x)} \, dx$ is convergent.

7. The integral $\int_0^\infty \left(\frac{1}{e^x - 1} - \frac{1}{x} + \frac{1}{2} \right) \frac{e^{-kx}}{x} dx$ converges.

8. The integral $\int_0^\infty \log(1 + 2\,\text{sech}\, x) dx$ is convergent.

9. The integral $\int_0^\infty \left(\frac{1}{1+x} - e^{-x} \right) \frac{dx}{x}$ is convergent.

10. The integral $\int_0^\infty \frac{x \log x}{(1+x^2)^2} dx$ converges to 0.

11. The integral $\int_1^\infty \frac{\sin x}{x^p} dx$ is convergent for $p > 0$.

12. The integral $\int_2^\infty \frac{\cos x}{\log x} dx$ is conditionally convergent.

13. The integral $\int_0^{n\pi} \frac{x}{1+x^4 \cos^2 x} dx$ is

14. The integral $\int_0^\infty \frac{dx}{1+x^4 \cos^2 x}$ is convergent.

15. The integral $\int_0^\infty x^{-\alpha} \sin^{1-\beta} dx$ is convergent if α lies between β and $2-\beta$.

16. The integral $\int_0^\infty \frac{\cosh at}{\cosh bt} dt$ converges iff $|b| > |a|$.

17. The integral $\int_0^\infty (\log x)^m \frac{\sin x}{x} dx$ converges when $m \geq 0$.

18. The integral $\int_0^\infty \frac{x^a}{a^x} dx$ converges only if $a > 1$.

19. If $f(x)$ is monotonic and $\int_0^\infty f(x) dx$ converges, then

$$x f(x) \to 0 \text{ as } x \to \infty.$$

20. The value of the integral $\int_0^\infty \frac{\sin ax}{x} dx = \frac{\pi}{2}$, 0 or $-\frac{\pi}{2}$ according as $a >, =, < 0$.

Chapter 9

Uniform Convergence of Sequence and Series of Functions

9.1 INTRODUCTION

In this chapter, we shall discuss the uniform convergence of sequence and series of functions. Mainly, we want to draw attention on the most important aspects of the probelms of sequence and series of functions.

9.2 POINTWISE CONVERGENCE

Let $\langle f_n \rangle$ be a sequence of real valued functions on a metric space (X, d). Let the function f_n be tends to a definite limit for all values of $x \in X$ as $n \to \infty$. Therefore, to each point $t \in X$, there corresponds a sequence of numbers $\langle f_n(t) \rangle$ with terms $f_1(t), f_2(t), f_3(t), \ldots$

Let this sequence $\langle f_n(t) \rangle$ converge to $f_n(t)$. Then pointwise converges can be defined as follows:

Definition. *Let $X \neq \phi$ and f be a function from X to R. Also, each $n \in N$ let $f_n : X \to R$. Then, the sequence of functions $\langle f_n \rangle$ converges pointwise to the function f, if for each $x \in X$, the sequence of real numbers $\langle f_n(x) \rangle$ converges to the real number $f(x)$.*

Therefore, $\langle f_n \rangle$ converges pointwise to f if $\lim\limits_{n \to \infty} f_n(x) = f(x) \forall x \in X$

☞ ILLUSTRATIONS

(1) For each $n \in N$. Let us define $f_n : R \to R$ by $f_n(x) = \dfrac{x}{n}, \forall x \in R$

Then $\langle f_n(x) \rangle$ converges to $f(x) = 0, \forall x \in R$

(2) The sequence $\langle f_n(x) \rangle = \langle x^n \rangle$ converges pointwise to the function $f : [0, 1] \to R$ defined

by $f(x) = \begin{cases} 0 & \text{if } x \in]0,1[\\ 1 & \text{if } x = 1 \end{cases}$

(3) The sequence $\langle x(1-x)^n \rangle$ converges pointwise to the function f that vanish identically.

(4) The sequence $\langle f_n = 1 + \dfrac{x}{1+nx} \rangle$ converges, pointwise to the function f defined by

$f(x) = 1, \forall x \in]0, \infty[$.

(5) The geometric series $1+x+x^2+x^3+\cdots$ converges to $(1-x)^{-1}, \forall x \in]-1, 1[$.

REMARKS

- If the sequence f_n converges to f, then, we can write $\lim\limits_{n \to \infty} f_n(x) = f(x)$ for $x \in X$
- Point wise convergent of a sequence of functions depends on the metric space as well as on the functions.

Definition. *Let (X, d) be a metric space and f be a function from X to R and $f_n : X \to R$, $\forall n \in N$. The sequence of function $\langle f_n \rangle$ converges pointwise to f if and only if for each $x \in X$ and for each positive real number ε, $\exists$ a positive integer m such that*

$$n \geq m \Rightarrow |f_n(x) - f(x)| < \varepsilon$$

Solved Examples

Example 1. *Let $< f_n >$ be the sequence defined by $f_n : R \to R$ such that*

$$f_n(x) = \frac{x}{n} \forall x \in R, n \in N$$

Show that the sequence converges pointwise to the zero function.

Solution. Here, we want to show that given sequence converges pointwise to the zero function *i.e.*, $f(x) = 0, x \in R$, then we must show that given $\varepsilon > 0$, we can find $m \in N$ such that

$$\forall n \geq m \Rightarrow \left| \frac{x}{n} - 0 \right| = \frac{|x|}{n} \qquad \qquad ..(1)$$

Let us choose $m > \dfrac{|x|}{\varepsilon}$

Then (1) gives $\qquad \forall n \geq m \Rightarrow \left| \dfrac{x}{n} - 0 \right| = \dfrac{|x|}{n} < \varepsilon$

Here, the given sequence converges pointwise to the zero function.

9.3 UNIFORM CONVERGENCE OF SEQUENCES

Let us suppose the sequence $\langle f_n(x) \rangle$ converges for every point x in X. Therefore, f_n tends to a definite limit as $n \to \infty$ for every $x \in X$. The limit is also a function of x.

Then by definition of limit, we must have for every $\varepsilon > 0$ $\exists$ a positive integer m such that

$$n \geq m \Rightarrow |f_n(x) - f(x)| < \varepsilon.$$

Here, it must be noted that the integer m depends upon x as well as ε.

Definition. *The sequence $\langle f_n(x) \rangle$ of functions is said to converge uniformly on X to a function f, if for every $\varepsilon > 0$, we can find a positive integer m such that*

$$n \geq m \Rightarrow |f_n(x) - f(x)| < \varepsilon, \forall x \in X$$

The convergence of a sequence $\langle f_n(x) \rangle$ at every point $x \in X$ (*i.e.*, pointwise convergence) does not necessarily ensure its uniform convergence on X. A sequence of functions may be convergent at every point of X, yet it may not be uniformly convergent on X.

Let us consider the following examples :

(1) The sequence of function $\langle f_n \rangle$ defined on R such that $f_n(x) = \dfrac{x}{n}, \forall n \in N$ converges pointwise to the zero function (*i.e.*, $f(x) = 0$) while, this sequence does not converges uniformly to this function.

We will prove that convergence is not uniform.

Let us suppose the sequence $\left\langle \dfrac{x}{n} \right\rangle$ converges uniformly to the zero function on R, then there is some $m \in N$ (m depending only on $\varepsilon = 1$) such that

$$n \geq m \Rightarrow |f_n(x) - f(x)| = \frac{|x|}{n} < 1, \forall x \in R$$

which is not true for all $x \in$ R for if $n = m$ and $x = m$, then $\dfrac{|x|}{m} = 1$

(2) Let $\langle f(x) \rangle = \langle x^n \rangle$ be the sequence of function defined on $[0, 1]$. Then we can easily verify that the given sequence $\langle f_n(x) \rangle$ converges pointwise to the limit function f,

defined by
$$f(x) = \begin{cases} 0 & \text{if} \quad 0 \le x < 1 \\ 1 & \text{if} \quad x = 1 \end{cases} \quad \text{for every } x \in [0, 1].$$

To check that the convergence is uniform, we consider the interval $[0, 1]$. Let $\varepsilon > 1$ be given. Then, we have
$$| f_n(x) - f(x) | < \varepsilon \Rightarrow | x^n - 0 | < \varepsilon \Rightarrow x^n < \varepsilon$$

$$\Rightarrow \qquad \frac{1}{x^n} > \frac{1}{\varepsilon} \Rightarrow n \log \frac{1}{x} > \log \frac{1}{\varepsilon}$$

i.e., $\qquad\qquad n > \dfrac{\log(1 / \varepsilon)}{\log(1 / x)}$ ... (1)

Therefore, when $x \ne 1$, $m \in$ N such that $m > \dfrac{\log(1 / \varepsilon)}{\log(1 / x)}$

In particular, when $x = 0$, $m = 1$.

Now as x increases from 0 to 1, it is clear from (1) that $n \to \infty$.

Therefore, it is not possible to find $m \in$ N such that
$$n \ge m \Rightarrow | f_n(x) - f(x) | < \varepsilon \text{ for all } x \in [0,1]. \text{ Hence, the}$$
given sequence is not uniformly convergent on $[0, 1]$.

REMARK

- If we consider the interval $]0, k[$, where $0 < k < 1$, then the greatest value of $\log (1/\varepsilon)/ \log (1/x)$ is $\log (1/ \varepsilon)/ \log (1/k)$ so that if we take $m > (\log 1/\varepsilon)/\log(1 / k) \in$ N, we have
$$n \ge m \Rightarrow |f_n (x) - f(x)| < \varepsilon \ \forall \ x \in [0, k].$$
Therefore, $\langle f_n(x) \rangle$ converges uniformly on $[0, k]$.

(3) The sequence of functions $< 1/(1 + nx) >$ does not converges uniformiy on R to the function f defined by
$$f(x) = \begin{cases} 0, & \text{if} \quad x \ne 0 \\ 1, & \text{if} \quad x = 0 \end{cases}$$

(4) Let a be any positive real number and for each $n \in$ N.

Define $\qquad f_n(x) = \dfrac{1}{1 + nx^2} \ \forall x \in [a, \infty]$

The sequence $\langle f_n(x) \rangle$ converges uniformly to the zero function *i.e.,* $f(x) = 0$ on $[a, \infty[$, because of $m \in$ N, $m > (1 - \varepsilon)/a^2$, then
$$n \ge m \Rightarrow | f_n(x) - 0 | = \frac{1}{1 + nx^2} \le \frac{1}{1 + mx^2}$$

$$\le \frac{1}{1 + ma^2} < \varepsilon \ \forall x \in [a, \infty[$$

9.3.1 POINT OF NON-UNIFORM CONVERGENCE

A point such that the sequence does not converge uniformly in any neighbourhood of it, however small, is said to be a point of non-uniform converges of the sequence.

9.3.2 SUM FUNCTION OF A SERIES

Consider the series $\sum_{n=1}^{\infty} u_n(x) = u_1(x) + u_2(x) + \ldots + u_n(x) + \ldots, x \in X$ of real valued function defined on a set X. This series gives rise to a sequence of function $\langle f_n(x) \rangle$ where $f_n(x) = u_1(x) + u_2(x) + \ldots + u_n(x)$

The series $\Sigma u_n(x)$ is said to be convergent on X if the corresponding sequence $\langle f_n(x) \rangle$ is convergent on X and the limit function $f(x)$ of the sequence is said to the sum function or the sum of the series.

9.4 UNIFORM CONVERGENCE OF A SERIES OF FUNCTIONS

The series $\sum_{n=1}^{\infty} u_n(x)$ is said to converge uniformly on X if the sequence $\langle f_n(x) \rangle$, where $f_n(x) = u_1(x) + u_2(x) + \ldots + u_n(x)$ converges uniformly on X.

9.5 CAUCHY'S GENERAL PRINCIPLE OF UNIFORM CONVERGENCE

THEOREM 1. *Let $\langle f_n \rangle$ be a sequence of real valued function defined on X. Then $\langle f_n \rangle$ converges uniformly on X if and only if for every $\varepsilon > 0$, there exists a positive integer m such that*

$$n \geq m, \, p \geq m, \, x \in X \Rightarrow |f_n(x) - f_p(x)| < \varepsilon.$$

Proof. **The only if part.** Let us first suppose, the sequence $\langle f_n \rangle$ converges uniformly to the function f on X. Then, by definition, for given $\varepsilon > 0 \, \exists$, a positive integer m such that

$$| f_n(x) - f(x)| < \varepsilon/2 \, \forall n \geq m, \, \forall x \in X \qquad \ldots(1)$$

Therefore, if $p, n \geq m$ we have for any $x \in X$.

$$| f_n(x) - f_p(x)| = | f_n(x) - f(x) + f(x) - f_p(x)|$$

$$\leq | f_n(x) - f(x)| + | f(x) - f_p(x)|$$

$$< \varepsilon/2 + \varepsilon/2 = \varepsilon.$$

Hence (1) holds for this m.

The if part. Let $\langle f_n \rangle$ be a sequence of function from X to R such that for given $\varepsilon > 0$, $\exists$ a positive integer m such that (1) holds.

To show $\exists$ a function f on X such that the sequence $\langle f_n \rangle$ converges uniformly to f on X.

Now, for each fixed $x \in X$, the sequence of real numbers $\langle f_n(x) \rangle$ is a Cauchy sequence and therefore $\lim_{n \to \infty} f_n(x)$ exists for every $x \in X$.

(∵ Every Cauchy sequence of real numbers is convergent.)

Define $f : X \to $ R by $f(x) = \lim_{n \to \infty} f_n(x), \forall x \in X$

We want to show that the sequence $<f_n>$ converges uniformly to f.

If $x \in $ R, $\varepsilon > 0$, then there is some $m \in $ N such that

$$n, p \geq m \Rightarrow | f_n(x) - f_p(x)| < \varepsilon/2 \text{ for all } x \in X.$$

For any fixed $p, p \geq m$ and fixed $x \in X$, consider that sequence $\left\langle | f_n(x) - f_p(x) | : n \in N \right\rangle$

Since

$$\lim_{n \to \infty} f_n(x) = f(x) \text{ and } | f_n(x) - f_p(x) | < \varepsilon / 2 \text{ for } n \geq p,$$

we have $\lim_{n \to \infty} | f_n(x) - f_p(x) | = |f_n(x) - f_p(x)| < \varepsilon / 2$

Therefore, if $p \geq m$, then

$$|f(x) - f_p(x)| < \varepsilon, \ \forall \ x \in X.$$

Hence, the sequence $\langle f_n(x) \rangle$ converges uniformly to f on X.

THEOREM 2. *The series $\Sigma u_n(x)$ converges uniformly on X if and only if for every $\varepsilon > 0$, $\exists$ a positive integer m such that*

$$n \geq m \Rightarrow | u_{n+1}(x) + u_{n+2}(x) + \dots + u_{n+p}(x) | < \varepsilon, p = 1, 2. \text{ for all } x \in X.$$

Proof. Let $s_n(x)$ denotes the sequence of partial sum of the given series such that

$$s_n(x) = u_1(x) + u_2(x) + \dots + u_n(x), x \in X.$$

Then, $s_{n+p}(x) - s_n(x) = u_{n+1}(x) + u_{n+2}(x) + \dots + u_{n+p}(x)$

The series $\sum_{n=1}^{\infty} u_n(x)$ converges uniformly to $f(x)$ on X if and only if $\langle f_n \rangle$ converges uniformly on X. But $\langle s_n(x) \rangle$ converges to $s(x)$ on X if and only if for given $\varepsilon > 0$, $\exists$ a positive integer m such that $n \geq m \Rightarrow | s_{n+p}(x) - s_n(x) | < \varepsilon, p = 1, 2$ for all $x \in X$.

Therefore,

$$n \geq m \Rightarrow | u_{n+1}(x) + u_{n+2}(x) + \dots + u_{n+p}(x) | < \varepsilon, \ p = 1, 2,$$

for all $x \in X$.

9.6 DINI'S CRITERION FOR UNIFORM CONVERGENCE OF A SEQUENCE OF CONTINUOUS FUNCTIONS

THEOREM 1. *Let $\langle f_n \rangle$ be a sequence of continuous real valued function defined on the compact metric space (X, d) such that*

$$f_1(x) \geq f_2(x) \geq \dots \geq f_n(x) \geq \qquad \dots (1)$$

for every $x \in X$. If $\langle f_n \rangle$ pointwise converges on X to the continuous function f on X, then $\langle f_n \rangle$ converges uniformly to f on X.

Proof. Let $g_n = f_n - f$ for each $n \in N$ Then, from (1), we get

$$g_1(x) \geq g_2(x) \geq g_n(x) \geq \dots \geq 0 \qquad \dots (2)$$

Also, since $\langle f_n \rangle$ converges to f on X, we have

$$\lim_{n \to \infty} g_n(x) = 0 \forall x \in X \qquad \dots (3)$$

To show, $\langle g_n \rangle$ converges uniformly to 0 on X. Let $\varepsilon > 0$ be given.

If $x \in X$, then from (3), $\exists$ a positive integer $m(x)$ such that $0 \leq g_m(x) \leq \varepsilon / 2$

Since $g_m(x)$ is continuous at x, therefore, $\exists$ an open sphere $S(x, r)$ such that $y \in S(x, r) \Rightarrow g_{m(x)}(y) < \varepsilon$. Therefore, the collection $C = \{ S(x, r) : x \in X, r > 0 \}$ forms an open cover of X. Since X is compat, therefore, by definition $\exists$

a finite subcover of C i.e., $\exists$ a finite number of open spheres $S(x, r)$ say $S(x_1, r_1), S(x_2, r_2),\dots S(x_k, r_k)$, which also cover X.

Now, let $m = \max\{m(x_1), m(x_2), \dots, m(x_k)\}$,

If y is any point of X, then $y \in S(x_i, r)$ for some $i = 1, 2,\dots, k$.

Therefore, $\qquad g_{m(x_i)}(y) < \varepsilon$

But since $m(x_i) \le m$, therefore from (2), we have $g_m(y) = g_m(x_i)(y)$

$\Rightarrow \qquad\qquad 0 \le g_m(y) < \varepsilon$, $\forall y \in X$

Thus, from (2), we have $0 \le g_n(y) < \varepsilon$, $\forall\ n \ge m, y \in X$.

Hence, $\langle g_n \rangle$ converges uniformly to 0 on X. This implies that $\langle f_n \rangle$ converges uniformly on Y to the function f.

9.7 TESTS FOR UNIFORM CONVERGENCE

THEOREM 1. **(M$_n$-test).** *Let $\langle f_n \rangle$ be a sequence of function defined on a metric space X. Let $\lim\limits_{n\to\infty} f_n(x) = f(x)$ for all $x \in X$ and let $M_n = \sup\{\,|f_n(x) - f(x)|:x \in X\}$*

Then $\langle f_n \rangle$ converges uniformly to f if and only if $M_n \to 0$ as $n \to \infty$.

Proof. **Necessary condition.** Let us suppose, the sequence $\langle f_n \rangle$ of functions converges uniformly to f on X. Then by definition, for a given $\varepsilon > 0$, $\exists$ a positive integer m (independent of x) such that $n \ge m \Rightarrow |f_n(x) - f(x)| < \varepsilon \ \forall x \in X$.

Also, M_n is the supremum of $|f_n(x) - f(x)|$.

Therefore

$$|f_n(x) - f(x)| < \varepsilon \,\forall n \ge m, \forall x \in X$$

$$\Rightarrow \qquad M_n = \sup_{x \in X} |f_n(x) - f(x)| < \varepsilon \forall n \ge m$$

$$\Rightarrow \qquad M_n \to 0, \text{as } n \to \infty$$

Sufficient condition. Let us assume that $M_n \to 0$ as $n \to \infty$. Then for a given $\varepsilon > 0$, $\exists$ a positive integer m such that

$$|M_n - 0| < \varepsilon \forall n \ge m, \forall x \in X$$

$$\Rightarrow \qquad M_n < \varepsilon \forall n \ge m, \forall x \in X$$

$$\Rightarrow \qquad \sup_{x \in X} |f_n(x) - f(x)| = M_n < \varepsilon, \forall n \ge m$$

$$\Rightarrow \qquad |f_n(x) - f(x)| \le M_n < \varepsilon \forall n \ge m, \forall x \in X|$$

$\Rightarrow \quad \langle f_n \rangle$ converges uniformly to f on X.

THEOREM 2. **(T$_n$-test for uniform convergence).** *Let $\langle f_n(x) \rangle$ be a sequence of functions defined on $X \subset R$ and let on X*

$$\lim_{n\to\infty} f_n(x) = f(x) \qquad\qquad \dots(1)$$

Let for $n \ge p \in N$, $T_n > 0$ exists such that

$$|f_n(x) - f(x)| \le T_n, \forall x \in X \text{ and } n > p \qquad\qquad \dots(2)$$

Then $\langle f_n(x) \rangle$ converges uniformly to $f(x)$ on X if

$$\lim_{n\to\infty} T_n = 0$$

$$\dots(3)$$

Proof. Let $\varepsilon > 0$ be given. Then by (3), $\exists\, m = m\,(\varepsilon) \in N$ such that

$$|T_n - 0| < \varepsilon \Rightarrow T_n < \varepsilon, \forall\, n \geq m \qquad \qquad ...(4)$$

Let $n_0 = \max\{m, p\}$, then by (2), we get

$$|f_n(x) - f_n| \leq T_n, \forall x \in X \text{ and } n \geq n_0 \qquad \text{[Using(4)]}$$
$$< \varepsilon$$

Hence, the sequence $\langle f_n(x) \rangle$ converges uniformly to $f(x)$ on X.

REMARKS

- Let $\langle f_n(x) \rangle$ be a sequence of functions defined on $X \subset R$ and let on X, $\lim\limits_{n \to \infty} f_n(x) = f(x)$

  If $\lim\limits_{n \to \infty} T_n \neq 0$ where $T_n = |f_n(x_n) - f(x_n)| \ \forall n \geq n_0 \in N, x_n = x(n) \in X$

  Then $\langle f_n(x) \rangle$ does not converges uniformly to $f(x)$.
- The choice of T_n is not unique as that of M_n.

Solved Examples

Example 1. Show that the sequence $\langle f_n \rangle$ where $f_n(x) = nx\,(1 - x)^n$ does not converge uniformly on $[0, 1]$

Solution. Here, we have

$$f(x) = \lim_{n \to \infty} f_n(x)$$

$$= \lim_{n \to \infty} \frac{nx}{(1 - x)^{-n}}$$

$$\left| \text{Form } \frac{\infty}{\infty} \right.$$

$$= \lim_{n \to \infty} \frac{x}{-(1-x)^{-n} \log(1-x)}$$

[Using L-Hospital rule]

$$= \lim_{n \to \infty} \left(-\frac{x(1-x)^n}{\log(1-x)} \right) = 0$$

$$[\because (1-x)^n \to 0 \,\forall x \in [0,1]]$$

$$\Rightarrow f(x) = 0, \forall x \in [0,1]$$

Now

$$M_n = \sup\{|f_n(x) - f(x)|\} : x \in [0,1]$$

$$= \sup\{nx(1-x)^n : x \in [0,1]\}$$

$$= \sup\,(1-x)^n nx, x \in [0, 1]$$

Therefore,

$$M_n \geq n \cdot \frac{1}{n}\left(1 - \frac{1}{n}\right)^n$$

Based on the following Results

- If the Sequence $\langle f_n \rangle$ converges to f then we can write $\lim\limits_{n \to \infty} f_n(x) = f(x) \forall x$ This function f is called the limit of the sequence $\langle f_n \rangle$.

- A sequence, $\langle f_n(x) \rangle$ of functions is said to converges uniformly to a function f if for every $\varepsilon > 0$ we can find a positive integer m such that
  $$n \geq m = |f_n(x) - f(x)| < \varepsilon \forall x \in X$$

- A point such that the sequence does not converges uniformly in any neighbourhood of it, is said to be a point of non-uniform convergence of the sequence.

- **(Cauchy General Principle of Convergence).** Let $\langle f_n \rangle$ be a sequence of real valued function defined on X. Then $\langle f_n \rangle$ converges uniformly on X if and only if for every $\varepsilon > 0 \exists$ a positive integer m such that
  $$|f_n(x) - f_p(x)| < \varepsilon \ \forall n, p \geq m, \forall x \in X.$$

- **(M_n- test).** Let $\langle f_n \rangle$ be a sequence of functions defined on X. Let $\lim\limits_{n \to \infty} f_n(x) = f(x)$ and $M_n = \sup\{|f_n(x) - f(x)| : x \in X\}$. Then $\langle f_n \rangle$ converges uniformly to f iff $M_n \to 0$ as $n \to \infty$.

- **(T_n-test).** Let $\langle f_n(x) \rangle$ be a sequence of functions defined on $X \subset R$ and let $\lim\limits_{n \to \infty} f_n(x) = f(x)$. Let for $n \geq p \in N$, $T_n > 0$ exists such that $|f_n(x) - f(x)| \leq T_n$. Then $\langle f_n(x) \rangle$ converges uniformly to $f(x)$ if $\lim\limits_{n \to \infty} T_n = 0$

$$\left(\text{Taking } x = \frac{1}{n} \in [0,1] \right)$$

$$= \left(1 - \frac{1}{n}\right)^n \to \frac{1}{e} \text{ as } n \to \infty$$

Hence, by M_n-test, $\langle f_n \rangle$ does not converge uniformly on [0, 1]. Therefore, 0 is a point of non-uniform convergence, since $x = \frac{1}{n} \to 0$ as $n \to \infty$.

Example 2. *Test for unifrom and non-uniform convergence the sequence of functions $\langle f_n(x) \rangle$ where $f_n(x) = e^{-nx}, x \ge 0$*

Solution. Here, we have $\quad f(x) = \lim_{n \to \infty} f_n(x) = \begin{cases} 1, & x = 0 \\ 0, & x > 0. \end{cases}$

Clearly, f is discontinuous at $x = 0$. Now, let $\delta > 0$ be any small number and let $X = \{x \in R : 0 \le x \le \delta\}$.

For $n \ge n_0 \in N$, let us take $x_n = \frac{1}{n}$ in such a way that $0 < \frac{1}{n} < \delta$, then for $n \ge n_0$,

$x_n \in X$ and $x_n \ne 0$. Therefore $\forall n \ge n_0 \Rightarrow T_n = f_n(x_n) - f(x_n) = e^{-n/n} = e^{-1}$

$\Rightarrow \qquad \qquad \lim_{n \to \infty} T_n \ne 0$

Hence, by T_n-test $\langle f_n(x) \rangle$ does not converge uniformly on X. We now discuss on the interval $0 < \delta \le x$, where δ is fixed. Then for $x \ge \delta$

$$|f_n(x) - f(x)| = e^{-n\delta} \ge e^{-n\delta} = T_n \text{ (say)}.$$

Also, $\qquad \qquad \lim_{n \to \infty} T_n = \lim_{n \to \infty} e^{-ns} = 0$

Hence, by T_n-test, the given sequence converges uniformly for every $x \ge \delta > 0$.

THEOREM 3. **(Weierstrass's M-test).** *A series $\sum_{n=1}^{\infty} u_n(x)$ of functions will converge uniformly on X if there exists a convergent series $\sum_{n=1}^{\infty} M_n$ of positive constant such that*

$$|u_n(x)| \le M_n \forall n \text{ and } \forall x \in X.$$

Proof. Since ΣM_n is convergent, therefore, by definition, for a given $\varepsilon > 0$ we can find a positive integer m such that

$$n \ge m \Rightarrow M_{n+1} + M_{n+2} + \ldots + M_{n+p} < \varepsilon \qquad \qquad \ldots (1)$$
$$\text{(for } p = 1, 2, 3, \ldots)$$

Since $\qquad \qquad |u_n(x)| \le M_n \forall n \text{ and } \forall x \in X \qquad \qquad \ldots (2)$

From (1) and (2), we conclude that

$$|u_{n+1}(x) + u_{n+2}(x) + \ldots + u_{n+p}(x)|$$

$$\le |u_{n+1}(x)| + |u_{n+2}(x)| + \ldots + |u_{n+p}(x)|$$
$$\le M_{n+1} + M_{n+2} + \ldots + M_{n+p}$$

$$< \varepsilon, \text{ for every } n \ge m \text{ and } \forall x \in X.$$

Hence, $\Sigma u_n(x)$ converges uniformly on X.

REMARKS

- The above series is also converges absolutely.
- It is clear that same proof would hold if M_n were a function of x and if the series $\sum M_n(x)$ is uniformly convergent on X.
- The series which satisfy M-test are called normally convergent.

Solved Examples

Example 1. *Show that the series*

$$\frac{\cos x}{1^p} + \frac{\cos 2x}{2^p} + \frac{\cos 3x}{3^p} + \dots + \frac{\cos nx}{n^p} +$$

converges uniformly on R *if* $p > 1$.

Solution.   Here, we have

$$\left| \frac{\cos nx}{n^p} \right| \le \frac{1}{n^p} \forall x \in \text{R}$$

Also, the series $\sum \dfrac{1}{n^p}$ is known to be convergent for $p > 1$.

Hence, by Weierstrass's M-test the given series converges uniformly on R for $p > 1$.

THEOREM 4.   **(Abel's Test).** *The series* $\sum u_n(x) v_n(x)$ *will converges uniformly in* $[a, b]$ *if*

(i) $\sum u_n(x)$ *is uniformly convergent in* $[a, b]$

(ii) *the sequence* $\langle v_n(x) \rangle$ *is monotonic for every* $x \in [a, b]$

(iii) *the sequence* $\langle v_n(x) \rangle$ *is uniformly bounded in* $[a, b]$ *by* k *i.e.,*

$$|v_n(x)| < k, \forall x \in [a,b] \text{ and } \forall n \in \text{N}.$$

Proof.   Let $R_{n,p}(x)$ be the partial remainder of the series $\sum u_n(x) v_n(x)$ and $r_{n,p}(x)$ that of the series $\sum u_n(x)$. Then

$$R_{n,p}(x) = u_{n+1}(x)v_{n+1}(x) + u_{n+2}(x)v_{n+2}(x) + \dots + u_{n+p}(x)v_{n+p}(x)$$

$$= r_{n,1}(x)v_{n+1}(x) + \{r_{n,2}(x) - r_{n,1}(x)\}v_{n+2}(x)$$

$$+ \{r_{n,3}(x) - r_{n,2}(x)\}v_{n+3}(x) + \dots + \{r_{n,p}(x) - r_{n,p-1}(x)\}v_{n+p}(x)$$

$$= r_{n,1}(x)\{v_{n+1}(x) - v_{n+2}(x)\} + r_{n,2}(x)\{v_{n+2}(x) - v_{n+3}(x)\} +$$

$$+ \{r_{n,p-1}(x)\{v_{n+p-1}(x) - v_{n+p}(x)\} + r_{n,p}(x)v_{n+p}(x). \qquad \dots(A)$$

Given that $\langle v_n(x) \rangle$ is monotonic, therefore,

$$\{v_{n+1}(x) - v_{n+2}(x)\}, \{v_{n+2}(x) - v_{n+3}(x)\}, \dots, \{v_{n+p-1}(x) - v_{n+p}(x)\} \qquad \dots(1)$$

all have the same sign for fixed value of x in $[a, b]$.

Also, given that $\langle v_n(x) \rangle$ is uniformly bounded by k, therefore

$$|v_n(x)| < k \text{ for all } x \in [a, b] \text{ and } \forall n \in \text{N}. \qquad \dots(2)$$

Also, since the given series $\sum u_n(x)$ is uniformly convergent in $[a, b]$, for a given $\varepsilon > 0$, $\exists$ a positive integer m, independent of x such that for $n \geq m$.

$$|r_{n,p}(x)| = |u_{n+1}(x) + u_{n+2}(x) + ... + u_{n+p}(x)| < \frac{\varepsilon}{3k} \qquad \text{... (3)}$$

From (1) and (3), we have

$$|R_{n,p}(x)| < \frac{\varepsilon}{3k}|v_{n+1}(x) - v_{n+2}(x)| + \frac{\varepsilon}{3k}|v_{n+2}(x) - v_{n+3}(x)| + ...$$

$$+ \frac{\varepsilon}{3k}|v_{n+p-1}(x) - v_{n+p}(x)| + \frac{\varepsilon}{3k}|v_{n+p}(x)|$$

$$= \frac{\varepsilon}{3k}|v_{n+1}(x) - v_{n+p}(x)| + \frac{\varepsilon}{3k}|v_{n+p}(x)|. \qquad \text{... (4)}$$

Using (A), we have

$$|v_{n+1}(x) - v_{n+2}(x)| + |v_{n+2}(x) - v_{n+3}(x)| + ... + |v_{n+p-1}(x) - v_{n+p}(x)|$$

$$= |v_{n+1}(x) - v_{n+2}(x) + v_{n+2}(x) - v_{n+3}(x) + ...$$

$$+ v_{n+p-1}(x) - v_{n+p}(x)|$$

$$= |v_{n+1}(x) - v_{n+p}(x)|.$$

Now $|v_{n+1}(x) - v_{n+p}(x)| \leq |v_{n+1}(x)| + |-v_{n+p}(x)|$

$$\leq k + k < 2k. \qquad \text{... (5)}$$

Then (4) can be written as

$$|R_{n,p}(x)| < \frac{\varepsilon}{3k}.2k + \frac{\varepsilon}{3k}.k = \varepsilon \qquad \text{... (6)}$$

i.e., $\quad |u_{n+1}(x).v_{n+1}(x) + ... + u_{n+p}(x)v_{n+p}(x)| < \varepsilon \; \forall n \geq m \; \forall x \in [a,b].$

Hence, from (6), the given series $\sum u_n(x)v_n(x)$ converges uniformly on $[a, b]$.

Solved Examples

Example 1. *Test the series* $\sum \frac{(-1)^{n-1}}{n}.x^n$ *for uniform convergence in* $[0, 1]$.

Solution. Let us suppose $v_n(x) = x^n$ and $u_n(x) = \frac{(-1)^{n-1}}{n}$

Clearly, the sequence $\langle v_n(x) \rangle$ is uniformly bounded and monotonically decreasing on $[0, 1]$.

Also, the series $\sum u_n(x) = \frac{\sum(-1)^{n-1}}{n}$ is convergent. Hence, by Abel's test the series

$$\sum u_n(x)v_n(x) = \frac{\sum(-1)^{n-1}}{n}.x^n$$

is uniformly convergent on $[0, 1]$.

THEOREM 5. **(Dirichlet's Test).** *The series $\sum u_n(x)v_n(x)$ will be uniformly convergent on $[a, b]$ if*

(i) *The sequence $\langle v_n(x) \rangle$ is a positive monotonic decreasing sequence converging uniformly to zero for all $x \in [a, b]$.*

(ii) $f_n(x) = \sum\limits_{r=1}^{n} u_r(x)$ *is uniformly bounded in $[a, b]$ i.e.,*

$$| f_n(x) | = \left| \sum_{r=1}^{n} u_r(x) \right| < k$$

for every value of x in $[a, b]$ and for all positive integral values of n, where k is a fixed number, independent of x.

Proof.   Proceed as in previous theorem, we have

$$R_{n, p}(x) = u_{n+1}(x)v_{n+1}(x) + u_{n+2}(x)v_{n+2}(x) + \ldots + u_{n+p}(x)v_{n+p}(x)$$

$$= [s_{n+1}(x) - s_n(x)]v_{n+1}(x) + [s_{n+2}(x) - s_{n+1}(x)]v_{n+2}(x) + \ldots$$
$$+ [s_{n+p}(x) - s_{n+p-1}(x)]v_{n+p}(x)$$

$$= s_{n+1}(x)[v_{n+1}(x) - v_{n+2}(x)] + s_{n+2}(x)[v_{n+2}(x) - v_{n+3}(x)] + \ldots$$
$$+ s_{n+p-1}(x)[v_{n+p-1}(x) - v_{n+p}(x)]$$
$$+ s_{n+p}(x)v_{n+p}(x) - s_n(x)v_{n+1}(x). \qquad \ldots(1)$$

Now, since $\langle v_n(x) \rangle$ is a positive monotonic decreasing sequence, therefore, $v_1(x)$, $v_2(x)$, $v_3(x), \ldots$ are all positive and

$$v_1(x) > v_2(x) > v_3(x) > \ldots > v_n(x) > \ldots$$

Also $| f_n(x) | < k$ for all x in $[a, b]$ and for all $n \in \mathbb{N}$. $\therefore$ From (1), we have

$$| R_{n,p}(x) | \le | f_{n+1}(x)[v_{n+1}(x) - v_{n+2}(x)] + \ldots$$
$$+ | f_{n+p-1}(x) | [v_{n+p-1}(x) - v_{n+p}(x)]$$
$$+ | f_{n+p}(x) | v_{n+p}(x) | + | s_n(x) | v_{n+1}(x) |$$
$$< k[v_{n+1}(x) - v_{n+p}(x) + v_{n+p}(x) + v_{n+1}(x)]$$

$$= 2k\, v_{n+1}(x). \qquad \ldots (2)$$

Also, since $\langle v_{n+1}(x) \rangle$ converges to zero, we have

$$| r_n(x) | < \frac{\varepsilon}{2k} \; \forall n \ge m \qquad \text{i.e.,} \qquad v_n(x) < \frac{\varepsilon}{2k} \; \forall n \ge m \quad \ldots (3)$$

From (2) and (3), we conclude that

$$| R_{n,p}(x) | < 2k . \frac{\varepsilon}{2k} \; \text{ for } n \ge m$$

$$\Rightarrow \qquad | R_{n,p}(x) | < \varepsilon \text{ for } n \ge m \; \forall \, x \in [a, b]$$

Hence, the series $\sum u_n(x)v_n(x)$ is uniformly convergent in $[a, b]$.

Solved Examples

Example 1. *Show that the series*

$$\sum_{n=1}^{\infty} (-1)^{n-1}.x^n$$

converges uniformly in $0 \le x \le k < 1$.

Solution. Let $u_n = (-1)^{n-1}, v_n(x) = x^n$

Since

$$s_n(x) = \sum_{r=1}^{n} u_r = \begin{cases} 0 & \text{if } n \text{ is even} \\ 1 & \text{if } n \text{ is odd} \end{cases}$$

$\Rightarrow s_n(x)$ is bounded for all $n \in \mathbb{N}$.

Also $\langle v_n(x) \rangle$ is positive monotonic-decreasing sequence, converging to zero for all values of x in $0 \le x \le k < 1$.

Hence, by Dirichlet's test, the given series is uniformly convergent in $0 \le x \le k < 1$.

Example 2. *Show that the sequence $\langle f_n \rangle$, where $f_n(x) = \dfrac{nx}{1+n^2 x^2}$ does not converges uniformly on* $\mathbb{R}$.

Solution. Here, we have

$$f(x) = \lim_{n \to \infty} f_n(x)$$

$$= \lim_{n \to \infty} \frac{nx}{1+n^2 x^2} = 0 \ \forall x \in \mathbb{R}$$

Based on the following Results

- **Weierstrass M-test.** A series $\displaystyle\sum_{n=1}^{\infty} u_n(x)$ of functions will converges uniformly on X if there exists a convergent series $\displaystyle\sum_{n=1}^{\infty} M_n$ of positive constant such that $|u_n(x)| \le M \ \forall n \ \forall \ x \in X$

- **Abel's test.** The series $\Sigma u_n(x)v_n(x)$ will converges uniformly in $[a, b]$ if
  (i) $\Sigma u_n(x)$ is uniformly convergent in $[a, b]$
  (ii) the sequence $\langle v_n(x) \rangle$ is monotonic for every $x \in [a, b]$
  (iii) the sequene $\langle v_n(x) \rangle$ uniformly bounded in $[a, b]$ by k i.e., $|v_n(x)| \le k \ \forall x \in [a,b]$

- **Dirichlet's test.** The series $\Sigma u_n(x)v_n(x)$ will be uniformly convergent on (a, b) if
  (i) the sequence $\langle v_n(x) \rangle$ is a positive monotonic decreasing sequence converges uniformly to zero $\forall x \in (a, b)$
  (ii) $f_n(x) = \displaystyle\sum_{r=1}^{n} u_r(x)$ is uniformly bounded in (a, b) i.e., $f_n(x)| = \left| \displaystyle\sum_{r=1}^{n} u_r(x) \right| < k$ for every value of x in (a, b).

Let if possible, the sequence converges uniformly on $\mathbb{R}$, then for a given $\varepsilon > 0$, $\exists$ a positive integer m such that

$$n \ge m, x \in \mathbb{R} \Rightarrow |f_n(x) - f(x)| = \frac{nx}{1+n^2 x^2} < \varepsilon \qquad \ldots (1)$$

If we take $\varepsilon = \dfrac{1}{3}$ and $x = \dfrac{1}{n}(n = 1,2,3,\ldots)$, then

$$|f_n(x) - f(x)| = \frac{n\dfrac{1}{n}}{1+n^2\dfrac{1}{n^2}} = \frac{1}{2} < \frac{1}{3} = \varepsilon$$

Thus, there is no single m such that (1) holds simultaneously for all $x \in \mathbb{R}$. For if, such an m exist, we would have (on taking $n = m$), $|f_m(x) - f(x)| < \dfrac{1}{3} \forall \ x \in \mathbb{R}$

but if we take $x = \dfrac{1}{m}$, we get a contradiction $\left(\because \text{ in this case } \dfrac{1}{2} < \dfrac{1}{3} \right)$ and therefore, the sequence is not uniformly convergent on $\mathbb{R}$. Also since $\dfrac{1}{m} \to 0$ therefore, 0 is a point of non-uniform convergence.

Example 3. *Discuss the uniform convergence of the series*

$$\sum_{n=1}^{\infty}\left[\frac{nx}{1+n^2x^2}-\frac{(n-1)x}{1+(n-1)^2x^2}\right]$$

Solution. Here, we have

$$u_1(x)=\frac{x}{1+x^2}-0$$

$$u_2(x)=\frac{2x}{1+2^2x^2}-\frac{x}{1+x^2}$$

$$\cdots \qquad \cdots \qquad \cdots \qquad \cdots$$

$$u_n(x)=\frac{nx}{1+n^2x^2}-\frac{(n-1)x}{(1+(n-1)^2x^2)}$$

On adding, we get

$$f_n(x)=\frac{nx}{1+n^2x^2}$$

Now do same as example (2).

Example 4. *Show that the sequence (f_n) where $f_n(x)=\dfrac{x}{1+nx^2}$ converges uniformly on R.*

Solution. Here, we have

$$y=\lim_{n\to\infty}\frac{x}{1+nx^2}=0\ \forall\,x\in R$$

Let

$$y=f_n(x)-f(x)=\frac{x}{1+nx^2}$$

For maxima and minima of y, we must have $\dfrac{dy}{dx}=0$

$$\Rightarrow \qquad \frac{(1+nx^2)-2nx^2}{(1+nx^2)^2}=0 \qquad \Rightarrow \qquad \frac{1-nx^2}{(1+nx^2)^2}=0$$

$$\Rightarrow \qquad\qquad x=\pm\frac{1}{\sqrt{n}}$$

Clearly, $\dfrac{d^2y}{dx^2}$ is negative when $x=\dfrac{1}{\sqrt{n}}$

$$\therefore \qquad \text{Maximum value of } y=\frac{1/\sqrt{n}}{1+n\left(\dfrac{1}{n}\right)}=\frac{1}{2\sqrt{n}}$$

Also, $\dfrac{1}{2\sqrt{n}}-|y|=\dfrac{1}{2\sqrt{n}}-\dfrac{|x|}{1+nx^2}=\dfrac{1+nx^2-2\sqrt{n}\,|x|}{2\sqrt{n}(1+nx^2)}$

$$=\frac{(1-|x|\sqrt{n})^2}{2\sqrt{n}(1+nx^2)}\geq 0$$

Now, $$M_n=\sup_{x\in R}|f_n(x)-f(x)|=\sup_{x\in R}\left|\frac{x}{1+nx^2}\right|=\sup_{x\in R}|y|$$

$$=\max.y=\frac{1}{2\sqrt{n}}\to 0\,\text{as}\,n\to\infty$$

Hence, by M_n-test the sequence is uniformly convergent on R.

Example 5. *Show that 0 is a point of non-uniform convergence of the sequence $\langle f_n(x) \rangle$ where*

$$f_n(x) = n\,x\,e^{-nx^2}, x \in R$$

Solution. Here, we have

$$f(x) = \lim_{n \to \infty} f_n(x) = \lim_{n \to \infty} nx\,e^{-nx^2}$$

$$= \lim_{n \to \infty} \frac{nx}{e^{nx^2}} \qquad \left| \text{Form} \frac{\infty}{\infty} \right.$$

$$= \lim_{n \to \infty} \frac{x}{x^2 e^{nx^2}} \qquad \text{(By L-Hospital rule)}$$

$$= 0.$$

Let if possible, the sequence be uniformly convergent in a neighbourhood $]0, k[$ of 0, where $k \in N$.

Then, for a given $\varepsilon > 0$, $\exists$ a positive integer m such that

$$n \geq m, x \in]0, k[\Rightarrow |f_n(x) - f(x)| = nxe^{-nx^2}$$

$$< \varepsilon \qquad \qquad \dots(1)$$

In particular, the inequality (1) must be true for $x = \dfrac{1}{\sqrt{n}}$, where n is a positive integer greater than m such that $0 < \dfrac{1}{\sqrt{n}} < k$.

Then (1) gives, $\qquad \dfrac{\sqrt{n}}{e} > \varepsilon$

Now, since $x \to 0$, when $n \to \infty$, we see that on taking x sufficiently near 0, we can take n so large that $\dfrac{\sqrt{n}}{e} > \varepsilon$, which is a contradiction.

Hence, 0 is a point of non-uniform convergence of the sequence.

Aliter. Let $\qquad \qquad y = f_n(x) - f(x) = nxe^{-nx^2} = 0$

For maxima and minima of y, we must have

$$\frac{dy}{dx} = 0 \Rightarrow ne^{-nx^2} - 2n^2x^2e^{-nx^2} = 0$$

$$\Rightarrow x = +\frac{1}{\sqrt{2n}}$$

Also, $\qquad \qquad \dfrac{d^2y}{dx^2} = -ve,$ when $x = \dfrac{1}{\sqrt{2n}}$

Therefore, $\qquad$ maximum $y = n \cdot \dfrac{1}{\sqrt{2n}} e^{-n \cdot \frac{1}{2n}} = \sqrt{\dfrac{n}{2e}}$

$$\Rightarrow \qquad M_n = \sup_{x \in R} |f_n(x) - f(x)|$$

$$= \sup_{x \in R} n\,|x|\,e^{-nx^2}$$

$$= \sup |y| = \text{Max. } y$$

$$= \sqrt{\dfrac{n}{2e}} \to \infty \text{ as } n \to \infty$$

$\Rightarrow$ M_n does not tends to zero as $n \to \infty$.

Hence by M_n-test, the given sequence is not uniformly convergent.

Also $x \to 0$ as $n \to \infty$, therefore, 0 is a point of non-uniform convergence.

Example 6.   *Show that if δ is any fixed positive number less than unity, the series $\sum\limits_{n=1}^{\infty} \dfrac{x^n}{n+1}$ is uniformly convergent in $[-\delta, \delta]$.*

Solution.   Let $u_n(x) = x^n$ and $v_n = \dfrac{1}{n+1}$

Since for all $x \in [-\delta, \delta]$, we have $|x| \le \delta < 1$

Therefore,
$$|s_n(x)| = |u_1(x) + u_2(x) + \dots + u_n(x)|$$
$$\le |x + x^2 + \dots x^n|$$
$$\le \delta + \delta^2 + \dots + \delta^n = \frac{\delta(1-\delta^n)}{1-\delta} < \frac{\delta}{1-\delta}$$

Also, $\langle v_n \rangle$ is a positive, monotonic decreasing sequence converging to zero. Hence, by Dirichlet's test, the given series $\sum \dfrac{x^n}{n+1}$ is uniformly convergent in $[-\delta, \delta]$.

Example 7.   *Show that the sequence $\langle f_n \rangle$ where $f_n = x^{n-1}(1-x)$ converges uniformly in the interval $[0, 1]$.*

Solution.   Here, we have
$$f(x) = \lim_{n \to \infty} f_n(x) = \lim_{n \to \infty} x^{n-1}(1-x) = 0 \, \forall x \in [0, 1]$$

Let
$$y = |f_n(x) - f(x)| = x^{n-1}(1-x)$$

For maxima or minima of y, we must have $\dfrac{dy}{dx} = 0$

$\Rightarrow (n-1)x^{n-2}(1-x) - x^{n-1} = 0 \qquad \Rightarrow \qquad x^{n-2}[(n-1)(1-x) - x] = 0$

$\Rightarrow \qquad\qquad x = 0, \dfrac{n-1}{n}$

Also, we can see that $\dfrac{d^2 y}{dx^2}$ is negative, when $x = \dfrac{n-1}{n}$

Now
$$M_n = \sup_{x \in [0,1]} |f_n(x) - f(x)| = \sup_{x \in [0,1]} |x^{n-1}(1-x)|$$
$$= \sup_{x \in [0,1]} \cdot |y| = \text{Max. } y$$
$$= \left(1 - \frac{1}{n}\right)^{n-1} \left(1 - \frac{n-1}{n}\right)$$
$$\to \frac{1}{e} \times 0 = 0 \text{ as } n \to \infty.$$

Hence, by M_n-test, the sequence is uniformly convergent on $[0,1]$.

Example 8.   *Show that the sequence $\langle f_n \rangle$ where $f_n(x) = \tan^{-1} nx, x \ge 0$ is uniformly convergent in any interval $[a, b], a > 0,$ but is only pointwise convergent in $[a, b]$.*

Solution. We have
$$f(x) = \lim_{n\to\infty} f_n(x) = \begin{cases} \pi/2 , & x > 0 \\ 0 , & x = 0 \end{cases}$$

Let $\varepsilon > 0$ be given, so that for $x > 0$, we have
$$|f_n(x) - f(x)| = \left| \tan^{-1} nx - \frac{\pi}{2} \right| < \varepsilon$$

$$\Rightarrow \qquad \cot^{-1} nx < \varepsilon$$

i.e., if $n > \dfrac{\cot\varepsilon}{x}$ which clearly decreases with x, the maximum value being $\dfrac{\cot\varepsilon}{a}$ in $[a, b]$.

Let m be any integer $\geq \dfrac{\cot\varepsilon}{a}$ Therefore, for a given $\varepsilon > 0$, $\exists\, m \in N$ such that $\forall x \in [a,b]$.

However, it is clear that $\dfrac{\cot\varepsilon}{x} \to \infty$ as $x \to 0$ so that no such integer m exists such that $n \geq m \Rightarrow |f_n(x) - f(x)| < \varepsilon$

Hence, the sequence is not uniformly convergent on $[0, b]$.

Also, $\qquad\qquad\qquad n \geq m \Rightarrow |f_n(x) - f(x)| < \varepsilon$

Hence, the sequence converges uniformly on $[a, b]$.

Example 9. *Show that the sequence $\langle f_n \rangle$ where $f_n(x) = \dfrac{n}{n+x}, x \geq 0$ is uniformly convergent in any finite interval.*

Solution. Here, we have $f(x) = \lim_{n\to\infty} f(x) = \lim_{n\to\infty} \dfrac{n}{n+x}$ $\qquad$ $\left| \text{Form } \dfrac{\infty}{\infty} \right.$

$$= \lim_{n\to\infty} \frac{1}{1 + x/n} = 1 \,\forall\, x \geq 0$$

For an arbitrary choosen positive number ε, we have
$$|f_n(x) - f(x)| < \varepsilon$$

if $\qquad\qquad \left| \dfrac{n}{n+x} - 1 \right| < \varepsilon$ or i.e., if $\left| \dfrac{-x}{n+x} \right| < \varepsilon$

or i.e., if $\qquad\qquad \dfrac{x}{n+x} < \varepsilon$ or i.e., if $\quad n > x\left(\dfrac{1}{\varepsilon} - 1 \right)$

Obviously, n increase with x and tends to ∞ as $x \to \infty$.

Therefore, convergence is not uniform in $[0, \infty[$.

But if $]0, k[$ is any finite interval, where $k > 0$, however large then m is any positive

integer $\geq k\left(\dfrac{1}{\varepsilon} - 1 \right)$ such that $n \geq m, x \in [0, k] \Rightarrow |f_n(x) - f(x)| < \varepsilon$

Hence, the sequence is uniformly convergent on $[0, k]$.

Example 10. *Let $g_n(x) = \dfrac{1}{n} e^{-nx} (0 \leq x < \infty)$. Show that $\langle g_n \rangle$ converges uniromly to 0 on $[0, \infty[$.*

Solution. Here, we have $g(x) = \lim_{n\to\infty} g_n(x) = \lim_{n\to\infty} \dfrac{e^{-nx}}{n} = 0$

Define $\qquad\qquad y = |g_n(x) - g(x)| = \left| \dfrac{e^{-nx}}{n} - 0 \right| = \dfrac{e^{-nx}}{n}$

Since e^{-nx} is monotonic decreasing with x $(0 \leq x \leq \infty)$, therefore,

$$\max . y = \frac{1}{n} \qquad \text{(where } x = 0\text{)}$$

Thus

$$M_n = \sup_{x \in [0, \infty[} |g_n(x) - g(x)| = \sup y = \max . y = \frac{1}{n}$$

$$\to 0 \text{ as } n \to \infty$$

Hence, by M_n-test, the given sequence $\langle g_n \rangle$ is uniformly convergent on $[0, \infty[$.

Example 11. *Show that the series*

$$\frac{\cos x}{1^2} + \frac{\cos 2x}{2^2} + \frac{\cos 3x}{3^2} + \dots$$

converges uniformly on R. Give the interval of uniform convergence.

Solution. Let

$$\sum_{n=1}^{\infty} u_n(x) = \sum_{n=1}^{\infty} \frac{\cos nx}{n^2}$$

Then, we have

$$|u_n(x)| = \left| \frac{\cos nx}{n^2} \right| \leq \frac{1}{n^2} \forall x \in R$$

Taking $M_n = \dfrac{1}{n^2}$, the series $\sum M_n = \sum \dfrac{1}{n^2}$ is convergent. Hence, by Weierstrass's M-test, the given series converges uniformly on R.

Also, the interval of uniform convergence is $a \leq x \leq b$, where a and b are any finite unequal real numbers.

Example 12. *Test the series* $\sum \dfrac{\sin nx}{n^p}$ *for uniform convergence in any interval.*

Solution. Let $\sum u_n(x) = \sum \dfrac{\sin nx}{n^p}$

Now, there are following two cases:

Case (i) Suppose $p > 1$, then we have for all $n \in N$.

$$|u_n(x)| = \left| \frac{\sin nx}{n^p} \right| \leq \frac{1}{n^p} = M_n \text{(say)}$$

Now, since $\sum M_n = \sum \dfrac{1}{n^p}$ is convergent if $p > 1$, therefore, by Weierstrass

M-test, the series $\sum \dfrac{\sin nx}{n^p}$ is uniformly convergent if $p > 1$.

Case (ii) Suppose $0 < p \leq 1$.

Take $$u_n(x) = \sin nx, v_n = \frac{1}{n^p}$$

Then $$f_n(x) = \sum_1^n u_n(x) = \sin x + \sin 2x + \dots + \sin nx$$

$$= \frac{\sin \left\{ x + \dfrac{(n-1)}{2} . x \right\} . \sin \left(\dfrac{nx}{2} \right)}{\sin \dfrac{x}{2}}$$

Therefore, $|f_n(x)| \leq \operatorname{cosec} \dfrac{x}{2}$. Also, $\operatorname{cosec} \dfrac{x}{2}$ becomes infinite if $x = 2m\pi$ for $m = 0, 1, 2, \ldots$. Evidently, with the exclusion of these points, $|f_n(x)|$ is bounded uniformly for the remaining points. Also, $\langle v_n \rangle = \left\langle \dfrac{1}{n^p} \right\rangle$ is monotonic decresing sequence converging to zero.

Hence, by Dirichlet's test, the given series is uniformly convergent in every interval from which the points $x = 2m\pi$ for $m = 0, 1, 2, 3, \ldots$ are excluded in case of $0 < p \leq 1$.

Example 13. *Show that the series* $\sum \dfrac{a_n x^n}{1 + x^{2n}}$ *converges uniformly for all real values of x if $\sum a_n$ is absolutely convergent.*

Solution. Let
$$u_n(x) = \frac{a_n x^n}{1 + x^{2n}}$$

For maxima or minima of $u_n(x)$, we must have
$$\frac{du_n}{dx} = 0 \Rightarrow \frac{a_n \left[nx^{n-1}(1 + x^{2n}) - 2nx^{2n-1} \cdot x^n \right]}{(1 + x^{2n})^2}$$

$$\Rightarrow \qquad 1 + x^{2n} - 2x^{2n} = 0 \Rightarrow x^n = 1.$$

We can easily verify that $\dfrac{d^2 u_n}{dx^2} < 0$, when $x^n = 1$ and

$$M_n = \max |u_n| = \max \left| \frac{a_n x^n}{1 + x^{2n}} \right| = \frac{|a_n|}{1+1} = \frac{|a_n|}{2}$$

Now, $|u_n(x)| \leq M_n$ and $\sum M_n = \dfrac{1}{2} \sum |a_n|$ is convergent as $\sum a_n$ is given to be absolutely convergent. Hence, by Weierstrass M-test, the series $\sum |u_n(x)|$ is uniformly convergent for all $x \in \mathbb{R}$.

Example 14. *Show that the series* $\sum\limits_{n=1}^{\infty} \dfrac{1}{1 + n^2 x}$ *converges uniformly in* $[1, \infty[$

Solution. Let $\sum u_n(x) = \sum \dfrac{1}{1 + n^2 x}$. Then, $u_n(x) = \dfrac{1}{1 + n^2 x}$

$$\Rightarrow \qquad |u_n(x)| \leq \frac{1}{1 + n^2} \ \forall x \in [1, \infty]$$

$$< \frac{1}{n^2} = M_n \text{ (say)}.$$

Since, the series $\sum\limits_{n=1}^{\infty} M_n = \sum\limits_{n=1}^{\infty} \dfrac{1}{n^2}$ is convergent. Hence, by Weierstrass M-test the given series is uniformly convergent on $[1, \infty[$.

Example 15. *Test for uniform convergence of the series* $\sum\limits_{n=1}^{\infty} xe^{-nx}$ *in the closed interval.*

Solution. Here, we have

$$f_n(x) = \sum_{n=1}^{n-1} xe^{-nx} = x + xe^{-x} + xe^{-2x} + ... + xe^{-(n-1)x}$$

$$= \frac{x(1-1/e^{nx})}{1-e^{-x}} = \frac{xe^x}{e^x-1}\left(1-\frac{1}{e^{nx}}\right)$$

$$\Rightarrow \qquad f(x) = \lim_{n\to\infty} f_n(x) = \begin{cases} 0 & , \text{ when } x = 0 \\ \dfrac{xe^x}{e^x-1} & , \text{ when } 0 < x \le 1 \end{cases}$$

For $0 < x < 1$, we have $M_n = n \sup\limits_{x\in]0,1]} |f_n(x)-f(x)| = \sup\limits_{x\in]0,1]} \dfrac{xe^x}{(e^x-1)e^{nx}}$

$$\ge \frac{\dfrac{1}{n}e^{1/n}}{(e^{1/n}-1)e} \qquad \left(\text{Taking } x = \frac{1}{n} \in]0, 1]\right)$$

Now, $\lim\limits_{n\to\infty} \dfrac{(1/n)e^{1/n}}{(e^{1/n}-1).e}$ $\qquad \left|\text{Form } \dfrac{0}{0}\right.$

$$= \lim_{n\to\infty} \frac{(1/n).e^{1/n}\left(-\dfrac{1}{n^2}\right)+\left(-\dfrac{1}{n^2}\right)e^{1/n}}{e.e^{1/n}\left(-\dfrac{1}{n^2}\right)} \qquad \text{(By L' Hospital Rule)}$$

$$= \lim_{n\to\infty} \frac{\left(\dfrac{1}{n}+1\right)}{e} = \frac{(0+1)}{e} = \frac{1}{e}$$

Therefore, M_n does not tends to zero as $n \to \infty$. Hence, by M_n-test the sequence $\langle f_n(x)\rangle$ is non-uniformly convergent. Also, 0 is a point of non-uniform convergence.

Example 16. *The sum to n terms of a series is*

$$f_n(x) = \frac{n^2x}{1+n^4x^2}$$

Show that it converges non-uniformly in the interval [0, 1].

Solution. Here, we have

$$f(x) = \lim_{n\to\infty} f_n(x) = \lim_{n\to\infty} \frac{n^2x}{1+n^4x^2} = 0 \forall x \in [0, 1],$$

Let if possible, the sequence $\langle f_n(x)\rangle$ converges uniformly on [0,1]. Then, by definition for a given $\varepsilon > 0$, $\exists\, m \in \mathbb{N}$ such that

$$n \ge m, x \in [0,1] \Rightarrow |f_n(x)-f(x)| = \frac{n^2|x|}{1+n^4x^2} < \varepsilon \qquad ...(1)$$

If $x = \dfrac{1}{n^2} (n \in N)$, then $|f_n(x) - f(x)| = \dfrac{n^2 \cdot \dfrac{1}{n^2}}{1 + n^4 \cdot \dfrac{1}{n^4}} = \dfrac{1}{2}$

If we take $\varepsilon = \dfrac{1}{2}$, there is no single m such that (1) holds simultaneously for all $x \in [0,1]$. For if such m exists, we would have

$$|f_m(x) - f(x)| < \dfrac{1}{2}, x \in [0,1]$$

In particular, when $x = \dfrac{1}{m^2}$, we get a contradiction

$$\left(\because \text{ in this case we would have } \dfrac{1}{2} < \dfrac{1}{2} \right)$$

Hence, convergence is non-uniform on [0,1].

Example 17 *Show that the series*

$$\dfrac{x}{1+x^2} + \left(\dfrac{2^2 x}{1+2^3 x^2} - \dfrac{x}{1+x^2} \right) + \left(\dfrac{3^2 x}{1+3^3 x^2} - \dfrac{2^2 x}{1+2^3 x^2} \right) + \cdots$$

does not converge uniformly on [0,1].

Solution. Here, we have $u_1(x) = \dfrac{x}{1+x^2}$

$$u_2(x) = \dfrac{2^2 x}{1+2^3 x^2} - \dfrac{x}{1+x^2}$$

$$\cdots \quad \cdots \quad \cdots \quad \cdots \quad \cdots \quad \cdots \quad \cdots \quad \cdots$$

$$u_n(x) = \dfrac{n^2 x}{1+n^3 x^2} - \dfrac{(n-1)^2 x}{1+(n-1)^3 x^2}$$

On adding, we get $f_n(x) = \dfrac{n^2 x}{1+n^3 x^2}$

Therefore, $f(x) = \lim\limits_{n \to \infty} f_n(x) = \lim\limits_{n \to \infty} \dfrac{n^2 x}{1+n^3 x^2}$ $\quad \left| \text{Form} \dfrac{\infty}{\infty} \right.$

$$= \lim\limits_{n \to \infty} \dfrac{x}{\dfrac{1}{n^2} + nx^2} = 0, \forall x \in [0,1]$$

Now, $M_n = \sup\limits_{x \in [0,1]} |f_n(x) - f(x)| = \sup\limits_{x \in [0,1]} \dfrac{n^2 x}{1+n^3 x^2}$

$$= \dfrac{n^2 \cdot \dfrac{1}{n^{3/2}}}{1+n^3 \dfrac{1}{n^3}} = \dfrac{\sqrt{n}}{2} \quad \left(\text{Taking } x = \dfrac{1}{n^{3/2}} \in]0,1[\right)$$

$$\to \infty \text{ as } n \to \infty.$$

Since M_n does not tend to zero as $n \to \infty$, the series is non-uniformly convergent on [0,1] by M_n-test. Also, 0 is a point of non-uniform convergence.

Example 18. *Show that 0 is a point of non-uniform convergence of the series*

$$\sum_{n=1}^{\infty} \frac{x}{[(n-1)x+1][nx+1]}$$

Solution.   Here, we have

$$u_n(x) = \frac{x}{[(n-1)x+1][nx+1]} = \frac{1}{(n-1)x+1} - \frac{1}{nx+1}$$

$$\Rightarrow \qquad u_1(x) = 1 - \frac{1}{x+1}$$

$$u_2(x) = \frac{1}{x+1} - \frac{1}{2x+1}$$

$$\dots \quad \dots \quad \dots \quad \dots \quad \dots \quad \dots \quad \dots \quad \dots$$

$$u_n(x) = \frac{1}{(n-1)x+1} - \frac{1}{nx+1}$$

On adding, we get $f_n(x) = 1 - \dfrac{1}{nx+1}$

Therefore, $f(x) = \lim_{n\to\infty} f_n(x) = \begin{cases} 0 & , \quad \text{when } x = 0 \\ 1 & , \quad \text{when } x \neq 0 \end{cases}$

If $x \neq 0$, then we have

$$M_n = \sup_{x \in R_0} |f_n(x) - f(x)| = \sup_{x \in R_0} \frac{1}{|nx+1|}$$

$$\geq \frac{1}{\left| n.\dfrac{1}{n}+1 \right|} = \frac{1}{2} \qquad\qquad \left(\text{Taking } x = \frac{1}{n} \right)$$

$\Rightarrow \quad M_n$ does not tends to zero $n \to \infty$.

Therefore, by M_n-test the given series is not uniformly convergent. Also zero is a point of non-uniform convergence.

Example 19. *Test for uniform convergence of the series*

$$\sum x \left(\frac{n}{1+n^2 x^2} - \frac{n+1}{1+(n+1)^2 x^2} \right)$$

Solution.   Here, we have $u_n(x) = x \left\{ \dfrac{n}{1+n^2 x^2} - \dfrac{n+1}{1+(n+1)^2.x^2} \right\}$

$$\Rightarrow \qquad u_1(x) = \frac{x}{1+x^2} - \frac{2x}{1+2^2 x^2}$$

$$u_2(x) = \frac{2x}{1+2^2 x^2} - \frac{3x}{1+3^2 x^2}$$

$$\dots \quad \dots \quad \dots \quad \dots \quad \dots \quad \dots \quad \dots \quad \dots \quad \dots \quad \dots$$

$$u_n(x) = \frac{nx}{1+n^2 x^2} - \frac{(n+1)x}{1+(n+1)^2 x^2}$$

On adding, we get

$$f_n(x) = \frac{x}{1+x^2} - \frac{(n+1).x}{1+(n+1)^2.x^2}$$

Therefore,

$$f(x) = \lim_{n \to \infty} f_n(x) = \begin{cases} 0 & , \text{ when } x = 0 \\ \dfrac{x}{1+x^2} & , \text{ when } 0 < x \le 1 \end{cases}$$

If $0 < x \le 1$, then, we have

$$M_n = \sup_{x \in]0,1]} |f_n(x) - f(x)| = \sup_{x \in]0,1]} \frac{(n+1).x}{1+(n+1)^2 x^2}$$

$$\ge \frac{(n+1).\dfrac{1}{n+1}}{1+(n+1)^2.\dfrac{1}{(n+1)^2}} \qquad \left(\text{Taking } x = \frac{1}{n+1} \in]0,1] \right)$$

$$= \frac{1}{2}.$$

Since M_n does not tends to zero as $n \to \infty$ and also $x \to 0$ as $n \to \infty$ therefore, the sequence $\langle f_n(x) \rangle$ and so the series $\sum u_n(x)$ does not converges uniformly in $[0, 1]$. Also, zero is a point of non-uniform convergence.

Example 20 *If (x) denotes the pointwise or negative excess of x over the greatest integer and if x is midway between two integers, let (x) be zero. Test the uniform convergence of the series*

$$\sum \frac{(nx)}{n^2} = \frac{(x)}{1} + \frac{(2x)}{2^2} + \frac{(3x)}{3^2} + \dots + \frac{(nx)}{n^2} + \dots$$

Solution. We can define

$$(x) = \begin{cases} x & , \text{ for } -\dfrac{1}{2} < x < \dfrac{1}{2} \\ 0 & , \text{ for } x = \dfrac{1}{2} \\ x-1 & , \text{ for } \dfrac{1}{2} < x < 1\dfrac{1}{2} \\ 0 & , \text{ for } x = 1\dfrac{1}{2} \\ x-2 & , \text{ for } 1\dfrac{1}{2} < x < 2\dfrac{1}{2} \text{ and so on.} \end{cases}$$

Similarly $(2x) = \begin{cases} 2x & , \text{ for } -\dfrac{1}{4} < x < \dfrac{1}{4} \\ 0 & , \text{ for } x = \dfrac{1}{4} \\ 2x-1 & , \text{ for } \dfrac{1}{4} < x < \dfrac{3}{4} \\ 0 & , \text{ for } x = \dfrac{3}{4} \\ 2x-2 & , \text{ for } \dfrac{3}{4} < x < \dfrac{5}{4} \text{ and so on.} \end{cases}$

Similarly, we can write the expression for $(3x)$, $(4x)$,..., etc.

Clearly $\quad\quad\quad\quad |(nx)| < \dfrac{1}{2}$

We, then have

$$|u_n(x)| = \left|\dfrac{(nx)}{n^2}\right| < \dfrac{1}{2n^2} < \dfrac{1}{n^2}.$$

Since, $\Sigma \dfrac{1}{n^2}$ is convergent. Hence by Weierstrass's M-test the series $\Sigma \dfrac{(nx)}{n^2}$ is uniformly convergent.

Example 21. *Test for uniform convergence the series* $\Sigma \dfrac{x}{(n+x^2)^2}$

Solution. We have

$$u_n(x) = n^{th} \text{ term of the series } = \dfrac{x}{(n+x^2)^2}$$

For the maxima and minima of $u_n(x)$, we must have $\dfrac{du_n(x)}{dx} = 0$

$$\Rightarrow \quad \dfrac{(n+x^2)^2 - 4x^2(n+x^2)}{(n+x^2)^4} = 0$$

$$\Rightarrow \quad 3x^4 + 2nx^2 - n^2 = 0 \text{ i.e., } x = \pm\sqrt{\dfrac{n}{3}}$$

We can easily show that at $x = \sqrt{\dfrac{n}{3}}, u_n(x)$ is maximum.

Hence, $\quad\quad \max. u_n(x) = \dfrac{\sqrt{\dfrac{n}{3}}}{\left(n+\dfrac{n}{3x}\right)^2} = \dfrac{3\sqrt{3}}{16n^{3/2}} = M_n \text{ (say)}$

$$\Rightarrow \quad\quad\quad |u_n(x)| \le M_n$$

Since, ΣM_n is convergent, therefore by Weierstrass's M-test, the given series is uniformly convergent for all values of x.

EXERCISE 9.1

1. Test the series $\displaystyle\sum_{n=1}^{\infty} \dfrac{(-1)^{n-1}}{n+x^2}$ for uniform convergence for all values of x.

2. Show that 0 is the point of non-uniform convergence of the sequence $\langle f_n(x)\rangle$ when

$$f_n(x) = e^{-nx}, x \ge 0$$

3. Show that 0 is a point of non-uniform convergence of the sequence $\langle f_n\rangle$ where

$$f_n(x) = 1 - (1 - x^2)^n$$

4. Test for uniform convergence the series

$$\sum_{n=0}^{\infty} a^n \cos nx$$

$$= 1 + a\cos x + a^2\cos 2x + \dots + a^n \cos nx \dots$$

5. Show that the sequence $\langle f_n(x)\rangle$ on $X = [0, 1]$ is convergent on every point of the metric space X but is not uniformly convergent on X, when $f_n(x) = x^n$ and

$$\lim_{n\to\infty} x^n = 0, \text{ when } 0 < x \le 1$$

$$\text{and } \lim_{n\to\infty} x^n = 1, \text{ when } x = 1$$

6. Show that the sequence $\langle f_n \rangle$ where
$$f_n(x) = x^n (1-x)$$
converges uniformly in $[0, 1]$.

7. Show that the series
$$\frac{1}{a} - \frac{2a}{a^2-1}\cos x + \frac{2a}{a^2-2^2}\cos 2x - \dots \text{ is uniformly}$$
convergent in any finite interval.

8. Show that the series
$$1 + \frac{e^{-2x}}{2^2-1} - \frac{e^{-4x}}{4^2-1} + \frac{e^{-6x}}{6^2-1} - \dots \quad \text{converges}$$
uniformly for all $x \geq 0$.

9. Show that the series
$$\sum_{n=0}^{\infty} \frac{(-1)^n x^{2n+1}}{(2n+1)!} = x - \frac{x^3}{3!} + \frac{x^5}{5!} - \dots \text{ is uniformly}$$
convergent in every interval.

10. Show that the series
$$\frac{1}{1+x^2} - \frac{1}{2+x^2} + \frac{1}{3+x^2} - \dots \quad \text{converges}$$
uniformly in the interval $x \geq 0$.

11. Show that the series $\sum \dfrac{x}{n(n+1)}$ is non-uniformly convergent in $]0, \infty[$.

12. Show that the series
$$\frac{x^2}{1+x} + \left(\frac{2x^2}{1+2x} - \frac{x^2}{1+x}\right) + \dots$$
$$+ \left(\frac{nx^2}{1+nx} - \frac{(n-1)x^2}{1+(n-1).x}\right) + \dots$$
converges uniformly on $[0, 1]$.

13. Show that the sequence $\left\langle (\sin x)^{1/n} \right\rangle$ converges but not uniformly on $[0, \pi]$.

14. Show that the sequence $\left\langle \left(\dfrac{\sin x}{x}\right)^{1/n} \right\rangle$ converges but not uniformly on $[0, \pi]$.

15. Show that the series $\sum\limits_{n=1}^{\infty} x^n(1-x^n)$ is not uniformly convergent on $[0, 1]$

16. Show that the series $\sum\limits_{n=1}^{\infty} \dfrac{1}{n^3+x^3}$ converges uniformly in $[0, k]$ for $k>0$

17. Show that 0 is a point of non-uniform convergence of the series
$$\sum \left\{ \frac{2n^2x^2}{e^{n^2}x^2} - \frac{2(n-1)^2.x^2}{e^{(n-1)^2}. x^2} \right\}$$

18. Show that the series
$$\frac{2x}{1+x^2} + \frac{4x^3}{1+x^4} + \frac{8x^7}{1+x^6} + \dots \quad \text{is uniformly}$$
convergent in $-1 < x < 1$.

19. Show that $\sum \dfrac{1}{n^p+n^qx^2}$ is uniformly convergent for all values of x if $p+q>2$.

20. Show that the given series $\sum \dfrac{x}{n(1+nx^2)}$ is uniformly convergent for all x.

21. Show that the series $\sum\limits_{n=1}^{\infty} \dfrac{nx^2}{n^3+x^3}$ is uniformly convergent in $[0, k]$ for any $k>0$.

22. Show that the sequence $\left\langle \dfrac{nx}{1+n^2x^2} \right\rangle$ converges to zero on closed interal $[0, 2]$, but does not uniformly convergent in $]0, 2[$.

23. Show that the series
$$\frac{1}{x+1} - \frac{1}{(x+1)(x+2)} - \frac{1}{(x+2)(x+3)} -$$
$$\dots - \frac{1}{(x+n-1)(x+n)}$$
converges uniformly in $[0, 1]$.

24. Show that the following series converges uniformly in $[-1, 1]$
(i) $\sum \dfrac{x^n}{n^2}$ \qquad (ii) $\sum \dfrac{x^2}{n(n+1)}$
(iii) $\sum \dfrac{x^{2n}}{n+x^{2n}}$

25. Show that the following series converges uniformly in $[-k, k]$ for all real k.
(i) $\sum \dfrac{1}{n^4+n^2x^2}$ \qquad (ii) $\sum \dfrac{1}{n^2+n^4x^2}$

========================= **Answers** =========================

1. Uniformly convergent for all x.
4. Uniformly convergent for $0 < a < 1$.

9.8 UNIFORM CONVERGENCE AND CONTINUITY

THEOREM 1. *Let $f_n\{n = 1, 2, 3, ...,\}$ be the real valued function defined on a metric space (X,d) and let the sequence $\langle f_n \rangle$ converges uniformly to f on X. Let x_0 be the limit point of X, and suppose that $\lim_{x \to x_0} f(x) = c_n (n = 1, 2, 3, ...)$*

Then, the sequence $\langle c_n \rangle$ of real constants converges, and $\lim_{x \to x_0} f(x) = \lim_{n \to \infty} c_n$

or $\lim_{x \to x_0} \lim_{n \to \infty} f_n(x) = \lim_{n \to \infty} \lim_{x \to x_0} f_n(x).$

Proof. Let $\varepsilon > 0$ be given. Since $\langle f_n \rangle$ converges uniformly on X, therefore, there exists a positive integer m such that

$$n \ge m, p \ge m, x \in X \Rightarrow |f_n(x) - f(x)| < \varepsilon \qquad \qquad ... (1)$$

Letting $x \to x_0$ in (1), we get $|c_n - c_p| < \varepsilon \ \forall \ n \ge m, p \ge m.$
Then by Cauchy's general principle of convergence, we have
The sequence $\langle c_n \rangle$ converges to c i.e., $\lim_{n \to \infty} c_n = c$
Now, since $\langle c_n \rangle$ converges to c and $\langle f_n \rangle$ converges uniformly to f therefore, there exists a positive integer k such that

$$|c_k - c| < \varepsilon / 3 \qquad \qquad ...(2)$$

and $$|f_k(x) - f(x)| < \varepsilon / 3 \ \forall x \in X. \qquad \qquad ... (3)$$

Also, $\lim_{x \to x_0} f_k(x) = c_k$ therefore $\exists \ \delta > 0$ such that

$$d(x, x_0) < \delta \Rightarrow |f(x) - c| < \varepsilon$$

$\Rightarrow$ $\lim_{x \to x_0} f(x) = c = \lim_{n \to \infty} c_n$

$\Rightarrow$ $\lim_{x \to x_0} \lim_{n \to \infty} f_k(x) = \lim_{n \to \infty} \lim_{x \to x_0} f_n(x)$

REMARK

- Let $\sum_{n=1}^{\infty} u_n(x)$ be a series of real valued functions defined on a metric space (X, d) and let

$\lim_{x \to x_0} u_n(x)$ exist $(n = 1, 2, 3, ...)$ where x_0 is a limit point of X. If the series $\sum u_n(x)$ converges

uniformly on X, then $\lim_{x \to x_0} \sum_{n=1}^{\infty} u_n(x) = \sum_{n=1}^{\infty} \left[\lim_{x \to x_0} u_n(x) \right]$

THEOREM 2. (Continuity of limit function). *Let (f_n) be a sequence of real valued function on a set $X \subseteq R$ which converges uniformly to the function f on X. If each f_n $(n=1, 2, 3 ...)$ is continuous on X, then f is also continuous on X.*

Proof. Let $a \in X$ be arbitrary. We shall show that f is continuous at a.

Since each f_n is continuous on $X \Rightarrow f_n$ is continuous at $x = a$.

Also, since $\langle f_n \rangle$ converges uniformly to f on X, therefore, for a given $\varepsilon > 0 \; \exists \; m \in N$ such that $n \geq m \Rightarrow |f_n(x) - f(x)| < \varepsilon / 3 \; \forall \; x \in X$

In particular, we have

$$|f_m(x) - f(x)| < \varepsilon / 3 \qquad \qquad \text{... (1)}$$

and $\qquad \quad |f_m(a) - f(a)| < \varepsilon / 3 \qquad \qquad \text{... (2)}$

Again, since f_m is continuous at a, $\exists \; \delta > 0$ such that

$$d(x, a) < \delta \Rightarrow |f_m(x) - f_m(a)| < \varepsilon / 3 \qquad \qquad \text{... (3)}$$

Hence, if $d(x, a) < \delta$, we have

$$|f(x) - f(a)| = |f(x) - f_m(x) + f_m(x) - f_m(a) + f_m(a) - f(a)|$$
$$\leq |f(x) - f_m(x)| + |f_m(x) - f_m(a)| + |f_m(a) - f(a)|$$
$$< \varepsilon / 3 + \varepsilon / 3 + \varepsilon / 3 = \varepsilon.$$

Therefore, for a given $\varepsilon > 0 \; \exists \; \delta > 0$ such that

$$d(x, a) < \delta \Rightarrow |f(x) - f(a)| < \varepsilon$$

Hence, f is continuous at a.

REMARK

- The above theorem can also be restated as
  'The limit function of a uniformly convergent sequence of continuous function is itself continuous.'

THEOREM 3. **(Continuity of sum function).** *Let* $\sum\limits_{n=1}^{\infty} u_n(x)$ *be a series of real valued continuous function defined on a set $X \subseteq R$. If the series converges uniformly to the function $f(x)$ on X, then $f(x)$ is continuous on X.*

Proof. Let $\qquad \qquad f_n(x) = u_1(x) + u_2(x) + \ldots + u_n(x)$

By definition of the uniform convergence of the series $\sum\limits_{n=1}^{\infty} u_n(x)$ depend upon the uniform convergence of the sequence $\langle f_n(x) \rangle$. Since $f_n(x)$ is the sum of finite number of continuous function, therefore, it is continuous. Now, since $\langle f_n(x) \rangle$ is a sequence of continuous function which converges uniformly to $f(x)$. Therefore, for given $\varepsilon > 0$, $\exists \; m \in N$ such that

$$n \geq m \Rightarrow |f_n(x) - f(x)| < \varepsilon / 3 \; \forall \; x \in X \qquad \qquad \text{... (1)}$$

Let $a \in X$ be arbitrary. Then by (1)

$$n \geq m \Rightarrow |f_n(a) - f(a)| < \varepsilon / 3 \qquad \qquad \text{... (2)}$$

Also, by continuity of f_n at $x = a$, we have for a given $\varepsilon > 0 \; \exists \; \delta > 0$ such that

$$d(x, a) < \delta \Rightarrow |f_n(x) - f_n(a)| < \varepsilon / 3 \qquad \qquad \text{...(3)}$$

Therefore,

$$|f(x) - f(a)| = |f(x) - f_n(x) + f_n(x) - f_n(a) + f_n(a) - f(a)|$$
$$\leq |f(x) - f_n(x)| + |f_n(x) - f_n(a)| + |f_n(a) - f(a)|$$
$$< \varepsilon / 3 + \varepsilon / 3 + \varepsilon / 3 = \varepsilon$$

Hence, $\qquad \qquad d(x, a) < \delta \Rightarrow |f(x) - f(a)| < \varepsilon$

Therefore, $f(x)$ is continuous at $x = a$.

REMARKS

- The above theorem can be restated as
  'The sum function of a uniformly convergent series of continuous function is itself continuous.'
- Uniform convergence of $\langle f_n \rangle$ is a sufficient but not a necessary condition for the continuity of the sum function *i.e.*, if the limit function is continuous on X, it is not necessary that $\langle f_n \rangle$ converges uniformly on X.
- If each f_n is continuous on X and if limit function is not continuous on X, then $\langle f_n \rangle$ does not converge uniformly on X.
- The above theorem also shows that if the series of continuous function defined on a metric space X has discontinuous sum, it cannot be uniformly convergent on a subset Y of X which contains a point of discontinuity.

THEOREM 4. *Let E be a compact set, and let*

    (i) $\langle f_n \rangle$ *be a sequence of continuous function on E,*

    (ii) $\langle f_n \rangle$ *converges pointwise to a continuous function f on E*

    (iii) $f_n(x) \geq f_{n+1}(x) \ \forall \ x \in E, \ n = 1, 2, 3, \ldots$

Then, f_n converges uniformly to f on E.

Proof.   Define $g_n(x) = f_n(x) - f(x)$

Given that f_n and f are continuous, therefore, being the difference of two continuous function, $g_n(x)$ is also continuous.

Also, since $f_n \to f$, we have $g_n = f_n - f \to 0$

Morever, $g_n - g_{n+1} = (f_n - f) - (f_{n+1} - f) = f_n - f_{n+1}$

and therefore $f_n \geq f_{n+1} \Rightarrow g_n \geq g_{n+1}$

Now, to show g_n converges uniformly on E. Let $\varepsilon > 0$ be arbitrary.

Since $g_n \to E$. Therefore, for each $x \in E \ \exists n_x \in N$, such that $|g_{n_x}(x)| < \varepsilon / 2$

Since g_{n_x} is continuous, $\exists$ an open nbd G_x, of $x \in E$ such that for every $y \in G_x$, we have,

$$|g_{n_x}(y) - g_{n_x}(x)| < \varepsilon / 2 \Rightarrow |g_{n_x}(y)| - |g_{n_x}(x)| < \varepsilon / 2$$

$\Rightarrow$ $|g_{n_x}(y)| < |g_{n_x}(x)| + \varepsilon / 2 < \varepsilon / 2 + \varepsilon / 2$ [Using (1)]

Therefore $|g_{n_x}(y)| < \varepsilon$ ... (2)

Now, since $g_n \geq g_{n+1} \ \forall \ n \in N$, we have

$$n \geq n_x \Rightarrow g_{n_x}(x) \geq g_n(x)$$ .... (3)

From (2) and (3), we conclude that $|g_n(y)| < \varepsilon$ for every $y \in G_x$ and $n \geq n_x$.

Clearly, the collection $\{G_x : x \in E\}$ is an open cover of E. Since E is compact, there exists a finite set of points $x_1, x_2, \ldots x_m$ such that $E \subset \overset{m}{\underset{i=1}{\cup}} G_{x_i}$

Let $n_0 = \max \{n_{x_1}, n_{x_2}, \ldots n_{x_m}\}$

Therefore from (4) and (5), we conclude that

$$|g_n(y)| < \varepsilon \text{ for every } y \in E \text{ and } n \geq n_0.$$

$\therefore$ g_n converges uniformly on E. Hence f_n converges uniformly to f on E.

Solved Examples

Example 1. Let $f_n(x) = n^2 x(1-x)^n, x \in R$ for each $n \in N$. Show that the limit function f is continuous, but $\langle f_n \rangle$ does not converge to f uniformly.

Solution. We have

$$\lim_{n \to \infty} n^2 x(1-x)^n = 0, \text{ if } x \in [0, 1]$$

Therefore, $\langle f_n \rangle$ converges on $[0, 1]$.

For maxima and minima of $f_n(x)$. we must have $\dfrac{df_n(x)}{dx} = 0$ which implies

$$-n^3 x(1-x)^{n-1} + n^2(1-x)^n = 0$$

$$\Rightarrow \qquad n^2(1-x)^{n-1}[-nx + (1-x)] = 0 \Rightarrow x = 1, \frac{1}{n+1}$$

Obviously, $\dfrac{d^2 f_n(x)}{dx^2}$ is negative when $x = \dfrac{1}{n+1} \in [0,1]$.

Therefore, $\qquad \text{max.} f_n(x) = f_n\left(\dfrac{1}{n+1}\right) = n\left(\dfrac{n}{n+1}\right)^{n+1}$

Now $\qquad\qquad M_n = \sup_{x \in [0,1]} |f_n(x) - f(x)| = \sup_{x \in [0,1]} n^2 x(1-x)^n$

$$= \text{max.}\{n^2 x(1-x)^n\} = n\left(\dfrac{n}{n+1}\right)^{n-1}$$

and $\qquad \lim_{n \to \infty} M_n = \lim_{n \to \infty} n\left(1 - \dfrac{1}{n+1}\right)^{n+1} = \lim_{n \to \infty} \dfrac{n}{e} = \infty \neq 0$

Therefore, by Weierstrass's M-test, the sequence $\langle f_n \rangle$ is not uniformly convergent in $[0,1]$, Hence, $\langle f_n \rangle$ converges to a continuous limit function on $[0,1]$, but $\langle f_n \rangle$ does not converge uniformly on $[0, 1]$.

Example 2. Show that the sum of the series

$$\Sigma\left(\frac{nx}{1+n^2 x^2} - \frac{(n-1)x}{1+(n-1)^2.x^2}\right)$$

is continuous for all values of x, although 0 is a point of non-uniform convergence of the series.

Solution. Here, we have

$$u_n(x) = \frac{nx}{1+n^2 x^2} - \frac{(n-1)x}{1+(n-1)^2.x^2}$$

Then, $\qquad f_n(x) = u_1(x) + u_2(x) + ... + u_n(x)$

$$= \left(\frac{x}{1+x^2} - 0\right) + \left(\frac{2x}{1+2x^2} - \frac{x}{1+x^2}\right) + ... +$$

$$\left(\frac{nx}{1+n^2 x^2} - \frac{(n-1).x}{1+(n-1)^2.x^2}\right)$$

$$= \frac{nx}{1+n^2 x^2}$$

Therefore, $$f(x) = \lim_{n \to \infty} f_n(x) = \lim_{n \to \infty} \frac{nx}{1 + n^2 x^2} \qquad \left|\text{Form } \frac{\infty}{\infty}\right.$$

$$= \lim_{n \to \infty} \frac{x}{(1/n) + nx^2} = 0, \forall x$$

Hence, the sum function $f(x)$ is continuous for all values of x.

Example 3. *Test for uniform convergence and continuity of the sum function of the series for which* $f_n(x) = nx(1 - x)^n$, *for* $0 \le x \le 1$

Solution. Here, we have

$$f_n(x) = nx(1 - x)^n, 0 \le x \le 1$$

Therefore, when $0 < x < 1$, we have

$$f(x) = \lim_{n \to \infty} nx(1 - x)^n = \lim_{n \to \infty} \frac{nx}{(1 - x)^{-n}}$$

$$= \lim_{n \to \infty} \frac{x}{(1 - x)^{-n} \log(1 - x)}$$

$$= \lim_{n \to \infty} \frac{x.(1 - x)^n}{\log(1 - x)} = 0$$

Therefore, $$f(x) = \lim_{n \to \infty} f_n(x) = 0, \text{when } 0 < x < 1$$

Also, $$f_n(x) = 0 \forall x \in [0,1]$$

Hence, the sum function $f(x)$ is continuous for all $x \in [0, 1]$, but the sequence $\langle f_n(x) \rangle$ is not uniformly convergent on $[0,1]$.

Example 4. *If the series* $\sum_{n=0}^{\infty} a_n$ *is convergent and has the sum s, then the series* $\sum_{n=0}^{\infty} a_n x^n$ *is uniformly convergent for* $0 \le x \le 1$ *and* $\lim_{x \to 1^-} \sum_{n=0}^{\infty} a_n x^n = s$

Solution. Given that the series $\sum_{n=0}^{\infty} a_n$ converges to s.

Therefore, by definition, for given $\varepsilon > 0 \ \exists \ m = m \ (\varepsilon) \in N$ such that

$$\left| \sum_{k=n+1}^{p} a_k \right| < \varepsilon \ \forall \, p > n \ge m$$

Since $\langle x^n \rangle$ is a monoionic decreasing sequence, then we have

$$\left| \sum_{k=n+1}^{p} a_k x^k \right| < \varepsilon . x^{n+1} < \varepsilon \ \forall x \in [0,1]$$

Therefore, the series $\sum_{n=0}^{\infty} a_n x^n$ is uniformly convergent on $[0,1]$.

Now, for each $n \in N$, $a_n x^n$ is continuous in $[0, 1]$ and $\sum_{n=0}^{\infty} a_n x^n$ is uniformly convergent on $[0, 1]$. Hence, the sum function is also continuous in $[0, 1]$. Therefore

$$\lim_{x \to 1^-} \left(\sum_{n=0}^{\infty} a_n x^n \right) = \sum_{n=0}^{\infty} a_n \left(\lim_{x \to 1^-} x^n \right) = \sum_{n=0}^{\infty} a_n = s$$

Example 5. Examine the continuity of the sum function of the series $\sum x^2(1-x^2)^{n-1}$ in the interval $-\sqrt{2} < -\delta \le x \le \delta < \sqrt{2}$.

Solution. Here, we have

$$f_n(x) = \sum_1^n u_n(x) = x^2 \sum_1^n (1-x^2)^{n-1}$$

$$= x^2[1+(1-x^2)+(1-x^2)^2 +...+ n \text{ terms}]$$

$$= \frac{x^2[1-(1-x^2)^n]}{1-(1-x^2)} = 1-(1-x^2)^n, \text{ if } x \ne 0$$

and $\qquad f(x) = \lim_{n \to \infty} f_n(x) = 0, \text{if } x = 0$

$$= 1 \text{ if } 0 < |x| < \sqrt{2}$$

$\Rightarrow f(x)$ is discontinuouns at $x = 0$. Hence, 0 is a point of non-uniform convergence.

9.9 UNIFORM CONVERGENCE AND INTEGRATION

THEOREM 1. *Let $\langle f_n \rangle$ be a sequence of real valued function defined on a closed and bounded interval $[a, b]$ and let $f_n \in R\ [a, b]$, for $n \in N$. If $\langle f_n \rangle$ converges uniformly to the function f on $[a, b]$, then $f \in R\ [a, b]$ and*

$$\int_a^b f(x)\,dx = \lim_{n \to \infty} \int_a^b f_n(x)\,dx$$

Proof. Let $\varepsilon > 0$ be given. Since f_n converges uniformly to f on $[a, b] \exists$ an integer $m > 0$ such that

$$n \ge m, x \in [a,b] \Rightarrow |f_n(x) - f(x)| < \frac{3}{3(b-a)} \qquad \qquad ...\ (1)$$

In particular, for $n = m$, we have

$$|f_m(x) - f(x)| < \frac{\varepsilon}{3(b-a)}$$

$$\Rightarrow \qquad f_m(x) - \frac{\varepsilon}{3(b-a)} < f(x) < f_m(x) + \frac{\varepsilon}{3(b-a)} \qquad \qquad ...\ (2)$$

Since f_m is R-integrable on $[a, b]$, $\exists$ a partition $P = \{a = x_0, x_1,..., x_n = b\}$ of $[a, b]$ such that

$$U(P, f_m) - L\ (P, f_m) < \varepsilon/3 \qquad \qquad ...\ (3)$$

Let $m_r^{(m)}, M_r^{(m)}$ and m_r, M_r denote the infima and suprema of f_m and f on $[x_{r-1}, x_r]$ respectively, then from (2), we have

$$f_m(x) < f(x) + \frac{\varepsilon}{3(b-a)} \ \forall x \in [a,b]$$

In particular, for all $x \in [x_{r-1}, x_r]$

$$m_r^{(m)} \le m_r + \frac{\varepsilon}{3(b-a)}$$

$$\Rightarrow \qquad \sum_{r=1}^{n} m_r^{(m)} \Delta x_r \le \sum_{r=1}^{n} m_r \Delta x_r + \frac{\varepsilon}{3(b-a)} \sum_{r=1}^{n} \Delta x_r$$

$$\Rightarrow \qquad L(P, f_m) \le L(P, f) + \varepsilon/3 \qquad\qquad \text{... (4)}$$

Similarly, from (2), we, have

$$f(x) < f_m(x) + \frac{\varepsilon}{3(b-a)}$$

$$\Rightarrow \qquad M_r \le M_r^{(m)} + \frac{\varepsilon}{3(b-a)}$$

$$\Rightarrow \qquad \sum_{r=1}^{n} M_r \Delta x_r \le \sum_{r=1}^{n} M_r^{(m)} \Delta x_r + \frac{\varepsilon}{3(b-a)} \sum_{r=1}^{n} \Delta x_r$$

$$\Rightarrow \qquad U(P, f) \le U(P, f_m) + \varepsilon/3 \qquad\qquad \text{... (5)}$$

Adding (4) and (5), we get

$$U(P, f) + L(P, f_m) \le L(P, f) + U(P, f_m) + \frac{2\varepsilon}{3}$$

$$\Rightarrow \qquad U(P, f) - L(P, f) \le U(P, f_m) - L(P, f_m) + \frac{2\varepsilon}{3}$$

$$< \varepsilon/3 + 2\varepsilon/3$$

$$= \varepsilon$$

which is the necessary and sufficient condition for a function to R-integrable

$$\Rightarrow \quad f \in R[a, b]$$

Now, we have for all $n \ge m$

$$\left| \int_a^b f_n - \int_a^b f \right| = \left| \int_a^b (f_n - f) \right| \le \int_a^b | f_n - f |$$

$$< \frac{\varepsilon}{3(b-a)} \int_a^b dx$$

$$= \frac{\varepsilon}{3(b-a)} \cdot (b-a) = \varepsilon/3$$

Hence, $\lim\limits_{n \to \infty} \left[\int_a^b f_n(x) dx \right] = \int_a^b f(x) dx$

THEOREM 2. Let $\langle f_n \rangle$ be a sequence of real valued continuous functions defined on [a, b] such that $f_n \to f$ uniformly on [a, b] then $f \in R[a, b]$

and $\qquad \lim\limits_{n \to \infty} \int_a^b f_n(x) dx = \int_a^b f(x) dx$

Proof. Since $f_n(n \in N)$ is continuous and therefore, $f_n \in R[a, b]$

Also, f is continuous on [a, b], therefore $f \in R[a, b]$. (∵ Every continuous function in a closed interval is R-integrable). The proof is same as in theorem 1.

THEOREM 3. **(Term by term integration).**

Let $\sum\limits_{n=1}^{\infty} u_n(x)$ be a series of real valued function defined on [a, b] such that $u_n(x) \in R[a, b]$,

for $n = 1, 2, 3, ...$. If the series converges uniformly to f on [a, b] then $f \in R[a, b]$

$$and \int_a^b \left[\sum_{n=1}^{\infty} u_n(x) \right] dx = \sum_{n=1}^{\infty} \int_a^b u_n(x)\,dx$$

Proof.　　Let　　　　　　　$f_n(x) = u_1(x) + u_2(x) + ... + u_n(x)$

Since the sum of a finite number of R-integrable functions is R-integrable. Therefore, $f_n \in R(a, b)$ for each fixed n.

We know that the uniform convergence of the series $\sum u_n(x)$ is the same thing as the uniform convergence of the sequence $\langle f_n \rangle$, so that $f_n \rightarrow f$ uniformly on $[a, b]$. Therefore, by above theorem $f \in R[a, b]$.

Also　　$\int_a^b \left[\sum_{n=1}^{\infty} u_n(x) \right] dx = \int_a^b f(x)\,dx = \lim_{n \to \infty} \int_a^b f_n(x)\,dx$

$$= \lim_{n \to \infty} \int_a^b \left[\sum_{m=1}^{n} u_m(x) \right] dx$$

$$= \lim_{n \to \infty} \sum_{n=1}^{\infty} \int_a^b u_m(x)\,dx = \sum_{n=1}^{\infty} \left[\int_a^b u_n(x)\,dx \right]$$

9.10 UNIFORM CONVERGENCE AND REIMANN-STIELTJES INTEGRATION

THEOREM 1. *Let α be monotonically increasing function on $[a, b]$ and $\langle f_n \rangle$ be a sequence of real valued function defined on $[a, b]$ such that $f_n \in RS(\alpha)$ on $[a, b]$ for $n = 1,2,3,....$ If $f_n \rightarrow f$ uniformly on $[a, b]$, then $f_n \in RS(\alpha)$ and*

$$\int_a^b f\,d\alpha = \lim_{n \to \infty} \int_a^b f_n\,d\alpha$$

Proof.　　Let $\varepsilon > 0$ be given. Choose $\eta > 0$ such that

$$\eta[\alpha(b) - \alpha(a)] \le \varepsilon / 3 \qquad \qquad ... (1)$$

Now, since $f_n \rightarrow f$ uniformly, there exists a positive integer m such that

$$|f_m(x) - f(x)| < \eta \; \forall x \in [a,b] \qquad \qquad ... (2)$$

Since $f_m \in RS(\alpha)$ on $[a, b]$, therefore, by definition, $\exists$ a partition $P = \{a = x_0, x_1, ... x_n = b\}$ of $[a, b]$ such that

$$U(P, f_m, \alpha) - L(P, f_m, \alpha) < \varepsilon / 3 \qquad \qquad ... (3)$$

Inequality (2) can be written as

$$f_m(x) - \eta < f(x) < f_m(x) + \eta$$

$\Rightarrow$　　　　$f_m(x) < f(x) + \eta \Rightarrow L(P, f_m, \alpha) \le L(P, f, \alpha) + \eta[\alpha(b) - \alpha(a)]$

Then, by (1), we have

$$L[P, f_m, \alpha] < L[P, f, \alpha] + \varepsilon / 3 \qquad \qquad ... (4)$$

Similarly,　　$f(x) < f_m(x) + \eta \Rightarrow U(P, f, \alpha) \le U(P, f_m, \alpha) + \varepsilon / 3 \qquad ... (5)$

Adding (4) and (5), we get

$$U(P, f, \alpha) - L(P, f_m, \alpha) \le U(P, f_m, \alpha) - L(P, f, \alpha) + \frac{2\varepsilon}{3}$$

$$\Rightarrow \quad U(P,f,\alpha) - L(P,f_m,\alpha) \le U(P,f_m,\alpha) - L(P,f,\alpha) + \frac{2\varepsilon}{3}$$

$$< \frac{\varepsilon}{3} + \frac{2\varepsilon}{3} = \varepsilon \qquad \text{[Using (3)]}$$

$$\Rightarrow \qquad f \in RS\,(\alpha)\ \text{on}\ [a, b]$$

Again for, $\varepsilon > 0$, by uniform convergence of $\langle f_n \rangle$ to f, $\exists$ a positive integer m such that

$$n \ge m, x \in [a,b] \Rightarrow |f_n(x) - f(x)| < \varepsilon \qquad \text{... (6)}$$

Therefore, for $n \ge m$, we have

$$\left| \int_a^b f d\alpha - \int_a^b f_n d\alpha \right| = \left| \int_a^b (f - f_n) d\alpha \right|$$

$$\le \int_a^b |f - f_n|\, d\alpha < \varepsilon \int_a^b d\alpha \qquad \text{[By (6)]}$$

$$= \varepsilon[\alpha(b) - \alpha(a)]$$

Hence, $\qquad \int_a^b f d\alpha = \lim_{n \to \infty} \int_a^b f_n d\alpha$

REMARK

- The condition of uniform convergence of the series $\sum u_n(x)$ is only sufficient but not necessary for the validity of term by term integration.

Solved Examples

Example 1. *Examine for term by term integration, the series for which $f_n(x) = nxe^{-nx^2}, x \in R$ Also indicate the interval for which your conclusion holds.*

Solution. Here, we have

$$f(x) = \lim_{n \to \infty} f_n(x) = \lim_{n \to \infty} \frac{nx}{e^{nx^2}}$$

$$= \lim_{n \to \infty} \frac{x}{e^{nx^2} \cdot x^2} = 0, \quad \text{for all finite values of } x.$$

Now, consider the interval $0 \le x \le 1$, we have

$$\int_0^1 f(x)dx = \int_0^1 0.dx = 0$$

Also $\qquad \int_0^1 f_n(x)dx = \int_0^1 nx\,e^{-nx^2}\,dx = \left[-\frac{1}{2}e^{-nx^2} \right]_0^1 = \frac{1}{2}[1 - e^{-n}] = \frac{1}{2}$ as $n \to \infty$

Hence, term by term integration over the interval $0 \le x \le 1$ is not justified. In fact, 0 is a point of non-uniform convergence.

REMARK

- Term by term integration is justified over the interval $[0, 1]$ where $0 \le c \le 1$. For this, we have

$$\int_c^1 f_n(x)dx = \int_c^1 nx\,e^{-nx^2}\,.dx = \frac{1}{2}[e^{-nx^2} - e^{-n}] = 0 \text{ as } n \to \infty$$

Hence, $\qquad \int_c^1 f(x)dx = \lim_{n \to \infty} \int_0^1 f_n(x)\,dx$

Example 2. *Test for uniform convergence and term by term integration of the series*

$$\sum \frac{x}{(n+x^2)^2}$$

Solution. Clearly, series is uniformly convergent for all real values of x, by Weierstrass's M-test. Hence, the series is integrable term by term between any finite limits. Therefore,

$$\int_0^1 f(x)\,dx = \lim_{n\to\infty}\sum_1^n \int_0^1 u_n(x)\,dx = \lim_{n\to\infty}\int_0^1 x(n+x^2)^{-2}\,dx$$

$$= \lim_{n\to\infty}\sum_1^n \left[\frac{(n+x^2)^{-1}}{-2}\right]_0^1 = \lim_{n\to\infty}\sum_1^n\left[-\frac{1}{2}\left\{\frac{1}{n+1}-\frac{1}{n}\right\}\right]$$

$$= \frac{1}{2}\lim_{n\to\infty}\left[\left(1-\frac{1}{2}\right)+\left(\frac{1}{2}-\frac{1}{3}\right)+\dots+\left(\frac{1}{n}-\frac{1}{n+1}\right)\right] = \frac{1}{2}\lim_{n\to\infty}\left[1-\frac{1}{n+1}\right] = \frac{1}{2}$$

Example 3. *Examine for term by term integration the series, the sum of whose first n terms is* $n^2x(1-x)^n$ $(0 \le x \le 1)$.

Solution. Here, we have

$$f_n(x) = n^2x(1-x)^n$$

$\therefore$

$$f(x) = \lim_{n\to\infty} f_n(x) = 0,\ \text{for}\ 0 \le x \le 1.$$

$\because$ When $0 < x < 1$, we have

$$\lim_{n\to\infty} f_n(x) = \lim_{n\to\infty}\frac{n^2x}{(1-x)^{-n}}$$

$$= \lim_{n\to\infty}\frac{2nx}{-(1-x)^{-n}.\log(1-x)} \qquad \text{(By L-Hospital Rule)}$$

$$= \lim_{n\to\infty}\frac{2x}{(1-x)^{-n}[\log(1-x)]^2} = 0$$

Then, $\qquad \int_0^1 f(x)\,dx = 0$

But $\qquad \int_0^1 f(x)\,dx = \int_0^1 n^2x(1-x)^n\,dx$

$$= n^2\left[\frac{-x(1-x)^{n+1}}{n+1} - \frac{(1-x)^{n+2}}{(n+1)(n+2)}\right]_0^1$$

$$= \frac{n^2}{(n+1)(n+2)} \to 1 \text{ as } n\to\infty.$$

Therefore, term by term integration is not justified in $0 \le x \le 1$. Also, the series is non-uniformly convergent in $0 \le x \le 1$, because if possible let it be uniformly convergent, then for a given ε, we have

$$|f_n(x)-f(x)| = n^2x(1-x)^n < \varepsilon\,\forall\,n \ge m\,\text{and}\,x \in [0,1] \qquad \dots (1)$$

Taking $x = \dfrac{1}{n}$ we have

$$| f_n(x) - f(x) | = n^2 . \frac{1}{n}\left(1 - \frac{1}{n}\right)^n = n\left(1 - \frac{1}{n}\right)^n$$

$$\to \infty \text{ as } n \to \infty \ (i.e., \text{ as } x \to 0)$$

which is a contradiction of (1). Hence, the series is non-uniformly convergent in $0 \le x \le 1$. Here, 0 is a point of non-uniform convergence.

Example 4. *Show that the series $\sum\limits_{1}^{\infty} u_n(x)$ for which the nth partial sum*

$$f_n(x) = \sum_{1}^{\infty} u_n(x) = nx(1-x)^n \text{ is term by term integrable.}$$

Solution. Here, we have

$$f_n(x) = nx(1-x)^n$$

Clearly $f_n(x) = 0$, when $x = 0$ or 1.
Also, when $0 < x < 1$, we have

$$\lim_{n\to\infty} f_n(x) = \lim_{n\to\infty} \frac{nx}{(1-x)^{-n}} = \lim_{n\to\infty} \frac{x}{-(1-x)^{-n}\log(1-x)} = 0$$

Therefore, for $0 \le x \le 1, f(x) = 0$

$$\therefore \qquad \int_0^1 f(x)\,dx = 0 \text{ and} \int_0^1 f_n(x)\,dx = \int_0^1 nx(1-x)^n dx$$

$$= n\left[-\frac{x(1-x)^{n+1}}{n+1} - \frac{(1-x)^{n+2}}{(n+1)(n+2)} \right]_0^1$$

$$= \frac{n}{(n+1)(n+2)} \to 0 \text{ as } n \to \infty$$

Hence, the. series is term by term integrable in $0 \le x \le 1$, although 0 is a point of non-uniform convergence of the series.

Example 5. *Test the series $\sum x^{n-1}(1 - 2x^n)$ in the interval $0 \le x \le 1$ for term by term integration.*

Solution. Here, we have

$$f_n(x) = \sum_{1}^{n} x^{n-1}(1-2x^n) = \sum_{1}^{n} x^{n-1} - 2\sum_{1}^{n} x^{2n-1}$$

$$= \frac{1-x^n}{1-x} - 2\frac{x(1-x^{2n})}{1-x^2}$$

Therefore,

$$f(x) = \lim_{n\to\infty} f_n(x) = \begin{cases} \dfrac{1}{1+x} & , \quad \text{when } 0 < x < 1 \\ 1 & , \quad \text{when } x = 0 \end{cases}$$

and $f(x) = -\infty$, when $x = 1$, because $M_n = -1$
Therefore, the series diverges to $-\infty$ at $x = 1$
Now, for $0 < x < 1$, we have

$$\int_0^1 f(x)\,dx = \int_0^1 \frac{1}{1+x}\,dx = \left[\log_e(1+x)\right]_0^1 = \log_e 2$$

and

$$\sum \int_0^1 u_n(x) = \int_0^1 (x^{n-1} - 2x^{2n-1})\,dx = \sum\left(\frac{1}{n} - \frac{2}{2n}\right) = 0$$

Hence, term by term integration is not justified.

EXERCISE 9.2

1. Show that the series

$$x^4 + \frac{x^4}{1+x^4} + \frac{x^4}{(1+x^4)^2} + \frac{x^4}{(1+x^4)^3} + \dots \text{ is not}$$

uniformly convergent on [0, 1].

2. Show that the series $\sum f_n(x)$ is non-uniformly convergent, where

$$f_n(x) = \frac{1}{1+nx} \text{ for } 0 \le x \le 1$$

3. Show that the series $\sum xe^{-nx}$ is non-uniformly convergent in any interval which includes $x = 0$.

4. Test for term-by-term integration of the series

$$\sum_{n=1}^{\infty} \frac{x^n}{n^2} \text{ on } [0, 1] \text{ and show that}$$

$$\int_0^1 \left(\sum_1^{\infty} \frac{x^n}{n^2} \right) dx = \sum_1^{\infty} \frac{1}{n^2(n+1)}.$$

5. Show that the series $1 - x + x^2 - x^3 + \dots$
$$(0 \le x \le 1)$$

admits of integration term by term in $0 \le x \le 1$ although it is not uniformly convergent in [0, 1].

6. Show that the series $\sum_1^{\infty} u_n(x)$ for which the nth partial sum $f_n(x)$ is given by

$$f_n(x) = \sum_1^n u_n(x) = \frac{1}{1+nx}$$

7. Examine for the continuity of the sum function and for term by term integration the series whose n^{th} terms is

$$n^2 xe^{-n^2 x^2} - (n-1)^2 xe^{-(n-1)^2 x^2}, 0 \le x \le 1.$$

8. Show that 0 is a point of non-uniform convergence of the series $\sum u_n(x)$ where

$$u_n(x) = \frac{1-(1+x)^n}{1+(1+x)^n} - \frac{1-(1+x)^{n-1}}{1+(1+x)^{n-1}}$$

9. Show that near $x = 0$, the series $\sum_{n=1}^{\infty} u_n(x)$ where $u_1(x) = x, u_n(x) = x^{1/(2n-1)} - x^{1/(2n-3)}$ and real values of x are concerned is discontinuous and non-uniformly convergent. Can the series be integrated term by term?

10. Test for term by term integration of the series $\sum u_n(x)$ for which n^{th} partial sum $f_n(x)$ is given by

$$f_n(x) = \frac{nx}{1+n^2x^2}, 0 \le x \le 1.$$

11. Show that the series $\sum u_n(x)$, where

$$u_n(x) = 2x \left[\frac{e^{-x^2/n^2}}{n^2} - \frac{e^{-x^2/(n+1)^2}}{(n+1)^2} \right]$$

can be integrated term-by-term in any interval [a, b].

9.11 UNIFORM CONVERGENCE AND DIFFERENTIATION

THEOREM 1. *Let $\langle f_n \rangle$ be a sequence of real valued functions defined on [a,b] such that*

   (i) f_n is differentiable on [a, b] for $n \in \mathbf{N}$

   (ii) the sequence $\langle f_n(c) \rangle$ converges for some point c of [a, b]

   (iii) the sequence $\left\langle f_n' \right\rangle$ converges uniformly on [a, b]

then the sequence $\langle f_n \rangle$ converges uniformly to a differentiable limit f and

$$\lim_{n \to \infty} f_n'(x) = f'(x), a \le x \le b$$

Proof.   Let $\varepsilon > 0$ be given. Since $\langle f_n(c) \rangle$ converges for some $c \in [a, b]$ and $\left\langle f_n' \right\rangle$ converges uniformly on [a, b], therefore, for both, Cauchy convergence criterian is satisfied.

Therefore, we can choose n such that $n \geq N, m \geq N$

$$|f_n(c) - f_m(c)| < \varepsilon/2 \qquad \text{... (1)}$$

and $\qquad |f_n'(t) - f_m'(t)| < \dfrac{\varepsilon}{2(b-a)} \ (a \leq t \leq b) \qquad \text{... (2)}$

By mean value theorem of differential calculus, we have

$$[f_n(x) - f_m(x)] - [f_n(t) - f_m(t)] = (x-t)[f_n'(\xi) - f_m'(\xi)]$$

for any x and t on $[a, b]$ and for some $\xi \in [x, t]$, if $n \geq N, m \geq N$

Now,

$$|f_n(x) - f_m(x) - f_n(t) + f_m(t)| \leq |x-t||f_n'(\xi) - f_m'(\xi)| < \dfrac{|x-t| \cdot \varepsilon}{2(b-a)}$$

$$< \varepsilon/2 \qquad \text{... (3)}$$

Using (1) and (3), we get

$$|f_n(x) - f_m(x)| = |f_n(x) - f_m(x) - f_n(c) + f_m(c) + f_n(c) - f_m(c)|$$

$$< \varepsilon/2 + \varepsilon/2 = \varepsilon$$

Therefore, $<f_n>$ converges uniformly on $[a, b]$

Now, let $\qquad f(x) = \lim\limits_{n\to\infty} f_n(x) \ (a \leq x \leq b)$

Fixed a point x on $[a, b]$ and define

$$\phi_n(t) = \dfrac{f_n(t) - f_n(x)}{t-x}, \phi(t) = \dfrac{f(t) - f(x)}{t-x} \text{ for } a \leq t \leq b, t \neq x \qquad \text{... (4)}$$

Then $\qquad \lim\limits_{t\to x} \phi_n(t) = f_n'(x) \ (n = 1,2,3,...) \qquad \text{... (5)}$

Now (3) gives

$$|\phi_n(t) - \phi_m(t)| < \dfrac{\varepsilon}{2(b-a)} \ (n \geq N, m \geq N)$$

Therefore $\langle \phi_n \rangle$ converges uniformly for $t \neq x$. Since $\langle f_n \rangle$ converges to f then from (4) we have $\qquad \lim\limits_{n\to\infty} \phi_n(t) = \phi(t) \qquad \text{... (6)}$

uniformly for $a \leq t \leq b, \ t \neq x$.

From (5) and (6), we conclude that

$$\lim_{t\to x} \phi(t) = \lim_{n\to\infty} f_n'(x) \text{ or } \lim_{t\to x} \dfrac{f(t)-f(x)}{t-x} = \lim_{n\to\infty} f_n'(x)$$

or $\qquad f'(x) = \lim\limits_{n\to\infty} f_n'(x)$

9.11.1 TERM BY TERM DIFFERENTIATION

THEOREM 1. *Let* $\displaystyle\sum_{n=1}^{\infty} u_n(x)$ *be a series of real valued differentiable function on* $[a, b]$ *such that*

$\displaystyle\sum_{n=1}^{\infty} u_n(c)$ *converges for some point c of $[a, b]$ and* $\displaystyle\sum_{n=1}^{\infty} u_n'(x)$ *converges uniformly* on $[a, b]$. Then the series $\sum\limits_{n=1}^{\infty} u_n(x)$ converges uniformly on $[a, b]$ to a differentiable sum function f and $f'(x) = \lim\limits_{n \to \infty} \sum\limits_{m=1}^{\infty} u'_m(x), (a \le x \le b)$

In other words if $a \le x \le b$, then

$$\frac{d}{dx}\left(\sum_{n=1}^{\infty} u_n(x)\right) = \sum_{n=1}^{\infty}\left[\frac{d}{dx} u_n(x)\right]$$

Proof. Let $\quad f_n(x) = u_1(x) + u_2(x) + ... + u_n(x).$

Clearly, $f_n'(x) = u_1'(x) + u_2'(x) + ... + u_n'(x).$

Therefore, the series $\sum\limits_{n=1}^{\infty} u_n(x)$ and $\sum\limits_{n=1}^{\infty} u_n'(x)$ are respectively equivalent to the sequences $\langle f_n \rangle$ and $\langle f_n' \rangle$. Now by previous theorem, we can obtained

$$\frac{d}{dx}\left(\sum_{n=1}^{\infty} u_n(x)\right) = \sum_{n=1}^{\infty}\left[\frac{d}{dx} u_n(x)\right]$$

THEOREM 2. *Let $F : [a, b] \to R$ be a mapping and let $F'(x_0)$ exist for $a < x_0 < b$. If $a < \alpha_n < x_0 < \beta_n < b$ for $n = 1, 2, 3...$ and $\beta_n \to x_0, \alpha_n \to x_0,$ as $n \to \infty,$ then*

$$\lim_{n \to \infty} \frac{F(\beta_n) - F(\alpha_n)}{\beta_n - \alpha_n} = F'(x_0).$$

Proof. Let us define $\qquad \lambda_n = \dfrac{\beta_n - x_n}{\beta_n - \alpha_n}$

Then clearly $0 < \lambda_n < 1$

Now $\dfrac{F(\beta_n) - F(\alpha_n)}{\beta_n - \alpha_n} - F'(x_0) = \lambda_n \left\{ \dfrac{F(\beta_n) - F(x_0)}{\beta_n - x_0} - F'(x_0) \right\}$

$$-(1 - \lambda_n)\left\{ \frac{F(\alpha_n) - F(x_0)}{\alpha_n - x_0} - F'(x_0) \right\} \qquad \text{... (1)}$$

By definition $\dfrac{F(\beta_n) - F(x_0)}{\beta_n - x_0}$ both $\dfrac{F(\alpha_0) - F(x_0)}{\alpha_n - x_0}$ both tends to $F'(x_0)$ as $n \to \infty$.

Also, since $\langle \lambda_n \rangle$ and $\langle 1 - \lambda_n \rangle$ are bounded sequence, therefore from (1), we conclude that

$$\lim_{n \to \infty}\left[\frac{F(\beta_n) - F(\alpha_n)}{\beta_n - \alpha_n} - F'(x_0) \right] = 0$$

$\Rightarrow \qquad\qquad \lim\limits_{n \to \infty} \dfrac{F(\beta_n) - F(\alpha_n)}{\beta_n - \alpha_n} = F'(x_0)$

THEOREM 3. *(Everywhere continuous but no where differentiable functions).*

There exists a real valued function F such that F is continuous on R but $F'(c)$ does not exist for any $c \in R$.

Proof.   Define a function

$$\phi(x) = |x| \quad (-1 \le x \le 1) \qquad \qquad \text{... (1)}$$

and we extend the definition of $\phi(x)$ to entire real axis by periodicity 2, such that

$$\phi(x+2) = \phi(x) \qquad \qquad \text{... (2)}$$

Then clearly ϕ is continuous on R *i.e.*, $|\phi(s) - \phi(t)| = \big|\,|s| - |t|\,\big| \le |s - t| \qquad \text{... (3)}$

Define, $F(x) = \displaystyle\sum_{n=0}^{\infty} \left(\frac{1}{4}\right)^n \phi(4^n.x) = \sum_{n=0}^{\infty}\left(\frac{1}{4}\right)^n .F_n(x) \text{(say)}$

Where $F_n(x) = \left(\dfrac{1}{4}\right)^n \phi(4^n.x)$

From (1), $0 < \phi \le 1$, so that

$$|F_n(x)| = \left|\left(\frac{1}{4}\right)^n .\phi(4^n.x)\right| \le \left|\frac{1}{4}\right|^n = M_n \text{ (say)}$$

Since $\sum M_n$ is a geometric series with common ratio less than 1.

Therefore, $\sum M_n$ is convergent.

Then, by Weierstrass's theorem for uniform convergence, we have $\sum F_n(x)$ converges uniformly.

$\Rightarrow \quad F$ is continuous on x.

Let m be a fixed positive integer and let x be a fixed real number.

Put $\delta_m = \pm\dfrac{1}{2}4^{-m}$ where the sign is so chosen that no integer lies between $4^m.x$ and $4^m(x+\delta_m)$.

This can be done because $4^m|\delta_m| = \left|\pm\dfrac{1}{2}\right| = \dfrac{1}{2}$

Now, define $\gamma_n = \dfrac{\phi(4^n(x+\delta_m)) - \phi(4^n.x)}{\delta_m} \qquad \text{... (6)}$

If $0 \le n \le m$, then (3) gives

$$|\gamma_n| = \frac{|\phi(4^n(x+\delta_m)) - \phi(4^n.x)|}{|\delta_m|} \le \frac{|4^n(x+\delta_m) - 4^n.x|}{|\delta_m|} = \frac{|4^n.\delta_m|}{|\delta_m|} = 4^m$$

Since $|\gamma_m| = \left|\dfrac{\phi(4^m(x+\delta_m)) - \phi(4^m.x)}{\delta_m}\right| = \dfrac{\left|\phi\left(4^m\left(x\pm\dfrac{1}{2}\right)\right) - \phi(4^m.x)\right|}{\delta_m}$

$$= \frac{\left|\phi\left(4^m\left(x\pm\dfrac{1}{2}\right)\right) - \phi(4^m.x)\right|}{\delta_m} = \frac{\left|\,\left|4^m\left(x\pm\dfrac{1}{2}\right)\right| - |4^m.x|\,\right|}{|\delta_m|}$$

$$= \frac{\dfrac{1}{2}}{\dfrac{1}{2}.4^m} = 4^m$$

Therefore, we conclude that

$$\left|\frac{F(x+\delta_m)-F(x)}{\delta_m}\right| = \left|\sum_{n=0}^{m}\left(\frac{3}{4}\right)^n \frac{\phi(4^n(x+\delta_m))-\phi(4^n.x)}{\delta_m}\right|$$

$$= \left|\sum_{n=0}^{m}\left(\frac{3}{4}\right)^n \gamma_n + \sum_{n=m+1}^{\infty}\left(\frac{3}{4}\right)^n .\gamma_n\right|$$

$$= \left|\sum_{n=0}^{m}\left(\frac{3}{4}\right)^n \gamma_n\right|, \text{ since } \gamma_n = 0 \text{ for } n>m$$

$$= \left|\left(\frac{3}{4}\right)^m .4^m - \left(-\sum_{n=0}^{m-1}\left(\frac{3}{4}\right)^n .\gamma_n\right)\right| \qquad [\because \gamma_m = 4^m]$$

$$\geq \left|3^m\right| - \left|\sum_{n=0}^{m}\left(\frac{3}{4}\right)^n .\gamma_n\right| \geq 3^m - \sum_{n=0}^{m-1}\left(\frac{3}{4}\right)^n .4^n$$

$$= 3^m - \sum_{n=0}^{m-1} 3^n = \frac{1}{2}(3^m + 1)$$

Since $\delta_m \to 0$ as $m \to \infty$, it follows that F is not differentiable at x.

9.12 WEIRSTRASS'S NON-DIFFERENTIABLE FUNCTIONS

Define
$$f(x) = \sum_{n=0}^{\infty} a^n \cos(b^n.\pi x)$$

where $0<a<1$, b is an odd positive integer. We shall show that f has no derivative at any point x, if $ab > 1+\frac{2}{3}\pi$. The continuity of function f follows from the fact that the series converges uniformly in any interval and the cosine function is continuous for all x. For a fixed value of x, we write

$$\phi_n(h) = \cos b^n.\pi(x+h) - \cos b^n \pi x$$

$$S_m = \sum_{n=1}^{m-1} a^n.\phi_n(h)/h \text{ and } R_m = \sum_{n=m}^{\infty} a^n \phi_n(b)/h$$

so that
$$\frac{f(x+h)-f(x)}{h} = S_m + R_m$$

By mean value theorem of differential calculus, we have

$$\phi_n(h) = -b^n \pi h \sin b^n \pi \sin b^n \pi(x+\theta h) \qquad (0 < \theta < 1)$$

Therefore,
$$\left|\frac{\phi_n(h)}{h}\right| \leq b^n.\pi \text{ and } |S_m| \leq \sum_{n=1}^{m-1} \pi a^n b^n$$

$$= \frac{\pi(a^m b^m - 1)}{ab-1} < \frac{\pi a^m.b^m}{ab-1}$$

Now, we find the limit of R_m by assigning to h a particular sequence of values as follows :

Suppose $b^m . x = r + \gamma$, where r is an integer nearest to $b^m x$ and $\dfrac{1}{2} \le \gamma < \dfrac{1}{2}$

Setting $\hspace{2cm} h = \dfrac{1-t}{b^m}$, so that $0 < h < \dfrac{3}{2} b^{-m}$

Then $\hspace{2cm} b^n \pi (x + h) = b^{n-m} . b^m \pi (x + h) = \pi b^{n-m} (r + 1), n \ge m.$

Also, $\hspace{2cm} \cos b^n \pi x = \cos \{ b^{n-m} (r + \gamma) \pi \}$

$$= \cos(b^{n-m} . r\pi) \cos(b^{n-m} \gamma \pi) = (-1)^r \cos(b^{n-m} \gamma \pi)$$

$\Rightarrow \hspace{2cm} \phi_n(h) = (-1)^{r+1} [(1 + \cos(b^{n-m} . \gamma \pi))]$

and $\hspace{2cm} R_m = (-1)^{r+1} \sum_{n=m}^{\infty} \dfrac{a^n [1 + \cos(b^{n-m} . \gamma \pi)]}{h}$

Since, none of the terms of the series is negative, we have on taking only the first one

$$|R_m| \ge \dfrac{a^m (1 + \cos \gamma \pi)}{h} \ge \dfrac{a^m}{h} \ge \dfrac{2}{3} a^m b^m$$

Hence, $\hspace{1cm} \left| \dfrac{f(x+h) - f(x)}{h} \right| = |R_m - S_m| = |R_m - (-S_m)| \ge |R_n| - |S_m|$

$$\ge \left(\dfrac{2}{3} - \dfrac{\pi}{ab - 1} \right) a^m . b^m$$

Let us suppose $ab > 1 + \dfrac{3}{2} \pi$. Then, we have

$$\dfrac{ab - 1}{\pi} > \dfrac{3}{2} i.e., \quad \dfrac{\pi}{ab - 1} < \dfrac{2}{3} \text{ so that } \dfrac{2}{3} - \dfrac{\pi}{ab - 1} > 0$$

Hence, f is not differentiable at x. Now, since x is arbitrary, therefore, we conclude that f is no where differentiable an any interval.

9.13 EQUICONTINUOUS FAMILY OF FUNCTIONS

Definition 1. *Let $\langle f_n \rangle$ be a sequence of function defined on a set E. Then $\langle f_n \rangle$ is said to be pointwise bounded on E, if the sequence $\langle f_n(x) \rangle$ is bounded for every $x \in E$ i.e., if there exists a finite-valued function ϕ defined on E such that*

$$| f_n(x) | < \phi(x), x \in E, n \in N$$

Definition 2. *The sequence of functions $\langle f_n \rangle$ defined on a set E is said to be uniformly bounded on E if there exists a number M such that $| f_n(x) | < M, x \in E, n \in N$*

REMARKS

- If $\langle f_n \rangle$ is pointwise bounded on E and F a countable subset of E then it is always possible to find a subsequence $\langle f_{n_k}(x) \rangle$ converges for every $x \in E$.

- If $\langle f_n \rangle$ is uniformly bounded sequence of continuous function on E, then it is not necessary that there exists a subsequence which converges pointwise on E.

Definition 3. *Let $X \subseteq R$ be non empty. A family F of complex functions f defined on a set E in X is said to be equicontinuous on E if for given $\varepsilon > 0$. $\exists\, \delta > 0$ such that*

$$x \in E, y \in E, d(x,y) < \delta, f \in F \Rightarrow |f(x) - f(y)| < \varepsilon$$

THEOREM 1. *Let K be a compact set $f_n \in F(K)$ for $n = 1, 2, 3 \ldots$ and let $\langle f_n \rangle$ converges uniformly on K, Then $\langle f_n \rangle$ is equicontinuous on K.*

Proof. Let $\varepsilon > 0$ be given. Then by definition of uniformly convergent of $\langle f_n \rangle$, there exists a positive integer N such that

$$n > N \Rightarrow \|f_n - f_N\| < \varepsilon / 3 \qquad \ldots (1)$$

We know that on the compact metric space, every continuous function is uniformly continuous, therefore $\exists\, \delta > 0$ such that

$$1 \le i \le N, d(x,y) < \delta \Rightarrow |f_i(x) - f_i(y)| < \varepsilon / 3 \qquad \ldots (2)$$

If $n > N$ and $d(x, y) < \delta$, if follows that

$$|f_n(x) - f_n(y)| = |f_n(x) - f_N(x) + f_N(x) - f_N(y) + f_N(y) - f_n(y)|$$

$$\le |f_n(x) - f_N(x)| + |f_N(x) - f_N(y)| + |f_N(y) - f_n(y)|$$

$$< \varepsilon / 3 + \varepsilon / 3 + \varepsilon / 3$$

$$= \varepsilon \qquad \text{[By (1) and (2)]} \qquad \ldots (3)$$

From (2) and (3), we conclude that the sequence $\langle f_n \rangle$ is equicontinuous on K.

THEOREM 2. **(Arzele Ascoli Theorem).** *Let K be a compact metric space, $f_n \in F(K)$ for $n = 1, 2, 3, \ldots$ and let $\langle f_n \rangle$ be pointwise bounded and equicontinuous on K. Then*

(i) $\langle f_n \rangle$ is uniformly bounded on K.

(ii) $\langle f_n \rangle$ contains a uniformly convergence subsequence.

Proof. (i) Let ε be given. Now, since $\langle f_n \rangle$ is equicontinuous on K, we can choose $\delta > 0$ such that

$$d(x,y) < \delta, x \in K, y \in K \Rightarrow |f_n(x) - f_n(y)| < \delta\, \forall n \qquad \ldots (1)$$

Since, K is compact, therefore there are finitely many points $p_1, p_2, \ldots, p_r$ in K such that to every $x \in K$, there corresponds at least one p_i with $d(x, p_i) < \delta$. Also, since $\langle f_n \rangle$ is pointwise bounded, $\exists M_i < \infty$ such that

$$|f_n(p_i)| < M_i \ \forall\ n = 1, 2, 3, \ldots$$

Set $M = \max\{M_1, M_2, \ldots M_r\}$. Then, we have

$$|f_n(x)| < M + \varepsilon\ \forall\ x \in K, n \in N$$

$\Rightarrow \langle f_n \rangle$ is uniformly bounded on K.

(ii) Since (X, d) is a compact metric space, there always exist a countable dense subset. Now, for every positive integer n, there are finitely many neighbourhoods of radius $\dfrac{1}{n}$ whose union covers K. The collection of such neighbourhoods is a countable base for K. Choose one point of K from each member of this countable base. This set is countably dense in K.

Let E be such countable dense subset of K. Then evidently $\langle f_n \rangle$ has a subsequence $\langle f_{n_i} \rangle$ such that $\langle f_{n_i}(x) \rangle$ converges for every $x \in E$.

Put, $f_{n_i} = g_i$ Now, we shall show that $\langle g_i \rangle$ converges uniformly on K.

Let $\varepsilon > 0$ and choose $\delta > 0$ as in (1). Let $V(x, \delta)$ denote the set of all $y \in K$ with $d(x, y) < \delta$. Since E is a dense subset in K and K is compact, there exists finitely many points $x_1, x_2, \ldots x_m$ in E such that

$$K \subset V(x_1, \delta) \cup V(x_2, \delta) \cup \ldots \cup V(x_m, \delta) \qquad \ldots(2)$$

Since $\langle g_i(x) \rangle$ converges for every $x_s (s = 1, 2, \ldots m)$, there exist finitely many points $x_1, x_2, \ldots x_m$ in E such that

$$i \geq N, j \geq N, 1 \geq s \geq m \Rightarrow |g_i(x_s) - g_j(x_s)| < \varepsilon / 3. \qquad \ldots (3)$$

If $x \geq K$, then (2) shows that $x \in V(x_s, \delta)$ for some s. Therefore

$$|g_i(x) - g_i(x_s)| < \varepsilon / 3 \text{ for every } i.$$

If $i \geq N, j \geq N$, then from (3), we have

$$|g_i(x) - g_j(x)| = |g_i(x) - g_i(x_s) + g_i(x_s) - g_j(x_s) + g_j(x_s) - g_j(x)|$$

$$\leq |g_i(x) - g_i(x_s)| + |g_i(x_s) - g_j(x_s)| + |g_j(x_s) - g_j(x)|$$

$$< \varepsilon / 3 + \varepsilon / 3 + \varepsilon / 3 = \varepsilon$$

We have shown that for a given $\varepsilon > 0$, $\exists$ a positive integer N such that

$$i \geq N, j \geq N, x \in E \Rightarrow |f_{n_i}(x) - f_{n_j}(x)| < \varepsilon$$

$\Rightarrow \langle f_n \rangle$ converges uniformly on K.

Hence, $\langle f_n \rangle$ contains a uniformly convergent subsequence.

9.13.1 BERNSTEIN POLYNOMIALS

If n is a positive integer and k is an integer such that $0 \leq k \leq n$, then the polynomial B_n of each n is defined by

$$B_n(x) = \sum_{k=0}^{n} (^n c_k) x^k (1-x)^{n-k} f(^k c_n) \, ; x \in [0,1]$$

are called the Bernstein polynomials associated with the real function f

REMARK

- Bernstein polynomials continuously converge to f on [0, 1].

9.14 WEIRSTRASS'S APPROXIMATION THEOREM

Let f be a real valued continuous function defined on a closed interval [a, b]. Then there exists a sequence of real polynomials $\langle P_n \rangle$ which converges uniformly to $\langle f_n \rangle$ on [a, b] i.e.,

$$\lim_{n \to \infty} p_n(x) = f_n(x) \text{ uniformly on } [a, b].$$

Proof. If $a = b$, then taking $P_n(x)$ to be a constant polynomial, result is obvious. Now assume that $a < b$.

Consider a linear transformation $h(x) = \dfrac{x-a}{b-a}$

Clearly, the transfromation is continuous mapping of [a, b] onto [0, 1]. Without loss of any generality, we may assume $a = 0, b = 1$.

Consider, $\qquad g(x) = f(x) - f(0) - x [f(1) - f(0)]$

Then $\qquad\qquad g(0) = g(1) = 0$

Also, g can be obtained as the limit function of a uniformly convergent sequence of polynomials, it is clear that same is true for f, since $f - g$ is a polynomial.

We define $f(x) = 0$ for x outside $[0, 1]$, Then f would be uniformly continuous on R.

Let us define a polynomial (non-negative for $|x| \leq 1$)

$$Q_n(x) = c_n(1 - x^2)^n \qquad (n=1, 2, 3,....) \qquad \text{...(1)}$$

where c_n independent of x, is so chosen that

$$\int_{-1}^{1} Q_n(x)\, dx = 1 \qquad \text{...(2)}$$

Therefore,
$$1 = \int_{-1}^{1} c_n(1 - x^2)^n\, dx = 2c_n \int_0^1 (1 - x^2)^n\, dx$$

$$\geq 2c_n \int_0^{1/\sqrt{n}} (1 - x^2)^n\, dx$$

$$\geq 2c_n \int_0^{1/\sqrt{n}} (1 - nx^2)\, dx$$

($\because$ By Bernaulli's inequality if $y > -1$, then $(1+y)^n \geq 1 + ny \; \forall \; n \in N$)

$$= 2c_n \left[x - \frac{nx^3}{3} \right]_0^{1/\sqrt{n}}$$

$$= \frac{4c_n}{3\sqrt{n}} > \frac{c_n}{\sqrt{n}}$$

$$\Rightarrow \qquad\qquad c_n < \sqrt{n} \qquad \text{... (3)}$$

Therefore, for any $\delta > 0$, equation (3) gives

$$Q_n(x) \leq \sqrt{n}(1 - \delta^2)^n \text{ when } \delta \leq |x| \leq 1 \qquad \text{... (4)}$$

so that $Q_n \to 0$ uniformly in $\delta \leq |x| \leq 1$

Let
$$P_n(x) = \int_{-1}^{1} f(x + t)Q_n(t)\, dt$$

$$= \int_{-1}^{-x} f(x + t)Q_n(t)\, dt + \int_{-x}^{1-x} f(x + t)Q_n(t)\, dt + \int_{1-x}^{1} f(x + t)Q_n(t)\, dt$$

For $|x| \leq 1, -1 + x \leq x + t \leq 0$, for $-1 \leq t \leq -x$ so that $x+t$ lies outside $[0, 1]$ and therefore $f(x+t) = 0$ and hence the first integral on R.H.S vanishes. Similarly, the third integral is also vanishes.

$$\therefore \qquad P_n(x) = \int_{-x}^{1-x} f(x + t)Q_n(t)\, dt$$

$$= \int_0^1 f(t)Q_n(t - x)\, dt, \text{ which is a polynomial in } x.$$

Therefore $\langle P_n \rangle$ is a sequence of real polynomials. Now, to show that $\langle P_n(x) \rangle$ converges uniformly to f on $[0,1]$. Since the continuous function defined on a compact set $[0,1]$ is bounded and uniformly continuous, therefore, f is uniformly continuous on $[0,1]$

$\Rightarrow \exists\, M$ such that

$$M = \sup_{x \in [0,1]} |f(x)| \qquad \text{...(5)}$$

and for any given $\varepsilon > 0$, we can choose $\delta > 0$ such that for any two points $x_1, x_2 \in [0,1]$

$$|f(x_1) - f(x_2)| < \varepsilon / 2, \quad \text{whenever } |x_1 - x_2| < \delta \leq 1. \qquad \dots (6)$$

For, $0 \leq x \leq 1$, we have

$$|P_n(x) - f(x)| = \left| \int_{-1}^{1} f(x+t) Q_n(t)(dt) - f(x) \right|$$

$$= \left| \int_{-1}^{1} \{f(x+t) - f(x)\} Q_n(t)(dt) \right| \qquad \text{[Using (2)]}$$

$$\leq \int_{-1}^{1} |f(x+t) - f(x)| Q_n(t)(dt) \qquad (\because Q_n(t) \geq 0)$$

$$= \int_{-1}^{-\delta} |f(x+t) - f(x)| Q_n(t)dt + \int_{-\delta}^{\delta} |f(x+t) - f(x)| Q_n(t)dt$$

$$+ \int_{\delta}^{1} |f(x+t) - f(x)| Q_n(t)dt$$

$$\leq 2M \int_{-1}^{-\delta} Q_n(t)dt + \frac{\varepsilon}{2} \int_{-\delta}^{\delta} Q_n(t)dt + 2M \int_{\delta}^{1} Q_n(t)dt$$

$$\qquad \text{[By (5) and (6)]}$$

$$\leq 2M \sqrt{n}(1-\delta^2)^n \left\{ \int_{-1}^{-\delta} dt + \int_{\delta}^{1} dt \right\} + \frac{\varepsilon}{2} \qquad \text{[By (2) and (4)]}$$

$$\leq 4M \sqrt{n}(1-\delta^2)^n + \frac{\varepsilon}{2} + \varepsilon, \quad \text{for large } n.$$

Therefore, for any given $\varepsilon > 0 \ \exists\, N$, such that

$$|P_n(x) - f(x)| < \varepsilon \ \forall \ n \geq N$$

$$\Rightarrow \qquad \lim_{n \to \infty} P_n(x) = f(x) \text{ uniformly on } [0, 1].$$

Solved Examples

Example 1. *Show that the series, for which*

$$f_n(x) = \frac{nx}{1+n^2 x^2}, 0 \leq x \leq 1$$

cannot be differentiated term by term at $x=0$.

Solution. For $0 \leq x \leq 1$

$$f(x) = \lim_{n \to \infty} f_n(x) = \lim_{n \to \infty} \frac{nx}{1+n^2 x^2} = 0$$

Also,
$$f_n'(0) = \lim_{h \to 0} \frac{f_n(0+h) - f_n(0)}{h} = \lim_{h \to 0} \frac{\frac{nh}{1+n^2 h^2} - 0}{h}$$

$$= \lim_{h \to 0} \frac{n}{1+n^2 h^2} = n \to \infty \text{ as } n \to \infty$$

Also, $f'(0) = 0$

Hence $f'(0) \neq \lim_{n \to \infty} f_n'(0)$

which shows that the given series cannot be differentiated term by term at $x=0$.

Example 2. *Show that the series*

$$\sum_{n=1}^{\infty} \frac{1}{n^3 + n^4 x^2} . x \in R$$

uniformly convergent for all values of x and is differentiable term by term.

Solution. Let
$$u_n(x) = \frac{1}{n^3 + n^4 x^2}$$

Since, for all values of x,

$$u_n(x) = \frac{1}{n^3} \cdot \left(\frac{1}{1 + nx^2} \right) \le \frac{1}{n^3}$$

and $\sum \frac{1}{n^3}$ is convergent, then by Weierstrass M-test the series $\sum u_n$, converges uniformly.

Now
$$u_n'(x) = \frac{1}{n^3} \cdot \frac{-2nx}{(1 + nx^2)^2} = -\frac{2}{n^2} \cdot \frac{x}{(1 + nx^2)^2}$$

we know that $\dfrac{2x}{n^2(1 + nx^2)^2}$ maximum if $\quad 2(1 + nx^2) - 2x.2(1 + nx^2).2nx = 0$

or if
$$1 + nx^2 - 4nx^2 = 0 \Rightarrow x = \frac{1}{\sqrt{3n}}$$

Therefore, maximum value of $\dfrac{2x}{n^2(1 + nx^2)^2} = \dfrac{2 \cdot \dfrac{1}{\sqrt{3n}}}{n^2 \left(1 + \dfrac{1}{3} \right)^2} = \dfrac{3\sqrt{3}}{8n^{5/2}}$

$$\Rightarrow \qquad\qquad\qquad |u_n'(x)| \le \frac{3\sqrt{3}}{8} \cdot \sum \frac{1}{n^{5/2}}$$

$$\le M_n, \text{ where } M_n = \frac{3\sqrt{3}}{8} \cdot \frac{1}{n^{5/2}}$$

But $\displaystyle\sum_{n=1}^{\infty} M_n = \frac{3\sqrt{3}}{8} \cdot \sum \frac{1}{n^{5/2}}$ is absolutely convergent.

Hence, by Weierstrass's M-test, series $\sum u_n'(x)$ is uniformly convergent in all finite intervals. Therefore given series is term by term differentiable.

Example 3. *If $f_n(x) = \dfrac{x^2}{x^2 + (1 - nx)^2}, (0 \le x \le 1), n = 1, 2, 3 \ldots$ show that*

    (i) the sequence $\langle f_n \rangle$ is uniformly bounded on $[0, 1]$
    (ii) no subsequence of $\langle f_n(x) \rangle$ can converge uniformly on $[0, 1]$
    (iii) the sequence $\langle f_n(x) \rangle$ is not equicontinuous on $[0, 1]$

Solution.   (i) Since, we have $x^2 + (1 - nx)^2 \ge x^2$

Therefore, $|f_n(x)| = \dfrac{x^2}{x^2 + (1 - nx)^2} \le 1 \ (0 \le x \le 1, n = 1, 2, 3 \ldots)$

$\Rightarrow \quad \langle f_n \rangle$ is uniformly bounded on $[0, 1]$.

(ii) Let $\langle f_{n_k} \rangle$ be any subsequence of $\langle f_n \rangle$, then

$$f_{n_k}(x) = \frac{x^2}{x^2 + (1 - n_k x^2)} = \frac{1}{1 + \left(\dfrac{1}{x} - n_k\right)^2} \quad (0 \le x \le 1, k = 1,2,3,....)$$

Now, we have

$$\lim_{k \to \infty} f_{n_k}(x) = 0, 0 \le x \le 1$$

Also

$$f_{n_k}(x) = 0 \ \forall k$$

∴

$$\lim_{k \to \infty} f_{n_k}(x) = 0, \forall \ k$$

Further if

$$f(x) = \sum_{n=1}^{\infty} \frac{1}{n^3 + n^4 x^2}$$

Then

$$f'(x) = \sum_{n=1}^{\infty} u_n'(x) = -2x \sum_{n=1}^{\infty} \frac{1}{n^2 (1 + nx^2)^2}$$

Hence, the subsequence $\langle f_{n_k} \rangle$ converges pointwise to 0 on [0, 1] but

$$f_{n_k}\left(\frac{1}{n_k}\right) = 1, k = 1,2,3,...$$

⇒ $\langle f_{n_k} \rangle$ cannot converge uniformly to 0 on [0, 1]. If we take $\varepsilon = \dfrac{1}{2}$,

then for $x = \dfrac{1}{n_k}$ we have $\left| f_{n_k}\left(\dfrac{1}{n_k}\right) - 0 \right| = |1 - 0| = 1 > \varepsilon \quad (k = 1, 2, 3, ...)$

(iii) To show that $\langle f_n \rangle$ is not equicontinuous on [0,1]. For this, we must find an $\varepsilon > 0$ corresponding to which there exists no $\delta > 0$ satisfying the definition of equicontinuity.

Let us take $\varepsilon = \dfrac{1}{4}, x = \dfrac{1}{n}$ and $y = \dfrac{1}{n+1}$. Then

$$|x - y| = \left| \frac{1}{n} - \frac{1}{n+1} \right| = \frac{1}{n(n+1)}$$

Since

$$f_n(x) = \frac{x^2}{x^2 + (1 - nx)^2} = \frac{1}{1 + \left(\dfrac{1}{x} - n\right)^2}$$

⇒

$$f_n\left(\frac{1}{n}\right) = 1 \qquad \text{and} \qquad f_n\left(\frac{1}{n+1}\right) = \frac{1}{1 + \{(n+1) - n\}^2} = \frac{1}{2}$$

⇒ $\left| f_n\left(\dfrac{1}{n}\right) - f_n\left(\dfrac{1}{n+1}\right) \right| = \left| 1 - \dfrac{1}{2} \right| = \dfrac{1}{2}$

⇒ For $\delta > 0$, we can always find a positive integer n such that $\dfrac{1}{n(n+1)} < \delta$.

Therefore, in this ease, we have $|x - y| < \delta$ but $|f_n(x) - f_n(y)| < \varepsilon$

Hence, the sequence $\langle f_n \rangle$ is not equicontinuous on [0, 1].

1. Show by an example that for term by term differentiation, the condition of uniform convergence is sufficient but not necessary.

2. Show that the series $\sum \dfrac{1}{n^2 + n^4 . x^2}$ is uniformly convergent for all real values of x and that it can be differentiated term by term.

3. Show that the series

$$a^x = 1 + \frac{\log a}{1!} . x + \frac{(\log a)^2}{2!} . x^2 + \dots$$

$$+ \frac{(\log a)^{n-1}}{(n-1)!} x^{n-1} + \dots$$

can be integrated and differentiated term by term.

4. Let $f_n(x) = \dfrac{1}{2n^2} \log(1 + n^4 x^2)$.

Show that the series $\sum u_n'(x)$ does not converges uniformly but the given series can be differentiated term by term.

5. Show that the function represented by $\displaystyle\sum_{n=1}^{\infty} \dfrac{\sin nx}{n^3}$ is differentiable for every x and its derivative is $\displaystyle\sum_{n=1}^{\infty} \dfrac{\cos nx}{n^2}$

6. Let $\left\langle f_n(x) = \dfrac{x}{1 + nx^2} \right\rangle$, where $n = 1,2,3,\dots$ and x is real. Show that $\langle f_n \rangle$ converges uniformly to a function f and the equation

$$f'(x) = \lim_{n \to \infty} f_n'(x) \quad \text{is correct if } x \neq 0 \text{ and}$$

false if $x = 0$.

7. Show that $f(x) = \displaystyle\sum \dfrac{1}{n^p + n^q . x^2}, p > 1$ is uniformly convergent for all values of x and can be differentiated term by term if $q < 3p - \alpha$.

8. Show that the series

$$\sin x = x - \frac{x^3}{3!} + \frac{x^5}{5!} - \frac{x^7}{7!} + \dots \text{ can be differented}$$

to obtained the expansion of $\cos x$.

9. If f is continuous on $[0, 1]$ and if $\int_0^1 x^n f(x) dx = 0$ for $n = 0, 1, 2, \dots$ show that $f(x) = 0$ on $[0, 1]$.

10. Show that, if f is continuous on R, then there exists a sequence $\langle P_n \rangle$ of polynomials converging uniformly to f each bounded subset of R.

CHAPTER REVIEW: A COMPETITIVE APPROACH

SELECTED TERMS AND RESULTS

TERMS

- **Pointwise convergence:** Let f be a function from X to R. Let for each $n \in$ N, f_n: $X \to$ R. Then the sequence of functions $\langle f_n \rangle$ converges pointwise to the function f, if for each $x \in X$, the sequence of real numbers $\langle f_n(x) \rangle$ converges to the real number $f(x)$.

- **Uniform convergence of a sequence:** The sequence $\langle f_n(x) \rangle$ of function is said to converge uniformly on X to a function f if for every $\varepsilon > 0$, we can find a positive integer m such that

$$n \geq m \Rightarrow |f_n(x) - f(x)| < \varepsilon \; \forall \; x \in X$$

- **Point of non-uniform convergence:** A point such that the sequence does non converge uniformly in any *nbd* of it, however small is said to be a point of non-uniform convergence of the sequence.

- **Uniform convergence of a series of functions:** The series $\sum_{n=1}^{\infty} u_n(x)$ is said to converge uniformly on X if the sequence $\langle f_n(x) \rangle$

where $\quad f_n(x) = u_1(x) + u_2(x) + \ldots + u_n(x)$

converges.

RESULTS

- Every uniformly convergent sequence is point-wise convergent and the uniform limit function is same as the pointwise limit function.

- Uniform convergence implies pointwise converegence but not vice-versa.

- A sequence which is not pointwise convergent cannot be uniformly convergent *i.e.*, non-pointwise convergence implies non-uniform convergence.

- (*Cauchy criterion for uniform convergence*). A sequence of functions $\langle f_n(x) \rangle$ converges uniformly on $[a, b]$ if and only if for given $\varepsilon > 0$ and for all $x \in [a, b]$ $\exists$ an integer m such that
$|f_{n+p}(x) - f_n(x)| < \varepsilon \; \forall n \geq m, p > 1$

- Let $\langle f_n \rangle$ be a sequence of functions such that $\lim_{n \to \infty} f(x) = f(x)$, $x \in [a, b]$ and let $M_n = \sup_{x \in [a,b]} |f_n(x) - f(x)|$. Then $f_n \to f$ uniformly on $[a, b]$ if and only if $M_n \to 0$ as $n \to \infty$.

- (*Weirstrass's M-test*): A series of function $\sum f_n$ will converge uniformly and absolutely on $[a, b]$ if there exists a convergent series $\sum M_n$ of positive numbers such that for all $x \in [a, b]$, $|f_n(x)| \leq M_n \; \forall \; n$.

- *Abel's test:* The series $\sum u_n(x) \, v_n(x)$ will converges uniformly in $[a, b]$ if
  (i) $\sum u_n(x)$ is uniformly convergent in $[a, b]$
  (ii) The sequence $\langle v_n(x) \rangle$ is monotonic for every $x \in [a, b]$
  (iii) The sequence $\langle v_n(x) \rangle$ is uniformly bounded in $[a, b]$ by k *i.e.*, $|v_n(x)| < k \; \forall x \in (a, b)$

- A uniformly convergent series $\sum u_n(x)$ remains uniformly convergent on $[a, b]$ if its each term is multiplied by a function $a_n(x)$, $a \leq x \leq b$ provided that the sequence $<a_n(x)>$ is uniformly bounded on (a, b).

- If $\sum_{n=1}^{\infty} a_n x^n$ is a series which converges for all valued of x when $|x| < R$ then $\sum_{n=1}^{\infty} a_n x^n$ is uniformly convergent in $[0, R]$ if and only if $\sum a_n R^n$ is convergent.

- *Dirichlet's test:* The series $\sum u_n(x) \, v_n(x)$ will be uniformly convergent on $[a, b]$ if
  (i) The sequence $\langle v_n(x) \rangle$ is a positive monotonic decreasing sequence converging uniformly to zero for all $x \in (a, b)$.
  (ii) $f_n(x) = \sum_{r=1}^{n} u_r(x)$ is uniformly bounded in (a, b).

- If $v_n(x)$ is a monotonic function of n for each fixed value of $x \in [a, b]$ and $b_n(x)$ tends to zero for $a \leq x \leq b$ and if $\Sigma u_n(x)$ either uniformly converges or oscillate finitely in $[a, b]$ then the series $\Sigma u_n(x) V_n(x)$ is uniformly convergent on $[a, b]$.

- If $\langle v_n \rangle$ is a monotonic sequence of real numbers that converges to zero, then each of the series $\sum v_n \sin n\theta, \sum v_n \cos n\theta$ is uniformly convergent with regard to θ in the interval $[\alpha, 2\pi - \alpha]$ where α is any fixed positive number less than π.

- If a sequence $\langle f_n(x) \rangle$ converges uniformly on $[a, b]$ and $x_0 \in [a, b]$ such that $\lim_{x \to x_0} f_n(x) = a_n, n = 1, 2, 3, ...$

  then (i) $\langle a_n \rangle$ converges

  and (ii) $\lim_{x \to x_0} f(x) = \lim_{n \to \infty} a_n$

- The limit of the sum function of a series = the sum of the series of limit of functions.

- If $\langle f_n \rangle$ is a sequence of continuous function on an interval $[a, b]$ and if $f_n \to f$ uniformly on $[a, b]$ then f is continuous on $[a, b]$.

- If a series $\sum f_n$ converges uniformly to f in an interval $[a, b]$ and its term f_n are continuous at a point x_0 of the interval then the sum function f is also continuous at x_0.

- If the sum function of a series or limit function of a sequence of continuous terms is not continuous on an interval the convergence can not be uniform.

- *Dini's theorem :*

  (i) If a sequence of continuous function $\langle f_n \rangle$ defined on $[a, b]$ is monotonic increasing and converges pointwise to a continuous function f then the convergence is uniform on $[a, b]$.

  (ii) If the sum function of a series $\sum f_n$ with non-negative continuous terms defined on an interval $[a, b]$ is continuous on $[a, b]$ then the series is uniformly convergent on $[a, b]$.

- *Weierstrass Approximation theorem:* If f is a real valued continuous function defined on a closed interval $[a, b]$ then there exists a sequence of real polynomials $\langle P_n \rangle$ which converges uniformly to $f(x)$ on $[a, b]$ i.e., $\lim_{x \to 0} P_n(x) = f(x)$ converges uniformly on $[a, b]$.

- For any interval $[-a, a]$ there is a sequence of real polynomials P_n such that $P_n(0) = 0$ and that uniformly on $[-a, a]$.

Difference between Pointwise Convergence and Uniform Convergence :

1. Pointwise convergence (or simple convergence) is a local property (to be satisfied at a point) whereas uniform convergence is essential associated with an interval, *i.e.*, uniform convergence is a global property.

2. The interval of uniform convergence is always closed, or the functions are bounded above, *i.e.*, $a < x < b$. This is a convenient, though not a necessary assumption.

3. In case of convergence, a positive integer m depend on both ε and x does not remain bounded, but in case of uniform convergence, m solely dependent on ε but independent on x remains bounded.

4. Convergence at each point of a closed interval does not mean uniform convergence in that interval.

5. Each term of the series is the function of x for uniform convergence but it is not necessary for convergence.

6. Uniform convergence is a far reaching *sufficient* condition for the validity of interchanging certain limits, but it does not provide the complete answer of the general question of whether we can reverse the order of two limit processes. That is to say, the conditions of uniform convergence is merely *sufficient* for the truth of theorems regarding continuity of a sum, term by term integration, and termwise differentiation, but it is by no means a necessary condition (Bromwich).

REVIEW QUESTIONS AND PROJECT WORK

1. Show that the sequences $\langle nx(1-x^2)^n \rangle$ and $\langle n^2x(1-x^2)^n \rangle$ are not uniformly convergent on $[0, 1]$.

2. Show that the sequence $\langle f_n \rangle$ where $f_n(x) = \dfrac{x}{n+x}$ is uniformly convergent in $[0, k]$, $k < \infty$ but only pointwise convergent when the interval extends to ∞.

3. Show that the series
$$\frac{x}{1+x} + \frac{x}{(1+x)(1+2x)} + \frac{x}{(1+2x)(1+3x)} + \dots$$
is uniformly convergent on $[a, b]$ $a > 0$ but only pointwise in $[0, b]$.

4. Show that the series $\displaystyle\sum \frac{x}{n^p + x^2 n^q}$ converges uniformly over any finite interval $[a, b]$ for

(i) $p > 1, q \geq 0$ (ii) $0 < p \leq 1, p + q > 2$

5. Show that the series $\displaystyle\sum (-1)^n \frac{x^2 + n}{n^2}$

converges uniformly in every bounded interval but does not converge absolutely for any value of x.

6. Show that $\displaystyle\sum \frac{\log n}{n^x}$ converges uniformly for all real $x > 1 + \alpha > 1$.

7. Show that $\displaystyle\sum \frac{1}{n^x}$ converges uniformly for all real $x > 1$.

8. Show that the series
$e^x + e^{2x} + e^{3x} + \dots, |x| \leq \dfrac{1}{4}$ converges uniformly.

9. Show that the sequence $\left\langle (\sin x)^{1/n} \right\rangle$ converges but not uniformly on $[0, \pi]$.

10. If f is continuous on $[0, 1]$ and if $\int_0^1 x^n f(x)dx = 0$ for $n = 0, 1, 2\dots$ then show that $f(x) = 0$ on $[0, 1]$.

OBJECTIVE TYPE QUESTIONS

FILL IN THE BLANKS

1. The term of a sequence need not be _____ .

2. Every convergent sequence in a _____ cauchy.

3. The sequence $\langle f_n \rangle$ where $f_n = nx(1-x)^n$ _____ converges uniformly.

4. The series which satisfy M_n-test is called _____ convergent.

TRUE/ FALSE

Write 'T' for true and F for False Statements

1. A point such that the sequence does not converges uniformly in any nbd of it, however small is said to be a point of non-uniform convergence.   **(T/F)**

2. Uniform convergence of a sequence $\langle f_n(x) \rangle$ is a property associated with a domain of x comprising more than one point whereas simple convergence is meaningful at each point of that domain.   **(T/F)**

3. The series which satisfy M_n-test is called normally convergent.   **(T/F)**

MULTIPLE CHOICE QUESTIONS

Choose the most appropriate one.

Problem Set- 1

1. A sequence $\langle f_n \rangle$ of real valued functions is said to be uniformly bounded if :

(a) $f_n(x) < M$ (b) $|f_n(x)| \leq M$
(c) $f_n(x) \geq M$ (d) none of the above

2. If a sequence $\langle f_n \rangle$ is convergent then limit function f of $< f_n (x) >$ is called

(a) sum function
(b) product function

(c) convergent function

(d) none of the above

3. The convergence of the sequence ensures it:
   (a) uniform convergence
   (b) not necessarily uniform convergence
   (c) pointwise convergence
   (c) both (a) and (b) are true

4. A point such that the sequence does not converge uniformly in any neighbourhood is called :
   (a) limit point
   (b) peak point
   (c) point of non-uniform convergence
   (d) none of the above

5. If $\qquad \lim_{n\to\infty} f_n(x) = f(x) \forall\ x \in X$ and
   $M_n = \sup\{|f_n(x) - f(x)| : x \in X\}$, then $\langle f_n \rangle$
   converges uniformly to f if and only if
   (a) $M_n = 0\ \forall\ n$
   (b) $M_n \to 0$ as $n \to 0$
   (c) $M_n \to 0$ as $n \to \infty$
   (d) none of the above

6. A series $\sum u_n(x)$ of functions will converge uniformly on x if there exists a convergent series ΣM_n of positive constant such that:
   (a) $u_n(x) = M_n$
   (b) $|u_n(x)| \geq M_n$
   (c) $|u_n(x)| \leq M_n$
   (d) none of the above

7. If the given series $\sum u_n(x)$ converges uniformly, then :
   (a) f is integrable
   (b) $\int_a^b \left[\sum_{n=1}^{\infty} u_n(x) \right] dx = \sum_{n=1}^{\infty} \int_a^b u_n(x) dx$
   (c) both (a) and (b) are true
   (d) none of the above

8. The condition of uniform convergence of the series $\sum u_n(x)$ for the validity of term by term integration is:
   (a) only necessary condition
   (b) only sufficient condition
   (c) necessary and sufficient both
   (d) none of the above

Problem Set- 2

1. The function defined by $u(x) = \lim_{n\to\infty} u_n(x)$ is called :
   (a) sum function   (b) limit function
   (c) both (a) and (b)   (d) none of these

2. If $s_n(x) = u_1(x) + u_2(x) + ... + u_n(x)$ and $s(x) = \lim_{n\to\infty} s_n(x)$ then infinite series $\sum_{n=1}^{\infty} u_n(x)$ is said to be convergent on a set F if the sequence $< s_n(x) >$ is:
   (a) convergent on F
   (b) bounded on F
   (c) divergent on F
   (d) none of the above

3. If $R_n(x) = |s(x) - s_n(x)|$ then it is called:
   (a) limit function
   (b) sum function
   (c) partial remainder
   (d) none of the above

4. A sequence of functions $<u_n>$ is said to converge uniformly to u on the set F if for each $\varepsilon > 0\ \exists$ a positive integer $m\ (\varepsilon)$ such that $n \geq m \Rightarrow$

   (a) $u_n(x) - u(x) < \varepsilon$
   (b) $|u_n(x) - u(x)| < \varepsilon$
   (c) $|u_n(x) - u(x)| > \varepsilon$
   (d) none of the above

5. If $u_n \to u$ uniformly and each u_n is bounded on F then $\langle u_n \rangle$ is said to be :
   (a) uniformly bounded
   (b) bounded
   (c) uniformly unbounded
   (d) none of the above

6. If $u_n \to u$ uniformly on a function f. If each u_n is continuous at a point x_0 of f then limit function u is:
   (a) continuous at x_0
   (b) uniformly continuous
   (c) discontinuous at x_0
   (d) none of the above

7. The series
   $$\frac{x}{x+1} + \frac{x}{(x+1)(2x+1)} + \frac{x}{(2x+1)(3x+1)} + ...$$
   (a) converges for each x in $0 \leq x \leq 1$
   (b) converges uniformly to 1
   (c) both (a) and (b) are true
   (d) none of the above

8. The series

$$s(x) = \frac{x^2}{(1+x)} + \left(\frac{2x^2}{1+2x} - \frac{x^2}{1+x} \right)$$

$$+ \ldots + \left(\frac{nx^2}{1+nx} - \frac{(n-1)x^2}{1+(n-1)x} \right) + \ldots$$

(a) converges
(b) converges uniformly to 0
(c) converges uniformly to 1
(d) none of the above

9. The series

$$\frac{1}{x+1} - \frac{1}{(x+1)(x+2)} + \ldots + \frac{1}{(x+n-1)(x+n)} \ldots$$

for which $s_n(x) = \frac{1}{x+n}$

(a) convergent at 0
(b) convergent at 1
(c) uniformly convergent in the interval (0,1]
(d) all are true

10. The series

$$s_n(x) = 1 + x + x^2 + \ldots + x^{n-1} \text{ is}$$

(a) converges pointwise in] 0,1 [
(b) does not converges uniformly on]0, 1[
(c) converges uniformly for $-\frac{1}{2} \le x \le \frac{1}{2}$
(d) all are true

11. The sequence for which

$$s_n(x) = \frac{x}{1+nx}, 0 \le x \le \infty$$

(a) converges uniformly on] 0, ∞ [
(b) converges uniformly on [0, ∞ [
(c) does not converges uniformly
(d) none of the above

12. The point of non-uniform convergence of the sequence $\langle s_n (x) \rangle$ where $s_n(x) = 1 - (1-x^2)^2$ is given by:

(a) $x = \infty$　　　　(b) $x = 0$
(c) $x = 1$　　　　(d) $x = n$

13. The sequence $\langle s_n (x) \rangle$ for which $s_n(x) = \frac{n}{x+n}$ for $x \ge n$ is:

(a) non-uniformly convergent on [0, ∞ [
(b) uniformly convergent on (0, λ]
(c) both (a) and (b) are true
(d) none of the above

14. The exponential series:

$$1 + x + \frac{x^2}{2!} + \frac{x^3}{3!} + \ldots \text{ is :}$$

(a) not converges uniformly
(b) converges uniformly for a particular subset of R.
(c) converges uniformly on every subset of R
(d) none of the above

15. The point of non-uniform convergence of the sequence $\langle s_n(x) \rangle$ where $s_n(x) = \frac{nx}{1+n^2x^2}$ is :

(a) $x = 0$　　　　(b) $x = \infty$
(c) $x = 1$　　　　(d) $x = n$

16. The point of non-uniform convergence of the sequence $\langle s_n(x) \rangle$ where $s_n(x) = nx \exp.(-nx^2)$ is given by :

(a) $x = 0$　　　　(b) $x = 1$
(c) $x = \infty$　　　　(d) $x = n$

17. The series $\sum\limits_{n=0}^{\infty} x(1-x)^n$ is :

(a) convergent on [0, 2]
(b) not uniformly convergent for $0 \le x \le p \le 2$
(c) both (a) and (b) are true
(d) none of the above

18. The sequence $\langle s_n(x) \rangle$ where $s_n(x) = \frac{nx}{1+n^2x^2}$ for $x \ge 0$

(a) not uniformly convergent
(b) uniformly convergent for $x \ge n$
(c) both (a) and (b) are true
(d) none of the above

19. The series $\sum \left(\frac{2n^2x^2}{e^{n^2x^2}} - \frac{2(n-1)^2x^2}{e^{(n-1)^2} \cdot x^2} \right)$ is :

(a) uniformly convergent on any interval containing $x = 0$
(b) not uniformly convergent on any interval containing $x = 0$
(c) both (a) and (b) are true
(d) none of the above

20. The sequence $\langle s_n(x) \rangle$ where $s_n(x) = nx (1-x)^n$:

(a) converges uniformly on [0, 1]
(b) does not converges uniformly on [0, 1]
(c) $x=0$ is not a point of non-uniform convergence
(d) none of the above

21. Let $f_n(x) = \dfrac{x}{1+nx^2}, n \in N$ and x is a real number. Also $f = \lim_{n\to\infty} f_n(x)$ then $\langle f_n(x) \rangle$:

(a) converges uniformly
(b) converges uniformly to f
(c) both (a) and (b) are true
(d) none of the above

22. The series $u_1(x) = x$ and $u_n(x) = x^{\frac{1}{2n-1}} x^{\frac{1}{2x-3}}$ is :

(a) convergent near $x = 0$
(b) uniformly convergent near $x = 0$
(c) non-uniformly convergent near $x = 0$
(d) none of the above

23. For a sequence $\langle s_n(x) \rangle$ where $s_n(x) = nx\,e^{-nx^2} \,\forall\, x \in R$ the point of non-uniform convergence is:

(a) $x = 0$ (b) $x = 1$
(c) $x = n$ (d) none of the above

24. The sequence $\langle s_n(x) \rangle$ where $s_n(x) = x^{n-1}(1-x)$:

(a) converges uniformly on $(0, 1)$
(b) does not converges uniformly
(c) both (a) and (b) are true
(d) none of the above

25. The point of non-uniform convergence of the series $\sum_{n=0}^{\infty} x\,e^{-nx}$ in the interval $[0, 1)$ is:

(a) $x = 0$ (b) $x = 1$
(c) $x = 2$ (d) none of the above

26. The series $\sum 3^n \sin\dfrac{1}{4^n x}$:

(a) converges uniformly on $] a, \infty [, a > 0$
(b) converges absolutely on $] a, \infty [, a > 0$
(c) both (a) and (b) are true
(d) none of the above

27. The series $\sum \dfrac{x}{(n+x^2)^2}$ is:

(a) uniformly convergent on R
(b) not uniformly convergent on R
(c) not convergent for a particular $x \in R$
(d) none of the above

28. The series $\sum_{n=1}^{\infty} \dfrac{\sin(x+nx)}{n(n+1)}$ is :

(a) uniformly convergent for all real x
(b) not uniformly convergent for all real x
(c) not convergent for a particular value of $x \in R$
(d) none of the above

29. The series $\sum \dfrac{\cos nx}{n^2}$:

(a) converges uniformly on a finite interval $[a, b]$
(b) not converges uniformly on a finite interval $[a, b]$
(c) not convergent for a particular value of $x \in R$
(d) none of the above

30. The series $\sum \dfrac{2}{(n+x)^2}$ is:

(a) converges uniformly
(b) not converges uniformly
(c) $x = 0$ is a point of non-uniform convergence
(d) none of the above

31. The series is $\sum \dfrac{1}{1+n^2x}$ is

(a) converges uniformly on $(1, \infty[$
(b) not converges uniformly
(c) $x = 0$ is a point of non-uniform convergence
(d) none of the above

32. The series $\sum \dfrac{x}{n(1+n^2x)}$:

(a) not converges uniformly
(b) converges uniformly $\forall\, x$,
(c) $x = 0$ is a point of non-uniform convergence
(d) none of the above

33. The series $\sum_{n=1}^{\infty} \{u_n(x)\}^2$ is :

(a) converges absolutely
(b) converges uniformly on $[a, b]$
(c) both (a) and (b) are true
(d) none of the above

34. The series of real functions $\sum_{n=1}^{\infty} \dfrac{(n+1)^3}{n^5}\left(\dfrac{x}{3}\right)^n$ is :

(a) not uniformly convergent
(b) uniformly convergent on $[-3, 3]$
(c) $x = 3$ is a point of non-uniform convergence
(d) none of the above

35. The series $\sum \left\{ \dfrac{2n^2x^2}{e^{n^2x^2}} - \dfrac{2^{(n-1)^2}.x^2}{3^{(n-1)^2}.x^2} \right\}$ is :

(a) converges uniformly
(b) not converges uniformly
(c) converges at $x = 0$
(d) none of the above

36. The series $\sum\limits_{n=1}^{\infty} \dfrac{1}{1+n^2x}$ converges uniformly in:

(a) $[1, \infty[$ (b) $]-\infty, \infty[$
(c) $[0, 1]$ (d) none of the above

37. The series $\sum \dfrac{x}{(n+x^2)^2}$ is:

(a) converges uniformly for all x
(b) not-converges uniformly
(c) $x = 0$ is a point of non-uniform convergence
(d) none of the above

38. The series $x^2 + \dfrac{x^2}{1+x^2} + \dfrac{x^2}{(1+x^2)^2} + ...$
$+ \dfrac{x^2}{(1+x^2)^n} + ...$ is:

(a) non-Uniformly convergent series on any interval containing $x = 0$
(b) absolutely convergent series on any interval containing $x = 0$
(c) both (a) and (b) are true
(d) none of the above

39. The series $\sum\limits_{n=1}^{\infty} \dfrac{x}{\{1+(n-1)x\}\{1+nx\}}$ is:

(a) uniformly convergent
(b) not uniformly convergent
(c) may or may not be convergent
(d) none of the above

40. The series for which $s_n(x) = \dfrac{1}{1+nx} (0 \le x \le 1)$ and $u_1(x) = \dfrac{1}{1+x}$ is:

(a) uniformly convergent on any interval containing $x=0$
(b) not uniformly convergent on any interval containing $x=0$
(c) converges absolutely
(d) none of the above

41. The series is $\sum\limits_{n=0}^{\infty} xe^{-nx}$ is:

(a) non-uniformly convergent series in the interval includes 0.
(b) Its sum function has a discontinuous limit in the *nbd* of $x=0$
(c) both (a) and (b) are true
(d) none of the above

42. For the series $u_n(x)$ for which $s_n(x) = \dfrac{1}{2n^2}\log(n^4x^2+1)$ the differeniated series $u_n'(x)$.

(a) does not converges uniformly on any finite interval
(b) term by term differentiation of $u_n(x)$ is possible
(c) both (a) and (b) are true
(d) none of the above

43. For the series $u_n(x)$ for which $s_n(x) = \log(n^3x^2+1)$, differentiated series $u_n'(x)$

(a) does not converges uniformly on any finite interval
(b) term by term differentiation of $u_n(x)$ is possible
(c) both (a) and (b) are true
(d) none of the above

44. The series for which $s_n(x) = \dfrac{nx}{1+n^2x^2}$ $0 \le x \le 1$

(a) can be differential termwise at $x = 0$
(b) cannot be differentialed termwise at $x = 0$
(c) both (a) and (b) are true
(d) none of the above

45. For the series such that $s(x) = \sum\limits_{n=1}^{\infty} \dfrac{1}{n^3+n^4x^2}$:

(a) series converges uniformly for each x
(b) term by term differentiation can be possible
(c) both (a) and (b) are true
(d) none of the above

46. The series $x + \dfrac{1}{2}\sin 2x + \dfrac{1}{2}\sin 3x + ...$

(a) converges uniformly $0 < a \le x \le b \le 2\pi$
(b) does not converges uniformly
(c) pointwise convergent
(d) none of the above

47. The series $\sum_{n=1}^{\infty} (-1)^{n-1} x^n$:

(a) converges uniformly in $0 \le x \le \lambda < 1$

(b) does not converges uniformly

(c) does not converges pointwise

(d) none of the above

48. The series $\sum \frac{x^n}{n+1}$ in the interval $(-\delta, \delta)$, where δ is a fixed positive number less than unity is:

(a) uniformly convergent on $(-\delta, \delta)$

(b) does not converges uniformly

(c) both (a) and (b) are true

(d) none of the above

49. The series $\sum \frac{(-1)^{n-1}}{n+x^2}$ is:

(a) converges uniformly for all real x

(b) does not converges uniformly

(c) both (a) and (b) are true

(d) none of the above

50. If Σu_n is a convergent series of positive constant then series $\Sigma a_n x^n$

(a) converges uniformly on [0, 1]

(b) does not converges uniformly

(c) converges uniformly for R

(d) none of the above

51. The series $\sum \frac{(-1)^n . x^n}{n}$ is:

(a) converges uniformly on [0, 1]

(b) converges uniformly on R

(c) does not converges uniformly

(d) none of the above

Answers

FILL IN THE BLANKS

1. distinct   **2.** not necessarily   **3.** does not   **4.** normally

TRUE/FALSE

1. T   **2.** T   **3.** T

MULTIPLE CHOICE QUESTIONS

Problem Set-1

1. (b) **2.** (a) **3.** (d) **4.** (c) **5.** (c) **6.** (c) **7.** (c) **8.** (b)

Problem Set-2

1. (c) **2.** (a) **3.** (c) **4.** (b) **5.** (a) **6.** (a) **7.** (c) **8.** (b) **9.** (d)
10. (d) **11.** (b) **12.** (b) **13.** (c) **14.** (c) **15.** (a) **16.** (a) **17.** (c) **18.** (c)
19. (b) **20.** (b) **21.** (c) **22.** (c) **23.** (a) **24.** (a) **25.** (a) **26.** (c) **27.** (a)
28. (a) **29.** (a) **30.** (a) **31.** (a) **32.** (b) **33.** (c) **34.** (b) **35.** (b) **36.** (a)
37. (a) **38.** (c) **39.** (b) **40.** (b) **41.** (b) **42.** (c) **43.** (c) **44.** (b) **45.** (c)
46. (a) **47.** (a) **48.** (a) **49.** (a) **50.** (a) **51.** (a)

HINTS TO SELECTED PROBLEMS

Problem Set-2

7. The series can be put in the more convenient from as

$$\left(1 - \frac{1}{x+1}\right) + \left(\frac{1}{x+1} - \frac{1}{2x+1}\right) + \dots$$

$$+ \left(\frac{1}{(n-1)x+1} - \frac{1}{nx+1}\right) + \dots$$

This implies that the sum $s_n(x)$ of n terms is

$$s_n(x) = 1 - \frac{1}{nx+1} = \frac{nx}{nx+1}$$

Hence

$$\lim_{n \to \infty} s_n(x) \equiv s(x) = \lim_{n \to \infty} \frac{nx}{nx+1} = 1, \text{if } x \neq 0$$

Therefore, if $x \neq 0$,

$$R_n(x) = s(x) - s_n(x) = \frac{1}{nx+1}$$

Now we wish to determine the magnitude of m, corresponding to a prescribed ε, such that for all $\qquad n \geq m$

$$|R_n(x)| < \varepsilon.$$

In fact, $\qquad R_n(x) = \frac{1}{nx+1}$

The condition $|R_n(x)| < \varepsilon$ becomes

$$R_n(x) = \frac{1}{nx+1} < \varepsilon$$

8. Obviously $s_n(x) = \frac{nx^2}{1+nx}$

Then $\qquad n > \frac{1}{x}\left(\frac{1}{\varepsilon} - 1\right) \qquad \dots(2)$

from which it is clear that m depends not only on ε, but also on the magnitude of x. For clarity of ideas, take $\varepsilon = 0.01$. Then (2) gives

$n > 99 / x$ and for $x = \frac{1}{2}$, m can be chosen to be 199. If $\varepsilon = 0.01$

but for $x = \frac{1}{4}$, m must be atleast $4 \times 99 + 1$

$= 397$. Clearly, if ε is fixed and x is allowed to assume values nearer and nearer to zero, then the value of m must be chosen larger and larger.

We see that for the series, m does not remain bounded (ε fixed) when x is allowed to assume various values in the interval $[0, 1]$. The series (1) converges for each x in $0 \leq x < 1$, but it is impossible to find a single m which will satisfy the inequality (2) uniformly well, that is for each x in $]0, 1]$. This implies that the series (1) is not uniformly convergent in $[0, 1]$.

9. $s(x) = \lim_{n \to \infty} s_n(x) = 0$

The condition of convergence provides :

$$|s_n(x) - s(x)| < \varepsilon$$

for each $n > m\ (\varepsilon, x)$,

i.e. $\qquad \left|\frac{1}{x+n} - 0\right| < \varepsilon$

Solving this inequality for n, we find $n > \frac{1}{\varepsilon} - x$.

This implies that $m(\varepsilon, x)$ is the integral part in $\left\{\frac{1}{\varepsilon} - x\right\}$. Thus if we set $m - 1$ as the greatest integer in $1/\varepsilon - x$, the condition of convergence holds for given $\varepsilon > 0$ whenever $n > m$ and clearly m is a function of ε and x. It is true that $1/\varepsilon > 1/\varepsilon - x$ since by hypothesis x is positive. Therefore if we set $m(\varepsilon) = 1/\varepsilon$, which is independent of x, and corresponding to given $\varepsilon > 0$, it is finite. That is, m remains bounded. Hence we have

$$\left|\frac{1}{x+n} - 0\right| < \varepsilon$$

for each $n > 1/\varepsilon = m(\varepsilon)$ independent of x, but dependent on ε. That is to say, the convergence is uniform by definition.

10. The nth partial sum

$$s_n(x) = \frac{1-x^n}{1-x}$$

Hence, $\qquad s(x) = \frac{1}{1-x}$ and

$$R_n(x) = \frac{x^n}{1-x} \text{ where } x \in]0,1[$$

The condition of uniform convergence

$$|R_n(x)| = |s_n(x) - s(x)| < \varepsilon$$

gives : $|x^n| < \varepsilon(1-x)$ for $x \in \,]\,0, 1[$.

Taking logarithm and solving for n, we have

$$n > \frac{\log \varepsilon(1-x)}{\log|x|} \qquad \ldots(1)$$

Clearly, the choice of m depends on both x and ε implies that the sequence $<s_n(x)>$ converges pointwise in $]\,0, 1[$. Now it is possible to choose an m that will serve for all values of $x \in \left[-\frac{1}{2}, \frac{1}{2}\right]$ Given a small $\varepsilon > 0$, the ratio $\log \varepsilon(1-x)/\log x$ assumes

its maximum value when $x = \pm\frac{1}{2}$. Then if m is

chosen so that $m > \dfrac{\log \varepsilon/2}{\log \dfrac{1}{2}} = 1 - \dfrac{\log \varepsilon}{\log 2}$, the

inequality $n > \dfrac{\log \varepsilon(1-x)}{\log|x|}$ will be satisfied

for all $n \geq m$. In view of uniform convergence, the series Σx^n converges uniformly for $-\frac{1}{2} \leq x \geq \frac{1}{2}$. However, the series does not converge uniformly in $[0,1]$, because in this interval the ratio in (1) will increase indefinitely as $x \to \pm 1$.

11. Clearly $0 \leq x < \infty \Rightarrow 0 \leq s(x) \leq \dfrac{1}{n}$. Hence, for $n > 0$ the statement

$$|s_n(x) - s(x)| = |\,s_n(x) - 0\,| < \varepsilon, n \geq m \text{ true}$$

for all x in $[0, \infty\,[$ simultaneously, provided only that $n_0 > 1/\varepsilon$, It is clear that

$$|s_n(x) - 0| \leq \frac{1}{n} \leq \frac{1}{m} < \varepsilon$$

for all x in $[0, \infty\,[$. Therefore for this sequence $< s_n(x) >$, an m can be determined such that the inequality $|s_n(x) - 0| < \varepsilon, n \geq m$ holds for all x in $[0, \infty\,[$. This m depends on ε and not on x. This shows that the sequence $<s_n(x)>$ is uniformly convergent on $[0, \infty\,[$.

12. Infact, $s(x) = 0$ at $x = 0$, and 1 at x in $0 < |x| < \sqrt{2}$

Now, $|s_n(x) - s(x)| = (1 - x^2)^n$ for all x in $0 < |x| < \sqrt{2}$

But $\lim_{x \to 0}(1 - x^2)^n = 1$. Hence the inequality $|s_n(x) - s(x)| < \varepsilon$ cannot hold good in any neighbourhood of $x = 0$. Therefore, the sequence is non-uniformly convergent in any neighbourhood of $x = 0$.

13. The sequence $< s_n(x) >$ is pointwise convergent for each $x \geq 0$ and the pointwise limit s is given by $s(x) = 1$ for each $x \geq 0$. Suppose $\varepsilon > 0$ is given.

Then $|s_n(x) - s(x)| = \left|\dfrac{n}{x+n} - 1\right| = \dfrac{x}{x+n} < \varepsilon$

if $\qquad n > x(1/\varepsilon - x)$,

Define $m(\varepsilon, x)$ as an integer just greater than $x(1/\varepsilon - x)$. Then $m(\varepsilon, x)$ increases as x increases and tends to infinity as $x \to \infty$. Hence it is not possible to choose any number m such that $n \geq m \Rightarrow |s_n(x) - s(x)| < \varepsilon$ for all $x \geq 0$. Therefore convergence is not uniform on $[0, \infty\,[$. Now we consider the interval, $[0, \lambda]$, Assume λ as an integer greater than $\lambda(1/\varepsilon - \lambda)$. Then each $n \geq m$ and each x in $[0, \lambda]$

$\Rightarrow \quad |s_n(x) - s(x)| < \varepsilon.$

Thus convergence is uniform on $[0, \lambda]$.

14. We have $\qquad u_n(x) = \dfrac{x^n}{n!}$

Now for each $\lambda > 0$ and $\varepsilon > 0$ there exists $m \in N$ such that for any $p \in N$ and $x \in [-\lambda, \lambda]$.

$$\left|\frac{x^n}{n!}\right| \leq \left|\frac{\lambda^n}{n!}\right| < \frac{\varepsilon}{p} \text{ for each } n \geq m.$$

Therefore for $\forall\, n \geq m$ and $p \in N$,

$|R_n(x)|$

$$= \left|\frac{x^{n+1}}{(n+1)!} + \ldots + \frac{x^{n+p}}{(n+p)!}\right| \leq \left|\frac{x^{n+1}}{(n+1)!}\right| + \ldots +$$

$$\left|\frac{x^{n+p}}{(n+p)!}\right| < \frac{\varepsilon}{p}.p = \varepsilon$$

Hence the given series converges uniformly on $[-\lambda, \lambda]$ for any $\lambda > 0$. Therefore the series converges uniformly on every bounded subset of R, However, it does not converge uniformly on R, for each $m \in N$ there is an $x \in R$

such that $\dfrac{x^m}{m!} > 1$

and hence for each $\varepsilon > 0$, $1 < \dfrac{x^m}{m!} \nleq \varepsilon$

15. Assume $\varepsilon > 0$ is given. Then

$$|s_n(x) - s(x)| < \varepsilon \Rightarrow \left|\frac{nx}{1+n^2x^2} - 0\right| < \varepsilon$$

$$\Rightarrow \frac{n|x|}{1+n^2x^2} < \varepsilon \Rightarrow n|x| < \varepsilon + n^2x^2\varepsilon$$

$$\Rightarrow n^2x^2\varepsilon - n|x| + \varepsilon > 0$$

$$\Rightarrow n > \frac{|x| + \sqrt{(x^2 - 4x^2\varepsilon^2)}}{2|x|\varepsilon}$$

Now $x \to 0 \Rightarrow n \to \infty$. Hence it is not possible to choose m such that $|s_n(x) - s(x)| < \varepsilon$ for each $n \geq m$ and each x in R.

17. We have

$$u_n(x) = x(1-x)^n$$

$$\Rightarrow \quad s_n(x) = x[1 + (1-x) + (1-x)^2 + ...]$$

$$= \frac{x[1 - (1-x)^n]}{1 - (1-x)} = 1 - (1-x)^n$$

$$\Rightarrow \quad s(x) = 0 \text{ if } x = 0 \text{ and } s(x) = 1 \text{ if}$$

$-1 < 1 - x < 1$ *i.e.*, if $0 \leq x < 2$

$\Rightarrow \sum u_n$ converges on $[0, 2[$, which is the interval of convergence of the series.
Now

$$|R_n(x)| = |s_n(x) - s(x)| = |1 - (1-x)^n - 1|$$

$$= \frac{1}{(1-x)^n} > \frac{1}{\varepsilon} \Rightarrow n > \frac{\log(1/\varepsilon)}{\log 1/(1-x)}$$

Define m (ε, x) an integer just greater

$$\frac{\log(1/\varepsilon)}{\log 1/(1-x)} \Rightarrow m = m(\varepsilon, x) \Rightarrow \sum u_n(x)$$

is not uniformly convergent on $[0, 2[$.

18. Our hypothesis reads that

$$s_n(x) = \frac{nx}{1+n^2x^2}$$

$$\Rightarrow \quad s_n(x) = \lim_{n \to \infty} s_n(x) = 0 \forall x$$

Now assume

$$M_n = \max|s_n(x) - s(x)| = \frac{n|x|}{1+n^2x^2}$$

$$\Rightarrow \quad \frac{dM_n}{dx} = 0 \text{ if } (1+n^2x^2)n - 2n^2xnx = 0,$$

or if $\quad n^2x^2 = n$, or if $\quad x = \pm\dfrac{1}{n}$

Also $\dfrac{d^2M_n}{dx^2}$ is positive for $x < \dfrac{1}{n}$ and negative for $x > \dfrac{1}{n}$, and zero for $x = \dfrac{1}{n} \Rightarrow x = \dfrac{1}{n}$ is a local maximum so that

$$M_n = \frac{nx}{1+n^2x^2}\bigg|_{x=1/n} = \frac{1}{2} \neq 0$$

Hence M_n-test asserts that the given sequence is not uniformly convergent on any interval containing $x = 0$.

For $x \geq \lambda$, the value of M_n is always zero as $n \to \infty$

This proves that $<s_n(x)>$ is uniformly convergent for $x \geq \lambda$ by M_n-test.

19. Clearly,

$$s_n(x) = u_1(x) + u_2(x) + ... + u_n(x)$$

$$= 2n^2x^2e^{-n^2x^2}$$

$\Rightarrow s(x) = 0$ by L' Hospital rule, for

$$s(x) = \lim_{n \to \infty} \frac{2n^2x^2}{e^{n^2x^2}}$$

$$= \lim_{n \to \infty} \frac{4nx^2}{e^{n^2x^2}.2nx^2}$$

$$= \lim_{n \to \infty} \frac{2}{e^{n^2x^2}} = 0$$

Also, $s_n(0) = 0$. Hence $s(x) = 0$ for all x.
Set $M_n(x) = \max|s_n(x) - s(x)|$

$$\Rightarrow \quad M_n(x) = \frac{2n^2x^2}{e^{n^2x^2}}$$

$$\Rightarrow \quad \frac{dM_n}{dx} = 0 \text{ at local maximum}$$

$$x = \frac{1}{n} \Rightarrow M_n = \frac{2}{e} \neq 0$$

⇒ the series is not uniformly convergent on any interval containing $x = 0$ in view of M_n-test of uniform convergence.

24. By hypothesis,

$$s_n(x) = x^{n-1}(1-x) \Rightarrow s(x) = 0 \,\forall\, x \in [0,1]$$

Set $\quad y = |s_n(x) - s(x)| = x^{n-1}(1-x)$

$$\Rightarrow \quad \frac{dy}{dx} = (n-1)x^{n-2}(1-x) - x^{n-1},$$

and $\dfrac{dy}{dx} = 0$ produces.

$$x^{n-2}\{(n-1)(1-x) - x\} = 0$$

$$\Rightarrow \qquad x^{n-2}(n-1-xn) = 0$$

$$\Rightarrow \quad x = 0 \text{ or } \frac{n-1}{n}$$

Also $\dfrac{d^2y}{dx^2}$ is negative if $x = \dfrac{n-1}{n}$

and $M_n = \max y$

$$= \left(1 - \frac{1}{n}\right)^{n-1}\left(1 - \frac{n-1}{n}\right)$$

$$= \left(1 - \frac{1}{n}\right)^{n}\left(1 - \frac{1}{n}\right)^{-1}\frac{1}{n}$$

$$= e^{-1}(1)0 = 0 \text{ as } n \to \infty$$

Hence, M_n-test proves that $<s_n(x)>$ converge uniformly on $[0, 1]$.

25. We have

$$s_n(x) = \sum_{n=0}^{\infty} xe^{-nx} = \frac{x(1-e^{-nx})}{1-e^{-x}}$$

$$= \frac{xe^x}{e^x - 1}\left(1 - \frac{1}{e^{nx}}\right)$$

$$\Rightarrow \quad s(x) = \lim_{n\to\infty} s_n(x)$$

$$= \begin{cases} 0 & , \text{ if } x = 0 \\ \dfrac{xe^x}{e^x - 1}, & \text{ if } 0 < x < 1 \end{cases}$$

Now, $M_n = \max |s_n(x) - s(x)| \,\forall\, x \in [0,1]$

$$= \max\left|\frac{xe^x}{(e^x - 1)e^{nx}} : x \in [0,1]\right| \geq 1 \Big/ \frac{ne^{1/n}}{(e^{1/n} - 1)e}$$

on setting $\quad x = \dfrac{1}{n} \in [0,1]$

$$= e^{-1}\frac{\dfrac{1}{n}e^{1/n}\left(-\dfrac{1}{n^2}\right) + \left(-\dfrac{1}{n^2}\right)e^{1/n}}{\left(-\dfrac{1}{n^2}\right)e^{1/n}}$$

when $n \to \infty$.

$$= e^{-1}\left(\frac{1}{n} + 1\right) \to e^{-1} \text{ is the limit.}$$

Now $M_n \neq 0 \Rightarrow$ by M_n-test, the series in not uniformly convergent on any interval containing $x = 0 \Rightarrow x = 0$ is a point of non-uniform convergence of the given series.

26. For any $x : 0 < x \in [a, \infty]$ there exists $m \in N$ such that $4^n x \geq 1 \,\forall\, n \in N$. We observe that the series after a finite number of terms consists of positive terms.

Now $\lim\left|\dfrac{u_{n+1}}{u_n}\right| \equiv 3\lim \dfrac{\sin\dfrac{1}{4^{n+1}x}}{\sin\dfrac{1}{4^n x}}$

$$= 3\lim \frac{\sin\dfrac{1}{4^{n+1}x}}{\dfrac{1}{4^{n+1}x}} \cdot \frac{1/4^n x}{\sin(1/4^n x)} \cdot \frac{1/4^{n+1}x}{1/4^n x}$$

$$= \frac{3}{4} < 1$$

⇒ the series converges absolutely on $]a, \infty[$ for $a > 0$.

For $n \geq m, \sin\dfrac{1}{4^n x} < \dfrac{1}{4^n x} < \dfrac{1}{4^{n-m}}$

$$\Rightarrow \quad |3^n \sin(1/4^n x)| \leq 4^m(3/4)n \,\forall\, n > m$$

$$\Rightarrow \quad \sum 4^m(3/4)^n \text{ is a series of non-negative}$$
terms and is also convergent. Hence by Weierstrass M-test, the given series converges uniformly on $]a, \infty[$ where $a > 0$.

27. Here, $u_n(x) = \dfrac{x}{(n + x^2)^2}$,

which is maximum if

$$\frac{du_n}{dx} = 0 \Rightarrow (n + x^2)^2 - 4x^2(n + x^2) = 0$$

$$\Rightarrow \qquad (n + x^2)(n + x^2 - 4x^2) = 0$$

$$\Rightarrow \qquad x = \sqrt{(n/3)}$$

$$\Rightarrow \qquad u_n(x) \leq \sqrt{(n/3)}\Big/\left(n + \frac{n}{3}\right)^2$$

$$= \frac{3\sqrt{3}}{16n^{3/2}} < \frac{1}{n^{3/2}}.$$

But convergence of $\sum 1/n^{3/2}$ implies uniform

convergence $\sum x/(n+x^2)^2$ of M-test.

29. Here

$$u_n(x) = \frac{\cos nx}{n^2} \Rightarrow |u_n(x)| < \frac{1}{n^2}$$

convergence of $\sum \frac{1}{n^2} \Rightarrow$ implies uniform

convergence of $\quad \sum u_n(x)$ on $[a,b]$.

30. $u_n(x) = \frac{1}{(n+x)x^2} \Rightarrow |u_n(x)| < \frac{1}{n^2}$

convergence of $\quad \sum \frac{1}{n^2} \Rightarrow$ uniform

convergence of $\quad \sum \frac{1}{(n+x)^2}.$

31. $u_n(x) = \frac{1}{1+n^2 x} \Rightarrow |u_n(x)| \le \frac{1}{1+n^2}$

$$\forall x \in [1, \infty] \le \frac{1}{n^2}$$

Now $\sum \frac{1}{n^2}$ converges $\Rightarrow \sum \frac{1}{1+n^2 x}$ converges

uniformly by M-test.

32. $u_n(x) = \frac{x}{n(1+nx^2)}$ and it has a maximum or

a minimum if

$$\frac{d}{dx} u_n(x) = 0 \Rightarrow n(1+nx^2) - 2n^2 x^2 = 0$$

$$\Rightarrow x = \pm \frac{1}{\sqrt{n}} \Rightarrow u_n(x) \text{ has a local maximum at}$$

$$x = \frac{1}{\sqrt{n}} \text{ and hence}$$

$$\max u_n(x) = \frac{1/\sqrt{n}}{n(1+1)} = \frac{1}{2n^{3/2}}$$

Convergence of $\sum M_n = \frac{1}{2} \sum \frac{1}{n^{3/2}} \Rightarrow$ uniform

convergence $\sum u_n(x)$ by M-test.

33. Weierstrass's M-test gives:

$$|u_n(x)| \le M_n \forall x \in [a,b] \text{ and } n \in N.$$

$$\Rightarrow \quad |u_n^2(x)| \le M_n^2 \forall x \text{ in } [a, b]$$

But $\sum M_n$ is convergent

$$\Rightarrow \lim_{n \to \infty} \frac{M_{n+1}}{M_n} < 1 \Rightarrow \lim_{n \to \infty} \frac{M_{n+1}^2}{M_n^2} < 1$$

$\Rightarrow$ the series $\sum M_n^2$ is convergent

$\Rightarrow$ by M-test, $\sum u_n^2(x)$ converges absolutely

and uniformly on $[a, b]$

34. We have

$$|u_n(x)| = \left| \frac{(n+1)^3}{n^5} \left(\frac{x}{3} \right)^n \right| \le \frac{(n+1)^3}{n^5} = M_n$$

$$\forall n \in N \text{ and } \forall x \in [-3,3]$$

But $\quad \sum M_n = (n^2 + 3n^2 + 3n + 1)\frac{1}{n^5}$

$$= \frac{1}{n^2} + \frac{3}{n^3} + \frac{3}{n^4} + \frac{1}{n^5}$$

$$\Rightarrow \quad \sum M_n = \sum \frac{1}{n^2} + 3\sum \frac{1}{n^2} + 3\sum \frac{1}{n^4} + \sum \frac{1}{n^5}$$

Now $\sum M_n$ is a convergent series since each
term on the right converge.

By M-test, the given series converges
absolutely and uniformly.

35. We have

$$u_1(x) = \frac{2x^2}{e^{x^2}} - 1$$

$$u_2(x) = \frac{2.2^2 x^2}{e^{2x^2}} - \frac{e x^2}{e^{x^2}}$$

.. ..

... ...

$$u_n(x) = \frac{2n^2 x^2}{e^{n^2 x^2}} - \frac{2(n-1)^2 . x^2}{e^{(n-1)^2 x^2}}$$

On adding all these we get

$$f_n(x) = \Sigma u_i(x)$$

$$= \frac{2n^2x^2}{e^{n^2x^2}} - 1$$

$\therefore \qquad f(x) = \lim_{n\to\infty} f_n(x) = 1 \forall x$

Now $\qquad M_n = \sup\{\,|\,f_n(x) - f(x)\,|: x \in R\}$

$$= \sup\left\{\frac{2n^2x^2}{e^{n^2x^2}} : x \in R\right\}$$

$$= \frac{2n^2 \cdot \dfrac{1}{n^2}}{e^{n^2 \cdot \frac{1}{n^2}}}$$

(Let us take $x = \dfrac{1}{n}$)

36. Here , $\quad u_n(x) = \dfrac{1}{1 + n^2 x}$

$\Rightarrow \qquad |\,u_n(x)\,| \le \dfrac{1}{1 + n^2 . x} \,\forall\, x \in [1, \infty]$

$$< \frac{1}{n^2}$$

But $\displaystyle\sum_{n=1}^{\infty} \frac{1}{n^2}$ is known to be convergent.

Hence, by Weierstrass M-test the given series is uniformly convergent on $[1, \infty]$.

37. We have $u_n(x) = \dfrac{x}{(n + x^2)^2}$

For the maxima and minima of $u_n(x)$, we must have $\dfrac{d u_n(x)}{dx} = 0$

$\Rightarrow \qquad (n + x^2)^2 - 4x^2(n + x^2) = 0$

$\Rightarrow \qquad 3x^4 + 2nx^2 - n^2 = 0$

$\Rightarrow \qquad x^2 = \dfrac{n}{3} \Rightarrow x = \sqrt{\dfrac{n}{3}}$

Clearly at $x = \sqrt{\dfrac{n}{3}}, \dfrac{d^2 u_n(x)}{dx^2}$ is negative.

Therefore,

$$\text{maximum } [u_n(x)] = \frac{\sqrt{\dfrac{n}{3}}}{\left(n + \dfrac{n}{3}\right)^2} = \frac{3\sqrt{3}}{16 n^3 / 2} = M_n$$

$\Rightarrow \qquad |\,u_n(x)\,| \le M_n$

But $\sum M_n$ is convergent.

Hence, by Weirstrass M-test the given series is uniformly convergent for all x.

38. Assume $x \ne 0$. Then the series is a geometric one with ratio $\dfrac{1}{1 + x^2}$ and the limit function

$s(x)$ is $\dfrac{x^2}{\left(1 - \dfrac{1}{1 + x^2}\right)}$, i.e., $1 + x^2$. In fact

$$\lim_{n\to\infty} \lim_{x\to0} s_n(x) = 0$$

Whereas

$$\lim_{x\to0} \lim_{n\to\infty} s_n(x) = \lim_{x\to0}(1 + x^2) = 1$$

Thus $s(x)$ possess a discontinuous limit at $x=0$. Hence $\sum u_n(x)$ is not uniformly convergent on any interval containing the origin.

39. We have $\quad u_n(x) = \dfrac{1}{1 + (n-1)x} - \dfrac{1}{1 + nx}$

$\Rightarrow \qquad s_n(x) = 1 - \dfrac{1}{1 + nx} \Rightarrow s_n(0) = 0$

$\Rightarrow \qquad s(0) = 0$

Also $\qquad s(x) = \lim_{n\to\infty} s_n(x) = 1 \qquad$ gives

$$\lim_{x\to0} s(x) = 1$$

Hence the requirement

$$\lim_{x\to0} \lim_{n\to\infty} s_n(x) = \lim_{n\to\infty} \lim_{x\to0} s_n(x)$$

is violated. Therefore, the series in question is not uniformly convergent including the origin.

40. Here

$$u_n(x) = \frac{1}{1 + nx} - \frac{1}{1 + (n-1)x}$$

$$= \frac{-x}{(1 + nx)\{1 + (n-1)x\}}$$

Now $\displaystyle\lim_{x\to0} s(x) = \lim_{x\to0} \lim_{n\to\infty} s_n(x) = 0$ for $x > 0$

and $\displaystyle\lim_{x\to0} \lim_{n\to\infty} s_n(x) = 1$ for all n.

41. $s(x) = \lim\limits_{n\to\infty} s_n(x) = \begin{cases} \dfrac{x}{1-e^{-x}}, & x \neq 0 \\ 0, & x = 0 \end{cases}$

$\Rightarrow \lim\limits_{x\to 0}\lim\limits_{n\to\infty} s_n(x) = 1$ by L' Hospital's rule,

$\neq \lim\limits_{n\to\infty}\lim\limits_{x\to 0} s_n(0) = 0$

42. $s_n(x) = x \cdot \dfrac{1-e^{-nx}}{1-e^{-x}}$

and $s_n(x) = \begin{cases} \dfrac{x}{1-e^{-x}} & \text{for } x \neq 0 \\ 0 & \text{for } x = 0 \end{cases}$

Now $\lim\limits_{x\to 0} s(x) = \lim\limits_{x\to 0} \dfrac{x}{1-e^{-x}}$

$= \lim\limits_{x\to 0} \dfrac{1}{e^{-x}} = \dfrac{1}{1} = 1$

Also,

$\lim\limits_{n\to\infty}\lim\limits_{x\to 0} s(x) = 0 \neq \lim\limits_{x\to 0}\lim\limits_{n\to\infty} s_n(x) = 1$

This shows that the sum function has a discontinuous limit. Hence the series cannot be uniformly convergent on any interval including $x=0$

43. We have $s_n(x) = \dfrac{1}{2n^2}\log(n^4 x^2) + 1$

$\Rightarrow \qquad s(x) = \lim\limits_{n\to\infty} \dfrac{\log(n^4 x^2 + 1)}{2n^2}$

$= \lim \dfrac{4n^3 x^2 / (n^4 x^2 + 1)}{4n}$

by L' Hospital's rule

$= \lim \dfrac{n^2 x^2}{n^4 x^2 + 1} = \lim \dfrac{n^2 x^2}{n^4 x^2} = \lim \dfrac{1}{n^2} = 0$

$\Rightarrow \dfrac{d}{dx}\lim\limits_{n\to\infty} s_n(x) = \dfrac{d}{dx} 0 = 0$

Also $\lim\limits_{n\to\infty} \dfrac{d}{dx} s_n(x)$

$= \lim\limits_{n\to\infty} \dfrac{d}{dx}\left[\dfrac{1}{2n^2}\log(n^4 x^2 + 1)\right]$

$= \lim\limits_{n\to\infty} \dfrac{2n^4 x}{2n^2(n^4 x^2 + 1)} = \lim\limits_{n\to\infty} \dfrac{n^2 x}{1 + n^4 x^2} = 0$

Hence, $\dfrac{d}{dx}\lim\limits_{n\to\infty} s_n(x) = \lim\limits_{n\to\infty} \dfrac{d}{dx} s_n(x)$

This proves that termwise differentiation is permissible. But we see that

$$\sum u'_n(x) = \dfrac{n^2 x}{1 + n^4 x^2} = f_n(x)\,\text{say,}$$

is not uniformly convergent on any interval containing $x = 0$. To show this, we wish to apply M_n-test.

$M_n = \max |f_n(x) - f(x)| = \max y,\text{(say)}$

$= \max\left|\dfrac{n^2 x}{1 + n^4 x^2} : x \in [0,1]\right|$

$\geq \dfrac{n^2 \cdot \dfrac{1}{n^2}}{1 + n^4 \cdot \dfrac{1}{n^4}}$

on letting $x = \dfrac{1}{n^2} \in [0,1]$

$M_n = \dfrac{1}{2} \neq 0$

By M_n-test, $f_n \nrightarrow f$ uniformly.

44. We have $s(x) = \lim\limits_{n\to\infty} s_n(x) = 0$ for $0 \leq x \leq 1$

$s'_n(0) = \lim\limits_{n\to\infty} \dfrac{s_n(0+h) - s_n(0)}{h},\ 0 \leq x \leq 1$

$= \lim\limits_{h\to 0} \dfrac{1}{h}\left(\dfrac{nh}{1 + n^2 h^2} - 0\right)$

$= \lim\limits_{h\to 0} \dfrac{n}{1 + n^2 h^2} = n$

$\Rightarrow \qquad s'_n(0) \to \infty \text{ as } n \to \infty$

But $s'(0) = 0 \text{ for all } x$

$\Rightarrow \qquad s'(0) \neq \lim\limits_{n\to\infty} s_n(0)$

This shows that termwise differentiation is invalid at $x=0$.

45. We have $\sum \dfrac{1}{n^3 + n^4 x^2} < \sum \dfrac{1}{n^3}$

Convergence of $\sum 1/n^3$ implies uniform

convergence of $\sum \dfrac{1}{n^3 + n^4 x^2}$ for all x in view of Weierstrass M-test.

To test whether $s'(x)$ is obtained by termwise

differentiation, we consider the series $\sum u_n'(x)$ where

$$u_n'(x) = \frac{d}{dx}\left(\frac{1}{n^3} + n^4 x^2\right)$$

$$= \frac{-2xn^4}{(n^3 + n^4 x^2)^2}$$

$$= -\frac{2x}{n^2(1+n^2 x)^2} = y, \text{say}$$

Now $\dfrac{dy}{dx} = 0$

$\Rightarrow \quad 2n^2(1+nx^2)^2 - 2xn^2.2(1+nx^2).2nx = 0$

$\Rightarrow \quad 1+nx^2 - 4nx^2 = 0 \Rightarrow x \pm \dfrac{1}{\sqrt{(3n)}}$

$\Rightarrow$ the function $u_n'(x)$ has a local maximum

at $x = \dfrac{1}{\sqrt{(3n)}}$ and has the value

$$\frac{-2.\dfrac{1}{\sqrt{(3n)}}}{\dfrac{(1+1/3)^2}{n^2}}, \text{i.e., } \frac{-3\sqrt{3}}{8n^{5/2}}$$

$\Rightarrow \qquad \sum u_n'(x) = \dfrac{-3\sqrt{3}}{8}\sum\dfrac{1}{n^{5/2}}$

and $\qquad |u_n'(x)| = \dfrac{-3\sqrt{3}}{8}\sum\dfrac{1}{n^{5/2}}$

Now $\sum\dfrac{1}{n^{5/2}}$ is a convergent series. By M-test

$\sum u_n'(x)$ converges uniformly, and

$$s'(x) = \sum_{n=1}^{\infty} u_n'(x)$$

46. Set $u_n(x) = \sin nx$ and $g_n = 1/x$

Now $f_n(x) = \sin x + \sin 2x + ... + \sin nx$

$$= \frac{\sin\left(x + \dfrac{n-1}{2}\right)\sin\dfrac{nx}{2}}{\sin\dfrac{x}{2}}$$

$$= \frac{1}{\sin\dfrac{x}{2}}\left(\sin\dfrac{n+1}{2}x\sin\dfrac{nx}{2}\right)$$

$\Rightarrow |f_n(x)| = \dfrac{1}{\left|\sin\dfrac{x}{2}\right|}\left|\sin\dfrac{n+2}{2}x\right|\left|\sin\dfrac{nx}{2}\right|$

$$\leq \frac{1}{\left(\sin\dfrac{x}{2}\right)}$$

$\Rightarrow |f_n(x)| \leq |\cosec(x/2)|$

In fact, coses $x/2$ is bounded for each x in $0 < a \leq x \leq b \leq 2\pi$. Define λ as the upper bound of cosec $x/2$ in this interval.

Hence $|f_n(x)| < \lambda$ for each x in the interval. Also the sequence $<g_n>$ is a monotonic decreasing function converging to zero. Hence by Dirichlet's test, the given series is uniformly convergent in $0 < a \leq x \leq b \leq 2\pi$.

47. Set $u_n = (-1)^{n-1}$ and $g_n(x) = x^n$

$\Rightarrow \sum u_n = f_n(x)$, say $=0$ or 1 as n is even or odd

$\Rightarrow f_n(x)$ is bounded for each n.

Also $<g_n(x)>$ is a positive monotonic decreasing sequence converging to zero for each x in $0 \leq x \leq \lambda \leq 1$.

Hence by Dirichlet's test, the given series is uniformly convergent in $0 \leq x \leq \lambda \leq 1$.

48. Define $u_n(x) = x^n$ and $g_n = \dfrac{1}{n+1}$

Now $|x| \leq \delta < 1$

$|f_n(x)| = |u_1(x) + ... + u_n(x)|$

$$= |x + x^2 + ... + x^n|$$

$$\leq |x| + |x|^2 + ... + |x|^n$$

$$\leq \delta + \delta^2 + ... + \delta^n = \frac{\delta(1-\delta^n)}{1-\delta} < \frac{\delta}{1-\delta}$$

$\Rightarrow f_n(x)$ is bounded for each x in $|x| \leq \delta < 1$. Moreover, $<g_n>$ is a monotonic decreasing sequence converging to zero. In view of Dirichlet's test the given series is uniformly convergent on $[-\delta, \delta]$.

49. Set $u_n = (-1)^{n-1}, g_n(x) = \dfrac{1}{n+x^2} \Rightarrow \sum u_n = 0$ or 1 according as n is even or odd. Set

$f_n(x) = \sum u_n \Rightarrow f_n(x)$ is bounded for each n.

50. Set $g_n(x) = x^n$. Then the sequence $<g_n(x)>$ is uniformly bounded on $[0, 1]$.

But $|g_n(x)| < |x^n| < 1$ for each n since $x \in [0, 1]$. Also $g_n(x) > g_{n+1}(x)$,

i.e., $<g_n(x)>$ is a monotonic decreasing real valued function. Convergence of Σa_n on $[0, 1]$ implies uniform convergence of Σa_n on $[0,1]$. The Abel's test asserts that $\Sigma a_n x^n$ converges.

uniformly on $[0, 1]$.

51. Set $f_n(x) = \dfrac{1}{n}(-1)^{n-1}$, and $g_n(x) = x^n$, $<g_n(x)>$ is uniformly bounded and monotonic decreasing on $[0, 1]$.

Also $\Sigma \dfrac{1}{n}(-1)^{n-1}$ is convergent by alternating series test.

Hence by Abel's test, the given series in uniformly convergent on $[0, 1]$.

SELF ASSESSMENT TEST

Verify the following :

1. The sequence $\left\langle e^{-nx} \right\rangle$ is uniformly convergent in $[a, b]$ where a and b are positive number but only pointwise in $[o, b]$

2. The sequence $\left\langle \dfrac{nx}{1+n^3 x^2} \right\rangle$ converges uniformly to zero for $0 \le x \le 1$.

3. The series $\sum f_n$ the sum of whose n terms is $s_n(x) = \dfrac{x}{1+nx^2}$ converges uniformly for all real x.

4. The sequence, where $f_n(x) = \dfrac{n^2 x}{1+n^3 x^2}$ is not uniformly convergent on $[0, 1]$.

5. The series $\sum \dfrac{a_n x^n}{1+x^{2n}}$ converges uniformly $\forall x \in \mathbb{R}$, if $\sum a_n$ is absolutely convergent.

6. The series $\sum \dfrac{\cos n\theta}{n^p}$ is uniformly and absolutely convergent for all real values of θ, $p>1$

7. $\sum n^{-x}$ is uniformly convergent in $[1+\delta, \infty[, \delta>0$

8. The series $\sum \dfrac{(-1)^n . x^{2n}}{n^p (1+x^{2n})}$ converges absolutely and uniformly for all real x if $p>1$.

9. Each of the series $\sum \dfrac{\sin nx}{n}, \sum \dfrac{\cos nx}{n}$, converges uniformly w.r.t. x in $(a, 2\pi -a)$, where $a<\pi$ is any fixed number.

10. The sequence $<\tan^{-1} nx>$ of continuous function has a discontinuous limit function on $[0, 1]$ and the convergence is not uniform on $[0, 1]$.

11. The sum function $(1+x)$ of the series $\sum (1-x^2)x^n$ is continuous on $[0, 1]$ although the convergence is not uniform.

12. The sum function of the series is $\sum\limits_{n=0}^{\infty} (1-x)x^n$ is $f(x) = \begin{cases} 1 & x \ne 1 \\ 0 & x = 1 \end{cases}$ which is discontinuous on $[0, 1]$

Also, the series is not uniformly convergent on $[0, 1]$.

13. Sequences $\left\langle \dfrac{n+x+1}{(n+1)x} \right\rangle$ and $\left\langle \dfrac{(n+1)x^2}{1+n^2 x^2} \right\rangle$

converges uniformly to $\dfrac{1}{x}$ and 0 respectively on $]0, 1[$ but their product sequence $\left\langle \dfrac{nx+x^2+x}{1+n^2 x^2} \right\rangle$ does not converge uniformly on $]0, 1[$.

14. The point 0 is the only point of non-uniform convergence of $e^{-n|x|}$ on $\mathbb{R}$.

15. The series $\sum 2^n x^{2^n}$ converges uniformly in $[-a, a]$ where $1>a>0$ but not on $]-1,+1[$.

16. The series $\sum 3^n \sin \dfrac{1}{4^n x}$ converges absolutely and uniformly on $]a, \infty[, a > 0$.

17. If $\sum f_n(x)$ converges uniformly on a set S then $<f_n(x)>$ converges uniformly to 0 on S.

18. The series $\dfrac{1}{1+x^2} - \dfrac{1}{2+x^2} + \dfrac{1}{3+x^2} - ....$ converges uniformly on $\mathbb{R}$.

19. The series $\sum \dfrac{1}{n^3 + n^4 x^2}$ can be differentiated term by term on any bounded interval.

20. The infinite product $\prod\limits_{n=1}^{\infty} \left\{ 1 + \dfrac{x^2}{(1+x^2)^n} \right\}$ is convergent for all values of x but, is not uniformly convergent on any interval I which either contains 0 as an interior point or has it as an end points.

10 Function of Several Variables

10.1 INTRODUCTION

Let f be a function from a set of ordered pair of real numbers to a set of real numbers; then f is said to be a real valued function of two real variables or, briefly, a real function of two variables. The value that f assumes at the arguments (x, y) is naturally written as $f(x, y)$. Let us suppose this value is called z. Then we write $z = f(x, y)$, where x and y are the independent variables and z is the dependent variable.

We shall write $z = z(x, y)$, which means that we are considering some function of two variables, where the independent variables are x and y and the dependent variable is z.

If to each pair of values of x and y there exists only one value of z, then the function is said to be single valued function. On the other hand, if there are two or more values of z correspond to some x and y or all of the values assigned to x and y, the function is called multiple valued.

Definition 1. *Let $f(x, y)$ be a function of two variables x and y, then we say* $\displaystyle\lim_{\substack{x \to x_0 \\ y \to y_0}} f(x, y)$

exists and is equal to l, if for every $\varepsilon > 0$, $\exists$ a $\delta > 0$ such that

$$|f(x, y) - l| < \varepsilon$$

for all values of x and y in the neighbourhood of (x_0, y_0) defined by

$$|x - x_0| < \delta, \ |y - y_0| < \delta.$$

10.2 LIMIT

Let $f(x, y)$ is a function of two variables x and y, we define several kind of limits. If (x_0, y_0) is the limiting point of a set of values on two dimensional space, then we have the following limits

$$\lim_{\substack{x \to x_0 \\ y \to y_0}} f(x, y), \ \lim_{x \to x_0} \lim_{y \to y_0} f(x, y), \ \lim_{y \to y_0} \lim_{x \to x_0} f(x, y).$$

Then limit of the first kind is known as simultaneous limit and the last two types are known as iterated limits.

REMARK

- It is not necessary that, the two repeated limits, even if they exist, may be equal *e.g.*, consider a function

$$f(x, y) = \frac{x^2 - y^2}{x^2 + y^2}; \ (x, y) \neq (0, 0)$$

We have $\qquad \lim\limits_{x\to 0} f(x,y) = -1 \, \forall \, y \qquad \Rightarrow \qquad \lim\limits_{y\to 0}\left[\lim\limits_{x\to 0} f(x,y)\right] = -1$

Now $\qquad \lim\limits_{y\to 0} f(x,y) = 1 \, \forall x \qquad \Rightarrow \qquad \lim\limits_{x\to 0}\left[\lim\limits_{y\to 0} f(x,y)\right] = 1$

Thus, while the two repeated limit exist in this case, they are unequal. It may also be seen that

$$\lim\limits_{(x,y)\to(0,0)} f(x,y) \text{ does not exist.}$$

10.2.1 NON-EXISTENCE OF A LIMIT

To determine whether a simultaneous limit exists or not, it is a difficult matter but a simple consideration, which we describe, says us to decide about the non-existence of a limit.

If $\quad \lim\limits_{(x,y)\to(a,b)} f(x,y) = l$ and if ϕ is any function of a single real variable such that

$$\lim\limits_{x\to a} \phi(x) = b$$

Then $\qquad \lim\limits_{x\to a} f[x,\phi(x)] = l$

Thus, we can determine two functions ϕ_1 and ϕ_2 such that

$$\lim\limits_{x\to a} f[x,\phi_1(x)] \ne \lim\limits_{x\to a} f[x,\phi_2(x)]$$

Then, we can say that the simultaneous limit $\quad \lim\limits_{(x,y)\to(a,b)} f(x,y)$ does not exist.

Solved Examples

Example 1. Show that $\lim\limits_{(x,y)\to(0,0)} f(x,y)$, where

$$f(x,y) = \frac{x^2 - y^2}{x^2 + y^2} \text{ does not exist.}$$

Solution. Here $f(x,y) = \dfrac{x^2 - y^2}{x^2 + y^2}$

Now taking $\phi(x) = m(x)$, then we have

$\lim\limits_{x\to 0} f(x,mx) = \dfrac{1-m^2}{1+m^2}$, which depend upon m. Therefore it is not unique.

Since, $\lim\limits_{x\to 0} f(x,mx)$ is not unique. Hence

$\lim\limits_{(x,y)\to(0,0)} f(x,y)$ does not exist.

Based on the following Results

- Simultaneous limit: $\lim\limits_{\substack{x\to x_0 \\ y\to y_0}} f(x,y)$

- Iterated limit: $\lim\limits_{x\to x_0} \lim\limits_{y\to y_0} f(x,y)$ and $\lim\limits_{y\to y_0} \lim\limits_{x\to x_0} f(x,y)$

- Two iterated limits, if they exist may be equal or not.

- To establish non-existence of a limit, we must find two methods of approaches to the limit point which gives different limiting values.

Example 2. Show that the simultaneous limit,

$$\lim\limits_{\substack{x\to 0 \\ y\to 0}} \frac{xy^3}{x^2 + y^6} \text{ does not exist.}$$

Solution. Let (x, y) tends to $(0, 0)$ through the line $y = x$, which is a line through the origin.

Put $y = x$, in the given function, we get

$$\lim\limits_{x\to 0} \frac{x^4}{x^2 + x^6} = \lim\limits_{x\to 0} \frac{x^2}{1 + x^4} = 0$$

Again, let $(x, y) \to (0, 0)$ through the curve $x = y^3$.

Put $x = y^3$, in the given function, we obtain $\lim\limits_{y \to 0} \dfrac{y^6}{y^6 + y^6} = \dfrac{1}{2}$

$\Rightarrow$ The limit obtained by two different methods are different.

Hence, the simultaneous limit does not exist.

Example 3. *Let* $f(x, y) = \dfrac{y - x}{y + x} \cdot \dfrac{1 + x}{1 + y}, (x, y) \ne (0, 0)$ *Show that two repeated limits exist at origin but are unequal.*

Solution.   We have

$$\lim\limits_{x \to 0}\lim\limits_{y \to 0} f(x, y) = \lim\limits_{x \to 0} -\left(\frac{1 + x}{1}\right) = -1$$

Again,

$$\lim\limits_{y \to 0}\lim\limits_{x \to 0} f(x, y) = \lim\limits_{y \to 0}\left(\frac{1}{1 + y}\right) = 1$$

Therefore, the two repeated limits exist at the origin are not equal.

Example 4. *Show that* $\lim\limits_{(x, y) \to (0, 0)} \dfrac{3x - 2y}{2x - 3y}$ *does not exist.*

Solution.   When $(x, y) \to (0, 0)$ along the straight line $y = x$, we have

$$\lim\limits_{(x, y) \to (0, 0)} f(x, y) = \lim\limits_{x \to 0} \frac{3x - 2x}{2x - 3x} = \lim\limits_{x \to 0} \frac{x}{-x} = \lim\limits_{x \to 0} -1 = -1$$

Now , again when $(x, y) \to (0, 0)$ along the straight line $y = 0$, we have

$$\lim\limits_{(x, y) \to (0, 0)} f(x, y) = \lim\limits_{x \to 0} \frac{3x - 0}{2x - 0}$$

$$= \lim\limits_{x \to 0} \frac{3x}{2x} = \lim\limits_{x \to 0} \frac{3}{2} = \frac{3}{2}$$

Since two methods of approach to the limiting point give different limiting values, hence the limit does not exist.

Example 5. *Show that* $\lim\limits_{(x, y) \to (0, 0)} \dfrac{2xy^2}{x^2 + y^4}$ *does not exist.*

Solution.   Let (x, y) tends $(0, 0)$ along the curve $x = my^2$, we put $x = my^2$ in the function and then allow y to tends to zero. Thus, we have

$$\lim\limits_{(x, y) \to (0, 0)} \frac{2xy^2}{x^2 + y^4} = \lim\limits_{y \to 0} \frac{2my^4}{(m^2 + 1)y^4} = \lim\limits_{y \to 0} \frac{2m}{1 + m^2} = \frac{2m}{1 + m^2}$$

which is different for different values of m.

e.g., if $m = 1$, then the limt $= \dfrac{2 \cdot 1}{1 + 1^2} = 1$ and if $m = 2$, then this limit $= \dfrac{2 \cdot 2}{1 + 2^2} = \dfrac{4}{5}$.

Therefore, the two methods of tends to the limiting point give different limiting values, hence the limit $\lim\limits_{(x, y) \to (0, 0)} \dfrac{2xy^2}{x^2 + y^4}$ does not exist.

Example 6. *Find that* $\lim\limits_{(x,y)\to(0,0)} \dfrac{x^3+y^3}{x-y}$ *does not exist.*

Solution. Let (x, y) tends to $(0, 0)$ through the curve $y = x - mx^3$, we have

$$\lim_{(x,y)\to(0,0)} \frac{x^3+y^3}{x-y} = \lim_{x\to 0} \frac{x^3+(x-mx^3)^3}{x-(x-mx^3)}$$

$$= \lim_{x\to 0} x^3 \left[\frac{1+(1-mx^2)^3}{mx^3} \right]$$

$$= \lim_{x\to 0} \frac{2-3mx^2+3m^2x^4-m^3x^6}{m} = \frac{2}{m}$$

which is different for different value of m.

Hence, the simultaneous limit $\lim\limits_{(x,y)\to(0,0)} \dfrac{x^3+y^3}{x-y}$ does not exist.

Example 7. *Show that the limit, when* $(x, y) = (0, 0)$, *exist in each case and equal to 0.*

$$(i)\ \lim \frac{x^3y^3}{x^2+y^2} \qquad\qquad (ii)\ \frac{x^4+y^4}{x^2+y^2}$$

Solution. (i) Here, we show that $\lim\limits_{(x,y)\to(0,0)} \dfrac{x^3y^3}{x^2+y^2} = 0$

Take any given $\varepsilon > 0$, for all $(x, y) = (0, 0)$, we have

$$\left| \frac{x^3y^3}{x^2+y^2} - 0 \right| = \left| \frac{x^3y^3}{x^2+y^2} \right|$$

$$= |r^4 \cos^3\theta \sin^3\theta| \qquad\qquad [\text{Putting } x=r\cos\theta,\ y = r\sin\theta]$$

$$= r^4 |\cos\theta|^3 |\sin\theta|^3$$

$$\leq r^4 \qquad\qquad [\because |\cos\theta| \leq 1 \text{ and } |\sin\theta| \leq 1]$$

$$= (x^2+y^2)^2$$

$$< \varepsilon \text{ if } x^2 < \sqrt{\varepsilon}\,/2 \text{ and } y^2 < \sqrt{\varepsilon}\,/2$$

$$i.e.,\ \text{if } |x| < \left(\frac{\sqrt{\varepsilon}}{2} \right)^{1/2} \text{ and } |y| < \left(\frac{\sqrt{\varepsilon}}{2} \right)^{1/2}$$

Now if we take $\delta = \dfrac{\sqrt{\varepsilon}}{2}$, then for any given $\varepsilon > 0$, there exists $\delta > 0$ such that

$$\left| \frac{x^3y^3}{x^2+y^2} - 0 \right| < \varepsilon \text{ whenever } |x| < \delta \text{ and } |y| < \delta$$

Hence, $\lim\limits_{(x,y)\to(0,0)} \dfrac{x^3y^3}{x^2+y^2} = 0$

(ii) Take any given $\varepsilon > 0$ and for all $(x, y) \neq (0, 0)$, we have

$$\left| \frac{x^4+y^4}{x^2+y^2} - 0 \right| = \frac{x^4+y^4}{x^2+y^2}$$

$$= r^2 (\cos^4\theta + \sin^4\theta) \quad [\text{Putting } x=r\cos\theta,\ y = r\sin\theta]$$

$$\le 2r^2 \qquad\qquad [\because \cos^4\theta \le 1 \text{ and } \sin^4\theta \le 1]$$
$$= 2(x^2 + y^2)$$
$$< \varepsilon \text{ if } x^2 < \varepsilon/4 \text{ and } y^2 < \varepsilon/4 \text{ i.e., if } |x| < \sqrt{\varepsilon}/2 \text{ and } |y| < \sqrt{\varepsilon}/2$$

Now, if we take $\delta = \dfrac{\sqrt{\varepsilon}}{2}$, then for any given $\varepsilon > 0$, there exists $\delta > 0$, such that

$$\left| \frac{x^4 + y^4}{x^2 + y^2} - 0 \right| < \varepsilon , \text{ whenever } |x| < \delta \text{ and } |y| < \delta$$

Hence, $$\lim_{(x,y)\to(0,0)} \frac{x^4 + y^4}{x^2 + y^2} = 0$$

Example 8.   Prove that $$\lim_{(x,y)\to(0,0)} \frac{\sqrt{x^2 y^2 + 1} - 1}{x^2 + y^2} = 0$$

Solution.   Since x, y are small in absolute values, we have

$$\frac{\sqrt{x^2 y^2 + 1} - 1}{x^2 + y^2} = \frac{(1 + x^2 y^2)^{1/2} - 1}{x^2 + y^2} \approx \frac{\frac{1}{2} x^2 y^2}{x^2 + y^2}$$

Take any given $\varepsilon > 0$, we have

$$\left| \frac{\sqrt{x^2 y^2 + 1} - 1}{x^2 + y^2} - 0 \right| = \frac{\sqrt{x^2 y^2 + 1} - 1}{x^2 + y^2} = \frac{\frac{1}{2} x^2 y^2}{x^2 + y^2}$$

$$= \frac{1}{2} \frac{r^4 \cos^2\theta \sin^2\theta}{r^2} \qquad\qquad [\because x = r\cos\theta, y = r\sin\theta]$$

$$= \frac{1}{2} r^2 \cos^2\theta \sin^2\theta$$

$$\le \frac{1}{2} r^2 \qquad\qquad [\because \cos^2\theta < 1 \text{ and } \sin^2\theta \le 1]$$

$$\le \frac{1}{2}(x^2 + y^2) \qquad\qquad [r^2 = x^2 + y^2]$$

$$< \varepsilon \text{ if } x^2 < \varepsilon \text{ and } y^2 < \varepsilon \text{ i.e., } |x| < \sqrt{\varepsilon} \text{ and } |y| < \sqrt{\varepsilon}.$$

Now if we take $\delta = \sqrt{\varepsilon}$, then for any given $\varepsilon > 0$, there exists $\delta > 0$ such that

$$\left| \frac{\sqrt{x^2 y^2 + 1} - 1}{x^2 + y^2} - 0 \right| < \varepsilon \text{ whenever } |x| < \delta \text{ and } |y| < \delta.$$

Hence, $$\lim_{(x,y)\to(0,0)} \frac{\sqrt{x^2 y^2 + 1} - 1}{x^2 + y^2} = 0$$

REMARKS

- If the simultaneous limit, $\lim\limits_{\substack{x\to x_0 \\ y\to y_0}} f(x, y)$ exist, then the single limit, $\lim\limits_{x\to x_0} f(x, y_0)$ $\lim\limits_{y\to y_0} f(x_0, y)$

  also exist. But it is necessary that the single limits, $\lim\limits_{x\to x_0} f(x, y)$, $\lim\limits_{\substack{x\to x_0 \\ y\to y_0}} f(x, y)$, exist for $y \ne y_0$, $x \ne x_0$ respectively.

- If two iterated limits are equal such that $\lim\limits_{x\to 0}\lim\limits_{y\to 0} f(x, y) = \lim\limits_{y\to 0}\lim\limits_{x\to 0} f(x, y)$

  then the limit of the funciton need not exist.

10.2.2 SQUARE NEIGHBOURHOOD OF A POINT (a, b)

Let neighbourhood of a point (a, b) in the xy-plane determined by a positive number δ be the square bounded by the lines.

$$x = a - \delta, \quad x = a + \delta,$$
$$y = b - \delta, \quad y = b + \delta.$$

If a point (x, y) lies in the neighbourhood, we have

$$a - \delta < x < a + \delta \Rightarrow |x - a| < \delta$$
$$b - \delta < y < b + \delta \Rightarrow |y - b| < \delta$$

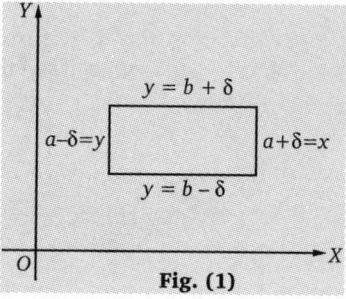

Fig. (1)

The centre of the square is at the point (a, b). This square is called the neighbourhood of the point (a, b). For every value of δ, we will get a neighbourhood.

10.2.3 CIRCULAR NEIGHBOURHOOD OF A POINT (a, b)

A circular neighbourhood of a point (a, b) in R^2 is the set of all points (x, y) whose distance from the point (a, b) is less than some given $\delta > 0$ i.e., the set of all point (x, y) such that

$$\sqrt{(x - a)^2 + (y - b)^2} < \delta$$

i.e.,
$$|(x, y) - (a, b)| < \delta$$

Here, $|(x, y) - (a, b)|$ stands for distance betwen the points (x, y) and (a, b) i.e.,

$$\sqrt{(x - a)^2 + (y - b)^2}$$

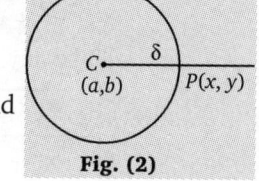

Fig. (2)

10.2.4 ALGEBRA OF LIMITS

If $\lim\limits_{(x,y)\to(a,b)} f(x, y) = l_1$ and $\lim\limits_{(x,y)\to(a,b)} g(x, y) = l_2$, then

(i) $\lim\limits_{(x,y)\to(a,b)} [f(x,y) + g(x,y)] = l_1 + l_2$ (ii) $\lim\limits_{(x,y)\to(a,b)} [f(x,y) - g(x,y)] = l_1 - l_2$

(iii) $\lim\limits_{(x,y)\to(a,b)} [f(x,y).g(x,y)] = l_1.l_2$ (iv) $\lim\limits_{(x,y)\to(a,b)} \left[\dfrac{f(x,y)}{g(x,y)}\right] = \dfrac{l_1}{l_2}$ (provided $l_2 \neq 0$)

THEOREM 1. *Let $z = f(x, y)$ be a function, then* $\lim\limits_{(x,y)\to(a,b)} f(x, y)$ *if exists, is unique.*

Proof. Let $z = f(x, y)$ be a function.

Let if possible

$$\lim\limits_{(x,y)\to(a,b)} f(x, y) = l_1 \text{ and } \lim\limits_{(x,y)\to(a,b)} f(x, y) = l_2$$

Now, to prove $\qquad l_1 = l_2$

Let us first suppose $\lim\limits_{(x,y)\to(a,b)} f(x, y) = l_1$, then by the definition of limit, we have

"for given $\varepsilon > 0 \; \exists \; \delta_1 > 0$ such that

$$|f(x, y) - l_1| < \varepsilon/2 \text{ whenever } |(x, y) - (a, b)| < \delta_1" \qquad \dots (1)$$

Now suppose $\lim\limits_{(x,y)\to(a,b)} f(x, y) = l_2$

"for given $\varepsilon > 0 \; \exists \; \delta_2 > 0$ such that

$$|f(x, y) - l_2| < \varepsilon/2 \text{ whenever } |(x, y) - (a, b)| < \delta_2" \qquad \dots (2)$$

Let $\delta = \min \{\delta_1, \delta_2\}$

Hence, we have

$|f(x, y) - l_1| < \varepsilon/2$ and $| f(x, y) - l_2| < \varepsilon/2$ whenever $|(x, y) - (a, b)| < \delta$

Now, consider

$$|l_1 - l_2| = |l_1 - f(x, y) + f(x, y) - l_2|$$
$$\leq |l_1 - f(x, y)| + | f(x, y) - l_2| \qquad \text{[By triangular inequality]}$$
$$\leq |f(x, y) - l_1| + |f(x, y) - l_2|$$
$$< \varepsilon/2 + \varepsilon/2 = \varepsilon$$

Since ε is arbitrary and small, therefore $l_1 - l_2 = 0 \Rightarrow l_1 = l_2$.

Hence, limit of a function is unique.

Solved Examples

Example 1. *Let $f : R^2 \to R$ be defined by*

$$f(x,y) = \begin{cases} \dfrac{xy}{x^2 + y^2} & , \quad (x, y) \neq (0,0) \\ 0 & , \quad (x, y) = (0,0) \end{cases}$$

Prove that $\lim\limits_{(x,y)\to(0,0)} f(x,y)$ *does not exist.*

Solution. Since, if $\lim\limits_{(x,y)\to(0,0)} f(x,y)$ exists, then this limit is independent of the path along which we approach the point (a, b). Let $(x, y) \to (0, 0)$ along the path $y = mx$ where $m \in R$. As $x \to 0$, from $y = mx$, we have $y \to 0$.

Consider

$$\lim_{(x,y)\to(a,b)} f(x,y) = \lim_{(x,y)\to(0,0)} \frac{xy}{x^2 + y^2}$$

$$= \lim_{x\to 0} \frac{x\,mx}{x^2 + m^2 x^2} = \lim_{x\to 0} \frac{mx^2}{x^2(1 + m^2)} \qquad \text{(Putting } y = mx)$$

$$= \lim_{x\to 0} \frac{m}{1 + m^2} = \frac{m}{1 + m^2}$$

which will be different for different values of m.

Therefore, $\lim\limits_{(x,y)\to(0,0)} f(x,y)$ does not exist.

Example 2. *Let $A = \{(x, y) : 0 < x < 1, 0 < y < 1, x, y \in R\}$. If $f : A \to R$ defined by $f(x, y) = x + y$, Show that*

$$\lim_{(x,y)\to\left(0,\frac{1}{2}\right)} f(x,y) = \frac{1}{2} \text{ where } x, y \in A.$$

Solution. Let $\varepsilon > 0$ be given (an arbitrary positive small number)

Also, let $|x - 0| < \varepsilon/2$ and $\left| y - \dfrac{1}{2} \right| < \varepsilon/2$

Consider $\left| f(x,y) - \dfrac{1}{2} \right| = \left| x + y - \dfrac{1}{2} \right|$

$$\leq |x| + \left| y - \frac{1}{2} \right| < \varepsilon/2 + \varepsilon/2 = \varepsilon$$

Then, by the definition of limit, we have

$$\lim_{(x,y)\to\left(0,\frac{1}{2}\right)} f(x,y) = \frac{1}{2}$$

Example 3. If $f(x,y) = y \sin\dfrac{1}{x} + x \sin\dfrac{1}{y}$ where $x \neq 0$, $y \neq 0$. Then prove that

$$f(x, y) \to 0 \ as \ (x, y) \to (0, 0)$$

Solution. Let ε be any given arbitrary small positive number since, $\varepsilon > 0$ let us take $\delta = \varepsilon$. Also, let $|x - 0| < \varepsilon/2$, $|y - 0| < \varepsilon/2$

$$\therefore \quad |(x, y) - (0, 0)| = \sqrt{(x-0)^2 + (y-0)^2} \leq |x - 0| + |y - 0|$$
$$< \varepsilon/2 + \varepsilon/2 = \varepsilon$$

$$\Rightarrow \quad |f(x, y) - 0| = \left| y \sin\frac{1}{x} + x \sin\frac{1}{y} \right| \leq \left| y \sin\frac{1}{x} \right| + \left| x \sin\frac{1}{y} \right|$$

$$\leq |y| \left| \sin\frac{1}{x} \right| + |x| \left| \sin\frac{1}{y} \right|$$

$$\leq |y| + |x| \qquad \left[\because \left| \sin\frac{1}{x} \right| \leq 1 \text{ and } \left| \sin\frac{1}{y} \right| \leq 1 \right]$$

$$< \varepsilon/2 + \varepsilon/2 = \varepsilon$$

$$\Rightarrow \quad |f(x, y) - 0| < \varepsilon. \ \text{Hence,} \quad \lim_{(x,y)\to(0,0)} f(x,y) = 0$$

Example 4. Show that $\displaystyle\lim_{(x,y)\to(0,1)} \tan^{-1}\dfrac{y}{x}$ does not exist.

Solution. Let $y = mx$

$$\therefore \quad \lim_{(x,y)\to(0,1)} \tan^{-1}\frac{y}{x} = \lim_{x\to0} \tan^{-1}\left(\frac{mx}{x}\right) = \tan^{-1} m$$

i.e., limit of the function depends on m; which means limit of the function at $(0, 1)$ is not unique. Hence, the limit of the function does not exist.

Example 5. If $f(x,y) = \begin{cases} 2xy & , if \ (x,y) \neq (1,2) \\ 0 & , if \ (x,y) = (1,2) \end{cases}$ then, show that $\displaystyle\lim_{(x,y)\to(1,2)} f(x,y) = 4$

Solution. By definition of limit, we know that for $\varepsilon > 0$, we have to find $\delta > 0$ such that

$$|2xy - 4| < \varepsilon \quad \text{whenever} \quad \sqrt{(x-1)^2 + (y-2)^2} < \delta$$
$$\Rightarrow \quad |2xy - 4| < \varepsilon \quad \text{whenever} \quad |x - 1| < \delta \text{ and } |y - 2| < \delta$$
$$\Rightarrow \quad 1 - \delta < x < 1 + \delta \text{ and } 2 - \delta < y < 2 + \delta$$
Hence, $(1 - \delta)(2 - \delta) < xy < (1+\delta)(2+\delta)$
$$\Rightarrow \quad 2(1 - \delta)(2 - \delta) < 2xy < 2(1 + \delta)(2 + \delta)$$
$$\Rightarrow \quad (2 - 2\delta)(2 - \delta) < 2xy < (2 + 2\delta)(2 + \delta)$$
$$\Rightarrow \quad 2\delta^2 - 6\delta + 4 < 2xy < 2\delta^2 + 6\delta + 4$$
$$\Rightarrow \quad 2\delta^2 - 6\delta < 2xy - 4 < 2\delta^2 + 6\delta$$

$\Rightarrow$ $\qquad$ $|2xy - 4| < 8\delta$ if $\delta \le 1$

Let $\qquad \delta = \dfrac{\varepsilon}{8}$. Then $|2xy - 4| < \varepsilon$ whenever $|x - 1| < \delta$ and $|y - 2| < \delta$

Hence, we have for given $\varepsilon > 0$, there exists a $\delta > 0$ such that $|2xy - 4| < \varepsilon$

$\Rightarrow \qquad \lim_{(x,y) \to (1,2)} (2xy) = 4$

Example 6. *Give an example to show that the order of iterated limits can be interchanged although the simultaneous limit does not exist.*

Solution. Consider the function $f(x,y) = \dfrac{xy}{x^2 + y^2}$ for iterated and simulaneous limit

at $(0, 0)$. Let us first suppose the variables approach the origin along the line $y = x$. Then putting $y = x$, we have

$$\lim_{x \to 0} \frac{x^2}{x^2 + x^2} = \frac{1}{2}$$

Now consider

$$\lim_{y \to 0} \lim_{x \to 0} \frac{xy}{x^2 + y^2} = \lim_{y \to 0} \frac{0}{0 + y^2} = 0$$

Since, the results obtained by two methods of approaches are different. Hence, the simultaneous limit does not exist.

Now, for iterated limits, we have

$$\lim_{x \to 0} \left[\lim_{y \to 0} \frac{x.y}{x^2 + y^2} \right] = \lim_{x \to 0} \left[\frac{x.0}{x^2 + 0} \right] = 0$$

and $\qquad \lim_{y \to 0} \left[\lim_{x \to 0} \dfrac{x.y}{x^2 + y^2} \right] = \lim_{y \to 0} \left[\dfrac{0.y}{0 + y^2} \right] = 0$

Hence, the order of iterated limits can be interchanged.

10.3 CONTINUITY OF A FUNCTION OF TWO VARIABLES

(i) A function $f(x, y)$ is said to be continuous at the point (a, b) if $\lim_{(x,y) \to (a,b)} f(x,y)$ exists and equal to $f(a, b)$.

(ii) A function $f(x, y)$ is said to be continuous at (a, b), if for every $\varepsilon > 0 \; \exists \; \delta > 0$ such that $|f(x, y) - f(a, b)| < \varepsilon$ whenever $|x - a| < \delta$, $|y - b| < \delta$.

Solved Examples

Example 1. *Show that the function $f : R^2 \to R$ defined by*

$$f(x,y) = \begin{cases} \dfrac{xy(x^2 - y^2)}{x^2 + y^2} & , \quad (x,y) \ne (0,0) \\ 0 & , \quad otherwise \end{cases}$$

is continuous at $(0, 0)$

Solution. Let $\varepsilon > 0$ be given. Now, let us suppose $|x - 0| < \sqrt{\varepsilon}$ and $|y - 0| < \sqrt{\varepsilon}$

Consider,

$$|f(x,y)-f(0,0)|=\left|\frac{xy(x^2-y^2)}{x^2+y^2}-0\right|\le|xy|\left|\frac{x^2-y^2}{x^2+y^2}\right|$$

$$\le|xy|\left[\because|x^2-y^2|\le x^2+y^2|\Rightarrow\left|\frac{x^2-y^2}{x^2+y^2}\right|\le1\right]$$

$\Rightarrow$ $\quad|f(x,y)-f(0,0)|\le|x||y|$

$\Rightarrow$ $\quad|f(x,y)-f(0,0)|<\sqrt{\varepsilon}\cdot\sqrt{\varepsilon}$

$\Rightarrow$ $\quad|f(x,y)-f(0,0)|<\varepsilon.$

Hence, we have $\lim_{(x,y)\to(0,0)}f(x,y)$ exists and equal to $(0,0)$.

Example 2. *Show that the function, $f(x,y)=x^2+3y$ is continuous at $(1,2)$.*

Solution. Here, we have

$$f(x,y)=x^2+3y$$

$\Rightarrow$ $\quad f(1,2)=1+3\times2=7$

We know, that if limit of the function $f(x,y)=x^2+3y$ at $(1,2)$ is also 7, then function is continuous at $(1,2)$.

Now to show $\lim_{(x,y)\to(1,2)}f(x,y)=7$

For given $\varepsilon>0$, there must exist $\delta>0$, such that

$$0<\sqrt{(x-1)^2+(y-2)^2}<\delta\Rightarrow|x^2+3y|<\varepsilon$$

Therefore

$$|x^2+3y-7|=|(x^2-1)+3(y-2)|$$
$$=|(x-1)(x+1)+3(y-2)|$$
$$\le|(x-1)|.|(x+1)|+3|(y-2)| \quad\quad ...(1)$$

But $\quad|y-2|<\sqrt{(x-1)^2+(y-2)^2}<\dfrac{\delta}{6} \quad\quad ...(2)$

and $\quad|x-1|<\sqrt{(x-1)^2+(y-2)^2}<\dfrac{\delta}{6} \quad\quad ...(3)$

$\therefore\quad 0<x-1<1\Rightarrow0<x<2\Rightarrow1<x+1<3$

Hence, from (1)

$$|x^2+3y-7|\le3|x-1|+3|y-2|$$

Using (1) and (2), we get

$$|(x^2+3y)-7|\le3|\sqrt{(x-1)^2+(y-2)^2}|+3|\sqrt{(x-1)^2+(y-2)^2}|$$
$$=6\sqrt{(x-1)^2+(y-2)^2}$$
$$<\frac{\delta}{6}=\varepsilon$$

$\Rightarrow\quad|(x^2+3y)-7|<\varepsilon$

$\Rightarrow\quad\lim_{(x,y)\to(1,2)}(x^2+3y)=7$

Hence, the limit and value of the function at point $(1,2)$ are equal therefore the function is continuous at $(1,2)$.

Example 3.   *Show that the function*
$$f(x,y) = \frac{xy^3}{x^2 + y^6}, x \neq 0, y \neq 0 \ and \ f(0,0) = 0$$

is not continuous at (0, 0) in (x, y).

Solution.   Here $f(0, 0) = 0$ (given)

Let us suppose $(x, y) \to (0, 0)$ through the curve $x = y^3$.

Then $\lim\limits_{(x,y)\to(0,0)} f(x,y) = \lim\limits_{y\to 0} \frac{y^6}{y^6 + y^6} = \frac{1}{2}$

Aganin, let $(x, y) \to (0, 0)$ through the line $y = x$, then

$$\lim\limits_{(x,y)\to(0,0)} f(x,y) = \lim\limits_{x\to 0} \frac{x.x^3}{x^2 + x^6} = \lim\limits_{x\to 0} \frac{x^2}{1 + x^4} = 0$$

Since, the limit obtained by two different approaches are different, hence $\lim\limits_{(x,y)\to(0,0)} f(x,y)$ does not exist. Hence, the given function is not continuous.

Example 4.   *Let $X = \{(x, y) : 0 < x < 1\}$ and $f : X \to R$ be defined by $f(x, y) = x+y$. Prove that f is continuous at every point of the domain X.*

Solution.   Let $X = \{(x, y) : 0 < x < 1, 0 < y < 1\}$

Let (a, b) be any point of X.

To prove, that $f(x, y)$ is continuous at (a, b).

Let $\varepsilon > 0$ be given.

Also, let $|x - a| < \varepsilon/2$ and $|y - b| < \varepsilon/2$ ... (1)

         Therefore $|f(x, y) - (a, b)| = |x+y - (a+b)|$

                           $= |(x - a) + (y - b)|$

                           $\leq |x - a| + |y - b|$ [By triangular inequality]

                           $< \varepsilon/2 + \varepsilon/2$ [From (1)]

                           $= \varepsilon$

$\Rightarrow$ $|f(x, y) - (a, b)| < \varepsilon$

$\Rightarrow$ $\lim\limits_{(x,y)\to(a, b)} f(x, y)$ exist and hence equal to (a, b).

Therefore f is continuous at (a, b) and f is continuous at every point of X.

Example 5.   *Show that the function*
$$f(x,y) = \frac{xy}{\sqrt{(x^2 + y^2)}}, x \neq 0, y \neq 0 \ and \ f(0,0) = 0$$

is continuous at the origin in (x, y) together.

Solution.   Let $\varepsilon > 0$ be given

Let us suppose $x = r \cos \theta, y = r \sin \theta$.

Then $f(r \cos \theta, r \sin \theta) = \frac{r^2 \cos\theta \sin\theta}{r\sqrt{\sin^2 \theta + \cos^2 \theta}} = r \cos \theta \sin \theta$

                                      $= \frac{1}{2} r \sin 2\theta$

Now, consider
$$| f(r \cos \theta, r \sin \theta) - f(0, 0)| = |f(r \cos \theta, r \sin \theta)|$$
$$= \left| \frac{1}{2} r \sin 2\theta \right|$$
$$= \frac{1}{2} r \, | \sin 2\theta \, |$$
$$\leq \frac{1}{2} r \qquad\qquad\qquad [\because \ |\sin 2\theta| \leq 1]$$

Now, if we choose $r = 2\varepsilon$.

Therefore we have $\varepsilon > 0$ such that
$$| f \, (r \cos \theta, r \sin \theta)| < \varepsilon \text{ for all values of } \theta \qquad \dots (1)$$
Equation (1) is true for all points within a circle about the origin and radius $r = 2\varepsilon$. Therefore $f(r \cos \theta, r \sin \theta)$ is uniformly continuous in r for all values of θ. Hence $f(x, y)$ is continuous in (x, y) at the origin.

Example 6. *Check the continuity at (1, 2) of the function*
$$f(x,y) = \begin{cases} x^2 + 4y &, \quad when \quad (x,y) \neq (1,2) \\ 0 &, \quad when \quad (x,y) = (1,2) \end{cases}$$

Solution. We have $\displaystyle\lim_{(x,y)\to(1,2)} x^2 + 4y = (1)^2 + 4\cdot2 = 9$

Therefore, the limit exists and is equal to 9.

Since it is given that $f(1, 2) = 0$ and $\displaystyle\lim_{(x,y)} f(x,y) = 9$

Thus $\displaystyle\lim_{(x,y)\to(1,2)} f(x,y) \neq f(1,2)$

Hence, the function is not continuous at (1, 2).

Example 7. *Show that the function f defined as follows has a removable discontinuity at (2, 3).*
$$f(x,y) = \begin{cases} 3xy &, \quad (x,y) \neq (2,3) \\ 6 &, \quad (x,y) = (2,3) \end{cases}$$

Again redefine the function to make it continuous at (2, 3)

Solution. We have
$$\lim_{(x,y)\to(2,3)} f(x,y) = \lim_{(x,y)\to(2,3)} 3xy = 3 \times 2 \times 3 = 18 \qquad \dots (1)$$
Since, given that $f(2, 3) = 6$ $\qquad\qquad \dots (2)$

Therefore, from equation (1) and (2), we get
$$\lim_{(x,y)\to(2,3)} f(x,y) \neq f(2, 3).$$
Hence, the function $f(x, y)$ is discontinuous at (2, 3).

Since $\displaystyle\lim_{(x,y)\to(2,3)} f(x,y)$ exists but is not equal to $f(2, 3)$ therefore the discontinuity is removable.

We can remove the discontinuity by redefining the function as follows
$$f(x,y) = \begin{cases} 3xy &, \quad (x,y) \neq (2,3) \\ 18 &, \quad (x,y) = (2,3) \end{cases}$$

Example 8. *Show that the following function are discontinuous at the origin*

$$f(x,y) = \begin{cases} \dfrac{x^4 - y^4}{x^4 + y^4} & , \ (x,y) \neq (0,0) \\ 0 & , \ (x,y) = (0,0) \end{cases}$$

Solution. Since, we know that the function $f(x,y)$ is continuous at a point (a,b) of its domain if $\lim\limits_{(x,y)\to(a,b)} f(x,y)$ exists and is equal to $f(a,b)$ otherwise is discontinuous at (a,b).

Now, let $(x,y) \to (0,0)$ along the straight line $y = mx$. Then

$$\lim_{(x,y)\to(0,0)} f(x,y) = \lim_{(x,mx)\to(0,0)} \frac{x^4 - y^4}{x^4 + y^4}$$

Putting $\qquad\qquad y = mx$

$$= \lim_{x\to 0} \frac{x^4 - m^4 x^4}{x^4 + m^4 x^4} = \lim_{x\to 0} \frac{1 - m^4}{1 + m^4} = \frac{1 - m^4}{1 + m^4}$$

which is different for different values of m. Therefore, $\lim\limits_{(x,y)\to(0,0)} f(x,y)$ does not exist. Hence, $f(x,y)$ is discontinuous at the origin.

Example 9. *Show that the following function is discontinuous at $(0,0)$*

$$f(x,y) = \begin{cases} \dfrac{x^3 + y^3}{x - y} & , \ x \neq y \\ 0 & , \ x = y \end{cases}$$

Solution. Since we know that a function $f(x,y)$ is continuous at a point (a,b) of its domain if $\lim\limits_{(x,\,y)\to(a,\,b)} f(x,y)$ exists and is equal to $f(a,b)$, otherwise $f(x,y)$ is discontinuous at (a,b).

Now, we shall show that $\lim\limits_{(x,\,y)\to(0,\,0)} f(x,y)$ does not exist.

Let (x,y) tends to $(0,0)$ through the curve $y = x - mx^3$. Then

$$\lim_{(x,\,y)\to(0,\,0)} f(x,y) = \lim_{(x,\,y)\to(0,\,0)} \frac{x^3 + y^3}{x - y}$$

$$= \lim_{x\to 0} \frac{x^3 + (x - mx^3)^3}{x - (x - mx^3)} \qquad\qquad [\because \ y = x - mx^3]$$

$$= \lim_{x\to 0} \frac{x^3[1 + (1 - mx^3)^3]}{mx^3} = \lim_{x\to 0} \frac{2 - 3mx^2 + 3m^2 x^4 - m^3 x^6}{m}$$

$$= \frac{2}{m}, \quad \text{(which is different for different values of } m.)$$

Therefore, $\lim\limits_{(x,y)\to(0,0)} f(x,y)$ does not exist.

Hence, the function $f(x,y)$ is discontinuous at the origin.

Example 10. *Show that the function is continuous at the origin*

$$f(x,y) = \begin{cases} \dfrac{x^3 y^3}{x^2 + y^2} & , \ (x,y) \neq (0,0) \\ 0 & , \ (x,y) = (0,0) \end{cases}$$

Solution. We shall show that
$$\lim_{(x,\ y)\to(0,\ 0)} f(x,y) = 0$$
Let $\varepsilon>0$. For all $(x,y) \neq (0, 0)$, we have
$$| f(x,y) - 0 | = \left| \frac{x^3 y^3}{x^2 + y^2} \right|$$
$$= \frac{x^3 y^3}{x^2 + y^2} \qquad \text{[Putting } x = r\cos\theta, y = r\sin\theta]$$
$$= r^4 \cos^3 \theta \sin^3 \theta \qquad [\because \cos^3\theta \leq 1, \sin^3\theta \leq 1]$$
$$\leq r^4 = (x^2 + y^2)^2$$
$$< \varepsilon \text{ if } x^2 < \sqrt{\varepsilon/2} \text{ and } y^2 < \sqrt{\varepsilon/2}.$$

Now, if we take $\delta = (\varepsilon/2)^{1/4}$, then, we see that for any given $\varepsilon > 0$, there exists $\delta>0$, such that
$$|f(x,y) - 0| < \varepsilon, \text{ whenever } |x| < \delta, |y| < \delta.$$

Therefore, $\lim_{(x,\ y)\to(0,\ 0)} f(x,y) = 0$

Since given that the $f(0, 0) = 0$, therefore $\lim_{(x,\ y)\to(0,\ 0)} f(x,y) = f(0,0)$
Hence, $f(x, y)$ is continuous at the origin.

Example 11. *Let $f : R^2 \to R$ be a function, defined by*
$$f(x,y) = \begin{cases} \dfrac{xy}{x^2 + y^2} & \text{when } (x,y) \neq (0,0) \\ 0 & \text{when } (x,y) = (0,0) \end{cases}$$
Show that f is not continuous at $(0, 0)$ but is continuous in each variable seperately.

Solution. For point (x, y) on the x-axis, we have $y = 0$ and $f(x,y) = f(x, 0) = 0$, so the function has the constant value, 0, every where on the x-axis, which gives that $f(x, y)$ is continuous at $x = 0$. Similarly $f(x, y)$ has the constant value, 0, at all points on the y-axis, so if we put $x = 0$, the function $f(x, y)$ is continuous at $y = 0$. Now, we shall show that $f(x, y)$ is not continuous at origin.

Let $y = x$ Then $f(x,y) = f(x,x) = \dfrac{x^2}{2x^2} = \dfrac{1}{2}$

Also $f(0,0) = 0$ (given)

Since there are points on the line arbitrarily close to the origin and since $f(0, 0) \neq \dfrac{1}{2}$, the function of two variable $f(x, y)$ is not continuous at the origin.

Example 12. *Show that the function $f(x, y) = xy$ is continuous at the point $(2, 3)$.*

Solution. Here, we have $f(x, y) = xy \Rightarrow f(2, 3) = 2 \times 3 = 6$
Now, we shall find the limit of $f(x, y) = xy$.
Let $\varepsilon > 0$ be given.
For a given $\varepsilon>0$, we want to find $\delta>0$ such that
$$0 < \sqrt{(x-2)^2 + (y-3)^2} < \delta \Rightarrow |(xy) - 6| < \varepsilon$$

$$|x - 2| \le \sqrt{(x-2)^2 + (y-3)^2} < \delta \qquad \dots (1)$$

and $\qquad |y - 3| \le \sqrt{(x-2)^2 + (y-3)^2} < \delta \qquad \dots (2)$

Now, consider

$$|xy - 6| = |xy - 2y + 2y - 6| = |y(x-2) + 2(y-3)|$$
$$\le |y|.|x-2| + 2|y-3| \qquad \dots (3)$$

Now, from (2) and (3), we have

$$|y - 3| < \delta < 1 \Rightarrow 2 < |y| < 4$$

$\therefore \qquad |xy - 6| < 4.|x-2| + 2|y-3| < 4\delta + 2\delta = 6\delta$

If $\delta = \dfrac{\varepsilon}{6}$, then

$$|xy - 6| < 6\frac{\varepsilon}{6} = \varepsilon$$

$\Rightarrow \qquad |xy - 6| < \varepsilon$

$\Rightarrow \qquad \lim\limits_{(x,y)\to(2,3)} xy = 6$

Since, $\qquad f(2, 3) = \lim\limits_{(x,y)\to(2,3)} f(x,y)$

Hence, the function $f(x, y) = xy$ is continuous at $(2, 3)$.

Example 13. *Let $f : R^2 \to$ R be a function defined by*

$$f(x,y) = \begin{cases} \dfrac{xy(x^2 - y^2)}{x^2 + y^2} & , \quad (x,y) \ne (0,0) \\ 0 & , \quad (x,y) = (0,0) \end{cases}$$

show that the function $f(x, y)$ is continuous at $(0, 0)$.

Solution. we know that the function $f(x, y)$ will be continuous at $(0, 0)$ if limit of the function = value of the function at $(0, 0)$. It is given that, the value of the function $f(x, y)$ at $(0, 0)$ is equal to zero *i.e.*, $f(0, 0) = 0$

Now, we shall find the limit of $f(x, y)$.

Let $\varepsilon > 0, \exists\ \delta > 0$, such that

$$|f(x, y) - f(0, 0)| < \varepsilon \text{ whenever } |(x, y) - (0, 0)| < \delta$$

or $\qquad \left| \dfrac{xy(x^2 - y^2)}{x^2 + y^2} - 0 \right| < \varepsilon$

whenever $\sqrt{(x-0)^2 + (y-0)^2} = \sqrt{x^2 + y^2} < \delta \qquad \dots (1)$

Now $\qquad \sqrt{x^2 + y^2} < \delta \Rightarrow |x| < \delta < \sqrt{\varepsilon}$ and $|y| < \delta < \sqrt{\varepsilon}$

The function (a, b), but naturally the point itself, must belong to the domain.

$\therefore \quad$ from (1) $\quad \left| \dfrac{xy(x^2 - y^2)}{x^2 + y^2} \right| \le \left| \dfrac{xy(x^2 + y^2)}{x^2 + y^2} \right| = |x|.|y| < \sqrt{\varepsilon}.\sqrt{\varepsilon} = \varepsilon$

which gives that

$$\left| \frac{xy(x^2 - y^2)}{x^2 + y^2} - 0 \right| < \varepsilon \text{ whenever } |(x, y) - (0, 0)| < \delta.$$

$$\Rightarrow \quad \lim_{(x,y)\to(0,0)} \left(\frac{xy(x^2 - y^2)}{(x^2 + y^2)} \right) = 0.$$

Hence, $f(x, y)$ is continuous at $(0, 0)$.

REMARKS

- For functions of two variables, continuous at some point (a, b), it is true that addition, subtraction, multiplication and division, produce new functions which are also continuous at this point [if only the domains of the resulting functions satisfy the condition, mentioned above in (1)].

- Every polynomial and every rational function of two variables will be continuous in all of its domain.

- For a function of two variables defined and continuous in a closed region it can be shown that the range of the function is a closed interval. Such a function thus has an absolute maximum value and an absolute minimum value.

- In order that function of two variables, $f(x, y)$ shall be continuous in both the variables together, it must have the same limiting value by all possible approaches to the critical point. Therefore, a necessary and sufficient condition involves the condition that the function is not only continuous in each direction, but the continuity is uniform for all direction.
  If we put $x = a + r \cos \theta, y = b + r \sin\theta$, then by the definition of continuity, we have
  $$|f(a + r \cos \theta, b + r \sin\theta) - f(a, b)| < \varepsilon$$
  which must hold for all values of r, which is independent of θ i.e., the transformed function must be uniformly continuous in r for all values of $|\theta| \le 2\pi$.

- If $\lim_{r\to 0} f(a + r \cos \theta, b + r \sin \theta) = f(a, b)$ for every value of θ, then, it is not necessary that the function is continuous at (a, b).

For example, Let $f(x, y) = \dfrac{xy^3}{x^2 + y^6}, x \ne 0, y \ne 0$ and $f(0, 0) = 0$ does not exist as x and y approaches zero. Thus function is therefore discontinuous at the origin. However, putting $x = r \cos\theta, y = r \sin\theta$, we have

$$\lim_{r\to 0} \frac{r^4 \cos\theta \sin^3\theta}{r^2 \cos^2\theta + r^6 \sin^6\theta} = \lim_{r\to 0} r^2 . \frac{\cos\theta \sin^3\theta}{\cos^2\theta + r^4 \sin^6\theta}$$

$$= 0 = f(0, 0) \text{ for each constant value of } \theta.$$

EXERCISE 10.1

1. Let $f : R^2 \to R$ be defined by $f(x, y) = x^2 + y^2$, show that
$$\lim_{(x,y)\to(0,0)} f(x, y) = 0$$

2. Show that $\lim_{(x,y)\to(0,0)} \dfrac{2x^3 - y^3}{x^2 + y^2} = 0$.

3. Show that $\lim_{(x,y)\to(0,0)} \dfrac{2x - y}{x^2 + y^2}$ does not exist.

4. Prove that $\lim_{(x,y)\to(0,0)} f(x, y)$ does not exist, where
$$f(x, y) = \frac{xy^2}{x^2 + y^4}, (x, y) \ne (0, 0)$$

5. Let
$$f(x, y) = \frac{x^2 y^2}{x^2 y^4 + (x - y)^2} (x^2 y^2 + (x - y)^2 \ne 0)$$

Show that

$$\lim_{x\to 0}\left[\lim_{y\to 0} f(x,y)\right] = \lim_{y\to 0}\left[\lim_{x\to 0} f(x,y)\right] = 0$$

6. Show that $f(x,y) = \dfrac{2y}{x}$ exists at $(0, 0)$.

7. Show that the iterated limit of the function

$$f(x,y) = y\sin\frac{1}{x}$$

exist at $(0, 0)$, but the limit of the function does not exist.

8. Show that $\displaystyle\lim_{(x,y)\to(0,0)}\left[\log\frac{a(1-e^x)}{(a-y^x)}\right]$ does not exist.

9. Show that iterated limit of the function

$$f(x,y) = y\sin\frac{1}{x}$$

exists at $(0, 0)$ but the limit of the function does not exist.

10. Show that $\displaystyle\lim_{(x,y)\to(0,0)}\left[\frac{x-y}{\sqrt{x^2-y^2}}\right]$ does not exist.

11. Let

$$f(x,y) = \begin{cases} \dfrac{xy}{\sqrt{x^2-y^2}} & \text{,when} \quad (x,y) \neq (0,0) \\ 0 & \text{, when} \quad (x,y) = (0,0) \end{cases}$$

Show that $f(x, y)$ is continuous at $(0, 0)$

12. Let $f(x,y) = \begin{cases} x\sin\dfrac{1}{y} & \text{, if} \quad y \neq 0 \\ 0 & \text{, if} \quad y = 0 \end{cases}$

is continuous at $(0, 0)$.

13. Show that the function xy is continuous at each point in the x-y plane.

14. Find the point where the function

$f(x,y) = \dfrac{x^2}{x^2-2y}$ is discontinuous.

15. Show that $f(x,y) = \sqrt{|xy|}$ is continuous at $(0, 0)$.

16. Let $f : R^2 \to R$ be defined as

$$f(x,y) = \begin{cases} 1 & \text{, if } x \text{ is irrational} \\ 0 & \text{, if } x \text{ is rational} \end{cases}$$

Show that for any point (a, b), $\displaystyle\lim_{(x,y)\to(a,b)} f(x,y)$ does not exist.

17. Show that the function $f(x, y)$ defined by

$$f(x,y) = \begin{cases} \dfrac{x^2-y^2}{x^2+y^2} & \text{if} \quad x \neq 0, y \neq 0 \\ 0 & \text{if} \quad x = y = 0 \end{cases}$$

is discontinuous at that point $(0, 0)$.

18. Show that the function defined by

$$f(x,y) = \frac{xy^2}{x^2+y^4}, f(0,0) = 0$$

is discontinuous in (x, y) at the origin. Also, show that this function is continuous along the radius vector $\theta = \pi/2$.

19. Let $f : R^2 \to R$ be a continuous function. Define a function $g : R^2 \to R$ as

$$g(x,y) = \begin{cases} f(x,y) & \text{if} \quad (x,y) \neq (0,0) \\ f(x,y)+1 & \text{if} \quad (x,y) = (0,0) \end{cases}$$

Show that g is not continuous at $(0, 0)$.

20. If $f : R^2 \to R$ be a function defined by

$$f(x,y) = \begin{cases} \dfrac{xy^2+x^2y}{x^3+y^3} & \text{if} \quad (x,y) \neq (0,0) \\ 0 & \text{if} \quad (x,y) = (0,0) \end{cases}$$

is not continuous at $(0, 0)$.

HINTS TO SELECTED PROBLEMS

2. $\displaystyle\lim_{\substack{x\to 0 \\ y\to 0}}\frac{2x^3-y^3}{(x^2+y^2)} = \lim_{x\to 0}\frac{2x^3-m^3x^3}{x^2+m^2x^2}$

$$= \lim_{x\to 0} x\left(\frac{2-m^3}{1+m^2}\right) = 0$$

9. $\displaystyle\lim_{x\to 0}\lim_{y\to 0} f(x,y) = \lim_{x\to 0}\lim_{y\to 0} y\sin\frac{1}{x}$

$$= \lim_{x\to 0} .0 = 0$$

Also

$$\lim_{\substack{x\to 0 \\ y\to 0}} f(x,y) = \lim_{\substack{x\to 0 \\ y\to 0}} y\sin\frac{1}{x} = \lim_{x\to 0} mx\sin\frac{1}{x}$$

13. $\displaystyle\lim_{\substack{x\to a \\ y\to b}} f(x,y) = \lim_{\substack{x\to a \\ y\to b}} xy = \lim_{x\to a} bx = ab = f(a,b)$

17. Taking the simultaneous limit along the line $y = mx$.

19. $\displaystyle\lim_{\substack{x\to 0 \\ y\to 0}} g(x,y) = \lim_{\substack{x\to 0 \\ y\to 0}} f(x,y)$

continuous and since $\displaystyle\lim_{\substack{x\to 0 \\ y\to 0}} f(x,y) = f(0,0)$.

Therefore, $\displaystyle\lim_{\substack{x\to 0 \\ y\to 0}} g(x,y)$ exists because $f(x,y)$ is

But $g(0,0) = f(0,0)+1$

$\therefore\ \displaystyle\lim_{\substack{x\to 0 \\ y\to 0}} g(x,y) = \lim_{\substack{x\to 0 \\ y\to 0}} f(x,y) = f(0,0) \neq g(0,0)$

10.4 PARTIAL DIFFERENTIATION OF FUNCTION OF TWO VARIABLES

Let $z = f(x,y)$ be a continuous function of two independent variables x and y. If we treat y as constant and only x changes, then the partial differential coefficient of z with respect to x, is denoted by

$$\frac{\partial z}{\partial x} \text{ or } \frac{\partial f}{\partial x} \text{ or } f_x \text{ or } D_x f.$$

Thus, $\dfrac{\partial f}{\partial x} = \displaystyle\lim_{\partial x\to 0} \dfrac{f(x+\delta x, y) - f(x,y)}{\partial x}$ provided this limit exists and is unique.

Similarly, if x is regarded as constant and only y-varies, then the differential coefficient of z with respect to y is denoted by

$$\frac{\partial f}{\partial y} \text{ or } \frac{\partial z}{\partial y} \text{ or } f_y \text{ or } D_y f.$$

Thus, $\dfrac{\partial z}{\partial y} = \displaystyle\lim_{\partial y\to 0} \dfrac{f(x, y+\delta y) - f(x,y)}{\partial y}$ provided this limit exists and is unique.

10.4.1 SECOND ORDER PARTIAL DIFFERENTIAL COEFFICIENTS

Partial differential coefficients $\dfrac{\partial f}{\partial x}$ and $\dfrac{\partial f}{\partial y}$ can be differentiated further, partially with respect to x and y.

Thus, partial differential coefficients of $\dfrac{\partial f}{\partial x}$ with respect to x and y are $\dfrac{\partial}{\partial x}\left(\dfrac{\partial f}{\partial x}\right)$ and $\dfrac{\partial}{\partial y}\left(\dfrac{\partial f}{\partial x}\right)$ respectively.

or $\qquad \dfrac{\partial^2 f}{\partial x^2}$ and $\dfrac{\partial^2 f}{\partial y \partial x}$ respectively

By means of the limit, the derivative of the second order are defined by

$$\frac{\partial^2 f}{\partial x^2} = \lim_{\partial x\to 0} \frac{f(x_0 + 2\delta x, y_0) - 2f(x_0 + \delta x, y_0) + f(x_0, y_0)}{(\delta x)^2}$$

$$\frac{\partial^2 f}{\partial y^2} = \lim_{\partial y\to 0} \frac{f(x_0, y_0 + 2\delta y) - 2f(x_0, y_0 + \delta y) + f(x_0, y_0)}{(\delta y)^2}$$

$$\frac{\partial^2 f}{\partial y \partial x} = \lim_{\delta y\to 0}\lim_{\delta x\to 0} \frac{f(x_0 + \delta x, y_0 + \delta y) - f(x_0, y_0 + \delta y) - f(x_0 + \delta x, y_0) + f(x_0, y_0)}{\delta y\, \delta x}$$

and $\dfrac{\partial^2 f}{\partial x \partial y} = \displaystyle\lim_{\delta x\to 0}\lim_{\delta y\to 0} \dfrac{f(x_0 + \delta x, y_0 + \delta y) - f(x_0 + \delta x, y_0) - f(x_0 y_0 + \delta y) + f(x_0, y_0)}{\delta x\, \delta y}$

REMARK

- It is clear that

$$\frac{\partial^2 f}{\partial x \partial y} = \frac{\partial^2 f}{\partial y \partial x}$$

Solved Examples

Example 1. Let $f(x, y)$ be a function, defined by

$$f(x, y) = x \sin \frac{1}{x} + y \sin \frac{1}{y},$$

$$x \neq 0, y \neq 0$$

$$f(0, y) = y \sin \frac{1}{y}, y \neq 0$$

$$f(x, 0) = x \sin \frac{1}{x}, x \neq 0$$

$$f(0, 0) = 0$$

Examine the existence of f_x and f_{yx} at $x = 0, y = 0$.

Solution. Here, we have

$$f_x(0, 0) = \lim_{\delta x \to 0} \frac{f(\delta x, 0) - f(0, 0)}{\delta x}$$

$$= \lim_{\delta x \to 0} \frac{\delta x \sin \frac{1}{\delta x} - 0}{\delta x}$$

$$= \lim_{\delta x \to 0} \sin \frac{1}{\delta x}$$

Since, limit does not exist

$\Rightarrow$ $f_x(0, 0)$ does not exists.

Now,

$$f_{yx} = \lim_{\delta y \to 0} \lim_{\delta x \to 0}$$

$$\frac{f(\delta x, \delta y) - f(0, \delta y)}{- f(\delta x, 0) + f(0, 0)}}{\delta x \delta y}$$

$$= \lim_{\delta y \to 0} \lim_{\delta x \to 0} \frac{\delta x \sin \frac{1}{\delta x} + \delta y \sin \frac{1}{\delta y} - \delta y \sin \frac{1}{\delta y} - \delta x \sin \frac{1}{\delta x}}{\delta x \delta y}$$

$$= \lim_{\delta y \to 0} \lim_{\delta x \to 0} \frac{0}{\delta x \delta y} = 0$$

Based on the following Results

- $$\frac{\partial f}{\partial x} = \lim_{\delta x \to 0} \frac{f(x + \delta x, y) - f(x, y)}{\delta x}$$

- $$\frac{\partial f}{\partial y} = \lim_{\delta y \to 0} \frac{f(x, y + \delta y) - f(x, y)}{\delta y}$$

- $$\frac{\partial^2 f}{\partial x^2} = \lim_{\delta x \to 0} \frac{f(x_0 + 2\delta x, y_0) - 2f(x_0 + \delta x, y_0) + f(x_0, y_0)}{\delta x}$$

- $$\frac{\partial^2 f}{\partial y^2} = \lim_{\delta y \to 0} \frac{f(x_0, y_0 + 2\delta y) - 2f(x_0, y_0 + \delta y) + f(x_0, y_0)}{\delta y}$$

- $$\frac{\partial^2 f}{\partial y \partial x} = \lim_{\delta y \to 0} \lim_{\delta x \to 0} \frac{f(x_0 + \delta x, y_0 + \delta y) - f(x_0, y_0 + \delta y) - f(x_0 + \delta x, y_0) + f(x_0, y_0)}{\delta y \delta x}$$

- $$\frac{\partial^2 f}{\partial x \partial y} = \lim_{\delta x \to 0} \lim_{\delta y \to 0} \frac{f(x_0 + \delta x, y_0 + \delta y) - f(x_0 + \delta x, y_0) - f(x_0, y_0 + \delta y) + f(x_0, y_0)}{\delta x \delta y}$$

In spite of the fact that limit is zero, the derivative $f_{yx}(0, 0)$ cannot be said to exist, since $f_y(0, 0)$ does not exist.

REMARK

☞ The existence of higher derivatives implies the existence of the corresponding derivatives of lower order. Thus, in order that f_{xy} should exist at a point, it is necessary that the partial derivatives f_x should exist in the neighbourhood of that point. However, it is possible for the limit, defining f_{xy} to exist without the partial derivative f_x existing. In such cases higher derivatives cannot be said to exist.

Example 2. Let $f(x,y) = \dfrac{x^2 y}{x^4 + y^2}$, *for* $x \neq 0$, $y \neq 0$ *and* $f(0, 0) = 0$

Show that the partial derivatives f_x, f_y exist everywhere in the region $-1 \leq x \leq 1$, $-1 \leq y \leq 1$, although $f(x, y)$ is discontinuous in (x, y) at the origin.

Solution. Here, we have

$$f(x,y) = \frac{x^2 y}{x^4 + y^2} \qquad\qquad (x \neq 0, y \neq 0)$$

$$\Rightarrow \qquad f_x = 2xy . \frac{y^2 - x^4}{(x^4 + y^2)^2} \qquad\qquad (x \neq 0, y \neq 0)$$

and $$f_y = x^2 . \frac{x^4 - y^2}{(x^4 + y^2)^2} \qquad\qquad (x \neq 0, y \neq 0)$$

Also, for $x = y = 0$, we obtain

$$f_x = \lim_{\delta x \to 0} \frac{\delta x . 0}{(\delta x)^4 + 0} = 0 \text{ and } f_y = 0$$

Similarly, the following results can be prove

$$f_x(x, y) = 0 \text{ for } x = 0, y \neq 0$$

$$f_x(x, y) = 0 \text{ for } x \neq 0, y = 0$$

$$f_y(x, y) = 0 \text{ for } x = 0, y \neq 0$$

$$f_y(x,y) = \frac{1}{x^2} \text{ for } x \neq 0, y = 0$$

Hence, the partial derivative f_x, f_y exist at all points of the given region.
Now, we shall check the continuity of the given function.
The limiting value of $f(x)$, along the line $y = 0$, is given by $\lim_{\delta x \to 0} .0 = 0$.

and, limiting value of $f(x)$, along the line $y = x^2$ is given by $\lim_{x \to 0} \dfrac{x^4}{x^4 + x^4} = \dfrac{1}{2}$.

Now, since limit obtained by two different approaches are different. Hence $f(x)$ is discontinuous in (x, y) at the origin.

Example 3. If $u = f(y / x)$, then prove that $x \dfrac{\partial u}{\partial x} + y \dfrac{\partial u}{\partial y} = 0$

Solution. Here, we have $u = f\left(\dfrac{y}{x}\right)$

$$\therefore \qquad \frac{\partial u}{\partial x} = f'\left(\frac{y}{x}\right)\left(-\frac{y}{x^2}\right)$$

$$\Rightarrow \qquad x\frac{\partial u}{\partial x} = -\frac{y}{x} f'\left(\frac{y}{x}\right) \qquad\qquad\qquad .... (1)$$

Also $$\frac{\partial u}{\partial y} = f'\left(\frac{y}{x}\right).\frac{1}{x}$$

$$\therefore \qquad y\frac{\partial u}{\partial y} = \frac{y}{x}f'\left(\frac{y}{x}\right) \qquad \qquad \dots (2)$$

Now, from (1) and (2), we get

$$x\frac{\partial u}{\partial x} + y\frac{\partial u}{\partial y} = 0$$

Example 4. If $z = x^2\tan^{-1}\left(\frac{y}{x}\right) - y^2\tan^{-1}\left(\frac{x}{y}\right)$, then prove that $\dfrac{\partial^2 z}{\partial y \partial x} = \dfrac{x^2 - y^2}{x^2 + y^2}$

Solution. Here, we have

$$z = x^2\tan^{-1}\left(\frac{y}{x}\right) - y^2\tan^{-1}\left(\frac{x}{y}\right)$$

$$\therefore \qquad \frac{\partial z}{\partial x} = 2x\tan^{-1}\frac{y}{x} + x^2.\frac{1}{1+\frac{y^2}{x^2}}\left(-\frac{y}{x^2}\right) - y^2\frac{1}{1+\frac{x^2}{y^2}}.\frac{1}{y}$$

$$= 2x\tan^{-1}\frac{y}{x} - \frac{x^2 y}{x^2 + y^2} - \frac{y^3}{x^2 + y^2}$$

Now, $\dfrac{\partial^2 z}{\partial y \partial x} = \dfrac{\partial}{\partial y}\left(\dfrac{\partial z}{\partial x}\right)$

$$= \frac{\partial}{\partial y}\left[2x\tan^{-1}\frac{y}{x} - \frac{x^2 y}{x^2 + y^2} - \frac{y^3}{x^2 + y^2}\right]$$

$$= 2x.\frac{1}{1+\frac{y^2}{x^2}}.\frac{1}{x} - x^2\left[\frac{(x^2 + y^2).1 - y.2y}{(x^2 + y^2)^2}\right] - \frac{(x^2 + y^2)3y^2 - y^3.2y}{(x^2 + y^2)^2}$$

$$= \frac{2x^2}{x^2 + y^2} - x^2\left[\frac{x^2 - y^2}{(x^2 + y^2)^2}\right] - \frac{y^2(3x^2 + 3y^2 - 2y^2)}{(x^2 + y^2)^2}$$

$$= \frac{2x^2}{x^2 + y^2} - x^2\left[\frac{x^2 - y^2}{(x^2 + y^2)^2}\right] - y^2\left[\frac{3x^2 + y^2}{(x^2 + y^2)^2}\right]$$

$$= \frac{2x^2(x^2 + y^2) - x^2(x^2 - y^2) - y^2(3x^2 + y^2)}{(x^2 + y^2)^2}$$

$$= \frac{2x^4 + 2x^2 y^2 - x^4 + x^2 y^2 - 3x^2 y^2 - y^4}{(x^2 + y^2)^2}$$

$$= \frac{x^4 - y^4}{(x^2 + y^2)^2} = \frac{(x^2 - y^2)(x^2 + y^2)}{(x^2 + y^2)^2}$$

$$= \frac{x^2 - y^2}{(x^2 + y^2)}$$

Example 5. If $x^x y^y z^z = c$, then show that $\dfrac{\partial^2 z}{\partial x \partial y} = -[x \log(ex)]^{-1}$ when $x=y=z$.

Solution. Here, we have

$$x^x y^y z^z = c$$

Taking logarithm, we get

$$x \log x + y \log y + z \log z = \log c.$$

Now, differentiating partially w.r.t. x, we get

$$\left[x.\frac{1}{x} + 1.\log x \right] + \left[z.\frac{1}{z} + 1.\log z \right]\frac{\partial z}{\partial x} = 0 \qquad [\because z \text{ is a function of } x \text{ and } y]$$

$$\Rightarrow \quad (1 + \log x) + (1 + \log z)\frac{\partial z}{\partial x} = 0$$

$$\Rightarrow \qquad\qquad \frac{\partial z}{\partial x} = -\frac{1 + \log x}{1 + \log z}$$

Similarly, we get

$$\frac{\partial z}{\partial y} = -\frac{1 + \log y}{1 + \log z}$$

Now

$$\frac{\partial^2 z}{\partial x \partial y} = \frac{\partial}{\partial x}\left(\frac{\partial z}{\partial y} \right) = \frac{\partial}{\partial x}\left[-\frac{1 + \log y}{1 + \log z} \right] = -(1 + \log y)\frac{\partial}{\partial x}\left(\frac{1}{1 + \log z} \right)$$

$$= -(1 + \log y)\left[\frac{-1}{(1 + \log z)^2}\cdot\frac{1}{z}\cdot\frac{\partial z}{\partial x} \right] = \frac{(1 + \log y)}{(1 + \log z)^2}\cdot\frac{1}{z}\cdot\frac{\partial z}{\partial x}$$

$$= \frac{(1 + \log y)}{(1 + \log z)^2}\left[-\frac{1}{z}\left(\frac{1 + \log x}{1 + \log z} \right) \right] = -\frac{(1 + \log y)(1 + \log x)}{z(1 + \log z)^3}$$

$$= -\frac{(1 + \log x)(1 + \log x)}{z(1 + \log x)^3}, \text{at } x = y = z$$

$$= -\frac{1}{x(1 + \log x)} = -\frac{1}{x[\log e + \log x]}$$

$$= -\frac{1}{x \log[ex]} = -[x \log(ex)]^{-1}$$

Example 6. If $u = x^y$, then show that $\dfrac{\partial^2 u}{\partial y \partial x} = \dfrac{\partial^2 u}{\partial x \partial y}$.

Solution. Here, we have $\quad u = x^y$

$$\Rightarrow \qquad\qquad \frac{\partial u}{\partial y} = x^y.\log x$$

$$\frac{\partial}{\partial x}\left(\frac{\partial u}{\partial y} \right) = \frac{\partial^2 u}{\partial x \partial y} = y x^{y-1}\log x + x^{y-1}x$$

$$= y x^{y-1}\log x + x^{y-1} \qquad\qquad\qquad\qquad \text{... (1)}$$

Now $\dfrac{\partial u}{\partial x} = yx^{y-1}$

$\therefore$ $\dfrac{\partial^2 u}{\partial y \partial x} = \dfrac{\partial}{\partial y}\left(\dfrac{\partial u}{\partial x}\right) = \dfrac{\partial}{\partial y}(yx^{y-1})$

$= x^{y-1} + yx^{y-1}\log x$ ... (2)

From (1)and (2), we conclude that

$$\dfrac{\partial^2 u}{\partial x\, \partial y} = \dfrac{\partial^2 u}{\partial y\, \partial x}.$$

Example 7. *Let $f(x, y)$ be a function defined by*

$$f(x,y) = \begin{cases} xy\dfrac{x^2 - y^2}{x^2 + y^2} & \text{for} \quad (x,y) \neq (0,0) \\ 0 & \text{for} \quad (x,y) = (0,0) \end{cases}$$

Prove that

(i) *f, f_x, f_y are continuous in (x, y),*

(ii) *f_{xy} and f_{yx} exist at every point (x, y) and are continuous except at $(0, 0)$*

(iii) *$f_{xy}(0, 0) = 1$ and $f_{yx}(0, 0) = -1$.*

Solution. Firstly, we examine the continuity of f at $(0, 0)$

Putting $x = r\cos\theta, y = r\sin\theta$, we get

$$f(r\cos\theta, r\sin\theta) = r^2\cos\theta\sin\theta\left(\dfrac{r^2\cos^2\theta - r^2\sin^2\theta}{r^2\cos^2\theta + r^2\sin^2\theta}\right)$$

$$= r^2\sin\theta\cos\theta\,(\cos^2\theta - \sin^2\theta)$$

$$= \dfrac{1}{2}r^2\sin 2\theta\cos 2\theta = \dfrac{1}{4}r^2\sin 4\theta$$

$$\leq \varepsilon \text{ where } r < 2\sqrt{(\varepsilon\,\text{cosec}\,4\theta)}$$

Since, $\sqrt{(\text{cosec}\,4\theta)}$ is never less than one, it follows that if we put $r = 2\sqrt{\varepsilon}$ then ε may be chosen such that $f(r\cos\theta, r\sin\theta)$ is always less than ε for all values of θ and r. Thus, the transformed function is uniformly continuous in r for all values of θ.

Hence, the function f is continuous in (x, y) together, at the origin.

Now, we show that f_x and f_y are continuous at $(0, 0)$.

For $(x, y) \neq (0, 0)$, we have

$$f_x = y\left(\dfrac{x^2 - y^2}{x^2 + y^2} + \dfrac{4x^2 y^2}{(x^2 + y^2)^2}\right), f_y = x\left(\dfrac{x^2 - y^2}{x^2 + y^2} - \dfrac{4x^2 y^2}{(x^2 + y^2)^2}\right)$$

For $(x, y) = (0, 0)$ we have

$$f_x(0,0) = \lim_{\delta x \to 0}\dfrac{\delta x.0}{\delta x}\left[\dfrac{(\delta x)^2 - 0}{(\delta x)^2 + 0}\right] = 0$$

$$f_y(0,0) = \lim_{\delta y \to 0}\dfrac{0.\delta y}{\delta y}\left[\dfrac{0 - (\delta y)^2}{0 + (\delta y)^2}\right] = 0$$

Now, we shall show that f_x and f_y are continuous in (x, y) at the origin. We only prove the continuity of f_x at the origin. Now, putting $x = r \cos \theta, y = r \sin \theta$, we get

$$f_x(r\cos\theta, r\sin\theta) = r\sin\theta \left[\frac{r^2\cos^2\theta - r^2\sin^2\theta}{r^2\cos^2\theta + r^2\sin^2\theta} + \frac{4r^4\cos^2\theta\sin^2\theta}{(r^2\cos^2\theta + r^2\sin^2\theta)^2} \right]$$

$$= r\sin\theta(\cos 2\theta + \sin^2 2\theta)$$

$$\leq 2r, \text{ for all values of } \theta.$$

Hence, if we take $r = \dfrac{\varepsilon}{2}$ then ε may be chosen such that

$$f_x(r\cos\theta, r\sin\theta) < \varepsilon \text{ for } r \text{ and for all values of } \theta.$$

Hence, f_x is continuous at the origin.

Similarly, we can shown that f_y is continuous at the origin.

Now,

$$f_{xy}(0,0) = \lim_{\delta x \to 0} \lim_{\delta y \to 0} \left[\frac{f(0+\delta x, 0+\delta y) - f(0+\delta x, 0) - f(0, 0+\delta y) + f(0,0)}{\delta x . \delta y} \right]$$

$$= \lim_{\delta x \to 0} \lim_{\delta y \to 0} \left[\frac{(\delta x)^2 - (\delta y)^2}{(\delta x)^2 + (\delta y)^2} \right]$$

$$= \lim_{\delta x \to 0} \left[\frac{(\delta x)^2 - 0}{(\delta x)^2 + 0} \right] = 1$$

and $f_{yx}(0,0) = \lim_{\delta y \to 0} \lim_{\delta x \to 0} \left[\frac{(\delta x)^2 - (\delta y)^2}{(\delta x)^2 + (\delta y)^2} \right]$

$$= \lim_{\delta y \to 0} \left[\frac{0 - (\delta y)^2}{0 + (\delta y)^2} \right] = -1$$

Hence, $f_{xy}(0, 0) \neq f_{xy}(0, 0)$, in the case. It may be shown that the order of differentiation can be interchanged at every other point of the finite region.

Example 8. Let $f(x,y) = \begin{cases} \dfrac{1}{4}(x^2 + y^2)\log(x^2 + y^2) & , \quad when(x,y) \neq (0,0) \\ 0 & , \quad when(x,y) = (0,0) \end{cases}$

Show that $f_{xy} = f_{yx}$ at all points (x, y). Also, show that neither of the derivatives is continuous in (x, y) at the origin.

Solution. For $x \neq 0 \, y \neq 0$, we have

$$f_x = \frac{1}{2}x\{1 + \log(x^2 + y^2)\} \text{ and } f_y = \frac{1}{2}y\{1 + \log(x^2 + y^2)\}$$

and $f_{xy} = f_{yx} = \dfrac{xy}{x^2 + y^2}$

For $x = 0, y = 0$, we have

$$f_x = f_y = f_{xy} = f_{yx} = 0$$

Hence, $f_{xy} = f_{yx}$ at every point.

Now, we show that $f_{xy} = f_{yx}$ is not continuous at $(0, 0)$.

Since $\lim_{(x,y) \to (0,0)} \dfrac{xy}{x^2 + y^2}$ does not exist.

(Because, if we put $y = mx$, then limit of the function, depends upon m, i.e., limit is not unique). Hence, the limit does not exist. It follows that $f_{xy} = f_{yx}$ is not continuous at the origin.

Example 9. Let $f(x,y) = \begin{cases} x^2 y^2 \cdot \cos\dfrac{1}{x} , & \text{for all values of } y \text{ so long as } x \neq 0 \\ 0 , & \text{for } x = 0 \end{cases}$

Show that

     (i) $f_{xy} = f_{yx}$ at all points (x, y),

     (ii) neither f_{xy} nor f_{yx} is continuous in x at $x = 0$, if $y \neq 0$.

and (iii) both f_{xy} and f_{yx} are continuous in (x, y), together at the origin.

Solution.   (i) For $x \neq 0$, $y \neq 0$, we have

$$f_{xy} = f_{yx} = 4xy\cos\frac{1}{x} + 2y\sin\frac{1}{x}$$

For $(x = 0, y \neq 0)$, $(x = 0, y = 0)$ and $(x \neq 0, y = 0)$, it can be easily shown that
$$f_{xy} = f_{yx} = 0$$
Hence, $f_{xy} = f_{yx}$ at every point.

(ii) If $y \neq 0$, then

$$\lim_{x\to 0}\left(4xy\cos\frac{1}{x} + 2y\sin\frac{1}{x}\right) \text{does not exist.}$$

Hence, neither f_{xy} nor f_{yx} is continuous in x at $x = 0$ if $y \neq 0$.

(iii) Both f_{xy} and f_{yx} are continuous in (x, y) together at the origin. Since, the simultaneous limit

$$\lim_{(x,y)\to(0,0)}\left\{4xy\cos\frac{1}{x} + 2y\sin\frac{1}{x}\right\} = 0 = f_{xy}(0,0) = f_{yx}(0,0)$$

Example 10. Find $\dfrac{\partial f}{\partial x}, \dfrac{\partial f}{\partial y}$ at $(1, 2)$ if $f(x, y) = 2x^2 - xy + 2y^2$

Solution.   We have

$$\left(\frac{\partial f}{\partial x}\right)_{(1,2)} = \lim_{h\to 0}\frac{f(1+h,2) - f(1,2)}{h}$$

$$= \lim_{h\to 0}\frac{\{2(1+h)^2 - (1+h).2 + 2(2)^2\} - \{2(1)^2 - 1.2 + 2(2)^2\}}{h}$$

$$= \lim_{h\to 0}\frac{2h^2 + 2h}{h} = \lim_{h\to 0}(2h+2) = 2$$

Similarly $$\left(\frac{\partial f}{\partial y}\right)_{(1,2)} = \lim_{k\to 0}\frac{f(1,2+k) - f(1,2)}{k}$$

$$= \lim_{k\to 0}\frac{\{2 - (2+k) + 2(2+k)^2\} - \{2 - 2 + 8\}}{k}$$

$$= \lim_{k\to 0}\frac{2k^2 + 7k}{k} = \lim_{k\to 0}2k + 7 = 7 .$$

EXERCISE 10.2

1. If $z = \left[\dfrac{x^2 + y^2}{x + y}\right]$, show that

$$\left(\dfrac{\partial z}{\partial x} - \dfrac{\partial z}{\partial y}\right)^2 = 4\left(1 - \dfrac{\partial z}{\partial x} - \dfrac{\partial z}{\partial y}\right).$$

2. If $u = f\left(\dfrac{y}{x}\right)$, show that

$$x\dfrac{\partial u}{\partial x} + y\dfrac{\partial u}{\partial y} = 0.$$

3. If $u = \sin^{-1}\dfrac{x}{y} + \tan^{-1}\dfrac{y}{x}$ show that

$$x\dfrac{\partial u}{\partial x} + y\dfrac{\partial u}{\partial y} = 0.$$

4. If $u = \sin^{-1}\dfrac{\sqrt{x} - \sqrt{y}}{\sqrt{x} + \sqrt{y}}$, show that

$$x\dfrac{\partial u}{\partial x} + y\dfrac{\partial u}{\partial y} = 0$$

5. If $u = \tan^{-1}\dfrac{xy}{\sqrt{1 + x^2 + y^2}}$, show that

$$\dfrac{\partial^2 u}{\partial x\,\partial y} = \dfrac{1}{(1 + x^2 + y^2)^{3/2}}.$$

6. If $u = \sin(\sqrt{x} + \sqrt{y})$, show that

$$x\dfrac{\partial u}{\partial x} + y\dfrac{\partial u}{\partial y} = \dfrac{1}{2}(\sqrt{x} + \sqrt{y})\cos(\sqrt{x} + \sqrt{y}).$$

7. If $u = f(r)$, where $r^2 = x^2 + y^2$, show that

$$\dfrac{\partial^2 u}{\partial x^2} + \dfrac{\partial^2 u}{\partial y^2} = f''(r) + \dfrac{f'(r)}{r}.$$

8. If $f(x,y) = \begin{cases} \dfrac{xy}{x^2 + y^2} & , \quad x \neq 0, y \neq 0 \\ 0 & , \quad x = 0, y = 0 \end{cases}$

Then, show that f_x, f_y exist at $(0, 0)$ and examine the continuity of f_x, f_y with respect to x, y and (x, y) together.

9. In any given region R, where $f(x, y)$ is continuous in x and y together and $f_x = 0$, $f_y = 0$ for all values of x and y in R. Show that $f(x, y)$ is constant in R.

10. Show that $f(x, y) = |x| + |y|$ is continuous at $(0, 0)$.

11. If $f(x,y) = (x^2 + y^2)\tan^{-1}\dfrac{y}{x}$ when $x \neq 0$

and $f(0, y) = \dfrac{\pi y^2}{2}$

Show that $f_{xy}(0,0) \neq f_{yx}(0,0)$.

12. If $f(x,y) = \begin{cases} xy & \text{if } |y| \leq |x| \\ -xy & \text{if } |y| > |x| \end{cases}$

Show that $f_{xy}(0, 0) \neq f_{yx}(0, 0)$

HINTS TO SELECTED PROBLEMS

1. $\dfrac{\partial z}{\partial x} = \dfrac{x^2 - y^2 + 2xy}{(x + y)^2}, \dfrac{\partial z}{\partial y} = \dfrac{y^2 - x^2 + 2xy}{(x + y)^2}$

2. $\dfrac{\partial u}{\partial x} = \left[f'\left(\dfrac{y}{x}\right)\right]\left(-\dfrac{y}{x^2}\right)$

$\Rightarrow x\dfrac{\partial u}{\partial x} = \left(-\dfrac{y}{x}\right)f'\left(\dfrac{y}{x}\right)$ ... (1)

Similarly, $\qquad y\dfrac{\partial u}{\partial y} = \dfrac{y}{x}f'\left(\dfrac{y}{x}\right)$ ... (2)

Adding (1) and (2) to get the required results.

10. The function $f(x, y)$ will be continuous at $(0, 0)$ if the simultaneous limit of $f(x, y)$ as $x \to 0, y \to 0$ exists.

$$\therefore \quad \lim_{\substack{x \to 0 \\ y \to 0}} f(x,y) = \lim_{\substack{x \to 0 \\ y \to 0}} (|x| + |y|)$$

$$= \lim_{\substack{x \to 0 \\ y \to 0}} (|x| + |mx|)$$

$$= \lim_{x \to 0} |x|.(1 + m) = 0 = f(0,0).$$

Answers

8. $f_x(0, 0) = f_y(0, 0) = 0, f_x$ is continuous in x, discontinuous in y at $(0, 0)$ and discontinuous in (x, y) at $(0, 0)$; f_y is continuous in y, discontinuous in x at $(0, 0)$ and discontinuous in (x, y) at $(0, 0)$.

10.5 HOMOGENEOUS FUNCTIONS

A function, in which every term is of the same degree say n, is known as a homogeneous function of degree n.

Let $f(x, y) = a_0x^n + a_1x^{n-1}y + a_2x^{n-2}y^2 + \ldots + a_{n-1}x.y^{n-1} + a_ny^n$ be a function of x and y. Here, we see that every term of the function has the same degree *i.e.* n. Hence, $f(x, y)$ is a homogeneous function of x and y of degree n.

The homogeneous function $f(x, y)$ can be written as

$$f(x, y) = x^n\left[a_0 + a_1\left(\frac{y}{x}\right) + a_2\left(\frac{y}{x}\right)^2 + \ldots + a_{n-1}\left(\frac{y}{x}\right)^{n-1} + a_n\left(\frac{y}{x}\right)^n\right]$$

$$= x^n F\left(\frac{y}{x}\right).$$

REMARK

- Let $f(x, y)$ be a function of x and y. Then $f(x, y)$ is said to be a homogeneous function of x and y of degree n if $f(tx, ty) = t^n f(x, y)$.

10.5.1 EULER'S THEOREM ON HOMOGENEOUS FUNCTIONS

If u is a homogeneous function of x and y of degree n, then $x\dfrac{\partial u}{\partial x} + y\dfrac{\partial u}{\partial y} = nu$

Proof. Since, u is a homogeneous function of x and y of degree n, then, we can write

$$u = x^n F\left(\frac{y}{x}\right) \qquad \ldots (1)$$

$$\therefore \quad \frac{\partial u}{\partial x} = nx^{n-1}F\left(\frac{y}{x}\right) + x^n F'\left(\frac{y}{x}\right)\left(-\frac{y}{x^2}\right) \quad \Rightarrow \quad x\frac{\partial u}{\partial x} = nx^n F\left(\frac{y}{x}\right) - x^{n-1}yF'\left(\frac{y}{x}\right)$$

$$\Rightarrow \quad x\frac{\partial u}{\partial x} = nu - x^{n-1}yF'\left(\frac{y}{x}\right) \qquad \therefore \qquad x\frac{\partial u}{\partial x} = nu - x^{n-1}yF'\left(\frac{y}{x}\right) \qquad \ldots (2)$$

Now, $\qquad \dfrac{\partial u}{\partial y} = x^n F'\left(\dfrac{y}{x}\right).\dfrac{1}{x} \qquad \Rightarrow \qquad y\dfrac{\partial u}{\partial y} = x^{n-1}yF'\left(\dfrac{y}{x}\right) \qquad \ldots (3)$

Adding (2) and (3), we get $x\dfrac{\partial u}{\partial x} + y\dfrac{\partial u}{\partial y} = nu$

REMARKS

- This theorem can be extended to a homogeneous function of any number of variables. Thus, if $f(x_1, x_2, \ldots x_n)$ be a homogeneous function of n variables, say $x_1, x_2, \ldots x_n$, then

$$x_1\frac{\partial f}{\partial x_1} + x_2\frac{\partial f}{\partial x_2} + \ldots + x_n\frac{\partial f}{\partial x_n} = nf$$

- If u is a homogeneous function of degree n, then

(a) $x\dfrac{\partial^2 u}{\partial x^2} + y\dfrac{\partial^2 u}{\partial x\partial y} = (n-1)\dfrac{\partial u}{\partial y}$, \qquad (b) $x^2\dfrac{\partial^2 u}{\partial x\partial y} + y\dfrac{\partial^2 u}{\partial y^2} = (n-1)\dfrac{\partial u}{\partial y}$,

and (c) $x^2\dfrac{\partial^2 u}{\partial x^2} + 2xy\dfrac{\partial^2 u}{\partial x\partial y} + y^2\dfrac{\partial^2 u}{\partial y^2} = n(n-1)u$

10.6 DIFFERENTIABILITY OF FUNCTION OF TWO VARIABLES

A function $f(x, y)$ defined on an open interval $]a, b[$ is said to be totally differentiable or simply differentiable in the open interval $]a, b[$, if there exist two constant α and β depending on f and $]a, b[$ such that

$$\lim_{\substack{\delta x \to 0 \\ \delta y \to 0}} \frac{f(a + \delta x, b + \delta y) - f(a,b) - \alpha \delta x - \beta \delta y}{\sqrt{(\delta x)^2 + (\delta y)^2}} = 0$$

or

$$\lim_{\substack{h \to 0 \\ k \to 0}} \frac{f(a + h, b + k) - f(a,b) - \alpha h - \beta k}{\sqrt{h^2 + k^2}} = 0$$

or

$$\lim_{\substack{h \to 0 \\ k \to 0}} \frac{f(a + h, b + k) - f(a,b) - h f_x(a,b) - k f_y(a,b)}{\sqrt{h^2 + k^2}} = 0$$

THEOREM 1. *If the function $f(x, y)$ is totally differentiable at the point (a, b), then it is continuous at (a, b).*

Proof. Since $f(x, y)$ is differentiable at (a, b) we have

$$\lim_{\substack{h \to 0 \\ k \to 0}} \frac{f(a + h, b + k) - f(a,b) - h f_x(a,b) - k f_y(a,b)}{\sqrt{h^2 + k^2}} = 0$$

$$\Rightarrow \quad \lim_{\substack{h \to 0 \\ k \to 0}} [f(a + h, b + k) - f(a,b) - h f_x(a,b) - k f_y(a,b)] = 0$$

$$\Rightarrow \quad \lim_{\substack{h \to 0 \\ k \to 0}} f(a + h, b + k) = f(a,b).$$

Hence, $f(x, y)$ is continuous at $[a, b]$.

THEOREM 2. *If a function $f(x, y)$ is totally differentiable, then the partial derivatives f_x and f_y both exist and finite.*

Proof. Let $f(x, y)$ be a function, which is totally differentiable , then there exist constants α and β such that

$$\lim_{\substack{h \to 0 \\ k \to 0}} \frac{f(a + h, b + k) - f(a,b) - \alpha h - \beta k}{\sqrt{h^2 + k^2}} \text{ exists.}$$

$$\Rightarrow \quad \lim_{k \to 0} \frac{f(a, b + k) - f(a,b) - \beta k}{k} \text{ exists.}$$

$$\Rightarrow \quad f_y(a, b) = \beta$$

Similarly, we get $f_x(a, b) = \alpha$

REMARKS

- Continuity in two variables is a necessary condition for total differentiability. It is not a sufficient condition.

- If $z = f(x, y)$, then $z + \delta z = f(x + \delta x, y + \delta y)$ so that $\delta z = f(x + \delta x, y + \delta y) - f(x, y)$. If z is differentiable, then the value of dz is given by

$$\delta z = \frac{\partial z}{\partial x} \delta x + \frac{\partial z}{\partial y} \delta y + \varepsilon \delta(x, y) \text{ where } \varepsilon \to 0 \text{ as } \delta x \text{ and } \delta y \text{ approaches zero simultaneously.}$$

- The increment $\{f(x, y) - f(a, b)\}$ itself may depend on x and y, the differentiability expresss that as an approximation we can replace the increment by the very simple linear function $A(x - a) + B(y - b)$ i.e., the differentiable of dz. The error, thus, introduce as $\varepsilon\delta(x, y)$, which is vanishing of a higher order than the linear expression, or in general small in comparision with this.

- Geometrically, the approximation , discussed above means that, the given surface $z = f(x, y)$ is replaced by the plane. $z' - z_0 = A(x-a) + B(y - b)$
  which is called the tangent plane to the surface at the point $P[x_0, y_0, f(x_0, y_0)]$. A normal vector to the surface (*i.e.*, to its tangent plane) at the point P, is therefore
  $$(A, B, -1) = (f_x'(a, b), f_y'(a, b), -1)$$

10.6.1 CONDITION FOR DIFFERENTIABILITY IN POLAR COORDINATES

Let $f(x, y)$ be a differentiable function at a point (a, b), then
$$\lim_{(h,k)\to(0,0)} \frac{f(a + h, b + k) - f(a,b) - hf_x(a,b) - kf_y(a,b)}{\sqrt{h^2 + k^2}} = 0$$

Put $h = r\cos\theta$, $y = r\sin\theta$ and take limit $r \to 0$, then we have
$$\lim_{r\to0}\left[\frac{f(a + r\cos\theta, b + r\sin\theta) - f(a,b)}{r} - \cos\theta f_x(a,b) - \sin\theta f_y(a,b) \right] = 0$$
$$\Rightarrow \lim_{r\to0}\left[\frac{f(a + r\cos\theta, b + r\sin\theta) - f(a,b)}{r} \right] = \cos\theta f_x(a,b) + \sin\theta f_y(a,b)$$

THEOREM 1. *Let $f(x, y)$ be a function such that $f_x(x, y)$, $f_y(x, y)$ exist at the point (x_0, y_0) and let one of those derivative say $f_y(x, y)$ exist for all values of x, y in the neighbourhood of (x_0, y_0) and be continuous at that point in two variables together. Then, $f(x, y)$ is totally differentiable at (x_0, y_0).*

Proof.   Let $\varepsilon > 0$ be given. Since, the partial derivative $f_x(x, y)$ exist at (x_0, y_0), there exist a $\delta_1 > 0$ such that for $|\delta x| < \delta_1$, we have
$$\left| \frac{f(x_0 + \delta x, y_0) - f(x_0, y_0)}{\delta x} - f(x_0 y_0) \right| < \frac{\varepsilon}{2}$$

Let $\dfrac{f(x_0 + \delta x, y_0) - f(x_0, y_0)}{\delta x} - f(x_0 y_0) = k$,   where $|k| < \dfrac{\varepsilon}{2}$ ... (1)

Since, $f_y(x, y)$ exists for all values of x and y in the neighbourhood of (x_0, y_0) and is continuous in (x, y) together at this point, then by mean value theorem, we have
$$f(x_0 + \delta x, y_0 + \delta y) - f(x_0 + \delta x, y_0) = \delta y\, f_y(x_0 + \delta x, y_0 + \theta\delta y) \qquad \text{... (2)}$$
Therefore,
$$\left| \frac{1}{\delta(x,y)}\left[f(x_0 + \delta x, y_0 + \delta y) - f(x_0, y_0) - \delta x f_x(x_0, y_0) - \delta y \cdot f_y(x_0, y_0) \right] \right.$$

$$= \left| \frac{1}{\delta(x,y)}[f(x_0 + \delta x, y_0 + \delta y) - f(x_0 + \delta x, y_0) + f(x_0 + \delta x, y_0) \right.$$

$$\left. - f(x_0, y_0) - \delta x f_x(x_0, y_0) - \delta y f_y(x_0, y_0) \right|$$

$$= \left| \frac{1}{\delta(x,y)} [\delta y\, f_y(x_0 + \delta x, y_0 + \theta \delta y) + \delta x\, f_x(x_0, y_0) + k\,\delta x - \delta x\, f_x(x_0, y_0) \right.$$
$$\left. - \delta_y f_y(x_0, y_0)] \right|$$

$$= \left| \frac{\delta y}{\delta(x,y)} \{ f_y(x_0 + \delta x, y_0 + \theta \delta y) - f_y(x_0, y_0) \} + \frac{\delta x}{\delta(x,y)}.k \right|$$

$$\le \left| \frac{\delta y}{\delta(x,y)} \right| |\, f_y(x_0 + \delta x, y_0 + \theta \delta y) - f_y(x_0, y_0)\,| + \left| \frac{\delta x}{\delta(x,y)}.k \right| \qquad \dots (3)$$

Now, since $f_y(x, y)$ is continuous in (x, y) at (x_0, y_0), there exist a $\delta_2 > 0$ such that

$$|\, f_y(x_0 + \delta x, y_0 + \theta \delta y) - f_y(x_0, y_0)\,| < \varepsilon / 2 \qquad \dots(4)$$

For $\qquad\qquad |\delta x| < \delta_2$ and $|\,\delta y\,| < \delta_2$.

Since, $\delta x \ne 0$, $\delta y \ne 0$, we have

$$\left| \frac{\delta x}{\delta(x,y)} \right| < 1, \left| \frac{\delta x}{\delta(x,y)} \right| < 1 \qquad \dots (5)$$

Let $\delta_3 = \min\{\delta_1, \delta_2\}$, then for $|\delta x| < \delta_3$ and $|\delta y| < \delta_3$ the inequality (3), with the help of (4) and (5), becomes

$$\left| \frac{1}{\delta(x,y)} [f(x_0 + \delta x, y_0 + \delta y) - f(x_0, y_0) - \delta x f_x(x_0, y_0) - \delta y f_y(x_0, y_0)] \right|$$
$$< \frac{\varepsilon}{2} + \frac{\varepsilon}{2} = \varepsilon$$

$$\therefore \quad \lim_{\substack{\delta x \to 0 \\ \delta y \to 0}} \frac{f(x_0 + \delta x, y_0 + \delta y) - f(x_0, y_0) - \delta x\, f_x(x_0, y_0) - \delta y\, f_y(x_0, y_0)}{\delta(x,y)} = 0$$

It follows that, $f(x, y)$ is totally differentiable at (x_0, y_0).

Solved Examples

Example 1. Let

$$f(x,y) = \begin{cases} \dfrac{xy}{\sqrt{x^2 + y^2}} & , (x,y) \ne (0,0) \\ 0 & , (x,y) = (0,0) \end{cases}$$

Show that $f(x, y)$ is continuous but not differentiable at $(0, 0)$.

Solution. (i) *Test for continuity.* Let us suppose (x, y) approaches $(0, 0)$ through the line $y = mx$, then

$$\lim_{\substack{x \to 0 \\ y \to 0}} \frac{xy}{\sqrt{x^2 + y^2}}$$

$$= \lim_{x \to 0} \frac{x\, mx}{\sqrt{x^2 + m^2 x^2}} = 0 \quad \dots (1)$$

Based on the following Results

- A function in which every term is of the same degree say n is known as homogeneous functions.
  It can also be written as $f(x,y) = x^n F\left(\dfrac{y}{x}\right)$

- **(Euler's Theorem for Homogeneous Function).** If u is a homogeneous function of x and y of degree n then $x\dfrac{\partial u}{\partial x} + y\dfrac{\partial u}{\partial y} = nu$

- $\delta z = \dfrac{\partial z}{\partial x}\,\partial x + \dfrac{\partial z}{\partial y}\,\partial y$ where $z = f(x, y)$

- $f_x(0,0) = \lim_{h \to 0} \dfrac{f(0+h) - f(0,0)}{h}$

- $f_y(0,0) = \lim_{k \to 0} \dfrac{f(0,0+k) - f(0,0)}{k}$

Again, let (x, y) approaches through the path $x = y^3$, then

$$\lim_{\substack{x \to 0 \\ y \to 0}} \frac{xy}{\sqrt{x^2 + y^2}} = \lim_{y \to 0} \frac{y^4}{\sqrt{y^6 + y^2}} = 0 \qquad \qquad \dots (2)$$

Now, from (1) and (2), it is obvious that

$$\lim_{\substack{x \to 0 \\ y \to 0}} \frac{xy}{\sqrt{x^2 + y^2}} \text{ exists and equal to 0.}$$

Since $f(0, 0) = 0$ (given)

Hence, the function $f(x, y)$ is continuous at $(0, 0)$.

(ii) *Test for differentiability.* Here, we have

$$f_x(0,0) = \lim_{h \to 0} \frac{f(0+h,0) - f(0,0)}{h}$$

$$= \lim_{h \to 0} \frac{f(h,0) - f(0,0)}{h} = 0$$

and $\qquad f_y(0,0) = \lim_{k \to 0} \frac{f(0,0+k) - f(0,0)}{k} = \lim_{k \to 0} \frac{f(0,k) - f(0,0)}{k} = 0$

Now $\quad \lim_{(h,k) \to (0,0)} \dfrac{f(0+h, y+k) - f(0,0) - hf_x(0,0) - kf_y(0,0)}{\sqrt{h^2 + k^2}}$

$$= \lim_{(h,k) \to (0,0)} \frac{f(h,k)}{\sqrt{h^2 + k^2}} \qquad [\because \ f_x(0, 0) = f_y(0, 0) = 0]$$

$$= \lim_{(h,k) \to (0,0)} \frac{hk}{h^2 + k^2}$$

$$= \lim_{(h,k) \to (0,0)} \frac{mh^2}{h^2(1 + m^2)}$$

$$=. \frac{m}{1 + m^2} \qquad \qquad [\text{Let } (h, k) \to (0, 0) \text{ along } k = mh]$$

which implies that the required limit is depend upon m i.e., it depends upon the path along with $(h, k) \to (0, 0)$, so this limit does not exist.

Hence, $f(x, y)$ is not differentiable at $(0, 0)$.

Example 2. *Examine the continuity and differentiablility of the function*

$$f(x, y) = \begin{cases} \dfrac{xy^2}{x^2 + y^2} & \text{when} \quad (x, y) \neq (0, 0) \\ 0 & \text{when} \quad (x, y) = (0, 0) \end{cases}$$

Solution. (i) *Test for continuity.* Firstly, we test the continuity of $f(x, y)$ at $(0, 0)$.

Let $\varepsilon > 0$ be given.

Let $x = r \cos\theta$ and $y = r \sin\theta$, then

$$f(r \cos\theta, y \sin\theta) = \frac{r \cos\theta \cdot r^2 \sin^2\theta}{r^2(\cos^2\theta + \sin^2\theta)} = r \cos\theta \sin^2\theta$$

Now, consider

$$| f(r\cos\theta, r\sin\theta) - f(0,0)| = |r\cos\theta\sin^2\theta - 0|$$

$$= |r\cos\theta| \sin^2\theta$$

$$\leq r \text{ for all values of } \theta.$$

If we set $r_0 = \varepsilon$, then for all values of θ and for $r < r_0$, we have

$$| f(r\cos\theta, r\sin\theta)| - f(0,0)| < \varepsilon$$

∴ the transformed function is uniformly continuous in r for all values of θ.

Hence, $f(x, y)$ is continuous in (x, y) together at the origin.

(ii) *Test for differentiability.* Now, we discuss the differentiability of function $f(x, y)$ at the origin

$$f_x(0,0) = \lim_{h\to 0} \frac{f(h,0) - f(0,0)}{h}$$

$$= \lim_{h\to 0} \frac{1}{\delta x}\left[\frac{\delta x.0}{(h)^2 + 0} - 0\right] = 0$$

Similarly $f_y(0, 0) = 0$

Now, consider

$$\lim_{\substack{h\to 0 \\ k\to 0}} \frac{f(h,k) - f(0,0) - hf_x(0,0) - kf_y(0,0)}{\delta(x,y)}$$

$$= \lim_{\substack{h\to 0 \\ k\to 0}} \frac{1}{\sqrt{h^2 + k^2}}\left[\frac{hk^2}{h^2 + k^2} - 0 - h.0 - k.0\right]$$

$$= \lim_{\substack{h\to 0 \\ k\to 0}} \frac{hk^2}{(h^2 + k^2)^{3/2}}$$

Now, if we put $k = mh$, then

$$\lim_{\substack{h\to 0 \\ k\to 0}} \frac{h.k^2}{(h^2 + k^2)^{3/2}} = \lim_{h\to 0} \frac{h.m^2h^2}{h^3(1 + m^2)^{3/2}} = \frac{m^2}{(1 + m^2)^{3/2}}$$

Since, this limit is not unique (it depends upon m), thus the limit does not exist. Hence, the given function $f(x, y)$ is not differentiable at $(0,0)$.

Example 3. *Show that the function $f(x, y) = \sin x + \cos y$ is differentiable every where.*

Solution. Let (a, b) be any arbitrary point of R^2, then

$$f_x(a,b) = \lim_{h\to 0} \frac{f(a+h,b) - f(a,b)}{h}$$

$$= \lim_{h\to 0} \frac{\sin(a+h) + \cos b - \sin a - \cos b}{h}$$

$$= \lim_{h\to 0} \frac{\sin(a+h) - \sin a}{h}$$

$$= \lim_{h\to 0} \frac{2\cos(a + h/2)\sin h/2}{h}$$

$$= \lim_{h \to 0} \cos\left(a + \frac{h}{2}\right) \frac{\sin h/2}{h/2} = \cos a$$

Similarly $\qquad f_y(a,b) = -\sin b$

Now, consider

$$\lim_{(h,k)\to(0,0)} [f(a+h,b+k) - f(a,b) - hf_x(a,b) - kf_y(a,b)]$$

$$= \lim_{(h,k)\to(0,0)} [\sin(a+h) + \cos(b+k)]$$

$$- \sin a - \cos b - h\cos a + k\sin b]$$

$$= \lim_{(h,k)\to(0,0)} \left[h\left\{ \frac{\sin(a+h) - \sin a}{h} - \cos a \right\} \right.$$

$$\left. + k\left\{ \frac{\cos(b+k) - \cos b}{k} + \sin b \right\} \right] = 0$$

Hence, the given function is differentiable at every point.

Example 4. *Let* $f(x, y) = xy + x + y^2$. *Show that f is differentiable at the origin.*

Solution. Here, we have

$$f(x, y) = xy + x + y^2.$$

It can be easily verified that

$$f_x(0, 0) = 1 \text{ and } f_y (0, 0) = 0$$

$\therefore \qquad \lim_{(h,k)\to(0,0)} [f(h,k) - f(0,0) - hf_x(0,0) - kf_y(0,0)]$

$$= \lim_{(h,k)\to(0,0)} (hk + k^2) = 0$$

Hence, $f(x, y)$ is differentiable at $(0, 0)$.

Example 5. *Let*

$$f(x,y) = \begin{cases} \dfrac{x^3 - y^3}{x^2 + y^2} & when \quad (x,y) \neq (0,0) \\ 0 & when \quad (x,y) = (0,0) \end{cases}$$

Show that the function f is continuous but not differentiable at the origin.

Solution. (i) *Test for continuity.* If we put $x = r\cos\theta$ and $y = r\sin\theta$, then

$$| f(r\cos\theta, r\sin\theta) - f(0,0) | = \left| \frac{r^3(\cos^3\theta - \sin^3\theta)}{r^2(\cos^2\theta + \sin^2\theta)} - 0 \right|$$

$$= r | (\cos^3\theta - \sin^3\theta) |$$

$$\leq r(| \cos^3\theta | + | \sin^3\theta |)$$

$$\leq 2r , \text{ for all values of } \theta$$

Now, set $r_0 = \dfrac{\varepsilon}{2}$, then for all values of θ and $r < r_0$, we have

$$| f(r\cos\theta, r\sin\theta) - f(0,0) | < \varepsilon$$

$\therefore$ The transformed function is uniformly continuous in r for all values of θ.

Hence $f(x, y)$ is continuous in (x, y) together at the origin.

(ii) *Test for differentiability.* Here, we have

$$f_x(0,0) = \lim_{h \to 0} \frac{f(0+h,0) - f(0,0)}{h} = \lim_{h \to 0} \frac{f(h,0) - f(0,0)}{h}$$

$$= \lim_{h \to 0} \frac{1}{h} \left[\frac{(h)^3 - 0}{(h)^2 + 0} - 0 \right] = 1$$

and $f_y(0,0) = \lim_{k \to 0} \frac{f(0,0+k) - f(0,0)}{k} = \lim_{k \to 0} \frac{f(0,k) - f(0,0)}{k}$

$$= \lim_{k \to 0} \frac{1}{k} \left[\frac{0 - (k)^3}{0 + (k)^2} - 0 \right] = -1$$

Now, consider

$$\lim_{(h,k) \to (0,0)} \frac{f(h,k) - f(0,0) - h f_x(0,0) - k f_y(0,0)}{\sqrt{h^2 + k^2}}$$

$$= \lim_{(h,k) \to (0,0)} \frac{1}{\sqrt{h^2 + k^2}} \left[\frac{h^3 - k^3}{h^3 + k^3} - h + k \right]$$

$$= \lim_{(h,k) \to (0,0)} \frac{(h-k)[(h)^2 + hk + (k)^2 - h^2 - k^2]}{[(h)^2 + (k)^2]^{3/2}}$$

$$= \lim_{(h,k) \to (0,0)} \left[\frac{(h-k)(h+k)}{(h^2 + k^2)^{3/2}} \right]$$

If we put $k = mh$, then the above limit

$$= \lim_{(h \to 0)} \frac{(h - mh)(hmh)}{(h)^3(1 + m^2)^{3/2}} = \frac{(1-m)m}{(1+m^2)^{3/2}}$$

Hence, this limit does not exist, since it depends upon m, It follows that the given function is not differentiable at $(0, 0)$.

Example 6. *Discuss the differentiability of f defined by*

$$f(x,y) = \begin{cases} x^2 \sin\dfrac{1}{x} + y^2 \sin\dfrac{1}{y} & when \quad (x,y) \neq (0,0) \\ 0 & when \quad (x,y) = (0,0) \end{cases}$$

Solution. We first calculate the partial derivatives at $(0, 0)$, we have

$$f_x(0,0) = \lim_{h \to 0} \frac{f(0+h,0) - f(0,0)}{h}$$

$$= \lim_{h \to 0} \frac{\left[(h)^2 \sin\dfrac{1}{h} - 0 \right]}{h} = \lim_{h \to 0} h \sin\frac{1}{h} = 0$$

Similarly $f_y(0, 0) = 0$

Since, we know that, the function will be differentiable or not at $(0, 0)$ according as

$$\frac{1}{r} \{ f(r\cos\theta, r\sin\theta) - f(0,0) \}$$

converges uniformly to the limit $\cos\theta f_x(0,0) + \sin\theta f_y(0,0) = 0$

Now, $\dfrac{1}{r}\{f(r\cos\theta, r\sin\theta) - f(0,0)\}$

$$= \frac{1}{r}\left\{r^2\cos^2\theta\sin\left(\frac{1}{r\cos\theta}\right) + r^2\sin^2\theta\sin\left(\frac{1}{r\sin\theta}\right) - 0\right\}$$

$$= \frac{1}{r}\left\{\cos^2\theta\sin\left(\frac{1}{r\cos\theta}\right) + \sin^2\theta\sin\left(\frac{1}{r\sin\theta}\right)\right\}$$

which converges uniformly to 0 in the closed interval $[0, \pi]$ as r tends to 0.

Hence, $f(x, y)$ is differentiable at (0, 0).

Example 7. *Show that the function*

$$f(x,y) = \begin{cases} x^2\sin(1/x) + y^2\sin(1/y) &, & x, y \neq 0 \\ x^2\sin(1/x) &, & x \neq 0, y = 0 \\ y^2\sin(1/y) &, & x = 0, y \neq 0 \\ 0 &, & x = y = 0 \end{cases}$$

is differentiable at the origin.

Solution. Since we know that function $f(x, y)$ is said to be differentiable at any point (a, b) of its domain if $f(a+h, b+k) - f(a, b)$ can be expressed in the form of

$$f(a+h, b+k) - f(a,b) = h.f_x(a,b) + k.f_y(a,b) + \sqrt{h^2 + k^2}.g(h,k)$$

where $g(h, k) \to 0$ as $(h, k) \to (0, 0)$

Hence, $f_x(0,0) = \lim\limits_{h\to 0}\dfrac{f(0+h,0) - f(0,0)}{h}$

$$= \lim\limits_{h\to 0}\frac{f(h,0) - f(0,0)}{h} = \lim\limits_{h\to 0}\frac{h^2\sin(1/h) - 0}{h}$$

$$= \lim\limits_{h\to 0}h\sin(1/h) = 0$$

Again $f_y(0,0) = \lim\limits_{k\to 0}\dfrac{f(0,0+k) - f(0,0)}{k} = \lim\limits_{k\to 0}\dfrac{f(0,k) - f(0,0)}{k}$

$$= \lim\limits_{k\to 0}\frac{k^2\sin 1/k}{k} = 0$$

Now, again

$$f(0+h, 0+k) - f(0,0) = f(h,k) - f(0,0)$$

$$= h^2\sin 1/h + k^2\sin 1/k - 0$$

$$= 0.h + 0.k + \sqrt{h^2 + k^2}\left[\frac{h^2}{\sqrt{h^2 + k^2}}\sin\frac{1}{h} + \frac{k^2}{\sqrt{h^2 + k^2}}\sin\frac{1}{k}\right]$$

$$= f_x(0,0).h + f_y(0,0).k + \sqrt{h^2 + k^2}.g(h,k)$$

where
$$g(h,k) = \frac{h^2}{\sqrt{h^2+k^2}}\sin\frac{1}{h} + \frac{k^2}{\sqrt{h^2+k^2}}\sin\frac{1}{k}$$

We have

$$\lim_{(h,k)\to(0,0)} g(h,k) = \lim_{(h,k)\to(0,0)}\left[\frac{h}{\sqrt{h^2+k^2}}.h\sin\frac{1}{h} + \frac{k}{\sqrt{h^2+k^2}}.k\sin\frac{1}{k}\right]$$

$$= 0 \qquad \left[\because \lim_{h\to 0} h\sin\frac{1}{h} = 0, \lim_{k\to 0} k\sin\frac{1}{k} = 0\right.$$

$$\left.\text{and both } \frac{h}{\sqrt{h^2+k^2}} \text{ and } \frac{k}{\sqrt{h^2+k^2}} \text{ are bounded}\right]$$

Hence, $f(x,y)$ is differentiable at $(0, 0)$.

EXERCISE 10.3

1. Show that the function $f(x, y)$ defined by
$$f(x,y) = \begin{cases} \dfrac{x^3-y^3}{x^3+y^3} & , (x,y) \neq (0,0) \\ 0 & , (x,y) = (0,0) \end{cases}$$
is continuous but not differentiable at $(0, 0)$.

2. Show that the function $f(x, y) = \{|xy|\}^{1/2}$ is not totally differentiable at $(0, 0)$ but that $\dfrac{\partial f}{\partial x}$ and $\dfrac{\partial f}{\partial y}$ both exist at the origin and have the value 0.

3. Show that the function
$$f(x,y) = \begin{cases} \dfrac{x^2y^2}{x^2+y^2} & \text{when } (x,y) \neq (0,0) \\ 0 & \text{when } (x,y) = (0,0) \end{cases}$$
is not totally differentiable at the origin.

4. Show that the function $f(x, y)$ defined by
$$f(x,y) = \begin{cases} \dfrac{2xy}{x^2+y^2} & , (x,y) \neq (0,0) \\ 1 & , (x,y) = (0,0) \end{cases}$$
is not differentiable at $(0, 0)$.

5. Show that the function $f(x, y)$ defined by
$$f(x,y) = \begin{cases} \dfrac{x^3+y^3}{x-y} & , \text{if } x \neq y \\ 1 & , \text{if } x = y \end{cases}$$
is not differentiable at $(0, 0)$.

6. If $u = x^2+y^2$, $v = x^3+y^3$, prove that, if x is considered as a function of u and v, then

$$\frac{\partial x}{\partial u} = -\frac{y}{2x(x-y)}, \frac{\partial x}{\partial v} = \frac{1}{3x(x-y)}.$$

7. Show that the function, $f(x, y)$ defined by
$$f(x,y) = \begin{cases} \dfrac{x^2y^2}{x^4+y^4} & \text{for } x \neq y \neq 0 \\ 0 & \text{for } x = y = 0 \end{cases}$$
is not totally differentiable at origin.

8. Verify the following function for differentiability
   (a) $f(x, y) = e^x + y$ at $(1, 3)$
   (b) $f(x, y) = \cos x + \sin y$ at $(0, 0)$
   (c) $f(x, y) = \cos(xy)$ at $\left(\dfrac{\pi}{4}, \dfrac{\pi}{4}\right)$
   (d) $f(x, y) = |x^2 - y^2|$ at origin.
   (e) $f(x, y) = y^2 \sin\dfrac{x}{y}$ at origin.

9. Examine the following function for total differentiability at the origin.
$$f(x,y) = \begin{cases} xy.\sqrt{\dfrac{x^2-y^2}{x^2+y^2}}, & \text{when } (x,y) \neq (0,0) \\ 0 & , \text{when } (x,y) = (0,0) \end{cases}$$

10. If $f(x, y, z) = 0$, where f is a differentiable function of x, y, z prove that
$$\left(\frac{\partial z}{\partial y}\right)_x \left(\frac{\partial x}{\partial z}\right)_y \left(\frac{\partial y}{\partial x}\right)_z = -1$$
where $\left(\dfrac{\partial z}{\partial y}\right)_x$ denotes the derivative of z with respect to y when x is constant.

HINTS TO SELECTED PROBLEMS

1. Along $y = mx$

$$\lim_{\substack{x \to 0 \\ y \to 0}} f(x,y) = \frac{1-m^3}{1+m^3} \text{ , not unique.}$$

Also $f_x(0,0) = \infty, f_y(0,0) = -\infty$

3. $f_x(0,0) = 0, f_y(0,0) = 0$

Also among $k = mh$, simultaneous double limit exists.

6. Since

$$\frac{\partial x}{\partial u} = -\frac{y}{2x(x-y)}, \frac{\partial x}{\partial v} = \frac{1}{3x(x-y)}$$

Now using the result

$$\frac{\partial x}{\partial x} = 1 = \frac{\partial x}{\partial u} \cdot \frac{\partial u}{\partial x} + \frac{\partial x}{\partial v} \cdot \frac{\partial v}{\partial x}$$

and

$$\frac{\partial x}{\partial y} = \frac{\partial x}{\partial u} \cdot \frac{\partial u}{\partial y} + \frac{\partial x}{\partial v} \cdot \frac{\partial v}{\partial y}$$

10.7 DIRECTIONAL DERIVATIVES OF A FUNCTION OF TWO VARIABLES

Let $f(x, y)$ be a function of two variables x and y and a line S is inclined at an angle θ, with x-axis, then the directional derivative of $f(x, y)$ along this line at a point $P(x, y)$ is defined as

$$\frac{\partial f}{\partial S} = f_x \cos\theta + f_y \sin\theta$$

REMARK

- The existence of all the directional derivatives at a point may not imply the continuity of the function at the point *e.g.* the function $f(x,y) = \dfrac{xy}{x^2 + y^2}$ for $x \neq y \neq 0$ and $f(0, 0) = 0$ then $f(x, y)$ admits of every directional derivatives at the origin $(0, 0)$, but $f(x, y)$ is not continuous at $(0, 0)$.

10.8 COMPOSITE FUNCTIONS

If z is a function of two variables x and y and these variables themselves are given to be the function of the variable t, then z is said to be the composite function of t.

i.e., the relation $\begin{cases} z = f(x,y) \\ x = \phi(t) \\ y = \Psi(t) \end{cases}$, define z as composite function of t.

Here, $\dfrac{dz}{dt}$ is called the total differential coefficients of z with respect to t.

THEOREM 1. *If z is a composite function of t, defined by the relations*

$$z = f(x, y), \quad x = \phi(t), \quad y = \psi(t)$$

where z, possesses first order partial derivatives with respect to x and y, and also x and y possesses continuous derivatives w.r.t., 't', then

$$\frac{dz}{dt} = \frac{\partial z}{\partial x} \cdot \frac{dx}{dt} + \frac{\partial z}{\partial y} \cdot \frac{dy}{dt}$$

Proof. Let $z = f(x, y)$ ... (1)

Let us suppose δt, δx and δy be the corresponding changes in t, x and y respectively, then

$$z + \delta z = f(x + \delta x, y + \delta y) \qquad \text{... (2)}$$

From (1) and (2), we get

$$\delta z = f(x + \delta x, y + \delta y) - f(x, y)$$

$$= [f(x + \delta x, y + \delta y) - f(x, y + \delta y) + f(x, y + \delta y) - f(x, y)]$$

$$\Rightarrow \quad \frac{\delta z}{\delta t} = \frac{f(x + \delta x, y + \delta y) - f(x, y + \delta y)}{\delta x} \cdot \frac{\delta x}{\delta t} + \frac{f(x, y + \delta y) - f(x, y)}{\delta y} \cdot \frac{\delta y}{\delta t} \quad \ldots (3)$$

Now, proceeding to limit $\delta t \to 0$ and consequently δx and δy also tends to zero, we have

$$\lim_{\delta t \to 0} \frac{\delta z}{\delta t} = \frac{dz}{dt}, \quad \lim_{\delta t \to 0} \frac{\delta x}{\delta t} = \frac{dx}{dt}, \quad \lim_{\delta t \to 0} \frac{\delta y}{\delta t} = \frac{dy}{dt}$$

and

$$\lim_{\delta x \to 0} \frac{f(x + \delta x, y + \delta y) - f(x, y + \delta y)}{\delta x} = \frac{\delta z}{\delta x}$$

[Because, while x change to $x + \delta x$, $y + \delta y$ remains unchanged.]

Similarly

$$\lim_{\delta y \to 0} \frac{f(x, y + \delta y) - f(x, y)}{\delta y} = \frac{\partial z}{\partial y}$$

Hence, (3) gives

$$\frac{dz}{dt} = \frac{\partial z}{\partial x} \cdot \frac{dx}{dt} + \frac{\partial z}{\partial y} \cdot \frac{dy}{dt}$$

REMARK

- In general, if $z = f(x_1, x_2, \ldots, x_n)$ where $x_1, x_2, \ldots, x_n$ all are functions of t, we have

$$\frac{dz}{dt} = \frac{\partial z}{\partial x_1} \cdot \frac{dx_1}{dt} + \frac{\partial z}{\partial x_2} \cdot \frac{dx_2}{dt} + \ldots + \frac{\partial z}{\partial x_n} \cdot \frac{dx_n}{dt}$$

10.9 MEAN VALUE THEOREM FOR TWO VARIABLES

Let $f(x, y)$ be a function, such that

(i) *$f(x, y)$ is continuous in the closed domain D*

(ii) *first order differential coefficient of f(x, y) exists in the open domain D'.*

Then $\quad f(a+h, b+k) = f(a, b) + h f_x (a+\theta h, b+\theta k) + k f_y(a+\theta h, b+\theta k)$

where, $\quad (a, b), (a+h), (b+k) \in D$ and $0 < \theta < 1$.

Proof. Let us define a new variable t as $x = a + ht, y = b + kt, 0 < t \leq 1$

where, h and k are constants, then we get the function of single variable t as

$$F(t) = f(x, y) = f(a + ht, b + kt) \quad \ldots(1)$$

By Lagrange's mean value theorem of single variable, we have

$$\frac{F(1) - F(0)}{1 - 0} = F'(\theta), \ 0 < \theta < 1$$

$$\Rightarrow \quad F(1) = F(0) + F'(\theta), \ 0 < \theta < 1 \quad \ldots (2)$$

But, from (1)

$$\therefore \quad F(1) = f(a + h, b + k) \quad \text{and} \quad F(0) = f(a, b)$$

$$f(a + h, b + k) - f(a, b) = F'(\theta), 0 < \theta < 1 \quad \ldots (3)$$

If $\ x = a + ht \quad$ and $\quad y = b + kt$, then $F(t) = f(x, y)$

$$\Rightarrow \quad F'(t) = \frac{d}{dt} F(t) = \frac{d}{dt} f(x, y)$$

$$= \frac{\partial f}{\partial x} \cdot \frac{dx}{dt} + \frac{\partial f}{\partial y} \cdot \frac{dy}{dt}$$

[By definition of directional derivatives]

$$= hf_x(x, y) + k f_y(x, y) \qquad \left[\because \frac{dx}{dt} = h, \frac{dy}{dt} = k \right]$$

or $\qquad F'(t) = hf_x(a + ht, b + kt) + k f_y(a + ht, b + kt) \qquad \ldots (4)$

Now substitute θ in place of t, in (4), we get

$$F'(\theta) = hf_x(a + \theta h, b + \theta k) + k f_y(a + h\theta, b + k\theta) \qquad \ldots (5)$$

Now, from (3) and (5), we have

$$f(a+h, b+k) - f(a, b) = hf_x(a+\theta h, b+\theta k) + k f_y(a+\theta h, b+\theta k)$$

$\Rightarrow \qquad f(a+h, b+k) = f(a, b) + hf_x(a+\theta h, b+\theta k) + kf_y(a+\theta h, b+\theta k)$

10.10 TAYLOR'S THEOREM FOR FUNCTION OF TWO VARIABLES

If $f(x, y)$ possesses continuous partial derivatives upto nth order inclusive for all points (x, y) in the region $a \le x \le a - k, b \le y \le b + k$

Then $\qquad f(x + h, y + k) = f(x, y) + \left(h\frac{\partial}{\partial x} + k\frac{\partial}{\partial y} \right) f(x, y)$

$$+ \frac{1}{2!}\left(h\frac{\partial}{\partial x} + k\frac{\partial}{\partial y} \right)^2 f(x, y) + \ldots + \frac{1}{(n-1)!}\left(h\frac{\partial}{\partial x} + k\frac{\partial}{\partial y} \right)^{n-1} f(x, y)$$

$$+ \frac{1}{n!}\left(h\frac{\partial}{\partial x} + k\frac{\partial}{\partial y} \right)^n f(x + \theta h, y + \theta k) \text{where } 0 < \theta < 1.$$

Proof. Let us suppose that x varies while y remains constant.

Thus, regarding $f(x+h, y - k)$ as a function of one variable only, say that of x. Then using Taylor's theorem for one variable we have

$$f(x + h, y + k) = f(x, y - k - h)\frac{\partial f(x, y + k)}{\partial x} + \frac{h^2}{2!}\frac{\partial^2 f(x, y + k)}{\partial x^2} + \ldots \qquad \ldots (1)$$

Further, expanding each term of R.H.S. of (1) by Taylor's regarding y as variable and x as constant, we have

$$f(x + h, y + k) = f(x, y) + k\frac{\partial(x, y)}{\partial y} + \frac{k^2}{2!}\frac{\partial^2 f(x, y)}{\partial y^2} +$$

$$+ h\frac{\partial}{\partial x}\left\{ f(x, y) + k\frac{\partial f(x, y)}{\partial y} + \ldots \right\} + \frac{h^2}{2!}\frac{\partial^2}{\partial x^2}\left\{ f(x, y) + k\frac{\partial f(x, y)}{\partial y} + \ldots \right\} + \ldots$$

Hence, $f(x + y, y + k) = f(x, y) + \left[h\frac{\partial f}{\partial x} + k\frac{\partial f}{\partial y} \right] + \frac{1}{2!}\left[h^2\frac{\partial^2 f}{\partial x^2} + 2hk\frac{\partial^2 f}{\partial x \partial y} + k^2\frac{\partial^2 f}{\partial y^2} \right] + \ldots$

or $\qquad f(x + h, y + k) = f(x, y) + \left(h\frac{\partial}{\partial x} + k\frac{\partial}{\partial y} \right) f(x, y)$

$$+ \frac{1}{2!}\left(h\frac{\partial}{\partial x} + k\frac{\partial}{\partial y} \right) f(x, y) + \ldots + \frac{1}{n!}\left(h\frac{\partial}{\partial x} + k\frac{\partial}{\partial y} \right)^n f(x, y) + \ldots$$

__THEOREM 1.__ *Suppose E is an open set in R^n, f maps E into R^m, f is differentiable at $x_0 \in E$. g maps an open set containing f(E) into R^k, and g is differentiable at $f(x_0)$. Then mapping F of E into R^k defined by F(x) = g(f(x)) is differentiable at t_0 and that*

$$F'(x_0) = g'(x_0))f'(x_0) \qquad \text{... (1)}$$

__Proof.__　Suppose that $y_0 = f(x_0), A = f'(x_0), B = g(y_0)$, and define

$$u(h) = f(x_0 + h) - f(x_0) - A.h,$$

$$v(k) = g(h_0 + k) - g(y_0) - B.k,$$

for all $h \in R^n$ and $k \in R^m$ for which $f(x_0+h)$ and $g(y_0+k)$ are defined. By definition

$$\lim_{h \to 0} \frac{|f(x_0 + h) - f(x_0) - Ah|}{|h|} = 0$$

and

$$\lim_{h \to 0} \frac{|g(y_0 + k) - g(y_0) - Bk|}{|k|} = 0$$

It follows that

$$\lim_{h \to 0} \frac{|u(h)|}{|h|} = 0 \text{ and } \lim_{k \to 0} \frac{|v(k)|}{|k|} = 0. \text{ Then}$$

$$|u(h)| = \varepsilon(h)|h|, |v(k)| = \eta(k)|k|, \qquad \text{...(2)}$$

where $\varepsilon(h) \to 0$, as $h \to 0$ adn $\eta(k) \to 0$ as $k \to 0$.

For given h, put $k = f(x_0 + h) - f(x_0)$, then

$$|k| = |Ah + u(h)| \le |Ah| + |u(h)|$$

$$\le [\|A\| + \varepsilon(h)]|h|, \text{ by (2)}$$

and $F(x_0 + h) - F(x_0) - BAh = g(f(x_0 + h)) - g(f(x_0)) - BAh$

$$= g([f(x_0) + f(x_0 + h) - f(x_0)] - g(f(x_0)) - BAh$$

$$= g(y_0 + k) - g(y_0) - BAh = v(k) + Bk - BAh$$

$$= B(k - Ah) + v(k) = Bu(h) + v(k)$$

Hence (2) an (3) imply, for $h \ne 0$ that

$$\frac{|F(x_1 + h) - F(x_0) - BAh|}{|h|} = \frac{|Bu(h) + v(k)|}{|h|}$$

$$\le \frac{|Bu(h)| + |v(k)|}{|h|} \le \frac{\|B!\| |u(h)| + \eta(k)|(k)}{|h|}$$

$$\le \|B\|\varepsilon(h) + [\|A\| + \varepsilon(h)|\eta(k)]$$

Let $h \to 0$ then $\varepsilon(h) \to 0$. Also $k \to 0$, by (3), so that $\eta(k) \to 0$. Thus

$$\lim_{h \to 0} \frac{|f(x_0 + h) - F(x_0) - BAh|}{|h|} = 0$$

It follows that $F'(x_0) = BA$, i.e.,

$$F'(x_0) = g'(y_0) + f'(x_0) = g'(f(x_0))f'(x_0)$$

Solved Examples

Example 1. *If* $f(x) = x^2 + y^2 + 2x + 3y$, *then find the directional derivatives of* $f(x)$ *at* $(2, 1)$ *in the direction of the vector* $2i + j$.

Solution.   Directional derivatives $= \dfrac{\partial f}{\partial x} \cos\alpha + \dfrac{\partial f}{\partial y} \sin\beta$.

It is obvious that $\cos\alpha = \dfrac{2}{\sqrt{5}}$ and $\sin\beta = \dfrac{1}{\sqrt{5}}$

and $\dfrac{\partial f}{\partial x} = 2x + 2, \quad \dfrac{\partial f}{\partial y} = 2y + 3$

$$\left(\dfrac{\partial f}{\partial x}\right)_{(2,1)} = 6, \left(\dfrac{\partial f}{\partial y}\right)_{(2,1)} = 5$$

Hence, the directional derivatives of $f(x)$ at $(2, 1)$ in the given direction

$$= 6 \times \dfrac{2}{\sqrt{5}} + 5 \times \dfrac{1}{\sqrt{5}} = \dfrac{17}{\sqrt{5}}$$

Example 2. *If* $f = xy^2 + 2xy$, *then find the directional derivatives of* f *at* $(1, 2)$ *in the direction* $\theta = \dfrac{\pi}{2}$.

Solution.   Let $f = xy^2 + 2xy$

$\Rightarrow$ $\dfrac{\partial f}{\partial x} = y^2 + 2y, \quad \dfrac{\partial f}{\partial y} = 2xy + 2x$

$\therefore$ $\left(\dfrac{\partial f}{\partial x}\right)_{(1,2)} = 8$ and $\left(\dfrac{\partial f}{\partial y}\right)_{(1,2)} = 6$

Hence, the directional derivative at $(1, 2)$ in the direction $\theta = \dfrac{\pi}{2}$

$$= \dfrac{\partial f}{\partial x} \cos\dfrac{\pi}{2} + \dfrac{\partial f}{\partial y} \sin\dfrac{\pi}{2}$$

$$= 8\cos\dfrac{\pi}{2} + 6\sin\dfrac{\pi}{2} = 6 .$$

Example 3. *Check the continuity of the function* $x^3 + y^3 - 3xy + y = 0$, *near the point* $(0, 0)$.

Solution.   Here, we have

$$f(x, y) = x^3 + y^3 - 3xy + y = 0, \text{ at } (0, 0)$$

Now check the continuity of f_x and f_y at $(0, 0)$

(i) *Continuity of* f_x *at* $(0, 0)$

The simultaneously limit $\lim_{\substack{x \to 0 \\ y \to 0}} (3x^2 - 3y) = 0$, exists.

$\therefore$ f_x is continuous.

The simultaneous limit $\lim_{\substack{x \to 0 \\ y \to 0}} (3y^2 - 3x + 1) = 1$, exists.

$\therefore$ f_y is continuous.

Example 4. *Check the continuity of the derivative of the following function:*

(a) $xy \sin x + \cos y = 0$ at $\left(0, \dfrac{\pi}{2}\right)$ (b) $y^3 \cos x + y^2 \sin^2 x = 7$ at $\left(\dfrac{\pi}{3}, 2\right)$.

Solution.   (a) $f(x, y) = xy \sin x + \cos y$, $\left(0, \dfrac{\pi}{2}\right)$

$\therefore$ $f\left(0, \dfrac{\pi}{2}\right) = 0 + \cos \dfrac{\pi}{2} = 0$

Now $f_x = y[x \cos x + \sin x]$, $f_y = x \sin x - \sin y$.

Further we check the continuity of f_x and f_y at $\left(0, \dfrac{\pi}{2}\right)$.

Continuity at $\left(0, \dfrac{\pi}{2}\right)$ *of* f_x.

The simultaneous limit of $f_x(x, y)$.

i.e., $\displaystyle \lim_{\substack{x \to 0 \\ y \to (\pi/2)}} (xy \cos x + \sin x) = 0$, exists

and the simultaneous limit of $f_y(x, y)$, i.e.,

$\therefore$ $\displaystyle \lim_{\substack{x \to 0 \\ y \to (\pi/2)}} (x \sin x - \sin y) = \lim_{y \to (\pi/2)} (-\sin y) = -1$, exists.

Hence both f_x and f_y are continuous

and $f_y\left(0, \dfrac{\pi}{2}\right) = \displaystyle\lim_{k \to 0} \dfrac{f\left(0, \dfrac{\pi}{2} + k\right) - f\left(0, \dfrac{\pi}{2}\right)}{k}$

$= \displaystyle\lim_{k \to 0} \dfrac{\cos\left(\dfrac{\pi}{2} + k\right)}{k} = \lim_{k \to 0} \dfrac{-\sin k}{k} = -1$

$\therefore$ $f_y\left(0, \dfrac{\pi}{2}\right) \neq 0$

(b) $f(x, y) = y^3 \cos x + y^2 \sin^2 x - 7 = 0$

$\therefore$ $f\left(\dfrac{\pi}{3}, 2\right) = 8 \cos \dfrac{\pi}{3} + 4 \sin^2 \dfrac{\pi}{3} - 7 = 4 + 3 - 7 = 0$

Now $f_x(x, y) = -y^3 \sin x + 2y^2 \sin 2x \cos x$

and $f_y(x, y) = 3y^2 \cos x + 2y \sin^2 x$

Next, we have to examine the continuity of f_x and f_y.

(i)   *Continuity of* $f_x (x, y)$ *at* $\left(\dfrac{\pi}{3}, 2\right)$.

The simultaneous limit is

$\displaystyle \lim_{\substack{x \to (\pi/3) \\ y \to 2}} [-y^3 \sin x + 2y^2 \sin x \cos x] = \lim_{y \to 2} \left[-y^3 \sin \dfrac{\pi}{3} + 2y^2 \sin \dfrac{\pi}{3} \cos \dfrac{\pi}{3}\right]$

$= -8 \cdot \dfrac{\sqrt{3}}{2} + 8 \cdot \dfrac{\sqrt{3}}{2} \cdot \dfrac{1}{2}$

$= -4\sqrt{3} + 4\sqrt{3} = 0$, exists

∴ The simultaneous limit is given by

$$\lim_{\substack{x \to (\pi/3) \\ y \to 2}} [3y^2 \cos x + 2y \sin^2 x] = \lim_{y \to 2}\left[3y^2 \cos\frac{\pi}{3} + 2y \sin^2\frac{\pi}{3}\right]$$

$$= 12.\frac{1}{2} + 4.\frac{3}{4} = 6+3 = 9 \text{ exists}$$

∴ f_x and f_y both are continuous.

Example 5. *Examine the following equations for the existence of a unique implicit function near the point indicated:*

(a) $y^2 + 2x^2 y + x^5 = 0$ *at* $(1, -1)$ (b) $x^2 + xy + y^2 - 1 = 0$ *at* $(1, 0)$

(c) $y^2 - yx^2 - 2x^5 = 0$ *at* $(0, 0)$

Solution.   (a) Let $f(x, y) = y^2 + 2x^2 y + x^5$

Then $f_x = 4xy + 5x^4$

$f_y = 2y + 2x^2$

Also, we have $f(1, -1) = 1 - 2 + 1 = 0$

and $f_y (1, -1) = -2 + 2 = 0$

which implies that both the partial derivatives f_x and f_y are continuous functions in a neighbourhood of $(1, -1)$.

(b) Let $f(x, y) = x^2 + xy + y^2 - 1$

Then, we have

$f_x = 2x + y , f_y = x + 2y$

Also, $f(1, 0) = 1 + 0 + 0 - 1 = 0$

$f_y(1, 0) = 1 \neq 0$

Clearly, both the derivatives f_x and f_y are continuous in the nbd of $(1, 0)$.

(c) Let $f(x, y) = y^2 - yx^2 - 2x^5$

Then, we have

$f_x = -2xy - 10x^4, f_y = 2y - x^2$

Also, $f(0, 0) = 0 \quad \Rightarrow f_y(0, 0) = 0$

We have both the partial derivatives f_x and f_y are continuous functions in a neighbouhood of $(0, 0)$.

EXERCISE 10.4

1. If $(x, y) = (x^2 + y^2) \log (x^2 + y^2)$ for $(x, y) \neq (0, 0)$ and $f(0, 0) = 0$; show that f_{xy} and f_{yx} are not continuous at $(0, 0)$ but $f_{xy}(0, 0) = f_{yx}(0, 0)$.

2. Find the first derivative of the equation.
$2xy - \log xy = 2$ at $(1, 1)$.

10.11 SCHWARZ'S THEOREM

Let f be a real valued function defined on a domain $D \subset R^2$ and (a, b) be a point of the domain D, such that

(i) f_{xy} *is continuous at* (a, b).

(ii) f_x *exists in a certain neighbourhood of* (a, b)

then, $f_{yx}(a, b)$ *exists and equal to* f_{xy} (a, b).

Proof. Let $f(x, y)$ be a real valued function in a domain D. Condition (i) and (ii) implies that there exist a certain neighbourhood of (a, b) at every point (x, y) of which $f_x(x, y), f_y(x, y)$ and $f_{xy}(x, y)$ exists.

Let us suppose $(a+h, b+k)$ be any point of this neighbourhood, then we may define two functions ϕ and g such that

$$\phi(h, k) = f(a+h, b+k) - f(a+h, b) - f(a, b+k) + f(a, b)$$

and
$$g(y) = f(a+h, y) - f(a, y)$$

so that
$$\phi(h, k) = g(b+k) - g(b) \qquad \dots (1)$$

Since, f_y, exists in a neighbourhood of (a, b), therefore the function g of one variable is differentiable in $[b, b + k]$, then by Lagrange's mean value theorem, we have

$$\frac{g(b+k) - g(b)}{b+k-b} = g'(b+\theta k), \ 0 < \theta < 1 \qquad \dots (2)$$

Using (1) and (2), we have

$$\phi(h, k) = kg'(b+\theta k)$$
$$= k[f_y(a+h, b+\theta k) - f_y(a, b+\theta k)] \qquad \dots (3)$$

Now, since, f_{xy} exist in the neighbourhood of (a, b), then again apply the Lagrange's mean value theorem to the right of (3), we get

$$\phi(h,k) = hk\, f_{xy}(a+\theta'h, b+\theta k)$$
$$(0 < \theta' < 1)$$

$$\Rightarrow \frac{1}{k}\left[\frac{f(a+h,b+k) - f(a,b+k)}{h} - \frac{f(a+h,b) - f(a,b)}{h} \right] = f_{xy}(a+\theta'h, b+\theta k)$$

Since, f_{xy} exist in a neighbourhood of (a, b), this gives when $h \to 0$

$$\frac{f_x(a,b+k) - f_x(a,b)}{k} = \lim_{h \to 0} f_{xy}(a+\theta'h, b+\theta k)$$

Letting $k \to 0$, we get

$$\lim_{k \to 0} \frac{f_x(a,b+k) - f_x(a,b)}{k} = \lim_{k \to 0}\left[\lim_{h \to 0} f_{xy}(a+\theta'h, b+\theta k) \right]$$

$$\Rightarrow \qquad f_{yx}(a,b) = \lim_{k \to 0}\lim_{h \to 0} f_{xy}(a+\theta'h, b+\theta k)$$
$$= f_{xy}(a, b)$$

Hence, we get

$$f_{yx}(a, b) = f_{xy}(a, b)$$

REMARK

- If f_{xy} and f_{yx} are both continuous at (a, b), then $f_{xy}(a, b) = f_{yx}(a, b)$

10.12 YOUNG'S THEOREM

Let f be a real valued function defined on a domain $D \subset R^2$ and (a, b) be any point of the domain D, such that f_x and f_y are both differentiable at (a, b), then $f_{xy}(a, b) = f_{yx}(a, b)$.

Proof. Since, f_x and f_y are both differentiable at (a, b) therefore, the derivatives of f_x and f_y exist in a certain neighbourhood of (a, b) and given by $f_{xx}, f_{yx}, f_{xy}, f_{yy}$.

Let $(a+h, b+h)$ be a point of this neighbourhood.

Then, we may define two functions ϕ and g such that

$$\phi(h, h) = f(a + h, b + h) - f(a + h, b) - f(a, b + h) + f(a, b)$$

$$g(y) = f(a+h, h) - f(a, y) \qquad \qquad \text{...(1)}$$

$$\Rightarrow \qquad \phi(h, h) = g(b+h) - g(b)$$

Since, f_y exist in a neighbourhood of (a, b), apply the mean value theorem to the expression on the right of (1), we get

$$\frac{g(b+h) - g(b)}{b+h-b} = g'(b+\theta h) ; \quad 0 < \theta < 1$$

$$\Rightarrow \qquad g(b+h) - g(b) = hg'(b+\theta h) ; \quad 0 < \theta < 1$$

$$\Rightarrow \qquad \phi(h,h) = hg'(b+\theta h) ; \quad 0 < \theta < 1$$

$$= h[f_y(a+h, b+\theta h) - f_y(a, b+\theta h)] \qquad \text{... (2)}$$

Since, f_y is differentiable at (a, b), we have by definition,

$$f_y(a+h, b+\theta h) - f_y(a,b) = hf_{xy}(a,b) + \theta h f_{yy}(a,b) + |h|\sqrt{(1+\theta^2)}\phi_1(h,h) \qquad \text{... (3)}$$

where $\phi_1 \to 0$ as $h \to 0$

and $\qquad f_y(a, b+\theta h) - f_y(a,b) = \theta h f_{yy}(a,b) + \theta |h| \phi_2(h,h) \qquad \text{... (4)}$

where $\phi_2 \to 0$ as $h \to 0$

Now, from (2), (3) and (4), we get

$$\frac{\phi(h,h)}{h^2} = f_{xy}(a,b) + \sqrt{(1+\theta^2)}\phi_1(h,h) - \theta\phi_2(h,h) \qquad \text{...(5)}$$

In a similar way considering $F(x) = f(x, b+h) - f(x, b)$, we can easily show that

$$\frac{\phi(h,h)}{h^2} = f_{yx}(a,b) + \sqrt{(1+\theta'^2)}.\psi_1(h,h) - \theta'\psi_2(h,h) \qquad \text{... (6)}$$

where $\psi_1, \psi_2 \to 0$ as $h \to 0$.

Now equating the right hand sides of (5) and (6) and letting $h \to 0$, we get

$$f_{xy}(a, b) = f_{yx}(a, b).$$

THEOREM 1. *Suppose f maps a convex open set $E \subset R^n$ into R^m, f is differentiable in E, and there is a real number M such that : $\|f'(x)\| \le M$, for every x. Then, for a, $b \in E$,*

$$|f(b) - f(a)| \le M|b-a|$$

What would happen if $f'(x) = 0$ for all x?

Proof. Let $a \in E$, $b \in E$ and define

$$\gamma(t) = (1-t)a + tb \qquad \text{... (1)}$$

for all $t \in R'$ such that $\gamma(t) \in E$.

If $0 \le t \le 1$, then by convexity of E, $\gamma(t) \in E$

Putting $\qquad g(t) = f(r(t))$. Then by chain rule, we have

$$g'(t) = f'(\gamma(t')(\gamma')(t)) = f'\gamma(t)(b-a), \text{by}(1)$$

So that

$$|g'(t)| = |f'\gamma(t)(b-a)| \le \|f'\gamma(t)\||(b-a)|$$

$$\le M|(b-a)| \text{ for all } t \in [0, 1].$$

Now, by the mean value theorem for vector valued function, we get

$$|g(1) - g(0)| \le (1-0)|g'(t)|$$

or $\qquad |g(1) - g(0)| \le M\,|\,b-a\,|$ by (2).

But $g(1) = f(\gamma)\,(1) = f(b)$ and $g(0) = f\gamma(0) = f(a)$. Thus $|f(b) - f(a)| \le M\,(b-a)$

Further $f' = 0$ for all $x \in E$, then f is constant.

By (2) if $f'(x) = 0$ for all $x \in E$, we have $g'\,(t) = 0$

So that $\qquad\qquad g'\,(t) = 0 \qquad \Rightarrow |g(1) - g(0)| \le 0$

$\Rightarrow \qquad\qquad\qquad g(1) = g(0)$

$\Rightarrow \qquad\qquad\qquad f(b) = f(a) \quad \Rightarrow f$ is constant.

Solved Examples

Example 1. *Explain the inequality*

$$f_{xy}(0,0) \ne f_{yx}(00)$$

for the function $f(x,y) = \begin{cases} \dfrac{xy(x^2 - y^2)}{x^2 + y^2} & , \quad (x,y) \ne (0,0) \\ 0 & , \quad (x,y) = (0,0) \end{cases}$

in view of the Schwariz's and Young's theorems.

Solution. Here we have $\quad f(x,y) = \dfrac{xy(x^2 - y^2)}{x^2 + y^2}; (x,y) \ne (0,0)$

$$f(0,0) = 0$$

According to Schwarz's iheorem. If $f_x(0,0)$ exists and f_{xy} is continuous at $(0,0)$. Then $f_{yx}(0,0)$ exists and equal to $f_{xy}\,(0,0)$.

First we check that $f_x(0,0)$ exists or not

$$f_x(0,0) = \lim_{\delta x \to 0} \frac{f(\delta x, 0) - f(0,0)}{\delta x} = \lim_{\delta x \to 0} \frac{0-0}{\delta x} = 0$$

$\therefore \quad f_x(0,0)$ exists

Now, $\quad f_x = y\left[\dfrac{x^2 - y^2}{x^2 + y^2} + \dfrac{4x^2 y^2}{(x^2 + y^2)^2} \right]$

$\therefore \quad f_{xy} = \dfrac{x^2 - y^2}{x^2 + y^2} + \dfrac{4x^2 y^2}{(x^2 + y^2)^2} + y\left[\dfrac{(x^2 + y^2)(-2y) - (x^2 - y^2)2y}{(x^2 + y^2)^2} \right.$

$$\left. \dfrac{+(x^2 + y^2)^2 8x^2 y - 4x^2 y^2 . 2(x^2 + y^2).2y}{(x^2 + y^2)^4} \right]$$

$= \dfrac{x^2 - y^2}{x^2 + y^2} + \dfrac{4x^2 y^2}{(x^2 + y^2)^2} + y\left[\dfrac{-4x^2 y}{(x^2 + y^2)^2} + \dfrac{(x^2 + y^2)8x^2 y - 16x^2 y^3}{(x^2 + y^2)^3} \right]$

$= \dfrac{x^2 - y^2}{x^2 + y^2} + \dfrac{4x^2 y^2}{(x^2 + y^2)^2} + y\left[\dfrac{-4x^2 y}{(x^2 + y^2)^2} + \dfrac{8x^4 y + 8x^2 y^3 - 16x^2 y^3}{(x^2 + y^2)^3} \right]$

$$= \frac{x^2 - y^2}{x^2 + y^2} + \frac{y(8x^4 y - 8x^2 y^3)}{(x^2 + y^2)^3}$$

$$\Rightarrow \qquad f_{xy} = \frac{x^2 - y^2}{x^2 + y^2} + \frac{8x^2 y^2 (x^2 - y^2)}{(x^2 + y^2)^3}$$

Now find the simultaneous limit of f_{xy} i.e.,

$$\lim_{\substack{x \to 0 \\ y \to 0}} \left[\frac{x^2 - y^2}{x^2 + y^2} + \frac{8x^2 y^2 (x^2 - y^2)}{(x^2 + y^2)^3} \right]$$

Taking $y = mx$, we get

$$\lim_{\substack{x \to 0 \\ y \to 0}} \left[\frac{x^2 - y^2}{x^2 + y^2} + \frac{8x^2 y^2 (x^2 - y^2)}{(x^2 + y^2)^3} \right] = \lim_{x \to 0} \left[\frac{1 - m^2}{1 + m^2} + \frac{8m^2 (1 - m^2)}{(1 + m^2)^3} \right]$$

$$= \left[\frac{1 - m^2}{1 + m^2} + \frac{8m^2 (1 - m^2)}{(1 + m^2)^3} \right]$$

Thus the limit does not exist since it depends upon m.

Hence, f_{xy} is not continuous at $(0, 0)$, and hence

$$f_{xy}(0, 0) \neq f_{yx}(0, 0)$$

According to Young's theorem, if f_x and f_y are both differentiable at $(0, 0)$ then $f_{xy}(0, 0) = f_{yx}(0, 0)$. Therefore first we check the differentiability of f_x and f_y at $(0, 0)$.

Since
$$f_x = y \left[\frac{x^2 - y^2}{x^2 + y^2} + \frac{4x^2 y^2}{(x^2 + y^2)^2} \right] = g(x, y) \qquad \text{(say)}$$

and
$$f_y = x \left[\frac{x^2 - y^2}{x^2 + y^2} - \frac{4x^2 y^2}{(x^2 + y^2)^2} \right] = h(x, y) \qquad \text{(say)}$$

Now find
$$g_x(0,0) = \lim_{\delta x \to 0} \frac{g(\delta x, 0) - g(0, 0)}{\delta x} = \lim_{\delta x \to 0} \frac{0 - 0}{\delta x} = 0$$

$$g_y(0,0) = \lim_{\delta y \to 0} \frac{g(0, \delta y) - g(0, 0)}{\delta y} = \lim_{\delta y} - \frac{\delta y}{\delta y} = -1$$

$$\therefore \quad \lim_{\substack{\delta x \to 0 \\ \delta y \to 0}} \frac{g(\delta x, \delta y) - g(0, 0) - \delta x \, g_x(0, 0) - \delta y g_y(0, 0)}{\sqrt{(\delta x)^2 + (\delta y)^2}}$$

$$= \lim_{\substack{\delta x \to 0 \\ \delta y \to 0}} \frac{\delta y \left[\dfrac{(\delta x)^2 - (\delta y)^2}{(\delta x)^2 + (\delta y)^2} + \dfrac{4(\delta x)^2 (\delta y)^2}{[(\delta x)^2 + (\delta y)^2]^2} \right] + \delta y}{\sqrt{(\delta x)^2 + (\delta y)^2}}$$

If we put $\delta y = m\delta x$, then above limit

$$= \lim_{\delta x \to 0} \frac{m\delta x\left[\dfrac{1-m^2}{1+m^2} + \dfrac{4m^2}{(1+m^2)^2}\right] + m\delta x}{\delta x\sqrt{1+m^2}}$$

$$= \frac{m\left[\dfrac{1-m^2}{1+m^2} + \dfrac{4m^2}{(1+m^2)^2}\right] + m}{\sqrt{1+m^2}}$$

This limit depends upon m. Hence the limit does not exist.

Therefore, $g(x,y) = f_x(x,y)$ is not differentiable at $(0,0)$.

Hence, $f_{xy}(0,0) \neq f_{yx}(0,0)$

Example 2. Let $f(x,y) = \dfrac{x^2 y^2}{x^2 + y^2}, (x,y) \neq (0,0)$ and $f(0,0) = 0$

Verify that f_{xy} and f_{yx} exist in a neighbourhood of $(0,0)$ but are not continuous at $(0,0)$ and yet are equal at $(0,0)$.

Solution. We have

$$f_{xy}(0,0) = \lim_{\delta x \to 0}\lim_{\delta y \to 0} \frac{f(\delta x, \delta y) - f(\delta x, 0) - f(0, \delta y) + f(0,0)}{\delta x \delta y}$$

$$= \lim_{\delta x \to 0}\lim_{\delta y \to 0} \frac{\dfrac{(\delta x)^2(\delta y)^2}{(\delta x)^2 + (\delta y)^2} - 0 - 0 + 0}{\delta x \delta y}$$

$$= \lim_{\delta x \to 0}\lim_{\delta y \to 0} \frac{\delta x \delta y}{(\delta x)^2 + (\delta y)^2}$$

$$= \lim_{\delta x \to 0}\left[\frac{\delta x . 0}{(\delta x)^2 + 0}\right] = 0$$

$\therefore$ f_{xy} exists and $f_{xy}(0,0) = 0$

and $$f_{yx}(0,0) = \lim_{\delta y \to 0}\lim_{\delta x \to 0} \frac{f(\delta x, \delta y) - f(0, \delta y) - f(\delta x, 0) + f(0,0)}{\delta x \delta y}$$

$$= \lim_{\delta y \to 0}\lim_{\delta x \to 0} \frac{\delta x \delta y}{(\delta x)^2 + (\delta y)^2}$$

$$= \lim_{\delta y \to 0}\left[\frac{0 . \delta y}{0 + (\delta y)^2}\right] = 0$$

$\therefore$ f_{yx} exist and $f_{yx}(0,0) = 0$

$\therefore$ $f_{xy}(0,0) = f_{yx}(0,0) = f_{xy}(0,0)$

Now find $$f_x = \frac{(x^2 + y^2).2xy^2 - x^2 y^2(2x)}{(x^2 + y^2)^2}$$

$$= \frac{2x^3 y^2 + 2xy^4 - 2x^3 y^2}{(x^2 + y^2)^2} = \frac{2xy^4}{(x^2 + y^2)^2}$$

$$\therefore \qquad f_{yx} = \frac{(x^2 + y^2)^2.(8xy^3) - 2xy^4.2(x^2 + y^2).2y}{(x^2 + y^2)^4}$$

$$= \frac{8x^3y^3 + 8xy^5 - 8xy^5}{(x^2 + y^2)^3} = \frac{8x^3y^3}{(x^2 + y^2)^3}$$

Similarly $\qquad f_{xy} = \dfrac{8x^3y^3}{(x^2 + y^2)^3}$

Now we check the continuity at $(0, 0)$.

The simulaneous limit $= \lim\limits_{\substack{x \to 0 \\ y \to 0}} \dfrac{8x^3y^3}{(x^2 + y^2)^3}$

Put $y = mx$.

$$\therefore \qquad \lim_{x \to 0} \cdot \frac{8m^3x^6}{x^6(1 + m^2)^3} = \frac{8m^3}{(1 + m^2)^3}$$

Thus the limit depends upon m hence $f_{xy} = f_{yx}$ is not continuous at $(0, 0)$.

10.13 INVERTIBLE FUNCTION

Let f be a real valued function with domain D and range E as subset of R^n.
Here, we can write
$$y = f(x), \ x \in D \quad and \ y \in E \qquad \qquad ... (1)$$
Let y_n can be written as $y_n = f_n (x_1, x_2, ... x_n)$

Then, the function f is a transformation, which transforms the set D to the set E. Now, to each point of D, there corresponds a point of E. Now there are two cases :
  (i) One-one transformation
  (ii) Many-one transformation.

Then, a one-one function f with domain D and range E is called invertible.
If $y = f(x) \Leftrightarrow x = g(y)$, then the function g is called the inverse of the function f.

10.14 LINEAR TRANSFORMATION

Let V_1 and V_2 be two vector spaces. A mapping $f : V_1 \to V_2$ is said to be linear transformation.
if $\qquad\qquad f_1 (x_1 + x_2) = f(x_1) + f(x_2)$
and $\qquad\qquad f(cx) = cf(x) \ \forall \ x, x_1, x_2 \in V_1, c \in R$

REMARKS

- Linear transformation V_1 into V_2 will be called linear operator on V_1.
- Linear operator A on V_1 is said to be invertiable if A is
  (i) one to one
  (ii) f maps V_1 onto V_2
- If A is invertiable, then an operator A^{-1} can be defined by setting
  $$A^{-1} (Ax) = x \ \forall x \in V_1$$

THEOREM 1. *If A is a linear operator on a finite dimensional vector space X. Then A is one to one iff range of A is full of X.*

Proof . Let $B = \{x_1, ... x_n\}$ be a basis of X. Let $R (A)$ denote the range of A. We shall show that the set $Q = (Ax_1, \ldots Ax_n)$ spans $R(A)$. Let $y \in R(A)$. Then $y = Ax$. For some $x \in X$. Since B spans X, therefore there exists scalars $c_1, \ldots c_n$ such that

$$x = c_1 x_1 + c_2 x_2 + \ldots + c_n x_n$$

Therefore, $y = A(c_1 x_1 + \ldots + c_n x_n) = c_n Ax_n$, by linearity of A. This shows that Q spans $R(A)$, then $R(A) = X$ if and only if Q is linearly independent. We have to prove that this happens if and only if A is one-to-one. Let us suppose Q is linearly independent and let $x \in X$ be arbitrary. Since B is a basis. Therefore,

$$x = \sum_{i=1}^{n} c_i x_i \text{ for some scalars } c_i, i=1, 2, \ldots n$$

Then $Ax = 0 \qquad \Rightarrow \quad A\left(\sum_{i=1}^{n} c_i x_i\right) = 0$

$\Rightarrow \quad \sum c_i Ax_i = 0$, by linearity of A

$\Rightarrow \quad c_1 = c_2 = \ldots = c_n = 0$, since Q is linearly independent

$\Rightarrow \quad x = 0$

Therefore,

$\qquad Ax = 0 \quad \Rightarrow \quad x = 0$

So, $Ax = Ay \quad \Rightarrow \quad Ax - Ay = 0$ \hfill ...(1)

$\Rightarrow \quad A(x - y) = 0$, since A is linear.

$\Rightarrow \quad x - y = 0$ by (1)

$\Rightarrow \quad x = y$

$\Rightarrow \quad A$ is one-to-one.

Conversely, suppose A is one-to-one.
Then

$$\sum_{i=1}^{n} c_i Ax_i = 0 \Rightarrow A\left(\sum_{i=1}^{n} c_i x_i\right) = 0 \text{ since } A \text{ is linear.}$$

$$\Rightarrow \sum_{i=1}^{n} c_i x_i = 0 \text{, since } A \text{ is one-to-one}$$

$$\Rightarrow c_1 = c_2 = \ldots = c_n = 0, \text{ since } B \text{ is independent.}$$

Thus we conclude that Q is linearly independent.

THEOREM 2. *Let W be the set of all invertible operators on R_n. Then, if $A \in W$ and $B \in L(R^n)$ such that $\|B - A\| \cdot \|A^{-1}\| < 1$, then $B \in w$.*

Proof. Put $\|A^{-1}\| = \dfrac{1}{\alpha}$ and $\|B - A\| = \beta$

Then $\|B - A\| \|A^{-1}\| < 1$ implies that $\beta < \alpha$. For every $x \in R^n$. we have

$$|x| = |A^{-1} Ax| \le \|A^{-1}\| |Ax| = \frac{1}{\alpha} |Ax| \text{ so that}$$

$$(\alpha - \beta)|x| = \alpha |x| - \beta |x| \le |Ax| - \|B - A\| |x|$$

$$\le |Ax| - |(B - A)x|, \text{since } |Ax| \le \|A\| |x|$$

$$= |Ax| - |Bx - Ax| = |Ax| \le \|Ax - Bx\|$$

$$\le |Ax| - (|Ax| - |Bx|), \text{ since } |y_1| - |y_2| \le |y_1 - y_2|$$

$$\text{for all } y_1, y_2 \in R^n$$

$$=\mid Bx \mid (x \in R^n)$$

Thus, $\mid Bx \mid \geq (\alpha - \beta) \mid x \mid$ for all $x \in R^n$ ...(2)

Note that $(\alpha - \beta) \mid x - y \mid$ cannot be negative since $\alpha - \beta > 0$.

Now $Bx = By \Rightarrow Bx - By = 0$

$$\Rightarrow \quad B(x - y) = 0 \Rightarrow \mid B(x - y) \mid = 0$$
$$\Rightarrow \quad (\alpha - \beta) \mid x - y \mid = 0 \text{ by (1)}$$
$$\Rightarrow \quad \mid x - y \mid = 0 \text{ since } \alpha \neq \beta$$
$$\Rightarrow \quad x - y = 0 \Rightarrow x = y$$

This show that B is one-to-one. B is also onto. Hence B is an invertible operator, and so $B \in W$.

10.14.1 INVERSE FUNCTION THEOREM

Let f be a vector valued function with its domain D and range E as subset of R^n. Let $a \in D$, $f(a) = b \in E$. Let f admit of continuous first order partial derivatives in a neighbourhood of a and let the Jacobian $J_f(a)$, $\neq 0$ then the function f is locally invertible at a. Also the local inverse g of f admits of continuous first order partial derivatives in a neighbourhood of b.

Proof.   Let f be a vector valued function with its domain D and range E as subset of R^n, thus, we show that, under the given conditions on f, there exist neighbourhoods P and Q of a and b respectively such that

(i) $f(P) = \{f(x) : x \in P\} = Q$

(ii)   No two different points of P correspond to the same point of V.

(iii) $f(a) = b$.

If g is the inverse of f with domain Q and range P so that

$$y = f(x) \Leftrightarrow x = g(y); \quad x \in P \quad \text{and} \quad y \in Q$$

the function g admits of continuous first order partial derivative in Q.

Now consider the determinant

$$\begin{vmatrix} D_1 f_1(x_1) & D_2 f_1(x_1) & \cdots & D_n f_1(x_1) \\ D_1 f_2(x_2) & D_2 f_2(x_2) & \cdots & D_n f_2(x_2) \\ \cdots & \cdots & \cdots & \cdots \\ \cdots & \cdots & \cdots & \cdots \\ D_1 f_n(x_n) & D_2 f_n(x_n) & \cdots & D_n f_n(x_n) \end{vmatrix}$$

where $x_1, x_2, ..., x_n$ are points in a neighbourhood of a. We see upon this determinant as a function of n^2 variables viz. The n^2 co-ordinates of the n-points $x_1, x_2, ..., x_n$, such that its value when $x_1, x_2, ..., x_n$ all take the same value a is non-zero. Also the determinant being a polynomial in its elements is a continuous function. Thus $\exists$ a neighbourhood of the point a such that if $P_1, P_2, P_3 .... P_n$ are any n arbitrary points of this neighbourhood, we have $\mid D_j f_i(P_i) \mid \neq 0$

Again there exist $s > 0$ s.t. the closed sphere $S(a, s)$ is contained in the neighbourhood.

(i)   Let $x \neq y$ be two points of the sphere $S(a, s)$. We shall show that

$$x \neq y \Rightarrow f(x) \neq f(y)$$

Let if possible, $f(x) = f(y)$

$\Leftrightarrow$ $f_i(x) = f_i(y), 1 \leq i \leq n$

By the mean value theorem, there exist θ_i, $0 < \theta_i < 1$ s.t.

$$0 = f_i(x) - f_i(y) = \sum_{j=1}^{n} (y_i - x_j) D_j f_i[x + \theta_i(y - x)]$$

we write $x + \theta_i(y - x) = P_i$ therefore, $\sum_{j=1}^{n} (y_j - x_j) D_j f_i(P_i) = 0$ thus we obtain a system of n linear equations

$$(y_1 - x_1) D_1 f_1(P_1) + (y_2 - x_2) D_2 f_1(P_1) + ... + (y_n - x_n) D_n f_1(P_1) = 0$$

$$\cdots \quad \cdots \quad \cdots \quad \cdots \quad \cdots \quad \cdots \quad \cdots \quad \cdots \quad \cdots$$
$$\cdots \quad \cdots \quad \cdots \quad \cdots \quad \cdots \quad \cdots \quad \cdots \quad \cdots \quad \cdots$$
$$\cdots \quad \cdots \quad \cdots \quad \cdots \quad \cdots \quad \cdots \quad \cdots \quad \cdots \quad \cdots$$

$$(y_1 - x_1) D_1 f_n(P_n) + (y_2 - x_2) D_2 f(P_n) + ... + (y_n - x_n) D_n f_n(P_n) = 0$$

whose determinant

$$|D_j f_i(P_i)| \neq 0, 1 \le i \le n, 1 \le j \le n$$

$\Rightarrow \quad y_1 - x_1 = 0, ..., y_n - x_n = 0 \Rightarrow x = y$

which is a contradiction.

Thus, we prove that $x \neq y \Rightarrow f(x) \neq f(y)$.

(ii) Now. we shall show that $f(\bar{S})$ contains, a neighbourhood of $b = f(a)$.

Here $f(S) = \{f(x) : x \in S\}$

Let T denote the boundary of the sphere $\bar{S}$ i.e., the set $\{x : \|x - a\| \le s\}$.

Now consider $\|f(x) - b\| : x \in T$.

Here, we have a real valued function with values $\|f(x) - f(a)\| = \|f(x) - b\|$ with domain T, as the function f is one-one with domain T and $a \notin T$, we see that $\|f(x) - b\|$ is positive $\forall x \in T$.

Since, T is a compact set and the real valued function is continuous $\forall x \in T$. We see that the infimum of the function is necessarily positive and be denoted by $2k$ so that we have

$$2k \le \|f(x) - b\| \, x \, \forall \, x \in T.$$

We now define a set Q s.t.

$$Q = \{y : \|y - P\| < k\} \text{ and show that } Q \text{ is contained in } f(\bar{S}).$$

Let $y \in Q$. We have to show that there exist $x \in \bar{S}$ s.t. $y = f(x)$. we have

$$\alpha k \le \|f(x) - b\| \le \|f(x) - y\| + \|y - b\|$$
$$\le \|f(x) - y\| + k$$

so that $\quad \|f(x) - y\| > k \, \forall \, y \in Q, \forall x \in T$ ... (1)

Also $\quad \|f(a) - y\| = \|y - b\| < k$ ... (2)

Taking y fixed, we consider the real valued function with value $\|f(x) - y\|^2$, $x \in \bar{S}$

From (1) and (2) we see that the infimum of .this function is assumed at an interior point of $\bar{S}$ and as such this infimum is a minimum value.

We have $\quad \sum_{i=1}^{n} [f_i(x) - f_i]^2 = \|f(x) - y\|$

Thus, we see that

$$\sum_{i=1}^{n} [f_i(x) - y_i] D_i f_i(x) = 0, \, 1 \le j \le n$$

Rewritting these equations, we obtain

$$[f_1(x) - y_1]D_1 f_1(x) + ... +, \; [f_n(x) - y_n]D_1 f_n(x) = 0$$

$$\begin{matrix} ... & ... & ... & ... & ... & ... \end{matrix}$$

$$[f_1(x) - y_1]D_n f_1(x) + ... +, \; [f_n(x) - y_n]D_n f_n(x) = 0$$

the determinant of this set of linear equations being non-zero.

We see that

$$f_1(x) - y_1 = 0, f_2(x) - y_2 = 0, \; ... f_n(x) - y_n = 0$$

which imply $f(x) = y$.

Thus, Q is contained in $f\{S\}$.

(iii) Let P be the set of points x such that $f(x) \in P$.

The set P is open, therefore f is continuous and Q is an open set.

Thus, we see that there exist an open set P and Q containing a and $f(a) = b$ respectively such that f is a one-one function from P onto Q. i.e., $x \in P \Rightarrow f(x) \in Q$ and to $y \in Q$ there corresponds one and only one $x \in P$ s.t. $f(x) = y$.

Thus the function f with domain P and range Q is invertible. Let g denote the inverse of the function f so that Q is its domain and P its range.

Thus, we have $y = f(x) \Leftrightarrow x = g(y), x \in P, y \in Q$.

(iv) Now, we have only show that function g itself have continuous first order partial derivatives. Let $y = f(x) \Leftrightarrow x = g(y)$.

For sufficiently small values of $\lambda, y + \lambda u_k \in Q$.

Let $x' = g(y + \lambda u_k) \Leftrightarrow y + \lambda u_k = x'$

$\therefore \qquad f(x') - f(x) = \lambda u_k$

$\Rightarrow f_i(x') f_i(x) = \delta_{ik}, 1 \le i \le n.$

Now by mean value theorem, we get

$$f_i(x') - f_i(x) = \sum_{j=1}^{n} (x'_j - x_j) D_j f_i [x + \theta_i(x' - x)]$$

so that we obtain

$$\sum_{j=1}^{n} (x'_j - x_j) D_j f_i [x + \theta_i(x' - x)] = \lambda \delta_{ik}$$

$$\Leftrightarrow \sum_{j=1}^{n} \frac{[g_j(y + \lambda u_k) - g_i(y)]}{\lambda} D_i [x + \theta_i(x' - x) = \delta_{ik}]$$

The determinant of this system of linear equations is non-zero. We see that

$$\frac{g_j(y + \lambda u_k) - g_j(y)}{\lambda} \text{ is determined; } 1 \le j \le n$$

Consider these equations, solved and letting $x' \to x$ which is equivalent to λ tending to 0, we get

$$\sum_{j=1}^{n} D_k g_j(y) D_j f_i(x) = \delta_{ik}$$

Thus $D_k g_j(y)$ exist $\forall \; 1 \le k \le n, \; 1 \le j \le n$ and $\forall y \in Q$ so that g possess of partial derivatives in Q.

Because the partial derivatives $D_k g_j(y)$ can be expressed as linear combinations of $D_j f_i(x)$, the partial derivatives $D_k g_j(y)$ are continuous in Q.

10.15 IMPLICIT FUNCTION

Let f be a real valued function of two variables so that its domain D is a subset of R^2. Let us define E as follows : $E = \{(x, y) : (x, y) \in D \text{ and } f(x, y) = 0\}$.

Now, there may or may not exist a real valued function g of single variable with domain $A \subset R$ such that $E = \{(x, g(x)) : x \in A\}$

If g exists, then $y = g(x)$ is a solution of $f(x, y) = 0$.

(i) *First differential coefficient of an implicit function.* If $u = f(x, y) = $ constant and x and y, both are the functions of t and r, then

$$\frac{dy}{dx} = -\frac{p}{q}$$

where

$$p = \frac{\partial u}{\partial x} \text{ and } q = \frac{\partial u}{\partial y}$$

$$\frac{\partial u}{\partial x} = \frac{\partial u}{\partial t} \cdot \frac{\partial t}{\partial x} + \frac{\partial u}{\partial r} \cdot \frac{\partial r}{\partial x}$$

and

$$\frac{\partial u}{\partial y} = \frac{\partial u}{\partial t} \cdot \frac{\partial t}{\partial y} + \frac{\partial u}{\partial r} \cdot \frac{\partial r}{\partial y}$$

(ii) *Second differential coefficient of an implicit function.* Let $u = f(x, y) = c$, be the implicit function. Then,

$$\frac{d^2 y}{dx^2} = -\left[\frac{q^2 r - 2pqs + p^2 t}{q^3} \right]$$

where $p = \dfrac{\partial f}{\partial x}$, $q = \dfrac{\partial f}{\partial y}$, $r = \dfrac{\partial^2 f}{\partial x^2}$, $s = \dfrac{\partial^2 f}{\partial x \, \partial y}$ and $t = \dfrac{\partial^2 f}{\partial y^2}$

Solved Example

Example 1.   *If $x^y + y^x = c$, find $\dfrac{dy}{dx}$.*

Solution.   Let $f(x, y) = x^y + y^x - c$

$$\therefore \qquad \frac{\partial f}{\partial x} = yx^{y-1} + y^x \log y$$

and $\dfrac{\partial f}{\partial y} = x^y \log x + xy^{x-1} = x^y \log x + xy^{x-1}$

Hence, $\dfrac{dy}{dx} = -\dfrac{\partial f / \partial x}{\partial f / \partial y} = -\left[\dfrac{yx^{y-1} + y^x \log y}{x^y \log x + xy^{x-1}} \right]$

10.16 IMPLICIT FUNCTION THEOREM

Let $f(x, y)$ be a function of two variables x and y and let (a, b) be a point of its domain such that
    (i) $f(a, b) = 0$
    (ii) f possesses continuous partial derivatives f_x and f_y in a certain neighbourhood of (a, b)
and (iii) $f_y (a, b) \neq 0$
then, there exist a rectangle $[a - h, a + h; b - k, b + k]$ about (a, b) such that $\forall x \in [a-h, a+h]$,

the equation $f(x, y) = 0$ has one and only one solution $y = g(x)$, lying in the interval $[b - k, b + k]$, which have the folowing properties :

    (i) $b = g(a)$

    (ii) $f(x, g(x)) = 0 \; \forall \, x \in [a–h, a+h]$

    (iii) *g is differentiable. and both g and g' are continuous in* $[a - h, a + h]$.

    Proof.   Let us suppose $f_y(a, b) > 0$ (If $f_y(a, b) \neq 0$, then replace $f(x, y)$ by $–f(x, y)$). Consider the figure (3):

    (i) *Existence and Uniqueness of the Solution.*   Let f_x, f_y be continuous in a neighbourhood R_1 of (a, b), where

$$R_1 = [a - h_1, a + h_1; b - k_1, b + k_1]$$

Now,   since f_x, f_y are continuous in R_1, therefore, f is also continuous in R_1.

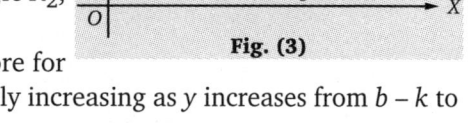

Also, since f_y is continuous at (a, b) and $f_y(a,b) > 0$, there exists a rectangle

$$R_2 = [a - h_2, a + h_2; b - k, b+k], \; h_2 < h_1, \; k < k_1$$

such that for every point (x, y) of the rectangle R_2, and $f_y (x, y) > 0$.

Now, since $f_y (x, y) > 0 \; \forall \; (x, y) \in R_2$, therefore for all $x \in [a - h_2, a + h_2]$ the function f of y strictly increasing as y increases from $b - k$ to $b+k$.

In particular, since $f(a, b) = 0$, we have

$$f(a, b - k) < 0, \; f(a, b + k) > 0 \qquad [\because (b - k) < 0 \text{ and } (b + k) > 0]$$

Since, f is continuous, and $f(a, b - k) < 0, \; f(a, b + k) > 0$, therefore, there exists an interval $[a - h, a + h], \; (h < h_2)$ such that for every x of this interval, we have

$$f(x, b - k) < 0, \; f(x, b+k) > 0, \text{ in some neighbourhood of } (a, b).$$

Now, for every fixed value of x in $[a - h, a+h]$, the continuous function of y strictly increases from a negative to positive value as y increases from $b - k$ to $b + k$ and therefore, there exists one and only one value of y for which $f(x, y) = 0$.

Hence, for each value of x in $[a - h, a + h]$, there is a unique value $y = g(x)$ for which $f(x, y) = 0$. Hence we can say the equation $f(x, y) = 0$ has one and only one solution $y = g(x)$ lying in the interval $[b - k, b + k]$ such that

        (a) $b = g(a)$

and   (b) $f(x, g(x)) = 0 \quad \forall \, x \in [a - h, a + h]$

   (ii) *Test for Continuity.*   Now, we shall prove g is continuous in $[a - h, a + h]$. Let x_0 be any point such that $x_0 \in [a - h, a + h]$.

Let $y_0 = g(x_0)$

Let $\varepsilon > 0$ be given. Now, consider a rectangle R', lying within the interval $R = [a - h, a + h; b - k, b + k]$ such that

$$R' = [x_0 - \delta_1, x_0 + \delta_1; y_0 - \varepsilon, y_0 + \varepsilon]$$

Since, $y = g(x)$ is the solution of $f(x, y) = 0$ in R which encloses R' . therefore $y = g(x)$ is also the solution of $f(x, y) = 0$ in R'.

Therefore, there exists an interval $]x_0 - \delta, x_0 + \delta[, (\delta \leq \delta_1)$ such that for every value of x in the interval $]x_0 - \delta, x_0 + \delta[, g(x)$ lies between $y_0 - \varepsilon$ and $y_0 + \varepsilon$ i.e.,

$$|y - y_0| = |g(x) - g(x_0)| < \varepsilon \text{ whenever } |x - x_0| < \delta$$

Hence, g is continuous at x_0 and therefore in $[a - h, a + h]$.

(iii) *Test for Differentiability.* Let x be any point of $[a- h, a + h]$ and let $x+p$ be another point of $[a - h, a + h]$.

Let
$$y = g(x), y + q = g(x+p)$$
$$\Rightarrow \qquad f(x, y) = 0, f(x+p, y+q) = 0$$
$$\Rightarrow f(x+p, y+q) - f(x, y) = 0.$$
$$\Rightarrow f(x+p, y+q) - f(x+p, y) + f(x+p, y) - f(x, y) = 0$$
$$\Rightarrow qf_y(x + p, y+\theta_1 q) + pf_x(x + \theta_2 p, y) = 0$$

(By mean value theorem, $0 < \theta_1 < 1$ and $0 < \theta_2 < 1$)

Since, $f_y \neq 0$ in R and $(x + p, y + \theta_1 q)$ is a point of R, we have

$$\frac{q}{p} = -\frac{f_x(x + \theta_2 p, y)}{f_y(x + p, y + \theta_1 q)}$$

Since, g is continuous and $q \to 0$ as $p \to 0$, therefore f_x and f_y being continuous, we get

$$g'(x) = -\frac{f_x(x, y)}{f_y(x, y)} \text{ when } (p, q) \to (0, 0)$$

Thus g is differentiable and $g'(x) = -\dfrac{f_x(x, y)}{f_y(x, y)}$

Also, since f_x and f_y are both continuous, therefore $g'(x)$ is continuous.

REMARKS

- Here, the function $y = g(x)$ is said to be the unique implicit function, determined by $f(x, y) = 0$.
- The implicit function theorem states that if a function of two variables satisfying certain assumptions of continuity and differentiability in a neighbourhood of a point (a, b) and If $(a, b) = 0$, then there exist a neighbourhood $[a - h, a + h; b - k, b+ k]$

of (a, b), such that $x \in [a - h, a + h]$, there exist a unique y belonging to $[b - k, b + k]$ such that $f(x, y) = 0$. Hence, we can say that $f(x, y) = 0$ defines a functions $y = g(x)$ in $[a - h, a + h]$; $y \in [b - k, b+k]$, such that $g(x)$ is differentiable.

Solved Examples

Example 1. Show that the equation $x^3 + y^3 - 3xy + y = 0$, determine unique solution near the point $(0, 0)$. Also find the first derivative of the solution.

Solution. Here, we have

$$f(x, y) = x^3 + y^3 - 3xy + y = 0 \text{ at } (0, 0)$$

(i) $f(0, 0) = 0$

(ii) $f_x = 3x^2 - 3y, f_y = 3y^2 - 3x + 1$

Now check the continuity of f_x and f_y at $(0,0)$.

(i) *Continuity of f_x at $(0, 0)$.*

The simultaneously limit is $\lim\limits_{\substack{x \to 0 \\ y \to 0}} (3x^2 - 3y) = 0$, exists.

∴ f_x is continuous.

The simultaneous limit is $\lim\limits_{\substack{x\to 0 \\ y\to 0}} (3y^2 - 3x + 1) = 1$, exists.

∴ f_y is continuous.

and $f_y(0,0) = \lim\limits_{k\to 0} \dfrac{f(0,k) - f(0,0)}{k}$

$$= \lim\limits_{k\to 0} \dfrac{k^3 + k}{k}$$

$$= \lim\limits_{k\to 0} k^2 + 1 = 1 \neq 0$$

∴ $f_y(0, 0) \neq 0$

Thus the equation $f(x, y) = 0$ determine a unique solution

$$y = \phi(x) \qquad\qquad \text{(By implicit function theorem)}$$

The unique solution of $f(x, y) = 0$ is given by

$$x^3 + y^3 - 3xy + y = 0$$

$$y^3 - y(3x - 1) + x^3 = 0$$

Let $y = \phi(x)$, therefore $[\phi(x)]^3 - \phi(x)[3x - 1] + x^3 = 0$

Since $\phi'(0) = 0$ ...(1)

Differentiating (1) , we get

$$3[\phi(x)]^2.\phi'(x) - 3\phi(x) - (3x - 1)\phi'(x) + 3x^2 = 0$$

$$0 - 3\times 0 - (-1)\phi'(0) + 0 = 0$$

$\Rightarrow$ $\phi'(0) = 0$

Example 2. *Show that the following equation determine unique solution near the point indicated.*

Also, find the derivative of the solution

(a) $xy \sin x + \cos y = 0$ at $\left(0, \dfrac{\pi}{2}\right)$ *(b) $y^3 \cos x + y^2 \sin^2 x = 7$ at $\left(\dfrac{\pi}{3}, 2\right)$*

Solution.   (a) $f(x,y) = xy \sin x + \cos y = 0$ Point $= \left(0, \dfrac{\pi}{2}\right)$

∴ $f\left(0, \dfrac{\pi}{2}\right) = 0 + \cos\dfrac{\pi}{2} = 0$

Now $f_x = y[x \cos x + \sin x], f_y = x \sin x - \sin y$

Further we check the continuity of f_x and f_y at $\left(0, \dfrac{\pi}{2}\right)$.

Continuity of f_x at $\left(0, \dfrac{\pi}{2}\right)$.

The simultaneous limit of $f_x(x, y)$. i.e., $\lim\limits_{\substack{x\to 0 \\ y\to(\pi/2)}} (xy \cos x + \sin x) = 0$ exists.

and the simultaneous limit of f_y (x, y) i.e,

$$\lim\limits_{\substack{x\to 0 \\ y\to(\pi/2)}} (x \sin x - \sin y) = \lim\limits_{y\to(\pi/2)} (-\sin y) = -1, \text{ exists}$$

Hence both f_x and f_y are continuous.

and
$$f_y\left(0,\frac{\pi}{2}\right) = \lim_{k\to 0}\frac{f\left(0,\frac{\pi}{2}+k\right) - f\left(0,\frac{\pi}{2}\right)}{k}$$

$$= \lim_{k\to 0}\frac{\cos\left(\frac{\pi}{2}+k\right)}{k} = \lim_{k\to 0}\frac{-\sin k}{k} = -1$$

$$\therefore \qquad f_y\left(0,\frac{\pi}{2}\right) \neq 0$$

Then by implicit function theorem, the equation $f(x, y) = 0$ determines a unique solulion. Now we have to find its first derivative

$$f(x, y) = 0$$

$$\Rightarrow \qquad xy \sin x + \cos y = 0$$

$$\therefore \qquad \phi'(x) = -\frac{f_x(x,y)}{f_y(x,y)} = -\frac{xy\cos x + y\sin x}{x\sin x - \sin y}$$

$$\phi'(x) = -\frac{0}{-1} = 0 \qquad\qquad \left[\begin{array}{l} \because\ f_x\left(0,\frac{\pi}{2}\right) = 0 \\[2mm] \text{and}\ f_y\left(0,\frac{\pi}{2}\right) = -1 \end{array}\right]$$

(b) $f(x,y) = y^3 \cos x + y^2 \sin^2 x - 7$; point $=\left(\dfrac{\pi}{3},2\right)$

$$\therefore \qquad f\left(\frac{\pi}{3},2\right) = 8\cos\frac{\pi}{3} + 4\sin^2\frac{\pi}{3} - 7 = 4 + 3 - 7 = 0$$

Now $\qquad f_x(x,y) = -y^3 \sin x + 2y^2 \sin x \cos x$

and $\qquad f_y(x,y) = 3y^2 \cos x + 2y \sin^2 x$

Next, we have to examine the continuity of f_x and f_y

Continuity of $f_x(x, y)$ at $\left(0,\dfrac{\pi}{2}\right)$

The simultaneous limit is

$$\lim_{\substack{x\to(\pi/3)\\y\to 2}}[-y^3\sin x + 2y^2\sin x\cos x] = \lim_{y\to 2}\left[-y^3\sin\frac{\pi}{3} + 2y^2\sin\frac{\pi}{3}\cos\frac{\pi}{3}\right]$$

$$= -8.\frac{\sqrt{3}}{2} + 8.\frac{\sqrt{3}}{2}.\frac{1}{2}$$

$$= -4\sqrt{3} + 2.\sqrt{3} = -2\sqrt{3} \neq 0,\ \text{exists}$$

$\therefore$ The simultaneous limit is given by

$$\lim_{\substack{x\to(\pi/3)\\y\to 2}}[3y^2\cos x + 2y\sin^2 x] = \lim_{y\to 2}\left[3y^2\cos\frac{\pi}{3} + 2y\sin^2\frac{\pi}{3}\right]$$

$$= 12.\frac{1}{2} + 4.\frac{3}{4} = 6 + 3 = 9,\ \text{exists}$$

$\therefore$ f_x and f_y both are continuous

$$f_y\left(\frac{\pi}{3}.2\right) = 12 \times \frac{1}{2} + 4 \times \frac{3}{4} = 6 + 3 = 9 \neq 0 \cdot$$

Then by implicit function theorem, $f(x, y) = 0$ determines a unique solution. And its derivative is given by

$$\phi'(x) = -\frac{f_x(x, y)}{f_y(x, y)}$$

$$\phi'\left(\frac{\pi}{3}\right) = -\left[\frac{-8\sin\frac{\pi}{3} + 8\sin\frac{\pi}{3}.\cos\frac{\pi}{3}}{9}\right] = -\left[\frac{-8\frac{\sqrt{3}}{2} + 2\sqrt{3}}{9}\right] = \frac{2\sqrt{3}}{9}$$

Let $\qquad y = \phi(x)$

$$\therefore \qquad \phi\left(\frac{\pi}{3}\right) = 2$$

$$[\phi(x)]^3 \cos x + [\phi(x)]^2 \sin^2 x - 7 = 0$$

$$3[\phi(x)]^2.\phi'(x).\cos x - [\phi(x)]^3 \sin x + 2\phi'(x)\phi(x)\sin^2 x + 2[\phi(x)]^2 \sin x \cos x = 0$$

$$3\left[\phi\left(\frac{\pi}{3}\right)\right]^2 \phi'\left(\frac{\pi}{3}\right).\cos\frac{\pi}{3} - \left[\phi\left(\frac{\pi}{3}\right)\right]^3 .\sin\frac{\pi}{3}$$

$$+ 2\phi'\left(\frac{\pi}{3}\right).\phi\left(\frac{\pi}{3}\right).\sin^2\frac{\pi}{3} + 2\left[\phi\left(\frac{\pi}{3}\right)\right]^2 \sin\frac{\pi}{3}\cos\frac{\pi}{3} = 0$$

$$12.\frac{1}{2}\phi'\left(\frac{\pi}{3}\right) - 8.\frac{\sqrt{3}}{2} + 4.\frac{3}{4}\phi'\left(\frac{\pi}{3}\right) + 8.\frac{\sqrt{3}}{2}.\frac{1}{2} = 0$$

$$9\phi'\left(\frac{\pi}{3}\right) - 4\sqrt{3} + 2\sqrt{3} = 0$$

$$\phi'\left(\frac{\pi}{3}\right) = \frac{2\sqrt{3}}{9}$$

EXERCISE 10.5

1. If $f(x, y) = (x^2 + y^2) \log (x^2 + y^2)$ for $(x, y) \neq (0, 0)$ and $f(0, 0) = 0$; show that f_{xy} and f_{yx} are not continuous at $(0, 0)$ but $f_{xy}(0, 0) = f_{yx}(0, 0)$.

2. Show that the following equation determine unique solutions near the point indicated. Find also the first derivative of the solution

$2xy - \log xy = 2$. at $(1, 1)$.

3. If f is a continuous function of each variable x and y separately in a certain neighbourhood of (a, b), and $f(a, b) = 0$, f_y is continuous at (a, b) and $f_y(a, b) \neq 0$, then the equation $f(x, y) = 0$ determines a unique continuous implicit solution $y = \phi(x)$ near (a, b).

4. Show that the least positive root of $xy = \tan y$ is a continuous function of x throughout the interval $[1, \infty[$ and increases from 0 to $\frac{\pi}{2}$ as y increases from 1 towards ∞.

10.17 CONCEPT OF MAXIMA AND MINIMA

If $y = f(x)$ be a continuous function. At a point $x = x_1$, if the function $f(x)$ does not increase and begins to decrease then $f(x)$ has its maximum value at $x = x_1$ and if at a point $x = x_2$, $f(x)$ does not decrease and begins to increase, then $f(x)$ has its minimum value at $x = x_2$.

If $f(x)$ is maximum at a point $x = x_1$ then $f(x)$ is an increasing function for the preceding values of x_1 and is a decreasing function for those value of x which the just below x_1 or we can say derivative of the function $\left(i.e., \dfrac{dy}{dx}\right)$ will be positive before $x = x_1$ and will be negative after $x = x_1$. But $\dfrac{dy}{dx}$ is a continuous function and $\dfrac{dy}{dx}$ changes the sign from positive to negative. So, $\dfrac{dy}{dx}$ will be zero at any point.

Therefore, for a maximum value of $y = f(x)$ at a point, we have $\dfrac{dy}{dx} = 0$ and $\dfrac{dy}{dx}$ changes the sign from positive to negative. On the other hand, for a minimum value of $y = f(x)$ at point we have $\dfrac{dy}{dx} = 0$ and $\dfrac{dy}{dx}$ changes the sign negative to positive.

REMARKS

- If $\dfrac{dy}{dx}$ changes the sign positive to negative; it means that $f(x)$ is a decreasing function of x i.e., $\dfrac{d^2y}{dx^2} < 0$.

- If $\dfrac{dy}{dx}$ changes the sign from negative to positive, it means that the $f(x)$ is an increasing function of x, i.e., $\dfrac{d^2y}{dx^2} > 0$.

- A function may have more than one maximum and minimum value.
- Any minimum value of the function $f(x)$ can be greater than any maximum value.
- Maximum and minimum values of the function occur alternately.
- Maximum and minimum values of the function are sometimes known as extreme value.
- From the definition of maxima and minima, it is clear that $\dfrac{dy}{dx} = 0$ is the necessary condition for maximum or minimum.
- $\dfrac{d^2y}{dx^2} < 0$ is sufficient condition for maximum and $\dfrac{d^2y}{dx^2} > 0$ is sufficient condition for minimum.

10.17.1 GRAPHICAL REPRESENTATION

In Fig. (4), we see that the A_2P_2 is the maximum point and the function has the maximum value A_2P_2 at $x = OP_2$ and has a minimum value A_4P_4 at $x = OP_4$.

Again, the function has a minimum value A_1P_1 at $x = OP_1$ and a minimum value A_3P_3 at $x = OP_3$.

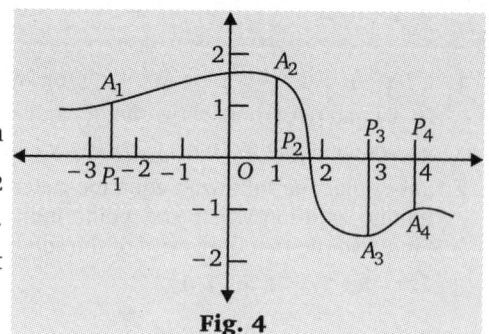

Fig. 4

WORKING PROCEDURE

Step 1. Find the derivative of the given function i.e., $\dfrac{dy}{dx}$.

Step 2. Put $\dfrac{dy}{dx} = 0$ and find all the real values of x. (say $x_1, x_2, x_3 \ldots$).

Step 3. Find $\dfrac{d^2y}{dx^2}$.

Step 4. Put $x = x_i$ in $\dfrac{d^2y}{dx^2}$ and find the result. If result is negative then the function $f(x)$ is maximum at $x = x_i$ and max. $f(x) = f(x_i)$. On the other hand, if result is positive then the function $f(x)$ is minimum at $x = x_i$ and minimum $f(x) = f(x_i)$.

REMARKS

- In a continuous function, maxima and minima values occur alternately, *i.e.*, between two successive maxima there is one minimum and between two successive minima, there is one maximum.

- If $\dfrac{d^2y}{dx^2}$ is equal to 0 at any point $x = x_i$ then find $\dfrac{d^3y}{dx^3}, \dfrac{d^4y}{dx^4}$, and find the values of these derivatives at $x = x_i$ successively and cheek the sign.

Solved Examples

Example 1. Find the value of x for which $f(x) = y = x^4 + 2x^3 - 3x^2 - 4x + 4$ is maximum or minimum and also find those value of $f(x)$.

Solution. Here, the given function is $y = f(x) = x^4 + 2x^3 - 3x^2 - 4x + 4$ ...(1)

So

$$\frac{dy}{dx} = 4x^3 + 6x^2 - 6x - 4$$

$$= 2(x+2)(2x+1)(x-1)$$

Now, put $\dfrac{dy}{dx} = 0$, we have

$$2\,(x+2)(2x+1)(x-1) = 0$$

So, $x = -2, -\dfrac{1}{2}, 1$

Again differentiating (2) *w.r.t.* to x, we get

Based on the following Results

- For a maximum value of $y = f(x)$ at point, we have $\dfrac{dy}{dx} = 0$ and $\dfrac{dy}{dx}$ change the sign from positive to negative. For a minimum value of $y = f(x)$ at a ponit we have $\dfrac{dy}{dx} = 0$ and $\dfrac{dy}{dx}$ changes the sign negative to positive.

- $\dfrac{dy}{dx} = 0$ is a necessary condition for the maxima and minima of $y = f(x)$.

- $\dfrac{d^2y}{dx^2} < 0$ is sufficient condition for maxima and $\dfrac{d^2y}{dx^2} > 0$ is for minima.

$$\frac{d^2y}{dx^2} = 12x^2 + 12x - 6$$

At $x = -2$, we have

$$\frac{d^2y}{dx^2} = 12(-2)^2 + 12(-2) - 6 = 48 - 24 - 6 = 18 > 0$$

Since, $\dfrac{d^2y}{dx^2} > 0$ (*i.e.*, positive). So $f(x)$ is minimum at $x = -2$. The minimum value of $f(x)$ at $x = -2$ is given by $f(-2) = (-2)^4 + 2(-2)^3 - 3(-2)^2 - 4(-2) + 4 = 0$

Now, at $x = -\dfrac{1}{2}$, we have

$$\frac{d^2y}{dx^2} = 12\left(-\frac{1}{2}\right)^2 + 12\left(-\frac{1}{2}\right) - 6 = 3 - 6 - 6 = -9 < 0.$$

Since, $\dfrac{d^2y}{dx^2} < 0$ (i.e., negative). So, $f(x)$ is maximum at $x = -\dfrac{1}{2}$ and maximum

value of $f(x)$ at $x = -\dfrac{1}{2}$ is

$$f\left(-\frac{1}{2}\right) = \left(-\frac{1}{2}\right)^4 + 2\left(-\frac{1}{2}\right)^3 - 3\left(-\frac{1}{2}\right)^2 - 4\left(-\frac{1}{2}\right) + 4$$

$$= \frac{1}{16} - \frac{1}{4} - \frac{3}{4} + 2 + 4 = \frac{81}{16}$$

Similarly, at $x = 1$, we have

$$\frac{d^2y}{dx^2} = 12(1)^2 + 12(1) - 6 = 12 + 12 - 6 = 18 > 0$$

Since, $\dfrac{d^2y}{dx^2} > 0$ (i.e., positive). So $f(x)$ is minimum at $x = 1$ and minimum value of $f(x)$ at $x = 1$ is

$$f(1) = (1)^4 + 2(1)^3 - 3\,(1)^2 - 4(1) + 4 = 1 + 2 - 3 - 4 + 4 = 0.$$

Example 2. *Find the maximum and minimum value of the function*

$$y = f(x) = x^3 - 12x^2 + 36x + 21$$

Solution. Here, the given function is $y = x^3 - 12x^2 + 36x + 21$

Now, differentiating *w.r.t. x*, we get

$$\frac{dy}{dx} = 3x^2 - 24x + 36$$

Puting $\dfrac{dy}{dx} = 0$, we get

$$3x^2 - 24x + 36 = 0 \text{ or } x^2 - 8x + 12 = 0$$

or $\qquad (x-2)(x-6) = 0 \quad$ or $\qquad\qquad x = 2, 6$

Again, differentiating w.r.t. x, we get $\dfrac{d^2y}{dx^2} = 6x - 24$

At $x = 2$, we have

$$\frac{d^2y}{dx^2} = 6(2) - 24 = -12 < 0$$

Since, $\dfrac{d^2y}{dx^2} < 0$ so $f(x)$ is maximum at $x = 2$. The maximum value of $f(x)$ at

$x = 2$ is given by

$$f(2) = (2)^3 - 12(2)^2 + 36(2) + 21 = 8 - 48 + 72 + 21 = 53.$$

Similarly, at $x = 6$, we have

$$\frac{d^2y}{dx^2} = 6 \times 6 - 24 = 36 - 24 = 12 > 0$$

Since, $\dfrac{d^2y}{dx^2} > 0$ so, $f(x)$ is minimum at $x = 6$ and minimum value of $f(x)$ at $x = 6$ is

$$f(6) = (6)^3 - 12(6)^2 + 36(6) + 21$$

$$= 216 - 432 + 216 + 21 = 453 - 432 = 21.$$

Example 3. *Find the value of x for which $y = x^3 - x^2 - 16x + 16$ is maximum or minimum. Also find those value of y.*

Solution. Here, the given function is $y = f(x) = x^3 - x^2 - 16x + 16$. Differentiating w.r.t. x, we get

$$\dfrac{dy}{dx} = 3x^2 - 2x - 16$$

Put $\dfrac{dy}{dx} = 0$, we get $3x^2 - 2x - 16 = 0$ or $(x + 2)(3x - 8) = 0$ or $x = -2, \dfrac{8}{3}$

Again, differentiating w.r.t. x, we get $\dfrac{d^2y}{dx^2} = 6x - 2$

At $x = -2$, we have $\dfrac{d^2y}{dx^2} = 6(-2) - 2 = -14 < 0$

Since, $\dfrac{d^2y}{dx^2} < 0$ so $f(x)$ is maximum at $x = -2$ and the maximum value of $f(x)$ at

$x = -2$ is

$$f(-2) = (-2)^3 - (-2)^2 - 16(-2) + 16 = -8 - 4 + 32 + 16 = 36$$

Similarly, at $x = \dfrac{8}{3}$ we have $\dfrac{d^2y}{dx^2} = 6\left(\dfrac{8}{3}\right) - 2 = 14 > 0$

Since, $\dfrac{d^2y}{dx^2} > 0$, so $f(x)$ is minimum at $x = \dfrac{8}{3}$ and the minimum value of $f(x)$ at

$x = \dfrac{8}{3}$ is given by

$$f\left(\dfrac{8}{3}\right) = \left(\dfrac{8}{3}\right)^3 - \left(\dfrac{8}{3}\right)^2 - 16 \times \dfrac{8}{2} + 16$$

$$= \dfrac{512}{27} - \dfrac{64}{9} - \dfrac{128}{3} + 16 = -\dfrac{400}{27}$$

Example 4. *Find the maximum and minimum values of the function $y = 2x^3 - 9x^2 + 12x - 1$*

Solution. Here, the given function is $y = 2x^3 - 9x^2 + 12x - 1$

So, $\dfrac{dy}{dx} = 6x^2 - 18x + 12$

Now, put $\dfrac{dy}{dx} = 0$, we have $6x^2 - 18x + 12 = 0$ or $6(x - 2)(x - 1) = 0$ or $x = 2, 1$

Again, differentiating w.r.t. x, we get $\dfrac{d^2y}{dx^2} = 12x - 18$

At $x = 2$, we have $\dfrac{d^2y}{dx^2} = 12 \times 2 - 18 = 6 > 0$

Since, $\dfrac{d^2y}{dx^2} > 0$. So, $f(x)$ is minimum and the minimum value of $f(x)$ at $x = 2$ is

$$f(2) = 2.(2)^3 - 9(2)^2 + 12(2) - 1 = 16 - 36 + 24 - 1 = 3$$

Similarly, at $x = 1$, we have

$$\frac{d^2y}{dx^2} = 12(1) - 18 = -6 < 0$$

Since, $\frac{d^2y}{dx^2} < 0$ so $f(x)$ is maximum at $x = 1$ and maximum value of $f(x)$ at $x = 1$ is given by

$$f(1) = 2(1)^3 - 9(1)^2 + 12(1) - 1 = 2 - 9 + 12 - 1 = 4$$

Example 5. *In a particular process the average cost $C = 80 - 12x + x^2$ where x is the number of units produced. Find the minimum average cost, and the corresponding number of units to be produced.*

... (1)

Solution. Here, average cost C is given by $C = 80 - 12x + x^2$

Differentiating w.r.t. x, we get

$$\frac{dC}{dx} = 2x - 12$$

Now $\frac{dC}{dx} = 0 \Rightarrow 2x - 12 = 0 \Rightarrow x = 6$

Again, differntiating, w.r.t. x, we get $\frac{d^2C}{dx^2} = 2$

So, at $x = 6, \frac{d^2C}{dx^2} = 2$ which is positive. So, C is minimum at $x = 6$ and the minimum value of C at $x = 6$ is given by

$$C = 80 - 12 \times 6 + (6)^2 = 80 - 72 + 36 = 44$$

and the number of units is $x = 6$.

Example 6. *The sale of fixed price product depend upon the cost of material C and the amount of labour L, according to the following relationship $S = 10CL - 2L^2$.*

Also the manufacturer plans his expenses such that $L + C = 12$

Find the maximum sale obtainable.

Solution. Here, it is given that $S = 10CL - 2L^2$ ...(1)

where $L + C = 12 \Rightarrow C = 12 - L$

On putting the value of C in (1), we get

$$S = 10(12 - L)L - 2L^2 \text{ or } S = 120L - 12L^2 \qquad ... (2)$$

Now, differentiating (2) w.r.t. L, we get $\frac{dS}{dL} = 120 - 24L$ ...(3)

For a maximum or minimum, put $\frac{dS}{dL} = 0$

$\Rightarrow \qquad 120 - 24L = 0 \Rightarrow L = 5$

Again, differentiating (2) w.r.t. L, we get $\frac{d^2S}{dL^2} = -24$

At $L = 5$, we have $\frac{d^2S}{dL^2} = -24$

Since, $\dfrac{d^2S}{dL^2}$ is negative. So, S is maximum at $L = 5$ and the maximum value of S at $L = 5$ is given by

$$S = 120 \times 5 - 12(5)^2 = 600 - 300 = 300$$

Hence, maximum sales is 300 units.

Example 7. *For a particular process, the average cost is $C = 56 - 8x + x^2$, where C is average cost and x is the number of units produced. Find the minimum value of the average cost, and the corresponding number of units to be produced.*

Solution. Here, the average cost is $C = 56 - 8x + x^2$ ...(1)

Differentiating (1) w.r.t. x, we get $\dfrac{dC}{dx} = -8 + 2x$ ... (2)

Now, put $\dfrac{dC}{dx} = 0 \Rightarrow 2x - 8 = 0 \Rightarrow x = 4$

Again differentiating (2) w.r.t. x, we get

$$\dfrac{d^2C}{dx^2} = 2$$

At $x = 4$, we have

$$\dfrac{d^2C}{dx^2} = 2$$

Since, $\dfrac{d^2C}{dx^2}$ is positive. So, C is minimum at $x = 4$ and the minimum value of C at $x = 4$ is

$$C = 56 - 8 \times 4 + (4)^2 = 56 - 32 + 16 = 40.$$

Hence, the minimum average cost is 40 and the number of units produced is 4.

Example 8. *Find the value of x for which $f(x) = x^3 - 3x + 4$ is maximum or minimum. Also, find the maximum or minimum value of $f(x)$.*

Solution. Here, $f(x) = x^3 - 3x + 4$

Differentiating the given equation, w.r.t. x, we get $f'(x) = 3x^2 - 3$

Put $f'(x) = 0 \Rightarrow 3x^2 - 3 = 0 \Rightarrow x^2 = 1 \Rightarrow x = \pm 1$

Again differentiating w.r.t. x, we get $f''(x) = 6x$

At $x = 1$, we have $f''(x) = 6$

Since $f''(x)$ is positive. So, $f(x)$ is minimum at $x = 1$ and minimum value of $f(x)$ at $x = 1$ is

$$f(1) = (1)^3 - 3 \times 1 + 4 = 1 - 3 + 4 = 2$$

Now at $x = -1$, we have

$$f''(x) = 6(-1) = -6$$

Since $f''(x)$ is negative at $x = -1$. So $f''(x)$ is maximum at $x = -1$ and maximum value of $f(x)$ at $x = -1$ is given by

$$f(-1) = (-1)^3 - 3(-1) + 4 = -1 + 3 + 4 = 6.$$

Example 9. Find the value of x for which $y = \dfrac{2}{3}x^3 + \dfrac{1}{2}x^2 - 6x + 8$ is maximum or minimum. Also find those value of y.

Solution. Here,
$$y = \dfrac{2}{3}x^3 + \dfrac{1}{2}x^2 - 6x + 8 \qquad \dots (1)$$

Differentiating (1) w.r.t. x, we get
$$\dfrac{dy}{dx} = 2x^2 + x - 6$$

Now, put $\dfrac{dy}{dx} = 0 \qquad \Rightarrow 2x^2 + x - 6 = 0$

$\Rightarrow \quad (2x - 3)(x + 2) = 0 \qquad \Rightarrow \qquad x = \dfrac{3}{2}, -2$

Again, differentiating (2) w.r.t. x, we get $\quad \dfrac{d^2y}{dx^2} = 4x + 1$

At $x = \dfrac{3}{2}$ we have $\dfrac{d^2y}{dx^2} = 4 \cdot \dfrac{3}{2} + 1 = 7 > 0$

Since, $\dfrac{d^2y}{dx^2}$ is positive. So, y is minimum at $x = \dfrac{3}{2}$ and the minimum value of y at $x = \dfrac{3}{2}$ is given by

$$f\left(\dfrac{3}{2}\right) = \dfrac{2}{3} \cdot \left(\dfrac{3}{2}\right)^3 + \dfrac{1}{2}\left(\dfrac{3}{2}\right)^2 - 6 \cdot \dfrac{3}{2} + 8 = \dfrac{9}{4} + \dfrac{9}{8} - 9 + 8 = \dfrac{19}{8}$$

Similarly, at $x = -2$, we have

$$\dfrac{d^2y}{dx^2} = 4(-2) + 1 = -7 < 0$$

Since, $\dfrac{d^2y}{dx^2}$ is negative. So, y is maximum at $x = -2$ and maximum value at $x = -2$ is given by
$$f(-2) = \dfrac{2}{3}(-2)^3 + \dfrac{1}{2}(-2)^2 - 6(-2) + 8$$

$$= -\dfrac{16}{3} + 2 + 12 + 8 = -\dfrac{16}{3} + 22 = \dfrac{50}{3}$$

Example 10. Investigate for maximum and minimum values, the function $(\sin x + \cos 2x)$.

Solution. Let $y = \sin x + \cos 2x$,

$\Rightarrow \quad \dfrac{dy}{dx} = \cos x - 2\sin 2x = \cos x - 4\sin x \cos x$

For stationary point

$\dfrac{dy}{dx} = 0 \quad \Rightarrow \qquad \cos x(1 - 4\sin x) = 0$

$\Rightarrow \quad \cos x = 0 \text{ or } 1 - 4\sin x = 0$

$\Rightarrow \qquad x = \dfrac{\pi}{2} \text{ or } \sin x = \dfrac{1}{4}$

For maxima or minima $\dfrac{d^2y}{dx^2} = -\sin x - 4\cos 2x$

$$= -\sin x - 4(1 - 2\sin^2 x)$$

$$= -\sin x - 4 + 8\sin^2 x$$

(i) At $x = \dfrac{\pi}{2}$,

$$\left(\dfrac{d^2y}{dx^2}\right) = -1 - 4 + 8 = 3 \text{ (which is positive.)}$$

So, given function is minimum at $x = \dfrac{\pi}{2}$.

(ii) At $\sin x = \dfrac{1}{4}$, $\left(\dfrac{d^2y}{dx^2}\right) = -\dfrac{1}{4} - 4 + 8 \cdot \dfrac{1}{16} = -\dfrac{1}{4} - 4 + \dfrac{1}{2}$

$$= \dfrac{-1 - 16 + 2}{4} = \dfrac{-15}{4} \text{(which is negative)}.$$

Hence, given function is minimum at $x = \sin^{-1}\dfrac{1}{4}$.

Example 11. *Find the maximum value of* $(x-1)(x-2)(x-3)$.

Solution. Let $f(x) = (x-1)(x-2)(x-3) = x^3 - 6x^2 + 11x - 6$

then $f'(x) = 3x^2 - 12x + 11$

For a maximum or minimum value of $f(x)$, we must have $f'(x) = 0$

$\Rightarrow$ $3x^2 - 12x + 11 = 0$

i.e., $$x = \dfrac{12 \pm \sqrt{144 - 4 \times 3 \times 11}}{6} = \dfrac{6 \pm \sqrt{(36-33)}}{3} = 2 \pm \dfrac{1}{\sqrt{3}}.$$

Also $f''(x) = 6x - 12$

Now $f''[2+(1/\sqrt{3})] = +$ve , therefore $f(x)$ has minimum value at $x = 2+(1/\sqrt{3})$. Again $f''[2-(1/\sqrt{3})] = -$ve   therefore $f(x)$ has a maximum value at $x = 2-(1/\sqrt{3})$. The maximum value of $f(x)$ is $= f[2-(1/\sqrt{3})]$

$$= [1-(1/\sqrt{3})](-1/\sqrt{3})[-1-(1/\sqrt{3})]$$

$$= (1 - 1/3)(1/\sqrt{3}) = 2/3\sqrt{3}$$

Example 12. *Find the maximum values of* $y = (x-1)(x-2)^2$.

Solution. Let $y = (x-1)(x-2)^2$

Then $\dfrac{dy}{dx} = (x-2)^2 + 2(x-1)(x-2) = (x-2)(3x-4)$

For maximum or minimum of y, we must have $\dfrac{dy}{dx} = 0$

i.e., $(x-2)(3x-4) = 0 \Rightarrow x = 2, 4/3$

Also $\dfrac{d^2y}{dx^2} = 6x - 10$

Now $\left(\dfrac{d^2y}{dx^2}\right)_{\text{at }x=2} = 2$ i.e., $+$ve $\Rightarrow y$ has a minimum value at $x = 2$.

Again $\left(\dfrac{d^2y}{dx^2}\right)_{x=4/3} = -2$ i.e., $-$ve $\Rightarrow y$ has a maximum value at $x = 4/3$

Example 13. *Show that* $\dfrac{x}{1+x\tan x}$ *is maximum when* $x = \cos x.$

Solution. Let $f(x) = \dfrac{x}{1+x\tan x}$

Then $f'(x) = \dfrac{(1+x\tan x)-x(x\sec^2 x+\tan x)}{(1+x\tan x)^2} = \dfrac{1-x^2\sec^2 x}{(1+x\tan x)^2}$

For a maximum or minimum of $f(x)$, we must have $f'(x) = 0$

$$1-x^2\sec^2 x = 0 \quad\Rightarrow\quad x = \pm\cos x$$

Now $f''(x) = (1-x^2\sec^2 x)\dfrac{d}{dx}\dfrac{1}{(1+x\tan x)^2}$

$$+\dfrac{1}{(1+x\tan x)^2}(-2x\sec^2 x - 2x^2\sec^2 x\tan x)$$

$$= (1-x^2\sec^2 x)\dfrac{d}{dx}\dfrac{1}{(1+x\tan x)^2} - \dfrac{2x\sec^2 x}{(1+x\tan x)^2}.(1+x\tan x)$$

$$= (1-x^2\sec^2 x)\dfrac{d}{dx}\dfrac{1}{(1+x\tan x)^2} - \dfrac{2x\sec^2 x}{(1+x\tan x)}$$

when $\quad x = \cos x, f''(x) = 0 - \dfrac{2}{\cos x(1+\sin x)}$ *i.e.*, $f''(x) = -$ve

Therefore $f(x)$ has a maximum value when $x = \cos x.$

Example 14. *Show that* $\sin x(1 + \cos x)$ *is a maximum at* $x = \pi/3.$

Solution. Let $f(x) = \sin x(1+\cos x) = \sin x + \dfrac{1}{2}\sin 2x$

Then $f'(x) = \cos x + 2\cos 2x$

For a maximum or a minimum value of $f(x), f'(x) = 0$

i.e., $\cos x + \cos 2x = 0 \Rightarrow 2\cos^2 x + \cos x - 1 = 0$

$$(2\cos x - 1)(\cos x + 1) = 0$$

$\therefore$ $\cos x = 1/2, -1 \Rightarrow x = \pi/3, \pi$

Now $f''(x) = -\sin x - 2\sin 2x$

$\therefore$ $f''\left(\dfrac{\pi}{3}\right) = -\sin\left(\dfrac{\pi}{3}\right) - 2\sin\left(\dfrac{2\pi}{3}\right) = -$ve

Hence $f(x)$ is maximum at $x = \pi/3.$

Example 15. *Find the maximum value of* $(\log x)/x$ *in* $0 < x < \infty.$

Solution. Let $f(x) = (\log x)/x$

Then $f'(x) = \dfrac{x.(1/x)-\log x}{x^2} = \dfrac{(1-\log x)}{x^2}$

For a maximum or a minimum value of $f(x), f'(x) = 0$

$$1-\log x = 0 \Rightarrow x = e$$

Now $f''(x) = (1-\log x)(-2/x^3) - 1/x^3$

$\therefore$ $f''(e) = -1/e^3 = -$ve. Therefore $f(x)$ is maximum at $x = e.$

Hence, the maximum value of $f(x)$ at $x = e$ is given by

$$f(e) = \frac{\log_e e}{e} = \frac{1}{e}$$

Example 16. *Find the maximum value of $(1/x)^x$.*

Solution. Let

$$y = (1/x)^x$$

$\Rightarrow \qquad \log y = x(\log 1 - \log x) = -x \log x$

$\Rightarrow \qquad \dfrac{1}{y}\dfrac{dy}{dx} = -1\log x - x(1/x) = -(1 + \log x)$

$\Rightarrow \qquad \dfrac{dy}{dx} = -y(1 + \log x) = -(1/x)^x(1 + \log x)$

For a maximum or a minimum of y, we must have $\dfrac{dy}{dx} = 0$

$\Rightarrow \qquad -(1/x)^x(1 + \log x) = 0$

$\Rightarrow \qquad 1 + \log x = 0 \Rightarrow x = 1/e$

$\Rightarrow \qquad \dfrac{d^2y}{dx^2} = -\dfrac{dy}{dx}(1 + \log x) - y(1/x)$

$$= -\dfrac{dy}{dx}(1 + \log x) - (1/x)^x.(1/x)$$

Therefore, when $x = 1/e$,

$$\dfrac{d^2y}{dx^2} = 0 - (e)^{1/e}.e = -\text{ve} \Rightarrow y \text{ is maximum at } x = 1/e.$$

Thus the maximum value of y is given by $e^{1/e}$.

Example 17. *The strength of a rectangular beam varies as the product of its breadth and the square of its depth. Find the dimensions of the strongest beam that can be cut from a round log of diameter $2a$.*

Solution. Let x be the breadth and y the depth of the beam, then as given, we have

$$x^2 + y^2 = 4a^2 \qquad \qquad ... (1)$$

If S denotes the strength of the beam, then

$$S \propto xy^2 \Rightarrow S = kxy^2$$

where k is constant.

Now $S = kx(4a^2 - x^2)$, from (1).

$\therefore \qquad \dfrac{dS}{dx} = k(4a^2 - 3x^2)$

For a maximum or a minimum of S, we must have

$\dfrac{dS}{dx} = 0$

i.e., $\qquad 4a^2 - 3x^2 = 0$ i.e., $x = 2a/\sqrt{3}$

Now $\dfrac{d^2S}{dx^2} = -6kx = \text{negative}$, when $x = 2a/\sqrt{3}$.

Therefore S is maximum when $x = 2a/\sqrt{3}$. The maximum value, when $x = 2a/\sqrt{3}$ is given by

$$y = \sqrt{\left(4a^2 - \dfrac{1}{3}.4a^2\right)} = 2a\sqrt{2/3}.$$

Hence, for the strongest beam, depth $= \sqrt{2/3}.a$ and breadth $= (1/\sqrt{3})a$

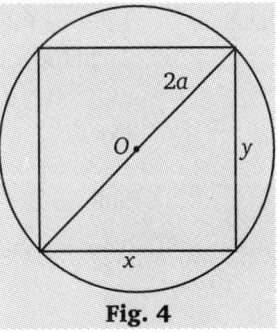

Fig. 4

Example 18. *Show that the semi-vertical angle of the cone of maximum volume and of given slant height is* $\tan^{-1}\sqrt{2}$.

Solution. Let Q be the semi-vertical angle of the cone and $OA = l$ be its slant height.

Then, V = volume of the cone $= \dfrac{1}{3}\pi r^2 h$

$$= \frac{1}{3}\pi(l\sin\theta)^2(l\cos\theta)$$

$$= \frac{1}{3}\pi l^3 \sin^2\theta\cos\theta$$

Since l is given, therefore V is a function of θ. Then

Fig. (5)

$$\frac{dV}{d\theta} = \frac{1}{3}\pi l^3(2\sin\theta\cos^2\theta - \sin^3\theta)$$

$$= \frac{1}{3}\pi l^3 \sin\theta(2\cos^2\theta - \sin^2\theta)$$

For a maximum or minimum of V, we must have $\dfrac{dV}{d\theta} = 0$

i.e., $\sin\theta(2\cos^2\theta - \sin^2\theta) = 0 \Rightarrow \sin\theta = 0$ or $\tan 2\theta = 2$

$\therefore \theta = \tan^{-1}\sqrt{2}$, because the values $\theta = 0$ and $\tan\theta = -\sqrt{2}$ are inadmissible.

Now $\dfrac{d^2V}{d\theta^2} = \dfrac{1}{3}\pi l^3 \cos\theta(2\cos^2\theta - \sin^2\theta)$

$$+ \frac{1}{3}\pi l^3 \sin\theta(-4\cos\theta\sin\theta - 2\sin\theta\cos\theta)$$

< 0 when $\tan\theta = \sqrt{2}$ i.e., when $2\cos^2\theta = \sin^2\theta$

Hence, V is maximum when $\theta = \tan^{-1}\sqrt{2}$.

Example 19. *Show that the semi-vertical angle of the right circular cone of given total surface (including area of the base) and maximum value is* $\sin^{-1}(1/3)$.

Solution. Let x be the radius of the base, h be the height and y the slant height of the cone. Then the total surface of the cone = constant

$$\Rightarrow \quad \pi x^2 + \pi xy = \text{constant} \qquad \qquad \dots (1)$$

Now $\quad V$ = volume of the cone $= \dfrac{1}{3}\pi x^2 h = \dfrac{1}{3}\pi x^2(y^2 - x^2)^{1/2}$

Since, $h = \sqrt{y^2 - x^2}$, therefore $V^2 = \dfrac{1}{9}\pi^2 x^4(y^2 - x^2)$.

Now, V is maximum or minimum according as V^2 or $\dfrac{9V^2}{\pi^2}$ is maximum or minimum.

Let $\quad S = \dfrac{9V^2}{\pi^2} = x^4(y^2 - x^2)$.

Then S can be regarded as a function of x because y is connected with x by (1).

We have $\quad \dfrac{dS}{dx} = 4x^3(y^2 - x^2) + x^4\left\{2y\left(\dfrac{dy}{dx}\right) - 2x\right\}$ $\qquad \dots(2)$

Differentiating (1) w.r. to x, we get

$$\pi\left(2x + y + x\frac{dy}{dx}\right) = 0 \text{ or } \frac{dy}{dx} = -\frac{2x + y}{x}$$

Substituting this value of $\dfrac{dy}{dx}$ in (2), we get

$$\frac{dS}{dx} = 4x^3 y^2 - 4x^5 + x^4 \left[-2y \frac{(2x+y)}{x} - 2x \right]$$

$$= 2x^3 y^2 - 6x^5 - 4x^4 y$$

For a maximum or a minimum of S, we must have $\dfrac{dS}{dx} = 0$

Now $\qquad \dfrac{dS}{dx} = 0 \Rightarrow 2x^3(y^2 - 2xy - 3x^2) = 0$

$$\Rightarrow 2x^3(y - 3x)(y + x) = 0$$

i.e., $\qquad y = 3x$ since $x \neq 0$ and $y \neq -x$.

Again $\qquad \dfrac{d^2 S}{dx^2} = 6x^2 y^2 + 4x^3 y \dfrac{dy}{dx} - 30x^4 + 16x^3 y - 4x^4 \dfrac{dy}{dx}$

When $y = 3x, \dfrac{dy}{dx} = -5$, so when $y = 3x$, we have

$$\frac{d^2 S}{dx^2} < 0$$

Therefore S is maximum when $y = 3x$.

Hence, V is maximum when $y = 3x$ or $x/y = 1/3$ or $\sin \theta = 1/3$ where θ is the semi-vertical angle of the cone.

Example 20. *Prove that least perimeter of an isosceles triangle in which a circle of radius r can be inscribed is $6r/\sqrt{3}$.*

Solution. Let α be the semi-vertical angle of an isoscales triangle in which a circle of radius r can be inscribed.

Then, $\qquad AQ = r \cot \alpha \quad$ and $\quad OA = r \operatorname{cosec} \alpha$

$\therefore \qquad AP = r + r \operatorname{cosec} \alpha = r(1 + \operatorname{cosec} \alpha)$ and $BP = AP \tan \alpha$

$$= r(1 + \operatorname{cosec} \alpha) \tan \alpha$$

$$= r(\tan \alpha + \sec \alpha)$$

Now if S is the perimeter of the triangle ABC, then

$\qquad S = 2AQ + 4PB \qquad\qquad$ [$\because AQ = AR, BQ = BP, CP = CR$]

$$= 2r \cot \alpha + 4r (\tan \alpha + \sec \alpha)$$

$$= 2r (\cot \alpha + 2 \tan \alpha + 2 \sec \alpha)$$

$\therefore \qquad \dfrac{dS}{d\alpha} = 2r(-\operatorname{cosec}^2 \alpha + 2\sec^2 \alpha + 2\sec \alpha \tan \alpha)$

For S to be a maximum or a minimum, we have $\dfrac{dS}{d\alpha} = 0$

$$- \operatorname{cosec}^2 \alpha + 2 \sec^2 \alpha + 2 \sec \alpha \tan \alpha = 0$$

or $\qquad\qquad -\dfrac{1}{\sin^2 \alpha} + \dfrac{2}{\cos^2 \alpha} + \dfrac{2 \sin \alpha}{\cos^2 \alpha} = 0$

or $\qquad\qquad \sin^3 \alpha + 2 \sin^2 \alpha - \cos^2 \alpha = 0$

or $\qquad\qquad \sin^3 \alpha + 2 \sin^2 \alpha - \cos^2 \alpha = 0$

or $\qquad\qquad 2\sin^3 \alpha + 3 \sin^2 \alpha - 1 = 0$

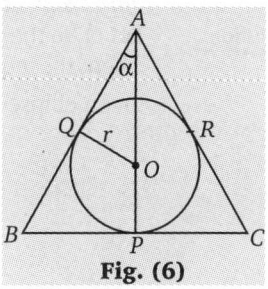

Fig. (6)

On solving the equation, we get $\sin \theta = -1$ or $\sin \theta = 1/2$

But $\sin \theta = -1$ is inadmissible. Therefore $\sin \theta = -1/2$ or $\theta = 30°$.

Also $\dfrac{d^2 S}{d\alpha^2} = 2r[2\operatorname{cosec}^2 \alpha \cot \alpha + 4\sec^2 \alpha \tan \alpha + 2\sec^3 \alpha + 2\sec \alpha \tan^2 \alpha]$

$= +ve$ for $\theta = 30°$.

Hence, $\theta = 30°$ gives least perimeter, which is given by

$$= 2r\left[\sqrt{3} + \frac{2}{\sqrt{3}} + 2.\frac{2}{\sqrt{3}}\right] = \frac{18r}{\sqrt{3}} = 6r\sqrt{3}.$$

Example 21. *Show that the triangle of maximum area that can be inscribed in a circle of radius a is an equilateral triangle.*

Solution. We observe that among all the inscribed triangles having AB as base, the area of that triangle is greatest for which the altitude of C w.r. to AB is greatest. Such a triangle is an isoscales triangle. Let α be the semi-vertical angle of such a triangle ABC inscribed in a given circle of radius r.

From Fig. (7), S = area of the triangle ABC

$$= 2\Delta\, AOC + \Delta\, AOB$$

$$= 2.\frac{1}{2}r^2 \sin(\pi - 2\alpha) + \frac{1}{2}r^2 \sin 4\alpha$$

$$= r^2 \sin 2\alpha + \frac{1}{2}r^2 \sin 4\alpha$$

$\therefore \qquad \dfrac{dS}{d\alpha} = 2r^2 \cos 2\alpha + 2r^2 \cos 4\alpha$

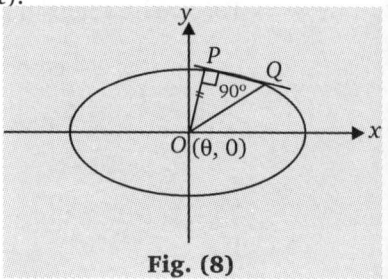

Fig. (7)

For a maximum or minimum of S, we must have $\dfrac{dS}{d\alpha} = 0$

$\cos 2\alpha + \cos 4\alpha = 0 \Rightarrow 2\cos 3\alpha \cos \alpha = 0$

i.e., $\quad 3\alpha = \pi/2$ or $\alpha = \pi/2$ i.e., $\alpha = \pi/6$ or $\alpha = \pi/2$

But $\quad \alpha = \pi/2$ is not possible. Hence, $\alpha = \pi/6$.

Also $\dfrac{d^2 S}{d\alpha^2} = -4r^2 \sin 2\alpha - 8r^2 \sin 4\alpha$, which is negative for $\alpha = \pi/6$.

Hence, area is maximum when $\alpha = \pi/6$ or $2\alpha = \pi/3$.

Thus the triangle is equilateral.

Example 22. *P is the foot of the perpendicular drawn from the centre O onto the tangent at a variable point Q on the ellipse*

$$x^2 / a^2 + y^2 / b^2 = 1, \qquad\qquad (where\ a > b)$$

Find the maximum length of QP and also the maximum area of the triangle OQP.

Solution. Let Q be the point with coordinates $(a \cos t,\ b \sin t)$.

Then the equation of the tangent at Q to the given ellipse is

$$\frac{x}{a}\cos t + \frac{y}{b}\sin t = 1$$

or $\qquad bx \cos t + ay \sin t = ab$

Let $\qquad OP = p$, then

$$p = \frac{ab}{\sqrt{(b^2 \cos^2 t + a^2 \sin^2 t)}}$$

Fig. (8)

$\therefore$
$$\frac{a^2b^2}{p^2} = b^2\cos^2 t + a^2\sin^2 t = b^2(1-\sin^2 t) + a^2(1-\cos^2 t)$$

$$= a^2 + b^2 - (a^2\cos^2 t + b^2\sin^2 t)$$
$$= a^2 + b^2 - OQ^2 \qquad (\text{because } OQ^2 = a^2\cos^2 t + b^2\sin^2 t)$$

$\therefore \qquad OQ^2 = a^2 + b^2 - (a^2b^2 / p^2)$

We have $\qquad QP^2 = OQ^2 - OP^2 = a^2 + b^2 - \left(\dfrac{a^2b^2}{p^2}\right) - p^2 = (z)\ \text{(say)} \qquad \ldots (1)$

Now QP is maximum or minimum according as QP^2 is maximum or minimum. But QP^2 i.e.,z is a function of p. For a maximum or a minimum of z, we have

$$\frac{dz}{dp} = \frac{2a^2b^2}{p^3} - 2p = 0 \ i.e., \qquad p^4 = a^2b^2 \ i.e.,\ p^2 = ab$$

Also $\qquad \dfrac{d^2z}{dp^2} = -\dfrac{6a^2b^2}{p^4} - 2 = -\text{ve, when } p^2 = ab.$

Therefore z is maximum when $p^2 = ab$. Putting $p^2 = ab$ in (1), we get

$$QP^2 = a^2 + b^2 - \frac{a^2b^2}{ab} - ab = (a-b)^2 \qquad (\because PN = a-b)$$

Hence, the maximum length of $QP = a$–b.

Now, let S be the area of the triangle OQP. Then $S = \dfrac{1}{2} OP \times QP$

Then $\qquad S^2 = \dfrac{1}{4} P^2 \times QP^2 = \dfrac{1}{4} p^2 \left(a^2 + b^2 - \dfrac{a^2b^2}{p^2} - p^2 \right) \qquad \text{[From (1)]}$

Let $\quad u = S^2 = \dfrac{1}{4}\{p^2(a^2+b^2) - a^2b^2 - p^4\}$

Now S is maximum or minimum according as S^2 i.e., u is maximum or minimum. For a maximum or minimum of u, we have

$$\frac{du}{dp} = \frac{1}{4}\{2p(a^2+b^2) - 4p^3\} = 0 \, i.e., p^2 = \frac{a^2+b^2}{2},\ \text{ since } p \ne 0$$

Also $\qquad \dfrac{d^2u}{dp^2} = \dfrac{1}{4}\{2(a^2-b^2) - 12p^2\} = -\text{ve, when } p^2 = \dfrac{a^2+b^2}{2}$

Hence u is maximum, when $p = \dfrac{a^2+b^2}{2}.$ Putting $p^2 = \dfrac{a^2+b^2}{2}$ in the value of S^2, we get

$$S^2 = \frac{1}{4}\left\{\frac{1}{2}(a^2+b^2)^2 - a^2b^2 - \frac{1}{4}(a^2+b^2)^2\right\}$$

$$= \frac{1}{16}\{(a^2+b^2)^2 - 4a^2b^2\} = \frac{1}{16}(a^2-b^2)^2$$

Hence, the maximum area of the triangle $OQP = \dfrac{1}{4}(a^2 - b^2).$

Example 23 *Investigate the maxima and minima of $ax + by$ when $xy = c^2$.*

Solution. Let $\qquad u = ax + by \qquad \ldots (1)$

Given that $xy = c^2$ or $y = c^2 / x$. Substituting this value of y in (1), we get

$$u = ax + bc^2 / x$$

Now, for maximum or minimum value of u, we must have

$$\frac{du}{dx} = 0 \quad \Rightarrow \quad \frac{du}{dx} = a - \frac{bc^2}{x^2} = 0$$

or $\qquad a = bc^2 / x^2$ or $x^2 = \dfrac{bc^2}{a}$ or $x = \pm c\sqrt{(b/a)}$

Again $\qquad \dfrac{d^2u}{dx^2} = \dfrac{2bc^2}{x^3} = +ve$ for $x = c\sqrt{(b/a)}$

and $\qquad \dfrac{d^2u}{dx^2} = -ve$ for $x = -c\sqrt{b/a}$

Hence u is maximum, when $x = -c.\sqrt{b/a}$ and minimum, when $x = c.\sqrt{b/a}$.

Therefore, maximum value of $u = -ac\sqrt{(a/c)} - bc^2 / \{c\sqrt{b/c}\}$

$$= -c\sqrt{(ab)} - c\sqrt{(ab)} = -2c\sqrt{(ab)}$$

And minimum value of $u = ac\sqrt{(b/a)} + bc^2 / \{c\sqrt{(b/a)}\}$

$$= c.\sqrt{ab} + c.\sqrt{(ab)} = 2c\sqrt{ab}$$

Example 24. *In a submarine telegraph cable the speed of signalling varies as* $\log x^2 \log(1/x)$, *where x is the ratio of the radius of the core to that of the covering. Show that the greatest speed is attained when this ratio is* $1: \sqrt{e}$.

Solution. Let S be the speed of signalling.

Then $\quad S = \mu x^2 \log(1/x) = -\mu x^2 \log x$ where μ is a constant.

For a maximum or a minimum of S, we have $\dfrac{dS}{dx} = 0$

i.e., $x(2\log x + 1) = 0 \Rightarrow x = 0$ or $\log x = -1/2$

But $x = 0$ is inadmissible. Therefore $\log x = -1/2$ or $x = e^{-1/2} = 1/\sqrt{e}$

Now $\qquad \dfrac{d^2S}{dx^2} = -\mu(2\log x + 1) - \mu x(2/x) = -\mu(2\log x + 1) = -2\mu$

When $x = 1/\sqrt{e}$, we have $2\log x + 1 = 0$, when $x = 1/\sqrt{e}$,

we have $\qquad \dfrac{d^2S}{dx^2} = -2\mu$ which is negative.

Hence, S is maximum, when $x = 1/\sqrt{e}$.

Example 25. *Show that maximum rectangle that can be inscribed in a circle is square.*

Solution. Let $ABCD$ be the rectangle inscribed in circle with centre O and radius a. Also, let $AB = 2x$ and $BC = 2y$. Then

$$a^2 = x^2 + y^2 \qquad \qquad ...(1)$$

Area of rectangle $ABCD$

$$A = (2x)(2y) = 4xy$$

$$= 4x\sqrt{a^2 - x^2} \quad \text{[From (1)]}$$

For maximum or minimum area, $\dfrac{dA}{dx} = 0$

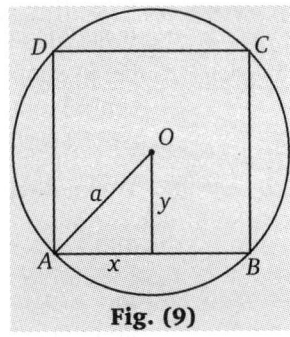

Fig. (9)

$$\Rightarrow \quad 4\left\{\sqrt{a^2 - x^2} - \frac{x^2}{\sqrt{a^2 - x^2}}\right\} = 0$$

$$\Rightarrow \quad 4\left\{\frac{a^2 - 2x^2}{\sqrt{a^2 - x^2}}\right\} = 0$$

$$\Rightarrow \quad a^2 - 2x^2 = 0 \Rightarrow x = \frac{a}{\sqrt{2}}$$

Now $\quad \dfrac{d^2 A}{dx^2} = 4\left\{(-4x)a^2 - x^2)^{1/2} + (a^2 - 2x^2)\left(-\frac{1}{2}\right)(a^2 - x^2)^{-3/2}(-2x)\right\}$

$$= 4\left[\frac{-4x}{\sqrt{a^2 - x^2}} \cdot \frac{x(a^2 - 2x^2)^2}{(a^2 - x^2)^{3/2}}\right]$$

$$\Rightarrow \quad \left(\frac{d^2 A}{dx^2}\right)_{x = a/\sqrt{2}} = -16 \text{ (which is negative.)}$$

Thus A is max. when $x = \dfrac{a}{\sqrt{2}}$.

From (1), $x = \dfrac{a}{\sqrt{2}}$ .Therefore $x = y = \dfrac{a}{\sqrt{2}}$

Hence, area is maximum when $x = y = \dfrac{a}{\sqrt{2}}$ *i.e.*, rectangle is square.

Example 26. *Show that the height of the closed cylinder of given surface and greatest volume is equal to its diameter.*

Solution. Let r be radius of base and h the height of a closed cylinder of given surface S, then

$$S = 2\pi r^2 + 2\pi rh \Rightarrow h = \frac{S - 2\pi r^2}{2\pi r} \qquad \dots (1)$$

If V be volume of cylinder then

$$V = \pi r^2 h = \pi r^2 \left(\frac{S - 2\pi r^2}{2\pi r}\right) = \frac{rS - 2\pi r^3}{2}$$

$$\Rightarrow \quad \frac{dV}{dr} = \frac{S}{2} - 3\pi r^2 \qquad \dots (2)$$

For max or min we have $\dfrac{dV}{dr} = 0$

$$\frac{S}{2} - 3\pi r^2 = 0 \Rightarrow S = 6\pi r^2$$

$$\Rightarrow \quad 2\pi r^2 + 2\pi rh = 6\pi r^2$$

$$\Rightarrow \quad h = 2r$$

From (2) $\dfrac{d^2 V}{dr^2} = -6\pi r$, (–ve) for any positive value of r.

Hence V is maximum when $h = 2r$ *i,e.*, when the height of cylinder is equal is diameter of base.

Example 27. *Prove that a conical tent of a given capacity will required the least amount of canvas when the height is $\sqrt{2}$ times the radius of the base.*

Solution. Let us suppose h be the height, r be the radius of the base l the slant height of the conical tent. Let V be the given capacity (*i.e.* volume) and S denote the area of the curved surface of the tent.

We know that

$$V = \frac{1}{3}\pi r^2 h \qquad \qquad \text{...(1)}$$

and

$$S = \pi l r = \pi(\sqrt{h^2 + r^2})r$$

$\Rightarrow$

$$S^2 = \pi^2 r^2(h^2 + r^2) = u \,(\text{say}) \qquad \qquad \text{... (2)}$$

From (1) and (2), we get

$$u = \pi^2 r^2\left[\frac{9V^2}{\pi^2 r^4} + r^2\right] = \frac{9V^2}{r^2} + \pi^2 r^4$$

$\therefore$

$$\frac{du}{dr} = -\frac{18V^2}{r^3} + 4\pi^2 r^3$$

and

$$\frac{d^2 u}{dr^2} = \frac{54V^2}{r^4} + 12\pi^2 r^2$$

Now

$$\frac{du}{dr} = 0 \Rightarrow V = \left(\frac{2}{3\sqrt{2}}\right)\pi r^3$$

for

$$V = \left(\frac{2}{3\sqrt{2}}\right)\pi r^3, \frac{d^2 u}{dr^2} > 0$$

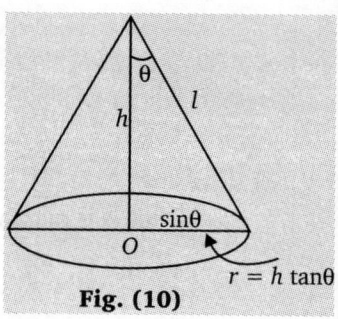

Fig. (10)

i.e., u is minimum when $V = \dfrac{2}{3\sqrt{2}}\pi r^3$

i.e., when $\dfrac{2}{3\sqrt{2}}\pi r^3 = \dfrac{1}{3}\pi r^2 h$ *i.e.*, $h = r\sqrt{2}$

$\Rightarrow$ u is minimum, when $h = r\sqrt{2}$.

Example 28. *Show that the radius of the right circular cylinder of greatest curved surface which can be inscribed in a given cone is half that of the cone.*

Solution. Suppose r is the radius and H is the height of the given cone

i.e.,

$$OB = r, OA = H$$

where O is the centre of the base circle.

Suppose x is the radius and h is the height of the cylinder inscribed in the given cone.

Now triangles AOB and ADE are similar, therefore

$$\frac{AD}{AO} = \frac{DE}{OB} \text{ or } \frac{H-h}{H} = \frac{x}{r}$$

$\Rightarrow$

$$1 - \frac{h}{H} = \frac{x}{r} \Rightarrow \frac{H}{h} = 1 - \frac{x}{r}$$

$\Rightarrow$

$$H = h\left(1 - \frac{x}{r}\right)$$

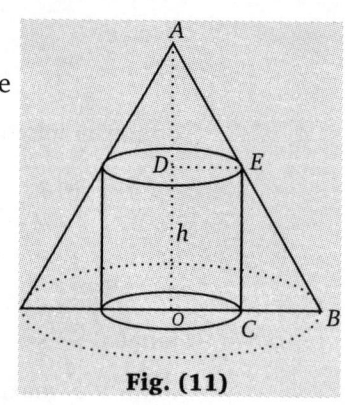

Fig. (11)

Now the curved surface of the cylinder

$\Rightarrow$

$$S = 2\pi.x.h = 2\pi rH\left(1 - \frac{x}{r}\right)$$

$$S = \frac{2\pi H}{r}(rx - x^2)$$

$\Rightarrow$

$\Rightarrow$ $$\frac{dS}{dx} = \frac{2\pi H}{r}(r - 2x)$$

so $\frac{dS}{dx} = 0$, we get $r - 2x = 0$ or $x = \frac{r}{2}$

Also $\frac{d^2S}{dx^2} = \frac{2\pi h}{r}(-2) < 0 \Rightarrow S$ is greatest, when $x = \frac{r}{2}$

i.e., when radius of the cylinder is half of that can be inscribed in a sphere of radius *a* is $2a / \sqrt{3}$.

Example 29. *Show that the height of the closed cylinder of given surface and of maximum volume is equal to the diameter of its base.*

Solution. Let $OO' = h$ be the height of the cylinder and radius $OB = r$.

Also let *V* be the volume of the cylinder then

$$V = \pi r^2 h \qquad \qquad ...(1)$$

and S be the surface area of the given cylinder, then

$$S = 2\pi rh + 2\pi r^2 \qquad \qquad ...(2)$$

Now from (1), $h = \frac{V}{\pi r^2} \qquad \qquad ...(3)$

Fig. (12)

Putting this value in (2), we get

$$S = 2\pi r \frac{V}{\pi r^2} + 2\pi r^2 = \frac{2V}{r} + 2\pi r^2. \qquad \qquad ...(4)$$

Differentiating (4) w.r. to *r* on both sides, we get

$$\frac{dS}{dr} = \frac{4V}{r^3} + 4\pi \qquad \qquad ...(5)$$

Again differentiating (5) w.r. to *r* on both sides, we get

$$\frac{d^2S}{dr^2} = \frac{4V}{r^3} + 4\pi \qquad \qquad ...(6)$$

For max. and min. $\frac{dS}{dr} = 0$, , then from (5) $-\frac{2V}{r^2} + 4\pi r = 0$

or $-\frac{2V}{r^2} = -4\pi r$ or $V = 2\pi r^3$.

Putting this value in (6), we get

$\frac{d^2S}{dr^2} = \frac{4}{r^3} + 4\pi = 8\pi + 4\pi = 12\pi$, which is positive.

i.e., $\frac{d^2S}{dr^2} > 0$. Hence S will be minimum for $V = 2\pi r^3$.

Then putting $V = 2\pi r^3$ in (3), we get $h = \frac{V}{\pi r^2} = \frac{2\pi r^3}{\pi r^2} = 2r$

$h = 2r =$ diameter of the base of the cylinder.

Example 30. *Show that the right circular cylinder of the given surface (including two ends) and maximum volume is such that it height equal to the diameter of the base.*

Solution. Let $OO' = h$ be the height of the cylinder and radius $OB = r$. Alse let V be the volume of the cylinder then

$$V = \pi r^2 h \qquad \ldots(1)$$

and S be the surface area of the given cylinder, then

$$S = 2\pi r h + 2\pi r^2 \qquad \ldots(2)$$

or $\qquad 2\pi r h = S - 2\pi r^2 \Rightarrow h = \dfrac{S - 2\pi r^2}{2\pi r}.$

Putting this value in (1), we get

$$V = \pi r^2 \left[\dfrac{S - 2\pi r^2}{2\pi r} \right] = \dfrac{1}{2}[Sr - 2\pi r^3]. \qquad \ldots(3)$$

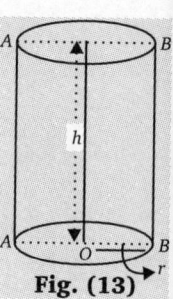

Fig. (13)

Differentiating (3) w.r.t. to r on both sides, we get

$$\dfrac{dV}{dr} = \dfrac{1}{2}[S - 6\pi r^2]$$

Again differentiating, we get

$$\dfrac{d^2V}{dr^2} = \dfrac{d}{dr}\left[\dfrac{dV}{dr} \right] = \dfrac{d}{dr}\left[\dfrac{1}{2}(s - 6\pi r^2) \right] \qquad \text{[From (4)]}$$

$$= -\dfrac{12\pi r}{2} = -6\pi r. \qquad \ldots(5)$$

For max., $\qquad \dfrac{dV}{dr} = 0,$ then from (4), we get

$$\dfrac{1}{2}[S - 6\pi r^2] = 0 \text{ or } S - 6\pi r^2 = 0 \text{ or } S = 6\pi r^2$$

or $\qquad r^2 = \dfrac{S}{6\pi} \Rightarrow r = \sqrt{\dfrac{S}{6\pi}}.$

Therefore V will be maximum

Now putting $S = 6\pi pr^2$ in (2), we get

$$6\pi r^2 = 2\pi rh + 2\pi r^2 \text{ or } 2\pi rh = 4\pi r^2$$

$$h = 2r.$$

i.e., height of the cylinder is equal to diameter of the base.

EXERCISE 10.6

1. Find the maximum or minimum value of $y = x^3 - 6x^2 + 9x + 15.$

2. Show that $y = x^5 - 5x^4 + 5x^3 - 10$ has a maximum value when $x = 1$ and a minimum value when $x = 3$ and neither maximum nor minimum when $x = 0.$

3. The cost C of manufacturing a certain product is given by $C = x^2 - 4x + 100,$ when x is the number of product manufactured. Find the minimum value of $C.$

4. Find the maximum or minimum value of the function $y = x^2 - 5x + 7.$

HINTS TO SELECTED PROBLEMS

1. $y = x^3 - 6x^2 + 9x + 15$

$$\frac{dy}{dx} = 3x^2 - 12x + 9.$$

Put $\frac{dy}{dx} = 0$

$\Rightarrow \quad 3x^2 - 12x + 9 = 0$

$\Rightarrow \quad x^2 - 4x + 3 = 0$

$\Rightarrow \quad (x - 3)(x - 1) = 0$

$$x = 1, 3$$

Now $\left(\frac{d^2y}{dx^2}\right) = 6x - 12$

$\left(\frac{d^2y}{dx^2}\right)_{at\,x=1} = 6 \times 1 - 12 = -6 < 0$

y is maximum at $x = 1$

$\left(\frac{d^2y}{dx^2}\right)_{at\,x=3} = 6 \times 3 - 12 = 6 > 0$

y is minimum at $x = 3$.

2. $y = x^5 - 5x^4 + 5x^3 - 10$

$$\frac{dy}{dx} = 5x^4 - 20x^3 + 15x^2.$$

Put $\frac{dy}{dx} = 0$

$\Rightarrow \qquad 5x^4 - 20x^3 + 15x^2 = 0$

$\Rightarrow \qquad 5x^2(x^2 - 4x + 3) = 0$

$\Rightarrow \qquad\qquad x = 0, 1, 3$

Now, $\frac{d^2y}{dx^2} = 20x^3 - 60x^2 + 30x$

$\left(\frac{d^2y}{dx^2}\right)_{at\,x=1} = 20(1)^3 - 60(1)^2 + 30(1) = -10 < 0$

y is maximum at $x = 1$

$\left(\frac{d^2y}{dx^2}\right)_{at\,x=3} = 20(3)^3 - 60(3)^2 + 30(3) > 0$

y is minimum at $x = 3$

$\left(\frac{d^2y}{dx^2}\right)_{at\,x=0} = 0 - 0 + 0 = 0 \Rightarrow y$

$\Rightarrow y$ is neither maximum nor minimum at $x = 0$.

Answer

1. maximum at $x = 1$, minimum at $x = 3$ **3.** at $x = 2$, minimum $C = 96$

4. minimum at $x = \dfrac{5}{2}$.

10.18 MAXIMA AND MINIMA OF A FUNCTION OF SEVERAL INDEPENDENT VARIABLES

Let $f(x, y, z, ...)$ be a function of several independent variables $x, y, z....$ If f is continuous and finite for all values of $x, y, z, ...$ in the neighbourhood of $x = a, y = b, z = c, ...$ respectively, then the value of $(a, b, c, ...)$ is said to be a maximum or minimum if $f(a+h, b+k, c+l, ...)$ is less than or greater than $f(a, b, c, ...)$ for all values of $h, k, l, ...$ (where $h, k, l, ...$) are sufficiently small, may be positive or negative provided they are not all zero.

In other words we can say, the value of $f(a, b, c, ...)$ is said to be a maximum or minimum if $f(a+h, b + k, c + l, ...) - f(a, b, c, ...)$ maintain an invariant sign (may be positive or negative) for all values of $h, k, l, ...$ positive or negative provided they are taken sufficiently small and finite.

A point $(a_1, a_2, ..., a_n)$ is called a stationary point, if all the first order partial derivative of the function $f(x_1, x_2, ..., x_n)$ vanish at the point. A stationary point, if it is maximum or minimum is known as extreme point and the value of the function at an extreme point is known as an extreme value.

REMARK

- A stationary point may be a maximum or minimum or neither of these two.

10.19 NECESSARY CONDITION FOR THE EXISTENCE OF MAXIMA OR MINIMA

Let $f(x, y, z, ...)$ be a function of several independent variables $x, y, z,...$ It is clear from the definition of maxima and minima that maximum or minimum of $f(x, y, z, ..)$ will occur for those values of $x, y, z, ...$, for which the expression $f(x+h, y+k, z+l, ...) - f(x, y, z, ...)$ maintain an invariant sign for all sufficiently small and finite values of $h, k, l, ...$ positive or negative.

Now, expanding $f(x+h, y+k, z+l, ...)$ by Taylor's theorem, we have

$$f(x+h, y+k, z+L...) = f(x, y, z) + \left(h\frac{\partial f}{\partial x} + k\frac{\partial f}{\partial y} + l\frac{\partial f}{\partial z} + ... \right)$$

+ terms of second and higher order.

$$\Rightarrow f(x+h, y+k, z+l...) - f(x, y, z,...) = \left(h\frac{\partial f}{\partial x} + k\frac{\partial f}{\partial y} + l\frac{\partial f}{\partial z} + ... \right)$$

+ terms of second and higher orders.   ...(1)

Now, since $h, k, l, ...$ are sufficiently small, the first degree expression

$$\left(h\frac{\partial f}{\partial x} + k\frac{\partial f}{\partial y} + l\frac{\partial f}{\partial z} + ... \right)$$

of the equation (1) can be made to govern the sign of right hand side and hence, of the left hand side as well as. Thus, by changing the sign of the left hand side of the equation (1) will also change.

Since, left hand side is to preserve an invariable sign for maxima or minima, therefore, as a necessary condition for maximum and minimum values, we must have

$$h\frac{\partial f}{\partial x} + k\frac{\partial f}{\partial y} + l\frac{\partial f}{\partial z} + ... = 0 \qquad ...(2)$$

Now, since $h, k, l, ...$ are arbitrary and independent of each other, we must have

$$\frac{\partial f}{\partial x} = 0, \frac{\partial f}{\partial y} = 0, \frac{\partial f}{\partial z} = 0, \text{ etc.} \qquad ...(3)$$

If the number of independent variables be n, we shall get n simultaneous equations in these n variables, which will give the values $a, b, c, ...$ of the n variables $x, y, z, ...$ respectively for which $f(x, y, z, ...)$ will have a maximum or a minimum values.

REMARKS

- The necessary condition for a function $f(x, y, z, ...)$ of the independent variables $x, y, z, ...$ to be maximum or minimum is given by

$$\frac{\partial f}{\partial x} = 0, \frac{\partial f}{\partial y} = 0, \frac{\partial f}{\partial z} = 0,$$

- The conditions given above is only a necessary condition for the maxima and minima of the function $f(x, y, z, ...)$. These conditions are not sufficient.

10.19.1 MAXIMA AND MINIMA FOR A FUNCTION OF TWO INDEPENDENT VARIABLES

(1) *To find the condition which governs the sign of a quadratic expression.*

Consider, a binary expression

$$I = ax^2 + 2hxy + by^2$$

of two variables x and y. Then I can be written as

$$I = ax^2 + 2hxy + by^2$$
$$= \frac{1}{a}[(ax + hy)^2 + (ab - h^2)y^2].$$

If $(ab - h^2)$ is positive, the sign of I will be the same as that of a.

But if $(ab - h^2)$ is negative, then, the expression within the brackets may be positive or negative and therefore we cannot say anything about the sign of expression I.

(2) *Stationary and extreme points (For the function of two independent variables)*:

Let $f(x, y)$ be a function of two independent variables x and y. A point (a, b) is called a stationary point, if both the first order partial derivatives $\left(\frac{\partial f}{\partial a} \text{ and } \frac{\partial f}{\partial b} \right)$ of the function $f(x, y)$ at (a, b) vanish.

A stationary point which is either a maximum or minimum is called an extreme point.

REMARKS

- A stationary point is not necessarily an extreme point, hence a stationary point may be a maximum or a minimum or neither of these two.
- The value of the function at extreme point is called extreme value.
- A point at which function is neither maximum nor minimum, is known as saddle point.

10.20 NECESSARY CONDITION FOR MAXIMA OR MINIMA

Let $f(x, y)$ be a function of two independent variables x and y. Then, we have the maximum or minimum of $f(x, y)$ at $x = a$ and $x = b$ if the expression $f(a + h, b + k) - f(a, b)$ is of invariable sign for all sufficiently small independent variables h and k provided both of them are not equal to zero.

We observe that,

(i) If the sign of $f(a+h, b+k) - f(a, b)$ is negative, then we have a maximum of $f(x, y)$ at $x = a, y = b$.

(ii) If the sign of $f(a+h, b+k) - f(a, b)$ is positive, we have a minimum of $f(x, y)$ at $x = a$, $y = b$.

Expand $f(a+h, b+k)$ by Taylor's theorem, we have

$$f(a+h,b+k) = f(a,b) + \left(h\frac{\partial f}{\partial x} + k\frac{\partial f}{\partial y} \right)_{\substack{x=a \\ y=b}}$$

$$+ \frac{1}{2!}\left(h^2\frac{\partial^2 f}{\partial x^2} + 2hk\frac{\partial^2 f}{\partial x\,\partial y} + k^2\frac{\partial^2 f}{\partial y^2} \right)_{\substack{x=a \\ y=b}} + \dots$$

$$\Rightarrow \quad f(a+h,b+k) - f(a,b) = h\left(\frac{\partial f}{\partial x}\right)_{\substack{x=a \\ y=b}} + k\left(\frac{\partial f}{\partial y}\right)_{\substack{x=a \\ y=b}}$$

+ term of the second and higher orders in h and k.

Now, since h and k are sufficiently small, the expression $h\left(\dfrac{\partial f}{\partial x}\right)_{\substack{x=a \\ y=b}} + k\left(\dfrac{\partial f}{\partial y}\right)_{\substack{x=a \\ y=b}}$ of the equation (1) can be made to govern the sign of right hand side and hence of the left hand side as well. Thus by changing the sign of h and k, the sign of the left hand side of the equation (1) will also change.

Since L.H.S. is to preserve an invariable sign for maximum or minimum, therefore as a necessary condition for maximum and minimum values, we must have

$$h\left(\frac{\partial f}{\partial x}\right)_{\substack{x=a \\ y=b}} + k\left(\frac{\partial f}{\partial y}\right)_{\substack{x=a \\ y=b}} = 0. \qquad\qquad \dots(2)$$

If $k = 0$, we find that if $\left(\dfrac{\partial f}{\partial x}\right)_{\substack{x=a \\ y=b}} \neq 0$, the R.H.S. of (2) changes sign when h changes sign. Therefore $f(x, y)$ cannot have a maximum or minimum at $x = a$, $y = b$ if $\left(\dfrac{\partial f}{\partial x}\right)_{\substack{x=a \\ y=b}} \neq 0$.

Similarly, taking $h = 0$, we see that $f(x, y)$ cannot have a maximum or a minimum at $x = a$, $y = b$ if $\left(\dfrac{\partial f}{\partial y}\right)_{\substack{x=a \\ y=b}} \neq 0$. Thus, a set of necessary conditions that $f(x, y)$ should have a maximum or minimum at $x=a$, $y=b$ is that

$$\left(\frac{\partial f}{\partial x}\right)_{\substack{x=a \\ y=b}} = 0 \text{ and } \left(\frac{\partial f}{\partial y}\right)_{\substack{x=a \\ y=b}} = 0.$$

10.21 SUFFICIENT CONDITION FOR MAXIMA OR MINIMA: THE LAGRANGE'S CONDITION

Let $f(x, y)$ be a function of two variables x and y.

Let $\qquad r = \dfrac{\partial^2 f}{\partial x^2},\ s = \dfrac{\partial^2 f}{\partial x\,\partial y},\ t = \dfrac{\partial^2 f}{\partial y^2}$ at $x = a$ and $y = b$.

As a set of necessary conditions for a maximum or minimum at (a, b) we have

$$\frac{\partial f}{\partial x} = 0 \text{ and } \frac{\partial f}{\partial y} = 0 \text{ at } (a, b)$$

then $\qquad f(a + h, b+ k) - f(a, b) = \dfrac{1}{2!}[rh^2 + 2shk + tk^2] + R$ $\qquad$ ...(1)

Where R consists of terms of third and higher order of small quantities h and k .

Now, by taking h and k sufficiently small, the second degree terms in R.H.S. of (1) may be made to govern the sign of R.H.S. and therefore of the L.H.S. also *i.e.*, for sufficiently small values of h and k, the sign of $\dfrac{1}{2}(rh^2 + 2shk + tk^2) + R$ is same as that of $rh^2 + 2shk + tk^2$.

If the sign is negative, then the function is maximum at (a, b) and if the sign is positive, then the function is minimum at (a, b).

Case (i) *If* $(rt - s^2) > 0$.

Then, neither r nor t can be zero. Hence, we can write

$$rh^2 + 2shk + tk^2 = \frac{1}{2}[r^2h^2 + 2rshk + rtk^2]$$

$$= \frac{1}{2}[(rh + sk)^2 + (rt - s^2)k^2]$$

since $rt - s^2 > 0$, therefore $(rh + sk)^2 + (rt - s^2)k^2 > 0$ for all values of h and k except when $rh + sk = 0, k = 0$ *i.e.*, at $h = 0, k = 0$, which is not possible.

Hence, in this case the expression $rh^2 + 2shk + tk^2$ will have the same sign for all values of h and k, and the sign is determined by the sign of r.

Thus, the function $f(x, y)$ will have a maximum or minimum at $x = a$ and $y = b$. If $rt - s^2 > 0$. The function $f(x, y)$ is maximum or minimum according as r is negative or positive.

Case (ii) *If* $(rt - s^2) < 0$.

If $rt - s^2$ is negative, we are not sure about the sign of second degree term of R.H.S. of (1) and hence there is neither a maximum nor a minimum value.

Case (iii) *If* $rt - s^2 = 0$.

If $rt = s^2$, then quadratic expression $rh^2 + 2shk + tk^2$ becomes $\dfrac{1}{r}(hr + ks)^2$.

So that, the quadratic expression will be of the same sign as that of r or t unless

$$\frac{h}{k} = -\frac{s}{r} = \alpha \text{ (say) } i.e., \ rh + sk = 0.$$

If this condition is sasfisied, then the second degree expression in R.H.S. of (1) vanishes and hence, the sign of the R.H.S. of (1) depends upon third degree expression in h and k, which change sign with the change of sign of h and k and hence, the sign of L.H.S. of (1) will also change and hence, there will be neither maximum nor minimum.

Thus, the necessary condition for the existence of maxima and minima now is that the cubic terms must vanish collectively in R.H.S. of (1) when $\dfrac{h}{k} = -\dfrac{s}{r} = \alpha;$ and then the biquadratic terms of R.H.S. of (1) must collecctively be of the same sign as r and t, when

$$\frac{h}{k} = -\frac{s}{r} = \alpha$$

i.e., $\qquad hr + ks = 0$

Hence, the case is doubtful.

Thus, if $rt - s^2 = 0$, the case is doubtful and further, investigation is needed to determine the maxima and minima of $f(x, y)$ at (a, b).

WORKING PROCEDURE

To discuss the maxima and minima at $x = a, y = b$, we must find

$$r = \left(\frac{\partial^2 u}{\partial x^2}\right)_{\substack{x=0 \\ y=0}}, \; s = \left(\frac{\partial^2 u}{\partial x \partial y}\right)_{\substack{x=a \\ y=b}}, \; t = \left(\frac{\partial^2 u}{\partial y^2}\right)_{\substack{x=a \\ y=b}}$$

Then, calculate $rt - s^2$.

Now following cases arise :

(i) If $rt - s^2 > 0$, then

    (A) If r is negative then, $f(x, y)$ is maximum at $x = a, y = b$.

    (B) If r is positive then, $f(x, y)$ is minimum at $x = a, y = b$.

(ii) If $rt - s^2 < 0$, $f(x, y)$ is neither maximum nor minimum at $x = a, y = b$.

(iii) If $rt - s^2 = 0$, the case is doubtful, and further investigation will be required.

REMARK

- While solving problems, we frequently used the identity, given by Lagrange.

$$\{(a^2 + b^2 + c^2)(p^2 + q^2 + r^2) - (ap + bq + cr)^2\}$$
$$= \{(br - cq)^2 + (cp + ar)^2 + (aq - bp)^2\}.$$

Solved Examples

Example 1. *Find all maximum or minimum values of the function :*
$$f(x, y) = y^2 + x^2 y + x^4.$$

Solution. Since, we have
$$f(x, y) = y^2 + x^2 y + x^4.$$

$\therefore$ $\dfrac{\partial f}{\partial x} = 2xy + 4x^3$ and $\dfrac{\partial f}{\partial y} = 2y + x^2.$

For a maximum or minimum of $f(x, y)$, we must have

$$\frac{\partial f}{\partial x} = 0 \text{ and } \frac{\partial f}{\partial y} = 0$$

$\therefore$ $\dfrac{\partial f}{\partial x} = 0 \Rightarrow 2xy + 4x^3 = 0$

         $\Rightarrow \quad 2x(y + 2x^2) = 0$ ...(1)

     $\dfrac{\partial f}{\partial y} = 0 \Rightarrow 2y + x^2 = 0$

Solving (1) and (2), we get $x = 0, y = 0$.

Thus $(0, 0)$ is the only point of maximum or minimum.

Now $r = \left(\dfrac{\partial^2 f}{\partial x^2}\right)_{(0,0)} = [2y + 12x^2]_{(0,0)} = 0$

         $s = \left(\dfrac{\partial^2 f}{\partial x \partial y}\right)_{(0,0)} = [2x]_{(0,0)} = 0$

and $t = \left(\dfrac{\partial^2 f}{\partial y^2}\right)_{(0,0)} = [2]_{(0,0)} = 2$

$$\therefore \qquad rt - s^2 = 0\,(2) - 0^2 = 0.$$

Thus, the case is doubtful and further investigation will be required.

Example 2. *Find the maximum or minimum values of the function $x^3 y^2 (1 - x - y)$.*

Solution.   Let $\qquad u = x^3 y^2 (1 - x - y)$

$$\Rightarrow \qquad \frac{\partial u}{\partial x} = 3x^2 y^2 (1 - x - y) - x^3 y^2$$

and $\qquad \dfrac{\partial u}{\partial y} = 2x^3 y (1 - x - y) - x^3 y^2.$

For a maximum or minimum of u, we must have $\dfrac{\partial u}{\partial x} = 0$ and $\dfrac{\partial u}{\partial y} = 0$

$$\Rightarrow \qquad 3x^2 y^2 (1 - x - y) - x^3 y^2 = 0 \qquad \qquad \text{...(1)}$$

and $\qquad 2x^3 y (1 - x - y) - x^3 y^2 = 0. \qquad \qquad \text{...(2)}$

Now, subtracting (2) from (1), we have $\quad x^2 y (1 - x - y)(3y - 2x) = 0$

which gives $\qquad y = \dfrac{2}{3} x.$

Putting the value of y in (1), we get $\quad x = \dfrac{1}{2}$

So $\left(\dfrac{1}{2}, \dfrac{1}{3} \right)$ be the point of maxima or minima.

Now $\qquad\qquad r = \dfrac{\partial^2 u}{\partial x^2} = 6xy^2 - 12x^2 y^2 - 6xy^3 \; = -\dfrac{1}{9}, \text{ at } \left(\dfrac{1}{2}, \dfrac{1}{3} \right)$

$$t = \frac{\partial^2 u}{\partial y^2} = 2x^3 - 2x^4 - 6x^3 y \; = -\frac{1}{8}, \text{ at } \left(\frac{1}{2}, \frac{1}{3} \right)$$

$$s = \frac{\partial^2 u}{\partial x \partial y} = 6x^2 y - 8x^3 y - 9x^2 y^2 = -\frac{1}{12} \text{ at } \left(\frac{1}{2}, \frac{1}{3} \right).$$

Now, $\qquad\qquad rt - s^2 = $ positive.

Also, r is negative, hence the function u has a maximum at $x = \dfrac{1}{2}, \, y = \dfrac{1}{3}.$

The maximum value is $= \left(\dfrac{1}{2} \right)^3 \left(\dfrac{1}{3} \right)^2 \left(1 - \dfrac{1}{2} - \dfrac{1}{3} \right) = \dfrac{1}{432}.$

Example 3. *Discuss the maximum or minimum values of u, where*

$$u = 2a^2 xy - 3ax^2 y - ay^3 + x^3 y + xy^3.$$

Solution.   We have $\qquad u = 2a^2 xy - 3ax^2 y - ay^3 + x^3 y + xy^3$

which gives

$$\frac{\partial u}{\partial x} = 2a^2 y - 6axy + 3x^2 y + y^3$$

and $\qquad \dfrac{\partial u}{\partial y} = 2a^2 x - 3ax^2 - 3ay^2 + x^3 + 3xy^2$

For a maximum and minima of u, we have

$$\frac{\partial u}{\partial x} = 0, \frac{\partial u}{\partial y} = 0$$

which gives,

$$y(2a^2 - 6ax + 3x^2 + y^2) = 0 \qquad \text{...(1)}$$

and $\qquad 2a^2x - 3ax^2 - 3ay^2 + x^3 + 3xy^2 = 0 \qquad \text{...(2)}$

Equation (1) and (2) gives the following values of x and y :

$$x = 0, \; y = 0; \; x = a, y = 0; \; x = 2a, \; y = 0; \; x = \frac{3}{2}a, \; y = \pm\frac{1}{2}a;$$

$$x = a, y = a, \; x = \frac{1}{2}, \; y = \frac{1}{2}a; x = a, y = -a; \; x = \frac{1}{2}a, \; y = -\frac{1}{2}a.$$

Then, we get the following pairs of values of x and y which make the function u stationary.

$$(0,0), \, (a,0), \, (2a,0), \, \left(\frac{3}{2}a, \frac{1}{2}a\right), \, \left(\frac{3}{2}a, -\frac{1}{2}a\right), \, (a,a), \, \left(\frac{1}{2}a, \frac{1}{2}a\right), \, (a,-a), \, \left(\frac{1}{2}a, -\frac{1}{2}a\right).$$

Also
$$r = \frac{\partial^2 u}{\partial x^2} = -6ay + 6xy,$$

$$s = \frac{\partial^2 u}{\partial x \, \partial y} = 2a^2 - 6ax + 3x^2 + 3y^2,$$

and
$$t = \frac{\partial^2 u}{\partial y^2} = -6ay + 6xy.$$

For (0, 0).

$$r = 0, \, s = 2a^2, \, t = 0$$

$\Rightarrow rt - s^2$, is negative.

Therefore, we have neither maximum nor a minimum of u at (0, 0).

Similarly, we can easily shown that u has neither a maximum nor a minimum at $(a, 0)$, $(2a, 0)$, (a, a), $(a, -a)$.

For $\left(\dfrac{3a}{2}, \dfrac{a}{2}\right)$.

$$r = \frac{3}{2}a^2, \, s = \frac{1}{2}a^2, \, t = \frac{3}{2}a^2, \quad \Rightarrow \quad rt - s^2 \text{ is positive.}$$

Here, since r is positive, therefore u has minimum at $\left(\dfrac{3a}{2}, \dfrac{a}{2}\right)$.

Similarly, we can check the maxima and minima at all other points.

Example 4.
Solution.

Find the maximum and minimum values of $xy(a - x - y)$.

Let $\qquad u = xy(a - x - y)$

Then $\qquad \dfrac{\partial u}{\partial x} = ay - 2xy - y^2 \;$ and $\; \dfrac{\partial u}{\partial y} = ax - x^2 - 2xy.$

For a maximum or minimum of u, we have

$$\frac{\partial u}{\partial x} = 0 \text{ and } \frac{\partial u}{\partial y} = 0.$$

Thus, we have

$$ay - 2xy - y^2 = 0 \Rightarrow y(a - 2x - y) = 0 \qquad \text{...(1)}$$

$$ax - x^2 - 2xy = 0 \Rightarrow x(a - x - 2y) = 0 \qquad \text{...(2)}$$

Solving (1) and (2), we get the following pairs of values x and y which makes the function stationary

$$(0,0),\ (0,a),\ (a,0),\ \left(\frac{1}{3}a,\frac{1}{3}a\right).$$

Here
$$r = \frac{\partial^2 u}{\partial x^2} = -2y,\quad s = \frac{\partial^2 u}{\partial x\,\partial y} = a - 2x - 2y,$$

and
$$t = \frac{\partial^2 u}{\partial y^2} = -2x.$$

For (0, 0). $\qquad r = 0,\ s = a,\ t = 0 \qquad \Rightarrow\ rt - s^2$ is negative.

∴ We have neither a maximum nor a minimum of u at $(0, 0)$.

For (0, a). $\qquad r = -2a,\ s = -a,\ t = 0 \ \Rightarrow\ rt - s^2$ is negative.

∴ We have neither a maximum nor a minimum of u at $(a, 0)$.

Similarly, we have neither a maximum nor a minimum of u at $(\alpha, 0)$.

For $\left(\dfrac{1}{3}a,\ \dfrac{1}{3}a\right)$.

$$\circ\quad r = -\frac{2}{3}a,\ s = -\frac{1}{3}a,\ t = -\frac{2}{3}a \qquad \Rightarrow\ rt - s^2 \text{ is positive.}$$

Since $\qquad rt - s^2 > 0.$

∴ $\quad u$ has an extreme value at $\left(\dfrac{1}{3}a,\dfrac{1}{3}a\right)$

$\Rightarrow\quad u$ has a maximum if r is negative, *i.e.*, if a is positive and u has a minimum if r is positive, *i.e.*, if a is negative.

Example 5. *Find a point within a triangle such that the sum of the squares of its distances from the vertices is a minimum.*

Solution. Let us suppose $[(x_r, y_r) : r = 1, 2, 3]$ be the vertices of the triangle and (x, y) be any point inside the triangle.

Now, let us define a function
$$u = \sum_{r=1}^{3} [(x - x_r)^2 + (y - y_r)^2].$$

Then, we have
$$\frac{\partial u}{\partial x} = \Sigma\, 2(x - x_r) = 2[(x - x_1) + (x - x_2) + (x - x_3)]$$

and
$$\frac{\partial u}{\partial y} = \Sigma\, 2(y - y_r) = 2[(y - y_1) + (y - y_2) + (y - y_3)].$$

For a maximum or minimum of u, we must have
$$\frac{\partial u}{\partial x} = 0 \Rightarrow (x - x_1) + (x - x_2) + (x - x_3) = 0 \Rightarrow x = \frac{x_1 + x_2 + x_3}{3}$$

and
$$\frac{\partial u}{\partial y} = 0 \Rightarrow (y - y_1) + (y - y_2) + (y - y_3) = 0 \Rightarrow y = \frac{y_1 + y_2 + y_3}{3}.$$

Thus, we have $\left(\dfrac{x_1 + x_2 + x_3}{3},\ \dfrac{y_1 + y_2 + y_3}{3}\right)$ is the only point at which u have a maximum or minimum.

Now $$r = \frac{\partial^2 u}{\partial x^2} = 6, s = \frac{\partial^2 u}{\partial x \, \partial y} = 0, t = \frac{\partial^2 u}{\partial y^2} = 6.$$

Now, at $\left[\dfrac{x_1 + x_2 + x_3}{3}, \dfrac{y_1 + y_2 + y_3}{3} \right]$

$$r = 6, s = 0, t = 6$$

$\Rightarrow \qquad rt - s^2 = 36 > 0.$

Also, since $\qquad r > 0.$

Therefore u have a minimum value at $\left[\dfrac{x_1 + x_2 + x_3}{3}, \dfrac{y_1 + y_2 + y_3}{3} \right].$

Hence, the point $\left(\dfrac{x_1 + x_2 + x_3}{3}, \dfrac{y_1 + y_2 + y_3}{3} \right)$ is the required point at which u is minimum.

REMARK

- The point $\left(\dfrac{x_1 + x_2 + x_3}{3}, \dfrac{y_1 + y_2 + y_3}{3} \right)$ is the centroid of the given triangle.

Example 6. *Show that the minimum value of* $u = xy + \left(\dfrac{a^3}{x} \right) + \left(\dfrac{a^3}{y} \right)$ *is* $3a^2$.

Solution. We have

$$u = xy + \left(\frac{a^3}{x} \right) + \left(\frac{a^3}{y} \right)$$

$\Rightarrow \qquad \dfrac{\partial u}{\partial x} = y - \dfrac{a^3}{x^2}$ and $\dfrac{\partial u}{\partial y} = x - \dfrac{a^3}{y^2}.$

For a maximum or minimum of u, we have

$$\frac{\partial u}{\partial x} = 0 \text{ and } \frac{\partial u}{\partial y} = 0$$

Now, $\qquad \dfrac{\partial u}{\partial x} = 0 \Rightarrow y - \dfrac{a^3}{x^2} = 0 \qquad \qquad \dots (1)$

and $\qquad \dfrac{\partial u}{\partial y} = 0 \Rightarrow x - \dfrac{a^3}{y^2} = 0. \qquad \qquad \dots (2)$

Solving (1) and (2), we get

$$x = a, y = a$$

Now $r = \dfrac{\partial^2 u}{\partial x^2} = \dfrac{2a^3}{x^3}, s = \dfrac{\partial^2 u}{\partial x \, \partial y} = 1$ and $t = \dfrac{\partial^2 u}{\partial y^2} = \dfrac{2a^3}{y^3}.$

At $x = y = a$, we have

$$r = 2, s = 1, t = 2$$

$\Rightarrow \qquad rt - s^2 = 3 > 0.$

Thus, at (a, a), $rt - s^2 > 0$ and $r > 0$. Therefore u is minimum at $x = a$, $y = a$.

The minimum value of $\qquad u = a.a + \left(\dfrac{a^3}{a} \right) + \left(\dfrac{a^3}{a} \right) = 3a^2.$

Example 7. *Determine the points where a function* $x^3 + y^3 - 3axy$ *has maximum or minimum.*

Solution. Here, we have

$$u = x^3 + y^3 - 3axy$$

$\Rightarrow$ $\qquad \dfrac{\partial u}{\partial x} = 3x^2 - 3ay$ and $\dfrac{\partial u}{\partial y} = 3y^2 - 3ax.$

For a maximum or minimum of u, we must have

$$\dfrac{\partial u}{\partial x} = 0 \text{ and } \dfrac{\partial u}{\partial y} = 0$$

which gives, $\qquad x^2 - ay = 0$ $\qquad\qquad$ ... (1)

and $\qquad\qquad y^2 - ax = 0$ $\qquad\qquad$ ... (2)

Solving (1) and (2), we get

$$x = 0, y = 0; x = a, y = a.$$

Thus $(0, 0)$ and (a, a) are the stationary points of u.

Now $\qquad\qquad r = \dfrac{\partial^2 u}{\partial x^2} = 6x, s = \dfrac{\partial^2 u}{\partial x \partial y} = -3a, t = \dfrac{\partial^2 u}{\partial y^2} = 6y.$

For $x = 0, y = 0$

$$r = 0, s = -3a \text{ and } t = 0$$

$\therefore$ $\qquad\qquad rt - s^2 = -9a^2 < 0$, for all values of a.

$\Rightarrow$ $\quad u$ is neither maximum nor minimum at $x = 0, y = 0$.

For $x = a, y = a$

$$r = 6a, s = -3a \text{ and } t = 6a$$

$\Rightarrow$ $\qquad\qquad rt - s^2 = 27a^2 > 0$, for all values of a.

Also $r = 6a$, which is positive if $a > 0$.

Thus (i) u is maximum at $x = a, y = a$ if $a < 0$

and (ii) u is minimum at $x = a, y = a$ if $a > 0$.

Example 8. *Discuss the maxima and minima of the function* $u = \sin x \sin y \sin (x+y).$

Solution. Here, we have

$$u = \sin x \sin y \sin (x + y)$$

$\Rightarrow$ $\qquad \dfrac{\partial u}{\partial x} = \sin y [\sin x \cos(x + y) + \cos x \sin (x + y)]$

and $\qquad \dfrac{\partial u}{\partial y} = \sin x [\sin y \cos(x + y) + \cos y \sin (x + y)].$

For a maxima and minima of u, we must have

$$\dfrac{\partial u}{\partial x} = 0 \text{ and } \dfrac{\partial u}{\partial y} = 0.$$

$\Rightarrow$ $\qquad \sin y [\sin x \cos (x + y) + \cos x \sin (x+y)] = 0$

and $\qquad \sin x [\sin y \cos (x + y) + \cos y \sin (x+y)] = 0.$

Equation (1) and (2) gives

$$\tan (x + y) = -\tan x \quad ...(1) \Rightarrow \tan x = \tan y$$

and $\qquad$ $\tan(x+y) = -\tan y$ $\quad$ ...(2) $\Rightarrow$ $\qquad$ $x = y$

From (1) and (2), we have

$$\tan 2x = -\tan x = \tan(\pi - x)$$

$\Rightarrow$ $\qquad$ $2x = \pi - x$ $\qquad \Rightarrow \qquad$ $3x = \pi$ $\qquad \Rightarrow \qquad$ $x = \dfrac{\pi}{3} = y.$

Moreover, $\qquad$ $\dfrac{\partial u}{\partial x} = 0$, gives $\sin y = 0 \Rightarrow y = 0$

and $\qquad$ $\dfrac{\partial u}{\partial y} = 0$, gives $\sin x = 0 \Rightarrow x = 0.$

Thus, we get the following pair of values, which makes the function u stationary

$$(0,0),\ \left(\frac{\pi}{3}, \frac{\pi}{3}\right).$$

Now $\qquad$ $r = \dfrac{\partial^2 u}{\partial x^2} = 2\sin y \cos(2x+y),$

$$s = \dfrac{\partial^2 u}{\partial x\,\partial y} = \sin 2(x+y),$$

and $\qquad$ $t = \dfrac{\partial^2 u}{\partial y^2} = 2\sin x \cos(2y+x).$

For (0, 0). $\qquad r = 0, s = 0, t = 0$

$\Rightarrow \qquad\qquad rt - s^2 = 0.$

$\therefore$ $\quad$ this case is doubtful and need further investigation.

For $\left(\dfrac{\pi}{3}, \dfrac{\pi}{3}\right).$

$$r = 2\sin\frac{1}{3}\pi.\cos\pi = -\sqrt{3},$$

$$s = \sin\left(\frac{4\pi}{3}\right) = -\sin\frac{\pi}{3} = -\frac{\sqrt{3}}{2},$$

and $\qquad$ $t = 2\sin\dfrac{1}{3}\pi\cos\pi = -\sqrt{3}.$

$\therefore \qquad\qquad rt - s^2 = \dfrac{9}{4} = $ positive.

Also $\qquad\qquad r = -\sqrt{3}.$

Hence, u has a maximum value at $\left(\dfrac{\pi}{3}, \dfrac{\pi}{3}\right).$

Example 9. *Discuss the maxima and minima of the function* $u = x^2 y^2 - 5x^2 - 8xy - 5y^2.$

Solution. $\quad$ Here, we have

$$u = x^2 y^2 - 5x^2 - 8xy - 5y^2$$

$\Rightarrow \qquad\qquad \dfrac{\partial u}{\partial x} = 2xy^2 - 10x - 8y$

and $\qquad\qquad \dfrac{\partial u}{\partial y} = 2x^2 y - 8x - 10y.$

For a maximum or minimum of u, we must have $\dfrac{\partial u}{\partial x} = 0$ and $\dfrac{\partial u}{\partial y} = 0.$

which implies $\quad 2xy^2 - 10x - 8y = 0,$ ... (1)

and $\qquad\qquad 2x^2y - 8x - 10y = 0.$ ... (2)

From equation (2) we have $\quad y = \dfrac{4x}{x^2 - 5}$

Put this value of y in equation (1), we get

$$x \cdot \frac{16x^2}{(x^2 - 5)^2} - 5x - \frac{16}{x^2 - 5} = 0$$

$$\Rightarrow \qquad x[-5x^4 + 50x^2 - 45] = 0$$

$$\Rightarrow \qquad x[x^4 - 10x^2 + 9] = 0$$

$$\Rightarrow \qquad x = 0, \pm 1, \pm 3.$$

Also from (2), $\qquad$ for $x = 0, \qquad y = 0$,

$\qquad\qquad\qquad$ for $x = 1, \qquad y = -1$,

$\qquad\qquad\qquad$ for $x = -1, \quad y = 1$,

$\qquad\qquad\qquad$ for $x = 3, \qquad y = 3$,

and $\qquad\qquad\quad$ for $x = -3, \quad y = -3.$

Hence, the function u is stationary at the points $(0, 0)$, $(1, -1)$, $(-1, 1)$, $(3, 3)$ and $(-3, -3)$.

Now $\qquad\qquad r = \dfrac{\partial^2 u}{\partial x^2} = 2y^2 - 10, \quad s = \dfrac{\partial^2 u}{\partial y \, \partial x} = 4xy - 8,$

and $\qquad\qquad t = \dfrac{\partial^2 u}{\partial y^2} = 2x^2 - 10.$

For (0, 0). $\qquad r = -10, s = -8, t = -10$

$$\Rightarrow \qquad rt - s^2 = 36 = + \text{ve}.$$

Since $r = -10 < 0$. Hence, u is maximum at $(0, 0)$.

For (1, –1). $\qquad r = -8, s = -12, t = -8$

$$\Rightarrow \qquad rt - s^2 = -80 < 0.$$

Hence, the stationary value of u at $(1, -1)$ is neither maximum nor minimum.

Similarly at $(-1, 1)$, $(3, 3)$ and $(-3, 3)$ the function u is neither maximum nor minimum.

Example 10. *Find the maximum value of* $x^2 + y^2 + z^2$ *when* $ax + by + cz = p.$

Solution. $\quad$ Here, $\qquad\qquad u = x^2 + y^2 + z^2$ ...(1)

Also $\qquad\qquad ax + by + cz = p$

$$\Rightarrow \qquad\qquad z = \frac{p - ax - by}{c}.$$

Put this value of z in equation (1), we get

$$u = x^2 + y^2 + \frac{(p - ax - by)^2}{c^2}$$

$$\Rightarrow \qquad \frac{\partial u}{\partial x} = 2x - \frac{2a}{c^2}(p - ax - by)$$

$$\Rightarrow \qquad \frac{\partial u}{\partial y} = 2y - \frac{2b}{c^2}(p - ax - by).$$

For a maxima and minima of u, we must have $\dfrac{\partial u}{\partial x} = 0$ and $\dfrac{\partial u}{\partial y} = 0$

$$\Rightarrow \qquad x = \frac{ap}{a^2 + b^2 + c^2} \text{ and } y = \frac{bp}{a^2 + b^2 + c^2}.$$

Now, $\qquad r = \dfrac{\partial^2 u}{\partial x^2} = 2 + \dfrac{2a^2}{c^2}, \quad s = \dfrac{\partial^2 u}{\partial x \, \partial y} = \dfrac{2ab}{c^2}$

and $\qquad t = \dfrac{\partial^2 u}{\partial y^2} = 2 + \dfrac{2b^2}{c^2}$

$$\Rightarrow \qquad rt - s^2 = 4\left(1 + \frac{a^2}{c^2}\right)\left(1 + \frac{b^2}{c^2}\right) - \frac{4a^2 b^2}{c^4}$$

$$= 4\left(1 + \frac{a^2}{c^2} + \frac{b^2}{c^2}\right) = \text{positive.}$$

Since r is positive and $rt - s^2 > 0$, therefore u is minimum for the above values of x and y.

The minimum value is $\dfrac{p^2}{a^2 + b^2 + c^2}$.

Example 11. *Find the stationary point of $x^4 + y^4 - 2x^2 + 4xy - 2y^2$ and determine their nature.*

Solution. Here, we have $\qquad u = x^4 + y^4 - 2x^2 + 4xy - 2y^2$

$$\Rightarrow \qquad \frac{\partial u}{\partial x} = 4x^3 - 4x + 4y$$

and $\qquad \dfrac{\partial u}{\partial y} = 4y^3 + 4x - 4y.$

For, a maxima and minima of u, we must have

$$\frac{\partial u}{\partial x} = 0 \Rightarrow 4x^3 - 4x + 4y = 0 \qquad \qquad ...(1)$$

and $\qquad \dfrac{\partial u}{\partial y} = 0 \Rightarrow 4y^3 + 4x - 4y = 0. \qquad \qquad ...(2)$

Solving (1) and (2), we get

$$4x^3 + 4y^3 = 0 \Rightarrow x^3 + y^3 = 0$$

$$\Rightarrow \qquad (x + y)(x^2 - xy + y^2) = 0$$

$$\Rightarrow \qquad \text{either } x + y = 0 \text{ or } x^2 - xy + y^2 = 0 \qquad \qquad ...(3)$$

$$x + y = 0 \Rightarrow y = -x.$$

Put $y = -x$ in equation (1), we get $4x^3 - 8x = 0$

$$\Rightarrow \qquad x(x^2 - 2) = 0$$

$$\Rightarrow \qquad x = 0, \sqrt{2}, -\sqrt{2}.$$

Then $\qquad y = 0, -\sqrt{2}, \sqrt{2}$

Also from (3) $x = 0, y = 0$ (only real solution)

Hence, the stationary points u are given by $(0,0),(\sqrt{2}, -\sqrt{2}), (-\sqrt{2},\sqrt{2})$.

Now, we have $r = \dfrac{\partial^2 u}{\partial x^2} = 12x^2 - 4, \quad s = \dfrac{\partial^2 u}{\partial x \partial y} = 4,$

and $t = \dfrac{\partial^2 u}{\partial y^2} = 12y^2 - 4.$

For (0, 0). $r = -4, s = 4, t = -4$

$\Rightarrow$ $rt - s^2 = 0.$

$\Rightarrow$ At the point $(0, 0)$ the case is doubtful, and there is a need of further investigation.

For $(\sqrt{2}, -\sqrt{2})$.

$r = 20, s = 4, t = 20$

$\Rightarrow$ $rt - s^2 = 400 - 16 = 384 > 0.$

Also $r > 0.$

$\Rightarrow$ u has a minimum value at $(\sqrt{2}, -\sqrt{2})$.

Similarly u has a minimum value at $(-\sqrt{2}, \sqrt{2})$.

Example 12. *Prove that the maxima or minima of the function*

$$u = \left[\frac{ax^2 + by^2 + 2hxy + 2gx + 2fy + c}{a'x^2 + b'y^2 + 2h'xy + 2g'x + 2f'y + c'} \right]$$

are given by the roots of the equation

$$\begin{vmatrix} a - a'u & h - h'u & g - g'u \\ h - h'u & b - b'u & f - f'u \\ g - g'u & f - f'u & c - c'u \end{vmatrix} = 0.$$

Solution.   Here, we have

$$u = \left[\frac{ax^2 + by^2 + 2hxy + 2gx + 2fy + c}{a'x^2 + b'y^2 + 2h'xy + 2g'x + 2f'y + c'} \right]$$

$\Rightarrow$ $u[a'x^2 + b'y^2 + 2h'xy + 2g'x + 2f'y + c']$

$$= [ax^2 + by^2 + 2hxy + 2gx + 2fy + c].   \qquad \text{...(1)}$$

Differentiating (1) partially w.r.t. x and y, we have

$$\frac{\partial u}{\partial x}[a'x^2 + b'y^2 + 2h'xy + 2g'x + 2f'y + c'] + u[2a'x + 2h'y + 2g']$$

$$= 2ax + 2hy + 2g   \qquad \text{...(2)}$$

and $\dfrac{\partial u}{\partial y}[a'x^2 + b'y^2 + 2h'xy + 2g'x + 2f'y + c'] + u[2b'y + 2h'x + 2f']$

$$= 2by + 2hx + 2f.   \qquad \text{...(3)}$$

For the maxima and minima of u, we must have

$$\frac{\partial u}{\partial x} = 0 \Rightarrow u[a'x + h'y + g'] = ax + hy + g   \qquad \text{...(4)}$$

and $\dfrac{\partial u}{\partial y} = 0 \Rightarrow u[h'x + b'y + f'] = hx + by + f.$ ...(5)

Now, multiplying (4) by x, (5) by y and adding, we have

$$u[a'x^2 + b'y^2 + 2h'xy + g'x + f'y] = ax^2 + by^2 + 2hxy + gx + fy \qquad \dots(6)$$

Subtracting (6) from (1), we get

$$u(g'x + f'y + c') = gx + fy + c \qquad \dots(7)$$

Now, from (4), (5) and (7), we have

$$(a - a'u)x + (h - h'u)y + (g - g'u) = 0 \qquad \dots(8)$$

$$(h - h'u)x + (b - b'u)y + (f - f'u) = 0 \qquad \dots(9)$$

$$(g - g'u)x + (f - f'u)y + (c - c'u) = 0. \qquad \dots(10)$$

By eliminating x and y from (8), (9) and (10), we get

$$\begin{vmatrix} a - a'u & h - h'u & g - g'u \\ h - h'u & b - b'u & f - f'u \\ g - g'u & f - f'u & c - c'u \end{vmatrix} = 0.$$

which is a cubic equation in u. The roots of this equation gives the required maxima and minima.

REMARK

- If there is a function of two variables x and y connected by a relation $g(x, y) = 0$. Then we find the maxima and minima of the function in the following manner.

Let $\qquad\qquad u = f(x, y) \qquad \dots(1)$
and $\qquad\qquad g(x, y) = 0. \qquad \dots(2)$

Generally, it is possible to eliminate one of the variables x and y from (1) and (2), then u is expressed in terms of a single variable and we can proceed in the usual way. But if it is not convenient to take the value of one variable in terms of the other from (2), then we should proceed as follows :

From (2), we get $\qquad\qquad \dfrac{dg}{dx} = -\dfrac{\partial g / \partial x}{\partial g / \partial y} \qquad \dots(3)$

Now, differentiating (1) with respect to x, we get

$$\dfrac{du}{dx} = \dfrac{\partial f}{\partial x} + \dfrac{\partial f}{\partial y}\dfrac{dg}{dx} \qquad \dots(4)$$

Now, from (3) and (4), we get

$$\dfrac{du}{dx} = 0 \qquad \dots(5)$$

Solve (5) with the help of (2), and get the required values of x and y for which u will have maximum or minimum values.

Example 13. *Test the function* $u = x^2y - y^2x - x + y$ *for maximum and minimum.*

Solution. For maximum of u, we must have

$$\dfrac{\partial u}{\partial x} = 2xy - y^2 - 1 = 0$$

and $\qquad\qquad \dfrac{\partial u}{\partial y} = x^2 - 2xy + 1 = 0.$

Solving these equations, we get $x = 1, y = 1, x = -1, y = -1$.

Now

$$r = \dfrac{\partial^2 u}{\partial x^2} = 2y, \quad s = \dfrac{\partial^2 u}{\partial x \, \partial y} = 2x - 2y,$$

$$t = \frac{\partial^2 u}{\partial y^2} = -2x.$$

For $x = 1, y = 1$, we have $r = 2, s = 0, t = -2$ so that $rt - s^2 = -4$, which is negative.

Hence, u has neither a maximum nor a minimum at $(1, 1)$. Thus $(1, 1)$ is a saddle point.

For $x = -1, y = -1$. We have $r = -2, s = 0, t = 2$. so that $rt - s^2 = -4$, which is negative.

Hence, u has neither a maximum nor a minimum at $(-1, -1)$.

Thus $(-1, -1)$ is a saddle point.

Example 14. *Show that distance l of any point (x, y, z) on the plane $2x + 3y - z = 12$ from the origin is given by*

$$l = \sqrt{[x^2 + y^2 + (2x + 3y - 12)^2]}.$$

Hence, find the point on the plane that is nearest to the origin.

Solution. If l is the distance from $(0, 0, 0)$ of any point (x, y, z) then $l = \sqrt{(x^2 + y^2 + z^2)}$. If the point (x, y, z) lies on the plane $2x + 3y - z = 12$, then

$$l = \sqrt{[x^2 + y^2 + (2x + 3y - 12)^2]}$$

$$[\because z = 2x + 3y - 12, \text{ from the equation of the plane}]$$

$$\therefore \quad l^2 = x^2 + y^2 + (2x + 3y - 12)^2$$

$$= 5x^2 + 10y^2 + 12xy - 48x + 72y + 144 = u \,(\text{say}).$$

Now l is maximum or minimum according as l^2 i.e., u is maximum or minimum.

For a maximum or minimum of u, we get

$$\frac{\partial u}{\partial x} = 10x + 12y - 48 = 0$$

and $$\frac{\partial u}{\partial y} = 20y + 12x - 72 = 0$$

Solving these equations, we get $x = \frac{12}{7}$ and $y = \frac{18}{7}$.

Also $$r = \frac{\partial^2 u}{\partial x^2} = 10, \, s = \frac{\partial^2 u}{\partial x \partial y} = 12 \text{ and } t = \frac{\partial^2 u}{\partial y^2} = 20.$$

Therefore $rt - s^2 = 10 \times 20 - (12)^2 = +\text{ve}$, since $rt - s^2 > 0$

and $r > 0$, then u is minimum and hence l is minimum.

When $x = \frac{12}{7}$ and $y = \frac{18}{7}$. Putting these values of x and y in the equation of the plane, we get

$$z = 2 \cdot \left(\frac{12}{7}\right) + 3 \cdot \left(\frac{18}{7}\right) - 12 = -\frac{6}{7}.$$

Hence, the required point is $\left(\frac{12}{7}, \frac{18}{7}, -\frac{6}{7}\right)$.

Example 15. *Find the points on $z^2 = xy + 1$ nearest to the origin.*

Solution. Let l be the distance from the origin $(0, 0, 0)$ of any point (x, y, z) on the surface

$$z^2 = xy + 1 \qquad \qquad ...(1)$$

Then $\qquad l = \sqrt{x^2 + y^2 + z^2} = \sqrt{(x^2 + y^2 + xy + 1)}$ $\qquad$ [Using equation (1)]

Since l is always greater than zero, therefore l is maximum or minimum according as l^2, *i.e.*, u is maximum or minimum, where $u = l^2$.

For a maximum or minimum of u, we must have

$$\frac{\partial u}{\partial x} = 2x + y = 0 \qquad \qquad ...(2)$$

and $\qquad \dfrac{\partial u}{\partial y} = 2y + x = 0. \qquad \qquad ...(3)$

Solving the equation (2) and (3), we get

$$x = 0, y = 0$$

Also $\qquad r = \dfrac{\partial^2 u}{\partial x^2} = 2, \; s = \dfrac{\partial^2 u}{\partial x \, \partial y} = 1, \; t = \dfrac{\partial^2 u}{\partial y^2} = 2.$

$\therefore \qquad rt - s^2 = 2.2 - 1 = 3 > 0.$

Since at $x = 0, y = 0$, then $rt - s^2 > 0$ and $r > 0$.

Therefore u is minimum at $x = 0, y = 0$. Hence l is minimum, when $x = 0, y = 0$.

Putting $x = 0, y = 0$ in the equation (1), we get $z^2 = 1$ *i.e.*, $z = \pm 1$.

Hence, the required points are $(0, 0, 1)$ and $(0, 0, -1)$.

EXERCISE 10.7

1. Find the points (x, y) where the function $f(x, y) = xy(1-x-y)$ is maximum or minimum. Also find the maximum value of $f(x, y)$.

2. Discuss the maxima and minima of the function $f(x, y) = x^2 + y^2 + \dfrac{2}{x} + \dfrac{2}{y}$.

3. Find the values of x and y for which the expression
$$(a_1 x + b_1 y + c_1)^2 + (a_2 x + b_2 y + c_2)^2 + ...$$
$$+ (a_n x + b_n y + c_n)^2$$
is minimum.

4. Discuss the maxima and minima of the function $f(x, y) = x^4 + 2x^2 y - x^2 + 3y^2$.

5. Examine for maximum and minimum values of the function $f(x, y) = x^2 - 3xy + y^2 + 2x$.

6. Examine the function $f(x, y) = x^2 y - y^2 x - x + y$ for maxima and minima.

7. Discuss the maxima and minima of the function
$$f(x, y) = 2\sin\frac{1}{2}(x + y)\cos\frac{1}{2}(x - y) + \cos(x + y).$$

8. Find points on $z^2 = xy + 1$ nearest to the origin.

9. Show that the distance l of any point (x, y, z) on the plane $2x + 3y - z = 12$ from the origin is given by
$$l = \sqrt{(x^2 + y^2) + (2x + 3y - 12)^2}.$$

10. Find the maximum and minimum values of $u = 6xy + (47 - x - y)(4x + 3y)$.

HINTS TO SELECTED PROBLEMS

1. $\dfrac{\partial f}{\partial x} = 0 \Rightarrow y - 2xy - y^2 = 0$

$\dfrac{\partial f}{\partial y} = 0 \Rightarrow x - 2xy - x^2 = 0.$

On solving above two equations, we get $(0, 0)$, $(1, 0)$, $(0, 1)$ and $(1/3, 1/3)$ are the extreme points.

At (0, 0), $rt - s^2$ is negative $\Rightarrow f(x, y)$ is neither maximum nor minimum.

At (1, 0), $rt - s^2$ is negative $\Rightarrow f(x, y)$ is neither maximum nor minimum.

At (0, 1), $rt - s^2$ is negative $\Rightarrow f(x, y)$ is neither maximum nor minimum.

At $(1/3, 1/3)$, $rt - s^2$ is positive and $r \left(= -\dfrac{2}{3} \right)$ is negative.

Hence, at $\left(\dfrac{1}{3}, \dfrac{1}{3} \right)$, $f(x, y)$ is maximum.

7. $\dfrac{\partial f}{\partial x} = \cos x - \sin(x + y), \quad \dfrac{\partial f}{\partial y} = \cos y - \sin(x + y)$

$\dfrac{\partial f}{\partial x} = 0, \dfrac{\partial f}{\partial y} = 0$, we get $\cos x = \sin(x + y)$, and $\cos y = \sin(x + y)$.

The extreme points are given by $\left(-\dfrac{\pi}{2}, \dfrac{\pi}{2} \right), \left(\dfrac{3\pi}{2}, \dfrac{\pi}{2} \right)$ and $\left(\dfrac{\pi}{2}, \dfrac{\pi}{2} \right)$.

Answers

1. $f(x, y)$ is maximum at the point $\left(\dfrac{1}{3}, \dfrac{1}{3} \right)$; maximum value $= \dfrac{1}{27}$.

2. $f(x, y)$ is minimum at $(1, 1)$.

3. $f(x, y)$ is minimum for the value of x and y which are obtained by

$$\Sigma (a_1^2) x + (a_1 b_1) y + a_1 c_1 = 0$$

and $\quad \Sigma (a_1 b_1) x + (b_1^2) y + b_1 c_1 = 0.$

4. $f(x, y)$ is minimum for $\left(\dfrac{\sqrt{3}}{2}, \dfrac{-1}{4} \right)$ and $\left(-\dfrac{\sqrt{3}}{2}, -\dfrac{1}{4} \right)$.

5. Stationary point is $x = \dfrac{4}{5}, y = \dfrac{6}{5}$. The function $f(x, y)$ is neither maximum nor minimum at $\left(\dfrac{4}{5}, \dfrac{6}{5} \right)$.

6. At $(1, 1)$ and $(-1, -1)$ function is neither maximum nor minimum.

7. $x = y = 2n\pi \pm \pi/2;$ neither maximum nor minimum
$x = y = n\pi + (-1)^n \pi/6;$ f is maximum.

8. $(0, 0, 1)$ and $(0, 0, -1)$.

10. Maximum value is 3384.

10.22 MAXIMA AND MINIMA OF THE FUNCTION OF THREE INDEPENDENT VARIABLES

(1) *To find the condition, which governs the sign of the quadratic equation of three independent variables.*

Let I be the expression of three independent variables x, y and z given by
$$I = ax^2 + by^2 + cz^2 + 2fyz + 2gzx + 2hxy$$

I can be written as
$$I = \dfrac{1}{a} \left[a^2 x^2 + aby^2 + acz^2 + 2afyz + 2agzx + 2ahxy \right] (a \neq 0)$$

$$= \frac{1}{a}\left[a^2x^2 + 2ax(gz+hy) + aby^2 + acz^2 + 2afyz\right]$$

$$= \frac{1}{a}\left[(ax+hy+gz)^2 + aby^2 + acz^2 + 2afyz - (gz+hy)^2\right]$$

$$= \frac{1}{a}\left[(ax+hy+gz)^2 + \left(ab-h^2\right)y^2 + 2yz(af-gh) + \left(ac-g^2\right)z^2\right]$$

Here, we observe that I be of the same sign as provided the expression within the square brackets is positive which will of course be so if $ab-h^2$ and $\{(ah-h^2)(ac-g^2)-(af-gh)^2\}$ are positive *i.e.,* if

$$ab-h^2 \quad \text{and} \quad a[abc+2fgh-af^2-bg^2-ch^2] \text{ are both positive.}$$

Hence, I will be positive if

$$a, \begin{vmatrix} a & h \\ h & b \end{vmatrix}, \begin{vmatrix} a & h & g \\ h & b & f \\ g & f & c \end{vmatrix}$$

be all positive and will be negative if these three expression are alternately negative and positive.

10.23 MAXIMA AND MINIMA FOR A FUNCTION OF THREE INDEPENDENT VARIABLES: THE LAGRANGE'S CONDITION

Let $f(x,y,z)$ be a given function of three independent variables x, y and z.

Let A, B, C, F, G, H stand for

$$\frac{\partial^2 f}{\partial x^2}, \frac{\partial^2 f}{\partial y^2}, \frac{\partial^2 f}{\partial z^2}, \frac{\partial^2 f}{\partial y \partial z}, \frac{\partial^2 f}{\partial z \partial x}, \frac{\partial^2 f}{\partial x \partial y} \text{ respectively.}$$

Let a set of the values of x, y, z obtained by solving the equations

$$\frac{\partial f}{\partial x} = \frac{\partial f}{\partial y} = \frac{\partial f}{\partial z} = 0 \text{ be } a, b, c.$$

By Taylor's theorem, we have

$$f(a+h,b+k,c+l),-f(a,b,c)$$

$$= \frac{1}{2!}\left[Ah^2 + Bk^2 + Cl + 2Fkl + 2Glh + 2Hhk\right] + R \qquad \ldots(1)$$

where, remainder term R consist of third and higher order of same quantity (*i.e.,* h, k, l).

Now, by taking h, k, l sufficiently small the second term of R.H.S. of (1) can be made to govern the sign of R.H.S. and therefore of L.H.S. also.

If for all such values of h, k and l, these terms be of permanent sign, then we shall have a maximum or minimum of $f(x,y,z)$ according as that sign is negative or positive.

Hence, the function will be minimum if the expression

$$A, \begin{vmatrix} A & H \\ H & B \end{vmatrix}, \begin{vmatrix} A & H & G \\ H & B & F \\ G & F & C \end{vmatrix} \text{ be all positive.}$$

The function will have a maximum value, if the above three quantities are alternately negative and positive. If these conditions are not satisfied, we have neither a maximum nor a minimum.

WORKING PROCEDURE

Let us $f(x, y, z)$ be a function of three independent variables x, y and z. Find the values of triads (a,b,c) of the value x, y and z by putting $\frac{\partial f}{\partial x} = 0, \frac{\partial f}{\partial y} = 0, \frac{\partial f}{\partial z}$. The values of triads (a,b,c) will give the stationary values of $f(x, y, z)$.

Now, to discuss maximum and minimum values, at (a, b, c) we find the following six partial derivatives of second order

$$A = \frac{\partial^2 f}{\partial x^2}, B = \frac{\partial^2 f}{\partial y^2}, C = \frac{\partial^2 f}{\partial z^2}, F = \frac{\partial^2 f}{\partial y \partial z}, G = \frac{\partial^2 f}{\partial z \partial x}, \text{and } H = \frac{\partial^2 f}{\partial x \partial y}$$

Now, we have the following cases :

Case (i) The function $f(x,y,z)$ will be minimum at (a,b,c) if the expressions

$$A, \begin{vmatrix} A & H \\ H & B \end{vmatrix}, \begin{vmatrix} A & H & G \\ H & B & F \\ G & F & C \end{vmatrix} \text{ be all positive at } (a, b, c).$$

Case (ii) The function $f(x, y, z)$ will be maximum at (a, b, c) if the expressions

$$A, \begin{vmatrix} A & H \\ H & B \end{vmatrix}, \begin{vmatrix} A & H & G \\ H & B & F \\ G & F & C \end{vmatrix}$$

be alternately negative and positive.

Case (iii) If the expression, using in case (i) and (ii) neither be all positive nor having alternately negative and positive sign at (a,b,c). Then $f(x, y, z)$ is neither maximum nor minimum at (a,b,c).

REMARK

- To find the maximum and minimum of the function at stationary point, it is sufficient to find the value of a second order partial derivative of function with respect to any of the independent variables. Then, the value of the function is maximum or minimum according as the value of this second order partial derivative at the stationary point under consideration is negative or positive.

Solved Examples

Example 1. *Find the maximum value of u, where* $u = \dfrac{xyz}{(a+x)(x+y)(y+z)(z+b)}$.

Solution. We have

$$u = \frac{xyz}{(a+x)(x+y)(y+z)(z+b)}$$

Taking, log of both the sides, we have

$$\log u = \log x + \log y + \log z - \log(a+x) - \log(x+y) - \log(y+z) - \log(z+b).$$

Differentiating w.r.t. x, we have

$$\frac{1}{u}\frac{\partial u}{\partial x} = \frac{1}{x} - \frac{1}{a+x} - \frac{1}{x+y} = \frac{ay - x^2}{x(a+x)(x+y)}$$

$$\Rightarrow \qquad \frac{\partial u}{\partial x} = \frac{\left(ay - x^2\right)u}{x(a+x)(x+y)}$$

Similarly $\quad \dfrac{\partial u}{\partial y} = \dfrac{\left(xz - y^2\right)u}{y(x+y)(y+z)}$ and $\dfrac{\partial u}{\partial z} = \dfrac{\left(by - z^2\right)u}{z(y+z)(z+b)}$

For, a maxima and minima of u, we must have

$$\frac{\partial u}{\partial x}=0 \quad \Rightarrow \quad ay-x^2=0 \; ; \; \frac{\partial u}{\partial y}=0 \; \Rightarrow \; xz-y^2=0$$

and $\qquad \dfrac{\partial u}{\partial z}=0 \quad \Rightarrow \quad by-z^2=0$

Here, we observe that $x^2=ay, y^2=xz, z^2=by$ which implies that a, x, y, z and b are in G.P. Let r be the common ratio of this G.P.

Then $\qquad ar^4=b \quad$ or $\quad r=\left(\dfrac{b}{a}\right)^{1/4}$

Also $\qquad x=ar, y=ar^2, z=ar^3$.

Hence, we have

$$u=\frac{ar.ar^2.ar^3}{a(1+r)ar(1+r)ar^2(1+r)ar^3(1+r)}$$

$$=\frac{1}{a(1+r)^4} = \frac{1}{\left[a\left[1+\left(\dfrac{b}{a}\right)^{1/4}\right]\right]^4} = \frac{1}{\left(a^{1/4}+b^{1/4}\right)^4}$$

which gives a stationary value of u. Now, to decide whether this value of u is a maximum or a minimum, we proceed to find the second order partial derivative of u.

Here $\qquad \dfrac{\partial^2 u}{\partial x^2} = \dfrac{-2ux}{x(a+x)(x+y)} + \left(ay-x^2\right)\dfrac{\partial}{\partial x}\left[\dfrac{u}{x(a+x)(x+y)}\right]$

When $x=ar, y=ar^2, z=ar^3$, we have

$$A=\frac{\partial^2 u}{\partial x^2} = -\frac{2u}{a^2 r(1+r)^2} < 0$$

Hence, the above stationary value of u is maximum.

Example 2. *Find the maxima and minima value of the function*

$$u = \sin x \sin y \sin z$$

where x,y and z are the vertex angles of a triangle.

Solution. Here, we have

$$u= \sin x \sin y \sin z \; ; \; \text{where } x+y+z= \pi \qquad \ldots(1)$$

$\therefore \qquad u= \sin x \sin y \sin [\pi-(x+y)] = \sin x \sin y \sin(x+y)$

$\therefore \qquad \dfrac{\partial u}{\partial x} = \cos x \sin y \sin(x+y)+\sin x \sin y \cos(x+y)$

$$= \sin y \sin(2x+y). \qquad \ldots(2)$$

Similarly $\dfrac{\partial u}{\partial y} = \sin x \sin(2y+x). \qquad \ldots(3)$

For a maxima and minima, we must have

$$\frac{\partial u}{\partial x} = 0, \frac{\partial u}{\partial y}=0$$

So, $\dfrac{\partial u}{\partial x}=0 \Rightarrow \sin y \sin(2x+y)=0$

$$\Rightarrow \quad \sin y=0 \quad \text{or} \quad \sin(2x+y)=0$$

$$\Rightarrow \quad y=0 \quad \text{or} \quad \sin(x+x+y)=0$$

$$\Rightarrow \quad y=0 \text{ or } \sin x \cos(x+y)+\cos x \sin (x+y)=0$$

$$\Rightarrow \quad \tan (x+y)= -\tan x$$

$$\Rightarrow \quad \tan(x+y)= \tan(- x) = \tan(\pi - x) \qquad \qquad ...(4)$$

$$\Rightarrow \quad x+y=\pi -x$$

$$\Rightarrow \quad 2x+y=\pi \qquad\qquad ...(5)$$

Similarly, from (3)

$$x=0 \quad \text{or} \quad \tan(x+y)=-\tan y. \qquad\qquad ...(6)$$

Now, by (4) and (6), we have

$$\tan x=\tan y \quad \Rightarrow \quad x = y.$$

Hence, by (5), we have

$$3y=\pi \quad \Rightarrow \quad y=\frac{\pi}{3} \text{ and } x=\frac{\pi}{3}$$

Therefore, the stationary points are $\left(\dfrac{\pi}{3},\dfrac{\pi}{3}\right)$ and $(0, 0)$.

For (0,0): $u=0$.

For $\left(\dfrac{\pi}{3},\dfrac{\pi}{3}\right)$

$$r=\frac{\partial^2 u}{\partial x^2} = 2\sin y \cos(2x + y)= 2\sin\frac{\pi}{3}\cos\left(\frac{2\pi}{3}+\frac{\pi}{3}\right)=-\sqrt{3} < 0$$

$$s=\frac{\partial^2 u}{\partial x \partial y}=\sin(2x + 2y)=\sin\left(\frac{2\pi}{3}+\frac{2\pi}{3}\right)= \sin\left(\frac{4\pi}{3}\right)=-\frac{\sqrt{3}}{2}<0$$

and $t=\dfrac{\partial^2 u}{\partial y^2} = 2\sin x \cos(x + 2y)= 2\sin\dfrac{\pi}{3}\cos \pi = -\sqrt{3} < 0$

Now $rt-s^2=\left(-\sqrt{3}\right)\left(-\sqrt{3}\right)-\left(\dfrac{\sqrt{3}}{2}\right)^2=\dfrac{9}{4}> 0$

Thus $rt-s^2>0$ and $r<0$.

Hence, the function u will be maximum at $\left(\dfrac{\pi}{3},\dfrac{\pi}{3}\right)$.

Example 3.   *Show that the points such that the sum of the squares of its distances from n given points shall be minimum, is the centre of the mean position of the given points.*

Solution .   Let n given points be (a_1, b_1, c_1), (a_2, b_2, c_2),...,(a_n, b_n, c_n) and let (x, y, z) be the coordinates of the required point.

If u denotes the sum of the squares of the distances of (x, y, z) from the n given points, then

$$u=\Sigma[(x-a_1)^2+(y-b_1)^2+(z-c_1)^2]$$

$$=\Sigma(x-a_1)^2+\Sigma(y-b_1)^2+\Sigma(z-c_1)^2$$

$$\Rightarrow \qquad \frac{\partial u}{\partial x} = 2\Sigma\left(x - a_1\right) = 2nx - 2\Sigma a_1$$

$$\frac{\partial u}{\partial y} = 2\Sigma\left(y - b_1\right) = 2ny - 2\Sigma b_1 \qquad \text{...(1)}$$

and $\qquad \dfrac{\partial u}{\partial z} = 2\Sigma\left(z - c_1\right) = 2nz - 2\Sigma c_1$

For the maxima and minima of u, we must have

$$\frac{\partial u}{\partial x} = 0, \frac{\partial u}{\partial y} = 0 \text{ and } \frac{\partial u}{\partial z} = 0 \qquad \text{...(2)}$$

Now from (1) and (2), we have

$$x = \frac{\Sigma a_1}{n}, y = \frac{\Sigma b_1}{n}, z = \frac{\Sigma c_1}{n}$$

Now $\qquad A = \dfrac{\partial^2 u}{\partial x^2} = 2n, B = \dfrac{\partial^2 f}{\partial y^2} = 2n, C = \dfrac{\partial^2 f}{\partial z^2} = 2n,$

$$F = \frac{\partial^2 f}{\partial y \partial z} = 0, G = \frac{\partial^2 f}{\partial z \partial x} = 0, H = \frac{\partial^2 f}{\partial x \partial y} = 0.$$

Here, we have $\qquad A = 2n, \begin{vmatrix} A & H \\ H & B \end{vmatrix} = \begin{vmatrix} 2n & 0 \\ 0 & 2n \end{vmatrix} = 4n^2$

and $\qquad \begin{vmatrix} A & H & G \\ H & B & F \\ G & F & C \end{vmatrix} = \begin{vmatrix} 2n & 0 & 0 \\ 0 & 2n & 0 \\ 0 & 0 & 2n \end{vmatrix} = 8n^3$

Since, these expressions are all positive, therefore u is minimum when

$$x = \frac{\Sigma a_1}{n}, y = \frac{\Sigma b_1}{n}, z = \frac{\Sigma c_1}{n}.$$

Hence, the function u is minimum when the point (x, y, z) is the centre of the mean position of n given points.

Example 4. *Show that the function $u=(x+y+z)^3-3(x+y+z)-24xyz+a^3$ has minimum at $(1,1,1)$ and maximum at $(-1,-1,-1)$.*

Solution. Here we have

$$u = (x+y+z)^3 - 3(x+y+z) - 24xyz + a^3$$

$$\Rightarrow \qquad \frac{\partial u}{\partial x} = 3(x+y+z)^2 - 3 - 24yz \qquad \text{...(1)}$$

$$\frac{\partial u}{\partial y} = 3(x+y+z)^2 - 3 - 24xz \qquad \text{...(2)}$$

and $\qquad \dfrac{\partial u}{\partial z} = 3(x+y+z)^2 - 3 - 24xy \qquad \text{...(3)}$

For the maxima and minima of u, we must have

$$\frac{\partial u}{\partial x} = 0, \frac{\partial u}{\partial y} = 0 \text{ and } \frac{\partial u}{\partial z} = 0$$

The equations (1), (2) and (3) are satisfied when $x=y=z$.

Putting $y=x$ and $z=x$ in (1), we get
$$27.\, x^2 - 3 - 24x^2 = 0$$
$$\Rightarrow \qquad x = \pm 1$$
$\Rightarrow \quad x = y = z = 1$ and $x = y = z = -1$ are the solutions of (1), (2) and (3).

Hence, the stationary points are $(1,1,1)$ and $(-1,-1,-1)$.

Now,
$$A = \frac{\partial^2 u}{\partial x^2} = 6(x+y+z), \quad B = \frac{\partial^2 u}{\partial y^2} = 6(x+y+z), \quad C = \frac{\partial^2 u}{\partial z^2} = 6(x+y+z),$$
$$F = \frac{\partial^2 u}{\partial y \partial z} = 6(x+y+z) - 24x, \quad G = \frac{\partial^2 u}{\partial z \partial x} = 6(x+y+z) - 24y,$$
$$H = \frac{\partial^2 u}{\partial x \partial y} = 6(x+y+z) - 24z.$$

For (1,1,1). $A=18, B=18, C=18, F=-6, G=-6, H=-6.$

$\therefore$ At the point $(1,1,1)$, we have $A=18. >0$

$$\begin{vmatrix} A & H \\ H & B \end{vmatrix} = \begin{vmatrix} 18 & -6 \\ -6 & 18 \end{vmatrix} = 288 > 0$$

and
$$\begin{vmatrix} A & H & G \\ H & B & F \\ G & F & C \end{vmatrix} = \begin{vmatrix} 18 & -6 & -6 \\ -6 & 18 & -6 \\ -6 & -6 & 18 \end{vmatrix} = 3426 > 0$$

Since, all these three expressions are positive, therefore u is minimum at the point $(1,1,1)$.

For (-1,-1,-1).
$$A=-18, B=-18, C = -18, F=6, G=6, H=6.$$

$\therefore$ At the point $(-1,-1,-1)$, we have $A = -18 < 0$

$$\begin{vmatrix} A & H \\ H & B \end{vmatrix} = \begin{vmatrix} -18 & 6 \\ 6 & -18 \end{vmatrix} = 288 > 0$$

and
$$\begin{vmatrix} A & H & G \\ H & B & F \\ G & F & C \end{vmatrix} = \begin{vmatrix} -18 & 6 & 6 \\ 6 & -18 & 6 \\ 6 & 6 & -18 \end{vmatrix} = -3426 < 0$$

Here, the above three expressions are alternately negative and positive. Hence, u is maximum at the point $(-1,-1,-1)$.

EXERCISE 10.8

1. Prove that the function $u = x^2 + y^2 + x - 2z - xy$ is minimum at $\left(-\dfrac{2}{3}, -\dfrac{1}{3}, 1\right)$.

2. Find the maximum and minimum values of $u = y^2 + 2z^2 - 5x^4 + 4x^5$.

3. Find the maximum or minimum values of the function u, where
$$u = axy^2z^3 - x^2y^2z^3 - xy^3z^3 - xy^2z^4$$

4. Find the maximum value of
$$\frac{e^{-\left(\alpha^2.x^2 + \beta^2 y^2 + \gamma^2 z^2\right)}}{(ax+by+cz)}.$$

5. A rectangle box is placed on x-y plane. The one end of the box is at the origin. If the vertex opposite to the origin be on the plane $6x+4y+3z=24$, then find the maximum value of this box.

6. In a plane triangle xyz, find the maximum value of $\sin x \sin y \sin z$.

=== **Answers** ===

2. Minimum at (1,0,0), neither maximum nor minimum at (0,0,0).

3. Maximum at $\left(\dfrac{a}{7}, \dfrac{2a}{7}, \dfrac{3a}{7}\right)$, max. value $= \dfrac{108a^7}{7^7}$

4. Maximum at $\left(\dfrac{a}{2\alpha^2 k}, \dfrac{b}{2\beta^2 k}, \dfrac{c}{2\gamma^2 k}\right)$ where $k = \sqrt{\left\{\dfrac{1}{2}\left(\dfrac{a^2}{\alpha^2} + \dfrac{b^2}{\beta^2} + \dfrac{c^2}{\gamma^2}\right)\right\}}$

Maximum value $= \sqrt{\left\{\dfrac{1}{2e}\left(\dfrac{a^2}{\alpha^2} + \dfrac{b^2}{\beta^2} + \dfrac{c^2}{\gamma^2}\right)\right\}}$

5. Maximum at $\left(\dfrac{4}{3}, 2\right)$. maximum value $= \dfrac{64}{9}$ cube units. Neither maximum nor minimum at (0,0).

6. Maximum at $\left(\dfrac{\pi}{3}, \dfrac{\pi}{3}, \dfrac{\pi}{3}\right)$, value $= \dfrac{3\sqrt{3}}{8}$

10.24 LAGRANGE'S METHOD OF UNDETERMINED MULTIPLIERS

Let $u = f(x_1, x_2, ..., x_n)$ be a function of n variables $x_1, x_2, ..., x_n$.

Let us suppose these variables $x_1, x_2, ..., x_n$ are connected by k equations

$$g_1(x_1, x_2, ..., x_n) = 0$$
$$g_2(x_1, x_2, ..., x_n) = 0$$
$$... \quad ... \quad ... \quad ... \quad ...$$
$$g_k(x_1, x_2, ..., x_n) = 0$$

so, that there are $n-k$ independent variables out of these n variables. For the maxima and minima of u, we find

$$du = \frac{\partial u}{\partial x_1} dx_1 + \frac{\partial u}{\partial x_2} dx_2 + ... + \frac{\partial u}{\partial x_n} dx_n = 0 \qquad ...(1)$$

Also

$$dg_1 = \frac{\partial g_1}{\partial x_1} dx_1 + \frac{\partial g_1}{\partial x_2} dx_2 + ... + \frac{\partial g_1}{\partial x_n} dx_n = 0 \qquad ...(2)$$

$$dg_2 = \frac{\partial g_2}{\partial x_1} dx_1 + \frac{\partial g_2}{\partial x_2} dx_2 + ... + \frac{\partial g_2}{\partial x_n} dx_n = 0 \qquad ...(3)$$

$$\vdots \qquad \vdots \qquad \vdots \qquad \vdots \qquad \vdots \qquad \vdots$$

$$dg_k = \frac{\partial g_k}{\partial x_1} dx_1 + \frac{\partial g_k}{\partial x_2} dx_2 + ... + \frac{\partial g_k}{\partial x_n} dx_n = 0 \qquad ...(k+1)$$

Multiplying equation (1),(2),(3)...(k+1) by $1, l_1, l_2, ..., k$ respectively and adding, we get the result, which can be written as

$$P_1 dx_1 + P_2 dx_2 + P_3 dx_3 + ... + P_n dx_n = 0 \qquad ...(4)$$

where

$$P_k = \frac{\partial u}{\partial x_k} + l_1 \frac{\partial g_1}{\partial x_k} + l_2 \frac{\partial g_2}{\partial x_k} + ... + l_k \frac{\partial g_k}{\partial x_k}$$

Now we have at our choice k multiple viz $l_1, l_2, ..., l_k$ and can be chosen such that

$$P_1 = 0, P_2 = 0, ..., P_k = 0$$

Then, the equation (4) reduces to

$$P_{k+1} dx_{k+1} + P_{k+2} dx_{k+2} + P_{k+3} dx_{k+3} + ... + P_n dx_n = 0 \qquad ...(5)$$

Now, let us suppose that out of n variables, the $(n-k)$ variables $x_{k+1}, x_{k+2}, ...,x_n$ are independent.

Then, since $n-k$ quantities $dx_{k+1}, dx_{k+2}, ..., dx_n$ are independent so their coefficients must be separately zero. Hence, we have

$$P_{k+1}=0, P_{k+2}=0, ..., P_n=0$$

Thus, we have $k+n$ equations

$$P_1=0, P_2=0, ..., P_n=0$$

and

$$g_1=0, g_2=0, ..., g_k=0.$$

Hence, we get $(n+k)$ equations which determine the k multipliers $l_1,l_2,...,l_k$ and get the possible value of u.

REMARKS

- The Lagrange's method of undetermined multipliers is very convenient to apply. It gives the maximum and minimum values of the function without actually determining the values of the multipliers $l_1,l_2,...,l_k$.
- It does not determine the nature of stationary point, which is the only drawback of this method.

10.24.1 APPLICATIONS OF THE METHOD OF UNDETERMINED MULTIPLIERS

The Lagrange's method of undetermined multipliers can be applied to determine the extreme values of the given functions, it does not detemine the nature of stationary point. Now, it is more convenient to find out the extreme values of a function F with the help of new function, given by

$$V=g+l_1f_1+l_2f_2+...+l_mf_m$$

and use the following method. Here, we give the method for four variables x,y,u,v connected by the following two relations.

Let $F=g(x, y, u, v)$ be subjected to the conditions

$$f_1(x,y,u,v)=0 \qquad \text{...(1)}$$

and

$$f_2(x,y,u,v)=0. \qquad \text{...(2)}$$

For the maxima and minima of F, we have

$$dF = \frac{\partial g}{\partial x}dx + \frac{\partial g}{\partial y}dy + \frac{\partial g}{\partial u}du + \frac{\partial g}{\partial v}dv = 0 \qquad \text{...(3)}$$

Now, from (1) and (2), we have

$$df_1 = \frac{\partial f_1}{\partial x}dx + \frac{\partial f_1}{\partial y}dy + \frac{\partial f_1}{\partial u}du + \frac{\partial f_1}{\partial v}dv = 0 \qquad \text{...(4)}$$

and

$$df_2 = \frac{\partial f_2}{\partial x}dx + \frac{\partial f_2}{\partial y}dy + \frac{\partial f_2}{\partial u}du + \frac{\partial f_2}{\partial v}dv = 0 \qquad \text{...(5)}$$

Multiplying (4) by l_1, (5) by l_2 and adding their sum to (3), we get

$$\left(\frac{\partial g}{\partial x}+l_1\frac{\partial f_1}{\partial x}+l_2\frac{\partial f_2}{\partial x}\right)dx + \left(\frac{\partial g}{\partial y}+l_1\frac{\partial f_1}{\partial y}+l_2\frac{\partial f_2}{\partial y}\right)dy$$
$$+\left(\frac{\partial g}{\partial u}+l_1\frac{\partial f_1}{\partial u}+l_2\frac{\partial f_2}{\partial u}\right)du + \left(\frac{\partial g}{\partial v}+l_1\frac{\partial f_1}{\partial v}+l_2\frac{\partial f_2}{\partial v}\right)dv = 0 \qquad \text{...(6)}$$

Here, we have l_1 and l_2 are arbitrary, therefore we can choose them to satisfy the two linear equations

$$\frac{\partial g}{\partial x} + l_1 \frac{\partial f_1}{\partial x} + l_2 \frac{\partial f_2}{\partial x} = 0 \qquad \text{...(7)}$$

and

$$\frac{\partial g}{\partial y} + l_1 \frac{\partial f_1}{\partial y} + l_2 \frac{\partial f_2}{\partial y} = 0 \qquad \text{...(8)}$$

Using (7) and (8), equation (6) reduces to

$$\left(\frac{\partial g}{\partial u} + l_1 \frac{\partial f_1}{\partial u} + l_2 \frac{\partial f_2}{\partial u} \right) du + \left(\frac{\partial g}{\partial v} + l_1 \frac{\partial f_1}{\partial v} + l_2 \frac{\partial f_2}{\partial v} \right) dv = 0$$

Since, the given function contains four variables (namely x, y, u and v) and we are given two equations of conditions, therefore, only two of the variables are independent and it is immaterial which two of the four variables are regarded as independent. Let them be u and v then du and dv are also independent, therefore, their coefficients must be zero separately. Thus

$$\frac{\partial g}{\partial u} + l_1 \frac{\partial f_1}{\partial u} + l_2 \frac{\partial f_2}{\partial u} = 0 \qquad \text{...(9)}$$

$$\frac{\partial g}{\partial v} + l_1 \frac{\partial f_1}{\partial v} + l_2 \frac{\partial f_2}{\partial v} = 0 \qquad \text{...(10)}$$

Now, we have six equations namely (1),(2),(7),(8),(9) and (10) to determine the two multipliers l_1, l_2 and values of the four variables x, y, u and v for which maximum and minimum values of F are possible.

Now, defined a new function $V(x, y, u, v)$ such that

$$V(x,y,u,v) = g(x, y, u, v) + l_1 f_1(x, y, u, v) + l_2 f_2(x, y, u, v).$$

Assuming that x, y, u, v are now all independent variables. Hence, for the maxima and minima of V, we must have

$$\frac{\partial V}{\partial x} = \frac{\partial g}{\partial x} + l_1 \frac{\partial f_1}{\partial x} + l_2 \frac{\partial f_2}{\partial x} = 0 \qquad \text{...(11)}$$

$$\frac{\partial V}{\partial y} = \frac{\partial g}{\partial y} + l_1 \frac{\partial f_1}{\partial y} + l_2 \frac{\partial f_2}{\partial y} = 0 \qquad \text{...(12)}$$

$$\frac{\partial V}{\partial u} = \frac{\partial g}{\partial u} + l_1 \frac{\partial f_1}{\partial u} + l_2 \frac{\partial f_2}{\partial u} = 0 \qquad \text{...(13)}$$

and

$$\frac{\partial V}{\partial v} = \frac{\partial g}{\partial v} + l_1 \frac{\partial f_1}{\partial v} + l_2 \frac{\partial f_2}{\partial v} = 0 \qquad \text{...(14)}$$

Equations (11), (12), (13) and (14) are exactly the same as the equations (7). (8), (9) and (10). Hence, the maxima and minima of $V(x, y, u, v)$ are same as those of $F(x, y, u, v)$ assuming that $V(x, y, u, v)$ the variables x, y, u, v are now all independent.

Now, we proceed to find whether the values of F obtained with the help of above equations are maximum or minimum. For this, adopt the procedure, which is discussed ahead.

From (3), we get

$$d^2F = \left(\frac{\partial}{\partial x} dx + \frac{\partial}{\partial y} dy + \frac{\partial}{\partial u} du + \frac{\partial}{\partial y} dy \right)^2 g + \left(\frac{\partial g}{\partial x} d^2x + \frac{\partial g}{\partial y} d^2y + \frac{\partial g}{\partial u} d^2u + \frac{\partial g}{\partial y} d^2v \right) \cdots \quad \text{...(15)}$$

Also

$$d^2f_1 = \left(\frac{\partial}{\partial x} dx + \frac{\partial}{\partial y} dy + \frac{\partial}{\partial u} du + \frac{\partial}{\partial v} dv \right)^2 f_1 + \frac{\partial f_1}{\partial x} d^2x + \frac{\partial f_1}{\partial y} d^2y + \frac{\partial f_1}{\partial u} d^2u + \frac{\partial f_1}{\partial v} d^2v = 0 \qquad \text{...(16)}$$

and $d^2 f_2 = \left(\dfrac{\partial}{\partial x} dx + \dfrac{\partial}{\partial y} dy + \dfrac{\partial}{\partial u} du + \dfrac{\partial}{\partial v} dv \right)^2 f_2 + \dfrac{\partial f_2}{\partial x} d^2 x + \dfrac{\partial f_2}{\partial y} d^2 y + \dfrac{\partial f_2}{\partial u} d^2 u + \dfrac{\partial f_2}{\partial v} d^2 v = 0$ ...(17)

Multiplying (16) by l_1 and (17) by l_2 and adding their sum to (15) and using the result (11), (12),(13) and (14), we have

$$d^2 F = \left(\dfrac{\partial}{\partial x} dx + \dfrac{\partial}{\partial y} dy + \dfrac{\partial}{\partial u} du + \dfrac{\partial}{\partial v} dv \right)^2 (g + l_1 f_1 + l_2 f_2)$$

$$= \left(\dfrac{\partial}{\partial x} dx + \dfrac{\partial}{\partial y} dy + \dfrac{\partial}{\partial u} du + \dfrac{\partial}{\partial v} dv \right)^2 V$$

$$= d^2 V.$$

Hence $d^2 F$ is equal to $d^2 V$, where $d^2 V$ is obtained by assuming all the variables x, y, u and v as independent. Therefore, it is clear that $d^2 F$ and $d^2 V$ have the same sign. Hence, F will be minimum or maximum according as V is minimum or maximum.

REMARK

• This method has the advantage over the Lagrange's methods that it enables us to decide whether the values are maximum or minimum.

Solved Examples

Example 1. *Find the maxima and minima of $x^2 + y^2 + z^2$ subject to the conditions :*

$$ax^2 + by^2 + cz^2 = 1$$

and $$lx + my + nz = 0$$

Solution. Here, we have

$$u = x^2 + y^2 + z^2 \qquad \text{...(1)}$$

where, the relations between the variables x, y and z are given by

$$ax^2 + by^2 + cz^2 = 1 \qquad \text{...(2)}$$

and $$lx + my + nz = 0 \qquad \text{...(3)}$$

For the maxima and minima of u, we must have

$$du = 0$$

$\Rightarrow \qquad 2x dx + 2y dy + 2z dz = 0$

$\Rightarrow \qquad x dx + y dy + z dz = 0 \qquad \text{...(4)}$

From (2) and (3), we get

$$ax\,dx + by\,dy + cz\,dz = 0 \qquad \text{...(5)}$$

$$l\,dx + m\,dy + n\,dz = 0 \qquad \text{...(6)}$$

Now, multiplying (4) by 1, (5) by l_1 and (6) by l_2 and adding, we get

$$(x\,dx + y\,dy + z\,dz) + l_1(ax\,dx + by\,dy + cz\,dz) + l_2(l\,dx + m\,dy + n\,dz) = 0$$

$\Rightarrow \quad (x + al_1x + ll_2)dx + (y + bl_1y + ml_2)dy + (z + cl_1z + nl_2)dz = 0$

Now equating the coefficient of dx, dy, dz to zero, we get

$$x + l_1 ax + l_2 l = 0 \qquad \text{...(7)}$$

$$y + bl_1 y + ml_2 = 0 \qquad \text{...(8)}$$

and
$$z + cl_1 z + nl_2 = 0 \qquad \qquad \text{...(9)}$$

Multiplying the equations (7), (8) and (9) by x, y and z respectively, and adding we get

$$x^2 + y^2 + z^2 + l_1(ax^2 + by^2 + cz^2) + l_2(lx + my + nz) = 0$$

or
$$u + l_1.1 + l_2.0 = 0 \qquad \qquad \text{[By using (1), (2) and (3)]}$$

$$\Rightarrow \qquad l_1 = -u$$

Substituting for l_1 in the equations (7), (8) and (9), we get

$$x = \frac{l_2 l}{au - 1}, y = \frac{l_2 m}{bu - 1}, z = \frac{l_2 n}{cu - 1} \qquad \qquad \text{...(10)}$$

Now from (10) and (3), we get

$$\frac{l_2 l^2}{au - 1} + \frac{l_2 m^2}{bu - 1} + \frac{l_2 n^2}{cu - 1} = 0$$

or
$$\frac{l^2}{au - 1} + \frac{m^2}{bu - 1} + \frac{n^2}{cu - 1} = 0 \qquad \qquad \text{...(11)}$$

which gives the maximum and minimum of $u = x^2 + y^2 + z^2$.

REMARKS

- Equation (11) is a quadratic in u. So it gives two stationary values of u.
- Geometrically, the surface $ax^2 + by^2 + cz^2 = 1$ represents an ellipsoid whose centre is origin, and $lx + my + nz = 0$ represents a plane passing through the origin. The points (x, y, z) satisfying both the conditions (2) and (3) lies on the conic in which (2) and (3) intersect. $x^2 + y^2 + z^2$ gives the square of the distance (x, y, z) from the origin, which is also the centre of the conic of intersection. The maximum value of this distance is the major axis of this conic, and the minimum value of this distance is the minor axis of this conic. Hence, equation (11) gives the squares of the lengths of the semi-axis of the conic of intersection.

Example 2. *Find the maxima and minima of $x^2 + y^2 + z^2$, where*
$$ax^2 + by^2 + cz^2 + 2fyz + 2gzx + 2hxy = 1.$$

Solution. Let
$$u = x^2 + y^2 + z^2 \qquad \qquad \text{...(1)}$$

where the relation between the variables x, y and z is
$$ax^2 + by^2 + cz^2 + 2fyz + 2gzx + 2hxy = 1. \qquad \qquad \text{...(2)}$$

For a maximum or minima of u, we must have
$$du = 0$$

$$\Rightarrow \qquad x\,dx + y\,dy + z\,dz = 0. \qquad \qquad \text{...(3)}$$

From (2), we have
$$2ax\,dx + 2by\,dy + 2cz\,dz + 2fy\,dz + 2fz\,dy + 2gz\,dx + 2gx\,dz + 2hx\,dy + 2hy\,dx = 0$$

$$\Rightarrow \quad (ax + hy + gz)dx + (hx + by + fz)dy + (gx + fy + cz)dz = 0. \qquad \text{...(4)}$$

Now, multiplying (3) by 1 and (4) by l_1, adding, and then equating the coefficient of dx, dy, dz to zero, we have

$$x + l_1(ax + hy + gz) = 0. \qquad \qquad \text{...(5)}$$
$$y + l_1(hx + by + fz) = 0. \qquad \qquad \text{...(6)}$$

$$z+l_1(gx+fy+cz)=0. \qquad \qquad ...(7)$$

Multiplying (5) by x, (6) by y, (7) by z and adding, we get

$$x^2+y^2+z^2+l_1(ax^2+by^2+cz^2+2fyz+2gzx+2hxy)=0$$

$\Rightarrow \qquad \qquad u+l_1.1=0 \qquad \qquad$ [From (1) and (2)]

$\therefore \qquad \qquad l_1 = -u.$

Hence, from (5), we have

$$x-u(ax+hy+gz)=0$$

$\Rightarrow \qquad \qquad \qquad \qquad \qquad \qquad \qquad \qquad ...(8)$

$$\left(a-\frac{1}{u}\right)x+hy+gz=0$$

Similarly from (6) and (7), we get

$$hx+\left(b-\frac{1}{u}\right)y+fz=0 \qquad \qquad ...(9)$$

and

$$gx+fy+\left(c-\frac{1}{u}\right)z=0 \qquad \qquad ...(10)$$

Eliminating x, y, z from (8), (9) and (10), we get

$$\begin{vmatrix} \left(a-\dfrac{1}{u}\right) & h & g \\ h & \left(b-\dfrac{1}{u}\right) & f \\ g & f & \left(c-\dfrac{1}{u}\right) \end{vmatrix} = 0 \qquad \qquad ...(11)$$

Hence, the maximum or minimum values of u are the roots of the equation (11).

Example 3. *Find the maximum and minima of $u=x^2+y^2$ subject to the condition*

$$ax^2+2hxy+by^2=1.$$

Solution. Here, we have $\qquad \qquad u = x^2+y^2 \qquad \qquad ...(1)$

where the relation between the variables x and y is

$$ax^2+2hxy+by^2 =1. \qquad \qquad ...(2)$$

For the maxima and minima of u, we must have

$$du = 0$$

$\Rightarrow \qquad \qquad 2x\, dx+2y\, dy = 0$

$\Rightarrow \qquad \qquad x\, dx+y\, dy = 0. \qquad \qquad ...(3)$

Now, from (2), we get

$$2ax\, dx+2hx\, dy+2hy\, dx+2by\, dy=0$$

$\Rightarrow \qquad \qquad (ax+hy)dx+(hx+by)dy=0 \qquad \qquad ...(4)$

Now, multiplying (3) by 1, (4) by l_1, adding and then equating the coefficients of dx, dy to zero, we have

$$x+l_1(ax+hy)=0 \qquad \qquad ...(5)$$

and $\qquad \qquad y+l_1(hx+by)=0 \qquad \qquad ...(6)$

Multiplying (5) by x, (6) by y and adding, we get

$$x^2+y^2+l_1(ax^2+2hxy+by^2)=0$$

$$\Rightarrow \qquad u+l_1.1=0 \qquad\qquad\text{[using (1) and (2)]}$$
$$\Rightarrow \qquad u=-l_1$$

Therefore, from (5), we have

$$x-u(ax+hy)=0$$

$$\Rightarrow \qquad \left(a-\frac{1}{u}\right)x+hy=0 \qquad\qquad ...(7)$$

Similarly from (6), we have

$$hx+\left(b-\frac{1}{u}\right)y=0 \qquad\qquad ...(8)$$

Eliminating x and y from (7) and (8), we get

$$\begin{vmatrix} a-\dfrac{1}{u} & h \\ h & b-\dfrac{1}{u} \end{vmatrix}=0$$

Hence, the maximum or minimum values of u are the roots of the equation (9).

Example 4. *Find the maximum value of $u=x^m y^n z^p$ subject to the condition $x+y+z=a$.*

Solution . Here, we have

$$u=x^m y^n z^p \qquad\qquad ...(1)$$

and x, y, z connected by the relation given by

$$x+y+z=a \qquad\qquad ...(2)$$

Taking log of both the sides of (1), we get

$$\log u = m\log x + n\log y + p\log z.$$

On differentiating, we get

$$\frac{1}{u}du = \frac{m}{x}dx + \frac{n}{y}dy + \frac{p}{z}dz$$

For the maxima and minima of u, we must have

$$du=0$$

$$\Rightarrow \qquad \frac{m}{x}dx + \frac{n}{y}dy + \frac{p}{z}dz = 0 \qquad\qquad ...(3)$$

Now, differentiating (2), we get

$$dx+dy+dz=0. \qquad\qquad ...(4)$$

Now, multiplying (3) by 1 and (4) by l, and equating the coefficient of dx, dy, dz to zero (after adding), we get

$$\frac{m}{x}+l=0, \quad \frac{n}{y}+l=0 \text{ and } \frac{p}{z}+l=0$$

which implies $\qquad x=-\dfrac{m}{l}, y=-\dfrac{n}{l}, z=-\dfrac{p}{l}$

Putting the values of x, y and z in (2), we get

$$l=-\left(\frac{m+n+p}{a}\right)$$

therefore, we can say that, u is stationary when

$$x=\frac{am}{m+n+p}, y=\frac{an}{m+n+p}, z=\frac{ap}{m+n+p}$$

Now, we find the nature of this stationary value of u.

Let us regard x and y as independent variable and z is a function of x and y given by (2) [It is justify, because the variables x, y and z are connected by the relation (2), any two of them may be regarded as independent].

Now from (1), we get

$$\log u = m \log x + n \log y + p \log z$$

$$\therefore \quad \frac{1}{u}\frac{\partial u}{\partial x} = \frac{m}{x} + \frac{p}{z}\frac{\partial z}{\partial x}$$

Now, differentiating (2) partially w.r.t x (treating y as constant), we get

$$1 + \frac{\partial z}{\partial x} = 0 \quad \Rightarrow \quad \frac{\partial z}{\partial x} = -1$$

Put this value in (5), we get

$$\frac{1}{u}\frac{\partial u}{\partial x} = \frac{m}{x} - \frac{p}{z}$$

$$\Rightarrow \quad \frac{1}{u}\frac{\partial^2 u}{\partial x^2} - \frac{1}{u^2}\left(\frac{\partial u}{\partial x}\right)^2 = -\frac{m}{x^2} + \frac{p}{z^2}\frac{\partial z}{\partial x} = -\frac{m}{x^2} - \frac{p}{z^2}$$

At stationary point $\dfrac{\partial u}{\partial x} = 0$

Therefore, $$\frac{1}{u}\frac{\partial^2 u}{\partial x^2} = \frac{-m}{x^2} - \frac{p}{z^2}$$

$$\Rightarrow \quad \frac{\partial^2 u}{\partial x^2} = u\left[-\frac{m}{x^2} - \frac{p}{z^2}\right] = -x^m y^n z^p \left[-\frac{m}{x^2} - \frac{p}{z^2}\right]$$

which is negative for the obtained values of x, y and z.

Hence, at the stationary point, u is maximum and maximum value is

$$= \left(\frac{am}{m+n+p}\right)^m \left(\frac{an}{m+n+p}\right)^n \left(\frac{ap}{m+n+p}\right)^p$$

Example 5. *Find the maximum and minimum value of* $u = \dfrac{5xyz}{(x+2y+4z)}$ *subject to the condition* $xyz = 8$.

Solution. Here, we have

$$u = \frac{5xyz}{(x+2y+4z)} \qquad \qquad ...(1)$$

The variables x, y, z are connected by the relation

$$xyz = 8. \qquad \qquad ...(2)$$

From (1) and (2), we get

$$u = \frac{40}{(x+2y+4z)} \Rightarrow du = \frac{-40}{(x+2y+4z)^2}(dx + 2dy + 4dz)$$

For the maxima or minima of u, we must have $du = 0$

$$\Rightarrow \quad dx + 2dy + 4dz = 0 \qquad \qquad ...(3)$$

From (2), we get

$$\log x + \log y + \log z = \log 8.$$

On differentiating, we get

$$\frac{1}{x}dx + \frac{1}{y}dy + \frac{1}{z}dz = 0 \qquad\qquad ...(4)$$

Now, multiplying (3) by 1, (4) by l, adding and then equating to zero the coefficients of dx, dy and dz, we get

$$1 + \frac{l}{x} = 0, 2 + \frac{l}{y} = 0, 4 + \frac{l}{z} = 0$$

Now using (2), we get $l = -4$

$\therefore$ u is stationary at the point given by $x=4$, $y=2$, $z=1$.

Regard x and y as independent variables and z is a function of x and y given by (2).

From (1)

$$\frac{\partial u}{\partial x} = -\frac{40}{(x+2y+4z)^2}\left[1 + 4\frac{\partial z}{\partial x}\right]$$

From (2), we get

$$\log x + \log y + \log z = \log 8$$

$\therefore$

$$\frac{1}{x} + \frac{1}{z}\frac{\partial z}{\partial x} = 0$$

$\Rightarrow$

$$\frac{\partial z}{\partial x} = -\frac{z}{x}$$

$\therefore$

$$\frac{\partial u}{\partial x} = -\frac{40}{(x+2y+4z)^2}\left[1 - 4\frac{z}{x}\right]$$

$\Rightarrow$

$$\frac{\partial^2 u}{\partial x^2} = \frac{80}{(x+2y+4z)^3}\left[1 + 4\frac{\partial z}{\partial x}\right]\left[1 - 4\frac{z}{x}\right] - \frac{40}{(x+2y+4z)^2}\left[\frac{4z}{x^2} - \frac{4}{x}\frac{\partial z}{\partial x}\right]$$

Now using $x = 4, y = 2, z = 1$. We get $\dfrac{\partial^2 u}{\partial x^2} = -ve$

$\therefore$ u is maximum at the point given by $x=4$, $y=2$, $z=1$.

The maximum value is given by $u = \dfrac{5 \times 4 \times 2 \times 1}{(4 + 2 \times 2 + 4 \times 1)} = \dfrac{40}{12} = \dfrac{10}{3}$.

Example 6. *In a plane triangle ABC, find the maximum value of* $u = \cos A \cos B \cos C$.

Solution . Here, we have

$$u = \cos A \cos B \cos C \qquad\qquad ...(1)$$

Since, we know that the sum of the angles of a triangle is always 180°.

$\therefore$ The variables A, B and C are connected by the relation

$$A + B + C = \pi \qquad\qquad ...(2)$$

From (1), we get

$$\log u = \log \cos A + \log \cos B + \log \cos C$$

$\Rightarrow$

$$\frac{1}{u}du = -\tan A\, dA - \tan B\, dB - \tan C\, dC.$$

For the maxima and minima of u, we must have $du=0$

$\Rightarrow$

$$\tan A\, dA + \tan B\, dB + \tan C\, dC = 0 \qquad\qquad ...(3)$$

Also from (2), $dA + dB + dC = 0$ ...(4)

Now, multiply (3) by 1, (4) by l, adding, and equating the coefficients of dA, dB and dC to zero, we get

$$\tan A + l = 0$$
$$\tan B + l = 0$$
$$\tan C + l = 0$$

$\Rightarrow$ $l = -\tan A = -\tan B = -\tan C \Rightarrow A = B = C.$

Now from (2), $A = B = C = \dfrac{\pi}{3}$ i.e., the triangle is equilateral.

Now to show that the stationary value of u given by $A = B = C = \dfrac{\pi}{3}$ is maximum. Let C be a function of A and B, regarding A and B as independent variables. From (1),

$$\log u = \log \cos A + \log \cos B + \log \cos C$$

$\Rightarrow$ $\dfrac{1}{u}\dfrac{\partial C}{\partial A} = -\tan A - \tan C \dfrac{\partial C}{\partial A}$

Now, differentiating (2), partially w.r.t. A, we get

$$1 + \dfrac{du}{dA} = 0 \quad \Rightarrow \quad \dfrac{\partial C}{\partial A} = -1$$

$\therefore$ $\dfrac{1}{u}\dfrac{\partial u}{\partial A} = -\tan A + \tan C$

$\Rightarrow$ $\dfrac{1}{u}\dfrac{\partial^2 u}{\partial^2 A} - \dfrac{1}{u^2}\left(\dfrac{\partial u}{\partial A}\right)^2 = -\sec^2 A + \sec^2 C.\dfrac{\partial C}{\partial A} = -\left(\sec^2 A + \sec^2 C\right)$

At stationary point $\dfrac{\partial u}{\partial A} = 0$

$\therefore$ $\dfrac{\partial^2 u}{\partial^2 A} = -u\left(\sec^2 A + \sec^2 C\right) = -\text{ve for } A=B=C=\dfrac{\pi}{3}.$

Hence, u is maximum at $A=B=C=\dfrac{\pi}{3}$ and the maximum value is given by

$$u = \left(\cos\dfrac{\pi}{3}\right)^3 = \left(\dfrac{1}{2}\right)^3 = \dfrac{1}{8}.$$

10.25 HIGHER ORDER PARTIAL DERIVATIVES

Definition. *Suppose $f{:}E \subset R^n \to R$ is a real valued function of n variables defined in an open set $E{\subset}R^n$, with partial derivatives D_1f, D_2f,... D_nf. If the functions D_if are differentiable then the second order partial derivatives of f be defined by*

$$D_{ij}f = D_iD_jf \ (i, j=1,2,...,n)$$

For $i = j$ we have the second order partial derivative $D_i^2 f$

For example

$$D_3D_1^2D_2^3 f = \left(D_3D_1\right)D_1D_2^3 f = D_1\left(D_3D_1\right)D_2^3 f$$
$$= D_1^2\left(D_1D_3\right)D_2^3 f = D_1^2\left(D_3D_2\right)D_2^3 f$$
$$= D_1^2\left(D_1D_3\right)D_2^2 f = D_1^2 D_2\left(D_3D_2\right)D_2 f$$
$$= D_1^2 D_2\left(D_2D_3\right)D_2 f = D_1^2 D_2^2\left(D_2D_3\right)f\, D_2 f$$
$$= D_1^2 D_2^2\left(D_2D_3\right)f = D_1^2 D_2^3 D_3 f$$

THEOREM 1. *Suppose f is a function defined in an open set $E \subset R^2$, if $D_1 f$, $D_{21} f$ and $D_2 f$ exist at every point of E and $D_{21} f$ is continuous at some point $(a, b) \in E$. then $D_{12} f$ exists at (a, b) and $(D_{12} f)(a, b) = (D_{21} f)(a, b)$.*

Proof. Let $(a, b) \in E$ be arbitrary but fix. Since E is open, we can obtain an open ball $B(0, \delta)$ such that $(a,b) + (h,k) \in E$ for every $(a,b) \in B(0,\delta)$. Define $F:B(0,\delta) \to R$ by

$$F(h,k) = f(a+h,b+k) - f(a,b+k) - f(a+h,b) + f(a,b) \text{ for } (h,k) \text{ in } B(0, \delta).$$

Let us consider the function g defined in a neighbourhood of x in R by

$$g(t) = f(t,b+k) - f(t,b)$$

By the mean value theorem, we have

$$g(a+h) - g(a) = f(a+h,b+k) - f(a+h,b) - f(a,b+k) + f(a,b)$$
$$= hg'(a+\theta_1 h) \neq \theta_1 \in (0,1)$$
$$= h\{D_1 f(a+\theta_1 h,b+k) - D_1 f(a+\theta_1 h,b)\}$$

Again by mean value theorem

$$F(h,k) = hk(D_2 D_1) f(a+\theta_1 h, b+\theta_1 k), \theta_2 \in (0,1)$$

Using the definition of continuity of $(D_{21} f)(a,b)$, we have

$$\lim_{(h,k) \to (0,0)} \frac{F(h,k)}{hk} = (D_{21} f)(a,b) \qquad \text{...(1)}$$

Similarly, by considering the function v defined in a neighbourhood of b in R by $v(s) = f(a+h, s) - f(a, s)$ and using the continuity of $(D_{12} f)(a, b)$ we get

$$\lim_{(h,k) \to (0,0)} \frac{F(h,k)}{hk} = (D_{12} f)(a,b) \qquad \text{...(2)}$$

Combining (1) and (2), we obtain the required identity given by

$$D_{12} f(a, b) = (D_{21} f)(a, b), (a, b) \in E.$$

EXERCISE 10.9

Using Lagrange's method of undetermined multiplirers:

1. Find the maximum and minimum values of

$$\frac{x^2}{a^4} + \frac{y^2}{b^4} + \frac{z^2}{c^4}$$

where $lx+my+nz=0$ and $\frac{x^2}{a^2} + \frac{y^2}{b^2} + \frac{z^2}{c^2} = 1$.

2. Find the maximum and minimum values of
$$f = a^2 x^2 + b^2 y^2 + c^2 z^2$$
where $x^2+y^2+z^2=1$ and $lx+my+nz=0$.

3. Show that the maximum and minimum values of $u=x^2+y^2+z^2$ subject to the conditions

$$px+qy+rz = 0 \text{ and } \frac{x^2}{a^2} + \frac{y^2}{b^2} + \frac{z^2}{c^2} = 1$$

are given by $\frac{a^2 p^2}{u-a^2} + \frac{b^2 q^2}{u-b^2}$.

4. Find the minimum value of $u=x+y+z$ subject to the condition $\frac{a}{x} + \frac{b}{y} + \frac{c}{z} = 1$.

5. Find the minimum value of $u=x^2+y^2+z^2$, subject to the condition $ax+by+cz=p$.

6. Find the minimum value of $x+y+z$ where $xyz=c^3$.

7. Find the extreme values of $x^p y^q z^r$ subject to the condition $\frac{a}{x} + \frac{b}{y} + \frac{c}{z} = 1$.

8. Show that the maximum and minimum values of the radii vectors of the sections of the surface

$$(x^2+y^2+z^2)^2 = \frac{x^2}{a^2} + \frac{y^2}{b^2} + \frac{z^2}{c^2}$$

by the plane $\lambda x+\mu y+vz=0$

are given by $\dfrac{a^2\lambda^2}{1-a^2r^2}+\dfrac{b^2\mu^2}{1-b^2r^2}+\dfrac{c^2v^2}{1-c^2r^2}=0$

9. Find the stationary points of the function $u=ax^p+by^q+cz^r$ subject to the condition
$$x^l+y^m+z^n=k.$$

10. If two variables x and y are connected by the relation $ax^2+by^2=ab$, show that the maximum and minimum values of the function $u=x^2+y^2+xy$ will be the roots of the equation
$$4(u-a)(u-b)=ab.$$

11. Prove that of all rectangular parallelopipeds of the same volume, the cube has the least surface.

12. Prove that if $x+y+z=1$, $ayz+bzx+cxy$ has an extreme value equal to
$$\frac{abc}{2bc+2ca+2ab-a^2-b^2-c^2}$$
Also, prove if a, b, c are all positive and c lies between $a+b-2\sqrt{ab}$ and $a+b+2\sqrt{ab}$ this value is true maximum and that if a, b, c are all negative and c lies between $a+b\pm2\sqrt{ab}$. It is true minimum.

13. Find the maximum value of u, when
$$u=\sin x \sin y \sin z$$
and x,y,z are the angles of a triangle.

14. Find the triangle of maximum area inscribed in a circle.

15. Prove that the rectangular solid of maximum volume which can be inscribed in a sphere is a cube.

16. Find a plane triangle ABC such that
$$u=\sin^a A \sin^b B \sin^c C$$
has maximum value.

17. Find the rectangular parallelopiped of maximum volume that can be inscribed in the ellipsoid
$$\frac{x^2}{a^2}+\frac{y^2}{b^2}+\frac{z^2}{c^2}=1$$

18. Divide a number n into three parts x, y, z such that $ayz+bzx+cxy$ shall have maximum or minimum and determine which it is.

19. Prove that a rectangular solid of maximum volume which can be inscribed in a sphere is a cube.

20. Find the maximum or minimum value of $x^p y^q z^r$ subject to the condition $ax+by+cz=p+q+r$.

21. Show that the maximum and minimum value of
$$u=ax^2+by^2+cz^2+2fyz+2gzx+2hxy$$
subject to the conditions $lx+my+nz=0$
and $x^2+y^2+z^2=1$
are given by the equation
$$\begin{vmatrix} a-u & h & g & l \\ h & b-u & f & m \\ g & f & c-u & n \\ l & m & n & 0 \end{vmatrix}=0$$

22. Show that of the perimeter of a triangle is constant, its area is maximum when it is equilateral.

Answers

1. The maximum and minimum values of the given function is given by the equation
$$\frac{l^2a^4}{a^2u-1}+\frac{m^2b^4}{b^2u-1}+\frac{n^2c^4}{c^2u-1}=0$$

2. The maximum and minimum values of the given function is given by the equation
$$\frac{l^2}{u-a^2}+\frac{m^2}{u-b^2}+\frac{m^2}{u-c^2}=0$$

4. Stationary points are
$$x=\sqrt{a}\left(\sqrt{a}+\sqrt{b}+\sqrt{c}\right), y=\sqrt{b}\left(\sqrt{a}+\sqrt{b}+\sqrt{c}\right), z=\sqrt{c}\left(\sqrt{a}+\sqrt{b}+\sqrt{c}\right)$$
minimum value is $\left(\sqrt{a}+\sqrt{b}+\sqrt{c}\right)^2$.

5. Minimum value is $\dfrac{p^2}{\left(a^2+b^2+c^2\right)}$.

6. u is minimum at the point $x=y=z=c$. Value is $=3c^4$.

7. u is stationary when $\dfrac{px}{a}=\dfrac{qy}{b}=\dfrac{rc}{c}=p+q+r$

Minimum value is $\dfrac{a^p b^q c^r}{p^p q^q r^r}(p+q+r)^{p+q+r}$.

9. Stationary points are given by $\dfrac{x^{p-1}}{l/pa}=\dfrac{y^{q-m}}{m/qb}=\dfrac{z^{r-n}}{n/rc}$

13. u is maximum, when $x=y=z=\dfrac{\pi}{3}$. Maximum value is $\dfrac{3\sqrt{3}}{8}$.

14. Equilateral.

16. u is maximum when, A, B, C are given by $\dfrac{\tan A}{a}=\dfrac{\tan B}{b}=\dfrac{\tan C}{c}$.

17. Stationary points are $x=\dfrac{a}{\sqrt{3}}, y=\dfrac{b}{\sqrt{3}}, z=\dfrac{c}{\sqrt{3}}$

Maximum value $=\dfrac{8abc}{3\sqrt{3}}$.

CHAPTER REVIEW: A COMPETITIVE APPROACH

SELECTED TERMS AND RESULTS

TERMS

- **Limit:** A function f is said to tends to a limit l as a point (x,y) tends to the point (a,b) if for every arbitrary small positive number ε there correspond a positive number δ such that
$$|f(x, y) - l| < \varepsilon$$
for every point (x, y), different from (a, b) such that $|(x, y) - (a, b)| < \delta$.

- **Repeated limit:** Let f be a function in the nbd of (a,b) then the limit $\lim_{y \to b} f(x,y)$, if exists is a function of x say $g(x)$. If $\lim_{x \to a} g(x)$ exists and is equal to λ, then we write
$$\lim_{x \to a} \lim_{y \to b} f(x,y) = \lambda$$
and say that λ is a repeated limit of f as $y \to a$, $x \to a$.

- **Continuity:** A function f is said to be continuous at a point (a, b) if $\lim_{(x,y) \to (a,b)} f(x,y) = f(a,b)$.

- **Homogeneous function:** A function in which every term is of the same degree say n is called the homogeneous function of degree n.

- **Differentiability:** A function $f(x,y)$ defined on $]a,b[$ is said to be totally differentiable in $]a,b[$ if there exists two constants α and β depending on f on $]a,b[$ such that
$$\lim_{\substack{\delta x \to 0 \\ \delta y \to 0}} \frac{f(a + \delta x, b + \delta y) - f(a,b) - \alpha \delta x - \beta \delta y}{\sqrt{(\delta x)^2 + (\delta y)^2}} = 0$$

- **Directional derivatives:** Let $f(x, y)$ be a function of two variables x and y and a line S inclined at an angle θ with x-axis then the directional derivative of $f(x,y)$ along this line at a point $P(x,y)$ is defined as
$$\frac{\partial f}{\partial S} = f_x \cos\theta + f_y \sin\theta$$

- **Maximum value:** A function $f(x, y)$ is said to have a maximum value at a point (a, b) if there exists a neighbourhood N of (a, b) such that
$$f(a+h, b+k) < f(a, b) \ \forall (a+h, b+k) \in N$$

- **Minimum value:** A function $f(x, y)$ is said to have a maximum value at a point (a, b) if there exists a nbd N of (a, b) such that
$$f(a+h, b+k) > f(a, b) \ \forall (a+h, b+k) \in N$$

- **Extreme value:** A function $f(x, y)$ is said to have extreme value at (a, b) if it has either a maximum or minimum at (a, b).

- **Absolute maximum:** A function $f(x, y)$ is said to have an absolute maximum at (a, b) if $f(a, b) > f(x, y) \ \forall (x, y)$ in the domain of f.

- **Absolute minimum:** A function $f(x, y)$ is said to have an absolute minimum at (a, b) if $f(a, b) < f(x, y) \ \forall (x, y)$ in the domain of f.

- **Critical point:** A point (a,b) is said to be a critical point of the function $f(x, y)$ if $f_x(a, b) = 0$ and $f_y(a, b) = 0$.

RESULTS

- If the simultaneous limit exists then two repeated limits if they exists are necessarily equal but converse is not true.
- If the repeated limits are not equal, the simultaneous limit can not exist.
- If f_x exists throughout a nbd of a point (a,b) and $f_y(a,b)$ exists then for any point $(a+h,b+k)$ of this neighbourhood
$$f(a + h, b + k) = f(a,b) + h f_x(a + \theta h, b + k)$$
$$+ k[f_y(a,b) + \eta]$$

where $0 < \theta < 1$ and η is a function of k, tending to zero with k.

- A sufficient condition that a function f be continuous at (a,b) is that one of the partial derivatives exists and is bounded in a neighbourhood of (a,b) and the other exists at (a,b).
- A sufficient condition that a function be continuous in a closed region is that both the partial derivatives exist and bounded throughout the region.

- A function which is differentiable at a point possesses the first order partial derivatives at that point and necessarily continuous at that point.

- If (a, b) be a point of domain of a function f such that

(i) f_x is continuous at (a, b)

(ii) f_y exists at (a, b)

then f is differentiable at (a, b)

- A function f is differentiable at (a, b) if f_x exists and f_y is continuous at (a, b) i.e., one of the partial derivatives is to be continuous and the other merely to exist at that point.

- (Young's theorem): If f_x and f_y are both differentiable at a point (a, b) of the domain of function f then $f_{xy}(a, b) = f_{yx}(a, b)$.

- (Schwarz's theorem): If f_y exists in a certain nbd of a point (a, b) of the domain of f and f_{yx} is continuous at (a, b) then $f_{xy}(a, b)$ exists and is equal to $f_{yx}(a, b)$.

- If f_{xy} and f_{yx} are both continuous at (a, b) then $f_{xy}(a, b) = f_{yx}(a, b)$.

- A necessary condition for $f(x, y)$ to have an extreme value at (a, b) is that $f_x(a, b) = 0$, $f_{yx}(a, b) = 0$ provided they exist.

- The necessary condition for a function $f(x, y)$ to be an extremum at (a, b) are that $f_x(a,b)=0$, $f_y(a,b)=0$.

- Lagrange's method of undetermined multipliers is used to find the extreme values of a function of three or more variables when the variables are not independent but are connected by some relation. It gives only the extreme points and not distinguish whether the point is a maxima or minima.

REVIEW QUESTIONS AND PROJECT WORK

1. Show that the function

$$f(x,y)=\begin{cases} x\sin\dfrac{1}{x}+y\sin\dfrac{1}{y} & \text{when } x, y \neq 0 \\ 0 & \text{when } x, y = 0 \end{cases}$$

is continuous at $(0,0)$.

2. Show that the function

$$f(x, y)=\begin{cases} x^3\sin\dfrac{1}{x^2}+y^3\sin\dfrac{1}{y^2} & x, y \neq 0 \\ x^3\sin\dfrac{1}{x^2} & x \neq 0, y = 0 \\ y^3\sin\dfrac{1}{y^2} & x = 0, y \neq 0 \end{cases}$$

is differentiable at $(0,0)$ whereas none of f_x, f_y is continuous at $(0,0)$.

3. Show that for

$$f(x,y)=\begin{cases} (x^2+y^2)\log(x^2+y^2), & x^2+y^2 \neq 0 \\ 0 & \text{when } x = y = 0 \end{cases}$$

we have $f_{xy}(0,0)=f_{yx}(0,0)$ even though the conditions of Young's and the Schwarz theorems at $(0,0)$ are not satisfied.

4. Show that $(|x|+|y|+|z|)$ is continuous but not differentiable at $(0,0,0)$.

5. Show that for $k>0$, $\{|x+y|+(x+y)\}^k$ is everywhere differentiable on the finite xy-plane.

6. Show that $f(x,y)=\sin x \sin y \sin(x+y)$ has minima at $(0,0)$ and maxima at $\left(\dfrac{\pi}{3},\dfrac{\pi}{3}\right)$ in the positive quadrant when $x^2+y^2\leq\dfrac{\pi^2}{2}$.

7. Show that $f(x,y)=(y-x)^4+x^4$ has minimum at origin.

8. Show that the function $f(x,y,z)=(x+y+z)^3-12(x+y+z)-24xyz^3$ has maxima at $(-2,-2,-2)$ and minima at $(2,2,2)$.

9. If $lx+my+nz=d$, show that

$$\min\{(x-a)^2+ (y-b)^2+(z-c)^2\}$$

$$= \dfrac{(al + bm + cn - d)^2}{l^2 +m^2 +n^2}.$$

10. Show that the function $f=x^2+y^2+2z^2$ when $xyz^2=1$ has four stationary points given by $(1,1,\pm1),(-1,-1,\pm1)$ giving a minima at each of these points.

11. Show that the largest length of the semi-axis of the ellopsoid $ax^2 +by^2 +cz^2 + 2dxy + 2exz + 2fyz = 1$ is given by the largest real root of the equation.

$$\begin{vmatrix} a-1/r^2 & d & e \\ d & b-1/r^2 & f \\ e & f & c-1/r^2 \end{vmatrix} = 0$$

12. Show that $xy + yz + zx$ has no extreme value when it is considered as a function of the independent variables x, y, z but it has a maximum value when $ax+by+cz=1$ where a,b,c are positive and

$$2(ab + bc + ca) > (a^2 + b^2 + c^2).$$

OBJECTIVE TYPE QUESTIONS

FILL IN THE BLANKS

1. A function of two variables x and y may be writen as _____ .

2. $\lim_{(x,y)\to(a,b)} f(x,y)$, if exists, is _____ .

3. If a function $f(x, y)$ is totally differentiable, then the partial derivatives f_x and f_y _____ .

4. If $u = f\left(\dfrac{y}{x}\right)$, then $x\dfrac{\partial u}{\partial x} + y\dfrac{\partial u}{\partial y} = $ _____ .

5. If u is a homogeneous function of x and y of degree n, then

$$x\frac{\partial u}{\partial x} + y\frac{\partial u}{\partial y} = \underline{\qquad}.$$

6. The statement given in (5) is known as _____ .

7. If u is a function of variables x and y, and x and y both are the functions of the variable t, then u is said to be _____ .

8. If $f_{xy}(a, b)$ and $f_{yx}(a, b)$, both are continuous, then f_{xy} _____ .

TRUE/FALSE

Write 'T' for true and 'F' for false statement.

1. The limit of a function is unique. **(T/F)**

2. $\lim_{(x,y)\to(1,1)} f(x,y)$ does not exist, where $f(x, y) = (x^2 + 2y)$. **(T/F)**

3. The two iterated limits obtained by reversing the order of limits, are always equal. **(T/F)**

4. The function $f(x, y) = xy$ is not continuous at $(2, 3)$. **(T/F)**

5. In case of a function of two variables, the continuity is a necessary condition for differentiability. **(T/F)**

MULTIPLE CHOICE QUESTIONS

Choose the most appropriate one:

Problem Set-1

1. Simultaneous limits are also called :
   (a) repeated limit   (b) double limit
   (c) both (a) and (b)   (d) none of these

2. Iterated limits are also called :
   (a) repeated limit   (b) double limit
   (c) both (a) and (b)   (d) none of these

3. The limit
   $$\lim_{\substack{x\to a \\ y\to b}} f(x,y) = \lim_{(x,y)\to(a,b)} f(x,y) \text{ is called :}$$
   (a) iterated limit   (b) simultaneous limit
   (c) both (a) and (b)   (d) none of these

4. The limit, $\lim_{x\to a}\left[\lim_{y\to b} f(x,y)\right]$ or $\left[\lim_{y\to b} f(x,y)\right]$
   is called:
   (a) iterated limit   (b) simultaneous limit
   (c) both (a) and (b)   (d) none of these

5. The simultaneous limit, $\lim_{\substack{x\to a \\ y\to b}} \dfrac{xy^3}{x^2 + y^6} = $
   (a) 0   (b) 2
   (c) 1   (d) does not exist

6. The value of $\lim_{(x,y)\to(0,0)} \dfrac{x^2 - y^2}{x^2 + y^2} = $
   (a) 0   (b) 1
   (c) 2   (d) does not exist

7. Let $f : R^2 \to R$ be defined by $f(x, y) = x^2 + y^2$ then value of $\lim_{(x,y)\to(0,0)} f(x,y) = $
   (a) 1   (b) 0
   (c) 2   (d) does not exist

8. The value of $\lim_{(x,y)\to(0,0)} \dfrac{2x^3 - y^3}{x^2 + y^2} = $
   (a) 1   (b) 0
   (c) 2   (d) does not exist

9. The value of $\lim\limits_{(x,y)\to(0,0)} \dfrac{xy^2}{x^2+y^4} =$

(a) 1   (b) 2

(c) 0   (d) does not exist

10. If $f(x,y) = \dfrac{x^2 y^2}{x^2 y^2 + (x-y)^2}$, where

$x^2 y^2 + (x-y)^2 \neq 0$. Then $\lim\limits_{x\to 0}\left[\lim\limits_{y\to 0} f(x,y)\right] =$

(a) 1   (b) 2

(c) 0   (d) does not exist

11. $\lim\limits_{(x,y)\to(0,0)}\left[\log\dfrac{a(1-e^x)}{(a-y^x)}\right] =$

(a) 1   (b) 0

(c) 2   (d) does not exist

12. $\lim\limits_{(x,y)\to(0,0)}\left[\dfrac{x-y}{\sqrt{x^2-y^2}}\right] =$

(a) 1   (b) 0

(c) 2   (d) does not exist

13. The value of $\lim\limits_{(x,y)\to(0,0)} f(x,y)$, where

$f(y) = \begin{cases} \dfrac{xy}{x^2+y^2}, & (x,y)\neq(0,0) \\ 0, & (x,y)=(0,0) \end{cases}$ is given by

(a) 1   (b) 0

(c) 2   (d) does not exist

14. If $f(x,y) = \begin{cases} 2xy, & \text{if } (x,y)\neq(1,2) \\ 0, & \text{if } (x,y)=(1,2) \end{cases}$. Then

$\lim\limits_{(x,y)\to(1,2)} f(x,y) =$

(a) 1   (b) 2

(c) 3   (d) 4

15. $\lim\limits_{(x,y)\to(0,0)} \dfrac{2x-y}{x^2+y^2} =$

(a) 0   (b) 1

(c) 2   (d) does not exist

16. $\lim\limits_{(x,y)\to(0,0)}\left(y\sin\dfrac{1}{x}+x\sin\dfrac{1}{y}\right) =$

(a) 1   (b) 0

(c) 2   (d) does not exist

17. $\lim\limits_{(x,y)\to(0,1)} \tan^{-1}\left(\dfrac{y}{x}\right) =$

(a) $\dfrac{\pi}{2}$ (b) 0

(c) 1   (d) does not exist

18. A function $f(x,y)$ is said to be continuous at the point (a,b) if :

(a) $\lim\limits_{(x,y)\to(a,b)} f(x,y)$ exists and equal to $f(a,b)$

(b) if for $\varepsilon > 0 \; \exists \; \delta > 0$ such that $|f(x,y) - f(a,b)| < \varepsilon, \; |x-a| < \delta$

(c) both (a) and (b) are true

(d) none of the above

19. The function $f(x,y)$ is said to be continuous in a domain D if it is continuous at :

(a) one point of D

(b) at each point of D

(c) at least two points of D

(d) none of the above

20. Which of the following is not true ?

(a) Every polynomial function of two variables is continuous

(b) Every rational functions of two variables is continuous

(c) Both (a) and (b) are true.

(d) None of the above

21. At $(1, 2)$ the function $f(x, y) = x^2 + 3y$ is :

(a) continuous

(b) discontinuous

(c) limit does not exist

(d) none of the above

22. The function

$f(x,y) = \begin{cases} \dfrac{xy(x^2-y^2)}{x^2+y^2}, & (x,y)\neq(0,0) \\ 0, & \text{otherwise} \end{cases}$

is:

(a) continuous everywhere

(b) continuous at (0,0)

(c) not continuous

(d) none of the above

23. The function $f(x,y) = \dfrac{xy}{\sqrt{x^2+y^2}}, x\neq 0, y\neq 0$ and $f(0, 0)$ is :

(a) continuous everywhere
(b) continuous at origin
(c) continuous at (0,0)
(d) none of the above

24. The function $f(x,y)=\dfrac{xy^3}{x^2+y^6}$, $x\neq0,y\neq0$ and $f(0,0)$ is :

(a) continuous at origin
(b) continuous everywhere
(c) discontinuous at origin
(d) none of the above

25. The function
$$f(x,y)=\begin{cases}\dfrac{xy}{\sqrt{x^2+y^2}},(x,y)\neq(0,0)\\ 0 \quad ,(x,y)=(0,0)\end{cases}\text{ is:}$$
(a) continuous everywhere
(b) discontinuous
(c) continuous at (0, 0)
(d) none of the above

26. The function
$$f(x,y)=\begin{cases}\dfrac{xy}{\sqrt{x^2+y^2}},(x,y)\neq(0,0)\\ 0 \quad ,(x,y)=(0,0)\end{cases}\text{ then } f \text{ is :}$$
(a) not continuous at (0,0)
(b) continuous at each variable separately
(c) both (a) and (b) are true
(d) none of the above

27. The function
$$f(x,y)=\begin{cases}\dfrac{xy\left(x^2-y^2\right)}{\sqrt{x^2+y^2}},(x,y)\neq(0,0)\\ 0 \quad ,(x,y)=(0,0)\end{cases}\text{ is :}$$
(a) continuous everywhere
(b) continuous at (0,0)
(c) discontinuous
(d) none of the above

28. The function $f(x,y)=\sqrt{|x,y|}$ is:
(a) continuous everywhere
(b) continuous at (0,0)
(c) discontinuous
(d) none of the above

29. The function
$$f(x,y)=\begin{cases}x\sin\dfrac{1}{y};y\neq0\\ 0 \quad \text{if}\,(x,y)=(0,0)\end{cases}\text{ is:}$$

(a) continuous everywhere
(b) continuous at (0,0)
(c) discontinuous
(d) none of the above

30. The function
$$f(x,y)=\dfrac{xy^2}{x^2+y^4}\,;f(0,0)=0\text{ is :}$$
(a) discontinuous at origin
(b) continuous along the radius vector $\theta=\dfrac{\pi}{2}$
(c) both (a) and (b) are true
(d) none of the above

31. The function
$$f(x,y)=\begin{cases}\dfrac{xy^2+x^2y}{x^3+y^3};(x,y)\neq(0,0)\\ 0 \quad ;(x,y)=(0,0)\end{cases}\text{ is:}$$
(a) continuous at (0, 0)
(b) not continuous at (0, 0)
(c) discontinuous everywhere
(d) none of the above

32. The function
$$f(x, y)=\begin{cases}0;\text{if }x\text{ is irrational}\\ 1;\text{if }x\text{ is rational}\end{cases}$$
(a) discontinuous at any point
(b) continuous at (0,0)
(c) continuous everywhere
(d) none of the above

33. If $f(x,y)=x^3+y^3-3axy$ then the value of $\dfrac{\partial f}{\partial x}=$
(a) $3x^2+3ay$ 　　　(b) $3x^2-3ay$
(c) $3y^2-3ax$ 　　　(d) $3x^2$

34. If $f(x,y)=x\cos y+y-\cos x$ then which of the following is true ?

(a) $\dfrac{\partial^2 f}{\partial x\partial y}\neq\dfrac{\partial^2 f}{\partial y\partial x}$ 　　(b) $\dfrac{\partial^2 f}{\partial x\partial y}=\dfrac{\partial^2 f}{\partial y\partial x}$

(c) $\dfrac{\partial f}{\partial x}=\dfrac{\partial f}{\partial y}$ 　　(d) None of these

35. If $u=x^3+y^3-3xy^2$ then value of $x\dfrac{\partial u}{\partial u}+y\dfrac{\partial u}{\partial y}=$
(a) 4 　　　(b) 24
(c) 34 　　　(d) 0

36. If $u=\tan^{-1}\left(\dfrac{x^3+y^3}{x-y}\right)$ then value of $x\dfrac{\partial u}{\partial x}+y\dfrac{\partial u}{\partial y}=$

(a) $\sin u$ (b) $\sin 2u$

(c) $\sin 3u$ (d) $\cos 3u$

37. If $u = \tan^{-1}\dfrac{y}{x}$ then value of $x\dfrac{\partial u}{\partial x} + y\dfrac{\partial u}{\partial y} =$

(a) 0   (b) 4

(c) 24   (d) 34

38. Let $f(x,y) = \sqrt{|x,y|}$ then :

(a) $f(x,y)$ is continuous at origin

(b) f_x exists at origin and equal to zero

(c) f_y exists at origin and equal to zero

(d) all are true

39. For $f(x,y) = \begin{cases} \dfrac{2xy}{x^2+y^2} & ;(x,y) \neq (0,0) \\ 0 & ;(x,y) = (0,0) \end{cases}$ we have:

(a) f_x and f_y exist at $(0, 0)$ and f is continuous at $(0, 0)$

(b) f_x and f_y exist at $(0, 0)$ and f is discontinuous at $(0, 0)$

(c) f_x and f_y do not exist

(d) none of the above

40. Let $f : R^2 \rightarrow R$ be a function defined by

$$f(x,y) = \begin{cases} \dfrac{x^2 y}{x^4 + y^2} & ;(x,y) \neq (0,0) \\ 0 & ;(x,y) = (0,0) \end{cases}$$

then the directional derivative of f at $(0,0)$ in the direction of the vector $\left(\dfrac{1}{\sqrt{2}}, \dfrac{1}{\sqrt{2}}\right)$ is :

(a) $\dfrac{1}{2}$ (b) $\dfrac{1}{\sqrt{2}}$

(c) $\dfrac{1}{2\sqrt{2}}$ (d) none of these

41. Let $f : R^2 \rightarrow R$ be a function defined by

$$f(x,y) = \begin{cases} x^2 + y^2 & ;(x,y) \neq (0,0) \\ 0 & ;(x,y) = (0,0) \end{cases}$$

(a) f is discontinuous at $(0,0)$

(b) f is continuous and differentiable at $(0,0)$

(c) f is differentiable everywhere

(d) none of the above

42. Let $f : R^2 \rightarrow R$ be a function such that $\dfrac{\partial f}{\partial x}$ and $\dfrac{\partial f}{\partial y}$ exist at all points. Then:

(a) total derivative of f exists at all points of R^2

(b) the function $f(x,y)$ as a function of x for every fixed y and $f(x,y)$ as a function of y for every fixed x are continuous

(c) f is continuous on R^2

(d) none of the above

43. Which of the following is true ?

(a) If a function $f(x,y)$ is differentiable at (a,b) it must be continuous at (a,b)

(b) If a function is continuous at (a,b) then it is differentiable at (a,b)

(c) Both (a) and (b) are true

(d) None of the above

44. If $f(x,y) = \begin{cases} \dfrac{x^2 - xy}{x+y} & ;(x,y) \neq (0,0) \\ 0 & ;(x,y) = (0,0) \end{cases}$ then

(a) $f_x(0,0) = 1$

(b) $f_y(0,0) = 0$

(c) both (a) and (b) are true

(d) none of the above

45. If $f(x,y) = \begin{cases} x^2 + 2y & ;(x,y) \neq (1,2) \\ 0 & ;(x,y) = (1,2) \end{cases}$ then

(a) $f(x, y)$ is discontinuous at $(1, 2)$

(b) $f(x, y)$ has a removable discontinuity at $(1, 2)$

(c) both (a) and (b) are true

(d) none of these

46. If $f(x,y) = \begin{cases} \dfrac{xy}{\sqrt{x^2+y^2}} & ;(x,y) \neq (0,0) \\ 0 & ;(x,y) = (0,0) \end{cases}$ then

(a) $f(x,y)$ is discontinuous at $(0,0)$

(b) both the functions of single variable are continuous at origin

(c) both (a) and (b) are true

(d) none of the above

47. If $f(x,y) = 2x^2 - xy + 2y^2$ then at $(1,2)$

(a) $\dfrac{\partial f}{\partial x} = 2$

(b) $\dfrac{\partial f}{\partial y} = 7$

(c) both (a) and (b) are true

(d) none of the above

48. If $f(x,y) = \begin{cases} \dfrac{2xy}{x^2+y^2} & ;(x,y) \neq (0,0) \\ 0 & ;(x,y) = (0,0) \end{cases}$ then:

(a) f_x exists at origin
(b) f_y exists at origin
(c) both (a) and (b) are true
(d) none of the above

49. If $f(x,y) = \begin{cases} \dfrac{xy}{x^2 + y^2} & ;(x,y) \neq (0,0) \\ 0 & ;(x,y) = (0,0) \end{cases}$ then:

(a) f_x and f_y exists at origin
(b) $f(x,y)$ is discontinuous at origin
(c) both (a) and (b) are true
(d) none of the above

50. If $f(x,y) = \begin{cases} \dfrac{xy}{\sqrt{x^2 + y^2}} & ;(x,y) \neq (0,0) \\ 0 & ;(x,y) = (0,0) \end{cases}$ then at $(0,0)$:

(a) f is continuous
(b) f possess partial derivatives
(c) f is not differentiable
(d) all are true

Problem Set-2

1. If both the first order partial derivatives of the function $f(x,y)$ vanish at that point, then this point is called :
   (a) stationary point   (b) saddle point
   (c) maxima point   (d) minima point

2. The function $f(x,y)$ will have a maximum or minimum at $x=a, y=b$ if :
   (a) $rt > s^2$ (b) $rt < s^2$
   (c) $rt = s^2$ (d) none of these

3. If $rt - s^2 < 0$, then $f(x,y)$ is :
   (a) maximum
   (b) minimum
   (c) neither maximum nor minimum
   (d) none of the above

4. If at a point (a,b), $\dfrac{\partial f}{\partial x} = 0$, $\dfrac{\partial f}{\partial y} = 0$, then $f(x,y)$ is maximum at (a,b) if at (a,b) :
   (a) $rt - s^2 > 0$
   (b) $rt - s^2 > 0$ and $r < 0$
   (c) $rt - s^2 > 0$ and $r > 0$
   (d) $rt - s^2 < 0$ and $r < 0$

5. The condition $\dfrac{\partial f}{\partial x} = 0$, $\dfrac{\partial f}{\partial y} = 0$, $\dfrac{\partial f}{\partial z} = 0$ for the maxima and minima of a function is :
   (a) necessary condition
   (b) sufficient condition
   (c) necessary and sufficient both
   (d) none of the above

6. The volume of the greatest rectangular parallelopiped that can be inscribed in the ellipsoid is :
   (a) $8abc$ (b) $\dfrac{8abc}{3\sqrt{3}}$
   (c) $\dfrac{8abc}{3}$ (d) $\dfrac{8abc}{\sqrt{3}}$

7. The maximum and minimum values of $u = a^2x^2 + b^2y^2 + c^2z^2$, where $x^2 + y^2 + z^2 = 1$ and $lx + my + nz = 0$ are the roots of the equation :
   (a) $\dfrac{l^2}{u - a^2} + \dfrac{m^2}{u - b^2} + \dfrac{n^2}{u - c^2} = 0$
   (b) $\dfrac{l}{u - a} + \dfrac{m}{u - b} + \dfrac{n}{u - c} = 0$
   (c) $\dfrac{l}{u - a^2} + \dfrac{m}{u - b^2} + \dfrac{n}{u - c^2} = 0$
   (d) none of the above

8. If we divide a number a into three parts such that their product will be maximum then parts of this number are :
   (a) $\dfrac{a}{3}, \dfrac{a}{3}, \dfrac{a}{3}$ (b) $\dfrac{a}{2}, \dfrac{a}{2}$
   (c) $\dfrac{a}{\sqrt{3}}, \dfrac{a}{\sqrt{3}}, \dfrac{a}{\sqrt{3}}$ (d) none of these

9. If $u = x^2y^3z^4$ and $2x + 3y + 4z = a$ then maximum value of u is given by :
   (a) $\dfrac{a}{9}$ (b) $\left(\dfrac{a}{9}\right)^2$
   (c) $\left(\dfrac{a}{9}\right)^9$ (d) none of these

10. The function $u = \sin x \sin y \sin z$, where x, y, z are the angles of a triangle is stationary at the point :
    (a) $x = y = z = \dfrac{\pi}{3}$ (b) $x = \dfrac{\pi}{2}, y = \dfrac{\pi}{4}, z = \dfrac{\pi}{4}$
    (c) $x = 0, y = \dfrac{\pi}{2} = 2$ (d) none of the above

Problem Set-3

1. The maximum value of

$u=\sin x \sin y \sin (x+y)$ is :

(a) $\dfrac{3\sqrt{3}}{8}$

(b) $\dfrac{\sqrt{3}}{8}$

(c) $\dfrac{3}{8}$

(d) none of these

2. When $x=y=\dfrac{\pi}{6}$ then function

$u= 2\sin \dfrac{x+y}{2} \cos \dfrac{x-y}{2} + \cos(x+y)$ is :

(a) minimum

(b) maximum

(c) neither maximum nor minimum

(d) none of the above

3. The minimum value of

$u= xy +a^3 \left(\dfrac{1}{x} + \dfrac{1}{y} \right)$ is :

(a) a^2

(b) $2a^2$

(c) $3a^2$

(d) $4a^2$

4. The function $u=x^3y^2(1-x-y)$ attains its maximum value at :

(a) $x = \dfrac{1}{2}, y = \dfrac{1}{2}$

(b) $x = \dfrac{1}{2}, y = \dfrac{1}{3}$

(c) $x = \dfrac{1}{3}, y = \dfrac{1}{2}$

(d) $x = 1, y = 1$

5. The function $u = 2a^2xy - 3ax^2y - ay^3 + x^3y + xy^3$ is:

(a) maximum at $\left(\dfrac{3a}{2}, \dfrac{-a}{2} \right)$ and $\left(\dfrac{a}{2}, \dfrac{a}{2} \right)$

(b) minimum at $\left(\dfrac{a}{2}, -\dfrac{a}{2} \right)$

(c) both (a) and (b) are true

(d) none of the above

6. The function $u=x^3+y^3-3axy$ is

(a) minimum at $x=0$

(b) maximum at $x=2^{1/3}$

(c) both (a) and (b) are true

(d) none of the above

7. The function $u=x^2y^2-5x^2-8xy-5y^2$ is maximum at:

(a) $x=y=1$

(b) $x=y=0$

(c) $x=y=5$

(d) none of these

8. The function $u=x^4+2x^2y-x^2+3y^2$ is minimum at:

(a) $x = \pm \dfrac{\sqrt{3}}{2}, y = \dfrac{1}{4}$

(b) $x=0, y=0$

(c) $x = \dfrac{1}{2}, y = \dfrac{1}{4}$

(d) none of the above

9. The function $xy^2(3x+6y-2)$ is:

(a) maximum at $(0, 0)$

(b) minimum at $(0, 0)$

(c) neither minimum nor maximum at $(0, 0)$

(d) none of the above

10. The function $y^2+4xy+3x^2+x^3$ is:

(a) minimum at $\left(\dfrac{2}{3}, -\dfrac{4}{3} \right)$

(b) neither minimum nor maximum at $(0,0)$

(c) both (a) and (b) are true

(d) none of the above

11. The function $\dfrac{(x+y-1)}{x^2+y^2}$ is:

(a) maximum at $(1, 1)$

(b) minimum at $(1, 1)$

(c) neither minimum nor maximum at $(1, 1)$

(d) none of the above

12. The function $\dfrac{(x+y)}{x^2+2y^2+6}$ is:

(a) maximum at $(2, 1)$

(b) minimum at $(-2, -1)$

(c) both (a) and (b) are true

(d) none of the above

13. The maximum value of

$u = axy^2z^3 - x^2y^2z^3 - xy^3z^3 - xy^2z^4$ is:

(a) $108a^7$

(b) $\dfrac{107a^7}{7^7}$

(c) $\dfrac{a^7}{7^7}$

(d) none of these

14. The function $u=y^2+2z^2-5x^4+4x^5$ is:

(a) minimum at $(1, 0, 0)$

(b) neither minimum nor maximum at $(0, 0, 0)$

(c) both (a) and (b) are true

(d) none of the above

15. The function

$(x+y+z)^3-3(x+y+z)-24xyz+a^3$ has a:

(a) minimum at (1, 1, 1)

(b) maximum at (–1, –1, –1)

(c) both (a) and (b) are true

(d) none of the above

16. The function

$u=x^2+y^2+z^2+x-2z-xy$ is:

(a) minimum at $\left(-\dfrac{2}{3},-\dfrac{1}{3},1\right)$

(b) maximum at $\left(-\dfrac{2}{3},-\dfrac{1}{3},1\right)$

(c) neither   maximum   nor   minimum

at $\left(-\dfrac{2}{3},-\dfrac{1}{3},1\right)$

(d) none of the above

17. The maximum value of

$xy(z-h)\left[\dfrac{x^2}{a^2}+\dfrac{y^2}{b^2}-\dfrac{z^2}{c^2}\right]$ is

(a) $\left(\dfrac{2h}{5}\right)^5$ (b) $\dfrac{ab}{c^4}$

(c) $\left(\dfrac{2h}{5}\right)^5\left(\dfrac{ab}{c^4}\right)$ (d) none of these

18. The point such that the sum of the squares of its distance from n given points is:

(a) centroid

(b) centre of the mean position of the given points

(c) both (a) and (b) are true

(d) none of the above

19. The maximum value of

$-\alpha^2x^2-\beta^2y^2-\gamma^2z^2+(ax+by+cz).e$

(a) $\dfrac{1}{2e}\left(\dfrac{a^2}{\alpha^2}+\dfrac{b^2}{\beta^2}+\dfrac{c^2}{\gamma^2}\right)^{1/2}$

(b) $\left[\dfrac{1}{2e}\left(\dfrac{a^2}{\alpha^2}+\dfrac{b^2}{\beta^2}+\dfrac{c^2}{\gamma^2}\right)\right]^{1/2}$

(c) $\dfrac{1}{2e}$

(d) none of the above

20. The function $u=ax^3y^2-x^4y^2-x^3y^3$ is :

(a) maximum at $x=\dfrac{a}{2},y=\dfrac{a}{3}$

(b) minimum at $x=\dfrac{a}{2},y=\dfrac{a}{3}$

(c) neither   maximum   nor   minimum   at

$x=\dfrac{a}{2},y=\dfrac{a}{3}$

(d) none of the above

21. If x, y, z are the angles of a triangle, then the function $u=\sin x \sin y \sin z$ is :

(a) maximum at $x=y=z=\dfrac{\pi}{3}$

(b) minimum at $x=y=z=\dfrac{\pi}{3}$

(c) neither   maximum   nor   minimum   at

$x=y=z=\dfrac{\pi}{3}$

(d) none of the above

22. If $x^2+y^2+z^2=1$ and $lx+my+nz=0$ then maximum or minimum value of $u=a^2x^2+b^2y^2+c^2z^2$ is given by :

(a) $\dfrac{l^2}{u-a^2}+\dfrac{m^2}{u-b^2}+\dfrac{n^2}{u-c^2}=0$

(b) $\dfrac{l^2}{u-a^2}+\dfrac{m^2}{u-b^2}+\dfrac{n^2}{u-c^2}=1$

(c) either (a) or (b)

(d) none of the above

23. If $lx+y+nz=0$ and $\dfrac{x^2}{a^2}+\dfrac{y^2}{b^2}+\dfrac{z^2}{c^2}=1$ then maximum   and   minimum   value   of

$u=\dfrac{x^2}{a^4}+\dfrac{y^2}{b^4}+\dfrac{z^2}{c^4}$ is given by :

(a) $\dfrac{l^2a^4}{1-a^2u}+\dfrac{m^2b^4}{1-b^2u}+\dfrac{n^2c^4}{1-c^2u}=0$

(b) $\dfrac{l^2a^4}{1-a^2u}+\dfrac{m^2b^4}{1-b^2u}+\dfrac{n^2c^4}{1-c^2u}=1$

(c) either (a) or (b)

(d) none of the above

24. If $xyz=c^3$ then $u=x^4+y^4+z^4$ is :

(a) minimum at $x=y=z=c$

(b) maximum at $x=y=z=c$

(c) neither   maximum   nor   minimum at $x = y = z = c$

(d) none of the above

25. If $a^xb^yc^z = A$ then maximum value of $(z+1)$ $(y+1)(z+1)$ is:

(a) $\dfrac{(\log A\,abc)^3}{\log a^3.\log b^3.\log c^3}$

(b) $\left(\log A\, abc\right)^3$

(c) $\dfrac{\left(\log A\, abc\right)^3}{\log abc}$

(d) none of the above

26. The triangle of maximum area which can be inserted in a circle is:

(a) equilateral   (b) isosceles

(c) right angled   (d) none of the above

27. If the parameter of a triangle is constant, then its area is maximum when triangle is

(a) equilateral   (b) isosceles

(c) right angled   (d) none of the above

28. The function $f(x, y) = 2x^4 - 3x^2 y + y^2$ has

(a) maximum at $(0, 0)$

(b) minimum at $(0, 0)$

(c) neither maximum nor minimum at $(0, 0)$

(d) none of the above

29. The function $f(x, y) = x^2 - 2xy + y^2 + x^3 - y^3 + x^5$ has:

(a) maximum value at origin

(b) minimum value at origin

(c) neither maximum nor minimum at origin

(d) none of the above

30. The function $f(x,y) = x^3 + y^2 - 63(x+y) + 12xy$ has:

(a) four stationary points

(b) maximum at $(-7, -7)$

(c) minimum at $(3, 3)$

(d) all are true

Answers

FILL IN THE BLANKS

1. $f(x, y)$ **2.** unique   **3.** both exist and equal   **4.** 0   **5.** nu

6. Euler's theorem   **7.** composite function   **8.** f_{yx}

TRUE/FALSE

1. T   **2.** F   **3.** F   **4.** F   **5.** T

MULTIPLE CHOICE QUESTIONS

Problem Set-1

1. (b)	**2.** (a)	**3.** (b)	**4.** (a)	**5.** (d)	**6.** (d)	**7.** (b)	**8.** (b)	**9.** (d)
10. (a)	**11.** (d)	**12.** (d)	**13.** (d)	**14.** (d)	**15.** (d)	**16.** (a)	**17.** (d)	**18.** (c)
19. (b)	**20.** (c)	**21.** (a)	**22.** (b)	**23.** (b)	**24.** (c)	**25.** (c)	**26.** (c)	**27.** (b)
28. (b)	**29.** (b)	**30.** (c)	**31.** (b)	**32.** (a)	**33.** (b)	**34.** (b)	**35.** (c)	**36.** (a)
37. (a)	**38.** (d)	**39.** (b)	**40.** (b)	**41.** (b)	**42.** (b)	**43.** (a)	**44.** (c)	**45.** (c)
46. (c)	**47.** (c)	**48.** (c)	**49.** (c)	**50.** (d)				

Problem Set-2

1. (a)	**2.** (a)	**3.** (c)	**4.** (b)	**5.** (a)	**6.** (b)	**7.** (a)	**8.** (a)	**9.** (c)
10. (a)								

Problem Set-3

1. (a)	**2.** (b)	**3.** (c)	**4.** (b)	**5.** (c)	**6.** (c)	**7.** (b)	**8.** (a)	**9.** (c)
10. (c)	**11.** (a)	**12.** (c)	**13.** (b)	**14.** (c)	**15.** (c)	**16.** (a)	**17.** (c)	**18.** (c)
19. (b)	**20.** (a)	**21.** (a)	**22.** (a)	**23.** (a)	**24.** (a)	**25.** (a)	**26.** (a)	**27.** (a)
28. (c)	**29.** (c)	**30.** (d)						

HINTS TO SELECTED PROBLEMS

Problem Set-1

5. Consider the line $y=x$

Let $(x,y) \to (0,0)$

For $y=x$ we get

$$\lim_{x \to 0} \frac{x \cdot x^3}{x^2 + x^6} = \lim_{x \to 0} \frac{x^2}{1 + x^2} = 0$$

Further, let $(x,y) \to (0,0)$ throughout, the curve $x=y^3$, then for $y=x^3$, we get

$$\lim_{y \to 0} \frac{y^6}{y^6 + y^6} = \frac{1}{2}$$

$\Rightarrow$ Limits obtained by two different methods are different.

Hence, the simultaneous limit does not exist.

6. Let us take $y=mx$. Then

$$\lim_{\substack{x \to 0 \\ y \to 0}} \frac{x^2 - y^2}{x^2 + y^2} = \lim_{x \to 0} \frac{1 - m^2}{1 + m^2}$$

$$= \frac{1 - m^2}{1 + m^2}, \text{ which depend upon } m.$$

$\Rightarrow \lim_{x \to 0} f(x, mx)$ is not unique.

Hence $\lim_{(x,y) \to (0,0)} \frac{x^2 - y^2}{x^2 + y^2}$ does not exist.

7. We have $\lim_{(x,y) \to (0,0)} f(x,y)$

$$= \lim_{(x,y) \to (0,0)} \left(x^2 + y^2\right)$$

$$= \lim_{\substack{x \to 0 \\ y \to 0}} \left(x^2 + y^2\right) = \lim_{y \to 0} \left(0 + y^2\right) = 0$$

8. We have

$$\lim_{(x,y) \to (0,0)} \frac{2x^3 - y^3}{x^2 + y^2} = \lim_{\substack{x \to 0 \\ y \to 0}} \frac{2x^3 - y^3}{x^2 + y^2}$$

$$\lim_{x \to 0} \frac{2x^3 - m^3x^3}{x^2 + m^2x^2} \quad \text{(by letting } y=mx)$$

$$= \lim_{x \to 0} \frac{x^3 \left(2 - m^3\right)}{x^2 \left(1 + m^2\right)}$$

$$= \lim_{x \to 0} x \left(\frac{2 - m^3}{1 + m^2}\right) = 0$$

9. Let us take $y^2 = mx$. Then, we have

$$\lim_{\substack{x \to 0 \\ y \to 0}} f(x,y) = \lim_{\substack{x \to 0 \\ y \to 0}} \frac{xy^2}{x^2 + y^4}$$

$$= \lim_{x \to 0} \frac{x(mx)}{x^2 + m^2x^2}$$

$$= \lim_{x \to 0} \frac{m}{1 + m^2} = \frac{m}{1 + m^2}$$

which depends upon m.

Hence, $\lim_{(x,y) \to (0,0)} f(x,y)$ is not unique.

$\Rightarrow \lim_{(x,y) \to (0,0)} f(x,y)$ does not exist.

10. We have $f(x,y) = \dfrac{x^2 y^2}{x^2 y^2 + (x-y)^2}$

Now $\lim_{x \to 0} \left[\lim_{y \to 0} f(x,y) \right]$

$$= \lim_{x \to 0} \left[\lim_{y \to 0} \frac{x^2 y^2}{x^2 y^2 + (x-y)^2} \right]$$

$$= \lim_{x \to 0} \left[\frac{0}{x^2} \right] = 0$$

and $\lim_{y \to 0} \left[\lim_{x \to 0} f(x,y) \right]$

$$= \lim_{y \to 0} \left[\lim_{x \to 0} \frac{x^2 y^2}{x^2 y^2 + (x-y)^2} \right]$$

$$= \lim_{y \to 0} \left[\frac{0}{y^2} \right] = 0$$

Hence,

$$\lim_{x \to 0} \left[\lim_{y \to 0} f(x,y) \right] = \lim_{y \to 0} \left[\lim_{x \to 0} f(x,y) \right] = 0$$

11. We have $\lim_{(x,y) \to (0,0)} \left[\log \frac{a\left(1 - e^x\right)}{\left(a - y^x\right)} \right]$

$$= \lim_{\substack{x \to 0 \\ y \to 0}} \left[\log \frac{a\left(1 - e^x\right)}{\left(a - y^x\right)} \right]$$

$$= \lim_{x \to 0} \left[\log \frac{a\left(1 - e^x\right)}{a - k^x} \right]$$

(Taking the limit along the line $y = k$)

$$= \left(\log \frac{0}{a - 1} \right)$$

$$= \log 0$$

$$= -\infty$$

$$= \text{does not exist.}$$

12. Let $f(x, y) = \dfrac{x - y}{\sqrt{x^2 - y^2}}, (x, y) \neq (0, 0)$

Then $\lim_{(x,y) \to (0,0)} f(x, y) = \lim_{\substack{x \to 0 \\ y \to 0}} f(x, y)$

$$= \lim_{\substack{x \to 0 \\ y \to 0}} f(x, y) \frac{x - y}{\sqrt{x^2 - y^2}}$$

$$= \lim_{\substack{x \to 0 \\ y \to 0}} \frac{x - mx}{\sqrt{x^2 + m^2 x^2}}$$

(Taking the limit along the line $y = mx$)

$$= \lim_{\substack{x \to 0 \\ y \to 0}} \frac{1 - m}{\sqrt{1 + m^2}} = \frac{1 - m}{\sqrt{1 + m^2}}$$

$$= \lim_{\substack{x \to 0 \\ y \to 0}} f(x, y)$$

depends upon m, so it is not unique.

Hence, $\lim_{(x,y) \to (0,0)} f(x, y)$ does not exist.

13. Since, if $\lim_{(x,y) \to (a,b)} f(x, y)$ exists, then this

limit is independent of the path along which we approach the point (a, b).

Let $f(x, y) \to (0,0)$ along the path $y = mx$, $m \in R$

As $x \to 0$, from $y = mx$, we have $y \to 0$

Consider $\lim_{(x,y) \to (0,0)} f(x, y)$

$$= \lim_{(x,y) \to (0,0)} \frac{xy}{x^2 + y^2}$$

$$= \lim_{x \to 0} \frac{x.mx}{x^2 + m^2 x^2} \quad \text{(along the line } y = mx\text{)}$$

$$= \lim_{x \to 0} \frac{mx^2}{x^2\left(1 + m^2\right)}$$

$$= \lim_{x \to 0} \frac{m}{1 + m^2} = \frac{m}{1 + m^2}$$

which is not unique.

Hence, $\lim_{(x,y) \to (0,0)} f(x, y)$ does not exist.

21. We have $f(x,y) = x^2 + 3y$

$$\Rightarrow f(1, 2) = 1 + 3 \times 2 = 7$$

We know that if limit of the function $f(x, y) = x^2 + 3y$ at $(1, 2)$ is also 7, then $f(x)$ is continuous at $(1, 2)$

i.e., we have to show that

$$\lim_{(x,y) \to (1,2)} f(x, y) = 7.$$

Now for given $\varepsilon > 0$ there must exist $\delta > 0$ such that

$$0 < \sqrt{(x - 1)^2 + (y - 2)^2} < \delta$$

$$\Rightarrow |x^2 + \delta y| < \varepsilon$$

$$\therefore |x^2 + 3y - 7| = |(x^2 - 1) + 3(y - 2)|$$

$$= |(x - 1)(x + 1) + 3(y - 2)|$$

$$\leq |x - 1|.|x + 1| + 3|y - 2| \quad \text{...(1)}$$

But $|y - 2| < \sqrt{(x - 1)^2 + (y - 2)^2} < \dfrac{\delta}{6}$ ...(2)

and $|x - 1| < \sqrt{(x - 1)^2 + (y - 2)^2} < \dfrac{\delta}{6}$ ...(3)

$$\therefore \quad 0 < x - 1 < 1$$

$$\Rightarrow \quad 1 < x < 2$$

$$\Rightarrow \quad 2 < x + 1 < 3$$

Then, from (1)

$$|x^2 + 3y - 7| \leq 3|x - 1| + 3|y - 2|$$

Using (1) and (2), we get

$$|(x^2 + 3y) - 7| \leq 3 |\sqrt{(x - 1)^2 + (y - 2)^2}|$$

$$+ 3 |\sqrt{(x - 1)^2 + (y - 2)^2}|$$

$$= 6 |\sqrt{(x - 1)^2 + (y - 2)^2}| < \frac{\delta}{6} = \varepsilon$$

$$\Rightarrow \quad |(x^2 + 3y) - 7| < \varepsilon$$

$$\Rightarrow \quad \lim_{(x,y) \to (1,2)} f(x, y) = 7$$

⇒ The limit and value of the function at the point (1,2) are equal, Hence the function is continuous at (1,2).

22. Let $\varepsilon > 0$ be given

Now, suppose that $|x - 0| < \sqrt{\varepsilon}$ and $|y - 0| < \sqrt{\varepsilon}$.

Consider

$$|f(x, y) - f(0, 0)| = \left| \frac{xy\left(x^2 - y^2\right)}{x^2 + y^2} - 0 \right|$$

$$\Rightarrow \quad |xy|\left|\frac{x^2 - y^2}{x^2 + y^2}\right| \le |xy|$$

$$\left[\because \left|x^2 - y^2\right| \le \left|x^2 + y^2\right| \left|\frac{x^2 - y^2}{x^2 + y^2} - 0\right| \le 1 \right]$$

$$\Rightarrow |f(x, y) - f(0, 0)| \le |x| \cdot |y|$$

$$\Rightarrow |f(x, y) - f(0, 0)| < \sqrt{\varepsilon} \cdot \sqrt{\varepsilon}$$

$$\Rightarrow |f(x, y) - f(0, 0)| < \varepsilon$$

Hence, $\lim\limits_{(x,y)\to(0,0)} f(x,y)$ exists and equal to $f(0,0)$.

23. Let $\varepsilon > 0$ be given

Let $x = r\cos\theta, y = r\sin\theta$

Then $f(r\cos\theta, r\sin\theta)$

$$= \frac{r^2\cos\theta\sin\theta}{r\sqrt{\sin^2\theta + \cos^2\theta}} = r\cos\theta\sin\theta$$

$$= \frac{1}{2}r\sin 2\theta$$

Now consider $|f(r\cos\theta, r\sin\theta) - f(0,0)|$

$$= |f(r\cos\theta, r\sin\theta)|$$

$$= \left|\frac{1}{2}r\sin 2\theta\right| = \frac{1}{2}r|\sin 2\theta| \le \frac{1}{2}r$$

If we choose $r = 2\varepsilon$

Then for $\varepsilon > 0$ we have $|f(r\cos\theta, r\sin\theta)| < \varepsilon$ for all θ.

Since the above equation is true for all points within a circle about the origin and radius $r = 2\varepsilon$. Hence, $f(x, y)$ is continuous in (x, y) at the origin.

24. Given that $f(0,0) = 0$

Let us suppose $(x,y) \to (0,0)$ through the curve $x = y^3$

Then we have

$$\lim\limits_{(x,y)\to(0,0)} f(x,y) = \lim\limits_{y\to 0} \frac{y^6}{y^6 + y^6} = \frac{1}{2}$$

Further, let $(x, y) \to (0, 0)$ through the line $y = x$, then

$$\lim\limits_{(x,y)\to(0,0)} f(x,y) = \lim\limits_{x\to 0} \frac{x \cdot x^3}{x^2 + x^6}$$

$$= \lim\limits_{x\to 0} \frac{x^2}{1 + x^4} = 0$$

Clearly, the limit obtained by two different approaches are different.

Therefore, $\lim\limits_{(x,y)\to(0,0)} f(x,y)$ does not exist.

Hence, the given function is not continuous.

25. We know that if the simultaneous limit, $\lim\limits_{(x,y)\to(0,0)} f(x,y)$ exists and is equal to $f(0,0)$,

then $f(x,y)$ will be continuous at $(0,0)$.

Then, we have

$$\lim\limits_{(x,y)\to(0,0)} f(x,y) = \lim\limits_{\substack{x\to 0 \\ y\to 0}} f(x,y)$$

$$= \lim\limits_{\substack{x\to 0 \\ y\to 0}} \left(\frac{xy}{\sqrt{x^2 + y^2}} \right)$$

$$= \lim\limits_{x\to 0} \frac{mx^2}{\sqrt{x^2 + m^2 x^2}}$$

(Along the line $y = mx$)

$$= \lim\limits_{x\to 0} \frac{mx}{\sqrt{1 + m^2}} = 0$$

$$\Rightarrow \lim\limits_{(x,y)\to(0,0)} f(x,y) = f(0,0) = 0$$

Hence, $f(x,y)$ is continuous at $(0, 0)$.

26. For point (x, y) on the x-axis, we have $y = 0$ and $f(x, y) = f(x, 0) = 0$

⇒ The function has the constant value 0 everywhere on the x-axis which gives that $f(x, y)$ is continuous at $x = 0$.

Similarly, $f(x, y)$ has the constant value 0 at all points on the y-axis so, if we put $x = 0$, the function $f(x,y)$ is continuous at $y = 0$.

Now, we shall show that $f(x,y)$ is not continuous at origin.

Let $y=x$.

Then $f(x, y)=f(x, x)=\dfrac{x^2}{2x^2}=\dfrac{1}{2}$.

Also $f(0, 0)=0$ (given)

Since, there are points on the line arbitrarily close to the origin and since $f(0,0)\neq\dfrac{1}{2}$ the function of two variables $f(x,y)$ is not continuous at the origin.

28. The simultaneous limit of $f(x, y)$ at $(0, 0)$ is

$$\lim_{(x,y)\to(0,0)} f(x,y)= \lim_{\substack{x\to0\\y\to0}} \sqrt{|xy|}$$

$$= \lim_{x\to0} \sqrt{|mx^2|} \quad \text{(Along the line } y = mx\text{)}$$

$$= \lim_{x\to0} x\sqrt{|m|} = 0$$

Also $f(0, 0) = 0$

$$\Rightarrow \lim_{(x,y)\to(0,0)} f(x,y) =f(0,0)=0$$

Hence, $f(x,y)$ is continuous at $(0, 0)$.

29. The simultaneous limit of $f(x,y)$ at $(0,0)$ is

$$\lim_{\substack{x\to0\\y\to0}} f(x,y) = \lim_{\substack{x\to0\\y\to0}} x\sin\frac{1}{y}$$

$$= \lim_{x\to0} x\sin\frac{1}{x}$$
$$\text{(Along the line } y = x\text{)}$$

$$= 0 \times a \text{ finite quantity}$$

$$= 0$$

$$\Rightarrow \lim_{\substack{x\to0\\y\to0}} f(x,y) \text{ exists and is equal to } f(0,0).$$

Hence, $f(x, y)$ is continuous at $(0,0)$.

30. Taking the limit along the curve $y^2=mx$, we have

$$\lim_{\substack{x\to0\\y\to0}} f(x,y) = \lim_{\substack{x\to0\\y\to0}} \frac{xy^2}{x^2+y^2}$$

$$= \lim_{x\to0} \frac{mx^2}{x^2+m^2.x^2} \quad (\because y^2 = mx)$$

$$\lim_{x\to0} \frac{m}{1+m^2} = \frac{m}{1+m^2}$$

which depends upon m and hence not unique

$$\Rightarrow \lim_{(x,y)\to(0,0)} f(x,y) \text{ does not exist.}$$

Hence, $f(x,y)$ is discontinuous at $(0, 0)$.

Further, let $\varepsilon>0$ be given and $x=r\cos\theta$, $y=r\sin\theta$, then

$$f(r\cos\theta, r\sin\theta) = \frac{r^3\cos\theta\sin^2\theta}{r^2\cos^2\theta+r^4\sin^4\theta}$$

$$= \frac{r\cos\theta\sin^2\theta}{\cos^2\theta+r^2\sin^4\theta}$$

Now,

$$|f(r\cos\theta, r\sin\theta)-f(0,0)|$$

$$= |f(r\cos\theta, r\sin\theta)|$$

$$= \left|\frac{r\cos\theta\sin^2\theta}{\cos^2\theta+r^2\sin^4\theta}\right|$$

If $\theta=\dfrac{\pi}{2}$, then $|f(r\cos\theta, r\sin\theta)-f(0,0)=0|$

$\Rightarrow f(x, y)$ is continuous along the radius vector $\theta=\dfrac{\pi}{2}$.

31. We have

$$\lim_{\substack{x\to0\\y\to0}} f(x,y)= \lim_{\substack{x\to0\\y\to0}} \frac{xy^2+x^2y}{x^3+y^3}$$

$$= \lim_{x\to0} \frac{m^2x^3+mx^3}{x^3\left(1+m^3\right)} \quad \text{(along the line } y=mx\text{)}$$

$$= \frac{m^3+m}{1+m^3}$$

which depends upon m and hence not unique.

Therefore, $\displaystyle\lim_{\substack{x\to0\\y\to0}} f(x,y)$ does not exist.

Hence, $f(x, y)$ is not continuous at $(0,0)$.

32. Let a be rational, then

$$\lim_{\substack{x\to0\\y\to0}} f(x,y)= \lim_{\substack{x\to a\\y\to b}} 0$$

$$(\because f(x, y)=0 \text{ if } x \text{ is rational})$$

$$= 0.$$

Now, if a is irrational, then we have

$$\lim_{\substack{x\to0\\y\to0}} f(x,y)=1$$

$$(\because f(x, y)=1 \text{ if } x \text{ is irrational})$$

$\Rightarrow \lim\limits_{(x,y)\to(a,b)} f(x,y)$ is not unique.

Hence, $f(x, y)$ is not continuous at (a, b).

39. $\lim\limits_{x\to 0} f(x,y) = \lim\limits_{x\to 0} \dfrac{2.x.mx}{x^2 + m^2 x^2}$

$= \lim\limits_{x\to 0} \dfrac{2m}{1+m^2}$,depend upon m.

$\Rightarrow \lim\limits_{(x,y)\to(0,0)} f(x,y)$ does not exist.

$\Rightarrow f(x,y)$ is discontinuous at $(0,0)$

Also $f_x(0,0) = \lim\limits_{h\to 0} \dfrac{f(0+h,0) - f(0,0)}{h}$

$\qquad\qquad = \lim\limits_{h\to 0} \dfrac{f(h,0)}{h} = \lim\limits_{h\to 0} 0 = 0$

Similarly, $f_y(0,0) = 0$

41. $\lim\limits_{(x,y)\to(0,0)} f(x,y) = \lim\limits_{(x,y)\to(0,0)} x^2 + y^2 = 0$

Also $f(0,0) = 0$ (given)

$\Rightarrow \lim\limits_{(x,y)\to(0,0)} f(x,y) = f(0,0)$

$\Rightarrow f(x,y)$ is continuous at $(0,0)$.

Now $f(0+h, y+k) - f(0,0)$

$= 0.h + 0.k + \sqrt{h^2+k^2}.\sqrt{h^2+k^2}$

$\qquad \phi(h,k) = \sqrt{h^2 + k^2}$

and $\lim\limits_{(h,k)\to(0,0)} \phi(h,k)$

$= \lim\limits_{(h,k)\to(0,0)} \sqrt{h^2 + k^2} = 0$

$\Rightarrow f(x,y)$ is continuous as well as differentiable

at $(0,0)$.

45. $\lim\limits_{(x,y)\to(1,2)} f(x,y) = \lim\limits_{(x,y)\to(1,2)} x^2 + 2y$

but $f(1, 2) = 0$

$\Rightarrow \lim\limits_{(x,y)\to(1,2)} f(x,y) \neq (1,2)$

$\Rightarrow f(x,y)$ is discontinuous at $(1,2)$

Further if we define

$$f(x,y) = \begin{cases} x^2 + 2y & ;(x,y) \neq (1,2) \\ 5 & ;(x,y) = (1,2) \end{cases}$$

$\Rightarrow f(x,y)$ is continuous

$\Rightarrow f(x,y)$ has a removable discontinuity at $(1,2)$.

49. Putting $y=mx$, we have $\lim\limits_{x\to 0} f(x,y) = \dfrac{m}{1+m^2}$ which depand upon m.

$\Rightarrow f(x, y)$ is not continuous at $(0,0)$

Now $f_x(0,0) = \lim\limits_{h\to 0} \dfrac{f(0,0+h) - f(0,0)}{h}$

$\qquad\qquad = \lim\limits_{h\to 0} \dfrac{f(h,0)}{h} = \lim\limits_{x\to 0} = 0$

and $f_y(0,0) = \lim\limits_{k\to 0} \dfrac{f(0,0+k)}{k}$

$\qquad\qquad = \lim\limits_{k\to 0} \dfrac{f(0,0+k)}{k} = 0$

$\Rightarrow f_x$ and f_y exist at origin.

50. We have $\left| \dfrac{xy}{\sqrt{x^2+y^2}} - 0 \right| = \dfrac{|xy|}{\sqrt{x^2+y^2}}$

Also, $x^2 + y^2 - |xy|$

$= \left(|x| - \dfrac{1}{2}|y| \right)^2 + \dfrac{1}{2}|y^2| \geq 0$

$\Rightarrow \qquad x^2 + y^2 - |xy| \geq 0$

$\Rightarrow \qquad \dfrac{|xy|}{\sqrt{x^2+y^2}} \leq \sqrt{x^2+y^2} \quad \forall\ (x,y) \neq (0,0)$

Therefore, $\left| \dfrac{xy}{\sqrt{x^2+y^2}} - 0 \right| < \varepsilon$ if $\sqrt{x^2+y^2} < \varepsilon$

$\Rightarrow f$ is continuous at $(0,0)$

we can easily see that $f_x(0,0) = 0 = f_y(0,0)$

$\Rightarrow$ partial derivative f_x and f_y exist at origin similarly we can prove that f is not differentiable.

Problem Set-3

1. We have $\dfrac{\partial u}{\partial y}=0$

or $\cos x \sin y \sin (x+y)$

$\qquad +\sin x \sin y \cos (x+y)=0$ ...(1)

$\dfrac{\partial u}{\partial y}=0$ or $\sin x \cos y \sin(x+y)$

$\qquad +\sin x \sin y \cos(x+y)=0.$ ...(2)

From (1), $\tan(x+y)=-\tan x.$

From (2), $\tan(x+y)=-\tan y.$

$\therefore \quad \tan x = \tan y$ or $x = y.$

Hence $\tan 2x+\tan x=0,\ \dfrac{\sin(2x+x)}{\cos 2x \cos x}=0,$

$\therefore \quad \sin 3x=0,\quad$ or $\quad 3x=0$ or $\pi;$

or $\qquad x=0\quad$ or $\quad \dfrac{\pi}{3}.$

$\therefore$ For max. or min., we have

$$x = y = 0 \text{ or } x = y = \frac{\pi}{3}.$$

In order to discriminate for max. and min. we

have to find the sign of $rt-s^2$, when $x=\dfrac{\pi}{3}=y.$

$\dfrac{\partial u}{\partial x}=\sin y \sin (x+x+y),$

$\dfrac{\partial u}{\partial y}=\sin x.\sin(x+y+y).$

$\therefore \quad r=\dfrac{\partial^2 u}{\partial x^2}=2\sin y \cos (2x+y)$

$\qquad =2\sin \dfrac{\pi}{3}\cos \pi=-\sqrt{\varepsilon}$

$t=\dfrac{\partial^2 u}{\partial y^2}=2\sin x \cos (x+2y)$

$\qquad =2\sin \dfrac{\pi}{3}\cos \pi=-\sqrt{\varepsilon}$

$s=\dfrac{\partial^2 u}{\partial x \partial y}$

$\qquad =\cos y \sin(2x+y)+\sin y \cos (2x+y)$

$\qquad =\sin(2x+2y)$

$\qquad =\sin \dfrac{4\pi}{3}=\sin\left(\pi+\dfrac{\pi}{3}\right)=-\sin \dfrac{\pi}{3}=-\dfrac{\sqrt{3}}{2}$

$\therefore \qquad rt-s^2=3-\dfrac{3}{4}=\dfrac{9}{4}$

i.e., $+$ ive and $r=-\sqrt{\varepsilon}$ i.e.,$-$ive.

$\therefore \qquad u$ is max. at $x=y=\dfrac{\pi}{3}$ and its value

is $\sin \dfrac{\pi}{3}\sin \dfrac{\pi}{3}\sin \dfrac{2\pi}{3}=\dfrac{3\sqrt{(3)}}{8}.$

2. We can write $u=\sin x+\sin y+\cos(x+y).$

$\dfrac{\partial u}{\partial x}=\cos x-\sin(x+y)=0,$

$\dfrac{\partial u}{\partial y}=\cos y-\sin(x+y)=0$ ...(1)

$\therefore \qquad \cos x=\cos y,\qquad (\because x = y)$

Putting in (1), we get

$\qquad \cos x-\sin 2x=0$

or $\cos x.(1-2\sin x)=0.$

If $\cos x=0$, then $x=\dfrac{\pi}{2},\dfrac{3\pi}{2},\dfrac{5\pi}{2},...,(2n+1)\dfrac{\pi}{2}$

If $\sin x=\dfrac{1}{2}$, then $x=\dfrac{\pi}{6}$ and in general

$n\pi+(-1)^n\dfrac{\pi}{6}$

$r=\dfrac{\partial^2 u}{\partial x^2}=-\sin x-\cos(x+y).$

$t=\dfrac{\partial^2 u}{\partial y^2}=-\sin y-\cos(x+y),$

$s=\dfrac{\partial^2 u}{\partial x \partial y}=-\cos(x+y).$

Now when $x=\dfrac{\pi}{2}=y$, then $r=0, t=0,$

$\therefore \qquad$ neither max. nor min.

When $x=\dfrac{3\pi}{2}=y$, then $r=2, t=2, s=1.$

$\therefore \qquad rt-s^2=4-1=3$

i.e., $+$ive and also $r=\ +$ive, therefore min. at

$x=y=\dfrac{3\pi}{2}.$

When $x=y=\dfrac{\pi}{6}$ and in general $n\pi+(-1)^n\dfrac{\pi}{6}$,

then $r=-1,$

$t=-1, s=-\dfrac{1}{2}$ and $rt-s^2=1-\dfrac{1}{4}$, i.e., $+$ive.

Also r is $-$ive : therefore u is max.

3. We have $\dfrac{\partial u}{\partial x}=0$ and $\dfrac{\partial u}{\partial y}=0$ give $x=y=a$ for

which $u=3a^2.$

Also $r=2, s=1, t=2;$

$\therefore rt-s^2=+$ive and r being $+$ive, hence min.

4. We have

$\dfrac{\partial u}{\partial x}=3x^2y^2(1-x-y)-x^3y^2=0$ ...(1)

$$\frac{\partial u}{\partial y} = 2x^3 y(1-x-y) - x^3 y^2 = 0. \qquad \ldots(2)$$

Subtracting, we get

$$x^2 y(1-x-y)(3y-2x) = 0; \quad \therefore y = \frac{2}{3}x$$

Putting in (1), we get

$$3x^2 \frac{4}{9} x^2 \cdot \left(1 - \frac{5}{3}x\right) - \frac{4}{9}x^5 = 0$$

or $\dfrac{4}{9}x^4 (3 - 5x - x) = 0$

$\therefore x = \dfrac{1}{2}$ and hence $y = \dfrac{1}{3}$

$$r = \frac{\partial^2 u}{\partial x^2} = 6xy^2 - 12x^2 y^2 - 6xy^3$$

$$= 6xy^2 (1 - 2x - y) = -\frac{1}{9} \text{ at } \left(\frac{1}{2}, \frac{1}{3}\right)$$

$$t = \frac{\partial^2 u}{\partial y^2} = 2x^3 - 2x^4 - 4x^3 y - 2x^3 y$$

$$= x^3 (2 - 2x - 6y) = -\frac{1}{8}$$

$$s = \frac{\partial^2 u}{\partial x \partial y} = 6x^2 y - 8x^3 y - 9x^2 y^2 = -\frac{1}{12}$$

Now $rt - s^2$ is +ive and r –ive and; hence maximum value when

$$x = \frac{1}{2} \text{ and } y = \frac{1}{3};$$

$$\therefore \quad u = \frac{1}{2^3} \cdot \frac{1}{3^2} \left(1 - \frac{1}{2} - \frac{1}{3}\right) = \frac{1}{432} \text{ max.}$$

5. We have $\dfrac{\partial u}{\partial x} = 2a^2 y - 6axy + 3x^2 y + y^3 = 0$

or $2a^2 - 6ax + 3x^2 + y^2 = 0$ $\ldots(1)$

$$\frac{\partial u}{\partial y} = 2a^2 x - 3ax^2 - 3ay^2 + 3xy^2 = 0 \qquad \ldots(2)$$

Multiplying (1) by x and subtracting from (2), we get

$$3ax^2 - 3ay^2 - 2x^3 + 2xy^2 = 0.$$

or $(x^2 - y^2)(3a - 2x) = 0$

$\therefore \quad x = \dfrac{3a}{2}$ and $x = \pm y$.

If $x = \dfrac{3a}{2}$, then from (1), $y = \pm \dfrac{a}{2}$

If $x = y$, then from (1), $2a^2 - 6ay + 4y^2 = 0$

$\therefore \qquad y = a, \dfrac{a}{2}$

If $x = -y$, then from (1), $2a^2 + 6ay + 4y^2 = 0$

$\therefore \qquad y = -a, -\dfrac{a}{2}$

Hence we have the following stationary points:

$$\left(\frac{3a}{2}, \frac{a}{2}\right), \left(\frac{3a}{2}, -\frac{a}{2}\right), (a, a), \left(\frac{a}{2}, \frac{a}{2}\right),$$

$$(a, -a), \left(\frac{a}{2}, -\frac{a}{2}\right)$$

$$r = \frac{\partial^2 u}{\partial x^2} = -6ay + 6xy, \ t = \frac{\partial^2 u}{\partial y^2} = -6ay + 6xy$$

$$s = \frac{\partial^2 u}{\partial x \partial y} = 2a^2 - 6ax + 3x^2 + 3y^2$$

$$r = \frac{3}{2}a^2, s = \frac{a^2}{2}, t = \frac{3}{2}a^2 \text{ at } \left(\frac{3a}{2}, \frac{a}{2}\right) \text{ and}$$

$rt - s^2$ is +ive, $\therefore$ min.

Similarly max. at $\left(\dfrac{3a}{2}, -\dfrac{a}{2}\right)\left(\dfrac{a}{2}, \dfrac{a}{2}\right)$

Min. at $\left(\dfrac{a}{2}, \dfrac{-a}{2}\right)$.

Neither max. nor min. at (a, a) and $(a, -a)$ as $rt - s^2$ will be –ive.

6. For max. or min. of y we must have $\dfrac{dy}{dx} = 0$.

Differentiating the given equation, we get

$$3x^2 + 3y^2 \frac{dy}{dx} = 3ay + 3ax \frac{dy}{dx}. \qquad \ldots(1)$$

Putting $\dfrac{dy}{dx} = 0$ for max. or min. of y, we have

$$x^2 = ay \qquad \ldots(2)$$

Putting for y from (2) in the given equation

$$x^3 + \frac{x^6}{a^3} = 3ax \cdot \frac{x^2}{a} \text{ or } x^3 (x^3 - 2a^3) = 0.$$

$\therefore \qquad x = 0 \quad$ or $\quad 2^{1/3} a$.

In order to discriminate for max. or min. we have to find the value of $\dfrac{d^2 y}{dx^2}$ at the above points.

Differentiating (1) again w.r.t. x, we get

$$6x + 6y \left(\frac{dy}{dx}\right)^2 + 3y^2 \frac{d^2 y}{dx^2} = 6a \frac{dy}{dx} + 3ax \frac{d^2 y}{dx^2}$$

Put $\dfrac{dy}{dx} = 0$, as y is to be max. or min., we have

$$6x = (3ax - 3y^2) \frac{d^2 y}{dx^2}$$

or $\quad \dfrac{d^2y}{dx^2} = \dfrac{2x}{ax - y^2}$

or $\quad \dfrac{d^2y}{dx^2} = \dfrac{2x}{ax - \dfrac{x^2}{a^2}} = \dfrac{2a^2}{a^3 - x^3}$

At $x = 0$, $\dfrac{d^2y}{dx^2} = \dfrac{2}{a} = +$ive;

$\therefore$ y is min.

At $x = 2^{1/3}a$ or $x^3 = 2a^3$, $\dfrac{d^2y}{dx^2} = \dfrac{-2}{a} = -$ ive,

$\therefore$ y is max.

Also from (2) we can find

$$y = \dfrac{x^2}{a} = \dfrac{2^{2/3}a^2}{a} = 2^{2/3}a$$

7. We have $\dfrac{\partial u}{\partial x} = 2xy^2 - 10x - 8y = 0$,

$\dfrac{\partial u}{\partial y} = 2x^2y - 8x - 10y = 0$

Subtracting, we get

$\quad 2xy(y-x) + 2(y-x) = 0$

or $\quad (y-x)(xy+1) = 0$

Either $\quad y = x$ or $\quad y = -\dfrac{1}{x}$

when $y = x$ then $2x^3 - 18x = 0$;

$\therefore \qquad x = 0, \pm 3$

when $y = -x$ then $\dfrac{2}{x} - 10x + \dfrac{8}{x} = 0$

or $1 - x^2 = 0$ $\therefore$ $x = \pm 1$

when $x = 0, y = 0, x = \pm 3, y = \pm 3$ $\because$ $y = x$

when $x = \pm 1$ then $y = \mp 1$ $\because$ $y = -\dfrac{1}{x}$

$r = \dfrac{\partial^2 u}{\partial x^2} = 2y^2 - 10$, $\quad t = \dfrac{\partial^2 u}{\partial y^2} = 2x^2 - 10$,

$s = \dfrac{\partial^2 u}{\partial x \partial y} = 4xy - 8$,

At $(0,0)$, $r = -10$, $t = -10$, $s = -8$

$rt - s^2 = 100 - 64 = 36 = +$ive.

Also r is –ive and hence max. at $(0,0)$.

At $(\pm 3, \pm 3)$, $r = 8$, $s = 28$, $t = 8$

$\quad rt - s^2 = 64 - 28 \times 28 = -$ive.

Hence there is neither max. nor min. at $(\pm 3, \pm 3)$.

At $(\pm 1, \mp 1)$, $r = -8$, $s = -12$, $t = -8$

$rt - s^2 = 64 - 144 = -$ive.

Hence there is neither max. nor min. at $(\pm 1, \mp 1)$.

8. We have

$$\dfrac{\partial u}{\partial x} = 4(x-y) - 4x^3 = 0,$$

$$\dfrac{\partial u}{\partial y} = -4(x-y) - 4y^3 = 0,$$

Adding, we get $-4(x^3 + y^3) = 0$ which gives $y = -x$

Putting in $\dfrac{\partial u}{\partial x} = 0$, we get $8x - 4x^3 = 0$

$\therefore$ $x = 0$ or $\pm \sqrt{2}$ and hence $y = 0$ or $\mp \sqrt{2}$.

$\therefore$ We have to consider $(0,0)$, $\left(\sqrt{2}, \sqrt{2}\right)$

Now $r = -20$, $t = -20$, $s = -4$ and $rt - s^2 = +$ ive.

$\therefore$ Max. at $x = \pm \sqrt{2}$, $y = \mp \sqrt{2}$.

At $x = 0$, $y = 0$, $rt - s^2 = 0$, Hence a doubtful case.

9. $\dfrac{\partial u}{\partial x} = y^2(3x + 6y - 2) + 3xy^2 = 0$ ...(1)

$\dfrac{\partial u}{\partial y} = 2xy(3x + 6y - 2) + 6xy^2 = 0$. ...(2)

Multiplying (2) by y and putting

$y^2(3x + 6y - 2) = -3xy^2$ from (1), we get

$2x(-3xy^2) + 6xy^3 = 0$ or $6xy^2(x - y) = 0$.

$\therefore$ $x = y$ or $x = 0$, $y = 0$.

Putting $x = y$ in (1), we get $12x^3 - 2x^2 = 0$;

$\therefore \qquad x = \dfrac{1}{6}$, 0.

Hence $(0,0)$ and $\left(\dfrac{1}{6}, \dfrac{1}{6}\right)$ are the points where

there can be max. or min.

$r = \dfrac{\partial^2 u}{\partial x^2} = 3y^2 + 3y^2$,

$t = \dfrac{\partial^2 u}{\partial y^2} = 6x^2 + 36xy$,

$s = \dfrac{\partial^2 u}{\partial x \partial y} = 12xy + 136y^2 - 4y$.

$r = 0$, $t = 0$, $s = 0$ at $(0,0)$. Hence neither max. nor min.

$r = \dfrac{1}{6}, t = \dfrac{7}{6}, s = \dfrac{1}{36}$, $rt-s^2$ is clearly +ive and r

is also +ive. Hence min. at $\left(\dfrac{1}{6}, \dfrac{1}{6}\right)$.

10. Using (9) and solving $\dfrac{\partial u}{\partial x} = 0$, $\dfrac{\partial u}{\partial y} = 0$, we get

(0,0) and $\left(\dfrac{2}{3}, -\dfrac{4}{3}\right)$; $r=6, t=2, s=4$ at (0, 0),

but $rt-s^2$ is –ive and hence neither max. nor min.

$r=10, t=2, s=4$ at $\left(\dfrac{2}{3}, -\dfrac{4}{3}\right)$, and $rt-s^2$ is +ive

and r is +ive and hence min.

11. $\dfrac{\partial u}{\partial x} = \dfrac{\left(x^2+y^2\right)-(x+y-1).2x}{\left(x^2+y^2\right)^2}$

$= \dfrac{y^2-x^2-2xy+2x}{\left(x^2+y^2\right)^2} = 0$

$\dfrac{\partial u}{\partial y} = \dfrac{\left(x^2+y^2\right)-(x+y-1).2y}{\left(x^2+y^2\right)^2}$

$= \dfrac{x^2-y^2-2xy+2y}{\left(x^2+y^2\right)^2} = 0$

Subtracting, we get (0,0) and (1,1).

Calculate $r = \dfrac{\partial^2 u}{\partial x^2} = -\dfrac{1}{2}$ at (1,1)

Similarly find t and s then $rt-s^2=+$ive and also r –ive.

$\therefore$ Max. at (1,1)

13. We have

$\dfrac{\partial u}{\partial x} = y^2 z^3 (a-2x-y-z) = 0;$

$\therefore \qquad 2x+y+z = a$ ...(1)

$\dfrac{\partial u}{\partial y} = xyz^3 (2a-2x-2z-3y) = 0;$

$\therefore \qquad 2x+3y+2z = 2a$ ...(2)

$\dfrac{\partial u}{\partial z} = xy^2 z^2 (3a-3x-4z-3y)$

$\therefore \qquad 3x+3y+4z = 3a$ ...(3)

Adding, we get $7(x+y+z) = 6a$;

$\therefore \qquad x+y+z = \dfrac{6a}{7}$

Putting in the relations (1), (2) and (3), we get

$x + \dfrac{6a}{7} = a \qquad \therefore\ x = \dfrac{a}{7}$

2. $\dfrac{6a}{7} + y = 2a, \qquad \therefore\ y = \dfrac{2a}{7}$

3. $\dfrac{6a}{7} + z = 3a, \qquad \therefore\ z = \dfrac{3a}{7}$

$A = \dfrac{\partial^2 u}{\partial x^2} = -2y^2 z^2 = -\dfrac{216a^5}{7^5}$

$B = \dfrac{\partial^2 u}{\partial y^2} = -\dfrac{162a^5}{7^5}$, $H = \dfrac{108a^5}{7^5}$

Now A is –ive and $\begin{vmatrix} A & H \\ H & B \end{vmatrix}$, i.e., $AB-H^2$ is +ive

and it may be shown that $\begin{vmatrix} A & H & G \\ H & B & F \\ G & F & C \end{vmatrix}$ is –ive.

Therefore u is max. when x, y, z have values written above

$\therefore\ 108\dfrac{a^7}{7^7}$ is the max. value of u.

14. We have $\dfrac{\partial u}{\partial x} = -20x^3 + 20x^4 = 0;\ \therefore\ x = 0,1.$

$\dfrac{\partial u}{\partial y} = 2y = 0\ \therefore\ y = 0, \dfrac{\partial u}{\partial z} = 4z = 0;\ \therefore\ z = 0.$

Hence (0,0,0) and (1,0,0).

Consider (1,0,0),

$A = \dfrac{\partial^2 u}{\partial x^2} = -60x^2 + 80x^3 = 20,$

$B = \dfrac{\partial^2 u}{\partial y^2} = 2, \qquad C = \dfrac{\partial^2 u}{\partial z^2} = 4,$

$F = \dfrac{\partial^2 u}{\partial y \partial z} = 0,\ \ G = \dfrac{\partial^2 u}{\partial z \partial x} = 0,\ H = \dfrac{\partial^2 u}{\partial x \partial y} = 0.$

At (1,0,0), $A = 20$ i.e., +ive.

$\begin{vmatrix} A & H \\ H & B \end{vmatrix}$ i.e., $AB-H^2 = 40$ i.e., +ive

and $\begin{vmatrix} A & H & G \\ H & B & F \\ G & F & C \end{vmatrix} = 160$ i.e., +ive

Since all the three are +ive, hence there is Min. at (1,0,0).

At (0,0,0) all the three expressions are zero and hence there is neither max. nor min. at (0,0,0).

15. $\dfrac{\partial u}{\partial x}=0$ gives $8yz=(x+y+z)^2-1.$ ...(1)

$\dfrac{\partial u}{\partial y}=0$ gives $8zx=(x+y+z)^2-1.$

$\dfrac{\partial u}{\partial z}=0$ gives $8xy=(x+y+z)^2-1$

$\therefore$ $yz=zx=xy$ or $\dfrac{1}{x}=\dfrac{1}{y}=\dfrac{1}{z}$ or $x=y=z.$

Putting in (1), we get $8x^2=9x^2-1$;

$x=\pm1.$

Hence the critical values are $x=y=z=1$ or

$x=y=z=-1$

$A=\dfrac{\partial^2 u}{\partial x^2}=6(x+y+z)=18=B=C$ at $(1,1,1)$

$=-18=B=C$ at $(-1,-1,-1)$

$F=\dfrac{\partial^2 u}{\partial y\partial z}=6(x+y+z)=-24x=-6=G=H$

at $(1,1,1)$

$=6=H=G$ at $(-1,-1,-1).$

For the point $(1,1,1),$

$A=18,$ $\begin{vmatrix} A & H \\ H & B \end{vmatrix}=\begin{vmatrix} 18 & -6 \\ -6 & 18 \end{vmatrix}=288$ i.e., $+$ive

and $\begin{vmatrix} A & H & G \\ H & B & F \\ G & F & C \end{vmatrix}$ can be shown to be $+$ive

$=6912$

Since all the three are $+$ive, hence u is min. at $(1,1,1).$

Again for the point $(-1,-1,-1)$ it can be shown as above that the three expressions as found above are alternately $-$ive and $+$ive, i.e., $-18,$ 288 and $-1728.$ Therefore, u is max. at $(-1,-1,-1)$

16. Proceed as usual, we get $A=B=C=2,$ $F=0,$ $G=0$, $H=-1$ etc.

Here $A=2,$ $\begin{vmatrix} A & H \\ H & B \end{vmatrix}=\begin{vmatrix} 2 & -1 \\ -1 & 2 \end{vmatrix}=3$

$\begin{vmatrix} A & H & G \\ H & B & F \\ G & F & C \end{vmatrix}=\begin{vmatrix} 2 & -1 & 0 \\ -1 & 2 & 0 \\ 0 & 0 & 2 \end{vmatrix}=6$

Since all the three are $+$ive we have a minimum

at the point $\left(-\dfrac{2}{3},\dfrac{-1}{3},1\right).$

17. For max. or min., we have

$\dfrac{\partial u}{\partial x}=y(z-h)\left[\dfrac{x^2}{a^2}+\dfrac{y^2}{b^2}-\dfrac{z^2}{c^2}\right]$

$+xy(z-h)\dfrac{2x}{a^2}=0$

or $\quad y(z-h)\left[\dfrac{3x^2}{a^2}+\dfrac{y^2}{b^2}-\dfrac{z^2}{c^2}\right]=0$ ...(1)

$\dfrac{\partial u}{\partial y}=x(z-h)\left[\dfrac{x^2}{a^2}+\dfrac{3y^2}{b^2}-\dfrac{z^2}{c^2}\right]=0$ ...(2)

$\dfrac{\partial u}{\partial z}=xy\left[\dfrac{x^2}{a^2}+\dfrac{y^2}{b^2}-\dfrac{z^2}{c^2}\right]$

$+xy(z-h)\left(-\dfrac{2z}{c^2}\right)=0$

or $\quad xy\left[\dfrac{x^2}{a^2}+\dfrac{y^2}{b^2}-\dfrac{3z^2}{c^2}+\dfrac{2hz}{c^2}\right]=0$ ...(3)

Above gives either $x=0,y=0,$ $z=h$

or $\quad \dfrac{3x^2}{a^2}+\dfrac{y^2}{b^2}-\dfrac{z^2}{c^2}=0$ ...(4)

$\dfrac{x^2}{a^2}+\dfrac{3y^2}{b^2}-\dfrac{z^2}{c^2}=0$ ...(5)

$\dfrac{x^2}{a^2}+\dfrac{y^2}{b^2}-\dfrac{3z^2}{c^2}+\dfrac{2hz}{c^2}=0$ ...(6)

Subtracting (4) and (5), we get

$\dfrac{x^2}{a^2}=\dfrac{y^2}{b^2};$ $\therefore \dfrac{z^2}{c^2}=\dfrac{4x^2}{a^2}.$

Hence from (6),

$\dfrac{x^2}{a^2}+\dfrac{x^2}{a^2}-\dfrac{12x^2}{a^2}+\dfrac{2h}{c}\left(\dfrac{2x}{a}\right)=0$

or $\quad \dfrac{10x}{a}=\dfrac{4h}{c}$ or $\dfrac{x}{a}+\dfrac{2}{5}\dfrac{h}{c}$

$\therefore \quad \dfrac{x}{a}=\dfrac{2}{5}\dfrac{h}{c}=\dfrac{y}{b}=\dfrac{z}{2c}.$

Putting in $u,$ we get

$u=\left(\dfrac{2}{5}h\dfrac{a}{c}\right)\left(\dfrac{2}{5}h.\dfrac{b}{c}\right)\left(\dfrac{4}{5}h-h\right)$

$\left[\dfrac{4}{25}+\dfrac{4}{25}-\dfrac{16}{25}\right]\dfrac{h^2}{c^2}$

Above gives us max. value of u as

$$A = \frac{\partial^2 u}{\partial x^2} = y(z-h)\frac{6x}{a^2} = \text{-ive.}$$

At the other point $(0,0,h)$, A is zero. Hence neither max. nor min.

18. We have

$$u = \Sigma[(x-x_1)^2 + (y-y_1)^2 + (z-z_1)^2]$$

for max. or min.

$$\frac{\partial u}{\partial x} = 0,$$

$$\frac{\partial u}{\partial y} = 0,$$

$$\frac{\partial u}{\partial z} = 0$$

$$\therefore \quad x = \frac{\Sigma x_1}{n}, y = \frac{\Sigma y_1}{n}, z = \frac{\Sigma z_1}{n}$$

and $A = \dfrac{\partial^2 u}{\partial x^2} = 2n$, $B = 2n$, $C = 2n$,

$$F = \frac{\partial^2 u}{\partial y \partial z} = 0, G = 0, H = 0$$

$$\therefore \quad A, \begin{vmatrix} A & H \\ H & B \end{vmatrix} \text{ and } \begin{vmatrix} A & H & G \\ H & B & F \\ G & F & C \end{vmatrix} \text{ are all + ive}$$

and hence u is min.

Hence (x,y,z) is the centroid of given points.

19. If the given expression be u, then u will be max. when $\log u$ is max.

Let $V = \log u$.

$$\therefore \quad V = \log(ax+by+cz) - (\alpha^2 x^2 + \beta^2 y^2 + \gamma^2 z^2)$$

$$\frac{\partial V}{\partial x} = \frac{a}{ax+by+cz} - 2\alpha^2 x = 0 \qquad \text{...(1)}$$

$$\frac{\partial V}{\partial y} = \frac{b}{ax+by+cz} - 2\beta^2 y = 0 \qquad \text{...(2)}$$

$$\frac{\partial V}{\partial z} = \frac{c}{ax+by+cz} - 2\gamma^2 z = 0 \qquad \text{...(3)}$$

$$x(ax+by+cz) = \frac{a}{2\alpha^2}, y(ax+by+cz) = \frac{b}{2\beta^2}$$

$$z(ax+by+cz) = \frac{c}{2\gamma^2}$$

Multiplying both sides of above by a, b, and c respectively and adding, we get

$$(ax+by+cz)^2 = \frac{1}{2}\left(\frac{a^2}{\alpha^2} + \frac{b^2}{\beta^2} + \frac{c^2}{\gamma^2}\right) = A^2 \text{ say}$$

$$\text{...(4)}$$

$$\therefore \quad ax+by+cz = A.$$

Hence from (1), (2) and (3), we get

$$x = \frac{a}{2\alpha^2 A}, y = \frac{b}{2\beta^2 A}, z = \frac{c}{2\gamma^2 A} \qquad \text{...(5)}$$

Putting the values of x, y and z in V,

$$V = \log A - \left(\frac{a^2}{4\alpha^2 A^2} + \frac{b^2}{4\beta^2 A^2} + \frac{c^2}{4\gamma^2 A^2}\right)$$

or $\log u - \log A = -\dfrac{1}{4A^2}2A^2 = -\dfrac{1}{2}$ from (4)

$$\therefore \quad \log \frac{u}{A} = -\frac{1}{2}$$

$$\therefore \quad u = Ae^{-1/2} = \frac{A}{\sqrt{e}} = \left[\frac{A^2}{e}\right]^{1/2}$$

or $\quad u = \left[\dfrac{1}{2e}\left(\dfrac{a^2}{\alpha^2} + \dfrac{b^2}{\beta^2} + \dfrac{c^2}{\gamma^2}\right)\right]^{1/2}$ From (4)

Also $\dfrac{\partial^2 V}{\partial x^2} = \left[-\dfrac{a^2}{(ax+by+cz)^2} - 2\alpha^2\right]$ i.e., –ive

Hence V is max. or u is max.

20. $r = \dfrac{a^4}{8}, t = -\dfrac{a^4}{8}$ and $s = \dfrac{a^4}{12}$

at $\left(\dfrac{a}{2}, \dfrac{a}{3}\right)$ and $rt - s^2 = +$ ive, hence max. because r is –ive.

21. Since x, y, z are the angles of a triangle,

$$\therefore \quad x+y+z = \pi.$$

Hence $u = \sin x \sin y \sin z.$ ...(1)

$\therefore$ Differentiating (1) logarithmically, we have

$$\frac{1}{u}du = \cot x \, dx + \cot y \, dy + \cot z \, dz = 0 \qquad \text{...(3)}$$

and $dx + dy + dz = 0.$ ...(4)

Multiplying (3) by 1 and (4) by λ and adding and equating to zero the coefficients of dx, dy and dz, we get

$$\cot x + \lambda = 0, \cot y + \lambda = 0, \cot z + \lambda = 0. \quad \text{...(5)}$$

$\therefore$ cot x=cot y=cot z

or $x=y=z=\dfrac{\pi}{3}$ by (2)

22. $du=a^2x\,dx+b^2y\,dy+c^2z\,dz=0,$

$df_1=x\,dx+y\,dy+z\,dz=0.$

$df_2=l\,dx+m\,dy+n\,dz=0.$

Multiplying them by 1, λ_1 and λ_2 and adding and equating to zero the coefficients of dx, dy and dz, we get

$a^2x+x\lambda_1+l\lambda_2=0,$

$b^2y+y\lambda_1+m\lambda_2=0,$

$c^2z+z\lambda_1+n\lambda_2=0.$

Multiplying the above by x, y and z respectively and adding, we get

$u+\lambda_1.1+\lambda_2(0)=0;$ $\therefore \lambda_1=-u.$

Putting for λ_1 in the above, we get

$x(u-a^2)=l\lambda_2$

$\therefore$ $x=\dfrac{l\lambda_2}{u-a^2}, y=\dfrac{m\lambda_2}{u-b^2}, z=\dfrac{n\lambda_2}{u-c^2}.$

Putting in $lx+my+nz=0$, we get

$$\dfrac{l^2}{u-a^2}+\dfrac{m^2}{u-b^2}+\dfrac{n^2}{u-c^2}=0$$

Above gives the max. or min. values of u.

23. Proceeding as usual we have the following equations :

$$\left.\begin{array}{l}\dfrac{x}{a^4}+\lambda_1 l+\lambda_2\dfrac{x}{a^2}=0,\\[2mm]\dfrac{y}{b^4}+\lambda_1 m+\lambda_2\dfrac{y}{b^2}=0,\\[2mm]\dfrac{z}{c^4}+\lambda_1 n+\lambda_2\dfrac{z}{c^2}=0,\end{array}\right\}\quad\dots(1)$$

Multiplying the equations (1) by x, y and z and adding, we get

$u+\lambda_1(0)+\lambda_2(1)=0,$

$\therefore$ $\lambda_2=-u$

Hence from (1), we get

$$\dfrac{x}{a^4}\left(1+\lambda_2a^2\right)=-\lambda_1 l$$

$\therefore$ $x=-\dfrac{\lambda_1 la^4}{1-a^2u}$ $\because \lambda_2=-u$

Similarly, $y=\dfrac{-\lambda_1 mb^4}{1-b^2u}$

and $z=\dfrac{-\lambda_1 nc^4}{1-c^2u}$

Putting the values of x, y and z in $lx+my+nz=0$, we find that the max and min value of x, y and z are given by

$$\dfrac{l^2a^4}{1-a^2u}+\dfrac{m^2b^4}{1-b^2u}+\dfrac{n^2c^4}{1-c^2u}=0$$

24. Proceeding as usual, we shall prove that

$x=y=z=c.$

25. Let $u=(x+1)(y+1)(z+1)$

where $x\log a+y\log b+z\log c=\log A$ ...(1)

$du=(y+1)(z+1)dx+(z+1)(x+1)dy$

$\qquad\qquad +(x+1)(y+1)dz=0$ ...(2)

$\log a\,dx+\log b\,dy+\log c\,dz=0.$ ...(3)

Multiplying (2) and (3) by 1 and λ respectively and adding and equating to zero the coefficients of dx, dy and dz, we get

$$\left.\begin{array}{l}(y+1)(z+1)+\lambda\log a=0,\\(z+1)(x+1)+\lambda\log b=0,\\(x+1)(y+1)+\lambda\log c=0.\end{array}\right\}\quad\dots(4)$$

Multiplying the equation (4) by $(x+1)$, $(y+1)$ and $(z+1)$ respectively and adding we get

$3u+\lambda[(x+1)\log a+(y+1)\log b$

$\qquad\qquad +(z+1)\log c]=0$

or $3u+\lambda(\log A+\log a+\log b+\log c)=0,$

by (1)

or $3u+\lambda\log(Aabc)=0;$

$\therefore$ $-\lambda=\dfrac{3u}{\log(Aabc)}$

Again from (4), $(y+1)(z+1)=-\lambda\log a$ etc.

Multiplying, $(x+1)^2(y+1)^2(z+1)^2=-\lambda^3$

$\log a\log b\log c.$

or $\quad u^2 = \dfrac{27u^3}{\left(\log Aabc\right)^3}\log a \log b \log c$, by (5)

$\therefore \quad u = \dfrac{\left(\log Aabc\right)^3}{3\log a.3\log b.3\log c}$

$\qquad = \dfrac{\left[\log\left(Aabc\right)\right]^3}{\log a^3.\log b^3.\log c^3}$

26. Let the radius of the circumcircle be R. Now

area of triangle$= \dfrac{1}{2}$ product of the sides $\times$

$\qquad\qquad\qquad$ (sine of included angle).

$\therefore \quad \Delta ABC = \Delta OBC + \Delta OCA + \Delta OAB$

or $\qquad \Delta = \dfrac{1}{2}R^2(\sin x + \sin y + \sin z) \qquad ...(1)$

where $x+y+z=2\pi$.

$\qquad d\Delta = \dfrac{1}{2}R^2(\cos x + \cos y + \cos z) = 0$

or $\quad \cos x + \cos y + \cos z = 0$

$\qquad df = dx + dy + dz = 0.$

Multiplying above by 1 and λ respectively and adding and equating to zero the coefficients of dx, dy and dz, we get

$\qquad \cos x + \lambda = 0,\ \cos y + \lambda = 0,\ \cos z + \lambda = 0.$

$\therefore \qquad \cos x = \cos y = \cos z = -\lambda$

or $\qquad x = y = z = \dfrac{2\pi}{3}$

Hence the triangle is equilateral.

27. Here $\Delta = \sqrt{\left[s(s-a)(s-b)(s-c)\right]}$

where $2s = a+b+c$ (a, b, c being variables).

Proceeding as usual you will find that

$\qquad s-a = s-b = s-c = -\dfrac{1}{\lambda}$

$\therefore\ a = b = c$ and hence triangle is equilateral. Also regarding c as a function of a and b as usual, you will find that at $a = b = c$,

$\qquad r = \dfrac{-2}{(s-a)^2} = t$ and $s = \dfrac{-1}{(s-a)^2}$

Thus $rt-s^2$ is + ive and r is – ive showing that Δ is max. when $a=b=c$.

28. $f_x(x, y) = 8x^3 - 6xy = 0$ at $(0, 0)$

$f_y\ (x, y) = -3x^2 + 2y = 0$ at $(0,0)$

$f_{yy}\ (x, y) = 2$ at $(0, 0)$

$\therefore$ at $(0, 0)$

$f_x(0,0).f_y(0,0) - [f_{xy}(0,0)]^2 = 0$

$\Rightarrow$ case is doubtful

Further $f_{(x,y)} = (x^2-y)(2x^2-y)$,

$\qquad f(0,0) = 0$

$\Rightarrow\ f(x,y) - f(0,0)$

$\qquad = (x^2-y)(2x^2-y) > 0$

for $y<0$ or $x^2 > y > 0$

$\qquad < 0$ for $y > x^2 > \dfrac{y}{2} > 0$

$\Rightarrow\ f(x,y) - f(0,0)$ does not keep the same sign near the origin,

$\Rightarrow\ f$ has neither maxima nor minima at the origin.

29. We can easily show that at $(0,0)$

$f_x = 0, f_y = 0, f_{xx} = 2, f_{xy} = -2, f_{yy} = 0$

$\Rightarrow\ rt - s^2 = 0$

Now, we have $f(0,0) = 0$

Since $f(x, y)$

$= (x-y)^2 + (x-y)(x^2 + xy + y^2) + x^5$

for points along the line $y - x = 0$, we have

$f(x,y) = x^5$

and accordingly in every nbd of $(0, 0)$ there are points where for $f(x,y)$ is positive and there are points for $f(x, y)$ is negative.

$\Rightarrow f(0, 0)$ is neither a maximum nor a minimum value.

30. Clearly, we have

$$f_x(x, y) = 3x^2 - 63 + 12\,y = 0$$

$$f_y(x,y) = 3y^2 - 63 + 12\,x = 0$$

$$f_{xx}(x,y) = f_{xy}(x,y) = 12$$

$$f_{yy}(x,y) = 6y$$

on solving we get 4 pair of solutions given by

$$(-7,-7), (3,3), (5,-1), (-1,5)$$

At $(-7,-7)$, $f_{xx} = 42 < 0$

$rt - s^2 = 1620 > 0$

$\Rightarrow$ the function is maximum at $(-7,-7)$ further, at $(3,3)$

$f_{xx} = 18 > 0$

and $rt - s^2 = 180 > 0$

$\Rightarrow$ function is minimum at $(3, 3)$.

Chapter

11 Implicit Functions and Jacobian

11.1 INTRODUCTION

Sometimes a function is not exclusively defined in terms of the independent variable and we assumed that a functional equation $f(x,y)=0$ gives y as a function of x. Some times y cannot be expressed in terms of x then we say that function is implicit.

Definition. *Let $f(x,y)$ be a function of two variables and $y=g(x)$ be a function of x such that for every value of x for which $g(x)$ is defined, $f(x,g(x))$ vanish identically, i.e, $y=g(x)$ is a root of the equation $f(x,y)=0$. Then $y=g(x)$ is called the implicit function defined by the functional equation $f(x,y)=0$.*

REMARK

- A functional equation in general may or may not define an implicit function.

11.2 EXISTENCE AND DERIVABILITY OF IMPLICIT FUNCTIONS

If $f_x(x,y)$, $f_y(x,y)$ are continuous in a nbd of (a,b) and $f(a,b)\neq 0$ then $\exists$ a rectangle $R : [a-h,a+h,b-k,b+k]$ about (a,b) such that

(i) for each $x\in[a-h,a+h]$, the equation $f(x,y)=0$ determine unique solution $y=g(x)$ in $(b-k, b+k)$

and (ii) $g'(x)$ is continuous in $[a-h,a+h]$ and $f_y(x,g(x))\neq 0$ and $g'(x)=\dfrac{-f_x\left(x,g(x)\right)}{f_y\left(x,g(x)\right)}$.

11.2.1 GENERAL CASE

If $f(x_1,x_2,...,x_n;y)$ be a function of $(n+1)$ variables $x_1, x_2,...,x_n,y$ and $(a_1,a_2,...,a_n,b)$ be a point of its domain such that

(i) $f(a_1,a_2,...,a_n,b)=0$

(ii) the partial derivatives w.r.t. all $(n+1)$ variables exists and are continuous in a nbd of $(a_1,a_2,...,a_n,b)$

(iii) $f_y(a_1,a_2,...,a_n,b)\neq 0$

then $\exists$ a nbd $(a_1-h_1, a_1+h_1; a_2-h_2, a_2+h_2; ...; a_n-h_n, a_n+h_n; b-k, b+k)$ of $(a_1, a_2, ..., a_n, b)$ such that for every point $(x_1, x_2, ..., x_n)$ of the nbd $R:(a_1-h_1,a_1+h_1;a_2-h_2,a_2+h_2;...;a_n-h_n,a_n+h_n)$.

The equation $f(x_1,x_2,...,x_n,y)=0$ gives only one value $y=g(x_1,x_2,...,x_n)$ in $[b-k,b+k]$ satisfying the following conditions :

(i) $b=g(a_1,a_2,...,a_n)$

(ii) $f(x_1,x_2,...,x_n,g)=0$ for every point $(x_1,x_2,...,x_n)$ in R.

(iii) g is continuous and having continuous first order partial derivatives w.r.t. $x_1,x_2,...x_n$ in R.

If the equation $f(x,y)=0$ defines y as a function of x. Then derivative $\dfrac{dy}{dx}$ can be obtained simply by differentiating the equation w.r.t. x assuming y as a function of $g(x)$.

Therefore, $\qquad\qquad f_x + f_y \dfrac{dy}{dx} = 0$

$$f_{xx} + f_{xy}\frac{dy}{dx} + \left(f_{xy} + f_{yy}\frac{dy}{dx}\right)\frac{dy}{dx} + f_y\frac{d^2y}{dx^2} = 0$$

or $\qquad\qquad f_{xx} + 2f_{xy}\dfrac{dy}{dx} + f_{yy}\left(\dfrac{dy}{dx}\right)^2 + f_y\dfrac{d^2y}{dx^2} = 0 \qquad$ provided $f_y \neq 0$

In a similar manner, we may find the other higher order derivatives.

11.3 **JACOBIAN**

Here we shall discuss some important definitions related to Jacobian and its properties.

(i) If u and v are the functions of two independent variables x and y, then the determinant

$$\begin{vmatrix} \dfrac{\partial u}{\partial x} & \dfrac{\partial u}{\partial y} \\[2ex] \dfrac{\partial v}{\partial x} & \dfrac{\partial v}{\partial y} \end{vmatrix}$$

is called the Jacobian of u and v with respect to x and y. It is denoted by $\dfrac{\partial(u,v)}{\partial(x,y)}$ or $J(u,v)$.

i.e., $\qquad\qquad \dfrac{\partial(u,v)}{\partial(x,y)} = \begin{vmatrix} \dfrac{\partial u}{\partial x} & \dfrac{\partial u}{\partial y} \\[2ex] \dfrac{\partial v}{\partial x} & \dfrac{\partial v}{\partial y} \end{vmatrix}$

(ii) If u,v and w are the functions of three independent variables x, y and z, then the determinant

$$\begin{vmatrix} \dfrac{\partial u}{\partial x} & \dfrac{\partial u}{\partial y} & \dfrac{\partial u}{\partial z} \\[2ex] \dfrac{\partial v}{\partial x} & \dfrac{\partial v}{\partial y} & \dfrac{\partial v}{\partial z} \\[2ex] \dfrac{\partial w}{\partial x} & \dfrac{\partial w}{\partial y} & \dfrac{\partial w}{\partial z} \end{vmatrix}$$

is called the Jacobian of u, v and w with respect to x, y and z. It is denoted by $\dfrac{\partial(u,v,w)}{\partial(x,y,z)}$ or $J(u,v,w)$.

(iii) If $u_1, u_2, ..., u_n$ are n functions of independent variables $x_1, x_2, ..., x_n$, then the determinant

$$\begin{vmatrix} \dfrac{\partial u_1}{\partial x_1} & 0 & 0 & \cdots & 0 \\[2ex] \dfrac{\partial u_2}{\partial x_1} & \dfrac{\partial u_2}{\partial x_2} & 0 & \cdots & 0 \\[2ex] \vdots & \vdots & \vdots & & \vdots \\[2ex] \dfrac{\partial u_n}{\partial x_1} & \dfrac{\partial u_n}{\partial x_2} & \dfrac{\partial u_n}{\partial x_3} & \cdots & \dfrac{\partial u_n}{\partial x_n} \end{vmatrix}$$

is called the Jacobian of $u_1, u_2, ..., u_n$ with respect to $x_1, x_2, ..., x_n$. It is denoted by

$$\frac{\partial(u_1, u_2, ..., u_n)}{\partial(x_1, x_2, ..., x_n)} \text{ or } J(u_1, u_2, ..., u_n).$$

(iv) If the functions $u_1, u_2, ..., u_n$ of n independent variables $x_1, x_2, ..., x_n$ are of the following form

$$u_1 = f_1(x_1), u_2 = f_2(x_1, x_2), ..., u_n = f_n(x_1, x_2, ..., x_n)$$

Then $\dfrac{\partial(u_1, u_2, ..., u_n)}{\partial(x_1, x_2, ..., x_n)} = \begin{vmatrix} \dfrac{\partial u_1}{\partial x_1} & 0 & 0 & \cdots & 0 \\ \dfrac{\partial u_2}{\partial x_1} & \dfrac{\partial u_2}{\partial x_2} & 0 & \cdots & 0 \\ \vdots & \vdots & \vdots & & \vdots \\ \dfrac{\partial u_n}{\partial x_1} & \dfrac{\partial u_n}{\partial x_2} & \dfrac{\partial u_n}{\partial x_3} & \cdots & \dfrac{\partial u_n}{\partial x_n} \end{vmatrix}$

11.4 IMPORTANT THEOREMS ON JACOBIANS

THEOREM 1. $\dfrac{\partial(u, v)}{\partial(x, y)} \cdot \dfrac{\partial(x, y)}{\partial(u, v)} = 1$

Proof. Let $u = u(x, y)$ and $v = v(x, y)$ be the given functions.
Differentiating partially each of these functions, we get

$$1 = \frac{\partial u}{\partial x} \cdot \frac{\partial x}{\partial u} + \frac{\partial u}{\partial y} \cdot \frac{\partial y}{\partial u}$$

$$0 = \frac{\partial u}{\partial x} \cdot \frac{\partial x}{\partial v} + \frac{\partial u}{\partial y} \cdot \frac{\partial y}{\partial v}$$

and

$$0 = \frac{\partial v}{\partial x} \cdot \frac{\partial x}{\partial u} + \frac{\partial v}{\partial y} \cdot \frac{\partial y}{\partial u}$$

$$1 = \frac{\partial v}{\partial x} \cdot \frac{\partial x}{\partial v} + \frac{\partial v}{\partial y} \cdot \frac{\partial y}{\partial v}$$

Now $\dfrac{\partial(u, v)}{\partial(x, y)} \cdot \dfrac{\partial(x, y)}{\partial(u, v)} = \begin{vmatrix} \dfrac{\partial u}{\partial x} & \dfrac{\partial u}{\partial y} \\ \dfrac{\partial v}{\partial x} & \dfrac{\partial v}{\partial y} \end{vmatrix} \cdot \begin{vmatrix} \dfrac{\partial x}{\partial u} & \dfrac{\partial x}{\partial v} \\ \dfrac{\partial y}{\partial u} & \dfrac{\partial y}{\partial v} \end{vmatrix}$

$$= \begin{vmatrix} \dfrac{\partial u}{\partial x} & \dfrac{\partial u}{\partial y} \\ \dfrac{\partial v}{\partial x} & \dfrac{\partial v}{\partial y} \end{vmatrix} \cdot \begin{vmatrix} \dfrac{\partial x}{\partial u} & \dfrac{\partial y}{\partial u} \\ \dfrac{\partial x}{\partial v} & \dfrac{\partial y}{\partial v} \end{vmatrix}$$

$$= \begin{vmatrix} \dfrac{\partial u}{\partial x} \cdot \dfrac{\partial x}{\partial u} + \dfrac{\partial u}{\partial y} \cdot \dfrac{\partial y}{\partial u} & \dfrac{\partial u}{\partial x} \cdot \dfrac{\partial x}{\partial v} + \dfrac{\partial u}{\partial y} \cdot \dfrac{\partial y}{\partial v} \\ \dfrac{\partial v}{\partial x} \cdot \dfrac{\partial x}{\partial u} + \dfrac{\partial v}{\partial y} \cdot \dfrac{\partial y}{\partial u} & \dfrac{\partial v}{\partial x} \cdot \dfrac{\partial x}{\partial v} + \dfrac{\partial v}{\partial y} \cdot \dfrac{\partial y}{\partial v} \end{vmatrix}$$

$$= \begin{vmatrix} 1 & 0 \\ 0 & 1 \end{vmatrix} = 1$$

THEOREM 2. *If the function $u_1, u_2, ..., u_n$ of n independent variables $x_1, x_2, ..., x_n$ are of the following form*

$$u_1 = f_1(x_1)$$
$$u_2 = f_2(x_1, x_2)$$
$$\vdots \quad \vdots \quad \vdots$$
$$u_n = f_n(x_1, x_2, ..., x_n)$$

Then
$$\frac{\partial(u_1, u_2, ..., u_n)}{\partial(x_1, x_2, ..., x_n)} = \frac{\partial u_1}{\partial x_1} \cdot \frac{\partial u_2}{\partial x_2} \cdot \frac{\partial u_3}{\partial x_3} \cdots \frac{\partial u_n}{\partial x_n}$$

Proof. We know that

$$\frac{\partial(u_1, u_2, ..., u_n)}{\partial(x_1, x_2, ..., x_n)} = \begin{vmatrix} \dfrac{\partial u_1}{\partial x_1} & 0 & 0 & \cdots & 0 \\ \dfrac{\partial u_2}{\partial x_1} & \dfrac{\partial u_2}{\partial x_2} & 0 & \cdots & 0 \\ \vdots & \vdots & \vdots & & \vdots \\ \dfrac{\partial u_n}{\partial x_1} & \dfrac{\partial u_n}{\partial x_2} & \dfrac{\partial u_n}{\partial x_3} & \cdots & \dfrac{\partial u_n}{\partial x_n} \end{vmatrix} \qquad ...(1)$$

(i) u_1 is a function of x_1 only, therefore $\dfrac{\partial u_1}{\partial x_1}$ exists and

$$\frac{\partial u_1}{\partial x_2} = 0, \frac{\partial u_1}{\partial x_3} = 0, ..., \frac{\partial u_1}{\partial x_n} = 0$$

(ii) u_2 is a function of x_1 and x_2 only, therefore $\dfrac{\partial u_2}{\partial x_1}$ and $\dfrac{\partial u_2}{\partial x_2}$ exist

and
$$\frac{\partial u_2}{\partial x_3} = 0, \frac{\partial u_2}{\partial x_4} = 0, ..., \frac{\partial u_2}{\partial x_n} = 0$$

(iii) u_3 is a function of x_1, x_2 and x_3 only, therefore $\dfrac{\partial u_3}{\partial x_1}, \dfrac{\partial u_3}{\partial x_2}$ and $\dfrac{\partial u_3}{\partial x_3}$ exist

and
$$\frac{\partial u_3}{\partial x_4} = 0, \frac{\partial u_3}{\partial x_5} = 0, ..., \frac{\partial u_3}{\partial x_n} = 0$$

Preceding in the same manner, we have

u_n is a function of $x_1, x_2, ..., x_n$, therefore

$$\frac{\partial u_n}{\partial x_1}, \frac{\partial u_n}{\partial x_2}, ..., \frac{\partial u_n}{\partial x_n} \text{ all exist.}$$

Putting all these values in (1), we get

$$\frac{\partial(u_1, u_2, ..., u_n)}{\partial(x_1, x_2, ..., x_n)} = \begin{vmatrix} \dfrac{\partial u_1}{\partial x_1} & 0 & 0 & \cdots & 0 \\ \dfrac{\partial u_2}{\partial x_1} & \dfrac{\partial u_2}{\partial x_2} & 0 & \cdots & 0 \\ \vdots & \vdots & \vdots & \vdots & \vdots \\ \dfrac{\partial u_n}{\partial x_1} & \dfrac{\partial u_n}{\partial x_2} & \dfrac{\partial u_n}{\partial x_3} & \cdots & \dfrac{\partial u_n}{\partial x_n} \end{vmatrix}$$

Now expanding the determinant along the first row, we get

$$\frac{\partial(u_1,u_2,...,u_n)}{\partial(x_1,x_2,...,x_n)} = \frac{\partial u_1}{\partial x_1}\cdot\frac{\partial u_2}{\partial x_2}\cdot\frac{\partial u_3}{\partial x_3}\cdots\frac{\partial u_n}{\partial x_n}$$

THEOREM 3. *If u_1,u_2 are functions of y_1, y_2 and y_1,y_2 are functions of x_1,x_2 then*

$$\frac{\partial(u_1,u_2)}{\partial(x_1,x_2)} = \frac{\partial(u_1,u_2)}{\partial(y_1,y_2)}\cdot\frac{\partial(y_1,y_2)}{\partial(x_1,x_2)}$$

Proof. Since u_1,u_2 are functions of y_1,y_2. Also y_1,y_2 are functions of x_1,x_2; therefore, we get

$$\left.\begin{aligned}
\frac{\partial u_1}{\partial x_1} &= \frac{\partial u_1}{\partial y_1}\cdot\frac{\partial y_1}{\partial x_1} + \frac{\partial u_1}{\partial y_2}\cdot\frac{\partial y_2}{\partial x_1} \\
\frac{\partial u_1}{\partial x_2} &= \frac{\partial u_1}{\partial y_1}\cdot\frac{\partial y_1}{\partial x_2} + \frac{\partial u_1}{\partial y_2}\cdot\frac{\partial y_2}{\partial x_2} \\
\frac{\partial u_2}{\partial x_1} &= \frac{\partial u_2}{\partial y_1}\cdot\frac{\partial y_1}{\partial x_1} + \frac{\partial u_2}{\partial y_2}\cdot\frac{\partial y_2}{\partial x_1} \\
\frac{\partial u_2}{\partial x_2} &= \frac{\partial u_2}{\partial y_1}\cdot\frac{\partial y_1}{\partial x_2} + \frac{\partial u_2}{\partial y_2}\cdot\frac{\partial y_2}{\partial x_2}
\end{aligned}\right\} \qquad ...(1)$$

We have

$$\frac{\partial(u_1,u_2)}{\partial(y_1,y_2)}\cdot\frac{\partial(y_1,y_2)}{\partial(x_1,x_2)} = \begin{vmatrix} \dfrac{\partial u_1}{\partial y_1} & \dfrac{\partial u_1}{\partial y_2} \\[2ex] \dfrac{\partial u_2}{\partial y_1} & \dfrac{\partial u_2}{\partial y_2} \end{vmatrix} \times \begin{vmatrix} \dfrac{\partial y_1}{\partial x_1} & \dfrac{\partial y_1}{\partial x_2} \\[2ex] \dfrac{\partial y_2}{\partial x_1} & \dfrac{\partial y_2}{\partial x_2} \end{vmatrix}$$

$$= \begin{vmatrix} \dfrac{\partial u_1}{\partial y_1}\dfrac{\partial y_1}{\partial x_1} + \dfrac{\partial u_1}{\partial y_2}\dfrac{\partial y_2}{\partial x_1} & \dfrac{\partial u_1}{\partial y_1}\dfrac{\partial y_1}{\partial x_2} + \dfrac{\partial u_1}{\partial y_2}\dfrac{\partial y_2}{\partial x_2} \\[2ex] \dfrac{\partial u_2}{\partial y_1}\dfrac{\partial y_1}{\partial x_1} + \dfrac{\partial u_2}{\partial y_2}\dfrac{\partial y_2}{\partial x_1} & \dfrac{\partial u_2}{\partial y_1}\dfrac{\partial y_1}{\partial x_2} + \dfrac{\partial u_2}{\partial y_2}\dfrac{\partial y_2}{\partial x_2} \end{vmatrix}$$

Now, using relation (1), we get

$$\frac{\partial(u_1,u_2)}{\partial(y_1,y_2)}\cdot\frac{\partial(y_1,y_2)}{\partial(x_1,x_2)} = \begin{vmatrix} \dfrac{\partial u_1}{\partial x_1} & \dfrac{\partial u_1}{\partial x_2} \\[2ex] \dfrac{\partial u_2}{\partial x_1} & \dfrac{\partial u_2}{\partial x_2} \end{vmatrix} = \frac{\partial(u_1,u_2)}{\partial(x_1,x_2)}$$

THEOREM 4. *If u_1,u_2,u_3 are functions of y_1, y_2, y_3 and y_1,y_2,y_3 are functions of x_1,x_2,x_3, then*

$$\frac{\partial(u_1,u_2,u_3)}{\partial(x_1,x_2,x_3)} = \frac{\partial(u_1,u_2,u_3)}{\partial(y_1,y_2,y_3)}\cdot\frac{\partial(y_1,y_2,y_3)}{\partial(x_1,x_2,x_3)}$$

Proof. Since u_1,u_2 and u_3 are functions of y_1,y_2 and y_3. Also y_1,y_2 and y_3 are functions of x_1,x_2 and x_3 therefore, we get

$$\frac{\partial u_1}{\partial x_1} = \frac{\partial u_1}{\partial y_1}\cdot\frac{\partial y_1}{\partial x_1} + \frac{\partial u_1}{\partial y_2}\cdot\frac{\partial y_2}{\partial x_1} + \frac{\partial u_1}{\partial y_3}\cdot\frac{\partial y_3}{\partial x_1} = \sum_{i=1}^{3}\frac{\partial u_1}{\partial y_i}\cdot\frac{\partial y_i}{\partial x_1}$$

$$\frac{\partial u_1}{\partial x_2} = \frac{\partial u_1}{\partial y_1}\cdot\frac{\partial y_1}{\partial x_2} + \frac{\partial u_1}{\partial y_2}\cdot\frac{\partial y_2}{\partial x_2} + \frac{\partial u_1}{\partial y_3}\cdot\frac{\partial y_3}{\partial x_2} = \sum_{i=1}^{3}\frac{\partial u_1}{\partial y_i}\cdot\frac{\partial y_i}{\partial x_2}$$

Similarly, $\qquad \dfrac{\partial u_1}{\partial x_3} = \displaystyle\sum_{i=1}^{3}\dfrac{\partial u_1}{\partial y_i}\cdot\dfrac{\partial y_i}{\partial x_3}, \dfrac{\partial u_2}{\partial x_1} = \displaystyle\sum_{i=1}^{3}\dfrac{\partial u_2}{\partial y_i}\cdot\dfrac{\partial y_i}{\partial x_1},$

$$\frac{\partial u_2}{\partial x_2} = \sum_{i=1}^{3} \frac{\partial u_2}{\partial y_i} \cdot \frac{\partial y_i}{\partial x_2}, \frac{\partial u_2}{\partial x_3} = \sum_{i=1}^{3} \frac{\partial u_2}{\partial y_i} \cdot \frac{\partial y_i}{\partial x_3},$$

$$\frac{\partial u_3}{\partial x_1} = \sum_{i=1}^{3} \frac{\partial u_3}{\partial y_i} \cdot \frac{\partial y_i}{\partial x_1}, \frac{\partial u_3}{\partial x_2} = \sum_{i=1}^{3} \frac{\partial u_3}{\partial y_i} \cdot \frac{\partial y_i}{\partial x_2},$$

and

$$\frac{\partial u_3}{\partial x_3} = \sum_{i=1}^{3} \frac{\partial u_3}{\partial y_i} \cdot \frac{\partial y_i}{\partial x_3}$$

Now,

$$\frac{\partial(u_1, u_2, u_3)}{\partial(y_1, y_2, y_3)} \cdot \frac{\partial(y_1, y_2, y_3)}{\partial(x_1, x_2, x_3)} = \begin{vmatrix} \dfrac{\partial u_1}{\partial y_1} & \dfrac{\partial u_1}{\partial y_2} & \dfrac{\partial u_1}{\partial y_3} \\ \dfrac{\partial u_2}{\partial y_1} & \dfrac{\partial u_2}{\partial y_2} & \dfrac{\partial u_2}{\partial y_3} \\ \dfrac{\partial u_3}{\partial y_1} & \dfrac{\partial u_3}{\partial y_2} & \dfrac{\partial u_3}{\partial y_3} \end{vmatrix} \begin{vmatrix} \dfrac{\partial y_1}{\partial x_1} & \dfrac{\partial y_1}{\partial x_2} & \dfrac{\partial y_1}{\partial x_3} \\ \dfrac{\partial y_2}{\partial x_1} & \dfrac{\partial y_2}{\partial x_2} & \dfrac{\partial y_2}{\partial x_3} \\ \dfrac{\partial y_3}{\partial x_1} & \dfrac{\partial y_3}{\partial x_2} & \dfrac{\partial y_3}{\partial x_3} \end{vmatrix}$$

$$= \begin{vmatrix} \sum \dfrac{\partial u_1}{\partial y_i} \cdot \dfrac{\partial y_i}{\partial x_1} & \sum \dfrac{\partial u_1}{\partial y_i} \cdot \dfrac{\partial y_i}{\partial x_2} & \sum \dfrac{\partial u_1}{\partial y_i} \cdot \dfrac{\partial y_i}{\partial x_3} \\ \sum \dfrac{\partial u_2}{\partial y_i} \cdot \dfrac{\partial y_i}{\partial x_1} & \sum \dfrac{\partial u_2}{\partial y_i} \cdot \dfrac{\partial y_i}{\partial x_2} & \sum \dfrac{\partial u_2}{\partial y_i} \cdot \dfrac{\partial y_i}{\partial x_3} \\ \sum \dfrac{\partial u_3}{\partial y_i} \cdot \dfrac{\partial y_i}{\partial x_1} & \sum \dfrac{\partial u_3}{\partial y_i} \cdot \dfrac{\partial y_i}{\partial x_2} & \sum \dfrac{\partial u_3}{\partial y_i} \cdot \dfrac{\partial y_i}{\partial x_3} \end{vmatrix}$$

Putting the values of each element of the determinant from the above relations, we get

$$\frac{\partial(u_1, u_2, u_3)}{\partial(y_1, y_2, y_3)} \cdot \frac{\partial(y_1, y_2, y_3)}{\partial(x_1, x_2, x_3)} = \begin{vmatrix} \dfrac{\partial u_1}{\partial x_1} & \dfrac{\partial u_1}{\partial x_2} & \dfrac{\partial u_1}{\partial x_3} \\ \dfrac{\partial u_2}{\partial x_1} & \dfrac{\partial u_2}{\partial x_2} & \dfrac{\partial u_2}{\partial x_3} \\ \dfrac{\partial u_3}{\partial x_1} & \dfrac{\partial u_3}{\partial x_2} & \dfrac{\partial u_3}{\partial x_3} \end{vmatrix}$$

$$= \frac{\partial(u_1, u_2, u_3)}{\partial(x_1, x_2, x_3)}$$

Generalization. If $u_1, u_2, ..., u_n$ are functions of $y_1, y_2, ..., y_n$ and $y_1, y_2, ..., y_n$ are functions of $x_1, x_2, ..., x_n$, then

$$\frac{\partial(u_1, u_2, u_3, ..., u_n)}{\partial(x_1, x_2, x_3, ..., x_n)} = \frac{\partial(u_1, u_2, u_3, ..., u_n)}{\partial(y_1, y_2, y_3, ... y_n)} \cdot \frac{\partial(y_1, y_2, y_3, ..., y_n)}{\partial(x_1, x_2, x_3, ..., x_n)}$$

The proof may be easily extended as in the case of two and three variables.

THEOREM 5. *If the functions u, v, w of three independent variables x, y and z are not independent, then the Jacobian of u, v, w with respect to x, y, z vanishes.*

Proof. Since, the functions u, v and w (of three independent variables x, y and z) are not independent. Then there will be a relation

$$F(u,v,w) = 0 \qquad \qquad \text{...(A)}$$

which will connect these independent variables.

Differentiating (A), with respect to x, y and z, we get

$$\frac{\partial F}{\partial u}\cdot\frac{\partial u}{\partial x}+\frac{\partial F}{\partial v}\cdot\frac{\partial v}{\partial x}+\frac{\partial F}{\partial w}\cdot\frac{\partial w}{\partial x}=0 \qquad \qquad \text{...(1)}$$

$$\frac{\partial F}{\partial u}\cdot\frac{\partial u}{\partial y}+\frac{\partial F}{\partial v}\cdot\frac{\partial v}{\partial y}+\frac{\partial F}{\partial w}\cdot\frac{\partial w}{\partial y}=0 \qquad \qquad \text{...(2)}$$

$$\frac{\partial F}{\partial u}\cdot\frac{\partial u}{\partial z}+\frac{\partial F}{\partial v}\cdot\frac{\partial v}{\partial z}+\frac{\partial F}{\partial w}\cdot\frac{\partial w}{\partial z}=0 \qquad \qquad \text{...(3)}$$

Eliminating $\dfrac{\partial F}{\partial u}$, $\dfrac{\partial F}{\partial v}$ and $\dfrac{\partial F}{\partial w}$ from (1), (2) and (3), we get

$$\begin{vmatrix} \dfrac{\partial u}{\partial x} & \dfrac{\partial v}{\partial x} & \dfrac{\partial w}{\partial x} \\[2mm] \dfrac{\partial u}{\partial y} & \dfrac{\partial v}{\partial y} & \dfrac{\partial w}{\partial y} \\[2mm] \dfrac{\partial u}{\partial z} & \dfrac{\partial v}{\partial z} & \dfrac{\partial w}{\partial z} \end{vmatrix}=0 \Rightarrow \begin{vmatrix} \dfrac{\partial u}{\partial x} & \dfrac{\partial u}{\partial y} & \dfrac{\partial u}{\partial z} \\[2mm] \dfrac{\partial v}{\partial x} & \dfrac{\partial v}{\partial y} & \dfrac{\partial v}{\partial z} \\[2mm] \dfrac{\partial w}{\partial x} & \dfrac{\partial w}{\partial y} & \dfrac{\partial w}{\partial z} \end{vmatrix}=0 \text{ (Interchanging rows and columns)}$$

Hence, $\qquad \dfrac{\partial(u,v,w)}{\partial(x,y,z)}=0$

THEOREM 6. $\quad \dfrac{\partial(u,v,w)}{\partial(x,y,z)}\times\dfrac{\partial(x,y,z)}{\partial(u,v,w)}=1$

Proof. Let us suppose

$$u=f_1(x,y,z);\ v=f_2(x,y,z)\ \text{and}\ w=f_3(x,y,z).$$

Then we may write these equations as

$$x=\phi_1(u,v,w);\ y=\phi_2(u,v,w)\ \text{and}\ z=\phi_3(u,v,w)$$

Now, differentiating $u=f_1(x,y,z)$ partially w.r.t. u, v and w respectively, we get

$$\frac{\partial u}{\partial u}=\frac{\partial u}{\partial x}\cdot\frac{\partial x}{\partial u}+\frac{\partial u}{\partial y}\cdot\frac{\partial y}{\partial u}+\frac{\partial u}{\partial z}\cdot\frac{\partial z}{\partial u}$$

$$\Rightarrow \qquad 1=\frac{\partial u}{\partial x}\cdot\frac{\partial x}{\partial u}+\frac{\partial u}{\partial y}\cdot\frac{\partial y}{\partial u}+\frac{\partial u}{\partial z}\cdot\frac{\partial z}{\partial u} \qquad \qquad \text{...(1)}$$

and $\qquad \dfrac{\partial u}{\partial v}=\dfrac{\partial u}{\partial x}\cdot\dfrac{\partial x}{\partial v}+\dfrac{\partial u}{\partial y}\cdot\dfrac{\partial y}{\partial v}+\dfrac{\partial u}{\partial z}\cdot\dfrac{\partial z}{\partial v}$

$$\Rightarrow \qquad 0=\frac{\partial u}{\partial x}\cdot\frac{\partial x}{\partial v}+\frac{\partial u}{\partial y}\cdot\frac{\partial y}{\partial v}+\frac{\partial u}{\partial z}\cdot\frac{\partial z}{\partial v} \qquad \qquad \text{...(2)}$$

Similarly $\qquad 0=\dfrac{\partial u}{\partial x}\cdot\dfrac{\partial x}{\partial w}+\dfrac{\partial u}{\partial y}\cdot\dfrac{\partial y}{\partial w}+\dfrac{\partial u}{\partial z}\cdot\dfrac{\partial z}{\partial w} \qquad \qquad \text{...(3)}$

Now differentiating $v=f_2(x,y,z)$ and $w=f_3(x,y,z)$ partially with respect to u, v and w respectively, we get

$$
\left.
\begin{aligned}
0 &= \frac{\partial v}{\partial x}\cdot\frac{\partial x}{\partial u}+\frac{\partial v}{\partial y}\cdot\frac{\partial y}{\partial u}+\frac{\partial v}{\partial z}\cdot\frac{\partial z}{\partial u}\\[4pt]
1 &= \frac{\partial v}{\partial x}\cdot\frac{\partial x}{\partial v}+\frac{\partial v}{\partial y}\cdot\frac{\partial y}{\partial v}+\frac{\partial v}{\partial z}\cdot\frac{\partial z}{\partial v}\\[4pt]
0 &= \frac{\partial v}{\partial x}\cdot\frac{\partial x}{\partial w}+\frac{\partial v}{\partial y}\cdot\frac{\partial y}{\partial w}+\frac{\partial v}{\partial z}\cdot\frac{\partial z}{\partial w}
\end{aligned}
\right\} \qquad \ldots(4)
$$

and

$$
\left.
\begin{aligned}
0 &= \frac{\partial w}{\partial x}\cdot\frac{\partial x}{\partial u}+\frac{\partial w}{\partial y}\cdot\frac{\partial y}{\partial u}+\frac{\partial w}{\partial z}\cdot\frac{\partial z}{\partial u}\\[4pt]
0 &= \frac{\partial w}{\partial x}\cdot\frac{\partial x}{\partial v}+\frac{\partial w}{\partial y}\cdot\frac{\partial y}{\partial v}+\frac{\partial w}{\partial z}\cdot\frac{\partial z}{\partial v}\\[4pt]
1 &= \frac{\partial w}{\partial x}\cdot\frac{\partial x}{\partial w}+\frac{\partial w}{\partial y}\cdot\frac{\partial y}{\partial w}+\frac{\partial w}{\partial z}\cdot\frac{\partial z}{\partial w}
\end{aligned}
\right\} \qquad \ldots(5)
$$

We have

$$
\frac{\partial(u,v,w)}{\partial(x,y,z)}\times\frac{\partial(x,y,z)}{\partial(u,v,w)}=
\begin{vmatrix}
\dfrac{\partial u}{\partial x} & \dfrac{\partial u}{\partial y} & \dfrac{\partial u}{\partial z}\\[6pt]
\dfrac{\partial v}{\partial x} & \dfrac{\partial v}{\partial y} & \dfrac{\partial v}{\partial z}\\[6pt]
\dfrac{\partial w}{\partial x} & \dfrac{\partial w}{\partial y} & \dfrac{\partial w}{\partial z}
\end{vmatrix}
\begin{vmatrix}
\dfrac{\partial x}{\partial u} & \dfrac{\partial x}{\partial v} & \dfrac{\partial x}{\partial w}\\[6pt]
\dfrac{\partial y}{\partial u} & \dfrac{\partial y}{\partial v} & \dfrac{\partial y}{\partial w}\\[6pt]
\dfrac{\partial z}{\partial u} & \dfrac{\partial z}{\partial v} & \dfrac{\partial z}{\partial w}
\end{vmatrix}
$$

$$
=
\begin{vmatrix}
\sum\dfrac{\partial u}{\partial x}\cdot\dfrac{\partial x}{\partial u} & \sum\dfrac{\partial u}{\partial x}\cdot\dfrac{\partial x}{\partial v} & \sum\dfrac{\partial u}{\partial x}\cdot\dfrac{\partial x}{\partial w}\\[8pt]
\sum\dfrac{\partial v}{\partial x}\cdot\dfrac{\partial x}{\partial u} & \sum\dfrac{\partial v}{\partial x}\cdot\dfrac{\partial x}{\partial v} & \sum\dfrac{\partial v}{\partial x}\cdot\dfrac{\partial x}{\partial w}\\[8pt]
\sum\dfrac{\partial w}{\partial x}\cdot\dfrac{\partial x}{\partial u} & \sum\dfrac{\partial w}{\partial x}\cdot\dfrac{\partial x}{\partial v} & \sum\dfrac{\partial w}{\partial x}\cdot\dfrac{\partial x}{\partial w}
\end{vmatrix}
$$

$$
=
\begin{vmatrix}
1 & 0 & 0\\
0 & 1 & 0\\
0 & 0 & 1
\end{vmatrix}=1 \qquad\qquad \text{(Using the relations (1) to (5))}
$$

Hence

$$
\frac{\partial(u,v,w)}{\partial(x,y,z)}\times\frac{\partial(x,y,z)}{\partial(u,v,w)}=1
$$

11.5 JACOBIAN OF IMPLICIT FUNCTIONS

THEOREM 1. *If $u_1,u_2,$ are implicit functions of x_1,x_2 that is $F_1(u_1,u_2,x_1,x_2)=0$ and $F_2(u_1,u_2,x_1,x_2)=0$,*

then

$$
\frac{\partial(u_1,u_2)}{\partial(x_1,x_2)}=(-1)^2\left[\frac{\partial(F_1,F_2)}{\partial(x_1,x_2)}\Big/\frac{\partial(F_1,F_2)}{\partial(u_1,u_2)}\right]
$$

Proof. We have

$$
\left.
\begin{aligned}
F_1(u_1,u_2,x_1,x_2)&=0\\
F_2(u_1,u_2,x_1,x_2)&=0
\end{aligned}
\right\} \qquad \ldots(1)
$$

Differentiating relation (1), partially *w.r.t.* x_1 and x_2 respectively, we get

$$\left.\begin{array}{l}\dfrac{\partial F_1}{\partial x_1}+\dfrac{\partial F_1}{\partial u_1}\cdot\dfrac{\partial u_1}{\partial x_1}+\dfrac{\partial F_1}{\partial u_2}\cdot\dfrac{\partial u_2}{\partial x_1}=0\\[2mm]\dfrac{\partial F_1}{\partial x_2}+\dfrac{\partial F_1}{\partial u_1}\cdot\dfrac{\partial u_1}{\partial x_2}+\dfrac{\partial F_1}{\partial u_2}\cdot\dfrac{\partial u_2}{\partial x_2}=0\\[2mm]\dfrac{\partial F_2}{\partial x_1}+\dfrac{\partial F_2}{\partial u_1}\cdot\dfrac{\partial u_1}{\partial x_1}+\dfrac{\partial F_2}{\partial u_2}\cdot\dfrac{\partial u_2}{\partial x_1}=0\\[2mm]\dfrac{\partial F_2}{\partial x_2}+\dfrac{\partial F_2}{\partial u_1}\cdot\dfrac{\partial u_1}{\partial x_2}+\dfrac{\partial F_2}{\partial u_2}\cdot\dfrac{\partial u_2}{\partial x_2}=0\end{array}\right]\qquad\ldots(2)$$

We have

$$\frac{\partial(F_1,F_2)}{\partial(u_1,u_2)}\times\frac{\partial(u_1,u_2)}{\partial(x_1,x_2)}=\begin{vmatrix}\dfrac{\partial F_1}{\partial u_1}&\dfrac{\partial F_1}{\partial u_2}\\[2mm]\dfrac{\partial F_2}{\partial u_1}&\dfrac{\partial F_2}{\partial u_2}\end{vmatrix}\times\begin{vmatrix}\dfrac{\partial u_1}{\partial x_1}&\dfrac{\partial u_1}{\partial x_2}\\[2mm]\dfrac{\partial u_2}{\partial x_1}&\dfrac{\partial u_2}{\partial x_2}\end{vmatrix}$$

$$=\begin{vmatrix}\dfrac{\partial F_1}{\partial u_1}\cdot\dfrac{\partial u_1}{\partial x_1}+\dfrac{\partial F_1}{\partial u_2}\cdot\dfrac{\partial u_2}{\partial x_1}&\dfrac{\partial F_1}{\partial u_1}\cdot\dfrac{\partial u_1}{\partial x_2}+\dfrac{\partial F_1}{\partial u_2}\cdot\dfrac{\partial u_2}{\partial x_2}\\[3mm]\dfrac{\partial F_2}{\partial u_1}\cdot\dfrac{\partial u_1}{\partial x_1}+\dfrac{\partial F_2}{\partial u_2}\cdot\dfrac{\partial u_2}{\partial x_1}&\dfrac{\partial F_2}{\partial u_1}\cdot\dfrac{\partial u_1}{\partial x_2}+\dfrac{\partial F_2}{\partial u_2}\cdot\dfrac{\partial u_2}{\partial x_2}\end{vmatrix}$$

$$=\begin{vmatrix}-\dfrac{\partial F_1}{\partial x_1}&-\dfrac{\partial F_1}{\partial x_2}\\[2mm]-\dfrac{\partial F_2}{\partial x_1}&-\dfrac{\partial F_2}{\partial x_2}\end{vmatrix}\qquad\text{Using relation (2),}$$

$$=(-1)^2\,\frac{\partial(F_1,F_2)}{\partial(x_1,x_2)}$$

$$\Rightarrow\qquad\frac{\partial(u_1,u_2)}{\partial(x_1,x_2)}=(-1)^2\left[\frac{\partial(F_1,F_2)}{\partial(x_1,x_2)}\Bigg/\frac{\partial(F_1,F_2)}{\partial(u_1,u_2)}\right]$$

THEOREM 2. *If u_1,u_2 and u_3 be the implicit functions of x_1,x_2,x_3 that is*

$$F_1(u_1,u_2,\,u_3,\,x_1,x_2,x_3)=0$$
$$F_2(u_1,u_2,\,u_3,\,x_1,x_2,x_3)=0$$
$$F_3(u_1,u_2,\,u_3,\,x_1,x_2,x_3)=0$$

then

$$\frac{\partial(u_1,u_2,u_3)}{\partial(x_1,x_2,x_3)}=(-1)^3\left[\frac{\partial(F_1,F_2,F_3)}{\partial(x_1,x_2,x_3)}\Bigg/\frac{\partial(F_1,F_2,F_3)}{\partial(u_1,u_2,u_3)}\right]$$

Proof. We have

$$\left.\begin{array}{l}F_1\left(u_1,u_2,u_3,x_1,x_2,x_3\right)=0\\[1mm]F_2\left(u_1,u_2,u_3,x_1,x_2,x_3\right)=0\\[1mm]F_3\left(u_1,u_2,u_3,x_1,x_2,x_3\right)=0\end{array}\right]\qquad\ldots(1)$$

Differentiating (1), partially w.r.t. x_1, x_2 and x_3 respectively, we get

$$\frac{\partial F_1}{\partial x_1}+\frac{\partial F_1}{\partial u_1}\cdot\frac{\partial u_1}{\partial x_1}+\frac{\partial F_1}{\partial u_2}\cdot\frac{\partial u_2}{\partial x_1}+\frac{\partial F_1}{\partial u_3}\cdot\frac{\partial u_3}{\partial x_1}=0$$

$\Rightarrow$

Similarly

and

$$\left.\begin{array}{l} \displaystyle\sum_{r=1}^{3} \frac{\partial F_1}{\partial u_r}\cdot\frac{\partial u_r}{\partial x_1} = -\frac{\partial F_1}{\partial x_1}, \\[2ex] \displaystyle\sum_{r=1}^{3} \frac{\partial F_1}{\partial u_r}\cdot\frac{\partial u_r}{\partial x_2} = -\frac{\partial F_1}{\partial x_2}, \\[2ex] \displaystyle\sum_{r=1}^{3} \frac{\partial F_1}{\partial u_r}\cdot\frac{\partial u_r}{\partial x_3} = -\frac{\partial F_1}{\partial x_3}, \\[2ex] \displaystyle\sum_{r=1}^{3} \frac{\partial F_2}{\partial u_r}\cdot\frac{\partial u_r}{\partial x_1} = -\frac{\partial F_2}{\partial x_1}, \\[2ex] \displaystyle\sum_{r=1}^{3} \frac{\partial F_2}{\partial u_r}\cdot\frac{\partial u_r}{\partial x_2} = -\frac{\partial F_2}{\partial x_2}, \\[2ex] \displaystyle\sum_{r=1}^{3} \frac{\partial F_2}{\partial u_r}\cdot\frac{\partial u_r}{\partial x_3} = -\frac{\partial F_2}{\partial x_3}, \\[2ex] \displaystyle\sum_{r=1}^{3} \frac{\partial F_3}{\partial u_r}\cdot\frac{\partial u_r}{\partial x_1} = -\frac{\partial F_3}{\partial x_1}, \\[2ex] \displaystyle\sum_{r=1}^{3} \frac{\partial F_3}{\partial u_r}\cdot\frac{\partial u_r}{\partial x_2} = -\frac{\partial F_3}{\partial x_2}, \\[2ex] \displaystyle\sum_{r=1}^{3} \frac{\partial F_3}{\partial u_r}\cdot\frac{\partial u_r}{\partial x_3} = -\frac{\partial F_3}{\partial x_3}, \end{array}\right\} \qquad \ldots(2)$$

Now consider

$$\frac{\partial(F_1,F_2,F_3)}{\partial(u_1,u_2,u_3)} \times \frac{\partial(u_1,u_2,u_3)}{\partial(x_1,x_2,x_3)} = \begin{vmatrix} \dfrac{\partial F_1}{\partial u_1} & \dfrac{\partial F_1}{\partial u_2} & \dfrac{\partial F_1}{\partial u_3} \\[2ex] \dfrac{\partial F_2}{\partial u_1} & \dfrac{\partial F_2}{\partial u_2} & \dfrac{\partial F_2}{\partial u_3} \\[2ex] \dfrac{\partial F_3}{\partial u_1} & \dfrac{\partial F_3}{\partial u_2} & \dfrac{\partial F_3}{\partial u_3} \end{vmatrix} \times \begin{vmatrix} \dfrac{\partial u_1}{\partial x_1} & \dfrac{\partial u_1}{\partial x_2} & \dfrac{\partial u_1}{\partial x_3} \\[2ex] \dfrac{\partial u_2}{\partial x_1} & \dfrac{\partial u_2}{\partial x_2} & \dfrac{\partial u_2}{\partial x_3} \\[2ex] \dfrac{\partial u_3}{\partial x_1} & \dfrac{\partial u_3}{\partial x_2} & \dfrac{\partial u_3}{\partial x_3} \end{vmatrix}$$

$$= \begin{vmatrix} \displaystyle\sum \dfrac{\partial F_1}{\partial u_r}\cdot\dfrac{\partial u_r}{\partial x_1} & \displaystyle\sum \dfrac{\partial F_1}{\partial u_r}\cdot\dfrac{\partial u_r}{\partial x_2} & \displaystyle\sum \dfrac{\partial F_1}{\partial u_r}\cdot\dfrac{\partial u_r}{\partial x_3} \\[2ex] \displaystyle\sum \dfrac{\partial F_2}{\partial u_r}\cdot\dfrac{\partial u_r}{\partial x_1} & \displaystyle\sum \dfrac{\partial F_2}{\partial u_r}\cdot\dfrac{\partial u_r}{\partial x_2} & \displaystyle\sum \dfrac{\partial F_2}{\partial u_r}\cdot\dfrac{\partial u_r}{\partial x_3} \\[2ex] \displaystyle\sum \dfrac{\partial F_3}{\partial u_r}\cdot\dfrac{\partial u_r}{\partial x_1} & \displaystyle\sum \dfrac{\partial F_3}{\partial u_r}\cdot\dfrac{\partial u_r}{\partial x_2} & \displaystyle\sum \dfrac{\partial F_3}{\partial u_r}\cdot\dfrac{\partial u_r}{\partial x_3} \end{vmatrix}$$

Now, using (2), we get

$$= \begin{vmatrix} -\dfrac{\partial F_1}{\partial x_1} & -\dfrac{\partial F_1}{\partial x_2} & -\dfrac{\partial F_1}{\partial x_3} \\[2ex] -\dfrac{\partial F_2}{\partial x_1} & -\dfrac{\partial F_2}{\partial x_2} & -\dfrac{\partial F_2}{\partial x_3} \\[2ex] -\dfrac{\partial F_3}{\partial x_1} & -\dfrac{\partial F_3}{\partial x_2} & -\dfrac{\partial F_3}{\partial x_3} \end{vmatrix} = (-1)^3 \begin{vmatrix} \dfrac{\partial F_1}{\partial x_1} & \dfrac{\partial F_1}{\partial x_2} & \dfrac{\partial F_1}{\partial x_3} \\[2ex] \dfrac{\partial F_2}{\partial x_1} & \dfrac{\partial F_2}{\partial x_2} & \dfrac{\partial F_2}{\partial x_3} \\[2ex] \dfrac{\partial F_3}{\partial x_1} & \dfrac{\partial F_3}{\partial x_2} & \dfrac{\partial F_3}{\partial x_3} \end{vmatrix}$$

$$= (-1)^3 \frac{\partial(F_1,F_2,F_3)}{\partial(x_1,x_2,x_3)}$$

Hence,

$$\frac{\partial(u_1,u_2,u_3)}{\partial(x_1,x_2,x_3)} = (-1)^3 \left[\frac{\partial(F_1,F_2,F_3)}{\partial(x_1,x_2,x_3)} \Big/ \frac{\partial(F_1,F_2,F_3)}{\partial(u_1,u_2,u_3)} \right]$$

Generalization. Let $u_1,u_2,...,u_n$ be the implicit functions of $x_1,x_2,...,x_n$, that is

$$F_1\left(u_1,u_2,...,u_n,x_1,x_2,...,x_n\right)=0$$
$$F_2\left(u_1,u_2,...,u_n,x_1,x_2,...,x_n\right)=0$$
$$\vdots \quad \vdots \quad \vdots \quad \vdots \quad \vdots$$
$$F_n\left(u_1,u_2,...,u_n,x_1,x_2,...,x_n\right)=0$$

Then,

$$\frac{\partial(u_1,u_2,...,u_n)}{\partial(x_1,x_2,...,x_n)} = (-1)^n \left[\frac{\partial(F_1,F_2,...,F_n)}{\partial(x_1,x_2,...,x_n)} \Big/ \frac{\partial(F_1,F_2,...,F_n)}{\partial(u_1,u_2,...,u_n)} \right]$$

The proof may be easily extended as in case of two and three implicit functions. (Theorem 7 and Theorem 8).

11.6 NECESSARY AND SUFFICIENT CONDITIONS FOR A JACOBIAN TO BE VANISH

THEOREM 1. *If $v_1,v_2,...,v_n$ be the functions of n independent variables $x_1,x_2,...,x_n$ For $F(v_1,v_2,...,v_n)=0$, it is necessary and sufficient that the Jacobian*

$$\frac{\partial(v_1,v_2,...,v_n)}{\partial(x_1,x_2,...,x_n)} \text{ should vanish identically.}$$

Proof.

Necessary Condition. Suppose that there exists a relation of $v_1, v_2,...,v_n$ such that

$$F(v_1,v_2,...,v_n)=0 \qquad \qquad ...(1)$$

To show that Jacobian is necessarily zero.

Differentiating (1), partially w.r.t. $x_1, x_2,...,x_n$ respectively, we get

$$\frac{\partial F}{\partial v_1}\cdot\frac{\partial v_1}{\partial x_1}+\frac{\partial F}{\partial v_2}\cdot\frac{\partial v_2}{\partial x_1}+...+\frac{\partial F}{\partial v_n}\cdot\frac{\partial v_n}{\partial x_1}=0$$

$$\frac{\partial F}{\partial v_1}\cdot\frac{\partial v_1}{\partial x_2}+\frac{\partial F}{\partial v_2}\cdot\frac{\partial v_2}{\partial x_2}+...+\frac{\partial F}{\partial v_n}\cdot\frac{\partial v_n}{\partial x_2}=0$$

$$... \quad ... \quad ... \quad ... \quad ... \quad ...$$

$$\frac{\partial F}{\partial v_1}\cdot\frac{\partial v_1}{\partial x_n}+\frac{\partial F}{\partial v_2}\cdot\frac{\partial v_2}{\partial x_n}+...+\frac{\partial F}{\partial v_n}\cdot\frac{\partial v_n}{\partial x_n}=0$$

Now eliminating $\dfrac{\partial F}{\partial v_1},\dfrac{\partial F}{\partial v_2},...,\dfrac{\partial F}{\partial v_n}$ from these equations, we get

$$\begin{vmatrix} \dfrac{\partial v_1}{\partial x_1} & \dfrac{\partial v_2}{\partial x_1} & ... & \dfrac{\partial v_n}{\partial x_1} \\ \dfrac{\partial v_1}{\partial x_2} & \dfrac{\partial v_2}{\partial x_2} & ... & \dfrac{\partial v_n}{\partial x_2} \\ ... & ... & ... & ... \\ \dfrac{\partial v_1}{\partial x_n} & \dfrac{\partial v_2}{\partial x_n} & ... & \dfrac{\partial v_n}{\partial x_n} \end{vmatrix}=0$$

$$\Rightarrow \qquad \frac{\partial(v_1, v_2, ..., v_n)}{\partial(x_1, x_2, ..., x_n)} = 0$$

Sufficient Condition. If the Jacobian $J(v_1, v_2, ..., v_n)$ is zero, then to show that there must exist a relation between $v_1, v_2, ..., v_n$.

The equation connecting the functions $v_1, v_2, ..., v_n$ and the variables $x_1, x_2, ..., x_n$ can be written as

$$g_1(x_1, x_2, ..., x_n, v_1) = 0$$
$$g_2(x_2, x_3, ..., x_n, v_1, v_2) = 0$$
$$\cdots \qquad \cdots \qquad \cdots \qquad \cdots \qquad \cdots$$
$$g_k(x_k, x_{k+1}, ..., x_n, v_1, v_2, ..., v_k) = 0$$
$$\cdots \qquad \cdots \qquad \cdots \qquad \cdots \qquad \cdots$$
$$g_n(x_n, v_1, v_2, ..., v_n) = 0$$

Then, we have

$$J = \frac{\partial(v_1, v_2, ..., v_n)}{\partial(x_1, x_2, ..., x_n)} = (-1)^n \frac{\left[\dfrac{\partial(g_1, g_2, ..., g_n)}{\partial(x_1, x_2, ..., x_n)}\right]}{\left[\dfrac{\partial(g_1, g_2, ..., g_n)}{\partial(v_1, v_2, ..., v_n)}\right]} = (-1)^n \frac{\left(\dfrac{\partial g_1}{\partial x_1} \cdot \dfrac{\partial g_2}{\partial x_2} \cdots \dfrac{\partial g_n}{\partial x_n}\right)}{\left(\dfrac{\partial g_1}{\partial v_1} \cdot \dfrac{\partial g_2}{\partial v_2} \cdots \dfrac{\partial g_n}{\partial v_n}\right)}$$

If $J = 0$, then $\dfrac{\partial g_1}{\partial x_1} \cdot \dfrac{\partial g_2}{\partial x_2} \cdots \dfrac{\partial g_r}{\partial x_r} \cdots \dfrac{\partial g_n}{\partial x_n} = 0$

$\Rightarrow$ At least one of $\dfrac{\partial v_1}{\partial x_1}, \dfrac{\partial v_2}{\partial x_2}, ..., \dfrac{\partial v_n}{\partial x_n}$ is zero.

$\Rightarrow$ $\dfrac{\partial g_k}{\partial x_k} = 0$ for some value of k between 1 and n.

$\Rightarrow$ For that particular value of k, the function g_k must not contain x_k and hence

$$g_k(x_{k+1}, ..., x_n, v_1, v_2, ..., v_k) = 0 \qquad ...(2)$$

Now we may easily eliminate the variables $x_{k+1}, x_{k+2}, ..., x_n$ between (2) and $g_{r+1} = 0$, $g_{r+2} = 0, ..., g_n = 0$ and an equation between $v_1, v_2, ..., v_n$ alone, can be obtained.

Solved Examples

Example 1. If $x = r \cos \theta$, $y = r \sin \theta$, show that

(i) $\dfrac{\partial(x, y)}{\partial(r, \theta)} = r$ (ii) $\dfrac{\partial(r, \theta)}{\partial(x, y)} = \dfrac{1}{r}$

Solution. (a) We have

$$\frac{\partial(x, y)}{\partial(r, \theta)} = \begin{vmatrix} \partial x / \partial r & \partial x / \partial \theta \\ \partial y / \partial r & \partial y / \partial \theta \end{vmatrix} = \begin{vmatrix} \cos\theta & -r\sin\theta \\ \sin\theta & r\cos\theta \end{vmatrix}$$
$$= r \cos^2\theta + r \sin^2\theta = r$$

(b) From the given relations, we get

$$r^2 = x^2 + y^2 \text{ and } \tan \theta = y/x$$

Now differentiating partially w.r.t. x and y, we obtain

$$2r\frac{\partial r}{\partial x} = 2x \qquad \text{or} \qquad \frac{\partial r}{\partial x} = \frac{x}{r}$$

$$2r\frac{\partial r}{\partial y} = 2y \qquad \text{or} \qquad \frac{\partial r}{\partial y} = \frac{y}{r}$$

and $\qquad \tan\theta = y/x \qquad \Rightarrow \quad \sec^2\theta\frac{\partial\theta}{\partial x} = -\frac{y}{x^2}$

or $\qquad \frac{\partial\theta}{\partial x} = -\frac{y}{x^2\sec^2\theta} = -\frac{y}{r^2\cos^2\theta\sec^2\theta} = -\frac{y}{r^2}$

and $\qquad \sec^2\theta\frac{\partial\theta}{\partial y} = \frac{1}{x} \quad \text{or} \quad \frac{\partial\theta}{\partial y} = \frac{1}{x\sec^2\theta} = \frac{\cos^2\theta}{x} = \frac{x^2}{r^2}\cdot\frac{1}{x} = \frac{x}{r^2}$

$$\frac{\partial(r,\theta)}{\partial(x,y)} = \begin{vmatrix} \dfrac{\partial r}{\partial x} & \dfrac{\partial r}{\partial y} \\ \dfrac{\partial\theta}{\partial x} & \dfrac{\partial\theta}{\partial y} \end{vmatrix} = \begin{vmatrix} x/r & y/r \\ -y/r^2 & x/r^2 \end{vmatrix}$$

$$= \frac{x^2}{r^3} + \frac{y^2}{r^3} = \frac{x^2+y^2}{r^3} = \frac{r^2}{r^3} = \frac{1}{r}$$

Example 2. If $x=r\sin\theta\cos\phi$, $y=r\sin\theta\sin\phi$, $z=r\cos\theta$, show that

$$\frac{\partial(x,y,z)}{\partial(r,\theta,\phi)} = r^2\sin\theta$$

Solution. We know that

$$\frac{\partial(x,y,z)}{\partial(r,\theta,\phi)} = \begin{vmatrix} \dfrac{\partial x}{\partial r} & \dfrac{\partial x}{\partial\theta} & \dfrac{\partial x}{\partial\phi} \\ \dfrac{\partial y}{\partial r} & \dfrac{\partial y}{\partial\theta} & \dfrac{\partial y}{\partial\phi} \\ \dfrac{\partial z}{\partial r} & \dfrac{\partial z}{\partial\theta} & \dfrac{\partial z}{\partial\phi} \end{vmatrix}$$

$$= \begin{vmatrix} \sin\theta\cos\phi & r\cos\theta\cos\phi & -r\sin\theta\sin\phi \\ \sin\theta\sin\phi & r\cos\theta\sin\phi & r\sin\theta\cos\phi \\ \cos\theta & -r\sin\theta & 0 \end{vmatrix}$$

[expanding the determinant along the third row]

$$= \cos\theta\,(r^2\sin\theta\cos\theta\cos^2\phi + r^2\sin\theta\cos\theta\sin^2\phi)$$
$$+ r\sin\theta(r\sin^2\theta\cos^2\phi + r\sin^2\theta\sin^2\phi)$$
$$= r^2\sin\theta\cos^2\theta + r^2\sin^3\theta = r^2\sin\theta(\cos^2\theta + \sin^2\theta)$$
$$= r^2\sin\theta.$$

Example 3. If $x=c\cos u\cosh v$ and $y=c\sin u\sinh v$ prove that

$$\frac{\partial(x,y)}{\partial(u,v)} = \frac{1}{2}c^2(\cos 2u - \cosh 2v)$$

Solution. We have

$$x = c \cos u \cosh v \quad \text{and} \quad y = c \sin u \sinh v$$

$$\Rightarrow \quad \frac{\partial x}{\partial u} = -c \sin u \cosh v, \qquad \frac{\partial x}{\partial v} = c \cos u \sinh v$$

and $\quad \dfrac{\partial y}{\partial u} = c \cos u \sinh v, \qquad \dfrac{\partial y}{\partial v} = c \sin u \cosh v$

Now $\quad \dfrac{\partial(x,y)}{\partial(u,v)} = \begin{vmatrix} \dfrac{\partial x}{\partial u} & \dfrac{\partial x}{\partial v} \\ \dfrac{\partial y}{\partial u} & \dfrac{\partial y}{\partial v} \end{vmatrix}$

$$= \begin{vmatrix} -c \sin u \cosh v & c \cos u \sinh v \\ c \cos u \sinh v & c \sin u \cosh v \end{vmatrix}$$

$$= -c^2 \sin^2 u \cosh^2 v - c^2 \cos^2 u \sinh^2 v$$

$$= -\frac{c^2}{2} \left[2 \sin^2 u \cosh^2 v + 2 \cos^2 u \sinh^2 v \right]$$

$$= -\frac{c^2}{2} \left[(1 - \cos 2u) \cosh^2 v + (1 + \cos 2u) \sinh^2 v \right]$$

$$= -\frac{c^2}{2} \left[\cos 2u \left(\sinh^2 v - \cosh^2 v \right) + \cosh^2 v + \sinh^2 v \right]$$

$$= \frac{c^2}{2} \left[\cos 2u - \cos 2v \right]$$

Example 4. *If* $y_1 = r \sin \theta_1 \sin \theta_2, \, y_2 = r \sin \theta_1 \cos \theta_2, \, y_3 = r \cos \theta_1 \sin \theta_3$
$y_4 = r \cos \theta_1 \cos \theta_3$, *find the value of Jacobian.*

Solution . We have

$$y_1 = r \sin \theta_1 \sin \theta_2 \qquad \qquad \dots(1)$$

$$y_2 = r \sin \theta_1 \cos \theta_2 \qquad \qquad \dots(2)$$

$$y_3 = r \cos \theta_1 \sin \theta_3 \qquad \qquad \dots(3)$$

$$y_4 = r \cos \theta_1 \cos \theta_3 \qquad \qquad \dots(4)$$

Squaring and adding the given four relations, we get

$$y_1{}^2 + y_2{}^2 + y_3{}^2 + y_4{}^2 = r^2$$

$$\therefore \quad y_1 \frac{\partial y_1}{\partial r} + y_2 \frac{\partial y_2}{\partial r} + y_3 \frac{\partial y_3}{\partial r} + y_4 \frac{\partial y_4}{\partial r} = r$$

and $\quad y_1 \dfrac{\partial y_1}{\partial \theta_i} + y_2 \dfrac{\partial y_2}{\partial \theta_i} + y_3 \dfrac{\partial y_3}{\partial \theta_i} + y_4 \dfrac{\partial y_4}{\partial \theta_i} = 0, i = 1,2,3$ $\qquad \dots(5)$

Also $y_3{}^2 + y_4{}^2 = r^2 \cos^2 \theta_1$, so that

$$y_1 \frac{\partial y_3}{\partial \theta_1} + y_4 \frac{\partial y_4}{\partial \theta_1} = -r^2 \cos \theta_1 \sin \theta_1$$

$$y_3 \frac{\partial y_3}{\partial \theta_j} + y_4 \frac{\partial y_4}{\partial \theta_j} = 0, j = 1,2,3 \qquad \dots(6)$$

Now the required Jacobian

$$J=\begin{vmatrix} \partial y_1/\partial r & \partial y_1/\partial \theta_1 & \partial y_1/\partial \theta_2 & \partial y_1/\partial \theta_3 \\ \partial y_2/\partial r & \partial y_2/\partial \theta_1 & \partial y_2/\partial \theta_2 & \partial y_2/\partial \theta_3 \\ \partial y_3/\partial r & \partial y_3/\partial \theta_1 & \partial y_3/\partial \theta_2 & \partial y_3/\partial \theta_3 \\ \partial y_4/\partial r & \partial y_4/\partial \theta_1 & \partial y_4/\partial \theta_2 & \partial y_4/\partial \theta_3 \end{vmatrix}$$

Operating, $y_1 R_1 + (y_2 R_2 + y_3 R_3 + y_4 R_4)$ and using result (5), we obtain

$$J=\frac{1}{y_1}\begin{vmatrix} r & 0 & 0 & 0 \\ \partial y_2/\partial r & \partial y_2/\partial \theta_1 & \partial y_2/\partial \theta_2 & \partial y_2/\partial \theta_3 \\ \partial y_3/\partial r & \partial y_3/\partial \theta_1 & \partial y_3/\partial \theta_2 & \partial y_3/\partial \theta_3 \\ \partial y_4/\partial r & \partial y_4/\partial \theta_1 & \partial y_4/\partial \theta_2 & \partial y_4/\partial \theta_3 \end{vmatrix}$$

$$=\frac{r}{y_1}\begin{vmatrix} \dfrac{\partial y_2}{\partial \theta_1} & \dfrac{\partial y_2}{\partial \theta_2} & \dfrac{\partial y_2}{\partial \theta_3} \\ \dfrac{\partial y_3}{\partial \theta_1} & \dfrac{\partial y_3}{\partial \theta_2} & \dfrac{\partial y_3}{\partial \theta_3} \\ \dfrac{\partial y_4}{\partial \theta_1} & \dfrac{\partial y_4}{\partial \theta_2} & \dfrac{\partial y_4}{\partial \theta_3} \end{vmatrix}=\frac{r}{y_1 y_3}\begin{vmatrix} \dfrac{\partial y_2}{\partial \theta_1} & \dfrac{\partial y_2}{\partial \theta_2} & \dfrac{\partial y_2}{\partial \theta_3} \\ -r^2\cos\theta_1\sin\theta_1 & 0 & 0 \\ \dfrac{\partial y_4}{\partial \theta_1} & \dfrac{\partial y_4}{\partial \theta_2} & \dfrac{\partial y_4}{\partial \theta_3} \end{vmatrix}$$

[Adding $y_4 R_3$ to $y_3 R_2$ and using the result (6)]

$$=\frac{r}{y_1 y_3}\cdot r^2\cos\theta_1\sin\theta_1\left[\frac{\partial y_2}{\partial \theta_2}\cdot\frac{\partial y_4}{\partial \theta_3}-\frac{\partial y_4}{\partial \theta_2}\cdot\frac{\partial y_2}{\partial \theta_3}\right]$$

$$=\frac{r^3\cos\theta_1\sin\theta_1}{y_1 y_3}\left[(-r\sin\theta_1\sin\theta_2)(-r\cos\theta_1\sin\theta_3)-0\right]$$

$$=\frac{r^5\sin^2\theta_1\cos^2\theta_1\sin\theta_2\sin\theta_3}{r^2\sin\theta_1\cos\theta_1\sin\theta_2\sin\theta_3}=r^3\sin\theta_1\cos\theta_1.$$

Example 5. If $y_1=1-x_1, y_2=x_1(1-x_2), y_3=x_1 x_2(1-x_3)...y_n=x_1 x_2...x_{n-1}(1-x_n)$. *Prove that*
$J(y_1,y_2,...,y_n)=(-1)^n x_1^{n-1}.x_2^{n-2}...x_{n-1}.$

Solution. In the above relations

y_1 is a function of x_1

y_2 is a function of x_1, x_2

y_3 is a function of $x_1 x_2 x_3$

...   .

and y_n is a function of $x_1 x_2,...,x_n.$

$$\therefore \frac{\partial(y_1,y_2,...,y_n)}{\partial(x_1,x_2,...,x_n)}=\begin{vmatrix} \dfrac{\partial y_1}{\partial x_1} & 0 & 0 & \cdots & 0 \\ \dfrac{\partial y_2}{\partial x_1} & \dfrac{\partial y_2}{\partial x_2} & 0 & \cdots & 0 \\ \vdots & \vdots & \vdots & \vdots & \vdots \\ \dfrac{\partial y_n}{\partial x_1} & \dfrac{\partial y_n}{\partial x_2} & \dfrac{\partial y_n}{\partial x_3} & \cdots & \dfrac{\partial y_n}{\partial x_n} \end{vmatrix}$$

$$=\frac{\partial y_1}{\partial x_1}\cdot\frac{\partial y_2}{\partial x_2}\cdot\frac{\partial y_3}{\partial x_3}\cdots\frac{\partial y_n}{\partial x_n}$$

$$= (-1)(-x_1)(-x_1x_2)...(-x_1x_2.x_3...x_{n-1})$$
$$= (-1)^n x_1^{n-1} x_2^{n-2}...x_{n-1}.$$

Example 6. If $u^3+v^3=x+y$ and $u^2+v^2=x^3+y^3$, then prove that

$$\frac{\partial(u,v)}{\partial(x,y)} = \frac{1}{2}\frac{y^2-x^2}{uv(u-v)}$$

Solution. Here we can write above relations, as

$$F_1 = u^3+v^3-x-y=0$$
$$F_2 = u^2+v^2-x^3-y^3=0$$

Now

$$\frac{\partial(u,v)}{\partial(x,y)} = (-1)^2 \frac{\partial(F_1,F_2)}{\partial(x,y)} \Big/ \frac{\partial(F_1,F_2)}{\partial(u,v)} \qquad \text{...(1)}$$

We have

$$\frac{\partial(F_1,F_2)}{\partial(x,y)} = \begin{vmatrix} \dfrac{\partial F_1}{\partial x} & \dfrac{\partial F_1}{\partial y} \\ \dfrac{\partial F_2}{\partial x} & \dfrac{\partial F_2}{\partial y} \end{vmatrix} = \begin{vmatrix} -1 & -1 \\ -3x^2 & -3y^2 \end{vmatrix} \qquad \text{...(2)}$$

$$= 3y^2 - 3x^2 = 3(y^2-x^2)$$

and

$$\frac{\partial(F_1,F_2)}{\partial(x,y)} = \begin{vmatrix} 3u^2 & 3v^2 \\ 2u & 2v \end{vmatrix} = 6u^2v - 6uv^2 = 6uv(u-v) \qquad \text{...(3)}$$

From equations (1), (2) and (3), we get

$$\frac{\partial(u,v)}{\partial(x,y)} = \frac{3(y^2-x^2)}{6uv(u-v)} = \frac{1}{2}\frac{y^2-x^2}{uv(u-v)}$$

Example 7. If $u^3+v^3+w^3=x+y+z$; $u^2+v^2+w^2=x^3+y^3+z^3$ and $u+v+w=x^2+y^2+z^2$

Prove that $\dfrac{\partial(u,v,w)}{\partial(x,y,z)} = \dfrac{(y-z)(z-x)(x-y)}{(u-v)(v-w)(w-u)}$

Solution. Here, given relation can be written as

$$F_1 = u^3+v^3+w^3-x-y-z=0$$
$$F_2 = u^2+v^2+w^2-x^3-y^3-z^3=0$$
$$F_3 = u+v+w-x^2-y^2-z^2=0.$$

Now

$$\frac{\partial(u,v,w)}{\partial(x,y,z)} = (-1)^3 \frac{\partial(F_1,F_2,F_3)}{\partial(x,y,z)} \Big/ \frac{\partial(F_1,F_2,F_3)}{\partial(u,v,w)}$$

We have

$$\frac{\partial(F_1,F_2,F_3)}{\partial(x,y,z)} = \begin{vmatrix} -1 & -1 & -1 \\ -3x^2 & -3y^2 & -3z^2 \\ -2x & -2y & -2z \end{vmatrix}$$

$$= -6\begin{vmatrix} 1 & 1 & 1 \\ x^2 & y^2 & z^2 \\ x & y & z \end{vmatrix} = 6\begin{vmatrix} 1 & 1 & 1 \\ x & y & z \\ x^2 & y^2 & z^2 \end{vmatrix}$$

$$= 6\begin{vmatrix} 1 & 0 & 0 \\ x & y-x & z-x \\ x^2 & y^2-x^2 & z^2-x^2 \end{vmatrix} = 6(y-x)(z-x)\begin{vmatrix} 1 & 1 \\ y+x & z+x \end{vmatrix}$$

$$= 6(y-x)(z-x)(z-y)$$
$$= 6(x-y)(y-z)(z-x)$$

Also
$$\frac{\partial(F_1, F_2, F_3)}{\partial(u, v, w)} = \begin{vmatrix} 3u^2 & 3v^2 & 3w^2 \\ 2u & 2v & 2w \\ 1 & 1 & 1 \end{vmatrix} = \begin{vmatrix} 1 & 1 & 1 \\ u & v & w \\ u^2 & v^2 & w^2 \end{vmatrix}$$

$$= -6(u-v)(v-w)(w-u)$$

From equations (1), (2) and (3), we get

$$\frac{\partial(u, v, w)}{\partial(x, y, z)} = -\frac{6(x-y)(y-z)(z-x)}{-6(u-v)(v-w)(w-u)}$$

$$= \frac{(y-z)(z-x)(x-y)}{(u-v)(v-w)(w-u)}$$

Example 8. *Prove that*

$$\frac{\partial(y_1, y_2, ..., y_n)}{\partial(x_1, x_2, ..., x_n)} \cdot \frac{\partial(x_1, x_2, ..., x_n)}{\partial(y_1, y_2, ..., y_n)} = 1$$

Solution . Let

$$\left. \begin{array}{l} y_1 = f_1(x_1, x_2, ..., x_n), y_2 = f_2(x_1, x_2, ..., x_n), ... \\ y_n = f_n(x_1, x_2, ..., x_n) \end{array} \right] \qquad ...(1)$$

Above relations can be written as

$$x_1 = F_1(y_1, y_2, ..., y_n), x_2 = F_2(y_1, y_2, ..., y_n)...x_n = F_n(y_1, y_2, ..., y_n)$$

Differentiating (1) partially w.r.t. $y_1, y_2, ..., y_n$, we have

$$\left[\begin{array}{l} 1 = \dfrac{\partial y_1}{\partial x_1} \cdot \dfrac{\partial x_1}{\partial y_1} + \dfrac{\partial y_1}{\partial x_2} \cdot \dfrac{\partial x_2}{\partial y_1} + ... + \dfrac{\partial y_1}{\partial x_n} \cdot \dfrac{\partial x_n}{\partial y_1} = \sum \dfrac{\partial y_1}{\partial x_r} \cdot \dfrac{\partial x_r}{\partial y_1} \\[3mm] 0 = \dfrac{\partial y_1}{\partial x_1} \cdot \dfrac{\partial x_1}{\partial y_2} + \dfrac{\partial y_1}{\partial x_2} \cdot \dfrac{\partial x_2}{\partial y_2} + ... + \dfrac{\partial y_1}{\partial x_n} \cdot \dfrac{\partial x_n}{\partial y_2} = \sum \dfrac{\partial y_1}{\partial x_r} \cdot \dfrac{\partial x_r}{\partial y_2} \\ ... \quad ... \quad ... \quad ... \quad ... \quad ... \quad ... \\ ... \quad ... \quad ... \quad ... \quad ... \quad ... \quad ... \\ 0 = \dfrac{\partial y_1}{\partial x_1} \cdot \dfrac{\partial x_1}{\partial y_n} + \dfrac{\partial y_1}{\partial x_2} \cdot \dfrac{\partial x_2}{\partial y_n} + ... + \dfrac{\partial y_1}{\partial x_n} \cdot \dfrac{\partial x_n}{\partial y_n} = \sum \dfrac{\partial y_1}{\partial x_r} \cdot \dfrac{\partial x_r}{\partial y_n} \end{array} \right.$$

Similarly other relations from $y_2, y_3, ..., y_n$ can be obtained.

Now,
$$\frac{\partial(y_1, y_2, ..., y_n)}{\partial(x_1, x_2, ..., x_n)} \cdot \frac{\partial(x_1, x_2, ..., x_n)}{\partial(y_1, y_2, ..., y_n)}$$

$$= \begin{vmatrix} \dfrac{\partial y_1}{\partial x_1} & \dfrac{\partial y_1}{\partial x_2} & ... & \dfrac{\partial y_1}{\partial x_n} \\ \dfrac{\partial y_2}{\partial x_1} & \dfrac{\partial y_2}{\partial x_2} & ... & \dfrac{\partial y_2}{\partial x_n} \\ ... & ... & ... & ... \\ \dfrac{\partial y_n}{\partial x_1} & \dfrac{\partial y_n}{\partial x_2} & ... & \dfrac{\partial y_n}{\partial x_n} \end{vmatrix} \times \begin{vmatrix} \dfrac{\partial x_1}{\partial y_1} & \dfrac{\partial x_1}{\partial y_2} & ... & \dfrac{\partial x_1}{\partial y_n} \\ \dfrac{\partial x_2}{\partial y_1} & \dfrac{\partial x_2}{\partial y_2} & ... & \dfrac{\partial x_2}{\partial y_n} \\ ... & ... & ... & ... \\ \dfrac{\partial x_n}{\partial y_1} & \dfrac{\partial x_n}{\partial y_2} & ... & \dfrac{\partial x_n}{\partial y_n} \end{vmatrix}$$

$$= \begin{vmatrix} \sum \dfrac{\partial y_1}{\partial x_r}\cdot\dfrac{\partial x_r}{\partial y_1} & \sum \dfrac{\partial y_1}{\partial x_r}\cdot\dfrac{\partial x_r}{\partial y_2} & \cdots & \sum \dfrac{\partial y_1}{\partial x_r}\cdot\dfrac{\partial x_r}{\partial y_n} \\ \sum \dfrac{\partial y_2}{\partial x_r}\cdot\dfrac{\partial x_r}{\partial y_1} & \sum \dfrac{\partial y_2}{\partial x_r}\cdot\dfrac{\partial x_r}{\partial y_2} & \cdots & \sum \dfrac{\partial y_2}{\partial x_r}\cdot\dfrac{\partial x_r}{\partial y_n} \\ \cdots & \cdots & \cdots & \cdots \\ \sum \dfrac{\partial y_n}{\partial x_r}\cdot\dfrac{\partial x_r}{\partial y_1} & \sum \dfrac{\partial y_n}{\partial x_r}\cdot\dfrac{\partial x_r}{\partial y_2} & \cdots & \sum \dfrac{\partial y_n}{\partial x_r}\cdot\dfrac{\partial x_r}{\partial y_n} \end{vmatrix}$$

(Operating row by column multiplication)

$$= \begin{vmatrix} 1 & 0 & \cdots & 0 \\ 0 & 1 & \cdots & 0 \\ \cdots & \cdots & \cdots & \cdots \\ 0 & 0 & \cdots & 1 \end{vmatrix}$$

Example 9. *If x,y,z are connected by a functional relation $f(x,y,z)=0$ then prove that*

$$\frac{\partial(y,z)}{\partial(x,z)} = \left(\frac{\partial y}{\partial x}\right)_{z=\text{constant}}$$

Solution . We have $f(x,y,z)=0 \quad \Rightarrow \quad y$ is a function of x and z. Also from this relation z may be regarded as a function of x and z.

$$\therefore \quad \frac{\partial(y,z)}{\partial(x,z)} = \begin{vmatrix} \left(\dfrac{\partial y}{\partial x}\right)_{z=\text{const.}} & \left(\dfrac{\partial y}{\partial z}\right)_{x=\text{const.}} \\ \dfrac{\partial z}{\partial x} & \dfrac{\partial z}{\partial z} \end{vmatrix}$$

$$= \begin{vmatrix} \left(\dfrac{\partial y}{\partial x}\right)_{z=\text{const.}} & \left(\dfrac{\partial y}{\partial z}\right)_{x=\text{const.}} \\ 0 & 1 \end{vmatrix} \qquad \left[\because \dfrac{\partial z}{\partial x}=0, \dfrac{\partial z}{\partial z}=1\right]$$

$$= \left(\frac{\partial y}{\partial x}\right)_{z=\text{const.}}$$

Example 10. *Show that $ax^2+2hxy+by^2$ and $Ax^2+2Hxy+By^2$ are independent unless*

$$\frac{a}{A} = \frac{h}{H} = \frac{b}{B}.$$

Solution . Let $u= ax^2+2hxy+by^2, v= Ax^2+2Hxy+By^2$. If the functions u, v are not independent, then $\dfrac{\partial(u,v)}{\partial(x,y)} = 0$

or $\quad \begin{vmatrix} \dfrac{\partial u}{\partial x} & \dfrac{\partial u}{\partial y} \\ \dfrac{\partial v}{\partial x} & \dfrac{\partial v}{\partial y} \end{vmatrix} = 0$ or $\begin{vmatrix} 2(ax+hy) & 2(hx+by) \\ 2(Ax+Hy) & 2(Hx+By) \end{vmatrix} = 0$

or $\quad (ax+hy)(Hx+By)-(hx+by)(Ax+Hy)=0$

or $\quad (aH-Ah)x^2+(aB-Ab)xy+(Bh-bH)y^2=0$

Since the variables x, y are independent, the coefficients of x and y in above equation must be separately zero. Hence we have $aH-Ah=0$ and $Bh-bH=0$

Hence, $$\frac{a}{A} = \frac{h}{H} = \frac{b}{B}.$$

Example 11. *If $u=x+2y+z$, $v=x-2y+3z$ and $w=2xy-xz+4yz-2z^2$, then prove that they are not independent. Find the relation between u, v and w.*

Solution . We have $\dfrac{\partial(u,v,w)}{\partial(x,y,z)} = \begin{vmatrix} 1 & 2 & 1 \\ 1 & -2 & 3 \\ 2y-z & 2x+4z & -x+4y-4z \end{vmatrix}$

$$= \begin{vmatrix} 1 & 0 & 0 \\ 1 & -4 & 2 \\ 2y-z & -x+2y-6z & -x+2y-3z \end{vmatrix} \quad \text{by } c_2-2c_1 \text{ and } c_3-c_1$$

$$= -2\begin{vmatrix} 1 & 0 & 0 \\ 1 & 2 & 2 \\ 2y-z & -x+2y-3z & -x+2y-3z \end{vmatrix} = 0$$

Here last two columns are identical. So the Jacobian of the functions u, v, w is zero, therefore these functions are not independent so there must exist a relation between them.

We have $u^2-v^2 = (x+2y+z)^2-(x-2y+3z)^2$
$$=(2x+4z)(4y-2z)$$
$$=4(x+2z)(2y-z)$$

By simplification $= 4(2xy-xz+4yz-2z^2)=4w$

Therefore $u^2-v^2=4w$, which is the required relation between u, v and w.

Example 12. *Show that the functions*
$$u=x+y+z, \ v=xy+yz+zx, \ w=x^3+y^3+z^3-3xyz$$
are not independent. Also find the relation between u, v and w.

Solution . We have $\dfrac{\partial(u,v,w)}{\partial(x,y,z)} = \begin{vmatrix} 1 & 1 & 1 \\ y+z & z+x & x+y \\ 3(x^2-yz) & 3(y^2-zx) & 3(z^2-xy) \end{vmatrix}$

$$= 3\begin{vmatrix} 1 & 1 & 1 \\ y+z & z+x & x+y \\ x^2-yz & y^2-zx & z^2-xy \end{vmatrix}$$

$$= 3\begin{vmatrix} 1 & 0 & 0 \\ y+z & z+x & x+y \\ x^2-yz & -(y-x)(x+y+z) & (z-x)(x+y+z) \end{vmatrix}$$

operating c_2-c_1 and c_3-c_1

$$= 3(x-y)(x-z)\begin{vmatrix} 1 & 0 & 0 \\ y+z & 1 & 1 \\ x^2-yz & -(x+y+z) & -(x+y+z) \end{vmatrix}$$

[Last two columns being identical]

since Jacobian is zero, therefore functions are not independent, therefore, a relation set up between them.

We have

$$w=x^3+y^3+z^3-3xyz$$
$$=(x+y+z)(x^2+y^2+z^2-yz-zx-xy)$$
$$=(x+y+z)[(x+y+z)^2-3(yz+zx+xy)]$$
$$=u(u^2-3v)=u^3-3uv.$$

∴ $u^3=3uv+w$ is the required relation.

Example 13. *If $u=(x+y)/(1-xy)$ and $v=\tan^{-1}x+\tan^{-1}y$, find $\dfrac{\partial(u,v)}{\partial(x,y)}$. Are u and v functionally related? If so, find their relationship.*

Solution. We have

$$\frac{\partial u}{\partial x}=\frac{1.(1-xy)-(-y)(x+y)}{(1-xy)^2}=\frac{1+y^2}{(1-xy)^2}$$
$$\frac{\partial u}{\partial y}=\frac{1.(1-xy)-(-x)(x+y)}{(1-xy)^2}=\frac{1+x^2}{(1-xy)^2}$$
$$\frac{\partial v}{\partial x}=\frac{1}{(1+x^2)},\frac{\partial v}{\partial y}=\frac{1}{(1+y^2)}$$

Now

$$\frac{\partial(u,v)}{\partial(x,y)}=\begin{vmatrix} \partial u/\partial x & \partial u/\partial y \\ \partial v/\partial x & \partial v/\partial y \end{vmatrix}$$
$$=\frac{\partial u}{\partial x}\cdot\frac{\partial v}{\partial y}-\frac{\partial u}{\partial y}\cdot\frac{\partial v}{\partial x}$$
$$=\frac{1+y^2}{(1-xy)^2}\cdot\frac{1}{1+y^2}-\frac{1+x^2}{(1-xy)^2}\cdot\frac{1}{1+x^2}$$
$$=\frac{1}{(1-xy)^2}-\frac{1}{(1-xy)^2}=0$$

Since the Jacobian of the function u, v is zero, therefore these functions are not independent and so they must be functionally related.

Now, we have $y=\tan^{-1}x+\tan^{-1}y=\tan^{-1}\dfrac{x+y}{1-xy}=\tan^{-1}u.$

Thus $v=\tan^{-1}u$ or $\tan v=u$ which is the required relation, between u and v.

EXERCISE 11.1

1. If $u=\dfrac{y^2}{2x}$ and $v=\dfrac{x}{2}+\dfrac{y^2}{2x}$, find $\dfrac{\partial(u,v)}{\partial(x,y)}$.

2. (a) If $u_1=\dfrac{x_2x_3}{x_1}$, $u_2=\dfrac{x_3x_1}{x_2}$, $u_3=\dfrac{x_1x_2}{x_3}$,

then show that $J(u_1,u_2,u_3)=4$.

(b) If $x=u(1+v)$, $y=v(1+u)$, then show that
$$\dfrac{\partial(x,y)}{\partial(u,v)}=1+u+v.$$

3. If $x=\sin\theta\sqrt{1-a^2\sin^2\phi}$, $y=\cos\theta\cos\phi$, then show that

$$\dfrac{\partial(x,y)}{\partial(\theta,\phi)}=-\sin\phi\dfrac{\left[\left(1-a^2\right)\cos^2\theta+a^2\cos^2\phi\right]}{\sqrt{1-a^2\sin^2\phi}}$$

4. If $y_1=x_1(1-x_1)$, $y_2=x_1x_2(1-x_3),...,$ $y_{n-1}=x_1x_2...x_{n-1}(1-x_n),y_n=x_1x_2...x_n$. Then prove that

$$\dfrac{\partial(y_1,y_2,...,y_n)}{\partial(x_1,x_2,...,x_n)}=x_1{}^{n-1}.x_2{}^{n-2}.....\,x_{n-1}.$$

5. If $y_1=\cos x_1$, $y_2=\sin x_1\cos x_2$, $y_3=\sin x_1\sin x_2$ $\cos x_3,...,$ $y_n=\sin x_1\sin x_2\sin x_3...\sin x_{n-1}\cos x_n$. Then find the Jacobian of y_1, y_2, ..., y_n with respect to $x_1,x_2,...,x_n$.

6. Show that $\dfrac{\partial(u,v)}{\partial(x,y)}\cdot\dfrac{\partial(x,y)}{\partial(u,v)}=1.$

7. If $u^3=xyz$, $\dfrac{1}{v}=\dfrac{1}{x}+\dfrac{1}{y}+\dfrac{1}{z}$, $w^2=x^2+y^2+z^2$,

show that

$$\dfrac{\partial(u,v,w)}{\partial(x,y,z)}=\dfrac{-v(y-z)(z-x)(x-y)(x+y+z)}{3u^2w(yz+zx+xy)}$$

8. If $u=2xy$, $v=x^2-y^2$, and $x=r\cos\theta$, $y=r\sin\theta$, show that $\dfrac{\partial(u,v)}{\partial(r,\theta)}=-4r^3.$

9. (i) If $u_1=x_1+x_2+x_3+x_4$, $u_1u_2=x_2+x_3+x_4$, $u_1u_2u_3=x_3+x_4$, $u_1u_2u_3u_4=x_4$. Show that

$$\dfrac{\partial(x_1,x_2,x_3,x_4)}{\partial(u_1,u_2,u_3,u_4)}=u_1^3u_2^2u_3.$$

(ii) If $x+y+z=u$, $y+z=uv$, $z=uvw$, show that

$$\dfrac{\partial(x,y,z)}{\partial(u,v,w)}=u^2v.$$

10. If $y_1.(x_1-x_2)=0$, $y_2.(x_1^2+x_1x_2+x_2^2)=0$, show that

$$\dfrac{\partial(y_1,y_2)}{\partial(x_1,x_2)}=3y_1y_2\dfrac{x_1+x_2}{x_1^3-x_2^3}.$$

11. If l, m, n are the roots of the equation in k.

$$\dfrac{x}{a+k}+\dfrac{y}{b+k}+\dfrac{z}{c+k}=1$$

Prove that

$$\dfrac{\partial(x,y,z)}{\partial(l,m,n)}=-\left[\dfrac{(m-n)(n-l)(l-m)}{(b-c)(c-a)(a-b)}\right].$$

12. If the roots of the equation

$(\lambda-x)^3+(\lambda-y)^3+(\lambda-z)^3=0$ are u, v, w. Prove that

$$\dfrac{\partial(u,v,w)}{\partial(x,y,z)}=-2\dfrac{(y-z)(z-x)(x-y)}{(v-w)(w-u)(u-v)}.$$

13. Show that the functions

$u=x+y-z$, $v=x-y+z$, $w=x^2+y^2+z^2-2yz$

are not independent of one another. Also find the relation between them.

14. If $u=x^2+y^2+z^2$, $v=x+y+z$, $w=xy+yz+zx$.

Show that the Jacobian $\dfrac{\partial(u,v,w)}{\partial(x,y,z)}$ vanish identically. Also find the relation between u, v and w.

15. If $u=x+y+z+t$, $v=x+y-z-t$, $w=xy-zt$, $r=x^2+y^2-z^2-t^2$. Show that $\dfrac{\partial(u,v,w,r)}{\partial(x,y,z,t)}=0.$

16. If $f(0)=0$ and $f'(x)=\dfrac{1}{1+x^2}$. Then prove that

$$f(x)+f(y)=f\left(\dfrac{(x+y)}{(1-xy)}\right)\text{(without using the}$$

method of integration).

17. $u=x\left(1-r^2\right)^{-\frac{1}{2}}$, $v=y\left(1-r^2\right)^{-\frac{1}{2}}$,

$w=z\left(1-r^2\right)^{-\frac{1}{2}}$ where $r^2=x^2+y^2+z^2$, then

show that $\dfrac{\partial(u,v,w)}{\partial(x,y,z)}=\left(1-r^2\right)^{-5/2}.$

18. If $u = \dfrac{x+y}{z}, v = \dfrac{y+z}{x}, w = y\dfrac{(x+y+z)}{xz}$

then show that u, v, w are not independent.

19. If
$$x_1 + x_2 + \dots + x_n = y_1$$
$$x_2 + x_3 + \dots + x_n = y_1 y_2$$
$$x_3 + x_4 + \dots + x_n = y_1 y_2 y_3$$
$$\dots \quad \dots \quad \dots \quad \dots \quad \dots \quad \dots$$
$$x_n = y_1 y_2 y_3 \dots y_n$$

then show that

$$\frac{\partial(x_1, x_2, \dots, x_n)}{\partial(y_1, y_2, \dots, y_n)} = y_1^{n-1} y_2^{n-2} \dots y_{n-2}^2 y_{n-1}.$$

20. If $u_1 = \dfrac{x_1}{x_n}, u_2 = \dfrac{x_2}{x_n}, u_3 = \dfrac{x_3}{x_n}, \dots, u_{n-1} = \dfrac{x_{n-1}}{x_n}$

and $x_1^2 + x_2^2 + \dots x_{n-1}^2 + x_n^2 = 1$, find the value

of $\dfrac{\partial(u_1, u_2, \dots, u_n)}{\partial(x_1, x_2, \dots, x_n)}$.

HINTS TO SELECTED PROBLEMS

1. $\dfrac{\partial u}{\partial x} = \dfrac{-y^2}{2x^2}$, $\qquad \dfrac{\partial u}{\partial y} = \dfrac{y}{x}$

$\dfrac{\partial v}{\partial x} = \dfrac{1}{2} - \dfrac{y^2}{2x^2}$, $\qquad \dfrac{\partial v}{\partial y} = \dfrac{y}{x}$

2. $\dfrac{\partial u_1}{\partial x_1} = -\dfrac{x_2 x_3}{x_1^2}, \dfrac{\partial u_1}{\partial x_2} = \dfrac{x_3}{x_1}, \dfrac{\partial u_1}{\partial x_3} = \dfrac{x_2}{x_1}$

$\dfrac{\partial u_2}{\partial x_1} = \dfrac{x_3}{x_2}, \dfrac{\partial u_2}{\partial x_2} = -\dfrac{x_3 x_1}{x_2}, \dfrac{\partial u_2}{\partial x_3} = \dfrac{x_1}{x_2}$

$\dfrac{\partial u_3}{\partial x_1} = \dfrac{x_2}{x_3}, \dfrac{\partial u_3}{\partial x_2} = \dfrac{x_1}{x_3}.$

3. $\dfrac{\partial x}{\partial \theta} = \cos\theta \sqrt{\left(1 - a^2 \sin^2 \phi\right)}$

$\dfrac{\partial x}{\partial \phi} = \dfrac{-a^2 \sin\theta \sin\phi \cos\phi}{\sqrt{1 - a^2 \sin^2 \phi}}$

$\dfrac{\partial y}{\partial \theta} = -\sin\theta \cos\phi$

$\dfrac{\partial y}{\partial \phi} = -\cos\theta \sin\phi$

9. $u_1 = x_1 + x_3 + x_4, u_2 = x_2 + x_3 + x_4, u_1 u_2 u_3 = x_3 + x_4, u_1 u_2 u_3 u_4 = x_4$. Therefore $x_3 = u_1 u_2 u_3 (1-u_4)$, $x_2 = u_1 u_2 (1-u_3), x_1 = u_1 (1-u_2)$. Now find required partial derivative of x_1, x_2 and x_3 w.r.t. u_1, u_2 and u_3.

13. If $\dfrac{\partial(u, v, w)}{\partial(x, y, z)} = 0$ Then u, v, w are not independent.

Answers

1. $-\dfrac{y}{2x}$ $\qquad$ **5.** $(-1)^n \sin^n x_1 \sin^{n-1} x_2 \sin^{n-2} x_3 \dots \sin x_n$

13. $u^2 + v^2 = 2w$ $\quad$ **14.** $v^2 = u + 2w$ $\qquad$ **15.** $uv = r + 2w$ $\qquad$ **20.** $\dfrac{1}{x_n^n}$

CHAPTER REVIEW: A COMPETITIVE APPROACH

SELECTED TERMS AND RESULTS

TERMS

- **Implicit function:** If $f(x, y)$ be a function of two variables and $y = g(x)$ be a function of x such that $f(x, g(x)) = 0$ then $y = g(x)$ is called an implicit function defined by the functional equation $f(x, y) = 0$.

- **Jacobian:** Let $u_1, u_2, ..., u_n$ be n functions of n variables $x_1, x_2, ..., x_n$, possessing partial derivatives of the first order, then the determinant

$$\begin{vmatrix} \dfrac{\partial u_1}{\partial x_1} & \dfrac{\partial u_1}{\partial x_2} & \cdots & \dfrac{\partial u_1}{\partial x_n} \\ \dfrac{\partial u_2}{\partial x_1} & \dfrac{\partial u_2}{\partial x_2} & \cdots & \dfrac{\partial u_2}{\partial x_n} \\ \vdots & \vdots & \vdots & \vdots \\ \dfrac{\partial u_n}{\partial x_1} & \dfrac{\partial u_n}{\partial x_2} & \cdots & \dfrac{\partial u_n}{\partial x_2} \end{vmatrix}$$

is called the Jacobian.

RESULTS

- A functional equation may or may not define an implicit function.

- If the functions $u_1, u_2, ..., u_n$ of n independent variables $x_1, x_2, ..., x_n$ are of the following form $u_1 = f_1(x_1), u_2 = f_2(x_1, x_2), ..., u_n = f_n(x_1, x_2, ..., x_n)$ then

$$\frac{\partial(u_1, u_2, ..., u_n)}{\partial(x_1, x_2, ..., x_n)} = \frac{\partial u_1}{\partial x_1} \cdot \frac{\partial u_2}{\partial x_2} \cdot \frac{\partial u_3}{\partial x_3} \cdots \frac{\partial u_n}{\partial x_n}$$

- If u_1, u_2 are functions of y_1, y_2 and y_1, y_2 are functions of x_1, x_2 then

$$\frac{\partial(u_1, u_2)}{\partial(x_1, x_2)} = \frac{\partial(u_1, u_2)}{\partial(y_1, y_2)} \cdot \frac{\partial(y_1, y_2)}{\partial(x_1, x_2)}$$

- If $u_1, u_2, ..., u_n$ are functions of $y_1, y_2, ..., y_n$ and $y_1, y_2, ..., y_n$ are functions of $x_1, x_2, ..., x_n$, then

$$\frac{\partial(u_1, u_2, u_3, ..., u_n)}{\partial(x_1, x_2, x_3, ..., x_n)} =$$

$$\frac{\partial(u_1, u_2, u_3, ..., u_n)}{\partial(y_1, y_2, y_3, ... y_n)} \cdot \frac{\partial(y_1, y_2, y_3, ..., y_n)}{\partial(x_1, x_2, x_3, ..., x_n)}$$

- If the functions u, v, w of three variables x, y, z are not independent, then the Jacobian of u, v, w w.r.t. x, y, z vanishes.

- $\dfrac{\partial(u, v, w)}{\partial(x, y, z)} \times \dfrac{\partial(x, y, z)}{\partial(u, v, w)} = 1$

- If u_1, u_2, are implicit functions of x_1, x_2, i.e., $F_1(u_1, u_2, x_1, x_2) = 0$ and $F_2(u_1, u_2, x_1, x_2) = 0$ then

$$\frac{\partial(u_1, u_2)}{\partial(x_1, x_2)} = (-1)^2 \left[\frac{\partial(F_1, F_2)}{\partial(x_1, x_2)} \Big/ \frac{\partial(F_1, F_2)}{\partial(u_1, u_2)} \right]$$

- If $v_1, v_2, ..., v_n$ be the functions of n independent variables $x_1, x_2, ..., x_n$ such that $F(v_1, v_2, ..., v_n) = 0$ it is necessary and sufficient that the Jacobian

$$\frac{\partial(v_1, v_2, ..., v_n)}{\partial(x_1, x_2, ..., x_n)}$$ should vanish identically.

REVIEW QUESTIONS AND PROJECT WORK

1. Show that the least positive root of $xy = \sec y$ is a continuous function of $x \in [0, \infty[$ and it increases monotonically from 0 to $\dfrac{\pi}{2}$ as x increases from 0 to ∞.

2. Show that the equation $y^3 \sin x + y^2 \cos^2 x = 7$ determine unique implicit functions in a *nbd* of that point $\left(\dfrac{\pi}{6}, 0\right)$.

3. Show that none of x, y, z can be expressed as a function of the other two in a nbd of any point when

$$\sin x + \sin y + 2 \sec z = 0.$$

4. If $f(1) = 0$ and $f'(x) = \dfrac{1}{x}$ without using integration, prove that

$$f(x) + f(y) = f(x, y)$$

5. If $xyz = \lambda^3$ and a, b, c are positive, show that $(x + a)(y + b)(z + c)$ is minimum when

$$\frac{x}{a} = \frac{y}{b} = \frac{z}{c} = \frac{\lambda}{(abc)^{1/3}}.$$

6. Show that $u = x + y + z$, $v = x - y + z$ and $w = x^2 + y^2 + z^2 - 2yz$ are not functionally independent.

7. If roots of the equation in t
$(t-x)^3+(t-y)^3+(t-z)^3=0$ are u,v,w, show that

$$\frac{\partial(u,v,w)}{\partial(x,y,z)}=-2\frac{(x-y)(y-z)(z-x)}{(u-v)(v-w)(w-u)}$$

OBJECTIVE TYPE QUESTIONS

FILL IN THE BLANKS

1. If u and v are the functions of two independent variables x and y, then Jacobian of u and v with respect to x and y is denoted by _____ .

2. The function u,v and w of three independent variables x,y and z will not _____ if
$$\frac{\partial(u,v,w)}{\partial(x,y,z)}=0.$$

3. It is _____ that $\dfrac{\partial(u,v)}{\partial(x,y)}=\dfrac{\partial(x,y)}{\partial(u,v)}$.

4. If $x=r\cos\theta, y=r\sin\theta$, then the value of
$$\frac{\partial(x,y)}{\partial(r,\theta)}=\underline{\hspace{2cm}}.$$

5. The value of $\dfrac{\partial(u,v)}{\partial(x,y)}\cdot\dfrac{\partial(x,y)}{\partial(u,v)}=\underline{\hspace{2cm}}.$

TRUE/FALSE

Write 'T' for true and 'F' for false statement.

1. If u_1, u_2 are functions of y_1, y_2 and y_1, y_2 are functions of x_1, x_2, then,

$$\frac{\partial(u_1,u_2)}{\partial(x_1,x_2)}=\frac{\partial(u_1,u_2)}{\partial(y_1,y_2)}\cdot\frac{\partial(y_1,y_2)}{\partial(x_1,x_2)} \qquad \textbf{(T/F)}$$

2. $\dfrac{\partial(u_1,u_2)}{\partial(x_1,x_2)}\cdot\dfrac{\partial(x_1,x_2)}{\partial(u_1,u_2)}=0.$ **(T/F)**

3. If the functions u, v, w of three independent variables x, y, z are not independent then the Jacobian of u, v, w with respect to x, y, z vanishes.   **(T/F)**

4. If u_1, u_2 be the implicit function of x_1, x_2, i.e., $F_1(u_1, u_2, x_1, x_2) = 0$ and $F_2(u_1, u_2, x_1, x_2) =0$ then

$$\frac{\partial(F_1,F_2)}{\partial(u_1,u_2)}\times\frac{\partial(u_1,u_2)}{\partial(x_1,x_2)}=(-1)^2\frac{\partial(F_1,F_2)}{\partial(x_1,x_2)}. \quad \textbf{(T/F)}$$

5. The necessary and sufficient condition for the existence of a relation of the form $F(u_1, u_2, ..., u_n) = 0$ is that the Jacobian should vanish identically.   **(T/F)**

MULTIPLE CHOICE QUESTIONS

Choose the most appropriate one.

1. If the functions u,v,w of three independent variables x,y,z and $\dfrac{\partial(u,v,w)}{\partial(x,y,z)}=0$ then the functions are :
   (a) independent
   (b) not independent
   (c) may be independent
   (d) none of the above

2. The value of $\dfrac{\partial(u,v)}{\partial(x,y)}\cdot\dfrac{\partial(x,y)}{\partial(u,v)}$ is :
   (a) 1   (b) 0
   (c) ∞ (d) none of these

3. The necessary and sufficient condition for the existence of a relation $F(u_1,u_2,...,u_n)=0$ is that the Jacobian must be :
   (a) equal to 1   (b) equal to 2
   (c) vanish identically (d) none of the above

4. If x,y,z are connected by a functional relation $f(x, y, z) = 0$ then the value of $\dfrac{\partial(y,z)}{\partial(x,z)}$ is equal to :

   (a) $\dfrac{\partial x}{\partial y}$ (b) $\dfrac{\partial y}{\partial x}$

   (c) $\left(\dfrac{\partial y}{\partial x}\right)_{z=\text{constant}}$ (d) $\left(\dfrac{\partial x}{\partial y}\right)_{z=\text{constant}}$

5. If $x=r\cos\theta, y=r\sin\theta$, then the value of $\dfrac{\partial(r,\theta)}{\partial(x,y)}$:

   (a) 1   (b) r
   (c) $1/r$ (d) none of the above

Answers

FILL IN THE BLANKS

1. $\dfrac{\partial(u,v)}{\partial(x,y)}$ or $J(u,v)$ **2.** independent   **3.** necessary   **4.** r **5.** 1

TRUE/FALSE

1. T   **2.** F   **3.** T   **4.** T   **5.** T

MULTIPLE CHOICE QUESTIONS

1. (b)   **2.** (a)   **3.** (c)   **4.** (c)   **5.** (c)

SELF ASSESSMENT TEST

Verify each of the following :

1. Let $f(x,y)=x^2+y^2-1$ and a point $(0,1)$ so that $f(0,1)=0$ and $f_y(0,1)\neq0$

   Of the two possible solutions $y = \pm\sqrt{1-x^2}$

   (i) $y = +\sqrt{1-x^2}$ is the implicit function in a nbd of $]0,1[$ where $|x|<1, y>0$.

   (ii) $y = -\sqrt{1-x^2}$ is the implicit function in a nbd of $]0,-1[$ where $|x|<1, y<0$.

2. If $f(x,y,z)$ is a function subject to the constraints $g(x, y, z) = 0$ then at a stationary point $f_x g_y - f_y g_x = 0$.

3. The equation $f(x,y)=y^5+2x^2y^3-x^5+y=0$ determines a unique function $y=g(x)$ defined for all x and has continuous derivative for all x.

4. If $f(0)=0$ and $f(x)= \dfrac{1}{\sqrt{1-x^2}}$ then without using integration we have

   $$f(x)+f(y)= f\left(x\sqrt{1-y^2} + y\sqrt{1-x^2}\right)$$

5. If $u_1 = \dfrac{x_1}{x_n}, u_2 = \dfrac{x_2}{x_n},...,u_{n-1} = \dfrac{x_{n-1}}{x_n}$ and $x_1^2+x_2^2+...+x_n^2=1$ then

   $$\frac{\partial(u_1,u_2,...,u_n)}{\partial(x_1,x_2,...,x_n)} = \frac{1}{x_n^{n+1}}.$$

12 Indeterminant Forms

12.1 INTRODUCTION

When a function involves the independent variable in such a manner that for a certain assigned value of that variable, its value cannot be found by simply substituting that value of the variable, the function is said to take an indeterminate form.

The most common cases occuring is that of a fraction whose numerator and denominator both vanish for the value of the variable involved.

As $f(x) \to 0$ and $g(x) \to 0$ when $x \to a$, then the quotient $\dfrac{f(x)}{g(x)}$ is said to have attained the indeterminate form $\dfrac{0}{0}$. Similarly if $\lim\limits_{x \to a} f(x) = \infty$ and $\lim\limits_{x \to a} g(x) = \infty$, then the fraction $\dfrac{f(x)}{g(x)}$ is said to have attained the indeterminate form $\dfrac{\infty}{\infty}$.

The other important indeterminate forms are $0 \times \infty$, $\infty - \infty$, 0^0, 1^∞ and ∞^0.

REMARKS

- The limiting value of the indeterminate forms is also called the true value.
- The most standard form among all indeterminate forms is $\dfrac{0}{0}$. We reduce all other cases of limits to this form.
- It will always be assumed that $f(x)$, $g(x)$ etc. and their respective derivatives are all continuous functions.
- The true value of the indeterminate form $\dfrac{0}{0}$ and $\dfrac{\infty}{\infty}$ is determined by the application of L-Hospital Rule.

12.2 L'HOSPITAL RULE FOR INDETERMINATE FORM $\dfrac{0}{0}$

To find $\lim\limits_{x \to a} \dfrac{f(x)}{g(x)}$ *when* $\lim\limits_{x \to a} f(x) = 0 = \lim\limits_{x \to a} g(x)$.

Let us assume $f(x)$ and $g(x)$ be continuous at $x = a$, then, we have

$$f(a) = \lim\limits_{x \to a} f(x) = 0, \quad g(a) = \lim\limits_{x \to a} g(x) = 0$$

By Taylor's theorem, we have

$$f(a + h) = f(a) + hf'(a + \theta_1 h) = hf'(a + \theta_1 h), \quad 0 < \theta_1 < 1$$
$$g(a + h) = g(a) + hg'(a + \theta_2 h) = hg'(a + \theta_2 h), \quad 0 < \theta_2 < 1$$

Therefore

$$\lim_{x \to a} \frac{f(x)}{g(x)} = \lim_{h \to 0} \frac{f(a+h)}{g(a+h)} = \lim_{h \to 0} \frac{hf'(a+\theta_1 h)}{hg'(a+\theta_2 h)}$$

$$= \lim_{h \to 0} \frac{f'(a+\theta_1 h)}{g'(a+\theta_2 h)} = \frac{f'(a)}{g'(a)} \qquad \text{(provided } g'(a) \neq 0)$$

$$= \lim_{x \to a} \frac{f'(x)}{g'(x)}$$

$$\Rightarrow \qquad \lim_{x \to a} \frac{f(x)}{g(x)} = \lim_{x \to a} \frac{f'(x)}{g'(x)}, \text{ provided } g'(a) \neq 0.$$

If $g'(a) = 0$, then this argument fails. The case when $g'(a) = 0$ but $f'(0) \neq 0$.

$$\lim_{x \to a} \frac{f'(x)}{g'(x)} \to +\infty \text{ or } -\infty.$$

If $f'(a) = 0 = g'(a)$, then by Taylor's theorem, we have

$$f(a+h) = f(a) + hf'(a) + \frac{h^2}{2!} f''(a+\theta_3 h), \quad 0 < \theta_3 < 1$$

$$= \frac{h^2}{2!} f''(a+\theta_3 h)$$

$$g(a+h) = g(a) + hg'(a) + \frac{h^2}{2!} g''(a+\theta_4 h), \quad 0 < \theta_4 < 1$$

$$= \frac{h^2}{2!} g''(a+\theta_4 h)$$

$$\Rightarrow \qquad \lim_{x \to a} \frac{f(x)}{g(x)} = \lim_{h \to a} \frac{f(a+h)}{g(a+h)}$$

$$= \lim_{h \to a} \frac{f''(a+\theta_3 h)}{g''(a+\theta_4 h)} = \frac{f''(a)}{g''(a)}, \text{ provided } g''(a) \neq 0.$$

The case of failure, when $g''(a) = 0$, the limit can be determined as before.
Now, in general if

$$f(a) = f'(a) = f''(a) = \dots = f^{n-1}(a) = 0$$

$$g(a) = g'(a) = g''(a) = \dots = g^{n-1}(a) = 0$$

and $\qquad g^n(a) \neq 0.$

Then, by Taylor's theorem, we get

$$f(a+h) = f(a) + hf'(a) + \dots + \frac{h^{n-1}}{(n-1)!} f^{n-1}(a) + \frac{h^n}{n!} f^n(a+\theta_n h),$$
$$0 < \theta_n < 1$$

$$= \frac{h^n}{n!} f^n(a+\theta_n h).$$

and $\qquad g(a+h) = g(a) + hg'(a) + \dots + \frac{h^{n-1}}{(n-1)!} g^{n-1}(a) + \frac{h^n}{n!} g^n(a+\theta'_n h),$

$$= \frac{h^n}{n!} g^n(a+\theta'_n h). \qquad 0 < \theta'_n < 1$$

Therefore,
$$\lim_{x \to a} \frac{f(x)}{g(x)} = \lim_{h \to a} \frac{f(a+h)}{g(a+h)}$$

$$= \lim_{x \to a} \frac{f^n(a+\theta_n h)}{g^n(a+\theta_n' \, h)} = \frac{f^n(a)}{g^n(a)}, \text{ if } g^n(a) \neq 0$$

$\Rightarrow \qquad \lim_{x \to a} \frac{f(x)}{g(x)} = \lim_{x \to a} \frac{f^n(x)}{g^n(x)}, \text{ provided } g^n(a) \neq 0.$

12.3 L'HOSPITAL RULE FOR INDETERMINATE FORM $\frac{\infty}{\infty}$

If $\lim_{x \to a} f(x) = \infty$ and $\lim_{x \to a} g(x) = \infty$, then

$$\lim_{x \to a} \frac{f(x)}{g(x)} = \lim_{x \to a} \frac{f'(x)}{g'(x)} \text{ provided } \lim_{x \to a} \frac{f'(x)}{g'(x)} \text{ exists.}$$

Proof. Consider
$$\lim_{x \to a} \frac{f(x)}{g(x)} = \lim_{x \to a} \frac{\dfrac{1}{g(x)}}{\dfrac{1}{f(x)}} \qquad\qquad \left[\frac{0}{0}\text{form}\right]$$

$$= \lim_{x \to a} \left\{ \frac{-\dfrac{g'(x)}{[g(x)]^2}}{-\dfrac{f'(x)}{[f(x)]^2}} \right\} \qquad\qquad [\text{By L' Hospital rule}]$$

$\Rightarrow \qquad \lim_{x \to a} \frac{f(x)}{g(x)} = \lim_{x \to a} \frac{g'(x)}{f'(x)} . \lim_{x \to a} \left[\frac{f(x)}{g(x)} \right]^2 \qquad ...(1)$

Now, let $\qquad \lim_{x \to a} \frac{f(x)}{g(x)} = l.$ $\qquad\qquad\qquad\qquad ...(2)$

Then there are following three cases :
Case (i) if $l \neq 0$ and $l \neq \infty$.
In this case, (1) becomes
$$l = \lim_{x \to a} \frac{g'(x)}{f'(x)} . l^2$$

Dividing by l^2, we get
$$\frac{1}{l} = \lim_{x \to a} \frac{g'(x)}{f'(x)}$$

$\Rightarrow \qquad \lim_{x \to a} \frac{f'(x)}{g'(x)} = l = \lim_{x \to a} \frac{f(x)}{g(x)} \qquad\qquad [\text{Using (2)}]$

Case (ii) if $l = 0$.
In this case, adding 1 to each side of (2), we get
$$l + 1 = \lim_{x \to a} \frac{f(x)}{g(x)} + 1 = \lim_{x \to a} \frac{f(x) + g(x)}{g(x)}$$

$$= \lim_{x \to a} \frac{f'(x) + g'(x)}{g'(x)} \qquad\qquad [\text{By case (i)}]$$

$$= \lim_{x \to a} \frac{f'(x)}{g'(x)} + 1 \quad \Rightarrow \quad l = \lim_{x \to a} \frac{f'(x)}{g'(x)}$$

Case (iii) Let $l = \infty$

In this case, by reciprocating, we have

$$\lim_{x \to a} \frac{g(x)}{f(x)} = 0$$

By case (ii)

$$0 = \lim_{x \to a} \frac{g(x)}{f(x)} = \lim_{x \to a} \frac{g'(x)}{f'(x)}$$

Therefore, $\lim_{x \to a} \frac{f'(x)}{g'(x)} = \infty$

Hence, the result $\lim_{x \to a} \frac{f(x)}{g(x)} = \lim_{x \to a} \frac{g'(x)}{f'(x)}$ has been established in every case.

REMARKS

- The above result can be extended to the case when $x \to \infty$, i.e., we can show that

$$\lim_{x \to \infty} \frac{f(x)}{g(x)} = \lim_{x \to \infty} \frac{f'(x)}{g'(x)}$$

Let $x = \dfrac{1}{y}$ then

$$\lim_{x \to \infty} \frac{f(x)}{g(x)} = \lim_{y \to 0} \frac{f\left(\dfrac{1}{y}\right)}{g\left(\dfrac{1}{y}\right)} = \lim_{y \to 0} \frac{f'\left(\dfrac{1}{y}\right)\left(-\dfrac{1}{y^2}\right)}{g'\left(\dfrac{1}{y}\right)\left(-\dfrac{1}{y^2}\right)} = \lim_{y \to 0} \frac{f'\left(\dfrac{1}{y}\right)}{g'\left(\dfrac{1}{y}\right)} = \lim_{x \to \infty} \frac{f'(x)}{g'(x)}$$

- While evaluating $\lim_{x \to \infty} \frac{f(x)}{g(x)}$ when it is of the form $\dfrac{\infty}{\infty}$, care must be taken to change over to the form $\dfrac{0}{0}$ as early as possible, otherwise process of differentiating the numerator and denominator may never terminate.

- While appplying L' Hospital rule, we are not to differentiate $\dfrac{f(x)}{g(x)}$ by the rule for finding the differential coefficient of the quotient of two functions, but we are to differentiate the numerator and denominator separately.

- It must be remember that $\log 1 = 0$, $\log 0 = -\infty$, and $\log \infty = \infty$.

Solved Examples

Example 1. Find $\lim_{x \to 0} \dfrac{e^x - e^{\sin x}}{x - \sin x}$.

Solution. We have $\lim_{x \to 0} \dfrac{e^x - e^{\sin x}}{x - \sin x}$ $\qquad \dfrac{0}{0}$ form

$$= \lim_{x \to 0} \frac{e^x - e^{\sin x} \cdot \cos x}{1 - \cos x} \qquad \frac{0}{0} \text{ form}$$

$$= \lim_{x \to 0} \frac{e^x - [\cos x \cdot e^{\sin x} \cdot \cos x + e^{\sin x}(-\sin x)]}{\sin x}$$

$$= \lim_{x \to 0} \frac{e^x - e^{\sin x}[\cos^2 x - \sin x]}{\sin x} \qquad \frac{0}{0} \text{ form}$$

$$= \lim_{x \to 0} \frac{e^x - e^{\sin x}[2\cos x(-\sin x) - \cos x] - [(\cos^2 x - \sin x)e^{\sin x} \cos x]}{\sin x}$$

$$= \lim_{x \to 0} \frac{e^x - e^{\sin x}[-\sin 2x - \cos x + \cos^3 x - \sin x \cos x]}{\cos x}$$

$$= \frac{1 - 1(-1+1)}{1} = \frac{1}{1} = 1.$$

Example 2. Find $\lim_{x \to 0} \dfrac{x \cos x - \log(1+x)}{x^2}$

Solution. We have $\lim_{x \to 0} \dfrac{x \cos x - \log(1+x)}{x^2}$

$$= \lim_{x \to 0} \frac{x\left(1 - \dfrac{x^2}{2!} + \dfrac{x^4}{4!} - ....\right) - \left(x - \dfrac{x^2}{2} + \dfrac{x^3}{3} - ...\right)}{x^2} \qquad \frac{0}{0} \text{ form}$$

$$= \lim_{x \to 0} \left(\frac{\dfrac{x^2}{2} - \dfrac{5}{6}x^3 + ...}{x^2}\right) = \lim_{x \to 0}\left(\frac{1}{2} - \frac{5}{6}x + \text{terms containing } x\right) = \frac{1}{2}.$$

Example 3. Find $\lim_{x \to 0} \dfrac{\cosh x - \cos x}{x \sin x}$.

Solution. Since we have $\lim_{x \to 0} \dfrac{\cosh x - \cos x}{x \sin x}$ $\qquad \frac{0}{0} \text{ form}$

$$= \lim_{x \to 0}\left[\left(\frac{\cosh x - \cos x}{x^2}\right)\left(\frac{x}{\sin x}\right)\right]$$

$$- \lim_{x \to 0} \frac{\cosh x - \cos x}{x^2} . \lim_{x \to 0}\left(\frac{x}{\sin x}\right) \qquad \frac{0}{0} \text{ form}$$

$$= \lim_{x \to 0} \frac{\sinh x + \sin x}{2x} . 1 \qquad \frac{0}{0} \text{ form}$$

$$= \lim_{x \to 0} \frac{\cosh x + \cos x}{2} = \frac{1+1}{2} = 1.$$

Example 4. Find $\lim_{x \to 0} \dfrac{(1+x)^{1/x} - e}{x}$.

Solution. We have $\lim_{x \to 0} \dfrac{(1+x)^{1/x} - e}{x}$ $\qquad \frac{0}{0} \text{ form}$

Evaluating the limit of expansion for $(1+x)^{1/x}$ in ascending power of x, we get

$$\lim_{x \to 0} \frac{(1+x)^{1/x} - e}{x} = \lim_{x \to 0} \frac{e\left[1 - \frac{1}{2}x + \frac{11}{24}x^2 + \ldots\right] - e}{x}$$

$$= \lim_{x \to 0} \frac{e\left[-\frac{1}{2}x + \frac{11}{24}x^2 + \ldots\right]}{x} = -\frac{1}{2}e.$$

Example 5. *Find* $\lim\limits_{x \to 0} \dfrac{\log \sin 2x}{\log \sin x}$.

Solution. We have $\lim\limits_{x \to 0} \dfrac{\log \sin 2x}{\log \sin x}$ $\left|\dfrac{\infty}{\infty}\right.$ form

$$= \lim_{x \to 0} \frac{\left(\dfrac{2}{\sin 2x}\right)\cos 2x}{\left(\dfrac{1}{\sin x}\right).\cos x} = \lim_{x \to 0} \frac{2\cot 2x}{\cot x} \qquad \left|\frac{\infty}{\infty}\right. \text{ form}$$

$$= \lim_{x \to 0} \frac{-4\cosec^2 2x}{-\cosec^2 x} \qquad \left|\frac{\infty}{\infty}\right. \text{ form}$$

$$= \lim_{x \to 0} \frac{4\sin^2 x}{\sin^2 2x} \qquad \left|\frac{0}{0}\right. \text{ form}$$

$$= \lim_{x \to 0} \frac{4\sin^2 x}{(2\sin x \cos x)^2} = \lim_{x \to 0} \frac{1}{\cos^2 x} = 1.$$

Example 6. *Find* $\lim\limits_{x \to 0} \dfrac{\log \log (1 - x^2)}{\log \log \cos x}$.

Solution. We have $\lim\limits_{x \to 0} \dfrac{\log \log (1 - x^2)}{\log \log \cos x}$ $\left|\dfrac{\infty}{\infty}\right.$ form

$$= \lim_{x \to 0} \frac{\dfrac{1}{\log(1 - x^2)} \cdot \dfrac{1}{(1 - x^2)}(-2x)}{\dfrac{1}{\log \cos x} \cdot \dfrac{1}{\cos x} \cdot (-\sin x)}$$

$$= 2 \lim_{x \to 0} \frac{x \cos x \log \cos x}{\sin x (1 - x^2)\log(1 - x^2)}$$

$$= \left(2 \lim_{x \to 0} \frac{x}{\sin x}\right)\left(\lim_{x \to 0} \frac{\cos x}{1 - x^2}\right).\left(\lim_{x \to 0} \frac{\log \cos x}{\log(1 - x^2)}\right)$$

$$= 2 \times 1 \times 1 \times \lim_{x \to 0} \frac{\log \cos x}{\log(1 - x^2)} \qquad \left|\frac{0}{0}\right. \text{ form}$$

$$= 2 \lim_{x \to 0} \frac{\dfrac{1}{\cos x} \cdot (-\sin x)}{\dfrac{1}{(1 - x^2)} \cdot (-2x)} = 2 \times \frac{1}{2} \cdot \lim_{x \to 0} \left(\frac{\sin x}{x} \cdot \frac{1 - x^2}{\cos x}\right) = 1.$$

Example 7. Find $\lim\limits_{x \to \infty} \dfrac{x^n}{e^x}$, where n is a positive integer.

Solution. We have $\lim\limits_{x \to \infty} \dfrac{x^n}{e^x} = \lim\limits_{x \to \infty} \dfrac{nx^{n-1}}{e^x}$ $\left|\dfrac{\infty}{\infty}\right|$ form

$$= \lim\limits_{x \to \infty} \dfrac{n(n-1)x^{n-2}}{e^x}$$ $\left|\dfrac{\infty}{\infty}\right|$ form

Repeating this process, we get

$$= \lim\limits_{x \to \infty} \dfrac{[n(n-1)(n-2)...n\,\text{factors}]}{e^x}$$

$$= \lim\limits_{x \to \infty} \dfrac{n!}{e^x} = \dfrac{n!}{e^\infty} = \dfrac{n!}{\infty} = 0.$$

Example 8. Find $\lim\limits_{x \to \frac{\pi}{2}} \dfrac{\log\left(x - \dfrac{\pi}{2}\right)}{\tan x}$.

Solution. We have $\lim\limits_{x \to \frac{\pi}{2}} \dfrac{\log\left(x - \dfrac{\pi}{2}\right)}{\tan x} = \lim\limits_{x \to \frac{\pi}{2}} \left(\dfrac{\dfrac{1}{x - \pi/2}}{\sec^2 x}\right)$ $\left|\dfrac{\infty}{\infty}\right|$ form

$$= \lim\limits_{x \to \frac{\pi}{2}} \left(\dfrac{\dfrac{1}{x - \pi/2}}{\dfrac{1}{\cos^2 x}}\right) = \lim\limits_{x \to \frac{\pi}{2}} \left(\dfrac{\cos^2 x}{x - \pi/2}\right)$$ $\left|\dfrac{0}{0}\right|$ form

$$= \lim\limits_{x \to \frac{\pi}{2}} \left(\dfrac{-2\cos x \sin x}{1}\right) = -2\cos\dfrac{\pi}{2}.\sin\dfrac{\pi}{2} = 0.$$

Example 9. Find the following limits :

(i) $\lim\limits_{x \to 0} \dfrac{\log x}{\cot x}$ (ii) $\lim\limits_{x \to 0} \dfrac{\tan x - x}{x^2 \tan x}$

Solution. (i) We have $\lim\limits_{x \to 0} \dfrac{\log x}{\cot x}$ $\left|\dfrac{\infty}{\infty}\right|$ form

$$= \lim\limits_{x \to 0} \dfrac{1/x}{-\text{cosec}^2 x} = -\lim\limits_{x \to 0} \dfrac{\sin^2 x}{x}$$

$$= -\lim\limits_{x \to 0} \left(\dfrac{\sin x}{x}\right).\sin x$$ $\left[\because \lim\limits_{x \to 0}\left(\dfrac{\sin x}{x}\right) = 1\right]$

$$= -1 \times 0 = 0.$$

(ii) We have $\lim\limits_{x \to 0} \dfrac{\tan x - x}{x^2 \tan x}$ $\left[\because \tan x = x + \dfrac{x^3}{3} + \dfrac{2}{15}x^5 + ...\right]$

$$= \lim_{x \to 0} \frac{\left(x + \dfrac{x^3}{3} + \dfrac{2}{15}x^5 + \dots\right) - x}{x^2\left(x + \dfrac{x^3}{3} + \dfrac{2}{15}x^5 + \dots\right)}$$

$$= \lim_{x \to 0} \frac{\dfrac{x^3}{3} + \dfrac{2}{15}x^5 + \dots}{x^3 + \dfrac{x^5}{3} + \dfrac{2}{15}x^7 + \dots} = \lim_{x \to 0} \frac{x^3\left(\dfrac{1}{3} + \dfrac{2}{15}x^2 + \dots\right)}{x^3\left(1 + \dfrac{x^2}{3} + \dfrac{2}{15}x^4 + \dots\right)}$$

$$= \lim_{x \to 0} \frac{\dfrac{1}{3} + \dfrac{2}{15}x^2 + \dots}{1 + \dfrac{x^2}{3} + \dfrac{2}{15}x^4 + \dots} = \frac{\dfrac{1}{3}}{1} = \frac{1}{3}.$$

EXERCISE 12.1

1. Evaluate the following limits :

(i) $\displaystyle\lim_{x \to 0} \frac{x - \sin x}{x^3}$

(ii) $\displaystyle\lim_{x \to 0} \frac{1 - \cos x}{x^2}$

(iii) $\displaystyle\lim_{x \to 0} \frac{a^x - b^x}{x}$

(iv) $\displaystyle\lim_{x \to 1} \frac{\log x}{x - 1}$

(v) $\displaystyle\lim_{x \to 0} \frac{(1+x)^n - 1}{x}$

(vi) $\displaystyle\lim_{x \to 0} \frac{xe^x - \log(1+x)}{x^2}$

(vii) $\displaystyle\lim_{x \to a} \frac{a^x - x^a}{x^x - a^a}$

(viii) $\displaystyle\lim_{x \to 0} \frac{5\sin x - 7\sin 2x + 3\sin 3x}{\tan x - x}$

(ix) $\displaystyle\lim_{x \to 0} \frac{\sin 2x + a \sin x}{x^2}$

(x) $\displaystyle\lim_{x \to 0} \frac{[\cosh x + \log(1 - x) - 1 + x]}{x^2}$

(xi) $\displaystyle\lim_{x \to 1} \frac{x^5 - 2x^3 - 4x^2 + 9x - 4}{x^4 - 2x^3 + 2x - 1}$

2. Find $\displaystyle\lim_{x \to 0} \frac{\sin x \sin^{-1} x - x^2}{x^6}$.

3. Find $\displaystyle\lim_{x \to 0} \frac{(1+x)^{1/x} - e + \dfrac{1}{2}ex}{x^2}$.

4. Evaluate the following limits :

(i) $\displaystyle\lim_{x \to \infty} \frac{a^{1/x} - b^{1/x}}{\log\left(\dfrac{x}{x-1}\right)}$

(ii) $\displaystyle\lim_{x \to \pi/2} \frac{\left(\dfrac{\pi}{2} - x\right)^2 \sin x}{\cos^2 x}$

(iii) $\displaystyle\lim_{x \to 0} \frac{e^x + \log\left(\dfrac{1-x}{e}\right)}{\tan x - x}$

(iv) $\displaystyle\lim_{x \to 0+} \frac{3^x - 2^x}{\sqrt{x}}$

5. (i) If $\displaystyle\lim_{y \to 0} \frac{re^y - q\cos y + pe^{-y}}{y \tan y} = 3$, find the vlaues of p, q, and r.

(ii) Find the values of a and b in order that $\displaystyle\lim_{x \to 0} \frac{x(1 + a\cos x) - b \sin x}{x^3}$ may be equatl to 1.

(iii) If $\displaystyle\lim_{x \to 0} \frac{\sin 2x + a \sin x}{x^3}$ be finite, find the value of a and the limit.

6. Evaluate the following limits :

(i) $\lim\limits_{x \to 0} \dfrac{\log x^2}{\cot x^2}$

(ii) $\lim\limits_{x \to a} \dfrac{\log(x-a)}{\log(e^x - e^a)}$

(iii) $\lim\limits_{x \to 1-0} \dfrac{\log(1-x)}{\cot \pi x}$

(iv) $\lim\limits_{x \to \infty} \dfrac{\log x}{a^x}, a > 1$

(v) $\lim\limits_{x \to \pi/2} \dfrac{\tan x}{\tan 3x}$

HINTS TO SELECTED PROBLEMS

1. (i) $\lim\limits_{x \to 0} \dfrac{x - \sin x}{x^3}$

$$= \lim_{x \to 0} \dfrac{x - \left(x - \dfrac{x^3}{3!} + \dfrac{x^5}{5!} - \ldots\right)}{x^3}$$

$$= \lim_{x \to 0} \left\{ \dfrac{\dfrac{x^3}{3!} - \dfrac{x^5}{5!} + \ldots}{x^3} \right\}$$

$$= \lim_{x \to 0} \left[\dfrac{1}{3!} - \dfrac{x^2}{5!} + \ldots \right] = \dfrac{1}{3!} = \dfrac{1}{6}.$$

(vii) $\lim\limits_{x \to a} \dfrac{a^x - x^a}{x^x - a^a} = \lim\limits_{x \to a} \dfrac{a^x \log a - a \cdot x^{a-1}}{x^x(1 + \log x)}$

$$= \dfrac{a^a \log a - a^a}{a^a(1 + \log a)}$$

$$= \dfrac{\log a - 1}{\log a + 1}.$$

5. (i) If $r - q + p = 0$, ...(1)

we obtained $\dfrac{0}{0}$ form,

Then we have

$$\lim_{y \to 0} \dfrac{re^y + q \sin y + pe^{-y}}{y \sec^2 y + \tan y}$$

Again if $r - p = 0$, ...(2)

We obtained $\dfrac{0}{0}$ form.

Then we have

$$\lim_{y \to 0} \dfrac{re^y + q \cos y + pe^{-y}}{\sec^2 y + 2y \sec^2 y \tan y + \sec^2 y}$$

$$\Rightarrow \quad r + q + p = 6 \quad \ldots(3)$$

Now solving (1), (2) and (3).

6. (iv) $\lim\limits_{x \to \infty} \dfrac{\log x}{a^x}, a > 1$ $\quad\left| \text{form } \dfrac{\infty}{\infty} \right.$

$$= \lim_{x \to \infty} \dfrac{1/x}{a^x \log a} = \dfrac{1}{\log a} \lim_{x \to \infty} \dfrac{1}{x a^x}$$

$$= \dfrac{1}{\log a} \cdot 0 = 0.$$

Answers

1. (i) $\dfrac{1}{6}$ (ii) $\dfrac{1}{2}$ (iii) $\log \dfrac{a}{b}$ (iv) 1 (v) n (vi) $\dfrac{3}{2}$ (vii) $\dfrac{\log a - 1}{\log a + 1}$ (viii) -15

(ix) $\begin{cases} \infty \text{ if } a \neq -2 \\ 0 \text{ if } a = -2 \end{cases}$ (x) 0 (xi) 4

2. $\dfrac{1}{18}$ **3.** $\dfrac{11e}{24}$ **4.** (i) $\log \dfrac{a}{b}$ (ii) 1 (iii) $-\dfrac{1}{2}$ (iv) 0

5. (i) $p = \dfrac{3}{2}, q = 3, r = \dfrac{3}{2}$ (ii) $a = -\dfrac{5}{2}, b = -\dfrac{3}{2}$ (iii) $\begin{cases} -1 \text{ if } a = -2 \\ \infty \text{ if } a \neq -2 \end{cases}$

6. (i) 0 (ii) 1 (iii) 0 (iv) 0 (v) 3

12.4 THE INDETERMINATE FORM $0 \times \infty$

To find $\lim_{x \to a} [f(x) \cdot g(x)]$, when $\lim_{x \to a} f(x) = 0$ and $\lim_{x \to a} g(x) = \infty$.

To determine this limit, the product may be transformed into the form $\dfrac{0}{0}$ or $\dfrac{\infty}{\infty}$, using any one of the following relations

$$f(x) \cdot g(x) = \frac{f(x)}{\dfrac{1}{g(x)}} \quad \text{or} \quad f(x) \cdot g(x) = \frac{g(x)}{\dfrac{1}{f(x)}}$$

and then apply previous method.

Solved Exmaples

Example 1. *Evaluate* $\lim_{x \to 0^+} (x \log x)$.

Solution.
$$\lim_{x \to 0^+} (x \log x) = \lim_{x \to 0^+} \frac{\log x}{1/x} \qquad \left| \frac{\infty}{\infty} \text{ form} \right.$$

$$= \lim_{x \to 0^+} \frac{1/x}{-1/x^2} = \lim_{x \to 0^+} (-x) = 0.$$

Example 2. *Evaluate* $\lim_{x \to 0} x \log \sin x$.

Solution.
$$\lim_{x \to 0} x \log \sin x \qquad \qquad \qquad | \, 0 \times \infty \text{ from}$$

$$= \lim_{x \to 0} \left(\frac{\log \sin x}{1/x} \right) \qquad \left| \frac{\infty}{\infty} \text{ form} \right.$$

$$= \lim_{x \to 0} \frac{(1/\sin x) \cdot \cos x}{-1/x^2} \qquad \left| \frac{\infty}{\infty} \text{ form} \right.$$

$$= \lim_{x \to 0} \frac{-x^2 \cos x}{\sin x} \qquad \left| \frac{0}{0} \text{ form} \right.$$

$$= \lim_{x \to 0} \frac{x^2 \sin x - 2x \cos x}{\cos x} = 0.$$

12.5 THE INDETERMINATE FORM $\infty - \infty$

To determine $\lim_{x \to a} [f(x) - g(x)]$, when $\lim_{x \to a} f(x) = \infty = \lim_{x \to \infty} g(x)$.

Here, this can be reduced to the form $\dfrac{0}{0}$ by the relation

$$f(x) - g(x) = \left\{ \frac{\left[\dfrac{1}{g(x)} - \dfrac{1}{f(x)} \right]}{\dfrac{1}{f(x) \cdot g(x)}} \right\}$$

and then by previous method.

WORKING PROCEDURE

Step 1. *Change all trigonometric-ratio into sin x and cos x (if T-ratio are present)*
Step 2. *Take L.C.M.*
Now the indeterminate form is reduced into $\dfrac{0}{0}$ *form.*

Solved Exmaples

Example 1. *Evaluate* $\lim\limits_{x\to 0}\left(\dfrac{1}{x^2}-\dfrac{1}{\sin^2 x}\right)$.

Solution.
$$\lim_{x\to 0}\left(\frac{1}{x^2}-\frac{1}{\sin^2 x}\right) \qquad\qquad |\,\infty-\infty\text{ form}$$

$$= \lim_{x\to 0}\frac{\sin^2 x - x^2}{x^2\sin^2 x} \qquad\qquad \left|\frac{0}{0}\text{ form}\right.$$

$$= \lim_{x\to 0}\frac{\left(x-\dfrac{x^3}{3!}+...\right)^2 - x^2}{x^2\left(x-\dfrac{x^3}{3!}+...\right)}$$

$$= \lim_{x\to 0}\frac{-\dfrac{2x^4}{3!}+\text{terms containing higher powers of }x}{x^4+\text{terms containing higher power of }x}$$

$$= \lim_{x\to 0}\frac{-\dfrac{2}{3!}+\text{terms containing }x\text{ in the numerator}}{1+\text{terms containing }x\text{ in the numerator}}$$

$$= -\frac{2}{3!} = -\frac{1}{3}$$

Example 2. *Evaluate* $\lim\limits_{x\to \pi/2}(\sec x - \tan x)$.

Solution. We have $\lim\limits_{x\to \pi/2}(\sec x - \tan x)$ $\qquad\qquad |\,\infty-\infty\text{ form}$

$$= \lim_{x\to \pi/2}\left(\frac{1}{\cos x}-\frac{\sin x}{\cos x}\right) \qquad\qquad \left|\frac{0}{0}\text{ form}\right.$$

$$= \lim_{x\to \pi/2}\left(\frac{1-\sin x}{\cos x}\right) = \lim_{x\to \pi/2}\frac{-\cos x}{-\sin x} = \lim_{x\to \pi/2}\cot x = 0.$$

Example 3. *Evaluate* $\lim\limits_{x\to \pi/2}\left(\sec x - \dfrac{1}{1-\sin x}\right)$.

Solution. We have $\lim\limits_{x\to \pi/2}\left(\sec x - \dfrac{1}{1-\sin x}\right)$ $\qquad\qquad |\,\infty-\infty\text{ form}$

$$= \lim_{x\to \pi/2}\left(\frac{1}{\cos x}-\frac{1}{1-\sin x}\right) \qquad\qquad |\,\infty-\infty\text{ form}$$

$$= \lim_{x\to \pi/2}\left(\frac{1-\sin x-\cos x}{\cos x-\cos x\sin x}\right) \qquad\qquad \left|\frac{0}{0}\text{ form}\right.$$

$$= \lim_{x\to \pi/2}\frac{-\cos x+\sin x}{-\sin x+\sin^2 x-\cos^2 x} = \frac{-0+1}{-1+1-0} = \infty.$$

12.6 THE INDETERMINATE FORM $0°$, $1^∞$, $∞°$

To determinate $\lim\limits_{x \to a} [f(x)]^{g(x)}$ *when the limit is of the form* $0°$, $1^∞$, $∞°$

Let $$y = [f(x)]^{g(x)}$$

Taking logs; $$\log y = g(x) \log f(x)$$

The R.H.S. assumes the indeterminate forms $0 \times \infty$ in each of these above cases. The limit can, therefore, be determined by the method discussed in the article 12.4.

Suppose

$$\lim_{x \to a} [g(x) \log f(x)] = l \text{ (say)}$$

$\Rightarrow$ $\lim\limits_{x \to a} \log y = l$ $\Rightarrow$ $\lim\limits_{x \to a} [\log y] = l$

$\Rightarrow$ $\lim\limits_{x \to a} y = e^l$ $\Rightarrow$ $\lim\limits_{x \to a} [f(x)]^{g(x)} = e^l$.

WORKING PROCEDURE

Step 1. *Let the given limit = y.*

Step 2. *Take logs on both sides to get the forms* $0 \times \infty$ *and proceed by the method of the type* $0 \times \infty$.

Solved Examples

Example 1. *Evaluate* $\lim\limits_{\theta \to \frac{\pi}{2}} (\cos\theta)^{\cos\theta}$.

Solution. Let $$y = (\cos\theta)^{\cos\theta} \qquad\qquad | \; 0° \text{ form}$$

Taking logs, $\log y = \cos\theta \log \cos\theta$.

$\therefore$ $\lim\limits_{\theta \to \frac{\pi}{2}} (\log y) = \lim\limits_{\theta \to \frac{\pi}{2}} \cos\theta \log \cos\theta \qquad\qquad | \; 0 \times \infty \text{ form}$

$$= \lim_{\theta \to \frac{\pi}{2}} \frac{\log \cos\theta}{\sec\theta} \qquad\qquad \left|\frac{\infty}{\infty}\right. \text{ form}$$

$$= \lim_{\theta \to \frac{\pi}{2}} \frac{\dfrac{1}{\cos\theta} \times -\sin\theta}{\sec\theta \tan\theta}$$

$$= \lim_{\theta \to \frac{\pi}{2}} (-\cos\theta) = 0$$

$\Rightarrow \lim\limits_{\theta \to \frac{\pi}{2}} (\log y) = 0 \Rightarrow \log\left(\lim\limits_{\theta \to \frac{\pi}{2}} y\right) = 0 \Rightarrow \lim\limits_{\theta \to \frac{\pi}{2}} y = e^0 = 1$

$\Rightarrow \lim\limits_{\theta \to \frac{\pi}{2}} (\cos\theta)^{\cos\theta} = 1.$

Example 2.   *Find* $\lim\limits_{x\to 0}\left(\dfrac{\tan x}{x}\right)^{1/x^2}$.

Solution.   Let
$$y=\left(\frac{\tan x}{x}\right)^{1/x^2} \qquad\qquad |\ 1^{\infty}\text{ form for } x=0$$

$$\Rightarrow \qquad\qquad \log y=\frac{1}{x^2}\log\frac{\tan x}{x}$$

$$\Rightarrow \qquad\qquad \lim_{x\to 0}\log y=\lim_{x\to 0}\frac{1}{x^2}\log\frac{\tan x}{x}$$

$$=\lim_{x\to 0}\frac{\log\dfrac{\tan x}{x}}{x^2} \qquad\qquad \left|\frac{0}{0}\text{ form}\right.$$

$$=\lim_{x\to 0}\left\{\frac{1\left[\dfrac{x\sec^2 x-\tan x}{x^2}\right]}{\left(\dfrac{\tan x}{x}\right)}\middle/ 2x\right\}$$

$$=\lim_{x\to 0}\frac{x\sec^2 x-\tan x}{2x^3} \qquad\qquad \left|\ \therefore\ \lim_{x\to 0}\frac{\tan x}{x}=1\right|$$

$$=\lim_{x\to 0}\frac{x.2\sec x\sec x\tan x+\sec^2 x-\sec^2 x}{6x^2}$$

$$=\lim_{x\to 0}\frac{2x\tan x\sec^2 x}{6x^2}=\lim_{x\to 0}\frac{\tan x\sec^2 x}{3x}$$

$$=\lim_{x\to 0}\left(\frac{1}{3}.\frac{\tan x}{x}.\sec^2 x\right)=\lim_{x\to 0}\frac{1}{3}\times 1\times\sec^2 x=\frac{1}{3}$$

$$\therefore \qquad\qquad \lim_{x\to 0}y=e^{1/3}\Rightarrow\lim_{x\to 0}\left(\frac{\tan x}{x}\right)^{1/x^2}=e^{1/3}.$$

Example 3.   *Evaluate* $\lim\limits_{x\to 0}\left(\dfrac{\sin x}{x}\right)^{1/x}$.

Solution.   Let $\quad y=\lim\limits_{x\to 0}\left(\dfrac{\sin x}{x}\right)^{1/x}$

$$\therefore \qquad \log y=\lim_{x\to 0}\left(\frac{1}{x}\log\frac{\sin x}{x}\right)=\lim_{x\to 0}\frac{1}{x}\log\left\{\frac{x-\dfrac{x^3}{3!}+\dfrac{x^5}{5!}-\cdots}{x}\right\}$$

$$=\lim_{x\to 0}\frac{1}{x}\log\left(1-\frac{x^2}{3!}+\frac{x^4}{5!}-\cdots\right)=\lim_{x\to 0}\frac{1}{x}\log\left[1-\left(\frac{x^2}{6}-\frac{x^4}{120}+\cdots\right)\right]$$

$$= \lim_{x \to 0} \frac{1}{x} \log(1 - z) \quad \text{where } z = \frac{x^2}{6} - \frac{x^4}{120} + \ldots$$

$$= \lim_{x \to 0} \frac{1}{x}\left(-z - \frac{z^2}{2} - \ldots \right)$$

$$= \lim_{x \to 0} \frac{1}{x}\left[-\left(\frac{x^2}{6} - \frac{x^4}{120} + \ldots \right) - \frac{1}{2}\left(\frac{x^2}{6} - \frac{x^4}{120} + \ldots \right)^2 - \ldots \right]$$

$$= \lim_{x \to 0} \frac{1}{x}\left[-\frac{x^2}{6} + \left(\frac{x^4}{120} - \frac{x^4}{72} \right) + \ldots \right] = \lim_{x \to 0} \frac{1}{x}\left[-\frac{x^2}{6} + \frac{x^4}{180} + \ldots \right]$$

$$= \lim_{x \to 0}\left[-\frac{x}{6} + \frac{x^3}{180} + \ldots \right] = 0.$$

Hence, $\quad y = e^0 = 1.$

Example 4. *Evaluate* $\lim_{x \to 0} (\operatorname{cosec} x)^{1/\log x}$.

Solution. We have

$$y = \lim_{x \to 0} (\operatorname{cosec} x)^{1/\log x} \qquad \qquad |\infty^0 \text{ form}$$

$$\therefore \qquad \log y = \lim_{x \to 0} \frac{1}{\log x}(\log \operatorname{cosec} x) \qquad \qquad \left|\frac{\infty}{\infty} \text{ form}\right.$$

$$= \lim_{x \to 0} \frac{\left(\dfrac{1}{\operatorname{cosec} x} \right)(-\operatorname{cosec} x \cot x)}{1/x}$$

$$= \lim_{x \to 0} \left(-\frac{x}{\tan x} \right) \qquad \qquad \left|\frac{0}{0} \text{ form}\right.$$

$$= \lim_{x \to 0} \left(-\frac{1}{\sec^2 x} \right) = -1.$$

$$\Rightarrow \qquad y = e^{-1} = \frac{1}{e}$$

Example 5. *Evaluate* $\lim_{x \to 0} \left(\dfrac{\sin x}{x} \right)^{1/x^3}$.

Solution. Let

$$y = \lim_{x \to 0} \left(\frac{\sin x}{x} \right)^{1/x^3}$$

$$\Rightarrow \qquad \log y = \lim_{x \to 0} \left(\frac{1}{x^3} \log \frac{\sin x}{x} \right)$$

$$= \lim_{x \to 0} \frac{1}{x^3} \log \left\{ \frac{x - \dfrac{x^3}{3!} + \dfrac{x^5}{5!} - \ldots}{x} \right\}$$

$$= \lim_{x \to 0} \frac{1}{x^3} \log\left(1 - \frac{x^2}{3!} + \frac{x^4}{5!} - \ldots\right)$$

$$= \lim_{x \to 0} \frac{1}{x^3} \log(1 - z) \qquad \left[\text{where } z = \frac{x^2}{6} - \frac{x^4}{120} - \ldots\right]$$

$$= \lim_{x \to 0} \frac{1}{x^3}\left(-z - \frac{z^2}{2} - \ldots\right)$$

$$= \lim_{x \to 0} \frac{1}{x^3}\left[-\left(\frac{x^2}{6} - \frac{x^4}{120} + \ldots\right) - \frac{1}{2}\left(\frac{x^2}{6} - \frac{x^4}{120} + \ldots\right)^2 - \ldots\right]$$

$$= \lim_{x \to 0} \frac{1}{x^3}\left[-\frac{x^2}{6} + \left(\frac{x^4}{120} - \frac{x^4}{72}\right) + \ldots\right]$$

$$= \lim_{x \to 0} \frac{1}{x^3}\left[-\frac{x^2}{6} - \frac{x^4}{180} + \ldots\right]$$

$$= \lim_{x \to 0} \frac{1}{x^3}\left[-\frac{1}{6x} - \frac{x}{180} + \ldots\right] = \infty.$$

$$\Rightarrow \qquad y = e^{\infty} = \infty.$$

Example 6. (a) *Evaluate the following limits :*

    (i) $\displaystyle\lim_{x \to 0} (\cos x)^{\cot x}$ (ii) $\displaystyle\lim_{x \to 0} \frac{e^x - e^{-x} - 2\log(1+x)}{x \sin x}$

(b) *If* $f(x) = (x-1)(x-3)(x-5)$, $a = 0$, $b = 4$, *find the value of* c *such that* $f'(c)$
has the same value as the slope of the chord joining the points for which $x = 0$
and $x = 4$

Solution. (a) (i) Let $\qquad y = \displaystyle\lim_{x \to 0} (\cos x)^{\cot x}$

$$\log y = \lim_{x \to 0} \log(\cos x)^{\cot x} = \lim_{x \to 0} \cot \log \cos x$$

$$\log y = \lim_{x \to 0} \frac{\log \cos x}{\tan x}$$

$$= \lim_{x \to 0} \frac{(-\sin x)}{\cos x \cdot \sec^2 x} = \lim_{x \to 0} \frac{-\sin x}{\sec x}$$

$$= \lim_{x \to 0} -\sin x \cos x = \lim_{x \to 0} -\frac{\sin 2x}{2 \cdot x} \cdot x$$

$$= -\lim_{x \to 0}\left(\frac{\sin 2x}{2x}\right) \cdot \lim_{x \to 0} x = -1 \times 0.$$

$$\therefore \qquad \log y = 0 \Rightarrow y = e^0$$
$$= y = 1.$$

(ii) We have $\displaystyle\lim_{x\to 0}\frac{e^x - e^{-x} - 2\log(1-x)}{x\sin x}$ $\left|\dfrac{0}{0}\right|$ form

$$= \lim_{x\to 0}\frac{e^x + e^{-x} - \dfrac{2}{1+x}}{x\cos x + \sin x} \qquad \left|\dfrac{0}{0}\right| \text{ form}$$

$$= \lim_{x\to 0}\frac{e^x - e^{-x} + \dfrac{2}{(1+x)^2}}{2\cos x - x\sin x}$$

$$= \frac{e^0 - e^0 + \dfrac{2}{(1+0)^2}}{2\cdot\cos 0 - 0} = \frac{1 - 1 + \dfrac{2}{(1+0)^2}}{2\cdot\cos 0 - 0} = \frac{2}{2} = 1.$$

(b) Same as (a).

Example 7. Find $\displaystyle\lim_{x\to 0}\left(\frac{\tan x}{x}\right)^{1/x^3}$.

Solution. Let

$$y = \lim_{x\to 0}\left(\frac{\tan x}{x}\right)^{1/x^3}$$

$$\log y = \lim_{x\to 0}\frac{1}{x^3}\log_e\left(\frac{\tan x}{x}\right)$$

$$= \lim_{x\to 0}\frac{\log_e \tan x - \log_e x}{x^3} \qquad \left|\dfrac{0}{0}\right| \text{ form}$$

$$= \lim_{x\to 0}\frac{\dfrac{\sec^2 x}{\tan x} - \dfrac{1}{x}}{3x^2} = \lim_{x\to 0}\frac{\dfrac{2x}{\sin 2x} - \dfrac{1}{x}}{3x^2}$$

$$= \lim_{x\to 0}\frac{2x - \sin 2x}{3x^2 \sin 2x} \qquad \left|\dfrac{0}{0}\right| \text{ form}$$

$$= \lim_{x\to 0}\frac{2 - 2\cos 2x}{6x^2 \sin 2x + 6x^3 \cos 2x} \qquad \left|\dfrac{0}{0}\right| \text{ form}$$

$$= \lim_{x\to 0}\frac{2\sin 2x}{15x^2 \cos 2x + 6x\sin 2x - 6x^3 \sin 2x} \qquad \left|\dfrac{0}{0}\right| \text{ form}$$

$$= \lim_{x\to 0}\frac{4\cos 2x}{-30x^2 \sin 2x + 30x\cos 2x + 6\sin 2x + 12x\cos 2x}$$
$$-18x^2 \sin 2x - 12x^3 \cos 2x$$

$$= \frac{4}{0} = \infty.$$

Hence, $y = e^\infty = \infty$.

EXERCISE 12.2

1. Evaluate the following limits :

   (i) $\displaystyle\lim_{x\to 0} x\log\tan x$ (ii) $\displaystyle\lim_{x\to 0}\tan\left(\frac{\pi}{2}-x\right)$

   (iii) $\displaystyle\lim_{x\to\infty} 2^x\sin\frac{a}{2^x}$ (iv) $\displaystyle\lim_{x\to\infty}(a^{1/x}-1)\cdot x$

   (v) $\displaystyle\lim_{x\to 1}\sec\frac{\pi}{2x}\log x$

   (vi) $\displaystyle\lim_{x\to 0} x^m(\log x)^n \; m;n\in Z^+.$

2. Evaluate the following limits :

   (i) $\displaystyle\lim_{x\to 0}\left[\frac{1}{x}-\frac{1}{x^2}\log(1+x)\right]$

   (ii) $\displaystyle\lim_{x\to 2}\left(\frac{1}{x-2}-\frac{1}{\log(x-1)}\right)$

   (iii) $\displaystyle\lim_{x\to 0}\left[\frac{1}{x^2}-\cosec^2 x\right]$

   (iv) $\displaystyle\lim_{x\to 0}\left(\frac{1}{x^2}-\cot^2 x\right)$

   (v) $\displaystyle\lim_{x\to 0}\left(\frac{1}{x^2}-\frac{1}{x\tan x}\right)$

   (vi) $\displaystyle\lim_{x\to 0}\left(\cosec x-\frac{1}{x}\right).$

3. Evaluate the following limits :

   (i) $\displaystyle\lim_{x\to 0}\left(\frac{1}{x}\right)^{\tan x}$ (ii) $\displaystyle\lim_{x\to 0} x^x$

   (iii) $\displaystyle\lim_{x\to\infty}\left(\frac{\pi}{2}-\tan^{-1}x\right)^{1/x}$

   (iv) $\displaystyle\lim_{x\to 0}\left(\frac{\tan x}{x}\right)^{1/x}$

   (v) $\displaystyle\lim_{x\to\pi/2}(\sin x)^{\tan x}$

   (vi) $\displaystyle\lim_{x\to\pi/4}(\tan x)^{\tan 2x}$

   (vii) $\displaystyle\lim_{x\to 0}\left[\frac{2(\cosh x-1)}{x^2}\right]^{1/x^2}$

   (viii) $\displaystyle\lim_{x\to 0}(\cosec x)^{1/\log x}$

   (ix) $\displaystyle\lim_{x\to a}\left(2-\frac{x}{a}\right)^{\tan\left(\frac{\pi x}{2a}\right)}$

   (x) $\displaystyle\lim_{x\to\infty}\left(1+\frac{k}{x}\right)^x$

4. Evaluate the following limits :

   (i) $\displaystyle\lim_{x\to 0}\left[\frac{a^x+b^x}{2}\right]^{1/x}$

   (ii) $\displaystyle\lim_{x\to 0}\left[\frac{a_1^x+a_2^x+...+a_n^x}{n}\right]^{1/x}$

   (iii) $\displaystyle\lim_{x\to\infty}\left(1+\frac{a}{x}\right)^x.$

HINTS TO SELECTED PROBLEMS

1. (iii) $\displaystyle\lim_{x\to\infty} 2^x.\sin\frac{a}{2^x}=\lim_{x\to\infty}\frac{\sin\left(\dfrac{a}{2^x}\right)}{\dfrac{1}{2^x}}$

   $=a\displaystyle\lim_{x\to\infty}\left[\frac{\sin\left(\dfrac{a}{2^x}\right)}{\dfrac{a}{2^x}}\right]$

   $=a\cdot 1=a.$

   (v) $\displaystyle\lim_{x\to 1}\sec\left(\frac{\pi}{2x}\right).\log x=\lim_{x\to 1}\frac{\log x}{\cos\left(\dfrac{\pi}{2x}\right)}$

   $=\displaystyle\lim_{x\to 1}\frac{\dfrac{1}{x}}{-\sin\left(\dfrac{\pi}{2x}\right)\dfrac{\pi}{2x^2}}=\lim_{x\to 1}\frac{2x}{\pi\sin\left(\dfrac{\pi}{2x}\right)}$

   $=\dfrac{2\times 1}{\pi\cdot\sin\left(\dfrac{\pi}{2}\right)}=\dfrac{2}{\pi}.$

2. (v) $\displaystyle\lim_{x\to 0}\left(\frac{1}{x^2}-\frac{1}{x\tan x}\right)=\lim_{x\to 0}\frac{\tan x-x}{x^2\tan x}$

   $=\displaystyle\lim_{x\to 0}\frac{\left(x+\dfrac{x^3}{3}+\dfrac{2}{15}x^5+...\right)-x}{x^2\left(x+\dfrac{x^3}{3}+\dfrac{2}{15}x^5+...\right)}$

$$= \lim_{x \to 0} \frac{x^3 \left(\frac{1}{3} + \frac{2}{15} x^2 + \ldots \right)}{x^3 \left(1 + \frac{x^2}{3} + \frac{2}{15} x^4 + \ldots \right)}$$

$$= \lim_{x \to 0} \frac{\frac{1}{3} + \frac{2}{15} x^2 + \ldots}{1 + \frac{x^2}{3} + \frac{2}{15} x^4 + \ldots} = \frac{1}{3}$$

$$= -\lim_{x \to 0} \tan x \log x$$

$$= -\lim_{x \to 0} \frac{\log x}{\cot x}$$

$$= -\lim_{x \to 0} \frac{1/x}{-\csc^2 x} = \lim_{x \to 0} \frac{\sin^2 x}{x}$$

$$= \lim_{x \to 0} \left(\frac{\sin x}{x} \right) . \sin x = 1 \times 0 = 0$$

3. (i) $\quad \log y = \lim_{x \to 0} \left(\frac{1}{x} \right)^{\tan x}$

$\Rightarrow \quad \log y = \lim_{x \to 0} \tan x \log \left(\frac{1}{x} \right)$

$\Rightarrow \qquad y = e^0 = 1.$

Answers

1. (i) 0   (ii) ∞ (iii) a (iv) $\log a$ (v) $-\dfrac{2}{\pi}$ (vi) 0.

2. (i) $\dfrac{1}{2}$ (ii) $-\dfrac{1}{2}$ (iii) $-\dfrac{1}{3}$ (iv) $\dfrac{2}{3}$ (v) $\dfrac{1}{3}$ (vi) 0.

3. (i) 1   (ii) 1   (iii) 1   (iv) 1   (v) 1   (vi) $\dfrac{1}{e}$ (vii) $e^{1/12}$ (viii) $\dfrac{1}{e}$

(ix) $e^{2/\pi}$ (x) e^k.

4. (i) $\sqrt{ab}$ (ii) $(a_1 . a_2 \ldots a_n)^{1/n}$ (iii) e^a.

CHAPTER REVIEW: A COMPETITIVE APPROACH

SELECTED TERMS AND RESULTS

TERMS

- When a function involves the independent variables in such a manner that for a certain assigned value of that variable, its value cannot be found by simply substituting that value of the variable, the function is said to take an indeterminate form.

RESULTS

- Important indeterminate forms are $0 \times \infty$, $\infty - \infty$, 0^0, 1^∞ and ∞^0.

- If $\lim_{x \to a} f(x) = \lim_{x \to a} g(x) = 0$

Then $\lim_{x \to a} \dfrac{f(x)}{g(x)} = \lim_{x \to a} \dfrac{f'(x)}{g'(x)}$

- If $\lim_{x \to a} f(x) = \infty$ and $\lim_{x \to a} g(x) = \infty$, then

$\lim_{x \to a} \dfrac{f(x)}{g(x)} = \lim_{x \to a} \dfrac{f'(x)}{g'(x)}$, provided

$\lim_{x \to a} \dfrac{f'(x)}{g'(x)}$ exists.

- If $\lim_{x \to a} f(x) = 0$ and $\lim_{x \to a} g(x) = \infty$

$$\lim_{x \to a} [f(x) \cdot g(x)] = \lim_{x \to a} \dfrac{g(x)}{\dfrac{1}{f(x)}}.$$

- If $\lim_{x \to a} f(x) = \infty = \lim_{x \to a} g(x)$

then $\lim_{x \to a} [f(x) - g(x)] = \lim_{x \to a} \dfrac{\left[\dfrac{1}{g(x)} - \dfrac{1}{f(x)}\right]}{\dfrac{1}{f(x) \cdot g(x)}}$

If limits of the form 0^0, 1^∞, ∞^0, such that

$\lim_{x \to a} \cdot [f(x)]^{g(x)} = e^l$

where $l = \lim_{x \to a} [g(x) \cdot \log f(x)]$.

REVIEW QUESTIONS AND PROJECT WORK

1. Show that $\lim_{x \to 0} \dfrac{(\tan^{-1} x)^2}{\log(1 + x^2)} = 1$

2. Show that $\lim_{x \to \infty} \dfrac{a^{1/x} - b^{1/x}}{\log \dfrac{x}{x-1}} = \log\left(\dfrac{a}{b}\right)$

3. Show that $\lim_{x \to a^+} \dfrac{\log(x - a)}{\log(e^x - e^a)} = 1$

4. Show that $\lim_{x \to 1} \sec\dfrac{\pi}{2x} \log x = \dfrac{2}{\pi}$

5. Show that $\lim_{x \to 0} (\cos x)^{\cot^2 x} = e^{-1/2}$

6. Show that $\lim_{x \to 0} \left(\dfrac{\sinh x}{x}\right)^{1/x^2} = e$

7. Show that

$\lim_{x \to \infty} \dfrac{a_0 x^n + a_1 x^{n-1} + \dots + a_n}{b_0 x^m + b_1 x^{m-1} + b_2 x^{m-2} + \dots + b_m}$

$= \begin{cases} \infty ; \text{ for } n > m \\ \dfrac{a_0}{b_0} ; \text{ for } n = m \\ 0 \text{ ; for } n < m \end{cases}$

8. Show that $\lim_{x \to 0^+} (\cos x)^{1/x^3} = 0$

9. Show that $\lim_{x \to a} \left(2 - \dfrac{x}{a}\right)^{\tan\left(\dfrac{\pi x}{2a}\right)} = e^{2/\pi}$

10. Show that $\lim_{x \to 0} \dfrac{\log \log(1 - x^2)}{\log \log \cos x} = 1$

OBJECTIVE TYPE QUESTIONS

FILL IN THE BLANKS

1. $\lim\limits_{x\to 1} \dfrac{\log x}{x-1} =$ _____

2. $\lim\limits_{x\to 0} \dfrac{\sin ax}{\cos ax} =$ _____

3. $\lim\limits_{x\to \infty} \dfrac{x^2 + 2x}{5 - 3x^2} =$ _____

4. $\lim\limits_{x\to 0} \dfrac{\tan x}{x} =$ _____

5. $\lim\limits_{x\to 0} x \log x =$ _____

6. $\lim\limits_{x\to 1} \left(\sec \dfrac{\pi}{2x} \right) \log x =$ _____

7. $\lim\limits_{x\to 0} x^x =$ _____

8. $\lim\limits_{x\to \infty} \left(\dfrac{1}{x_1} \right)^{1/x} =$ _____

9. $\lim\limits_{x\to 1} x^{x-1} =$ _____

10. $\lim\limits_{x\to 0} (1+x)^{1/x} =$ _____

TRUE/FALSE

Write 'T' for true and 'F' for false statement.

1. The indeterminate form $\dfrac{\infty}{\infty}$ can be converted into the form $\dfrac{0}{0}$. **(T/F)**

2. 1° is not an indeterminate form. **(T/F)**

3. 0° is not an indeterminte form. **(T/F)**

4. $\infty + \infty$ and $\infty \times \infty$ are indeterminate form. **(T/F)**

5. By L' Hospital rule to find $\lim\limits_{x\to a} \dfrac{f(x)}{g(x)}$, if the form is $\dfrac{0}{0}$, we are differentiate $\dfrac{f(x)}{g(x)}$ is a fraction. **(T/F)**

6. $\lim\limits_{x\to 0} (1+nx)^{1/x}$ is e^n. **(T/F)**

7. $\lim\limits_{x\to 0} \dfrac{x^2 + 2x - 2}{x \sin^3 x}$ is $\dfrac{1}{6}$. **(T/F)**

8. If $\lim\limits_{x\to a} f(x) = \infty = \lim\limits_{x\to a} g(x)$ then $\lim\limits_{x\to a} \dfrac{f(x)}{g(x)} = \lim\limits_{x\to a} \dfrac{f'(x)}{g'(x)}$. **(T/F)**

MULTIPLE CHOICE QUESTIONS

Choose the most appropriate one.

1. $\lim\limits_{x\to 0} \dfrac{\tan x}{x}$ is :
   (a) 0   (b) ∞
   (c) 1   (d) -1

2. $\lim\limits_{x\to 0} (1+nx)^{1/x}$ is :
   (a) 1   (b) e^{-n}
   (c) e^2 (d) e^n

3. $\lim\limits_{x\to 1} \dfrac{x^5 - 2x^3 - 4x^2 + 9x - 4}{x^4 - 2x^3 + 2x - 1}$ is:
   (a) 1   (b) 2
   (c) 3   (d) 4

4. $\lim\limits_{x\to 0} \dfrac{\log(1 - x^2)}{\log \cos x}$ is :
   (a) 1   (b) 2
   (c) 3/2   (d) 2/3

5. $\lim\limits_{x\to 0} \dfrac{\sin ax}{\sin bx}$ is :
   (a) 1   (b) 0
   (c) a/b (d) b/a

6. $\lim\limits_{x\to 0} \dfrac{x - \sin x}{x^3}$ is :
   (a) $\dfrac{1}{2}$ (b) $\dfrac{1}{3}$
   (c) $\dfrac{1}{5}$ (d) $\dfrac{1}{6}$

7. $\lim\limits_{x\to 0} \dfrac{x^2 + 2\cos x - 2}{x \sin x}$ is :
   (a) $\dfrac{1}{12}$ (b) $\dfrac{1}{6}$
   (c) $\dfrac{2}{5}$ (d) 1

8. $\lim\limits_{x\to 1}\dfrac{\log x}{x-1}$ is :

(a) 0   (b) –1

(c) 1   (d) ∞

9. $\lim\limits_{x\to 1}\dfrac{\sin x\,\sin^{-1}x}{x^2}$ is :

(a) 0   (b) – 1

(c) 1   (d) ∞

10. $\lim\limits_{x\to 0}\dfrac{x\cos x-\log(1+x)}{x^2}$ is :

(a) 0   (b) 1

(c) 1/2   (d) 1/3

11. $\lim\limits_{x\to a}\left(\dfrac{a^x-b^x}{x}\right)$ is :

(a) 1   (b) ∞

(c) $\log b/a$ (d) $\log a/b$

12. $\lim\limits_{x\to 0}\dfrac{\log x}{\cot x}$ is :

(a) 1   (b) – 1

(c) 0   (d) ∞

13. $\lim\limits_{x\to \pi/2}\dfrac{\log(x-\pi/2)}{\tan x}$ is :

(a) 1   (b) 0

(c) – 1   (d) ∞

14. $\lim\limits_{x\to \pi/2}\dfrac{\tan 5x}{\tan x}$ is :

(a) 1   (b) 5

(c) 1/5   (d) –1

15. $\lim\limits_{x\to 1}(1-x)\tan\dfrac{\pi x}{2}$ is:

(a) 1   (b) 2π

(c) π (d) $2/\pi$

16. $\lim\limits_{x\to \infty}(a^{1/x}-1)x$ is :

(a) 1   (b) $\log a$

(c) a (d) 0

17. $\lim\limits_{x\to 0}x\log x$ is :

(a) 1   (b) –1

(c) 0   (d) ∞

18. $\lim\limits_{x\to 0}\left(\dfrac{1}{x}-\cot x\right)$ is :

(a) 1   (b) –1

(c) 0   (d) ∞

19. $\lim\limits_{x\to 0}\left[\dfrac{1}{e^x-1}-\dfrac{1}{x}\right]$ is :

(a) 1   (b) 1/2

(c) –1   (d) –1/2

20. $\lim\limits_{x\to 0}\left[\dfrac{\cot x-1/x}{x}\right]$ is :

(a) 1   (b) –1

(c) 3   (d) –1/3

21. $\lim\limits_{x\to \pi/2}\left[\sec x-\dfrac{1}{(1-\sin x)}\right]$ is :

(a) 0   (b) – 1

(c) ∞   (d) – ∞

22. $\lim\limits_{x\to \infty}\left[1+\dfrac{a}{x}\right]^x$ is :

(a) 0   (b) 1

(c) e^a (d) e^{-a}

23. $\lim\limits_{x\to 0}(1+x)^{1/x}$ is :

(a) 0   (b) 1

(c) 1/e   (d) e

24. $\lim\limits_{x\to \pi/2}(\sin x)^{\tan x}$ is :

(a) 0   (b) $e^{\pi/2}$

(c) 1   (d) –1

25. $\lim\limits_{x\to 1}(x)^{1/x-1}$ is :

(a) e^{-2} (b) e^2

(c) e (d) e^{-1}

26. $\lim\limits_{x\to 0}\left(\dfrac{\tan x}{x}\right)^{1/x}$ is :

(a) 0   (b) 1

(c) –1   (d) ∞

27. $\lim\limits_{x\to 0}(a^x+x)^{1/x}$ is :

(a) $2ae$ (b) a/e

(c) ae (d) $(ae)^2$

28. $\lim\limits_{x\to 0} x^x$ is :

    (a) 0   (b) 1

    (c) –1   (d) ∞

29. $\lim\limits_{x\to\infty} (1/x)^{1/x}$ is :

    (a) 0   (b) 1

    (c) –1   (d) ∞

30. $\lim\limits_{x\to 0} (\cot x)^{\sin x}$ is :

    (a) 0   (b) 1

    (c) –1   (d) ∞

Answers

FILL IN THE BLANKS

1. 1   **2.** $\dfrac{a}{b}$ **3.** $-\dfrac{1}{3}$ **4.** 1   **5.** 0   **6.** $\dfrac{2}{\pi}$ **7.** 1   **8.** 1   **9.** e^{-1}
10. e.

TRUE/FALSE

1. T   **2.** T   **3.** F   **4.** F   **5.** F   **6.** T   **7.** F   **8.** T

MULTIPLE CHOICE QUESTIONS

1. (c)   **2.** (d)   **3.** (d)   **4.** (b)   **5.** (c)   **6.** (d)   **7.** (a)   **8.** (c)   **9.** (c)
10. (c)   **11.** (d)   **12.** (c)   **13.** (b)   **14.** (c)   **15.** (d)   **16.** (b)   **17.** (c)   **18.** (d)
19. (d)   **20.** (d)   **21.** (c)   **22.** (c)   **23.** (d)   **24.** (c)   **25.** (c)   **26.** (b)   **27.** (c)
28. (b)   **29.** (b)   **30.** (b)

SELF ASSESSMENT TEST

Verify each of the following :

1. $\lim\limits_{x \to 0}\left(\dfrac{1}{\log(x + \sqrt{1+x^2})} - \dfrac{1}{\log(1-x)}\right) = -\dfrac{1}{2}$

2. $\lim\limits_{x \to 0}\dfrac{\log(1+x)}{\sin x} = 1$

3. $\lim\limits_{x \to 0}\dfrac{xe^{x^2}}{\int_0^x e^{t^2}\,dt} = 1$

4. $\lim\limits_{x \to \infty}\left(\dfrac{2^{n+1}+3^{n+1}}{2^n + 3^n}\right) = 3$

5. If f and g are twice differentiable functions and $f(p) = 3$, $f'(p) = -2$, $g(p) = -1$, $g'(p) = 4$, then

$\lim\limits_{x \to p}\dfrac{g(x)f(p) - g(p)f(x)}{x - p} = 10$

6. $\lim\limits_{x \to b}\dfrac{x^b - b^x}{x^x - b^b} = \dfrac{1 - \log b}{1 + \log b}$

7. $\lim\limits_{x \to a}\dfrac{a^x - x^a}{x^x - a^a} = \log\left(\dfrac{a}{e}\right)\log(ae)$

8. $\lim\limits_{a \to b}\dfrac{a^b - b^a}{a^a - b^b} = \log\left(\dfrac{e}{b}\right)\log(be)$

9. $\lim\limits_{x \to 1}\dfrac{x^x - x}{x - 1 - \log x} = 2$

10. If $f''(x)$ exists and is continuous in a neighbourhood of $x = a$, then

$\lim\limits_{h \to 0}\dfrac{f(a+h) - 2f(a) + f(a-h)}{h^2} = f''(a)$

11. If f and g are two functions such that when $x \to c$

(i) $\lim |g(x)| = \infty$ (ii) $\lim\left(\dfrac{f'(x)}{g'(x)}\right) = l$ then

$\lim\left(\dfrac{f(x)}{g(x)}\right) = l$ when $x \to c$

12. $\lim\limits_{x \to 1^-}(1 - x^2)^{\frac{1}{[\log(1-x)]}} = e$

AN IMPORTANT TABLE

Form	Given	To find	Method
1. $\dfrac{0}{0}$ form	$\lim\limits_{x\to a} f(x) = 0$ and $\lim\limits_{x\to a} g(x) = 0$	$\lim\limits_{x\to a} \dfrac{f(x)}{g(x)}$	$\lim\limits_{x\to a} \dfrac{f(x)}{g(x)} = \lim\limits_{x\to a} \dfrac{f'(x)}{g'(x)}$ provided $g'(x) \neq 0$ or $\lim\limits_{x\to a} \dfrac{f(x)}{g(x)} = \lim\limits_{x\to a} \dfrac{f^n(x)}{g^n(x)}$ provided $g^n(x) \neq 0$
2. $\dfrac{\infty}{\infty}$ form	$\lim\limits_{x\to a} f(x) = \infty$ and $\lim\limits_{x\to a} g(x) = \infty$	$\lim\limits_{x\to a} \dfrac{f(x)}{g(x)}$	$\lim\limits_{x\to\infty} \dfrac{f(x)}{g(x)} = \lim\limits_{x\to a} \dfrac{f'(x)}{g'(x)}$
3. $0 \times \infty$ form	$\lim\limits_{x\to a} f(x) = \infty$ and $\lim\limits_{x\to a} g(x) = \infty$	$\lim\limits_{x\to a} [f(x) \cdot g(x)]$	$\lim\limits_{x\to a} \dfrac{f(x)}{g(x)} = \lim\limits_{x\to a} \dfrac{g(x)}{\dfrac{1}{f(x)}}$ and proceed same as $\dfrac{\infty}{\infty}$
4. $\infty - \infty$ form	$\lim\limits_{x\to a} f(x) = \infty$ and $\lim\limits_{x\to a} g(x) = \infty$	$\lim\limits_{x\to a} [f(x) - g(x)]$	$\lim\limits_{x\to a} [f(x) - g(x)]$ $= \lim\limits_{x\to a} \dfrac{\left[\dfrac{1}{g(x)} - \dfrac{1}{f(x)}\right]}{\dfrac{1}{f(x) \cdot g(x)}}$ and proceed same as $\dfrac{0}{0}$ form
5. 0^0, 1^∞, ∞^0 form	(i) $\lim\limits_{x\to a} f(x) = 0$, $\lim\limits_{x\to a} g(x) = 0$ (ii) $\lim\limits_{x\to a} f(x) = 1$ $\lim\limits_{x\to a} g(x) = \infty$ (iii) $\lim\limits_{x\to a} f(x) = \infty$ and $\lim\limits_{x\to a} g(x) = 0$	$\lim\limits_{x\to a} [f(x)]^{g(x)}$	$\lim\limits_{x\to a} .[f(x)]^{g(x)} = e^l$ where $l = \lim\limits_{x\to a} [g(x) \cdot \log f(x)]$.

Chapter

13 Power Series

13.1 INTRODUCTION

A series of the form

$$\sum_{n=0}^{\infty} a_n (x - x_0)^n = a_0 + a_1 (x - x_0) + a_2 (x - x_0)^2 + \ldots + a_n (x - x_0)^n + \ldots$$

where x is a continuous variable and the constants $a_0, a_1, \ldots, a_n$ and x_0 are real and independent of x, is called a power series.

REMARKS

- If we change the variable $x = t + x_0$, then power series reduces to $\sum_{n=0}^{\infty} a_n t^n$.
- For $x=0$, every power series is convergent, whatever be the value of the coefficients.
- A power series is either
  (1) convergent for no value of x other then $x=0$, then it is said to be nowhere convergent.
  (2) convergent for all values of x and is called everywhere convergent.
  (3) convergent for some values of x and diverges for others.
- The totality of points x for which a power series converges is called the 'Region of Convergence'.

THEOREM 1. *If a power series* $\sum_{n=0}^{\infty} a_n x^n$ *converges for a particular value* x_0 *of* x *then it converges absolutely for all values of* x *for which* $|x| < |x_0|$.

Proof. Let the series $\sum a_n x_0^n$ be converges. Then n^{th} term of the given series $a_n x_0^n$ must tends to 0 as $n \to \infty$. Therefore, we can find a number $M > 0$ such that

$$| a_n x_0^n | \leq M \text{ for all } n.$$

Then
$$\left| a_n x^n \right| \leq M \left| \frac{x}{x_0} \right|^n \qquad \ldots (1)$$

Now, since $|x| < |x_0|$, then geometric series $\sum \left| \frac{x}{x_0} \right|^n$ converges. Hence, from (1), the series $\sum a_n x^n$ converges for all values of x for which $|x| < |x_0|$.

THEOREM 2. *The power series* $\sum a_n x^n$ *either*

    (i) converges absolutely for all x,

or (ii) converges for $x=0$ *only*

or (iii) there exists $R > 0$ *such that the series converges absolutely when* $|x| < R$ *and diverges when* $|x| > R$.

Proof. Let $x \in R^+$ and P be the set of all x for which the given power series $\sum a_n x^n$ converges. The given series is definitely converges to 0.

$$\therefore \qquad\qquad 0 \in P.$$

Then P may or may not contain numbers other than 0. If $x_0 \in P$ then by previous theorem, every x with $0 \le x \le x_0$ is also in P.

If all non-negative real numbers are in P, then $\sum a_n x^n$ absolutely converges for all x.

On the other hand, if P does not contain all non-negaiive numbers, then it,has a least upper bound R $(R>0)$. Let $R>0$ and $|x_0| < R$. Then, to show that $\sum a_n x_0^n$ absolutely converges.

Let us choose R_0 such that $|x_0| < R_0 < R$. Then $R_0 \in P$ and so the series converges for $x = R_0$. Then by previous theorem $\sum |a_n x_0^n|$ converges.

Finally, we prove if $|x'| > R \ge 0$, the series cannot converge for $x = x'$.

Select R_0 with $R < R_0 < |x'|$. If $\sum a_n x'^n$ was to converge, then by previous theorem $\sum a_n R_0^n$ also converges, which contradicts the fact $R = \sup P$.

13.2 RADIUS OF CONVERGENCE

The number R(obtained in above theorem) is called the radius of convergence of the power series $\sum\limits_{n=0}^{\infty} a_n x^n$ and the set of all x for which $|x| < R$ i.e., the open interval $]-R, R[$ is called interval of convergence.

A power series $\sum a_n x^n$ absolutely converges for values of x inside the circle of convergence and diverges outside the circle. For values of x on the circumference of the circle, the series may converge, diverge or oscillate.

For example:

(1) The series $\sum\limits_{n=0}^{\infty} a_n x^n$ converges for $|x| < 1$ and diverges for $|x| > 1$.

(2) The series $\sum\limits_{n=0}^{\infty} \dfrac{x^n}{n}$ converges for $-1 \le x < 1$ and diverges elsewhere.

(3) The series $\sum\limits_{n=0}^{\infty} \dfrac{x^n}{n^2}$ converges absolutely for $|x| \le 1$ and diverges for $|x| > 1$.

13.2.1 EXPRESSION OF THE RADIUS OF CONVERGENCE

We know that

$$\liminf_{n \to \infty} \left|\frac{a_{n+1}}{a_n}\right| \le \liminf_{n \to \infty} \sqrt[n]{|a_n|} \le \limsup_{n \to \infty} \sqrt[n]{|a_n|} \le \limsup_{n \to \infty} \left|\frac{a_{n+1}}{a_n}\right|$$

provided the first and the last limit exist. Therefore, if

$$\liminf_{n \to \infty} \left|\frac{a_{n+1}}{a_n}\right| = \limsup_{n \to \infty} \left|\frac{a_{n+1}}{a_n}\right|$$

Then, we have

$$\liminf_{n \to \infty} \sqrt[n]{|a_n|} = \limsup_{n \to \infty} \sqrt[n]{|a_n|}$$

i.e., if $\lim\limits_{n\to\infty}\left|\dfrac{a_{n+1}}{a_n}\right|$ exists, then $\lim\limits_{n\to\infty}\sqrt[n]{|a_n|}$ also exist and are equal.

Therefore, the radius of convergence can also be found by the relation

$$R=\lim\limits_{n\to\infty}\left|\frac{a_{n+1}}{a_n}\right| \text{ provided the limit exist.}$$

We can also find the radius of convergence by using the following formula

$$R=\frac{1}{\lim\limits_{n\to\infty}\sup\sqrt[n]{|a_n|}}$$

THEOREM 1. (*Uniform Convergence of Power Series*).

The power series $\sum a_n x^n$ is uniformly convergent for $|x|\le\rho<R$ where R is the radius of convergence.

Proof. Consider a number ρ' between ρ and R. Since, the series is convergent for $|x|=\rho'$, then by definition $\exists$ a number k, independent of n so that

$$|a_n\rho'^n|<k \,\forall\, n$$

$$\Rightarrow \quad \text{for}\quad |x|\le\rho, |a_n x^n|=\left|a_n\rho'^n\left(\frac{x}{\rho'}\right)^n\right|<k\left(\frac{\rho}{\rho'}\right)^n$$

which is independent of x.

But the series is geometric series with common ratio $\dfrac{\rho}{\rho'}<1$, therefore the series $k\sum\left(\dfrac{\rho}{\rho'}\right)^n$ is convergent. Thus, by Weierstrass's M-test the power series is uniformly convergent for $|x|\le\rho<R$.

Hence, every power series is uniformly convergent within its radius of convergence.

13.2.2 ALGEBRA OF POWER SERIES

(i) If $f(x)$ and $g(x)$ be two power series such that $f(x)=\sum\limits_{n=0}^{\infty}a_n x^n$ and $g(x)=\sum\limits_{n=0}^{\infty}b_n x^n$

then within the interval of convergence, we have $f(x)\pm g(x)=\sum\limits_{n=0}^{\infty}(a_n\pm b_n)x^n$

(ii) If the series $f(x)$ and $g(x)$ are absolutely convergent in $]-R,R[$ then for any $x\in]-R, R[$, we have

$$f(x).g(x)=\sum\limits_{n=0}^{\infty}C_n.x^n \text{ where } C_n=\sum\limits_{k=0}^{\infty}a_k b_{n-k}.$$

13.3 UNIQUENESS OF POWER SERIES

THEOREM 1. If two real power series $\sum a_n x^n$ and $\sum b_n x^n$ have a radius of convergence $R>0$ and converge to the same function in $]-R,R[$ then the two series are identical.

Proof. Let the two power series $\sum a_n x^n$ and $\sum b_n x^n$ converges to the same sum when $|x|<R$, then, we have

$$a_0+a_1 x+a_2 x^2+\ldots=b_0+b_1 x+b_2 x^2+\ldots$$

Since, we know that the function defined by the sum of a power series is continuous in its interval of convergence at $x=0$. Therefore, when $x\to 0$, we must have $a_0=b_0$.

Therefore, $a_1x+a_2x^2+...=b_1x+b_2x^2+...$ i.e., $a_1+a_2x+...=b_1+b_2x+...$

Again taking $x\to 0$, we have $a_1=b_1$

Proceeding in the same way, we get $a_n = b_n$, $n = 0, 1, 2, ...$

Hence, both the given series are identical.

13.4 ABEL'S SUMMABILITY

Definition 1. *The series* $\sum_{n=0}^{\infty} a_n$ *is Abel summable to a value s, if the associated power series* $\sum_{n=0}^{\infty} a_nx^n$ *converges for* $0\le x<1$ *to a function f and* $\lim_{x\to 1^-} f(x)=s$.

Definition 2. *A summabiliiy method T, for a sequence is said to be regular if, whenever the sequence* $\langle x_n\rangle$ *converges to s, then* $\langle x_n\rangle$ *is also T-summable to s.*

THEOREM 1. **(Abel's Theorem).** *If the series* $\sum_{n=0}^{\infty} a_n$ *is convergent and has the sum s, then the series* $\sum_{n=0}^{\infty} a_nx^n$ *is uniformly convergent for* $0\le x<1$ *and* $\lim_{x\to 1^-}\sum_{n=0}^{\infty} a_nx^n = s$.

Proof. Given that the series $\sum a_n$ is convergent, therefore, for $n\ge m$ we have
$|a_{n+1}+a_{n+2}+...+a_{n+p}|<\varepsilon$ for every integral value of $p>0$.

Also, since the sequence $\langle x_n\rangle$ is monotonically decreasing for all values of $x\in[0,1]$.

Then from Abel's inequality
$$\left|a_nx^n+a_{n+1}x^{n+1}+...+a_{n+p}x^{n+p}\right|\le \varepsilon x' \le \varepsilon (x\in[0,1]).$$

Therefore, the series $\sum a_nx^n$ is uniformly convergent for $0\le x\le 1$.

which implies $\sum_{n=0}^{\infty} a_nx^n$ is continuous function of $x\in[0,1]$ and therefore,

$$\lim_{x\to 1^-}\sum_{n=0}^{\infty} a_nx^n = \sum_{n=0}^{\infty}\lim_{h\to 0} a_n(1-h)^n = \sum_{n=0}^{\infty} a_n = s$$

THEOREM 2. *If* $\sum c_n$ *converges and for* $0<x<1$, *define* $f(x)=\sum_{n=0}^{\infty} c_nx^n$ *then* $\lim_{x\to 1} f(x)=\sum_{n=0}^{\infty} c_n$

Proof. Let S_n denote the nth partial sum of the series $\sum c_n$: let $c_n=a_n$ that is,
$$S_n=c_0+c_1+c_2+...+c_n, S_{-1}=0 \text{ and let } \sum_{n=0}^{\infty} a_n=S, \text{ then}$$

$$f(x) = \sum_{n=0}^{m} a_nx^n = \sum_{n=0}^{m}(S_n-S_{n-1})x^n$$
$$= \sum_{n=0}^{m-1} S_nx^n - \sum_{n=0}^{m} S_{n-1}x^n +S_mx^m$$
$$= \sum_{n=0}^{m-1} S_nx^n - x\sum_{n=0}^{m} S_{n-1}x^{n-1} +S_mx^m \qquad [\because S_{-1}=0]$$

$$= \sum_{n=0}^{m-1} S_n x^n - x \sum_{n=0}^{m-1} S_n x^n + S_m x^m \qquad \ldots(1)$$

For $|x|<1$, when $m\to\infty$, since $S_m\to S$, and $x^m\to 0$, we get from (1),

$$f(x)= \sum_{n=0}^{\infty} a_n x^n = (1-x) \sum_{n=0}^{\infty} S_n x^n, \text{ for } 0<x<1 \qquad \ldots(2)$$

Again, since $S_n\to S$ for any given $\varepsilon>0$ we can choose an integer N such that

$$\forall\, n\geq N \quad \Rightarrow \quad |S_n-S|<\frac{\varepsilon}{2} \qquad \ldots(3)$$

Also $(1-x) \sum_{n=0}^{\infty} x^n = 1 \quad |x|<1$ \qquad ...(4)

Hence for $n\geq N$ we have, for $0<x<1$

$$|f(x)-S| = \left|(1-x) \sum_{n=0}^{\infty} S_n x^n - S\right| \qquad \text{[Using (2)]}$$

$$= \left|(1-x) \sum_{n=0}^{\infty} (S_n - S)x^n\right| \qquad \text{[Using (4) and (3)]}$$

$$\leq (1-x) \sum_{n=0}^{N} |S_n - S|x^n < \frac{\varepsilon}{2}$$

But for a fixed N, $(1-x)\sum_{n=0}^{N} (S_n - S)x^n$ is a positive continuous function of x, having zero value at $x=1$. Therefore $\exists\, \delta>0$ such that for $1-\delta<x<1$,

$$|f(x)-s| \leq (1-x)\sum_{n=0}^{N} |S_n - S|x^n + \frac{\varepsilon}{2}$$

$\therefore \qquad |f(x)-s| < \frac{\varepsilon}{2}+\frac{\varepsilon}{2} = \varepsilon$ where $1-\delta<x<1$,

Hence $\qquad \lim_{x\to 1-0} f(x) = S = \sum_{n=0}^{\infty} a_n = \sum_{n=0}^{\infty} C_n \qquad$ by $[a_n=C_n]$

REMARKS

- **Abel's Inequality.** If the sequence $\langle s_n\rangle$ of positive terms is monotonic decreasing, then

$$hs_1 < \sum_{1}^{p} a_n s_n < Hs_1$$

where H and h are the upper and lower limits of the sums $a_1, a_1+a_2, a_1+a_2+a_3,..., a_1+a_2+...,a_p$.
- The converse of the above theorem is not necessarily true.

13.5 PROPERTIES OF POWER SERIES

(i) A power series $\sum a_n x^n$ is a continuous function of x within its interval of convergence.

(ii) A power series $\sum a_n x^n$ can be integrated term by term so long as the limits of integration lie strictly within the open interval $]-R,R[$.

(iii) The derived series of the power series $\sum a_n x^n$ is $\sum n a_n x^{n-1}$. If R' be the radius of convergence of the derived series $\sum n a_n x^{n-1}$, then we have

$$R' = \frac{1}{\lim\limits_{n \to \infty} \sup |n a_n|^{1/n}} = \lim\limits_{n \to \infty} \frac{1}{n^{1/n} |a_n|^{1/n}}$$

$$= \lim\limits_{n \to \infty} \frac{1}{|a_n|^{1/n}} \qquad \left[\because \lim\limits_{n \to \infty} n^{1/n} = 1 \right]$$

$$= R.$$

Therefore, the differentiated series has the same radius of convergence as the original series. Hence, a power series may also be differentiated term by term at any point $x \in]-R, R[$.

(iv) The series obtained by integrating or differentiating a power series term by term has the same radius of convergence as the original series.

13.6 TAUBER'S THEOREM

Here, we prove the modified converse of Abel's theorem, which is known as Tauber's theorem.

THEOREM 1. *If* $\lim\limits_{n \to \infty} n a_n = 0$ *and* $f(x) = \sum a_n x^n$ *with* $\lim\limits_{n \to 1^-} f(x) = s$ *then the series* $\sum a_n$ *converges to the sum s.*

Proof. By Cauchy's first theorem on limits, we know that

$$\lim\limits_{n \to \infty} \frac{|a_1| + 2|a_2| + \dots + n|a_n|}{n} = \lim\limits_{n \to \infty} n a_n = 0$$

Now, since $\lim\limits_{n \to 1^-} f(x) = s$, then by definition, for any given $\varepsilon > 0$, we can find a positive integer N such that for all $n \geq N$, we have

$$\left| f\left(1 - \frac{1}{n}\right) - s \right| < \varepsilon/3 \qquad \dots(1)$$

$$|n a_n| < \varepsilon/3 \text{ i.e., } |a_n| < \frac{\varepsilon}{3n} \qquad \dots(2)$$

and

$$\frac{|a_1| + 2|a_2| + \dots + n|a_n|}{n} < \frac{\varepsilon}{3} \qquad \dots(3)$$

Let s_n denote the nth partial sum of the series $\sum a_n$ i.e., $s_n = a_0 + a_1 + a_2 + \dots + a_n$

then

$$|s_n - s| = |f(x) - s + s_n - f(x)|$$

$$= \left| f(x) - s + a_0 + \sum_{k=1}^{n} a_k - a_0 - \sum_{k=1}^{\infty} a_k x^k \right|$$

$$= \left| f(x) - s + \sum_{k=1}^{n} a_k - \sum_{k=1}^{\infty} a_k x^k \right|$$

$$= \left| f(x) - s + \sum_{k=1}^{n} a_k - \sum_{k=1}^{n} a_k x^k - \sum_{k=n+1}^{\infty} a_k x^k \right|$$

$$= \left| f(x) - s + \sum_{k=1}^{n} a_k \left(1 - x^k \right) - \sum_{k=n+1}^{\infty} a_k x^k \right|$$

$$\leq \left| f(x) - s \right| + \left| \sum_{k=1}^{n} a_k \left(1 - x^k \right) \right| + \left| \sum_{k=n+1}^{\infty} a_k x^k \right|$$

Now since $(1-x^k) = (1-x)(1+x+x^2+...x^{k-1}) < k(1-x)$ as $0<x<1$.

Therefore, $$\left| s_n - s \right| \leq \left| f(x) - s \right| + (1-x) \sum_{k=1}^{n} k |a_k| + \sum_{k=n+1}^{\infty} |a_k| x^k \qquad \qquad ...(4)$$

From (2), we have

$$\sum_{k=n+1}^{\infty} |a_k| x^k < \frac{\varepsilon}{3n} \sum_{k=n+1}^{\infty} x^k < \frac{\varepsilon}{3n} \cdot \frac{x^{n+1}}{1-x} \quad \text{as } 0<x<1$$

$$< \frac{\varepsilon}{3n(1-x)}$$

Therefore (4) can be written as

$$\left| s_n - s \right| \leq \left| f(x) - s \right| + (1-x) \sum_{k=1}^{n} k |a_k| + \frac{\varepsilon}{3n(1-x)}$$

Pulling $x = 1 - \dfrac{1}{n}$ and using (1) and (3), we get

$$\left| s_n - s \right| < \left| f\left(1 - \frac{1}{n} \right) - s \right| + \frac{1}{n} \sum_{k=1}^{n} k |a_k| + \frac{\varepsilon}{3}$$

$$< \frac{\varepsilon}{3} + \frac{\varepsilon}{3} + \frac{\varepsilon}{3} = \varepsilon \ \forall n \in N$$

Hence $\lim\limits_{n \to \infty} s_n = s,$

$\Rightarrow$ $\sum a_n$ converges to the sum s.

REMARKS

- Let γ be an Euler's constant and let

$$1 + \frac{1}{2} + \frac{1}{3} + ... + \frac{1}{n} = \log n + \gamma_n \ \text{ then } \ \lim_{n \to \infty} \gamma_n = \gamma.$$

- If $\Delta a_n = a_n - a_{n+1}$ then for any sequence $\langle a_n \rangle$ and $\langle b_n \rangle$, $\Delta(a_n b_n) = (\Delta a_n) b_n + a_{n+1} (\Delta b_n)$

- (Cauchy products). For any sequence $\langle a_n \rangle$ and $\langle b_n \rangle$ $\sum_{n=0}^{\infty} \sum_{k=0}^{\infty} a_n b_k x^{n+k} = \sum_{n=0}^{\infty} \sum_{k=0}^{\infty} a_{n-k} b_k x^n$

- If the radius of convergence for the power series $\sum_{n=0}^{\infty} a_n x^n$ is R then the radius of convergence of

the power series $\sum_{n=0}^{\infty} n a_n x^{n-1}$ and $\sum_{n=0}^{\infty} \dfrac{a_n}{n+1} x^{n+1}$ is also R.

- Let R be the radius of convergence of the power series then $f(x) = \sum\limits_{n=0}^{\infty} a_n x^n$, $x \in R$.

  (i) f is differentiable in the interval $]-R, R[$ and for $x \in]-R, R[, f'(x) = \sum\limits_{n=0}^{\infty} n a_n x^{n-1}$

  (ii) f is integrable in the interval $]-R, R[$ and $\int_0^x f(x) dx = \sum\limits_{n=0}^{\infty} \dfrac{a_n}{n+1} x^{n+1}$, $0 \le x \le R$.

THEOREM 2. Let $f(x) = \sum\limits_{n=0}^{\infty} a_n x^n$ be a series converges in $|x| < R$ such that if $-R < a < R$, then f can be expanded in a power series about $x = a$ which converges $|x-a| < R - |a|$.

Proof.   We have

$$f(x) = \sum_{n=0}^{\infty} a_n x^n = \sum_{n=0}^{\infty} a_n \left[(x-a) + a\right]^n \qquad \text{where } [a_n = C_n]$$

$$= \sum_{n=0}^{\infty} a_n \sum_{m=0}^{n} \left({}^n C_m\right) a^{n-m} (x-a)^m,$$

i.e., $\quad f(x) = \sum\limits_{n=0}^{\infty} \left[\sum\limits_{m=0}^{n} \left({}^n C_m\right) a_n a^{n-m} \right] (x-a)^m;$ ...(1)

which is the desired expansion about $x = a$.

Now, we have to show that this is permissible if

$$\sum_{n=0}^{\infty} \sum_{m=0}^{\infty} \left| a_n \left({}^n C_m\right) a^{n-m} (x-a)^m \right| \qquad \qquad ...(2)$$

converges. But (2) is the same as

$$\sum_{n=0}^{\infty} |a_n| \cdot \left(|x-a||a|\right)^n \qquad \qquad ...(3)$$

We see that (2) converges, if $|x-a| + |a| < R$.

Now $\quad f^m(x) = \sum\limits_{n=0}^{\infty} n(n-1)...(n-m+1) a_n a^{n-m}$

Put $x = a$ we obtain

$$f^{(m)}(a) = \sum_{n=m}^{\infty} n(n-1)...(n-m+1) a_n a^{n-m}$$

or $\quad \dfrac{f^{(m)}(a)}{m!} = \sum\limits_{n=m}^{\infty} \dfrac{n!}{m!(n-m)!} a_n a^{n-m} = \sum\limits_{n=m}^{\infty} \left({}^n C_m\right) a_n a^{n-m}$ ...(4)

Hence (1) and (4) yields

$$f(x) = \sum_{m=0}^{\infty} \frac{f^{(m)}(a)}{m!}(x-a)^m = \sum_{n=0}^{\infty} \frac{f^{(n)}(a)}{n!}(x-a)^n \quad (|x-a| < R - |a|) \qquad ...(5)$$

Example 1. *Prove that the power series* $1+x+\dfrac{x^2}{2!}+\dfrac{x^3}{3!}+\dfrac{x^4}{4!}...$ *has infinite radius of convergence.*

It converges (absolutely) for all x.

Solution . Here, the given series

$$\sum_{n=0}^{\infty} \frac{x^n}{n!} = \sum_{n=0}^{\infty} a_n x^n \text{ (say) then } \quad a_n = \frac{1}{n!}$$

the radius of convergence R is given by the formula

$$R = \lim_{n\to\infty} \left|\frac{a_n}{a_{n+1}}\right| = \lim_{n\to\infty} \frac{(n+1)!}{n!} = \lim_{n\to\infty} (n+1) = \infty$$

therefore, the given power series convergence absolutely for all x.

Example 2. *Prove that the power series* $1+x+2!x^2+3!x^3+...$ *fails to converge for any value of x other than 0. Its radius of convergence R=0.*

Solution . Here, the given power series is
$$\sum a_n x^n = 1+x+2!x^2+3!x^3+... .$$
Here, $\qquad a_n = n!.$
Then, the radius of convergence is given by

$$R = \lim_{n\to\infty} \left|\frac{a_n}{a_{n+1}}\right| = \lim_{n\to\infty} \left|\frac{n!}{(n+1)!}\right| = \lim_{n\to\infty} \frac{1}{n+1} = 0$$

Therefore, the given power series fail to converges for any value of x other than 0.

Example 3. *Find the interval of absolute convergence for the power series* $x+\dfrac{x^2}{2^2}+\dfrac{x^3}{3^3}+... .$

Solution . Here, the given power series is

$$\sum_{n=1}^{\infty} \frac{x^n}{n^n} = \sum_{n=0}^{\infty} a_n x^n \text{ (say) then } \quad a_n = \frac{1}{n^n}$$

Now, the radius of converges R is

$$\frac{1}{R} = \lim_{n\to\infty} \sup |a_n|^{1/n} = \lim_{n\to\infty} \sup \left|\frac{1}{n^n}\right|^{1/n} = \lim_{n\to\infty} \frac{1}{n} = 0$$

$$\Rightarrow \qquad R = \frac{1}{0} = \infty$$

Therefore, the given power series converges absolutely for all $x \in]-\infty, \infty[$.

Example 4. *Find the radius of convergence of the following power series :*

(i) $\displaystyle\sum_{n=0}^{\infty} \frac{(2n)!}{(n!)^2} z^n.$ $\qquad$ (ii) $\displaystyle\sum_{n=0}^{\infty} \frac{n!}{n^n} z^n.$

Solution . (i) Here, the given power series is $\displaystyle\sum_{n=0}^{\infty} \frac{(2n)!}{(n!)^2} z^n.$

On comparing with $\sum\limits_{n=0}^{\infty} C_n z^n$, we have $C_n = \dfrac{(2n)!}{(n!)^2}$ then by ratio test,

$$R = \dfrac{1}{\lim\limits_{n\to\infty} \sup \dfrac{C_{n+1}}{C_n}}$$

$$R = \dfrac{1}{\lim\limits_{n\to\infty} \dfrac{(2n+2)!(n!)^2}{(2n)!((n+1)!)^2}} = \dfrac{1}{\lim\limits_{n\to\infty} \dfrac{\left(2+\dfrac{2}{n}\right)\left(2+\dfrac{1}{n}\right)}{\left(1+\dfrac{1}{n}\right)^2}} = \dfrac{1}{4}$$

(ii) Here, the given power series is $\sum\limits_{n=0}^{\infty} \dfrac{n!}{n^n} z^n$.

On comparing with $\sum\limits_{n=0}^{\infty} C_n z^n$, we obtain $C_n = \dfrac{n!}{n^n}$

so, radius of convergence $R = \dfrac{1}{\lim\limits_{n\to\infty} \sup \dfrac{C_{n+1}}{C_n}} = \lim\limits_{n\to\infty}\left(1+\dfrac{1}{n}\right)^n = e$

$\Rightarrow$ $R = e$

Example 5. *Find the radius of convergence of the following power series :*

(i) $\quad \sum\limits_{n=0}^{\infty} n^n.z^n$

(ii) $\quad \sum\limits_{n=0}^{\infty} \dfrac{z^n}{n!}$

(iii) $\quad \sum\limits_{n=0}^{\infty} \dfrac{2^n}{n^2} z^n$

(iv) $\quad \sum\limits_{n=0}^{\infty} \dfrac{n^3}{3^n} z^n$

Solution. (i) Here, the given power series is $\sum\limits_{n=0}^{\infty} n^n.z^n$.

On comparing with $\sum\limits_{n=0}^{\infty} C_n z^n$, we obtain $C_n = n^n$ then, the radius of convergence R is given by

$$\dfrac{1}{R} = \lim\limits_{n\to\infty} \sup \sqrt[n]{|C_n|} = \lim\limits_{n\to\infty} \sup n = +\infty.$$

$\Rightarrow$ $R = 0.$

(ii) Here, the given power series is $\sum\limits_{n=0}^{\infty} \dfrac{z^n}{n!}$

On comparing with $\sum\limits_{n=0}^{\infty} C_n z^n$, we obtain $C_n = \dfrac{1}{n!}$ then, the radius of convergence R is given by

$$\dfrac{1}{R} = \lim\limits_{n\to\infty} \sup \dfrac{C_{n+1}}{C_n} = \lim\limits_{n\to\infty} \dfrac{n!}{(n+1)!} = \lim\limits_{n\to\infty} \dfrac{1}{n+1} = 0$$

$\Rightarrow$ $R = +\infty$

(iii) Here, the given power series is $\sum\limits_{n=0}^{\infty} \dfrac{2^n}{n^2} z^n$.

On comparing with $\sum\limits_{n=0}^{\infty} C_n z^n$, we obtain $C_n = 2^n/n^2$, then the radius of convergence R is

$$\frac{1}{R} = \lim_{n \to \infty} \sup \sqrt[n]{|C_n|} = \lim_{n \to \infty} \frac{2}{n^{2/n}} = \frac{2}{\left(\lim\limits_{n \to \infty} n^{1/n}\right)^2} = 2$$

$$\Rightarrow \qquad R = \frac{1}{2}$$

(iv) Here, the given power series is $\sum\limits_{n=0}^{\infty} \dfrac{n^3}{3^n} z^n$.

On comparing with $\sum\limits_{n=0}^{\infty} C_n z^n$, we obtain $C_n = \dfrac{n^3}{3^n}$

then, the radius of convergence is

$$\frac{1}{R} = \lim_{n \to \infty} \sup \sqrt[n]{|C_n|} = \lim_{n \to \infty} \frac{n^{3/n}}{3} = \lim_{n \to \infty} \frac{\left(n^{1/n}\right)^3}{3} = \frac{1}{3} \qquad \left[\because \lim_{n \to \infty} n\sqrt{n} = 1\right]$$

$R = 3$.

Example 6. *Show that the radius of convergence of the power* $\sum\limits_{n=0}^{\infty} \dfrac{z^n}{(n+1)^\alpha} (\alpha > 1)$ *is* 1.

Solution . On comparing the given power series with $\sum\limits_{n=0}^{\infty} C_n z^n$, we obtain $C_n = \dfrac{1}{(n+1)^\alpha} (\alpha > 1)$,

then the radius of convergence is

$$\frac{1}{R} = \lim_{n \to \infty} \sup \frac{C_{n+1}}{C_n} = \lim_{n \to \infty} \frac{(n+1)^\alpha}{(n+2)^\alpha} = \lim_{n \to \infty} \frac{\left(1+\frac{1}{n}\right)^\alpha}{\left(1+\frac{2}{n}\right)^\alpha} = 1$$

$$\Rightarrow \qquad R = 1.$$

Example 7. *Prove that the following power series is convergent for* $-1 \le x \le 1$

(i) $\quad x - \dfrac{x^3}{3} + \dfrac{x^5}{5} - \dfrac{x^7}{7} + ...$

(ii) $\quad \dfrac{x^2}{2} - \dfrac{x^3}{3}\left(1+\dfrac{1}{2}\right) + \dfrac{x^4}{4}\left(1+\dfrac{1}{2}+\dfrac{1}{3}\right) - ...$

Solution . (i) Here, the given power series is $\sum\limits_{n=1}^{\infty} \dfrac{(-1)^{n-1} . x^{2n-1}}{2n-1}$

On comparing with $\sum\limits_{n=1}^{\infty} a_n x^{2n-1}$, we obtain $a_n = \dfrac{(-1)^{n-1}}{2n-1}$ then, the radius of convergence R is

$$\frac{1}{R} = \lim_{n \to \infty} \sup \left(\frac{a_{n+1}}{a_n}\right) = \lim_{n \to \infty} \frac{2n-1}{2n+1} = 1$$

$$\Rightarrow \qquad R=1$$

So, the power series $\displaystyle\sum_{n=1}^{\infty}\frac{(-1)^{n-1}}{2n-1}x^{2n-1}$ is convergent for $-1 < x < 1$.

Now for $x=1$, $\qquad \displaystyle\sum_{n=1}^{\infty}\frac{(-1)^{n-1}}{2n-1}x^{2n-1}=\sum_{n=1}^{\infty}\frac{(-1)^{n-1}}{2n-1}$

By Leibnitz test, $\displaystyle\sum_{n=1}^{\infty}\frac{(-1)^{n-1}}{2n-1}$ is convergent. Since $\dfrac{1}{2n-1}>0\,(n\geq 1)$

is monotonic decreasing and $\displaystyle\lim_{n\to\infty}\frac{1}{2n-1}=0.$ So the power series

$\displaystyle\sum_{n=1}^{\infty}\frac{(-1)^{n-1}}{2n-1}x^{2n-1}$ is convergent for $-1\leq x\leq 1$.

(ii) Here, the given power series is $\displaystyle\sum_{n=1}^{\infty}(-1)^{n-1}\left(1+\frac{1}{2}+\frac{1}{3}+...+\frac{1}{n}\right)x^{n+1}$.

On comparing with $\displaystyle\sum_{n=1}^{\infty}a_n x^{n+1}$ we obtain $a_n=(-1)^{n-1}C_n$

and $\qquad C_n=\dfrac{1}{n+1}\displaystyle\sum_{k=1}^{\infty}\frac{1}{k}$

then $\qquad \left|\dfrac{a_{n+1}}{a_n}\right|=\left|\dfrac{C_{n+1}}{C_n}\right|=\left(\dfrac{n+1}{n+2}\right)\dfrac{\left(1+\dfrac{1}{2}+\dfrac{1}{3}+...+\dfrac{1}{n+1}\right)}{\left(1+\dfrac{1}{2}+...+\dfrac{1}{n}\right)}$

$$=\left(1-\frac{1}{n+1}\right)\left\{1+\frac{1}{(n+1)\left(1+\dfrac{1}{2}+...+\dfrac{1}{n}\right)}\right\}$$

so $\qquad \dfrac{1}{R}=\displaystyle\limsup_{n\to\infty}\left|\frac{a_{n+1}}{a_n}\right|=1 \quad\Rightarrow\quad R=1$

Therefore, the given series is convergent for $-1<x<1$.

Now, for $x=1$, $\qquad \displaystyle\sum_{n=1}^{\infty}a_n x^{n+1}=\sum_{n=1}^{\infty}(-1)^{n-1}C_n$.

We have

$$C_n-C_{n+1}=\Delta C_n=\left(\frac{1}{n+1}\cdot\sum_{k=1}^{n}\frac{1}{k}\right)=\Delta\left(\frac{1}{n+1}\right)\sum_{k=1}^{n}\frac{1}{k}+\frac{1}{n+2}\Delta\left(\sum_{k=1}^{n}\frac{1}{k}\right)$$

$$=\left(\frac{1}{(n+1)(n+2)}\right)\sum_{k=1}^{n}\frac{1}{k}-\frac{1}{(n+1)(n+2)}=\left(\frac{1}{(n+1)(n+2)}\right)\sum_{k=2}^{n}\frac{1}{k}>0$$

So the series $\{C_n\}_1^{\infty}$ is monotonic decreasing, then, we have

$$\lim_{n\to\infty}C_n=\lim_{n\to\infty}\left[\frac{1}{n+1}(\log n+\gamma_n)\right]$$

$$= \lim_{n \to \infty} \frac{\log n}{n+1} + \lim_{n \to \infty} \frac{\gamma_n}{n+1} = 0$$

By Leibnitz test, the series $\sum_{n=1}^{\infty} (-1)^{n-1} c_n$ is convergent.   Hence   the

power series $\sum_{n=1}^{\infty} a_n x^{n+1}$ is covergent for $-1 \le x \le 1$.

EXERCISE 13.1

1. Show that the radius of convergence of the series $\sum_{n=1}^{\infty} n x^{n-1}$ is 1.

2. Find the radius of convergence of the series
$$x + \frac{x^2}{2^2} + \frac{2!}{3^3} x^3 + \dots .$$

3. Find the radius of convergence of the series
$$\frac{1}{2} x + \frac{1.3}{2.5} x^2 + \frac{1.3.5}{2.5.8} x^3 + \dots .$$

4. Show that the series $\sum n^2 x^n$ has radius of convergence 1. Also show that the series converges absolutely for $-1 < x < 1$, but it fail to converges for $x = \pm 1$.

5. Show that the power series
$$1 + a.\frac{b}{1.c}.x + \frac{a(a+1)b(b+1)}{1.2c(c+1)} x^2 + \dots$$
has unit radius of convergence.

6. Show that the series $\sum_{n=1}^{\infty} \frac{x^{n-1}}{n^2}$ converges absolutely for $-1 < x < 1$.

7. Show that the series $x - \frac{x^2}{2} + \frac{x^3}{3} - \frac{x^4}{4} + \dots$ has
unit radius of convergence. Also show that the series converges absolutely for $-1 < x < 1$ and it converges at $x = 1$ but not at $x = -1$.

8. Find the radius of convergence and exact interval of convergence of the power series
$$\sum \frac{(-1)^n . x^{2n}}{(n!)^2 \, 2^{2n}} .$$

9. If the power series $\sum a_n x^n$ has radius of convergence R, then show that for any positive integer α, $\sum a_n x^{\alpha n}$ has radius of convergence $R^{1/\alpha}$.

10. Prove that $\log(1+x) = x - \frac{x^2}{2} + \frac{x^3}{3} - \frac{x^4}{4} + \dots ,$
$-1 < x \le 1$.

11. Show that the following power series is uniformly convergent :

(i) $x - \frac{x^3}{3} + \frac{x^5}{5} - \frac{x^7}{7} + \dots , [-1,1]$

(ii) $1 + \frac{x}{2} + \frac{x^2}{3} + \frac{x^3}{4} + \dots , [-1,k], 0 < k < 1$

(iii) $1 + \frac{1}{2} \frac{x^3}{3} + \frac{1.3}{2.4} \frac{x^5}{5} + \frac{1.3.5}{2.4.6} \frac{x^7}{7} + \dots , [-1,1]$

12. Prove   that   the   power   series
$1 + 2x + 3x^3 + 4x^3 + \dots$ is uniformly convergent in $[-k, k]$, where $0 < k < 1$.

13. Show that the power series
$$x + \frac{x^2}{2^2} + \frac{x^3}{3^2} + \frac{x^4}{4^2} + \dots$$
is uniformly convergent in $[-1, 1]$.

14. Show that

(i) $\log 2 = 1 - \frac{1}{2} + \frac{1}{3} - \frac{1}{4} + \dots$

(ii) $\sin^{-1} x = x + \frac{1}{2} \frac{x^3}{3} + \frac{1.3}{2.4} \frac{x^5}{5}$
$$+ \frac{1.3.5}{2.4.6} \frac{x^7}{7} + \dots \quad -1 < x \le 1.$$

(iii) $\log(1 - x) = -x - \frac{x^2}{2} - \frac{x^3}{3} \dots , -1 \le x < 1.$

(iv) $\frac{\pi}{2} = 1 + \frac{1}{2} . \frac{1}{3} + \frac{1.3}{2.4} . \frac{1}{5} + \frac{1.3.5}{2.4.6} . \frac{1}{7} + \dots$

(v) $\frac{1}{2} [\log(1 - x)]^2 = \frac{x^2}{2} + \left(1 + \frac{1}{2}\right) \frac{x^3}{3}$
$$+ \left(1 + \frac{1}{2} + \frac{1}{3}\right) \frac{x^4}{4} + \dots , -1 \le x < 1.$$

15. Show that

$$\frac{\log(1+x)}{(1+x)} = x - \left(1+\frac{1}{2}\right)x^2 + \left(1+\frac{1}{2}+\frac{1}{3}\right)x^3$$
$$- \left(1+\frac{1}{2}+\frac{1}{3}+\frac{1}{4}\right)x^4 + ...,$$
$$-1 < x < 1.$$

16. Show that

$$\frac{\sin^{-1}x}{\sqrt{1-x^2}} = x + \frac{2}{3}x^3 + \frac{2.4}{3.5}x^5 + \frac{2.4.6}{3.5.7}x^7 + ...,$$
$$-1 < x < 1.$$

17. Show that $\int_0^k \frac{dt}{1+t^n} = x - \frac{x^{n+1}}{n+1} + \frac{x^{2n+1}}{2n+1} - ...,$

$$-1 < x < 1, n > 0.$$

18. Show that

$$\int_0^1 \frac{\sin^{-1}x}{x} dx = \sum_{n=0}^{\infty} \frac{1.3.5...(2n-1)}{2.4.6...2n} \cdot \frac{1}{(2n+1)^2}.$$

19. Show that

$$\frac{1}{2}\left(\sin^{-1}x\right)^2 = \frac{x^2}{2} + \frac{2}{3}\frac{x^4}{4} + \frac{2.4}{3.5}\frac{x^6}{6} - ...,$$
$$-1 < x \le 1$$

20. Show that

$$\frac{1}{2}\left(\tan^{-1}x\right)^2 = \frac{x^2}{2} - \frac{x^4}{4}\left(1+\frac{1}{3}\right) +$$
$$\frac{x^6}{6}\left(1+\frac{1}{3}+\frac{1}{5}\right) - ...$$
$$-1 < x < 1.$$

CHAPTER REVIEW: A COMPETITIVE APPROACH

SELECTED TERMS AND RESULTS

TERMS

- **Power Series:** An expression of the form

$$a_0 + a_1 x + a_2 x^2 + \ldots + a_n x^n + \ldots = \sum_{n=0}^{\infty} a_n x^n$$

is called a power series.

- **Radius of Convergence:** A power series

$\sum_{n=0}^{\infty} a_n x^n$ is said to have radius of convergence,

$0, r$ and ∞ according as it converges only for $x=0$ or for $|x| < r$ or for any finite x respetively.

- **Abel's Summability:** A series $\sum a_n$ is said to be Abel summable to a sum s, if the associated power series $\sum a_n x^n$ with radius of convergence $r > 0$ converges on $]-r, r[$ to a function $f(x)$ and $\lim_{x \to r^-} f(x) = s$.

- **Nowhere Convergent:** If a power series converges for no value other than $x=0$ then given series is called nowhere convergent.

- **Everywhere Convergent:** If a given series converges for all values of x, we say that the given power series is called everywhere convergent.

- **Region of Convergence:** If the given power series converges for some value of x and diverges for the other values of x then the set of all values of x for which it is convergent is called region of convergence.

RESULTS

- If the power series $\sum a_n x^n$ is such that $a_n \neq 0$

$\forall n$ and $\lim_{n \to \infty} \left| \dfrac{a_{n+1}}{a_n} \right| = \dfrac{1}{R}$ then $\sum a_n x^n$ is convergent for $|x| < R$ and divergent for $|x| > R$.

- If the power series $\sum a_n x^n$ is such that $a_n \neq 0 \ \forall n$

and $\lim_{n \to \infty} |a_n|^{1/n} = \dfrac{1}{R}$ then $\sum a_n x^n$ is convergent for $|x| < R$ and divergent for $|x| > R$.

- Every power series converges absolutely within its interval of convergence.

- If a power series $\sum a_n x^n$ converges for $x=x_0$ then it is absolutely convergent for every $x=x_1$ where $|x_1| < |x_0|$.

- A power series will be absolutely convergent within the interval of convergence while it will be just convergent at one or both the end points of the interval of convergence.

- If a power series diverges for $x=x_i$ then it diverges for every $x=x_i'$ where $|x_i'| > |x_i|$.

- A power series is absolutely convergent within its interval of convergence and divergence outside it.

- The series obtained by integrating and differentiating power series term by term has the same radius of convergence as the original series.

- The power series $\sum a_n x^n$ and its corresponding series of derivatives $\sum n a_n x^{n-1}$ have the same radius of convergence.

- **(Abel's theorem)** If a power series $\sum a_n x^n$ converges at the end point $x=R$ of the interval $[-R, R]$ then it is uniformly convergent in the closed interval $[0, R]$.

- The sum function of the power series is continuous at the end point of the interval of convergence of the power series.

1. If $\sum_{n=0}^{\infty} a_n x^n = \sum_{n=0}^{\infty} b_n x^n$ in same nbd of 0, show that $a_n = b_n \ \forall n = 0, 1, 2, ...$

2. If $f(x) = \sum_{n=0}^{\infty} a_n x^n$ in some nbd of 0, show that no other power series can represent $f(x)$ in this nbd *i.e.*, when exists the power series representation of a function is unique.

3. If the power series $\sum_{n=0}^{\infty} a_n x^n$ converges on $[0,r[$, show that supremum r is the radius of convergence of the power series.

4. If $\sum a_n$ is Abel's summable to s and $\lim na_n = 0$ then prove that $\sum a_n$ converges to s.

5. If the series $\sum_{n=0}^{\infty} u_n, \sum_{n=0}^{\infty} v_n$ and their product series $\sum_{n=0}^{\infty} \left(\sum_{n=0}^{\infty} u_r v_{n-r} \right)$ converges to the sums s, t and p respectively, then show that $p = st$.

6. If $\sum a_n$ converges, show that $\sum a_n x^n$ converges uniformly on $[0,1]$ and converges absolutely on $]-1,1[$.

7. Show that the radius of convergence of the power series

$$\sum_{n=0}^{\infty} \frac{1.3...(2n-1)}{2.4...2n}\left(1+\frac{1}{2}+...+\frac{1}{n}\right)x^n \text{ is 1.}$$

8. If the term a_n, b_n are non-negative, then show that $\overline{\lim} (a_n + b_n)^{1/n} = \max\left\{\overline{\lim} a_n^{1/n}, \overline{\lim} b_n^{1/n}\right\}$.

FILL IN THE BLANKS

1. For $x=$ _____ every power series is convergent whatever the value of the coefficients.

2. The totality of point x for which a power series is converges is called _____ of convergence.

3. If a power series $\sum_{n=0}^{\infty} a_n x^n$ converges for a particular value x_0 of x then it converges absolutely for all values of x for which $|x| <$ _____.

4. The power series $\sum a_n x^n$ is uniformly convergent for $|x| \le \rho < R$ where R is the _____ of convergence.

TRUE/FALSE

Write 'T' for true and 'F' for false statement.

1. Every power series is uniformly convergent within its radius of convergence. **(T/F)**

2. A power series $\sum a_n x^n$ absolutely converges for values of x inside the circle of convergence and diverges outside the circle. **(T/F)**

3. $R = \lim_{n\to\infty} \left|\frac{a_{n+1}}{a_n}\right|$. **(T/F)**

4. The series obtained by integrating or differentiating a power series term by term has the same radius of convergence as the original series. **(T/F)**

5. The power series $x + \frac{x^2}{2^2} + \frac{x^3}{3^3} + ...$ converges absolutely for all $x \in]-\infty, \infty[$. **(T/F)**

MULTIPLE CHOICE QUESTIONS

Choose the most appropriate one:

Problem Set-1

1. A power series within its circle of convergence is :
   (a) converges uniformly

   (b) converges absolutely
   (c) converges absolutely and uniformly
   (d) none of the above

2. If the power series $\sum a_n x^n$ is convergent but the series $\sum |a_n x^n|$ is not convergent then the

   series $\sum_{n=0}^{\infty} a_n x^n$ is
   (a) convergent
   (b) conditionally convergent
   (c) divergent
   (d) none of the above

3. The radius of the convergence of the series $\sum n^n x^n$ is :
   (a) 1   (b) 0
   (c) 2   (d) none of these

4. The product series of any two absolutely convergent series is :
   (a) absolutely convergent
   (b) convergent
   (c) divergent
   (d) none of the above

5. The radius of convergence of the series
   $\sum \dfrac{n!}{n^n} x^n$:

(a) 1   (b) e
(c) 0   (d) none of these

6. In a convergent series, Σu_n :
   (a) necessarily tends to 0
   (b) tends to a definite limit
   (c) tends to ∞
   (d) none of the above

7. The radius of convergence of the series
   $\sum \dfrac{n}{2^n} x^n$ is :
   (a) 1   (b) 2
   (c) 0   (d) ∞

8. The radius of convergence of the series
   $\sum \left(1 - \dfrac{1}{n}\right)^{n^2} x^n$ is :
   (a) e (b) $2e$

   (c) $\dfrac{1}{e}$ (d) 0

9. The radius of convergence of the series
   $\sum_{n=0}^{\infty} \dfrac{2^n}{n!} x^n$ is :
   (a) ∞ (b) e
   (c) 1   (d) 0

10. The radius of convergence of the series
    $\sum \dfrac{1}{n^p} x^n$ is :
    (a) 0   (b) e
    (c) 1   (d) ∞

Problem Set-2

1. The power series $\sum a_n x^n$ is said to be absolutely convergent if the series :

   (a) $\sum |a_n x^n|$ is convergent

   (b) $\sum |a_n| \cdot |x^n|$ is convergent

   (c) both (a) and (b) are true
   (d) none of the above

2. If $\sum_{n=0}^{\infty} u_n$ and $\sum_{n=0}^{\infty} v_n$ be two given series and if

   $\lim_{n \to \infty} \dfrac{u_n}{v_n} = 0$ then :

(a) both series are Convergent
(b) both series are divergent
(c) either (a) and (b)
(d) none of the above

3. If $|u_n| \le |v_n|$ then $\sum u_n$ is absolutely convergent if :
   (a) $\sum v_n$ is convergent

   (b) $\sum |v_n|$ is convergent

   (c) either (a) or (b)
   (d) none of the above

4. If $|u_n| \geq |v_n|$ and $\sum v_n$ is divergent then :

(a) $\sum |u_n|$ is divergent

(b) $\sum u_n$ may or may not be convergent

(c) both (a) and (b) are true

(d) none of the above

5. If $\lim\limits_{n \to \infty} |u_n|^{1/n} = l$. Then the series $\sum u_n$ is :

(a) converges absolutely if $l < 1$

(b) divergent if $l > 1$

(c) both (a) and (b) are true

(d) none of the above

6. If $\sum \dfrac{|u_{n+1}|}{u_n} = l$, then :

(a) $\sum u_n$ converges absolutely if $l < 1$

(b) $\sum u_n$ is divergent if $l > 1$

(c) both (a) and (b) are true

(d) none of the above

7. The auxiliary series $\sum \dfrac{1}{n^p}$ is :

(a) convergent if $p > 1$

(b) divergent if $p < 1$

(c) both (a) and (b) are true

(d) none of the above

8. The power series $\sum a_n x^n$ is :
(a) converges for all values of x
(b) converges for $x = 0$ only
(c) converges for x in some region
(d) any one of the above

9. If a power series $\sum a_n x^n$ converges for a particular value x_0 of x then it converges for all values of x for which :

(a) $x < x_0$ (b) $|x| < |x_0|$
(c) $|x_0| < |x|$ (d) none of these

10. For every power series $\sum a_n x^n$, there exists a number R such that $0 \leq R \leq \infty$ with the property that the series :
(a) converges absolutely for every x such that $|x| < R$
(b) diverges if $|x| > R$
(c) both (a) and (b) are true
(d) none of the above

11. The series $\sum a_n (x - a)^n$ is :

(a) converges for $|x-a| < R$
(b) converges for $|x-a| > R$
(c) converges for $|x-a| = R$
(d) none of the above

12. The series $\sum a_n (x - a)^n$ is :

(a) diverges for $|x-a| > R$
(b) diverges for $|x-a| < R$
(c) diverges for $|x-a| = R$
(d) none of the above

13. If $f(x) = \sum\limits_{n=0}^{\infty} a_n x^n$, then $f(x)$ is called the :

(a) product functions
(b) sum function
(c) both (a) and (b) are true
(d) none of the above

14. The radius of the convergence of the series $\sum \dfrac{1}{n^n} \cdot x^n$ is :

(a) 1   (b) 0
(c) 2   (d) ∞

15. The radius of convergence of the series $\sum \dfrac{n!}{n^n} x^n$ is :

(a) 1   (b) e
(c) 2   (d) ∞

16. The radius of convergences of the series $\sum \left(1 + \dfrac{1}{n}\right)^{n^2} x^n$ is :

(a) e (b) 1
(c) $\dfrac{1}{e}$ (d) 0

17. The radius of convergence of the series $\sum \dfrac{x^n}{n!}$ is :
(a) e (b) 1
(c) $\dfrac{1}{e}$ (d) ∞

18. The radius of convergence of the series $\sum \dfrac{(n!)^2}{(2n)!} x^n$ is :

(a) e (b) 4
(c) $\dfrac{1}{}$ (d) ∞

19. The radius of convergence of the series
$\sum \dfrac{(n+1)}{(n+2)(n+3)}x^n$ is :

(a) 1 (b) 0

(c) $\dfrac{1}{2}$ (d) ∞

20. The radius of convergence of the series
$\sum \dfrac{x^n}{2^n+1}$:

(a) 2 (b) $\dfrac{1}{2}$

(c) e (d) $\dfrac{1}{e}$

21. The radius of convergence of the series
$\sum 2^{\sqrt{n}}.x^n$ is :

(a) 2 (b) $\dfrac{1}{2}$

(c) 1 (d) ∞

22. The radius of convergence of the series
$1+\dfrac{a.b}{1.c}x+\dfrac{a(a+1)b(b+1)}{1.2c(c+1)}x^2+... \text{ is :}$

(a) $\dfrac{1}{2}$ (b) 1

(c) 0 (d) ∞

23. The radius of convergence of the series
$\dfrac{1}{2}x+\dfrac{1.3}{2.5}x^2+\dfrac{1.3.5}{2.5.8}x^3+... \text{ is :}$

(a) $\dfrac{3}{2}$ (b) $\dfrac{2}{3}$

(c) 1 (d) ∞

24. If R_1 and R_2 are the radius of convergence
of the power series $\sum a_n x^n$ and $\sum b_n x^n$
respectively, then radius of convergence of
the series $\sum a_n b_n x^n$ is:

(a) R_1 (b) R_2

(c) $R_1.R_2$ (d) none of these

25. For the power series $f(x)= \sum\limits_{n=0}^{\infty} \dfrac{x^n}{2^n+1}$ which of
the following is/are true :

(a) $R=2$

(b) $(2-x)f(x)-2\to 0$ as $x\to 2$

(c) Both (a) and (b) are true

26. The domain of convergence of the power
series $\sum \dfrac{1.3.5...(2n-1)}{n!}\left(\dfrac{1-x}{x}\right)^n$:

(a) $\dfrac{3}{2}$ (b) $\dfrac{2}{3}$

(c) 1 (d) 0

27. The region of convergence of the series
$\sum\limits_{n=1}^{\infty} \dfrac{(x+2)^{n-1}}{(n+1)^3.4^n}$ is :

(a) $|x+2|\le 4$ (b) $|x+2|\le 3$
(c) $|x+2|\ge 3$ (d) none of the above

28. The region of convergence of the series
$\sum\limits_{n=1}^{\infty} n!x^n$ is :

(a) $x=0$ (b) $x\ne 0$
(c) $x=\infty$ (d) none of the above

29. The region of convergence of the series
$\sum\limits_{n=1}^{\infty} \dfrac{(-1)^{n-1}x^{2n-1}}{(2n-1)!}$ is :

(a) all finite x
(b) all finite x except $x=0$
(c) all finite x including 0
(d) none of the above

30. For the series $\sum\limits_{n=1}^{\infty} \dfrac{1}{(x^2+1)^n}$, which of the foll-

owing is/are true ?
(a) Series is convergent if $|x^2+1|>1$
(b) Sum of the given series is $\dfrac{1}{x^2}$
(c) Both (a) and (b) are true
(d) None of the above

31. For the series $\sum\limits_{n=0}^{\infty} (-1)^n .\left(x^n+x^{n+1}\right)$, which of

the following is/are true ?
(a) The series is convergent if $|x|<1$ $(x\ne 0)$
(b) Sum of the series is 1
(c) Both (a) and (b) are true
(d) None of the above

32. The behaviour of the series $\sum \dfrac{x^n}{n}$ on the cir-

cle of convergence $a_n = \dfrac{1}{n}$ is/are given by :

(a) its sequence of product sums is bounded

(b) $\lim\limits_{n \to 0} u_n = 0$

(c) the series is convergent for every value of x other than 1

(d) all are true

33. The behaviour of the power series $\sum \dfrac{x^{4n}}{4n+1}$ on the circle of convergence is/are :

(a) sequence of partial sums is bounded.

(b) $\lim\limits_{n \to \infty} u_n = 0$

(c) the given series $\sum a_n u_n$ is convergent for every value of x other than $x = \pm 1, \pm i$

(d) all are true

34. The behaviour of the series $\sum\limits_{n=2}^{\infty} \dfrac{x^n}{n(\log n)^2}$ on the circle of convergence is given by :

(a) radius of circle of convergence is 1

(b) the given series is absolutely convergent

(c) both (a) and (b) are true

(d) none of the above

35. The domain of Convergence of the series $\sum\limits_{n=1}^{\infty} \dfrac{1.3...(2n-1)}{n!}\left(\dfrac{1-x}{x}\right)^n$ is :

(a) $|x-1| < \dfrac{2}{3}$

(b) $|x-4| < \dfrac{2}{3}$

(c) $\left|x-\dfrac{4}{3}\right| < \dfrac{2}{3}$

(d) none of the above

36. The behaviour of the power series $\sum\limits_{n=1}^{\infty} (-1)^n \cdot \dfrac{x^n}{n}$ on the circle of convergence is that :

(a) given series is not convergent for $x = -1$

(b) its sequence of partial sums is bounded

(c) both (a) and (b) are true

(d) none of the above

37. For the series $\sum\limits_{n=1}^{\infty} \dfrac{(x+2)^{n+1}}{(n+1)^3 \, 4^n}$ which of the following is/are true ?

(a) its radius of convergence is 4

(b) its circle of convergence is –2

(c) both (a) and (b) are true

(d) none of the above

38. The radius of convergence of the series $\sum\limits_{n=2}^{\infty} \dfrac{x^n}{\log n}$ is :

(a) 2

(b) $\dfrac{1}{2}$

(c) 1

(d) 10

39. The radius of convergence of the series $\sum\limits_{n=1}^{\infty} \dfrac{(-1)^n}{n} \cdot x^n$ is :

(a) 2

(b) $\dfrac{1}{2}$

(c) 1

(d) 0

40. The series $\sum \dfrac{x^n}{n}$ converges :

(a) at every point of the circle of convergence

(b) at every point of the circle of convergence other than 1

(c) at every point of the circle of convergence other than 0

(d) none of the above

41. The series $\sum\limits_{n=1}^{\infty} c^{n^2} . e^{nx}$ is absolutely convergent for every value of x if :

(a) $|c| < 1$

(b) $|c| > 1$

(c) $|c| = 1$

(d) none of these

FILL IN THE BLANKS

1. 0 　　　2. region 　　　3. $|x_0|$ 　　　4. radius

TRUE/FALSE

1. T 　　　2. T 　　　3. F 　　　4. T 　　　5. T

MULTIPLE CHOICE QUESTIONS

Problem Set-1

1. (c)	**2.** (b)	**3.** (b)	**4.** (a)	**5.** (b)
6. (a)	**7.** (b)	**8.** (a)	**9.** (a)	**10.** (c)

Problem Set-2

1. (c)	**2.** (c)	**3.** (b)	**4.** (c)	**5.** (c)
6. (c)	**7.** (c)	**8.** (d)	**9.** (b)	**10.** (c)
11. (a)	**12.** (a)	**13.** (b)	**14.** (a)	**15.** (b)
16. (c)	**17.** (d)	**18.** (b)	**19.** (a)	**20.** (a)
21. (a)	**22.** (b)	**23.** (a)	**24.** (a)	**25.** (a)
26. (b)	**27.** (a)	**28.** (b)	**29.** (b)	**30.** (c)
31. (c)	**32.** (a)	**33.** (b)	**34.** (b)	**35.** (a)
36. (a)	**37.** (a)	**38.** (b)	**39.** (c)	**40.** (c)
41. (d)	**42.** (d)	**43.** (c)	**44.** (c)	**45.** (c)
46. (c)	**47.** (c)	**48.** (c)	**49.** (b)	**50.** (a)

Verify each of the following :

1. The radius of convergence of the series $\sum_{n=1}^{\infty} n x^{n-1}$ is 1. Also the given series converges absolutely on $]-1,1[$ and the series does not converges at $x=\pm 1$.

2. The radius of convergence of the series $1+x^2+x^4+x^6+...$ is 1.

3. The series $1+x+2!x^2+3!x^3+...$ does not converge for any value of x (other than 0).

4. The series $1+x+\dfrac{x^2}{2!}+\dfrac{x^3}{3!}+\dfrac{x^4}{4!}...$ has infinite radius of convergence and it converges absolutely for all values of x.

5. The series $\sum_{n=0}^{\infty} \dfrac{x^{n-1}}{n}$ converges at $x=-1$ but not at $x=1$. Also it converges absolutely on $]-1,1[$.

6. The series $x-\dfrac{x^2}{2}+\dfrac{x^3}{3}-\dfrac{x^4}{4}+...$ converges absolutely on $]-1,1[$. It converges at $x=1$ but not at $x=-1$.

7. The series $\dfrac{1}{2}x+\dfrac{1.3}{2.5}x^2+\dfrac{1.3.5}{2.5.8}x^3+...$ converges absolutely for all $|x|<\dfrac{3}{2}$.

8. If the series $\sum(-1)^n a_n$ converges then
$$\lim_{x\to-1+0} f(x)=\sum(-1)^n a_n .$$

9. If the power series $\sum a_n x^n$ has radius of convergence R, then for any positive integer k, $\sum a_n x^{kn}$ has radius of convergence $R^{1/k}$.

10. The power series
$$1+\dfrac{a.b}{1.c}x+\dfrac{a(a+1)b(b+1)}{1.2c(c+1)}x^2+...$$
has unit radius of convergence.

14 Double and Fourier Series

14.1 INTRODUCTION

In this chapter we shall study about double series which can be looked upon an extention of infinite series. Also, in the second part of this chapter we shall discuss the fourier series in details.

14.2 FUNDAMENTAL CONCEPTS OF DOUBLE SEQUENCE

Definition 1. *A function whose domain is the set $N \times N$, (where N is the set of natural numbers) and range is the subset of R, the set of real numbers is called double sequence.*

It is denoted by $<f_{nm}>$ or $<a_{nm}>$

Definition 2. *Let $<a_{mn}>$ be the double sequence then $<a_{mn}>$ tends to a definite limit a when m and n tends to ∞ independently then we say that $a_{mn} \to a$ as $m, n \to \infty$ independently*

It can be written as $\quad \lim\limits_{(m,n) \to \infty} a_{mn} = a$

Here, a is called the limit of the sequence.

Definition 3. *A double sequence $<a_{mn}>$ is said to converge to a real number a if for given $\varepsilon > 0 \; \exists \; n_0 > 0, \; n_0 \in N$ such that $|a_{mn} - a| < \varepsilon \quad \forall m, n \in N, m, n \geq n_0$.*

Definition 4. *(Cauchy's general principle of convergence)*

The double sequence $<a_{mn}>$ converges if and only if for given $\varepsilon > 0 \; \exists \; a$ positive number $n_0 \in N$ such that

$$|a_{pq} - a_{mn}| < \varepsilon \; \forall \; p \geq m \geq n_0 \text{ and } q \geq n \geq n_0.$$

14.3 DOUBLE SERIES

Let $< a_{mn}>$ be a double sequence. Then the double series is defined by a double summation extended over all terms a_{mn} for $m, n \in N$, where the terms are to be summed up in any arbitrary order. It is denoted by $\sum\limits_{m,n=1}^{\infty} a_{mn}$.

REMARK

- There are infinitely many ways of double summation of the terms a_{mn}. The summation by the method of rectangles, squares, rows and columns are respectively denoted by

$$\lim_{m,n \to \infty} s_{mn}, \; \lim_{n,m \to \infty} s_{mn}, \; \lim_{n \to \infty} s_{mn} \left(\lim_{m \to \infty} s_{mn} \right) \text{ and } \lim_{n \to \infty} \left(\lim_{m \to \infty} s_{mn} \right).$$

14.4 CONVERGENCE OF A DOUBLE SERIES

Let $\sum\limits_{m,n=1}^{\infty} a_{mn}$ be the given double series, then it is said to be convergent iff there is a number s such that the double sequence $<s_{mn}>$ of partial sums of the terms a_{mn} converges to s as $m, n \to \infty$.

Here, the number 's' is called the sum of the double series $\sum\limits_{m,n=1}^{\infty} a_{mn}$ such that $\sum\limits_{m,n=1}^{\infty} a_{mn} = s$

THEOREM 1. *A necessary condition for the convergence of a double series* $\sum\limits_{m,n=1}^{\infty} a_{mn}$ *is that* $\lim\limits_{m,n\to\infty} a_{mn} = 0$.

Proof. Let us suppose the given double series $\sum\limits_{m,n} a_{mn}$ converges. Then $\lim\limits_{m,n\to\infty} s_{mn}$ exists finitely, such that $\lim\limits_{m,n\to\infty} s_{mn} = s$

Then for given $\varepsilon > 0 \; \exists \; n_0 \in N$ such that $|s_{mn} - s| < \dfrac{\varepsilon}{4} \forall m, n \geq n_0$ ... (1)

Now consider

$$|s_{mn} + s_{m-1,n-1} - s_{m-1,n} - s_{m,n-1}|$$

$$= |s_{mn} - s + s_{m-1,n-1} - s - s_{m-1,n} + s - s_{m,n-1} + s|$$

$$\leq |s_{mn} - s| + |s_{m-1,n-1} - s| + |s_{m-1,n} - s| + |s_{m,n-1} - s|$$

$$< \frac{\varepsilon}{4} + \frac{\varepsilon}{4} + \frac{\varepsilon}{4} + \frac{\varepsilon}{4} = \varepsilon \; \forall m, n \geq n_0 + 1$$

$$\Rightarrow |a_{mn}| < \varepsilon \; \forall m, n \geq n_0 + 1$$

Hence $\lim\limits_{m,n\to\infty} a_{mn} = 0$

14.5 CAUCHY'S GENERAL PRINCIPLE OF CONVERGENCE

A necessary and sufficient condition for a double series $\sum\limits_{mn} a_{mn}$ to be convergent is that for its partial sum s_{mn}, given $\varepsilon > 0 \; \exists \; n_0 \in N$

$$|s_{pq} - s_{mn}| < \varepsilon \; \forall \; p \geq m \geq n_0 \text{ and } q \geq n \geq n_0$$

IMPORTANT RESULTS

If $\sum a_{mn}$ converges to s_1 and $\sum b_{mn}$ converges to s_2 then

(i) $\sum (a_{mn} \pm b_{mn})$ converges to $s_1 \pm s_2$
(ii) $\sum k a_{mn}$ converges to ks_1, k is independent of m, n.

THEOREM 1. *If a double series is convergent to the sum s and its sum exists then each of these sum is equal to s.*

Proof. Let the given double series be $\sum\limits_{m,n} a_{mn}$ converges to the sum s then

$$s_{mn} = \sum_{i=1}^{m} \sum_{j=1}^{n} a_{ij} \text{ converges to } s.$$

Then by definition, for given $\varepsilon > 0 \; \exists \; n_0 \in N$ such that $|s_{mn} - s| < \varepsilon \forall \, m, n \geq n_0$

$\Rightarrow \qquad \left| \lim_{n \to \infty} s_{mn} - s \right| < \varepsilon \; \forall m \geq n_0 \qquad\qquad (\because \lim_{n \to \infty} s_{mn} \text{ exists})$

$\Rightarrow \qquad \left| \lim_{m \to \infty} \left(\lim_{n \to \infty} s_{mn} \right) - s \right| < \varepsilon$

$\Rightarrow \qquad \lim_{m \to \infty} \left(\lim_{n \to \infty} s_{mn} \right) = s \qquad \Rightarrow \qquad \sum_{m=1}^{\infty} \left(\sum_{n=1}^{\infty} a_{mn} \right) = s$

In a similar way we may get $\displaystyle\sum_{n=1}^{\infty} \left(\sum_{m=1}^{\infty} a_{mn} \right) = s$.

THEOREM 2. *A double series $\sum a_{mn}$ with $a_{mn} \geq 0$ either converges to a finite sum s or else it converges to ∞.*

Proof. Let us suppose $\displaystyle\sum_{m,n} a_{mn}$ be a double series and s_{mn} denote the sequence of partial sums of the terms a_{mn}.

Since $a_{mn} \geq 0 \Rightarrow \; < s_{mn} >$ is monotonically non-decreasing.

$\Rightarrow \qquad < s_{mn} >$ is either convergent to a finite limit or divergent to ∞.

$\Rightarrow \qquad < s_{mn} > \to s$ or diverges to ∞.

$\Rightarrow \qquad$ The series $\displaystyle\sum_{m,n} a_{mn}$ with $a_{mn} \geq 0$ either converges to a finite

sum s or diverges to ∞.

THEOREM 3. *(Comparison test) If $\sum a_{mn}$ and $\sum b_{mn}$ are two double series such that $0 \leq a_{mn} \leq b_{mn} \; \forall \; m, n \in N$, then.*

(i) *$\sum b_{mn}$ converges $\Rightarrow \sum a_{mn}$ converges.*

(ii) *$\sum a_{mn}$ diverges $\Rightarrow \sum b_{mn}$ diverges.*

Proof. Let $<s_{mn}>$ and $<r_{mn}>$ be the sequence of partial sums of terms of the double series $\sum a_{mn}$ and $\sum b_{mn}$ respectively.

It is given that $0 \leq a_{mn} \leq b_{mn} \qquad \Rightarrow \qquad 0 \leq s_{mn} \leq r_{mn} \; \forall \; m, n \in N$

(i) It is given that $\sum b_{mn}$ is convergent

$\Rightarrow \qquad$ for given $\varepsilon > 0 \; \exists \; n_0 \in N$ such that

$\qquad\qquad |r_{pq} - r_{mn}| < \varepsilon \forall \, p \geq m \geq n_0$ and $q \geq n \geq n_0$

$\Rightarrow \qquad |s_{pq} - s_{mn}| \leq |r_{pq} - r_{mn}| < \varepsilon$

$\Rightarrow \qquad |s_{pq} - s_{mn}| < \varepsilon$

$\Rightarrow \qquad \sum a_{mn}$ is convergent

(ii) Let us suppose that $\sum a_{mn}$ is divergent $\qquad \Rightarrow \qquad < s_{mn} >$ is unbounded

$\Rightarrow \qquad \lim_{m,n \to \infty} s_{mn} = \infty$

Now, since $s_{mn} \leq r_{mn} \; \forall m, n \in N \qquad \Rightarrow \qquad \lim_{m,n \to \infty} r_{mn} = \infty$

$\Rightarrow \qquad <r_{mn}>$ is not bounded above

$\Rightarrow \qquad \displaystyle\sum_{m,n} b_{mn}$ diverges

REMARKS

- The series $\sum \dfrac{1}{c_n}$ and the double series $\sum a_{mn}$ where $a_{mn} = (c_m c_n)^{-1}$ converges or diverges together.

- The double series $\sum\limits_{m,n} \dfrac{1}{(m+n)b_p}$ converges if $\sum\limits_{m,n} \dfrac{1}{b_p}$ converges and $\sum\limits_{m,n} \dfrac{1}{(m+n)b_p}$ diverges if $\sum \dfrac{1}{b_p}$ diverges where $p = m+n$.

14.6 ABSOLUTE CONVERGENCE OF A DOUBLE SERIES

A double series $\sum\limits_{m,n} a_{mn}$ is said to be absolutely convergent if the double series $\sum\limits_{m,n} |a_{mn}|$ converges.

THEOREM 1. *Every absolutely convergent double series is convergent.*

Proof. Let $\sum\limits_{m,n} a_{mn}$ be an absolutely convergent double series $\Rightarrow \sum\limits_{m,n} |a_{mn}|$ converges.

By comparison test $\sum\limits_{m,n} \dfrac{1}{2}(|a_{mn}|+a_{mn})$ and $\sum\limits_{m,n} \dfrac{1}{2}(|a_{mn}|-a_{mn})$ also converges

$\Rightarrow$ The series $\sum\limits_{m,n} a_{mn} = \sum\limits_{m,n} \left\{ \dfrac{1}{2}(|a_{mn}|+a_{mn}) - \dfrac{1}{2}(|a_{mn}|-a_{mn}) \right\}$ is convergent.

$\Rightarrow \sum\limits_{m,n} a_{mn}$ is convergent.

REMARK

- If one of the series $\sum\limits_{m}\sum\limits_{n} a_{mn}, \sum\limits_{n}\sum\limits_{m} a_{mn}$ and $\sum a_{mn}$ is absolutely convergent then the remaining two and the sum of all the three series are the same.

Solved Examples

Example 1. *Show that the double series* $\sum\limits_{m,n=1}^{\infty} \dfrac{1}{2^m 3^n}$ *is convergent*

Solution. Let us suppose

$$\sum\limits_{m,n=1}^{\infty} \dfrac{1}{2^m 3^n} = \sum\limits_{m,n=1}^{\infty} a_{mn} \text{ where } a_{mn} = \dfrac{1}{2^m 3^n}$$

Now sequence of partial sum $s_{mn} = \sum\limits_{i=1}^{m}\sum\limits_{j=1}^{n} a_{ij} = \sum\limits_{i=1}^{m}\sum\limits_{j=1}^{n} \dfrac{1}{2^i 3^j}$

$$= \sum\limits_{i=1}^{m} \left[\dfrac{1}{2^i} \left\{ \dfrac{1}{3} + \dfrac{1}{3^2} + ... + \dfrac{1}{3^n} \right\} \right]$$

$$= \left(\dfrac{1}{2} + \dfrac{1}{2} + ... + \dfrac{1}{2^m} \right) \left(\dfrac{1}{3} + \dfrac{1}{3^2} + ... + \dfrac{1}{3^m} \right)$$

Now, since $\sum\limits_{m=1}^{\infty} \dfrac{1}{2^m}$ and $\sum\limits_{n=1}^{\infty} \dfrac{1}{3^n}$ are convergent being the G.P. of common ratio

$\dfrac{1}{2} < 1$ and $\dfrac{1}{3} < 1$ respectively, therefore, there exists real number $k > 0$ such that

$$\left(\frac{1}{2} + \frac{1}{2^2} + \dots + \frac{1}{2^m} \right)\left(\frac{1}{3} + \frac{1}{3^2} + \dots + \frac{1}{3^n} \right) \le k \ \forall m, n \in \mathbb{N}$$

$\Rightarrow \quad 0 \le s_{mn} \le k$

$\Rightarrow \quad < s_{mn} >$ is bounded and monotonic

$\Rightarrow \quad < s_{mn} >$ is convergent.

Example 2. *Test the convergence of the double series* $\sum\limits_{m,n} \dfrac{1}{m^{\alpha} n^{\beta}} - \alpha, \beta \in \mathbb{R}$

Solution. Let $\sum\limits_{m,n} \dfrac{1}{m^{\alpha} n^{\beta}} = \sum\limits_{m,n} a_{mn}$, where $a_{mn} = \dfrac{1}{m^{\alpha} n^{\beta}}$

Now $\quad s_{mn} = \sum\limits_{i=1}^{m} \sum\limits_{j=1}^{n} a_{ij} = \sum\limits_{i=1}^{m} \sum\limits_{j=1}^{n} \dfrac{1}{i^{\alpha} j^{\beta}}$

$$= \sum\limits_{i=1}^{m} \frac{1}{i^{\alpha}} \left\{ \frac{1}{1^{\beta}} + \frac{1}{2^{\beta}} + \dots + \frac{1}{n^{\beta}} \right\}$$

$$= \left(\frac{1}{1^{\alpha}} + \frac{1}{2^{\alpha}} + \dots + \frac{1}{m^{\alpha}} \right)\left(\frac{1}{1^{\beta}} + \frac{1}{2^{\beta}} + \dots + \frac{1}{n^{\beta}} \right)$$

$\Rightarrow \quad s_{mn}$ and so $\sum\limits_{m,n} \dfrac{1}{m^{\alpha} n^{\beta}}$ converges iff $\sum \dfrac{1}{m^{\alpha}}$ and $\sum \dfrac{1}{n^{\beta}}$ converges.

But $\sum \dfrac{1}{m^{\alpha}}$ and $\sum \dfrac{1}{n^{\beta}}$ converges for $\alpha > 1$ and $\beta > 1$ respectively and diverges for

$\alpha \le 1$ and $\beta \le 1$

Example 3. *Test the convergence of the double series* $\sum\limits_{m,n} \dfrac{1}{(m+n)^{\alpha}}$, $\alpha \in \mathbb{R}$

Solution. Let $a_{mn} = \dfrac{1}{(m+n)^{\alpha}} = \dfrac{1}{(m+n)(m+n)^{\alpha-1}}$

Let us write $r_p = (m+n)^{\alpha-1}$

Therefore, the given double series becomes

$$\sum\limits_{m,n} a_{mn} = \sum\limits_{m,n} \frac{1}{(m+n)^{\alpha}} = \sum\limits_{m,n} \frac{1}{(m+n)r_p}$$

$\Rightarrow$ The given double series converges iff the series $\sum\limits_{m,n} \dfrac{1}{(m+n)^{\alpha}}$ converges.

But the series $\sum\limits_{m,n} \dfrac{1}{(m+n)^{\alpha}}$ converges iff the single series $\sum\limits_{m,n} \dfrac{1}{r_p} = \sum\limits_{m,n} \dfrac{1}{(m+n)^{\alpha-1}}$

converges only if $\alpha - 1 > 1$ *i.e.*, $\alpha > 2$.

1. Show that the double series $\displaystyle\sum_{m,n} \sin \frac{1}{2^m.3^n}$ is convergent.

2. If $\displaystyle\sum_{m,n} a_{mn}$ is convergent, prove that

$$\lim_{m,n\to0} a_{mn} = 0.$$

3. Show that $\displaystyle\sum_{m=2}^{\infty} \sum_{n=1}^{\infty} \frac{1}{(2n)^m} = \log 2.$

4. Show that the double series $\displaystyle\sum_{m,n} \frac{1}{(m^2+n^2)^\alpha}$ converges iff $\alpha > 1$.

5. Show that the double series

$$\sum_{m,n=0}^{\infty} \frac{(m+n)!}{m!.n!} \left(\frac{x}{2}\right)^{m+n} \quad \forall x \in]-2,1[\, .$$

14.7 FOURIER SERIES

In this section, we shall study a special type of functional series extensively studied by Joseph Fourier. Joseph Fourier represented expansions in trigonometrical series in connection with boundary value problem in conduction of heat. Although such expansions had been studied earlier, these series bear the name 'Fourier series'because of the major contributions of Fourier in this field.

14.8 PERIODIC FUNCTIONS

A function $f(x)$ which satisfies the relation $f(x + T) = f(x)$ for all real x and some fixed T is called a periodic function. The smallest positive number T, for which this relation holds, is called the period of $f(x)$.

If T is the period of $T(x)$. Then

$$f(x) = f(x + T) = f(x + 2T) =... = f(x + nT) = ...$$

Also, $\qquad f(x) = f(x - T) = f(x - 2T) =... = f(x - nT) = ...$

$\therefore \qquad f(x) = f(x \pm nT)$, where n is a positive integer.

For example: Consider the function $f(x) = \sin x$. We have

$$\sin x = \sin (x + 2\pi) = \sin (x + 4\pi) = ...$$

Here, $f(x) = \sin x$ is a periodic function with period 2π. This function is also called sinusoidal periodic function.

We have studied about the Macluarian's theorem which is used to expand a function provided the function's derivative are continuous. Now, the need arise to expand functions which have discontinuities in their derivatives. By Fourier series, we can expand both types of functions under certain conditions as an infinite series of sine and cosine of x and it's integral multiple of a function $f(x)$ is defined in the interval $c < x < c + 2\pi$.

Then, Fourier series of $f(x)$ is given by

$$f(x) = \frac{a_0}{2} + \sum_{n=1}^{\infty} a_n \cos nx + \sum_{n=1}^{\infty} b_n \sin nx \qquad ...(1)$$

where a_0, a_n and b_n are called Fourier coefficient of $f(x)$ and their values are given as :

$$a_0 = \frac{1}{\pi} \int_{c}^{c+2\pi} f(x)dx \qquad ...(2)$$

$$a_n = \frac{1}{\pi} \int_{c}^{c+2\pi} f(x)\cos nx \, dx \qquad ... (3)$$

$$b_n = \frac{1}{\pi} \int_c^{c+2\pi} f(x)\sin nx\,dx \qquad \qquad \dots (4)$$

The series (1) with coefficients a_0, a_n and b_n given by (2), (3) and (4) respectively is called the Fourier series of $f(x)$ and the coefficients a_0, a_n and b_n are called the Fourier coefficients corresponding to $f(x)$.

(i) When $c = 0$, the interval becomes $0 < x < 2\pi$ and formula for a_0, a_n, b_n is obtained by putting $c = 0$.

(ii) When $c = -\pi$, then interval becomes $-\pi < x < \pi$. In this interval, the formula for a_0, a_n and b_n becomes as under :

(a) When $f(x)$ is an odd function, then

$$a_0 = \frac{1}{\pi} \int_{-\pi}^{\pi} f(x)dx = 0$$

$$a_n = \frac{1}{\pi} \int_{-\pi}^{\pi} f(x)\cos nx\,dx = 0 \qquad \text{[By property of definite integral]}$$

$$b_n = \frac{1}{\pi} \int_{-\pi}^{\pi} f(x)\sin nx\,dx = \frac{2}{\pi} \int_0^{\pi} f(x)\sin x\,dx$$

Hence, if function $f(x)$ is odd, its Fourier expansion contains only sine series,

i.e., $\qquad f(x) = \sum_{n=1}^{\infty} b_n \sin nx, \text{where } b_n = \frac{2}{\pi} \int_0^{\pi} f(x)\sin nx\,dx.$

(b) When $f(x)$ is even function, then formula for a_0, a_n and b_n are given by

$$a_0 = \frac{1}{\pi} \int_{-\pi}^{\pi} f(x)dx = \frac{2}{\pi} \int_0^{\pi} f(x)dx$$

$$a_n = \frac{1}{\pi} \int_{-\pi}^{\pi} f(x)\cos nx\,dx = \frac{2}{\pi} \int_0^{\pi} f(x)\cos nx\,dx$$

and $\qquad b_n = \frac{1}{\pi} \int_{-\pi}^{\pi} f(x)\sin nx\,dx = 0 \qquad \qquad [\because f(x) \sin nx \text{ is odd.}]$

Hence, if a periodic function $f(x)$ is even, its Fourier expansion contains only cosine terms,

i.e., $f(x) = \frac{a_0}{2} + \sum_{n=1}^{\infty} \int_0^{\pi} f(x)dx$, where

$$a_0 = \frac{2}{\pi} \int_0^{\pi} f(x)dx \text{ and } a_n = \frac{2}{\pi} \int_0^{\pi} f(x).\cos nx\,dx$$

14.9 SOME IMPORTANT RESULTS

The following results are useful in the Fourier series :

(i) $\sin n\pi = 0, \cos n\pi = (-1)^n, \cos\left(n+\frac{1}{2}\right)\pi = 0$, where $n \in Z$.

(ii) $\int uv = uv_1 - u'v_2 + u''v_3 - u'''v_4 +..., \text{ where } u' = \frac{du}{dx}, u'' = \frac{d^2u}{dx^2},...$
$v_1 = \int v\,dx, v_2 = \int v_1 dx,...$

(iii) $\int_0^{2\pi} \sin nx\, dx = 0$

(iv) $\int_0^{2\pi} \cos nx\, dx = 0$

(v) $\int_0^{2\pi} \sin^2 nx\, dx = \pi$

(vi) $\int_0^{2\pi} \cos^2 nx\, dx = \pi$

(vii) $\int_0^{2\pi} \sin nx.\sin mx\, dx = 0$

(viii) $\int_0^{2\pi} \cos nx.\cos mx\, dx = 0$

(ix) $\int_0^{2\pi} \sin nx.\cos nx\, dx = 0$

(x) $\int_0^{2\pi} \sin nx.\cos nx\, dx = 0$

(xi) $\int e^{ax}\sin bx\, dx = \dfrac{e^{ax}}{a^2+b^2}(a\sin bx - b\cos bx)+c$

(xii) $\int e^{ax}\cos bx\, dx = \dfrac{e^{ax}}{a^2+b^2}(a\cos bx - b\sin bx)+c$

14.10 DETERMINATION OF FOURIER COEFFICIENTS: EULER'S FORMULAE

The fourier series is given by

$$f(x) = \frac{a_0}{2} + a_1\cos x + a_2\cos 2x + \dots + a_n\cos nx$$
$$+ b_1\sin x + \dots + b_2\sin 2x + \dots + b_n\sin nx + \dots \qquad \dots(i)$$

or $\qquad f(x) = \dfrac{a_0}{2} + \sum_{n=1}^{\infty} a_n\cos nx + \sum_{n=1}^{\infty} b_n\sin nx.$

To find a_0 : Integrating both sides of equation (1) from $x = c+0, x = c+2\pi$

$$\int_c^{c+2\pi} f(x)dx = \frac{a_0}{2}\int_c^{c+2\pi} dx + \int_c^{c+2\pi}\left(\sum_{n=1}^{\infty} a_n\cos nx\right)dx + \int_c^{c+2\pi}\left(\sum_{n=1}^{\infty} b_n\sin nx\right)dx$$

$$= \frac{a_0}{2}(c+2\pi-c)+0+0 = a_0\pi$$

$\Rightarrow \qquad a_0 = \dfrac{1}{\pi}\int_c^{c+2\pi} f(x)\,dx.$

To find a_n : Multipling each side of equation (1) by $\cos nx$ and integrate w.r.t. x, between the limit c to $c+2\pi$.

$$\int_c^{c+2\pi} f(x)\cos nx\, dx = \frac{a_0}{2}\int_c^{c+2\pi}\cos nx\, dx + \int_c^{c+2\pi}\left(\sum_{n=1}^{\infty} a_n\cos nx\right)\cos nx\, dx$$

$$+ \int_c^{c+2\pi}\left(\sum_{n=1}^{\infty} b_n\sin nx\right)\cos nx\, dx$$

$$= 0 + a_n\pi + 0 = a_n\pi$$

$\Rightarrow \qquad a_n = \dfrac{1}{\pi}\int_c^{c+2\pi} f(x)\cos nx\, dx.$

To find b_n : Multiplying each side of equation (1) by $\sin nx$ and integrate w.r.t. x between the limit c to $c + 2\pi$.

$$\int_c^{c+2\pi} f(x)\sin nx\, dx = \frac{a_0}{2}\int_c^{c+2\pi}\sin nx\, dx + \int_c^{c+2\pi}\left(\sum_{n=1}^{\infty} a_n\cos nx\right)\sin nx\, dx +$$

$$+ \int_c^{c+2\pi}\left(\sum_{n=1}^{\infty} b_n\sin nx\right)\sin nx\, dx$$

$$= 0 + 0 + b_n\pi = b_n\pi$$

$$\Rightarrow \qquad b_n = \frac{1}{\pi}\int_c^{c+2\pi} f(x)\sin nx\, dx$$

These values of a_0, a_n and b_n are called Euler's formulae.

14.11 DIRICHLET'S CONDITIONS

Any function $f(x)$ can be expressed as a Fourier series

$\frac{a_0}{2} + \sum_{n=1}^{\infty} a_n\cos nx + \sum_{n=1}^{\infty} b_n\sin nx$, where a_0, a_n and b_n are constants.

(i) $f(x)$ is finite and single valued in the interval $c < x < c + 2\pi$.
(ii) $f(x)$ is periodic with period 2π
(iii) $f(x)$ and $f'(x)$ are piecewise continuous in the interval $c < x < c + 2\pi$.
The Fourier series with its coefficients converge to
(a) $f(x)$ if x is a point of continuity.
(b) $\dfrac{f(x+0)+f(x-0)}{2}$, if x is a point of discontinuity.
The conditions (i), (ii) and (iii) imposed on $f(x)$ are sufficient but not necessary. *i.e.*, if the conditions are satisfied, the convergence is guranteed. However, if they are not satisfied the series may or may not converge.

Solved Examples

Example 1. *Expand the function $f(x) = x\sin x$ as a Fourier series in interval $-\pi \le x \le \pi$. Deduce that* $\dfrac{1}{1.3} - \dfrac{1}{3.5} + \dfrac{1}{5.7} - \dfrac{1}{7.9} + ... = \dfrac{\pi-2}{4}$

Solution. Since $x\sin x$ is an even function of x, so $b_n = 0$, then Fourier series is given by

$$f(x) = x\sin x = \frac{a_0}{2} + \sum_{n=1}^{\infty} a_n\cos nx,$$

where

$$a_0 = \frac{2}{\pi}\int_0^\pi x\sin x\, dx = \frac{2}{\pi}\left[-x\cos x + \sin x\right]_0^\pi = \frac{2}{\pi}(-\pi\cos\pi) = 2$$

$$a_n = \frac{2}{\pi}\int_0^\pi x\sin x\cos nx\, dx = \frac{1}{\pi}\int_0^\pi x.2\cos nx\sin x\, dx$$

$$= \frac{1}{\pi}\int_0^\pi x\{\sin(n+1)x - \sin(n-1)x\}\, dx$$

$$= \frac{1}{\pi}\left[x\left\{\frac{-\cos(n+1)x}{n+1} + \frac{\cos(n+1)x}{n-1}\right\}\right.$$

$$\left. -1\left\{\frac{-\sin(n+1)x}{(n+1)^2} + \frac{\sin(n-1)x}{(n+1)^2}\right\}\right]_0^\pi$$

$$= \frac{1}{\pi}\left[x\left\{\frac{-\cos(n+1)\pi}{n+1} + \frac{\cos(n-1)\pi}{n-1}\right\}\right]$$

$$= \frac{\cos(n-1)\pi}{n-1} - \frac{\cos(n+1)\pi}{n+1}; n \neq 1$$

$$= \begin{cases} \dfrac{1}{n-1} - \dfrac{1}{n+1} = \dfrac{2}{n^2-1} & \text{if } n \text{ is odd } n \neq 1 \\ \dfrac{-1}{n-1} + \dfrac{1}{n+1} = \dfrac{-2}{n^2-1} & \text{if } n \text{ is even} \end{cases}$$

When $n=1$, then $a_1 = \dfrac{2}{\pi}\int_0^\pi x \sin x \cos x\, dx = \dfrac{1}{\pi}\int_0^\pi x \sin 2x\, dx$

$$= \frac{1}{\pi}\left[x\left(\frac{-\cos 2x}{2}\right) - \left(\frac{-\sin 2x}{4}\right)\right]_0^\pi = \frac{1}{\pi}\left[\frac{-\pi\cos 2\pi}{2}\right] = -\frac{1}{2}$$

∴ $$x \sin x = 1 - \frac{1}{2}\cos x - 2\left[\frac{\cos 2x}{2^2-1} - \frac{\cos 3x}{3^2-1} + \frac{\cos 4x}{4^2-1} - \frac{\cos 5x}{5^2-1} + \ldots\right]$$

Putting $x = \dfrac{\pi}{2}$, we get $\dfrac{\pi}{2} = 1 - 2\left(\dfrac{-1}{2^2-1} + \dfrac{1}{4^2-1} - \dfrac{1}{6^2-1} + \ldots\right)$

$\Rightarrow$ $\dfrac{\pi}{2} - 1 = 2\left(\dfrac{1}{3} - \dfrac{1}{15} + \dfrac{1}{35} - \ldots\right) \Rightarrow \dfrac{\pi-2}{4} = \left(\dfrac{1}{1.3} - \dfrac{1}{3.5} + \dfrac{1}{5.7} - \ldots\right).$

Example 2. *Find the Fourier series to represent e^{ax} in interval $-\pi < x < \pi$.*

Solution. Let $f(x) = e^{ax} = \dfrac{a_0}{2} + \sum_{n=1}^\infty a_n \cos nx + \sum_{n=1}^\infty b_n \sin nx$

$$a_0 = \frac{1}{\pi}\int_{-\pi}^\pi f(x)\, dx = \frac{1}{\pi}\int_{-\pi}^\pi e^{ax}\, dx = \frac{1}{\pi}\left[\frac{e^{ax}}{a}\right]_{-\pi}^\pi = \frac{1}{a\pi}(e^{a\pi} - e^{-a\pi}) = \frac{2\sinh a\pi}{\pi a}$$

$$a_n = \frac{1}{\pi}\int_{-\pi}^\pi f(x)\cos nx\, dx = \frac{1}{\pi}\int_{-\pi}^\pi e^{ax}\cos nx\, dx$$

$$= \left[\frac{e^{ax}}{\pi(a^2+n^2)}a\cos nx + a\sin nx\right]_{-\pi}^\pi$$

$$= \frac{a\cos n\pi(e^{a\pi} - e^{-ia\pi})}{\pi(a^2+n^2)} = \frac{2a(-1)^n \sinh a\pi}{\pi(a^2+n^2)}$$

Similarly, we can set

$$b_n = \frac{2n(-1)^n \sinh a\pi}{\pi(a^2+n^2)}$$

$$\therefore \qquad e^{ax} = \frac{\sinh a\pi}{a\pi} + \sum_{n=1}^{\infty} \frac{2a(-1)^n \sinh a\pi}{\pi(a^2+n^2)} \cos nx + \sum_{n=1}^{\infty} \frac{2n(-1)^n \sinh a\pi}{\pi(a^2+n^2)} \sin n\pi$$

$$= \frac{2\sinh a\pi}{\pi}\left[\frac{1}{2a} - a\left(\frac{\cos x}{a^2+1^2} - \frac{\cos 2x}{a^2+2^2} + \frac{\cos 3x}{a^2+3^2} - \cdots\right)\right.$$

$$\left. - \left(\frac{\sin x}{a^2+1^2} - \frac{2\sin 2x}{a^2+2^2} + \frac{3\sin 3x}{a^2+3^2} - \cdots\right)\right]$$

Example 3. *Obtain the Fourier series for the function $f(x) = x^2$, $-\pi < x < \pi$. Sketch the graph f function $f(x)$. Hence, show that*

(i) $\dfrac{1}{1^2} + \dfrac{1}{2^2} + \dfrac{1}{3^2} + \dfrac{1}{4^2} + \cdots = \sum_{n=1}^{\infty} \dfrac{1}{n^2} = \dfrac{\pi^2}{6}$

(ii) $\dfrac{1}{1^2} - \dfrac{1}{2^2} + \dfrac{1}{3^2} - \dfrac{1}{4^2} + \cdots = \dfrac{\pi^2}{12}$

(iii) $\dfrac{1}{1^2} + \dfrac{1}{3^2} + \dfrac{1}{5^2} + \cdots = \sum_{n=1}^{\infty} \dfrac{1}{(2n-1)^2} = \dfrac{\pi^2}{8}$

Solution. $f(x) = x^2$ is an even function, therefore $b_n = 0$

Now $f(x) = x^2 = \dfrac{a_0}{2} + \sum_{n=1}^{\infty} a_n \cos nx$. Then

$$a_0 = \frac{2}{\pi}\int_0^\pi f(x)\,dx = \frac{2}{\pi}\int_0^\pi x^2\,dx = \frac{2}{\pi}\left[\frac{x^3}{3}\right]_0^\pi = \frac{2}{3}\pi^2$$

$$a_n = \frac{2}{\pi}\int_0^\pi f(x)\cos nx\,dx = \frac{2}{\pi}\int_0^\pi x^2 \cos nx\,dx$$

$$= \frac{2}{\pi}\left[x^2\left(\frac{\sin nx}{n}\right) - 2x\left(\frac{-\cos nx}{n^2}\right) + 2\left(\frac{-\sin nx}{n^2}\right)\right]_0^\pi$$

$$= \frac{2}{\pi}\left[2x\frac{\cos n\pi}{n^2}\right] = 4\frac{(-1)^n}{n^2}$$

$$\therefore \qquad x^2 = \frac{\pi^2}{3} - 4\left(\frac{\cos x}{1^2} - \frac{\cos 2x}{2^2} + \frac{\cos 3x}{3^2} - \frac{\cos 4x}{4^2} + \cdots\right)$$

$$\Rightarrow \qquad x^2 = \frac{\pi^2}{3} + 4\sum_{n=1}^{\infty} \frac{(-1)^n}{n^2}\cos nx \qquad\qquad \cdots (1)$$

Put $x=\pi$ in (1), we get

$$\pi^2 = \frac{\pi^2}{3} - 4\left(-\frac{1}{1^2} - \frac{1}{2^2} - \frac{1}{3^2} - \frac{1}{4^2} - \cdots\right)$$

$$\Rightarrow \qquad \frac{2\pi^2}{3} = -4\left(-\frac{1}{1^2} - \frac{1}{2^2} - \frac{1}{3^2} - \frac{1}{4^2} - \cdots\right)$$

$$\therefore \quad \frac{1}{1^2} + \frac{1}{2^2} + \frac{1}{3^2} + \frac{1}{4^2} \cdots = \frac{\pi^2}{6} \qquad \ldots (2)$$

Put $x = 0$ in (1), we get

$$0 = \frac{\pi^2}{3} - 4\left(\frac{1}{1^2} - \frac{1}{2^2} + \frac{1}{3^2} - \frac{1}{4^2} - \cdots\right)$$

$$\therefore \quad \frac{1}{1^2} - \frac{1}{2^2} - \frac{1}{3^2} - \frac{1}{4^2} - \cdots = \frac{\pi^2}{12} \qquad \ldots (3)$$

Adding (2) and (3), we get

$$\frac{\pi^2}{4} = 2\left(\frac{1}{1^2} + \frac{1}{3^2} + \frac{1}{5^2} + \cdots\right)$$

$$\therefore \quad \frac{1}{1^2} + \frac{1}{3^2} + \frac{1}{5^2} + \cdots = \frac{\pi^2}{8}.$$

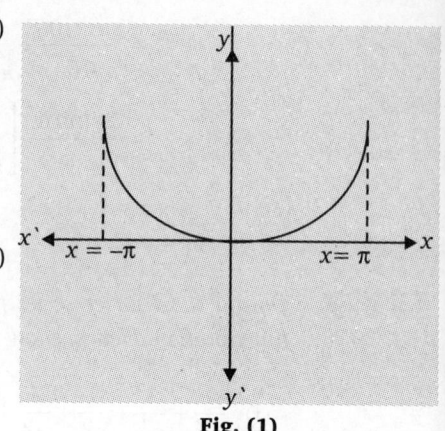

Fig. (1)

Example 4. *Obtain the Fouries series for $f(x) = e^{-x}$ in the interval $0 < x < 2\pi$.*

Solution. Let $f(x) = e^{-x}$. The Fourier series of $f(x)$ can be written as

$$f(x) = e^{-x} = \frac{a_0}{2} + \sum_{n=1}^{\infty} a_n \cos nx + \sum_{n=1}^{\infty} b_n \sin nx$$

Then,

$$a_0 = \frac{1}{2}\int_0^{2\pi} f(x)dx = \frac{1}{\pi}\int_0^{2\pi} e^{-x}dx = \frac{1}{\pi}\cdot\left[-e^{-x}\right]_0^{2\pi} = \frac{1-e^{-2\pi}}{\pi}$$

$$a_n = \frac{1}{\pi}\int_0^{2\pi} f(x)\cos nx\, dx = \frac{1}{\pi}\int_0^{2\pi} e^{-x}\cos nx\, dx$$

$$= \frac{1}{\pi(1+n^2)}[e^{-x}(-\cos nx + n\sin nx)]_0^{2\pi} = \frac{1-e^{-2\pi}}{\pi(1+n^2)}$$

$$b_n = \frac{1}{\pi}\int_0^{2\pi} f(x)\sin nx\, dx = \frac{1}{\pi}\int_0^{2\pi} e^{-x}\sin nx\, dx$$

$$= \frac{1}{\pi(1+n^2)}[-\sin nx - n\cos nx]_0^{2\pi} = \frac{1-e^{-2\pi}}{\pi}\cdot\frac{n}{1+n^2}$$

$$\therefore \quad e^{-x} = \frac{1-e^{-2\pi}}{\pi}\left[\frac{1}{2} + \left(\frac{1}{2}\cos x + \frac{1}{5}\cos 2x + \frac{1}{10}\cos 3x + \cdots\right)\right]$$

$$+ \left(\frac{1}{2}\sin x + \frac{2}{5}\sin 2x + \frac{3}{10}\sin 3x + \cdots\right)$$

$$= \frac{1-e^{-2\pi}}{2\pi} + \frac{1-e^{-2\pi}}{\pi}\sum_{n=1}^{\infty}\frac{\cos nx}{1+n^2} + \frac{1-e^{-2\pi}}{\pi}\sum_{n=1}^{\infty}\frac{n\sin nx}{1+n^2}$$

Example 5. *Obtain Fourier series for the function f(x), given by*

$$f(x) = \begin{cases} 1 + \dfrac{2x}{\pi}; & -\pi \le x \le 0 \\ 1 - \dfrac{2x}{\pi}; & 0 \le x \le \pi \end{cases}$$

Hence, deduce that $\dfrac{1}{1^2} + \dfrac{1}{3^2} + \dfrac{1}{5^2} + ... = \dfrac{\pi^2}{8}$

Solution. When $-\pi \le x \le 0 \Rightarrow 0 \le -x \le \pi$

$$\therefore \quad f(-x) = 1 - \frac{2(-x)}{\pi} = 1 + \frac{2x}{\pi} = f(x)$$

When $0 \le x \le \pi \Rightarrow -\pi \le -x \le 0$

$$\therefore \quad f(-x) = 1 + \frac{2(-x)}{\pi} = 1 - \frac{2x}{\pi} = 1 - \frac{2x}{\pi} = f(x)$$

Therefore, $f(x)$ is an even function of x in the interval $[-\pi, \pi]$. Hence $b_n = 0$.
Now, Fourier series of $f(x)$ is given by

$$f(x) = \frac{a_0}{2} + \sum_{n=1}^{\infty} a_n \cos nx$$

Then, $\quad a_0 = \dfrac{2}{\pi} \int_0^\pi f(x)\,dx = \dfrac{2}{\pi} \int_0^\pi \left(1 - \dfrac{2x}{\pi}\right) dx = \dfrac{2}{\pi}\left[x - \dfrac{x^2}{\pi}\right]_0^\pi = 0$

$$a_n = \frac{2}{\pi} \int_0^\pi f(x) \cos nx\,dx = \frac{2}{\pi} \int_0^\pi \left(1 - \frac{2x}{\pi}\right) \cos nx\,dx$$

$$= \frac{2}{\pi}\left[\left(1 - \frac{2x}{\pi}\right)\frac{\sin nx}{n} - \left(-\frac{2}{\pi}\right)\left(-\frac{\cos nx}{n^2}\right)\right]_0^\pi$$

$$= \frac{2}{\pi}\left[-\frac{2\cos n\pi}{\pi n^2} + \frac{2}{\pi n^2}\right] = \frac{4}{\pi^2 n^2}[1 - (-1)^n]$$

$$\Rightarrow \quad f(x) = \frac{4}{\pi^2} \sum_{n=1}^{\infty} [1-(-1)^n]\frac{\cos nx}{n^2}$$

$$= \frac{4}{\pi^2}\left(\frac{2\cos x}{1^2} + \frac{2\cos 3x}{3^2} + \frac{2\cos 5x}{5^2} + ...\right)$$

$$= \frac{8}{\pi^2}\left(\frac{\cos x}{1^2} + \frac{\cos 3x}{3^2} + \frac{\cos 5x}{5^2} + ...\right)$$

Putting $x = 0$, we get $\dfrac{1}{1^2} + \dfrac{1}{3^2} + \dfrac{1}{5^2} + ... = \dfrac{\pi^2}{8}$ [Since $f(0) = 1)$]

Example 6. *Find a Fourier series to represent $x - x^2$ from $x = -\pi$ to $x = \pi$.*

Deduce that $\dfrac{1}{1^2} - \dfrac{1}{2^2} + \dfrac{1}{3^2} - \dfrac{1}{4^2} + ... = \dfrac{\pi^2}{12}$.

Solution. The Fourier series for $f(x)$ in $(-\pi, \pi)$ is

$$f(x) = a_0 + \sum_{n=1}^{\infty} a_n \cos nx + \sum_{n=1}^{\infty} b_n \sin nx$$

Here, $a_0 = \dfrac{1}{2\pi}\int_{-\pi}^{\pi}(x - x^2)\,dx = \dfrac{1}{2\pi}\left[\dfrac{x^2}{2} - \dfrac{x^3}{3}\right]_{\pi}^{-\pi} = -\dfrac{\pi^2}{3}$

$a_n = \dfrac{1}{\pi}\int_{-\pi}^{\pi}(x - x^2)\cos nx\,dx$

$\quad = \dfrac{1}{\pi}\left[(x - x^2)\dfrac{\sin nx}{n} - (1 - 2x)\left(-\dfrac{\cos nx}{n^2}\right) + (-2)\left(-\dfrac{\sin nx}{n^3}\right)\right]_{-\pi}^{\pi} = \dfrac{-4(-1)^n}{n^2}$

and $\quad b_n = \dfrac{1}{\pi}\int_{-\pi}^{\pi}(x - x^2)\sin nx\,dx$

$\quad = \dfrac{1}{\pi}\left[(x - x^2)\left(-\dfrac{\cos nx}{n}\right) - (1 - 2x)\cdot\left(-\dfrac{\sin nx}{n^2}\right) + (-2)\left(\dfrac{\cos nx}{n^3}\right)\right]_{-\pi}^{\pi}$

$\quad = \dfrac{(-2)(-1)^n}{n}$

∴ The required Fourier series is

$$x - x^2 = -\dfrac{\pi^2}{3} + 4\left[\dfrac{\cos x}{1^2} - \dfrac{\cos 2x}{2^2} + \dfrac{\cos 3x}{3^2} - \dfrac{\cos 4x}{4^2} + \dots\right]$$
$$+ 2\left[\dfrac{\sin x}{1} - \dfrac{\sin 2x}{2} + \dfrac{\sin 3x}{3} - \dfrac{\sin 4x}{4} + \dots\right] \qquad \dots (1)$$

Deduction. Putting $x = 0$ in (1), we get $0 = -\dfrac{\pi^2}{3} + 4\left(\dfrac{1}{1^2} - \dfrac{1}{2^2} + \dfrac{1}{3^2} - \dfrac{1}{4^2} + \dots\right)$

or $\quad \dfrac{1}{1^2} - \dfrac{1}{2^2} + \dfrac{1}{3^2} - \dfrac{1}{4^2} + \dots = \dfrac{\pi^2}{12}$.

Example 7. *Find the Fourier series of the function defined as*

$$f(x) = \begin{cases} x + \pi & ; \quad 0 \le x \le \pi \\ -x - \pi & ; \quad -\pi \le x \le 0 \end{cases} \quad and\; f(x + 2\pi) = f(x)$$

Solution. Let $\quad f(x) = \dfrac{a_0}{2} + \sum_{n=1}^{\infty} a_n \cos nx + \sum_{n=1}^{\infty} b_n \sin nx$

Then, $\quad a_0 = \dfrac{1}{\pi}\int_{-\pi}^{\pi} f(x)\,dx = \dfrac{1}{\pi}\int_{-\pi}^{0} f(x)\,dx + \dfrac{1}{\pi}\int_{0}^{\pi} f(x)\,dx$

$\quad = \dfrac{1}{\pi}\int_{-\pi}^{0}(-x - \pi)\,dx + \dfrac{1}{\pi}\int_{0}^{\pi}(x + \pi)\,dx = \dfrac{1}{\pi}\left[\left(-\dfrac{x^2}{2} - \pi x\right)_{-\pi}^{0} + \left(\dfrac{x^2}{2} + \pi x\right)_{0}^{\pi}\right]$

$\quad = \dfrac{1}{\pi}\left\{\left(\dfrac{\pi^2}{2} - \pi^2\right) + \left(\dfrac{\pi^2}{2} + \pi^2\right)\right\} = \pi$

$\quad a_n = \dfrac{1}{\pi}\int_{-\pi}^{\pi} f(x)\cos nx\,dx = \dfrac{1}{\pi}\int_{-\pi}^{0} f(x)\cdot\cos nx\,dx + \dfrac{1}{\pi}\int_{0}^{\pi} f(x)\cdot\cos nx\,dx$

$$= \frac{1}{\pi}\int_{-\pi}^{0}(-x-\pi)\cos nx\, dx + \frac{1}{\pi}\int_{0}^{\pi}(x+\pi)\cos nx\, dx$$

$$= \frac{1}{\pi}\left[(-x-\pi)\frac{\sin nx}{n} - (-1)\left\{-\frac{\cos nx}{n^2}\right\}\right]_{-\pi}^{0}$$

$$+ \frac{1}{\pi}\left[(x+\pi)\frac{\sin nx}{n} - (-1)\left\{-\frac{\cos nx}{n^2}\right\}\right]_{0}^{\pi}$$

$$= \frac{1}{\pi}\left[-\frac{1}{n^2} + \frac{(-1)^n}{n^2}\right] + \frac{1}{\pi}\left[\frac{(-1)^n}{n^2} - \frac{1}{n^2}\right]$$

$$= \frac{2}{n^2\pi}[(-1)^n - 1] = \begin{cases} -\dfrac{4}{n^2\pi} & ; \quad \text{if } n \text{ is odd} \\ 0 & ; \quad \text{if } n \text{ is even} \end{cases}$$

Also, $$b_n = \frac{1}{\pi}\int_{-\pi}^{\pi} f(x)\sin nx\, dx = \frac{1}{\pi}\left\{\int_{-\pi}^{0} f(x).\sin nx\, dx + \int_{0}^{\pi} f(x).\sin nx\, dx\right\}$$

$$= \frac{1}{\pi}\left\{\int_{-\pi}^{0}(-x-\pi)\sin nx\, dx + \int_{0}^{\pi}(x+\pi)\sin nx\, dx\right\}$$

$$= \frac{1}{\pi}\left[(-x-\pi)\left(-\frac{\cos nx}{n}\right) - (-1)\left\{-\frac{\sin nx}{n^2}\right\}\right]_{-\pi}^{0}$$

$$+ \frac{1}{\pi}\left[(x+\pi)\left(-\frac{\cos nx}{n}\right) - (-1)\left\{-\frac{\sin nx}{n^2}\right\}\right]_{0}^{\pi}$$

$$= \frac{1}{\pi}\left[\frac{\pi}{n}\right] + \frac{1}{\pi}\left[\frac{-2\pi}{n}(-1)^n + \frac{\pi}{n}\right] = \frac{1}{n}[1 - 2(-1)^n + 1] = \frac{2}{n}[1 - (-1)^n]$$

$$= \begin{cases} \dfrac{4}{n} & , \quad \text{if } n \text{ is odd} \\ 0 & , \quad \text{if } n \text{ is even} \end{cases}$$

The required Fourier series is given by

$$f(x) = \frac{a_0}{2} + a_1\cos x + a_2\cos 2x + \ldots + b_1\sin x + b_2\sin 2x + \ldots$$

$$= \frac{\pi}{2} - \frac{4}{\pi}\left(\frac{\cos x}{1^2} + \frac{\cos 3x}{3^2} + \ldots\right) + 4\left(\frac{\sin x}{1} + \frac{\sin 3x}{3} + \ldots\right)$$

Example 8. *Find the Fourier series for the function* $f(x) = x + x^2$, $-\pi < x < \pi$. *Hence, show that*

(i) $\dfrac{\pi^2}{6} = 1 + \dfrac{1}{2^2} + \dfrac{1}{3^2} + \dfrac{1}{4^2} + \ldots$ (ii) $\dfrac{\pi^2}{12} = \dfrac{1}{1^2} - \dfrac{1}{2^2} + \dfrac{1}{3^2} - \dfrac{1}{4^2} + \ldots$

Solution. Let the Fourier series be

$$x + x^2 = \frac{a_0}{2} + \sum_{n=1}^{\infty} a_n\cos nx + \sum_{n=1}^{\infty} b_n\sin nx \qquad \ldots (1)$$

Here,
$$a_0 = \frac{1}{\pi} \int_{-\pi}^{\pi} (x+x^2)dx$$

$$= \frac{1}{\pi}\left[\int_{-\pi}^{\pi} x\,dx + \int_{-\pi}^{\pi} x^2\,dx \right] = \frac{2}{\pi}\int_0^{\pi} x^2 dx = \frac{2}{3}\pi^2$$

$$a_n = \frac{1}{\pi}\int_{-\pi}^{\pi} (x+x^2)\cos nx\,dx$$

$$= \frac{1}{\pi}\left[\int_{-\pi}^{\pi} x\cos nx\,dx + \int_{-\pi}^{\pi} x^2\cos nx\,dx \right] = \frac{2}{\pi}\int_0^{\pi} x^2\cos nx\,dx$$

$$= \frac{2}{\pi}\left[\left(x^2\,\frac{\sin nx}{n} \right)_0^{\pi} - \int_0^{\pi} 2x.\frac{\sin nx}{n}dx \right] = -\frac{4}{n\pi}\int_0^{\pi} x\sin nx\,dx$$

$$= -\frac{4}{n\pi}\left[\left\{ x\left(-\frac{\cos nx}{n} \right) \right\}_0^{\pi} - \int_0^{\pi} 1.\left(-\frac{\cos nx}{n} \right)dx \right]$$

$$= -\frac{4}{n\pi}\left(-\frac{\pi}{n}\cos nx \right) = \frac{4}{n^2}\cos n\pi = \frac{4}{n^2}(-1)^n$$

and
$$b_n = \frac{1}{\pi}\int_{-\pi}^{\pi} (x+x^2)\sin nx\,dx$$

$$= \frac{2}{\pi}\int_0^{\pi} x\sin nx\,dx + \frac{2}{\pi}\int_0^{\pi} x^2\sin nx\,dx \qquad \left[\because \int_0^{\pi} x^2\sin nx\,dx = 0 \right]$$

$$= \frac{2}{\pi}\left(-\frac{\pi}{n}\cos n\pi \right) = -\frac{2}{n}(-1)^n.$$

From (1),
$$x+x^2 = \frac{\pi^2}{3} + 4\sum_{n=1}^{\infty} \frac{(-1)^n}{n^2}\cos nx - 2\sum_{n=1}^{\infty} \frac{(-1)^n}{n}\sin nx$$

$$f(x) = \frac{\pi^2}{3} + 4\left[-\frac{1}{1^2}\cos x + \frac{1}{2^2}\cos 2x - \frac{1}{3^2}\cos 3x + \ldots \right]$$

$$-2\left[-\frac{1}{1}\sin x + \frac{1}{2}\sin 2x - \frac{1}{3}\sin 3x + \ldots \right]. \qquad \ldots (2)$$

We observe that the series on the R.H.S. given by equation (2), always represents $x+x^2$ for all values of x except the end points $-\pi$ or π.

At the point of discontinuity

$$f(-\pi) = \frac{1}{2}(\text{L.H.L.} + \text{R.H.L.})$$

$$= \frac{1}{2}[f(-\pi-0) + f(-\pi+0)] = \frac{1}{2}[f(\pi-0) + f(-\pi+0)]$$

$$= \frac{1}{2}[\pi + \pi^2 + (-\pi) + (-\pi)^2] = \pi^2.$$

Putting $x = -\pi$ in equation (2), we get

$$\pi^2 = \frac{\pi^2}{3} + 4\left[\frac{1}{1^2} + \frac{1}{2^2} + \frac{1}{3^2} + \frac{1}{4^2} + \ldots \right]$$

Therefore, $\dfrac{\pi^2}{6} = 1 + \dfrac{1}{2^2} + \dfrac{1}{3^2} + \dfrac{1}{4^2} + \ldots$ ... (3)

Again, putting $x = 0$ in equaton (2), we get

$$0 = \frac{\pi^2}{3} + 4\left[-\frac{1}{1^2} + \frac{1}{2^2} - \frac{1}{3^2} + \frac{1}{4^2} - \ldots \right]$$

$$\Rightarrow \quad \frac{\pi^2}{12} = \frac{1}{1^2} - \frac{1}{2^2} + \frac{1}{3^2} - \frac{1}{4^2} \ldots$$

Example 9. *Express $f(x) = |x|$, $-\pi < x < \pi$, as Fourier series. Hence, show that*

$$\frac{1}{1^2} + \frac{1}{3^2} + \frac{1}{5^2} + \ldots = \frac{\pi^2}{8}.$$

Solution. Here, $f(-x) = |-x| = |x| = f(x)$

$\therefore$ $f(x)$ is an even function and hence $b_n = 0$.

Let $f(x) = |x| = \dfrac{a_0}{2} + \displaystyle\sum_{n=1}^{\infty} a_n \cos nx$

Then, $a_0 = \dfrac{2}{\pi}\int_0^{\pi} f(x)dx = \dfrac{2}{\pi}\int_0^{\pi}|x|\,dx = \dfrac{2}{\pi}\int_0^{\pi}x\,dx = \dfrac{2}{\pi}\left[\dfrac{x^2}{2}\right]_0^{\pi} = \pi$

and $a_n = \dfrac{2}{\pi}\int_0^{\pi} f(x)\cos nx\,dx$

$$= \frac{2}{\pi}\int_0^{\pi}|x|.\cos nx dx = \frac{2}{\pi}\int_0^{\pi} x\cos nx dx$$

$$= \frac{2}{\pi}\left[x\left(\frac{\sin nx}{n}\right) - 1\left(-\frac{\cos nx}{n^2}\right)\right]_0^{\pi} = \frac{2}{\pi}\left[\frac{\cos nx}{n^2} - \frac{1}{n^2}\right]$$

$$= \frac{2}{\pi n^2}[(-1)^n - 1] = \begin{cases} 0 & , \quad \text{if } n \text{ is even} \\ -\dfrac{4}{\pi n^2} & , \quad \text{if } n \text{ is odd} \end{cases}$$

Hence, $|x| = \dfrac{\pi}{2} - \dfrac{4}{\pi}\left(\cos x + \dfrac{\cos 3x}{3^2} + \dfrac{\cos 5x}{5^2} + \ldots\right)$ ... (1)

Deduction. Putting $x = 0$, in equation (1), we get

$$\frac{1}{1^2} + \frac{1}{3^2} + \frac{1}{5^2} + \ldots = \frac{\pi^2}{8}$$

Example 10. *Find the Fourier series expansion of $f(x)$, if $f(x) = \begin{cases} -\pi & , \quad -\pi < x < 0 \\ x & , \quad 0 < x < \pi \end{cases}$*

Deduce that $\dfrac{1}{1^2} + \dfrac{1}{3^2} + \dfrac{1}{5^2} + \ldots = \dfrac{\pi^2}{8}.$

Solution. Let the Fourier series be

$$f(x) = \frac{a_0}{2} + \sum_{n=1}^{\infty} a_n \cos nx + \sum_{n=1}^{\infty} b_n \sin nx$$

Then, $a_0 = \dfrac{1}{2\pi} \int_{-\pi}^{\pi} f(x)dx = \dfrac{1}{2\pi}\left[\int_{-\pi}^{0}(-\pi)dx + \int_{0}^{\pi} x\, dx\right]$

$$= \dfrac{1}{\pi}\left[-\pi(x)_{-\pi}^{0} + \left(\dfrac{x^2}{2}\right)_{0}^{\pi}\right] = \dfrac{1}{\pi}\left[-\left(\pi^2 + \dfrac{\pi^2}{2}\right)\right] = -\dfrac{\pi}{2}$$

$a_n = \dfrac{1}{\pi}\int_{-\pi}^{\pi} f(x)\cos nx\, dx = \dfrac{1}{\pi}\left[\int_{-\pi}^{0}(-\pi)\cos nx dx + \int_{0}^{\pi} x\cos nx\, dx\right]$

$$= \dfrac{1}{\pi}\left[-\pi\left(\dfrac{\sin nx}{n}\right)_{-\pi}^{0} + \left(\dfrac{x\sin nx}{n} + \dfrac{\cos nx}{n^2}\right)_{0}^{x}\right]$$

$$= \dfrac{1}{\pi}\left[0 + \dfrac{1}{n^2}\cos n\pi - \dfrac{1}{n^2}\right] = \dfrac{1}{\pi n^2}(\cos n\pi - 1)$$

and $b_n = \dfrac{1}{\pi}\int_{-\pi}^{\pi} f(x)\sin nx\, dx = \dfrac{1}{\pi}\left[\int_{-\pi}^{0}(-\pi)\sin nx\, dx + \int_{0}^{\pi} x\sin nx\, dx\right]$

$$= \dfrac{1}{\pi}\left[\left(\dfrac{\pi \cos nx}{n}\right)_{-\pi}^{0} + \left(-\dfrac{\cos nx}{n} + \dfrac{\sin nx}{n^2}\right)_{0}^{\pi}\right]$$

$$= \dfrac{1}{\pi}\left[\dfrac{\pi}{n}(0 - \cos n\pi) - \dfrac{\pi}{n}\cos n\pi\right] = \dfrac{1}{n}(1 - 2\cos n\pi)$$

The required Fourier series is

$$f(x) = -\dfrac{\pi}{4} - \dfrac{2}{\pi}\left(\cos x + \dfrac{\cos 3x}{3^2} + \dfrac{\cos 5x}{5^2} + ...\right)$$

$$+ \left(3\sin x - \dfrac{\sin 2x}{2} + \dfrac{3\sin 3x}{3} - \dfrac{\sin 4x}{4} + ...\right) \qquad ... (1)$$

Deduction. Putting $x = 0$ in (1), we get

$$f(0) = \dfrac{\pi}{4} - \dfrac{2}{\pi}\left(1 + \dfrac{1}{3^2} + \dfrac{1}{5^2} + ...\right) \qquad ... (2)$$

But $f(x)$ is continuous at $x = 0$, and we have $f(0 - 0) = -\pi$ and $f(0+0) = 0$

$\therefore \qquad f(0) = \dfrac{1}{2}[f(0-0) + f(0+0)] = -(\pi/2) \qquad ... (3)$

Hence, from (2) and (3), we have

$$-\dfrac{\pi}{2} = -\dfrac{\pi}{4} - \dfrac{2}{\pi}\left[\dfrac{1}{1^2} + \dfrac{1}{3^2} + \dfrac{1}{5^2} + ...\right] \text{ or } \dfrac{1}{1^2} + \dfrac{1}{3^2} + \dfrac{1}{5^2} + ... = \dfrac{\pi^2}{8}.$$

EXERCISE 14.2

1. Express $f(x) = \dfrac{1}{2}(\pi - x)$ in a Fourier series in the interval $0 < x < 2\pi$.

2. Find the Fourier series to represent the function $f(x) = |\sin x|, -\pi < x < \pi$.

3. Obtain the Fourier series to represent $f(x) = \dfrac{1}{4}(\pi - x)^2, 0 < x < 2\pi$. Hence, obtain the following results :

(i) $\dfrac{1}{1^2} + \dfrac{1}{2^2} + \dfrac{1}{3^2} + \dfrac{1}{4^2} + ... = \dfrac{\pi^2}{6}$

(ii) $\dfrac{1}{1^2} - \dfrac{1}{2^2} + \dfrac{1}{3^2} - \dfrac{1}{4^2} + \ldots = \dfrac{\pi^2}{12}$

(iii) $\dfrac{1}{1^2} + \dfrac{1}{3^2} + \dfrac{1}{5^2} + \ldots = \dfrac{\pi^2}{8}$

4. Expand in a Fourier series the function $f(x) = x$ in the interval $0 < x < 2\pi$, sketch its graph from $x = -4\pi$ to $x = 4\pi$.

5. Show that for $-\pi < x < \pi$

$\sin ax =$

$\dfrac{2\sin a\pi}{\pi}\left(\dfrac{\sin x}{1^2 - a^2} - \dfrac{2\sin 2x}{2^2 - a^2} + \dfrac{3\sin 3x}{3^2 - a^2} - \ldots\right)$

6. Obtain a Fourier expansion for $\sqrt{1 - \cos x}$ in the interval $-\pi < x < \pi$.

7. Obtain a Fourier series to represent e^{-ax} from $x = -\pi$ to $x = \pi$. Hence derive the series for

$\dfrac{\pi}{\sinh \pi}$.

8. Find the Fourier series to represent the periodic function:

$f(x) = \begin{cases} x & , \quad -\pi/2 < x < \pi/2 \\ \pi - x & , \quad \pi/2 < x < 3\pi/2 \end{cases}$

9. Find a series of sines and cosines to multiples of x which will represent $\dfrac{\pi}{\sinh \pi} e^x$ in the interval $-\pi < x < \pi$.

10. Prove that

$x^2 = \dfrac{\pi^2}{3} + 4\sum_{n=1}^{\infty}(-1)^n\dfrac{\cos nx}{n^2}, -\pi < x < \pi$.

11. Prove that in the interval

$x\cos x = -\dfrac{1}{2}\sin x + 2\sum_{n=2}^{\infty}\dfrac{n(-1)^n}{n^2 - 1}\sin nx$

12. If $f(x) = \cos\omega x$, $-\pi < x < \pi$, where ω is a fraction as a fourier series, prove that

$\cot\theta = \dfrac{1}{\theta} + \dfrac{2\theta}{\theta^2 - \pi^2} + \dfrac{2\theta}{\theta^2 - 4\pi^2} + \ldots$

HINTS TO SELECTED PROBLEMS

3. $a_0 = \dfrac{1}{\pi}\int_0^{2\pi}\dfrac{1}{4}(\pi - x)^2 dx = \dfrac{1}{4\pi}\left[\dfrac{(\pi - 3x)^3}{-3}\right]_0^{2\pi}$

$= -\dfrac{1}{12\pi}[-\pi^3 - \pi^3] = \dfrac{\pi^2}{6}$

$a_n = \dfrac{1}{\pi}\int_0^{2\pi}f(x)\cos nx\, dx$

$= \dfrac{1}{\pi}\int_0^{2\pi}\dfrac{1}{4}(\pi - x)^2\cos nx\, dx$

$= \dfrac{1}{4\pi}\left[(\pi - x)^2\dfrac{\sin nx}{n} - \{-2(\pi - x)\}\right.$

$\left.\left(-\dfrac{\cos nx}{n^2}\right) + 2\left(\dfrac{-\sin nx}{n^3}\right)\right]_0^{2\pi}$

$= \dfrac{1}{4\pi}\left[\dfrac{2\pi}{n^2} + \dfrac{2\pi}{n^2}\right] = \dfrac{1}{n^2}$

and

$b_n = \dfrac{1}{\pi}\int_0^{2\pi}\dfrac{1}{4}(\pi - x)^2\sin nx\, dx$

$= \dfrac{1}{4\pi}\left[(\pi - x)^2\left(-\dfrac{\cos nx}{n}\right) - \{-2(\pi - x)\}\right.$

$\left.\left(-\dfrac{\sin nx}{n^2}\right) + 2\left(\dfrac{\cos nx}{n^3}\right)\right]_0^{2\pi}$

$= \dfrac{1}{4\pi}\left[\left(-\dfrac{\pi^2}{n} + \dfrac{2}{n^3}\right) - \left(-\dfrac{\pi^2}{n} + \dfrac{2}{n^3}\right)\right] = 0$

$\therefore f(x) = \dfrac{\pi^2}{12} + \sum_{n=1}^{\infty}\dfrac{\cos nx}{n^2}$

$= \dfrac{\pi^2}{12} + \dfrac{\cos x}{1^2} + \dfrac{\cos 2x}{2^2} + \dfrac{\cos 3x}{3^2} + \ldots \quad \ldots (1)$

(i) Putting $x = 0$ in equation (1), we get

$\dfrac{\pi^2}{4} = \dfrac{\pi^2}{12} + \left(\dfrac{1}{1^2} + \dfrac{1}{2^2} + \dfrac{1}{3^2} + \dfrac{1}{4^2} + \ldots\right)$

$\dfrac{\pi^2}{6} = \dfrac{1}{1^2} + \dfrac{1}{2^2} + \dfrac{1}{3^2} + \dfrac{1}{4^2} + \ldots \quad \ldots(2)$

(ii) Putting $x = \pi$ in equation (1), we get

$0 = \dfrac{\pi^2}{12} + \left[\left(\dfrac{-1}{1^2}\right) + \dfrac{1}{2^2} + \left(-\dfrac{1}{3^2}\right) + \dfrac{1}{4^2} + \ldots\right]$

$\dfrac{\pi^2}{12} = \dfrac{1}{1^2} - \dfrac{1}{2^2} + \dfrac{1}{3^2} - \dfrac{1}{4^2} + \ldots \quad \ldots (3)$

(iii) Adding equations (2) and (3), we get

$\Rightarrow \dfrac{\pi^2}{6} + \dfrac{\pi^2}{12} = 2\left(\dfrac{1}{1^2} + \dfrac{1}{3^2} + \dfrac{1}{5^2} + \ldots\right)$

$\Rightarrow \dfrac{\pi^2}{4} = 2\left(\dfrac{1}{1^2} + \dfrac{1}{3^2} + \dfrac{1}{5^2} + \ldots\right) \quad \ldots(4)$

$\Rightarrow \dfrac{\pi^2}{8} = \dfrac{1}{1^2} + \dfrac{1}{3^2} + \dfrac{1}{5^2} + \ldots$

5. Here, $a_0 = 0$, $a_n = 0$

$$[\because f(x) \text{ is an odd function}]$$

$$b_n = \frac{2}{\pi}\int_0^\pi \sin ax \sin nx\, dx$$

$$= \frac{1}{\pi}\int_0^\pi [\cos(n-a)x - \cos(n+a)x]\, dx$$

$$= \frac{1}{\pi}\left[\frac{\sin(n-a)x}{(n-a)} - \frac{\sin(n+a)x}{n+a}\right]_0^\pi$$

$$= \frac{1}{\pi}\left[\frac{\sin(n-a)\pi}{n-a} - \frac{\sin(n+a)\pi}{n+a}\right]$$

$$= \frac{1}{\pi}\left[\frac{(-1)^n(-\sin a\pi)}{n-a} - \frac{(-1)^n \sin a\pi}{n+a}\right]$$

$$= \frac{(-1)^n \sin a\pi}{\pi}\left[\frac{1}{n-a} + \frac{1}{n+a}\right]$$

$$= (-1)^{n+1}\frac{2n \sin a\pi}{\pi(n^2 - a^2)}$$

$$\therefore\ \sin ax = \frac{2\sin a\pi}{\pi}\sum_{n=1}^\infty \frac{(-1)^{n+1}n}{n^2 - a^2}\sin nx$$

$$= \frac{2\sin a\pi}{\pi}\left[\frac{\sin x}{1^2 - a^2} - \frac{2\sin 2x}{2^2 - a^2} + \frac{3\sin 3x}{3^2 - a^2} - \cdots\right]$$

9. $f(n) = \frac{1}{2l}\int_{-l}^l f(x)dx$

$$+ \frac{1}{l}\sum_{n=1}^\infty\left[\cos\frac{n\pi x}{l} - \int_{-l}^l f(x)\cdot\cos\frac{n\pi x}{l}\,dx\right.$$

$$\left. + \sin\frac{n\pi x}{l}\int_{-l}^l f(x)\sin\frac{n\pi x}{l}\,dx\right]$$

$$\frac{\pi}{2\sin n\pi}e^x = \frac{1}{2\pi}\int_{-\pi}^\pi \frac{\pi}{2\sin n\pi}e^x dx$$

$$+ \frac{1}{\pi}\sum_{n=1}^\infty \cos nx\int_{-\pi}^\pi \frac{\pi}{2\sin n\pi}e^x \cos nu\, du$$

$$+ \frac{1}{\pi}\sum_{n=1}^\infty \sin nx\int_{-\pi}^\pi \frac{\pi}{2\sin n\pi}e^x \sin nu\, du.$$

We have

$$\int_{-\pi}^\pi e^u du = \left[e^u\right]_{-\pi}^\pi = 2\sin n\pi$$

$$\int_{-\pi}^\pi e^u \cos nu\, du$$

$$= \left[e^u \frac{\sin nu}{n}\right]_{-\pi}^\pi - \frac{1}{n}\int_{-\pi}^\pi e^u \sin nu\, du$$

$$= \frac{1}{n^2}\left[e^u \cos nu\right]_{-\pi}^\pi - \frac{1}{n^2}\int_{-\pi}^\pi e^u \cos nu\, du$$

$$\left(1 + \frac{1}{n^2}\right)\int_{-\pi}^\pi e^u \cos nu\, du = \frac{1}{n^2}(e^\pi - e^{-\pi})\cos n\pi$$

$$\int_{-\pi}^\pi e^u \cos nu\, du = \frac{2}{1+n^2}\sin n\pi \cos n\pi$$

$$\int_{-\pi}^\pi e^u \sin nu\, du$$

$$= \left[\frac{-e^u \cos nu}{n}\right]_{-\pi}^\pi + \frac{1}{n}\int_{-\pi}^\pi e^u \cos nu\, du$$

$$= -\frac{1}{n}(e^\pi - e^\pi)\cos n\pi$$

$$+ \frac{1}{n}\left[\left\{e^u \frac{\sin nu}{n}\right\}_{-\pi}^\pi - \int_{-\pi}^\pi \frac{e^u \sin nu}{n}\,du\right]$$

$$\left(1 + \frac{1}{n^2}\right)\int_{-\pi}^\pi e^u \sin nu\, du$$

$$= -\frac{1}{n}(e^\pi - e^{-\pi})\cos n\pi$$

$$\int_{-\pi}^\pi e^u \sin nu\, du = \frac{2}{1+n^2}\sin n\pi \cos n\pi$$

$$\therefore\ \frac{\pi}{2\sin n\pi}e^x = \frac{1}{2} + \sum_{n=1}^\infty \frac{\cos n\pi}{1+n^2}\cos nx$$

$$- \sum_{n=1}^\infty\left\{\frac{n}{1+n^2}\cos n\pi \sin nx\right\}$$

$$= \frac{1}{2} - \left(\frac{1}{2}\cos x - \frac{1}{2}\cos 2x\right.$$

$$\left. + \frac{1}{10}\cos 3x - \frac{1}{17}\cos 4x + \cdots\right)$$

$$+ \left(\frac{1}{2}\sin x - \frac{2}{5}\sin 2x\right.$$

$$\left. + \frac{3}{10}\sin 3x - \frac{4}{17}\sin 4x + \cdots\right)$$

Answers

1. $f(x) = \sum_{n=1}^{\infty} \dfrac{\sin nx}{n}$ **2.** $|\sin x| = \dfrac{2}{\pi} \cdot \dfrac{4}{\pi}\left(\dfrac{\cos 2x}{3} + \dfrac{\cos 4x}{15} + ... + \dfrac{\cos 2nx}{4n^2 - 1} + ...\right)$

3. $f(x) = \dfrac{\pi^2}{12} + \sum_{n=1}^{\infty} \dfrac{\cos nx}{n^2} = \dfrac{\pi^2}{12} + \dfrac{\cos x}{1^2} + \dfrac{\cos 2x}{2^2} + \dfrac{\cos 3x}{3^2} + ...$ **4.** $f(x) = \pi - 2 \cdot \sum_{n=1}^{\infty} \dfrac{\sin nx}{n}$

5. $\sin ax = \dfrac{2\sin a\pi}{\pi} \sum_{n=1}^{\infty} \dfrac{(-1)^{n+1}}{n^2 - a^2} \sin nx$ **6.** $\sqrt{1 - \cos x} = \dfrac{2\sqrt{2}}{\pi} - \dfrac{4\sqrt{2}}{\pi} \sum_{n=1}^{\infty} \dfrac{\cos nx}{4n^2 - 1}$

7. $e^{-ax} = 2\dfrac{\sin h\, a\pi}{\pi}\left[\left(\dfrac{1}{2a} - \dfrac{a\cos x}{1^2 + a^2} + \dfrac{a\cos 2x}{2^2 + a^2} - ...\right) - \left(\dfrac{\sin x}{1^2 + a^2} - \dfrac{2\sin 2x}{2^2 + a^2} + \dfrac{3\sin 3x}{3^2 + a^2} ...\right)\right]$

$\dfrac{\pi}{\sinh \pi} = 2\left[\dfrac{1}{2^2 + 1} - \dfrac{1}{3^2 + 1} + \dfrac{1}{4^2 + 1} - ...\right]$

8. $f(x) = \dfrac{4}{\pi}\left[\dfrac{\sin x}{1^2} - \dfrac{\sin 3x}{3^2} + \dfrac{\sin 5x}{5^2} - ...\right]$

9. $\dfrac{\pi}{2\sin n\pi} e^x = \dfrac{1}{2} + \sum_{n=1}^{\infty} \dfrac{\cos n\pi}{1 + n^2} \cos nx - \sum_{n=1}^{\infty}\left\{\dfrac{n}{1 + n^2}\cos nx \sin n\pi\right\}$

$= \dfrac{1}{2} - \left(\dfrac{1}{2}\cos x - \dfrac{1}{2}\cos 2x + \dfrac{1}{10}\cos 3x - \dfrac{1}{17}\cos 4x + ...\right)$

14.12   FOURIER SERIES FOR DISCONTINUOUS FUNCTIONS

At the point of discontinuity, the value of function for Fourier series is obtained by the average of left hand limit and right hand limit of function at that point of discontinuity.

Solved Examples

Example 1.   *Obtain Fourier series for the function* $f(x) = \begin{cases} x & ; \quad -\pi < x < 0 \\ -x & ; \quad 0 < x < \pi \end{cases}$ *and hence show*

that $\dfrac{1}{1^2} + \dfrac{1}{3^2} + \dfrac{1}{5^2} + ... = \dfrac{\pi^2}{8}$.

Solution.   We know that $f(x) = \dfrac{a_0}{2} + \sum_{n=1}^{\infty} a_n \cos nx + \sum_{n=1}^{\infty} b_n \sin nx$ ... (1)

$$a_0 = \dfrac{1}{\pi}\int_{-\pi}^{\pi} f(x)\,dx = \dfrac{1}{\pi}\left[\int_{-\pi}^{0} x\,dx + \int_{0}^{\pi} -x\,dx\right]$$

$$= \dfrac{1}{\pi}\left[\left(\dfrac{x^2}{2}\right)_{-\pi}^{0} - \left(\dfrac{x^2}{2}\right)_{0}^{\pi}\right] = \dfrac{1}{\pi}\left[0 - \dfrac{\pi^2}{2} - \dfrac{\pi^2}{2}\right] = -\pi$$

$$a_n = \dfrac{1}{\pi}\int_{-\pi}^{\pi} f(x)\cos nx\,dx = \dfrac{1}{\pi}\left[\int_{-\pi}^{0} x\cos nx\,dx + \int_{0}^{\pi} -x\cos nx\,dx\right]$$

$$= \dfrac{1}{\pi}\left[\left(\dfrac{x\sin nx}{n}\right)_{-\pi}^{0} - \int_{-\pi}^{0}\dfrac{\sin nx}{n}dx + \left(-x\dfrac{\sin nx}{n}\right)_{0}^{\pi} - \int_{0}^{\pi}(-1)\dfrac{\sin nx}{n}dx\right]$$

$$= \frac{1}{\pi}\left[\frac{1}{n^2}(\cos nx)_{-\pi}^0 - \frac{1}{n^2}(\cos nx)_0^\pi\right] = \frac{1}{\pi}\left[\left\{\frac{1-(-1)^n}{n^2}\right\} - \left\{\frac{(-1)^n - 1}{n^2}\right\}\right]$$

$$= \frac{1}{\pi}\left[\frac{2\{1-(-1)^n\}}{n^2}\right] = \frac{2}{\pi n^2}[1-(-1)^n] = \begin{cases} 0 & ; \quad \text{if } n \text{ is even} \\ \dfrac{4}{\pi n^2} & ; \quad \text{if } n \text{ is odd} \end{cases}$$

and $\quad b_n = \dfrac{1}{\pi}\int_{-\pi}^\pi f(x)\sin nx\,dx = \dfrac{1}{\pi}\left[\int_{-\pi}^0 x\sin nx\,dx + \int_0^\pi -x\sin nx\,dx\right]$

$$= \frac{1}{\pi}\left[\left(x\frac{-\cos nx}{n}\right)_{-\pi}^0 - \int_{-\pi}^0 \frac{-\cos nx}{n}dx\right.$$

$$\left. + \left(x\frac{\cos nx}{n}\right)_0^\pi - \int_0^\pi (-1)\frac{-\cos nx}{n}dx\right]$$

$$= \frac{1}{\pi}\left[\frac{-\pi}{n}(-1)^n + \frac{1}{n}(-1)^n\right] = 0$$

From (1) $f(x) = -\dfrac{\pi}{2} + \dfrac{4}{\pi}\left(\dfrac{\cos x}{1^2} + \dfrac{\cos 3x}{3^2} + \dfrac{\cos 5x}{5^2} + ...\right)$

At the point of discontinuity $f(0) = \dfrac{1}{2}[f(0^-) + f(0^+)] = \dfrac{1}{2}[0-0] = 0$

Putting, $x = 0$ in (2), we get $\quad 0 = -\dfrac{\pi}{2} + \dfrac{4}{\pi}\left(\dfrac{1}{1^2} + \dfrac{1}{3^2} + \dfrac{1}{5^2} + ...\right)$

Hence, $\quad \dfrac{1}{1^2} + \dfrac{1}{3^2} + \dfrac{1}{5^2} + ... = \dfrac{\pi^2}{8}$

Example 2. *Obtain the Fourier series to represent $f(x)$ given as follows :*

$$f(x) = \begin{cases} x & ; \quad \text{for } 0 \le x \le \pi \\ 2\pi - x & ; \quad \text{for } \pi \le x \le 2\pi \end{cases}$$

Solution. Let $\quad f(x) = \dfrac{a_0}{2} + \sum_{n=1}^\infty a_n\cos nx + \sum_{n=1}^\infty b_n\sin nx, 0 \le x \le 2\pi$ ... (1)

where $\quad a_0 = \dfrac{1}{\pi}\int_0^{2\pi} f(x)\,dx$

$$= \frac{1}{\pi}\left[\int_0^\pi x\,dx + \int_\pi^{2\pi}(2\pi - x)dx\right] = \frac{1}{\pi}\left[\left(\frac{x^2}{2}\right)_0^\pi + \left(2\pi x - \frac{x^2}{2}\right)_\pi^{2\pi}\right]$$

$$= \frac{1}{\pi}\left[\frac{\pi^2}{2} + 2\pi(2\pi - x) - \frac{1}{2}(4\pi^2 - \pi^2)\right] = \frac{1}{\pi}(\pi^2) = \pi$$

$$a_n = \frac{1}{\pi}\int_0^{2\pi} f(x)\cos nx\,dx = \frac{1}{\pi}\left[\int_0^\pi x\cos nx\,dx + \int_0^{2\pi}(2\pi - x)\cos nx\,dx\right]$$

$$= \frac{1}{\pi}\left[\left\{x\frac{\sin nx}{n} + \frac{\cos nx}{n^2}\right\}_0^\pi + \left\{(2\pi - x)\frac{\sin nx}{n} - \frac{\cos nx}{n^2}\right\}_\pi^{2\pi}\right]$$

$$= \frac{1}{\pi}\left[\left(\frac{\cos n\pi - 1}{n^2}\right) - \left(\frac{1 - \cos n\pi}{n^2}\right)\right] = \frac{2}{n^2\pi}[(-1)^n - 1] = \begin{cases} 0 & , \text{ if } n \text{ is even} \\ -\dfrac{4}{n\pi^2} & , \text{ if } n \text{ is odd} \end{cases}$$

Again $\quad b_n = \dfrac{1}{\pi}\int_0^{2\pi} f(x).\sin nx\, dx = \dfrac{1}{\pi}\left[\int_0^{\pi} x\sin nx\, dx + \int_0^{2\pi}(2\pi - x)\sin nx\, dx\right]$

$$= \frac{1}{\pi}\left[\left\{-\frac{x\cos nx}{n} + \frac{\sin nx}{n^2}\right\}_0^{\pi} + \left\{-(2\pi - x)\frac{\cos nx}{n} - \frac{\sin nx}{n^2}\right\}_\pi^{2\pi}\right]$$

$$= \left[\frac{-\pi\cos n\pi}{n} + \frac{\pi\cos n\pi}{n}\right] = 0$$

Therefore, $f(x) = \dfrac{\pi}{2} - \dfrac{4}{\pi}\left[\cos x + \dfrac{\cos 3x}{3^2} + \dfrac{\cos 5x}{5^2} + ...\right], 0 \le x \le 2\pi$

which is the required Fourier series for $f(x)$.

Example 3. If $f(x) = \begin{cases} 0 & , & -\pi \le x \le 0 \\ \sin x & , & 0 \le x \le \pi \end{cases}$ *Prove that* $f(x) = \dfrac{1}{\pi} + \dfrac{1}{2}\sin x - \dfrac{2}{\pi}\sum_{n=1}^{\infty}\dfrac{\cos 2nx}{4n^2 - 1}.$

Hence, show that

(i) $\dfrac{1}{1.3} + \dfrac{1}{3.5} + \dfrac{1}{5.7} + ... = \dfrac{1}{2}$ (ii) $\dfrac{1}{1.3} - \dfrac{1}{3.5} + \dfrac{1}{5.7} - ... = \dfrac{\pi - 2}{4}$

Solution. Let

$$f(x) = \frac{a_0}{2} + \sum_{n=1}^{\infty} a_n \cos nx + \sum_{n=1}^{\infty} b_n \sin nx$$

Then, $a_0 = \dfrac{1}{\pi}\int_{-\pi}^{\pi} f(x)\, dx = \dfrac{1}{\pi}\left[\int_{-\pi}^{0} 0.dx + \int_0^{\pi}\sin x\, dx\right] = \dfrac{2}{\pi}$

$a_n = \dfrac{1}{\pi}\int_{-\pi}^{\pi} f(x).\cos nx\, dx = \dfrac{1}{\pi}\left[\int_{-\pi}^{0} 0.dx + \int_0^{\pi}\sin x\cos nx\, dx\right]$

$$= \frac{1}{2\pi}\int_0^{2\pi} 2\cos nx.\sin x\, dx = \frac{1}{2\pi}\int_0^{\pi}[\sin(n+1)x - \sin(n-1)x]\, dx$$

$$= \frac{1}{2\pi}\left[-\frac{\cos(n+1)x}{n+1} + \frac{\cos(n-1)x}{n-1}\right]_0^{\pi}, n \ne 1$$

$$= \frac{1}{2\pi}\left[-\frac{\cos(n+1)\pi}{n+1} + \frac{\cos(n-1)\pi}{n-1} + \frac{1}{n+1} - \frac{1}{n-1}\right]$$

$$= \frac{1}{2\pi}\left[-\frac{(-1)^{n+1}}{n+1} + \frac{(-1)^{n-1}}{n-1} + \frac{1}{n+1} - \frac{1}{n-1}\right]$$

$$= \begin{cases} \dfrac{1}{2\pi}\left(-\dfrac{1}{n+1} + \dfrac{1}{n-1} + \dfrac{1}{n+1} - \dfrac{1}{n-1}\right) & , \text{ when } n \text{ is odd} \\ \dfrac{1}{2\pi}\left(\dfrac{1}{n+1} - \dfrac{1}{n-1} + \dfrac{1}{n+1} - \dfrac{1}{n-1}\right) & , \text{ when } n \text{ is even} \end{cases}$$

$$= \begin{cases} 0 & , \quad \text{when } n \text{ is odd}, i.e., n = 3,5,7,... \\ -\dfrac{2}{\pi(n^2 - 1)} & , \quad \text{when } n \text{ is even} \end{cases}$$

When $n=1$, we have

$$a_1 = \frac{1}{\pi}\int_0^\pi \sin x \cos x\, dx = \frac{1}{2\pi}\int_0^\pi \sin 2x\, dx = \frac{1}{2\pi}\left[-\frac{\cos 2x}{2}\right]_0^\pi = 0$$

and $\quad b_n = \frac{1}{\pi}\int_{-\pi}^\pi f(x)\sin nx\, dx = \frac{1}{\pi}\left[\int_{-\pi}^0 0.dx + \int_0^\pi \sin x \sin nx\, dx\right]$

$$= \frac{1}{2\pi}\int_0^\pi 2\sin nx \sin x\, dx = \frac{1}{2\pi}\int_0^\pi [\cos(n-1)x - \cos(n+1)x]dx$$

$$= \frac{1}{2\pi}\left[\frac{\sin(n-1)x}{(n-1)} - \frac{\sin(n+1)x}{(n+1)}\right]_0^\pi = 0, n \neq 1$$

When $n = 1$, we have $b_1 = \frac{1}{\pi}\int_0^\pi \sin x \sin x\, dx$

$$= \frac{1}{2\pi}\int_0^\pi (1 - \cos 2x)dx = \frac{1}{2\pi}\left[x - \frac{\sin 2x}{2}\right]_0^\pi = \frac{1}{2}$$

$$\therefore \quad f(x) = \frac{1}{\pi} - \frac{2}{\pi}\left[\frac{\cos 2x}{2^2 - 1} + \frac{\cos 4x}{4^2 - 1} + \frac{\cos 6x}{6^2 - 1} + ...\right] + \frac{1}{2}\sin x$$

$$= \frac{1}{\pi} + \frac{1}{2}\sin x - \frac{2}{\pi}\sum_{n=1}^\infty \frac{\cos 2nx}{(2n)^2 - 1}$$

Putting $x = 0$ in equation (1), we have $0 = \frac{1}{\pi} - \frac{2}{\pi}\sum_{n=1}^\infty \frac{1}{4n^2 - 1}$

$$\frac{1}{2} = \sum_{n=1}^\infty \frac{1}{4n^2 - 1} = \sum_{n=1}^\infty \frac{1}{(2n-1)(2n+1)} = \frac{1}{1.3} + \frac{1}{3.5} + \frac{1}{5.7} + ...$$

Putting $x = \pi/2$ in equation (1), we have,

$$1 = \frac{1}{\pi} + \frac{1}{2} - \frac{2}{\pi}\sum_{n=1}^\infty \frac{\cos n\pi}{4n^2 - 1}$$

$$\Rightarrow \quad \frac{1}{2} - \frac{1}{\pi} = -\frac{2}{\pi}\sum_{n=1}^\infty \frac{(-1)^n}{4n^2 - 1}$$

$$\Rightarrow \quad \frac{\pi - 2}{4} = -\sum_{n=1}^\infty \frac{(-1)^n}{(2n-1)(2n+1)} = -\left(-\frac{1}{1.3} + \frac{1}{3.5} - \frac{1}{5.7} + ...\right)$$

$$\Rightarrow \quad \frac{1}{1.3} - \frac{1}{3.5} + \frac{1}{5.7} - ... = \frac{\pi - 2}{4}$$

EXERCISE 14.3

1. Find the Fourier series for the following function:

$$f(x) = \begin{cases} x^2 & , \quad 0 \le x \le \pi \\ -x^2 & , \quad -\pi \le x \le 0 \end{cases}$$

2. Find the Fourier series to represent the function:

$$f(x) = \begin{cases} -k & , \quad \text{when } -\pi < x < 0 \\ k & , \quad \text{when } 0 < x < \pi \end{cases}$$

Also deduce that $\quad \dfrac{\pi}{4} = 1 - \dfrac{1}{3} + \dfrac{1}{5} - \dfrac{1}{7} + ...$

3. Find the Fourier series for the function:

$$f(x) = \begin{cases} -1 & , \quad -\pi < x < -\pi/2 \\ 0 & , \quad -\pi/2 < x < \pi/2 \\ 1 & , \quad \pi/2 < x < \pi \end{cases}$$

4. Find the Fourier series expansion for $f(x)$ if

$$f(x) = \begin{cases} -\pi & , \quad -\pi < x < 0 \\ x & , \quad 0 < x < \pi \end{cases}$$

Deduce that $\dfrac{1}{1^2} + \dfrac{1}{3^2} + \dfrac{1}{5^2} + ... = \dfrac{\pi^2}{8}$

5. Find the Fourier expansion of the function defined in one period by the relations:

$$f(x) = \begin{cases} 1 & , \quad 0 < x < \pi \\ 2 & , \quad \pi < x < 2\pi \end{cases} \quad \text{and deduce that}$$

$\dfrac{\pi}{4} = 1 - \dfrac{1}{3} + \dfrac{1}{5} - \dfrac{1}{7} + ...$

6. An alternating current after passing through a rectifier has the form

$$i = \begin{cases} I_0 \sin x & \text{for} \quad 0 \le x < \pi \\ 0 & \text{for} \quad \pi \le x \le 2\pi \end{cases}$$

where I_0 is the maximum current and the period is 2π. Express i as a Fourier series.

HINTS TO SELECTED PROBLEMS

3. $a_0 = \dfrac{1}{\pi} \int_{-\pi}^{-\pi/2} (-1)\,dx$

$+ \dfrac{1}{\pi} \int_{-\pi/2}^{\pi/2} 0\,dx + \dfrac{1}{\pi} \int_{\pi/2}^{\pi} 1\,dx = 0$

$a_n = \dfrac{1}{\pi} \int_{-\pi}^{-\pi/2} (-1)\cos nx\,dx$

$+ \dfrac{1}{\pi} \int_{-\pi/2}^{\pi/2} (0)\cos nx\,dx$

$+ \dfrac{1}{\pi} \int_{\pi/2}^{\pi} (1)\cos nx\,dx = 0$

$b_n = \dfrac{1}{\pi} \int_{-\pi}^{\pi/2} (-1)\sin nx\,dx$

$+ \dfrac{1}{\pi} \int_{-\pi}^{\pi/2} (0)\sin nx\,dx$

$+ \dfrac{1}{\pi} \int_{\pi/2}^{\pi} (1)\sin nx\,dx = 0$

$= \dfrac{2}{n\pi} \left[\cos \dfrac{n\pi}{2} - \cos n\pi \right]$

$b_1 = \dfrac{2}{\pi}, b_2 = -\dfrac{2}{\pi}, b_3 = \dfrac{2}{3\pi}$

$f(x) = \dfrac{1}{\pi} \left[2\sin x - 2\sin 2x + \dfrac{2}{3}\sin 3x + ... \right]$

Answers

1. $f(x) = 2\left(\pi - \dfrac{4}{\pi}\right)\sin x - \pi \sin 2x + \dfrac{2}{3}\left(\pi - \dfrac{4}{9\pi}\right)\sin 3x - \dfrac{\pi}{2}\sin 4x + ..$

2. $f(x) = \dfrac{4k}{\pi}\left(\sin x + \dfrac{\sin 3x}{3} + \dfrac{\sin 5x}{5} + ...\right)$
3. $f(x) = \dfrac{2}{\pi}\left[\sin x - \sin 2x + \dfrac{\sin 3x}{3} + ...\right]$

4. $f(x) = -\dfrac{\pi}{4} - \dfrac{2}{\pi}\left(\cos x + \dfrac{\cos 3x}{3^2} + \dfrac{\cos 5x}{5^2} + ...\right) + \left(3\sin x - \dfrac{\sin 2x}{2} + \sin 3x - \dfrac{\sin 4x}{4} + ...\right)$

5. $f(x) = \dfrac{3}{2} - \dfrac{2}{\pi}\left(\sin x + \dfrac{\sin 3x}{3} + \dfrac{\sin 5x}{5} + ...\right)$

6. $i = \dfrac{I_0}{\pi} + \dfrac{I_0}{2}\sin x - \dfrac{2I_0}{\pi}\left(\dfrac{\cos 2x}{2^2 - 1} + \dfrac{\cos 4x}{4^2 - 1} + \dfrac{\cos 6x}{6^2 - 1} + ...\right)$

14.13 CHANGE OF INTERVAL

In many problems, the interval of Fourier expansion is $2l$ and not 2π. In order to apply this theory, this interval must be transformed into an interval of length 2π.

Consider a periodic function $f(x)$ defined in the interval $c < x < c + 2l$. To change the interval into one of length 2π, we put

$$\frac{x}{l} = \frac{z}{\pi} \text{ or } z = \frac{\pi x}{l} \text{ so that at } x = c, z = \frac{\pi c}{l} = d(\text{say})$$

When $x = c + 2l, z = \dfrac{\pi(c + 2l)}{l} = \dfrac{\pi c}{l} + 2\pi = d + 2\pi$

Thus, the function $f(x)$ of period $2l$ in $(c, c+2l)$ is transformed to the function $f\left(\dfrac{lz}{\pi}\right) = F(z)$

say, or period in $(d, d+2\pi)$ and then function $F(z)$ can be expressed as a Fourier series such that

$$F(z) = \frac{a_0}{2} + \sum_{n=1}^{\infty} a_n \cos nz + \sum_{n=1}^{\infty} b_n \sin nz \qquad \dots (1)$$

where,

$$a_0 = \frac{1}{\pi}\int_d^{d+2\pi} F(z)dz; a_n = \frac{1}{\pi}\int_d^{d+2\pi} F(z)\cos nz\, dz$$

and

$$b_n = \frac{1}{\pi}\int_d^{d+2\pi} F(z)\sin nz\, dz$$

Now, making the inverse substitution $z = \dfrac{\pi x}{l}, dz = \dfrac{\pi}{l}dx$, when $z = d$, $x = c$ and when $z = d + 2\pi, x = c + 2l$. The expression (1) becomes

$$F(z) = F\left(\frac{\pi x}{l}\right) = F(x) = \frac{a_0}{2} + \sum_{n=1}^{\infty} a_n \cos\frac{n\pi x}{l} + \sum_{n=1}^{\infty} b_n \sin\frac{n\pi x}{l} \qquad \dots (2)$$

The coefficient a_0, a_n, b_n in (2) becomes

$$a_0 = \frac{1}{l}\int_c^{c+2l} f(x)dx, \quad a_n = \frac{1}{l}\int_c^{c+2l} f(x)\cos\frac{n\pi x}{l}dx,$$

$$b_n = \frac{1}{l}\int_c^{c+2\pi} f(x)\sin\frac{n\pi x}{l}dx$$

REMARKS

• If $c = 0$, the interval become $0 < x < 2l$ and the a_0, a_n, b_n are given by

$$a_0 = \frac{1}{l}\int_0^{2l} f(x)dx, a_n = \frac{1}{l}\int_0^{2l} f(x)\cos\frac{n\pi x}{l}dx, \quad b_n = \frac{1}{l}\int_0^{2l} f(x)\sin\frac{n\pi x}{l}dx.$$

• If $c = -l$, the interval become $-l < x < l$ and a_0, a_n, b_n are given by

$$a_0 = \frac{1}{l}\int_{-l}^{l} f(x)dx, \quad a_n = \frac{1}{l}\int_{-l}^{l} f(x)\cos\frac{n\pi x}{l}dx, \quad b_n = \frac{1}{l}\int_{-l}^{l} f(x)\sin\frac{n\pi x}{l}dx.$$

Solved Examples

Example 1. *Find the Fourier series to represent $f(x) = x^2 - 2$ when $-2 \le x \le 2$.*

Solution. Here, $b_n = 0$ because $f(x)$ is an even function

Let $f(x) = x^2 - 2 = \dfrac{a_0}{2} + \sum_{n=1}^{\infty} a_n \cos\dfrac{n\pi x}{2}$ $[\because 2l = 4 \Rightarrow l = 2]$

Then, $a_0 = \frac{2}{2}\int_0^2 (x^2 - 2)dx = \left[\frac{x^3}{3} - 2x\right]_0^2 = \frac{8}{3} - 4 = -\frac{4}{3}$

and $a_n = \frac{2}{2}\int_0^2 (x^2 - x)\cos\frac{n\pi x}{2}dx$

$$= \left[(x^2 - 2)\frac{\sin n\pi x / 2}{(n\pi / 2)} - 2x\right]\left(-\frac{\cos\frac{n\pi x}{2}}{(n^2\pi^2 / 4)} + 2\left(\frac{\sin\frac{n\pi x}{2}}{(n^3\pi^3 / 8)}\right)\right)_0^2$$

$$= \frac{16\cos n\pi}{n^2\pi^2} = \frac{16(-1)^n}{n^2\pi^2}$$

$\therefore \quad f(x) = (x^2 - 2) = -\frac{2}{3} + \frac{16}{\pi^2}\sum\frac{(-1)^n}{n^2}\cos\frac{n\pi x}{2}$

$$= -\frac{2}{3} - \frac{16}{\pi^2}\left(\cos\frac{\pi x}{2} - \frac{1}{4}\cos\pi x + \frac{1}{9}\cos\frac{3\pi x}{2} - ...\right)$$

Example 2. Obtain the Fourier series for the function

$$f(x) = \begin{cases} \pi x & ; \quad 0 \le x \le 1 \\ \pi(2-x) & ; \quad 1 \le x \le 2 \end{cases}$$

Solution. Here, $2l = 2 \Rightarrow l = 1$.

Let $f(x) = \frac{a_0}{2} + \sum_{n=1}^{\infty} a_n \cos n\pi x + \sum_{n=1}^{\infty} b_n \sin n\pi x$

where $a_0 = \int_0^2 f(x)dx = \int_0^1 \pi x dx + \int_1^2 \pi(2-x)dx = \pi\left[\frac{x^2}{2}\right]_0^1 + \pi\left[2x - \frac{x^2}{2}\right]_1^2$

$$= \pi\left(\frac{1}{2}\right) + \pi\left[(4-2) - \left(2 - \frac{1}{2}\right)\right] = \pi$$

$a_n = \int_0^2 f(x)\cos n\pi x\, dx = \int_0^1 \pi x\cos n\pi x\, dx + \int_1^2 \pi(2-x)\cos n\pi x\, dx$

$$= \left[\pi x\frac{\sin n\pi x}{n\pi} - \pi\left(-\frac{\cos n\pi x}{n^2\pi^2}\right)\right]_0^1 + \left[\pi(2-x)\frac{\sin n\pi x}{n\pi} - (-\pi)\left(-\frac{\cos n\pi x}{n^2\pi^2}\right)\right]_1^2$$

$$= \left(\frac{\cos n\pi}{n^2\pi} - \frac{1}{n^2\pi}\right) + \left[-\frac{\cos 2n\pi}{n^2\pi} + \frac{\cos n\pi}{n^2\pi}\right] = \frac{2}{n^2\pi}(\cos n\pi - 1)$$

$$= \frac{2}{n^2\pi}[(-1)^n - 1] = \begin{cases} 0 & ; \quad \text{if } n \text{ is even} \\ -\frac{4}{n^2\pi} & ; \quad \text{if } n \text{ is odd} \end{cases}$$

and $b_n = \int_0^2 f(x)\sin n\pi x dx = \int_0^1 \pi x \sin n\pi x\, dx + \int_1^2 \pi(2-x)\sin n\pi x\, dx$

$$= \left[\pi x\left(\frac{-\cos n\pi x}{n\pi}\right) - \pi\left(-\frac{\sin n\pi x}{n^2\pi^2}\right)\right]_0^1$$

$$+ \left[\pi(2-x)\frac{\cos n\pi x}{n\pi} - (-\pi)\left(-\frac{\sin n\pi x}{n^2\pi^2}\right)\right]_1^2$$

$$= \left[-\frac{\cos n\pi}{n}\right] + \left[\frac{\cos n\pi}{n}\right] = 0$$

Hence, $f(x) = \dfrac{\pi}{2} - \dfrac{4}{\pi}\left(\dfrac{\cos \pi x}{1^2} + \dfrac{\cos 3\pi x}{3^2} + \dfrac{\cos 5\pi x}{5^2} + ...\right)$

Example 3. *Expand $f(x) = e^{-x}$ as a Fourier series in the interval $(-l, l)$.*

Solution. Let $f(x) = e^{-x} = \dfrac{a_0}{2} + \sum\limits_{n=1}^{\infty} a_n \dfrac{\cos n\pi x}{l} + \sum\limits_{n=1}^{\infty} b_n \dfrac{\sin n\pi x}{l}$

Then, $a_0 = \dfrac{1}{l}\int_{-l}^{l} e^{-x} dx = \dfrac{1}{l}\left[-e^{-x}\right]_{-l}^{l} = -\dfrac{1}{l}(e^l - e^{-l}) = \dfrac{2\sinh l}{l}$

$$a_n = \frac{1}{l}\int_{-l}^{l} e^{-x}\cos\frac{n\pi x}{l}dx = \frac{1}{l}\left[\frac{e^{-x}}{1+\left(\dfrac{n\pi}{l}\right)^2}\left(-\cos\frac{n\pi x}{l} + \frac{n\pi}{l}\sin\frac{n\pi x}{l}\right)\right]_{-l}^{l}$$

$$= \frac{1}{l^2+(n\pi)^2}[-e^{-l}\cos n\pi + e^l\cos n\pi]$$

$$= -\frac{2l\cos n\pi}{l^2+(n\pi)^2}\left(\frac{e^l - e^{-l}}{2}\right) = \frac{2l(-1)^n \sinh l}{l^2+(n\pi)^2}$$

$$b_n = \frac{1}{l}\int_{-l}^{l} e^{-x}\sin\frac{n\pi x}{l}dx$$

$$= \frac{1}{l}\left[\frac{e^{-x}}{1+\left(\dfrac{n\pi}{l}\right)^2}\left(-\sin\frac{n\pi x}{l} - \frac{n\pi}{l}\cos\frac{n\pi x}{l}\right)\right]_{-l}^{l}$$

$$= -\frac{1}{l^2+(n\pi)^2}\left[\frac{n\pi}{l}(e^{-l} - e^l)\cos n\pi\right]$$

$$= \frac{2n\pi\cos n\pi}{l^2+(n\pi)^2}\left(\frac{e^l - e^{-l}}{2}\right) = \frac{2n\pi(-1)^n \sinh l}{l^2+(n\pi)^2}$$

Hence, $e^{-x} = \sinh l\left[\dfrac{1}{l} - 2l\left(\dfrac{1}{l^2+\pi^2}\cos\dfrac{\pi x}{l} - \dfrac{1}{l^2+2^2\pi^2}\cos\dfrac{2\pi x}{l}\right.\right.$

$$\left. + \frac{1}{l^2+3^2\pi^2}\cos\frac{3\pi x}{l} - ...\right)$$

$$\left. - 2\pi\left(\frac{1}{l^2+\pi^2}\sin\frac{\pi x}{l} - \frac{2}{l^2+2^2\pi^2}\sin\frac{2\pi x}{l} + \frac{3}{l^2+3^2\pi^2}\sin\frac{3\pi x}{l} - ...\right)\right]$$

Example 4. *Prove that $\dfrac{l}{2} - x = \dfrac{l}{\pi}\sum\limits_{n=1}^{\infty}\dfrac{1}{n}\sin\dfrac{2n\pi x}{l}, 0 < x < l.$*

Solution. Let $f(x) = \dfrac{l}{2} - x, 0 < x < l$

The Fourier series for $f(x)$ in the interval $(0, l)$ is

$$f(x) = \frac{a_0}{2} + \sum_{n=1}^{\infty}\left[a_n \cos\frac{n\pi x}{l/2} + b_n \sin\frac{n\pi x}{l/2}\right]$$

Here,
$$a_0 = \frac{1}{(l/2)}\int_0^l f(x)\,dx = \frac{2}{l}\int_0^l\left(\frac{1}{2} - x\right)dx = \frac{2}{l}\left[\frac{lx}{2} - \frac{x^2}{2}\right]_0^l = 0$$

$$a_n = \frac{1}{(l/2)}\int_0^l f(x)\cos\frac{n\pi x}{(l/2)}\,dx = \frac{2}{l}\int_0^l\left(\frac{l}{2} - x\right).\cos\frac{2n\pi x}{l}\,dx$$

$$= \frac{2}{l}\left[\left(\frac{l}{2} - x\right)\frac{1}{2n\pi}\sin\frac{2n\pi x}{l} + (-1)\frac{l^2}{4n^2\pi^2}\cos\frac{2n\pi x}{l}\right]_0^l$$

$$= \frac{2}{l}\left[-\frac{l^2}{4n^2\pi^2}\cos 2n\pi + \frac{l^2}{4n^2\pi^2}\right] = \frac{2}{l}.\frac{l^2}{4n\pi^2}(-\cos 2n\pi + 1)$$

$$= \frac{1}{2n^2\pi^2}(-1 + 1) = 0$$

and
$$b_n = \frac{1}{(l/2)}\int_0^l f(x).\frac{\sin n\pi x}{(l/2)}\,dx = \frac{2}{l}\int_0^l\left(\frac{l}{2} - x\right)\sin\frac{2n\pi x}{l}\,dx$$

$$= \frac{2}{l}\left[\left(\frac{l}{2} - x\right).\left(-\frac{1}{2n\pi}\cos\frac{2n\pi x}{l}\right) - (-1)\left(-\frac{l^2}{4n^2\pi^2}\sin\frac{2n\pi x}{l}\right)\right]_0^l$$

$$= \frac{2}{l}\left[\frac{l}{2}.\frac{1}{2n\pi}\cos 2n\pi + \frac{l}{2}.\frac{1}{2n\pi}(l)\right] = \frac{2}{l}\left[\frac{l^2}{2n\pi}\right] = \frac{l}{n\pi}$$

The required Fourier series is

$$f(x) = \sum_{n=1}^{\infty}\frac{1}{n\pi}\frac{\sin 2n\pi x}{l} \quad \text{or} \quad \frac{1}{2} - x = \frac{1}{\pi}\sum_{n=1}^{\infty}\frac{1}{n}\sin\frac{2n\pi x}{l}.$$

Example 5. Find the Fourier expansion for the function $f(x) = x - x^2; -1 < x < 1$.

Solution. Let $f(x) = \dfrac{a_0}{2} + \sum_{n=1}^{\infty} a_n \cos n\pi x + \sum_{n=1}^{\infty} b_n \sin n\pi x$

Then, $a_0 = \int_{-1}^1 (x - x^2)\,dx = \int_{-1}^1 x\,dx - \int_{-1}^1 x^2\,dx = 0 - 2\int_0^1 x^2\,dx = -2\left[\frac{x^3}{3}\right]_0^1 = -\frac{2}{3}$

$$a_n = \int_{-1}^1 (x - x^2)\cos n\pi x\,dx$$

$$= \int_{-1}^1 x\cos n\pi x\,dx - \int_{-1}^1 x^2 \cos n\pi x\,dx = 0 - 2\int_0^1 x^2 \cos n\pi x\,dx$$

$$= -2\left[x^2 \frac{\sin n\pi x}{n\pi} - 2x\left(-\frac{\cos n\pi x}{n^2\pi^2}\right) + 2\left(-\frac{\sin n\pi x}{n^3\pi^3}\right)\right]_0^1$$

$$= -2\left[\frac{2\cos n\pi}{n^2\pi^2}\right] = -\frac{4(-1)^n}{n^2\pi^2}$$

and $b_n = \int_{-1}^{1}(x - x^2)\sin n\pi x\, dx = \int_{-1}^{1} x\sin n\pi x\, dx = -1\int_{-1}^{1} x^2 n\pi x\, dx$

$$= 2\int_0^1 x\sin n\pi x\, dx - 0$$

$$= 2\left[x\left(-\frac{\cos n\pi x}{n\pi}\right) - 1\left(-\frac{\sin n\pi x}{n^2\pi^2}\right)\right]_0^1 = 2\left[-\frac{\cos n\pi}{n\pi}\right] = -2\frac{(-1)^n}{n\pi}$$

$$\therefore x - x^2 = -\frac{1}{3} + \frac{4}{\pi^2}\left(\frac{\cos \pi x}{1^2} - \frac{\cos 2\pi x}{2^2} + \frac{\cos 3\pi x}{3^2} - \cdots\right)$$

$$+ \frac{2}{\pi}\left(\frac{\sin \pi x}{1} - \frac{\sin 2\pi x}{2} + \frac{\sin 3\pi x}{3} - \cdots\right)$$

EXERCISE 14.4

1. Develop $f(x)$ in a Fourier series in the interval $(0, 2)$ if $f(x) = \begin{cases} x & , \quad 0 < x < 1 \\ 0 & , \quad 1 < x < 2 \end{cases}$

2. Given $f(x) = \begin{cases} 0 & , \quad 0 < x < c \\ 1 & , \quad c < x < 2c \end{cases}$ expand $f(x)$ in a Fourier series of period $2c$.

3. Expand $f(x)$ in Fourier series in the interval $(-2, 2)$ when $f(x) = \begin{cases} 0 & , \quad -2 < x < 0 \\ 1 & , \quad 0 < x < 2 \end{cases}$

4. Find a Fourier series for the function given by

$$f(t) = \begin{cases} t & , \quad 0 < t < 1 \\ 1 - t & , \quad 1 < t < 2 \end{cases}$$

5. Find a Fourier series corresponding to the function $f(x)$ defined in $(-2, 2)$ as follows;

$$f(x) = \begin{cases} 2 & , \quad \text{if} \quad -2 \le x \le 0 \\ x & , \quad \text{if} \quad 0 < x < 2 \end{cases}$$

6. Find a Fourier series for the function

$$f(x) = \begin{cases} 0 & , \quad \text{when} \quad -2 < x < -1 \\ k & , \quad \text{when} \quad -1 < x < 1 \\ 0 & , \quad \text{when} \quad 1 < x < 2 \end{cases}$$

HINTS TO SELECTED PROBLEMS

2. $a_0 = \frac{1}{c}\int_0^{2c} f(x)dx = \frac{1}{c}\int_0^c 0.dx + \frac{1}{c}\int_c^{2c} 1.dx$

$$= \frac{1}{c}[x]_c^{2c} = 1$$

$a_n = \frac{1}{c}\int_0^{2c} f(x)\cos\frac{n\pi x}{c}dx$

$$= \frac{1}{c}\int_0^c 0.\cos\frac{n\pi x}{c}dx + \frac{1}{c}\int_c^{2c} 1.\cos\frac{n\pi x}{c}dx$$

$$= \frac{1}{c}\left[\frac{c}{n\pi}\sin\frac{n\pi x}{c}\right]_c^{2c}$$

$$= \frac{1}{n\pi}[\sin 2n\pi - \sin n\pi] = 0$$

$b_n = \frac{1}{c}\int_0^{2c} f(x).\sin\frac{n\pi x}{c}dx$

$$= \frac{1}{c}\int_0^c 0.\sin\frac{n\pi x}{c}dx + \frac{1}{c}\int_c^{2c} 1.\sin\frac{n\pi x}{c}dx$$

$$= \frac{1}{c}\left[-\frac{c}{n\pi}\cos\frac{n\pi x}{c}\right]_c^{2c}$$

$$= -\frac{1}{n\pi}[\cos 2n\pi - \cos n\pi] = -\frac{1}{n\pi}[1 - (-1)^n]$$

$$= \begin{cases} -\dfrac{2}{n\pi}, & \text{when } n \text{ is odd} \\ 0, & \text{when } n \text{ is even} \end{cases}$$

Then, $f(x) = \dfrac{1}{2} - \dfrac{2}{\pi}\left(\dfrac{1}{1}\sin\dfrac{\pi x}{c} + \dfrac{1}{3}\sin\dfrac{3\pi x}{c} + ...\right)$

5. $a_0 = \dfrac{1}{l}\int_{-l}^{l} f(x)dx = \dfrac{1}{2}\left[\int_{-2}^{0} 2dx + \int_0^2 x\,dx\right]$

$\qquad = \dfrac{1}{2}\left[(2x)_{-2}^0 + \left(\dfrac{x^2}{2}\right)_0^2\right] = 3$

$a_n = \dfrac{1}{l}\int_{-l}^{l} f(x)\cos\left(\dfrac{n\pi x}{l}\right)dx$

$\qquad = \dfrac{1}{2}\left[\int_{-2}^{0} 2\cos\dfrac{n\pi x}{2}dx + \int_0^2 x\cos\dfrac{n\pi x}{2}dx\right]$

$\qquad = \dfrac{1}{2}\left[\dfrac{4}{n\pi}\left(\sin\dfrac{n\pi x}{2}\right)_{-2}^0\right.$

$\qquad \left. + \left(x\dfrac{2}{n\pi}\sin\dfrac{n\pi x}{2} + \dfrac{4}{n^2\pi^2}\cos\dfrac{n\pi x}{2}\right)_0^2\right]$

$\qquad = \dfrac{1}{2}\left[\dfrac{4}{n^2\pi^2}\cos n\pi - \dfrac{4}{n^2\pi^2}\right]$

$\qquad = \dfrac{2}{n^2\pi^2}[(-1)^n - 1]$

$$= \begin{cases} -\dfrac{4}{n^2\pi^2}, & \text{when } n \text{ is odd} \\ 0, & \text{when } n \text{ is even} \end{cases}$$

$b_n = \dfrac{1}{l}\int_{-l}^{l} f(x)\sin\left(\dfrac{n\pi x}{l}\right)dx$

$\qquad = \dfrac{1}{2}\left[2\int_{-2}^{0}\sin\dfrac{n\pi x}{2}dx + \int_0^2 x\sin\dfrac{n\pi x}{2}dx\right]$

$\qquad = \dfrac{1}{2}\left[2\left(-\dfrac{2}{n\pi}\cos\dfrac{n\pi x}{2}\right)\right]_{-2}^0$

$\qquad + \dfrac{1}{2}\left[x\left(-\dfrac{2}{n\pi}\cos\dfrac{n\pi x}{2}\right) + (1)\dfrac{4}{n^2\pi^2}\sin\dfrac{n\pi x}{2}\right]_0^2$

$\qquad = \dfrac{1}{2}\left[-\dfrac{4}{n\pi} + \dfrac{4}{n\pi}\cos n\pi\right]$

$\qquad + \dfrac{1}{2}\left[-\dfrac{4}{n\pi}\cos n\pi + \dfrac{4}{n^2\pi^2}\sin n\pi\right]$

$\qquad = \dfrac{1}{2}\left[-\dfrac{4}{n\pi}\right] = -\dfrac{2}{n\pi}$

$f(x) = \dfrac{3}{2} - \dfrac{4}{\pi^2}\left\{\dfrac{1}{1^2}\cos\dfrac{\pi x}{2} + \dfrac{1}{3^2}\cos\dfrac{3\pi x}{2} + ...\right\}$

$\qquad - \dfrac{2}{\pi}\left\{\dfrac{1}{1}\sin\dfrac{\pi x}{2} + \dfrac{1}{2}\sin\dfrac{2\pi x}{2} + \dfrac{1}{3}\sin\dfrac{3\pi x}{2} + ...\right\}.$

Answers

1. $f(x) = \dfrac{1}{4} - \dfrac{2}{\pi^2}\left(\cos\pi x + \dfrac{\cos 3\pi x}{3^2} + \dfrac{\cos 5\pi x}{5^2} + ...\right) + \dfrac{1}{\pi}\left(\sin\pi x - \dfrac{\sin 2\pi x}{2} + \dfrac{\sin 3\pi x}{3} + ...\right)$

2. $f(x) = \dfrac{1}{2} - \dfrac{2}{\pi}\left\{\sin\dfrac{\pi x}{c} + \dfrac{1}{3}\sin\dfrac{3\pi x}{c} + ...\right\}$

3. $f(x) = \dfrac{1}{2} + \dfrac{2}{\pi^2}\left(\sin\dfrac{\pi x}{2} + \dfrac{1}{3}\sin\dfrac{3\pi x}{2} + \dfrac{1}{5}\sin\dfrac{5\pi x}{2} + ...\right)$

4. $f(t) = -\dfrac{4}{\pi^2}\left(\cos\pi t + \dfrac{\cos 3\pi t}{3^2} + \dfrac{\cos 5\pi t}{5^2} + ...\right) + \dfrac{2}{\pi}\left(\sin\pi t + \sin\dfrac{3\pi t}{3} + ...\right)$

5. $f(x) = \dfrac{3}{2} - \dfrac{4}{\pi^2}\left\{\dfrac{1}{1^2}\cos\dfrac{\pi x}{2} + \dfrac{1}{3^2}\cos\dfrac{3\pi x}{2} + ...\right\} - \dfrac{2}{\pi}\left\{\sin\dfrac{\pi x}{2} + \dfrac{1}{2}\sin\dfrac{2\pi x}{2} + \dfrac{1}{3}\sin\dfrac{3\pi x}{2} + ...\right\}$

6. $f(x) = \dfrac{k}{2} + \dfrac{2R}{\pi}\left(\cos\dfrac{\pi x}{2} - \dfrac{1}{3}\cos\dfrac{3\pi x}{2} + \dfrac{1}{5}\cos\dfrac{5\pi x}{5} - ...\right)$

14.14 HALF RANGE SERIES

When we require to expand a function $f(x)$ in the range $(0, \pi)$ in a Fourier series of period 2π or more generally in the range $(0, l)$ in a Fourier series of period $2l$, a function $f(x)$ defined over the interval $0 < x < l$ is capable of two distinct half range series.

The half range cosine series is $f(x) = \dfrac{a_0}{2l} + \sum\limits_{n=1}^{\infty} a_n \cos\dfrac{n\pi x}{l}$

where, $a_0 = \dfrac{2}{l}\int_0^l f(x).dx,$ where $a_n = \dfrac{2}{l}\int_0^l f(x)\cos\dfrac{n\pi x}{l}dx$

The half range sine series is

$$f(x) = \sum\limits_{n=1}^{\infty} b_n \sin\dfrac{n\pi x}{l},\ \text{where}\ b_n = \dfrac{2}{l}\int_0^l f(x)\sin\dfrac{n\pi x}{l}dx$$

Solved Examples

Example 1. If $f(x) = \begin{cases} x & ; & 0 < x < \pi/2 \\ \pi - x & ; & \pi/2 < x < \pi \end{cases}$ Show that

(i) $f(x) = \dfrac{4}{\pi}\left[\sin x - \dfrac{\sin 3x}{3^2} + \dfrac{\sin 5x}{5^2} - \ldots\right]$

(ii) $f(x) = \dfrac{\pi}{4} - \dfrac{2}{\pi}\left[\dfrac{\cos 2x}{1^2} + \dfrac{\cos 6x}{3^2} + \dfrac{\cos 10x}{5^2} + \ldots\right]$

Solution . (i) *Half range sine series, we have* $l = \pi$ *so*

$$f(x) = \sum\limits_{n=1}^{\infty} b_n \sin\dfrac{n\pi x}{\pi} = \sum\limits_{n=1}^{\infty} b_n \sin nx$$

$$b_n = \dfrac{2}{\pi}\int_0^\pi f(x)\sin nx\, dx = \dfrac{2}{\pi}\left[\int_0^{\pi/2} x\sin nx\, dx + \int_{\pi/2}^\pi (\pi - x)\sin nx\, dx\right]$$

$$= \dfrac{2}{\pi}\left[x\left(-\dfrac{\cos nx}{n}\right) - 1\left(-\dfrac{\sin nx}{n^2}\right)\right]_0^{\pi/2}$$

$$+ \dfrac{2}{\pi}\left[(\pi - x)\left(-\dfrac{\cos nx}{nx}\right) - (-1)\left(-\dfrac{\sin nx}{n^2}\right)\right]_0^\pi$$

$$= \dfrac{2}{\pi}\left[-\dfrac{\pi}{2n}\cos\dfrac{n\pi}{2} + \dfrac{1}{n^2}\sin\dfrac{n\pi}{2}\right] + \dfrac{2}{\pi}\left[\dfrac{\pi}{2n}\cos\dfrac{n\pi}{2} + \dfrac{1}{n^2}\sin\dfrac{n\pi}{2}\right]$$

$$= \dfrac{2}{\pi}\left[\dfrac{2}{n^2}\sin\dfrac{n\pi}{2}\right] = \dfrac{4}{\pi n^2}\sin\dfrac{n\pi}{2}$$

Hence, $f(x) = \dfrac{4}{\pi}\left[\sin x - \dfrac{\sin 3x}{3^2} + \dfrac{\sin 5x}{5^2} - \ldots\right]$

(ii) *Half range cosine series*

Let $f(x) = \dfrac{a_0}{2} + \sum\limits_{n=1}^{\infty} a_n \cos nx$

Then, $a_0 = \dfrac{2}{\pi}\int_0^\pi f(x)dx = \dfrac{2}{\pi}\int_0^{\pi/2} x\,dx + \int_{\pi/2}^\pi (\pi - x)dx$

$$= \dfrac{2}{\pi}\left[\dfrac{x^2}{2}\right]_0^{\pi/2} + \left[\pi x - \dfrac{x^2}{2}\right]_{\pi/2}^\pi = \dfrac{2}{\pi}\left[\dfrac{\pi^2}{8} + \left(\pi^2 - \dfrac{\pi^2}{2}\right) - \left(\dfrac{\pi^2}{2} - \dfrac{\pi^2}{8}\right)\right]$$

$$= \frac{2}{\pi}\left[\frac{\pi^2}{4}\right] = \frac{\pi}{2}$$

and $a_n = \frac{2}{\pi}\int_0^\pi f(x)\cos x\,dx = \frac{2}{\pi}\left[\int_0^{\pi/2} x\cos nx\,dx + \int_{\pi/2}^\pi (\pi-x)\cos nx\,dx\right]$

$$= \frac{2}{\pi}\left[\frac{x\sin nx}{n} - 1\left(-\frac{\cos nx}{n^2}\right)\right]_0^{\pi/2} + \frac{2}{\pi}\left[(\pi-x)\frac{\sin nx}{n} - (-1)\left(\frac{\cos nx}{n^2}\right)\right]_{\pi/2}^\pi$$

$$= \frac{2}{\pi}\left[\frac{\pi}{2n}\sin\frac{n\pi}{2} + \frac{1}{n^2}\cos\frac{n\pi}{2} - \frac{1}{n^2}\right] + \frac{2}{\pi}\left[\frac{\cos n\pi}{n^2} - \frac{\pi}{2n}\sin\frac{n\pi}{2} + \frac{1}{n^2}\cos\frac{n\pi}{2}\right]$$

$$= \frac{2}{\pi}\left[\frac{2}{n^2}\cos\frac{n\pi}{2} - \frac{\cos n\pi}{n^2} - \frac{1}{n^2}\right] = \frac{2}{\pi n^2}\left[2\cos\frac{n\pi}{2} - \cos n\pi - 1\right]$$

Put $n = 0, 1, 2, 3, \ldots$ in equation (1), we get

$$a_1 = 0, a_2 = \frac{2}{\pi \cdot 2^2}(2\cos\pi - \cos 2\pi - 1) = \frac{-2}{1^2 \cdot \pi}$$

$$a_3 = 0, a_4 = 0, a_5 = 0, a_6 = \frac{2}{6^2\pi}(2\cos 3\pi - \cos 6\pi - 1) = \frac{-2}{3^2\pi}$$

$$a_7 = a_8 = a_9 = 0, a_{10} = \frac{2}{10^2 \cdot \pi}(2\cos 5\pi - \cos 10\pi - 1) = \frac{-2}{5^2\pi}$$

Hence, $f(x) = \frac{\pi}{4} - \frac{2}{\pi}\left[\frac{\cos 2x}{1^2} + \frac{\cos 6x}{3^2} + \frac{\cos 10x}{5^2} + \ldots\right]$

Example 2. *Develop the* $\sin\frac{\pi x}{l}$ *in half range cosine series in range* $0 < x < l$.

Solution . Let $\sin\frac{\pi x}{l} = \frac{a_0}{2} + \sum_{n=1}^\infty a_n \cos\frac{n\pi x}{l}$

where, $a_0 = \frac{2}{l}\int_0^l \sin\frac{\pi x}{l}dx = \frac{2}{l}\left[-\frac{\cos(\pi x/l)}{\pi/l}\right]_0^l = \frac{2}{\pi}[\cos\pi - 1] = \frac{4}{\pi}$

and $a_n = \frac{2}{l}\int_0^l \sin\frac{\pi x}{l}\cos\frac{n\pi x}{l}dx = \frac{1}{l}\int_0^l\left[\sin(n+1)\frac{\pi x}{l} - \sin(n-1)\frac{\pi x}{l}\right]dx$

$$= \frac{1}{l}\left[-\frac{\cos(n+1)\frac{\pi x}{l}}{(n+1)\pi/l} + \frac{\cos(n+1)\frac{\pi x}{l}}{(n-1)\pi/l}\right]_0^l$$

$$= \frac{1}{\pi}\left[-\frac{(-1)^{n+1}}{n+1} + \frac{(-1)^{n-1}}{n+1} + \frac{1}{n+1} - \frac{1}{n-1}\right]$$

(i) When n is odd

$$a_n = \frac{1}{\pi}\left[-\frac{1}{n+1} + \frac{1}{n-1} + \frac{1}{n-1} - \frac{1}{n-1}\right] = 0$$

(ii) When n is even

$$a_n = \frac{1}{\pi}\left[\frac{1}{n+1} - \frac{1}{n-1} + \frac{1}{n-1} - \frac{1}{n-1}\right] = \frac{2}{\pi}\left[\frac{1}{n+1} - \frac{1}{n-1}\right]$$

$$= \frac{-4}{\pi(n+1)(n-1)}, n \neq 1$$

$$\therefore \quad \sin\frac{\pi x}{l} = \frac{2}{\pi} - \frac{4}{\pi}\left[\frac{\cos\dfrac{2\pi x}{l}}{1.3} + \frac{\cos\dfrac{4\pi x}{l}}{3.5} + \frac{\cos\dfrac{6\pi x}{l}}{5.7} + ...\right]$$

Example 3. *Obtain the half range sine series for function $f(x) = x^2$ in the interval $0 < x < 3$.*

Solution . The Fourier half range sine series in the interval $(0, c)$ is given by

$$f(x) = \sum_{n=1}^{\infty} b_n \sin nx \qquad \qquad ... (1)$$

where, $b_n = \dfrac{2}{c}\int_0^c f(x)\sin\dfrac{n\pi x}{c}dx$

Here, $c = 3$ and $f(x) = x^2$

$$\therefore \quad b_n = \frac{2}{3}\int_0^3 x^2 \sin\frac{n\pi x}{3}dx$$

$$= \frac{2}{3}\left[x^2\left(\frac{-3}{n\pi}\right)\left(\cos\frac{n\pi x}{3}\right) + 2x\left(\frac{3}{n\pi}\right)\left(\frac{3}{n\pi}\right)\sin\frac{n\pi x}{3}\right.$$

$$\left. -2\left(\frac{3}{n\pi}\right)\left(\frac{3}{n\pi}\right)\left(\frac{3}{n\pi}\right)\cos\frac{n\pi x}{3}\right]_0^3$$

$$= \frac{2}{3}\left[\left\{-\frac{27}{n\pi}(-1)^n - \frac{54}{n^3\pi^3}(-1)^n\right\} + \frac{54}{n^3\pi^3}\right]$$

$$= \frac{2}{3}\left[\frac{54}{n^3\pi^3}\{1-(-1)^n\} - \frac{27}{n\pi}(-1)^n\right] = \begin{cases} \dfrac{2}{3}\left(\dfrac{108}{n^3\pi^3} + \dfrac{27}{n\pi}\right) , & \text{if } n \text{ is odd} \\[4mm] -\dfrac{18}{n\pi} , & \text{if } n \text{ is even} \end{cases}$$

Hence, the required half range sine series is given by
$$f(x) = b_1 \sin x + b_2 \sin 2x + b_3 \sin 3x + ...$$

$$= \frac{2}{3}\left[\frac{108}{\pi^3}\left(\frac{\sin x}{1^3} + \frac{\sin 3x}{3^3} + \frac{\sin 5x}{5^3} + ...\right) + \frac{27}{\pi}\left(\frac{\sin x}{1} + \frac{\sin 3x}{3} + \frac{\sin 5x}{5} + ...\right)\right.$$

$$\left. -\frac{18}{\pi}\left(\frac{\sin 2x}{2} + \frac{\sin 4x}{4} + ...\right)\right]$$

Example 4. (i) *Express $f(x) = x$ as a half range sine series in $0 < x < 2$,*

(ii) *Express $f(x) = x$ as a half-range cosine series in $0 < x < 2$.*

Solution . (i) The Fourier since series for $F(x)$ in $(0, 2)$ is

$$f(x) = \sum_{n=1}^{\infty} b_n \sin\frac{n\pi x}{2}$$

where $$b_n = \frac{2}{2}\int_0^2 f(x)\sin\frac{n\pi x}{2}\,dx = \int_0^2 x\sin\frac{n\pi x}{2}\,dx$$

$$= \left[-\frac{2x}{n\pi}\cos\frac{n\pi x}{2} + \frac{4}{n^2\pi^2}\sin\frac{n\pi x}{2}\right]_0^2 = \frac{-4(-1)^n}{n\pi}$$

$$\Rightarrow \quad b_1 = 4/\pi_1, b_2 = -4/2\pi, b_3 = 4/3\pi, b_4 = -4/4\pi, \quad \text{etc.}$$

Required half range Fourier sine series is

$$f(x) = \frac{4}{\pi}\left[\sin\frac{\pi x}{2} - \frac{1}{2}\sin\frac{2\pi x}{2} + \frac{1}{3}\sin\frac{3\pi x}{2} - \frac{1}{4}\sin\frac{4\pi x}{2} + \ldots\right]$$

(ii) The Fourier cosine series for $f(x)$ in $(0, 2)$ is

$$f(x) = \frac{a_0}{2} + \sum_{n=1}^{\infty} a_n \cos\frac{n\pi x}{2}$$

where $$a_0 = \frac{2}{2}\int_0^2 f(x)\,dx = \int_0^2 x\,dx = 2$$

and $$a_n = \frac{2}{2}\int_0^2 f(x)\cos\frac{n\pi x}{2}\,dx = \int_0^2 x\cos\frac{n\pi x}{2}\,dx$$

$$= \left[\frac{2x}{n\pi}\sin\frac{n\pi x}{2} + \frac{4}{n^2\pi^2}\cos\frac{n\pi x}{2}\right]_0^2 = \frac{4}{n^2\pi^2}[(-1)^n - 1]$$

$$\Rightarrow \quad a_1 = -8/\pi^2, a_2 = 0, a_3 = -8/3^2\pi^2, a_4 = 0, a_5 = -8/5^2\pi^2$$

Required half range Fourier series is given by

$$f(x) = 1 - \frac{8}{\pi^2}\left[\frac{\cos\pi x/2}{1^2} + \frac{\cos 3\pi x/2}{3^2} + \frac{\cos 5\pi x/2}{5^2} + \ldots\right].$$

Example 5. *Find a series of cosines of multiples of x which will represent x sin x in the interval*
(0, π) and show that.

$$\frac{1}{1.3} - \frac{1}{3.5} + \frac{1}{5.7} - \ldots = \frac{\pi - 2}{4}.$$

Solution . Let $$x\sin x = \frac{a_0}{2} + \sum_{n=1}^{\infty} a_n \cos nx$$

Then $$a_0 = \frac{2}{\pi}\int_0^\pi x\sin x\,dx = \frac{2}{\pi}[x(-\cos x) - 1.(-\sin x)]_0^\pi = \frac{2}{\pi}[-\pi\cos x] = 2$$

and $$a_n = \frac{2}{\pi}\int_0^\pi x\sin x\cos nx\,dx = \frac{1}{\pi}\int_0^\pi x(2\cos nx\sin x)\,dx$$

$$= \frac{1}{\pi}\int_0^\pi x[\sin(n+1)x - \sin(n-1)x]\,dx$$

$$= \frac{1}{\pi}\left[x\left\{-\frac{\cos(n+1)x}{n+1} + \frac{\cos(n-1)x}{n-1}\right\}\right.$$

$$\left.-1\left\{-\frac{\sin(n+1)x}{(n+1)^2} - \frac{\sin(n-1)\pi}{(n-1)^2}\right\}\right]_0^\pi$$

$$= \frac{1}{\pi}\left[-\frac{\pi\cos(n+1)\pi}{n+1} + \frac{\pi\cos(n-1)\pi}{(n-1)}\right], \text{ when } n \neq 1$$

$$= \frac{-(-1)^{n+1}}{n+1} + \frac{(-1)^{n-1}}{n-1} = (-1)^n\left[\frac{1}{n-1} - \frac{1}{n+1}\right] = \frac{2(-1)^{n-1}}{(n-1)(n+1)}$$

When $n=1$, we have

$$a_1 = \frac{2}{\pi}\int_0^\pi x \sin x \cos x \, dx = \frac{1}{\pi}\int_0^\pi x \sin 2x \, dx$$

$$= \frac{1}{\pi}\left[x\left(-\frac{\cos 2x}{2}\right) - 1\left(-\frac{\sin 2x}{2^2}\right)\right]_0^\pi = \frac{1}{\pi}\left[-\frac{\pi\cos 2x}{2}\right] = -\frac{1}{2}$$

$$\therefore \quad x \sin x = 1 - \frac{1}{2}\cos x - 2\left(\frac{\cos 2x}{1.3} - \frac{\cos 3x}{2.4} + \frac{\cos 4x}{3.5} - \dots\right)$$

Putting $x = \dfrac{\pi}{2}$, we get $\dfrac{\pi}{2} = 1 - 2\left(-\dfrac{1}{1.3} + \dfrac{1}{3.5} - \dfrac{1}{5.7} - \dots\right)$

$$\therefore \quad 1 + \frac{2}{1.3} - \frac{2}{3.5} + \frac{2}{5.7} - \dots = \frac{\pi}{2} \quad \Rightarrow \quad \frac{2}{1.3} - \frac{2}{3.5} + \frac{2}{5.7} - \dots = \frac{\pi}{2} - 1$$

Hence, $\dfrac{1}{1.3} - \dfrac{1}{3.5} + \dfrac{1}{5.7} - \dots = \dfrac{\pi-2}{4}$.

Example 6. Obtain the half range sine series for e^x in $0 < x < 1$.

Solution. Let $e^x = \sum\limits_{n=1}^{\infty} b_n \sin n\pi x$ $\qquad\qquad [\because l = 1]$

Then, $b_n = 2\int_0^l e^x \sin n\pi x \, dx = 2\left[\dfrac{e^x}{1+(n\pi)^2}(\sin n\pi x - n\pi \cos n\pi x)\right]_0^l$

$$= 2\left[\frac{e}{1+(n\pi)^2}(-n\pi \cos n\pi x) - \frac{1}{1+(n\pi)^2}(-n\pi)\right]$$

$$= \frac{2}{1+n^2\pi^2}[-en\pi(-1)^n + n\pi] = \frac{2n\pi}{1+n^2\pi^2}[1 - e(-1)^n]$$

Hence, $e^x = 2\pi \sum\limits_{n=1}^{\infty} \dfrac{n[1 - e(-1)^n]}{1+n^2\pi^2}$

$$= 2\pi\left[\frac{1+e}{1+\pi^2}\sin \pi x + \frac{2(1-e)}{1+4\pi^2}\sin 2\pi x + \frac{3(1+e)}{1+9\pi^2}\sin 3\pi x + \dots\right]$$

Example 7. Expand $f(x) = \begin{cases} \dfrac{1}{4} - x & ,\text{if } 0 < x < \dfrac{1}{2} \\ x - \dfrac{3}{4} & ,\text{if } \dfrac{1}{2} < x < 1 \end{cases}$ as the Fourier series of sine terms.

Solution. The Fourier sine series for $f(x)$ in $(0, 1)$ is $f(x) = \sum\limits_{n=1}^{\infty} b_n \sin n\pi x$

where, $b_n = \dfrac{2}{l}\int_0^1 f(x) \sin n\pi x \, dx$

Double and Fourier Series

$$= 2\left[\int_0^{1/2}\left(\frac{1}{4}-x\right)\sin n\pi x\, dx + \int_{1/2}^1\left(x-\frac{3}{4}\right)\sin n\pi x\, dx\right]$$

$$= 2\left|-\left(\frac{1}{4}-x\right)\frac{\cos n\pi x}{n\pi}-\frac{\sin n\pi x}{n\pi}\right|_0^{1/2}+2\left|-\left(x-\frac{3}{4}\right)\frac{\cos n\pi x}{n\pi}+\frac{\sin n\pi x}{n^2\pi^2}\right|_{1/2}^1$$

$$= 2\left[\frac{1}{4n\pi}\cos\frac{n\pi}{2}+\frac{1}{4n\pi}-\frac{\sin n\pi/2}{n^2\pi^2}\right]$$

$$+2\left[-\frac{1}{4n\pi}\cos n\pi-\frac{1}{4n\pi}\cos\frac{n\pi}{2}-\frac{\sin n\pi/2}{n^2\pi^2}\right]$$

$$= \frac{1}{2n\pi}[1-(-1)^n]-\frac{4\sin n\pi/2}{n^2\pi^2}$$

$$\Rightarrow \qquad b_1 = \frac{1}{\pi}-\frac{4}{\pi^2}, b_2 = 0, b_3 = \frac{1}{3\pi}+\frac{4}{3^2\pi^2}, b_4 = 0, b_5 = \frac{1}{5}-\frac{4}{5^2\pi^2}, b_6 = 0 \text{ etc.}$$

Hence, the required Fourier series is

$$f(x) = \left(\frac{1}{\pi}-\frac{4}{\pi^2}\right)\sin \pi x+\left(\frac{1}{3\pi}+\frac{4}{3^2\pi^2}\right)\sin 3\pi x+\left(\frac{1}{5\pi}-\frac{4}{5^2\pi^2}\right)\sin 5\pi x+\dots .$$

Example 8. *Find the half range cosine series of the function*

$$f(t) = \begin{cases} 2t & ; \quad 0<t<1 \\ 2(2-t) & ; \quad 1<t<2 \end{cases}$$

Solution . The Fourier half range cosine series in interval $(0, C)$ is

$$f(t) = \frac{a_0}{2}+a_1\cos\frac{\pi t}{c}+a_2\cos\frac{2\pi t}{c}+a_3\cos\frac{3\pi t}{c}+\dots \qquad \dots (1)$$

Here, $c = 2$ we have

$$a_0 = \frac{2}{c}\int_0^c f(t)dt = \frac{2}{2}\int_0^1 2t\, dt + 2\int_1^2 2(2-t)dt$$

$$= [t^2]_0^1+\left[2\left(2t-\frac{t^2}{2}\right)\right]_1^2 = 1+[(4t-t)]_1^2 = 1+(8-4-4+1) = 2$$

and $$a_n = \frac{2}{c}\int_0^c f(t)\cos\frac{n\pi t}{c}dt = \frac{2}{2}\int_0^1 2t\cos\frac{n\pi t}{2}dt+\frac{2}{2}\int_1^2 2(2-t)\cos\frac{n\pi t}{2}dt$$

$$= \left[2t\left(\frac{2}{n\pi}\sin\frac{n\pi t}{2}\right)-(2)\left(-\frac{4}{n^2\pi^2}\cos\frac{n\pi t}{2}\right)\right]_0^1$$

$$+\left[(4-2t)\left(\frac{2}{n\pi}\sin\frac{n\pi t}{2}\right)-(2)\left(-\frac{4}{n^2\pi^2}\cos\frac{n\pi t}{2}\right)\right]_1^2$$

$$= \left[\frac{4}{n\pi}\sin\frac{n\pi}{2}+\frac{8}{n^2\pi^2}\cos\frac{n\pi}{2}-\frac{8}{n^2+\pi^2}\right]$$

$$+\left[0+\frac{8}{n^2\pi^2}\cos\frac{n\pi}{2}-\frac{4}{n\pi}\sin\frac{n\pi}{2}+\frac{8}{n^2\pi^2}\cos\frac{n\pi}{2}\right]$$

$$= \frac{8}{n^2\pi^2}\cos\frac{n\pi}{2} - \frac{8}{n^2\pi^2} - \frac{4}{n\pi}\sin\frac{n\pi}{2} = \frac{8}{n^2\pi^2}\left[\cos\frac{n\pi}{2} - 1 - \frac{n\pi}{2}\sin\frac{n\pi}{2}\right]$$

If $n=1$, $a_1 = \frac{8}{\pi^2}\left[0 - 1 - \frac{\pi}{2}\right] = \frac{-8}{\pi^2} - \frac{4}{\pi}$

If $n=2$, $a_2 = \frac{8}{4\pi^2}[-1-1] = \frac{-16}{4\pi^2} = -\frac{4}{\pi^2}$

If $n=3$, $a_3 = \frac{8}{9\pi^2}\left[0 - 1 + \frac{3\pi}{2}\right] = \frac{-8}{9\pi^2} + \frac{4}{3\pi}$

Putting these values of $a_0, a_1, a_2, a_3, \ldots$ in equation (1), we get

$$f(1) = 1 - \left(\frac{8}{\pi^2} + \frac{4}{\pi}\right)\cos\frac{\pi t}{2} - \frac{4}{\pi^2}\cos\frac{2\pi t}{2} + \left(-\frac{8}{9\pi^2} + \frac{4}{3\pi}\right)\cos\frac{4\pi t}{2} + \ldots.$$

EXERCISE 14.5

1. Find the Fourier half range even expansion of the function

$$f(x) = (-x/l) + 1, 0 \le x \le l.$$

2. Find a series of sines of multiples of x which will represent $f(x)$ in the interval $(0, \pi)$, where

$$f(x) = \begin{cases} \frac{1}{3}\pi &, \quad 0 < x < \frac{1}{3}\pi \\ 0 &, \quad \frac{1}{3}\pi < x < \frac{2}{3}\pi \\ -\frac{1}{3}\pi &, \quad \frac{2}{3} < x < \pi \end{cases}$$

Also, represent this function by a series of cosines of multiples of x as well. Draw graph of these series and find the sine and cosine series where $x = -\frac{1}{3}\pi, -\frac{2}{3}\pi, -\pi$

3. Find the half range cosine series for function $f(x) = (x-1)^2$ in the interval $0 < x < 1$. Hence show that

(i) $\frac{1}{1^2} + \frac{1}{2^2} + \frac{1}{3^2} + \frac{1}{4^2} + \ldots = \frac{\pi^2}{6}$,

(ii) $\frac{1}{1^2} - \frac{1}{2^2} + \frac{1}{3^2} - \frac{1}{4^2} + \ldots = \frac{\pi^2}{12}$,

(iii) $\frac{1}{1^2} + \frac{1}{3^2} + \frac{1}{5^2} + \frac{1}{7^2} + \ldots = \frac{\pi^2}{8}$.

4. If $f(x) = mx$, $0 \le x \le \pi/2$
$= m(\pi - x)$, $\pi/2 \le x \le \pi$

Then show that

$$f(x) = \frac{4m}{\pi}\left[\frac{\sin x}{1^2} - \frac{\sin 3\pi}{3^2} + \frac{\sin 5x}{5^2} - \ldots\right].$$

5. If $f(x) = \begin{cases} \frac{hx}{a} &, \quad 0 < x < a \\ \frac{h(l-x)}{l-a} &, \quad a < x < l \end{cases}$

Prove that for all values of x between 0 and l

$$f(x) = \frac{2hl^2}{a(l-a)\pi^2}\left[\sin\frac{\pi a}{l}\sin\frac{\pi x}{l}\sin\frac{\pi x}{l}\right.$$
$$\left. + \frac{1}{2^2}\sin\frac{2\pi a}{l}\sin\frac{2\pi x}{l} + \ldots\right].$$

6. Expand $\pi x - x^2$ as a half range sine series in the interval $(0, \pi)$ upto first three terms.

7. Obtain a half range cosine series for

$$f(x) = \begin{cases} kx &, \quad \text{for} \quad 0 \le x \le l/2 \\ k(l-x) &, \quad \text{for} \quad l/2 \le x \le l \end{cases}$$

Deduce the sum of the series $\frac{1}{1^2} + \frac{1}{3^2} + \frac{1}{5^2} + \ldots$.

8. Find the half range sine series for the function $f(t) = t - t^2$ in the interval $0 < t < 1$.

HINTS TO SELECTED PROBLEMS

1. $a_0 = \dfrac{2}{l}\int_0^l f(x)dx = \dfrac{2}{l}\int_0^l \left(-\dfrac{x}{l}+1\right)dx$

$= \dfrac{2}{l}\left[-\dfrac{x^2}{2l}+x\right]_0^l = \dfrac{2}{l}\left[-\dfrac{l^2}{2l}+l\right] = 1$

$a_n = \dfrac{2}{l}\int_0^l f(x)\cos\dfrac{n\pi x}{l}\,dx$

$= \dfrac{2}{l}\int_0^l \left(-\dfrac{x}{l}+1\right)\cos\dfrac{n\pi x}{l}\,dx$

$= \dfrac{2}{l}\left[\left(-\dfrac{x}{l}+1\right)\left(\dfrac{l}{n\pi}\sin\dfrac{n\pi x}{l}\right)\right.$

$\left.-\left(-\dfrac{1}{l}\right)\left(-\dfrac{l^2}{n^2\pi^2}\cos\dfrac{n\pi x}{l}\right)\right]_0^l$

$= \dfrac{2}{l}\left[0-\dfrac{l}{n^2\pi^2}\cos n\pi+\dfrac{l}{n^2\pi^2}\right]$

$= \dfrac{2}{n^2\pi^2}[1-(-1)^n]$

$= \begin{cases} \dfrac{4}{n^2\pi^2}, & \text{when } n \text{ is odd}\\ 0, & \text{when } n \text{ is odd}\end{cases}$

and $b_n = 0$

$\Rightarrow f(x) = \dfrac{1}{2}+\dfrac{4}{\pi^2}\left[\dfrac{1}{1^2}\cos\dfrac{\pi x}{l}\right.$

$\left.+\dfrac{1}{3^2}\cos\dfrac{3\pi x}{l}+\dfrac{1}{5^2}\cos\dfrac{5\pi x}{l}+...\right].$

2. $b_n = \dfrac{2}{\pi}\int_0^\pi f(v)\sin nv\,dv$

$= \dfrac{2}{\pi}\left[\int_0^{\pi/2}\dfrac{\pi}{3}\sin nv\,dv\right.$

$\left.+\int_{\pi/3}^{2\pi/3}0.\sin nv\,dv+\int_{2\pi/3}^{\pi}-\dfrac{\pi}{3}.\sin nv\,dv\right]$

$= \dfrac{2}{3}\left[-\dfrac{\cos nv}{n}\right]_0^{\pi/3}-\dfrac{2}{3}\left[-\dfrac{\cos nv}{n}\right]_{2\pi/3}^{\pi}$

$= \dfrac{2}{3n}\left[1-\cos\dfrac{n\pi}{3}+\cos n\pi-\cos\dfrac{2n\pi}{3}\right]$

$= -\dfrac{8}{3n}\sin\dfrac{n\pi}{6}\sin\dfrac{n\pi}{3}.\cos\dfrac{n\pi}{2}$

$f(x) = -\dfrac{8}{3}\sum_{n=1}^{\infty}\dfrac{1}{n}\sin\dfrac{n\pi}{6}\sin\dfrac{n\pi}{3}\cos\dfrac{n\pi}{2}\sin nx$

$= \dfrac{1}{2}\left[\dfrac{1}{2}\sin 2x+\dfrac{1}{2}\sin 4x\right.$

$\left.+\dfrac{1}{3}\sin 3x+\dfrac{1}{10}\sin 10x+...\right]$

$a_0 = \dfrac{1}{\pi}\int_0^\pi f(v)dv$

$= \dfrac{1}{\pi}\int_0^{\pi/3}\dfrac{\pi}{3}dv+\int_{\pi/3}^{2\pi/3}0.dv+\int_{2\pi/3}^{\pi}-\dfrac{\pi}{3}dv = 0$

$a_n = \dfrac{2}{\pi}\int_0^\pi f(v)\cos nv\,dv$

$= \dfrac{2}{\pi}\left[\int_0^{\pi/3}\dfrac{\pi}{3}.\cos nv\,dv\right.$

$\left.+\int_{\pi/3}^{2\pi/3}0.dv+\int_{2\pi/3}^{\pi}-\dfrac{\pi}{3}.\cos nv\,dv\right]$

$= \dfrac{2}{3n}\left[\sin\dfrac{n\pi}{3}+\sin\dfrac{2n\pi}{3}\right]$

$= \dfrac{4}{3n}\sin\dfrac{n\pi}{2}\cos\dfrac{n\pi}{6}$

$\Rightarrow f(x) = \dfrac{4}{3}\sum_{n=1}^{\infty}\dfrac{1}{n}\cos\dfrac{n\pi}{6}\sin\dfrac{n\pi}{2}.\cos nx$

$= \dfrac{2}{\sqrt{3}}\left[\cos x-\dfrac{1}{5}\cos 5x\right.$

$\left.+\dfrac{1}{7}\cos 7x-\dfrac{1}{11}\cos 11x+...\right]$

6. $b_n = \dfrac{2}{\pi}\int_0^\pi (\pi x-x^2)\sin x\,dx$

$= \dfrac{2}{\pi}\left[(\pi x-x^2)\left(-\dfrac{\cos nx}{n}\right)-(\pi-2x)\right.$

$\left.\left(-\dfrac{\sin nx}{n^2}\right)+(-2)\left(\dfrac{\cos nx}{n^3}\right)\right]_0^\pi$

$= \dfrac{2}{\pi}\left[-\dfrac{2\cos n\pi}{n^3}+\dfrac{2}{n^3}\right]$

$= \dfrac{4}{\pi n^3}[1-(-1)^n] = 0 \text{ or } \dfrac{8}{n\pi^3}$

according as n is even or odd

$\Rightarrow \pi x-x^2 = \dfrac{8}{\pi}\left(\sin x+\dfrac{\sin 3x}{3^3}+\dfrac{\sin 5x}{5^3}+...\right).$

7. $a_0 = \dfrac{2}{l}\int_0^l f(x)\,dx$

$= \dfrac{2}{l}\left[\int_0^{l/2}kx\,dx+\int_{l/2}^{l}k(l-x)dx\right] = \dfrac{kl}{2}$

$$= \frac{2}{l}\left[\int_0^{l/2} k.x \cos\frac{n\pi x}{l}\,dx\right.$$

$$\left.+\int_{l/2}^l k(l-x)\frac{\cos n\pi x}{l}\,dx\right]$$

$$= \frac{2kl}{n^2\pi^2}\left[2\cos\frac{n\pi}{2}-1-\cos n\pi\right]$$

When n is odd, $\cos\frac{n\pi}{2}=0$ and $\cos n\pi = -1$,

$\Rightarrow a_n = 0 \Rightarrow a_1 = a_3 = a_5 = \dots = 0$

When n is even,

$$a_2 = \frac{2kl}{2^2\pi^2}[2\cos\pi-1-\cos 2\pi] = -\frac{8kl}{2^2\pi^2}$$

$$a_4 = \frac{2kl}{4^2\pi^2}[2\cos 2\pi-1-\cos 4\pi] = 0$$

$$a_6 = \frac{2kl}{6^2\pi^2}[2\cos 3\pi - 1 - \cos 6\pi]$$

$$= \frac{2kl}{6^2\pi^2}(-2-1-1) = -\frac{8kl}{6^2\pi^2}\quad\text{and so on}$$

$$\Rightarrow f(x) = \frac{kl}{4} - \frac{8kl}{\pi^2}\left[\frac{1}{2^2}\cos\frac{2\pi x}{l}+\frac{1}{6^2}\cos\frac{6\pi x}{l}+\dots\right]$$

$$\dots\text{(1)}$$

Putting $x=1, f(x)=0$

From (1), we have $\quad 0 = \frac{kl}{4} - \frac{8kl}{\pi^2}\left(\frac{1}{2^2}+\frac{1}{6^2}+\dots\right)$

$$\frac{1}{2^2}+\frac{1}{6^2}+\dots = \frac{\pi^2}{32}$$

$$\Rightarrow \quad \frac{1}{2^2}\left(\frac{1}{1^2}+\frac{1}{3^2}+\dots\right)\dots = \frac{\pi^2}{32}.$$

Hence $\quad \frac{1}{1^2}+\frac{1}{3^2}+\dots = \frac{\pi^2}{8}$

Answers

1. $f(x) = \frac{1}{2} + \frac{4}{\pi^2}\left[\frac{1}{1^2}\cos\frac{\pi x}{l}+\frac{1}{3^2}\cos\frac{3\pi x}{l}+\frac{1}{5^2}\cos\frac{5\pi x}{l}+\dots\right]$

2. $f(x) = \frac{1}{2}\left[\frac{1}{2}\sin 2x+\frac{1}{2}\sin 4x+\frac{1}{8}\sin 8x+\frac{1}{10}\sin 10x\dots\right]$

and $\quad f(x) = \frac{2}{\sqrt 3}\left[\cos x-\frac{1}{5}\cos 5x+\frac{1}{7}\cos 7x-\frac{1}{11}\cos 11x+\dots\right]$

3. $\frac{1}{3}+\frac{4}{\pi^2}\left(\cos\pi x+\frac{\cos 2\pi x}{2^2}+\frac{\cos 3\pi x}{3^2}+\dots\right)$ **6.** $\pi x-x^2 = \frac{8}{\pi}\left(\sin x+\frac{\sin 3x}{3^2}+\frac{\sin 5x}{5^3}+\dots\right)$

7. $\frac{2kl}{n^2\pi^2}\left[2\cos\frac{n\pi}{2}-1-\cos n\pi\right]$ **8.** $f(t) = \frac{a}{2}+\frac{2a}{\pi}\left[\sin x+\frac{1}{3}\sin 3x+\frac{1}{5}\sin 5x+\frac{1}{7}\sin 7x+\dots\right]$

14.15 PARSEVEL'S IDENTITY FOR FOURIER SERIES

Consider the Fourier series $\frac{a_0}{2}+\sum\limits_{n=1}^{\infty}(a_n\cos nx+b_n\sin nx)$ .If $f(x)$ converges uniformly to $f(x)$ at every point of the interval $(0, 2\pi)$, then

$$\frac{1}{\pi}\int_0^{2\pi}\{f(x)\}^2 dx = \frac{a_0^2}{2}+\sum_{n=1}^{\infty}(a_n^2+b_n^2)$$

Proof. Let the series $\frac{a_0}{2}+\sum\limits_{n=1}^{\infty}(a_n\cos nx+b_n\sin nx)$ represent the Fourier series of $f(x)$.

Also, let this series converges uniformly to $f(x)$ at every point of the interval $(0, 2\pi)$ so that

$$f(x) = \frac{a_0}{2}+\sum_{n=1}^{\infty}(a_n\cos nx+b_n\sin nx)\qquad\dots\text{(1)}$$

and that term by term integration is possible.

To prove that $\dfrac{1}{\pi}\int_0^{2\pi}\{f(x)\}^2 dx = \dfrac{a_0^2}{2} + \sum(a_n^2 + b_n^2)$

We have $\qquad a_n = \dfrac{1}{\pi}\int_0^{2\pi} f(x).\cos nx\, dx \qquad (n = 0, 1, 2, 3, \ldots)$

$\qquad\qquad\quad b_n = \dfrac{1}{\pi}\int_0^{2\pi} f(x).\sin nx\, dx \qquad (n = 0, 1, 2, 3 \ldots)$

Multiplying (1) by $f(x)$ and then integrating from $x=0$ to $x=2\pi$, we get

$$\int_0^{2\pi}\{f(x)\}^2 dx = \dfrac{a_0}{2} + \int_0^{2\pi} f(x)\, dx$$

$$+ \sum_{n=1}^{\infty}\left(a_n \int_0^{2\pi} f(x).\cos nx dx + b_n \int_0^{2\pi} f(x)\sin nx dx\right)$$

$$= \dfrac{a_0}{2}.\pi a_0 + \sum_{n=1}^{\infty}(\pi a_n^2 + \pi b_n^2)$$

Dividing by π, we get $\dfrac{1}{\pi}\int_0^{2\pi}\{f(x)\}^2 dx = \dfrac{a_0^2}{2} + \sum_{n=1}^{\infty}(a_n^2 + b_n^2).$

Solved Examples

Example 1. *Obtain the Fourier series expansion of* $f(x) = x^2$ *in* $-\pi < x < \pi$ *and prove that*

$$\sum_{n=1}^{\infty}\dfrac{1}{n^4} = \dfrac{\pi^4}{90} \text{ by using Parsevel's theorem.}$$

Solution. Since $f(x) = x^2$ is even function so $b_n = 0$

Let the Fourier series expansion of $f(x)$ is given by

$$f(x) = x^2 = \dfrac{a_0}{2} + \sum_{n=1}^{\infty} a_n \cos nx \qquad\qquad \ldots (1)$$

where, $\qquad a_0 = \dfrac{2}{\pi}\int_0^{\pi} f(x)dx = \dfrac{2}{\pi}\int_0^{\pi} x^2 dx = \dfrac{2\pi^2}{3}$

$$a_n = \dfrac{2}{\pi}\int_0^{\pi} f(x)\cos nx\, dx = \dfrac{2}{\pi}\int_0^{\pi} x^2\cos nx\, dx$$

$\Rightarrow \qquad a_n = \dfrac{2}{\pi}\left[x^2 \dfrac{\sin nx}{n} + 2x.\dfrac{\cos nx}{n^2} - 2\dfrac{\sin nx}{n^2}\right]_0^{\pi} = \dfrac{\pi(-1)^2}{n^2}$

$\therefore\quad$ (1) becomes,

$$x^2 = \dfrac{\pi^2}{3} + 4\sum_{n=1}^{\infty}\dfrac{(-1)^n \cos nx}{n^2} \qquad\qquad \ldots (2)$$

which is the required Fourier expansion.

Now, by Parsevel's theorem, we have

$$\int_{-\pi}^{\pi}\{f(x)\}^2 dx = \pi\left[\dfrac{a_0^2}{2} + \sum_{n=1}^{\infty}(a_n^2 + b_n^2)\right]$$

Hence, $\int_{-\pi}^{\pi} x^4 dx = \pi\left[\dfrac{4\pi^4}{2.9} + \sum_{n=1}^{\infty} \dfrac{16}{n^4}\right]$

$\Rightarrow \quad \left(\dfrac{x^5}{5}\right)_{-\pi}^{\pi} = \dfrac{2\pi^5}{9} + \pi \sum_{n=1}^{\infty} \dfrac{16}{n^4}$ or $\dfrac{2\pi^5}{5} - \dfrac{2\pi^5}{9} = \pi \sum_{n=1}^{\infty} \dfrac{16}{n^4}$

$\Rightarrow \quad \dfrac{\pi^4}{90} = \sum_{n=1}^{\infty} \dfrac{1}{n^4} .$

Example 2. By using the sine series for $f(x) = 1$ in $0 < x < \pi$, show that

$$\dfrac{\pi^2}{8} = 1 + \dfrac{1}{3^2} + \dfrac{1}{5^2} + \dfrac{1}{7^2} + \dots .$$

Solution . The Fourier sine series for $f(x) = 1$ in $(0, \pi)$ is $f(x) = \Sigma\, b_n \sin nx$,

where, $b_n = \dfrac{2}{\pi}\int_0^{\pi} f(x)\sin nx\, dx = \dfrac{2}{\pi}\int_0^{\pi}(1).\sin nx\,dx = \dfrac{2}{\pi}\left(-\dfrac{\cos nx}{n}\right)_0^{\pi}$

$= -\dfrac{2}{n\pi}[\cos n\pi - 1] = -\dfrac{2}{n\pi}[(-1)^n - 1] = \begin{cases} \dfrac{4}{n\pi} &, \text{ if } n \text{ is odd} \\ 0 &, \text{ if } n \text{ is even} \end{cases}$

The Fourier sine series is

$$1 = \dfrac{4}{\pi}\sin x + \dfrac{4}{3\pi}\sin 3x + \dfrac{4}{5\pi}\sin 5x + \dfrac{4}{7\pi}\sin 7x + \dots$$

By Parsevel's formula, we get

$$\int_0^{\pi}[f(x)]^2 dx = \dfrac{c}{2}[b_1^2 + b_2^2 + b_3^2 + b_4^2 + b_5^2 + \dots]$$

$\Rightarrow \quad \int_0^{\pi}(1)^2 dx = \dfrac{\pi}{2}\left[\left(\dfrac{4}{\pi}\right)^2 + \left(\dfrac{4}{3\pi}\right)^2 + \left(\dfrac{4}{5\pi}\right)^2 + \left(\dfrac{4}{7\pi}\right)^2 + \dots\right]$

$\Rightarrow \quad [x]_0^{\pi} = \left(\dfrac{\pi}{2}\right)\left(\dfrac{16}{\pi^2}\right)\left[1 + \dfrac{1}{3^2} + \dfrac{1}{5^2} + \dfrac{1}{7^2} + \dots\right]$

$\Rightarrow \quad \pi = \dfrac{\pi}{2}\left(\dfrac{16}{\pi^2}\right)\left[1 + \dfrac{1}{3^2} + \dfrac{1}{5^2} + \dfrac{1}{7^2} + \dots\right]$

Hence, $\dfrac{\pi^2}{8} = 1 + \dfrac{1}{3^2} + \dfrac{1}{5^2} + \dfrac{1}{7^2} + \dots$

Example 3. If $f(x) = \begin{cases} \pi x &, \quad 0 < x < 1 \\ \pi(2 - x) &, \quad 1 < x < 2 \end{cases}$

Using half range cosine series, show that $\dfrac{1}{1^4} + \dfrac{1}{3^4} + \dfrac{1}{5^4} + \dots = \dfrac{\pi^4}{96}$

Solution The half range cosine series for $f(x)$ in $(0, c)$ is

$$f(x) = \dfrac{a_0}{2} + \sum_{n=1}^{\infty} a_n \cos\dfrac{n\pi x}{c}$$

Here, $a_0 = \dfrac{2}{c}\int_0^c f(x)dx = \dfrac{2}{2}\left[\int_0^1 \pi x\, dx + \int_1^2 \pi(2-x)dx\right]$

$$= \pi\left[\dfrac{x^2}{2}\right]_0^1 + \pi\left[2x - \dfrac{\pi^2}{2}\right]_0^1 = \dfrac{\pi}{2} + \pi\left[(4-2)-\left(2-\dfrac{1}{2}\right)\right] = \pi$$

$a_n = \dfrac{2}{c}\int_0^c f(x).\cos\dfrac{n\pi x}{c}dx$

$$= \dfrac{2}{2}\left[\int_0^1 \pi x\cos\dfrac{n\pi x}{2}dx + \int_1^2 \pi(2-x)\cos\dfrac{n\pi x}{2}dx\right]$$

$$= \pi\left[x\dfrac{\sin\dfrac{n\pi x}{2}}{\dfrac{n\pi}{2}} - \left(-\dfrac{\cos\dfrac{n\pi x}{2}}{\dfrac{n^2\pi^2}{4}}\right)\right]_0^1 + \pi\left[(2-x)\dfrac{\sin\dfrac{n\pi x}{2}}{\dfrac{n\pi}{2}} - (-1)\left(-\dfrac{\cos\dfrac{n\pi x}{2}}{\dfrac{n^2\pi^2}{4}}\right)\right]_1^2$$

$$= \pi\left[\dfrac{2}{n\pi}\sin\dfrac{n\pi}{2} + \dfrac{4}{n^2\pi^2}\cos\dfrac{n\pi}{2} - \dfrac{4}{n^2\pi^2}\right]$$

$$+ \pi\left[0 - \dfrac{4}{n^2\pi^2}\cos n\pi - \dfrac{2}{n\pi}\sin\dfrac{n\pi}{2} + \dfrac{4}{n^2\pi^2}\cos\dfrac{n\pi}{2}\right]$$

$$= \pi\left[\dfrac{8}{n^2\pi^2}\cos\dfrac{n\pi}{2} - \dfrac{4}{n^2\pi^2} - \dfrac{4}{n^2\pi^2}\cos n\pi x\right] = \dfrac{4}{n^2\pi}\left[2\cos\dfrac{n\pi}{2} - 1 - \cos n\pi\right]$$

Putting $n = 1, 2, 3, ...$, we get

$$a_1 = 0, a_2 = \dfrac{-4}{\pi}, a_3 = 0, a_4 = 0, a_5 = 0, a_6 = -\dfrac{4}{9\pi}$$

By Parsevel's formula, we get

$$\int_0^c \{f(x)\}^2 dx = \dfrac{c}{2}\left[\dfrac{a_0^2}{2} + a_1^2 + a_2^2 + a_3^2 + ...\right]$$

$$\int_0^1 (\pi x)^2 dx + \int_1^2 \pi^2(2-x)^2 dx = \dfrac{2}{2}\left[\dfrac{\pi^2}{2} + \dfrac{16}{\pi^2} + \dfrac{16}{81\pi^2} + ...\right]$$

$$\pi^2\left[\dfrac{x^3}{3}\right]_0^1 - \pi^2\left[\dfrac{(2-x)^3}{3}\right]_1^2 = \dfrac{\pi^2}{2} + \dfrac{16}{\pi^2} + \dfrac{16}{81\pi^2} + ...$$

$$\Rightarrow \qquad \dfrac{\pi^2}{3} - \pi^2\left(0 - \dfrac{1}{3}\right) = \dfrac{\pi^2}{3^2} + \dfrac{16}{\pi^2}\left[1 + \dfrac{1}{81} + ...\right]$$

$$\Rightarrow \qquad \dfrac{2\pi^2}{3} - \dfrac{\pi^2}{2} = \dfrac{16}{\pi^2}\left[1 + \dfrac{1}{3^4} + \dfrac{1}{5^4} + ...\right]$$

$$\Rightarrow \qquad \dfrac{\pi^2}{6} = \dfrac{16}{\pi^2}\left[1 + \dfrac{1}{3^4} + \dfrac{1}{5^4} + ...\right] \Rightarrow \dfrac{\pi^4}{96} = 1 + \dfrac{1}{3^4} + \dfrac{1}{5^4} + ...$$

CHAPTER REVIEW: A COMPETITIVE APPROACH

SELECTED TERMS AND RESULTS

TERMS

- **Double sequence :** Double sequence is a function whose domain is N×N and range can be any subset of real numbers R.

- **Convergence of double sequence :** A double sequence $<s_{mn}>$ is said to converge to a real number s if for given $\varepsilon > 0$ ∃ a positive number $n_0 \in N$ such that

-   $|s_{mn} - s| < \varepsilon \ \forall m, n \in N$ and $m, n \geq n_0$

- **Double series :** Let $<s_{mn}>$ be a double sequence, then the double series is defined as a double summation extended over all terms $s_{mn} \in N$ where the terms are to be summed up in any arbitrary order unless it is otherwise specified.

- **Convergence of a double series :** Let $\displaystyle\sum_{m,n=1}^{\infty} s_{mn}$ be a double series then it is said to be convergent iff there is a number s such that the double sequence $<a_{mn}>$ of partial sums of the terms $<s_{mn}>$ converges to s as $m, n \to \infty$.

- **Absolute convergence :** A double series $\displaystyle\sum_{m,n} s_{mn}$ is said to be absolutely convergent if the double series $\displaystyle\sum_{m,n} |s_{mn}|$ converges.

- **Fourier series :** Let f be a real valued bounded and integrable function defined on $[-\pi, \pi]$ such that

$$f(x) = \frac{a_0}{2} + \sum_{n=1}^{\infty} (a_n \cos nx + b_n \sin nx) \quad \dots (1)$$

where $a_n = \dfrac{1}{\pi}\displaystyle\int_{-\pi}^{\pi} f(x) \cos nx\, dx$ and

$b_n = \dfrac{1}{\pi}\displaystyle\int_{-\pi}^{\pi} f(x) \sin nx\, dx$ then (1) is called Fourier series.

- **Periodic function :** A functon $f(x)$ which satisfies the relation $f(x+\lambda) = f(x) \ \forall x$ and some fixed λ, is called periodic function with period λ.

- **Half range sine series :**

$$f(x) = \sum_{n=1}^{\infty} b_n \sin\frac{n\pi x}{l},$$

where $b_n = \dfrac{2}{l}\displaystyle\int_{0}^{2l} f(x) \sin\frac{n\pi x}{l}\, dx$

- **Half range cosine series :**

$$f(x) = \frac{a_0}{2l} + \sum_{n=1}^{\infty} a_n \cos\frac{n\pi x}{l}$$

where

$a_0 = \dfrac{2}{l}\displaystyle\int_{0}^{l} f(x)dx, a_n = \dfrac{2}{l}\displaystyle\int_{0}^{l} f(x)\cos\frac{n\pi x}{l}\, dx$

RESULTS

- The double sequence $<s_{mn}>$ converges iff for $\varepsilon > 0$ ∃ a positive number $n_0 \in N$ such that $|s_{pq} - s_{mn}| < \varepsilon \ \forall p \geq m \geq n_0, q \geq n \geq n_0$

- A double series $\displaystyle\sum_{m,n=1}^{\infty} s_{mn}$ is said to be convergent iff there is a number s such that the double sequence a_{mn} of partial sums of the terms $<s_{mn}>$ converges to s as $m, n \to \infty$

- The necessary and sufficient condition for a double series $\displaystyle\sum_{m,n} s_{mn}$ to be convergent is that for its partial sums a_{mn} and for given $\varepsilon > 0$ ∃

$n_0 \in N$ such that

$|s_{pq} - s_{mn}| < \varepsilon \forall p \geq m \geq n_0; q \geq n \geq n_0$

- If a double series is convergent to s and its sums by rows and columns each exist, then each of them sum is equal to s.

- A double series $\displaystyle\sum_{m,n} s_{mn}$ with $s_{mn} \geq 0$ either converges to a finite sum s or else it diverges to ∞.

- If $\sum a_{mn}$ and $\sum b_{mn}$ two double series such that $0 \leq a_{mn} \leq b_{mn} \forall m, n \in N$ then

(i) $\sum b_{mn}$ converges $\Rightarrow \sum a_{mn}$ converges

(ii) $\sum a_{mn}$ diverges $\Rightarrow \sum b_{mn}$ diverges

• Every absolutely convergent double series is convergent.

• If the series $\dfrac{a_0}{2} + \sum_{n=1}^{\infty} (a_n \cos nx + b_n \sin nx)$

converges uniformly to a function f on $[-\pi, \pi]$ then it is the Fourier series for f on $[-\pi, \pi]$.

• If f is bounded and integrable on $[-\pi, \pi]$ and if a_n, b_n are its Fourier coefficients then

$$\sum_{n=1}^{\infty} (a_n^2 + b_n^2) \text{ converges.}$$

• The inequality $\sum_{n=1}^{\infty} (a_n^2 + b_n^2) \le \dfrac{1}{\pi} \int_{-\pi}^{\pi} f^2 dx$ is called Bessel's inequality.

REVIEW QUESTIONS AND PROJECT WORK

1. Prove that

$$\sum_{n=1}^{\infty} \frac{\sin(2n-1)x}{(2n-1)^3} = \frac{\pi x}{8}(\pi - x) \forall x \in]-\pi, \pi].$$

2. Prove that $\displaystyle\sum_{n=1}^{\infty} \frac{\cos(2n-1)x}{(2n-1)^2} = \frac{\pi}{4}\left(\frac{\pi}{2} - x\right).$

3. Prove that

$$e^x = \frac{e^\pi - e}{\pi}\left\{\frac{1}{2} + \sum_{n=1}^{\infty} \frac{(-1)^n}{n^2+1}(\cos nx - n \sin nx)\right\}$$

for $-\pi < x < \pi$.

4. Prove that

$$\sum_{n=1}^{\infty} \frac{\cos nx}{n^2} = \frac{3x^2 - 6\pi x + 2\pi^2}{12} \forall x \in [0, \pi].$$

5. Prove that

$$\sum_{n=1}^{\infty} \frac{\sin 2nx}{n} = \begin{cases} -\dfrac{\pi}{2} - x & \text{when } x \in]-\pi, 0[\\[2mm] \dfrac{\pi}{2} - x & \text{when } x \in]0, \pi[\\[2mm] \dfrac{3\pi}{2} - x & \text{when } x \in]\pi, 2\pi[\end{cases}$$

6. Show that for $|x| < 1$

$$x + x^2 = \frac{1}{3} + \frac{2}{\pi}\sum_{n=1}^{\infty} (-1)^n \left(\frac{2 \cos n\pi x}{n^2 \pi} - \frac{\sin n\pi x}{n}\right).$$

7. Show that for $0 < x < 2\pi$

$$\frac{\pi - x}{2} = \frac{\sin x}{1} + \frac{\sin 2x}{2} + \frac{\sin 3x}{3} + \cdots .$$

8. Show that the fourier series for $x + x^2$ on $[-\pi, \pi]$ is

$$\frac{\pi^2}{3} + 4\sum_{n=1}^{\infty} (-1)^n \left(\frac{\cos nx}{n^2} - \frac{\sin nx}{2n}\right).$$

9. Show that every uniformly convergent trigonometric series is a Fourier series.

10. Show that following are not Fourier series

(i) $\displaystyle\sum_{n=1}^{\infty} \frac{1}{\sqrt{n}} \cos nx$

(ii) $\displaystyle\sum_{n=1}^{\infty} \left(\sin nx + \frac{1}{n}\cos nx\right)$

OBJECTIVE TYPE QUESTIONS

FILL IN THE BLANKS

1. The double sequence $\left\langle \dfrac{1}{mn}\right\rangle$ converges to ... .

2. The double sequence $\left\langle (-1)^{m+n}\right\rangle$ oscillates ...

3. The double sequence $\left\langle (-1)^{m+n}(m+n)\right\rangle$ oscillates ... .

4. A necessary condition for convergence of a double series $\displaystyle\sum_{m,n} a_{mn}$ that $\displaystyle\lim_{m,n\to\infty} a_{mn} = ...$

5. Every absolutely convergent double series is ...

TRUE / FALSE

Write 'T' for true and 'F' for false statement.

1. $\int_{-\pi}^{\pi} \sin nx \, dx = 0 = \int_{-\pi}^{\pi} \cos nx \, dx \, \forall n$. **(T/F)**

2. The inequality $\sum_{n=1}^{\infty} (a_n^2 + b_n^2) \le \frac{1}{\pi} \int_{-\pi}^{\pi} f^2 dx$ is called Bessel's inequality. **(T/F)**

3. If f is integrable on $[-\pi, \pi]$ and has period 2π then $\int_{-\pi}^{\pi} f(t)dt = \int_{-\pi}^{\pi} f(t+a)dt$ for any $a \in R$ **(T/F)**

4. $\int_{0}^{\infty} \frac{\sin t}{T} dt = \pi$ **(T/F)**

5. Fourier series has period 2π. **(T/F)**

MULTIPLE CHOICE QUESTIONS

Choose the most appropriate one

1. A function $f(x)$ is called periodic if it is defined for real $x \in R$ and:
   (a) if there is any positive number p such that $f(x+p) = f(x) + f(p)$
   (b) if there is any positive number p, such that $f(x+p) > f(x)$
   (c) if there is any positive number p such that $f(x+p) = f(r)$,
   (d) None of the above .

2. In Fourier series
$$f(x) = a_0 + \sum_{n=1}^{\infty} (a_n \cos nx + b_n \sin nx),$$
   Euler formula is:
   (a) $a_0 = \frac{1}{2\pi} \int_{-\pi}^{\pi} f(x) dx$
   (b) $a_0 = \frac{1}{2\pi} \int_{-\pi}^{\pi} f(x) dx$
   (c) $a_0 = \frac{1}{2\pi} \int_{-\pi}^{\pi} f(x) \sin x \, dx$
   (d) none of the above

3. In Fouries series
$$f(x) = a_0 + \sum_{n=1}^{\infty} (a_n \cos nx + b_n \sin nx),$$
   Euler formula is:
   (a) $a_n = \frac{1}{\pi} \int_{-\pi}^{\pi} f(x) dx$
   (b) $a_n = \frac{1}{\pi} \int_{-\pi}^{\pi} f(x) \sin nx \, dx$
   (c) $a_n = \frac{1}{\pi} \int_{-\pi}^{\pi} f(x) \cos nx \, dx$
   (d) none of the above

4. If a function $f(x)$ of period $p = 2l$ has Fourier series, then in series $f(x) = a_0 + \sum_{n=1}^{\infty} (a_n \cos nx + b_n \sin nx)$, the Fourier coefficients a_n is given by ;
   (a) $\int_{-l}^{l} f(x) . \frac{\sin n\pi x}{l} dx$
   (b) $\frac{1}{l} \int_{-l}^{l} f(x) . \frac{\cos n\pi x}{l} dx$
   (c) $\int_{-l}^{l} f(x) . \frac{\cos n\pi x}{l} dx$
   (d) none of the above

5. The Fourier series of an even function of period 2π is a:
   (a) Fourier cosine series
   (b) Fourier sine series
   (c) Fourier complex series
   (d) none of the above

6. The Fourier series of an odd function of period $2l$ is given by:
   (a) $f(x) = a_0 + \sum_{n=1}^{\infty} a_n \frac{\cos n\pi}{l}$, where
   $a_0 = \frac{1}{l} \int_{0}^{l} f(x) dx, \ a_n = \frac{2}{l} \int_{0}^{l} f(x) . \frac{\cos n\pi x}{l} dx$
   (b) $f(x) = \sum_{n=1}^{\infty} b_n \sin \frac{n\pi x}{l}$, where
   $b_0 = \frac{2}{l} \int_{0}^{l} f(x) \sin \frac{n\pi x}{l} dx$
   (c) $f(x) = \sum_{n=-\infty}^{\infty} C_n e^{inx}$, where
   $C_n = \frac{1}{2\pi} \int_{-\pi}^{\pi} f(x) e^{-inx} dx, n = 1, 2, \ldots$
   (d) none of the above

7. The Fourier series of an odd function of period $2l$ is given by:
   (a) Fourier cosine series
   (b) Fourier sine series
   (c) Fourier complex series
   (d) none of the above

8. The Fourier series of an even function of period $2l$ is given by:

(a) $f(x) = a_0 + \sum_{n=1}^{\infty} a_n \dfrac{\cos n\pi}{l}$, where

$a_0 = \dfrac{1}{l} \int_0^l f(x)\,dx$

and $a_n = \dfrac{2}{l} \int_0^l f(x) \cdot \dfrac{\cos n\pi x}{l}\,dx$

(b) $f(x) = \sum_{n=1}^{\infty} b_n \sin \dfrac{n\pi x}{l}$, where

$b_0 = \dfrac{2}{l} \int_0^L f(x) \sin \dfrac{n\pi x}{l}\,dx$

(c) $f(x) \sum_{n=1}^{\infty} c_n e^{inx}$, where

$c_n = \dfrac{1}{2\pi} \int_{-\pi}^{\pi} f(x) e^{-inx}\,dx, n = 1, 2, \ldots$

(d) none of the above.

=== **Answers** ===

FILL IN THE BLANKS

  1. 0   **2.** finitely   **3.** infinitely   **4.** 0   **5.** Convergent

TRUE/FALSE

  1. T   **2.** T   **3.** T   **4.** F   **5.** T

MULTIPLE CHOICE QUESTIONS

  1. (c)   **2.** (c)   **3.** (a)   **4.** (b)   **5.** (a)   **6.** (b)   **7.** (b)   **8.** (a)

SELF ASSESSMENT TEST

Verify each of the following:

1. $\lim \int_0^a g \dfrac{\sin nx}{\sin x} dx = \lim \int_0^a g \dfrac{\sin nx}{x} dx \ \ 0 \le a \le \pi$

2. Using $\int_0^\infty \dfrac{\sin x}{x} dx = \dfrac{\pi}{2}$. we get

$$\dfrac{1}{2}a_0 + \sum_{n=1}^\infty a_n = \dfrac{f(0^-) + f(0^+)}{2}$$

3. $f(x) = |x| = \dfrac{\pi}{2} - \dfrac{4}{\pi}\left(\dfrac{\cos x}{1^2} + \dfrac{\cos 3x}{3^2} + \dfrac{\cos 5x}{5^2} + \ldots\right)$

4. The function x^2 is periodic with period $2l$ on the interval $(-l, l)$ and its Fourier series is given by

$$\dfrac{l^2}{3} + \dfrac{4l^2}{\pi^2} \sum_{n=1}^\infty \dfrac{(-1)^n}{n^2} \cos\left(\dfrac{n\pi x}{l}\right).$$

5. $\left|\cos\left(\dfrac{\pi x}{l}\right)\right| = \dfrac{4}{\pi} + \sum_{n=1}^\infty \dfrac{2(-1)^{n+1}}{\pi(4n^2 - 1)} \cos\left(\dfrac{2n\pi x}{l}\right)$

6. $x - [x] - \dfrac{1}{2} = \sum_{n=1}^\infty \left(-\dfrac{1}{n\pi}\right) \sin(2n\pi x) \text{ in} \left[-\dfrac{1}{2}, \dfrac{1}{2}\right]$

7. For all values of x in $[-\pi, \pi]$, k is not an integer

$$\cos kx = \dfrac{\sin k\pi}{\pi}\left[\dfrac{1}{k} + \sum_{n=1}^\infty \dfrac{(-1)^n 2k \cos nx}{k^2 - n^2}\right]$$

8. The Fourier series of e^x in $[-l, l]$ is given by

$$\dfrac{\sinh l}{l} + 2l \sinh l \sum \dfrac{(-1)^n \cos \dfrac{n\pi x}{l}}{l^2 + n^2 \pi^2}$$

$$-2\pi \sinh l \sum \dfrac{(-1)^n n \sin \dfrac{n\pi x}{l}}{l^2 + n^2 \pi^2}$$

15 Metric Spaces

15.1 INTRODUCTION

The set of real numbers has two type of properties. The first type consist of the algebraic which deals with addition, multiplication etc. The second type consists of properties having to do with the notion of distance between two numbers and with the concept of limit. The second type of properties are called topological or metric property. In this chapter we shall study these properties in a general space in which the notion of distance is defined.

Definition. *Let X be a non-empty set. Then a mapping $d : X \times X \to R^+ \cup \{0\}$ is said to be metric, if it satisfies the following conditions :*

M(1) $d(x, y) > 0 \ \forall x, y \in X$

M(2) $d(x, y) = 0$ iff $x = y \ \forall \ x, y \in X$

M (3) $d(x, y) = d(y, x) \ \forall \ x, y \in X$ (*Symmetric property*)

M (4) $d(x, y) < d(x, z) + d(z, y) \forall x, y, z \in X$ (*Triangle inequality*).

If d is a metric on X, then ordered pair (X, d) is said to be metric space.

REMARKS

- In 1905, the French mathematician Maurice Frechet thought of generalizing the notion of distance and extending it to arbitrary set, which seems the begining of metric space.
- The conditions of metric space can be defined in terms of distance as follows :

    (i) $d(x, y) \geq 0 \Rightarrow$ The distance between any two points of X is a non-negative real number.

    (ii) $d(x, y) = 0$ iff $x = y \Rightarrow$ If two points coincides then the distance is zero and if the distance is zero, then two points are same.

   (iii) $d(x, y) = d(y, x) \Rightarrow$ The distance does not depend on the order of the points x and y.

   (iv) $d(x, y) \leq d(x, z) + d(z, y) \Rightarrow$ The sum of the length of two sides of a triangle is greater than or equal to the length of the third side. The sign of equality holds when three points lie on the straight line.

15.2 SOME PARTICULAR METRIC SPACES

(1) The real line R. Let R be the set of real numbers and let $d : R \times R \to R$ be the function defined by

$$d(x, y) = |x-y| \ \forall \ x, y \in R$$

Then clearly, we have

(i) d is a non-negative real valued function on $R \times R$

(ii) $d(x, y) = 0 \Leftrightarrow |x-y| = 0 \Leftrightarrow x = y \forall x, y \in R$

(iii) $d(x, y) = |x-y| = |-(y-x)| = |y-x| = d(y, x) \; \forall \; x, y \in R$

(iv) Let x, y, z be any three elements of R.

Then $\qquad d(x, y) = |x-y| = |(x-z) + (z-y)|$

$\Rightarrow \qquad\qquad \leq |x-z| + |z-y| = d(x, z) + d(z, y)$

$\Rightarrow \qquad\qquad d(x, y) < d(x, z) + d(z, y) \; \forall \; x, y, z \in R.$

$\Rightarrow$ d is a metric on R and the ordered pair (R, d) is a metric space.

REMARK

- The metric defined above is called usual metric on R.

(2) The Euclidean plane R^2. Let R^2 be the set of all ordered pairs of real numbers. Define a mapping $d : R^2 \times R^2 \to R$ such that

$$d(x, y) = \sqrt{\left\{ (x_1 - y_1)^2 + (x_2 - y_2)^2 \right\}} \; \forall \; x, y \in R^2.$$

where $\qquad\qquad x = (x_1, x_2), y = (y_1, y_2).$

Then, clearly we have

(i) d is a non-negative real valued function on $R^2 \times R^2$.

(ii) $d(x, y) = 0 \quad \Leftrightarrow \quad \sqrt{\left\{ (x_1 - y_1)^2 + (x_2 - y_2)^2 \right\}} = 0$

$\Leftrightarrow \qquad x_1 - y_1 = 0 \qquad \text{and} \qquad x_2 - y_2 = 0$

$\Leftrightarrow \qquad x_1 = y_1 \qquad \text{and} \qquad x_2 = y_2 \Leftrightarrow x = y$

(iii) For all $x, y \in R^2$

$$d(x, y) = \sqrt{\left\{ (x_1 - y_1)^2 + (x_2 - y_2)^2 \right\}}$$

$$= \sqrt{\left\{ (y_1 - x_1)^2 + (y_2 - x_2)^2 \right\}} = d(y, x).$$

(iv) Let $x = (x_1, x_2), y = (y_1, y_2), z = (z_1, z_2)$ be any three elements of R^2. Now, by triangle inequality for real numbers, we have

$$\sqrt{\left\{ (a_1 + b_1)^2 + (a_2 + b_2)^2 \right\}} \leq \sqrt{\left(a_1^2 + a_2^2 \right)} + \sqrt{\left(b_1^2 + b_2^2 \right)} \text{ for } a_1, b_1, a_2, b_2 \in R.$$

Put $a_1 = x_1 - z_1, a_2 = x_2 - z_2, b_1 = z_1 - y_1, b_2 = z_2 - y_2$ in above inequality, we have

$$\sqrt{\left\{ (x_1 - y_1)^2 + (x_2 - y_2)^2 \right\}} \leq \sqrt{\left\{ (x_1 - z_1)^2 + (x_2 - z_2)^2 \right\}} + \sqrt{\left\{ (z_1 - y_1)^2 + (z_2 - y_2)^2 \right\}}$$

$\Rightarrow \qquad\qquad d(x, y) < d(x, z) + d(y, z).$

Hence, d is a metric on R^2.

REMARK

- The metric space (R^2, d) defined above is called Eucledian plane.

(3) Discrete metric space. Let X be any non-empty set. The mapping $d : X \times X \to R$ defined by

$$d(x, y) = \begin{cases} 0 \text{ if } x = y \\ 1 \text{ if } x \neq y \end{cases}$$

is a metric on X, called the discrete metric on X.

Proof. $d(x, y)$ is a metric because it satisfy all the following properties such that

(i) $d(x, y) = 0$ or $d(x, y) = 1$

$\Rightarrow$ $d(x, y) \geq 0 \; \forall \; x, y \in X.$

(ii) If $x = y \Rightarrow$ $d(x, y) = 0.$

    Conversely, if $d(x, y) = 0$ then $x = y.$

    If it is not possible *i.e.,* $x \neq y$ Then $d(x, y) = 1.$

    It contradicts the fact that $d(x, y) = 0$ $\Rightarrow$ $d(x, y) = 0$ iff $x = y.$

(iii) If $x = y$ $d(x, y) = 0 = d(y, x)$

    If $x \neq y$ $d(x, y) = 1 = d(y, x) \; \forall \; x, y \in X.$

    Hence $d(x, y) = d(y, x).$

(iv) Let $x, y, z \in X.$

    If $x = y$ then $d(x, y) = 0$ ...(1)

                                $d(x, z) \geq 0 \Rightarrow d(z, y) \geq 0$

$\Rightarrow$ $d(x, z) + d(z, y) \geq 0$

$\Rightarrow$ $d(x, z) + d(z, y) \geq d(x, y)$

    Similarly, if $x \neq y,\; d(x, y) = 1$ [from equation (1)]

    $\Rightarrow \; \exists \, z \in X$ such that at least

                         $d(x, z) = 1$ and $d(z, y) = 1$

$\Rightarrow$ $d(x, y) \leq d(x, z) + d(z, y)$

    Hence d is metric and (X, d) is a metric space which is called Discrete metric space.

(d)   Let (X_1, d_1) and (X_2, d_2) be two metric spaces.

    Define $X = X_1 \times X_2$

    and $d(x, y) = d_1(x_1, y_1) + d_2(x_2, y_2) \forall x, y \in X$, when $x = (x_1, y_1)$ and $y = (x_2, y_2).$

    Then, (X, d) is a metric space. Hence, cartesian product of two metric spaces is a metric space.

REMARK

• The space (X, d) defined above is called Product metric space.

(5) The Postman metric for R^2. Consider a well-planned city, in which the roads are either parallel or perpendicular to each other and there are rectangular blocks of housing complexes. Suppose someone wants to go from point A to point B.
How do we find the minimum distance that he has to travel.
Since he can not go as a cross flies.
Therefore, the Eucledian metric is useless.
The product metric is also useless.
Then he has to go along one road till he reaches a road on which B is situated and then move along the perpendicular road till he reaches B.
If the co-ordinates of A and B with reference to a pair of rectangular axes, one of which is parallel to one set of roads, and the other is perpendicular to it, be (x_1, x_2) and (y_1, y_2) respectively, then he will have to move a distance

$$|x_1 - y_1| + |x_2 - y_2|$$

Therefore, we can define a metric as follows :

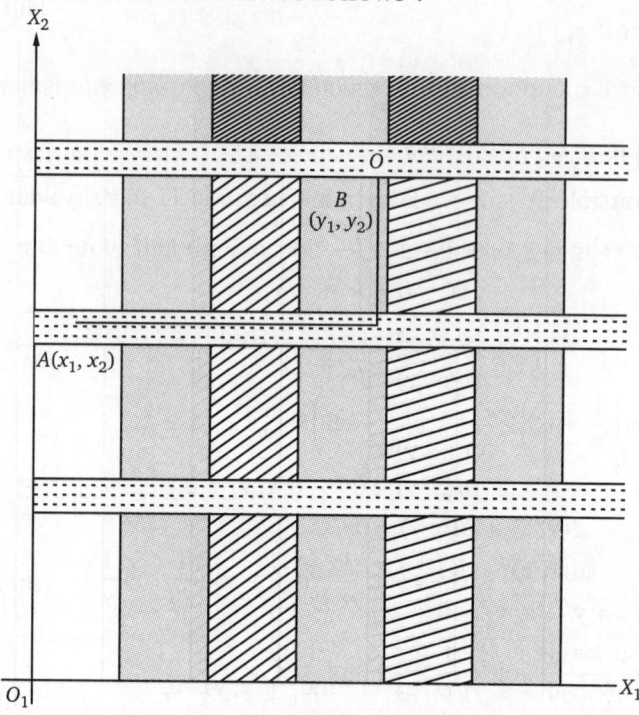

Fig. 1

Let R^2 be the set of all ordered pairs of real numbers and let $d : R^2 \times R^2 \to R$
defined by
$$d(x, y) = |x_1 - y_1| + |x_2 - y_2|$$
where
$$x = (x_1, x_2) \text{ and } y = (y_1, y_2).$$
Then, clearly we have

(i) d is a non-negative real-valued function on $R^2 \times R^2$.

(ii) $d(x, y) = 0 \iff |x_1 - y_1| + |x_2 - y_2| = 0$
$$\iff |x_1 - y_1| = 0 \text{ and } |x_2 - y_2| = 0$$
$$\iff x_1 = y_1 \text{ and } x_2 = y_2 \iff x = y.$$

(iii) For all $x, y \in R^2$
$$d(x, y) = |x_1 - y_1| + |x_2 - y_2|$$
$$= |y_1 - x_1| + |y_2 - x_2| = d(y, x).$$

(iv) Let $x = (x_1, x_2), y = (y_1, y_2), z = (z_1, z_2)$ be any three points of R^2.

Consider
$$d(x, y) = |x_1 - y_1| + |x_2 - y_2|$$
$$= |(x_1 - z_1) + (z_1 - y_1)| + |(x_2 - z_2) + (z_2 - y_2)|$$
$$\leq (|x_1 - z_1| + |z_1 - y_1|) + (|x_2 - z_2 + y_2 - z_2|)$$
$$= (|x_1 - z_1| + |x_2 - z_2|) + (|z_1 - y_1| + |z_2 - y_2|)$$
$$= d(x, z) + d(z, y)$$

$\Rightarrow \quad d(x, y) \leq d(x, z) + d(z, y).$

Hence, d is a metric on R^2.

15.2.1 SUBSPACE

Let (X, d) be a metric space and Y be a proper subset of X. Let d_1 be the restriction of d on Y i.e., $d_1(x, y) = d(x, y)$, $\forall x, y \in Y \times Y$.

Then (Y, d_1) is called subspace of (X, d).

15.2.2 PSEUDO-METRIC

A mapping $d : X \times X \to R^+ \cup \{0\}$ is called a pseudo-metric or semi-metric for X if and only if

(i) $d(x, y) \geq 0$, $\forall x, y \in X$

(ii) $d(x, x) = 0$, $\forall x \in X$

(iii) $d(x, y) = d(y, x)$; $\forall x, y \in X$

(iv) $d(x, y) \leq d(x, z) + d(z, y)$; $\forall x, y, z \in X$.

REMARKS

- The pseudo metric is said to be finite if $d(x, y) < \infty$; $\forall x, y \in X$.
- The Pseudo metric d differ from metric in the sense that
  - (a) $d(x, y)$ may be equal to zero even if $x \neq y$ i.e., distance between a pair of distinct points may be zero.
  - (b) $d(x, y) = \infty$, for some $x, y \in X$ i.e., ∞ is defined as a measure of a distance between a pair of points.

For Example :
(1) Consider a set X of real valued functions defined over the closed interval $[-1, 1]$. Let $f(x)$ and $g(x)$ be two arbitrary real valued functions defined over $[-1, 1]$. Let us define

$$d(f, g) = \int_{-1}^{1} \left\{ \left[f(x) - g(x) \right]^2 dx \right\}^{1/2}.$$

It is easy to verify that d is a pseudo metric on X.

(2) Let (X, d) be a pseudo metric and let '~' be a relation in X defined by setting

$$x \sim y \text{ iff } d(x, y) = 0.$$

The relation '~' is an equivalence relation in X. In fact,

(i) Since $\qquad\qquad d(x, x) = 0$, $\forall \ x \in X$

$\Rightarrow \qquad\qquad\qquad x \sim x$, $\quad \forall \ x \in X$.

Therefore, the relation '~' is reflexive.

(ii) $x \sim y \Leftrightarrow d(x, y) = 0 \Leftrightarrow d(y, x) = 0 \Leftrightarrow y \sim x \qquad \Rightarrow \qquad$ '~' is symmetric.

(iii) Since $x \sim y$ and $y \sim z \Rightarrow d(x, y) = 0$ and $d(y, z) = 0$

$\Rightarrow \qquad\qquad d(x, y) + d(y, z) = 0$

$\Rightarrow \qquad\qquad\qquad d(x, z) = 0$

$\Rightarrow \qquad\qquad\qquad\qquad x \sim z$

$\Rightarrow$ '~' is transitive.

15.2.3 NORM

The size of an element x is a real number denoted by $\|x\|$ and is called norm. Norm is a generalisation of the real valued functions, which satisfying the following conditions :

(i) $\|x\| \geq 0$ (ii) $\|x\| = 0 \Leftrightarrow x = 0$

(iii) $\|cx\| = |c| \, \|x\|$ (iv) $\|x + y\| \leq \|x\| + \|y\|$.

REMARK

- The metric defined with the help of norm as follows $d(x, y) = \|x-y\|$
  This metric is known as metric induced by norm.
  For Example :
  Consider the set of all bounded real valued functions defined on $[0, 1]$.

  Define norm of a function as follows :
  $$\|f\| = \sup \{|f(x)| : x \in [0, 1]\}.$$
  Here, it can be easily verified that
  $$d(f, g) = \|f-g\| = \sup |f(x)-g(x)|$$
  is a metric on $[0, 1]$. We denote this space by $C[0, 1]$.

15.3 DISTANCE BETWEEN TWO SETS : DIAMETER OF A SET

Definition 1. *Let (X, d) be a metric space and let A be a non-empty subset of X. Then diameter of the set A, denoted by $\delta(A)$, is defined by*
$$\delta(A) = \sup \{d(x, y) : x, y \in A\}.$$

Definition 2. *The distance between a point $x \in X$ and a set A is denoted by $d(x, A)$ and is defined by*
$$d(x, A) = \inf\{d(x, y) : y \in A).$$

Definition 3. *The distance between two non-empty subsets A and B of a metric space X is denoted and defined by*
$$d(A, B) = \inf\{d(x, y) : x \in A, y \in B\}.$$

REMARKS

- The diameter of a set is always non-negative
- The diameter of empty set is ∞.
- A set A is said to be bounded if $\delta(A) < \infty$.
- If A is a closed sphere of radius r, then $d(A) = 2r$.
- The diameter of a finite set is finite and diameter for an infinite set is infinite.
- For the empty set ϕ, the distance of ϕ from point x is denoted by $d(x, \phi) = \infty$.
- $d(x, A) = 0$ if $x \in A$.
- The distance of set A and empty set is ∞ i.e., $d(A, \phi) = \infty$.
- $d(A, B) = 0$ iff $A \cap B \neq \phi$.

THEOREM 1. *Let (X, d) be a metric space and let x, y, z be any three points of X, then*
$$d(x, y) \geq |d(x, z)-d(z, y)|.$$

Proof. By definition of metric space, we have
$$d(x, z) \leq d(x, y)+d(y, z)$$
$$= d(x, y)+d(z, y) \qquad \text{(By symmetric property)}$$
$$\Rightarrow \quad d(x, z)-d(z, y) \leq d(x, y). \qquad \qquad ...(1)$$
By M (iv), we have
$$d(z, y) \leq d(z, x)+d(x, y)$$
$$= d(x, z)+d(x, y) \qquad \qquad [d(x, z)=d(z, x)]$$

$$\Rightarrow \qquad d(z, y) - d(x, z) \le d(x, y). \qquad \text{...(2)}$$

From (1) and (2), we conclude that

$$d(x, y) \ge |d(x, z) - d(z, y)|.$$

REMARK

- The above inequality states that the difference of the lengths of any two sides of a triangle is less than or equal to the third side. Also, the sign of equality occurs, when three poirts lie on the straight line.

THEOREM 2. *Let (X, d) be a metric space, then*

$$|d(x, y) - d(x', y')| \le d(x, x') + d(y, y') \ \forall \ x, x', y, y' \in X.$$

Proof. By M(iv), we have

$$d(x, y) \le d(x, x') + d(x', y)$$
$$\qquad\qquad \le d(x, x') + d(x', y') + d(y', y) \qquad\qquad [\text{By } M(\text{iv})]$$
$$\qquad\qquad = d(x, x') + d(x', y') + d(y, y') \qquad\qquad [d(y, y') = d(y', y)]$$
$$\Rightarrow \quad d(x, y) - d(x', y') \le d(x, x') + d(y, y'). \qquad\qquad \text{...(1)}$$

Also, we have

$$d(x', y') \le d(x', x) + d(x, y')$$
$$\qquad\qquad \le d(x', x) + d(x, y) + d(y, y') \qquad [\because d(x, y') \le d(x, y) + d(y, y')]$$
$$\qquad\qquad = d(x, x') + d(x, y) + d(y, y') \qquad [\because d(x', x) = d(x', x)]$$
$$\Rightarrow \quad d(x', y') - d(x, y) \le d(x, x') + d(y, y'). \qquad\qquad \text{...(2)}$$

From (1) and (2), we conclude that

$$|d(x, y) - d(x', y')| \le d(x, x') + d(y, y')$$

THEOREM 3. *Let $X \ne \phi$. Then a mapping $d: X \times X \to R$ is a metric if and only if the following conditions hold*

(i) $d(x, y) = 0 \Leftrightarrow x = y; \ \forall \ x, y \in X.$

(ii) $d(x', y) \le d(x, y) + d(y, z); \ \forall \ x, y, z \in X.$

Proof. Let us first suppose d is a metric on R.

Then by M(iv), we have

$$d(x, y) \le d(x, z) + d(z, y) \qquad\qquad\qquad\qquad\qquad \text{...(1)}$$

Also $\qquad d(z, y) = d(y, z). \qquad\qquad\qquad\qquad\qquad\qquad\qquad \text{...(2)}$

From (1) and (2), we conclude that

$$d(x, y) \le d(x, z) + d(y, z); \ \forall \ x, y, z \in X.$$

Now, suppose condition (i) and (ii) holds : To show d is a metric.
Let x, y be any two elements of X, then by (ii), we have

$$d(x, x) \le d(x, y) + d(x, y)$$

[Replacing x, x, y for x, y, z respectively]

i.e., $\qquad\qquad 2d(x, y) \ge d(x, x). \qquad\qquad\qquad\qquad\qquad \text{...(3)}$

But from (i) $d(x, x) = 0$, therefore from (3), we have

$$d(x, y) \ge 0 \ \forall \ x, y \in X \Rightarrow M \ (\text{i}) \text{ is satisfied.}$$

Now apply condition (ii) for the points x, y, z, we have

$$d(x, y) \leq d(x, x) + d(y, x) = 0 + d(y, x) \qquad [\because d(x, y) = 0]$$

$$\Rightarrow \qquad d(x, y) \leq d(y, x). \qquad \qquad \ldots(4)$$

Again applying condition (ii) for the points y, x, y, we get

$$d(y, x) \leq d(y, y) + d(x, y) = 0 + d(x, y) \qquad [\because d(y, y) = 0]$$

$$= d(x, y). \qquad \qquad \ldots(5)$$

From (4) and (5), we conclude that

$$d(x, y) = d(y, x) \qquad \Rightarrow M\text{(iii) is satisfied.}$$

Finally, for any x, y, z, we have

$$d(x, y) \leq d(x, z) + d(y, z) = d(x, z) + d(z, y)$$

$$\Rightarrow \qquad d(x, y) \leq d(x, z) + d(z, y)$$

$$\Rightarrow \qquad M \text{ (iv) is satisfied.}$$

Hence, d is a metric on X.

THEOREM 4. *Let (X_1, d_1) and (X_2, d_2) be any two metric spaces defined by*

$$d[(x_1, x_2), (y_1, y_2)] = \sqrt{d_1^2(x_1, y_1) + d_2^2(x_2, y_2)}$$

where $x_1, y_1 \in X_1$, $x_2, y_2 \in X_2$. Show that d is a metric for $X_1 \times X_2$.

Proof. Since (X_1, d_1) and (X_2, d_2) are two metric spaces. Therefore, by definition, we have

(i) $d_1(x_1, y_1) \geq 0,\ d_2(x_2, y_2) \geq 0$

(ii) $d_1(x_1, y_1) = 0 \Leftrightarrow x_1 = y_1$ and $d_2(x_2, y_2) = 0 \Leftrightarrow x_2 = y_2$

(iii) $d_1(x_1, y_1) = d_1(y_1, x_1)$ and $d_2(x_2, y_2) = d_2(y_2, x_2)$

(iv) $d_1(x_1, y_1) \leq d_1(x_1, z_1) + d_1(z_1, y_1)$ and $d_2(x_2, y_2) \leq d_2(x_2, z_2) + d_2(z_2, y_2)$

To show $d[(x_1, x_2), (y_1, y_2)] = \sqrt{d_1^2(x_1, y_1) + d_2^2(x_2, y_2)}$

(a) Since $d_1(x_1, y_1) \geq 0$ and $d_2(x_2, y_2) \geq 0$

Therefore $d_1^2(x_1, y_1) \geq 0$ and $d_2^2(x_2, y_2) \geq 0$

$$\Rightarrow \qquad d_1^2(x_1, y_1) + d_2^2(x_2, y_2) \geq 0$$

$$\Rightarrow \qquad \sqrt{d_1^2(x_1, y_1) + d_2^2(x_2, y_2)} \geq 0$$

$$\Rightarrow \qquad d[(x_1, x_2), (y_1, y_2)] \geq 0.$$

(b) Here, we have

$$d_1(x_1, y_1) = 0 \Leftrightarrow x_1 = y_1$$

and $d_2(x_2, y_2) = 0 \Leftrightarrow x_2 = y_2$.

Now $\qquad d[(x_1, x_2), (y_1, y_2)] = 0$

$$\Leftrightarrow \quad \sqrt{d_1^2(x_1, y_1) + d_2^2(x_2, y_2)} = 0$$

$$\Leftrightarrow \quad d_1^2(x_1, y_1) = 0,\ d_2^2(x_2, y_2) = 0$$

$$\Leftrightarrow \quad d_1(x_1, y_1) = 0 \Leftrightarrow x_1 = y_1 \text{ and } d_2(x_2, y_2) = 0 \Leftrightarrow x_2 = y_2$$

$$\Rightarrow \quad (x_1, x_2) = (y_1, y_2)$$

(c) Here, we have

$$d_1(x_1, y_1) = d_1(y_1, x_1) \quad \text{and} \quad d_2(x_2, y_2) = d_2(y_2, x_2).$$

$$\Rightarrow \quad d_1^2(x_1, y_1) = d_1^2(y_1, x_1) \quad \text{and} \quad d_2^2(x_2, y_2) = d_2^2(y_2, x_2)$$

$$\Rightarrow \quad d_1^2(x_1, y_1) + d_2^2(x_2, y_2) = d_1^2(y_1, x_1) + d_2^2(y_2, x_2)$$

$$\Rightarrow \quad \sqrt{d_1^2(x_1, y_1) + d_2^2(x_2, y_2)} = \sqrt{d_1^2(y_1, x_1) + d_2^2(y_2, x_2)}$$

$$\Rightarrow \quad d[(x_1, x_2), (y_1, y_2)] = d[(y_1, y_2), (x_1, x_2)].$$

(d) Let $(x_1, x_2), (y_1, y_2), (z_1, z_2) \in X_1 \times X_2$.

Then, we have

$$d(x, y) = \sqrt{d_1^2(x_1, y_1) + d_2^2(x_2, y_2)}$$

$$\leq \sqrt{\left\{ \left[d_1(x_1, z_1) + d_1(z_1, y_1) \right]^2 + \left[d_2(x_2, z_2) + d_2(z_2, y_2) \right]^2 \right\}}$$

(By triangle inequality)

$$\leq \left[d_1^2(x_1, z_1) + d_2^2(x_2, z_2) \right]^{1/2} + \left[d_1^2(z_1, y_1)^2 + d_2^2(z_2, y_2)^2 \right]^{1/2}$$

$$\leq d(x, z) + d(z, y).$$

Hence, from (a), (b), (c) and (d), we conclude that

$$d(x, y) = \sqrt{\left[d_1^2(x_1, y_1) + d_2^2(x_2, y_2) \right]} \text{ is a metric on } X.$$

15.4 SOME IMPORTANT INEQUALITIES

(1) *Triangle Inequality.*

$$|z + w| \leq |z| + |w|, \ z, w \in \mathbb{C}$$

(2) *Holder's Inequality.*

(i) If a_i, b_i $(i = 1, 2, \ , n)$ are non-negative real numbers, then

$$\sum_{i=1}^{n} a_i b_i \leq \left(\sum_{i=1}^{n} a_i^P \right)^{1/P} \left(\sum_{i=1}^{n} a_i^q \right)^{1/q}$$

where $p > 1$ and $\dfrac{1}{p} + \dfrac{1}{q} = 1$.

(ii) *Holder's Inequality for integrals.* If f, g are non-negative real-valued integrable functions defined on $[a, b]$, then

$$\int_a^b f.g \, dx \leq \left\{ \int_a^b (f(x))^P \, dx \right\}^{1/P} \left\{ \int_a^b (g(x))^q \, dx \right\}^{1/q}.$$

(3) *Cauchy-Schwarz inequality.* Let $z = (z_1, z_2, ..., z_n)$ and $w = (w_1, w_2, ..., w_n)$ be two n-tuples of real or complex numbers, then

$$\sum_{i=1}^{n} |z_i w_i| \leq \left(\sum_{i=1}^{n} |z_i|^2 \right)^{1/2} \left(\sum_{i=1}^{n} |w_i|^2 \right)^{1/2}$$

REMARK

- In terms of norms, the above inequality can be written as $\displaystyle\sum_{i=1}^{n} |z_i w_i| \leq \|z\| \|w\|$.

(4) *Minkowski's inequality.*

(i) If $p \geq 1$ and a_i, b_i $(i=1, 2, \ldots, n)$ are non-negative real numbers then,

$$\left(\sum_{i=1}^{n}(a_ib_i)^p\right)^{1/p} \leq \left(\sum_{i=1}^{n}a_i^p\right)^{1/p} + \left(\sum_{i=1}^{n}b_i^p\right)^{1/p}$$

(ii) *Minkowski's inequality for integrals.* Let f, g be non-negative real-valued functions defined on $[a, b]$ then

$$\left\{\int_a^b (f+g)^p\, dx\right\}^{1/p} \leq \left\{\int_a^b (f(x))^p\, dx\right\}^{1/p} + \left\{\int_a^b (g(x))^p\, dx\right\}^{1/p}$$

(iii) *Minkowski's inequality in terms of norm.* Let $z = (z_1, z_2, \ldots, z_n)$ and $w = (w_1, w_2, \ldots, w_n)$ be two n-tuples of real or complex numbers. Then

$$\left(\sum_{i=1}^{n}|z_i + w_i|^2\right)^{1/2} \leq \left(\sum_{i=1}^{n}|z_i|^2\right)^{1/2} + \left(\sum_{i=1}^{n}|w_i|^2\right)^{1/2}$$

or $\|z+w\| \leq \|z\| + \|w\|$.

(iv) *Minkowski's inequality for complex numbers.* If z_i, w_i $(i=1, 2, \ldots, n)$ are complex numbers, then

$$\left(\sum_{i=1}^{n}|z_i + w_i|^p\right)^{1/p} \leq \left(\sum_{i=1}^{n}|z_i|^p\right)^{1/p} + \left(\sum_{i=1}^{n}|w_i|^p\right)^{1/p}, (p \geq 1).$$

Solved Examples

Example 1. *Show that the mapping $d: R^2 \times R^2 \rightarrow R^2$ defined by $d(x, y) = max.\{|x_1-y_1|, |x_2-y_2|\}$ where $x = (x_1, x_2)$, $y = (y_1, y_2) \in R$ is metric on R^2.*

Solution.

(i) Since $|x_1-y_1| \geq 0$
and $|x_2-y_2| \geq 0$
which implies
$max \{(x_1-y_1), (x_2-y_2)\} \geq 0$
$\Rightarrow$ $d(x, y) \geq 0$.

(ii) Since, we know that

$|x_1-y_1| = 0$ iff $x_1 = y_1$
and $|x_2-y_2| = 0$ iff $x_2 = y_2$.
Therefore,
$|x_1-y_1| + |x_2-y_2| = 0$ iff
$x_1 = y_1, x_2 = y_2$.
$\Rightarrow max\{|x_1-y_1| + |x_2-y_2|\} = 0$
iff $x_1 = y_1, x_2 = y_2$
$d(x, y) = 0$ iff $x = y$.

(iii) We know that
$|x_1-y_1| = |y_1-x_1|$
and $|x_2-y_2| = |y_2-x_2|$

Based on the following Results

● Let $X \neq \phi$. Then, a mapping $d: X \times X \rightarrow R^+ \cup \{0\}$ is said to be metric if following four conditions are satisfied:

(i) $d(x, y) \geq 0 \,\forall\, x, y \in X$

(ii) $d(x, y) = 0 \Leftrightarrow x = y \,\forall x, y \in X$

(iii) $d(x, y) = d(y, x) \,\forall\, x, y \in X$

(iv) $d(x, y) \leq d(x, z) + d(z, y) \,\forall x, y, z \in X$

● Diameter of a set : $\delta(A) = \sup\{d(x, y): x, y \in A\}$

● Distance between two sets A and B
$d(A, B) = \inf\{d(x, y): x \in A, y \in B\}$

● *Triangle Inequality.* $|x+y| \leq |x| + |y|$

● *Holder's Inequality.*

$$\sum_{i=1}^{n}a_ib_i \leq \left(\sum_{i=1}^{n}a_i^p\right)^{1/p}\left(\sum_{i=1}^{n}a_i^q\right)^{1/q}$$

● *Cauchy-Schwarz Inequality.*

$$\sum_{i=1}^{n}|z_iw_i| \leq \left(\sum_{i=1}^{n}|z_i|^2\right)^{1/2}\left(\sum_{i=1}^{n}|w_i|^2\right)^{1/2}$$

● *Minkowski's Inequality.*

$$\left(\sum_{i=1}^{n}(a_i+b_i)^p\right)^{1/p} \leq \left(\sum_{i=1}^{n}a_i^p\right)^{1/p} + \left(\sum_{i=1}^{n}b_i^p\right)^{1/p}$$

$$\Rightarrow \quad |x_1-y_1|+|x_2-y_2| = |y_1-x_1|+|y_2-x_2|$$
$$\Rightarrow \quad d(x, y) = d(y, x)$$

(iv) Consider

$$d(x, y) = \max.\{|x_1-y_1|, |x_2-y_2|\}$$
$$= \max\{|x_1-z_1+z_1-y_1|, |x_2-z_2+z_2-y_2|\}$$
$$\leq \max\{|x_1-z_1|+|z_1-y_1|, |x_2-z_2|+|z_2-y_2|\}$$
$$\leq \max\{|x_1-z_1|, |x_2-z_2|\}+\max[|z_1-y_1|+|z_2-y_2|]$$
$$\Rightarrow \quad d(x, y) \leq d(x, z)+d(z, y).$$

From (i), (ii), (iii) and (iv), we conclude that d is a metric on $\mathbb{R}^2$.

Example 2. *Let X be the set of real valued bounded continuous function defined on the closed interval [0, 1]. We define the norm of a function $f \in X$ by*

$$\|f\| = \int_0^1 |f(x)|\,dx.$$

Define a mapping $d: X \times X \to \mathbb{R}$ by $d(f, g) = \|f-g\| = \int_0^1 |f(x)-g(x)|\,dx \ \forall f, g \in X.$
Show that d is a metric on X.

Solution. (i) Since, we have

$$|f(x)-g(x)| \geq 0.$$

Therefore, $\int_0^1 |f(x)-g(x)|\,dx \geq 0$

$$\Rightarrow \quad \|f-g\| \geq 0$$
$$\Rightarrow \quad d(f, g) \geq 0.$$

(ii) Since

$$|f(x)-g(x)| = 0 \text{ iff } f(x)=g(x)$$

therefore, $\int_0^1 |f(x)-g(x)|\,dx = 0 \text{ iff } f(x)=g(x)$

$$\Rightarrow \quad \|f-g\| = 0 \text{ iff } f(x)=g(x)$$
$$\Rightarrow \quad d(f, g)=0 \text{ iff } f(x)=g(x)$$

(iii) Since $|f(x)-g(x)| = |g(x)-f(x)|.$

Therefore, $\int_0^1 |f(x)-g(x)|\,dx = \int_0^1 |g(x)-f(x)|\,dx$

$$\Rightarrow \quad \|f-g\| = \|g-f\|$$
$$\Rightarrow \quad d(f, g) = d(g, f).$$

(iv) Let $f, g, h \in X$. Then, we have

$$\|f-g\| = \int_0^1 |f(x)-g(x)|\,dx = \int_0^1 |f(x)-h(x)+h(x)-g(x)|\,dx$$
$$\leq \int_0^1 |f(x)-h(x)|+|h(x)-g(x)|\,dx$$
$$= \int_0^1 |f(x)-h(x)|\,dx + \int_0^1 |h(x)-g(x)|\,dx$$
$$= \|f-h\|+\|h-g\|$$
$$\Rightarrow \quad d(f, g) \leq d(f, h)+d(h, g).$$

Hence, from (i), (ii), (iii) and (iv), we conclude that d is a metric on $\mathbb{R}$.

Example 3. *Let L_∞ denote the set of all bounded sequence. If $x = \langle x_n \rangle$ and $y = \langle y_n \rangle$ are any two parts of L_∞, we define*

$$d(x, y) = \sup\{|x_n - y_n| : n \in N\}.$$

Show that l_∞ is a metric space under d.

Solution.　(i)　Since 　　　　$|x_n - y_n| \geq 0$. Therefore, $\sup|x_n - y_n| \geq 0$

$\Rightarrow$ 　　　　　　　$d(x, y) \geq 0$.

(ii)　We have 　　$|x_n - y_n| = 0$ iff $x_n = y_n$ 　$\Rightarrow$　$\sup.|x_n - y_n| = 0$ iff $x_n = y_n$

$\Rightarrow$ 　　　　　$d(x, y) = 0$ iff $x = y$.

(iii)　Since 　　$|x_n - y_n| = |y_n - x_n|$. Therefore $\sup|x_n - y_n| = \sup|y_n - x_n|$

$\Rightarrow$ 　　　　$d(x, y) = d(y, x)$.

(iv)　Let $z = \langle z_n \rangle$ be an element of l_∞. Then for any positive integer n, we have

$$|x_n - y_n| = |x_n - z_n + z_n - y_n|$$
$$\leq |x_n - z_n| + |z_n - y_n|$$
$$\leq \sup\{|x_n - z_n| : n \in N\} + \sup\{|z_n - y_n| : n \in N\}$$
$$= d(x, z) + d(z, y)$$

$\Rightarrow$ 　　　　$d(x, y) \leq d(x, z) + d(z, y)$.

Example 4. *Let (X, d) be a metric space and let M be a positive number, then there exists a metric d_1 on X such that the metric space (X, d_1) is bounded with $\delta(x) \leq M$.*

Solution.　Define d_1 by

$$d_1(x, y) = \frac{Md(x, x_1)}{1 + d(x, x_1)} ; \text{where } x, y \in X.$$

To show d_1 is a metric,

(i)　Since 　　$d(x, y) \geq 0$ and $M \geq 0$.

$$\therefore \quad \frac{Md(x, x_1)}{1 + d(x, x_1)} \geq 0 \quad\quad \Rightarrow \quad\quad d_1(x, y) \geq 0$$

(ii)　　$d_1(x, y) = 0 \Leftrightarrow \dfrac{Md(x, x_1)}{1 + d(x, x_1)} = 0$

$$\Leftrightarrow d(x, y) = 0 \Leftrightarrow x = y.$$

(iii)　$d_1(x, y) = \dfrac{Md(x, x_1)}{1 + d(x, x_1)} = \dfrac{Md(y, x)}{1 + d(y, x)} = d_1(y, x).$

(iv)　Let $x, y, z \in X$, therefore

$$d_1(x, y) = \frac{Md(x, y)}{1 + d(x, y)} = M - \frac{M}{1 + d(x, y)}$$

$$\leq M - \frac{M}{1 + d(x, z) + d(z, y)} + \frac{M[d(x, z) + d(z, y)]}{1 + d(x, z) + d(z, y)}$$

$$= \frac{Md(x, z)}{1 + d(x, z) + d(z, y)} + \frac{Md(z, y)}{1 + d(x, z) + d(z, y)}$$

$$\leq \frac{Md(x, z)}{1 + d(x, z)} + \frac{Md(z, y)}{1 + d(z, y)}$$

$$\leq d_1(x, z) + d_1(z, y)$$

$$\Rightarrow \qquad d_1(x, y) \le d_1(x, z) + d_1(z, y).$$

From (i), (ii), (iii) and (iv), we conclude that

$$d_1(x, y) = \frac{Md(x, x_1)}{1 + d(x, x_1)} \text{ is a metric on } X$$

also since $\quad d_1(x, y) = \dfrac{Md(x, x_1)}{1 + d(x, x_1)} \le M \quad$ for every points $x, y \in X$

therefore, d_1 is a bounded metric for X with $\delta(x) \le M$.

REMARKS

- A metric space (X, d) is said to be bounded if there exist a positive number M such that $d(x, y) \le M$ for every pair of points x, y of X.
- A metric space which is not bounded is said to be unbounded.
- A metric space X is said to be bounded if its diameter is finite.

Example 5. *Let (X, d) be a metric space and let $d_1(x, y) = min.\{1, d(x, y)\}$*

Then show that d_1 is a metric on X.

Solution. (i) Here, we have

$$d(x, y) = 1 \text{ or } d_1(x, y) = d(x, y).$$

Clearly $\qquad d(x, y) \ge 0$ $\qquad\qquad\qquad\qquad\qquad\qquad$ [d is a metric]

Therefore, in both the cases $d_1(x, y) \ge 0$.

(ii) If $d_1(x, y) = 0$, then $d_1(x, y) = d(x, y) = 0$.

Since d is a metric, therefore $d(x, y) = 0 \Leftrightarrow x = y$.

Hence $\qquad d_1(x, y) = 0 \Leftrightarrow x = y$

(iii) Since either

$$d_1(x, y) = 1 \text{ or } d_1(x, y) = d(x, y).$$

Therefore, if $d_1(x, y) = d(x, y)$, then $d(x, y) < 1$.

Hence, $\qquad d(y, x) = d(x, y) < 1$.

But $d(y, x) < 1$, gives $d_1(y, x) = d(y, x) = d(x, y) = d_1(x, y)$.

If $d_1(x, y) = 1$, then $d(x, y) \ge 1$ and therefore $d(y, x) \ge 1$.

But $d(y, x) \ge 1$ gives $d_1(y, x) = 1$.

Hence, $\qquad d_1(x, y) = d_1(y, x)$.

Therefore, in each case $d_1(x, y) = d_1(y, x)$.

(iv) Here, we want to prove that

$$d_1(x, y) \le d_1(x, z) + d_1(z, y). \qquad\qquad\qquad …(A)$$

If $d(x, y) \le 1$, then if either $d_1(x, z) = 1$ or $d_1(z, y) = 1$

$\Rightarrow \qquad$ inequality (A) holds good.

If both $d_1(x, z) \ne 1$ and $d_1(z, y) \ne 1$, then

$$d_1(x, z) = d(x, z) \text{ and } d_1(z, y) = d(z, y).$$

Then, we have

$$d(x, y) \le d(x, z) + d(z, y) = d_1(x, z) + d_1(z, y). \qquad …(B)$$

But $\qquad d_1(x, y) = min \{1, d(x, y)\} \le d(x, y). \qquad\qquad …(C)$

From (B) and (C), we have

$$d_1(x, y) \le d_1(x, z) + d_1(z, y).$$

Hence from (i), (ii), (iii) and (iv), we conclude that d_1 is a metric on X.

Example 6. *Let d be a metric for non-empty set X. Show that d_1 defined by*

$$d_1(x, y) = 2d(x, y) \text{ is also metric for X.}$$

Solution.   (i)   Since $d(x, y)$ is a metric on X, therefore $d(x, y) \geq 0$.

$\Rightarrow$ $\qquad\qquad$ $2d(x, y) \geq 0$ $\quad \Rightarrow$ $d_1(x, y) \geq 0$.

(ii)   Since $d(x, y)$ is a metric on X, therefore

$$d(x, y) = 0 \Leftrightarrow x = y$$

or $\qquad\qquad$ $2d(x, y) = 0 \Leftrightarrow x = y$

or $\qquad\qquad$ $d_1(x, y) = 0 \Leftrightarrow x = y$

(iii)   Since $d(x, y)$ is a metric on X, therefore $d(x, y) = d(y, x)$

$\Rightarrow$ $\qquad\qquad$ $2d(x, y) = 2d(y, x)$

$\Rightarrow$ $\qquad\qquad$ $d_1(x, y) = d_1(y, x)$.

(iv)   Since $\qquad\qquad$ $d(x, y) \leq d(x, z) + d(z, y)$

$\Rightarrow$ $\qquad\qquad$ $2d(x, y) \leq 2d(x, z) + 2d(z, y)$

$\Rightarrow$ $\qquad\qquad$ $d_1(x, y) \leq d_1(x, z) + d_1(z, y)$.

Hence, from (i), (ii), (iii) and (iv), we conclude that d_1 is a metric on X.

Example 7. *Show that the function d, defined by $d(p, q) = \|p-q\|$ where p, q are vectors in a normed vector space V is metric on V.*

Solution.   Let $p, q, r \in V$ and x is any scalar.

By definition of normed vector space, we have

(a) $\|p\| > 0$ $\qquad\qquad\qquad\qquad$ (b) $\|p\| = 0 \Leftrightarrow p = 0$

(c) $\|xp\| = |x| \|p\|$ $\qquad\qquad\qquad$ (d) $\|p+q\| \leq \|p\| + \|q\|$.

To show d is a metric on V.

(i)   Since $\qquad$ $d(p, q) = \|p-q\|$

Then by (a) $\|p-q\| \geq 0$.

(ii)   Using (b), $\quad d(p, q) = 0$ $\qquad \Leftrightarrow \|p-q\| = 0$

$\Leftrightarrow \quad p - q = 0 \Leftrightarrow p = q$

(iii)   Using (c), we have

$$d(p, q) = \|p-q\|$$

$\Rightarrow$ $\qquad$ $d(q, p) = \|q-p\| = \|-(p-q)\| = \|p-q\| = d(p, q)$.

(iv)   Consider

$$\|p-q\| = \|p-r+r-q\| \leq \|p-r\| + \|r-q\|$$

$\Rightarrow$ $\qquad$ $d(p, q) \leq d(p, r) + d(r, q)$.

Hence, from (i), (ii), (iii) and (iv), we conclude that $d(p, q)$ is a metric on V.

Example 8. *Let C be the set of all complex number then show that mapping, $d: C \times C \rightarrow R$ is a metric on C if d is defined as*

$$d(z_1, z_2) = |z_1 - z_2|, \ \forall z_1, z_2 \in C.$$

Solution.   Since d is defined by

$$d(z_1, z_2) = |z_1 - z_2|, \ \forall z_1, z_2 \in C.$$

We have to show that (X, d) is metric space $i, e., d$ is metric.

(i) Since $|z_1-z_2| \geq 0,\ \forall\ z_1, z_2 \in C$

$\Rightarrow \quad d(z_1, z_2) \geq 0,\ \forall\ z_1, z_2 \in C$

(ii) Since $|z_1-z_2|=0 \Leftrightarrow z_1-z_2=0 \Leftrightarrow z_1=z_2$

$\Rightarrow \quad d(z_1, z_2)=0$ iff $z_1=z_2$.

(iii) $|z_1-z_2|=|z_2-z_1|,\ \forall\ z_1, z_2 \in C$

$\Rightarrow \quad d(z_1, z_2)=d(z_2, z_1),\ \forall\ z_1, z_2 \in C.$

(iv) Let $z_1, z_2, z_3 \in C$

$$|z_1-z_2| = |(z_1-z_3)+(z_3-z_2)|$$
$$\leq |z_1-z_3|+|z_3-z_2|,\ \forall\ z_1, z_2, z_3 \in C$$

$\Rightarrow \quad d(z_1, z_2) \leq d(z_1, z_3)+d(z_3, z_2),\ \forall\ z_1, z_2, z_3 \in C.$

Hence by above properties d is a metric.

Example 9. *Let R be the set of real numbers and the mapping $d: R \times R \to R$ defined by*
$$d(x, y)=|x^2-y^2|,\ \forall x, y \in R$$
show that it is a Pseudo-metric on R which is not a metric on R.

Solution. Here 'd' is defined as
$$d(x, y)=|x^2-y^2|,\ \forall x, y \in R$$

To show it a metric.

(i) Since $|x^2-y^2| \geq 0,\ \forall\ x, y \in R$

$\Rightarrow \qquad d(x, y) \geq 0,\ \forall\ x, y \in R.$

(ii) Since $\qquad d(x, x)=|x^2-x^2|=0,\ \forall\ x \in R$

but here $\qquad d(x, y)=|x^2-y^2|=0$

$\Rightarrow \qquad x^2-y^2=0 \Rightarrow x=\pm y.$

$\Rightarrow \qquad d(x, y)=0$ if $x=y$ but not converse,

(iii) Since $\qquad d(x, y)=|x^2-y^2|=|y^2-x^2|$

$\Rightarrow \qquad d(x, y)=d(y, x),\ \forall\ x, y \in R.$

(iv) Let $x, y, z \in R$

$$|x^2-y^2|=|(x^2-z^2)+(z^2-y^2)|$$
$$\leq |(x^2-z^2)|+|(z^2-y^2)|$$

$\Rightarrow \qquad d(x, y) \leq d(x, z)+d(z, y),\ \forall\ x, y, z \in R$

By above properties $d(x, y)$ is Pseudo metric space.

Example 10. *Let d_1, d_2 be two metric for a non-empty set X. Show that the mapping d defined by*
$$d(x, y)=d_1(x, y)+d_2(x, y),\quad \forall x, y \in X$$
is also a metric for X.

Solution. Let d_1, d_2 are matrices on a non-empty set X which satisfying the following properties:

(i) $d_1(x, y) \geq 0,\ d_2(x, y) \geq 0,\ \forall\ x, y \in X$

(ii) $d_1(x, y)=0$ iff $x=y$ and $d_2(x, y)=0$ iff $x=y$

(iii) $d_1(x, y)=d_1(y, x)$ and $d_2(x, y)=d_2(y, x),\ \forall\ x, y \in X$

(iv) $d_1(x, y) \leq d_2(x, z)+d_1(z, y),\ \forall\ x, y \in X$

$\qquad d_2(x, y) \leq d_2(x, z)+d_2(z, y),\ \forall\ x, y \in X.$

Now we have to show that
$$d(x, y) = d_1(x, y) + d_2(x, y), \quad \forall\, x, y \in X$$
is metric on X.

(i) $d_1(x, y) \geq 0,\ d_2(x, y) \geq 0$

$\Rightarrow$ $d_1(x, y) + d_2(x, y) \geq 0\ \forall x, y \in X$

$\Rightarrow$ $d(x, y) \geq 0\ \forall\, x, y \in X.$

(ii)   Since $d_1(x, y) = 0$ iff $x = y$ and $d_2(x, y) = 0$ iff $x = y$

$\Rightarrow$ $d_1(x, y) + d_2(x, y) = 0$ iff $x = y$

$\Rightarrow$ $d(x, y) = 0$ iff $x = y$.

(iii) Since $d_1(x, y) = d_1(y, x)$ and $d_2(x, y) = d_2(y, x)\ \forall\, x, y \in X$

$\qquad\qquad d(x, y) = d_1(x, y) + d_2(x, y)$

$\qquad\qquad\qquad\quad = d_1(y, x) + d_2(y, x) = d(y, x)\ \forall x, y \in X$

(iv)   Let x, y, z be elements of X such that

$$d_1(x, y) \leq d_1(x, z) + d_1(z, y)$$

and $\qquad\qquad d_2(x, y) \leq d_2(x, z) + d_2(z, y)$

$\Rightarrow$ $d_1(x, y) + d_2(x, y) \leq d_1(x, z) + d_1(z, y) + d_2(x, z) + d_2(z, y)$

$\Rightarrow$ $d(x, y) \leq [d_1(x, z) + d_2(x, z)] + [d_1(z, y) + d_2(z, y)]$

$\Rightarrow$ $d(x, y) \leq d(x, z) + d(z, y)\ \forall\, x, y, z \in X.$

so $d(x, y)$ is a metric on X.

Example 11.   Let (X_1, d_1) and (X_2, d_2) be two metric spaces. For any pair of points $x = (x_1, x_2)$, $y = (y_1, y_2)$ in $X = X_1 \times X_2$ and is defined as
$$d(x, y) = d_1(x_1, y_1) + d_2(x_2, y_2)$$
then prove that d is metric on $X = X_1 \times X_2$.

Solution.   Since d_1 and d_2 are two metric spaces then it satisfying all properties of metric spaces. Proceeding same as above example, we have to show that d is metric for $X = X_1 \times X_2$ defined by $d(x, y) = d_1(x_1, y_1) + d_2(x_2, y_2)$.

(1)   Since $d_1(x_1, y_1) \geq 0,\ \forall\, x_1, y_1 \in X_1$ and $d_2(x_2, y_2) \geq 0,\ \forall\, x_2, y_2 \in X_2$

$\Rightarrow$ $d_1(x_1, y_1) + d_2(x_2, y_2) \geq 0,\ \forall\, x_1, y_1 \in X_1,\ \forall\, x_2, y_2 \in X_2$

$\Rightarrow$ $d(x, y) \geq 0,\ \forall\ x, y \in X$, where $x = (x_1, x_2),\ y = (y_1, y_2)$.

(2) $d_1(x_1, y_1) = 0$ iff $x_1 = y_1$ and $d_2(x_2, y_2) = 0$ iff $x_2 = y_2$

$\Rightarrow$ $d_1(x_1, y_1) + d_2(x_2, y_2) = 0$ iff $x_1 = y_1$ and $x_2 = y_2$.

$\Rightarrow$ $d(x, y) = 0$ iff $(x_1, x_2) = (y_1, y_2)$

$\Rightarrow$ $d(x, y) = 0$ iff $x = y$.

(3)   We have $d_1(x_1, y_1) = d_1(y_1, x_1),\ \forall\, x_1, y_1 \in X_1$

$\qquad\qquad\qquad d_2(x_2, y_2) = d_2(y_2, x_2),\ \forall\, x_2, y_2 \in X_2$

$\qquad\qquad\qquad\quad d(x, y) = d_1(x_1, y_1) + d_2(x_2, y_2)$

$\qquad\qquad\qquad\qquad\quad = d_1(y_1, x_1) + d_2(y_2, x_2)$

$\qquad\qquad\qquad\qquad\quad = d_1(y, x)\ \forall\, x, y \in X.$

(4) Let $\quad x=(x_1, x_2), y=(y_1, y_2), z=(z_1, z_2)$

and $\qquad\qquad X=X_1 \times X_2, x_1, y_1, z_1 \in X_1, x_2, y_2, z_2 \in X_2$

since $\qquad\qquad d_1(x_1, y_1) \le d_1(x_1, z_1)+d_1(z_1, y_1)$

and $\qquad\qquad d_2(x_2, y_2) \le d_2(x_2, z_2)+d_2(z_2, y_2)$

Now $\quad d_1(x_1, y_1)+d_2(x_2, y_2) \le d_1(x_1, z_1)+d_1(z_1, y_1)+d_2(x_2, z_2)+d_2(z_2, y_2)$

$\qquad d_1(x_1, y_1)+d_2(x_2, y_2) \le d_1(x_1, z_1)+d_2(x_2, z_2)+d_1(z_1, y_1)+d_2(z_2, y_2)$

$\qquad\qquad d(x, y) \le d(x, z)+d(z, y), \forall x, y, z \in X.$

Hence, d is metric space.

Example 12. *Let* R *be the set of all real number and let* R^2 *the set of all ordered pairs of real number. Then the function*

$$d : R^2 \times R^2 \to R$$

where d is defined as $d(x, y)=\left[(x_1-y_1)^2+(x_2-y_2)^2\right]^{1/2}$, *is a metric on* R.

Solution. Here we shall show that d is a metric space defined as

$$d : R^2 \times R^2 \to R$$

such that $\qquad d(x, y)=\left[(x_1-y_1)^2+(x_2-y_2)^2\right]^{1/2}$

(1) Since $\qquad (x_1-y_1)^2 \ge 0, (x_2-y_2)^2 \ge 0$

$\Rightarrow \qquad\qquad (x_1-y_1)^2+(x_2-y_2)^2 \ge 0$

$\Rightarrow \qquad \left[(x_1-y_1)^2+(x_2-y_2)^2\right]^{1/2} \ge 0$

$\Rightarrow \qquad\qquad d(x, y) \ge 0, \forall x, y \in R^2.$

(2) Let $\quad d(x, y) = 0 \Leftrightarrow \left[(x_1-y_1)^2+(x_2-y_2)^2\right]^{1/2}=0$

$\qquad\qquad \Leftrightarrow (x_1-y_1)^2+(x_2-y_2)^2=0$

$\qquad\qquad \Leftrightarrow (x_1-y_1)^2=0$ and $(x_2-y_2)^2=0$

$\qquad\qquad \Leftrightarrow x_1-y_1=0$ and $x_2-y_2=0$

$\qquad\qquad \Leftrightarrow x_1=y_1$ and $x_2=y_2 \Leftrightarrow x=y$

Hence $\qquad d(x, y)=0 \Leftrightarrow x=y \ \forall x, y \in R$

(3) $d(x, y)=\left[(x_1-y_1)^2+(x_2-y_2)^2\right]^{1/2}$

$\qquad =\left[(y_1-x_1)^2+(y_2-x_2)^2\right]^{1/2}=d(y, x)$

(4) $d(x, y)=\left[(x_1-y_1)^2+(x_2-y_2)^2\right]^{1/2}$

$\qquad =\left[\{(x_1-z_1)+(z_1-y_1)\}^2+\{(x_2-z_2)+(z_2-y_2)\}^2\right]^{1/2}$

$\qquad \le \left[(x_1-z_1)^2+(x_2-z_2)^2\right]^{1/2}+\left[(z_1-y_1)^2+(z_2-y_2)^2\right]^{1/2}.$

$\qquad = \quad d(x, z)+d(z, y)$

$\left[\because \text{By Minkowski's inequality, we have } \left[\sum_{i=1}^{n}(x_i+y_i)^p\right]^{1/p} \le \left(\sum_{i=1}^{n}x_i^p\right)^{1/p}+\left[\sum_{i=1}^{n}y_i^p\right]^{1/p}\right.$

$\therefore \qquad\qquad d(x, y) \le d(x, z)+d(z, y), \forall x, y, z \in R^2.$

Hence, we conclude that d is a metric on R^2.

EXERCISE 15.1

1. Give an example of a pseudo-metric which is not metric. Is every metric a pseudo metric?

2. Let (X, d) be a metric space and let x, y, z be any three points of X, then show that
$$d(x, y) \geq |d(x, z) - d(z, y)|$$

3. Let (X, d) be a metric space and let $x, x', y, y' \in X$. Then show that
$$|d(x, y) - d(x', y')| \leq d(x, x') + d(y, y').$$

4. Let A, B be subsets of a metric space (X, d), then show that
$$\delta(A \cup B) \leq \delta(A) + \delta(B) + d(A, B).$$

5. Define the diameter of subset A of the metric space X. What is the diameter of empty set? What do you mean by the distance between two non-empty subset A and B of metric space X. If x is a point of X and A is a subset of X, write the distance of x from A.

6. Let R[0, 1] denote the classes of all Reimann integrable function f from [0, 1] into R. Consider the mapping $d : R[0, 1] \times R[0, 1] \to$ R defined by
$$d(f,g) = \int_0^1 |f - g|(x)\,dx = \int_0^1 |f(x) - g(x)|\,dx$$
Show that d is a pseudo-metric but not metric on R.

7. Show that $d : R^2 \times R^2 \to R$ defined by
$$d(x, y) = |x_1 - y_1| + |x_2 - y_2|,$$
$$x = (x_1, x_2), y = (y_1, y_2)$$
where $x \in R^2$ is a metric on R^2.

8. Let $X = R^n$ denote the set of all ordered n-tuples of real numbers for a fixed $n \in N$. Let
$$x = (x_1, x_2, \ldots, x_n), y = (y_1, y_2, \ldots, y_n).$$
Define the mapping d_1, d_2 and d_3 of $R^n \times R^n$ into R by

(i) $d_1(x, y) = \left[\sum_{i=1}^{n} (x_i - y_i)^2 \right]^{1/2}.$

(ii) $d_2(x, y) = \sum_{i=1}^{n} |x_i - y_i|.$

(iii) $d_3(x,y) = \max.\{|x_1 - y_1|, |x_2 - y_2|, \ldots, |x_n - y_n|\}.$
Show that d_1, d_2, d_3 are metrics on R^n.

9. Show that the set C of all complex numbers is a metric space under

$$d(z_1, z_2) = \frac{|z_1 - z_2|}{\left[\left(1 + |z_1|^2\right)\left(1 + |z_2|^2\right) \right]^{1/2}}.$$

10. If (X, d) be a metric space, then show that d_1, defined by
$$d_1(x, y) = \frac{d(x, y)}{1 + d(x, y)}$$
is also a metric on X.

11. Show that sum of two metric spaces is again a metric space.

12. If $d(x, y) = 2|x - y|$. Then show that d is a metric on R.

13. Let (X, d) be a metric space and let A, B be subsets of X, then show that
$$A \subset B \Rightarrow \delta(A) \leq \delta(B).$$

14. Let $d(x, y) = \min \{2, |x - y|\}$. Show that d is a metric on R.

15. Let C[0, 1] denote the collection of all real valued bounded continuous functions defined on the closed interval [0, 1], we define the norm of $f \in C$ [0, 1] by
$$\|f\| = \sup\{|f(x)| : x \in [0, 1]\}$$
where d is defined as
$$d(f, g) = \|f - g\| = \sup\{|f(x) - g(x)| : x \in [0, 1]\}$$
then show that d is a metric for C[0, 1].

16. Let $x = \langle x_n \rangle$ be a sequence then $F = \{x : x = \langle x_n \rangle\}$ is said to be Frechet space, for which d is defined as
$$d(x, y) = \sum_{n=1}^{\infty} \frac{|x_n - y_n|}{2^n (1 + |x_n - y_n|)}, (x, y \in F)$$
show that d is metric on F.

17. Let R be the set of real numbers and let
$$d(x, y) = \frac{|x - y|}{1 + |x - y|}, \forall x, y \in R$$
show that d is a metric for R.

18. Let (X, d) be any metric space then show that function d^* defined by
$$d^*(x, y) = \min\{2, d(x, y)\}, \forall x, y \in X$$
is a metric on X.

HINTS TO SELECTED PROBLEMS

1. $d(x, y) = |x^2 - y^2|$ $\forall$ $x, y \in R$.

12. Since, we know that $d(x, y) = |x-y|$ is a metric. Use this result to prove $d(x, y) = 2|x-y|$ is a metric.

13. By definition of the diameter of a set, we have

$$d(A) = \sup\{d(x, y): x, y \in A\}$$

Since $A \subset B$

$$x, y \in A \Rightarrow x, y \in B.$$

$$\therefore \quad \{d(x, y) : x, y \in A\} \Rightarrow \{d(x, y) : x, y \in B)$$

$$\Rightarrow \sup\{d(x, y) : x, y \in A\}$$

$$\leq \sup|d(x, y) : x, y \in B\}$$

$$(A \subset B)$$

$$\delta(A) \leq \delta(B).$$

17. Do same as example 4.

18. Do same as example 5.

Answers

4. $d(x, y) = |x^2 - y^2|$, Yes

15.5 OPEN AND CLOSED SETS IN A METRIC SPACE

Definition. *Let (X, d) be a metric space. Let $p \in X$ and $r > 0$ be given. Then, the set of all points $x \in X$ such that $d(p, x) < r$ is called the open ball or open sphere of radius r and centre p in X and is denoted by $S(p, r)$ or $d(p, r)$*

i.e., $\qquad S(P, r) = \{x \in X: d(x, p) < r\}.$

The set of all points $x \in X$ such that $d(p, x) \leq r$ is called closed ball of radius r and centre p and is denoted by $S^*(p, r)$

$$S^*(P, r) = \{x \in X: d(x, p) \leq r\}.$$

☛ **ILLUSTRATIONS**

In the case of real line :

(1) $S(p, r)$ is the open interval $]p-r, p+r[$.

(2) The open interval $]a, b[$ is the open ball with centre at the points $p = \frac{1}{2}(a+b)$ and radius $r = \frac{1}{2}(b-a)$.

(3) $S(p, r)$ is the closed interval $[p-r, p+r]$.

(4) The closed interval $[a, b]$ is the closed ball $S^*(p, r)$ with centre $p = \frac{1}{2}(a+b)$ and radius $r = \frac{1}{2}(b-a)$.

REMARKS

- Open sphere is defined as spherical neighbourhood of a point p.
- An open (or closed) sphere is always non-empty, since it contains its centre at least.

15.5.1 OPEN SET

Definition. *Let (X, d) be a metric space. A subset A of X is said to be open iff to each $x \in A$, there exist $r > 0$ such that $S(x, r) \subseteq A$.*

For Example :

(1) The subset $]0, 3[$ is open in $X = [0, 3]$ under the metric d given by $d(x, y) = |x-y|$.

Since d is open set in X.

15.5.2 LIMIT POINT

Let (X, d) be a metric space and $A \subset X$. A point $x \in X$ is called a limit point or limiting point or accumulation point or cluster point if every open sphere centered at x contains a point of A, other than x

i.e., $x \in X$ is called limit point of A if $[S(x, r)-\{x\}] \cap A \neq \phi, r \in R^+$.

15.5.3 ADHERENT POINT

Let (X, d) be a metric space and A be any subset of X. A point $x \in X$ is said to be adherent point of A if every open sphere centered at x contains at least one point of A.

i.e. if $\{S(x, r)-\{x\}\} \cap A$ is empty or non-empty for every open sphere $S(x, r)$.

There are two type of adherent points :

(a) *Limit point.* A point $x \in X$ is called limiting point of A if $[S(x, r)-\{x\}] \cap A \neq \phi \ \forall \ r \in R^+$.

(b) *Isolated point.* A point $x \in X$ is called isolated point of A if
$$[S(x, r)-\{x\}] \cap A = \phi \ \forall \ r \in R^+.$$

i.e. if x is not the limit point of A.

15.5.4 DERIVED SET

Set of all limit points of a set A is called derived set of A and is denoted by $D(A)$.

15.5.5 CLOSED SET

Let (X, d) be a metric space and $A \subset X$. Then A is called closed set if the derived set of A contains in A i.e. $D(A) \subset A$. (or) if every limit point of A belongs to the set itself.

15.5.6 PERFECT SET

A closed set which has no isolated point is called perfect set.

15.5.7 DIFFERENCE BETWEEN LIMIT AND LIMITING POINT

The limit and limit point both are different terms. For example, consider the sequence $\langle 1,1,1,... \rangle$ in which every element is identically equal to 1. The limit of the sequence is 1, but the set of all points of this sequence is the singleton set [1] and hence its limit point does not exists, because finite set has no limit point.

Here, it is also possible that the set of points of a sequence may have a limit point, but cannot have a limit.

THEOREM 1. *In a metric space (X, d), the empty set ϕ and the whole space X, are open sets.*

Proof. Let (X, d) be a metric space. To show that ϕ and X are open sets.

To prove that ϕ is open in X, it is suffices to show that
$$x \in \phi \Rightarrow \exists \ \varepsilon > 0: S(x, \varepsilon) \subset \phi.$$

Since ϕ does not contain any element and hence this condition is automatically fulfilled.

Now to show X is open.

Since, corresponding to every point $x \in X \ \exists$ an open sphere with its centre at x which is contained in X. Hence, X is an open set.

THEOREM 2. *In a metric space (X, d), ϕ and X are closed sets.*

Proof. Let (X, d) be a metric space. To show ϕ and X are closed sets.

We know that $D(\phi) = \phi$.

Therefore $D(\phi) \subset \phi$.

$\Rightarrow$ ϕ contain all its limit points.

$\Rightarrow$ ϕ is closed.

Now to show X is closed.

Since, all the limit points of X belongs to X.   ($\because$ X is the universal set)

i.e. $\qquad\qquad\qquad\qquad x \in D(X) \Rightarrow x \in X.$

$\Rightarrow$ X contains all its limit points.

$\Rightarrow$ X is a closed set.

THEOREM 3. *In a metric space, every open sphere is an open set.*

Proof. Let (X, d) be a metric space and let $S(p, r)$ be any open sphere.

To show that to each point $x_0 \in S(p, r)$ there exists an open sphere centred at x_0 and contained in $S(p, r)$.

Let $x_0 \in S(p, r)$ be arbitrary.

Now, $\qquad\qquad x_0 \in S(p, r) \Rightarrow d(x_0, p) < r \Rightarrow r - d(x_0, p) > 0.$

Let us define $\qquad\qquad \rho = r - d(x_0, p).$ $\qquad$ Then clearly $\rho > 0$.

Define an open sphere $S(x_0, \rho)$.

To show $\qquad\qquad S(x_0, \rho) \subset S(p, r).$

Let $x \in S(x_0, \rho)$. Then by definition of open sphere, we have

$$d(x, x_0) < \rho = r - d(x_0, p).$$

Now, $\qquad\qquad d(x, p) < d(x, x_0) + d(x_0, p)$ $\qquad$ (By Triangle inequality)

$$< r - d(x_0, p) + d(x_0, p) = r$$

$\Rightarrow$ $\qquad\qquad d(x, p) < r \Rightarrow x \in S(p, r).$

Since x is arbitrary.

Hence, $\qquad\qquad S(x_0, \rho) \subset S(p, r)$ $\qquad \Rightarrow \qquad S(p, r)$ is an open set.

THEOREM 4. *In a metric space (X, d), arbitrary union of open sets is open.*

Proof. Let (X, d) be a metric space and let $[G_\lambda : \lambda \in \Lambda]$ be an arbitrary collection of open subset of X.

Define $\qquad\qquad G = \cup [G_\lambda : \lambda \in \Lambda].$

To show G is open.

Let $x \in G$ $\qquad \Rightarrow \qquad x \in G_\lambda$ for some λ.

Since G_λ is open, therefore there exist $r > 0$ such that

$$S(x, r) \subset G_\lambda. \qquad\qquad\qquad\qquad ...(1)$$

By definition of G, we have

$$G_\lambda \subset G. \qquad\qquad\qquad\qquad ...(2)$$

From (1) and (2), we have

$$S(x, r) \subset G_\lambda \subset G$$

Since x is arbitrary therefore, we can say that to each $x \in G$, there exists a number $r > 0$ such that $\qquad S(x, r) \subset G.$

$\Rightarrow$ G is open.

Hence, the arbitrary union of open sets in a metric space, is open.

THEOREM 5. *In a metric space* (X, d), *the intersection of a finite number of open sets is open.*

Proof. Let (X, d) be a metric space and G_i $(i=1, 2, ..., n)$ be a finite collection of open subsets of X.

Define $\qquad\qquad\qquad H = \cap[G_i : i = 1, 2, ..., n]$.

To show H is open.

Let $x \in H$. Then by definition of H, $x \in G_i$ for each $i = 1, 2, ..., n$.

Also each G_i is open, therefore there exist $r_i > 0$ such that

$$S(x, r_i) \subset G_i, \text{ for each } i.$$

Let $\quad r = \min\{r_1, r_2, ..., r_n\}$

Then $\qquad\qquad S(x, r) \subset S(x, r_i) \subset G_i, \forall\, i = 1, 2, ..., n$

$\Rightarrow \qquad\qquad S(x, r) \subset G_i, \qquad\qquad \forall\, i = 1, 2, ..., n$

$\Rightarrow \qquad\qquad S(x, r) \subset \cap [G_i : i = 1, 2, ..., n]$

Since x is arbitrary, therefore it is shown that to each $x \in H$, there exist $r > 0$, such that $S(x, r) \subset H$. Hence H is open.

THEOREM 6. *In a metric space* (X, d), *for every pair of distinct points* $x, y \in X$, *there exist disjoint open sets* U *and* V *such that* $x \in U$ *and* $y \in V$.

Proof. Let (X, d) be a metric space and x, y be two distinct point of X such that $x \neq y$.

Then clearly $\qquad d(x, y) > 0$.

Define $\varepsilon = \dfrac{1}{3} d(x, y)$, then clearly $\varepsilon > 0$.

Also, define $U = S(x, \varepsilon)$ and $V = S(y, \varepsilon)$, then $x \in U$ and $y \in V$.

Since every open sphere is an open set, therefore U and V both are open sets.

Now to show U and V are distinct i.e. $U \cap V = \phi$.

Let if possible $U \cap V \neq \phi$ and $p \in U \cap V$.

Therefore, $\qquad p \in U \cap V \Rightarrow p \in U$ and $p \in V$.

Now $\qquad\qquad p \in U \quad \Rightarrow d(p, x) < \varepsilon \qquad\qquad (\because U \text{ is an open sphere})$

and $\qquad\qquad p \in V \quad \Rightarrow d(p, y) < \varepsilon. \qquad\qquad (\because V \text{ is an open sphere})$

By triangle inequality, we have

$$d(x, y) \leq d(x, p) + d(p, y)$$
$$= d(p, x) + d(p, y) \qquad\qquad [\because d(p, x) = d(x, p)]$$
$$< \varepsilon + \varepsilon = 2\varepsilon.$$

$\Rightarrow \qquad\qquad d(x, y) < 2\varepsilon$

$\Rightarrow \qquad\qquad d(x, y) < 2 . \dfrac{1}{3} d(x, y)$

$\Rightarrow \qquad\qquad d(x, y) < \dfrac{2}{3} d(x, y)$ which is absurd.

Therefore, we have a contradiction $\qquad$ Hence, $\qquad\qquad U \cap V = \phi$

Hence, we can find two disjoint open sets U and V such that $x \in U$ and $y \in V$.

THEOREM 7. *A subset of a metric space is open if and only if it is the union of a family of open spheres.*

Proof.   Let (X, d) be a metric space and A be any subset of X.

Let us first suppose A is open. To show A can be written as the union of family of open spheres.

If $A = \phi$, then it can be written as the union of empty family of open sphere.

If $A \neq \phi$, let $x_1 \in A$. Now, since A is open, therefore, there exists an open sphere $S(x_1, r)$ such that $S(x_1, r) \subset A$

Similarly for $x_2 \in A$ There exists an open sphere $S(x_2, r)$ such that $S(x_2, r) \subset A$.

Proceeding in the same way, we can say that for each point x_i of A there exists an open sphere $S(x_i, r)$ such that $S(x_i, r) \subset A$

Therefore, $A \subset \cup \, [S(x_i, r) : x_i \in A] \subset A$

$\Rightarrow$ $A = \cup \, [S(x_i, r) : x_i \in A]$.

Hence A can be written as the union of a family of open spheres.

Conversely, let A can be written as the union of open spheres.

To show A is open set.

Since, we know that every open sphere in a metric space is an open set. Therefore $S(x_i, r)$ is an open set. Also, finite union of open sets is again open.

Since $A = \cup \, [S(x_i, r) : x_i \in A]$.

Also the right hand side is the union of open sets. Therefore, A is open.

THEOREM 8. *Every non-empty open set on the real line is the union of a countable collection of pairwise disjoint open intervals.*

Proof.   Let G be an open subset of R. Let $x \in G$.

Since G is open, there exists an open interval $S(x, r)$, centered at x such that

$$S(x, r) \subset G.$$

Let I_x be the union of all open intervals, which contain x and are contained in G. Then we have

(a) I_x is open interval containing x and contained in G.

(b) I_x contains each open interval which contains x and is contained in G.

(c)   If y is any other point in I_x, then $I_x = I_y$.

If x and y are two distinct points of G, then either $I_x = I_y$ or $I_x \cap I_y = \phi$.

If $z \in I_x \cap I_y$. Then we have $z \in I_x$ and $z \in I_y$

$\Rightarrow$ $I_x = I_z$ and $I_y = I_z$ $\Rightarrow$ $I_x = I_y$.

Now, let I be the collection of all distinct sets of the form I_x for points x belonging to G.

Then clearly, we have I is the collection of open intervals and G is the union of this collection.

Now, to show I is countable.

Let G_i be the set of all rational points in G. Then, obviously $G_i \neq \phi$.

Now, define a map f of G_i onto I such that for each $i \in G_i$, $f(i)$ be the unique interval in I to which i belongs.

$\Rightarrow$ G_i is countable $\qquad\qquad$ ($\because$ G_i is a non-empty subset of the countable set Q of all rational numbers)

$\Rightarrow$ I is countable.

Hence, we can say that every non-empty open set on the real line is the union of a countable collection of pairwise disjoint open intervals.

THEOREM 9. *Let (X, d) be a metric space and A, any subset of X. Then A is closed if and only if its complement (i.e. X–A) is open.*

Proof. Let (X, d) be a metric space. Let us first suppose A is closed. To show its complement $(X–A)$ is open.

Let $x \in X–A$, then $x \notin A$. Since A is closed and $x \notin A$.

$\Rightarrow$ x is not the limit point of A.

Then by definition of limit point $\exists$ an open sphere $S(x, r)$ such that

$$S(x, r) \cap A = \phi$$

$\Rightarrow$ $\qquad\qquad S(x, r) \subset X–A$ for some $r > 0$.

Since $x \in X–A$ is arbitrary, therefore each point of $X–A$ is the centre of some open sphere which is contained in $X–A$.

$\Rightarrow$ $X–A$ is open.

Conversely, let $X–A$ is open. Let $x \in X$ be any limit point of A

If $x \in A$, then A is closed $\qquad\qquad$ (By definition of closed set)

If $x \notin A$, then $x \in X–A$, also since $X–A$ is open therefore there exists an open sphere $S(x, r)$ which contained in $X–A$ i.e. $S(x, r) \subset X–A$

and $\qquad\qquad S(x, r) \cap A = \phi$ for some $r > 0$

$\Rightarrow$ x is not the limit point, which is a contradiction. Therefore, $x \in A$.

Hence, A is closed.

THEOREM 10. *In a metric space, every closed sphere is a closed set.*

Proof. Let (X, d) be a metric space. Consider a closed sphere $S(x_0, r)$ in X. To show $S(x_0, r)$ is a closed set. For this, we shall show that the complement $S'(x_0, r)$ of $S(x_0, r)$ is open.

Let $x \in S'(x_0, r)$. Then $x \notin S(x_0, r)$.

Then by definition of open sphere $d(x, x_0) > r$.

Define $\qquad\qquad \rho = d(x, x_0) - r$.

Obviously $\qquad\qquad \rho > 0$. $\qquad\qquad\qquad\qquad$ ...(1)

Now, we take an open sphere $S(x, \rho)$ of radius ρ centered at x.

To show $\qquad\qquad S(x, \rho) \subset S'(x_0, r)$.

Let $\qquad\qquad y \in S(x, \rho) \Rightarrow d(x, y) < \rho$. $\qquad\qquad\qquad$ ...(2)

Consider $\qquad\qquad d(x, x_0) \leq d(x, y) + d(y, x_0)$

or,
$$d(y, x_0) \geq d(x, x_0) - d(x, y)$$
$$> d(x, x_0) - r \qquad \text{[using (2)]}$$
$$= d(x, x_0) - [d(x, x_0) - r] \qquad \text{[using (1)]}$$
$$= r$$

$\Rightarrow \qquad d(y, x_0) > r$

$\Rightarrow \qquad y \in S'(x_0, r)$

$\Rightarrow \qquad S(x, p) \subset S'(x_0, r) \qquad (\because y \text{ is arbitrary})$

$\Rightarrow \qquad S'(x_0, r)$ is void

$\Rightarrow \qquad S'(x_0, r)$ is open

Hence, $S(x_0, r)$ is closed.

REMARK

- Let (X, d) be a metric space and S be any subset of X defined by
$$S = \{x \in X : d(x, x_0) = r\}$$
where $r > 0$ and $x \in X$. Then S is a closed set.

THEOREM 11. *In a metric space (X, d), the intersection of an arbitrary family of closed sets is closed.*

Proof. Let (X, d) be a metric space and let $(H_\lambda : \lambda \in \Lambda)$ be an arbitrary collection of closed subset of X. Then to show $\cap [H_\lambda : \lambda \in \Lambda]$ is also a closed set.

Since H_λ is closed for each $\lambda \in \Lambda$

$\Rightarrow \quad X - H_\lambda$ is open

$\Rightarrow \quad \cup (X - H_\lambda)$ is open $\qquad (\because$ Arbitrary union of open sets is open)

$\Rightarrow \quad X - \cap [H_\lambda : \lambda \in \Lambda]$ is open $\qquad$ (By De-morgan's law)

$\Rightarrow \quad \cap [H_\lambda : \lambda \in \Lambda]$ is closed $\qquad (\because$ Complement of an open set is closed)

Hence, the arbitrary intersection of closed sets is closed.

THEOREM 12. *In a metric space (X, d) the finite union of closed sets is closed.*

Proof. Let (X, d) be a metric space and $H_i : i = 1, 2, ..., n$ be closed subsets of X. To show $\bigcup_{i=1}^{n} H_i$ is closed.

Since each H_i is closed

$\Rightarrow \quad H_i'$ is open $\qquad (\because$ Complement of a closed set is open)

$\Rightarrow \quad \bigcap_{i=1}^{n} H_i'$ is open $\qquad (\because$ Finite intersection of open sets is open)

$\Rightarrow \quad \left[\bigcup_{i=1}^{n} H_i \right]'$ is open $\qquad$ (By De-morgan's law)

Hence, $\bigcup_{i=1}^{n} H_i$ is closed.

THEOREM 13. *In a discrete metric space, every set is open.*

Proof. Let A be any non-empty subset of a discrete metric space (X, d).

If $A = \phi \Rightarrow A$ is open.

If $A \neq \phi$ and $x \in A$, where x is any arbitrary point of A

and $\qquad S\left(x, \dfrac{1}{2}\right) = \{x\} \subseteq A$

$\Rightarrow \quad A$ is open.

Hence in a discrete metric space every set is open.

REMARKS

- The union of infinite number of closed sets may or may not be closed.

For Example: Let $\qquad F_x = \left[0, \dfrac{n}{n+1}\right]$ Then $\overset{n}{\underset{n=1}{\cup}} F_n = [0, 1[= $ semi open, which is not closed.

- The intersection of an infinite number of open sets is not necessarily open.

For Example : Consider an open interval collection $\left\{\left]-\dfrac{1}{n}, \dfrac{1}{n}\right[: n \in N\right\}$ in R with usual metric $d(x, y) = |x-y|$.

So the intersection of this type collection will be

$$\cap \left\{\left]-\dfrac{1}{n}, \dfrac{1}{n}\right[: n \in N\right\} = \{0\} \text{ which is not open.}$$

Hence in a metric space the intersection of an infinite collection of open sets is not open.

15.6 NEIGHBOURHOOD

Let (X, d) be a metric space. A set N of X is said to be neighbourhood (*nbd*) of a point $p \in X$ if there exists an $\varepsilon > 0$ such that $S(x, \varepsilon) \subset N$.

For Example :

(1) *On the real line* :

(i) The open interval $]a, b[$ is a nbd of each of its points

(ii) R is a nbd of each of its points.

(iii) The closed interval $[a, b]$ is a nbd of each of its points except the end points.

(iv) The set of integers Z is not a nbd of any of its points.

(v) The set of rational numbers Q is not a nbd of any of its points.

(2) Let (X, d) be a discrete metric space and $x \in X$. Then $\{x\}$ is a nbd of x. Further every superset of $\{x\}$ is also a nbd of x.

15.7 PROPERTIES OF NEIGHBOURHOOD

THEOREM 1. *Let (X, d) be a metric space and A be any subset of X. If N is the neighbourhood of A and $M \supset N$, then M is also a neighbourhood of A. i.e., 'every superset of a neighbourhood of A is also a neighbourhood of A ".*

Proof. Let (X, d) be a metric space and N be a nbd of $A \subset X$.

Since N is a nbd of A, therefore, by definition there exists an open set G such that

$\qquad \qquad A \subset G \subset N \qquad \qquad \qquad ...(1)$

Given that $\qquad \qquad N \subset M. \qquad \qquad \qquad ...(2)$

From (1) and (2), we conclude that

$\qquad \qquad A \subset G \subset M$

$\Rightarrow \quad M$ is a nbd of A.

THEOREM 2. *The intersection of a finite number of neighbourhood of A is also a neighbourhood of A.*

Proof.   Let (X, d) be a metric space and A is any subset of X.

Also let $N_1, N_2, ..., N_k$ are a finite number of neighbourhoods of A, then to show that
$$\cap \{N_i : i = 1, 2,..., k] \text{ is also a neighbourhood of } A$$

Since each N_i $(i = 1, 2,..., n)$ is a nbd of A, then by definition there exist open sets G_i $(i = 1, 2,..., k)$ such that
$$A \subset G_i \subset N_i \; ; i=1, 2,..., k.$$

Since $A \subset G_i$ for each $i = 1, 2,..., k$.

Therefore $A \subset \cap \{G_i : i=1, 2,..., k\}$

Also, $G_i \subset N_i \;\; \forall\, i = 1, 2, ..., k$

$\Rightarrow$ $\cap \{G_i : i = 1, 2, ..., k\} \subset \cap \{N_i : i = 1, 2,..., k\}$

$\Rightarrow$ $A \subset \cap \{G_i : i = 1, 2,..., k\}$
$$\subset \cap \{N_i : i = 1, 2, ..., k\}. \qquad\qquad ...(1)$$

Since each G_i is open, therefore $\cap \left[(G_i : i = 1, 2,..., k\right]$ is open.

$$[\because \text{Finite intersection of open sets is again open}]$$

Hence, from (1), we conclude that $\cap \{N_i : i = 1, 2,..., k\}$ is also a neighbourhood of A.

REMARK

- The intersection of an infinite number of neighbourhood of a set A is not necessarily the nbd of A. For example, in the metric space (R, d) where $d(x, y) = |x-y|$, (*i.e.* usual metric)

$$\left]-\frac{1}{n}, \frac{1}{n}\right[\text{ is a nbd of } A \text{ for each } n \in N \text{ but } \cap \left\{ \left]-\frac{1}{n}, \frac{1}{n}\right[: n \in N \right\} = \{0\}, \text{ which is not a nbd of } A.$$

THEOREM 3. *In a metric space, every open sphere is a neighbourhood of each of its points.*

Proof.   Let (X, d) be a metric space and let $S(x_0, r)$ be an open sphere centred at x_0, and of radius r; Let $y \in S(x_0, r)$.

Now to show that $S(x_0, r)$ is a neighbourhood of y. We have to show that there exists an open sphere centred at p, which is contained in $S(x_0, r)$.

Now we have
$$y \in S(x_0, r) \Rightarrow d(y, x_0) < r$$

$\Rightarrow$ $r - d(y, x_0) > 0.$

Let $\varepsilon = r - d(y, x_0)$ where $\varepsilon > 0.$

Now we have to show that $S(y, \varepsilon)$ contained in $S(x_0, r)$. Let $x \in S(y, \varepsilon)$

$\Rightarrow$ $d(x, y) < \varepsilon \Rightarrow d(x, y) < r - d(y, x_0)$
$$d(x, x_0) \le d(x, y) + d(y, x_0) < r - d(y, x_0) + d(y, x_0) < r$$

$\Rightarrow$ $d(x, x_0) < r \Rightarrow x \in S(x_0, r)$

$\Rightarrow$ $S(y, \varepsilon) \subseteq S(x_0, r)$ $(\because x \text{ is arbitrary})$

$\Rightarrow$ $S(x_0, r)$ is a neighbourhood of y

Since, y is an arbitrary point of $S(x_0, r)$, therefore, $S(x_0, r)$ is a neighbourhood of each of its points.

THEOREM 4. *A subset in a metric space is open if and only if it is a neighbourhood of each of its points.*

Proof. Let (X, d) be a metric space and A is any subset of X i.e, $A \subset X$.

Let us first suppose A is open. To show it is a nbd of each of its points.

Let x be any arbitrary point of A i.e. $x \in A$.

Since A is open, we can write $x \in A \subset A$ $\Rightarrow$ A is a nbd of x.

Conversely, let A is a nbd of each of its points. To show A is open.

Let $x \in A$, and A is a nbd of x. Then by definition of nbd, there exists an open set G_x such that

$$x \in G_x \subset A$$

Let $G = \cup \{G_x : x \in A\}$.

We claim that $G = A$

If $x \in A$ then by definition $x \in \cup \{G_x : x \in A\} = G$

$\Rightarrow$ $A \subset G.$ ...(1)

Now, if $y \in G$, then $y \in G_x$ for some $x \in A$. But $G_x \subset A$ and hence $y \in A$

$\Rightarrow$ $G \subset A.$ ...(2)

From (1) and (2) we conclude that $A = G$.

Now since G is open   (Being the union of collection of open sets)

Therefore, A is open.

THEOREM 5. *Let (X, d) be a metric space and let $x \in X$. If $\{N_i : i = 1, 2, ..., k\}$ are finite number of neighbourhood of X, then $\cap \{N_i : i = 1, 2, ..., k\}$ is also a nbd of x.*

Proof. Given that N_i $(i = 1, 2, ..., k)$ is a nbd of x for each i.

Therefore by definition of neighbourhood there exist open sets G_i $(i = 1, 2, ..., k)$ such that

$$x \in G_i \subset N_i : i = 1, 2, ..., k$$

$\Rightarrow$ $x \in \cap \{G_i : i = 1, 2, ..., n\} \subset \cap \{N_i : i = 1, 2, ..., k\}.$ ...(1)

Since, each G_i is open, therefore $\cap \{G_i : i = 1, 2, ..., n\}$ is open

                              ($\because$ Finite intersection of open sets is open)

Hence, from (1) we conclude that $\cap [N_i : i = 1, 2, ..., k]$ is a nbd of x.

THEOREM 6. *Let (X, d) be a metric space and A be a subset of X. A point $x \in X$ is a limit point of A if and only if every open sphere $S(x, r)$ centered at x contains infinitely many points of A.*

Proof. Let us first suppose every open sphere $S(x, r)$ centered at x contains infinitely many points of A. Then clearly, we can say that every open sphere $S(x, r)$ centered at x contains at least one point of A other than x. Therefore x is the limit point of A.

Conversely, let x is the limit point of A. To show every open sphere $S(x, r)$ centered at x contains infinitely many points of A.

Let if possible there exists a sphere $S(x, r)$ which contains only a finite number of points of A. Let $x_1, x_2, ..., x_n$ be those points of $S(x, r) \cap A$ which are distinct from x.

Define $r_1 = \min\{d(x, x_m): 1 \le m \le n\}$

Then clearly $r_1 > 0$.

Then open sphere $S(x, r_1)$ contains no points of A distinct from x.

$\Rightarrow$ x is not the limit point of A; which is a contradiction.

Hence, every open sphere centered at x must contain infinitely many points of A.

THEOREM 7. *Let (X, d) be a metric space and $A \subset X$. A point $x \in X$ is an adherent point of A if and only if $d(x, A) = 0$.*

Proof. Let us first suppose $d(x, A) = 0$

Then $d(x, A) = \inf. \{d(x, y): y \in A\} = 0$.

Let $S(x, r)$ be any sphere centered at x.

By definition of infimum, there exists a point $y_0 \in A$ such that $0 \le d(x, y_0) \le r$

$\Rightarrow$ $y_0 \in S(x, r)$

$\Rightarrow$ x is an adherent point of A.

Conversely, let us suppose x be an adherent point of A. To show $d(x, A) = 0$.

Since x is an adherent point of A therefore, every open sphere centered at x must contain a point of A.

$\Rightarrow$ to each $r > 0$ there exists a $y \in A$ such that $0 \le d(x, y) < r$.

Since r is arbitrary, letting $r \to 0$ (By taking r very small)

$\Rightarrow$ $\inf \{d(x, y): y \in A\} = 0$

$\Rightarrow$ $d(x, A) = 0$.

THEOREM 8. *A subset A of a metric space X is closed if and only if $D(A) \subset A$ i.e. iff A contains all its limit points.*

Proof. Let (X, d) be a metric space and $A \subset X$.

Let us first suppose A is closed. To show A contains all its limit point.

Since A is closed $\Rightarrow A'$ is open.

Let $x \in A'$, since A' is open, therefore by definition of open set there exists a nbd N of x such that $N \subset A$.

Now, since $A \cap A' = \phi$. $\Rightarrow$ N contains no point of A.

$\Rightarrow$ x is not a limit point of A.

$\Rightarrow$ No point of A' can be a limit point of A.

$\Rightarrow$ A contains all its limit points.

$\Rightarrow$ $D(A) \subseteq A$

Conversely, let A contains all its limit point *i.e.* $D(A) \subset A$. To show A is closed.

For this, we shall show that A' is open.

Let $x \in A'$, then $x \notin A$. Since $D(A) \subset A$ and $x \notin A$ therefore $x \notin D(A)$

$\Rightarrow$ x is not the limit point of A

$\Rightarrow$ there exists a nbd N of x such that $N \cap A = \phi$

$\Rightarrow$ $N \not\subset A$

$\Rightarrow$ $\qquad\qquad\qquad\qquad$ $N \subset A'$

$\Rightarrow$ $\quad$ A' contains a neighbourhood of x.

Since x is arbitrary, therefore, we can say A' is a nbd of each of its points

$\Rightarrow$ $\quad$ A' is open.

Hence, A is closed.

THEOREM 9. *Let A be any subset of a metric space (X, d) then derived set $D(A)$ of A is a closed set.*

Proof. $\quad$ To show $D(A)$ is closed, we show that $D(A)$ contains all its limit point.

Let x be a limit point of $D(A)$. Then for all $r>0$, the open sphere $S(x, r)$ contains infinitely many points of $D(A)$.

We know that each point of $D(A)$ is a limit point of A

$\Rightarrow$ $\quad$ every open sphere $S(x, r)$ must contain infinitely many points of A

$\Rightarrow$ $\quad$ x is a limit point of A.

$\Rightarrow$ $\qquad\qquad\qquad\qquad$ $x \in D(A)$

$\Rightarrow$ $\quad$ $D(A)$ contains all its limit points and so $D(A)$ is closed.

THEOREM 10. *Let A and B be subset of a metric space X. Then*

$\quad$ *(i)* $A \subset B \Rightarrow D(A) \subset D(B)$ *(ii)* $D(A \cap B) \subset D(A) \cap D(B)$

$\quad$ *(iii)* $D(A \cup B) = D(A) \cup D(B)$.

Proof. $\quad$ (i) $\quad A \subset B$. To show $D(A) \subset D(B)$.

$\qquad$ Let $\qquad\qquad\qquad$ $x \in D(A)$

$\qquad$ $\Rightarrow$ $\quad$ x is the limit point of A

$\qquad$ $\Rightarrow$ $\quad$ every nbd of x contains at least one point of A, other than x

$\qquad$ $\Rightarrow$ $\quad$ every nbd of x contains at least one point of B, other than $x (\because A \subset B)$

$\qquad$ $\Rightarrow$ $\quad$ x is the limit point of B

$\qquad$ $\Rightarrow$ $\qquad\qquad\qquad$ $x \in D(B)$

$\qquad$ Since x is arbitrary, therefore $D(A) \subset D(B)$.

$\quad$ (ii) $\quad$ To show $D(A \cap B) \subset D(A) \cap D(B)$.

$\qquad$ Since we know that $A \cap B \subset A \Rightarrow D(A \cap B) \subset D(A)$ $\qquad\qquad$ ...(1)

$\qquad\qquad\qquad\qquad\qquad\qquad\qquad\qquad\qquad\qquad\qquad$ [Using (1)]

$\qquad$ and $\qquad\qquad$ $A \cap B \subset B \Rightarrow D(A \cap B) \subset D(A)$

$\qquad$ From (1) and (2), we conclude that $D(A \cap B) \subset D(A) \cap D(B)$.

$\quad$ (iii) $\quad$ To show $\quad D(A \cup B) = D(A) \cup D(B)$

$\qquad$ Let $\qquad\qquad\qquad$ $x \notin D(A) \cup D(B)$

$\qquad$ $\Rightarrow$ $\quad$ $x \notin D(A)$ and $x \notin D(B)$

$\qquad$ $\Rightarrow$ $\quad$ x is not the limit point of A and x is not the limit point of B

$\qquad$ $\Rightarrow$ $\quad$ $x \notin D(A \cup B)$

$\qquad$ Since x is arbitrary, therefore

$\qquad\qquad\qquad$ $D(A \cup B) \subset D(A) \cup D(B)$. $\qquad\qquad$ ...(1)

$\qquad$ Also $\qquad\qquad$ $A \subset A \cup B \Rightarrow D(A) \subset D(A \cup B)$ $\qquad\qquad$ ...(2)

$\qquad$ and $\qquad\qquad$ $B \subset A \cup B \Rightarrow D(B) \subset D(A \cup B)$ $\qquad\qquad$ ...(3)

Now, (2) and (3) gives

$$D(A) \cup D(B) \subset D(A \cup B). \qquad \qquad ...(4)$$

From (1) and (4) we conclude that

$$D(A) \cup D(B) = D(A \cup B).$$

15.7.1 EQUIVALENT METRICS

Two metrices d and d^* on the same set X are said to be equivalent iff every d-open set is d^*-open and every d^*-open set is d-open.

Example. *Let (X, d) be a metric space and d^* a mapping such that $d^*: X \times X \to R$*

defined by $\qquad d^*(x, y) = \dfrac{Md(x,y)}{1+d(x,y)}, M > 0$

is also a metric for X. Also show that d and d^* are equivalent.

Solution. We have already proved that $d^*(x, y) = \dfrac{Md(x,x_1)}{1+d(x,x_1)}$ is metric for X as it follows the four properties of metric.

Now it remains only to show that d and d^* are equivalent.

For this we shows that d-open sphere centred at $x \in X$ contains a d^* open sphere centred at x and vice-versa.

Let $S(x, r), r>0$ be any d open sphere centred at $x \in X$.

Let $S(x, \rho), \rho>0$ be any d^* open sphere centred at $x \in X$ where $\rho = \dfrac{Mr}{1+r}$.

Now we have to show that $S^*(x, \rho) \subseteq S(x, r)$.

Here let $x_1 \in S^*(x, \rho) \qquad \Rightarrow \quad d^*(x, x_1) < \rho$

$\Rightarrow \qquad \dfrac{Md(x,x_1)}{1+d(x,x_1)} < \dfrac{Mr}{1+r}$

$\Rightarrow \qquad d(x, x_1) + rd(x, x_1) < r + rd(x, x_1)$

$\Rightarrow \qquad d(x, x_1) < r \Rightarrow x_1 \in S(x, r)$

$\Rightarrow \qquad S^*(x, \rho) \subseteq S(x, r)$

Now it remains to show that $S(x, r) \subseteq S^*(x, \rho)$.

Here $\qquad r = \dfrac{\rho}{M-\rho}$.

Let $\qquad x_1 \in S(x, r) \Rightarrow d(x, x_1) < r$

$\Rightarrow \qquad \dfrac{d^*(x,x_1)}{M-d^*(x,x_1)} < \dfrac{\rho}{M-\rho}$.

$\Rightarrow \quad Md^*(x, x_1) - \rho d^*(x, x_1) < \rho M - \rho d^*(x, x_1)$

$\Rightarrow \qquad d^*(x, x_1) < \rho \Rightarrow x_1 \in S^*(x, \rho)$

$\Rightarrow \qquad S(x, r) \subseteq S^*(x, \rho)$

Hence d and d^* are equivalent metric.

Solved Examples

Example 1. *Find the closed and open spheres for the usual metric for R.*

Based on the following Results

Solution. We know that the usual metric for R is defined by

$$d(x, y) = |x-y|$$

Let $x_0 \in R$. Then the open sphere $S(x_0, r)$, centered at x_0, with radius r is given by

$$S(x_0, r) = \{x \in R : |x-x_0| < r\}$$
$$= \{x \in R : x_0-r < x < x_0+r\}$$
$$=]x_0-r, x_0+r[.$$

Hence, the open spheres on the real line are open intervals.

Similarly, the closed sphere with centre x_0 and radius r is the closed interval $[x_0-r, x_0+r]$.

- A subset A of a metric space (X, d) is said to be open if and only if to each, $x \in A \; \exists \; r > 0$ such that
$$S(x, r) \subseteq A.$$

- For x to be limit point of A : $\{S(x, r) - \{x\}\} \cap A \neq \phi$

- For x to be an isolated point of A : $\{S(x, r) - \{x\}\} \cap A = \phi$

- If $D(A) \subset A$, then A is closed.

- A set N of a metric space $[X \, d]$ is said to be nbd of a point $p \in X$ if there exist an $\varepsilon > 0$ such that $S(x, \varepsilon) \subset N$.

- $A \subset B \Rightarrow D(A) \subset D(B)$

- $D(A \cap B) \subset D(A) \cap D(B)$

- $D(A \cup B) = D(A) \cup D(B)$

Example 2. *Consider, the usual metric $d(x, y) = |x-y|$ for $[0, 1]$*

describe $\qquad S\left(\left|\frac{1}{2}, 1\right|\right)$ *and* $S\left[\frac{1}{2}, 1\right]$

Solution. We have $S\left(\left|\frac{1}{2}, 1\right|\right) = \left\{x \in [0,1] : \left|x - \frac{1}{2}\right| < 1\right\} = \left\{x \in [0,1] : \frac{1}{2} - 1 < x < \frac{1}{2} + 1\right\}$

$$= \left[x \in [0,1] : -\frac{1}{2} < x < \frac{3}{2}\right]$$
$$= [0, 1] \qquad \qquad \text{[Don't go outside the given reason [0, 1]]}$$

Similarly, $\qquad S\left[\frac{1}{2}, 1\right] = [0, 1].$

Example 3. *Let R be the set of all real numbers with usual metric $d(x, y) = |x-y|$. Find whether or not the given sets are open*

    (i) $A = [0, 1[$ (ii) $B =]0, 1[$ (iii) $C =]0, 1]$

    (iv) $D = [0, 1]$ (v) $E = \{1\}$ (vi) $F = \{1, 2, 3\}$.

Solution. Since, we know that an open sphere about $x_0 \in R$ and 'r' is the radius of open interval or sphere $]x_0-r, x_0+r[$. To show whether a set is open or not we check that for each point of given set an open interval of type described above exists or not and also contained in the given set.

    (i) Here $A = [0, 1[$. Let us choose a positive number r. So the open interval $]0-r, 0+r[=]-r, r[\notin A$.

        $\Rightarrow$ No open sphere with centred '0' contained in A

        $\Rightarrow$ A is not open.

(ii) Let us take a point x of set $B=]0, 1[$ and let $r=\min\{x-0, 1-x\}$ so it is obvious that $]x-r, x+r[\subseteq B$

$\Rightarrow$ B is an open set.

(iii) Here $C=]0, 1]$. Let us choose a positive number r.

So the open interval $]1-r, 1+r[\notin A$.

Thus no open sphere contained in A having radius $(1+r)$.

$\Rightarrow$ C is not open set.

(iv) $D=[0, 1]$ is not an open set because D an interval $]0-r, 0+r[\not\subset D$.

(v) $E=\{1\}$ is not open set because it contains a single point 1 and so it is not possible to find $r>0$ such that $]1-r, 1+r[\subseteq F$.

(vi) F is not open because it consist elements $\{1, 2, 3\}$ but it is not possible to find $r>0$ such that $]1-r, 1+r[\subseteq F$ and other so it is not open.

Example 4. *Describe the open spheres of unit radius about (0, 0) for each of the following matrices for* $\mathbb{R}^2$.

(i) $d(z_1, z_2)=\sqrt{(x_1-x_2)^2+(y_1-y_2)^2}$

(ii) $d(z_1, z_2)=\max\{|x_1-x_2|, |y_1-y_2|\}$

where $z_1=(x_1, y_1)$, $z_2=(x_2, y_2)$ *are any two points of* $\mathbb{R}^2$.

Solution. (i) Let d be the usual metric on $\mathbb{R}^2$ and, here we are given that

$$d(z_1, z_2)=\sqrt{(x_1-x_2)^2+(y_1-y_2)^2}$$

where $z_1=(x_1, y_1)$ and $z_2=(x_2, y_2)$ be any two points of $\mathbb{R}^2$.

The open space with centre z_0 and radius r is given by

$$S(z_0, r)=\left\{\sqrt{(x_1-x_0)^2+(y_1-y_0)^2}\right\}<r$$

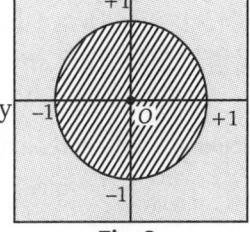

Here, the open spheres $S(z_0, r)$ consists of all points of the cartesian plane which lie within the circle.

$$(x-x_0)^2+(y-y_0)^2=r^2.$$

Fig. 2

(ii) Here, we have

$S(z_0, 1)=\{(x, y) \in \mathbb{R}^2 : d(z, z_0)<1\}$

$=\{(x,y)\in\mathbb{R}^2:\max[\{|x-0|, |y-0|\}<1]\}$

$=\{(x, y) \in \mathbb{R}^2 : \max[\{|x|, |y|\}<1]\}.$

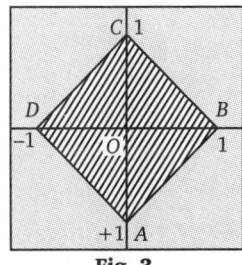

$\Rightarrow$ The open spheres $S(z_0, 1)$ is the interior space of the square bounded by the lines

$$x=1, x=-1, y=1 \text{ and } y=-1.$$

Fig. 3

Example 5. *Let (X, d) be any discrete metric space. Describe open sphere for d.*

Solution. The discrete metric space is defined by

$$d(x, y)=\begin{cases}0 & \text{if } x=y \\ 1 & \text{if } x \neq y\end{cases}$$

Let $x_0 \in X$ and $r>1$, then we have

$$S(x_0, r) = \{x \in X : d(x, x_0 < r] = X$$

[Since $d(x, x_0)=0$ or 1, each of which is less than r so that $x \in X \Rightarrow x \in S(x_0, r)$].

If $r \leq 1$ then, we have

$$S(x_0, r) = \{x \in X : d(x, x_0) < r\} = \{x_0\}$$

$$[\because d(x_0, x_0) = 0 < r \text{ and } d(x, x_0) = 1 \ \forall \ r \text{ if } x \neq x_0]$$

Example 6. *Show that every singleton set in* R *is closed for the usual metric d for* R.

Solution. Let $a \in$ R. To show $\{a\}$ is closed.

Consider $\quad$ R$-\{a\} =]-\infty, a[\ \cup \ ] a, \infty[$

$$= \text{union of two open sets} = \text{open}$$

$\Rightarrow \quad$ R$-\{a\}$ is open

$\Rightarrow \quad \{a\}$ is closed.

Example 7. *Show that every closed interval is a closed set for the usual metric for* R.

Solution. Consider a closed interval [a, b] where $a \in$ R, $b \in$ R.

Also, consider

$$\text{R} - [a, b] = [x \in \text{R} : a < x < b]$$
$$= [x \in \text{R} : x < a, x > b]$$
$$= [x \in \text{R} : x < a] \cup [x \in \text{R} : x > b] =]-\infty, a \ [\ \cup \ ] \ b, \infty[$$

$$(\because \text{ Union of two open sets is again open})$$

which is open.

Hence, [a, b] is closed.

Example 8. *Give an example of two closed subsets A and B of real line* R *such that*

$$d(A, B) = 0, \ A \cap B = \phi.$$

Solution. Let d be the usual metric on R.

Define $\quad N_1 = [n+1 : n \in \text{N}]$

$$A = [n+1 : n \in N_1] \text{ and } B = \left[n + \frac{1}{n} : n \in N_1 \right].$$

Now if $n \in A \Rightarrow n + \frac{1}{n} \in B \Rightarrow B - A = \frac{1}{n}$ which tends to 0 as $n \to \infty$.

Now $\qquad d(A, B) = \inf \{d(x, y) : x \in A, y \in B\}$

$$= \inf. \ \{ |x-y| : x \in A, y \in B \}$$

$$= \inf. \ \left\{ \frac{1}{n} : n \in A, n + \frac{1}{n} \in B \right\} = 0$$

Hence $\qquad d(A, B) = 0$ and $A \cap B = \phi$.

Example 9. *Give an example of a set which has*

(i) *no limit point* $\qquad\qquad$ (ii) *exactly one limit point*

(iii) *exactly two limit points* $\qquad$ (iv) *infinite number of limit points*

(v) *every point of the set as its limit points.*

Solution. (i) The set of rational number Q, has no limit, point.

(ii) The set $S = \left[\frac{1}{n} : n \in \text{N} \right]$ has exactly one limit point namely 0.

(iii) The set $S = \left\{ \frac{1}{2} - \frac{1}{2}, \frac{2}{3}, -\frac{2}{3} \right\}$ has exactly two limit points 1 and -1.

(iv) The open interval]1, 2[has infinite number of limit points.

(v) The closed interval [1, 2] is a set in which every point is its limit point.

Example 10. *Given an example to show that in a metric space the union of an infinite collection of closed sets is not necessarily closed.*

Solution. Let us consider an infinite collection $F_n = \left[\dfrac{1}{n}, 1\right], n \in N$ of closed intervals for usual metric space (R, d) and we know that in usual metric every closed interval is a closed set

$\Rightarrow$ F_n is a closed set in (R, d).

Now $\cup\{f_n : n \in N\} = \{1\} \cup \left[\dfrac{1}{2}, 1\right] \cup \left[\dfrac{1}{3}, 1\right] \cup \left[\dfrac{1}{4}, 1\right] \cup \ldots =]0, 1]$

$\Rightarrow$ which is not closed

Hence, union of an infinite collection of closed sets is not necessarily closed.

Example 11. *If A and B are closed sets of a metric space (X, d) then show that*

(a) *A ∪ B is closed*   (b) *A ∩ B is closed.*

Solution.   (a)   Since A and B are closed set of a metric space X

$\Rightarrow$ $A' = X{-}A$ and $B' = X{-}B$ both are open set of X

$\Rightarrow$ $A' \cap B'$ is an open set of X

$\Rightarrow$ $(A' \cap B')'$ is closed set of X

$\Rightarrow$ $(A')' \cup (B')'$ is closed set of X (By De-Morgan's law)

                                                                ($\because (A')' = A$, $(B')' = B$)

$\Rightarrow$ $A \cup B$ is closed set of X.

(b)   Since A and B are closed set of X

$\Rightarrow$ A' and B' both are open sets of X

$\Rightarrow$ $A' \cup B'$ is an open sets of X

$\Rightarrow$ $(A' \cup B')'$ is closed set of X

$\Rightarrow$ $(A')' \cap (B')'$ is closed set of X (By De-Morgan's law)

                                                                ($\because (A')' = A$, $(B')' = B$)

$\Rightarrow$ $A \cap B$ is closed set of X.

Example 12. *Let (X, d) be a metric space and $S(x_0, r)$ the open sphere with centre x_0 and radius r. Let A be a subset of X which intersects $S(x_0, r)$ and has diameter less than r. Show that $A \subseteq S(2r, x_0)$.*

Solution. Since given that $S(r, x_0)$ is open sphere with centre x_0 and radius r and A is any non-empty subset of X which intersects $S(x_0, r)$

Therefore, $A \subseteq S(x_0, r) \neq \phi$.

Let $y \in A \cap S(x_0, r)$

Then $y \in A$ and $y \in S(x_0, r) \Rightarrow d(y, x_0) < r$.

Since $x \in A$ and x is arbitrary point of A and also $y \in A$

$\Rightarrow$ $d(x, y) < r$

Now $d(x, x_0) \leq d(x, y) + d(y, x_0)$ (By triangular inequality)

$$<r+r=2r$$

$$\Rightarrow \qquad d(x, x_0)<2r \Rightarrow x \in S(x_0, 2r)$$

$$\Rightarrow \qquad A \subseteq S(x_0, 2r).$$

Example 13. Let (X, d) be a metric space and let $p \notin S(x_0, r)$ where $x_0 \in X$ and $r>0$ then show that $\qquad d(p, S(x_0, r)) \geq d(x_0, p)-r.$

Solution. Let $x \in S(x_0, r)$ where x is any arbitrary point of $S(x_0, r)$, then

$$d(x_0, p) \leq d(x_0, x)+d(x, p)$$

$$d(x, p) \geq d(x_0, p)-d(x_0, x) \qquad\qquad \text{...(1)}$$

But we consider that $x \in S(x_0, r)$

$$\Rightarrow \qquad d(x_0, x)<r \qquad\qquad \text{...(2)}$$

Using equation (2) in equation (1) we have $d(x, p) \geq d(x_0, p)-r$

This condition holds for all $x \in S(x_0, r)$

$$\Rightarrow \qquad d(p, S(x_0, r)) \geq d(x_0, p)-r$$

Example 14. Let (X, d) be a metric space and let $p \notin S(x_0, r)$, where $x_0 \in X$ and $r>0$, then show that $\qquad d(p, S[x_0, r])>d(x_0, p)-r.$

Solution. Let x be an arbitrary point of $S(x_0, r)$

Now, $\qquad d(x_0, p) \leq d(x_0, x)+d(x, p)$

$\qquad\qquad$ ($\because$ in a metric space, $d(x, y) \leq d(x, z)+d(z, y)$ by triangle inequality)

$$d(x, p) \geq d(x_0, p)-d(x_0, x)$$

But $\qquad\qquad x \in S(x_0, r) \qquad \Rightarrow \qquad d(x_0, x)<r$

So $\qquad\qquad d(x, p) \geq d(x_0, p)-r$

Since $\qquad\qquad x \in S(x_0, r) \qquad \Rightarrow \qquad d(p, S(x_0, r)) \geq d(x_0, p)-r.$

Example 15. Let (X, d) be a metric space and let $p \notin S[x_0, r]$ and $x_0 \in X$ and $r>0$. Then show that $\qquad d(p, S[x_0, r])>d(x_0, p)-r.$

Solution. We have $d(p, S[x_0, r])$ is infimum of $\{d(x, p) : x \in S[x_0, r]\}$ that is

$$d(p, S[x_0, r])= \inf \{d(x, p): x \in S[x_0, r]\}$$

Let $x \in S[x_0, r]$ then we have

$$d(x_0, p) \leq d(x_0, x)+d(x, p) \qquad\qquad \text{(By triangular inequality)}$$

or $\qquad\qquad d(x, p) \geq d(x_0, p)-d(x_0, x).$

But $\qquad\qquad x \in S[x_0, r]$

$$\Rightarrow \qquad d(x_0, x) \leq r \qquad \text{so} \qquad d(x, p) \geq d(x_0, p)-r$$

$$\Rightarrow \qquad \inf\{d(x, p): x \in S[x_0, r]\} \geq d(x_0, p)-r$$

$$\Rightarrow \qquad d(p, S[x_0, r]) \geq d(x_0, p)-r.$$

Example 16. In the usual metric space (R, d) find the derived set of the set Q of all rational numbers.

Solution. Here we have to show that every real number is a limit point of the set Q.

Let x be any real number and for given $\varepsilon>0$, then $x-\varepsilon$ and $x+\varepsilon$ are two distinct real numbers and exist infinitely many points between them thus for given $\varepsilon>0$ open interval $]x-\varepsilon, x+\varepsilon[$ contain at least one point of rational set Q other than x

$\Rightarrow$ x is a limit point of Q

$\Rightarrow$ every real number is a limit point of Q

$\Rightarrow$ the set of the limits of Q is the set of all real number R so $D(Q)=R$.

Example 17   *On the real line R, the set of integers, Z has no limit point.*

Solution.   Let $x \in R.$

Consider $]x-\varepsilon, x+\varepsilon[.$

Then $]x-\varepsilon, x+\varepsilon[\subseteq Z$

(By Denseness property of real numbers)

EXERCISE 15.2

1. Consider the usual metric

$$d(x, y) = |x-y|, \text{ for } [0, 1]$$

Describe $S(]1/4, 1/4[), S[1/4, 1/4]$, S
$$\left(\left]0, \frac{1}{8}\right]\right) \text{ and } S\left(\left[\frac{1}{16}, \frac{1}{16}\right[\right).$$

2. Show that the Cantor set C is not open.

3. Show that on the real line, every open interval is an open set.

4. Show that on the real line, every closed interval is a closed set.

5. Show that in a metric space, the intersection of two open spheres need not be an open sphere but that it will always contain another open sphere.

6. If A and B are open sets of a metric space X. Then show that

   (i) $A \cap B$ is open set   (ii) $A \cup B$ is also open.

7. Let x_1 and x_2 be two distinct elements in the metric space (X, d), show that two disjoint open spheres will exists, which are centred at x_1 and x_2 respectively.

8. If A_1 is open set in metric space (X_1, d_1) and A_2 is open set in metric space (X_2, d_2) then if $X = X_1 \times X_2$, show that $A_1 \times A_2$ is open.

9. Show that the right half open inerval $[a, b[$ is neither closed nor open (that is say clopen) for the usual metric on R.

10. In a metric space (X, d) the empty set ϕ and the whole space X are closed as well as open.

11. Show that every finite subset of R is closed with respect to the usual metric for R.

12. Show that every subset of X containing x is a neighbourhood of x, where (X, d) is discrete space defined as

$$d(x, y) = \begin{cases} 0 \text{ if } x = y \\ 1 \text{ if } x \ne y \end{cases}$$

13. Let (X, d) be any metric space and let

   (a) $d^*(x, y) = \dfrac{d(x, y)}{1 + d(x, y)} \quad \forall x, y \in X.$

   Show that d^* is also a metric on X and d and d^* are equivalent.

   (b) If $d^*(x, y) = \min \{1, d(x, y)\}$ then show that $d^*(x, y)$ is a metric on X and also d and d^* are equivalent.

14. Show that in a discrete metric space, every set is open as well as closed.

15. Show that the closed open interval $[a, b[$ is neither closed nor open for the usual metric on R.

16. Give an example of a set which (i) is both open and closed (ii) is neither open nor closed.

17. If A' denote the derived set of A, then find a set A such that

   (i) $A \cap A' = \phi$ (ii) $A = A'$

   (iii) $A \subseteq A$ (iv) $A \subseteq A'$.

18. On the real line, show that the set of rational numbers Q, has no limit point.

HINTS TO SELECTED PROBLEMS

1. Since we know that the Cantor set is the intersection of closed sets, therefore, being the intersection of closed set, it is again closed.

4. Using the following result

$R-[a, b] = \{x \in R: x<a \text{ or } x>b\}$
$= \{x \in R: x<a\} \cup \{x \in R: x>b\}$
$=]\infty, a[\cup]b, \infty[$, union of two open sets.

16. (i) ϕ and R   (ii) Set of rational numbers.

18. Using the denseness property of real numbers.

Answers

1. $]0, \frac{1}{2}[, [0, \frac{1}{2}], [0, \frac{1}{8}[,]0, \frac{1}{8}[$

8. (i) ϕ and R   (ii) Q, the set of rationals

9. (i) $A = [1, \frac{1}{2}, \frac{1}{3}, ..., \frac{1}{n}]$ (ii) $A = \{$set of closed interval$\}$.

15.8 CLOSURE, INTERIOR AND EXTERIOR

Definition. *In a metric space (X, d), the set of all adherent points of a set $A \subseteq X$ is called the closure of A and is denoted by $C(A)$ or $\bar{A}$.*

For Example :

(1) On the real line, every real number is an adherent point of the set R~Q of all irrational numbers. For, if p be any real number whatever any $\varepsilon>0$ be given, then $]p-\varepsilon, p+\varepsilon[$ contain infinitely many irrational numbers and hence $]p - \varepsilon, p+\varepsilon[\cap (R\sim Q) \neq \phi$. Hence, p is an adherent point of R~Q and $[R\sim Q] = R$.

(2) On the real line, the closure of each of the sets $]0, 1[,]0, 1]$ and $[0, 1]$ is $[0, 1]$.

(3) The closure of the set of integers is Z itself.

(4) Let $A = \{\frac{1}{n} : n \in Z^+\}$. Then closure $A = \bar{A} = \{\frac{1}{n} : n \in Z^+\} \cup \{0\}$.

THEOREM 1. *In a metric space (X, d), the closure of a set $A \subset X$, is a closed superset of A.*

Proof.   Let p be any point of A and N be any nbd of p. Then $p \in N \cap A$ i.e. $N \cap A \neq \phi$. Since every nbd of p intersects A, therefore $p \in \bar{A}$. Since

$$p \in A \Rightarrow p \in \bar{A} \Rightarrow A \subset \bar{A}.$$

Now, $\bar{A}$ is a closed set $\Leftrightarrow X - \bar{A}$ is an open set.

$$\Leftrightarrow X - \bar{A} \text{ is a nbd of each of its points.}$$

Therefore, to show that $\bar{A}$ is a closed set, it is enough to show that $X - \bar{A}$ is a nbd of each of its points. Consider an arbitrary point $q \in X - \bar{A}$

Now, $q \in X - \bar{A} \Rightarrow q \notin \bar{A}$

$\Rightarrow$ there exists an $\varepsilon>0$ such that $S(q, \varepsilon)$ contains no point of A.

It can be easily seen that no point of $S(q, \varepsilon)$ can be in $\bar{A}$. In fact if $r \in S(q, \varepsilon)$, then $S(q, \varepsilon)$ is a nbd of r containing no point of A and therefore $r \notin \bar{A}$. Since r is arbitrary, it follows that no point of $S(q, \varepsilon)$ is in $\bar{A}$. Thus $S(q, \varepsilon) \subset X - \bar{A}$, showing that $X - \bar{A}$ is a nbd of q. Again since q is arbitrary point of $X - \bar{A}$, therefore it follows that $X - \bar{A}$ is a nbd of each of its points. Hence $\bar{A}$ is closed.

THEOREM 2. *In a metric space (X, d), the closure of a set $A \subset X$ is the smallest closed set containing A.*

Proof. Since we have already shown that the closure of a set A is a closed superset of A, it only remains to show that $\bar{A}$ is the smallest among all closed superset of A i.e. if B is any closed set containing A, then $\bar{A} \subset B$.

Now, B is closed superset of $A \Rightarrow X - B$ is an open set disjoint from A

$\Rightarrow$ no point of $X - B$ is an adherent point of A

$\Rightarrow$ $\qquad\qquad\qquad \bar{A} \subset B$

REMARK

- The properties of $\bar{A}$, expressed in the above two theorems is sometimes taken as definition of closure of A. It gives us a description of $\bar{A}$ without using the notion of an adherent point of A. Therefore, we can say :

   (i) The closure of A i.e. $\bar{A}$ is the smallest closed set containing A.

   (ii) The closure of $\bar{A}$ i.e., $\bar{A}$ is the intersection of all closed sets containing A i.e.
   $$\bar{A} = \cap \; [F : F \text{ is closed and } A \subset F].$$

THEOREM 3. *A subset A of a metric space is closed if and only if $\bar{A} = A$.*

Proof. Let (X, d) be a metric space.

Let us first suppose A is closed, then A is the smallest closed set containing A

$\Rightarrow$ $\qquad\qquad\qquad \bar{A} = A$

Conversely if $\bar{A} = A$, then since $\bar{A}$ is closed, therefore A is closed.

THEOREM 4. *Let A be a subset of a metric space, then*
$$\bar{A} = A \cup D(A).$$

Proof. Let p be the limit point of A then p is an adherent point of A also.

i.e. $\qquad\qquad\qquad A' \subset \bar{A}.$ ...(1)

Now every point of A is an adherent point of A

i.e. $\qquad\qquad\qquad A \subset \bar{A}.$ ...(2)

From (1) and (2), we conclude that

$\qquad\qquad\qquad A \cup A' \subset \bar{A}.$ ...(3)

Now let p be any point of $\bar{A}$.

If $p \in A$, then clearly $\bar{A} \subset A \cup A'$.

Now, if $\qquad\qquad p \in \bar{A}, p \notin A$. Let N be any nbd of p. Since $p \in \bar{A}, N \cap A \neq \phi$ and $p \notin A$. It follows that $p \notin N \cap A$ so that $N \cap A$ is a non-empty set, containing a point other than p. Since N is any nbd of p, it follows that $p \in A'$.

Therefore $\qquad\qquad \bar{A} \subset A \cup A'.$ ...(4)

From (3) and (4), we conclude that

$\qquad\qquad\qquad \bar{A} = A \cup A' \Rightarrow \bar{A} = A \cup D(A)$

THEOREM 5. *Let (X, d) be a metric space and A, B be subsets of X. Then*

(i). $\bar{\phi} = \phi$.   (ii) $A \subset \bar{A}$ (iii) $A \subset B \Rightarrow \bar{A} \subset \bar{B}$

(iv) $\overline{A \cup B} = \bar{A} \cup \bar{B}$ (v) $\overline{A \cap B} \subset \bar{A} \cap \bar{B}$ (vi) $\overline{\bar{A}} = \bar{A}$

Proof.

(i) Since ϕ is closed, therefore using theorem (3), we have $\bar{\phi} = \phi$.

(ii) By definition, $\bar{A}$ is the smallest closed set containing A i.e. $A \subset \bar{A}$

(iii) From (ii), we have $B \subset \bar{B}$.

Since $A \subset B$. Therefore $A \subset B \subset \bar{B}$. $\Rightarrow$ $A \subset \bar{B}$.

But $\bar{B}$ is closed $\Rightarrow$ $\bar{B}$ is the closed set containing A

Also, $\bar{A}$ is the smallest closed set containing A. Therefore $\bar{A} \subset \bar{B}$.

(iv) Since we know that

$$A \subset A \cup B \text{ and } B \subset A \cup B.$$

Therefore from (iii)

$$\bar{A} \subset \overline{A \cup B} \text{ and } \bar{B} \subset \overline{A \cup B}$$
$$\bar{A} \cup \bar{B} \subset \overline{A \cup B}. \qquad \qquad ...(1)$$

Now, $\bar{A}$ and $\bar{B}$ are closed sets, $\bar{A} \cup \bar{B}$ is also closed (being the union of two closed sets). Also $A \subset \bar{A}$ and $B \subset \bar{B}$

$\Rightarrow$ $A \cup B \subset \bar{A} \cup \bar{B}$

$\Rightarrow$ $\bar{A} \cup \bar{B}$ is the closed set containing $A \cup B$.

But $\overline{A \cup B}$ is the smallest closed set containing $A \cup B$.

Therefore $\overline{A \cup B} \subset \bar{A} \cup \bar{B}$.   ...(2)

From (1) and (2), we conclude that $\overline{A \cup B} = \bar{A} \cup \bar{B}$.

(v) Since we know that

$$A \cap B \subset A \Rightarrow \overline{A \cap B} \subset \bar{A} \quad \text{and} \quad A \cap B \subset B \Rightarrow \overline{A \cap B} \subset \bar{B}$$

Hence $\overline{A \cap B} \subset \bar{A} \cap \bar{B}$.

(vi) Since $\bar{A}$ is a closed set, therefore using theorem (3), we have

$$\left(\overline{\bar{A}} \right) = \bar{A}.$$

REMARK

• The inclusion in part (v) of the above theorem can not be replaced by an equality.

For Example. Let $A =]0, 1[$ and $B =]1, 2[$ Then $\bar{A} = [0, 1]$ and $\bar{B} = [1, 2]$

$\Rightarrow$ $\bar{A} \cap \bar{B} = \{1\}$.

Also $\bar{A} \cap \bar{B} = \phi \Rightarrow \left(\overline{A \cap B} \right) = \phi$.

we find that in this case $\left(\overline{A \cap B} \right)$ is a proper subset of $\bar{A} \cap \bar{B}$.

THEOREM 6. *A finite set in a metric space has no limit point.*

Proof.   Let (X, d) be a metric space and A be any finite subset of X.

As $p \in X$ is a limit point of any set B if every open sphere $S(x, r)$ contains an infinite number of points of B, other then p.

But A has finite number of points. Hence, A has no limit point.

THEOREM 7. *Let (X, d) be a metric space and $A \subset X$. Then $\bar{A} = \{x \in X : d(x, A) = 0\}$.*

Proof. Let (X, d) be a metric-space and A be any subset of X.

To show

$$\bar{A} = \{x \in X : d(x, A) = 0\} \text{ where } \bar{A} = A \cup D(A).$$

Let $d(x, A) = \varepsilon > 0$, then $x \notin A$

and $\qquad \{S(x, \varepsilon/3) - [x]\} \cap A = \phi.$

Let $x \notin D(A)$, then

$$d(x, A) = \varepsilon \quad \Rightarrow x \notin A, x \notin D(A) \Rightarrow x \notin A \cup D(A)$$
$$\Rightarrow x \notin \bar{A}. \qquad\qquad [\because \bar{A} = A \cup D(A)]$$

Now $\qquad d\{x, A\} = 0 \Rightarrow x \in A$ and $\{S(x, \varepsilon) - [x]\} \cap A \neq \phi$
$$\Rightarrow x \in D(A).$$

Finally, $\qquad d(x, A) = 0 \Rightarrow x \in A$ or $x \in D(A)$
$$\Rightarrow x \in A \cup D(A)$$
$$\Rightarrow x \in \bar{A}. \qquad\qquad [\because \bar{A} = A \cup D(A)]$$

and $\qquad d(x, A) = \varepsilon > 0 \Rightarrow x \notin \bar{A}.$

Further more $\qquad d(x, A) \geq 0$ and hence the result follows.

THEOREM 8. *In a metric space (X, d), all finite sets are closed.*

Proof. Since we know that every finite set can be written as the finite union of singletons and each singleton is closed. Also, finite union of closed sets is again closed.

Hence, finite set is always closed.

THEOREM 9. (Bolzano-Weirstrass Theorem) *Every bounded sequence in a metric space has a limit point.*

Proof. Let $<x_n>$ be a bounded sequence in a metric space (X, d). If there are only a finite number of distinct elements in $\{x_n : n \in N\}$ then at least one of them must occur infinitely often. If this element be x, then x is the limit point of $<x_n>$.

If the sequence $<x_n>$ contains infinitely many distinct elements in X, then $S = \{x_n : n \in N\}$ is an infinite bounded set and therefore has a limit point, say y. Since y is the limit point of S, therefore every nbd of y must contains infinitely many points of S. This implies that the sequence $<x_n>$ must be frequently in every nbd of y. Hence, y is a limit point of $<x_n>$.

15.9 INTERIOR POINT AND INTERIOR OF A SET

Definition 1. *Let (X, d) be a metric space and A be any subset of X. A point $p \in A$ is said to be an interior point of A, if A is a nbd of p.*

Definition 2. *Let (X, d) be a metric space and $A \subset X$. A point $p \in A$ is said to an interior point of A if there exist $\varepsilon > 0$ such that $S(p, \varepsilon) \subset A$.*

Definition 3. *The set of all interior points of a set $A \subset X$, is called the interior of A and is denoted by $A°$.*

☞ **ILLUSTRATIONS**

On the real line :

(1) Every point of $]a, b[$ is an interior point of $]a, b[$ i.e. int($]a, b[$)=$]a, b[$.

(2) No point of Z, is an interior point of Z i.e. $Z° = \phi$.

(3) The interior of the closed ray $[0, \infty[$ is the open ray $]0, \infty[$.

(4) 0 is not an interior point of $[0, 1[$. Every point of $]0, 1[$ is an interior point of $[0, 1[$. The interior of $[0, 1[$ is $]0, 1[$.

(5) The set of rational numbers Q has no interior point i.e, $Q° = \phi$.

(6) Let d be the discrete metric on a non-empty set X, and let $p \in X$. Then int $\{p\} = \{p\}$. In fact, for each set $A \subset X$, int $A = A$.

15.10 PROPERTIES OF THE INTERIOR

THEOREM 1. *Let (X, d) be a metric space and A be a subset of X then*

     (i) $A°$ is an open set.

     (ii) $A°$ is the largest open set contained in A.

     (iii) A is open if and only if $A° = A$.

Proof.   (i)   Let $x \in A°$. Then x is an interior point of A

         $\Rightarrow$ A is a nbd of x (By definition of interior point)

         $\Rightarrow$ $\exists$ an open set G such that $x \in G \subset A$ (By definition of neighbourhood)

         Since G is open, therefore G is a nbd of each of its points. Also, Since $G \subset A$, therefore A is also a nbd of each point of G

         $\Rightarrow$ Every point of G is an interior point of A

         $\Rightarrow$ $G \subset A°$

         $\Rightarrow$ to each $x \in A°$, there exists an open set G such that $x \in G \subset A°$

         $\Rightarrow$ $A°$ is the nbd of each of its points

    (ii) Let G be an open subset of A and let $x \in G$. Then $x \in G \subset A$

         Now, since G is open, A is a nbd of x

         $\Rightarrow$ x is an interior point of A $\Rightarrow$ $x \in A°$

         Therefore, $x \in G \Rightarrow x \in A°$ ($\because G \subset A°$)

         $\Rightarrow$ $A°$ contains every open subset of A and it is, therefore, the largest open subset of A .

    (iii) Let us first suppose $A = A°$.

         Since $A°$ is open, therefore A is also open.

         Conversely, let A is open. To show $A = A°$.

         Since, G is open, therefore A is the largest open set contained in A.

         Then A is surely identical with $A°$.

         Also, $A°$ is the largest open subset of A. Hence, $A = A°$.

THEOREM 2. *Let (X, d) be a metric space and A be a subset of X. Then $A°$ is equals the set of all those points of A which are not limit point of A'.*

Proof.   Let x be a point of A which is not a limit point of A'. Then there exists a nbd N of x which contains no point of A'.

       $\therefore$ $N \subset A$

       $\Rightarrow$ A is also a nbd of x.

$\Rightarrow$ $x \in A^{\circ}$.

Conversely, let $x \in A^{\circ}$.

Since A° is open   (By definition)

$\Rightarrow$ A° is a nbd of x ($\because$ Every open set is a nbd of each of its points)

Also, A° contains no point of A' ($\because$ $A^{\circ} \subset A$ and A contains no point of A')

$\Rightarrow$ x is not a limit point of A'

$\Rightarrow$ no point of A° is a limit point of A'.

Hence, A' consists exactly those points of A which are not limit points of A'.

15.11 EXTERIOR POINT AND EXTERIOR OF A SET

Definition 1. *Let A be a subset of a metric space (X, d). A point $x \in X$ is said to be exterior point of A if it is an interior point of the complement of A.*

Definition 2. *The set of all exterior points of A is called the exterior of A and is denoted by ext.(A) or A^e.*

REMARKS

- Clearly ext.$(A) = (A')^{\circ}$ and ext.$(A') = (A'')^{\circ} = A^{\circ}$. Also $A \cap$ ext $(A) = \phi$
- Since ext. (A) is the interior of A' therefore ext.(A) is the largest open set containing A'.

15.12 FRONTIER POINT AND FRONTIER OF A SET

Definition 1. *Let (X, d) be a metric space. A point x of a metric space is said to be frontier point of a subset $A \subset X$ if and only if it is neither an interior nor exterior point of A.*

Definition 2. *The set of all frontier points of A is called the frontier of A and is denoted by $F_r(A)$.*

15.13 BOUNDARY POINT AND BOUNDARY OF A SET

Definition 1. *Let (X, d) be a metric space. A point $x \in X$ is said to be boundary point of a subset $A \subset X$ if it is a frontier point of A and belongs to A*

Definition 2. *The set of all boundary points of A is called the boundary of A and it is denoted by $b(A)$.*

REMARK

- $b(A) \subset F_r(A)$

15.13.1 SOME MORE DEFINITIONS

Let (X, d) be a metric space and A, B be subsets of X. Then

(1) A is said to be dense in B if $B \subset \bar{A}$

(2) A is said to be dense in X or everywhere dense if $\bar{A} = X$.

(3) A is said to be nowhere dense in X if $(\bar{A})^{\circ} = \phi$ *i.e.* interior of the closure of A is empty.

(4) A is said to be dense in itself if $A \subset D(A)$.

(5) A is said to be perfect iff A is dense-in-itself and closed.

REMARKS

- A is said to be every where dense iff every point of X is an adherent point of A.
- A is said to be perfect if $A = D(A)$.

THEOREM 1. *Let (X, d) be a metric space and A be any subset of X. Then*

        (i) $A° = \cup [G : G \text{ is open, } G \subset A]$

        (ii) ext. $(A) = \cup [G : G \text{ is open, } G \subset A']$.

Proof.   (i) Let $x \in A°$, then by definition of interior point.

        A is a nbd of x. $\Rightarrow$ there exists an open set G such that $x \in G \subset A$.

        $\Rightarrow$ $x \in \cup [G : G \text{ is open, } G \subset A]$.

        Conversely, let $x \in \cup \{G : G \text{ is open, } G \subset A\}$. Then $x \in G$ for some open set G such that $G \subset A$

    $\Rightarrow$ A is a neighbourhood of x

    $\Rightarrow$ $x \in A°$.

        Therefore, $x \in A°$ if and only if $x \in \cup \{G : G \text{ is open, } G \subset A\}$.

        Since, x is arbitrary, therefore

    $\Rightarrow$ $A° = \cup \{G : G \text{ is open, } G \subset A\}$.

THEOREM 2. *Let A be a subset of a metric space (X, d). Then a point $x \in X$ is an exterior point of A if and only if x not an adherent point of A i.e. $x \in (\bar{A})'$.*

Proof.   Let us first suppose x be an exterior point of A.

    $\Rightarrow$ x is an interior point of A'.

    $\Rightarrow$ A' is a nbd of x containing no point of A

    $\Rightarrow$ x is not an adherent point of A

    $\Rightarrow$ $x \in (\bar{A})'$.

        Conversely, let $x \in (\bar{A})'$. To show x is an exterior point of A

        Now, $x \in (\bar{A})' \Rightarrow x$ is not an adherent point of A

    $\Rightarrow$ there exists a nbd N of x which contains no points of A

    $\Rightarrow$ $x \in N \subset A'$

    $\Rightarrow$ A' is a nbd of x.

    $\Rightarrow$ x is an interior point of A'

    $\Rightarrow$ x is an exterior point of A.

THEOREM 3. *Let (X, d) be a metric space and $A \subset X$. Then, a point $x \in X$ is a frontier point of A if and only if every nbd of x intersects both A and A'.*

Proof.   Here, we have

$$x \in F_r(A) \Leftrightarrow x \notin A° \text{ and } x \notin \text{ext}(A) = (A')°$$

                  $\Leftrightarrow$ neither A nor A' is a nbd of x

                  $\Leftrightarrow$ no nbd of x can be contained in A or in A'.

                  $\Leftrightarrow$ every nbd of x intersects both A and A'.

REMARK

- $F_r(A) = F_r(A')$

THEOREM 4. *Let A be any subset of a metric space (X, d). Then $A°$, $ext(A)$ and $F_r(A)$ are mutually disjoint and $X = A° \cup ext(A) \cup F_r(A)$. Also, $F_r(A)$ is a closed set.*

Proof.   Let (X, d) be a metric space. We know that $ext(A) = (A')°$

Also, $A° \subset A$, $(A')° \subset A'$ and $A \cap A' = \phi$

$\Rightarrow$ $A° \cap ext (A) = A° \cap (A')° = \phi$.

Now, $x \in F_r(A) \Leftrightarrow x \notin A°$ and $x \notin ext. (A)$

                        $\Leftrightarrow x \notin A° \cup ext (A)$

                        $\Leftrightarrow x \in [A° \cup ext(A)]'$.

Therefore, $F_r(A) = [A° \cup ext(A)]'$

$\Rightarrow$ $F_r(A) \cap A° = \phi$ and $F_r(A) \cap ext(A) = \phi$

and $x = A° \cup ext (A) \cup F_r (A)$.

Finally, since $A°$ and $ext (A)$ both are open.
Therefore, $(A°)'$ and $(ext (A))'$ both are closed.

Hence, $F_r(A)$ is closed.

THEOREM 5. *Let (X, d) be a metric space and A, B be any subsets of X. Then*

   (i) $X° = X$, $\phi° = \phi$ (ii) $A° \subset A$ (iii) $A \subset B \Rightarrow A° \subset B°$

   (iv) $(A \cap B)° = A° \cap B°$ (v) $A° \cup B° \subset (A \cup B)°$ (vi) $A°° = A°$.

Proof.   (i) Since we know that X and ϕ both are open sets.

                Therefore $X° = X$ and $\phi° = \phi$ $(\because A$ is open iff $A° = A)$

        (ii) Let $x \in A° \Rightarrow x$ is an interior point of A

                                $\Rightarrow A$ is a nbd of x

                                $\Rightarrow x \in A \Rightarrow A° \subset A$

        (iii) Let $A \subset B$ and let $x \in A°$

                                $\Rightarrow x$ is an interior point of A

                                $\Rightarrow x$ is an interior point of B $(\because A \subset B)$

                                $\Rightarrow x \subset B° \Rightarrow A° \subset B°$. $(\because x$ is arbitrary$)$

        (iv) Since we know that

$$A \cap B \subset A \Rightarrow (A \cap B)° \subset A° \left. \right|$$
$$\text{and}\quad A \cap B \subset B \Rightarrow (A \cap B)° \subset B° \right]$$
                                                                    [using (iii)]

                This implies $(A \cap B)° \subset A° \cap B°$.   ...(1)

                Now let $x \in A° \cap B°$

                $\Rightarrow$ $x \in A°$ and $x \in B°$

                $\Rightarrow$ x is an interior point of A and x is an interior point of B

                $\Rightarrow$ x is an interior point of $A \cap B$

                Therefore $(A° \cap B°) \subset (A \cap B)°$.   ...(2)

                From (1) and (2), we conclude that

                        $A° \cap B° = (A \cap B)°$.

        (v) We know that

                        $A \subset A \cup B \Rightarrow A° \subset (A \cup B)°$

and $\qquad B \subset A \cup B \Rightarrow B° \subset (A \cup B)°.$

This implies that

$$A° \cup B° \subset (A \cup B)°.$$

(vi) Since $A°$ is always open and we know that A is open if and only if $A°=A$.

Apply the above result for $A°$, we get

$$(A°)°=A°$$

$$\Rightarrow \qquad A°\,° =A°.$$

REMARK

- In result (v) $A° \cup B° \neq (A \cup B)°.$

  For Example : If $\qquad A=[0, 1[$ and $B=[1, 2[$

  $\Rightarrow \qquad\qquad A°=]0, 1[$ and $B°=]1, 2[$

  $\Rightarrow \qquad A° \cup B°=]0, 1[\cup]1, 2[=]0, 2[-[1]$

  also $\qquad A \cup B=[0, 2[\Rightarrow (A \cup B)°=]0, 2[.$

  Therefore, $(A \cup B)° \neq A° \cup B°$

THEOREM 6. *Let (X, d) be a metric space and let A, B be subsets of X. Then*

$\qquad$ (i) *ext $(X)=\phi$, ext $(\phi)=X$* $\qquad\qquad$ (ii) *ext$(A) \subset A'$*

$\qquad$ (iii) *ext $(A)=ext[(ext(A))\,']$* $\qquad\qquad$ (iv) *$A \subset B \Rightarrow$ ext $[B] \subset$ ext $[A]$*

$\qquad$ (v) *$A° \subset ext[ext(A)]$* $\qquad\qquad$ (vi) *ext $(A \cap B)=$ ext $(A) \cap$ ext (B).*

Proof. $\qquad$ (i) $\quad$ ext $(X)=(X')°=\phi°=\phi$ $\qquad\qquad\qquad\qquad\qquad\qquad$ $(\because X'=\phi)$

$\qquad\qquad\qquad$ ext $(\phi)=(\phi')°=X°=X$ $\qquad\qquad\qquad\qquad\qquad\qquad\quad$ $(\because \phi'=X)$

$\qquad$ (ii) $\quad$ ext$(A) = (A')° \subset A'.$

$\qquad$ (iii) $\quad$ Here, we have

$$\text{ext } [(\text{ext } (A))'\,] = \text{ext } (A'\,°)\,']=\text{ext } (A'\,°\,')$$

$$=[(A'\,°\,')\,']° = (A'\,°'')°=(A'\,°)° \qquad\qquad (\because A''=A)$$

$$= A'\,°\,° =A'\,°=(A')°=\text{ext } (A) \qquad\qquad (\because A°°=A°)$$

$\qquad$ (iv) $\quad$ Now $\quad A \subset B$ $\qquad\qquad \Rightarrow B' \subset A'$

$\qquad\qquad\qquad\qquad\qquad\qquad\qquad\quad \Rightarrow (B')° \subset (A')°$

$\qquad\qquad\qquad\qquad\qquad\qquad\qquad\quad \Rightarrow \text{ext}(B) \subset \text{ext}(A).$

$\qquad$ (v) $\quad$ We have ext $(A) \subset A' \Rightarrow \text{ext}(A') \subset \text{ext}(\text{ext}(A))$

$\qquad\qquad\qquad\qquad\qquad\qquad\qquad\quad \Rightarrow A° \subset \text{ext}(\text{ext}(A)) \qquad\qquad (\because A°=\text{ext } (A'))$

$\qquad$ (vi) $\quad$ We have

$$\text{ext } (A \cup B)=[(A \cup B)']°=[A' \cap B']°$$

$$=A'° \cap B'°=\text{ext } (A) \cap \text{ext } (B).$$

THEOREM 7. *Let (X, d) be a metric space and $A \subset X$. Then*

$\qquad$ (i) *closure of the complement of A is the complement of the interior of A.*

$\qquad$ (ii) *the interior of A is the complement of the closure of the complement of A.*

$\qquad$ (v) *the closure of A is the complement of the interior of the complement of A.*

Proof. $\qquad$ (i) $\quad$ Since we know that

$$A° = A'^{-'}.$$

Therefore $\quad (A^\circ)' = (A'^-)=A'^{-"}=A^{\circ\prime}=A'^-.$

(ii) Using (i) $\qquad A^{\circ\prime} = A'^-$

$\Rightarrow \qquad\qquad A^{\circ\prime\prime} = A'^{-\prime}.$ $\qquad\qquad$ (By taking complement)

$\Rightarrow \qquad\qquad A^\circ = A'^{-\prime}.$

(iii) Using (i) $\qquad A''^- = A^{\circ\prime}.$

Replacing A by A', we get

$$(A')'^- = (A')^{\circ\prime}$$

$$A'^- = A'^{\circ\prime} \Rightarrow \bar{A}=A^{\circ\prime}.$$

THEOREM 8. *Let (X, d) be a metric space and $A\ X$. Then* $\qquad A=A^\circ \cup F_r(A)$

Proof. $\qquad$ Since we know that

$$\bar{A} = \cap\ [F: F \text{ is closed and } A \subset F]$$

$\Rightarrow \qquad\qquad (-\bar{A})' = \cup\ [F': F' \text{ is open and } F' \subset A'] = \text{ext }(A)$

$\Rightarrow \qquad\qquad (\bar{A})'' = [\text{ext }(A)]'$ $\qquad\qquad$ (By taking complement)

$\Rightarrow \qquad\qquad \bar{A} = A^\circ \cup F_r(A).$

Hence, we have $\qquad \bar{A} = A^\circ \cup F_r(A).$

THEOREM 9. *Let (X, d) be a metric space and let A, B be subsets of X. Then*

(i) $F_r(A)= \bar{A} \cap A'^- = \bar{A}-A^\circ$ $\qquad$ (ii) $A^\circ=A - F_r(A)$

(iii) $[F_r(A)]' = A^\circ \cup A'^\circ$ $\qquad$ (iv) $F_r(A^\circ) \subset F_r(A)$

(v) $F_r(\bar{A}) \subset F_r(A)$ $\qquad$ (vi) $F_r(A \cup B) \subset F_r(A) \cup F_r(B)$

Proof. $\qquad$ Let (X, d) be a metric space and A and B be any two subset of X.

(i) We know that

$$F_r (A)= [A^\circ \cup \text{ext }(A)]' = A'^\circ \cap [\text{ext }(A)]'$$

$\qquad\qquad\qquad\qquad\qquad$ (By De-Morgan's law)

$$= A'^{-"} \cap A^{-"}=A'^- \cap \bar{A} \qquad\qquad (\because A'' =A)$$

Now $\qquad A \cap A'^- = \bar{A}-A'^{-\prime}$ $\qquad\qquad [\because A \cap B' = A-B]$

$$= \bar{A}-A^\circ$$

which implies $\quad F_r (A) = \bar{A} \cap A'^- = A'-A^\circ.$

(ii) Since $\qquad F_r (A)= A' - A^\circ$

$\Rightarrow \qquad A-F_r(A) = A' -(A'-A^\circ)=A^\circ.$

(iii) $\qquad [F_r(A)]' = (\bar{A} \cap A'-)' = A-' \cup A'-'$

Now using $\qquad A'^\circ = (A')^\circ = (A')'^{-\prime}=A''-'=A-'$

(iv) Now $\qquad Fr(A^\circ) = A^{\circ-} \cap A^{\circ\prime-} = A^{\circ-} \cap A'^{-"-}$ $\qquad [\because A'^{-\prime} = A^\circ]$

$$= A^{\circ-} \cap A' \subseteq F_r(A). \ \phi$$

(v) $\qquad F_r(\bar{A}) = \bar{\bar{A}} \cap A^{-\prime-} = \bar{A} \equiv A^{-\prime-}$ $\qquad \left[\because \bar{\bar{A}} = \bar{A}\right]$

Now, $\qquad A \subset \bar{A} \Rightarrow (\bar{A})' \subset A'$

$$\Rightarrow A^{-\prime-} \subset A'^-$$

Hence $\qquad F_r(\bar{A}) \subset \bar{A} \cap A'^- = F_r(A)$

$$\Rightarrow F_r(\bar{A}) \subset F_r(A).$$

(vi) Consider

$$F_r(A \cup B) = \overline{(A \cup B)} \cap (A \cup B)'^{-}$$

$$= (\overline{A} \cup \overline{B}) \cap (A' \cup B')^{-} \subset (\overline{A} \cup \overline{B}) \cap (A'^{-} \cup B'^{-})$$

$$= \left[\overline{A} \cap (A'^{-} \cap B')\right] \cup \left[\overline{B} \cap (A'^{-} \cap B'^{-})\right]$$

$$= \left[(\overline{A} \cup A'^{-}) \cap B'\right] \cup \left[(\overline{B} \cap B'^{-}) \cap A'\right] \subset F_r(A) \cup F_r(B).$$

THEOREM 10. *Let (X, d) be a metric space and A be any subset of X. Then*

(i) *If A is open, $F_r(A) = \overline{A} - A$.*

(ii) *$F_r(A) = \phi$ if and only if A is open as well as closed.*

(iii) *A is open if and only if $A \cap F_r(A) = \phi$.*

(iv) *A is closed if and only if $F_r(A) \subset A$.*

Proof.   (i)   Since we know that

$$F_r(A) = \overline{A} - A.$$

If A is a open then $A = A^\circ$.

$\therefore$ $F_r(A) = \overline{A} - A^\circ$.

(ii)   Let us first suppose $F_r(A) = \phi$. Then, we have

$$F_r(A) = \phi \Rightarrow \overline{A} - A^\circ = \phi \Rightarrow \overline{A} \subset A^\circ \Rightarrow \overline{A} \subset A \qquad (\because A^\circ \subset A)$$

$$\Rightarrow D(A) \subset A \qquad (\because \overline{A} = A \cup D(A))$$

$$\Rightarrow A \text{ is closed.}$$

Also, $F_r(A) \Rightarrow \phi = \overline{A} - A^\circ = \phi \Rightarrow \overline{A} \subset A^\circ$

$$\Rightarrow A \cup D(A) \subset A^\circ \Rightarrow A \subset A^\circ.$$

But $A^\circ \subset A$. Therefore $A^\circ = A$

$\Rightarrow$ A is open   ($\because A$ is open if and only if $A^\circ = A$)

Hence $Fr(A) = \phi$ then A is closed as well as open.

Conversely, let A is open as well as closed. To show $F_r(A) = \phi$.

Since we know that $F_r(A) = \overline{A} - A^\circ$.   ...(1)

Also if A is closed, then $\overline{A} = A$ and if A is open, then $A^\circ = A$.

Put the above values in (1), we get $F_r(A) = \phi$

(iii)   We know that

$$F_r(A) = \overline{A} \cap A'^{-}$$

If A is open, then A' is closed ,

$\Rightarrow$ $A'^{-} = A'$.

Now $A \cap F_r(A) = A \cap [\overline{A} \cap A'^{-}] = A \cap [\overline{A} \cap A']$

$$= [A \cap \overline{A}] \cap A' = A \cap A' = \phi$$

Conversely, let $A \cap F_r(A) = \phi$.

Then, $A \cap F_r(A) = \phi \Rightarrow A \cap (\overline{A} \cap A'^{-}) = \phi$

$$\Rightarrow (A \cap A') \cap A'^{-} = \phi.$$

$$\Rightarrow A \cap A'^{-} = \phi \Rightarrow A \subset A'^{-'} \Rightarrow A \subset A^0.$$

Now $A^\circ \subset A$ gives $A^\circ = A$

$\Rightarrow$ A is open.

(iv) Let A is closed. Then $\bar{A} = A$ $(\because A$ is closed iff $\bar{A} = A)$

$\Rightarrow$ $F_r(A) = \bar{A} \cap A^{'-} = A \cap A^{'-} \subset A.$

Conversely, let $F_r(A) \subset A$ Then $A \cup F_r(A) = A$

But $A \cup F_r(A) = \bar{A} \Rightarrow A = \bar{A} \Rightarrow A$ is closed. $(\because A$ is closed iff $\bar{A} = A)$

Solved Examples

Example 1. *Consider the usual metric space (R, d).*
Find the closure of the following sets

(i) $A = \left[\dfrac{1}{n} : n \in N\right]$ (ii) Z

(iii) Q   (iv)]0, 1[.

Solution. (i) Since 0 is the only limit point
of A

$\therefore$ $D(A) = [0]$

and so

$$\bar{A} = \left\{\dfrac{1}{n} : n \in N\right\} \cup [0].$$

Based on the following Results

- The smallest closed set containing A is called closure of A. It is denoted by $\bar{A}$.

- A is closed $\Leftrightarrow \bar{A} = A.$

- $\bar{\phi} = \phi; A \subset \bar{A}; A \subset B \Rightarrow \bar{A} \subset \bar{B}; \overline{A \cup B} = \bar{A} \cup \bar{B}$

- $\overline{A \cap B} \subseteq \bar{A} \cap \bar{B}; \bar{\bar{A}} = \bar{A}.$

- The-interior of A is the largest open set contained in A. It is denoted by $A°$.

- A is open $\Leftrightarrow A = A°.$

(ii) Since $D(Z) = \phi$, therefore $\bar{Z} = Z \cup \phi = Z;$

(iii) $D(Q) = R.$

$\therefore$ $\bar{Q} = Q \cup D(Q) = Q \cup R = R.$

(iv) $\overline{]0,1[} = [0,1].$

Example 2. *Given an example to show that in a metric space it is not necessary that $\overline{(A \cap B)} = \bar{A} \cap \bar{B}$.*

Solution. Let us consider the usual metric space (R, d)

and $A = [0, 1[$ and $B =]1, 2]$

Clearly, $A \cap B = \phi \Rightarrow \overline{(A \cap B)} = \bar{\phi} = \phi$

and $\bar{A} = [0,1]$ and $\bar{B} = [1,2]$

$\Rightarrow$ $\bar{A} \cap \bar{B} = \{1\}.$

Hence $\overline{(A \cap B)} \neq \bar{A} \cap \bar{B}.$

Example 3. *Define a dense set and give an example.*

Solution. By definition of dense set we have

A subset A of metric space (X, d) is said to be dense in X iff $\bar{A} = X$.

Example of dense set: let (R, d) be a usual metric space on R and Q be set of rational number then $Q \subseteq R$

$$\bar{Q} = Q \cup D(Q) = Q \cup R = R$$

$\Rightarrow$ $\bar{Q} = R$

$\Rightarrow$ Q is dense in R.

Example 4. *Define a nowhere dense set by giving a suitable example.*

Solution. Let A be non-empty subset A of X, then A is said to be nowhere dense in X iff $(\overline{A})^\circ = \phi$.

Example. Let (R, d) be usual metric on R and Z is the set of integers, $Z \subseteq R$.

$$\overline{Z} = Z \cup D(Z) = Z \cup \phi = Z$$

$$\Rightarrow \qquad (\overline{Z})^\circ = Z^\circ = \phi \Rightarrow Z \text{ is nowhere dense in } R.$$

Example 5. *Show that Cantor's set E is a non-dense set.*

Solution. Since we know that Cantor set is always a closed set.

$$\therefore \qquad\qquad \overline{E} = E \qquad\qquad\qquad\qquad (\because A \text{ is closed iff } \overline{A} = A)$$

$$\Rightarrow \qquad (\overline{E})^\circ = (E)^\circ = \cup [G \subset [0, 1] : G \subset E] = \phi$$

$$\Rightarrow \qquad (\overline{E})^\circ = \phi$$

$\Rightarrow$ E is non-dense.

Example 6. *In any metric space, show that $(\overline{A})' = (A')^\circ$.*

Solution. Let (X, d) be a metric space and A be any subset of X.

Now $(\overline{A})' = X - \overline{A} = X - $ intersection of all closed set F_i

$$= X - \cap F_i \quad \text{where } F_i \text{ is closed and } A \subset F_i$$

$$= \cup [X - F_i] \quad \text{where } X - F_i \text{ is open.}$$

Since $X - F_i \subset X - A$

Therefore $(\overline{A})' = $ Union of open subsets of $X - A = A' = (A')^\circ$.

Example 7. *Consider the usual metric $d(x, y) = |x - y|$ on R and find (i) interior, (ii) exterior, (iii) frontier, (iv) and boundary of each of the following subsets of R*

(a) $A =]0, 1[$ (b) $B = [0, 1[$ (c) $C = \left\{\dfrac{1}{n} : n \in N\right\}$ (d) N

Solution. (a) (i) Since A is an open set

$\Rightarrow$ A is a nbd of each of its points

$\Rightarrow$ every point of A is an interior point of A.

$\Rightarrow$ $A^\circ = A =]0, 1[$.

(ii) $A' =]-\infty, a[\cup [1, \infty[$

Here A' is a nbd of each of its points except 0 and 1

$\Rightarrow$ ext $(A) = (A')^\circ =]-\infty, 0 [\cup] 1, \infty[$.

(iii) $F_r(A) = [A^\circ \cup \text{ext }(A)]' = \{0, 1\}$

$\Rightarrow$ $F_r (A)$ contains two points 0 and 1.

(iv) $b(A) = \phi$. $\qquad\qquad\qquad\qquad$ ($\because$ no frontier point is a point of A)

(b) (i) $B^\circ =]0, 1[$ $\qquad\qquad\qquad\qquad\qquad\qquad$ (same as in (a))

(ii) $B' =]-\infty, 0[\cup [1, \infty[$

$\Rightarrow$ ext$(B) = (b')^\circ =]-\infty, 0[\cup]1, \infty[$

(iii) $F_r(B) = \{0, 1\}$

(iv) $b(B) = \{0\}$ $\qquad\qquad$ ($\because$ 0 is the only frontier point which belongs to B)

(c) (i) Since C can not be a nbd of any of its point $\dfrac{1}{n}$, $n = 1, 2, \ldots$, since $\exists \varepsilon > 0$ such that

$$\left]\dfrac{1}{n} - \varepsilon, \dfrac{1}{n} + \varepsilon\right[\subset D$$

$\Rightarrow$ no point of D can be its interior point so that $D^\circ = \phi$.

(ii) Clearly $D' = R - D$ is a nbd of each of its point except 0. Hence
$$\text{ext}(D) = (D')^\circ = R - \{D \cup (0)\}.$$

(iii) $F_r(D) = [D^\circ \cup \text{ext}(D)]' = D \cup \{0\}.$

(iv) $b(D) = D.$ Since all the points of D are frontier point of $D.$

(d) (i) Since N is not a nbd of any of its point $\Rightarrow N^\circ = \phi.$

(ii) Since N has no limit points, therefore we have $\overline{N} = N.$

$\Rightarrow$ N is a closed set

$\Rightarrow$ N' is open set. $\Rightarrow$ $\text{ext}(N) = (N')^\circ = N' = R - N.$

(iii) $F_r(N) = [N^\circ \cup \text{ext}(N)]' = [\phi \cup \text{ext}(N)]' = [\text{ext}(N)]' = (N')' = N.$

(iv) $b(N) = N$ \qquad (because every point of N is a frontier point of N).

Example 8. *Show that a subset F of a metric space X is closed iff $\{x \in X : d(x, F) = 0\} \subseteq F.$*

Solution. Let (X, d) be a metric space and $F \subseteq X.$

We know that $\qquad \overline{F} = \{x \in X : d(x, F) = 0\}.$

Let F be closed. Now we have to show that $\{x \in X : d(x, f) = 0\} \subseteq F.$

Since F is closed $\quad \Rightarrow \quad \overline{F} = F \Rightarrow \overline{F} \subseteq F$ \qquad (by the definition of closure)

$\Rightarrow \quad \{x \in X : d(x, F) = 0\} \subseteq F.$

Conversely let $\{x \in X: d(x, F)\} = 0 \subseteq F$ then we have to prove that F is closed.

Since we have

$$\{x \in X: d(x, F)\} = 0 \subseteq F \quad \Rightarrow \quad \overline{F} \subseteq F$$

But $\qquad\qquad\qquad F \subseteq \overline{F} \Rightarrow F = \overline{F}$

$\Rightarrow \quad F$ is closed.

15.14 SOME MORE DEFINITIONS

(1) Separable spaces. A metric space (X, d) is said to be separable if X contains a countable dense subset *i.e.* there exists a countable subset A of X such that $\overline{A} = X.$

For Example: The usual metric space (R, d) is separable, because the set Q of rational numbers is a countable dense subset of R.

(2) Bases for the neighbourhood system of a point. Let (X, d) be a metric space and $N(x)$ denotes the family of all neighbourhoods of a point $x \in X.$ A sub family $\beta(x)$ of $N(x)$ is said to be a base for $N(x)$ if to each $N \in N(x)$ there exist $B \in \beta(x)$ such that $B \subset N.$

REMARKS

- Here, $\beta(x)$ is said to be local base at x or a fundamental system of neighbourhoods of $x.$
- The set of all open interval in R form a base for the family of open subsets of R.

(3) Bases for the open sets of a metric space.

Let (X, d) be a metric space and let G be the family of all open subsets of a metric space $(X, d).$ Then a subfamily β of G is said to be base if for each point $x \in X$ and each nbd N of $x \ni B \in \beta$ such that $x \in B \subset N.$

(4) First and second countable spaces.

(a) A metric space (X, d) is said to satisfy the first axiom of countability if each point $x \in X$ possesses a countable local base. Such a space is said to be *first countable.*

(b) A metric space (X, d) is said to satisfy the second axiom of countability if there exist a countable local base for G, where G denotes the family of all open subsets of X. Such a space is said to be *completely separable space*.

15.14.1 SOME MORE RESULTS

- A second countable space is also called completely separable space.
- The usual metric space (R, d) is a first as well as a second countable spaces.
- Every metric space (X, d) is first countable.
- A metric space is separable if and only if it is second countable.
- Let (X, d) be a metric space and Y be a proper subset of X. Let d^* denote the restriction of d

on $Y \times Y$ i.e., $d^*(x, y) = d(x, y)$ whenever x and y are points of Y. Then d^* is a metric for Y called the induced metric. Then (Y, d^*) is said to be subspace of (X, d).

- A property of a metric space is said to be heriditary if and only if every sub-space of that space has that property.
- Every subspace of a separable metric space is separable *i.e.* separability is a heriditary properly in a metric space.

EXERCISE 15.3

1. Let (X, d) be a metric space and A be any subset of X. Then show that the following statements are equivalent:

   (a) A is closed.

   (b) A contains all its limit points.

   (c) $\bar{A} = A$.

2. Let (X, d) be a metric space and G be any open set in X, Show that G is disjoint from A iff G is disjoint from $\bar{A}$.

3. Let (X, d) be a metric space and $A \subset X$. Then show that $A°$ equals the set of all those points of A, which are not limit point of A'.

4. Show that the frontier of a subset of a metric space is closed.

5. Let (X, d) be a metric space and $A \subset X$. Find (i) $A°$, (ii) ext (A), (iii) $F_r(A)$, (iv) $b(A)$.

6. Are the following subsets of R, d-nbds of 3 when d denotes the usual metric defined by $d(x, y) = |x-y|$ for R?

   (i) $]2, 4[$ (ii) $[1, 3[$

   (iii) $[3, 4[$ (iv) N.

7. Show that every subspace of a discrete metric space is discrete.

8. Consider the following subset of R. Find their

closures relative to usual metric $d(x, y) = |x-y|$

   (i) $A = \{1, 2, 3, 4\}$ (ii) $B = [1]$

   (iii) $C =]2, \infty[$ (iv) $D =]1, 2[\cup]3, 4[$

   (v) $E = \left\{ \dfrac{n+1}{n} : n \in N \right\}$

9. Show that the diameter of a subset of a metric space is equal to the diameter of its closure.

10. Show that a closed set is nowhere dense if and only if its complement is everywhere open.

11. Let A be subset of a metric space (X, d). Prove that A is non-dense in X if and only if $X - \bar{A}$ is dense in X.

12. Let (X, d) be any metric space and let A be any subset of X. Prove that

    (i) $\overline{X - A} = X - A°$

    (ii) $X - \bar{A} = (X-A)°$.

13. In the metric space (R, d) where d is the usual metric on R, find the boundary of the set of integers Z.

14. Give an example of two subsets A and B of R such that

$$D(A \cap B) \neq D(A) \cap D(B)$$

the metric on R is the usual metric.

15. In a metric space prove that $(\bar{A})' = (A')°$

HINTS TO SELECTED PROBLEMS

9. Define $f(A) = \sup\{d(x, y) : x, y \in A\}$.

   By definition of closure, we have $\bar{A}$ is the smallest closed set containing A.

   Also A is closed $\Rightarrow \bar{A} = A \Rightarrow d(\bar{A}) = d(A)$.

10. Since A is dense in X if $\bar{A} = X$ and A is nowhere dense in X if int $(\bar{A}) = \phi$.

$\therefore$ A is non-dense in X

$\Leftrightarrow$ int $(\overline{A}) = \phi$

$\Leftrightarrow$ $\exists$ no open neigbourhood of any point of $\overline{A}$ such that $N \subset \overline{A}$

$\Leftrightarrow$ $\overline{A}$ contains no nbd

$\Leftrightarrow$ $S(r, x) \subset \overline{A}, r > 0 \; \forall \, x \in X$

$\Leftrightarrow$ $S(r, x) \cap (\overline{A})' \neq \phi$

$\Leftrightarrow$ x is abherent point of $(\overline{A})' \; \forall \, x \in X$.

$\Leftrightarrow$ $\overline{\left((A')\right)} = X \Leftrightarrow \overline{(A)}'$ is dense in X.

Answers

5. (i) A (ii) A' (iii) ϕ (iv) ϕ

6. (i) Yes   (ii) No   (iii) No   (iv) No

8. (i) A (ii) B (iii) $[2, \infty[$ (iv) $[1, 2] \cup [3, 4]$

    (v) $E \cup [1]$.

15.15 COMPLETE METRIC SPACE

Definition 1. *A metric space (X, d) is said to be complete if every Cauchy sequence in X converges.*

A metric space, which is not complete is said to be incomplete.

Solved Examples

Example 1. *Show that the real line is a complete metric space.*

Solution. Let $\langle s_n \rangle$ be a Cauchy sequence in R. Then by definition given any $\varepsilon > 0$ there exist a positive integer k such that

$$|x_n - x_k| < \varepsilon, \; n \geq k.$$

Let $\varepsilon = 1$.

Then $|x_n - x_{k_1}| < 1, \forall \, n \geq k_1$ for some k_1

$\Rightarrow$ $x_{k_1} - 1 < x_n < x_{k_1} + 1, \; \forall \, n \geq k_1.$

Let us define

$$p = \min\{x_1, x_2, \; \dots \; , x_{k_1} - 1, \, x_{k_1} + 1\}$$

and

$$q = \min\{x_1, x_2, \dots \quad , x_{k_1} - 1, \, x_{k_1} + 1\}$$

Then $p \leq x_n \leq q, \forall \; n \in N$

$\Rightarrow$ $\langle x_n \rangle$ is bounded sequence of real numbers.

Let $x = \lim \inf \langle x_n \rangle$ and $y = \lim \sup \langle x_n \rangle$.

Let if possible $x \neq y$ and let $y - x = s$ then $s \geq 0$.

Now since $\langle x_n \rangle$ is a Cauchy sequence, therefore, there exists a positive integer n such that

$$|x_l - x_m| < \varepsilon/2, \forall \; l, m \geq n. \tag{1}$$

Since x is the limit inferior of $\langle x_n \rangle$, therefore $x \leq x_n < x + s/4$ for infinitely many values of n. In particular, we can find a positive integer $u > n$ such that

$$x \leq x_u < x + s/4. \tag{2}$$

Again since y is the limit superior of $\langle x_n \rangle$, therefore

$$y - s/4 < x_n \leq y \text{ for infinitely many values of } n.$$

In particular, we can find a positive integer $v > n$ such that

$$y - s/4 < x_v \leq y.$$ ...(3)

From (2) and (3), we conclude that there exist positive integers $u, v > n$ such that

$$|x_u - x_v| > s/2.$$ ...(4)

This contradicts (1) which says that for all positive integers $l, m \geq n$

$$|x_l - x_m| < s/2.$$ ...(5)

In view of the above contradiction, we must have $s \not> 0$. Thus, $s = 0$ and therefore $x = y$ which gives

$$\lim \inf \langle x_n \rangle = \lim \sup \langle x_n \rangle.$$

Therefore $\lim_{n \to \infty} \langle x_n \rangle$ exists.

Hence, the real line is complete.

Example 2. *The set Z of integers with the usual metric $(d(x, y) = |x-y| \; \forall \, x, y \in Z)$ is a complete metric space.*

Solution. Let $\langle x_n \rangle$ be any Cauchy sequence in Z. Take $\varepsilon = \dfrac{1}{2}$.

Then we can find a positive integer p such that $|x_m - x_n| < \dfrac{1}{2} \; \forall \, m, n \geq p$.

Since $x_n \in Z \; \forall \, n \in N$, therefore $|x_m - x_n|$ must be a non-negative integer $\forall \, m, n \in N$. Therefore, we have

(i) $|x_m - x_n| \; \forall \, m, n \geq p$.

(ii) $|x_m - x_n|$ is a non-negative integer $\forall \, m, n \in N$.

Hence, we find that

$$|x_m - x_n| = 0 \; \forall \, m, n \geq p \quad \text{i.e.} \quad x_m = x_n \; \forall \, m, n \geq p.$$

Consequently $x_n = x_p \; \forall \, n \geq p$

$\Rightarrow$ $\langle x_n \rangle$ is eventually a constant sequence

$\Rightarrow$ $\langle x_n \rangle$ must converges to x_p.

Since, an arbitrary Cauchy sequence in Z converges.

Therefore, the set Z of integers equipped with the usual metric is a complete metric space.

Example 3. *Let $X =]0, 1[$ and let $d(x, y) = |x-y|$ for all $x, y \in X$. Then show that (X, d) is not complete.*

Solution. Let $\langle x_n \rangle$ be the sequence in X defined by setting $x_n = \dfrac{1}{n}$ for all $n \in N$. Let ε be any arbitrary positive number.

Now $d(x_m, x_n) = |x_m - x_n|.$

$$= \left| \frac{1}{m} - \frac{1}{n} \right| \leq \frac{1}{m} + \frac{1}{n}.$$ ...(1)

Let p be any positive integer greater than $\dfrac{2}{\varepsilon}$, then

$$\frac{1}{m} + \frac{1}{n} < \frac{\varepsilon}{2} + \frac{\varepsilon}{2} \; \forall \, m, n \geq p.$$

Therefore, from (1) we can find that

$$d(x_m, x_n) < \varepsilon \quad \forall \, m, n \geq p$$

$\Rightarrow$ $\langle x_n \rangle$ is a Cauchy sequence in X.

Now let $x \in X$. Then by the Archimedean property of real numbers, we can find a positive integer n such that

$$n \le \frac{1}{n} < n+1 \implies \frac{1}{n} \ge x > \frac{1}{n+1}$$

Now there are two different cases :

Case (i) If $x = \frac{1}{n}$, then $\left] \frac{1}{(n-1)}, \frac{1}{(n+1)} \right[$ is a nbd of x in $]0, 1[$ which contains only one element of the sequence $\left\langle \frac{1}{n} \right\rangle$.

Case (ii) If $x \ne \frac{1}{n}$ then $\left] \frac{1}{n+1}, \frac{1}{n} \right[$ is a nbd of x in $]0, 1[$ which does not contain any element of the sequence.

In either case the sequence can not converge.

Since x is any point of X, therefore the sequence $\langle x_n \rangle$ can not converge to any point of X. Hence (X, d) is an incomplete metric space.

THEOREM 1. *Let (X, d) be a complete metric space and Y be a subspace of X. Then Y is complete if and only if Y is closed.*

Proof. Let (X, d) be a complete metric space. Let us first suppose Y be a complete subspace of X. To show Y is closed in X.

Let $p \in X$ be any limit point of Y. Then, by definition of limit point, for every positive integer n, the open sphere $s\left(p, \frac{1}{n} \right)$ must contains a point q_n of Y such that the sequence $\langle q_n \rangle$ converges to p

$\implies \langle q_n \rangle$ is Cauchy in Y ($\because$ Every convergent sequence is Cauchy)

Since Y is complete therefore $p \in Y$.

Now since $p \in Y$ is arbitrary, therefore we can say Y contains all its limit points and hence Y is closed.

Conversely, let Y is closed. To show Y is complete.

Let $\langle s_n \rangle$ be any Cauchy sequence in Y.

$\implies \langle s_n \rangle$ is a Cauchy sequence in X.   ($\because Y$ is a subspace of X)

Also, since X is complete, therefore $\langle s_n \rangle$ must converge to a point $s_0 \in X$. We want to show $s_0 \in Y$. If the range set of $\langle s_n \rangle$ consist of finite number of distinct points then $\langle s_n \rangle$ must be of the form $\langle s_1, s_2, ..., s_n, s_0, s_0, s_0, ... \rangle$ where n is finite and hence $s_0 \in Y$. If the range set of $\langle s_n \rangle$ contains infinite many points, then s_0 is a limit of the range set of $\langle s_n \rangle$.

> ($\because$ If the range set of a convergent sequence in a metric space consists of infinitely many distinct points, then the limit of the sequence is a limit point of the range set of the sequence)

$\implies s_0$ is also a limit point of Y

$\implies s_0 \in Y$ ($\because$ Y is closed)

$\implies$ every Cauchy sequence in Y converges in Y. Hence, Y is complete.

THEOREM 2. (Canter's Intersection Theorem). *Let (X, d) be a metric space and let $\langle F_n \rangle$ be a nested sequence of non-empty closed subset of X such that $\delta(F_n) \to 0$ as $n \to \infty$, Then X is complete if and only if $\overset{\infty}{\underset{n=1}{\cap}} F_n$ consist of exactly one point.*

Proof.

(i) Necessary condition. Let (X, d) be a complete metric space and let $\langle F_n \rangle$ be the sequence of non-empty closed subsets of X such that $F_{n+1} \subset F_n \ \forall n \in N$ (by definition of nested sequence) and $\delta(F_n) \to 0$ as $n \to \infty$. To show $\underset{n \in N}{\cap} F_n$ consist of exactly one point.

Step 1. For each $n \in N$, choose a point $x_n \in F_n$. It can be easily seen that $\langle x_n \rangle$ is a Cauchy sequence of points of X. In fact, given $\varepsilon > 0$, we can find a positive integer m such that

$$\delta(F_n) < \varepsilon \ \forall \ n \geq m. \qquad \qquad ...(1)$$

Since the sequence $\langle F_n \rangle$ is nested, therefore $F_n \subseteq F_m \ \forall \ n \geq m$ and consequently

$$x_n \in F_m \ \forall \ n \geq m. \qquad \qquad ...(2)$$

Since we know that for any non-empty bounded set S in a metric space (X,d)

$$\delta(S) = \sup\{d(x, y) : x, y \in S\}$$

so that $\quad\quad d(x, y) \leq \delta(S) \ \forall \ x, y \in S \qquad \qquad ...(3)$

From (2) and (3), we conclude that

$$d(x_n, x_m) \leq \delta(F_m) \ \forall \ n \geq m \qquad \qquad ...(4)$$

Now (1) and (4) gives

$$d(x_n, x_m) < \varepsilon \ \forall \ x \geq m$$

$\Rightarrow \quad \langle x_n \rangle$ is a Cauchy sequence in X.

Step 2. Since the metric space (X, d) is complete, therefore the Cauchy sequence $\langle x_n \rangle$ converges *i.e.* there exist $x_0 \in X$ such that $x_n \to x_0$

We shall show that $x_0 \in F_n \ \forall \ x \in N$ and consequently $\underset{n \in N}{\cap} F_n$ is non-empty.

Let $k \in Z$ be fixed. Now from (2) $x_n \in F_k \ \forall \ n \geq k$

i.e. $\langle x_k, x_{k+1}, x_{k+2}, ... \rangle$ is a sequence in F_k.

The sequence $\langle x_k, x_{k+1}, ... \rangle$ converges to x_0 and therefore $x_0 \in F_k$.

Since F_k is closed $\Rightarrow X \in F$.

Now $x_0 \in F_k$ and k is arbitrary, therefore $x_0 \in F_n \ \forall \ n \in N$

$\Rightarrow \quad\quad\quad\quad\quad x_0 \in \underset{n \in N}{\cap} F_n$.

Step 3. (Uniqueness). Let if possible x_0, y_0 be two points in the intersection of F_n's. To show $x_0 = y_0$

Suppose if possible $x_0 \neq y_0$ and let $d(x_0, y_0) = \varepsilon > 0$.

Since $\delta(F_n) \to 0$, therefore we can find a positive integer m such that

$$\delta(F_n) < \varepsilon/2 \ \forall \ n \geq m.$$

Since $x_0, y_0 \in F_n \ \forall \ n \in N$, therefore in particular $x_0, y_0 \in F_m$

$\Rightarrow \quad\quad\quad\quad \delta(F_n) < \varepsilon/2 \ \forall \ n \geq m.$

Since $d(x_0, y_0) = \varepsilon$, we have $\varepsilon = d(x_0, y_0) < \varepsilon/2$

which is a contradiction and therefore $x_0 = y_0$ and so $\underset{n \in N}{\cap} F_n$ consist of exactly one point.

(ii) **Condition is sufficient.** Let us suppose that every nested sequence of closed sets with diameter tending to zero has non-empty intersection.

To show X is complete.

Let $\langle x_n \rangle$ be a Cauchy sequence in X.

Correspond to $\varepsilon = 1/2$, we can find $n_1 \in N$ such that

$$d(x_{n_1}, x_n) < \frac{1}{2} \ \forall \ n \geq n_1 \text{ and for } n_2 > n_1$$

$$d(x_{n_2}, x_n) < \frac{1}{2^2} \ \forall \ n \geq n_2 .$$

Proceeding in the same manner, we can construct a strictly increasing sequence $\langle n_1, n_2, n_3, \ldots \rangle$ of positive integers such that

$$d\left(x_{n_1}, x_n\right) < \frac{1}{2}, \forall \, n \geq n_1$$

$$d\left(x_{n_2}, x_n\right) < \frac{1}{2^2}, \forall \, n \geq n_2$$

$$\ldots \quad \ldots \quad \ldots \quad \ldots \quad \ldots$$

$$d\left(x_{n_k}, x_n\right) < \frac{1}{2^k}, \forall \, n \geq n_k$$

$$\ldots \quad \ldots \quad \ldots \quad \ldots$$

Let us write $\qquad F_k = S^*\left(x_{n_k}, 2^{-k+1}\right), k = 1, 2, 3, \ldots$

Then $\langle F_n \rangle$ is a sequence of closed sets with diameters tending to zero.

To see that $\langle x_k \rangle$ is nested *i.e.* $F_{k+1} \subset F_k \ \forall \ k \geq 1$, take a point $y \in F_{k+1}$.

Then $\qquad d\left(x_{n_k}, y\right) \leq d\left(x_{n_k}, x_{n_{k+1}}\right) + d\left(x_{n_{k+1}}, y\right) \qquad$ (By triangle inequality)

$$\leq 2^{-k} + 2^{-k} = 2^{-k+1}$$

$$\Rightarrow \qquad y \in S\left(x_{n_k}, 2^{-k+1}\right) \subset F_k$$

$$\Rightarrow \qquad F_{k+1} \subset F_k.$$

Thus, we find that $\langle F_n \rangle$ is a nested sequence of non-empty closed sets with diameters tending to zero therefore there exist $x_0 \in X$ such that $x_0 \in F_k \ \forall \ N$.

Consider the sequence $\langle x_{n_1}, x_{n_2}, x_{n_3}, \ldots \rangle$.

Since $n_1 < n_2 < n_3$ therefore the above sequence is a subsequence of $\langle x_n \rangle$.

Also, $\qquad \left(x_{n_k}, x_0\right) < 2^{-k} \quad \forall \, k \in N$

Since $\langle x_n \rangle$ is a Cauchy sequence in x and $\langle x_{n_k} \rangle$ is a subsequence of $\langle x_n \rangle$ converging to x_0, therefore $\langle x_n \rangle$ converges to x_0.

Hence (X, d) is complete.

15.16 METRIC SPACE OF FIRST AND SECOND CATEGORY

Definition 1. *Let (X, d) be a metric space. A subset of a metric space is said to be of first category if and only if it can be written as the union of a countable family of nowhere dense sets: otherwise it is said to be the second category.*

For example. Set of rational numbers Q is of second category.

Definition 2. (Contracting Mapping). *Let (X, d) be a complete metric space.*

A mapping $f : X \rightarrow X$ is called a contracting, mapping (or contraction on X) if there exist a real number α with $0 \leq \alpha < 1$ such that

$$d[f(x), f(y)] \leq \alpha \, d(x, y) < d(x, y) \ \forall \, x, y \in X.$$

Definition 3. *Let X be a non-empty set and let f:X→X be a mapping. A point x ∈ X is said to be a fixed point of f if f(x)=x.*

For example. Let $f: X \to X$ be the identity mapping on X such $f(x)=x$
Therefore every point of X is a fixed point.

<u>THEOREM 1.</u> *Let A be a subset of a metric space X, then following statements are equivalent:*

      (i) *A is non-dense in X.*

      (ii) *A contains no neighbourhoods*

      (iii) *$(\bar{A})'$ is dense in X.*

<u>Proof.</u> Let (X, d) be a metric space and A be any subset of X.

Here, we first prove (i) ⟺ (ii).

A is non-dense in $X \Leftrightarrow (\bar{A})° = \phi$

                ⟺ No point of X is an interior point of $\bar{A}$

                ⟺ $\bar{A}$ is not a nbd of any of its points

                ⟺ $\bar{A}$ contains no nbd.

We now prove (ii) ⟺(iii)

$\bar{A}$ contains no nbd ⟺ For every $x \in X$, $S(x_0, r) \not\subset \bar{A}, r > 0$

                ⟺ $S(x, r) \cap (\bar{A})' \neq \phi$ for every $x \in X$ and every $r>0$

                ⟺ every nbd of x contain a point of $(\bar{A})'$ for every $x \in X$

                ⟺ x is an adherent point of $(\bar{A})'$ for every $x \in X$

                ⟺ $\left[\overline{(A')}\right] = X$ [By definition of adherent point]

                ⟺ $(\bar{A})'$ is dense in X.

REMARKS

- Since $(\bar{A})'$=ext (A), it follows, from that A is non-dense if and only if ext (A) is everywhere dense.
- If A is non-dense in X, then A' is dense in X.
- If A is nowhere dense, then $\bar{A}$ is not the entire space X.

<u>THEOREM 2.</u> *The union of a finite number of nowhere dense sets is no where dense.*

<u>Proof.</u> Let A and B be two nowhere dense subsets of a metric space (X, d).

Let us write $G = \left(\overline{A \cup B}\right)°$ so that $G \subset \overline{A \cup B} = \bar{A} \cup \bar{B}$.

It follows that

$$G \cap (\bar{B})' \subset (\bar{A} \cup \bar{B}) \cap (\bar{B})'$$

$$= [\bar{A} \cap (\bar{B})'] \cup [\bar{B} \cap (\bar{B})'] \qquad \text{(By distributivity)}$$

$$= \bar{A} \cap (\bar{B})' \qquad\qquad [\because \bar{B} \cap (\bar{B})' = \phi]$$

$$\subset \bar{A}$$

$$\Rightarrow \qquad [G \cap (\bar{B})']° \subset (\bar{A})° = \phi \qquad\qquad\qquad\qquad ...(1)$$

                               $[\because$ A is non-dense *i.e.* $(A)°=\phi]$

But $[G \cap (\bar{B})']°=G \cap (\bar{B})' \qquad\qquad\qquad\qquad ...(2)$

                               $[\because G \cap (\bar{B})'$ is open so use $A°=A]$

From (1) and (2), we conclude that

$$G \cap (\overline{B})' = \phi$$

$$\Rightarrow \qquad G \subset \overline{B} \Rightarrow G° = (\overline{B})° = \phi \qquad [\because B \text{ is non-dense in } X \text{ so } (\overline{B})° = \phi]$$

But $\qquad G° = (A \cup B)°°$

$$= \overline{(A \cup B)}°$$

so that $\qquad (A \cup B)° = \phi.$

Hence, $A \cup B$ is a non-dense.

In general, the union of a finite number of no where dense set is no where dense.

THEOREM 3. **(Bair's Category Theorem).** *Every complete metric space is of second category.*

Proof. Let (X, d) be a complete metric space. To show X is of second category. Let if possible, X is not of second category *i.e.* X is of first category.

$\Rightarrow$ X can be expressed as a countable union of nowhere dense sets arranged in a sequence $\langle A_n \rangle$. Now, since A_1 is non-dense and so there exists a closed sphere K_1 of radius $r_1 < \dfrac{1}{2}$ s.t. $K_1 \cap A_1 = \phi.$

Let the open sphere, with same centre and redius as r_1, be denoted by S_1.

In S_1, we can find a closed sphere K_2 of radius $r_2 < \left(\dfrac{1}{2}\right)^2$ such that

$$K_2 \cap A_2 = \phi \text{ and so } K_2 \cap A_1 = \phi.$$

Continuing this process, we constitute a nested sequence $\langle K_n \rangle$ of closed spheres which have the following properties :

(i) For each positive integer, n, K_n does not intersect $A_1, A_2, ..., A_{n...}$

(ii) The radius of K_n tends to zero as $n \to \infty.$ $\qquad \left(\text{For } \dfrac{1}{2^n} \to 0 \text{ as } n \to \infty \right)$

Since (X, d) is complete, therefore by Cantor's intersection theorem $\underset{n}{\cap} K_n$ contains a single point x_0, so that

$$x_0 \in K_n, \forall n.$$

By (i), $x_0 \notin A_n \ \forall \ n$. By assumption, it is not possible, because X is the union of A_n's. Hence, X is not of first category, which is a contradiction.

Hence, X is of second category.

REMARKS

- The Bair's category theorem can also be stated as follows :

"If $\langle A_n \rangle$ is a sequence of nowhere dense sets in a complete metric space (X, d) ∃ a point in X, which is not in A_n's. or "If a complete metric space is the union of a sequence of its subsets, then the closure of at least one set in the sequence must have non-empty interior.

THEOREM 4. *Every contracting mapping is continuous.*

Proof. Let f be a contracting mapping on a metric space (X, d), therefore there exists a positive real number $\alpha < 1$ such that

$$f:(X, d) \to (X, d), \forall \ x, y \in X$$

and $\qquad d\,[f(x),f(y)] \le \alpha\,d(x,y) < d(x,y).$ ...(1)

Taking $d(x,y)<\varepsilon$, we get $d[f(x),f(y)]<\varepsilon.$

Given $\varepsilon>0\ \exists\ \delta>0$ such that $d[f(x),f(y)]<\varepsilon$, whenever $d(x,y)<\delta.$

Here $\qquad\qquad\qquad \varepsilon = \delta.$

Hence, f is a continuous mapping.

THEOREM 5. **(Banach Fixed Point Theorem).** *If f is a contracting mapping on a complete metric space (X,d), then there exists a unique point p in X such that $f(p)=p$.*

Proof. Let (X,d) be a complete metric space. Define a contracting mapping $f:(X,d)\to(X,d)$. Then by definition of contracting mapping there exists a positive real number $\alpha<1$ such that

$$d\,[f(x),f(y)] \le \alpha\,d(x,y)\ \forall\ x,y \in X.$$ ...(1)

Then $\qquad d[f^2(x),f^2(y)]=d[f(f(x)),f(f(y))]$

$$\le \alpha\,d[f(x),f(y)] \qquad\qquad \text{[using (1)]}$$

$$\le \alpha.\alpha.d(x,y) \qquad\qquad \text{[using (1)]}$$

$$\Rightarrow \qquad d[f^2(x),f^2(y)] \le \alpha^2\,d(x,y).$$

In general, we get

$$d[f^n(x),f^n(y)] \le \alpha^n.d(x,y),\ n \in \mathbb{N}.$$ ...(2)

Next, we suppose that $x_0 \in X$ and let

$$x_1=f(x_0),\ x_2=f(x_1)=ff(x_0)=f^2(x_0)$$
$$x_n=f(x_{n-1})=f^n(x_0).$$

Thus $\qquad x_n=f^n(x_0),\ x_{n+1}=f^{n+1}(x_0)=f^n[f(x_0)]=f^n(x_1)$

$$\Rightarrow \qquad x_n=f^n(x_0),\ x_{n+1}=f^n(x_1).$$ ...(3)

We claim that $\langle x_n\rangle$ is a Cauchy sequence.

Let $m,n \in \mathbb{N}$ be arbitrary such that $m>n$ and $m=n+p$, p being a positive integer ≥ 1.

Now $\qquad d(x_n,x_m)=d(x_n,x_{n+p})$

$$\le d(x_n,x_{n+1})+d(x_{n+1},x_{n+2})+...+d(x_{n+p-1},x_{n+p})$$

$$\text{(By triangle inequality)}$$

Now using (3)

$$d(x_n,x_n) \le d[f^n(x_0),f^n(x_1)]+d[f^{n+1}(x_0),f^{n+1}(x_1)]$$
$$+d[f^{n+p-1}(x_0),f^{n+p-1}(x_1)]$$

On using (2)

$$d(x_n,x_m) \le \alpha^n d(x_0,x_1)+\alpha^{n+1}d(x_0,x_1)+...+\alpha^{n+p-1}d(x_0,x_1)$$

$$= \alpha^n d(x_0,x_1)[1+\alpha+\alpha^2+...+\alpha^{p-1}]$$

$$= d(x_0,x_1).\alpha^n\frac{\left(1-\alpha^p\right)}{(1-\alpha)} \qquad\qquad \text{[being the sum of G.P.]}$$

$$\le d(x_0,x_1)\frac{\alpha^n}{1-\alpha} \qquad \text{for } 0<\alpha<1.$$

Thus $\qquad d(x_n,x_m) \le \dfrac{\alpha^n}{1-\alpha}d(x_0,x_1).$

Therefore $\qquad 0<\alpha<1 \Rightarrow \lim_{n\to\infty} \alpha^n = 0$

Now (4) gives $\qquad d(x_n, x_m)<\varepsilon.$

$\Rightarrow \quad \langle x_n \rangle$ is a Cauchy sequence.

Now since (X, d) is complete, therefore by definition, every Cauchy sequence in X converges to some point in X

$\Rightarrow \quad \exists\, p \in X\, s.t.\, x_n \to p$ as $n \to \infty.$

$\Rightarrow \qquad\qquad\qquad \lim x_n = p.$

Also, f is contracting mapping $\Rightarrow f$ is continuous.

$\qquad\qquad\qquad\qquad\qquad\qquad$ ($\because$ Every contracting mapping is continuous)

Therefore $\qquad\qquad x_n \to p \Rightarrow f(x_n) \to f(p) \Rightarrow \lim f(x_n) = f(p).$

But $\qquad\qquad\qquad x_{n+1} = f(x_n).$

$\therefore \qquad\qquad\qquad f(p) = \lim f(x_n) = \lim x_{n+1} = \lim x_n = p$

or $\qquad\qquad\qquad f(p)=p$

$\Rightarrow \quad p$ is a fixed point.

Now to show the uniqueness of the point p.

Let if possible $\exists$ another fixed point $q \in X$ such that $p \neq q$ and $f(q) = q.$

Consider $\qquad d[f(p), f(q)] \leq \alpha\, d(p, q)$

$\Rightarrow \quad d(p, q) \leq \alpha\, d(p, q).$ But $d(p, q)\neq 0$

$\Rightarrow \quad \alpha \geq 1$, which is a contradiction $\qquad\qquad\qquad\qquad (0< \alpha< 1)$

Hence, p is unique.

15.16 COMPLETION OF A METRIC SPACE

Definition 1. *A metric space (X, d) is said to be isometric to a space (Y, d_1) if there exist one-one mapping $f : X \to Y$ which preserves the distance i.e.*

$$d(x, y) = d_1[f(x), f(y)] \ \forall\, x, y \in X.$$

Definition 2. *If (X, d) is not complete metric space, then we extend the metric space (X, d) to (X_1, d_1) which is complete and (X, d) is a dense subspace.*

A metric space (X_1, d_1) is called completion of a metric space (X, d) if X_1 is complete and X is isometric to a dense subset of X.

THEOREM 1. *The metric space (R, d) is complete where d denotes usual metric for R.*

Proof. Let $\langle s_n \rangle$ be a Cauchy sequence of real numbers. Now, we define a sequence $\langle n_k \rangle$ of positive integers as n_{k+1} is the smallest integer greater than n_k such that

$$n, m \geq n_k \Rightarrow |s_n - s_m| < \frac{1}{2^{k+1}}$$

$\qquad\qquad\qquad\qquad\qquad$ (Since $\langle s_n \rangle$ in Cauchy sequence)

Now let I_k be the closed interval $\left[s_{n_k} - 2^{-k}, s_{n_k} + 2^{-k} \right].$

Here, it is easy to see that $I_{k+1} \subset I_k$, we have

$$\left| s_{n_k} - s_{n_{k+1}} \right| < \frac{1}{2^{k+1}}$$

Also, the length of $I_k \to 0$ as $k \to \infty$. So by nested interval theorem, $\bigcap_{k=1}^{\infty} I_k$ consists of exactly one point, say $a \in R$.

Thus $a \in I_k$, $\forall k \in N$ so that $\left| a - s_{n_k} \right| < \frac{1}{2^k}$ for all $k \in N$. ...(2)

Now for $n \geq n_k$, we have from (1)

$$\left| s_{n_k} - s_n \right| < \frac{1}{2^{k+1}} < \frac{1}{2^k}.$$ ...(3)

Hence, for all $n \geq n_k$.

$$\left| a - s_n \right| = \left| a - s_{n_k} + s_{n_k} - s_n \right|$$
$$\leq \left| a - s_{n_k} \right| + \left| s_{n_k} - s_n \right|$$
$$< \frac{1}{2^k} + \frac{1}{2^k} = \frac{1}{2^{k-1}} \qquad \text{[by (2) and (3)]}$$

It follows that $\lim_{n \to \infty} s_n = a$.

Thus every Cauchy sequence in R converges to a point in R.

Hence, R is complete.

THEOREM 2. *The set R^n of all n-tuples $x = (x_1, x_2, \ldots, x_m)$ of real numbers is a complete metric space with respect to the usual metric*

$$d(x, y) = \left[\sum_{i=1}^{n} (x_i - y_i)^2 \right]^{1/2}.$$

Proof. Here, we shall denote the elements of R^n by a functional notation. Thus, an element of R^n will be a real function defined on the set $\{1, 2, 3, \ldots, n\}$. Let $f(m)$ stand for the element $[f_m(1), f_m(2) \ldots f_m(n)]$ of R^n. Now we shall show that R^n is complete.

Let $\langle f_m \rangle$ be a Cauchy sequence in R^n, then for a given $\varepsilon > 0$, $\exists$ a positive integer $n(\varepsilon)$ such that

$$p, q \geq n(\varepsilon) \Rightarrow d(f_p, f_q) < \varepsilon$$
$$\Rightarrow d^2(f_p, f_q) < \varepsilon^2$$
$$\Rightarrow \sum_{i=1}^{n} \left[f_p(i) - f_q(i) \right]^2 < \varepsilon^2$$
$$\Rightarrow [f_p(i) - f_q(i)]^2 < \varepsilon^2 \qquad (i = 1, 2, 3, \ldots, n)$$
$$\Rightarrow [f_p(i) - f_q(i)] < \varepsilon \qquad (i = 1, 2, 3, \ldots, n)$$

So $\langle f_m(i) \rangle$ is a Cauchy sequence of real numbers. Since R is complete, so every Cauchy sequence in R converge to a point in R.

i.e. The sequence $\langle f_m \rangle$ converges pointwise to a limit function f defined by

$$\lim f_m(i) = f(i).$$

Since the set $\{1, 2, \ldots, n\}$ is finite, this convergence is uniform. Hence there exists a positive integer n_0 such that

$$\left| f_m(i) - f(i) \right| < \frac{\varepsilon}{\sqrt{n}}, \text{ for all } m \geq n_0 \text{ and } \forall i.$$

Now, squaring and adding the above result for $i = 1, 2,..., n$, we get

$$\sum_{i=1}^{n} \left| f_m(i) - f(i) \right|^2 < \frac{\varepsilon^2}{n} \cdot n = \varepsilon^2$$

or, $\qquad d^2(f_m, f) < \varepsilon^2 \; \forall \; m \geq n_0$ thus $d(f_m, f) < \varepsilon, \; \forall \; m \geq n_0$

which shows that the Cauchy sequence $\langle f_m \rangle$ converges to the limit f.

Hence, R^n is complete.

THEOREM 3. *The Hilbert space (l_2, d) is complete.*

Proof. Let $\langle s_n \rangle$ be a Cauchy sequence in l_2. Since each $<s_n>$ is a sequence. So we have

$$s_n = \left\langle s_1^n, s_2^n, s_3^n \, ... \right\rangle.$$

So that s_k^n denotes the kth term of the sequence $<s_n>$. Now

$$d(s_m, s_n) = \left[\sum_{i=1}^{n} \left(s_i^m - s_i^n \right)^2 \right]^{1/2}.$$

Now, Since $\langle s_n \rangle$ in a Cauchy sequence, so for a given $\varepsilon > 0 \; \exists$ a positive integer m such that

$$m, n \geq m_0 \Rightarrow d(s_m - s_n) < \varepsilon \Rightarrow d^2(s_m - s_n) < \varepsilon^2$$

$$\Rightarrow \sum_{i=1}^{n} \left(s_i^m - s_i^n \right)^2 < \varepsilon^2 \Rightarrow \left(s_i^m - s_i^n \right) < \varepsilon^2, \forall \, i \in N$$

$$\Rightarrow \left| s_i^m - s_i^n \right| < \varepsilon \; \forall \; i \in N.$$

So, $\{ \left\langle s_i^n \right\rangle : i = 1,...,n \}$ is a Cauchy sequence of real numbers for all $i \in N$.

and it must converge to some real number x_i, let $X = \left\langle x_1, x_2, ... , \right\rangle$.

Now, we shall show that the sequence $\langle s_n \rangle$ converges to X, and that $X \in l_2$.

Now for a fixed integer M, we have

$$\sum_{i=1}^{m} \left(s_i^m - s_i^n \right) < \varepsilon^2, \forall \; m, n \geq m_0.$$

If we fix n and let $m \rightarrow \infty$, we get

$$\sum_{i=1}^{m} \left(s_i - s_i^n \right)^2 < \varepsilon^2. \qquad \qquad ...(1)$$

Since, (1) hold for all m, we have for $n \geq m_0$.

$$\sum_{i=1}^{\infty} \left(s_i - s_i^n \right)^2 < \varepsilon^2. \qquad \qquad ...(2)$$

i.e., $\qquad d^2(x, s_n) < \varepsilon^2 \Rightarrow d(x, s_n) < \varepsilon.$

Hence, $\qquad \lim_{n \to \infty} s_n = x.$

Now we shall show that $x \in l_2$, we have

$$x_i^2 = \left(x_i - s_i^{m_0} + s_i^{m_0} \right)^2$$

$$= \left(x_i - s_i^{m_0} \right)^2 + \left(s_i^{m_0} \right)^2 + 2 \left(x_i - s_i^{m_0} \right) \left(s_i^{m_0} \right)$$

$$\geq 2 \left(x_i - s_i^{m_0} \right)^2 + 2 \left(s_i^{m_0} \right)^2. \qquad \qquad (\because 2ab \leq a^2 + b^2)$$

So
$$\sum_{i=1}^{\infty} x_i^2 \le 2\sum_{i=1}^{\infty} \left(x_i - s_i^{m_0}\right)^2 + 2\sum_{i=1}^{\infty} \left(s_i^{m_0}\right)^2$$

$$< 2\varepsilon^2 + 2\sum_{i=1}^{\infty} \left(s_i^{m_0}\right)^2 \qquad \text{[by (2)]} \qquad \qquad ...(3)$$

Since $s_{m_0} \in l_2$, the series $\sum_{i=1}^{\infty} \left(s_i^{m_0}\right)^2$ converges so that for some $B>0$, we have

$$\sum_{i=1}^{\infty} \left(s_i^{m_0}\right)^2 < B \qquad\qquad ...(4)$$

Now from (2) and (4), we have
$$\sum_{i=1}^{\infty} x_i^2 < 2\varepsilon^2 + 2B.$$

So the series $\sum_{i=1}^{\infty} x_i^2$ is convergent and consequently $x \in l_2$.

Thus, every Cauchy sequence, $\langle s_n \rangle$ in l_2 converges to a point x in l_2.

Hence, (l_2, d) is complete.

Completeness of $C[a, b]$. Let $C[a, b]$ is the set of all real valued continuous funcions defined on I ($I=[a, b]$). For $f . g \in C[a, b]$. We define $d(f, g)=\sup\{[f(x)-g(x): x \in I\}$. We have already seen that d is matric for $C[a, b]$. So this space is known as the space of continuous functions on I

THEOREM 4. *The space $C[a, b]$ is complete.*

Proof.   Let $I=[a, b]$ and let $\langle f_m \rangle$ be a Cauchy sequence in $C[a, b]$.

Let for a given $\varepsilon>0$, there exist a positive integers m_0 such that
$$m, n \ge m_0 \Rightarrow d(f_m, f_n)<\varepsilon$$
$$\Rightarrow \sup\{|f_m(x)-f_n(x)| : x \in I\}<\varepsilon$$

So $|f_m(x)-f_n(x)|< \varepsilon, \forall m, n \ge m_0$ and for all $x \in l$.

This is the condition for uniform convergence. Thus the sequence $\langle f_m \rangle$ is a uniform convergent sequence of continuous functions and it must converge to a continuous function f on I.

Thus every Cauchy sequence $\langle f_m \rangle$ in $C[a, b]$ converges to a point f in $C[a, b]$. Hence, $C[a, b]$ is complete.

Solved Example

Example 1.   *Show that the set C of complex numbers with usual metric is a complete metric space.*

Solution.   Let C is the set of complex number and $z_1, z_2 \in C$ such that
$$z_1=x_1+iy_1 \text{ and } z_2= x_2+ iy_2 \text{ where } x_1, y_1, x_2, y_2 \in R$$
Let us definedan metric on C as
$$d(z_1, z_2)= |z_1-z_2|.$$
Let $\langle z_n \rangle$ be a Cauchy sequence in C.

To show that C is a complete metric space, we have to show that $\langle z_n \rangle$ converges to a point $z \in C$. Let for given $\varepsilon>0$, there exist $m \in N$ such that
$$|z_n-z_m|<\varepsilon, \forall n \ge m$$
$$\Rightarrow \qquad |(x_n+iy_n)-(x_m+iy_m)|<\varepsilon$$

$$\Rightarrow \quad |(x_n - x_m) + i(y_n - y_m)| < \varepsilon$$

$$\Rightarrow \quad |(x_n - x_m) + i(y_n - y_m)|^2 < \varepsilon^2$$

$$\Rightarrow \quad |x_n - x_m|^2 + |y_n - y_m|^2 < \varepsilon^2$$

$$\Rightarrow \quad |x_n - x_m| < \varepsilon \text{ and } |y_n - y_m| < \varepsilon, \forall\, n \geq m$$

$$\Rightarrow \quad \langle x_n \rangle \text{ and } \langle y_n \rangle \text{ are Cauchy sequence in R}$$

and we know that every Cauchy sequence converges to a point.

$$\Rightarrow \quad x_n \to x \text{ and } y_n \to y \text{ in R}$$

$$\Rightarrow \quad z_n \to x + iy = z \text{ in C}$$

$$\Rightarrow \quad \langle z_n \rangle \text{ converges to a point } z \text{ in C}$$

$$\Rightarrow \quad \text{C is convergent}$$

$$\Rightarrow \quad (C, d) \text{ is a complete metric space.}$$

THEOREM 5. *Let C[a, b] be the set of all continuous functions on [a, b]. For f, g ∈ C[a, b], define*

$$d(f, g) = \left\{ \int_a^b |f(x) - g(x)|^2\, dx \right\}^{1/2}$$

then ρ is a metric for C[a, b], which is not complete.

Proof. It is easy to see that ρ is a metric for C[a, b].

Now we shall show that the space is not complete. For this, consider the sequence of continuous functions $\langle f_n \rangle$ defined on [−1, 1] by

$$f_n(x) = \begin{cases} 0 & \text{if } -1 \leq x \leq 0 \\ nx & \text{if } 0 \leq x \leq \dfrac{1}{n} \\ 1 & \text{if } \dfrac{1}{n} \leq x \leq 1 \end{cases}$$

Then for $m > n$, we have

$$d^2(f_m, f_n) = (m-n)^2 \int_0^{1/m} x^2\, dx + \int_{1/m}^{1/n} (1 - nx)^2\, dx$$

$$= \frac{(m-n)^2}{3m^3} + \left[\frac{1}{3n} \left(1 - \frac{n}{m} \right)^3 \right]$$

$$= \frac{(m-n)^2}{3m^3} + \frac{(m-n)^3}{3nm^3} = \frac{(m-n)^2}{3nm^2}$$

$$< \frac{1}{3n} < \varepsilon \text{ if } n > \frac{1}{3\varepsilon}$$

Therefore, the sequence is a Cauchy sequence.

Suppose, if possible, this Cauchy sequence converges to a continuous function f so that

$$d(f_n, f) = \int_{-1}^{1} |f_n(x) - f(x)|^2\, dx \to 0.$$

This implies that the integral with any limits between ±1 also tends to 0. Thus

$$\int_{-1}^{0} |f_n(x) - f(x)|^2\, dx \to 0.$$

But $f_n(x) = 0$ when $x \leq 0$ and hence this interval is independent of n. So the continuous function f is such that $\int_{-1}^{0} |f(x)|^2\, dx = 0$.

It follows that $f(x) = 0$ when $x \le 0$. Again, if $c > 0$, then

$$\int_c^1 |f_n(x) - f(x)|^2\, dx \to 0 \text{ as } n \to \infty.$$

If we choose $n \to \dfrac{1}{c}$, we have

$$\int_c^1 |1 - f(x)|^2\, dx \to 0 \text{ as } n \to \infty.$$

As the interval is independent of n, is vanishes, and since $f(x)$ is continuous we have $f(x) = 1$ for $x \ge c$. But we can choose as near to zero as we want, thus there exists a continuous function which vanishes when $x \le 0$ which is equal to 1 when $x > 0$. So the Cauchy sequence does not converge to a point of $C[a, b]$. Hence the space is not complete.

THEOREM 6. *Let (X, d) and (Y, e) be two complete metric space then the product space $Z = X \times Y$ with metric*

$$\rho(z_1, z_2) = \sqrt{\left[d^2(x_1, x_2) + e^2(y_1, y_2) \right]}$$

is complete where $z_1 = (x_1, y_1)$ and $z_2 = (x_2, y_2)$.

Proof. We know that ρ is a metric for Z. Now we show that the product space (Z, ρ) is complete. Let $\langle z_n \rangle$ be a Cauchy sequence in Z. Then for a given $\varepsilon > 0$, there exists a positive integer m_0 such that

$$m, n \ge m_0 \Rightarrow \rho(z_m, z_n) < \varepsilon$$
$$\Rightarrow \rho^2(z_m, z_n) < \varepsilon^2$$
$$\Rightarrow d^2(x_m, x_n) + e^2(y_m, y_n) < \varepsilon^2$$
$$\Rightarrow d^2(x_m, x_n) < \varepsilon^2 \text{ and } e^2(y_m, y_n) < \varepsilon^2$$
$$\Rightarrow d(x_m, x_n) < \varepsilon \text{ and } e(y_m, y_n) < \varepsilon$$

It follows that $\langle x_n \rangle$ and $\langle y_n \rangle$ are Cauchy sequences in the space X and Y respectively. Since these space are complete, the sequences $\langle x_n \rangle$ and $\langle y_n \rangle$ converges respectively to point $x \in X$ and $y \in Y$. It follows that the sequence $\langle z_n \rangle$ converges to $z = (x, y) \in Z$ and consequently the product space $Z = X \times Y$ is complete.

REMARK

- The space defined in the above theorem is known as product metric space.

EXERCISE 15.4

1. Show that set of real numbers is complete.

2. Show that the Hilbert space is complete.

3. Define a complete metric space. Give an example of a complete metric space.

4. Define an incomplete metric space. Give an example of an incomplete metric space.

5. Show that set C^n of all ordered n-tuples $z = (z_1, z_2, ..., z_n)$ of complex number is a complete metric space with respect to the usual metric d defined by

$$d(z, u) = \left[\sum_{i=1}^n |z_i - u_i|^2 \right]^{1/2}$$

where $(z = z_1, z_2, ..., z_n)$ and $u = (u_1, u_2, ..., u_n)$

6. Show that the metric space of rational numbers with the usual metric is incomplete.

7. Let $D(a, b)$ denote the set of all functions f on $[a, b]$ which have continuous derivatives at all points of $I = [a, b]$. For $f, g \in D[a, b]$, define

$$d(f,g) = |f(a) - g(b)| + \sup\{|f'(x) - g'(x)| : x \in Z\}.$$

show that d is a metric for $D(a, b)$ and that the space $(D[a, b], d)$ is complete.

8. Show that the completeness is preserved under isometrics.

9. Let X consist of all ordered pairs $X = (x_1, x_2)$ be real numbers with metric

$d(x, y) = \max\{|x_1 - y_1|, |x_2 - y_2|\}$.
Prove that X is complete.

10. Let X consist of all bounded sequences $X = \langle x_n \rangle$ in R. Prove that

$$d(x, y) = \sup\{|x_i - y_i| : i \in N\}$$

is a metric on X and X is complete.

11. Which of the following are Cauchy sequence in R with respect to the usual metric d

(i) $\left\langle \dfrac{1}{n} \right\rangle$ (ii) $\left\langle \dfrac{n-1}{n+1} \right\rangle$

(iii) $\left\langle \dfrac{2^{n+1}}{2^n} + 1 \right\rangle$ (iv) $\left\langle 1 + \dfrac{(-1)^n}{n} \right\rangle$.

12. Let X consist of all sequence $X = \langle x_n \rangle$ in R. Show that

$$d(x, y) = \sum_1^\infty \dfrac{|x_n - y_n|}{[1 + |x_n - y_n|]}$$

is a metric on X and X is complete.

HINTS TO SELECTED PROBLEMS

10. Let $\langle x_n \rangle$ be a cauchy sequence in the given space X so that given $\varepsilon > 0$, $\exists\, n_0 \in N$ such that

$m, n \geq n_0 \Rightarrow d(x_m - x_n) < \varepsilon$

$\Rightarrow \sup\{|x_m - x_n| : n \in N\} < \varepsilon$

$\Rightarrow \langle x_n \rangle$ is uniformly convergent

$\Rightarrow \langle x_n \rangle$ is convergent.

$\Rightarrow X$ is complete.

9. Do same as above.

11. Let $s_n = \dfrac{1}{n}$

$$d(s_n, s_m) = |s_n - s_m| = \left|\dfrac{1}{n} - \dfrac{1}{m}\right|$$

$$\leq \dfrac{1}{n} + \dfrac{1}{m}$$

Now there may exists a positive integer p greater then $\dfrac{2}{\varepsilon}$ then

$$\dfrac{1}{n} + \dfrac{1}{m} < \dfrac{\varepsilon}{2} + \dfrac{\varepsilon}{2} \quad \forall\, m, n \geq p$$

$\therefore$ $d(s_n, s_m) \leq \dfrac{1}{n} + \dfrac{1}{m} < \dfrac{\varepsilon}{2} + \dfrac{\varepsilon}{2} \quad \forall\, m, n \geq p$

$\Rightarrow$ $d(s_n, s_m) < \varepsilon$

$\Rightarrow$ $\langle s_n \rangle$ is a Cauchy sequence.

CHAPTER REVIEW: A COMPETITIVE APPROACH

SELECTED TERMS AND RESULTS

TERMS

- **Metric :** Let X be a non-empty set and d be a function defined on $X \times X$ into the set R (*i.e.* set of reals) such that the image of any ordered pair (x, y) of $X \times X$ is denoted by d (x, y), then d is called a metric or distance function iff d satisfies the following axioms :

$M_1 : d\ (x, y) > 0\ \forall\ x, y \in X$ (*Non-Negativity*)

i.e. distance between any two points is a non-negative real number.

$M_2 : d(x, y) = 0 \Leftrightarrow x = y\ \forall\ x, y \in X.$

i.e. distance between two points is zero iff the points coincide.

$M_3 : d(x, y) = d(y, x)\ \forall\ x, y \in X$ (*Symmetry*)

i.e. distance between x and y is the same as the distance between y and x.

$M_4 : d(x, y) \leq d(x, z) + d(z, y)\ \forall\ x, y, z \in Y.$

(*Triangular Inequality*)

i.e. any side of a triangle is always less than or equal to the sum of the remaining two sides. The equality occurring when all the three points forming a triangle are collinear.

- **Metric Space:** The set X along with metric d defined as above is called metric space and is written as (X, d). By defining the function d in various manners we can have various metric spaces on a given set X.

- **Metrizable:** A non-empty set X is said to be metrizable iff a metric d can be defined on it satisfying the four axioms M_1, M_2, M_3, M_4.

- **Pseudo Metric:** The mapping $d : X \times X \to R$ is called a pseudo metric for X iff d satisfies M_1, M_3, M_4 axioms written above and in place of axiom M_2 it satisfies M_2' *i.e.* $d(x, x) = 0$

- **(Usual Metric on R).** If R be the set of real members then the mapping $d : R \times R \to R$ defined as $d(x, y) = |x - y|\ \forall\ x, y \in R$ is a metric on R (known as usual metric on R)

- **(Discrete or Trivial Metric).** Let X be a non-empty set and $d : X \times X \to R$ defined as

$$d(x, y) = \begin{cases} 0 \text{ if } x = y \\ 1 \text{ if } x \neq y \end{cases} \quad \forall x, y \in R$$

then d is a metric on X, (called the discrete or trivial metric.)

- **Distance between a point $p \in X$ and a subset A of X.**

$$d(p, A) = \inf.\{d(p, x) : x \in A\}$$

- **Diameter of subset A.**

$$\delta(A) = \sup.\{d(x, y) : x, y \in A\}$$

- **Open sphere $S(x_0, r)$:** Let (X, d) be a metric space and $x_0 \in X$ and r is any +ive real number then the set of all those points of X which are within a distance of r from x_0 is called an open sphere (ball) centred at x_0 and radius r. It is denoted by

$S\ (x_0, r)$ or $S_r\ (x_0)$ or $B(x_0, r)$ or $B_r\ (x_0)$.

$\therefore\quad S\ (x_0, r) = \{x \in X : d\ (x, x_0) < r\}$

- **Closed sphere $S[x_0, r]$:**

$S\ [x_0, r] = \{x \in X : d\ (x, x_0) \leq r\}$ (Note the sign of equality) is called a closed sphere or ball centred at x_0 and of radius r. It may be noted that a ball is non-empty as it always contains its centre x_0 as $d(x_0, x_0) = 0 \leq r$, as r is a +ive real number.

- **Subspace.** Let (X, d) be a metric space, so that $d : X \times X \to R$. And if Y be a non-empty subset of X then the same mapping d when restricted to point of Y *i.e.* $d : Y \times Y \to R$ will be a metric on Y and space (Y, d).

- **Range set of a sequence :** The range set X of a sequence may be finite or infinite.

- **Sub-sequence:** Let $<s>$ and $<t>$ be two sequences in a set then the sequence $<t>$ is said to be sub-sequence of $<s>$ if there exists a mapping $\phi : N \to N$ such that

(1) $t = s \circ \phi$

(2) For each n in N there exists an m in N such that $\phi\ (i) \geq n$ for every $i \geq m$ in N.

- **Cluster point of a sequence.** Let (X, d) be a metric space. A point $x \in X$ is said to be cluster point of sequence s in X if s is frequently in every neighbourhood of X.

- **Convergence of a sequence :** A sequence $\langle x_n \rangle = \langle x_1, x_2, x_n, ... \rangle$ in a metric space (X, d) is said to converge to a point x of X if and only if for each $\varepsilon > 0$, $\exists$ a +ive integer $n \in N$ such that

$$n \geq n(\varepsilon) \Rightarrow d(x_n, x) < \varepsilon.$$

- **First and Second Category :** A subset of a metric space is said to be of first category if and only if it can be written as a union of a countable number of nowhere dense sets, otherwise it is said to be of second category.

RESULTS

- In a metric space $d(x, y) = 0$ iff $x = y$ but in pseudo metric space $d(x, y)$ may be zero even though x may not be equal to y. In other words $x = y \Rightarrow d(x, y) = 0$ but $d(x, y) = 0$ does not necessarily imply that $x = y$. Then every pseudo metric space is not a metric space but every metric space is a pseudo metric space For example, if we consider the set R and $d : R \times R \to R$ such that $d(x, y) = |x^2 - y^2| \; \forall \; x, y \in R$ then d is a pseudo metric because $d(x, y) = 0 \Rightarrow |x^2 - y^2| = 0 \Rightarrow x^2 - y^2 = 0 \Rightarrow x = \pm y$. In other words $d(x, y) = 0$ does not necessarily imply that $x = y$.

- A metric space (X, d) is said to be bounded metric space if there exists a positive real number k such that $d(x, y) < k \; \forall \; x, y \in X$.

- A metric space which is not bounded is called unbounded.

- (*An alternative set of axioms for a metric*). Let X be a non-empty set and d be a real valued function of $X \times Y$ into R. then d is a metric if and only if the following two conditions hold
  (1) $d(x, y) = 0 \Leftrightarrow x = y$
  (2) $d(x, y) \leq d(x, z) + d(y, z)$.

- If $1 < p < \infty$ and $1 < q < \infty$ such that $\dfrac{1}{p} + \dfrac{1}{q} = 1$ and a, b be two non-negative real numbers then $a^{1/p} b^{1/q} \leq \dfrac{a}{p} + \dfrac{b}{q}$.

- **Holder's Inequality :** If x_i, y_i ($i = 1, 2, ...,$ n) are non-negative real numbers, then

$$\sum_{i=1}^{n} x_i y_i \leq \left(\sum_{i=1}^{n} x_i^p \right)^{1/p} \left(\sum_{i=1}^{n} y_i^q \right)^{1/q}$$

where $\dfrac{1}{p} + \dfrac{1}{q} = 1$.

- **Nowhere dense set :** A subset A of a metric space (X, d) is said to be nowhere dense if its closure has empty interior *i.e.*, $(\overline{A})^\circ = \phi$.

- In a metric space (X, d), a mapping $f : X \to X$ is called a contracting mapping or simply a contraction on X if and only if there exists a real number $0 \leq \alpha < 1$ such that

$$d(f(x), f(y)) \leq \alpha d(x, y) < d(x, y) \text{ for every } x, y \in X \qquad ...(1)$$

- *Cauchy Inequality.* If we choose $p = q = 2$ so that $\dfrac{1}{p} + \dfrac{1}{q} = \dfrac{1}{2} + \dfrac{1}{2} = 1$ in Holder's inequality then it becomes

$$\sum_{i=1}^{n} (x_i y_i) \leq \left(\sum_{i=1}^{n} x_i^2 \right)^{1/2} \left(\sum_{i=1}^{n} y_i^2 \right)^{1/2}$$

- *Cauchy-Schwarz Inequality :* If x_i, y_i ($i = 1, 2, ..., n$} are complex numbers, then

$$\left| \sum x_i y_i \right| \leq \left[\sum |x_i|^2 \right]^{1/2} \left[\sum |y_i|^2 \right]^{1/2}, \text{ where } \sum = \sum_{i=1}^{n}.$$

- *Minkowski Inequality :* If x_i, y_i ($i = 1, 2, ..., n$) are non-negative real numbers, then

$$\left[\sum (x_i + y_i)^p \right]^{1/p} \leq \left(\sum x_i^p \right)^{1/p} + \left(\sum y_i^p \right)^{1/p}$$

where $p > 1$ and $\sum = \sum_{i=1}^{n}$.

$$\sum (x_i + y_i)^p = \sum (x_i + y_i)(x_i + y_i)^{p-1}$$
$$= \sum x_i (x_i + y_i)^{p-1} + \sum y_i (x_i + y_i)^{p-1}$$

- Let (X, d) be a given metric space and A, B be subsets of X. Let $x \in A$ and $y \in B$ then we know that :

$$d(x, y) \geq 0.$$

Hence $\{d(x, y) : x \in A \text{ and } y \in B\}$ is a set of real numbers which is bounded below by zero and consequently this set must have a greatest lower bound *i.e.* infimum.
Thus if $A \cap B \neq \phi$ then $d(A, B) = 0$.

- Let A, B be two subsets of a metric space (X, d), then $\delta(A \cup B) \leq \delta(A) + \delta(B) + \delta(A, B)$ and if $A \cap B \neq \phi$, then

$$\delta(A \cup B) \leq \delta(A) + \delta(B)$$

- A is a non-empty subset of a metric space (X, d) and x, y are any two points of X, then
$$|d(x, A) - d(y, A)| \leq d(x, y).$$

- Let (X, d) be a metric space and B be the collection of all open spheres $S(x, r)$ where r is any +ive real number and x any point in X. Then B is a base for some topology on X.

- In a metric space (X, d) a non-empty subset G of X is open relative to d-metric topology for X if and only if for each $x \in G$ there exists an open sphere $S(x, r)$ centred at x and contained in G i.e. $S(x, r) \subset G$.

- In a metric space (X, d), $S = S(x_0, r)$ is an open sphere and p is any point of $S(x_0, r)$. Then there exists an open sphere $T = S(p, \in)$ centred at p and contained in $S(x_0, r)$ i.e. $T \subset S$.

- Let (X, d) be a metric space and let $p \in X$. If S and T are open spheres such that $p \in S \cap T$ then there exists an open sphere $S(p)$ centred at p such that $S(p) \subset S \cap T$.

- In a metric space (X, d) each open sphere is an open set.

- In a metric space (X, d), ϕ and X are open sets.

- Let X be a metric space then a subset G of X is open iff G is union of open spheres.

- In a metric space (X, d), Then

  (a) union of an arbitrary collection of open sets is open.

  (b) any finite intersection of open sets in X is open.

- Let (X, d) be a metric space, then every closed sphere in X is a closed set relative to the metric topology for X.

- In a metric space (X, d) all finite sets are closed.

- Let (X, d) be a metric space and f be a bijection from X to a set X^*. Then there exists a metric d^* for X^* such that the metric spaces (X, d) and (X^*, d^*) are isometric.

- Every metric is first countable.

- A metric space is separable if and only if it is second countable

- The function f is said to be continuous iff it is continuous at every point of x_1.

- If G_1 is an open set in (X_1, d_1) and G_2 is an open set in (X_2, d_2) then $G_1 \times G_2$ is open in the product space $(X_1 \times X_2, d)$.

- Let (X, d) be a metric space, then a sequence s in X is a function from set N of all +ive integers into X.

- Let s be a sequence in a non empty set X and Y be a sub set of X, then the sequence s is said to be eventually in Y if there exists a natural number m such that
$$s(n) \in Y \text{ for every } n \geq m$$

- The sequence s is said to be frequently in Y if for each natural number m, there exists a natural number $n \geq m$ such that $s(n) \in Y$.

- The limit of a sequence in a metric space if it exists is unique.

- Let x_0, y_0 be any two points of a metric space (X, d) and $<y_n>$ be a sequence converging to y_0 then $<d(x_n, y_n)>$ converges to $d(x_0, y_0)$.

- Let (X, d) be a metric space and $<x_n>$ and $<y_n>$ are sequences in X such that $x_n \to x$ and $y_n \to y$ then $d(x_n, y_n) \to d(x, y)$.

- Let x be a limit point of a subset A of a metric space (X, d) then there exists a sequence $<x_n>$ of points of A all distinct from x which converges to x.

- A sequence $<x_n>$ in a metric space (X, d) is said to be a Cauchy sequence if and only if for each $\varepsilon > 0$ there exists a +ive integer n_ε, such that $d(x_m, x_n) < \varepsilon \ \forall \ m, n \geq n_\varepsilon$.

- A Cauchy sequence is also called a fundamental sequence.

- In a metric space (X, d) every convergent sequence is a Cauchy sequence.

- If $<x_n>$ is a Cauchy sequence in metric space then any cluster point of $<x_n>$ is the limit of $<x_n>$.

- Let (X, d) be a metric space and $<x_n>$ be a Cauchy sequence in X. If $<y_n>$ be a sequence in X such that $d(x_n, y_n) < (1/n)$ for every positive integer n, then

  (a) $<y_n>$ is also a Cauchy sequence.

  (b) $<y_n>$ converges to a point $x \in X$ iff $<x_n>$ converges to x.

- A metric space (X, d) is said to be complete if every Cauchy sequence in X converges to a point in X.

- Let Y be a subspace of a complete metric space (X, d), then Y is complete $\Leftrightarrow Y$ is closed.

- Let Y be a subspace of a complete metric space (X, d), then Y is complete $\Leftrightarrow Y$ is closed.

- Let $<A_n>$ be a sequence of subsets of X in a metric space (X, d), then this sequence is called a decreasing sequence if $A_1 \supset A_2 \supset A_3 \supset \ldots$

This sequence is sometimes called Nested Sequence.

- (*Cantor's Intersection Theorem*). Let $<A_n>$ be a nested sequence, *i.e.* monotonic decreasing sequence of non-empty closed subsets of a metric space (X, d) such that $\delta(A_n) \to 0$ as $n \to \infty$ (*i.e.*, whose diameters tend to zero) then X is complete if and only if $\cap \{A_n : n \in N\} \neq \phi$ and contains exactly one point.

- (*Baire Category Theorem*). Every complete metric space is of second category.

- The metric space (R, d) is complete where d denotes the usual metric given by
$$d(x, y) = |x-y| \ \forall \ x, y \in R.$$

- The set R^n of all n-tuples $(x_1, x_2, ..., x_n)$ of real numbers is a complete metric space with respect to the usual metric d defined by
$$d(x, y) = \left[\sum_{i=1}^{n} (x_i - y_i)^2 \right]^{\frac{1}{2}}$$

- The set C of all complex numbers $z = x + iy$ is a complete metric space with respect to the usual metric d defined by $d(z_1, z_2) = |z_1 - z_2|$.

- The set C^n of all n-tuples $z = (z_1, z_2, ..., z_n)$ of complex numbers is a complete metric space with respect to the usual metric space defined by
$$d(z, u) = \left[\sum_{i=1}^{n} |z_i - u_i|^2 \right]^{\frac{1}{2}}$$
where $z = (z_1, z_2, ..., z_n)$ and $u = (u_1, u_2, ..., u_n)$.

- The Hilbert space (l_2, d) is complete.

- The space $C[a, b]$ is complete with the metric defined as
$$d(f, g) = \sup\{ |f(x) - g(x)| : x \in [a, b]\}$$

- Every contraction mapping is continuous.

- (*Banach Fixed Point Theorem*): Every contractive mapping in a complete metric space has one and only one fixed point such that $f(x) = x$.

REVIEW QUESTIONS AND PROJECT WORK

1. Let (X, d) be a discrete metric space and let $x \in X$. Show that every subset of X containing x is a nbd of x.

2. Show that every subset of a discrete metric space is closed.

3. Show that in a metric space the compliment of a finite set is open.

4. For any two subsets A and B of a metric space $[X, d]$ show that $\delta(A \cup B) = \delta(A) \cup \delta(B)$.

5. Show that the metric space (R^n, d) is complete where d is usual metric on R^n

OBJECTIVE TYPE QUESTIONS

FILL IN THE BLANKS

1. In a metric space (X, d), $d(x, y) = 0$ iff _____.

2. In a metric space (X, d), $x, y, z \in X$ then we have $d(x, y) \leq d(x, z) +$ _____.

3. Let (X, d) be a metric space and A be non-empty subset of X then diameter of A denoted as $\delta(A) =$ _____.

4. Let $X = R^n$, the set of all ordered n-tuples of real numbers let $x = (x_1, x_2, ..., x_n)$ and $y = (y_1, y_2, ..., y_n)$ then $d(x, y) =$ _____.

5. In a metric space the union of an arbitrary collection of open set is _____.

6. In a metric space intersection of finite number of open set is _____.

7. In a metric space intersection of infinite number of open set is _____.

8. In a metric space, a subset in it is open iff it is of each of its points.

9. A subset A of metric space is closed iff $D(A)$ _____.

10. Let A be a subset of metric space then $\overline{A} = A$ $\cup$ _____.

11. Let A and B are two closed sets then :
    (i) $A \cap B$ is _____.
    (ii) $A \cup B$ is _____.

12. Let A and B both are open set then :
    (i) $A \cap B$ is _____.
    (ii) $A \cup B$ is _____.

13. A subset A of X is said to dense set if $\overline{A} =$ _____.

14. A subset A of X is said to be nowhere dense set if $(\overline{A})°$ _____.

15. The open interval $]a, b[$ is a nbd of _____ of its point.

16. The set of real numbers R is a nbd of _____ of its point.

17. The set of integers Z is _____ a nbd of any of its points.

18. The empty set of ϕ is a nbd of _____ of its points.

19. Every open interval is an _____ set.

20. Every closed interval is a _____ set.

21. If A and B are two open sets in a metric space (X, d), then $A \cap B$ is _____ .

22. The set $[1, 2] \cup [3, 4]$ is a _____ set.

23. In a metric space, every finite set is _____ set.

24. On the real line, the set Z of integers has _____ limit points.

25. On the real line, the set $S = \left\{ \dfrac{1}{n} : n \in N \right\}$ has _____ limit points.

26. A subset of a metric space is _____ iff it contains all its limit point.

27. In a metric space, the derived set of a metric space is _____ .

28. If $S \subset T$ then $S' \subset$ _____ .

29. In a metric space, the set of all adherent points is called _____ .

30. On the real line, the closure of each of the

31. A set A is closed iff $\overline{A}$ _____ .

32. A set A is open iff A° _____ .

33. In a metric space (X, d) a point $x \in X$ is an _____ point of a set $S \subset X$ iff every open set containing x contain a point of S.

34. The _____ of a set A is the intersection of all closed supersets of A.

35. A sequence $\langle s_n \rangle$ is said to be _____ if $\exists$ a number $M > 0$ such that $|s_n| < M \ \forall \ n \in N$.

36. A sequence which is not convergent is said to be _____ .

37. A sequence $\langle s_n \rangle$ in X is said to converge to a point $s_0 \in X$ if for each $\varepsilon > 0$ of a positive integer m such that $d \langle s_n, s_0 \rangle < \varepsilon$ for all _____ .

38. In a metric space the convergent sequence has a _____ limit.

39. Every convergent sequence in a metric space is _____ sequence.

40. A metric space (X, d) is said to be _____ iff every Cauchy sequence in X converges to a point in X.

41. The open interval is _____ .

42. A subspace of a metric space is complete iff it is _____ .

TRUE/FALSE

Write 'T' for true and 'F' for false statement.

1. In case of real line, the open sphere $S(x, r)$ is the open interval $]x{-}r, x{+}r[$. **(T/F)**

2. For the metric space (X, d), the open sphere centerd at z_0 and having radius r is the set $\{z : |z - z_0| < r\}$. **(T/F)**

3. The closed interval $[a, b]$ is a nbd of each of its points. **(T/F)**

4. In a metric space (X, d), the intersection of the family of all nbds of a point $p \in X$ is $\{p\}$. **(T/F)**

5. A finite set in a metric space can be open. **(T/F)**

6. A finite non-empty set in a metric space can be open. **(T/F)**

7. The intersection of any arbitrary family of open sets in a metric space is always open. **(T/F)**

8. Union of an arbitrary family of open sets in a metric space is always open. **(T/F)**

9. Every infinite set in a metric space is open. **(T/F)**

10. In a discrete metric space, every subset is open. **(T/F)**

11. Every closed interval is a closed set. **(T/F)**

12. In a metric space, every singleton is a closed set. **(T/F)**

13. The arbitrary union of closed set in a metric space is necessarily closed. **(T/F)**

14. The arbitrary union of open set in a metric space is necessarily open. **(T/F)**

15. If S is open and T is closed then $S{-}T$ is open. **(T/F)**

16. If S is closed and T is open then $S{-}T$ is closed. **(T/F)**

17. A subset of a metric space is closed if it containing all its limit points. **(T/F)**

18. On the real line, the set of integers has no limit point. **(T/F)**

19. The closure of a set A is the smallest closed set. **(T/F)**

20. The interior of a set A is the largest closed set. **(T/F)**

21. In a metric space every convergent sequence is cauchy. **(T/F)**

22. In a metric space, every Cauchy sequence is convergent. **(T/F)**

23. In a complete metric space, every Cauchy sequence is convergent. **(T/F)**

24. A sequence in a metric space can converge to at most one point of that space. **(T/F)**

25. If a Cauchy sequence in a metric space has a convergent subsequences, then the sequence is convergent. **(T/F)**

26. Completeness is not preserved under isometrics. **(T/F)**

27. The real line is complete. **(T/F)**

28. The image of a Cauchy sequence under a uniformly continuous mapping is a Cauchy sequence. **(T/F)**

MULTIPLE CHOICE QUESTIONS

Choose the most appropriate one:

Problem Set-1

1. The interior of the set of rational numbers Q is:
   (a) ϕ
   (b) Q
   (c) R
   (d) None of these

2. The interior of the set of integers Z is:
   (a) Z
   (b) ϕ
   (c) R
   (d) None of these

3. Which of the following is not true:
   (a) int.$\phi = \phi$
   (b) int X=X
   (c) $S \subset T \Rightarrow$ int $S \subset$ int T
   (d) None of these

4. If p is an interior point of S then S is:
   (a) empty
   (b) nbd of p
   (c) exterior of p
   (d) None of these

5. Which is not true:
   (a) X–int $(A)=$ cl $(X$–$A)$
   (b) cl$(A)=X$~int$(X$–$A)$
   (c) X–cl$(A)=$int $(X$–$A)$
   (d) None of these

6. A complete subspace of a metric space is:
   (a) Closed
   (b) Open
   (c) Neither open nor closed
   (d) None of these

7. The set of integers with the usual metric is:
   (a) Incomplete
   (b) Complete
   (c) Both are true
   (d) None of these

8. Set of rational numbers with the usual metric is:
   (a) Always complete
   (b) May or may not be complete
   (c) Not complete
   (d) None of these

9. The set of irrational with the usual metric is:
   (a) Always complete
   (b) May or may not be complete
   (c) Not complete
   (d) None of these

10. Which space is complete, with the usual metric:
    (a) $X =]0, 1[$
    (b) $X = [0, 1[$
    (c) $X =]0, \infty[$
    (d) None of these

11. The triangle inequality in a metric space (X, d) holds equality sign when three points (x, y), (y, z) and (x, z) are :
    (a) Collinear
    (b) non-collinear
    (c) on the triangle
    (d) None of these

12. If $X=R$ and $d(x, y)=|x{-}y| \; \forall \, x, y \in X$ then d is a metric on X. This metric is known as:
    (a) Pseudo metric
    (b) usual metric
    (c) trival metric
    (d) None of these

13. If A is a closed subset of a complete metric space, then:
    (a) A is complete
    (b) A is incomplete
    (c) undefined
    (d) None of these

14. Let A be the subset of a metric space X. A point $x \in A$ is an interior point of A if A is a nbd of X. What is the correct option for the interior point of A i.e., A^o:

(a) A^o is the union of all open set contained in A
(b) A^o is the largest open set contained in A
(c) both (a) and (b)
(d) None of these

15. A closed sphere is defined by:

(a) $S(x_0, r) = \{x \in X : d(x_0, x) = r\}$
(b) $S(x_0, r) = \{x \in X : d(x_0, x) \le r\}$
(c) $S(x_0, r) = \{x \in X : d(x_0, x) < r\}$
(d) None of these

16. Definition of closure of a set states that:

(a) $\overline{A} = \cup [F : F \text{ is closed}, F \supset A]$
(b) $\overline{A} = \cap [F : F \text{ is closed}, F \subset A]$
(c) $\overline{A} = \cap [F : F \text{ is closed}, F \supset A]$
(d) None of these

17. If every cauchy sequence $<s_n>$ is also:

(a) Convergent   (b) Divergent
(c) Oscillatory   (d) None of these

18. Every subspace of X containing $x \in X$ in a discrete space (X, d) is:

(a) not a nbd of x (b) is a nbd of x
(c) is a closed set   (d) None of these

19. A sequence $<s_n>$ in a metric space (X, d) is a Cauchy sequence in X such that $\forall\ \varepsilon > 0, \exists$ a positive integer n such that:

(a) $m, n \ge n(\varepsilon) \Rightarrow d(s_m, s_n) < \varepsilon$
(b) $m, n \ge n(\varepsilon) \Rightarrow |(s_m, s_n)| < \varepsilon$
(c) $m, n \ge n(\varepsilon) \Rightarrow d(s_m, s_n) > \varepsilon$
(d) None of these

20. In the usual metric space R, the derived set Q of all rational numbers is:

(a) Q   (b) Z
(c) I   (d) R

21. Let $[X, d]$ be a metric space. If $\exists$ a positive integer M such that $d(x, y) \le M\ \forall\ x, y \in X$. Then X is called:

(a) Bounded metric space
(b) unbounded
(c) complete
(d) None of these

22. If $\overline{A}$ denote the closure of A in a metric space (X, d), which one of the following is false:

(a) $A \subset \overline{A}$
(b) $A \subset B \Rightarrow \overline{A} \subset \overline{B}$
(c) $\overline{A \cup B} = \overline{A} \cup \overline{B}$.
(d) $\overline{A \cap B} = \overline{A} \cap \overline{B}$

23. If (X, d) be a metric space and A be a non-empty subset of X, the diameter of A i.e., $\delta(A)$ is:

(a) $\delta(A) = \sup\{d(x, y) : x, y \in A\}$
(b) $\delta(A) = \inf\{d(x, y) : x, y \in A\}$
(c) $\delta(A) = \{d(x, y) : x, y \in A\}$
(d) None of these

24. The sequence $< \dfrac{1}{n} : n \in N>$ in metric space R with usual metric R is:

(a) Convergent   (b) divergent
(c) not cauchy   (d) None of these

25. A metric space X is bounded iff its diameter is:

(a) not-defined   (b) finite
(c) infinite   (d) 0

26. A metric space X satisfy Bolzano-Weirstrass property, then:

(a) every finite sequence $\langle x_n \rangle$ in Y has no cluster point
(b) every infinite sequence has a cluster point
(c) X is not sequentially compact
(d) None of these

27. A is said to be no where dense if :

(a) $(\overline{A})^c \ne \phi$ (b) $(\overline{A})^o = R$
(c) $(\overline{A})^o = \phi$ (d) None of these

28. By BWP, every bounded sequence in a metric space has a:

(a) limit   (b) limit point
(c) interior point   (d) None of these

29. If A is a non-empty subset of a metric space X is complete, then:

(a) A is closed   (b) A is open
(c) $A = \phi$ (d) None of these

30. The open sphere of radius 2 about $(1, 1)$ in R^2 with metric

$$d(z_1, z_2) = \sqrt{(x_1 - x_2)^2 + (y_1 - y_2)^2} :$$

$$z_1 = (x_1, y_1), z_2 = (x_2, y_2) \text{ is :}$$

(a) $(x-1)^2 + (y-1)^2 = 4$
(b) $(x-1)^2 + (y-1)^2 = 2$
(c) $(x-1)^2 + (y-1)^2 = 1$
(d) None of these

Problem Set-2

1. A set $E \subset R$ is compact, then it is :
  (a) closed
  (b) bounded
  (c) closed and bounded
  (d) none of these

2. A compact subset A of a metric space (X, d) implies :
  (a) A is closed
  (b) A is bounded
  (c) Closure of B is compact for every $B \subseteq A$
  (d) all are true

3. If S and T are subsets of R, then which one of the following is not true :
  (a) $S° \cap T° = (S \cap T)°$
  (b) $S° \cup T° \subseteq (S \cup T)°$
  (c) $\bar{S}$ is closed in R
  (d) none of these

4. If every sequence $<x_n>$ in a metric space (X, d) is convergent then every cauchy sequence in $<x_n>$ is :
  (a) convergent
  (b) divergent
  (c) oscillatory
  (d) none of these

5. A collection F of sets have finite intersection property of :
  (a) any finite sub collection of F has empty, intersection
  (b) any finite sub collection of F has non empty intersection
  (c) any finite sub collection of F has empty intersection
  (d) none of these

6. If A is an open subset of complete metric space X then :
  (a) A is closed
  (b) A is open
  (c) A′ is closed
  (d) none of these

7. The union of any finite collection of non empty closed sets is :
  (a) closed
  (b) open
  (c) null
  (d) none of these

8. If A is a closed subset of complete metric space, then :
  (a) A is a complete
  (b) A is a not complete
  (c) Can't say
  (d) none of these

9. If X is a complete metric space, E is a non empty open subset of X, then E is of :
  (a) first category
  (b) second category
  (c) Can't say
  (d) none of these

10. A set E is said to be of first category if :
  (a) E is the union of collection of nowhere dense sets
  (b) E is the union of countable collection of nowhere dense sets
  (c) Can't say
  (d) none of these

11. A set is said to be of second category if it is :
  (a) of first category
  (b) not of first category
  (c) empty
  (d) none of these

12. The empty set of a metric space is :
  (a) open
  (b) closed
  (c) both open and closed
  (d) none of these

13. The function $d : X \rightarrow R$ is said to be pseudo metric if :
  (a) $d(x, y) = 0 \Rightarrow x = y$
  (b) $d(x, y) = 0$ for some $x \neq y$
  (c) $d(x, y) \neq 0$ for $x = y$
  (d) none of these

14. A set of metric space (X, d) is said to be closed if it :
  (a) contains all its limit points
  (b) it is not open
  (c) its compliment is open
  (d) all are true

15. A function of mapping $f : (X, d_1) \rightarrow (Y, d_2)$ is said to be homomorphism if :
  (a) $f(x) = y$
  (b) $f(x) \subset Y$
  (c) $f(x) \neq y$
  (d) none of these

16. If X and Y are metric spaces and for every open set O of Y, $O \subset Y$, $f^{-1}(O) \subset X$ is open then :
  (a) f is continuous
  (b) f is constant
  (c) f is discontinuous
  (d) none of these

17. Which of the following is/are true ?
  (a) Q is of first category in R w.r.t. usual metric
  (b) If X is of second category and if $X = A \cup B$, then either A or B must be of second category.
  (c) every countable subspace of R is of first category in R
  (d) all are true

18. Let (X, d) be a metric space and x, y, z be points of X then $|d(x, y) - d(y, z)|$:

    (a) $\leq d(x, z)$ (b) $\geq d(x, z)$

    (c) $= d(x, z)$ (d) none of these

19. Let (X, d) be a metric space and $d^*(x, y) = \min\{1, d(x, y)\}$ then d^* is :

    (a) a metric on x (b) pseudo metric

    (c) not a metric   (d) none of these

20. If d is metric an X and $d^* : X \times X \to R$ defined as

    $$d^*(x, y) = \frac{d(x, y)}{1 + d(x, y)} \text{ then } d^* \text{ is :}$$

    (a) a metric on X (b) pseudo metric

    (c) not a metric   (d) none of these

21. Let X be the set of all real valued continuous function defined on the closed unit interval $(0, 1)$ and d be a mapping of $X \times X \to R$ defined by $d(f, g) =$

    (a) a metric on X (b) pseudo metric

    (c) not a metric   (d) none of these

22. Let $C[0, 1]$ denote the collection of all real valued bounded continuous function defined on $[0, 1]$. Let d be defined by

    $$d[f, g] = \{|f(x) - g(x) : x \in [0, e]|\}$$

    then d is :

    (a) a metric on x (b) pseudo metric

    (c) not a metric   (d) none of these

23. If A and B be two subsets of a metric space (X, d) then :

    (a) $\delta(A \cup B) \leq \delta(A) + \delta(B) + \delta(A, B)$
    (b) $\delta(A \cup B) \leq \delta(A) + \delta(B)$ if $A \cap B \neq \phi$
    (c) both (a) and (b) are true
    (d) none of these

24. If A is a non empty subset of a metric space (X, d) and x, y are any two points of X then $|d(x, A) - d(y, A)|$:

    (a) $\leq d(x, y)$ (b) $\geq d(x, y)$

    (c) $= d(x, y)$ (d) none of these

25. The closure of A can be defined by :

    (a) $\overline{A} = \{x \in X : d(x, A) = 0\}$

    (b) $\overline{A} = \{x \in X : d(x, A) = 1\}$

    (c) $\overline{A} = \{x \in X : d(x, A) \neq 0\}$

    (d) none of these

26. A subset A of metric space (X, d) is closed if and only if $\{x \in X : d(x, A) = 0\}$:

    (a) $< A$ (b) $> A$

    (c) $= A$ (d) none of these

27. The metric space (R, d) is :

    (a) complete

    (b) not complete

    (c) may or may not be complete

    (d) none of these

28. The set R^n of all n tuples $(x_1, x_2, ..., x_n)$ of real numbers is :

    (a) complete
    (b) not complete
    (c) may or may not be complete
    (d) none of these

29. Let C be the set of complex numbers and $d(z_1, z_2) = |z_1 - z_2|$ then d is :

    (a) complete
    (b) not complete
    (c) may or may not be complete
    (d) none of these

30. If $d[f, g] = \{|f(x) - g(x) : x \in [a, b]|\}$ then d is :

    (a) complete
    (b) not complete
    (c) may or may not be complete
    (d) none of these

31. There is a countable family $\{O_i\}$ of open sets such that for any open set $O \subset X$, $O = \cup O_i$ then metric space is :

    (a) separable   (b) not separable

    (c) can't say   (d) none of these

32. Let (X, d) be a complete metric space. Then "no non empty open subset of X is of first category *i.e.* the union of a countable collection of no where dense subsets" is called :

    (a) Bair's category theorem

    (b) Bolzano-Weirstrass theorem

    (c) both (a) and (b) are true

    (d) none of these

33. Which one of the following is not correct :

    (a) $S \subset R^n$ is closed, if it contains all its limit points

    (b) $S \subset R^n$ is closed if it does not contains all its limit points

    (c) $S \subset R^n$ is closed if it contains all isolated points

    (d) none of these

34. Canter ternary set is :

    (a) closed
    (b) uncountable
    (c) closed and uncountable
    (d) none of these

35. If x is an accumulation point of $S \subset R^n$, then :

    (a) every open sphere contains infinitely many points

(b) every open sphere contains finitely many points
(c) every open sphere contains no points
(d) none of these

36. Which of the following is not true :
(a) The union of any collection of open sets is closed
(b) The union of any collection of open sets is not open
(c) The union of any collection of open sets is open
(d) none of these

37. Which one of the following is true :
(a) The Intersection of an arbitrary collection of closed sets is closed.
(b) The intersection of an arbitrary collection of closed sets is open.
(c) The intersection of an arbitrary collection of closed sets is empty.
(d) none of these

38. Which of the following is not true :
(a) $S \subset R^n$ is closed $\Rightarrow S = \bar{S}$
(b) $S = \bar{S} \Rightarrow S \subset R^n$ is closed
(c) $S = \bar{S} \Rightarrow S \subset R^n$ is open
(d) none of these

39. Let A and B be two sets of positive real numbers bounded above. Let $a = \sup(A)$ and $b = \sup(B)$ and $C = [xy : x \in A, y \in B]$ then :
(a) $x > 0, y > 0 \Rightarrow xy < ab$
(b) $ab = \sup(C)$
(c) both (a) and (b) are true
(d) none of these

40. If d is an extended metric on the set X, then for some $x, y \in X$:
(a) $d(x, y) > \infty$ (b) $d(x, y) < \infty$
(c) $d(x, y) = \infty$ (d) none of these

Answers

FILL IN THE BLANKS

1. $x = y$ **2.** $d(z, y)$ **3.** $\sup\{d(x, y) : x \in A, y \in A\}$ **4.** $\left[\sum (x_i - y_i)^2\right]^{1/2}$

5. open   **6.** open   **7.** not necessarily open   **8.** neighbourhood
9. $\subseteq A$ **10.** $D(A)$ **11.** (i) closed   (ii) closed   **12.** (i) open   (ii) open
13. X **14.** ϕ **15.** each   **16.** each   **17.** not   **18.** each
19. open   **20.** closed   **21.** open   **22.** closed   **23.** closed   **24.** no
25. one   **26.** closed   **27.** closed   **28.** T' **29.** closure   **30.** $[0, 1]$
31. A **32.** A **33.** adherent   **34.** closure   **35.** Bounded   **36.** Divergent
37. $n \geq m$ **38.** Unique   **39.** Cauchy   **40.** Complete   **41.** Complete   **42.** Closed

TRUE/FALSE

1. T   **2.** T   **3.** F   **4.** F   **5.** T   **6.** T   **7.** F   **8.** F   **9.** F
10. T   **11.** T   **12.** T   **13.** F   **14.** T   **15.** T   **16.** T   **17.** T   **18.** T
19. T   **20.** T   **21.** T   **22.** F   **23.** T   **24.** T   **25.** T   **26.** F   **27.** T
28. T

MULTIPLE CHOICE QUESTIONS

Problem Set-1

1. (a)   **2.** (b)   **3.** (d)   **4.** (b)   **5.** (d)   **6.** (a)   **7.** (b)   **8.** (c)   **9.** (c)
10. (d)   **11.** (a)   **12.** (b)   **13.** (a)   **14.** (c)   **15.** (b)   **16.** (b)   **17.** (a)   **18.** (c)
19. (a)   **20.** (d)   **21.** (a)   **22.** (d)   **23.** (a)   **24.** (a)   **25.** (b)   **26.** (b)   **27.** (c)
28. (b)   **29.** (a)   **30.** (a)

Problem Set-2

1. (c)   **2.** (d)   **3.** (d)   **4.** (a)   **5.** (a)   **6.** (c)   **7.** (a)   **8.** (a)   **9.** (a)
10. (b)   **11.** (b)   **12.** (c)   **13.** (b)   **14.** (d)   **15.** (a)   **16.** (a)   **17.** (d)   **18.** (a)
19. (a)   **20.** (a)   **21.** (a)   **22.** (a)   **23.** (a)   **24.** (a)   **25.** (a)   **26.** (a)   **27.** (a)
28. (a)   **29.** (a)   **30.** (a)   **31.** (a)   **32.** (a)   **33.** (a)   **34.** (c)   **35.** (a)   **36.** (c)
37. (a)   **38.** (c)   **39.** (c)   **40.** (a)

SELF ASSESSMENT TEST

Verify each of the following :

1. The function d defined by $d(z_1, z_2) = |z_1 - z_2|$: $z_1, z_2 \in C$ is a metric on the set C of complex numbers.

2. Let $X = \left\{ 1, \dfrac{1}{2}, \dfrac{1}{3}, ..., \dfrac{1}{n}, ... \right\}$ and d is the usual metric defined on X.

   Let $A = \left\{ 1, \dfrac{1}{3}, \dfrac{1}{5}, ..., \dfrac{1}{2n-1}, ... \right\}$ and

   $B = \left\{ \dfrac{1}{2}, \dfrac{1}{4}, \dfrac{1}{6}, ..., \dfrac{1}{2n}, ... \right\}$ then $d(A, B) = 0$.

3. The function $d : C \times C \rightarrow R$ defined by

   $d(x, y) = \dfrac{2|x - y|}{\sqrt{1 + |x|^2} \cdot \sqrt{1 + |y|^2}}$ is a metric on

   the set of complex numbers.

4. Let $R[0, 1]$ be the set of all R-integrable function defined on $[0, 1]$ and $d(f, g) = \int_0^1 |f(x) - g(x)| dx \quad \forall \ f, g \in R[0, 1]$ then d is not a metric on $R[0, 1]$.

5. In the metric space (R, d) of real numbers with the usual metric d, the open sphere $S(r, a)$ is the open interval $]a-r, a+r[$ and the closed sphere $S[r, a]$ is the closed interval $[a-r, a+r], a \in R, r > 0$.

6. The subset $A = [0, 1]$ of the metric space $[X, d]$ where $X = [0, 2[$ and d is the usual metric, is an open set.

7. The subset $A = [(x, y) : y^2 < x, x, y \in R]$ of R^2 with the Euclidean metric is an open set.

8. The derived set of every subset of a discrete space is empty.

9. Every real number is a limit point of the set of rationals.

10. A finite set and the set of integers has no limit point.

11. The set $\left[\dfrac{1+i}{n} : n \in N \right]$ is neither open nor closed with respect to the usual metric in the complex plane.

12. Let $X = R$ and d is the usual metric, $A = Q$ then $\text{int}(A) = \phi$ and $\text{ext}(A) = \phi$, $F_r(A) = R$ and $bd(A) = \phi$.

13. Let $[X, d]$ be a discrete metric space and $A \subseteq X$, then $\text{int}(A) = A$, $\text{ext}(A) = A'$, $F_r(A) = bd(A) = \phi$.

14. The Euclidean space R^n is separable.

15. The Canter set is nowhere dense.

16. The discrete space $[X, d]$ and the space (R, d) are complete metric space.

17. Any closed interval with the usual metric space is compact.

18. The open interval $]0, 1[$ with the usual metric is not compact.

19. For any non-empty subset A of a metric space $[X, d]$ the infimum $f : X \rightarrow R$ given by $f(x) = d(x, A) \ \forall \ x \in X$ is uniformly continuous. Also $f(x) = 0 \Leftrightarrow x \in \overline{A}$.

20. If $n \geq 2$ then there are open subset of R^n which cannot be expressed as the union of a countable family of pairwise disjoint open spheres in R^n.

Chapter

16 The Lebesgue Integrals

16.1 INTRODUCTION

We know that, the length of an interval is defined as the difference of end points of the interval. Thus if intervals are $[a, b]$, (a, b), $[a, b)$, $(a, b]$ i.e., closed, open, closed-open, open-closed respectively, then by definition each interval will be of length equal to $b - a$ and be denoted by $|I|$. In fact the empty set and every singleton subset of R are intervals of zero length.

Definition 1. *If G is any open subset of $[a, b]$, then there will exist a countable family of disjoint open intervals say $\{I_n\}$ s.t.* $G = \underset{n}{\cup} I_n$

The length $|G|$ of the open set G is defined as the sum of the lengths of the intervals of this family i.e.,
$$|G| = \sum_n |I_n|$$

REMARKS

- If $G_1 \supseteq G_2$, then $|G_1| \geq |G_2|$
- If $G_1 \cap G_2 = \phi$, then $G_1 \cup G_2 = |G_1| + |G_2|$

Definition 2. *If F is any closed subset of $[a, b]$, then the length $|F|$ of the closed set F is defined as $|F| = |G| - |G - F|$ where G is any open subset of $[a, b]$ s.t. $F \subseteq G$.*

REMARK

- $|F| = (b - a) - |F^c|$.
- If G_1 and G_2 are any open subsets of (a, b), then $|G_1| + |G_2| + = |G_1 \cup G_2| + |G_1 \cap G_2|$.

16.2 OUTER LEBESGUE MEASURE OF A SET

The outer Lebesgue measure of a set $A \subset R$ is defined by
$$m^*(A) = \begin{cases} 0 & \text{if } A = \phi, \\ \inf. & \{x : x = \Sigma |I_i|\} \end{cases}$$

where $\{I_i\}$ is countable family of open intervals s.t. $\cup I_i \supseteq A$, when $A \neq \phi$ and is denoted by $m^*(A)$. It is also known as Lebesgue exterior measure or more briefly the outer measure and is also denoted by $m_e(A)$ or $\overline{m}(A)$.

IMPORTANT PROPERTIES

(1) $m^*(A) \geq 0$, $\forall A$. It implies that $m^*(A)$ is non-negative. Here, $|I_i|$ is positive and hence x is positive and hence inf $\{x\}$ is also positive.

(2) Infimum gives that $m^*(A)$ is the least length to cover the set A from outside.

(3) For each $\varepsilon > 0$, however small, $\exists$ at least one countable family of open intervals such that $\underset{n}{\cup} I_n \supseteq A$ and $m^*(A) + \varepsilon > \underset{n}{\sum} |I_n|$.

(4) If $A \subseteq B$, then $m^*(A) \leq m^*(B)$.

(5) If I is an interval of the real line, $m^*(I) = $ length of I.

(6) If $A \subseteq [a, b] \Rightarrow m^*(A) \leq (b - a)$.

(7) If $A = \{a\} = a$ singleton set, then $m^*(A) = 0$.

(8) For every set A, $m^*(A)$ is unique.

(9) If G is an open set of real numbers then $m^*(G) = |G|$.

(10) The outer measure m^* is a set function from the set P(R) to the set of all non-negative extended real numbers.

THEOREM 1. If $\{A_1, A_2, A_3, ...\}$ is a countable family of subsets of R. then $m^*\left(\overset{\infty}{\underset{n=1}{\cup}} A_n\right) \leq \overset{\infty}{\underset{n=1}{\sum}} m^*(A_n)$

Proof. If one of the sets A_n has an infinite measure then right hand side sum is infinity and the theorem holds good. Suppose $m^*(A_n)$ is finite for all A_n. Then for $\varepsilon = \dfrac{\lambda}{2^n} > 0$, we can have a countable family $\{I_{nk}\}$ of open intervals s.t.

$$A_n \subseteq \underset{k}{\cup} I_{nk} \text{ and } m^*(A_n) + \frac{\lambda}{2^n} > \overset{\infty}{\underset{k=1}{\sum}} |I_{nk}|,$$

$$\therefore \quad \overset{\infty}{\underset{n=1}{\sum}} m^*(A_n) + \lambda > \overset{\infty}{\underset{n=1}{\sum}}\left(\overset{\infty}{\underset{k=1}{\sum}} |I_{nk}|\right) \quad\quad ...(1)$$

since $\overset{\infty}{\underset{n=1}{\sum}} \dfrac{\lambda}{2^n} = \lambda \overset{\infty}{\underset{n=1}{\sum}} \dfrac{1}{2^n} = \lambda$.

Now, since each $m^*(A_n)$ is finite, so $\Sigma m^*(A_n)$ is also finite, so $\Sigma\Sigma|I_{nk}|$ is finite and hence convergent. Therefore the terms may be added in any order. Let us write these intervals in a sequence $I_1, I_2, I_3,...$ Then

$$A = \underset{n\in N}{\cup} A_n \subseteq \cup I_n,$$

$$\therefore \quad m^*(A) \leq \Sigma|I_n| = \Sigma\Sigma|I_{nk}| < \Sigma m^*(A_n) + \lambda \text{ being arbitrary.} \quad\text{by (1)}$$

Therefore,

$$m^*(A) \leq \Sigma m^*(A_n), \Rightarrow m^*\left(\overset{\infty}{\underset{n=1}{\cup}} A_n\right) \leq \overset{\infty}{\underset{n=1}{\sum}} m^*(A_n).$$

THEOREM 2. If A is countable then $m^*(A) = 0$.

Proof. Let $A = \{a_1, a_2, a_3,...\}$. This is obviously countable. Evidently, $A = \underset{i}{\cup}\{a_i\} = $ countable union of singleton sets $\{a_i\}$.

$$\Rightarrow m^*(A) = m^*(\cup\{a_i\}) \leq \overset{\infty}{\underset{i=1}{\sum}} m^*(\{a_i\}),$$

Hence, $m^*(A) \leq 0$, as outer measure of every singleton set is zero i.e., $m^*(\{a_i\}) = 0$, $\forall i$.

But $m^*(A) \geq 0$, Hence, $m^*(A) = 0$.

REMARKS

- Since the set of all rational numbers is countable, its outer Lebesgue measure is zero.
- The converse of the above theorem is not always true *i.e.*, a set with outer measure zero may or may not be countable.
- A set with outer measure different from zero is uncountable.

THEOREM 3. *If A and B are any two disjoint subsets of R, then*
$$m^* (A \cup B) = m^*(A) + m^*(B)$$

Proof. By theorem 1, we have $m^*(A \cup B) \leq m^*(A) + m^*(B)$. So we are only to show that $m^* (A) + m^* (B) \leq m^* (A \cup B)$. Let $C = \{I_n\}$ be the family of open intervals such that $m^*(A \cup B) = \Sigma |I_n|$. Since $A \cap B = \phi$ we can split this family into two disjoint subfamilies. Now let $C_1 = \{I'_k : k \in N\}$ and $C_2 = \{I''_k : k \in N\}$ by the two subfamilies of C which cover A and B respectively.

Clearly,
$$C = C_1 \cup C_2 \text{ and } C_1 \cap C_2 = \phi \text{ as } A \cap B = \phi,$$
$$\Rightarrow \qquad |C| = \sum_n |I_n| = \sum_k |I'_k| + \sum_k |I''_k|. \qquad \qquad ...(1)$$

Now, since $C_1 = \{I'_k : k \in N\}$ is a family of open intervals covering A, so
$$m^*(A) \leq \sum_k |I'_k|. \qquad \qquad ... (2)$$

Similarly
$$m^*(B) \leq \sum_k |I''_k|. \qquad \qquad ... (3)$$

Adding (2) and (3), we have
$$m^*(A) + m^*(B) \leq \sum_k |I'_k| + \sum_k |I''_k| = \sum_n |I_n|. \qquad \qquad ... (4)$$

By (1), we have
$$m^*(A) + m^*(B) \leq m^*(A \cup B)$$
and hence,
$$m^*(A \cup B) = m^*(A) + m^*(B).$$

THEOREM 4. *The outer measure of an interval is equal to its length.*

Proof. **Case I.** If the interval $I = (a, b)$ *i.e.*, open then result is obvious.

Case II. If the interval I is closed interval $[a, b]$, then obviously
$$I = [a,b] \subset \left(a - \frac{\varepsilon}{2}, b + \frac{\varepsilon}{2} \right)$$
where, $\varepsilon > 0$ being arbitrary, however small.
$$\Rightarrow \qquad m^*(I) < \left| \left(a - \frac{\varepsilon}{2}, b + \frac{\varepsilon}{2} \right) \right| < (b - a) + \varepsilon$$

But ε is arbitrary, so we must have
$$m^*(I) \leq (b - a) \qquad \qquad ...(1)$$

Now we have to show that $m^*(I) \geq b - a$. For let $\{I'_k\}$ be a countable family of open intervals covering $[a, b]$. Then by the Heine-Borel theorem, $\exists$ a finite sub-collection $\{I_n\}$ of open intervals covering I *i.e.*, $\cup I_n \supseteq I$
$$\Rightarrow \qquad |I| \leq \Sigma |I_n| \Rightarrow \Sigma |I_n| \geq (b - a).$$

This result is true for all families $\{I'_k\}$ and hence for all families $\{I_n\}$ s.t., $\Sigma |I_n| \leq \Sigma |I'_k|$ which cover I.

Therefore, taking min of $\Sigma|I_n|$, we get

$$\min \Sigma\,|\,I_n\,| \ge (b-a) \text{ or } \inf (\Sigma|I_n|) \ge (b-a)$$

$$\Rightarrow \qquad m^*\,(I) \ge b - a$$

(1) and (2) $\Rightarrow \qquad m^*\,(I) = m^*[a, b] = (b-a)$.

Case III. Let $I = (a, b)$ and $a > -\infty$. If $a = b$, then $I = \phi$

$\Rightarrow m*(I) = 0 = b - a$

Suppose $a < b$. Let $0 < \varepsilon < b - a$, then

$$I' = [a + \varepsilon, b] \subset I \Rightarrow m^*(I) \ge m^*\,(I')$$

$$\Rightarrow \qquad m^*\,(I) \ge b - a - \varepsilon = |I| - \varepsilon$$

Also $\qquad I \subset [a, b + \varepsilon] \Rightarrow m*(I) < m* [a, b + \varepsilon] = |[a, b + \varepsilon]|$

$$\Rightarrow m*(I) < b - a + \varepsilon = |I| + \varepsilon$$

$$\Rightarrow |I| - \varepsilon \le m*(I) < |I| + \varepsilon.$$

Since ε is arbitrary small quantity, we get, $|I| \le m*(I) \le |I| \Rightarrow m*(I) = |I|$

Case IV. Suppose I is an infinite interval. Four types of such intervals exist.

Let $I = (-\infty, a]$. This is an infinite interval. In this case if M is any number >0 however large, we can get a number $k \le a$ s.t. the interval $[k, k + M)$ is contained in I i.e., $[k, k + M) \subseteq I$, M being chosen arbitrarily.

$$\Rightarrow \qquad m^*(I) > M$$

But M is arbitrarily chosen large number

$$\Rightarrow \qquad m^*(I) = \infty = |I| = l(I).$$

Similarly, we can establish the result for other types of infinite intervals.

16.3 LEBESGUE INNER MEASURE

The inner measure of a set A, denoted by $m_*(A)$, or $m_i(A)$ is defined by

$$m_*(A) = b - a - m^*\,(A'),$$

where A' is the complement of A relative to interval $[a, b]$ s.t. $A \subseteq [a, b]$.

REMARKS

- $m_*(\phi) = 0$
- $m_*(A) = b - a - \inf \{|G| : G \supseteq A' \text{ and is open}\}$

    $= \sup \{|H| : H \subseteq A \text{ and is closed}\}$
- $m_*(A) \ge 0$, since $A' \subseteq [a, b] \Rightarrow m^*(A') \le b - a$.
- $m_*(A) \ge |H|$, for any closed set $H \subseteq A$.
- If H is closed set, then $m_*(H) = |H|$.
- For each $\varepsilon > 0$, we can find a closed set $H \subset A$ and satisfying the condition $|H| > m_*(A) - \varepsilon$.

THEOREM 1. $m^*(A) \ge m_*(A)$ i.e., $m_e(A) \ge m_i(A)$ for any set $A \subset R$.

Proof. Let H be any closed set s.t. $H \subseteq A$, and G be an open set s.t. $A \subseteq G$.

$$\Rightarrow \qquad H \subseteq A \subseteq G \text{ or } H \subseteq G \quad \Rightarrow |H| \le |G|.$$

Taking the infimum over all such G, we get

$$|H| \leq \inf |G| \text{ s.t. } G \supseteq A \Rightarrow |H| \leq m^*(A).$$

Taking supremum over all such H, we get

$$\sup |H| \leq m^*(A) \Rightarrow m_*(A) \leq m^*(A).$$

16.4 LEBESGUE MEASURABLE SETS

Definition 1. *The set A is said to be Lebesgue measurable if for each set $E \subseteq R$ we have* $m^*(E) = m^*(E \cap A) + m^*(E \cap A').$

Definition 2. *The set A is said to be Lebesgue measurable, if $m^*(A) = m_*(A)$ and the common value is called its measure and is denoted by $m(A)$. Thus, if A is Lebesgue measurable, then*

$$m^*(A) = m_*(A) = m(A).$$

THEOREM 1. *A linear set A of outer measure zero is Lebesgue measurable.*

Proof.   We know that $m^*(A) \geq m^*(A)$ ...(1)

Given $m^*(A) = 0$ then from (1) $m_*(A) \leq 0$

But $m_*(A) \geq 0$, so $m_*(A) = 0$

$\therefore$ $m^*(A) = m_*(A) = 0.$

Hence, set A is lebesgue measurable.

REMARKS

- Any subset of A (whose outer measure is zero) is also measurable.
- The necessary and sufficient condition for a set E to the measurable with measure zero is that
$$m^*(E) = 0.$$

THEOREM 2. *Every countable set is L-measurable, and its measure is zero.*

Proof.   We know that every countable set has its outer measure zero. So by theorem 1, it is L-measurable and has its measure zero.

Theorem 3. *Let A and B be two sets such that $A \subseteq B$; then*

(i) $m^*(A) \leq m^*(B)$ (ii) $m_*(A) \leq m_*(B)$.

Proof.   (i) By deflnition, we have

$m^*(A) = \inf \{|G| : G$ is an open set containing $A\}$,

$m^*(B) = \inf \{|G| : G$ is an open set containing $B\}$,

Let $m^*(B) = |G'|$, s.t. G' is an open set and $G' \supseteq B$.

Now $B \subseteq G' \Rightarrow A \subset G', (\because A \subseteq B)$

$\Rightarrow$ G' is open superset of A.

$\Rightarrow$ $m^*(A) \leq |G'| = m^*(B).$

(ii) We have $A \subseteq B \subseteq [a, b]$ (suppose) $\Rightarrow B' \subseteq A' \Rightarrow m^*(B') \leq m^*(A')$

$\Rightarrow$ $-m^*(B') \geq -m^*(A')$

Adding $(b - a)$ to both the sides, we get

$(b - a) - m^*(B') \geq (b - a) - m^*(A')$

$\Rightarrow$ $m_*(B) \geq m_*(A).$

THEOREM 4. *Every bounded open set, and every bounded closed set is measurable.*

Proof.   Let F be a closed set which is bounded; then we have

$$m_*(F) = \sup \{|H| : H \text{ is closed and } H \subseteq F\}$$

which implies that $\quad m_*(F) \geq |F| = m^*(F) \Rightarrow m_*(F) \geq m^*(F)$ ...(1)

But $\qquad\qquad\qquad m_*(F) \leq m^*(F).$ ...(2)

From (1) and (2), $m^*(F) = m_*(F)$ *i.e.*, F is measurable. Again, we know that open sets are the complements of closed sets and so when a bounded closed set is measurable, its complement will also be measurable *i.e.*, bounded open sets will also be measurable.

THEOREM 5. *Any set A is measurable, iff an open set G containing A and a closed set H contained in A can be so determined, that $|G| - |H| < \varepsilon$, where ε is an arbitrary positive real number (however small).*

Proof. For each $\varepsilon/2 > 0$ we can find an open set $G \supseteq A$ and a closed Set $H \subseteq A$, s.t.

$$|G| < m^*(A) + (\varepsilon/2)$$ ...(1)

and $\qquad\qquad m_*(A) < |H| + (\varepsilon/2).$ ...(2)

Since A is measurable $\quad \Rightarrow m^*(A) = m_*(A).$ ...(3)

From (1), (2) and (3), we have

$$|G| - (\varepsilon/2) < |H| + \varepsilon/2 \text{ or } |G| - |H| < \varepsilon.$$

Conversely, let $|G| - |H| < \varepsilon$, where $\varepsilon > 0$. Again, by definition of Lebesgue outer measure of a set A, we know that

$$m^*(A) \leq |G| \text{ and } m_*(A) \geq |H| \text{ or } -m_*(A) \geq -|H|,$$

which implies (on adding both) that $m^*(A) - m_*(A) \leq |G| - |H|$,

But $\qquad\qquad |G| - |H| < \varepsilon$, (given)

So the last inequality gives $m^*(A) = m_*(A)$ *i.e.*, $< \varepsilon$.

But ε is arbitrary. Hence, $\quad m^*(A) = m_*(A)$ *i.e.*, A is measurable.

THEOREM 6. *Union of two measurable sets is also measurable.*

Proof. If set A is measurable, we can find a closed set H_1 and an open set G_1 s.t. $H_1 \subseteq A \subseteq G_1$ and $|G_1| - |H_1| < \varepsilon/2$. Similarly for measurable set B, there exists closed set H_2 and open set G_2 s.t.

$$|G_2| - |H_2| < \varepsilon/2 \text{ s.t. } H_2 \subseteq B \subseteq G_2.$$

Let $G = G_1 \cup G_2$ and $H = H_1 \cup H_2$. Then G is an open set being the finite union of open sets. Similarly, H is closed.

Again, $\qquad A \subseteq G_1, B \subseteq G_2 \Rightarrow A \cup B \subseteq G_1 \cup G_2 \Rightarrow A \cup B \subseteq G$ ... (1)

Similarly, $\qquad H_1 \cup H_2 \subseteq A \cup B$ ... (2)

Let $\qquad\qquad E = A \cup B$

From (1) and (2), we have

$$H \subseteq E \subseteq G \Rightarrow m_*(H) \leq m_*(E) \text{ and } m^*(E) \leq m^*(G).$$

But H and G are measurable. So

$$m^*(H) = m_*(H) = m(H),$$ ...(3)

and $\qquad\qquad m^*(G) = m_*(G) = m(G),$ ...(4)

So, we have $\quad m(H) \leq m_*(E)$

and $\qquad\qquad m^*(E) \leq m(G)$

or $\qquad -m_*(E) \le -m(H)$

Adding the above two inequalities, we get

$$m^*(E) - m_*(E) \le m(G) - m(H). \qquad \qquad ...(5)$$

Again, $\qquad\qquad G = (G-H) \cup H \text{ and } (G-H) \cap H = \phi$

$\therefore \qquad\qquad m(G) = m(G-H) + m(H)$

$\Rightarrow \qquad\qquad m(G-H) = m(G) - m(H). \qquad\qquad ...(6)$

Using (6) in (5), we have

$$m^*(E) - m_*(E) \le m(G-H) = m((G_1 - H_1) \cup (G_2 - H_2))$$
$$\le m(G_1) - m(H_1) + m(G_2) - m(H_2)$$
$$= |G_1| - |H_1| + |G_2| - |H_2|.$$

But $\qquad |G_1| - |H_1| < \varepsilon/2 \text{ and } |G_2| - |H_2| < \varepsilon/2.$

Hence,

$$m^*(E) - m_*(E) < \varepsilon/2 + \varepsilon/2 = \varepsilon.$$

But ε is arbitrary. Hence it can be made as small as we please. Making $\varepsilon \to 0$, in the last inequality $m^*(E) - m_*(E) \le 0$, but $m^*(E) - m_*(E) \ge 0$ is always true. Combining these two, we get the desired result that $m^*(E) - m_*(E) = 0$, $m_*(E) = m^*(E)$ i.e., E is measurable.

REMARKS

- Countable union of measurable sets is measurable.
- Intersection of two measurable sets is also measurable.
- The finite intersection of measurable sets is also measurable.
- Difference of two measurable sets is also measurable.
- Symmetric difference $A \Delta B$ is measurable as $A \Delta B = (A - B) \cup (B - A)$.
- If A and B are disjoint measurable sets, then $m(A \cup B) = m(A) + m(B)$.

THEOREM 7. *The union of finite number of measurable sets is also measurable.*

Proof. Let $E = \overset{n}{\underset{r=1}{\cup}} E_r$ where $E_1, E_2, ..., E_n$ are measurable sets.

We are to prove that the set E is measurable. Let $\varepsilon > 0$ be an arbitrary small number. Since E_r is measurable, $\exists$ closed set H_r and an open set G_r, s.t. $H_r \subset E_r \subset G_r$ and

$$m(G_r) - m(H_r) < \varepsilon/n \qquad\qquad ...(1)$$

If we take $H = \overset{n}{\underset{r=1}{\cup}} H_r$ and $G = \overset{n}{\underset{r=1}{\cup}} G_r$ then being the finite unions of closed and open sets, the set H is closed and the set G is open such that $H \subset E \subset G$.

$\Rightarrow \qquad\qquad m_*(H) \le m_*(E) \text{ and } m^*(E) \le m^*(G).$

But H and G are closed and open sets, therefore H and G are measurable.

$\Rightarrow \qquad\qquad m^*(H) = m_*(H) = m(H) \text{ and } m^*(G) = m_*(G) = m(G).$

Thus, we get $\qquad m(H) \le m^*(E) \text{ and } m^*(E) \le m(G).$

$\Rightarrow \qquad\qquad -m^*(E) \le -m(H) \text{ and } m^*(E) \le m(G)$

Adding these two, $m^*(E) - m_*(E) \le m(G) - m(H). \qquad\qquad ...(2)$

Now $G - H = G \cap H'$. But G is open and H' is also open $\Rightarrow G - H$ is measurable.

Now $G = (G - H) \cup H$ where $(G - H) \cap H = \phi$.

Similarly, $m(G_r - H_r) = m(G_r) - m(H_r)$.

Also

$$(G - H) \subset \bigcup_{r=1}^{n} (G_r - H_r) \Rightarrow m(G - H) \leq \sum_{r=1}^{n} m(G_r - H_r)$$

$$\Rightarrow m(G) - m(H) \leq \sum_{r=1}^{n} [m(G_r) - m(H_r)] < n.\frac{\varepsilon}{n} = \varepsilon, \text{ by } (1)$$

Therefore (2) $\Rightarrow m^*(E) - m_*(E) < \varepsilon.$

But ε is arbitrary small positive quantity. Therefore making $\varepsilon \to 0$, the above inequality yields that $m^*(E) - m_*(E) \leq 0$

$$\Rightarrow \qquad\qquad m^*(E) \leq m_*(E) \qquad\qquad\qquad\qquad …(3)$$

But $\qquad\qquad\qquad m^*(E) \geq m_*(E) \qquad\qquad\qquad\qquad …(4)$

Hence, from (3) and (4) $\quad m^*(E) = m_*(E)$

$$\Rightarrow \quad E = \bigcup_{i=1}^{n} E_i \text{ is measurable.}$$

REMARK

- The arbitrary union of measurable sets is also measurable.

THEOREM 8. *If a measurable set G_1 contains another measurable sets G_2, then the set $G_1 - G_2$ is measurable and its measure is $m(G_1) - m(G_2)$.*

Proof. Let G_1 and G_2 be two measurable sets such that $G_2 \leq G_1$.

$\Rightarrow \quad G_1'$ and G_2' are also measurable sets.

$\Rightarrow \quad G_1 \cap G_2'$ is also measurable,

But $G_1 \cap G_2' = G_1 - G_2$, as $G_2 \subset G_1 \Rightarrow G_1 \cap G_2 = G_2$ and $G_1 \subseteq (a, b)$.

$\Rightarrow \quad G_1 - G_2$ is measurable.

Again $\qquad G_1 = (G_1 - G_2) \cup G_2$ and $(G_1 - G_2) \cap G_2 = \phi$.

$\therefore \qquad m(G_1) = m(G_1 - G_2) + m(G_2)$ or $m(G_1 - G_2) = m(G_1) - m(G_2)$.

THEOREM 9. (First Fundamental Theorem). *If $E_1, E_2, E_3, \ldots$ are pairwise disjoint measurable sets and $E = E_1 \cup E_2 \cup \ldots$, then E is measurable and $m(E) = \sum_{k=1}^{\infty} m(E_k)$ i.e., Lebesgue measure is countable additive.*

Proof. Given $\quad E_i \cap E_j = \phi, \forall i \neq j \qquad\qquad\qquad\qquad …(1)$

we have

$$m^*(\cup E_k) \leq \sum m^*(E_k) \text{ or } m\left(\bigcup_{k=1}^{\infty} E_k\right) \leq \sum_{k=1}^{\infty} m(E_k) \qquad …(2)$$

as $\cup E_k$ is measurable.

Now $\qquad\qquad \bigcup_{k=1}^{\infty} E_k \supseteq \bigcup_{k=1}^{n} (E_k)$, for all n

$$\Rightarrow \qquad m\left(\bigcup_{k=1}^{\infty} E_k\right) \geq m\left(\bigcup_{k=1}^{n} E_k\right) \qquad\qquad\qquad …(3)$$

But we know that if $E_1 \cap E_2 = \phi$, then $m(E_1 \cup E_2) = m(E_1) \cap m(E_2)$.

Extending, by induction, the result for finite number of pairwise disjoint sets $E_1, E_2, \ldots E_n$, we get

$$m\left(\bigcup_{k=1}^{n} E_k\right) = \sum_{k=1}^{n} m(E_k).$$

(3) $\Rightarrow$
$$m\left(\bigcup_{k=1}^{\infty} E_k\right) \geq \sum_{k=1}^{n} m(E_k), \forall n$$

Letting $n \to \infty$

$$m\left(\bigcup_{k=1}^{\infty} E_k\right) \geq \sum_{k=1}^{\infty} m(E_k). \qquad \ldots(4)$$

(2) and (4) $\Rightarrow$ $m\left(\bigcup_{k=1}^{\infty} E_k\right) = \sum_{k=1}^{\infty} m(E_k).$

$\Rightarrow$ Lebesgue measure is countably additive for pairwise disjoint sets.

THEOREM 10. *If $E_1, E_2, E_3, \ldots$ are measurable sets of real numbers, then $\bigcup_{n=1}^{\infty} E_n$ is measurable and*

$$m\left(\bigcup_{n=1}^{\infty} E_n\right) \leq \sum_{n=1}^{\infty} m(E_n).$$

Proof.   We have
$$\bigcup_{n=1}^{\infty} E_n = E_1 \cup [E_2 - E_1] \cup [E_3 - (E_1 \cup E_2)] \cup \ldots$$
$$\ldots \cup [E_n - (E_1 \cup E_2 \cup \ldots \cup E_{n-1})] \cup \ldots$$

Clearly the sets $E_1, E_2 - E_1, E_3 - (E_1 \cup E_2) \ldots$, are disjoint measurable sets and hence, their union *i.e.*, the set $\bigcup_{n=1}^{\infty} E_n$ is measurable.

REMARK

- Interval (a, ∞) is measurable.

THEOREM 11. **(Second fundamental theorem).** *If $E_1, E_2, \ldots$ are measurable sets, then $\bigcup_{k=1}^{\infty} E_k$ is measurable.*

Proof.   Let
$$G = \bigcap_{k=1}^{\infty} E_k.$$

Then
$$G' = \left[\bigcap_{k=1}^{\infty} E_k\right]' = \bigcup_{k=1}^{\infty} E_k'$$

Now given that E_k is measurable, $\forall k \in N$

$\Rightarrow$ E_k' is measurable

$\Rightarrow$ $\bigcup_{k=1}^{\infty} E_k'$ is measurable   (By first fundamental theorem)

$\Rightarrow$ G' is measurable

$\Rightarrow$ G is measurable.

THEOREM 12. *If $\langle E_n \rangle$ is monotonically increasing (or non-decreasing) sequence of measurable sets and $E = \bigcup_{k=1}^{\infty} E_k$; then $m(E) = \lim_{n \to \infty} m(E_n).$*

Proof. Given that $E_1 \subseteq E_2 \subseteq E_3 \ldots$, so we may write

$$E = E_1 \cup (E_2 - E_1) \cup (E_3 - E_2) \cup \ldots \cup (E_r - E_{r-1}) \cup \ldots$$

i.e., $$E = E_1 \overset{\infty}{\underset{k=1}{\cup}} (E_{k+1} - E_k)$$

Clearly, $(E_{k+1} - E_k)$ and hence $\overset{\infty}{\underset{k=1}{\cup}} (E_{k+1} - E_k)$ are measurable sets. Since E is a disjoint union of measuratble sets, we have

$$m(E) = m(E_1) + \sum_{k=1}^{\infty} m(E_{k+1} - E_k)$$

or $$m(E) = m(E_1) + \sum_{k=1}^{\infty} [m(E_{k+1}) - m(E_k)], (\because \ E_k \subseteq E_{k+1})$$

$$= m(E_1) + \lim_{n\to\infty} [\{m(E_2) - m(E_1)\} + \ldots + \{m(E_n) - m(E_{n-1})\}]$$

$$= m(E_1) + \lim_{n\to\infty} [-m(E_1) + m(E_n)]$$

$$= m(E_1) - m(E_1) + \lim_{n\to\infty} m(E_n) = \lim_{n\to\infty} m(E_n)$$

THEOREM 13. *If* $<E_1, E_2, \ldots>$ *is a monotonically decreasing (or non-increasing) sequence of measurable sets s.t.* $m(E_1) < \infty$ *and* $E = \overset{\infty}{\underset{k=1}{\cap}} E_k$, *then* $m(E) = \lim_{n\to\infty} m(E_n)$.

Proof. Let $$E = \overset{\infty}{\underset{k=1}{\cap}} E_k$$

Obviously, $$E_1 = E \cup (E_1 - E_2) \cup (E_2 - E_3) \cup \ldots$$

Then $$m(E_1) = m(E) + \sum_{k=1}^{\infty} [m(E_k) - m(E_{k+1})]$$

$$= m(E) + \lim_{n\to\infty} [m(E_1) - m(E_n)]$$

$\Rightarrow$ $$m(E_1) = m(E) + m(E_1) - \lim_{n\to\infty} m(E_n)$$

$\Rightarrow$ $$m(E) = \lim_{n\to\infty} m(E_n).$$

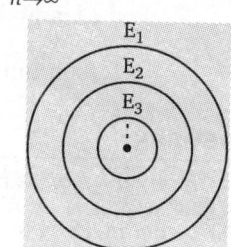

Fig. (1)

16.5 CANTOR TERNARY SET

Consider the closed interval $[0, 1]$. Divide this closed interval into three equal parts and remove the middle one *i.e.*, remove the open interval $\left(\dfrac{1}{3}, \dfrac{2}{3}\right)$. This is our first step of construction of Cantor ternary set.

$$\dfrac{1}{3^2} \quad \dfrac{2}{3^2} \quad \dfrac{1}{3} \qquad \text{Removed} \qquad \dfrac{2}{3} \quad \dfrac{7}{3^2} \quad \dfrac{8}{3^2} \qquad 1$$

First step

Second step — Removed — — Removed —

Again divide the each of the remaining two intervals into three equal parts and remove the middle one *i.e.*, remove the open intervals $\left(\dfrac{1}{3^2}, \dfrac{2}{3^2}\right)$ and $\left(\dfrac{7}{9}, \dfrac{8}{9}\right)$. This is our second step of construction. Continuing in this way infinitely many times, we have that at p^{th} step. 2^{p-1} open intervals are removed each of length $\dfrac{1}{3^p}$. The remaining set constitutes Cantor ternary set. Let us denote this set by F. Then clearly F' is open being enumerable union of mutually disjoint open intervals $\Rightarrow$ Being the complement of an open set, set F is closed.

Here $F = [0,1] - F'$, where $F' = \left(\dfrac{1}{3},\dfrac{2}{3}\right) \cup \left(\dfrac{1}{9},\dfrac{2}{9}\right) \cup \left(\dfrac{7}{9},\dfrac{8}{9}\right) \cup ....$.

Thus Cantor ternary set surely contains the points

$$0,1,\frac{1}{3},\frac{2}{3},\frac{1}{9},\frac{2}{9},\frac{7}{9},\frac{8}{9} \text{ etc.}$$

Definition. *The Cantor set F is the set of all numbers in the interval $[0, 1]$ which have a ternary expansion without the digit 1 i.e., ternary expansion involves only two digits 0 or 2.*

REMARKS

- Cantor set contains no open interval.
- Cantor ternary set is measurable and its measure is zero.
- The cantor set and its each subset is measurable and has measure zero. For if $S \subseteq F$. Then $m(S) \le m(F) = 0 \Rightarrow m(S) = 0$ and hence S is measurable.
- Cantor set is bounded as $F \subset [0, 1]$.
- By Heine-Borel theorem, being closed and bounded, set F is compact also.
- Cantor ternary set is perfect.
- Cantor set is uncountable.
- There exists uncountable set of measure zero. (For example, cantor set).
- The cardinal number of the Cantor set is the cardinal number c of the linear continuum.
- The Cantor set can be put into a one-to-one correspondence with the interval $[0, 1]$.
- The family M of all measurable sets has the cardinal number 2^c.

16.6 SOME MORE RESULTS

1. A set A is Lebesgue measurable iff its complement A' is measurable.
2. If $A \subset (a, b)$ is Lebesgue measurable set, then $m(A) + m(A') = b - a$.
3. If E_1 and E_2 are subsets of $[a, b]$, then
$$m^*(E_1) + m^*(E_2) \ge m^*(E_1 \cup E_2) + m^*(E_1 \cap E_2)$$
and $\qquad m_*(E_1) + m_*(E_2) \le m_*(E_1 \cup E_2) + m_*(E_1 \cap E_2)$.

4. If $E_1, E_2, ...$ are pairwise disjoint subsets of $[a, b], m^*\left(\bigcup_{n=1}^{\infty} E_n\right) \ge \sum_{n=1}^{\infty} m^*(E_n)$.

5. For the above sets, we have $m\left(\bigcup_{n=1}^{\infty} E_n\right) = \sum_{n=1}^{\infty} m(E_n)$,

where E_n's are measurable sets.

__THEOREM 1.__ *If E_1 and E_2 are measurable subsets of $[a, b]$ then*
$$m(E_1) + m(E_2) = m(E_1 \cup E_2) + m(E_1 \cap E_2)$$

Proof.   Since, E_1 and E_2 are measurable therefore $m_*(E_1) = m^*(E_1) = m(E_1)$

and $\qquad m_*(E_2) = m^*(E_2) = m(E_2)$.   ...(1)

Again E_1 and E_2 are measurable $\Rightarrow$ $E_1 \cup E_2$ and $E_1 \cap E_2$ are measurable

$\Rightarrow \quad m_*(E_1 \cup E_2) = m^*(E_1 \cup E_2) = m(E_1 \cup E_2)$

and $\quad m_*(E_1 \cap E_2) = m^*(E_1 \cap E_2) = m(E_1 \cap E_2).$ ...(2)

Now, we know that

$$m^*(E_1) + m^*(E_2) \geq m^*(E_1 \cup E_2) + m^*(E_1 \cap E_2)$$

and $m_*(E_1) + m_*(E_2) \leq m_*(E_1 \cup E_2) + m_*(E_1 \cap E_2)$

In view of (1) and (2), above inequalities become

$$m(E_1) + m(E_2) \geq m(E_1 \cup E_2) + m(E_1 \cap E_2)$$ ...(3)

and $\quad m(E_1) + m(E_2) \leq m(E_1 \cup E_2) + m(E_1 \cap E_2)$ ...(4)

Combining (3) and (4), we get

$$m(E_1) + m(E_2) = m(E_1 \cup E_2) + m(E_1 \cap E_2)$$

THEOREM 2. *If E_1 and E_2 are measurable subsets of $[2, 3]$ s.t. $m(E_1)=1$, then $m(E_1 \cap E_2)=m(E_2)$*

Proof. We have $\quad m(E_1 \cap E_2) = m(E_1) + m(E_2) - m(E_1 \cup E_2)$ ...(1)

Now $\quad E_1 \cup E_2 \supseteq E_1 \Rightarrow m(E_1 \cup E_2) \geq m(E_1) \geq 1$

also $E_1 \subseteq [2, 3], E_2 \subseteq [2, 3] \Rightarrow E_1 \cup E_2 \subseteq [2, 3]$

$\Rightarrow m(E_1 \cup E_2) \leq m[2,3] = 3 - 2 = 1$

$\Rightarrow m(E_1 \cup E_2) = 1$

Therefore (1) $\Rightarrow m(E_1 \cap E_2) = 1 + m(E_2) - 1 = m(E_2)$

THEOREM 3. *If $m^*(A) = 0$, then $m^*(A \cup B) = m^*(B)$*

Proof. We know that $m^*(A \cup B) \leq m^*(A) + m^*(B)$

But $\quad m^*(A) = 0 \Rightarrow m^*(A \cup B) \leq m^*(B).$

Also $\quad B \subseteq A \cup B \Rightarrow m^*(B) \leq m^*(A \cup B)$

$\Rightarrow m^*(A \cup B) = m^*(B)$

REMARK

- If set A is Lebesgue measurable and $m^*(A \Delta B) = 0$, then B is also measurable and $m^*(B)=m^*(A)$.

THEOREM 4. *If A, B, C are any three measurable subsets of $[a, b]$, then*

$$m(A) + m(B) + m(C) = m(A \cup B \cup C) + m(B \cap C)$$
$$+ m(C \cap A) + m(A \cap B) - m(A \cap B \cap C).$$

Proof. We know that if A and B are measurable sets

then $\quad m(A) + m(B) = m(A \cup B) + m(A \cap B).$ ...(1)

Now, A, B are measurable $\Rightarrow A \cup B$ is measurable.

So from (1), we have $m(A \cup B) + m(C) = m(A \cup B \cup C) + m[(A \cup B) \cap C].$ ...(2)

Form (1), we have $\quad m(A \cup B) = m(A) + m(B) - m(A \cap B).$

Using this in (2), we have

$$m(A) + m(B) - m(A \cap B) + m(C) = m(A \cup B \cup C) + m[(A \cap C) \cup (B \cap C)]$$
$$= m(A \cup B \cup C) + m(A \cap C) + m(B \cap C)$$
$$- m(A \cap B \cap C)$$

or $\quad$ $m(A) + m(B) + m(C) = m(A \cup B \cup C) + m(B \cap C)$
$$+ m(C \cap A) + m(A \cap B) - m(A \cap B \cap C).$$

REMARKS

- If M is a measurable set, then prove that for any set E, $m^*(E) = m^*(EM) + m^*(E - EM)$ where $EM = E \cap M$.
- A is any measurable set and $A_1, A_2, \ldots A_n$ are pairwise disjoint measurable sets, then

$$m\left(A \cap \left[\bigcup_{i=1}^{n} A_i\right]\right) = \sum_{i=1}^{n} m(A \cap A_i), \forall n$$

Solved Examples

Example 1. Show that the Lebesgue measure of the following set is zero:

$\{x \in R : 0 < x < 1 \text{ and } x \text{ has a decimal expansion not using the digit 7}\}$.

Solution. Divide the open interval $(0, 1)$ into 10 equal parts and remove the interior of 8th part $\left(\text{length}\dfrac{1}{10}\right)$. Again subdivide each of the remaining 9 intervals into 10 equal parts and remove the interior of 8th part of each of them. All the 9 deleted open intervals are of length $\dfrac{1}{10^2}$. Continue this process indefinitely. The remaining subset of $(0, 1)$ constitute the required set. Let us donote this set by A.

Number of step	Number of intervals deleted	Number of remaining intervals	Length of each deleted interval	Total length of open intervals deleted
1	1	9	$\dfrac{1}{10}$	$\dfrac{1}{10}$
2	9	9^2	$\dfrac{1}{10^2}$	$\dfrac{9}{10^2}$
3	9^2	9^3	$\dfrac{1}{10^3}$	$\dfrac{9}{10^3}$
—	—	—	—	—
—	—	—	—	—

Clearly, A' is an open set. So sum of dropped intervals *i.e.*, $m(A')$

$$= \frac{1}{10} + 9.\frac{1}{10^2} + 9^2.\frac{1}{10^3} + \ldots = \frac{1/10}{1 - 9/10} = 1$$

Now $\quad$ $m(A') + m(A) = 1 \Rightarrow m(A) = 1 - 1 = 0$

REMARK

- Above set A is formed by the same pattern as Cantor ternary set was formed. In general a set obtained by dividing $[0, 1]$ in n parts and then removing some particular interval, is known as Cantor's *n*-ary set.

Example 2. *Let x be a number in the interval (0, 1) written in the scale of 10 as x = 0.$x_1 x_2 x_3$ ... where none of the x_i's is 3 or 5. Find the measure of the set so defined.*

Solution. Here every time we have to delete two subintervals (interior of 4th and 6th parts at each step of construction). Hence if the required set is A then removed intervals will will constitute the A'.

Number of step	Number of intervals deleted	Number of remaining intervals	Length of each deleted interval	Total length of open intervals deleted
I	2	8	$\dfrac{1}{10}$	$2 \times \dfrac{1}{10}$
II	2×8	8^2	$\dfrac{1}{10^2}$	$2 \times \dfrac{8}{10^2}$
III	2×8^2	8^3	$\dfrac{1}{10^3}$	$2 \times \dfrac{8}{10^3}$
—	—	—	—	—
—	—	—	—	—

$$\Rightarrow \qquad m(A') = 2 \times \frac{1}{10} + 2 \times \frac{8}{10^2} + \frac{2 \times 8^2}{10^3} + ... = \frac{\dfrac{2}{10}}{1 - \dfrac{8}{10}} = 1$$

$$\Rightarrow \qquad m(A) = |(0,1)| - m(A') = 1 - 1 = 0$$

Example 3. *Construct a non-dense perfect set in the interval [0, 1], whose measure is $\dfrac{1}{2}$.*

Solution. Let $\alpha \in [0,1]$. Remove, from the interval [0, 1] and also in succession from the remaining free sub-intervals, a part of length $l_n (n = 1, 2, 3, ...)$ satisfying

$$l_1 + 2l_2 + 2^2 l_3 + ... + 2^{n-1} l_n = \alpha - \frac{\alpha}{n+2}.$$

Suppose A denotes the remaining set which is obviously perfect and non-dense as in the case of Cantor ternary set.

We have $m(A) = 1 - \alpha$, for the sum of the lengths of intervals dropped upto n^{th} stage of construction.

$$= l_1 + 2l_2 + 2^2 l_3 + ... + 2^{n-1} l_n = \alpha - \frac{\alpha}{n+2}.$$

So limiting sum of the dropped intervals

$$= \lim_{n \to \infty} \left[\alpha - \frac{\alpha}{n+2} \right] = \alpha$$

Choosing $\alpha = \dfrac{1}{2} \in [0,1]$, we have $m(A) = \dfrac{1}{2}$.

Example 4. *Show that if $<E_n>$ is a sequence of measurable sets, all contained in a set of finite measure and limit E_n exists then limit E_n is measurable and $m(\lim E_n) = \lim m(E_n)$.*

Solution. First of all we shall show that $\overline{\lim} E_n$ and $\underline{\lim} E_n$ are measurable.

Recall that $\overline{\lim E_n} = \bigcap\limits_{n=1}^{\infty}\left(\bigcup\limits_{k=n}^{\infty} E_k\right)$ and $\underline{\lim E_n} = \bigcup\limits_{n=1}^{\infty}\left(\bigcap\limits_{k=n}^{\infty} E_k\right)$

We are given that sets E_n are measurable. Also we know that countable union and intersection of measurable sets are also measurble. Hence, $\overline{\lim E_n}$ and $\underline{\lim E_n}$ are measurable.

Further if the $\lim E_n$ exists, we get

$$\underline{\lim E_n} = \overline{\lim E_n} = \lim E_n \text{ and hence } \lim E_n \text{ is also measurable.}$$

Now $\bigcup\limits_{k=n}^{\infty} E_k = A_n$ and $\bigcap\limits_{k=n}^{\infty} E_k = B_n$.

Then $<A_n>$ is a monotonically non-increasing (or say decreasing) sequence of measurable sets. Then, we get

$$m\left(\overline{\lim E_n}\right) = m\left(\bigcap\limits_{n=1}^{\infty} A_n\right) = \lim_{n\to\infty} m(A_n) = \lim_{n\to\infty} (E_n) \qquad \text{...(1)}$$

Again $<B_n>$ is a monotonically non-decreasing (or say increasing sequence of measurable sets, we get,

$$m\left(\overline{\lim E_n}\right) = m\left(\bigcap\limits_{n=1}^{\infty} B_n\right) = \lim_{n\to\infty} m(B_n)$$

$$= \lim_{n\to\infty} m(E_n) \qquad \left[\because \quad \text{when } n\to\infty, B_n = \bigcap\limits_{k=n}^{\infty} E_k = E_n\right]$$

Since limit exists, so we get $\overline{(\lim E_n)} = m(\underline{\lim E_n}) = m(\lim E_n)$

$\Rightarrow \qquad m(\lim E_n) = \lim m(E_n)$

REMARK

- If E_1, E_2 are any subsets of real numbers s.t. the measure of the symmetric difference of E_1 and E_2 is zero and E_1 is measurable then E_2 is measurable and $m(E_2) = m(E_1)$.

16.7 BOREL SETS AND THEIR MEASURABILITY

The set which can be obtained by taking countable unions or intersectons of open and closed sets, are called Borel sets. Therefore Borel sets are measurable sets.

For every class A of subsets of a space X, there always exists a smallest σ-algebra V containing the class A. If $\{V_\alpha, a \in \Delta\}$, is a collection of all σ-algebras containing A, then $\bigcap\limits_{\alpha\in\Delta} V_\alpha$ is the smallest σ-algebra containing A. This smallest algebra is called σ-algebra generated by A.

Definition. *The σ-algebra generated by the class of all open intervals of the form (a, b), is called the class of Borel sets in R and denoted by B. The members of Borel class B are called Borel sets of R.*

THEOREM 1. *Every Borel set of R is measurable.*

Proof.   We know that an interval is always measurable and hence every open set is measurable. Also B is the smallest algebra containing open sets, we have $B \subset M \Rightarrow$ every Borel set is Lebesgue measurable.

16.7.1 BOREL MEASURABLE SET

All Borel sets are also known as Borel measurable sets.

16.8 F_σ AND G_δ SETS

We know that arbitrary intersecton of closed sets is closed and the finite union of closed sets is also closed, while the infinite union of closed sets may not be closed. Similarly arbitrary union of open sets is open as well the finite intersection of open sets is open while the infinite intersection of open sets may not be open.

Definition-1 (F_σ-set). *A set A is said to be a F_σ-set if it is the countable (may be finite or infinite) union of closed sets.*

i.e.,
$$A = \underset{i \in N}{\cup} F_i; F_i \text{ are closed.}$$

Definition-2 (G_δ-set). *A set A is said to be a G_δ-set if it is countable intersection of open sets i.e.,* $A = \underset{i \in N}{\cup} G_i, G_i$ *being open sets.*

REMARKS

- Complement of a G_δ-set is a F_σ-set and vice versa
  For Example : An open interval (a, b) is a F_σ-set as
  $$(a,b) = \underset{n=1}{\overset{\infty}{\cup}} \left[a + \frac{1}{n}, b - \frac{1}{n} \right]$$
  $$= \text{ a countable union of closed sets.}$$
- F_σ and G_δ sets are Borel sets.
- All that F_σ and G_δ sets are measurable sets.
- A closed-interval $[a, b]$ is G_δ-set.

16.9 NON-MEASURABLE SET

In analysis most of the sets, not all, are measurable. There exists sets which are non-measurable in the Lebesgue sence.

THEOREM 1. *There exists a none-measurable set in the interval $[0, 1[$.*

Proof. Define a relation '~' in the set $[0, 1[$ as follows :

$x \sim y$ iff $x - y$ is a rational number.

Obviously (i) $x - x = 0$, a rational number $\Rightarrow$ '~' is a reflexive relation.

(ii) If $x \sim y \Rightarrow x - y$ is a rational number

$\Rightarrow y - x$ is also a rational number

$\Rightarrow y \sim x \Rightarrow$ '~' is a symmetric relation.

(iii) If $x \sim y$ and $y \sim z \Rightarrow x - y = r_1$ and $y - z = r_2$, where r_1, r_2 are both rationals

$\Rightarrow x - z = r_1 + r_2$, is also rational number

$\Rightarrow x - z \Rightarrow$ '~' is transitive relation also.

$\Rightarrow$ relation '~' is an equivalence relation.

$\Rightarrow$ this relation '~' partitions the set $[0, 1[$ into disjoint equivalence classes such that any two elements of one class will differ by a rational number while two elements taken from different classes will differ by an irrational number.

Now, let P be the set formed by choosing one element from each equivalence class induced by the '$\sim$' relation in the set $[0, 1[$ Obviously $P \subset [0, 1[$.

We shall show that this set P is non-measurable.

Let $\langle r_i \rangle_0$ be a sequence of the rationals in $[0, 1[$ with $r_0 = 0$. Let us define

$$P_i = P + r_i, \forall i = 0, 1, 2, \ldots \infty.$$

Obviously $\qquad\qquad P_0 = P + r_0 = p + 0 = P.$

To prove that $P_i \cap P_j = \phi$ when $i \neq j$.

If $x \in P_i \cap P_j \; i \neq j$ then $x \in P + r_i$ and $x \in P + r_j$

$\Rightarrow \qquad x = p_i + r_i$ and $x = p_j + r_j$ where $p_i, p_j \in P$

$\Rightarrow \qquad P_i + r_i = P_j + r_j \Rightarrow P_i - P_j = r_j - r_i$ which is a rational number.

$\Rightarrow \qquad P_i \sim P_j$ i.e., P_i and P_j are the elements of P belonging to the same equivalence class which is a contradiction as P contains only one element from each equivalence class.

$\Rightarrow \qquad \exists$ no, $x \in P_i \cap P_j \Rightarrow P_i, P_j$ are disjoint

$\Rightarrow \qquad <P_i>$ is a sequence of pairwise disjoint sets.

To prove that $\bigcup_i P_i \subseteq [0,1[$.

On the other hand each element $x \in [0, 1[$ is in some equivalence class and so is $\sim$ related to some element y of P. If x differs from y by rational number r_i. Then $x \in p_i$ and hence $[0,1[\subseteq \bigcup P_i$.

Also each $P_i \subseteq [0,1[$ therefore $\bigcup_i P_i \subseteq [0,1[$

$\Rightarrow \qquad\qquad \bigcup_i P_i = [0,1[$.

To show that set P is non-measurable.

Let if possible, suppose P is a measurable set.

Then, since each P_i is a 'translation modulo 1 or P' so each each P_i is also measurable and $m(P_i) = m(P)$

$$\Rightarrow \qquad m(\bigcup_i P_i) = \sum_{i=0}^{\infty} m(P_i) = \sum_{i=0}^{\infty} m(P) = \begin{cases} 0 & , \quad \text{if} \quad m(P) = 0 \\ \infty & , \quad \text{if} \quad m(P) > 0 \end{cases} \qquad \ldots(1)$$

In this case $m\left(\bigcup_{i=0}^{\infty} P_i\right) = m([0,1]) = 1$ which is a contradiction to (1). Hence our supposition that P is measurable is not true.

$\Rightarrow \qquad$ set P is a non-measurable set.

REMARK

- Every set of positive measure contains a non-measurable subset.

16.10 REGULARITY OF A MEASURE

An extended real-valued, non-negative, countably additive set function m defined on a σ-ring R i.e., $R \to R^*$ is said to be a regular measure if for every $\varepsilon > 0$, and $E \subseteq R$, ∃ an open set $O \supseteq E$ and a closed set $G \subseteq E$ in R s.t.

$$m(O - E) \leq \varepsilon \text{ and } m(E - G) - \varepsilon.$$

EXERCISE 16.1

1. Let E be any set. Then show that for given $\varepsilon > 0$, however small ∃ an open set G containing E such that

   $$m^*(G) \leq m^*(A) + \varepsilon.$$

2. Construct a set which is not measurable in the sence of Lebesgue.

3. Show that Lebesgue measure is invariant under translation modulo I.

4. Prove that

   $$m^*(E_1 \cup E_2) \leq m^*(E_1) + m^*(E_2).$$

5. Prove that the set of all irrational numbers in [0, 1] is measurable and find its measure.

6. Prove that $m^*(E) = 0$, for a countable set E. Hence deduce that the closed set [0, 1] is not countable.

7. Show that the Lebesgue measure of the following set is zero :

   $\{x \in R : 0 < x < 1 \text{ and } x \text{ has a decimal expansion not using the digit 5}\}$.

8. Prove that every Borel set is Lebesgue measurable.

   In particular every open set and closed set is measurable.

9. Show that $m^*(S) = \inf[m(T) : S \subset T]$, for all measurable sets T containing S.

10. Show that a set S is measurable iff for every T,

    $$m^*(T) = m^*(S \cap T) + m^*(S' \cap T).$$

16.11 CARATHEODORY'S POSTULATES FOR OUTER MEASURE

An extended real valued set function m_0 (or m_0^*) defined on a hereditory σ-ring R is called an Caratheodory's outer measure, if it satisfies the following postulates:

(1) $m_0(A_r) \geq 0, \forall A_i \in R$

(2) $m_0(\phi) = 0$

(3) $A_r, A_s \in R$ and $A_r \subseteq A_s \Rightarrow m_0(A_r) \leq m_0(A_s)$

(4) $m_0\left(\underset{r \in N}{\cup} A_r\right) \leq \underset{r \in N}{\sum} m_0(A_r)$ i.e., m_0 is σ-subadditive.

16.12 MEASURABLE SET

If μ denotes an outer measure function defined on a hereditory σ-ring $R = \{A_i : i \in \Delta\}$, then any member $A \in R$ is said to be measurable w.r.t the outer measure function μ (i.e., μ-measurable), *if*

$$\mu(E) = \mu(A \cap E) + \mu(A' \cap E), \forall E \in R \qquad \qquad ...(1)$$

REMARK

- Since $E = (A \cap E) \cup (A' \cap E)$ for any E, by postulate (iii) of outer measure, we have

  $$\mu(E) \leq \mu(A \cap E) + \mu(A' \cap E)$$

THEOREM 1. *A set A is μ-measurable, iff A' is μ-measurable, μ being an outer measure.*

Proof. Let A' be outer measurable, so that

$$\mu(E) = \mu(A' \cap E) + \mu[(A')' \cap E)], \text{ for any } E \subseteq R$$

$$= \mu(A' \cap E) + \mu(A \cap E) \qquad\qquad [\because (A')' = A]$$

$$= \mu(A \cap E) + \mu(A' \cap E), \forall E \subseteq R$$

This shows that A is m-measurable.

Conversely, suppose that A is μ-measurable. So for any set E,

$$\mu(E) = \mu(A \cap E) + \mu(A' \cap E) = \mu(A' \cap E) + \mu[(A')' \cap E]$$

Hence A' is μ-measurable.

THEOREM 2. *The union of two m_0-measurable (or outer measurable) sets is also m_0-measurable (outer measurable).*

Proof. Let A and B be the two m_0-measurable sets, then we have

$$m_0(E) = m_0(A \cap E) + m_0(A' \cap E) \qquad\qquad \text{... (1)}$$

and $$m_0(F) = m_0(B \cap F) + m_0(B' \cap F) \qquad\qquad \text{... (2)}$$

for any E and F.

Taking $F = E \cap A'$ and applying the criterion (2) of measurability to set B, we have

$$m_0(E \cap A') = m_0(B \cap E \cap A') + m_0(B' \cap E \cap A')$$

or $$m_0(E \cap A') = m_0(E \cap A' \cap B) + m_0[E \cap (A \cup B)']. \qquad\qquad \text{... (3)}$$

Putting $E = E \cap (A \cup B)$ in (1), we get

$$m_0[E \cap (A \cup B)] = m_0[A \cap (E \cap (A \cup B))] + m_0[A' \cap E \cap (A \cup B)]$$

On simplificaton, we get

$$m_0[E \cap (A \cup B)] = m_0(A \cap E) + m_0(E \cap A' \cap B)$$

Putting the value of $m_0(E \cap A' \cap B)$ from (3), we have

$$m_0(E \cap (A \cup B)) = m_0(A \cap E) + m_0(E \cap A') - m_0[E \cap (A \cup B)']$$

or $m_0[(E \cap (A \cup B)] + m_0[(E \cap (A \cup B')] = m_0(A \cap E) + m_0(A' \cap E) = m_0(E).$

Hence, $A \cup B$ is outer measurable *i.e.*, m_0-measurable.

REMARK

- By induction we may show that $\overset{n}{\underset{r=1}{\cup}} A_r$ is also m_0-measurable.

THEOREM 3. *If the outer meausre of a set is zero, then the set is outer measurable.*

Proof. Let $m_0(A) = 0$, for some set A. In order to prove that the set A is measurable, we have simply to prove that for any set E,

$$m_0(E) \geq m_0(E \cap A) + m_0(E \cap A')$$

$\because E \cap A' \subset E, \therefore m_0(E \cap A') \leq m_0(E)$ \qquad\qquad ... (1)

Again, since $E \cap A \subset A$

$\Rightarrow \qquad\qquad m_0(E \cap A) \leq m_0(A)$

$\Rightarrow \qquad\qquad m_0(E \cap A) \leq 0$ as $m_0(A) = 0 \text{(given)}$

But $\qquad\qquad m_0(E \cap A) \geq 0$ therefore $m_0(E \cap A) = 0$ \qquad\qquad ... (2)

From (1) and (2), we get

$$m_0(E) \geq m_0(E \cap A) + m_0(E \cap A')$$

THEOREM 4. *Let A be an outer measurable set. If B is any set, then*

$$m_0(A \cup B) + m_0(A \cap B) = m_0(A) + m_0(B).$$

Proof. Les us suppose that $m_0(B)$ is finite. For if $m_0(B) = \infty$, the given result is obviously satisfied. Again, given that A is measurable. So, we have

$$m_0(E) = m_0(E \cap A) + m_0(E \cap A'), \forall E \in R \qquad \qquad ...(1)$$

Suppose $E = A \cup B$. Then from (1), we have

$$m_0(A \cup B) = m_0[(A \cup B) \cap A] + m_0[(A \cup B) \cap A']$$

$$= m_0(A) + m_0(B \cap A') \qquad \qquad ...(2)$$

Let $E = B$. Then from (1)

$$m_0(B) = m_0(B \cap A) + m_0(B \cap A')]. \qquad \qquad ...(3)$$

Putting the value of $m_0[B \cap A']$ from (2) in (3), we have

$$m_0(B) = m_0(B \cap A) + m_0(A \cup B) - m_0(A)$$

or $\qquad m_0(A \cup B) + m_0(A \cap B) = m_0(A) + m_0(B).$

REMARK

- This result also holds good even if B is also outer measurable.

THEOREM 5. *If A and B are disjoint, then* $m_0(A \cup B) = m_0(A) + m_0(B)$.

Proof. We know that $m_0(\phi) = 0$. Now, given that $A \cap B = \phi$ so that $m_0(A \cap B) = 0$. Here putting the value of $m_0(A \cap B)$ in the result of previous theorem, we get

$$m_0(A \cup B) = m_0(A) + m_0(B).$$

REMARK

- By induction the result can easily be extended for finite number of sets *i.e.*,

$$m_0\left(\bigcup_{i=1}^{n} A_i \right) = \sum_{i=1}^{\infty} m_0(A_i) \text{ provided } A_i \cap A_j = \phi, \text{ for } i \neq j.$$

THEOREM 6. *If* $< A_r >$ *is a sequence of pairwise disjoint measurable subsets of* R *then*

$$m_0\left(\bigcup_{i=1}^{\infty} A_i \right) = \sum_{i=1}^{\infty} m_0(A_i).$$

Proof. Since $\bigcup_{r=1}^{n} A_r \subseteq \bigcup_{r=1}^{\infty} A_r$, we have $m_0\left(\bigcup_{r=1}^{n} A_r \right) \leq m_0\left(\bigcup_{r=1}^{\infty} A_r \right)$.

But $m_0\left(\bigcup_{r=1}^{n} A_r \right) = \sum_{r=1}^{n} m_0(A_r)$; therefore

$$\sum_{r=1}^{n} m_0(A_r) \leq m_0\left(\bigcup_{r=1}^{\infty} A_r \right). \qquad \qquad ... (1)$$

This inequality is true for all n and therefore, letting $n \to \infty$, we have

$$\sum_{r=1}^{\infty} m_0(A_r) \le m_0\left(\bigcup_{r=1}^{\infty} A_r\right). \qquad \qquad ...(2)$$

Again, we know that

$$m_0\left(\bigcup_{r=1}^{\infty} A_r\right) \le \sum_{r=1}^{\infty} m_0(A_r). \qquad \qquad ...(3)$$

Combining (2) and (3), we get

$$m_0\left(\bigcup_{r=1}^{\infty} A_r\right) = \sum_{r=1}^{\infty} m_0(A_r)$$

REMARK

- Let (X, R, m) be a measurable space. Then

  (i) $A, B \in R$ s.t. $A \subseteq B \Rightarrow m(A) \le m(B)$ (ii) $m\left(\bigcup_{r=1}^{\infty} A_r\right) \le \sum_{r=1}^{\infty} m(A_r)$, where $A_r \in R \ \forall i = 1, 2, 3, ...$

THEOREM 7. *Let $< A_r >$ be a monotonically increasing (or non-decreasing) sequence of m_0-measurable sets; then*

$$m_0(A) = \lim_{r \to 0} m_0(A_r), \text{where } A = \bigcup_{r=1}^{\infty} A_r$$

Proof. Let us suppose that $m_0(A_i) < \infty, \forall \ i$, because if $m_0(A_i) = \infty$, for some i, then the equality holds good automatically.

Now, since $A_1 \subseteq A_2 \subseteq A_3 \subseteq ....$ Therfore, we write

$$A = A_1 \cup (A_2 - A_1) \cup (A_3 - A_2) \cup ....$$

which is a union of pairwise disjoint set, as

$$(A_2 - A_1) \cap A_1 = \phi$$
$$(A_3 - A_2) \cap (A_2 - A_1) = \phi, \text{etc.}$$

Therefore, we have

$$m\left(\bigcup_{r=1}^{\infty} A_r\right) = m_0(A_1) + m_0(A_2 - A_1) + m_0(A_3 - A_2) + ... \qquad ...(1)$$

But $m_0(A_1) + m_0(A_2 - A_1) + m_0(A_3 - A_2) + ... = \lim_{r \to \infty}\left[m_0(A_1) + \sum_{i=1}^{r-1} m_0(A_{i+1} - A_i)\right]$

$$... (2)$$

Again $\qquad \qquad A_{i+1} = (A_{i+1} - A) \cup A_i, \text{ where } (A_{i+1} - A_i) \cap A_i = \phi$

$$\Rightarrow \qquad \qquad m_0(A_{i+1}) = m_0(A_{i+1} - A_i) + m_0(A_i)$$

or $\qquad m_0(A_{i+1}) - m_0(A_i) = m_0(A_{i+1} - A_i).$ $\qquad \qquad ...(3)$

Using (3) in R.H.S. of (2), we get

$$m_0(A_1) + m_0(A_2 - A_1) + m_0(A_3 - A_2) + ...$$

$$= \lim_{r \to \infty}\left[m_0(A_1) + \sum_{i=1}^{r-1}[m_0(A_{i+1}) - m_0(A_i)]\right]$$

$$= \lim_{r \to \infty}[m_0(A_1) + \{-m_0(A_1) + m_0(A_r)\}]$$

$$= \lim_{r \to \infty} m_0(A_r)$$

Therefore $(1) \Rightarrow m\left(\bigcup_{r=1}^{\infty} A_r\right) = \lim_{r \to \infty} m_0(A_r)$

THEOREM 8. *If $<A_n>$ is a monotonically decreasing (or non-increasing) sequence of outer measurable sets s.t. $m_0(A_n) < \infty$ for at least one n, then*

$$m_0(A_1) < \infty \quad or \quad m_0\left(\bigcap_{n=1}^{\infty} A_n\right) = \lim_{n \to \infty} m_0(A_n)$$

Proof.   Let for $n = k, m_0(A_k) < \infty$

Since $A_1 \supseteq A_2 \supseteq A_3 \supseteq \dots \supseteq A_k \supseteq A_{k+1} \supseteq \dots$ we have $\quad m_0(A_r) < \infty, \forall r > k$

and also $\quad \lim_{n \to \infty} m_0(A_n) < \infty$

Now $\qquad\qquad A_k = E \cup (A_k - A_{k+1}) \cup (A_{k+1} - A_{k+2}) \cup \dots$

where $\qquad\qquad E = \bigcap_{n=1}^{\infty} A_n$ and $(A_k - A_{k+1}) \cap (A_{k+1} - A_{k+2}) = \phi$ etc.

$\Rightarrow \quad A_k$ is the union of pairwise disjoint outer measurable sets.

Hence

$$m_0(A_k) = m_0(E) + \sum_{r=k}^{\infty} m_0(A_r - A_{r+1})$$

$$= m_0(E) + \lim_{p \to \infty} \sum_{r=k}^{p} m_0(A_r - A_{r+1})$$

$$= m_0(E) + \lim_{p \to \infty} \sum_{r=k}^{p} [m_0(A_r) - m_0(A_{r+1})]$$

$$= m_0(E) + \lim_{p \to \infty} [m_0(A_k) - m_0(A_{p+1})]$$

$\Rightarrow \qquad\qquad m_0(A_k) = m_0(E) + m_0(A_k) - \lim_{p \to \infty} m_0(A_{p+1})$

$\Rightarrow \qquad\qquad m_0(E) = \lim_{p \to \infty} m_0(A_{p+1})$

$\Rightarrow \qquad\qquad m_0\left(\bigcap_{n=1}^{\infty} A_n\right) = \lim_{n \to \infty} m_0(A_n).$

REMARKS

- If $<A_n>$ is a sequence of outer measurable sets, then $m_0(\underline{\lim} A_n) \leq \underline{\lim}(m_0 A_n)$.
- If $<A_n>$ is a convergent sequence of outer measurable sets then $m_0(\lim A_n) = \lim(m_0(A_n))$
- If A is an outer measurable subset of a set B and E be any set such that $m_0(E) < \varepsilon$ then
$$m_0(E \cap B \cap A') = m_0(E \cap B) - m_0(E \cap A).$$
- If $<E_i>$ is a sequence of pairswise disjoint outer measurable sets then for any set A
$$m_0(A \cap E) = \sum_{i=1}^{\infty} m_0(A \cap E_i) \text{ where } E = \bigcup_{i=1}^{\infty} E_i$$
- If m_0 is an outer measure then class of m_0-measurable sets is σ-algebra.

Solved Example

Example 1. *If E is a subset of real numbers such that $0 < m_0(E) < \infty$ and if μ is a real valued set function defined as $\mu(A) = m_0(A \cap E)$, $\forall A \in R$, then show that m is an outer measure on the ring R, m_0 being an outer measure.*

Solution. We have

  (i) $m_0(A \cap E) \geq 0$ implies that $\mu(A) = m_0(A \cap E) \geq 0$

  (ii) If $A = \phi$, then $A \cap E = \phi$ implies that $m_0(A \cap E) = 0$

  and so $\mu(A) = m_0(A \cap E) = 0$

  Thus $\mu(A) = 0$, when $A = \phi$.

  (iii) Let $A \cap B = \phi$. Then $\mu(A \cup B) = m_0[(A \cap E) \cup (B \cap E)]$

  $$[\because (A \cup B) \cap E = (A \cap E) \cup (B \cap E)]$$

  Now, $A \cap B = \phi \Rightarrow (A \cap E) \cap (B \cap E) = \phi;$

  So, we have $m_0[(A \cap E) \cup (B \cap E)] = m_0(A \cap E) + m_0(B \cap E)$

  or $\mu(A \cup B) = m_0(A \cap E) + m_0(B \cap E) = \mu(A) + \mu(B)$.

  (iv) $A \subseteq B \Rightarrow (A \cap E) \subseteq (B \cap E)$

  $$\Rightarrow m_0(A \cap E) \leq m_0(B \cap E) \Rightarrow \mu(A) \leq \mu(B).$$

  (v) $\mu(\cup A_r) \leq \sum \mu(A_r),$

  as $\mu(\cup A_r) = m_0[(\cup A_r) \cap E] = m_0[\cup(A_r \cap E)]$

  $$\leq \sum m_0(A_r \cap E) = \sum \mu(A_r)$$

  $\therefore$ $\mu(\cup A_r) \leq \sum \mu(A_r)$.

  Hence, μ is an outer measure (Caratheodory's Outer Measure) on R and is called the measure induced by the outer measure m_0.

Example 2. *Show that the intersection of two outer measurable sets is outer measurable.*

Solution. Let A, B be any two outer measurable sets. Now $A \cap B = (A' \cup B')'$ (by Demorgan's law for complements).

A, B are outer measurable

$\Rightarrow$ A' and B' are outer measurable

$\Rightarrow$ $A' \cup B'$ is outer measurable

$\Rightarrow$ $(A' \cup B')'$ is outer measurable

$\Rightarrow$ $((A \cap B)')'$ is outer measurable

$\Rightarrow$ $A \cap B$ is outer measurable.   (By Demorgan's law)

REMARK

- By induction, we can show that $\overset{n}{\underset{i=1}{\cap}} A_i$ is also outer measurable.

Example 3. (a) *If A and B are outer measurable sets, then show that (A–B) and A Δ B are also outer measurable.*

(b) *Prove that if $B \subset A$, then $m_0(A - B) = m_0(A) - m_0(B)$.*

Solution. (a) B is outer measurable $\Rightarrow$ B' is outer measurable

$\Rightarrow A \cap B'$ is outer measurable.

Now, $\qquad A \cap B' = A - B$.

But $A \cap B'$ is measurable $\Rightarrow A - B$ is measurable.

Similarly $B - A$ is outer measurable.

$\Rightarrow \qquad A \triangle B = (A - B) \cup (B - A)$ is also measurable.

(b) If $B \subseteq A$, then

$$A = (A - B) \cup B \text{ and } (A - B) \cap B = \phi$$

$$\Rightarrow \qquad m_0(A) = m_0(A - B) + m_0(B)$$

$$\Rightarrow \qquad m_0(A - B) = m_0(A) - m_0(B).$$

Example 4. *Show that the class M of m_0-measurable sets is a Boolean algebra.*

Solution. If $A, B \in M \Rightarrow A, B$ are m_0-measurable

$$\Rightarrow \quad A', A \cap B \text{ are also measurable}$$

$$\Rightarrow \quad M \text{ is a Boolean algebra.}$$

Example 5. *Show that every subset of a set of measure zero is measurable and its measure is also zero.*

Solution. Let $B \subset A$ s.t. $m_0(A) = 0$.

Now $B \subset A \Rightarrow m_0(B) \le m_0(A) = 0$

$$\Rightarrow \quad m_0(B) \le 0 \qquad\qquad\qquad [\because m_0(A) = 0]$$

$$\Rightarrow \quad m_0(B) = 0 \text{ since we also have } m_0(B) \ge 0.$$

Hence, B is m_0-measurable with outer measure zero.

Example 6. *Show that an open set in a metric space is measurable w.r.t. any metric outer measure.*

Solution. Let E be a subset of an open set G.

Define $E_n = \left\{ x \in E : d(x, G') \ge \dfrac{1}{n} \right\}$.

$\Rightarrow \quad \langle E_n \rangle$ is a monotonically increasing sequence and $\lim\limits_{n \to \infty} E_n = E$

Now we know that

$$m_0(B) \le m_0(B \cap G) + m_0(B \cap G'), \forall B \in R. \qquad \ldots(1)$$

Set $B \cap G = E$. Then

$$B = E \cup (B \cap G') \supset E_n \cup (B \cap G').$$

$$\therefore \qquad m_0(B) \ge m_0[E_n \cup (B \cap G')].$$

Now by our definition of E_n, we have

$$d(E_n, B \cap G') = 0 \text{ i.e., } E_n \cap (B \cap G') = \phi.$$

$$\therefore \qquad m_0(B) \ge m_0(E_n) + m_0(B \cap G')$$

Making $\quad n \to \infty$

$$\lim_{n \to 0} m_0[E_n \cup (B \cap G')] = \lim_{n \to 0} m_0(E_n) + \lim_{n \to 0} m_0(B \cap G')$$

$$= m_0(E) + m_0(B \cap G')$$

$$= m_0(B \cap G) + m_0(B \cap G'), \qquad (\because E = B \cap G)$$

Hence, $\qquad m_0(B) \ge m_0(B \cap G) + m_0(B \cap G')$. $\qquad \ldots(2)$

From (1) and (2), we get
$$m_0(B) = m_0(B \cap G) + m_0(B \cap G').$$
Thus G is outer measurable.

Example 7. *Show that an outer measure is monotonic and σ-subadditive*

Solution. Let m_0 be an outer measure defined on a σ-ring R, containing subsets of X.

We are to prove that $A \subset B \subset C \subset D \subset \ldots \quad \Rightarrow \quad m_0(A) \le m_0(B) \le m_0(C) \le \ldots$

Let $\qquad A \subset B \subset C \subset D \subset \ldots \subseteq X$

Since $A \subset B$, we get $B = (B - A) \cup A$,

$\Rightarrow \quad m_0(B) = m_0(B - A) + m_0(A) \ge m_0(A)$ [∵ $(B - A) \cap A = \phi$ and $m_0(B - A) \ge 0$]

$\Rightarrow \quad m_0(A) \le m_0(B)$

Similarly, $B \subset C \Rightarrow m_0(B) \le m_0(C)$; $C \subset D \Rightarrow m_0(C) \le m_0(D)$ etc.

$\Rightarrow \quad m_0(A) \le m_0(B) \le m_0(C) \le m_0(D) \le \ldots$

$\Rightarrow \quad m_0$ is monotonic.

To prove that m_0 is σ-subadditive.

Let $<A_n>$ be a sequence of sets in σ-ring R and let $B_n = A_n - \bigcup_{i=1}^{n-1} A_i$.

Then $\qquad B_1 = A_1, B_2 = A_2 - A_1, B_3 = A_3 - (A_1 \cup A_2), \ldots$

$\Rightarrow \qquad \bigcup_{i=1}^{\infty} B_i = \bigcup_{i=1}^{\infty} A_i =$ and $B_n \subset A_n, \forall n$ and $B_i \cap B_j = \phi$ whenever $i \ne j$

$\Rightarrow \qquad m_0\left(\bigcup_{i=1}^{\infty} B_i \right) = \sum_{i=1}^{\infty} m_0(B_i)$

$\Rightarrow \qquad m_0\left(\bigcup_{i=1}^{\infty} A_i \right) = \sum_{i=1}^{\infty} m_0(B_i) \le \sum_{i=1}^{\infty} m_0(A_i)$, as $B_i \subset A_i$

$\Rightarrow \qquad m_0\left(\bigcup_{i=1}^{\infty} A_i \right) \le \sum_{i=1}^{\infty} m_0(A_i)$.

EXERCISE 16.2

1. Show that ϕ and whole space X are measurable.

2. Show that an open set in a metric space is measurable with respect to any metric outer measure.

3. A bounded set E is said to be measurable if for each bounded set A, $\mu^*(A) = \ldots$

4. If $<A_n>$ is a convergent sequence of μ-measurable sets and $\mu(A_1) < \infty$, (in general $\mu(A_i) < \infty$, for at least one i) prove that

$$\mu\left(\lim_{n \to \infty} A_n \right) = \mu\left(\bigcap_{n \to N} A_n \right) = \lim_{n \to \infty} \mu(A_n)$$

and by an example show that finiteness condition is essential.

5. Let X be a non-empty set and let $x_0 \in X$ be a fixed element. Define a set function μ s.t.

$$\mu(\phi) = 0. \text{ and } \mu(A) = \begin{cases} 0 & \text{if } x_0 \notin A \\ 1 & \text{if } x_0 \in A \end{cases}$$

where A is a non-empty subset of X. Prove that m is an outer measure on X.

6. If $E_1, E_2, E_1 \cap E_2$ and $E_1 \cup E_2$ are all bounded, measurable sets. Prove that

$$\mu(E_1 \cup E_2) + \mu(E_1 \cup E_2) = \mu(E_1) + \mu(E_2).$$

16.12 MEASURABLE FUNCTIONS

We know that a real valued function may be Lebesgue integrable even if the function is not continuous. For the existence of a Lebesgue integral, function must satisfy a less restrictive condition than that of continuity. This requirement give rise to a new class of functions, known as measurable functions.

Definition 1. *An extended real valued function f defined over a measurable set E is said to be measurable in the sense of Lebesgue, if set $E(f> a) = \{x \in E : f(x) > a\}$ is a measurable for all extended real numbers a.*

This definition states that f is a measurable function if for every real number a, the inverse image of $]a, \infty]$ under f i.e., $f^{-1}]a, \infty]$ is a measurable set.

REMARK

- Measure of set $E(f > a)$ may be finite or infinite.

THEOREM 1. *Let f be a measurable function defined over a measurable set E_r, $\forall r \in N$: then f is measurable on E, where $E = \bigcup_{r=1}^{\infty} E_r$.*

Proof. Since E is an enumerable union of measurable sets, so measurable. Also, the set $\{x \in E_r : f(x) > a\}$ is measurable for all real numbers a and $\forall r \in$ N.

Now
$$E(f > a) = \left(\bigcup_{r=1}^{\infty} E_r \right)(f > a) = \bigcup_{r=1}^{\infty} [E_r(f > a)]$$

But f is measurable on each E_r; so that the set $E_r(f > a)$ is measurable. Hence, we conclude that $E(f > a)$ is measurable being enumerable union of measurable sets. Therefore, f is measurable on E.

THEOREM 2. *The characteristic function of a set A is measurable, iff A is measurable.*

Proof.
(i) Since $A = \{x : \phi_A(x) > 0\}$ therefore if the characteristic function ϕ_A is measurable, the set $\{x : \phi_A(x) > 0\}$ is measurable i.e., A is a measurable set.

(ii) Conversely, Let A be measurable and a be any real number. Then

$$E(\phi_A > a) = \begin{cases} \phi & \text{if } a \geq 1, \\ A & \text{if } 0 \leq a < 1, \\ A \cup A' (= E) & \text{if } a < 0. \end{cases}$$

Every set on R.H.S. is measurable. It follows that $E(\phi_A > 0)$ is measurable. Hence, ϕ_A is measurable on A.

REMARK

- The characteristic functions of non-measurable sets A are also non-measurable even if the domain set E is measurable.

16.12.1 SIMPLE FUNCTION

A real valued function $\psi : E \to R$ is said to be a simple function if there exist a finite collection $\{A_1, A_2, ..., A_n\}$ of disjoint measurable subsets of E s.t. $\bigcup_{i=1}^{n} A_i = E$, and n non-zero real numbers $k_1, k_2, ..., k_n$ such that

$$\psi(x) = k_i, \ \forall x \in A_i, \ (i = 1, 2, ... n).$$

If ψ is a simple function and assumes the values $k_1, k_2, ..., k_n$, then we can write

$$\psi = \sum_{i=1}^{n} k_i \phi_{A_i}, \text{ where } \phi_{A_i}, \text{ is the charactristic function of the set } A_i = \{x : \psi(x) = k_i\}$$

This representation of ψ is called canonical representation and such a representation of ψ is not unique.

For example, the characteristic function of a set A is a simple function as it takes only two values (0 or 1) on two disjoint sets A and A'.

16.12.2 STEP FUNCTION

A real valued function s defined on an interval $[a, b]$ is said to be a step function if there exists a partition $a = x_0 < x_1 < ... < x_n = b$ s.t. the function assumes one and only one value in each interval i.e., $s(x) = c_i, \forall x \in] x_{i-1}, x_i], i = 1, 2, ..., n$.

REMARKS

- The step functions also assumes finite number of values like simple functions but here the sets $\{x : s(x) = c_i\}$, are intervals for each i.
- Every step function is also a simple function but the converse is not true as the

$$f : R \to R \text{ s.t. } f(x) = \begin{cases} 1 & , & x \text{ is rational} \\ 0 & , & x \text{ is irrational} \end{cases}$$

is a simple function but not step as the sets of rational and irrationals are not intervals.

For Example : The function $f : R \to [0, 1[$ s.t. $f(x) = 5 - [x]$ is a step function where $[x]$ denotes the greatest integer $\leq x$.

THEOREM 1. *The definition for the measurable function can be replaced by other three equivalent definitions.*

Proof. The four definitions for a function f to be Lebesgue measurable are

$\left. \begin{array}{ll} (1) & E(f > a) \\ (2) & E(f \geq a) \\ (3) & E(f < a) \\ (4) & E(f \leq a) \end{array} \right\}$ if any of these sets is measurable, $\forall a \in R$,

where R is the extended set of real numbers.

To prove their equivalence.

(1) $\Rightarrow$ (2). Since $E(f \geq a) = \bigcap_{n=1}^{\infty} \left[E\left(f > a - \frac{1}{n} \right) \right]$.

But according to (1), $E\left(f > a - \frac{1}{n} \right)$ is measurable for all n

therefore $E(f \geq a)$ is measurable being an enumerable intersection of measurable sets.

(2) $\Rightarrow$ (1). Since $E(f > a) = \bigcup_{n=1}^{\infty} \left[E\left(f \geq a + \frac{1}{n} \right) \right]$,

therefore $E(f > a)$ is measurable, being a countable union of measurable sets.

(2)$\Leftrightarrow$(3). Obviously $E(f < a) = [E(f \geq a)]'$ and $[E(f \geq a)] = [E(f < a)]'$.

Also, we know that a set is measurable, iff its complement is measurable. Thus, if set of one of the sides is measurable then set of other side will also be measurable.

Similarly (1)$\Leftrightarrow$(4).

(3) ⇒ (4). Since $E(f \le a) = \overset{\infty}{\underset{n=1}{\cap}}\left[E\left(f < a+\frac{1}{n}\right)\right]$.

therefore being an enumerable intersection of measurable sets, $E(f \le a)$ is measurable.

(4) ⇒ (3). Since $E(f < a) = \overset{\infty}{\underset{n=1}{\cup}}\left[E\left(f \le a-\frac{1}{n}\right)\right]$

therefore, being an enumerable union of measurable sels, $E(f < a)$ is measurable.

THEOREM 2. *A constant function over a measurable set E is measurable over that set E.*

Proof.   Let $f(x) = c, \forall\, x \in E$ where E is a measurable set.

Now $E(f < a) = \begin{cases} E & , & \text{if } c > a \\ \phi & , & \text{if } c \le a \end{cases}$

The sets E and ϕ are measurable and hence $E(f > a)$ is measurable for all a and hence the function f is measurable on E.

THEOREM 3. *If f is measurable over a measurable set E, then f is also measurable over any measurable subset A of E.*

Proof.   We have

$$A(f > a) = [E(f > a)] \cap A .$$

Thus $A(f>a)$ is measurable, being intersection of two measurable sets. Note that $E(f > a)$ is measurable since f is measurable on E.

⇒ f is also measurable over A.

THEOREM 4. *If f(x) and g(x) are measurable functions, then max (f(x), g(x)) and min (f(x), g(x)) are measurable.*

Proof.   We have $E[max\ (f, g) < a] = [E(f < a)] \cap E[g < a]$

since f, g are measurable, $[E(f < a)]$, $E[g < a]$ are measurable, so

$E\ [max\ (f, g) < a]$ is measurable and hence max $(f(x), g(x))$ is measurable, Similarly, min $(f(x).g(x))$ is measurable.

Alternatively,

$$min\ (f, g) = -\ max\ (-f, -g).$$

But $-f, -g$ are measurable and hence $-max\ (-f, -g)$ is also measurable

⇒ $min(f, g)$ is measurable.

THEOREM 5. *Let f be a measurable function defined over a measurable set E. Then cf, $-f$, $f+c$, $|f|$, $f^2, \frac{1}{f}$ (if f vanishes nowhere on E) are also measurable functions.*

Proof.   (i) To prove that cf is measurable on E. We have

$$E(cf > a) = \begin{cases} E\left(f > \dfrac{a}{c}\right) & , & \text{if } c > 0 \\[2mm] E\left(f < \dfrac{a}{c}\right) & , & \text{if } c < 0 \end{cases}$$

In both the cases, $E(c\,f > a)$ is measurable as (a/c) is also a real number. Also, if $c = 0$ then $cf(x) = 0, \forall x \in E$. In this case cf is a constant function. Therefore cf is measurable.

Particularly taking $c = -1$, we get $-f$ is also measurable.

(ii) We have, $E(c + f > a) = E(f > a - c) = a$ measurable set, since $a - c$ is a real number.

Hence $f + c$ is a measurable function.

(iii) We have

$$E(|f| > a) = \begin{cases} E & , \quad \text{if } a < 0 \\ [E(f > a)] \cup [E(f < -a)] & , \quad \text{if } a \geq 0 \end{cases}$$

Thus $E(|f| > a)$ is measurable for all a, since E is measurable and the union of two measurable sets is also meaurable. Hence $|f|$ is measurable.

(iv) We have

$$E(f^2 > a) = \begin{cases} E & , \quad \text{if } a < 0 \\ E(|f| > \sqrt{a}) & , \quad \text{if } a \geq 0 \end{cases}$$

But $E(|f| > \sqrt{a}) = [E(f > \sqrt{a})] \cup [E(f < -\sqrt{a})]$,

$\therefore$ $$E(f^2 > \sqrt{a}) = \begin{cases} E & , \quad \text{if } a < 0 \\ E(f > \sqrt{a}) \cup [E(f < -\sqrt{a})] & , \quad \text{if } a \geq 0 \end{cases}$$

But f is measurable over $E \Rightarrow E(f > \sqrt{a})$ and $[E(f < -\sqrt{a})]$ are measurable sets

$\Rightarrow E(f > \sqrt{a}) \cup [E(f < -\sqrt{a})]$ is a measurable set

$\Rightarrow E(f^2 > a)$ is a measurable set and hence, function f^2 is also measurable.

(v) Given that, $f(x) \neq 0$, $\forall x \in E$ and thus $\dfrac{1}{f}$ exists. So, we have

$$E\left(\frac{1}{f} > a\right) = \begin{cases} E(f > 0) & , \quad \text{if } a = 0 \\ [E(f > 0)] \cap E\left(f < \dfrac{1}{a}\right) & , \quad \text{if } a > 0 \\ \left[E(f < 0) \cap E\left(f < \dfrac{1}{a}\right)\right] \cup [E(f > 0)] & , \quad \text{if } a < 0 \end{cases}$$

Thus $E\left(\dfrac{1}{f} > a\right)$ is the finite union and intersection of measurable sets and hence, measurable and so is $\dfrac{1}{f}$.

THEOREM 6. *Let f and g be measurable functions defined over a measurable sets E. Then $f + g$, $f - g$, fg, are measurable functions over E.*

Proof. We shall first prove that if f and g are measurable over E then the set $E(f > g)$ is also measurable.

Now $f > g \Rightarrow \exists$ a rational number r, s.t. $f(x) > r > g(x)$.

Thus $E(f > g) = \underset{r \in Q}{\cup} [E(f > r)] \cap [E(g < r)]$

$\Rightarrow$ $E(f > g)$ is a measurable set ($\because$ A countable union of measurable sets as rational numbers are countable).

(i)   To prove that $f + g$ is measurable over E. Let a be any real number. Now,

$$E(f + g > a) = E(f > a - g). \qquad \qquad \dots(1)$$

Again, g is measurable $\Rightarrow cg$ is measurable, c is a constant.

$\Rightarrow a + cg$ is measurable, $\forall a \in R$

$\Rightarrow a - g$ is measurable, by taking $c = -1$.

Since, f and $a - g$ are measurable

$\Rightarrow E(f > a - g)$ is measurable

$\Rightarrow E(f + g > a)$ is a measurable set

$\Rightarrow f + g$ measurable function.

(ii) g is measurable $\Rightarrow -g$ is measurable

$\Rightarrow f + (-g)$ is measurable over E

$\Rightarrow f - g$ is measurable over E.

(iii)   Since, $f + g$ and $f - g$ are measurable functions over E

$\Rightarrow (f + g)^2, (f - g)^2$ are measurable functions over E

$\Rightarrow (f + g)^2 - (f - g)^2$ is a measurable function over E

$\Rightarrow \frac{1}{4}[(f + g)^2 - (f - g)^2]$ is a measurable function over E

$\Rightarrow fg$ is a measurable function over E.

REMARK

- If $f_1, f_2, ..., f_n$ are measurable functions then $\sum_{i=1}^{n} f_i$ and also $\sum_{i=1}^{n} k_i f_i$ are also measurable functions, k_i being numbers.

THEOREM 7. *If f and g are measurable functions defined over a measurable set E and if g vanishes nowhere on the set E, then the function f/g is also measurable over E.*

Proof.   Since, $g(x) \neq 0, \forall x \in E$ implies that $1/g$ exists. Again g is measurable $\Rightarrow 1/g$ is measurable function over E.

Now f and $1/g$ are measurable functions over E

$\Rightarrow$ $f. 1/g$ is a measurable function over E

$\Rightarrow$ f/g is a measurable function over E.

THEOREM 8.   *A continuous function defined over a measurable set E is measurable.*

Proof.   Firstly, we shall prove that the set $B = E(f \geq a)$ is a closed set *i.e.,* $D(B) \subseteq B$.

Let $x_0 \in D(B)$ be arbitrary; then for every nbd. G of x_0, we have

$$(G - \{x_0\}) \cap B \neq \phi.$$

Let $x \in (G - \{x_0\}) \cap B \Rightarrow x \in G, x \neq x_0$ and $x \in B$

$$\Rightarrow f(x) \geq a \qquad\qquad (\because x \in B)$$

Thus, for all $x \in G$ s.t. $x \neq x_0 \Rightarrow f(x) \geq a$.

Now f is continuous, therefore $f(x_0) \geq a \Rightarrow x_0 \in B$.

Thus,   any $x_0 \in D(B) \Rightarrow x_0 \in B$

$\Rightarrow$ every limit point of set B belongs to B

$\Rightarrow$ B is a closed set.

But, every closed set is measurable; it follows that B is measurable, which implies that $B = E(f \geq a)$ is measurable. Hence, f is measurable over E.

REMARKS

- Converse of the above theorem is not necessarily true,
- Measurability is not the sufficient condition for the continuity of a function. It also shows that the criterion for measurability of a function is somewhat weaker than the criterion for continuity.

THEOREM 9. *If f and g are measurable real valued functions defined on X and F $[f(x), g(x)] = h(x)$, $x \in X$ be real and continuous function on the Euclidean plane R^2, then h is measurable.*

Proof. For any real number a, let

$$G_a = \{(p,q) : F(p,q) > a\}$$

Obviously, G_a is an open subset of R^2 and so it can be expressed as countable union of open intervals.

Let

$$G_a = \overset{\infty}{\underset{n=1}{\cup}} I_n$$

where

$$I_n = \{(p,q) : p \in]a_n, b_n[\text{ and } q \in]c_n, d_n[\}$$

and

$$a_n, b_n \in R \text{ etc.}$$

Again, given that f is measurable, so the sets $X(f > a_n)$ and $X(f < b_n)$ are measurable sets. Thus

$$\{x \in X : f(x) \in]a_n, b_n[\} = X(f > a_n) \cap X(f < b_n)$$

$$= \text{an intersection of two measurable sets}$$

$$= \text{a measurable set.} \qquad \qquad \dots(1)$$

Similarly we can prove that the set

$$\{x \in X : g(x) \in]c_n, d_n[\} \text{ is a measurable set.} \qquad \qquad \dots(2)$$

Now, (1) and (2) implies that the set

$$\{x \in X : F(f(x), g(x)) \in I_n\} = \{x \in X : f(x) \in]a_n, b_n[\} \cap \{x \in X : g(x)\} \in]c_n, d_n[\}$$

is a measurable set.

Now, $\quad \{x \in X : h(x) > a\} = \{x \in X : (f(x), g(x)) \in G\}$

$$= \overset{\infty}{\underset{n=1}{\cup}} \{x \in X : (f(x), g(x)) \in I_n\}$$

$$= \text{an enumerable union of measurable sets}$$

$$= \text{a measurable set.}$$

$\Rightarrow h$ is measurable function on X.

THEOREM 10. *If $<f_n>$ is sequence of measurable functions defined over a measurable set E, then sup $<f_1, f_2, f_3, \dots >$ and inf $<f_1, f_2, f_3, \dots >$ are also measurable over E.*

Proof. Given that each f_n, $\forall n \in N$ is measurable over E. So $E(f_n > a)$ is a measurable subset of E for every $n \in N$ and $a \in R$.

Define

$$f(x) = \sup < f_n(x) : n = 1, 2, 3, \dots >, x \in E,$$

and

$$g(x) = \inf < f_n(x) : n = 1, 2, 3, \dots >, x \in E,$$

Observe that

$$g(x) = -\sup < -f_n(x) : n = 1, 2, 3, \dots >.$$

Thus it is enough to prove that sup $<f_n>$ i.e., f is measurable over E.

We have

$$E(f > a) = \bigcup_{n=1}^{\infty} [E(f_n > a)]$$

$$= a \text{ countable union of measurable sets}$$

$$= a \text{ measurable set.}$$

Hence, f is measurable over $E \Rightarrow \sup < f_n >$ and $\inf <f_n>$ are measurable over E.

THEOREM 11. *Let $<f_n>$ be a sequence of measurable functions defined over a measurable set E. Then $\overline{\lim} f_n$ and $\underline{\lim} f_n$ are measurable over E. Hence, $\lim f_n$ is measurable over E, if $\lim f_n$ exists.*

Proof.   Let us define $g_k(x) = \sup_{n \geq k} \langle f_n(x) \rangle$ and $h_k(x) = \inf_{n \geq k} \langle f_n(x) \rangle$

Then $\overline{\lim} f_n(x) = \inf_{k \geq 1} \langle g_k(x) \rangle$ and $\underline{\lim} f_n(x) = \sup_{k \geq 1} \langle h_k(x) \rangle$

Clearly, $g_k(x)$ and $h_k(x)$ are measurable functions defined over $E \; \forall \; x \in N$.
Again, $\inf_{k \geq 1} \langle g_k(x) \rangle$ and $\sup_{k \geq 1} \langle h_k(x) \rangle$ are measurble, and

Hence, $\overline{\lim} f_n(x) = \underline{\lim} f_n(x)$ are measurable.

To prove that $\lim f_n$ is measurable, when $\lim f_n$ exists.
Since,

$$\overline{\lim} f_n = \underline{\lim} f_n = \lim f_n .$$

But, we have already shown that $\overline{\lim} f_n$ and $\underline{\lim} f_n$ are measurable, so $\lim f_n$ is measurable over E.

16.13 BOREL MEASURABLE FUNCTIONS

A function f defined on a Borel set E is said to be a Borel measurable function or simply Borel function on E if for all real numbers a, the set $E(f > a)$ is a Borel set.

REMARK

- Since every Borel set is also a measurable set, so $E(f > a)$ is also Lebesgue measurable and hence, every Borel function is a measurable function. But converse is not necessarily, true *i.e..* every measurable function is not necessarily a Borel function.

  For example the characteristic function of a set of a non-Borel Lebesgue measurable set is Lebesgue measurable function but not Borel measurable.

Solved Examples

Example 1   *Show that a function f is measurable, iff the set $\{x : f(x) < r\}$ is measurable $r \in Q$ i.e., for all rational numbers r.*

Solution.   Let a be any real number and given that f be measurable over E, so that $E(f < a)$ is a measurable set. Now since, $Q \subset R$ implies that any $r \in Q \Rightarrow r \in R$. Therefore $E(f < r)$ is a measurable set, $\forall \; r \in Q$.

Conversely.   Let $\{x : f(x) < r)$ be measurable $\forall \; r \in Q$.

To prove that f is measurable. Let a be any real number.

Then $E(f < a) = (x \in E : f(x) < a\}$

$$= \underset{r \in Q}{\cup} \{x \in E : f(x) < r < a\}$$

$$= \underset{r < a}{\cup} \{x \in E : f(x) < r, r \in Q\}.$$

Thus $E(f < a)$ is an enumerable union of measurable sets (for Q is enumerable) and so a measurable, set. Hence f is measurable over E.

Example 2. *If f is measurable. Then show that max (f, 0) and max (–f, 0) are both measurable.*

Solution. We know that $\max(f, 0) \equiv f \cup 0 = f^+ \equiv \dfrac{1}{2}[f + |f|]$

and $\max(-f, 0) \equiv (-f) \cup 0 \equiv f^- = \dfrac{1}{2}[|f| - f].$

Since f is measurable, $|f|$ is also measurable and hence $\dfrac{1}{2}[f + |f|]$ and $\dfrac{1}{2}[|f| - f]$ are also measurable

$\Rightarrow$ $\max(f, 0)$ and $\max(-f, 0)$ also measurable.

Example 3. *Show that every function defined on a set of measure zero is measurable.*

Solution. Suppose function f is defined on a measurable set E s.t. $m(E) = 0$.
Then for any extended real number a,

$$E(f > a) = \{x \in E : f(x) > a\}.$$

Since $E(f > a) \subseteq E$, we get

$$m(E(f > \infty)) \le m(E)$$

$\Rightarrow$ $m(E(f > a)) = 0$ as $m(E) = 0$

$\Rightarrow$ $E(f > a)$ is measurable for all a and hence f is measurable.

Example 4. *Show that the signum function S is measurable.*

Solution. We know that $S(x) = sig(x) = \begin{cases} 1 & , \quad \text{when} \quad x > 0 \\ 0 & , \quad \text{when} \quad x = 0 \\ -1 & , \quad \text{when} \quad x < 0 \end{cases}$

Obyiously it is a step function, therefore measurable.

Example 5. *If f is a measurable function then show that $|f|$ is also measurable, but the converse is not true.*

Solution. We have already proved the first part of this example. For the second part, let A be a non-measurable subset of R. Then define the function

$$f(x) = 4, \text{ if } x \in A \text{ and } f(x) = -4, \text{ if } x \in A'.$$

Now $R(f > 0) = A$ which is non-measurable and hence f is not measurable.
But $|f|(x) = |f(x)| = 4, \forall x \in R$, which is a measurable function, being a constant function.

Example 6. *Let D and E be measurable sets and f is a function with domain $D \cup E$. Show that f is measurable iff its restrictions to D and E are measurable.*

Solution. We have

$$(D \cup E)(f > a) = D(f > a) \cup E(f > a).$$

By the definition of restriction of f, we get

$$f|D(x) = f(x), x \in D \text{ and } f|E(x) = f(x), x \in E.$$

So $f(x) > a$ iff $f|D(x) > a$, where $x \in D$

and $f(x) > a$ iff $f|E(x) > a$, when $x \in E$

$\Rightarrow D \cup E (f > a) = D(f | D > a) \cup E(f | E > a)$

Further, Let $f|D$ and $f|E$ be measurable

$\Rightarrow \quad D(f|D > a)$ and $E(f|E > a)$ are measurable

$\Rightarrow D \cup E(f > a)$ is measurable and hence, f is measurable.

Now, Let f be measurable over $D \cup E$ (which is a measurable set). Then f is also measurable over D (a measurable subset of $D \cup E$). Hence, $D(f > a)$ is measurable.

But $\forall x \in D, f(x) = f | D(x)$, so we get

$$D(f | D > a) = D(f > a) = a \text{ measurable set}$$

$\Rightarrow \quad f|D$ is measurable.

Similarly, we can prove that $f|E$ is measurable.

Example 7 *Show that a function is simple, iff it is measurable and assumes only a finite number of values.*

Solution.   Let f be any simple function. Then by Definition, we can write

$$f \equiv \sum_{i=1}^{n} k_i \phi_{A_i}$$

where $A_1, A_2, ..., A_n$ are measurable sets and ϕ_{A_i} is the characteristic function of A_i. To show that f is measurable and assume only finite number of values. For A_i is measurable.

$\Rightarrow \quad \phi_{A_i}$ is measurable $\Rightarrow \quad k_i \phi_{A_i}$ is measurable

$\Rightarrow \quad \sum_{i=1}^{n} k_i \phi_{A_i} \equiv f$ is a measurable function.

Further, $x \notin \cup A_i \Rightarrow x \notin A_i$, for all $i = 1, ..., n \Rightarrow f(x) = k_1 \phi_{A_1}(x) + ... + k_n \phi_{A_n}(x)$

$$= k_1.0 + ... + k_n.0 \text{ as } x \notin A_i, \forall i$$

$\Rightarrow \quad f(x) = 0.$

Now $x \in A_j$ $\Rightarrow f(x) = \sum_{i=1}^{n} k_i \phi_{A_i}(x)$

$$= k_1 \phi_{A_1}(x) + ... + k_j \phi_{A_j}(x) + ... + k_n \phi_{A_n}(x)$$

$$= k_1.0 + ... k_j.1 + ... + k_n.0 = k_j$$

$\Rightarrow \quad$ If x belongs to a single set A_j, then $f(x)$ can at most assume n values, namely $k_1, ..., k_n$.

$\Rightarrow \quad$ Number of values of $f(x)$ in this case is at most ${}^nC_1$.

Now $x \in A_j \cap A_p \Rightarrow f(x) = k_j + k_p$ (sum of two numbers).

But two numbers out of n can be chosen in ${}^nC_2$ ways. So $f(x)$ can have at most ${}^nC_2$ much values.

Now $x \in A_j \cap A_p \cap A_q \Rightarrow f(x) = k_j + k_p + k_q$ (sum of three numbers).

But such values can be chosen at most ${}^nC_3$ in number.

Continuing in this way, we observe that $f(x)$ can have at most
$$1 + {}^nC_1 + {}^nC_2 + \ldots + {}^nC_n (= 2^n)$$
values in number and thus $f(x)$ assumes only finite number of values.

On the other hand let f be a measurable function which can take the values $k_1, k_2, \ldots k_n$ only, then we can write
$$f(x) = \sum_{i=1}^{n} k_i \phi_{A_i}$$
Here $A_i = \{x : f(x) = k_i\}$

But f is a measurable function $\Rightarrow A_i$ is measurable
$$\Rightarrow f \text{ is a simple function.}$$

Example 8. *Determine whether the function, defined below, is measurable:*
$$\begin{aligned} f(x) &= x + 5 & , & \quad if \quad x < -1 \\ &= 2 & , & \quad if \quad -1 \le x < 0 \\ &= x^2 & , & \quad if \quad 0 \le x \end{aligned}$$

Solution. If a is any real number and R is the set of all real numbers, then we can write
$$R(f \le a) = \begin{cases}]-\infty, a-5] & , & \text{if} \quad a < 0 \\ ]-\infty, -5] \cup \{0\} & , & \text{if} \quad a = 0 \\ ]-\infty, a-5] \cup [0, \sqrt{(a)}] & , & \text{if} \quad 0 < a < 2 \\ ]-\infty, a-5] \cup [-1, \sqrt{(a)}] & , & \text{if} \quad 2 \le a < 4 \\ ]-\infty, \sqrt{(a)}] & , & \text{if} \quad 4 \le a \end{cases}$$

Clearly all the sets on the right side are measurable and hence the set $R(f \le a)$ is measurable $\forall a \in R$. Hence, the function f is measurable.

Example 9. *Let f be a function defined on $[0, 1/\pi]$ as follows:*
$$\begin{aligned} f(x) &= 0.1 & , & \quad if \quad x = 0 \\ &= 2x \sin\left(\frac{1}{x}\right) & , & \quad if \quad x > 0 \end{aligned}$$

Determine the measure of the set $\{x : f(x) \ge 0\}$.

Solution. We have $f(x) \ge 0$ for $x = 0, \dfrac{1}{\pi}$ and in the intervals $\left[\dfrac{1}{(2n+1)\pi}, \dfrac{1}{2n\pi}\right]$, $n \in N$.

Hence $\{x : f(x) \ge 0\} = \left\{0, \dfrac{1}{\pi}, \left[\dfrac{1}{(2n+1)\pi}, \dfrac{1}{2n\pi}\right], n \in N\right\}$

$\Rightarrow \quad m\{x : f(x) \ge 0\} = m\left\{0, \dfrac{1}{\pi}\right\} + \sum_{n=1}^{\infty} m\left[\dfrac{1}{(2n+1)\pi}, \dfrac{1}{2n\pi}\right]$

$$= 0 + \sum_{n=1}^{\infty} m\left[\dfrac{1}{2n\pi} - \dfrac{1}{(2n+1)\pi}\right] = \dfrac{1}{\pi} \sum_{n=1}^{\infty} \left(\dfrac{1}{2n} - \dfrac{1}{2n+1}\right)$$

$$= \dfrac{1}{\pi}\left[\dfrac{1}{2} - \dfrac{1}{3} + \dfrac{1}{4} - \dfrac{1}{5} + \ldots\right] = \dfrac{1}{\pi}\left[1 - 1 + \dfrac{1}{2} - \dfrac{1}{3} + \dfrac{1}{4} \ldots\right]$$

$$= \dfrac{1}{\pi}[1 - \log_e 2].$$

EXERCISE 16.3

1. Prove that if the functions f and g are finite and measurable, then

$$cf, f+g, f-g \text{ and } fg \text{ are also measurable}$$

functions.

2. If f is an extended real valued function s.t for each extended real number α, the set $\{x : f(x) = \alpha\}$ is measurable. Must f be measurable? Justify your answer.

3. Show that if $<f_n>$ is a sequence of measurable functions, then $\overline{\lim} f_n$ and $\underline{\lim} f_n$ and the $\lim f_n$, if exists, are measurable.

4. If I_1, I_2 are intervals of real numbers, and f is measurable function, then show that $f^{-1}(I_1 \cup I_2)$ is a messurable set.

5. If f is an extended real valued function with measurable domain D and $D_1 = \{x : f(x) = \infty\}$ and $D_2 = \{x : f(x) = -\infty\}$ then show that f is measurable on $D \Leftrightarrow D_1, D_2$ are measurable and the restriction of f to $D - (D_1 \cup D_2)$ is measurable.

6. Show that a function $f(x)$ is measurable, iff for every pair of real numbers a, b s.t. $a < b$ the set (i) $E = E[a < f < b]$ (ii) $E = E[a \le f \le b]$ is measurable.

7. Show that the set M of all measurable functions is an algebra. Also show that M is a real linear space.

8. Let f be an extended real valued function whose domain is measurable. Prove the equivalence of following statements:

(i) for each real number α, the set $\{x : f(x) > \alpha\}$ is measurable.

(ii) for each real number α, the set $\{x : f(x) \ge \alpha\}$ is measurable.

9. Show that if f is a measurable real valued function and g a continuous function defined on $(-\infty, \infty)$, then $g \circ f$ is measurable.

10. Show that the function $f(x) = [x]$, the greatest integer not greater than x, is measurable.

11. Let f be a non-negative measurable function and let $\{r_k : k = 1, 2, ...\}$ be an enumeration of positive rationals. Show that

$$\sqrt{[f(x)]} = \frac{1}{2} \inf_k \left(\frac{1}{r_k} f(x) + r_k \right)$$

and hence $f^{1/2}$ is a measurable function.

12. Let $f : R \to R$ s.t. $f(x) = \begin{cases} \dfrac{1}{x(x-1)} &, \quad x \ne 0, 1 \\ 2 &, \quad x = 1, 1 \end{cases}$

is f measurable?

13. Let $[f_n(x)]$ be sequence of measurable functions. For $x \in X$, where X is measurable space, put

$$g(x) = \sup \; f_n(x) \,; n = 1, 2, 3, ...$$

$$h(x) = \lim_{n \to \infty} \sup f_n(x)$$

Prove that g and h are measurable.

14. Show that a measurable functions of a continuous function is not necessarily always measurable.

15. Show that the function f defined on $E = [0, 1]$ is measurable, where

$$f(x) = \begin{cases} 3 &, \quad \text{if} \quad x = 0 \\ 1/x &, \quad \text{if} \quad 0 < x < 1, \\ 5 &, \quad \text{if} \quad x = 1 \end{cases}$$

16. Show that each characteristic function of a measurable set is a simple function.

17. Show that the product of two simple functions and any finite linear combination of simple functions are also simple functions.

18. Show that the function $f : R \to \{0, 1\}$ defined as $f(x) = 1$, if $0 \le x < 1$, and $f(x) = 0$, otherwise, is a measurable but not continuous.

19. If f is a measurable function on each of the sets in a countable collection $\{E_i\}$ of pairwise disjoint measurable sets, then show that f is measurable on $\cup E_i$ also.

20. Prove that an increasing function of a measurable function is also a measurable function.

16.14 LOWER AND UPPER LEBESGUE SUMS

Let $P = \{S_1, S_2,\ldots S_n\}$ be a measurable partition of the closed interval $[a, b]$ and f be a bounded function defined on $[a, b]$; then

$$U[f; P] = \text{Upper Lebesgue sum} = \sum_{r=1}^{n} M[f; S_r] m(S_r)$$

and $\qquad L[f; P] = \text{Lower Lebesgue sum} = \sum_{r=1}^{n} m[f; S_r] m(S_r)$

where $m(S_r)$ is the Lebesgue measure of S_r.

It is obvious that $L[f : P] \le U[f; P]$.

Also $\qquad L[-f, P] = -U[f, P]$ and $U[-f, P] = -L[f, P]$.

Now if $S_1, S_2, \ldots S_n$ are the interval components of a Riemann subdivision σ of the interval $[a, b]$, then the above defined upper-Lebesgue sum $U[f; P]$ becomes the same as the upper Riemann sum $U[f; \sigma] = \Sigma M [f; S_r]$ (length of interval S_r), since in case in interval "length of the interval = Lebesgue measure of the interval".

REMARK

- Every $U[f; \sigma]$ gives a Lebesgue upper sum $U[f; P]$ and hence, the set of numbers $U[f; P]$ for all the Riemann subdivision of $[a, b]$, is a subset of the set of numbers $U[f; P]$ for all measurable partitions P of the $[a, b]$. A similar conclusion also holds for $L[f; \sigma]$ and $L[f; P]$.

16.15 LEBESGUE INTEGRAL OF BOUNDED FUNCTIONS ON INTERVAL [a, b]

16.15.1 UPPER AND LOWER LEBESGUE-INTEGRALS

Let f be a bounded function defined on $[a, b]$. The supremum of $L[f : Q]$ is called the lower Lebesgue integral on $[a, b]$ and is denoted by $L\underline{\int}_a^b f(x)\, dx$, supremum is taken over all measurable partitions Q of $[a, b]$.

Similarly, the infimum of $U[f : P]$ is called the upper Lebesgue integral and is denoted by

$$L\overline{\int}_a^b f(x)\, dx .$$

Thus $L\underline{\int}_a^b f(x)\, dx = \sup\{L[f; Q] : Q \text{ is a measurable partition of } [a, b]\}$

and $L\overline{\int}_a^b f(x)\, dx = \inf\{U[f; P] : P \text{ is a measurable partition of } [a, b]\}$.

For simplicity, sometimes we denote the lower and upper integrals of f by $L\underline{\int}_a^b f$ and $L\overline{\int}_a^b f$.

Since $\qquad L[-f ; Q] = -U[f ; Q]$, and $U[-f , P] = -L[f , P]$,

it follows that $\qquad L\underline{\int}_a^b (-f) = -\left(L\overline{\int}_a^b f\right) \quad$ and $\quad L\overline{\int}_a^b (-f) = -\left(L\underline{\int}_a^b f\right)$.

REMARKS

- $L\int_{\underline{a}}^{b} f \geq L[f;Q]$, $\forall$ measurable partitions.

- $L\int_{a}^{\overline{b}} f \leq U[f;P]$ for all measurable partitions P.

- For every $\varepsilon > 0$, however small, there always exists at least one partition P' s.t. $L\int_{a}^{\overline{b}} f + \varepsilon > U[f;P']$.

  Similarly for every $\varepsilon > 0$, however small, there always exists at least one partition Q' s.t.
  $$L\int_{\underline{a}}^{b} f - \varepsilon < L[f;Q'].$$

16.15.2 LEBESGUE INTEGRAL

Let f be a bounded function defined on the interval $[a, b]$; we say that f is Lebesgue integrable on $[a, b]$, iff

$$L\int_{\underline{a}}^{b} f = L\int_{a}^{\overline{b}} f$$

and their common value is called the L-integral of f on $[a, b]$ and is denoted by $\int_{a}^{b} f$.

We shall denote the class of all bounded functions f which are Lebesgue integrable on $[a, b]$ by $L[a, b]$. Thus $f \in L[a,b] \Leftrightarrow f$ is L-integrable on $[a, b]$. The numbers a and b are called the lower and upper limits of integration respectively.

REMARKS

- The concept of integrability of a function over an interval as introduced here is subject to two very important limitations, viz,
  (i) the function is bounded.
  (ii) the interval of integration is finite so that neither of the end points is infinite.

- Every bounded function is not necessarily integrable $i.e.$, there may exist a bounded function f for which $L\int_{a}^{b} f \neq L\int_{a}^{\overline{b}} f$.

- The statement that $\int_{a}^{b} f$ exists means that the function f is bounded and integrable over $[a, b]$.

THEOREM 1. *Let f be a bounded function defined on $[a, b]$. Then every upper sum is greater than or equal to every lower sum for f.*

Proof. Let P, Q be any two measurable partitions of $[a, b]$. Then we have to show that
$$U[f;P] \geq L[f;Q]. \qquad \qquad ...(1)$$
Firstly, we shall prove that
$$L[f;Q] \leq L[f;Q^*] \text{ where } Q^* \text{ is a refinement of } Q. \qquad ...(2)$$
Let $S_r(r = 1, 2, 3, ... n)$ be the components of Q and let $S_1, S_2, ..., S_i^*, S_i^{**}, ..., S_n$ be the components of Q^* where $S_i^* \cap S_i^{**} = \phi$ and $S_i = S_i^* \cup S_i^{**}$ and thus Q^* is obtained from Q by breaking only one component of Q.

$$\Rightarrow \qquad m(S_i) = m(S_i^*) + m(S_i^{**}).$$

Now $\qquad S_i \supseteq S_i^* \qquad \Rightarrow \qquad m[f;S_i] \leq m[f;S_i^*]$

and $\qquad S_i \supseteq S_i^{**} \qquad \Rightarrow \qquad m[f;S_i^*] \leq m[f;S_i^{**}].$

We have, therefore

$$L[f,Q^*] = \sum_{\substack{j=1 \\ j \neq i}}^{n} m[f\,;S_j]m(S_j) + m[f\,;S_i^*]m(S_i^*) + m[f\,;S_i^{**}]m(S_i^{**})]$$

$$\geq \sum_{\substack{j=1 \\ j \neq i}}^{n} m[f\,;S_j]m(S_j) + m[f\,;S_i] \times [m(S_i^*) + m[S_i^{**}])]$$

$$\geq \sum_{\substack{j=1 \\ j \neq i}}^{n} m[f\,;S_j]m(S_j) + m[f\,;S_i].m(S_i) = L[f\,;Q]$$

Thus $\qquad L[f;Q^*] \leq L[f;Q]$

Similarly, we can prove that $U[f;P] \geq U[f;P*]$.

Let $P = \{S_1, S_2, ..., S_n\}$ and $Q = \{T_1, T_2, ..., T_m\}$. Consider the collections of subset of the form $S_i \cap T_j, (i = 1,2,...,n; j = 1,2,3,...,m)$ and denote it by PQ. Then PQ is the measurable partition and is refinement of both P and Q. So we have

$$L[f;Q] \leq L[f;PQ] \text{ and } U[f;PQ] \leq U[f;P]$$

$\Rightarrow \qquad U[f;P] \geq U[f;PQ] \geq L[f;PQ] \geq L[f;Q]$

or $\qquad U[f;P] \geq L[f;Q]$

THEOREM 2. *If f is a bounded function on [a, b] and if P is a measurable partition of [a, b], then* $\sup L[f;Q] \leq \inf U[f;P]$; *where supremum and infimum are taken over all measurable partitions Q and P of [a, b].*

Proof. We know that

$$U[f\,;P] \geq L\,[f\,;Q\,], \forall\, P, Q$$

Thus $L[f\,;Q]$ is a lower bound for the set of all upper sum $U[f\,;P]$, and hence

$$\inf U[f;P] = \inf\ U[f;P] \geq L[f;Q], \text{ for every } Q$$

$\Rightarrow \qquad \sup L\,[f\,;Q] \leq \inf\ U[f;P]$

$\Rightarrow \qquad \sup L[f;Q] \leq \inf U[f;P]$

REMARKS

- $L\underline{\int}_a^b f \leq L\overline{\int}_a^b f$.

- If for some partition $P, U[f\,;P] = L[f\,;P]$, then since $L[f,P] \leq L\underline{\int}_a^b f \leq L\overline{\int}_a^b f \leq U[f,P]$

  we have $\qquad L\underline{\int}_a^b f = L\overline{\int}_a^b f = U[f,P] = L[f,P] \qquad \Rightarrow \qquad$ f is L-integrable and $L\int_a^b f = U[f,P]$

THEOREM 3. *If f is a bounded function defined on [a, b] and f is R-integrable, on [a, b] then f is also L-integrable on [a, b] and*

$$L\int_a^b f = R\int_a^b f$$

Proof. Let $R\int_a^b f = L[f;\sigma_1]$ and $R\overline{\int_a^b} f = U[f;\sigma_2]$

where σ_1 and σ_2 are the Riemann subdivisons of $[a, b]$.

But since σ_1 and σ_2 will also give rise to measurable partitions P_1 and P_2 of $[a, b]$.

Therefore $\quad L\overline{\int_a^b} f \le U[f; P_1]$ and $L\int_{\underline{a}}^b f \ge L[f; P_2]$

$\Rightarrow \quad L[f; P_2] \le L\int_{\underline{a}}^b f \le L\overline{\int_a^b} f \le U[f;P_1]$

$\Rightarrow \quad L[f;\sigma_1] \le L\int_{\underline{a}}^b f \le L\overline{\int_a^b} f \le U[f;\sigma_2]$

$\Rightarrow \quad R\int_{\underline{a}}^b f \le L\int_{\underline{a}}^b f \le L\overline{\int_a^b} f \le R\overline{\int_a^b} f$ ...(1)

But given that f is R-integrable. So $R\int_{\underline{a}}^b f = R\overline{\int_a^b} f = R\int_a^b f$

Therefore (1) $\quad\Rightarrow\quad R\int_a^b f = L\int_{\underline{a}}^b f = L\overline{\int_a^b} f = R\int_a^b f$

$\quad\Rightarrow\quad f$ is also L-integrable over $[a, b]$ and $L\int_a^b f = R\int_a^b f$

Further note that if both the R- and L-integrals, exist, their values are equal. So in this case we may denote the integral by the symbols $\int_a^b f$ or $\int_a^b f(x)dx$.

Converse of this theorem is not necessarily true.

THEOREM 4. (First Mean Value Theorem) *Let f be a bounded measurable real valued function such that $a \le f(x) \le b$ on a measurable set $E[p,q] \subset R$. Then*

$$a.m(E) \le \int_E f(x)\, dx \le b.m(E).$$

Proof. Let $m \in N$ be so that

$$a \le f(x) \le b \Rightarrow \left(a - \frac{1}{m}\right) < f(x) < \left(b + \frac{1}{m}\right), \forall x \in E.$$

Taking $a - \frac{1}{m}\alpha, b + \frac{1}{m} = \beta$, we have $\alpha < f(x) < \beta, \forall x \in E$.

Divide the interval $[\alpha, \beta]$ by means of points

$$\alpha = y_0 < y_1 < y_2 < ... < y_n = \beta.$$

Define $\quad E_0 = \{x \in E : y_0 < f(x) < y_1\},$

and $\quad E_r = \{x \in E : y_r \le f(x) < y_{r+1}\}, r = 1,2,...,n-1.$

Clearly, $\quad E = \bigcup_{r=0}^{n-1} E_r$ and $E_r \cap E_s = \phi$ (For $r \ne s$).

Since f is measurable over E and hence measurability of the sets E_r is implied and

$$m(E) = \sum_{r=0}^{n-1} m(E_r)$$ ...(1)

Thus we have got a measurable partition $P = \{E_0, E_1, ... E_{n-1}\}$ of E.

Again, $\quad \alpha \le y_r \le \beta \Rightarrow \alpha.m(E_r) \le y_r.m(E_r) \le \beta.m(E_r).$

$\Rightarrow \quad \sum_{r=0}^{n-1} \alpha.m(E_r) \le \sum_{r=0}^{n-1} y_r.m(E_r) \le \sum_{r=0}^{n-1} \beta.m(E_r)$

i.e.,
$$\alpha.m(E) \le \sum_{r=0}^{n-1} y_r.m(E_r) \le \beta.m(E) \qquad ...(2)$$

Now, making max $(y_{r+1}- y_r)\to 0$, we get $y_{r+1}\to y_r$ and hence for this partition,

$$U[f,P] = \sum_{r=0}^{n-1} y_{r+1}.m(E_r) \to \sum_{r=0}^{n-1} y_r.m(E_r) = L[f,P]$$

So
$$\int_E f = \Sigma y_r.m(E_r)$$

Therefore,

(2) $\Rightarrow$
$$\alpha.m(E) \le \int_E f(x)dx \le \beta.m(E)$$

or
$$\left(a -\frac{1}{m}\right).m(E) \le \int_E f(x)dx \le \left(b +\frac{1}{m}\right).m(E)$$

Making $m \to \infty$, we get

$$a.m(E) \le \int_E f(x)dx \le b.m(E)$$

THEOREM 5. *Let f be a bounded function defined on [a, b] . Then the function f is L-integrable, iff for each $\varepsilon>0$, however small, there exists a measurable partition P of [a, b] such that*
$$U[f ; P]- L[f ; P]< \varepsilon$$

Proof. *The condition is necessary.* Let f be L-integrable. Then
$$L\overline{\int}_a^b f = g.l.b.U[f;P]= l.u.b.[f; Q]= L\underline{\int}_a^b f$$

For $\varepsilon > 0$, we may partitions P and Q such that

$$L\overline{\int}_a^b f +\frac{\varepsilon}{2}> U[f; P] \text{ or } U[f; P]< L\overline{\int}_a^b f +\frac{\varepsilon}{e} \qquad ...(1)$$

$$L\underline{\int}_a^b f -\frac{\varepsilon}{2}< L[f; Q] \Rightarrow -L\underline{\int}_a^b f +\frac{\varepsilon}{2}> -L[f; Q]$$

$\Rightarrow$
$$-L[f; Q]< -L\underline{\int}_a^b f +\frac{\varepsilon}{2}. \qquad ...(2)$$

Adding (1) and (2),

$$U[f;P]- L[f;Q]< L\overline{\int}_a^b f - L\underline{\int}_a^b f +\varepsilon = \varepsilon, \text{ as } L\overline{\int}_a^b f = L\underline{\int}_a^b f$$

$\Rightarrow$
$$U[f;P]- L[f;Q]< \varepsilon. \qquad ...(3)$$

Now consider the common refinement PQ of both P and Q, whose components are the pairwise intersections of the components of P and Q. Then, we get
$$L[f;PQ]\ge L[f;Q] \text{ or} - L[f; PQ]\le -L[f;Q] \qquad ...(4)$$

and
$$U[f;PQ]\le U[f : P] \qquad ... (5)$$

Adding (4) and (5), we get
$$U[f;PQ]- L[f : PQ]\le U[f;P]- L[f;Q]$$

$\Rightarrow$
$$U[f;PQ]- L[f : PQ]< \varepsilon, \text{ using (3)}$$

Thus we have obtained at least one measurable partition PQ for which the condition of the theorem holds good necessarily.

Conversely, let, for a given $\varepsilon > 0$, $\exists$ a measurable partition P of $[a, b]$ such that

$$U[f;P] - L[f;P] < \varepsilon.$$ ... (6)

Also, by definitions of upper and lower integrals, we have

$$L\overline{\int}_a^b f \le U[f; P] \text{ and } L\underline{\int}_a^b f \ge L[f; P]$$

or say $\qquad -L\underline{\int}_a^b f \le -L[f; P].$

On adding, $\quad \left(L\overline{\int}_a^b f - L\underline{\int}_a^b f\right) < U[f;P] - L[f;P] < \varepsilon$ [Using (6)]

or $\qquad L\overline{\int}_a^b f - L\underline{\int}_a^b f + \varepsilon.$

Since ε is arbitrary, it follows that

$$L\overline{\int}_a^b f \le L\underline{\int}_a^b f$$ ...(7)

But, $\qquad L\overline{\int}_a^b f \ge L\underline{\int}_a^b f$

Combining (7) and (8), we get

$$L\overline{\int}_a^b f = L\underline{\int}_a^b f, \text{ i.e., } f \text{ is L-integrable.}$$

REMARK

- If for some function f, $\exists$ a measurable partition P s.t $U[f; P] = L[f; P]$, then $f \in L[a, b]$.

THEOREM 6. *Every bounded measurable function f defined on $[a, b]$ is Lebesgue-integrable over $[a, b]$.*

Proof. Since f is bounded, $\exists$ a $k \in N$ s.t $f(x) \in [-k, k]$, $\forall x \in [a, b]$
i.e., $\qquad$ range of $f \subset [-k, k]$.

Let $\varepsilon > 0$ be given. Then we can divide $[-k, k]$ by means of finite number of points $\alpha_0, \alpha_1, \alpha_2, \dots \alpha_n$, such that

$$-k = \alpha_0 < \alpha_1 < \alpha_2 < \dots < \alpha_n = k$$

where $\qquad \alpha_i - \alpha_{i-1} < \dfrac{\varepsilon}{b-a}, i = 1, 2, \dots, n.$

Define $\qquad E_i = f^{-1}[\alpha_{i-1}, \alpha_i[, \forall i = 1, 2, 3, \dots, n.$

Thus $\qquad x \in E_i \Leftrightarrow \alpha_{i-1} \le f(x) < \alpha_i.$

Clearly, each E_i is measurable being the inverse image of any interval under a measurable function. Also these $E_i's$ are pairwise disjoint.

Thus, $\{E_1, E_2, \dots E_n\} = P$ is measurable partition of $[a, b]$.

Clearly, $\qquad M[f;E_i] = \sup.\{f(x) ; x \in E_i\} \le \alpha_i$

So, $\qquad U[f;P] = \sum_{i=1}^{n} M[f;E_i].m(E_i) \le \sum_{i=1}^{n} \alpha_i.m(E_i)$ ...(1)

Similarly, $\qquad L[f;P] = \sum_{i=1}^{n} \alpha_{i-1}.m(E_i)$

or $\qquad -L[f;P] \le -\sum_{i=1}^{n} \alpha_{i-1}.m(E_i).$ ... (2)

Adding (1) and (2) $U[f;P] - L[f;P] \le \sum_{i=1}^{n}(\alpha_i - \alpha_{i-1}).m(E_i) < \dfrac{\varepsilon}{b-a}\sum_{i=1}^{n} m(E_i).$ ...(3)

Since $E_i \cap E_j = \phi$ for $i \neq j$ and $\overset{n}{\underset{i=1}{\cup}} E_i = [a,b]$ we have

$$\sum_{i=1}^{n} m(E_i) = m\left[\overset{n}{\underset{i=1}{\cup}} E_i\right] = m([a,b]) = b - a.$$

Using this in (3), we get

$$U[f;P] - L[f;P] < \frac{\varepsilon}{b-a} \sum_{i=1}^{\infty} m(E_i) < \frac{\varepsilon}{b-a}.(b-a).$$

Thus　　　　$U[f;P] - L[f;P] < \varepsilon.$

Hence, f is L-integrable on $[a, b]$.

THEOREM 7. *If f is a bounded function in $L[a, b]$ and if g is a bounded function on $[a, b]$ such that $f(x) = g(x)$ almost everywhere in $[a, b]$, then*

$$g \in L[a,b] \text{ and } \int_a^b g = \int_a^b f.$$

Proof. Since $f(x) = g(x)$ almost everywhere in $[a, b]$, so g is also measurable on $[a, b]$. Again, g is given to be bounded, therefore $g \in L[a, b]$.

Let　$E = \{x : a \leq x \leq b\}$. Also let $E_0 = \{x : x \in E \text{ and } f(x) \neq g(x)\}$

$\Rightarrow$　　　　$m(E_0) = 0$ and $f(x) = g(x). \ \forall \, x \in E - E_0$

or　　say　$(f - g)(x) = 0, \ \forall \, x \in E - E_0$

Let $P = \{E_0, E_1, ... E_n\}$. This is obviously a measurable partition of $[a, b]$.

Hence　$U[f - g; P] = M[f - g; E_0].m(E_0) + M[f - g; E - E_0].m(E - E_0)$

$\Rightarrow$　　　$U[f - g; P] = M[f - g; E_0].0 + 0.m(E - E_0) = 0$

Similarly,　$L[f - g; P] = 0$

Since for the partition P,

$$U[f - g; P] = L[f - g; P], \text{ the function } f - g \in L[a,b]$$

$\Rightarrow$　　　$-(f - g) = g - f \in L[a,b]$

$\Rightarrow$　　　$(g - f) + f \in L(a,b)$

$\Rightarrow$　　　$g \in L[a,b]$

16.16 LEBESGUE INTEGRAL FOR BOUNDED FUNCTIONS OVER A SUBSET OF REAL NUMBERS

Definition 1. *Let $E \subset [a, b]$ be a measurable subset and f be a bounded function in $L[a, b]$. Then $\int_E f$ is defined as*

$$\int_E f = \int_a^b f\phi_E \text{ where } \phi_E \text{ is the characteristic function of } E.$$

Definition 2. *If function f is a simple function and has the canonical repres*

$f = \sum_{i=1}^{n} a_i \Psi_{E_i}$, *then we define*

$$\int_E f(x)dx = \sum_{i=1}^{n} a_i.m(E_i)$$

where $E = \overset{n}{\underset{r=1}{\cup}} E_i$ and $E_i \underset{i \neq j}{\cap} E_j = \phi$ and $E_i = \{x \in E : f(x) = a_i\}$ and

REMARK

- Every simple function f is L-integrable and its L-integral is the same as elementary integral and denoted by $\int_a^b f$.

THEOREM 1. *If A and B are disjoint measurable subsets of $[a, b]$ and if f is a bounded L-integrable function on $[a, b]$, then $\int_{A \cup B} f = \int_A f + \int_B f$.*

Proof.   Since $A \cap B = \phi$; therefore $\phi_{A \cup B} = \phi_A + \phi_B$

$\therefore$ $f \phi_{A \cup B} = f \phi_A + f \phi_B$

Now, by definition, we have

$$\int_{A \cup B} f = \int_a^b f \phi_{A \cup B} = \int_a^b (f \phi_A + f \phi_B)$$

$$= \int_a^b f \phi_A + \int_a^b f \phi_B = \int_A f + \int_B f .$$

THEOREM 2. *If f is bounded real valued measurable function defined on a measurble set E of finite measurable set s.t. $a \le f(x) \le b$, then*

$$a.m(E) \le \int_E f \le b.m(E)$$

(First mean theorem when set E is not necessarily an interval, may be any subset of real numbers).

Proof.   Consider a function $\psi_1(x) = a, \forall x \in E$.

Then $a \le f(x) \Rightarrow \psi_1(x) \le f(x), \forall x \in E$,

$\Rightarrow f(x) - \psi_1(x) \ge 0$, on E.

We know that

$$\int_E [f - \psi_1] \ge 0 \Rightarrow \int_E \psi_1 \le \int_E f \qquad \qquad ...(1)$$

But $\int_E \psi_1 = a.m(E)$,

Hence from (1) $a.m(E) \le \int_E f$ \qquad\qquad ...(2)

Similarly defining a function $\psi_2(x) = b, \forall x \in E$, we can obtain

$$\int_E f \le b.m(E). \qquad\qquad ...(3)$$

From, (2) and (3) we ge t $a.m.(E) \le \int_E f(x)dx \le b.m.(E)$

THEOREM 3. **(Countable additive property of integrals)**

If $E = E_1 \cup E_1 \cup ... \cup E_n$, where all $E_i's$ are pairwise disjoint and if f is a bounded measurable function (integrable) defined over E, then

$$ f + \int_{E_2} f + ... + \int_{E_n} f .$$

$$= \bigcup_{i=1}^{\infty} E_k \ s.t. E_k \cap E_r = \phi \ for \ k \ne r$$

rinciple of induction on the result of throrem 1.

expressed as

$$E = \left(\bigcup_{k=1}^{n} E_k \right) \cup \left(\bigcup_{k=n+1}^{\infty} E_k \right)$$

Let $$R_{n+1} = \bigcup_{k=n+1}^{\infty} E_K; \quad \text{then} \quad E = \left(\bigcup_{k=1}^{n} E_K\right) \cup R_{n+1}.$$

$$\Rightarrow \quad \int_E f(x)dx = \sum_{k=1}^{n} \int_{E_k} f(x)dx + \int_{R_{n+1}} f(x)dx \qquad \qquad ...(1)$$

Since f is bounded, $\alpha \le f(x) \le \beta$

On the set R_{n+1}, applying First mean value theorem, we get

$$\alpha.m(R_{n+1}) \le \int_{R_{n+1}} f(x)dx \le \beta.m(R_{n+1}) \qquad \qquad ...(2)$$

Again $$m(R_{n+1}) = m\left(\bigcup_{k=n+1}^{\infty} E_k\right), \text{where } E_k \cap E_r = \phi \text{ for } k=n.$$

$$\therefore \qquad m(R_{n+1}) = \sum_{k=n+1}^{\infty} m(E_k).$$

Letting $n \to \infty$, we get

$$\lim_{n\to\infty} m(R_{n+1}) = \lim_{n\to\infty} \sum_{k=n+1}^{\infty} m(E_k) = 0 \quad \text{(By general principle of convergence)}$$

$$\therefore \qquad \lim_{n\to\infty} m(R_{n+1}) = 0. \qquad \qquad ...(3)$$

Taking limit $n \to \infty$ in (2) and using (3) we get $\alpha.0 \le \lim_{n\to\infty} \int_{R_{n+1}} f(x)dx \le \beta.0$

$$\lim_{n\to\infty} \int_{R_{n+1}} f(x)dx = 0.$$

Now letting $n \to \infty$ in (1), we get

$$\int_E f(x)dx = \lim_{n\to\infty} \sum_{k=1}^{n} \int_{E_k} f(x)dx + 0.$$

Hence, $$\int_E f = \sum_{k=1}^{\infty} \int_{E_k} f.$$

16.17 GENERAL LEBESGUE INTEGRAL

Definition. *A measurable function f is Lebesgue integrable over E if f^+ and f^- are both Lebesgue integrable over E, In this case we define $\int_{E_k} f$ as*

$$\int_{E_k} f = \int_{E_k} f^+ - \int_E f^-$$

THEOREM 1. *If f is a bounded function, Lebesgue integrable on a measurable subset of [a, b] then $|f|$ is also L-inegrable on E and*

$$\left|\int_E f\right| \le \int_E |f| \text{ When does equality hold?}$$

Proof. If $f \in L[a, b]$, then its positive and negative parts $f^+, f^- \in L[a,b]$. Now $|f| = f^+ + f^-$

$$\Rightarrow \qquad |f| \in L[a,b]$$

Also $$\int_E |f| = \int_E (f^+ + f^-) = \int_E f^+ + \int_E f^-$$

and since f^+ and f^- are both non-negative, so we have $\int_E f^+ + \int_E f^- \geq 0$

Now $\qquad \int_E f = \int_E f^+ - \int_E f^- \leq \int_E f^+ + \int_E f^- = \int_E |f|$

$\Rightarrow \qquad\qquad \int_E f \leq \int_E |f| \qquad\qquad\qquad …(1)$

and $\qquad -\int_E f = -\int_E f^+ + \int_E f^- \leq \int_E f^+ + \int_E f^- = \int_E |f|$

$\Rightarrow \qquad\qquad -\int_E f \leq \int_E |f| \qquad\qquad\qquad …(2)$

From (1) and (2) we get $\quad \left| \int_E f \right| \leq \int_E |f|$

Now,

If $\int f \geq 0$. Then $\int |f| = \int$ and hence $\int (|f| - f) = 0$

$\Rightarrow \qquad\qquad |f| = f,\ a.e. \Rightarrow f \geq 0\ a.e.$

If $\int f \leq 0$. Then $\int |f| = \int (-f) = -f,\ a.e.$

$\Rightarrow \qquad\qquad f \leq 0\ a.e.,$

Thus equality will occur when either $f \geq 0$, almost everywhere, $f \leq 0$, almost everywhere.

THEOREM 2. *If f is measurable on a measurable set E, then f is Lebesgue integrable iff $|f|$ is Lebesgue integrable.*

Proof. If f is measurable over E, then f^+ and f^- both are integrable over E. Also $|f| = f^+ + f^-$, being the sum of two integrable functions, is also integrable.

Conversely, if $|f|$ is integrable the $\int_E |f|$ is finite. Also we have

$$\int_E |f| = \int_E f^+ + \int_E f^-$$

$\Rightarrow \quad \int_E f^+$ and $\int_E f^-$ and both also finite.

$\qquad\qquad\qquad$ (If any of these is infinite, then $\int_E |f|$ will become infinite)

$\Rightarrow \quad f^+$ and f^- both are integrable.

$\Rightarrow \quad f$ is integrable.

THEOREM 3. *Let f be a bounded function defined on measurable set E with $m(E) < \infty$. Then*

$$\inf_{f \leq \psi} \int_E \psi(x)\,dx = \sup_{f \geq \phi} \int_E \phi(x)\,dx$$

for all simple functions ϕ and ψ iff f is measurable.

Proof. *Sufficient part.* Let f be measurable.

Since, f is bounded. Let $|f| \leq M$ on $E \Rightarrow -M \leq f(x) \leq M$

$$E_k = \left\{ x \in E : \frac{(k-1)M}{n} < f(x) \leq \frac{kM}{n} \right\},\ \text{where } -n \leq k \leq n.$$

Then $\{E_k\}_{k=-n}^n$ is a family of disjoint measurable subsets of E s.t. $\cup E_k = E$

$\Rightarrow \quad m(E) = \sum_{k=-n}^{n} m(E_k) \qquad\qquad\qquad …(1)$

Let us define two simple functions ψ_n and ϕ_n on E as follows:

$$\psi_n \equiv \sum_{k=-n}^{n} \left(\frac{Mk}{n}\right)\psi_{E_k} \quad \text{and} \quad \phi_n \equiv \sum_{k=-n}^{n} \frac{(k-1)M}{n}\psi_{E_k} \, .$$

Then $\phi_n(x) \le f(x) \le \psi_n(x), \forall x \in E$

$\Rightarrow \quad \inf_{\psi \ge f}\int_E \psi(x)dx \le \int_E \psi_n(x)dx = \sum_{k=-n}^{n} \frac{Mk}{n}.m(E_k) = \frac{M}{n}\sum_{k=-n}^{n} k.m(E_k)$ ...(2)

and $\quad \sup_{\phi \le f}\int_E \phi(x)dx \ge \int_E \phi_n(x)dx = \frac{M}{n}\sum_{k=-n}^{n}(k-1).m(E_k)\,.$

$\Rightarrow \quad -\sup_{\phi \le f}\int_E \phi(x)dx \le -\frac{M}{n}\sum_{k=-n}^{n}(k-1).m(E_k)$ ... (3)

Adding (2) and (3), we get

$$0 \le \inf\int_E \psi(x)dx - \sup\int_E \phi(x)dx \le \frac{M}{n}\sum_{k=-n}^{n} m(E_k) = \frac{M}{n}m(E)\cdot$$ ... (4)

Letting $n \to \infty$, we get

$$0 \le \inf\int_E \psi(x)dx - \sup\int_E \phi(x)dx \le 0 \quad \Rightarrow \quad \inf_{\psi \ge f}\int_E \psi(x)dx = \sup_{\phi \le f}\int_E \phi(x)dx$$

Necessary part. Given that

$$\inf_{f \le \psi}\int \psi(x)dx = \sup_{\phi \le f}\int \phi(x)dx,$$

We have to prove that f is measurable

$\Rightarrow \quad$ For every n, $\exists$ simple functions ϕ_n and ψ_n s.t. $\phi_n(x) \le f(x) \le \psi_n(x)$

and $\quad \int \psi_n(x)dx - \int \phi_n(x)dx < \dfrac{1}{n}$

Let $\quad \psi^* = \inf\langle\psi_n\rangle$ and $\phi^* = \sup\langle\phi_n\rangle.$
Then ψ^* and ϕ^* are also measurable and

$$\phi^*(x) \le f(x) \le \psi^*(x)$$ ...(5)

Let F be the set s.t. $F = \{x : \phi^*(x) \ne \psi^*(x)\} = \{x : \phi^*(x) < \psi^*(x)\}$

Then $\quad F = \underset{r \in N}{\cup}\, F_r, \text{ where } F_r = \left\{x : \phi^*(x) < \psi^*(x) - \dfrac{1}{r}\right\}$

But $\quad F_r \subseteq \left\{x : \phi_n(x) < \psi_n(x) - \dfrac{1}{r}\right\}, \forall n\,.$

Also the measure of the right super set is less than $\dfrac{r}{n}\,.$

$\Rightarrow \quad m(F_r) \le \dfrac{r}{n}\,\forall\, n.$

Since n is arbitrary. letting $n \to \infty$, we get $m(F_r) = 0$ for all r and hence

$$m(F) = 0 \quad \Rightarrow \quad \phi^* = f = \psi \text{ almost everywhere.}$$

$\Rightarrow \quad f$ is measurable.

16.18 THE LEBESGUE INTEGRAL OF UNBOUNDED FUNCTIONS

Let $f(x)$ be an unbounded, measurable and non-negative real valued function defined on $[a, b]$. Let $n \in N$ be arbitrary. We define a function $[f(x)]_n$ on $[a, b]$, such that

$$[f(x)]_n = \begin{cases} f(x) & \text{when} \quad f(x) \leq n \\ n & \text{when} \quad f(x) > n \end{cases}$$

Thus
$$[f(x)]_n = \min(f(x), n)$$

THEOREM 1. $[f(x)]_n$ *as defined above is bounded and measurable over* $[a, b]$.

Proof.   Since $[f(x)]_n = \min(f(x), n), 0 \leq [f(x)]_n$ and hence $[f(x)]_n$ is bounded over $[a, b]$. Now, let $d \in R$ be arbitrary. Then

$$\{x \in [a,b] : [f(x)]_n > d\} = \begin{cases} \{x \in [a,b] : f(x) > d\} & \text{if} \quad n > d \\ \phi & \text{if} \quad n \leq d \end{cases}$$

But $f(x)$ is measurable over $[a, b]$

∴ $\{x \in [a, b] : f(x) > d\}$ is a measurable set .

Also ϕ is measurable set.

Hence $\{x \in [a, b] : [f(x)]_n > d\}$ is a measurable set, which implies that $[f(x)]_n$ is a measurable function on $[a, b]$.

Definition. *If E is any measurable subset of $[a, b]$, and if f is a non-negative real vlaued function $\in L[a, b]$, then $\int_E f$ is defined as*

$$\int_E f = \lim_{n \to \infty} \int_E [f(x)]_n$$

16.19 INTEGRAL OF NON-NEGATIVE FUNCTIONS

A non-negative measurable function f defined on a measurable set E is said to be integrable if $\int_E f < \infty$ where we define $\int_E f = \sup_{g \leq f} \int_E g(x)$, g being a bounded measurable function s.t. $m[E(x : g(x) \neq 0)] < \infty$.

THEOREM 1. *If f, g are non-negative measurable function defined on E, then $\int_E (f + g) = \int_E f + \int_E g$.*

Proof.   Let $h(x) \leq f(x)$ and $k(x) \leq g(x)$, then
$$h(x) + k(x) \leq f(x) + g(x) = (f+g)(x)$$

⇒ $\int_E h + \int_E k \leq \int_E (f + g)$.

Taking supremum, we get

$$\int_E f + \int_E g \leq \int_E (f + g). \tag{1}$$

Further let $F(x)$ be a bounded measurable function s.t. $F(x) \leq (f+g)(x)$ and $F(x)$ vanishes outside a set of finite measure. Now taking

$$h(x) = \min(f(x)), F(x))$$

and $k(x) = F(x) - h(x)$.

Then $h(x) + k(x) = F(x) \leq f(x) + g(x)$.

We have $\qquad h(x) \le f(x)$ and $k(x) \le g(x)$.

Also we have $|h|$ and $|k|$ are bounded by the bound of $|F|$ and vanish where F vanishes, so

$$\int_E F = \int_E h + \int_E k \le \int_E f + \int_E g$$

This is true for all $\quad F \le (f+g)$. So taking supremum, we get

$$\int_E (f+g) \le \int_E f + \int_E g. \qquad\qquad\qquad ...(2)$$

Finally (1) and (2) $\qquad \Rightarrow \int_E (f+g) = \int_E f + \int_E g$.

THEOREM 2. *If f and g are non-negative measurable functions defined on a measurable set E s.t. f>g on E and f is integrable over E, then g is also integrable on E.*

Proof. We have $\qquad \int_E f = \int_E (f - g + g) = \int_E (f - g) + \int_E g$. $\qquad ...(1)$

But given that f is integrable over $E \Rightarrow \int_E f < \infty$. $\qquad\qquad ...(2)$

and $f - g$, g both are non-negative measurable functions

$$\Rightarrow \qquad \int_E (f - g) \ge 0, \int_E g \ge 0$$

Hence, in view of (2), both $\int_E (f - g) < \infty$ and $\int_E g < \infty$

$\Rightarrow$ function g is integrable on E.

Also (1) $\Rightarrow \int_E (f - g) = \int_E f - \int_E g$.

THEOREM 3. *If f is a non-negative measurable functions over a set E, then for given $\varepsilon > 0$, $\exists$ a $\delta > 0$ s.t. for every set $A \subset E$ with $m(A) < \delta$, we get $\int_A f < \varepsilon$.*

Proof. **Case I.** If f is a bounded function on E i.e., $|f(x)| \le M$, $\forall x \in E$, then $\forall x \in A$ also $|f(x)| \le M$. Choosing $\delta = (\varepsilon/M) > 0$, for $A \subset E$, $m(A) < \delta$, we get

$$\int_A f \le M.m(A) < M.\delta = M.\frac{\varepsilon}{M} = \varepsilon$$

Case II. Suppose f is unbounded on E. Then define the new functions

$$f_n(x) = \begin{cases} f(x) & , \quad \text{if} \quad f(x) \le n \\ n & , \quad \text{if} \quad f(x) > n \end{cases}$$

Then $\langle f_n \rangle$ is an increasing sequence of bounded functions on E s.t. $f_n \to f$. Therefore, by monotone convergence theorem, given $\varepsilon > 0$, $\exists$ an integer $n_0 \in N$ s.t.

$$\int_E f_{n_0} > \int_E f - \frac{\varepsilon}{2} \quad \text{or} \quad \int_E f - \int_E f_{n_0} < \frac{\varepsilon}{2}.$$

But

$$\int_E (f - f_{n_0}) = \int_E f - \int_E f_{n_0} < \frac{\varepsilon}{2}$$

Now choosing $\delta < \frac{\varepsilon}{2n_0}$, for all $A \subset E$ s.t. $m(A) < \delta$, we get

$$\int_A f = \int_A (f - f_{n_0}) + \int_A f_{n_0} \le \int_E (f - f_{n_0}) + n_0.m(A) \text{ as } f_{n_0} \text{ is bounded and}(f_{n_0}) \le n_0$$

$$< \frac{\varepsilon}{2} + n_0.\delta < \frac{\varepsilon}{2} + n_0.\left(\frac{\varepsilon}{2n_0}\right) < \varepsilon$$

Solved Example

Example 1. Let f be a bounded measurable function defined on a measurable set E, such that $f(x) \geq 0$ and $\int_E f(x) dx = 0$, prove that $f(x) = 0$ in a.e. in E, i.e., $f(x)$ is equivalent to zero function on E.

Solution. Let

$$E_0 = E[f(x) = 0] = \{x \in E : f(x) = 0\}$$

$$E_r = E\left(\frac{M}{r+1} < f(x) \leq \frac{M}{r}\right), \forall r \in \mathbb{N}$$

where $f(x) \leq M$, $\forall x \in E$. i.e., M is an upper bound and $M > 0$

Obviously, $E = \bigcup_{r=0}^{\infty} E_r, E_r \cap E_s = \phi$ for $r \neq s$. ... (1)

Each E_r is measurable, since f measurable over E

$$\left(\because \text{The set } E_r = E\left(f(x) \leq \frac{M}{r}\right) \cap E\left(f(x) > \frac{M}{r+1}\right), \text{being intersection of} \atop \text{two measurable sets, is measurable.} \right)$$

From (1), we have $E - E_0 = \bigcup_{r=1}^{\infty} E_r$. ...(2)

$\therefore$ $m(E - E_0) = \sum_{r=1}^{\infty} m(E_r)$ ...(2)

Now $\frac{M}{r+1} < f(x), \forall x \in E_r$ and $n \in \mathbb{N}$.

Applying first mean value theorem, we get

$$\frac{M}{r+1}.m(E_r) < \int_{E_r} f(x) dx \leq \sum_{r=0}^{\infty} \int_{E_r} f(x) dx = \int_E f(x) dx = 0 \text{ (given)}$$

$\therefore$ $\frac{m}{r+1}.m(E_r) \leq 0, \forall r \in \mathbb{N}$. Also $\frac{M}{r+1} \geq 0, m(E_r) \geq 0 \Rightarrow \frac{m}{r+1} m(E_r) \geq 0$

or $m(E_r) = 0, \forall r \in \mathbb{N}$ ($\because$ Measure of any set is non-negative.)

$\therefore$ $\sum_{r=1}^{\infty} m(E_r) = 0$. ... (3)

Using (3) in (2), we have

$$m(E - E_0) = 0.$$

Thus, $f(x) = 0, \forall x \in E_0$ and $m(E - E_0) = 0$, i.e., where $f(x) \neq 0$ the measure of that set is zero.

Hence, $f(x) = 0$ a.e., on the set E.

REMARKS

- If E is any measurable subset of $[a, b]$ and if $f, g \in L[a, b]$ are arbitrary functions s.t. $f(x) = g(x)$, a.e. on E, then
$$\int_E f = \int_E g.$$
- $\int_E (f \pm g) = \int_E f \pm \int_E g$

- $\int_E (af) = a\int_E f$ and $\int_E (af + bg) = a\int_E f + b\int_E g$; a, b are real numbers.

- $\int_{E_1 \cup E_2} f = \int_{E_1} f + \int_{E_2} f$, if $E_1 \cap E_2 = \phi$

Example 2. *If $\int_A f dx = 0$ for every measurable subset A of a measurable set E, show that $f(x) = 0$ a.e. on E.*

Solution. Let if possible, suppose $m[B] = E\{f > 0\} > 0$.

$\Rightarrow$ $\exists$ a closed set $F \subset B$ s.t. $m(F) > 0$. If set $O = B - F$. Since $B \subset E$, we get

$$0 = \int_B f = \int_F f + \int_O f \Rightarrow \int_O f = -\int_F f \neq 0$$

But O being open set, it is a countable union of disjoint open intervals (a_n, b_n) and hence

$$\int_O f = \sum_{n \in N} \int_{a_n}^{b_n} f$$

$\Rightarrow$ for at least one $n, \int_{a_n}^{b_n} f \neq 0$, which is a contradiction

$\Rightarrow$ $m[E\{f > 0\}] > 0$ is not true.

Similarly, we can show that $m[E\{f < 0\}] > 0$ is also not true

$\Rightarrow$ $m[E\{f \neq 0\}] > 0$ is not true.

$\Rightarrow$ $m[E\{f \neq 0\}] = 0 \Rightarrow f = 0$ a.e on E.

Example 3. *If $f \in L[a, b]$ and if $f(x) \geq 0$ almost everywhere in $[a, b]$, then show that $\int_a^b f \geq 0$*

Solution. We may take $f(x) \geq 0$, $\forall x \in [a, b]$. Then obviously, $U[f ; P] \geq 0$ for every P.

So $$L\overline{\int_a^b} f = \inf .U[f; P] \geq 0 .$$

since $f \in L[a, b]$,

$$\int_a^b f = L\overline{\int_a^b} f \geq 0 \Rightarrow \int_a^b f \geq 0 .$$

Example 4. *If f, g∈L $[a, b]$ and if $f(x) \leq g(x)$ a.e. in $[a, b]$, then show that*

$$\overline{\lim} f_n(x) = \underline{\lim} f_n(x)$$

Solution. We know that $f \in L[a,b] \Rightarrow kf \in L[a,b]$ where $k \in R$ be arbitary. Taking $k = -1$, we have

$$f \in L[a,b] \Rightarrow -f \in L[a,b]$$

Again, we have

$$f \in L[a,b]; g \in L[a,b] \Rightarrow g - f \in L[a,b]$$

Since $f(x) \leq g(x)$ a.e. in $[a, b]$ $\Rightarrow$ $[g(x) - f(x)] > 0$ a.e. in $[a, b]$

Then, we have

$$\int_a^b (g - f) \geq 0 \text{ or } \int_a^b g + \int_a^b (-f) \geq 0$$

or $$\int_a^b g - \int_a^b f \geq 0 \text{ or } \int_a^b f \leq \int_a^b g$$

Example 5. *If E is a measurable subset of [a, b], and f is a bounded measurable function (or say $f \in L[a, b]$) s.t. $f(x) \geq 0$ a.e. on E, then show that $\int_E f \geq 0$.*

Solution. We know that

$$\int_E f = \int_E f \phi_E, \text{ where } \phi_E \text{ is the characteristic function of } E.$$

Since $f \geq 0$ a.e. on E and, $\phi_E \geq 0$ on $[a, b]$, we have $f \phi_E \geq 0$ a.e. on $[a, b]$.

Hence, we get

$$\int_a^b f \phi_E \geq 0 \text{ implying that } \int_E f \geq 0.$$

Example 6. *Show that if f is a bounded function in L[a, b], then $|f| \in L[a, b]$ and*

$$\left| \int_a^b f \right| \leq \int_a^b |f|$$

Solution. Since $f \in L[a,b], f$ is bounded and measurable. Therefore $|f|$ is also bounded measurable and hence $|f| \in L[a,b]$.

Further since $f(x) \leq |f(x)| = |f|(x), \forall x \in [a,b]$, therefore

$$\int_a^b f \leq \int_a^b |f| \qquad \dots (1)$$

Also since

$$-f(x) \leq |f(x)| = |f|(x)$$

we get

$$-\int_a^b f \leq \int_a^b |f| \qquad \dots(2)$$

Hence from (1) and (2) $\left| \int_a^b f \right| \leq \int_a^b |f|$.

Example 7. *By an example, show that a function which is Lebesgue integrable is not necessarily R-integrable.*

Solution. Let $h(x)$ be a function defined over [0, 1] such that

$$h(x) = \begin{cases} 1 & , \text{ if } x \text{ is irrational} \\ 0 & , \text{ if } x \text{ is rational} \end{cases}$$

Clearly, $h(x)$ is discontinuous and points of discontinuity form a measurable set. Since every subinterval of [0, 1] contains both rationals and irrationals, for all values of r and for all modes of subdivision.

Now, $M_r = 1, m_r = 0$.

Let $0 = x_0 < x_1 < x_2 < \dots < x_n = 1$ be any mode of subdivision.

Lower Riemann sum $= \sum_{r=0}^{n-1} m_r.(x_{r+1} - x_r) = 0$, since $m_r = 0$.

Upper Riemann sum $= \sum_{r=0}^{n-1} M_r.(x_{r+1} - x_r)$

$$= \sum_{r=0}^{n-1} 1.(x_{r+1} - x_r) = 1 \quad \left(\because \sum_{r=0}^{n-1} (x_{r+1} - x_r) = x_n - x_0 = 1 \right)$$

Thus, upper Riemann sum $\neq$ lower Riemann sum. Hence $h(x)$ is not R-integrable over [0, 1]

Now, let A = set of all points of irrational numbers in [0, 1], and B = set of all points of rational numbers in $[0, 1]$.

Then, $A \cup B = [0, 1]$ and $A \cap B = \phi$ *i.e.,* $m(A \cap B) = 0$

Clearly, $P = \{A, B\}$ is a measurable partition of $[0, 1]$

Also, h is identically 1 on A and 0 on B

$\Rightarrow$ $M[h; A] = m[h; A] = 1$

and $M[h; B] = m[h; B] = 0$.

Hence $U[h; P] = M[h; A]m(A) + M[h; B]m(B)$

$$= 1.\, m(A) + 0.\, m(B) = m(A) = 1$$

[For, $m(A) = $ Length$[0, 1] - m(A')$]

$$= 1 - m(B) = 1 - 0 = 1 \text{ as } B \text{ is countable and hence } m(B) = 0]$$

Similarly $L[h; P] = m[h; A]\, m(A) + m[h; B]\, m(B)$

$$= 1.\, m(A) + 0.\, m(B) = 1$$

$\Rightarrow$ $U[h; P] = L[h; P] = 1$

or $U[h; P] - L[h; P] = 0 < \varepsilon$

Hence $h \in L[0, 1]$

Thus h as defined above is a function which is Lebesgue-integrable but not Riemann integrable.

Example 8. *Show that, if* $f(x) = \dfrac{1}{x}$ $(0 < x \le 1), f(0) = 19$ *then f is not integrable on* $[0, 1]$.

Solution.   Clearly, $f(x)$ is an unbounded function. Let $m \in N$, then define

$$[f(x)]_m = \begin{cases} f(x) & \text{if} \quad 0 \le f(x) \le m \\ m & \text{if} \quad f(x) > m \end{cases}$$

Therefore,

$$[f(x)]_m = \frac{1}{x}, \text{ if } \frac{1}{m} \le x \le 1, \text{and} [f(x)]_m = m, \text{if } 0 < x < \frac{1}{m}$$

Now $\int_0^1 [f(x)]_m = \int_0^{1/m} [f(x)]_m + \int_{1/m}^1 [f(x)]_m = \int_0^{1/m} m\, dx + \int_{1/m}^1 \frac{1}{x} dx$

$$= [\log x]_{1/m}^1 + [mx]_0^{1/m} = -\log \frac{1}{m} + 1 = \log m + 1$$

Making $m \to \infty$, we have

$$\int_0^1 f(x)dx = 1 + \lim_{m \to \infty} \log m.$$

Since $\lim\limits_{m \to \infty} \log m$ does not exist, f is not Lebesgue integrable over $[0, 1]$.

Example 9. *Prove that the function*

$$f(x) = \frac{d}{dx}\left(x^2 \sin \frac{1}{x^2} \right) = 2x \sin \frac{1}{x^2} - \frac{2}{x} \cos \frac{1}{x^2}$$

is not L-integrable over $[0, 1]$

Solution. Given that
$$f(x) = 2x \sin \frac{1}{x^2} - \frac{2}{x} \cos \frac{1}{x^2}$$

Clearly, that function $2x \sin \frac{1}{x^2}$ is bounded and continuous over $[0, 1]$ and

hence measurable over $[0, 1]$. Again it is bounded and meaurable over $[0, 1]$; hence

L-integrable over $[0\ 1]$. Thus, in order to show that $f(x)$ is not L-integrable, it is

sufficient to prove that the function $-\frac{2}{x} \cos \frac{1}{x^2}$ is not L-integrable over $[0, 1]$.

Let
$$a_n = \left\{ \left(2n + \frac{1}{3}\right) \pi \right\}^{-1/2}, b_n \left\{ \left(2n - \frac{1}{3}\right) \pi \right\}^{-1/2}.$$

Then
$$a_n^2 \le x^2 \le b_n^2 \Rightarrow \left(2n - \frac{1}{3}\right) \pi \le \frac{1}{x^2} \le \left(2n - \frac{1}{3}\right) \pi$$

$$\Rightarrow \quad \left| \cos \frac{1}{x^2} \right| \ge \frac{1}{2} \qquad \qquad \ldots (1)$$

Therefore $\int_0^1 \frac{1}{x} \left| \cos \frac{1}{x^2} \right| dx = \sum_{n=1}^{\infty} \int_{a_n}^{b_n} \frac{1}{x} \left| \cos \frac{1}{x^2} \right| dx \ge \sum_{n=1}^{\infty} \int_{a_n}^{b_n} \frac{1}{2x} dx$ [Using (1)]

$$= \sum_{n=1}^{\infty} \frac{1}{4} \log \frac{b_n^2}{a_n^2} = \frac{1}{4} \sum_{n=1}^{\infty} \log \frac{6n+1}{6n-1} \to \infty.$$

Therefore, function $\frac{1}{x} \left| \cos \frac{1}{x^2} \right|$ is not and hence, $f(x)$ is also not L-integrable.

Example 10. *Show that the function defined on $E = [0, \infty[$ by $f(x) = \frac{\sin x}{x}$, for $x \ne 0$ and $f(0) = 0$, is not Lebesgue integrable on E.*

Solution. We have $\int_0^{n\pi} \left| \frac{\sin x}{x} \right| dx = \sum_{r=1}^{n} \int_{(r-1)x}^{\pi} \frac{|\sin x|}{x} dx$

$$= \sum_{r=1}^{n} \int_0^{\pi} \frac{|\sin\{t + (r-1)\pi\}|}{t + (r-1)\pi} dt \ge \sum_{r=1}^{n} \int_0^{\pi} \frac{|\sin\{t + (r-1)\pi\}|}{r\pi} dt$$

$$= \frac{1}{\pi} \sum_{r=1}^{n} \frac{1}{r} \int_0^{\pi} |\sin t| \, dt = \frac{1}{\pi} \sum_{r=1}^{n} \frac{1}{r} \int_0^{\pi} \sin t \, dt = \frac{2}{\pi} \sum_{r=1}^{n} \frac{1}{r}$$

$$\Rightarrow \quad \lim_{n \to \infty} \int_0^{n\pi} \left| \frac{\sin x}{x} \right| dx \ge \frac{2}{\pi} \sum_{r=1}^{\infty} \frac{1}{r}.$$

But series $\sum_{r=1}^{\infty} \frac{1}{r}$ is divergent, hence, $\sum_{r=1}^{\infty} \frac{1}{r} = \infty$

$\Rightarrow \quad \int_0^{\infty} |f(x)| dx \ge \infty \Rightarrow \int_0^{\infty} |f(x)| dx = \infty \Rightarrow |f|$ is not L-integrable and hence function $f(x)$ is also not L-integrable.

Example 11. *Give an example to show that the integral of a nowhere zero function can be zero.*

Solution. Let $f : Q \to R$ s.t $f(x) = 1, \forall x \in Q$

Obviously function $f(x)$ is nowhere zero.

But $m(Q) = 0$, being a countable set. By first mean value theorem,

$$1.m(Q) \leq \int_Q f \leq 1.m(Q) \qquad\qquad [\because 1 \leq f(x) \leq 1]$$

$$\Rightarrow \qquad\qquad 0 \leq \int_Q f \leq 0 \Rightarrow \int_Q f = 0.$$

Example 12. *If the Lebesgue integral of non-negative measurable function f over* [0, 1] *be zero, show that f=0 a.e. or* $\int f = 0 \Rightarrow f = 0$ *a.e.*

Solution. Given that $f(x) \geq 0$, $\forall x \in [0, 1]$ and $\int_0^1 f(x) = 0$.

Now $\int_0^1 f(x)dx = \inf_P U[f; P] = U[f; P^*]$, (say) where $P^* = \{A_1, A_2, ..., A_n\}$

$$\Rightarrow \quad \sum_{i=1}^{n} M[f; A_i].m(A_i) = 0 \qquad\qquad ...(1)$$

But $m(A_i) > 0$, a.e., on P^* and $M[f; A_i]$ is also non-negative. Hence, (1) can hold good only if $M[f; A_i] = 0$

$\Rightarrow$ max. value of $f(x)$ is zero on almost each A_i

$\Rightarrow$ max. $f(x) = 0$ a.e. on [0, 1]

$\Rightarrow$ $f(x) = 0$, a.e. on [0, 1].

Example 13. *If f is a non-negative integrable function, then show that the function* $F(x) = \int_{-\infty}^{x} f(t)dt$ *is continuous on R.*

Solution. We know that for given $\varepsilon > 0$, $\exists$ a $\delta > 0$, s.t. for all subsets $A \subset R$ s.t $m(A) < \delta$, we get

$$\int_A f < \varepsilon \text{ and hence } \left| \int_A f \right| < \varepsilon.$$

Let x_0 be arbitrary element belonging to R. Then for all $x \in R$ s.t $|x - x_0| < \delta$, we have $\left| \int_{x_0}^{x} f(t)dt \right| < \varepsilon$

$$\Rightarrow \qquad \left| \int_{-\infty}^{x} f(t)dt - \int_{-\infty}^{x_0} f(t)dt \right| < \varepsilon \text{ or } |F(x) - F(x_0)| < \varepsilon$$

$\Rightarrow$ $F(x)$ is continuous at $x_0 \in R$

$\Rightarrow$ $F(x)$ is continuous on R.

16.20 CONVERGENCE OF SEQUENCE OF MEASURABLE FUNCTIONS

16.20.1 CONVERGENCE ALMOST EVERYWHERE

Let $\langle f_n \rangle$ be a sequence of measurable functions defined over a measurable set E. Then $\langle f_n \rangle$ is said to converge almost everywhere in E if there exists a subset E_0 of E s.t.

(i) $f_n(x) \to f(x)$, $\forall x \in E - E_0$ and (ii) $m[E_0] = 0$.

16.20.2 POINTWISE CONVERGENCE

Let $\langle f_n \rangle$ be a sequence of measurable functions on a measurable set E. Then $\langle f_n \rangle$ is said to converge "pointwise" in E, if $\exists$ a measurable function f on E such that

$$f_n(x) \to f(x), \forall x \in E \text{ or } \lim_{n \to \infty} f_n(x) = f(x).$$

i.e., for arbitrarly chosen positive quantity ε, however small we must get a number $n_0(\varepsilon, x) \in N$ s.t $|f_n(x) - f(x)| < \varepsilon, \forall n \geq n_0$.

Thus the number $n_0(\varepsilon, n)$ will depend upon 'ε' and 'x' both i.e., for different $x \in E$, we may get different number n_0, but we must get n_0 for every $x \in E$.

16.20.3 CONVERGENCE IN MEASURE

Let $\langle f_n \rangle$ be a sequence of measurable functions defined over a measurable set E. Then the sequence $\langle f_n \rangle$ is said to converge in measure to the function f, written as $f_n \longrightarrow f$, if

(1) f is measurable function on E s.t. $f(x) < \infty$ a.e. on the set E

(2) $\lim m[E(|f_n - f| \geq \varepsilon)] = 0$, for $\varepsilon > 0$, however small i.e, for each $\varepsilon > 0$, $\exists$ a number $n_0(\delta)$ s.t. $\forall n \geq n_0(\delta)$, we have $m[E(|f_n - f| \geq \varepsilon)] < \delta$.

16.20.4 UNIFORM CONVERGENCE, ALMOST EVERYWHERE

Let $\langle f_n \rangle$ be a sequence of measurable functions defined over a measurable set E. Then the sequence $\langle f_n \rangle$ is said to converge uniformly a.e. to f, if $\exists$ a set $E_0 \subset E$ s.t.

(i) $m(E_0) = 0$ and

(ii) $\langle f_n \rangle$ converges uniformly to f on the set $E - E_0$.

i.e, for arbitrarily chosen small quantity $\varepsilon > 0$, we can find a number $n_0(\varepsilon) \in$ N, depending upon ε alone and independent of x, such that

$$\forall n \geq n_0 \Rightarrow |f_n(x) - f(x)| < \varepsilon, \forall x \in E - E_0$$

REMARK

- We must not the differnce between "pointwise convergence" and "uniform convergence". In both types of convergence we get a number $n_0 \in$ N, $\forall x$. But in pointwise convergence n_0 depends upon x also, i.e., n_0 may be different for different x, while in uniform covergence one number n_0 (independent of x) works for all elements x.

THEOREM 1. (F. Riesz Theorem) *Let $\langle f_n \rangle$ be a sequence of measurable functions which converges in measure to f, then $\exists$ a subsequence $\langle f_{n_k} \rangle$ of $\langle f_n \rangle$ which also converges to f almost everywhere.*

Proof. Let $\langle \sigma_n \rangle$ be a monotonic decreasing sequence of positive terms s.t. $\lim \sigma_n = 0$.

Now, since $\langle f_n \rangle$ converges in measure to f, we get $\lim_{n \to \infty} m[E(|f_n - f| \geq \sigma_k)] = 0$

$\Rightarrow$ corresponding to each σ_k we can find a number $n_k \in$ N s.t. $\forall n \geq n_k$

$$|m[E(|f_n - f| \geq \sigma_k)] - 0| < \sigma_k.$$

Particularly for $n = n_k$, we obtain

$$m[E(|f_{n_k} - f| \geq \sigma_n)] < \sigma_k.$$

This corresponding to the sequence $\langle \sigma_n \rangle$, we can construct a sequence $\langle n_k \rangle$ satisfying (1). Now we want to prove that

$$\lim_{k \to \infty} f_{n_k}(x) = f(x), \text{a.e. on } E.$$

Let $E_i = \bigcup_{k=i}^{\infty} [E(|f_{n_k} - f| \geq \sigma_k)]$ and $B = \bigcap_{i=1}^{\infty} E_i$.

Obviously, $\langle E_i \rangle$ is a monotonic decreasing sequence of measurable sets. Therefore

$$\lim_{n \to \infty} m(E_n) = m(B).$$

Again, by (1); $m(E_n) < \sum\limits_{k=n}^{\infty} \sigma_k$ or $|m(E_n) - 0| < \sum\limits_{k=n}^{\infty} \sigma_k$

$\Rightarrow \qquad \lim\limits_{n\to\infty} m(E_n) = 0 \Rightarrow m(B) = 0$.

Let $y \in E - B$ be arbitrary, then by $y \notin B \Rightarrow \exists$ at least one $n_0 \in N$ s.t. $y \notin E_{n_0}$

$\Rightarrow \qquad y \notin E(|f_{n_k} - f| \geq \sigma_k), \forall k \geq n_0$

$\Rightarrow \qquad |f_{n_k}(y) - f(y)| < \sigma_k, \forall k \geq n_0 |$

$\Rightarrow \qquad \lim\limits_{k\to\infty} f_{n_k}(y) = f(y), \forall y \in E - B$

$\Rightarrow \qquad \lim\limits_{k\to\infty} f_{n_k} = f, \text{a.e on } E.$

THEOREM 2. **(D. F. Egorffs Theorem).** *If a sequence of almost everywhere, finite valued measurable functions is convergent almost everywhere on a measurable set E, of finite measure, then for every $\varepsilon > 0$, $\exists$ a measurable set $E_0 \subset E$ such that*

(a) $m(E_0) < \varepsilon$,
(b) the sequence is uniformly convergent on $(E - E_0)$.
(Little wood's third principle of measurability.)

Proof. Let $\langle f_n \rangle$ be a sequence of finite valued measurable functions defined over E such that this sequence converges almost everywhere on E to a finite measurable function f. Let B be the set on which $\lim\limits_{n\to\infty} f_n(x) = f(x)$.

then obviously, $m(E - B) = 0$. Hence f is measurable on $E - B$ as every function is measurable on a set of measure zero. Thus f is measurable on E.

Let $\qquad E_n^p = \bigcap\limits_{i=n}^{\infty}\left[E\left(|f_i - f| < \dfrac{1}{p}\right)\right]$ where p is a positive integer.

Then $\qquad E_1^p \subset E_2^p \subset ... \subset E.$

Now $\qquad E - E_n^p = E - \bigcap\limits_{i=n}^{\infty}\left[E\left(|f_i - f| < \dfrac{1}{P}\right)\right]$

$$= \bigcup\limits_{i=n}^{\infty}\left[E - E\left(|f_i - f| < \dfrac{1}{P}\right)\right] = \bigcup\limits_{i=n}^{\infty} E\left(|f_i - f| \geq \dfrac{1}{P}\right).$$

Since f_i and f are measurable over E, we have each $E\left(|f_i - f| \geq \dfrac{1}{p}\right)$ is measurable and hence $E - E_n^p$ is measurable.

$\Rightarrow \quad \langle E - E_n^p \rangle$ is a non-increasing sequence of measurable sets and $\bigcap\limits_{n=1}^{\infty}(E - E_n^p) \subseteq B$.

Therefore $\quad \lim\limits_{n\to\infty} m(E - E_n^p) = m\left(\bigcap\limits_{n=1}^{\infty}(E - E_n^p)\right) \leq m(B) = 0$

$\Rightarrow \qquad \lim\limits_{n\to\infty} m(E - E_n^p) = 0$

$\Rightarrow \quad$ for given $\varepsilon > 0$, $\exists\, n_0 \in N$ (depending upon ε and p)s.t. $\forall\, n \geq n_0$
$$|m(E - E_n^p) - 0| < \varepsilon / 2^P$$

or $\qquad m(E - E_n^p) < \varepsilon / 2^P$ as $m(E - E_n^p) \geq 0$.

Particularly taking $n = n_0$, we get

$$m(E - E_n^p) < \varepsilon / 2^p . \qquad \text{... (1)}$$

Define

$$E_0 = \overset{\infty}{\underset{p=1}{\cup}} [E - E_{n_0}^p] .$$

So, $E_0 \subseteq E$. Also E_0 is a measurable set being the countable union of measurable sets.

$$\therefore \qquad m(E_0) = m \left[\overset{\infty}{\underset{p=1}{\cup}} (E - E_{n_0}^p) \right] \le \overset{\infty}{\underset{p=1}{\sum}} m(E - E_{n_0}^p) < \overset{\infty}{\underset{P=1}{\sum}} \frac{\varepsilon}{2^p} < \varepsilon \qquad \text{... (2)}$$

$$\text{[By (1)]}$$

Again, $\qquad E - E_0 = E - \overset{\infty}{\underset{p=1}{\cup}} (E - E_{n_0}^p) = E \cap \left(\overset{\infty}{\underset{p=1}{\cup}} E - E_{n_0}^p \right) .$

Thus $\forall n \ge n_0(p, \varepsilon)$ and $x \in E - E_0 \Rightarrow x \in E_{n_0}^p .$

$$\Rightarrow \qquad x \in \overset{\infty}{\underset{i=n_0}{\cap}} \left(E \mid f_i - f \mid < \frac{1}{p} \right) \Rightarrow \mid f_n(x) - f(x) \mid < \frac{1}{p}, \forall n \ge n_0 .$$

<u>THEOREM 3.</u> **(Lebesgue bounded convergence theorem).** *Let $\langle f_n \rangle$ be a sequence of bounded measurable functions defined on a set E of finite measure. If $\exists$ a positive real number M such that $\mid f_n(x) \mid \le M, \forall n \in N$ and $\forall x \in E$*

and $\langle f_n \rangle$ converges in measure to a measurable function f on the set E, then

$$\underset{n \to \infty}{\lim} \int_E f_n(x) = \int_E f_n dx .$$

Proof. Since f_n is bounded and measurable on E and therefore integrable on E. By hypothesis, for all $\delta > 0$. $\underset{n \to \infty}{\lim} m[E(\mid f_n - f \mid \ge \delta)] = 0$

and $\mid f_n(x) \mid \le M, \forall n \in N \mid$ and $x \in E$. So, $\mid f(x) \mid \le M \ \forall x \in E$.

Thus $f(x)$ is bounded and measurable on E, therefore integrable on E. Let $\lambda > 0$ be arbitrary and let $E_n = E(\mid f_n - f \mid \ge \lambda), E_n' = E(\mid f_n - f \mid < \lambda)$

Then $\quad E = E_n \cup E_n'$ and $E_n \cap E_n' = \phi .$

Since $\langle f_n \rangle$ is not convergent in E_n, by definition, we have

$$\underset{x \to 0}{\lim} m(E_n) = 0 . \qquad \text{... (2)}$$

By countable additivity property of the integrals, we have

$$\int_E \mid f_n - f \mid = \int_{E_n} \mid f_n - f \mid + \int_{E_n'} \mid f_n - f \mid \qquad \text{... (3)}$$

Now, $\mid f_n - f \mid < \lambda, \forall x \in E_n'$. Therefore by first mean vlaue theorem, we have

$$\int_{E_n'} \mid f_n - f \mid < \lambda . m(E_n') \le \lambda . m(E)$$

$$\Rightarrow \qquad \int_{E_n'} \mid f_n - f \mid < \lambda . m(E) . \qquad \text{... (4)}$$

For arbitrary value of $\varepsilon > 0$, choose λ such that $\lambda . m(E) < \dfrac{\varepsilon}{2}$ i.e., $\lambda < \dfrac{\varepsilon}{2 . m(E)}$

Then (4) $\Rightarrow \qquad \int_{E_n'} \mid f_n - f \mid < \dfrac{\varepsilon}{2} . \qquad \text{... (5)}$

Again, $\qquad \mid f_n - f \mid \le \mid f_n \mid + \mid f \mid \le M + M = 2M .$

Again by Ist mean value theorem, we have

$$\int_{E_n} |f_n - f| \le 2M.m(E_n).$$ ... (6)

From (2) for arbitrary $\varepsilon > 0$, $\exists\ m_0 \in N$ s.t. $\forall\ n \ge m_0$, we have $|m(E_n) - 0| < \dfrac{\varepsilon}{4M}$.

Thus $\qquad n \ge m_0 \Rightarrow m(E_n) < \dfrac{\varepsilon}{4M}$ ... (7)

Using (7) in (6), we get $\qquad\qquad\qquad$ $(\because m(E_n)$ is non-negative.)

$$\int_{E_n} |f_n - f| < \frac{1}{2}\varepsilon, \forall n \ge m_0$$ ... (8)

Using (5) and (8) in (3), we get

$$\int_{E_n} |f_n - f| < \frac{1}{2}\varepsilon + \frac{1}{2}\varepsilon = \varepsilon, \forall n \ge m_0.$$

But $\qquad\qquad \int_{E_n} |f_n - f| \le \int_E |f_n - f| < \varepsilon,$

$\Rightarrow \qquad\qquad \left| \int_{E_n} (f_n - f) - 0 \right| < \varepsilon, \forall n \ge m_0$

$\Rightarrow \qquad\qquad \lim\limits_{n\to\infty} \int_{E_n} (f_n - f) = 0$ or $\lim\limits_{n\to\infty} \int_E f_n(x)dx - \int_E f(x)dx = 0$

or $\qquad\qquad \lim\limits_{n\to\infty} \int_E f_n(x)dx = \int_E f(x)dx.$

THEOREM 4. **(Lebesgue Dominated convergence Theorem)** *Let $\langle f_n \rangle$ be a sequence of measurable functions defined over a measurable set E, s.t.*

$$|f_n(x)| < \psi(x), \forall x \in E \ and \ \forall\ n \in N$$

where $\psi(x)$ is an inegrable function over E. Let $\langle f_n \rangle$ converge in measure to a measurable function f over E. Then

$$\lim\limits_{x\to 0} \int_E f_n(x)dx = \int_E f(x)dx.$$

Proof. Given that $\langle f_n \rangle$ is a sequence of measurable functions. Also, $|f_n(x)| < \psi(x), \forall\ n \in N$, and $\forall\ x \in E$, where $\psi(x)$ is integrable over E

$\Rightarrow \quad f_n(x)$ is bounded, $\forall\ n \in N$, $\qquad (\because \psi(x)$ is bounded being integrable).

Thus $\langle f_n \rangle$ is a sequence of bounded and measurable functions over E and hence integrable (Lebesgue) over E.

Again, given $\langle f_n \rangle$ converges in measure to a measurable function f on E, so we have

$$\lim\limits_{n\to\infty} m[E(|f_n - f| \ge \sigma)] = 0 \text{ for every } \sigma > 0.$$

$\therefore\ \langle f_n \rangle$ is a sequence of integrable functions over E, $\forall\ n \in N$ and $f_n \to f$ in measure.

$\Rightarrow \quad f(x)$ is integrable over E.

Now $\langle f_n \rangle$ and f are integrable over E, $\Rightarrow (f_n - f)$ is integrable over E, $\forall\ n \in N$

Let $\delta > 0$ be arbitrary.

Also, let $\qquad\qquad E_n = E(f_n - f| \ge \delta),\ E_n' = E(|f_n - f| < \delta).$

Obviously, $\qquad \lim\limits_{n\to 0} m(E_n) = 0,$ and $E = E_n \cup E_n', E_n \cap E_n' = \phi.$

Using countable additive property of integrals, we get

$$\int_E |f_n - f|\,dx = \int_{E_n} |f_n - f|\,dx + \int_{E_n'} |f_n - f|\,dx$$ ... (1)

Now $\quad |f_n - f| < \delta, \forall x \in E_n'$.

$\therefore \qquad \int_{E_n} |f_n - f| dx < \delta.m(E_n') \le \delta.m(E)$

i.e., $\qquad \int_{E_n} |f_n - f| dx < \delta.m(E)$. $\qquad \qquad$ ... (2)

Let $\varepsilon > 0$ be arbitrary and choose δ such that $\delta.\, m(E) < \varepsilon/2$.

Using this in (2), we get

$$\int_{E_n} |f_n - f| dx < \varepsilon/2.$$ $\qquad \qquad$ ... (3)

Now, δ is fixed, so $\quad |f_n - f| \le |f_n| + |f| < \psi + \psi = 2\psi$ or $|f_n - f| < 2\psi$

or $\qquad |f_n(x) - f(x)| < 2\psi(x), \forall x \in E$.

Since, we have

$$\int_{E_n} |f_n - f| dx < 2\int_{E_n} \psi(x)dx.$$ $\qquad \qquad$ ...(4)

Since $\qquad |f_n| < \psi, \forall x \in E$

$\Rightarrow \qquad \psi(x) \ge 0, \forall x \in E \qquad \Rightarrow \int_{E_n} \psi(x) \ge 0$

$\Rightarrow \qquad \left| \int_{E_n} \psi(x)dx \right| = \int_{E_n} \psi(x)dx$.

Using this in (4), we have

$$\int_{E_n} |f_n - f| dx < 2 \left| \int_{E_n} \psi(x)dx \right|.$$

$\because \qquad \lim_{n \to \infty} m(E_n) = 0$.

$\Rightarrow \quad$ Given $\mu > 0, \exists\, m \in N$ s.t. $n > m \quad |m(E_n) - 0| < \mu \Rightarrow m(E_n) < \mu \qquad (\because m(E_n) \ge 0)$

Using absolute continuity of the integral, we get

$$\left| \int_{E_n} \psi(x)dx \right| < \frac{\varepsilon}{4}.$$ $\qquad \qquad$ ...(6)

Combining (5) and (6), we get $\int_{E_n} |f_n - f| dx < \frac{\varepsilon}{2}$.

Combining (3) and (7) and using in (1), we get

$\int_{E_n} |f_n - f| dx < \varepsilon \Rightarrow \lim_{n \to \infty} \int_E f_n(x)dx = \int_E f(x)dx$

THEOREM 5. **(Lebesgue Monotone Convergence Theorem).** *Let* $\langle f_n \rangle$ *be a non-decreasing sequence of integrable functions defined over a measurable set E. Let*

$\lim_{n \to \infty} f_n(x)dx = f(x)$ *be integrable over E, then* $\lim_{n \to \infty} \int_E f_n(x)dx = \int_E f(x)dx$.

Proof. Since $\langle f_n \rangle$ is non-decreasing sequence, then

$$f_1 \le f_2 \le f_3 \le ... \Rightarrow f_n - f_1 \ge 0, \forall n \in N$$

Write $\quad f_n - f = \psi_n$. So $\psi_n \ge 0, \forall n \in N$.

Hence $\langle \psi_n \rangle$ is a sequence of non-negative real valued functions.

Moreover $\langle f_n - f \rangle$ is a sequence of integrable functions

$\Rightarrow \quad \langle \psi_n \rangle$ is a sequence of integrable functions

Case I. Let ψ_n be bounded measurable.

Using Lebesgue bounded convergence theorem, we get

$$\lim_{n\to\infty} \int_E \psi_n(x)dx = \int_E \lim_{n\to\infty} \psi_n(x)dx .$$

Case II. If $\psi_n(x)$ is unbounded, then define

$$[\psi_n(x)]_m = \begin{cases} \psi_n(x) & , \quad \text{when}\,\psi_n(x) \le m \\ m & , \quad \text{when}\, \psi_n(x) > m \end{cases}$$

then $[\psi_n(x)]_m$ is bounded and measurable. Hence again by Lebesgue bounded convergence theorem, we get

$$\lim_{n\to\infty} \int_E [\psi_n(x)]_m dx = \int_E \lim_{n\to\infty} [\psi_n(x)]_m dx$$

Making $m \to \infty$, we get

$$\lim_{n\to\infty} \int_E \psi_n(x)dx = \int_E \lim_{n\to\infty} \psi_n(x)dx,$$

as when $m \to \infty$, $[\psi_n(x)]_m \to \psi_n(x)$

Thus, in either case I holds good.

$$\lim_{n\to\infty} \int_E (f_n - f_1)dx = \int_E \lim_{n\to\infty} (f_n - f_1)dx,$$

or $$\lim_{n\to\infty} \int_E f_n dx - \int_E f_1\,dx = \int_E \lim_{n\to\infty} f_n dx - \int_E f_1 dx$$

or $$\lim_{n\to\infty} \int_E f_n(x)dx = \int_E \lim_{n\to\infty} f_n dx$$

REMARKS

- This is also known as 'Beppo-Levis Theorem'.
- If E is a measurable set and $\langle f_n \rangle$ is a sequence of non-negative measurable functions and

$$f(x) = \sum_{n=1}^{\infty} f_n(x), x \in E, \quad \text{then} \int_E f = \sum_{n=1}^{\infty} \int_E f_n .$$

THEOREM 6. **(Fatou's Lemma).** *Let $\langle f_n \rangle$ be a sequence of non-negative integrable functions defined over a measurable set E s.t.*

(i) $\lim_{n\to\infty} \inf f_n = f, a.e$ on $E,(or \lim f_n \to f, a.e.$ on $E)$

(ii) $\lim_{n\to\infty} \inf \int_E f_n(x)dx < \infty$, $\lim_{n\to\infty} \inf \int_E f_n(x)dx \ge \int_E f(x)dx$ or $\lim_{n\to\infty} \int_E f_n \ge \int_E f$.

Proof.   Let $g_r(x) = \inf\{f_n(x) : n \ge r\}$

Then $g_r(x) \le f_n(x), \forall n \in N.$
Integrating, we get

$$\int_E g_n(x)dx \le \int_E f_n(x)dx, \forall n \in N$$

Hence $\lim_{n\to\infty} \int_E g_n(x)dx \le \lim_{n\to\infty} \inf \int_E f_n(x)dx.$ ... (1)

Obviously, $\langle g_n \rangle$ is an increasing sequence of non-negative integrable functions. Using 'Lebesgue monotone convergent theorem', we get

$$\lim_{n\to\infty} \int_E g_n(x)dx = \int_E \lim_{n\to\infty} g_n(x)dx$$

$$= \int_E \lim_{n\to\infty} \inf f_n(x)dx = \int_E f(x)dx \qquad \left[\because \lim_{n\to\infty} \inf f_n(x) = f(x)\right]$$

Thus $\qquad \lim_{n\to\infty} \inf \int_E g_n(x)dx = \int_E f(x).$

Using this in (1), we get

$$\lim_{n\to\infty} \inf \int_E f_n(x)dx \le \int_E f(x)dx.$$

16.21 LITTLE WOOD'S THREE PRINCIPLES

J.E. Little wood suggested following three principles:
(1) Every measurable set is nearly a finite union of intervals.
(2) Every measurable function is nearly continuous.
(3) Every convergent sequence of measurable functions is nearly uniformly convergent.

Solved Examples

Example 1. *Use Lebesgue dominated convergence theorem to evaluate*

$$\lim_{x\to 0} \int_0^1 f_n(x)\,dx, \text{ where } f_n(x) = \frac{n^{3/2}x}{1+n^2x^2}, n = 1,2,3,...,0 \le x \le 1.$$

Solution. We have $f_n(x) = \dfrac{n^{3/2}x}{1+n^2x^2} = \dfrac{1}{x}\cdot\dfrac{n^{3/2}x^2}{1+n^2x^2} = \dfrac{1}{x}\cdot\dfrac{\left(\dfrac{1}{\sqrt{n}}x^2\right)}{\dfrac{1}{n^2}+x^2} < \dfrac{1}{x}\cdot\dfrac{.1.x^2}{0+x^2} \le \dfrac{1}{x} = \psi(x) \text{ (say)}$

$\Rightarrow \qquad f_n(x) \le \psi(x) \text{ and } \psi(x)\in L[0, 1]$

Hence, by Legesgue dominated convergence theorem.

$$\lim_{n\to\infty} \int_0^1 f_n(x)dx = \int_0^1 \lim_{n\to\infty} f_n(x)dx = \int_0^1 \lim_{n\to\infty}\left(\frac{n^{3/2}x}{1+n^2x^2}\right)dx$$

$$= \int_0^1 \lim_{n\to\infty}\left(\frac{1}{\sqrt{n}}\right)\left(\frac{x}{\frac{1}{n^2}+x^2}\right)dx = \int_0^1 0\,dx = 0.$$

Example 2. *Show that the theorem of bounded convergence applies to*

$$f_n(x) = \frac{nx}{1+n^2x^2},0 \le x \le 1.$$

Solution. Let $f_n(x) = \dfrac{nx}{1+n^2x^2} = \dfrac{1}{\dfrac{1}{nx}+nx} = \dfrac{1}{\left[\dfrac{1}{\sqrt{nx}}-\sqrt{nx}\right]^2+2} \le \dfrac{1}{2}.$

Thus $\exists$ a number $\dfrac{1}{2}$ s.t.$|f_n(x)| \le \dfrac{1}{2}.$ Hence it satisfies the conditions of bounded convergence theorem. Now

$$\lim_{n\to\infty} \int_0^1 f_n(x)dx = \lim_{n\to\infty} \int_0^1 \frac{nx}{1+n^2x^2}dx = \lim_{n\to\infty} \frac{1}{2n}\log(1+n^2) \qquad \left(\frac{\infty}{\infty}\text{form}\right)$$

$$= \lim_{n\to\infty} \frac{[1/(1+n^2)]2n}{2} = \lim_{n\to\infty} \frac{n}{1+n^2} = 0$$

and $\int_0^1 \lim_{x\to 0} f_n(dx) = \int_0^1 \lim_{n\to\infty} \left(\frac{nx}{1+n^2x^2}\right) dx = \int_0^1 (0)dx = 0$

$$\Rightarrow \lim_{n\to\infty} \int_0^1 f_n(x)\,dx = \int_0^1 \lim_{n\to\infty} f_n(x)dx$$

This verifies the result of bounded convergence theorem.

Example 3. If $\alpha < 0$, prove that $\lim_{n\to\infty} \int_0^n \left(1-\frac{x}{n}\right)^n x^{\alpha-1}dx = \int_0^\infty e^{-x}.x^{\alpha-1}dx$, where the integrals are taken in the Lebesgue-sense.

Solution. If $f_n(x) = \left(1-\frac{x}{n}\right)^n .x^{\alpha-1} > 0$, then $f_n(x) \le g(x)$, where $g(x) = e^{-x}.x^{\alpha-1}$.

$$\left[\because \lim_{n\to\infty} \left(1-\frac{x}{n}\right)^n = e^{-x}\right]$$

Also $g(x) \in L[0, \infty]$. Hence by Lebesgue dominated convergence theorem, we get

$$\lim_{x\to 0} \int_0^n f_n(x)\,dx = \int_0^\infty \lim_{n\to\infty} f_n(x)dx$$

$$= \int_0^\infty \lim_{n\to\infty} \left(1-\frac{x}{n}\right)^n .x^{\alpha-1}dx = \int_0^\infty e^{-x}.x^{\alpha-1}dx.$$

EXERCISE 16.4

1. Let f be integrable over a set E and suppose that E is the union of a countable family of pairwise disjoint measurable set E_i where

$$E = E_1 \cup E_2 \cup ..., E_i \cap E_j = \phi, \text{ for } i \ne j.$$

Prove that $\int_E f = \sum_n \int_{E_n} f$

2. Show that $f: [a, b] \to R$ is Riemann integrable, iff the discontinuities of f form a set of Lebesgue measure zero or say f is continuous a.e., over $[a, b]$.

3. Prove that two bounded measurable functions which are equal a.e. have the same integral. Also prove that the converse is not necessarily true.

4. Prove that if $f(x) = 0$ at every point of Cantor's ternary set and $f(x) = p$ in each of the complementary intervals of length 3^{-p}, then

$\int_0^1 f(x)dx$ exists in the Lebesgue sense and is equal to 3.

5. If A and B are disjoint measurable subsets of $[a, b]$ and if f is a bounded and Lebesgue-integrable on $[a, b]$, then show that

$$\int_{A\cup B} f = \int_A f + \int_B f .$$

6. If f is a bounded measurable function on $[a, b]$ such that $f(x) \ge 0$ a.e. $x \in [a, b]$ and if $\int_a^b f = 0$, then prove that $f(x) = 0$ a.e. in $[a, b]$.

7. If A is a measurable set $B \subseteq A$ a s.t $m(A - B) = 0$ then show that

$$\int_A f = \int_B f .$$

8. Suppose f is measurable on E, $|f| < g$ and g is integrable on E, then f is also integrable on E.

9. If f is Lebesgue-integrable on $[a, b]$ then prove that

$$\int_a^b f(t)dt = \int_{-b}^{-a} f(-t)dt$$

10. If f is bounded and integrable in the Riemann sense in a closed interval $[a, b]$, then prove that f is measurable and also integrable in the Lebesgue sense and the Riemann integral of f over $[a, b]$ is equal to the Lebesgue integral over $[a, b]$.

11. If f is the function from $R \to R$ defined by

$$f(x) = \begin{cases} 0 & , \quad x \notin [0,1] \\ 1 & , \quad x \in [0,1] \text{ and rational} \\ -1 & , \quad x \in [0,1] \text{ and irrational} \end{cases}$$

Show that f is Lebesgue integrable.

12. If f is Lebesgue integrable, show that
$$\int f(x) = -\int f(-x)\,dx.$$

13. Prove that the integral of the sum of a finite number of bounded measurable functions is the sum of integrals of the separate functions.

14. Let a function f be defined on the interval $[a, b]$, as follows:
$$f(x) = \begin{cases} 0 & , \quad \text{if } x \text{ is irrational,} \\ 1 & , \quad \text{if } x \text{ is rational} \end{cases}$$

CHAPTER REVIEW: A COMPETITIVE APPROACH

SELECTED TERMS AND RESULTS

TERMS

- **Outer measure of a Set:**

$$m^*(A) = \inf_{A \subseteq UI_n} \sum_n l(I_n)$$

- **Inner Measure of a Set:** Inner measure of A is computed by closed set which are contained in A i.e., $m(A) \geq l(B)$

- **Measurable Set :** A is measurable if $m(A) = m^*(A) = m_*(A)$

- **Lebesgue Measurable Set:** The set A is said to be Lebesgue measurable if for each set $E \subseteq R$ we have $m^*(E) = m^*(E \cap A) + m^*(E \cap A')$.

- **Cantor Set:** The set of all numbers in the interval $[0, 1]$ which have a ternary expansion without the digit 1 i.e., ternary expansion involves only two digits 0 and 2.

- **Borel Set :** A set which can be obtained by taking countable unions or intersections of open and closed sets are called Borel set.

- **F_σ-Set:** A set is called F_σ-Set if it is countable union of closed sets.

- **G_δ-Set:** A set is called G_δ-set if it is countable intersection of open sets.

- **Measurable Function:** An extended real valued function f defined over a measurable set E is said to be measurable if the set $E[f > a] = \{x \in E : f(x) > a\}$ is measurable.

- **Borel Measurable Function:** A function f defined on a Borel set E is called Borel measurable function on E if for all real number a, the set $E[f > a]$ is a Borel set.

- **Lower and Upper Lebesgue Sums:** Let $P = \{S_1, S_2, \dots S_n\}$ be a measurable partition of the closed interval $[a, b]$ and f be a bounded function defined on (a, b) then

$U(f, P) =$ Upper Lebesgue sum

$$= \sum_{r=1}^{n} M[f : S_r] m[S_r]$$

and $L(f, P) =$ Lower Lebesgue sum

$$= \sum_{r=1}^{n} m[f : S_r] m[S_r]$$

where $m[S_r]$ is the Lebesgue measure of S_r.

- **Upper and Lower Lebesgue Intergrals:**

(i) $L\underline{\int}_a^b f(x)dx = \sup\{L[f,Q]: Q$ is a measurable partition of $(a, b)\}$.

(ii) $L\overline{\int}_a^b f(x)dx = \inf\{U[f,P]: P$ is a measurable partition of $(a, b)\}$

- **Lebesgue Integral:** Let f be a bounded function defined on the interval $[a, b]$ then f is Lebesgue integrable on $[a, b]$ if

$$L\underline{\int}_a^b fdx = L\overline{\int}_a^b fdx .$$

- **Lebesgue Integral of Unbounded functions:** Let $f(x)$ be unbounded, measurable and non-negative function defined on $[a, b]$. Let $n \in N$

Then $[f(x)]_n = \begin{cases} f(x) & ; \text{ when } f(x) \leq n \\ n & ; \text{ when } f(x) > n \end{cases}$

RESULTS

- For any two disjoint subset A and B of R
$m^*(A \cup B) = m^*(A) + m^*(B)$

- The outer measure of an interval is equal to its length.

- A linear set of outer measure zero is lebesgue measurable.

- Every countable set is (L-measurable and its measure is zero).

- Every bounded open set and every bounded closed set is measurable.

- Every countable set is measurable and its measure is zero.

- Countable union of measurable set is measurable.

- The finite intersection of measurable sets is measurable.

- Difference of two measurable sets is measurable.

- (*First Fundamental Theorem*) If E_1, E_2, ... are pairwise disjoint measurable sets and $E = E_1 \cup E_2$ ... then E is measurable and

$$m(E) = \sum_{k=1}^{\infty} m(E_k)$$

- (*Second Fundamental Theorem*). If E_1, E_2, ... are measurable sets then $\bigcap_{k=1}^{\infty} E_k$ is measurable.

- Cantor set is measurable and its measure is zero.

- Cantor set is perfect.

- A set is Lebesgue measurable $\Leftrightarrow$ its complement is L-measurable.
- $m(E_1) + m(E_2) = m(E_1 \cup E_2) + m(E_1 \cap E_2)$
- F_σ and G_δ sets are Borel sets.
- If the outer measure of a set is zero then the set is outer measurable.
- Let f be a measurable function defined over a measurable set E then cf, $-f$, $f+c$, $|f|$, f^2, $\frac{1}{f}(f \pm 0)$, $c \in R$ are also measurable.
- Let f and g be two measurable functions defined over a measurable set E then $f+g$, $f-g$, $f.g$, $\frac{f}{g}(g \neq 0)$ are measurable functions over E.
- A continuous function defined over a measurable set E is measurable (converse is not true).
- f is measurable $\Rightarrow |f|$ is measurable. Converse is not true.
- A function is simple if and only if it is measurable and assumes only a finite number of values.
- If f is measurable then any positive integral power of f is also measurable.
- An increasing function of a measurable function is also measurable.
- Each characteristic function of a measurable set is a simple function.
- Every bounded function is not necessarily integrable.
- (First mean value theorem): Let f be a bounded measurable real valued function such that $a \leq f(x) \leq b$ on a measurable set $E \subset R$ then

  a. $m(E) \leq \int_E f(x)dx \leq b.m(E)$
- Every bounded measurable function f defined on $[a, b]$ is lebesgue integrable over $[a, b]$.
- If f is measurable then f is Lebesgue integrable iff $|f|$ is Lebesgue integrable.
- If the lebesgue integral of non-negative measurable function f over $[0, 1]$ is zero then $f = 0$ almost every where.
- The integral of a nowhere zero function can be zero.

REVIEW QUESTIONS AND PROJECT WORK

1. Show that every countable set is a Borel set of measure zero.

2. Show that every enumerable set is measurable and its measure is zero.

3. If E is a measurable set, then show that for given $\varepsilon > 0$ $\exists$ a closed set $F \subset E$ such that $m^*(E - F) < \varepsilon$.

4. Show that the necessary and sufficient condition for set E to the measurable set of measure zero is that $m^*(E) = 0$.

5. Show that the Lebesgue measure is finitely additive.

6. Show that a σ-additive set function is a continuous set function.

7. If f is an extended real valued non-negative function of the points of a set S then show that the function is defined on the ring R of all finite subsets of S defined as

$$\mu(x_1 x_2 ... x_n) = \sum_{i=1}^{n} f(x_i)$$

8. If E_1, E_2, $E_1 \cap E_2$ and $E_1 \cup E_2$ are all bounded measurable sets, prove that

$$\mu(E_1 \cup E_2) + \mu(E_1 \cap E_2) = \mu(E_1) + \mu(E_2).$$

9. Show that a step function is a measurable functions.

10. Show that the space of measurable functions is closed under the usual operation of arithmetic.

11. Show that a function is measurable iff its positive and negative parts are measurable.

12. Let f be a function with measurable domain D. Show that f is measurable iff the function g defined by

$$g(x) = \begin{cases} f(x) & \text{if } x \in D \\ 0 & \text{if } x \notin D \end{cases}$$

is measurable.

13. Show that a necessary and sufficient condition for measurability of a function f is that $\{x ; a \leq f(x) \leq b\}$ should be measurable for all extended real numbers a and b such that $a < b$.

14. Show that the sum, difference and product of two simple functions is also a simple function.

15. Show that the set of all measurable functions forms a vector space.

OBJECTIVE TYPE QUESTIONS

FILL IN THE BLANKS

1. $m^*(A)$ _____ 0.

2. If $A \subset (a, b)$ then $m^*(A) \leq$ _____ .

3. For every set A, $m^*(A)$ is _____ .

4. If G is an open set of real numbers then $m^*(G)$ = _____ .

5. A set with outer measure different from 0 is _____ .

6. The outer measure of an interval is equal to _____ .

7. $m^*(A)$ _____ $m_*(A)$.

8. A linear set of outer measure _____ is Lebesgue measurable.

9. The measure of a countable set is _____ .

10. Any subset of A whose outer measure is zero is _____ .

TRUE /FALSE

Write 'T' for true and 'F' for false statement.

1. Every bounded open (or closed)set is measurable. **(T/F)**

2. Countable union of measurable sets is measurable. **(T/F)**

3. Difference of two measurable sets is not necessarily measurable . **(T/F)**

4. If $G_1 \cap G \neq \phi$ then $| G_1 \cup G_2 | = |G_1| + |G_2|$ **(T/F)**

5. Cantor set is bounded, measurable and perfect . **(T/F)**

6. The family of all measurable sets has the cardinality 2^C. **(T/F)**

7. If $A \in [a, b]$ then $m(A) + m(A^{'}) = b - a$ **(T/F)**

8. If E_1 and E_2 are messurable subsets of $[2, 3]$ such that $m(E_1) = 1$ then $m(E_1 \cup E_1) = m(E_2)$. **(T/F)**

9. Borel set is not always measurable. **(T/F)**

10. F_σ and G_δ sets are not necessarily measurable. **(T/F)**

MULTIPLE CHOICE QUESTIONS

Choose the most appropriate one

1. If $\langle E_i \rangle$ is a sequence of Lebesgue measurable sets then Lebesgue measure $m(\cup E_i)$:
   (a) $\leq \Sigma m(E_i)$
   (b) $= \Sigma m(E_i)$
   (c) $> \Sigma m(E_i)$
   (d) None of the above

2. If E is a measurable set of finite measure, $\langle f_n \rangle$ is a sequence of measurable functions defined on E such that $f_n(x) \to f(x)$ then given $\varepsilon > 0$ $\exists \delta > 0$ such that there is a measurable set $A \subset E$ such that:
   (a) $m(A) < \delta$
   (b) $m(A) = \delta$
   (c) $m(A) > \delta$
   (d) None of the above

3. Which one of the following is true?
   (a) $]-\delta, b[$ is Lebesgue measurable set
   (b) $]-\delta, b[$ is not Lebesgue measurable set
   (c) $]-\delta, b[$ may be a Lebesgue measurable set
   (d) None of the above

4. Which one of the following is true?
   (a) $]a, \infty[$ is not Lebesgue integrable
   (b) $]a, \infty[$ is Lebesgue integrable
   (c) $m^*(a, \infty) = 0$
   (d) None the the above

5. If $<E_i>$ is a sequence of disjoint measurable sets and A is any set, then:
   (a) $m^*(A \cap \cup E_i) = \Sigma m^*(A \cap E_i)$
   (b) $m^*(A \cap \cup E_i) < \Sigma m^*(A \cap E_i)$
   (c) both (a) and (b) are true
   (d) None the the above

6. If f is a measurable function and $f = g$ almost everywhere then:
   (a) g is not measurable
   (b) g is measurable
   (c) $m(g) = 0$
   (d) None of the above

7. If E_1 and E_2 are Lebesgue measurable sets, then $m(E_1 \cup E_2)$ is equal to:
   (a) $m(E_1) + m(E_2)$
   (b) $m(E_1) + m(E_2) - m(E_1 \cap E_2)$
   (c) both (a) and (b)are true
   (d) None of the above

8. If A is countable then $m^*(A)$ is equal to :
(a) 1
(b) ∞
(c) 0
(d) None of the above

9. If Legesgue outer measure of a set E, such that $m^*(E) = 0$ then :
(a) E is not measurable
(b) E is measurable
(c) E is always empty
(d) None of the above

10. Let $\langle A_n \rangle$ be a countable collection of sets of real numbers then:
(a) $m^*(\cup A_n) \le \Sigma m^*(A_n)$
(b) $m^*(\cup A_n) = \Sigma m^*(A_n)$
(c) $m^*(\cup A_n) > \Sigma m^*(A_n)$
(d) None of the above

11. If $f_n(x) = \begin{cases} \dfrac{1}{x} & ; \quad n < x < n+1 \\ 0 & ; \quad \text{other wise} \end{cases}$

and if $f(x) = \lim_{x \to 0} f_n(x)$ then

$\int_0^\infty f(x)dx = \lim_{n \to \infty} \int_0^\infty f_n(x)\,dx$ follows by
(a) bounded convergence theorem
(b) dominated convergence theorem
(c) both (a) and (b) are true
(d) None of the above

12. Any countable set of points on the real axis has:
(a) no measure
(b) zero measure
(c) measure one
(d) None of the above

13. If f is integrable on $[a, b]$ and

$\int_a^x f(t)dt = 0 \ \forall x \in (a,b)$ then $f(t)=0$:
(a) almost everywhere
(b) nowhere
(c) both (a) and (b) are true
(d) None of the above

14. Let $\langle g_n \rangle$ be a sequence of integrable functions which converges almost everywhere to an integrable function g. If $\langle f_n \rangle$ is a sequence of measurable functions such that $|f_n| \le g_n$ and $\langle f_n \rangle$ converges to f a.e. If $\int g = \lim \int g_n$

then:
(a) $\int f > \lim \int f_n$
(b) $\int f = \lim \int f_n$
(c) $\int f < \lim \int f_n$
(d) None of the above

15. If f is measurable function then f is integrable over E if :
(a) f^+ and f^- are both integrable over E
(b) $\int_E f(x)dx = \int_E f^+dx + \int_E f^-dx$
(c) both (a) and (b) are true
(d) None of the above

16. If $f^+(x) = \max\{f(x); 0\}$ and

$f^-(x) = \max\{-f(x); 0\}$ then:
(a) $|f| = f^+ + f^-$
(b) $|f| = f^+ - f^-$
(c) both (a) and (b) are true
(d) None of the above

17. Let $\langle u_n \rangle$ be a sequence of non-negative measurable functions and $f = \displaystyle\sum_{n=1}^{\infty} u_n$ then:
(a) $\int f = \int u_n$
(b) $\int f = \displaystyle\sum_{n=1}^{\infty} \int u_n$
(c) both (a) and (b) are true
(d) None of the above

18. If f is a non-negative function and $\langle E_i \rangle$ a disjoint sequence of measurable sets. Let $E = \cup E_i$ then
(a) $\int_E f = \cup \int_{E_i} f$
(b) $\int_E f = \Sigma \int_{E_i} f$
(c) both
(d) None of the above

19. If f is a measurable function, then for $c \in R$
(a) $\int_E cf = c\int_E f$
(b) $\int_E cf < c\int_E f$
(c) both (a) and (b) are true
(d) None of the above

20. If f is a non-negative function which is integrable over a set E then given $\varepsilon > 0 \ \exists \ \delta > 0$ such that for every set $A \subset E$ with $mA > \delta$ we have
(a) $\int_A f < \varepsilon$
(b) $\int_A f > \varepsilon$

(c) $\int_A f = \varepsilon$

(d) None of the above

21. A non-negative measurable function f is integrable over the measurable set E if:

(a) $\int_E f = \infty$

(b) $\int_E f < \infty$

(c) $\int_E f < \infty$

(d) None of the above

22. If f and g are measurable functions and $f \leq g$ almost everywhere then:

(a) $\int_E f = \int_E g$ (b) $\int_E f \leq \int_E g$

(c) $\int_E f \geq \int_E g$ (d) None of these

23. If f and g are two non-negative measurable functions. If f is integrable over E and $g(x) < f(x)$ on E then:

(a) $\int_E f - g < \int_E f - \int_E g$

(b) $\int_E (f - g) = \int_E f - \int_E g$

(c) both (a) and (b) are true

(d) None of the above

24. If $\langle s_n \rangle$ is a sequence of non-negative measurable functions and $f_n(x) \to f(x)$ almost everywhere on a set E, then by Fatou's lemma:

(a) $\int_E f \leq \lim \int_E f_n$

(b) $\int_E f \leq \int_E f_n$

(c) $\int_E f \geq \lim \int_E f_n$

(d) None of the above

25. Let $\langle f_n \rangle$ be a sequence of measurable functions then:

(a) inf $\{f_1, f_2, \ldots f_n\}$ is measurable

(b) inf f_n is measurable

(c) both (a) and (b) are true

(d) None of the above

26. If $<f_n>$ is a sequence of measurable functions with same domain then:

(a) $\sup\{f_1, f_2, \ldots f_n\}$ is measurable

(b) $\sup_n f_n$ is measurable

(c) both (a) and (b) are true

(d) None of the above

27. which of the following is/are true

(a) Closed set is L-integrable

(b) Open set is L-integrable

(c) Both (a) and (b) are true

(d) None of the above

28. Which of the following is/are true?

(a) the set [0,1] is not countable

(b) $m^*([0, 1]) = 1$

(c) both (a) and (b) are true

(d) None of the above

29. Which of the following is/are true?

(a) Cantor set is perfect

(b) Cantor set is nowhere dense

(c) Measure of cantor set is zero

(d) All are true

30. Which of the following is/are true?

(a) Singleton set is measurable with measure zero

(b) Countable set is meaurable

(c) Both (a) and (b) are true

(d) None of the above

Answers

FILL IN THE BLANKS

1. ≥ 0 2. $b - a$ 3. unique   4. $|G|$
5. uncountable   6. its length   7. ≥ 0 8. 0   9. 0
10. measurable

TRUE OR FALSE

1. T   2. T   3. F   4. F   5. T   6. T   7. T   8. T   9. T
10. F

MULTIPLE CHOICE QUESTIONS

1. (b)	2. (a)	3. (a)	4. (b)	5. (a)	6. (b)	7. (b)	8. (b)	9. (b)
10. (a)	11. (b)	12. (b)	13. (a)	14. (b)	15. (c)	16. (a)	17. (b)	18. (b)
19. (a)	20. (a)	21. (b)	22. (b)	23. (b)	24. (a)	25. (c)	26. (c)	27. (c)
28. (c)	29. (d)	30. (c)						

SELF ASSESSMENT TEST

Verify each of the following:

1. If f is measurable function on $[a, b]$ and $k \in R$ then $f+k$, and kf are measurable.

2. Constant functions are measurable.

3. If $f(x) = \begin{cases} \dfrac{1}{x} & ; \text{ if } \quad 0 < x \leq 1 \\ 0 & ; \text{ if } \quad x = 0 \end{cases}$

   then f is not Lebesgue integrable on $[0, 1]$.

4. Bounded convergence theorem is applicable for the sequence of function

   $$f(x) = \frac{1}{\left(1 + \dfrac{x}{n}\right)^n}; 0 \leq x \leq 1, n \in N$$

5. For the function $f_n(x) = 2n$ for $x \in \left] \dfrac{1}{2n}, \dfrac{1}{n} \right[$

$n \in N$. The Fatou's Lemma holds but the Lebesgue dominated convergence theorem does not.

6. If f is integrable then f is finite value almost everywhere.

7. The function

   $$f(x) = \begin{cases} 0 & ; \quad \text{when } x \text{ is rational} \\ n & ; \quad \text{when } x \text{ is irrational} \end{cases}$$

8. Every non-empty open set has positive measure.

9. Every subset of a set of measure zero is of measure zero.

10. Every set with positive outer measure contains a non-measurable subsets.

Chapter

17 Functions of Bounded Variation

17.1 INTRODUCTION

We know that a real-valued function f defined on $[a, b]$ is said to be absolutely continuous on $[a, b]$, if for any arbitrary $\varepsilon > 0$, however small, $\exists$ a, $\delta > 0$, such that

$$\sum_{r=1}^{n} |f(b_r) - f(a_r)| < \varepsilon, \text{ whenever } \sum_{r=1}^{n} (b_r - a_r) < \delta.$$

where $a_1 < b_1 \le a_2 < b_2 \le \ldots \le a_n < b_n$ i.e., a_i's and b_i's are forming finite collection $\{(a_i, b_i) : i = 1, 2, \ldots, n\}]$ of pairwise disjoint (non-overlapping) intervals.

REMARKS

- Every absolutely continuous function is continuous.
- If a function satisfies $\sum |f(b_r) - f(a_r)| < \varepsilon$, even then it is absolutely continuous.

- The condition $\sum_{\lambda=1}^{n} (b_r - a_r) < \delta$ means that total length of all the intervals must be less than δ.

17.2 MONOTONIC FUNCTION

A function f defined on an interval I is said to be monotonically decreasing, iff
$$x > y \Rightarrow f(x) \le f(y), \forall \ x, y \in I$$
and monotonically increasing, iff $x > y \Rightarrow f(x) \ge f(y), \forall \ x, y \in I$
Also f is said to be strictly decreasing, iff $x > y \Rightarrow f(x) < f(y)$
and strictly increasing, iff $x > y \Rightarrow f(x) > f(y)$.

17.3 FUNCTIONS OF BOUNDED VARIATION

Let f be a real valued function defined on $[a, b]$ which is divided by means of points
$$a = x_0 < x_1 < x_2 < \ldots < x_n = b.$$
Then the set $P = \{x_0, x_1, x_2, \ldots, x_n\}$ is termed as subdivision or partition of $[a, b]$.

Let us take $\quad \underset{a}{\overset{b}{V}}(f, p) = \sum_{r=0}^{n-1} |f(x_{r+1}) - f(x_r)|.$ and $\quad \underset{a}{\overset{b}{V}}(f) = \underset{a}{\overset{b}{V}}(f, P)$ for all possible

subdivisions P of $[a, b]$. (This is called total variation of f over $[a, b]$ and also denoted by $\underset{a}{\overset{b}{T}}(f)$.)

If $\underset{a}{\overset{b}{V}}(f)$ is finite, then f is called a function of bounded variation or function of finite variation over $[a, b]$.

Set of all the functions of bounded variation on $[a, b]$ is denoted by $BV[a, b]$.

REMARK

- If f is defined on R, then we define $\overset{\infty}{\underset{-\infty}{V}}(f) = \lim_{a \to \infty} \overset{a}{\underset{-a}{V}}(f)$.

17.3.1 SOME RESULTS ON FUNCTIONS OF BOUNDED VARIATION

Let $f: [a, b] \to R$ and P be any subdivision of $[a, b]$. Then

(1) $|f(x) - f(a)| \leq \overset{x}{\underset{a}{V}}(f), x \in [a, b]$.

(2) $\overset{a}{\underset{a}{V}}(f) = 0$.

(3) $P_1 \subset P_2 \Rightarrow \overset{b}{\underset{a}{V}}(f, P_1) \leq \overset{b}{\underset{a}{V}}(f, P_2)$ where P_1 and P_2 are any two subdivisions of $[a, b]$.

(4) $\overset{b}{\underset{a}{V}}(f, P) \leq \overset{b}{\underset{a}{V}}(f)$, for all subdivisions P of $[a, b]$.

(5) For each $\varepsilon > 0$, however small, at least one subdivision P' of $[a, b[$ s.t. $\overset{b}{\underset{a}{V}}(f, P') + \varepsilon > \overset{b}{\underset{a}{V}}(f)$.

(6) $\overset{b}{\underset{a}{V}}(f) \geq 0$.

(7) $a < b < c \Rightarrow \overset{b}{\underset{a}{V}}(f) < \overset{c}{\underset{b}{V}}(f)$

17.3.2 LIPSCHITZ CONDITION

A function f defined on $[a, b]$ is said to satisfy Lipschitz-condition, if $\exists$ a constant $M > 0$ s.t.
$$|f(x) - f(y)| \leq M|x - y|, \ \forall \ x, y \in [a, b].$$

17.3.3 LEBESGUE POINT OF A FUNCTION

A point x is said to be a Lebesgue point of the function $f(t)$, if
$$\lim_{h \to 0} \frac{1}{h} \int_x^{x+h} |f(t) - f(x)| dt = 0 \cdot$$

17.3.4 COVERING IN THE SENSE OF VITALI

A set E is said to be covered in the sense of Vitali by a family of intervals (may be open, closed or half open), M in which none is a singleton set, if every point of the set E is contained in some small interval of M i.e., for each $x \in E$, $\exists \ \varepsilon > 0$, an interval $I \in M$ s.t. $x \in I$ and $l(I) < \varepsilon$.

The family M is called the Vitali Cover of set E.

For example, if $E = \{q : q$ is a rational number in the interval $[a, b]\}$, then the family $\{I_{qi}\}$ where $I_{qi} = \left[q - \dfrac{1}{i}, q + \dfrac{1}{i}\right], i \in N$ is a vitali cover of $[a, b]$.

THEOREM 1. **(Vitali's covering theorem)** *Let E be a set of finite outer measure and M be a family of intervals which cover E in the sense of Vitali; then for a given $\varepsilon > 0$, it is possible to find a finite family of disjoint intervals $\{I_k, k = 1, 2, ..., n\}$ of M, such that*

$$m^* \left[E - \overset{n}{\underset{k=1}{\cup}} I_k \right] < \varepsilon.$$

Proof. Let us assume that every interval of family M is a closed interval, because if not we replace each interval by its closure and observer that the set of end points of $I_1, I_2, ... J_n$ has measure zero.

Suppose O is an open set containing E s.t. $m^*(O) < m^*(E) + 1 < \infty$ we assume that each interval in M is contained in O, if this can be achieved by discarding the intervals of M extending beyond O and still the family M will cover the set E in the sense of Vitali.

Now we shall use the induction method to determine the sequence $<I_k : k = 1, 2,...,n>$ of disjoint intervals of M as follows :

Let I_1 be any interval in M and let l_1 be the supremum of the lengths of the intervals in M disjoint from I_1. Then $l_1 < \infty$ as $I_1 \leq m(O) < \infty$.

Now select an interval I_2 from M, disjoint from I_1, such that $l(I_2) > \frac{1}{2} l_1$. Let l_2 be the supremum of lengths of all those intervals of M which do not have any point common with I_1 or I_2. Obviously $l_2 < \infty$. In general, suppose we have already chosen r mutually disjoint intervals $I_1, I_2,..I_r$. Let l_r be the supremum of the lengths of those intervals of M which do not have any point in common with $\underset{i=1}{\cup} I_i$. Then

$$l_r \leq m(0) < \infty.$$

Now if E is contained in $\underset{i=1}{\overset{r}{\cup}} I_i$, then theorem is proved.

Suppose $\underset{i=1}{\overset{r}{\cup}} I_i \subset E$. Then we can find interval I_{r+1}, s.t. $l(I_{r+1}) > \frac{1}{2} l_r$ which is disjoint from $I_1, I_2,...,I_r$. Thus at some finite iteration either the theorem will be established or we shall get an infinite sequence $<I_r>$ of disjoint intervals of M s.t.

$l(I_{r+1}) > \frac{1}{2} l_r$ and $I_r < \infty$, $n = 1,2,3,....$

Since, $<I_r>$ is a monotonically decreasing sequence of non-negative real numbers. Obviously, we have that $\underset{i=1}{\overset{r}{\cup}} I_r \subset O \Rightarrow \overset{\infty}{\underset{r=1}{\sum}} l(I_r) \leq m(0) < \infty.$

Hence for any arbitrary $\varepsilon > 0$, we can find an integer N s.t.

$$\overset{\infty}{\underset{r=N+1}{\sum}} l(I_r) < \frac{1}{5} \varepsilon.$$

Let us set
$$F = E - \overset{N}{\underset{r=1}{\cup}} I_r$$

Now, we show that $m^*(F) < \varepsilon$.

Let $x \in F$, then $x \notin \overset{N}{\underset{r=1}{\cup}} I_r \Rightarrow x$ is an element of E not belonging to the closed set $\overset{N}{\underset{r=1}{\cup}} I_r$

$\Rightarrow \exists$ an interval I in M s.t. $x \in I$ and $l(I)$ is so small that I does not meet the $\overset{N}{\underset{r=1}{\cup}} I_r$

i.e., $I \cap I_r = \phi$, $\forall r = 1, 2,..., N$.

Therefore we shall have $l(I) < l_N < 2l(I_{n+1})$. If also $I \cap I_{N+1} = \phi$, we should have $l(I) \leq l_{N+1}$.

If the interval I does not meet any of the intervals in the sequence $<I_r>$, then
$$l(I) \leq I_r, \forall r$$

which is not true as $I_r < 2l(I_{r+1}) \to 0$ as $r \to \infty$

$\Rightarrow$ f must meet at least one of the intervals of the sequence $<I_r>$.

Let p be the least integer such that I meets I_p. Then $p>N$ and $l(I) \le l_{p-1} < 2l(I_p)$. Further let $x \in I$ as well $x \in I_p$, then the distance of x from the mid point of I_p is at most

$$l(I) + \frac{1}{2} l(I_p) < 2l\ (I_p) + \frac{1}{2} l(I_p) = \frac{5}{2} l(I_p).$$

Therefore, if I_p is an interval having ihe same mid point as I_p but length 5 times the length of I_p i.e., $l(J_p) = 5l(I_p)$ Then $x \in J_p$ also.

$\Rightarrow$ For every $x \in F$, $\exists$ an integer $p > N$ s.t. $x \in J_p$ and $l(J_p)=5l(I_p)$.

Also

$$F \subset \bigcup_{N+1}^{\infty} J_p$$

$\Rightarrow$

$$m^*(F) \le \sum_{p=N+1}^{\infty} l\left(J_p\right) = 5 \sum_{p=N+1}^{\infty} l\left(I_p\right) < 5.\frac{\varepsilon}{5} = \varepsilon$$

17.4 INDEFINITE INTEGRAL

Let $f(x)$ be L-integrable over $[a,b]$; then the function $F(x)$ defined by

$$F(x) = \int_a^x f(t)dt + c, \ \forall \ x \in [a, b] \text{ and } c \text{ is any constant}$$

is called an indefinite integral of $f(x)$.

17.5 FOUR DINI'S DERIVATIVES

Dini's derivatives, may be defined even at the points where the function is not differentiable. Now,

(1) $D^+ f(x) = \overline{\underset{h \to 0+}{\text{Lim}}} \dfrac{f(x+h)-f(x)}{h}$, called upper right derivative

(2) $D_+ f(x) = \underset{h \to 0+}{\underline{\text{Lim}}} \dfrac{f(x+h)-f(x)}{h}$, called lower right derivative

(3) $D^- f(x) = \overline{\underset{h \to 0-}{\text{Lim}}} \dfrac{f(x+h)-f(x)}{h}, = \overline{\underset{h \to 0+}{\text{Lim}}} \dfrac{f(x-h)-f(x)}{-h}$, called upper left derivative.

(4) $D_- f(x) = \underset{h \to 0-}{\underline{\text{Lim}}} \dfrac{f(x+h)-f(x)}{h}, = \underline{\underset{h \to 0+}{\text{Lim}}} \dfrac{f(x-h)-f(x)}{-h}$, called lower left derivative.

REMARKS

- $D^+ f(x) \ge D_+ f(x)$ and $\overline{D} f(x) \ge \underline{D} f(x)$.
- If $D^+ f(x) = D_+ f(x)$, then we conclude that right hand derivative of $f(x)$ exists at the point x and denoted by $f'(x+)$. Similarly if $D^- f(x) = D_- f(x)$, we say that $f(x)$ is left differentiable at x and denote this common value by $f'(x-)$.
- The function is said to be differentiable at x if all the four Dini's derivatives are equal but different from $\pm\infty$ i.e., if $D^+ f(x) = D_+ f(x) = D^- f(x) = D_- f(x) \ne \pm\infty$ and their common value is denoted by $f'(x)$.

17.5.1 PROPERTIES OF DINI'S DERIVATIVES

(1) Dini's derivatives always exist, may be finite or infinite for every function f.

(2) $D^+(f+g) \le D^+f + D^+g$

(3) If f and g are continuous at a point 'x', then $D^+(f.g)(x) \le f(x)D^+g(x) + g(x)D^+f(x)$

(4) $D_+f(x) = -D^+(-f(x))$ and $D_-f(x) = -D^-(-f(x))$.

(5) If f is a continuous function on $[a, b]$ and one of its derivatives (say D^+) is non-negative on (a, b). Then f is non-decreasing on $[a, b]$ i.e., $f(x) \le f(y)$ whenever $x \le y$, $x, y \in [a, b]$.

(6) If f is any function on an interval $[a, b]$, then the four derivatives if exist are measurable.

THEOREM 1. *A monotonic function on $[a,b]$ is of bounded variation.*

Proof.   Let $[a, b]$ be the given interval. Divide the interval $[a, b]$ by means of points $a = x_0 < x_1 < x_2 < \dots < x_n = b$.

Let $f(x)$ be a decreasing function on $[a, b]$. Since, if f is a decreasing function, $-f$ is an increasing function and so by taking $-f = g$, we see that g is an increasing function and so we are allowed to consider only increasing functions. So,

$$x_r < x_{r+1} \implies f(x_r) \le f(x_{r-1})$$
$$\implies f(x_{r+1}) - f(x_r) \ge 0$$
$$\implies |f(x_{r+1}) - f(x_r)| = f(x_{r+1}) - f(x_r). \qquad \dots (1)$$

Now $V = \sum_{r=0}^{n-1} |f(x_{r+1}) - f(x_r)| = \sum_{r=0}^{n-1} \{f(x_{r+1}) - f(x_r)\}$ [Using (1)]

$\therefore$ $V = f(x_n) - f(x_0) = f(b) - f(a)$.

Here, $f(b)$ and $f(a)$ are finite quantities.

$\implies$ $V = a$ finite quantity. Hence f is of bounded variation.

REMARK

- If f is a monotonic function on $[a, b]$, then $\overset{b}{\underset{a}{T}}(f) = |f(b) - f(a)|$.

THEOREM 2. *Let V, P, N denote total, positive and negative variations of a bounded function f on $[a, b]$; then*
$$V = P + N, \quad and \quad P - N = f(b) - f(a).$$

Proof.   Divide the interval $[a, b]$ by means of points $a = x_0 < x_1 < x_2 < \dots < x_n = b$.

Let, $v = \sum_{r=0}^{n-1} |f(x_{r+1}) - f(x_r)|$.

If p denotes the sum of those differences $f(x_{r+1}) - f(x_r)$ which are positive and $-n$ for negative, then clearly,

$$v = p + n, f(b) - f(a) = p - n. \qquad \dots (1)$$

Suppose that $p = \sup p, V = \sup v, N = \sup n$, $\dots (2)$

From (1), we have

$$v = 2p + f(a) - f(b), \qquad \dots (3)$$

$$v = 2n + f(b) - f(a). \qquad \dots (4)$$

Taking supremum of (3) and (4) and using (2), we get

$$V = 2P + f(a) - f(b), \qquad \qquad ...(5)$$

$$V = 2N + f(b) - f(a). \qquad \qquad ...(6)$$

Using, (5) and (6) we get

$$V = P + N \text{ and } f(b) - f(a) = P - N.$$

THEOREM 3. *If f_1 and f_2 are non-decreasing functions on $[a,b]$, then $f_1 - f_2$ is of bounded variation on $[a, b]$.*

Proof. Let $f = f_1 - f_2$ defined on $[a, b]$. Then for any partition $P = [a = x_0, x_1, ..., x_n = b]$, we have

$$\Sigma |f(x_i) - f(x_{i-1})| \le [f_1(x_i) - f_1(x_{i-1})] + [f_2(x_i) - f_2(x_{i-1})]$$

$$\le [f_1(b) - f_1(a)] + [f_2(b) - f_2(a)]$$

$$(\because f_1 \text{ and } f_2 \text{ are monotonically increasing})$$

$$\Rightarrow \quad \overset{b}{\underset{a}{V}} \le f_1(b) + f_2(b) - f_2(a) - f_1(a), \text{ which is a finite quantity.}$$

$$\Rightarrow \quad \overset{b}{\underset{a}{V}}(f) < \infty \text{ and hence } f \text{ is of bounded variation.}$$

REMARKS

- If $f \in BV\,[a, b]$ and $c \in (a, b)$, then $f \in BV\,[a, c]$ and $f \in BV\,[c, b]$. Also

$$\overset{b}{\underset{a}{V}}(f) = \overset{c}{\underset{a}{V}}(f) + \overset{b}{\underset{c}{V}}(f).$$

- We can define a new function (called variation function) such that

$$v(x) = \overset{b}{\underset{a}{V}}(f), \ \forall \, x \in [a, b]$$

- If $x < y$ in $[a, b]$, then $\overset{y}{\underset{a}{V}}(f) = \overset{x}{\underset{a}{V}}(f) + \overset{y}{\underset{x}{V}}(f)$ i.e., $v(y) = v(x) + \overset{y}{\underset{x}{V}}(f)$

$$\Rightarrow v(x) \text{ is an increasing function.}$$

- If $a < c_1 < c_2 < ... < c_n < b$, then $\overset{b}{\underset{a}{V}}(f) = \overset{c_1}{\underset{a}{V}}(f) + \overset{c_2}{\underset{c_1}{V}}(f) + ... \overset{b}{\underset{c_n}{V}}(f)$.

- $f \in BV\,[a, b] \Leftrightarrow f \in BV\,[a, c], f \in BV\,[c, b]$ for each $c \in [a, b]$.

THEOREM 4. *Let f be a function of bounded variation in $[a,b]$ and is continuous at $c \in [a, b]$, then the function defined by $v(x) = \overset{x}{\underset{a}{V}}\,(f)$, is also continuous at $x = c$ and vice-versa.*

Proof. Let f be continuous at $x = c$. Then for arbitrary $\varepsilon > 0$, we can find a $\delta_1 > 0$ s.t.

$$a \le c - \delta_1 < x < c \text{ or } |x - c| < \delta_1 \Rightarrow |f(x) - f(c)| < \varepsilon/2. \qquad ...(1)$$

Now, we can get a subdivision

$$P = \{a = x_0, x_1, x_2, ..., x_n = c] \text{ of } [a, c] \text{ s.t } \overset{c}{\underset{a}{V}}(f) < \overset{c}{\underset{a}{V}}(f, P) + \varepsilon/2. \qquad ...(2)$$

Now choosing positive $\delta < \min\,[\delta_1, c - x_{n-1})$, we get for any x s.t. $c - \delta < x < c$, also we have $x_{n-1} < x < x_n$.

Now, (2) implies $\overset{c}{\underset{a}{V}}(f) < \sum_{r=1}^{n-1} |f(x_r) - f(x_{r-1})| + |f(x_n) - f(x_{n-1})| + \dfrac{1}{2}\varepsilon$

$$< \sum_{r=1}^{n-1} |f(x_r) - f(x_{r-1})| + |f(x_n) - f(x) + f(x) - f(x_{n-1})| + \dfrac{1}{2}\varepsilon$$

$$< \sum_{r=1}^{n-1} |f(x_r) - f(x_{r-1})| + |f(x) - f(x_{n-1})| + |f(x_n) - f(x)| + \dfrac{1}{2}\varepsilon$$

$$< \overset{x}{\underset{a}{V}}(f) + |f(c) - f(x)| + \dfrac{1}{2}\varepsilon,$$

$\Rightarrow \qquad \overset{c}{\underset{a}{V}}(f) < \overset{x}{\underset{a}{V}}(f) + \varepsilon,$ i.e., $0 \le \overset{c}{\underset{a}{V}}(f) - \overset{x}{\underset{a}{V}}(f) < \varepsilon$ By (1)

$\Rightarrow$ For $c - \delta < x < c$, $v(c) - v(x) < \varepsilon$.

$\Rightarrow \qquad \lim_{x \to c-0} v(x) = v(c).$

Hence, $v(x)$ is continuous on the left at $x = c$.

Similarly considering the partition of $[c, b]$, we can show that $v(x)$ is right continuous also at $x = c$ and hence $v(x)$ is also continuous at $x = c$.

Conversely, let $v(x)$ be coniinuous at $x = c$, then for arbitrary small $\varepsilon > 0$, $\exists$ a $\delta > 0$ s.t,

$$|v(x) - v(c)| < \varepsilon, x \in \,]c - \delta, c + \delta[. \qquad \qquad ...(1)$$

Now let $c < x < c + \delta$. Then, we get

$$\overset{x}{\underset{a}{V}}(f) = \overset{c}{\underset{a}{V}}(f) + \overset{x}{\underset{c}{V}}(f) \Rightarrow v(x) = v(c) + \overset{x}{\underset{c}{V}}(f)$$

$\Rightarrow \qquad v(x) - v(c) = \overset{x}{\underset{c}{V}}(f) \ge |f(x) - f(c)|$

$\Rightarrow \qquad |f(x) - f(c)| \le |v(x) - v(c)| < \varepsilon \qquad \qquad ...(2)$

Similarly we can show that $|f(c) - f(x)| < \varepsilon$, if $c - \delta < x < c$. $...(3)$

Hence, from (2) and (3), $f(x)$ is also continuous at $x = c$.

THEOREM 5. *Let f and g be functions of bounded variation on $[a, b]$; then $f+g$, $f-g$, fg and f/g ($|g(x)| \ge \sigma > 0$, $\forall x$) and cf are functions of bounded variation, c being constant.*

Proof. (i) Let $f + g = h$, then

$$|h(x_{r+1}) - h(x_r)| = |[f(x_{r+1}) + g(x_{r+1})] - [f(x_r) + g(x_r)]|$$

$$\text{where } a = x_0 < x_1 < x_2 < ... < x_n = b$$

$$= |[f(x_{r+1}) - f(x_r)] + [g(x_{r+1}) - g(x_r)]|$$

$$\le |f(x_{r+1}) - f(x_r)| + |g(x_{r+1}) - g(x_r)|.$$

$\Rightarrow \quad \sum_{r=0}^{n-1} |h(x_{r+1}) - h(x_r)| \le \sum_{r=0}^{n-1} |f(x_{r+1}) - f(x_r)| + \sum_{r=0}^{n-1} |g(x_{r+1}) - g(x_r)|$

or $\qquad \overset{b}{\underset{a}{V}}(h) \le \overset{b}{\underset{a}{V}}(f) + \overset{b}{\underset{a}{V}}(g).$

Now since, f, g are functions of bounded variation, therefore

$$\overset{b}{\underset{a}{V}}(f) \text{ and } \overset{b}{\underset{a}{V}}(g) \text{ are finite} \Rightarrow \overset{b}{\underset{a}{V}}(h) = a \text{ finite quantity.}$$

Hence $h=f+g$ is of bounded variation in $[a, b]$.

(ii) Let $h = f-g$. Then proceed same as above,

$$|h(x_{r+1})-h(x_r)| \le |f(x_{r+1})-f(x_r)| + |g(x_{r+1})-g(x_r)|.$$

$$\Rightarrow \qquad \overset{b}{\underset{a}{V}}(h) \le \overset{b}{\underset{a}{V}}(f) + \overset{b}{\underset{a}{V}}(g).$$

$$\Rightarrow \qquad \overset{b}{\underset{a}{V}}(h) = a \text{ finite quantity.}$$

Hence $f-g$ is of bounded variation in $[a, b]$.

(iii) Let $h(x) = f(x) \cdot g(x)$. Then

$$\begin{aligned}
|h(x_{r+1})-h(x_r)| &= |f(x_{r+1}).g(x_{r+1})-f(x_r).g(x_r)| \\
&= |f(x_{r+1}).g(x_{r+1}) - f(x_r).g(x_{r+1}) + f(x_r).g(x_{r+1}) \\
&\qquad\qquad\qquad\qquad\qquad\qquad -f(x_r).g(x_r)| \\
&\le |g(x_{r+1})| |[f(x_{r+1})-f(x_r)]| + |f(x_r)[g(x_{r+1})-f(x_r)]| \\
&\le |g(x_{r+1})| . |f(x_{r+1})-f(x_r)| + |f(x_r).[g(x_{r+1})-f(x_r)]|
\end{aligned}$$

Let $\qquad A = \sup\{|f(x)| : x \in [a, b]\}, B=\sup\{|g(x)| : x \in [a, b]\}.$

Then $|h(x_{r+1})-h(x_r)| \le B.|f(x_{r+1})-f(x_r)| +A.|g(x_{r+1})-g(x_r)|.$

$$\Rightarrow \sum_{r=0}^{n-1}|h(x_{r+1})-h(x_r)| \le B.\sum_{r=0}^{n-1}|f(x_{r+1})-f(x_r)| + A.\sum_{r=0}^{n-1}|g(x_{r+1})-g(x_r)|$$

$$\Rightarrow \qquad \overset{b}{\underset{a}{V}}(h) \le B.\overset{b}{\underset{a}{V}}(f) + A.\overset{b}{\underset{a}{V}}(g) = a \text{ finite quantity.}$$

Hence $f(x) \cdot g(x)$ is of bounded variation in $[a, b]$.

(iv) Firstly, we shall show that $1/g$ is of bounded variation, where $g(x) \ge \sigma > 0$, $\forall$ $x \in [a, b]$.

$$g(x) \ge \sigma > 0, \forall x \in [a, b] \Rightarrow \frac{1}{g(x)} \le \frac{1}{\sigma} > 0, \forall x \in [a, b].$$

Now, we have

$$\left|\frac{1}{g(x_{r+1})} - \frac{1}{g(x_r)}\right| = \left|\frac{g(x_r)-g(x_{r+1})}{g(x_r)g(x_{r+1})}\right| \le \frac{1}{\sigma^2}|g(x_r)-g(x_{r+1})|$$

$$\therefore \qquad \sum_{r=0}^{n-1}\left|\frac{1}{g(x_{r+1})} - \frac{1}{g(x_r)}\right| \le \frac{1}{\sigma^2} \sum_{r=0}^{n-1}|g(x_r)-g(x_{r+1})|$$

$$\Rightarrow \qquad \overset{b}{\underset{a}{V}}\left(\frac{1}{g}\right) \le \frac{1}{\sigma^2} \overset{b}{\underset{a}{V}}(g) = a \text{ finite quantity.}$$

Hence $\dfrac{1}{g}$ is of bounded variation in $[a, b]$.

Now f and $\dfrac{1}{g}$ are of bounded variation in $[a, b]$.

$$\Rightarrow \qquad f.\frac{1}{g} \text{ is bounded variation in } [a, b]. \qquad\qquad \text{[By case (iii)]}$$

$\Rightarrow \quad \dfrac{f}{g}$ is of bounded variation in $[a, b]$.

(v) Left for the reader

REMARK

- If p, q are any constants, then $f, g \leftarrow BV[a, b] \Rightarrow af+bg \in BV[a, b]$.

THEOREM 6. *Every absolutely continuous function f defined on $[a, b]$ is of bounded variation.*

Proof. Since f is absolutely continuous on $[a, b]$ then for $\varepsilon > 0, \exists$ a $\delta>0$

s.t. $$\sum_{i=1}^{n} |f(b_i)- f(a_i)| < 1. \qquad\qquad (\because \text{ take } \varepsilon = 1)$$

whenever $$\sum_{i=1}^{n} |(b_i)-(a_i)| < \delta,$$

and $$a=a_1 < b_1 \le a_2 < b_2 \le \dots \le a_n < b_n = b.$$

Now consider another subdivision of $[a, b]$ or refinement of P by adjoining additional points to P in such a way that all the intervals can be divided into r parts of total length less than δ.

Let the r sub-intervals be $[c_0, c_1], [c_1, c_2],.., [c_{r-1}, c_r]$ s.t.

$$a = c_0, c_r = b \text{ and } (c_{k+1}, c_k)<\delta, \forall\, k= 1, 2,..., r$$

Then, $$\sum_i |f(x_{i+1})- f(x_i)| < 1, , \text{ where } x_{i+1}, x_i \in [c_k, c_{k+1}]$$

or $$\overset{c_{k+1}}{\underset{c_k}{V}} (f)< 1,$$

Hence $$\overset{b}{\underset{a}{V}}(f) = \overset{c_1}{\underset{c_0}{V}}(f) + \overset{c_2}{\underset{c_1}{V}}(f) + \dots \overset{c_r}{\underset{c_{r-1}}{V}}(f) < 1+1+1+\dots+1 = r$$

$$= \text{A finite quantity}$$

Hence f is of bounded variation.

REMARK

- There extsts functions of bounded variation but not absolutely continuous.

THEOREM 7. **(Jordan Decomposition Theorem).** *A function f is of bounded variation, if and only if it can be expressed as a difference of two monotonic functions both non-decreasing.*

Proof. Let $f\colon [a, b]\to R$ be the function. We have the following cases :

Case I. $f \in BV [a, b]$. Then we can write $f = v-(v-f),$ ...(1)

so that $$f(x) = v(x)-(v(x)-f(x)), x \in [a, b].$$

Now, if $x, y \in [a, b]$ s.t. $x < y$, then, we get

$$\overset{y}{\underset{a}{V}}(f) = \overset{x}{\underset{a}{V}}(f) + \overset{y}{\underset{x}{V}}(f) \quad \Rightarrow \quad v(y)-v(x) = \overset{y}{\underset{x}{V}}(f) \ge 0$$

$$\Rightarrow \quad v(y) \le v(x) \text{ and hence } v \text{ is a non-decreasing function on } [a, b].$$

Again, if $x<y$ in $[a, b]$, then

$$v(y)-v(x) = \overset{y}{\underset{x}{V}}(f) \geq \left|f(y)-f(x)\right| \geq f(y)-f(x)$$

$$\Rightarrow \qquad v(y)-f(y) \geq v(x)-f(x) \quad \Rightarrow \qquad (v-f)y \geq (v-f)x$$

$\Rightarrow$ $v-f$ is also a non-decreasing function on $[a, b]$.

Hence we conclude that f is expressible as a difference of two monotonically non-decreasing functions.

Case II. If $g(x)$ and $h(x)$ are two increasing functions such that $f(x)=g(x)-h(x)$.

Divide the closed interval $[a, b]$ such that $a = x_0 < x_1 < x_2 < \ldots < x_n = b$.

Let
$$V = \sum_{r=0}^{n-1} \left|f(x_{r+1})-f(x_r)\right|$$

Now,

$$\begin{aligned}
|f(x_{r+1})-f(x_r)| &= |g(x_{r+1})-h(x_{r+1})-(g(x_r)-h(x_r))| \\
&= |[g(x_{r+1})-g(x_r)]+[h(x_r)-h(x_{r+1})| \\
&\leq |g(x_{r+1})-g(x_r)|+| h(x_r)-h(x_{r+1})| \\
&\leq |g(x_{r+1})-g(x_r)|+| h(x_{r+1})-h(x_r)|.
\end{aligned}$$

Since, $g(x)$ and $h(x)$ are monotonically increasing functions, so that $g(x_{r+1})-g(x_r) \geq 0$ and $h(x_{r+1})-h(x_r) \geq 0$; therefore

$$|g(x_{r+1})-g(x_r)|=g(x_{r+1})-g(x_r)$$

and
$$|h(x_{r+1})-h(x_r)|=h(x_{r+1})-h(x_r).$$

Hence,
$$|f(x_{r+1})-f(x_r)| \leq [g(x_{r+1})-g(x_r)]+[h(x_{r+1})-h(x_r)].$$

$\therefore$
$$\sum_{r=0}^{n-1}\left|f(x_{r+1})-f(x_r)\right| \leq \sum_{r=0}^{n-1}\left|g(x_{r+1})-g(x_r)\right|+\sum_{r=0}^{n-1}\left|h(x_{r+1})-h(x_r)\right|.$$

Now
$$\sum_{r=0}^{n-1}\left[g(x_{r+1})-g(x_r)\right]=[g(x_1)-g(x_0)]+[g(x_2)-g(x_1)]+\ldots+[g(x_n)-g(x_{n-1})]$$
$$=g(x_n)-g(x_0)=g(b)-g(a)$$

Similarly, $\displaystyle\sum_{r=0}^{n-1}\left[h(x_{r+1})-h(x_r)\right]= h(b)-h(a)$.

Therefore, $\displaystyle\sum_{r=0}^{n-1}\left|f(x_{r+1})-f(x_r)\right| \leq [\, g(b)-g(a)+h(b)-h(a)].$

Now, since f is finite in $[a, b]$ therefore $g(b), g(a), h(b), h(a)$ are finite.

Hence, $\displaystyle\sum_{r=0}^{n-1}\left|f(x_{r+1})-f(x_r)\right| <\infty \Rightarrow \overset{b}{\underset{a}{V}}(f) <\infty.$

$\Rightarrow$ f is a function of bounded variation.

REMARK

- A continuous function is of bounded variation iff it can be expressed as a difference of two continuous monotonically increasing functions.

THEOREM 8. *Let f be a function of bounded variation, then f′(x) exists a.e.*

Proof. Since f is of bounded variation, then it can be expressed as a difference of two monotonic non-decreasing functions. We know that a monotonic non-decreasing function is always differentiable a.e., i.e., if $f(x)=g(x)-h(x)$, the $g'(x)$ and $h'(x)$ exists a.e. and hence $f'(x)=g'(x)-h'(x)$, also exists a.e.

THEOREM 9. *If f is a finite-valued monotonic increasing function defined on* $[a, b]$, *then f is continuous on* $[a, b]$ *except atmost a countable number of points.*

Proof.   By definition of jump function $\delta f(x)$ of $f(x)$, we have
$$\delta f(x) = \inf f(x + h) - \sup f(x-h), \ h>0.$$
Since f is monotonic increasing, therefore $\delta f(x)>0$.

We know that f is continuous at x iff $\delta f(x) = 0$.

Also $\displaystyle\sum_{x_i\in[a,b]} \delta f(x_i) \le f(b) - f(a)$ where $\langle x_i \rangle$ is any sequence of points in $[a, b]$.

$\Rightarrow$ the set $E_n=\{x : \delta f(x)>1/n\}$ contains atmost $n\,[f(b)-f(a)]$ points.

$\Rightarrow$ set E_n is countable, for all n.

Since the set of points of discontinuity is $E = \displaystyle\bigcup_{n=1}^{\infty} E_n$ which is a countable union of countable sets and hence set E is countable.

REMARKS

- If $f \in BV[a, b]$, then f is continuous a.e. on $[a, b]$.
- An absolutely contiuous function is differentiable a.e.

THEOREM 10. *An indefinite integral is a function of bounded variation.*

Proof.   Let $f \in L[a, b]$ and $F(x)$ is indefinite integral of $f(x)$, i.e., $F(x)=\int_a^x f(t)dt$,then $F \in BV[a, b]$. And
$$\overset{b}{\underset{a}{V}}(F)\le \int_a^x |f|.$$
Since $f \in L\,[a, b]$,   also $|f|\in L[a,b]$.   Let $P=\{x_i, : i = 0, 1. 2, ... n\}$ be a subdivision of the interval $[a, b\}$. Then
$$\sum_{i=1}^{n}|F(x_i)-F(x_{i-1})| = \sum_{i=1}^{n}\left|\int_a^{x_i} f - \int_a^{x_{i-1}} f\right|$$
$$= \sum_{i=1}^{n}\left|\int_{x_{i-1}}^{x_i} f\right| \le \sum_{i=1}^{n}\int_{x_{i-1}}^{x_i} |f| = \int_a^b |f| < \infty.$$

$\Rightarrow$ $F \in BV[a, b]$ and $\overset{b}{\underset{a}{V}}(F, P) \le \int_a^b |f|$. which is true for any subdivision of P of $[a, b]$.

Hence taking supremum, we get $\overset{b}{\underset{a}{V}}(F)\le \int_a^b |f|$.

THEOREM 11. *If* $f \in B\,V[a, b]$, *then f is absolutely continuous on* $[a, b]$ *iff the variation function*
$$v(x)= \overset{x}{\underset{a}{V}}\,(f)\ \text{is absolutely continuous an}\ [a, b].$$

Proof.   Let us first suppose $v(x)$ is absolutely continuous.

Then, by definition for arbitrary $\varepsilon> 0, \exists\ \delta>0$ s.t.
$$\sum_{r=1}^{n}|v(b_r)-v(a_r)| <\varepsilon, \text{ whenever } \sum_{r=1}^{n}|b_r - a_r| < \delta$$

Also $\quad |f(x)-f(a)| \le \overset{x}{\underset{a}{V}}(f) = v(x)$

So, $\quad \displaystyle\sum_{r=1}^{n}\left|f(b_r)-f(a_r)\right| = \sum_{r=1}^{n}\left|f(b_r)-f(a)+f(a)-f(a_r)\right|$

$$= \sum_{r=1}^{n}\left|\left[f(b_r)-f(a)\right]+\left[f(a_r)-f(a)\right]\right|$$

$$\le \sum_{r=1}^{n}\left|v(b_r)+v(a_r)\right| < \varepsilon \text{ whenever } \sum_{r-1}^{n}\left|b_r-a_r\right| < \delta$$

$\Rightarrow$ f is also absolutely continuous on $[a, b]$.

Conversely let us suppose that f is absolutely continuous on $[a, b]$,

Then, for a given $\varepsilon > 0$, $\exists$ a $\delta > 0$ s.t. $\displaystyle\sum_{r=1}^{n}\left|f(b_i)-f(a_i)\right| < \varepsilon$, ...(1)

for every finite collection $P\{]a_i, b_i[, i=1, 2,..., n\}$ of pairwise disjoint sub-intervals of $[a, b]$ such that

$$\sum_{r=1}^{n}\left|(b_i-a_i)\right| < \delta.$$

Now, let $P_i=\{]x^i_{k-1}, x^i_k[, k=1,..., m_i\}$ be a finite collection of non-overlapping intervals of the interval $[a_i, b_i]$.

Then the collection $\{]x^i_{k-1}, x^i_k[:i=1, 2, ..., n, k=1,..., m_i\}$

is a finite collection of non-overlapping sub-intervals of $[a, b]$ s.t.

$$\sum_{i=1}^{n}\left[\sum_{k=1}^{m_i}\left(x^i_k - x^i_{k-1}\right)\right] = \sum_{i=1}^{n}(b_i-a_i) < \delta,$$

$\Rightarrow \quad \displaystyle\sum_{i=1}^{n}\sum_{k=1}^{m_i}\left|f\left(x^i_k\right)-f\left(x^i_{k-1}\right)\right| < \varepsilon.$ (Using (1))

Taking supremum over all collections of P_i of $[a_i, b_i]$ for $i = 2, ..., n$, we get

$$\sum_{i=1}^{n}\overset{b_i}{\underset{a_i}{V}}(f) < \varepsilon.$$

But $\quad \overset{b_i}{\underset{a}{V}}(f) = \overset{a_i}{\underset{a}{V}}(f) + \overset{b_i}{\underset{a_i}{V}}(f) \Rightarrow \overset{b_i}{\underset{a_i}{V}}(f) = \overset{b_i}{\underset{a}{V}}(f) - \overset{a_i}{\underset{a}{V}}(f)$

$\Rightarrow \quad \overset{b_i}{\underset{a}{V}}(f) = v(b_i) - v(a_i)$

Hence, from (2) $\displaystyle\sum_{i=1}^{n}\left|v(b_i)-v(a_i)\right| < \varepsilon \Rightarrow v(x)$ is absolutely continuous.

THEOREM 12. *Let $f(x)$ and $g(x)$ be absolutely continuous functions, then $f(x) \pm g(x)$ and $f(x).g(x)$ are also absolutely continuous functions. Hence, $\dfrac{f(x)}{g(x)}$ (if $|g(x)| > 0$, $\forall\ x$) is also absolutely continuous function.*

Proof. Given $f(x)$ and $g(x)$ are absolutely continuous functions on the closed interval $[a, b]$, therefore for each $\varepsilon > 0$, there exists $\delta > 0$ such that

$$\sum_{r=1}^{n} \left| f(b_r) - f(a_r) \right| < \varepsilon \text{ and } \sum_{r=1}^{n} \left| g(b_r) - g(a_r) \right| < \varepsilon,$$

whenever $\displaystyle\sum_{r=1}^{n} (b_r - a_r) < \delta$. for all the points $a_1, b_1, a_2, b_2, \ldots, a_n, b_n$ such that

$a_1 < b_1 \leq a_2 < b_2 \leq \ldots \leq a_n < b_n$.

(i) We have $\displaystyle\sum_{r=1}^{n} \left| \left[f(b_r) \pm g(b_r) \right] - \left[f(a_r) \pm g(a_r) \right] \right|$

$$\leq \sum_{r=1}^{n} \left| \left[f(b_r) - f(a_r) \right] \right| + \sum_{r=1}^{n} \left| g(b_r) - g(a_r) \right|$$

Now if $\displaystyle\sum_{r=1}^{n} (b_r - a_r) < \delta$. then

$$\sum_{r=1}^{n} \left| f(b_r) - f(a_r) \right| < \frac{1}{2}\varepsilon \text{ and } \sum_{r=1}^{n} \left| g(b_r) - g(a_r) \right| < \frac{1}{2}\varepsilon$$

$\therefore \quad \displaystyle\sum_{r=1}^{n} \left| \left[f(b_r) \pm g(b_r) \right] - \left[f(a_r) \pm g(a_r) \right] \right| < \frac{\varepsilon}{2} + \frac{\varepsilon}{2} = \varepsilon.$ whenever $\displaystyle\sum_{r=1}^{n} (b_r - a_r) < \delta$.

$\Rightarrow \quad [f(x) \pm g(x)]$ are also absolutely continuous functions over $[a, b]$.

(ii) Consider

$$\sum_{r=1}^{n} \left| f(b_r) g(b_r) - f(a_r) g(a_r) \right| = \sum_{r=1}^{n} \left| f(b_r) g(b_r) - f(b_r) g(a_r) + f(b_r) g(a_r) - f(a_r) g(a_r) \right|$$

$$= \sum_{r=1}^{n} \left| f(b_r) \left[g(b_r) - g(a_r) \right] + g(a_r) \left[f(b_r) - f(a_r) \right] \right|$$

$$\leq \sum_{r=1}^{n} \left| f(b_r) \left[g(b_r) - g(a_r) \right] \right| + \sum_{r=1}^{n} \left| g(a_r) \left[f(b_r) - f(a_r) \right] \right|$$

$$\leq \sum_{r=1}^{n} \left| f(b_r) \right| \left| g(b_r) - g(a_r) \right| + \sum_{r=1}^{n} \left| g(a_r) \right| \left| f(b_r) - f(a_r) \right|.$$

Since, every absolutely continuous function is bounded, therefore $f(x)$ and $g(x)$ are bounded in the closed interval $[a, b]$.

Let $\qquad |f(x)| \leq k_1, \; |g(x)| \leq k_2, \; \forall \, x \in [a, b]$.

Then we have

$$\sum_{r=1}^{n} \left| f(b_r) g(b_r) - f(a_r) g(a_r) \right| \leq k_1 \varepsilon + k_2 \varepsilon = \varepsilon \left(|k_1| + |k_2| \right). \text{ whenever } \sum_{r=1}^{n} (b_r - a_r) < \delta.$$

If $\varepsilon(|k_1| + |k_2|) = \varepsilon_1$, then $\displaystyle\sum_{r=1}^{n} \left| f(b_r) g(b_r) - f(a_r) g(a_r) \right| < \varepsilon_1$,

whenever $\displaystyle\sum_{r=1}^{n} (b_r - a_r) < \delta$, when $\qquad a_1 < b_1 \leq a_2 < b_2 \leq \ldots \leq a_n < b_n$,

$\Rightarrow$ product of two absolutely continuous functions is also absolutely continuous.

(iii) We have $\qquad |g(x)| > 0, \; \forall \, x \in [a, b]$; therefore

$$|g(x)| > \rho, \text{ where } \rho > 0, \; \forall \, x \in [a, b].$$

Now, $\sum_{r=1}^{n} \left| \dfrac{1}{g(b_r)} - \dfrac{1}{g(a_r)} \right| = \sum_{r=1}^{n} \left| \dfrac{g(a_r) - g(b_r)}{g(b_r)g(a_r)} \right| < \dfrac{\varepsilon}{\rho^2}$, whenever $\sum_{r=1}^{n} (b_r - a_r) < \delta$. Setting

$\dfrac{\varepsilon}{\rho^2} = \varepsilon_2$, we get $\sum_{r=1}^{n} \left| \dfrac{1}{g(b_r)} - \dfrac{1}{g(a_r)} \right| < \varepsilon_2$, where $\varepsilon_2 > 0$

$\Rightarrow \quad \dfrac{1}{g(x)}$ is absolutely continuous function over $[a, b]$.

Now $f(x), \dfrac{1}{g(x)}$ are absolutely continuous.

$\Rightarrow \quad f(x) \cdot \dfrac{1}{g(x)}$ is absolutely continuous.

$\Rightarrow \quad \dfrac{f(x)}{g(x)}$ is absolutely continuous over $[a, b]$.

REMARK

- The set of all absolutely continuous functions on $[a, b]$ is a proper subspace of the space $BV[a,b]$ of all functions of bounded variation on $[a, b]$.

THEOREM 13. *Let $<E_k>$ be a sequence of subsets of a measurable set E s.t. $\lim\limits_{k \to \infty} m(E_k) = 0$.*

Let f be integrable over E, then $\lim\limits_{k \to \infty} \int_{E_k} f(x)dx = 0$ or $\left| \int_{E_k} f(x)dx \right| < \varepsilon, k \geq k_0$, whenever $m(E_k) < \rho$.

Proof. Consider the case when $f(x)$ is bounded over E.

Since $E_k \subset E$, $\forall\, k$ then $f(x)$ is also bounded over each E_k.

Now let $\alpha \leq f(x) \leq \beta$, $\forall\, x \in E_k$.

Then $\quad \alpha.m(E_k) \leq \int_{E_k} f(x) \leq \beta.m(E_k)$

Letting $k \to \infty$ and using $\lim\limits_{k \to \infty} m(E_k) = 0$.

we get $\quad \alpha.0 \leq \lim\limits_{k \to \infty} \int_{E_k} f(x)dx \leq \beta.0$ or $\lim\limits_{k \to \infty} \int_{E_k} f(x)dx = 0$.

Now suppose that $f(x)$ be unbounded, then we define

$$[f(x)]_m = \begin{cases} f(x), \text{if } f(x) \leq m, \\ m, \text{if } f(x) > m, \end{cases}$$

where $m \in N$ is arbitrary, then $[f(x)]_m$ is L-integrable and we define

$$\int_E f(x)dx = \lim\limits_{m \to \infty} \int_E [f(x)]_m\, dx. \qquad \qquad \text{...(1)}$$

$\Rightarrow$ for each $\varepsilon > 0$, however small, $\exists\, n_0 \in N$, s.t.

$$m \geq n_0 \Rightarrow \left| \int_{E_k} \left\{ f(x) - [f(x)]_m \right\} dx \right| < \dfrac{1}{2}\varepsilon \qquad \text{...(2)}$$

Now $[f(x)]_m \leq f(x) \qquad \Rightarrow \qquad f(x) - [f(x)]_m \leq 0$,

$$\Rightarrow \qquad \left| \int_{E_k} \left\{ f(x) - [f(x)]_m \right\} dx \right| \geq 0 \qquad \text{...(3)}$$

Using (3), in relation (2), we get

$$\int_{E_k}\left[f(x)-\left[f(x)\right]_m\right]dx < \frac{1}{2}\varepsilon.$$

In particular, for $m = n_0$, we have

$$\int_{E_k}\left[f(x)-\left[f(x)\right]_{n_0}\right]dx < \frac{1}{2}\varepsilon. \qquad \qquad ...(4)$$

Now, $\lim\limits_{k\to\infty} m(E_k) = 0 \Rightarrow$ for each $\rho > 0$, however small, $\exists\ k_0 \in N$ s.t.

$$|m\ (E_k)| < \rho\ \forall\ k \geq k_0$$

$$\Rightarrow \qquad\qquad m(E_k) < \rho \qquad\qquad ...(5)$$

Since $[f(x)]_{n_0} \leq n_0$, we have $\int_{E_k}\left[f(x)\right]_{n_0} dx \leq n_0.m(E_k)$

Using (5), we get

$$\int_{E_k}\left[f(x)\right]_{n_0} dx < n_0.\rho\ , \text{ for all } k \geq k_0$$

or $\qquad \int_{E_k}\left[f(x)\right]_{n_0} dx \leq \frac{1}{2}\varepsilon\ , \text{ where } n_0\rho = \varepsilon/2, \text{ for all } k \geq k_0. \qquad ...(6)$

Using (4) and (6), we get

$$\int_{E_k}\left\{f(x)-\left[f(x)\right]_{n_0}\right\}dx + \int_{E_k}\left[f(x)\right]_{n_0} dx < \frac{1}{2}\varepsilon + \frac{1}{2}\varepsilon$$

$$\Rightarrow\qquad\qquad \int_{E_k} f(x)dx < \varepsilon, \text{ for all } k \geq k_0. \qquad\qquad ...(7)$$

Now suppose that $f(x) \geq 0$, since if the result is true for positive function then it is also applicable to negative functions. Thus (7) becomes

$$\int_{E_k}|f(x)|dx < \varepsilon\ , \text{ for all } k \geq k_0$$

or $\qquad \left|\int_{E_k} f(x)dx\right| \leq \int_{E_k}|f(x)|dx < \varepsilon\ , \text{ for } k \geq k_0. \text{ whenever } m(E_k) < \rho$

i.e., $\qquad \lim\limits_{k\to\infty}\int_{E_k} f(x)dx = 0.$

THEOREM 14. *A necessary and sufficient condition that a function should be an indefinite integral is that it should be absolutely continuous.*

Proof. **Condition is sufficient.** Let $f(x)$ be an absolutely continuous function over the closed interval $[a, b]$. Then f is of bounded variation, and hence we can express $f(x)$ as

$$f(x) = f_1(x) - f_2(x)$$

where $f_1(x)$ and $f_2(x)$ are monotonically increasing functions and hence both are differentiable.

$\Rightarrow\quad f'(x)$ exists and $|f'(x)| \leq f_1'(x) + f_2'(x)$

$$\Rightarrow\qquad\qquad \int|f'(x)| \leq f_1(b) + f_2(b) - f_1(a) - f_2(a) < \infty,$$

$\Rightarrow\quad f'(x)$ is integrable.

Let $F(x)$ be an indefinite integral of $f'(x)$ i.e.,

$$F(x) = F(a) + \int_a^x f'(t)dt, x \in [a, b]. \qquad\qquad ...(1)$$

Using Fundamental theorem of integral calculus, we get

$$F'(x) = f(x) \text{ a.e. or } F'(x) = f'(x) + c \qquad \text{...(2)}$$

From (1), we have $F(a) = f(a)$. Using this in (2), we get, $c = 0$ and so $F(x) = f(x)$. Hence, every absolutely continuous function $f(x)$ is an indefinite integral of its own derivative.

Condition is necessary. Let $F(x)$ be an indefinite integral of $f(x)$ defined on the closed interval $[a, b]$, so that $F(x) = \int_a^x f(t)\,dt + F(a)$, $\forall\, x \in [a, b]$. and $f(x)$ is integrable over $[a, b]$.

Corresponding to arbitrary small $\varepsilon > 0$, let $\delta > 0$ be such that if $m(A) < \delta$, then $\int_A |f| < \varepsilon$. Now select $2n$ real numbers such that $a_1 < b_1 \le a_2 < b_2 \le \ldots \le a_n < b_n$.

s.t. $\qquad A = \overset{n}{\underset{i=1}{\cup}} [a_i, b_i]$ and $\sum\limits_{i=1}^{n} (b_i - a_i) < \delta$.

Then $\qquad \sum\limits_{i=1}^{n} |F(b_i) - F(a_i)| = \sum\limits_{i=1}^{n} \left| \int_a^{b_i} f - \int_a^{a_i} f \right|$

$$= \sum\limits_{i=1}^{n} \left| \int_{a_i}^{b_i} f \right| \le \sum\limits_{i=1}^{n} \int_{a_i}^{b_i} |f| = \int_A |f| < \varepsilon.$$

$\Rightarrow \quad F$ is a absolutely continuous.

Hence, every indefinite integral is absolutely continuous.

THEOREM 15. *An integral is a continuous function.*

Proof. Let $F(x)$ be an indefinite integral of $f(x)$ defined over $[a, b]$ so that $f(x)$ is integrable over $[a, b]$ and

$$F(x) = \int_a^x f(t)\,dt + F(a), x \in [a, b].$$

Let x is any element of the open interval $]a, b[$ and $x_1, x_2 \in]x - \delta/2, x + \delta/2[$. If $E_k = [x_1, x_2]$, then $m(E_k) = m([x_1, x_2]) < \delta$.

Thus $\qquad \left| \int_{E_k} f(t)\,dt \right| < \varepsilon, \text{ for } k \ge k_0.$

But $E_k = [x_1, x_2]$, then

$$\left| \int_{x_1}^{x_2} f(t)\,dt \right| < \varepsilon, \text{ whenever } m([x_1, x_2]) < \delta$$

or $\qquad |F(x_2) - F(x_1)| < \varepsilon, \text{ whenever } |x_2 - x_1| = m([x_1, x_2]) < \delta.$

$\Rightarrow \quad F(x)$ is a continuous function.

REMARKS

- If a function f is absolutely continuous in an interval $[a, b]$ and if $f'(x) = 0$ a.e., in $[a, b]$, then f is constant.
- If the derivatives of two absolutely continuous functions are equivalent, then the functions differ by a constant.

THEOREM 16. *If f is a bounded and measurable function on $[a, b]$ and $F(x) = \int_a^x f(t)\,dt + F(a)$,*

then $\qquad F'(x) = f(x)$ *a.e. in $[a, b]$.*

Proof. Since every indefinite integral is a function of bounded variation therefore $F(x)$ is a function of bounded variation over $[a, b]$. Thus $F(x)$ can be expressed as a difference of two monotonic functions and since every monotonic function has a finite differential coefficient at every point of a set of non-zero measure, therefore $F(x)$ has a finite differential coefficient a.e. in $[a, b]$.

Now $f(t)$ is bounded, then

$$|f(t)| \leq M, \text{ say.} \qquad \qquad \text{...(1)}$$

Set $\qquad \qquad f_n(x) = \dfrac{F(x+h) - F(x)}{h}, \text{ where } h = \dfrac{1}{n}.$

Then $\qquad | f_n(x) | = \left| \dfrac{1}{h} \int_x^{x+h} f(t)dt \right| \leq \dfrac{1}{h} \int_x^{x+h} | f(t) | dt \leq \dfrac{M}{h} \int_x^{x+h} dt = M$

$\Rightarrow \qquad \qquad |f_n(x)| \leq M$

Since $f_n(x) \to F'(x)$ a.e., then by the Lebesgue-bounded convergence theorem, we have

$$\int_0^x F'(x)\,dx = \lim_{n \to \infty} \int_a^x f_n(x)\,dx = \lim_{h \to 0} \dfrac{1}{h} \int_a^x [F(x+h) - F(x)]\,dx$$

$$= \lim \left[\dfrac{1}{h} \int_x^{x+h} [F(x)\,dx - \dfrac{1}{h} \int_a^{a+h} F(x)]\,dx \right] = F(x) - F(a)$$

$$= \int_a^x f(t)\,dt,$$

or $\qquad \int_a^x F'(t) - f(t)]\,dt = 0, \forall\, x.$

$\Rightarrow \qquad \qquad F'(x) - f(x) = 0 \text{ a.e. in } [a, b].$

Hence, $\qquad \qquad F'(x) = f(x) \text{ a.e. in } [a, b].$

THEOREM 17. *If f is an integrable function on $[a, b]$ and if $F(x) = \int_a^x f(t)\,dt + F(a)$,*

$\qquad$ *then $\qquad \qquad F'(x) = f(x) \text{ a.e. in } [a, b].$*

Proof. $\qquad$ Let us suppose that $f(x) \geq 0, \forall\, x$. Now define a function $[f(x)]_m$ as follows :

$$[f(x)_m] = \begin{cases} f(x), \text{ if } f(x) \leq m, \\ m, \qquad \text{ if } f(x) > m. \end{cases}$$

$\qquad \qquad$ Clearly, $\qquad [f(x)]_m \leq f(x),$

$\Rightarrow \qquad \qquad f(x) - [f(x)]_m \geq 0.$

$\therefore \qquad \qquad G_m(x) = \int_a^x \left[f(t) - \left[f(t) \right]_m \right] dt$

is an increasing function of x, so that $G_m(x)$ has a non-negative differential coefficient. Using previuos theorem, we have

$$\dfrac{d}{dx} \int_a^x \left[f(t) \right]_m dt = \left[f(t) \right]_m \text{ a.e.}$$

Also $\qquad G_m(x) + \dfrac{d}{dx} \int_a^x \left[f(x) \right]_m = F'(x).$

Since $\dfrac{d}{dx} G_m(x)$ is non-negative, therefore $F'(x) \geq [f(x)]_m$ a.e.

But since m is arbitrary, we have $F'(x) \geq f(x)$

$\Rightarrow \qquad \qquad \int_a^b F'(x)\,dx \geq \int_a^b f(x)\,dx = F(b) - F(a),$

and
$$\int_a^b F'(x)\,dx = F(b) - F(a) = \int_a^b f(x)\,dx.$$

$$\int_a^b F'(x)\,dx = F(b) - F(a) = \int_a^b f(x)\,dx.$$

or
$$\int_a^b \left[F'(x) - f(x) \right] dx = 0.$$

Now since $F'(x) - f(x) \geq 0$, therefore we have

$$F'(x) - f(x) = 0 \text{ a.e.} \quad \text{or} \quad F'(x) = f(x) \text{ a.e.}$$

REMARKS

- Let f be an absolutely continuous function on $[a, b]$ and F be another function which satisfies Lipschitz condition in the segment $[c, d] \subset [a, b]$: then the composite function $F[f(x)]$ is absolutely continuous in $[c, d]$.
- Let x be a Lebesgue point of a function $f(t)$; then the indefinite integral $F(x) = F(a) + \int_a^x f(t)\,dt$ is differentiable at each point x and $F'(x) = f(x)$.
- Every point of continuity of an integrable function $f(t)$ is a Lebesgue point of $f(t)$.

THEOREM 18. **(Lebesgue Differentiation Theorem).** *Let $f: [a, b] \to R$ be a finite valued monotonically increasing function, then f is differentiable. Also $f : [a, b] \to R$ is L-integrable and*

$$\int_a^b f'(x)\,dx \leq f(b) - f(a).$$

Proof. Let $\langle f_n \rangle$ be a sequence of non-negative functions, where $f_n : [a, b] \to R$ s.t.

$$f_n(x) = n \left[f\left(x + \frac{1}{n}\right) - f(x) \right], \ \forall\, x \in [a, b] \qquad \ldots(1)$$

Set $f(x) = f(b)$, for $x \geq b$.

Since, $f: [a, b] \to R$ is an increasing function; therefore $f_n : [a, b] \to R$ is also an increasing function and hence Lebesgue integrable.

Again from (1), we have

$$\lim_{n \to \infty} f_n(x) = \lim_{1/n \to 0} \frac{f\{x + (1/n)\} - f(x)}{(1/n)}, \forall\, x \in [a, b]$$

$\Rightarrow$ the sequence $\langle f_n \rangle$ converges to $f'(x)$ a.e.

Using Fatou's Lemma, we have

$$\int_a^b f'(x)\,dx \leq \lim_{n \to \infty} \inf \left\{ \int_a^b f_n(x)\,dx \right\}. \qquad \ldots(2)$$

Also
$$\lim_{n \to \infty} \inf \int_a^b f_n(x)\,dx \leq \lim_{n \to \infty} \inf n \int_a^b \left[f\left(x + \frac{1}{n}\right) - f(x) \right] dx$$

$$= \lim_{n \to \infty} \inf n \left[\int_a^b f\left(x + \frac{1}{n}\right) dx - \int_a^b f(x)\,dx \right]. \qquad \ldots(3)$$

Putting $t = x + (1/n)$, we get

$$\int_a^b f\left(x + \frac{1}{n}\right) dx = \int_{a+(1/n)}^{b+(1/n)} f(x)\,dx.$$

$\therefore$
$$\lim_{n \to \infty} \inf \int_a^b f_n(x)\,dx = \lim_{n \to \infty} \inf n \left[\int_{a+(1/n)}^{b+(1/n)} f(x)\,dx - \int_a^b f(x)\,dx \right]$$

$$= \lim_{n \to \infty} \inf n \left[\int_b^{b+(1/n)} f(x)\,dx - \int_a^{a+(1/n)} f(x)\,dx \right] \qquad \ldots(3)$$

Now extend the definition of f by assuming $f(x) = f(b)$, $\forall\, x \in [b, b+1/n]$.

Also $f(a) \le f(x)$, for $x \in \left(a, a+\dfrac{1}{n}\right)$, Then

$$\int_a^{a+(1/n)} f(x)\,dx \ge \int_a^{a+(1/n)} f(a)\,dx = \frac{1}{n} f(a)$$

$$\Rightarrow \qquad -\int_a^{a+(1/n)} f(x)\,dx \le -\frac{1}{n} f(a)$$

Then, (3) $\Rightarrow$ $\displaystyle \lim_{n\to\infty} \inf \int_a^b f_n(x)\,dx = \lim_{n\to\infty} \inf\, n\left[\int_b^{b+(1/n)} f(b)\,dx + \left(-\int_a^{a+(1/n)} f(x)\,dx\right)\right]$

$$= \lim_{n\to\infty} \inf\, n\left[f(b)\cdot\frac{1}{n} + \left(-\frac{1}{n}\right)f(a)\right] \le f(b) - f(a)$$

Thus, from (2), we get

$$\int_a^b f'(x)\,dx \le f(b) - f(a).$$

$\Rightarrow$ $f'(x)$ is integrable and therefore finite a.e. Hence f is differentiable a.e.

Solved Examples

Example 1.   *Let f be a function defined by $f(0) = 0$ and $f(x) = x \sin (1/x)$ for $x \ne 0$. Find $D^+ f(0)$, $D_+ f(0)$, $D^- f(0)$, $D_- f(0)$.*

Solution.   We have $D^+ f(0) = \overline{\lim_{h\to 0+}} \dfrac{f(0+h) - f(0)}{h} = \overline{\lim_{h\to 0}} \dfrac{h\sin\dfrac{1}{h} - 0}{h}$

$$= \overline{\lim_{h\to 0}} \sin\frac{1}{h} = 1, \quad \text{as } -1 \le \sin\frac{1}{h} \le 1$$

Also, $D_+ f(0) = \underline{\lim_{h\to 0+}} \dfrac{f(0+h) - f(0)}{h} = \underline{\lim_{h\to 0+}} \sin\frac{1}{h} = -1$

$$D^- f(0) = \overline{\lim_{h\to 0-}} \dfrac{f(0-h) - f(0)}{0-h} = \overline{\lim_{h\to 0}} \dfrac{(-h)\sin\left(-\dfrac{1}{h}\right) - 0}{-h}$$

$$= \overline{\lim_{h\to 0}} -\sin\frac{1}{h} = 1$$

and $D_- f(0) = \underline{\lim_{h\to 0+}} \dfrac{f(0-h) - f(0)}{-h} = \underline{\lim_{h\to 0+}} \left(-\sin\frac{1}{h}\right) = -1.$

Example 2.   *Find the four Dini's derivatives of fucntion $f : [0, 1] \to R$ s.t. $f(x) = 0$, if $x \in Q$ and $f(x) = 1$ if $x \notin Q$.*

Solution.   Let $x \in Q$, then $D^+ f(x) = \overline{\lim_{h\to 0}} \dfrac{f(x+h) - f(x)}{h} = \overline{\lim_{h\to 0}} \dfrac{f(x+h) - 0}{h}.$

But $f(x+h)$ will have the value 0 or 1 depending upon h is rational or not. Therefore $\dfrac{f(x+h)}{h}$ will have the value 0 or $\dfrac{1}{h}$,

$$\Rightarrow \qquad D^+ f(x) = \overline{\lim_{h\to 0}} \left(0 \text{ or } \frac{1}{h}\right) = \infty.$$

Also $D_+f(x)=0$ and $\quad D^-f(x)= \varlimsup_{h\to 0} \dfrac{f(x-h)-f(x)}{-h}$

$$= \varlimsup_{h\to 0} \dfrac{f(x+h)-0}{-h} = \varlimsup_{h\to 0} \dfrac{0\,\text{or}\,1}{-h}=-\infty$$

and $\qquad\qquad D_-f(x)=-\infty.$

Similarly when $x \in Q$, then

$$D^+f(x)= \varlimsup_{h\to 0+} \dfrac{f(x+h)-f(x)}{h} = \varlimsup_{h\to 0+} \dfrac{f(x+h)-1}{h}$$

$$= \varlimsup_{h\to 0+} \dfrac{(0\,\text{or}\,1)-1}{h} = \varlimsup_{h\to 0+} \left(-\dfrac{1}{h}\,\text{or}\,0\right)= 0.$$

Similarly as above,

$$D_+f(x)= \varliminf_{h\to 0+} \left(-\dfrac{1}{h}\,\text{or}\,0\right)= -\infty$$

$$D^-f(x)= \varlimsup_{h\to 0-} \dfrac{f(x-h)-f(x)}{-h} = \varlimsup_{h\to 0-} \dfrac{(0\,\text{or}\,1)-1}{-h} = \varlimsup_{h\to 0-} \left(\dfrac{1}{h}\,\text{or}\,0\right)= \infty$$

and $\qquad\qquad D_-f(x)=0.$

Example 3. *If the function f assumes its maximum at c, show that $D^+f(c) \le 0$ and $D_-f(c) \ge 0$.*

Solution. Since $g(x)$ assumes its maximum at $x = c$

$\Rightarrow \qquad f(c+h) \le f(c)$ and $f(c-h) \le f(c)$

$\Rightarrow \qquad f(c+h)-f(c) \le 0$ and $f(c-h)-f(c) \le 0$

$\Rightarrow \qquad \dfrac{f(c+h)-f(c)}{h} \le 0$ for h^+ and $\dfrac{f(c-h)-f(c)}{-h} \ge 0$, for h^+

$$D^+f(c)= \varlimsup_{h\to 0+} \dfrac{f(c+h)-f(c)}{h} \le 0 \text{ and } D_-f(c) \ge 0.$$

Example 4. *Let f be an absolutely continuous monotone function on [a, b] and E a set of measure zero, then show that f(E) has measure zero.*

Solution. Let the function f be monotonically increasing. By the definition of absolute continuity of f, for $\varepsilon > 0$, $\exists\ \delta > 0$ and non-overlapping intervals $\{I_n = [a_n, b_n]\}$ s.t.

$$\Sigma(b_n-a_n) < \delta \Rightarrow \Sigma|f(b_n)-f(a_n)| < \varepsilon$$

or $\qquad\qquad \Sigma[f(b_n)-f(a_n)] < \varepsilon.$ $\qquad\qquad\qquad$ ...(1)

Now, $\qquad\qquad E \subseteq [a, b] \Rightarrow E \subseteq \cup I_n$

$\Rightarrow \qquad\qquad f(E) \subset f(\cup I_n) = \cup f(I_n)$

$\Rightarrow \qquad\qquad m^*(f(E)) \le \Sigma m^*(f(I_n)) \le \Sigma[\bar{f}(x_n)-\underline{f}(x_n)] < \varepsilon.$

where $\bar{f}(x_n)$ and $\underline{f}(x_n)$ are the maximum and minimum values of $f(x)$ in the interval $[a_n, b_n]$.

Also note that $\Sigma|\bar{x}_n - \underline{x}_n| \le \Sigma(b_n- a_n) < \delta$

$\Rightarrow \quad m^*(f(E)) \le \varepsilon,$

$\Rightarrow \quad m^*(f(E))=0 \Rightarrow m(f(E))=0.$

Example 5. *Show that the function f defined by*

$$f(x) = x^P \sin\frac{1}{x} \text{ for } 0 < x \leq 1, f(0) = 0, p \geq 2.$$

is of bounded variation [0,1].

Solution. Note that

$$Rf'(0) = \lim_{h \to 0} \frac{(0+h)^P \sin\frac{1}{h} - 0}{h} = \lim_{h \to 0} h^{(p-1)} \sin\frac{1}{h} = 0$$

and

$$Lf'(0) = \lim_{h \to 0} \frac{(-h)^P \sin\left(-\frac{1}{h}\right) - 0}{-h} = 0$$

$$\Rightarrow \quad f'(0) = 0 \text{ and } f'(x) = x^P \cos\frac{1}{x}\left(-\frac{1}{x^2}\right) + px^{P-1}\sin\frac{1}{x}$$

$$\Rightarrow \quad f'(x) = x^{P-2}\left[px\sin\frac{1}{x} - \cos\frac{1}{x}\right], \text{ for } 0 < x \leq 1$$

$$\Rightarrow \quad f'(x) \text{ is bounded for } 0 \leq x \leq 1.$$

Hence, $f \in BV [0, 1]$.

EXERCISE 17.1

1. Prove that if f is absolutely continuous, then $f'(x)$ exists almost everywhere.

2. Show that, if $\qquad F(x) = F(a) + \int_a^x f(t)dt$

   Then $\qquad F'(x) = f(x)$ a.e.

3. Give an example to show that a bounded function need not be of bounded variation.

4. Prove that a monotone function is differentiable almost everywhere.

5. Let $f : [a, b] \to R$ be such that

   $|f(y) - f(x)| \leq k(y-x)$, whenever $a \leq x \leq y \leq b$.
   Prove that

   $\int_a^b f'(x)dx = f(b) - f(a).$

6. Explain the statement that $f(x)$ is an absolulely continuous function on an interval $[a, b]$. If f is integrable on $[a, b]$, show that

   $F(x) = \int_a^x f(t)dt$, $(a < x < b)$

   is absolutely continuous on the given interval $[a, b]$. Give an example of a function which is continuous but not absolutely continuous on a given interval.

7. Show that the function f defined on $[0, 1]$ by

   $$f(x) = \begin{cases} x\cos\left(\dfrac{\pi x}{2}\right); & \text{for } 0 < x \leq 1, \\ 0 & ; \text{for } x = 0 \end{cases}$$

   is continuous but not of bounded variation on $[0, 1]$.

8. If f is integrable on $[a, b]$ and $\int_a^x f(t)dt = 0$ for all $x \in [a, b]$, show that $f(t) = 0$ a.e. in $[a, b]$.

9. Define absolute continuity for a real variable. Show that $F(x)$ is an indefinite integral, if F is absolutely continuous.

10. If $f, g : [0, 1] \to R$ are absolutely continuous, prove that $f+g$ and fg are also absolutely continuous.

11. If $\langle f_n \rangle$ is a sequence of real valued functions defined on $[a, b]$, which converges to f at each point of $[a, b]$, then show that

    $$\overset{b}{\underset{a}{V}}(f) \leq \varliminf \overset{b}{\underset{a}{V}}(f_n).$$

12. Show that the function f defined on $[0, 1]$ by

    $f(x) = x\sin\dfrac{1}{x}$ when $x \neq 0$ and $f(0) = 0$ is not of bounded variation.

13. If $f \in BV [a, b]$ and $\exists$ a m s.t, $0 < m \leq |f(a)|$, $\forall x \in [a,b]$, show that $\dfrac{1}{f} \in BV [a, b]$ and

    $$\overset{b}{\underset{a}{V}}\left(\frac{1}{f}\right) \leq \left(\frac{1}{m^2}\right)\overset{b}{\underset{a}{V}}(f).$$

14. If f is continuous at $x_0 \in [a, b]$, show that supremums of total positive and negative variations of f over $[a, b]$ are also continuous at $x_0 \in [a, b]$.

15. Show that $BV[a, b]$ is a normed linear space where norm is defined as

$$\|f\| = |f(a)| + \overset{b}{\underset{a}{V}}(f).$$

16. If f is differentiable on $[a, b]$ s.t. $|f'(x)| \leq M < \infty$, $\forall x \in [a, b]$, then $f \in BV[a, b]$ and

$$\overset{b}{\underset{a}{V}}(f) \leq M(b-a).$$

17. Show that the function f defined on $[0, 1]$ by $f(x) = x \sin(\pi/x)$ for $x > 0$, $f(0) = 0$ is continuous but is not of bounded variation on $[0, 1]$.

18. If $f \in BV[a, b]$, then show that $f' \in L[a, b]$.

19. Give an example of a function which is continuous but not absolutely.

20. Show that if a function f is Lipschitizian, then it is absolutely continuous.

21. Show that if f' exists and is bounded on $[a, b]$ then $f \in BV[a, b]$

22. Show that a continuous function may not be of bounded variations.

23. Show that a function of bounded variation may not be continuous.

CHAPTER REVIEW: A COMPETITIVE APPROACH

SELECTED TERMS AND RESULTS

TERMS

- **Absolutely Continuous Function:** A real valued function f defined on $[a, b]$ is said to be absolutely continuous on $[a, b]$, if for given $\varepsilon > 0$, $\exists\ \delta > 0$, such that

$$\sum_{r=1}^{n} |f(b_r) - f(a_r)| < \varepsilon,\ \text{whenever}\ \sum_{r=1}^{n} (b_r - a_r) < \delta.$$

- **Functions of Bounded Variation:** Let us define

$$\overset{b}{\underset{a}{V}}(f, p) = \sum_{r=0}^{n-1} |f(x_{r+1}) - f(x_r)|.$$

If $\overset{b}{\underset{a}{V}}(f)$ is finite, then f is called a function of bounded variation.

- **Lebesgue Point of a Function:** A point x is said to be a Lebesgue point of the function $f(t)$, if $\lim_{h \to 0} \dfrac{1}{h} \int_x^{x+h} |f(t) - f(x)|\, dt = 0$.

- **Covering in the Sense of Vitali:** A set E is said to be covered in the sense of Vitali by a family of intervals M in which none is a singleton set, if every point of the set E is contained in some small intervals of M i.e., for each $x \in E$ there exists $\varepsilon > 0$ and an interval $I \in M$ such that $x \in I$ and $l(I) < \varepsilon$.

- **Indefinite Integral:** Let $f(x)$ be L-integrable over $[a, b]$ then the function $F(x)$ defined by $F(x) = \int_a^x f(t)\, dt + c,\ \forall\ x \in [a, b]$ is called an indefinite integral of $f(x)$.

RESULTS

- Every monotonic function on $[a, b]$ is of bounded variation.

- If V, P, N denote total, positive and negative variations of a bounded function f on $[a, b]$ then $V = P + N$ and $P - N = f(b) - f(a)$.

- If a function f is of bounded variation in $[a, b]$ is continuous at $c \in [a, b]$ then the function defined by $v(x) = \overset{x}{\underset{a}{V}}(f)$ is also continuous at $x = c$ and vice-versa.

- Every absolutely continuous function f defined on $[a, b]$ is of bounded variations.

- A function f is of bounded variation if and only if it can be expressed as a difference of two monotonic functions both non-decreasing.

- A continuous function is of bounded variation iff it can be expressed as a difference of two continuous monotonically increasing functions.

- If f is a function of bounded variation then $f'(x)$ exists almost everywhere.

- Every absolutely continuous function is differentiable almost everywhere.

- The necessary and sufficient condition that a function should be an indefinite integral is that it should be absolutely continuous.

- An integral is a continuous function.

- If a function f is absolutely continuous in an interval $[a, b]$ and if $f'(x) = 0$ a.e. then f is constant.

- If the derivatives of two absolutely continuous functions are equivalent then the function differ by a constant.

- Every point of continuity of an integrable function $f(t)$ is a Lebesgue point of $f(t)$.

- Let $f: [a, b] \to R$ be a finite valued monotonically increasing function then f is differentiable.

REVIEW QUESTIONS AND PROJECT WORK

1. Show that the product of two functions of bounded variations is also of bounded variations.

2. Show that a function of bounded variation is necessarily bounded.

3. Show that the variation function of a function f of bounded variation is continuous if and only if f is a continuous function.

4. Show that the function $f(x) = x^2 \sin\left(\dfrac{1}{x}\right)$ if $x \neq 0$, and $f(0) = 0$ is of bounded variation on $[0, 1]$.

5. Show that $\sin x$ and $\cos x$ are the functions of bounded variation over a finite interval.

6. Show that a polynomial function is of bounded variations over any finite interval.

OBJECTIVE TYPE QUESTIONS

FILL IN THE BLANKS

1. Every absolutely continuous function is _____ .

2. A function $f(x)$ is said to be monotonically _____ iff $x > y \Rightarrow f(x) \leq f(y)$.

3. A function f defined on $[a, b]$ is said to satisfy Lipschitz-condition if $\exists$ a constant $M > 0$ such that $|f(x) - f(y)|$ _____ $M|x - y|$.

4. $\overset{b}{\underset{a}{V}}(f) = $ _____ .

5. $\overset{b}{\underset{a}{V}}(f)$ _____ 0.

TRUE/FALSE

Write 'T' for true and 'F' for false statement.

1. If $P_1 \subset P_2$ then $\overset{b}{\underset{a}{V}}(f, P_1) \leq \overset{b}{\underset{a}{V}}(f, P_2)$. **(T/F)**

2. $\overset{b}{\underset{a}{V}}(f, P) \leq \overset{b}{\underset{a}{V}}(f)$. **(T/F)**

3. If $F(x) = \int_a^x f(t)\,dt + c \; \forall \, x \in [a, b]$ then $F(x)$ is called the definite integral of $f(x)$. **(T/F)**

4. Dini's derivatives always exist, may be finite or infinite for every function f. **(T/F)**

5. Every bounded function on $[a, b]$ is of bounded variation. **(T/F)**

MULTIPLE CHOICE QUESTIONS

Choose the most appropriate one.

1. Every absolutely continuous function f defined on $[a, b]$ is :
   (a) of bounded variation
   (b) continuous
   (c) both (a) and (b) are true
   (d) none of the above

2. A function f is of bounded variation if and only if it can be expressed as a difference of two :
   (a) monotonic non-decreasing function
   (b) monotonic non-increasing function
   (c) both (a) and (b) are true
   (d) none of the above

3. Every absolutely continuous function is :
   (a) continuous
   (b) of bounded variation
   (c) differentiable a.e.
   (d) all are true

4. Let f be a function of bounded variation then :
   (a) $f'(x)$ may not exist
   (b) $f'(x)$ exists a.e.
   (c) $f'(x)$ exists always
   (d) none of the above

5. Every indefinite integral is :
   (a) absolutely continuous
   (b) continuous
   (c) both (a) and (b) are true
   (d) none of the above

6. If the derivatives of two absolutely continuous functions are equivalent then the function differ by :
   (a) constant   (b) 0
   (c) $\dfrac{\pi}{2}$ (d) none of these

7. Every continuous function is :
   (a) of bounded variation
   (b) not necessarily of bounded variation
   (c) never be of bounded variation
   (d) none of the above

8. Every function of bounded variation is :

(a) continuous

(b) not continuous

(c) may not be continuous

(d) none of the above

Answers

FILL IN THE BLANKS

1. continuous   **2.** decreasing   **3.** $\leq$ **4.** 0   **5.** $\geq$

TRUE/FALSE

1. T   **2.** T   **3.** F   **4.** T   **5.** T

MULTIPLE CHOICE QUESTIONS

1. (c)   **2.** (a)   **3.** (d)   **4.** (b)   **5.** (c)   **6.** (a)   **7.** (b)   **8.** (c)

SELF ASSESSMENT TEST

Verify each of the following :

1. If f is of bounded variation and one-one on $[a, b]$ to $[c, d]$ then f^{-1} is of bounded variation on $[c, d]$.

2. If f is a function of bounded variation on $[a, b]$ and there exists $k>0$ such that $f \geq k$ on $[a, b]$ then $\dfrac{1}{f}$ is also of bounded variation.

3. The function $f(x) = \begin{cases} x \cos \dfrac{1}{x}, & \text{when } x \neq 0 \\ 0, & \text{when } x = 0 \end{cases}$

   is not of bounded variation on any closed interval containing zero.

4. The function $f(x) = \begin{cases} x^2 \sin \dfrac{1}{x}, & \text{when } x \neq 0 \\ 0, & \text{when } x = 0 \end{cases}$ is

   of bounded variation on $[-1, 1]$.

5. The function $f(x) = \begin{cases} x; & \text{when } x \text{ is rational} \\ 0; & \text{when } x \text{ is irrational} \end{cases}$

   then $f(x)$ is not of bounded variation on $[0, 1]$.

6. The function $x^p \sin \dfrac{1}{x}$ and $x^p \cos \dfrac{1}{x}$ when $x \neq 0$ and each equal to zero when $x=0$ are of bounded variation on any closed interval $[0, a]$ where $a>0$ if and only if $p>0$.

18 Beta and Gamma Functions

18.1 INTRODUCTION

The definite integral $\int_0^\infty e^{-x} x^{n-1}\, dx$, for $n > 0$ is known as the gamma function and is denoted by $\Gamma(n)$ ['read as Gamma n']. Gamma function is also called the Eulerian integral of second kind. Weierstrass defined it as infinite product as

$$\frac{1}{\Gamma(z)} = z\,e^{2/z} \prod_{n=1}^{\infty}\left[\left(1+\frac{z}{n}\right)e^{-z/n}\right]$$

for non-zero and non-negative number z and n is an Euler's constant.

REMARK

- The integral is valid only for $n > 0$ because it is for just those values of m and n that the above integral are convergent.

18.2 PROPERTIES OF GAMMA FUNCTION

(1) $\Gamma(1) = 1$.

    Proof.   We have

$$\Gamma(n) = \int_0^\infty e^{-x} x^{n-1} dx,\ n > 0. \qquad \ldots(1)$$

Put $n = 1$ in equation (1), we get

$$= \int_0^\infty e^{-x}\, dx = \left[-e^{-x}\right]_0^\infty = 1.$$

    $\therefore$ $\Gamma(1) = 1$

(2) $\Gamma(n+1) = n\,\Gamma(n),\ n > 0$.

    Proof.   We have

$$\Gamma(n) = \int_0^\infty e^{-x} x^{n-1}\, dx,\ \text{for } n > 0$$

Replacing n by $(n+1)$, we have

$$\Gamma(n+1) = \int_0^\infty e^{-x} x^{n+1-1} dx = \int_0^\infty e^{-x} x^n\, dx$$

$$= \left[x^n \cdot (-e^{-x})\right]_0^\infty - \int_0^\infty (nx^{n-1})(-e^{-x})\, dx$$

$\therefore$ $$\Gamma(n+1) = -\lim_{x\to\infty}\frac{x^n}{e^x} + 0 + n\int_0^\infty e^{-x}x^{n-1}dx \qquad ...(1)$$

$$\left(\because \lim_{x\to\infty} x^n e^{-x} = 0 \text{ as } n > 0\right)$$

But
$$\lim_{x\to\infty}\frac{x^n}{e^x} = \lim_{x\to\infty}\frac{x^n}{1+\dfrac{x}{1!}+\dfrac{x^2}{2!}+...+\dfrac{x^n}{n!}+\dfrac{x^{n+1}}{(n+1)!}+...}$$

$$= \lim_{x\to\infty}\frac{1}{\dfrac{1}{x^n}+\dfrac{1}{1!x^{n-1}}+...+\dfrac{1}{n!}+\dfrac{x}{(n+1)!}+...}$$

$$= 0 \qquad ...(2)$$

Also, by definition, we have

$$\Gamma(n) = \int_0^\infty e^{-x}x^{n-1}\,dx . \qquad ...(3)$$

Using (2) and (3), (1) redcuces to

$$\Gamma(n+1) = n\,\Gamma(n).$$

REMARKS

- The formula $\Gamma(n+1) = n\,\Gamma(n)$ is known as a recurrence formula for gamma function.
- The gamma function can be generalized to $n < 0$ by recurrence formula in the form of $\Gamma(n) = \dfrac{\Gamma(n+1)}{n}$ This process is known as analytic continuation.

(3) *If n is a non-negative integer, then* $\Gamma(n+1) = n\,!$.

Proof. We know that for $n > 0$.

$$\Gamma(n+1) = n\,\Gamma(n)$$
$$= n\,\Gamma(n-1+1)$$
$$= n(n-1)\Gamma(n-1) \qquad \text{[By property 2]}$$
$$= n(n-1)(n-2)\,\Gamma(n-2)$$
$$= n(n-1)(n-2)\,...\,3.\,2.\,1.\,\Gamma(1)$$
$$= n\,! \qquad [\because \Gamma(1)=1]$$

REMARK

- Gauss's Pi-function is denoted by $\pi(n)$ and is defined by $\pi(n) = \Gamma(n+1)$, when n is +ve integer.

(4) $\Gamma(1/2) = \sqrt{\pi}$.

Proof. By definition, we have

$$\Gamma(n) = \int_0^\infty e^{-t}t^{n-1}\,dt, n > 0 \qquad ...(1)$$

Replacing n by $1/2$ in equation (1), we get

$$\Gamma(1/2) = \int_0^\infty e^{-t}t^{-1/2}dt = 2\int_0^\infty e^{-u^2}\,du \qquad ...(2)$$

[Putting $t = u^2$, i.e., $dt = 2u\,du$]

$\therefore$ $$\Gamma(1/2) = 2\int_0^\infty e^{-x^2}dx \text{ and } \Gamma(1/2) = 2\int_0^\infty e^{-y^2}dy \qquad ...(3)$$

(Limits remaining same)

Multiplying the corresponding sides of two equations of (3), we get

$$[\Gamma(1/2)]^2 = \left(2\int_0^\infty e^{-x^2}dx\right)\left(2\int_0^\infty e^{-y^2}dy\right)$$

$$= 4 \int_0^\infty \int_0^\infty e^{-(x^2+y^2)} dx\, dy.$$

Now, changing the variables to polar co-ordinates (r, θ) where $x = r\cos\theta$, $y = r\sin\theta$

$\Rightarrow$ $\qquad\qquad\qquad x^2 + y^2 = r^2$ and $dx\, dy = r\, d\theta\, dr$

we have $\qquad\qquad [\Gamma(1/2)]^2 = 4 \int_{\theta=0}^{\pi/2} \int_{r=0}^\infty e^{-r^2} r\, d\theta\, dr$.

The area of integration in the positive quadrant of plane is

$$= 2 \int_0^{\pi/2} \left\{ \int_0^\infty 2e^{-r^2} r\,.dr \right\} d\theta .$$

$$= 2 \int_0^{\pi/2} \left[-e^{-v} \right]_0^\infty d\theta = 2 \int_0^{\pi/2} d\theta = 2 [\theta]_0^{\pi/2} = \pi$$

(Putting $r^2 = v$, so that $2r\, dr = dv$)

Therefore, $[\Gamma(1/2)]^2 = \pi$ so that $\Gamma(1/2) = \sqrt{\pi}$.

(5) $\Gamma(n) = \int_0^1 (\log 1 / y)^{n-1} dy$.

Proof. By definition of gamma function, we have

$$\Gamma(n) = \int_0^\infty e^{-x} x^{n-1} dx, n > 0$$

Putting $x = \log (1/y)$ in gamma function, we get

$$\Gamma(n) = -\int_1^0 (\log 1 / y)^{n-1} dy = \int_0^1 (\log 1 / y)^{n-1} dy.$$

18.3 SOME TRANSFORMATIONS OF GAMMA FUNCTION

Gamma function is given by

$$\Gamma(n) = \int_0^\infty x^{n-1} e^{-x} dx .$$ ...(1)

(1) $\qquad\qquad \dfrac{\Gamma(n)}{a^n} = \int_0^\infty e^{-ay} y^{n-1} dy, n > 0, a > 0$

Proof. We have

$$\Gamma(n) = \int_0^\infty x^{n-1} e^{-x} dx , n > 0.$$

Put $x = ay$, so that $\qquad dx = a\, dy$.

When $x = 0$, $y = 0$ and when $x \to \infty$, $y \to \infty$.

$\therefore \qquad\qquad\qquad \Gamma(n) = \int_0^\infty e^{-ay} (ay)^{n-1}.a\, dy$.

Hence, $\qquad \int_0^\infty e^{-ay} y^{n-1} dy = \dfrac{\Gamma(n)}{a^n}$.

(2) $\qquad\qquad\qquad \Gamma(n) = \dfrac{1}{n} \int_0^\infty e^{-x^{1/n}} dx, n > 0$

Proof. We have $\qquad \Gamma(n) = \int_0^\infty e^{-x} x^{n-1} dx, n > 0$ ...(1)

Put $\qquad\qquad x^n = t$. i.e., $nx^{n-1} dx = dt$, then (1) gives

$$\Gamma(n) = \frac{1}{n} \int_0^\infty e^{-t^{1/n}} dt$$

$\Rightarrow \qquad\qquad \Gamma(n) = \dfrac{1}{n} \int_0^\infty e^{-x^{1/n}} dx$ $\qquad\qquad$ [By the property of definite integral]

(3) $\Gamma(n) = 2\int_0^\infty e^{-x^2} x^{2n-1} dx, n > 0$

Proof.   We have

$$\Gamma(n) = \int_0^\infty e^{-x} x^{n-1} dx, \qquad \qquad \ldots(1)$$

Put $x = t^2$, so that $dx = 2t\, dt$,

Therefore, $\Gamma(n) = \int_0^\infty e^{-t^2} (t^2)^{n-1} 2t\, dt$

or $\Gamma(n) = 2\int_0^\infty e^{-t^2} t^{2n-1} dt$

$\Rightarrow$ $\Gamma(n) = 2\int_0^\infty e^{-x^2} x^{2n-1} dx$.

Solved Examples

Example 1.   *Evaluate :*

(i) $\int_0^\infty e^{-x} x^4\, dx$

(ii) $\int_0^\infty x^6 e^{-2x}\, dx$

Based on the following Results

- $\Gamma(n) = \int_0^\infty e^{-x} x^{n-1} dx,\ n > 0$
- $\Gamma(n+1) = n\,\Gamma(n)$
- $\Gamma\left(\dfrac{1}{2}\right) = \sqrt{\pi}$
- $\Gamma(n+1) = n!$
- $\Gamma(1) = 1$

Solution.   (i) We have $\int_0^\infty e^{-x} x^4\, dx$

$$= \int_0^\infty e^{-x} x^{5-1}\, dx$$

[By definition of gamma function]
$$= \Gamma(5) = (4)! = 24.$$

(ii) Let $I = \int_0^\infty x^6 e^{-2x} dx$
$$\ldots(1)$$

Put $2x = t$, so that $dx = 1/2\, dt$ then

$$I = \int_0^\infty \left(\frac{t}{2}\right)^6 e^{-t} \cdot \frac{1}{2} dt = \frac{1}{2^7} \int_0^\infty e^{-t} t^{7-1} dt$$

$$= \frac{1}{2^7} \Gamma(7) \qquad \text{[By definition of gamma function]}$$

$$= \frac{1}{2^7} \times (6!) = \frac{45}{8}.$$

Example 2.   *Show that*

$$\int_0^1 \frac{dx}{\sqrt{(-\log x)}} = \sqrt{\pi}\ .$$

Solution.   We know that
$$\Gamma(n) = \int_0^1 (-\log x)^{n-1} dx$$

Putting $n = 1/2$, we have

$$\Gamma(1/2) = \int_0^1 (-\log x)^{(1/2)-1} dx$$

or $\sqrt{\pi} = \int_0^1 (-\log x)^{-1/2} dx$

or $\sqrt{\pi} = \int_0^1 \frac{dx}{\sqrt{(-\log x)}}.$

Example 3. *If n is a positive integer, prove that*
$$2n\ \Gamma(n+1/2) = 1.\ 3.\ 5.\ ...(2n-1)\ \sqrt{\pi}$$

Solution. We know that $\Gamma(n+1) = n\Gamma(n)$ ...(1)

Now $\Gamma(n + 1/2) = \Gamma(n - 1/2 + 1)$

$= (n - 1/2)\ \Gamma(n - 1/2)$ [Using (1)]

$= (n - 1/2)\ \Gamma(n - 3/2 + 1)$

$= (n - 1/2)\ (n - 3/2)\ \Gamma(n - 3/2)$

$= \dfrac{2n-1}{2} \cdot \dfrac{2n-3}{2} \cdot \Gamma\left(\dfrac{2n-3}{2}\right)$

$= \dfrac{2n-1}{2} \cdot \dfrac{2n-3}{2} \cdot \dfrac{5}{2} \cdot \dfrac{3}{2} \cdot \dfrac{1}{2} \cdot \Gamma\left(\dfrac{1}{2}\right)$

$= \dfrac{(2n-1)(2n-3)...5.3.1}{2^n}\sqrt{\pi}$ $[\because \Gamma(1/2) = \sqrt{\pi}]$

Hence, $2^n\Gamma(n + 1/2) = (2n-1)\ (2n-3)\ ...\ 5.3.1.\ \sqrt{\pi}$

Example 4. *Show that* $\int_0^\infty \exp(2ax - x^2)\,dx = \dfrac{1}{2}\sqrt{\pi}\exp a^2.$

Solution. Consider $\int_0^\infty \exp(2ax - x^2)\,dx$

$= \int_0^\infty e^{2ax-x^2}\,dx = \int_0^\infty e^{a^2-(x^2-2ax+a^2)}.\,dx$

$= \int_0^\infty e^{a^2-(x-a)^2}\,dx = e^{a^2}\int_0^\infty e^{-(x-a)^2}\,dx$

$= e^{a^2}\int_0^\infty e^{-t^2}\,dt.$ Put $x - a = t, \therefore dx = dt$

$\Rightarrow \int_0^\infty \exp(2ax - x^2)\,dx = \exp a^2 \int_0^\infty e^{-t^2}\,dt.$ ...(1)

Now $\Gamma(n) = \int_0^\infty e^{-u}u^{n-1}\,du.$ ...(2)

Putting $n=1/2$ in (2), we have

$\Gamma(1/2) = \int_0^\infty e^{-u}u^{-1/2}\,du$. ...(3)

Putting $u=t^2$ so that $du = 2t\,dt$ in (3), we get

$\Gamma(1/2) = \int_0^\infty e^{-t^2}t^{-1}.2t\,dt$

or $\sqrt{\pi} = 2\int_0^\infty e^{-t^2}\,dt$ or $\int_0^\infty e^{-t^2}\,dt = \sqrt{\pi}/2$ ...(4)

Using (4), (1) reduces to

$\int_0^\infty \exp(2ax - x^2)\,dx = \dfrac{1}{2}\sqrt{\pi}\exp a^2.$

Example 5. *Evaluate* $\int_0^\infty t^{-3/2}(1 - e^{-t})\,dt$.

Solution. We have $\int_0^\infty t^{-3/2}(1 - e^{-t})\,dt$

$= (1 - e^{-t})\left[\dfrac{t^{-1/2}}{-1/2}\right]_0^\infty - \int_0^\infty (e^{-t})\left(\dfrac{t^{-1/2}}{-1/2}\right)dt$

$$= 0 + 2\int_0^\infty e^{-t} t^{(1/2)-1} dt$$

$$= 2\Gamma(1/2) \qquad \text{[By definition of gamma function]}$$

$$= 2\sqrt{\pi}.$$

Example 6. *Prove that*

(i) $\int_0^\infty xe^{-\alpha x}\cos\beta x\, dx = \dfrac{\alpha^2 - \beta^2}{(\alpha^2 + \beta^2)^2}$, $\alpha > 0$ *(Remember)*

(ii) $\int_0^\infty xe^{-\alpha x}\sin\beta x\, dx = \dfrac{2\alpha\beta}{(\alpha^2 + \beta^2)^2}$, $\alpha > 0$ *(Remember)*

Solution. We know that

$$\int_0^\infty e^{-kx} x^{n-1} dx = \frac{\Gamma(n)}{k^n}, n > 0, k > 0. \qquad \text{...(1)}$$

Putting $k = \alpha - i\beta$ and $n = 2$ in (1), we get

$$\int_0^\infty e^{-(\alpha - i\beta)x} x\, dx = \frac{\Gamma(2)}{(\alpha - i\beta)^2}$$

or $\displaystyle\int_0^\infty xe^{-\alpha x} e^{i\beta x} dx = \dfrac{(\alpha + i\beta)^2}{(\alpha - i\beta)^2 (\alpha + i\beta)^2}$ [as $\Gamma(2)=1$]

$$\int_0^\infty xe^{-\alpha x} e^{i\beta x} dx = \frac{\alpha^2 - \beta^2 + 2i\alpha\beta}{[(\alpha + i\beta)(\alpha - i\beta)]^2}$$

$\Rightarrow \displaystyle\int_0^\infty xe^{-\alpha x}(\cos\beta x + i\sin\beta x)\, dx = \dfrac{\alpha^2 - \beta^2 + 2i\alpha\beta}{(\alpha^2 + \beta^2)^2}$

or $\displaystyle\int_0^\infty xe^{-\alpha x}\cos\beta x\, dx + i\int_0^\infty xe^{-\alpha x}\sin\beta x$

$$= \frac{\alpha^2 - \beta^2}{(\alpha^2 + \beta^2)^2} + i\frac{2\alpha\beta}{(\alpha^2 + \beta^2)^2}.$$

Equating real and imaginary parts of both sides, we get

$$\int_0^\infty xe^{-\alpha x}\cos\beta x\, dx = \frac{\alpha^2 - \beta^2}{(\alpha^2 + \beta^2)^2}$$

and $\displaystyle\int_0^\infty xe^{-\alpha x}\sin\beta x\, dx = \dfrac{2\alpha\beta}{(\alpha^2 + \beta^2)^2}$.

Example 7. *Show that*

$$\int_0^\infty \frac{x^c}{c^x} dx = \frac{\Gamma(c+1)}{(\log c)^{c+1}}, c > 0.$$

Solution. We have $\displaystyle\int_0^\infty \frac{x^c}{c^x} dx = \int_0^\infty x^c c^{-x}\, dx$

$$= \int_0^\infty x^c [e^{\log_e c}]^{-x} dx \qquad [\because c = e^{\log_e c} \text{ if } c \geq 0]$$

$$= \int_0^\infty x^{(c+1)-1} e^{-x \log_e c} dx$$

$$= \frac{\Gamma(c+1)}{(\log_e c)^{c+1}} \left[\because \int_0^\infty x^{n-1} e^{-kx} dx = \frac{\Gamma(n)}{k^n}; n > 0, k > 0\right]$$

Example 8. *With certain limitations on the values of a, b, m and n, prove that*

$$\int_0^\infty \int_0^\infty e^{-(ax^2+by^2)}x^{2m-1}y^{2n-1}dx\,dy = \frac{\Gamma(m)\Gamma(n)}{4a^m b^n}.$$

Solution. Let

$$I = \int_0^\infty \int_0^\infty e^{-(ax^2+by^2)}x^{2m-1}y^{2n-1}dx\,dy \qquad \text{...(1)}$$

$$\Rightarrow \qquad I = \int_0^\infty e^{-ax^2}x^{2m-1}\,dx \times \int_0^\infty e^{-by^2}y^{2n-1}\,dy = I_1 \times I_2 \qquad \text{...(2)}$$

where

$$I_1 = \int_0^\infty e^{-ax^2}x^{2m-1}\,dx \qquad \text{...(3)}$$

$$I_2 = \int_0^\infty e^{-by^2}y^{2n-1}\,dy \qquad \text{...(4)}$$

Put $ax^2 = t$, $\quad x = \left(\dfrac{t}{a}\right)^{1/2}$ so that $dx = \dfrac{dt}{2\sqrt{at}}$ then equation (3) becomes

$$I_1 = \int_0^\infty e^{-t}\left[\frac{t}{a}\right]^{\frac{(2m-1)}{2}}\frac{dt}{2\sqrt{at}} = \frac{1}{2a^m}\int_0^\infty e^{-t}t^{m-1}\,dt$$

$$= \frac{\Gamma(m)}{2a^m} \quad \text{[By definition of gamma function taking } n>0, a>0]$$

Then $\qquad I_2 = \dfrac{\Gamma(n)}{2b^n}$ if $n > 0, b > 0$.

$\therefore$ from (1) and (2), we get

$$I = I_1 \times I_2 = \frac{\Gamma(m)\Gamma(n)}{4a^m b^n}.$$

18.4 BETA FUNCTION

Definition. *The definite integral* $\int_0^1 x^{m-1}(1-x)^{n-1}dx,$ *for* $m > 0, n > 0$ *is known as the Beta function and denoted by B(m, n) which is read as "Beta m, n", where m, n are positive number or integers. Thus* $B(m,n) = \int_0^1 x^{m-1}(1-x)^{n-1}\,dx$.

REMARK

- Beta function is also called the Eulerian integral of first kind.

18.5 PROPERTIES OF BETA FUNCTION

(1) *Symmetry of beta function i.e.,* $B(m, n) = B(n, m)$

    Proof. By definition of beta function, we have

$$B(m,n) = \int_0^1 x^{m-1}(1-x)^{n-1}\,dx$$

$$= \int_0^1 (1-x)^{m-1}[1-(1-x)]^{n-1}dx \quad \left[\because \int_0^a f(x)dx = \int_0^a f(a-x)dx\right]$$

$$= \int_0^1 (1-x)^{m-1}x^{n-1}dx = \int_0^1 x^{n-1}(1-x)^{m-1}\,dx$$

$$= B(n,m) \qquad\qquad\qquad \text{[By definition of Beta function]}$$

$$B(m,n) = B(n,m)$$

i.e., the interchange of position of m and n does not change the value of beta function.

REMARK

- This is the fundamental property of beta function and also called symmetric property of beta function.

(2) *Beta function B(m, n) can be evaluated in an explicit form if m or n is a positive integer.*
Proof. Case I. *When 'n' is a positive integer.*
If $n = 1$, then by definition of Beta function, we have

$$B(m,n) = \int_0^1 x^{m-1}(1-x)^{n-1}\,dx \qquad\qquad \dots(1)$$

$\Rightarrow$

$$B(m,1) = \int_0^1 x^{m-1}(1-x)^{1-1}\,dx$$

$$= \int_0^1 x^{m-1}\,dx = \left[\frac{x^m}{m}\right]_0^1 = \frac{1}{m}. \qquad\qquad \dots(2)$$

Now, let $n > 1$, then from (1), we have

$$B(m,n) = \int_0^1 (1-x)^{n-1}x^{m-1}\,dx$$

$$= \left[(1-x)^{n-1}.\frac{x^m}{m}\right]_0^1 - \int_0^1 (n-1)(1-x)^{n-2}.(-1)\frac{x^m}{m}\,dx.$$

Integrating by parts taking x^{m-1} as second function, we have

$$B(m, n) = 0 + \frac{n-1}{m}\int_0^1 x^m(1-x)^{n-2}\,dx \qquad \left[\begin{array}{l} \because n > 1 \\[2mm] \text{and } \lim_{x\to 0}(1-x)^{n-1}\dfrac{x^m}{m} = 0 \end{array}\right]$$

$$= \frac{n-1}{m}\int_0^1 x^{(m+1)-1}(1-x)^{(n-1)-1}\,dx$$

$$= \frac{n-1}{m}B(m+1, n-1)$$

Thus

$$B(m,n) = \frac{n-1}{m}B(m+1, n-1) \qquad\qquad \dots(3)$$

Now replacing m by $m+1$ and n by $n-1$ in (3), we get

$$B(m+1, n-1) = \frac{n-1-1}{m+1}B(m+2, n-2) \qquad\qquad \dots(4)$$

Using equation (4), the equation (3) becomes

$$B(m,n) = \frac{n-1}{m}.\frac{n-2}{m+1}B(m+2, n-2) \qquad\qquad \dots(5)$$

After applying the above process successively, we get

$$B(m,n) = \frac{n-1}{m}.\frac{n-2}{m+1}.\frac{n-3}{m+2}\cdots\frac{1}{m+n-2}B(m+n-1, 1) \qquad \dots(6)$$

$$= \frac{n-1}{m}.\frac{n-2}{m+1}.\frac{n-3}{m+2}\cdots\frac{1}{m+n-2}\int_0^1 x^{m+n-2}(1-x)^0\,dx$$

$$= \frac{n-1}{m}.\frac{n-2}{m+1}.\frac{n-3}{m+2}\cdots\frac{1}{m+n-2}\left[\frac{x^{m+n-1}}{m+n-1}\right]_0^1$$

$$= \frac{n-1}{m} \cdot \frac{n-2}{m+1} \cdot \frac{n-3}{m+2} \cdots \frac{1}{m+n-2} \cdot \frac{1}{m+n-1}$$

$\Rightarrow \qquad B(m,n) = \dfrac{n-1}{m} \cdot \dfrac{n-2}{m+1} \cdot \dfrac{n-3}{m+2} \cdots \dfrac{1}{m+n-2} \cdot \dfrac{1}{m+n-1}$

$\therefore \qquad B(m,n) = \dfrac{(n-1)!}{m(m+1)(m+2)\ldots(m+n-2)(m+n-1)} \qquad \ldots(7)$

Case II. *When m is a positive integer.*
Since the beta function is symmetrical in m and n i.e., $B(m, n) = B(n, m)$ therefore by interchanging m and n in Case I we get

$$B(m,n) = \frac{(m-1)!}{n(n+1)(n+2)\ldots(n+m-2)(n+m-1)}. \qquad \ldots(8)$$

Case III. *When both m and n are positive integers.*
We have, by Case I

$$B(m,n) = \frac{(n-1)!}{m(m+1)(m+2)\ldots(m+n-2)(m+n-1)}.$$

$$= \frac{[1.2.3\ldots(m-1)](n-1)!}{1.2.3\ldots m(m+1)(m+2)\ldots(m+n-2)(m+n-1)}$$

Multiplying both numerator and denominator by 1. 2. 3. .. .(m–1) !, we get

$$B(m,n) = \frac{(m-1)!(n-1)!}{(m+n-1)!}$$

18.6 TRANSFORMATION OF BETA FUNCTION

The beta function

$$B(m,n) = \int_0^1 x^{m-1}(1-x)^{n-1}\, dx \qquad \ldots(A)$$

can be transformed into many forms given below :

(I) $B(m,n) = \displaystyle\int_0^\infty \frac{x^{n-1}}{(1+x)^{m+n}}\, dx = \int_0^\infty \frac{x^{m-1}}{(1+x)^{m+n}}\, dx$.

Proof. Put $x = \dfrac{1}{(1+y)}$ and $dx = -\dfrac{dy}{(1+y)^2}$ and $y \to 0$ when $x=1$, $y \to \infty$, when $x = 0$.

$\therefore \qquad B(m,n) = \displaystyle\int_\infty^0 \left(\frac{1}{1+y}\right)^{m-1}\left[1 - \frac{1}{1+y}\right]^{n-1}\left[\frac{-dy}{(1+y)^2}\right]$

$\qquad\qquad = \displaystyle\int_0^\infty \frac{(y)^{n-1}}{(1+y)^{m+1}}\left(\frac{1}{1+y}\right)^{n-1}.dy = \int_0^\infty \frac{y^{n-1}}{(1+y)^{m+n}}\, dy$

$\Rightarrow \qquad B(m,n) = \displaystyle\int_0^\infty \frac{x^{n-1}\, dx}{(1+x)^{m+n}}. \qquad \ldots(1)$

Since m and n are interchangeable in beta function by symmetric property therefore (1) gives

$$B(m,n) = \int_0^\infty \frac{x^{m-1}}{(1+x)^{m+n}}\, dx$$

thus $\qquad B(m,n) = \displaystyle\int_0^\infty \frac{x^{n-1}\, dx}{(1+x)^{m+n}} = \int_0^\infty \frac{x^{m-1}\, dx}{(1+x)^{m+n}}.$

(II) $B(m,n) = 2\int_0^{\pi/2} \cos^{2m-1}\theta \sin^{2n-1}\theta\, d\theta$.

Proof. Put $x = \sin 2\theta$ and $dx = 2\sin\theta\cos\theta\, d\theta$ and when $x = 0$, $\theta = \pi/2$ when $x = 1$ in (A) we get

$$B(m,n) = 2\int_0^{\pi/2} \sin^{2m-1}\theta \cos^{2n-1}\theta\, d\theta$$

$$= 2\int_0^{\pi/2} \cos^{2m-1}\theta \sin^{2n-1}\theta\, d\theta$$

[By symmetric property of beta function]

(III) $B(m,n) = \dfrac{1}{a^{m+n-1}}\int_0^a x^{m-1}(a-x)^{n-1}dx$.

Proof. Put $\quad x = \dfrac{y}{a}$, i.e., $dx = \dfrac{1}{a}dy$ and when $x \to 0$, then $y \to 0$, when $x = 1$ then $y \to a$.

So $\qquad B(m,n) = \dfrac{1}{a^{m+n-1}}\int_0^a y^{m-1}(a-y)^{n-1}dy$

$$= \dfrac{1}{a^{m+n-1}}\int_0^a x^{m-1}(a-x)^{n-1}dx.$$

(IV) $\dfrac{B(m,n)}{a^n(1+a)^m} = \int_0^1 \dfrac{x^{m-1}(1-x)^{n-1}dx}{(x+a)^{m+n}}$.

Proof. Let $\quad \dfrac{x}{1+a} = \dfrac{t}{t+a}$, (Remember)

$\Rightarrow \qquad\qquad dx = a(1+a)\dfrac{dt}{(t+a)^2}$

then we have

$$B(m,n) = \int_0^1 (1+a)^{m-1}\left(\dfrac{t}{t+a}\right)^{m-1} a^{n-1}\left(\dfrac{1-t}{a+t}\right)^{n-1}\dfrac{a(a+1)}{(t+a)^2}dt$$

$$= a^n(1+a)^m \int_0^1 \dfrac{t^{m-1}(1-t)^{n-1}}{(t+a)^{m+n}}dt$$

$$= a^n(1+a)^m \int_0^1 \dfrac{x^{m-1}(1-x)^{n-1}}{(x+a)^{m+n}}dx$$

Hence, $\qquad \dfrac{B(m,n)}{a^n(1+a)^m} = \int_0^1 \dfrac{x^{m-1}(1-x)^{n-1}}{(x+a)^{m+n}}dx$

(V) $B(m,n)(a-b)^{m+n-1} = \int_b^a (x-b)^{m-1}(a-x)^{n-1}dx$.

Proof. Put $\quad x = \dfrac{t-b}{a-b}$ so that $dx = \dfrac{dt}{a-b}$. in (A), we get

$$B(m,n) = \int_b^a \left(\dfrac{t-b}{a-b}\right)^{m-1}\left(\dfrac{a-t}{a-b}\right)^{n-1}\dfrac{dt}{a-b}$$

$$= \dfrac{1}{(a-b)^{m+n-1}}\int_b^a (t-b)^{m-1}(a-t)^{n-1}dx$$

$$= \dfrac{1}{(a-b)^{m+n-1}}\int_b^a (x-b)^{m-1}(a-x)^{n-1}dx.$$

$\therefore B(m,n)(a-b)^{m+n-1} = \int_b^a (x-b)^{m-1}(a-x)^{n-1}dx.$

(VI) $\dfrac{1}{a^n b^m} B(m,n) = \int_0^1 \dfrac{x^{m-1}(1-x)^{n-1}\,dx}{\{a+(b-a)x\}^{m+n}}$

Proof.   We put

$$\frac{a}{y} - \frac{b}{x} = a-b. \text{ (Remember)} \qquad \qquad \dots(1)$$

$\therefore \qquad\qquad\qquad \dfrac{b}{x} = \dfrac{a}{y} + (b-a) = \dfrac{a+(b-a)y}{y}$

$\Rightarrow \qquad\qquad\qquad x = \dfrac{by}{a+(b-a)y} \qquad\qquad\qquad\qquad \dots(2)$

$\Rightarrow \qquad\qquad\qquad dx = \dfrac{b[a+(b-a)y] - by(b-a)}{\{a+(b-a)y\}^2}\,dy$

i.e., $\qquad\qquad\qquad dx = \dfrac{ab\,dy}{[a+(b-a)y]^2}. \qquad\qquad\qquad \dots(3)$

Again from (1), we see that when $x = 1, y = 1$ and $x = 0, y = 0$.
and from (2), we have

$$1 - x = 1 - \frac{by}{a+by-ay} = \frac{a(1-y)}{a+(b-a)y}. \qquad\qquad \dots(4)$$

Using (2), (3) and (4), (1) gives

$$B(m,n) = \int_0^1 \left\{ \frac{by}{a+(b-a)y} \right\}^{m-1} \left\{ \frac{a(1-y)}{a+(b-a)y} \right\}^{n-1} \frac{ab\,dy}{\{a+(b-a)y\}^2}$$

$$= a^n b^m \int_0^1 \frac{y^{m-1}(1-y)^{n-1}}{\{a+(b-a)y\}^{m+n}}\,dy = a^n b^m \int_0^1 \frac{x^{m-1}(1-x)^{n-1}\,dx}{\{a+(b-a)x\}^{m+n}}$$

$$\Rightarrow \qquad \frac{1}{a^n b^m} B(m,n) = \int_0^1 \frac{x^{m-1}(1-x)^{n-1}\,dx}{\{a+(b-a)x\}^{m+n}}.$$

Solved Examples

Based on the following Results

Example 1.   *Express* $\int_0^1 x^m(1-x^p)^n\,dx$ *in terms of beta function and hence evaluate*

$$\int_0^1 x^5(1-x^3)^{10}\,dx.$$

Solution.   Put $x^p = t$ so that

$$dx = \left(\frac{1}{p}\right) t^{1/p-1}\,dt.$$

- $B(m,n) = \int_0^1 x^{m-1}(1-x)^{n-1}\,dx$

- $B(m,n) = \dfrac{(m-1)!(n-1)!}{(m+n-1)!}$

- $B(m,n) = 2\int_0^{\pi/2} \cos^{2m-1}\theta \sin^{2n-1}\theta\,d\theta$

- $B(m,n) = \dfrac{1}{(a-b)^{m+n-1}} \int_a^b (x-b)^{m-1}(a-x)^{n-1}\,dx$

$\therefore$

$$\int_0^1 x^m(1-x^p)^n\,dx = \int_0^1 t^{m/p}(1-t)^n(1/p)t^{1/p-1}\,dt$$

$$= \frac{1}{p}\int_0^1 t^{(m+1)/p-1}(1-t)^{n+1-1}\,dt$$

$$= \frac{1}{p} B\left(\frac{m+1}{p}, n+1\right).$$

Writing $m = 5$, $n = 10$ and $p = 3$, we have

$$\int_0^1 x^5 (1-x^3)^{10} dx = \frac{1}{3} B\left(\frac{5+1}{3}, 11\right) = \frac{1}{3} B(2,11)$$

$$= \frac{1}{3} \frac{\Gamma(2)\Gamma(11)}{\Gamma(13)} = \frac{1}{3} \frac{1.\Gamma(11)}{12.11\Gamma(11)} = \frac{1}{3.12.11} = \frac{1}{396}.$$

Example 2. *Evaluate the following integrals by expressing them in terms of beta function*

(i) $\int_0^1 x^m (1-x^2)^n \, dx$, $m > 1, n > -1$ (ii) $\int_0^1 \dfrac{x^2 dx}{\sqrt{(1-x^5)}}$.

Solution. (i) We have

$$\int_0^1 x^m (1-x^2)^n dx = \int_0^1 x^{m-1}(1-x^2)^n x \, . dx$$

$$= \int_0^1 y^{\frac{(m-1)}{2}} (1-y)^n . \frac{dy}{2}$$

(Putting $x^2 = y$ so that $2x \, dx = dy$)

$$= \frac{1}{2} \int_0^1 y^{\frac{(m-1)}{2}}(1-y)^n dy$$

$$= \frac{1}{2} \int_0^1 y^{\frac{m+1}{2}-1}(1-y)^{(n+1)-1} dy = \frac{1}{2} B\left[\frac{1}{2}(m+1), n+1\right].$$

(ii) We have

$$\int_0^1 \frac{x^2}{\sqrt{(1-x^5)}} dx = \int_0^1 x^2 (1-x^5)^{-1/2} dx$$

$$= \int_0^1 x^2 . \frac{1}{x^4}(1-x^5)^{-1/2} x^4 dx = \int_0^1 x^{-2}(1-x^5)^{-1/2} x^4 dx$$

$$= \int_0^1 y^{-2/5}(1-y)^{-1/2} \frac{1}{5} dy \quad \text{(Putting } x^5 = y, i.e., 5x^4 dx = dy)$$

$$= \frac{1}{5} \int_0^1 y^{-2/5}(1-y)^{-1/2} dy$$

$$= \frac{1}{5} \int_0^1 y^{(3/5)-1}(1-y)^{(1/2)-1} dy = \frac{1}{5} B\left(\frac{3}{5}, \frac{1}{2}\right).$$

Example 3. *Show that* $\int_0^1 \dfrac{x^{m-1}(1-x)^{n-1}}{(a+bx)^{m+n}} dx = \dfrac{1}{(a+b)^m a^n} B(m,n)$.

Solution. Let $I = \int_0^1 \dfrac{x^{m-1}(1-x)^{n-1}}{(a+bx)^{m+n}} dx$

$$= \int_0^1 \left(\frac{x}{a+bx}\right)^{m-1} . \left(\frac{1-x}{a+bx}\right)^{n-1} \frac{1}{(a+bx)^2} dx.$$

Put $\dfrac{x}{a+bx} = \dfrac{y}{a+b}$ i.e. $\dfrac{(a+bx)-x.b}{(a+bx)^2} dx = \dfrac{dy}{a+b}$ $\Rightarrow$ $\dfrac{1}{(a+bx)^2} dx = \dfrac{dy}{a(a+b)}$

$$\Rightarrow \qquad \frac{1-x}{a+bx} = \frac{1}{a}\left(\frac{a-ax}{a+bx}\right) = \frac{1}{a}\left[\frac{a+bx-ax-bx}{a+bx}\right]$$

$$= \frac{1}{a}\left[1 - \frac{x(a+b)}{a+bx}\right] = \frac{1-y}{a}.$$

Also when $x = 0$, $y = 0$, and when $x = 1$, $y = 1$. Therefore,

$$I = \int_0^1 \left(\frac{y}{a+b}\right)^{m-1}\left(\frac{1-y}{a}\right)^{n-1} \cdot \frac{dy}{a(a+b)}$$

$$= \frac{1}{(a+b)^m \cdot a^n}\int_0^1 y^{n-1}(1-y)^{n-1}dy = \frac{B(m,n)}{(a+b)^m \cdot a^n}.$$

Example 4. *Evaluate $\int_0^\infty x^m e^{-ax^n}dx$, when m, n and a are all positive constant.*

Solution. Let $ax^n = y \quad \Rightarrow \quad nax^{n-1} = dy$ and at $x = 0, y = 0$ and at $x = \infty, y = \infty.$

$$\therefore \qquad \int_0^\infty x^m e^{-ax^n}dx = \frac{1}{na}\int_0^\infty \left(\frac{y}{a}\right)^{m/n}\left(\frac{y}{a}\right)^{1/n-1}e^{-y}dy$$

$$= \frac{1}{na^{(m+1)/n}}\int_0^\infty y^{\left(\frac{m+1}{n}\right)-1}\cdot e^{-y}dy$$

$$= \frac{1}{na^{(m+1)/n}}\Gamma\left(\frac{m+1}{n}\right).$$

Example 5. *Prove that*

$$\int_0^\infty \frac{x^{m-1}-x^{n-1}}{(1+x)^{m+n}}dx = 0, m > 0, n > 0.$$

Solution. We have

$$\int_0^\infty \frac{x^{m-1}}{(1+x)^{m+n}}dx - \int_0^\infty \frac{x^{n-1}}{(1+x)^{m+n}}dx$$

$$= B(m,n) - B(n,m) = B(m,n) - B(m,n) = 0.$$

18.7 RELATION BETWEEN BETA AND GAMMA FUNCTION

We have $\qquad B(m,n) = \dfrac{\Gamma(m)\Gamma(n)}{\Gamma(m+n)}, m > 0, n > 0$

$$= \int_0^\infty \frac{y^{n-1}dy}{(1+y)^{m+n}}.$$

Proof. We have $\int_0^\infty y^{n-1}e^{-xy}dy = \dfrac{\Gamma(n)}{x^n}$

or $\qquad \Gamma(n) = \int_0^\infty x^n y^{n-1}e^{-xy}dy$ ...(1)

Also $\qquad \Gamma(m) = \int_0^\infty x^{m-1}e^{-x}dx$ ...(2)

Multiplying both sides of (1) by $x^{m-1}e^{-x}$, we have

$$\Gamma(n).x^{m-1}e^{-x} = \int_0^\infty x^{n+m-1}y^{n-1}e^{-(y+1)x}dy \cdot$$

Integrating both sides with respect to x within limits $x = 0$ to $x = \infty$, we have

$$\Gamma(n)\int_0^\infty x^{m-1}e^{-x}dx = \int_0^\infty \left[\int_0^\infty x^{n+m-1}e^{-(y+1)x}dx\right]y^{n-1}dy \qquad \ldots(3)$$

But $\quad \int_0^\infty x^{(n+m)-1}e^{-(y+1)x}dx = \dfrac{\Gamma(n+m)}{(1+y)^{m+n}}.$

Hence with the help of this result and (2), we get from (3)

$$\Gamma(n)\Gamma(m) = \int_0^\infty \Gamma(n+m)\frac{y^{n-1}}{(1+y)^{n+m}}dy$$

$$= \Gamma(n+m)\int_0^\infty \frac{y^{n-1}}{(1+y)^{n+m}}dy = \Gamma(n+m)B(m,n)$$

or $\qquad\qquad B(m,n) = \dfrac{\Gamma(m)\Gamma(n)}{\Gamma(n+m)}.$

Deduction 1. $\Gamma(n)\Gamma(1-n) = \dfrac{\pi}{\sin n\pi}$, where $0 < n < 1.$

Proof. We have

$$B(m,n) = \int_0^\infty \frac{x^{n-1}dx}{(1+x)^{m+n}}, \ m > 0, \ n > 0.$$

Therefore the relation between beta and gamma functions becomes

$$\int_0^\infty \frac{x^{n-1}dx}{(1+x)^{m+n}} = \frac{\Gamma(m)\Gamma(n)}{\Gamma(m+n)}.$$

Taking $m + n = 1$, so that $\quad m = 1 - n$, we get

$$\int_0^\infty \frac{x^{n-1}}{1+x}dx = \frac{\Gamma(1-n)\Gamma(n)}{\Gamma(1)}, \ 0 < n < 1.$$

$$[\because m > 0 \Rightarrow 1 - n > 0 \Rightarrow n < 1. \text{ Also } n > 0]$$

But also we know that

$$\int_0^\infty \frac{x^{n-1}}{1+x}dx = \frac{\pi}{\sin n\pi} \ \text{ and } \Gamma(1) = 1$$

$\therefore \qquad\qquad \dfrac{\pi}{\sin n\pi} = \Gamma(1-n)\Gamma(n), \ 0 < n < 1.$

Deduction 2. $\Gamma(1/2) = \sqrt{\pi}.$

Proof. We have just proved that

$$\Gamma(n)\Gamma(1-n) = \frac{\pi}{\sin n\pi} \qquad \ldots(1)$$

Putting $n = 1/2$ in (1), we obtain

$$\Gamma\left(\frac{1}{2}\right)\Gamma\left(1-\frac{1}{2}\right) = \frac{\pi}{\sin \pi/2} \ \text{ or } \ \left[\Gamma\left(\frac{1}{2}\right)\right]^2 = \pi$$

$$\Gamma\left(\frac{1}{2}\right) = \sqrt{\pi}.$$

Aliter. We know $\qquad B(m,n) = \dfrac{\Gamma(m)\Gamma(n)}{\Gamma(m+n)}.$

Putting $m = n = 1/2$, we get

$$B(1/2,1/2) = \frac{\Gamma(1/2)\Gamma(1/2)}{\Gamma(1/2+1/2)} = \frac{\{\Gamma(1/2)\}^2}{\Gamma(1)} \qquad [\because \Gamma(1)=1]$$

or

$$\{\Gamma(1/2)\}^2 = B(1/2,1/2) = \int_0^1 x^{(1/2)-1}(1-x)^{(1/2)-1}dx$$

$$= \int_0^1 x^{-1/2}(1-x)^{-1/2}dx = \int_0^1 \frac{dx}{\sqrt{x}\sqrt{1-x}}$$

$$= \int_0^{\pi/2} \frac{2\sin\theta\cos\theta\, d\theta}{\sin\theta\sqrt{(1-\sin^2\theta)}} \qquad \text{(By putting } x = \sin^2\theta)$$

$$= 2\int_0^{\pi/2} d\theta = 2[\theta]_0^{\pi/2} = 2\left(\frac{\pi}{2}\right) = \pi$$

$\Rightarrow$

$$\left\{\Gamma\left(\frac{1}{2}\right)\right\}^2 = \pi \Rightarrow \Gamma\left(\frac{1}{2}\right) = \sqrt{\pi}.$$

Deduction 3. $\int_0^1 e^{-x^2}dx = \dfrac{1}{2}\sqrt{\pi}.$

Proof. We have $\int_0^1 e^{-x^2}dx = \int_0^1 e^{-y}.\dfrac{1}{2\sqrt{y}}dy$, putting $x^2 = y$, $2x\,dx = dy$

$$= \frac{1}{2}\int_0^\infty e^{-y}y^{-1/2}dy = \frac{1}{2}\int_0^1 e^{-y}y^{(1/2)-1}dy$$

$$= \frac{1}{2}\Gamma\left(\frac{1}{2}\right) \qquad\qquad \left[\because \int_0^\infty e^{-x}.x^{n-1}dx = \Gamma n\right]$$

$$= \frac{1}{2}\sqrt{\pi} \qquad\qquad\qquad\qquad\qquad \left[\because \Gamma\left(\frac{1}{2}\right) = \sqrt{\pi}\right]$$

$\Rightarrow$

$$\int_0^\infty e^{-x^2}dx = \frac{1}{2}\sqrt{\pi}.$$

Deduction 4.

$$\int_0^{\pi/2} \cos^m\theta\sin^n\theta\, d\theta = \frac{\Gamma\left(\dfrac{m+1}{2}\right)\Gamma\left(\dfrac{n+1}{2}\right)}{2\Gamma\left(\dfrac{m+n+2}{2}\right)}$$

for all values of m and n such that m > –1, n > –1.

Proof. We put $\sin^2\theta = x,$ $\Rightarrow$ $2\sin\theta\cos\theta\, d\theta = dx$

$\Rightarrow$ $2\sin\theta.\sqrt{(1-\sin^2\theta)}\, d\theta = dx$ $\Rightarrow$ $2x^{1/2}\sqrt{1-x}\, d\theta = dx$

$\Rightarrow$

$$d\theta = \frac{dx}{2x^{1/2}(1-x)^{1/2}}.$$

Also, when $\theta = \pi/2$, $x = 1$ and $\theta = 0$, $x = 0$.

Putting these values in L.H.S. of the given equation, we get

$$\int_0^{\pi/2} \cos^m\theta\sin^n\theta\, d\theta = \int_0^{\pi/2}(1-\sin^2\theta)^{m/2}.\sin^n\theta\, d\theta$$

$$= \int_0^1 (1-x)^{m/2}.x^{n/2}.\frac{dx}{2x^{1/2}(1-x)^{1/2}}$$

$$= \frac{1}{2}\int_0^1 x^{\frac{(n-1)}{2}} (1-x)^{\frac{(m-1)}{2}} \, dx$$

$$= \frac{1}{2}\int_0^1 x^{\left\{\frac{(n+1)}{2}\right\}-1} (1-x)^{\left\{\frac{(m+1)}{2}\right\}-1} \, dx$$

$$= \frac{1}{2} B\left(\frac{m+1}{2}, \frac{n+1}{2}\right)$$

$$= \frac{\frac{1}{2}\Gamma\frac{1}{2}(m+1)\Gamma\frac{1}{2}(n+1)}{\Gamma\frac{1}{2}(m+n+1+1)} \qquad \left(\text{Because } B(m,n) = \frac{\Gamma(m)\Gamma(n)}{\Gamma(m+n)}\right)$$

$$= \frac{\Gamma\frac{1}{2}(m+1)\Gamma\frac{1}{2}(n+1)}{2\Gamma\frac{1}{2}(m+n+2)}.$$

Deduction 5. $\int_0^{\pi/2} \sin^{p-1}\theta \cos^{q-1}\theta \, d\theta = \dfrac{\Gamma(p/2)\Gamma(q/2)}{2\Gamma\left(\dfrac{p+q}{2}\right)}.$

Proof. By definition of beta function, we have

$$B(m,n) = \int_0^1 x^{m-1}(1-x)^{n-1} dx$$

$$= 2\int_0^{\pi/2} \cos^{2m-1}\theta \sin^{2n-1}\theta \, d\theta = \frac{\Gamma(m).\Gamma(n)}{\Gamma(m+n)} \qquad \qquad \dots(1)$$

Let $2m = p$ and $2n = q$. So that $m = p/2$ and $n = q/2$
Put in equation (1), we get

$$\int_0^{\pi/2} \sin^{p-1}\theta \cos^{q-1}\theta \, d\theta = \frac{\Gamma(p/2)\Gamma(q/2)}{2\Gamma\left(\dfrac{p+q}{2}\right)}.$$

Deduction 6. $\int_0^{\pi/2} \sin^{p-1}\theta \, d\theta = \int_0^{\pi/2} \cos^{p-1}\theta \, d\theta$

$$= \frac{\Gamma\left(\dfrac{p}{2}\right)\Gamma\left(\dfrac{1}{2}\right)}{2\Gamma\left(\dfrac{p+1}{2}\right)} = \frac{\sqrt{\pi}}{2} \frac{\Gamma(p/2)}{\Gamma\left(\dfrac{p+1}{2}\right)}. \qquad \left(\because \Gamma\left(\dfrac{1}{2}\right) = \sqrt{\pi}\right)$$

Proof. Replacing q by 1 in deduction 5, we get

$$\int_0^{\pi/2} \sin^{p-1}\theta \, d\theta = \frac{\Gamma\left(\dfrac{p}{2}\right)\Gamma\left(\dfrac{1}{2}\right)}{2\Gamma\left(\dfrac{p+1}{2}\right)}.$$

Next, replacing p by 1 and q by p in deduction 5, we have

$$\int_0^{\pi/2} \cos^{p-1}\theta \, d\theta = \frac{\Gamma\left(\dfrac{1}{2}\right)\Gamma\left(\dfrac{p}{2}\right)}{2\Gamma\left(\dfrac{1+p}{2}\right)} = \frac{\sqrt{\pi}}{2}\cdot\frac{\Gamma\left(\dfrac{p}{2}\right)}{\Gamma\left(\dfrac{p+1}{2}\right)}$$

Solved Examples

Example 1. Evaluate the following integrals:

(i) $\int_0^1 x^4(1-x^2)dx$

(ii) $\int_0^a y^4\sqrt{a^2-y^2}\,dy$

(iii) $\int_0^2 x(8-x^3)^{1/3}dx$

(iv) $\int_0^\infty \dfrac{x\,dx}{1+x^6}$

Based on the following Results

- $B(m,n)=\dfrac{\Gamma(m)\Gamma(n)}{\Gamma(m+n)}, \; m>0, n>0$

- $\Gamma(n)\Gamma(1-n)=\dfrac{\pi}{\sin n\pi}, 0<n-1$

- $\int_0^\infty e^{-x^2}dx = \dfrac{1}{2}\sqrt{\pi}$

- $\int_0^{\pi/2}\cos^m\theta\sin^n\theta\,d\theta = \dfrac{\Gamma\left(\dfrac{m+1}{2}\right)\Gamma\left(\dfrac{n+1}{2}\right)}{2\Gamma\left(\dfrac{m+n+2}{2}\right)}$

Solution.   (i) We have
$$\int_0^1 x^4(1-x)^2 dx$$
$$= \int_0^1 x^{5-1}(1-x)^{3-1}dx$$
$$= \frac{\Gamma(5)\Gamma(3)}{\Gamma(5+3)} = \frac{4!\,2!}{7!}$$
$$= \frac{4!\times 2}{7\times 5\times 4!\times 6} = \frac{1}{105}.$$

(ii) $\int_0^a y^4\sqrt{a^2-y^2}\,dy$.

Let $y^2 = a^2 t$ so that $dy = \dfrac{a^2\,dt}{2y} = \dfrac{a\,dt}{2\sqrt{t}}$, then

$$I = \int_0^1 (a^2t)^2\sqrt{(a^2-ta^2)}\left(\frac{a\,dt}{2\sqrt{t}}\right)$$

$$= \frac{a^6}{2}\int_0^1 t^{3/2}(1-t)^{1/2}dt$$

$$= \frac{a^6}{2}\int_0^1 t^{(5/2)-1}(1-t)^{(3/2)-1}dt$$

$$= \frac{a^6}{2}\,\frac{\Gamma\left(\dfrac{5}{2}\right)\Gamma\left(\dfrac{3}{2}\right)}{\Gamma\left(\dfrac{5}{2}+\dfrac{3}{2}\right)} = \frac{a^6}{2}\,\frac{\dfrac{3}{2}\cdot\dfrac{1}{2}\sqrt{\pi}\cdot\dfrac{1}{2}\sqrt{\pi}}{3!} = \frac{\pi a^6}{32}.$$

(iii) Let $\int_0^2 x(8-x^3)^{1/3}dx = I$

Put $x^3 = 8t$ or $x = 2t^{1/3}$ so that $dx = \dfrac{2}{3}t^{-2/3}dt.$, we get

$$I = \int_0^1 (2t^{1/3})(8-8t)^{1/3}\left(\frac{2}{3}t^{-2/3}\,dt\right)$$

$$= \frac{8}{3}\,\frac{\Gamma\left(\dfrac{2}{3}\right)\Gamma\left(\dfrac{4}{3}\right)}{\Gamma\left(\dfrac{2}{3}+\dfrac{4}{3}\right)} = \frac{8}{3}\,\frac{\Gamma\left(1-\dfrac{1}{3}\right)\Gamma\left(1+\dfrac{1}{3}\right)}{\Gamma(2)}$$

$$= \frac{8}{3}\Gamma\left(1-\frac{1}{3}\right).\frac{1}{3}\Gamma\left(\frac{1}{3}\right) = \frac{8}{9}\frac{\pi}{\sin\pi/3} = \frac{16\pi}{2\sqrt{3}}.$$

$$(\because \Gamma 2 = 1! = 1, \Gamma(1+p) = p\Gamma(p) \text{ and } \Gamma(1-n)\Gamma n = \frac{\pi}{\sin n\pi})$$

(iv) Let $I = \int_0^\infty \frac{x\,dx}{1+x^6}$.

Put $x^6 = y$ or $x = y^{1/6}$. $\Rightarrow$ $dx = \frac{1}{6}.y^{-5/6}\,dy$

$$\therefore \qquad I = \frac{1}{6}\int_0^\infty \frac{y^{1/6}.y^{-5/6}}{1+y}\,dy = \frac{1}{6}\int_0^\infty \frac{y^{-2/3}}{1+y}\,dy$$

$$= \frac{1}{6}\int_0^\infty \frac{y^{(1/3)-1}}{(1+y)^{2/3+1/3}}\,dy = \frac{1}{6}B\left(\frac{1}{3},\frac{2}{3}\right)$$

$$= \frac{1}{6}\frac{\Gamma\left(\frac{1}{3}\right)\Gamma\left(\frac{2}{3}\right)}{\Gamma\left(\frac{1}{2}+\frac{2}{3}\right)} = \frac{1}{6}\frac{\Gamma\left(\frac{1}{3}\right)\Gamma\left(1-\frac{1}{3}\right)}{\Gamma 1} = \frac{1}{6}\frac{\pi}{\sin\dfrac{\pi}{3}}$$

$$\left[\because \Gamma(n)\Gamma(1-n) = \frac{\pi}{\sin n\pi}\right]$$

$$= \frac{1}{6}.\frac{\pi}{\left(\sqrt{3}/2\right)} = \frac{1}{6}.\frac{2\pi}{\sqrt{3}} = \frac{\pi}{3\sqrt{3}}.$$

Example 2. Prove that $\int_0^2 (8-x^3)^{-1/3}dx = \dfrac{2\pi}{3\sqrt{3}}$

Solution. Let $x^3 = 8t$, then $x = 2t^{1/3} \Rightarrow dx = \dfrac{2}{3}t^{-2/3}dt$ and when $x = 0$ to $x = 2, t = 0$ to $t = 1$

So $\int_0^2 (8-x^3)^{-1/3}dx = \int_0^1 (8-8t)^{-1/3}.\dfrac{2}{3}t^{-2/3}dt$

$$= (8)^{-1/3}.\frac{2}{3}\int_0^1 t^{-2/3}(1-t)^{-1/3}dt$$

$$= \frac{1}{3}\int_0^1 t^{(1/3)-1}(1-t)^{(2/3)-1}dt$$

$$= \frac{1}{3}B\left(\frac{1}{3},\frac{2}{3}\right) = \frac{1}{3}\frac{\Gamma\left(\frac{1}{3}\right)\Gamma\left(\frac{2}{3}\right)}{\Gamma\left(\frac{1}{3}+\frac{2}{3}\right)}$$

$$= \frac{1}{3}\frac{\Gamma\left(\frac{1}{3}\right)\Gamma\left(1-\frac{1}{3}\right)}{\Gamma(1)} = \frac{1}{3}\frac{\pi}{\sin\dfrac{\pi}{3}} = \frac{2\pi}{3\sqrt{3}}.$$

Example 3. Show that $\int_0^1 \dfrac{dx}{(1-x^n)^{1/2}} = \dfrac{\sqrt{\pi}\,\Gamma(1/n)}{n\Gamma\left(\dfrac{1}{n}+\dfrac{1}{2}\right)}$.

Solution. Let $x^n = \sin^2\theta \Rightarrow x = \sin^{2/n}\theta$ so that $dx = 2.\dfrac{1}{n}.\sin^{\left(\frac{2}{n}-1\right)}\theta\cos\theta\,d\theta$

then
$$\int_0^1 \frac{dx}{\sqrt{(1-x^n)}} = \frac{2}{n}\int_0^{\pi/2}\frac{\sin^{(2/n)-1}\theta\cos\theta\,d\theta}{\cos\theta}$$

$$= \frac{2}{n}\int_0^{\pi/2}\sin^{(2/n)-1}\theta\cos^0\theta\,d\theta$$

$$= \frac{2}{n}.\frac{\Gamma\left(\dfrac{1}{n}\right)\Gamma\left(\dfrac{1}{2}\right)}{2\Gamma\left(\dfrac{1}{n}+\dfrac{1}{2}\right)} = \frac{\sqrt{\pi}}{n}.\frac{\Gamma\left(\dfrac{1}{n}\right)}{\Gamma\left(\dfrac{1}{n}+\dfrac{1}{2}\right)}.$$

Example 4. *Evaluate*
$$\int_0^\infty \frac{x^8(1-x^6)}{(1+x)^{24}}\,dx .$$

Solution. We have
$$\int_0^\infty \frac{x^8(1-x^6)}{(1+x)^{24}}\,dx = \int_0^\infty \frac{x^8\,dx}{(1+x)^{24}} - \int_0^\infty \frac{x^{14}}{(1+x)^{24}}\,dx$$

$$= \int_0^\infty \frac{x^{9-1}}{(1+x)^{9+15}}\,dx - \int_0^\infty \frac{x^{15-1}}{(1+x)^{15+9}}\,dx$$

$$= B(9,15) - B(15,9) = 0 \qquad [\because B(m,n) = B(n,m)]$$

Example 5. *Prove that* $\int_0^\infty \dfrac{y^{n-1}}{1+x}\,dx = \dfrac{\pi}{\sin n\pi}.$

Solution. We know that
$$B(m,n) = \int_0^\infty \frac{x^{n-1}}{(1+x)^{m+n}}\,dx$$

Put $m = 1 - n$, we get
$$B(1-n,n) = \int_0^\infty \frac{x^{n-1}}{1+x}\,dx$$

$\Rightarrow \qquad \int_0^\infty \dfrac{x^{n-1}}{1+x}\,dx = B(n,1-n)$ $\qquad\qquad [\because B(m,n) = B(n,m)]$

$$= \frac{\Gamma(n)\Gamma(1-n)}{\Gamma(n+1-n)} = \Gamma(n)\Gamma(1-n)$$

$$= \frac{\pi}{\sin n\pi}.$$

Example 6. *Prove that*

(a) $\int_0^{\pi/2} \sqrt{\tan\theta}\,d\theta = \dfrac{1}{2}\Gamma\left(\dfrac{1}{4}\right)\Gamma\left(\dfrac{3}{4}\right) = \dfrac{\pi\sqrt{2}}{2}.$

(b) $\int_0^{\pi/2} \tan^n x\,dx = \dfrac{\pi}{2}\sec\dfrac{n\pi}{2}, \; -1 < n < 1.$

Solution. (a) We have

$$\int_0^{\pi/2} \sqrt{\tan\theta}\, d\theta = \int_0^{\pi/2} \left(\frac{\sin\theta}{\cos\theta}\right)^{1/2} d\theta = \int_0^{\pi/2} \sin^{1/2}\theta \cos^{-1/2}\theta\, d\theta$$

$$= \frac{\Gamma\left(\dfrac{1+\dfrac{1}{2}}{2}\right)\Gamma\left(\dfrac{1-\dfrac{1}{2}}{2}\right)}{2\Gamma\left(\dfrac{\dfrac{1}{2}-\dfrac{1}{2}+2}{2}\right)}$$

$$\left[\because \int_0^{\pi/2} \sin^n\theta\cos^m\theta\, d\theta = \frac{\Gamma\left(\dfrac{n+1}{2}\right)\Gamma\left(\dfrac{m+1}{2}\right)}{2\Gamma\left(\dfrac{n+m+2}{2}\right)}, n>-1, m>-1\right]$$

$$= \frac{\Gamma\left(\dfrac{3}{4}\right)\Gamma\left(\dfrac{1}{4}\right)}{2\Gamma(1)} = \frac{1}{2}\Gamma\left(\frac{3}{4}\right)\Gamma\left(\frac{1}{4}\right)$$

$$= \frac{1}{2}\Gamma\left(\frac{1}{4}\right)\Gamma\left(1-\frac{1}{4}\right) = \frac{1}{2}\frac{\pi}{\sin\left(\dfrac{\pi}{4}\right)} = \frac{\pi\sqrt{2}}{2}.$$

(b) Consider

L.H.S. $\quad \int_0^{\pi/2} \tan^n x\, dx = \int_0^{\pi/2} \sin^n x \cos^{-n} x\, dx$

$$= \frac{\Gamma\left(\dfrac{1+n}{2}\right)\Gamma\left(\dfrac{1-n}{2}\right)}{2\Gamma\left(\dfrac{n-n+2}{2}\right)} \quad \left[\begin{array}{l}\text{Here } \dfrac{1+n}{2}>0, \dfrac{1-n}{2}>0 \\ \Rightarrow \quad n>-1 \text{ and } n<1\end{array}\right]$$

$$= \frac{1}{2}\Gamma\left(\frac{1+n}{2}\right)\Gamma\left(1-\frac{1+n}{2}\right)$$

$$= \frac{1}{2}\frac{\pi}{\sin\left(\dfrac{1+n}{2}\right)\pi} = \frac{\pi}{2\sin\left(\dfrac{\pi}{2}+\dfrac{n\pi}{2}\right)} = \frac{\pi}{2\cos\dfrac{n\pi}{2}}$$

$$= \frac{\pi}{2}\sec\frac{n\pi}{2}, \quad \text{where } -1<n<1.$$

Example 7. Evaluate $\int_{-1}^{1} \left(\dfrac{1+x}{1-x}\right)^{1/2} dx$.

Solution. Let $\quad I = \int_{-1}^{1}\left(\dfrac{1+x}{1-x}\right)^{1/2} dx \qquad \qquad \text{...(1)}$

Putting $t = \dfrac{1}{2}(1+x)$ so that $x = 2t-1$, $dx = 2\,dt$ in (1), we get

$$I = \int_0^1 \left(\frac{1+2t-1}{1-2t+1}\right)^{1/2} 2\,dt = 2\int_0^1 \left(\frac{t}{1-t}\right)^{1/2} dt$$

$$= 2\int_0^1 t^{(3/2)-1}(1-t)^{(1/2)-1}dt = 2B\left(\frac{3}{2},\frac{1}{2}\right)$$

$$= \frac{2\Gamma\left(\frac{3}{2}\right)\Gamma\left(\frac{1}{2}\right)}{\Gamma\left(\frac{3}{2}+\frac{1}{2}\right)} = 2\frac{\frac{1}{2}\sqrt{\pi}\sqrt{\pi}}{\Gamma(2)} = \pi.$$

Example 8. *Simplify*

$$\int_0^1 \frac{35x^3\,dx}{32\sqrt{1-x}}.$$

Solution. Let $\quad I = \int_0^1 \frac{35x^3\,dx}{32\sqrt{1-x}}$ ...(1)

Putting $x = \sin^2\theta$ so that $dx = 2\sin\theta\cos\theta\,d\theta$ then (1) gives

$$I = \int_0^{\pi/2}\frac{35\sin^6\theta\,2\sin\theta\cos\theta}{32\cos\theta} = \frac{35}{16}\int_0^{\pi/2}\sin^7\theta\,d\theta$$

$$= \frac{35}{16}\cdot\frac{6}{7}\cdot\frac{4}{5}\cdot\frac{2}{3} = 1.$$

Example 9. *If $p > 0, q > 0, m+1 > 0, n+1 > 0$, then prove that*

$$\int_0^p x^m(p^q - x^q)^n\,dx = \frac{p^{nq+m+1}}{q}\cdot B\left(n+1,\frac{m+1}{q}\right).$$

Solution. Let $\quad I = \int_0^p x^m(p^q - x^q)^n\,dx$ ...(1)

Putting $x^q = p^q t \Rightarrow dx = \left(\frac{p}{q}\right)(t)^{(1/q)-1}dt$ in (1) then we have

$$I = \int_0^1 (pq^{1/q})^m(p^q - p^q t)^n\left(\frac{p}{q}\right)t^{1/q-1}dt$$

$$= \frac{p^m\cdot p^{nq}\cdot p}{q}\int_0^1 t^{(m/q)+(1/q)-1}(1-t)^{(n+1)-1}dt$$

$$= \frac{p^{nq+m+1}}{q}B\left(\frac{m+1}{q},n+1\right)$$

$$= \frac{p^{nq+m+1}}{q}B\left(n+1,\frac{m+1}{q}\right).$$

Example 10. *Show that* $B(n,n+1) = \frac{1}{2}\frac{\Gamma(n)^2}{\Gamma(2n)}$ *and hence deduce that*

$$\int_0^{\pi/2}\left(\frac{1}{\sin^3\theta} - \frac{1}{\sin^2\theta}\right)^{1/4}\cos\theta\,d\theta = \frac{\left\{\Gamma\left(\frac{1}{4}\right)\right\}^2}{2\sqrt{\pi}}.$$

Solution. We have

$$B(n,n+1) = \frac{\Gamma(n)\Gamma(n+1)}{\Gamma(n+n+1)} = \frac{\Gamma(n).n\Gamma n}{(2n)\Gamma(2n)} \qquad ...(1)$$

$$\therefore \qquad B(n,n+1) = \frac{1}{2}\frac{\{\Gamma(n)\}^2}{\Gamma(2n)} \qquad [\because \ \Gamma(p+1)=p\Gamma(p)]$$

Let

$$I = \int_0^{\pi/2} \left(\frac{1}{\sin^3\theta} - \frac{1}{\sin^2\theta}\right)^{1/4} \cos\theta\, d\theta \qquad ...(2)$$

Putting $x = \sin\theta$, so that $dx = \cos\theta\, d\theta$ in (2), we get

$$I = \int_0^1 \left(\frac{1}{x^3} - \frac{1}{x^2}\right)^{1/4} dx = \int_0^1 \left(\frac{1-x}{x^3}\right)^{1/4} dx$$

$$= \int_0^1 x^{-3/4}(1-x)^{1/4}\, dx$$

$$= \int_0^1 x^{(1/4)-1}(1-x)^{(5/4)-1}\, dx$$

$$= B\left(\frac{1}{4},\frac{5}{4}\right)$$

$$= B\left(\frac{1}{4},\frac{1}{4}+1\right) = \frac{1}{2}\frac{\left\{\Gamma\left(\frac{1}{4}\right)\right\}^2}{\Gamma\left(\frac{1}{2}\right)} = \frac{\left\{\Gamma\left(\frac{1}{4}\right)\right\}^2}{2\sqrt{\pi}}.$$

Example 11. Show that $I = \int_0^{\pi/2}\sqrt{\sin n\theta}\, d\theta . \int_0^{\pi/2}\frac{d\theta}{\sqrt{\sin\theta}} = \pi.$.

Solution. We know that

$$\int_0^{\pi/2}\sin^p d\theta = \frac{\Gamma\left(\frac{p+1}{2}\right)\sqrt{\pi}}{2\Gamma\left(\frac{p+2}{2}\right)} \qquad ...(1)$$

Now,

$$I = \int_0^{\pi/2}\sin^{1/2}\theta.d\theta.\int_0^{\pi/2}\sin^{-1/2}\theta\, d\theta$$

$$= \frac{\Gamma\left(\frac{1/2+1}{2}\right)\sqrt{\pi}.\Gamma\left(\frac{-1/2+1}{2}\right)\sqrt{\pi}}{2\Gamma\left(\frac{1/2+2}{2}\right).2\Gamma\left(\frac{-1/2+2}{2}\right)}$$

$$= \frac{\Gamma\left(\frac{3}{4}\right)\sqrt{\pi}\, \Gamma\left(\frac{1}{4}\right)\sqrt{\pi}}{2\Gamma\left(\frac{5}{4}\right)\, 2\Gamma\left(\frac{3}{4}\right)} = \frac{\pi\Gamma\left(\frac{1}{4}\right)}{4\Gamma\left(1+\frac{1}{4}\right)} = \frac{\pi\Gamma\left(\frac{1}{4}\right)}{4.\frac{1}{4}.\Gamma\left(\frac{1}{4}\right)} = \pi.$$

Example 12. Prove that

(a) $\int_0^{\pi}\frac{\sin^{n-1}x\, dx}{(a+b\cos x)^n} = \frac{2^{n-1}}{(a^2-b^2)^{n/2}}B\left(\frac{n}{2},\frac{n}{2}\right).$

(b) $\int_0^\pi \dfrac{\sqrt{\sin x}}{[5 + 3\cos x]^{3/2}} = \dfrac{\left[\Gamma\left(\dfrac{3}{4}\right)\right]^2}{2\sqrt{2}\,\pi}.$

Solution.   (a)   Let $I = \int_0^\pi \dfrac{\sin^{n-1}x\,dx}{(a + b\cos x)^n} = \int_0^\pi \dfrac{(\sin x)^{n-1}\,dx}{(a + b\cos x)^n}$ ...(1)

$$= \int_0^\pi \dfrac{\left(2\sin\dfrac{x}{2}\cos\dfrac{x}{2}\right)^{n-1}dx}{\left[a\left\{\cos^2\left(\dfrac{x}{2}\right) + \sin^2\left(\dfrac{x}{2}\right)\right\} + b\left\{\cos^2\left(\dfrac{x}{2}\right) - \sin^2\left(\dfrac{x}{2}\right)\right\}\right]^n} \qquad \text{[by (1)]}$$

$$= 2^{n-1}\int_0^\pi \dfrac{\sin^{n-1}\left(\dfrac{x}{2}\right)\cos^{n-1}\left(\dfrac{x}{2}\right)dx}{\left[(a+b)\cos^2\left(\dfrac{x}{2}\right) + (a-b)\sin^2\left(\dfrac{x}{2}\right)\right]^n}$$

$$= \dfrac{2^{n-1}}{(a+b)^n}\int_0^\pi \dfrac{\sin^{n-1}\left(\dfrac{x}{2}\right)\cos^{n-1}\left(\dfrac{x}{2}\right)dx}{\cos^{2n}\left(\dfrac{x}{2}\right)\left[1 + \dfrac{a-b}{a+b}\tan^2\dfrac{x}{2}\right]^n}$$

$$= \dfrac{2^{n-1}}{(a+b)^n}\int_0^\infty \dfrac{\tan^{n-1}\left(\dfrac{x}{2}\right)\sec^2\left(\dfrac{x}{2}\right)dx}{\left[1 + \dfrac{a-b}{a+b}\tan^2\dfrac{x}{2}\right]^n}$$

$$= \dfrac{2^{n-1}}{(a+b)^n}\int_0^\infty \dfrac{\left[\dfrac{a+b}{a-b}t\right]^{\frac{(n-2)}{2}}\cdot\dfrac{a+b}{a-b}dt}{(1+t)^n}$$

Put $\dfrac{a-b}{a+b}\tan^2\dfrac{x}{2} = t$, i.e., $2\dfrac{a-b}{a+b}\tan\dfrac{x}{2}\sec^2\dfrac{x}{2}\cdot\dfrac{dx}{2} = dt$, we get

$$I = \dfrac{2^{n-1}}{[(a+b)(a-b)]^{n/2}}\int_0^\infty \dfrac{t^{(n/2)-1}}{(1+t)^{n/2 + n/2}}\,dt$$

$$= \dfrac{2^{n-1}}{(a^2 - b^2)^{n/2}}B\left(\dfrac{n}{2}, \dfrac{n}{2}\right).$$

(b)   Taking $n = 3/2$, $a = 5$ and $b = 3$, in part (a), we get

$$\int_0^\pi \dfrac{\sin^{(3/2)-1}x\,dx}{(5 + 3\cos x)^{3/2}} = \dfrac{(2)^{(3/2)-1}}{(25-9)^{3/4}}B\left(\dfrac{3}{4}, \dfrac{3}{4}\right)$$

$$= \frac{\sqrt{2}}{2^3} \cdot \frac{\Gamma\left(\frac{3}{4}\right)\Gamma\left(\frac{3}{4}\right)}{\Gamma\left(\frac{3}{4}+\frac{3}{4}\right)} = \frac{\sqrt{2}\left\{\Gamma\left(\frac{3}{4}\right)\right\}^2}{8\Gamma\left(\frac{3}{2}\right)}$$

$$= \frac{\sqrt{2}\left\{\Gamma\left(\frac{3}{4}\right)\right\}^2}{8 \cdot \frac{1}{2}\sqrt{\pi}} = \frac{\left\{\Gamma\left(\frac{3}{4}\right)\right\}^2}{2\sqrt{2\pi}}.$$

EXERCISE 18.1

1. Show that $\int_0^\infty e^{-4x} x^{3/2}\, dx = \frac{3\sqrt{\pi}}{128}$.

2. Show that $\int_0^\infty e^{-x^2} \cdot x^2\, dx = \frac{\sqrt{\pi}}{4}$.

3. Show that $\int_0^1 \frac{dx}{\sqrt{(-\log x)}} = \sqrt{\pi}$.

4. Show that $\Gamma\left(-\frac{15}{2}\right) = \frac{2^8\sqrt{\pi}}{1.3.5.7.9.11.13.15}$.

5. Show that $\int_0^1 x^{n-1}\left(\log\frac{1}{x}\right)^{m-1} dx = \frac{\Gamma(m)}{n^m}$

$$m > 0, n > 0.$$

6. Show that $\int_0^{\pi/2} \sin^m\theta \cos^n\theta\, d\theta$

$$= \frac{\Gamma\left(\frac{m+1}{2}\right)\Gamma\left(\frac{n+1}{2}\right)}{2\Gamma\left(\frac{m+n+2}{2}\right)}$$

$$= \frac{1}{2} B\left(\frac{m+1}{2},\frac{n+1}{2}\right).$$

7. Show that $\int_0^\infty 3^{-4x^2} dx = \frac{\sqrt{\pi}}{4\sqrt{(\log 3)}}$.

8. Show that

$$\int_0^a (a-x)^{m-1} x^{n-1} dx = \frac{a^{m+n-1}\Gamma(m)\Gamma(n)}{\Gamma(m+n)}.$$

9. Show that

$$\int_0^{\pi/2} \sin^p\theta \cos^q\theta\, d\theta = \frac{1}{2} B\left(\frac{p+1}{2},\frac{q+1}{2}\right),$$

$$p > -1, q > -1.$$

Deduce that

$$\int_0^2 x^4 (8-x^3)^{-1/3} dx = \frac{16}{3} B\left(\frac{5}{3},\frac{2}{3}\right).$$

10. Show that

$$\int_0^1 \left(\frac{1}{x}-1\right)^{1/4} dx = B\left(\frac{5}{4},\frac{3}{4}\right) = \frac{\pi}{2\sqrt{2}}.$$

11. Prove that
(i) $B(l, m)\, B(l+m, n) = B(m, n)B(m+n, l)$
$$= B(n, l)\, B(n+l, m)$$
(ii) $B(l, m)B(l+m, n)\, B(l+m+n, p)$
$$= \frac{\Gamma(l)\Gamma(m)\Gamma(n)\Gamma(p)}{\Gamma(l+m+n+p)}.$$

12. Show that

$$\int_0^1 x^m (1-x^n)^p dx = \frac{1}{n} B\left(\frac{m+1}{n}, p+1\right).$$

13. Show that $\int_0^1 \left(1-x^n\right)^{1/n} dx = \frac{1\left[\Gamma(1/n)\right]^2}{n\,2\Gamma(2/n)}$.

14. Show that

$$\int_0^1 \frac{x^2 dx}{(1-x^4)^{1/2}} \times \int_0^1 \frac{dx}{(1+x^4)^{1/2}} = \frac{\pi}{4\sqrt{2}}.$$

15. Show that $B(m,n) = B(m+1, n) + B(m, n+1)$
for $m > 0, n > 0$.

HINTS TO SELECTED PROBLEMS

1. $I = \int_0^\infty e^{-4x} x^{3/2} dx = \int_0^\infty e^{-4x} x^{(5/2)-1} dx$

$$= \frac{\Gamma\left(\frac{5}{2}\right)}{4^{5/2}} = \frac{\frac{3}{2} \cdot \frac{1}{2} \sqrt{\pi}}{2^5} = \frac{3\sqrt{\pi}}{128}.$$

3. $I = \int_0^1 \dfrac{dx}{\sqrt{\log \dfrac{1}{x}}} = \int_0^1 \left[\log\left(\dfrac{1}{x}\right)\right]^{-1/2} dx \cdot$

Now put $\log(1/x) = t$.

5. $I = \int_0^1 x^{n-1} \left(\log \dfrac{1}{x}\right)^{m-1} dx.$

Put $\dfrac{1}{x} = t, i.e., x = e^{-t} \Rightarrow dx = -e^{-t} dt.$

6. The given integral can be written as

$$I = \int_0^{\pi/2} (\sin^2 \theta)^{(m-1)/2} (1-\sin^2)^{(n-1)/2} \sin\theta \cos\theta\, d\theta$$

Now put $\sin^2\theta = x$ i.e., $dx = 2\sin\theta\cos\theta\, d\theta$.

7. Put $3^{-4x^2} = e^{-t}$ i.e., $x = \dfrac{\sqrt{t}}{2\sqrt{\log 3}}.$

8. Put $x = at$ in the LHS.

9. Do same as (6).

12. Put

$$x^n = t \ i.e., \ x = (t)^{1/n} \Rightarrow \ dx = \frac{1}{n}(t)^{1/n-1} dt.$$

18.8 DUPLICATION FORMULA

We have

$$\Gamma(n)\Gamma\left(n+\frac{1}{2}\right) = \frac{\sqrt{\pi}}{2^{2n-1}}\Gamma(2n), \ n > 0.$$

Proof. $\qquad\qquad B(m,n) = \dfrac{\Gamma(m)\Gamma(n)}{\Gamma(m+n)}$ where $m > 0, n > 0.$ ...(1)

Now putting $m = n$ in equation (1), we get

$$B(n,n) = \frac{[\Gamma(n)]^2}{\Gamma(2n)}.$$ ...(2)

By definition of beta function, we get

$$B(n,n) = \int_0^1 x^{n-1}(1-x)^{n-1} dx.$$

Putting $x = \sin^2\theta$ so that $dx = 2\sin\theta\cos\theta\, d\theta$ in (1), we get

$$B(n,n) = \int_0^{\pi/2} (\sin^2\theta)^{n-1}(1-\sin^2\theta)^{n-1}.2\sin\theta\cos\theta\, d\theta$$

$$= 2\int_0^{\pi/2}(\sin\theta\cos\theta)^{2n-1} d\theta = 2\int_0^{\pi/2}\left(\frac{\sin 2\theta}{2}\right)^{2n-1} d\theta$$

$$= \frac{1}{2^{2n-2}}\int_0^{\pi/2}\sin^{2n-1} 2\theta\, d\theta$$

$$= \frac{1}{2^{2n-2}}\int_0^{\pi}\sin^{2n-1}\phi\,\frac{d\phi}{2} \ (\text{By putting } 2\theta=\phi \Rightarrow d\theta = \frac{1}{2}d\phi)$$

$$= \frac{1}{2^{2n-1}}\int_0^{\pi}\sin^{2n-1}\phi\, d\phi$$

$$= \frac{1}{2^{2n-2}}\int_0^{\pi/2}\sin^{2n-1}\phi\, d\phi$$

$$\left[\because \int_0^{2a} f(x)\,dx = 2\int_0^a f(x)\,dx \text{ when } f(2a-x) = f(x)\right]$$

$$= \frac{1}{2^{2n-2}} \int_0^{\pi/2} \sin^{2n-1} \phi (\cos \phi)^0 \, d\phi$$

$$= \frac{1}{2^{2n-2}} \frac{\Gamma\left(\dfrac{2n-1+1}{2}\right)\Gamma\left(\dfrac{0+1}{2}\right)}{2\Gamma\left(\dfrac{2n-1+0+2}{2}\right)}$$

$$\therefore \qquad B(n,n) = \frac{1}{2^{2n-1}} \cdot \frac{\Gamma(n)\sqrt{\pi}}{\Gamma\left(n+\dfrac{1}{2}\right)} \text{ as } \Gamma\left(\frac{1}{2}\right) = \sqrt{\pi}. \qquad \ldots(3)$$

Equating two values of $B(n, n)$ given by (2) and (3), we obtain

$$\frac{[\Gamma(n)]^2}{\Gamma(2n)} = \frac{1}{2^{2n-1}} \frac{\Gamma(n)\sqrt{\pi}}{\Gamma\left(n+\dfrac{1}{2}\right)}$$

or $\qquad \Gamma(n)\Gamma\left(n+\dfrac{1}{2}\right) = \dfrac{\sqrt{\pi}}{2^{2n-1}}\Gamma(2n).$ $\qquad \ldots(4)$

Deduction 1. *For all positive real value of p, we have* $2^p\Gamma\left(\dfrac{p+1}{2}\right)\Gamma\left(\dfrac{p+2}{2}\right) = \sqrt{\pi}\,\Gamma(p+1)$
Proof. We know that

$$\Gamma(n)\Gamma\left(n+\frac{1}{2}\right) = \frac{\sqrt{\pi}}{2^{2n-1}}\Gamma(2n). \qquad \ldots(1)$$

Putting $2n - 1 = p$, so that $n = \dfrac{1}{2}(p+1)$ in equation (1), we get

$$\Gamma\left(\frac{p+1}{2}\right)\Gamma\left(\frac{p+1}{2}+\frac{1}{2}\right) = \frac{\sqrt{\pi}}{2^p}\Gamma(p+1)$$

$$\Rightarrow \qquad 2^p\Gamma\left(\frac{p+1}{2}\right)\Gamma\left(\frac{p+2}{2}\right) = \sqrt{\pi}\,\Gamma(p+1).$$

Deduction 2. *For any positive integer n, we have* $\Gamma\left(n+\dfrac{1}{2}\right) = \dfrac{(2n)!}{2^{2n}.n!}\sqrt{\pi}.$

Proof. Let n be positive integer, then we have

$$\frac{\Gamma(2n)}{\Gamma(n)} = \frac{(2n-1)!}{(n-1)!} = \frac{(2n)(2n-1)!}{2.n.(n-1)!} = \frac{(2n)!}{2.(n)!} \qquad \ldots(1)$$

Now, from the duplication formula (4), we have

$$\Gamma\left(n+\frac{1}{2}\right) = \frac{\sqrt{\pi}}{2^{2n-1}} \cdot \frac{\Gamma(2n)}{\Gamma(n)} = \frac{\sqrt{\pi}}{2^{2n-1}} \cdot \frac{(2n)!}{2.n!} \qquad \text{[By (1)]}$$

$$= \frac{(2n)!}{2^{2n}.n!}\sqrt{\pi}.$$

Deduction 3. *For any integer n, we have* $\Gamma\left(\dfrac{1}{n}\right)\Gamma\left(\dfrac{2}{n}\right)\Gamma\left(\dfrac{3}{n}\right)...\Gamma\left(\dfrac{n-1}{n}\right) = \dfrac{(2\pi)^{(n-1)/2}}{n^{1/2}}.$

Proof. Let $\quad X = \Gamma\left(\dfrac{1}{n}\right)\Gamma\left(\dfrac{2}{n}\right)\Gamma\left(\dfrac{3}{n}\right)...\Gamma\left(\dfrac{n-2}{n}\right)\Gamma\left(\dfrac{n-1}{n}\right).$ ...(1)

Writing the above expression in the reversed order, we get

$$X = \Gamma\left(\dfrac{n-1}{n}\right)\Gamma\left(\dfrac{n-2}{n}\right)...\Gamma\left(\dfrac{2}{n}\right)\Gamma\left(\dfrac{1}{n}\right)$$

$$X = \Gamma\left(1-\dfrac{1}{n}\right)\Gamma\left(1-\dfrac{2}{n}\right)...\Gamma\left(1-\dfrac{n-2}{2}\right)\Gamma\left(1-\dfrac{n-1}{n}\right). \qquad ...(2)$$

Multiplying (1) and (2) and arranging in products of terms in the $\Gamma(n)\,\Gamma(1-n)$, we have

$$X^2 = \left[\Gamma\left(\dfrac{1}{n}\right)\Gamma\left(1-\dfrac{1}{n}\right)\right]\cdot\left[\Gamma\left(\dfrac{2}{n}\right)\Gamma\left(1-\dfrac{2}{n}\right)\right]....$$

$$...\left[\Gamma\left(\dfrac{n-2}{n}\right)\Gamma\left(1-\dfrac{n-2}{n}\right)\right]\left[\Gamma\left(\dfrac{n-1}{n}\right)\Gamma\left(1-\dfrac{n-1}{n}\right)\right]$$

$$= \dfrac{\pi.\pi}{\sin\dfrac{\pi}{n}.\sin\dfrac{2\pi}{n}}...\dfrac{\pi}{\sin\dfrac{n-2}{n}\pi\sin\dfrac{n-1}{n}\pi} \qquad \left[\because \Gamma(m)\Gamma(1-m) = \dfrac{\pi}{\sin m\pi}\right]$$

$$\therefore \qquad X^2 = \dfrac{\pi^{n-1}}{\sin\left(\dfrac{\pi}{n}\right)\sin\left(\dfrac{2\pi}{n}\right)...\sin\left(\dfrac{(n-1)\pi}{n}\right)}.$$

Now, using the following trigonometrical identity :

$$2^{n-1}\sin\left(\theta+\dfrac{\pi}{n}\right)\sin\left(\theta+\dfrac{2\pi}{n}\right)\sin\left(\theta+\dfrac{n-1}{n}\pi\right) = \dfrac{\sin n\theta}{\sin\theta} \qquad ...(4)$$

and, taking limit as $\theta \to 0$, equation (4) gives

$$2^{n-1}\sin\dfrac{\pi}{n}\sin\dfrac{2\pi}{n}...\sin\left(\dfrac{n-1}{n}\pi\right) = \lim_{\theta\to 0}\dfrac{\sin n\theta}{\sin\theta}$$

$$= n\lim_{\theta\to 0}\left[\dfrac{\sin n\theta}{n\theta}.\dfrac{\theta}{\sin\theta}\right] = n.$$

$$\therefore \qquad \sin\left(\dfrac{\pi}{n}\right)\sin\left(\dfrac{2\pi}{n}\right)...\sin\left\{\dfrac{(n-1)\pi}{n}\right\} = \dfrac{n}{2^{n-1}}. \qquad ...(5)$$

Using (5), (3) reduces to

$$X^2 = \dfrac{\pi^{n-1}}{(n/2)^{n-1}} = \dfrac{(2\pi)^{n-1}}{n} \quad\text{or}\quad X = \dfrac{(2\pi)^{(n-1)/2}}{n^{1/2}} \qquad ...(6)$$

From (1) and (6), we get

$$\Gamma\left(\frac{1}{n}\right)\Gamma\left(\frac{2}{n}\right)\ldots\Gamma\left(\frac{n-1}{n}\right) = \frac{(2\pi)^{\frac{(n-1)}{2}}}{n^{1/2}}$$

Deduction 4.

(i) $\int_0^\infty e^{-ax}\cos bx\, x^{m-1}dx = \dfrac{\Gamma(m)}{r^m}\cos m\theta,$

(ii) $\int_0^\infty e^{-ax}\sin bx\, x^{m-1}dx = \dfrac{\Gamma(m)}{r^m}\sin m\theta$

where $r = (a^2+b^2)^{1/2}$ and $\theta = \tan^{-1}\left(\dfrac{b}{a}\right).$

Proof. We know that

$$\int_0^\infty e^{-kx}x^{m-1}dx = \frac{\Gamma(m)}{k^m},\ m>0,\ k>0 \qquad \ldots(1)$$

Putting $k = a - ib$ in both sides of (1), we get

$$\int_0^\infty e^{-(a-ib)x}x^{m-1}dx = \frac{\Gamma(m)}{(a-ib)^m}$$

or $\qquad \int_0^\infty e^{-ax}e^{ibx}x^{m-1}dx = \dfrac{\Gamma(m)(a+ib)^m}{[(a+ib)(a-ib)]^m}$

$\Rightarrow \qquad \int_0^\infty e^{-ax}x^{m-1}(\cos bx + i\sin bx)dx = \dfrac{\Gamma(m)(a+ib)^m}{(a^2+b^2)^m} \qquad \ldots(2)$

Let $\qquad\qquad a+ib = r(\cos\theta + i\sin\theta) \qquad \ldots(3)$

Equating real and imaginary parts of both sides, we get

$$r^2 = a^2 + b^2, \qquad \tan\theta = \frac{b}{a} \qquad \ldots(4)$$

Now $\qquad\qquad (a+ib)^m = [r(\cos\theta + i\sin\theta)]^m \qquad$ [By (3)]

$\qquad\qquad\qquad (a+ib)^m = r^m(\cos m\theta + i\sin m\theta) \qquad \ldots(5)$

[By De'Moivre's theorem]

Using (4) and (5), (2) reduces to

$$\int_0^\infty e^{-ax}x^{m-1}(\cos bx + i\sin bx)dx = \frac{\Gamma(m)r^m(\cos m\theta + i\sin m\theta)}{r^{2m}}.$$

Equating real and imaginary parts of both sides, we get

$$\int_0^\infty e^{-ax}x^{m-1}\cos bx\, dx = \frac{\Gamma(m)}{r^m}\cos m\theta$$

$\qquad\qquad\qquad\qquad\qquad\qquad\qquad\qquad\qquad\qquad \ldots(6)$

and $\qquad \int_0^\infty e^{-ax}x^{m-1}\sin bx\, dx = \dfrac{\Gamma(m)}{r^m}\sin m\theta. \qquad \ldots(7)$

Deduction 5. Let $m = 1$, then $\Gamma(m) = \Gamma(1) = 1$, so (6), (7) reduces to

$$\int_0^\infty e^{-ax} \cos bx \, dx = \frac{\cos\theta}{r} \qquad \qquad ...(8)$$

$$\int_0^\infty e^{-ax} \sin bx \, dx = \frac{\sin\theta}{r}. \qquad \qquad ...(9)$$

But $\tan\theta = \dfrac{b}{a}$, so that $\sin\theta = \dfrac{b}{\sqrt{a^2+b^2}}$ and $\cos\theta = \dfrac{a}{\sqrt{(a^2+b^2)}}$

Also, $r^2 = (a^2 + b^2)$. Hence (8) and (9) becomes

$$\int_0^\infty e^{-ax} \cos bx \, dx = \frac{a}{a^2+b^2} \qquad \qquad ...(10)$$

and $$\int_0^\infty e^{-ax} \sin bx \, dx = \frac{b}{a^2+b^2} \qquad \qquad ...(11)$$

Solved Examples

Example 1. *Express $\Gamma(1/6)$ in terms of $\Gamma(1/3)$.*

Solution.   By duplication formula, we have

$$\Gamma(n)\Gamma\left(n+\frac{1}{2}\right) = \frac{\sqrt{\pi}}{2^{n-1}}\Gamma(2n) \qquad \qquad ...(1)$$

Put $n = 1/6$ in (1), we get

$$\Gamma\left(\frac{1}{6}\right)\Gamma\left(\frac{2}{3}\right) = \frac{\sqrt{\pi}\,\Gamma\left(\frac{1}{3}\right)}{2^{-2/3}} \Rightarrow \Gamma\left(\frac{1}{6}\right) = \frac{\sqrt{\pi}\,\Gamma\left(\frac{1}{3}\right)}{2^{-2/3}\Gamma\left(\frac{2}{3}\right)}. \qquad \qquad ...(2)$$

Also, we know that

$$\Gamma(n)\Gamma(1-n) = \frac{\pi}{\sin n\pi}. \qquad \qquad ...(3)$$

Putting $n = 1/3$ in (3), we get

$$\Gamma\left(\frac{1}{3}\right)\Gamma\left(\frac{2}{3}\right) = \frac{\pi}{\sin(\pi/3)} = \frac{2\pi}{\sqrt{3}}$$

$$\Gamma\left(\frac{2}{3}\right) = \frac{2\pi}{\sqrt{3}\,\Gamma(1/3)} \qquad \qquad ...(4)$$

Substituting the value of $\Gamma(2/3)$ given by (4) in (2), we get

$$\Gamma\left(\frac{1}{6}\right) = \frac{\sqrt{\pi}\,\Gamma\left(\frac{1}{3}\right)}{2^{-2/3}} \cdot \frac{\sqrt{3}\,\Gamma\left(\frac{1}{3}\right)}{2\pi} = \frac{\sqrt{3}}{2^{1/3}\sqrt{\pi}}\left[\Gamma\left(\frac{1}{3}\right)\right]^2.$$

Example 2. *Find the value of $\Gamma\left(\dfrac{1}{9}\right)\Gamma\left(\dfrac{2}{9}\right)\Gamma\left(\dfrac{3}{9}\right)...\Gamma\left(\dfrac{8}{9}\right)$.*

Solution.   We know that
$$\Gamma\left(\frac{1}{n}\right)\Gamma\left(\frac{2}{n}\right)\Gamma\left(\frac{3}{n}\right)...\Gamma\left(\frac{n-1}{n}\right) = \frac{(2\pi)^{(n-1)/2}}{n^{1/2}}.$$

Putting $n = 9$, in the above relation, we get

$$\Gamma\left(\frac{1}{9}\right)\Gamma\left(\frac{2}{9}\right)\Gamma\left(\frac{3}{9}\right)... = \frac{(2\pi)^{(9-1)/2}}{9^{1/2}} = \frac{(2\pi)^4}{3} = \frac{16}{3}\pi^4.$$

Example 3. *Prove that* $B(m,m)\,B\left(m+\frac{1}{2}, m+\frac{1}{2}\right) = \frac{\pi m^{-1}}{2^{4m-1}}.$

Solution. $\text{L.H.S.} = \dfrac{\Gamma(m)\Gamma(m)}{\Gamma(m+m)} \cdot \dfrac{\Gamma\left(m+\dfrac{1}{2}\right)\Gamma\left(m+\dfrac{1}{2}\right)}{\Gamma\left(m+\dfrac{1}{2}+m+\dfrac{1}{2}\right)}$

$$= \frac{[\Gamma(m)\Gamma(m+1/2)]^2}{\Gamma(2m)\Gamma(2m+1)} = \frac{[\Gamma(m)\Gamma(m+1/2)]^2}{\Gamma(2m).2m\Gamma(2m)} \qquad [\because \Gamma(p+1)=p\Gamma(p)]$$

$$= \frac{1}{2m}\left[\frac{\Gamma(m)\Gamma(m+1/2)}{\Gamma(2m)}\right]^2 = \frac{1}{2m}\cdot\left(\frac{\sqrt{\pi}}{2^{2m-1}}\right)^2 \qquad \text{(By duplication formula)}$$

$$= \frac{\pi}{2m.2^{4m-2}} = \frac{\pi m^{-1}}{2^{4m-1}}.$$

Example 4. *Prove that* $\int_{-\infty}^{\infty} \cos\dfrac{\pi}{2}x^2\,dx = 1.$.

Solution. Let $\qquad I = \int_{-\infty}^{\infty} \cos\dfrac{1}{2}\pi x^2\,dx$. $\qquad\qquad\qquad\qquad\qquad$ ...(1)

Since $\cos\dfrac{1}{2}\pi x^2$ is an even function therefore (1) gives

$$I = 2\int_0^{\infty} \cos\frac{1}{2}\pi x^2\,dx \ . \qquad\qquad\qquad\qquad\qquad ...(2)$$

Putting $x^2 = t$ so that $x = t^{1/2}$ and $dx = \left(\dfrac{1}{2}\right)t^{-1/2}dt$ then equation (2) reduces to

$$I = 2\int_0^{\infty} \cos\frac{1}{2}\pi t \cdot \frac{1}{2}t^{-1/2}dt$$

$$= \int_0^{\infty} (t)^{1/2-1}\cos\frac{1}{2}\pi t\,dt = \frac{\Gamma\left(\dfrac{1}{2}\right)}{\left(\dfrac{\pi}{2}\right)^{1/2}}\cos\left(\frac{1}{2}\cdot\frac{\pi}{2}\right)$$

$$\left[\because \int_0^{\infty} x^{m-1}\cos bx\,dx = \frac{\Gamma(m)}{b^m}\cos\frac{m\pi}{2}\right]$$

$$= \frac{\Gamma(1/2)}{(\pi/2)^{1/2}}\cos\left(\frac{1}{2}\cdot\frac{\pi}{2}\right) = \frac{\sqrt{\pi}}{\sqrt{\pi/2}}\cdot\frac{1}{\sqrt{2}} = 1.$$

Example 5. *Show that*

(i) $2^n \Gamma \left(n + \dfrac{1}{2} \right) = 1.3.5....(2n-1)\sqrt{\pi}$, *where n is a positive integer.*

(ii) $\Gamma \left(\dfrac{3}{2} - x \right) \Gamma \left(\dfrac{3}{2} + x \right) = \left(\dfrac{1}{4} - x^2 \right) \pi \sec \pi x$, *provided* $-1 < 2x < 1$.

Solution. (i) We have

$$\Gamma \left(n + \dfrac{1}{2} \right) = \left(n - \dfrac{1}{2} \right) \Gamma \left(n - \dfrac{1}{2} \right) = \left(n - \dfrac{1}{2} \right) \left(n - \dfrac{3}{2} \right) \Gamma \left(n - \dfrac{3}{2} \right)$$

$$= \left(n - \dfrac{1}{2} \right) \left(n - \dfrac{3}{2} \right) \left(n - \dfrac{5}{2} \right) ... \dfrac{3}{2} . \dfrac{1}{2} \Gamma \left(\dfrac{1}{2} \right)$$

$$= \dfrac{2n-1}{2} . \dfrac{2n-3}{2} . \dfrac{2n-5}{2} ... \dfrac{3}{2} . \dfrac{1}{2} . \sqrt{\pi}$$

$$= \dfrac{1}{2^n} (2n-1)(2n-3)(2n-5)...3.1.\sqrt{\pi}.$$

Hence, $2^n \Gamma \left(n + \dfrac{1}{2} \right) = 1.3.5....(2n-1)\sqrt{\pi}$.

(ii) We have

$$\Gamma \left(\dfrac{3}{2} - x \right) \Gamma \left(\dfrac{3}{2} + x \right) = \left(\dfrac{1}{2} - x \right) \Gamma \left(\dfrac{1}{2} - x \right) . \left(\dfrac{1}{2} + x \right) \Gamma \left(\dfrac{1}{2} + x \right)$$

$$= \left(\dfrac{1}{4} - x^2 \right) \Gamma \left(\dfrac{1-2x}{2} \right) \Gamma \left(\dfrac{1+2x}{2} \right)$$

$$= \left(\dfrac{1}{4} - x^2 \right) \Gamma \left(\dfrac{1-2x}{2} \right) \Gamma \left(1 - \dfrac{1-2x}{2} \right)$$

$$= \left(\dfrac{1}{4} - x^2 \right) \dfrac{\pi}{\sin \left(\dfrac{1-2x}{2} \pi \right)} = \left(\dfrac{1}{4} - x^2 \right) \dfrac{\pi}{\sin \left(\dfrac{\pi}{2} - \pi x \right)}$$

$$= \left(\dfrac{1}{4} - x^2 \right) . \dfrac{\pi}{\cos \pi x} = \left(\dfrac{1}{4} - x^2 \right) \sec \pi x . \pi.$$

Example 6. *Prove that* $\int_0^\infty \cos(bx^{1/n}) dx = \dfrac{\Gamma(n+1)}{b^n} \cos \dfrac{n\pi}{2}$.

Solution. Let $I = \int_0^\infty \cos(bx^{1/n}) dx$. ...(1)

Putting $x = t^n$ so that $dx = nt^{n-1} dt$ then (1) gives

$$I = n \int_0^\infty \cos(bt) . t^{n-1} dt = \dfrac{n\Gamma(n)}{b^n} \cos \dfrac{n\pi}{2}$$

$$= \dfrac{\Gamma(n+1)}{b^n} \cos \dfrac{n\pi}{2}.$$

Aliter. $\int_0^\infty \cos(bz^{1/n}) dz = \dfrac{1}{b^n} \Gamma(n+1) \cos \dfrac{n\pi}{2}$.

Put $z^{1/n} = x \Rightarrow z = x^n$. so that $dz = nx^{n-1} dx$.

$$\therefore \qquad \int_0^\infty \cos(bz^{1/n})dz = \int_0^\infty \cos(bx).nx^{n-1}dx$$

$$= n\int_0^\infty x^{n-1}\cos(bx)dx$$

$$= \text{real part of } n\int_0^\infty e^{-bxi}x^{n-1}dx$$

$$= \text{real part of } n\frac{\Gamma(n)}{(bi)^n}$$

$$= \text{real part of } \frac{n\Gamma(n)}{b^n}\left(\cos\frac{\pi}{2}+i\sin\frac{\pi}{2}\right)^{-n}$$

$$= \text{real part of } \frac{\Gamma(n+1)}{b^n}\left(\cos\frac{n\pi}{2}-i\sin\frac{n\pi}{2}\right)$$

$$= \frac{1}{b^n}\Gamma(n+1)\cos\left(\frac{n\pi}{2}\right).$$

Example 7. *Evaluate*

(i) $\int_0^\infty \cos(c^2x^2)dx$ \qquad (ii) $\int_0^\infty \sin x^2\, dx.$

Solution. \qquad (i) Let $\qquad I = \int_0^\infty \cos(c^2x^2)\,dx$ \hfill ...(1)

Putting $\quad x^2 = t$ so that $x = t^{1/2}$ $\quad\Rightarrow\quad$ $dx = \left(\dfrac{1}{2}\right)t^{-1/2}dt$

Then, (1) reduces to

$$I = \int_0^\infty \cos(c^2t)\frac{1}{2}t^{-1/2}dt = \frac{1}{2}\int_0^\infty (t)^{(1/2)-1}\cos(c^2t)\,dt$$

$$= \frac{1}{2}\frac{\Gamma(1/2)}{(c^2)^{1/2}}\cos\left(\frac{1}{2}\cdot\frac{\pi}{2}\right) = \frac{\sqrt{\pi}}{2c}\cdot\frac{1}{\sqrt{2}} = \frac{1}{2c}\sqrt{\frac{\pi}{2}}.$$

(ii) Let $\qquad I = \int_0^\infty \sin x^2\, dx$ \hfill ...(2)

Putting $x^2 = t$, so that $x = t^{1/2}$ and $dx = \left(\dfrac{1}{2}\right)t^{-1/2}dt$, then (2) reduces to

$$I = \frac{1}{2}\int_0^\infty t^{(1/2)-1}\cos t\, dt \quad = \frac{1}{2}\frac{\Gamma(1/2)}{1}\sin\left(\frac{1}{2}\cdot\frac{\pi}{2}\right)$$

$$\left[\because \int_0^\infty x^{m-1}\sin bx\, dx = \frac{\Gamma(m)}{b^m}\sin\frac{m\pi}{2}\right]$$

$$= \frac{\sqrt{\pi}}{2}\cdot\frac{1}{\sqrt{2}} = \frac{1}{2}\sqrt{\frac{\pi}{2}}.$$

EXERCISE 8.2

1. Show that

$$\Gamma(n)\Gamma\left(\frac{1-n}{2}\right) = \frac{\sqrt{\pi}\,\Gamma(n/2)}{2^{1-n}\cos\left(\dfrac{n\pi}{2}\right)}, 0 < n < 1\,.$$

2. Show that $\int_0^1 \dfrac{dx}{\sqrt{(1-x^6)}} = \dfrac{\sqrt{3}}{2}\int_0^1 \dfrac{dx}{\sqrt{1-x^3}}$

$$= \frac{1}{2^{7/3}\pi}\left[\Gamma\left(\frac{1}{3}\right)\right]^3.$$

3. Show that

$$\Gamma(0.1).\Gamma(0.2).\Gamma(0.3)...\Gamma(0.9) = \frac{(2\pi)^{9/2}}{\sqrt{10}}.$$

4. Show that $\int_0^1 \frac{dx}{\sqrt{(1-x^4)}} = \frac{\sqrt{2}}{8\sqrt{\pi}}\left[\Gamma\left(\frac{1}{4}\right)\right]^2.$

5. Show that if $m > -1$ then

$$\int_0^\infty x^m e^{-n^2 x^2} dx = \frac{1}{2n^{m+1}}\Gamma\left(\frac{m+1}{2}\right).$$

HINTS TO SELECTED PROBLEMS

2. Let $A = \int_0^1 \frac{dx}{\sqrt{1-x^6}}.$

Now putting $x^6 = \sin^2\theta$ i.e., $x = \sin^{1/3}\theta \Rightarrow dx = \frac{1}{2}\sin^{-2/3}\theta\cos\theta\,d\theta.$

$$\therefore \quad A = \frac{1}{3}\int_0^{\pi/2}\sin^{-2/3}\theta\cos^0\theta\,d\theta = \frac{1}{3}\frac{\Gamma\left(\frac{1}{6}\right)\Gamma\left(\frac{1}{2}\right)}{2\Gamma\left(\frac{2}{3}\right)} = \frac{1}{6}\frac{\Gamma\left(\frac{1}{6}\right).\sqrt{\pi}}{\Gamma\left(\frac{2}{3}\right)}.$$

Now find the values of $\Gamma\left(\frac{1}{6}\right)$ and $\Gamma\left(\frac{2}{3}\right)$ separately.

4. Let $I = \int_0^1 \frac{dx}{\sqrt{1-x^4}}.$ Then solve after putting $x^4 = t.$

5. Let $I = \int_0^\infty x^m e^{-n^2 x^2} dx.$ Then solve after putting $n^2 x^2 = t.$

CHAPTER REVIEW: A COMPETITIVE APPROACH

SELECTED TERMS AND RESULTS

TERMS

- **Gamma Function** : The definite integral $\int_0^\infty e^{-x} x^{n-1} dx, n > 0$ is called the gamma function.

- **Beta Function** : The definite integral $\int_0^1 x^{m-1}(1-x)^{n-1} dx, \ m > 0, n > 0$ is called beta function.

RESULTS

- $\Gamma(n+1) = n\Gamma(n), \quad \Gamma(1) = 1$
- $\Gamma(n+1) = n!$
- $\Gamma\left(\dfrac{1}{2}\right) = \sqrt{\pi}$
- $B(m, n) = B(n, m)$
- $B(m,n) = 2\int_0^{\pi/2} \cos^{2m-1}\theta \sin^{2n-1}\theta \, d\theta$
- $B(m,n) = \dfrac{\Gamma(m)\Gamma(n)}{\Gamma(m+n)}, \ m > 0, n > 0$
- $\Gamma(n)\Gamma(1-n) = \dfrac{\pi}{\sin n\pi}, 0 < n < 1$

- $\int_0^{\pi/2} \cos^m\theta \sin^n\theta \, d\theta = \dfrac{\Gamma\left(\dfrac{m+1}{2}\right)\Gamma\left(\dfrac{n+1}{2}\right)}{2\Gamma\left(\dfrac{m+n+2}{2}\right)}$

- **Duplication formula:**

$$\Gamma(n)\Gamma\left(n+\dfrac{1}{2}\right) = \dfrac{\sqrt{\pi}}{2^{2n-1}}\Gamma(2n), n > 0$$

- $\Gamma\left(\dfrac{1}{n}\right)\Gamma\left(\dfrac{2}{n}\right)\Gamma\left(\dfrac{3}{n}\right)...\Gamma\left(\dfrac{n-1}{n}\right)$

$$= \dfrac{(2\pi)^{(n-1)/2}}{n^{1/2}}, \ n \in Z.$$

REVIEW QUESTIONS AND PROJECT WORK

1. If $p > 0, q > 0$ then show that

$$\int_{-1}^{1} \dfrac{(1+x)^{2p-1}(1-x)^{2q-1}}{(1+x^2)^{p+q}} dx$$

$$= 2^{p+q-2} B(p,q) \cdot$$

2. Show that $\Gamma(p) = \lim_{n\to\infty} n^p B(p,n), p > 0$

3. Show that

$$\sqrt{\pi}.\Gamma(2p) = 2^{2p-1}\Gamma(p)\Gamma\left(\pi + \dfrac{1}{2}\right), p > 0 \cdot$$

4. Show that $\int_0^1 \dfrac{(1-x^4)^{3/4}}{(1+x^4)^2} dx = \dfrac{B\left(\dfrac{7}{4}, \dfrac{1}{4}\right)}{2^{9/4}}.$

5. Show that for $x > 0$, $\lim_{n\to\infty} \dfrac{\Gamma(n+x)}{n^x \Gamma(n)} = 1$.

6. Show that the gamma function is continuous on $]0, \infty[$.

7. Show that

$$\int_0^{\pi/2} \sin^p x \, dx \int_0^{\pi/2} \sin^{p+1} x \, dx = \dfrac{\pi}{2(p+1)}$$

if $p > -1$.

8. Show that the beta function $B(x, y)$ is a convex function of x on $]0, \infty[$ for each fixed $y > 0$.

OBJECTIVE TYPE QUESTIONS

FILL IN THE BLANKS

1. $\Gamma(1) = $ _____ .

2. $\Gamma(n+1) = $ _____ $\Gamma(n), n > 0.$

3. $\Gamma\left(\dfrac{1}{2}\right) = $ _____ .

4. $\int_0^\infty x^{n-1}e^{-x}dx =$ _____ .

5. $\int_0^1 x^{m-1}(1-x)^{n-1}dx =$ _____
$m>0, n>0$.

6. Beta function is also called _____ integral of first kind.

7. $\dfrac{\Gamma(m)\Gamma(n)}{\Gamma(m+n)} =$ _____ .

TRUE/FALSE

Write 'T' for true and 'F' for false statement.

1. Beta function is also called Eulerian integral of first kind. **(T/F)**

2. Beta function is also called Eulerian integral of second kind. **(T/F)**

3. $\int_0^\infty e^{-ax}\sin bx\, dx = \dfrac{b}{a^2+b^2}$. **(T/F)**

4. $\int_0^{\pi/2} e^{-ax}\sin bx\, dx = \dfrac{b}{a^2+b^2}$. **(T/F)**

5. $\int_{-\infty}^\infty \cos\dfrac{\pi}{2}x^2.dx = 1$. **(T/F)**

MULTIPLE CHOICE QUESTIONS :

Choose the most appropriate one :

1. The value of $\Gamma\left(\dfrac{1}{2}\right)$ is equal to :

(a) π (b) $\sqrt{\pi}$
(c) 1 (d) 2

2. $\Gamma(x) = \int_0^\infty e^{-x}x^{n-1}dx$ if :

(a) $n>0$ (b) $n<0$
(c) $n=0$ (d) none of these

3. The value of $\Gamma(n+1)$, if $n>0$ is:

(a) n (b) $\Gamma(n)$

(c) $n\,\Gamma(n)$ (d) none of these

4. If n is a non-negative integer then value of $\Gamma(n+1)$ is equal to :

(a) n (b) $n!$
(c) $(n+1)!$ (d) none of these

5. $\Gamma(m)\,\Gamma(1-m) =$ _____ , $(0<m<1)$.

(a) $\dfrac{1}{\sin m\pi}$ (b) $\dfrac{\pi}{\sin m\pi}$
(c) $\dfrac{m}{\sin m\pi}$ (d) none of these

Answers

FILL IN THE BLANKS

  1. 1   **2.** n **3.** $\sqrt{\pi}$ **4.** $\Gamma(n)$ **5.** $B(m,n)$ **6.** Eulerian   **7.** $B(m,n)$

TRUE/FALSE

  1. T   **2.** F   **3.** T   **4.** F   **5.** T

MULTIPLE CHOICE QUESTIONS

  1. (b)   **2.** (a)   **3.** (c)   **4.** (b)   **5.** (b)

SELF ASSESSMENT TEST

Verify each of the following :

1. If $p > 0$, $B(p,p) = B\left(p+\dfrac{1}{2}, p+\dfrac{1}{2}\right) = \dfrac{\pi}{2^{4p-1}\cdot p}$.

2. $\int_0^1 x^{-1/3}(1-x)^{-2/3}(1-2x)^{-1}\,dx = \dfrac{1}{9^{1/3}} B\left(\dfrac{2}{3}, \dfrac{1}{3}\right)$

3. $\int_{-\infty}^{\infty} \dfrac{e^{pt}}{1+e^t}\,dt = \Gamma(p)\Gamma(1-p), 1 > p > 0$

4. If $p > 0$, $q > 1$ then $\sum\limits_{n=0}^{\infty} B(p+r,q)$ converges to $B(p, q-1)$.

5. $\sin\dfrac{\pi}{2n} \cdot \sin\dfrac{2\pi}{2n} \cdot \sin\dfrac{3\pi}{2n} \cdots \sin\dfrac{(n-1)\pi}{2n} = \sqrt{n}\cdot 2^{-n+1}$.

6. If $p > 0$, $q > 0$, $a > 0$, $b > 0$, then $\int_0^{\pi/2} \dfrac{\cos^{2p-1}\theta \sin^{2q-1}\theta}{(a\cos^2\theta + b\sin^2\theta)^{p+q}}\,d\theta = \dfrac{B(p,q)}{2a^p\cdot b^q}$.

7. If $n > 0$, then

$$\dfrac{\Gamma\left(n+\dfrac{1}{2}\right)}{\Gamma(n)} = \dfrac{(2n)!}{2^{2n}\cdot n!}.$$

8. $\int_0^{\pi/2} \dfrac{d\theta}{\sqrt{\sin\theta}} \times \int_0^{\pi/2} \sqrt{\sin\theta}\,d\theta = \pi$.

9. $\int_0^1 \sqrt{(1-x^4)}\,dx = \dfrac{1}{12}\sqrt{\dfrac{2}{\pi}}\left[\Gamma\left(\dfrac{1}{4}\right)\right]^2$.

10. $\int_0^{\infty} x^{m-1}\cos bx = \dfrac{\Gamma(m)}{b^m}\cos\left(\dfrac{m\pi}{2}\right)$.

Chapter

19 Multiple Integrals

19.1 INTRODUCTION

Double integral is an extension of a definite integral in two-dimensional space. Let (x, y) be a single valued function of x and y, bounded and defined in the region R of XY-plane, and A be the area of region R and let R be divided in any manner into n-sub regions α_1, $\alpha_2,...,\alpha_n$, whose areas are $\delta s_1, \delta s_2,... \delta s_n$ respectively. If $p_r(\xi_r, \eta_r)$ is any point inside the region α_n. $\beta_n = f(\xi_1)$.

Let $B_n = \sum_{r=1}^{n} f(\xi_r, \mu_r)\delta s_r$ then the limits of B_n

which is assumed to exists as $n \to \infty$ such that every $\alpha_r \to 0$ in all its dimensions is known as double integral of $f(x, y)$ over the region R and is denoted by

$$\int_R f(x,y)ds$$

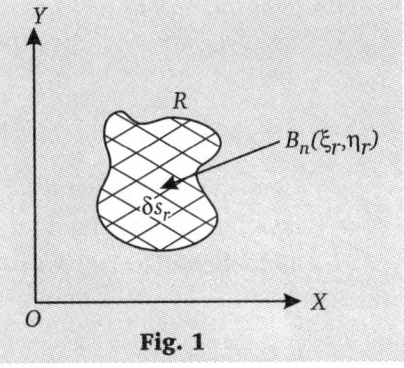

Fig. 1

or

$$\iint_R f(x,y)dx\,dy .$$

Hence, the area R is called the region or field of integration for the double integral and ds is called element of area.

REMARK

- Let the region A be divided into the rectangular partitions and dx be the length of a sub-rectangular and dy be its width, so that the $dx\,dy$ is an element of area in cartesian co-ordinates, then the integral $\iint f(x, y)\,ds$ is written as $\iint_A f(x,y)dx\,dy$ and is called the double integral of $f(x, y)$ over the region R.

19.2 PROPERTIES OF DOUBLE INTEGRALS

(1) When the region R is partitioned into two parts say R_1 and R_2 then
$$\iint_R f(x,y)dx\,dy = \iint_{R_1} f(x,y)dx\,dy + \iint_{R_2} f(x,y)dx\,dy$$

Similarly, we divide the region into three or more parts.

(2) The double integral of a algebraic sum of a fixed number of functions is equal to the algebraic sum of double integrals taken for each term separately. Thus
$$\iint_R [f_1,(x,y) + f_2(x,y) + f_3(x,y) + ...]dx\,dy$$
$$= \iint_R f_1(x,y)dx\,dy + \iint_R f_2(x,y)dx\,dy + \iint_R f_3(x,y)dx\,dy + ...$$

(3) A constant factor may be taken outside the integral sign. Thus
$$\iint_R mf(x,y)dx\,dy = m\iint_R f(x,y)dx\,dy \text{ where } m \text{ is a constant.}$$

19.3 EVALUATION OF DOUBLE INTEGRALS

(i) *Over a rectangular region R.* If the region R be given by the inequalities $a \leq x \leq b, c \leq y \leq d$, then the double integral

$$\iint_R f(x,y)dx\,dy = \int_a^b \int_c^d f(x,y)dx\,dy$$

$$= \int_a^b \left[\int_c^d f(x,y)dy \right] dx . \qquad \ldots (1)$$

We first evaluate $\int_c^d f(x,y)dy$ i.e., integrate $f(x, y)$ with respect to y regarding x as constant and then resulting function of x is to be integrated with respect to x between the limits a and b

or $$\iint_R f(x,y)dx\,dy = \int_c^d \int_a^b f(x,y)dx\,dy$$

$$= \int_c^d \left[\int_a^b f(x,y)dx \right] dy . \qquad \ldots (2)$$

Now, we integrate $\int_a^b f(x,y)dx$ and then integrate with respect to y.

(ii) *Over the regions which are not rectangular.* Let the region R be described by $a \leq x \leq b$ and $\phi_1(x) \leq y \leq \phi_2(x)$ so that $y = \phi_1(x)$ and $y = \phi_2(x)$ respectively, the boundary of R then

$$\iint_R f(x,y)dx\,dy = \int_a^b \left[\int_{\phi_1(x)}^{\phi_2(x)} f(x,y)dy \right] dx$$

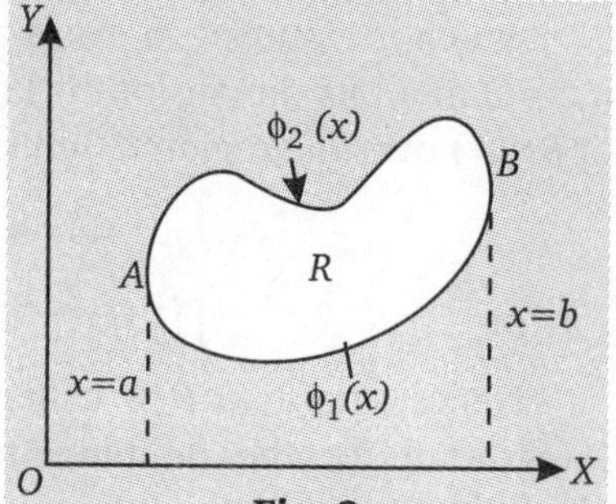

Here, the inner integral $\int_{\phi_1(x)}^{\phi_2(x)} f(x,y)dy$ is integrated first and in this integral the result of integration is a function of x, say $\phi_1(x)$ then $\phi_1(x)$ is integrated with respect to x between the limits a and b to obtain the value of double integral.

In a similar way, if R can be described by

$$c \leq y \leq d, \quad \phi_3(y) \leq x \leq \phi_4(y)$$

then we get

$$\iint_R f(x,y)dx\,dy = \int_c^d \left[\int_{\phi_3(y)}^{\phi_4(y)} f(x,y)dx \right] dy .$$

Here, the result of integration

$$\int_{\phi_3(y)}^{\phi_4(y)} f(x,y)dx .$$

which is evaluated first, is a function of y say $\phi_2(y)$, then $\phi_2(y)$ is integrated with respect to y between the limits c to d.

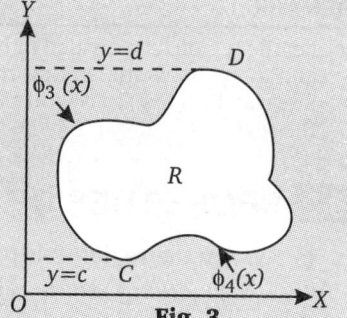

Fig. 2

Fig. 3

WORKING PROCEDURE

While evaluating double integrals, first integrate with respect to variable having variable limits and treating the other variable as constant and then integrate with respect to variable with constant limits. In case the limits of integration of both the variables are constants.

17.3.1 CONVERSION OF CARTESIAN CO-ORDINATES TO POLAR

The transformation formula required is $x = r \cos \theta, y = r \sin\theta$ and elementary area $\delta A = r\delta\theta . \delta r$ so that

$$\iint f(x,y)dx\,dy = \iint f(x,y)dA = \iint f(r,\theta)r\,d\theta\,dr.$$

Solved Examples

Example 1. Evaluate $\int_1^2 \int_0^{y/2} y \, dy \, dx$.

Solution. We have $\int_1^2 \int_0^{y/2} y \, dy \, dx = \int_1^2 y(x)_0^{y/2} \, dy = \int_1^2 y\left(\frac{1}{2}y\right) dy$

$$= \frac{1}{2}\int_1^2 y^2 dy = \frac{1}{2}\left[\frac{1}{3}y^3\right]_1^2 = \frac{1}{6}(2^3 - 1^3) = 7/6 \; .$$

Example 2. Evaluate $\int_1^2 \int_0^x \frac{1}{x^2 + y^2} \, dx \, dy$.

Solution. We have $\int_1^2 \int_0^x \frac{dx \, dy}{x^2 + y^2} = \int_1^2 \left[\int_0^x \frac{dy}{x^2 + y^2}\right] dx = \int_1^2 \left[\frac{1}{x}\tan^{-1}\frac{y}{x}\right]_{y=0}^x dx$

$$= \int_1^2 \left[\frac{1}{x}(\tan^{-1} 1 - \tan^{-1} 0)\right] dx = \int_{\frac{\pi}{4}}^{\frac{\pi}{4}}\int_1^2 \frac{dx}{x} = \frac{\pi}{4}[\log x]_1^2$$

$$= \frac{\pi}{4}\cdot[\log 2 - \log 1] = \frac{1}{4}\pi \log 2.$$

Example 3. Show that $\int_1^2 \int_0^{y/2} y \, dy \, dx = \int_1^2 \int_0^{x/2} x \, dx \, dy$.

Solution. We have $\int_1^2 \int_0^{y/2} y \, dy \, dx = \int_1^2 [y]\left[\int_0^{y/2} dx\right] dy$

$$= \int_1^2 y[x]_0^{y/2} dy = \int_1^2 y[y/2 - 0] dy$$

$$= \frac{1}{2}\int_1^2 y^2 dy = \frac{1}{2}\left[\frac{y^3}{3}\right]_1^2 = \frac{7}{6} \; .$$

Again $\int_1^2 \int_0^{x/2} x \, dx \, dy = \int_1^2 x\left[\int_0^{x/2} dy\right] dx = \int_1^2 x[y]_0^{x/2} dx$

$$= \int_1^2 x\left[\frac{x}{2} - 0\right] dx = \frac{1}{2}\int_1^2 x^2 dx$$

$$= \frac{1}{2}\left[\frac{x^3}{3}\right]_1^2 = \frac{1}{6}(8 - 1) = \frac{7}{6} \; .$$

Hence, $\int_1^2 \int_0^{y/2} y \, dy \, dx = \int_1^2 \int_0^{x/2} x \, dx \, dy$.

Example 4. Evaluate the double integral of $x^2 y^3$ over the rectangle bounded by $x=2$, $x=3$, $y=2$, $y=4$.

Solution. The required integral $= \int_2^3 \int_0^4 x^2 y^3 dx \, dy = \int_2^3 \left[\frac{1}{4}y^4\right]_2^4 x^2 dx$

$$= \frac{1}{4}(4^4 - 2^4)\int_2^3 x^2 dx = 60\left[\frac{1}{3}x^3\right]_2^3 = 20[27 - 8] = 380 \; .$$

Example 5. Evaluate $\int_0^3 \int_1^2 xy(1 + x + y) \, dx \, dy$

Solution. We have $\int_0^3 \int_1^2 xy(1 + x + y) \, dx \, dy$

$$= \int_0^3 \left[x\frac{y^2}{2} + x^2\frac{y^2}{2} + x\frac{y^3}{3}\right]_{y=1}^2 dx$$

$$= \int_0^3 \left[\frac{x}{2}(4-1) + \frac{x^2}{2}(4-1) + \frac{x}{3}(8-1) \right] dx$$

$$= \int_0^3 \left[\left(\frac{3}{2} + \frac{7}{3} \right) x + \frac{3}{2} x^2 \right] dx = \left[\frac{23}{6} \cdot \frac{x^2}{2} + \frac{3}{2} \cdot \frac{x^2}{3} \right]_0^3$$

$$= \frac{23}{6} \cdot \frac{9}{2} + \frac{27}{2} = \frac{123}{4} .$$

Example 6. *Evaluate $\iint_A (x^2 + y^2)\,dx\,dy$, where A is the region bounded by $x=0$, $y=0$, $x+y=1$.*

Solution. Let R be the region of integration $x+y=1$ and the limit of itegration can be expressed as $0 \le x \le 1, 0 < y < 1-x$.

From the equation $x+y=1$, we have $x = 1$ for $y = 0$ and for the positive quadrant x varies from 0 to 1 and for y which varies from $y = 0$ to $y = 1-x$. First integrate with respect to y, treated x as constant and then integrate with respect to 'x'.

Hence, the integral $= \int_0^1 \int_0^{1-x} (x^2 + y^2)\,dx\,dy = \int_0^1 \left(x^2 y + \frac{1}{3} y^3 \right)_0^{1-x} dx$

$$= \int_0^1 \left[x^2(1-x) + \frac{1}{3}(1-x)^3 \right] dx$$

$$= \int_0^1 (1-x) \left\{ x^2 + \frac{1}{3}(1-x)^2 \right\} dx$$

$$= \int_0^1 \frac{1}{3} [1 - 3x + 6x^2 - 4x^3]\,dx$$

$$= \frac{1}{3} \left[x - \frac{3}{2} x^2 + 2x^3 - x^4 \right]_0^1 = \frac{1}{3} \left[1 - \frac{3}{2} + 2 - 1 \right] = \frac{1}{6}$$

Example 7. *Find the area by double integration the region bounded by circle $x^2 + y^2 = a^2$.*

Solution. The area of a small element at any point (x, y) is $dx\,dy$. Now to find the area bounded by the circle $x^2 + y^2 = a^2$, the region of integration R can be expressed as

$$-a \le y \le a, -\sqrt{a^2 - y^2} \le x \le \sqrt{(a^2 - y^2)}.$$

Now, first integration is to be performed w.r. to x regarding y as constant.

∴ The required area

$$= \iint_R dx\,dy = \int_{y=-a}^a \int_{x=-\sqrt{(a^2-y^2)}}^{\sqrt{(a^2-y^2)}} 1.dy\,dx$$

$$= \int_{-a}^a \left[2\int_0^{\sqrt{(a^2-y^2)}} 1.dx \right] dy, \text{ by property of definite integral}$$

$$= 2\int_{-a}^a [x]_0^{\sqrt{(a^2-y^2)}} dy = 2\int_{-a}^a \sqrt{(a^2 - y^2)}dy = 2.2\int_0^a \sqrt{(a^2 - y^2)}dy$$

$$= 4 \left[\frac{y\sqrt{(a^2 - y^2)}}{2} + \frac{a^2}{2} \sin^{-1} \frac{y}{a} \right]_0^a$$

$$= 4 \left[0 + \frac{a^2}{2} \sin^{-1} 1 \right] = 4 \cdot \frac{1}{2} a^2 \cdot \frac{1}{2} \pi = \pi a^2 .$$

Example 8. *Evaluate $\iint (x+y)^2 dx\,dy$ over the region bounded by ellipse $\dfrac{x^2}{a^2} + \dfrac{y^2}{b^2} = 1$. Hence*

find the mass of an elliptic plate whose density per unit area is given by $\rho = k(x+y)^2$.

Solution. Since the region is bounded by ellipse $\dfrac{x^2}{a^2} + \dfrac{y^2}{b^2} = 1$, we expressed it as:

$$x = -a \text{ and } x = a \quad y = -b\sqrt{(1 - x^2/a^2)}, y = b\sqrt{(1 - x^2/a^2)} \ .$$

$$\therefore \quad \iint (x+y)^2 dx\,dy = \int_{-a}^{a} \int_{-b\sqrt{(1-x^2/a^2)}}^{b\sqrt{(1-x^2/a^2)}} (x^2 + y^2 + 2xy)\,dx\,dy$$

$$= \int_{-a}^{a} 2\int_{0}^{b\sqrt{(1-x^2/a^2)}} (x^2 + y^2)\,dx\,dy$$

[∵ 2xy being an odd function of f, its integration under the given limits of y is 0]

$$= 2\int_{-a}^{a} \left[x^2 y + \frac{y^3}{3} \right]_{0}^{b\sqrt{1-x^2/a^2}} dx$$

$$= 2\int_{-a}^{a} \left\{ x^2 b \sqrt{\left(1 - \frac{x^2}{a^2}\right)} + \frac{b^3}{3}\left(1 - \frac{x^2}{a^2}\right)^{3/2} \right\} dx$$

$$= 2 \times 2\int_{0}^{a} \left\{ x^2 b \sqrt{\left(1 - \frac{x^2}{a^2}\right)} + \frac{b^3}{3}\left(1 - \frac{x^2}{a^2}\right)^{3/2} \right\} dx$$

$$= 4b\int_{0}^{\pi/2} \left\{ a^2 \sin^2\theta \cos\theta + \frac{b^2}{3}\cos^3\theta dx \right\} a\cos\theta\, d\theta$$

(By putting $x = a\sin\theta$ so that $dx = a\cos\theta\, d\theta$)

$$= 4ab\int_{0}^{\pi/2} \left[a^2 \sin^2\theta \cos^2\theta + \frac{b^2}{3}\cos^4\theta \right] d\theta$$

$$= 4ab \left[a^2 \int_{0}^{\pi/2} \sin^2\theta \cos^2\theta\, d\theta + \frac{b^2}{3}\int_{0}^{\pi/2}\cos^4\theta\, d\theta \right]$$

$$= 4ab \left[\frac{1}{16}\pi a^2 + \frac{1}{16}\pi b^2 \right] = \frac{1}{4}\pi ab(a^2 + b^2)$$

= the mass of elliptic plate whose density is given by $\rho = k(x+y)^2$

$$= \iint_R k(x+y)^2 dx\,dy$$

(where integration is to be performed over the area *A* of ellipse.)

$$= k.\frac{1}{4}\pi ab(a^2 + b^2)$$

Example 9. *Evaluate $\iint xy(x+y)\,dx\,dy$ over the region between $y = x^2$ and $y = x$.*

Solution. When we draw the given curve, the parabola $y = x^2$ and line $y = x$ intersect at the point $(0, 0)$ and $(1, 1)$, $x^2 = x$ or $x(x-1) = 0$ i.e., $x=0$ or 1, when $x =0, y = 0$, and $x=1, y=1$].

So the area of integration for x is from $x=0$ to

$x = 1$ and for y from x^2 to x.

∴ Given integral $= \int_0^1 \int_{x^2}^x xy(x+y)dx\,dy$

$= \int_0^1 \int_{x^2}^x x(yx + y^2)dx\,dy$

$= \int_0^1 x\left[x.\dfrac{y^2}{2} + \dfrac{1}{3}y^3\right]_{x^2}^x dx$

$= \int_0^1 x.\left[\left(x.\dfrac{x^2}{2} + \dfrac{1}{3}x^3\right) - \left(x.\dfrac{x^4}{2} + \dfrac{1}{3}x^6\right)\right]dx$

$= \int_0^1 x\left[\dfrac{5x^3}{6} - \dfrac{1}{2}x^5 - \dfrac{1}{3}x^6\right]dx = \int_0^1\left[\dfrac{5}{6}x^4 - \dfrac{1}{2}x^6 - \dfrac{1}{3}x^7\right]dx$

$= \left[\dfrac{1}{6}x^5 - \dfrac{1}{14}x^7 - \dfrac{1}{24}x^8\right]_0^1 = \dfrac{1}{6} - \dfrac{1}{14} - \dfrac{1}{24} = \dfrac{3}{56}$

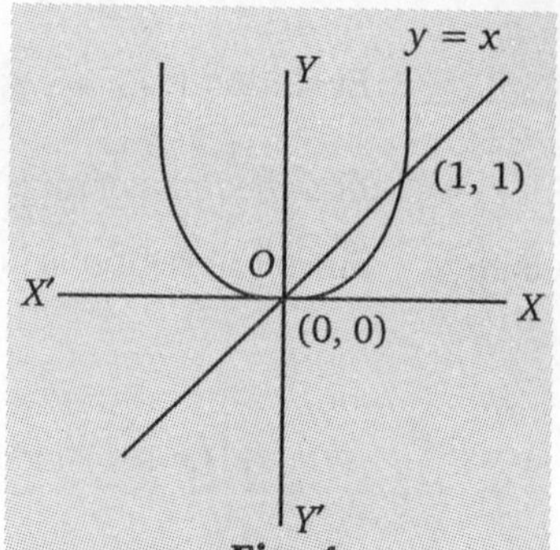

Fig. 4

Example 10. *When the region of integration A is the triangle given by $y=0$, $y=x$ and $x=1$, show that* $\iint_A \sqrt{4x^2 - y^2}\,dx\,dy = \dfrac{1}{3}\left(\dfrac{\pi}{3} + \dfrac{\sqrt{3}}{2}\right)$.

Solution. Here we draw straight lines $y = 0$, $y = x$ and $x = 1$. We can express the region of integration as $0 \le y \le x$, $0 \le x \le 1$.

∴ $\iint_A \sqrt{4x^2 - y^2}\,dx\,dy = \int_{x=0}^1 \int_{y=0}^x \sqrt{(4x^2 - y^2)}\,dx\,dy$

$= \int_0^1\left[\dfrac{y}{2}\sqrt{4x^2 - y^2} + 2x^2 \sin^{-1}\dfrac{y}{2x}\right]_{y=0}^x dx$

Integrating w.r. to y treating x as constant.

$= \int_0^1\left[\dfrac{x}{2}\sqrt{4x^2 - x^2} + 2x^2 \sin^{-1}\dfrac{1}{2} - 0\right]dx$

$= \int_0^1\left[\dfrac{\sqrt{3}}{2}x^2 + \dfrac{\pi}{3}x^2\right]dx = \left[\dfrac{\sqrt{3}}{2}.\dfrac{x^3}{3} + \dfrac{\pi}{3}.\dfrac{x^3}{3}\right]_0^1$

$= \dfrac{1}{3}\left(\dfrac{\sqrt{3}}{2} + \dfrac{\pi}{3}\right)$.

EXERCISE 19.1

1. Evaluate $\int_2^3 dx\int_0^1(x^2 + 3y^2)dy$.

2. Evaluate $\int_0^2\int_0^{\sqrt{4+x^2}} \dfrac{dx\,dy}{(4+x^2+y^2)}$.

3. Evaluate $\int_0^{\pi/2}\int_{\pi/2}^\pi \cos(x+y)dx\,dy$.

4. Evaluate $\int_0^2\int_0^{\sqrt{2x-x^2}} x\,dx\,dy$.

5. Evaluate $\int_0^1\int_0^{x^2} e^{y/x}\,dx\,dy$.

6. Evaluate $\int_0^1\int_0^1 \dfrac{dx\,dy}{\sqrt{(1-x^2)(1-y^2)}}$

7. Evaluate $\iint e^{2x+3y}dx\,dy$ over the triangle bounded by $x=0$, $y=0$ and $x+y=1$.

8. Evaluate $\iint_p x\sin(x+y)dx\,dy$, where p is a rectangle $[0 \le x \le \pi, 0 \le y \le \pi/2]$.

9. Show that

$$\int_1^2 \int_3^4 (xy + e^y) dx\, dy = \int_3^4 \int_1^2 (xy + e^y) dy\, dx$$

10. Evaluate $\iint x^2 y^2 dx\, dy$ over the region bounded by $x = 0, y = 0$, where A is the region bounded by $x^2 + y^2 = 1$.

11. Find the area of the ellipse $\dfrac{x^2}{a^2} + \dfrac{y^2}{b^2} = 1$ by double integration.

12. Show that by double integration that the area between the parabolas $y^2 = 4ax$ and $x^2 = 4.b.y$ is $(16/3)\,ab$.

13. Find by double integration the region included between the parabola $x^2 = 4ay$ and the curve $y = 8a^3/(x^2 + 4a^2)$.

14. Evaluate $\iint y\, dx\, dy$ over the region between the parabolas $y^2 = 4x$ and $x^2 = 4y$.

15. Find the double integration the region lying between the parabola $y = 4x - x^2$, and the line $y = x$.

16. Find by double integration the area of the region lying between the parabola $y^2 = 4ax$, and line $y = mx$.

17. Find by double integration the area of the region lying between the semi-cubical parbola $y^2 = x^3$, and line $y = mx$.

18. Find by double integration the area of the region lying between the circle $x^2 + y^2 = a^2$, and line $x + y = a$ (in first quadrant)

19. Find by double integration the area of the region lying between the curves $(x^2 + 4a^2) y = 8a^3$, $2y = x$ and $x = 0$.

20. Evaluate

(i) $\displaystyle\int_0^1 \int_0^{\sqrt{1+x^2}} \dfrac{dx\, dy}{1 + x^2 + y^2}$

(ii) $\displaystyle\int_0^a \int_0^{\sqrt{a^2 + y^2}} (a^2 - x^2 - y^2) dx\, dy$

HINTS TO SELECTED PROBLEMS

5. $I = \displaystyle\int_0^1 \int_0^{x^2} e^{y/x} dx\, dy = \int_0^1 x [e^{y/x}]_{y=0}^{x^2} dx$

$= \displaystyle\int_0^1 x(e^x - 1) dx = \int_0^1 xe^x dx - \int_0^1 x\, dx$

$= (xe^x)_0^1 - [e^x]_0^1 - \left(\dfrac{x^2}{2}\right)_0^1$

$= (e - 0) - (e - 1) - \dfrac{1}{2} = \dfrac{1}{2}$

10. $I = \displaystyle\iint_A x^2 y^2 dx\, dy = \int_0^1 \int_0^{\sqrt{1-x^2}} x^2 y^2 dx\, dy$

$= \displaystyle\int_0^1 x^2 \left[\dfrac{y^2}{3}\right]_0^{\sqrt{1-x^2}} dx = \dfrac{1}{2} \int_0^1 x^2 (1 - x^2)^{3/2} dx$

Now put $x^2 = t$.

11. $I = \displaystyle\iint_A dx\, dy = \int_{-a}^a \int_{-b/a\sqrt{a^2-x^2}}^{b/a\sqrt{a^2-x^2}} dx\, dy$

12. $A = \displaystyle\int_0^{4a^{1/3}b^{2/3}} \int_{x^2/4a}^{\sqrt{4ax}} dx\, dy$.

13. $A = \displaystyle\int_{-2a}^{2a} \int_{x^2/4a}^{8a^3/x^2 + 4a^2} dx\, dy$.

14. The curves $y^2 = 4a$ and $x^2 = 4y$ intersect at the points where $x = 0$ and $x = 4$. Also, when

$$0 < x < 4, \sqrt{4x} > \dfrac{x^2}{4}.$$

$\therefore \quad I = \displaystyle\int_0^4 \int_{x^2/4}^{\sqrt{4x}} y\, dx\, dy.$

16. Since, the two corners cut at the point where

$$x = 0 \text{ and } x = \dfrac{4a}{m^2}$$

$\therefore \quad I = \displaystyle\int_0^{4a/m^2} \int_{mx}^{\sqrt{4ax}} dx\, dy$

Answers

1. $\dfrac{22}{3}$ 2. $\dfrac{\pi}{4}\log(1 + \sqrt{2})$ 3. -2 4. $\dfrac{\pi}{2}$ 5. $\dfrac{1}{2}$ 6. $\dfrac{\pi^2}{4}$ 7. $\dfrac{1}{6}(e-1)^2(2e+1)$

8. $\pi + 2$ 9. $\dfrac{21}{4} + e^4 - e^3$ 10. $\pi/96$ 11. πab 13. $\left(2\pi - \dfrac{4}{3}\right)a^2$ 14. $48/5$ 15. $9/2$

16. $8a^2/3m^2$ 17. $1/10\, m^5$ 18. $\dfrac{1}{4}(\pi - 2)a^2$ 19. $(\pi - 1)a^2$ 20. (i) $\dfrac{\pi}{4}\log(1 + \sqrt{2})$ (ii) $\dfrac{\pi a^4}{8}$.

19.4 DOUBLE INTEGRAL IN POLAR CO-ORDINATES

Let us consider a function $f(r, \theta)$ of polar co-ordinates r, θ over a certain area A with whose boundary is also given in terms of polar co-ordinates. We divide the area into n parts of elementary areas $\delta A_1, \delta A_2, \delta A_3, \ldots \delta A_n$ and let

$$s_n = \sum_{r=1}^{n} f(r, \theta) \delta A$$

where (r_1, θ_1) is a point inside the elementary area δA_1, the dobule integral of $f(r, \theta)$ is then defined as

$$\iint_A f(r, \theta) dA = \lim_{\substack{n \to \infty \\ \delta A_i \to 0}} \sum_{i=1}^{n} f(r_i, \theta_i) \delta A_i \,,$$

provided limit toward right hand side exists.

REMARK

- In case of cartesian co-ordinates when the double integral $\iint_A f(x, y) dA$ is expressed in the form of repeated integral, dA representes the area of the rectangle with sides dx and dy and hence $dA = dx\, dy$.

If the radius vector of OS and OP are r and $r + \delta r$ respectively and
$$\angle POQ = d\theta,$$

$$RS = r\, d\theta$$

as RS and PQ are arcs of circles. Then
$$dA = RS \times SP$$
$$= r\, d\theta . dr = r\, dr\, d\theta$$

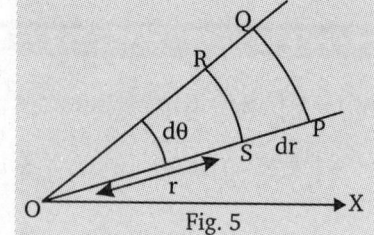

Fig. 5

Solved Examples

Example 1. Evaluate the double integral
$$\int_0^{\pi/2} \int_0^{2a\cos\theta} r^2 \sin\theta . \cos\theta . d\theta\, dr\,.$$

Solution. We have $\int_0^{\pi/2} \int_0^{2a\cos\theta} r^2 \sin\theta . \cos\theta . d\theta\, dr$

$$= \int_0^{\pi/2} \int_0^{2a\cos\theta} (r^2 \sin\theta . \cos\theta)\, dr\, d\theta$$

$$= \int_0^{\pi/2} \left[\frac{r^3}{3} . \sin\theta . \cos\theta \right]_0^{2a\cos\theta} d\theta$$

$$= \frac{1}{3} \int_0^{\pi/2} (2a\cos\theta)^3 . \sin\theta . \cos\theta\, d\theta$$

$$= \frac{8a^3}{3} \int_0^{\pi/2} \sin\theta . \cos^4\theta\, d\theta$$

$$= -\frac{8a^3}{3} \int_0^{\pi/2} \cos^4\theta\, d(\cos\theta) = -\frac{8a^3}{3} \left[\frac{\cos^5\theta}{5} \right]_0^{\pi/2}$$

$$= -\frac{8a^3}{3} \left[0 - \frac{1}{5} \right] = \frac{8a^3}{15}\,.$$

Example 2. *Evaluate* $\iint \dfrac{r\,d\theta\,dr}{\sqrt{a^2+r^2}}$ *over one loop of the lemniscate* $r^2 = a^2\cos 2\theta$

Solution.   In lemniscate, there are two loops.

We see that when $-\pi/4 < \theta < \pi/4$ or $3\pi/4 < \theta < 5\pi/4$, where r is real.

We want to evaluate the given integral over the right loop of the leminscate.

Therefore, $\iint \dfrac{r\,d\theta\,dr}{\sqrt{a^2+r^2}}$

$$= \int_{-\pi/4}^{\pi/4}\int_{-a\sqrt{\cos 2\theta}}^{a\sqrt{\cos 2\theta}} \dfrac{r}{\sqrt{r^2+a^2}}\,dr\,d\theta$$

as $r^2 = a^2\cos 2\theta.$

$\therefore$ $r = \pm a\sqrt{\cos 2\theta}$

Thus, r varies from $-a\sqrt{\cos 2\theta}$ to $a\sqrt{\cos 2\theta}$.

Since, there is a symmetry about X-axis, we should evaluate the double integral over half of the right loop as follows:

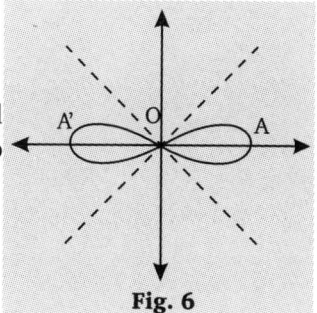

Fig. 6

$$\iint \dfrac{r\,d\theta\,dr}{\sqrt{a^2+r^2}} = \int_0^{\pi/4}\int_0^{a\sqrt{\cos 2\theta}} \dfrac{r}{\sqrt{a^2+r^2}}\,dr\,d\theta$$

$$= \dfrac{1}{2}\int_0^{\pi/4}\int_0^{a\sqrt{\cos 2\theta}} \dfrac{2r}{\sqrt{a^2+r^2}}\,dr\,d\theta$$

$$= \dfrac{1}{2}\int_0^{\pi/4}\int_0^{a\sqrt{\cos 2\theta}} \dfrac{d(a^2+r^2)}{\sqrt{a^2+r^2}}\,d\theta$$

$$= \dfrac{1}{2}\int_0^{\pi/4}\left[2\sqrt{a^2+r^2}\right]_0^{a\sqrt{\cos 2\theta}} d\theta$$

$$= \int_0^{\pi/4}\left[\sqrt{a^2+a^2\cos 2\theta} - \sqrt{a^2+0}\right]d\theta$$

$$= \int_0^{\pi/4}[\sqrt{2}.a\cos\theta - a]d\theta$$

$$= \sqrt{2}.a\int_0^{\pi/4}\cos\theta\,d\theta - a\int_0^{\pi/4} d\theta$$

$$= \sqrt{2}.a[\sin\theta]_0^{\pi/4} - a[\theta]_0^{\pi/4}$$

$$= \sqrt{2}.a.\dfrac{1}{\sqrt{2}} - a.\dfrac{\pi}{4} = a - \dfrac{a}{4}\pi$$

$\therefore$ Value of double integral over the complete right loop

$$= 2a - \dfrac{2a\pi}{4} = 2a(1-\pi/4) \ .$$

REMARK

- If we evaluate the double integral as $\int_{-\pi/4}^{\pi/4}\int_{-a\sqrt{\cos 2\theta}}^{a\sqrt{\cos 2\theta}} \dfrac{r}{\sqrt{a^2+r^2}}\, d\theta\, dr$

  then it will become zero due to oddness of function $\dfrac{r}{\sqrt{a^2+r^2}}$ therefore we must not calculate the double integral over the complete loop.

Example 3. *Evaluate $\iint r^2 d\theta\, dr$ over the area of circle $r = a\cos\theta$.*

Solution. In the given region of circle, θ varies from $-\pi/2$ to $\pi/2$ and r varies from 0 to a.

$$\therefore \iint r^2 d\theta\, dr = \int_{-\pi/2}^{\pi/2}\int_0^{a\cos\theta} r^2 d\theta\, dr$$

$$= \int_{-\pi/2}^{\pi/2}\left[\frac{r^3}{3}\right]_0^{a\cos\theta} d\theta$$

$$= \frac{a^3}{3}\int_{-\pi/2}^{\pi/2}\cos^3\theta\, d\theta$$

$$= \frac{a^3}{3}\int_{-\pi/2}^{\pi/2}\left(\frac{3}{4}\cos\theta + \frac{1}{4}\cos 3\theta\right)d\theta$$

$$= \frac{a^3}{3}\times\frac{3}{4}\int_{-\pi/2}^{\pi/2}[\cos\theta\, d\theta] + \frac{a^3}{12}\int_{-\pi/2}^{\pi/2}\cos 3\theta\, d\theta$$

$$= \frac{a^3}{4}[\sin\theta]_{-\pi/2}^{\pi/2} + \frac{a^3}{12}\left[\frac{\sin 3\theta}{3}\right]_{-\pi/2}^{\pi/2}$$

$$= \frac{a^3}{4}.2 + \frac{a^3}{12}\times\frac{1}{3}\left[\sin\frac{3\pi}{2} + \sin\frac{3\pi}{2}\right] = \frac{a^3}{4} - \frac{a^3}{18} = \frac{4}{9}a^3.$$

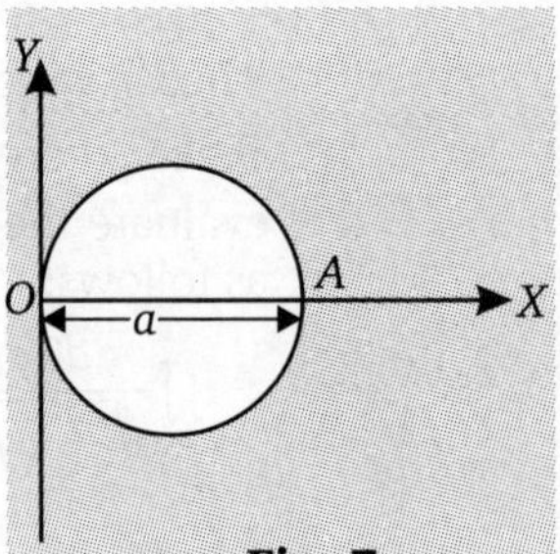

Fig. 7

Example 4. *Evaluate $\int_0^\pi \int_0^{a(1+\cos\theta)} r^2 \cos\theta\, d\theta\, dr$.*

Solution. We have $\int_0^\pi \int_0^{a(1+\cos\theta)} r^2 \cos\theta\, d\theta\, dr$

$$= \int_0^\pi \cos\theta\left[\frac{r^3}{3}\right]_0^{a(1+\cos\theta)} d\theta = \frac{1}{3}\int_0^\pi \cos\theta.a^3(1+\cos\theta)^3 d\theta$$

$$= \frac{a^3}{3}\int_0^\pi \cos\theta(1+3\cos\theta+3\cos^2\theta+\cos^3\theta)d\theta$$

$$= \frac{a^3}{3}\int_0^\pi [\cos\theta+3\cos^2\theta+3\cos^3\theta+\cos^4\theta]d\theta$$

$$= 2.\frac{a^3}{3}\int_0^\pi [3\cos^2\theta+\cos^4\theta]d\theta \qquad \left[\because \int_0^\pi \cos^n\theta\, d\theta = 0, \text{since } n \text{ is odd}\right]$$

$$= \frac{2a^3}{3}\left[3.\frac{1}{2}.\frac{\pi}{2} + \frac{3}{4}.\frac{1}{2}.\frac{\pi}{2}\right] = \frac{2a^3}{3}.\frac{3\pi}{4}\left[1+\frac{1}{4}\right]$$

$$= \frac{2a^3}{3}.\frac{3\pi}{4}.\frac{5}{4} = \frac{5\pi a^3}{8}$$

Example 5. *Evaluate* $\int_0^{\pi/2} \int_0^{\sin\theta} r\, d\theta\, dr$.

Solution. We have

$$\int_0^{\pi/2} \int_0^{\sin\theta} r\, d\theta\, dr = \int_0^{\pi/2} d\theta \int_0^{\sin\theta} r\, dr = \int_0^{\pi/2} \left(\frac{1}{2}r^2\right)_0^{\sin\theta} d\theta$$

$$= \frac{1}{2}\int_0^{\pi/2} \sin^2\theta\, d\theta = \frac{1}{4}\int_0^{\pi/2}(1-\cos 2\theta)\, d\theta$$

$$= \frac{1}{4}\left[\theta - \frac{1}{2}\sin 2\theta\right]_0^{\pi/2} = \frac{1}{4}\left[\frac{1}{2}\pi - \frac{1}{2}(0)\right] = \frac{\pi}{8} \ .$$

Example 6. *Evaluate* $\int_0^{\pi/2} \int_{a(1-\cos\theta)}^{a\sin\theta} r\, d\theta\, dr$.

Solution. We have

$$\int_0^{\pi/2} \int_{a(1-\cos\theta)}^{a\sin\theta} r\, d\theta\, dr = \int_0^{\pi/2}\left[\frac{1}{2}r^2\right]_{a(1-\cos\theta)}^{a\sin\theta} d\theta$$

$$= \frac{1}{2}a^2 \int_0^{\pi/2}[\sin^2\theta - (1-\cos\theta)^2]\, d\theta$$

$$= \frac{1}{2}a^2 \int_0^{\pi/2}[2\cos\theta - \cos^2\theta - (1-\sin^2\theta)]\, d\theta$$

$$= a^2 \int_0^{\pi/2}(\cos\theta - \cos^2\theta)\, d\theta$$

$$= a^2 \int_0^{\pi/2}\left[\cos\theta - \frac{1}{2}(1+\cos 2\theta)\right]\, d\theta$$

$$= a^2\left[\sin\theta - \frac{1}{2}\theta - \frac{1}{4}\sin 2\theta\right]_0^{\pi/2} = a^2\left[1 - \frac{\pi}{4}\right] = \frac{1}{4}a^2(4-\pi) .$$

EXERCISE 19.2

1. Integrate $r\sin\theta$ over the area of cardiod $r = a(1+\cos\theta)$ lying above the initial line.
2. Find by double integration that the area lying inside the cardiod $r = a(1+\cos\theta)$ and outside the circle $r = a$.
3. Find by double integration the area lying inside the cardiod $r = 1+\cos\theta$ and outside the parabola $r(1+\cos\theta) = 1$.
4. Find by double integration the area lying inside the circle $r = a\sin\theta$ and outside the cardioid $r - a(1-\cos\theta)$.

Answers

1. $\dfrac{4a^3}{3}$ 2. $\dfrac{1}{4}a^2(\pi+8)$ 3. $\dfrac{9\pi+16}{12}$ 4. $\dfrac{a^2}{4}(4-\pi)$.

19.5 APPLICATIONS OF DOUBLE INTEGRATION

Double integration is generally used to find the area of curves, volume and surface of solids of revolution.

(1) *Area of curves.* Let AD be an arc of curve $y=f(x)$.

Let area $ABCD$ be divided into sub-area by drawing lines parallel to X and Y axis respectively such that distance between two adjoining lines drawn parallel to Y-axis be δx and those drawn parallel to X-axis be δy.

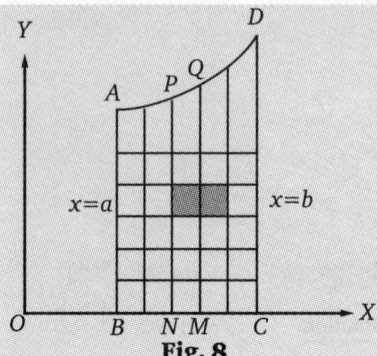

Fig. 8

(i) Let $P(x,y)$ and $Q(x+\delta x, y+\delta y)$ be two neighbouring points on the curve AD. PN and QM are the co-ordinates at P and Q respectively. Then the area of element shown by shaded lines is $\delta x\, \delta y$. Therefore, the area of strip PN

$$= \int_{y=0}^{f(x)} dx\, dy \text{ where } y = f(x).$$

The required area

$$ABCD = \int_{x=a}^{b} \int_{y=0}^{f(x)} dx\, dy .$$

(ii) We can find the area bounded by the two curves $y = f_1(x)$ and $y = f_2(x)$ and the ordinates $x=a$ and $x=b$

$$\int_{x=a}^{\beta} \int_{y=f_2(x)}^{f_1(x)} dx\, dy .$$

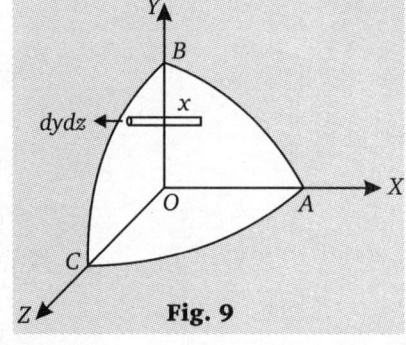

Fig. 9

(iii) *In polar co-ordinates.* The area bounded by curve $r = f(\theta)$ where $f(\theta)$ is a single valued function of θ in the domain (α, β) and the radii vector $\theta = \alpha$ and $\theta = \beta$ is

$$\int_{\theta=\alpha}^{\beta} \int_{r=0}^{f(\theta)} r\, d\theta\, dr .$$

(2) *Volume of a solid.* Consider the area $dy\, dz$ on the plane $x=0$ through each point on the boundary of this small area. Draw the lines parallel to X-axis and thus construct a small cylinder whose base is area to X-axis. This cylinder cuts the given surface, and volume of this cylinder

$$= x\, dy\, dz .$$

∴ Volume of solid $= \iint x\, dy\, dz .$

REMARKS

- By considering area $dx\, dy$ on plane $z=0$ the volume of solid $= \iint z\, dx\, dy .$
- By considering area $dx\, dz$ on plane $y=0$ the volume of solid $= \iint y\, dx\, dz .$

(3) *Area of surface of a solid.* Let the equatioin of surface be $z = f(x, y)$. Consider a point $P(x, y, z)$ on this surface surrounding this point P. Consider an element of area δs of the surface. Let $\delta x\ \delta y$ be the projection of this area δs on the plane $z = 0$, then we have

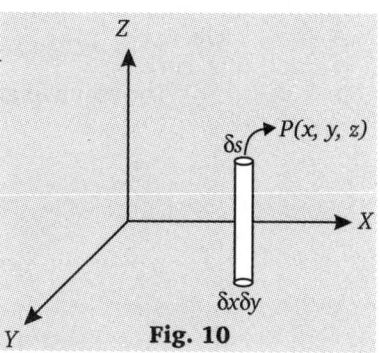

$$\delta x\ \delta y = \delta s \cos\alpha \qquad \dots (1)$$

where α is the angle between the tangent plane to the given surface at $P(x, y, z)$ and the plane $z=0$ then by co-ordinate geometry, we have

Fig. 10

$$\sec\alpha = \sqrt{\left[1+\left\{\frac{\partial z}{\partial x}\right\}^2+\left\{\frac{\partial z}{\partial y}\right\}^2\right]} \qquad \dots(2)$$

From (1) we have $\delta s = \delta x\ \delta y \sec\alpha$

$$= \delta x \delta y \sqrt{\left[1+\left\{\frac{\partial z}{\partial x}\right\}^2+\left\{\frac{\partial z}{\partial y}\right\}^2\right]} \qquad \text{[From (2)]}$$

$\therefore$ The required area of surface

$$= \iint \sqrt{\left[1+\left\{\frac{\partial z}{\partial x}\right\}^2+\left\{\frac{\partial z}{\partial y}\right\}^2\right]}.dx.dy \ .$$

Solved Examples

Example 1. *Find the area of ellipse* $\dfrac{x^2}{a^2}+\dfrac{y^2}{b^2}=1.$

Solution. Required area of ellipse

$$= 4 \text{ (area of quadrants } OABO \text{ of ellipse)}$$

$$= 4\int_{x=0}^{a}\int_{y=0}^{f(x)}dx\,dy, \text{where } y = f(x) = \frac{b}{a}\sqrt{(a^2-x^2)}$$

$$= 4\int_0^a [y]_0^{f(x)}dx = 4\int_0^a f(x)dx = 4\int_0^a \frac{b}{a}\sqrt{(a^2-x^2)}\,dx$$

$$= \frac{4b}{a}\left[\frac{1}{2}x\sqrt{(a^2-x^2)}+\frac{1}{2}a^2\sin^{-1}\left(\frac{x}{a}\right)\right]_0^a$$

$$= \frac{2b}{a}[0+a^2\sin^{-1}(1)] = \frac{2b}{a}a^2.\frac{\pi}{2} = ab\pi$$

Example 2. *Find the whole area of curve* $a^2x^2 = y^3(2a-y).$

Solution. The shape of curve is shown in fig. 11.

The required area $= 2\times$ area OAB

$$= 2\int_{y=0}^{2a}\int_{x=0}^{f(y)}dy\,dx$$

where $x = f(y)$ i.e., $x = y^{3/2} \dfrac{\sqrt{2a-y}}{a}$ is the equation of curve.

$\therefore$ The required area

$$= 2\int_{y=0}^{2a} [x]_0^{f(y)} dy$$

$$= 2\int_0^{2a} f(y)\, dy$$

$$= 2\int_0^{2a} \frac{y^{3/2}\sqrt{2a-y}}{a}\, dy$$

$$\left[\because f(y) = x = y^{3/2}\frac{\sqrt{2a-y}}{a} \right]$$

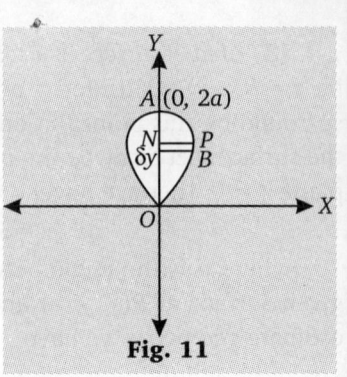

Fig. 11

Put $\qquad y = 2a\sin^2\theta$

$\Rightarrow \qquad dy = 4a\sin\theta\cos\theta\, d\theta$

at $\qquad y = 0, \theta = 0$ and $y = 2a, \theta = \pi/2$

$\therefore$ Required area $= \dfrac{2}{a}\int_0^{\pi/2}(2a\sin^2\theta)^{3/2}\sqrt{(2a-2a\sin^2\theta)}\,4a\sin\theta\cos\theta\, d\theta$

$$= 32a^2\int_0^{\pi/2}\sin^4\theta\cos^2\theta\, d\theta$$

$$= \frac{32a^2\Gamma(5/2)\Gamma(3/2)}{2\Gamma 4} = \frac{32a^2.(3/2).(1/2).\sqrt{\pi}.(1/2).\sqrt{\pi}}{2.3.2.1} = \pi a^2.$$

Example 3. *Find by double integration the area between* $y = \dfrac{3x}{(x^2+2)}$ *and* $4y = x^2$

Solution. We have

$$4y = x^2, \qquad \text{and} \qquad y = \frac{3x}{(x^2+2)}$$

$$\Rightarrow \qquad 4y = \frac{12x}{(x^2+2)}, \qquad \text{and} \qquad 4y = x^2$$

$$\Rightarrow \qquad x^2 = \frac{12x}{(x^2+2)} \qquad \Rightarrow \quad x^4 + 2x^2 - 12x = 0$$

$$\Rightarrow \quad x(x^3 + 2x - 12) = 0$$

$$x = 0, 2$$

$\therefore$ Required area $= \int_{x=0}^2 \int_{y=x^2/4}^{3x/(x^2+2)} dx\, dy$

$$= \int_0^2 [y]_{x^2/4}^{3x/(x^2+2)} dx = \int_0^2 \left[\frac{3x}{x^2+2} - \frac{x^2}{4} \right] dx$$

$$= \frac{3}{2}\int_0^2 \frac{2x\, dx}{x^2+2} - \frac{1}{4}\int_0^2 x^2\, dx = \frac{3}{2}\left[\log(x^2+2) \right]_0^2 - \frac{1}{4}\left(\frac{1}{3}x^3 \right)_0^2$$

$$= \frac{3}{2}[\log(6) - \log(2)] - \frac{1}{12}(8-0) = \frac{3}{2}\log 3 - \frac{2}{3}.$$

Example 4. *Find the area of curve r=a (1+cos θ).*

Solution.   The required area

$$= 2\times area\ OABO = 2\int_{\theta=0}^{\pi}\int_{r=0}^{f(\theta)} r\,d\theta\,dr,$$

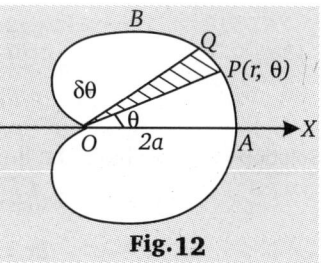

(where $r = f(\theta)$ and $r = a(1+\cos\theta)$ is the equation of the curve.)

$$= 2\int_{\theta=0}^{\pi}\left[\frac{1}{2}r^2\right]_{r=0}^{f(\theta)} d\theta = \int_{0}^{\pi}[f(\theta)]^2\,d\theta$$

$$= \int_{0}^{\pi} a^2(1+\cos\theta)^2\,d\theta = a^2\int_{0}^{\pi}(2\cos^2\theta/2)^2 d\theta$$

$$= 4a^2\int_{0}^{\pi}\cos^4\frac{\theta}{2}d\theta$$

$$= 8a^2\int_{0}^{\pi/2}\cos^4\phi\,d\phi, \qquad\qquad \text{(Putting } \theta = 2\phi)$$

$$= 8a^2.\frac{3}{4}.\frac{1}{2}.\frac{1}{2}\pi = (3/2)a^2\pi.$$

Fig. 12

Example 5. *Find the volume bounded by co-ordinates planes and the plane $\dfrac{x}{a}+\dfrac{y}{b}+\dfrac{z}{c}=1$.*

Solution.   The plane cuts X, Y and Z- axis at point $(a, 0, 0)$, $(0, b, 0)$ and $(0, 0, c)$ respectively. The surface ABCD of co-ordinates planes will be equal to

$$\int_{0}^{a}\int_{0}^{b(1-x/a)}\int_{0}^{c(1-x/a-y/b)} dx\,dy\,dz$$

$$= \int_{0}^{a}\int_{0}^{b(1-x/a)} c\left(1-\frac{x}{a}-\frac{y}{b}\right)dy\,dx$$

$$= c\int_{0}^{a}\int_{0}^{b(1-x/a)}\left(1-\frac{x}{a}-\frac{y}{b}\right)dy\,dx$$

$$= c\int_{0}^{a}\left[y-\frac{x}{a}.y-\frac{y^2}{2b}\right]_{0}^{b(1-x/a)} dx$$

$$= c\int_{0}^{a}\left[b\left(1-\frac{x}{a}\right)-\frac{x}{a}.b\left(1-\frac{x}{a}\right)-\frac{1}{2b}b^2\left(1-\frac{x}{a}\right)^2\right] dx.$$

Example 6. *Find the volume bounded by the cylinder $x^2+y^2=4$ and the hyperboloid $x^2+y^2-z^2=-1$.*

Solution.   Here, surfaces $x^2+y^2=4$ and $x^2+y^2-z^2=-1$ are symmetrical about all the three axes. Therefore, volume

$$V = \iint z\,dx\,dy = 8\int_{0}^{2}\int_{0}^{\sqrt{4-x^2}}\sqrt{(x^2+y^2+1)}\,dx\,dy.$$

Change to polar co-ordinates and change the limits of integrations for the region of quadrant of circle $r = 2$ and $\theta = 0$ to $\pi/2$

$$V = 8\int_{0}^{\pi/2}\int_{0}^{2}\sqrt{(r^2+1)}r\,d\theta\,dr = 8\int_{0}^{\pi/2}\left[\frac{1}{3}(r^2+1)^{3/2}\right]_{0}^{2} d\theta$$

$$= 8\int_{0}^{\pi/2}\frac{1}{3}(5\sqrt{5}-1)d\theta = \frac{4\pi}{3}(5\sqrt{5}-1) .$$

Example 7. *Transform the integral*

$$\int_0^2 \int_0^{\sqrt{2x-x^2}} \frac{x\,dx\,dy}{\sqrt{(x^2+y^2)}}$$

by changing into polar co-ordinates and hence evaluate it.

Solution. We have the limit of integration be

$$y = 0, y = \sqrt{(2x-x^2)} \text{ and } x = 0, x = 2.$$

$$x^2 + y^2 - 2x = 0 \text{ which is change into}$$

$$r^2(\cos^2\theta + \sin^2\theta) - 2r\cos\theta = 0$$

or $r = 2\cos\theta.$

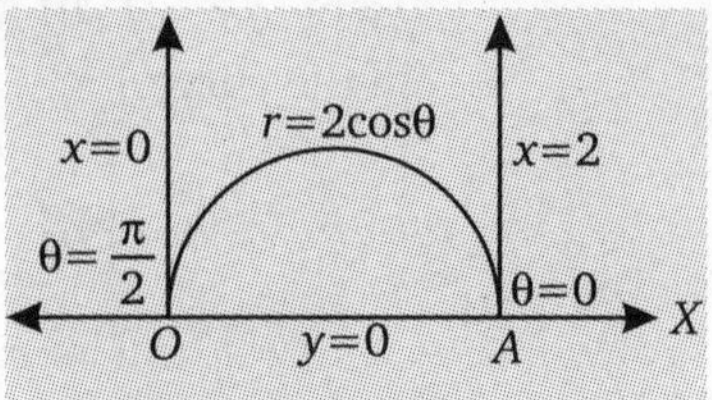

Fig. 13

Now r varies form 0 to $2\cos\theta$ and θ varies from 0 to $\pi/2$.

Note that at the point A of the circle, $\theta = 0$ and at point O, $r = 0$ and so from $r = 2\cos\theta$, we get

$$\theta = \frac{\pi}{2} \text{ at } O$$

the polar equivalent of the elementary area $dx\,dy$ is $r\,d\theta\,dr$.

$$\therefore \quad \iint_A f(x,y)dx\,dy = \iint_A f(r\cos\theta, r\sin\theta)r\,d\theta\,dr$$

where A is the region of integration.

Therefore, transforming to polar co-ordinates, the given double integral

$$= \int_{\theta=0}^{\pi/2}\int_{r=0}^{2\cos\theta} \frac{r\cos\theta}{r}r\,d\theta\,dr = \int_0^{\pi/2}\cos\theta\left[\frac{r^2}{2}\right]_0^{2\cos\theta}d\theta$$

$$= \int_0^{\pi/2}\frac{1}{2}\cos\theta.4\cos^2\theta\,d\theta = 2\int_0^{\pi/2}\cos^3\theta\,d\theta$$

$$= 2.\frac{2}{3} = \frac{4}{3}.$$

Example 8. *Find the area of the surface $z^2 = 2xy$ included between planes $x=0$, $x=a$, $y=0$, $y=b$.*

Solution. The given surface is $z^2 = 2xy$.

$$\therefore \quad 2z\frac{\partial z}{\partial x} = 2y \text{ or } \frac{\partial z}{\partial x} = \frac{y}{z}$$

Similarly $\quad \dfrac{\partial z}{\partial y} = \dfrac{x}{z}$

Then required area of the surface

$$= \iint \sqrt{\left[1 + \left(\frac{\partial z}{\partial x}\right)^2 + \left(\frac{\partial z}{\partial y}\right)^2\right]}dx\,dy = \int_{x=0}^a\int_{y=0}^b \sqrt{\left\{1 + \left(\frac{y}{z}\right)^2 + \left(\frac{x}{z}\right)^2\right\}}dx\,dy$$

$$= \int_{x=0}^a\int_{y=0}^b \sqrt{\left(\frac{z^2 + y^2 + x^2}{2xy}\right)}dx\,dy = \int_{x=0}^a\int_{y=0}^b \sqrt{\left(\frac{x^2 + y^2 + z^2}{2xy}\right)}dx\,dy$$

$$= \int_{x=0}^a\int_{y=0}^b \sqrt{\frac{x^2 + y^2 + 2xy}{2xy}}dx\,dy = \int_{x=0}^a\int_{y=0}^b \frac{(x+y)}{\sqrt{2}\sqrt{(xy)}}dx\,dy$$

$$= \frac{1}{\sqrt{2}} \int_{x=0}^{a} \int_{y=0}^{b} \left(\sqrt{x} \cdot \frac{1}{\sqrt{y}} + \sqrt{y} \cdot \frac{1}{\sqrt{x}} \right) dx \, dy$$

$$= \frac{1}{\sqrt{2}} \int_{x=0}^{a} \sqrt{x} (2\sqrt{y})_0^b \, dx + \frac{1}{\sqrt{2}} \int_{x=0}^{a} \frac{1}{\sqrt{x}} \left(\frac{2}{3} y^{3/2} \right)_0^b dx$$

$$= \sqrt{(2b)} \int_0^a \sqrt{x} \, dx + \frac{\sqrt{2}}{3} b^{3/2} \int_0^a \frac{1}{\sqrt{x}} dx$$

$$= \sqrt{2b} \left[\frac{2}{3} x^{3/2} \right]_0^a + \frac{1}{3} \sqrt{2b^3} (2x^{1/2})_0^a$$

$$= \frac{2}{3} \sqrt{2} \sqrt{(ab)} (a+b) \ .$$

Example 9. *Show that the area of the surface of paraboloid $x^2 + y^2 = a^2$ which lies between the planes $z = 0$ and $z = a$ is $(\pi/6)(5 \cdot \sqrt{5} - 1)a^2$.*

Solution. The projection of given suface between the planes $z=0$ and $z=a$ on the x-y planes is circle $x^2 + y^2 = a^2, z = 0$.

Also, $$\frac{\partial z}{\partial x} = \frac{2x}{a}, \frac{\partial z}{\partial y} = \frac{2y}{a}$$

$\therefore$ $$S = \iint_A \sqrt{1 + \left(\frac{\delta z}{\delta x} \right)^2 + \left(\frac{\partial z}{\partial y} \right)^2} \, dx \, dy$$

$$= \iint_A \sqrt{1 + \frac{4x^2}{a^2} + \frac{4y^2}{a^2}} \, dx \, dy = \frac{1}{a} \iint_A \sqrt{a^2 + 4x^2 + 4y^2} \, dx \, dy$$

where A is the circle $x^2 + y^2 = a^2$ in the xy-plane.

The equation of the circle $x^2 + y^2 = a^2$ in polar co-ordinates is $r = a$. Hence ,transforming the above double integral into polar co-ordinates, we have

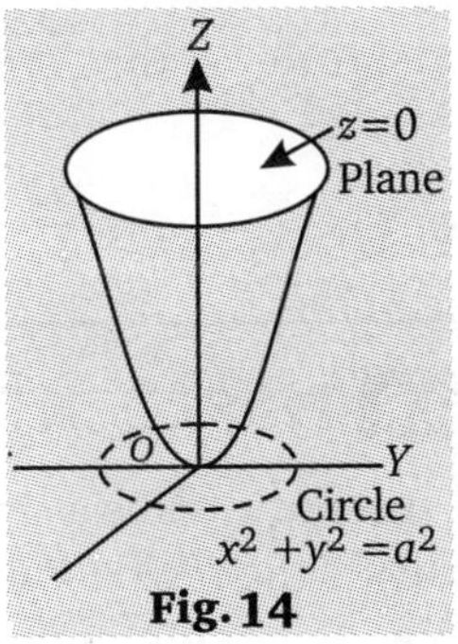

$$S = \frac{1}{a} \int_0^{2\pi} \int_0^a \sqrt{a^2 + 4r^2} (r \, d\theta \, dr)$$

$$= \frac{1}{a} \int_0^{2\pi} \left[\int_0^a r \sqrt{a^2 + 4r^2} \, dr \right] d\theta$$

$$= \frac{1}{8a} \int_0^{2\pi} \left[\int_0^a \sqrt{a^2 + 4r^2} \, d(a^2 + 4r^2) \right] d\theta$$

$$= \frac{1}{8a} \int_0^{2\pi} \left[\frac{2}{3} . (a^2 + 4r^2)^{3/2} \right]_0^a d\theta$$

$$= \frac{1}{8} \times \frac{2}{3} \times \frac{1}{a} \int_0^{2\pi} [(5a^2)^{3/2} - (a^2)^{3/2}] d\theta$$

$$= \frac{1}{12a} \int_0^{2\pi} (5\sqrt{5}a^3 - a^3) d\theta = \frac{1}{12a} (5\sqrt{5} - 1)a^3 \int_0^{2\pi} d\theta$$

$$= \frac{1}{12a} (5\sqrt{5} - 1)a^3 . 2\pi = \frac{\pi}{6} (5\sqrt{5} - 1)a^2 \ .$$

Fig. 14

1. Find by double integration, the area of the region enclosed by curves
   (a) $y = 4x - x^2, y = x$
   (b) $(x^2 + 4a^2)y = 8a^3, 2y = x$ and $x = 0$
   (c) $y = \dfrac{3x}{(x^2 + 2)}, 4y = x^2$.

2. Show that by double integration that the area between the parabolas $y^2 = 4ax$ and $x^2 = 4.b.y$ is $(16/3)ab$.

3. Find by double integration the area included between the parabola $x^2 = 4ay$ and the curve
   $$y = \frac{8x^3}{(x^2 + 4a^2)}.$$

4. Find the volume of the region bounded by $z = x^2 + y^2$ and $z = 2x$.

5. Find the volume cut off the sphere $x^2 + y^2 + z^2 = a^2$ by the cone $x^2 + y^2 = z^2$

6. Transforms the following double integrals to polar co-ordinates and hence, evaluate them
   (a) $\int_{y=0}^{a} \int_{x=0}^{\sqrt{a^2 - y^2}} (a^2 - x^2 - y^2) dx\, dy$
   (b) $\int_0^1 \int_x^{\sqrt{2x - x^2}} (x^2 + y^2) dx\, dy$
   (c) $\int_0^a \int_0^{\sqrt{a^2 - x^2}} y^2 \sqrt{(x^2 + y^2)} dx\, dy$

7. Evaluate $\iint r^2 d\theta\, dr$ over the area of circle $r = a\cos\theta$.

2. Required area $= \int_{x=0}^{4a^{1/3}b^{2/3}} \int_{y=x^2/4b}^{2\sqrt{ax}} dx\, dy$

3. Given that $x^2 = 4ay$ and $y = \dfrac{8a^3}{x^2 + 4a^2}$

   After simplification, we get
   $$x = \pm i2\sqrt{2}a, \pm 2a$$
   $\therefore$ Required area $= \int_{x=-2a}^{2a} \int_{y=x^2/4a}^{8a^3/x^2 + 4a^2} dx\, dy$

4. $I = \int_0^\pi \int_0^{a(1+\sin\theta)} r\sin\theta.rd\theta\, dr$

5. Since the two curves intersect at the points where $\cos\theta = 0$ i.e., $\theta = \dfrac{\pi}{2}$
   $\therefore\ I = \int_{\theta = -\pi/2}^{\pi/2} \int_{r=a}^{a(1+\cos\theta)} rd\theta\, dr$.

7. The required volume is given by
   $$V = \int_0^{2\pi} \int_0^{a/\sqrt{2}} \int_r^{\sqrt{a^2 - r^2}} dz(r\, d\theta\, dr).$$

1. (a) $= \dfrac{9}{2}$ (b) $(\pi - 1)a^2$ (c) $\dfrac{3}{2}\log 3 - \dfrac{2}{3}$ 3. $\left(2\pi - \dfrac{4}{3}\right)a^2$ 4. $\dfrac{\pi^3}{-2}$ 5. $\dfrac{(2 - \sqrt{2})\pi a^3}{3}$

6. (a) $\dfrac{\pi a^4}{8}$ (b) $\dfrac{3\pi}{8} - 1$ (c) $\dfrac{\pi a^5}{20}$ 7. $\dfrac{4a^3}{9}$.

19.6 TRIPLE INTEGRAL

Let $f(x, y, z)$ be a single-valued function of the independent variables x, y, z in finite region V. Divide the region V into n subregions $\delta V_1, \delta V_2, \delta V_3, \dots$ Let P be any point on the boundary or inside.

Take a point in each part and form the sum
$$s_n = f(x_1, y_1, z_1)\delta V_1 + f(x_2, y_2, z_2)\delta V_2 + \dots + f(x_n, y_n, z_n)\delta V_n$$
$$= \sum_{r=1}^{n} f(x_r, y_r, z_r)\delta V_r \qquad \dots(1)$$

when n tends to infinity. The limit of sum (1) tends to zero is called the triple integral of function $f(x, y, z)$ over the region V and is denoted by

$$\iiint_V f(x,y,z)dv .$$

The triple integral can be utilised in evaluating a number of physical quantities like, $f(x, y, z) = 1$

We find volume, $V = \iiint_V dV$ and putting $f(x, y, z) = \rho$

We get, $mass = \iiint_V \rho dV .$

19.6.1 EVALUATION OF TRIPLE INTEGRALS

The region V divide into elementary cuboids by drawing parallel co-ordinate planes. The volume V can then be considered as the sum of number of columns parallel to z-axis extending from the lower surface of V say $z=z_1(x, y)$ to the upper surface of V say $z = z_2 (x, y)$ the bases of these as column (only one column has been shown in fig. 15) are the elementary area δs_r, which cover a certain area S in x-y plane *i.e.* plane $z=0$.

$\therefore$ Summing up over the elementary cuboids in the same column first and then taking the sum of all such columns

we can write

$$\sum_{r=1}^{n} f(x_r, y_r, z_r) \text{ as } \sum_r \sum_m f(x_r, y_r, z_m)\delta_z]\delta s_r$$

where (x_r, y_r, z_r) is a point in the *mth* cuboid.

When δS_r and δz tend to zero this becomes equal to

$$\iint_S \left\{ \int_{z=z_1(x,y)}^{z_2(x,y)} f(x,y,z)dz \right\} ds$$

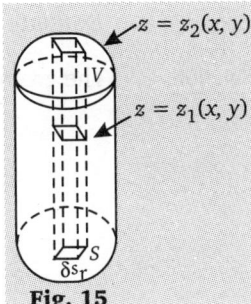

Fig. 15

(a) If the region V be specified by inequalities $a \leq x \leq b, c \leq y \leq d, e \leq z \leq f$ then triple integral

$$\iiint_V f(x,y,z)dx\,dy\,dz = \int_a^b \int_c^d \int_e^f f(x,y,z)dx\,dy\,dz$$

$$= \int_a^b dx \int_c^d dy \int_e^f f(x,y,z)dz.$$

Here, we integrate first with respect to z keeping x and y constant and then the remaining integration is done as in the case of doube integrals.
The integration with respect to z is performed first regarding x and y as constant then integration w.r to y regarding x as a constant and then integrate w.r to x.

(b) If the limits of z are function of x and y and y as function of x and x takes the constant values as from $x = a$ to $x = b$.

$$\iiint_V f(x,y,z)dx\,dy\,dz = \int_a^b dx \int_{y_1(x)}^{y_2(x)} f(x,y,z)dz$$

The integration with respect to z perform first regarding x and y as constant then integral w.r.t. y regarding x as a constant and then integrate w.r.t. x.

Solved Examples

Example 1.   *Evaluate* $\int_0^1 \int_{y^2}^1 \int_0^{1-x} x\,dy\,dx\,dz$

Solution.   We have

$$I = \int_0^1 \int_{y^2}^1 (z)_0^{1-x} x\,dy\,dx$$

$$= \int_0^1 \int_{y^2}^1 x(1-x)\,dy\,dx$$

$$= \int_0^1 \int_{y^2}^1 (x - x^2)\,dy\,dx = \int_0^1 \left[\frac{1}{2}x^2 - \frac{1}{3}x^3 \right]_{y^2}^1 dy$$

$$= \int_0^1 \left[\left\{ \frac{1}{2}(1)^2 - \frac{1}{3}(1)^3 \right\} - \left\{ \frac{1}{2}(y^2)^2 - \frac{1}{3}(y^2)^3 \right\} \right] dy$$

$$= \int_0^1 \left[\left(\frac{1}{2} - \frac{1}{3} \right) - \left(\frac{1}{2}y^4 - \frac{1}{3}y^6 \right) \right] dy$$

$$= \int_0^1 \left(\frac{1}{6} - \frac{1}{2}y^4 + \frac{1}{3}y^6 \right) dy = \left(\frac{1}{6}y - \frac{1}{10}y^5 - \frac{1}{21}y^7 \right)_0^1$$

$$= \frac{1}{6} - \frac{1}{10} + \frac{1}{21} = \frac{4}{35}.$$

Example 2. *Evaluate* $\int_{x=0}^1 \int_{y=0}^{\sqrt{1-x^2}} \int_{z=0}^{\sqrt{1-x^2-y^2}} xyz\,dx\,dy\,dz$

Solution. The given integral

$$I = \int_{x=0}^1 \int_0^{\sqrt{1-x^2}} xy \left(\frac{1}{2}z^2 \right)_0^{\sqrt{1-x^2-y^2}} dx\,dy$$

$$= \frac{1}{2} \int_{x=0}^1 \int_{y=0}^{\sqrt{1-x^2}} xy(1-x^2-y^2)\,dx\,dy$$

$$= \frac{1}{2} \int_{x=0}^1 \int_{y=0}^{\sqrt{1-x^2}} x[y(1-x^2) - y^3]\,dx\,dy$$

$$= \frac{1}{2} \int_{x=0}^1 x \left[\frac{1}{2}(1-x^2)y^2 - \frac{1}{4}y^4 \right]_0^{\sqrt{1-x^2}} dx$$

$$= \frac{1}{2} \int_0^1 x \left[\frac{1}{2}(1-x^2)(1-x^2) - \frac{1}{4}(1-x^2)^2 \right] dx$$

$$= \frac{1}{2} \int_0^1 x \left(\frac{1}{2} - \frac{1}{4} \right)(1-x^2)^2 dx$$

$$= \frac{1}{8} \int_0^1 (x - 2x^3 + x^5)\,dx = \frac{1}{8} \left[\frac{1}{2}x^2 - \frac{1}{2}x^4 + \frac{1}{6}x^6 \right]_0^1$$

$$= \frac{1}{8} \left(\frac{1}{2} - \frac{1}{2} + \frac{1}{6} \right) = \frac{1}{48}.$$

Example 3. *Evaluate* $\int_0^4 \int_0^{2\sqrt{z}} \int_0^{\sqrt{4z-x^2}} dz\,dx\,dy$.

Solution. The given triple integral

$$I = \int_0^4 \int_0^{2\sqrt{z}} \left[\int_0^{\sqrt{4z-x^2}} dy \right] dz\,dx$$

$$= \int_0^4 \int_0^{2\sqrt{z}} [y]_0^{\sqrt{4z-x^2}} dz\,dx = \int_0^4 \left[\int_0^{2\sqrt{z}} \sqrt{4z - x^2}\,dx \right] dz$$

$$= \int_0^4 \left[\frac{x}{2}\sqrt{4z-x^2} + \frac{4z}{2}\sin^{-1}\frac{x}{2\sqrt{z}} \right]_0^{2\sqrt{z}} dz$$

$$= \int_0^4 \left[0 + \frac{4z}{2}\sin^{-1}\frac{2\sqrt{z}}{2\sqrt{z}} \right] dz \quad = \int_0^4 2z.\frac{\pi}{2}dz = \int_0^4 \pi z\, dz$$

$$= \pi \left[\frac{z^2}{2} \right]_0^4 = \frac{\pi}{2}[16] = 8\pi.$$

Example 4.   *Evaluate* $\int_0^{\log a} \int_0^x \int_0^{x+y} e^{x+y+z} dx\, dy\, dz.$

Solution.   Let

$$I = \int_0^{\log a} \int_0^x \int_0^{x+y} e^{x+y} e^z dx\, dy\, dz \; = \int_0^{\log a} \int_0^x e^{x+y} (e^z)_0^{x+y} dx\, dy$$

$$= \int_0^{\log a} \int_0^x e^{x+y} [e^{x+y} - 1] dx\, dy$$

$$= \int_0^{\log a} \int_0^x e^{2(x+y)} dx\, dy - \int_0^{\log a} \int_0^x e^{x+y} dx\, dy$$

$$= \int_0^{\log a} \int_0^x e^{2x}.e^{2y} dx\, dy - \int_0^{\log a} \int_0^x e^{x+y} dx\, dy$$

$$= \int_0^{\log a} e^{2x} \left(\frac{1}{2}e^{2y} \right)_0^x dx - \int_0^{\log a} e^x (e^y)_0^x dx$$

$$= \frac{1}{2}\int_0^{\log a} e^{2x} (e^{2x} - e^0) dx - \int_0^{\log a} e^x (e^x - e^0) dx$$

$$= \frac{1}{2}\int_0^{\log a} (e^{4x} - e^{2x}) dx - \int_0^{\log a} (e^{2x} - e^x) dx$$

$$= \frac{1}{2}\int_0^{\log a} (e^{4x} - 3e^{2x} + 2e^x) dx = \frac{1}{2}\left[\frac{1}{4}e^{4x} - \frac{3}{2}e^{2x} + 2e^x \right]_0^{\log a}$$

$$= \frac{1}{8}(e^{4\log a} - e^0) - \frac{3}{4}(e^{2\log a} - e^0) + (e^{\log a} - e^0)$$

$$= \frac{1}{8}(a^4 - 1) - \frac{3}{4}(a^2 - 1) + (a - 1)$$

$$= \frac{1}{8}a^4 - \frac{3}{4}a^2 + a - \frac{3}{8} = \frac{1}{8}[a^4 - 6a^2 + 8a - 3] \; .$$

Example 5.   *Evaluate* $\iiint_V (x^2 + y^2 + z^2) dx\, dy\, dz$, *where V is the volume of cube bounded by the co-ordinates planes and the planes* $x=y=z=a$.

Solution.   Here, the limits of x, y and z are varies from 0 to a.

Therefore, the given integral

$$I = \int_0^a \int_0^a \int_0^a (x^2 + y^2 + z^2) dx\, dy\, dz$$

$$= \int_0^a \int_0^a \left[x^2 z + y^2 z + \frac{1}{3}z^3 \right]_0^a dx\, dy$$

$$= \int_0^a \int_0^a \left(x^2 a + y^2 a + \frac{1}{3}a^3 \right) dx\, dy$$

$$= \int_0^a \left[x^2 ay + \frac{1}{3}y^3 a + \frac{1}{3}ya^3 \right]_0^a dx$$

$$= \int_0^a \left(x^2 a^2 + \frac{1}{3}a^4 + \frac{1}{3}a^4 \right) dx$$

$$= \left[\frac{1}{3}x^3 a^2 + \frac{1}{2}a^4 x + \frac{1}{3}a^4 x \right]_0^a = a^5.$$

Example 6. *Evaluate the volume of tetrahedron bounded by the co-ordinate planes and the planes $x+y+z=1$.*

Solution. The volume of tetrahedron can be expressed as

$$0 \le x \le 1, 0 \le y \le 1-x, 0 \le z \le 1-x-y.$$

∴ The integral $I = \int\int\int dx\, dy\, dz = \int_0^1 \int_0^{1-x} \int_0^{1-x-y} dx\, dy\, dz$

$$= \int_0^1 \int_0^{1-x} [z]_0^{1-x-y} dx\, dy$$

$$= \int_0^1 \int_0^{1-x} (1-x-y) dx\, dy = \int_0^1 \left[(1-x)y - \frac{y^2}{2} \right]_0^{1-x} dx$$

$$= \int_0^1 \left[(1-x)^2 - \frac{(1-x)^2}{2} \right] dx = \int_0^1 \frac{1}{2}(1-x^2) dx$$

$$= \frac{1}{2}\left[\frac{(1-x)^3}{3.(-1)} \right]_0^1 = -\frac{1}{6}(0-1) = \frac{1}{6}.$$

Example 7. *Evaluate $\int\int\int_V zy^2 dx\, dy\, dz$, where V is the region bounded between the xy plane and the sphere, $x^2 + y^2 + z^2 = 1$.*

Solution. Here the column parallel to z-axis is bounded by the plane $z=0$ and the surface of sphere $x^2 + y^2 + z^2 = 1$

i.e., $z = \sqrt{1-x^2-y^2}$.

The region S above which the volume V stands, is the area of circle of intersection of sphere $x^2 + y^2 + z^2 = 1$ by the xy plane.

Hence, the region S is the circle $x^2 + y^2 = 1$.

It is clear from the figure that limits of integration for y are $-\sqrt{1-x^2}$ to $\sqrt{1-x^2}$ and for x are -1 to 1.

Hence, the given integral.

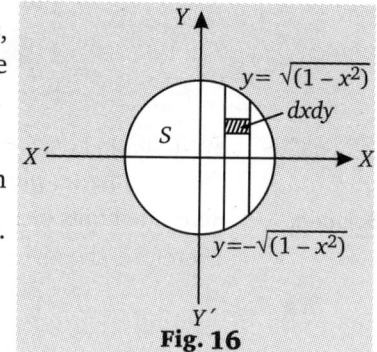

Fig. 16

$$I = \int_{x=-1}^1 \int_{y=-\sqrt{1-x^2}}^{\sqrt{1-x^2}} \int_{z=0}^{\sqrt{1-x^2-y^2}} zy^2 dx\, dy\, dz$$

$$= \int_{x=-1}^1 \int_{y=-\sqrt{1-x^2}}^{\sqrt{1-x^2}} y^2 \left(\frac{1}{2}z^2 \right)_0^{\sqrt{1-x^2-y^2}} dx\, dy$$

$$= \frac{1}{2}\int_{x=-1}^{1}\int_{y=-\sqrt{1-x^2}}^{\sqrt{1-x^2}} y^2(1-x^2-y^2)\,dx\,dy$$

$$= \frac{1}{2}\int_{x=-1}^{1}\int_{y=-\sqrt{1-x^2}}^{\sqrt{1-x^2}} (y^2-x^2y^2-y^4)\,dx\,dy$$

$$= \frac{1}{2}\int_{x=-1}^{1}\left(\frac{1}{3}y^3-\frac{1}{3}x^2y^3-\frac{1}{5}y^5\right)_{-\sqrt{1-x^2}}^{\sqrt{1-x^2}}dx$$

$$= \frac{1}{2}\int_{x=-1}^{1}\left[\left(\frac{2}{3}\right)(1-x^2)^{3/2}-\frac{2}{3}x^2(1-x^2)^{3/2}-\frac{2}{5}(1-x^2)^{5/2}\right]dx$$

$$= \int_{x=-1}^{1}\left[\frac{1}{3}(1-x^2)^{5/2}-\frac{1}{5}(1-x^2)^{5/2}\right]dx$$

$$[\because (1-x^2)^{3/2}-x^2(1-x^2)^{3/2}=(1-x^2)^{5/2}]$$

$$= \frac{2}{15}\int_{x=-1}^{1}(1-x^2)^{5/2}dx = \frac{4}{15}\int_{0}^{1}(1-x^2)^{5/2}dx$$

$$= \frac{4}{15}\int_{0}^{\pi/2}(1-\sin^2\phi)^{5/2}\cos\phi\,d\phi \qquad\qquad (\text{putting } x=\sin\phi)$$

$$= \frac{4}{15}\int_{0}^{\pi/2}\cos^6\phi\,d\phi = \frac{4}{15}\cdot\frac{5}{6}\cdot\frac{3}{4}\cdot\frac{1}{2}\cdot\frac{\pi}{2} = \frac{\pi}{24}.$$

Example 8. *Find the volume of region bounded by the cylinder $x^2+y^2=16$ and point $z=0$ to $z=3$.*

Solution. Here the limits are given from $z=0$ to $z=3$.

Also, from the equation of cylinder $x^2+y^2=16$ we find that the limits of y are from $-\sqrt{16-x^2}$ to $+\sqrt{16-x^2}$ and limit of x are from $-\sqrt{16}$ to i.e., from -4 to $+4$.

$\therefore$ Required volume $v=\int_{x=-4}^{4}\int_{y=-\sqrt{16-x^2}}^{\sqrt{16-x^2}}\int_{z=0}^{3}n\,dx\,dy\,dz$

$$= 4\int_{0}^{4}\int_{0}^{\sqrt{16-x^2}}\int_{0}^{3}dx\,dy\,dz = 4\int_{0}^{4}\int_{0}^{\sqrt{16-x^2}}(z)_{0}^{3}dx\,dy$$

$$= 12\int_{0}^{4}\int_{0}^{\sqrt{16-x^2}}dx\,dy$$

$$= 12\int_{0}^{4}(y)\int_{0}^{\sqrt{16-x^2}}dx = 12\int_{0}^{4}\sqrt{4^2-x^2}\,dx$$

$$= 12\left[\frac{1}{2}x\sqrt{4^2-x^2}+\frac{1}{2}4^2\sin^{-1}\left(\frac{x}{4}\right)\right]_{0}^{4}$$

$$= 12.\left[\frac{1}{2}.16\sin^{-1}(1)\right] = 96(\pi/2) = 48\pi.$$

EXERCISE 19.4

1. Evaluate $\int_{x=0}^{1}\int_{y=0}^{2}\int_{z=1}^{2}x^2\,yz\,dx\,dy\,dz$.

2. Evaluate $\int_{-a}^{a}\int_{-b}^{b}\int_{-c}^{c}(x^2+y^2+z^2)\,dx\,dy\,dz$.

3. Evaluate $\int_{-1}^{1}\int_{0}^{z}\int_{x-z}^{x+z}(x+y+z)\,dy\,dx\,dz$.

4. Evaluate $\int_{0}^{1}\int_{0}^{1-x}\int_{0}^{1-x-y}\dfrac{dy\,dx\,dz}{(1+x+y+z)^3}$.

5. Evaluate $\int_0^{\pi/2} d\theta \int_0^{a\sin\theta} dr \int_0^{(a^2-r^2)/a} r\, dz$.

6. Evaluate $\int_0^a \int_0^{a-x} \int_0^{a-x-y} x^2\, dx\, dy\, dz$.

7. Evaluate $\int_0^2 \int_0^x \int_0^{x+y} e^x (y+2z)\, dx\, dy\, dz$.

8. Evaluate $\int_0^{\log 2} \int_0^x \int_0^{x+\log y} e^{x+y+z}\, dx\, dy\, dz$.

9. Evaluate the integral $\int\int\int xyz\, dx\, dy\, dz$ over the the volume enclosed by three co-ordinates plane and the plane $x+y+z=1$

10. Evaluate $\int\int\int \dfrac{dx\, dy\, dz}{(x+y+z+1)^2}$ over the region $x \geq 0, y \geq 0, z \geq 0, x+y+z \leq 1$.

11. Evaluate $\int\int\int (z^5 + z)\, dx\, dy\, dz$ over the sphere $x^2 + y^2 + z^2 = 1$.

12. Evaluate $\int\int\int_R u^2 v^2 w\, du\, dv\, dw$, where R is the region $u^2 + v^2 \leq 1, 0 \leq w \leq 1$.

13. Find the volume of the tetrahedron bounded by the plane $\dfrac{x}{a} + \dfrac{y}{b} + \dfrac{z}{c} = 1$ and $(x+z = a)$ and coordinate plane.

14. Evaluate $\int_0^2 \int_0^x \int_0^{x+y} e^x (y+2z)\, dx\, dy\, dz$.

================ **Answers** ================

1. 1 **2.** $\dfrac{2}{3} abc(a^2 + b^2 + c^2)$ **3.** 0 **4.** $\dfrac{1}{2}\left(\log 2 - \dfrac{5}{8}\right)$ **5.** $\dfrac{5a^3\pi}{64}$ **6.** $\dfrac{a^5}{60}$ **7.** $19[(1/3)e^2 + 1]$

8. $\dfrac{8}{3}\log 2 - \dfrac{19}{9}$ **9.** 0 **10.** $\dfrac{1}{2}\left(\log 2 - \dfrac{5}{8}\right)$ **11.** 0 **12.** $\pi/48$ **13.** $abc/5$ **14.** $\dfrac{19}{3}(e^2 + 3)$

19.7 DIRICHLET'S THEOREM FOR THREE VARIABLES

Statements. *Let V be the region given by $x \geq 0, y \geq 0, z \geq 0, x+y+z \leq 1, l, m, n$ are positive.* Then

$$\int_V x^{l-1} y^{m-1} z^{n-1}\, dx\, dy\, dz = \frac{\Gamma(l)\Gamma(m)\Gamma(n)}{\Gamma(l+m+n+1)}.$$

Proof. We evaluate the given integral over the volume enclosed by the three co-ordinates planes and the plane $x+y+z=1, x=0, y=0, z=0$. The limits of integration for this region can be expressed as $0 \leq x \leq 1, 0 \leq y \leq 1-x, 0 < z \leq 1-x-y$.

Hence we may write the given triple integral as

$$\int_0^1 \int_0^{1-x} \int_0^{1-x-y} x^{l-1} y^{m-1} z^{n-1}\, dx\, dy\, dz$$

$$= \int_0^1 \int_0^{1-x} x^{l-1} y^{m-1} [z^n / n]_0^{1-x-y}\, dx\, dy$$

$$= \frac{1}{n} \int_0^1 \int_0^{1-x} x^{l-1} y^{m-1} (1-x-y)^n\, dx\, dy$$

$$= \frac{1}{n} \int_0^1 \int_0^1 x^{l-1} \{(1-x)t\}^{m-1} [1-x-(1-x)t]^n (1-x)\, dx\, dt$$

(Putting $y = (1-x)t, \Rightarrow dy = (1-x)dt$))

$$= \frac{1}{n} \int_0^1 \int_0^1 x^{l-1} (1-x)^{m-1} t^{m-1} (1-x)^n (1-t)^n (1-x)\, dx\, dt$$

$$= \frac{1}{n} \int_0^1 \int_0^1 x^{l-1} (1-x)^{m+n} t^{m-1} (1-t)^n\, dx\, dt$$

$$= \frac{1}{n} \int_0^1 x^{l-1} (1-x)^{m+n}\, dx \times \int_0^1 (t)^{m-1} (1-t)^n\, dt$$

$$= \frac{1}{n}B(l,m+n+1)B(m,n+1) \quad \text{(By definition of Beta function)}$$

$$= \frac{1}{n} \cdot \frac{\Gamma(l)\Gamma(m+n+1)}{\Gamma(l+m+n+1)} \cdot \frac{\Gamma(m)\Gamma(n+1)}{\Gamma(m+n+1)}$$

$$\left[\because B(m,n) = \frac{\Gamma(m)\Gamma(n)}{\Gamma(m+n)} \right]$$

$$= \frac{\Gamma(l)\Gamma(m)}{\Gamma(l+m+n+1)} \cdot \frac{n\Gamma(n)}{n} \qquad\qquad [\because \Gamma(n+1) = n\,\Gamma\,(n)]$$

$$= \frac{\Gamma(l)\Gamma(m)\Gamma(n)}{\Gamma(l+m+n+1)} \, .$$

Another proof. Here we solve it by first consider the double integral

$$I_2 = \iint x^{l-1} y^{m-1} dx\, dy$$

where, integral extended to all positive values of variables. The condition is $x+y \le 1$.

Here, 2-dimensional Euclidean space, (*i.e.*, the region of integration of I_2), is bounded by the straight lines $x=0, y = 0$ and $x+y = 1$, and the region expressed as $0 \le x \le 1, 0 \le y \le 1-x$.

$$\therefore \qquad I_2 = \int_{x=0}^{1}\int_{y=0}^{1-x} x^{l-1} y^{m-1} dx\, dy$$

$$= \int_0^1 x^{l-1}\left[\frac{y^m}{m} \right]_0^{1-x} dx = \int_0^1 \frac{1}{m} x^{l-1}(1-x)^m dx$$

$$= \frac{1}{m}\int_0^1 x^{l-1}(1-x)^{m+1-1} dx = \frac{1}{m}B(l,m+1) \qquad\qquad \text{[By Beta function]}$$

$$= \frac{1}{m}\frac{\Gamma(l)\Gamma(m+1)}{\Gamma(l+m+1)} = \frac{1}{m}\frac{\Gamma(l)\Gamma(m)\Gamma(m)}{\Gamma(l+m+1)}$$

$$= \frac{\Gamma(l)\Gamma(m)}{\Gamma(l+m+1)} \, . \qquad\qquad\qquad\qquad\qquad \dots (1)$$

It is for two variables.

Now, we consider the double integral as follows

$$U_2 = \iint x^{l-1} y^{m-1} dx\, dy, x + y \le h \, .$$

We have $\qquad x+y \le h \Rightarrow \dfrac{x}{h}+\dfrac{y}{h} \le 1$

So putting $\dfrac{x}{h} = u$ and $\dfrac{y}{h} = v$ so that $dx = h\,du$ and $dy = h\,dv$ the integrals U_2 becomes

$$U_2 = \iint (hu)^{l-1}(hv)^{m-1} h^2 du\, dv$$

$$= h^{l+m}\iint u^{l-1} v^{m-1} du\, dv, \text{where}\, v + u \le 1$$

$$= h^{l+m}\frac{\Gamma(l)\Gamma(m)}{\Gamma(l+m+1)} \qquad\qquad \text{[By (1)]} \qquad\qquad \dots (2)$$

Now, we consider the triple integral

$$I_3 = \iiint x^{l-1} y^{m-1} z^{n-1} dx\, dy\, dz \, .$$

Condition $x + y + z \leq 1 \, i.e., y + z \leq 1 - x$. We have

$$I_3 = \int_{x=0}^{1} \left[\iint y^{m-1} z^{n-1} dy \, dz \right] x^{l-1} dx,$$

where $y + z \leq 1 - x$

$$= \int_0^1 (1-x)^{m+n} \frac{\Gamma(m)\Gamma(n)}{\Gamma(m+n+1)} x^{l-1} dx$$

[By (2)]

$$= \frac{\Gamma(m)\Gamma(n)}{\Gamma(m+n+1)} \int_0^1 x^{l-1} (1-x)^{m+n+1-1} dx$$

$$= \frac{\Gamma(m)\Gamma(n)}{\Gamma(m+n+1)} B(l, m+n+1)$$

$$= \frac{\Gamma(m)\Gamma(n)}{\Gamma(m+n+1)} \cdot \frac{\Gamma(l)\Gamma(m+n+1)}{\Gamma(l+m+n+1)} = \frac{\Gamma(l)\Gamma(m)\Gamma(n)}{\Gamma(l+m+n+1)}.$$

REMARKS

- Dirichlet's theorem holds good even if the conditons is taken as $x+y+z<1$ in place of $x+y+z \leq 1$.
- The triple integral $\iiint x^{l-1} y^{m-1} z^{n-1} dx \, dy \, dz = h^{l+m+n} \dfrac{\Gamma(l)\Gamma(m)\Gamma(n)}{\Gamma(l+m+n+1)}$

where the integral is extended to all positive values of the variables x, y and z, when $x+y+z \leq h$.

19.8 DIRICHLET'S THEOREM FOR n VARIABLES

Statement. *If the integral is extended to all positive values of the variables* $x_1, x_2, .., x_n$ *subject to the condition* $x_1 + x_2 + ... + x_n \leq 1$. *Then*

$$\int\int...\int x_1^{l_1-1} x_2^{l_2-1} ... x_n^{l_n-1} dx_1 dx_2 ... dx_n = \frac{\Gamma(l_1)\Gamma(l_2)...\Gamma(l_n)}{\Gamma(1+l_1+l_2+...+l_n)}$$

Proof. We shall prove this theorem by mathematical induction.

First we prove the theorem for 2-variables *i.e.*, $n=2$

Let us consider the integral

$$I_2 = \iint x_1^{l_1-1} x_2^{l_2-1} dx_1 dx_2 \text{ such that } x_1 + x_2 \leq 1.$$

Now, using previous theorem, we have

$$I_2 = \frac{\Gamma(l_1)\Gamma(l_2)}{\Gamma(1+l_1+l_2)}$$

... (2)

Equation (1) is true for two variables. Now assume that theorem is true for n variables. Therefore

$$I_n = \iint...\int x_1^{l_1-l} x_2^{l_2-l} ... x_n^{l_n-l} dx_1 . dx_2 .... dx_n$$

$$= \frac{\Gamma(l_1)\Gamma(l_2)...\Gamma(l_n)}{\Gamma(1+l_1+l_2+...+l_n)}$$

...(2)

with condition $x_1 + x_2 + ... + x_n \leq 1$.

If the condition $x_1 + x_2 + ... + x_n \leq h$, then putting

$\dfrac{x_1}{h} = u_1, \dfrac{x_2}{h} = u_2 ... \dfrac{x_n}{h} = u_n$ so that $dx_1 = h \, du_1, dx_2 = h \, du_2, ... dx_n = h \, du_n$

We have $\iint...\int x_1^{l_1-1} x_2^{l_2-1} ... x_n^{l_n-1} dx_1 dx_2 ... dx_n$

$$= h^{l_1+l_2+...+l_n}\int\int...\int u_1^{l_1-1}u_2^{l_2-1}...u_n^{l_n-1}du_1 du_2...du_n$$

subject to the condition $u_1 + u_2 + ... + u_n \le 1$

$$= h^{l_1+l_2+...+l_n}\frac{\Gamma(l_1)\Gamma(l_2)...\Gamma(l_n)}{\Gamma(1+l_1+l_2+...+l_n)} \qquad ...(3)$$

(Using the assumed result (2))

Now for $n+1$ variables the conditions are

$x_1 + x_2 +...+ x_n + x_{n+1} \le 1$ i.e., $x_2 + x_3 +...+ x_n + x_{n+1} \le 1 - x_1$ and $0 \le x_1 \le 1$.

We have

$$\int\int...\int x_1^{l_1-1}x_2^{l_2-1}...x_n^{l_n-1}x_{n+1}^{l_{(n+1)}-1}dx_1 dx_2...dx_n dx_{n+1}$$

where $x_1 + x_2 +...x_{n+1} \le 1$

$$= \int_{x_1=0}^{1} x_1^{l_1-1}\left[\int\int...\int x_2^{l_2-1}...x_{n+1}^{l_{n+1}-1}dx_2...dx_{n+1}\right]dx_1$$

$$= \frac{\Gamma(l_2)\Gamma(l_3)...\Gamma(l_{n+1})}{\Gamma(l_1+1+l_2+...l_n+l_{n+1})}\cdot\int_0^1 x_1^{l_1-1}(1-x_1)^{(1+l_2+l_3+...+l_{n+1})-1}dx_1$$

Using (3)

$$= \frac{\Gamma(l_2)\Gamma(l_3)...\Gamma(l_{n+1})}{\Gamma(1+l_2+...+l_{n+1})}\cdot\frac{\Gamma(1+l_2+...+l_{n+1})}{\Gamma(1+l_1+l_2+...+l_n+l_{n+1})}$$

$$= \frac{\Gamma(l_1)\Gamma(l_2)...\Gamma(l_{n+1})}{\Gamma(1+l_1+l_2+...+l_{n+1})} \qquad ...(4)$$

The result (4) shows that the theorem hold for $(n+1)$ variables. Hence, by principle of mathematical induction, theorem is true for all values of n.

Solved Examples

Example 1.   Evaluate $\int\int\int x^{l-1}y^{m-1}z^{n-1}dx\,dy\,dz$ in which $x \ge 0, y \ge 0, z = 0$ and

$$(x/a)^{1/2} + (y/b)^{1/2} + (z/c)^{1/2} \le 1.$$

Solution.   Let $(x/a)^{1/2} = u$, $(y/b)^{1/2} = v$ and $(z/b)^{1/2} = w$

Then $x = au^2, y = bv^2, z = cw^2$

$$dx = 2au\,du;\ dy = 2bv\,dv;\ dz = 2cw\,dw,$$

$$u \ge 0, v \ge 0, w \ge 0 \text{ and } u+v+w \le 1$$

Hence, $\int\int\int (au^2)^{l-1}(bv^2)^{m-1}(cw^2)^{n-1}.2au.2bv.2cw\,du\,dv\,dw$

$$= 8a^l b^m c^n \int\int\int u^{2l-1}v^{2m-1}w^{2n-1}\,du\,dv\,dw$$

$$= 8a^l b^m c^n \frac{\Gamma(2l)\Gamma(2m)\Gamma(2n)}{\Gamma(2l+2m+2n-1)}$$

Example 2. *Evaluate $\iint dx\, dy\, dz$ over the region in the positive quadrant for which $x+y \le 1$.*

Solution. We have $x+y \le 1$ and $x \ge 0, y \ge 0$ and so by Dirichlet's theorem, we get

$$I = \iint x^{2-1} y^{2-1} dx\, dy = \frac{\Gamma(2)\Gamma(2)}{\Gamma(2+2+1)} = \frac{1}{4}\cdot\frac{1}{\Gamma(3)}$$

$$= \frac{1}{4.3.2\Gamma(1)} = \frac{1}{24\times 1} = \frac{1}{24}.$$

Example 3. *Show that the integral $\iiint x^{l-1} y^{m-1} z^{n-1} \, dx\, dy\, dz$ integrand over the region in the first octant below the surface $(x/a)^p + (y/b)^q + (z/c)^r = 1$*

is

$$\frac{a^l b^m c^n}{pqr}\cdot\frac{\Gamma(l/p)\Gamma(m/q)\Gamma(n/r)}{\Gamma[(l/p)+(m/q)+(n/r)+1]}.$$

Solution. Putting

$$\left(\frac{x}{a}\right)^p = u \ \text{or}\ x = au^{1/p} \qquad \Rightarrow \qquad dx = a(1/p)u^{(1/p)-1}.du.$$

Similarly putting $(y/b)^q = v$ and $(z/c)^r = w$, we get

$$dy = b(1/q)v^{(1/q)-1} dv \ \text{and}\ dz = c(1/r)w^{(1/r)-1} dw.$$

∴ $$x^{l-1} dx = a^{l-1} u^{(l-1)/p} a(1/p)u^{(1-p)/p} du = a^l(1/p)u^{(l/p)-1} du$$

Similarly, $y^{m-1} dy = b^m(1/q)v^{(m/q)-1} dv \, ; \ z^{n-1} dz = c^n(1/r)w^{(n/r)-1} dw$

Hence, subject to the condition $u+v+w \le 1$, the given integral

$$= \frac{a^l b^m c^n}{pqr}\iiint u^{(l/p)-1} v^{(m/q)-1} w^{(n/r)-1} du\, dv\, dw$$

$$= \frac{a^l b^m c^n}{pqr}\cdot\frac{\Gamma(l/p)\Gamma(m/q)\Gamma(n/r)}{\Gamma[(l/p)+(m/q)+(n/r)+1]}$$

Example 4. *Find the value of $\iint\ldots\int dx_1\, dx_2\ldots dx_n$ extended to all positive values of variables subject to the condition $x_1^2 + x_2^2 + \ldots + x_n^2 < R^2$.*

Solution. To find the value of integral I extended to all positive values of $x_1, x_2, \ldots x_n$ subject to the condition

$$\frac{x_1^2}{R^2} + \frac{x_2^2}{R^2} + \ldots + \frac{x_n^2}{R^2} = 1$$

We put $$\left(\frac{x_1}{R}\right)^2 = u_1 \Rightarrow x_1 = Ru_1^{1/2} \quad \Rightarrow \quad dx_1 = (1/2)Ru_1^{-1/2} du_1$$

$$\left(\frac{x_2}{R}\right)^2 = u_2 \Rightarrow x_2 = Ru_2^{1/2} \ \text{so that}\ dx_2 = (1/2)Ru_2^{-1/2} du_2 \ \text{and so on.}$$

Then the required integral

$$I = \iint\ldots\int (1/2)^n R^n u_1^{-1/2}\ldots u_n^{-1/2} du_1.du_2\ldots du_n$$

$$= \left(\frac{R}{2}\right)^2 \int\int ... \int u_1^{(1/2)-1} u_2^{(1/2)-1} ... u_n^{(1/2)-1} du_1\, du_2 ... du_n$$

Condition is $u_1 + u_2 + ... + u_n < 1$.

$$= \left(\frac{R}{2}\right)^2 \frac{\{\Gamma(1/2)\}^n}{\Gamma(1+n/2)} \qquad \text{(By Dirichlet's theorem)}$$

$$= \left(\frac{R}{2}\right)^2 \cdot \frac{\pi^{n/2}}{\Gamma\left(1+\dfrac{n}{2}\right)} \qquad [\because \Gamma(1/2) = \sqrt{\pi}]$$

Example 5. *Evaluate* $\int\int\int dx\, dy\, dz$ *where* $\dfrac{x^2}{a^2} + \dfrac{y^2}{b^2} + \dfrac{z^2}{c^2} \le 1$

or find the volume of $(x^2/a^2) + (y^2/b^2) + (z^2/c^2) = 1$.

Solution. Let

$$\frac{x^2}{a^2} = u, x = au^{1/2} \text{ so that } dx = \frac{1}{2}au^{-1/2}du .$$

Similarly, putting $\dfrac{y^2}{b^2} = v$ and $\dfrac{z^2}{c^2} = w$, we get

$$dy = \frac{1}{2}bv^{-1/2}dv \quad \text{and} \quad dz = \frac{1}{2}cw^{-1/2}dw .$$

$$\therefore \quad \int\int\int dx\, dy\, dz = \int\int\int \frac{1}{2}au^{-1/2}du . \frac{1}{2}bv^{-1/2}.dv . \frac{1}{2}cw^{-1/2}dw$$

$$= \frac{1}{8}abc \int\int\int u^{1/2-1}v^{1/2-1}w^{1/2-1}du\, dv\, dw$$

$$= \frac{1}{8}abc \frac{\Gamma(1/2)\Gamma(1/2)\Gamma(1/2)}{\Gamma(1/2+1/2+1/2+1)}$$

$$= \frac{1}{8}abc \frac{\sqrt{\pi}\sqrt{\pi}\sqrt{\pi}}{\Gamma(5/2)} = \frac{\pi\sqrt{\pi}\, bca}{8 \cdot 3/2 \cdot 1/2\sqrt{\pi}} = \frac{1}{6}\pi abc$$

Example 6. *Find the volume enclosed by the surface* $(x/a)^{2n} + (y/b)^{2n} + (z/c)^{2n} = 1$.

Solution. The given surface is symmetrical in all the eight octants. Now we want to find the volume V in the positive octant. Clearly

$$V = \int\int\int dx\, dy\, dz$$

where the integral is extended to all positive values of the variables x, y, z subject to the condition $(x/a)^{2n} + (y/b)^{2n} + (z/c)^{2n} \le 1$.

Now put $(x/a)^{2n} = u, (y/b)^{2n} = v, (z/c)^{2n} = w$

$$\Rightarrow \qquad x = au^{1/2n}, y = bv^{1/2n}, z = cw^{1/2n}$$

So that $$dx = \frac{a}{2n}u^{(1/2n)-1}du$$

$$\therefore \qquad V = \frac{abc}{8n^3} \int\int\int u^{(1/2n)-1}v^{(1/2n)-1}w^{(1/2n)-1}du\, dv\, dw$$

$$= \frac{abc}{8n^3} \frac{[\Gamma(1/2n)]^3}{\Gamma\{(3/2n)+1\}} = \frac{abc}{8n^3} \frac{[\Gamma(1/2n)]^3}{(3/2n).\Gamma(3/2n)}$$

$$= \frac{abc}{12n^2} \frac{[\Gamma(1/2n)]^3}{\Gamma(3/2n)} \cdot$$

Hence, the total volume enclosed by given surface

$$= 8V = \frac{2}{3}.\frac{abc}{n^2} \frac{[\Gamma(1/2n)]^3}{\Gamma(3/2n)} \cdot$$

Example 7. *Find the volume of the tetrahedron bounded by the plane* $\frac{x}{a}+\frac{y}{b}+\frac{z}{c}=1$ *and the co-ordinate planes.*

Solution. The volume of a small element at a point $(x, y, z) = dx\,dy\,dz$

$\therefore$ the volume of the given tetrahedron $= \iiint dx\,dy\,dz$ where the integral is extended to all positive values of variables x, y, z.

Put $x/a = u, y/b = v, z/c = w$ subject to the condition so that $\frac{x}{a}+\frac{y}{b}+\frac{z}{c} \leq 1$

$$dx = a.du, dy = b.dv \text{ and } dz = c.dw$$

then   the required volume $= \iiint abc\,du\,dv\,dw$ where $u+v+w \leq 1$

$$= abc\iiint u^{1-1}v^{1-1}w^{1-1}\,du\,dv\,dw$$

$$= abc\,\frac{[\Gamma(1)]^3}{\Gamma(1+1+1+1)} \qquad \text{[By Dirichlet's theorem]}$$

$$= abc\,\frac{1}{\Gamma(4)} = \frac{abc}{3.2.1} = \frac{abc}{6}$$

19.9 LIOUVILLE'S EXTENSION OF DIRICHLET'S THEOREM

Statement. *If x, y, z are all positive and such that* $h_1 < x+y+z \leq h_2$ *then*
$$\iiint f(x+y+z)x^{l-1}y^{m-1}z^{n-1}dx\,dy\,dz$$

$$= \frac{\Gamma(l)\Gamma(m)\Gamma(n)}{\Gamma(l+m+n)}\int_{h_1}^{h_2} f(u)u^{l+m+n-1}du \,.$$

Proof. From Dirichlet's theorem, we have
$$I = \iiint x^{l-1}y^{m-1}z^{n-1}dx\,dy\,dz = \frac{\Gamma(l)\Gamma(m)\Gamma(n)}{\Gamma(l+m+n)}u^{(l+m+n)} \qquad \dots (1)$$

subject to the condition that $x, y, z \geq 0$ and $x+y+z \leq u$.

Now if $x, y, z \geq 0$ and $x+y+z \leq u+\delta u$, then we have
$$I = \iiint x^{l-1}y^{m-1}z^{n-1}dx\,dy\,dz$$

$$= \frac{\Gamma(l)\Gamma(m)\Gamma(n)}{\Gamma(l+m+n+1)}(u+\delta u)^{(l+m+n)} \qquad \dots (2)$$

So the value of integral given above extended to all such positive value of x, y, z such that $x+y+z$ lies between u and $u+\delta u$, is given by
$$I = \iiint x^{l-1}y^{m-1}z^{n-1}dx\,dy\,dz$$

$$= \frac{\Gamma(l)\Gamma(m)\Gamma(n)}{\Gamma(l+m+n+1)}[(u+\delta u)^{l+m+n} - u^{l+m+n}]$$

$$= \frac{\Gamma(l)\Gamma(m)\Gamma(n)}{\Gamma(l+m+n+1)}u^{l+m+n}\left[\left(1+\frac{\delta u}{u}\right)^{l+m+n} - 1\right]$$

$$= \frac{\Gamma(l)\Gamma(m)\Gamma(n)}{\Gamma(l+m+n+1)}u^{l+m+n}\left[1+(l+m+n)\frac{\delta u}{u}+\ldots-1\right]$$

[On expanding by Taylor's series]

$$= \frac{\Gamma(l)\Gamma(m)\Gamma(n)}{\Gamma(l+m+n+1)}(l+m+n)u^{(l+m+n-1)}\delta u$$

[Neglecting the second and higher degree terms of δu]

$$= \frac{\Gamma(l)\Gamma(m)\Gamma(n)}{\Gamma(l+m+n)}u^{(l+m+n-1)}\delta u$$

$$[\because \Gamma(l+m+n+1) = (l+m+n)\Gamma(l+m+n)]$$

Now, consider the intergral $\iiint f(x+y+z)x^{l-1}y^{m-1}z^{n-1}dx\,dy\,dz$.

Since $u \le x+y+z \le \delta u$, so the function $f(x+y+z)$ will differ by a small quantity of same order of solution. Hence, the integral

$$\iiint f(x+y+z)x^{l-1}y^{m-1}z^{n-1}dx\,dy\,dz = \frac{\Gamma(l)\Gamma(m)\Gamma(n)}{\Gamma(l+m+n)}F(u)u^{(l+m+n-l)}\delta u$$

subject to the condition that $x, y, z \ge 0$ and $u \le x+y+z \le u+\delta u$, to the first approximation. So finally for the given condition that for positive x, y, z such that $h_1 < x+y+z \le h_2$,

we get $\iiint f(x+y+z)x^{l-1}y^{m-1}z^{n-1}dx\,dy\,dz = \frac{\Gamma(l)\Gamma(m)\Gamma(n)}{\Gamma(l+m+n)}\int_{h_1}^{h_2} f(u)^{(l+m+n-1)}du$

Solved Examples

Example 1. Evaluate $\iiint e^{x+y+z}dx\,dy\,dz$ taken over the positive octant such that $x+y+z \le 1$.

Solution. In the positive octant x, y, z are all positive and therefore

$$0 < (x+y+z) \le 1.$$

Therefore, we have

$$\iiint e^{x+y+z}dx\,dy\,dz = \frac{\Gamma(1)\Gamma(1)\Gamma(1)}{\Gamma(1+1+1)}\int_0^1 e^h h^{1+1+1-1}dh \qquad \text{[By Liouville's theorem]}$$

$$= \frac{1}{\Gamma(3)}\int_0^1 h^2 e^h dh = \frac{1}{2!}\left[(h^2 e^h)_0^1 - \int_0^1 2h e^h dh\right]$$

$$= \frac{1}{2}\left[e - 2\left\{(he^h)_0^1 - \int_0^1 e^h dh\right\}\right]$$

$$= \frac{1}{2}\left[e - 2\left\{e - (e^h)_0^1\right\}\right] \qquad = \frac{1}{2}[e - 2\{e - e + 1\}]$$

$$= \frac{1}{2}(e-2).$$

Example 2. *Evaluate* $\iiint \log(x + y + z)\,dx\,dy\,dz$ *taken over all positive values of x, y, z subject to the condition* $x+y+z \le 1$.

Solution. Since x, y, z are to be taken positive value only, we have $0 < (x+y+z) \le 1$.
Therefore, we have

$$\iiint \log(x + y + z)\,dx\,dy\,dz$$

$$= \iiint \log(x + y + z)x^{1-1}y^{1-1}z^{1-1}\,dx\,dy\,dz$$

$$= \frac{\Gamma(1)\Gamma(1)\Gamma(1)}{\Gamma(1+1+1)} \int_0^1 (\log h)h^{1+1+1-1}\,dh, \qquad \text{[By Liouville's theorem]}$$

$$= \frac{1}{\Gamma(3)} \int_0^1 h^2 (\log h)\,dh$$

$$= \frac{1}{2!}\left[\left\{(\log h)\frac{1}{3}h^3\right\}_0^1 - \int_0^1 \frac{1}{h}\cdot\frac{1}{3}h^3\,dh\right]$$

$$= \frac{1}{6}\left[h^3 \log h - \frac{h^3}{3}\right]_0^1 = -\frac{1}{18}.$$

Example 3. *Evaluate* $\iiint (x + y + z)\,dx\,dy\,dz$ *over the tetrahedron* $x = 0, y = 0, z = 0$ *and* $x+y+z \le 1$

Solution. We are given that : $0 \le x+y+z \le 1$ therefoe, by Liouville's extension of Dirichlets theorem $\iiint (x + y + z)\,dx\,dy\,dz$

$$= (x + y + z)x^{1-1}y^{1-1}z^{1-1}\,dx\,dy\,dz$$

$$= \frac{\Gamma(1)\Gamma(1)\Gamma(1)}{\Gamma(1+1+1)} \int_0^1 u.u^{1+1+1-1}\,du = \frac{1}{\Gamma(3)} \int_0^1 u^3\,du$$

$$= \frac{1}{2!}\left[\frac{u^4}{4}\right]_0^1 = \frac{1}{2}\left[\frac{1}{4}\right] = \frac{1}{8}.$$

Example 4. *Prove that*

$$\iiint \frac{dx\,dy\,dz}{\sqrt{(1 - x^2 - y^2 - z^2)}} = \frac{\pi^2}{8}; \text{ the integral being extended to a positive values of}$$

variables for which the expression is real.

Solution. Since x, y, z are to be taken positive values only, we have $0 < x^2 + y^2 + z^2 < 1$.
Put $x^2 = u$ or $x = u^{1/2}$ so that $dx = 1/2u^{-1/2}\,du$.
Similarly putting $y^2 = v$ and $z^2 = w$, we get $dy = 1/2\,v^{-1/2}\,dv$, $dz = \frac{1}{2}w^{-1/2}dw$.
Now, given integral

$$= \iiint \frac{(1/2)u^{-1/2}(1/2)v^{-1/2}(1/2)w^{-1/2}}{\sqrt{1-(u+v+w)}}\,du\,dv\,dw \text{ where } 0<u+v+w<1$$

$$= \frac{1}{8}\iiint \frac{u^{1/2-1}v^{1/2-1}w^{1/2-1}}{\sqrt{1-(u+v+w)}}\,du\,dv\,dw$$

$$= \frac{1}{8} \frac{\Gamma(1/2)\Gamma(1/2)\Gamma(1/2)}{\Gamma(1/2+1/2+1/2)} \int_0^1 \frac{1}{\sqrt{1-h}} h^{1/2+1/2+1/2-1} dh$$

$$= \frac{1}{8} \frac{\sqrt{\pi}\sqrt{\pi}\sqrt{\pi}}{\Gamma(3/2)} \int_0^1 \sqrt{\left(\frac{h}{1-h}\right)} dh$$

$$= \frac{1}{4}\pi \int_0^{\pi/2} \sqrt{\left(\frac{\sin^2\theta}{\cos^2\theta}\right)} 2\sin\theta\cos\theta d\theta$$

                                                   Putting $h = \sin^2\theta$

$$= \frac{1}{4}\pi \int_0^{\pi/2} 2\sin^2\theta d\theta = \frac{1}{4}\pi \int_0^{\pi/2} (1-\cos 2\theta) d\theta$$

$$= \frac{1}{4}\pi \left[\theta - \frac{1}{2}\sin 2\theta\right]_0^{\pi/2} = \frac{1}{4}\pi(1/2\pi) = \frac{1}{8}\pi^2 .$$

Example 5.   *Evaluate $\iiint x^\alpha y^\beta z^\gamma (1-x-y-z)^\lambda \, dx \, dy \, dz$ over the interior of tetrahedron formed by the co-ordinate plane and the plane $x+y+z=1$.*

Solution.   The region of integration is bounded by the plane $x = 0, y = 0, z = 0$ and $x+y+z=1$. So, the variable x, y, z take all positive values subject to the condition

         $0<x+y+z<1$.

Therefore the given integral

$$= \iiint x^{(\alpha+1)-1} y^{(\beta+1)-1} z^{(\gamma+1)-1} [1-(x+y+z)]^\lambda \, dx \, dy \, dz$$

$$= \frac{\Gamma(\alpha+1)\Gamma(\beta+1)\Gamma(\gamma+1)}{\Gamma(\alpha+\beta+\gamma+3)} \int_0^1 u^{\alpha+1+\beta+1+\gamma+1-1}(1-u)^\lambda \, du$$

                                     [By Liouville's extension of Dirichlet's theorem]

$$= \frac{\Gamma(\alpha+1)\Gamma(\beta+1)\Gamma(\gamma+1)}{\Gamma(\alpha+\beta+\gamma+3)} \int_0^1 u^{(\alpha+\beta+\gamma+3)-1}(1-u)^{(\lambda+1)-1} \, du .$$

$$= \frac{\Gamma(\alpha+1)\Gamma(\beta+1)\Gamma(\gamma+1)}{\Gamma(\alpha+\beta+\gamma+3)} B(\alpha+\beta+\gamma+3, \lambda+1)$$

$$= \frac{\Gamma(\alpha+1)\Gamma(\beta+1)\Gamma(\gamma+1)}{\Gamma(\alpha+\beta+\gamma+3)} \cdot \frac{\Gamma(\alpha+\beta+\gamma+3)\Gamma(\lambda+1)}{\Gamma(\alpha+\beta+\gamma+4)}$$

$$= \frac{\Gamma(\alpha+1)\Gamma(\beta+1)\Gamma(\gamma+1)\Gamma(\lambda+1)}{\Gamma(\alpha+\beta+\gamma+\lambda+4)}$$

Example 6.   *Evaluate $\iiint \sqrt{(a^2b^2c^2 - b^2c^2x^2 - c^2a^2y^2 - a^2b^2z^2)} dx \, dy \, dz$ taken throughout the ellipsoid*

$$\frac{x^2}{a^2} + \frac{y^2}{b^2} + \frac{z^2}{c^2} = 1 .$$

Solution.   Let us first evaluate the given integral over the region of ellipsoid which lie in the positive octants the given ellipsoid $\dfrac{x^2}{a^2} + \dfrac{y^2}{b^2} + \dfrac{z^2}{c^2} = 1$ is symmetrical in all the eight octants.

Put $\dfrac{x^2}{a^2} = u, \dfrac{y^2}{b^2} = v, \dfrac{z^2}{c^2} = w$ then $x = au^{1/2}$ $\Rightarrow$ $dx = 1/2 au^{-1/2} du$

$$y = bv^{1/2} \qquad \Rightarrow \qquad dx = \frac{1}{2}bv^{-1/2}dv$$

and $\qquad z = cw^{1/2} \qquad \Rightarrow \qquad dz = \frac{1}{2}cw^{-1/2}dw$

The given integral

$$I = abc\iiint \sqrt{\left(1 - \frac{x^2}{a^2} - \frac{y^2}{b^2} - \frac{z^2}{c^2}\right)}\,dx\,dy\,dz$$

where $0 < x^2/a^2 + y^2/b^2 + z^2/c^2 \le 1$

$$= abc\iiint \sqrt{1-u-v-w}\,(1/8)abcu^{-1/2}v^{-1/2}w^{-1/2}du\,dv\,dw$$

where $0 < u+v+w \le 1$

$$= \frac{a^2b^2c^2}{8}\iiint u^{(1/2)-1}v^{(1/2)-1}w^{(1/2)-1}\sqrt{1-(u+v+w)}\,du\,dv\,dw$$

$$= \frac{a^2b^2c^2}{8}\frac{[\Gamma(1/2)]^3}{\Gamma(3/2)}\int_0^1 \sqrt{1-t}.t^{1/2+1/2+1/2-1}dt \qquad \text{[By Liouville's theorem]}$$

$$= \frac{a^2b^2c^2}{8}\frac{(\sqrt{\pi})^3}{1/2\sqrt{\pi}}\int_0^1 (1-t)^{(3/2)-1}t^{(3/2)-1}dt$$

$$= \frac{a^2b^2c^2}{8}.2\pi\frac{\Gamma(3/2)\Gamma(3/2)}{\Gamma(3)} = \frac{\pi^2 a^2b^2c^2}{32}.$$

Hence, if the integration is extended throughout the ellipsoid then the given integral

$$= 8I = 8.\frac{\pi^2 a^2b^2c^2}{32} = \frac{\pi^2 a^2b^2c^2}{4}.$$

Example 7. *Evaluate* $\iiint_R (x+y+z+1)^2 dx\,dy\,dz$ *where R defined by* $x \ge 0,\ y \ge 0,\ z \ge 0,$ $x+y+z \le 1.$

Solution. As given x, y, z are all positive such that $0 \le x+y+z \le 1$.

$\therefore \quad \iiint (x+y+z+1)^2 dx\,dy\,dz$

$$= \iiint x^{1-1}y^{1-1}z^{1-1}\{(x+y+z)+1\}^2 dx\,dy\,dz$$

$$= \frac{\Gamma(1)\Gamma(1)\Gamma(1)}{\Gamma(1+1+1)}\int_0^1 (u+1)^2.u^{1+1+1-1}du$$

[By Liouville's extension of Dirichlet's theorem]

$$= \frac{1}{2}\int_0^1 (u^2 + 2u + 1)u^2 du = \frac{1}{2}\left[\frac{u^2}{5} + \frac{3u^4}{4} + \frac{u^3}{3}\right]_0^1$$

$$= \frac{1}{2}\left[\frac{1}{5} + \frac{1}{2} + \frac{1}{3}\right] = \frac{1}{2}\frac{(6+15+10)}{5\times 2\times 3} = \frac{1}{2}.\frac{31}{30} = \frac{31}{60}.$$

Example 8. *Evaluate*

$$\iiint x^{-1/2}y^{-1/2}z^{-1/2}(1-x-y-z)^{1/2}dx\,dy\,dz$$

extended to all positive values of variable subject to the condition $x+y+z<1$

Solution. The given condition is $0 < x+y+z < 1$

∴   the given integral

$$= \iiint x^{1/2-1} y^{1/2-1} z^{1/2-1} [1 - (x+y+z)^{1/2} dx\, dy\, dz]$$

$$= \frac{\Gamma(1/2)\Gamma(1/2)\Gamma(1/2)}{\Gamma(1/2+1/2+1/2)} \int (1-h)^{1/2} h^{1/2+1/2+1/2} \, dh$$

$$= \frac{\sqrt{\pi}\sqrt{\pi}\sqrt{\pi}}{\Gamma(3/2)} \int_0^1 h^{1/2}(1-h)^{1/2} dh$$

$$= \frac{\pi\sqrt{\pi}}{1/2\sqrt{\pi}} \int_0^1 h^{3/2-1}(1-h)^{3/2-1} dh \quad = 2\pi B(3/2, 3/2)$$

$$= \frac{2\pi.\Gamma(3/2)\Gamma(3/2)}{\Gamma(3/2+3/2)} = \frac{2.\pi.(1/2\sqrt{\pi}).(1/2\sqrt{\pi})}{\Gamma(3)} = \frac{\pi^2}{2\Gamma(3)} = \frac{\pi^2}{2.2.1} = \frac{\pi^2}{4}.$$

Example 9.   *Prove that when x and y are positive and x+y<h*

$$\iint f'(x+y) x^{l-1} y^{-1} dx\, dy = \frac{\pi}{\sin \pi l}[f(h) - f(0)]$$

Solution.   The given integral

$$I = \iint f'(x+y) x^{l-1} y^{(1-l)-1} dx\, dy \text{ where } 0 < x+y < h$$

$$= \frac{\Gamma(l)\Gamma(l-1)}{\Gamma(l+1-l)} \int_0^h f'(u) u^{l+(1-l)-1} du \qquad \text{[By Liouville's extension]}$$

$$= \frac{\Gamma(l)\Gamma(l-1)}{\Gamma(1)} \int_0^h f'(u) du = \frac{\pi}{\sin \pi.l}[f(u)]_0^h$$

$$= \frac{\pi}{\sin \pi l}[f(h) - f(0)]$$

Example 10.   *Show that* $\displaystyle \iint \left(\frac{1-x^2-y^2}{1+x^2+y^2}\right)^{1/2} dx\, dy = \frac{\pi}{8}(\pi - 2)$

over the positive quadrant of circle $x^2+y^2=1$.

Solution.   The given integral is to be extended to all positive values of *x* and *y* such that

$$0 \le x^2 + y^2 \le 1 \qquad\qquad \text{... (1)}$$

Put $x^2 = u, y^2 = v \implies x = u^{1/2}, y = v^{1/2}$ so that

$$dx = \frac{1}{2} u^{-1/2} \, du, \quad dy = \frac{1}{2} v^{-1/2} dv$$

With these substitution, the condition (1) become $0 \le u \le v \le 1$
Therefore the integral

$$= \iint \left[\frac{1-(u+v)}{1+(u+v)}\right]^{1/2} \frac{1}{4} u^{-1/2} v^{-1/2} du\, dv$$

$$= \frac{1}{4} \iint \left[\frac{1-(u+v)}{1+(u+v)}\right]^{1/2} u^{(1/2)-1} v^{(1/2)-1} du\, dv \qquad \text{where } 0 \le u+v \le 1$$

$$= \frac{1}{4}\frac{\Gamma(1/2)\Gamma(1/2)}{\Gamma(1/2+1/2)}\int_0^1\left[\frac{1-h}{1+h}\right]^{1/2}.h^{(1/2)+(1/2)-1}dh$$

[By Liouville's extension of Dirichlet's theorem]

$$= \frac{1}{4}\frac{\sqrt{\pi}\sqrt{\pi}}{\Gamma(1)}\int_0^1\frac{1-h}{\sqrt{(1-h^2)}}dh$$

$$= \frac{\pi}{4}\int_0^1\frac{(1-\sin\theta)}{\cos\theta}\cos\theta\,d\theta \qquad \text{Putting } h=\sin\theta, \text{ so that } dh = \cos\theta\,d\theta$$

$$= \frac{\pi}{4}[\theta + \cos\theta]_0^{\pi/2} = \frac{\pi}{4}\left[\frac{\pi}{2}-1\right] = \frac{\pi}{8}(\pi-2).$$

Example 11. *Prove that* $I = \iiint dx\,dy\,dz\,dw$, *for all positive values of the variables for which* $x^2 + y^2 + z^2 + w^2$ *is not less than* a^2 *and not greater than* b^2, *is* $\pi^2(b^4 - a^4)/32$.

Solution. As per given, we have $a^2 < x^2 + y^2 + z^2 + w^2 < b^2$.

Put $x^2 = u_1$ or $x = u_1^{1/2}, \Rightarrow dx = 1/2u_1^{-1/2}du_1$.

Similarly, putting $y^2 = u_2$, $z^2 = u_3$, $w^2 = u_4$, we get

$$dy = \frac{1}{2}u_2^{-1/2}du_2, dz = \frac{1}{2}u_3^{-1/2}du_3, dw = \frac{1}{2}u_4^{-1/2}du_4$$

∴ Then

$$I = \iiiint \frac{1}{2}u_1^{-1/2}\frac{1}{2}u_2^{-1/2}.\frac{1}{2}u_3^{-1/2}\frac{1}{2}u_4^{-1/2}du_1du_2du_3du_4$$

$$= \frac{1}{16}\iiiint u_1^{1/2-1}u_2^{1/2-1}u_3^{1/2-1}u_4^{1/2-1}du_1\,du_2\,du_3\,du_4$$

$$= \frac{1}{16}\frac{\Gamma(1/2)\Gamma(1/2)\Gamma(1/2)\Gamma(1/2)}{\Gamma(1/2+1/2+1/2+1/2)}\int_{a^2}^{b^2}h^{\frac{1}{2}+\frac{1}{2}+\frac{1}{2}+\frac{1}{2}-1}dh$$

$$= \frac{1}{16}\frac{(\sqrt{\pi})^4}{\Gamma(2)}\int_{a^2}^{b^2}h\,dh \qquad\qquad [\because \Gamma(1/2) = \sqrt{\pi}]$$

$$= \frac{\pi^2}{16}\left(\frac{1}{2}h^2\right)_{a^2}^{b^2} = \frac{\pi^2}{32}(b^4 - a^4).$$

EXERCISE 19.5

1. Show that if l, m, n are all positive, then

$$\iiint x^{l-1}y^{m-1}z^{n-1}dx\,dy\,dz$$

$$= \frac{a^l b^m c^n}{8}.\frac{\Gamma(l/2)\Gamma(m/2)\Gamma(n/2)}{\Gamma(l/2+m/2+n/2+1)}$$

where the triple integral is taken throughout the part of the ellipsoid $\dfrac{x^2}{a^2}+\dfrac{y^2}{b^2}+\dfrac{z^2}{c^2}=1$ which lies in the positive octant.

2. Evaluate $\iiint x^p y^q z^r (1-x-y-z)^s dx\,dy\,dz$ over the interior of the tetrahedron formed by four planes $x = 0, y = 0, z = 0, x+y+z = 1$.

3. Find the volume in the positive octant of the ellipsoid $\dfrac{x^2}{a^2}+\dfrac{y^2}{b^2}+\dfrac{z^2}{c^2}=1$.

4. Find the volume of ellipsoid $\dfrac{x^2}{a^2}+\dfrac{y^2}{b^2}+\dfrac{z^2}{c^2}=1$.

5. Evaluate $\iint x^{2l-1}y^{2m-1}dx\,dy$ such that $x^2+y^2\le c^2$ for all positive values of x and y.

6. Find the volume of solid surrounded by the surface
$$\left(\frac{x}{a}\right)^{2/3}+\left(\frac{y}{b}\right)^{2/3}+\left(\frac{z}{c}\right)^{2/3}=1.$$

7. Evaluate the double integral
$$\iint_P x^{1/2}y^{1/2}(1-x-y)^{2/3}dx\,dy$$
over the domain D bounded by lines $x=0$, $y=0$, $x+y=1$.

8. Evaluate $\iint_T x^{1/2}y^{1/2}(1-x-y)^{3/2}dx\,dy$, where T is the region bounded by $x\ge 0, y\ge 0$, $x+y\le 1$.

9. Find the volume of the tetrahedron bounded by $\dfrac{x}{a}+\dfrac{y}{b}+\dfrac{z}{c}=1$ and the co-odinates axes.

10. Evaluate $\iiint\sqrt{\left(\dfrac{1-x^2-y^2-z^2}{1+x^2+y^2+z^2}\right)}dx\,dy\,dz$, integral being taken over all positive values of x, y, z such that $x^2+y^2+z^2\le 1$.

11. Evaluate $\iint\sqrt{\left[\dfrac{1-(x^2/a^2-y^2/b^2)}{1+(x^2/a^2+y^2/b^2)}\right]}dx\,dy$ where $\dfrac{x^2}{a^2}+\dfrac{y^2}{b^2}\le 1$.

12. Evaluate $\iint_R\sqrt{(x^2+y^2)}dx\,dy$, where R is the region in the xy plane bounded by $x^2+y^2=4$ and $x^2+y^2=9$.

13. Prove that $\iiint\dfrac{dx\,dy\,dz}{(x+y+z+1)^2}=\dfrac{1}{2}\left[\log 2-\dfrac{5}{8}\right]$ throughout the volume bounded by the co-ordinates planes and plane $x+y+z=1$.

14. Evaluate the integral
$$\iiint_R(ax^2+by^2+cz^2)dx\,dy\,dz$$
where R is the region given by $x^2+y^2+z^2\le d^2$

15. Find the value of
$$\iiint xyz\,\sin(x+y+z)\,dx\,dy\,dz,$$
the integral being extended to all positive values of variables subject to the condition $x+y+z\le \pi/2$.

16. Evaluate $\iiint_R x^2y^2z^2\,dx.dy.dz$ where R is the region given by $x^2+y^2<1, 0\le z\le 1$.

17. Find the value of
$$\iint x^{l-1}y^{-1}e^{x+y}\,dx,dy, 0<l<1$$
to all positive values subject to $x+y<h$.

HINTS TO SELECTED PROBLEMS

1. Put $\dfrac{x^2}{a^2}=u$ i.e., $x=au^{1/2},\dfrac{y^2}{b^2}=v$ i.e., $y=bv^{1/2}$

and $z=cw^{1/2}$

Then apply Dirichlet's theorem.

3. Do same as (1).

5. Put $x^2=c^2u, y^2=c^2v$

6. Put $\left(\dfrac{x}{a}\right)^{2/3}=u,\left(\dfrac{y}{b}\right)^{2/3}=v,\left(\dfrac{z}{c}\right)^{2/3}=w$.

7. The given integral can be written as
$$I=\iint x^{(3/2)-1}y^{(3/2)-1}[1-(x+y)]^{2/3}dx\,dy\ .$$
Then apply Liouville's extension of Dirichlet's theorem.

9. Put $\dfrac{x}{a}=u,\dfrac{y}{b}=v,\dfrac{z}{c}=w$ and apply Liouville's extension of Dirichlet's theorem.

10. Put $x^2=u, y^2=v, z^2<w$ and apply Liouville's extension of Dirichlet's theorem.

11. Put $\dfrac{x^2}{a^2}=u$ and $\dfrac{y^2}{b^2}=v$ and apply Liouville's extension of Dirichlet's theorem.

12. Put $x^2=u, y^2=v$ and Liouville's extension of Dirichlet's theorem.

14. Put $x^2=d^2u, y^2=d^2v, z^2=d^2w$.

2. $\dfrac{\Gamma(p+1)\Gamma(q+1)\Gamma(r+1)\Gamma(s+1)}{\Gamma(p+q+r+s+4)}$ **3.** $\dfrac{\pi abc}{6}$ **4.** $\dfrac{\pi abc}{6}$ **5.** $\dfrac{c^{2l+2m}}{4}\dfrac{\Gamma(l)\Gamma(m)}{\Gamma(l+m+1)}$ **6.** $\dfrac{4}{35}\pi abc$

7. $\dfrac{27\pi}{1760}$ **8.** $\dfrac{2\pi}{315}$ **9.** $\dfrac{abc}{6}$ **10.** $\dfrac{\pi}{8}\left[B\left(\dfrac{3}{4},\dfrac{1}{2}\right)-B\left(\dfrac{5}{4},\dfrac{1}{2}\right)\right]$ **11.** $\pi ab\left[\dfrac{\pi}{2}-1\right]$

12. $\dfrac{38\pi}{3}$ **14.** $\dfrac{4}{15}\pi(a+b+c)d^5$ **15.** $\dfrac{1}{384}[\pi^4-48\pi^2+384]$ **16.** $\dfrac{\pi}{48}$ **17.** $\dfrac{\pi}{\sin l\pi}(e^n-1)$.

19.10 CHANGE OF VARIABLES

Some time we change the variables from one system to another system for more convenient way to find the double integrals. The variables x, y in $\iint_R f(x,y)dx\,dy$ are changed to u, v by means of the relations $x= f_1(u, v), y = f_2(u, v)$ then the double integral is transformed into

$$\iint f\{f_1(u,v), f_2(u,v)\}\,|J|\,du\,dv$$

where $J = \begin{vmatrix} \dfrac{\partial x}{\partial u} & \dfrac{\partial x}{\partial v} \\ \dfrac{\partial y}{\partial u} & \dfrac{\partial y}{\partial v} \end{vmatrix}$ and R' is the region in the u-v plane corresponding to region R in the x-y plane.

WORKING PROCEDURE

Replace x, y by their equivalent in terms of u and v, the element of area $dx\,dy$ by $(J)\,du\,dv$ and the region R of integration in xy plane by the region R', in the uv plane.

19.10.1 CHANGE TO POLAR CO-ORDINATES

To change the variable from cartesian to polar form we put $x = r\cos\theta, y = r\sin\theta$.

Then $J = \begin{vmatrix} \dfrac{\partial x}{\partial r} & \dfrac{\partial x}{\partial \theta} \\ \dfrac{\partial y}{\partial r} & \dfrac{\partial y}{\partial \theta} \end{vmatrix} = \begin{vmatrix} \cos\theta & -r\sin\theta \\ \sin\theta & r\cos\theta \end{vmatrix} = r$

$$\iint_R f(x,y)dx\,dy = \iint_{R'} f(r\cos\theta, r\sin\theta)\,|J|\,dr\,d\theta$$

$$= \iint_{R'} f(r\cos\theta, r\sin\theta)\,|r\,dr\,d\theta$$

Solved Examples

Example 1. *Transform $\iint f(x,y)dx\,dy$, by the substitution $x+y= u, y=vu$.*

Solution. We have $x+y= u$ and $y = uv$ and $y=uv$ therefore,

$$x = u - y = u - uv \text{ and } y = uv.$$

$\therefore \quad \dfrac{\partial x}{\partial u} = 1-v, \dfrac{\partial x}{\partial v} = -u, \dfrac{\partial y}{\partial u} = v \text{ and } \dfrac{\partial y}{\partial v} = u$ ...(1)

$\therefore \quad J = \dfrac{\partial(x,y)}{\partial(u,v)} = \begin{vmatrix} \dfrac{\partial x}{\partial u} & \dfrac{\partial x}{\partial v} \\ \dfrac{\partial y}{\partial u} & \dfrac{\partial y}{\partial v} \end{vmatrix} = \begin{vmatrix} 1-v & -u \\ v & u \end{vmatrix} = u$

∴ $dx\,dy = J\,du\,dv = u\,du\,dv.$

Hence, the given integral transforms to $\iint F(u, vu)du\,dv$

Example 2.   *Transform to polar co-ordinate and integrate*

$$\iint \sqrt{\left(\frac{1-x^2-y^2}{1+x^2+y^2}\right)}\,dx\,dy$$

the integral being extended over all positive values of x and y subject to $x^2 + y^2 \le 1$.

Solution.   Here, x varies from 0 to 1 and y varies from 0 to $\sqrt{1-x^2}$ in the first quadrant where x and y are both positive.

∴   Given integral $I = \int_0^1 \int_0^{\sqrt{1-y^2}} \sqrt{\frac{(1-x^2-y^2)}{1+x^2+y^2}}\,dx\,dy$

Now change it into polar form by putting $x = r\cos\theta, y = r\sin\theta$
then the circle $x^2 + y^2 = 1$ transform into $r^2 = 1$ to $r=1$ and its first quadrant θ varies from 0 to $\pi/2$ and r varies from 0 to 1, then integral

$$I = \int_{\theta=0}^{\pi/2}\int_{r=0}^{1} \sqrt{\left(\frac{1-r^2}{1+r^2}\right)}\,r\,d\theta\,dr = \int_0^{\pi/2}d\theta\int_0^1 \sqrt{\left(\frac{1-r^2}{1+r^2}\right)}\,r\,dr$$

$$= [\theta]_0^{\pi/2}\left[\frac{1}{2}\left(\frac{\pi}{2}-1\right)\right] = \frac{1}{2}\pi.\frac{1}{2}\left(\frac{1}{2}\pi-1\right) = \frac{1}{4}\pi\left(\frac{1}{2}\pi-1\right).$$

Example 3.   *Evaluate* $\iint \sqrt{(a^2-x^2-y^2)}\,dx\,dy$, *over the semi-circle* $x^2 + y^2 = ax$ *in the positive quadrant.*

Solution.   The region of integration is a semi-circle, we change it into the polar co-ordinate by putting $x = r\cos\theta$, and $y = r\sin\theta$ in $x^2 + y^2 = ax$, then we have

$$r^2\cos^2\theta + r^2\sin^2\theta = ar\cos\theta$$
$$r^2(\sin^2\theta + \cos^2\theta) = aR\cos\theta$$
$$r = a\cos\theta.$$

The equation $r = a\cos\theta$ represent a circle which passing through the pole for the given region where r varies from 0 to $a\cos\theta$ and θ varies from 0 to $\pi/2$.

∴ $\iint\sqrt{(a^2-x^2-y^2)}dx\,dy = \int_0^{\pi/2}\int_0^{a\cos\theta}\sqrt{a^2-r^2}.r\,d\theta\,dr$

$$[\because x^2+y^2 = r^2 \text{ and } dx\,dy = r\,d\theta\,dr]$$

$$= \int_0^{\pi/2}\left[\int_0^{a\cos\theta}\left\{-\frac{1}{2}(a^2-r^2)^{1/2}(-2r)\right\}dr\right]d\theta$$

$$= \int_0^{\pi/2}\left[-\frac{1}{2}.\frac{2}{3}(a^2-r^2)^{3/2}\right]_0^{a\cos\theta}d\theta$$

$$= \frac{-1}{3}\int_0^{\pi/2}(a^3\sin^3\theta - a^3)d\theta$$

$$= \frac{-a^3}{3}\left[\frac{2}{3.1}-\frac{\pi}{2}\right] = \frac{1}{3}a^3\left(\frac{\pi}{2}-\frac{2}{3}\right).$$

Example 4. *By using the transformation $x + y = u$, $y = vu$, show that*

$$\int_0^1 \int_0^{1-x} e^{y/(x+y)} dx\, dy = \frac{1}{2}(e-1)$$

Solution. We have $\qquad dx\, dy = u\, du\, dv$

The region of integration is bounded by the lines $y = 0$, $y = 1-x$, $x = 0$ and $x = 1$
Changing these equations into new variables u and v by using the relation

$$x = u - y = u - uv = u(1-v)$$

and $y = uv$, we have $uv = 0$, $uv = 1 - u(1-v)$, $u(1-v) = 0$ and $u(1-v) = 1$
giving $\qquad v = 0$ to $v = 1$, $u = 0$ to $u = 1$.
Therefore for the given region v varies from 0 to 1 and u varies from 0 to 1.
and $\qquad e^y / (x+y) = e^{uv/u} = e^v$.
Changing the variables to u, v the given integral becomes

$$I = \int_0^1 \int_0^1 e^v u\, du\, dv = \int_0^1 [e^v]_0^1 u\, du$$

$$= \int_0^1 (e^1 - e^0) u\, du$$

$$= (e-1)\int_0^1 u\, du = (e-1)\left[\frac{u^2}{2}\right]_0^1 = \frac{1}{2}(e-1) .$$

Example 5. *Evaluate the integral $\int_0^a \int_0^{\sqrt{a^2-y^2}} (x^2 + y^2) dy\, dx$ by changing into polar co-ordinates.*

Solution. Putting $x = r\cos\theta$, $y = r\sin\theta$, we have

$$J = \begin{vmatrix} \dfrac{\partial x}{\partial r} & \dfrac{\partial x}{\partial \theta} \\[2mm] \dfrac{\partial y}{\partial r} & \dfrac{\partial y}{\partial \theta} \end{vmatrix} = \begin{vmatrix} \cos\theta & -r\sin\theta \\ \sin\theta & r\cos\theta \end{vmatrix} = r$$

$\therefore \quad dx\, dy$ is to be replaced by $J\, dr\, d\theta$.

$$x^2 + y^2 = r^2 \cos^2\theta + r^2 \sin^2\theta = r^2 .$$

Again we find that in the upper limit $x = \sqrt{a^2 - y^2}$, y varies from 0 to a and x varies from 0 to any point on the circle $x^2 + y^2 = a^2$.
In the polar form of the circle $r^2 = a^2$ i.e., $r = a$ we find that r varies from 0 to a and θ varies from 0 to $\pi/2$.

$$\therefore \quad \int_0^a \int_0^{\sqrt{a^2-y^2}} (x^2 + y^2) dy\, dx = \int_{\theta=0}^{\pi/2} \int_{r=0}^{a} r^2 . r\, d\theta\, dr$$

$$= \int_{\theta=0}^{\pi/2} \left[\frac{r^4}{4}\right]_0^a d\theta$$

$$= \frac{1}{4} a^4 (\theta)_0^{\pi/2} = \left(\frac{1}{8}\right)\pi a^4 .$$

Example 6. *Prove that $\iint_D e^{-x^2-y^2} dx\, dy = \dfrac{\pi}{4}(1 - e^{-R^2})$*

where D is the region defined by $x \geq 0$, $y \geq 0$ and $x^2 + y^2 \leq R^2$.

Solution.   Let $x = r\cos\theta$, $y = r\sin\theta$, then

$$dx\,dy = \begin{vmatrix} \dfrac{\partial x}{\partial r} & \dfrac{\partial x}{\partial \theta} \\[2mm] \dfrac{\partial y}{\partial r} & \dfrac{\partial y}{\partial \theta} \end{vmatrix} dr\,d\theta = \begin{vmatrix} \cos\theta & -r\sin\theta \\ \sin\theta & r\cos\theta \end{vmatrix} dr\,d\theta = r\,dr\,d\theta$$

$$\iint_D e^{-x^2-y^2}\,dx\,dy = \int_{\theta=0}^{\pi/2}\int_{r=0}^{R} r\,dr\,d\theta$$

$$= \int_0^{\pi/2}\left[-\frac{e^{-r^2}}{2}\right]_0^R d\theta = \frac{1}{2}(1-e^{-R^2})\int_0^{\pi/2} d\theta$$

$$= \frac{1}{2}(1-e^{-R^2})[\theta]_0^{\pi/2} = \frac{1}{2}(1-e^{-R^2})\left[\frac{\pi}{2}\right]$$

$$= \frac{\pi}{4}(1-e^{-R^2})$$

EXERCISE 19.6

1. Transform $\int_0^a\int_0^{a-x} f(x,y)\,dx\,dy$,   by   the substitution $x+y = u$, $y=uv$.

2. By using the transformation $x+y = u$, $y = uv$ show that

$$\iint \{xy(1-x-y)\}^{1/2}dx\,dy$$

taken over the area of the triangle bounded by lines $x=0$, $y=0$, $x+y =1$ is $\dfrac{2\pi}{105}$.

3. Transform the integral

$$\int_0^a\int_0^{\sqrt{a^2-x^2}} y\sqrt{x^2+y^2}\,dx\,dy$$

by changing to polar co-ordinates and hence solve it.

4. Evaluate $\int_0^2\int_0^{\sqrt{2x-x^2}} \dfrac{x\,dx\,dy}{\sqrt{x^2+y^2}}$, by changing

to polar co-ordinates.

5. Evaluate $\iint (x^2+y^2)^{7/2}dx\,dy$, over the circle $x^2+y^2 = 1$.

6. Evaluate $\iint xy(x^2+y^2)^{3/2}dx\,dy$, over the positive axes of circle $x^2+y^2 = 1$.

7. Transform the integral

$$\int_0^{\pi/2}\int_0^{\pi/2} \sqrt{\frac{\sin\phi}{\sin\theta}}\,d\phi\,d\theta$$

by the substitutions $x = \sin\phi\cos\theta$, $y = \sin\phi\sin\theta$ and show that its value is π.

HINTS TO SELECTED PROBLEMS

1. $x+y = u$ and $y = uv \Rightarrow x= u-uv$ and $y = uv$

Now,

$$J = \frac{\partial(x,y)}{\partial(u,v)} = \begin{vmatrix} \dfrac{\partial x}{\partial u} & \dfrac{\partial x}{\partial v} \\[2mm] \dfrac{\partial y}{\partial u} & \dfrac{\partial y}{\partial v} \end{vmatrix} = \begin{vmatrix} 1-v & -u \\ v & u \end{vmatrix} = u$$

$\Rightarrow$ $dx\,dy = u\,du\,dv$.

2. Proceed same as (1), we have $dx\,dy = u\,du\,dv$

$\therefore$ $\{xy\,(1-x-y)\}^{1/2} = [u(1-v)uv(1-u)]^{1/2}$
$= u(1-u)^{1/2}\,v^{1/2}\,(1-v)^{1/2}$.

3. $x = r\cos\theta$, $y = r\sin\theta$, $J =r$

$\Rightarrow$ $dx\,dy = r\,d\theta\,dr$

$\therefore$ $\int_0^a\int_0^{\sqrt{(a^2-x^2)}} y\sqrt{(x^2+y^2)}dx\,dy$

$= \int_{\theta=0}^{\pi/2}\int_{r=0}^{a} r\sin\theta.r\,d\theta\,dr$

Answers

1. $\int_0^a \int_0^1 F(u,v) u \, du \, dv$ **3.** $\dfrac{a^4}{4}$ **4.** $\dfrac{4}{3}$ **5.** $\dfrac{2\pi}{9}$

6. $\dfrac{1}{14}$ **7.** $\int_0^1 \int_0^{\sqrt{1-y^2}} \dfrac{dx\,dy}{\sqrt{y-y(x^2+y^2)}}$

19.11 CHANGE OF ORDER OF INTEGRATION

If the limits of integration are constants in the double integration then the value of integration can be obtained by integrating with respect to any independent variable.

When the limits are not constant but are the function of x and y then firstly we integrate with respect to first independent variable and then with respect to second.

In this case the limits of integration are determined in the given region by drawing the strips parallel to Y-axis or X-axis.

In this case limits of y are function of x then we find the new limits of x as function of y and new constant.

WORKING PROCEDURE

Step 1. If we perform the integration first with respect to y, we take the elementary strip parallel to y-axis and determine the limits of y and add up the vertical strip from extreme left to the extreme right of the region.

Step 2. If the order of integration is performed first with respect to x, we take the elementary strip parallel to x-axis and proceed.

Solved Examples

Example 1. *Evaluate the following integral by changing the order of integration*

$$\int_0^{2a} \int_0^{\sqrt{2ax-x^2}} a - \sqrt{(a^2-y^2)}\,dx\,dy \, .$$

Solution. Here, we have the figure (17).

The limits of y are from $y = 0$ to $y = \sqrt{2ax-x^2}$ or between x-axis and semicircle

$$x^2 + y^2 - 2ax = 0$$

or $\qquad (x-a)^2 + y^2 = a^2$

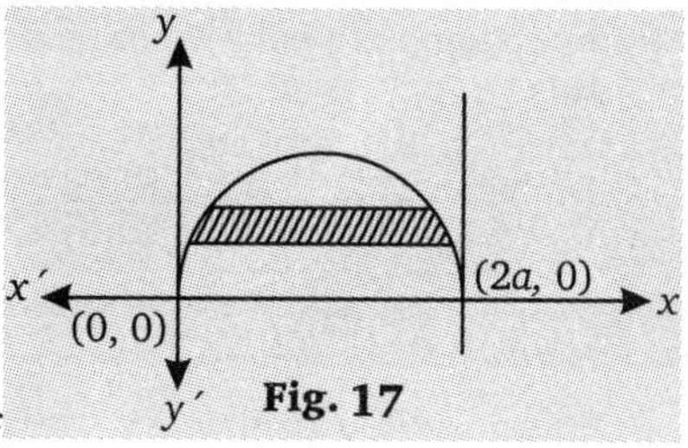

Fig. 17

In the figure, the region x varies from one end of the circle to other and

i.e., $\qquad (x-a)^2 = a^2 - y^2 \Rightarrow x = a \pm \sqrt{a^2-y^2}$.

The strips are taken parallel to x-axis and y varies from 0 to a. So the order of integration is changed as

$$I = \int_0^{2a} \int_0^{\sqrt{2ax-x^2}} (a - \sqrt{a^2-y^2})\,dx\,dy$$

$$= \int_0^a \int_{a-\sqrt{a^2-y^2}}^{a+\sqrt{a^2-y^2}} (a - \sqrt{a^2-y^2})\,dx\,dy.$$

Example 2. *Change the order of integration in the integral $\int_0^a \int_0^x f(x,y)\,dx\,dy$.*

Solution. The given limits shows that the region of integration is bounded by the curve $y = 0$, $y = x, x = 0, x = a$.

Hence $y = 0$ represent X-axis and $y=x$ represent a straight line through the origin. Also $x = 0$ and $x = a$, represent straight lines parallel to y-axis therefore the region of the integration is the triangle OAB in the figure 18 and B is (a, a).

In the given integral, the limits of integration of y being variable, we are required to integrate first w.r. to y regarding x as constant and then w.r. to x. To change the order of integration drawn parallel strip along x-axis, straight from the line OB and terminating on the line AB.

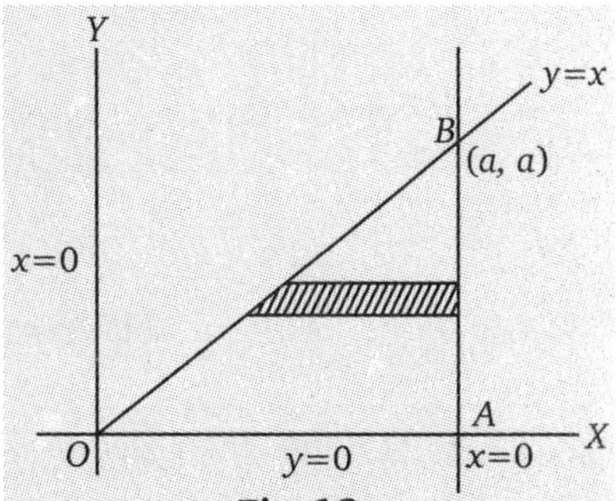

Thus in region OBA, x varies from y to a and y varies from 0 to a.

Hence, by changing the order of integration we have

Fig. 18

$$\int_0^a \int_0^x f(x,y)\,dx\,dy = \int_0^a \int_y^a f(x,y)\,dy\,dx.$$

Example 3. *Evaluate $\iint xy(x+y)\,dx\,dy$ over the area between $y=x^2$ and $y = x$.*

Solution. Here $x^2 = y$ represents a parabola whose vertex is the origin and axis is the axis of y. The equation $y=x$ is a line through origin making an angle of $45°$ with x-axis. Solving $y=x^2$ and $y = x$ we find that the parabola $y = x^2$ and the line $y = x$ intersect at the point $(0, 0)$ and $(1, 1)$. When we integrate with respect to x-along a strip parallel to x-axis the strip starts from the line $y = x$ and ends on the parabola $y =x^2$ and A is $(1, 1)$.

$$\therefore \qquad \text{Required value} = \int_{x=0}^{1}\int_{y=x}^{x^2} xy(x+y)\,dx\,dy$$

$$= \int_{x=0}^{1}\int_{y=x}^{x^2} (x^2 y + xy^2)\,dx\,dy$$

$$= \int_{x=0}^{1}\left(\frac{1}{2}x^2 y^2 + \frac{1}{3}xy^3\right)_0^{x^2} dx$$

$$= \int_0^1 \left(\frac{1}{2}x^6 + \frac{1}{3}x^7\right) - \left(\frac{1}{2}x^4 + \frac{1}{3}x^4\right) dx$$

$$= \int_0^1 \left(\frac{1}{2}x^6 + \frac{1}{3}x^7 - \frac{5}{6}x^4\right) dx$$

$$= \left[\left(\frac{1}{14}\right)x^7 + \left(\frac{1}{24}\right)x^8 - \left(\frac{1}{6}\right)x^5\right]_0^1$$

$$= \left(\frac{1}{14}\right) + \left(\frac{1}{24}\right) - \left(\frac{1}{6}\right) = \frac{3}{56}.$$

Example 4. *Change the order of integration and evaluate $\int_0^1 \int_{e^x}^e \frac{dx\,dy}{\log y}$.*

Solution. The region of integration is bounded by $e^x = y, y = e, x = 0$ and $x =1$.
Here $y = e^x$ represents a curve. Putting $x = 0$ and $x=1$ in $y = e^x$, then we get $y =1$

and $y = e$. So $A(0, 1)$ and $B(1, e)$ are the points on this on this curve.

When we integrate with respect to x first drawn a strip parallel to x-axis.

The strip starts from $x = 0$ and exends upto the curve $y = e^x$ i.e., $x = \log y$. Also for the given region y varies from $y = 1$ to $y = e$. On changing the order of integration, the given integral

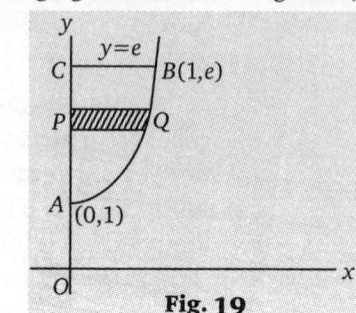

$$I = \int_1^e \int_0^{\log y} \frac{dy \, dx}{(\log y)}$$

$$= \int_1^e \frac{1}{\log y} (x)_0^{\log y} \, dy$$

$$= \int_1^e \frac{1}{\log y} (\log y - 0) dy$$

$$= \int_1^e dy = (y)_1^e = e - 1.$$

Fig. 19

Example 5. *Change the order of integration in $\int_0^a \int_{\sqrt{a^2-x^2}}^{x+2a} f(x,y) dx dy$.*

Solution. The area of integration is bounded by the curves $y = \sqrt{a^2 - x^2}$ i.e., $x^2 + y^2 = a^2$.

This is the equation of the circle with centre $(0, 0)$ and radius a. Also $y = x + 2a$ represents a straight line which passing through $(0, 2a)$ i.e., the Y-axis and the line $x=a$ which is parallel to Y-axis.

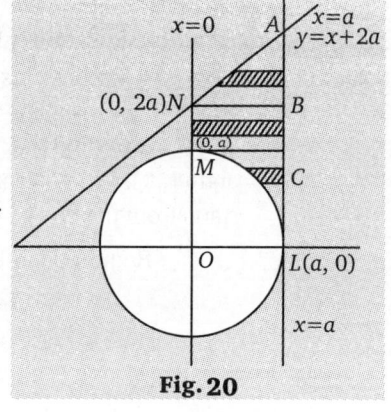

We draw the curves $x^2 + y^2 = a^2$, $y = x + 2a$, $x = 0$ and $x = a$. We observe that the region of integration is the area $MLANM$.

To change the order of integration we draw a strip parallel to x-axis. Draw the lines MC and MB parallel to X-axis. So the area of integration is divided into three portions $MLEC$, $NNCB$ and NAB.

Fig. 20

For region MLC, x varies from $x^2 + y^2 = a^2$ circle's arc to line $x = a$ or $x = \sqrt{a^2 - y^2}$ to a and y varies from 0 to a.

For region $NMCB$, x varies from 0 to a and y varies from a to $2a$.

For region NBA, x varies from $y - 2a$ to a and y varies from $2a$ to $3a$.

So, the given integral transform to

$$\int_0^a \int_{\sqrt{a^2-y^2}}^a f(x,y) dy \, dx + \int_a^{2a} \int_0^a f(x,y) dy \, dx + \int_{2a}^{3a} \int_{y-2a}^a f(x,y) dy \, dx.$$

Example 6. *Change the order of integration in the integral*
$$\int_0^{\pi/2} \int_0^{2a \cos\theta} f(r,\theta) d\theta \, dr.$$

Solution. The limits are given by $\theta = 0$ to $\theta = \pi/2$ and $r = 0$ to $r = 2a \cos\theta$. Also the curve $r = 2a \cos\theta$ is the circle.

The region of integration is the area *OABO* of circle. In the given integral the limits of integration of r is variable while limit of θ are constant. Now draw a strip parallel to θ (pole) such strip extends from the points O to the point A i.e., $r = 0$ to $r = 2a$ and for a particular circular strip of this type we observe that θ varies from $\theta = 0$ to θ of curve i.e., $\theta = \cos^{-1}(r/2a)$. Hence, the given integral

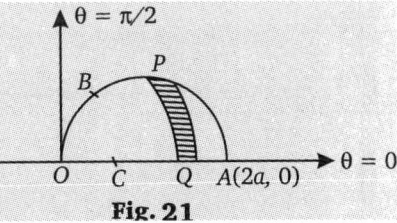

$$= \int_{r=0}^{2a} \int_{\theta=0}^{\cos^{-1}(r/2a)} f(r,\theta)\,dr\,d\theta.$$

Example 7. *Change the order of integration in*

$$\int_0^a \int_x^{a^2/x} \phi(x,y)\,dx\,dy.$$

Solution. We observe that the region is bounded by $y = x$ and $y = a^2/x$ and $x = 0$ to $x = a$. Clearly the region is *OAC*. Now draw a line *AB* parallel to x-axis, which divides the given region into two parts, *OAB* and *BAC*, Therefore, we draw a strip parallel to x-axis in *OAB*, where the left end of the strip is at y-axis and the right end is at $y = x$ and y takes the values from 0 to a. Also in region *BAC*, the left end of this strip is at y-axis whereas the right end is at the curve $y = a^2/x$, and y takes the values from $y = a$ to $y = \infty$. Hence, the given integral becomes

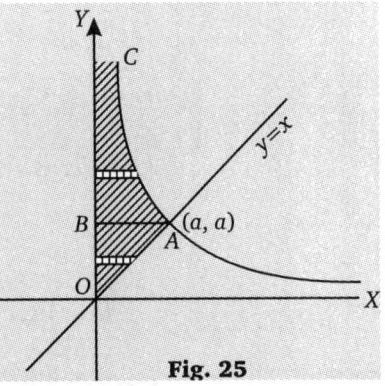

Fig. 25

$$\int_0^a \int_x^{a^2/x} \phi(x,y)\,dx\,dy = \int_0^a \int_0^y \phi(x,y)\,dx\,dy + \int_a^\infty \int_0^{a^2/y} \phi(x,y)\,dx\,dy.$$

EXERCISE 19.7

Change the order of integration in the following integral (Ques. 1-9)

1. $\int_0^{a/2} \int_{x^2/2}^{x-x^2/a} dy\,dx.$

2. $\int_0^a \int_x^{a^3/x} f(x,y)\,dx\,dy.$

3. $\int_0^1 \int_x^{x(2-x)} f(x,y)\,dx\,dy.$

4. $\int_0^{a\cos\alpha} \int_{x\tan\alpha}^{\sqrt{a^2-x^2}} f(x,y)\,dx\,dy.$

5. $\int_0^a \int_{-mx}^{lx} f(x,y)\,dx\,dy.$

6. $\int_0^a \int_{(b/a)\sqrt{a^2-x^2}}^{b} f(x,y)\,dx\,dy,$ where $c < a.$

7. $\int_0^{2a} \int_{\sqrt{2a-x^2}}^{\sqrt{2ax}} f(x,y)\,dx\,dy.$

8. $\int_0^{\pi/2} \int_0^{2a\cos\theta} f(r,\theta)\,rd\theta\,dr.$

9. $\int_0^a \int_0^{b/(b+x)} f(x,y)\,dx\,dy.$

10. Change the order of integration in
$$\int_0^\infty \int_0^\infty \frac{e^{-y}}{y} dx\,dy$$ and hence find its value.

11. Change the order of integration in
$$\int_0^\infty \int_x^\infty f(x,y)\,dx\,dy.$$

12. Change the order of integration in
$$\int_0^{2a} \int_{x(2a-x)/2a}^{\sqrt{2a-x^2}} f(x,y)\,dxdy.$$

13. Change the order of integration and evaluate
$$\int_0^\infty \int_0^x xe^{-x^2/y}\,dx\,dy.$$

14. Change the order of integration in
$$\int_0^a \int_{\sqrt{ax-x^2}}^{\sqrt{ax}} f(x,y)\,dx\,dy.$$

===== **Answers** =====

1. $\int_0^{a/4} \int_{\frac{1}{2}\left[a-\sqrt{a^2-4ay}\right]}^{\sqrt{ay}} f(x,y)\,dy\,dx.$ **2.** $\int_0^a \int_0^y f(x,y)\,dy\,dx + \int_a^\infty \int_0^{a^2/y} f(x,y)\,dy\,dx$

3. $\int_0^1 \int_{1-\sqrt{1-y}}^y f(x,y)\,dy\,dx.$ **4.** $\int_0^{a\sin\alpha} \int_0^{y\cot\alpha} f(x,y)\,dy\,dx + \int_{a\sin\alpha}^a \int_0^{\sqrt{a^2-y^2}} f(x,y)\,dy\,dx$

5. $\int_0^{am} \int_{y/l}^{y/m} f(x,y)\,dy\,dx + \int_{am}^{al} \int_{y/l}^a f(x,y)\,dy\,dx$

6. $\int_0^{b\sqrt{1-(c^2/a^2)}} \int_{a\sqrt{1-y^2/b^2}}^a f(x,y)\,dy\,dx + \int_b^b \int_{b\sqrt{1-c^2/a^2}}^a \int_c^a f(x,y)\,dy\,dx$

7. $\int_0^a \int_{y^2/2a}^{a-\sqrt{a^2-y^2}} f(x,y)\,dy\,dx + \int_0^a \int_{a+\sqrt{a^2-y^2}}^{2a} f(x,y)\,dy\,dx + \int_a^{2a} \int_{y^2/2a}^{2a} f(x,y)\,dy\,dx$

8. $\int_0^{2a} \int_0^{\cos^{-1}(r/2a)} f(r,\theta)\,dr\,d\theta$ **9.** $\int_0^{b/(a+b)} \int_0^a f(x,y)\,dy\,dx + \int_{b/(a+b)}^1 \int_0^{b(1-y)/y} f(x,y)\,dy\,dx$

10. 1 **11.** $\int_0^\infty \int_0^y f(x,y)\,dy\,dx$

12. $\int_0^{a/2} \int_{a-\sqrt{a^2-y^2}}^{a-\sqrt{a^2-2ay}} f(x,y)\,dx\,dy + \int_{a/2}^a \int_{a-\sqrt{a^2-y^2}}^{a+\sqrt{a^2-y^2}} f(x,y)\,dx\,dy + \int_a^{a/2} \int_{a+\sqrt{a^2-2ay}}^{a+\sqrt{a^2-y^2}} f(x,y)\,dx\,dy$

13. $\int_0^\infty \int_0^y xe^{-x^2/y}\,dx\,dy$

14. $\int_0^{a/2} \int_{y^2/a}^{a/2-\sqrt{(a^2/4)-y^2}} f(x,y)\,dx\,dy + \int_{a/2}^a \int_{y^2/a}^a f(x,y)\,dx\,dy + \int_0^{a/2} \int_{a/2+\sqrt{a^2/4-y^2}}^a f(x,y)\,dx\,dy$

19.12 DIFFERENTIATION UNDER THE SIGN OF INTEGRATION

Let $\quad \phi(x) = \int_{u_1}^{u_2} f(x,\alpha)\,dx$ where $a \le \alpha \le b$ ...(1)

and where u_1 and u_2 depend on the parameter α. We have to show that

$$\frac{d\phi}{d\alpha} = \int_{u_1}^{u_2} \frac{\partial f}{\partial \alpha} + f(u_2,\alpha)\frac{du_2}{d\alpha} - f(u_1,\alpha)\frac{du_1}{d\alpha} \qquad ...(2)$$

For $\alpha \in [a, b]$, $f(x, \alpha)$, $\dfrac{\partial f}{\partial \alpha}$ are continuous at both x and α in $x\alpha$-plane's region including $u_1 \le x \le u_2$, $a \le \alpha \le b$ and u_1, u_2 are continuous and having continuous derivatives in open interval $]a, b[$. Also if u_1, u_2 are constants, then we have

$$\frac{\partial \phi}{\partial \alpha} = \int_{u_1}^{u_2} \frac{\partial f}{\partial \alpha}\,dx \qquad ...(3)$$

This result is established as follows :

Let $\quad \phi(\alpha) = \int_{u_1(\alpha)}^{u_2(\alpha)} f(x,\alpha)\,dx \quad$ then $\Delta\phi = \phi(\alpha + \Delta\alpha) - \phi\alpha$

or $\quad \Delta\phi = \int_{u_1(\alpha+\Delta\alpha)}^{u_2(\alpha+\Delta\alpha)} [f(x,\alpha+\Delta\alpha)\int_{u_1(\alpha)}^{u_2(\alpha)} f(x,\alpha)dx + \int_{u_1(\alpha)}^{u_2(\alpha+\Delta\alpha)} f(x,\alpha+\Delta\alpha)\,dx$

$$- \int_{u_1(\alpha)}^{u_2(\alpha)} f(x,\alpha)\,dx$$

$$= \int_{u_1(\alpha)}^{u_2(\alpha)} f(x,\alpha+\Delta\alpha) - f(x,\alpha)]\,dx + \int_{u_1(\alpha)}^{u_2(\alpha+\Delta\alpha)} f(x,\alpha+\Delta\alpha)dx$$

$$- \int_{u_1(\alpha)}^{u_1(\alpha+\Delta\alpha)} f(x,\alpha+\Delta\alpha)\,dx \qquad ...(4)$$

Now by derivatives of mean value theorem, we have

$$\int_{u_1(\alpha)}^{u_2(\alpha+\Delta\alpha)}\left[f(x,\alpha+\Delta\alpha)dx = \int_{u_1(\alpha)}^{u_2(\alpha)}\Delta\alpha.f_\alpha(x,t)dx\right.$$

where $f_\alpha(x, t)$ is the partial derivatives of f with respect to α and t lies between α and $\alpha+\Delta\alpha$.

So, $$\int_{u_1(\alpha)}^{u_2(\alpha)}\left[f(x,\alpha+\Delta\alpha)dx - f(x,\alpha)\right]dx = \Delta\alpha\int_{u_1}^{u_2}\frac{\partial f}{\partial\alpha}dx \qquad ...(5)$$

Also by mean value theorem we have

$$\int_{u_1(\alpha)}^{u_1(\alpha+\Delta\alpha)} f(x,\alpha+\Delta\alpha)dx = f(t,\alpha+\Delta\alpha)\left[u_1(\alpha+\Delta\alpha)-u_1(\alpha)\right]$$

$$= f(t_1,\alpha+\Delta\alpha).\Delta u_1 \qquad ...(6)$$

where t_1 lies between $u_1(\alpha)$ and $u_1(\alpha+\Delta\alpha)$

$$= f(t_2,\alpha+\Delta\alpha).\Delta u_2 \qquad ...(7)$$

where t_2 lies between $u_2(\alpha)$ and $u_2(\alpha+\Delta\alpha)$.

Now substituting the values of (5), (6), (7) in (4) and divided by $\Delta\alpha$, we get

$$\frac{\Delta\phi}{\Delta\alpha} = \int_{u_1}^{u_2}\frac{\partial f}{\partial\alpha}dx + f(t_2,\alpha+\Delta\alpha)\frac{\Delta u_2}{\Delta\alpha} - f(t_1,\alpha+\Delta\alpha)\frac{\Delta u_1}{\Delta\alpha}$$

Now taking the limit $\Delta\alpha\to0$, we get

$$\frac{d\phi}{d\alpha} = \int_{u_1}^{u_2}\frac{\partial f}{\partial\alpha}dx + f\left[u_2(\alpha),\alpha\right]\frac{du_2}{d\alpha} - f\left[u_1(\alpha),\alpha\right]\frac{du_1}{d\alpha}$$

$$= \int_{u_1}^{u_2}\frac{\partial f}{\partial\alpha}dx + f(u_2,\alpha)\frac{du_2}{d\alpha} - f(u_1,\alpha)\frac{du_1}{d\alpha}.$$

If u_1 and u_2 are constant then it reduces to $\dfrac{d\phi}{d\alpha} = \int_{u_1}^{u_2}\dfrac{\partial f}{\partial\alpha}dx$

REMARKS

- The result of the above theorem is also called Leibnitz's rule.
- This result is known as differentiation under the integral sign because in the R.H.S., integrand is differentiation under the integral sign.

19.13 INTEGRATION UNDER THE INTEGRAL SIGN

Let $$\phi(\alpha)= \int_{u_1}^{u_2} f(x,\alpha)dx \;; a \le \alpha \le b \qquad ...(1)$$

and $f(x, \alpha)$ is continuous in x and α is a region including $u_1 \le x_1 \le u_2$, $a \le \alpha \le b$, then if u_1 and u_2 are constants then

$$\int_a^b \phi(\alpha)d\alpha = \int_a^b\left[\int_{u_1}^{u_2} f(x,\alpha)dx\right]d\alpha = \int_{u_1}^{u_2}\left\{\int_a^b f(x,\alpha)d\alpha\right\}dx.$$

This result can be established as follows :

Suppose $$\psi(\alpha) = \int_{u_1}^{u_2}\left\{\int_a^\alpha f(x,\alpha)d\alpha\right\}dx \qquad ...(2)$$

Now by Leibnitz's rule, we have

$$\frac{d\psi}{d\alpha} = \int_{u_1}^{u_2}\frac{\partial f}{\partial\alpha}\left\{\int_a^b f(x,\alpha)d\alpha\right\} \qquad \text{[∵ } u_1 \text{ and } u_2 \text{ are constant.]}$$

$$= \int_{u_1}^{u_2} f(x,\alpha)dx = \phi(\alpha) \qquad \begin{matrix}\text{[By (1)]}\\...(3)\end{matrix}$$

Integrating equation (3) with respect to α under the limit a to α, we get,

$$\psi(\alpha) = \int_a^\alpha \phi(\alpha)d\alpha + c \text{ ; where } c \text{ is a constant of integration.} \qquad ...(4)$$

Now if $\alpha = a$, then $\psi(a) = c$ $\qquad\qquad\qquad\qquad\left[\because \int_a^a \phi(\alpha)d\alpha = 0\right]$

But $\qquad\qquad\qquad\qquad \psi(a)=0$ $\qquad\qquad\qquad\qquad\qquad\qquad\qquad$ [from (2)]

$\therefore \qquad\qquad\qquad c=0$ and $\psi(\alpha) = \int_a^\alpha \phi(\alpha)d\alpha = 0$ $\qquad\qquad\qquad$ ...(5)

So by (2) and (5), we get

$$\int_a^\alpha \phi(\alpha)d\alpha = \int_{u_1}^{u_2}\left\{\int_a^\alpha f(x,\alpha)d\alpha\right\}dx$$

or $\qquad \int_a^\alpha\left\{\int_{u_1}^{u_2} f(x,\alpha)dx\right\}d\alpha = \int_{u_1}^{u_2}\left\{\int_a^\alpha f(x,\alpha)d\alpha\right\}dx$

Now putting $\alpha = b$ in upper limit, we get

$$\int_a^b\left\{\int_{u_1}^{u_2} f(x,\alpha)dx\right\}d\alpha = \int_{u_1}^{u_2}\left\{\int_a^b f(x,\alpha)d\alpha\right\}dx \qquad\qquad\qquad ...(6)$$

This shows that the order of Integration can be interchanged if the limits are independent of variables.

REMARK

- It is not applicable in improper integrals because the validity of above results (6) depends on limits.

Solved Examples

Example 1. Evaluate $I= \int_0^{\pi/2}\dfrac{\log(1+\cos\alpha\cos x)}{\cos x}dx$.

Solution. We have $I= \int_0^{\pi/2}\dfrac{\log(1+\cos\alpha\cos x)}{\cos x}dx$

$$= \int_0^{\pi/2}\frac{1}{\cos x}\cdot\frac{-\sin\alpha\cos x}{1+\cos\alpha\cos x}dx = -\int_0^{\pi/2}\frac{\sin\alpha}{1+\cos\alpha\cos x}dx$$

$$= -\int_0^{\pi/2}\frac{\sin\alpha\sec^2 x/2\,dx}{\sec^2 x/2+\cos\alpha\left(1-\tan^2 x/2\right)} \qquad\qquad\qquad \text{[By Leibnitz rule]}$$

$$= -\sin\alpha\int_0^{\pi/2}\frac{\sec^2 x/2}{(1+\cos\alpha)+(1-\cos\alpha)\tan^2 x/2}dx$$

$$\therefore \qquad \frac{dI}{d\alpha} = -\frac{\sin\alpha}{1-\cos\alpha}\int_0^1\frac{2dt}{\dfrac{1+\cos\alpha}{1-\cos\alpha}+t^2}$$

$\qquad\qquad\qquad\qquad\qquad\qquad\left(\because \text{If } \tan\dfrac{x}{2}=t \text{ then } \dfrac{1}{2}\sec^2\dfrac{x}{2}dx = dt\right.$

$\qquad\qquad\qquad\qquad\qquad\qquad\left.\text{also } x = 0 \Rightarrow t = 0 \text{ and } x = \dfrac{\pi}{2}\Rightarrow t = 1\right)$

$$= -\frac{2\times 2\sin\dfrac{\alpha}{2}\cos\dfrac{\alpha}{2}}{2\sin^2\dfrac{\alpha}{2}}\int_0^1\frac{dt}{\cot^2\dfrac{\alpha}{2}+t^2} = -2\cot\frac{\alpha}{2}\cdot\frac{1}{\cot\dfrac{\alpha}{2}}\left[\tan^{-1}\frac{t}{\cot\dfrac{\alpha}{2}}\right]_0^1$$

$$= -2\left[\tan^{-1}\tan\frac{\alpha}{2}-\tan^{-1}0\right] = -\alpha$$

So, $I = -\int \alpha \, d\alpha = -\dfrac{\alpha^2}{2} + C$

When $\alpha = \dfrac{\pi}{2}, I = 0, \text{So } 0 = -\dfrac{1}{2} \cdot \dfrac{\pi^2}{4} + c \text{ or } c = \dfrac{\pi^2}{8}$

$\therefore$ $I = -\dfrac{1}{2}\alpha^2 + \dfrac{\pi^2}{8} = \dfrac{1}{8}\left[\pi^2 - 4\alpha^2\right].$

Example 2. Find the value of $\int_0^\infty \dfrac{\sin ax}{x} dx \ (a > 0)$ by the method of differentiation and integration under the sign of integration.

Solution.   Let $\phi(a) = \int_0^\infty e^{-bx} \dfrac{\sin ax}{x} dx, \ a > 0$

So, $\dfrac{d\phi}{da} = \int_0^\infty \dfrac{\partial}{\partial a}\left[e^{-bx} \dfrac{\sin ax}{x}\right] dx = \int_0^\infty e^{-bx} \cos ax \, dx$

$= \dfrac{e^{-bx}}{a^2 + b^2}[-b \cos ax + a \sin ax]_0^\infty = 0 - \dfrac{-b}{a^2 + b^2} = \dfrac{b}{a^2 + b^2}$

So, $\phi(a) = \int \dfrac{b}{a^2 + b^2} da = \tan^{-1}\left(\dfrac{a}{b}\right) + c$

If $a = 0 \ \Rightarrow \ \phi(a) = 0 \ $ or $\ c = 0$

$\therefore$ $\phi(a) = \tan^{-1}\left(\dfrac{a}{b}\right)$ or $\int_0^\infty e^{-bx} \dfrac{\sin ax}{x} dx = \tan^{-1}(\infty) = \dfrac{\pi}{2}$

Taking $b = 0$, we get

$\int_0^\infty \dfrac{\sin ax}{x} dx = \tan^{-1}\left(\dfrac{a}{0}\right) = \tan^{-1}(\infty) = \dfrac{\pi}{2}.$

Example 3. Show that $I = \int_0^\infty e^{-x^2} dx = \dfrac{\sqrt{\pi}}{2}.$

Solution.   Writing ax in place of x, we get

$I = \int_0^\infty e^{-(ax)^2} . d(ax) = \int_0^\infty e^{-a^2 x^2} dx . a$

Multiplying both sides, by e^{-a^2} we get

$Ie^{-a^2} = \int_0^\infty e^{-a^2(1+x^2)} . a . dx$

Now integrating with respect to a on both sides under the limits 0 to ∞, we get

$I = \int_0^\infty e^{-a^2} da = \int_0^\infty \left\{\int_0^\infty e^{-a^2(1+x^2)} . a \, da\right\} dx$

or $I^2 = \int_0^\infty \left\{\dfrac{e^{-a^2(1+x^2)}}{-2(1+x^2)}\right\}_0^\infty dx = \int_0^\infty -\dfrac{1}{2(1+x^2)}(0-1)dx$

$= \dfrac{1}{2}\left[\tan^{-1} x\right]_0^\infty = \dfrac{1}{2}\left[\dfrac{\pi}{2} - 0\right] = \dfrac{\pi}{4}$

Hence, $I = \dfrac{\sqrt{\pi}}{2}.$

Example 4. *Verify the following :*

$$\int_0^1\left\{\int_1^2\left(\alpha^2-x^2\right)dx\right\}d\alpha = \int_1^2\left\{\int_0^1\left(\alpha^2-x^2\right)d\alpha\right\}dx.$$

Solution. We have

$$\text{L.H.S.}= \int_0^1\left\{\int_1^2\left(\alpha^2-x^2\right)dx\right\}d\alpha = \int_0^1\left[\alpha^2 x-\frac{x^3}{3}\right]_1^2 d\alpha$$

$$= \int_0^1\left[\left(2\alpha^2-\frac{8}{3}\right)-\left(\alpha^2-\frac{1}{3}\right)\right]d\alpha$$

$$= \int_0^1\left[\alpha^2-\frac{7}{3}\right]d\alpha = \left[\frac{\alpha^3}{3}-\frac{7}{3}\alpha\right]_0^1 = \frac{1}{3}-\frac{7}{3}=-2.$$

$$\text{R.H.S.}= \int_1^2\left\{\int_0^1\left(\alpha^2-x^2\right)d\alpha\right\}dx = \int_1^2\left[\frac{\alpha^3}{3}-\alpha x^2\right]_0^1 dx$$

$$= \int_1^2\left(\frac{1}{3}-x^2\right)dx = \left[\frac{1}{3}x-\frac{x^3}{3}\right]_1^2$$

$$= \left(\frac{2}{3}-\frac{8}{3}\right)-\left(\frac{1}{3}-\frac{1}{3}\right)=-2.$$

Example 5. *Show that* $I=\int_0^\infty \dfrac{\tan^{-1}\alpha x\tan^{-1}\beta x}{x^2}dx = \dfrac{\pi}{2}\log\left\{\dfrac{(\alpha+\beta)^{\alpha+\beta}}{\alpha^\alpha\beta^\beta}\right\}$ *where* $\alpha,\beta\geq 0$.

Solution. We have $\dfrac{\partial I}{\partial\alpha} = \int_0^\infty \dfrac{\partial}{\partial\alpha}\left\{\dfrac{\tan^{-1}\alpha x.\tan^{-1}\beta x}{x^2}\right\}dx$

$$= \int_0^\infty \frac{1}{x^2}\tan^{-1}\beta x.\frac{1.x}{1+\alpha^2 x^2}dx = \int_0^\infty \frac{\tan^{-1}\beta x}{x\left(1+\alpha^2 x^2\right)}dx$$

Also, $\dfrac{\partial^2 I}{\partial\beta\partial\alpha} = \int_0^\infty \dfrac{\partial}{\partial\beta}\left\{\dfrac{1}{x\left(1+\alpha^2 x^2\right)}\tan^{-1}\beta x\right\}dx = \int_0^\infty \dfrac{1}{x\left(1+\alpha^2 x^2\right)}.\dfrac{x}{1+\beta^2 x^2}dx$

$$= \int_0^\infty \frac{1}{\left(1+\alpha^2 x^2\right)\left(1+\beta^2 x^2\right)}dx = \frac{1}{\alpha^2-\beta^2}\int_0^\infty\left[\frac{\alpha^2}{1+\alpha^2 x^2}-\frac{\beta^2}{1+\beta^2 x^2}\right]dx$$

$$= \frac{1}{\alpha^2-\beta^2}\left[\alpha\tan^{-1}\alpha x-\beta\tan^{-1}\beta x\right]_0^\infty$$

$$= \frac{1}{\alpha^2-\beta^2}\left[\alpha\frac{\pi}{2}-\beta\frac{\pi}{2}\right] = \frac{1}{\alpha+\beta}\frac{\pi}{2}.$$

Now integrating with respect to β, we get

$$\frac{\partial I}{\partial\alpha} = \frac{\pi}{2}\log(\alpha+\beta)+c,$$

Initially when $\beta=0, \dfrac{\partial I}{\partial\alpha}=0 \Rightarrow c=-\dfrac{\pi}{2}\log\alpha$

$$\therefore \qquad \frac{\partial I}{\partial\alpha} = \frac{\pi}{2}\log(\alpha+\beta)-\frac{\pi}{2}\log\alpha.$$

Integrating again with respect to α, we get

$$I = \frac{\pi}{2}\int 1.\log(\alpha+\beta)\,d\alpha - \frac{\pi}{2}\int 1.\log\alpha\,d\alpha + D$$

$$= \frac{\pi}{2}\left[\alpha\log(\alpha+\beta) - \int\alpha.\frac{1}{\alpha+\beta}\,d\alpha - \alpha\log\alpha + \int\alpha.\frac{1}{\alpha}\,d\alpha\right] + D$$

$$= \frac{\pi}{2}\left[\alpha\log(\alpha+\beta) - \int\frac{\alpha+\beta-\beta}{\alpha+\beta}\,d\alpha - \alpha\log\alpha + \alpha\right] + D$$

$$= \frac{\pi}{2}\left[\alpha\log(\alpha+\beta) - \alpha\log\alpha + \alpha - \alpha + \beta\log(\alpha+\beta)\right] + D$$

$$= \frac{\pi}{2}\left[(\alpha+\beta)\log(\alpha+\beta) - \alpha\log\alpha\right] + D$$

Initially when $\alpha = 0, I = 0 \Rightarrow D = -\frac{\pi}{2}[\beta\log\beta]$

$$\therefore \qquad I = \frac{\pi}{2}\left[(\alpha+\beta)\log(\alpha+\beta) - \alpha\log\alpha\right] - \frac{\pi}{2}.\beta\log\beta$$

$$= \frac{\pi}{2}\left[(\alpha+\beta)\log(\alpha+\beta) - \alpha\log\alpha - \beta\log\beta\right]$$

$$= \frac{\pi}{2}\log\left\{\frac{(\alpha+\beta)^{\alpha+\beta}}{\alpha^{\alpha}\beta^{\beta}}\right\}.$$

EXERCISE 19.8

1. Find the value of $\int_0^{\pi}\frac{dx}{a+b\cos x}, a>0, |b|<a$ and deduce that

$$\int_0^{\pi}\frac{dx}{(a+b\cos x)^2} = \frac{\pi a}{(a^2-b^2)^{3/2}}$$

and $\int_0^{\pi}\frac{\cos x}{(a+b\cos x)^2}\,dx = -\frac{\pi b}{(a^2-b^2)^{3/2}}.$

2. If $\int_0^{\infty}e^{-\alpha x}\,dx = \frac{1}{\alpha}$ show that

$$\int_0^{\infty}e^{-\alpha x}x^n\,dx = \frac{n!}{a^{n+1}}.$$

3. Show that $\int_0^{a}\frac{\log(1+x)}{1+x^2}\,dx = \frac{\pi}{8}\log 2.$

4. Evaluate $\int_0^{\pi/2}\log\left(\frac{a+b\sin x}{a-b\sin x}\right)\frac{dx}{\sin x}, a>b.$

5. Evaluate

$$\int_0^{\pi/2}\log\left(a^2\cos^2\theta + b^2\sin^2\theta\right)d\theta; a,b>0.$$

6. Show that

$$\int_0^{\pi}\frac{\log(1+\sin\alpha\cos x)}{\cos x}\,dx = \frac{1}{2}\left(\pi\alpha - \alpha^2\right).$$

7. Show that $\int_0^{\infty}\frac{\log(1+a^2x^2)}{(1+b^2x^2)}\,dx = \frac{\pi}{6}\log\left(\frac{a+b}{b}\right).$

HINTS TO SELECTED PROBLEMS

1. $I = \int_0^{\pi}\frac{dx}{a+b\cos x}$

$$= \int_0^{\pi}\frac{dx}{a+b\cos(\pi-x)} \Rightarrow \int_0^{\pi}\frac{dx}{a-b\cos x}$$

$$= I$$

$$\Rightarrow \quad 2I = \int_0^{\pi}\frac{2a}{a^2-b^2\cos^2 x}\,dx$$

$$\Rightarrow \quad I = 2a\int_0^{\pi/2}\frac{\sec^2 x}{a^2+\tan^2 x+a^2-b^2}\,dx$$

Now putting $a\tan x = V$.

2. $\Gamma(n) = \int_0^\infty e^{-y} y^{n-1} dy$.

Putting $y = \alpha x$, we get $\dfrac{\Gamma(n)}{\alpha^n} = \int_0^\infty e^{-\alpha x} x^{n-1} dx$

$\Rightarrow \quad \dfrac{1}{\alpha} = \int_0^\infty e^{-\alpha x} dx$.

3. $I = \int_0^1 \dfrac{\log(1+x)}{(1+x^2)} dx$.

Now solve it after putting

$x = \tan\theta \Rightarrow dx = \sec^2\theta\, d\theta$.

4. Let $F(a, b) = \int_0^{\pi/2} \log\left(\dfrac{a+b\sin x}{a-b\sin x}\right) \dfrac{dx}{\sin x}, a > b$.

$\Rightarrow \dfrac{\partial F}{\partial b} =$

$\int_0^{\pi/2} \dfrac{\partial}{\partial b} \left[\log(a+b\sin x) - \log(a-b\sin x) \right] \dfrac{dx}{\sin x}$

$= 2a \int_0^{\pi/2} \dfrac{\cos ec^2 x}{a^2 - b^2 + a^2 \cot^2 x} . dx$.

Now put $\cot x = t$ i.e., $-\text{cosec}^2 x\, dx = dt$ and then solve.

5. Let $V = \int_0^{\pi/2} \log\left(a^2 \cos^2\theta + b^2 \sin^2\theta\right) d\theta$

$\Rightarrow \quad \dfrac{dV}{d\theta} = \int_0^{\pi/2} \dfrac{2a\cos^2\theta}{a^2\cos^2\theta + b^2\sin^2\theta} d\theta$.

Now putting $t = \tan\theta$ and then solve.

6. Let $F(\alpha) = \int_0^{\pi/2} \dfrac{\log(1+\sin\alpha\cos x)}{\cos x} dx$

$\Rightarrow \quad \dfrac{\partial F}{\partial\alpha} = \int_0^{\pi/2} \dfrac{\partial}{\partial\alpha}\left[\log(1+\sin\alpha\cos x)\right]\dfrac{dx}{\cos x}$

$= \cos\alpha \int_0^{\pi/2} \dfrac{dx}{1+\sin\alpha\cos x}$

Now putting $\cos x = \dfrac{1-\tan^2 x/2}{1+\tan^2 x/2}$ and then solve.

Answers

4. $\pi \sin^{-1}\left(\dfrac{b}{a}\right)$

5. $\pi \log\left[\dfrac{1}{2}(a+b)\right]$

CHAPTER REVIEW: A COMPETITIVE APPROACH

SELECTED TERMS AND RESULTS

TERMS

- **Double Interval:** Double integral is an extension of a definite integral in two-dimensional space.

- **Triple Interval:** Let $f(x, y, z)$ be a single valued function of the independent variables x, y, z in a finite region V.

RESULTS

- The double integral of a algebraic sum of a fixed number of functions is equal to the algebraic sum of double integrals taken for each term separately.

- A constant factor may be taken outside the integral sign.

- $\int_V x^{l-1} y^{m-1} z^{n-1} dx\, dy\, dz = \dfrac{\Gamma(l)\Gamma(m)\Gamma(n)}{\Gamma(l+m+n+1)}$

- $\underbrace{\int \int \ldots \int}_{n\ \text{times}} x_1^{l_1-1} . x_2^{l_2-1} \ldots x_n^{l_n-1} dx_1 dx_2 \ldots dx_n$

  $= \dfrac{\Gamma(l_1)\Gamma(l_2)\ldots\Gamma(l_n)}{\Gamma(1+l_1+l_2+\ldots+l_n)}$

- If x, y, z are all positive and such that $h_1 < x+y+z \le h_2$ then

  $\int\int\int f(x+y+z) x^{l-1} y^{m-1} z^{n-1} dx\, dy\, dz$

  $= \dfrac{\Gamma(l)\Gamma(m)\Gamma(n)}{\Gamma(l+m++n)} \int_{h_1}^{h_2} f(u) u^{l+m+n-1} du$

- If the limits of integration are constants in the double integration then the value of integration can be obtained by integrating with respect to any independent variable.

- If $\phi(x) = \int_{u_1}^{u_2} f(x,\alpha)\, d\alpha,\ a \le \alpha \le b$ and where u_1 and u_2 depend on the parameter α then

  $\dfrac{d\phi}{d\alpha} = \int_{u_1}^{u_2} \dfrac{\partial f}{\partial \alpha} + f(u_2,\alpha)\,\dfrac{du_2}{d\alpha} - f(u_1,\alpha)\dfrac{du_1}{d\alpha}$

- If $\phi(\alpha) = \int_{u_1}^{u_2} f(x,\alpha)\, dx,\ a \le \alpha \le b$ and $f(x,\alpha)$ is continuous in x and α is a region including $u_1 \le x \le u_2,\ a \le \alpha \le b$ then if u_1 and u_2 are constants then

  $\int_a^b \phi(\alpha)\, d\alpha = \int_a^b \left[\int_{u_1}^{u_2} f(x,\alpha)\, dx \right] d\alpha$

  $= \int_{u_1}^{u_2} \left\{ \int_a^b f(x,\alpha)\, d\alpha \right\} dx$

REVIEW QUESTIONS AND PROJECT WORK

1. Prove that

$$\int_1^2 \int_3^4 \left(xy + e^y\right) dy\, dx = \int_3^4 \int_1^2 \left(xy + e^y\right) dx\, dy.$$

2. Show that the value of $\int\int \sqrt{a^2 - x^2 - y^2}\, dx\, dy$ over the semi-circle $x^2 + y^2 = ax$ in the positive quadrant is $\dfrac{a^3}{3}\left[\dfrac{\pi}{2} - \dfrac{2}{3}\right]$.

3. Show that

$$\iiint_{x^2+y^2+z^2 \le 1} (x^2 + y^2 + z^2)\, dx\, dy\, dz = \frac{4\pi}{5}.$$

4. If n is a positive integer, show that

$$\iiint_{x^2+y^2+z^2 \le 1} (x^2 + y^2 + z^2)^n\, dx\, dy\, dz = \frac{4\pi}{2n+3}.$$

5. Show that

$$\iint_D x^{m-1} y^{n-1} dx\, dy = \frac{\Gamma(m)\Gamma(n)}{\Gamma(1+m+n)} . h^{m+n}$$

where D is the region $x \ge 0, y \ge 0$ and $x+y \le h$.

6. Show that the volume of the solid whose surface is represented by the equation

$$\frac{x^4}{a^4} + \frac{y^4}{b^4} + \frac{z^4}{c^4} = 1 \text{ is } \frac{abc}{6\sqrt{2}.\pi}\left[\Gamma\left(\frac{1}{4}\right)\right]^2$$

7. Show that the volume enclosed by the surface

$$\left(\frac{x}{a}\right)^{2n} + \left(\frac{y}{b}\right)^{2n} + \left(\frac{z}{c}\right)^{2n} = 1,$$

is $\dfrac{2abc\left(\Gamma\left(\dfrac{1}{2n}\right)\right)^3}{3n^2\Gamma\left(\dfrac{3}{2n}\right)} \cdot n$ being an integer.

8. Show that mass of an octant of the ellipsoid

$\dfrac{x^2}{a^2} + \dfrac{y^2}{b^2} + \dfrac{z^2}{c^2} = 1$, the density at any point being $r = kxyz$ is $\dfrac{ka^2b^2c^2}{48}$.

OBJECTIVE TYPE QUESTIONS

FILL IN THE BLANKS

1. Double integral is an extension of a definite integral in _____ dimensional space.

2. The element of the area is generally denoted by _____.

3. To change the integral from cartesian to polar, we put $x = r \cos \theta$ and $y =$ _____.

4. The value of $\int_0^{\pi/2} \int_0^{\sin\theta} r \, d\theta \, dr =$ _____.

5. The value of the double integration of the region bounded by the circle $x^2 + y^2 = a^2$ is _____.

6. If we consider the area $dx \, dy$ on the plane $z = 0$ then the volume of the solids $= \iint$ _____.

7. The whole area of the curve $a^2 x^2 = y^3 (2a - y)$ is given by _____.

8. The volume bounded by the cylinder $x^2 + y^2 = 4$ and the hyperbola $-x^2 - y^2 + z^2 = 1$ is _____.

9. The transformation formula required as $x = r \cos \theta$, $y = r \sin \theta$ and elementary area $dA =$ _____.

10. The volume cut off by paraboloid $\dfrac{y^2}{b} + \dfrac{y^2}{c} = 2x$ and the plane $x = a$ is given by _____.

TRUE/FALSE

Write 'T' for true and 'F' for false statement.

1. The transformation formula required as $x = r \cos \theta$, $y = r \sin \theta$ and elementary area $\delta A = r \, \delta\theta \, \delta r$. **(T/F)**

2. The volume inside the paraboloid $x^2 + 4z^2 + 8y = 16$ and on the positive side of xz-plane is π. **(T/F)**

3. In evaluating the double integral, if we perform the integration first with respect to y, we take the elementary strip parallel to y-axis and determine the limits for y and add up the vertical strips from extreme left to the extreme right of the region A. **(T/F)**

4. If the order of integration is performed first with respect to x, we take the elementary strip parallel to x-axis. **(T/F)**

5. The value of $\dfrac{\partial(x,y)}{\partial(u,v)}$ is known as Jacobian of x and y with respect to u and v. **(T/F)**

6. The value of $\dfrac{\partial(x,y)}{\partial(u,v)}$ is known as Jacobian of u and v with respect to x and y. **(T/F)**

MULTIPLE CHOICE QUESTIONS
Choose the most appropriate one.

1. The value of integration
$\int_0^{\pi/2} \int_0^{2a\cos\theta} r \sin\theta \, d\theta \, dr$ is :
 (a) a^2
 (b) $2a^2$
 (c) $\dfrac{2a^2}{3}$
 (d) $4a^2$

2. If we consider the integration $\int_R f(x,y) \, ds$ then R is called :

 (a) region
 (b) field of integration

 (c) both (a) and (b)
 (d) none of these

3. In question (2), ds is called :
 (a) elementary strip
 (b) element of area
 (c) volume
 (d) none of these

4. If A be a region bounded by the areas $y = f_1(x)$ and $y = f_2(x)$, $x = a$ and $x = b$, then $\iint_A f(x,y) \, dA$ is equal to :

 (a) $\int_a^b \left\{ \int_{f_1}^{f_2} f(x,y) \, dy \right\} dx$

 (b) $\int_a^b \left\{ \int_{f_1}^{f_2} f(x,y) \, dx \right\} dy$

 (c) both are true

 (d) none of these

5. To transform the cartesian equation into polar, we put $x = r \cos \theta$, $y = r \sin \theta$, then the value of $\int\int f(x,y)\,dx\,dy$ is :

(a) $\int\int f(r,\theta)r\,d\theta\,dr$ (b) $\int\int f(r,\theta)d\theta\,dr$

(c) $\int\int f(r,\theta)r\theta\,d\theta\,dr$ (d) none of these

───────────────────── **Answers** ─────────────────────

FILL IN THE BLANKS

1. Two   **2.** dS **3.** $r \sin \theta$ **4.** $\dfrac{\pi}{8}$ **5.** πa^2

6. $z\,dx\,dy$ **7.** πa^2 **8.** $\dfrac{4\pi}{3}\left(5\sqrt{5}-1\right)$ **9.** $r\,d\theta\,dr$ **10.** $a^2\left(1-\dfrac{\pi}{4}\right)$.

TRUE/FALSE

1. T   **2.** F   **3.** T   **4.** T   **5.** T   **6.** F

MULTIPLE CHOICE QUESTIONS

1. (c)   **2.** (c)   **3.** (b)   **4.** (a)   **5.** (a)

SELF ASSESSMENT TEST

Verify each of the following :

1. $\int_0^{a/\sqrt{2}} \int_y^{\sqrt{a^2-y^2}} \log\left(x^2 + y^2\right) dx\, dy \,(a > 0)$

$$= \frac{\pi a^2}{4}\left(\log a - \frac{1}{2}\right).$$

2. $\int_0^a \int_{x^2/a}^{2a-x} xy\, dy\, dx = \frac{3}{8}a^4.$

3. $\iiint_V \sqrt{1-\left(x^2 + y^2 + z^2\right)}dx\, dy\, dz = \frac{\pi^2}{3^2},$

where V is the region interior to the sphere $x^2+y^2+z^2=1.$

4. $\iiint\int dx\, dy\, dz\, dw = \frac{\pi^2}{32}\left(b^4 - a^4\right)$, where $a^2 < x^2 + y^2 + z^2 + w^2 < b^2, a < b.$

5. The volume of the solid bounded by the surface .

$$\left(\frac{x}{a}\right)^{2/3} + \left(\frac{y}{b}\right)^{2/3} + \left(\frac{z}{c}\right)^{2/3} = 1 \text{ is } \frac{4\pi abc}{35}$$

6. The volume determined by the surface $x^n+y^n+z^n=a^n$, $n>0$ in the positive octant is

$$\frac{a^3\left[\Gamma\left(1+\frac{1}{n}\right)\right]^3}{\sqrt{\Gamma\left(1+\frac{3}{n}\right)}}.$$

7. The volume bounded by the surface $\frac{x^2}{a^2} + \frac{y^2}{b^2} + \frac{z^2}{c^2} = 1$ is $\frac{8\pi abc}{5}.$

8. The volume of solid whose surface is represent by the equation

$$\frac{x^4}{a^4} + \frac{y^4}{b^4} + \frac{z^4}{c^4} = 1 \text{ is } \frac{abc}{6\sqrt{2}.\pi}\left[\Gamma\left(\frac{1}{4}\right)\right]^4.$$

9. The volume of the ellipsoid $\frac{x^2}{a^2} + \frac{y^2}{b^2} + \frac{z^2}{c^2} = 1 \text{ is } \frac{4}{3}\pi abc.$

10. The area of the ellipse $\frac{x^2}{9} + \frac{y^2}{4} = 1 \text{ is } 6\pi.$

11. The volume of the tetrahedron bounded by the co-ordinate planes and the plane $x+y+z=1 \text{ is } \frac{1}{6}.$

12. The area of ellipse $\frac{x^2}{a^2} + \frac{y^2}{b^2} = 1 \text{ is } \pi ab.$

13. $\iiint(x + y + z)dx\, dy\, dz = \frac{1}{8}$ over the tetrahedron bounded by the planes $x=0$, $y=0$, $z=0$ and $x+y+z=1.$

14. The value of $\iint\sqrt{a^2 - x^2 - y^2}dx\, dy$ over the semi-circle $x^2+y^2=ax$ in the positive quadrant is $\frac{a^3}{3}\left(\frac{\pi}{2} - \frac{2}{3}\right).$

15. If R is a region bounded by the curves $x=f_1(y)$, $x=f_2(y), y=c\, y=d$ then

$$\iint_R f(x,y)dA = \int_c^d\left[\int_{f_1(y)}^{f_2(y)} f(x,y)dx\right]dy.$$

Bibliography

1.	**Aposotol, T.M.**	*Mathematical Analysis* Addison-Wasley
2.	**Berman, G.N.**	*A Problem Book in Mathematical Analysis, MIR, Moscow*
3.	**Chatterjee, D**	*Real Analysis* PHI, India
4.	**Gupta, SC, Rani, H**	*Fundamental Real Analysis* Vikas Publishing House, New Delhi
5.	**Halmos, P.R.**	Measure theory, Van Hostrand, Princeton
6.	**Krishanan, V.K.**	*Fundamental of Real Analysis* Pearson- New Delhi
7.	**Lang, S.**	*Analysis-I* Addition wesley
8.	**Malik, S.C.**	*Mathematical Analysis,* New Age International, New Delhi
9.	**Narayan S.**	*A course of Mathematical Analysis* S. Chand and Company, New York
10.	**Ray, W.O.**	*Real Analysis* Prentice Hall
11.	**Ross K.A.**	*Elementry Analysis* Springer-Verlag
12.	**Royden, H.L.**	*Real Analysis* Macmillan, New York
13.	**Roydin, W**	*Real and Complex Analysis* TMH, New Delhi
14.	**Saxena, S.C.**	*Introduction to Real Variable Theory* PHI, India
15.	**Stirling D.S.G.**	*Mathematical Analysis* Ellis Hor wood Ltd.
16.	**White A.J.**	*Real Analysis : An introduction* Addition Wesley Publishing Company
17.	**Wider, D.V.**	*Advanced Calculas,* PHI India
18.	**Wilson, E.B.**	Advanced Calculas, Dover, New York
19.	**Zannen, A.C.**	*Integration* North Holland, Amesterdam.

Index